D1568685

Internal Combustion Engine Handbook

Basics, Components, Systems, and Perspectives

Other SAE titles of interest:

Direct Injection Systems:
The Next Decade in Engine Technology
By Cornel Stan
(Order No. R-347)

Engine Combustion Instrumentation and Diagnostics
By Hua Zhao and Nicos Ladommatos
(Order No. R-264)

Introduction to Internal Combustion Engines
By Richard Stone
(Order No. R-278)

The Romance of Engines
By Takashi Suzuki
(Order No. R-188)

For more information or to order a book, contact SAE at
400 Commonwealth Drive, Warrendale, PA 15096-0001;
phone 724-776-4970; fax 724-776-0790;
e-mail CustomerService@sae.org;
website http://store.sae.org.

Internal Combustion Engine Handbook

Basics, Components, Systems, and Perspectives

Edited by

Richard van Basshuysen

and

Fred Schäfer

SAE International™
Warrendale, Pa.

For permission and licensing requests, contact:

SAE Permissions
400 Commonwealth Drive
Warrendale, PA 15096-0001 USA
E-mail: permissions@sae.org
Tel: 724-772-4028
Fax: 724-772-4891

Library of Congress Cataloging-in-Publication Data

Handbuch Verbrennungsmotor. English
Internal combustion engine handbook: basics, components, systems, and perspectives / edited by Richard van Basshuysen and Fred Schäfer.
p. cm.
Includes bibliographical references and index.
ISBN 0-7680-1139-6.
1. Internal combustion engines. I. van Basshuysen, Richard, 1932– II. Schäfer, Fred, 1948– III. Title.

TJ755.H2513 2004
621.43—dc22 2004048172

Translated from the German language edition:
Handbuch Verbrennungsmotor by Richard van Basshuysen and Fred Schäfer

SAE
400 Commonwealth Drive
Warrendale, PA 15096-0001 USA
E-mail: CustomerService@sae.org
Tel: 877-606-7323 (inside USA and Canada)
724-776-4970 (outside USA)
Fax: 724-776-1615

ISBN 0-7680-1139-6

SAE Order No. R-345

Printed in Canada.

Foreword

The complexity of a modern internal combustion engine is certainly one of the reasons why one person is no longer able to comprehensively present all the important interplays in their full depth. Perhaps it is also one of the reasons why there has been no complete work on this subject to date. Although a large number of technical books deal with certain aspects of the internal combustion engine, there has been no publication until now that covers all of the major aspects of the topic.

The more than 100-year development of the internal combustion engine has resulted in an enormous amount of important information and detailed knowledge on the different demands, the large number of components, and their interaction. With a volume of almost 950 pages, more than 1250 illustrations, and nearly 700 bibliographical references, we believe that with this book we have now succeeded in covering all the main technical aspects of the internal combustion engine.

It was, therefore, a particular endeavor of the publishers to place emphasis in all the right places and thus to present a work that closes a significant gap in the technical literature. Of particular note is the fact that this book was produced in just eighteen months and, therefore, effectively reflects the current high status of the present-day technical development.

Apart from illustrating the latest level of knowledge in engine development, the editors were extremely keen to present theory and practice in a balanced ratio. This was achieved, in particular, by winning the cooperation of more than 90 authors from science and industry. With their help, a publication has been created that is a valuable source of information and advice in the day-to-day work of education, research, and practice.

It is aimed, in particular, at specialists involved in science and practice in the automotive, engine, mineral oil, and accessories industry and at students for whom it is designed to provide valuable help throughout their studies. Furthermore, it is intended to be a useful advisor for patent lawyers, the motor vehicle trade, government offices, journalists, and interested members of the general public.

The question of the future of the internal combustion engine is reflected in many new approaches to the solution of the problems concerning fuel consumption and environmental compatibility. Particularly under these aspects, by comparison with the alternatives, it is not difficult to predict that the reciprocating piston engine as the driving power for cars will probably remain with us in its fundamental elements for many years to come. New drive systems always have the problem of having to compete with more than 100 years of development with enormous development capacities worldwide. Starting from the present-day status of motor development, it is important to answer the questions: In what direction is the internal combustion engine developing? What is its potential after more than 100 years of development? How is the fuel situation of the future to be assessed? Are there competing systems that could replace it in the coming years and decades? This book tries to give conclusive answers to these questions.

Even though the main focus of the book is on the car engine, certain basic aspects also relate to the commercial vehicle engine. It is also new that the different aspects of the gasoline engine as compared with the diesel engine in many areas are illustrated in this book. Will there be any fundamental difference between the gasoline and the diesel engines in a few years? We have only to look at the growing approximation between gasoline and diesel engines: Gasoline engines with direct injection—in the future perhaps diesel engines with homogeneous combustion.

Our special thanks go to all the authors for their collaboration and for their appreciation of this difficult task. With their discipline they made it easy for us to coordinate more than 90 authors. Of particular note is the punctuality of the authors that enabled the book to appear six months ahead of schedule—an almost unique occurrence.

The editors know that the work on this book has often been at the expense of partners and families, and so we express our thanks to them, too, for their understanding.

A few improvements have been made to the second edition: Numerous illustrations have been enlarged, and the formulas improved.

Thanks also to the Society of Automotive Engineers for its constructive and understanding cooperation.

Last but not least we thank Siemens VDO Automotive for the technical and material support in the creation of this work, without whose cooperation this book could never have been published.

Bad Wimpfen/Hamm, June 2002

Richard van Basshuysen
Fred Schäfer

Chapters, Articles and Authors

Index of Companies and Universities

Index of Companies

Federal Mogul, Nuremberg	Dipl.-Phys. Jürgen Goller Dipl.-Ing. Jens Scholz
Federal Mogul, Wiesbaden	Philippe Damour Dr.-Ing. Uwe Lehmann
FEV Motorentechnik GmbH, Aachen	Dipl.-Ing. Bernd Haake Dr.-Ing. Franz Koch Dr.-Ing. Peter Wolters
Freudenberg & Co. KG, Weinheim	Dr.-Ing. Uwe Meinig
Gates GmbH, Aachen	Dr.-Ing. Manfred Arnold Dipl.-Ing. Matthias Farrenkopf
Georg Fischer Fahrzeugtechnik AG, Schaffhausen (CH)	Dr. Leopold Kniewallner
Hydraulik-Ring, Nürtingen	Dipl.-Ing. Andreas Knecht Dipl.-Ing. Wolfgang Stephan
IAV GmbH, Berlin	Dipl.-Ing. Reinhold Bals Dr.-Ing. Michael Fischer
INA Motorenelemente, Hirschaid	Dipl.-Ing. Michael Haas
IWIS GmbH & Co. KG, Munich	Dr.-Ing. Peter Bauer
Mahle Kolben und Motorkomponenten GmbH, Stuttgart	Dipl.-Ing. Kay Brodesser Dr.-Ing. Alfred Elsäßer Dipl.-Ing. Rolf Kirschner Dr.-Ing. Martin Lechner Dr.-Ing. Uwe Mohr Dipl.-Ing. Wolfgang Schilling Dipl.-Ing. Jan Schmidt
Mann + Hummel GmbH, Ludwigsburg	Dr.-Ing. Olaf Weber
Matter Engineering, Wohlen (CH)	Dr. Markus Kasper
Miba Gleitlager AG, Laarkirchen (A)	Dipl.-Ing. Ulf G. Ederer
Mubea Muhr & Bender, Attendorn	Dr.-Ing. Rudolf Bonse
Peiner Umformtechnik, Peine	Dipl.-Ing. Siegfried Jende

Dr. Ing. h.c. F. Porsche AG, Stuttgart	Dipl.-Ing. Günter Helsper Dipl.-Ing. Karl B. Langlois
Schatz Thermo Engineering, Erling-Andechs	Dr.-Ing. Oskar Schatz
SHW GmbH, Bad Schussenried	Christof Härle Dr.-Ing. Christof Lamparski Bernd Schreiber
Siemens VDO Automotive, Regensburg	Dr.-Ing. Anton Grabmeier Dipl.-Ing. Friedrich Graf Dipl.-Ing. Stefan Klöckner Dipl.-Ing. Achim Koch Dr.-Ing. Robert Rehbold Dipl.-Ing. Rainer Riecke Dipl.-Ing. Peter Skotzek Dr. Michael Ulm Dr. Klaus Wenzlawski
Tenneco Automotive Heinrich Gillet, Edenkoben	Dipl.-Ing. Hubert Neumaier
TRW Deutschland GmbH, Barsinghausen	Dr.-Ing. Klaus Gebauer
TTM, Niederrohrdorf (CH)	Dipl.-Ing. Andreas Mayer
WiTech Engineering GmbH, Brunswick	Prof. Dr.-Ing. Ulrich Seiffert

Index of Universities

Aargauische Fachhochschule, Windisch	Prof. Dr. Heinz Burtscher
University of Applied Sciences, Esslingen	Prof. Dipl.-Ing. Dr. techn. Hartmut Bathelt
University of Applied Sciences, Giessen/Friedberg	Prof. Dr.-Ing. Stefan Zima
University of Applied Sciences, Südwestfalen, Iserlohn	Prof. Dr.-Ing. Wilhelm Hannibal Prof. Dr.-Ing. Fred Schäfer
Technical University of Dresden	Prof. Dr.-Ing. Hans Zellbeck
Technical University of Vienna	ao. Univ.-Prof. Dipl.-Ing. Dr. techn. Ernst Pucher

University of Cottbus	Dipl.-Ing. Dirk Goßlau Prof. Dr.-Ing. Peter Steinberg
University of Hanover	Univ.-Prof. Dr.-Ing. habil. Günter P. Merker
University of Karlsruhe	Prof. Dr.-Ing. Ulrich Spicher
University of Magdeburg	Dr.-Ing. Hanns Erhard Heinze Dr.-Ing. Detlef Hieber Prof. Dr.-Ing. Helmut Tschöke

Index of Authors

Author	Affiliation
Dr. rer. nat. Dipl.-Phys. Manfred Adolf	Manager of Ignition Engineering, Beru AG, Ludwigsburg, Germany
Dr.-Ing. Manfred Arnold	Business Unit Manager, Gates GmbH, Aachen, Germany
Dipl.-Ing. Reinhold Bals	Department Manager of Engines and Drive Systems, IAV Berlin, Germany
Dipl.-Ing. Matthias Banzhaf	Manager of Product and Process Development, Behr, Stuttgart, Germany
Prof. Dipl.-Ing. Dr.-techn. Hartmut Bathelt	Professor at University of Applied Sciences, Esslingen and former Manager of the Acoustics Department at AUDI AG, Neckarsulm, Germany
Dr.-Ing. Peter Bauer	Key Account Manager, IWIS, Munich, Germany
Dr. Andrée Bergmann	Emitec, Lohmar, Germany
Dr. rer. nat. Tilman Beutel	Senior Chemist, Engelhard Technologies, Hannover, Germany
Dr.-Ing. Erich Blümcke	Department I/EK-6, Audi AG, Ingolstadt, Germany
Dr.-Ing. Rudolf Bonse	Manager of the Valve Train Business Unit, Muhr & Bender, Attendorn, Germany
Dipl.-Ing. Stefan Brandt	Project Manager of Lean S.I. Engines, Engelhard Technologies, Hannover, Germany
Dipl.-Ing. Thomas Breier	Former Head of Development for the Business Unit Special Seals, Elring Klinger, Dettingen, Germany
Dipl.-Ing. Kay Brodesser	Division Manager for Testing and Advanced Design, Mahle, Stuttgart, Germany
Prof. Dr. Heinz Burtscher	Instructor at the Institute for Signals and Sensors at Aargau Polytechnic in Windisch, Switzerland

Dipl.-Ing. Uwe Dahle	Project Engineer for Lean S.I. Engines, Engelhard Technologies, Hannover, Germany
Philippe Damour	Chief Engineer for Bearing Shells and Connecting Rods, Federal Mogul, Wiesbaden, Germany
Dipl.-Ing. Armin Diez	Development Manager for the Business Unit Cylinder Head Seals, Elring Klinger, Dettingen, Germany
Dr.-Ing. Werner Dirschmid	Department I/EK-61, Audi AG, Ingolstadt, Austria
Dipl.-Ing. Ulf G. Ederer	Manager of Advanced Development of New Technologies, Miba Gleitlager, Laakirchen, Austria
Dr.-Ing. Alfred Elsäßer	Project Manager for Air Cycle Valves, Mahle, Stuttgart, Germany
Dipl.-Ing. Matthias Farrenkopf	Technical Director, Gates GmbH, Aachen, Germany
Dr.-Ing. Michael Fischer	Divisional Manager for S.I. Engines, IAV, Berlin, Germany
Dr.-Ing. Klaus Gebauer	Director of the Research and Development Department, TRW, Barsinghausen, Germany
Dipl.-Ing. Dirk Goßlau	Scientific Assistant to the Chair for Automotive Engineering and Vehicle Propulsion Systems at the Brandenburg University at Cottbus, Germany
Dr.-Ing. Anton Grabmeier	Siemens VDO Automotive, Regensburg, Germany
Dipl.-Ing. Friedrich Graf	Siemens VDO Automotive, Regensburg, Germany
Dipl.-Ing. Eberhard Griesinger	Manager of Applied Engineering, Elring Klinger, Dettingen, Germany
Dipl.-Phys. Jürgen Goller	Manager of Technical Calculations for Pistons, Federal Mogul, Nuremberg, Germany

Dipl.-Ing. Bernd Haake	Project Manager, FEV, Aachen, Germany
Dipl.-Ing. Michael Haas	Manager of Variable Valve Trains, INA Motorenelemente, Hirschaid, Germany
Prof. Dr.-Ing. Wilhelm Hannibal	Manager of the Laboratory for Construction and CAE Applications and the CAD Laboratory at South Westfalia University of Applied Sciences, Iserlohn, Germany
Christof Härle	Manager of Testing, SHW, Bad Schussenried, Germany
Dr.-Ing. Hanns Erhard Heinze	Scientific Assistant at the Institute for Machine Metrology and Piston Machines at the University of Magdeburg, Germany
Dipl.-Ing. Günter Helsper	Developer of Engine Drive Systems, Dr. Ing. h.c. F., Porsche AG, Stuttgart, Germany
Dr.-Ing. Detlef Hieber	Scientific Assistant at the Institute for Machine Metrology and Piston Machines at the University of Magdeburg, Germany
Dr.-Ing. Rolf Jakobs	Manager of Research and Development for Piston Rings, Federal Mogul, Burscheid, Germany
Dipl.-Ing. Siegfried Jende	Plant Management at Peiner Umformtechnik, Peine, Germany
Dr. Markus Kasper	Managing Director, Matter Engineering, Wohlen, Switzerland
Dipl.-Ing. Rolf Kirschner	Advanced Design and System Design for Valve Train Systems, Mahle, Stuttgart, Germany
Dipl.-Ing. Stefan Klöckner	Siemens VDO Automotive, Regensburg, Germany
Dipl.-Ing. Uwe Georg Klump	FEM Department, Elring Klinger, Dettingen
Dipl.-Ing. Andreas Knecht	Manager of Development for Engine Technology, Hydraulik-Ring GmbH, Nürtingen, Germany

Dr.-Ing. Leopold Kniewallner	Manager of the Research and Development Department, Fischer Fahrzeugtechnik, Schaffhausen
Dr.-Ing. Franz Koch	Department Manager for Construction and Engine Mechanics, FEV, Aachen, Germany
Dipl.-Ing. Achim Koch	Siemens VDO Automotive, Regensburg, Germany
Dr.-Ing. Gerd Krüger	Manager of the Development Department, Bleistahl, Wetter, Germany
Dipl.-Ing. Wilhelm Kullen	Developmental Engineer for the Business Unit Special Seals, Elring Klinger, Dettingen
Dr.-Ing. Christof Lamparski	Manager of Development, SHW, Bad Schussenried
Dipl.-Ing. Karl B. Langlois	Development of Engine Propulsion Systems, Dr. Ing. h.c. F. Porsche AG, Stuttgart, Germany
Dr.-Ing. Martin Lechner	Manager of Advanced Design, Mahle, Stuttgart, Germany
Dr.-Ing. Uwe Lehmann	Manager of Technical Calculations for Friction Bearings, Federal Mogul, Wiesbaden, Germany
Dipl.-Ing. Andreas Mayer	Manager of Technology, TTM, Niederrohrdorf, Switzerland
Dr.-Ing. Uwe Meinig	Manager of Development, Freudenberg Spezialdichtungsprodukte, Weinheim, Germany
Univ. Prof. Dr.-Ing. habil. Günter P. Merker	Manager at the Institute for Technical Combustion at the University of Hannover, Germany
Dr.-Ing. Uwe Mohr	Manager of the Research and Development Center, Mahle, Stuttgart, Germany
Dipl.-Ing. Markus Müller	Manager of Cylinder Product Development, Federal Mogul, Burscheid, Germany

Dipl.-Ing. Hubert Neumaier	Manager of Product Development, Tenneco Automotive-Heinrich Gillet, Edenkoben, Germany
Dr. Walter Piock	Team Manager for S.I. Engine Drive Systems for Passenger Cars, AVL List, Graz, Austria
ao. Univ.-Prof. Dipl.-Ing. Dr. techn.	Institute for Internal Combustion Engines and Automobile Construction at the
Ernst Pucher	Technical University of Vienna, Austria
Dipl.-Ing. Alfred Punke	Manager of New Technologies, Engelhard Technologies, Hannover, Germany
Dr.-Ing. Robert Rehbold	Siemens VDO Automotive, Regensburg, Germany
Dipl.-Ing. Rainer Riecke	Siemens VDO Automotive, Regensburg, Germany
Prof. Dr.-Ing. Fred Schäfer	Professor at University of Applied Sciences, South Westfalia for Motor Engines and Machines
Dr.-Ing. Oskar Schatz	Manager at Schatz Thermo Engineering, Gauting, Germany
Dr. rer. nat. Peter Scherm	Project Manager for Diesel Oxidation Catalytic Converters, Engelhard Technologies, Hannover, Germany
Dipl.-Ing. Wolfgang Schilling	Electronics Developer, Mahle, Stuttgart, Germany
Dipl.-Ing. Jan Schmidt	Control Software Developer, Mahle, Stuttgart, Germany
Dipl.-Ing. Heinz-Georg Schmitz	Developer of Cold-Start Systems, Beru AG, Ludwigsburg, Germany
Dipl.-Ing. Jens Scholz	Principal Scientist for Materials, Federal Mogul, Nuremberg, Germany
Dipl.-Ing. Johann Schopp	Department Manager for Construction and Mechanical Testing, BMW AG, Munich, Germany

Bernd Schreiber	Manager of Marketing, SHW, Bad Schussenried, Germany
Dipl.-Ing. Günter H. Seidel	Former Manager of Development, ARAL AG, Bochum, Germany
Prof. Dr.-Ing. Ulrich Seiffert	Managing Member of WiTech Engineering, Professor at Braunschweig Technical University
Dr. Stephan Siemund	Project Manager of Three-Way Catalytic Converters, Engelhard Technologies, Hannover, Germany
Dipl.-Ing. Peter Skotzek	Siemens VDO Automotive, Regensburg, Germany
Prof. Dr.-Ing. Ulrich Spicher	Manager of the Institute for Piston Machines at the University of Karlsruhe, Germany
Prof. Dr.-Ing. Peter Steinberg	Chair of Automotive Technology and Drive Systems at the Brandenburg University of Cottbus, Germany
Dipl.-Ing. Wolfgang Stephan	Manager of Development, Hydraulik-Ring GmbH, Nürtingen, Germany
Dr.-Ing. Susanne Stiebels	Project Engineer for Three-Way Catalytic Converters, Engelhard Technologies, Hannover, Germany
Dr. Rüdiger Teichmann	Business Segment Manager of Indicating Systems and Sensors, AVL List, Graz, Austria
Prof. Dr.-Ing. Helmut Tschöke	Manager at the Institute for Machine Metrology and Piston Machines at the University of Magdeburg, Germany
Dr. Michael Ulm	Siemens VDO Automotive, Regensburg, Germany
Dr.-Ing. E.h. Richard van Basshuysen	Publisher of ATZ/MTZ
Dr.-Ing. Olaf Weber	Manager of Central Development, Mann + Hummel, Ludwigsburg, Germany
Dr. Klaus Wenzlawski	Siemens VDO Automotive, Regensburg, Germany

Dr. rer. nat. Hans-Peter Werner	Manager of Testing and Simulation, Elring Klinger, Dettingen, Germany
Dr. Erich Winklhofer	Team Manager for Optical Procedures, Measuring and Testing Systems, AVL List, Graz, Austria
Dr.-Ing. Hans-Walter Wodtke	Manager of NVH, AFT Atlas Fahrzeugtechnik, Werdohl, Germany
Dr.-Ing. Peter Wolters	Division Head of Spark Ignition Engine Processes, FEV, Aachen, Germany
Prof. Dr.-Ing. Hans Zellbeck	Chair of Internal Combustion Engines at the Technical University of Dresden, Germany
Prof. Dr.-Ing. Stefan Zima	Professor of Mechanical Engineering, Foundry Technology and Materials Engineering at the University of Applied Sciences, Giessen/Friedberg, Germany
Dipl.-Ing. Frank Zwein	Project Manager, Federal Mogul, Burscheid, Germany

Contents

Note to the reader

Bibliographic references, given in square brackets at the end of a section, that are not marked in the text by superscripts indicate additional literature. This may serve to provide the reader with more in-depth information on the material covered in the respective section.

1 Historical Review

Motor vehicles have been built for more than a century.[4–8] The advancements in vehicle appearance, even to the technical layman, are astonishing. Advancements in basic engine *appearance,* on the other hand, have been relatively minimal. The similarity in dimensions and layout (and a few other details) between engines of the past and current models hide just how much has also been done in engine technology over the years (Fig. 1-1).

The origins of motor vehicle engines lie ultimately in the needs of the craftsmen and small traders who could not afford the expensive and complex steam engines as power generators. The costly steam engines were subject to strict regulations and were primarily owned by larger companies who could afford them. Thus, the first internal combustion engines (gasoline-powered stationary motors for driving machines of all kinds) were produced because of the need for an affordable and simple source of power.

Work on such drive systems had been done in various parts of the world. In 1876 Nikolaus August Otto successfully implemented the four-stroke process patented by the Frenchman Beau de Rochas. This engine had a decisive advantage when compared to the gasoline engines already being built by the Frenchman Jean Joseph Etienne Lenoir; it utilized precompression of the mixture. The British engineer Dougald Clerk "shortened" the four-stroke process to the two-stroke process by eliminating the charge cycle strokes. In 1886 Karl Benz and Gottlieb Daimler (with Wilhelm Maybach) simultaneously (and independently) developed the light, high-speed engine from which most modern gasoline engines would descend. Similar engines would also power airships and airplanes in the years that followed.

Rudolf Diesel's "rational heat engine" 1893–1897 could initially be used only for stationary applications; the same applied to its predecessors, the motors designed by George Bailey Brayton and Herbert Akroyd Stuart. It was to be decades before the diesel engine finally "hit the road."

The fundamental design of the internal combustion engine was duplicated from the steam engine: the crank drive controls the sequence of the thermodynamic process and converts the vapor pressure first into an oscillating and then into a rotary movement. The high development level of the steam engine at the end of the 19th century formed the foundation for the engines. The level of mastery in casting, forging, and precise machining of automotive components also increased as a result of the steam engine. It was the one-piece self-tensioning piston ring from John Ramsbottom (1854) that enabled the high working pressures in the combustion chamber of internal combustion engines to be maintained. The piston ring was, therefore, just as much a precondition for the control of the engine process as the knowledge and experience of engine bearings and their lubrication.

One of the initial developmental issues with the internal combustion engine was a question of presenting central engine functions. The most difficult problem of the early engines was the ignition. The flame ignition (Otto) and uncontrolled glow tube ignition (Maybach/Daimler) presented an obstacle to the engine development that was overcome only with the advent of electric ignition methods. These ignition types included snapper ignition (Otto), vibrator ignition (Benz), the Bosch magnetic low-voltage ignition with contact-breaking spark, and, finally, the high-voltage magnetic ignition (Bosch). Next, the quality and quantity of the mixture formation had to be improved. Wick-surface and brush carburetors allowed only the low-boiling fractions of the gasoline (final boiling point approximately 100°C) to be used. The fuel particles that *could* be used did not vaporize simultaneously, creating another problem. In the Wilhelm Maybach nozzle carburetor the fuel was atomized and no longer "vaporized." Now, it was possible to use a higher percentage of the gasoline (final boiling point around 200°C) productively.[9] The spectrum of fuels that could be used was significantly extended. In particular, the mixture could be formed in practically any quantity (a precondition for a further increase in performance and power). Carburetors with automatic auxiliary air control from Krebs, Claudel (Zenith) as well as Menesson and Goudard (Solex) improved the operating behavior of the engines and reduced the fuel consumption.

With the increase in power, more heat had to be dissipated with the coolant. Now, it was the simple evaporation cooling that proved to be the power-limiting factor. Heat dissipation was too low with the cooling system of the time. It required a large amount of water to be stored (and transported) on the vehicle in order to work effectively. Critical components could not be adequately and reliably cooled with a natural water circulation (thermal siphon) creating another problem. The Wilhelm Maybach *honeycomb cooler* offered the physically "workable" solution that allowed for the intensification of the heat transfer on the side of the weak thermal transition (on the air side).

Once these basics had been established on the engine side of vehicle technology, the motor vehicle industry developed rapidly. Advances on the engine side inspired the advances on the vehicle side (and vice versa). More and more companies took up the production of motor vehicles and engines.

In order to increase power and enhance the smoothness of running, the number of cylinders was increased—from one to two and then to four, as in the Mercedes Simplex engine. The splitting of the combustion chamber into several cylinders enabled higher speeds and a better utilization of the combustion chamber, i.e., higher specific work (effective mean pressure). The construction of motor vehicles and engines had also started in other countries

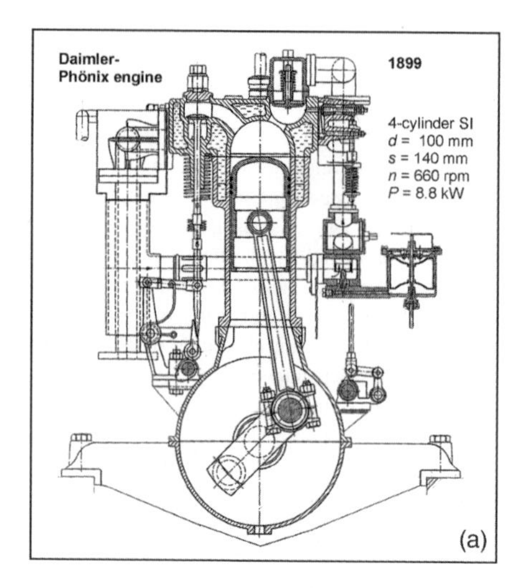

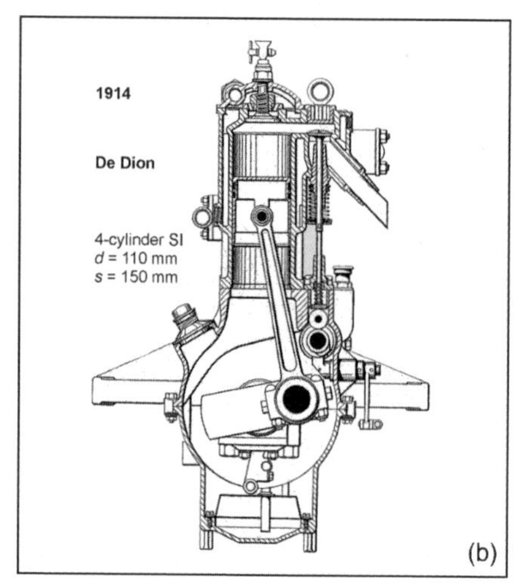

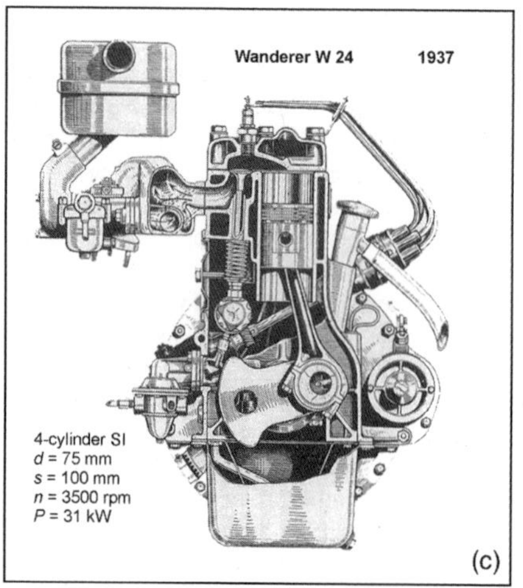

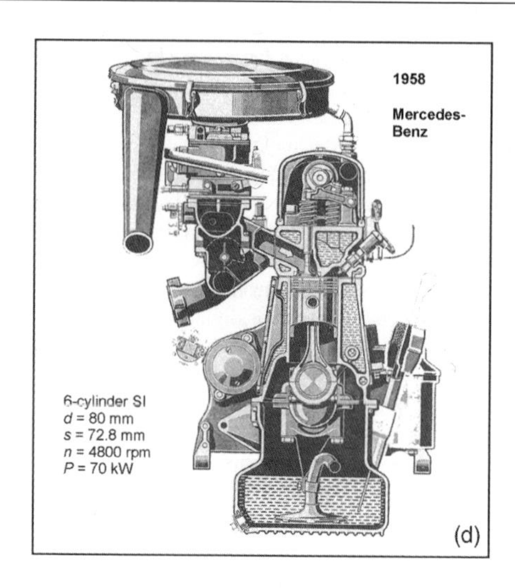

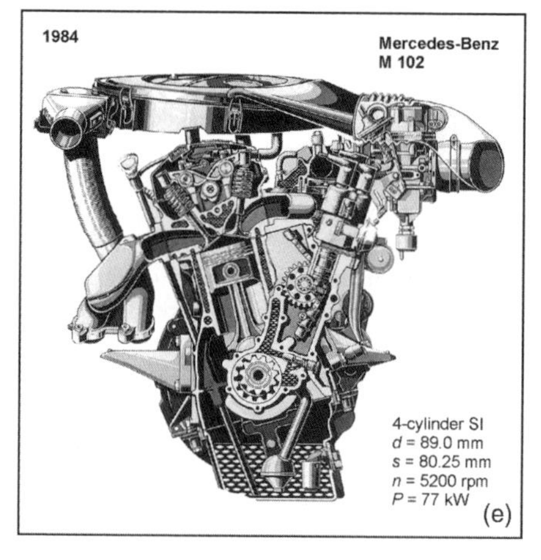

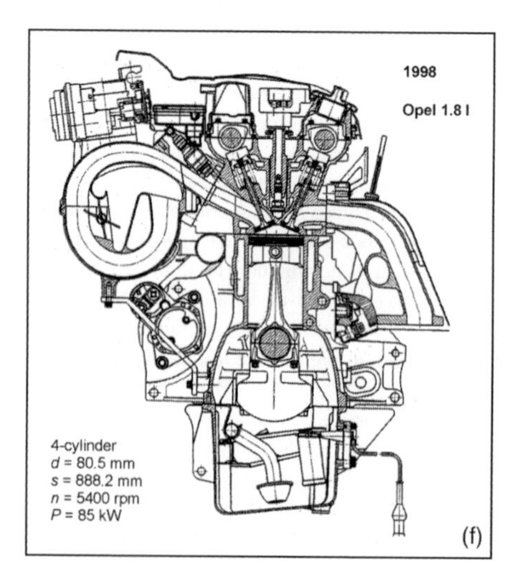

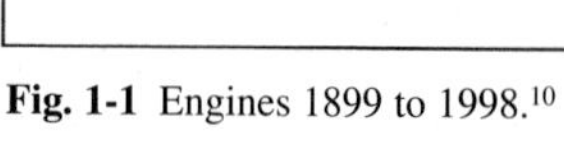
Fig. 1-1 Engines 1899 to 1998.[10]

(France, Italy, England, and, later, the United States). These engines were initially modeled using the German design, but soon other manufacturers began to create their own designs. The engine technology enjoyed an enormous boost from the aircraft development, from which the motor vehicle engines also benefited. Experience was shared, so that the errors made in the aircraft engine development (and recognized as such) could be avoided from the outset in the motor vehicle engines. Nevertheless, there was competition between several drive concepts, including the technically mature steam engine. This design had benefits as a power source for road vehicles, including the fact that the engine was self-starting, had an elastic operating curve to match the required tractive power of the vehicle, and was smooth running. The electric drive appeared to offer even greater benefits, but the disadvantages of this drive concept quickly became apparent.

As the engine power increased, so did the speed and weight of the vehicles. Now, it was a question of adapting the engine functions such as mixture composition, ignition timing, lubrication, and cooling to the conditions of road operation. The complex technical system engine had to be made controllable even for untrained personnel (namely, the vehicle owner). Fuel and oil consumption had to be reduced, the latter not only for cost reasons but also because the exhaust gases enriched with fully and partially combusted oil were a cause of public annoyance.

This mixture of demands, faults, experience, and new findings led to the development of engine concepts with different but also with similar design elements. W-type, radial-type, single-shaft reciprocating piston, and rotary piston engines were only occasionally built for motor vehicles. The standard design was the inline engine with four, six, and eight cylinders. V-engines with 8, 12, or even 16 cylinders were also built. The "typical" engine consisted of a low crankcase with mounted single or twin cylinders. The cylinder and cylinder head were cast in one piece, and the upright valves were driven by the camshaft(s) mounted low in the crankcase. The crankshaft was suspended in bearing brackets with bearings after only every second or even third throw. Although the automatic intake valves had been replaced by driven valves, the valve timing still presented several problems: valves burned through, valve springs broke, and the noise level became high. For this reason, the smooth running Knight slide valve gear appeared to be superior at the time. Knight sleeve valve engines were built in England by Daimler Co., in Belgium by Minerva, in the United States by Willys, and in Germany by Daimler-Motorengesellschaft. But ultimately the valve timing system with its simpler design and operation was preferred.

In the United States, the personal vehicle changed from a leisure pastime of the wealthy to an article of daily use before World War I. In 1909 Henry Ford started production of the Model T (Tin Lizzie). By 1927 more than 15 million of these vehicles had been manufactured. In Europe the widespread use of motor vehicles (predominantly commercial vehicles) started during World War I. The mass production necessitated a certain unification and standardization of parts. Operation under the extreme conditions at the front mercilessly revealed design errors. The operation, maintenance, and repair of so many vehicles necessitated the training and qualification of the operating personnel. The development of the aircraft engines driven by the war gave powerful impetus to the improvement of motor vehicle engines in the early 1920s, and this applies to both the design (basic construction) and to the details of individual parts. Alongside upright valves with L- and T-shaped cylinder heads, engines with suspended valves and compact combustion chambers were built enabling higher compression ratios—a precondition for more power and lower consumption.

With the piston competition of 1921 organized by the German Imperial Ministry of Transport, the German engine industry quickly discovered the benefits of the light alloy piston compared with the cast iron piston. As a result, the engines of the 1920s were changed to light alloy pistons. In spite of numerous setbacks, this resulted in a significant increase in power and efficiency. The controlled piston enabled piston knock to be reduced and ultimately eliminated. In the early 1920s, there had been significant problems with the conrod bearings of the aircraft engines; they had reached the limits of their load-bearing capacity. The steel leaded bronze bearing, developed by Norman Gilmann at Allison (United States), provided the remedy. These bearings were first used in the diesel engines for commercial vehicles and later in high-performance car engines. The next step in development was the three-material bearing, consisting of a steel supporting shell, a leaded bronze intermediate layer, and a babbit metal running layer; they had been developed by Clevite in the United States.

Higher speeds and increased demands on the reliability of the engines required better engine lubrication.

This development advanced from wick and pot lubrication (lubrication from storage vessels) and lubrication with hand pumps. Consumers were supplied with lubricant, and by immersion of engine parts or by special scoop mechanisms were able to lubricate various components. This solution was followed by the forced circulation lubrication as was the method commonly used in aircraft engines. Two-stroke engines operated with mixture lubrication, i.e., by adding oil to the fuel.

Thermal siphon cooling did not allow sufficient heat to be dissipated from the parts subject to high thermal loads. As a result, forced circulation cooling was introduced.

Piston knock had become a power-limiting criterion in gasoline engines even during World War I. In 1921 Thomas Midgley, Jr. and T.A. Boyd in the United States discovered the effectiveness of tetraethyl lead (TEL) as an "antiknock additive." The addition of TEL to the fuel reduced the knock, permitted higher compression ratios, and resulted in higher efficiencies.

In the 1920s, a large number of small automobiles were developed whose engines had to be light, simple, and cheap. The two-stroke system with its high power density was an obvious choice. There were two mutually exclusive arguments in favor of this solution: high power density and design simplicity. Valveless, two-stroke engines with crankcase scavenging were suitable for motorbikes and small automobiles. The development of Schnürle reverse or loop scavenging from DKW was an important advancement compared to the cross-flow scavenging method because it permitted better scavenging of the cylinder. This method also enabled flat pistons to replace stepped pistons (with high thermal load). The "Roaring Twenties" heralded the era of the "great" Mercedes, Horch, Stöhr, and Maybach with eight-cylinder inline and 12-cylinder V-engines. In England there were Rolls Royce, Bentley, and Armstrong-Siddeley, in France Delage and Bugatti, and in the United States Pierce Arrow, Duesenberg, Auburn, Cord, Cadillac, and Packard.

Influenced by the development in aircraft engine construction, the engine builders started to turbocharge the engines with displacement-type fans (Roots blowers) that could be switched on and off, depending on the power requirements. The air cooling of the aircraft engines also appeared to offer benefits, but this proved to be far more difficult with motor vehicle engines because of the low vehicle speed and less favorable operating conditions. A pioneer of air cooling was the Franklin Mfg. Co. from the United States. This company manufactured an air-cooled six-cylinder inline engine even before World War I. General Motors also tried air cooling with a Chevrolet (Chevrolet copper engine), where the cooling fins were made of copper to improve the heat dissipation. Because of technical problems, however, this engine never went into mass production. In Europe air-cooled motor vehicle engines were also developed and built in the 1920s and 1930s. Commercial vehicle engines from Krupp and Phänomen, and car engines from Tatra and Ferdinand Porsche for the new Volkswagens were produced. The air-cooled opposed-cylinder (boxer) engine from Volkswagen became a synonym for reliability and sturdiness (first in the jeep and the amphibian vehicle and later in the "Beetle").

In the 1920s, a highly efficient accessory industry was built up in symbiosis with the automotive and engine industry. It served as a development center that united not only knowledge and experience in the various areas but also enabled more cost-effective production. This industry produced for several (or even all) of the engine manufacturers and thus was able to offer proven, more or less standardized, and inexpensive accessories such as pistons, bearings, radiators, carburetors, electrical equipment, and diesel injection systems. The motor vehicle development promoted and enhanced the construction and expansion of long-distance highways. Better roads permitted higher speeds and wheel loads. The traffic density increased slowly but surely. Operation of the engines was simplified, particularly by the electric starter introduced by Charles F. Kettering at General Motors that made starting not only easier but also safer. Ignition timing *(advance-retard)* and mixture composition *(lean-rich)* no longer had to be adjusted by the driver and were controlled automatically. In the 1930s, cars were increasingly driven during the winter months. Up to this point, many cars had not been used in winter. The year-round operation of vehicles required different oils depending on the outside temperature *(i.e., summer oil–winter oil)*. Consideration had to be given to the outdoor temperatures by controlling the coolant temperature, first by covering the radiator with leather blankets, then by using adjustable radiator shutters, and finally by using a thermostat to control the coolant temperature.

In the 1930s, alternative concepts were developed for vehicle engines. In Europe the steam engine was used in commercial vehicles (Foden, Sentinel, Leyland, and Henschel) to cut fuel costs and to achieve higher power outputs than were possible at the time with vehicle diesel engines. Even the thought of a cost-effective independent operation played a role in the development of these engines. In the United States, Doble automobiles powered by steam engines had become known for their quiet running. Despite the favorable tractive force curve, the steam engine ultimately failed to assert itself against the internal combustion engine. Commercial vehicles were operated with gas from an accumulator or with generator gas.

During World War II and in the time period thereafter, automobile engines had to be converted to generator gas because of the shortage of fuel (Fig. 1-2).

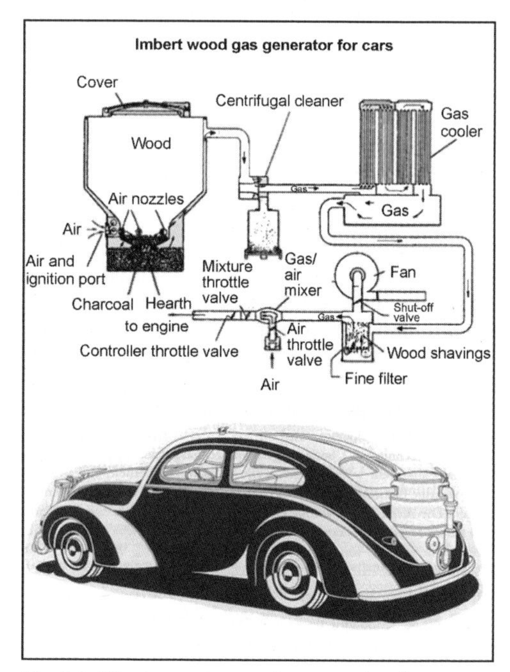

Fig. 1-2 Wood gas generator for car engines.[3]

Fuel injection using compressed air ("air injection") had been an obstacle to the use of diesel engines in the motor vehicle. In the early 1920s, intensive work on a "compressorless (airless) injection" was carried out in various areas. Based on the preliminary work conducted before and during World War I (L'Orange, Leissner), compressorless (airless) diesel engines for motor vehicles were developed—in Germany by MAN, Benz (later, Mercedes-Benz), and Junkers. On the basis of Acro patents, Robert Bosch developed complete fuel injection systems for vehicle diesel engines. The fuel injection pumps had helix and overflow control. But since direct fuel injection had not been mastered for motor vehicle engines with their wide speed range, indirect injection (prechamber and whirl chamber, air accumulator) was preferred. The diesel engine proved to be effective in heavy commercial vehicles and was increasingly used in light commercial vehicles and, ultimately, also in the automobile (Mercedes-Benz, Hanomag, Oberhänsli, Colt, Cummins, etc.). One of the first automobiles with a diesel engine was a Packard with a Cummins engine. In order to demonstrate the suitability of the diesel engine for cars, specially modified vehicles were entered in races. In 1930 a Packard Roadster powered by a Cummins diesel engine achieved a speed of 82 miles per hour (132 km/h) on the Daytona Beach racetrack in Florida. In Germany a Hanomag streamlined vehicle with a diesel engine reached 97 miles per hour (155.6 km/h); in 1978 it was a Mercedes-Benz C 111 that set the record at 197 miles per hour (316.5 km/h).

Despite the benefits of diesel engines, large gasoline engines were used to drive commercial vehicles. In the United States and in Germany, the 12-cylinder engine of the Maybach-Zeppelin powered omnibuses, fire engines, and half-track vehicles. The Opel Blitz commercial vehicle (with the six-cylinder inline engine of the Opel Admiral) became the standard vehicle of the German Wehrmacht. Small delivery vehicles (Tempo, Goliath, and Standard) were also driven by gasoline engines. Gradually, the diesel engine also broke into the automobile sector. The most common automobile powered by a diesel engine was the taxi.

During World War II, the development of automobile engines stagnated worldwide because other things now had priority. After the war, the production of prewar engines started again. In the United States car owners could afford large engines, and six-cylinder inline and eight-cylinder V-type engines were common. In Europe a large number of compact and subcompact cars were built with air and water-cooled two-stroke and four-stroke engines. German manufacturers included Gutbrod, Lloyd, Goliath, and DKW. France also had several manufacturers (Dyna-Panhard, Renault 4 CV, and Citroën 2 CV). England had Austin and Morris, and Italy had Fiat. To avoid the high fuel consumption of two-stroke gasoline engines due to the scavenging losses, Gutbrod and Goliath engines had a mechanical fuel injection system. During the "economic boom" in Germany, the demand for small cars fell, preventing the two-stroke engine from establishing itself in the automobile (with the exception of the Wartburg and Trabant cars, which were equipped with this engine type until the end of the 1980s in the German Democratic Republic).

In the early 1950s, many four-stroke car engines still had side valves, and the crankshaft rested in bearings only after every second throw. After this time period, engines started to show a more modern design: crankcase drawn down well under the middle of the crankshaft, bearings for the crankshaft after every throw, compact combustion chambers with overhead valves (OHV), bucket tappets with overhead camshafts (OHC) at higher engine speeds, and increased piston displacement. Mercedes-Benz successfully competed in races again; the engines of the Silver Arrows had a gasoline injection system and positive-closing valves [desmodromic (positive) control] derived from aircraft engines.

The economic upswing in the western world allowed prosperity to rise in general, and broad segments of the population could afford automobiles. As a result, vehicle production increased. There was plenty of opportunity for vehicle development. In Japan a new producer appeared on the world market that revolutionized automobile production with a high standard of quality, a reduction of the manufacturing depth, the splitting of production, improved assembly and development processes, and just-in-time delivery. Global competition necessitated even tighter cost control; the engines were produced in much larger quantities and were built with cost-effective production in mind. These engines required only simple maintenance and repair after production. Electronic data processing (EDP) started to establish itself in research and development in the 1970s. This practice utilized computer-aided design (CAD) to simulate engine processes using a technique called *finite elements method* (FEM). FEM resulted in rationalized, accelerated, and higher precision development.

The concept of the reciprocating piston engine was questioned time and again. At the end of the 1940s, Rover in England had developed a vehicle with a gas turbine engine (Fig. 1-3).

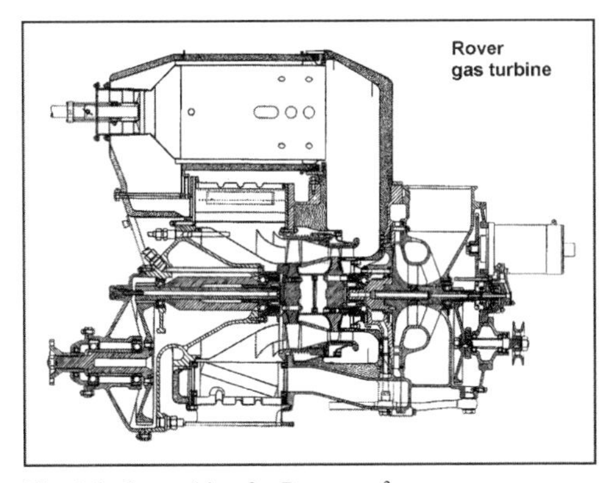

Fig. 1-3 Gas turbine for Rover car.[2]

High power density, compact design, a small number of moving parts, no free mass effects (hence smooth running), good pollution control (thanks to smoke-free exhaust), and good cold-starting properties are major points in favor of the gas turbine. However, it was discovered that gas turbines are not suitable for the low powers and operating conditions of automobile engines. The gap losses are too high, resulting in poor efficiency.

In the 1960s, the rotary piston engine of Felix Wankel developed by NSU (Fig. 1-4) appeared to offer an alternative to the reciprocating piston engine. Its kinematics, power density, and compact design are benefits compared with the reciprocating piston engines. However, the disadvantages outweighed the benefits: limited compression ratio, unfavorable combustion chamber, combustion with high constant pressure ratio, "late" combustion into the expansion phase, and problematical sealing of the combustion chamber led to high fuel consumptions and poor exhaust emission values. Only *Mazda* managed to build sporty vehicles with rotary piston engines with any degree of success.

The energy crises in the 1970s and the heightened public awareness of environmental problems led to a call for more economical engines with lower exhaust emissions. Starting from mechanical injection, a low-pressure fuel injection system with electronically controlled fuel metering was designed (much of the work done by Bosch). Despite the high development level of carburetor technology (twin carburetors, two-stage carburetors, constant-pressure carburetor), fuel injection quickly became the established solution. Electronics became more and more involved in the engine control. A common microprocessor-controlled electronic system with map storage controls ignition and mixture formation.

As measures inside the engine were no longer sufficient to reduce pollutant emissions to legally specified limits, three-way catalytic converters were employed that demanded precise control of the stoichiometric excess-air factor (lambda). Continuous measurement of the oxygen content in the exhaust gases using the lambda sensor allows the pollutant emissions to be reduced. An additional improvement is achieved with controlled exhaust gas recirculation (EGR).

Exhaust gas turbocharging as a means of increasing power and reducing consumption began to be employed in commercial vehicle engines from the 1960s. With increasing development levels, exhaust gas turbochargers could be "miniaturized" to such an extent that automobile gasoline engines could also be equipped. Since the fluid mechanics-based exhaust gas turbocharger and the reciprocating piston-powered internal combustion engine exhibited different operating behaviors, the "air supply" of the turbocharger and the "air demand" of the engine had to be balanced in order for these two machines to work together

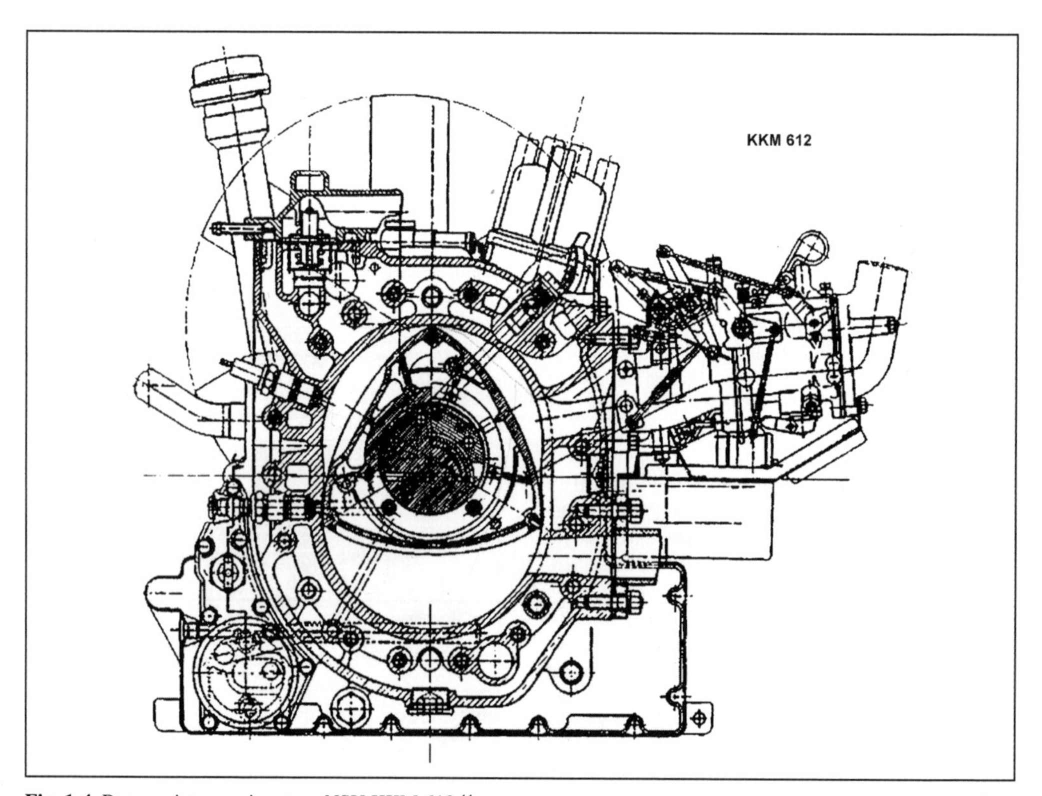

Fig. 1-4 Rotary piston engine, type NSU KKM 612.[11]

properly. Bypassing the turbine with part of the exhaust gas stream (waste gate control) and, for diesel engines, using variable turbine geometry were some of the initial advancements. A further improvement was achieved by cooling the charge air in the intercooler. As far as their response behavior for automobiles is concerned, mechanically powered turbochargers are at an advantage. Volkswagen developed a spiral turbocharger (G charger), and Mercedes-Benz uses Roots blowers for its "sporty" vehicle engines. An outstanding concept for turbocharging is the pressure wave supercharger (Comprex charger) from BBC, in which the energy from the exhaust gas is transferred dynamically directly to the charge air, i.e., without exhaust gas turbine and without a turbo compressor. Despite enormous development efforts, however, this principle has been unable to establish itself in the automotive industry (one of the reasons is because of high cost). Another drawback to this process is the fact that the exhaust gas temperature of gasoline engines is too high for proper operation.

The automotive diesel engine was at series-production maturity as early as the 1930s. It found an admittedly limited, but loyal, group of fans in the 1950s among taxi drivers and high-mileage drivers who attached less importance to sporty driving and more to low fuel consumption and long service life. Apart from the Mercedes-Benz and Borgward engines, Peugeot and Fiat were the only other diesel engine manufacturers at the time. In the 1970s, VW introduced an automotive diesel engine shortly followed by other German manufacturers (Opel, BMW, Ford, and Audi). The distributor injection pump arrived on the scene in the 1960s/1970s and proved to be ideal, particularly for the small injection volumes required by automotive diesel engines. Direct fuel injection offered significant consumption benefits for these engines, thus helping them to establish themselves for commercial vehicle engines in the 1960s. By the late 1980s, Ford had already equipped a delivery van with an engine using direct fuel injection. Audi was also in the process of delivering low-pollution car engines with direct injection around this time. Other companies followed suit, making direct injection a standard for diesel engines that still exists today. Turbochargers and intercoolers are becoming more and more common in diesel engines. High injection pressures are achieved with the unit injector system (UIS) and, more recently, with the accumulator injection system (common rail). In order to reduce the thermal load on the diesel engine pistons, they are cooled either by spraying the undersides of the pistons or by using cooling channels.

In the 1980s and 1990s, the charge cycle became one of the major developmental focal points. Flow coefficients and volumetric efficiency were improved with the multivalve technology. Further improvements came from variable valve timings and valve strokes as well as variable-configuration intake manifolds. The development trend is now towards electromagnetically actuated and controlled valves. As a result, the intake cycle can be dethrottled reducing one of the major problems with this engine type. The direct injection into the cylinders of gasoline engines results in higher performance, reduced pollutant emissions, and lower consumption.

On the engine side, the fuel consumption has been reduced by means of a whole range of measures: smaller dimensions and weights of the engine (downsizing), roller tappets rather than sliding tappets in the control, low-viscosity oils (that demand controlled operation of blowers and pumps), etc.

Increasing engine speeds of the five-cylinder inline and V-6 engine designs demanded measures to improve the machine dynamics of the engines. Differentials ensure the desired smoothness of running, as do rotational oscillation dampers.

Limited resources and generally higher pollutant emissions are driving the search for different drive concepts. On the one hand, it is a question of finding a substitute for crude oil, and, on the other, of relieving the environment. A solution strongly favored by politicians for a time was the use of regenerative energies in the form of vegetable oil (rape oil methyl ester). The rape growing areas is not sufficient for an adequate supply of fuel (quite apart from the ecological problems associated with monocultures), nor is it technically expedient to replace mineral oils in motor vehicle engines.

Another development is aimed at the use of hydrogen as fuel. Hydrogen, in conventional reciprocating piston engines as well as in fuel cells, can help to alleviate the pollutant situation. On the downside, hydrogen is difficult to produce. It has to be "generated" either by reverse electrolysis that requires a great deal of energy or by converting methanol or gasoline (not a good method for conserving resources). A feasible scenario lies in the increased use of natural gas-powered engines. This solution would ensure the energy supply with ever-decreasing resources of crude oil while preparing the way for the advent of a gas technology using hydrogen.

Bibliography

[1] Robert Bosch GmbH [eds.], Bosch und die Zündung, in Bosch-Schriftenreihe Folge 5, Stuttgart, 1952.

[2] Bussien, R. [ed.], Automobiltechnisches Handbuch, 18, Aufl., Technik Verlag H. Cram, 1965.

[3] Eckermann, E., Alte Technik mit Zukunft (Hrsg. Deutsches Museum), R. Oldenbourg, Munich, 1986.

[4] von Fersen, O. [ed.], Ein Jahrhundert Automobiltechnik—Personenwagen, VDI-Verlag, Düsseldorf, 1986.

[5] von Frankenberg, R., and M. Mateucci, Geschichte des Automobils, Siegloch, Künzelsau, 1988.

[6] Kirchberg, P., Plaste, Bleche und Planwirtschaft, Die Geschichte des Automobilbaus in der DDR, Nicolasche Verlagsbuchhandlung, Berlin, 2000.

[7] Krebs, R., 5 Jahrtausende Radfahrzeuge, Springer, Berlin, 1994.

[8] Sass, F., Geschichte des deutschen Verbrennungsmotorenbaues, Springer, Berlin, 1962.

[9] Pierburg, Vom Docht zur Düse, Ausgabe 8/1979, Fa. Pierburg, Neuss, 1979.

[10] Zima, S., Kurbeltriebe, 2, Aufl., Vieweg, Wiesbaden, 1999.

[11] ATZ 69 (1967), 9, pp. 279–284.

2 Definition and Classification of Reciprocating Piston Engines

2.1 Definitions

Piston machines are machines in which energy is transferred from a fluid (a gas or a liquid) to a moving displacer (e.g., a piston) or from the piston to the fluid.[1,2] They are thus part of the category of fluid energy machines that, as driven machines, absorb mechanical energy in order to increase the energy of the conveyed fluid. In drive machines, on the other hand, mechanical energy is released in the form of useful work at the piston or at the crank mechanism.

The occurrence of a periodically changing working chamber as a result of the motion of the displacer (piston) is characteristic of the manner of operation of piston engines. One differentiates between reciprocating displacer engines and rotary displacer engines depending on the nature of the displacer's movement. In reciprocating piston engines, the displacer takes the form of a cylindrical piston that moves between two extreme positions, the "dead centers," in a cylinder. The term "piston" is also frequently applied to noncylindrical displacers. In rotary piston engines, a rotating displacer is normally responsible for varying the working chamber.

Combustion engines are machines in which chemical energy is converted to mechanical energy as a result of the combustion of an ignitable mixture of air and fuel. The best-known combustion engines are internal combustion engines and gas turbines. Figure 2-1 provides an overview.

Internal combustion engines are piston engines. One differentiates between reciprocating piston engines (featuring oscillating piston movement) and rotary piston engines (featuring rotating piston movement) depending on the geometry of the gastight, changing working chamber and on the type of piston motion.[7] Rotary piston engines are, for their part, subclassified again into rotary engines (featuring an internal and an external rotor with purely rotary motion about fixed axes) and planetary rotary engines (that feature an internal rotor, the axis of which describes a circular motion). Figure 2-2 shows the differing working principles. Only the Wankel engine, a planetary piston engine, has achieved any significance.

It is also necessary, depending on the type of working process, to differentiate between combustion engines with

Type of working process		Open process				Closed process	
		Internal combustion				External combustion	
		Combustion gas = working fluid				Combustion gas $\neq$ working fluid	
						Change of phase of the working fluid	
						No	Yes
Type of combustion		Cyclical combustion			Continuous combustion		
Type of ignition		Auto-ignition	Supplied ignition				
Machine type	Engine	Diesel	Hybrid	Gasoline	Rohs[4]	Stirling[5]	Steam[6]
	Turbine	—	—	—	Gas	Superheated steam	Steam
Mixture type		Heterogeneous		Homogeneous (heterogeneous)	Heterogeneous		
		(in the combustion chamber)			(in the continuous flame)		

Fig. 2-1 The classification of combustion engines, after Ref. [3].

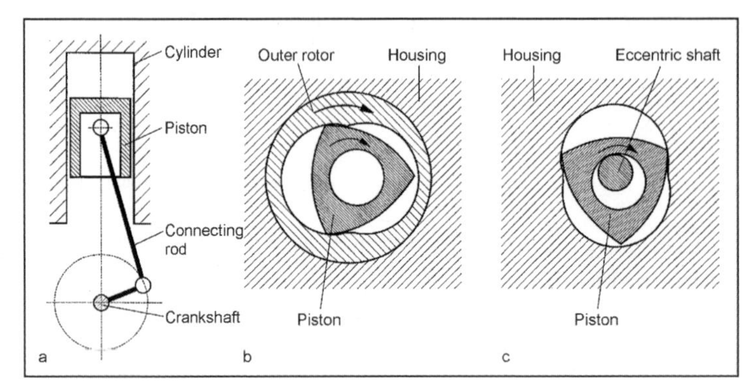

Fig. 2-2 The working principles of reciprocating piston engines, rotary engines, and planetary piston engines. (a) Trunk piston engine. (b) Rotary piston engine; power-yielding external rotor with epitrochoid internal contour and internal rotor as a sealing element. (c) Planetary piston engine (Wankel engine): Housing with epitrochoid internal contour and power-yielding internal rotor that rotates eccentrically around a pinion and seals simultaneously.

internal combustion and those with external combustion. In engines featuring internal combustion, the working fluid (air) is simultaneously the source of the oxygen necessary for combustion. Combustion of the fuel fed produces waste gas, which must be replaced in a "gas exchange" cycle prior to every working cycle. Combustion is therefore cyclical, differentiation being made between gasoline, diesel, and hybrid engines, depending on the combustion process.

In the case of external combustion engines (such as the Stirling engine, for example), the heat produced outside the working chamber as a result of continuous combustion is transferred to the working fluid. This permits a closed-circuit working process and the use of any fuel.

Only reciprocating piston engines featuring internal, cyclical combustion is examined from this point on.

2.2 Potentials for Classification

The potentials for the classification of reciprocating engines are extremely diverse because of the complex interrelationships involved. Internal combustion reciprocating engines[8] can be differentiated by their

- Combustion process
- Fuel
- Working cycle
- Mixture generation system
- Gas exchange control system
- Charging system
- Configuration

Further differentiating features may take the form of the[9,10]

- Ignition system
- Cooling system
- Load-adjustment system
- Application
- Speed and output graduations

A number of differentiating features are currently only of historical significance, however.

2.2.1 Combustion Processes

Among the combustion processes, differentiation is made primarily between the Otto cycle and the diesel cycle. Hybrid engines exhibit characteristics of both the Otto cycle and the diesel cycle.

The gasoline engine is a combustion engine in which combustion of the compressed fuel + air mixture is initiated by means of synchronized extraneous ignition. In the diesel engine, on the other hand, the liquid fuel injected into the combustion chamber ignites on the air charge after this has previously been heated, by means of compression, to a temperature sufficiently high to initiate ignition.[8]

In the case of hybrid engines, one differentiates between engines featuring charge stratification and multi-fuel engines.[3]

2.2.2 Fuel

Gaseous, liquid, and solid fuels can be combusted in combustion engines:

- Gaseous fuels: Methane, propane, butane, natural gas (CNG), generator, blast furnace, biogas (sewage treatment and landfill gas), and hydrogen
- Liquid fuels:
  Light liquid fuels: Gasoline, kerosene, benzene, alcohols (methanol, ethanol), acetone, ether, liquefied gases (LNG, LPG)
  Heavy liquid fuels: Petroleum, gas oil (diesel fuel), fatty-acid methyl esters (FAME), and, primarily in Europe, rape-seed[56] methyl esters (RME), also referred to as "biodiesel," vegetable oils, heavy fuel oils, and marine fuel oil (MFO)
  Hybrid fuels: Diesel + RME, diesel + water, and gasoline + alcohol
- Solid fuels: Pulverized coal

2.2.3 Working Cycles

In the field of working cycles, differentiation is made between four-stroke and two-stroke processes. Common to both is the compression of the charge (air, or a fuel vapor + air mixture) in the first step (stroke) by the reduction of the working chamber and ignition occurring shortly before the reversal of piston motion. Also, combustion associated with an increase in pressure up to the maximum cylinder pressure and the expansion of the working gas in the subsequent stroke, during which work is applied to the piston, is similar in both processes.

The four-stroke process requires two further strokes in order to remove the combustion gas from the working chamber by means of displacement and to fill the working chamber with a fresh charge by means of natural induction (normal aspiration).

In the two-stroke process, gas exchange occurs in the vicinity of bottom dead center as a result of expulsion of the combustion gases by the fresh charge with only a slight change in the working volume, with the result that the complete stroke is not exploited for compression and expansion. An additional scavenging blower is necessary for the scavenging process.

2.2.4 Mixture Generation

Combustion engines can be differentiated in terms of their type of mixture generation:

- External mixture generation: Formation of the fuel-air mixture in the inlet system
- Internal mixture generation: Formation of the mixture in the working chamber

on the basis of the quality of mixture generation:

- Homogeneous mixture generation: Carburetor and intake manifold injection in the case of the gasoline engine, or gasoline direct injection during the induction stroke
- Nonhomogeneous mixture generation: Injection at extremely short intervals in the diesel engine and in gasoline engines with gasoline direct injection (GDI)

and on the basis of the location of mixture generation:

- Direct injection into the working chamber in the case, for example, of DI diesel engines and GDI engines. Injection may be air-directed, jet-directed, or wall-directed.
- Indirect injection into a subsidiary chamber, such as antechamber, swirl-chamber, and air-chamber diesel engines.
- Intake manifold injection (in gasoline engines).

2.2.5 Gas Exchange Control

Valve, port, and slide-valve timing systems are used for control of the gas exchange.

In the case of valve timing mechanisms, one differentiates between overhead and side-actuated engines.[8] The overhead-actuated engine has overhead valves; i.e., the closing movement of the valves occurs in the same direction as the movement of the piston toward top dead center (TDC). The side-actuated engine, on the other hand, has vertical valves, and closure of the valves occurs in the same direction as the movement of the piston toward bottom dead center (BDC).

Only the OHV arrangement, with overhead valves located in the cylinder head, is used in modern four-stroke engines. The camshaft may be located in the cylinder head or in the crankcase.

Two-stroke engines mainly employ port-based timing systems (slots, or "ports" in the cylinder sleeve, with the piston acting as a slide valve), and also bevel slide valves, disk valves, slide valves, and diaphragm timing systems in individual cases. In addition, a valve timing system (an exhaust valve in many cases) is also used in some recent motor-car and large marine engine developments.

2.2.6 Supercharging

In a normally aspirated engine, the fresh charge (air or mixture) is drawn into the cylinder by the working piston (natural aspiration).

Supercharging enlarges the quantity of the charge as a result of precompression; a supercharger conveys the fresh charge into the cylinder. The primary aims of supercharging are the enhancement of power and torque output and the reduction of fuel consumption and exhaust gas emissions.

Figure 2-3 shows an overview of possible types of supercharging (after Ref. [11]).

The most widely used and effective variant in practice is self- or auto-supercharging, using a compressor:

- Mechanical supercharging: The compressor is driven directly by the engine.
- Exhaust turbo-supercharging: A turbine (exhaust turbine) powered by the engine exhaust drives the compressor.

Processes without a compressor, which exploit the gas-dynamic processes in the intake and exhaust systems to increase the charge, are also used.

2.2.7 Configuration

Numerous variants of cylinder arrangement have been suggested in the more than 120-year history of the internal combustion engine. Only a few standard configurations have stood the test of time.[9,10]

Starting from the single-cylinder engine, the number of cylinders selected can range up to as high as 12 in the case of vehicle engines. Aircraft engines with up to 28, or even as many as 48 cylinders, and high-performance engines with up to 56 cylinders have also been constructed.

There are numerous possible combinations for the cylinder arrangement, some of which are identified in a self-explanatory manner by letters. Figure 2-4 shows a selection of possible cylinder arrangements and configurations.

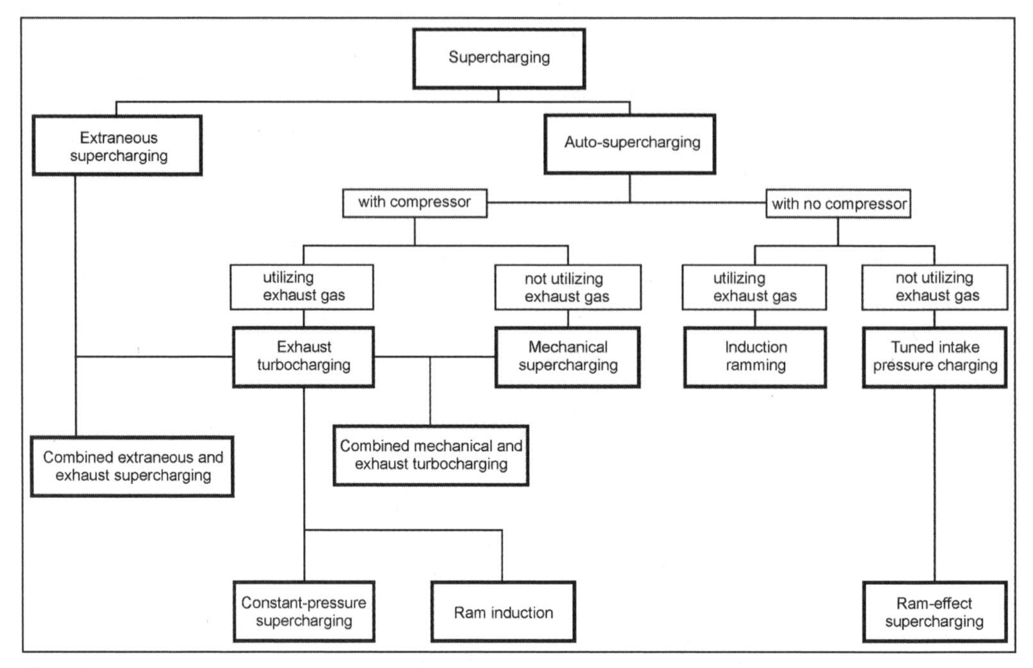

Fig. 2-3 Various supercharging methods, after Ref. [11].

The following are presently of significance:

- The inline engine (one bank of cylinders and one crankshaft).
- The V-engine (two banks of cylinders and one crankshaft): Two connecting rods are coupled to each crank pin. Common V-angles are 45°, 60°, 90°, and 180°. The VR engine[12] has a V-angle of 15°, the crankshaft having a separate crank pin for each connecting rod.
- The W-engine (three banks of cylinders and one crankshaft): Three connecting rods are connected in each case to one crank pin. A V-engine consisting of two VR banks is referred to as a V-VR engine, or also as a W-engine.[12]
- The boxer (flat-opposed) engine: Unlike the 180° V-engine, each connecting rod is connected to a separate crank pin.[13]

The crank mechanism has proven its value in engine design. Trunk piston engines and crosshead engines may be differentiated as variants. Slider crank mechanisms and cam engines are also described in the relevant literature, as are crankshaftless engines (curved-plate, curved-track, and swash-plate engines).[9]

Single- and double-acting engines can be differentiated according to their manner of action, depending on whether the combustion gases act on only one side or on both sides of the piston. The double-piston engine has two pistons to each combustion chamber, the pistons being arranged either opposing (opposed-piston engine) or concurrent (U-piston engine).

Vertical, horizontal, and overhead engines are differentiated on the basis of the location of the cylinder axis, and overhead-actuated and side-actuated engines by the location of the timing mechanism.

2.2.8 Ignition

The fuel-air mixture may be ignited by means of supplied ignition or compression ignition:

- Supplied ignition (gasoline engine): An electrical spark ignites the mixture in the cylinder (spark ignition).
- Autoignition (diesel engine): The fuel injected ignites spontaneously in the air heated by compression in the cylinder (compression ignition).

2.2.9 Cooling

In view of the high temperatures that occur, the combustion engine needs to be cooled, in order to protect its components and the lubricating oil. It is necessary to differentiate between direct and indirect engine cooling.

Direct cooling is accomplished using air (air cooling) either with or without the assistance of a fan.

In the case of indirect cooling, the engine is cooled with a mixture of water, antifreeze, and corrosion inhibitors, or with oil (liquid cooling). Removal of heat to the environment is accomplished via a heat exchanger arrangement. One differentiates between evaporative, recirculating, once-through, and hybrid cooling.

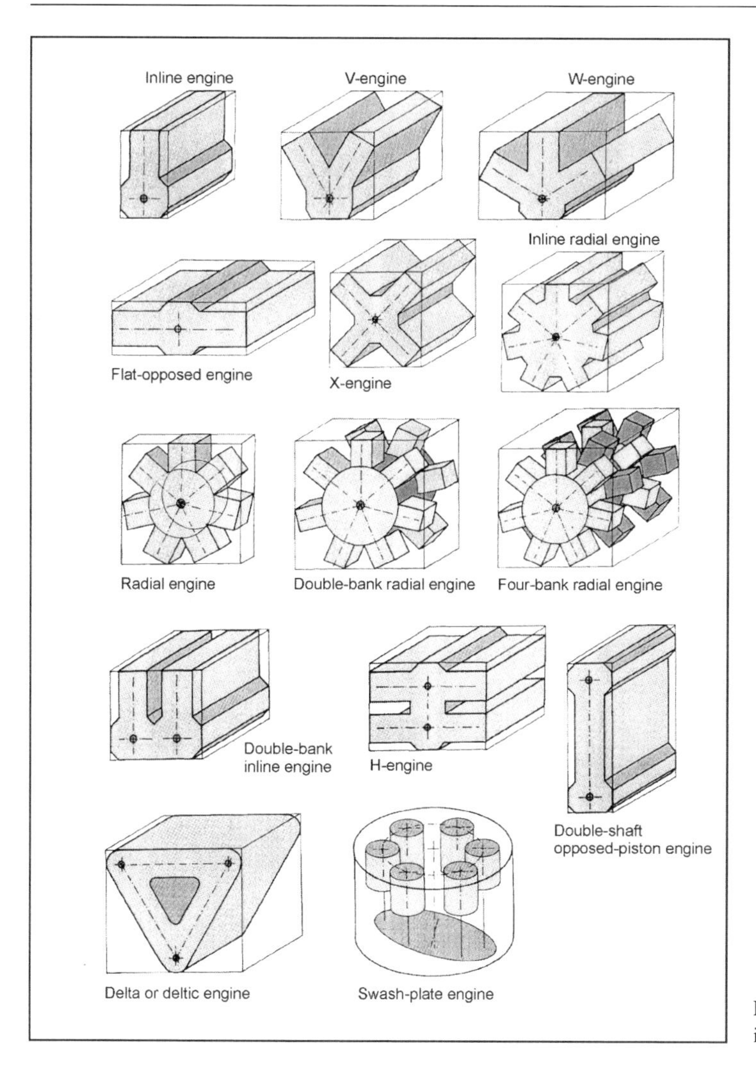

Fig. 2-4 Cylinder arrangements in reciprocating piston engines.

2.2.10 Load Adjustment

Motor output P

$$P = M \cdot \omega = M \cdot 2 \cdot \pi \cdot n \tag{2.1}$$

can be matched to the power requirement by modifying both speed n and torque M (load). In the context of load adjustment, it is necessary to differentiate between

- Quantity control and filling control: With an approximately constant air ratio λ, a throttling element (butterfly, rotary disk, slide, or other valve) controls the quantity of mixture that flows into the cylinder (conventional gasoline engine).
- Quality control: In diesel engines, and in GDI gasoline engines in certain operating ranges, the fuel is metered in as required. The injection flow is varied, with a practically constant flow of air (variable air ratio λ).

2.2.11 Applications

A number of examples of the use of combustion engines:

- Land-based vehicles: Road vehicles (motorcycles, automobiles, buses, commercial vehicles), off-road vehicles, and rail vehicles
- Marine craft: Boats, inland, coastal, and ocean-going ships
- Aircraft: Airplanes and airships
- Agricultural machines and vehicles: Tractors, harvesting machines
- Commercial and industrial applications: Construction machines, handling, conveying, and lifting equipment, tugs, and tractors
- Stationary engine installations: Engine-powered generating plants, unit-type cogeneration plants (UCPs), electrical generating sets, emergency-power sets, and supply systems.

2.2.12 Speed and Output Graduations

An extremely broad range of combustion engine speeds and outputs are used. Power ranges extend from model engines of 0.1 kW up to large-scale commercial installations of as much as 50 000 kW. An engine's speed range also defines its output and size.

The following can be differentiated by their speed[1]:

- Low-speed engines used, for example, in ships (60 to 200 rpm, in the case of diesel engines)
- Medium-speed engines (200 to 1000 rpm in diesel engines, maximum speed <4000 rpm in gasoline engines)
- High-speed engines, for use, for example, in motorcars (maximum speed >4000 rpm in diesel engines and >4000 rpm in gasoline engines).

Engines for sports and racing vehicles reach speeds of up to 22 000 rpm.

Bibliography

[1] Beitz, W., and K.-H. Grote [eds.], Dubbel—Taschenbuch für den Maschinenbau, 20th edition, Springer, Heidelberg, 2001.

[2] Kleinert, H.-J. [ed.], Taschenbuch Maschinenbau—Bd. 5 Kolbenmaschinen, Strömungsmaschinen, 1st edition, Verlag Technik, Berlin, 1989.

[3] Robert Bosch GmbH [eds.]: Kraftfahrtechnisches Handbuch, 23, Aufl., Braunschweig, Vieweg, Wiesbaden, 1999.

[4] Rohs, U., Kolbenmotor mit kontinuierlicher Verbrennung, Offenlegungsschrift DE 199 09 689 A 1, published 07.09.2000.

[5] Werdich, M., and K. Kübler, Stirling-Maschinen, Grundlagen-Technik-Anwendung, 7th edition, Ökobuch, Staufen, 1999.

[6] Buschmann, G., et al., Zero Emission Engine—Der Dampfmotor mit isothermer Expansion, in MTZ 61, 2000, Volume 5, pp. 314–323.

[7] Bensinger, W.-D., Rotationskolben—Verbrennungsmotoren, Springer, Berlin, Heidelberg, 1973.

[8] DIN Deutsches Institut für Normung [eds.], DIN 1940: Verbrennungsmotoren-Hubkolbenmotoren-Begriffe, Formelzeichen, Einheiten, Beuth, Berlin, 1976.

[9] van Basshuysen, R., and F. Schäfer, Shell Lexikon Verbrennungsmotoren, Vieweg, Wiesbaden, 1995–2001 (Supplement to ATZ/MTZ).

[10] Beier, R., et al., Verdrängermaschinen, Part II: Hubkolbenmotoren, TÜV Rheinland, Cologne, 1983.

[11] DIN Deutsches Institut für Normung [eds.], DIN 6262: Verbrennungsmotoren-Arten der Aufladung-Begriffe, Beuth, Berlin, 1976.

[12] Braess, H.-H., and U. Seiffert [eds.], Vieweg Handbuch Kraftfahrzeugtechnik, Vieweg, Braunschweig, Wiesbaden, 2000.

[13] Zima, S., Kurbeltriebe, 2nd edition, Vieweg, Braunschweig, Wiesbaden, 1999.

3 Characteristics

Engine characteristics[1,3,5] serve the developers, designers, and users of internal combustion engines as important aids in designing the fundamental dimensions, assessing engine power and consumption, and evaluating and comparing different engines. A distinction is made between *engine* characteristics such as stroke, bore, piston displacement, and compression ratio and *operating* characteristics such as power, torque, engine speed, mean pressure, volumetric efficiency, and fuel consumption.

3.1 Piston Displacement and Bore-to-Stroke Ratio

Piston Displacement

The piston displacement or swept volume V_h for an engine cylinder is the distance traveled by the piston during one piston stroke from BDC to TDC.

$$V_H = V_h \cdot z = \frac{\pi \cdot d_K^2}{4} \cdot s \cdot z \tag{3.1}$$

where

s = Piston stroke
d_K = Piston diameter or cylinder bore
V_h = Piston displacement for one cylinder
V_H = Total piston displacement of the engine
z = Number of cylinders

Calculation of Stroke and Piston Displacement from the Crankshaft Position; see Fig. 3-1

$$s_\alpha = r + l - x = r + l - r \cdot \cos\alpha - l \cdot \cos\beta \tag{3.2}$$

where
r = Crank radius
l = Connecting rod length

Between the crank offset α and the connecting rod sweep angle β (connecting rod offset), we have the relationship

$$l \cdot \sin\beta = r \cdot \sin\alpha \tag{3.3}$$

$$\beta = \arcsin\left(\frac{r}{l} \cdot \sin\alpha\right) \tag{3.4}$$

Allowing for

$$\cos\beta = \sqrt{1 - \sin^2\beta} = \sqrt{1 - (r/l)^2 \cdot \sin^2\alpha} \tag{3.5}$$

and inserting the connecting rod ratio

$$\lambda_s = \frac{r}{l} \tag{3.6}$$

we obtain the equation for the piston stroke:

$$s_\alpha = r \cdot \left(1 + \frac{l}{r} - \cos\alpha - \frac{l}{r} \cdot \sqrt{1 - (r/l)^2 \cdot \sin^2\alpha}\right) \tag{3.7}$$

$$s_\alpha = r \cdot \left[(1 - \cos\alpha) + \frac{1}{\lambda_s} \cdot (1 - \sqrt{1 - \lambda_s^2 \cdot \sin^2\alpha})\right] \tag{3.8}$$

or

$$s_\alpha = r \cdot f(\alpha) \tag{3.9}$$

where

$f(\alpha)$ = Stroke function

The connecting rod ratio λ_s for car engines normally lies in the range from 0.2 to 0.35. It is difficult to work with the equation for the piston travel, particularly when piston speed or piston acceleration is to be calculated. An approximation equation can normally be used for simplicity in which the radical of a power series (MacLaurin series) is developed:

$$\sqrt{1 - \lambda_s^2 \cdot \sin^2\alpha} = 1 - \frac{1}{2} \cdot \lambda_s^2 \cdot \sin^2\alpha - \frac{1}{8} \cdot \lambda_s^4 \cdot \sin^4\alpha - \frac{1}{16} \cdot \lambda_s^6 \cdot \sin^6\alpha - \ldots \tag{3.10}$$

Because of the values of $\lambda_s \approx 0.2$ to 0.35, the 3rd term is already very small compared with the 1st term (1) so that

$$\sqrt{1 - \lambda_s^2 \cdot \sin^2\alpha} \approx 1 - \frac{1}{2} \cdot \lambda_s^2 \cdot \sin^2\alpha \tag{3.11}$$

can be assumed.

Using the trigonometric function

$$\sin^2\alpha = \frac{1}{2} \cdot (1 - \cos 2\alpha) \tag{3.12}$$

we then obtain for the piston travel s_α

$$s_\alpha \approx r \cdot \left[(1 - \cos\alpha) + \frac{1}{\lambda_s} \cdot \left(1 - 1 + \frac{1}{2} \cdot \lambda_s^2 \cdot \sin^2\alpha\right)\right] \tag{3.13}$$

$$s_\alpha \approx r \cdot \left[(1 - \cos\alpha) + \frac{1}{2} \cdot \lambda_s \cdot \frac{1}{2} \cdot (1 - \cos 2\alpha)\right] \tag{3.14}$$

$$s_\alpha \approx r \cdot \left[1 - \cos\alpha + \frac{\lambda_s}{4} - \frac{\lambda_s}{4} \cdot \cos 2\alpha\right] \tag{3.15}$$

For the momentary combustion chamber volume V_α we obtain

$$V_\alpha = V_c + A_K \cdot s_\alpha \tag{3.16}$$

where

V_C = Compression ratio (see Section 3.2)
A_K = Piston surface area

We thus obtain

$$V_\alpha = V_c + A_K \cdot r \cdot \left[1 - \cos\alpha + \frac{1}{4} \cdot \lambda_s \cdot (1 - \cos 2\alpha)\right] \tag{3.17}$$

3.2 Compression Ratio

The compression ratio is defined as the quotient of the maximum and minimum cylinder volumes: The maximum cylinder volume is when the piston is at BDC. When the piston is in TDC position, the volume is minimal and is referred to as compression or dead volume.

The compression volume is made up of the combustion chamber volume of the cylinder head, the valve pockets in the piston, a piston recess, and the top land volume up to the upper compression ring. Compression volume and piston displacement can be determined by gauging in liters.

Figure 3-1 shows the swept volume and compression volume schematically.

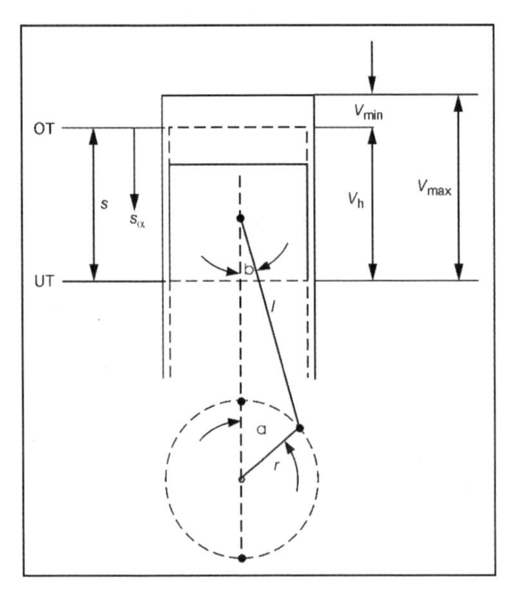

Fig. 3-1 Swept volume and compression ratio.

For the compression ratio of a four-stroke engine we thus obtain

$$\varepsilon = \frac{V_{max}}{V_{min}} = \frac{V_h + V_c}{V_c} \tag{3.18}$$

$V_C = V_{min}$ = Compression volume or dead volume

The compression ratio of a spark ignition (SI) engine is limited by the knock and by autoignition.

In spark ignition engines with direct fuel injection, an increase in the compression ratio is possible because of the improved internal cooling by the internal mixture preparation. This gives them a higher efficiency compared to the spark ignition engine with intake manifold injection.

For the diesel engine, the compression ratio has to be selected so as to ensure reliable starting when cold. In general, the thermodynamic efficiency increases with increasing compression ratio. An excessively high compression ratio, however, results in a decrease in the effective efficiency at full load due to the sharply increasing friction forces. In part-load operation, a high compression ratio has a positive effect on the efficiency. Irrespective of that, the peak pressure that is limited by the material strength limits the compression ratio that can be achieved in practice.

Figure 3-2 shows the influence of compression ratio on the effective efficiency and on the mean effective pressure in a spark ignition engine during full-load operation. The ignition timing was set to maximum torque. The increase in efficiency up to a compression ratio of approximately 17:1 is clearly seen. The efficiency then drops, in this case because of increasing frictional forces and a less favorable combustion chamber form because of increasing percentages of quench areas.

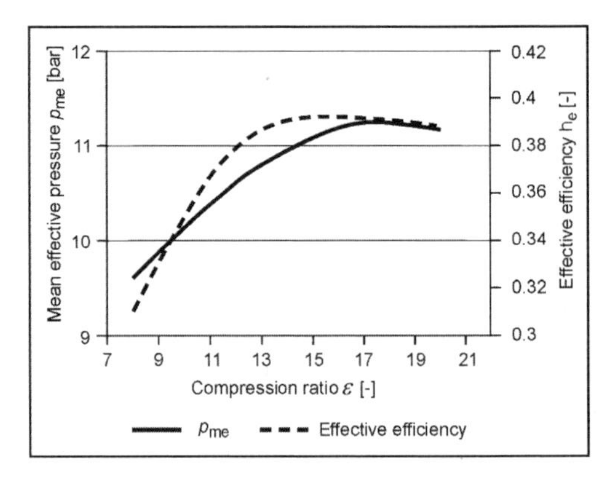

Fig. 3-2 Influence of compression ratio on mean effective pressure and effective efficiency at full load of a gasoline engine.[2]

With increasing compression, the NO_x and HC emissions initially continue to increase. The nitrous oxides rise because of the increased combustion temperatures in the combustion chamber, and the HC emissions rise because of the greater splitting of the combustion chamber (larger relative proportion of gaps) and the increase in the ratio of combustion chamber surface area to combustion chamber volume (surface-to-volume ratio). In order to avoid this, combustion chambers must be designed as compactly as possible. With increasing compression, the exhaust gas temperature also drops because of the better efficiency so that postreactions of unburned hydrocarbons and carbon monoxide in the exhaust system are prevented. At the same time, however, an increase in compression results in a better lean-off capability and allows the ignition to be retarded due to the faster combustion. This enables the HC and NO_x emissions to be further reduced.

In two-stroke engines with slot control, a distinction is made between the geometric compression ratio ε and the effective compression ratio ε'. Figure 3-3 shows the difference. The effective compression begins only after the

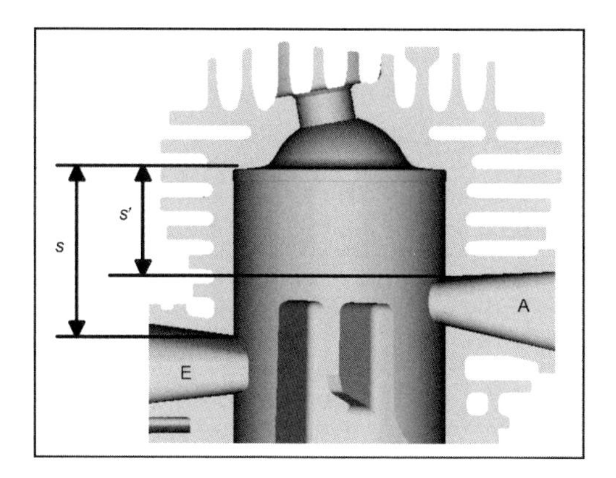

Fig. 3-3 Geometric and effective compression ratio of the two-stroke engine.

piston has closed the intake and exhaust slots. The effective compression ratio is calculated as

$$\varepsilon' = \frac{V'_h + V_c}{V_c} \tag{3.19}$$

where

$$V'_h = \frac{\pi \cdot d_K^2}{4} \cdot s' \tag{3.20}$$

V_h = Dead volume above the slots
s' = Residual stroke above the slots

Figure 3-4 shows the possible ranges of the compression ratios for common engines.

New developments are geared to varying the compression ratio according to the operating point while the engine is running. In the SI engine, the compression ratio is selected for optimum efficiency in part-load operation, whereas in full-load operation, the compression ratio is reduced to prevent knock. In the diesel engine, the compression ratio is limited by the maximum cylinder pressure (because of the component load). For diesel engines, the geometric compression ratio for full load can be optimally selected between a high efficiency and a maximum component load. For reliable cold starting, the compression ratio is set as high as possible.

3.3 Rotational Speed and Piston Speed

Rotational Speed

$$n = \frac{\text{Number of crankshaft revolutions}}{\text{Time}} \tag{3.21}$$

Angular Velocity

$$\omega = 2 \cdot \pi \cdot n \tag{3.22}$$

Piston Speed

The piston speed as a function of the crank angle is determined by the temporal derivation from the equation of the movement of the crank drive together with the angular velocity.

$$\dot{s}_\alpha = \frac{ds_\alpha}{dt} = \frac{ds_\alpha}{d\alpha} \cdot \frac{d\alpha}{dt} \tag{3.23}$$

$$\frac{d\alpha}{dt} = \omega = 2 \cdot \pi \cdot n \tag{3.24}$$

Consequently,

$$\dot{s}_\alpha = \omega \cdot \frac{ds_\alpha}{d\alpha} \approx \omega \cdot r \cdot \left[\sin \alpha + \frac{1}{2} \cdot \lambda_s \cdot \sin 2\alpha\right] \tag{3.25}$$

With increasing piston speed, the

- Mass forces
- Wear
- Flow resistance during intake
- Friction
- Noise

also increase. The maximum permissible mass forces, in particular, limit the piston speed and hence the maximum

Engine type	ε		Limited by
	From	To	
Two-stroke SI engine	7.5	10	Autoignition
SI engine (two-valve)	8	10	Knock, autoignition
SI engine (four-valve)	9	11	Knock, autoignition
Direct injection SI engine	11	14	Knock, autoignition
Diesel (indirect injection)	18	24	Loss of efficiency at full load, component load
Diesel (direct injection)	17	21	Loss of efficiency at full load, component load

Fig. 3-4 Compression ratio of modern engines.

Engine type	Max. speed [rpm] approx.	Mean piston speed [m/s] approx.
Racing engine (Formula 1)	18 000	25
Small engines (two-stroke)	20 000	19
Motorcycle engines	13 500	19
Car SI engine	7500	20
Car diesel engines	5000	15
Truck diesel engines	4200	14
Larger high-speed diesel engines	2200	13
Medium high-speed engines (diesel)	1200	10
Crosshead engines (two-stroke diesel)	200	8

Fig. 3-5 Maximum rotational speed and mean piston speed at rated revs of modern engines.

rotational speed. On engines with internal mixture formation, i.e., diesel engines and SI engines with direct injection, the rotational speed is additionally limited by the time necessary for the mixture formation. In diesel engines, this is one of the reasons for the significantly lower maximum revs compared with an SI engine of a similar size.

Mean Piston Speed

$$c_m = 2 \cdot s \cdot n \tag{3.26}$$

The mean piston speed is a measure for comparing the drives of various engines. It provides information on the load on the sliding partners and indications of the power density of the engine.

Figure 3-5 lists rotational speeds and piston speeds of modern engines for orientation.

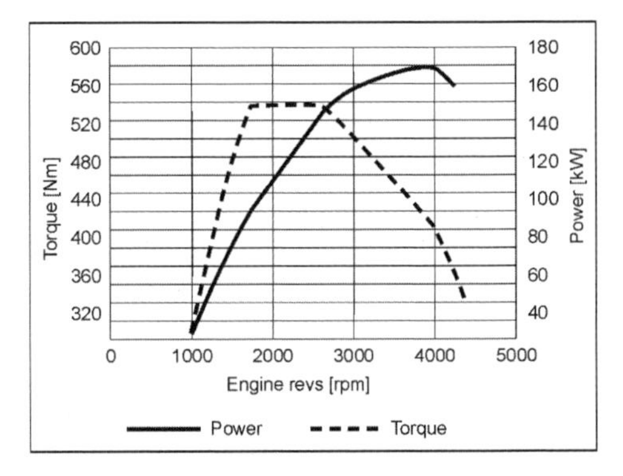

Fig. 3-6 Power and torque curves for a turbocharged diesel engine.[6]

3.4 Torque and Power

The power at any working point of the engine is calculated from the torque and engine revs:

$$P_e = M_d \cdot \omega = M_d \cdot 2 \cdot \pi \cdot n \tag{3.27}$$

According to this equation, an increase in power can be achieved by increasing the rotational speed or the torque. Both are subject to certain limits (see Chapter 3.3).

As an example, Fig. 3-6 shows motor characteristics of a diesel engine. The maximum torque and the maximum power are each plotted against the engine revs. The maximum power is not necessarily always achieved at the maximum engine revs. Not only the peak values for power and torque but also their curves against the engine revs are critical for the assessment of the interplay between engine and vehicle or engine and machine (see also Chapter 3.6: Gas work and mean pressure).

If the effective power P_e is related to the swept volume V_H, we speak of the specific power output P_l or power output per liter displacement.

$$P_l = \frac{P_e}{V_H} \tag{3.28}$$

If the engine weight m_M is referred to the power, then we obtain the power-to-weight ratio m_G:

$$m_G = \frac{m_M}{P_e} \tag{3.29}$$

Empirical values for this are shown in Fig. 3-7.

Engine type	Specific power output [kW/l] up to	Power-to-weight ratio [kg/kW] up to	At engine speed [rpm]
Racing engine (Formula 1)	200	0.4	($n \approx$ 18 000 rpm)
Car SI engine	70	2.0	($n \approx$ 6500 rpm)
Turbocharged car SI engine	100	3.0	($n \approx$ 6000 rpm)
Car diesel engine (naturally aspirated)	45	5.0	($n \approx$ 4500 rpm)
Turbocharged car diesel engine	64	4.0	($n \approx$ 4500 rpm)
Commercial vehicle diesel engine	30	5.5	($n \approx$ 3000 rpm)
High-speed diesel engine	15.0	11.0	($n \approx$ 4500 rpm)
Medium-speed diesel engine	7.5	19.0	($n \approx$ 500 rpm)
Slow large diesel engine (two-stroke)	3.0	55.0	($n \approx$ 100 rpm)

Fig. 3-7 Empirical values for specific power output and power-to-weight ratio.

3.5 Fuel Consumption

The energy admitted with the fuel is calculated as

$$E_K = m_K \cdot H_u \tag{3.30}$$

where

m_K = Weight of fuel admitted
H_u = Net calorific value of the fuel

The fuel consumption is measured as a volumetric flow or as a mass flow

$$\dot{m}_K = \frac{m_K}{t} = \rho_K \cdot \dot{V}_K \tag{3.31}$$

where

ρ_K = Density of the fuel

For better comparability, the fuel consumption can also be referred to the indicated or effective power.

Indicated specific fuel consumption:

$$b_i = \frac{\dot{m}_K}{P_i} = \frac{1}{\eta_i \cdot H_u} \tag{3.32}$$

where

η_i = Indicated efficiency

Effective specific fuel consumption:

$$b_e = \frac{\dot{m}_K}{P_e} = \frac{1}{\eta_e \cdot H_u} \tag{3.33}$$

where

η_e = Effective efficiency

The equation

$$b_e = \frac{1}{\eta_e \cdot H_u} \tag{3.34}$$

shown graphically in Fig. 3-8 illustrates the relationship between effective efficiency and effective specific fuel consumption.

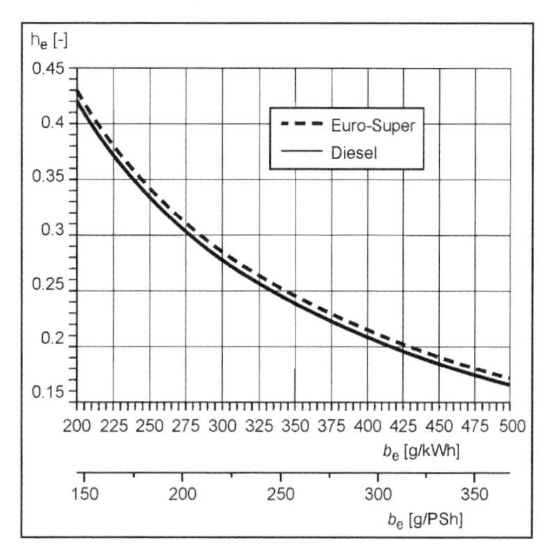

Fig. 3-8 Efficiency of the fuel consumption ($H_{U,\text{ Euro-Super}}$ = 42.0 MJ/kg; $H_{U,\text{ Diesel}}$ = 42.8 MJ/kg).

Figures 3-9 to 3-11 show examples of power and fuel consumption curves for a car SI engine, a car diesel

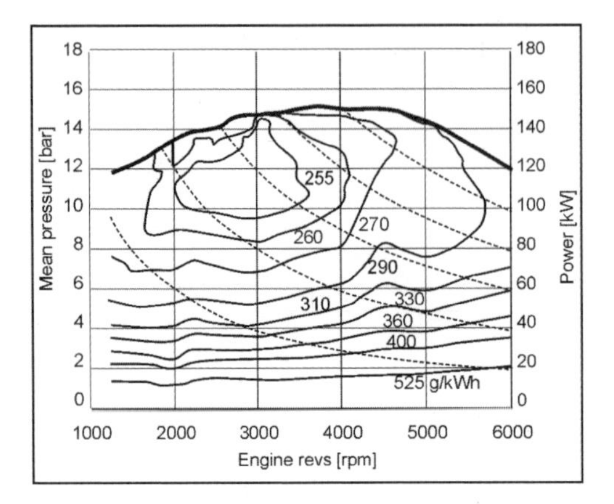

Fig. 3-9 Power and consumption curves for a car SI engine.[7]

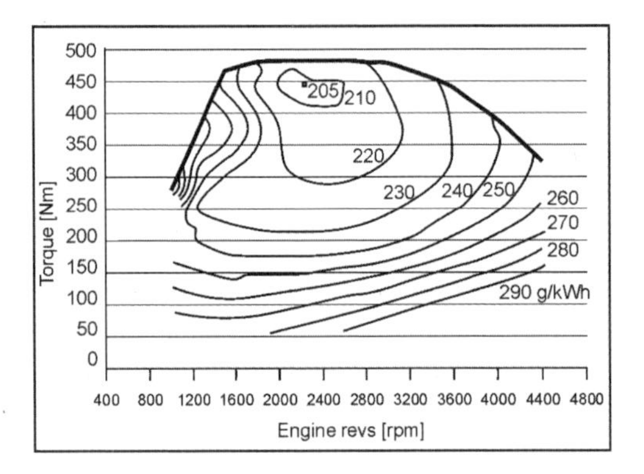

Fig. 3-10 Consumption curve, car diesel engine V8-TDI.[8]

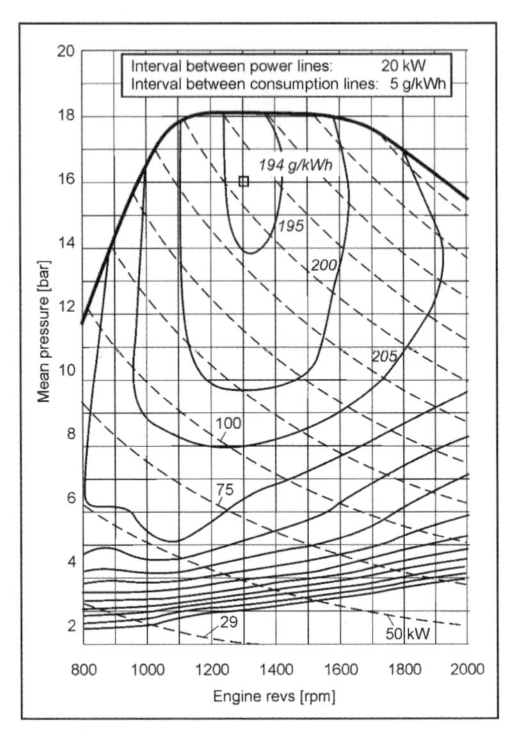

Fig. 3-11 Power and consumption curves for a commercial vehicle engine with $V_H = 12\ l$.[1]

engine, and a commercial vehicle diesel engine. The isolines (bell-shaped curves) indicate working points of equivalent fuel consumption. In order to assess the fuel consumption of an engine, the fuel consumption not only at the best point but also at all the working points has to be taken into consideration.

Figure 3-12 shows empirical values for the specific fuel consumption.

3.6 Gas Work and Mean Pressure

Gas work is the work done by the cylinder pressure at the piston. With the mean pressure, we distinguish between indicated and effective mean pressure and the frictional mean pressure.

Indicated Mean Pressure

The indicated mean pressure p_{mi} is equivalent to the specific work acting on the piston.

The indicated mean pressure is determined from the cylinder pressure curve and the swept volume (Fig. 3-13). The indicated mean pressure can be determined from the *p-V* diagram by planimetry (measurement of the area). If the surface enclosed by the curve is surrounded in a clockwise direction, we have a positive indicated mean pressure; if it is surrounded in a counterclockwise direction, we have a negative indicated mean pressure. Therefore, a distinction can be made between an indicated mean pressure of the high-pressure section and an indicated mean pressure of the gas exchange cycle. The sum of these two portions gives the indicated mean pressure of the engine p_{mi} (Fig. 3-14). The indicated mean pressure of the gas exchange cycle p_{miGW} comprises the intake and exhaust work and can, therefore, be regarded as a measure of the quality of the gas exchange.[9] For naturally aspirated engines, the p_{miGW} is generally negative, i.e., a work loss. For turbocharged engines this portion is normally positive.

The indicated mean pressure, Fig. 3-14, can be derived from the work of the gas force transmitted to the piston during a working cycle.

$$dW_{KA} = p \cdot A_K \cdot ds_\alpha \tag{3.35}$$

Engine type	Specific fuel consumption [g/kWh] up to	Efficiency [%] up to
Small engines (two-stroke)	350	25
Motorcycle engines	270	32
Car SI engines	250	35
Indirect injection car diesel engines	240	35
Turbocharged DI car diesel engines	200	42
Turbocharged truck diesel engines	190	45
Crosshead engines (two-stroke diesel)	156	54

Fig. 3-12 Empirical values for fuel consumption and efficiency at the best point.

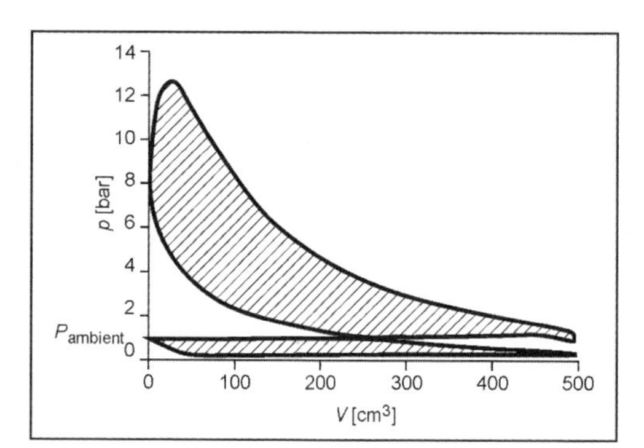

Fig. 3-13 Cylinder pressure over swept volume (2000 rpm, $p_{mi} = 2$ bar, $V_h = 500$ cm^3).

where

p = Combustion pressure or cylinder pressure
A_K = Piston or cylinder surface area
s_α = Piston travel $= f$(crank angle α)
W_{KA} = Gas work at the piston per working cycle

with the change in volume, depending on the piston travel:

$$A_K \cdot ds_\alpha = dV_\alpha \tag{3.36}$$

dV_α = Change in volume $= f$(crank angle α)

and integration over the whole working cycle gives

$$W_{KA} = \oint p \cdot dV_\alpha \tag{3.37}$$

The indicated power P_{iZ} of a cylinder is hence calculated as

$$P_{iZ} = n_A \cdot W_{KA} \tag{3.38}$$

where

n_A = Working cycles per unit of time $= i \cdot n$
n = Engine revolutions per unit of time
i = Working cycles per revolution
For four-stroke engines: $i = 0.5$
For two-stroke engines: $i = 1$

The cylinder power is calculated as

$$P_{iZ} = i \cdot n \cdot W_{KA} \tag{3.39}$$

The gas work W_{KA} referred to the swept volume V_h per working cycle is defined as the indicated mean pressure p_{mi}:

$$p_{mi} = \frac{W_{KA}}{V_h} \tag{3.40}$$

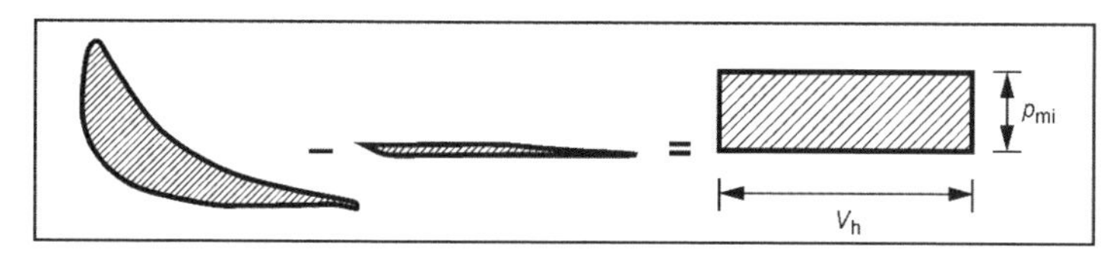

Fig. 3-14 Determination of the indicated mean pressure from the areas over the swept volume.

or

$$p_{mi} \cdot V_h = W_{KA} \tag{3.41}$$

The indicated cylinder power can be expressed as

$$P_{iZ} = i \cdot n \cdot p_{mi} \cdot V_h \tag{3.42}$$

This equation is true for one cylinder. An engine with several cylinders (z = number of cylinders) has the indicated power:

$$P_i = i \cdot n \cdot p_{mi} \cdot V_h \cdot z = i \cdot n \cdot p_{mi} \cdot V_H \tag{3.43}$$

The indicated mean pressure of several consecutive cycles is used to assess the regularity of the combustion, e.g., by calculation of the variance. Irregular combustion and misfiring can be determined in this way. These are criteria for hydrocarbon emissions, power, and smooth running of the engine. For well-designed engines, the variance of the indicated mean pressure is less than 1%, whereby the variance increases with increasing engine revs.

The variance is calculated as follows:

$$COV = \frac{\sigma_{p_{mi}}}{\overline{p}_{mi}} \tag{3.44}$$

$$\sigma_{pmi} = \sqrt{\frac{1}{n-1}\sum_{i=1}^{n}(p_{mii} - \overline{p}_{mi})^2} \tag{3.45}$$

where

COV = Variance (coefficient of variation)
$\sigma_{p_{mi}}$ = Standard deviation of the indicated mean pressure
$\overline{p}_{mi}$ = Mean value of the indicated mean pressure

By analogy with the indicated mean pressure p_{mi}, we also define the effective mean pressure p_{me} and the friction mean pressure p_{mr}.

Effective Mean Pressure

The effective mean pressure can be determined from the torque M_d:

$$p_{me} = \frac{M_d \cdot 2\pi}{V_H \cdot i} \tag{3.46}$$

M_d = Torque of the engine
i = Working cycles per revolution (0.5 for four-stroke, 1 for two-stroke engines)
V_H = Total swept volume of the engine

Figure 3-15 shows examples of the effective mean pressure of modern engines.

Friction Mean Pressure

The friction mean pressure is the difference between indicated mean pressure and effective mean pressure:

$$p_{mr} = p_{mi} - p_{me} \tag{3.47}$$

The friction mean pressure according to Society of Automotive Engineers (SAE) is the power loss due to mechanical friction in the engine and the pump losses in the crankcase. The friction in the engine is primarily dependent on the engine revs and hence on the piston speed, where the friction increases with increasing engine revs. The cylinder pressure, i.e., engine load and engine temperature, and the oil viscosity have a lesser effect on the friction. The friction losses according to DIN (German Industry Standard) also include the drive powers for auxiliary components of the engine such as the alternator, air conditioning compressor, or servo pump.

3.7 Efficiency

In the internal combustion engine, a distinction is made among the indicated, effective, and mechanical efficiencies.

Engine type	Effective mean pressure [bar]
	up to
Motorcycle engines	12
Racing engines (Formula 1)	16
Car SI engines (without turbocharger)	13
Car SI engines (with turbocharger)	17
Truck diesel engines (with turbocharger)	22
Car diesel engines (with turbocharger)	20
Larger high-speed diesel engines	30
Medium-speed diesel engines	25
Crosshead engines (two-stroke diesel)	15

Fig. 3-15 Effective mean pressure of modern engines.

The indicated and the effective efficiencies are essentially determined from the energy stored in the fuel.

The energy admitted with the fuel per unit of time is calculated as

$$\frac{E_K}{t} = \dot{m}_K \cdot H_u \tag{3.48}$$

where

$\dot{m}_K$ = Admitted mass of fuel per unit of time
H_u = Net calorific value of the fuel

If we consider the engine power P as the output of the engine process and the admitted fuel energy per unit of time as the input, then the efficiency η can be calculated as

$$\eta = \frac{\text{Power output}}{\text{Fuel input}} = \frac{P}{\frac{E_K}{t}} = \frac{P}{\dot{m}_K \cdot H_u} \tag{3.49}$$

Indicated Efficiency

$$\eta_i = \frac{P_i}{\dot{m}_K \cdot H_u} \tag{3.50}$$

Effective Efficiency

$$\eta_e = \frac{P_e}{\dot{m}_K \cdot H_u} \tag{3.51}$$

The ratio of effective efficiency to indicated efficiency is described by the mechanical efficiency.

Mechanical Efficiency

$$\eta_m = \frac{\eta_e}{\eta_i} = \frac{P_e}{P_i} \tag{3.52}$$

Figure 3-16 shows the breakdown of the admitted fuel energy into thermal losses and useful and frictional work. It also shows the breakdown of the frictional work or inertia work into the various portions.

3.8 Air Throughput and Cylinder Charge

The power of an engine is dependent on the cylinder charge. The air expenditure λ_a and the volumetric efficiency λ_l are used to assess and characterize the cylinder charge.

Air Expenditure

The air expenditure is a measure of the fresh charge admitted to the engine. It is assumed that the charge is in gaseous form. For the air expenditure, we have the relationship

$$\lambda_a = \frac{m_G}{m_{\text{th}}} = \frac{m_G}{V_h \cdot \rho_{\text{th}}} \quad \text{or} \quad \lambda_a = \frac{m_{G\,\text{ges}}}{V_H \cdot \rho_{\text{th}}} \tag{3.53}$$

m_G = Total fresh charge mass admitted to a cylinder per working cycle
$m_{G\,\text{ges}}$ = Total fresh charge mass admitted to the engine per working cycle
m_{th} = Theoretical charge mass per working cycle (cylinder or complete engine)
ρ_{th} = Theoretical charge density

The total fresh charge mass admitted consists of
SI engine:

$$m_G = m_K + m_L \quad \text{or} \quad m_{G\,\text{ges}} = m_{K\,\text{ges}} + m_{L\,\text{ges}} \tag{3.54}$$

Diesel engine:

$$m_G = m_L \quad \text{or} \quad m_{G\,\text{ges}} = m_{L\,\text{ges}} \tag{3.55}$$

The theoretical fresh charge mass is calculated from the geometric swept volume and the ambient state of the charge. For turbocharged engines, the thermodynamic state up line of the intake organs is used instead of the ambient state. For engines with internal mixture formation, the charge consists of air, and for engines with external mixture formation, the charge consists of air and fuel.

The gas equation gives us

$$p_u \cdot V_h = m_{\text{th}} \cdot R \cdot T_u \text{ or } p_u \cdot V_H = m_{\text{th ges}} \cdot R \cdot T_u \tag{3.56}$$

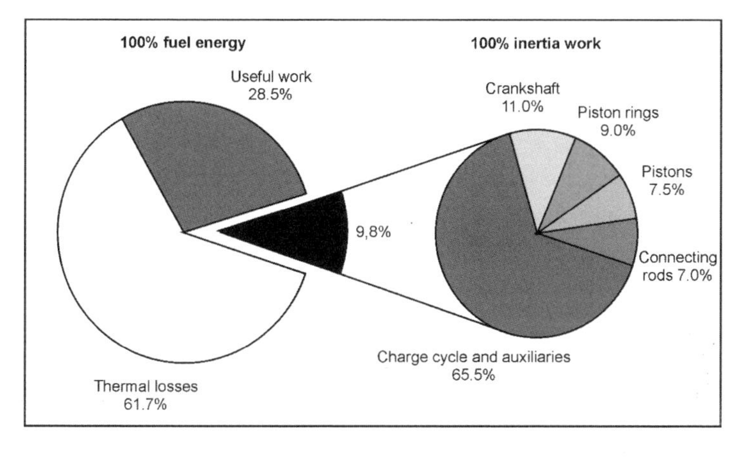

Fig. 3-16 Classification of the efficiency in a four-stroke SI engine.[4]

where

$R = R_G$ (Gas constant of the mixture) in the SI engine
$R = R_L$ (Gas constant of air) in the diesel engine or direct injection SI engine
T_u = Ambient temperature
p_u = Ambient pressure

If the density of the mixture or air taken in is assumed to be equal to the theoretical charge density ρ_{th}, the air expenditure can also be calculated using volumetric parameters:

$$m_G = V_G \cdot \rho_G \text{ or } m_{G\,\text{ges}} = V_{G\,\text{ges}} \cdot \rho_G \tag{3.57}$$

where

V_G = Volumetric charge input per working cycle of a cylinder
$V_{G\,\text{ges}}$ = Volumetric charge input per working cycle of the engine

SI engine:

$$\lambda_a = \frac{V_G}{V_h} \quad \text{or} \quad \lambda_a = \frac{V_{G\,\text{ges}}}{V_H} \tag{3.58}$$

Diesel engine:

$$\lambda_a = \frac{V_L}{V_h} \quad \text{or} \quad \lambda_a = \frac{V_{L\,\text{ges}}}{V_H} \tag{3.59}$$

In order to determine the air expenditure empirically at the engine, the intake air volume or air mass is measured. In addition, the pressure and temperature of the air and the ambient conditions, as well as the fuel consumption with the SI engine, have to be recorded.

Volumetric Efficiency

The volumetric efficiency is a measure of the fresh charge remaining in the cylinder at the end of the charge cycle. As with the air expenditure, this is referred to the theoretical charge density.

$$\lambda_l = \frac{m_Z}{m_{th}} = \frac{m_Z}{V_h \cdot \rho_{th}} \quad \text{or} \quad \lambda_l = \frac{m_{Z\,\text{ges}}}{V_H \cdot \rho_{th}} \tag{3.60}$$

The cylinder fresh charge is calculated as m_Z or $m_{Z\,\text{ges}}$:
For the SI engine:

$$m_Z = m_{ZL} + m_{ZK} \quad \text{or} \quad m_{Z\,\text{ges}} = m_{ZL\,\text{ges}} + m_{ZK\,\text{ges}} \tag{3.61}$$

For the diesel engine:

$$m_Z = m_{ZL} \quad \text{or} \quad m_{Z\,\text{ges}} = m_{ZL\,\text{ges}} \tag{3.62}$$

where

m_{ZL} = Air mass in one cylinder
$m_{ZL\,\text{ges}}$ = Air mass in all the engine cylinders
m_{ZK} = Fuel mass in one cylinder
$m_{ZK\,\text{ges}}$ = Fuel mass in all the engine cylinders

The charge mass remaining in the cylinder or in all the engine cylinders cannot be calculated or measured directly. The following method is employed as an approximation:

(a) Cylinder pressure indication in one or all the engine cylinders
(b) Assumption that the cylinder charge temperature at the moment the "intake valve closes" is roughly the same as the temperature in the intake duct upline of the intake valve (measurement of this temperature using a thermocouple)
(c) Application of the gas equation at the moment the "intake valve closes"

$$p_{ZEs} \cdot V_{Es} = m_Z \cdot R \cdot T_{ZEs}$$

R_G or R_L is assumed again for the gas constant R.

With four-stroke SI engines, the crank angle range of the valve overlap (the time during which both intake and exhaust valves are open at the same time during the charge cycle) is relatively small. For the case of the small valve overlap, $\lambda_a \approx \lambda_l$ can be assumed as a good approximation.

For engines without a turbocharger, λ_a and λ_l are always smaller than 1, as flow resistance during the intake and exhaust prevents a complete scavenging of the geometric swept volume. Turbocharged engines and engines with ram-effect supercharging are examples of engines that have operating states in which λ_a and λ_l are larger than 1.

Diesel engines, particularly those with a turbocharger, have large valve overlaps in order to achieve internal cooling and a better scavenging of the remaining gas out of the combustion chamber. Here λ_a can become $\cong \lambda_l$.

With slot-controlled two-stroke engines, a considerable difference exists between air expenditure and volumetric efficiency because of the overflow losses. The quotient of volumetric efficiency and air expenditure gives the retention rate that is a measure of the fresh charge remaining in the cylinder.

3.9 Air-Fuel Ratio

During combustion in the engine, the ratio of the air mass actually in the cylinder m_L to the stoichiometric air mass $m_{L,St}$ is referred to as the excess-air factor λ.

The stoichiometric air requirement L_{St} is defined as the quotient of the air mass and the fuel mass under stoichiometric conditions:

$$L_{St} = \frac{m_{L,St}}{m_K} \tag{3.63}$$

$$\lambda = \frac{m_L}{m_{L,St}} = \frac{m_L}{m_K \cdot L_{St}} \tag{3.64}$$

where

$m_{L,St}$ = Air mass under stoichiometric conditions
m_K = Fuel mass

The stoichiometric air requirement can be calculated from the percentage by weight of the chemical elements contained in the fuel, whereby the combustion products (exhaust gases) resulting from the combustion also have to be taken into consideration. The combustion process proper covers a large number of intermediate reactions in which numerous, but also predominantly short-lived, compounds or "radicals" are involved. The most important combustion products with complete combustion are carbon dioxide (CO_2), water (H_2O), and sulfur dioxide (SO_2), as well as the air nitrogen (N_2, inert gas) that is practically unchanged by the combustion. For complete combustion of a fuel with the composition $C_xH_yS_qO_z$, we thus obtain the chemical reaction equation:

$$C_xH_yS_qO_z + \left(x + \frac{y}{4} + q - \frac{z}{2}\right) \cdot O_2 \Rightarrow x \cdot CO_2 + \frac{y}{2} \cdot H_2O + q \cdot SO_2 \quad (3.65)$$

with the stoichiometric components

$$x = \frac{M_K}{M_C} \cdot c \qquad y = \frac{M_K}{M_H} \cdot h$$

$$q = \frac{M_K}{M_S} \cdot s \qquad z = \frac{M_K}{M_O} \cdot o$$

where

c, h, s, o = Percentages by weight of the elements carbon (c), hydrogen (h), sulfur (s), and oxygen (o) contained in the fuel

M_C, M_H, M_S, M_O = Molar weights of the elements in the fuel

M_K = Molar weight of the fuel

Allowing for the percentage by weight of oxygen in the air $\xi_{O_2,L}$ we obtain for the stoichiometric air requirement

$$L_{St} = \frac{1}{\xi_{O_2,L}} \cdot \frac{m_{O_2,St}}{m_K} = \frac{1}{\xi_{O_2,L}} \cdot \frac{M_{O_2}}{M_K} \cdot \frac{n_{O_2,St}}{n_K} \quad (3.66)$$

where

M_{O_2} = Molar weight of oxygen

$n_{O_2}; n_K$ = Volumes of oxygen and fuel

With the relations $n_{O_2,St} = x + \frac{y}{4} + q - \frac{z}{2}$ and $n_K = 1$ from the chemical reaction equations we obtain

$$L_{St} = \frac{1}{\xi_{O_2,L}} \cdot \left(\frac{M_{O_2}}{M_C} \cdot c + \frac{1}{4} \cdot \frac{M_{O_2}}{M_H} \cdot h + \frac{M_{O_2}}{M_S} \cdot s - o\right) \quad (3.67)$$

$$L_{St} = \frac{1}{0.232} \cdot (2.664 \cdot c + 7.937 \cdot h + 0.988 \cdot s - o) \quad (3.68)$$

Figure 3-17 shows exemplary data of a fuel analysis.

The fuel metering during engine operation is influenced by the stoichiometric air requirement. For this reason, the mixture forming system has to be adapted accordingly when using different fuels (e.g., gasoline and alcohol-based fuels).

During combustion in the engine, the mixture ratio deviates more or less from the stoichiometric ratio.

A mixture with excess air ($\lambda > 1$) is referred to as a "lean mixture" (lean operation), while a mixture with an air deficiency ($\lambda < 1$) is referred to as a "rich mixture." SI engines with intake manifold injection are operated today in wide program map ranges almost exclusively with a

	Unit	Value	
Mean molar mass of the fuel	G/mol	99.1	
Composition of the fuel specimen	wt.%	87.08	Carbon
	wt.%	12.87	Hydrogen
	wt.%	0.05	Oxygen
Theoretical total formula	—	7.2	Carbon
	—	12.6	Hydrogen
	—	0.0	Oxygen
Gross calorific value (Ho)	MJ/kg	45.72	
Net calorific value (Hu)	MJ/kg	42.88	
Theoretical stoichiometric air demand	kg air / kg fuel	14.47	

Fig. 3-17 Example of a fuel analysis, Euro-Super.

stoichiometric mixture ($\lambda = 1$). SI engines with direct injection can be operated homogeneously with $\lambda = 1$, homogeneous lean ($\lambda > 1$), and also stratified lean (on average for the combustion chamber $\lambda \gg 1$, but partially also with $\lambda = 1$). Diesel engines are always operated with excess air ($\lambda > 1$), and small two-stroke engines are predominantly operated in the air deficiency range ($\lambda < 1$).

Bibliography

[1] Mollenhauer, K. [ed.], Handbuch Dieselmotoren, Springer, Berlin, 1997, ISBN 3-540-62514-3.

[2] Heywood, John B., Internal Combustion Engine Fundamentals, McGraw-Hill, New York, 1988, ISBN 0-07-100499-8.

[3] Spicher, Ulrich, Umdruck zur Vorlesung Verbrennungsmotoren, University, Karlsruhe, 1996.

[4] N.N. [ed.], Einflussgrössen auf die Reibleistung der Kolbengruppe, Technische Information Nr. 7148 Mahle GmbH, Stuttgart, 1994.

[5] Robert Bosch GmbH [eds.], Kraftfahrtechnisches Taschenbuch, 23, Aufl., Vieweg, Braunschweig, 1999, ISBN 3-528-03876-4.

[6] Anisizs, F., K. Borgmann, H. Kratochwill, and F. Steinparzer, Der erste Achtzylinder-Dieselmotor mit Direkteinspritzung von BMW, in MTZ 60 (1999), Heft Nr. 6, pp. 362–371.

[7] Fortnagel, M., B. Heil, J. Giese, M. Mürwald, H.-K. Weining, and P. Lückert, Technischer Fortschritt durch Evolution: Neue Vierzylinder Ottomotoren von Mercedes-Benz auf der Basis des erfolgreichen M111, in MTZ 61 (2000), Heft Nr. 9, pp. 582–590.

[8] Bach, M., R. Bauder, H. Endress, H.-W.Pölzl, and W. Wimmer, Der neue TDI-Motor von Audi: Teil 3 Thermodynamik, in MTZ 60 (1999), Sonderausgabe 10 Jahre TDI-Motor von Audi, pp. 40–46.

[9] Kuratle, R., Motorenmesstechnik, 1, Aufl., Vogel, Würzburg, 1995, ISBN 3-8023-1553-7

4 Maps

The working point of an internal combustion engine is defined by its speed and its torque. The full range of all possible working points in a two-dimensional presentation gives the "engine map." In this map, the working range of the internal combustion engine is limited by the full-load curve and by the minimum and the maximum engine revs (Fig. 4-1). The power output by the engine at any particular working point is calculated from the equation $P_e = 2 \cdot \pi \cdot M \cdot n$. Lines of constant power are referred to in the engine map as power hyperbolas.

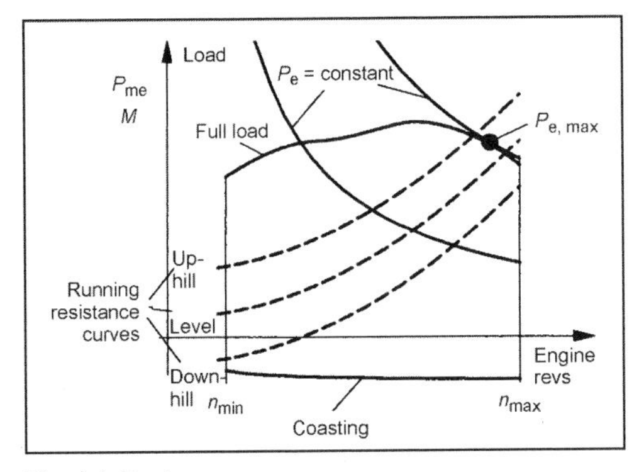

Fig. 4-1 Engine map.

The engine map is used to document certain engine characteristics as a function of the working point. This representation can consist of an indication of discrete values in individual points. When a large number of individual values are available over the whole working range of the engine, lines of equal value for the respective engine characteristic, the "isolines," can be created from these individual values by interpolation. The most common map representation is of the specific fuel consumption whose isolines are presented in the map as the "conchoids" (see also Fig. 4-3 below).

Apart from the engine characteristics, the characteristics of the vehicle and its powertrain can also be displayed in the map. This normally takes the form of the running-resistance curve. These curves show the relationship between the engine revs and the torque drawn by the powertrain for each gear during constant travel on a level road. Uphill or downhill travel results in a parallel shifting of the running-resistance curve (see Fig. 4-1).

If the working point of the engine is above the running-resistance curve, the vehicle accelerates; if it is below the curve, the vehicle brakes. The surplus power available for acceleration results from the current engine revs and the surplus torque corresponding to the distance between the running-resistance curve and the full-load curve. A gear shift results in a different torque for the same travel speed because of the change in engine revs with an approximately equal power requirement; i.e., the working point is shifted along the power hyperbola up to the intersection with the running-resistance curve corresponding to the gear shift. In this way, the engine map allows the changes in the operating or emission behavior to be assessed in relation to the boundary conditions of the vehicle and the method of operation.

For operating conditions with a low power requirement, such as is the case for large portions of the emissions cycles for type testing of a vehicle or in town traffic, operating points with low to medium engine revs/load combinations are of greater relevance. The typical load collectives for highway driving, on the other hand, lie in the top right-hand area of the engine map.

For reasons of comparability of engines with different swept volumes, the specific parameters of the load, the specific mean pressure, or the specific work referred to the swept volume are frequently used instead of the torque.

Maps are used both for documentation of operating parameters, such as ignition timing, injection timing, or excess air factor to illustrate the operating strategy, and for evaluation of the resulting measured and calculated parameters, such as emissions, fuel consumption, or temperatures. Figure 4-2 shows how an engine map is used to illustrate the operating strategy of the engine, taking as an example an SI engine with direct injection.

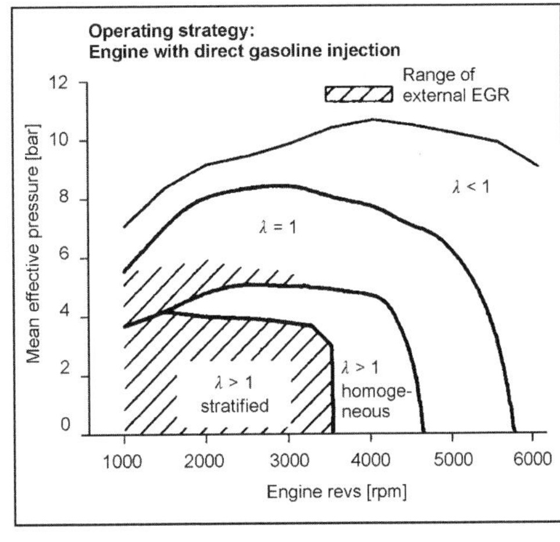

Fig. 4-2 Overview map.

To clarify the operating strategy, characteristic areas of the map are marked differently. In this example, the engine is operated below a load of P_{me} = 4 bar and up to engine speeds of 3500 rpm by injection during the compression stroke with a stratified fuel-air mixture with a large excess of air. In the rest of the map, the fuel is injected during the intake phase with the consequence that the longer mixture preparation time results in a homogeneous mixture. Even during homogeneous operation, there is a phase with an excess of air in the lower load range for engine speeds between 3500 and 4500 rpm.

In the remaining load and engine speed range above the stratified and homogeneous lean operation, this engine is operated like a conventional SI engine with a stoichiometric mixture. As full load is approached (particularly at higher engine revs), the mixture is enriched to protect the catalytic converter against any excessive exhaust gas temperatures and to achieve a higher power.

This operating map also shows by appropriate marks that an external exhaust gas recirculation takes place in the whole stratified and in part of the stoichiometric range. Further features characteristic of the engine operating strategy can be illustrated in the operating map in the same way. These include, for example, the balance of a camshaft adjustment or of a controlled intake manifold.

For the experienced engineer, the engine map represents a source of highly compacted information from which he/she can derive an assessment of the engine in question. When comparing and evaluating maps on the basis of specific engine parameters, it must be remembered that in practice, design criteria such as that for swept volume or stroke-to-bore ratio, the compression ratio or the design and arrangement of the injection valves are reflected in only minor differences. On the other hand, operative measures such as the setting of variable systems (controlled intake manifolds, camshaft adjusters) and of the engine control as well as measures for exhaust gas posttreatment (e.g., catalytic converter systems, thermal insulation of the exhaust system up to the catalytic converter) result in very significant differences in the operating behavior even for similar engines. One such example is with SI engines with direct injection. On the Japanese market, these engines exhibit a similar behavior to that of the engine described above, whereas the adjustment of the same basic engine for the European market exhibits no stratified or homogeneous lean range in the whole map. This shows clearly that the measures for exhaust gas posttreatment for different markets or even just for more stringent emissions certification levels result in more significant differences in the engine map than, for example, manufacturer-specific or design differences suggest.

4.1 Consumption Maps

Figure 4-3 shows a typical consumption map for conventional SI engines with intake manifold injection. As already mentioned, the lines of constant specific fuel consumption are also called conchoids due to their form. The minimum specific fuel consumption is found in the lower engine rev range in the range of high load. Only a flat gradient of the consumption increase is seen in a wider range around the minimum consumption. The gradient rises sharply toward the low load range. One of the main reasons for this is because of increasing throttle losses in the SI engine and the increasing proportion of friction in relation to the useful torque output. These two factors also lead to the visible increase in consumption at constant load and increasing engine revs. Toward the full-load range, the mixture has to be enriched, on the one hand, to counter the knock tendency of the engine and, on the other, to keep the exhaust gas temperature below a critical limit temperature for catalytic converter aging. This leads to a sharper gradient of the consumption increase.

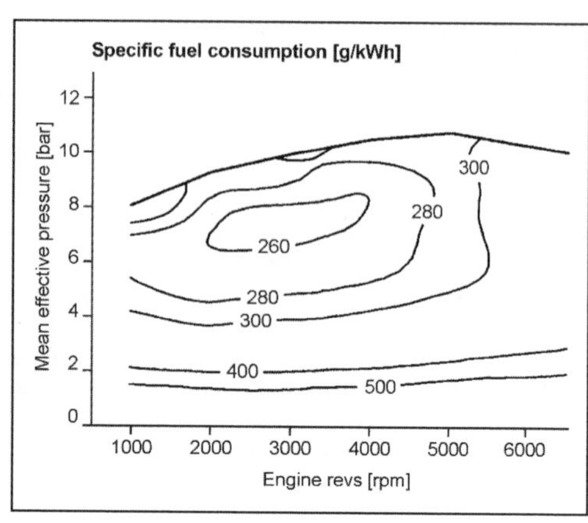

Fig. 4-3 Consumption map (MPI SI engine).

Figure 4-4 shows the typical consumption map for a diesel engine with direct injection and turbocharger. Of particular note is the slighter increase in consumption with decreasing load, as the quality control of the diesel engine is not related to throttle losses. Despite the far more favorable part-load consumption values as compared with the SI engine, the consumptions achieved with the vehicle calibration lie above those for a consumption-optimized setting, particularly in the relevant map area for the European driving cycle. One reason for this is the retarded setting of the injection timing necessary to comply with the permissible NO_x and particulate emissions.

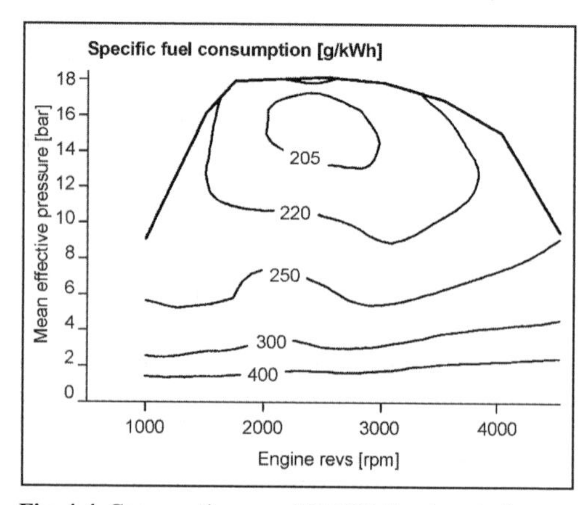

Fig. 4-4 Consumption map (DI-TCI diesel engine).

If the main vehicle-specific data such as the running resistances and gear ratios are known, the consumption map of an engine can also be used to calculate the fuel consumption of the vehicle. In order to calculate the consumption in the nonstatic test cycle, the operating curve is broken down into a sequence of static working points as a function of the vehicle-specific parameters, each characterized by the engine revs and the torque. The load points are then entered into the calculation of the cycle consumption and weighted temporally according to their relevance for the driving schedule. The models necessary for the exact calculation of the consumption take into account not only the vehicle-specific data, but also consumption-influencing processes such as the engine warm-up, gear shifting, and other nonstatic effects. These models allow vehicle-related effects on the consumption and emissions behavior of the engine in the vehicle to be assessed. Examples of the application of such methods are the transmission setting or the control strategy of a continuously variable transmission (CVT).

4.2 Emission Maps

The subjects of the emission maps are generally the raw emissions of the legally limited pollutant components, hydrocarbons, nitrous oxides, and carbon monoxide. Normally, these maps show the work-related specific values (in g/kWh) or mass flows (in g/h). For diesel engines and for SI engines with direct injection, the maps of the particulate emissions are also of significance. Apart from the raw emissions maps, the emission values down line of the catalytic converters are also often shown. These values permit an evaluation of the conversion in the catalytic converter as well as enable an estimation in the volumes of pollutants emitted by the vehicle in a driving cycle.

Figures 4-5 to 4-9 show characteristic maps for conventional SI engines and selected maps of operative parameters relevant to the emission behavior. The engines on which the illustrated maps are based are all equipped with three-way catalytic converters for efficient exhaust gas treatment. The maps relate to the running of the engine at operating temperature. This and the λ control employed with the selected engines to obtain an exact stoichiometric mixture guarantee a high conversion rate of all the pollutant components in the three-way catalytic converter. The air-fuel ratio shown in Fig. 4-5 clearly illustrates the large map area of active λ control. As in the example of the SI engine with direct injection shown above, a mixture enrichment is employed here again both in the full-load range and at high engine revs. In the area of the rated power, the minimum air-fuel ratios are calibrated with values of around $\lambda = 0.80$.

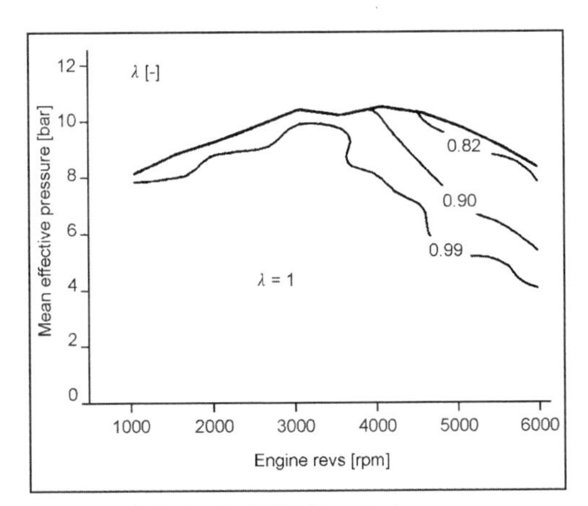

Fig. 4-5 Air-fuel ratio (MPI SI engine).

The CO concentration is predominantly a function of the excess-air factor, as the maps in Figs. 4-5 and 4-6 show. In the map area with active λ control, the concentrations generally lie in an uncritical order of between 0.5 and 0.8 vol.%. At full load, the combustion takes place with an air deficiency due to the mixture enrichment. The maximum CO concentration of 7.5 vol.% occurs at the maximum enrichment rates in the area of the rated power output. This relationship between the CO concentration and the air-fuel ratio illustrated in Fig. 4-6 can be regarded as typical for modern SI engines with high specific powers.

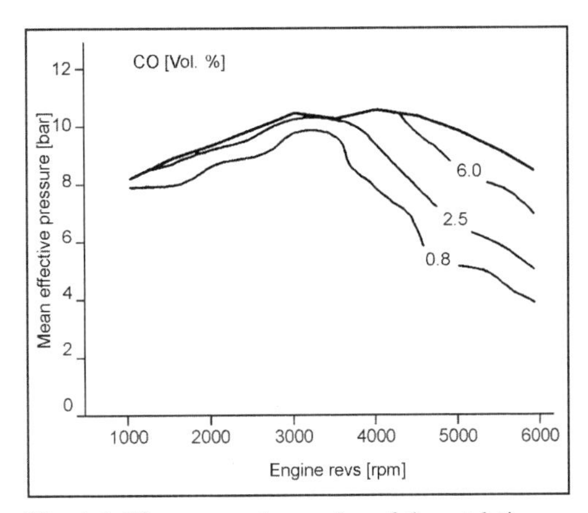

Fig. 4-6 CO concentration up line of the catalytic converter (MPI SI engine).

The level of the NO_x raw emissions can also be influenced in stoichiometric operation by adjustment of the operative parameters. When calibrating the engine in the map, a retarded calibration of the ignition angle—within limits—is selected. However, this measure does result in a reduced efficiency so that it also has to be taken into consideration when evaluating the consumption map. On the other hand, EGR in part-load operation offers a significant potential for reducing the NO_x raw emissions while at the same time improving efficiency due to the related dethrottling of the engine. Exhaust gas recirculation can be performed either externally via a valve or internally as exhaust gas recirculation by modifying the ignition timing. The map of the specific NO_x emissions of an SI engine with external exhaust gas recirculation in

Fig. 4-7 and the corresponding map of the EGR rates calibrated with the EGR valve in Fig. 4-8 show an example of the practical use of exhaust gas recirculation. The minimum NO_x emissions are achieved at the working point with the maximum EGR rate. Outside the map area of external EGR, we obtain a typical behavior of the NO_x emissions. The sharp reduction in NO_x emissions recognizable at full load and at high engine revs is a result of the mixture enrichment.

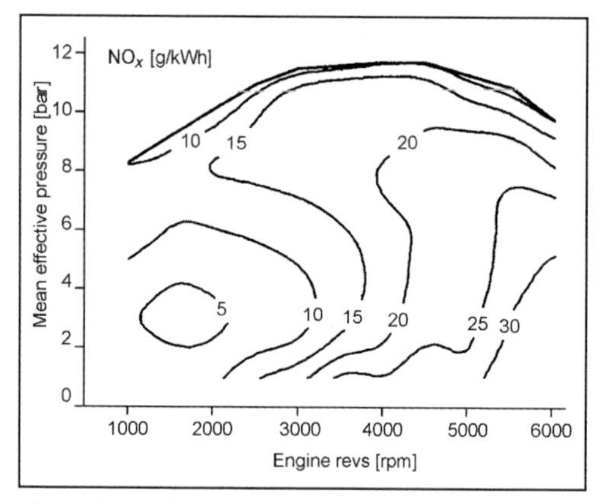

Fig. 4-7 Specific NO_x emissions up line of the catalytic converter (MPI SI engine).

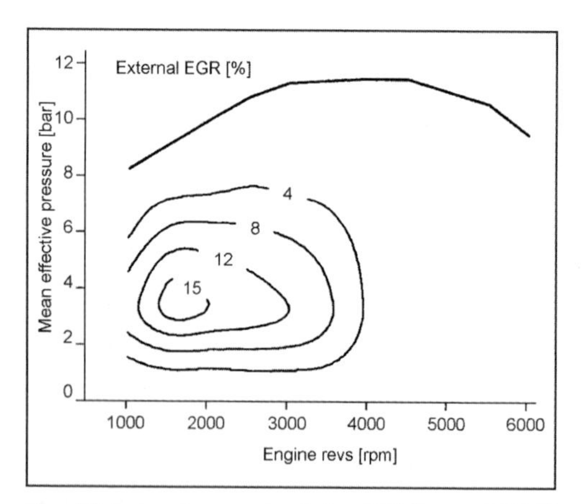

Fig. 4-8 Exhaust gas recirculation (EGR) rate (MPI SI engine).

Continuously operating systems for camshaft adjustment are frequently used in mass-produced engines not only to achieve an internal exhaust gas recirculation but also to improve the torque behavior at full load. Optimization of the ignition timing as a function of the engine revs allows air expenditure benefits to be achieved as a result of the improved torque curve.

In contrast to the NO_x and CO emissions, the level of the HC raw emissions is influenced far more strongly by design parameters. The first aspect here is the form of the combustion chamber, with the surface-to-volume ratio representing a characteristic parameter. Although the HC emissions are sensitive to operative parameters with the engine at operating temperature, it is of subordinate significance in the normal range of variations. An internal EGR using variable valve timing can have a positive effect on HC emissions as the typical HC peak observed toward the end of the exhaust cycle is returned to the combustion. A typical HC emissions map of an SI engine with single-stage intake camshaft adjustment is shown in Fig. 4-9.

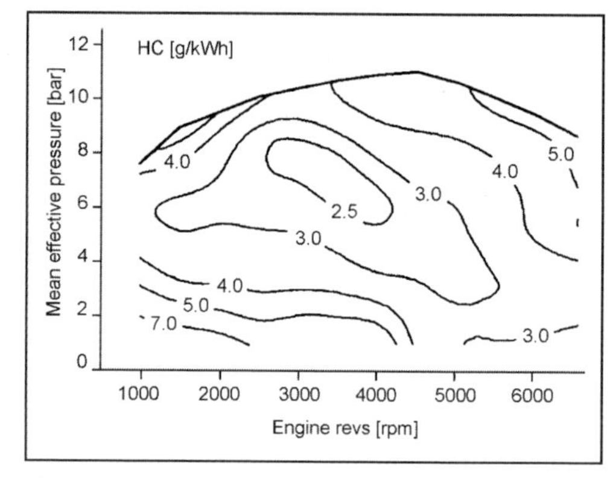

Fig. 4-9 Specific HC emissions (MPI SI engine).

The maps of the emissions down line of the catalytic converter for modern SI engines with a three-way catalytic converter not illustrated here are characterized by the practically complete conversion of the pollutants. Deviations from the extremely low emission levels occur in the map areas with substoichiometric operation where the catalytic oxidation of the HC and CO contents remains limited due to the oxygen insufficiency.

Because of the combustion with excess air typical for the diesel engine, the carbon monoxide and HC emissions are significantly lower compared with the SI engine (Figs. 4-10 and 4-11). The residual oxygen that always

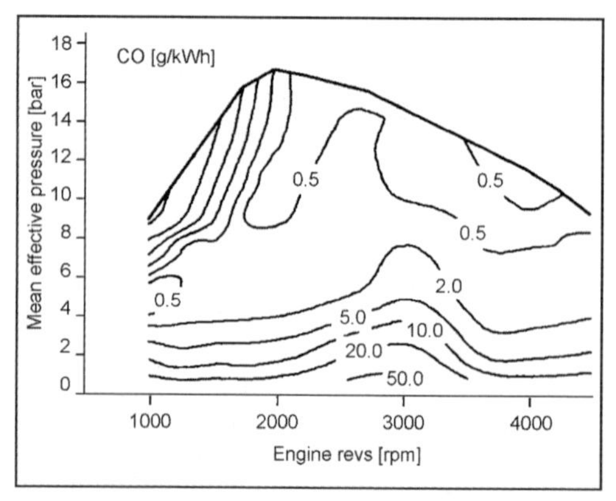

Fig. 4-10 Specific CO emissions (DI-TCI diesel engine).

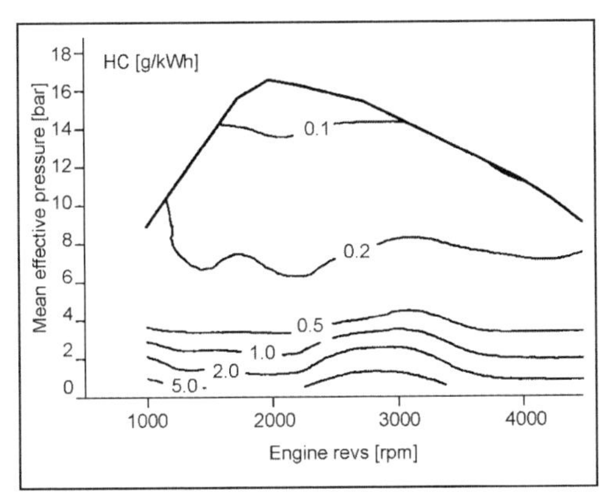

Fig. 4-11 Specific HC emissions (DI-TCI diesel engine).

exists in the exhaust gases from diesel engines permits a further reduction of these pollutant components in oxidation catalytic converters.

More critical for diesel engines, however, are the NO_x raw emissions (Fig. 4-12). Since catalytic posttreatment with excess air is not effective here, the primary solution pursued is to limit the occurrence of NO_x by influencing the combustion process. The measures employed here are the same as with the SI engine, exhaust gas recirculation, and the retarding of the injection process that is more or less the equivalent of the retarded ignition of the SI engine.

To increase the reduction effect of the EGR for the emitted nitrous oxides, the recirculated exhaust gases of the diesel engine are cooled.

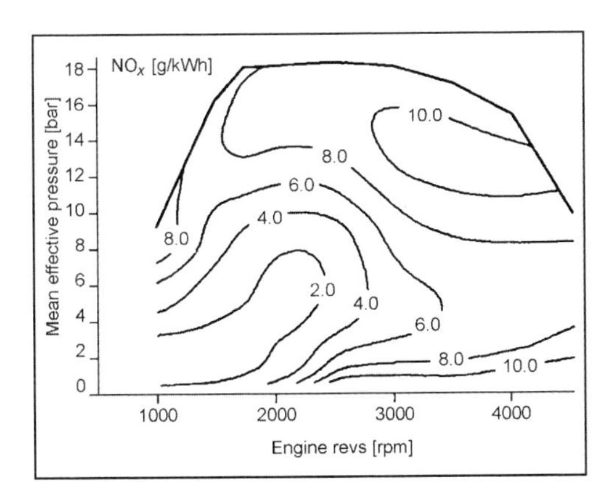

Fig. 4-12 Specific NO_x emissions (DI-TCI diesel engine).

The map of the EGR rates in Fig. 4-13 shows that in this example, the exhaust gas recirculation is essentially calibrated for the emissions-relevant map area. The exhaust gas recirculation rates can be as high as 50% and thus lie far higher than those for an SI engine. In contrast with the SI engine, the possibilities of exhaust gas recirculation are not limited here by the occurrence of combustion misses. It must be remembered here that combustion takes place

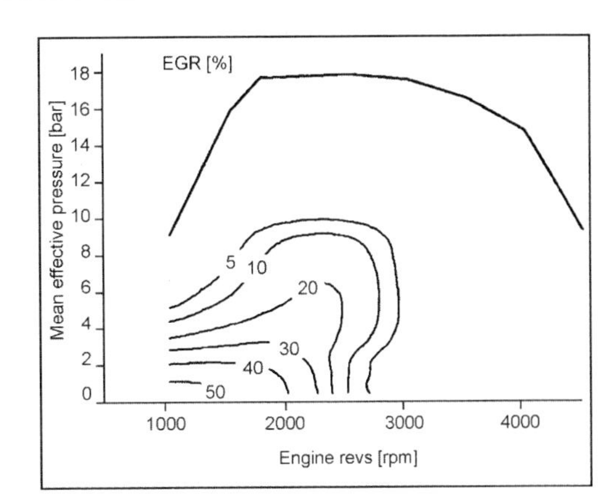

Fig. 4-13 Exhaust gas recirculation rate (DI-TCI diesel engine).

with a large air surplus and that the oxygen concentration in the exhaust gas is still as high as 15 vol.%.

In addition, for diesel vehicles, the law regulates the amount of particulate emissions that can come out of the exhaust. A common method for assessing the particulate emissions from diesel engines is the Bosch smoke number. The increased black smoke values in the emissions-relevant map area (Fig. 4-14) are indicative of the relationship between particulate formation and EGR. This relationship also draws attention to the known conflict of goals between NO_x and particulate emissions. Outside the area of the map geared to EGR, the level of the smoke numbers is relatively low and increases significantly only as full load approaches, particularly at low engine revs because of the lower air-fuel ratio prevailing here.

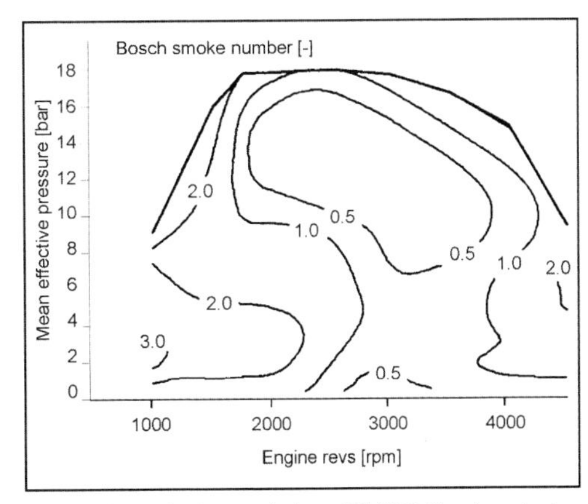

Fig. 4-14 Particulate emissions (DI-TCI diesel engine).

Particulate formation has to be countered by good preparation of the injected diesel fuel. That is why the high-pressure injection with high-quality atomization represents one of the major development directions of modern diesel engines. With a further toughening of the particulate emission limits, the use of particulate filter systems will permit major development steps to be taken in addition to the internal motor measures. Despite the relatively high cost, this technology is already in use in mass-produced vehicles. In contrast with the stationary calibration documented in the engine maps, the intermittent regeneration of the particulate filter necessitates intervention in the calibration of the engine that serves to temporarily increase the exhaust gas temperatures in certain map areas in order to promote the burn-off of the particles collected on the filter surface.

4.3 Ignition and Injection Maps

The typical calibration of the ignition angle in conventional engines with λ control exhibits a strong dependence on the operating point. In the middle of the part load, the ignition angle is generally calibrated in the area of optimum efficiency. Figure 4-15 shows a fundamental trend to an increasing need for advanced ignition with increasing engine revs and decreasing load. This behavior is superimposed by further effects. In the lower load range, a significant advance adjustment of the ignition is seen (even at low engine revs). For the engine shown, an external exhaust-gas recirculation is calibrated in this area. The recirculated exhaust gas that acts as an inert gas delays the combustion process that therefore must be initiated correspondingly earlier. Furthermore, a retarding of the ignition is seen close to full load at engine revs of around 4500 rpm. This behavior is attributable to the frequently observed tendency to knock in the area of the highest air expenditure. In line with the latest state of the art, the disadvantages resulting from this measure can be minimized with the use of dynamic knock control systems. These permit a torque-optimized preignition angle without the risk of engine damage caused by knocking combustion.

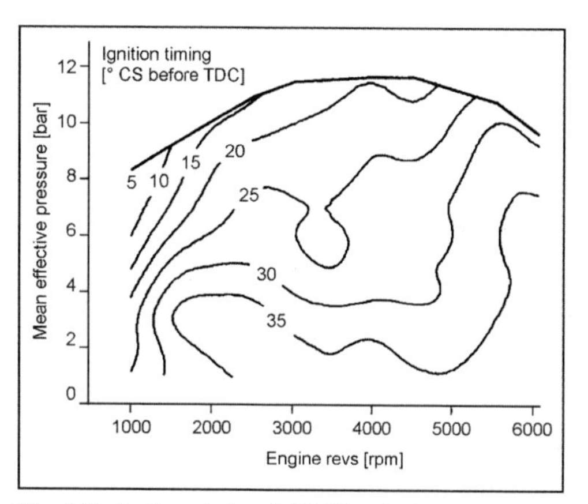

Fig. 4-15 Ignition timing (MPI SI engine).

In diesel engines, the combustion is primarily controlled by the fuel injection process. The start of injection therefore has a significance comparable to that of the ignition angle in the SI engine. With the transition to direct injection that predominates today, the fast combustion with sharp gradients of the cylinder pressure curve leads to acoustic problems. An effective measure for reducing the cylinder pressure gradients with modern electronic diesel control (EDC) engine control systems is the preinjection. With preinjection, the combustion is first triggered by a smaller injection of fuel. Then the remaining volume of fuel is admitted to the process during the main injection. Figures 4-16 and 4-17 show the maps for the start of injection for the preinjection and main injection in a modern car diesel engine. It can clearly be seen how the preinjection is limited to a specific engine speed range by the EDC engine control system.

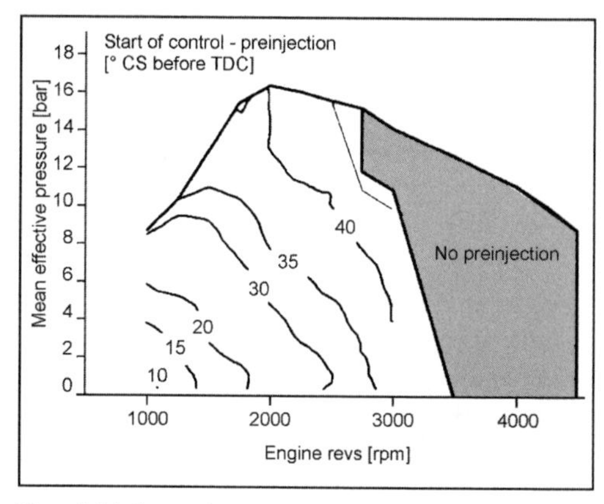

Fig. 4-16 Start of preinjection control (DI-TCI diesel engine).

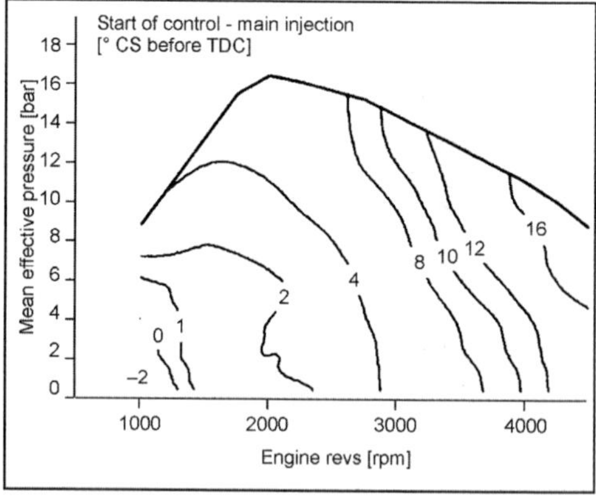

Fig. 4-17 Start of main injection control (DI-TCI diesel engine).

Furthermore, the relatively late positions of the start of injection of the main injection indicate the use of the measures to reduce the NO_x emissions described above. In the map area without preinjection, on the other hand, the main injection is shifted earlier.

4.4 Exhaust Gas Temperature Maps

The behavior of the exhaust gas temperature of an SI engine is shown in Fig. 4-18. The sharp increase in exhaust gas temperature to high loads necessitates specific measures to protect the exhaust gas catalytic converter from thermal aging or even destruction. Both design measures and the calibration of the engine operating parameters are employed here. For engines with exhaust gas turbocharging, the gas temperature at the turbine inlet is also critical for component protection. For SI engines, an enrichment of the fuel-air mixture is therefore employed as an effective component protection measure in the map area with critical exhaust gas temperatures as described above.

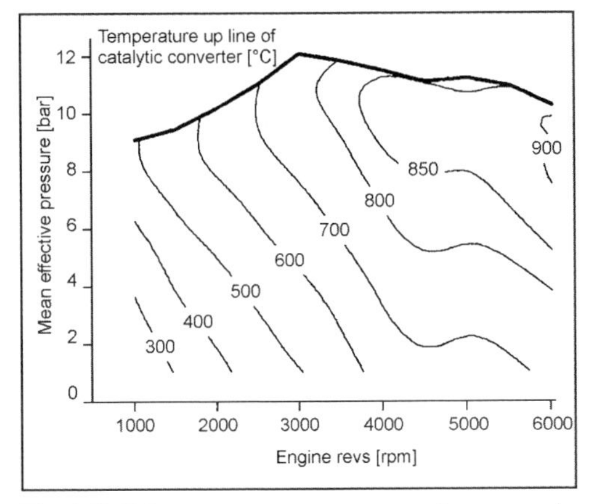

Fig. 4-18 Exhaust gas temperature map at the entry to the catalytic converter (MPI SI engine).

For operation with low load points, on the other hand, an excessively low exhaust gas temperature must be avoided so that the catalytic converter does not cool down. For this reason, a relatively retarded ignition timing can be necessary. In addition to these measures recognizable in the static maps, deviating control parameters for the ignition angle and EGR rates are normally calibrated after the engine cold start so that the catalytic converter quickly reaches the light-off temperature necessary for a conversion of the raw emissions into harmless components.

5 Thermodynamic Fundamentals

Internal combustion engines are heat engines in which chemically bound energy is converted into mechanical energy.[1–3] This is done by means of a reaction, the combustion process, in which energy is released. A part of this heat released in the combustion chamber of the cylinder is converted into mechanical energy by the crankshaft drive, and the remaining energy is carried away with the exhaust and released to a coolant via the walls neighboring the combustion chamber as well as directly to the environment.

The goal of the process of converting chemical energy into mechanical energy is to attain the greatest possible process effectiveness (strongly dependent on the thermodynamics).

These conversion processes are very complex, especially the combustion process with its energy-substance exchange processes and the chemical processes of the gas in the cylinder.[4] In addition, the process of the transfer of heat from the gas to the wall directly surrounding the combustion chamber, the neighboring engine components, and the coolant or oil can be approximated only with great effort.[5–8]

Since the fuels for spark-ignition and diesel engines are mixtures consisting of various hydrocarbons, it is practically impossible to describe the reaction kinetics of the numerous reactions. Frequently, pure substances such as methanol, methane, and hydrogen are used that have a sufficiently precise reaction mechanism with all the associated substance data. Depending on the methodology, it is sufficient to use specific reaction processes such as formation of NO[9] or the simplifying assumption of an O-H-C equilibrium at the flame front.[10]

If the process is considered from a locally multidimensional, nonstationary perspective with all of the transport mechanisms that actually exist in the gas, complex mathematical models result that yield a somewhat imprecise substance data (if any at all) for the physical-chemical description.

Hence, to obtain qualitative information on the relationship of certain process variables to predetermined parameters, more or less simple model calculations are used. This allows basic conclusions to be made regarding the effect of the conversion of energy based on engine-related parameters that are much less complex.

A series of methodologies were produced in the past that extend from a simple closed process control to more or less complicated open multizone models.[9,11–13]

5.1 Cyclical Processes

To obtain basic information, simplified models are created and described as cyclical processes. Cyclical processes are sequential state changes of a fuel in which the fuel is returned to its initial state. They are described as closed cyclical processes with the supply and removal of heat (Fig. 5-1).

This type of model does not treat the conversion of the initial products of combustion such as air and fuels into exhausts (CO, HC, NO_x, CO_2, HCO, H_2, N_2, etc.).

The four cycles of the combustion engines are compression, supply of heat as a "replacement" for the combustion process, expansion, and heat removal as a replacement of the charge cycle. The state of the medium, for example, at the beginning of compression and at the end of heat removal is identical.

State diagrams for internal combustion engines are

- Pressure-Volume diagram (*PV* diagram): The contained area represents work that is termed the indicated work.
- Temperature-Entropy diagram (*TS* diagram): The areas represent heat. The cyclical process work is the difference between the supplied and the removed heat. The area enclosed by the lines of the state changes is a measure of the useful work of the cyclical process.

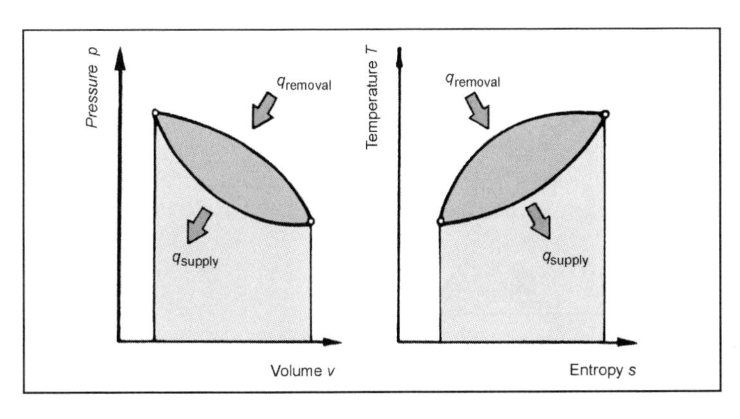

Fig. 5-1 State changes and work in a cyclical process.[14]

Essential information on the engine process attainable with such cyclical processes relates to the process efficiency.

A definition of one such type of efficiency, thermal efficiency, is

$$\eta_{th} = \frac{q_{zu} - q_{ab}}{q_{zu}} = 1 - \frac{q_{ab}}{q_{zu}} \tag{5.1}$$

where

q_{zu} = supplied quantity of heat

q_{ab} = removed quantity of heat.

The theory of cyclical processes originates from the French officer Sadi Carnot (1796–1832) who recognized that to convert heat into work there must be a temperature gradient. He also noticed that the thermal efficiency of a heat engine increases with the increase of temperature where the heat is supplied and with the decrease of temperature where it is removed. This becomes particularly clear with the optimum cyclical process that he described, the Carnot process (Fig. 5-2).

The state changes of the Carnot process are

- Isothermic compression
- Isentropic compression
- Isothermic expansion
- Isentropic expansion

In the *TS* diagram, the Carnot process is portrayed as a rectangle. The thermal efficiency results as a ratio of useful work to supplied heat.

$$\eta_{th} = \frac{q_{zu} - q_{ab}}{q_{zu}} = 1 - \frac{q_{ab}}{q_{zu}} \tag{5.2}$$

$$\eta_{th_c} = 1 - \frac{T_{min} \cdot (s_1 - s_2)}{T_{max} \cdot (s_4 - s_3)} = 1 - \frac{T_{min}}{T_{max}} \tag{5.3}$$

The thermal efficiency assumes the highest attainable value at a given temperature ratio in the Carnot process. In the *PV* diagram, the diagram area of the Carnot process is so small that the temperatures and pressures would have to be raised to an unacceptable level to obtain acceptable useful work (corresponding to the area in the *PV* diagram). This was realized by Rudolf Diesel when he wanted to implement the Carnot process with his rational heat engine. A rectangular process in the *PV* diagram yields the greatest amount of work, but is much less efficient because of the small area in the *TS* diagram. A rectangular process is therefore not suitable in practice.

The cyclical processes that are technically feasible with a heat engine are subject to the restrictions of the geometry and kinematics of the respective machine type, the conditions of energy conversion, and the state of the art. The evaluative criteria for comparative processes that are described in the following are

- Efficiency
- Work yield
- Technical feasibility

5.2 Comparative Processes

5.2.1 Simple Model Processes

The cyclical processes of an engine describe the energy conversion where the individual state changes of the fuel most closely approximate the actual behavior in the engine. With this in mind, internal combustion engines represent closed systems in which the energy conversion is discontinuous. A characteristic of the cyclical processes of internal combustion engines is that state changes occur in a work area whose size changes as a result of the movement of the crankshaft drive over the course of the combustion cycle. Compression and expansion can be described by simple state changes. The combustion and the charge cycle are replaced by heat addition and heat removal.

Ideal cyclical processes for internal combustion engines are differentiated according to the type of heat supply. A general process can be represented by the heat supply at a constant volume (isochor) and at constant pressure (isobar), as described by Myron Seiliger (1874–1952) as the Seiliger process. Borderline cases can be derived from this such as pure constant volume (only an isochoric supply of heat) and pure constant pressure (only isobaric heat supply) cycles.

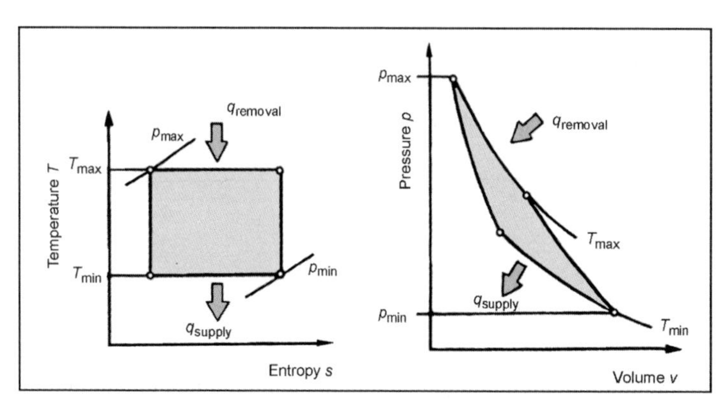

Fig. 5-2 State changes in a Carnot process.[14]

5.2.1.1 Constant Volume Cycle

Figure 5-3 presents the state change during the constant volume cycle. The sequence of state changes in this process is

- Isentropic compression
- Isochoric supply of heat
- Isentropic expansion
- Isochoric heat removal

This is the thermodynamically best process that can occur in a machine with a periodically changing working chamber with a reasonable amount of engineering.[1] Given the same compression ratio, the resulting thermal efficiency is greater than that of the Seiliger cycle and the constant pressure cycle. The efficiency depends on the type of gas (isentropic exponent) and the compression ratio. The constant volume cycle increases with the compression ratio and is calculated with

$$\eta_{\text{th}} = \frac{q_{\text{zu}\,v} - q_{\text{ab}}}{q_{\text{zu}\,v}} \tag{5.4}$$

5.2.1.2 Constant Pressure Cycle

The state changes of the constant pressure cycle are in Fig. 5-4. The sequence of the state changes in this process is

- Isentropic compression
- Isobaric supply of heat
- Isentropic expansion
- Isochoric heat removal

It can then be used as a comparative process when, for reasons of component load, the maximum pressure must be limited. The thermal efficiency is calculated as follows:

$$\eta_{\text{th}} = \frac{q_{\text{zu}\,p} - q_{\text{ab}}}{q_{\text{zu}\,p}} \tag{5.5}$$

The efficiency of this process depends on the gas type (isentropic exponent), the compression ratio, and the supplied quantity of heat at a constant pressure. It rises as the compression ratio increases and falls as the supply of heat increases. Of the three considered process controls, the constant pressure cycle has the least efficiency.

5.2.1.3 Seiliger Process

The state changes of the Seiliger process are shown in Fig. 5-5. In particular, these are

- Isentropic compression
- Isochoric supply of heat
- Isobaric supply of heat
- Isentropic (adiabatic reversible) expansion
- Isochoric heat removal

At a given compression ratio, a maximum pressure limit must be specified. The heat supply is partly isochoric and partly isobaric. The thermal efficiency from this process control is

$$\eta_{\text{th}} = \frac{q_{\text{zu}\,v} + q_{\text{zu}\,p} - q_{\text{ab}}}{q_{\text{zu}\,v} + q_{\text{zu}\,p}} \tag{5.6}$$

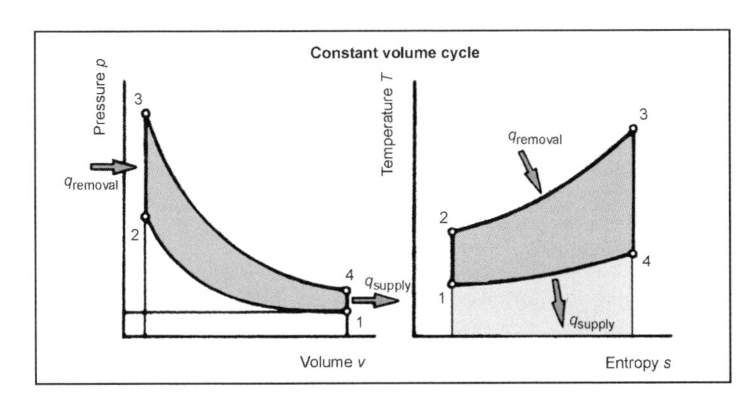

Fig. 5-3 State changes in the constant volume cycle.[14]

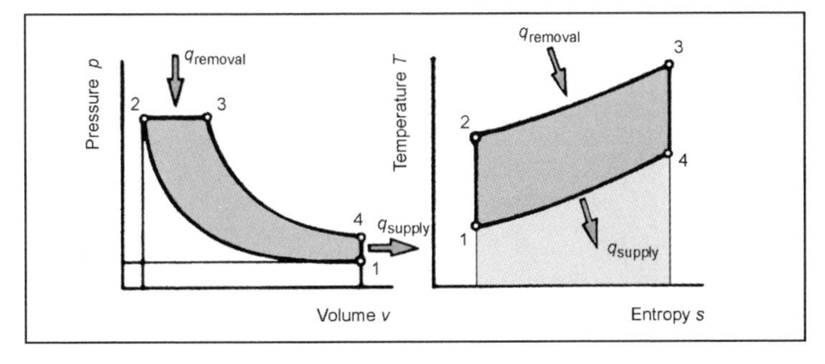

Fig. 5-4 State changes in the constant pressure cycle.[14]

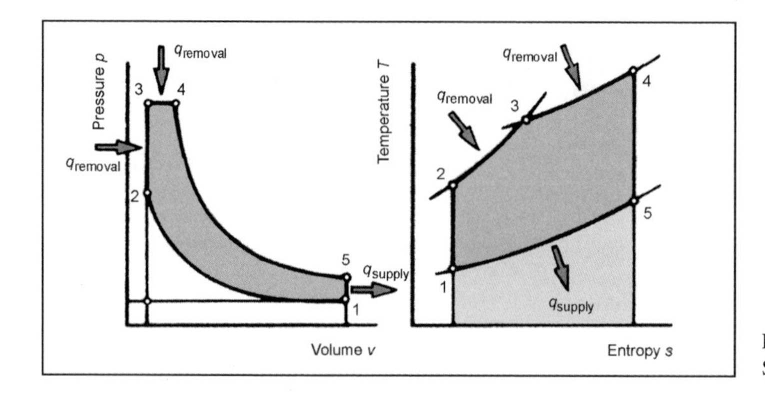

Fig. 5-5 State changes in the Seiliger process.[14]

It must be noted that the quantity of heat $q_{zu\,v}$ is supplied at a constant volume, and therefore it is measured from the temperature difference with reference to the specific heat at a constant *volume* (c_v). The supply of heat at a constant pressure $q_{zu\,p}$ is measured from the temperature difference with reference to the specific heat at a constant *pressure* (c_p).

Depending on the distribution of the supplied quantity of heat between the isochoric and isobaric state changes, the thermal efficiency results as a limit curve that would exist with constant volume and constant pressure.

Applying this process to a supercharged engine yields the relationships shown in Fig. 5-6.

In principle, supercharging does not change the process in the engine; only the pressure level rises. The compression in the engine is upstream from the compression in the compressor, and the expansion in the turbine follows the expansion in the engine and expansion in the exhaust pipe.

- Isentropic compression in the compressor
- Isentropic compression in the engine
- Isochoric supply of heat in the engine
- Isobaric supply of heat in the engine
- Isentropic expansion in the engine
- Isochoric heat removal from the engine
- Isobaric supply of heat to the turbine
- Isentropic expansion in the turbine
- Isobaric heat removal from the turbine

The work of the exhaust turbine and the compressor are correspondingly represented as the areas in the *PV* diagram (Fig. 5-6).

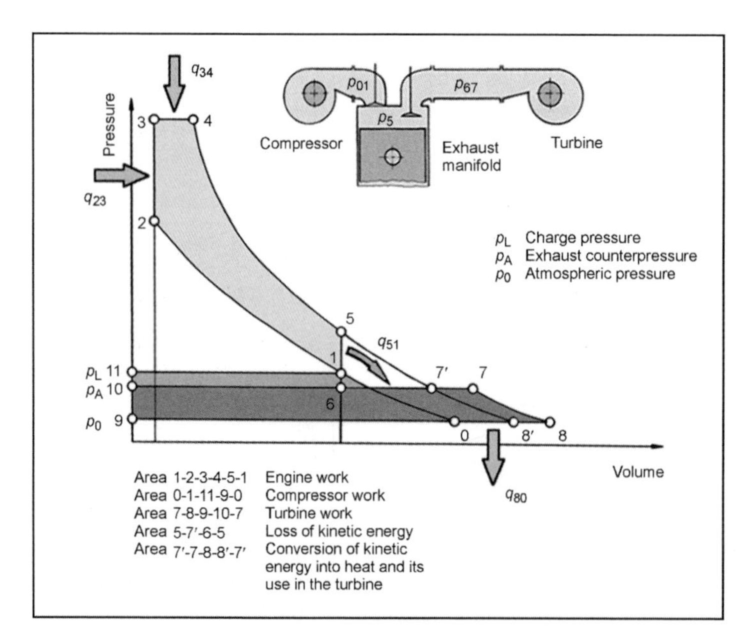

Fig. 5-6 State changes in the Seiliger process of an exhaust-gas turbocharged engine.[14]

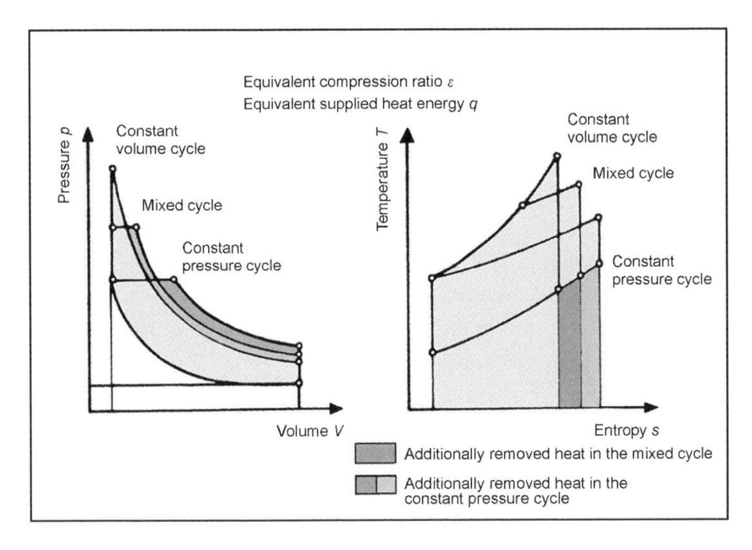

Fig. 5-7 Comparative engine processes.[14]

5.2.1.4 Comparison of the Cyclical Processes

Figure 5-7 compares the three considered processes in the *PV* and *TS* diagrams. The efficiency of the constant volume cycle is the maximum attainable given an equivalent compression ratio. This is because of the low quantity of heat that is removed given an equivalent compression ratio and the same amount of supplied heat in comparison to the two other process controls.

5.2.2 Energy Losses

The exergetical perspective of the discussed process controls shows that the exergy of the supplied energy can be only partially converted into mechanical work. Exergy is the energy that can be converted into any other form of energy in a predetermined environment. Anergy is the part of the energy that cannot be converted into exergy.[1]

An illustration of this in an example of the constant volume process is provided in Fig. 5-8. In the *PV* diagram, two types of process loss can be illustrated:

- If the medium is expanded from point 4 to point 5, i.e., to the initial pressure, the work (area 4-5-1-4) would be useful.
- The area 5-6-1-5 would be useful if the medium is expanded to the initial pressure as well as to the initial temperature. This must be followed by isothermic compression to the initial pressure.

However, in a real engine, this would require a substantial amount of additional engineering that would be out of proportion to the gain.

The third loss arises from the anergy of the supplied energy. It is not directly attributable to the process control. If a medium reaches the environmental temperature and environmental pressure, it is in a thermal and mechanical equilibrium with the environment. The second law of thermodynamics prevents the conversion of internal energy into exergy or useful work.[1]

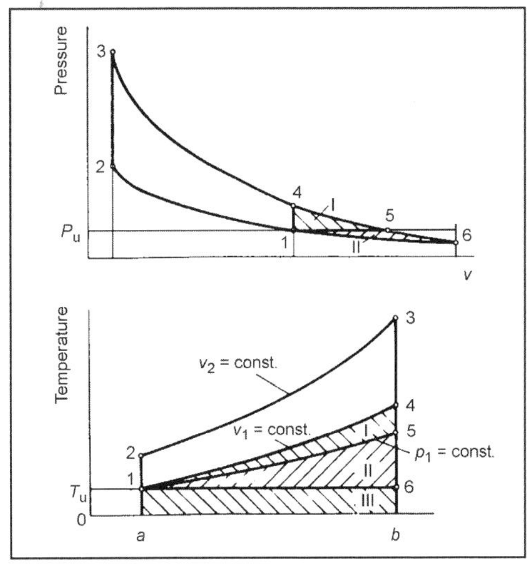

Fig. 5-8 Thermodynamic loss with the example of the constant volume cycle.

5.3 Open Comparative Processes

5.3.1 Work Cycle of the Perfect Engine

The ideal cyclical processes are only a crude approximation that can be used to arrive at a few basic conclusions. In regard to efficiency, they yield satisfactory values in comparison with reality: The work yield is greater and the efficiency is better than in real engines since the properties of the working gas, air, is treated as a real gas. Further, the heat loss, charge cycle loss, friction loss, and chemical reactions are not included.

To obtain more detailed information on the process cycle and answers regarding the optimum process control, further processes have been defined that allow for a better approximation of real engines. This is possible with open comparative processes. A helpful and frequently used comparative process is the "perfect" engine process.

The parameters under which this process occurs are as follows:

- The charge in the combustion chamber has no residual exhaust gas.
- The air-fuel ratio is the same as the actual engine.
- There is a loss-free charge cycle (no flow and leakage loss).
- Combustion occurs according to set laws.
- Heat-insulating walls are present.
- Isentropic compression and expansion occur with specific heats c_p and c_v depending on the temperature.
- The combustion products are in chemical equilibrium.

With the process defined in this manner, we can determine the influences of the parameters of compression and air-fuel ratio on average pressure, process effectiveness, and a few concentrations of substance components (Fig. 5-9).

Depending on the methodology, a process control can be selected that uses simple cyclical processes. This can be an isochoric (constant volume combustion), an isobaric (constant pressure compression), or a mixed isochoric-isobaric cycle.

5.3.1.1 Elements of Calculation

The calculation of the cycle of the perfect engine can be divided into the following steps:

(a) Isentropic compression of the fresh mixture. The initial state is described by the pressure p_0, the temperature

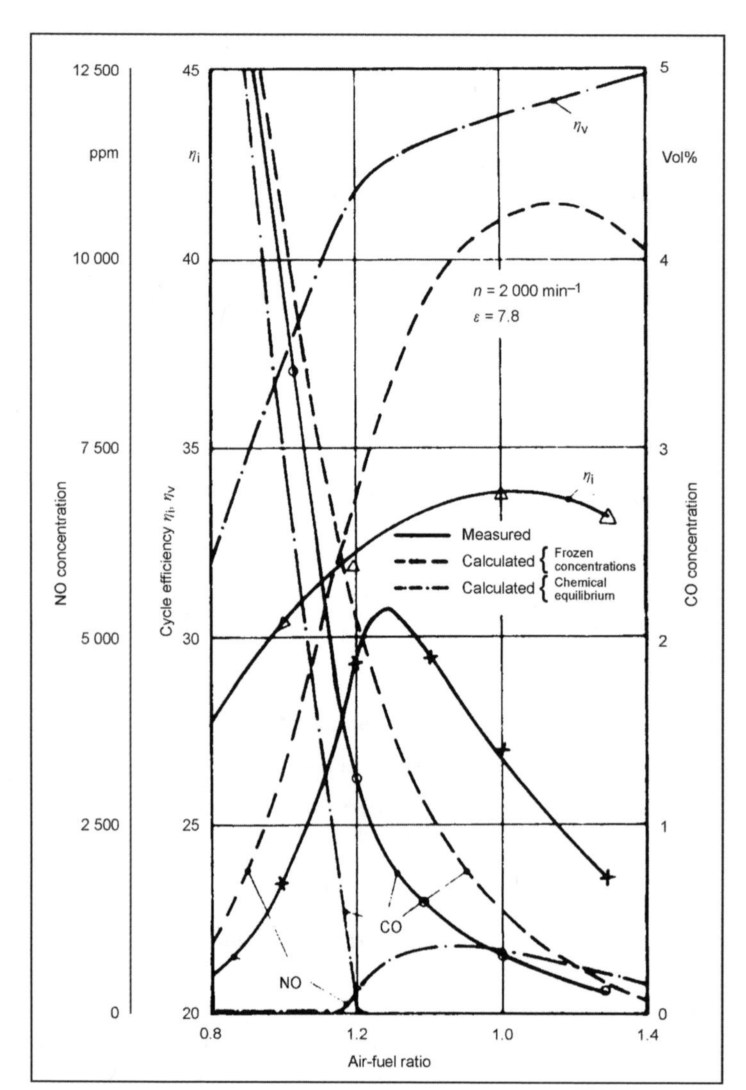

Fig. 5-9 Quantities calculated with the working cycle of the perfect engine, and quantities measured on a test bench.[15]

T_0, and the composition of the fresh gas characterized by the air-fuel ratio λ. This is defined by

$$\lambda = \frac{\dot{m}_{\text{Air}}}{\dot{m}_{\text{Kr}} \cdot m_{\text{Air, stoich}}} \tag{5.7}$$

where $\dot{m}_{\text{Air}}$ refers to the air mass, $\dot{m}_{\text{Kr}}$ the fuel mass, and $m_{\text{Air, stoich}}$ the stoichiometric air mass of the corresponding fuel. The compression ratio can be taken from the corresponding test engine. As a representative of gasoline, isooctane (C_8H_{18}) can be used since it more or less approximates the physical and chemical properties of commercially available fuels.

It is assumed that the gas composition during compression remains constant. With p = pressure, v = specific volume, T = temperature, R_m = general gas constant, and σ_i = specific moles of component i, the final compression state can be calculated with the aid of isentropic relationship $S_{1,T1} = S_{2,T2}$ and the thermal state equation for ideal gases $p \cdot v = \sum_i \sigma_i \cdot R_m \cdot T$, using the following equation:

$$\sum_i \sigma_{i,1} \cdot \left(s^0_{i,T1} - R_m \cdot \ln \frac{p_1}{p^0} \right) = \sum_i \sigma_{i,2} \cdot \left(s^0_{i,T2} - R_m \cdot \ln \frac{p_2}{p^0} \right) \tag{5.8}$$

$s^0_{i,T1}$ is the entropy of component i at standard pressure p^0 and temperature T.

The solution to the equation can, for example, be found using an iterative process.

(b) Isochoric adiabatic combustion. It is assumed that there is total chemical equilibrium. The combustion products consist, for example, of the components:

CO, CO_2, N_2, NO, NO_2, NH_3, O_2, O, H, N, H_2, H_2O, and OH.

The state of the gas mixture in the cylinder after combustion is characterized by the pressure p_3, the temperature T_3, and the specific moles of the participating (in this case, 13) components. To determine these quantities, 15 independent equations are necessary. These equations are

1. First law of thermodynamics for closed systems

If we assume that no heat is supplied or removed during combustion and no work is done, it follows that $du = 0$; i.e., there is no change in the internal energy. The following accordingly results:

$$\sum_i \sigma_{i,2} \cdot u_{i,T2} = \sum_i \sigma_{i,3} \cdot u_{i,T3} \tag{5.9}$$

2. Thermal state equation

This is expressed by the following:

$$p_3 \cdot v_3 = \sum_i \sigma_i \cdot R_m \cdot T_3 \tag{5.10}$$

3. Chemical equilibrium

The 13 gas components that chemically react with each other consist of the basic elements oxygen, nitrogen, hydrogen, and carbon. To describe the chemical equilibrium, nine independent reaction equations are therefore required with the stoichiometric coefficients $\tau_{j,i}$ (j = 1 to 9):

$$\sum_i \mu_i \cdot \tau_{j,i} = 0.$$

μ_i is the chemical potential of component i and is defined as

$$\mu_i = g^0_{i,T} + R_m \cdot T \cdot \ln \frac{p_i}{p^0} \tag{5.11}$$

where $g^0_{i,T}$ represents the molar free enthalpy of component i in a standard state.

4. Material balances

The remaining equations for determining the state after combustion are provided by the material balances. During combustion, the amount of the four basic materials j = 1–4 O, H, N, and C does not correspondingly change so that the material balances are expressed as follows:

$$\sigma_{B,j} = \sum_i \alpha_{j,i} \cdot \sigma_i \tag{5.12}$$

$\alpha_{j,i}$ is the number of atoms of the basic material j in component i.

The nonlinearity equation system that thereby arises consisting of 15 equations can (for example) be solved using a Newtonian method.

(c) Expansion. The parameters to represent the expansion state are the chemical equilibrium and constant gas composition. The state change is isentropic. The following equation results:

$$\sum_i \sigma_{i,3} \cdot \left(s^0_{i,T3} - R_m \cdot \ln \frac{p_3}{p^0} \right) = \sum_i \sigma_{i,3} \cdot \left(s^0_{i,T4} - R_m \cdot \ln \frac{p_4}{p^0} \right) \tag{5.13}$$

5.3.1.2 Work of the Perfect Engine

The work W_{VM} of the perfect engine results from the difference of the internal energy as expressed by the following:

$$W_{\text{VM}} = U_4 - U_1 \tag{5.14}$$

or by

$$W_{\text{VM}} = m \cdot \left(\sum_i \sigma_{i,1} \cdot u_{i,T1} - \sum_i \sigma_{i,4} \cdot u_{i,T4} \right) \tag{5.15}$$

where U and u_i represent internal work.

5.3.1.3 Effectiveness of the Perfect Engine

The effectiveness η_{VM} of the perfect engine is basically defined as

$$\eta_{\text{VM}} = \frac{W_{\text{VM}}}{m_{\text{Kr}} \cdot H_u} \tag{5.16}$$

with H_u as the bottom calorific value of the fuel, and m_{Kr} as the fuel mass. If the effectiveness is defined as the ratio of the obtained process work W_{VM} and the maximum theoretically obtainable work, $m_{Kr} \cdot H_u$ must be replaced by the term $W_{theoretisch}$. The quantity $W_{theoretisch}$ then can be defined as the maximum obtainable work in a reversible process control, or as the reversible reaction work. This results from the difference of the free enthalpy from the state of the fresh mixture and the exhaust gas in the equation:

$$W_{theoretisch} = \frac{H_{T0}^{n} - H_{T0}^{nn} - T_0 \cdot \left(\sum_i S_{i,p0,T0}^{n} - \sum_i S_{i,p0,T0}^{nn} \right)}{m_{Kr}} \approx H_u \tag{5.17}$$

where H_{T0}^{n} and H_{T0}^{nn} are the enthalpy of the material flows of the combusted and noncombusted material in reference to the environmental state; $S_{i,p0,T0}^{n}$ and $S_{i,p0,T0}^{nn}$ represent the entropy of component i in the combusted and uncombusted material in reference to the environmental state.

The differences from the reversible reaction work and bottom caloric value are very low for a few substances defined as substitute fuels such as C_7H_{14}, C_8H_{18}, or methanol so that $W_{theoretisch}$ is approximately the same as H_u. For hydrogen, the difference is approximately 6%.[15]

5.3.1.4 Exergy Loss in the Perfect Cycle

From the basic characteristic of the effectiveness in a perfect engine, we can see that the effectiveness rises with the air-fuel ratio. To further discuss these results, we need to look at exergy loss. The specific exergy for a closed system is defined by the following:

$$e_{T,p} = u_T - u_{0,T0} - T_0 \cdot (s_{T,p} - s_{0,T0,p0}) + p_0 \cdot (v - v_0) \tag{5.18}$$

u_T and $s_{T,p}$ mean the specific internal energy or entropy at temperature T and pressure p, and $u_{0,T0}$ and $s_{0,T0,P0}$ are quantities that result when the combustion gases are in a thermodynamic equilibrium with the environment.

The relative energy loss E_V of combustion can be defined by the following equation:

$$E_V = \frac{E_2 - E_3}{E_1} \tag{5.19}$$

and the relative exhaust exergy loss is defined by

$$E_A = \frac{E_4}{E_1} \tag{5.20}$$

Figure 5-10 shows the characteristic of the relative exergy loss in a perfect spark-injection cycle.

The relative exergy loss of the exhaust *falls* as the air-fuel ratio increases, whereas the relative exergy loss of combustion *rises* as the air-fuel ratio increases. The overall result is an increase of the effectiveness with the air-fuel ratio.

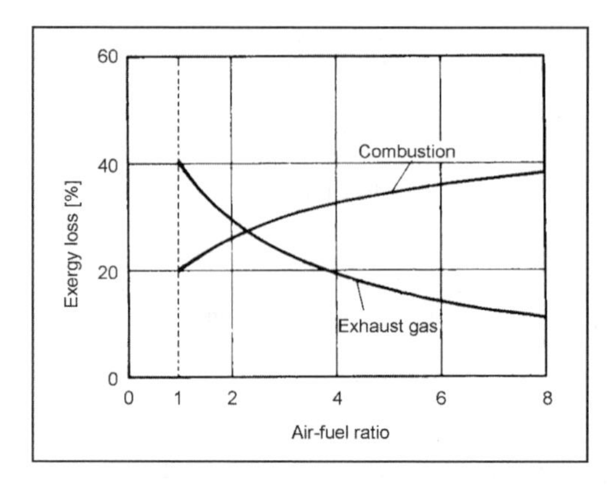

Fig. 5-10 Exergy loss from combustion and exhaust (according to Ref. [15]).

5.3.2 Approximation of the Real Working Cycle

The simple cyclical processes, as well as the process of the perfect engine provide only limited information on the real processes occurring in the engine. Models are therefore necessary to further approximate the real process. In particular, information on the indicated average pressure, internal effectiveness, combustion processes (combustion functions), combustion temperatures, pollutant formation, etc., is desirable. Such information is obtained from models that, for example, can be described as two-zone models.

Additional model calculations are possible that are based on the specified injection rate that can be used to gain information on the combustion and NO emissions,[8,16,17] or that use single-zone models with a set substitute combustion characteristic.[18]

Many of these models do not include the reaction process but instead use suitable functions that describe the energy released from combustion[19] such as the die Vibe function.[20]

More extensive thermodynamic analyses can yield models that use local coordinates in addition to the progress over time of parameters. However, because of their multidimensionality, these require a large amount of computing.

5.3.2.1 Models to Determine Combustion Behavior

Since it is practically impossible to directly determine the conversion of material over time during combustion in the engine, model calculations are used. Despite the simplification, experience shows that they can at least yield very good qualitative information.

We now discuss a model based on thermodynamics that is defined as follows:

- The use of the pressure characteristic is measured in the engine for calculating the cycle.
- At the time of ignition, the contents within the cylinder consist of residual exhaust gas and a fresh mixture.

- The mass flowing into the cylinder remains completely in the cylinder (no mass loss).
- During compression, no chemical reactions occur.
- The charge in the cylinder during combustion consists of two homogeneous areas in reference to pressure, temperature, and composition (area I = noncombusted material; area II = combusted material).
- The two homogeneous areas are separated by an infinitesimally thin flame front and exchange mass but no heat.
- The state change of area I occurs at constant enthalpy.
- The gas leaving the flame front is conveyed into area II and mixes with it to form a new state of equilibrium.
- The transfer of heat between the respective areas (combusted, noncombusted) to the combustion chamber wall occurs according to fixed laws.
- The composition in area I does not change during combustion.

The goals are to determine temperature as a function of time in the combusted and uncombusted materials, the specific moles in the combusted material, and the so-called combustion function that expresses the ratio of combusted fuel mass to the overall fuel mass. In the uncombusted material, the specific moles do not change by definition. These quantities can yield information on the combustion speed, the length of combustion, and the combustion delay. The process is then calculated with the following steps as a function of time or the crankshaft angle α:

1. Cylinder charge and the beginning of the reaction

The temperature can be determined with the thermal state equation

$$p \cdot v = \sum_i \sigma_i \cdot R_m \cdot T \tag{5.21}$$

and the empirically determined quantities of combustion chamber pressure, the volume above the piston, and fresh gas composition.

2. Combustion process

Zone I of noncombusted material: The thermal state equation and the first law of thermodynamics for open systems yield

$$p \cdot v_{\mathrm{I}} = \sum_i \sigma_{i\mathrm{I}} \cdot R_m \cdot T_{\mathrm{I}} \tag{5.22}$$

and

$$\frac{dT_{\mathrm{I}}}{d\alpha} = \frac{1}{\sum_{i=1}^{k_{\mathrm{I}}} \sigma_{i\mathrm{I}} \cdot c_{p\,mi}(T_{\mathrm{I}})} \cdot \left(\frac{dq_{\mathrm{I}}}{d\alpha} + \frac{R_m \cdot T_{\mathrm{I}}}{p} \cdot \frac{dp}{d\alpha} \cdot \sum_{i=1}^{k_{\mathrm{I}}} \sigma_{i\mathrm{I}} \right) \tag{5.23}$$

Zone II (combusted material): The unknown quantities are the k_{II}-specific moles in the combusted material $\sigma_{i\,\mathrm{II}}$, the temperature T_{II}, and the converted mixture mass. It is useful to use the components CO_2, CO, OH, H, O, O_2, H_2O, H_2, and N_2 as inert components for the gas composition in zone II. r-independent equations for the chemical equilibrium and b equations from the basic material balances ($k_{\mathrm{II}} = r + b$) serve to determine the k_{II}-specific moles. The equation system is completed with an independent equation for the temperature in the combusted material and the material conversion. This is characterized by the combustion function that is defined as follows:

$$x_B = \frac{m_{\mathrm{II}}}{m_{\text{overall fuel mass}}}$$

Accordingly, r equations result in the following form:

$$\sum_{i=1}^{k_{\mathrm{II}}} v_{i,j} \cdot \left(S^0_{\mathrm{mi}}(T_{\mathrm{II}}) - R_m \cdot \ln \frac{\sigma_{i\mathrm{II}}}{\sum_{i=1}^{k_{\mathrm{II}}} \sigma_{i\,\mathrm{II}}} \cdot \frac{p}{p^0} \right) \cdot \frac{dT_{\mathrm{II}}}{d\alpha}$$
$$= R_m \cdot T_{\mathrm{II}} \cdot \sum_{i=1}^{k_{\mathrm{II}}} v_{i,j} \cdot \left(\frac{v_{i,j}}{\sigma_{i\mathrm{II}}} - \frac{\sum_{i=1}^{k_{\mathrm{II}}} v_{i,j}}{\sum_{i=1}^{k_{\mathrm{II}}} \sigma_{i\mathrm{II}}} \right) \cdot \frac{d\sigma_{i\mathrm{II}}}{d\alpha}$$
$$- \frac{dp}{d\alpha} \cdot \frac{R_m \cdot T_{\mathrm{II}}}{p} \cdot \sum_{i=1}^{k_{\mathrm{II}}} v_{i,j} \tag{5.24}$$

and b equations from the basic material balance:

$$\sum_{i=1}^{k_{\mathrm{II}}} a_{i,l} \cdot \frac{d\sigma_{i,\mathrm{II}}}{d\alpha} = 0 \qquad \text{with } l = 1 \ldots b \tag{5.25}$$

The equations for the temperature of the noncombusted material include

$$\frac{dT_{\mathrm{II}}}{d\alpha} = \frac{1}{x_B \cdot \sum_{i=1}^{k_{\mathrm{II}}} \sigma_{i,j} \cdot cp_{mi}(T_{\mathrm{II}})}$$
$$\times \left[\left(\sum_{i=1}^{k_{\mathrm{II}}} h_{i,\,\mathrm{Flame}} - \sum_{i=1}^{k_{\mathrm{II}}} \sigma_{i,j} \cdot H_{mi}(T_{\mathrm{II}}) \right) \frac{dX_B}{d\alpha} + x_B \cdot \frac{dq_{\mathrm{II}}}{d\alpha} + \frac{x_B}{p} \cdot R_m \cdot T_{\mathrm{II}} \cdot \frac{dp}{d\alpha} \cdot \sum_{i=1}^{k_{\mathrm{II}}} \sigma_{i,j} \right]$$
$$- x_B \cdot \sum_{i=1}^{k_{\mathrm{II}}} H_{mi}(T_{\mathrm{II}}) \cdot \frac{d\sigma_{i,\mathrm{II}}}{d\alpha} \tag{5.26}$$

The equation for the percent fuel conversion is

$$\frac{dx_B}{d\alpha} = \frac{1}{\frac{R_m}{p} \cdot \left(T_{\mathrm{II}} \cdot \sum_{i=1}^{k_{\mathrm{II}}} \sigma_{i,\mathrm{II}} - T_{\mathrm{I}} \cdot \sum_{i=1}^{k_{\mathrm{I}}} \sigma_{i,\mathrm{I}} \right)}$$
$$\times \left[\frac{dV}{m \cdot d\alpha} - \frac{x_B \cdot R_m}{p} - \frac{(1 - x_B) \cdot R_m}{p} \times \left(\frac{T_{\mathrm{II}}}{d\alpha} \cdot \sum_{i=1}^{k_{\mathrm{I}}} d\sigma_{i,\mathrm{I}} - \frac{T_{\mathrm{I}}}{p} \cdot \frac{dp}{d\alpha} \sum_{i=1}^{k_{\mathrm{I}}} d\sigma_{i,\mathrm{I}} \right) \right]$$
$$\times \left(\frac{dT_{\mathrm{II}}}{d\alpha} \cdot \sum_{i=1}^{k_{\mathrm{II}}} \sigma_{i,\mathrm{II}} + \mathrm{T}_{\mathrm{II}} \cdot \sum_{i=1}^{k_{\mathrm{II}}} \frac{d\sigma_{i,\mathrm{II}}}{d\alpha} - \frac{T_{\mathrm{II}}}{p} \times \frac{dp}{d\alpha} \sum_{i=1}^{k_{\mathrm{II}}} d\sigma_{i,\mathrm{II}} \right) \tag{5.27}$$

There are accordingly $k_{II} + 3$ equations for determining the combustion function x_B, the temperature in the noncombusted material T_I, the temperature of the combusted material T_{II}, and the composition of the combustion material $\sigma_{1,II} \ldots \sigma_{8,II}$.

Typical definitions that can be represented with such models are shown in Figs. 5-11 and 5-12.

A better approximation of the real process, especially the real combustion in the engine, can be obtained by including in the model calculation transport processes such as diffusion and heat conduction in the gas.

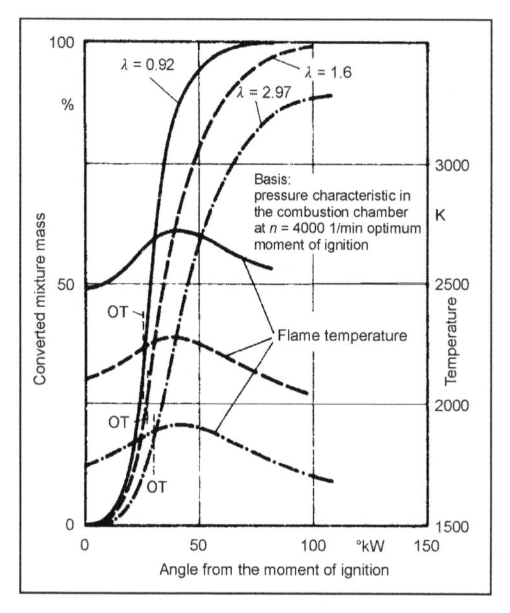

Fig. 5-11 Calculated combustion function and flame temperatures using a two-zone model (methanol-H_2).

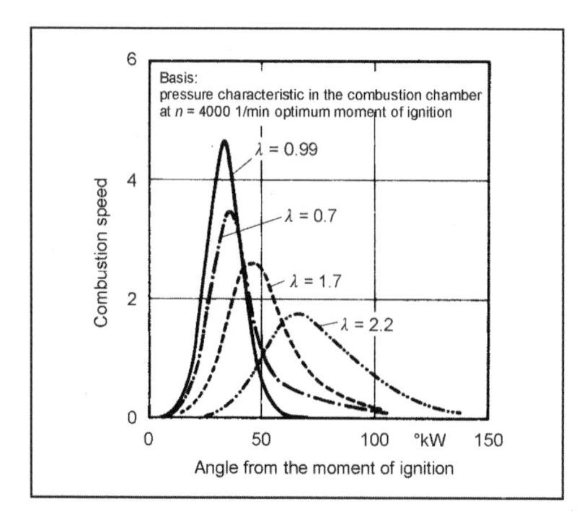

Fig. 5-12 Calculated combustion rates using a two-zone model (methanol-H_2).

This requires the description of both the temporal and local behavior of important process quantities. The necessary formulation of the balance equations is based on the thermodynamics of irreversible processes. Continuous systems are considered; i.e., the intensive state variables such as the temperature, pressure, and density are always functions of time and place. The balance equations describe the local changes in each volume element. In addition to the source therm for production or decomposition of the permitted components, there is an exchange of energy and material with the neighboring element.[4] If friction influences and the temporal and local pressure gradients are not included, the essential equations to describe such systems are the quantity balance and energy balance.

(a) Quantity balance: Taking into account chemical reactions and diffusion, the following results for the change of specific moles σ_i:

$$\rho \frac{\partial \sigma_i}{\partial t} = -v \cdot \rho \frac{\partial \sigma_i}{\partial x} - \frac{\partial I_i}{\partial x} + \sum_{j=1}^{r} (v_{j,i}^{n} - v_{j,i}^{nn}) \cdot J_j \tag{5.28}$$

(b) Energy balance: Not included are external force fields, friction influences, and local and temporal pressure gradients.

$$\begin{aligned}\sum_{i=1}^{k} \sigma_i \cdot \rho \frac{\partial H_{m,i}}{\partial t} = &-\sum_{i=1}^{k} H_{m,i} \sum_{j=1}^{r} \cdot (v_{j,i}^{n} - v_{j,i}^{nn}) \cdot J_j \\ &- \sum_{i=1}^{k} \vec{I_i} \cdot \mathrm{gradH}_{m,i} \\ &- \mathrm{div}\vec{I_Q} \sum_{i=1}^{k} \sigma_i \cdot \rho \cdot v \cdot \mathrm{grad}\, H_{m,i}\end{aligned} \tag{5.29}$$

i means the number of permitted components in the gas, j means the number of permitted chemical reactions, nn characterizes the combusted materials, n is the noncombusted materials, I_j is the diffusion flow density, J_j is the reaction speed of reaction j, $\vec{I_Q}$ characterizes the heat flow, and $H_{m,i}$ is the partial molar enthalpy of component i.

5.4 Efficiency

The consideration of the simple cyclical processes (Section 5.2) yields efficiency defined as thermal efficiency η_{th}, which can be evaluated as the maximum possible efficiency depending on the selected process. Given the previously cited prerequisites, the "perfect engine" yields efficiency η_v that is less efficient than η_{th} with the same process control.

As the computational models grow closer in their approximation of the real process, we grow increasingly distant from the ideal. The obtained efficiency continually falls and more closely approximates reality.

The deviations in efficiency of the perfect engine from the internal efficiency η_i of a real engine are determined by the following:

- Incomplete combustion and combustion process. The exhaust still contains components that can be further oxidized and hence represents a calorific value that is not exploited in the process. In addition, the real combustion process deviates from the comparative process.
- Leaks, heat loss, and charge cycle loss.

The internal efficiency η_i of a real engine can be determined from the indication of the high-pressure and low-pressure loops. The additional step to obtain effective efficiency η_e is to consider additional loss such as friction loss (powertrain friction, accessories, auxiliary drives, etc.).

5.5 Energy Balance in the Engine

If an engine is operated while stationary, i.e., with a fixed operating point, the process is a stationary flow process in which technical work is accomplished. To portray an energy balance, a system limit is defined, and the material and energy flows that go beyond this limit are considered (Fig. 5-13).

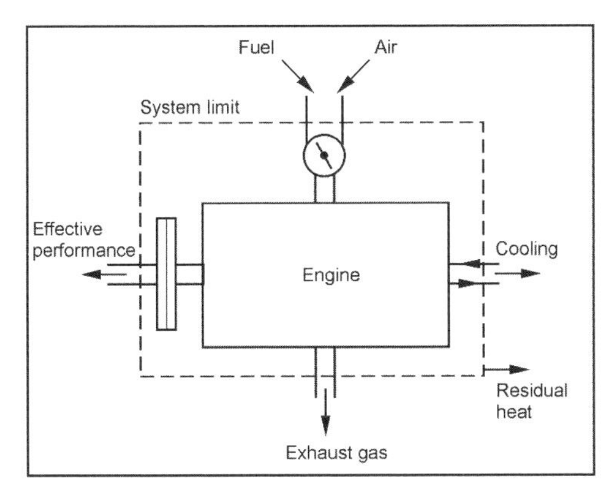

Fig. 5-13 Material and energy flows in the engine.

In particular, the following flows go beyond the system limit:

P_e Effective power

$\dot{Q}_{\text{Rest}}$ Residual heat (flow of heat into the environment due to heat radiation, heat conduction, and convection)

$\dot{H}_{\text{Air}}$ Enthalpy flow of the air

$\dot{H}_{\text{Kr}}$ Enthalpy flow of the fuel

$\dot{H}_{\text{KWE}}$ Enthalpy flow of the cooling water (entrance)

$\dot{H}_{\text{KWA}}$ Enthalpy flow of the cooling water (exit)

$\dot{H}_{\text{exhaust}}$ Enthalpy flow of the exhaust gas

5.5.1 Balance Equation

If the material and energy flows that pass through the control chamber are balanced, the following results:

$$\dot{H}_{\text{Kr}} + \dot{H}_{\text{Air}} + \dot{H}_{\text{KWE}} = \dot{H}_{\text{KWA}} + P_e + \dot{Q}_{\text{Rest}} + \dot{H}_{\text{exhaust } T_2} \tag{5.30}$$

The energy difference from different gas flow speeds between entering and leaving the engine is not considered. The air and fuel are converted by means of a chemical process into exhaust. For calculation, the definition of the calorific value is used:

$$H_u = \frac{\dot{H}'_1 - \dot{H}''_1}{\dot{m}_{\text{Kr}}} \tag{5.31}$$

where $\dot{H}'_1$ is the enthalpy flow of the noncombusted materials at temperature T_1, and $\dot{H}''_1$ is the enthalpy flow of the combusted materials (exhaust) at temperature T_1. The temperature T_1 of the combusted materials is attained by cooling the combusted materials to the initial temperature. The enthalpy flows are defined as

$$\dot{H}'_1 = \dot{H}_{\text{Air}} + \dot{H}_{\text{Kr}} \quad \text{and} \quad \dot{H}''_1 = \dot{H}_{\text{exhaust } T_1} \tag{5.32}$$

It accordingly follows that:

$$\dot{H}_{\text{KWA}} - \dot{H}_{\text{KWE}} + P_e + \dot{Q}_{\text{Rest}} + \dot{H}_{\text{exhaust } T_2} = H_u \dot{m}_{\text{Kr}} + \dot{H}_{\text{exhaust } T_1} \tag{5.33}$$

or

$$H_u \dot{m}_{\text{Kr}} = \Delta\dot{H}_{\text{KW}} + P_e + \dot{Q}_{\text{Rest}} + \Delta\dot{H}_{\text{exhaust}} \tag{5.34}$$

It must be remembered that $\Delta\dot{H}_{\text{exhaust}}$ is the enthalpy difference between the exhaust at the respective exhaust temperature T_2 and temperature T_1.

From the preceding equation, we can clearly see the distribution of the energy supplied by the fuel or the calorific value. It is divided into effective power, residual heat, the enthalpy difference of the cooling water, and the enthalpy difference of the exhaust gas.

The enthalpy of the cooling water is calculated with the following equation:

$$\Delta\dot{H}_{\text{KW}} = \dot{m}_{\text{KW}} \cdot c_{\text{W}} \cdot (T_{\text{KWA}} - T_{\text{KWE}}) \tag{5.35}$$

with

$\dot{m}_{\text{KW}}$ = Flow rate of cooling water

c_W = Specific heat of the water (4.185 kJ/kg K)

T_{KW_A} = Temperature of the cooling water upon exit

T_{KW_E} = Temperature of the cooling water upon entrance

The enthalpy difference of the exhaust is found with the equation

$$\Delta\dot{H}_{\text{exhaust}} = \dot{m}_{\text{exhaust}} \cdot (c_{p\,\text{exhaust}}|_0^{T_2}\, T_2 - c_{p\,\text{exhaust}}|_0^{T_1}\, T_1) \tag{5.36}$$

with

$\dot{m}_{\text{exhaust}}$ = Mass flow of the exhaust

$c_{p\,\text{exhaust}}|_0^{\text{T}}$ = Average specific heat of the exhaust

The exhaust mass flow is $\dot{m}_{\text{exhaust}} = \dot{m}_{\text{L}} + \dot{m}_{\text{Kr}}$.

The residual heat that essentially consists of the radiated heat, conducted heat, and convection can accordingly be calculated since all other quantities can be calculated from the measured data

$$\dot{Q}_{\text{Rest}} = H_u \cdot \dot{m}_{\text{Kr}} - P_e - \Delta\dot{H}_{\text{KWA}} - \Delta\dot{H}_{\text{exhaust}} \quad (5.37)$$

Bibliography

[1] Behr, H.D, Thermodynamik, Springer, Berlin, Heidelberg, New York, 1989.

[2] Pischinger, R., G. Kraβnig, G. Taucar, and Th. Sams, Thermodynamik der Verbrennungskraftmaschine, Die Verbrennungskraftmaschine, Reissue Volume 5, Springer, Vienna, 1989.

[3] Heywood, J.B., Internal Combustion Engine Fundamentals, New York, McGraw-Hill International Editions, 1988.

[4] Schäfer, F., Thermodynamische Untersuchung der Reaktion von Methanol-Luft-Gemischen unter der Wirkung von Wasserstoffzusatz, VDI Progress Reports, Series 6, Energietechnik/Wärmetechnik No. 120, VDI Verlag, Düsseldorf, 1983.

[5] Eiglmeier, C., and G.P. Merker, neue Ansätze zur phänomenologischen Modellierung des gasseitigen Wandwärmeübergangs im Dieselmotor, MTZ 61 (2000) 5.

[6] Bargende, M., Ein Gleichungsansatz zur Berechnung der instationären Wandwärmeverluste im Hochdruckteil von Ottomotoren, Dissertation, TH Darmstadt, 1990.

[7] Woschni, G., Die Berechnung der Wandwärmeverluste und der thermischen Belastung der Bauteile von Dieselmotoren, MTZ 31 (1970).

[8] Mollenhauer, K., Handbuch Dieselmotoren, Springer, Berlin, 1997.

[9] Heider, G., G. Woschni, and K. Zeilinger, 2-Zonen Rechenmodell zur Vorausberechnung der NO-Emission von Dieselmotoren, MTZ 59 (1998) 11.

[10] Torkzadeh, D.D., W. Längst, and U. Kiencke, Combustion and Exhaust Gas Modeling of a Common Rail Diesel Engine—An Approach, SAE 2001-01-1243.

[11] Jungbluth, G., and G. Noske, Ein quasidimensionales Modell zur Beschreibung des ottomotorischen Verbrennungsablaufs, Parts 1 and 2, MTZ 52 (1991).

[12] Stiech, G., Phänomenologisches Multizonen-Modell der Verbrennung und Schadstoffbildung im Dieselmotor, VDI Fortschrittberichte, Reihe 12, Verkehrstechnik/Fahrzeugtechnik, No. 399, VDI Verlag, Düsseldorf, 1999.

[13] Ohyama, Y., and O. Yoshishige, Engine Control Using a Real Time Combustion Model, SAE 2001-01-0256.

[14] Zima, S., Unpublished statements.

[15] Jordan, W., Erweiterung des ottomotorischen Betriebsbereiches durch Verwendung extrem magerer Gemische unter Einsatz von Wasserstoff als Zusatzkraftstoff, Dissertation, University of Kaiserslautern, 1977.

[16] Chmela, F., G. Orthaber, and W. Schuster, Die Vorausberechnung des Brennverlaufs von Dieselmotoren with direkter Einspritzung auf der Basis des Einspritzverlaufs, MTZ 59 (1998) 7.

[17] Sams, T., G. Regner, and F. Chmela, Integration von Simulationswerkzeugen zur Optimierung von enginekonzepten, MTZ 61 (2000) 9.

[18] Barba, C., C. Burkhard, K. Boulouchos, and M. Bargende, Empirisches Modell zur Vorausberechnung des Brennverlaufs bei Common-Rail-Dieselmotoren, MTZ 60 (1999) 4.

[19] Codan, E., Ein Programm zur Simulation des thermodynamischen Arbeitsprozesses des Dieselmotors, MTZ 57 (1996) 5.

[20] Vibe, I., Brennverlauf und Kreisprozess von Verbrennungsmotoren, VEB Verlag Technik, Berlin, 1970.

6 Crank Gears

6.1 Crankshaft Drive

6.1.1 Design and Function

The crank gear, a colloquial term for the crankshaft drive, is a functional group that not only efficiently transforms oscillating movement (back-and-forth movement) into rotary movement (and vice versa), but is also excellent at converting thermodynamic processes to yield the maximum work, efficiency, and technical feasibility. These advantages are gained at the cost of serious disadvantages, however:

- Limitation of speed—and hence the development of power—due to free inertia
- Uneven force transmission that requires special measures in the form of multiple cylinder crank gears, a suitable throw and firing sequence, mass balancing, and mass balancing gears
- Excitation of rotational oscillations that place a great deal of stress on the crankshaft and the drivetrain
- High fluctuations in the force characteristics in comparison to the nominal values for these forces
- Problematic component geometry in regard to the flow of force with high stress peaks
- Tribological problems

The crankshaft drive in automotive engines consists of pistons with rings, piston pins, conrods (connecting rods), a crankshaft with countermass(es) (counterweights), bearings (connecting rod bushing, connecting rod bearing, crankshaft main bearing), and the lubricant (Fig. 6-1).

In the following discussion, we refer to the kinematically relevant parts of the crankshaft drive. The individual parts of the crankshaft drive execute various movements:

- The piston oscillates in the cylinder (back and forth).
- The conrod
  (a) is articulated to the small conrod eye by the piston pin and also moves back and forth.
  (b) with the large conrod eye—articulated to the crank pin—also rotates.
  (c) with the conrod shaft swings within the plane of the crank circle.
- The crankshaft rotates (Fig. 6-2).

During a single rotation of the crankshaft, the piston moves from top to bottom and returns to top dead center; it thereby executes two strokes. It accelerates and decelerates while executing this movement. The crank gear movement, i.e., the respective position of the piston, is described by the crankshaft angle φ—the angle between the cylinder axis and the crankshaft throw. The crankshaft angle is a measure of both path and time since it indicates the time in which the crank gear has reached a certain

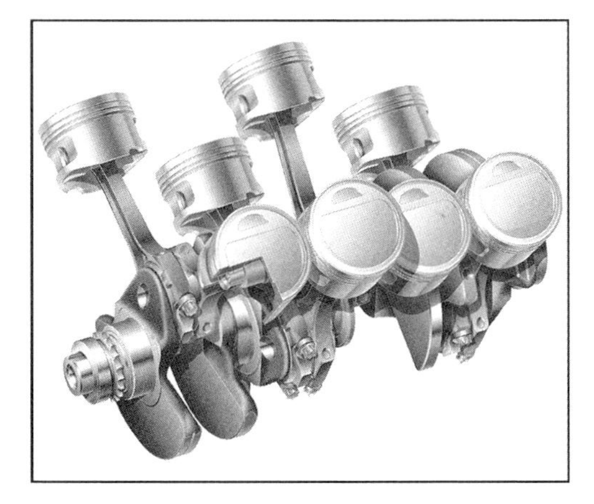

Fig. 6-1 Crank gear of a V-8 passenger car spark-ignition engine.

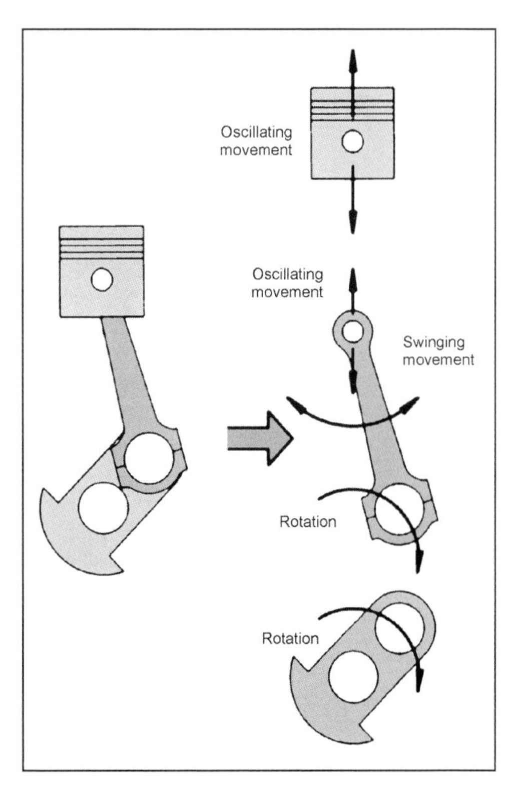

Fig. 6-2 Movements of the crank gear parts.

position independent of the respective speed. The following numerical value equation applies:

$$\varphi[°\text{crankshaft angle}] = 6 \cdot n\,[\text{min}^{-1}] \cdot t\,[\text{s}] \qquad (6.1)$$

The piston movement is calculated with the piston travel equation, i.e., by the relationship of the piston travel to the crankshaft angle, $s = f(\varphi)$; it results from the geometric relationships (Fig. 6-3).

r = Crankshaft radius
s = Piston travel
l = Conrod length
v = Piston speed

$\lambda = \frac{r}{l}$ Conrod ratio

a = Piston acceleration

$$s_0 = l + r \qquad (6.2)$$

$$s_x = r \cdot \cos\varphi + l \cdot \cos\psi \qquad (6.3)$$

$$s = s_0 - s_x \qquad (6.4)$$

$$s = l + r - (r \cdot \cos\varphi + l \cdot \cos\psi) \qquad (6.5)$$

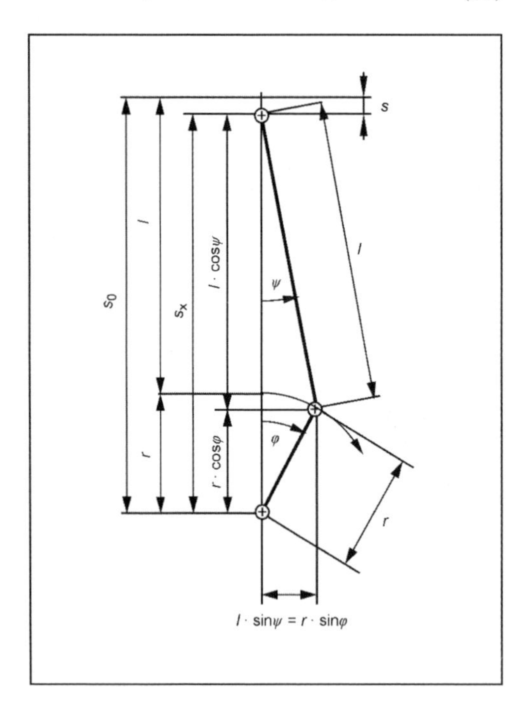

Fig. 6-3 Geometric relationships in the crankshaft drive.

The relationship between the crankshaft angle φ and the conrod angular travel ψ can be represented as follows:

$$\psi = \arctan = \frac{\lambda \cdot \sin\varphi}{\sqrt{1 - \lambda^2 \cdot \sin^2\varphi}} \qquad (6.6)$$

$$s = r \cdot \left[1 - \cos\varphi + \frac{1}{\lambda} \cdot \left(1 - \sqrt{1 - \lambda^2 \cdot \sin^2\varphi}\right)\right] \qquad (6.7)$$

Since it is difficult to use the radical in the piston travel equation, it is replaced by a quickly converging series that can be terminated after the second element because both λ and $\sin\varphi$ are less than 1, and their exponents or products are much less. (Another possibility is to develop the piston travel equation in a Fourier series and correspondingly truncate it according to the desired precision.)

$$\sqrt{1 + x} = 1 + \frac{1}{2}x - \frac{1}{8}x^2 + \frac{1}{16}x^3 - \ldots \qquad (6.8)$$

$$x = -\lambda^2 \cdot \sin^2\varphi \qquad (6.9)$$

The simplified piston travel equation is accordingly (Fig. 6-4)

$$s = r \cdot \left(1 - \cos\varphi + \frac{1}{2} \cdot \lambda \cdot \sin^2\varphi\right) \qquad (6.10)$$

By including time, we gain the piston speed (Fig. 6-5):

$$v = r \cdot \omega \cdot \left(\sin\varphi + \frac{1}{2} \cdot \lambda \cdot \sin 2\varphi\right) \qquad (6.11)$$

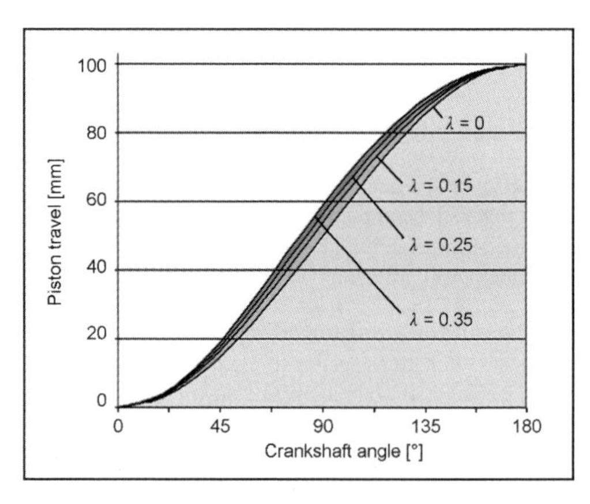

Fig. 6-4 Piston travel as a function of the crankshaft angle for different conrod ratios.

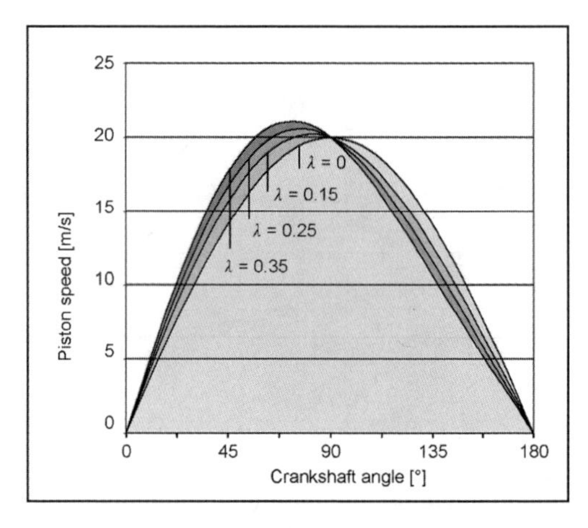

Fig. 6-5 Piston speed as a function of crankshaft angle for different conrod ratios.

The average piston speed is the path of two strokes traveled during a rotation in reference to the associated time $t = 1/n$

$$v_{Km} = 2 \cdot s \cdot n \qquad (6.12)$$

By squaring the time in the piston travel equation, we obtain the piston acceleration (Fig. 6-6):

$$a = r \cdot \omega^2 \cdot (\cos \varphi + \lambda \cdot \cos 2\varphi) \qquad (6.13)$$

The piston travel, speed, and acceleration are influenced by the conrod ratio λ. For a connecting rod ($\lambda = 0$) of infinite length, we can dispense with the perturbation function in the piston travel equation $\frac{1}{2} \cdot \lambda \cdot \sin^2 \varphi$ [expressed otherwise: $\frac{\lambda}{4} \cdot (1 - \cos 2\varphi)$]; the piston movement hence corresponds to harmonic movement. In general, the bigger the conrod ratio λ, the larger the deviation from the harmonic movement. Large conrod ratios, i.e., relative to the stroke of short connecting rods, reduce the engine height, yet they produce greater friction because of the stronger angle of the connecting rods. The λ values of German passenger car engines (from 1990–2000) are between 0.22 (Mercedes-Benz A-class) and 0.35 (Opel). Different conrod ratios are used for the same engine type when, for example, different variations of an engine are built with different strokes but the same conrod length.

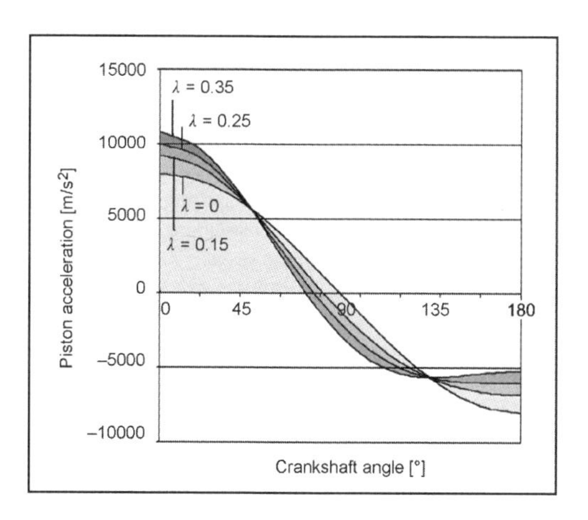

Fig. 6-6 Piston acceleration as a function of crankshaft angle for different conrod ratios.

Different approaches are used to try and reduce the oscillating masses or keep them from increasing despite an increase in output:

- Piston: Less compression height (example: BMW 2.5 l: reduced from 9.0 to 4.0 mm[1]), drawn-in bolt eyes, reduced eye spacing
- Conrod: smaller conrod ratio for lower second-order inertial forces, stepped conrod.

The oscillating masses in passenger car engines ranging from 1.25 to 1.6 dm^3 stroke volume yield 370 to 460 g (Fig. 6-7).

The masses increase substantially with the engine size and load; for example, the mass of the "naked" piston of the V8 Audi spark-ignition engine is 355 g,[4] and that of the complete piston of the Porsche Carrera is 650 g.[5]

By transposing or "deaxising" the crankshaft drive, i.e., shifting the articulation point of the conrod to the piston and crank pin out of the cylinder axis by an amount y,

Example[2,3]		Ford Fiesta 1.4 l		Opel Astra
Year		1996	1998	1998
Nominal piston diameter	mm	76	76	80.5
Piston	g	265	225	222
1st ring	g	5.6	5.6	
2nd ring	g	6.5	6.5	19
Oil control ring	g	5.6	5.6	
Piston pin	g	80	67.5	69
Piston, complete	g	362.7	310.2	310
Conrod mass	g	96	83	
Total osc. mass	g	458.7	393.2	

Fig. 6-7 Comparison of the oscillating masses in a Ford Fiesta and Opel Astra.

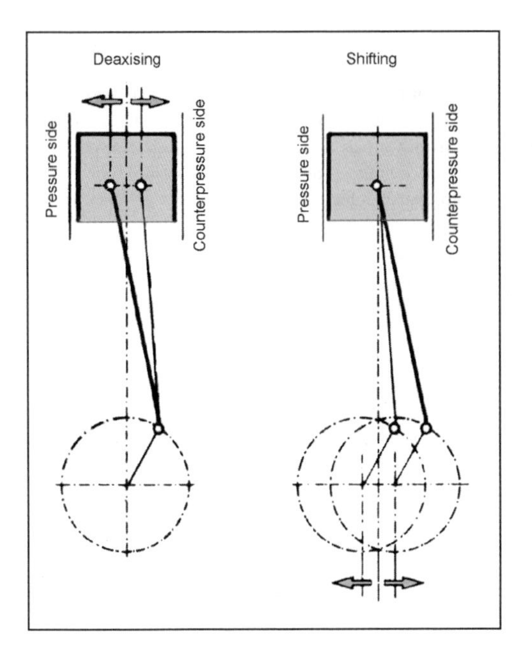

Fig. 6-8 Shifting and deaxising the crankshaft drive.

the movement of the crankshaft drive can be changed as desired (Fig. 6-8). There are

- Shifted crankshaft drives in which the crankshaft middle is displaced from the middle of the cylinder
- Axially offset crankshafts drive in which the piston pin is moved from the middle of the cylinder

It is also possible to combine shifting and deaxising. By shifting, the movement is changed so that the elongated positions of the crank gear no longer lie in the cylinder axis, the piston travel is no longer symmetrical with BDC, and the piston speeds in the advance and return strokes assume different values. The piston travel, speed, and acceleration of the shifted crankshaft drive can be determined with the shift y that refers to the conrod length:

$$e = \frac{y}{l} \quad \text{for [Refs. 6, 7]} \tag{6.14}$$

$$s = r \cdot \left[\cos\varphi + \frac{1}{\lambda} \cdot \sqrt{1 - (\lambda \cdot \sin\varphi + e)^2} \right] \tag{6.15}$$

$$v = -r \cdot \omega \left[\sin\varphi + \frac{\cos\varphi \cdot (\lambda \cdot \sin\varphi + e)}{\sqrt{1 - (\lambda \cdot \sin\varphi + e)^2}} \right] \tag{6.16}$$

$$\begin{aligned} a = -r \cdot \omega^2 \cdot \Bigg[\cos\varphi &+ \frac{\lambda \cdot \cos^2\varphi \cdot (\lambda \cdot \sin\varphi + e)}{[1 - (\lambda \cdot \sin\varphi + e)^2]^{2/3}} \\ &+ \frac{\lambda \cdot \cos^2\varphi - \sin\varphi \cdot (\lambda \cdot \sin\varphi + e)}{\sqrt{[1 - (\lambda \cdot \sin\varphi + e)^2]}} \\ - r \cdot \omega^2 \cdot \Bigg[\sin\varphi &+ \frac{\cos\varphi \cdot (\lambda \cdot \sin\varphi + e)}{\sqrt{1 - (\lambda \cdot \sin\varphi + e)^2}} \Bigg] \end{aligned} \tag{6.17}$$

There are different reasons for shifting and deaxising. In the early period of engine construction, the crankshaft drive was limited to $1/10$ of the stroke.[8] This was to align the connecting rod with the cylinder axis when it passed through TDC to reduce the normal force (piston-side force) around the ignition and hence reduce the load and wear. Today, shifting is used with VR engines (V-engines with V-angles between 10° and 20°) to allow for the necessary free travel of the opposing cylinder.[9,10]

Axially offsetting in the direction of pressure (direction in which the piston contacts the cylinder barrel in the expansion stroke) causes an earlier contact change for the piston when the normal force on the piston is weaker. The tilting movement of the piston causes it to first contact the cylinder with the "soft" bottom part (piston skirt), which reduces impact. One therefore speaks of deaxising to reduce noise. The optimum amount of axially offsetting

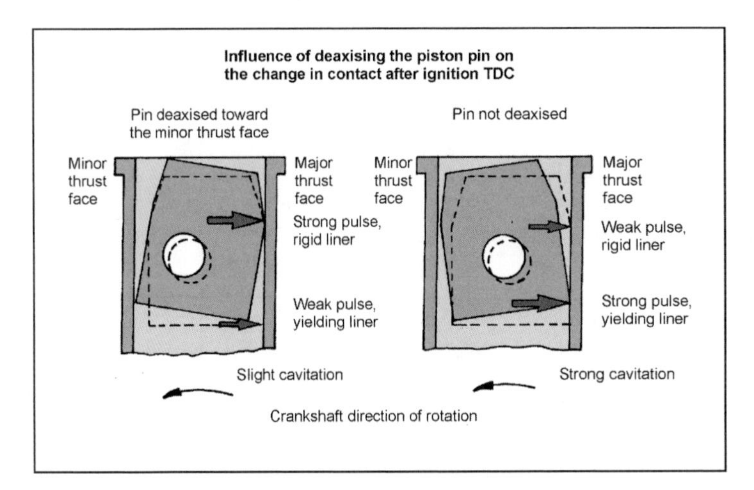

Fig. 6-9 Piston pin deaxising.

has been determined experimentally. For an opposed cylinder engine, this is, e.g., 0.9 mm. In automotive diesel engines, thermal deaxising is used—axially offsetting to the counterpressure side. This allows the piston (within the piston play) to stay more in the middle of the cylinder, which has a positive effect on the seal of the piston rings and counteracts the collection of carbon deposits on the fire land (Fig. 6-9).

6.1.2 Forces Acting on the Crankshaft Drive

The forces in the crankshaft drive of an internal combustion engine arise from the gas pressure in the combustion chamber and from inertial forces (Fig. 6-10).

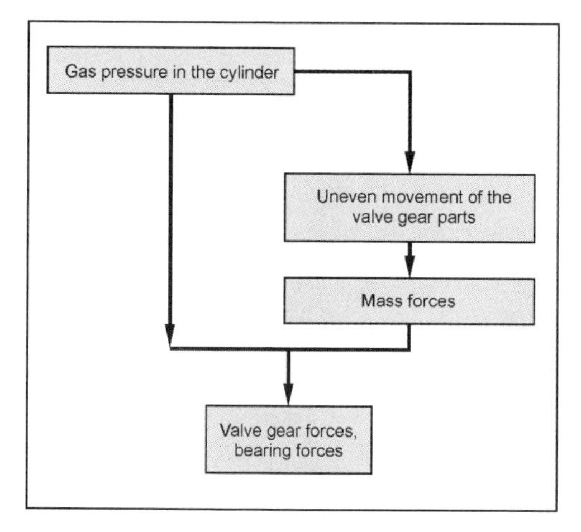

Fig. 6-10 Diagram of crank gear forces.

The shares of gas and inertial forces within the crank gears forces depend on the

- Thermodynamic process: spark-ignition engine/diesel engine
- Design of the engine: naturally aspirated engine/ supercharged engine
- Load level in the program map, e.g.,
  (a) High gas force, low inertial force
  (b) Low gas force, high inertial force

Because of the nonuniform processes of work and movement in the reciprocating-piston engine, the size and direction of the forces in the crank gear change during a work cycle.

To make it easier to conceptualize the processes in the engine, the forces are viewed statically as if in a snapshot. In the precomputer days, this approach corresponded to the procedures in engine development. Today, computer-supported calculations allow the computation of the movement, deformation, and strength behavior of complex mechanical structures with a high degree of precision and reliability. Forces are accordingly no longer viewed as vectors that attack at a point; rather, the transmission of force from crank gear part to crank gear part occurs in space under the pressure of a lubricating film with consideration of structural rigidity and flexibility and angled pin position.

The following act on the crank gear:

- Gas force
- (Oscillating) inertial force of the piston
- Oscillating inertial force of the conrod
- Rotating inertial force of the conrod
- Rotating inertial force of the crankshaft throw
- Rotating inertial force of the countermass

The inertial force from the rotating motion of the conrod is not included. In the following discussion, the cited forces are those that occur briefly after ignition TDC with a crankshaft angle of 30° after TDC (Fig. 6-11).

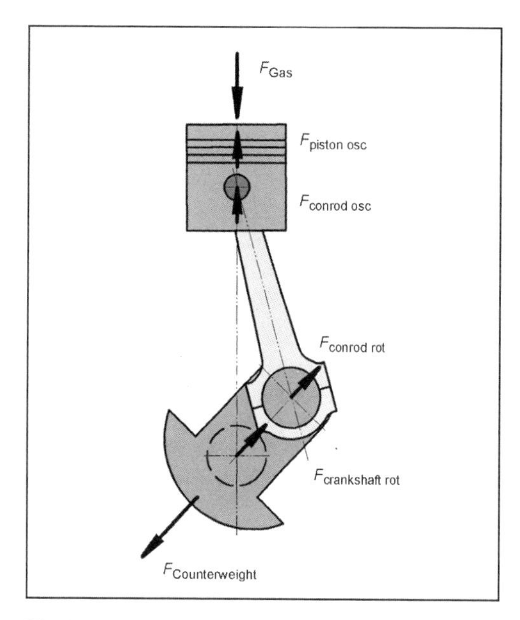

Fig. 6-11 Forces acting on the crank gear.

The gas pressure that arises from the combustion of the mixture depends on the amount and change of different influences, such as the

- Thermodynamic process
- Combustion process
- Power
- Operating point in the program map at which the engine is driven

The gas pressure is determined with a process calculation or by measurement (indication) (Fig. 6-12).

To express it simply as is often done, the oscillating inertial forces are summarized as a single force F_{osc}. This

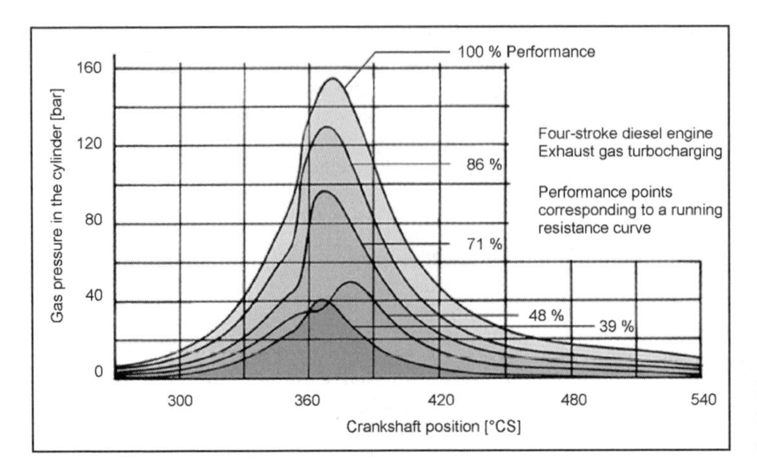

Fig. 6-12 Gas pressure characteristics of a supercharged diesel engine with direct fuel injection.

counters the gas force applied to the piston. The gas and inertial forces together yield the piston force F_K.

$$F_K = F_{Gas} + F_{Piston} + F_{conrod\ osc}$$

$$F_{Gas} = p(\varphi) \cdot A_K \qquad A_K = \frac{\pi}{4} \cdot d^2$$

$$F_{osc} = -m_{osc} \cdot r \cdot \omega^2 \cdot (\cos\varphi + \lambda \cdot \cos 2\varphi) \tag{6.18}$$

$$m_{osc} = (m_{Piston} + m_{conrod\ osc})$$

$$F_K = p(\varphi) \cdot A_{Piston} - r \cdot \omega^2 \cdot m_{osc} \cdot (\cos\varphi + \lambda \cdot \cos 2\varphi)$$

Since the connecting rod, apart from the dead centers, assumes a position that deviates from the direction of the cylinder axis, the piston force F_K must be correspondingly diverted. This results in the rod force F_{ST} and the normal force, i.e., perpendicular to the cylinder wall, F_N (also termed the sliding path force or piston-side force) (Figs. 6-13 and 6-14).

$$F_{ST} = \frac{F_K}{\cos\psi} \tag{6.19}$$

$$F_N = -F_K \cdot \tan\psi \tag{6.20}$$

Results become clearer when we look separately at the paths of the individual forces from the combustion chamber to the crankshaft bearing or engine suspension.[11]

Forces acting on the piston: The gas pressure acting on the piston produces the gas force; it is counteracted by the (oscillating) inertial force of the pistons. The sum of these two forces produces the piston force F'_K. The piston force alternates between positive and negative several times over the course of the power cycle of a four-stroke engine and subjects the piston to a dynamic load.

$$F'_K = F_{Gas} + F_{Piston} \tag{6.21}$$

$$F_{Piston} = -m_{Piston} \cdot r \cdot \omega^2 \cdot (\cos\varphi + \lambda \cdot \cos 2\varphi) \tag{6.22}$$

Forces acting on the piston pin: The piston force acting via the top of the piston and the bolt eyes on the

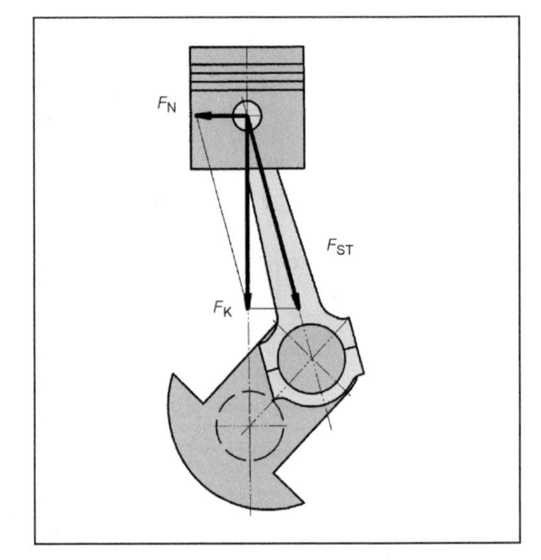

Fig. 6-13 Division of the piston force.

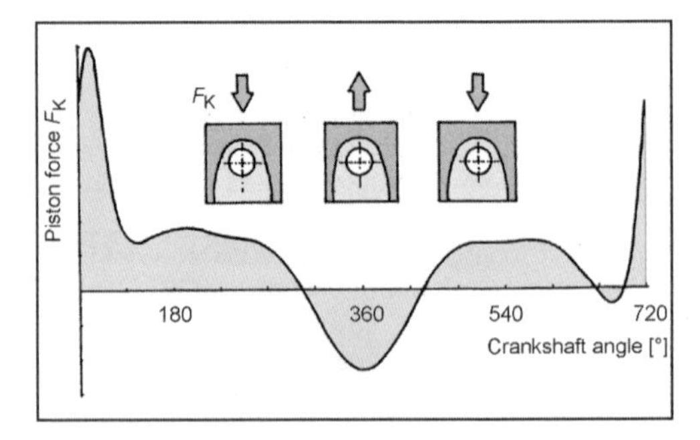

Fig. 6-14 Characteristic of the piston force of a fast-running four-stroke diesel engine over a power cycle.

piston pin F'_K is diverted toward the connecting rod. A force parallelogram results from the piston force F'_K, the rod force F'_{ST} in the direction of the conrod, and a normal force F'_K that is normal (perpendicular) to the cylinder barrel. The piston pin receives this rod force F'_{ST}.

$$F'_{ST} = \frac{F'_K}{\cos \psi} = \frac{F'_K}{\sqrt{1 - \lambda^2 \cdot \sin^2 \varphi}} \tag{6.23}$$

$$F'_N = -F'_K \cdot \tan \psi = -F'_K \cdot \frac{\lambda \cdot \sin \varphi}{\sqrt{1 - \lambda^2 \cdot \sin^2 \varphi}} \tag{6.24}$$

Forces acting on the connecting rod: The piston force F'_K divides as described above into the rod force F'_{ST} and the normal force F'_N. The oscillating conrod force $F_{\text{conrod osc}}$ acts in the direction of the cylinder axis;

$$F_{\text{conrod osc}} = -m_{\text{conrod osc}} \cdot r \cdot \omega^2 \cdot (\cos \varphi + \lambda \cdot \cos 2\varphi) \tag{6.25}$$

It divides into a component in the direction of the conrod and into a normal component. The first component reduces the conrod force from F'_{ST} to F_{ST}, the latter component reduces the normal force from F'_N to F_N.

$$F_{ST} = F'_{ST} - F_{\text{conrod osc}} \cdot \frac{1}{\cos \psi} = \frac{F_K}{\cos \psi} \tag{6.26}$$

$$F_N = -F'_N + F_{\text{conrod osc}} \cdot \tan \psi = -F_K \cdot \tan \psi \tag{6.27}$$

The sign change of the normal force F_N means that this occurs several times during a power cycle (Fig. 6-15).

The piston is pushed from one side of the cylinder barrel to the other with undesirable consequences:

- In a cold engine, it becomes manifested in light metal pistons by an annoying noise and piston rattle. Control pistons are developed at great effort to suppress the noise (pistons with cast steel strips).
- The cylinder bushing is excited to execute vibrations that the coolant cannot follow, and cavitation may occur.

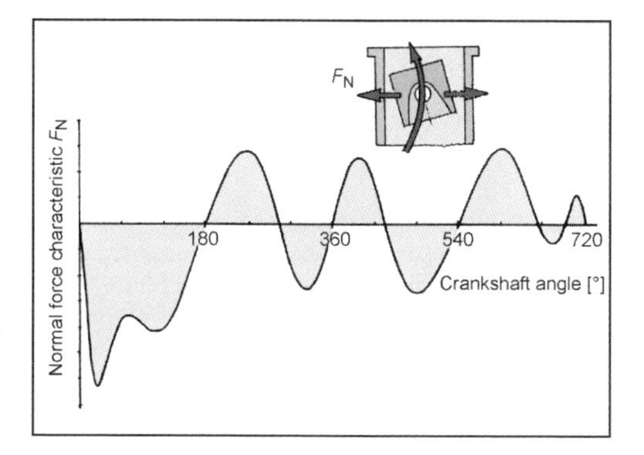

Fig. 6-15 Characteristic of the normal force of a fast-running diesel engine over the power cycle.

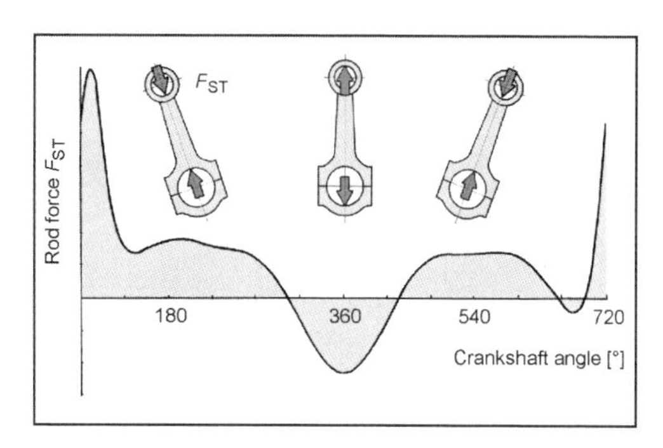

Fig. 6-16 Characteristic of the rod force of fast-running four-stroke diesel engines over a power cycle.

The rod force F_{ST} is directed to the crank pin (Fig. 6-16).

The crank pin rotates under the effect of the rod force along the circle of rotation of the crankshaft radius. Combined with the tangential component of the rod force (F_T), the crankshaft radius yields the torque M (Figs. 6-17 and 6-18).

$$F_T = F_{ST} \cdot \sin(\varphi + \psi) = F_K \cdot \frac{\sin(\varphi + \psi)}{\cos \psi} \tag{6.28}$$

The radial component, the radial force F_R, does not contribute to the engine torque; it is applied only to the crankshaft throw upon bending (Fig. 6-19), and it is a powerless or blind force.

$$F_R = F_{ST} \cdot \cos(\varphi + \psi) = F_K \cdot \frac{\cos(\varphi + \psi)}{\cos \psi} \tag{6.29}$$

Since every action yields an equal and opposite reaction, a torque opposite the useful torque on the engine block necessarily arises: the reaction torque. This results

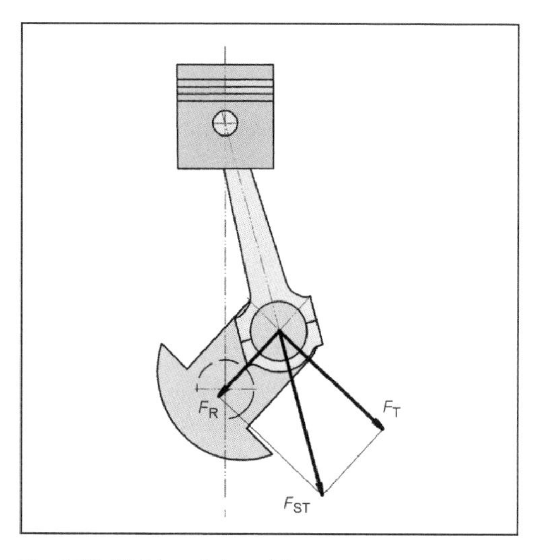

Fig. 6-17 Division of the rod force.

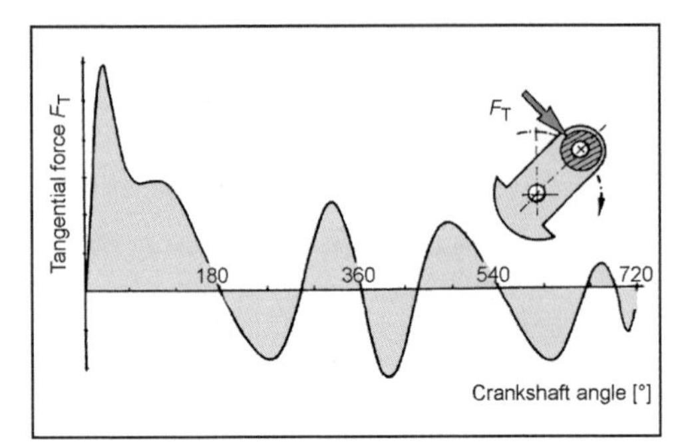

Fig. 6-18 Tangential force characteristic of a fast-running diesel engine.

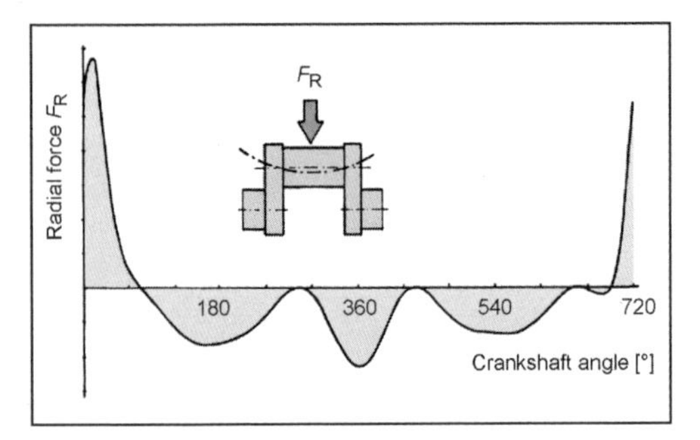

Fig. 6-19 Radial force characteristic of a fast-running diesel engine.

from the normal force F_N and distance of the normal force b that changes with the piston position.

$$M = F_T \cdot r \; M_{Reaction} = F_N \cdot b$$

$$b = r \cdot \cos\varphi + l \cdot \cos\psi \tag{6.30}$$

Hence the supporting forces F_A and F_B result from the reaction torque and their distance a (Fig. 6-20):

$$F_A = -\frac{M_{Reaction}}{a} \tag{6.31}$$

$$F_B = \frac{M_{Reaction}}{a} \tag{6.32}$$

The crank pin is subject to the rod force F_{ST} and the rotating inertial force of the conrod $F_{PL\,rot}$. Added geometrically, these forces yield the crank pin force F_{HZ}.

$$F_{HZ} = \sqrt{F_{ST}^2 + F_{conrod\,rot}^2 - 2 \cdot F_{ST} \cdot F_{conrod\,rot} \cdot \cos(\varphi + \psi)} \tag{6.33}$$

As a reaction to the crank pin force F_{HZ}, the conrod bearing force F_{PL} acts on the conrod bearing (Fig. 6-21).

$$F_{PL} = -F_{HZ} \tag{6.34}$$

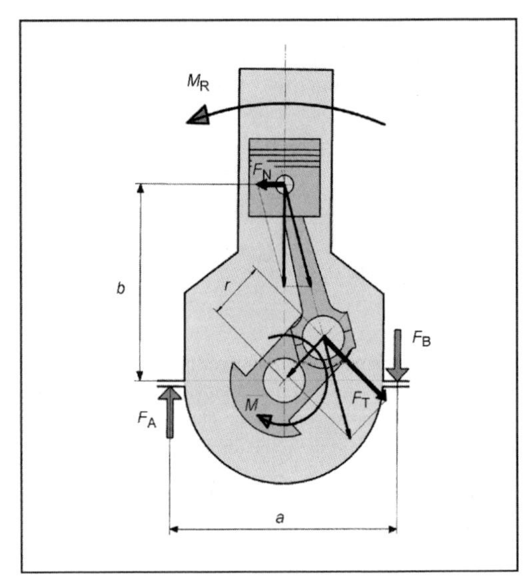

Fig. 6-20 Action torque, reaction torque, and supporting forces.[30]

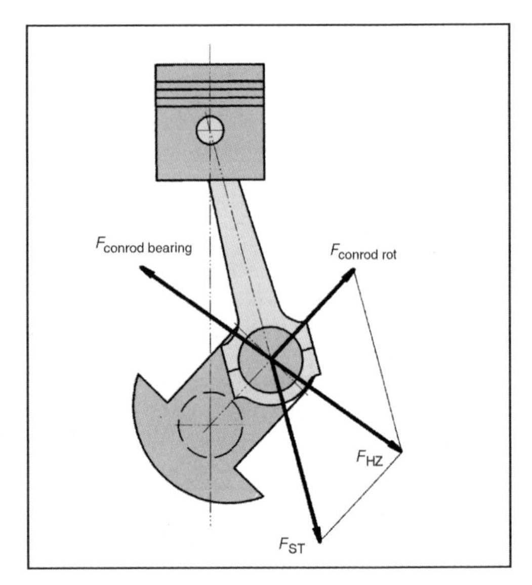

Fig. 6-21 Crank pin force.

When the size and direction of forces change during a power cycle, as is the case with the conrod bearing force, these forces are represented in polar diagrams by plotting them under the respective angle of their force-transfer direction in the sequence of the crankshaft angle (Fig. 6-22).

The crankshaft angle and angle of force are not identical. In order to follow the change of the force over time, the crankshaft angle must be given for the individual

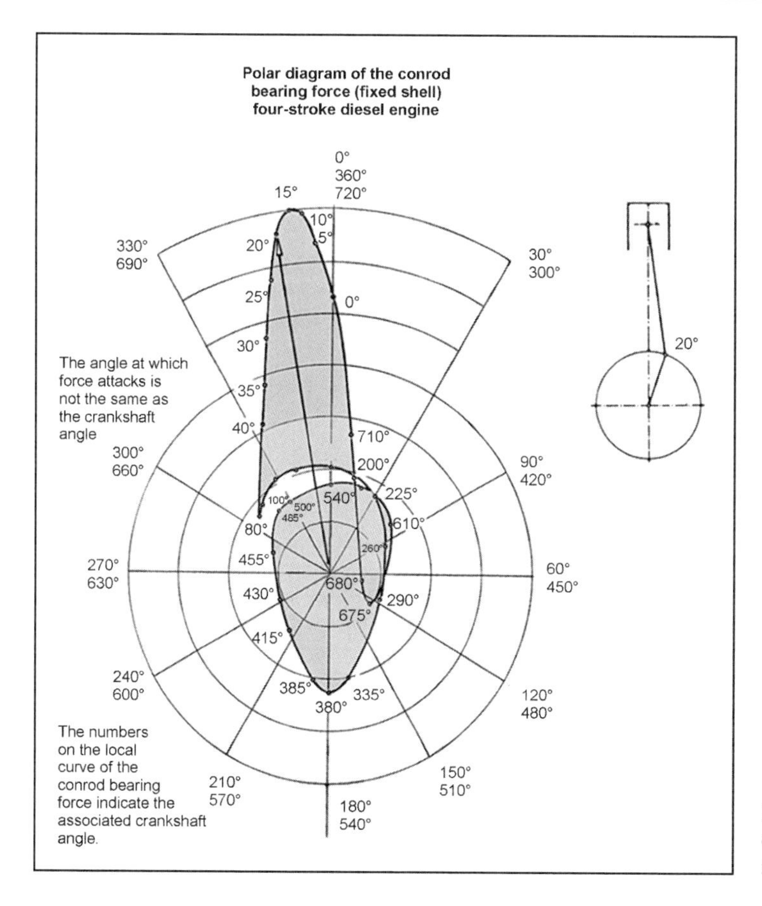

Fig. 6-22 Polar diagram of the conrod bearing force of a fast-running diesel engine.

points of the force characteristic. It is frequently useful to refer the forces to different coordinate systems (Fig. 6-23).

- Fixed spatial (or fixed housing) system (such as main bearing forces)
- Fixed pin system (such as the effect of the forces on rotating pins)
- Fixed rod system (such as the effect of forces on the conrod bearing)

The crank gear forces are transmitted via the main bearing pin and main bearing to the crankcase. The rotating inertial force of the crankshaft throws $F_{\text{KR rot}}$, the crank pin force F_{HZ} or its components F_{T}, F_{R}, and $F_{\text{conrod rot}}$, and the forces of the countermass $F_{\text{countermass}}$ ("counterweights") together to form the main bearing force F_{GL} (Fig. 6-24).

$$F_{\text{GL}} = \sqrt{(F_{\text{crankshaft}_{\text{rot}}} + F_{\text{R}} + F_{\text{conrod rot}} - F_{\text{countermass}})^2 + F_{\text{T}}^2} \tag{6.35}$$

The rotating masses of the throw are related to the crank pin axis.

$$m_{\text{throw}} = m_{\text{crank pin}} + 2 \cdot m_{\text{crank web}} \tag{6.36}$$

$$m_{\text{crank web}} = m_{\text{crank}} \cdot \frac{r_{\text{center of gravity}}}{r} \tag{6.37}$$

As a reaction to the main bearing force F_{GL}, the equal and opposite main bearing pin force F_{GZ} arises. The main bearing force F_{GL} is divided into the two main bearings neighboring the crankshaft throw.

Apart from single cylinder engines, the crankshaft has more than two bearings and represents a statically indeterminate system. In view of the fluctuating gas pressure from work cycle to work cycle, the tolerances of the masses, the deformation of the crankshaft and the oil film, and the flexibility of the bearing, the supporting forces are frequently not determined with (apparent) precision. The crankshaft is viewed as consisting of individual throws that are articulated to each other. The difference between the results of the statically indeterminate system and the statically determinate system is slight. The partial supporting forces resulting from each throw are added to yield the overall bearing force.

It is useful to calculate the crank gear force by dividing the forces into their X and Y components, totaling the X and Y components—taking into account whether they are positive or negative—and geometrically adding these

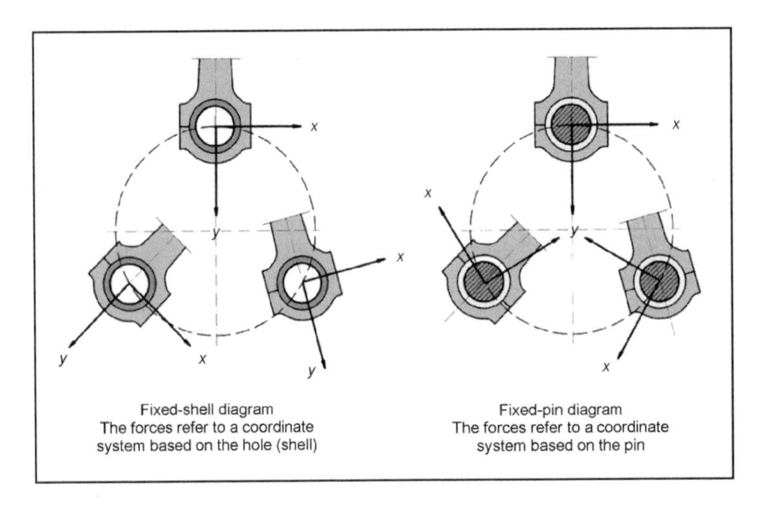

Fig. 6-23 Coordinate systems.

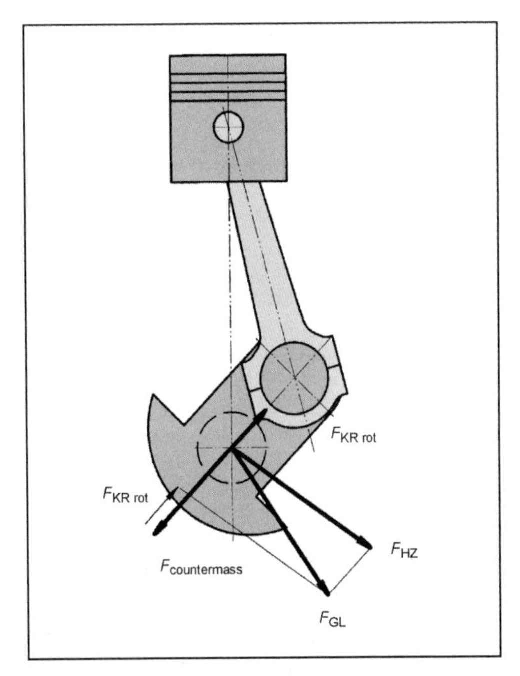

Fig. 6-24 Main bearing force.

sums. The direction of the resultant force is obtained from the quotients of the *X* and *Y* components. The tangent is periodic with π. The quadrants in which the angles lie are obtained from the sign of the individual components.

$$Z = \sqrt{(\sum X)^2 + (\sum Y)^2} \tag{6.38}$$

$$\gamma = \arctan\frac{\sum X}{\sum Y} \tag{6.39}$$

The gas force that presses the piston downward also attempts to lift the cylinder head. This is prevented by the cylinder head screws that hold the cylinder head on the cylinder crankcase. On the other hand, the gas force acts via the piston, conrod, and crankshaft on the crankshaft main bearings. These are held by the main bearing bridges (main bearing cap) and the main bearing screws. This closes the flow of force, and the crankcase intermediate wall is (dynamically) stressed from tension (Fig. 6-25).

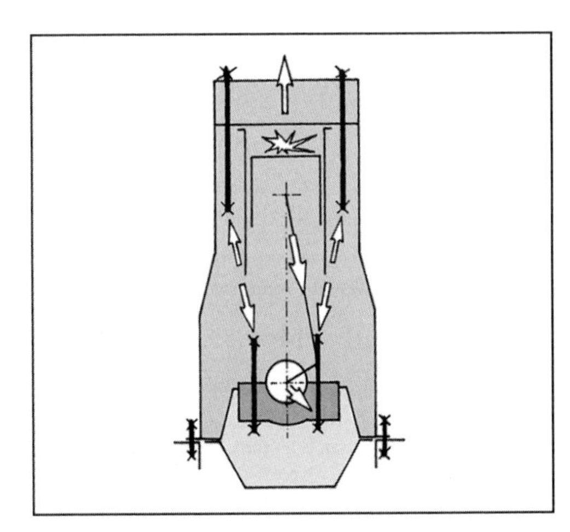

Fig. 6-25 Flow of force in the crankcase.

6.1.3 Tangential Force Characteristic and Average Tangential Force

The tangential force (torsional force) also fluctuates with the periodically changing gas and inertial forces. The average tangential force is calculated from the tangential force characteristic over a power cycle. The area enclosed by the tangential force and the diagram axes is a measure of the (indicated or internal) work W_i. If this work is related to the length of the power cycle, we get the *aver-*

age tangential force F_{Tm}. This is only a fraction of the maximum tangential force (Fig. 6-26).

$$F_{Tm} = \frac{1}{\varphi_p} \cdot \int_0^{\varphi_p} F_T(\varphi) \cdot d\varphi \tag{6.40}$$

$$F_{Tm} = \frac{(\text{Sum of positive areas}) + (\text{Sum of negative areas})}{\varphi_p} \cdot m_F \cdot m_\varphi \tag{6.41}$$

m_F = Measure of force
m_φ = Measure of the angle
φ_p = Length of the power cycle in crankshaft degrees

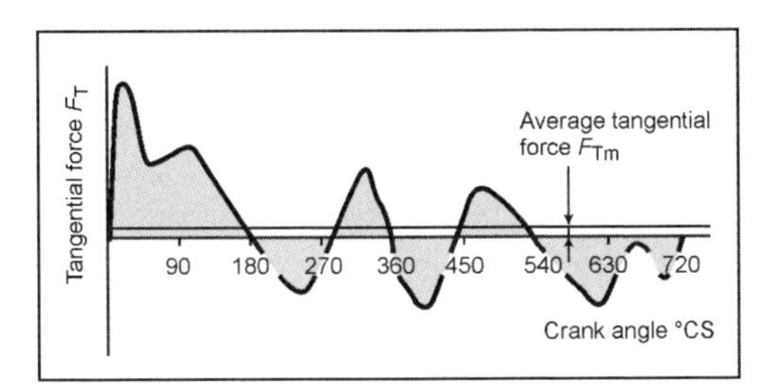

Fig. 6-26 Tangential force characteristic and average tangential force.

To even out the tangential force characteristic and increase the power, engines are built with multiple cylinders, apart from certain exceptions. The tangential forces (torsional forces) of the individual cylinders add up in displaced phases corresponding to the angular ignition spacing over the crankshaft to form the overall torsional force on the clutch side of the engine. This evens out the tangential force so that the fluctuations in the tangential force drop to a fraction of that of a single-cylinder crank gear even in a six-cylinder inline crank gear (Fig. 6-27).

The irregular torsional force characteristic results in fluctuations in the speed because torsional force $F_T(\varphi)$ above the average F_{Tm} accelerates the crank gear, and decelerates it when the force falls below the average. The fluctuation of the energy supplied to the crank gear is termed work fluctuation W_S. Given the moment of inertia I of the crank gear, we obtain the following:

$$W_S = \frac{1}{2} \cdot I \cdot (\omega_{max}^2 - \omega_{min}^2) = \frac{1}{2} \cdot I \cdot (\omega_{max} - \omega_{min}) \cdot (\omega_{max} + \omega_{min}) \tag{6.42}$$

$$\omega_m = 2 \cdot \pi \cdot n \approx \frac{1}{2} \cdot (\omega_{max} + \omega_{min}) \tag{6.43}$$

The speed fluctuation can be reduced with a flywheel. The flywheel acts as an energy accumulator that stores

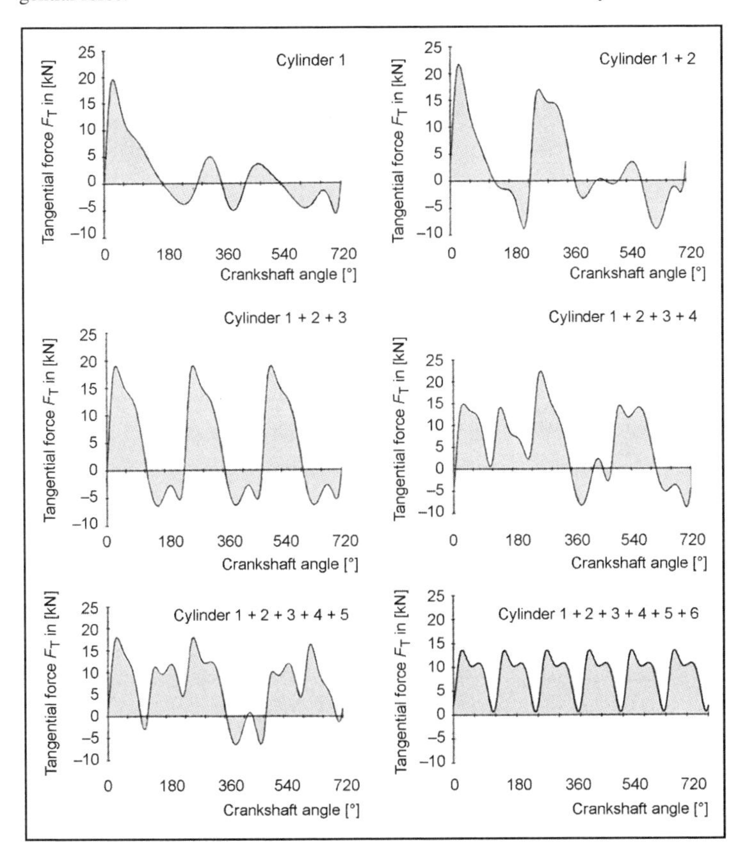

Fig. 6-27 Overlapping tangential forces of a four-stroke six-cylinder inline engine.

excess energy. Depending on the type of machine driven by the engine, different requirements are placed on the constant velocity. The speed fluctuation is indicated by the cyclic irregularity δ. The smoother the engine is to run, the lower the cyclic irregularity δ has to be; in particular, when revving the engines under a load, the cyclic irregularity is unpleasant since it causes the engine accessories to vibrate.

$$\delta = \frac{\omega_{\max} - \omega_{\min}}{\omega_m} \tag{6.44}$$

$$W_S = I \cdot \delta \cdot \omega_m^2 \tag{6.45}$$

$$\delta = \frac{W_S}{I \cdot \omega_m^2} \quad \text{or} \quad I = \frac{W_S}{\delta \cdot \omega_m^2} \tag{6.46}$$

The average tangential force can be derived from the internal power of the engine:

$$P_i = A_K \cdot s \cdot z \cdot w_i \cdot n \cdot i \tag{6.47}$$

$$P_i = M_i \cdot \omega \qquad \omega = 2 \cdot \pi \cdot n \tag{6.48}$$

$$M_i = F_{Tm} \cdot r \qquad r = \frac{s}{2} \tag{6.49}$$

$$F_{Tm} = \frac{A_K \cdot z \cdot w_i \cdot i_i}{\pi} \tag{6.50}$$

$$w_i = w_e \cdot \frac{1}{\eta_m} \tag{6.51}$$

$$P_e = F_{Tm} \cdot r \cdot 2 \cdot \pi \cdot n \cdot \eta_m \tag{6.52}$$

A_K = Piston surface
r = Crankshaft radius
s = Stroke
z = Number of cylinders
P_e = Effective power
w_i = Indicated specific work
w_e = Effective specific work
i = Cycles
η_m = Mechanical efficiency

The area enclosed by the tangential force line above the line of the average tangential force F_{Tm} corresponds to the acceleration work of the crank gear, and the area below corresponds to the deceleration work. If these areas are represented by arrows or pointers preceding from the corresponding quantity from the F_{Tm} line—the pointer for the areas above the F_{Tm} line directed upward, and the pointer for the areas beneath directed downward—then the difference A_s between the maximum and minimum of these pointers is a measure for the maximum work fluctuation W_S (Fig. 6-28).

$$W_S = A_s \cdot m_F \cdot m_\varphi \cdot \frac{\pi}{180} \cdot r \tag{6.53}$$

The crankshaft is subject to a load by the following (Fig. 6-29):

- The useful torque or working torque from the average tangential force that adds up from throw to throw.
- The pulsating torque results from the strongly fluctuating characteristic of the tangential force. The torsional forces of the individual cylinder add up corresponding to their phase shift (angular ignition spacing). On the clutch side, the pulsating torque

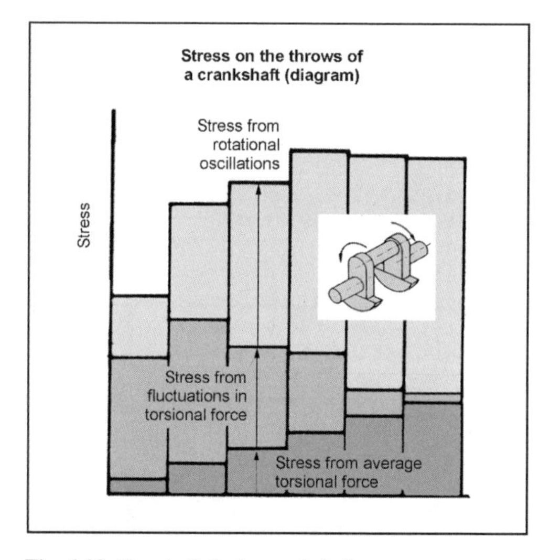

Fig. 6-29 “Load pile” of a crankshaft.

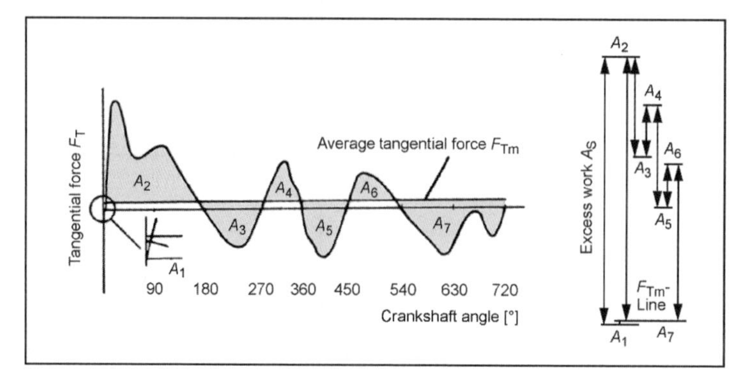

Fig. 6-28 Determining the excess work from the tangential force characteristic.

evens out; however, the primary factor of the load on the crankshaft is the range of fluctuation in the individual throws.

- The rotational oscillations cause additional torque in the crankshaft. This vibrational torque can be a multiple of the other types of torque.

6.1.4 Inertial Forces

In reciprocating-piston engines, inertial effects arise that originate from the movement of the crank gear parts. The inertial forces have both positive and negative effects:

- On the one hand, they are undesirable since they generate additional loads and impair the development of power of the reciprocating-piston engine.
- On the other hand, they even out the release of force of the crank gear by compensating for the force arising from gas pressure peaks and, hence, reduce force and load.

The crank gear executes rotating, oscillating, and swinging motions. To simplify the calculation, the crank gear is reduced to two mass points (Fig. 6-30) in which the oscillating and rotating masses are viewed as concentrated:

- On the articulation point of the conrod on the piston (piston bolt axis)
- On the articulation point of the conrod on the crankshaft (crank pin axis)

The conrod also executes a swinging motion (Fig. 6-31) that results in inertial forces. In fast-running engines, this cannot be ignored.

The mass of the conrod is divided into an oscillating and a rotating part inversely proportional to the respective spacing of the centers of gravity (a, b) so that the center of gravity of the conrods is retained. In connecting rods for automotive engines, this approximately corresponds to a ratio of 1/3 (oscillating mass) to 2/3 (rotating mass).

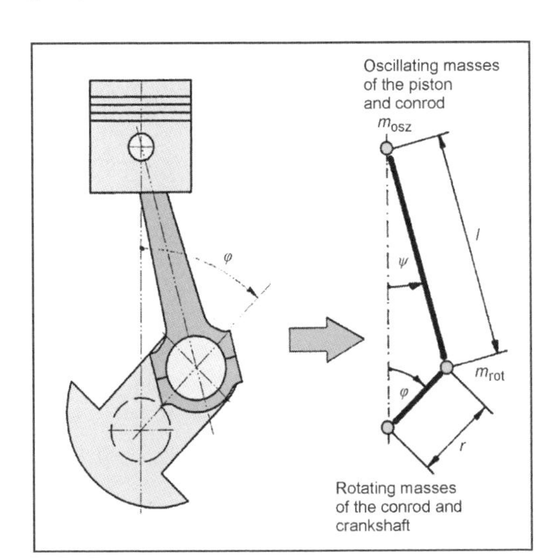

Fig. 6-30 Reduction of the crankshaft drive to two mass points.

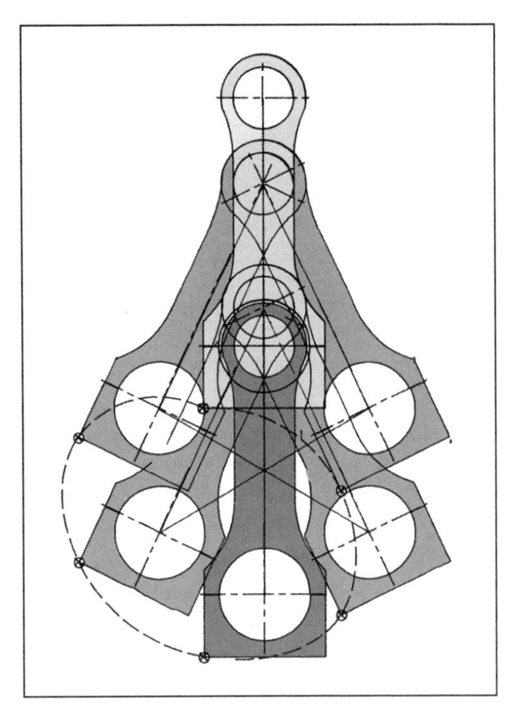

Fig. 6-31 Conrod pattern: Envelope curve of all conrod positions during a crankshaft rotation.

$$m_{\text{conrod osc}} = \frac{a}{l} \cdot m_{\text{conrod}} \tag{6.54}$$

$$m_{\text{conrod rot}} = \frac{b}{l} \cdot m_{\text{conrod}} \tag{6.55}$$

These inertial forces and the inertial torque that they generate proceed in an outward direction as free forces and free torques that try and move the crankcase back and forth in a horizontal and perpendicular direction. In addition, they cause the engine axes to tip. These free forces and torques can be more or less compensated (even completely compensated with a corresponding effort) by countermasses (counterweights) and/or by a corresponding number and arrangement of throws to make the engine externally stable.

6.1.4.1 Inertial Forces in Single-Cylinder Crank Gears

In crank gears, rotating inertial force arises as well as oscillating inertial forces of the first order and higher. If the demands of precision are not particularly high, only the oscillating inertial forces up to and including those of the second order are taken into account.

- Rotating inertial force

The rotating inertial force is a centrifugal force; it stays the same at a constant engine speed, but its direction

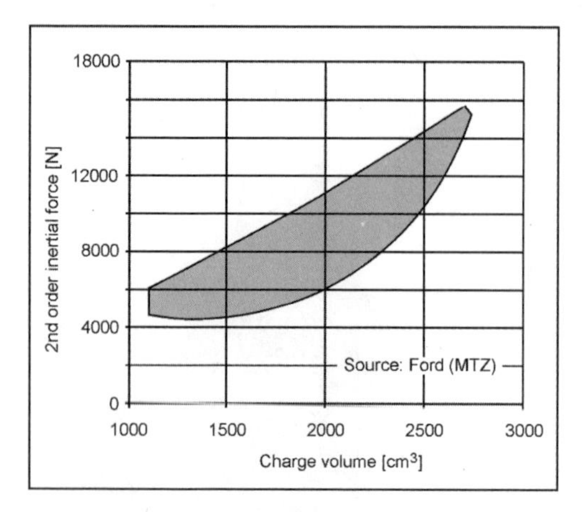

Fig. 6-32 Size of second-order inertial forces within inline engines.

changes with the crankshaft angle. The rotating inertial force rotates at the crankshaft frequency. Its locus diagram is a circle.

$$F_{\text{rot}} = m_{\text{rot}} \cdot \omega^2 \cdot r \tag{6.56}$$

- Oscillating inertial forces

The oscillating inertial forces act in the direction of the cylinder axis, and their size and sign (direction) changes over the course of the piston stroke:

$$F_{\text{osc}} = m_{\text{osc}} \cdot \omega \cdot r \cdot (\cos \varphi + \lambda \cdot \cos 2\varphi) \tag{6.57}$$

$$F_{\text{osc}} = m_{\text{osc}} \cdot \omega^2 \cdot r \cdot \cos \varphi + m_{\text{osc}} \cdot \omega^2 \cdot r \cdot \lambda \cdot \cos 2\varphi \tag{6.58}$$

- First-order inertial force

To be understood as an "order" in this context is "the frequency at which an event occurs in relationship to the crankshaft speed." The amount of the first-order inertial force changes with the crankshaft frequency—hence "first order"—and changes direction twice per rotation.

$$F_{\text{I osc}} = m_{\text{osc}} \cdot \omega^2 \cdot r \cdot \cos \varphi \tag{6.59}$$

- Second-order inertial force

The maximum is only the lth part of the oscillating first-order inertial force (Fig. 6-32); its amount changes at twice the crankshaft frequency, and it changes direction four times per rotation.

$$F_{\text{II osc}} = m_{\text{osc}} \cdot \omega^2 \cdot r \cdot \lambda \cdot \cos 2\varphi \tag{6.60}$$

One can conceive of the oscillating inertial forces as two oppositely rotating vectors that are one-half their maximum, the vectors of the first order rotating at the same speed as the crankshaft, and those of the second order rotating at twice the crankshaft speed. The sum of the two perpendicular components of these vectors yields the momentary inertial force; the horizontal components cancel each other out (Fig. 6-33).

The characteristics of the oscillating inertial forces of the first order and of the second order add to form the resulting oscillating inertial force (Fig. 6-34).

This overall inertial force for a cylinder results from the vectoral addition of the rotating and oscillating inertial forces of the first and second orders, and possibly the forces of a higher order (Fig. 6-35).

6.1.4.2 Inertial Forces in a Two-Cylinder V Crank Gear

If two cylinders at the angle δ to each other act together on a crankshaft throw (V-engine), the inertial forces of both cylinders are added as vectors (Fig. 6-36).

The locus diagram of the rotating inertial forces of both cylinders is a circle, and the locus diagrams of the oscillating inertial forces (depending on the V-angle δ and the order of the force under consideration) can be circles, ellipses, and straight lines (Fig. 6-37).

- Rotating inertial force

As is the case with a single-cylinder crank gear, the resulting rotating inertial force is constant with a vector that revolves at the crankshaft speed. The rotating mass is composed of the rotating masses of the two conrods and the rotating mass of the crankshaft throw; its locus diagram is a circle.

$$F_{\text{V2rot}} = m_{\text{V2}} \cdot \omega^2 \cdot r \tag{6.61}$$

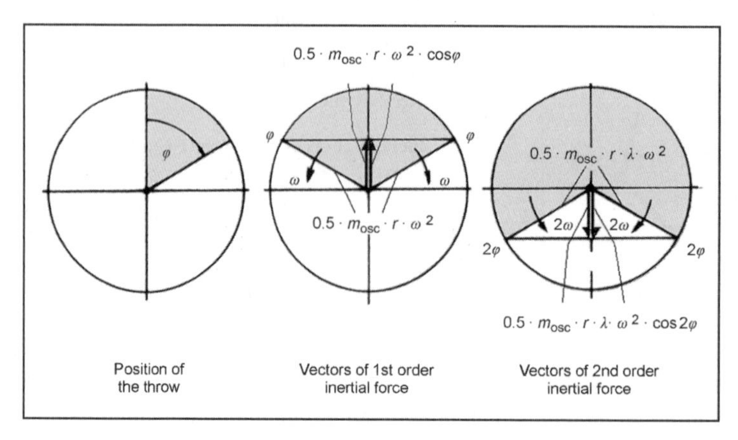

Fig. 6-33 Representation of the vectors of oscillating inertial forces.

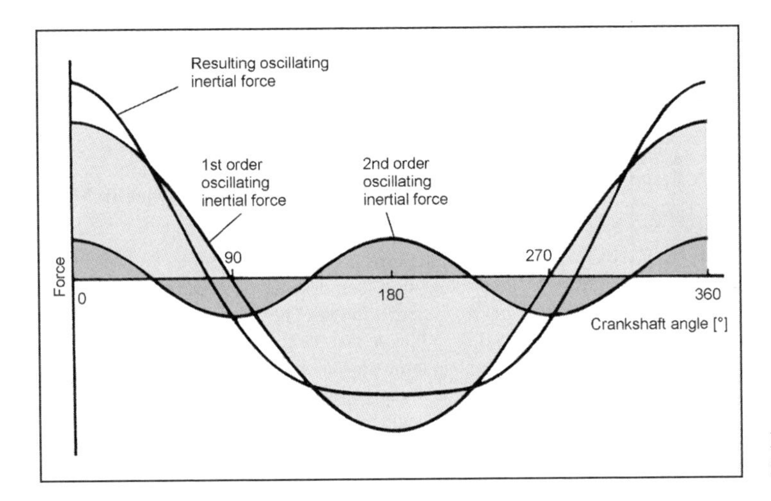

Fig. 6-34 Resulting oscillating inertial force.

$$m_{V2} = 2 \cdot m_{\text{conrod rot}} + (m_{\text{crankshaft}_{\text{rot}}} - m_{\text{countermass}}) \tag{6.62}$$

- Oscillating first-order inertial force

The resulting oscillating first-order inertial force results from the vectoral addition of the inertial forces of the two cylinders *A* and *B*. If the crankshaft angle φ of the crankshaft throw is measured from the bisector of the V-angle, then the crankshaft angle of the cylinder *A* (for right rotation) is $\varphi_A = \varphi + (\delta/2)$, and the angle of cylinder *B* is $\varphi_B = \varphi - (\delta/2)$. Between the oscillating inertial forces of cylinders *A* and *B*, there is an operating time difference equal to the V-angle δ.

$$F_{\text{I osc}A} = F_{\text{I}} \cdot \cos\left(\varphi + \frac{\delta}{2}\right) \tag{6.63}$$

$$F_{\text{I osc}B} = F_{\text{I}} \cdot \cos\left(\varphi - \frac{\delta}{2}\right) \tag{6.64}$$

$$F_{\text{I}} = m_{\text{osc}} \cdot r \cdot \omega^2 \tag{6.65}$$

$$F_{\text{I osc res}} = 2 \cdot F_{\text{I}} \cdot \sqrt{\cos\delta \cdot \cos^2\varphi + \sin^4\frac{\delta}{2}} \tag{6.66}$$

The resulting force can be graphically determined by representing the crankshaft throw in its respective position by a vector with the quantity F_{I}.

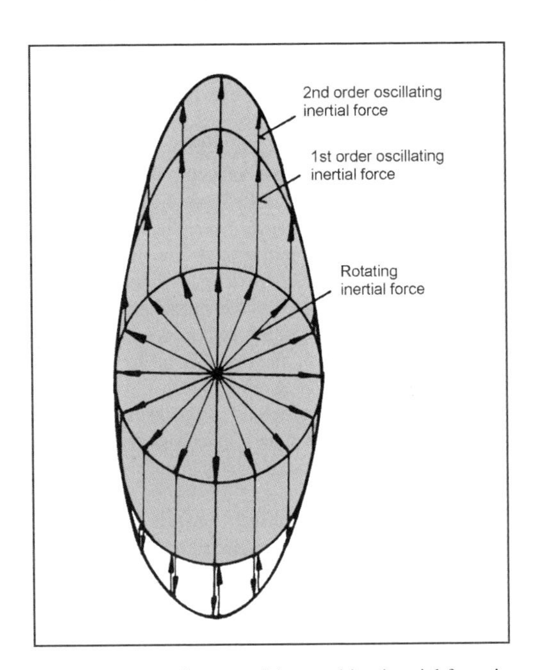

Fig. 6-35 Locus diagram of the resulting inertial force in a single-cylinder crank gear.

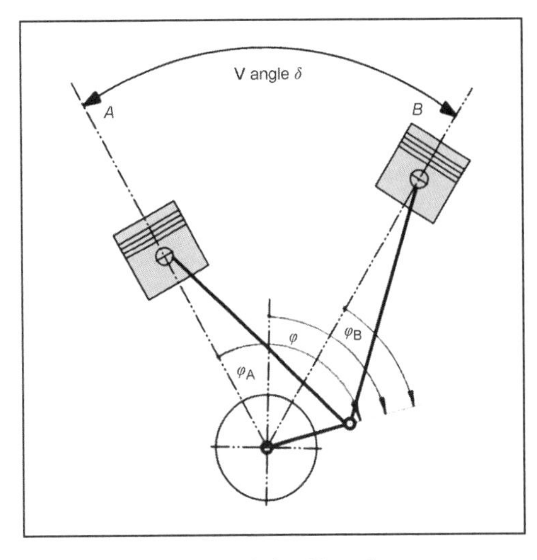

Fig. 6-36 Crankshaft angle in a V crank gear.

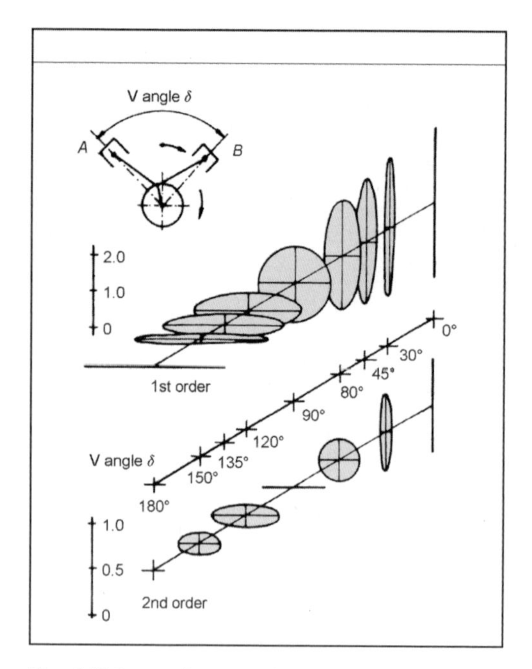

Fig. 6-37 Locus diagrams of the free inertial forces of V-crank gears depending on the V-angle.

These vectors are projected onto cylinder axes A and B. The momentary values of the inertial forces of the two cylinders determined in this manner are added vectorally, and they yield the resulting inertial force vector of the first order ($F_{\text{osc I res}}$) (Fig. 6-38).

- Oscillating second-order inertial force

The resulting oscillating second-order inertial force is also composed of the inertial forces of cylinders A and B. Since the oscillating second-order inertial force changes at twice the crankshaft frequency, the vectoral rotary angle is twice that of the first order. This amount is the lth of the first-order inertial force.

$$\varphi_A = 2 \cdot \varphi + \delta \tag{6.67}$$

$$\varphi_B = 2 \cdot \varphi - \delta \tag{6.68}$$

$$F_{\text{II osc} A} = F_{\text{II}} \cdot \cos (2\varphi + \delta) \tag{6.69}$$

$$F_{\text{II osc} B} = F_{\text{II}} \cdot \cos (2\varphi - \delta) \tag{6.70}$$

$$F_{\text{II}} = \lambda \cdot m_{\text{osc}} \cdot \omega^2 \cdot r \tag{6.71}$$

$$F_{\text{II osc res}} = \sqrt{2} \cdot \text{F}_{\text{II}} \times \sqrt{\cos^2 2\varphi \cdot (\cos 2\delta + \cos \delta) + \sin^2 \delta \cdot (1 - \cos \delta)} \tag{6.72}$$

The resulting force can be graphed by determining the momentary values of the oscillating second-order inertial forces for cylinders A and B and adding them vectorally. The momentary value for cylinder A is determined by plotting from the cylinder axis A the inertial force vector F_2 at the angle $\varphi_A = 2\varphi + \delta$ and projecting it onto cylinder axis A. The momentary value for cylinder B is obtained by plotting the vector F_2 at angle $\varphi_B = 2\varphi - \delta$ but counting from cylinder axis B and projecting it on the axis of cylinder B.

6.1.4.3 Inertial Forces and Inertial Torque in Multi-cylinder Crank Gears

The inertial forces in the individual throws produce torque corresponding to their distance from the engine's center of gravity—inertial torque. The forces and torques are vectoral quantities so that the force and torque vectors of the individual throws are shifted in the plane of gravity of the engine, and can be added to form resulting forces and torques. V-engines are two inline crank gears separated by the V-angle. Therefore, the mass effect of one line of crank gears can be determined and added to the other, phase shifted by the V-angle, or the resultant force of the crank gears opposing each other across the V can be added like the inline crank gear. Available computation programs allow the locus diagram to be graphically displayed in addition to analytical calculations. The inertial effects are determined by the position of the respective throws (Fig. 6-39).

- **Inertial forces.** The rotating forces act in the direction of throw, while vectors rotating in the opposite direction represent the oscillating forces. By projecting the crankshaft throws on the plane of gravity of the engine, i.e., the throw or crank diagram (also termed phase direction diagram), the directions of the inertial force vectors are represented. As a reference, the first throw (depending on whether you are counting from the force transmission side or the counterforce transmission) is in the TDC position. The position of the following throws is determined by the respective throw spacing (throw angle).

  For the oscillating second-order inertial force, the throw diagram of the second order (phase direction diagram of the second order) is used that is obtained by placing the throws under twice the throw angle.
- **Inertial torque.** The torque vector is perpendicular to its plane of action. The sign depends on the position of the relevant throw in reference to the engine's center of gravity; it therefore must be correspondingly taken into consideration. If the throw is to the left of the center of gravity, the vector is positive; if it is to the right, the counting proceeds in a negative direction. From the perspective of the torque diagram, the vectors illustrate the torque that originates from the forces to the left of the engine's center of gravity proceeding from the midpoint of the crankshaft, and proceeding toward the midpoint for torque to the right of the center of gravity. Because the torque vector is perpendicular to its plane of action, i.e., perpendicular to its throw, the torque diagram follows the crank diagram by 90°. The torque vectors can therefore be drawn in the direction of throw, and the vector of the resulting torque can be set back 90° counterclockwise. In V-engines, the iner-

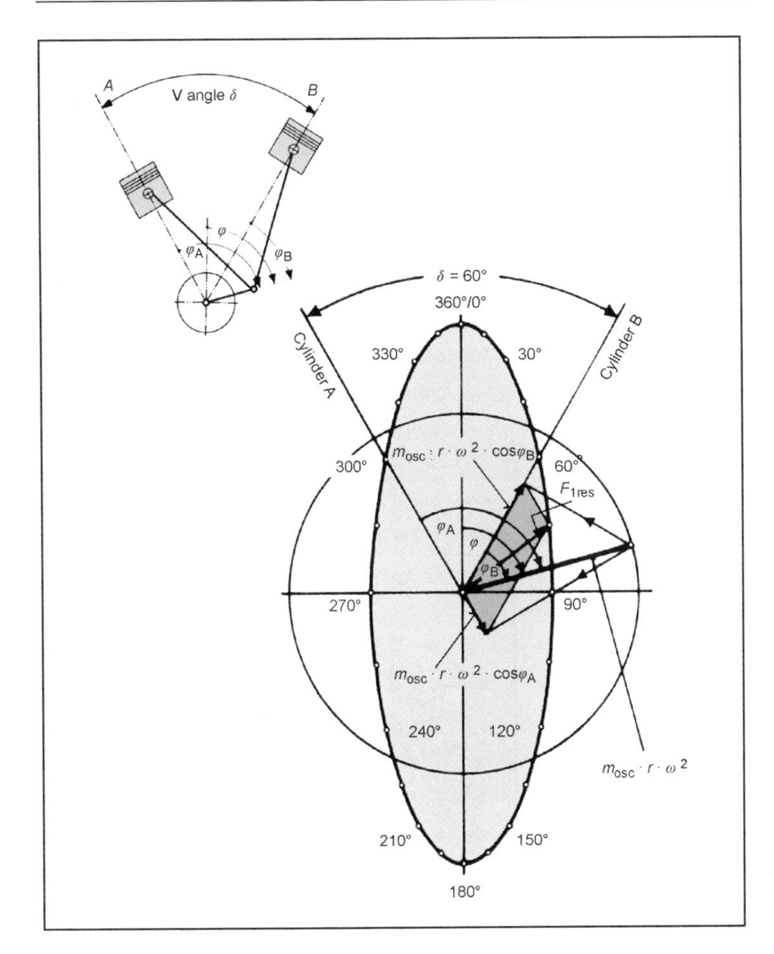

Fig. 6-38 Locus diagram of the oscillating first-order inertial force of a 2-V 60° crank gear.

tial forces of two cylinders acting on a throw are combined and used to determine the inertial torque.

- **Rotating inertial torque.** The inertial torque results from the rotating inertial force and the respective distance from the plane of gravity. It is correspondingly geometrically added to the throw diagram.

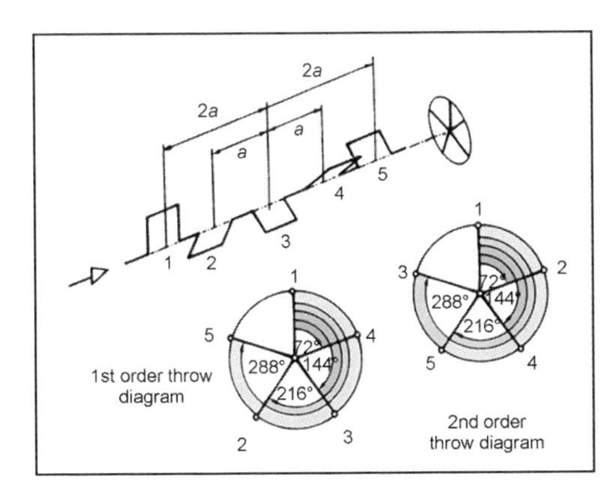

Fig. 6-39 Throw diagrams.

- **Oscillating inertial torque.**
  (a) Oscillating first-order inertial torque

  The vectors of oscillating first-order inertial torque are plotted in the direction of the throw diagram of the first order. After adding the vectors, the resulting torque vector is projected onto the cylinder axis because the oscillating forces act only in the direction of the cylinder axis. The projection is rotated 90° counterclockwise; this is then the resulting oscillating first-order inertial torque.

  (b) Oscillating second-order inertial torque

  The same procedure is used for the oscillating second-order inertial torque, except the throw diagram of the second order is used as a basis.

6.1.4.4 Example

To illustrate these relationships, the functions of a five-stroke shaft are graphed and analyzed. We assume the following:

- Equivalent masses of the crank gears in all throws
- Equivalent cylinder spacing

- The engine's center of gravity is in the middle of the engine in the crankshaft axis
- The first crankshaft throw is in TDC position

Rotating inertial torque
The throw spacing in the throw diagram of the first order is

$\alpha_1 = 0 \qquad \alpha_2 = 216°$

$\alpha_3 = 144°$ (not used since the throw is in the center of gravity)

$\alpha_4 = 72° \qquad \alpha_5 = 288°$

Taking into consideration the sign of the torque of the individual throws, we get the effective directions of the torque (Fig. 6-40):

$\varphi_1 = 0 \qquad \varphi_2 = 216°$

φ_3 not used $\quad 72° \, (+\,180°) = 252°; \qquad \varphi_4 = 252°$

$288° \, (+180°) - (360°) = 108; \qquad \varphi_5 = 108°$

$$F_{\text{rot}} = m_{\text{rot}} \cdot r \cdot \omega^2 \tag{6.73}$$

$$\sum M_X = a \cdot F_{\text{rot}} \cdot (2 \cdot \sin 0° + \sin 216° + \sin 252° + 2 \cdot \sin 108°)$$

$$\sum M_X = a \cdot F_{\text{rot}} \cdot 0.363$$

$$\sum M_Y = a \cdot F_{\text{rot}} \cdot (2 \cdot \cos 0° + \cos 216° + \cos 252° + 2 \cdot \cos 108°)$$

$$\sum M_Y = a \cdot F_{\text{rot}} \cdot 0.264$$

$$M_{\text{rot res}} = a \cdot F_{\text{rot}} \cdot \sqrt{0.363^2 + 0.264^2} = a \cdot F_{\text{rot}} \cdot 0.4488 \tag{6.74}$$

$$\tan \delta = \frac{0.363}{0.264} = 1.375 \qquad \delta = 54°$$

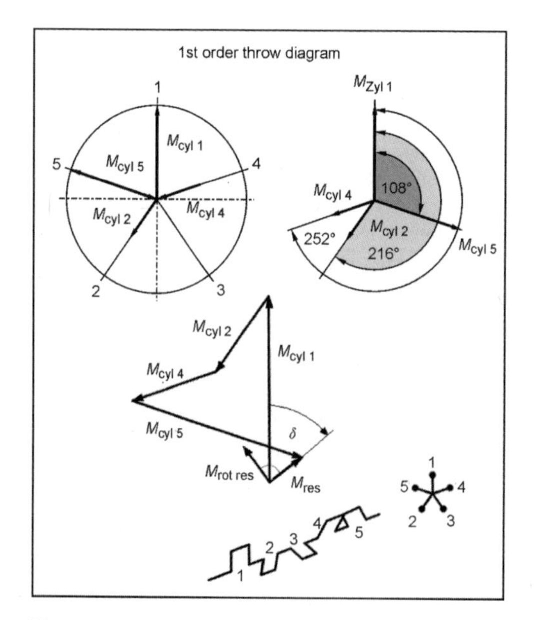

Fig. 6-40 Determining the resulting rotating inertial torque.

Oscillating first-order inertial torque
The effective directions of the vectors are the same as for the rotating inertial torque (Fig. 6-41).

$$F_1 = m_{\text{osc}} \cdot r \cdot \omega^2 \tag{6.75}$$

$$\sum M_X = a \cdot F_1 \cdot (2 \cdot \sin 0° + \sin 216° + \sin 252° + 2 \cdot \sin 108°)$$

$$\sum M_Y = a \cdot F_1 \cdot (2 \cdot \cos 0° + \cos 216° + \cos 252° + 2 \cdot \cos 108°)$$

$$\sum M_Y = a \cdot F_1 \cdot 0.264$$

$$M_{\text{osc 1 max}} = a \cdot F_1 \cdot \sqrt{0.363^2 + 0.264^2} = a \cdot F_1 \cdot 0.4488 \tag{6.76}$$

$$\tan \delta = \frac{0.363}{0.264} = 1.375$$

$$\delta = 54°$$

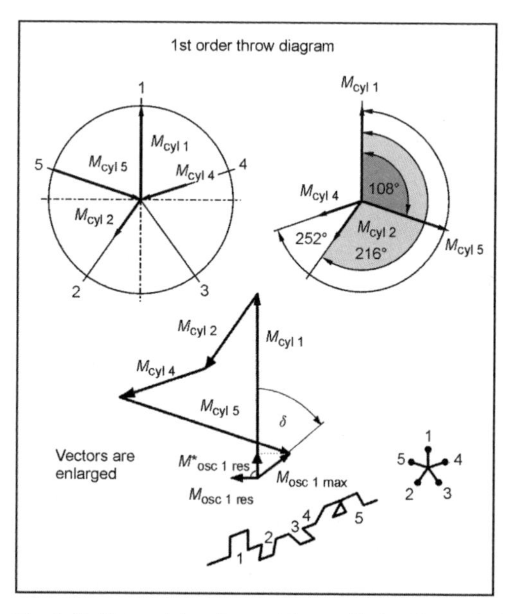

Fig. 6-41 Determining the resulting oscillating first-order inertial torque.

Oscillating second-order inertial torque
The throw spacing in the throw diagram of the second order is

$\alpha_1 = 0 \qquad \alpha_2 = 72°$

$\alpha_3 = 288°$ not used $\qquad \alpha_4 = 144°$

$\alpha_5 = 216°$

Taking into account the sign of the torque of the individual throws, we get the effective directions (Fig. 6-42).

$\varphi_1 = 0 \qquad \varphi_2 = 216$

φ_3 not used $\qquad \varphi_4 = 144° \, (+180) = 324°$

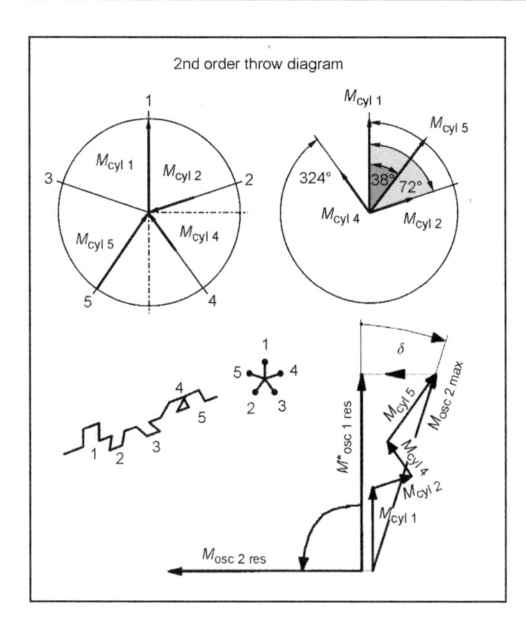

Fig. 6-42 Determining the resulting second-order inertial torque.

$$\varphi_5 = 216° (+180) - (360°) = 36°$$

$$F_2 = \lambda \cdot m_{osc} \cdot r \cdot \omega^2 \qquad (6.77)$$

$$\sum M_X = a \cdot F_2 \cdot (2 \cdot \sin 0° + \sin 72° + \sin 324° + 2 \cdot \sin 36°)$$

$$\sum M_Y = a \cdot F_2 \cdot (2 \cdot \cos 0° + \cos 72° + \cos 324° + 2 \cdot \cos 36°)$$

$$\sum M_X = a \cdot F_2 \cdot 1.539$$

$$\sum M_Y = a \cdot F_2 \cdot 4.736$$

$$M_{osc\ 2\ max} = a \cdot F_2 \cdot \sqrt{1.539^2 + 4.736^2} = a \cdot F_{rot} \cdot 4.98 \qquad (6.78)$$

$$\tan \delta = \frac{1.539}{4.736} = 0.325$$

$$\delta = 18°$$

6.1.5 Mass Balancing

To be understood as mass balancing is the compensation of imbalances due to construction. The balancing of manufacturing-related imbalances is merely termed balancing.

6.1.5.1 Balancing Single-Cylinder Crank Gears

The rotating inertial force can be balanced by countermass(es) where the condition must be fulfilled that the static torque (product of the mass and distance from the rotary axis) of the rotating masses and the balancing mass(es) must correspond.

$$F_{balance} = F_{rot} \qquad (6.79)$$

$$m_{balance} \cdot r_{balance} = m_{rot} \cdot r \qquad (6.80)$$

$$m_{balance} = m_{rot} \cdot \frac{r}{r_{balance}}$$

By dividing the balancing mass into two counterweights, we obtain the following:

$$m_{balance} = \frac{1}{2} \cdot m_{rot} \cdot \frac{r}{r_{balance}} \qquad (6.81)$$

To keep the balancing mass small, it must be affixed at the greatest possible distance from the rotary axis (crankshaft axis); this is greatly limited by the constructive conditions. Basically, mass balancing should include a large static torque and a small moment of inertia.

Oscillating inertial forces can also be compensated by revolving countermasses since their force vector is composed of components in the direction of the cylinder axis (*Y* direction) and perpendicular to the cylinder axis (*X* direction). The balancing mass is selected so that the component in the direction of the cylinder axis corresponds to the oscillating inertial force; this is balanced, but at the price of a free component perpendicular to the cylinder axis (Fig. 6-43).

$$F_{balance} = m_{balance} \cdot r \cdot \omega^2 \qquad (6.82)$$

$$X_{balance} = m_{balance} \cdot r \cdot \omega^2 \cdot \sin \varphi \qquad (6.83)$$

$$Y_{balance} = m_{balance} \cdot r \cdot \omega^2 \cdot \cos \varphi \qquad (6.84)$$

Better conditions result when the oscillating first-order inertial force is not completely balanced. Since the crankcase along its height (*Y* direction) is more rigid than in the transverse direction (*X* direction), the oscillating

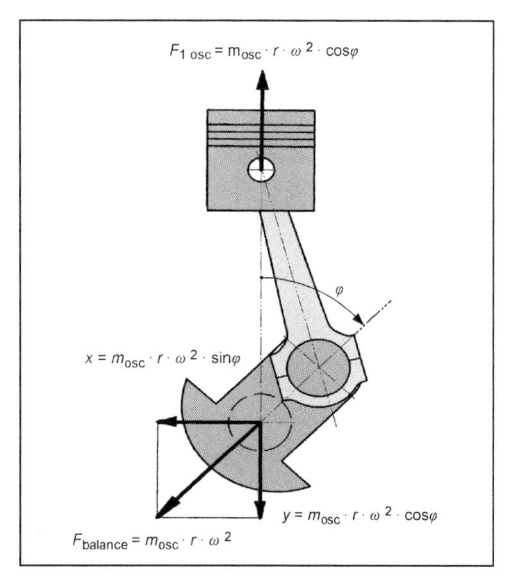

Fig. 6-43 Balance of oscillating forces using a revolving mass.

first-order inertial force is not completely compensated so that the free X component does not become too large, and it is only 50% balanced. Completely balancing the rotating inertial force F_{rot} and the 50% balance of the oscillating first-order inertial force is termed a normal balance—it was used even in the 19th century for drivetrains of steam locomotives. The mass balancing of designed passenger car engines is 50% to 60% of the oscillating inertial force and 80% to 100% of the rotating inertial force.

$$m_{balance} \cdot r_{balance} = (\alpha_1 \cdot m_{rot} + \alpha_2 \cdot m_{osc}) \cdot r \quad (6.85)$$

$$m_{normal\ balance} = (1 \cdot m_{rot} + 0.5 \cdot m_{osc}) \cdot \frac{r}{r_{balance}} \quad (6.86)$$

Another method for balancing oscillating inertial force is to use the so-called foot balance in which additional mass on the large connecting rod eye moves the conrod center of gravity toward the crank pin.[12] The oscillating first-order inertial force is completely balanced when two balancing masses revolving in the opposite direction that are half the oscillating crank gear masses are symmetrically arranged in relation to the vertical engine axis. Then the two components in the direction of the cylinder axis compensate the oscillating inertial force; the two components perpendicular to the cylinder axis cancel each other out (Fig. 6-44).

To obtain a balance of the second order, the countermass must rotate at twice the crankshaft speed (Fig. 6-45).

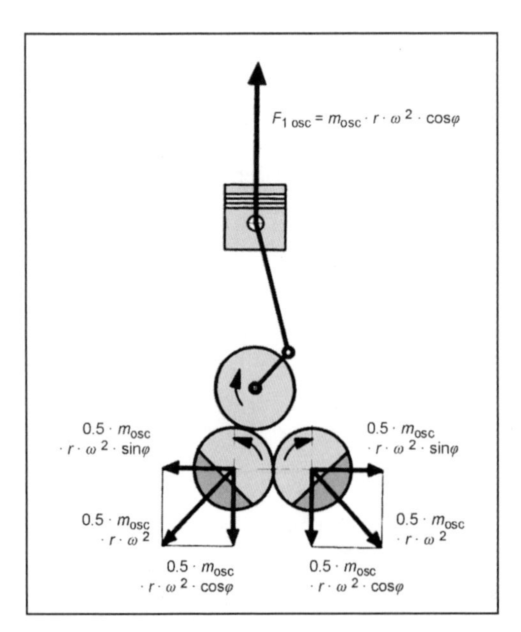

Fig. 6-44 Complete balance of the inertial forces of the first order.

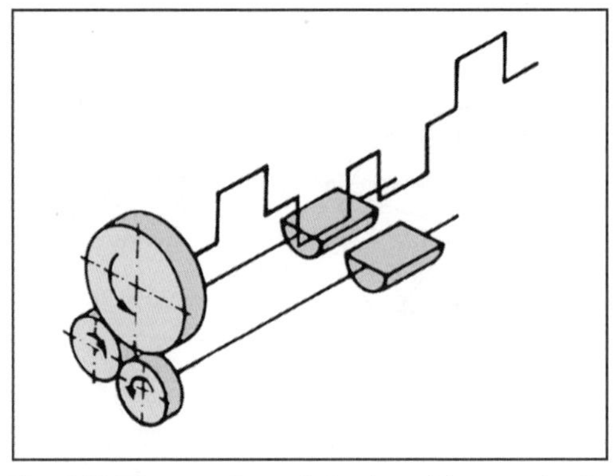

Fig. 6-45 Diagram of mass balancing of the second order in a four-stroke crank gear.

6.1.5.2 Balancing Multicylinder Crank Gears

Automobile engines are built with multiple cylinders, i.e., with 3 to 12 (16) cylinders, as three-, four-, five-, and six-cylinder inline engines and V6, V8, and V12 (V16) engines, and as VR5 and VR6 engines. Earlier, there was also a V-4 engine (Ford 12 M). Recently, three-row engines (W-engines) with 12 cylinders have been developed. These engines have three-, four-, five-, and six- (eight-) stroke crankshafts so that with a corresponding arrangement, the mass effects of the individual throws cancel each other out (self-balance). For this purpose, the throws are to be distributed evenly in the peripheral direction and lengthwise direction:

- With centrally symmetrical shafts (equal to the throw spacing across the perimeter), the free forces cancel each other out.
- Centrally and longitudinally symmetrical arrangements of the throws of a four-stroke engine shaft have no free forces and torques of the first order; starting with six strokes, the shafts are completely force-free and torque-free.

The criteria for the throw sequence are

- No or very low free mass effects. A simple rule of thumb for throw sequences with favorable mass balances was presented by O. Kraemer in Refs. [13, 14].
- Additional torque may not arise from mass balancing, and no additional inertial forces may arise from torque balancing.
- Even angular ignition spacing.

Free first-order inertial torque can be balanced by a shaft rotating in the opposite direction at the crankshaft speed with two countermasses of a corresponding size and lengthwise spacing (torque differential). The arrangement in the engine can be freely selected. Gears or chains provide the drive; frequently the oil pump drive is connected. To balance torque of the second order, the differential rotates at twice the crankshaft speed (Fig. 6-46).

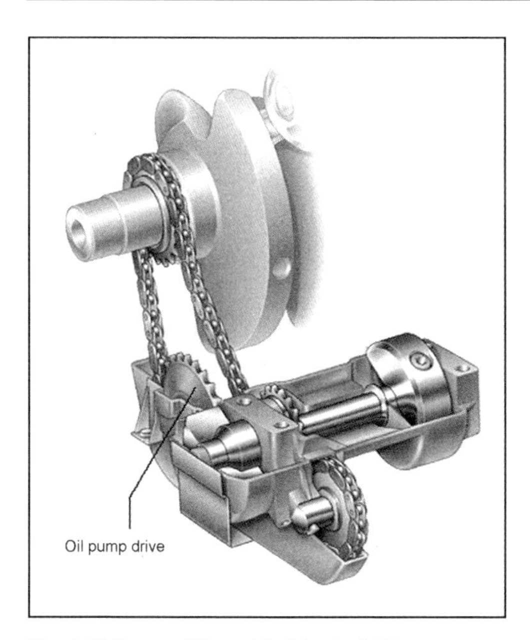

Fig. 6-46 Torque differential of the Audi V6.

The following holds true for crankshafts of four-stroke engines:

- Three-stroke shaft: free torque of the first and second orders occurs. The torque of the first order is compensated—especially in V-engines—with a torque differential.
- Four-stroke shaft: In four-cylinder, four-stroke inline engines, the inertial forces of second order are additive. These forces are balanced by two oppositely rotating shafts with countermasses (differential). Earlier, this was done only with tractor engines since the engine, the transmission, and rear axle housing form the bearing element of the vehicle.

Today, such differentials are also used for passenger car engines since beginning at 4000 min^{-1}, the free second-order inertial forces are noticeable. The vertical accelerations are guided into the body and cause an "unpleasant humming." [15]

Because of the high peripheral speeds of the bearing pin of this differential—up to 14 m/s—the bearing and drive must be carefully designed. The balance shafts are driven by a gear on a crankshaft web where the tooth face play of the drive must be harmonized to the shifts and rotational oscillations of the crankshaft (Figs. 6-47 and 6-48).

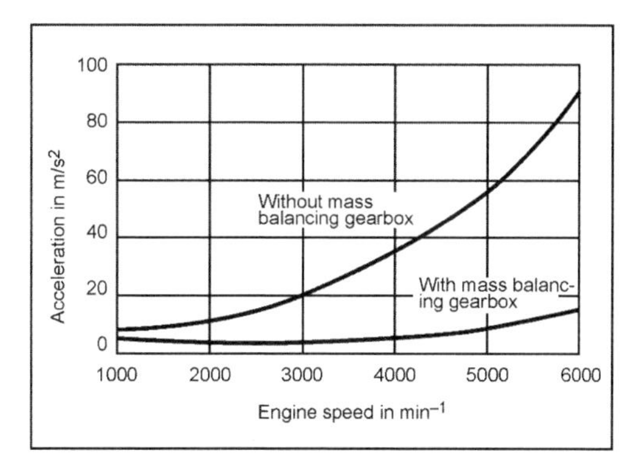

Fig. 6-48 Effect of the mass differential in a four-cylinder inline engine.

By offsetting the height of the balance shafts, an additional oscillating torque of the second order can be generated that can also balance with gas force components of the oscillating torque. Since this is only slightly effective, it is not used[16] (Fig. 6-49).

- Five-stroke shaft: Free inertial torque arises, especially a large oscillating moment of inertia of the second order (see example). Passenger car and truck engines

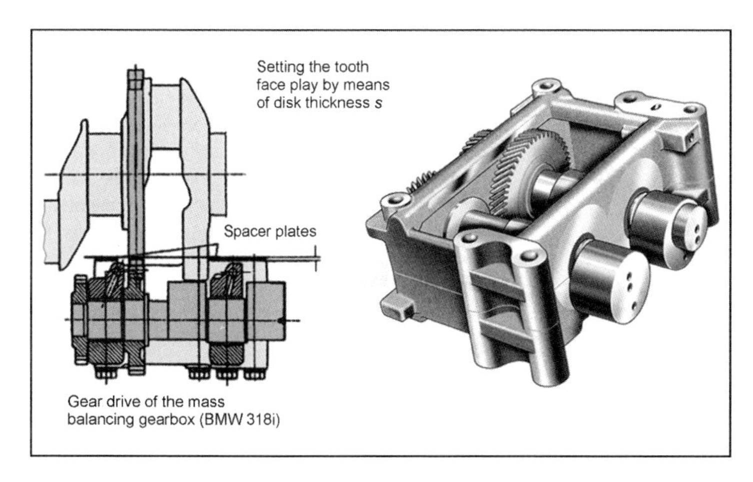

Fig. 6-47 Differential for inertial forces of the second order.

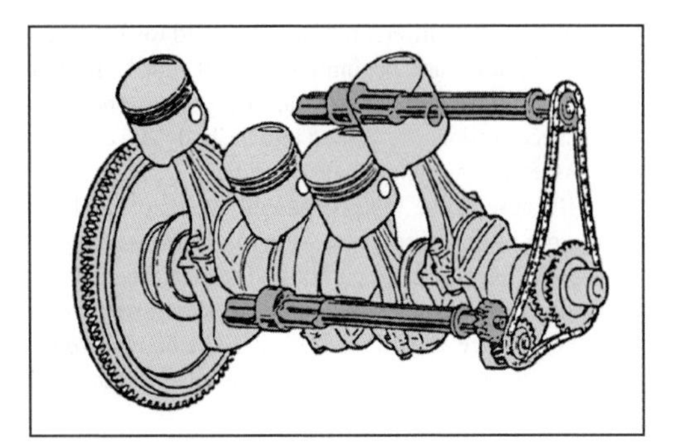

Fig. 6-49 Mass balancing of the second order with height-offset balance shafts (Mitsubishi).

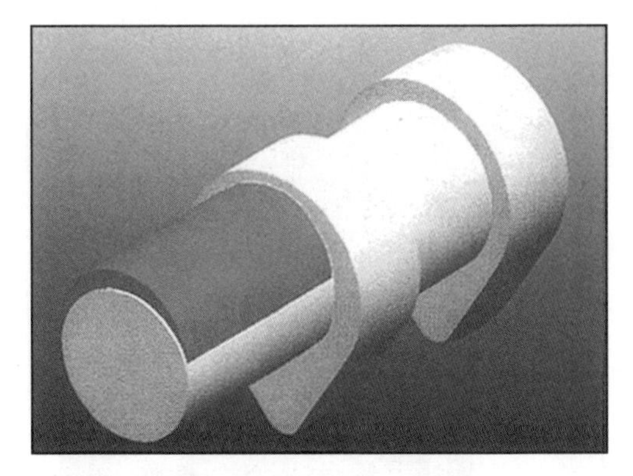

Fig. 6-50 Camshaft with balancing mass.

are built both with and without separate torque balance. In passenger car diesel engines, torque balance is not used, and the engine movement is captured by elastic bearings and shock absorbers.[17] In five-cylinder truck engines, the torque differential is optional depending on the installation in the vehicle (engine systems with a flange-mounted gearbox and retarder, or busses because of the resonance behavior of the vehicle body).

- Six-stroke shaft: Centrally symmetrical and longitudinally symmetrical shafts (starting at six shafts) are balanced by themselves; they do not have any free mass effects.

The most important considerations in designing the mass balancing system are

- Complexity of assembly (differential)
- Operating behavior at high speed (second order): bearing, lubrication, etc.
- Increasing or decreasing the load on the crank gear bearing
- Balance of the gas force
- Rotational oscillation behavior
- Inertia
- Friction behavior

The free forces and torques of the different cylinder configurations are summarized in tables in the relevant literature.

Mass balancing is used not only on the crankshaft drive but also on the valve gear, i.e., camshafts:

- The body is drilled eccentrically so that the manufactured imbalance can largely compensate for the free valve mass forces.
- Balancing masses are placed directly on the camshaft (Fig. 6-50).

6.1.6 Internal Torque

In addition to imbalanced inertial forces and inertial torques that are perceptible free inertial effects, internal torque also exists in engines. This includes the bending moment that arises in the (freely floating) crankshaft[18] (Fig. 6-51).

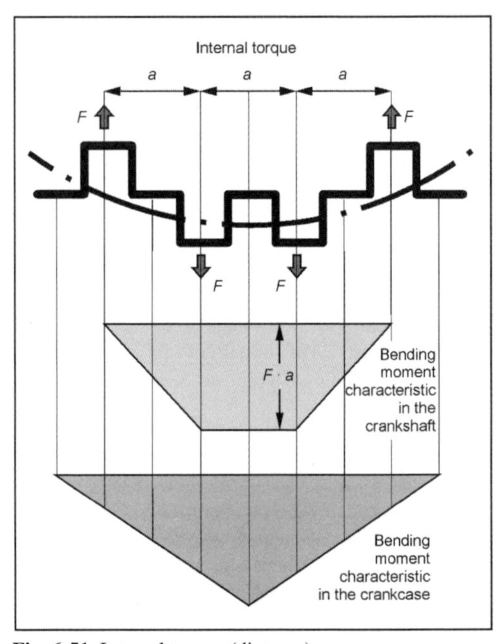

Fig. 6-51 Internal torque (diagram).

These internal torques provide an additional load on the crankshaft main bearing and subject the crankcase to flexural stress. As the engine speed increases, the internal torque places higher demands on the construction of the engine, especially V-12 and V-16 engines. The internal moment increases from the crankshaft ends to the middle of the engine. With longitudinally symmetrical shafts, the average bearing is subject to high loads from the inertial forces of the neighboring throws in the same direction, which can be prevented by internal mass balancing, i.e., balancing the inertial forces at their origin, i.e., at every throw (Fig. 6-52).

The advantages of complete internal mass balancing need to be weighed against the disadvantages of increasing the mass, moment of inertia, and cost.

Fig. 6-52 Four-cylinder spark-ignition engine (Opel-Ecotec) balanced on all cheeks.

6.1.7 Throw and Firing Sequences

To obtain a very even torque characteristic, the ignition of the individual cylinders must be evenly distributed over the power cycle. A requirement is that the throws must be evenly distributed over the perimeter. Hence, the following throw spacing results:

Four-stroke engines 720° crankshaft angle/cylinder number

Two-stroke engines 360° crankshaft angle/cylinder number

The firing sequence is also determined by the direction of rotation of the crankshaft. For automotive engines, the direction of rotation is established in DIN 73021.

- Clockwise rotation: when viewed from the counterforce transmission side; the cylinders are counted from the counterforce transmission side.
- Counterclockwise rotation: counterclockwise looking at the counterforce transmission side; the cylinders are counted from the counterforce transmission side.

The cylinders in V-engines are (viewed from the counterforce transmission side) counted from the right row starting from the left engine row that starts with $z/2 + 1$ from the right row. In V-engines, the same angular ignition spacing can be kept only when the V-angle corresponds to the power cycle (720° or 360° crankshaft angle) divided by the cylinder number. Other factors for the firing sequence are

- No or very small free inertial effects
- Favorable rotational oscillation behavior
- Good supercharging conditions

In two-stroke engines with a power cycle length corresponding to a 360° crankshaft angle, the throw sequence corresponds to the firing sequence; four-stroke engines have two dead centers with a 720° crankshaft angle power cycle:

- Ignition TDC
- Charge cycle TDC

Hence, for each throw sequence there are several firing sequences because of

- Short angular ignition spacing (V-angle δ)
- Long angular ignition spacing (depending on the direction of rotation: V-angle $\delta + 360°$ or V-angle $\delta - 360°$)

The number of possible firing sequences for inline engines with k = throw number[19] is

- Fully symmetrical shaft (four-stroke engines)

  $$2^{\left(\frac{k}{2}-1\right)}$$

- Partially symmetrical shafts (four-stroke engines with an uneven number of throws; two-stroke engines)

  $$\cdot\, k!/2 \cdot k$$

V-engines represent a good compromise between high power density and a compact basic design. The V-engine is therefore a preferred design in passenger car engines as well. A small V-angle requires a longer conrod (smaller conrod ratios $\lambda = r/l$) and possibly a shifting of the crankshaft drive to provide the necessary free travel of the cylinder. This yields a higher crankcase with reduced piston-side forces since the angular travel of the connecting rod is shorter. For vehicle engines, the 90° V-angle is preferred since it allows the first-order inertial forces to be completely balanced with rotating counterweights; in addition, in eight-cylinder V-90° four-stroke engines, the V-angle corresponds to the even angular ignition spacing. If the number of cylinders and V-angle do not correspond, an even ignition spacing is still attained by "spreading" the crank pins by the difference between the V-angle and angular ignition spacing (offset crank pin, stroke offset, split-pin crankshaft). Accordingly, six-cylinder passenger car and truck engines are being built today with a V-angle of 90° (such as Audi, Deutz, DaimlerChrysler), 60° (Ford), and even 54° (Opel), eight-cylinder engines with a 75° angle (DaimlerChrysler), which requires a total crank offset of 30°, 60°, 66°, and 15°. To select the V-angle, the clearance space of the engine and the harmonization of the engine program must be considered in addition to the crank gear mechanics.

Determining the firing sequences:

In two-stroke engines, the firing sequence corresponds to the throw sequence.

In four-stroke engines, the two crankshaft rotations of a power cycle are reduced to one rotation. This yields a 0.5-order phase diagram. The ignitions are evenly distributed over the perimeter and in the lengthwise direction. Viewed in terms of crank gear mechanics, V-engines are two inline engines offset from each other by the V-angle δ with half the number of cylinders. The ignition spacing of the cylinders that act together on a throw is

- $\delta°$ (short angular ignition spacing)
- $(\delta + 360)°$ (long ignition spacing)

The phase diagrams of the two partial inline engines are superposed for the short angular ignition spacing by $(\delta/2)°$, and for the long angular ignition spacing by $([\delta + 360]/2)°$ from which the ignition intervals can be determined.

6.2 Rotational Oscillations

6.2.1 Fundamentals

The crank gear is a spring-mass system that is excited to vibrate (oscillating rotational movement of the sequential individual masses on the shaft) by the periodic torsional forces (tangential forces) that overlap the actual rotational movement of the crankshaft. The rotational movement of the crankshaft therefore comprises three components:

- Even rotation corresponding to the speed
- Speed fluctuation as a result of the uneven torsional force characteristic (tangential force characteristic) over a power cycle ("static speed fluctuation")
- Vibration over the displacement angle caused by the torsional force ("dynamic speed fluctuation")

The movement of the system is described by the angle of twist of the moments of inertia in comparison to the initial position.

The kinetic energy stored in the moments of inertia is released to the coil springs and converted into potential energy in order to be reconverted back into kinetic energy. Given loss-free energy conversion, the free vibrations would last forever; the natural frequency depends exclusively on the system properties of spring rigidity and mass. Because of the resistance to the movement, energy is withdrawn from the system and converted into heat: The vibration is suppressed and slows at a greater or lesser rate depending on the damping.

If a periodic force acts on the system from the outside, then it forces the system to assume different vibration behavior; the system vibrates—after a transient phase—at the frequency of the exciting force. If the natural and exciting forces correspond, resonance occurs. Without damping, the vibration amplitude would assume an infinite value. However, the always-present damping limits the amplitude, and the size of the amplitude depends on the strength of the damping. This situation is illustrated by the magnification function V as a function of frequency ratio Ω/ω. [The magnification function is the ratio of the (maximum) vibration amplitude of the system to the amplitude that would result if the spring of the system were under a static load from the exciting force.]

If the path of the vibration amplitudes of the individual masses is represented over the length of the shaft as a curve trace, we get the mode of vibrations with the zero transition points of this curve as vibration nodes in which two neighboring masses vibrate in the opposite direction.

No rotational oscillation movement occurs at these points (certainly rotational oscillation stress, however) (Fig. 6-53).

For each possible form of vibration, there is a natural frequency that the system can use to execute free vibrations in the relevant mode of vibration. The mode of vibrations and the natural frequencies depend on the size and distribution of the torsional rigidities and the moments of inertia in the system.

Since the resonance can lead to vibration amplitudes that can destroy the crankshaft (Fig. 6-54), it is important to identify such dangerous conditions beforehand and undertake corresponding measures to eliminate them.

The properties of the crank gear are therefore calculated in this regard. Since it is a complex system, the crank gear must be conceptually simplified (reduced) so that it can be computed with a reasonable amount of effort. The basis of such a simplification (reduction) is the harmonization of the dynamic properties of the reduced system with those of the actual system. The calculation of the rotational oscillation consists of

- Reducing the machine system
- Calculating the natural frequencies and modes of natural vibration
- Calculating the exciting forces and amplitudes
- Calculating the crankshaft excursions in the case of resonance

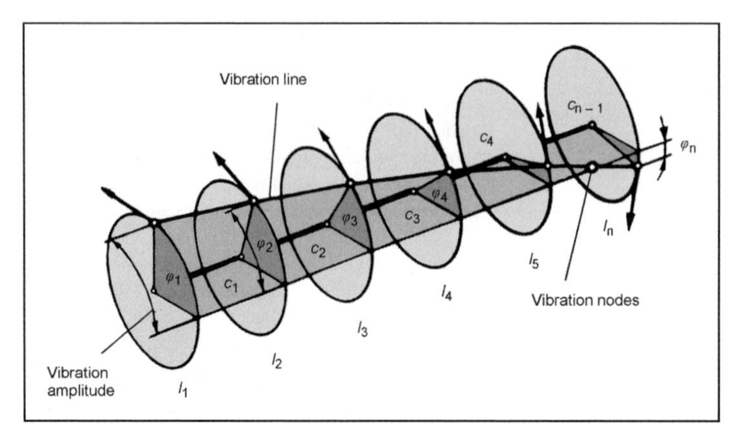

Fig. 6-53 Diagram of a rotational oscillation system.

Fig. 6-54 Torsion break of a passenger car crankshaft made of GGG 70.

- Calculating the crankshaft stress from the vibration excursions in the case of resonance
- Calculating the critical speeds

6.2.2 Reduction of the Machine System

The crank gear with the coupled masses (flywheel, crank wheel mechanism, valve gear, belt drive, etc.) is reduced to a simple geometrical model so that potential and kinetic energies of the actual and reduced systems correspond.

- Mass reduction: The crankshaft with the conrod, piston, and masses that it drives (crank wheel, flywheel, damper, etc.) is replaced by regular cylindrical disks with a constant moment of inertia. Although the moments of inertia of the crankshaft drives change from the piston and conrod movement, for the calculation, constant moments of inertia are assumed.
- Length reduction: The crankshaft throw is replaced with a straight, inertia-less shaft piece with the same diameter as the crankshaft main bearing (or the crank pin) whose length is such that the throw and shaft pieces have the same torsional rigidity (spring constant). There is a series of reduction formulas to accomplish this.

For passenger car engines, the BICERA formula is used. Since the shape of the crankshaft throw impairs its rotation, its reduced length is generally greater than the length of the throw (Fig. 6-55).

6.2.3 Natural Frequencies and Modes of Natural Vibration

The crank gear consists of coupled moments of inertia and torsional rigidities with mutually influential vibration behavior.

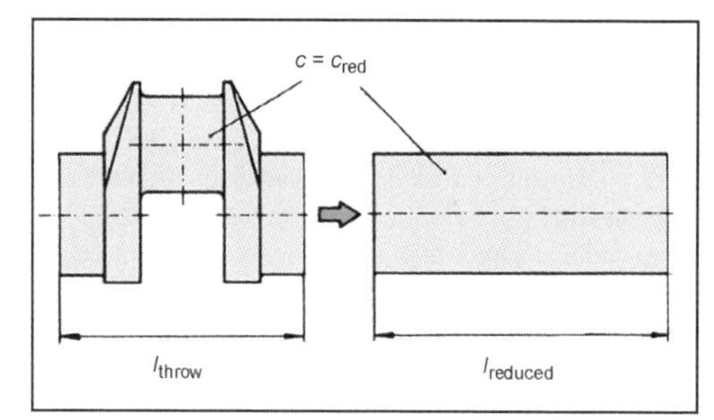

Fig. 6-55 Length reduction of a crankshaft throw.

Movement equations are created for the individual moments of inertia.

$$I_k \cdot \ddot{\varphi} + c_{k-1} \cdot (\varphi_k - \varphi_{k-1}) + c_k \cdot (\varphi_k - \varphi_{k+1}) = 0 \tag{6.87}$$

I = Mass moment of inertia
φ = Angle of twist of the moment of inertia
c = Torsional rigidity of the shaft piece
k = Counter for the moments of inertia

A system is obtained of homogeneously coupled linear differential equations with constant coefficients that describe the equilibrium between (Fig. 6-56)

- Moments of acceleration from the moment of inertia arising from the inertial torque and the angular acceleration
- Returning torque from the spring rigidity and difference between the angles of twist on both sides of the examined mass.

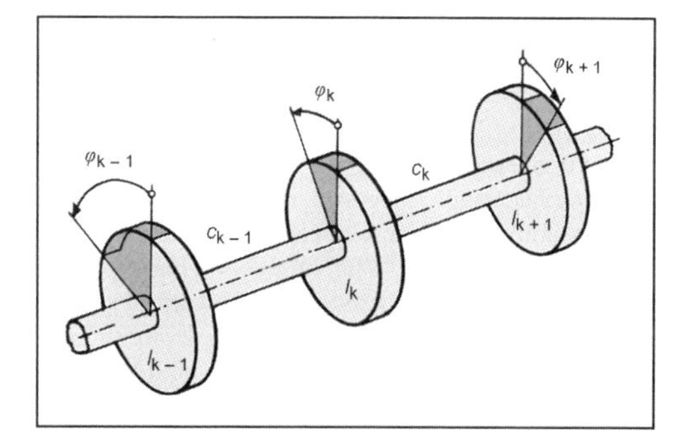

Fig. 6-56 Opposite rotation of the moments of inertia of the reduced crank gears.

The damping moment can be ignored when determining the natural frequencies since the natural frequencies are only slightly influenced by the damping when it is weak. The integration of these equations yields the natural frequencies of the system. To solve these differential equations, a model in the form of harmonic movement is created. Systems with more than three moments of inertia make the equation systems overly complex and difficult to deal with; for this reason, different experimental procedures have been developed. Of these, the procedure by Gümbel-Holzer-Tolle has gained broad acceptance. It provides insight into the physical behavior of the vibration processes and can be carried out using a simple and clear computational approach in which the results of one calculation step are used in the other as a pattern. The basic concept is as follows.

An oscillating torque is imagined that acts on the end of a system capable of vibration so that the system executes forced (undampened) vibrations; the amplitude of this oscillating torque (exciter moment amplitude) is set so that the vibration excursion of the first mass assumes the value 1. If the exciter frequency is then changed, the exciter moment M (residual exciter moment) also changes, which is necessary to maintain the vibration excursion 1 of the first mass. If the exciter frequency corresponds to one of the natural frequencies of the system, the amplitude M_k of the necessary exciter moment M is zero.

$$M_{k+1} = -M_k + I_k \cdot \Omega^2 \cdot u_k \quad (6.88)$$

$$u_{k+1} = u_k - \frac{M_{k+1}}{c_{k+1}} \quad (6.89)$$

$$u_1 = 1 \qquad M_1 = 0 \quad (6.90)$$

M = Exciter moment
u_k = Relative excursion
I = Mass moment of inertia
c = Spring rigidity
Ω = Exciter frequency
k = Counter for the moments of inertia

When doing the calculation, the residual exciter moment is calculated for the different exciter frequencies that are necessary to maintain vibration excursion 1 of the first mass, and the residual exciter moment is plotted over the exciter frequency. The intersections of the residual exciter moment curve with the abscissa yield the desired natural frequencies (Fig. 6-57).

If the calculation is repeated with the natural frequencies found in this manner, we obtain the respective modes of natural oscillation ("Sum of the amplitudes of all moments of inertia that define the deformational state of the oscillating system for each frequency.") However, only the relative excursions, i.e., the excursions of the individual moments of inertia in reference to the excursion of the first moment of inertia (Fig. 6-58), are determined.

We are therefore dealing with a problem of intrinsic value whose solution is only for one common factor. To determine the absolute amplitudes, we need the exciting forces. Another solution corresponding to the Gümbel-Holzer-Tolle method is a matrix calculation. The relationships derived from the motion equations between the amplitudes of the rotational oscillation excursions and the return torques provide an equation system that can be represented with matrices and can be solved with a computer.

$$I \cdot \ddot{\varphi} + D \cdot \dot{\varphi} + c \cdot \varphi = M(t) \quad (6.91)$$

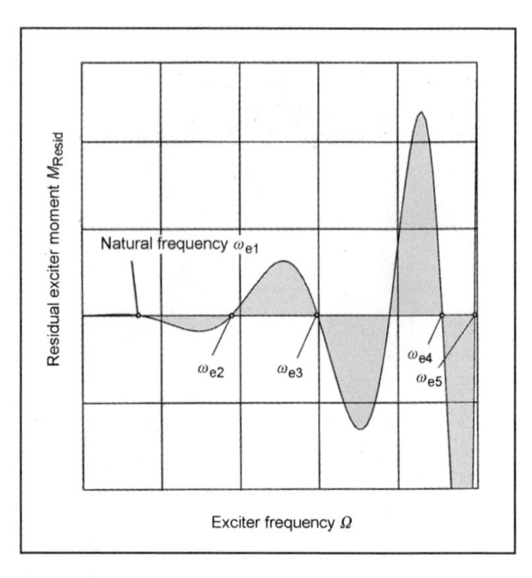

Fig. 6-57 Residual exciter moment curve.

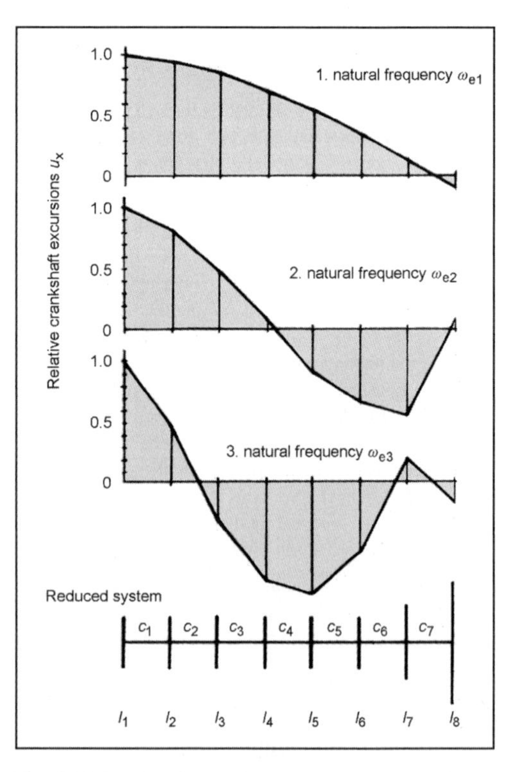

Fig. 6-58 Modes of natural oscillation for the three initial natural frequencies of a six-stroke crank gear with crank wheel and clutch.

6.2.4 Exciter Forces and Exciter Work

The vibration-exciting torsional force (tangential force) is composed of

- Gas torsional force (Fig. 6-59)
- Torsional force of the oscillating inertial forces (the rotating inertial forces do not participate in the excitation) (Fig. 6-60)

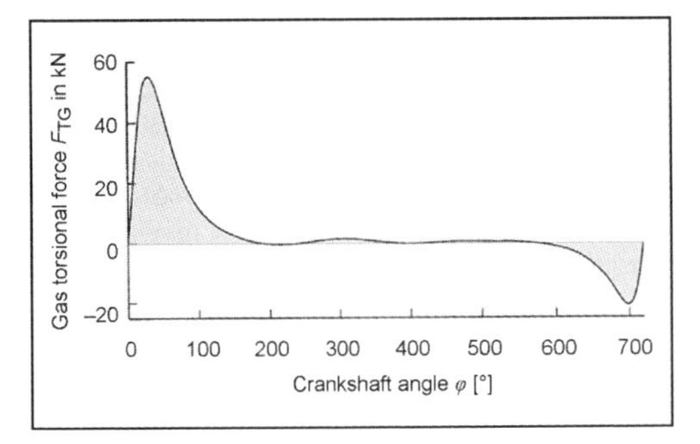

Fig. 6-59 Gas torsional force characteristic of a four-stroke diesel engine.

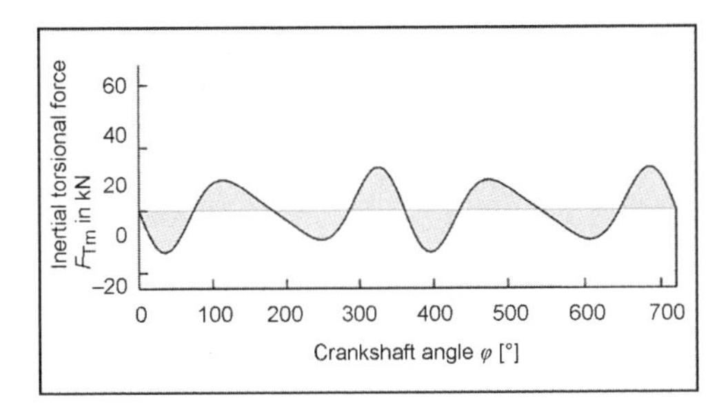

Fig. 6-60 Inertial torsional force characteristic of a four-stroke diesel engine.

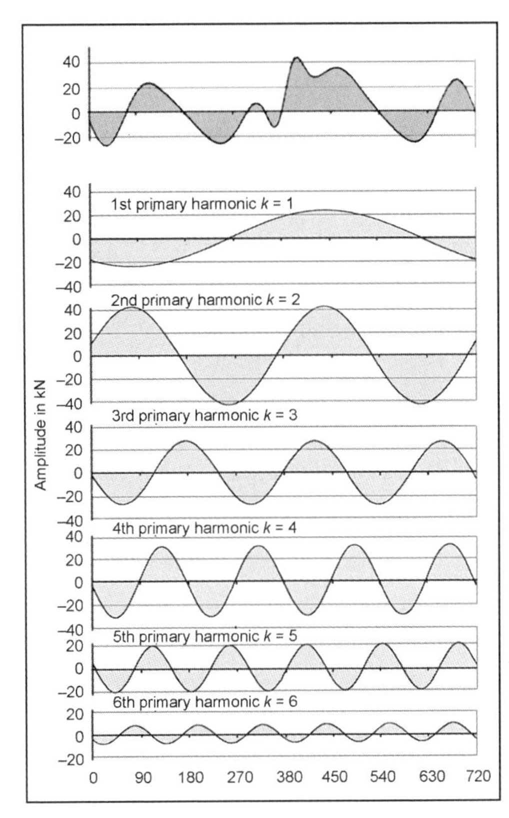

Fig. 6-61 Fourier analysis of a tangential force diagram: The tangential force curve is composed of the first six harmonics.

Since the gas torsional force is a function of the load (specific work), and the inertial torsional force is a function of the square of the rpm, their influence is investigated separately.

The gas torsional force cannot be described by a closed function and is therefore subject to a Fourier analysis; this is composed of a static component (nominal load torque) and a dynamic component (a basic vibration and overlapping harmonics). The exciting frequencies are, hence, the basic frequency (number of work cycles per unit time) and their integral multiples. They are proportional to the crankshaft speed. All of these exciting frequencies can resonate with one of the natural frequencies (Fig. 6-61).

The exciter work is the essential determinant in exciting vibration. An exciter force (resulting exciter force amplitude from the amplitudes of the gas and inertial torsional forces for the individual exciter frequencies) generates a greater excursion the farther it acts from the oscillating nodes (exciter work = exciter force × vibration amplitude). The phase angle of the exciter forces, i.e., their sequence over time, is represented in phase direction diagrams. The phase direction diagrams of the individual orders result from the order throw diagram of the 0.5 order (four-stroke) and of the first order (two-stroke) (Fig. 6-62).

Taking into consideration the vibration amplitude of the individual throws and the phase shift (firing sequence), we get the effective exciter force of the engine.

The relative crankshaft excursions of the individual cylinders are added geometrically in the direction of the rays of the phase direction diagrams. This shows us that certain orders are particularly dangerous because their geometric sum becomes very large. The geometric sum is described as the specific exciter work, i.e., exciter work of the engine in reference to force 1. Depending on the order and phase angle, the specific exciter work assumes different values.

The amplitude—the absolute excursion—of mass 1 is calculated from the equilibrium of the excitation work and damping work (per vibration). This allows us to determine

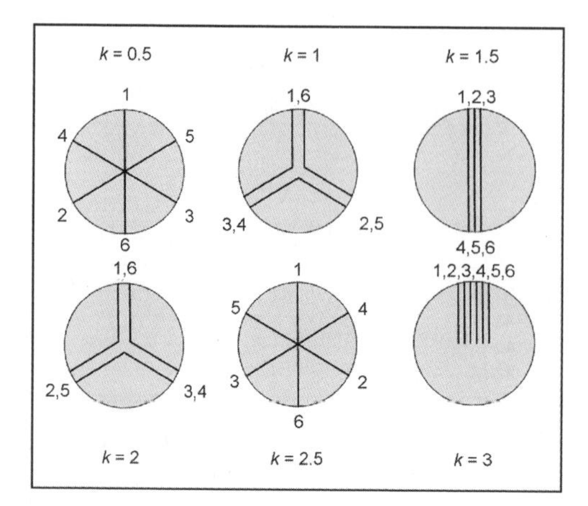

Fig. 6-62 Phase direction diagrams up to the 6th order for an inline six-cylinder four-stroke-crank gear.

the absolute excursions A of the individual masses of the substitute system:

$$A_1 = \frac{F_{Tk} \cdot \sum_1^z u_x}{\omega_e \cdot \sum_1^z \beta_x \cdot (u_x)^2} \tag{6.92}$$

$$A_x = u_x \cdot A_1 \tag{6.93}$$

F_{Tk} = Resulting exciter force amplitude from the amplitudes of the gas and inertial torsional forces (assumed to be the same for all cylinders)
u_x = Relative crankshaft excursions
ω_e = Natural frequency
β_x = Damping coefficient of the xth cylinder; usually the same damping coefficient is assumed for all cylinders
A_1 = Amplitude (absolute excursion) of the first mass of the system
u_x = Geometric sum of the relative crankshaft excursions
Index x = Number of cylinders
Index k = Order

The relative twist $\Delta\varphi$ of the masses x and $x + 1$ from the rotational oscillation stresses the crankshaft in addition to static torsional force.

$$\Delta\varphi = (u_x - u_{x+1}) \cdot A_1 \tag{6.94}$$

$$\tau = \frac{M_d}{W_p} = \frac{c_x \cdot A_1 \cdot (u_x - u_{x+1})}{W_p} \tag{6.95}$$

In particular, the gas forces excite vibrations of an order that are an integral multiple of the number i of ignitions within a crankshaft rotation.

- Four-stroke engine: $i = z/2$ ignitions per crankshaft rotation
- Two-stroke engine: $i = z$ ignitions per crankshaft rotation

All integral multiples of $z/2$ (four-stroke) or z (two-stroke engine) are dangerous since the exciters of all the cylinders are aligned for these orders. The critical speeds result from the intersections of the main harmonics with the exciter frequencies. The extent of the danger to the engine at the individual critical speeds can be found by calculating the resonance excursions of the crankshaft.

6.2.5 Measures to Reduce Crankshaft Excursions

Without damping, the excursions of the crankshaft would become increasingly larger until the shaft breaks. In practice, however, damping always exists: Material damping, friction damping, and damping from the lubrication film. However, these are usually insufficient in today's highly stressed crank gears so that additional measures must be taken. To avoid hazardous rotational oscillation states, one can

- Influence the exciter work by varying the firing sequence
- Shift the natural vibration frequency by changing the mass and spring rigidity

The feasibility and effectiveness of these measures is limited, however. An apparently simple measure is to increase the moment of inertia of the flywheel. This lowers the natural frequency, but at the same time the oscillating nodes are displaced toward the flywheel, and the shaft load is increased.

For these reasons, the only possibility is to reduce the rotational oscillations to a safe level. There are basically two options for this:

- Damping: Convert the vibration energy into heat. In the case of stationary forced vibrations and speed-proportional damping, there is an equilibrium between the moments of mass inertia, damping, return force, and excitation. The greater the damping moment, the smaller the vibration amplitude.
- Absorption: That is, "extinguishing" resonances by detuning the system, or, more precisely, shifting the natural frequencies into other speed ranges by counteracting with a mass: By coupling an additional mass, the absorber, the system is given one more degree of freedom. The original natural frequency splits into two natural frequencies that lie closely above and below the original. If the system is excited in the original natural frequency, then it remains unexcited while the absorber vibrates. Such absorbers are effective only for a single frequency. A pendulum attuned to a specific vibration frequency and articulated to the oscillating system enters a reverse phase when this vibration arises and, hence, counteracts the exciting moment. The resonance speed is split and shifted upward or downward. Centrifugal force absorbers are speed dependent.

The effect of vibration dampers in passenger vehicle engines are based on both damping and absorbing. With regards to spring rigidity, damping behavior, and mass inertia, they are designed to continuously reduce rotational oscillation excursions of the system.

For passenger car engines, rubber vibration dampers are used: An annular damper mass (secondary part) connected to the primary-side L-shaped driving disk is elastically coupled via a vulcanized rubber layer. The vibration energy is converted by the material damping (hysteresis) of the rubber into heat. The resonance peak is divided into two resonances whose peaks are reduced by the damping. Depending on the design, the damper mass is affixed radially and/or axially to the primary part. Two-stage dampers are also used in which two damper masses are tuned to two different frequencies[20] (Fig. 6-63). An example of this is with the two-mass rubber vibration damper for a five-cylinder diesel engine (2.5 L) in which both masses are harmonized to the torsion.

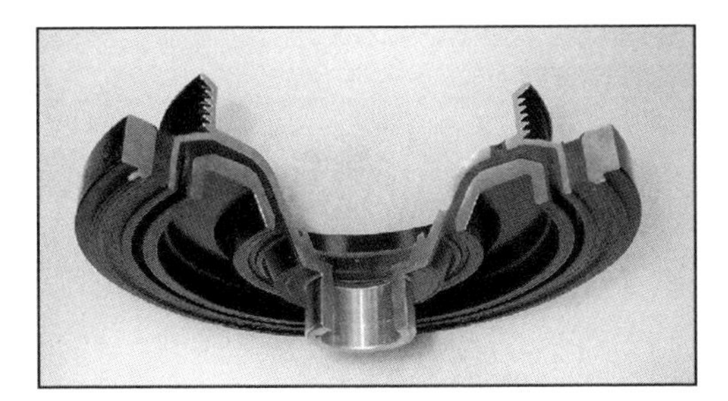

Fig. 6-63 Two mass rubber vibration damper (by Palsis) with vulcanized strips of rubber, V-belt strip on the primary side, primary side with shaft sealing flange made of St24W, secondary side made of GGG 40, primary-side moment of inertia $\Theta = 0.008$ kg m^2, secondary side 0.012 kg m^2/220 Hz and 0.006 kg m^2/360 Hz, rubber AEM (Vamac) (source: Palsis).

By reducing the rotational oscillation amplitude (Fig. 6-64), not only are the crankshaft and camshaft mechanically relieved, the play-induced noise of the engine and the excitation of the accessories to vibrate are reduced.[21]

Passenger car engines increasingly require vibration dampers to deal with large engine dimensions (stroke volume) and greater specific work (effective average pressure) because of the stronger excitation. These are also used to lower natural frequencies as a result of greater crank gear masses. (The natural frequencies of passenger car crank gears range from 300 to 700 Hz.)

Recently, viscous dampers like the ones that have been used for larger engines have also been used (Fig. 6-65).

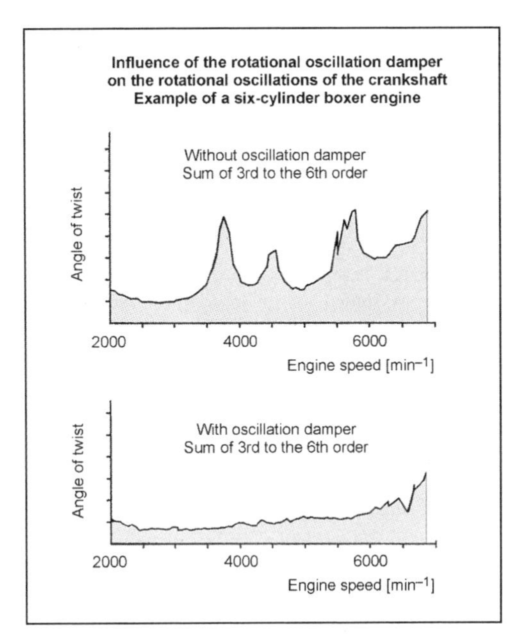

Fig. 6-64 Effect of a vibration damper.

Fig. 6-65 Viscose vibration damper with a decoupled belt pulley (torsionally elastic rubber coupling) for inline six-cylinder diesel engines (Palsis).

6.2.6 Two-Mass Flywheels

The drivetrain of a vehicle consists of an engine, a transmission, and the vehicle itself. The vibrations excited by the engine are also transmitted to the other components of the drivetrain. Engine-induced vibrations of the transmission are manifested as

- Bucking: The engine excites the system with 0.5-order vibrations that vibrate against the vehicle
- Chatter: The engine excites the transmission primarily with four- to six-order vibrations so that gears and synchronizer rings that do not lie in the flow of force vibrate against each other at comparatively large amplitudes.

In addition, the drivetrain is twisted during load changes and swings, which is only slightly dampened. These vibrations are noticeable, impair driving comfort, and additionally put stress on the components. To improve the vibration and noise behavior of the drive, two-mass flywheels are used: The mass of the engine flywheel is divided into a primary part rigidly fixed to the crankshaft and a secondary part articulated to the primary part. The primary and secondary parts are connected by torsionally elastic springs. This isolates the vibration; i.e., the operating range is shifted to the supercritical range of the enlargement function. Since different rigidities and damping properties are required to suppress the transmission chatter in the different operating ranges (traction, thrust, idling), the characteristics of the springs must be correspondingly engineered. This is accomplished, for example, by a series of springs with different rigidities. With correspondingly adjusted feather key systems, friction provides the desired damping[22] (Fig. 6-66).

The rotational oscillation behavior of the engine drivetrain changes because of the lower moment of inertia of the primary part of the flywheel (Fig. 6-67).

With two-mass flywheels, not only is the driving comfort improved, but the transmission is freed from additional oscillating torque. They are primarily used in passenger car engines with piston displacement ≥2 l, especially for diesel engines.[23] Three-mass flywheels are now also being used.

Bibliography

[1] Albrecht, F. u.a., Die Technik der neuen BMW Sechszylindermotoren, MTZ 61 (2000) 9.
[2] Der neue Ford Focus, Special addition ATZ/MTZ, 1/1999.
[3] Kemmann, H.K., Der neue Motor mit 1,8 l Hubraum, MTZ 59 (1998) 4.
[4] Bauder, A., W. Krause, M. Mann, R. Pischke, and H.-W. Pölzl, Die neuen V8-Ottomotoren von Audi mit Fünfventiltechnik, MTZ 60 (1999) 1, p. 16.
[5] Dorsch, H., H. Körkemeier, S. Peiters, S. Rutschmann, and P. Zwickwolf, Der 3,6-Liter-Doppelzündungsmotor des Porsche Carrera 4, MTZ 50 (1989) 2.
[6] Biezeno, C.B., and R. Grammel, Technische Dynamik, Springer, Berlin, 1953 (2nd reprint 1982).
[7] NN, Kolben für Pkw- und Nkw-Motoren, Grundlagen Pub., Kolbenschmidt, Chapter 1.
[8] Riedl, C., Konstruktion und Berechnung moderner Automobil- und Kraftradmotoren, 3rd edition, R.C. Schmidt, Berlin, 1937, p. 224–231.

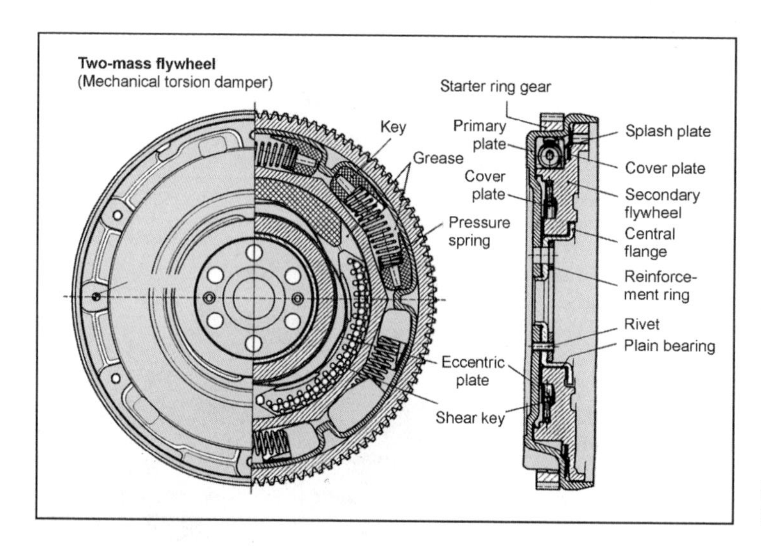

Fig. 6-66 Two-mass flywheel (GAT).

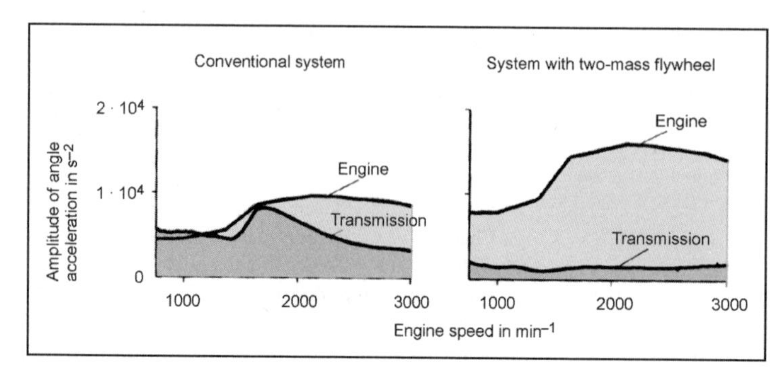

Fig. 6-67 Effect of a two-mass flywheel.

[9] Krüger, H., Sechszylindermotoren mit kleinem V-Winkel, MTZ 51 (1990) 10.

[10] Krüger, H., Der Massenausgleich des VR6-Motors, MTZ 54 (1993) 2.

[11] Küntscher, Kraftfahrzeugmotoren, 3, Aufl., Verlag Technik Berlin, Berlin, 1995, p. 134–156.

[12] Krüger, H., Massenausgleich durch Pleuelgegenmassen, MTZ 53 (1992) 2.

[13] Kraemer, O., Kurbelfolge günstigsten Massenausgleichs 1, Ordnung, ZVDI 81 (1937) 51, p. 1476.

[14] Kraemer, O., Bau und Berechnung der Verbrennungsmotoren, 4, Aufl., Springer, Berlin, 1963, p. 74.

[15] Breitwieser, K., M. Hofmann, H. Jacobs, and U. Vetter, Der neue Vierventilmotor mit 2.2 l Hubraum für den Opel Sintra, MTZ 57 (1996) 9.

[16] Flierl, R., and R. Jooß, Ausgleichswellensystem für den BMW-Vierzylindermotor im neuen 316i und 318i, MTZ 60 (1999) 5.

[17] Eberhard, A., and O. Lang, Der Fünfzylinder-Reihenmotor und seine triebwerksmechanischen Eigenschaften, MTZ 36 (1975) 4.

[18] Benz, W., Innere Biegemomente und Gegengewichtsanordnungen bei mehrfach gekröpften Kurbelwellen, MTZ 13 (1952) 1.

[19] Lang, O.R., Triebwerke schnelllaufender Verbrennungsmotoren, Springer, Berlin, 1966, p. 66.

[20] Anisits, F., K. Borgmann, H. Kratochwill, and F. Steinparzer, Der neue BMW Sechszylinder Dieselmotor, MTZ 59 (1998) 11.

[21] Pilgrim, R., and K. Gregotsch, Schwingungstechnisch-akustische Entwicklung am Sechszylinder-Triebwerk des Porsche Carrera 4, MTZ 50 (1989) 3.

[22] Nissen, P.-J., D. Heidingsfeld, and A. Kranz, Der MTD–Neues Dämpfungssystem für Kfz-Antriebsstrange, MTZ 61 (2000) 6.

[23] Reik, W., R. Seebacher, and A. Kooy, Das Zweimassenschwungrad, LuK 6th Colloquium Mach 19/20, 1998.

7 Engine Components

7.1 Pistons / Wristpins / Wristpin Circlips

7.1.1 Pistons

7.1.1.1 Requirements and Functions

The functions carried out by the piston include accepting the pressures created by the ignition of the fuel and air mixture, transferring these forces via the wristpin and the connecting rod to the crankshaft, and, in addition, providing guidance for the small conrod eye.

As a moving wall that, working in conjunction with the piston rings, transfers power, the piston has to reliably seal the combustion chamber against gas escaping and lubricant oil flowing by in all operating situations. Increases in engine performance have caused parallel increases in the demands on the pistons.

One example for piston loading: When a gasoline engine is running at 6000 rpm, every piston (D = 90 mm) at peak cylinder pressure of 75 bar, 50 times a second, is subjected to a load of about 5 tons.

Satisfying the various functions—such as adaptability to various operating situations, security against the pistons seizing while at the same time achieving smooth running, low weight at sufficient strength, low oil consumption, and low pollutant emissions—results in requirements for engineering and materials that in some cases are contradictory. These criteria have to be weighed carefully against each other for each type of engine. Consequently, the solution that is ideal in any particular instance may be quite different.

Compiled in Fig. 7-1 are the operating situation for the pistons, the resultant requirements for their design, and the requirements in terms of engineering and materials.

7.1.1.2 Engineering Designs

We find that, given the operational requirements of the various internal combustion engine designs (two-cycle, four-cycle, gasoline, and diesel engines), the aluminum-silicon alloys are as a rule the most suitable piston materials. Steel pistons are used in special cases, but they then require special cooling measures.

In the interest of weight reduction, a carefully worked out engineering design for the pistons is necessary, combined with the requirement for good piston cooling. Important terms and dimensions used to describe the geometry are shown in Figs. 7-2 and 7-3.

The increase in engines' specific performance is affected in part by increasing engine speed. The strong rise in the mass inertias that results in the reciprocating engine components is largely compensated for by reducing the compression height and optimizing the weight in the piston engineering design.

Particularly in smaller, high-speed engines the total length of the piston (GL), referenced to piston diameter, is shorter than in larger engines running at medium speeds.

The compression height influences overall engine height and most decisively the weight of the piston. The engineer thus strives to keep this dimension as small as possible. Consequently, the compression height is always a compromise between demands for a short piston and for high operational reliability.

The values given in Fig. 7-2 for the head thickness s apply generally for pistons with a flat and level head, as well as for those with a convex or concave crown. In the case of pistons for diesel engines with direct injection, with deep recesses, the head thicknesses, depending on maximum cylinder pressure, lie between 0.16 and 0.23 times the maximum recess diameter (D_{Mu}).

We learn from the guideline values in Fig. 7-2, in regard to the wristpin diameter, that the higher working pressures in diesel engines require larger wristpin diameters. The piston ring zone, together with the piston rings themselves, represents moving seals between the combustion chamber and the crankcase. The length of this zone depends on the number and thickness of the piston rings used and the lengths of the lands between the rings. The compression ring set, with just a few exceptions, comprises two compression rings and an oil control ring. The three-ring piston is the standard design today.

The length of the first ring land is selected in accordance with the ignition pressure occurring in the engine and the temperature of the land. The lengths of the lands located below are shorter, which is because of the falling temperature and loading due to gas pressure.

The piston skirt is used to guide the piston within the cylinder. It transfers to the cylinder wall, in sliding fashion, the lateral forces occurring because of the deflection of the conrod. With sufficient skirt length and close guidance the so-called "piston slapping," occurring at the moment when contact shifts from the one side of the piston to the opposite side (secondary piston motion), is kept to a minimum. This is important for smooth engine running and to reduce wear at all the piston's sliding surfaces.

The piston bosses must transfer all longitudinal forces from the piston to the wristpin and must therefore be well supported against the head and the skirt. Sufficient distance between the upper face of the boss bore and the inside of the piston head favors a more uniform distribution of stresses at the cross section for the support area. At high loads particularly careful design of the support area is thus required. To avoid fissures forming at the bosses, the mean calculated surface pressure in the boss bore (dependent on the boss and wristpin configuration and particularly dependent on the boss temperature) should not exceed values of between 55 and 75 N/mm^2. Attaining higher values is possible only by adopting special measures to increase the strength at the boss bore.

Operating conditions	Requirements for the piston	Engineering solution	Materials solution
Mechanical loading (a) Piston head / combustion recess Gasoline engines: Ignition pressures 50 to 90 bar Diesel engines: Ignition pressures 80 to 180 bar (b) Piston skirt: Lateral force: approx. 6% to 8% of max. ignition pressure (c) Piston boss: Permissible surface pressure, temperature-dependent	High static and dynamic strength at high temperatures. High surface pressure in the bores in the bosses. Little plastic deformation.	Sufficient wall strength, stable engineering design, uniform power flow, and heat flow Boss bushing, Ferrotherm piston heads made of steel	Various Al-Si casting alloys, with heat exposure (T5) or hardening by precipitation (T6), cast or forged special brass, bronze
High temperature in combustion chamber: Mean gas temperature approx. 1000°C At piston head / edge of recess: 200 to 400°C for ferrous materials: approx. 350 to 500°C At the wristpin boss: 150 to 260°C At the piston skirt: 120 to 180°C	Strength must be maintained even at higher temperatures. Indicator values: Hot hardness, permanent strength, high thermal conductivity, resistance to scale (steel)	Sufficient thermal convection cross sections, cooling channels	As above
Acceleration of piston and conrod at higher speeds: In some cases far above 25 000 m/s^2	Low weight, resulting in small inertial forces and moments of inertia	Lightweight construction with maximum utilization of material capabilities	Al-Si alloy, compacted
Sliding friction in the ring grooves, at the skirt, in the wristpin bearings. Unfavorable lubrication situation in some cases.	Low friction resistance, high wear resistance (influences service life), low tendency to seize	Sliding surfaces of sufficient size, uniform pressure distribution. Hydrodynamic piston shapes in the skirt area. Armored grooves	Al-Si alloys, skirt tinned, graphited, coated; groove reinforcement by ring carriers cast in place
Change of contact from one side of the cylinder to the other (above all at top dead center).	Low noise, no "piston slapping" with engine cold and warm, little susceptibility to cavitation, no impact pulses	Low play when running, elastic skirt design with an optimized piston shape, offset bores in the bosses	Low coefficient of thermal expansion. Eutectic or supereutectic Al-Si alloys

Fig. 7-1 Operating conditions and the resulting demands on the piston as well as for solutions based on the engineering design and materials selection.

The distance between the two bosses AA depends on the width of the small-end eye. This value has to be optimized in the interest of lower deformation values for the piston and wristpin. Only with the smallest possible boss clearances can ideal support be achieved and the reciprocating masses kept small.

	Gasoline engines		Diesel engines (four-cycle)
	Two-cycle	Four-cycle	Passenger car diesel
Diameter D (mm)	30 to 70	65 to 105	65 to 95
Overall length GL/D	0.8 to 1.0	0.6 to 0.7	0.8 to 0.95
Compression height KH/D	0.4 to 0.55	0.30 to 0.45	0.5 to 0.6
Wristpin diameter BO/D	0.20 to 0.25	0.20 to 0.26	0.32 to 0.40
Fire land F (mm)	2.5 to 3.5	2 to 8	4 to 15
First ring land St/D*	0.045 to 0.06	0.040 to 0.055	0.05 to 0.09
Groove height for first ring (mm)	1.2 and 1.5	1.0 to 1.75	1.75 to 3.0
Skirt length SL/D	0.55 to 0.7	0.4 to 0.5	0.5 to 0.65
Boss clearance AA/D	0.25 to 0.35	0.20 to 0.35	0.20 to 0.35
Head thickness s/D or s/D_{Mu}	0.055 to 0.07	0.06 to 0.10	0.15 to 0.22**

* Values for diesel engines are applicable to pistons with ring carriers (groove inserts), depending on peak combustion pressure.
** For direct injection models ~0.2 × combustion recess diameter (D_{Mu}).

Fig. 7-2 Major dimensions for lightweight metal pistons and passenger cars.

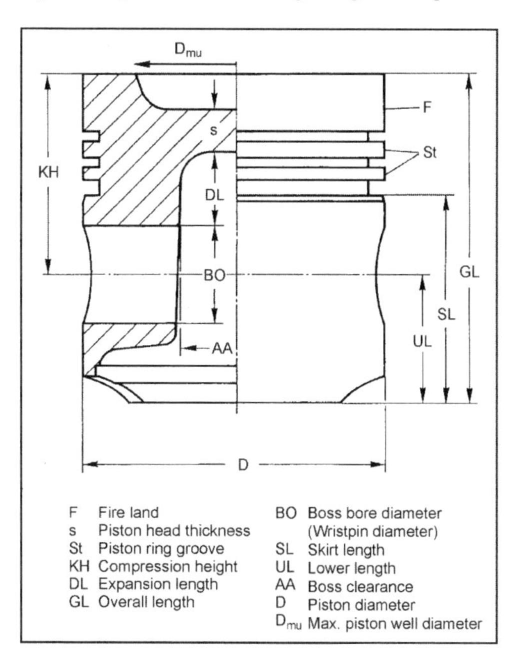

Fig. 7-3 Important terms and dimensions at the piston.

7.1.1.3 Offsetting the Boss Bore

Offsetting the axis of the wristpin in relation to the piston's longitudinal axis optimizes the contact properties for the piston at the change of sides. The impact pulses can be influenced decisively with this measure. The location and amount of offset to the piston's longitudinal axis can be optimized by calculating for the piston movement. Thus a reduction of the piston running noise and minimization of cavitation hazard at the cylinder liner is achieved.

7.1.1.4 Installation Play and Running Play

One attempts to keep installation play at the piston skirt as small as possible so that uniformly smooth running is achieved in all operating situations. When working with light-alloy pistons, this objective can be achieved only with special engineering efforts. This is because of the high coefficient of thermal expansion for lightweight alloys. In the past steel strips were often cast in place to influence expansion in response to heat ("regulating piston").

Figure 7-4 provides an overview of the amount of play found at the skirt and fire land for various piston designs.

The amount of play at the wristpin, inside the wristpin boss, is important for smooth piston running and low wear at these bearing points. When determining the minimum play (Fig. 7-5), it is necessary, in the case of gasoline engines, to determine whether a floating wristpin is used or whether it is fixed in the small-end eye by shrink fit. The floating wristpin is the standard design and the version that can handle the highest loads in the piston bosses. The "shrink-fit" conrod, which according to statements by some engine builders is more economical, is used only in gasoline engines. The shrink-fit conrod design is not suitable for modern diesel engines and for turbocharged gasoline engines.

Piston designs	Regulating piston		Without regulating strips		
	Hydrothermik®	Hydrothermatik®	Al piston		Modern light-weight pistons
Operating principle	Gasoline	Gasoline and diesel	Gasoline (two-cycle)	Diesel	Gasoline (four-cycle)
Installation examples (nominal dimension range)	0.3 to 0.5		0.6 to 1.3	0.7 to 1.3	0.3 to 0.5
Upper end of skirt	0.6 to 1.2	1.8 to 2.2	1.4 to 4.0*	1.8 to 2.4	1.7 to 2.2

* Only for single-ring designs and maximum performance engines (end of skirt near the fire land)

Fig. 7-4 Normal installation play dimensions for light-alloy pistons in vehicular engines (as ‰ of nominal diameter; installation in gray cast engine block).

Floating wristpin	Shrink-fit wristpin (fixed pin)
0.002 to 0.005	0.006 to 0.012

Fig. 7-5 Minimum wristpin play in gasoline engines, in mm (not for racing engines).

7.1.1.5 Piston Masses

The piston and its accessories (rings, wristpin, circlips) form, together with the reciprocating share of the conrod, the reciprocating masses. Depending on the engine design, free mass inertias and/or free moments occur; in some cases these can no longer be compensated for or may be compensated for only with considerable effort. It is because of this phenomenon that, above all in the case of high-speed engines, the need to achieve the lowest possible reciprocating masses arises. The piston and the wristpin account for the largest share of the reciprocating masses. Consequently, weight optimization has to start here.

About 80% of the piston weight is located between the center of the wristpin and the upper surface of the head. The remaining 20% is located between the center of the wristpin and the end of the skirt. Of the major dimensions previously discussed, the determination of the compression height obtains decisive significance; with the determination of the compression height, about 80% of the piston weight is predetermined.

When dealing with direct-injection gasoline engines the piston head is used to deflect the stream and is shaped accordingly; see Fig. 7-6. The pistons are both taller and heavier. The center of gravity shifts upwards.

Fig. 7-6 Piston for a gasoline engine with direct injection.

The piston's masses G_N can best be compared when one references them to the comparison volume $V \sim D^3$ (without piston rings and the wristpin). It should be noted here, however, that the length of the compression height is always to be included in any analyses of the engine.

The mass indices G_N/D^3 (without rings and the wristpin) for proven piston designs are shown in Fig. 7-7.

Material	Operating principle	G_N/D^3 (g/cm^3)
Aluminum alloys	Four-cycle gasoline engines*	0.40 to 0.55
	Two-cycle gasoline engines*	0.5 to 0.7
	Four-cycle diesel engines	0.80 to 1.10*

* Intake manifold injection

Fig. 7-7 Mass indices for passenger car pistons <100 mm diameter.

7.1.1.6 Operating Temperatures

An important factor regarding operational reliability and safety and service life is the component temperature for both the pistons and the cylinders. The piston head, exposed to the hot combustion gases, absorbs varying amounts of heat, depending on the operating situation (engine speed, torque). These volumes of heat, where the pistons are not oil cooled, are given off to the cylinder wall primarily through the first piston ring and, to a far lesser degree, through the piston skirt. When piston cooling is affected, by contrast, a major part of the heat volume is transferred to the motor oil. Because of the material cross sections determined by the engineering, there appear heat flows that result in characteristic temperature fields. Figures 7-8 and 7-9 show typical temperature distributions at pistons for gasoline and diesel engines.

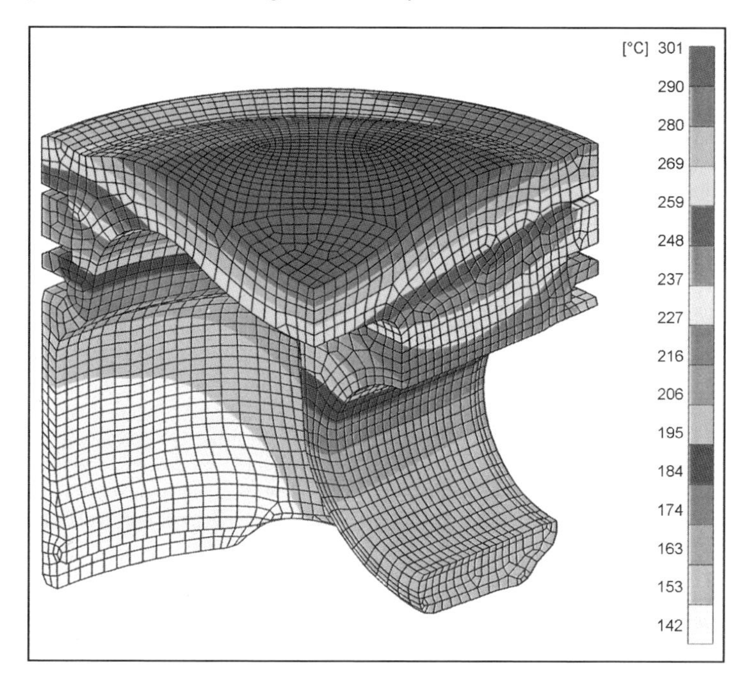

Fig. 7-8 Temperature distribution at a piston for a gasoline engine. *(See color section.)*

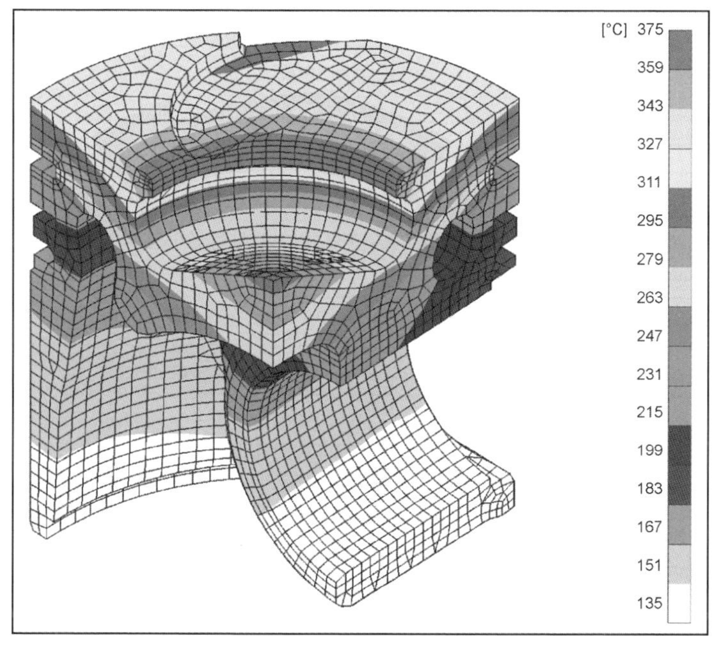

Fig. 7-9 Temperature distribution at a piston with cooling channel for a diesel engine. *(See color section.)*

Severe thermal loading, on the one hand, reduces the durability of the material from which the piston is made. The critical points in this regard are the zenith of the boss and the edge of the recess in direct-injection diesel engines, and the transitional area between the hub connection point and the piston head in gasoline engines.

On the other hand, the temperatures in the first piston ring groove are significant in regard to oil carbonization. Whenever certain limit values are exceeded, the piston rings tend to stick and as a result are limited in their functioning. In addition to the maximum temperatures, the dependency of piston temperatures on engine operating conditions (such as engine speed, mean pressure, ignition angle, and volume injected) is of significance. Figure 7-10 shows typical values for gasoline and diesel engines used in passenger cars, in the area around the first piston ring groove, depending on the operation conditions.

7.1.1.7 Piston Cooling

Spray Cooling

One version often found is a nozzle located at the lower end of the cylinder, through which motor oil is sprayed onto the inside contours of the piston. The cooling effect is dependent upon the volume of cooling oil and the surface area available for heat transfer. In this way temperature reductions of up to 30°C can be attained at the first groove and the boss. A simpler version is a bore through the big-end eye, which is provided with oil from the conrod bearing lubrication system. In addition to a lesser cooling effect, the part of the stream of oil that meets the cylinder running surfaces provides better lubrication, which in turn offers greater security against fuel friction.

Pistons with Cooling Oil Cavities

A more complex but more effective option for piston cooling is to provide cavities in those areas at the piston head and the ring grooves that are subjected to severe thermal loading. An annular cooling channel is supplied with oil, through a feed opening, by a spray nozzle; after taking on heat (ΔT up to about 40°C) the oil passes through a discharge opening on the opposite side of the piston and returns to the oil sump. The recommended specific masses for cooling oil come to about 5 kg/kWh. A cooling channel cast directly at the ring carrier ("cooled ring carrier") provides ideal effectiveness in regard to groove cooling.

Figure 7-11 shows the typical application ranges for various piston designs.

7.1.1.8 Piston Designs

Ongoing piston development has produced a large number of designs, the most important of which, having proven themselves in practice, are presented here. In addition, new directions for development are being pursued, for example, pistons for engines with an extremely low profile, pistons made of composites with local reinforcing elements, or pistons with a variable compression height, which permit variable compression ratios.

Engine conditions	Change in engine conditions	Change in piston temperature at groove 1
Water cooling	Water temperature 10°C	4 to 8°
	50% antifreeze	+5 to 10°C
Lubricating oil temperature (without piston cooling)	10°C	1 to 3°C
Piston cooling with motor oil	Injection nozzle in conrod big end	–8 to 15°C on one side
	Normal injection nozzle (stationary nozzle)	–10 to 30°C
	Cooling channel	–25 to 50°C
	Cooling oil temperature 10°C	4 to 8°C (also at edge of recess)
Mean pressure (n = constant)	0.1 MPa	5 to 10°C (15 to 20°C at edge of recess)
Engine speed (p_e = constant)	100 1 rpm	2 to 4°C
Ignition point, start of injection	1 crankshaft degree	1.5 to 3.5°C
Fuel-to-air ratio, lambda	Lambda = 0.8 to 1.0	Little influence

Fig. 7-10 Influence of engine operating conditions on the piston groove temperatures.

Operating principle	Loading		
Gasoline	No piston cooling	Piston with spray cooling	Forged piston with spray cooling
	Low ≈40 kW/l	Medium ≈65 kW/l	High ≥60 kW/l°
Passenger car diesel	Spray cooling	Piston with cooling channel	Cooled ring carrier
	Low ≤35 kW/l	Medium 35 to 45 kW/l	High >45 kW/l

Fig. 7-11 Survey of cooling variants.

Modern gasoline engines use lightweight designs with symmetrical or asymmetrical oval skirt shapes (cam ground pistons) and, if indicated, differing wall thicknesses for the contact side and the opposite side. These piston designs are distinguished by optimized weight and particular flexibility in the center and lower skirt areas. It is for the reasons mentioned here that the regulating piston is becoming less and less common. Older designs are also discussed briefly in the interest of completeness.

Pistons with Strip Inserts to Regulate Thermal Expansion, for Installation in Gray Cast Iron Engine Blocks

The primary objective in regulating piston design and for many inventions in this sector was and is the effort to reduce the relatively large differences in the coefficients of thermal expansion between gray cast engine blocks and aluminum pistons. Known solutions range from Invar strip pistons to the Hydrothermik® or Hydrothermatik® pistons.

Hydrothermik® Piston

Hydrothermik® pistons, Fig. 7-12, are designs with a skirt profile formed in accordance with hydrodynamic aspects. They are installed in gasoline engines for passenger cars. The pistons are slotted at the transition from the piston head to the skirt, at the level of the third groove. These pistons are characterized by particularly smooth running and long service lives. The strips cast in place between the skirt and the wristpin bosses, made of nonalloyed steel, in conjunction with the lightweight metal that surrounds them, form regulation elements that reduce the thermal expansion of the skirt in the direction that is important for guidance within the cylinder.

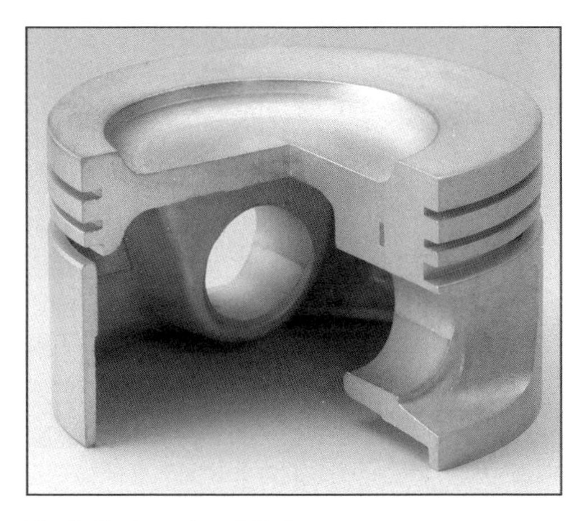

Fig. 7-12 Hydrothermik® piston.

Hydrothermatik® Pistons

Hydrothermatik® pistons, Fig. 7-13, operate on the same expansion regulation principle as the Hydrothermik® pistons.

In the Hydrothermatik® piston, the transition from the head area to the skirt is not slotted; the transitional cross sections are dimensioned so that, on the one hand, the flow of heat from the piston head to the skirt remains relatively unhindered while, on the other hand, the effect of the steel strips, because of the connection of the skirt with the rigid head section, is not affected in any essential way. Thus, this piston design joins the high strength of the nonslotted piston with the advantages of the design using regulation strips. The Hydrothermatik® piston is also suitable for use on naturally aspirated diesel engines.

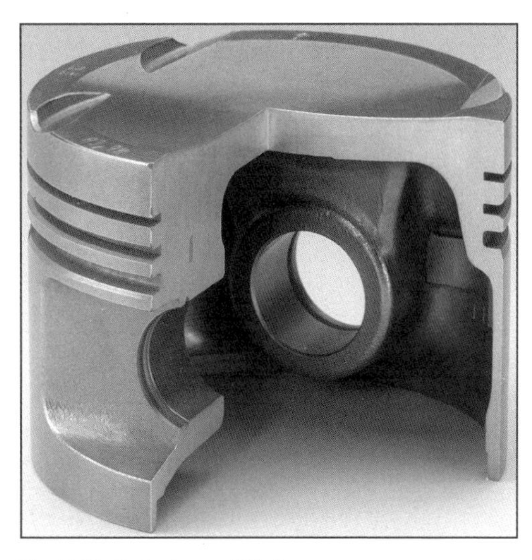

Fig. 7-13 Hydrothermatik® piston.

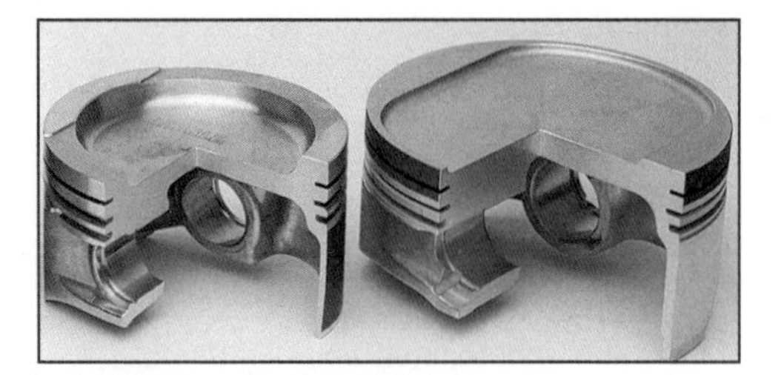

Fig. 7-14 Asymdukt® piston.

To extend the service life and to reduce wear the pistons used in diesel engines are fitted with a ring carrier (groove insert) made of austenitic cast iron.

Asymdukt® Piston

The Asymdukt® piston, Fig. 7-14, is a modern piston design distinguished by very low weight, optimized support, and a boxlike, oval-shaped skirt section. It is excellently suited for use in modern gasoline engines for passenger cars. It is suitable both for aluminum engine blocks and for gray cast engine blocks. With the flexible skirt design the differences in thermal expansion between the gray cast block and the aluminum pistons can be excellently compensated within the elastic range. The pistons may be either cast or forged. The forged version is used above all in high-performance sport engines or in heavily loaded, turbocharged gasoline engines.

Piston for Race Cars

These are always special designs, Fig. 7-15. The compression height (KH in the illustrations) is very short, and the piston as a whole is superbly optimized for weight. Only forged pistons are used here. Weight optimization and piston cooling are decisive criteria for the design of these pistons. In Formula 1 engines' specific output of more than 200 kW/l and engine speeds exceeding 18 000 rpm are common. The service life of the pistons is matched to the extreme operating conditions.

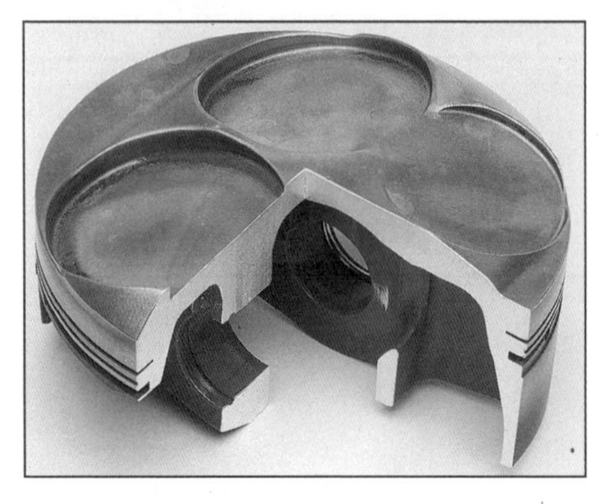

Fig. 7-15 Formula 1 piston, forged.

Fig. 7-16 Piston and cylinder for a two-cycle engine.

Pistons for Two-Cycle Engines

In the two-cycle piston, Fig. 7-16, the thermal loading is particularly high because of the more frequent exposure to heat; there is one ignition event for each rotation of the crankshaft. In addition, it has to close or open the inlet and outlet channels in the cylinder during its upward and downward strokes. This means that it has to control the exchange of gases. The result is severe thermal and mechanical loading.

Two-cycle pistons are equipped with one or two piston rings and, with regard to their outward design, can vary from the open "windowed" piston to a version with a full skirt. This depends upon the design of the overflow channels (long or short channels). In this case the pistons are normally manufactured from the MAHLE 138 supereutectic Al-Si alloy.

Ring Carrier Piston

In the case of ring carrier pistons, Fig. 7-17, introduced to mass production as early as 1931, the topmost ring

Fig. 7-17 Ring carrier piston.

groove and, in some cases, the second ring groove lie in a so-called ring carrier or groove insert that is joined permanently with the piston material by a metallic bond.

The ring carrier material is made of a nonmagnetic cast iron with a coefficient of thermal expansion similar to the material used for the piston itself. This material is particularly resistant to friction and impact wear. The groove that is most seriously endangered and the piston ring seated in it are effectively protected in this way against excessive wear. This is particularly advantageous where high operating temperatures and pressures are encountered, such as those found in diesel engines in particular.

Cooled Pistons

There are various types of cooling channels and cooling spaces to achieve particularly effective heat dissipation in the area near the combustion chamber and to combat the elevated temperatures resulting from performance increases. The cooling oil is generally delivered through the so-called fixed nozzles mounted in the crankcase.

In the cooling channel piston, Fig. 7-18, the ring-shaped cavities are created by inserting salt cores during casting. These cores are dissolved and removed with water introduced at very high pressure.

Fig. 7-18 Cooling channel piston with ring carrier for a diesel engine for passenger car use.

Piston with Cooled Ring Carrier

A new piston design is the piston with a cooled ring carrier, Fig. 7-19. It is used in diesel engines for both passenger cars and utility vehicles. The cooled ring carrier permits much improved cooling of the first ring groove and the edge of the combustion recess, which is subjected to extreme thermal loading. The intensive cooling of the first ring groove makes it possible to use a rectangular ring instead of the dual-trapezoid (double bevel) ring normally employed.

Fig. 7-19 Passenger car piston with cooled ring carrier.

Piston with Bushing in the Boss Bore

One of the most heavily loaded areas of the piston is the wristpin bearing area. There the piston material is subjected to thermal loads of up to 240°C and thus enters the temperature range in which the strength of the aluminum alloys declines.

When dealing with extremely heavily loaded pistons, measures such as shaped bores, relief pockets, and oval boss bores are no longer sufficient to increase the load-carrying capacities of the boss. That is why reinforcing was developed for boss bores into which shrink-fit bushings made of a higher-strength material (e.g., CuZn31Si1) are inserted.

Ferrotherm® Piston

In the Ferrotherm® piston, Fig. 7-20, the guidance and sealing functions are separated one from another. The two sections—the piston head and the piston skirt—are joined one with another in movable fashion by the wristpin.

The piston head, made of forged steel, transfers the ignition pressure through the wristpin and the conrod to the crankshaft.

The lightweight aluminum skirt handles only the lateral forces that are created by the angular positions of the conrod and, being of the appropriate shape, guarantees the oil cooling necessary for the piston head. In addition to this "shaker cooling" via the skirt, enclosed cooling cavities can also be integrated into the piston head. For this purpose the outer cooling cavity for the steel piston head is closed off with split tabs made of spring steel, Fig. 7-20.

With this design the Ferrotherm® piston offers not only greater strength and temperature resistance but also low wear values. Its constant, low oil consumption, its

Fig. 7-20 Ferrotherm® piston.

small dead space, and its relatively high surface temperature offer good prerequisites for complying with low exhaust emission limit values.

Monotherm® Piston

The Monotherm® piston 4, Fig. 7-21, grew out of development work for the Ferrotherm® piston. This new design is a one-piece piston made of forged steel and highly optimized for weight. At smaller compression heights and with machining above the clearance for the eye (on the inside) the piston weight, with the wristpin, can almost match the weight of a comparable aluminum piston with its wristpin. In the interest of improving piston cooling, the external cooling cavity is closed with two halves of a spring steel plate. The Monotherm® piston is used primarily for utility vehicle engines subjected to heavy loading.

Fig. 7-21 Monotherm® piston for utility vehicle engines.

7.1.1.9 Piston Manufacture

The latest in casting and machining equipment, in conjunction with an integrated quality management system, guarantees maximum quality across the entire product range.

Die Casting

Pistons made of aluminum alloys are manufactured in the main using the gravity die-casting process. The molds, made of ferrous materials, cause quick solidification of the molten metal; a fine-grained structure with good strength properties is formed at short casting cycle times. Optimized mold casting in conjunction with carefully designed riser and gating technology is necessary in order to achieve the most error-free and dense casting possible, this through graduated solidification aligning with the differences in wall thickness from the thin skirt to the thick piston head as mandated by the design. Multipart casting forms and casting cores provide great latitude in laying out the piston geometry so that even undercuts—inside the piston, for example—can be realized. To increase wear resistance at the ring grooves, ring carriers made of austenitic cast iron with intermetallic bonding (Al-fin bonding) can be cast in place with only as little trouble as for expansion-regulating steel struts or other engineering elements. By casting around cores made of compressed salt, which are then dissolved and removed with water, hollow cavities can be formed for piston cooling purposes. In order to do justice to high demands for quality and economy, multicavity molds and casting robots are used in mass-production operations.

Centrifugal Casting

The centrifugal casting (spin casting) process is used to manufacture the ring carriers used to reinforce the piston ring grooves. Tubes made of austenitic cast iron with flaked graphite are cast in rotating molds, and the ring carrier rings are then made up from the tubes.

Continuous Casting

This process is known for use with wrought alloys—primarily for bars, ingots, and blocks. MAHLE has further refined this process, in which the extrusion is cooled with water immediately after leaving the mold, so that it can be used with standard piston alloys. The high solidification speed has beneficial effects on the internal structure.

The extrusions are cast in various diameters and serve as the feedstock material for forged pistons or piston components.

Forging (Pressing)

Forging or warm flow pressing is used to manufacture pistons and piston skirts (assembled pistons) from aluminum alloys for engines subject to heavy loading. Sections of extrusion castings are normally used as the feedstock material. Reforming results in much higher and much more uniform strength values than can be achieved with casting. A further option is found in using semifinished products made of blast-compacted materials or those made up in a powder-metallurgical process. This process technology makes it possible to employ extremely heat-resistant materials for high-performance (racing) pistons, which could not be manufactured with hot metal technology.

Liquid Pressing (Liquostatik®, Squeeze Casting)

Squeeze casting differs from gravity die casting by the pressure applied to the molten material (up to and beyond 100 MPa), which is maintained until the casting has fully solidified. The extremely good contact of the molten material with the mold walls as it solidifies makes for very fast solidification. In this way, a very fine structure, advantageous in terms of material strength, is created.

Squeeze casting makes it possible to manufacture pistons that are reinforced locally with ceramic fibers or porous metallic materials at the piston head or in the areas around the ring grooves or bosses. These cast-in-place components are penetrated completely by the piston alloy owing to the pressure applied to the molten metal.

Tempering

Lightweight alloy pistons, depending on their alloy and the manufacturing process used, are subjected to single-stage or multistage heat treatment. In this way, the hardness and strength of most alloys can be increased. In addition, the remaining changes in volume ("growing") and the dimensional changes that would otherwise occur under the influence of operating temperature are preempted.

Machining

Leading piston makers themselves develop manufacturing concepts and special equipment for machining pistons. The distinguishing features are found in

- Complex shapes at the exterior of the pistons and close tolerances in piston diameter
- Complex piston head shapes (round, oval, or special shapes) and close boss bore tolerances
- High surface quality and geometry in rectangular and trapezoidal grooves in aluminum piston alloys as well as in ring carriers made of Niresist
- Close compression height tolerances

Thus, complex exterior piston shapes are machined on user-programmable shaping lathes whose CNC controls guarantee great flexibility and high quality. Irregular piston shapes that may, for example, be discovered empirically in engine test series can easily be manufactured in volume.

The same applies to the machining of the boss bore. Using a precision drill press, which is also user-programmable, differing boss bore shapes are possible along the direction of the boss bore axis and at the circumference of the boss bore.

Machining the piston grooves in the ferrous material making up ring carrier type pistons places particularly high demands on machinery capabilities.

7.1.1.10 Protection of Running Surfaces/Surfaces

The materials that have been highly developed to date and the precision machining processes used for pistons ensure high wear resistance and good running properties. In spite of applying protective coatings to the piston skirt, offering special emergency running properties is advantageous for the break-in phase and unfavorable operating conditions—dry running following frequent cold start attempts, temporary loading, insufficient lubrication. Under certain circumstances wear protection finishes may be required in the groove area. Severe thermal loading at the piston head must be counteracted with additional local protective measures. The coatings and finishes described below have proven their suitability for the various tasks in many applications.

With the use of automated machinery engineered especially for surface treatment, pistons may be finished in various ways:

- Tin plating the entire piston surface
- Applying phosphate and graphite (spray or spatter process)
- Applying graphite (screen printing) with and without phosphate
  (a) Piston skirt
  (b) Piston shaft and ring section
- Partial iron plating of the piston skirt (in conjunction with cylinder running surfaces made of aluminum)
- Hard anodized finishing
  (a) First groove
  (b) Piston head (complete or partial).

Improving Slip Properties

A thin plating of tin, which is applied by a chemical process to the lightweight metal piston, protects against seizing during cold starts and during break-in at unfavorable lubrication conditions. The layer is about 1 μm thick.

Where there are narrow installation tolerances and very high requirements for protection against seizure, the GRAFAL® running surface is used to an ever greater extent. This finish comprises a graphite-filled synthetic resin that adheres permanently to the piston running surface. This layer is generally 20 μm thick. Pistons for passenger car engines are typically finished with the GRAFAL® 255 version, applied in a screen printing process, while the sprayed GRAFAL® 240 or the screen printed GRAFAL® 255 version is used on pistons for utility vehicle engines and industrial engines.

In aluminum pistons the pairing of the wristpin and the boss is normally not critical in terms of sliding processes, and they require no special coatings—assuming the correct shapes and tolerances. In Ferrotherm® pistons, on the other hand, special protective measures are required. As an alternate to boss bushings, a slip phosphate coating for the upper section of the piston becomes more important here.

Increasing Wear Protection

FERROSTAN® pistons are paired with noncoated SILUMAL® cylinders or other noncoated, Al-Si-based cylinder materials. The skirt of FERROSTAN® pistons are iron plated to a thickness of 6 μm and hardness of HV 350 to 600. The iron layer is precipitated out, to precise dimensions, from special electrolytes. To conserve and improve slip properties, the iron-plated piston is finished with an additional layer of tin, 1 μm thick. Something new in technology is the application of layers, containing iron particles, using a screen printing process. Known as FERROPRINT® layers, they have been introduced successfully into mass production.

Owing to increased thermal and mechanical loads, wear and fretting effects are more frequently seen along the flanks of the first groove in gasoline engine pistons. Hard anodizing for the endangered area has been introduced in volume production as an effective countermeasure. When hard anodizing aluminum alloys, a zone near the surface of the aluminum substrate is transformed by electrolytic means into aluminum oxide. The layer created here is ceramic in nature, with hardness of about 400 HV. In this application a layer about 15 μm is specified, and the process parameters are optimized so that layer roughness is relatively moderate, eliminating the need for subsequent machining of the groove flanks.

Using Thermal Protection

Pistons for diesel engines are subjected to severe temperature alternation loading in the area at the top and in the combustion recess. The result may be fissures resulting from temperature alternation. A hard oxide layer at the top of the aluminum piston, shown in Fig. 7-22, typically about 80 μm thick, improves resistance to the effects of temperature alternation and thus prevents fissuring at the edge of the recess and/or in the top. Cutouts along the direction of the wristpin make sense in order to avert notch effects in the area where maximum tensile strain occurs.

7.1.1.11 Piston Materials

Aluminum Alloys

Pure aluminum is too soft and too susceptible to wear for use in pistons and for many other purposes. That is why alloys have been developed that are matched particularly to the requirements found in piston engineering. They combine, at low specific weight, good heat strength properties with a low tendency to wear, high thermal conductivity, and, in most cases, a low coefficient of thermal expansion as well.

Fig. 7-22 Hard anodized piston heads.

Two groups of alloys have come into being, depending on the primary additive—silicon or copper:

Aluminum-Silicon Alloys:

- Eutectic alloys containing from 11% to 13% silicon and smaller amounts of Cu, Mg, Ni, and the like. Included in this group of piston alloys, the ones used most frequently in engine construction, is MAHLE 124, which is also used for cylinders. For most applications they offer an ideal combination of mechanical, physical, and technological properties. The MAHLE 142 alloy, with a greater proportion of copper and nickel, was developed for use particularly at high temperatures. It is distinguished by better thermal stability and considerably improved strength when heated. A further step in this direction is the nearly eutectic MAHLE 174 alloy.
- Supereutectic alloys contain from 15% to 25% of silicon and use copper, magnesium, and nickel as additives to deal with high temperatures; examples include MAHLE 138 and MAHLE 145. They are used wherever a need for reduced thermal expansion and greater wear resistance is in the foreground. The MAHLE 147 (SILUMAL®) alloy is used for cylinders and/or engine blocks without any special treatment for the running surfaces.

Figures 7-23 and 7-24 show characteristic values for the materials.

Aluminum-Copper Alloys: To a lesser extent, alloys containing copper but almost no silicon and just a small amount of nickel as an additive are used for their good heat strength. In comparison with the Al-Si alloys, they exhibit greater thermal expansion and less wear resistance. While the Al-Si alloys can be both cast and reformed when warm, the Al-Cu alloys are more suitable for warm reforming.

Designation		MAHLE 124	MAHLE 138	MAHLE 142
Young's modulus E [N/mm²]	20°C 150°C 250°C 350°C	80 000 77 000 72 000 65 000	84 000 80 000 75 000 71 000	84 000 79 000 75 000 70 000
Thermal conduction coefficient λ [W/mk]	20°C 150°C 250°C 350°C	155 156 159 164	143 147 150 156	130 136 142 146
Mean, linear thermal expansion α [1/k × 10^{-6}]	20 to 100°C 20 to 200°C 20 to 300°C 20 to 400°C	20 21 21.9 22.8	18.6 19.5 20.2 20.8	19.2 20.5 21.1 21.8
Density ρ [g/cm³]	20°C	2.70	2.68	2.77

Fig. 7-23 Physical properties of MAHLE aluminum piston alloys.

Strength values are applicable to test bars made up separately.					
Designation		MAHLE 124 G	MAHLE 124 P	MAHLE 138 G	MAHLE 142
Tensile strength R_m [N/mm²]	20°C 150°C 250°C 350°C	200 to 250 180 to 230 100 to 150 40 to 65	300 to 370 250 to 300 110 to 170 40 to 70	180 to 220 170 to 210 100 to 140 60 to 80	200 to 280 180 to 240 100 to 160 50 to 70
Elongation limit $R_{p0.2}$ [N/mm²]	20°C 150°C 250°C 350°C	190 to 230 180 to 220 70 to 110 20 to 30	280 to 340 230 to 280 90 to 120 10 to 30	170 to 200 150 to 190 80 to 120 20 to 40	190 to 250 180 to 220 80 to 120 40 to 60
Ductile yield A [%]	20°C 150°C 300°C 400°C	0.1 to 1.5 1.0 to 1.5 2 to 4 9 to 15	1 to 3 2.5 to 4.5 8 to 10 31 to 35	0.2 to 1.0 0.3 to 1.2 1.0 to 2.2 5 to 7	0.1 to 0.5 0.2 to 1.0 1 to 3.5 5 to 13
Fatigue strength at reversed bending stresses σ_{bw} [N/mm²]	20°C 150°C 250°C 350°C	80 to 120 70 to 110 50 to 70 15 to 30	110 to 140 90 to 120 60 to 70 15 to 25	80 to 110 60 to 90 40 to 60 15 to 30	90 to 130 70 to 110 50 to 70 30 to 50
Relative wear index		1		0.9	0.95
Brinell hardness HB 2.5/62.5		90 to 130			100 to 150

Fig. 7-24 Mechanical properties of MAHLE aluminum piston alloys.

Lightweight Alloy Bonded Materials

The introduction of bonded materials technology opened a number of different options for significantly increasing the load-bearing capacities of lightweight metal pistons. Here reinforcement elements such as ceramics, carbon fibers, or porous metallic materials are arranged in closely defined positions in regions of the piston that are subject to particularly high loading. The bonded material is manufactured by infiltrating the reinforcing elements with lightweight metals such as aluminum or magnesium using the squeeze casting process. High price and unfavorable creep properties are the primary reasons magnesium is not yet used in mass production.

Among the many options available, reinforcing aluminum pistons with short ceramic fibers made of aluminum oxide is the one most widely adopted for mass production. Following a washing process to remove components that are not fiber shaped, the fibers are processed to create mold components that can be cast (preforms) with fiber content of between 10% and 20% by volume. In this way considerable improvements in strength can be achieved at the edge of the recess in direct-injection diesel pistons, for instance.

A reinforcing element made of porous sintered steel with uniform porosity from 30% to 50% was developed for ring grooves. The Porostatik® material offers favorable wear properties and a sure bond with the surrounding aluminum material. It is suitable, for example, for reinforcing ring grooves that are at an extremely high location, leaving hardly any room to cast around it on the side toward the piston head.

7.1.2 Wristpins

7.1.2.1 Functions

The wristpin makes the connection between the piston and the connecting rod. It is subjected to the extreme, alternating loads exerted by the pressure of the exploding gas and the mass inertias. Because of the small relative motions (rotary motions) between the piston and the wristpin and between the wristpin and the conrod, the lubrication situation is unfavorable.

7.1.2.2 Designs

The wristpin with cylindrical inside and outside contours has been successful in most applications. To reduce weight and with it the mass inertias, the outer ends of the wristpins' inside bore may be conical since the load is less there.

Wristpins in passenger car gasoline engines are often held in the conrod with a tension due to shrinkage (shrink wristpin). In more heavily loaded gasoline and diesel engines, the wristpin "floats" in the conrod. It is secured with circlips to keep it from wandering laterally and out of the piston (see Section 7.1.3).

7.1.2.3 Requirements and Dimensioning

Under the influence of the forces described above, loading on the wristpins is very complex and is influenced, in addition, by deformation of the piston and wristpin.

The essential aspects for the design of the wristpin are

- Sufficient wristpin strength (operating safety)
- Reverse effect on piston loading
- Weight (mass inertia)
- Surface quality, dimensional accuracy (running properties)
- Surface hardness (wear)

Today the wristpin is usually dimensioned with the aid of 3-D FE calculations, in some cases taking into account the shape of the lubricating oil film (pressure distribution) in the boss and the conrod. Solid knowledge on the dynamic behavior of the material is required to evaluate a material's dynamic properties. Guideline values for selecting the wristpin diameter for the various application ranges can be found in Fig. 7-25.

7.1.2.4 Materials

The materials that are used primarily today are 17Cr3 and 16MnCr5 case-hardened steels. Nitrated steel alloy 31CrMoV9 can be used where higher loading is anticipated. Figure 7-26 shows characterizing values for the materials used in wristpins.

Wristpins for racing use are manufactured in an electroslag remelting process to ensure a higher degree of purity in the material.

7.1.3 Wristpin Snap Rings

Since the wristpin is not held in the connecting rod by shrink fit, it has to be secured against wandering laterally from the holes in the boss and making contact with the cylinder wall. Used almost exclusively for this purpose, inside snap rings (made of spring steel) are installed in grooves at the outer edge of the boss holes (see Fig. 7-27).

Where the wristpin diameters are small, wound rings made from round wire are normally used. In engines that run at slower speeds, the ends of the snap rings may be bent inward to form a hooklike shape to facilitate installation. Such rings, when made up for racing use, are often bent outward at one end to keep them from rotating. If, in isolated cases, greater axial thrust is encountered in the wristpins, outside snap rings may also be used. These snap rings are mounted in grooves at the ends of the wristpins.

Application		Ratio of wristpin outside diameter to piston diameter	Ratio of wristpin outside diameter to wristpin inside diameter
Gasoline engines	Small two-cycle engines	0.20 to 0.25	0.60 to 0.75
	Passenger cars	0.20 to 0.26	0.55 to 0.70
Diesel engines	Passenger cars	0.32 to 0.40	0.48 to 0.52

Fig. 7-25 Wristpin dimensions (guideline values).

Material class		L (17Cr3) tool steel	M (16MnCr5) tool steel	N (31CrMoV9) nitriding steel
Chemical composition in % by weight	C	0.12 to 0.20	0.14 to 0.19	0.26 to 0.34
	Si	0.15 to 0.40	0.15 to 0.40	0.15 to 0.35
	Mn	0.40 to 0.70	1.00 to 1.30	0.40 to 0.70
	P	Max. 0.035	Max. 0.035	Max. 0.025
	S	Max. 0.035	Max. 0.035	Max. 0.25
	Cr	0.40 to 0.90	0.80 to 1.10	2.3 to 2.7
	Mo	—	—	0.15 to 0.25
	V	—	—	0.10 to 0.20
Surface hardness HRC		59 to 65 (vol. const. 57 to 65)	59 to 65	59 to 65
Core strength in N/mm^2		From 700 to 1500, depending on wall thickness	From 850 to 1350, depending on wall thickness	1000 to 1400
Mean linear thermal expansion $1/K \times 10^{-6}$, 20 to 200° C		12.8	12.7	13.1
Heat conductivity index $W/m \cdot K$	20° C 200° C	51.9 48.2	50.0 48.7	46.4 45.5
Young's modulus N/mm^2		210 000	210 000	210 000
Density kg/dm^3		7.85	7.85	7.85
Use		Standard material for wristpins	For heavily loaded loaded wristpins	For heavily loaded wristpins (special cases)

Fig. 7-26 Wristpin steels DIN 73 126.

Fig. 7-27 Wristpin snap rings.

Bibliography

[1] Zima, S., Kurbeltriebe, Konstruktion, Berechnung und Erprobung von den Anfängen bis heute, 2nd edition, Vieweg Publishers, Braunschweig, Wiesbaden, 1999.

[2] Junker, H., and W. Issler, "Kolben für hochbelastete Diesel-Motoren mit Direkteinspritzung," 8th Aachen Colloquium on Automotive and Engine Technology, Aachen, 1999.

[3] Röhrle, M., Kolben für Verbrennungsmotoren, Verlag moderne Industrie AG, Landsberg, 1994.

[4] Kemnitz, P., O. Maier, and R. Klein, "Monotherm, a new forged steel piston design for heavily loaded diesel engines," SAE 2000-01-0924.

7.2 Connecting Rod

The power system for reciprocating internal combustion engines uses a crank drive in which the connector rod end or the connecting rod joins the piston with the crankshaft.

The conrod converts the reciprocating movement of the piston into rotary motion. Moreover, the conrod transfers forces from the piston to the crankshaft. A further function of the conrod is to accept channels used to supply lubricating oil to the piston bushing in cases where the wristpin is of a floating design.

The weight and design of the conrod have a direct influence on the power-to-weight ratio, power output, and smooth engine operation. This is why conrods that have been optimized in terms of weight are gaining more importance in terms of engine running quality.

Corresponding to the inverted attitude of the conrod in the early engines, those built in the 19th century, the lower section (at the piston) is sometimes referred to as the conrod foot while the upper end (at the crankshaft) is called the conrod head.

7.2.1 Design of the Connecting Rod

The connecting rod has two so-called conrod ends.[1]

It is at the small conrod eye that the connection to the piston is made by the wristpin. Because of the lateral deflection of the connecting rod as the crankshaft turns, the rod end has to be attached to the piston in a way that allows it to rotate. This is done with the help of a sliding bearing. For this purpose a bearing bushing is pressed into the small conrod eye during assembly (Fig. 7-28). Alternately, the bearing may be integrated into the piston. In this case the wristpin is held in the small connecting rod eye with shrink fit.

The split, large connecting rod eye is located at the crankshaft end of the rod. Proper functioning is ensured with a sliding bearing (rolling bearings are used less often) and by fixing and screwing down the conrod bearing cap.

The connecting rod shaft joins the two connecting rod eyes. This section may have a special cross section, depending on the requirements at hand, e.g., *I shaped* or *H shaped*.

The connecting rod has to ensure sufficient slip in the bearings at both the small and the large ends.

Grooves may be machined in the ends to improve lubrication at the large end and/or to lubricate the cylinder and the wristpin; these grooves facilitate lubricant feed.

The wristpin bearing may be lubricated by means of a hole along the longitudinal axis of the shaft, through which oil is fed from the large end. This channel interferes with the structural relationships within high-performance conrods. That is why, as an alternate to a longitudinal channel through the shaft, one or more holes may be drilled at the small end in the surface facing the piston (Fig. 7-28). This solution is more economical.

7.2.2 Loading

The connecting rod is subject to a load exerted by the gas forces inside the cylinder and the inertia of the moving masses. Figure 7-29 shows the kinematic relationships in the crankshaft drive.

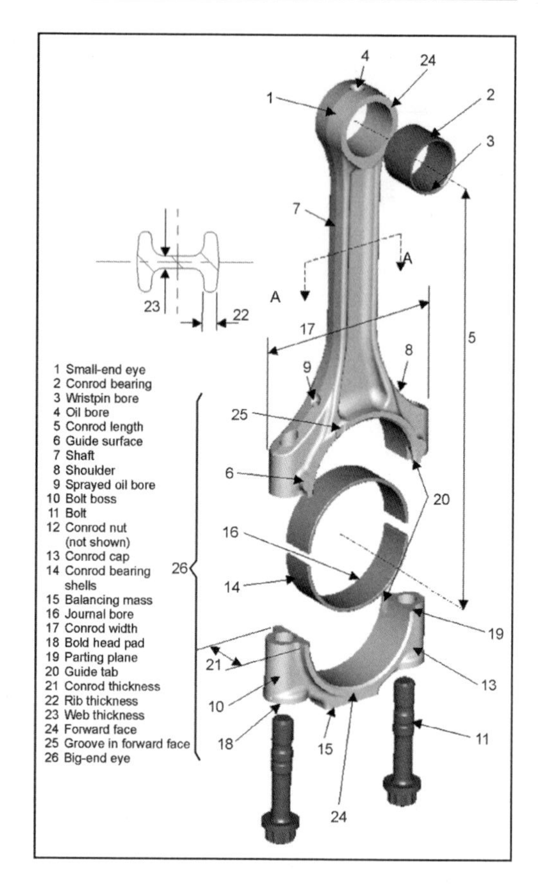

Fig. 7-28 Geometry and designations for a connecting rod with straight split (Federal Mogul).

The lateral deflection in the conrod oscillation plane generates centrifugal forces leading to bending that, however, can be neglected in the first approximation.

The accelerated and decelerated motion of the masses in the conrod and piston causes tensile strain in the shaft and at the transition from the shaft to the large eye. Thus, the conrod is subjected to alternating tensile and compressive forces; in diesel and turbocharged gasoline engines the magnitude of the compressive force exceeds that for the tensile force. For this reason resistance to buckling has to be examined carefully when engineering the conrod.

The tensile forces are also decisive in today's high-speed gasoline engines.

The inertial forces generated during accelerated and decelerated motion within a reciprocating engine's working cycle are influenced by the masses of the piston, the wristpin, and the conrod.

To simplify the determination of the resulting forces, the mass of the conrod is divided into rotating and reciprocating portions, assuming that the overall mass and the center of gravity for the conrod are retained unchanged.

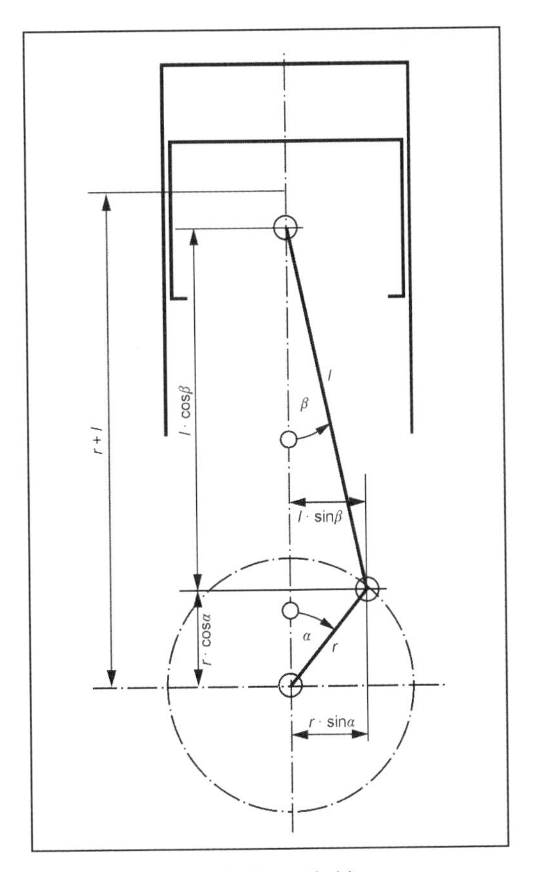

Fig. 7-29 Kinematics for the crank drive.

The masses concentrated in the large eye are assigned exclusively to rotational movement; those concentrated in the small eye are assigned to reciprocating motion.

To determine the various shares of the overall mass, it is first necessary to find the center of gravity (SP in the equation) for the conrod. The share of mass for the small eye results from

$$m_{\text{conrod, small eye}} = m_{\text{conrod total}} \cdot \frac{\text{SP}}{1} \quad (7.1)$$

with l as the distance between the centers of the conrod eyes, which is defined as the conrod length. The difference between this and overall weight gives the share of mass for the large eye.[2]

The reciprocating masses for the conrod (and the piston with the wristpin and piston rings) influence, by the inertial forces they generate, the loading and the smooth running of the engine. These reciprocating forces can be fully compensated only by providing additional compensating shafts.

Thus, it is necessary to reduce the conrod mass and/or the conrod's share of reciprocating mass. This can be done by optimizing the shape of the conrod shaft and, for example, by using a trapezoidal design for the small eye.

The true movement situation for a particle of mass in a conrod and thus the force effects are far more complex than what is reflected in the breakdown described above, this being only an approximation. Essentially, each particle of mass between the small and large conrod eyes executes a reciprocating and a rotating movement. The reciprocal component declines in the direction of the large conrod eye.

Suitable FEM calculation processes make it possible to attain a more or less exact calculation of the mass forces that are exerted on each element of mass.

This makes it possible to optimize design and the deformations under dynamic loading, and any play in the screw connection can be examined.

The masses for various conrods are shown in Fig. 7-30.

The bearings are the interface for the transfer of power from and to the wristpins and the crankshaft journals. Thus the conrod design has a major influence on the performance of the piston bushing and the conrod bearing,

7.2.3 Conrod Bolts

The upper and lower halves of the big end are held together with the conrod bolts.

These threaded connections must fulfill two functions:[3]

- The conrod bolts must prevent any gap forming in the separation plane between the lower half and the upper half of the big end. The forces that are effective on the conrod bolts include the inertial forces of the conrod

Application	Mass	Material
Mass production truck diesel	1.6 to 5 kg	Forged steel
Mass production passenger car gasoline engine	0.4 to 1 kg	Forged steel, gray casting, sintered steel
Sport use	0.4 to 0.7 kg	Steel, titanium
Racing engine/F1	0.3 to 0.4 kg	Titanium, carbon fiber
Compressor	0.2 to 0.6 kg	Aluminum

Fig. 7-30 Conrod masses for various applications.

and the piston along with a transverse force resulting from off-center loading and the forces resulting from "crushing" the bearing shell protrusion. During engine assembly, the bolts, opposing the effective inertial force, are normally preloaded by controlled tightening to the 0.2 offset limit or rotary torque plus rotation angle.[4, 5]

- The conrod and the cap have to be moved toward one another precisely and secured against shifting (offset). There are several options to choose from:
  a. Guidance by the conrod bolts, the shoulder or grooves of which are in the parting plane and thus prevent the upper and lower halves from shifting.
  b. Guidance by small pins next to the bolts or the bushings that surround the bolts (Fig. 7-31).
  c. Milling toothed ridges into the parting plane.
  d. Guidance by the separation surface at split (cracking) (Fig. 7-32).

Fig. 7-31 Fitting bearings and expanding bolt.

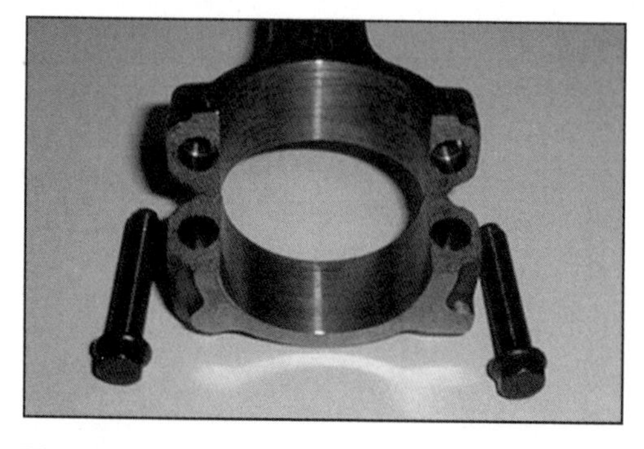

Fig. 7-32 Fracture-split conrod.

If pins, bushings, or fracture-split conrods are used, one may do without body-fit bolts. In this case, the structure at the parting surface or the pins and bushings offer sufficient resistance to relative motion between the upper and lower halves.

7.2.4 Design

The following aspects are of significance in regard to conrod design:

- Dimensional stability of the areas that accept the two bearing shells
- Oil channels for lubricating the small end eye may be required (unusual in modern designs)
- Separation of the big end bearing for mounting on the crank shaft journal.
- Fixing and securing the conrod cap.
- Engineering the conrod web to optimize design and/or reduce masses
- Design of critical zones in accordance with loading

Figure 7-33 shows the stress analysis for a conrod with an angular split.

Fig. 7-33 Stress analysis for a conrod with an angular split, with a trapezoidal small end (half model) (Federal Mogul). *(See color section.)*

To reduce the mass of the piston and/or the conrod, the small end may be flattened toward the top, creating a trapezoidal shape. This shape, for reasons associated with the loading (in turbocharged engines, for example), is advantageous since it permits close spacing to the wristpin bosses and thus reduced wristpin flexure.

The big end of the conrod is split to permit assembly on the crankshaft and is held together with two bolts.

The big end is normally split perpendicular to the long axis of the conrod. As an alternative, to reduce the maximum width of the conrod, the big end can also be split at an angle. This angular version makes it possible to pass the conrod (without the cap mounted at the big end) through the cylinder for assembly. The disadvantage of the angled split for the big end is that the blind hole for the conrod bolt terminates in the area subjected to the most severe loading and that great lateral forces have to be handled in the separation plane. Conrods with an angled split are used above all in V-block engines and in large diesel engines, which, because of the loading involved, have large-diameter crankshaft journals.

The big end and small end are joined by the conrod web, which has an I or H cross section. This makes it possible to satisfy requirements for reduced weight at a high section modulus.

7.2.4.1 Conrod Ratio

The conrod ratio is a comparative geometric magnitude, based on the crank radius r and the distance l between the centers of the small-end and big-end eyes (Fig. 7-29). It is defined as

$$\lambda = r/l \tag{7.2}$$

In passenger car engines, this value is normally between 0.28 and 0.33 with the lower values applicable to diesel engines. The selection of conrod length is influenced by many factors such as the stroke/bore ratio, piston speed, engine speed, peak combustion chamber pressure, engine block height, piston design, etc.

The lateral forces on the piston rise with the conrod ratio. This can, for instance, result in modified specifications for the piston's engineering design. As the conrod ratio falls, the overall height of the engine rises as a result of the increase in cylinder block height. Finally, restrictions imposed by the manufacturing process (cylinder block height) may prohibit a change in the conrod.

7.2.5 Conrod Manufacture

7.2.5.1 Manufacturing the Blank

The blank for the conrod may be manufactured in any of a number of different ways, depending on the particulars of the application:

a. **Drop forging.** The feedstock material for making up the blank is a steel bar with a round or rectangular cross section, which is heated to a temperature of between 1250 and 1300°C. A roll-forging process is used to effect a preliminary redistribution of the masses toward the big and small ends. As an alternative to roll forging, cross-wedge rolling may also be employed, improving the preliminary geometry for the blank.

   The major reforming process takes place in a press or a hammer unit. Excess material flows into flash, which is removed in a subsequent operation. Simultaneous with flash removal, the big eye and, in the case of larger conrods, the small eye are punched.

   To achieve the required structural and strength characteristics, the conrod requires various treatment processes, the choice depending on the steel alloy used.
   - Hardening with the forging heat (VS)
   - Controlled cooling in an air stream (BY)
   - Conventional hardening

   Then the scale on the blank is removed by blasting; here compression stresses of 200 MPa are generated near the surface. Additional procedures such as fissure inspections follow.

   In most cases, the conrod web and the end are cast as a unit and then separated during later machining. Depending on the conrod and the capacity of the available equipment, productivity can be boosted by tandem forging, i.e., shaping two conrods simultaneously.

b. **Casting.** The starting point for making up the blank is a model made of plastic or metal, comprising two halves that, when put together, create a positive image of the conrod. Several such identical halves are mounted on a model plate and joined with the model for the casting and gating system. In a process that can be reproduced many times, the two model plates are imaged by compacted green sand. The sand molds represent a negative image of the corresponding model plate. Placed one above the other, they form a hollow cavity in the shape of the conrod being manufactured. This is filled with liquid casting iron that is melted in a cupola blast furnace or electric furnace with steel scrap used as the feedstock material. The metal solidifies slowly inside the mold.

c. **Sintering.** The manufacturing process begins with servohydraulic pressing of the powder, in its final alloy, to create a powder preform. Weighing follows to ensure that this preform is within narrow weight tolerances of ±0.5%.

   The sintering process, illustrated in Fig. 7-34, takes place at about 1120°C in an electrically heated,

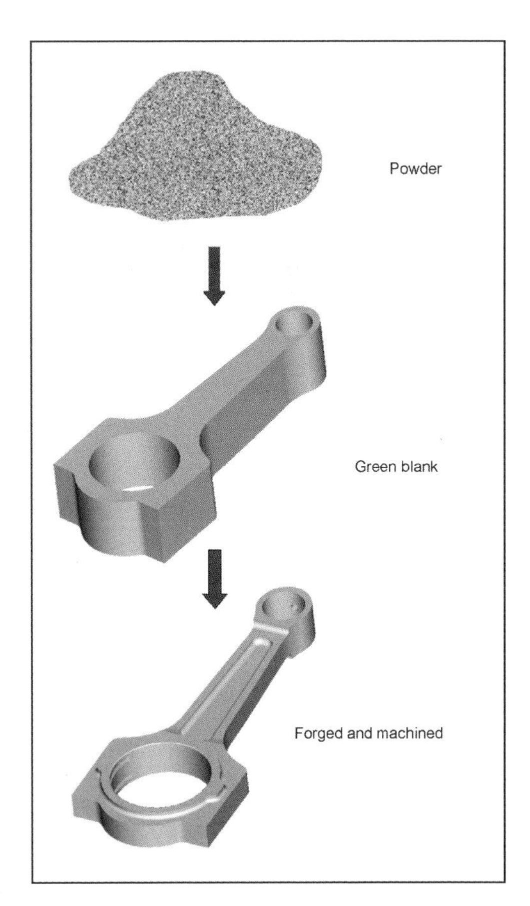

Fig. 7-34 Process–sinter-forged conrod.

continuous charge furnace. The parts remain here for about 15 min.

Subsequent forging merely reduces the height of the component in order to increase component density to the maximum theoretical limit. Then ball blasting is used to relieve the strain in the surface to the desired level.

Since the forging procedure in this manufacturing process is costly, developments are currently being pursued with the goal of eliminating this by using new powder technologies.[6,7]

7.2.5.2 Machining

The blanks are machined down to the final dimensions. In mass production this is done in fully automatic lines that are integrated into the engine manufacturing process. Machining centers with a lower degree of automation are available for smaller production runs. After machining, the finished part is weighed and classified. Conrods in a particular weight class are then installed in any given engine. If the blank was already manufactured to close weight tolerances, then it may be possible to do without this classification step.

In order to achieve the specified weight for the finished conrod, tabs can be provided at the small and/or big end of the blank (Fig. 7-35). During mechanical finishing, these tabs are ground down far enough that the specified weight value is attained.

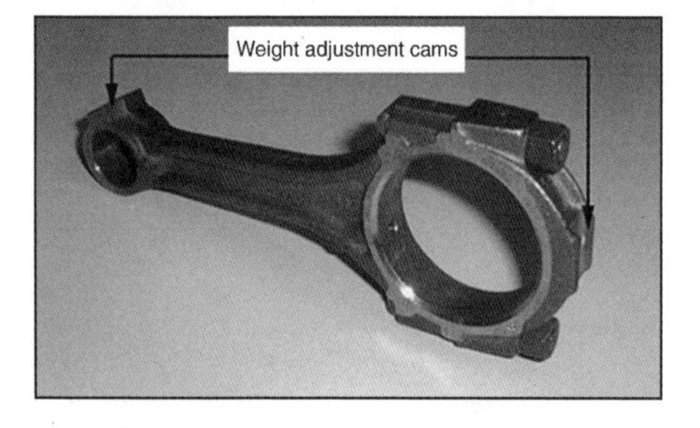

Fig. 7-35 Traditional conrod; body and big end cap forged separately and secured with nuts and bolts.

In more modern manufacturing processes, the manufacturing parameters can be monitored exactly so that blanks can be made within adequate weight tolerances.

Thus, grinding to remove excess material provided deliberately for this purpose is seen only rarely today.

The processing steps are described below, by example, for conrods that are split after manufacture (cracking):

- Grinding the faces of the big and small ends
- Prespindling the big and small ends
- Drilling and tapping the bolt holes
- Cracking
- Bolting the cap to the upper half of the big end and—if necessary—inserting the guide bushing
- Finishing final grinding
- Drilling out the small eye
- Spindling the big end and optional honing

The term "cracking" of "fracture splitting" describes the separation of the conrod web and the cap by breaking the latter away during processing. The prerequisites for this process are, in terms of the materials, a coarse-grain structure and, in terms of equipment, a cracking unit that can apply the required breaking energy at high speed. If the material exhibits a ratio of tensile strength to tensile yield strength (0.2 offset limit) that is near 2:1, then cracking can be carried out without any major deformation of the part. Blanks made with any of the modern manufacturing processes can be split by cracking.[8] The difference in the design of the conrod is shown in Fig. 7-36.

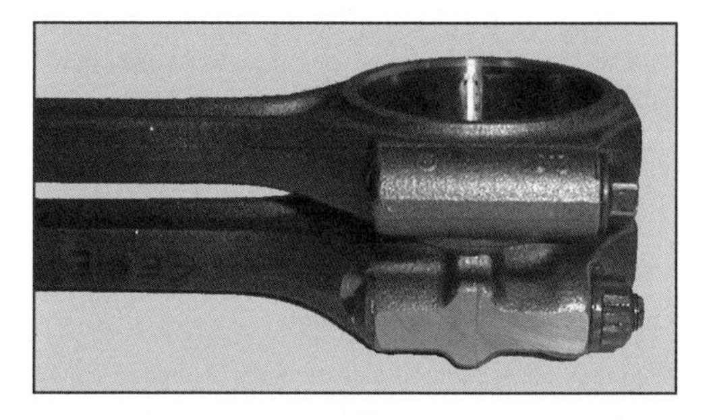

Fig. 7-36 Design differences between a fracture-split conrod (above) and a sawed conrod.

In preparation for cracking, notches are made in the side surfaces of the big-end eye by laser or broaching to achieve a deep notch effect at the desired separation plane (see Fig. 7-37). The large eye is positioned over a two-part breaker drift and fixed in place. The breaker drift is spread at high speed, and the stresses created in the workpiece initiate breaks within the notches. These breaks then propagate radially outward. If this process runs optimally to conclusion, then the out-of-roundness following cracking will be 30 μm at the most.

The advantage offered by fracture splitting is found above all in reducing the number of processing steps. Machining the separation surfaces, which used to be standard, can be done away with. The two halves fit together exactly after cracking and, with the irregular surface, are secured against relative movement, eliminating the need for any additional guide elements. A further benefit is found in the use of a simplified conrod bolt since it does not need to carry out guidance or lateral fixing functions.[9]

Fracture-split conrods are an economical alternative to conrods separated in a conventional fashion.

7.2.6 Conrod Materials

Depending on the particulars of the application and the resultant loads, any of a number of different materials may be used for connecting rods.

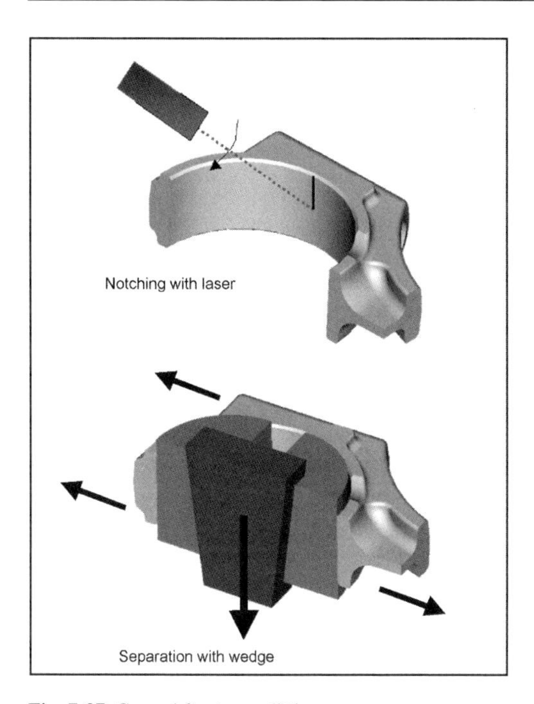

Fig. 7-37 Conrod fracture splitting.

Cast materials. The casting materials used most widely for connecting rods are nodular cast iron (GGG-70) and black malleable cast iron (GTS-70). GGG-70 has both technical and economic advantages when compared with malleable cast iron. In particular, the specific oscillation resistance, which is important for conrods, is considerably greater for GGG-70.

GGG-70 is an iron-carbon casting material; graphite inclusions that are largely spherical are introduced into a basic structure that is primarily pearlitic. The compact shape of the graphite gives the material an optimum strength and ductility. At the same time, the graphite is also responsible for the good casting properties. The required structure is created during the casting process without additional heat treatment.

In the case of malleable iron, which is also an iron-carbon material, the structure is determined by applying heat subsequent to casting.

Forging steel. The large majority of all conrods are manufactured from steel in the drop forge process. In most cases, microalloyed steel such as 27MnVS6 BY or carbon manganese steels like C40 mod BY are used. Steel with high carbon content (C70 S6 BY) is used for forged and fracture-split conrods. These materials attain tensile strength of Rm = 1000 MPa.[10]

Available for high-performance conrods is 34CrNi-Mo6 V (or 42CrMo4), a steel alloy that achieves tensile strength of 1200 MPa. In this case, additional heat treatment (hardening) is required.

New developments in steel have reached tensile strengths—even in materials used for cracking—of up to 1300 MPa at 0.2 offset limits in excess of 700 MPa. These steels are identified with the designation "C70+" in the table of materials.[11]

Powdered metal. Materials such as Sint F30 and Sint F31 are available for manufacturing conrods from powdered metal. They achieve tensile strengths of up to 900 MPa.[12]

Alternate materials. In addition to the materials used for conrods in mass production, explorations into using alternate materials pursue above all the objective of reducing conrod weight while maintaining load-handling capabilities. Carbon fiber reinforced aluminum or carbon fiber reinforced plastic is used for this purpose.

Widely used in racing are titanium conrods, with which a considerable weight reduction is achieved. The disadvantage of the titanium conrods is the strong tendency for bores to expand during operations, which has a deleterious effect on the tightness of the seat for the bearing shells. Another drawback is the fact that titanium is not a good "friction partner" for steel. Consequently, slip coatings on the mating surfaces are needed to protect against scuffing (friction-induced damage) and/or on the bearing's steel backing to prevent fretting.

Common to all conrods made of these alternate materials and fabricated for individual engines are the high manufacturing costs that hinder greater use in mass-production engines.

The most important materials and their properties are summarized in Fig. 7-38.

Material Name	NCI	C70	C70+	PMF	PMF	C38	42Cr	Al	TiAl4V4
Process comment	cast	forged & fractured car/truck		closed die	open die	forged BY	forged HT	cast	forged aircraft
Young Modulus (MPa)	170000	210000	210000	190000	199500	210000	210000	68900	128000
Fatigue Strength (pull) (MPa)	200	320/300	365/340	320	360	420	480	50	225
Fatigue Strength (push) (MPa)	200	320/300	365/340	320	360	420	480	50	309
Rp 0.2% Yield Strength (MPa)	410	550/500	750/700	685	550	550	>800	130	1000
Compressive Yield Str. (MPa)	–	600/550	–750/700		–620	–620	–850	–150	
Rm : Tensile Strength (MPa)	750	900/850	1050/950	900	850	900	1050	200	1080
Conrod Material Density	7.2	7.85	7.85	7.6	7.8	7.85	7.85	2.71	4.51

Fig. 7-38 Properties of conrod manufacturing materials.

Bibliography

[1] Küntscher, Kraftfahrzeug Motoren, Verlag Technik, Berlin, 1995.
[2] Greuter and Zima, Motorschäden, Vogel Fachbuch.
[3] Fisher, S., "Berechnungsbeispiel einer Pleuellagerdeckelverschraubung," VDI-Berichte, No. 478, 1983.
[4] VDI 2230, Systematische Berechnung hochbeanspruchter Schraubenverbindungen, Beuth-Verlag, Berlin and Cologne, 1986.
[5] Thomala, W., Commentary on VDI Guideline 2230, Sheet 1, 1986; RIBE-Blauheft, No. 40, 1986.
[6] Ohrnberger, V., and M. Hähnel, "Bruchtrennen von Pleueln erlangt Serienreife," Werkstatt und Betrieb, No. 125, Aalen (1992) 3.
[7] Adlof, W.W., "Bruchgetrennte Pleuelstangen aus Stahl," Schmiede-Journal, September, 1996.
[8] Herlan, Th., Optimierungs und Innovationspotential stahlgeschmiedeter Pleuel, VDI, Schwelm, 1996 or 1997.
[9] Moldenhauer, F., "Verbesserungen bei bruchtrennfähigen Pleuelstangen durch neuen mikrolegierten Stahl," in MTZ Motortechnische Zeitschrift, Vol. 61, 2000, No. 4.
[10] Weber, M., "Comparison of Advanced Procedures and Economics for Production of Connecting Rods," Powder Metallurgy International, Vol. 25, 1993, No. 3, pp. 125–129.
[11] Richter, K., E. Hoffann, Aüsselsheimm, K. Lipp, and C.M. Sonsino, "Single-Sintered Con Rods—An Illusion?" Metal Powder Report, Darmstadt, Vol. 49, 1994, No. 5, pp. 38–45.
[12] Skoglund, P., S. Bengtsson, A. Bergkvist, J. Sherborne, and M. Gregory, "Performance of High Density P/M Connecting Rods," Powdered Metal Applications (SP-1535).

7.3 Piston Rings

Piston rings are metallic gaskets whose functions are to seal the combustion chamber against the crankcase, to transmit heat from the piston to the cylinder wall, and to regulate the amount of oil present on the cylinder sleeve, a function of the oil control ring in particular.

It is necessary for this purpose that the piston rings be in close contact with both the cylinder wall and the flank of the groove machined into the piston. Contact with the cylinder wall is ensured by the spring action inherent to the ring itself, which expands the ring radially. Figure 7-39 shows the forces at a piston ring.

Oil control rings are usually given further support with an additional spring.

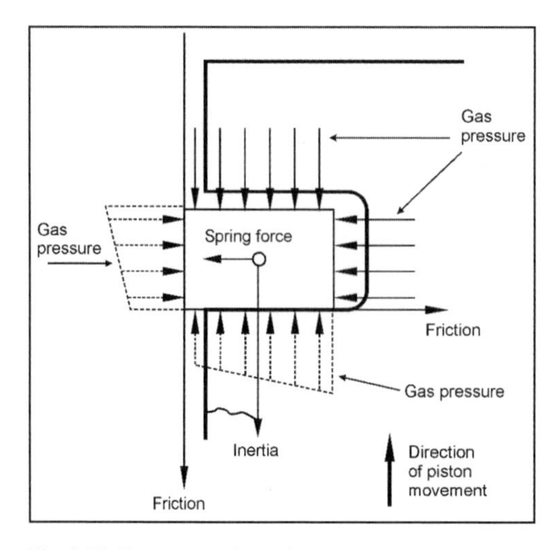

Fig. 7-39 Forces at a piston ring.

The gas pressure in the combustion chamber tends to reinforce both radial and axial contact in the ring groove in the piston. Axial contact may alternate between the lower and the upper flanks of the groove due to the influences of both mass and friction.

Trouble-free piston ring functioning depends on the thermal and dynamic loads generated by combustion, the engineering details, machining quality, and the choice of materials for the piston, piston rings, and cylinder.

The quality of the rings themselves has a decisive influence on their operating properties.

The number of rings per piston influences the friction losses inside the engine. The rings' masses represent a part of the reciprocating mass forces.

These reasons have driven the trend to fewer rings per piston. A three-ring configuration is standard: two compression rings and one oil control ring. Two-ring arrangements to reduce friction losses are also found in mass-production models. Here it is necessary to take account of the risk that if one ring should fail, the sealing effect for the complete ring set would be lost.

Figure 7-40 shows the most important terms and concepts.

7.3.1 Embodiments

One may differentiate among the types of piston rings on the basis of their functions:

- Sealing rings to keep the expanding gases inside the combustion chamber
- Oil control rings to strip away excess lubricating oil

7.3.1.1 Compression Rings

Among the types of compression rings available (Fig. 7-41) one differentiates the following:

Rectangular ring (Fig. 7-41a) with its rectangular cross section. This ring is used for sealing purposes at normal operating conditions.

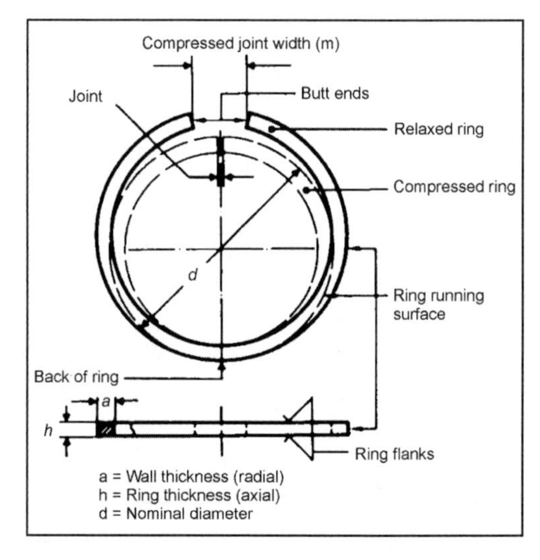

Fig. 7-40 Piston ring terms.

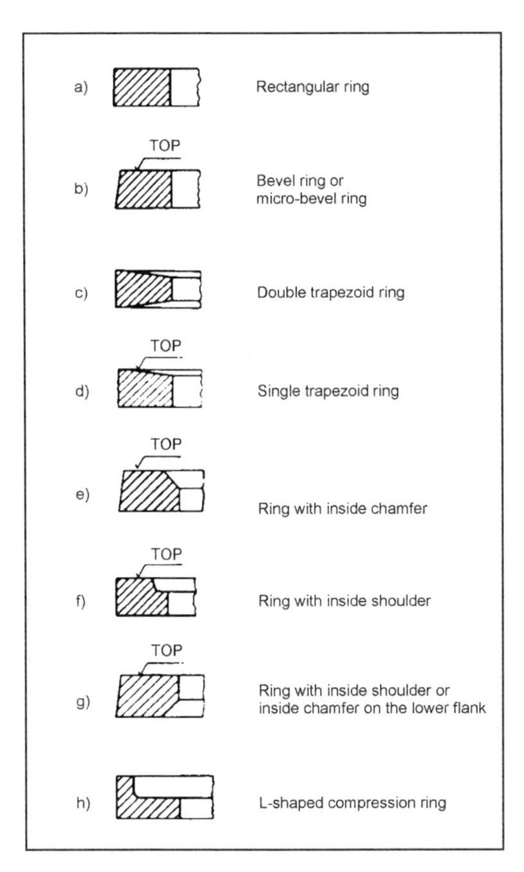

Fig. 7-41 Compression rings.

Bevel edge ring (Fig. 7-41b) with a conical running surface that shortens the wear-in period. Because of its oil stripper effect, it also supports oil consumption control.

Double trapezoid ring (Fig. 7-41c). The conical flanks of the ring significantly reduce "sticking" at the rings, since they are continuously freed of soot and combustion residues. This design is used only in diesel engines.

Single-sided trapezoid ring (Fig. 7-41d). This ring has a sloped flank at the top. Like the double trapezoid ring, it reduces sticking and is used primarily in diesel engines.

Ring with inner chamfer (Fig. 7-41e) **or inner groove** (Fig. 7-41f). The effect of the interruption in the cross section by the inner chamfer or groove is that the ring is deformed slightly when it is installed, thus creating a concave shape and, as a result, a conical running surface. Just like the beveled ring, it has an oil stripping effect.

Ring with inner chamfer or inner groove at the lower flank (Fig. 7-41g), the so-called negative torsion ring. This interruption in the cross section causes a negative twist after installation.[1]

L-shaped compression ring (Fig. 7-41h). This design is used primarily in small two-cycle engines as the so-called "head-land" ring; the end of the vertical arm of the "L" is flush with the piston head upper surface.[2] Because of the gas pressure effective behind the vertical arm of the "L," this ring seals tightly even when in contact with the upper flank of the piston ring groove. Along with use in two-cycle engines, it has been employed occasionally in automotive diesel engines to minimize the dead spaces in the combustion chamber.[3]

7.3.1.2 Oil Control Rings

Oil control rings are of particular significance in managing the engine's oil supply and consumption; they are subdivided into the following:

- Self-expanding cast iron rings (e.g., to support oil control action in high-speed engines and as the only oil control ring on a piston)
- Spring-expanded and spring-backed oil control rings, manufactured as castings, for gasoline and diesel engines
- Spring-expanded, oil control rings made of profiled steel, for gasoline and diesel engines
- Spring-expanded, oil control rings made of steel strip, for gasoline engines.

The fundamental versions of oil control rings are the following:

The **shoulder ring** (Fig. 7-42a) and **shoulder/bevel ring** (Fig. 7-42b) are, for all intents and purposes, compression rings with oil control properties.

The slotted ring (Fig. 7-42c) has an oil control effect because of the high surface pressure at the edges of the two rails. The slots at the circumference facilitate the return flow of the oil stripped off the cylinder wall.

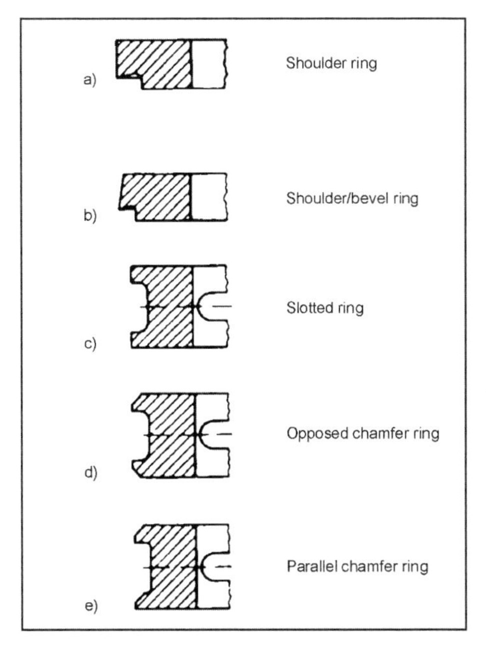

Fig. 7-42 Oil control ring (self-expanding cast rings).

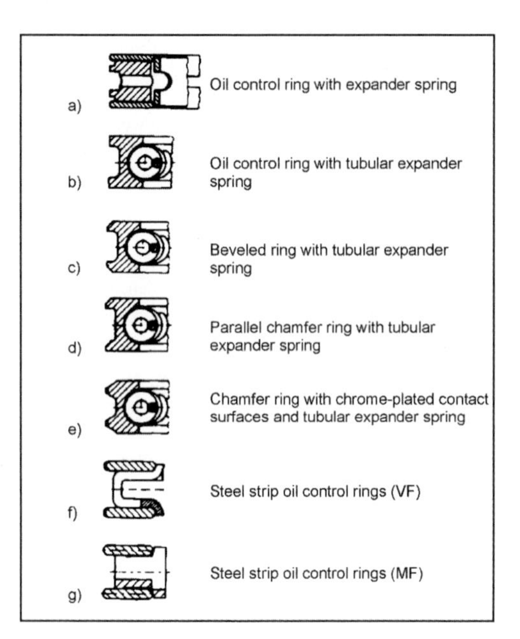

Fig. 7-43 Spring-expanded and spring-supported oil control rings.

Opposed chamfer (Fig. 7-42d) and **parallel chamfer** (Fig. 7-42e): Chamfering the contact surfaces of the slotted ring additionally increases the contact pressure at the edges, in turn enhancing stripping action.

Spring-expanded and spring-supported oil control rings (Fig. 7-43) are high-flexibility sealing rings with improved capacity for filling available space in the groove. They can adapt to compensate for cylinder warping and ensure particularly low oil consumption in the engine.

Oil control ring with expander spring (Fig. 7-43a): This version is found primarily in piston ring sets used for repairs.

Oil control ring with tubular expander spring (Figs. 7-43b–e): In this ring, the surface pressure and shape compensation capability are reinforced with a coiled, cylindrical compression spring (tubular spring).

Strip steel oil control rings (Figs. 7-43f and g) are used primarily in gasoline engines for passenger vehicles. They comprise two steel rails and a steel spacer spring.

7.3.2 Ring Combinations

The combination of rings for gasoline engines shown in Fig. 7-44 illustrates an example of current trends in the selection of rings.

1st groove: Steel ring 1.2 mm high, with crowned running surface, nitrided on all sides.

2nd groove: Shoulder/bevel ring 1.5 mm high, made of standard gray cast iron.

3rd groove: Strip steel oil control ring, 3.0 mm high, with chrome-plated steel rails at the contact surface or with nitrided spacer springs and steel rails nitrided on all sides.

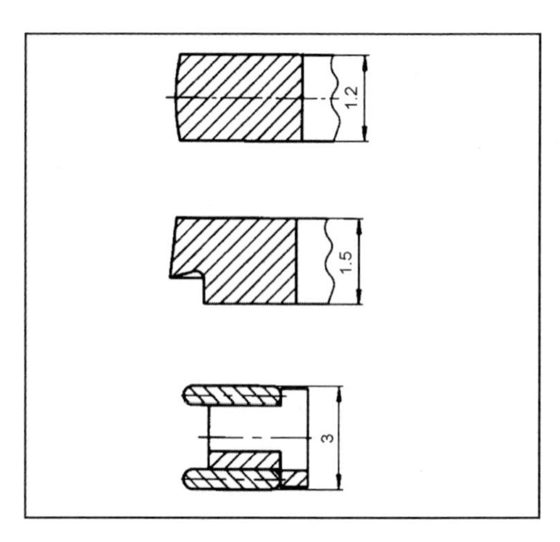

Fig. 7-44 Ring combinations for gasoline engines.

A typical combination for passenger-car diesel engines is shown in Fig. 7-45.

1st groove: Chrome-ceramic coated, rectangular ring, 2.5 mm high, spheroid casting, asymmetrically crowned contact surface, sharp lower running edge.

2nd groove: Negative-twist beveled ring, 2.0 mm high, made of gray cast iron, hardened.

3rd groove: Chrome-plated oil control ring with tubular spring, 3.0 mm high, made of standard gray cast iron, with running edge ground to form a profile, centerless grind at the butt joint for the tightly wound tubular spring.

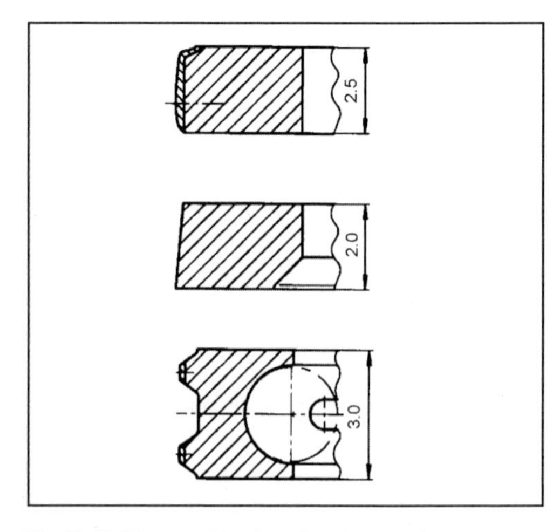

Fig. 7-45 Ring combinations for diesel engines.

At higher thermal loading, the first ring is a double trapezoid ring, 3.0 mm high, which otherwise exhibits identical characteristics.

7.3.3 Characterizing Features

Tangential force. The tangential force F_t is that force which must be present at the ends of the ring, at the outside diameter, in order to compress the piston ring to the specified gap at the joint (Fig. 7-46).

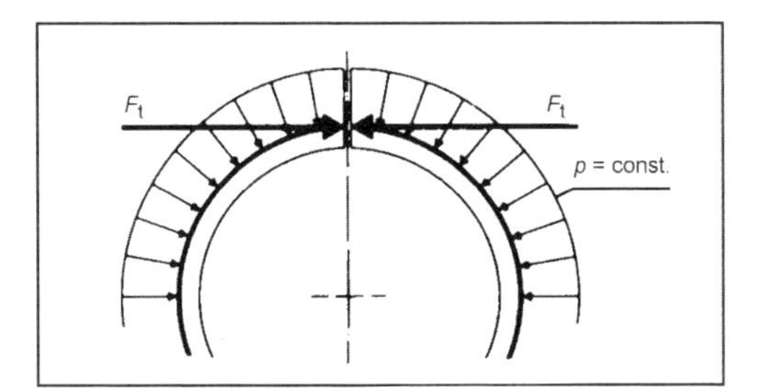

Fig. 7-46 Tangential force at a piston ring.

This is the determinant magnitude for the contact pressure. The contact pressure influences the sealing function and is the force with which the piston ring presses against the cylinder wall. It is calculated as shown below, where p = contact pressure, d = nominal diameter, h = ring height:

$$p = \frac{2 \cdot F_t}{d \cdot h} \text{ [N/mm}^2\text{]} \tag{7.3}$$

Radial pressure characteristics. Contact pressure can be set up to be constant around the circumference of the ring or to correspond to specified graduations in radial pressure. Figure 7-47 shows three typical forms for radial pressure characteristics.

The four-cycle characteristic (positive oval) (Fig. 7-47a) with increased radial pressure at the ends of the rings helps to "damp" piston ring flatter, which, in general, starts at the ends of the rings.

Rings with this characteristic show greater wear at the gap than those with uniform distribution of pressure (circular characteristic) (Fig. 7-47b).

Diesel engines, which do not run at such high speeds but which develop greater pressures, are thus equipped with rings that exhibit uniform radial pressure characteristics.

Where wear near the gap is to be reduced even further, rings with two-cycle characteristics (negative oval) (Fig. 7-47c) may also be used. Here the radial pressure at the ends of the rings is greatly reduced.[4]

Installed flexure tension. This is the flexure load to which the piston ring is subjected when installed in the cylinder. Maximum tension is found at the back of the ring and is calculated as follows:

Rectangular ring:

$$\delta_b = \frac{a \cdot E}{d - a} \cdot 2 \cdot k \text{ [N/mm}^2\text{]} \tag{7.4}$$

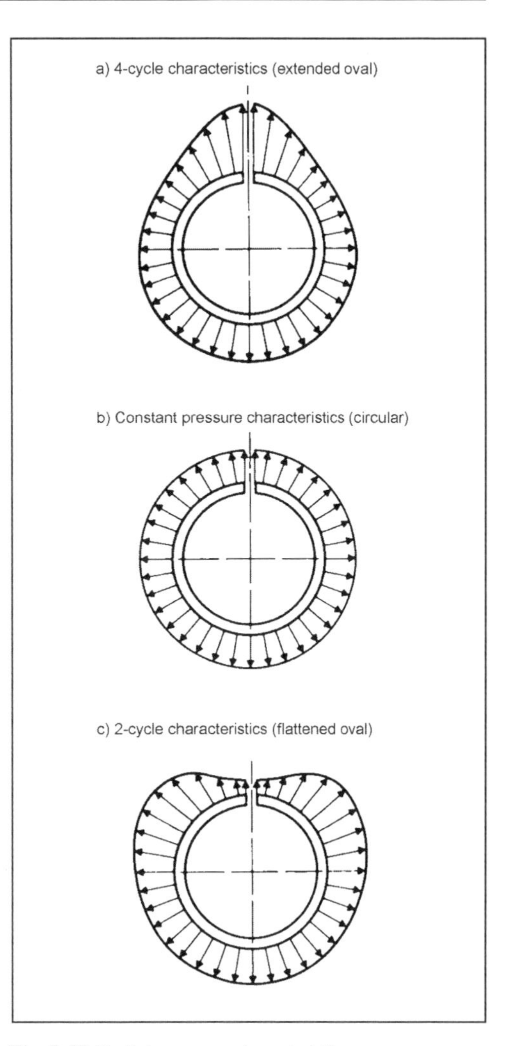

Fig. 7-47 Radial pressure characteristics.

Slotted ring:

$$\delta_b = \frac{x_1 \cdot E}{d - a} \cdot 2 \cdot k \cdot \frac{I_m}{I_s} \text{ [N/mm}^2\text{]}. \tag{7.5}$$

where

a = Ring wall thickness
d = Nominal diameter
E = Young's modulus for the ring material
k = Piston ring parameter
x_1 = Twice the distance of the center of gravity to the outside diameter

$$I_m = \frac{I_u + I_s}{2}$$

I_u = Geometrical moment of inertia for the cross section without slotting
I_s = Geometrical moment of inertia for the slotted oil control ring

The **piston ring parameter** k characterizes the elastic property of the piston ring. For rings with a rectangular cross section, for instance, it is defined as

$$k = 3\frac{(d-a)^2}{h \cdot a^3} \cdot \frac{F_t}{E} \quad (7.6)$$

where tangential force F_t is used, or

$$k = \frac{2}{3 \cdot \pi} \cdot \frac{m}{d-a} \quad (7.7)$$

when using gap width m (Fig. 7-40).

The **capacity to fill the space** is the capability of the piston ring to adapt even to cylinders that are out of round. Good shape-filling capability ensures correct sealing against the gas and the lubricating oil and thus good performance and low oil consumption.

Taking u_i as the radial, harmonic deformation of the cylinder of the ith order (Fourier analysis), the shape-filling capacity at which the ring is just in contact with the cylinder wall, exerting radial pressure of $p = 0$, is calculated as follows:

$$u_i = \frac{k \cdot r}{(i^2-1)^2} = \frac{1}{8}\frac{(d-a)^3}{(i^2-1)^2} \cdot \frac{F_t}{E \cdot I} \quad (7.8)$$

a = Ring wall thickness
k = Piston ring parameter
d = Nominal diameter
E = Young's modulus for the piston ring material, $r = d - a/2$
I = Geometrical moment of inertia for the cross section of the ring

$$I = \frac{h \cdot a^3}{12}$$

Gas pressure behind the ring improves space-filling properties at

$$u_{iz} = u_i\left(1 + \frac{p_z}{p}\right) \quad (7.9)$$

u_{iz} = Radial irregularity of the cylinder to the ith order, taking into account the gas pressure p_z
p = Contact pressure of the piston ring without gas pressure[5,6]

Ring gap. The ring gap is the space left between the ends of the ring after installation; this space is necessary to allow for thermal expansion in the piston ring. It is to be laid out for temperature differentials of at least 100°C between the compression rings and the cylinder and for 80° at oil control rings. If the ring gap is too large, then gas loss (blow-by) will result; if it is too narrow, then ring expansion can exert pressure on the ends of the rings and cause ring failure.

Butt joints with straight end surfaces are normally used. Bevel joints and lap joints are not used for passenger car engines and do not offer any advantages in regard to the tightness of the seal. Ring gaps with increased sealing quality (roll-shaped or beveled)[7] improve the sealing quality of the rings in comparison with the butt joint. These joint designs are recommended for use in two-ring piston concepts and have been employed with differing degrees of success.

7.3.4 Manufacturing

The performance and service lives of modern internal combustion engines can be ensured only with components that satisfy the highest quality requirements. In piston rings, the determinant magnitudes that must meet these requirements are the material and the shape of the part.

Piston rings made of cast iron are manufactured in a single casting process as single, double, or multiple blanks, on mold plates following a mathematically determined model, and are cast in stack molding. Another manufacturing option is to make up cast bushings in stationary or centrifugal casting.

Cold-drawn, profiled steel is preferred for manufacturing steel piston rings. Here not only the simple profiles may be selected for the compression rings but special profiles for oil control rings as well.

7.3.4.1 Shaping

While conventional processes (face-milling, lapping) are used to work the flanks of the rings, the outside contour, which determines piston ring characteristics, is shaped using the special processes tandem turning and winding.

Tandem turning is the manufacturing process most frequently used to lend the piston ring the desired shape. Here the blank, the flanks of which have been ground, is worked simultaneously on the inside and outside using a copying lathe, ensuring uniform wall thickness all around the circumference of the ring. Once the section of the ring corresponding to the width of the gap (Fig. 7-40) has been removed, the ring exhibits the uncompressed shape that will develop the desired degree of radial pressure distribution once it has been inserted into the cylinder. The shape of the copying cam is determined mathematically, separately for each radial pressure distribution pattern.

Winding is the process used for steel piston rings. The steel wire, having been drawn to the appropriate profile, is wound around a mandrel; the coil that is thus created is split lengthwise, separating the turns, and the rings that result are then mounted on a shaping mandrel and annealed to set the shape. The outside contour of the drift corresponds to the shape of an open, uncompressed ring, with this shape based on a certain radial pressure characteristic.

7.3.4.2 Wear-Protection Layers

To diminish piston ring and cylinder wear (to extend service life), the ring running surfaces, in particular, are provided with wear-reducing protective layers. They, nonetheless, must provide great resistance to burns[8] and cause the least possible wear to the cylinders at TDC in diesel engines.[9, 10] The following types of protective finishes are used:

Chrome plating. Layers of hard chrome, applied to the running surfaces by electroplating, exhibit very high resistance to abrasive and corrosive wear and are less susceptible to burns than unfinished running surfaces. Experience has shown that using a chrome-plated ring in the first piston ring groove reduces wear at the entire ring set to about 30% of the values for rings that are not chrome plated. Wear at the cylinder running surface is reduced by 50%.[11] Additional "special lapping" may be used for chrome-plated ring contact surfaces[12] to eliminate roughness formed by plateaus and valleys in the surface while supplementary "channel chrome plating" creates a channel-shaped network of fissures[7] in the chrome layer by porous etching. These techniques have been used with good results to cover the break-in phase for the engines.

Molybdenum coating. It is used above all because of its great resistance to burns. Molybdenum is applied to the piston ring running surface as a thermal spray layer, usually in a flame spatter process. The molybdenum layer's great resistance to burns can be traced hypothetically to the material's high melting point (about 2600°C) and its porous structure.

Plasma spatter layers. Plasma spatter technology makes it possible to apply mixed metallic and/or metal-ceramic layers whose component materials exhibit particularly high melting points. The wear protection layers created in this way have even higher wear resistance than molybdenum layers and higher resistance to burns than chrome layers.[9]

Chrome-ceramic layer. The good wear characteristics of the hard chrome layer are improved even further in the chrome-ceramic layer. The inclusion of ceramic particles (aluminum oxide) in the electrodeposited chrome layer not only improves its wear resistance across the entire service life of the layer but its thermal loading capacity, and thus its resistance to burns, is also increased.

The schematic structure of the chrome-ceramic layer is depicted in Fig. 7-48. It shows that the layer is created in an application process that is repeated several times to distribute the ceramic particles throughout the layer. Chrome-ceramic layers have been used with great success in diesel engines for automotive service.[13]

Nitriding and nitrocarburizing. Here thermochemical treatment (diffusion) is used to introduce nitrogen and, in some cases, carbon into the surface of the piston rings (primarily in rings made of steel). This diffusion process creates extreme surface hardness (approximately 1300 HV 0.025), which imparts high wear resistance to the layer. Layer hardness and thickness rise with the amount of alloying elements that form nitrides in the ring material (largely steel containing 13% or 18% chrome). In gasoline engines, this is used as an alternate to electroplated chrome layers and in part also to thermal spray layers, particularly at ring thicknesses of $\leq$1.2 mm. Additional advantages are dimensional trueness, which makes it possible to create sharp running edges at the piston ring, and coating on all the surfaces, providing additional protection against wear at the flanks. The burn resistance of these layers is similar to the chrome layers deposited with normal electroplating processes while that found in thermal spray layers is not reached.[14, 15]

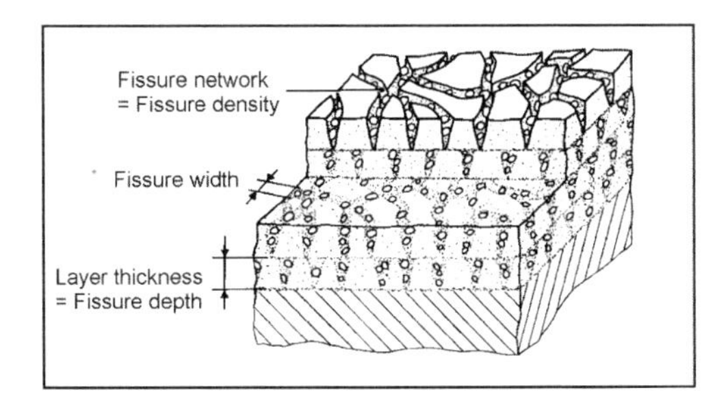

Fig. 7-48 Structure of the chrome-ceramic layer.

PVD layers (Physical Vapor Deposition). Employing the modern technology used to vapor deposit hard materials such as TiN or CrN gives wear protection layers that replicate exactly the contour of the substrate. In this way, one can treat only the functional surfaces of the paired wearing materials, which may be advantageous. PVD layers are characterized in part by great wear resistance, high burn resistance, and low TDC wear at cylinders in diesel engines. The layer thicknesses that can be created (5 to 50 μm, depending on the type of layer) do, however, limit the range of uses. Application is currently restricted to isolated instances in racing use and a few mass-production applications.

HVOF layers (High Velocity Oxy-Fuel). HVOF coating, a high-velocity flame spray, is based on the plasma coating's superior resistance to burns, further reducing inherent wear values and cylinder wear values. In HVOF coating, a supersonic flame is used to accelerate and heat the sprayed material. This creates a layer that is considerably denser and stronger than one applied with plasma spatter. These fundamental advantages for the engine, when compared with plasma, can be realized only when the coating materials are ideally matched to the properties of the process being used. The materials most frequently employed are metals with high carbide content.

7.3.4.3 Surface Treatments

The surface treatments listed below are employed with piston rings primarily to protect against corrosion during storage, to cover up minor surface defects, to improve break-in properties, secondarily to reduce wear at the running surfaces and flanks, and not at all to increase burn resistance during the run-in period.

Phosphating (zinc-phosphate and/or manganese-phosphate layers). The surface of the piston ring is transformed into phosphate crystals with chemical treatment. This phosphate layer is softer than the substrate material and thus wears away more easily, which accelerates ring wear-in.

Tin and copper plating. Both these metallic layers are applied by electroplating. Because of their softness they act somewhat like lubricants.

7.3.4.4 Contact Surface Shapes for Piston Rings

The contact surface shapes exert a major influence on the running properties of the piston rings. Symmetrically crowned, asymmetrically crowned, and optimized, asymmetrically crowned shapes such as those shown in Fig. 7-49 represent the state of the art. They have proven their suitability in practice, and this has been confirmed in many tribologic examinations.

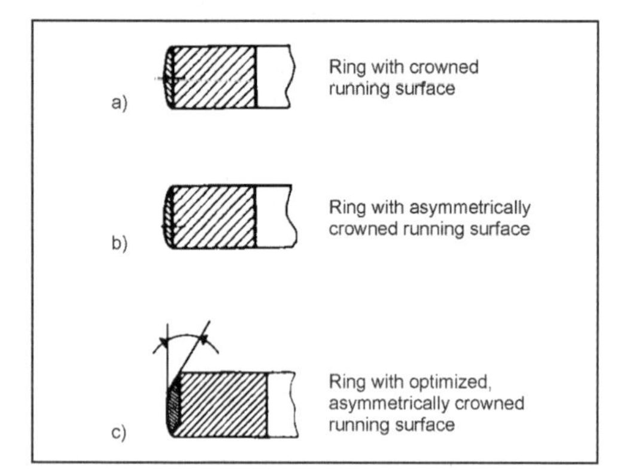

Fig. 7-49 Contact surface shapes.

In the optimized asymmetrical geometry (Fig. 7-49c), the upper third of the running surface is rendered conical, at an angle of less than 3°, so that the gas pressure in front of the ring counteracts in part the pressure behind the ring, thus avoiding excessive surface pressure at the running surfaces. Additionally, the amount of oil swept upward into the combustion chamber is minimized. Optimized crowning is used only in piston rings destined for use in diesel engines running at medium and high speeds.

Of great importance for the functional properties of the piston rings (oil consumption and blow-by) is the sharpness of the lower edge of the running surface.

7.3.4.5 Materials for Piston Rings

Determinants for the selection of the piston ring materials are demands for good running properties in both normal and emergency situations, for good elastic properties, and for good heat conductivity.

Great strength is required whenever extreme conditions such as high engine speeds or sharp rates of rise in the combustion pressure are present. The following materials may be used:

- **Cast iron with flaked graphite, nonhardened.** This is the "standard material" for piston rings, with good break-in and emergency running properties and satisfactory wear properties. The hardness of this material ranges from 210 to 290 HB, and the values for resistance to flexure—350 N/mm^2 at minimum—are relatively low. The standard material is used today only for the ring in the second piston ring groove and for oil control rings.
- **Cast iron with flaked graphite, alloyed, hardened.** The low shape retention properties of the standard material are improved by hardening. Resistance to flexures is at least 450 to 800 N/mm^2, and hardness reaches 320 to 470 HB.
- **Cast iron with spheroidal graphite (nodular cast iron), alloyed, and hardened.** This type of cast iron is distinguished particularly by great resistance to flexure, of at least 1300 N/mm^2, and hardness of from 310 to 470 HB. Because of its high resistance to flexure, nodular cast iron is given preference for rings mounted in the first piston ring groove.
- **Steel.** Because of its great breaking strength, steel is used, for example, at low ring heights ($h \leq 1.2$ mm) for gasoline engines and in diesel engines with steep rates of pressure rise. Steel is also used for the rails and spacer springs in oil control rings as well as in profiled oil control rings.

7.3.5 Loading, Damage, Wear, Friction

The piston rings are loaded by outward stresses when they are stretched to pass over the cylinder and, when installed, by the inward bending stresses imparted when the ring is compressed so it can enter the cylinder.[7] When the rings are properly engineered, the outward stress is equal to the strain imparted when the ring is passed over the cylinder.

Dynamic loading occurs in addition, namely, an axial motion of the piston ring caused by the interactions of gas, mass, and friction forces. In critical situations this leads to ring chatter and vibrations, which in turn can cause ring failure.[16, 17] Extraordinarily high loading on the ring can arise from soot collecting in the piston ring groove, which can cause sticking and ring failure. In addition to the axial motion, the ring also rotates around the circumference. Additional ring damage includes burn traces[8] and seizure.

The service life of the seal made at the piston rings is determined to a large degree by the amount of wear. This includes radial wear (wear on the running surface[18]), axial wear (wear at the flanks, "microwelding,"[19] piston groove wear), and secondary wear at oil control rings (wear between the ring and the tubular spring and between the rails and the spacer spring). The tribologic system surrounding the seal created by the piston ring is extremely complex since virtually all the normal types of wear—abrasive, adhesive, and corrosive—occur to a greater or lesser extent and effect.[8]

The piston group accounts for about 40% of all the friction in the engine. The piston rings cause a bit more than half of this friction.

The factors that influence piston ring friction include the surface pressure, ring thickness (width of the running surface), the rail height in oil control rings, the shape (crowning) of the contact surface, the coefficient of friction for the running surface layer (only in mixed friction areas at TDC and BDC, where the piston speed is very slow), and the number of rings per piston.[20] Measures taken to reduce piston ring friction must not interfere with ring functioning. The sealing effect of the ring set for both gases and the lubricating oil have to be maintained undiminished.

Bibliography

[1] N.N., Kolbenringhandbuch, AE Goetze GmbH, Burscheid, 1995.
[2] Jakobs, R., "Ein Beitrag zur Wirkungsweise von negativ vertwistenden Minutenringen in der zweiten Nut von Pkw-Dieselmotoren," Fachschrift K 41, der Goetze AG, 1988.
[3] Furuhama, S., and H. Ichikawa, "L-Ring Effect on Air-Cooled Two-Stroke Gasoline Engines," SAE Paper 730188, 1973.
[4] McLean, D.H., *et al.*, "Development of Headland Ring and Piston for a Four-Stroke Direct Injection Diesel Engine," SAE Paper 860164, 1986.
[5] Arnold, H., and F. Florin, "Zur Berechnung selbstspannender Kolbenringe von konstanter Stärke," Konstruktion, Vol. 1, 1949, No. 9.
[6] Gintsburg, B.I., "Splittles-type Piston Rings," Russian Engineering Journal, Vol. XLVIII, No. 7.
[7] Mierbach, A., "Radialdruckverteilung und Spannbandform eines Kolbenringes," in MTZ, Vol. 55, 1994, No. 2.
[8] Wiemann, L., "Die Bildung von Brandspuren auf den Laufflächen der Paarung Kolbenring-Zylinder in Verbrennungsmotoren," in MTZ, Vol. 32, 1971, No. 2.
[9] Buran, U., Chr. Mader, and M. Morsbach, "Plasmaspritzschichten für Kolbenringe: Stand und Einsatzmöglichkeiten," Fachschrift K 35, Goetze AG, Burscheid, 1983.
[10] Jakobs, R., "Einflussgrößen beim Zylinder(Zwickel)-Verschleiß von Pkw-Dieselmotoren," in MTZ, Vol. 44, 1983, No. 12.
[11] Charlsworth, W.H., and W.L. Brown, "Wear of Chromium Piston Rings in Modern Automotive Engines," SAE Paper 670 042, 1967.
[12] Plankert, H.W., and F. Stecher, "Oberflächengestaltung von Kolbenringlaufflächen—ein Ergebnis tribologischer Untersuchungen," Fachschrift K 24, Goetze AG, Burscheid, 1979.
[13] Buran, U., "Chrom-Keramik-Kombinationsschichten für Kolbenringe," MTZ/ATZ special edition, "Werkstoffe im Automobilbau 1996."
[14] Neuhäuser, H.J., *et al.*, "Steel Piston Rings—State of Development and Application Potential," T&N Symposium 1995, Paper 16.
[15] Brauers, B., and H.J. Neuhäuser, "Nitrierschichten als Verschleißschutz für Kolbenringoberflächen: Werkstoffe, Erprobungsstand, Einsatzmöglichkeiten," Fachschhrift K 46, Goetze AG, Burscheid, 1989.
[16] Wachtmeister, G., and K. Zeilinger, "Einfluss der Druckanstiegsgeschwindigkeit auf die Bauteilebelastung," FVV Research Report No. 413, 1988.
[17] Jöhren, P., "Gestaltfestigkeitsuntersuchungen an Kolbenringen im Ottomotor," in MTZ, Vol. 43, 1982, No. 4.
[18] Morsbach, M., and P. Jöhren, "Praxisrelevante Verschleißermittlung an Kolbenring- und Zylinderlaufflächen," in MTZ, Vol. 52, 1991, No. 3.
[19] Ishaq, R., and F. Grunow, "Wege zur Optimierung des Reibsystems Kolbenring und Ringnut," in MTZ, Vol. 60, 1999, No. 9.
[20] Jakobs, R., "Zur Reibleistung der Kolbenringe bei Personenwagen-Ottomotoren," in MTZ, Vol. 49, 1988, No. 7/8, and Fachschriften K 34 (1983), Fachschriften K 39 (1985), Fachschriften K 40 (1933), Goetze AG, Burscheid.

7.4 Engine Block

The engine block is the component that encloses the cylinders, the cooling jacket, and the engine block shell.

7.4.1 Assignments and Functions

The primary functions that the engine block fulfills are

- Absorbing the gas and mass forces in the crankshaft bearings and at the cylinder head bolts.
- Accepting the energy conversion assembly, comprising the pistons, conrods, crankshaft, and flywheel.
- Accepting and connecting the cylinders or, in the case of multisection engine blocks, connections to the individual cylinders or to the cylinder bank block or blocks.
- Carrying the crankshaft and (only rarely today) the camshaft.
- Accepting channels to convey operating media, primarily lubricants and coolant. Lubricants to supply the crankshaft bearings and conrod bearings, including the oil feeds and drains serving the cylinder head and in some cases serving as the mounting point for the spray nozzles used for oil cooling of the pistons.
- Transporting the coolant through cavities and channels contained in the engine block in liquid-cooled engines. This is also the mounting point for the coolant pump. Feed and drain channels guarantee that coolant reaches the cylinder head.
- Integrating a system for crankcase venting.
- Connecting to the transmission and to the valve actuators (with cover) and carrying and guiding power transmission elements such as chains.
- Connecting to and mounting position for various auxiliary assemblies, such as the engine mounts, coolant preheating components, oil-to-water heat exchanger, oil filter, oil separator for crankcase venting, and a variety of sensors for oil pressure and temperature, crankshaft speed, knock detection, etc.
- Isolating the crankcase from the outside world with the oil pan and—by way of radial shaft seals—at the point where the crankshaft passes through the engine block.

Because of the variety of functions to be carried out, the engine block is subjected to differing types of loads that are superimposed one upon another. It is exposed to tensile and compression loading, bending, and torsion as a result of mass and gas forces. Taken individually, these are

- Ignition gas forces, which have to be absorbed by the cylinder head bolts and the crankshaft bearings
- Internal mass moments (flexural moments), resulting from rotating and reciprocating mass forces
- Internal torsion moments (tipping moments) between individual cylinders
- Crankshaft torque and the resulting reactive forces in the engine mounts
- Free mass forces and moments, resulting from reciprocating mass forces, which have to be borne by the engine mounts

The effect of forces and the resulting moments, both inside the engine block and outside (engine mounts, mechanical vibrations, noise emission), depend on the engineering design for the engine.

The major parameters in the engine design that have effects on engine block loading are the number and arrangement of the cylinders, the arrangement of the crankshaft throws, and the ignition sequence. The loads occurring in the engine block influence the type of engine block selected and its specific design in view of achieving

Materials
(common materials for crankcases)

Material group:	Aluminum						Iron		
Material:	AlSi6 Cu4		AlSi17 Cu4Mg		AlSi9 Cu3		GG 25	GG 30	GGV
Remarks:	hypoeutectic		hypereutectic, heat treated	hypereutectic	hypoeutectic		cast iron with lamellar graphite	cast iron with lamellar graphite	vermicular graphite
Casting technique:	Sand and chill casting	Die casting	Sand and chill casting	Die casting	Sand and chill casting	Die casting			
Proof stress R_{p02} (N/mm_)	100–180	150–220	190–320	150–210	100–180	140–240	165–228	195–260	240–300
Tensile strength R_m (N/mm_)	160–240	220–300	220–360	260–300	240–310	240–310	250	300	300–500
Elongation at fracture A_6 (%)	0.5–3	0.5–3	0.1–1.2	0.3	0.5–3	0.5–3	0.8–0.3	0.8–0.3	2–6
Brinell hardness HB	65–110	70–100	90–150	25	65–110	80–120	180–250	200–275	160–280
Bending fatigue strength (N/mm_) NG = 25*108	60–80	70–90	90–125	70–95	60–95	70–90	87.5–125	105–150	160–210
E modulus (kN/mm_)	73–76	75	83–87	83–87	74–78	75	103–118	108–137	130–160
Thermal expansion coeff. (20–200° C) (10^{-4}/K)	21–22.5	22.5	18–19.5	18–19.5	21–22.5	21	11.7	11.7	11–14
Thermal conductivity (W/mK)	105–130	110–130	117–150	117–150	105–130	110–130	48.5	47.5	42–44
Density (kg/dm_)	2.75	2.75	2.75	2.75	2.75	2.75	7.25	7.25	7.0–7.7

Source: Kolbenschmidt AG, Neckarsulm, Handbuch Aluminium – Gussteile, Volume 18
DIN 1691 Cast iron with lamellar graphite (gray cast iron)
Porsche Technical Terms of Delivery 2002
Vermicular graphite cast iron (GGV) – A new material for the internal combustion engine,
Aachen Colloquium "Automobile and Engine Technology" 1995,
Prof. Dr. techn. F. Indra, Dipl.-Ing. M. Tholl, Adam Opel AG, Rüsselsheim

Fig. 7-50 Materials for engine blocks.

sufficient strength, minimum deformations, economical manufacturing, recyclability, noise emissions, engine block weight, and, with it, the total engine weight.

The strength of the engine block is determined by the material used, by the choice of heat treatment (which depends on the material and the casting process), and by the engineering design (characterized by the type of engine block, ribs or fins, wall thickness, etc.).

Common engine block materials, in comparison with vermicular graphite cast iron, and the most important material properties are shown in Fig. 7-50.

Engine blocks are characterized by the following major dimensions, which depend on the engine configuration, such as inline, V-block, or boxer (pancake) engine (Fig. 7-51):

- Length, measured from the front edge of the engine block to the transmission flange
- Width, as maximum overall width

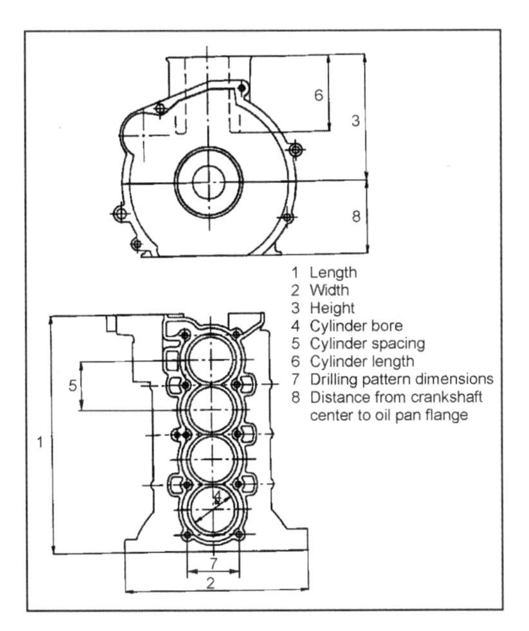

Fig. 7-51 Major dimensions of the engine block.

- Height, measured from the center of the crankshaft, along the axis of a cylinder, to the top plate plane
- Cylinder bore, expressed as the nominal inside diameter of the cylinders. Cylinder spacing, given as the distance between the centers of two adjacent cylinders
- Cylinder offset in V-block, W-block, and boxer engines, specified as the distance between the centers of two cylinders located opposite each other in adjacent banks of cylinders
- Cylinder length, measured from the top plate to the lower end of the cylinder. Dimensions and drilling pattern for cylinder head bolts: depending on their design, e.g., four or six per cylinder.
- Vertical distance from the center of the crankshaft to the oil pan flange:
  (a) Equal to zero where the oil pan attachment plane is level with the center of the crankshaft
  (b) Height of the deep skirts where the engine block side walls extend downward
  (c) Height of the lower engine block section

Figure 7-51 shows the most important dimensions.

The conrod executes a swinging motion with each revolution of the crankshaft. The path that it follows, determined by the outside contour of the conrod and the cranking radius, has a shape similar to the body of a guitar (Fig. 7-52).

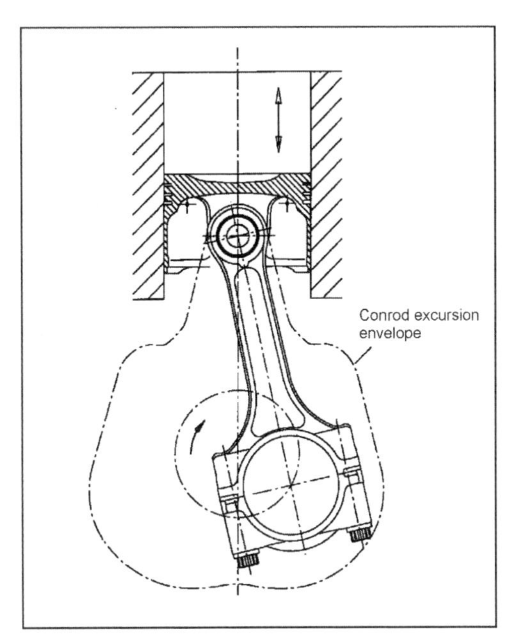

Fig. 7-52 Conrod envelope.

It is necessary, when laying out an engine block, to ensure that there is sufficient clearance for this outline. The most critical close clearances between the engine block and the envelope for the conrod are normally

- Lower surface of the cylinder and in V-block, W-block, and boxer engines that of the opposite cylinder, too
- Engine block sidewalls with channels located next to the conrod for oil return or for crankcase venting

The clearance is, as a rule, between 3.5 and 4.5 mm and is determined after having taken into account all the tolerances for the components involved, to include the casting tolerances for the engine block itself.

7.4.2 Engine Block Design

7.4.2.1 Types of Engine Blocks

The types of engine blocks can be classified according to the engineering design in the areas at the

- Top plate
- Main bearing pedestals
- Cylinders

Since a separate section is devoted to the cylinders, they are not dealt with here.

Top Plate

A basic engineering feature, one that limits the selection of the casting process, is the engine block top plate.

Here, one differentiates between closed-deck and open-deck designs.

(a) Closed-deck designs. In this version, the top of the engine block is largely closed in the area around the cylinders. In the top plate there are always, depending upon the specifics of the design, openings for the cylinders, openings for the tapped holes for the cylinder head bolts, and bores and channels for oil feed and return (for coolant circulation and for crankcase venting) (Fig. 7-53).

Here, with the exception of the cylinders, the top plate is penetrated essentially only by the smaller openings of appropriate cross sections to allow for coolant passage. These openings join the water jacket surrounding the cylinders (with the water jacket inside the cylinder head) through specified channel cross sections in the head gasket and at openings in the cylinder head combustion chamber plate. This design suffers disadvantages regarding cylinder cooling in the TDC area.

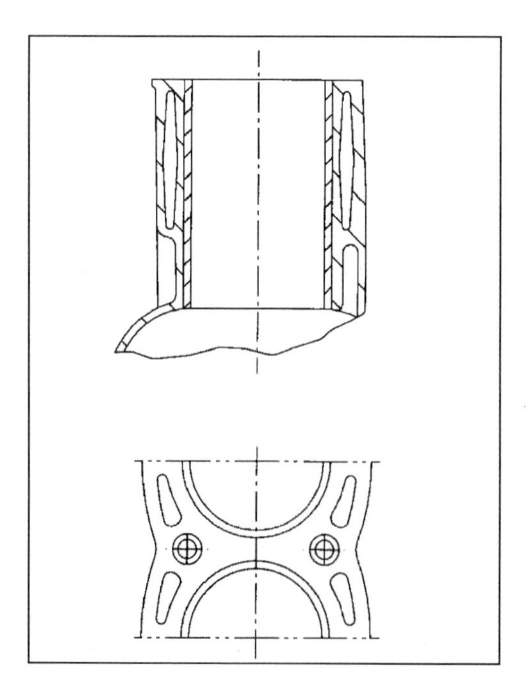

Fig. 7-53 Closed-deck design.

Producing the engine block water jacket requires a sand core in the closed-deck version because the water jacket, in the upper area of the engine block, is largely sealed off by the top plate. Consequently, the water jacket cannot be created as a feature of the external casting mold for the engine block upper section; a core has to be inserted inside the casting mold. These bearing points are generally found in the finished engine block as casting "eyes" in the engine block side walls. The openings for the core inserts are closed off with sheet metal plugs. Once the engine is assembled, core insertion points such as this are an indication that it is a closed-deck engine block.

The advantage of the closed-deck design in comparison to the open-deck version is the greater stiffness of the top plate. This has positive effects on top plate deformation, cylinder warping, and acoustic properties.

Selecting an engine block with closed-deck design does, however, limit the casting processes that can be used. The sand core required for the water jacket makes it possible to fabricate the closed-deck type only in sand casting and die-casting processes.

Engine blocks made of gray cast iron, made in a sand casting process, are almost exclusively of closed-deck design.

Engine blocks made of aluminum-silicon alloys in a closed-deck design are made in mass production primarily as die castings, as low-pressure castings, and, more recently, in a sand casting process.

(b) Open-deck designs. In the open-deck version, the water jacket surrounding the cylinders is open at the top as shown in Fig. 7-54. From the casting technology viewpoint, this means that no sand core and, thus, no core inserts are required to form the water jacket. The casting core for the water jacket requires no undercuts and may be made up as a steel mold.

The water jacket open at the top enables better cooling of the cylinder's hot upper section (compared to the closed-deck version).

There is less stiffness in the top plate of the open-deck design than that of the closed-deck version. A metallic head gasket is used to compensate for the increase in negative influence on the top plate by deformation resistance and cylinder warping. The metallic head gasket permits a lower preload value for the head bolts, thus reducing top plate deformation and cylinder warping.

Manufacturing open-deck engine blocks enables the use of essentially all types of casting processes.

In rare cases, gray iron cast engine blocks with an open-deck design are manufactured using a sand casting technique.

The open-deck design makes it possible to manufacture engine blocks from an Al-Si alloy using the economical die-casting process. Over and above this, it enables the realization of special techniques for the cylinders and cylinder sleeves.

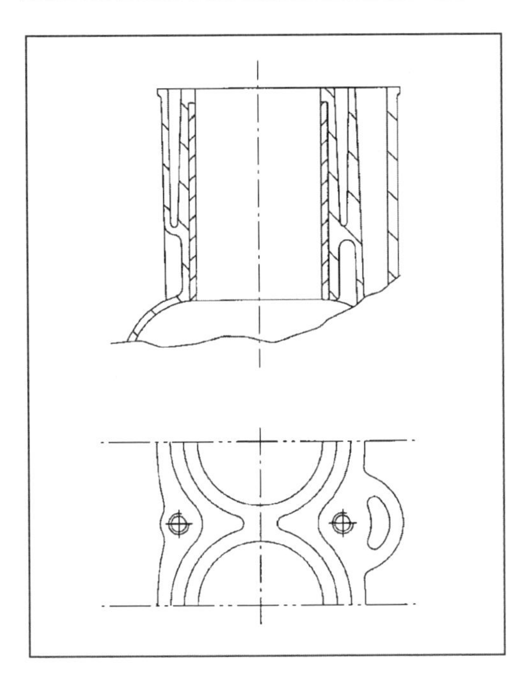

Fig. 7-54 Open-deck design.

Main Bearing Pedestal Area

The main bearing pedestal area in engine blocks is the area around the crankshaft bearings. The engineering design for this area is of particular importance because the forces acting on the crankshaft bearings have to be taken up here.

Options for further structuring the design of the engine block include selecting the location for the separation plane between the engine block and the oil pan, and the engineering of the main bearing caps.

One uses this separation plane to distinguish between an oil pan with the flange level at the center of the crankshaft and one that is below the center of the crankshaft.

In designing the main bearing caps, one distinguishes between individual main bearing caps, their integration into a longitudinal frame unit, and integration into the engine block lower section.

Main bearing cap

The main bearing caps represent the lower boundary of the main bearing pedestals; the caps are affixed and bolted to the main bearing pedestals. The main bearing caps and the main bearing pedestals have essentially the same function, i.e., absorbing the forces and torques imposed upon the crankshaft, accepting the corresponding bearings including the thrust bearing (collar bearing or thrust washers), as well as accepting a radial shaft sealing ring at the transmission output end, at the final main bearing, to seal the rear end of the crankshaft.

The main bearing caps and main bearing pedestals in the engine block are machined together and are joined during postmachining assembly procedures. The normal methods used for fixing these items are surfaces broached at the side in the main bearing pedestals or bores for guide bushings.

Main bearing caps are manufactured exclusively as gray castings and are combined with engine blocks made both of gray cast iron and of aluminum alloys. Working the aluminum main bearing pedestal and the gray cast bearing cap simultaneously is not without its difficulties because of the differences in ideal cutting speeds (specific to the materials). This is the procedure used in mass production today. The combination of an aluminum main bearing pedestal and a cast iron main bearing cap has advantages resulting from the gray cast iron: the low coefficient of thermal expansion in the main bearing cap made of gray cast iron limits the amount of play in the crankshaft bearings that develops during operation. This reduces the amount of oil that passes through the main crankshaft bearings. Reduced main bearing play and greater stiffness in the cast iron bearing cap (Young's modulus for gray cast iron is higher than that for aluminum) reduce noise generation and emissions in the area around the main bearing pedestals.

The version most widely used in mass production is the engine block made of gray casting with main bearing caps of the same material. The engine blocks are engineered either with the oil pan flange level with the center of the crankshaft or as an engine block with side walls or skirts that extend downward.

In V-block engines one most commonly finds an aluminum engine block combined with individual cast iron main bearing caps.

Main bearing pedestal

The upper section of the crankshaft-bearing surface in the engine block is referred to as the main bearing pedestal. Regardless of the engineering design of an engine block in the area around the crankshaft bearings, the main bearing pedestals are always a part of the casting for the engine block or for the upper section of the engine block (Fig. 7-55).

The number of main bearing pedestals for an engine block depends on the engine type and, in particular, on the number of cylinders and their arrangement. Today, for reasons associated with vibration phenomena, engine blocks are almost always made with a full set of bearings for the crankshaft. Crankshafts such as these have a main bearing journal next to each crankshaft throw. A four-cylinder inline engine thus has five main bearing pedestals, six-cylinder inline and boxer engines have seven main bearings, V-6 engines have four main bearings, V-8 engines have five main bearings, etc.

The major functions of the main bearing pedestals are

- Accepting axial and radial forces and moments impinging upon the crankshaft bearing system
- Accepting the upper sliding bearing shell for the crankshaft radial bearings along with accepting the

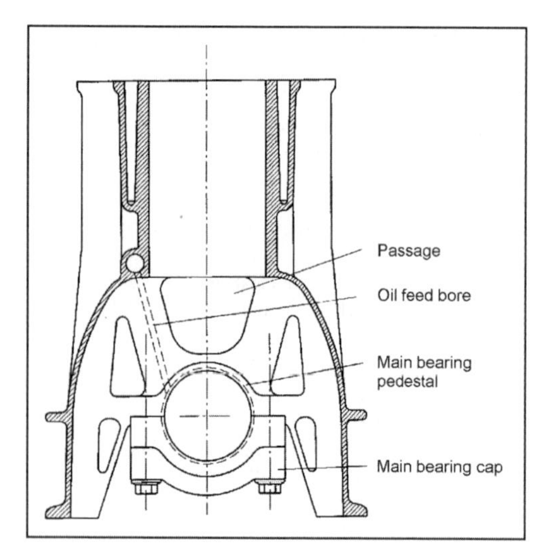

Fig. 7-55 Main bearing pedestal/main bearing cap.

collar bearings or thrust washers in a main bearing pedestal, the so-called thrust bearing, for axial control of the crankshaft

- Accepting the threads, fixing holes, or fixing bushings used in attaching and fixing main bearing caps or longitudinal frames or the lower section of the engine block
- Accepting oil feed bores and oil grooves used to supply the crankshaft main bearings with oil
- Depending on the engine design, accepting the radial shaft-sealing ring in the last main bearing pedestal, used to seal the rear end of the crankshaft

The main bearing pedestals often exhibit passageways to equalize pressures in the individual chambers in the crankcase area and, thus, reduce losses due to internal engine friction.

Vertical holes or channels for oil return from the cylinder head or for crankcase venting through the main bearing pedestals are commonly found.

These many functions require great care in the engineering and design of the main bearing pedestals and the components that interface with them—the main bearing caps or longitudinal frame or lower engine block section. Engineering for these assemblies is carried out today almost exclusively with the engineering aids now available, such as FEM calculations.

Engine block lower section

Just as in the longitudinal frame design, the individual main bearing caps in the engine block lower section are combined into a single component. In contrast to the longitudinal frame, the engine block lower section does not lie within the engine. Instead, the sidewalls of the engine block lower section form the outer limits of the crankcase; the lower plane forms the flange to the oil pan.

An engine block lower section offers essentially the same engineering design options as those described for the longitudinal frame concept. Since engine block lower sections are mass produced almost exclusively from aluminum alloys and in a die-casting process, additional functions can be integrated into it:

- Oil removal, i.e., radial stripping of the motor oil around the envelope for the crankshaft counterweights and the conrods
- Parts of the motor oil circuit such as the oil intake channel between the oil pump and the oil sump, the oil channel between the oil filter head and the oil pump, the oil filter head itself, the oil return channels, the main oil channel and oil channels to the individual main bearing points, partial integration of the oil pump housing
- Accepting shaft seal rings to seal the crankshaft

Engine block lower sections are used in mass-production, all-aluminum engines, and in racing engines.

Longitudinal frame concept

Similar to the situation where an engine block lower section is used, in the longitudinal frame concept the individual main bearing caps are consolidated into a single component, Fig. 7-56. In contrast to the engine block lower section, the longitudinal frame has no flange plane interfacing the oil pan. Rather, the longitudinal frame lies inside the engine and, thus, is enclosed by the oil pan in the version where the oil pan flange is centered on the crankshaft or by the deep sidewalls of engine blocks that incorporate the same. The advantages of a longitudinal frame are

- Greater stiffness in comparison to individual main bearing caps and thus better acoustic properties, easier and faster installation

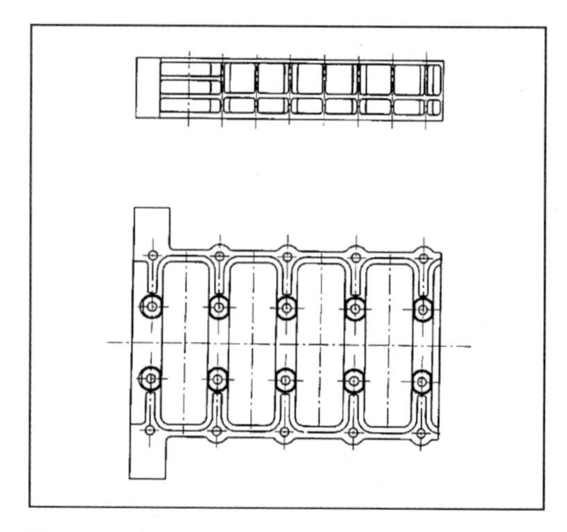

Fig. 7-56 Longitudinal frame design.

- Almost the same degree of engineering freedom as the engine block lower section in regard to integrating functions
- More economical and lighter in weight than an engine block lower section.

Longitudinal frames made of aluminum alloys can be manufactured using die casting. This also allows the integration of cast oil grooves to supply oil to the main bearings.

In the areas around the individual bearing points, insets made of cast iron with spherical graphite (e.g., GGG 60) can be cast in place. This yields the same advantages (reducing the bearing play at the crankshaft, increasing stiffness of the longitudinal frame, and reducing noise radiation in the main bearing pedestal area) as for the combination of aluminum engine block and main bearing cap made of gray cast iron.

In existing engine block designs with individual gray cast iron main bearing caps, these may be replaced with a longitudinal frame construction to increase stiffness and/or to improve acoustic properties without having to completely reengineer the block. Also possible are combined solutions in which individual bearing caps are joined by bolting them to a separate cast part shaped like a ladder.

The most widely used mass-production version of engine blocks with longitudinal frames is the combination of an aluminum engine block and an aluminum longitudinal frame; here the engine block may be laid out so that the oil pan flange is level with the center of the crankshaft or as an engine block with a long skirt.

Oil pan flange level with the center of the crankshaft

A distinction in engineering still often found today is the separation plane between the engine block and the oil pan being level with the center of the crankshaft, Fig. 7-57. In this version the upper halves of the crankshaft bearing seats are integrated into the casting for the engine block as main bearing pedestals. The lower halves of the crankshaft bearing seats are engineered either as individual main bearing caps or as a longitudinal frame.

The seal between the engine block and the oil pan is between the two flanges congruent with the separation plane.

The seal for the crankshaft at the front and rear ends depends on the particular engine design. The front end of the crankshaft may be sealed by a radial shaft seal in the oil pump housing or in the front-end cover. The rear end of the crankshaft may be sealed by a radial shaft seal in the last main bearing pedestal or in a separate cover.

Gray cast iron engine blocks in which the separation plane for the oil pan is level with the center of the crankshaft and with individual main bearing caps are often used for small-displacement (to about 1.8 liters) four-cylinder, inline engines (and in some V-6 and V-8 engines).

The advantages of this design are found in favorable manufacturing costs. The disadvantages of this design, in comparison to engine blocks with deep skirts or a lower engine block section, are less stiffness and less favorable acoustic properties.

Oil pan flange below the center of the crankshaft

With the separation plane between the crankshaft and the oil pan in this location, one differentiates between two types of engine block construction:

(a) Design with upper engine block section and lower engine block section (Fig. 7-58a). In this version,

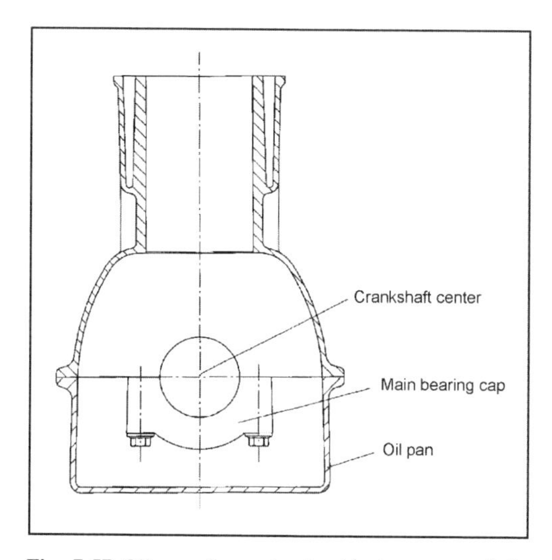

Fig. 7-57 Oil pan flange level with the center of the crankshaft.

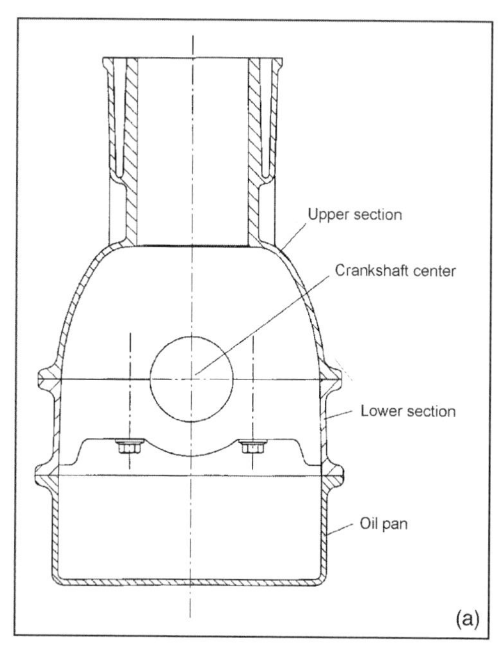

Fig. 7-58 (a) Version with upper and lower engine block sections.

the main bearing caps are joined to form a bearing case, the so-called engine block lower section. The separation plane between the upper and lower sections of the engine block is level with the center of the crankshaft. This means that here the component designated as the engine block upper section corresponds to the engine block of the type where the oil pan flange is level with the center of the crankshaft.

The lower face of the lower engine block section forms the flange surface with which the oil pan mates. Depending on the engine design, the crankshaft is sealed at the rear end (toward the transmission) by a radial shaft seal in the last main bearing pedestal and at the front end with another radial shaft seal (located in the oil pump housing or front-end cover).

The advantages of this concept are great stiffness, good acoustic properties, and the engineering design options available for the lower engine block section as elucidated at the description of the lower engine block section and longitudinal frame design (e.g., casting in place for inserts made of cast iron with spherical graphite in the area of the individual bearing points for lower engine block sections made of aluminum alloys and manufactured in a die-casting process). The disadvantages are higher manufacturing costs and, in some cases, slightly greater weight than if individual main bearing caps are used.

This concept is built in mass production with the upper and lower engine block sections made of aluminum alloys. Since racing engines are often integrated into the frame as a load-bearing component in the overall concept for the vehicle, racing engine blocks (because of the high degree of stiffness required) are designed almost exclusively using this engineering principle.

(b) Engine block with long side walls (Fig. 7-58b). In this version, the outside walls of the engine block are extended to below the middle of the crankshaft and end at the flange interfacing the oil pan. The separation of the main bearing pedestals continues to be centered on the crankshaft for reasons associated with the machining.

Designs that have been realized exhibit both individual main bearing caps and main bearing caps that have been combined to form a longitudinal frame.

The benefits of using a longitudinal frame are stiffness and acoustic properties similar to the concept with separate upper and lower engine block sections. The manufacturing costs for this method may be slightly lower, depending on the manufacturing volume.

Gray cast iron engine blocks for engines produced in volume are often engineered with a deep skirt and individual gray cast main bearing caps. Designs of aluminum engine blocks with extended sidewalls and aluminum longitudinal frame components have recently been used in mass production.

7.4.3 Optimizing Acoustic Properties

Complying with noise emission regulations and satisfying owners' expectations for quiet operation are key areas of attention in acoustic development for drive components.

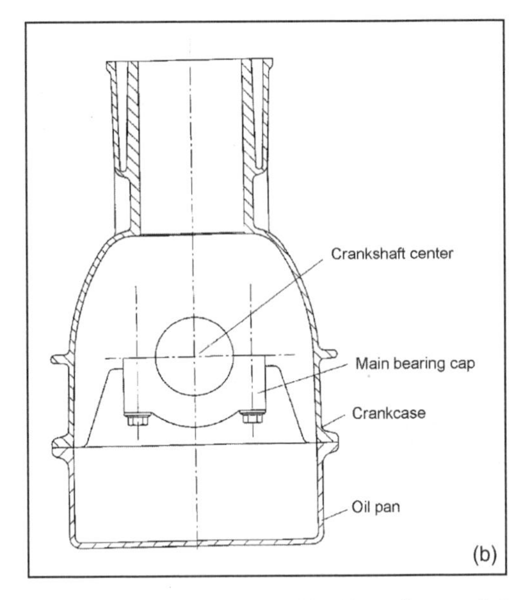

Fig. 7-58 (b) Engine block with side walls extended downward.

The acoustic properties and smooth running of an internal combustion engine depend on many parameters and are predetermined to a great degree by the selection of the design for the engine and engine block.

Optimizing the acoustic properties for the engine block structure, such as increasing stiffness at the engine block sidewalls, taking account of the many and varied functional requirements, is an important development target. This is achieved by low noise radiation, avoiding natural frequencies, and damping resonance-inducing vibrations.

The loading on the engine block resulting from the nonuniform progression of torques in the crankshaft due to the free mass forces and moments causes mechanical vibration. Their exciter frequency is in a certain relationship to the rotational speed of the crankshaft, according to the orders of excitation for the free effects of ignition gas and mass. Mechanical vibrations are caused by low exciter orders, are at a low frequency, and are found primarily in the area of the main bearing pedestals and the crankcase.

High-frequency vibrations in the engine block walls are induced by the combustion process itself, and are in part because of pulselike power transmission in the valve actuators and by forces induced at the pistons. The high frequencies are in the audible spectrum and are referred to as acoustic vibrations. A part of the high-frequency acoustic vibrations is radiated from the sidewalls of the engine block.

Low- and high-frequency vibrations exert their effects through the interface of the engine block with the engine mounts in the vehicle. Depending on the type of engine mount used, vibrations and structural noise may be transferred to the vehicle.

To be taken into account in the acoustic optimization of an engine are the following:

- The above-mentioned causes for initiation of structural noise
- The structural noise propagation paths in the cylinder head, cylinders, pistons, wristpins, conrods, and crankshaft
- The design of the engine mounts and their connection to the engine block or to other engine and drive train components
- The structure of the engine block in conjunction with the engine block engineering concept selected

Modern engine block development is undertaken in a closed CAE process chain. The 3-D CAD depiction and networking for the housing structure form the foundation for FEM calculations of strength, stiffness, and dynamic and acoustic properties.

An experimental model analysis at the finished engine block provides additional information on the forms of its own vibrations.

Both experience and the engineering calculation and analysis options available today support the basic claim that noise-optimized engine block design requires the stiffest possible engine block and the stiffest possible combination of engine and transmission.

This is achieved by measures that are independent of the selected engine block design and by exploiting advantages specific to a particular design, such as

- Manufacturing engine block surface structures with reinforced areas and ribs or fins to reduce airborne noise propagation.
- Stiff top plate and a force engagement point for the head bolts that is well below the top surface of the top plate. They minimize deformations at the sealing services and cylinders. The latter is a prerequisite for low piston play and, thus, low piston noise.
- Stiffness at the crankshaft's main bearing pedestal configuration, which permits only slight bearing play.
- Stiff flanges interfacing with the oil pan and the transmission as a prerequisite for a stiff engine and transmission assembly.

The various engine block designs have differing specific acoustic advantages:

- The closed-deck design has a stiff top surface with benefits, in comparison with the open-deck design, in regard to deformation at the sealing surfaces and cylinders.
- A design comprising upper and lower engine block sections gives a stiff engine and transmission group in comparison to an engine block with side walls extended below the center of the crankshaft in combination with individual main bearing caps. In the latter design, stiffness is increased by joining the individual main bearing caps to form a longitudinal frame.
- In solid aluminum engine blocks, composed of upper and lower sections, gray cast iron components that are cast into the engine block at the main bearing points reduce thermal expansion and, in turn, bearing play.
- Using a cast aluminum oil pan with a flange interfacing with the transmission provides a stiff engine and transmission group.

7.4.4 Minimizing Engine Block Mass

Important objectives in engine development are the reduction of pollutants, the lowering of fuel consumption, and an improvement in performance. This target requires, in addition to other measures, consistent implementation of lightweight engineering techniques for all vehicle components. Reducing the engine block weight is one contribution to reducing weight for the entire drive train.

Depending on the engine's size, design, combustion principle, and engine block design, the engine block accounts for between 25% and 33% of the overall engine weight (as per DIN 70020 A). Reducing the engine block weight thus makes a vital contribution to reducing vehicle engine weight.

The measures undertaken to reduce engine block weight can be subdivided into weight reductions attained by optimizing the structures and weight reductions specific to the materials.

Reducing weight by optimizing the structure. The design of the engine block has a critical influence on total engine block weight. The engineering and calculation methods (such as CAD and FEM) that are commonplace today enable more closely targeted optimization of the design needs, along with loading and functional needs, than could be achieved in the past.

This means that the wall cross sections required to carry out important functions such as the exact position, number, and geometry of ribs (which increase stiffness and improve acoustic properties) can be designed using minimal amounts of material.

Cylinders that are cast together and the integration of many functions into the engine block also contribute to reducing overall engine weight.

Weight reductions through material selection. Presently, the majority of the engine blocks in mass production are gray castings. The necessity to reduce weight has resulted in using aluminum silicon alloys more frequently for the engine block in small-displacement engines. Engine blocks of comparable design, but using Al-Si alloys are not lighter than cast iron engine blocks in exactly the same ratio as that for the specific weights of the materials. Figure 7-59 shows the data for some materials used for engine blocks. In addition to the density, additional materials properties (such as fatigue strength under reversed bending stress and Young's modulus) have to be taken into account. In comparison to gray cast iron engine blocks, the weight of aluminum engine blocks may be reduced by 40% to 60%, depending on the size of the engine.

Material	0.2% offset limit N/mm²	Density g/cm³	Young's modulus kN/mm²	Fatigue strength under reversed bending stress N/mm²
Die-cast magnesium alloy	140 to 160	1.8	45	
Die-cast Al-Si alloy	140 to 240	2.75	74 to 78	70 to 90
Gray cast iron GG 25		7.2 to 7.7	115 to 135	120 to 145

Fig. 7-59 Materials for engine blocks.

In engine blocks made of gray cast iron, it is possible to reduce weight by a combination of optimizing the structure and thin-wall casting. With this casting technique, wall thickness of as little as about 3 mm is generally possible. In comparison, the walls of cast iron engine blocks are normally in a range of from 4.0 to 5.5 mm thick.

Using vermicular graphite cast iron (GGV), a casting material with great strength, enables weight reductions by about 30% in comparison to conventional casting materials such as GG 25. Weight reduction to this extent requires engineering for the engine block, taking into account the particular needs of the material.

Vermicular graphite cast iron engine blocks have not yet entered mass production, but trials are currently being conducted.

The advantages of substituting GGV for GG 25 in engine block manufacture include weight-saving potential, greater stiffness, and better acoustic properties. The costs for the material are detrimental, estimated to be from 20% to 28% higher. These costs can be offset by the weight savings and longer service lives for machining tools.

Magnesium is a material that exhibits even lower density than aluminum. In the past, engine blocks made of magnesium alloy have been used in air-cooled engines. Examples include the four-cylinder boxer engines used in the Volkswagen Beetle and the six-cylinder boxer engines in the Porsche 911; from the end of the 1960s to the beginning of the 1970s, their engine blocks were made of a magnesium alloy.

Today magnesium blocks are used only in racing engines. The low specific weight supports using magnesium alloys for engine blocks. Disadvantages in comparison with the Al-Si alloys normally used today in mass production are the high costs for the material, the lower material strength, and the lower resistance to corrosion.

It is also true that engine blocks that are made of magnesium alloys cannot be lighter at a proportion corresponding to the ratio of their specific weights. In working out an engineering design that is in line with the loading, the differences in the material properties must be taken into account. In comparison to an engine block made of an Al-Si alloy, using a magnesium alloy in a comparable design can cause weight savings on an order of magnitude of 25%. The lower strength and the lower Young's modulus for the magnesium alloys must be compensated for, in large part, by higher design strength for the component. Thus, for example, the longitudinal cooling channels for the cross-stream cooling, integrated into the magnesium engine block in racing engines, contribute to increased design stiffness.

The reasons magnesium engine blocks are not currently mass-produced are many and varied.

The cost advantage for Al-Si alloys in comparison to magnesium alloys is an order of magnitude of about the factor three and results essentially from the absence of a recycling market for magnesium. While Al-Si alloys are available at low cost in the form of secondary alloys from components that have been melted down, it is necessary to draw upon the costly primary alloys for magnesium alloys. The higher costs for magnesium alloy materials are, however, to be set off, case by case, against the lower costs resulting from weight savings and by shorter processing times and longer service lives for die-casting molds and machining tools.

The corrosion resistance of magnesium alloy components, where no additional protective measures are adopted, is lower than that for components made from Al-Si parts; their natural surface or skin after casting already provides sufficient corrosion resistance. It is necessary to differentiate between contact corrosion and surface corrosion.

Contact corrosion arises when parts made of magnesium alloys come into contact with components made of other metals or alloys. It results from the differing positions of the various metals along the electrochemical series. Contact corrosion may arise at threaded connectors and at holes for fixing elements such as alignment bushings and pins. To achieve satisfactory corrosion protection, it is necessary to adopt measures that increase costs: Using washers made of an Al-Si alloy and special surface protection for bolts and guide bushings by galvanizing and chrome plating.

To avoid surface corrosion at the outermost surfaces of components made of standard magnesium alloys, it is necessary to apply surface treatments such as chrome plating before machining the component and, after machining, to apply wax or powder coating. Some components made of high-purity magnesium alloys offer sufficient protection against surface corrosion even without the above-mentioned surface treatments, while sufficient protection has to be seen in conjunction with the amount of corrosion exposure.

For mass production use of magnesium engine blocks today it is necessary, to achieve the engine life expectancy now required at 160 000 km or 100 000 mi, to provide sufficient resistance to surface corrosion in the water jacket.

Using pistons made of Al-Si alloys directly in magnesium cylinders is precluded by the tribologic properties of the magnesium alloy. Magnesium engine blocks require production-ready development of a cylinder running surface technology compatible with the basic magnesium alloy analogous to the cylinder running surface technologies described earlier for engine blocks made of Al-Si alloys (gray cast iron or aluminum bushings, bonding technologies).

7.4.5 Casting Processes for Engine Blocks

Engine blocks for automotive engines are manufactured from cast iron or aluminum-silicon alloys. The costs, numbers produced, and engineering design are the main criteria applied when selecting the casting process.

7.4.5.1 Die Casting

Permanent molds made of hardened hot-work steels are used in the pressure die-casting process. The sections of the mold have to be treated with a parting agent before each casting is made.

In contrast to sand casting and die casting, no cores can be inserted into the mold since the lightweight metal melt is introduced into the casting form at high pressure and high speed.

The pressure level depends on the size of the casting and is between 400 bar and about 1000 bar. The pressure is maintained during solidification. In larger castings the two halves of the form are cooled, allowing a directional solidification of the cast component.

In contrast to sand casting and die casting, pressurized die casting provides the most precise reproduction of the hollow cavity in the mold and thus for the greatest precision in the cast component. Thin-walled castings with close dimensional tolerances, great exactness of shape, and superb surface quality can be fabricated. Casting eyes, holes, passages, and lettering to exact dimensions eliminates the need for subsequent machining and casting in place bushings such as the cylinder sleeves made of gray cast iron. Pressure die casting, when compared with sand casting or die casting, offers the highest productivity since almost all the casting and mold movement processes are fully automated.

The drawbacks are the limited engineering freedom for the cast component, since undercuts are not possible. Air or gas bubbles that might be trapped in the casting preclude double heat treatment, as for sand casting and die casting.

Engine blocks made of aluminum-silicon alloys, in particular with special cylinder sleeve technologies, are produced to an even greater extent in pressure die casting.

7.4.5.2 Die Casting

A die is a permanent metal mold made of gray cast iron or hot-worked steels and is used to manufacture cast components from lightweight metal alloys. Just as in sand casting, sand cores are inserted into the casting mold, offering the benefit of greater freedom in the engineering design. Undercut areas are possible, this in contrast to pressure die casting. The die-casting process makes it possible to use each mold for many casting cycles, unlike in sand casting where new sand cores are required for each cycle.

Again, in contrast to the sand casting mold, solidification of the metal melt in the die is fast and directional. Closely defined cooling of the die is possible, and this option is often used.

The die has to be protected against the lightweight metal melt by applying a parting agent.

In comparison to sand casting, castings taken from the die exhibit a finer inner structure, greater strength, greater dimensional accuracy, and better surface quality.

Double heat treatment is possible for die-cast components. In addition to carefully defined control of cooling for the cast component inside the die, the first heat treatment undertaken is often a further heat treatment, controlled cooling.

In die casting one differentiates between gravity die casting and low-pressure die casting. The difference lies essentially in the way in which the melt is introduced to the mold.

In the low-pressure casting process the molten metal is introduced into the die from below at a gauge pressure of from 0.2 to 0.5 bar, which is then maintained during solidification. The almost perfectly directional solidification of the casting that results is one of the fundamental reasons for the high quality of low-pressure cast components.

In gravity die casting, by contrast, the mold is filled at atmospheric pressure, using the force of gravity acting upon the molten metal.

7.4.5.3 Lost-Foam Process

This is a special variation of the sand casting process. A plastic model is made of the piece to be cast, using EPS (expanded polystyrene), by foaming in place and, if necessary, by gluing individual segments together. The expanded polystyrene model is coated with a water-based parting agent. The model, coated and dry, is placed in a casting shell in which pure quartz sand (without any binders) is filled using vibratory compaction. In this fast casting procedure (taking 15 to 20 s), the molten metal is directed to the plastic model as a so-called full-mold casting. The heat in the molten metal degrades the plastic model: its liquid and gaseous components are absorbed by the casting sand. Following cooling and deforming a flash-free casting is obtained.

The particular advantages of this process are found in the capabilities for making up plastic models that replicate casting geometries not possible with conventional sand casting processes because of technical limitations on the latitude for mold fabrication.

The lost-foam process is suitable for making up both gray castings and lightweight metal alloy castings.

7.4.5.4 Sand Casting

Sand casting is the process traditionally used for engine blocks made of gray cast iron in mass production. Models and core boxes made of hardwood, metal, or plastic are used to replicate the later engine block casting inside the sand mold. The casting molds are normally made of quartz sand (either natural or synthetic sand) and binders (synthetic resin, CO_2). The sand is introduced, using "sand shooting machines," to make the cores. Combining individual cores to form a core package and assembling this core package and the outer casting mold is handled mechanically and fully automatically, even when producing only moderate numbers of castings.

Model, core, and mold parting in various planes, and inserting cores in the casting mold make it possible to produce complex cast components with undercut areas.

During the casting process, the hollow cavities between the outside mold and the cores are filled with molten metal.

Following the casting process and after the metal has solidified, the casting is removed from the sand mold. The mold is destroyed when doing so. The casting is then reworked to remove traces of the gating, sprues, casting skin, and flash.

In sand cast components made of Al-Si alloys, double heat treatment to increase strength is possible. The first heat treatment phase is found in the controlled cooling period for the casting inside the sand mold. The second heat treatment occurs during time- and temperature-controlled storage of the casting in a kiln.

The sand mold can be used to produce only a single casting.

Recently engine blocks in Al-Si alloys have also been produced in large numbers using a precision sand casting process.

Further applications for sand casting are creating prototypes and performing short production runs.

7.4.5.5 Squeeze Casting

The squeeze casting process represents a combination of low-pressure die casting and the pressure casting process. Permanent metal molds are filled from below with molten lightweight metal at a gauge pressure of from 0.2 to 0.5 bar. This is followed by solidification under high pressure at about 1000 bar.

The excellent density attained when filling the mold also makes it possible to use high-strength alloys with less favorable properties.

The solidification of the melt while under high pressure imparts a very fine internal structure to the cast component.

Slow filling of the mold and solidification under high pressure give a structure virtually free of pores. As a result, the material is capable of enduring high strength against alternating loads along with great resistance to temperature changes in comparison to both low-pressure casting and die casting.

As in die casting, the use of sand cores is not possible in squeeze casting. Since undercuts cannot be created, the same engineering restrictions apply to squeeze casting as for die casting.

In contrast to die casting, double heat treatment is possible, since there are virtually no pores in the structure.

Thus, squeeze casting joins the advantages of die casting, low-pressure casting, and pressure casting.

7.5 Cylinders

The piston group is mounted in the cylinders. With their surface and the material used—and working in concert with the piston rings—the cylinders also support slip and sealing functions. Over and above this they contribute, depending on the design, to heat dissipation via the engine block or directly into the coolant.

7.5.1 Cylinder Design

Both engineering and materials aspects have to be taken into account when designing the cylinder and the cylinder running surface. Both aspects are linked one with another. Taking the materials as the starting point, the designs for the cylinders and engine block may be subdivided as follows:

- Monolithic design
- Insert technology
- Bonding technology

7.5.1.1 Monolithic Design

Typical representatives for a monolithic (monometal) design are engine blocks made of cast iron alloys in which the cylinders are an integral part of the engine block. The required surface quality is achieved by machining in several steps, including preliminary and precision reaming and honing. Monolithic engine blocks made of Al-Si alloys are found in two versions:

- Manufacturing the engine block casting from a hypereutectic Al-Si alloy. Al-Si alloys are deemed to be hypereutectic if their silicon content exceeds 12%. The primary silicon precipitated out in the cast component, following the machining of the engine block at the cylinder running surfaces, is exposed by a chemical etching process or with special mechanical honing. A hard, wear-resistant cylinder running surface (referred to as nonreinforced) is created; it has to be mated with an iron-plated piston.

  Because of the higher share of silicon in hypereutectic Al-Si alloys, workpieces made of this material cannot be machined as readily as cast components made of hypoeutectic alloys. The primary silicon crystals precipitated out in the cast part are damaged and splinter during mechanical processing. This results in undesirably short chips.

In hypereutectic Al-Si alloys and a closed-deck design, this monolithic cylinder block or engine block design can be manufactured in a low-pressure process, and in hypoeutectic Al-Si alloys and open-deck design, in a pressure casting process. When using this latter process, the primary silicon grains are found, which are far smaller than with low-pressure casting processes. This improves machining properties significantly. Because of the reduced tendency to splinter, the smaller silicon crystals can be worked faster while better cutting results are achieved at the same time.

- Manufacturing the engine block from a hypoeutectic Al-Si alloy in combination with a finish for the cylinder running surface. The finish may be applied either by electroplating or with a thermal spatter process. In the meantime cylinder running surfaces that are remelted or plated using lasers are in the development phase.

  Used exclusively in mass production to date is the quasimonolithic engine block design in which a nickel dispersion layer is electrodeposited on the cylinder running surface. This layer comprises a nickel matrix into which silicon-carbide particles are inserted at uniform distribution. Cylinder surfaces finished in this way exhibit excellent running properties and low wear. Moreover, they may be combined with pistons and piston rings made of conventional materials. The layer is, however, to a certain extent sensitive to cold corrosion when using fuels that contain sulfur, so that its application in diesel engines is precluded.

  The nickel dispersion layer is better known under trademarked names such as NIKASIL®, GALNICAL®, and GILNISIL®. The nickel dispersion layer can be combined with both closed-deck and open-deck designs.

  Since even miniscule porosity in the cylinder running surface can cause plating problems in that the plated layer spalls off, the selection of the casting processes that may be used for Al-Si alloy engine blocks is restricted. The conventional pressure die-casting process, for example, cannot be used without special techniques such as vacuum support.

  Nickel dispersion plated cylinders are used frequently for single-cylinder motorcycle engines, as well as in the very few automotive engines that are made up of separate cylinders. Multicylinder engine blocks for automotive use, incorporating nickel dispersion finished cylinders, are mass-produced only to a very limited extent.

Cylinder Engineering for Monolithic Designs

One differentiates between cylinders that are cast as a single unit along the engine block's longitudinal axis and those that are not cast together.

In the past, engine blocks in both closed-deck and open-deck designs made of either cast iron or aluminum-silicon alloys were executed with cylinders that were not joined together along the engine block's longitudinal axis. This was done to achieve the most uniform possible temperature distribution in the cylinders (by coolant present between the cylinders) and the smallest possible degree of cylinder warping (by preventing mutual influencing of neighboring cylinders). Detrimental here was the greater length of the engine block that this involved.

Today, suitable engineering measures can be employed to ensure that cylinders that are cast as a single, solid unit along the engine's longitudinal axis can exhibit almost uniform temperature distribution in spite of the absence of coolant between the cylinders. This eliminates any appreciable warping problems and the concomitant functional problems such as excess oil consumption or blowby. The advantages of unitized cylinders are greater engine block strength, a shorter engine block, and lower engine weight.

Today, the reduced engine length is a dominant criterion in view of transverse engine mounting and in light of the ever declining amount of space available for installing drivetrain components. Depending on the particular engine design (inline, V-block, or boxer engine), designing the engine block with cylinders cast as a unit results in differing degrees of reduction in length and weight. The lower limit for joining cylinders is represented by the web remaining between the cylinders. Regardless of the material used, engine blocks manufactured in mass production incorporate cylinder webs thinner than 5.5 mm.

This was made possible by employing metal head gaskets with low compression and setting properties, thus requiring lower preload values at the head bolts. In addition to perfect sealing at the cylinder web, cylinder deformation is reduced to a minimum due to the lower preload values in the aggregate comprising the cylinder head and the engine block.

7.5.1.2 Insertion Technique

Insertion techniques are normally used for cylinder sleeves in automotive engines in conjunction with aluminum engine blocks. Sleeves made of any of a variety of materials are inserted into the engine block in any of a number of ways. Following differentiation by function into wet and dry cylinders, one distinguishes whether the sleeve is cast in place, pressed in place, shrink fit, or slid in place in the engine block. Moreover, one may distinguish according to the material used for the sleeve, possibly cast iron or aluminum.

Wet Cylinder

Wet cylinders are slid into the engine block, mating with mounting areas machined and prepared accordingly. The water jacket around the cylinder is formed between the engine block and the sleeve, Fig. 7-60.

The hanging cylinder sleeve features a collar at its upper end; this collar is clamped between the engine block and the head gasket or cylinder head. The sleeve is centered in the engine block at the collar itself or at a diameter below the collar. Using the collar for centering offers the advantage of good cooling for the top end of

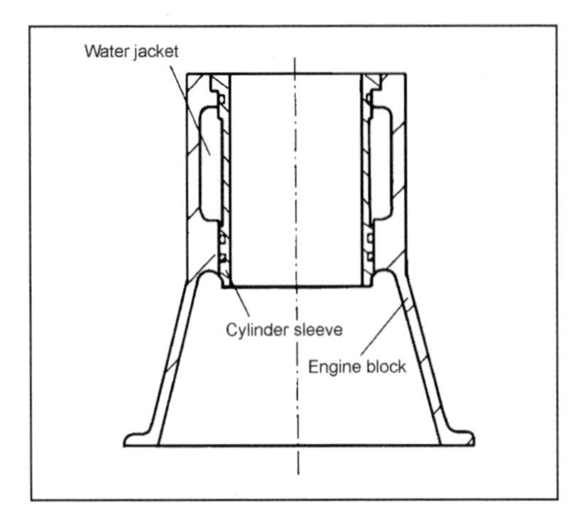

Fig. 7-60 Wet cylinder.

the cylinder sleeve, which is subjected to severe thermal loading. The disadvantage is the heavy loading at the fillet in the engine block.

Centering the sleeve at a point below the collar causes less satisfactory cooling at the upper end of the sleeve but does relieve this fillet. O rings are used at wet sleeves that are suspended from the top to seal against coolant at the top and against oil from the crankcase at the bottom.

In the standing wet cylinder, the support and centering functions are at the lower end of the sleeve. This sleeve concept requires particularly careful engineering to keep down cylinder deformation. Sealing is by the head gasket at the top and a flat gasket at the bottom below the sleeve support surface or by O rings. Misalignment of wet cylinder sleeves at the top plate, resulting in protrusion or depression, can be a problem. This has a negative effect both on the surface pressure applied by the head gasket around the cylinder and on cylinder deformation. Consequently, sleeve protrusion or depression can be reduced to the unavoidable minimum.

Inserting fully machined, wet cylinder sleeves into an engine block after the top plate has been finished is done by imposing extremely close tolerances on the relevant sleeve dimensions. When installing standing sleeves, shimming is a commonly used technique. A further option is final machining of the engine block top plate and the sleeves after the latter have been installed.

Wet sleeves are normally manufactured from gray cast iron. The less common aluminum sleeves may be made of either hypereutectic or hypoeutectic Al-Si alloys. As has already been described, in the case of hypereutectic Al-Si alloys the cylinder running surface is treated by chemical etching while a nickel dispersion layer is applied to hypoeutectic Al-Si alloys.

Wet sleeves, regardless of their design and material, can be used for both open-deck and closed-deck designs and can be combined with all casting processes that are normally used for engine blocks. They are given preference, for cost reasons, in die-cast engine blocks with open-deck design.

The advantages in using wet sleeves are freedom in selecting the material for the sleeve, flexibility in regard to the cylinder bore, and, thus, displacement specified by combining the appropriate sleeves with one and the same engine block. Further benefits are simple interchangeability and repairs. Unfavorable are the higher manufacturing costs when compared with monolithic concepts.

Wet aluminum sleeves are found almost exclusively in lightweight metal engines for sports cars or racing cars, where lower weight and better heat transfer are given preference over cost considerations.

Dry Sleeves

Dry sleeves are pressed, shrink fit, or cast in place in the engine block (Fig. 7-61). When cast in place, the sleeves are inserted in the engine block mold, and the molten aluminum alloy is cast around them. In contrast to wet sleeves, the water jacket is not between the sleeve material and the engine block casting but, as in the monolithic design, is a component of the engine block casting. Consequently, no sealing is required between the sleeve and the engine block.

Any protrusion of the dry sleeve—pressed or cast in place—in relation to the top plate level is corrected by machining the deck plate and the inserted sleeve together. Dry sleeves may be made of either gray cast iron or (hypereutectic) aluminum alloys; sintered sleeves made of powdered metal materials are another option.

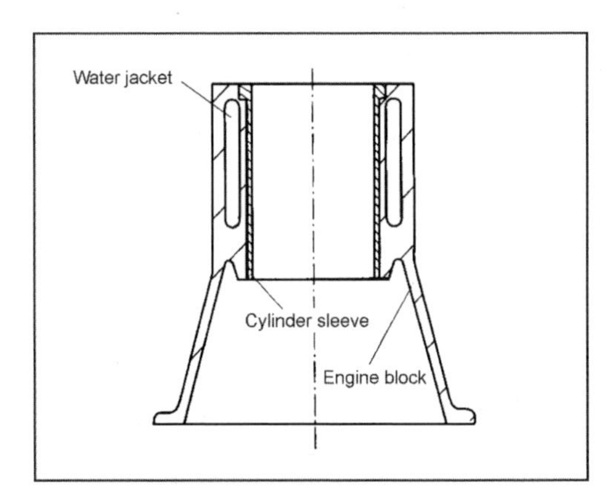

Fig. 7-61 Dry sleeves.

The running surfaces for dry sleeves made of gray cast iron or aluminum are treated the same way as wet sleeves and, thus, exhibit the properties listed there.

Dry sleeves, regardless of the material, may be used for both open-deck and closed-deck designs and can be combined with all the casting processes that are normally used for engine blocks. Aluminum engine blocks are found in mass production and are characterized by the closed-deck design, made in die casting or low-pressure casting, with gray cast iron cylinder sleeves pressed in place; also seen is the open-deck design made with pressure casting where gray cast iron sleeves are cast in place.

The advantages of dry cylinder sleeves are freedom in the selection of the materials, the easy repair option (in the case of gray cast sleeves) by reaming out to oversize dimensions, separate manufacture (in aluminum sleeves) of the cylinder running surface, and the option for combining sleeves with an engine block made of a different aluminum alloy. One embodiment of this is the cast-in-place sleeve made of a hypereutectic, spray-compacted aluminum alloy with the trademarked designation SILITEC®. A disadvantage in this concept, inherent to the concept, is poorer heat transfer between the cylinder running surface and the water jacket.

Regarding manufacturing costs, there may be advantages or disadvantages when compared to a monolithic design depending on the number of units produced, the casting process selected, and the engineering details of the engine block and sleeve. Particularly when manufacturing large numbers of units in pressure die casting or automated sand casting processes, cast-in-place gray cast sleeves can be very economical in terms of overall costs.

7.5.1.3 Bonding Technology

Bonding technology can be used only in engine blocks made of aluminum alloys. In aluminum engine blocks incorporating bonding technology—and in contrast to aluminum engine blocks in classical monolithic design—an inseparable unit comprising the engine block and the cylinder running surface is created by special manufacturing processes. This may be designated as "local material engineering."

There are two fundamental embodiments. In the first, shaped cylindrical bodies, so-called preforms, made of a bond of suitable metallic and ceramic materials, are inserted in the casting molds and are infiltrated by the molten aluminum alloy at high pressure during the casting process. In the second variant, sleeves made up of several layers and/or of several metallic materials are joined with the engine block by an intermetal bond during the casting operation. The bonding technology limits the choice of casting process to pressure die casting and processes derived from pressure die casting, such as squeeze casting or the new die-casting process developed by Honda.

The limitations that the technology places on die casting and related processes make it necessary to adopt an open-deck design when implementing a bonding process. Cylinders that are cast together as a unit and those that are cast separately can be realized.

Regarding the use of preforms, one may differentiate between two bonding technology processes:

- Honda MMC process. This metal matrix composite process has been in volume production for some years now. It is similar in principle to the Lokasil® process. Fiber preforms are inserted into the mold prior to casting. The preforms comprise a bond of Al_2O_3 fibers and carbon fibers and, in the Honda new die-casting processes, are infiltrated by the molten aluminum alloy.
- Lokasil® process by KS ATAG. A high-porosity, cylindrical body made of silicon is infiltrated by a liquid aluminum alloy at high pressure during the squeeze casting process. The cylinder running surface is prepared with three honing phases. In preliminary honing using diamond strips, many of the silicon crystals in the surface are damaged. Intermediate honing using silicon carbide removes this damaged silicon crystal layer. The third honing phase, using grains bound up elastically in the honing strips, exposes the silicon grains. Similar to the silicon crystals that are exposed in the monolithic version by etching the hypereutectic aluminum alloy, these crystals form a hard and wear-resistant cylinder running surface. An iron-plated piston is required for use as the mating material. As a rule, a set of piston rings similar to that used with cylinders made of gray cast iron is sufficient.

When using bonding technology with metallic sleeves, it is possible, in one embodiment, to cast in place (during die casting) a sleeve made up completely by thermal spraying and comprising various materials in multiple layers (GOEDEL® technology). The intermetallic bond between the sleeve and the molten aluminum alloy is ensured by the appropriate choice of materials (normally an Al-Si alloy similar to that used for the engine block) and the special surface at the outside face of the thermal spray sleeve. Regarding the material used for the cylinder running surface, with its influence on tribologic properties, the thermal spray process allows a broad choice of alloys, ranging from iron-based alloys and Fe-Mo alloys to hypereutectic Al-Si alloys. Machining for the cylinder running surface in each case (as a rule, by a two- or three-step honing process) is chosen to suit the material selected. The same applies to the selection of the mating materials in the pistons and piston rings.

In another embodiment, the desired intermetallic bond can be achieved by applying the thermally sprayed outer layer of the GOEDEL® sleeve on a conventional gray cast iron sleeve (referred to by the manufacturer as the HYBRID sleeve). Thus, the usual situations apply to machining the cylinder running surface and selecting the pistons and piston rings for the gray cast iron running surfaces.

This option is thus particularly economical and has been used for mass production. This is different from other bonding technology solutions that are limited essentially to sports cars and other high-performance engines.

7.5.2 Machining Cylinder Running Surfaces

The cylinder running surface in internal combustion engines is the tribologic mating material and sealing surface for the pistons and piston rings. The properties of the cylinder running surfaces have a determinant influence on establishing and distributing an oil film between the mating components. There is a strong interrelation between cylinder roughness, oil consumption, and wear inside an engine. Cylinder roughness values at Ra <0.3 μm are state of the art.

Final machining of the cylinder running surface is effected by precision boring or turning and subsequent honing. During the honing process, rotational and alternating translatory motions are superimposed upon each other to create a cutting motion. In this way, deviations in cylinder shape of less than 10 μm and uniform surface roughness can be achieved. The scoring arising from the cutting motion includes the so-called honing angle.

This processing, as shown in Fig. 7-62, should be as gentle on the material as possible in order to avoid breakout, pinching at the edges, and the formation of burrs. The material is cut with the assistance of honing strips running under a water-based coolant/lubricant or special honing oil. At the prescribed surface pressure or advancing speed, material removal of 100 μm in diameter is achieved in less than a minute.

7.5.2.1 Machining Processes

In standard honing in a single-stage or multistage machining process, a surface structure exhibiting normal distribution is created, so that in the roughness profile there are as many valleys as peaks.

Plateau honing, on the other hand, levels peaks with a supplementary machining step, creating a plateaulike slide surface with deep scoring that retains oil.

Helical slide honing is a further refinement of plateau honing. It differs from plateau honing primarily by the reduced roughness (and, in particular, the peak roughness) and a very large honing angle of from 120° to 150° for the deep scoring. Very uniform surface roughness is achieved using special honing strips that follow the shape of the bore.

Laser texturing offers almost unlimited freedom in surface design through carefully defined removal of material by the laser beam. The cylinder running surface is textured in the TDC area and is otherwise made as smooth as possible. Textures and structures such as helically arranged slots and pits, as well as cupping are possible in addition to conventional, uniform cross-scored textures.

The roughness profiles for various honing processes are shown in Fig. 7-63.

A complex variation of honing, in which free grinding grains are used, is lapping. Here loose grain is used to give the cylinder running surface a random high-and-low structure. Solid strips press this hard lapping agent in part into the surface, and a plateau surface is created.

In brush honing the surface texture is rounded and deburred following standard honing; a brush coated with a carbide material is used for this purpose. Fluid blasting is yet another process used to remove the metal "frost" (sometimes also referred to as spangle) from the surface and to flush out pores present in the surface. In this process, the entire cylinder running surface is blasted with a water-based coolant/lubricant at a pressure of about 120 bar.

Exposure honing for aluminum cylinder running surfaces employs specially designed honing strips to depress the soft aluminum matrix in comparison with the reinforcing fibers or particles. The particles can also be exposed by etching. The purpose is to depress the aluminum, which has a tendency to weld, by 0.5 to 1 μm. The oil retention spaces created by the suppression of the aluminum improve the running properties of the surface.

Plasma or flame-sprayed cylinder sleeves can be smoothed ideally, similar to inductance-hardened gray casting. The oil retention spaces created by the material's porosity guarantee good running properties.

Other elaborate special processes are nitriding and phosphatizing the honed cylinder running surfaces. Nitriding creates a very rough and hard layer that is not suitable for use as a cylinder running surface without supplementary treatment such as phosphatizing. Phosphatizing is also used without nitriding and has a smoothing effect while also acting as a solid lubricant.

The surface images after honing gray cast iron and aluminum cylinder running surfaces are shown in Figs. 7-64 and 7-65.

7.5.3 Cylinder Cooling

7.5.3.1 Water Cooling

With just a very few exceptions today's automotive engines are water cooled. In contrast to air-cooled cylinders, which are fitted with cooling fins, the cylinders are surrounded by a water-filled cavity, the water jacket or cooling jacket. An important engineering dimension is the water jacket depth, defined as the distance from the top

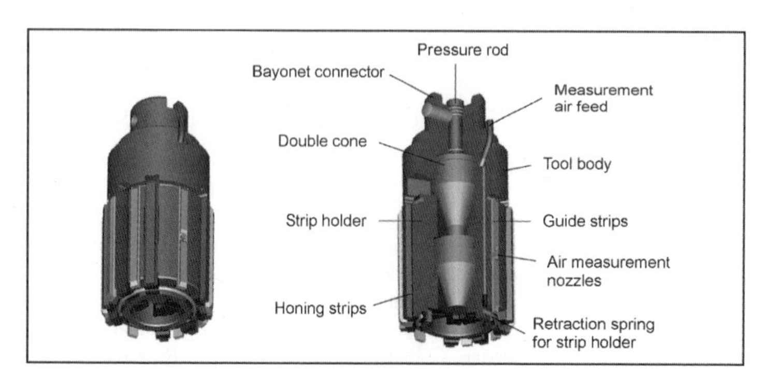

Fig. 7-62 Multistrip honing tool with air measurement system (mfg. Nagel).

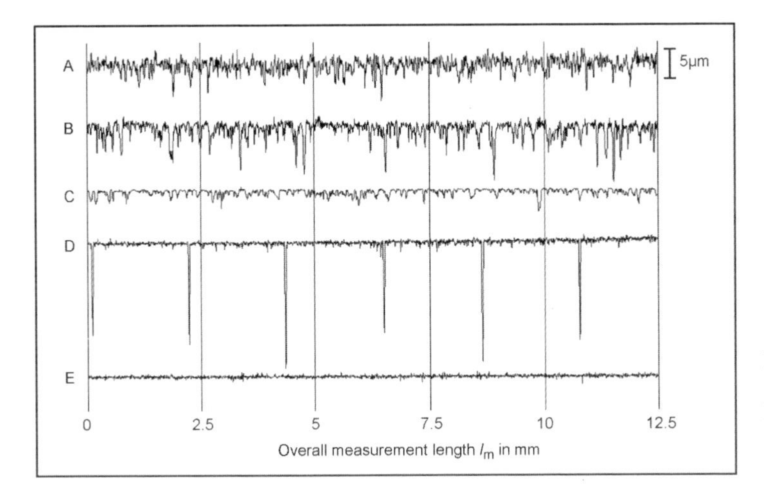

Fig. 7-63 Roughness profile for standard honing (A), plateau honing (B), helical slide honing (C), laser-imparted texture (D), and smooth standard honing (E).

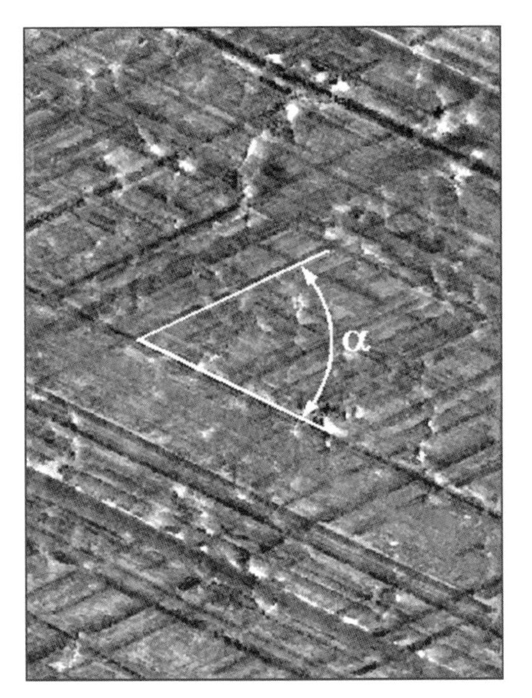

Fig. 7-64 The 3-D surface image for a honed, gray cast iron cylinder running surface with white marbling and the honing angle $\alpha = 47°$ sketched in.

Fig. 7-65 The 3-D surface image of an aluminum cylinder running surface with exposed reinforcing particles.

plate plane to the lowest point in the water jacket. In earlier gray cast engine blocks, this dimension was as much as 95% of the length of the cylinder running surface. In modern cast iron engine block designs, the water jacket ends in the area swept by the lower piston ring, i.e., in the area between the first compression ring and the oil control ring when the piston is at BDC.

The water jacket is even shorter in modern aluminum engine blocks. The water jacket depth corresponds to about one third of the length of the cylinder running surface. This is made possible by the greater thermal conductivity of aluminum alloys in comparison with cast iron materials and by pistons with ever shorter compression heights. A short water jacket reduces the coolant volume in the engine and, thus, the engine weight. The smaller coolant volume and thermal capacitance shortens the engine's warm-up phase, with positive effects in terms of unburned hydrocarbon emissions and the response time for the catalytic converter.

7.5.3.2 Air Cooling

Only a very few manufacturers still use air-cooled cylinders in automotive engines today. Heat dissipation in air-cooled cylinders is dependent upon the thermal conductivity of the cylinder fins and of the cylinder materials, shape of the cooling fins, and the way in which cooling air passes across the fins.

Shape of the Cooling Fins

In air-cooled cylinders, cooling fins are located on the cylinder outside walls to increase the effective surface area for heat transmission. In theory, fins with a triangular cross section are the most effective. In cast cylinders the particulars of the process result in slightly trapezoidal fins with rounded edges being formed; these are hardly any less effective than fins with a triangular cross section. Heat convection at cooling fins can be increased by

- Increasing the fin surface area by, for example, lengthening and by greater fin height
- Increasing cooling air velocity
- Converting from random to directed cooling air flow by installing air baffles and deflectors, for example
- Using a material with the highest possible heat transmission capacity for the cylinder and fins, such as aluminum alloys instead of gray cast iron

The fins on motor blocks carry out other functions in addition to heat dissipation:

- Increasing the stiffness of the engine block side walls, which improves acoustic properties
- Optimizing force transmission from less stiff areas into load-bearing areas of the engine block structure by, for example, making connections between casting eyes, both among eyes and with load-bearing areas such as the top plate, the oil pan flange, the transmission flange, and the like.
- Optimizing the casting process to achieve better flow of the molten metal to areas in the engine block.

Modern calculation methods make it possible to optimize fins in regard to weight, structural strength, and heat dissipation.

The thermal conductivity of aluminum alloys is almost three times that of cast iron materials. That is why cast iron cylinders, once suitable running surface finishing technologies had been developed for aluminum cylinders, were replaced by cylinders made of aluminum alloys.

In air-cooled gasoline and diesel engines that are in severe service the dimensional stability of the pure lightweight metal cylinder may in some cases not be sufficient. It was for that reason that cylinders were used in which a cast iron or steel liner is surrounded by a jacket of fins made of lightweight metal. These so-called bonded cast cylinders were made up in two processes: Casting around a gray iron cylinder sleeve (with a roughened outer surface) a rib jacket made of a lightweight metal alloy in a pressure casting process and casting around a steel or cast iron sleeve to which, prior to the casting process, a thin iron-aluminum coat had been applied. This gives an intermetal bond between the cylinder sleeve and the rib jacket and results in a uniform heat flow.

In engines subjected to less severe loading, cylinder designs were also used in which a lightweight metal rib jacket was cast around a cast iron sleeve without any particular bonding, or a cast iron sleeve was shrink fit into a prepared aluminum fin jacket.

Cooling Air Flow

In air-cooled automotive engines forced or positive cooling using a fan was and is implemented without exception. Here the cooling air is routed from the fan housing, through baffle plates that surround the cylinder and cylinder heads, and is thus directed onto and between the cooling fins. The more favorable the flow characteristics and the thermal conductance values in each case, the lower the amount of cooling air and fan output required.

Bibliography

[1] Flores, G., Grundlagen und Anwendungen des Honens, Vulkan-Verlag, Essen, 1992.
[2] Klink, U., "Laser-Honing für Zylinderlaufbahnen," in MTZ Motortechnische Zeitschrift, Vol. 58, Verlag-Vieweg, Wiesbaden, 1997.
[3] Robota, A., and F. Zwein, "Einfluss der Zylinderlaufflächentopografie auf den Ölverbrauch und die Partikelemissionen eines DI-Dieselmotors," in MTZ Motortechnische Zeitschrift, Vol. 60, Verlag-Vieweg, Wiesbaden, 1999.
[4] Weigmann, U.-P., "Neues Honverfahren für umweltfreundliche Verbrennungsmotoren," Werkstatt und Betrieb, Vol. 132, Carl Hanser Verlag, Munich, 1999.

7.6 Oil Pan

The oil supply for passenger engines today is provided almost exclusively with a wet sump lubrication design. In such engines the oil pan forms the bottom termination for the engine block, Fig. 7-66.

The most important of the oil pan functions are

- Serving as a container to receive the motor oil when oil is first installed and as the collecting basin for motor oil returning from the bearings and lubrication points.
- Enclosing the crankcase it serves and, in specially engineered oil pan types, serving at the same time to stiffen the engine and transmission assembly.
- Taking the threads for the oil drain plug and the dipstick guide tube and often housing, in addition, an oil level gauge showing the oil fill in the vehicle.

7.6.1 Oil Pan Design

In mass production engines the oil pan is normally a single-layer, deep-drawn component made of sheet steel. To improve acoustic properties, a design has recently been introduced that incorporates two layers of sheet steel with a plastic film between them.

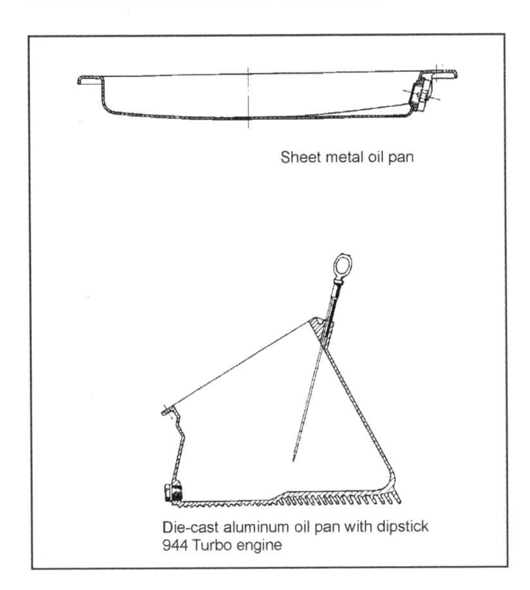

Fig. 7-66 Oil pan.

Used in conjunction with large-displacement engines incorporating cast iron or aluminum engine blocks are oil pans made of Al-Si alloys, manufactured by die casting or pressure casting. These oil pans are usually an integral component in a stiff engine and transmission assembly. This is achieved with a stiff design for the oil pan side walls and, primarily, with an integral flange at the clutch end of the engine, as the connection to the transmission flange. This design makes a significant contribution to stiffening the engine and transmission group and, consequently, to better acoustic properties.

Oil pans made of aluminum alloys are made in single- and two-component versions. Two-part oil pans compose an upper section made of a lightweight metal and a lower section made of sheet steel and bolted to the upper section.

The steel component can be changed more economically in case of deformation (if the car bottoms out). In comparison, an oil pan made entirely of aluminum would have to be completely replaced if deformation occurs.

Today, this advantage is only of subordinate significance because of the underbody claddings being used more frequently to enclose the engine.

7.7 Crankcase Venting

During reciprocating engine operation, gases (the so-called blowby gases) from the combustion chambers pass through the gap between the cylinder and the piston and/or piston rings and into the crankcase.

The blowby gases contain, in addition to unburned fuel components, the complete spectrum of emissions, identical to the exhaust gas. The hydrocarbon (HC) concentration in the blowby gases may, depending on the engine's loading situation, be many times the concentrations contained in the exhaust gases. The blowby gases mix, inside the crankcase, with motor oil, which is present there in the form of oil vapor.

The pressurized blowby gases (the volume of which depends on engine loading) and the reciprocating motion of the pistons create overpressure (proportional to engine speed) below the pistons in the crankcase. Since the crankcase is joined with the cylinder head (by channels for oil return, crankcase ventilation, and a timing chain case, which may be present), the overpressure also prevails at these points inside the engine.

In the early years of engine construction, this pressure was relieved by venting the mix of blowby and untreated motor oil into the atmosphere. In response to newer legal requirements, controlled, closed crankcase ventilation systems have been in use for some time now.

Positive crankcase ventilation passes blowby gases, largely free of motor oil, into the engine's fuel intake system, ensuring that virtually no overpressure is present inside the engine.

7.7.1 Conventional Crankcase Ventilation

A conventional crankcase ventilation system (Fig. 7-67) carries out the following essential functions.

The blowby gases, mixed with oil, pass out of the crankcase, through one or more channels, to the highest point in the engine (normally inside the cylinder head).

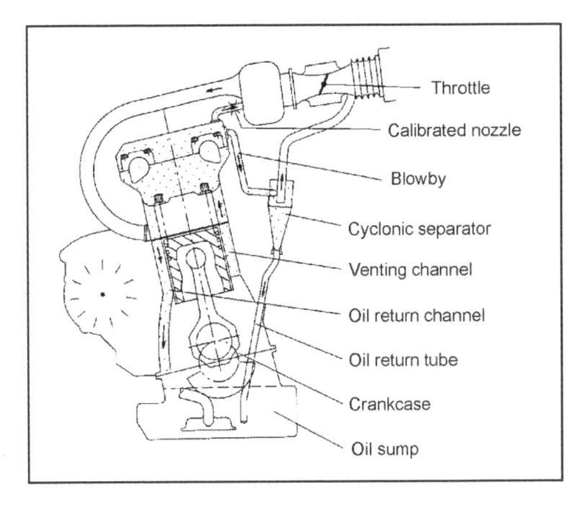

Fig. 7-67 Conventional crankcase ventilation.

These channels may be integrated into the engine block casting or may take the form of a channel (hose, tubing) located outside. The gas is introduced into the cylinder head at one or more points that are protected against oil spray. Oil separation is affected: Separating the motor oil swept up with the blowby gases.

One distinguishes among various arrangements and types of oil separators: those that are mounted in the cylinder head, those that are integrated into the valve cover, or those that are realized as a separate settling space in the crankcase. The effective surface area available for oil separation is often enlarged. This was done in the past with steel mesh and today with expanded sheet metal. Another option is to locate the oil separator, in the form of a cyclonic oil separator, outside the engine.

The arrangement and type depends on many criteria, including the engine design, the available installation space, and, finally, the engineering principles adopted by the engine manufacturer. Thus in V-block engines, for example, oil separators may be integrated either partially or completely into the engine block, in the space between the cylinder banks, or situated in this space as a separate, external oil separator.

The blowby gases, from which oil has been removed, are then introduced to the fuel intake system at a point where a vacuum will prevail at virtually all the engine's operating states—upstream from the throttle, for instance.

In the past—in engines fitted with carburetors, for example—the crankcase ventilation tube terminated in the air filter case, whereas today, in fuel injection engines, the crankcase ventilation tube terminates just upstream from the throttle in order to avoid soiling the air volume sensor and the idle speed stabilizer.

The crankcase is joined with the fuel intake system, downstream from the throttle, by a tube to create a vacuum in the crankcase housing. An integrated, calibrated choke limits the effective vacuum level.

Excessive vacuum or pressure in the engine block can lead to failure of the engine sealing system (crankshaft end seals, oil pan gasket, etc.). Where vacuum is excessive, unfiltered air can enter the engine, triggering accelerated aging of the oil due to oxidation and sludge formation. Excessive pressure can cause engine leaks.

Ensuring the integrity of system functions depends upon the calibration of the choke (affected during optimization work) used to limit the vacuum present in the engine.

When the engine is not yet at normal operating temperature, the blowby gases contain fuel and water vapors, which can lead to engine icing. Icing is avoided by routing hoses in a suitable configuration to avoid creating a siphon and by other measures as indicated.

7.7.2 Positive Crankcase Ventilation (PCV) System

In this system a controlled continuous or a load-dependent flow of fresh air is introduced into the engine, tapping this air downstream from the intake filter. The air is then blended with the mixture of blowby gas and motor oil. The system is regulated with matched and harmonized chokes and valves. Oil separation is, in principle, the same as for conventional or vacuum-regulated systems.

The water and fuel vapors contained in the blowby gases are replaced by the supplementary fresh air and extracted continuously from the crankcase.

The disadvantage of the PCV system, in addition to the higher effort involved in its construction, is the hazard of accelerated motor oil aging by oxidation and oil sludge formation. Motor oil oxidation, which is always present, is aggravated by the oxygen in the air added to the crankcase. Residual particles of grime in the fresh air, even though it has largely been cleaned at the air filter, can cause oil sludge.

7.7.3 Vacuum-Regulated Crankcase Ventilation

In this system the blowby gas exiting the oil separator is introduced—via a pressure differential valve—into the air intake manifold downstream from the throttle; see Fig. 7-68. In comparison with the conventional system, the connection between the oil separator and the intake system upstream of the throttle is eliminated. The vacuum line with its integrated choke, connecting the engine with the intake system at a point downstream from the throttle, is also removed.

The pressure differential valve is a spring-loaded, diaphragm-type valve with a matched bypass. It regulates the vacuum inside the engine to an acceptable maximum value during almost all engine loading situations.

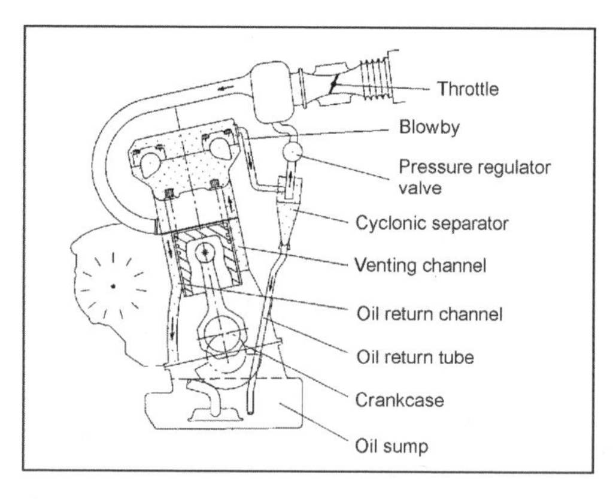

Fig. 7-68 Vacuum-regulated crankcase ventilation.

This system makes it possible to maintain a vacuum inside the crankcase across the entire engine map. This concept, when compared with the conventional crankcase ventilation system, requires fewer components (hoses, hose clamps), and the hazard of icing inside the hoses is reduced. Introducing blowby gases into the intake system downstream from the throttle largely precludes any accumulation of grime at the air flow sensor and at the idle speed stabilizer.

7.8 Cylinder Head

Great importance is attached to cylinder head design and engineering during engine development. The cylinder head determines, like no other subassembly in the engine, operating properties such as performance level, torque, exhaust emissions, fuel consumption, and acoustic properties.

The section that follows provides insights into development work and on current cylinder head design. The key issues dealt with during cylinder head development and in the manufacturing processes are discussed in sequence below. Because of the scope of the material involved, the discussion is limited to passenger car engines; two-cycle engines will not be included.

7.8.1 Basic Design for the Cylinder Head

The engineering designs for the cylinder head have been continuously developed and refined over the past 100 years of engine history. Even today new developments require decisions on what shape and which cylinder head components should be used in the new design. Current technologies such as variable valve actuation or direct-injection combustion concepts in gasoline and diesel engines take a prominent place in the discussion accompanying the new development of any engine. Not every company in the automotive industry follows the same path, because of differing requirements and the "signposts" set accordingly within the firms. As was the case about 100 years ago, this is the reason for employing a variety of designs in passenger car engines.

The cylinder head contains the fundamental elements used for mechanical control of gas exchange and combustion. The valve timing concept is of particular importance here. In the last 20 years it is in this sector, in particular, that the techniques and components used in valve timing have become far more sophisticated. The two-valve engines, in which two valves are used for each combustion chamber, have been largely displaced by the more modern, multivalve engines. Particularly the great increase in volumetric efficiency achieved in recent years demands refined geometries for the charge exchange process. The features inherent to multivalve technology, such as the use of two camshafts, provide greater freedom in engine management. Variable valve timing is used in almost all modern gasoline engines.[1]

7.8.1.1 Layout of the Basic Geometry

A number of technical requirements have to be satisfied when laying out the basic geometry for the cylinder head. At the beginning of new development for a cylinder head it is still possible to influence the individual parameters for a gasoline engine such as valve angle, cylinder head exterior dimensions, location of the gas ports, and the location of the spark plugs. Once the main geometries for these items have been established, the developer's choices with regards to the remaining cylinder head geometries are limited.

Shown in Fig. 7-69 are the factors that influence the shape of the cylinder head. If, at the beginning of a new development project, only the overall engine type has been determined (in-line or V-block), then it is necessary to find a compromise that takes several factors into consideration. These factors are the space available in the engine compartment, installation of the complete engine in that space, other influencing factors (such as the valve train components and their dimensions, the shape of the gas ports), and requirements stemming from manufacturing, such as the technologies available for casting and mechanical machining. A great deal of experience is required to identify the compromise that culminates in improvements in the objectives set for the engine, such as reducing consumption and exhaust emissions.

Not all the paths taken while developing a new cylinder head lead toward the defined goal. This may be the reason why the engines produced in the course of a series exhibit differing cylinder head designs. For example, the number of valves per multivalve cylinder may vary between three and five in mass-production gasoline engines.

Traditionally, the two-valve cylinder head is the most economical solution. Its valve train components are limited to a minimum, with just one intake valve and one exhaust valve. The number of moving parts is small, and

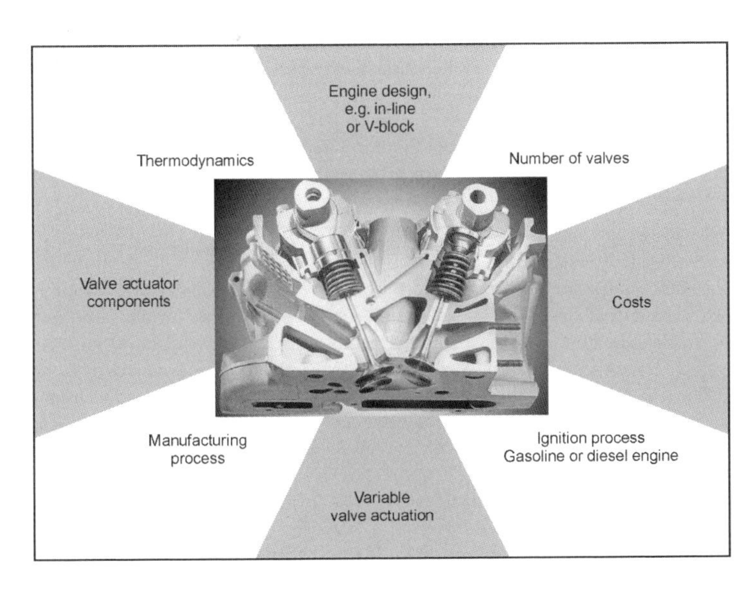

Fig. 7-69 Factors influencing cylinder head design.

its contribution to friction loss is commensurately low. The cylinder head can be kept compact in regard to its outside dimensions. There is great latitude in selecting the shapes for the gas exchange ports. In addition to and because of this design freedom, the component geometries can be better controlled in mass production with respect to the casting models and the shape of the cores. That is why the two-valve engines continue to be widely used in the standard engines, both gasoline and diesel, offered by many car makers.

The exact design of the cylinder head varies widely, in accordance with the general design—inline, boxer, or V-block engine—because of the fact that many engine components are mounted on the cylinder block and the cylinder head. These include the intake manifold, exhaust system, camshaft drive, and vacuum and pressure pumps. Only rarely are engineers successful in using one cylinder head, with all its complexity, for a four-cylinder or for a V-8 engine. As a rule, a unique cylinder head has to be developed for each engine variant. It is thus in the pursuit of controlling costs that an attempt is made to use as many identical parts as possible in assembling the various cylinder heads.

7.8.1.2 Determining the Manufacturing Processes

The casting process used for the cylinder head is established very early in the proceedings. It is advisable, once the casting technique has been selected, to take account of the knowledge and expertise available in the model shop and casting department when laying out the basic cylinder head design. Not all the geometries that the engineer might want can be realized with each and every casting process. The development team often faces a daunting challenge in its attempt to boost product quality in the highly complex cylinder head casting while at the same time realizing the complex geometries in the head. In this scenario it is important to continuously refine the casting processes suitable for producing cylinder heads.

Also taken into consideration early in the development phase is choosing the techniques to be used for machining the cylinder head; this depends in part on the numbers to be produced. Here, new designs, in particular, are subject to severe cost pressures.

7.8.1.3 Layout of the Gas Exchange Components

The shape and location of the intake and exhaust ports and the shape of the combustion chamber determine overall cylinder head geometry in part. Many studies on this are carried out either empirically, through experimentation and trials, or calculated based on 3-D simulations. Flow trials in the ports, carried out using rapid prototyping models, serve to determine flow values. Fabricating single-cylinder engines during preliminary development work makes it possible to respond flexibly to developments at the ports. Depending upon the fuel, either gasoline or diesel, a wide range of basic investigations are conducted prior to defining the geometries. These basic studies are also performed in parallel to cylinder head development. The concept for a diesel should, for example, identify a favorable shape for a swirl-inducing intake port. When exploring a new combustion concept, such as is the case when developing a direct-injection, multivalve diesel engine, it is necessary to test many versions. Only in the course of overall cylinder head development are all the geometries for the components in the cylinder head determined.

7.8.1.4 Variable Valve Control

Implementing variable valve control, as a rule, makes it necessary to develop new cylinder head concepts. Using camshaft shifters in modern gasoline engines requires adaptation work only at the camshaft drive and for the oil management concept in the cylinder head. Fully variable valve control such as that implemented by BMW in its "Valvetronic" system[2] makes it necessary to use cylinder heads developed entirely from the ground up. The components needed to vary valve stroke length are novel, and extensive adaptations have to be made in the cylinder head geometry. The amount of development work associated with this concept is considerable; several cylinder head construction stages have to be tested before the overall concept can go into volume production. The parameter studies required to optimize gas exchange ports, valve diameters, combustion chamber variations, and timing, as well as the control of valve stroke lengths, are very extensive.

7.8.2 Cylinder Head Engineering

The cylinders' bore and spacing determine the basic layout for the cylinder head. As a rule, the number of valves per combustion chamber has already been specified for new engineering. The minimum wall thickness required by manufacturing constraints and the necessary degree of stability narrows the space available for installing valve train components. Since the number of camshafts is specified at the outset of engineering work, it is then necessary to specify the locations and arrangement of the valve train components, taking the geometry of the gas exchange elements such as the ports and combustion chamber into account. Studies then follow to determine how the rough dimensions of the cylinder head change when parameters such as the valve angle, unrestricted valve flow area, or design of the gas exchange ports are modified.

7.8.2.1 Laying out the Rough Dimensions

One way to establish the basic cylinder head geometry is to prepare rough engineering sketches for the valve train components. This is done with CAD support. Parameters can be assigned to the components' individual geometric dimensions while doing so. Varying certain dimensions such as the valve angle, valve spring installation dimension, location of the camshafts, or spark plug length enables a rough evaluation of the overall concept. Depicted in Fig. 7-70 are rough dimensions for a parameter study used in engineering a five-valve cylinder head with pushrods.[3] This cylinder head incorporates three

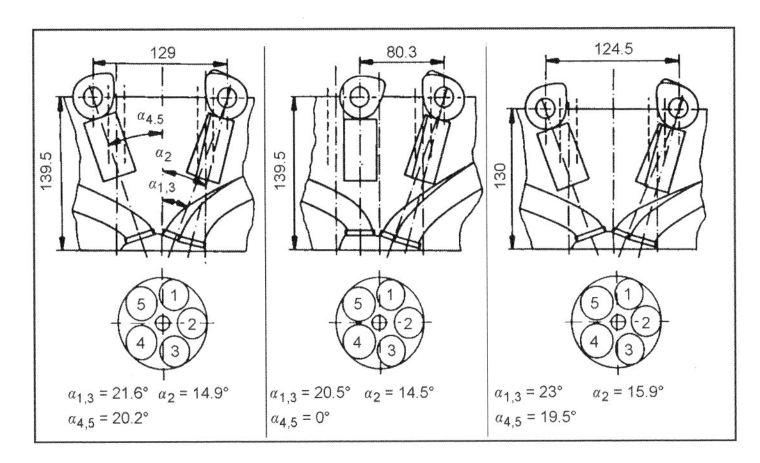

Fig. 7-70 Study on the basic geometric design of a five-valve cylinder head.[3]

intake and two exhaust valves. The spark plug is shown at the center of the combustion chamber. Indicated beneath the cam geometry shown there is the installation space required for the pushrods. The locations of the head bolts, which also require a certain amount of free space for installation, restrict the latitude for varying the valve angle. Accessibility to the head bolts after the head has been completely assembled is mandatory for almost all engines because of manufacturing and maintenance requirements. Illustrated in the figure at the center, for example, is the situation in which, with a vertically suspended exhaust valve having a valve angle of 0°, the head bolts are located outside the camshaft axis for accessibility. In a V-block engine this type of cylinder head design provides more space on the exhaust valve side for the design of the exhaust components. Exhaust routing in the manifolds could be optimized. These studies help in cylinder head development by allowing better evaluation of the overall effects on the engine. Using parametrized assumptions in the CAD system can, particularly in this development phase, make it possible to examine the basic cylinder head geometry with regard to its effects on the engine as a whole. Concept comparisons between pushrod and cam follower designs can also be carried out very well in this way.

One criterion for selecting the valve angle and the location and size of the valves is the determination of the unrestricted flow area around the valve disk. This is the unrestricted area available for gas exchange, as a function of the valve stroke, as described by Dong.[4] To influence engine breathing, an attempt is made, in coordination with the remaining potential geometric configurations for valve train components and gas exchange runners, to make this area as large as possible. Structural requirements and values resulting from experience—such as the width of webs between the runners—have to be maintained. In basic examinations of the geometric layout to preassess the situation regarding valve angle geometry, it is possible to compare variations with one another both quickly and simply.[5] Concept studies using various numbers of valves can be carried out quickly and easily. To ensure that these studies can be completed quickly, in the early phase of cylinder head conceptualization, simple PC programs should be used, as is mentioned in Refs. [3, 5]. Depicted in Fig. 7-71 are examples of the parameters that are pertinent to the basic design of a six-valve cylinder head. Minimum web widths between the valves have to be maintained for both cooling and cylinder head strength. One objective here is to incorporate the largest possible valve diameters. The results of such examinations are geometric magnitudes such as the utilization of available surface areas. This term is understood to be the quotient of the total intake or exhaust surface to the surface area for the cylinder bore. The results vary in dependency on the cylinder bore; when interpreted, this information gives differing numbers of valves. This phase of cylinder head development is particularly exciting since specifying the number of valves at a predetermined cylinder bore has decisive impact on cylinder head design.

7.8.2.2 Combustion Chamber and Port Design

The geometry of the combustion chamber is of major significance in cylinder head engineering. Technical calculations for this purpose are carried out simultaneously during the early development phase, which is why, before finalizing the concept, the geometries to be developed for the combustion chamber variants are determined. In coordination with the portion of the combustion chamber volume accounted for by the bowl at the top of the piston, extensive basic examinations are performed. Concepts such as charge stratification in direct-injection gasoline engines are assessed in conjunction with the port and combustion chamber geometries and are tested on real-world models. Three examples for the development of a two-valve concept with various combustion chamber designs are shown in Fig. 7-72. The rough geometry of the combustion chamber is also determined by the variation of the valve angle. In this example, in the interest of better comparison, the same cam follower design was used for all three embodiments. Among the matters examined was the extent to which the charge volume available burned most favorably. The overall influence becomes

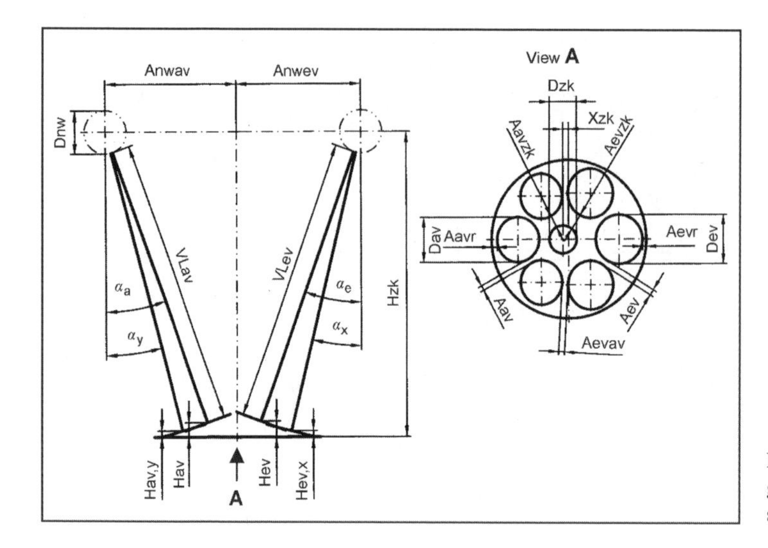

Fig. 7-71 Study to establish geometry for the valve cross section.[5]

clear in specific consumption values, the amenability to leaning out the mix, and, in particular, the raw NO_x and hydrocarbon emissions in the exhaust gases. The situation shown in the right-hand depiction proved to be advantageous. The spark plug, extending well into the combustion chamber, is arranged so that it is fully surrounded by the mix drawn into the combustion chamber. In the design selected here, about 70% of the combustion chamber volume is inside the cylinder head and 30% is in the piston. The interdependencies described here between combustion chamber geometry and the effects on the engine are to be found again in direct-injection gasoline engines currently under development where fully variable valve control is used. The development effort required there is considerable. The parameter studies to be defined for combustion chamber trials demand a great deal of experience and development discipline by the thermodynamics engineers.

Four-valve cylinder heads with the spark plug at the center offer the fundamental advantage of short combustion paths in the combustion chamber. Because of the valve head's large share of the total surface that defines the combustion chamber, the casting contour has only a slight influence on the volumetric tolerances, which can be kept very narrow at, in one example, 0.5 cm^3. To reduce thermodynamic losses during combustion, one strives to achieve the lowest possible ratio of combustion chamber surface area to combustion chamber volume. One key thrust in development is optimizing the geometry of the squish surface. Here the location varies in relationship to

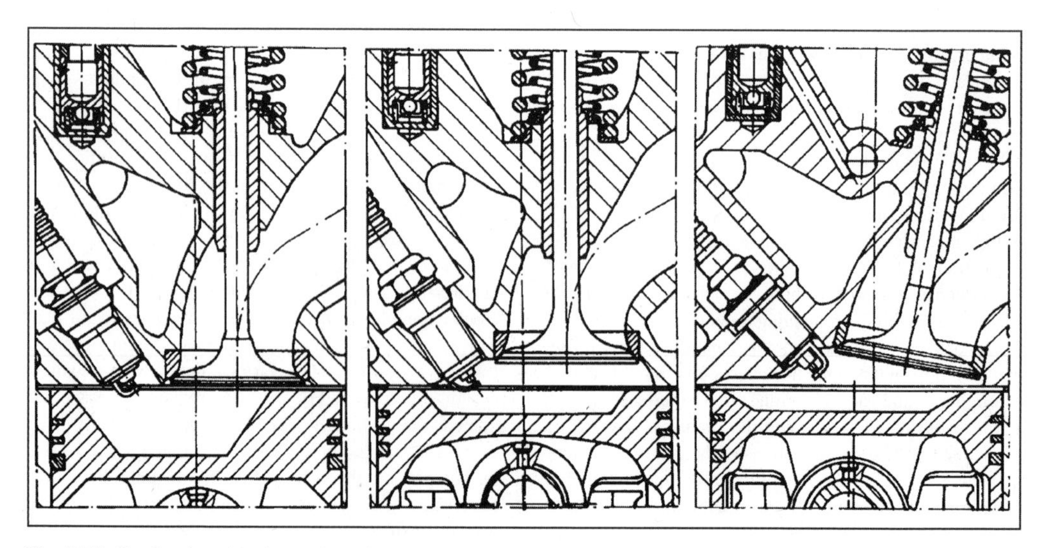

Fig. 7-72 Combustion chamber variants for a two-valve cylinder head.

the valves, shape, and size. An excessive share of squish area has proved to be detrimental because of the increase of the surface-to-volume ratio and the associated heat losses. Using the example of the four-cylinder engine shown here, a squish surface share of 7% proved to be favorable. In modern, four-cylinder engines with external fuel mix blending, the trend is toward flat piston heads with the bulk of the combustion chamber located inside the cylinder head.

With the development of new ignition concepts such as direct-injection gasoline and diesel engines, the development of the ports has become a science in itself. Attaining specific, reproducible charge flow is the subject of many basic research projects that are taking place parallel to overall cylinder head development. The design of the port has to be seen in conjunction with the designs for the intake and exhaust manifolds. This topic is dealt with primarily through trials and flow simulations. Here the engineer pays attention to finalizing these geometries early in the work since changes at the port can often trigger major changes in the cylinder head. Often so many thermodynamic interactions occur while defining the geometries for ports and combustion chambers that it is difficult to estimate how much time engineering will take. Potential port arrangements for a direct-injection diesel engine are shown in Fig. 7-73.[6] In diesel engines, swirl is imposed on the incoming air in order to intensify blending of the fuel and air mixture. There are two basic options for intake port design that may be drawn upon here:

- Helical (swirl or helical port)
- Sloped port configuration (tangential port)

In selecting the shape for the port, one pursues the objective of achieving the required swirl characteristic and the best possible flow throughput. This effect is to be preserved to mass production. In the swirl port shape, the port imparts the swirling motion on the incoming air. This results in smaller swirl deviation at relatively less favorable flow throughput values. In the tangential design, in contrast to the above, the incoming air is set in rotation by the cylinder wall, because of the port's off-center location. Typical here are high throughputs at good cylinder fill. Combining a swirl chamber with a downstream tangential port is thus a very good compromise in the conflict of goals between throughput and swirl stability.

The "helical port design, oriented vertically from the top to the combustion chamber," as shown in Fig. 7-74, improves port quality when compared with an arrangement at the side. Additionally, the glow plugs can be situated on the colder side of the cylinder head, where the thermal load is less. The short run for the exhaust port inside the cylinder head keeps heating to a minimum.[6] The port configuration described here makes a symmetrical valve arrangement with beneficial effects on the location of the valve train possible, Fig. 7-74.

7.8.2.3 Valve Train Design

The discussion as to which valve train concept is the best for a particular engine is one that we will not go into here. The engines' requirements profile—which depend on its use—results in differing engineering strategies and, in turn, in differing valve train concepts. One does observe, however, a trend toward roller-actuated cam followers or rocker arms. These designs have the lowest friction valves for the individual valve trains. But these solutions, in comparison with sliding cam follower concepts, are heavier; consequently, they are not used in sports car engines, for example. The goal here is to keep the masses in motion as small as possible and to minimize elasticity, which is why concepts using mechanical valve play adjustment are used in such engines.

The design of the valve train takes high priority in cylinder head development. In new developments pushrod concepts have proven their superiority to cam follower concepts. The installation situations for the valves are different. Different valve guide lengths have been worked out through time for pushrod and cam follower heads. Cam follower timing requires a better, and thus longer, valve stem guide than a pushrod concept since the pushrod itself has a guide. The valve length, in turn, results from the installation length required for the valve spring. During new developments, these mutual interdependencies

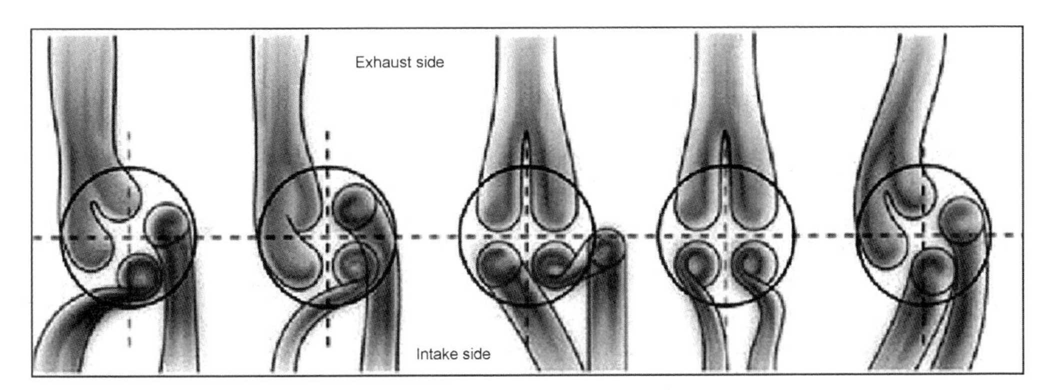

Fig. 7-73 Intake and exhaust valve variations for a four-valve diesel engine.[6]

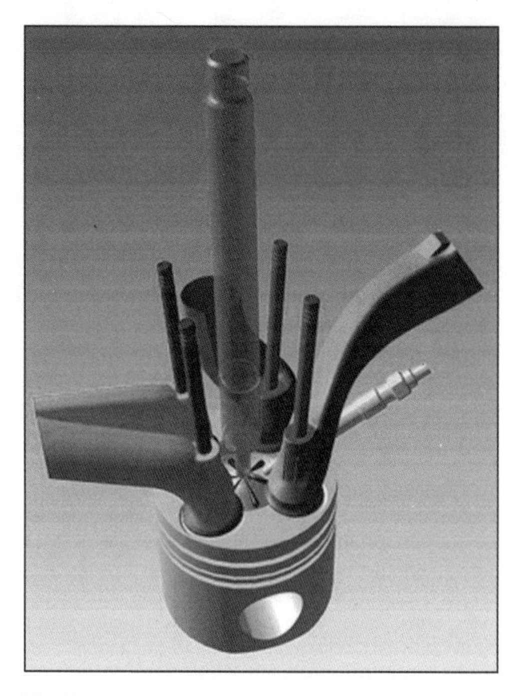

Fig. 7-74 Arrangement of the gas exchange ports in the cylinder head.[6]

result in increased employment of simulation techniques during the predevelopment phase in order to keep the number of prototypes required for testing as small as possible.

Taking the refinement of the valve train for a BMW six-cylinder engine as an example, Fig. 7-75 shows the development steps undertaken over several model years in efforts to reduce the weight of the valve train.

To keep the valves from lifting at high engine speeds, the valve spring has to be built for a minimum force of F_1, and the shape of the cam lobe has to be selected to suit. The required spring force and the associated spring geometry determine the minimum installation space for the springs. To limit spring force F_2 at maximum valve stroke, the primary thrust in valve train engineering is to keep down the masses that act on the valve.

7.8.2.4 Cooling Concepts

In discussing cooling for the cylinder head, differentiation is made where water cooling is used, among cross-flow cooling, longitudinal flow cooling, and a combination of these two types. In cross-flow cooling the coolant flows from the hot exhaust valve side to the intake valve side; in longitudinal flow cooling the coolant flows parallel to the long axis of the cylinder head. The objective in cooling is to equalize temperature distribution within any cylinder head segment at a low level and to create uniform cooling conditions for all the cylinder segments. Moreover, the top of the combustion chamber and the valve webs are to be generously supplied while at the same time keeping pressure loss throughout the cylinder head flow pattern as small as possible. The coolant passes from the engine block, through several transfer ports and the head gasket, into the lower face of the cylinder head. The shape, location, and size of these transfer ports have to be harmonized appropriately. The coolant flow calculations described in Section 7.8.2.9 represent the state of the art. Only by simulation can problem areas such as the webs between the exhaust ports or the area around the spark plugs be engineered for complete reliability.

7.8.2.5 Lubricating Oil Management

Motor oil under pressure, used to lubricate the cylinder head, is generally delivered by the oil pump in the engine block, through transfer ports in the head gasket. The oil passes through lateral bores or special supplementary lines to the points served, such as the camshaft bearings, hydraulic valve lifters, hydraulic valve play compensating elements, camshaft shifters, or oil spray nozzles above the cams. The pressurized motor oil supply for the cylinder head is managed by the cross sections of the supply tubing and specially provided choke points to keep the oil volume to an absolute minimum. To keep the hydraulic valve play adjusters and the camshaft shifters from running dry, check valves are provided in the lines supplying these elements. Multivalve cylinder heads, because of their greater number of lubrication points, are more difficult to coordinate and involve greater oil requirements. A more powerful oil pump is often required where camshaft shifters are employed. In spite of this, it has been possible in recent years to keep the total oil volume, even

	Model year 1990	Model year 1993	Model year 1995
Lifter	35 mm diam, 80 g	35 mm diam, 65 g	33 mm diam, 48 g
Spring cup	15 g	11.1 g	7.9 g
Valve keeper	1.5 g	1 g	1.0 g
Spring	Cylindrical Double spring 69 g x ½ = 34.5 g	Cylindrical Single spring 51 g x ½ = 25.5 g	Conical Single spring 40.5 g x 1/3 = 13.5 g
Valve stem	7 mm diam, 58 g	6 mm diam, 46 g	6 mm diam, 46 g
Total	**189.5 g**	**148.6 g**	**116.4 g**

Basis: 2.5-liter engine, intake valve side

Fig. 7-75 Development steps for reducing weight in valve train components.[7]

in multivalve engines, within reasonable limits. This goal was met with higher precision in machining to minimize play, through more precise tuning of the oil circuit and through technical calculations.

The oil flows back to the sump through return bores of appropriate size, located between the cylinder head and the engine block. These returns are situated at the lowest possible point, which depends in part on the engine's attitude when mounted in the engine compartment. The rotation of the camshafts in some cases slings the oil so severely that it foams. Accordingly, sufficient cross sections are also provided in the area below the camshafts to ensure draining toward the engine block. Particularly in boxer or V-block engines, it is necessary, because of the installation attitude for the cylinder head, to engineer the design to ensure sufficiently large drain cross sections.

7.8.2.6 Engineering Design Details

The cylinder head bolts are normally bolts with collars. Here the collar, because of the surface pressure to be transferred between the bolt contact surface area and the cylinder head, is broader than the bolt head itself. In monolithic cylinder heads, this can impose limitations on the camshaft arrangement. The diameter of the tool used to tighten the bolts or the outside diameters of the bolts themselves thus determines the locations of the camshafts if the latter are to remain in place inside the cylinder head while the cylinders are being installed. In some cases the cylinder heads are made in two or more sections, and the valve timing elements are borne by one or two separate cast components. In this case the design of the lower cylinder head section is simpler, as is the casting technology. Because of cost considerations monolithic cylinder heads are used in the majority of all passenger car engines.

Depending on the combustion process, appropriate space must be provided in the cylinder head to accommodate spark plugs, glow plugs, or injection nozzles and the diameters of the tools used to install and remove them. Wherever possible, spark plugs should be selected that use commonplace thread diameters and wrench sizes. In diesel engines or direct-injection gasoline engines, the arrangement of the cylinder head components is tight, particularly where a multivalve concept is used. It is for these reasons that the number of valves per combustion chamber is limited to four. The space required for these components can be modeled by assigning parameters using 3-D CAD when defining the basic layout for the cylinder head. This makes it easy to depict potential geometric arrangements. The wall thicknesses required around these components in the rough cylinder head casting reduce the overall installation space for the valve assembly or the camshafts. The cross sections required for cooling are also limited in the same fashion.

Modern multivalve engines incorporate camshaft shifters in the cylinder head. The systems in mass production are all located on the camshaft drive and are driven by the crankshaft by either a timing belt or a timing chain. Suitable oil supply lines have to be built into the cylinder head to serve the shifter. This is simpler when a cylinder head is developed from scratch. The space required by the shifter is not particularly great for the vane-type system normally used today. With these shifters the shift angle for the camshafts can be rotated steplessly in relation to the crankshaft.[1]

The diameters of the camshaft drive gears determine the minimum clearance between the camshafts. Particularly where the camshafts are driven directly by the crankshaft, this distance has great influence on the cylinder head design. Often, and in multivalve engines, too, the camshafts are driven by intermediate gearing. When using camshaft shifters, however, drive directly at the end of the cylinder head is the most economical. In this camshaft drive concept the clearance between the camshafts is of appropriate size or an intermediate gear is used between the crankshaft and the camshaft. The most widely used arrangement is with the camshaft drive at the forward end of the engine, i.e., at the end opposite the clutch. Drives centered between the cylinders are seldom used in passenger car engines, while they are being seen more frequently in motorcycle engines. Drives at the clutch end of the engine are also unusual.

7.8.2.7 Engineering in Construction Steps

It is impossible to predict all the influences that will be encountered while engineering the cylinder head, particularly when new combustion processes are being developed. Computer assistance in the basic design or the calculation processes used in simulation technologies do, indeed, help to generate a great deal of information in advance. The mutually influencing factors on cylinder head development are very complex, however, so that there is much to be said for using several construction stages in cylinder head development. Moreover, testing the engine's thermodynamic and mechanical properties delivers many findings that also cannot be predicted in advance (Fig. 7-76).

When developing entirely new cylinder head concepts, it may make good sense to obtain cylinder heads as prototypes—quickly and economically—for use in preliminary development. When building these prototypes, it is often advisable to use manufacturing techniques that differ from those used in mass production. Thus, small numbers of cylinder heads to be used as prototypes may be built in a low-pressure sand-casting process. Figure 7-77 shows an example of the flowchart for fabricating cylinder head prototypes by this procedure. Smaller companies have specialized in this particular field and are able to deliver initial prototypes as quickly and economically as possible.

To reduce overall cylinder head development time, the goals to be met in any given construction stage must be defined exactly. The project management work required here is of vital importance. As a rule, the development of the second construction step is commenced while the first construction step is still being tested. Here the manufacturing processes foreseen for mass-production use should

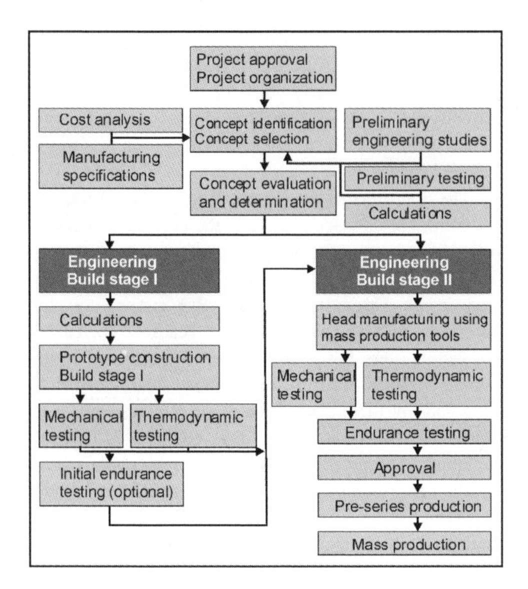

Fig. 7-76 Example of development steps for cylinder heads in two construction steps.

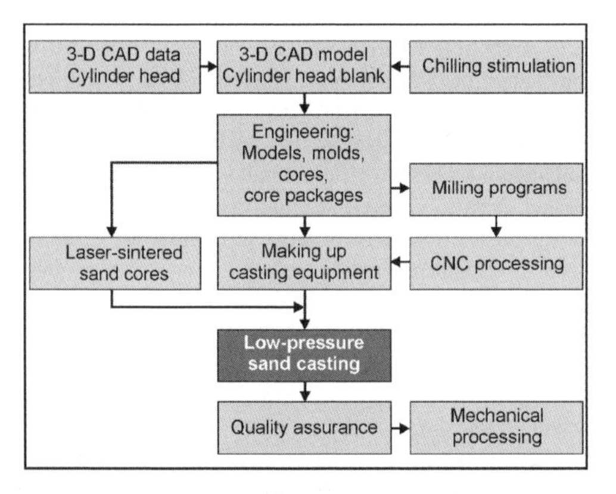

Fig. 7-77 Example of a flowchart for making up cylinder head prototypes, after Becker.[8]

be employed. Particularly, the rough cylinder head casting should be made up using the casting process selected for volume production.

The development of a cylinder head to readiness for mass production in a single step is possible for designs based on existing heads and exhibiting only minor modifications.

7.8.2.8 Using CAD in Engineering

Because of the multiple uses of CAD data, cylinder heads are modeled in the CAD systems in complete, three-dimensional renderings. The specifications for the model and the casting equipment can be derived from this data. The geometries can also be used for simulation calculations. When engineering a new cylinder head, interdependencies among its components can be parametrized; refer also to Section 8.5.2.1. This makes it possible to carry out basic studies simply and quickly. Model builders and casting specialists should be consulted continuously during detailed engineering work, beginning as soon as the rough cylinder head concept has been finalized with the definition of the internal components and the major dimensions. In this way, manufacturing considerations are accounted for in the process early. Engineering methods vary, depending on which CAD system is used. It makes sense, for example, to limit parametrization of the cylinder head to a few parameters to maintain flexibility when changes are made to the model. All the engineers involved in the project should use identical software with identical default settings. Because of the complexity of the CAD methods, one person on the development team should be responsible for adherence to the methods. Since the cylinder head involves many interfaces to adjacent components, transfer conditions to these components have to be defined.

The consistency of the CAD process provides many advantages. Data become more reproducible, can be used more easily for series of cylinder heads, and largely preclude any inaccuracies between engineering and manufacturing. Cylinder head engineers who prepare the overall concept for a new component need a great deal of practical experience. Today the designs are generated completely using CAD.

7.8.2.9 Computer-Assisted Design

A large number of calculation techniques are used today to dimension cylinder head geometries.[9] With the early employment of calculations—even prior to the concept phase—calculation findings can be utilized in the initial cylinder head prototypes. This makes the steering of subsequent development steps more effective, and in this way the number of components used in testing can be reduced. Ongoing verification of calculations against test results continues to be necessary. Computer support ranges from rough component dimensions and detailed design to optimization and simulation calculations. The target criteria for new engines—improved environmental compatibility, reduced exhaust emissions and fuel consumption, and improved performance, product quality, and ride—can be better satisfied through technical calculations.

Before the first prototypes are fabricated, the calculations are devoted primarily to specifying the valve, combustion chamber, and gas exchange port geometries. To a greater extent 3-D CAD data for the head geometry, once it has been prepared, can be used directly for technical calculations. During the development of a cylinder head in construction stages—understood to be differing cylinder head component development stages—technical calculations start right at the outset of development. During the course of development, the largest share of calculations are performed in the first construction stage. The goal here is to provide support in identifying and defining the concept for the main cylinder head geometries. During testing in subsequent construction steps, technical calculations are used more to lend precision to the concept and to specify details. Calculation activities decline the closer the design gets to mass-production launch.

At this point we mention briefly only a few activities that play a vital part in dimensioning the cylinder heads. Technical calculations contribute to making it possible to interpret, in a more understandable fashion, the complex processes involved in cylinder head development.

The PROMO[10] program is used to calculate the gas charge exchange. Here dynamic gas flows in the intake and exhaust systems of aspirated and turbocharged systems are calculated. The gas exchange components in an engine, with its intake and exhaust systems, are assembled to form a virtual model. Events associated with flow, such as pressure fluctuations or mass flows, can be analyzed at various points in the engine. The program provides information on the characterizing values to be expected for the engine, such as charging efficiency, maximum torque, or power output for a particular engine configuration. The core for calculations is embedded in an interactive graphic user interface from which data record conditioning and result evaluation are undertaken. By establishing the geometry for the ports in the cylinder head, the PROMO program is particularly well suited for initial dimensioning of the gas exchange components in the early concept definition phase, and, in particular, for laying out the timing. In this way it is possible, when developing cylinder head concepts with variable valve control, for example, to minimize the scope of costly trials

In engine development the program also delivers findings on

- Intake manifold dimensions
- Concepts for switching and resonance intake manifolds
- Evaluation of cam lobe contours and timing
- Estimating the potentials of various concepts for variable valve timing
- Evaluating different port shapes
- Exhaust manifold design in regard to length and diameter

In addition to this, three-dimensional flow simulations are conducted to design the intake and exhaust ports and the combustion chambers in the cylinder head and pistons. The charge motions are simulated on the basis of the CAD description of the port and the combustion chamber surfaces. The calculations provide insights into the flow situation in the intake and exhaust ports as well as for the charge as it flows into the cylinder. Solving the equations makes it possible to simulate the complex flow processes for static situations and for those which change through time. When dealing with transient calculations (i.e., for states that change through time) the calculation network to be prepared is modified at each timing phase in accordance with momentary valve and piston positions. The results of the simulation—which include pressures, velocities, turbulence, and blending values—have to be assessed with an eye toward perfect combustion. Shown in Fig. 7-78 as an example of calculation results is that for an intake valve at center stroke position; reproduced there is the velocity distribution for the charge as it flows into the cylinder (here 90° after intake TDC). The three-dimensional flow simulation is helpful, particularly when developing new combustion processes. Swirl or tumble effects can be better analyzed and further refined in accordance with the findings.

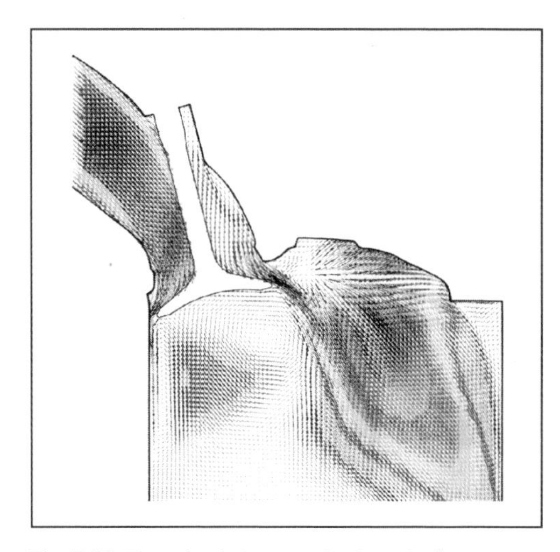

Fig. 7-78 Flow simulation at an intake valve.[9]

The design of the valve lifting lobes and the simulation of the valve train dynamics take a prime position in cylinder head development. The findings here have a direct impact on cylinder head design. Geometries such as the pushrod diameter, valve length, valve stem diameter, valve spring dimensions, and cam follower geometry are determined by these calculations. Imaging of the entire valve train in mechanical models also makes it possible to determine precisely the dynamic properties. The findings are reflected in the camshaft geometry and/or valve drive components.[11]

A major contribution to designing the cylinder head coolant cavities is made by the three-dimensional flow simulation for the complete coolant circuit.[9] This method is integrated into a larger calculation scheme that optimizes the entire cooling system, including the design of the water pump and the radiator. The geometry of the cylinder block and head cavities through which coolant flows is modeled and then compiled in a calculation matrix. Figure 7-79 shows the section of the water jacket as an example of coolant flow simulation in a five-valve cylinder head with cross-flow cooling. The cylinder head receives the coolant through transfer bores in the cylinder head gasket. Their graduated diameters ensure nearly identical distribution of coolant to the various cylinders. About two-thirds of the coolant passes into the cylinder head at the exhaust valve side. The coolant flow passes across the top of the combustion chamber and past the exhaust ports to the spark plug well. Behind the spark

Fig. 7-79 Section of the water jacket for coolant flow simulation.[9] *(See color section.)*

plug the flow continues along a center coolant collector channel that runs longitudinally through the cylinder head. Presented in Fig. 7-79, as an example of the results of a simulation calculation, is the depiction of the convective heat transfer coefficients in the area around the exhaust port, which is subjected to severe thermal loading. The dark areas correspond to a high thermal transfer coefficient, a result that is achieved by the optimized position and selection of the diameters for the transfer bores in the head gasket. By optimizing the cylinder head cooling pattern, with the support of simulation calculations, the temperature level at all the cylinders can be kept constant, with only minor deviations. This method makes a contribution to cylinder head development, which could be achieved using conventional techniques only with extensive effort in trials and testing.

Strength calculations represent a major area where technical calculations are used in engine development to determine the dimensioning and geometries of cylinder heads and their components. In order to make cylinder heads as light as possible and nonetheless sufficiently stiff, finite element calculations are carried out for the entire cylinder head.[9,12] The structural strength of the camshafts and their bearings can, for example, be examined for the design and position of the camshaft bearings. Wall thicknesses can be minimized by using strength analysis. Stiffening ribs are provided to increase structural strength. Thus, designs with a favorable effect on force flow can be predetermined in detail. A section from the FEM model for a complete cylinder head is shown in Fig. 7-80.[12] The loading magnitudes for the calculation are the spring and mass forces of the valve train, the belt and chain forces at the end of the camshaft, and the forces applied by the cylinder head bolts. Shown in Fig. 7-80, after Mises, are the comparative stresses at the deformed cylinder head when subjected to thermal loading at nominal power.

Because of extreme demands for reliability and smooth running at the valve train, the design of the lobe

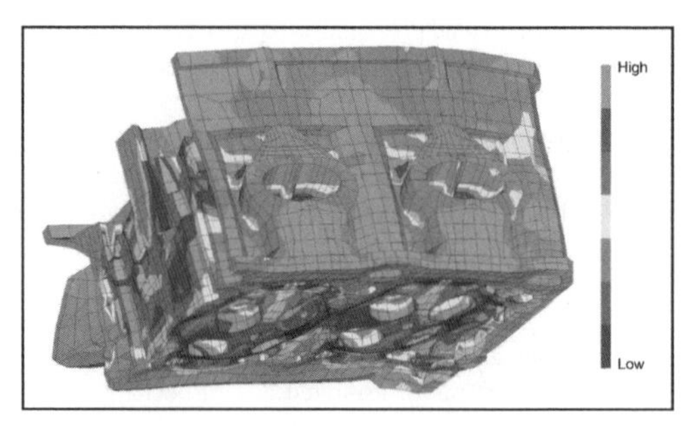

Fig. 7-80 Strength analysis at the cylinder head.[12] *(See color section.)*

contour assumes great importance. In addition to the purely kinematic design of cam contours, various computer programs are used to ensure good dynamic behavior in the valve train. To conduct the simulation calculations, the valve train structure is expressed as a multibody oscillating system with adjustable coupling conditions for friction, stiffness, damping, and degrees of freedom in movement. The dynamic simulation for the entire valve train is obtained by calculating the design of individual valve systems to better evaluate the interactions of individual components with one another. The valve train is actuated by the lobe contour. Stiffness is determined on the basis of measurements made at the actual components or by using FEM calculations. The damping values are primarily experience values that are determined by comparing calculations and measurements. The valve spring, as the main vibrating or oscillating element, is broken down into many oscillating subsystems. One goal in dynamic calculations is demonstrating rotation speed strength for the valve springs at the smallest possible valve spring forces, in order to keep overall valve train friction as low as possible. Simulation calculations make it possible to estimate even at a very early development stage the interactions among individual components. Closely defined changes in component properties make it possible to influence the overall structure of the cylinder head and its components in such a way that the components' own shape properties are manageable within the valve train's excitation spectrum. Suitable tuning for the actuation itself, which is determined primarily by the cam contour, can also bring about a marked reduction in the dynamic effects at the valve train.

Oil circuit calculations can be conducted to fine-tune oil management in the cylinder head.[9] Calculations for subsystems, such as for oil management at the cylinder head, make it possible, by simulating the entire motor oil supply system, to minimize the amount of oil required. This, in turn, keeps the amount of power consumed by the oil pump as low as possible. To do this, all the components in the engine in which oil is found are modeled in a virtual hydraulic system. The objective is to optimize by simulating the oil using points in the cylinder head, such as the push rods, camshaft bearings, camshaft shifter, and oil spray nozzles. The calculation models are further refined, incorporating the results of basic experiments. These preliminary calculations make it possible to predetermine with considerable accuracy the cross sections for the oil passages; this reduces the number of costly trials that would otherwise have to be undertaken using the complete engine.

7.8.3 Casting Process

Cylinder heads for internal combustion engines place considerable demands on the mechanical properties of the materials in a temperature range beyond 150°C. The design latitude for the geometries in the cylinder head is severely limited by the components to be used in the cylinder head. Particularly when developing new cylinder heads for direct-injection diesel engines the complexity in the type and magnitude of the stresses occurring during operation has risen considerably. To satisfy these more exacting requirements, the materials available for use have to be optimized and further developed. Any of a variety of materials may be used for cylinder heads, depending on the requirements profile and the casting process used. In addition to aluminum, cast iron materials are also used for industrial engines and utility vehicle power plants. In passenger car engines aluminum is used almost exclusively, with just a few exceptions. Cylinder heads may be manufactured both from primary alloys—aluminum extracted from ore at the refinery—and from recovered alloys—recycled aluminum following melting and purification; these may be delivered as ingots or as liquids. Aluminum casting alloys are also used for heavily loaded diesel engines with direct injection, but not all the available casting techniques may be used with these cylinder heads.

At ignition pressures exceeding 150 bar it is necessary to use alloys that satisfy the following stringent demands:

- High tensile strength and high creep resistance between room temperature and elevated temperatures of about 250°C
- Great thermal conductivity
- Low porosity
- High ductility and elasticity at great resistance to thermal shock
- Good casting properties at low susceptibility to heat fissuring

The central area of the cylinder head near the combustion chamber and, in particular, all the webs located near the exhaust ports are subjected to severe thermal loading in a range of from 180 to 220°C, this in addition to mechanical loading.[13] The casting technique should be determined as soon as the concept for a new cylinder head is finalized. An early evaluation by the model shop and the casting department helps to avoid errors in the engineering phase. The job of the casting department is to influence cylinder head design to optimize casting for the rough component. The filling and solidification processes in the casting procedure are assessed largely from simulation. These 3-D calculations give the casting department valuable information on problematic areas that might be anticipated right from the conceptualization phase. The geometry of the cylinder head can be modified to accommodate these areas before the first prototype is built. Considerable cost savings can be realized in the development process in this way.

The casting techniques used for engine blocks can also be used for cylinder heads. A brief review of the most commonly used casting techniques is provided below.

7.8.3.1 Sand Casting

Models and core boxes made of hardwood, metal, or plastic are used to replicate the later cylinder head casting inside the sand mold. The casting molds are normally made of quartz sand (either natural or synthetic sand) and binders (synthetic resin, CO_2). The sand cores are formed in core casting machines into which the sand is introduced under pressure; the mix of sand and resin is compacted to create the core by applying heat. It is advisable to use the laser sintering processes when making sand cores in the prototype phase. Combining individual cores to form a core package and assembling this core package and the outer casting mold are handled mechanically and fully automatically, even when producing only limited numbers of castings. Parting for the model, core, and mold in various planes and inserting cores in the casting mold make it possible to produce complex cast components with undercut areas. During the casting process, the hollow cavities between the outside mold and the cores are filled with molten metal. Following the filling process and after the metal has solidified, the casting is removed from the sand mold. The sand mold is destroyed here (which is why this is referred to as a "lost mold" process). Following casting the rough part is cleaned, and the gate and risers are separated. In mass production operations, these steps are fully automated. Sand-cast components made of Al-Si alloys permit double heat treatment. The first heat treatment phase is found in controlled cooling of the casting while still inside the sand mold. The second heat treatment takes place during time- and temperature-controlled exposure of the casting to heat in a kiln. These heat treatments increase the strength of the cast component and relieve inherent stresses created during the cooling process. The geometry of the components may include undercuts since the lost mold is used for only a single casting.

One advantage of sand casting is that the fabrication equipment can be set up quickly and economically when making small numbers of units. Cylinder heads for special types of engines, such as sports car engines, can be quickly realized; implementing changes during development is relatively simple and economical since plastic positives are used.

The low-pressure sand-casting process is suitable for prototypes and short production runs. Here the melt is introduced from below, through a riser, and into the sand mold; pressure at about 0.1 to 0.5 bar is applied to the molten metal (Fig. 7-81). This pressure is maintained

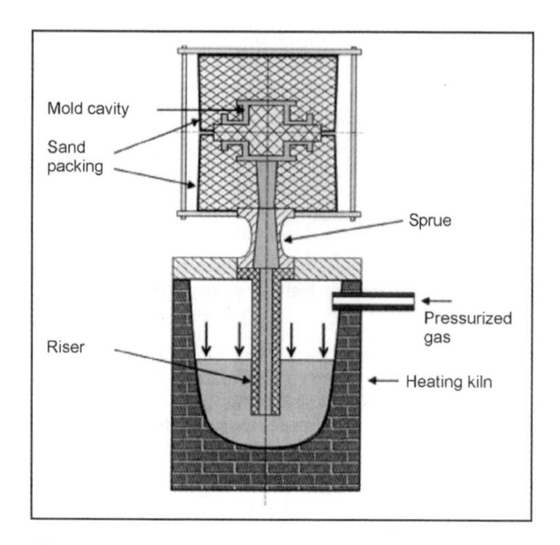

Fig. 7-81 Low-pressure sand-casting process.

during casting. Since solidification under pressure is almost directional, the structures in the cylinder heads are very fine.

The Cosworth low-pressure sand-casting process is used for cylinder heads, too, because of great dimensional accuracy and strength, a compact structure, and freedom from pores. In accordance with the specifications for the process, an aluminum alloy, in the form of assayed ingots, is melted in a resistance electric furnace under a blanket of inert gas, Fig. 7-82. The melts are buffered in a generously dimensioned holding kiln, once again blanketed with inert gas. Casting is affected with an electromagnetic pump that moves the molten aluminum upward to the sand mold, where it flows from below into the mold cavity. Just as in the low-pressure die-casting process, the pressure on the molten metal is maintained during solidification. Programmable regulation of pump output makes it possible to set a delivery rate suitable to the particular shape of the cavity. Casting can be automated to a great extent; the finished molds are moved one after another to the casting station, above the electromagnetic pump.

The core package process has been used to manufacture cylinder heads for about 20 years now. In this sand-casting process a closed sand core package is assembled from several individual sand cores. Adhesives are normally used to hold these together, but screws may also be used. Core packages are used for cores of complex design, which cannot be made up in a single piece. In its original embodiment the core package process, based on the low-pressure die-casting principle using an electromagnetic pump, was limited to short production runs for cylinder heads because of its low productivity. The latest approaches also point out perspectives for using this process in mass production once manufacturing facilities have been modified appropriately. The cast components do not fall below a temperature of about 500°C after casting through complete removal of the sand. Thus, they are cast virtually free of strains, giving the parts superior dimensional accuracy. Since each part is cast in a new, cold mold, practically no dimensional deviations are found such as those that occur in die casting, where the permanent molds are subject to wear.

7.8.3.2 Die Casting

About 90% of the cylinder heads made in Europe are manufactured by the die-casting process. The dies are permanent metal molds made of gray cast iron or hot-work tool steels and used to manufacture cast parts from lightweight alloys. Just as in sand casting, the sand cores are positioned inside the casting mold. Die casting can be subdivided into the gravity and low-pressure processes.

In gravity casting the mold is filled solely with the force of gravity acting on the molten metal and at atmospheric pressure. The casting process is used in partially or fully automated casting systems. In this casting process,

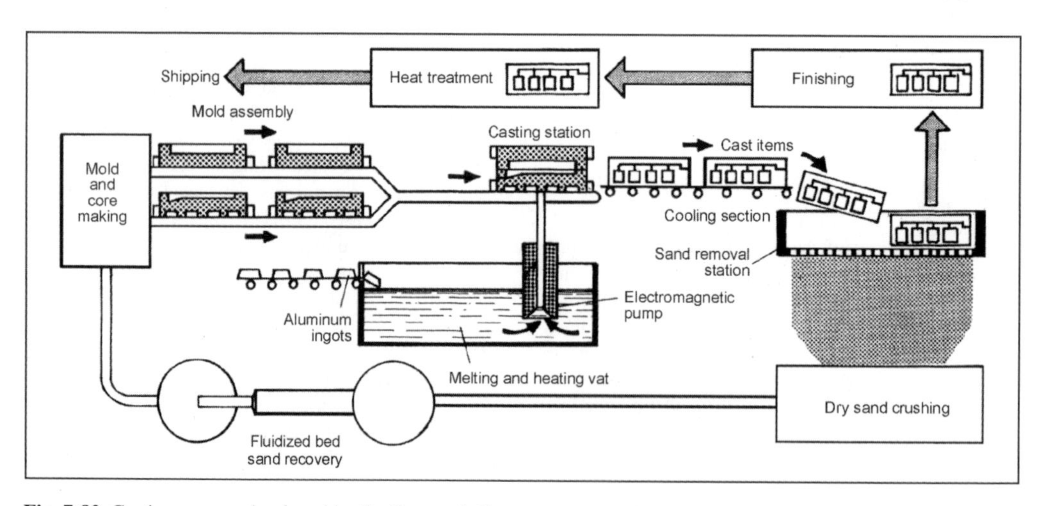

Fig. 7-82 Casting process developed by the Cosworth Company.

in contrast to sand casting, the dies can be used many times. It is necessary only to make new sand cores for each casting cycle, which is referred to as lost-core casting. Because of the use of sand cores, die casting, like sand casting, offers the advantage of greater freedom in the engineering design. Undercut areas are possible, in contrast to pressure die casting. Using steel as the die promotes fast and directional solidification of the molten metal; this is not the case in sand casting. The die is protected against the lightweight metal melt by applying a parting agent, also referred to as a refractory coating. In comparison with sand casting, the die cast components exhibit a finer internal structure, greater strength, improved dimensional accuracy, and better surface quality. Both die castings and sand castings can be further processed with double heat treatment. In addition to the advantages of carefully defined control of cooling inside the die, which is the first heat treatment, additional heat treatment is often implemented. As opposed to sand casting, there may be no undercuts in the permanent molds since they are used over and over.

Most of the cylinder heads at the VW Corporation, for instance, are manufactured using this process. The combustion chamber side of the cylinder head is cooled by inserting one steel die per cylinder. The sprue is at the upper side of the cylinder head, and the molten metal fills the mold as it flows downward from this point. The area around the combustion chamber cools faster because of the cooled combustion chamber dies, and this increases strength in that specific area. The casting process takes place on a turntable system with several stations; this reduces mass-production manufacturing costs to a minimum. The standard alloy used for this purpose is G-AlSi7MgCu0.5. Smaller runs are outsourced to suppliers. Similar processes are used there, and in some cases the cylinder heads are cast from below using special runners. The results are comparable in terms of the quality found in the final product.

A large number of cylinder heads is also produced with low-pressure casting, as is the case at the HONSEL Company in Meschede. This is one of the processes the casting department at BMW uses for their diesel engines and a majority of their gasoline engines. In much the same way as described above, the inductance-heated melt is pressed into the mold through a riser at a pressure of about 0.1 to 0.3 bar. The combustion chamber, at the bottom of the mold, is filled from below. Here, too, the combustion chamber plate is cooled with air or water. The cavities for water and lubricating oil and the geometry required for the camshaft timing chain are formed with sand cores. The remainder of the geometric forms in the cylinder head are shaped with dies. Thanks to the low-pressure casting process, the surfaces at the cylinder head are densely compacted. This process is particularly good for diesel cylinder heads that are subjected to heavy loading.

A technique developed by the VAW Mandl&Berger Company is known as the Rotacast process. The entire mold is rotated during the casting process. This process is intended to achieve turbulence-free mold filling. The form is filled from below and, during filling, is rotated through 180° within a period of 15 s. The charge passes into the mold through several, variable openings. Metallurgical studies have revealed that—with this process and the G-AlSi7Mg0.5 containing 0.19% iron—very good and highly reproducible structures are achieved, particularly in the area around the combustion chamber. When using the "LM Rotacast T6" alloy the mechanical properties with the 0.2 offset limit (Rm) in the combustion chamber area, at 272 MPa, are better than for the G-AlSi7MgCu0.5 alloy (gravity casting) at 260 MPa. The exact values depend upon the casting process used and subsequent heat treatments. The Isuzu Company, for example, manufactures cylinder heads using the Rotacast technique.

7.8.3.3 Lost-Foam Process (Full Mold Process)

The full mold (or lost-foam) process is used for mass production in the United States. At BMW's Landshut plant this process was used for the first time in a six-cylinder, inline gasoline engine. The lost-foam process may also be considered a special form of the sand-casting process. The key steps in manufacturing a cylinder head are shown schematically in Fig. 7-83.

First, the polystyrene granulate is warmed, expanded to about 30 times its original volume, dried, and stored. In the first step in the casting process the contours, from which the cylinder head is assembled in various layers, are foamed using the polystyrene material. For dimensional stability the foaming tools are cooled with water. Grippers remove the foam blank, which then cures on a conveyor belt. The sum of the foamed contours represents the exact geometry of the cylinder head, taking into account thermal shrinkage. The individual contours are now joined at two stations with hot-melt adhesive. The positive model of a cylinder head comprises five polystyrene layers glued together in this way. Two cylinder head models

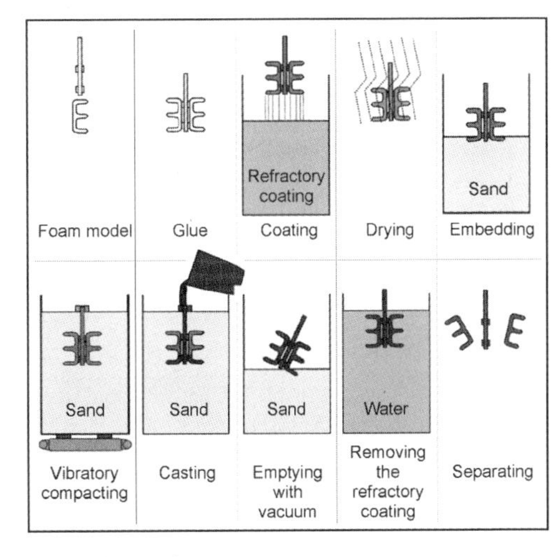

Fig. 7-83 Lost-foam process.

are glued together with the sprue and runners to form a cluster. At the third station, this cluster is immersed in a water-soluble ceramic refractory coating. The unit is rotated to better homogenize the application of the refractory coating. At the fourth station, the cluster is dried in a stream of dry, heated air. This extracts the water to form a dense, gas-permeable refractory coating. In the next step the cluster is inserted into the casting frame, and unbonded quartz sand is filled loosely around it. The sand is compacted at the sixth station by vibration. Casting then follows. A charge of molten aluminum is prepared and poured into the mold automatically using a casting ladle. The polystyrene retracts and gasifies during filling. At the eighth station, the mold is removed, and the sand is taken out of the casting frame. The refractory coating is removed in a water bath, and in the final step the individual cylinder heads are separated from the cluster.

The casting process itself demands familiarity with this technique, too. There is a great range of freedom in the engineering design for the cylinder head. Bores in the cylinder head, down to a minimum wall thickness of 4 mm, can be cast directly with the cylinder head. Changes in the course of the production run can be implemented at the tooling relatively easily and thus at favorable costs, since the tooling is made of aluminum. At a cycle time of four heads in 3 min, this system offers production capacity of about 330 000 cylinder heads per year, Fig. 7-84. In consideration of the high strength requirements for direct-injection diesel engines, this process has not been used to date in mass production for this particular application.

Shown in Fig. 7-84 is the polystyrene casting cluster used by BMW for the first time in Europe to cast a cylinder head in the lost-foam process. The material used here is G-AlSi6Cu4 (aluminum alloy 226). A thermally decoupled, secondary air channel was integrated into the exhaust side for the U.S. versions.

This process makes it possible

- To cast oil channels in virtually any desired shape
- To obtain water cavities with elaborately shaped flow control fins
- To implement curved intake and outlet ports
- To achieve markedly narrower tolerances in the combustion chamber area
- To use only a single foaming die for the duration of the production run
- To reduce significantly the amount of postcasting machining work required for the cylinder head

7.8.3.4 Pressure Die-Casting Process

Permanent molds made of hardened, hot-work tool steels are used in the pressure die-casting process. The plug has to be coated with a parting agent or refractory coating before each casting cycle, also known as a "charge." In contrast to sand casting and die casting, no cores can be inserted into the mold since the lightweight metal melt is introduced into the casting form at high pressure and high

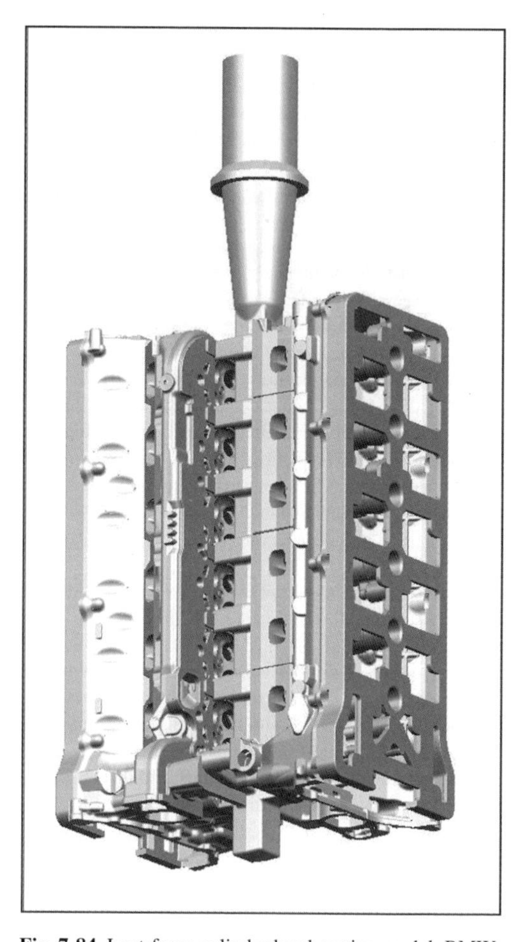

Fig. 7-84 Lost-foam cylinder head casting model, BMW.

speed. The pressure level depends upon the size of the cast component and as a rule ranges from 400 bar to about 1000 bar. As in low-pressure casting, this pressure is maintained during solidification. When casting larger components, the two halves of the casting mold are cooled for directional solidification and quicker cooling of the casting. Once the casting has solidified the mold, comprising fixed and moving components and possibly moving sliders, is opened and the casting is demolded with ejector pins. This process can be used only for air-cooled cylinder heads such as those used with small engines.

In contrast to sand casting and die casting, pressurized die casting provides the most precise reproduction, as well as the greatest precision in the cylinder head geometry. Thin-walled castings with close dimensional tolerances, great exactness of shape, and superb surface quality can be fabricated. Exact casting of eyes, bores, mating surfaces, and other surfaces is often possible without subsequent machining. Pressure die casting, when compared with sand casting, die casting, and low-pressure die casting, offers the highest productivity since almost all the casting and mold movement processes are fully automated. The drawbacks are the limited engineering freedom for the cast component, since undercuts are not possible. Air or gas bubbles that might be trapped in the casting preclude double heat treatment, as for sand casting, die casting, and low-pressure die casting. This process is not suitable for mass production of water-cooled passenger car engines.

7.8.4 Model and Mold Construction

When making the casting models, cores, dies, and all the casting tooling, almost all the parts are generated as models on the basis of 3-D CAD data throughout the CAD/CAM process chain. Thus, the geometry data are more reproducible and the response to change is more flexible. With the creation of the cylinder head design all the CAD models required for model making can be derived, from the CAD rough casting model to the fully machined component. Here a carefully designed data management system is required to maintain transparency so that everyone participating in the project is kept informed of changes and so that changes at the CAD cylinder head component are reflected in all the data records required for model and tool making. The model shop specifies all the traditional details such as mold sectioning, drafts, casting shrinkage, supplements for manufacturing, and any deformations that might be expected in casting; these are taken into account in the CAD model. An early and lively exchange of experience with the cylinder head engineers pays off in the long run. The model building activities vary, depending upon whether designing for prototypes or mass production and upon the choice of casting processes.

The low-pressure sand-casting process used by the Becker Company[8] is superbly suited for small production runs and prototypes. Figure 7-85 shows a core plug (above) and the package used for a water jacket core (below). The rough casting contour plus the allowance for shrinkage (of the metal during solidification) serves as the starting point for forming the model. Here areas of a cast component that align with a given demolding axis are represented as a positive model in the so-called core mold tool. These areas in the cylinder head include, for instance, the crowned head of the combustion chamber, the ends, the intake and exhaust channel sides, the intake and exhaust channels, the camshaft bearing area, and the interior contours for water and oil flow. All the core tools are fitted with sealing surfaces and the so-called core markers that make possible exact alignment and sealing of the cores. These cores are CNC milled in a special plastic resin, in only a few days, on the basis of the 3-D data. In the casting department these cores are filled with sand to which bonding resin has been added; this cures in a short period of time without further treatment. The sand core thus removed from the reusable core mold tool now exhibits the negative contour of the ultimate cast part. A special version here is the so-called sand laser sinter core, which can be made, layer by layer, directly from 3-D CAD data. No core mold tools are required here. Cores for detailed interior contours such as the water jacket or oil-carrying cavities are especially suitable for this process since manufacturing a core mold tool for these cores is both costly and time consuming. Finally, all these core segments (both conventional and sand laser sintered) are assembled to create the core package, and molten material is then poured around it in the low-pressure casting process. A core package can be used for only a single casting.

Fig. 7-85 Core mold tool and package for a water jacket core, by Becker.[8]

A section of the overall core is shown in Fig. 7-86 for a cylinder head made by BMW for an eight-cylinder engine, using the low-pressure casting process. All the cores are made of sand. The core frames required for this purpose are made of steel for mass production work. The spaces between the core segments are filled with molten aluminum. During the development stage, the sand cores are made as rapid prototyping models, used to evaluate the overall geometry. In the lower section of the illustration

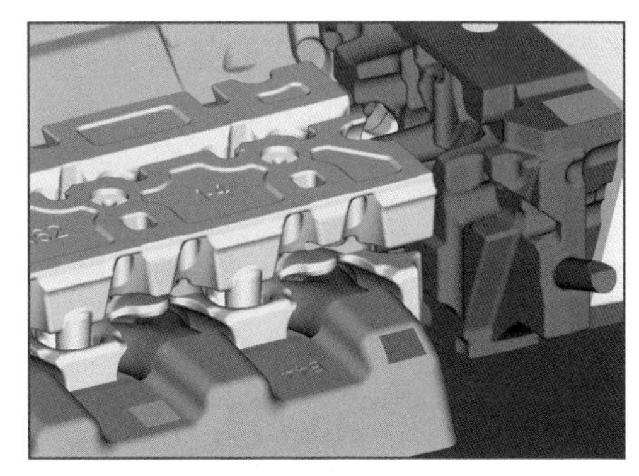

Fig. 7-86 Model of a cylinder head for an eight-cylinder BMW engine.

one sees the combustion chamber plate, shown in dark gray. To the right of that is the core for the timing chain case. In the foreground is the exhaust channel core package, which projects into the water jacket core. Located above this is the core for the oil cavity.

7.8.5 Machining and Quality Assurance

7.8.5.1 Mass-Production Manufacture

Cylinder heads are machined in mass-production operations on transfer lines or at linked machining centers, which make it possible to respond more flexibly to changes. A trend toward machining at sequential machining centers is emerging. Here the rough component passes through several machining stations, one following another. It is necessary for each station to adhere to the prescribed cycle time. To limit the high overall investment costs, as many machining phases as possible are implemented at any given station. When developing a cylinder head, manufacturing planners should be integrated into the project following the tenets of simultaneous engineering in order to take into account the needs of manufacturing at an early date, all with the goal of achieving economical realization. Changes to the cylinder head that have to be implemented retroactively at transfer lines are expensive and time consuming since the entire manufacturing process has to be interrupted. Because of the needs found in mass production, it is often necessary to adopt compromises at cylinder heads that restrict developers' design latitude.

7.8.5.2 Prototype Manufacturing

Machining centers are normally used to work small production runs and prototypes. Often these individual stations are standardized machine tools that can be flexibly programmed and allow changes in the cylinder head to be implemented quickly. The machining costs are higher in comparison with mass production. The combustion chambers are in some cases machined to achieve better uniformity in the combustion processes. It is possible to machine the transitional areas from the gas exchange ports to the combustion chamber and the complete port shapes.

7.8.5.3 Quality Assurance for Cylinder Heads

Failure of a cylinder head in the field often results in complete destruction of the engine. The goal for both the casting and the machining is to achieve a high quality standard for the customer, so the entire cylinder head is tested 100% for leaks. Spot checks by measuring components are standard procedures in quality control. It is imperative to minimize the reject rate in manufacturing. Computer-assisted tomography, known from the field of medicine, can be used to examine cylinder heads and to check the wall thicknesses, slice by slice, for compliance with the specified shapes and dimensions. Such examinations are standard, particularly for thin walls in a range of

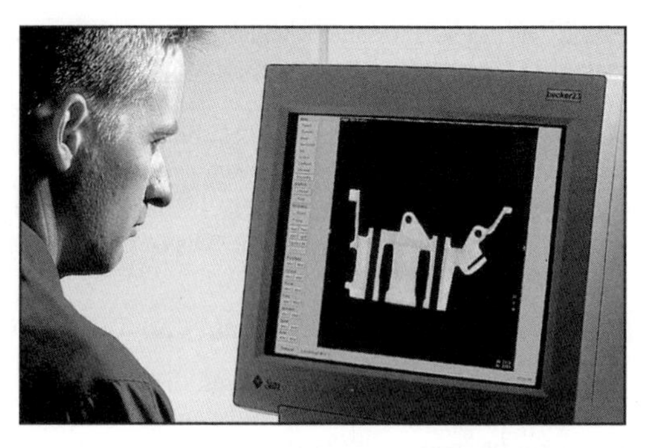

Fig. 7-87 Computer tomograph section of a cylinder head.[8]

about 2.5 mm, as required in racing engines for reducing weight; see Fig. 7-87.

Figure 7-88 shows verification measurements for a cylinder head using a coordinate measurement unit. This makes it possible to measure channel inside geometry, too. The channel surface can be traced point by point to form clusters of individual points. Deviations from the geometry described in the CAD data records can be detected. Using the points transmitted to CAD systems makes it possible to apply reverse engineering methods to establish surface areas based on the cluster of points, which can also be used for three-dimensional flow simulations. These

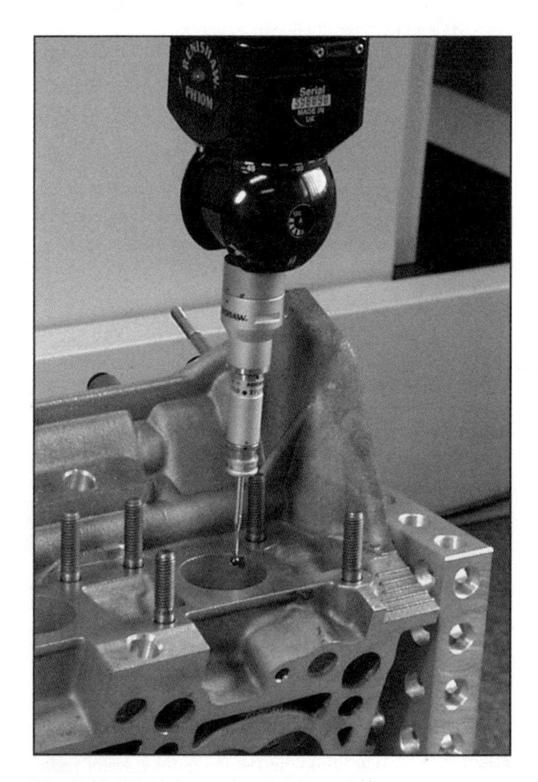

Fig. 7-88 Digitizing an intake channel.[14]

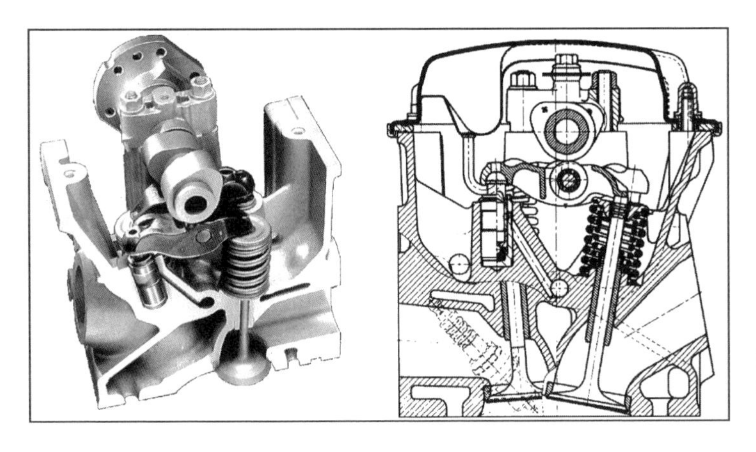

Fig. 7-89 Two-valve cylinder head for the BMW V-12 engine with roller-type cam follower.

techniques are particularly valuable in association with direct-injection engines since here even slight dimensional deviations can have considerable effects on the engine.

7.8.6 Shapes Implemented for Cylinder Heads

7.8.6.1 Cylinder Heads for Gasoline Engines

Four-cycle engines are discussed here. The cylinder heads illustrated here provide a selection from the multitude of valve train concepts found on the market, which have considerable influence on head geometry. The first example in Fig. 7-89 shows a two-valve cylinder head with roller cam followers made by BMW. This compact cylinder head concept is used in four- and twelve-cylinder engines. The head for the V-12 engine shown here is designed to be reversible and thus is identical for both cylinder blocks. To minimize friction, roller cam followers made of precision castings are used. This choice reduces friction in the valve train by as much as 70% when compared with the cylinder head without rollers, previously used. For weight restrictions, a hollow camshaft was developed using the process devised by the Süko Company.

Pushrods with hydraulic adjustment are often used in mass-production engines. Figure 7-90 shows as an example a four-valve cylinder head crafted by BMW for use in a V-8 engine. Longitudinal bores are provided in the unitized cylinder head to supply oil to the valve lifters; after casting these channels are drilled into, from the outside, near the valve lifter bores. In V-block engines with hydraulic pushrods the oil requirements in the cylinder head and the danger of oil foaming due to camshaft rotation are considerable so that drains of sufficient cross section have to be provided for oil to return through the engine block and to the oil pan. In this cylinder head, six return ports are provided for each bank of cylinders. The diameters of the intake valve disks are 32 mm for the three-liter engine and 35 mm for the four-liter engine; the exhaust valves measure 28.5 and 30.5 mm in diameter, respectively. Valve stem diameter is just 6 mm. The angle between the port and the valve is 39°45′ on the intake side

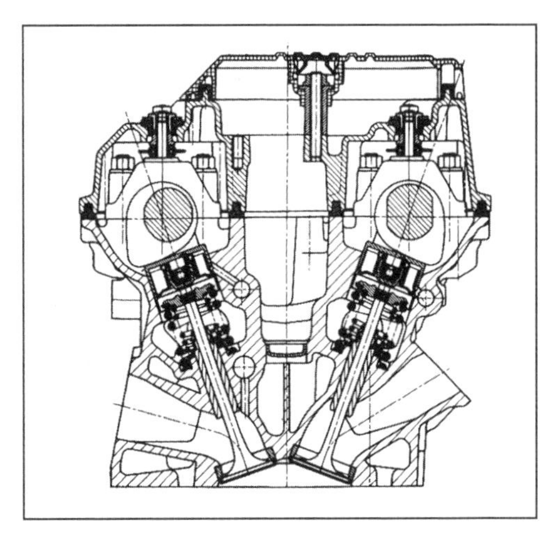

Fig. 7-90 Four-valve cylinder head with pushrods made by BMW.

and 55°45′ on the exhaust side. The intake and exhaust valves form an included angle of 39°30′ and thus make possible a very compact, crowned combustion chamber. The spark plug is located at the center, between the valves. The valve cover is mounted elastically and thus is largely acoustically decoupled. The combustion chambers inside the cylinder head are machined throughout to maintain close tolerances for the volume. The longitudinal-flow cylinder head is cast from aluminum alloy 226. For weight limits, the head is not designed to be reversible in this eight-cylinder engine. Both variants of the cylinder head are manufactured at a single production line and arrive fully assembled at the final installation point.

Figure 7-91 shows a four-valve cylinder head concept using push rods in a multisection design. Separate bearing strips are provided for the camshafts and the pushrods, both on the intake and on the exhaust sides. Thus, the cylinder head, when in mass production, can be made using die casting since there are no undercuts in the upper area of the cylinder head.

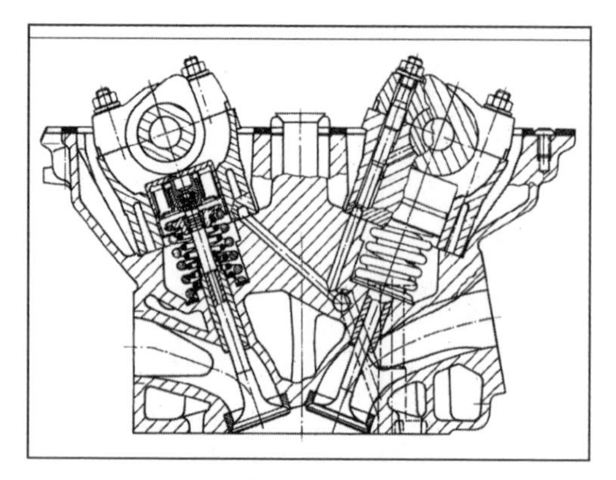

Fig. 7-91 Multisection, four-valve cylinder head made by BMW.

An example of a four-valve cylinder head with roller cam followers is depicted in Fig. 7-92. This cylinder head, made by BMW, is a further refinement of the head illustrated in Fig. 7-91. The objective in reworking the valve train was to reduce friction in the cylinder head, which was previously fitted with pushrods. Hydraulic compensation is affected here by static adjustment elements. Positioning the play adjustment unit in the stationary part of the valve train makes possible lower spring forces, because of the reduced oscillating masses, even though the valve stroke and opening period are retained. At the start of engineering, manufacturing operations had specified that the existing production line was to be retained. Thus, the valve angles and positions and the camshaft bearings were kept from the previous design. The scope of changes is thus limited to eliminating the bearing strips with the pushrod bores, the mounting bores for the compensators, which were arranged in a cloverleaf pattern around the spark plug, and the oil supply. Casting the camshaft bearings in place also lent stiffness to the cylinder head. The intake and exhaust ports and the combustion chamber were taken over without modification from the previous cylinder head.

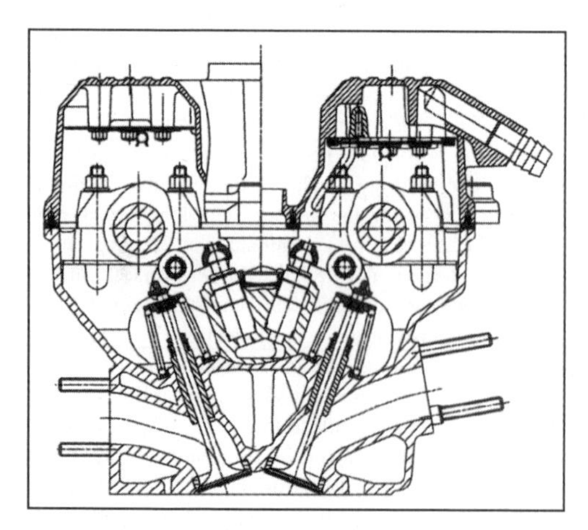

Fig. 7-92 Four-valve BMW cylinder head with roller cam followers.

Three-valve cylinder head concepts are used on the V-block engines made by DaimlerChrysler, Fig. 7-93. These cylinder heads use an overhead camshaft and roller rocker arms for valve actuation. Two spark plugs are used in each combustion chamber for faster burn propagation. In its eight- and twelve-cylinder engines, DaimlerChrysler incorporates cylinder cutout in this rocker arm concept, in the interest of reducing fuel consumption. Four cylinders are shut down in the eight-cylinder version and six in the twelve-cylinder version. Positioning a camshaft shifter is made more difficult by this single-camshaft solution. Because of the relatively heavy rocker arm, this cylinder head concept is not suitable for concepts involving high engine speeds. The overall concept is, however, more economical than a four-valve arrangement with two camshafts.

In 1994, with the introduction of its A4 series, Audi built for the first time a five-valve cylinder head in passenger car engines. This cylinder head has been adopted throughout the VW Corporation for four-, six-, and eight-cylinder engines, Fig. 7-94. With the exception of the eight-cylinder engine that uses roller cam followers, these engines employ pushrods with hydraulic compensation. For geometric reasons (the valve stem centerline would

Fig. 7-93 Three-valve cylinder head made by DaimlerChrysler.[15]

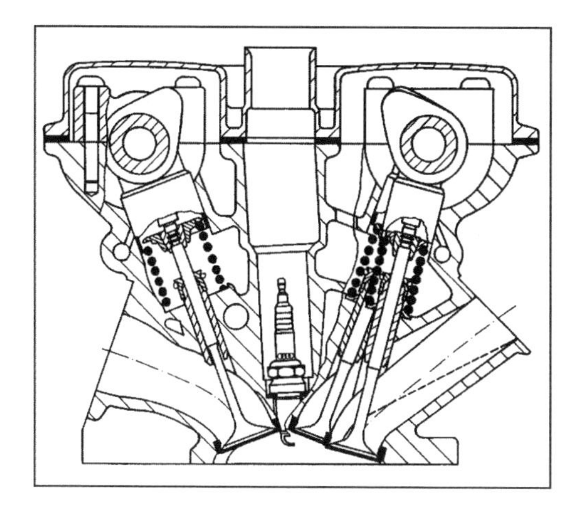

Fig. 7-94 Five-valve cylinder head made by Audi.[3]

otherwise intersect the camshaft) the angle for the center intake valve differs from the other two. The valve angle for the outer intake valves is 21.6°, that for the center valve is 14.9°, and the exhaust valve angle is 20.2°. To improve force transfer at the head bolts, a bushing is screwed into the cylinder head; thus, the collar on the head screws can be kept small. This effect helps alleviate the tight geometric situation at the cylinder head. In addition, the camshaft clearance can be kept at 129 mm since the bolts pass close by the camshafts. This is a one-piece cylinder head made up in gravity die casting. Similar five-valve designs had been used prior to their debut at Audi in one-, two-, and four-cylinder motorcycle engines made by Yamaha.

7.8.6.2 Cylinder Heads for Diesel Engines

Presented as the first example of an engineering design is the cylinder head for a two-valve engine with swirl chamber. These diesel engine concepts have dictated cylinder head design ever since diesel power plants were introduced in passenger cars. Seen in the cross section through the cylinder head in Fig. 7-95 is the prechamber with the injection valve and the glow plug. The hollow-cast camshaft actuates the intake and exhaust valves (with diameters of 36 and 31 mm, respectively) via pushrods with hydraulic compensation. In passenger car engineering, this design has been used in mass production at BMW since 1983.

With the introduction of direct-injection diesel engines by Audi in 1989, the share of diesel power plants in passenger cars has risen distinctly, primarily in Europe. Four-valve technology was introduced to a greater extent to achieve even higher power densities in diesel engines, too. Because of the greatly increased ignition pressures, maximum demands for strength and durability are made on today's diesel engine cylinder heads. Roller cam followers can be employed to minimize friction losses in the cylinder head, Fig. 7-96.

This example shows a six-cylinder engine made by BMW, which employs this head technology in its four- and eight-cylinder engines, too. The cylinder head is fitted with swirl ports; here the air is introduced from above, through the cylinder head. The cylinder head is cast from an alloy produced in primary refining. The timing chain case is cast into the front end of the cylinder head. This lends the component significant additional strength. An exhaust gas return channel is integrated into the rear section. The camshafts are driven by straight-tooth spur pinions, while the intake camshafts are driven by chains. The common rail injection technology used here requires two rails attached at the side of the cylinder head to supply fuel to the injection valves, which are positioned at the center of the cylinder head. The coolant flows inside the cylinder head from the exhaust side to the intake side. To ensure that crosswise flow is maintained, the cylinder

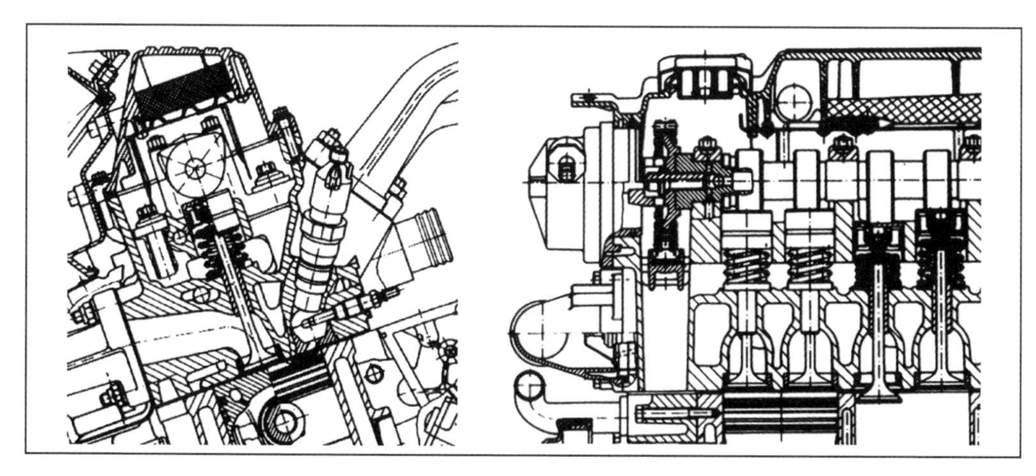

Fig. 7-95 Lateral and longitudinal sections with the installation situation for a two-valve diesel cylinder head made by BMW.

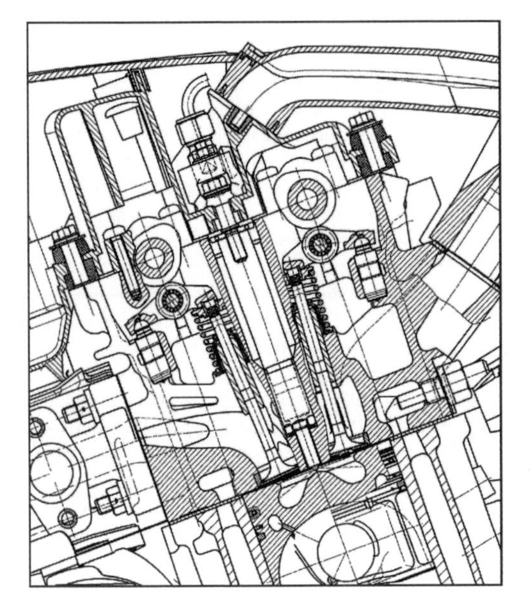

Fig. 7-96 Four-valve cylinder head with roller cam follower for a six-cylinder engine.

units are separated one from another by partitions inside the cooling cavity and have a water collector manifold cast as a unit on the intake side.

A further process used for diesel direct injection is the pump nozzle technique developed by Volkswagen. A separate injection pump, actuated by the camshaft, is provided for each cylinder, and this naturally has a major impact on the overall cylinder head concept, Fig. 7-97.

This two-valve cylinder head is equipped with pushrods with hydraulic valve play compensation. Located to the side, above the camshaft, is a bearing axis for rocker arm actuation of the pump nozzle elements. Used as the timing element is a synchronous belt that has to be fabricated from a high-strength material since the moments that the pump nozzle drive induces at the camshaft are very high. Fuel is supplied to the pump nozzle elements inside the cylinder head by one each of supply and return rails.

A vane pump driven by the camshaft delivers the required feed pressure. These pump nozzle elements can currently achieve injection pressures exceeding 2000 bar. This makes it possible to resolve the conflict of interest between low pollutant emissions and higher specific output since, even with small nozzle orifices and high injection pressures, it is possible to achieve a short injection period and rated output. Eliminating the distributor-type injection pump with its mounting bracket, drive components, and injection lines allows unification of auxiliary component arrangements on gasoline engines.

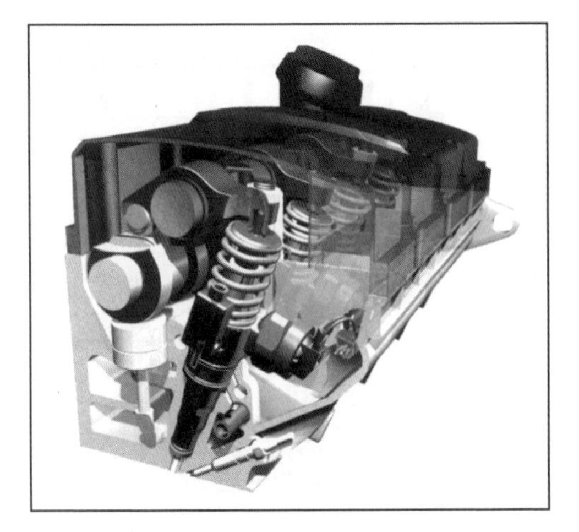

Fig. 7-97 Pump nozzle cylinder head made by Volkswagen.[16]

7.8.6.3 Special Cylinder Head Designs

In the VR series of engines made by VW five- and six-cylinder engines are made with a V-angle of 15° for a very compact configuration, in fact, a sort of synthesis of the inline and V-block engines. The one-piece cylinder heads are quite wide.[17] Placing the intake and exhaust runners on either side of the cylinder head mandates differing intake and outlet port lengths for the two cylinder banks. Concepts with symmetrical gas exchange ports are also possible, but they require a minimum of three camshafts instead of the two used here.[18] Figure 7-98 shows two sections through the mass-production, four-valve cylinder

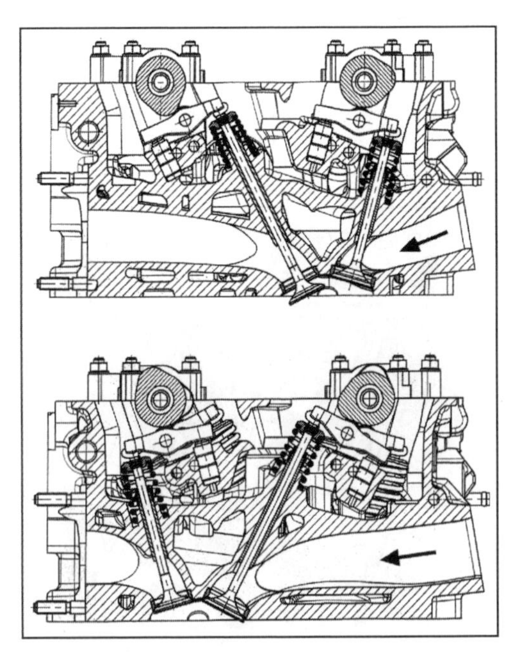

Fig. 7-98 Sections of the type VR four-valve cylinder head made by Volkswagen.

head, illustrating the various lengths of the gas ports. The cylinder head is provided with a camshaft shifter and uses precision-cast roller cam followers in the valve train. The design selected here, with its two camshafts, permits the spark plugs to be located at the center by adjusting the valve lengths. The difference in valve lengths is 33.9 mm. The valve diameters are 31 mm for the intake valves and 27 mm for the exhaust valves; the valve stem diameter is 6 mm. The combustion chambers of the two cylinder banks are almost mirror images of one another. The angle between the intake and exhaust valves is 42.5°. The cross-flow concept used in the VR cylinder head requires differing angles for the valves in relation to the cylinder centerline: 34.5° for the long ports and 8.0° for the short ports. In addition, the angles differ in relation to the port axes. To achieve uniform combustion in both cylinder banks, the short and long intake ports have to be tuned to achieve uniform flow-through and tumble effects.

Air-cooled cylinder heads are very rare in passenger cars. The two-valve cylinder head for a six-cylinder boxer engine made by Porsche and shown in Fig. 7-99 has been supplanted in the current series by water-cooled, four-valve cylinder heads. To handle the degree of heat dissipation required at the cylinder head, cooling fins with large surface areas are needed in addition to the existing cooling fan. In this example, a ceramic port liner is cast in place in the cylinder head. Its insulating effect limits the amount of heat transferred into the cylinder head. In addition, this keeps the exhaust temperature high, accelerating catalytic converter warm-up following a cold start.

High engine speeds and with them very lightweight valve train components are required for sports engines with their extremely high volumetric efficiency. The masses in motion are kept as small as possible. Here it is advisable to eliminate heavy hydraulic valve play compensation elements. An example of such a design has been realized by BMW in a six-cylinder engine with precision-cast cam followers and mechanical valve play adjustment. These sliding-type cam followers are very lightweight and rest on a shaft inserted into the cylinder head. In selecting the lever ratio for the cam follower, component stiffness was given precedence over minimizing the amount of installation space required. The rocker arm drive uses a 1:1 lever ratio in order not to induce any bending loads, Figure 7-100. The four-valve cylinder head used here is a one-piece unit cast in a steel die. A cross-flow cooling concept is used here. Integrated into the cylinder head is an air distribution line into which supplementary air is blown. Splitting off from this line, which is 12 mm in diameter, are 4 mm bores leading directly into the exhaust port, next to each exhaust valve.

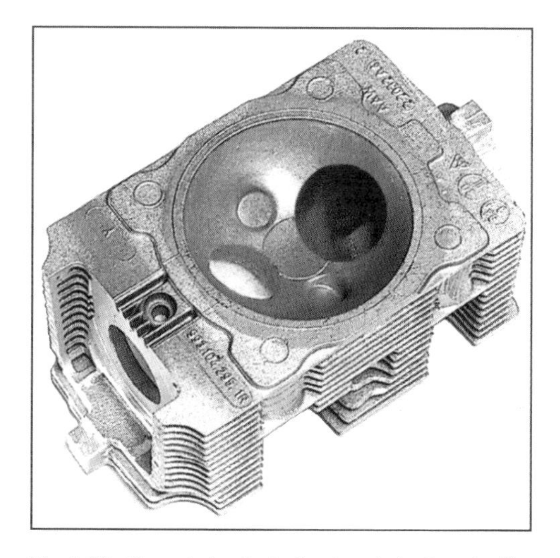

Fig. 7-99 Air-cooled cylinder head made by Porsche.[19]

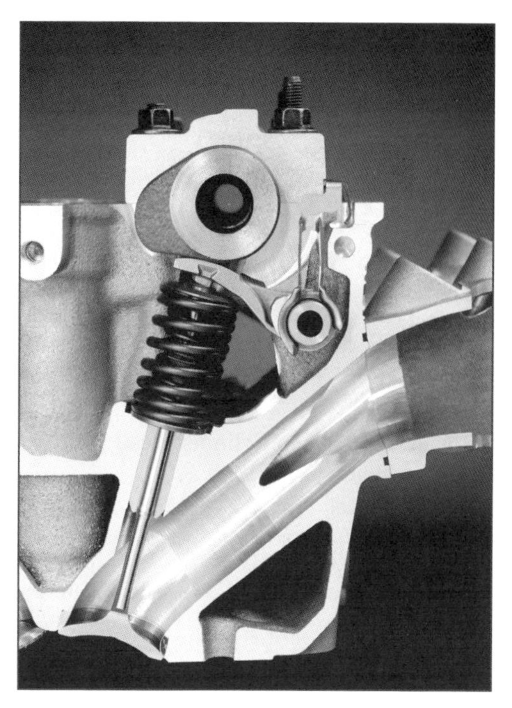

Fig. 7-100 Four-valve BMW cylinder head with sliding cam followers.

Direct-injection gasoline engines are currently being developed to readiness for mass production all over the world. It is necessary to free up space in the cylinder head, next to the spark plug, to accommodate the injection nozzle, a situation similar to that seen in direct-injection diesel engines. Space is tight even for a four-valve concept. Developing and fine-tuning the locations and shapes of the gas exchange ports is extremely difficult because of the charge stratification associated with the combustion process. Figure 7-101 shows a section of a four-valve cylinder head used in mass production for a direct-injection engine built by Mitsubishi. The introduction of the air through the intake port is effected, from the intake manifold, at the top surface of the cylinder head to induce

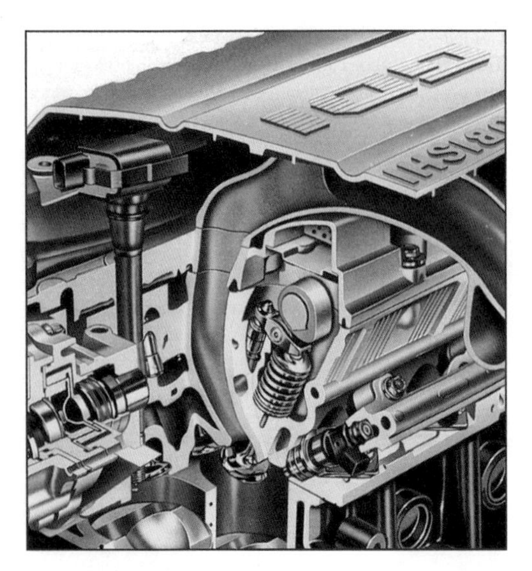

Fig. 7-101 Cylinder head for a direct-injection gasoline engine made by Mitsubishi.[20]

a defined tumble effect in the flow, matched to the shape of the bowl in the piston crown. The valves are actuated by roller cam followers. The injection nozzle is located toward the side, at the cylinder head. The spark plug is located at the center of the cylinder head.

7.8.7 Perspectives in Cylinder Head Technology

The gas charge, the exhaust gas, and the combustion process itself are controlled in the cylinder head. Further development in cylinder head technology is targeted on lightweight construction, higher-strength materials, and more economical manufacturing processes while at the same time improving all the engine targets.

Multivalve cylinder heads have made a breakthrough on all fronts to include diesel engines. Their introduction and better gas control make it possible to realize higher specific outputs per cylinder. Adopting advanced cylinder head concepts leads to downsizing concepts that result in high-performance engines with favorable emissions and consumption levels in a range of engines available for the customer's choice depending on power requirements.

Cylinder head development has literally taken on a new dimension with the employment of fully variable timing, initially in gasoline engines. Setting power output without a butterfly throttle has reduced fuel consumption considerably. At present purely mechanical concepts have been implemented in mass-production engines.[2] It will be exciting to observe the degree to which throttleless control will provide greater degrees of freedom in valve timing and thus in engine control in the future. The demands on cylinder heads will change with future systems which might incorporate electromechanical or other innovative actuation principles.

Similar to 100 years ago, the diversity in cylinder head concepts will continue to be great. This presents a major challenge to engine developers, who must keep pace with demands as they continue to develop.

Bibliography

[1] Hannibal, W., and K. Meyer, "Patentrecherche und Überblick zu variablen Ventilsteuerungen," Lecture, Haus der Technik, 2000.

[2] Flierl, R., R. Hofmann, C. Landerl, T. Melcher, and H. Steyer, "Der neue BMW-Vierzylinder-Ottomotor mit VALVETRONIC," in MTZ Motortechnische Zeitschrift, Vol. 62, 2001, No. 6.

[3] Hannibal, W., and F. Lukas, "Rechnergestützte Auslegung des Audi-Fünfventil Zylinderkopfkonzeptes," in MTZ Motortechnische Zeitschrift, Vol. 55, 1994, No. 12.

[4] Dong, X., "Öffnungsquerschnitt von Ventilen," in MTZ Motortechnische Zeitschrift, Vol. 46, 1985, No. 6.

[5] Schäfer, F., S. Barte, and M. Bulla, "Geometrische Zusammenhänge an Zylinderköpfen," in MTZ Motortechnische Zeitschrift, Vol. 58, 1997, Nos. 7/8.

[6] Eidenböck, T., R. Ratzberger, J. Stastny, and W. Stütz, "Zylinderkopf in Vierventiltechnik für den BMW DI-Dieselmotor," in MTZ Motortechnische Zeitschrift, Vol. 59, 1998, No. 6.

[7] Krappel, A., W. Riedl, D. Schmidt-Troje, and J. Schopp, "Der neue BMW Sechszylindermotor in neuer Hubraumstaffelung und innovativer Leichtbauweise," in MTZ Motortechnische Zeitschrift, Vol. 56, 1995, No. 6.

[8] N.N., "Becker CAD CAM CAST," Becker GmbH, Steffenberg-Quotshausen, 2001.

[9] Hannibal, W., "Begleitende Entwicklung der Audi-Fünfventil-Technologie mittels Rechnereinsatz," Lecture at the 1995 Wiener Motoren Symposium.

[10] Seifert, H., "20 Jahre erfolgreiche Entwicklung des Programmsystems PROMO," in MTZ Motortechnische Zeitschrift, Vol. 51, 1990, No. 11.

[11] Dirschmid, W., and M. Schober, "Computersimulation in der Ventiltriebsauslegung," in MTZ Motortechnische Zeitschrift, Vol. 57, 1996, No. 4.

[12] Nefischer, P., S. Blumenschein, A. Keber, and B. Seli, "Verkürzter Entwicklungsablauf beim neuen Achtzylinder-Dieselmotor von BMW," in MTZ Motortechnische Zeitschrift, Vol. 60, 1999, No. 10.

[13] Scheeren, H.W., A. Koreneef, and H. Fuchs, "Herstellung von Zylinderköpfen für hochbeanspruchte Diesel- und Ottomotoren," Lecture, Haus der Technik, Essen, 2000.

[14] Hannibal, W., and A. Metzlaw, "Von der Idee zum Produkt, in Digitalisierung und Flächenrückführung in der CAD-Prozesskette," QZ, Qualität und Zuverlässigkeit, Vol. 46, 2001, No. 7.

[15] Fortnagel, M., G. Doll, K. Kollmann, and H.-K. Weining, "Aus Acht mach Vier: Die neuen V8-Motoren mit 4,3 und 5 l Hubraum," in MTZ Motortechnische Zeitschrift, Mercedes-Benz S-Class, Special edition, 1998.

[16] Dorenkamp, R., J. Hadler, B. Simon, and D. Neyer, "Der Vierzylinder-Pumpe-Düse-Motor von Volkswagen," in MTZ Motortechnische Zeitschrift, Special edition, 1999.

[17] Aschoff, G., B. Ebel, S. Eissing, and F. Metzner, "Der neue V6-Vierventilmotor von Volkswagen," in MTZ Motortechnische Zeitschrift, Vol. 60, 1999, No. 11.

[18] Fuoss, K., W. Hannibal, and M. Paul, "Mehrzylinder-Brennkraftmaschine. Patentanmeldung," DE 34 44 501, German Patent Office, Munich, 1993.

[19] Klos, R., "Aluminium Gusslegierungen," Die Bibliothek der Technik, No. 116, Verlag Moderne Industrie, 1995.

[20] N.N., "Ansicht eines Mitsubishi-Galant-Motors," Brochure published by Mitsubishi Japan, 2001.

7.9 Crankshafts

7.9.1 Function in the Vehicle

The internal combustion engine continues to be the prevalent power plant in motor vehicles, largely in the form of a reciprocating engine. This will continue to hold true in coming years. The development goal through 2008 is to reduce average CO_2 emissions from the current 190 g/km to 140 g/km, which corresponds to fuel consumption of about 5.7 liters per 100 km.

7.9.1.1 The Crankshaft in the Reciprocating Piston Engine

The piston's linear movements are converted, with the intervention of the conrod, into rotational movement at the crank of the crankshaft, thus making torque available for use at the wheels.

Because of the strains involving forces that change in both time and location, with rotational and flexural torques and the resulting excitation for vibrations, the crankshaft is subject to very high, very complex loads.

7.9.1.2 Requirements

The crankshaft's service life is influenced by

- Resistance to flexural loading (weak points at the transition from the bearing seat to the web)
- Resistance to alternating torsion (the oil bores are often weak points)
- Torsion alternation behavior (stiffness, noise)
- Wear resistance, at the main bearings, for example
- Wear at shaft seals (leaks, motor oil escaping)

For ecological reasons the trend is toward high-torque engines that will develop high moments even at low engine speeds. In these engines the crankshaft is subjected to far greater loading in all the respects mentioned above than is the case in conventional, aspirated engines.

7.9.2 Manufacturing and Properties

About 16 million crankshafts are required every year in Europe. The large majority (about 15.5 million) is used in passenger cars and light utility vehicles.

7.9.2.1 Processes and Materials

Crankshafts are either cast or forged. The shares accounted for by the individual manufacturing processes are shown in Fig. 7-102.

The necessity for reducing CO_2 emissions (and fuel consumption) is leading increasingly to turbocharged gasoline engines, which at present are preferably fitted with forged crankshafts.

	Vehicles	Forged	Cast	Cast
Western Europe	11.3	3.1	8.2	73%
United States	6	0.35	5.65	94%
Japan	8.5	4.75	3.75	44%

Fig. 7-102 Passenger car crankshafts, by manufacturing process (in millions of units, 1993).

Process	Attitude in the mold	Mold process
Green sand IMD	Horizontal	Automatic system with mold frames
Green sand	Horizontal	Automatic system with mold frames
Shell mold	Vertical	Croning sand shells in frames, backed with steel pellets
Shell carrier	Vertical	Croning sand shells in steel backing shells
Waterglass-CO_2 process	Vertical	Double-sided molds from the automatic mold unit, shaped in the horizontal = 1 packet. Gassing in vertical attitude
SF process with cold box sand	Horizontal	Automatic mold machine with frame; lost sand is replaced
Lost foam	Vertical	Styroprene model in frames, backed with sand

Fig. 7-103 Survey of casting processes used to manufacture crankshafts.

Casting

There are several processes available for casting crankshafts; they are listed in Fig. 7-103.

Based on the evaluation of the various processes, we find that, due to better dimensional stability, there are advantages for the green sand IMD process.[5] But the technique most commonly used in practice is the shell mold process.

Forging

Two companies in Germany concentrate on making forged crankshafts for road vehicles;[1] see Fig. 7-104. There were 13 such companies 30 years ago. Because of technological considerations, the trend toward forged crankshafts is continuing.

Advantages and Disadvantages of Forged and Cast Crankshafts

Advantages of Cast Crankshafts over Forged Crankshafts

- Cast crankshafts are considerably more economical than forged units.

Manufacturer	Market share 98	
Company A	12%	Germany
Company B	12%	Germany
Others (some overseas)	6%	
Total	30%	Europe/United States

Fig. 7-104 Market shares held by manufacturers of forged crankshafts.

- Casting materials respond well to surface finishing processes used to boost oscillation resistance. Thus, for example, the resistance to flexural loading can be increased considerably by rolling the radii at the transition between the journals.
- Cast crankshafts can be hollow and thus may be as much as 1.5 kg lighter in weight.
- Cast crankshafts of the same design offer a weight advantage of about 10% compared with steel, which is because of the lower density of the nodular cast iron.
- Machining the cast crankshaft is in general simpler. It is possible to work with smaller supplements for later machining, the mold parting flash is reduced and no longer needs be removed, and the slopes in the webs can be specified more closely. In fact, it is often possible to do without any machining of the webs at all.

Disadvantages of Cast Crankshafts Compared with Forged Crankshafts

- Casting materials have a lower Young's modulus than steel. As a consequence, cast crankshafts are less stiff and exhibit different vibration properties.
- Measures implemented to increase drive train stiffness are even more necessary when aluminum blocks and crankcases are used, as this material has a far lower Young's modulus and thus less material stiffness.
- Cast crankshafts, when compared with steel, may exhibit less favorable wear characteristics at the bearing journals, which is because of the microvoids in the surface (exposed spheroliths), lower fundamental hardness, and less enhancement of hardness in the usual hardening processes.

It is possible, however, to compensate for these advantages:

- By larger diameters in the bearing area, which is not possible for existing engine concepts and which is not desirable in new concepts because of greater friction losses and the associated rise in fuel consumption.
- With complex vibration dampers, which increase system costs, however, and can offset the cost advantage of cast crankshafts in comparison with forged versions.
- With a very stiff design for the engine block, joined with crankshaft bearing bridges, cast oil pans, and a stiff linkage with the transmission.
- With the ISAD (integrated starter alternator damper) system[4] currently under development.

By eliminating the separate starter and alternator, this system offers the option of damping engine oscillations using "alternating reactive power." Here any crankshaft overspeeding is braked by kicking in generator action and any lag is compensated by applying the energy stored in capacitors.

In torque-optimized engines the cast crankshafts have more physical problems with torsional oscillations, because of Young's modulus and less stiffness, and this can lead to an unacceptable noise level in the vehicle. The Young's modulus for steel is 210 kN/mm^2 while that for GJS is 180 kN/mm^2. Consequently, forged crankshafts are used at present in engines that develop high torques at low engine speeds. This applies, in particular, to engines with more than four cylinders as well as to diesel engines.

7.9.2.2 Materials Properties for Crankshafts

Crankshaft properties are shown in Figs. 7-105 and 7-106.

Steel	Status	Tensile strength [N/mm^2]	0.2% offset limit [N/mm^2]	Elongation at failure [%]	Hardness [HB]
Ck 45	Hardened	600–720	360	18	210
37Cr4	Hardened	800–950	550	14	220
Today: 38 MnS 6	BY*	780–930	450	12	235–280

*BY with controlled cooling from melt temperature.

Fig. 7-105 Properties of forged crankshafts.

Casting material	Status	Tensile strength [N/mm²]	0.2% offset limit [N/mm²]	Elongation at failure [%]	Hardness [HB]
GJS-700-2	Casting status	700	420	2	230–280
GJS-800-2	Casting status	800	500	2	250–300
ADI	Double hardened	800–900	600	5	260–310

Fig. 7-106 Properties of nodular cast iron (GJS); minimum values.

GJS-700-2 is the material normally desired for crankshafts. The engine manufacturers sometimes have their own company specifications for spread of hardness values.

7.9.3 Lightweight Engineering and Future Trends

Design changes for crankshafts can be incorporated only to a very limited extent, since the space available in the crankcase does not offer any additional room.

7.9.3.1 Hollow Cast Crankshafts

In general, cast crankshafts of a comparable design weigh about 10% less than a forged unit because of the lesser density. Hollow cast crankshafts offer a further weight reduction of up to 1.5 kg (Fig. 7-107).

7.9.3.2 ADI Austempered Ductile Iron

This material, made in a complex heat treatment process, has a bainitic-ferritic structure offering high strength, good elongation properties, and great hardness. The material, however, has poor amenability to machining. The heat treatment causes warping, which in turn makes it necessary to straighten the crankshaft.

Aside from the considerably higher costs, "nearly finished"[2] manufacturing does nothing to alleviate the basic problem associated with nodular graphite casting materials: The Young's modulus cannot be raised to above the values for normal GJS even with the extensive heat treatment processes employed to impart greater strengths.

7.9.3.3 Increasing Component Strength through Postcasting Treatment

The static properties say little about crankshaft service life. Component strength, heavily influenced by sufficient vibration resistance, is achieved only through supplementary treatment processes; this is true for both castings and steel (Fig. 7-108).

Radius Rollers

Rolling the radii is the standard process[9] used to enhance fatigue strength under reversed bending stress for both cast and steel crankshafts. Here pressured-induced self stresses are created at the transitions from the bearing journals to the webs; this improves long-term strength considerably in this heavily loaded area.

Inductive Hardening, Radii with/without Journals

This process is used in some cases on crankshafts for diesel engines in order to increase the bearing journals' resistance to oscillation and wear.

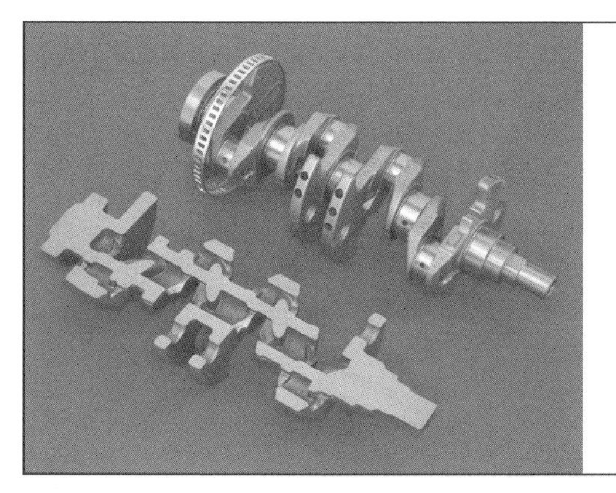

Fig. 7-107 Cast crankshaft for a four-cylinder motor using GJS-600-3 (hollow version weighing 10.6 kg at the left, solid version weighing 12 kg at the right).

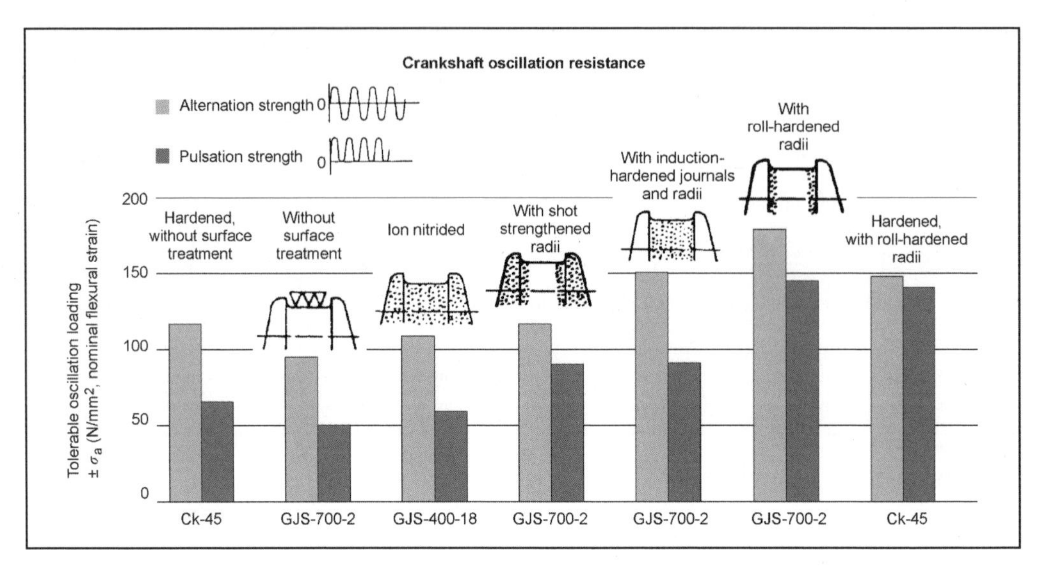

Fig. 7-108 Influence of postcasting treatment on crankshaft vibration resistance.

Nitriding

In this process, too, pressure-induced self stresses are induced in the journals and radii areas with a positive effect on enduring resistance to vibration and wear. But nitriding is used decreasingly since it cannot be integrated into the manufacturing line and disposal of the salts is difficult.

Ball Calibration

This process is employed to boost resistance to torsional vibration by strengthening the oil bores in the bearing journals. It has not yet found its way into practical use since it can hardly be integrated into mass production operations at reasonable costs.

Bibliography

[1] Adlof, W., "Wer an Leichtbau denkt, kommt an einer Stahlkurbelwelle nicht vorbei," in Schmiede-Journal, March 1994, pp. 13–16.

[2] Heck, K., *et al.*, "Innovative gießtechnologische Entwicklung zur Herstellung von Endnah-Guss-Kurbelwellen," Gießerei, No. 85, February 1998.

[3] IMC Consultants, "IMC Report for Georg Fischer/DISA, Analysis of alternative strategies designed to increase market share of the magnesium converter," May 1998.

[4] "ISAD der integrierte Starter–Alternator–Dämpfer," 1998 Motors and Environment Convention, AVL Graz.

[5] Becker, E., and K. Hornung, "Projekt 78274: Kurbelwellenfertigung im Masken- oder Grünsandverfahren," F&E Berichte, August 1985, September 1985.

[6] "Gusseisen mit Kugelgraphit," Technical Bulletin, Georg Fischer Company.

[7] "Lagerverhalten von Gusskurbelwellen," Technical Bulletin, Georg Fischer Company.

[8] "Beanspruchungsgerechte Gestaltung und anwendungsbezogene Eigenschaften von Gussteilen," Technical Bulletin, Georg Fischer Company.

[9] "Steigerung der Schwingfestigkeit von Bauteilen aus Gusseisen mit Kugelgraphit," Technical Bulletin, Georg Fischer Company.

7.10 Valve Train Components

7.10.1 Valve Train

There has emerged in recent years a trend in passenger car engines toward overhead camshafts (OHC) and double overhead camshafts (DOHC) while engines with camshafts located below [overhead valve (OHV)] continue to be used, particularly in large-displacement, V-block engines. Engines with overhead camshafts are developed so valve trains can be engineered to withstand the high speeds required in higher-performance engines. DOHC concepts give the engineer the option of mutually independent timing for the intake and exhaust camshafts using camshaft shifters. OHV and OHC concepts are characterized by compact shapes and sizes and by economy in their manufacture.

For diesel engines for utility vehicles, one sees a trend toward four-valve concepts. Rocker arms or double rocker arms fitted with mechanical valve play adjustment and driven by pushrods and camshafts located below—as is the case in two-valve designs—affect valve lift.

OHC concepts are employed, in addition to OHV versions, in smaller utility vehicle engines that utilize engine braking effect, and hydraulic valve lifters are used increasingly to compensate for valve lash.

7.10.1.1 Direct Drive Valve Trains

This category embraces valve trains with hydraulic (Fig. 7-109) or mechanical valve lifters as well as so-called "bridge" solutions in which components, guided by columns, lift multiple valves by direct actuation with a single camshaft. A subgroup within the latter solution is represented by the bridge that interfaces with two hydraulic valve lifters (Opel direct-injection diesel engines).

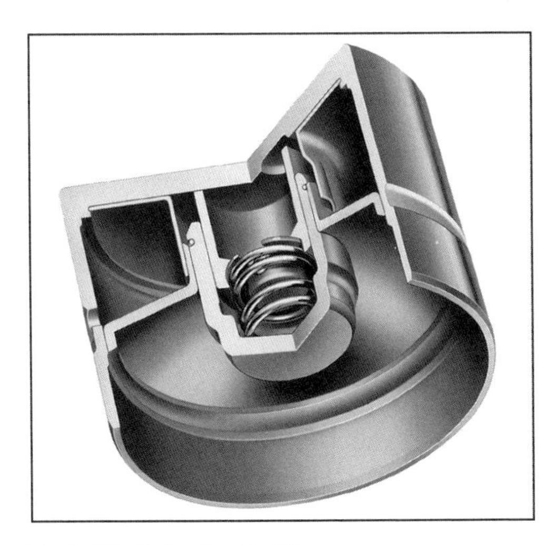

Fig. 7-109 Hydraulic valve lifters.

Direct drive always offers very good stiffness values with relatively modest masses in motion. This is the prerequisite for trouble-free valve train operation even at very high speeds (loss of contact force, premature valve seating). Thus, efficient, high-speed engines can be realized particularly through employing valve lifters.

In the interest of reducing the masses in motion, preference among mechanical valve lifters is given to those with graduated crown thickness (Fig. 7-110) or those with adjustment shims located at the bottom.

For service work (adjusting valve play), rods with an adjustment shim at the top (Fig. 7-111) are preferable since with this version it is not absolutely necessary to remove the camshaft. Such units are, however, considerably heavier and require more installation space than the other version (at identical valve lift). The basic body

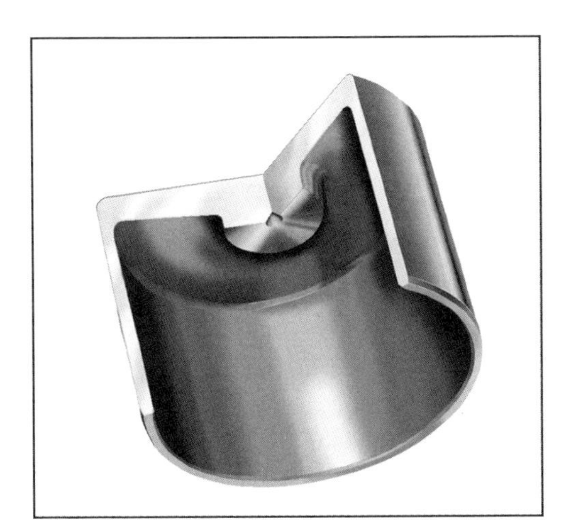

Fig. 7-110 Mechanical valve lifters with graduated crown thickness.

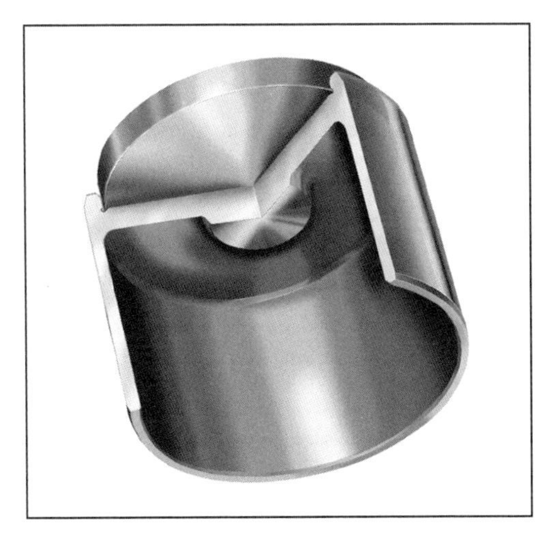

Fig. 7-111 Mechanical valve lifters with adjustment shim at the top.

for the valve lifter is made of ductile steel. Aluminum is found in only two applications (Toyota Lexus V-8 and Jaguar V-6 and V-8). The shims are usually made of steel that can be hardened. When bodies made from deep-drawn sheet steel and small hydraulic elements (11 mm O.D.) are used, hydraulic valve lifters achieve very low masses that, at identical lobe contact diameters, are far lighter than mechanical valve lifters with shims at the top.

The sliding contact with the lobe requires careful machining at the camshaft—stone finishing following cam lobe grinding has proved to be the most favorable. Over and above this, the camshaft material has to be matched to the loading situation to avoid wear. The versions that have been found to be particularly advantageous are hard-cast camshafts and camshafts made of gray cast iron with a remelted surface. The valve lifters and shims should rotate in order to achieve uniform wear for the cam lobe contact surfaces. This is achieved by shifting the cam in relation to the shim (toward the camshaft centerline) or with offsetting and an angular grind for the cam lobe at the point where the lobe contacts the valve lifter. Valve trains with valve lifters and the mechanical versions in particular offer the advantage of lower cylinder head height in DOHC designs. Valve lifters are found in many different applications, e.g., two- and four-valve gasoline engines and diesel engines.

Volkswagen uses a valve lifter incorporating a special hydraulic element designed to prevent increases in contact force while it sweeps the lobe's circular segment in all of its pump-nozzle diesel engines.

7.10.1.2 Indirect Drive Valve Trains

Included in this group of valve trains are

- Cam follower valve trains with stationary valve play compensation elements; the cam follower rests on the spherical upper end of the hydraulic element.

- Rocker arms that pivot on a shaft.
- OHV concepts comprising the cam follower (flat or roller valve lifter), pushrods, and rocker arms.

There is a clear trend in cam follower drive trains toward cam followers that are made of sheet metal and are fitted with a rolling bearing at the point of contact with the camshaft. Cam followers made from cast steel in a precision-casting process give the engineer greater design leeway (stiffness, moment of inertia). The cost advantages for the sheet metal cam are so great, however, that precision cast cam followers are used only in exceptional cases (Fig. 7-112). When compared with plain cam followers or valve trains with valve lifters, the use of the rolling bearing effects a reduction in friction, particularly in the lower speed range that is so relevant to reducing fuel consumption. This reduction in friction losses is, however, paid for with a significant reduction in damping of torsional vibrations at the camshaft, which has consequences for the timing chain or belt. Moments of inertia and stiffness are highly dependent on the shape of the lever. Short levers cause low moments of inertia, with masses on the valve side that are lower than for valve lifters. Seen as a whole, roller cam followers are inferior to valve lifters in regard to stiffness.

Fig. 7-112 Roller cam followers with hydraulic element.

The profiles for the lobes in valve trains incorporating roller cam followers differ significantly from those in valve trains that use valve lifters (greater radius at the apex, shorter lobe stroke—depending on the lever ratio, and concave flanks). In order to keep the concavities of the cam narrow enough that they can still be ground with mass production technology, preference is given to valve train geometries in which the roller is positioned approximately at the center between the valve and the hydraulic element. Here the camshaft is located above the roller. This arrangement makes it possible to keep the hazard of "pumping up" under control (see hydraulic valve play compensation).

This configuration, with the cam lobe offset from the valve stem centerline, makes the cam follower concept interesting for four-valve, direct-injection diesel engines since in these units the valve stems either are parallel or have only a very small included angle (Fig. 7-113). Only with the use of cam followers is there sufficient clearance between the camshafts. Using cam followers also makes it possible to serve "inverted" valve arrangements (e.g., DCC OM 668).

Fig. 7-113 Valve train for a direct-injection diesel engine with cam followers.[1]

As opposed to cam followers, rocker arms are mounted on shafts. One differentiates between rocker arms in which the pivot point is toward the center of the lever (Fig. 7-114) and those that pivot at one end; the latter are also known as cam followers.

The camshaft is located below one end of the rocker arm while cam motion is transferred via either a plain, sliding surface or a cam roller. To achieve low friction losses, needle-bearing cam rollers are used in most modern rocker arms. The valve is lifted at the opposite end of the lever, via a hydraulic valve play compensating element or a setting screw used for mechanical valve play adjustment (Fig. 7-115).

The contact surface at the rocker arm has to be angled to maintain unbroken contact between the adjustment

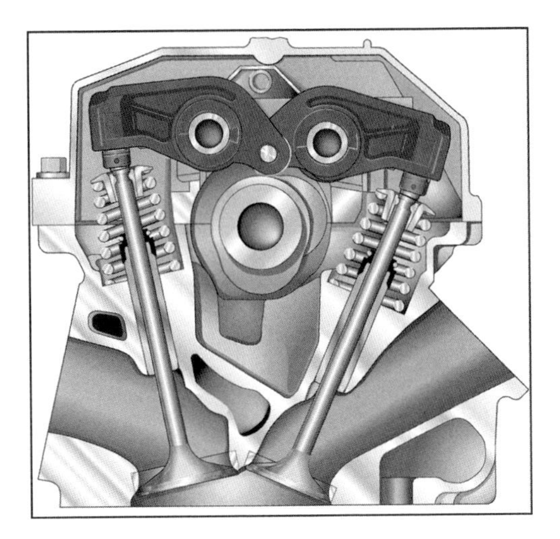

Fig. 7-114 Typical rocker lever valve train.

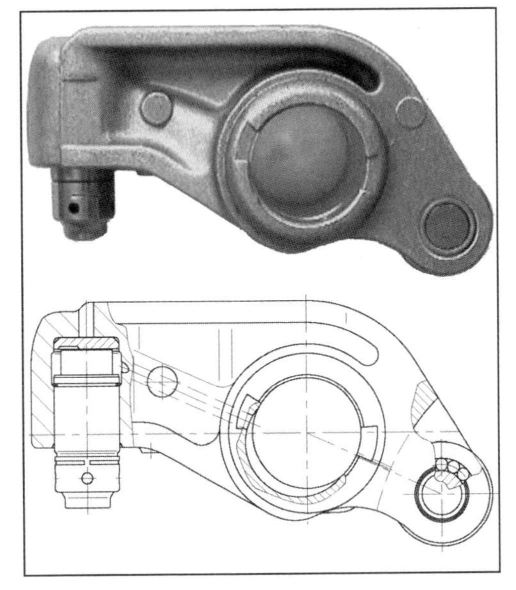

Fig. 7-115 Side view and section through an aluminum rocker arm.

element and the end of the valve stem as the arm executes its rocking motion. Since neither the hydraulic compensation element nor the mechanical adjustment screw is mounted in the rocker arm to specify a direction, the contact surface at the valve actuation element is crowned. This geometric design leads to relatively high surface pressures at the end of the valve stem. Where surface pressures are excessive, hydraulic elements are employed, which incorporate a pivoted foot at the point of contact with the valve. The contact itself is at a virtually flat surface, while the pivoted foot executes a movement around a ball mounted on the hydraulic element (Fig. 7-116).

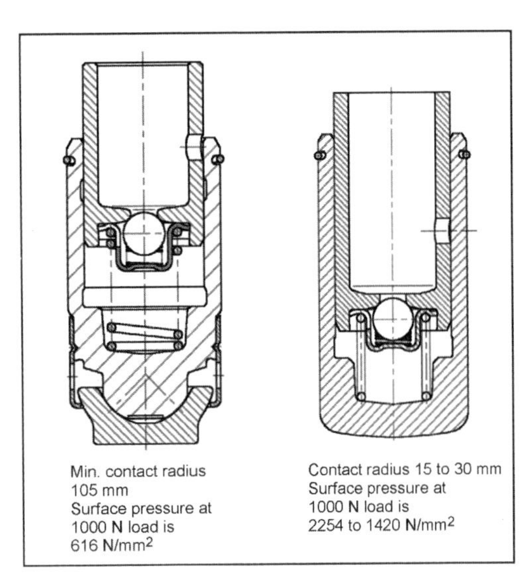

Fig. 7-116 Hydraulic elements for rocker arms, 11 mm O.D.

Aluminum, preferably manufactured in a die-casting process, or steel is used here.

Oil is supplied to the hydraulic compensator elements from the rocker arm shaft. From this point, bores in the rocker arm lead to the hydraulic elements. Support shims with a little play in the guide, which are always used in aluminum rocker arms, permit the escape of air that, for instance, can get into the hydraulic element when the engine is started. Either shims such as this or very tiny bores are used to vent steel rocker arms.

Starting at the oil supply bores in the rocker arm shaft, bores in the rocker arm can be used to spray the cam roller or the cam sliding surface.

Rocker arms of this shape and design are found in diesel and gasoline engines. Using rocker arms makes it possible to set up two-, three-, or four-valve arrangements with just a single camshaft. Where valve trains with two intake or exhaust valves are used, double or twin rocker arms can be used to lift two valves simultaneously with a single cam. Valve play is compensated individually, however, with the aid of hydraulic elements.

It is even possible to actuate three valves (Fig. 7-117). Audi uses a triple cam follower in the valve train for its V-8 engines incorporating three intake valves. Actuation force flows from two cam lobes to two rollers in the rocker arm and then to three hydraulic compensators.

In addition to the solutions previously mentioned, where the rocker arm actuates the valve directly, there are also rocker-arm valve trains that use bridges, either guided on posts or free-moving, to lift two valves simultaneously. In four-valve diesel engines, including those with an inverted valve arrangement, it is possible to actuate all the valves with just a single camshaft while at the same time maintaining the space needed for the injection nozzles.

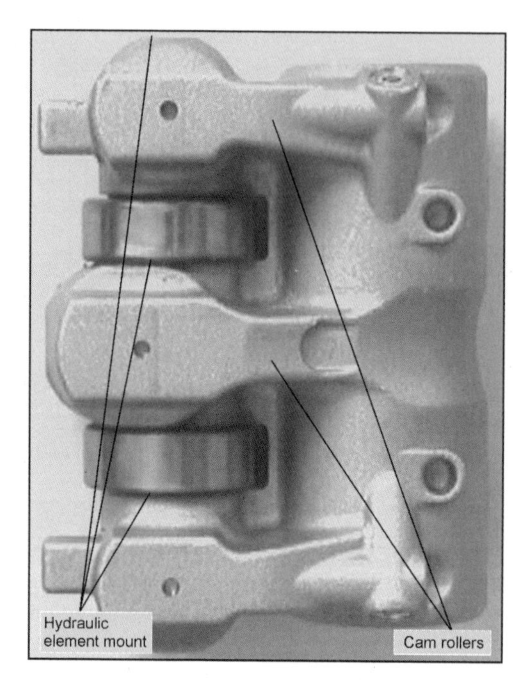

Fig. 7-117 Triple cam follower for the Audi V-8 engine.

Stiffness values in rocker arms are low, because of the geometry and, particularly, the great distance between the cam contact point and valve contact point, the relatively large number of contact points and the shaft, which have to be taken into account, in addition. The much more direct force flow in cam follower designs produces far better stiffness values.

7.10.1.3 Hydraulic Valve Play Compensation

For many years now one goal of engine builders has been to keep the adjustment and service work of the engine to a minimum. Thus, it is hardly a surprise that the first engines with hydraulic—and thus automatic—valve play compensation were produced well before World War II. These were, however, large-displacement engines that ran at moderate speeds. Higher engine speeds were attained in the 1970s in the Mercedes Benz V-8 engines with hydraulic screw-in elements (cam follower system). A further milestone reached in the 1970s was the introduction of hydraulic valve lifters in the V-8 engine used in the Porsche 928. Today hydraulic valve play compensation is employed in all engine classes and even in high-speed engines such as those used by Ferrari and Porsche.

The hydraulic elements consist of an outer casing in which a plunger with an integrated check valve is installed. These two parts can slide one inside the other and, at the contact surface, form a leak gap only a few micrometers wide. A spring on the inside keeps the two components apart.

During the valve stroke the valve spring and mass forces impose load on the hydraulic element. High pressure is developed in the space defined by the casing and the plunger (with the check valve closed). A small amount of oil escapes through the very narrow gap and is passed to the reserve space inside the plunger. In the following phase, while contact is made with the lobe's circular segment (valve closed), the inside spring pushes the hydraulic element apart until the valve play is once again fully compensated. The differential pressure thus arising causes the check valve to open; the amount of oil required for compensation can flow in. Thus, the length of the hydraulic element can change in both directions.

The advantages of hydraulic compensation for valve play include

- Simple mounting of the cylinder head (no measurement or adjustment work since the hydraulic element compensates for all tolerances)
- Freedom from service requirements
- Constant timing at all throttle settings and at all times (no need to adjust time to account for thermal effects or wear in valve train components)
- Low noise level (thanks to low opening and closing ramps at the camshaft and low opening and closing speeds).

Achieving this places certain demands on the oil circuit (oil pressure, foaming). It is also necessary to observe close shape tolerances when machining the circular lobe segment. The elements could become compressible in the event of a deficiency in the oil supply (air in the high pressure chamber), which would result in insufficient valve lift and consequently would induce noise or changes in dynamic response at high engine speeds. The hydraulic element recognizes loss of contact force as valve play, and this could result in an undesired lengthening of the element, with the result that the valves would not close completely.

7.10.1.4 Mechanical Valve Play Adjustment

Valve play is adjusted with

- Screws
- Adjusting shims of graduated thicknesses
- Valve lifters with graduated crown thickness (only for valve trains incorporating valve lifters)

Common to all three options is finite adjustment precision, which needs to be taken into account in the design of the lobe ramps for opening and closing the valves. It is necessary to measure and adjust for valve play when mounting the cylinder head. The increase in valve play resulting from wear at valve train components can be corrected by adjustments made during service work; changes in play resulting from temperature development in the engine cannot be automatically corrected. The effects enumerated here harbor the potential for a wide spread in the amount of play and necessitate steep ramps with great opening and closing speeds. This wide spread implies critical changes in timing and thus has negative effects on exhaust gas quality; rapid closing causes valve train noise.

The advantages of mechanical valve play adjustment (compared to comparable hydraulic valve train components) include

- Greater stiffness
- Lower friction losses (by eliminating friction at the lobe's circular segment and through modified valve spring characteristics)
- Lower component costs

7.10.1.5 Future Trends

Variable Valve Drive Trains with Single-Step and Multistep Variability

Building upon the systems explained in Section 7.10.1.1, it is possible to respond to the needs of engine designers and the desire of thermodynamics engineers to apply differing lift curves, selectively, to an engine valve. This is done by introducing a shifting capability into the transmission path for the valve train.

Valve lift cutout and switching systems using defeatable force transmission elements such as rocker arms, cam followers, and valve lifters have already been implemented in small production runs (Fig. 7-118). A separate cam has to be provided to initiate the stroke for each additional and alternate valve stroke length—unless the alternate stroke is no lift at all.

When valves are simply disengaged (to shut down specific cylinders, for example), it is then possible to do without a second cam for each cam follower.

Here the element that follows the cam lobe is decoupled from the engine valve. This "lost" motion lends its name to what is sometimes called a "lost motion" stroke; the negative mass forces here have to be absorbed by a lost motion spring, since the valve spring is no longer actuated. The section of the valve train for which no valve cutout or cylinder shutdown is planned then executes the stroke motion without any effect on valve stroke length.

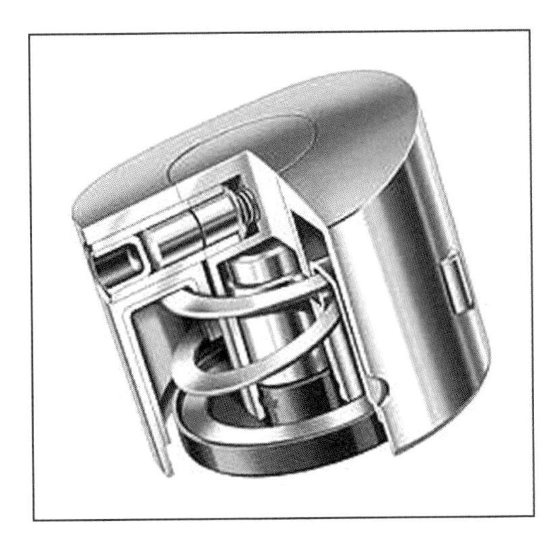

Fig. 7-118 Switchable valve lifter.

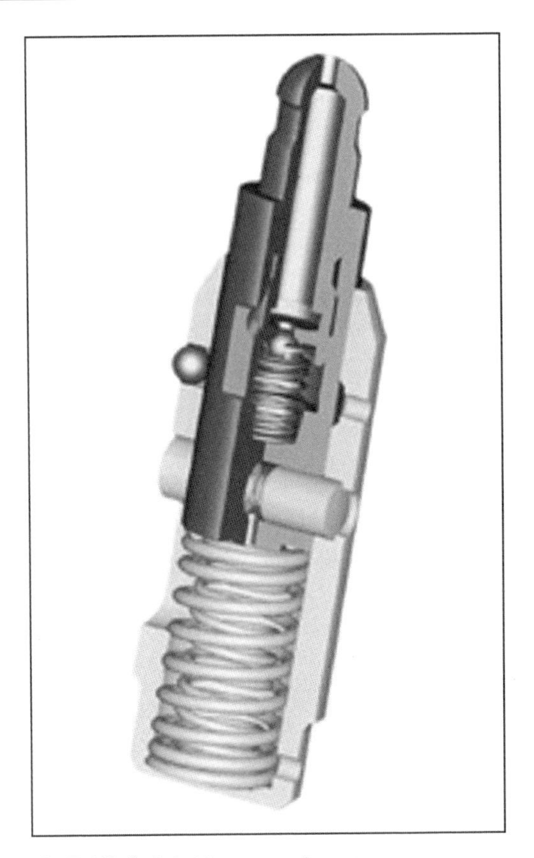

Fig. 7-119 Switchable support element.

In the case of defeatable cam follower units, which may also be fitted with hydraulic elements, this lost motion can be absorbed in supporting elements (Fig. 7-119); alternately a compression spring between two sections of the lever absorbs the mass forces for the part of the lever that continues to be moved (Fig. 7-120).

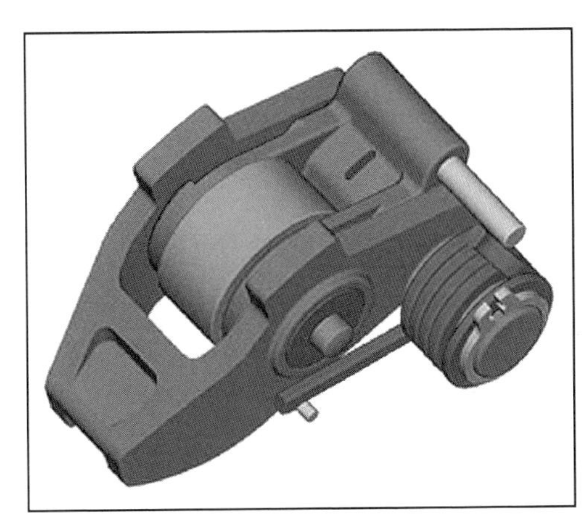

Fig. 7-120 Switchable cam follower. (*See color section.*)

The situation is similar for cam follower and rocker arm valve trains. Here the separation of the physical connection in the lever (or lever system) and/or in the switching mechanism is the normal configuration since, because the levers are borne on shafts, decoupling at the support presents difficult engineering problems (Fig. 7-121).

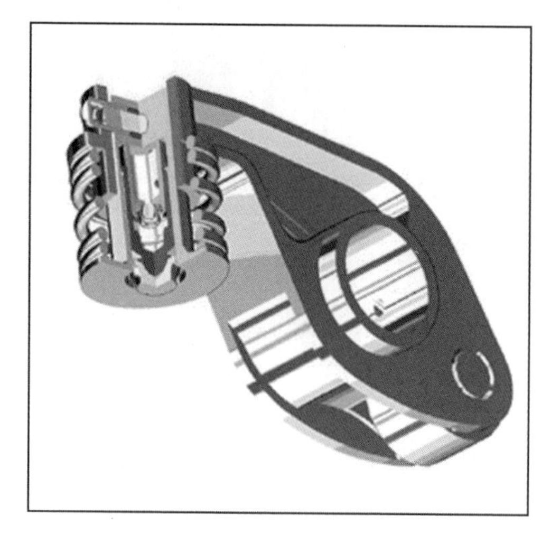

Fig. 7-121 Switchable rocker arm.

In OHV valve trains, used primarily in older types of large-displacement engines, shutting down the valves is simple. Here it makes sense to interrupt the physical connection at a point near the cam—such as at the (roller-type) push rod—to keep the masses in a motion as small as possible when in the deactivated state. The switchable, roller-type push rod shown in Fig. 7-122 is designed for cutout operation and as a consequence is fitted with only one rolling bearing to follow the cam contour. This has such a great effect on the engineering design that it differs severely from the switchable valve lifter shown in Fig. 7-118 even though the two units are similar in function.

The latter, with two cam contact surfaces (sliding surfaces in this case) and operating in conjunction with a cam packet, can also be employed as a true selector between two differing valve stroke curves. This stroke selector is used to activate different valve lift curves, the choice depending on the momentary operating situation.

In addition to the two-stage concepts, multistage concepts have also been implemented (Fig. 7-123); they approach valve trains with stepless, fully variable lift stroke. The fully variable valve drives, engineered without a throttle if at all possible, involve considerably greater space requirements and considerable complexity in both engineering and control technology. Consequently, one could well conceive of systems that use cam stroke selectors with separate changeover for the individual valves in multivalve engines in order to achieve a multistep effect with less effort than that required for fully variable systems.

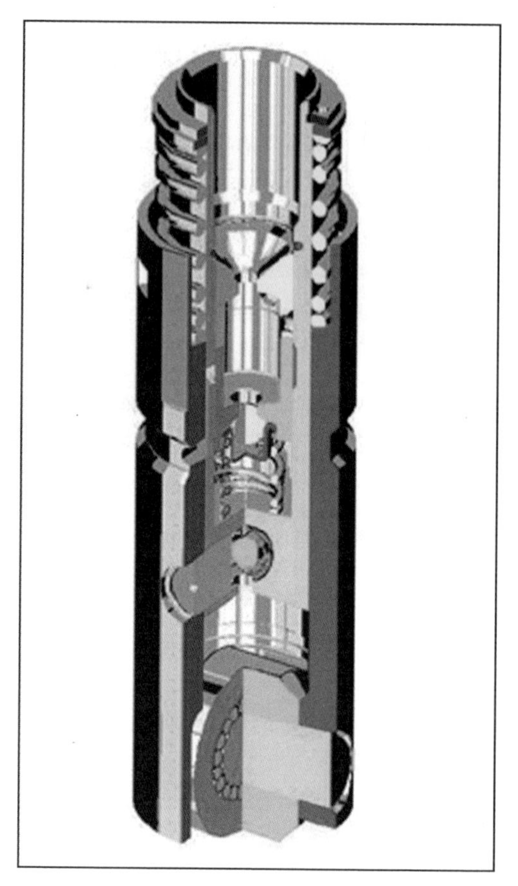

Fig. 7-122 Switchable OHV valve train.

Enhanced variability, particularly in regard to the intake valve actuation train, can be achieved with less effort in combination with camshaft adjustment or shifting systems. Here the results achieved by optimizing intake valve lift (as dictated by the operating situation) by varying the lift cycle and by shifting phases do, in fact, approach the fully variable valve train while at the same time utilizing familiar, rugged components (Fig. 7-124).

The coupling mechanism can be actuated either hydraulically or mechanically. Examples of mechanical actuation elements are linear and rotational electromagnets that activate the coupling and lockout mechanism via a physical connection. A hydraulic control concept (Fig. 7-125) uses the oil circuit already on hand in the cylinder head. The assignment of the switching states (coupled/decoupled) is made here with changes in oil pressure. Here the engineering implementation of both potential versions (coupling at zero pressure or decoupling at zero pressure) has been successful, and this broadens freedom in thermodynamic design.

The mechanical switching period that takes into account the coupling element excursion only may be in the

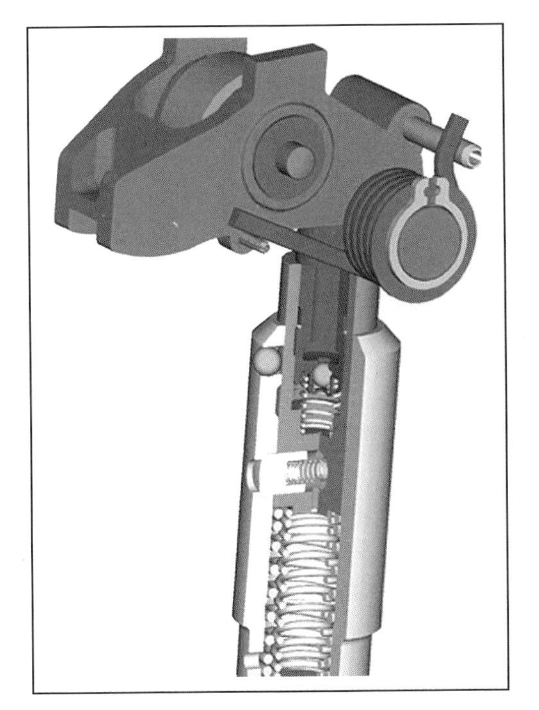

Fig. 7-123 Switchable support element and switchable cam follower. *(See color section.)*

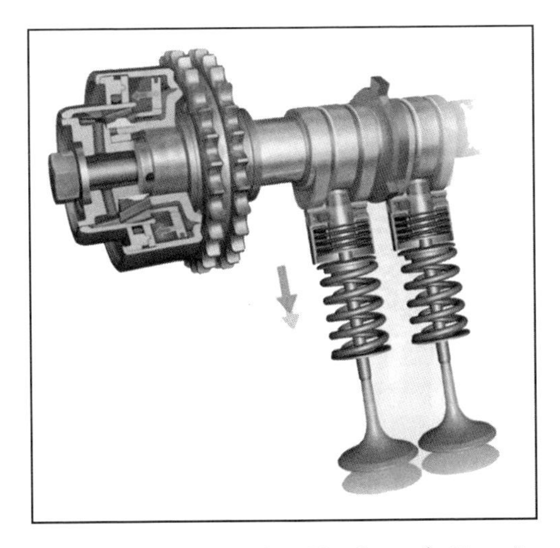

Fig. 7-124 Porsche VarioCam Plus System.[2] *(See color section.)*

range of about 10 to 20 msec where peripheral conditions are good. Since the other influences such as electrical and hydraulic dead times can be largely eliminated by the engine electronics, it becomes clear that switching from one operational status to the next is possible within a single camshaft revolution, up to high rotation speeds.

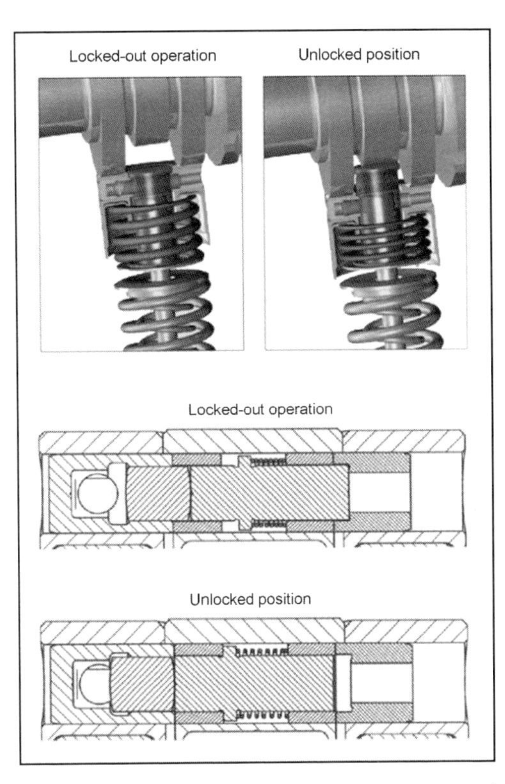

Fig. 7-125 Coupling mechanism (switching positions).[2] *(See color section.)*

Cylinder Shutdown

A method for implementing one of the variabilities described above is cylinder shutdown, which is used primarily in large-displacement engines (with 8, 10, or 12 cylinders, for example). The purpose of cylinder shutdown is to minimize the gas exchange losses (pumping and throttle losses, Fig. 7-126) and/or to shift the operating point. Reduction of friction loss is achieved with lower spring forces at the deactivated cylinders. Here the camshaft works only against the lost-motion spring forces that are less than comparable valve spring forces, by a factor of 4 to 5. Equidistant ignition sequences make it possible to "convert" standard V-8 and V-12 engines to V-4 or six-cylinder inline engines, respectively. Trials carried out using a V-8 engine at a test bed showed that employing cylinder shutdown made it possible to achieve fuel savings potentials of from 8% to 15% in normal driving cycles.

Stroke Changeover

A second way to implement variability in the valve stroke is to change the length of the valve lifting stroke. This concept aims to increase thermodynamic efficiency particularly by reducing the losses associated with the gas charge change. Positive effects are also expected here in regard to friction losses since the lost-motion springs used

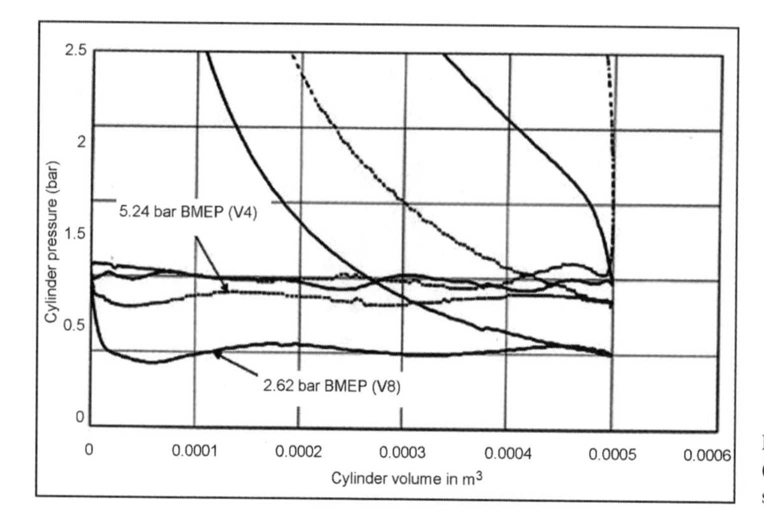

Fig. 7-126 Gas charge change (p-v chart) with/without cylinder shutdown.

here are also relatively weak; thus the total effective valve spring and lost-motion spring forces in partial stroke operations are less than the valve spring forces effective in full stroke operation. Implementing this system together with camshaft shifting lets us achieve thermodynamic optimization at many of the engine's operating points, and this will be reflected in a significant drop in fuel consumption.

This technology has achieved full maturity for volume production in the new Porsche 911 Turbo. Because of the four-valve technology widely used today, this system can achieve a multistep effect that represents a great advance toward full variability (Figs. 7-127 and 7-128).

Fully Variable Valve Trains

Among the fully mature embodiments of fully variable valve trains is the BMW Valvetronic concept. It offers great benefits in terms of consumption as well as in retaining stoichiometric operation with all its advantages and, in addition, can be used all around the world, regardless of fuel formulations (sulfur content).

The Valvetronic achieves engine operation without the need for a butterfly throttle valve. Cylinder fill at partial load is regulated by the intake valve lifting stroke and opening period. The intake and exhaust camshafts are driven by variable cam adjustment.

To achieve stepless adjustment of intake valve stroke, an intermediate lever, backed against an eccentric shaft, is inserted between the camshaft and the cam follower. The contour of the contact surface between this intermediate lever and the roller cam follower defines the valve lifting curve. Rotating the eccentric shaft moves the fulcrum for the intermediate lever and thus—steplessly—changes the lever ratio and, consequently, the relationship between the cam lobe stroke and the valve stroke. In this way, it is

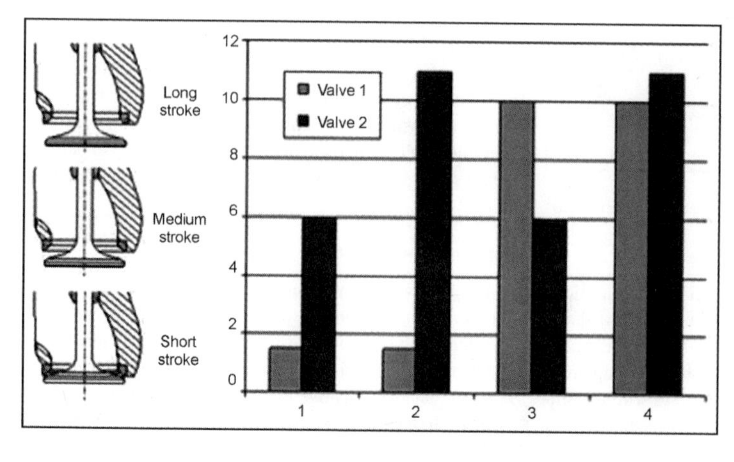

Fig. 7-127 Multistep.

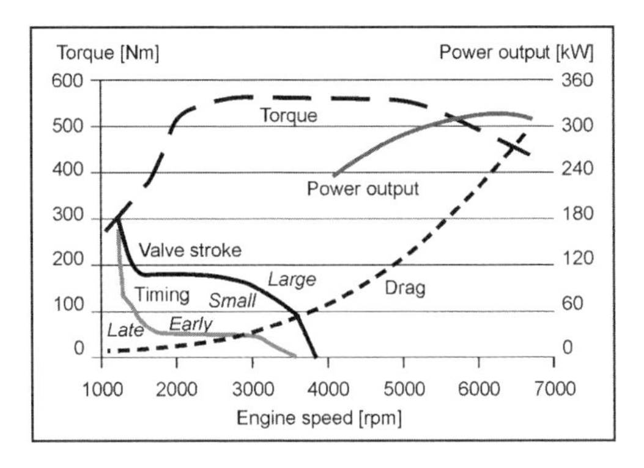

Fig. 7-128 Porsche 911 Turbo engine map.[2]

possible to achieve valve excursion from about 0.3 mm at idle to 9.7 mm at full throttle.[3]

7.10.2 Belt Tensioning Systems, Idler and Deflection Pulleys

7.10.2.1 Introduction

For 40 years now synchronous belts have been used successfully in mass production to drive camshafts and balancer shafts in internal combustion engines. The first application was in a four-cylinder engine made by Glas without any auxiliary components such as idler or deflection pulleys. In later designs the preload in the toothed belt was generated either with a component that was driven by the toothed belt (such as the water pump) and mounted on an eccentric bracket or by fixed idlers (eccentric tensioning pulleys or the like). Ideal adjustment of belt tension is not possible in such systems since they cannot compensate for belt tension fluctuations caused by temperature changes or aging; neither is it possible to compensate for dynamic effects (belt vibration, reverse influences from the valve train, etc.). Compensating for such fluctuations and effects is absolutely necessary in modern synchronous belt drives since only in this way can one achieve the targeted system service lives (corresponding to engine service life) of at least 240 000 km for gasoline engines and at least 160 000 km for diesel engines.

The influence that a fixed idler (tensioner) pulley has on static belt preload is illustrated in Fig. 7-129.

Using an automatic belt tensioning system makes it possible to considerably reduce the spread in preload values at initial assembly and to keep preload values nearly constant across the engine's full operating temperature range.

Automatic tensioners have been used with synchronous belt drives for internal combustion engines since the beginning of the 1990s and, for the reasons mentioned above, have forced fixed systems almost completely off the market.

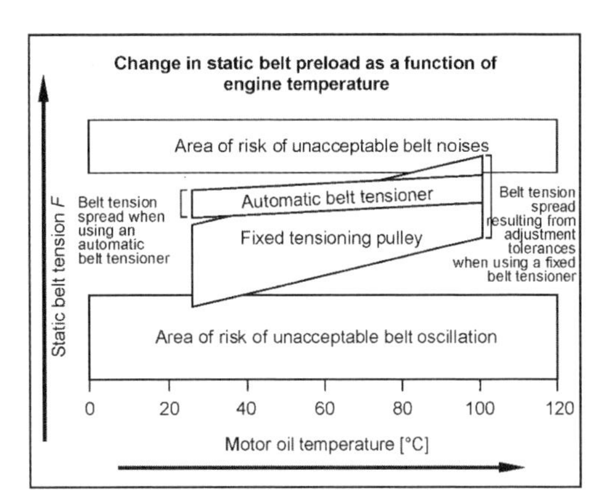

Fig. 7-129 Changing the static belt preload value using motor oil temperature as the lead variable; comparison between fixed tensioning idler and automatic tensioning system.

7.10.2.2 Automatic Belt Tensioning System for Synchronous Belt Drives

The primary requirements for automatic tensioning systems are derived from the conditions enumerated above and are the following:

- Setting specified belt tension at initial installation and after service (compensating for belt, diameter, and positioning tolerances).
- Maintaining the most constant belt tension possible at all operating states across the entire required system service life (compensation for thermal elongation, belt stretch, and wear, taking account of crankshaft and camshaft dynamics).
- Ensuring ideal noise levels while at the same time reducing belt vibration.
- Preventing tooth jump.

The parameters shown in Fig. 7-130 have to be taken into consideration when specifying the working range for a tensioning system such as this.

Of the various styles for synchronous belt tensioning systems (with hydraulic damping, linear action with

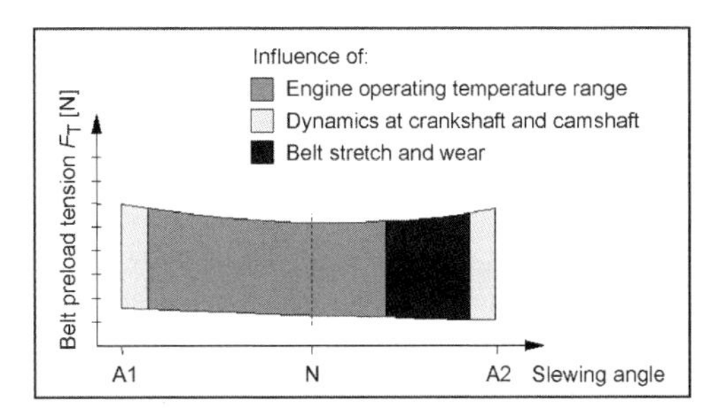

Fig. 7-130 Mechanical synchronous belt tensioning unit—sample operational chart with influencing parameters.

reversing lever; with hydraulic damping, rotating; with mechanical damping, rotating), the rotating mechanical systems are most widely used for reduced costs and less space required. The temperature-based tensioning systems using wax thermostats, employed in some engines in the past, never made a breakthrough. The basic design of a mechanical tensioning system such as this, using the so-called double-eccentric principle, is shown in Fig. 7-131.

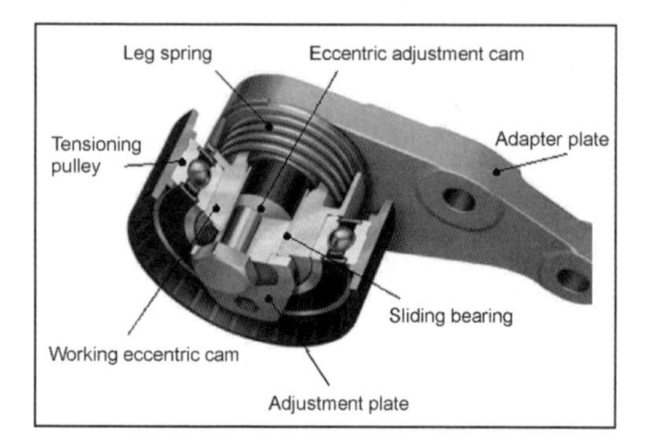

Fig. 7-131 Mechanical synchronous belt tensioning unit with double-eccentric design.

Here the adjustment eccentric compensates for the tolerances in all the components present in the belt drive; its setting is fixed after initial adjustment. The working eccentric mounted movably on the adjustment eccentric compensates for the temperature-induced changes in length at all the components used in the belt drive, for belt stretch and wear, and for dynamic effects originating at the crankshaft and camshaft. The lever spring is designed in accordance with ideal belt preload. Damping is affected with the slide bearing, and the tensioner's geometry is matched to the requirements of the belt drive.

7.10.2.3 Idler and Deflection Pulleys for Synchronous Belt Drives

It is for the foregoing reasons that fixed tensioning pulleys are found only rarely in modern engines. Deflection pulleys used, for instance, to calm critical sections of the belt, to avoid collisions with adjacent components, or to increase the wrap angle at neighboring pulleys have to satisfy the same requirements regarding service life and noise development. High-precision, single-row ball bearings with enlarged grease reserve spaces have proven their suitability for this application; if necessary, double-row angular ball bearings, also with optimized reserve grease spaces, may be used. These bearings are normally packed with high-temperature rolling bearing grease and fitted with suitable sealing rings; standard bearings are less suitable for this purpose. These bearings serve as the centers for pulleys that match the prevailing geometric requirements. Exemplary embodiments with plastic and steel pulleys are depicted in Fig. 7-132; these pulleys may

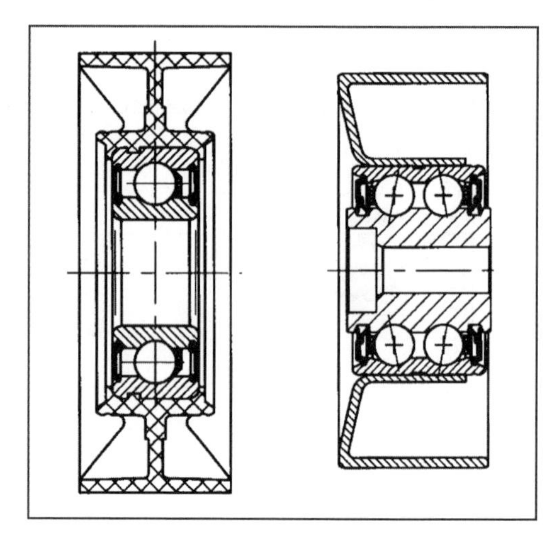

Fig. 7-132 Deflector pulley with single- and double-row ball bearings and pulleys in plastic and steel.

be equipped with flanges on one or both sides to guide the belt.

7.10.2.4 Prospects for the Future

Modern synchronous belt drives for internal combustion engines are no longer conceivable without automatic tensioning systems since only with their help can the required system service life of 240 000 km and more be achieved. Because of cost and space considerations, hydraulically damped systems are being replaced to greater extent by those that are mechanically damped. Key areas of emphasis in development work at the present are mechanical tensioning systems for the heavily loaded synchronous belt systems used in diesel engines, systems designed to facilitate installation, and mechanical tensioning systems (with either open-loop or closed-loop control) for ideal matching of the preload force to engine operating conditions.

7.10.3 Chain Tensioning and Guide Systems

7.10.3.1 Introduction

In addition to the synchronous belt tensioning and deflection systems described in Section 7.10.2, timing chains have long been used to drive camshafts in internal combustion engines. In addition, they are found in conjunction with the balancer shaft drives more frequently and as the power connection between the crankshaft and the oil pump. Chain tensioning and compensation for chain wear are normally affected with a chain tensioning element; any of a number of systems may be chosen, depending on the installation location.

In contrast to belt drives, the use of free span lengths in chain drives is very limited so that the tensioning and guide rails used to guide the chain attain great significance. Some of the fundamentals for the individual

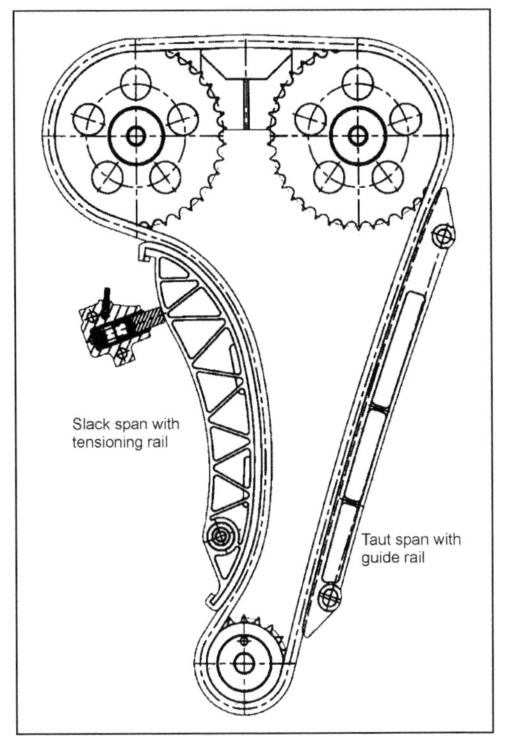

Fig. 7-133 DOHC timing scheme to explain the nomenclature.[4]

components are to be discussed in greater detail below. The terms used here are explained in Fig. 7-133 on the basis of a DOHC timing drive configuration.

7.10.3.2 Chain Tensioning Element

Hydraulic chain tensioning equipment is used in most of the timing chain drives found on the market today. They are, as shown, situated at the slack span of the chain drive and are connected with the motor oil circuit by supply bores.

They must satisfy the following main requirements:

- Preloading the chain drive to keep the chain from "climbing" or jumping on the sprockets
- Compensating for the chain wear occurring during the engine's service life
- Damping the oscillations induced by the chain at the tensioning rail
- Reducing tooth jump, particularly in chains stretched as a consequence of wear

In some cases the tensioning devices, into which oil spray nozzles are integrated, also handle the lubrication function needed to ensure proper chain drive functioning.

Figure 7-134 provides a schematic depiction of a tensioning element when installed.

This is a speed-proportional, leakage-gap damping unit that, because of the inclusion of the check valve, exhibits directional (single-sided) damping. The motor oil passes initially through the supply bore and a system of grooves in the tensioning element and into the reserve chamber. If, when the load is relieved on the slack span, excess pressure is created in the high-pressure chamber (located between the plunger and the housing) because of the spring-induced extension of the tensioning plunger, then the check valve is opened and oil passes from the reserve chamber into the high-pressure chamber. When a load is placed on the plunger, the valve spring and the pressure building up in the high-pressure chamber close

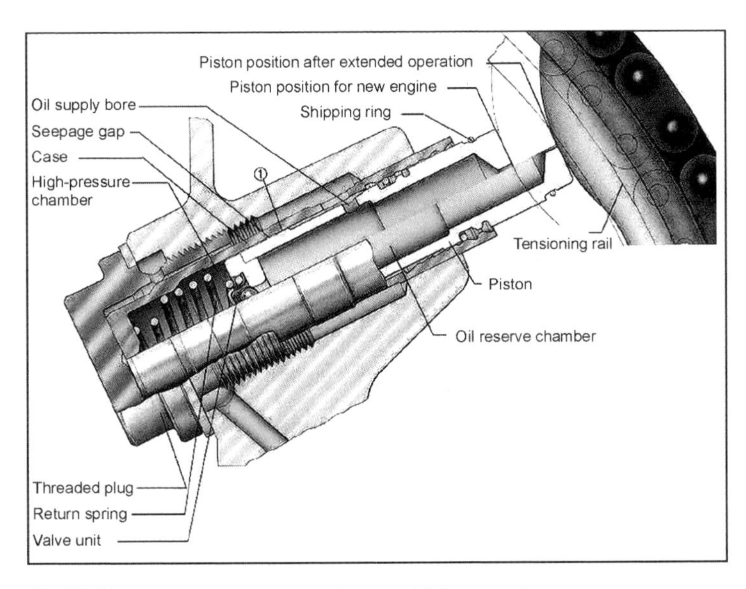

Fig. 7-134 Leakage gap tensioning element with integrated retraction stop.

the check valve. Oil is forced through the leakage gap and out of the high-pressure chamber. Thus, a damping effect is created, the amount being a factor of the width of the gap between the plunger and the housing. The pressure prevailing in the high-pressure chamber during the damping process can be as much as 80 bar and even more in isolated cases.

Because of the sensitivity of the damping characteristics to the size of the leakage gap, manufacturing precision in the individual components has a major influence on tensioner quality.

Some of the tensioner versions that have been engineered incorporate retraction stops to keep the tensioner from collapsing when the load is reversed at engine standstill. This effectively prevents tooth jump when the engine is restarted.

When designing the tensioner components, particular attention is paid to those segments of the excursion stroke that are active while the engine is running. This is shown schematically in Fig. 7-135. The portion of the stroke that compensates for thermal expansion has to be determined carefully, taking account of the materials used in building the engine. If this portion is underdimensioned, then there is a hazard that the tensioner will bottom out and, as a consequence, the chain will be overtightened.

Simple tensioners are used, especially in oil pump drives, because it is often possible to do without closely defined damping. In some cases plastic elements are even chosen instead of the steel components normally employed.

7.10.3.3 Tensioning and Guide Rails

Tensioning and guide rails are used to guide the chain along the spans. They normally consist of a slip-promoting plastic surface and a backing element adapted to suit the geometry of the space in which it is installed. Backing elements made of die-cast aluminum are superseded more and more today by injection-molded plastic parts. Glass reinforcing fibers may be added to the injection resin for the backing unit in order to stiffen the relatively soft plastic, which is also highly sensitive to heat.

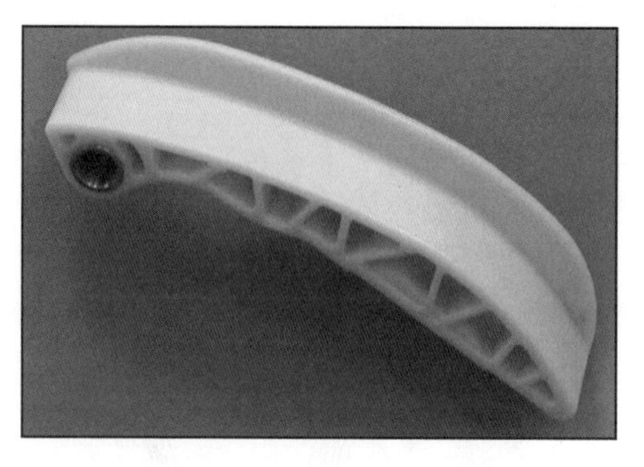

Fig. 7-136 Tensioning rail made of solid plastic.[4]

Design work here often uses finite element calculations to determine the loading that will be encountered. Figure 7-136 shows a tensioning rail manufactured completely from nonreinforced plastic that incorporates metal reinforcement only in the area around the pivot point.

7.10.3.4 Sprockets

Sprockets are used to transmit chain forces to the various shafts in the engine. The tooth geometry for these sprockets follows applicable standards without, however, making full use of the tolerances allowed there. Either precision stamped or sintered components are employed, depending on the sprocket geometry. In isolated cases, and particularly where multiweb chain is used, these sprockets may be manufactured so as to induce internal stresses.

Bibliography

[1] Eidenböck, T., R. Ratzberger, J. Stastny, and W. Stütz, "Zylinderkopf in Vierventiltechnik für den BMW DI-Dieselmotor," in MTZ, 1998, No. 6, p. 372.

[2] Proceedings, ÖVK, Institute for Combustion Engines and Vehicle Engineering at the Technical University of Vienna (organizer), 21st International Vienna Engine Symposium, Vienna, 2000, VDI, Düsseldorf, 2000.

[3] Proceedings, ÖVK, Institute for Combustion Engines and Vehicle Engineering at the Technical University of Vienna (organizer), 22nd

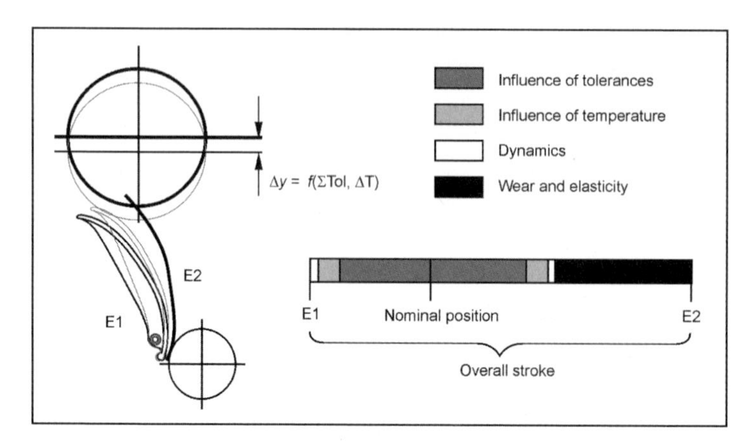

Fig. 7-135 Subdivision of chain tightener stroke.

International Vienna Engine Symposium, Vienna, 2001, VDI, Düsseldorf, 2001.
[4] Proceedings 200, ASK Altmann (organizer), 4th Plastic Engine Components Forum, Spitzingsee, 2001.

7.11 Valves

7.11.1 Functions and Explanation of Terms and Concepts

Intake and exhaust valves are precision engine components used to block gas flow ports and to control the exchange of gases in internal combustion engines. They are intended to seal the working space inside the cylinder against the manifolds. An example of a valve in place in the engine is shown in Fig. 7-137.

The intake valves, which are not subjected to such extreme thermal loading, are cooled by incoming gases, by thermal transmission at the seat, and by other means. Exhaust valves, by contrast, are exposed to severe thermal loads and chemical corrosion.

These two types of valves are manufactured using different materials, matched to the functions they perform. It may be assumed that during an engine's service life the valves will execute about 300 million operating cycles, many at very high temperatures. The most important terms used to describe valves are depicted in Fig. 7-138.

7.11.2 Types of Valves and Manufacturing Techniques

Valves may be subdivided essentially into three major groups: Monometallic valves, bimetallic valves, and hollow valves.

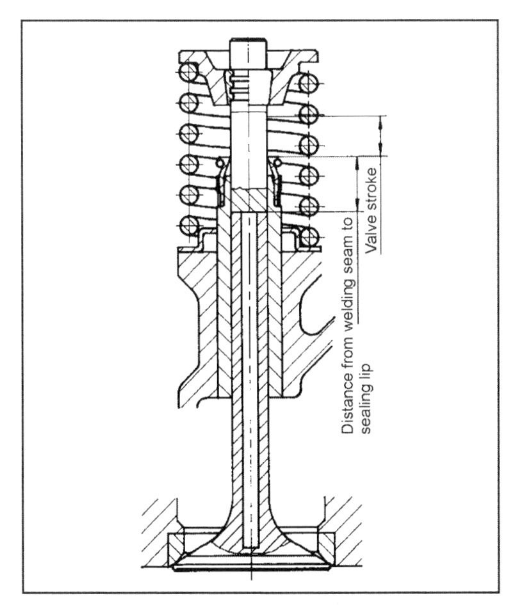

Fig. 7-137 Hollow valve in place in the engine.

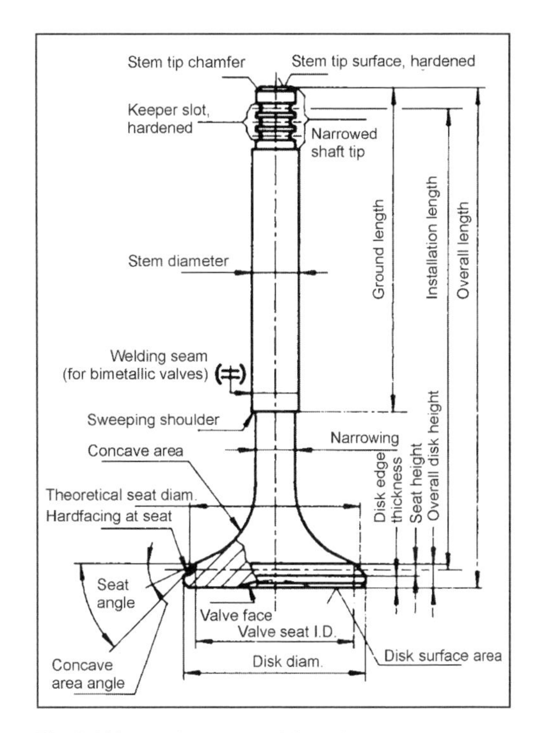

Fig. 7-138 Terminology used for valves.

7.11.2.1 Monometallic Valves

Monometallic valves may be manufactured either in the hot extrusion process or by upsetting.

The starting point in the hot extrusion process is a section of rod whose diameter is about two-thirds of the final disk diameter; its length corresponds to the volume of the blank to be manufactured. This rod is heated and reformed to make the blank in two forging steps.

During the upsetting process a ground section of rod, the diameter of which is slightly greater than the ultimate valve stem diameter, is first heated at the end; the rod is then forced forward to form a "pear," which is then reformed in a die to create the valve head.

7.11.2.2 Bimetallic Valves

Bimetallic valves permit the ideal combination of materials, each matched exactly to the needs of the valve stem and the valve head. Here again one works on the basis of a heat re-formed head that is made in the process described above and then attached to the stem by friction welding, Fig. 7-139.

The pairs of materials normally given preference are X53CrMnNiN219, X50CrMnNiNb219, X60CrMnMoV-NbN2110, and NiCr20TiAl for the head and X45CrSi93 for the stem.

It makes good sense to position the welding seam at a point along the valve stem so that, when the valve is closed, the seam is inside the guide by half of the valve

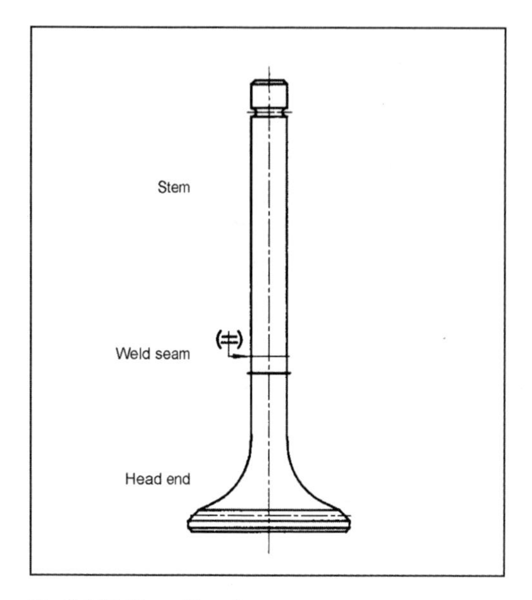

Fig. 7-139 Bimetallic valve.

lifting stroke and/or about 6 mm above the sweeping shoulder. Here, for reasons dictated by the manufacturing technology, it is necessary to ensure that the length of the cylindrical section on the head itself is at least 1.5 times the stem diameter. The seats for bimetallic valves can, of course, also be hardfaced.

7.11.2.3 Hollow Valve

This version is used primarily for the exhaust valves and, in certain special circumstances, on the intake side as well, to lower the temperatures primarily in the concave area at the back of the head and at the disk area and for weight reduction.

The sodium used for heat transmission is located in the hollow cavity in the valve stem and can move freely. A portion of the heat impinging on the concave area at the back of the head and the valve face is passed by the liquid sodium to the valve guide and, from there, to the coolant circuit, Fig. 7-140.

If one employs hollow valves to reduce temperatures, then about 60% of the volume inside the hollow space will be filled with metallic sodium. This liquid sodium (melting point 97.5°C) is shaken inside the valve cavity to an extent corresponding to the engine speed. It transports heat from the valve head into the valve stem. The degree of temperature reduction at perfect thermal energy flow and the smallest possible working clearances are in a range of from 80 to 150°C.

Hollow valve variants:

- "Tube on solid metal" version: The head, which is drilled from the stem end (forming the tube), is attached by friction welding to a (solid) stem end section, which is alloyed so it can be hardened.

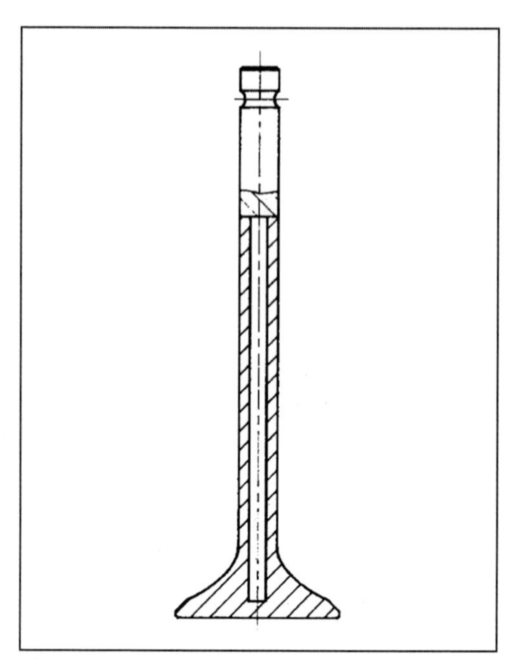

Fig. 7-140 Hollow valve.

- "Closed-off" version: This version is far more elaborate in its manufacture than the version described above. The workpiece is also drilled from the stem end. The bore is closed with inductive heating and subsequent forging. The stem end section is attached with friction welding. Such closed-off hollow-stem valves are used primarily in high-performance engines and aviation applications.
- Hollow valve: This valve represents a further measure taken to reduce weight and enhance heat transfer away from the center of the valve disk. These valves, in contrast to the above-mentioned designs, are drilled and machined from the disk end. The opening is closed by inserting a capping plate, using a special process to do so. These valves, more expensive to manufacture, are used primarily in racing engines, Fig. 7-141.

Hollow valves may be made at stem diameters of 5 mm and upwards. The diameter of the internal bore is about 60% of the stem diameter.

To avoid exposing the valve stem seals to excess temperatures, the bore inside the valve has to end about 10 mm away from the contact range for the sealing lip. Any change in clearance between the valve stem and the valve guide, different from that found in solid valves, is also observed. Valve sticking is reduced by tapering the stems slightly to compensate for the temperature gradient.

Hollow valves may be of a single metal, but bimetallic valves with the following combinations of materials are more common: X53CrMnNiN219, X50CrMnNiNb219, and NiCr20TiAl for the head section and X45CrSi93 for the stem section.

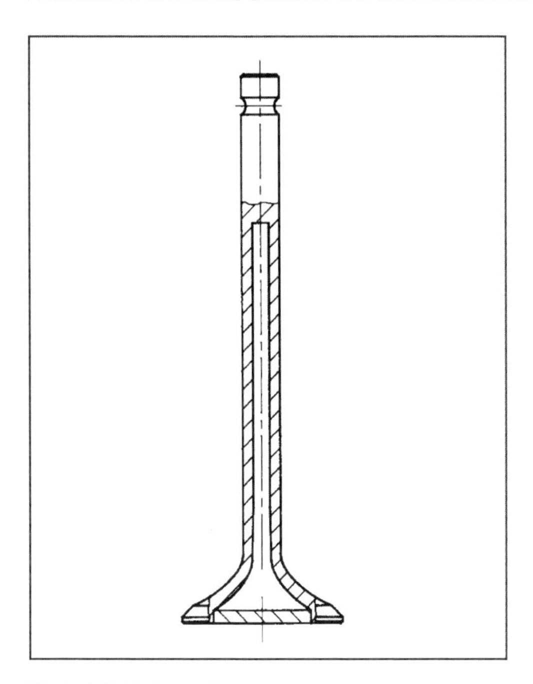

Fig. 7-141 Hollow valve.

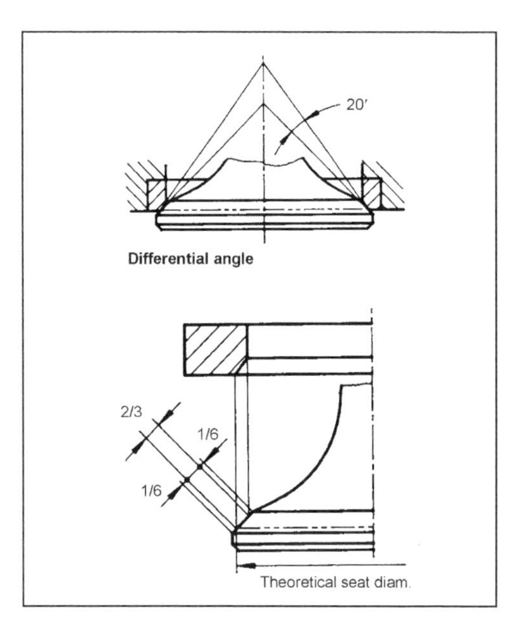

Fig. 7-142 Differential angle and valve seat width.

7.11.3 Embodiments

7.11.3.1 Valve Head

The theoretical diameter of the valve seat is the basis for the engineering design of the valve.

Overall disk height depends upon the combustion pressure and the average valve component temperature. This height is established by finite element analysis. Practice has shown that values of from 7% to 10% of valve head diameter are common.

The thickness of the edge of the disk determines the stiffness of the valve head and is coordinated with the valve seat angle; at 45° it is about 50% of overall disk height, at 30° approximately 55% to 60% of overall disk height.

The valve seat angle is generally 45°. Seat angles of 30° and 20° may, however, also be selected to reduce valve seat wear. Small seat angles are indispensable in gas-fired engines. Manufacturing technology requires a difference of at least 5° between the valve seat angle and the valve face angle, Fig. 7-142.

The differential angle between the valve seat and the seating ring achieves initial sealing along a line of contact, thus creating better seal of the face against the combustion chamber. Attention is needed to ensure that the valve seat width is greater than the seating ring contact width.

Curved depressions on the valve face are provided to reduce valve weight, to influence combustion chamber shape, and to distinguish between intake and outlet valves or similar valves.

The ideal shape for the transition from the concave area at the back of the head to the valve stem can be identified only with the appropriate engine trials.

7.11.3.2 Valve Seat

The seat for the exhaust valve is heavily impacted by heat and corrosion, which is why, as a rule, it is hardfaced with special alloys. In isolated cases, this is also done for the intake valve even though here martensitic hardening is normally used because of the material selected. Hardfacing can be used to reduce wear and enhance the sealing effect. The following processes are used for valve hardfacing:

- Fusion welding, in which the hardfacing material in rod form is melted and applied by means of an oxy-acetylene flame.
- Electrical PTA process (plasma-transferred arc) in which the pulverized hardfacing material is melted in a plasma arc and applied to the workpiece.

These hardfacing techniques are used for hollow valves, bimetallic valves, and occasionally for monometallic valves, as well. To keep any reduction in hardness at the inductively hardened valve seat within acceptable limits, it is necessary to ensure that valve temperatures do not exceed a maximum of from 550 to 600°C.

7.11.3.3 Valve Stem

This component is used to guide the valve inside the valve guide and is defined by the first keeper slot provided to mate with the conical keeper and by the sweeping shoulder and/or the transition to the concave area at the back of the head.

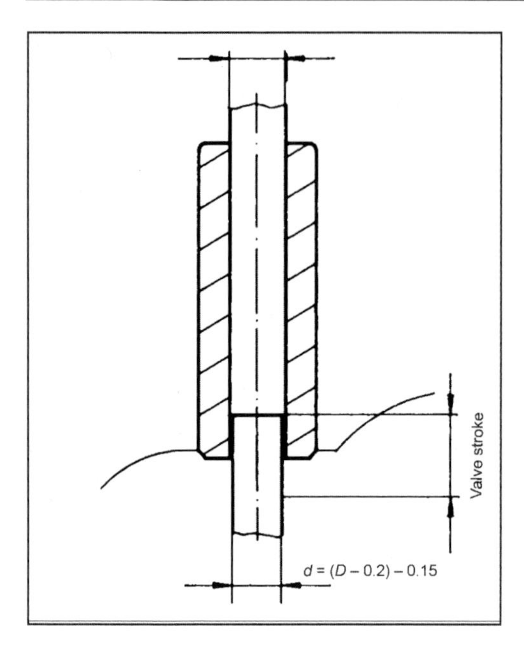

Fig. 7-143 Valve stem with narrowing and sweeping shoulder.

To limit the formation of soot on the end toward the gas port, a sweeping shoulder is made by narrowing the stem diameter, Fig. 7-143. When the valve is closed, this shoulder should be inside the valve guide by about one-half of the valve lifting stroke.

If, during the valve closing phase, bending is induced because of cylinder head warping or noncongruence of the centerlines, then it is desirable for the welding seam to be inside the valve guide. This is why the friction welding seam for bimetallic valves is moved to at least one-half stroke length inside the valve guide.

Depending on the tribologic situation, it is necessary to protect the valve stem surface against wear by using chrome plating or nitriding. There is a specific ratio between valve stem diameter and the valve disk diameter. The disk-to-stem ratio for intake valves is 6:1; for exhaust valves it is 5.5:1.

As a rule, the valve stem is cylindrical in shape. To take account of the variations in expansion due to the temperature gradation the valve stem may be tapered between 10 and 15 μm, depending on the length and diameter of the stem.

Valve stem ends featuring multiple keeper slots, the purpose of which is to support unrestricted valve rotation, is always inductively hardened in the area where the keeper makes contact in order to avoid wear. It is for the same reason that, where the valve actuators exert very high surface pressures, shims made of tungsten carbide or a hardenable stem material are welded into the end of the shaft, Fig. 7-144. Valves with a single keeper slot are seldom hardened, but here, too, shaft end shims made of tungsten carbide or other materials that can be hardened may be used for wear protection.

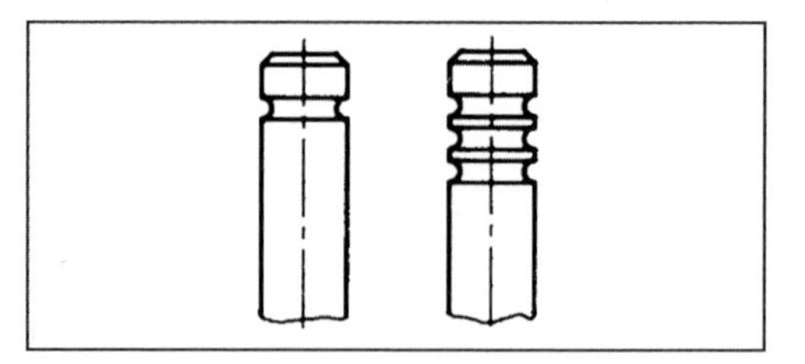

Fig. 7-144 Types of keeper slots in valve stems.

The distance between the end of the stem and the middle of the keeper slot may not exceed 2.5 mm. The sharp edges at the end of the stem are smoothed either by chamfering at less than 45° or 30° or by rounding, which assists in automatic valve installation.

Valve Guide

The valve guide ensures that the valve centers in the valve seat and that heat can be dissipated from the valve head, through the valve stem and to the cylinder head. This necessitates an ideal clearance valve between the guide bore and the valve stem. If there is insufficient clearance, then the valve tends to stick. Too much clearance interferes with heat dissipation. One should strive to achieve the smallest possible valve guide clearance. In addition, it is necessary to ensure that the end of the valve guide does not protrude unprotected into the exhaust port as otherwise there is a danger of the valve guide dilating and combustion residues entering the valve guide. As a rule of thumb, the length of the valve guide should be at least 40% of the length of the valve.

To ensure perfect valve functioning, it is necessary that the offset between the centerlines of the valve shaft and the seating ring be kept within certain limits (0.02 to 0.03 mm in a new engine). Excessive misalignment can cause above all serious bending of the valve disk in relationship to the stem. This excessive loading can lead to premature failure; other consequences may also be leaks, poor heat transmission, and high oil consumption.

7.11.4 Valve Materials

The demands made on a valve include endurance strength at elevated temperatures, wear resistance, resistance to high-temperature corrosion, and oxidation and corrosion resistance.

The standard valve materials are the following:

- Ferritic-martensitic valve steels: X45CrSi93 is the standard choice for monometallic intake valves and is used exclusively as the material for the stem in bimetallic valves. X85CrMoV182 is a higher alloy and is used as an intake valve material where the thermal and mechanical loading does not permit the use of the Cr-Si material.
- Austenitic valve steels: Here the austenitic Cr-Mn steels have proven to be an economical solution. A

widely used choice is the X53CrMnNiN21-9 (21-4N) alloy, which is deemed to be the classic exhaust valve material—for hollow valves, too.

- Valve materials with high nickel content: If the Cr-Mn steels no longer satisfy thermal requirements, then a transition to materials with high nickel content is the correct remedy. They are necessary where maximum operational reliability, and that means resistance to spalling and corrosion, are needed (in aviation engines, for racing use, in highly turbocharged diesel engines, and for using heavy oil as the fuel).

Valve steels prepared in a powder metallurgical (PM) process are available as special materials. In this way, material qualities are achieved that have a positive effect on strength and on resistance to hot corrosion.

7.11.4.1 Heat Treatment

Closely defined heat treatment makes it possible to further improve the technical characteristics of the valve steels. In many cases this can obviate the need for going to higher-quality alloys.

Martensitic valve steels are generally hardened. The hardness and strength of austenitic steels can be boosted by so-called structural (precipitation) hardening.

7.11.4.2 Surface Finishing

The following techniques may be used:

- Hard chrome plating for the valve stem: The manufacturing process, choice of materials, and operating conditions may make it necessary to chrome plate standard valves at the contact area along the stem. In standard bimetallic valves the chrome layer, from 3 to 7 μm thick, covers both valve materials. Thicker applications of chrome, up to 25 μm, may be employed in truck or industrial engines where there are high load levels or where there is more severe wear.
- Abrasive polishing: In all cases the stem has to be polished whenever the valve is chrome plated in order to remove any chrome nodules still present and to level out any unevenness. Roughness after the polishing operation is a maximum of Ra 0.2 (maximum Ra 0.4 for nonplated), which has a very favorable effect on valve guide wear and thus permits engineering for minimum clearance.
- Nitriding the valves: Bath immersion and plasma nitriding are used. The nitriding layers, approximately 10 to 30 μm in thickness, are extremely hard at the surface (approximately 1000 HV 0.025) and are particularly insensitive to wear. Like chrome-plated valves, immersion-nitrided valves are abrasively polished to finish them.

7.11.5 Special Valve Designs

Racing imposes the severest demands on the valve, and here it is a matter of withstanding extreme loading for relatively short periods of time.

Achieving engine speeds of some 18 000 rpm requires a very free-running valve train and a lightweight valve.

The next step toward weight reduction, in addition to adopting hollow valves, is to choose more exotic materials—such as titanium. This material permits components that are about 40% lighter when compared with steel. It must be kept in mind, however, that titanium does not offer very good high-temperature strength. That is why, when using titanium for exhaust valves, it is essential to ensure particularly effective heat dissipation. This is done with hollow valves in conjunction with seat rings exhibiting high thermal conductivity.

7.11.5.1 Exhaust Control Valves

Turbocharger regulation valves (overrun control valve): The overrun control valve (also referred to as a "waste gate") limits the charging pressure developed by the exhaust turbocharger and, in gasoline engines, can be intermittently exposed to temperatures of about 1000°C; the thermal load in the diesel engine is about 850°C. This is the criterion used when selecting engineering materials. Diesel engines can usually get along with the 21-4N alloy (X53CrMnNiN21-91), while a material that can withstand high temperatures, such as Nimonic 80A (NiCr20TiA1), is used in gasoline engines. The overrun control valve is secured with screws or rivets. Typical embodiments are shown in Fig. 7-145.

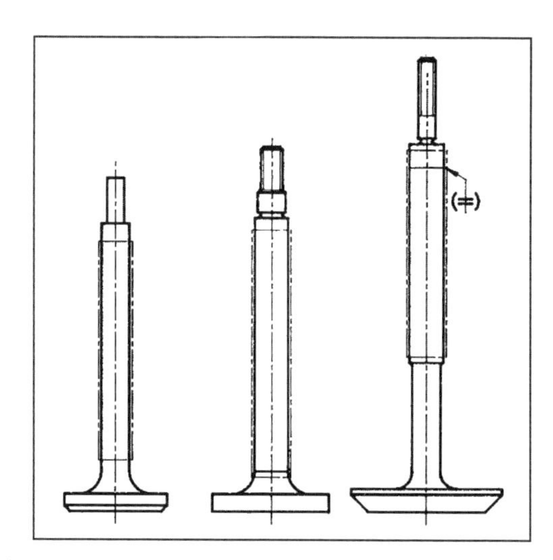

Fig. 7-145 Configurations for overrun control valves.

Exhaust gas return (EGR) valve: EGR valves have to cope with temperatures of up to about 800°C. Of the valve materials available for use, the 21-4N alloy (X53CrMnNiN21-9) has been found to be sufficient for this application since the valves are subjected to thermal stress only; they have moderate exposure to corrosive effects and very little mechanical loading.

7.11.6 Valve Keepers

7.11.6.1 Tasks and Functioning

The purpose fulfilled by valve keepers is joining the valve spring collar with the valve in such a way that the valve spring always keeps the valve in the required position.

Cold-embossed valve keepers are state of the art for valve stems up to 12.7 mm in diameter. The C10 and/or SAE1010 qualities are used.

The valve keepers are classified according to their function as follows:

- Clamping connection creating a frictional connection among the valve, the valve keeper, and the valve spring collar
- Nonclamping connection, which allows for unrestricted valve rotation

Clamping connections. Clamping valve keepers transfer force through a frictional connection. To achieve this, it is necessary that a narrow gap be maintained between the two halves of the valve keeper. That is why valve keepers with conical angles of 14°, 15°, and 10° are used. Valve keepers with smaller conical angles bring about far more intensive clamping action. They are suitable particularly for engines that run at extremely high speeds. Where the clamped connections are heavily loaded, the use of case-hardened (480 to 610 HV 1) or nitrided (≥400 HV 1) valve keepers is recommended.

Figure 7-146 shows an example of a clamping valve keeper in its installed position.

Nonclamping connections A nonclamping connection is achieved by using valve keepers with a conical angle of 14° 15′. Because of the fact that the two halves of the valve keeper, when installed, rest against each other at flat surfaces, they provide clearance between the valve keeper halves and the valve stem.

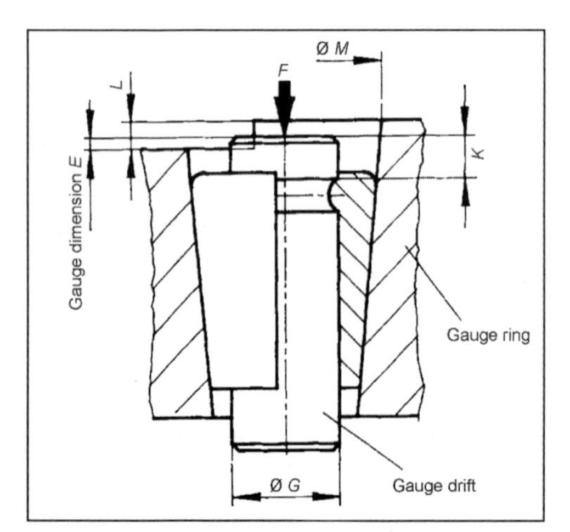

Fig. 7-146 Installation principle for valve keepers.

This allows the valve to rotate in the spring collar. Rotation is supported by vibration, by eccentric contact between the rocker arm and the end of the valve stem, and by the impetus provided by valve lifter rotation.

When a nonclamping connection is used, the forces along the axial direction are transferred by the three or four beads inside the valve keeper. That is why case hardening the valve keeper is indispensable. Figure 7-147 shows an example of a clamping valve keeper when installed.

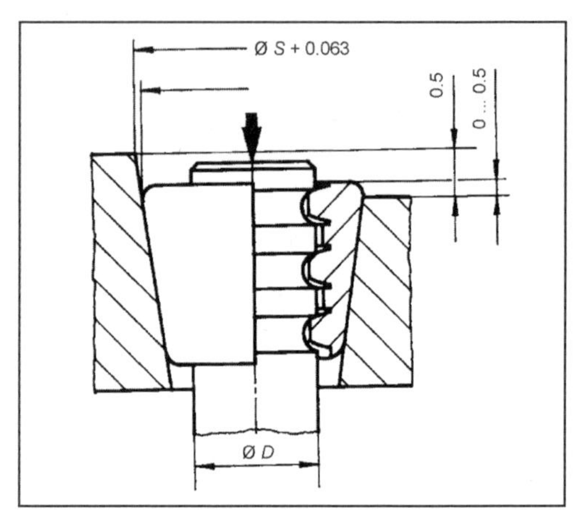

Fig. 7-147 Installation principle for valve keepers with clearance to prevent clamping.

7.11.6.2 Manufacturing Techniques

Valve keepers are cold pressed from profiled strip steel. Multislot valve keepers are always case hardened and ground at their mating planes. Other versions may be used without hardening or with case hardening, or they may be nitrided, as desired. Manufacturing may require that the outside jacket, about halfway down the side, be made concave by as much as 0.06 mm, the amount depending on the exact design. The outside jacket may never be convex.

In free-rotation, multislot valve keepers' correct valve stem clearance is achieved by dimensioning 0.06 mm smaller than the nominal diameter.

The conical section in the spring collar has to be long enough that the valve keeper does not hang over at either end when installed. The conical jacket may in no case be convex and should serve as the reference surface for the dimensional and positioning tolerances in the spring collar.

7.11.7 Valve Rotation Devices

7.11.7.1 Function

Regular rotation of the valve is of critical importance to its perfect functioning. In this way, valve head temperatures are stabilized and leaks due to warping are avoided. Carbon deposits on the valve seat are prevented as well.

Positive-action rotation devices are used in industrial engines, for example, wherever natural valve rotation is not sufficient.

7.11.7.2 Designs and Functioning

Valve rotators function according to one of two principles:

- Rotation during the valve opening stroke: The system comprises a round base featuring several oblong slots along its circumference. Mounted in each slot are a ball and a coil spring that forces the ball to the upper end of an inclined race. A flexible washer is located around the base's center hub, and this is topped by a collar, Fig. 7-148.

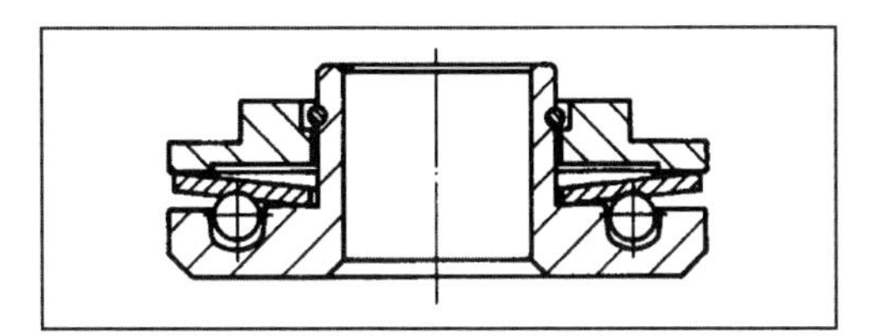

Fig. 7-148 Valve rotation during the opening stroke.

When the valve opens, the stroke is transferred to the collar (due to rising valve spring force), and the collar then flattens the flexible washer. This washer forces the balls in the slots to roll downward along the inclined races; the washer itself rolls downward on those balls. The contact with the balls causes the pressure exerted by the flexible washer on the hub to be reduced, causing slippage. The collar and the flexible washer, however, are joined one with the other by friction, thus preventing rotation. When the rotator is located below, the relative rotation between the base and the unit formed by the collar and the flexible washer is transferred to the valve via the collar, valve spring, flexible washer, and keeper. When the valve closes, the flexible washer is relieved and the coil springs move the balls, which do not roll in this phase, back into their initial position at the top of the inclined races.

It is a known fact that when a coil spring is compressed, the two ends of the spring rotate in opposite directions, and, when relieved, the ends rotate back into their original position. This rotation effect is preserved; the balls are mounted in the slots in the base in such a way that when the valve opens, the rotator action and valve spring rotation are aggregated, while, when the valve closes, only the reverse rotation of the valve spring is effective. The difference between the two values gives the actual valve rotation angle per stroke.

- Rotation during the valve closing stroke: If at all possible, the rotator should be located at the top, since in this location its functioning is less likely to be affected by grime, Fig. 7-149.

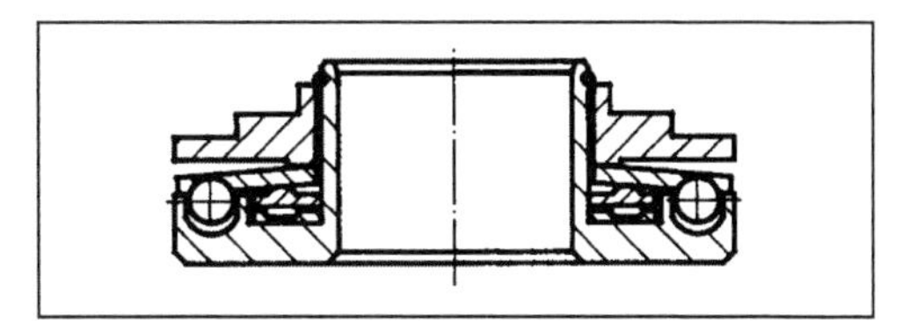

Fig. 7-149 Valve rotation during the closing stroke.

Rotator function here is the reverse of the situation where the valve is rotated during the opening stroke.

Either type may, in principle, be employed in the versions situated above or below. In high-speed engines, preference is given to location at the bottom to avoid increasing the masses in the valve train.

In the version at the top, the rotator replaces the spring collar. It is used in slow-running engines as well as when the version at the bottom cannot be used because of space limitations. What is important here is continuous valve rotation in dependency on engine speed.

Bibliography

[1] TRW Thompson GmbH & Co. KG. *Handbuch.* 7th edition, 1991.

[2] R. Milbach. *Ventilschäden und ihre Ursache.* TRW Thompson GmbH & Co. KG, 5th edition 1989.

7.12 Valve Springs

The purpose of the valve spring is to close the valve in a controlled fashion. This requires maintaining constant contact among valve train components during valve movement. In the "valve closed" state, the spring force F_1 must be great enough to keep the valve from bouncing on the valve seat immediately after closing. In the "valve open" state, it is necessary to prevent "fly-over," i.e., the valve stem lifting off and breaking contact with the cam at maximum deceleration. The kinematics are such that the required spring force F_2 is the product of the valve's mass and the maximum valve deceleration a_{max}.[1]

When engineering the valve springs, additional, and sometimes conflicting, objectives are to be achieved:

- Reducing spring forces: Among other factors, fuel consumption can be influenced by the engine's internal friction. The friction losses occurring in the valve train are proportional to the required spring forces. The maximum required spring forces are determined by the inertia of the moving valve train components, from the cam lobe to the valve. Consequently, the mass of the spring, the cam lobe contour, and maximum camshaft rotation speed are influencing factors. A reduction in spring mass can be influenced by increasing vibration resistance and optimizing the shape of the valve spring.
- Reducing height: Reducing the height of the assembly can also have a positive effect on fuel consumption. On the one hand, this provides greater latitude for the

design of the hood and improving vehicle aerodynamics. On the other hand, reducing the height of the assembly is another key to reducing engine weight. The design of the valve spring and an increase of its fatigue limit can have a favorable influence on assembly height.

- Ensuring minimum failure rates: The increased demands on the valve springs unavoidably lead to an increase in operational strength. In the course of an engine's service life, at about 200 000 km, the spring has to withstand up to 300 million loading cycles. At the same time, only a miniscule spring failure rate is acceptable. The use of multivalve technology makes it necessary to further reduce the failure rate for individual springs. If one assumes, for example, a failure rate of just 1 ppm for the valve springs and if one is building a 24-valve engine, then the result is that, at maximum, only one engine in 40 000 will fail as a result of valve spring failure. Ensuring low failure rates imposes stiff demands on valve spring design, materials, and production.
- Economy of product improvement: The demands presented here have to be economically justifiable; i.e., the benefit associated with any given measure has to be greater than any additional costs that might be incurred. This challenge has been taken up by valve spring manufacturers in the face of increasingly tougher competition.

Determining Strain under Load

Fundamentally, the loading on a coiled compression spring is that of a rod subjected to torsion. When torsional moment M_t is applied as is shown in Fig. 7-150, two shearing strains τ are induced in the longitudinal and transverse sections. According to Mohr's circle, these shearing strains can be assigned to two primary direct stresses σ_1 and σ_2 at less than 45°

Whereas pure shear stress load is induced in the torsion bar, the situation in a coiled spring is different. Because of the spring's geometry and the potential deviation of the effective force axis from the spring's centerline, the bending moment M_b, the lateral force Q, and the standard force N can generate additional stresses under load. Moreover, because of the curvature of the wire, the strains along the circumference are not uniform. Maximum load tensions thus occur on the inside of springs made of round wire.

The equations used to calculate helical compression springs are given in DIN 2089. The following situations apply to the spring rate R, the force F, and the torsional strain τ:

$$R = \frac{G}{8} \cdot \frac{d^4}{D_m^3 n} \tag{7.10}$$

$$F = s \cdot R \tag{7.11}$$

$$\tau = \frac{8}{\pi} \cdot \frac{D_m}{d^3} \cdot F \tag{7.12}$$

The approximation formula developed by Bergsträsser is among the techniques used to correct

$$k = \frac{w + 0.5}{w - 0.75} \tag{7.13}$$

the strain values resulting from the curvature of the wire. The strains on the inside of the spring when under load are thus as follows: [2]

$$\tau = k \cdot \frac{8}{\pi} \cdot \frac{D_m}{d^3} \cdot F \tag{7.14}$$

The shear stresses determined analytically do not take into account the additional load strains previously mentioned, which result from the bending moment and the

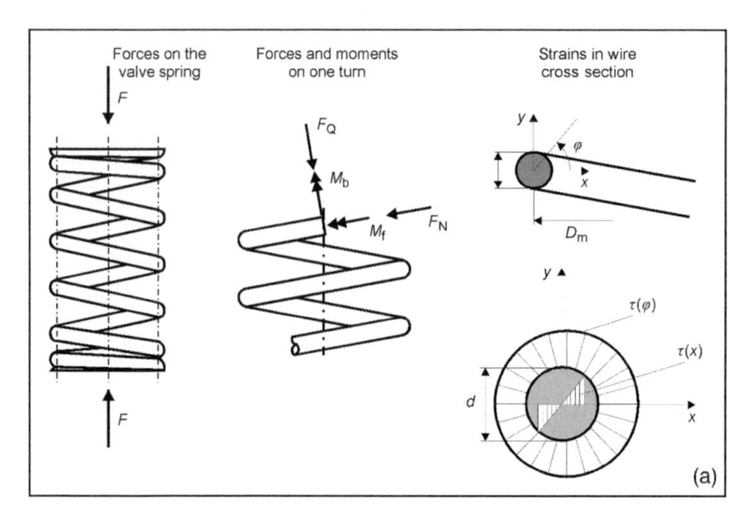

Fig. 7-150 **(a)** Forces, moments, and strains in valve springs.

transverse and standard forces. In addition, the spring's natural vibrations at high engine speeds cause dynamic overshooting to values that can exceed by as much as 50% the load strains determined in static testing. These dynamic effects can be ascertained either with multibody simulation programs or by metrologically using strain gauges. The experiments are usually performed on specially prepared engine mockups.[3] The resulting tracing shows the load strain plotted against engine speed and crankshaft angle.

Depending on the loading and the limitations imposed by available installation space, the shapes shown in Fig. 7-150b have been developed. The standard shape is the symmetrical, cylindrical spring. In this spring the distances between the turns are symmetrical at both ends of the spring, and the diameter of the turns is constant. Progression in the spring characteristics is achieved by the partial contact of the turns across the spring deflection path. Depending on the progression engineered into the spring, the spring rate and the spring's natural frequency may change across the spring deflection path. The dynamic excitation of the spring thus becomes broader in spectrum, and dynamic overshooting is reduced.

The spring may be wound asymmetrically in order to keep the spring masses in motion as small as possible. This means that the closely spaced turns required for progression are located toward the cylinder head. The disadvantage of the asymmetrically wound spring is that additional measures have to be implemented to ensure that the spring is properly oriented for mounting in the cylinder head.

The conical valve spring offers the advantage that, on the one hand, the moving masses are smaller than for a cylindrical spring and, on the other hand, the fully compressed height is slightly shorter. Furthermore, a conical spring permits the use of a smaller spring collar at the valve, which in turn has a positive influence on the masses in motion. A disadvantage is that a conical spring often exhibits less progression than a cylindrical spring.

The so-called "beehive spring" comprises a cylindrical section fixed in place and a conical section in contact with the spring collar. This shape is always used when the integrated valve stem seal precludes using a strictly conical spring shape. In this way, the masses in motion can be reduced significantly by employing a spring collar smaller than the one used with a cylindrical spring. The required degree of progression can be determined in the cylindrical section.

Round and multiarc ("egg-shaped") wires are the shapes normally used. With the multiarc wire one has, in addition, the benefits of reduced installation height and more uniform distribution of strains across the wire's cross section. This is in contrast to round wire, which, as mentioned above, is subjected to the greatest stress on the inside of the spring. Ideal utilization of the material's properties is achieved with the wire cross sections as analyzed by Yamomoto.[4] This cross section provides, on the one hand, the equivalent diameter of a round wire and the axial ratio of the two primary axes. Thus "3.8 MA 25" designates a multiarc wire whose axial ratio is 1:1.25 and whose polar geometrical moment of inertia corresponds to that of a round wire 3.8 mm in diameter.

The low failure rates required here place the maximum demands on the material used to make valve springs. Primary reasons for valve failure are found in nonmetallic inclusions in the spring wire or in mechanical damage to the surface. The Cr-V steels which were often used in the past can no longer satisfy the demands for tensile strength as found in heavily loaded valve springs. They have largely been supplanted in Europe by Cr-Si alloys. Cr-Si steels, in comparison with Cr-V steels, exhibit fewer nonmetallic inclusions and greater tensile strength. Being used to an even greater extent is HT (high-tensile) wire alloyed with Cr-Si-V or Cr-Si-Ni-V. The wire rod is

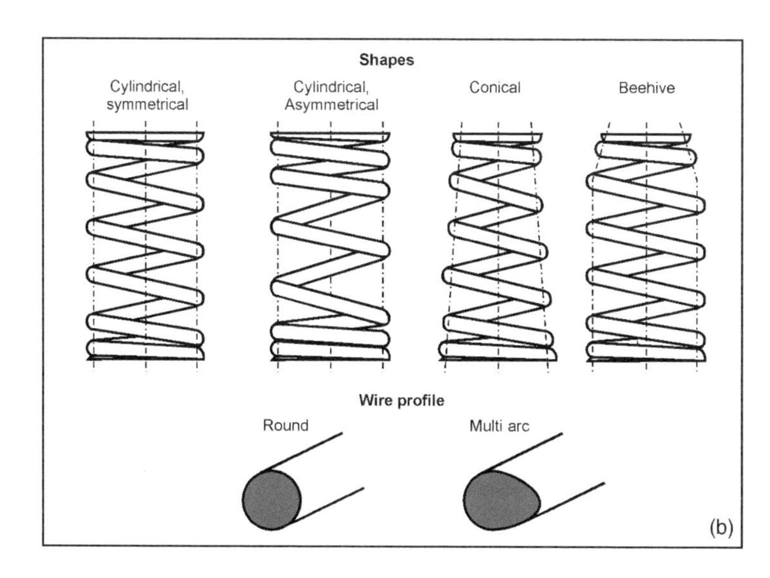

Fig. 7-150 (b) Valve spring shapes and wire profiles.

Product type Manufacturing step		Factors in fatigue strength (c)				
		Purity	Surface	Key mech. values	Microstructure	Intrinsic stress
Molten steel	Smelting and refining	•		•		
Slab ingot/Block	Casting	•				
Billets	Hot rolling	•	•			
Rolled rod wire	Hot rolling	•	•		•	
	Peeling		•			
	Patenting			•	•	
Valve spring wire	Cold drawing		•	•		
	Oil tempering			•	•	
Valve spring	Turning		•			•
	Stress-relief annealing			•	•	•
	Grinding spring ends flat					
	Shot peening		•	(•)		•
	Heat setting					•

Fig. 7-150 **(c)** Factors affecting valve spring fatigue strength.

peeled prior to cold drawing in order to achieve a wire that is free of surface defects. The required degree of strength is attained by a hardening process; this is usually oil tempering, but inductive hardening may also be used. Following hardening, eddy current sensors are used to check the wire for surface defects. Any faulty areas are marked, and the wire is rejected before it goes to the spring manufacturing process.

After the spring has been turned, it is stress-relief annealed to reduce the internal strains in the turns. Finally, the ends of the spring are ground flat to ensure that they are parallel with their mating surfaces. The spring may be chamfered, depending on the specifics of the application. The ball blasting process may be used to compact the surface and to introduce residual compressive force in the areas near the surface. The tensile stresses occurring during operation are superimposed on these intrinsic compression stresses and prevent fissure propagation.

To further increase fatigue strength, springs subject to severe loading are hardened as well. In this way, the amount of stress that can be handled is increased significantly, by about 10%, when compared with conventional springs.

Moreover, valve springs are nitrided for some applications and then shot peened once again. Because of the costs associated with this process, it has not yet been used in either Europe or North America.

Bibliography

[1] Muhr, T., "Zur Konstruktion von Ventilfedern in hochbeanspruchten Verbrennungsmotoren," Dissertation, RWTH Aachen, 1992.

[2] Deutsches Institut für Normung e.V [ed.], Zylindrische Schraubendruckfedern aus runden Stäben, 7th edition, DIN 2089 (Part 1), Beuth Publishers, DIN Pocket Books, Berlin, 1984.

[3] Niepage, P., "Messstellenermittlung und Messwertkalibrierung zur Spannungmessung an Ventilfedern mittels Dehnungsmessstreifen," Draht, Vol. 41, 1990, No. 3, pp. 333–336.

[4] Yamamoto, "Valve Spring Made by Sankos Multi-Arc Wire," Sanko Senzai Kogyo Co. Ltd., Kyoto, 1989.

7.13 Valve Seat Inserts

7.13.1 Introduction

Valve seat inserts and valve guides are important components within the valve train and are essential to perfect ignition and combustion within the cylinder. Together with the valve, these components must ensure complete sealing of the combustion chamber so that the required compression and ignition pressures can be generated inside the cylinder. Excessive wear causes changes in the

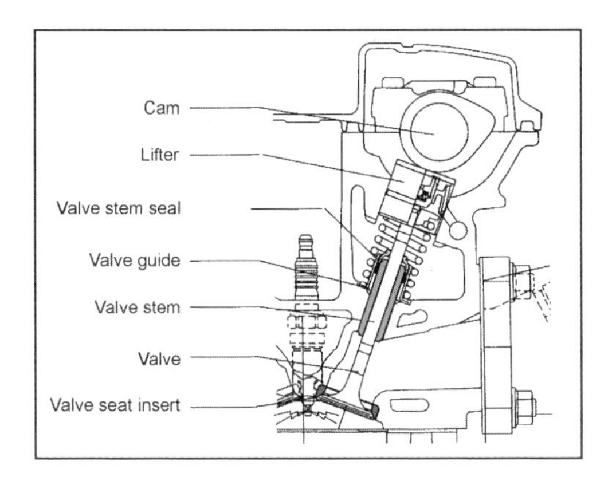

Fig. 7-151 Drive train with hydraulic valve lifters and overhead camshaft.

combustion parameters and thus degrades engine performance and emission data.

Figure 7-151 shows a drive train with hydraulic valve lifters and an overhead camshaft. The valve seat and the valve guide are components that are typically produced in large numbers. Figure 7-152 provides a survey of the passenger car engines built in 1998 and 1999.[1] This represents demand for between 900 million and one billion components. Around the world 13 companies manufacture valve seat inserts, and they can be subdivided by materials groups into cast materials and powdered metal materials, which account for a 90% share of the market.

Market	Passenger car engines manufactured	
	1998	1999
Europe	14 511 410	14 743 841
NAFTA	7 989 249	8 185 106
Mercosur	1 607 090	1 328 285
Asia	11 335 091	12 343 175
Rest of the world	541 321	492 295
Total	38 211 745	39 453 822

Fig. 7-152 Worldwide production of vehicular engines.

7.13.2 Demands Made on Valve Seat Inserts

More than 99% of all aluminum cylinder heads are fitted with separate valve seat inserts because the properties of aluminum and its alloys are not adequate for making up valve seats. The valve seat insert, together with the valve itself, forms a tribologic system that has to ensure sealing capacity even after several million operating cycles. Thus, specifications for modern engines mandate maintenance-free operation of the mechanical valve train without compensation for clearance, for mileage of up to 300 000 km (<2 μm/1000 km). All this takes place in an extremely demanding operating setting. The major factors influencing wear at valve seat inserts are discussed below.

7.13.2.1 Loading on Valve Seat Inserts

Varying loads are encountered in the valve seat contact area, depending on the specific engine design. The method used for adding fuel, the compression ratio, the ignition pressure, and the associated specific forces, as well as the temperatures prevailing in the contact area, all vitally influence wear and deformation in the tribologic system comprising the valve and valve seat insert. The wear factors that thus arise are summarized below.

(a) **Mechanical loading at the valve seat area.** This loading comprises the spring preload (F_f), the valve's closing force (F_B), and the pressure exerted by combustion (F_P). Figure 7-153 provides a survey of the percentages for the various types of loading imposed on the valve seat in an overhead camshaft engine.

	Share in overall loading
Spring preload	1% to 3%
Closing force (maximum acceleration 1500 to 7900 m/s^2)	2% to 17%
Combustion pressure	80% to 97%

Fig. 7-153 Distribution of loading at valve seat.[2]

This loading is subdivided into forces exerted perpendicular and parallel to the seating surface; the split varies with the valve seat angle. The parallel forces are the primary factor in wear and deformation at the valve seat. The sizes of the forces and the distribution of the loads they generate depend on the engine design and the current operating status (e.g., electromagnetic valve operation, engine braking).

(b) **Dynamic loads exerted on the valve seat due to valve motion relative to the valve seat insert.** One portion of the motion is the rotation of the valve. This depends on engine speed and, in valves actuated conventionally, may be as much as 10 rpm or, when using the so-called Rotocaps, up to 45 rpm. This motion is desirable since, on the one hand, it ensures uniform valve temperature and, on the other hand, it has a cleaning effect on the valve seat. A further dynamic load on the seat results from valve disk deflection, which occurs automatically when the pressure in the combustion chamber impinges upon the valve head. This effect is reinforced by a differential

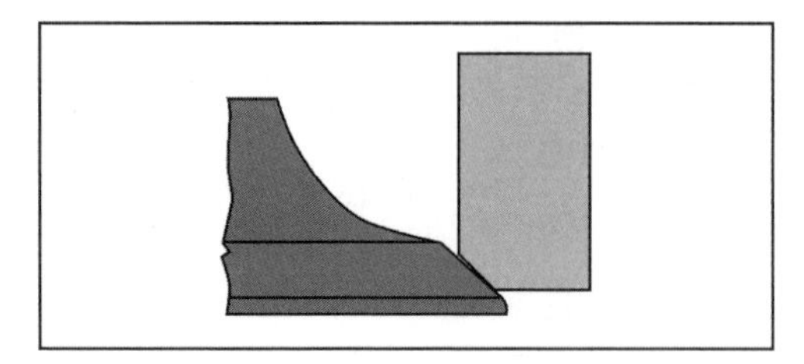

Fig. 7-154 Differential angle at the valve.

between the contact angles at the valve and valve seat, between 0.5° and 1°, which is referred to as the differential angle (Fig. 7-154). In this way, a narrower seat diameter and thus higher pressure at the sealing surface, with enhanced sealing effect, is achieved where ignition pressures are low. When the pressure is increased, the bearing portion of the contact surface increases because of bending at the valve disk, resulting in reduced surface pressure at the valve seat.

(c) **Lubricating the seat contact area.** The wear rates at the tribologic system formed by the valve and valve seat insert are greatly influenced by intermediate lubricating layers. The effects at the intake and exhaust sides differ, depending on the composition of the fuel mixture. Figure 7-155 compares the influence of the types of fuels on wear between the valve and the valve seat insert.

These effects are essentially subordinated by further, superimposed phenomena. Mentioned here, in particular, is the potential enrichment of the mix resulting from introducing crankcase vapors at the intake. Additionally, oily components can pass through the valve stem seal and along the valve stem to the seat contact area.

(d) **The partner in wear—the valve.** When designing the valve train, it is important to ensure that the valve contact surface is harder than the mating surface at the valve seat insert. This is necessary to achieve proper distribution of wear—one-third at the valve and two-thirds at the valve seat insert. This wear ratio is necessary since, in the opposite case, the valve disk could gradually be weakened. Consequently, the valve could slip into the valve seat, causing engine damage. Typical hardness values are summarized in Fig. 7-156.

	Intake		Exhaust	
Gasoline Wear rate 1 to 5 μm/1000 km	++ –	Liquid lubrication in aspirated and turbocharged engines. No lubrication in diesels using Otto cycle since only the combustion air passes through the intake port	+	Solid lubrication with deposits from the combustion gases
Diesel fuel Wear rate 1 to 5 μm/1000 km	–	No lubrication by fuel since only the combustion air passes through the intake port	++	Solid lubrication with deposits from the combustion gases
Alcohol Wear rate 1 to 10 μm/1000 km	o	Liquid lubrication in aspirated and turbocharged engines but with corrosive components; effect will vary with alcohol content	o	Little solid lubrication, increased water content, effect will vary with alcohol content
CNG Wear rate 2 to 50 μm/1000 km	–	No lubrication since only a gas blend passes through the intake port	– –	Little solid lubrication due to minimal combustion residues
LPG Wear rate 20 to 70 μm/1000 km	– –	No lubrication since only the gas blend passes through the intake port	– –	Little solid lubrication due to minimal combustion residues
Hydrogen Wear rate 20 to 70 μm/1000 km	– –	No lubrication since only a gas blend passes through the intake port	– –	No lubrication as there are no combustion residues; increased corrosion due to water vapor
Evaluation: ++ Very good; + Good; o Medium; - Poor; - - Very poor				

Fig. 7-155 Influence of type of fuel on wearing action at the valve and valve seat insert.

	Valve	Valve seat insert
Intake	270 to 370 HBW 2.5/187.5 Hardened > 48 HRC	220 to 320 HBW 2.5/187.5
Exhaust (hardfaced)	30 to 50 HRC	30 to 46 HRC

Fig. 7-156 Comparison of hardness for the valve and the valve seat insert.

7.13.2.2 Materials and Their Properties

Materials

Casting alloys. Various production methods, including die or sand casting and centrifugal casting, are employed to form components from these alloys. The components are manufactured as follows:

- **Cast iron**:[3] Low-alloyed gray casting material is used at both the intake and the exhaust ports for engines developing low internal loading. The high share of free graphite in the material ensures good emergency (dry) running properties. The material's properties can be further improved by heat treatment, e.g., to enhance ductility, which is necessary when using titanium valves. Austenitic cast iron is used to harmonize with the coefficients of thermal expansion found in aluminum cylinder heads. Increasing the amount of carbide increases wear resistance in this material.
- **Martensitic steel castings:**[3] These materials are based on tool steels and rust-free martensitic steels. They are generally employed as hardened qualities for intake and exhaust valve seat inserts in utility vehicle engines involving moderate and high loading, at temperatures of up to about 600°C. Good corrosion resistance is achieved by adding chrome.
- **Nonferrous alloys**:[3] This group of materials comprises high-alloy nickel- or cobalt-based alloys. Such alloys are used particularly at the exhaust side in engines where high loading occurs. Characteristic of this group of materials is the high shares of carbides and the intermetallic phases. Excellent high-temperature characteristics, capable of handling up to 875°C are attained. Disadvantageous are the high costs for the materials, their low thermal conductivity, and the difficulties in machining. In high-performance engines (racing and Formula 1) copper-based alloys doped with beryllium are used because of their great thermal conductivity.

Powdered metal materials: Here a powder mixture is compacted at pressure of up to 900 MPa inside a mold that is close to the final contours. The resulting blanks, the so-called green bodies (powder preforms), are sintered at high temperatures (1000 to 1200°C for ferrous alloys) and then subjected to heat treatment. Mechanical machining—turning and polishing—concludes the production process. Additional manufacturing steps may, however, be required, depending on the type of material used. The goal of modern powdered metal development is to keep down the number of manufacturing steps in the interest of achieving major cost savings.[4]

Powdered metal materials are subdivided into several groups:

- **Low-alloy steels:** Low-alloy steels are used primarily for intake valve seat inserts in gasoline engines. These materials are based on a Fe-Cu-C system. The structure is ferritic/pearlitic in nature, with a share of cementite. Small amounts of nickel or molybdenum are used to improve wear properties. Solid lubricants (such as MnS, Pb, MoS_2, CaF_2, or graphite) are often used to improve amenability to machining by cutting (free-cutting properties). Overall, the amount of alloying material is less than 5%.
- **Medium-alloy steels:** These materials are generally used as the valve seat inserts for gasoline engines and at both the intake and the exhaust ports in diesel engines. This group of materials is the one most widely used and provides a broad range of variants, of which the three most common groups are worthy of mention.

  In the martensitic steels the microstructure is essentially a martensitic tempered structure with finely divided carbides, solid lubricants, and, if appropriate, hard metal phases (intermetallic phases of great hardness and temperature resistance such as Co-Mo-Cr-Si Laves phases and Co-Cr-W-C phases[5]). High-speed steels derive their superior wear strength from a martensitic matrix with a fine distribution of specially formulated M_6C or MC type carbides, which can be made by alloying elements such as Cr, W, V, Mo, and/or Si. Taking the standard high-speed steel alloys (such as M2, M4, and M35) as the basis and using alloying technology modifications, such as diluting with iron powder, adding solid lubricants, or adding other hard-phase elements, finally culminate in the valve seat material. In contrast to the other two materials groups, bainitic steels do not have any tempered structure but instead a thermally more stable bainitic basic structure. The addition of solid lubricants, carbide-forming agents, and hard phases in combination with the fundamental structure produces good wearing properties when hot. Typical alloying elements include Co, Ni, and Mo.

  The medium-alloyed steel groups can also be purchased as copper-infiltrated qualities. Here the open volume between the pores in the sintered body is filled with liquid copper during the sintering process. The advantage of this alloy, in addition to improved heat conductivity, is found in better machinability.
- **High-alloy steels:** This group includes martensitic and austenitic materials. They are used in engines with higher demands for resistance to high-temperature oxidation and corrosion. Typical alloying elements include Ni, Cr, and Co. Because of the high alloying element content, these materials are very costly when compared with the other materials groups. It is for this reason that dual-layer technology is often used in which the valve seat insert is made of two layers of different materials—a high-alloy material at the valve seat and a low-alloy material facing the port.[6]
- **Nonferrous alloys:** The basic Ni and Co alloys, in contrast to the casting alloys, are encountered only very seldom in powdered metal technology. Copper-based materials are particularly interesting for racing applications. One objective of modern materials development efforts is to identify substitutes for toxic beryllium as an alloying element. Adding ceramic particles (such as Al_2O_3) has already made it possible to achieve wear values comparable to those in the standard applications.[7]

Properties

Valve seat insert materials must exhibit certain properties to satisfy the material technology requirements. The key properties are enumerated below:

- **Hot hardness:** A material's hardness generally corresponds to its wear resistance. For this reason hot hardness is used as an indicator of a material's wear resistance at elevated temperatures. Severe drops in hardness at rising temperatures may point to potential temperature limits for a given material (Fig. 7-157).

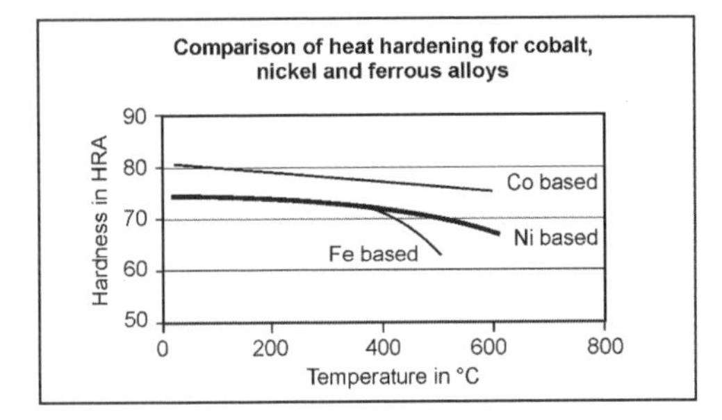

Fig. 7-157 Hot hardness comparison.[8]

- **Structural stability at elevated temperatures:** Structural stability at elevated temperatures identifies changes in the material due to the influence of heat. Figure 7-158 summarizes various effects. One must assume that there are diffusion-related changes, in particular, for materials with tempered structures when they are subjected to thermal stress.
- **Coefficient of thermal expansion:** The coefficients of thermal expansion for valve seat inserts and cylinder head materials are of considerable significance when mounting the inserts in the cylinder head with a press fit. It is beneficial if the materials used for both items exhibit similar coefficients of thermal expansion. If this is not the case, then a reduction in the holding force may occur when the system heats up (which is the case when combining ferrous valve seat inserts and aluminum cylinder heads). This can cause the valve seat to be dislodged from the cylinder head bore and result in damage to or destruction of the engine. Figure 7-159 shows typical values for coefficients of thermal expansion.
- **Thermal conductivity:** To keep valve temperature within reasonable limits it is necessary to ensure good transfer of heat from the valve disk, via the valve seat insert, to the cylinder head. This is achieved, in addition to engineering good heat transmission interfaces, by selecting materials with high thermal conductivity. Figure 7-160 depicts the theoretical heat flows at the valve.

Theoretical calculations[8] have revealed that an increase in conductivity from 20 to 40 W/mK reduces the operating temperature at the valve seat insert by 50 K and that at the valve by 30 K. Measurements in various engines have confirmed this reduction in valve head tem-

Temperature	Process	Effect
−190 to 21°C	Conversion of residual austenite into martensite	Increase in hardness Dimensional changes
250 to 900°C	Reduction of intrinsic stresses Diffusion processes Precipitation processes	Hardness changes Changes in properties Structural changes

Fig. 7-158 Effects due to thermal loading.

		Thermal expansion [10^{-6} K]
Cylinder head	Cast iron	9 to 11
	Aluminum	23 to 27
Valve seat insert	Ferrous (martensitic)	9 to 13
	Ferrous (austenitic)	17 to 19
	Ni basis	12 to 16
	Co basis	12 to 14

Fig. 7-159 Coefficients of thermal expansion.

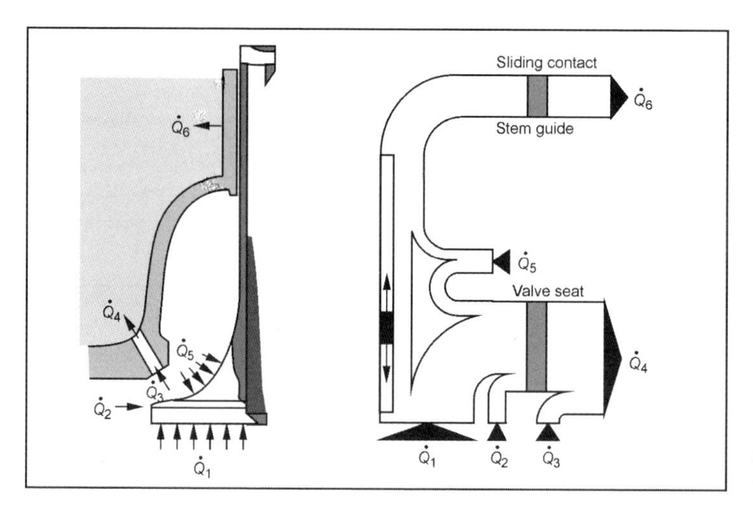

Fig. 7-160 Thermal flow at the valve.[7]

	Thermal conductivity [W/mK]
Ferrous	17 to 35
Ferrous (Cu infiltration)	40 to 49
Ni basis	16 to 18
Co basis	14 to 15
Cu basis	100 to 200

Fig. 7-161 Thermal conductivity.

perature.[9] One method commonly used to achieve these values is to infiltrate the medium-alloyed materials used on the exhaust side with copper. Figure 7-161 summarizes some representative values. When engineering the cylinder head, one must take into account the fact that the increased injection of heat into the aluminum comprising the cylinder head around high-conductivity valve seat inserts causes a loss of strength in the aluminum. Fissuring in the web area is the result of this type of thermal overloading.

- **Density:** In order to keep the stresses on materials as low as possible, materials with higher density are favorable because of their higher specific contact area at any given loading level. This also keeps the notching effect of the pores from initiating fatigue, culminating in material breaking away. In contrast to cast valve seat inserts, one must expect a certain volume of pores in powdered metal products.
- **Resistance to oxidation and corrosion:** Because of the extreme operating situation, valve seat inserts must be able to withstand corrosion and oxidation resulting from exposure to the hot exhaust gases. This can be achieved either with the chemical composition of the material or by a carefully defined passivation of the component's surfaces by preoxidation, for instance.
- **Wear resistance:** The following wear-inducing mechanisms are effective here:
  Adhesion: Local microwelds with subsequent failure at the contact points. Material is transferred from one interface surface to the other, and pitting will take place.
  Abrasion: Material removal due to grinding and cutting mechanisms in the microscopic range. Material transfer is found to only a limited extent.
  Oxidation: Forming brittle, loose oxide layers, which will spall off under load.
  Corrosion: The formation of reaction phases, such as the nickel-sulfur eutectic with its low melting temperature, can cause high nickel content material to weaken and break away.
- **Machining properties:** Good machining properties are an important criterion when evaluating materials for valve seat inserts since the final machining of the valve seat has to be effected after the insert has been mounted, which is because of the close tolerances for the cylinder and the valve seat insert. The nature of the microstructure, the highest possible density, and the addition of solid lubricants can have a positive effect on tool lives.

7.13.2.3 Geometry and Tolerances

Valve seat inserts, in general, exhibit a simple ring shape. Special shapes with contoured exterior surfaces are used for components that are cast in place during cylinder head manufacture. These contours are intended to create a positive connection, to keep the valve seat inserts from being dislocated.[10] Figure 7-162 shows a typical contour for a valve seat insert. Figure 7-163 summarizes common tolerance values.

- **Valve seat:** The valve seat in the insert is the actual functional area for this component. As a rule, final finishing by milling is carried out only after the

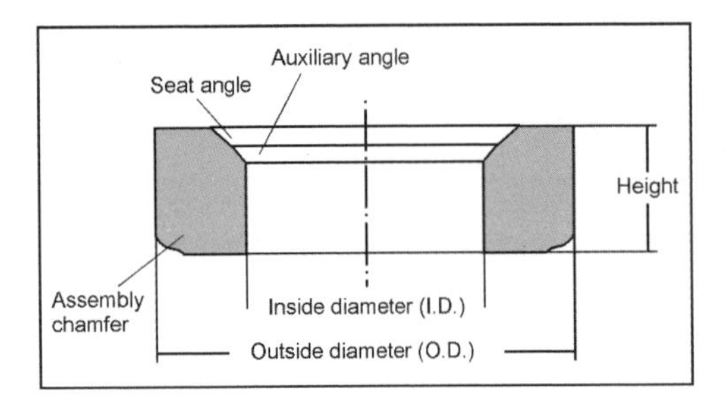

Fig. 7-162 Typical valve seat insert contour.

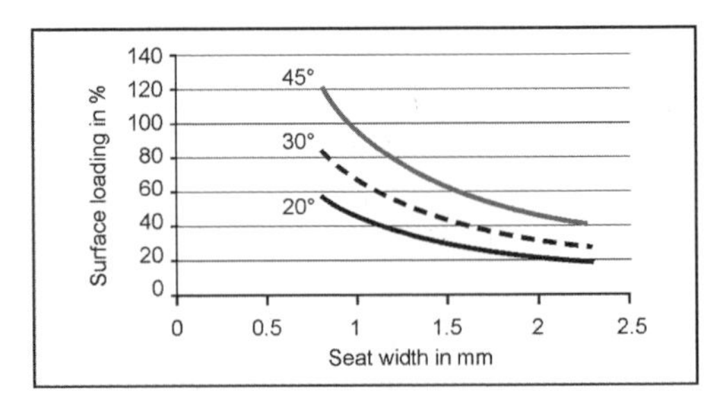

Fig. 7-164 Comparison of surface loading depending on valve seat angle and width.

component has been mounted in the cylinder head so as to achieve exact congruence of the valve axis and the valve seat insert axis (centerline offset of 0.02 to 0.03 mm in new engines). Two engineering options available to reduce wear at the valve seat are found in reducing the valve seat angle and increasing the width of the valve seat. Reducing the valve seat angle or widening the valve seat reduces the loads that are effective parallel to the seating surface, as is depicted in Fig. 7-164. Investigations have revealed that reducing longitudinal surface loading results in a reduction of the wear rate. Common values for the valve seat angle and valve seat widths are given in Fig. 7-165.

- **Installation chamfer:** The chamfer positions the valve seat insert and lowers the forces required for pressing it in place, prior to and while mounting in the cylinder head. Turned chamfers are normally a simple sloped area with an angle of from 10° to 45°. When valve seat inserts are formed in a powder metallic production process, the chamfers that are imparted are often rounded, with radii of from 0.4 to 1.4 mm, and with an area sloped by 10° to 15° on the outside surface. It may be assumed in principle that smaller angles for the sloped areas result in lower assembly forces. In addition, it is necessary to ensure that no burrs are created at the assembly surface during milling. This is prevented by fine grinding of the components.

	Valve seat width in mm		Valve seat angle
	Intake	Exhaust	
Gasoline engine	1.2 to 1.6	1.4 to 1.8	45°
Diesel engine			
Passenger cars	1.6 to 2.2	1.6 to 2.2	45°
Utility vehicles	2.0 to 3.0	2.0 to 3.0	20° to 45°
Gas engine	1.8 to 2.5	1.8 to 2.5	20° to 45°

Fig. 7-165 Valve seat widths and angles.

- **Inside diameters:** The inside diameter of valve seat inserts are generally not machined. To optimize gas flow patterns, the inner surfaces of intake valve seating rings in certain families of motors are specially shaped to impart Venturi contours, for example.

Outside diameter	*Da* < 45 mm	± 0..013 mm
	Da > 45 mm	± 0.010 mm
	Perpendicularity	0.03 referenced to chamfer side
	Surface	Ra = 1.25
Inside diameter	Cylinder dimension	± 0.1
	Taper dimension	± 0.15
	Surface	*Ra* = 3.2
	Center line congruence	0.2
Seat	Angle	± 1°
	Surface	*Ra* = 3.2
Height	Dimension	± 0.05
	Parallel	0.04
	End surfaces	*Ra* = 1.6
Assembly chamfer	Tolerance at radius	± 0.15 to ± 0.3
	Tolerance at taper	± 2°

Fig. 7-163 Tolerance ranges in valve seat engineering.

To improve run-in conditions and to achieve constant valve seat widths following final machining of the valve seat insert (in the cylinder head), subordinate angles are often provided at the valve seat area. The normal value for such angles is 30° (Fig. 7-166).

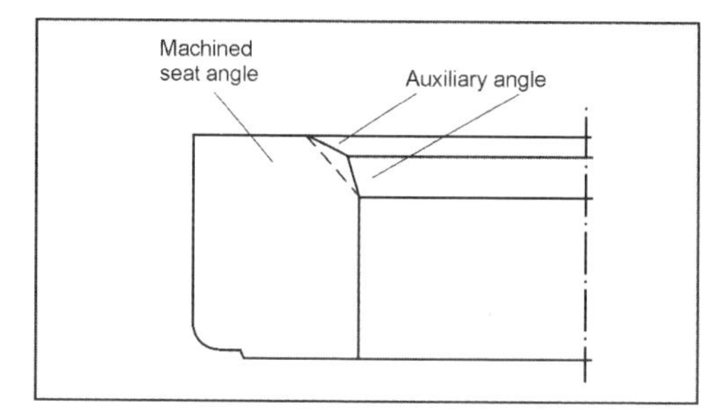

Fig. 7-166 Subordinate angles.

- **Wall thickness:** More compact designs for modern engines impose demands for thinner walls at the valve seat inserts. This is limited by the mechanical loading on the valve seat insert and by aspects associated with production reliability. The wall thicknesses normally produced in mass production exceed 1.8 mm. The ratio of height to wall thickness should be as is shown in Fig. 7-167.

Insert height H	Height ÷ Wall thickness
5 to 6 mm	≤2.5
6 to 9 mm	≤3.0
>9 mm	≤4.0

Fig. 7-167 H:W ratio.

- **Outside diameter**: To achieve a sufficiently tight press fit in the cylinder head, the insert is usually from 0.05 to 0.13 mm shorter than the bore in the cylinder head.[8] A further orientation value for the design of aluminum cylinder head assemblies is calculated as follows: Insert length differential = 0.3% to 0.4% of the diameter of the bore in the cylinder head. The amount of excess length should always be selected to suit the particulars of the application. Transferring heat to the cylinder head requires good contact with the inside diameter of the cylinder head bore, particularly at the face toward the combustion chamber, since it is here that the greatest amount of heat transfer takes place. Figure 7-168 shows the temperature distribution inside a valve seat insert at the exhaust port. When valve seat inserts are made using powdered metal technology it is necessary to ensure that the ratio of the outside diameter to the wall thickness is in a range of from 10 to 13. This is necessary to ensure sufficient "green-body" stability in the powder blanks before they are sintered. No such limitation is imposed on cast parts. The roughness of the outside surface has an influence on the forces required to press the valve seat insert into the cylinder head.

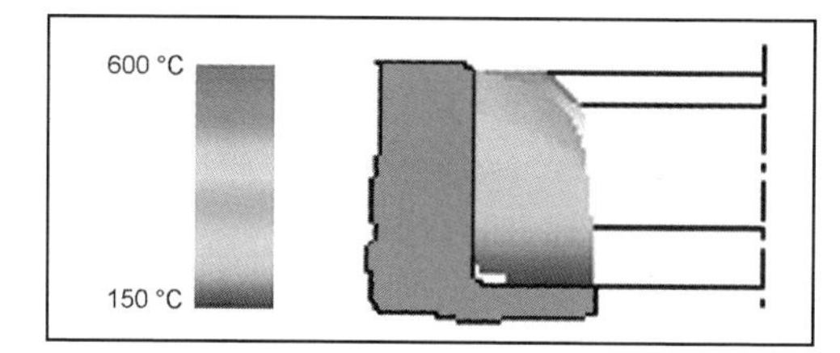

Fig. 7-168 Temperature distribution inside a valve seat insert at the exhaust port. (*See color section.*)

7.13.2.4 Cylinder Head Geometry and Assembly

The geometry of the cylinder head has a significant influence on the functioning of the valve seat inserts. The temperatures inside the insert can be influenced, particularly with appropriate engineering and assembly procedures. Good contact between the insert's outside surface and the inside of the bore in the cylinder head is critical. Consequently, perfect roundness and an exact 90° angle between the outside surfaces and the bore are important factors, as is the tendency of the cylinder head to warp. When using valve seat insert materials with enhanced thermal conductivity, it is necessary to remember that this will cause increased thermal loading at the web area in the cylinder head. This can lead in turn to fissuring in this area, particularly in higher-performance engines.

When installing valve seat inserts in the cylinder head at room temperature, there is a danger that—because of the slight differential between the lengths of the insert and the bore—plastic deformation of the cylinder head material with a displacement of material can occur during assembly. This impedes creating a satisfactory contact surface. Preinstallation chilling with liquid nitrogen offers the advantages of reducing slightly the length differential and lowering the insertion forces. Disadvantageous is the fact that the valve seat insert material is more brittle at low temperatures. In addition, exact process design is absolutely necessary since any delays during assembly immediately change the thermal insertion conditions, where the consequences are increased insertion forces and the risk of inexact seating.

Bibliography

[1] VDA-Mitteilungen 1999, www.vda.de.
[2] Dolenski, T., "Konstruktion eines Hochtemperatur-Stift-Scheibe-Verschleissprüfstandes," Thesis, Bochum Technical College, 1998.
[3] "SAE Valve Seat Information Report," Society for Automotive Engineers, Inc., SAE J 1692, Warrendale, PA, 1993.

[4] Rodrxigues, H., "Sintered Valve Seat Inserts and Valve Guides: Factors Affecting Design, Performance & Machinability," Proceedings of the International Symposium on Valve Train System Design and Materials, ASM, 1997.
[5] Dooley, D., T. Trudeau, and D. Bancroft, "Materials and Design Aspects of Modern Valve Seat Inserts," Proceedings of the International Symposium on Valve Train System Design and Materials, ASM, 1997.
[6] Motooka, N., *et al.*, "Double-Layer Seat Inserts for Passenger Car Diesel Engines," SAE Technical Paper Series 850455, 1985.
[7] "Valve seat insert information report," SAE J 1692, October 30, 1993.
[8] Richmond, J., D.J.S Barrett, and C.V. Whimpenny. ImechE, C389/057, 1992, pp. 121–128.
[9] Dalal, K., G. Krüger, U. Todsen, and A. Nadkarni, "Dispersion strengthened copper valve seat inserts and guides for automotive engines," SAE Technical Paper Series 980327, 1998.
[10] Rehr, A., Published application DE 3937402 A1, German Patent Office, 1991.

7.14 Valve Guides

Valve guides, just like the valves and valve seat inserts, are essential components in the valve train. Consequently, annual demand is comparable to that for the mating components, coming to between 900 million to one billion units per year (see Fig. 7-152).

In terms of the materials, the market is divided into powdered metal, reformed brass, and cast iron qualities.

7.14.1 Requirements for Valve Guides

The function of the valve guide is to stabilize the reciprocating valve in such a way that it is always perfectly positioned at the sealing surface inside the valve seat insert. The tribologic system is formed by the valve stem and the valve guide. Lubrication occurs when motor oil seeps through the gap between the valve stem and the valve guide. In some materials, certain alloying additives and/or components in the microstructure contribute to lubrication. Because of the increasingly stringent exhaust emission laws, it will become more important to reduce oil seepage rates in the future. Required here are combinations of materials that permit running dry, i.e., without additional lubricating oil. Increased abrasive or adhesive wear, particularly at the ends of the valve guides, will result in poorer performance and emission values for the engine. Adhesive wear can, in fact, cause seizure. As at the valve seat inserts, there are various influencing factors that have to be taken into account when engineering and using valve guides.

7.14.1.1 Loading on Valve Guides

The loads encountered inside the valve guide are reactions to forces that the valve stem introduces into the tribologic system represented by the valve guide and the valve itself; these forces tend to tip the valve. They consist of the following:[1]

- Friction action at the end of the valve (F_q)
- The lateral forces exerted by the valve spring (F_f)
- The standardized eccentric force on the end of the valve (F_n)
- The forces exerted by gases on the valve disk (F_{gas})

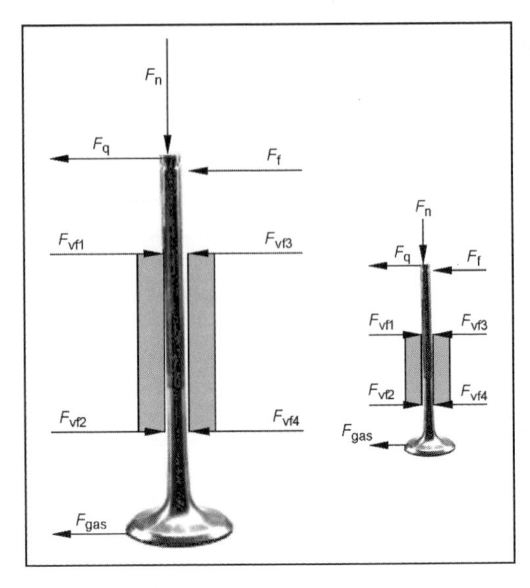

Fig. 7-169 Forces at the valve and valve guide.

The moments thus generated are neutralized by opposing forces at both ends of the valve guide. Figure 7-169 illustrates this equalization of forces.

$$\sum F = 0 = F_q + F_n + F_f + F_{gas} + \sum_{n=1}^{N=4} F_{vfn} \quad (7.15)$$

When running dry, the loading at the ends of the valve guides causes metal-to-metal contact with the valve stem. Oil inside the valve guide forms a hydrodynamic lubricating film as a result of the valve's reciprocating motion; pressure is developed at the ends of the valve guide. This lubricating film separates the mating surfaces through to the point that the motion is reversed. Then there is a brief period of direct contact between the surfaces' solid bodies, which then reverts again from adhesive sliding to sliding action. In principle, the contact between the valve stem and the guide cycles continuously through the friction situations described in the so-called Stribeck curve, depending on the sliding velocity. The following items influence loading inside the valve guide:

(a) Valve train: The forces occurring at the ends of the valve guide vary, depending on the type of valve train that is used. Consequently, the lateral forces for rocker arm valve trains are as much as five times as great as those found in valve lifter designs. Figure 7-170 shows the typical cycle of lateral forces in a rocker arm valve train.

(b) Valve clearance: Dynamic processes in valve lifting induce additional forces (Fig. 7-171). Increasing valve clearance by 0.1 mm increases the lateral force by 22%.[1]

(c) Valve stem seal: Creating a hydrodynamic lubricating film in the contact area between the valve stem and the valve guide requires both a sufficient quantity

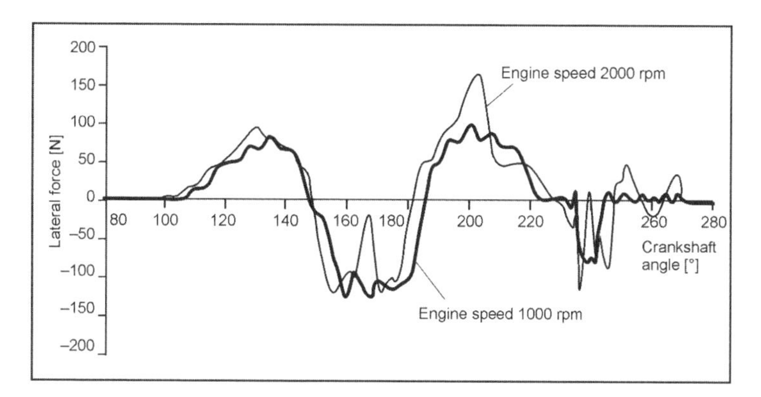

Fig. 7-170 Lateral forces at a valve guide, at varying speeds[1] (engine driven, valve play 0.1 mm, valve guide play 45 μm, oil temperature 50°C, rocker arm valve train).

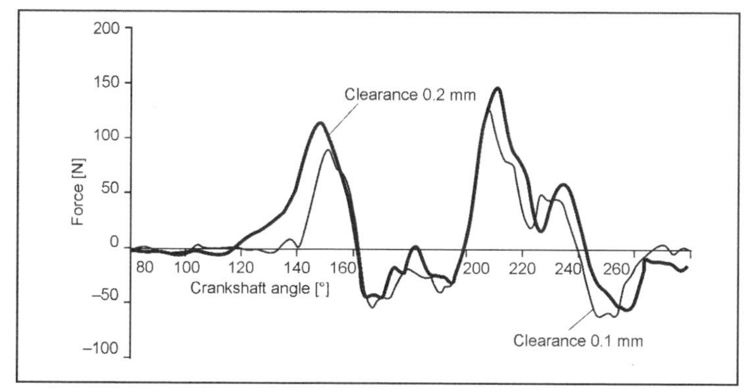

Fig. 7-171 Lateral forces at a valve guide, at varying clearances[1] (engine driven, engine speed 1000 rpm, valve guide play 45 μm, oil temperature 60°C, rocker arm valve train).

of oil and an adequate valve sliding velocity. This is achieved with valve stem seals that allow defined volumes of oil to pass through the stem sealing area. Normal values lie in a range of from 0.007 to 0.1 $cm^3/10$ h. When using turbochargers or engine braking in utility vehicles, the pressure situation on the port side of the valve guide can fluctuate, thus influencing the oil seepage rate. Investigations have revealed that gauge pressure of 0.8 bar on the port side can cause the oil to be forced out of the valve guide, resulting in insufficient lubrication with increased wear and the potential for seizure.[1] Specially engineered shapes for the valve stem seals can eliminate this problem.

(d) Valve guide clearance: The valve guide is responsible for exact positioning of the valve in the seat at the valve seat insert. To ensure that this task is fulfilled, the valve guide bore and the outside diameter of the valve stem have to be sized to match one another, always striving to achieve the smallest possible amount of play at the valve guide. In addition to improved heat transfer, the hazard of the valve's tipping is reduced. Moreover, this geometric matching of the mating components supports the establishment of the hydrodynamic lubricating film. The lower limits for the difference between the diameters are determined by the divergent coefficients of thermal expansion for the guide and the valve stem. Figure 7-172 provides some basic valves for valve guide clearances.

Stem diameter [mm]	Intake [μm]	Exhaust [μm]
6 to 7	10 to 40	25 to 55
8 to 9	20 to 50	35 to 65
10 to 12	40 to 70	55 to 85

Fig. 7-172 Basic valves for valve guide clearance.[2]

(e) Valve: As the component mating with the valve guide, the valve itself has a critical influence on wear phenomena by two factors.

(1) ***The heat applied via the valve stem:*** Theoretical calculations assume that some 10% to 25% of all the heat impinging on the valve is dissipated through the valve guide. This effect depends on the thermal conductivity of the valve stem material (12 to 21 W/mK), while the engineering design of the valve is also of decisive importance. Hollow valves filled with liquid sodium serve to lower (by between 80 and 150 K) the temperature at the critical curved area at the back of the valve head. Cooling is achieved by the liquid sodium inside the valve transporting heat from the head to the stem area. The higher thermal loading thus imposed upon the guide makes particular demands on the material and system tuning.

(2) ***The material for the stem:*** Distinction is made here among the following groups of materials:
Ferrous alloys: Valve stems are made up primarily of martensitic or austenitic qualities. Surface roughness is $R_a < 0.4$. The surface finish can be improved by chrome plating or nitriding. Typical thickness values for chrome plating are from 3 to 15 μm and from 10 to 30 μm where nitriding is employed.[2] Post-treatment of the finished surfaces by polishing is indispensable since residues from the production process (chrome nodules or nitride needles) have to be removed completely to prevent increased wear at the valve guides. The target value for surface roughness is $R_a < 0.2$.
Nickel-based alloys: This group of materials is used, in particular, wherever exhaust valves are exposed to high thermal and mechanical loading. In general, this group of materials is known as "nimonic" alloys. When compared with the ferrous alloys, there are no particular factors of interest regarding the tribologic system comprising the valve stem and valve guide.
Lightweight metal alloys: To reduce the masses in motion in the valve train, current research activities are focussing on the use of titanium and aluminum alloys for valves.
Nonmetallic materials: The types of ceramic materials now in use exhibit good wear-resistance properties. No special adaptive measures are required when such stems are used in conjunction with conventional valve guide materials. The reason is found in the excellent surface quality of the ceramic valves.

7.14.2 Materials and Properties

7.14.2.1 Materials

Powdered metal materials: This group of materials, which accounts for a continuously rising share of the market, can be used in all types of passenger cars and utility vehicles.

- **Ferrous materials:** The microstructures of these steel qualities, containing small amounts of alloying elements Cu, P, and Sn, are generally ferritic or pearlitic. Copper, when used as an alloying element, assumes a variety of tasks. On the one hand, it improves dimensional stability during the sintering process; moreover, it has a positive influence on thermal conductivity and mechanical properties such as hardness and strength. When tin is also present, there are reactions with the copper, including the formation of a bronze phase with a low melting point. This gives rise to liquid phases even at relatively low sintering temperatures, culminating in greater density in the sintered component. Phosphorous, together with iron and carbon, forms the Fe-P-C hard phase known in materials used for casting. Solid lubricants—such as MnS, MoS_2, graphite, CaF_2, and BN—improve dry running properties in case lubrication is interrupted.
  Powdered metal valve guides are relatively porous, and this is reflected in a density value of from 6.2 to 7.1 g/cm^3. These pores are often filled with oil in order to provide basic lubrication between the valve stem and the valve guide when engines are first started. The pores can be charged with oil by immersing the component in a heated oil bath. Capillary action and surface tensions cause the oil to enter the open pores in the sintered part. This process is very sensitive to outside influences including oil condition, component cleanliness, temperatures, oil viscosity, etc. Another process with far better reproducibility is impregnation with oil. Here the valve guides are first placed in a vacuum chamber to evacuate the air from the pores. Then the chamber is flooded with heated oil that enters the pore under ambient pressure. In this way, one can be sure that almost all the open pores are filled with oil.
- **Nonferrous materials.** In this context, application is restricted to copper-based materials. In addition to special materials such as dispersion-strengthened copper,[3] various powdered metal brass qualities have been tested. Market introduction has, however, not been seen since; when compared with current materials, neither cost nor functional advantages have been demonstrated.

Nonferrous metals: Copper-based wrought alloys (Cu-CN compounds) are often specified for use in valve guides for vehicular engines. These materials are purchased as drawn tubular or bar material, which is then turned to make the valve guides. The microstructure comprises two main phases,

- The cubic, surface-centered α phase: This is characterized by good cold reforming capability and is thus characteristic for all wrought brass alloys. The values for hardness and tensile strength are relatively low. This phase dominates where the tin content in the alloy is less than 37.5%.
- The cubic, body-centered β phase: The presence of this phase permits increases in both hardness and tensile strength. Toughness is reduced. An increase in the share for this phase is attained by raising the tin content from 38% to about 46%.

The heterogeneity of Cu-Zn alloys offers the ability to modify their properties to suit the particular application while increasing the materials' amenability to cutting operations. Adding aluminum boosts strength without any adverse influence on warm reforming capacities. At the same time, the slip properties are improved.[4] The material used for valve guides is primarily the $CuZn_{40}Al_2$ alloy. Various additions of further alloying elements such as Mn and Si serve to improve wear resistance. In addition to the superior machinability in comparison with other valve guide materials, high thermal conductivity is a further beneficial property of this material.

Cast iron/Cast steel: Valve guides made of ferrous casting alloys are widely used, particularly in the utility vehicles sector. The microstructure comprises a ferritic/pearlitic fundamental structure with free graphite elements (at sizes of about 4 to 7 μm). These act as an "integral" solid lubricant. The share of ferrite here is generally less than 5%. When phosphorous is present, phosphide compounds, individual and finely distributed structural components, and distinct networks may be formed. When the demands on the component are more severe, careful addition of alloying elements (Si, P, Cu, Mo, or Mn) can increase wear resistance. To be mentioned here, in particular, is the ternary Fe-P-C compound, which is often found as the hard phase in casting alloys. Cr is of rather lesser significance as an alloying element and is used in special materials chosen where good corrosion resistance at high temperatures is required. Sand casting is the manufacturing process of choice. The manufacturers indicate that these materials are compatible with all types of fuels. Maximum operating temperature is 600°C.

7.14.2.2 Materials Properties

To satisfy the requirements in the application technology, it is necessary that valve guides exhibit certain key properties, discussed below.

- **Wear resistance:** The main loads on valve guides occur at the ends, wherein the end toward the port generally exhibits more severe wear than the end toward the camshaft; refer to Fig. 7-173. This is because of higher thermal loading at this end. The wear mechanisms effective here are both abrasion and adhesion. The latter, in borderline cases, can cause seizure between the valve guide and the valve stem and, consequently, result in engine failure. Austenitic stem materials show a greater propensity for adhesive wear. When using chrome-plated or nitrided valve surfaces, wear appears mainly in the valve guide. The increased wear at the port end on the exhaust side poses a problem. The increase in the clearance between the valve and its guide can allow exhaust gas components to enter and be deposited in the sliding contact area.

  In extreme cases this can cause the valve stem to block inside the guide; engine failure will result.
- **Density:** Nonporous metals such as the reformed nonferrous metals and casting materials, because of their high specific contact area, have the advantage that the loading on the material is kept down at a given load level. This reduces susceptibility to wear.

  This also allows avoidance of fatigue phenomena and the associated fissuring and propagation of fissuring, which can appear because of the notch effect induced by the pores. When dealing with powdered metal products, one must always assume a certain amount of porosity. When making powdered metal valve guides, there appears, because of the compaction of the powder at each end, a density gradient; the area of greatest porosity is at the center of the valve guide (Fig. 7-174). This type of density distribution and pore distribution is a beneficial property for this category of valve guide since the greatest density is found in the area of greatest loading. The center section of the powdered metal guides can, because of greater porosity, take on a larger amount of oil and thus serve as an oil reservoir. Refer to Fig. 7-175 for the density values for the various groups of materials used in valve guides.
- **Thermal conductivity:** Thermal conductivity is a critical value for exhaust valve guides. On the one hand, a part of the heat in the valve has to be transferred to the cylinder head through the valve guide.

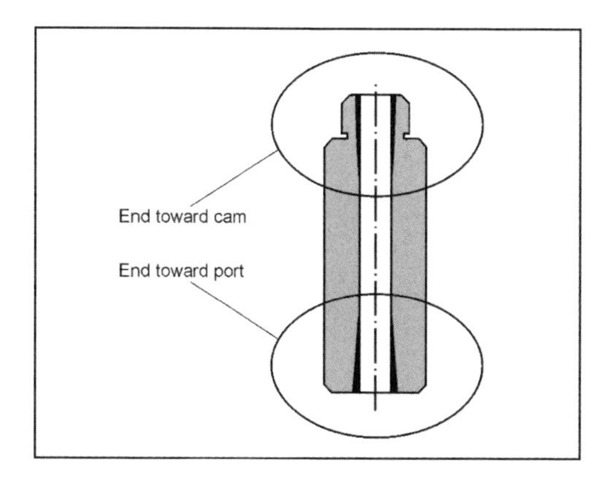

Fig. 7-173 Wear-prone areas in the valve guide.

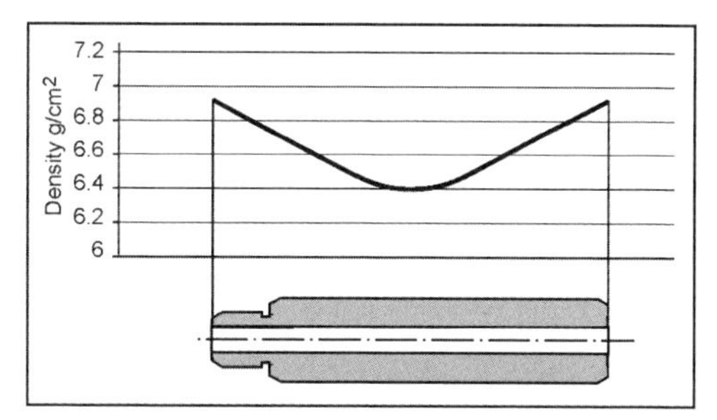

Fig. 7-174 Density distribution inside a powdered metal valve guide.

	Density [g/cm³]
Nonferrous materials (based on $CuZn_{40}A_{12}$)	> 8.0
Powdered metal materials (ferrous)	6.2 to 7.0
Casting materials (ferrous)	7.1

Fig. 7-175 Density values.

Measurements at test beds have shown that the temperature at the valve head can be lowered by up to 8%, depending on the thermal conductivity of the material used for the guide. On the other hand, the exhaust valve guides are exposed to hot exhaust gases. Consequently, good thermal conductivity reduces the thermal loading on the component itself. The temperature at the end of the valve guide toward the camshaft should not exceed 150°C as otherwise the functioning of the valve stem seal is endangered. Figure 7-176 shows temperature development at exhaust valve guides. Quite apparent here is the divergence of thermal loading at the two ends of the valve guide; the physical processes associated with dissipating heat to the cylinder head take place in the lower half of the valve guide, at the end toward the valve port (positions A to D). Above this area the various thermal conductance capacities are of lesser significance. Figure 7-177 summarizes some typical values for thermal conductivity.

- **Thermal expansion:** Like valve seat inserts, valve guides are held in the cylinder head by a press fit. Because of the lower temperature level and the larger mating surfaces, the danger of loosening due to differences in thermal expansion is low. If one observes the tribologic system consisting of the valve stem and guide, then one sees that there are combinations of materials that can narrow, or in extreme cases even eliminate, a preestablished valve guide clearance due to outside temperature influences causing the valve to seize. This is always the case when

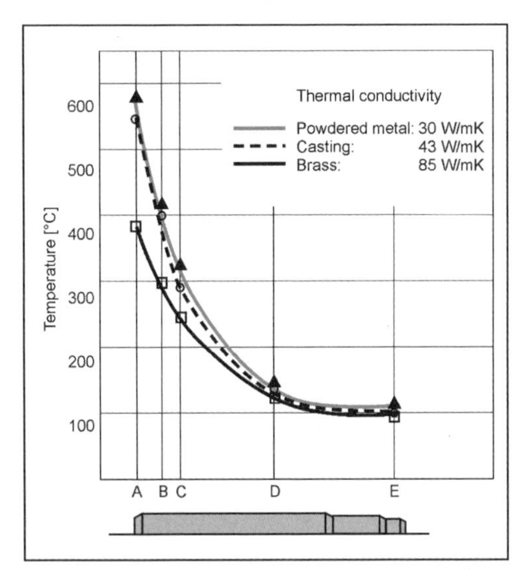

Fig. 7-176 Temperature distribution in valve guides at differing thermal conductivity values.[5]

	Thermal conductivity [W/mK]
Nonferrous metals (based on $CuZn_{40}A_{12}$)	46 to 100
Powdered metal materials (ferrous)	21 to 48
Casting materials (ferrous)	38 to 45

Fig. 7-177 Thermal conductivity values.

$$\lambda_{\text{valve shaft}} \geq \lambda_{\text{valve guide}} \tag{7.16}$$

λ = Coefficient of thermal expansion

If this relationship is inverted, then the end toward the port will dilate, causing an increase in the clearance inside the valve guide. This opens the possibility for exhaust gas contaminants to enter the valve guide and be deposited on the sliding surfaces. The result is that the valve seizes. Any hard particles entering the gap between the stem and its guide promotes abrasive wear. Figure 7-178 summarizes some coefficients of thermal expansion.

- **Hardness:** The requirements for hardness in the valve guides are relatively low. This can be traced back to the fact that the loading on this valve train component is not extremely high. In addition, the polished and (in some cases) coated surfaces on the valve stems do not provide much opportunity for abrasive attack. Figure 7-179 shows the normal hardness ranges for valve guide materials.
- **Oil content:** Oil content is a characteristic that is found only in valve guides made by powdered metal sintering. This figure indicates the amount of oil (in percent by weight) held in the component's pores. The characteristic values are at an order of magnitude of from 0.5% to 1.2% by weight.
- **Machining:** Final machining of valve guides is undertaken with the guides mounted in the cylinder head, parallel to machining the seat in the valve seat insert. This ensures that the centerline offset between the valve guide and the valve seat insert is kept within certain limits. Values for a new engine lie in a range of from 0.02 to 0.03 mm.[2]

The inside diameter of the valve guides is set by reaming. To do this, broaches with from one to six blades made of TiN-coated hard metal qualities are used. Machining tools made of cubic boronitride or polycrystalline diamonds are used only in exceptional cases. Tool life depends on a variety of influencing factors. Narrow tolerances for the centerline offset between the guide and the valve seat insert have a beneficial effect. Burr-free reaming and a homogenous microstructure also extend tool lives. Hard phases or martensitic components in the microstructure have an adverse effect because of their extreme hardness. Small inside diameters for long valve guides should also be avoided, as this generates high torsional torques in the broaching tool. Common values for the inside diameter, as a function of length, are shown in Fig. 7-180.

		Thermal expansion $[10^{-6} K]$
Valve guides	Nonferrous metals (based on $CuZn_{40}A_{12}$)	18 to 22
	Powdered metal materials (ferrous)	9 to 13
	Casting materials (ferrous)	9 to 11
Valves	Ferrous (martensitic)	9 to 13
	Ferrous (austenitic)	17 to 19
	Nickel based	12 to 16

Fig. 7-178 Coefficients of thermal conductivity.

	Brinell hardness 2.5	Loss of hardness, in percent up to 250°C
Nonferrous metals (based on $CuZn_{40}A_{12}$)	150 to 170	~20%
Powdered metal materials (ferrous)	120 to 200	0%
Casting materials (ferrous)	190 to 250	0%

Fig. 7-179 Hardness ranges for valve guide materials.

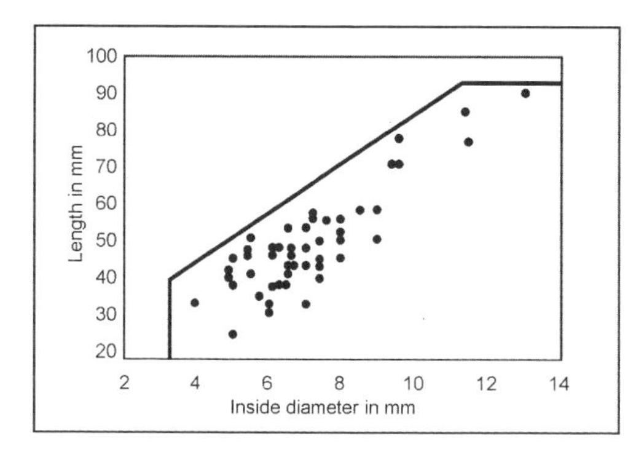

Fig. 7-180 Ratios of valve guide inside diameter to length.[6]

7.14.3 Geometry of the Valve Guide

Valve guides are typically cylindrical; the ends may assume any of a number of shapes, depending on the exact design. At the side toward the port simple chamfering may be found, serving as an aid in press-fit installation. There is greater variety at the end toward the camshaft, which depends in each case on the type of valve stem seal used in the particular instance. Over and above this, there are versions with a collar at the outside which forms a stop when pressing the valve guide in place (refer to Fig. 7-181 for examples). Figure 7-182 provides some standard tolerance values for valve guides.

- **Outside diameter**: The valve guide outside diameter has to be matched carefully to the bore in the cylinder head as this is responsible for a perfect press fit in the cylinder head. The standard values for the difference in length between the cylinder head bore and the valve guide are from 0.02 to 0.05 mm for cast iron and from 0.04 to 0.08 mm for aluminum cylinder heads.[6] The following ratio should be maintained when manufacturing valve guides in a powdered metal process:

  Length ÷ O.D. $\leq$ 4 (powdered metal valve guide);
  6 (cast iron valve guide) (7.17)

- **Wall thickness:** The minimum wall thickness for powdered metal valve guides is 1.8 mm (this is affected by the flow properties for the specific powder used and restrictions imposed by the compression technology). If stem seal seats are turned into the guide, then the initial wall thickness should be no less than 2.6 mm, since this is reduced by turning operations. The shorter the valve guide, the thicker the wall

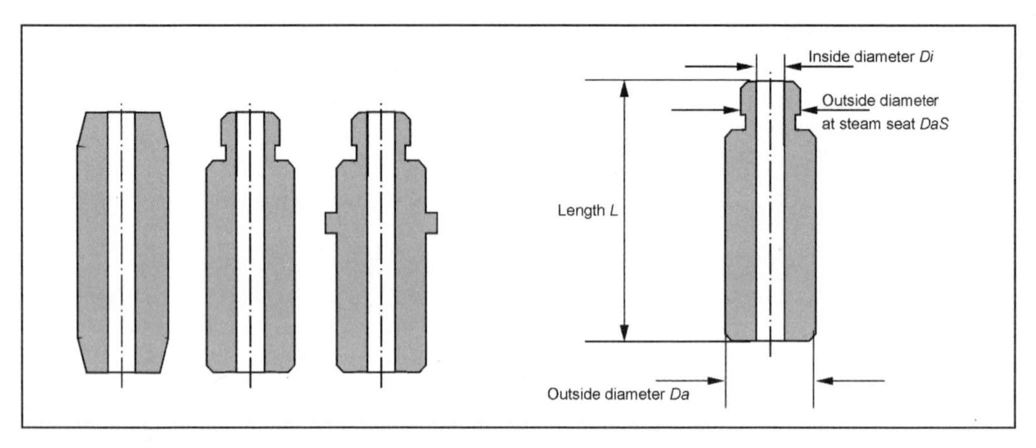

Fig. 7-181 Valve guide contours.

Outside diameter	Da	±0.01 mm
	Cylindrical shape	0.01
	Surface	R_a = 1.6
Inside diameter	Di	±0.1 mm
	Surface (before machining the cylinder head)	R_a = not machined
	Surface (after machining the cylinder head)	R_a = 2.0
	Centering on the outside surface (coaxiality)	0.15
	Cylindrical shape	0.1
Height	Dimension	±0.25 mm
	Surface at the ends	R_a = 6.3
Assembly chamfer	Tolerance for radii	±0.15 mm to ±0.3 mm
	Angular tolerance	±1°

Fig. 7-182 Tolerance ranges for valve guide engineering.

should be since—because of the shorter lever arm—the reaction forces involved in guiding the valve stem are increased. This induces greater loading at the ends of the valve guides. Figure 7-183 shows standard values for wall thicknesses as a function of the length of cylindrical valve guides.

- **Inside diameter:** The inside surface of the valve guide is not machined before it is mounted in the cylinder head.
- **Length:** Using valve guides with the maximum possible installed length is essentially advantageous in order to keep the tipping angle of the valve as small as possible. The length of the valve guide should be at least 40% of the length of the valve.[2]

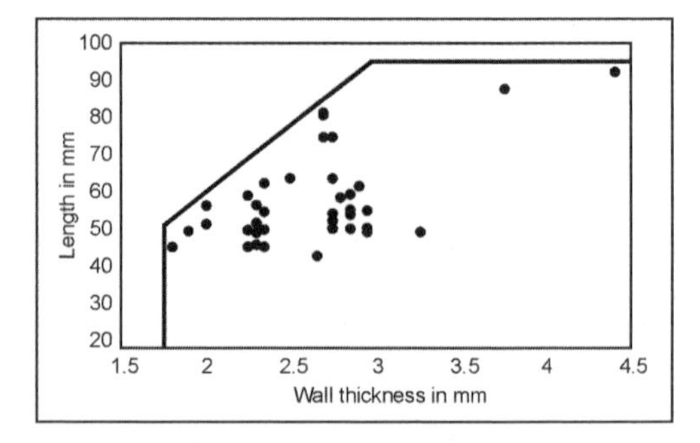

Fig. 7-183 Wall thicknesses for cylindrical, powdered metal valve guides as a function of the length.[6]

7.14.4 Installing in the Cylinder Head

Valve guides are installed by pressing them into the bores in the cylinder head, generally with both the guide and the cylinder head at ambient temperature.

The following engineering notes should be observed:

- The length of the valve guide outside surface should be as long as the bore in the cylinder head so that the ends of the valve guides, exposed to greater loading, is backed by the cylinder head material.
- The end toward the port should not protrude into the intake or exhaust port. This would have an adverse influence on gas flow, and the end of the valve guide would be subjected to extreme thermal loads. This could, under certain circumstances, cause increased wear; in the event of improper matching with the valve stem material disturbances in valve train operation and even engine failure could ensue.

Bibliography

[1] Meinecke, M., "Öltransportmechanismen an den Ventilen von 4-Takt-Dieselmotoren," FVV Final Report, Project No. 556, Institut für Reibungstechnik und Maschinenkinetik, Technische Universität Clausthal, 1994.

[2] Linke, A., and F. Ludwig, Handbuch TRW Motorenteile, TRW Motorkomponenten GmbH, 7th edition, 1991.

[3] N.N., "Kupfer-Zink-Legierungen, Messing und Sondermessing," Informational brochure No. I 005, Deutsches Kupfer-Institut.

[4] Todsen, U., "Interne Schulungsunterlagen Motorentechnik," Hannover Technical College, Piston Engine Laboratory, 1996.

[5] Funabashi, N., *et al.*, "U.S.-Japan PM Valve Guide History and Technology," Proceedings of the International Symposium on Valve Train Systems Design and Materials, ASM, 1998.

[6] Rehr, A., Published application DE3937402 A1, German Patent Office, 1991.

7.15 Oil Pump

The oil pump plays a central role in modern internal combustion engines. Increases in power density and tremendous torques—even at low speeds and especially in turbocharged diesels as well as in gasoline engines—make it necessary to enlarge the oil pump and to achieve greater oil throughput. This is necessary since component temperatures rise, while at the same time the bearings are subjected to heavier loading. On the other hand, an oil pump with optimal efficiency has to be utilized in order to reduce fuel consumption. If we realize that oil pump drag at certain operating states can be as much as 8% of that for the engine as a whole, then we can immediately recognize the importance of this factor.

7.15.1 Overview of Oil Pump Systems

There are many types of oil pumps, but not all systems are suitable for use in internal combustion engines. The primary selection criteria are size, costs, and efficiency in a given situation. Also, the ability to use the system in a wide range of applications is important. The only types of pumps found in mass production are rotor-type pumps and dual gear pumps—the so-called internal and external gear pumps, but also a variation of the sliding-vane pump, i.e., the rotary vane pump.

7.15.1.1 Internal Gear Pump

As mentioned above, the internal gear pump is a member of the double-gear pump family. These are double gears because two interlocking elements, the inner and outer gears, together execute the rotary motion. The inner rotor always turns inside the outer rotor. The outer rotor is always positioned eccentric to the inner rotor by one-half of tooth height so that the teeth mesh when engaged, but on the opposite side a seal is formed between the tips of the teeth. As a rule, the system is driven by the inner rotor. The outer rotor is driven by teeth contact. One differentiates between two basic designs, one with and one without a crescent.

Oil Pumps Without a Crescent

Oil pumps without a crescent usually exhibit a tooth ratio of $z_I = z_A - 1$. Typical tooth counts lay between 4/5 and 13/14 teeth. The way they work is described below. The engaged teeth at the driving inner rotor and the driven outer rotor move fluid under pressure from the intake to the outlet areas. They are separated on the one hand by contact at the flanks where the teeth are engaged and, on the other hand, by the two tips of the gears passing one by the other in the tip sealing area. This group of oil pumps is sealed by the teeth themselves, wherein there is a seal only along a narrow line at mating teeth. This low sealing capability makes it clear why these pumps can be used only in the low-pressure range, particularly since there is no compensation for changes in the gap due to thermal effects. Thus, there is always a small gap that, in turn, causes hydraulic losses. On the other hand, a gap such as this is naturally desirable as it helps to reduce friction losses, particularly when one is dealing with higher engine speeds and higher speeds at the circumference. The displacement process begins as soon as the two circles describing the tips of the teeth make contact, i.e., immediately following the tooth "pocket" with the greatest volume in the area where the tips meet to form a seal. One may also refer to continuous displacement here. One of three tooth geometries is used in general. One is a profile constructed from continuous circular arcs, called the gerotor (see Fig. 7-184); the other is one made of noncontinuous circular arcs, called the Duocentric® (see Fig. 7-185). The gerotor is used primarily in the North American market, where it was developed, while Duocentric® toothing is found primarily in Europe. The more or less unlimited selection of the geometry for the sealing and driving flanks makes possible a more compact design when using Duocentric® toothing. It also makes it possible to engineer higher teeth, giving better utilization of the pump's physical size, which can be used to reduce overall pump size. As a rule, the size advantages when compared with the gerotor concept is between 8% and 12%, depending on the particulars of the situation. In addition to the tooth designs mentioned above, a cycloid

Fig. 7-184 Gerotor pump.

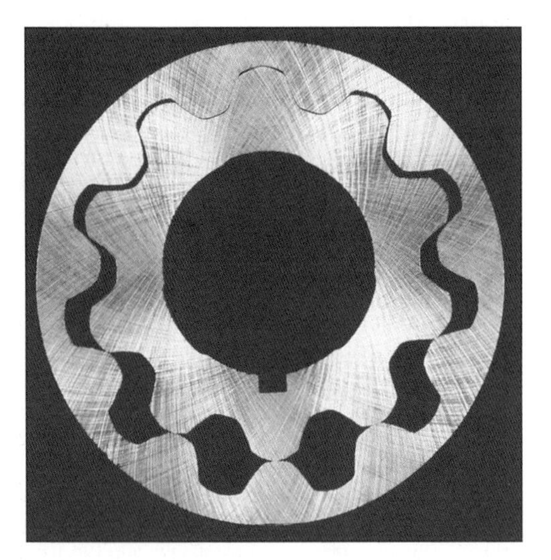

Fig. 7-185 Duocentric® pump.

tooth design has recently appeared on the market. This tooth profile is made up of hypocycloid and epicycloid elements, and it is called the IC®; see Fig. 7-186. The advantage of this tooth design is lower noise generation, resulting from the quieter running of the cycloid profile when compared with toothing engineered using circular arcs. This pump system is found both on the crankshaft and in auxiliary drive trains, e.g., in the oil sump. Pumps driven directly by the crankshaft typically exhibit from 8/9 teeth to 13/14 teeth. At smaller numbers of teeth, the teeth are too high; the opposite is the case for larger numbers of teeth. In pumps running in the oil sump, with diameters that are reduced accordingly and thus with lower speeds at the circumference, common values are between 4/5 and 7/8, while 6/7 is typical here.

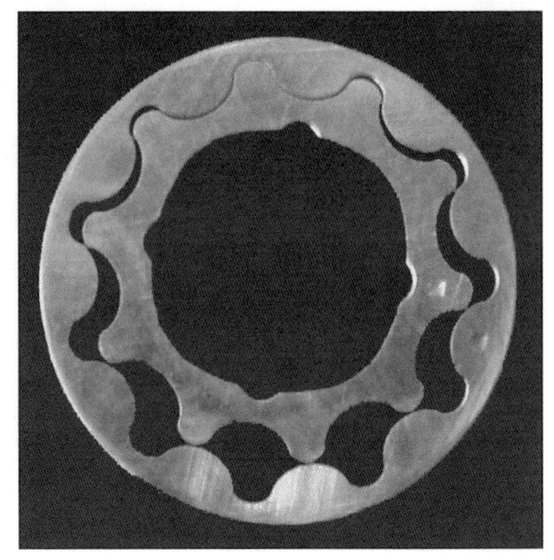

Fig. 7-186 Duocentric® IC® pump.

Crescent-Type Oil Pumps

As the name says, a crescent serves as the sealing element in this type of pump. This crescent is used to create a sealing surface that extends across several teeth. Consequently, this type can also be employed for higher pump pressures. The disadvantage of this design is the greater amount of space that it occupies. Two toothing systems are used, one with involute toothing and a second with a profile formed from trochoid toothing. In the first, one finds tooth counts of 19/24; see Fig. 7-187. In the case of the Trochocentric® toothing with a trochoid profile, this ratio is 11/13; see Fig. 7-188. These systems are used exclusively in pumps driven directly by the crankshaft. They are preferably mounted in the timing gear cover at

Fig. 7-187 Involute toothing.

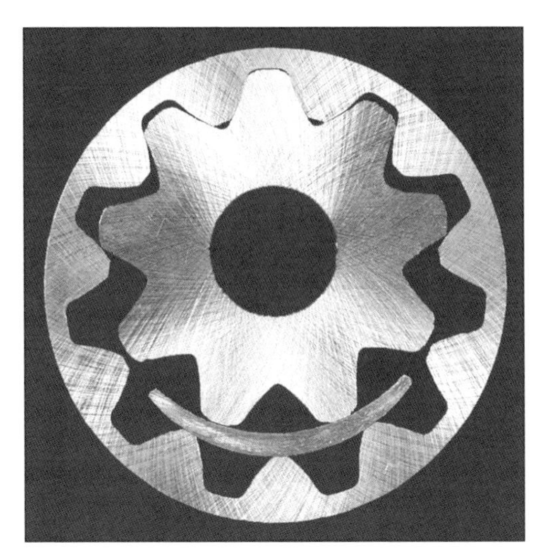

Fig. 7-188 Trochoid toothing.

the front of the engine. The space requirements and manufacturing constraints for the crescent do not permit positioning these pumps at an auxiliary drive such as in the oil sump.

7.15.1.2 External Gear Pump

The external gear pump comprises two or more spur gears, one of which is driven. The way they work is described below. The teeth that are not engaged sweep the circumference and act as displacing vanes, moving oil from the intake to the outlet port, where the teeth move oil under pressure out of the pockets between the teeth; see Fig. 7-189. One problem with this type of pump is the presence of "squished" oil at the roots of the teeth; it can assume very high pressures. These peak pressures can be limited with appropriate relief grooves, which make it difficult to use this design across a broad speed range. Considering that costs are high and that high system pressures are not developed, these pumps are built without any kind of clearance compensation. With the possibility of radial filling not available in internal gear pumps, this design principle gives special advantages where higher displacement volumes are involved.

Fig. 7-189 External gear pump.

7.15.1.3 Vane Pumps

The vane pumps are members of the rotary pump family but are not tandem-gear systems (such as the internal gear pumps) but rather belong to the so-called moving pusher systems; see Fig. 7-190.

Fig. 7-190 Vane pump.

7.15.1.4 Benefits and Drawbacks of Individual Pump Systems

Referring to the key values for the pump systems see Figs. 7-191, 7-192, and 7-193.

In principle, they operate as follows: The displacement vanes are mounted so that they can move in radial slots in the rotor. They are guided by an eccentric ring along a circular curve. The displacement cavity (or pocket) formed by two vanes (the rotor ring and the side plates) moves oil under pressure from the intake to the outlet in response to the rotary motion. Here the volume of the cavity changes continuously.

System	Max. drive speed [rpm]	Eff. displacement volume at 1500 rpm [L/min]	Acceptable operating pressure [bar]	Acceptable operating temperature [°C]	Kin. viscosity [mm²/s]
Crescent pump	1200 to 5000	5.6 to 576	63 to 250	−20 to +80	20 to 100
Gear pump without crescent	1500 to 1800	4 to 50	120	−10 to +80	16 to 150
External gear pump	800 to 3000	6.5 to 280	120	−15 to +80	22 to 90
Vane pump	500 to 3000	2.7 to 42	100	−10 to +80	10 to 52

Fig. 7-191 Typical key values for pump systems, taken from the literature.

System	Max. drive speed [rpm]	Typical operating pressure [bar]	Typical installed width [mm]	Acceptable operating temperature [°C]	Kin. viscosity [mm²/s]
Crescent pump	650 to 6500	1 to 13	8 to 14	−35 to +160	5 to...
Gear pump without crescent	600 to 7500 (crankshaft) 350 to 5000 (auxiliary drive)	1 to 13	8 to 14 (crankshaft) 20 to 32	−35 to +160	5 to...
External gear pump	350 to 5000	1 to 13	25 to 60	−35 to +160	5 to...
Vane pump	400 to 5000	1 to 13	15 to 30	−35 to +160	5 to...

Fig. 7-192 Typical values for oil pumps for internal combustion engines.

7.15.2 Regulation Principles

Because of discrepancies between effective displacement and the actual oil requirements for the internal combustion engine (see Fig. 7-194), it is necessary to integrate some kind of regulation into the system. The purpose is to limit the maximum system pressure in the engine. This can be done by limiting maximum pressure with a preloaded regulation spring, throttling down the displacement volume; alternately, the pump speed can serve as the leading variable.

7.15.2.1 Direct Regulation

Direct regulation is referred to when the regulation valve is located inside the oil pump and the pump pressure itself is the leading variable. Once this regulation system has gone into action, a virtually constant pressure is achieved down line from the pump, regardless of speed and temperature. Depending on the number of flow restrictions and using units before the main gallery, the system pressure is reduced by the time the oil reaches the main bearings. The disadvantage of this regulation system is found in the fact that the pressure level at the main bearings can fluctuate considerably as a factor of engine speed and temperature, the extent depending on the engineering for the oil circuit. It is necessary as a rule to set a relatively high regulation pressure in order to achieve sufficient pressure at the main bearings whenever oil viscosity is high. The result is that at lower temperatures, pressure tends to be too high, and this causes unnecessary hydraulic losses. The advantage of this system is the simplicity of the design. The preferred configuration is to divert the excess oil into an internal bypass in the pump. This slightly increases the pressure at the intake, which in turn helps to reduce cavitation. Much more important, however, is the fact that the internal oil return does not contribute to additional foaming in the sump.

7.15.2.2 Indirect Regulation

One refers to indirect regulation when the regulation valve itself is located inside the pump, in this case the lead variable is not pump pressure but a pressure tapped elsewhere in the engine. As a general rule, this is the pressure at the main gallery. With this regulation system, one can achieve a nearly constant system pressure in the main gallery, regardless of engine speed and temperature, as soon as the system begins regulating. At low temperatures this type of regulation tends to set overly high pump pressure since the throttling losses at high viscosities in the lines and the

Criterion	External gear pump	Duocentric®	Duocentric IC®	Gerotor	Trochocentric®	Involute	Vane pump
Installed size	+	++	++	+	+	–	+
Noise	–	+	++	+	++	++	+
Attached at crankshaft	Not possible	++	++	+	+	+	Not possible
Volumetric efficiency	–	+	+	+	++	+	+
Mechanical efficiency	+	++	++	+	–	– –	–
Regulation capacities	++	+	++	+	+	+	++
Sensitivity to grime	–	+	+	+	+	+	– –
Pulsation	–	+	+	+	++	++	++
Costs	–	+	+	+	–	–	– –

Fig. 7-193 Comparison of pump systems, evaluated from – very poor to ++ very good.

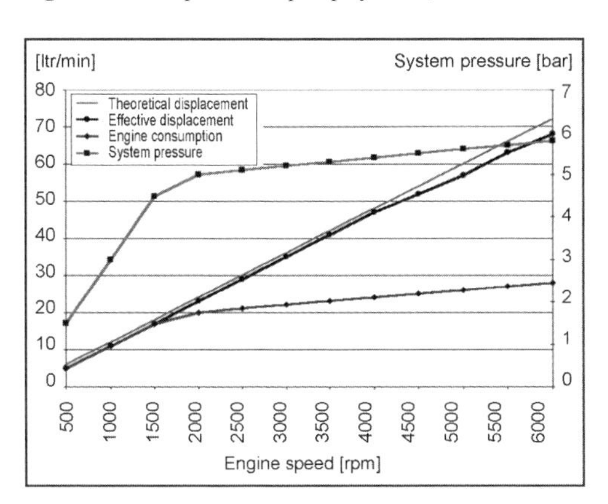

Fig. 7-194 Engine consumption curve, system pressure, theoretical and effective displacements.

filter are also high. The behavior of this system is particularly critical during cold starts. In this situation, because of the small amount of oil required by the engine and the lag in the system, excessive pressure may be applied to the oil filter, causing damage there. To prevent this, one can install a pressure relief valve at the oil pump, limiting maximum pressure at that point. Safety valves such as these are usually set to open at between 10 and 13 bar.

7.15.2.3 Regulation in the Clean Oil Stream

A disadvantage in both the regulation systems described above is that they are installed up line of the oil filter and thus are exposed to contaminated oil. The result may be that one or more particles of grime carried in the oil stream can cause the regulation valve to seize up. This can provoke one of the following two basic states: The regulating valve sticks in the closed position; maximum pressure can no longer be regulated, and there is damage to the filter or the hydraulic valve lifter may be dilated. The other situation is that the valve sticks in the open position, the result being that the oil pump no longer draws any oil since the regulation system diverts the flow to the bypass. This can be counteracted by having the regulation system externally blow off into the crankcase—with the disadvantages described above, but at least there is a chance that the regulation piston will work itself free.

More effective, however, is a system that responds to the cleaned oil stream. Here the regulation valve itself, or at least the oil that comes into contact with the regulation piston, has been cleaned by passing through the primary filter. The major drawback to this process is the extreme pressure losses in the system, since the entire output of the oil pump has to pass through the filter. This disadvantage is the reason why this system is not widely used. With skilled engineering of the regulating valve, it is also possible to effectively keep a regulation valve from sticking.

7.15.2.4 Two-Stage or Multistage Regulation

The classical regulation unit design has a response point at which the valve opens, thus stabilizing the pressure level when this value is reached. The resulting system pressure corresponds to the solid line shown in the chart

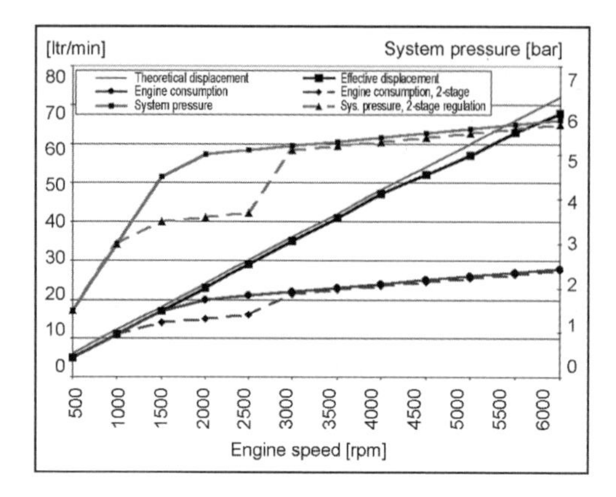

Fig. 7-195 Two-step control.

in Fig. 7-195. It has been found that it is not necessary in every situation to have a pressure curve like this in the engine. Rather, it is quite possible to use other pressure curves, such as pressure reduced in dependency on speed, at medium and lower engine speeds; see Fig. 7-195 (dashed line). By selecting this curve, it is possible to reduce the hydrostatic output at the oil pump and, nonetheless, maintain minimum oil pressure in the engine.

7.15.2.5 Two-Stage Regulation Pump

In the wake of further growing displacement volumes in internal combustion engines, it became impossible in some cases to provide the required flow with a single-stage pump. Other solutions have to be identified when the maximum possible displacement, as shown in Fig. 7-192, is achieved. One such solution is to use a two-stage or multistage oil pump. Here two or more pumping stages are connected in parallel. This parallel configuration makes it possible to achieve at least twice the delivery volume. The advantage of this parallel concept is that when a certain system pressure is exceeded and the regulation valve responds, the second and/or further stages proceeds to idling operation. If the system pressure again falls below this regulation point, any additional stages then are reactivated. In this way only the temporary increase in oil needs is met without generating full pump output at all times. Most of the mechanical losses in the system continue to be encountered. At present only two-stage oil pumps are used for large-displacement engines. Since there are only two operating states—delivery or nondelivery—there is always a jump in the pressure in the engine when the additional stage or stages are switched in and out; in some cases this can have adverse effects on certain engine components.

7.15.2.6 Regulated Internal Gear Pump

In order to better manage the discrepancy between oil pump delivery volume and engine consumption (depicted in Fig. 7-194), it is possible to match the actual delivery volume to engine requirements without blowing off through a relief valve, with all the associated losses. Known among the internal gear pumps is a system that varies the delivery volume by rotating an eccentric cam. To achieve quick regulation, this is done by toothing located outside the pump; see Fig. 7-196.

7.15.2.7 Regulated External Gear Pump

When using external gear pumps, the delivery volume is regulated by shifting the gears axially, one toward the other, where the width of the teeth involved in pumping is changed steplessly.

7.15.2.8 Regulated Vane Pump

In the vane pump, shifting the central eccentric ring relative to the outside contour causes a change in the geometric volume of the oil displacement cavities.

7.15.3 Engineering Basics

At this point one should first go into the theoretical basics so that later the relationships among the dimensions for the gear sets used in sump and crankshaft pumps are clear.

In the ideal situation, there are neither space nor cost specifications at the beginning of a project. There are engine oil requirements and pressure curves for various temperatures. These may be calculated or drawn upon measurements made with comparable engines.

It is on the basis of these measured and calculated key values that the theoretical delivery volume for the oil pump is figured, taking the volumetric efficiency into account. If

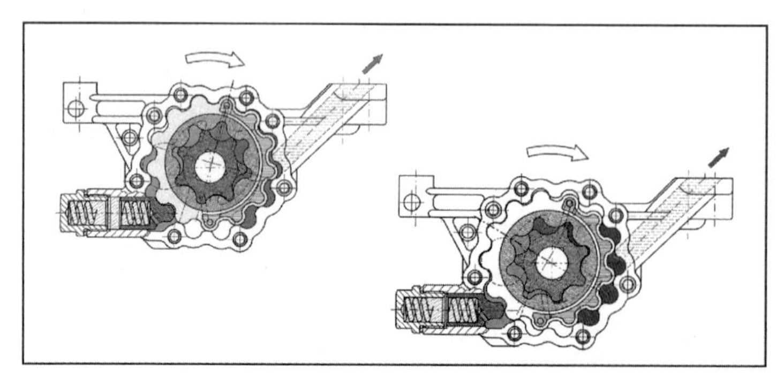

Fig. 7-196 Internal gear pumps without crescent.

it has not been determined whether the oil pump is to be located in the sump or at the crankshaft, then a set of gears should be worked out for each variation.

Theoretical Design of Gearing

If the amount of oil required by the engine has been closely estimated, then the critical points or situations have to be identified. This is normally idling when the engine is hot. This means that the oil pump has to be matched to this situation so that sufficient oil pressure is available during operation.

The size of the gears is ascertained first during engineering. One calculates backwards, from the known or assumed delivery volume to the theoretical pumping surface and the required root and outside diameters for the inner rotor. To do this, it is necessary to know the gear width, which is taken from the amount of installation space available. If the installation space has not been defined, then one may use values based on prior experience. If the above-mentioned dimensions are known, then it is possible to determine the outside diameter, taking appropriate wall thicknesses into account.

The formulas given below are employed in calculations.

$$q_{\text{th}} = n \cdot \frac{Q_{\text{eff}}}{\eta_{\text{vol}} \cdot 1000} \quad (7.18)$$

q_{th} = The oil pump delivery per revolution [cm^3]
Q_{eff} = Effective delivery volume, taken from the engine consumption curve [dm^3/min]
η_{vol} = Volumetric efficiency
n = Engine speed at Q_{eff} [rpm]

Once this value is known, then the theoretical delivery surface can be determined by the gear width.

$$A = \frac{q_{\text{th}}}{\text{RB}} \quad (7.19)$$

A = Delivery surface for the gear pair [cm^2]
RB = Gear width [cm]

Knowing the delivery surface it is possible to use the following equation:

$$A = (d_{k1}^2 - d_{f1}^2) \cdot \frac{\pi}{4} \quad (7.20)$$

to calculate

d_{k1} = Outside diameter for the inner rotor [cm]
d_{k1} = Root diameter for the inner rotor [cm]

while this formula

$$d_t = mo \cdot z \quad (7.21)$$

d_t = Pitch circle
mo = Modulus
z = Number of teeth (inner and outer rotors)

is used to calculate the dimensions for the toothing.

Now the geometric dimensions for the pairs of gears for the crankshaft and the pump in the sump are available. Using the gear dimensions as the basis, one can calculate the depths required for the kidney-shaped ports used to fill and empty the pair of gears.

In the next step, the intake speeds and the critical circumferential speeds of the pair of gears are determined in conjunction with engine and pump speeds. Ascertaining the gear width in this way permits an estimate of the space required for the oil pumps.

Once the designs for both pumps—at the crankshaft and in the sump—have been determined, it is necessary to decide at which location the pump is to be installed. The following selection criteria are available, from a technical viewpoint, when making this decision:

- Installed size
- Drive power and output
- Noise and pulsation

The cost factor has to be considered from the commercial point of view.

The following sections provide details on the variants available for crankshaft and sump pumps.

7.15.3.1 Crankshaft Pump

Crankshaft pumps, Fig. 7-197, are used today above all by the automobile industries in North America and in Japan. In Europe they are used primarily by companies that are influenced by parent firms in North America.

The structure of the crankshaft pumps is usually as follows:

- Die-cast case with shaft seal pressed in place and integral regulation valve
- Die-cast or steel cover
- Inner rotor centered directly on the crankshaft
- Outer rotor driven by the inner rotor

Internal gear pumps in the Duocentric®, DuoIC®, or gerotor designs are normally used.

Crankshaft pumps are normally employed for reasons of costs. Since every engine requires a front cover to take the shaft seal, it is logical to integrate the oil pump into this cover.

Because of the crankshaft diameter and the required sealing space between the cover and the crankshaft bore in the housing, on the one hand, and the root diameter for the inner rotor, on the other hand, a certain root diameter in the inner rotor results automatically. This is the determinant factor in the geometric dimensions.

As described in Section 7.15, the selection of the numbers of teeth for the inside toothing gives a theoretical delivery surface at the appropriate root diameter. Selecting a suitable width for the gear, assuming a suitable volumetric efficiency, gives the dimensions for the crankshaft-mounted oil pump.

In practice, crankshaft diameters lie between 35 mm (in small three- to four-cylinder engines) and about 50 mm (V6 and V8 engines). It is quite conceivable that at crankshaft diameters greater than 40 mm the outer gear will have a very large outside diameter.

Fig. 7-197 Crankshaft pump.

In practice, it is often necessary to strike compromises in crankshaft pumps since the overall length of the engine is to be kept down. This means that there is often an insufficient cross section available for oil intake. Moreover, for cost reasons, a steel cover is used to enclose the gears. This is normally a stamped component 4 to 5 mm wide. This cover, however, makes it impossible to fill the oil pump very easily from the side at the cover if the gear sets are large and wide. Consequently, intake is possible only on the casing side, and this will result in the filling, cavitation, and noise problems at medium and high speeds.

Normally the inner rotors are attached to the crankshaft. It is less often that designs are chosen that exhibit a bearing collar on the inner rotor. The housing also has a bearing bore to accept the inner rotor. The disadvantages are higher costs and somewhat more friction. The benefit of this design is that the inner rotor no longer needs to follow the movements of the crankshaft. The outer gear clearance and the distance between the tips of the teeth can be less than that in the version attached directly to the crankshaft. Here the offset between the pump and the crankshaft centerline and the crankshaft motion has to be taken up by outer gear play and play between the tips of the teeth so that there are no clashes. To ensure the most precise alignment of the pump on the crankshaft centerline, the crankshaft pumps are normally centered on the engine using centering pins, bushings, or tabs.

Power is normally transferred to the inner rotor via two flats, hex heads, or inside gearing with a variety of tooth geometries. Special designs such as polygonal transfer journals are also found.

The pickup tube is attached with a threaded intake flange and gasket. The outlet is direct from the oil pump to the engine block, with a seal or O ring.

Since the crankshaft pump often serves as the terminating cover, there is a gasket between the engine block and the oil pump. In this case the shaft seal is mounted in the oil pump case. The oil pan is often flange mounted to the bottom of the oil pump, and this is another area requiring sealing.

7.15.3.2 Sump Pump

Sump pumps, Fig. 7-198, are used primarily by German car makers. The goal is to buy power savings at somewhat higher costs. It is also possible to shorten engines in this way.

Sump pumps normally comprise the following components:

- Die-cast housing with bearing bore for the drive shaft and integrated regulation or cold-start valve
- Die-cast cover with bearing bore for the drive shaft
- Inner rotor, pressed onto the drive shaft
- Outer rotor driven by the inner rotor

Sump pumps are usually driven directly by the crankshaft, via a chain and sprocket.

Multistage pumps are also often found in the oil sump. These comprise one or more pumping stages and may also include one or more scavenger stages. An arrangement such as this is not possible on the camshaft.

Since these pumps are located in the oil sump, the long pickup tube with sieve is often eliminated. In the ideal design, this can be integrated directly into the case or the cover, as a die-cast component.

Since the oil pumps are in the sump, no additional sealing is required to retain any leaking oil.

The oil pumps are mounted on the engine block by centering pins or bushings to ensure that the chain is properly aligned. The oil is transferred between a cast bore in the housing and a machined bore in the engine block.

Internal gear pumps in the Duocentric®, DuoIC®, or gerotor designs are normally used. External gear pumps are also used from time to time.

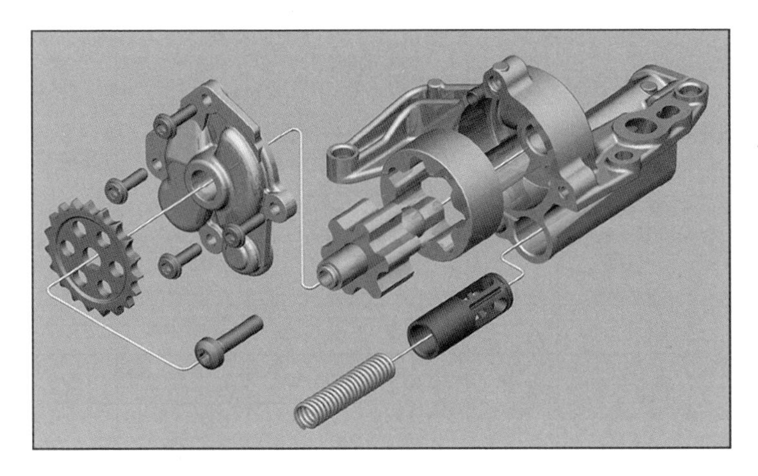

Fig. 7-198 Sump pump.

A subset of the sump pumps comprises oil pumps that are driven by auxiliary drives. The following drive types and locations are common:

- Oil pump driven via a shaft from the distributor shaft
- Oil pump integrated into the engine block and driven via a chain or spur gears
- Oil pump integrated into the balancer shaft gearing and driven via a chain or spur gears

7.15.3.3 Key Oil Pump Values Taken from Practice

Figure 7-199 shows key oil pump values found in practice. Compared here are four-, six- and eight-cylinder engines. The four-cylinder, all-aluminum engine with the crankshaft pump has the most oil using points; consequently, a large oil pump is used. The four-cylinder engine with a sump pump, the six-cylinder engine, and the eight-cylinder engine have been on the market for some time now.

7.15.3.4 Comparison between Crankshaft and Sump Pumps

Every pump system offers advantages and disadvantages. Experience has shown that compromises have to be reached on the following items:

- Build size (engine height and length)
- Costs
- Drive power and delivery
- Pulsation and noise

Build size:

If a crankshaft pump is to be used it is normally limited by the first engine bearing and the path for the timing chain or belt. Normally attention is paid to attaining the shortest possible engine length. Given this objective, there is little room for a wide set of gears and good filling properties.

In contrast to crankshaft pumps, sump pumps can usually be very wide in design. In most engine designs there is enough space to install a sump pump between the engine block, the outline described by the conrod, the crankshaft webs, and the oil pan.

Costs:

The sump pump is normally more expensive. This is primarily because of the additional costs for the chain or drive pinion. If sump pump engineering is carried out with careful attention to costs, then it can come very close to the costs for the crankshaft pump.

Technical comparison:

A comparison between a crankshaft and a sump pump is to be made below. Both pumps are designed for motor oil consumption of 5.2 l/min at 750 rpm engine speed, 1.5 bar, and 120°C. This means that both pumps can deliver the same flow volume to the engine when the hot engine is idling. Both pumps were run on a component test bed, at conditions relevant to engine operation and at a motor oil consumption curve for oil temperature at 100°C. Similar pressure levels (see Fig. 7-200) were developed by both versions.

The technical data for the pumps are shown in Fig. 7-201.

Drive power and delivery:

The basic assumption for a pump with low drive power requirements is an ideal design of the gear set (ratio of diameters to width). As is seen in Fig. 7-202, the drive power required for the crankshaft pump is significantly greater than that for the sump pump. This is essentially for two reasons. First, with the wider design for the sump pump the outside diameter can be reduced in comparison with the crankshaft pump, the advantage being that friction at the pair of gears is lower. Second, lowering the step-down ratio for the sump pump can further reduce the amount of friction in this type of pump. Stepping down can be disadvantageous when the hot engine is idling. At hot idle conditions a certain delivery volume is required to maintain oil pressure in the engine. Because of the temperature, however, the oil is quite thin, and thus

Designation	4-cyl. crankshaft	4-cyl. sump	6-cyl. crankshaft	6-cyl. sump	8-cyl. crankshaft	8-cyl. sump
Number of teeth	8/9	6/7	9/10	6/7	9/10	6/7
Theor. delivery volume referenced to engine speed [cm³] (including step-down for sump pumps)	20.6	10.3	17.0	12.8	15.72	21.3
Outside diameter [mm]	84.5	58.2	91	58.2	90.0	65.2
Gear width [mm]	14	20	10.8	25	10.7	31.2
Max. pump speed [rpm]	6800	4300	7000	4500	5800	4450
Output at 6000 rpm* engine speed [watts]	2560	830	2330	970	2100	1830
Key values for motor oil using units						
Hydr. valve clearance compensation	X	X	X	X		X
Camshaft shifter	X					
Turbocharger		X		X		
Timing chain	X				X	
Balancing shaft(s)	X					
All-aluminum engine	X					

*Power values at 80°C oil temperature, 5 bar oil pressure, 5W30 Shell Helix Ultra oil.

Fig. 7-199 Typical oil pump values.

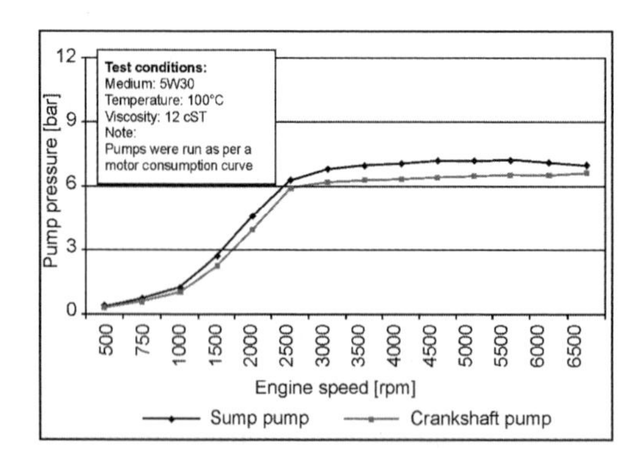

Fig. 7-200 Pump pressures for crankshaft and sump pumps.

there are relatively high losses through the leakage gap. These losses are nearly the same at all speeds. If there is a high step-down ratio, then the percentage leakage losses are high in comparison to the volume of oil delivered to the engine. Here one refers to poor volumetric efficiency at lower speeds and higher temperatures (see Fig. 7-203).

If, however, the step-down ratio and the clearances are selected properly, then the sump pump has a comparable delivery volume in the lower speed range. Thanks to stepping down, the oil pump is filled well at high engine speeds, and the oil delivery volume does not drop off so severely as is the case with the crankshaft pump (see Fig. 7-204). The result is better cavitation behavior.

7.15.3.5 Cavitation and Noise Emissions

The noise properties in the hydrostatic gear pumps are determined essentially by the following influencing factors, all of which are dependent on speed:

- Mechanical characteristics of the toothing
- Cavitation and the formation of vapor bubbles
- Changes in flow speed due to periodic fluctuations in delivery volume
- Sudden pressure equalization when cavities at differing pressures meet

At the lower speed range the mechanical properties of the toothing have a major impact on noise. Deviations in toothing because of the manufacturing process, changing tooth spring stiffness, and load-induced disturbances

Designation	Crankshaft pump	Sump pump
Number of teeth	9/10	6/7
Theor. delivery volume referenced to engine speed [cm^3] (including step-down for sump pump)	8.1	8.1
Gear width [mm]	8.1	20
Outside diameter [mm]	72	58.2
Root diameter [mm]	46.1	29.4
Shaft diameter [mm]	35	16
Gear ratio	1	2 (step-down)

Fig. 7-201 Technical data for crankshaft and sump pump.

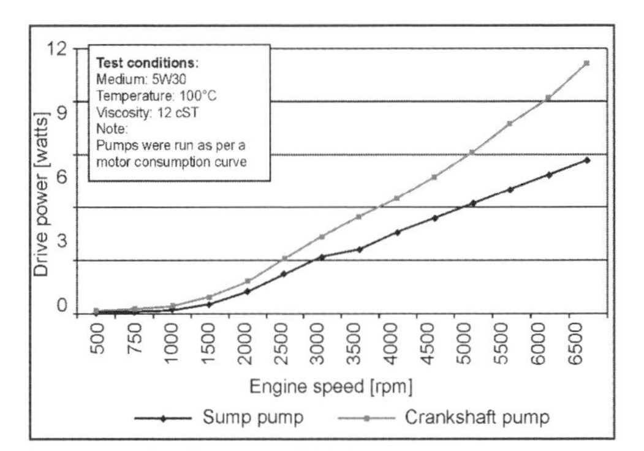

Fig. 7-202 Drive power for crankshaft and sump pumps.

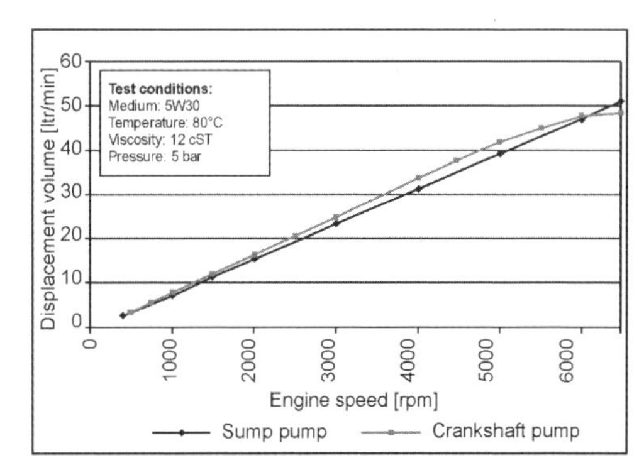

Fig. 7-204 Delivery volumes for crankshaft and sump pumps.

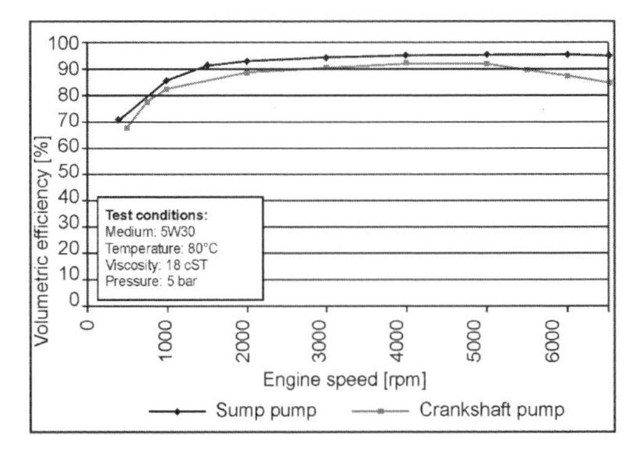

Fig. 7-203 Volumetric efficiencies for crankshaft and sump pumps.

during engagement are visible in the acoustic spectrum with harmonics up to many multiples of the basic frequency.

Inaccuracies in machining the teeth on the pinion and the gear with inside teeth appear in addition with a so-called repeat rate, referred to as the "hunting tooth frequency." This frequency can be determined using the following equation. Here f_{GMF} stands for tooth pitch, and z_1 and z_2 represent the numbers of teeth at the gears, while n_a is the number of potential installation attitudes.

$$f_{\mathrm{HT}} = \frac{f_{GMF}}{z_1 \cdot z_2} \cdot n_a \tag{7.22}$$

The mechanical noises in a toothed system, with the exception of the disturbances in engagement for outside teeth as a factor of load, are receding ever further into the background in view of the high quality found in sintered or erosion machined gear sets. In the involute gears of an external gear pump, selecting angular toothing and proper profile corrections allows a reduction in the noise level.

With decreasing pressure in the cavity between teeth during the intake phase, the gas dissolved in the oil is liberated in the form of bubbles. The mix of gas and liquid thus created increases in the noise propagated through the air. If the pressure in the flowing liquid falls below a critical level, then oil vapor bubbles will occur, Fig. 7-205.

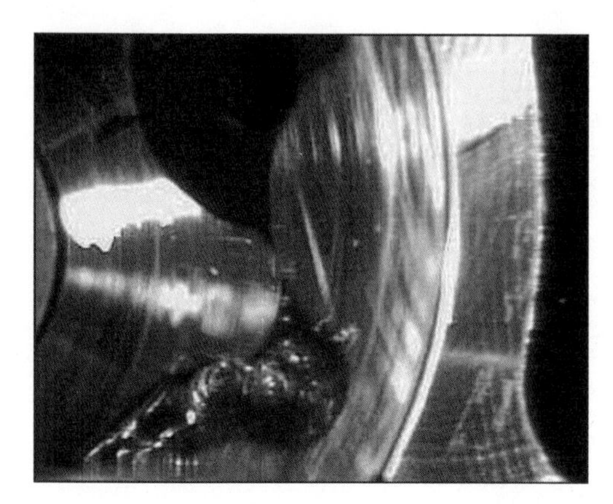

Fig. 7-205 Formation of oil vapor bubbles in an internal gear pump, photographed at 500 images per second.

If the pressure in the cavity between the teeth rises sharply because of the rapid meeting of the pumping cavities with the outlet, then the vapor bubbles implode suddenly. This process is accompanied by very high, localized pressure peaks that contribute to considerable noise generation. This typical phenomenon is known as cavitation. Important for this process is the presence of cavitation "seeds" in the form of the air bubbles previously mentioned. The start of the increase in noise as a result of cavitation is determined not only by the flow speed, which is proportional to engine speed, but also by the number of cavitation seeds in the liquid. With increasing oil foaming, one may observe an increase in the number of cavitation bubbles at otherwise constant pressure and constant flow velocity. At high air content cavitation starts even at low speeds. The associated early rise in the noise level then continues only gradually. If, on the other hand, the air content in the oil is very low, then oil vapor bubbles start to form only at very high circumferential speeds. In this case, the noise level rises suddenly.

A further source of noise is the alternating pressure (pulsation) in the liquid, which is because of the discontinuous oil pumping in the individual cavities between the teeth. When the pumped volume in the cavity meets the high pressure side, the volume is first compressed by the greater liquid pressure at the outlet side. The reduction in the size of the cavities during continuing transport ensures that they are emptied. The pulsation in pressure thus arising causes fluctuating speeds for the liquid particles, which is superimposed on smooth flow.

Figure 7-206 shows the results of the analysis for a gear pump. Quite apparent is the rise of the acoustic pressure level toward higher speeds. At about 4000 rpm the levels rise steeply, up to the third order. This effect is traceable to the cavitation of the vapor bubbles previously described, which appear more frequently with the increase of the speed and associated pressure drop in the liquid.

A further phenomenon—which appears in the pump examined here at a speed range of between 2000 and

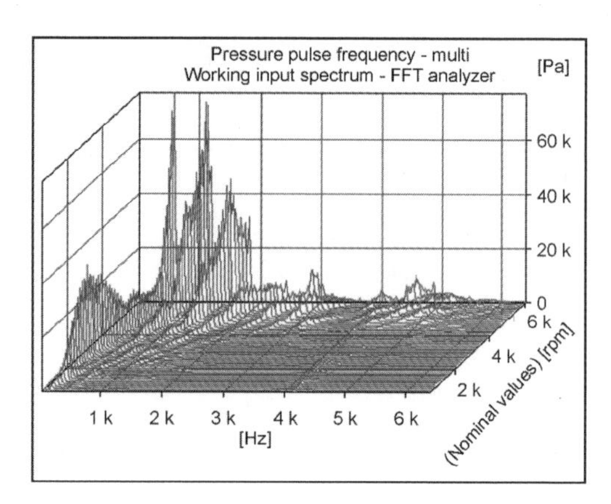

Fig. 7-206 Analysis as gear pump speed rises.

4000 rpm—is because of torsion resonance in the drive train. A good prediction of the acoustic pressure level, Figs. 7-207 and 7-208, is provided by the equation below, taken from VDI Guideline 3743, applicable to the cavitation-free range. Measurements have confirmed their validity for external gear pumps.

$$L_{WA} = 78 + 11 \cdot \log\left(\frac{P}{P_0}\right) \pm 3 \text{ [dB]} \qquad (7.23)$$

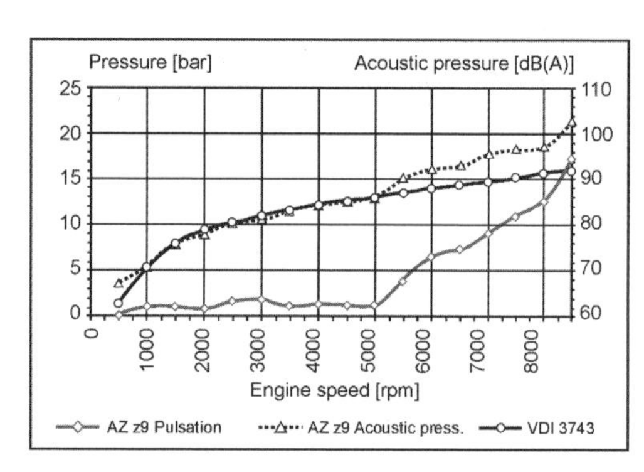

Fig. 7-207 Acoustic pressure level for an external gear pump, referenced to surface area.

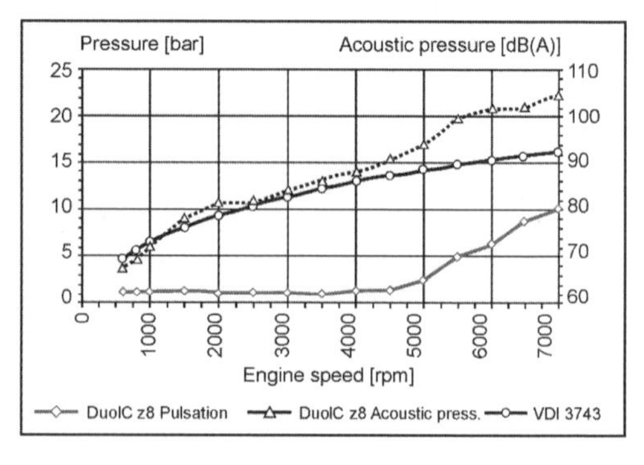

Fig. 7-208 Acoustic pressure level for an internal gear pump, referenced to surface area.

One achieves a better approximation for internal gear pumps with a modified form of the previous equation.

$$L_{\text{WA}} = 78 + 17 \cdot \log\left(\frac{P}{P_0}\right) \pm 5 \text{ [dB]} \qquad (7.24)$$

In both cases the calculated noise level corresponds closely to the level measured, through to the start of cavitation, indicated by the sharp rise in pulsation.

7.15.4 Calculation

7.15.4.1 Numerical Simulation of Flow—CFD

Formulation of the motion equations for multidimensional flow results in partial differential equations that are dependent on both time and location. The complex geometry of the flow because of the small amount of space available renders a unified solution to this equation impossible. It is thus helpful to simplify the spaces used for information so that they can be broken down into volumes that can be calculated more simply. With the appropriate transfer and peripheral conditions, such complex flow channels can be depicted as the sum of the individual, simple spatial elements. This idea is the basis for modern flow simulation (CFD—computational fluid dynamics). Today not only the maintenance equations for the mass, pulse energy, and species are calculated, but also the heat transport in laminar and turbulent flows.

In this way, it is possible, right from the engineering phase, to determine the most favorable flow patterns—in the pickup tube of an oil pump, for example. With the assistance of moving grid techniques, it is possible to simulate the movement of the flow cavities as well. Calculating the behavior of regulation pistons and optimizing flow in the valve area and reverse flow into the pumping channel can contribute to both increasing the efficiency of the pump and at the same time delivering important information on the tendency of valves to oscillate.

Figure 7-209 shows an example of a study of the filling cycle to obtain an optimized contour for the crescent-shaped area. The high-performance CFD products now available make it possible to calculate the filling process for the pocket formed by the gear and mating gear, dependent on the speed.

Complete simulation of gear pumps today is possible only to a limited extent. Although the "flexible grid" technology has made great advances, it is restricted at present to special solutions. Thus, for example, the change of gases in a cylinder can already be depicted. The geometry of the pockets between the teeth changes, however, is far more difficult to describe. This is aggravated by the fact that the volumes are theoretically reduced to zero, at least theoretically, during the course of a revolution. This results in numerical singularities. A simulation for the pump can be carried out only if one applies an equidistant offset to the toothing profile of one of the gears.

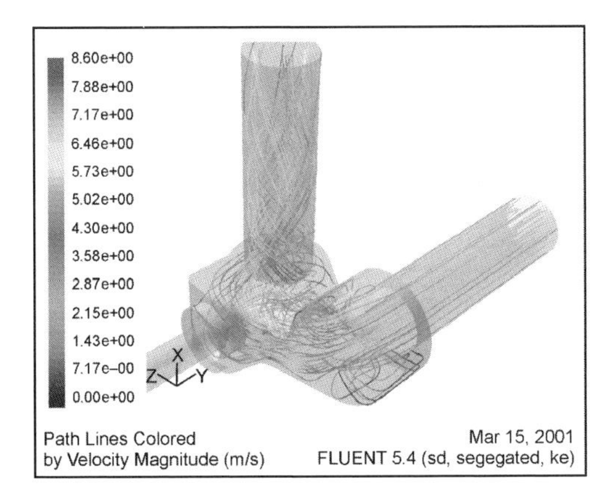

Fig. 7-209 Flow simulation near the valve using FLUENT V5. (*See color section.*)

7.15.4.2 One-Dimensional Simulation of Flow Grids

In addition to the classical multidimensional simulation of flow, knowledge-based programs used to calculate branched flow grids have been available for some years now. These programs are based on the flow string theory. They are capable of describing static and dynamic flow events in compressible gases and noncompressible liquids. The particular strength of these programs is found in their modular structure, while the individual modules—such as those for oil pumps, valves, bearings, camshaft shifter systems, manifolds, and many more—are knowledge based. Equations tailored individually to the particular component, usually backed up by test results, reflect the flow properties for these components. Often the database can also be expanded with one's own experience or test results. Using the modules mentioned above, it is possible to construct and examine multistrand grids such as the oil management concept for an internal combustion engine. These procedures are often used in an initial projection of oil requirements at various engine operating states.

Bibliography

[1] Findeisen, D. and F., Ölhydraulik, 4th edition, Springer Verlag.
[2] Lips, W., Strömungsakustik in Theorie und Praxis, 2nd edition, Expert Verlag.
[3] Emissionskennwerte technischer Schallquellen, VDI Guideline 3743, Sheets 1 and 2.
[4] Heckl, M., and H.A. Müller, Taschenbuch der technischen Akustik, 2nd edition, Springer Verlag.

7.16 Camshaft

The internal combustion engine is a machine that works intermittently. A fresh fuel mix flows through an open intake port and into the cylinder where it is compressed and ignited; it expands and passes through the open exhaust port into the exhaust system. Cam-actuated valves are normally used in four-cycle engines, less often in two-cycle engines, to open and close the ports.

In Wankel and two-cycle engines, the piston itself normally takes care of opening and closing the ports. Other potential embodiments such as rotating or reciprocating sleeves are no longer used in mass production.

7.16.1 Camshaft Functions

The primary function of the camshaft is to open and close the intake and exhaust valves so that gases can be exchanged; these actions are synchronized with the position of the piston and thus with the crankshaft.

Normally the valves are opened by transferring force from the cam to the cam follower, to other actuation elements where required, and ultimately to the valve, opening (or lifting) the valve against the force of the valve spring. During the closing cycle, the valve spring closes the valve. When the follower is in contact with the cam's base circle (with the cam exerting no lift), the valve spring keeps the valve closed against any gas pressure in the port (turbocharger pressure or exhaust gas counterpressure). During engineering it is particularly important to pay attention to the dynamics of all the peripheral conditions.

Desmodromic systems employed to increase potential engine speed (both the opening *and* closing phases are cam driven) are rarely used in mass production because of reduced valve train masses in multivalve engines and because improved valve springs have brought about an improvement in performance.

In the four-cycle engine, the camshaft is driven by the crankshaft and rotates at half the crankshaft speed. The valve timing for each individual valve is determined by the geometry and the phase rotation angle of the individual cams, normally separate for intake and exhaust valves and for the cylinders that are located along one or more camshafts. In multivalve engines it is possible to actuate several valves using a single cam with the intervention of linkages or forked levers. In special designs, the valves of multiple cylinders or the intake and exhaust valves are activated by the same cam.

In addition to the movements of the intake and exhaust valves required to control gas flow, the camshaft can also be used to generate the additional valve movements required for engine braking systems used in medium- and heavy-duty utility vehicles. Here existing or additional cams are employed so that engine drag is increased during overrun or coast down; the exhaust valve might, for example, be opened briefly around dead center in the compression stroke.

A further function of the camshaft, in addition to supplying power to auxiliary units (such as vacuum, hydraulic, fuel, or injection pumps), is actuating individual injection pumps in the engine block (pump-line nozzle) or pump nozzles in the cylinder head. Here, in addition to the cams that actuate the valves at the cylinders, further cams are provided to generate the stroke motion in the injection pump(s). Because of the additional loading encountered here, the cams usually have to be considerably more stable in design.

Torque, power output, fuel consumption, and pollutant emissions are influenced decisively by valve timing. The high specific power desired by the customer, smooth torque development, and low fuel consumption and pollutant emissions all across the speed range are difficult to achieve with conventional valve trains (see also the sections on camshaft shifting systems and variable valve actuation).

In every application the valve stroke length, velocity, and acceleration are the products of compromises between the fastest possible opening and closing for the individual valves and the forces and surface pressures created thereby. The friction and friction losses at the camshaft and the valve train as a whole are also important criteria in engineering.

7.16.2 Valve Train Configurations

When using overhead valves (OHV) the camshaft is located in the engine block, with the lift motion transferred to the valve by tappets or cam followers, push rods, and rocker arms. The configuration used for this type of drive train is usually simpler, but the stiffness is markedly lower than in systems with an overhead camshaft (OHC) or double overhead camshafts (DOHC). In the latter designs, the camshaft or camshafts are located in the cylinder head and driven off the crankshaft by gears, chains, or belts (and in a few cases toothed chains). The valves are actuated by rocker arms, cam followers, or valve lifters. The various types of valve trains used in passenger cars and utility vehicles and their application ranges are shown in Fig. 7-210. The materials listed here for the cams and cam followers are discussed later.

When the lift stroke is transferred to the cam follower (rocker arm, tappet, or valve lifter), one may differentiate between sliding contact and rolling contact. Current development trends are toward rolling contact in order to reduce drive losses and increase the tolerable loading. Another trend of simple valve lifter drives is toward sliding contact (without hydraulic clearance compensation) to reduce costs.

In addition to reduced friction losses (which means greater engine efficiency), the improved tribologic characteristics can also reduce wear. Where rolling contact is used, the tolerable surface pressure between the cam and the cam follower is considerably greater than for sliding contact. In the same comparison, Hertzian pressure rises because of the transition from sliding to rolling contact and the curved radii.

Materials with adequate rolling fatigue strength have to be selected when engineering for rolling contact; hardened steel (such as antifriction bearing steel) is normally used.

Two variants in the camshaft bearing concept are "open bearings" and "tunnel bearings." In the open bearing concept the bearing races are part of the camshaft; split bearings have to be used to support the camshaft. In tunnel bearings the camshaft has bearing races with a diameter greater than the maximum cam height. The camshaft can thus be slid completely into solid bearing races in the cylinder head or the engine block, Fig. 7-211.

	OHV Push rod	OHC Rocker arm	OHC Cam follower	OHC Valve lifter
• Trend:	Now only for basic engine output $V_H < 1.3$ ltr and simple V-block engines	Not widely used, Constant	Increasing	Standard
• Variants:	Sliding contact Rolling contact With/w/o hydraulic valve lifters	Sliding contact Rolling contact With/w/o hydraulic valve lifters	Sliding contact Rolling contact With/w/o hydraulic valve lifters	Sliding contact With/w/o hydraulic valve lifters
• Cam follower (cam contact)	Steel(Rolling contact) Cast iron(Sliding contact) (GG, CCI)	(GG, CCI) Steel(Rolling contact) Steel, cast iron(Sliding contact) (GG, GGG)	Steel(Rolling contact) Steel, cast iron(Sliding contact) CCI (GG, GGG)	Steel(Sliding contact)
• Cam material Rolling contact Sliding contact	Steel Cast iron GG/GGG, CCI (GG/GGG)	Steel Cast iron GG/GGG, CCI (GG/GGG)	Steel, powdered metal Cast iron GG/GGG, CCI (GG/GGG)	Cast iron GG/GGG, CCI (GG/GGG)

Fig. 7-210 Valve train configurations for passenger car, motorcycle, and utility vehicle engines.

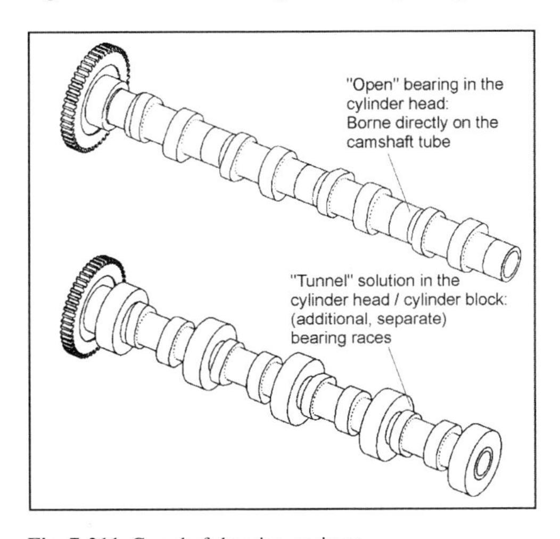

Fig. 7-211 Camshaft bearing variants.

7.16.3 Structure of a Camshaft

The basic camshaft design is shown in Fig. 7-212.

The main component is the cylindrical shaft (either hollow or solid), upon which the individual valve actuation cams are located. As was previously mentioned, additional cams for the injection system may also be included. The actuation forces are backed at camshaft bearings, most of which are axial bearings that stabilize the camshaft along the longitudinal direction. The crankshaft is driven by a drive sprocket that is attached either permanently or detachably to the drive flange at the end of the camshaft. As an alternative to this arrangement, the second camshaft in DOHC engines may be driven by the first camshaft. In this case, the first camshaft is fitted with an additional driving wheel (usually a sprocket or gear).

Auxiliary units are driven with an additional driving flange or takeoff at the free end of the camshaft or, for example, by an eccentric or lift profile at some point along the camshaft. A trigger wheel (generating one or more pulses per revolution) may also be mounted on the camshaft in order to ascertain the angular position of a camshaft.

The cam comprises one section with a constant radius (base circle) and the lifting area (run-up and run-down ramps, cam flank, and cam nose). The difference between the base circle and the highest point on the cam represents the cam lift stroke, which is selected to be proportional to the desired kinematic valve stroke.

Systems with mechanical clearance adjustment faults in the cam's base circle (deviations of the base circle from constant radius) have no effect on operational properties. A system with hydraulic valve play adjustment, by contrast, responds to every change in the base circle. Where there is a fault opposite the direction of movement, the hydraulic valve lifter compensates for this fault as valve play; in this case the valve stroke increases. If there is an error in the cam base circle in the lift direction, then the valve is already opened in the base circle segment because of the associated rise in force. This "pumping up" can, in extreme cases, result in complete loss of combustion chamber compression and engine failure.

7.16.4 Technologies and Materials

Camshafts made of cast iron are very widely used and differ in terms of the microstructure and hardness. Figure 7-213 provides an overview of the technologies and materials used.

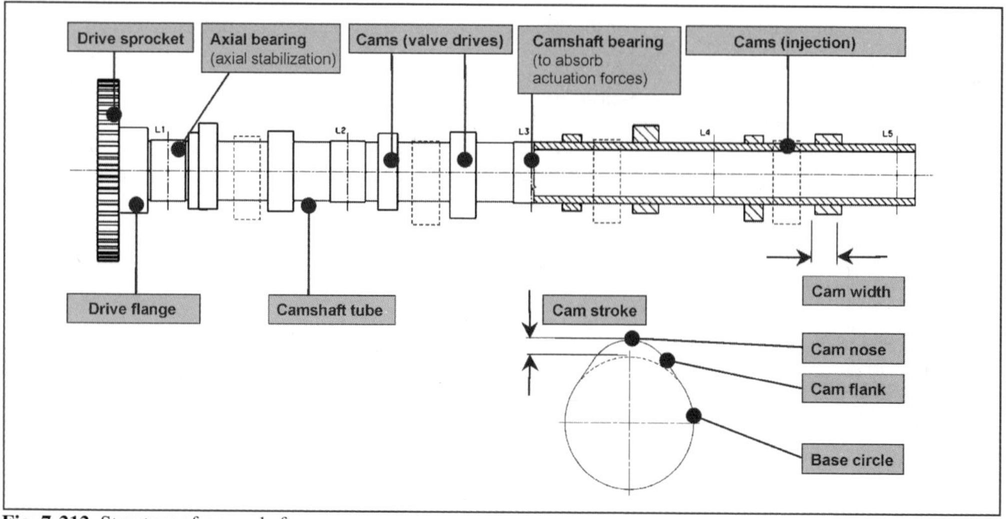

Fig. 7-212 Structure of a camshaft.

Technology:	(Cam) materials:	Mass production for passenger cars / utility vehicles
Cast camshaft	Cast iron with nodular graphite (GGG), inductance hardened	Passenger cars
	Cast iron with laminar graphite (GG), refluxing hardened (WIG)	Passenger cars
	Chilled cast iron, cast iron with laminar graphite (CCI, GG)	Passenger cars / utility vehicles
	Chilled cast iron, cast iron with nodular graphite (CCI, GGG)	Passenger cars / utility vehicles
	Cast steel (GS)	Under development
Assembled camshaft	Steel	Passenger cars (utility vehicles under development)
	Powdered metal materials	Passenger cars
	Powdered metal materials (precision cams)	Passenger cars
Forged camshaft	Steel	Passenger cars / utility vehicles
Worked from bar material	Steel	Utility vehicles

Fig. 7-213 Camshaft technologies and materials.

Assembled camshafts are made up of individual components (tube, cams, drive flange, etc.) that have been assembled. The materials can thus be matched exactly to the particular requirements.

When demands are extreme, camshafts forged from steel or machined from solid material (bar material) are used. A new manufacturing technology, cast steel camshafts, is currently under development.

7.16.4.1 Cast Camshaft

A camshaft made of cast iron with nodular or laminar graphite is often the ideal tribologic match for sliding contact and low-load rolling contact in many applications. With proper alloying and closely defined hardening of the cams, tolerable pressure levels of well over 1000 MPa can be attained.

In the case of chilled cast iron the cam area is cooled quickly following casting to create a wear-resistant carbide structure (ledeburite) with great hardness and good tribologic compatibility. A gray casting with good machining properties is available for use in the core area and the camshaft bearing points, Fig. 7-214.

7.16.4.2 Assembled Camshaft

Serving as the basis for an assembled camshaft is a tube to which individual cams are attached by shrink fit, press fit, interior high-pressure forming, or a comparable joining process. It is possible to distinguish between camshafts in which the tube and all the attached components are present as finished parts when they are attached and require no further machining and those processes in which the camshaft following assembly is available as a rough component (either in whole or in part), which has to be ground like conventional (unitized) camshafts.

Steel or sintering material (powdered steel) is used for the cams.

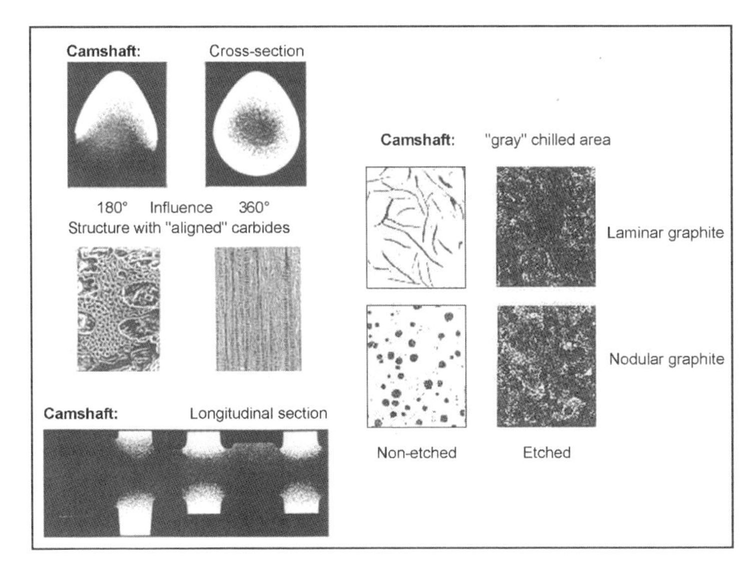

Fig. 7-214 Chilled cast iron in cross section.

Steel cams are normally forged as a rough part; the inner bore is then machined, and the cam is mounted on the tube. To attain the required materials properties, the cam can be hardened and tempered before or after attachment.

Using sintered material at rolling contact points makes it possible, since the cam geometry can be sintered more exactly than the required manufacturing tolerances, to build a camshaft that need not be further worked once the inside bore has been machined and the cam has been mounted on a tube with final geometry.

Figure 7-215 shows some examples of cam materials for assembled camshafts.

A high-alloy, liquid-phase sintered, powdered-metal steel was developed for use as a sintering material for sliding contact.

7.16.4.3 Steel Camshaft

Used for almost all applications with rolling contact in utility vehicles and in many passenger cars are forged steel camshafts or steel camshafts that are machined from solid material. When there are high demands in terms of torsional and/or tensile strength, steel shafts also have to be used for sliding contact.

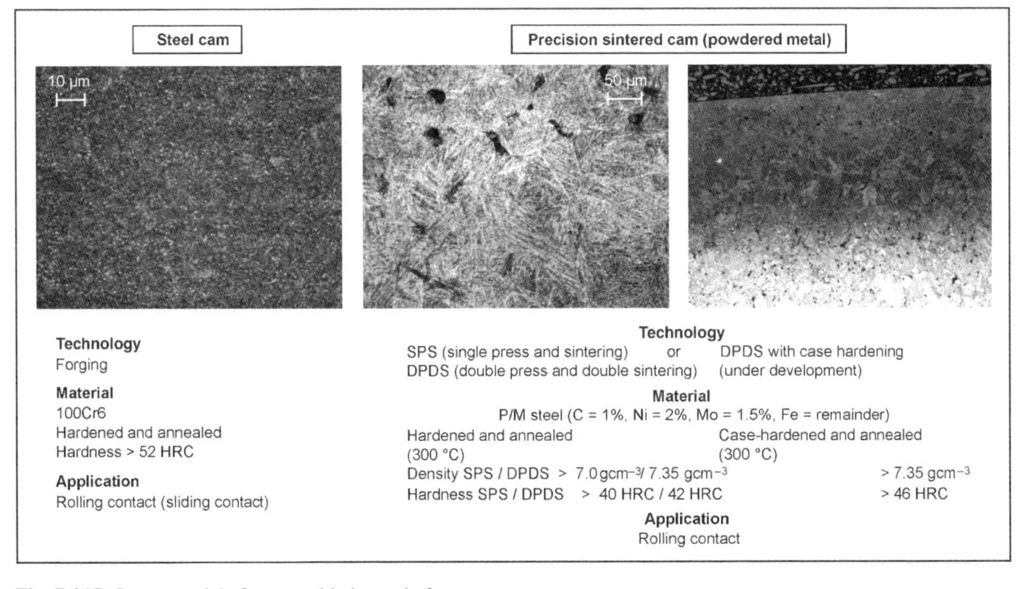

Fig. 7-215 Cam materials for assembled camshafts.

With the high tolerable pressure levels and the good mechanical properties of the material, these camshafts can be used for maximum demands, provided that correct tribologic mating materials are used.

7.16.4.4 Materials Properties and Recommended Matches

Figure 7-216 shows, for example, the spreads for torsional and tensile strengths for various cast materials. Various potential matches for rolling and sliding contact and the tolerable Hertzian pressures in each case are shown in Fig. 7-217 and in the summary of trends in Fig. 7-210.

Starting with the simplest gray cast camshaft, with cast tappets as the cam followers for sliding contact, it is possible to cover the entire range with the pairs of materials depicted, through to high-load rolling contact with cams and rollers made of roller bearing steel (100Cr6).

7.16.5 Reduction of Mass

Similar to the situation for the vehicle as a whole or for the overall valve train, the camshaft as an individual component is subject to the necessity to reduce masses. On the one hand, the engine's static mass is minimized, while, on the other hand, the moving (rotating) masses have great influence on the dynamics of the total system.

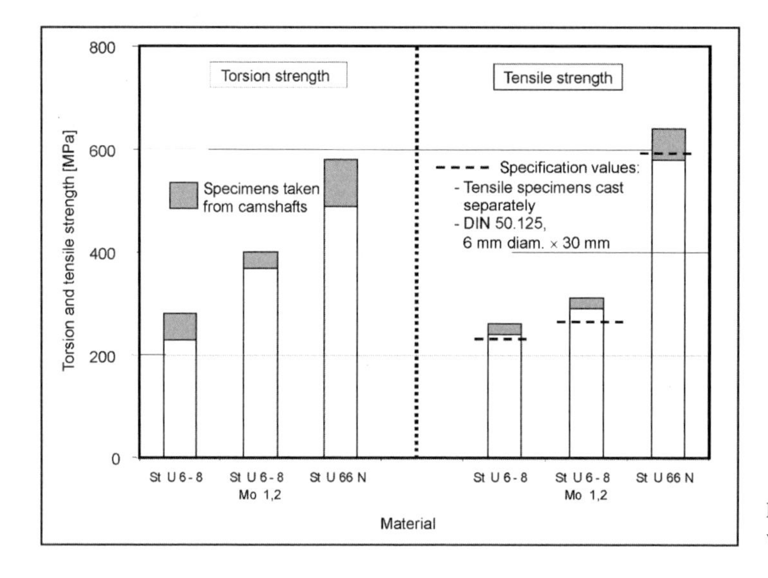

Fig. 7-216 Strength values for various casting materials.

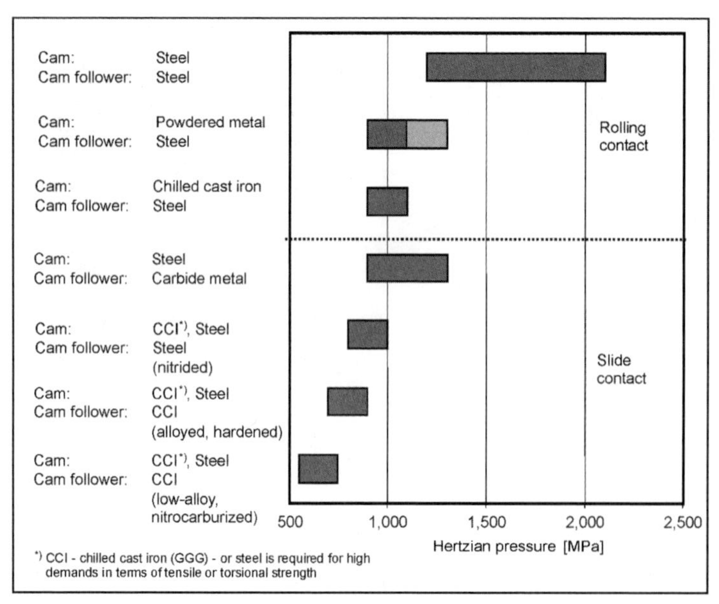

Fig. 7-217 Pairs of materials and Hertzian pressures.

At the same time, it is always necessary to reach a compromise among technical feasibility (minimum wall thickness, etc.), costs (materials, machining steps, etc.), and functioning (cam width, diameter of the base circle, torsional stiffness etc.).

The simplest possible option is hollow drilling or cylindrical hollow casting for camshafts (20% reduction in mass). When using hollow casting techniques with a graduated inner contour (profiled cavity), the mass is reduced even further. Figure 7-218 shows some examples of mass reductions in comparison with camshafts made of solid material and examples of chilled cast iron and steel cast camshafts with hollow profiles.

The assembled camshaft today presents the greatest potential for reducing masses. The steel tube's wall thickness can be reduced further than the wall thickness in the casting process. Integrating the camshaft bearing into the camshaft itself (tube diameter = inner race diameter) permits additional savings in masses. An important design criterion for such shafts is the joint between the cams and the tube, with its influence on the moment, which can be transferred.

7.16.6 Factors Influencing Camshaft Loading

The kinematics of the valve drive is the primary determinant for camshaft loading. The peripheral geometric conditions such as the step-down ratio or cam profile (e.g., high acceleration rates) are decisive here, in particular. Moreover, the camshaft is loaded by the valve train masses in motion and the total forces exerted by the valve springs and exhaust gas counterpressure. An integrated engine braking system can impose further and usually very significant loading on the camshaft (five to ten times the forces encountered during normal changes of gas charges). Figure 7-219 shows some of the influencing factors for camshaft loading.

The contact forces created between the cam and the camshaft induce both torsional and flexural moments in the camshaft which, together with the drive moment for auxiliary units, give the total torsional and flexural loads for the camshaft. In addition to the loading, the Young's modulus for the cam and the cam follower and the crowning of the components in the contact area are decisive for pressures and deformations.

7.16.7 Designing Cam Profiles

The progress of the valve stroke required, which is usually specified by the engine manufacturer, is a compromise for ideal filling across the entire speed range (high moment in the lower and medium speed ranges and at the same time high maximum power output). Here the peripheral geometric conditions such as valve diameter, valve stroke, and valve clearance to the piston at TDC are most important on one hand, while the demands in terms of functioning and manufacturing (such as jerk-free transitions in the entire valve stroke cycle or thermal loading of the exhaust valve while opening) are the most important parameters, on the other hand.

This specified and targeted advance for the valve stroke, depending on the type of valve train and its kinematics, is recalculated to form a cam profile matched to the cam follower.

If mechanical valve clearance adjustment is implemented, there is always some play in the total system between the cam and the valve. This play causes

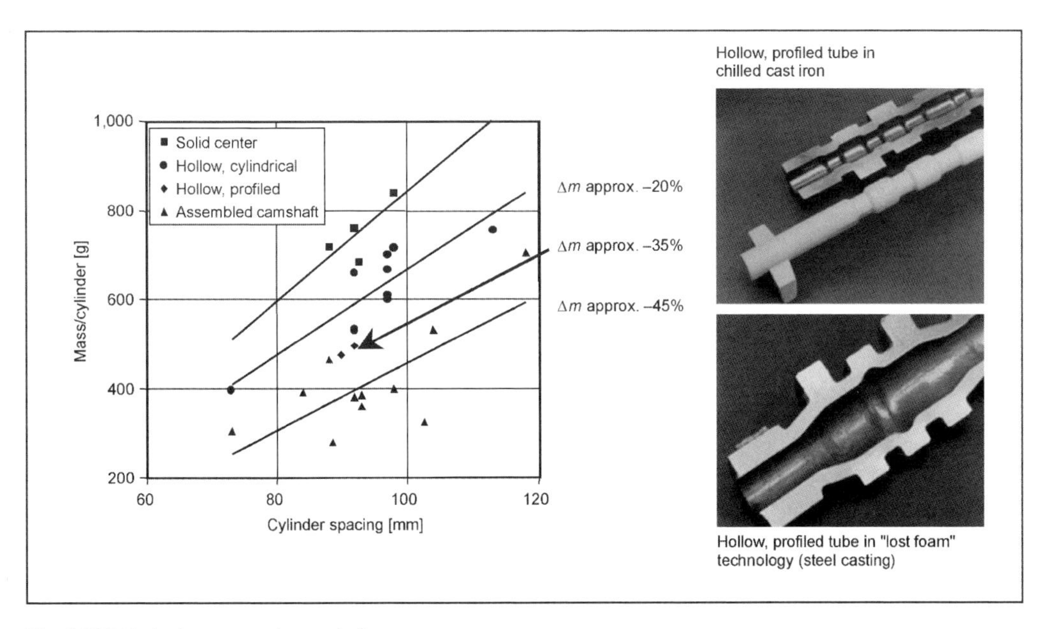

Fig. 7-218 Reducing masses in camshafts.

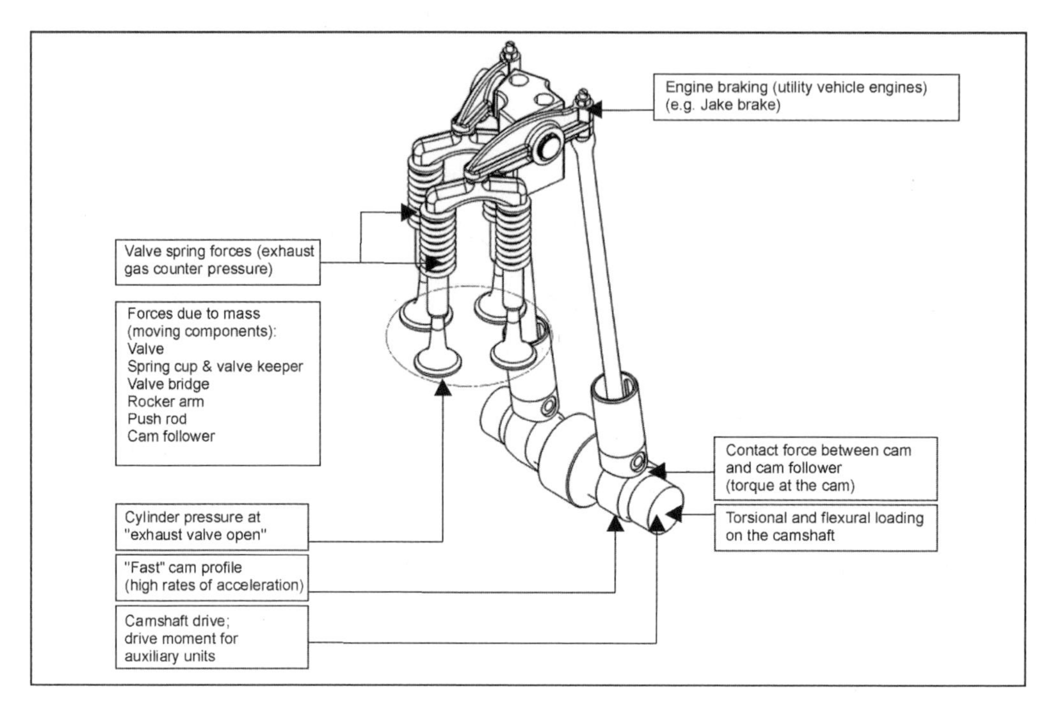

Fig. 7-219 Factors influencing camshaft loading.

inconsistency at the start of the stroke and thus always creates a sudden load. During the closing cycle there is also a "bump" because the valve contacts the valve seat before the cam stroke is completed. In order to limit the seating velocities and sudden accelerations for the valve train components involved, it is necessary to provide the appropriate opening and closing ramps. Variances in the valve stroke occur in systems with mechanical valve clearance compensation, the extent depending on wear and temperature; valve overlap also varies (phase during which the intake and exhaust valves are both open). In valve trains incorporating hydraulic valve lifters, these ramps are far flatter, Fig. 7-220; the valve stroke and overlap are nearly constant.

An important criterion for design is Hertzian pressure. This indicator value describes the compressive load on the mating components. Using the maximum tolerable Hertzian pressure allows us to preselect potential materials for cams and cam followers. The dynamics calculation usually shows, in comparison with the basic kinematic design, more realistic values for the location and size of maximum pressure values, Fig. 7-221.

When a roller is used as the cam follower (roller tappet, roller lever) there are often concave radii in the flanks of the cams. Here it is necessary to consider manufacturing limitations in reference to grinding. It may be necessary under certain circumstances to accept deviations from the specified valve stroke curve. When using sintered cams, the outer contours of which require no additional machining, any concave radius can be realized (at least in principle).

When using an assembled camshaft, it is necessary to pay attention to the moments that are transferred, depending on the system, as a decisive magnitude. During engineering, one must ensure that the maximum dynamic moments can be transferred with the required degree of confidence.

7.16.8 Kinematics Calculation

In the kinematic (quasistatic) calculation, the moving masses in the individual valve train are reduced to one single mass and one spring (the valve spring). A targeted motion (corresponding to the progression of the valve stroke) is imposed upon this individual mass. The mass and spring forces are considered in this way; additional outside forces such as gas forces coming into play when the exhaust valve is opened can be taken into account.

The most important results of kinematic calculations include the hydrodynamically effective speed for sliding contact, roller speed for rolling contact, and/or Hertzian pressures between the cam and its follower (as well as bearing loads for the valve train components), loading, and relative motion of the driving element at the end of the valve shaft or a valve link (e.g., valve finger radius, elephant foot, etc.).

The hydrodynamically effective speed (total speed, lubrication index) is a measure of the cohesion of the lubricating film between the components in contact, Fig. 7-222. In the case of sliding contact there are two "zero intersections" (change of sign) in this curve during

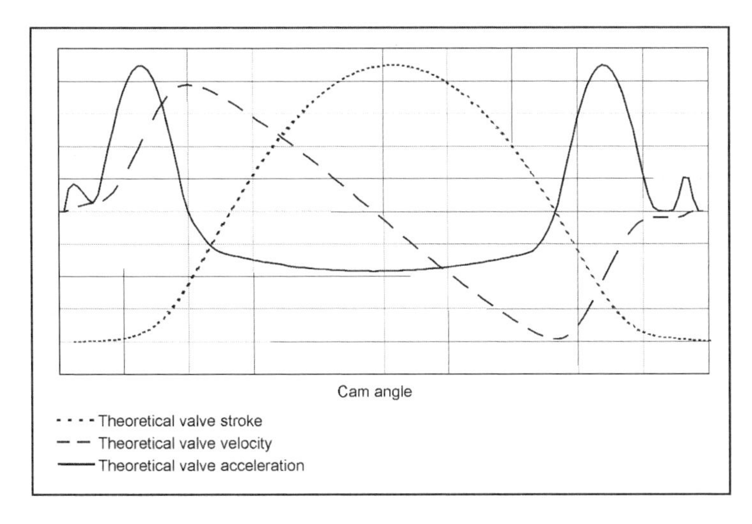

Fig. 7-220 Valve stroke, speed, and acceleration plotted against the cam angle for a roller cam follower valve train with hydraulic valve lifters.

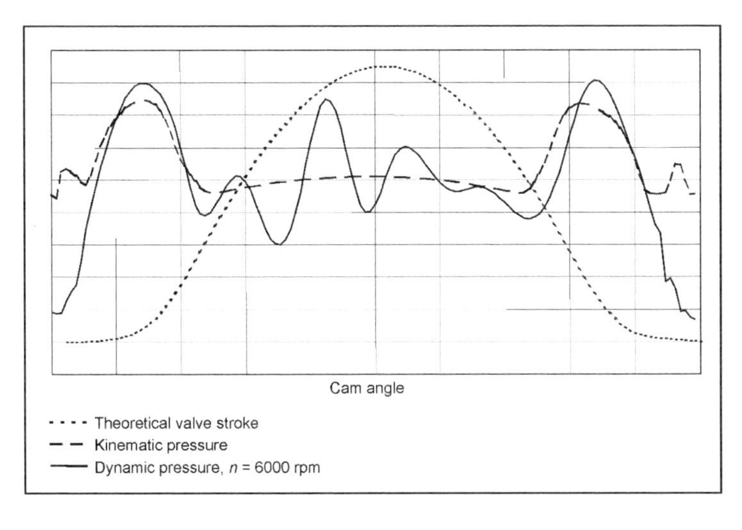

Fig. 7-221 Theoretical valve stroke and Hertzian pressure (kinematic and dynamic) for a roller cam follower drive train with hydraulic valve lifters.

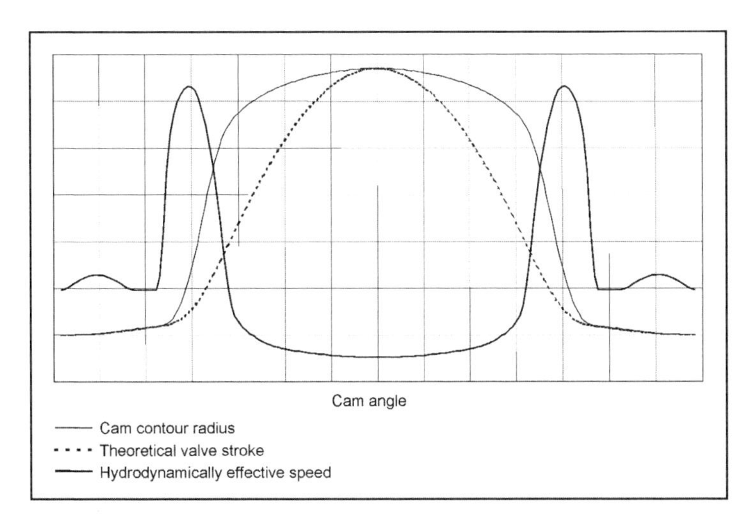

Fig. 7-222 Cam contour, theoretical valve stroke, and hydrodynamically effective speed plotted against the cam angle at contact between cam and flat tappet.

each cam revolution. Since at that particular moment the load-bearing capacity of the lubricating film collapses, the risk of wear can be reduced by suitable design.

When there is rolling contact with roller bearings (such as a needle bearing in the case of a roller cam follower), it is possible to analyze the service life (taking into account various loading populations).

7.16.9 Dynamics Calculations

Calculations of the dynamics supply a far more accurate image of real system behavior than does the relatively simple kinematics model. Accordingly, greater effort is required for modeling. Multibody simulation is the tool used for dynamics calculations. Common to all such programs is that the mechanical systems being assessed are broken down into individual masses and that they are then coupled one with another by means of spring and damping elements corresponding to the stiffness of the components and their damping properties. In addition to integrating hydraulic subsystems (hydraulic valve lifters) into the simulation, it is also possible to use the results from FEM calculations, e.g., force- or path-dependent stiffnesses for components.

The degree of detail for the dynamics calculation is virtually as desired and is limited only by the ratio of benefit to effort.

With all these elements and peripheral conditions, there arises a model capable of oscillation that, in addition to the stiffness, also depicts the *eigen* frequencies for the system being observed. The output depicts the motions of the individual components and the forces and pressures effective upon them.

One sees in Fig. 7-223 that the force between the cam and roller deviates distinctly from the progress determined kinematically, which is the result of the oscillations superimposed on the targeted motion. Particularly in valve trains with hydraulic clearance compensation, loss of contact can result in grave problems (pumping up the hydraulic valve lifters). A dynamic analysis of the valve train can identify critical components right in the engineering stage (long before parts are developed for measurements and engine operation) and thus shorten the development process considerably.

7.16.10 Camshaft Shifter Systems

To comply with future exhaust gas regulations and to reduce fuel consumption, elements that influence valve timing are used more often in gasoline engines. The camshaft shifter is one such device. It enables continuous change in the timing for a camshaft, across a wide angular range. This makes possible a change in valve overlap in DOHC engines and thus influences the residual gas content in the combustion chamber. In addition it is possible, above all at idle and full throttle, to tune timing for maximum comfort and/or maximum torque and highest performance. Camshaft shifters have been used in vehicles since the mid-1980s, initially as two-state shifters with simple controls but today more often as continuously adjustable systems operating under closed-loop regulation.

In DOHC engines, camshaft shifters are used mostly on the intake shaft; typical adjustment angles lie between 40 and 60 crankshaft degrees. There are, however, also shifters in mass production, used on the exhaust side, preferably in turbocharged engines. Both degrees of freedom may be combined where there are maximum demands regarding performance and exhaust gas quality.

In some DOHC engines camshaft shifters are used for dethrottling, i.e., reducing consumption by closing the intake valve late. In this concept, however, neither an increase in performance nor an improvement of comfort at idle can be achieved since the valve overlap is not changed.

Continuous camshaft shifting operates in a closed-loop regulation circuit and today is hydraulically powered in all cases.

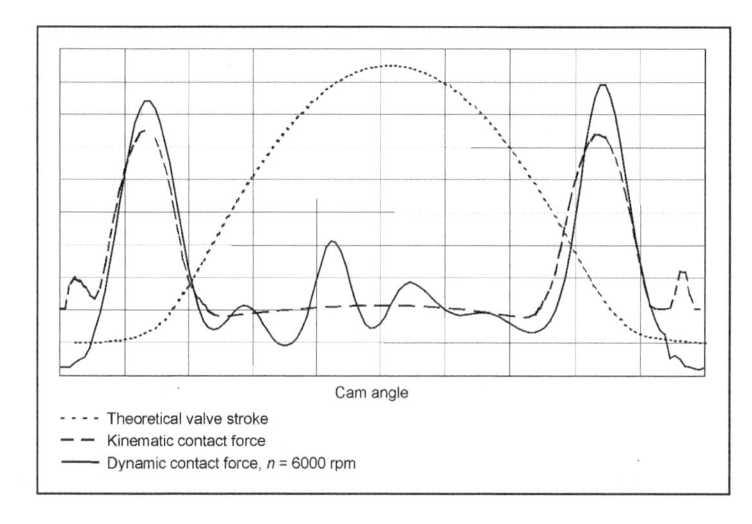

Fig. 7-223 Theoretical valve stroke, kinematic contact force, and dynamic contact force plotted against the cam angle for a rocker arm valve train with hydraulic valve lifters.

In the engine management system, the required set-point angle for timing adjustment is taken from an engine map dependent on load and speed. This is compared with the actual, measured angle. Deviations between the set-point and actual angles are evaluated with a regulation algorithm and cause a change in the electrical power applied to the control valve. Thus, the valve diverts oil into the chamber at the valve shifter in a fashion corresponding to the desired adjustment direction, while oil is allowed to escape from the opposite chamber. The angular position of the camshaft changes in accordance with the degree of fill at the oil chambers in the shifting unit. Sensors scan the trigger wheels at the camshaft and crankshaft; the actual value is calculated on the basis of these signals. This regulation process runs continuously, at high frequency, and thus leads to good response characteristics when there are rapid changes in the set-point angle, giving high angular accuracy in maintaining the set-point angle. The system generally uses the motor oil circuit as its power supply; systems with a separate high-pressure supply are also found in sports engines.

The following components are needed to implement camshaft shifting:

- The **hydraulic shifter unit**, mounted on the drive end of the camshaft. This component sets the adjustment angle in response to alternating filling of two oil chambers. Low leak rates and sufficiently large piston surfaces ensure good stiffness under load. The shifter unit is built in various styles—with a linear piston and helical toothing or with a rotary piston.
- The **regulation valve**, built into the cylinder head or an attached component, should be located near the point at which oil is transferred to the camshaft. This valve is controlled electrically, usually with a pulse-width modulated signal; it regulates the flow of oil into and out of the chambers in the shifter unit. A high flow rate during adjustment phases and precise regulation capacities to fix the angle are the most important features of the valve.
- The **regulation circuit** for continuous adjustment comprises suitable software and a power output stage in the engine management unit as well as trigger wheels and sensors at the crankshaft and camshaft. Components already present in the engine can be used for this purpose, although the trigger wheel at the camshaft has to be modified.

The overall system for continuous camshaft shifting and the components described above are shown in Fig. 7-224.

Two concepts for the hydraulic shifter unit have become commonplace. A brief review of their basic design is provided below. The camshaft shifter with helical toothing comprises these main functional components: the drive sprocket (joined with the crankshaft), adjustment piston, and output hub (bolted to the camshaft). These components are joined one with another in pairs, via

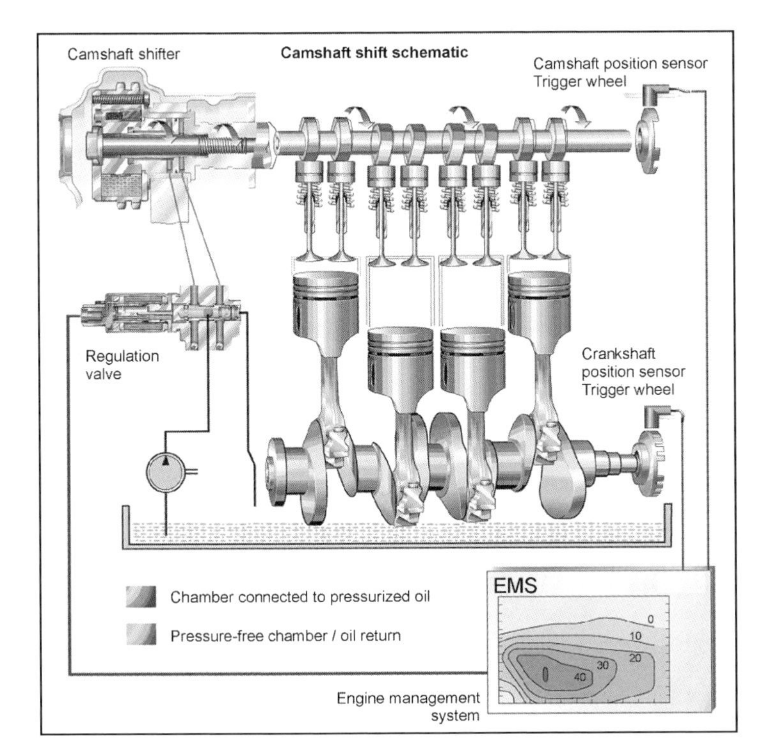

Fig. 7-224 Continuous camshaft shifting. *(See color section.)*

helical inner toothing, so that an axial shift of the adjustment cylinder causes the drive hub to rotate in relationship to the drive wheel. The transfer of the torque using inner toothing is very rugged. The design shown in Fig. 7-225 is completely sealed, for use in toothed belt drives.

When the engine starts, the spring shown in the illustration keeps the shifting piston in its home position. Both chambers are filled with oil during regulated operation; good sealing between the two chambers provides good stiffness under load. Quick responses demanded by the engine are achieved with engine oil at a pressure of about 1.5 bar.

Shown in Fig. 7-226 is the slewing motor or vane shifter in a version for chain drive. This version of the camshaft shifter is more compact and economical than the version with helical toothing; it comprises only the drive gear and the output hub. Rotary torque is transferred during operations by the oil fill in the chambers. Only during engine starting does a locking element normally ensure a fixed mechanical link between the drive and output elements. This locking element is unlatched hydraulically once the camshaft shifter has filled with oil. The locked end position here is, as a rule, "late" timing when adjusting the intake camshaft and the "early" timing setting when adjusting at the exhaust camshaft.

The regulation valve comprises a hydraulic section and a solenoid. The hydraulic slider is located in a bore with connections for oil supply, actuator chambers for the camshaft shifter, and oil return. A spring moves the slider toward the home position. When power is applied to the solenoid, the slide is shifted against the force of the spring. This changes the flow of oil into and out of the two chambers; in the so-called regulated position all the oil ports are largely closed. This achieves stiff holding of the adjustment piston in the camshaft shifter. In accordance with the givens of the particular application, the regulation valve either is integrated directly into the cylinder head or is attached by an intermediate housing. The regulation valve is connected electrically to the engine management unit.

Bibliography

[1] Bensinger, W.-D., Die Steuerung des Gaswechsels in schnelllaufenden Verbrennungsmotoren, Konstruktionsbücher, Vol. 16, Springer-Verlag, 1967.

[2] Holland, J., "Die instationäre Elastohydrodynamik," Konstruktion, Vol. 30, No. 9, 1978.

[3] Ruhr, W., "Nockenverschleiss—Auslegung und Optimierung von Nockentrieben hinsichtlich des Verschleissverhaltens," FVV Research Project No. 285, 1985.

[4] Holland, J., "Nockentrieb Reibungsverhältnisse—Untersuchung zur Verminderung der Reibung am Nocken-Gegenläufer-System unter Verwendung von Gleit- und Rollengegenläufern," FVV Research Project No. 341, 1986.

[5] Brands, Ch., "Dynamische Ventilbelastung–Rechnergestützte Simulation der Beanspruchung des Ventiltriebs," FVV Research Project No. 614, 1998.

[6] Dachs, A., Beitrag zur Simulation und Messung von Tassenstösselventiltrieben mit hydraulischem Ventilspielausgleich, Dissertation, Technical University of Vienna, 1993.

[7] Ruhr, W., Nockentriebe mit Schwinghebel, Dissertation, Technical University of Clausthal, 1985.

[8] Rahnejat, H., Multi-Body Dynamics, Vehicles, Machines and Mechanisms, SAE International, 1998.

[9] Beitz, W., and K.-H. Küttner, *Dubbel* Taschenbuch des Maschinenbau, Springer-Verlag

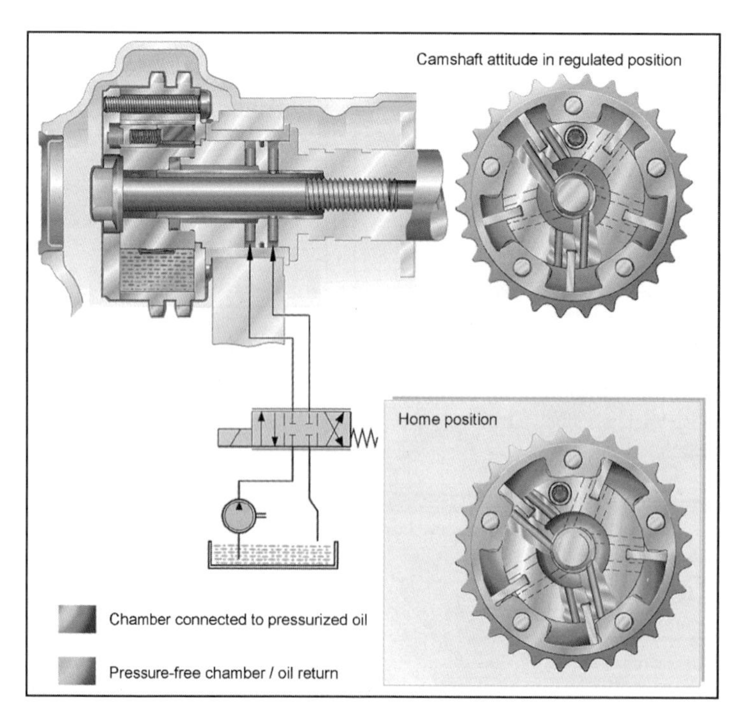

Fig. 7-225 Slewing motor or vane shifter. (*See color section.*)

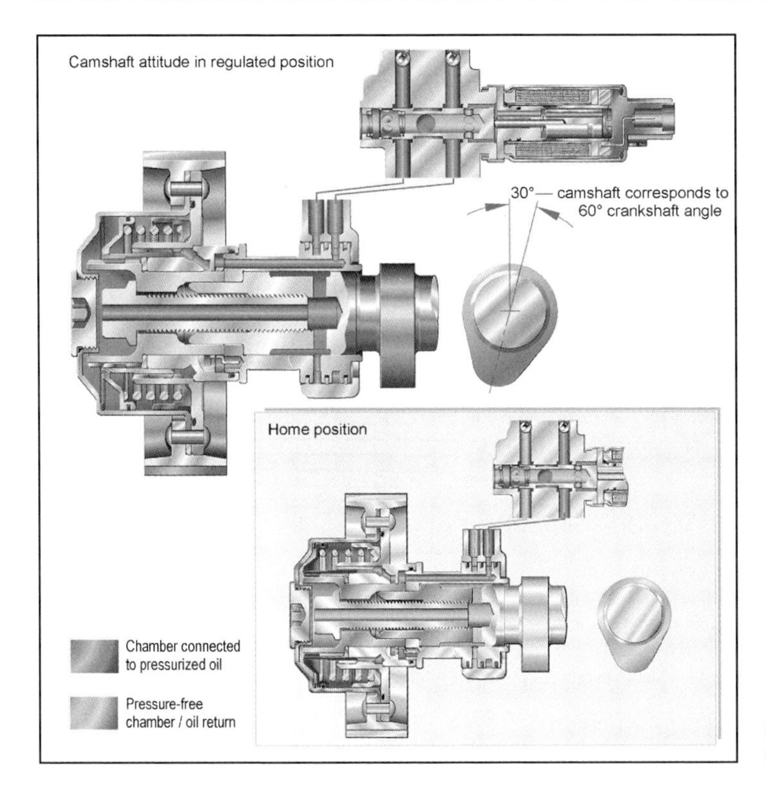

Fig. 7-226 Camshaft shifter with helical toothing. (*See color section.*)

7.17 Chain Drive

The primary function of the camshaft is to ensure that the valves open and close at the correct times. In modern overhead valve engines this is done by power transmission from the crankshaft. In most cases, toothed (synchronous) belts, or roller toothed (silent) or bushed roller chains[1, 2] of various weights are used. The selection of the design depends on the engine maker's philosophy.

The most important criteria in the choice of drive concept are the costs, the amount of space occupied, maintainability, service life, and noise generation.

A comparative evaluation of a timing chain and synchronous timing belt is shown in Fig. 7-227.

In modern engines, these power transmission systems often serve not only the camshaft, but other components such as the oil pump, water pump, and fuel injection pump as well. Figure 7-228 shows examples of potential arrangements.

Since neither the camshaft nor the crankshaft runs entirely smoothly and since the power required by the injection pump is subject to severe, periodic fluctuations, this drive system is exposed to very complex dynamic loading.[3, 4]

In the course of decades of experience, certain dimensions for the roller and bushed roller chains used in timing drives have proven to be particularly suitable.

	Timing chain	Synchronous belt
Installed size	+	O
Service life	++	O
Costs	O	+
Maintainability	O	O
Noise generation	O	O

Legend: ++ Very good + Good O Adequate

Fig. 7-227 Comparative evaluation of timing chains and belts.

7.17.1 Chain Designs

Among the standard chains, one differentiates between roller and bushed roller chains. In addition, there are both simplex and duplex chains, Fig. 7-229. A special form of the chain is the toothed chain, Fig. 7-230, also referred to as a silent chain.

The plates in toothed chains are shaped so as to enable direct force transfer between the chain and the sprocket, while in roller and bushed roller chains the interface with

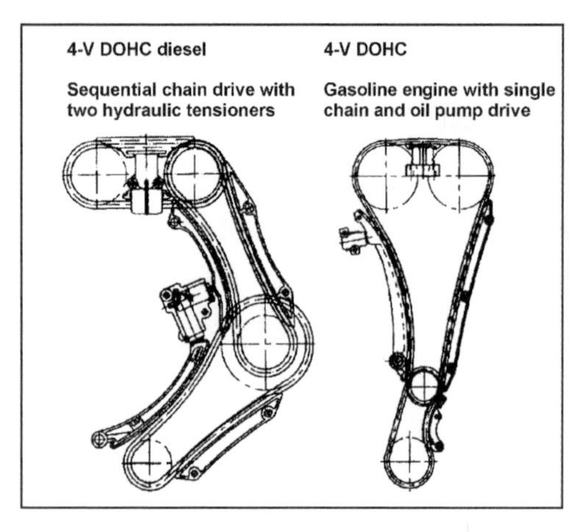

Fig. 7-228 Timing chain drive.

the sprocket takes place at the pivot joint via pins, bushes, or rollers. Silent chains can be made of any conceivable width without any fundamental change in design. Guide plates are provided to keep the chain from wandering off the sprocket; they may be located either at the center or on either of the outside edges.

The rollers, rotating over the bushes in a roller chain, encounter a small amount of friction when rolling along the sprocket's teeth. Thus, the contact point at the circumference changes continuously. The lubricant between the rollers and the bushes contributes to noise and impact damping. In a bushed roller chain, by contrast, the fixed bushes always mate with sprocket teeth at the same point. Thus, perfect lubrication for such drives is particularly important.

At the same pitch and failure strength, a bushed roller chain exhibits a larger joint surface than corresponding roller chains. A larger joint surface causes lower pressures at this surface area and thus less wear.

Bushed roller chains have proven their value, particularly for heavily loaded camshaft drives in high-rpm diesel engines. Whenever the transfer of a given torque at a certain maximum sprocket diameter using a simplex chain requires a number of teeth greater than 18, it is advisable to go to a multiple chain with the same or smaller pitch.

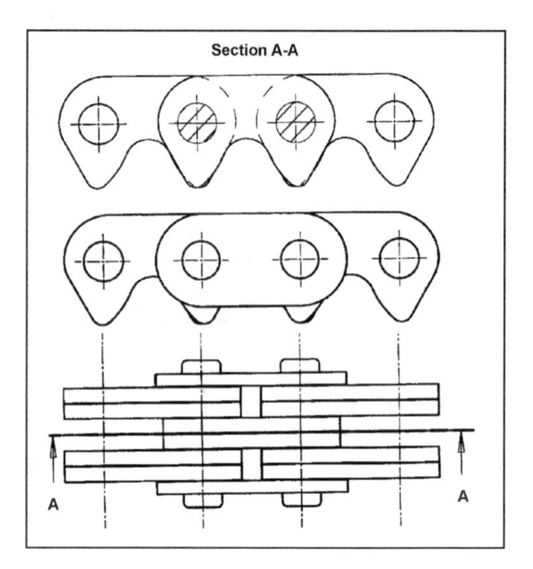

Fig. 7-230 Silent chain.

7.17.2 Typical Chain Values

Three essential factors characterize a chain's suitability for use as a timing chain:

- Breaking strength
- Endurance, Fig. 7-231
- Wear resistance

One cause that might be responsible for failure is exceeding the static or dynamic breaking load.

Particularly in timing drives, one does not encounter uniform loading. Pulsating loading on the chain results from fluctuating torques at the camshaft and the injection pump (in diesel engines, for example), nonuniform camshaft rotation, and pulsating longitudinal chain forces caused by the polygonal effect. Here the chain's fatigue strength must never be exceeded since the number of load alterations during an engine's service life is in all cases greater than 10^8.

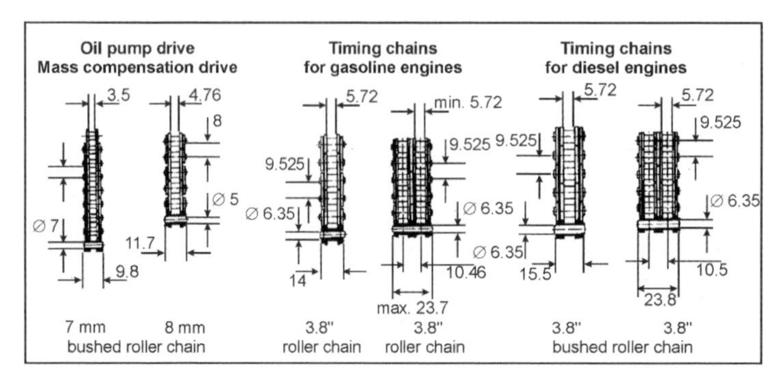

Fig. 7-229 Chain designs.

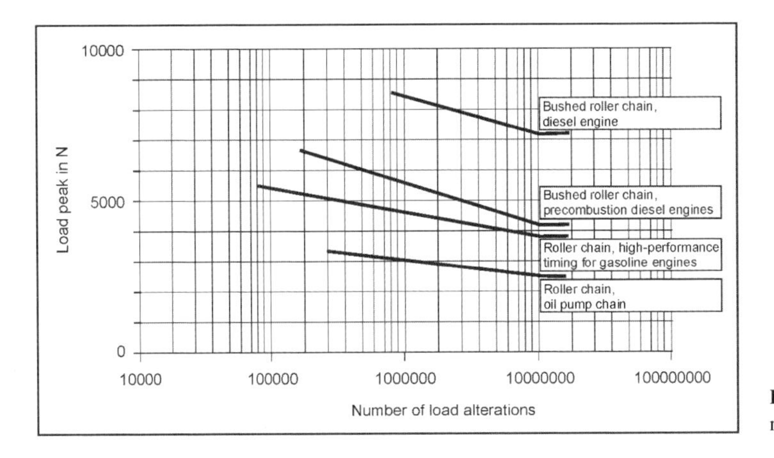

Fig. 7-231 Fatigue strengths for roller and bushed roller chains.

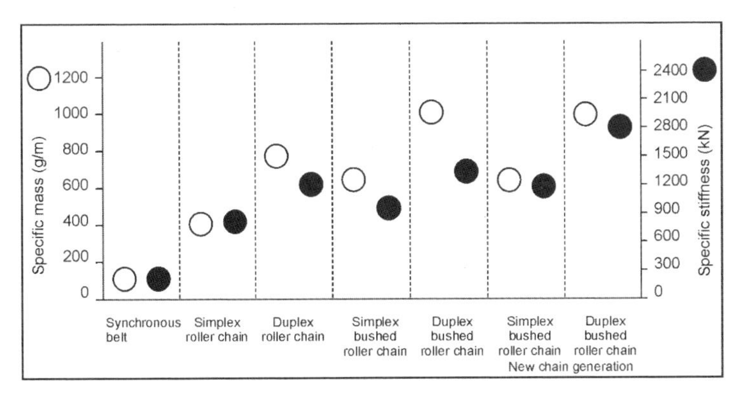

Fig. 7-232 Typical chain values: Stiffness and mass.

In today's engines with their precise timing, minimal stretching due to wear, at from 0.2% to 0.5% of chain length at up to 250 000 km in service, can be attained.

A chain timing system with its mass, stiffness, and damping represents a system capable of oscillation, having several degrees of freedom (Fig. 7-232). In response to excitation by the camshaft, crankshaft, injection pump, etc., this can cause resonance effects that result in extreme loading of the timing drive system.

Engineering measures make it possible to increase stiffness in the chain while retaining its specific mass. This shifts resonance points toward the higher frequencies.

7.17.3 Sprockets

The shapes of the teeth in sprockets intended for use with roller chains, bushed roller chains, and silent chains are standardized (DIN 8196). Proper tooth profile is just as important to reliable operation of the timing system as, for example, the chains' wear resistance.

Usually sprockets with the widest tooth gap are used. This makes possible, because of the short teeth and the wider gap between teeth, uninterrupted engagement and disengagement of the chain even at higher chain speeds.

Depending on the amount of space available and the particulars of the application, pulleys or sprockets with one or two rows of teeth may be used (Fig. 7-233). The selection of the materials depends on the timing drive system parameters, the operating conditions, and the amount of power to be transferred.

Carbon steel, alloyed steels, and sintered materials are used for the sprockets.

The materials used for precision punched sprockets include C 10 or 16MnCr5 for sprockets made with cutting processes, and D 11 for sintering processes, together with the heat treatment suitable for each particular material.

7.17.4 Chain Guide Elements

With the introduction of continuous-action tensioning and guide elements that are matched exactly to the particular engine, the drive can be optimized to such an extent that its service life equals that of the engine, without any special care being required beyond the prescribed engine maintenance.

The chain tensioner, Fig. 7-234, assumes a number of functions in the timing drive. First, the timing chain is preloaded (along the slack span) to a defined value under all operating conditions, even where stretching due to wear has occurred. A damping element, using either friction or viscous damping, reduces oscillations to an acceptable amount.

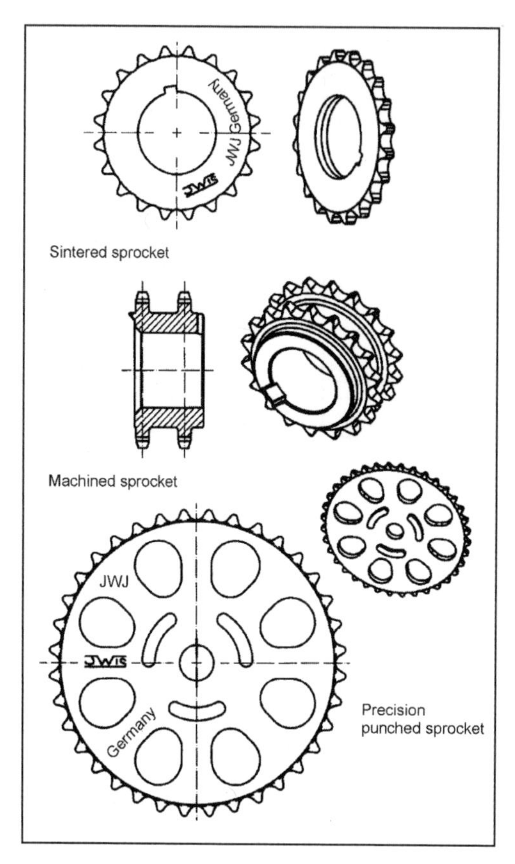

Fig. 7-233 Sprockets.

Simple rails made of plastic or metal are used as guide elements. They usually have a plastic surface and are either flat or curved to fit the chain's path, Fig. 7-235. The newer versions of these rails are usually injection molded plastic.

As regards the tensioner rails, a slip-promoting covering made of PA 46 is injected or clipped onto a backing element made of PA 66 with 50% glass fiber content for reinforcement purposes. The slip rails are usually manufactured as a unitized component.

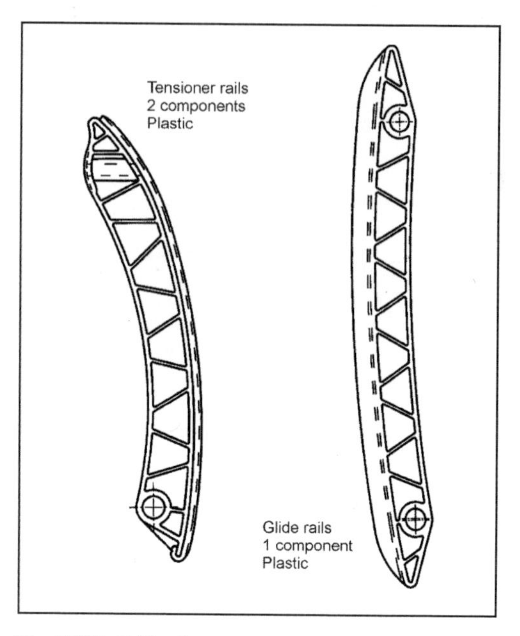

Fig. 7-235 Guide elements.

Bibliography

[1] Arnold, M., M. Farrenkopf, and S. McNamar, "Zahnriementriebe mit Motorlebensdauer für zukünftige Motoren," 9th Aachen Colloquium on Vehicle and Engine Technology, Aachen, 2000 ika/VKA.

[2] IWIS-Ketten Handbuch Kettentechnik, Munich.

[3] Fritz, P., "Dynamik schnellaufender Kettentriebe," VDI Fortschrittsberichte, Series 11: Schwingungstechnik, No. 253, VDI-Verlag GmbH, Düsseldorf, 1998.

[4] Fink, T., and V. Hirschmann, "Kettentriebe für den Einsatz in modernen Verbrennungsmotoren," in MTZ, Vol. 62, 2001, No. 10, pp. 796–806.

7.18 Belt Drives

This section provides an overview of the demands and functions of today's belt drives in internal combustion engines, synchronous belt drives used to drive the camshafts, and Micro-V® belt drives used to run auxiliary components.

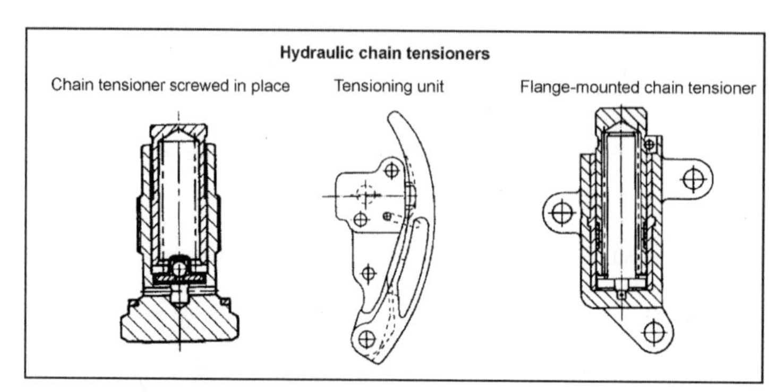

Fig. 7-234 Chain tensioner.

7.18.1 Belt Drives Used to Drive Camshafts

Synchronous belt camshaft drives today hold a 75% market share in European engines. This can be traced essentially to advantages found in the simplicity of the drive concept, flexibility in belt guidance, low friction, and cost advantages when compared with other drive systems. Moreover, auxiliary units such as oil or water pumps can be integrated into the drive concept.

7.18.1.1 Synchronous Belt Drive

Design of the Synchronous Belt

The synchronous belt is a bonded system made of three components (Fig. 7-236):

- Nylon fabric
- Rubber blend
- Tensile member

The facing fabric is made of high-strength nylon and is coated to reduce wear. It protects the rubber teeth against wear and against their shearing off. The rubber blend is a high-strength polymer. Polychloropene (CR) was used in early versions. Because of stringent requirements in terms of dynamic strength and resistance to temperature and aging, HNBR (hydrogenated nitrile rubber) materials are used exclusively today.

The cords in the tensile member are made of glass fiber—a material distinguished by its great tensile strength and amenability to bending. Consequently, it is particularly well suited for camshaft drives in which the crankshaft sprockets are small in diameter. The manufacturing process is such that the strands in the tensile member are twisted, clockwise and counterclockwise, in pairs, in order to achieve largely neutral running properties for the belt.

The synchronous belt is manufactured using a vulcanization process. Specific coatings for the fabric and the tensile cords ensure bonds between the materials that will endure for the life of the engine.

Synchronous Belt Profile

There has been a significant evolution in the profiles used for synchronous belts since they were initially employed as timing belts. A wide variety of profiles are in use today. The various profiles and their properties are discussed below.

The first camshaft drive belts were based on the classical Power Grip® design with its trapezoidal teeth, at that time already in widespread use in industrial applications. In response to increasing demands regarding power transmission, ratcheting resistance, and quiet running, curvilinear profiles (Power Grip® HTD/High Torque Drive) were developed. When compared with the trapezoidal shape, the forces are introduced more smoothly to the tooth with the rounder profiles, and this in turn reduces the possibility of tension peaks (Fig. 7-237). Rounded profiles are used exclusively today.

In the first generation of synchronous timing belts—with the trapezoidal teeth—there were two different tooth shapes, the smaller "C tooth" for gasoline engines and the larger "B tooth" for diesel engines, each with a pitch of 9.525 mm (Fig. 7-238). This differentiation is no longer made in the newly developed HTD tooth profiles.

When the HTD profile was introduced to the market, it was necessary to take into account the fact that some car makers continued to use the existing trapezoidal tooth sprockets.

To suit these applications, the profiles were optimized in regard to the radius at the root, flank shape, and tooth height (power function profile) so that they could be used with the existing trapezoidal sprockets. The associated sprockets, type ZA (C or CF tooth) and type B (B or BF tooth) are defined in ISO 9011.

HTD stands for the "high torque drive," which was developed and patented by Gates. This curvilinear profile represented a considerable improvement in noise reduction, in power transmission, and, in turn, in terms of service life.

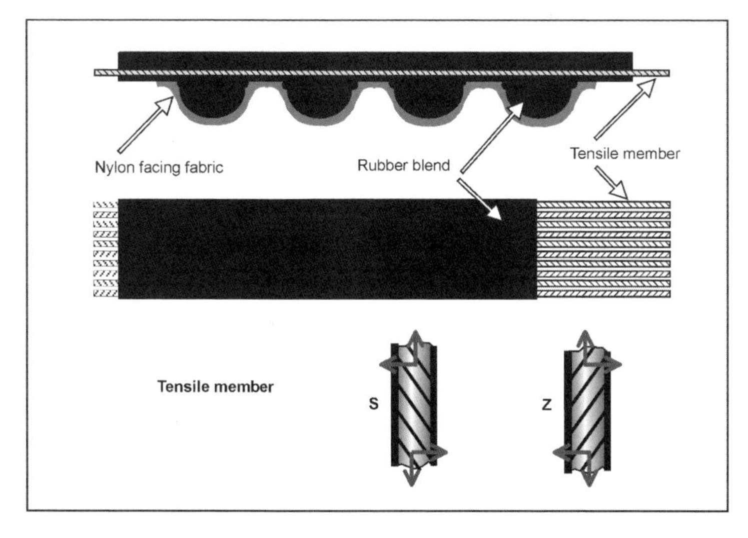

Fig. 7-236 Structure of the synchronous belt.

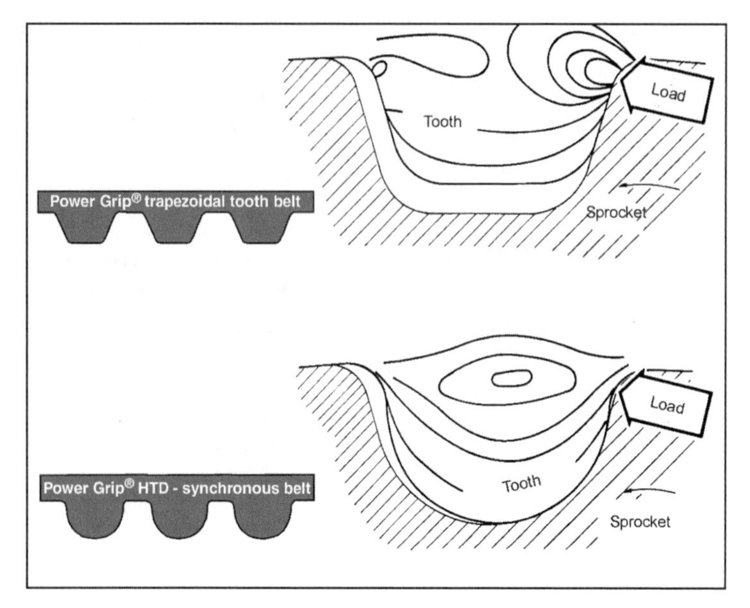

Fig. 7-237 Development of tooth profiles.

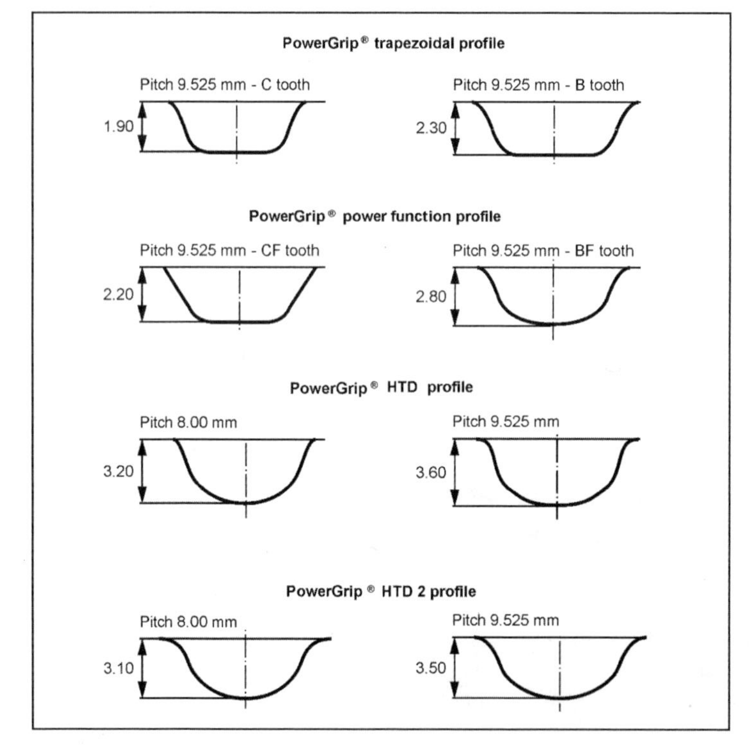

Fig. 7-238 Tooth profiles.

With the introduction of the succeeding HTD 2 generation, the existing advantages of HTD profiles were further enhanced. Here the radii at the root and the flank angles are once again enlarged.

Unique sprocket profiles are used for both types of profiles. The exact data for the profiles are available from Gates. Two pitch values are used for the two profiles: 9.525 and 8.00 mm. The smaller pitch has benefits in regard to noise and, because of the smaller sprocket diameter, permits a more compact design.

Both of the above-mentioned profiles can also be used in a double-sided synchronous belt (Fig. 7-239). Double-sided synchronous belts are used, for example, to drive balancer shafts.

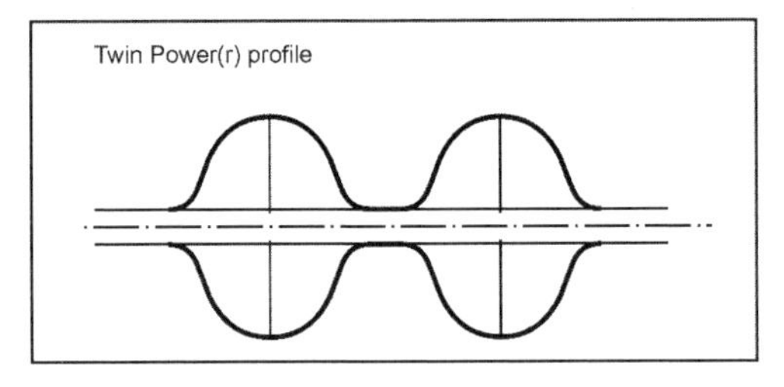

Fig. 7-239 Double-sided synchronous belts.

Key Values—Synchronous Belts and Sprockets

The most important values for the synchronous belt are shown in Fig. 7-240. The height of the tooth and depth of the backing together give the overall thickness of the belt. The pitch line distance (the distance from the root of the tooth to the center of the tensile member) depends on the belt design, the thickness of the fabric, and the diameter of the tensile cords. The width of the synchronous belt is selected in accordance with the alternating dynamic loading; in internal combustion engines it normally lies between 20 and 28 mm and, in isolated applications, is as much as 32 mm.

7.18.1.2 Synchronous Belt Drive System

The most important demand placed on the toothed belt system is synchronizing the camshaft over the engine's entire lifetime. This is an important criterion for maintaining emission values even after extended periods in service. With proper selection of the materials for the belt, the use of an automatic tensioning system, and the use of optimized system dynamics, stretch in the synchronous belt can be kept to less than 0.1% of belt length. In four-cylinder engines this represents a timing deviation of from 1 to 1.5 crankshaft degrees.

The usual requirements in engine building continue to apply, regarding engine life (currently 240 000 km), temperatures of about 120°C, the smallest possible build size, and minimum weight.

Bothersome noises generated by the belt drive are not acceptable.

Design Criteria

Complex synchronous drives are engineered with computer support. A survey of the most important parameters considered in the design and some general design criteria is provided here.

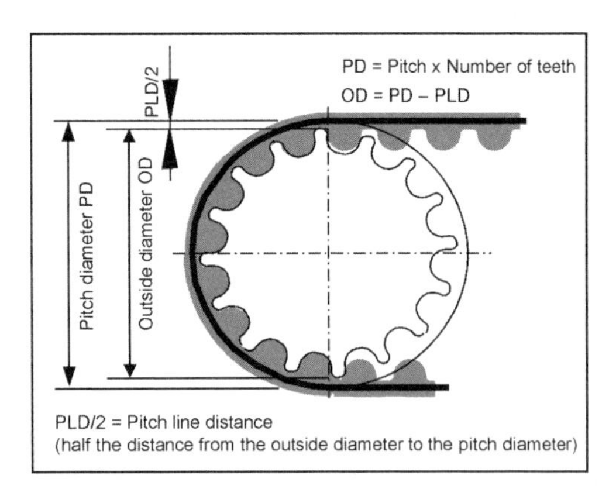

Fig. 7-241 Key values for the sprocket.

The profile of the sprocket has to be selected to match the diameter. The effective diameter is the product of the number of teeth and the pitch; the outside diameter of the sprocket is reduced by a value corresponding to the pitch line distance (Fig. 7-241).

The design of complex synchronous belt drives is computer supported. The most important parameters in design as well as a few general design criteria are discussed here. Important input data include the arrangement of the components, i.e., the drive configuration, torque development at the components, and the dynamic circumferential forces calculated from them, along with the data for the belt itself. With these data at hand, it is possible to calculate and optimize not only the span lengths and wrap angles, but also the belt's lifetime in reference to various failure modes. The dynamic forces and oscillations are used to calculate in the same way the other components in the system, such as the design of the reversing pulleys and the idler pulleys.

Given below are a few general design criteria that must be observed in synchronous belt systems in order to engineer a functional system that will achieve the 240 000 km lifetime required today:

Recommended Minimum Wrap Angle

Crankshaft	150°
Crankshaft/Injection pump	100°
Auxiliary unit sprocket	90°
Tensioning pulley (smooth or toothed) min. 30° and better	>70°
Deflection pulley (smooth or toothed)	30°

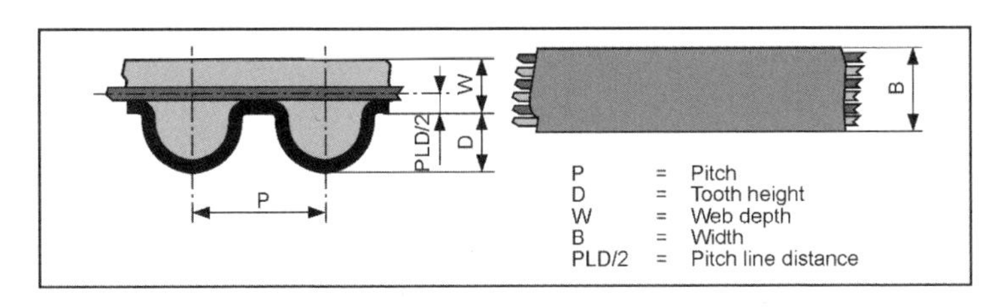

Fig. 7-240 Key values for toothed belts.

Periodic Tooth Engagement

Periodic tooth engagement means that a given tooth always engages with the same sprocket groove. This is to be avoided so as to preclude irregular belt wear and the belt damage that it may cause. The appearance of periodicity is calculated as follows:

X.nnn = Number of teeth at the belt ÷ Number of teeth at the sprocket

Here the following values for X.nnn are to be avoided:

X.nnn = X.0, X.5 (must in all cases be avoided)
X.nnn = X.25, X.333, X.666, X.75 (ought to be avoided)

Span Lengths

In order to avoid resonance-induced noise at idle, unsupported span lengths should not lie in a range of from 75 to 130 mm.

Minimum Diameters for Sprockets and Deflector Pulleys

Pitch 9.525 mm 18 teeth (54.57 mm diam.)
Pitch 8.00 mm 21 teeth (53.48 mm diam.)
Smooth deflector pulleys 52 mm diam.

Tolerances for Sprockets and Deflector Pulleys

Run-out / Lateral run-out; Diam. 50 to 100 mm ± 0.1 mm
Diam. > 100 mm ± 0.001 mm per mm diam.
Outside circumference taper: ≤0.001 mm per mm of pulley width
Parallel alignment of bore and toothing: ≤0.001 mm per mm of pulley width
Surface roughness: $R_a \leq 1.6\ \mu m$
Pitch error < 100 mm diam. ± 0.03 mm groove/groove/0.10 mm through 90°
100 to 180 mm diam. ± 0.03 mm groove/groove/0.13 mm through 90°
>180 mm diam. ± 0.03 mm groove/groove/0.15 mm through 90°

Axial Guidance

A synchronous belt has to be guided on at least one sprocket by flanges to keep the belt from wandering out of alignment. As a rule, guide flanges for the belt are located at the crankshaft (driving) sprocket. In this case, the crankshaft damper often serves as the forward flange. The rear flange is attached to or integrated into the crankshaft sprocket. Additional flanges may be required in complex, multivalve trains, depending on the number of sprockets and deflector pulleys. In these cases, it is advisable to locate the flanges at sprockets and not at deflector pulleys. In general, it is important to ensure that sprockets with flanges are aligned exactly with the other pulleys and sprockets to avoid deflecting the belt from its prescribed path. Sprockets and pulleys with just a single flange or without a flange are made wider than the belt itself in order to ensure that the belt runs stably on the sprocket or pulley. The width of the sprockets and the geometric design of the axial guide flanges are depicted in Fig. 7-242.

Belt Tensioning Systems

Fixed Tensioning Pulleys

In the past, tensioning pulleys were always fixed. Deflection pulleys mounted on an eccentric were most often used (Fig. 7-243). Preload was set mechanically on the line and was checked with suitable measurement instruments (span frequency measurement). One disadvantage for fixed tensioning pulleys is the increase in tension that results from the greater expansion of the engine in comparison to belts when the engine heats up. Another problem is that they cannot compensate for the loss of belt tension through the service life due to stretch and wear.

Automatic Tensioner Pulleys

Because of the drawbacks associated with fixed tensioning pulleys and because of the increased dynamic forces in camshaft drives, accompanied at the same time by increased expectations regarding lifetime, automatic tensioning idlers are used to a greater extent. This technology compensates both for the temperature-related rise in tension and for belt stretch. It also keeps constant the high tension required for dependable operation at high engine dynamics. The most widely used is the mechanical, friction-damped, compact-design tensioner. Hydraulic tensioning pulleys are used in some applications where very high dynamic forces are found in the belt drive system. With

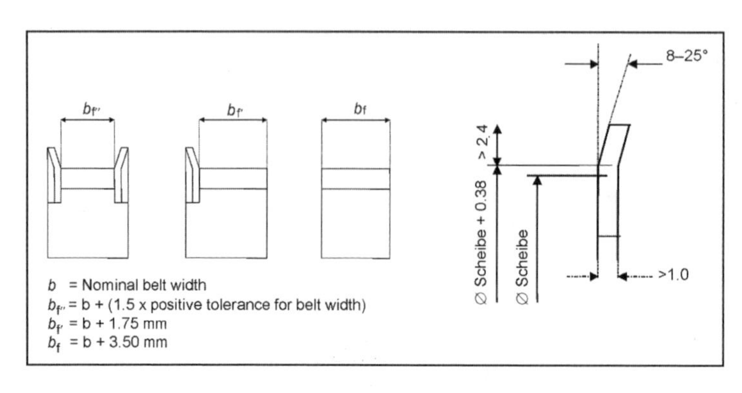

Fig. 7-242 Sprocket width and belt guidance.

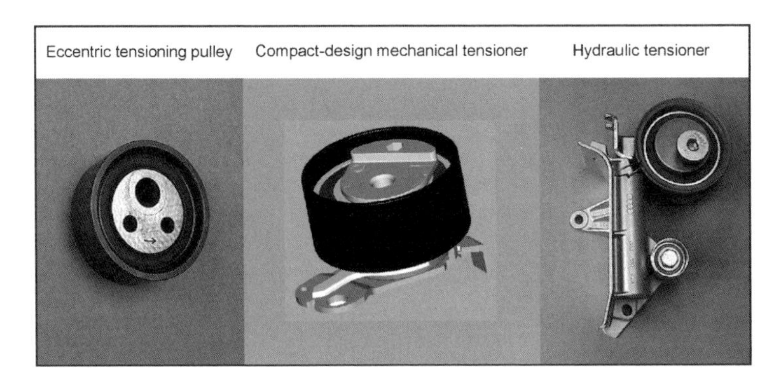

Fig. 7-243 Belt tensioning systems.

their asymmetrical damping they exhibit very good damping properties even at low preload values.

7.18.1.3 Synchronous Belt Dynamics

Optimizing system dynamics is an important step along the road to synchronous belt drives promising long engine life since forces and loads can be minimized and at the same time monitored. Here it is important to ensure that all the components in the system reach the targeted lifetime under these conditions.

Dynamic loading on the drive, rotational oscillations, dynamic forces, and oscillations along the spans are optimized as a whole. To do this, numerous parameters are optimized to minimize dynamic loading on the system. These parameters include the tensioner response characteristics, preload and damping, belt values, belt stiffness and damping, the belt profile, and the moments of inertia for the sprockets at the camshaft. Figure 7-244 shows two important values for the dynamics in the synchronous belt drive—the alternating load at the crankshaft and the rotary oscillations at the camshaft. System resonance, here at 4000 rpm if possible, is reduced to a minimum with optimized system design and has to be monitored over the service life of this drive. At the same time, the loads on other system components such as deflection and tensioning pulleys are also minimized.

7.18.1.4 Application Examples

Depicted in Fig. 7-245 are typical application examples for two engines. In both cases the water pump is integrated into the drive system. In many diesel engines the injection pumps (distributor injection pump or common rail pump) are integrated into the primary belt drive. The service lives of today's drives are 160 000 km for gasoline engines and 120 000 for diesels. To be anticipated for future engines are belt drive lives of 240 000 km thanks to optimized systems and improved belt designs.[1]

7.18.2 Toothed V-Belt Drive to Power Auxiliary Units

Auxiliary units were driven in the past with simple V-belts. Because of the increased complexity triggered by owners' increased demands in terms of comfort, integrating the alternator, water pump, power steering pump, and air

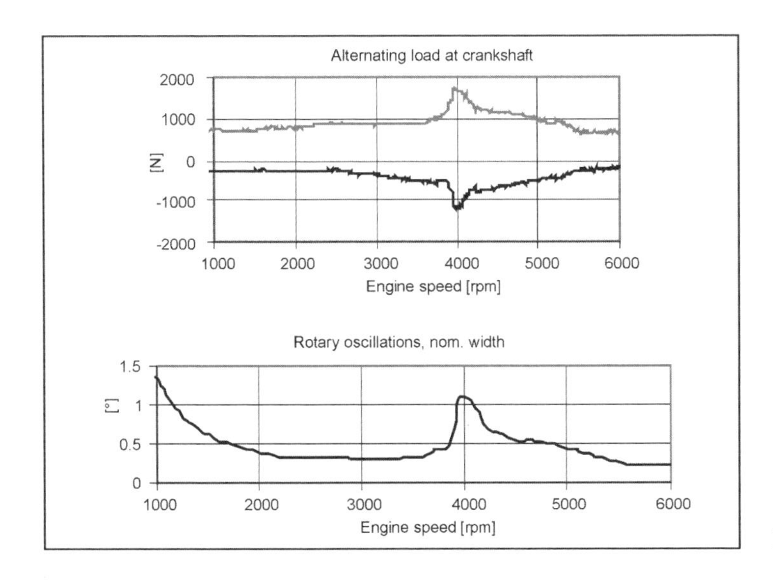

Fig. 7-244 System resonance.

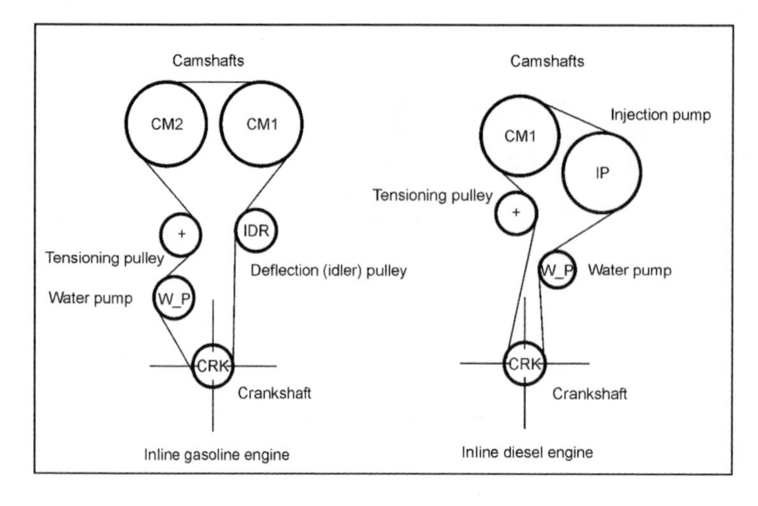

Fig. 7-245 Application examples.

conditioning compressor into this drive system is now state of the art. The complexity of the drives is further increased by additional units such as the fan and mechanical turbochargers or pumps used for secondary air injection. Today the auxiliary units are driven in a serpentine configuration with multirib V-belts (Micro-V® belts). The major benefits that the Micro-V® belt offers when compared with V-belt drives are greater power transmission and reduced installation space in complex drives.

7.18.2.1 Micro-V® Drive Belts

Structure of the Micro-V® Belt

The Micro-V® belt is a bonded system made of three components (Fig. 7-246):

- Fiber-reinforced rubber blend
- Tensile cords
- Overcord or rubber backing

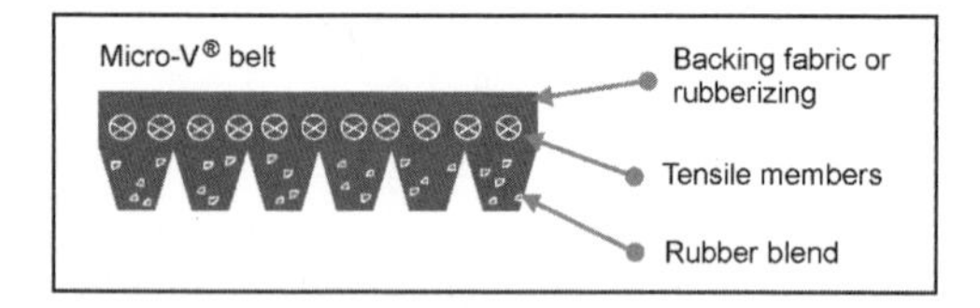

Fig. 7-246 Structure of the Micro-V® belt.

The tensile cords transmit drive power from the crankshaft to the auxiliary units, absorb dynamic loads at low stretch, and provide good resistance to alternating flexure.

The cords are made of nylon, polyester, or aramid; the widely differing moduli of elasticity for the tensile cords enable optimized tuning of system dynamics. The rubber forms the V ribs and transfers the drive forces from the pulley into the tensile cords. Chloroprene or EPDM is used as the material; fiber material is added to the rubber blend to stiffen the product.

The overcord can either use a backing fabric or be made through rubberizing. During the manufacturing process, the cords in the tensile member are twisted, clockwise and counterclockwise in pairs, in order to achieve largely neutral running properties for the belt.

The Micro-V® belt is manufactured in a vulcanization process. The V ribs are either molded from the very outset or are cut into the belt after the vulcanization. In double-sided belts this grinding process is carried out on both sides.

Micro-V® Belt Profile

It is the PK profile (as per ISO standard) that is normally used for automotive applications. The groove spacing is 3.56 mm. The designation for the belt, such as 6 PK 1270, means six ribs, PK profile, 1270 mm reference length. When components that draw a great deal of power—such as the alternator, power steering pump, or air conditioning compressor—are driven with the back of the belt, the belt can also be designed as a double-sided Micro-V® belt, with ribs on both sides (Fig. 7-247).

Fig. 7-247 Double-sided Micro-V® belt.

Characteristic Values for Micro-V® Belts and Sprockets

The most important key values for the Micro-V® belt are shown in Fig. 7-248. The belt width is calculated by multiplying the number of ribs by 3.56 mm (PK profile). Belt thickness, depending on the design, is between 4.3 and

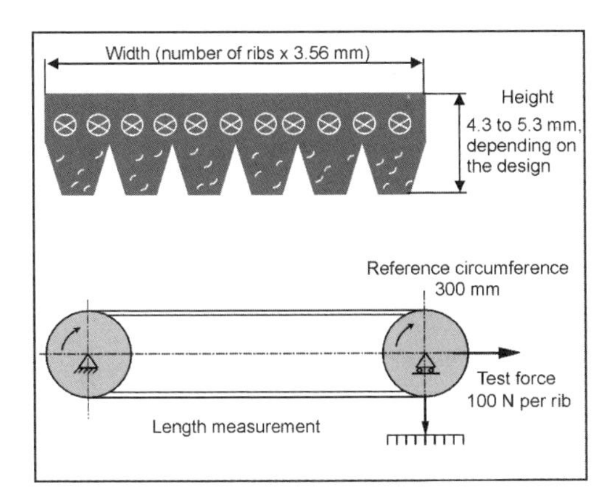

Fig. 7-248 Characteristic values for the Micro-V® belt.

5.3 mm. The reference belt length is determined on a two-pulley test bed at a defined preload (ISO 2790). The reference circumference of the pulleys used here is 300 mm.

The standardized profile used for the pulleys is shown in Fig. 7-249. The outside diameter of the flanges is one dimension used to describe the pulley. More important for the design and determination of belt length, however, is the pulley diameter across the test balls (2.5 mm diam.). With this measurement technique, the profile of the pulley and thus the groove angle is also taken into account. The groove angle is matched to the belt profile running (and deformed) in the wrap arc and deformed, dependent on the diameter of the pulley. Normal groove angles lie in a range of from 40° to 44°. The effective diameter is then calculated, in accordance with the belt design, using the diameter measured across the test balls. The effective diameter is congruent with the center of the tensile cords in Micro-V® belts. Characteristic values for common belt designs are defined in DIN 7876 and ISO 9981. During detailed design work, however, it is necessary to draw upon the characteristic values published by the belt and/or pulley manufacturer.

The pulleys are made of either steel or plastic.

7.18.2.2 Auxiliary Component Drive System

The most important demand on any auxiliary unit drive system is slip-free drive for all auxiliary units, at all loading states, for the length of the engine's useful life. In modern engines with full drives, it is thus possible, using the Micro-V® belts in a five- or six-rib design, to transfer maximum torques of up to 30 Nm and maximum power of from 15 to 20 kW with all the auxiliary units running at full load. The ambient temperatures at 80 to 100°C on average are somewhat lower than in a synchronous belt drive. It is important to avoid, in particular, noises such as the well-known belt squeal caused in cold and damp weather by slippage between the belt and the pulley. This is achieved with optimum system design in regard to the geometry and dynamics. It is also necessary to avoid belt noises caused by misalignment of pulleys, doing so right from the engineering stage. For auxiliary units, too, 240 000 km is taken today as the desired life expectancy in current engineering development work.

Design Criteria

Auxiliary unit drives are engineered with computer support.

A survey of the most important parameters to be considered in the design and some general engineering criteria are to be provided here. Important input data include the arrangement of the components (i.e., the drive configuration), torque development at the components, and the moments of inertia for the components as well as the data for the belt itself. With these data at hand, it is possible to calculate and optimize not only the span lengths and wrap angles, the system's *eigen* frequencies, and the limit values for slip, but the belt's lifetime as well.

Discussed below are a few general design criteria that must be observed in Micro-V® belt systems in order to engineer a functional system meeting today's longevity expectations:

Recommended Minimum Wrap Angles

Crankshaft	150°
Alternator	120°
Power steering pump, A/C compressor	90°
Tensioning pulley	60°

Alignment Error/Run-In Angle

In order to avoid unacceptable belt wear and noise, the belt's run-in angle into the grooved pulleys should not exceed 1°.

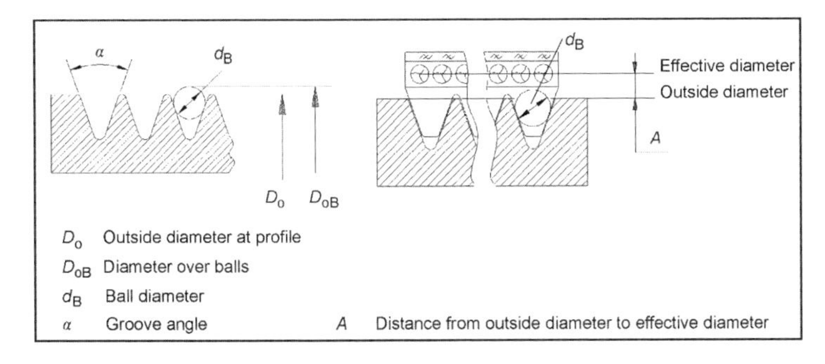

Fig. 7-249 Key values for Micro-V® belt pulleys.

System *eigen* Frequency

The system's *eigen* frequency should not be in the engine's idle range (second engine order).

Minimum Diameters for Pulleys and Deflector Pulleys

In practice, the smallest pulley is often found at the alternator, which is needed to achieve the high rotation speeds required there. Typical alternator pulleys have a diameter of from 50 to 56 mm. Belt fatigue rises exponentially when small pulleys are used; this has to be taken into account when engineering the belt. It is advisable to use diameters of no less than 70 mm for deflection pulleys.

Belt Tensioning Systems

Belt tensioning in auxiliary unit drives is normally handled today with automatic tensioning pulleys. The tensioning pulleys ensure constant tension throughout the service life and compensate for belt stretch and belt wear. The design of the tensioning pulleys is determined essentially by the available installation area (Fig. 7-250). In long-arm tensioners the spring-and-damping system lies in the same plane as the belt drive system; where Z-type tensioners are used the tensioner housing is recessed into the area behind the belt drive. Preload is generated by a leg spring; the tensioner is friction damped at the same time. The preloads for 6 PK belts normally lie in a range of from 250 to 400 N, the exact value depending on the system's dynamics.

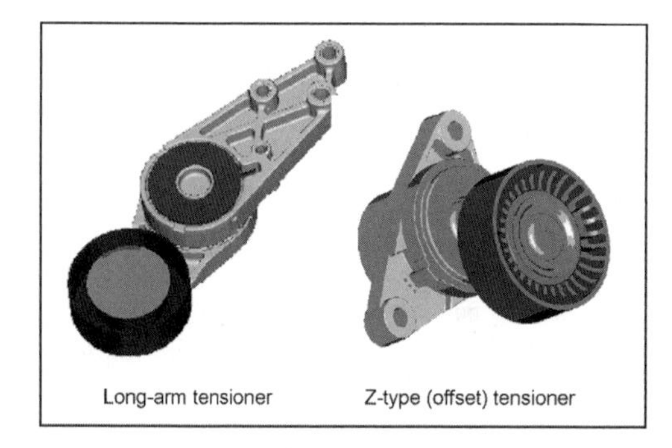

Fig. 7-250 Automatic belt tensioning systems.

7.18.2.3 Application Examples

Figure 7-251 shows a typical Micro-V® belt drive. In many drive concepts the power steering pump and the air conditioning compressor have already been integrated into the standard belt drive design. Particularly when the drive configurations are complex, additional deflector pulleys are required in order to ensure the required wrap angle at all driven units and thus slip-free operation.

Bibliography

[1] Arnold, M., M. Farrenkopf, and S. McNamara, "Zahnriementriebe mit Motorlebensdauer für zukünftige Motoren," in MTZ, Vol. 62, 2001, No. 2.

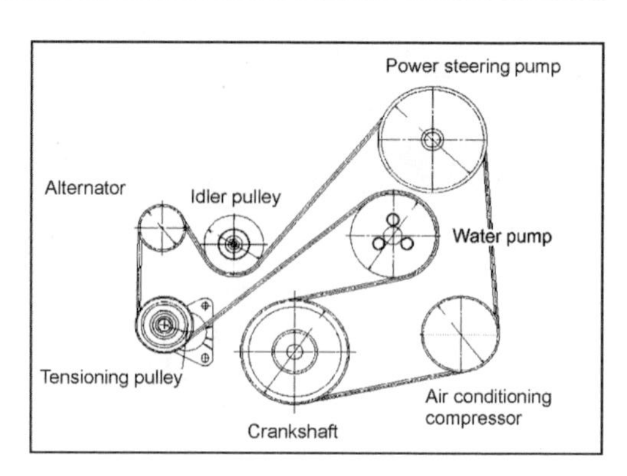

Fig. 7-251 Example of auxiliary component drive system.

7.19 Bearings in Internal Combustion Engines

The shafts found in multicylinder reciprocating engines—the crankshaft, valve train, and balancers—generally run in plain (sliding or friction) bearings. The reasons for selecting this type include their great ability to withstand shock and their damping properties, easy division for assembly around the crankshaft or camshaft, low space requirements, insensitivity to grime, and, last but not least, the low costs when compared with rolling bearings. The fundamental disadvantage of plain bearings compared to rolling bearings is the higher friction level and the resulting greater oil requirements.

Rolling bearings are employed in engines in some cases wherever the advantages of the plain bearing are not fully exploited: At the crankshaft for small, single-cylinder engines, at the bearings for the sprocket drive, and, to an increasing extent, at the valve train (roller tappets).

7.19.1 Fundamentals

7.19.1.1 Radial Bearing

Constant Loading

The lubricant is drawn into a plain, radial bearing by adhesion, filling the lubrication gap between the surfaces that move relative onto the other; this causes a buildup of pressure that keeps external forces in balance and that keeps the mating components—journal and bearing—separated by an oil film, Fig. 7-252.

The dimensionless Sommerfeld number describes the interrelationships in a cylindrical radial bearing.

$$\mathrm{So}_D = \frac{\overline{p} \cdot \psi^2}{\eta \cdot \omega} = f(b/d, \varepsilon) \qquad (7.25)$$

The terms in the above formula are as follows:

$\overline{p}$	$\mathrm{N/m^2}$	Specific bearing loading $F/(b \cdot d)$
ω	$\sec^{-1}$	Angular velocity
ψ	–	Relative bearing clearance, s/d

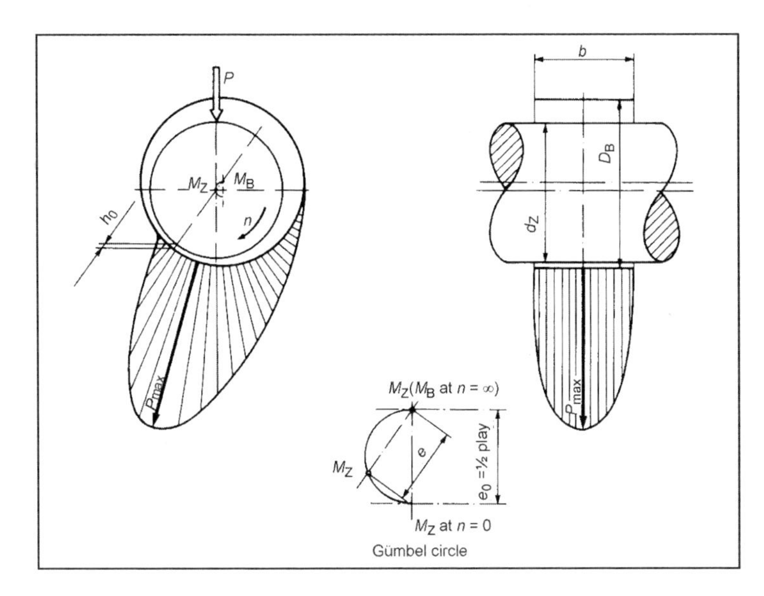

Fig. 7-252 Buildup of hydrodynamic pressure as a result of rotation.

η Nm^2 · sec   Dynamic viscosity
ε –   Relative eccentricity (displacement) of the journal's centerline within the bearing clearance

Every load and velocity value corresponds to a certain eccentric equilibrium situation for the journal in the bearing.

$\varepsilon = 0 \rightarrow \mathrm{So}_D = 0; \quad \varepsilon = 1 \rightarrow \mathrm{So}_D = \infty$

Dynamic Loading

A characteristic feature for the bearings used in engines is loading, which alternates periodically in both magnitude and direction; this results, for example, from the ignition and inertial forces at the crankshaft and from the pulsating loads resulting from the camshaft's actuating the valves.

The change in force causes an imbalance that causes the shaft's centerline to shift in the radial and circumferential directions. This eccentricity rises with rising loads; resistance to the displacement of the lubricant damps the radial motion. The high shock resistance of the plain bearing is the result.

The resultant additional bearing capacity is defined by the Sommerfeld number for lubricant displacement:

$$\mathrm{So}_V = \frac{\overline{p} \cdot \psi}{\eta \cdot (\partial \varepsilon / \partial t)} = f(b/d, \varepsilon) \tag{7.26}$$

The overall force at the bearing results from vectoral addition of both effects, Fig. 7-253.

Friction

If a continuous and complete separation of the sliding surfaces were to be achieved by the oil film, then no bearing material would be required; the bearing would run entirely in accordance with hydrodynamic principles. Friction, in this case, is determined only by the oil's shear strength and is very low, on an order of magnitude of $\mu = 0.002 - 0.005$. In real-world operations, however, there is contact between the mating surfaces since the bearing cannot form a sufficient hydrodynamic lubricating film for every operational state. This "mixed lubrication" situation is associated with far greater friction levels, increasing by as much as a factor of ten. The familiar, generalized Stribeck curve describes the interactions (Fig. 7-254).

The system becomes thermally unstable if the friction energy thus generated cannot be dissipated. The probability that a thermally unstable situation is reached in a plain bearing, i.e., the susceptibility of the bearing to malfunctions, is dependent on the energy density in the bearing system (load, velocity).

Following dynamic loading, the shaft centerline describes periodically within the bearing a certain displacement path (see also Fig. 7-257 below) with the smallest lubrication gap changing in size and location. The results are, on the one hand, that a far higher degree of direct material contact can be handled and that the dimensions of the bearing can be far smaller than one that is under constant loading; on the other hand, every area is subject to pulsation loading and the material's endurance becomes an issue.

7.19.1.2 Axial Bearing

Axial bearings are used to stabilize the shafts longitudinally and absorb the axial thrust generated by helical toothing and by any angular positioning. Higher loads may occur briefly, emanating from the clutch or resulting from shock triggered by acceleration.

Axial bearings may be engineered as thrust washers or combined with a radial bearing to form a so-called locating bearing. These bearings are simple, plane surfaces made of bearing metal. They work in the mixed lubrication sector; i.e., no hydrodynamic pressure is established.

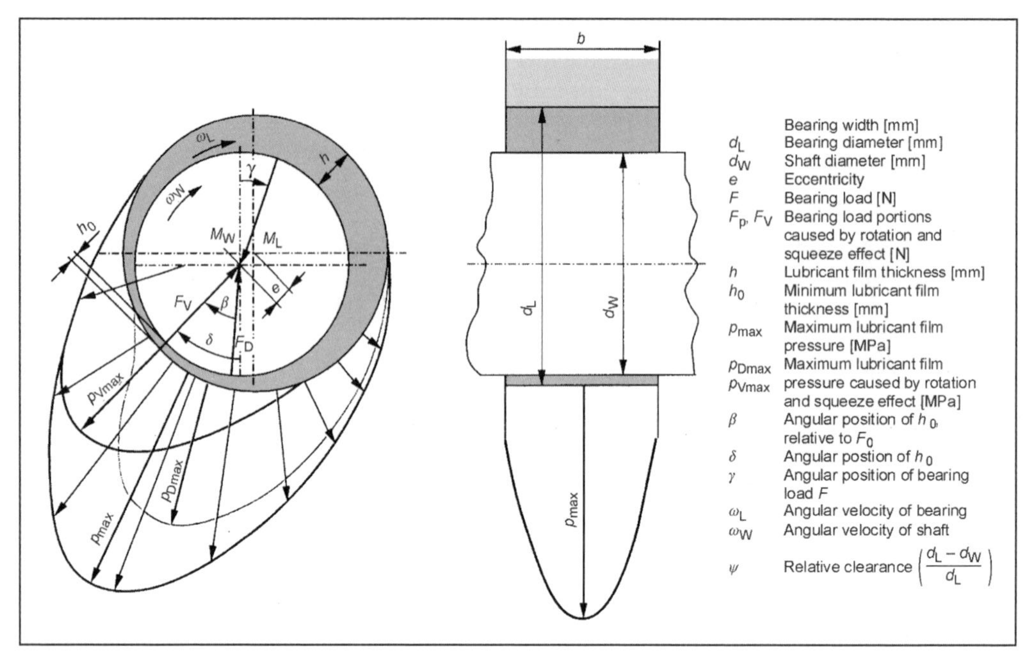

Fig. 7-253 Buildup of hydrodynamic pressure due to rotation and displacement.

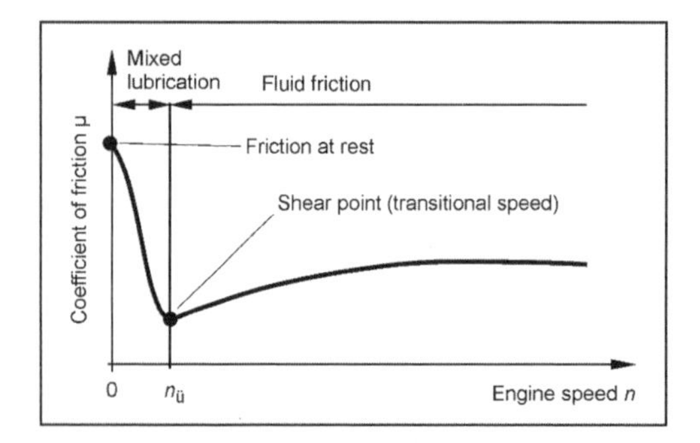

Fig. 7-254 Stribeck curve.

It is important that the wetting of the surfaces with lubricating oil is assured.

Overheating is generally the reason for axial bearing failure; failure resulting from overloading due to shock or vibration is not likely.

7.19.2 Calculating and Dimensioning Engine Bearings

A bearing is dimensioned in several steps during engine design work. The major dimensions, diameter and width, are determined primarily on the basis of the design parameters for the engine and mating components.

Once bearing loading has been calculated, it becomes possible, during the concept phase, to use specific bearing loading ($F/b \cdot d$) as a rough reference value. Because of the great influence exerted by load characteristics, the ratio of width to diameter, bearing clearance, oil viscosity, and engineering details, exact calculations for bearing dimensioning have to be made as early as possible.

The primary results of calculation work are the selection of the appropriate type of bearing for the application and establishing the bearing dimensions, in conjunction with the acceptable boundary values.

7.19.2.1 Loading

The loading on engine bearings changes cyclically. The forces effective at the crankshaft are illustrated by a representative example in Fig. 7-255. These forces are made up of cylinder pressure and the reciprocating and rotating inertial forces.

Figure 7-256 uses polar coordinates to show the progress of bearing forces—in both magnitude and vector—at the camshaft bearing in a diesel engine over a complete operating cycle and running at maximum torque. At higher speeds with lower loading, the peaks triggered by ignition decline and the ellipse representing inertial forces increases.

When designing the crankshaft system, the bearing loads are normally calculated together with the stiffness and oscillation situation in the crankshaft, taking account of elastic deformations. Thus for the main bearings above all (statistically indeterminate bearing), one can ascertain more exactly the distribution of loading across the individual bearing points. Having calculated cyclical loading in this way, one may then calculate the hydrodynamic pressures that are generated and the widths of lubrication gaps. The most commonly used method here is to calculate the bearing journal displacement path.

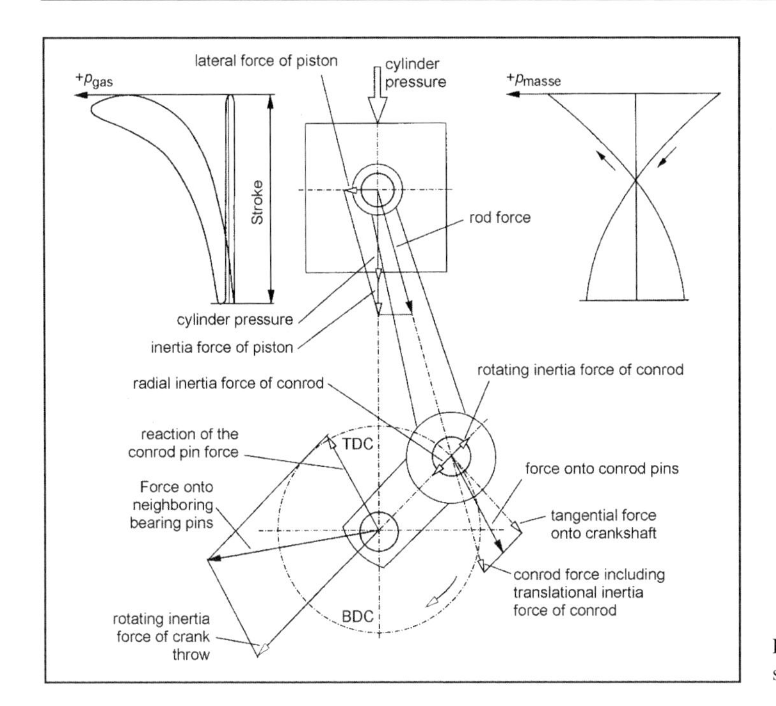

Fig. 7-255 Forces in the crankshaft system.

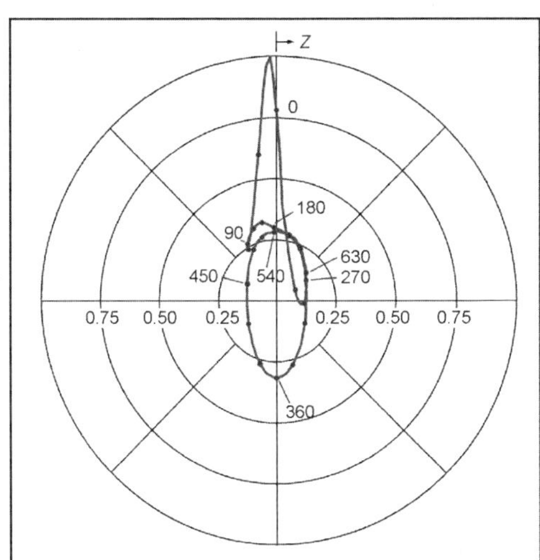

Fig. 7-256 Polar chart for forces at the conrod bearing in a diesel engine.

7.19.2.2 Bearing Journal Displacement Path

The displacement path that the journal executes in each full cycle, shown in Fig. 7-257, can be calculated with relatively simple means. The results are strongly influenced by the nature of the model (method after Holland-Lang or the mobility method after Booker), by the peripheral conditions for the pressure curve, and by the assumptions of oil viscosity. Thus, it is possible to compare the results delivered by different programs only if these assumptions are identical. The acceptable boundary values, determined by applying experience from practical operations and drawing on test results for the calculated data, apply only for comparable calculation models.

The path is iterated across the full cycle through to convergence in steps of a few degrees of crankshaft angle. Calculations are carried out separately for each loading situation. As a rule, the values are ascertained for operation at nominal load and at maximum torque with low engine speed.

The most important results from the calculation are as follows:

- Smallest lubrication gap
- Maximum lubricating film pressure

Additional information is obtained, and this includes the oil throughput rate, hydrodynamic friction, and the resultant oil heating. The period through which the smallest lubrication gap remains in a certain area provides information on the concentration of friction energy and thus on the amount of wear to be expected.

The calculation of the displacement path is suitable particularly for parameter studies in an early stage of motor engineering, e.g., to determine the ideal layout for the balancer in view of the crankshaft bearings and/or the influence of design parameters such as the ratio of width to diameter or bearing play. Calculations for loading and the displacement path are often integrated.

7.19.2.3 Elastohydrodynamic Calculation

Elastohydrodynamic calculation is a more precise method, one developed in recent years, to calculate engine bearings.

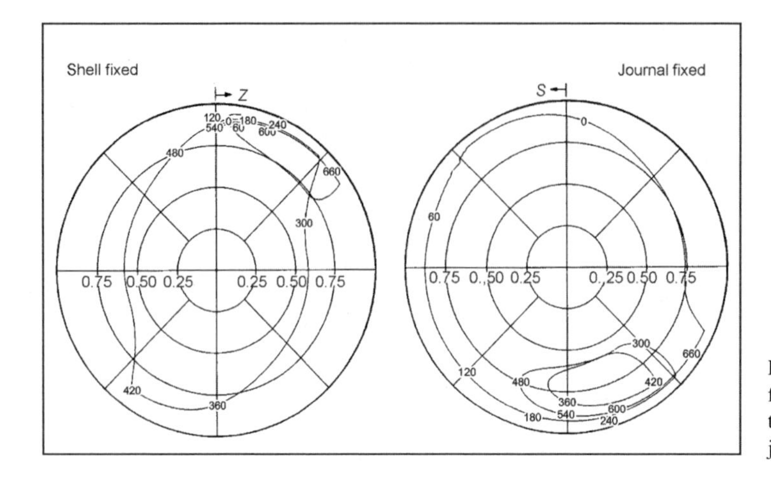

Fig. 7-257 Displacement path for a conrod bearing (viewed relatively from the bearing and the journal).

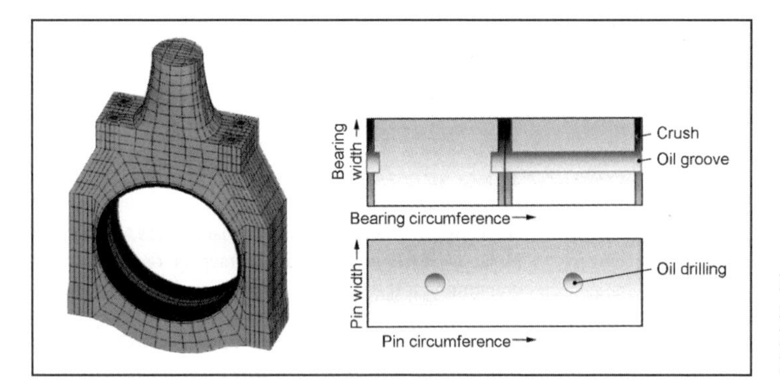

Fig. 7-258 Housing and bearing model (developed) used to calculate for the elastohydrodynamic lubricating film.

Here the distribution of the lubrication film in the bearing is calculated locally, taking elastic deformations into account, Fig. 7-258.

With the numerical solution for the Reynolds differential equation, one can take into account the stiffness of the bearing's environment and the influence of local geometry characteristics for the bearing and the journal. In addition, this calculation method makes it possible to examine and verify values, such as the degree of fill at the gap, which are taken as givens in the global observation.

Figure 7-259 shows the results for a conrod bearing with a slightly asymmetrical impingement of load, shortly after ignition.

This method requires far more detailed data and significantly greater calculation effort than calculating the displacement path. Consequently, it makes sense to employ this at an advanced stage in engineering and to examine local influences.

An estimate of the service life supported by cumulative damage models may follow if the load population and

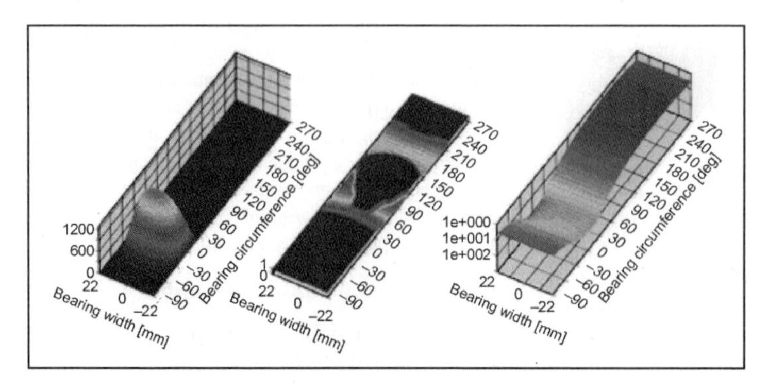

Fig. 7-259 Results of hydrodynamic calculations.

the necessary materials data are known. As a rule, service life and operational reliability are verified today by field testing and accompanying component testing.

7.19.2.4 Major Dimensions: Diameter, Width

The bearing diameter and width are defined within narrow limits by the engine design and the dynamic forces at the shafts. It is possible to influence specific bearing loading within these limits, and this may be decisive for the selection of the bearing design.

The ratio of width to diameter is usually from 0.25 to 0.35. At the same specific loading $F/(W \cdot D)$ a relatively smaller diameter and greater width causes a larger lubrication gap, lower peak pressures, and smaller friction losses. Because of the low circumferential velocity, sensitivity to contact with foreign objects, and malfunction falls, this situation is one that should be targeted. The minimum journal diameter required for sufficient crankshaft stiffness imposes limits on optimization.

7.19.2.5 Oil Feed Geometry

Additional information on the distribution of lubricating and cooling oil is provided in Chapter 8. Here only the features that affect the bearing directly are described.

The establishment of the hydrodynamic lubrication film is greatly influenced by the grooves and bores required for lubricating oil supply—at the main bearings, for example. An annular groove in the main bearings is ideal for continuous supply to the conrod bearings; this does, however, in otherwise identical conditions, reduce the smallest lubrication gap to about 30%. This is compensated in part by better oil delivery to the bearing point so that the load-bearing capacity declines to only about one-half.

Thus, one must attempt to achieve sufficient oil supply with bores and grooved sections in those areas around the bearing that exhibit lower loading and large lubricating gap widths. The displacement path described above provides information on the most favorable location for grooves (bearings) and bores (shafts).

In passenger car engines, a semicircular groove in the upper shell of the main bearing and a bore in the crankshaft, exiting at the conrod journal about 45° before the apex in the direction of rotation, has been established as standard.

To avoid inconstancies in oil flow, which could cause oil starvation and cavitation, it is often necessary to eliminate coarse inconstancies in the oil feed geometry. This is done by rounding bores and with tapered run-out at the grooves.

When planning the lubricating oil supply, it is necessary to pay attention not only to sufficient delivery but also to adequate cross sections in drain channels. This is true particularly for thrust bearings where continuous, radial grooves in the running surface ensure both wetting of the axial bearing surface and a slight restriction of flow out of the radial bearing.

Grooves are often required in the bearing housing for oil distribution, and here it is important that there be no hollows behind the bearing shells in zones that are subjected to loading since the shells could bend in response to lubricating film pressure and breaks in the bearing metal could occur.

7.19.2.6 Precision Dimensions

The actual bearing engineering work concentrates not only on the selection of the bearing type, but also on precision dimensioning:

- Tight seating, overhang
- Bearing play
- Progress of bearing thickness around the circumference, gap at the bearing shell ends
- Surface properties, shapes, and positioning tolerances at the ends

Tight Seating, Overhang

The bearing force has to be transferred to the housing. To do so, it is necessary that the bearing be seated firmly in the housing, to reliably suppress the relative motions resulting from the pulsating load. This tight seating in the radial bearing is affected by the excess diameter and the so-called "crush height," S_n, at the ends of the bearing shells, Fig. 7-260. The tabs or pins normally used to ensure correct positioning of the bearing are not suitable for fixing the bearing.

The limits are, on the one hand, sufficient radial pressure (see Fig. 7-260) and, on the other hand, tangential strain that can be tolerated by the bearing shell without great plastic deformation. All standard bearing metals are overtaxed at the low bearing thicknesses prevailing today; a steel backing shell is required to provide sufficiently tight seating. An engine bearing thus is made of a composite material incorporating steel and the actual bearing metal, with or without any additional coating, depending on the composition and particulars of the employment. Only in isolated cases, such as in large wrist pin bushings, can a single, solid material be used.

The low-alloy steels required in bearing manufacture have a maximum compression yield point of 360 N/mm^2; this dictates a lower limit for bearing thickness, at about 2.5% of the diameter.

Of particular significance is temperature development in aluminum housings. Because of the differing degrees of thermal expansion for steel and aluminum, there is a reduction as temperatures rise; this can even lead to a loss of the preload. At lower outside temperatures, by contrast, the strength limits for the bearing shell and/or the bore may be exceeded.

In such cases the area immediately around the bearing is stiffened by sintered or cast steel components that are cast in situ. The global models for press fit calculations are no longer adequate when engineering composite bearings; local strains and deformations have to be ascertained using finite element methods.

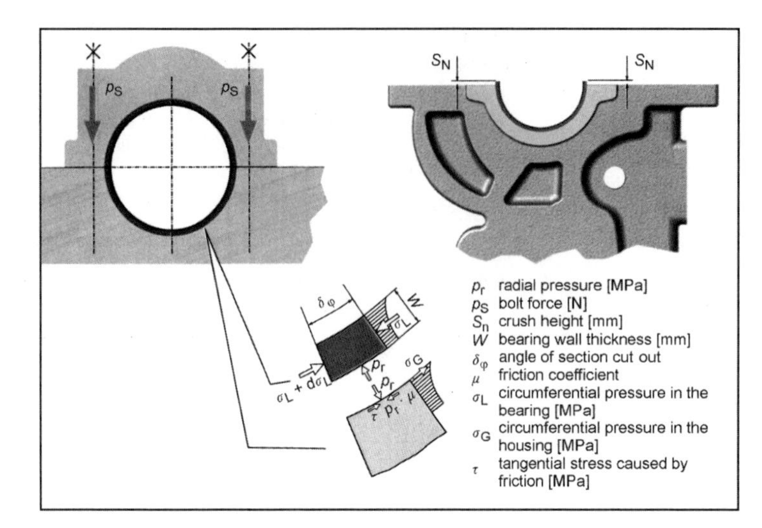

Fig. 7-260 Excess diameter and installation strains.

Bearing Play

Bearing play is the most important, user-definable magnitude in bearing design. A smaller amount of clearance, nominally, creates greater hydrodynamic load-bearing capacity and—because of to the greater damping to counter displacement—better acoustic conditions. In contrast, at larger bearing play the lubricating oil throughput rises excessively (more than with the square of the clearance); the bearing becomes more tolerant of deformations and disturbances. Thus, one sets the value for minimum play to be as small as possible while still ensuring operational reliability. The maximum play results from the manufacturing tolerances for bearing wall thickness (6 to 12 μm) and for the adjoining components and can become unacceptably high for small engines where $D < 60$ mm. Classification of bearing thickness is often a more favorable method than more exact manufacturing to limit the tolerance in play.

As for the press fit, mastering bearing play for aluminum housings is difficult. Across the operating temperature range there is an unacceptable amount of change from, for instance, 15 μm at -30°C to as much as 120 μm at 130°C (at 50 mm diameter). Limiting maximum play requires more exact classification in which the bore, shaft, and play are associated one with another.

Wall Thickness, Clearance at Ends

An undisturbed, perfectly cylindrical bore is ideal for the bearing function. The strains resulting from bearing installation and inertial forces usually, however, cause a bore that is not a true circle; this is compensated by a continuous change in bearing shell thickness, from the center to the ends. When bearings are split into two semicircular shells, a gap some millimeters in length and about 5 to 15 μm in depth equalizes the differing thicknesses for the shells. Figure 7-261 shows typical values.

Also essential to uninterrupted bearing function is the correct design of the bore and journals in regard to alignment, roundness, crowning, waviness, and surface roughness. Reference is made here to the applicable engineering guidelines and standards.

Acceptable boundary values are applied when selecting the type of bearing, which is done in consideration of loading and other peripheral conditions. The loading limits and the characteristics for use in the normal bearing designs are described in greater detail at Section 7.19.4. It is important that simultaneous development be carried out by the bearing maker right from the engine's draft design stage.

7.19.3 Bearing Materials

In addition to its primary function, transferring load during relative motions, the bearing has the additional important task of concentrating any disturbances in the system upon itself. The engine block, the crankshaft, and the conrod should be protected for as long as possible from any consequences of a fault in the system. Bearing metals are thus constructed so that they can absorb the adverse consequences of mixed friction largely without damage to themselves or to adjacent components. As a rule, they are made up of a harder matrix (e.g., CuSn, AlCu) into which are embedded the soft, immiscible phases that melt at lower temperatures (primarily Pb, Sn). This produces an error-tolerant alloy with good heat transfer properties, a low coefficient of friction, and a reduced tendency to weld to the steel.

Every good bearing material is a compromise between the contradictory requirements for strength and good tribologic properties. The best composition takes account of the weighting for the particular application.

In spite of the multitude of different but in some cases very similar materials made by various bearing manufac-

Bearing point	Operating conditions				Engineering magnitudes		
	Type of movement	Type of loading	U [m/sec]	P_{max} [N/mm²]	ψ_{min} [%]	W/D	$p_{r\,min}$ [N/mm²]
Crankshaft drive: Wrist pin bushing	Slewing	Pulsating load from cylinder pressure, reciprocating masses	2 to 3	70 to 120	0.8	<1.0	9
Conrod bearing	Nonuniform rotation, ~n	Pulsating load from wrist pin force and rotating masses	10 to 18	50 to 90	0.5	0.28 to 0.35	10
Main bearing	Rotating, n	Pulsating load from adjacent conrod bearings	12 to 20	40 to 60	0.8	0.25 to 0.32	8
Axial bearing	Sliding	Thrust, coupling force, impact load	15 to 24	<2 Permanent <5 Brief <12 Impact	—	—	—
Valve train: Rocker arm bearing	Slewing, >0	Spike load		60 to 90	0.7	0.5 to 0.8	9
Camshaft bearing	Rotating, $n/2$	Pulsating		20 to 50			8
Balancer	Rotating, $2n$	Planetary		20 to 40	1.2	0.3 to 0.4	>10
Sprocket drive, sprockets, auxiliary units	Rotating	Uniform	Dictated by the engineering design				

Fig. 7-261 Characteristic values and typical guideline values for the most important bearing locations.

turers, one may categorize those that are most important for use in internal combustion engines in three groups of bearing metals and two groups of overlays (Fig. 7-262).

The exact definitions, tolerances for the material composition, and mechanical properties are listed in Ref. [1] and in the above-mentioned standards.

7.19.3.1 Bearing Metals

Babbit Metals

Steel and babbit metals are currently found only (on rare occasions) in passenger car engine designs, in bearings that are subjected only to low loads (camshaft bearings, sprockets). The SnSb8Cu and PbSn8 alloys have superb running properties, but their long-term strength is insufficient to handle the pulsating loads occurring in the drive train in modern engines.

The composite material incorporating steel is manufactured in stationary sand casting or centrifugal casting for thick-walled bearings and in strip casting for thin-walled bearings of smaller dimensions.

Aluminum Alloys (Fig. 7-263)

Alloys based on aluminum have proven their utility as main bearings and camshaft bearings across a broad range of applications. When used as a two-material bearing without an overlay, they represent a very economical solution for moderate loads; as a three-material bearing and grooved bearing, they are in direct competition with leaded bronze compounds. Aluminum alloys are not suited, as per today's standards, for heavily loaded bushings where a slewing motion is encountered, e.g., in the small conrod end and the rocker arm; neither do they provide a satisfactory basis for sputter bearings.

AlSn alloys are the ones most frequently used. Upwards of about 15% tin content, these alloys exhibit good slip characteristics; their excellent corrosion resistance makes it possible above all to use them in gas-fired

Bearing metals

Babbit metals, cast (DIN-ISO 4381, SAE 12 - 17)

$PbSb_{14}Sn_9Cu$

$SnSb_8Cu_4$, $SnSb_{12}Cu_5$

Aluminum alloys, roll-bonded (SAE 770 - 788)

$AlSn_{40}Cu$, $AlSn_{20}Cu$, $AlSn_6Cu$

$AlSn_{12}Si_4$, $AlSn_{10}NiMn$

$AlZn_{4,5}SiPb$

Leaded bronzes, cast, sintered
(DIN 1716; DIN-ISO 4382,4383; SAE 790-798) :

$CuPb_{30}$

$CuPb_{25}Sn_4$, $CuPb_{20}Sn_2$

$CuPb_{15}Sn_7$, $CuPb_{10}Sn_{10}$

Running surfaces

Babbit metal, electroplated (SAE 19):

$PbSn_8$, $PbSn_{10}Cu_2$, $PbSn_{16}Cu_3$

$PbIn_9$, $SnSb_{12}Cu$

Aluminum alloy, sputtered: $AlSn_{20}Cu$

Fig. 7-262 The most important bearing metals for composite bearings.

engines and large four-cycle engines fired with heavy oil. Both AlSiSn materials and AlPb alloys are used in the Anglo-Saxon regions and in Japan.

AlZn4 and 5SiPb are used when dealing with heavy loads such as those found in conrod bearings. This material does not have an embedded soft phase and thus is suitable for use as the substrate for three-material or grooved bearings only when an overlay is applied.

The manufacture of aluminum bearing alloys is affected in a continuous or semicontinuous casting process; the process windows are limited by the formation of separations (liquidation) in the soft phase and by the appearance of fissures in the hard phase. The stronger the matrix and the higher the tin content, the narrower the processing window.

The method used most widely today is horizontal extrusion casting, which is noncritical for AlSn materials but which, however, cannot produce any higher-strength microstructures. A somewhat more homogenous structure can be achieved with vertical extrusion casting although the process is more sensitive to interference since the cooling conditions are more difficult to control.

Belt casting, the newest technological development, permits a broader bandwidth in the process and, beyond that, the combination of a high share of matrix-strengthening elements and higher soft phase content. Since here the ingot—in contrast to the other two processes—is actually a belt that runs simultaneously, the chilling and solidification parameters are better tuned to suit the particular material composition.

After casting, the strips are rolled out in several steps and heat treated; AlSn alloys are then joined with a thin aluminum bonding layer and, depending on the thickness of the finished bearing, are wound into coils or stored as strips.

The join with the steel is made by roll bonding, which is essentially a friction welding process (Fig. 7-264). The surfaces of the two strips are cleaned and activated; they are heated and rolled together, then they are rolled down by 20% to 35%. The finished strip is then coiled up again. In smaller batches, plating is more economical in strips that are several meters in length; the process is essentially the same.

The newer AlSn alloys are also roll bonded with alloyed intermediate layers such as AlZn so that their higher strength can also be utilized in the composite.

Copper Alloys

The copper-based materials used for bearings are many and varied. CuPbSn type alloys are used almost exclusively for composite materials. Other alloys such as CuAl or CuZn are used as solid materials only in special cases.

Leaded bronze comprises a fixed CuSn matrix in which the lead is embedded. From 1% to 10% tin and from 10% to 30% lead is alloyed in. The higher the tin content, the stronger the material. The higher the lead content, the better the slip properties. Two groups are formed:

- CuPb(18–23)Sn(1–3) for higher slip speeds as found in conrod bearings and main bearings and
- CuPb(10–15)Sn(7–10) for rocking movements as found in rocker arms and wrist pin bushings

In rotating applications, leaded bronzes are suitable only with an additional electroplated or sputtered overlay. Wristpin and rocker arm bushings may be used with or without an overlay, the choice depending on their size.

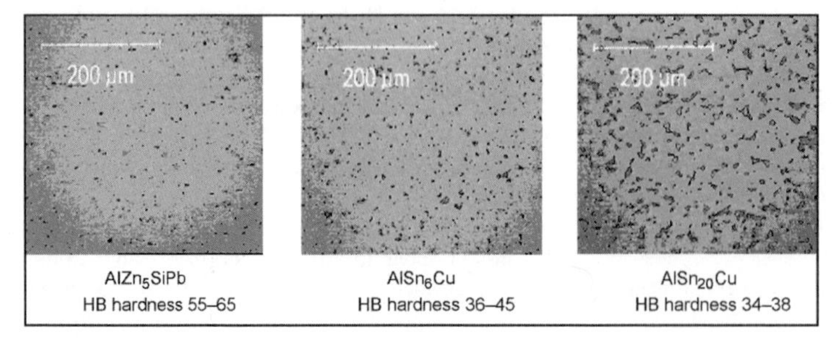

Fig. 7-263 Comparison of microstructures for Al bearing alloys.

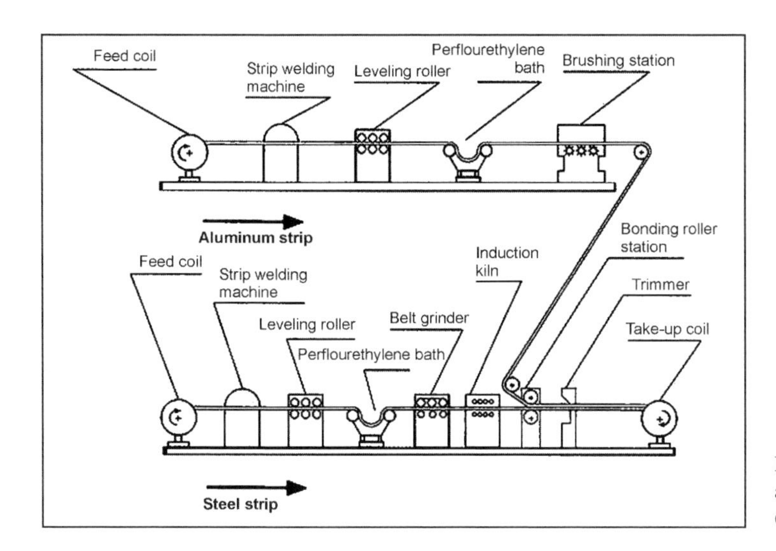

Fig. 7-264 Making the steel and aluminum bonded material (taken from Ref. [1]).

A major disadvantage of leaded bronze is lead's sensitivity to corrosive attack by sulfur and chlorine compounds. Consequently, aluminum alloys are given preference when running with heavy oil and in gas-fired engines.

The bronze/steel composite material is made by casting or sintering.

Strip casting is a suitable process for composite material of up to about 6 mm thick; centrifugal casting is used for thicker bearings.

In the strip casting process used most widely for passenger car bearings, the edges of the pretreated lead strip are bent upward and the molten metal is cast into the "trough" thus formed. After cooling, the surface is milled down and the edges are trimmed. Stretching the strip slightly during these last two steps ensures stable steel strength. An optional, subsequent rolling step boosts the strength of both the steel and the bronze in heavy-duty bearings (sputter bearings). The strip is coiled up again for intermediate storage (Fig. 7-265).

When sintering, the sheet metal strip is pretreated, and then bronze powder is spread over it. The sintering process proper (sintering and rolling) is carried out in two steps in order to achieve a structure with only a very few, very small pores.

The microstructures differ markedly (Fig. 7-266) and the strength of cast bronze is, without having to take any additional steps, greater than that of sintered bronze.

7.19.3.2 Overlays

Overlays have to be applied to all higher-strength bearing materials in order to achieve running properties of adequate quality and insensitivity to disturbances. Basically there are two fundamentally different types of coatings:

- Babbit metals deposited electrochemically
- AlSn alloys applied with the PVD (physical vapor deposition or sputter) process.

Surface modifications such as zinc phosphating are found in certain application niches but have not made a broad breakthrough.

An intermediate layer is required to ensure good bonding with the substrate and/or to suppress diffusion effects; nickel or NiSn is normally used for this purpose.

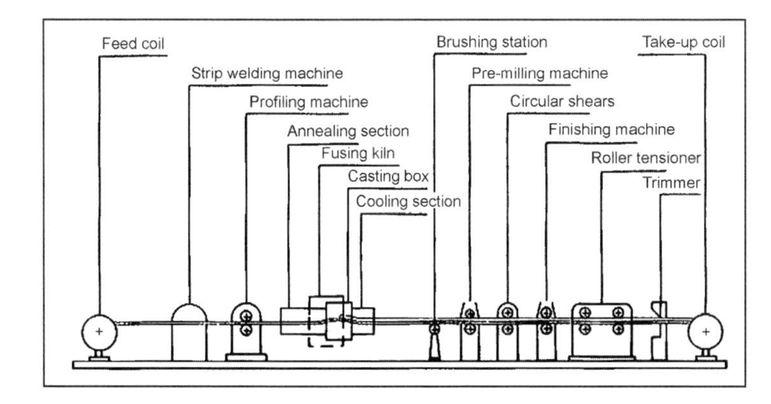

Fig. 7-265 Making up the leaded bronze composite material in strip casting (taken from Ref. [1]).

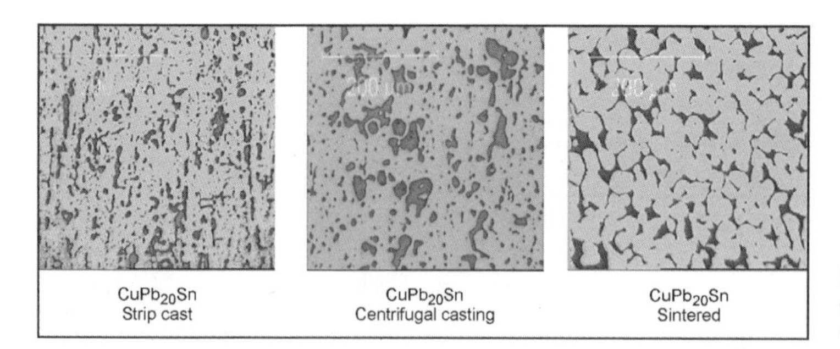

Fig. 7-266 Structures of the CuPb20Sn2 alloy made up in various manufacturing processes.

Nickel is not a material that offers good slip properties; consequently, the thickness of this layer should be considerably less than the surface roughness. Common are from 1 to 3 μm as otherwise larger, contiguous Ni areas appear on the running surface, and the bearing responds aggressively to disturbances where the overlay is worn.

Electroplated Overlays

These overlays are, from the alloy technology viewpoint, similar to the cast babbit metals but exhibit less hardness and a finer structure since they are precipitated out at temperatures below the melting point, sort of in a "frozen" state (see Fig. 7-270 below, three-material bearing). They are very insensitive to mixed lubrication but also wear very quickly because of their low hardness level, from 14 to 22 HV.

The most widespread is the PbSn(8–18)Cu(0–8) system, where the share of tin reduces corrosion sensitivity and the copper increases durability. Tin content in excess of 16% leads to faster diffusion and thus to long-term instability, while more than 6% copper can cause brittleness so that the strength-enhancing effect is negated. PbSn has a certain degree of significance in the Anglo-Saxon regions as does SnSb7 for bearings in large industrial engines, without their having made any widespread breakthrough.

These layers are applied in galvanic baths with the application of current. This is done in a four-stage process encompassing pretreatment, applying and activating the intermediate layer, precipitating the overlay, and using subsequent heat treatment to stabilize the structure and to induce a sufficient diffusion bond.

Overlay thickness is limited for several reasons:

- Durability drops rapidly with increasing thickness.
- The geometry of the lubrication gap must not change unacceptably as a result of wear.
- A concentration of electrical voltages causes the layers to be thicker at corners and edges.

For economic reasons, too, mass electroplating is to be targeted if at all possible.

As a rule, overlays of from 15 to 35 μm thick are applied in mass electroplating processes; where thicker layers are necessary—in large bearings, for instance—they have to be reworked retroactively.

Sputtered Overlay

A development that has made considerable advances in mass production only in recent decades is the use of the sputter process to deposit AlSn layers on plain bearings.

Sputtering (cathode ionization) is a coating process in which a working gas (argon) is ionized in a high vacuum. An electrical field accelerates the ions to the cathode, the "target," and atoms are dislodged from the target by the impact of the atoms. These atoms condense on the bearing running surface and form the slip-promoting film, Fig. 7-267.

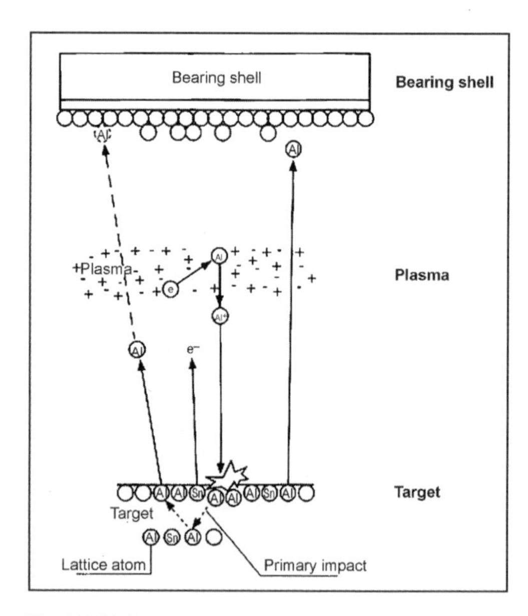

Fig. 7-267 Sputter process, schematic.

Atomic deposition creates a strong structure with an extremely fine distribution of the soft phase, which, in spite of high hardness at about 90 HB, gives good running properties (Fig. 7-268).

A further advantage of the process is the increase in bonding strength achieved by precleaning the substrate by sputter etching under vacuum, producing a highly active surface.

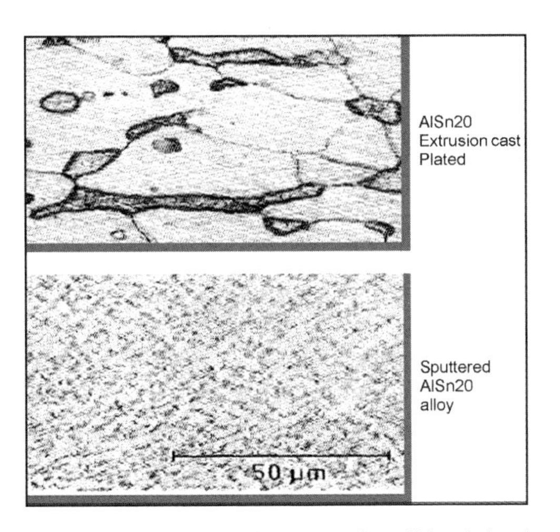

Fig. 7-268 Comparison of structures for roll bonded and sputtered AlSn20 layers.

Today AlSn20Cu is used almost exclusively as the sputter layer for heavy-duty bearings, but the process is, in principle, quite flexible and enables deposition of a very much broader range of alloys than is possible with conventional electrochemical processes. The only major drawback is the high cost for the coating.

7.19.4 Types of Bearings—Structure, Load-Bearing Capacity, Use

For cost reasons, one strives to satisfy the requirements for the particular application with the simplest possible bearing design. But contradictory demands for strength, tight seating, and good running properties ultimately lead to a "division of labor" and to a multilayer structure for the bearing.

The use properties of the bearings and above all their dynamic load-bearing capacities are influenced not only with the selection of materials but also with the structure and thickness of the layers and other engineering measures. Thus, there are, beyond classical multilayer concepts, newer types that optimize the bearing's utility with a closely defined sequence of layers and/or design of the running surface.

The fundamental advantages and disadvantages have already been mentioned in the discussion of the materials. Figure 7-269 provides a survey of the types of bearings most commonly used for a particular application range.

7.19.4.1 Solid Bearings

Solid material is used primarily in large, industrial engines in the form of hard bronzes for thick-walled bushings and AlSn6 for thrust washers (axial bearings). The advantage is simple manufacture and in thrust rings

	Backing shell	Bearing metal	Overlay	Max. p_{transv}	Primary use
Solid bearing	None	CuPb15Sn7 AlSn6	None	60	Wristpin bushings Thrust washers, camshaft bearings
Two-material bearing	Steel	CuPb10Sn10 CuPb15Sn7 AlSn6 AlSn20 SnSb12Cu	None	120 45 40 20	Wristpin bushings, rocker arm bushings Thrust washers, camshaft bearings Main bearings, conrod bearings Camshaft bearings
Three-material bearing	Steel	CuPb10Sn10 CuPb20Sn2 AlZn4.5	PbSnCu PbSn16Cu PbSn10 Cu PbSn10 PbSn10 ceramic SnSb 7 CuPb30 PbSn16Cu2	90 55 65 80 50	Large wristpin bushings Conrod bearing, main bearing Conrod bearing
Grooved bearing	Steel	CuPb20Sn2 AlSn6Cu AlZn4.5	SnSb7 PbSn16Cu2 PbSn10 ceramic PbSn16Cu2	50 50 65 55	Main bearings for large engines Main bearing, conrod bearing
Sputter bearing	Steel	CuPb20Sn2 CuPb10Sn10	AlSn20	>100	Conrod bearing

Fig. 7-269 Most important types of bearings and application ranges.

the additional option of enabling use at both ends with proper engineering.

In passenger car engines the slow-running camshafts are borne directly in the aluminum cylinder head. Although these alloys are not bearing metals, proper functioning is reliable because of the low energy density in the bearings.

7.19.4.2 Two-Material Bearing (Fig. 7-270)

Here there are two essentially different application areas:

- Suitable for use at wristpin and rocker arm bearings are rolled bushings made of leaded bronzes, since the high specific loading of up to 120 N/mm^2 requires endurance strength and the disadvantage of low running capacity, because of the low sliding speed, is of little significance. When there is an insufficient oil supply, these bushings tend to liberate lead from the material and cause oil carbonization.
- Bearings based on AlSn, because of their excellent ratio of performance to cost, are the preferred solution for moderately loaded applications involving rotational movement and thus primarily the main and conrod bearings in gasoline engines and industrial diesels. Their wear is low but there are limits to their adaptability. The low wear also harbors a risk: the appearance of the bearings changes hardly at all. Consequently, evaluating their condition with a visual evaluation is difficult. This makes necessary adequate statistical confidence for service life during the testing stage.

The continuously rising loads found in new developments and advances in engines have resulted in the development of two-material bearings using higher-strength

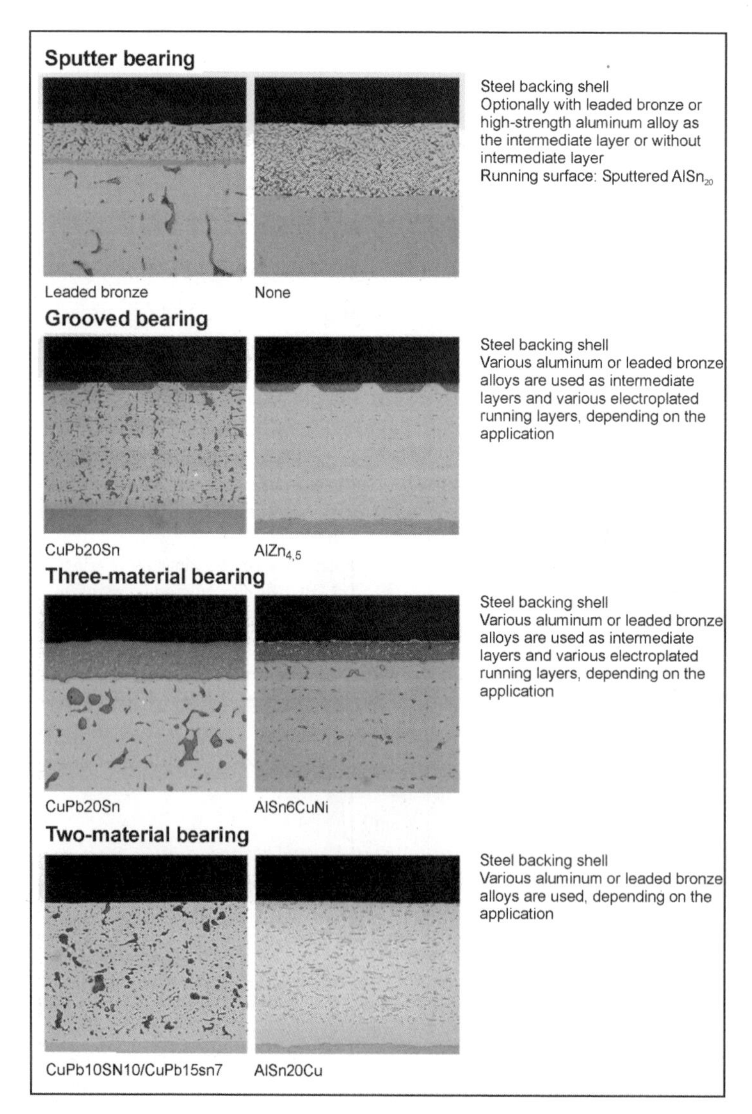

Fig. 7-270 Material structure (examples).

AlSn alloys. The above applies to them in principle, but these bearings, because of the smaller lubrication gap and the greater energy density, are at greater risk of friction and wear damage. The increase of strength achieved by reducing the tin content is thus not an option, which helps meet development objectives. And the bonding layer made of pure aluminum can become a weak point.

7.19.4.3 Three-Material Bearing (Fig. 7-270)

Three-material bearings with an overlay applied by electroplating and above all on a leaded bronze basis are the type used predominantly for crankshaft bearings. They represent a fully mature technology, are available worldwide, and offer a good ratio of cost to benefit. They are distinguished by good adaptability and are tolerant of grime and error for as long as the soft overlay is present. In larger engines three-material bearings based on aluminum are also used.

Three-material bearings are suited only with some limitations where high loading situations are encountered, above all in the conrod bearings for modern direct-injection engines (both gasoline and diesel). Their weak point is faster wear at the overlay as loading increases. Corrosion resistance, too, which becomes more important at longer oil change intervals, is not high. Wear at the overlay, from 15 to 30 μm thick, has in and of itself only an insignificant effect on the bearing function; exposure of the substrate, however, leads to a drastic increase in sensitivity to disturbances. The classical three-material bearing with a PbSnCu overlay is therefore supplanted to an even greater extent by higher-strength, two-material aluminum bearings in the lower load range and by the true high-performance concepts—grooved bearings for industrial engines and sputter bearings for passenger car and utility vehicle engines.

7.19.4.4 Miba™ Grooved Bearings (Figs. 7-270 and 7-271)

The grooved bearings developed by Miba™ almost 20 years ago and shown in Fig. 7-271 delay the degradation of the running layer with a special geometry for the surface. The overlay is embedded in very fine grooves in the running direction; between them are lands made of the harder bearing material. The ratio of materials at the running surface is about 75% overlay to 25% bearing metal. With this geometry, it is possible to continue to determine the tribologic properties by selecting the overlay material but to protect that layer against wear with the harder lands. Thus the good running properties are retained for a much longer time than in three-material bearings.

The grooved bearing today finds its primary application in diesel engines with greater specific power and is used to drive locomotives and ships; in the passenger car and utility vehicle engine segment it has been supplanted in recent years to an increasing extent by the sputter bearing, because of the continuously rising loads.

7.19.4.5 Sputter Bearing (Fig. 7-270)

The bearings that can stand the most extreme loading and are produced in large number today are three-material, leaded bronze bearings with a sputter overlay. Because of their greater load-handling ability, up to more than 100 N/mm^2, and with good running properties at the same time, they are installed in engines with high power density and used for passenger cars, utility vehicles, and drives for fast ships. Today, hardly any other type of bearing can be considered, above all for conrod bearings in direct-injection diesel engines for passenger cars.

The only major drawback to the sputter bearing is its price. Because of the complex vacuum coating process, a sputter bearing is five to eight times as expensive as a three-material bearing. Thus, in the conrod and main bearings a sputter shell on the side subjected to heavy loads is combined with a three-material or grooved bearing shell on the side with less loading. This combination offers the additional advantage that tiny particles of grime cannot become embedded in the soft overlay.

The application limits and costs for the various bearing designs are shown in Fig. 7-272.

7.19.5 Bearing Failure

7.19.5.1 Progress of Damage

Bearing damage, Fig. 7-273, in the narrower sense is always an interference in the geometry of the slip space to an extent that precludes stable operation of the bearing

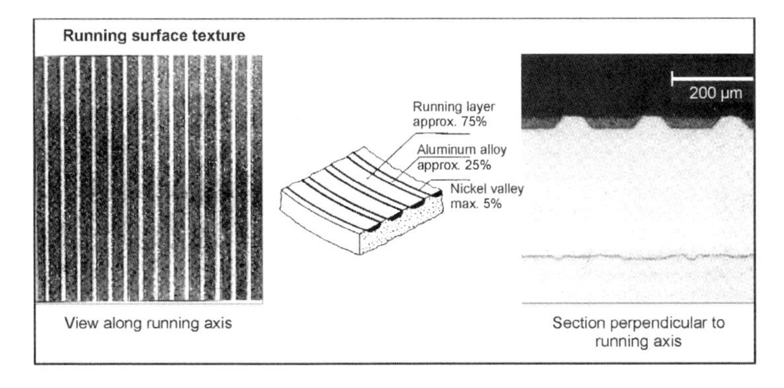

Fig. 7-271 Miba™ grooved bearings.

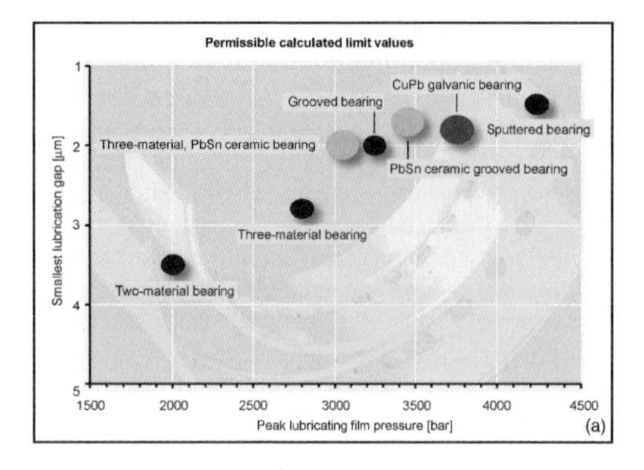

a

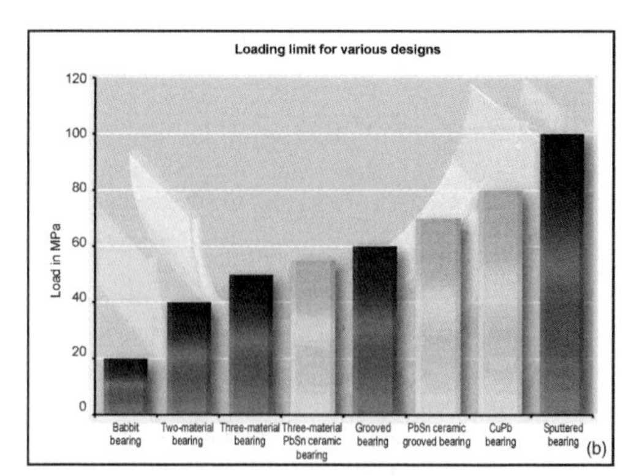

b

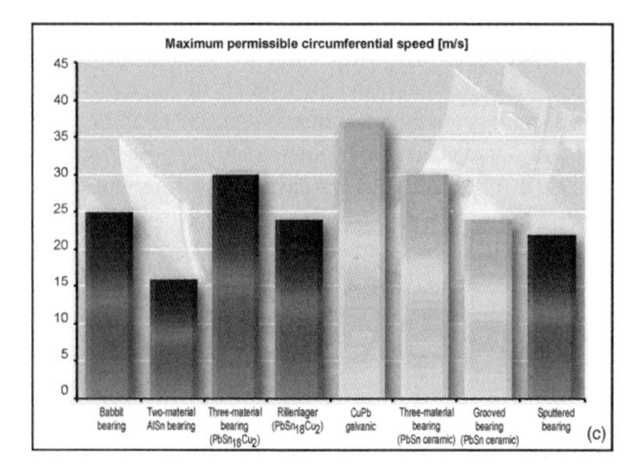

c

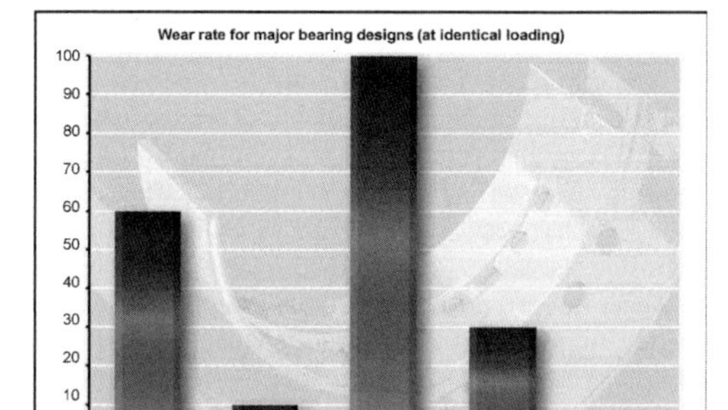

d

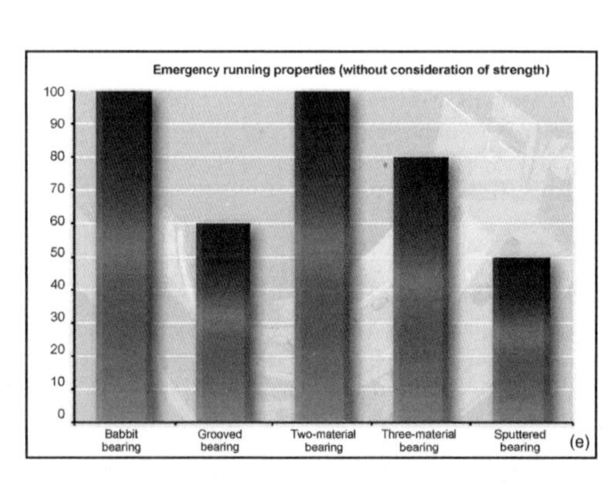

e

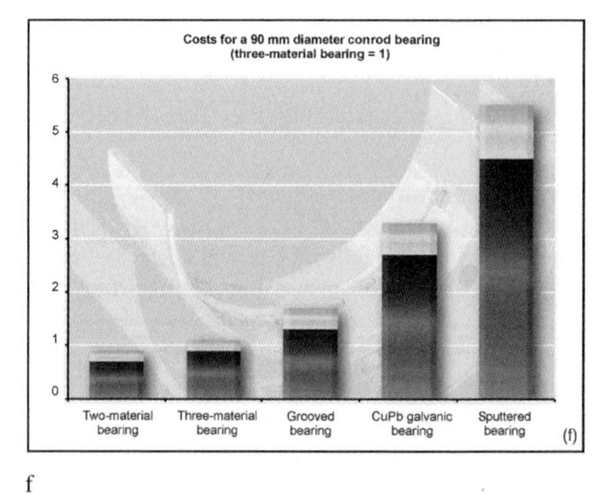

f

Fig. 7-272 Guideline values for various bearing designs.

location. The results are great friction and with it local overheating and destruction of the bearing and adjoining components, right through to complete failure of the engine.

In internal combustion engines where, in contrast to mechanical engineering bearings, the sizes of the loads and the direction change cyclically, the course of damage depends on the location, time, and loading level, and thus is to be seen statistically. This can cause total damage to one bearing while in the adjacent bearing almost no damage at all can be found. Disturbances can be covered for a period of time and even corrected if the cause is

Fig. 7-273 Total failure of a leaded bronze bearing.

eliminated (e.g., excess temperature, oil shortage, etc.); an error in the geometry may be attenuated by wear or adaptation.

Because of the serious subsequent damage that can follow bearing failure, even events that, seen on their own, do not cause a failure are deemed to be bearing damage. Events such as these are seen as early warning signals for impending bearing damage and thus are very important when diagnosing the status of the system.

7.19.5.2 Types of Bearing Damage

DIN-ISO standard 7146 and trade literature published by plain bearing manufacturers describe the most frequent types of bearing damage; consequently, only a brief overview is given here. This depiction follows the organization used in DIN-ISO 7146.

Damage to the Running Surface

Foreign Objects, Grime

Foreign objects that are swept into the bearing with the lubricating oil continue to represent the most frequent cause of bearing failure, particularly in main bearings and in spite of great efforts to maintain cleanliness during assembly, as well as in operation. The problem found in such events, in addition to the permanent disturbance, which reduces service life, is the scoring or embedding process itself. While this is happening, extremely high friction is generated locally.

Broad-Surface Wear

At high loads, with the wrong oil (too thin) or the selection of an unsuitable bearing design, premature wear can appear in the zone where the narrowed lubrication gap is found. As a rule, wear is not a problem in normal operations; three-material bearings do, however, become more susceptible to disturbances when the tolerant overlay is no longer present.

Edge Collar, Local Overloading, Overheating

Deficiencies in the geometry, localized contact points due to elastic deformations, and minor assembly errors can be attenuated by localized wear at the soft layer. This process, however, leads to an increased degree of mixed lubrication, corresponding to a local increase in temperature and, in an extreme situation, to instability and damage.

Fatigue Fracture

The bearing material has to exhibit sufficient durability so that the pulsating loads can be reliably transferred throughout the required service life. If this is not the case, then fine fissures appear and later particles spall off. The hazard potential represented by fatigue fracture is dependent on the thickness of the layer affected: spalling at the running layers seldom leads directly to bearing failure. Fractures in the bearing material, about ten times as thick, have an enduring adverse effect on slip gap geometry.

Cavitation

Cavitation is the result of vapor bubbles in lubricating oil, which arise when the lubrication oil pressure at some points falls below the vapor pressure. These bubbles implode when they again enter an area of higher pressure. The pressure surge thus created tears particles out of the bearing surface and in serious cases right through the bearing metal and into the steel in the backing shell.

Cavitation is quite often a design problem (groove shape, bearing play, etc.). In addition to changes in the geometry of the oil flow, its prevalence can also be reduced by measures that raise the oil pressure in the system.

Corrosion

Of the materials commonly used in bearing technology, the lead in the electroplated overlay and in the leaded bronze is most often affected by reactions with sulfur and chlorine. In those cases where corrosion is to be anticipated during operation, e.g., where industrial engines are run on heavy oil or landfill gas, an increase in the tin contact in the CuPb materials or the use of AlSn instead of CuPbSn is necessary

Damage at the Back of the Bearing

Insufficiently Tight Seating

The second important functional surface of the radial bearing is the outside diameter. Sufficient friction is necessary to transfer the force. The tight seating of the bearing in the housing bore is achieved by sufficient overage of the diameters and bearing halves by excess length, the so-called "crush height." Because of elastic deformations resulting from the operating forces, there is thrust loading at the interface between the bearing and the housing; insufficiently tight seating can result in relative movements between the bearing and the housing.

The consequences are material displacement, fretting corrosion, material transfer (pitting), and, in serious cases, shell movement.

These relative movements can be suppressed by greater crush height. The limit is imposed by the tangential stresses in the steel shell, which may not exceed the creep limit. Increasing the operating speeds for existing engines thus frequently necessitates engineering modifications.

Assembly Errors

In addition to the operating loads and geometric deficiencies, errors in assembling the bearings are often the reason behind serious bearing damage. Thus, bearings should be designed in such a way that incorrect positioning, interchanging, and the like can be positively avoided.

7.19.6 Prospects for the Future

Rapid developments in engine technology, which are further accelerated with the introduction of direct-injection engines, are flanked by component development and in some cases made possible by such developments.

The major driving forces behind new developments in bearings include

- Loading capacity (higher ignition pressures, mean pressures, service periods)
- Costs (heavy-duty, multilayer bearings are expensive)
- Environmental aspects (lead, cleaning, manufacturing processes)

Even if today all the requirements are covered from the technical side, there are combinations of loading capacity, running properties, and manufacturing costs that are not ideal. The goals of new developments are above all an improvement in the ratio of costs to utility.

For economic reasons the use of bearings without an overlay is also targeted at higher loading levels so that most materials developments in recent years are aimed at increasing strength with the least possible limitations in regard to running capacities. Developments are going essentially in two directions:

- Improving the load-bearing capacities of two-material bearings so that they can replace three-material bearings in some applications. This is done by developing new aluminum alloys in conjunction with advanced casting technology. Newer developments are, for example, AlSn10NiMn, AlSn12Si4, and AlSn25CuCoZr—the first, because of the significant reduction of the tin content, is more suited for use in smaller (passenger car) engines at moderate loads.
- Increasing the wear resistance and durability of electroplated overlays by new materials, on the one hand (CuPb system), and by hardening the PbSn layer by means of microscopic ceramic particles, on the other hand. This leads to improved three-material and grooved bearings.

Several new developments of this type are close to introduction into mass production. They will certainly not supplant conventional bearings—and particularly not the sputter bearings—but will provide a sensible complement for areas that are not covered ideally today.

Bibliography

[1] Affenzeller, J., and H. Gläser, Lagerung und Schmierung von Verbrennungsmotoren (includes extensive bibliography), Springer, 1996.
[2] Lang, O.R., and W. Steinhilper, Berechnung und Konstruktion von Gleitlagern mit dynamischer Belastung, Springer, 1978.
[3] N.N., Gleitlager-Handbuch, Miba Gleitlager AG, 2000.
[4] Ederer, U.G., and R. Aufischer, "Schadenswahrscheinlichkeit und Grenzen der Lebensdauer," Esslingen Technical Academy, 1992.
[5] Arnold, O., and R. Budde, "Konstruktive Gestaltung von Lagerungen in Verbrennungsmotoren," HdT Essen, 1999.
[6] Ederer, U.G., "Werkstoffe, Bauformen und Herstellung von Verbrennungsmotoren-Gleitlagern," HdT Essen, 1999.
[7] Schäden an Gleitlagern, DIN-ISO 7146.

7.20 Intake Systems

The air intake systems in modern internal combustion engines serve a number of functions in addition to routing and filtering the combustion air. The demands placed on intake systems continue to rise with increasing engine complexity. Two major trends are emerging.

System Competence

The entire air routing configuration is seen as a system extending from the intake opening to the cylinder head; it is engineered and manufactured by the supplier and delivered ready for installation. This presumes that the supplier will fully understand the system, going beyond the air supply system proper and including the exhaust system in view of the exchange of gases in the cylinder.

Modularization

A second trend is the increasing modularity of the intake system. A modular design makes good sense since the air supply system is spread out all around the engine and, simply because of its size, lends itself to the attachment of discrete components. These components are not necessary components of the air management system proper. One example is locating the motor management circuitry inside the air filter; passing air is used to cool the electronics. Modularization requires, in addition to an understanding of the system as a whole, increased competency in manufacturing and integration.

Figure 7-274 schematically shows the air path in a four-cylinder engine together with the main functions and some of the attached components. The thermodynamic situation along the air path is explained below along with the proximity to the fields of acoustics and filtration.

7.20.1 Thermodynamics in Air Intake Systems

The thermodynamics in the air supply system depends on the combustion process (gasoline or diesel) and on the charging principle. The external air piping and air filter are similar in all the variants. The systems differ markedly, however, downstream from the air filter, depending on the charging principle; see Fig. 7-275.

Turbocharged engines are fitted with an elaborate clean-air section with a compressor and aftercooler, while

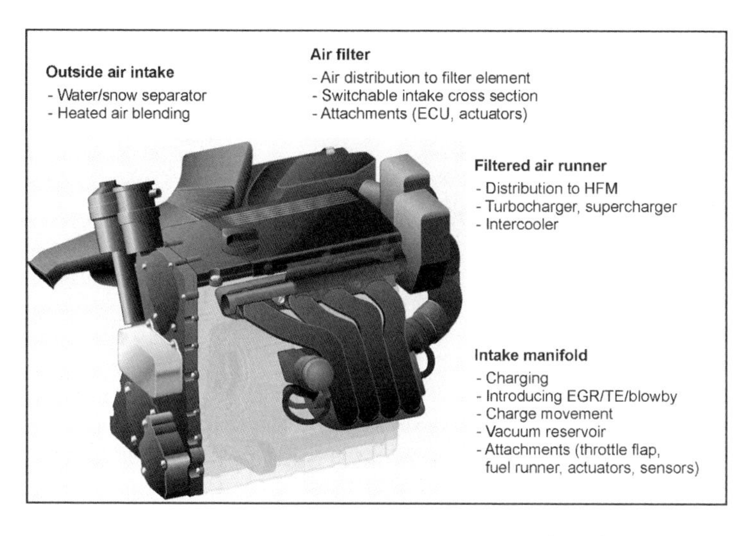

Fig. 7-274 Air flow system for an internal combustion engine (schematic).

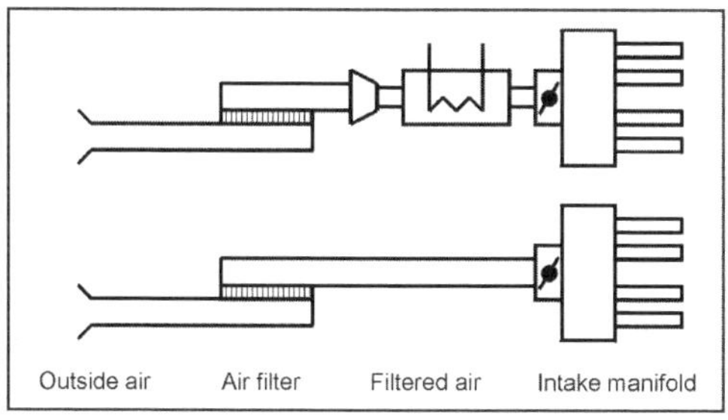

Fig. 7-275 Air path for naturally aspirated engines (below) and for turbocharged engines (above).

the intake manifold is simple in design, as it serves primarily to distribute air to the cylinders.

Naturally aspirated engines, by contrast, have a simple clean-air section but in most cases complex, active intake manifolds to improve cylinder fill.

External Air Section

The external air path, i.e., the section of the intake system between the intake opening and the air filter, not only guides the air but also is used for the addition of warm air and the elimination of dirt. Blending in warm air influences the engine's operating properties, particularly in the cold starting phase. This function grows in significance in the future as more stringent limit values are adopted in exhaust gas legislation. Drying the filter element and thawing snow are further reasons for adding warm air. Fuel consumption can be favorably influenced by intelligent temperature regulation for the intake air. Warm air is drawn in through a second take-up point near the exhaust manifold; it is activated by flaps. The flaps are actuated by thermostat elements or by vacuum or electrical actuators.

A suitable external air routing system also separates coarse particles (droplets, snow, dirt) with minimal pressure loss by bends. This preliminary separation helps to keep down the amount of grime collected at the air filter and protects the filter element against moisture. Particle separation and pressure loss at diversion points are determined in advance with the help of CFD.

Air Filter Body

What might colloquially be referred to as the "air filter" or "air cleaner" comprises the filter element, the air cleaner body, and the cover. In addition to its acoustic effect, the body serves to optimize the airflow path and air distribution around the filter proper. The ideal here is the most uniform distribution possible. Air velocity perpendicular to the filter element has to be homogeneous over the entire filter surface area. When the arrival of the air is not uniform, there is greater pressure loss at the filter element, and engine efficiency is degraded. The dirt and dust trapping capacity of the filter material is also optimized with homogeneous airflow.

Three-dimensional CFD techniques are employed at a very early stage to engineer the airflow in air filter bodies, Fig. 7-276. This makes it possible to determine the ideal geometry very early and with a minimum of effort for physical testing.

In one example, it was possible to reduce physical size by 30% with the same pressure loss level and dust capacity. It was possible to attain air distribution that was within 3% of the ideal test bed value.

Clean-Air Channel

The air impinging on the mass flow meter that measures the intake air on the clean-air side is analyzed for new intake systems, using CFD simulation, in order to achieve uniform flow. In view of more stringent emission limits, reliable functioning of this meter, in all operating states and for the life of the vehicle, is specified. Gradual degradation of the meter resulting from deposits on the sensor (oil droplets from the crankcase or from the exhaust gas return system) can also be reduced dramatically by applying CFD simulation to the air flow path.

Fig. 7-276 Airflow to the filter elements: Not uniform and with great pressure drop (left); nearly ideal (right).

The gas pulsations generated by the engine become more intensive downstream on the clean-air side. If thermodynamics and acoustics are not seen as a whole, then this has to be done at the latest in the clean-air runner since both disciplines exert an effect on air routing. In the area associated with the clean-air runner one finds acoustic components (shunt resonators, $\lambda/4$ tubes) that also have an influence on gas exchange in the cylinders. Today simulation tools are used for such components. The airflow rate and noise at the inlet tube are thus calculated at a very early point in the design phase. The effort required for modeling can be significantly reduced since a single calculation model delivers both results.

Supercharged and turbocharged engines have a longer airflow path than naturally aspirated engines. In engines with a turbocharger, the intake air passes from the forward module and through the air filter to the compressor located near the exhaust manifold. The compressed air is then returned to the forward module, where the aftercooler is located. Finally, the clean-air runner terminates at the intake manifold at the engine.

Intake Systems

Engines with mechanical superchargers or exhaust-driven turbochargers require intake manifolds and runners to distribute the combustion air to the cylinders. The aim here is to achieve short intake runners with little pressure loss and good uniformity of distribution to the cylinders.

Naturally aspirated engines use the wave effects initiated by the piston to compress intake air. The procedure known as "resonance tube charging" is described in Fig. 7-277.

When the intake valve is opened, the piston, as it moves downward, creates a vacuum "wave" that moves opposite the direction of airflow, away from the combustion chamber and along the resonance tube. The vacuum wave is reflected at the collector, because of a change in cross section. The pressure wave moving back toward the combustion chamber can be utilized to improve cylinder filling, provided that it arrives before the intake valve closes.

Ideal tube length, at constant speed of sound a, is inversely proportional to engine speed n. To achieve good cylinder fill over a broad engine speed range, all vehicle classes are seeing an increasing use of intake manifolds that can alternate between short and long resonance tubes. A typical response curve for an active intake manifold with two resonance tube lengths is shown in Fig. 7-278.

With increasing intake system complexity, the increase in the airflow rate depends more and more on the quality of the manufacturing and materials. Figure 7-279 uses gasket quality to illustrate the sensitivity of airflow rate to leakage. A gasket that permits an increase in the airflow rate in a two-stage active manifold can in three- and four-stage manifolds quite conceivably lead to a reduction of the airflow rate.

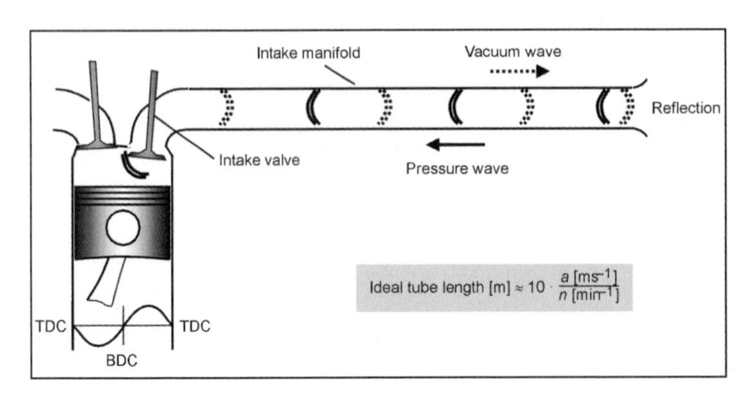

Fig. 2-277 Principle of resonance tube charging.

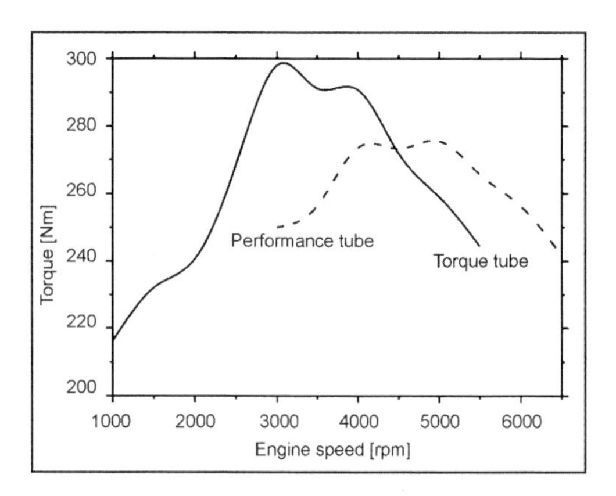

Fig. 7-278 Torque progression in a six-cylinder engine with active intake manifold.

In addition to leakage through the switching elements, there is a series of other variables that influence the change of gases in the intake manifold. Figure 7-280 provides a survey of potential sources of loss.

This makes it necessary for suppliers of modern intake systems to define the entire intake system in both thermodynamic and mechanical terms at a very early point in development work. This requires linking and networking all the CAE tools right from the outset of the development project.

7.20.2 Acoustics

Sound is understood to be mechanical oscillations and waves in an elastic medium. Section 7.20.1 used Fig. 7-277 to illustrate how piston movement, after the intake valve was opened, triggered a vacuum wave moving against the gas flow direction. These pressure fluctuations are propagated as sound through the air filter intake opening (intake noise). Moreover, the pulsation inside the components induces wall vibration (structural noise), which is then again propagated as airborne noise. Those in the vicinity do not always perceive this sound as pleasant, which is why restrictions have been imposed; every vehicle must meet these limits (see also Chapter 27).

Legislation

The noise problems associated with motor vehicles can be subdivided into two areas:

- Interior noises
- Pass-by noise

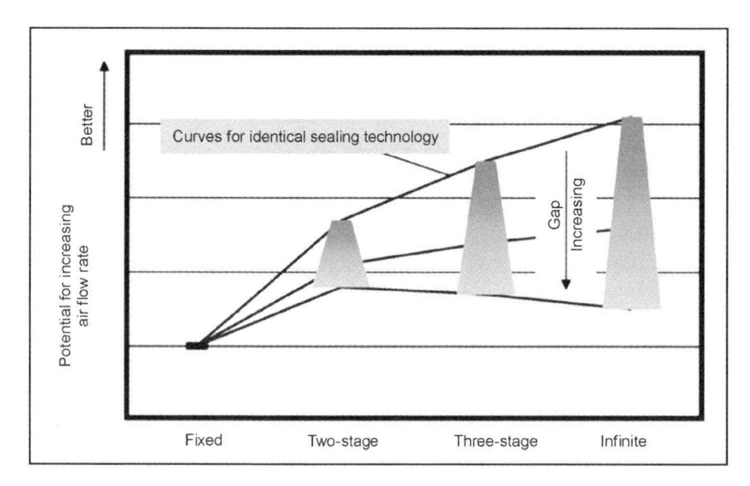

Fig. 7-279 The influence of selector flap gasket technology on air flow rate.

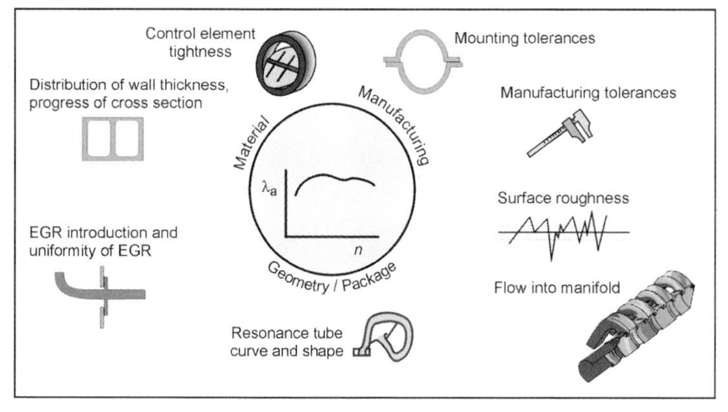

Fig. 7-280 Factors influencing the air flow rate for active intake manifolds.

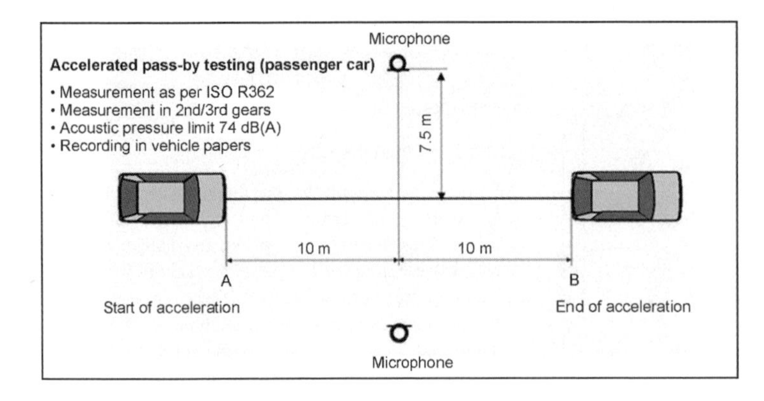

Fig. 7-281 Motor vehicle noise.

While the reduction of interior noise goes hand in hand with passengers' increasing comfort expectations and thus is a factor in the make's image, a legislated limit value applies to noises generated by vehicles accelerating past a given point. The procedure used to measure this value is illustrated in Fig. 7-281. Since October 1, 1995, passenger cars may not exceed a value of 74 dB (A) in accelerated pass-by testing.

The overall acoustic level is the sum of just a few individual noise sources. Regarding pass-by noise, these are engine noise, intake and exhaust noise, tire noise, and wind noise.

Noise Creation

In an internal combustion engine, the pistons, with their reciprocal motion, create fluctuations in air pressure (air pulsations) and airborne noise that results from them. The pistons thus act as an air-pulsation noise source. Disturbances in airflow along the intake system can also act as aerodynamic sources, contributing to intake noise.

This noise is emitted primarily through the intake opening and thus passes directly into the environment. A second portion of the pulsation energy inside the intake system incites structural noise oscillations in the elastic structure. These are then transmitted from the exterior surfaces to the surrounding air or through attachment points to the body. This situation is illustrated schematically in Fig. 7-282.

Optimization Measures

The objective of the measures undertaken to optimize intake noise is consistent acoustical development wherein the noise is to be reduced right from the draft stage. The work carried out to optimize noise is subdivided into primary efforts and secondary efforts.

Primary efforts: These exert an influence on the sound source. The noise created by airborne sound means a reduction in the alternating pressures, while the noises excited by structural sound require a reduction in the exciting forces and a change in structural noise behavior and in propagation (admittance and degree of propagation).

Secondary efforts: These retroactively reduce the airborne sound generated and reduce noise emissions with mufflers and/or encapsulation.

The pulsed-air noise source in the intake system is the engine; any influence on this source often conflicts with the objectives set in the thermodynamics analysis. That is why one employs secondary measures such as damping filters and shunt resonators to reduce intake noise. The effect of acoustic corrections on the throat noise and on the gas exchange is shown in the example in Fig. 7-283.

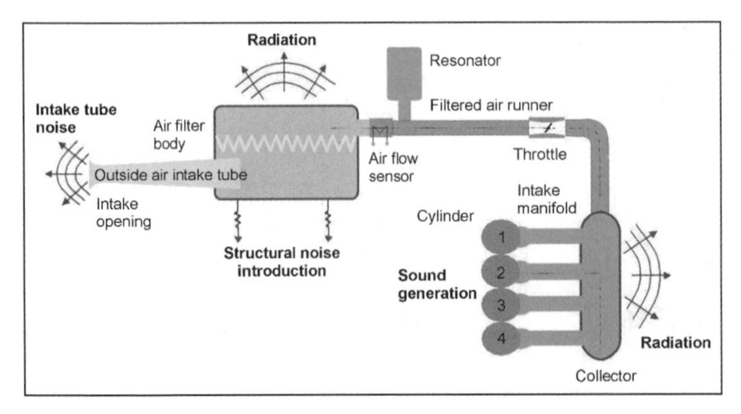

Fig. 7-282 Noise sources in an intake system.

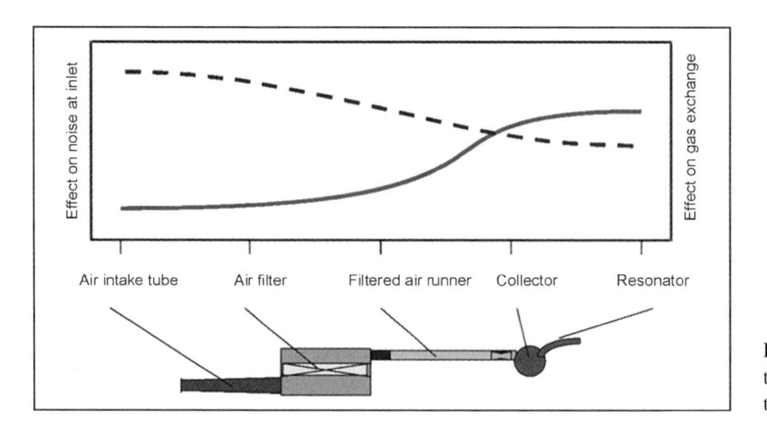

Fig. 7-283 Effect on the intake tube noise and gas exchange in the cylinders.

Acoustic Elements in Runner Systems

A variety of acoustic principles can be used to attenuate intake noises; see Fig. 7-284. The most important damper design is constructed in principle like a so-called series resonator. This is a system taking the form of a Helmholtz resonator, in which a damping chamber is connected to a section of tubing. A resonator such as this functions in principle like a sprung mass system in which the spring is represented by the compressible air in the chamber, the mass, on the other hand, by the air pulses in the pipe. Depending on its dimensions, a resonance frequency f_0 can be calculated at which a resonator such as this amplifies the sound introduced. The following formula is used to calculate the frequency:

$$f_0 = \frac{c}{2 \cdot \pi} \sqrt{\frac{A_w}{l_{\text{acoust}} \cdot V}} \qquad (7.27)$$

where A_w is the mean cross section of the resonator throat, l_{acoust} the effective length of the throat, and V the volume of the chamber. Inversely, frequencies upward of $f_0 \cdot \sqrt{2}$ are damped. The objective is to use this phenomenon in the damper filter. In order to achieve the best possible damping, f_0 has to be as low as possible, meaning well below the frequencies occurring during operation. This can be achieved by increasing the volume of the air filter body, by reducing the intake cross section or by lengthening the intake snout. Because of the fact that the installation size is usually limited, the volume of the body cannot be increased at will. A severely reduced intake cross section also has undesirable side effects since the flow of intake air is throttled. Increased pressure loss always means a loss of engine performance, which is why in practice the pressure loss in the intake tube is kept in bounds by designing a diffuserlike intake opening, similar to a Venturi tube. Lengthening the intake tube also runs into system-imposed limits, while a corrective measure such as this also harbors the hazard of tube resonances that can counteract damping at certain frequencies. That is why exact tuning of the entire system is needed to identify the ideal compromise between expense and profitability.

Acoustic Measurement and Simulation Tools

Many tools are available for use in designing an intake system; the simulation tools, in particular, have grown in significance in recent years since they can be used to predict acoustic properties even in a very early development stage. In addition to the finite element method, 1-D calculation programs based on the transfer matrix method or the finite differences method have become established here. The latter offers the advantage that, in addition to the acoustic values, thermodynamic values can also be calculated. The calculation results can be validated at simple component test beds as soon as initial samples are available. Final optimization with parts close to the mass-production components is then carried out on the engine acoustic test bed or in the vehicle.

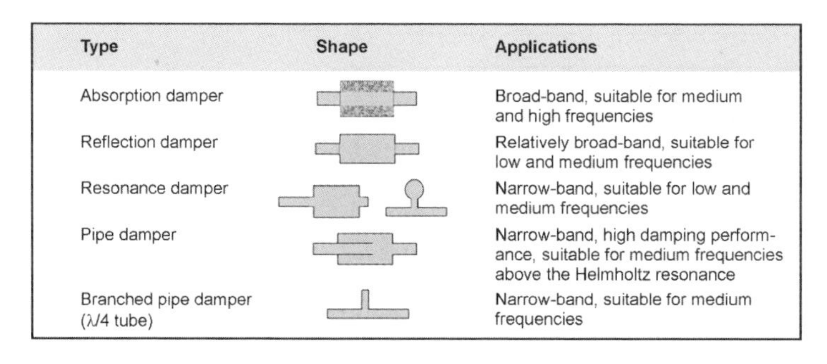

Type	Shape	Applications
Absorption damper		Broad-band, suitable for medium and high frequencies
Reflection damper		Relatively broad-band, suitable for low and medium frequencies
Resonance damper		Narrow-band, suitable for low and medium frequencies
Pipe damper		Narrow-band, high damping performance, suitable for medium frequencies above the Helmholtz resonance
Branched pipe damper (λ/4 tube)		Narrow-band, suitable for medium frequencies

Fig. 7-284 Construction forms for acoustic dampers and their application range.

In addition to the pure reduction of the acoustic pressure level, the quality of the noise is playing an even more important part in development. Here “dummy head” recordings are used to log the noises that are then evaluated subjectively by test persons in comparison listening tests. These tools are shown in Fig. 7-285.

Future Systems

In addition to passive measures, adaptive measures are employed more and more in intake systems. Active manifolds are used primarily to increase the airflow rate. But in the air routing systems, too, these components can be employed to optimize acoustic properties. Thus, for example, by using a single-stage, active manifold in the lower speed range, when the engine does not require its full volumetric flow, a smaller intake cross section can be used in order to achieve low frequency tuning of the Helmholtz resonator. Figure 7-286 shows such a configuration as an example.

The advent of electronics in the intake system has paved the way for entirely different system designs, such as the use of inverted or dephased sound to cancel out noise. If the noise arriving from the engine is countered by a wave of the same amplitude but 180° out of phase, then these two waves cancel each other out. This principle is also known as active noise control and is depicted in Fig. 7-287.

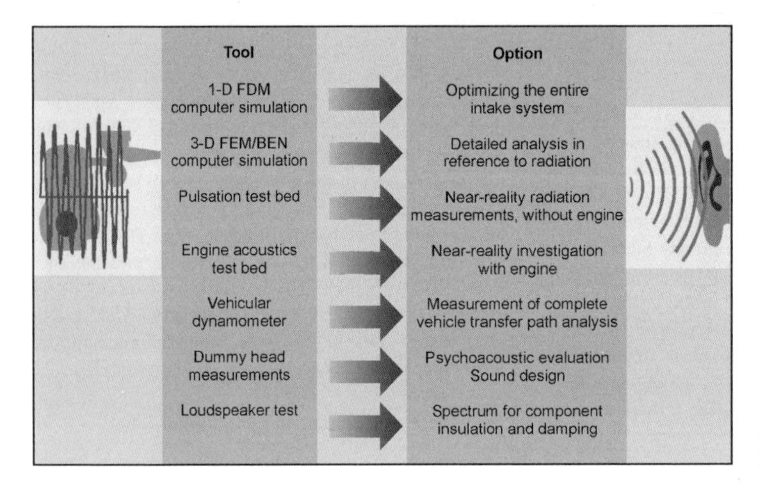

Fig. 7-285 Acoustic measurement and simulation tools.

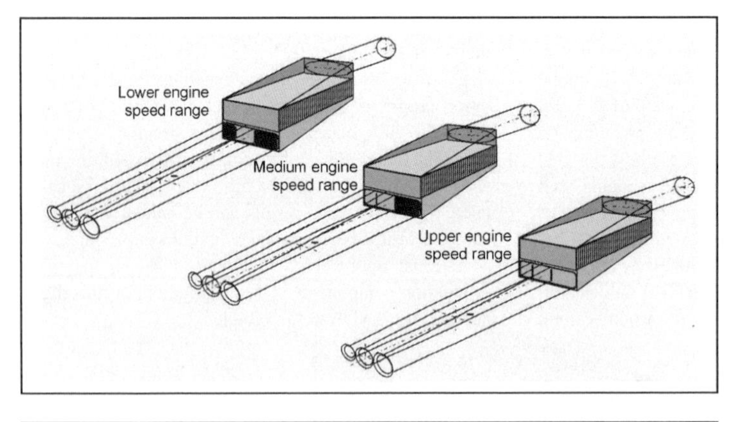

Fig. 7-286 Multistage, active intake manifold.

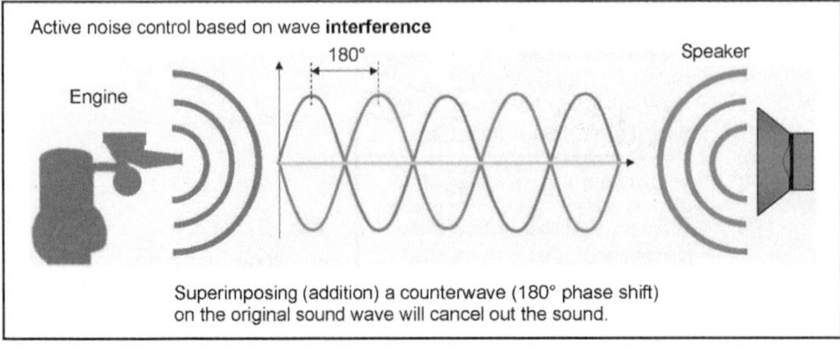

Fig. 7-287 Active noise control.

Bibliography

[1] Müller, K., and W. Mayer, Einfluss der Ventilgeometrie auf das Einströmverhalten in den Brennraum, 3rd edition, Vieweg, Wiesbaden, 1999.
[2] Wild, S., "Torque vs. Power—No Conflict with Highly Variable Resonance Runners," Global Powertrain Congress, Detroit, 2001.
[3] Weber, O., and St. Wild, "Leistung plus Drehmoment–optimierte Sauganlage mit voll variablen Resonanzrohren," 22nd International Vienna Engine Symposium, 2001, VDI Fortschrittsberichte, Series 12, No. 455, Vol. 2, pp. 320–332.
[4] Alex, M., Akustikoptimierung bei der Filterentwicklung, Haus der Technik, Essen, 1996.
[5] Weber, O., "topsys–A New Concept for Intake Systems," SAE 98 "Merra."
[6] Weber, O., H. Paffrath, H. Beutnagel, and W. Cedzich, "Thermodynamische und akustische Auslegung von Ansaugsystemen für Fahrzeugmotoren unter Berücksichtigung fertigungstechnischer Belange," 19th International Vienna Engine Symposium, May 1998.
[7] Paffrath, H., K.-E. Hummel, and M. Alex, "Technology for Future Intake Air Systems," SAE, March 1999.
[8] Weber, O., R. Vaculik, R. Füsser, and F. Pricken, "Qualitativ hochwertige Akustik von Ansaugsystemen und Kunststoffen–ein Widerspruch?; High Quality Acoustics of Plastic Intake Systems—Vision or Contradiction?" 20th International Vienna Engine Symposium, May 1999.
[9] Pricken, F., "Active noise cancellation in future air intake systems," SAE, 2000.
[10] Pricken, F., "Sound Design in the Passenger Compartment with Active Noise Control in the Air Intake System," SAE, 2001.
[11] DIN ISO 362, Akustik, Messung des von beschleunigten Strassenfahrzeugen abgestrahlten Geräusches; Verfahren der Genauigkeitsklasse 2 (ISO 362 AMD 1-1985, ISO 382; 1981).

7.21 Sealing Systems

Many different types of seals and gaskets, and almost as many different materials, are found in internal combustion engines. One normally becomes aware of these inconspicuous engineering elements only when they fail. In such cases, however, functioning of the entire system is endangered. The great importance of seals and gaskets is clear right from the early stages of engine development. A lack of properly functioning seals makes it virtually impossible to undertake component testing.

Modern sealing systems are extremely reliable. Great development effort has been devoted to devising solutions that ensure long and dependable service life even under critical conditions such as aggressive media, high pressures, and extreme temperatures.

This section is intended to provide the reader with an overview of the various types of seals, their uses, and basic information on how they function.

7.21.1 Cylinder Head Sealing Systems

The head gasket is becoming more important in modern engines. In addition to sealing off the combustion chamber, the cooling system, and the oil passages, the head gasket also serves to transmit forces between the cylinder head and the engine block. Thus, it exerts considerable influence on force distribution within the entire assembly system and the associated deformations in elastic components.

More stringent requirements for fuel consumption and emissions have given rise to engine designs with optimized weights and, particularly in diesel engines, higher ignition pressures. The use of aluminum and the reduction of wall thicknesses in castings are reasons to anticipate further reductions in component stiffness. In order to further reduce cylinder warping, which is detrimental to exhaust gas composition, engineers are striving to reduce bolt forces. These efforts result in a considerably greater load on the head gasket in the form of dynamic fluctuations at the sealing gap. The combustion chamber seal must be able to ensure the minimum sealing force, permanently and at all operating states. This causes very high demands on the durability of the sealing system selected for use here.

7.21.1.1 Ferrolastic Elastomer Head Gaskets

The head gasket made of asbestos-free ferrolastic elastomers, Fig. 7-288, is the system most widely used after converting to materials containing no asbestos at the end of the 1980s. The structure consists of a notched metal substrate with elastomers rolled onto both sides.

The sealing effect is distributed over the entire surface area, and that requires high bolt forces. The disadvantages of this system are found in the relatively low elastic resilience. Great dynamic fluctuations in sealing gap width or changes in pressure due to thermal effects cannot be

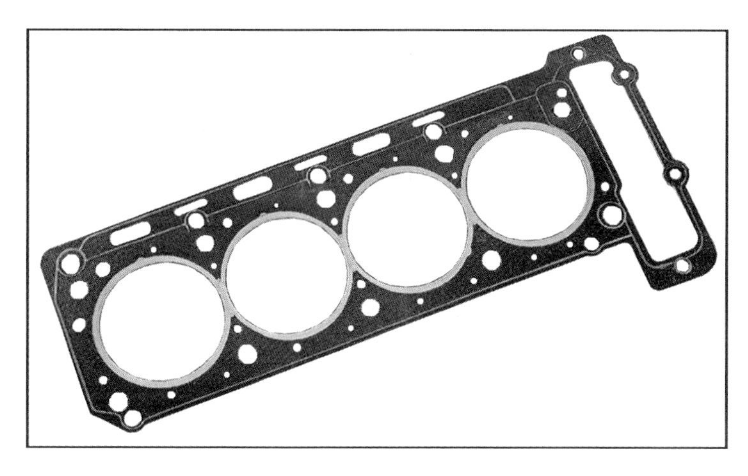

Fig. 7-288 Ferrolastic elastomer head gasket.

compensated and can be neutralized only in part by greater bolt forces. Engines with high thermal loads, narrow web widths, and wide oscillations in the sealing gap mark this system's limits, triggering the development of higher-performance systems.

7.21.1.2 Metal-Elastomer Head Gaskets

Metal-elastomer head gaskets, Fig. 7-289, are used today primarily in heavy-duty utility vehicle engines. The principle behind this design (Fig. 7-290) is distinguished by the separation of functions (separate sealing for the combustion chamber and the liquid circuits) and the system's great potential in each case. Not only are bead concepts with purely plastic properties used to seal the combustion chamber, but elastic systems as well. The passageways for liquids sealed with elastomer sealing lips exhibit great adaptability and elastic resilience. Selecting a suitable elastomer material ensures suitable aging resistance when exposed to fuel, coolant, and oil. Depending on the overall concept for the gasket, the elastomer lips may be injected onto the end of the sealing plate or on the surface. As an alternative, so-called inserts, i.e., metal substrates with a sealing lip vulcanized in place, may be used.

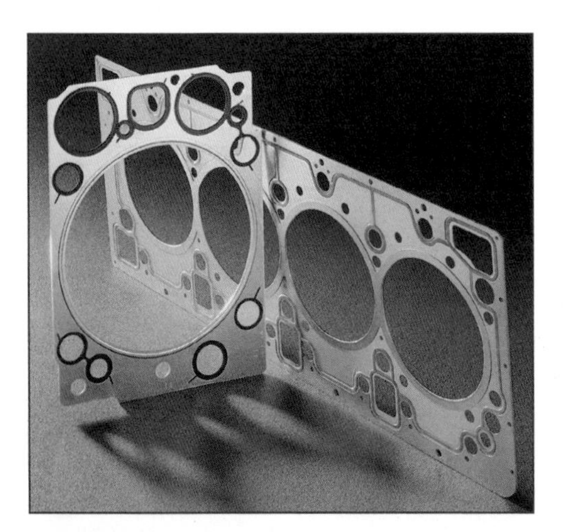

Fig. 7-289 Metal-elastomer head gasket.

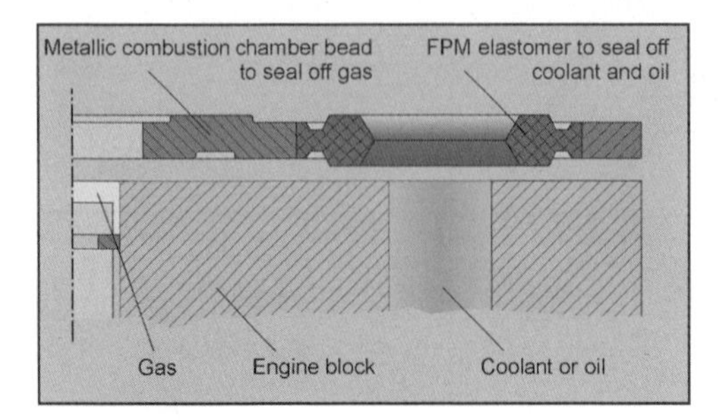

Fig. 7-290 Cross section of combustion chamber, through a metal-elastomer head gasket.

To avoid component warping and to achieve closely defined introduction of pressure into adjacent components, support elements may optionally be provided at the outer edge of the gasket.

Since the elastomer elements require only insignificant sealing forces in relationship to the bolt force, almost all the bolt force can be devoted to combustion chamber sealing and, if indicated, to supporting the components. In this way, the available bolt force is very efficiently utilized.

7.21.1.3 Metaloflex® Layered Metal Head Gaskets

Multilayer steel gaskets have been used as head gaskets (Fig. 7-291) in mass production since 1992. Particularly in modern diesel engines and in high-performance gasoline engines, extreme effort is required to devise a solution suitable for mass production when using the elastomer seals employed up to that time. The essential advantage of the layered metal head gaskets from the developer's viewpoint is that the gasket design can be matched precisely to the engine's technical requirements. As a result, cost-intensive and, above all, time-consuming iteration steps can be avoided. The metal head gasket is composed of one or more layers, depending on the application.

Function

The sealing function of the layered metal head gasket is essentially dependent on the beads in the spring steel layers. The deformation characteristics permit plastic adapta-

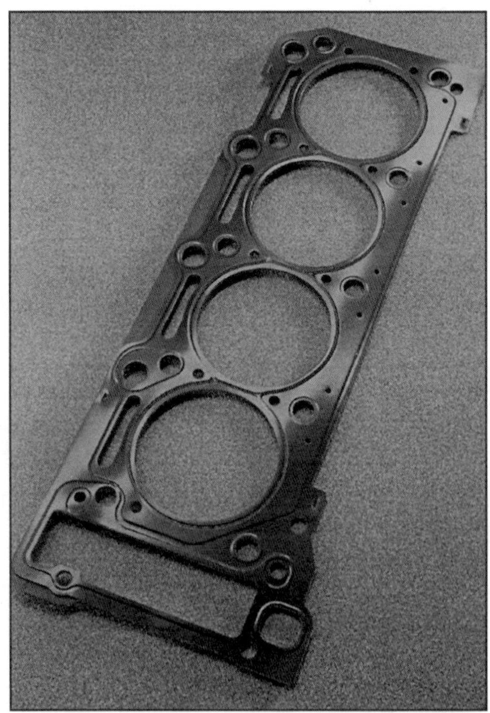

Fig. 7-291 Layered metal head gasket.

tion to component stiffness, on the one hand, and, on the other hand, great resilience to compensate for dynamic fluctuations in the sealing gap and for thermally induced component deformations. With the use of half-beads in the liquid sealing areas and full beads at the combustion chambers, the compression levels along a given line, required for sealing in each case, are achieved (Fig. 7-292).

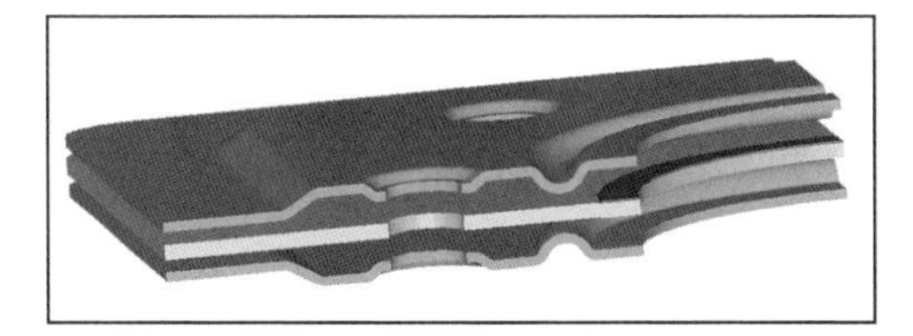

Fig. 7-292 A 3-D section through a layered metal head gasket.

The stopper induces an elastic preload in the components around the edge of the combustion chamber. In this way, the fluctuations in the sealing gap resulting from the gas forces are reduced, while at the same time unacceptable deformation of the full beads is prevented. Normal stopper heights lie within a range of from 100 to 150 μm. An intermediate layer may be inserted to achieve the required installation thickness or to accommodate differing thickness adjustments for diesel engines; this intermediate layer has no influence on the sealing function. At $3^1/_2$-layer and multilayer gaskets the stopper effect has to be split into the functional layers to protect the full beads. This means, for example, that in $3^1/_2$-layer gaskets the intermediate layer in the area near the stopper has to be cropped. In this way, the stopper is centered inside the head gasket. Without this distribution of the stopper effect to both functional layers, protection is not ensured for that full bead that does not lie on the stopper side.

Application Examples

Multifunction, Layered Design, Fig. 7-293

Excessive sealing gap fluctuations cause dynamic overloading of the beads; the full beads at the combustion chamber are especially endangered. Relaxation—a reduction in bead force and resilience—occurs and fissures may even appear in the beads. The functional layers in the Metaloflex® head gaskets that are provided with beads use their resilience to compensate for the sealing gap fluctuations occurring in the engine. With the use of multiple functional layers, the overall amplitude can be distributed to the individual layers and thus reduced to an acceptable level. The total resilience of the gasket rises with the number of functional layers used. In this way, it is possible to ensure function and durability even at low bolt forces and high peak pressures.

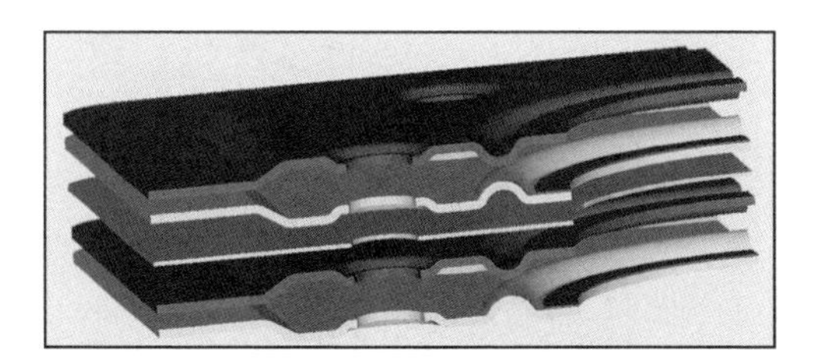

Fig. 7-293 A 3-D view of a head gasket in a multiple functional layer design.

Variable Stopper Thickness

Proper stopper design makes it possible to exert a closely defined influence on the sealing gap fluctuation. The gasket is normally between 0.10 and 0.15 mm thicker in the area around the stopper, the exact amount depending on engine stiffness. This causes an increase in pressure and elastic preload in the sealing system. Where the stiffness in adjacent engine components is not uniform, it may be necessary to graduate the thickness of the stopper. This allows more uniform distribution of pressure on the stopper and thus more uniform preload at the head gasket and engine (or cylinder) block. In this way weak points in components, characterized by low stiffness values, can be preloaded. The available bolt force is directed exactly to the desired areas and thus utilized ideally.

Two embodiments are used in principle: the plastic stopper (Fig. 7-294) and the graduated-height stopper (Fig. 7-295).

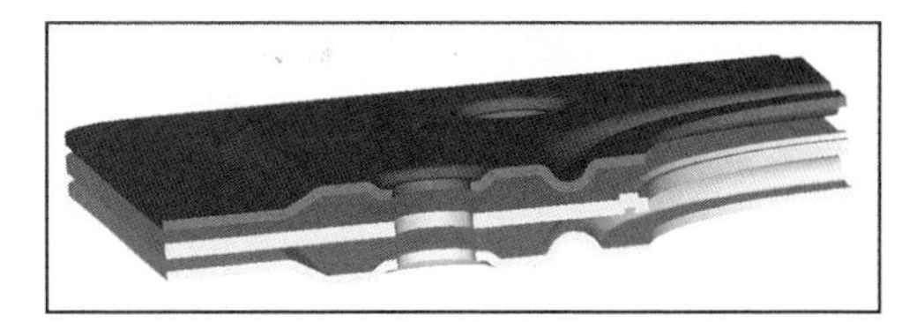

Fig. 7-294 A 3-D view of a head gasket with plastic stopper.

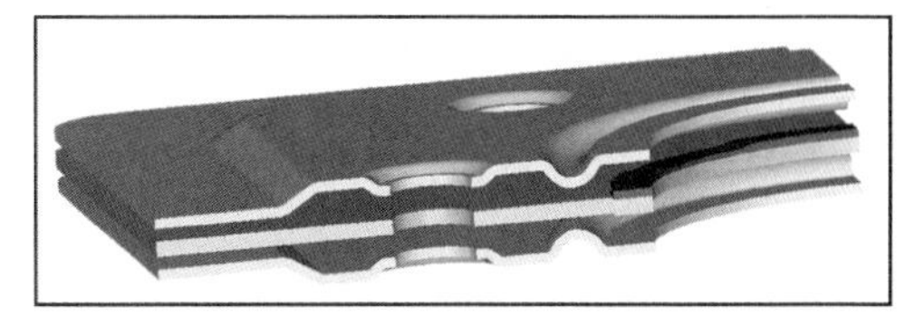

Fig. 7-295 A 3-D view of a head gasket with graduated-height stopper.

In the plastic stopper the vertical profile is achieved with a plastic adaptation in the engine, while in the graduated-height stopper the profiling is created during gasket manufacture. Using the graduated-height stopper makes it possible, in contrast to the plastic stopper, to achieve higher profiling even at low installation thicknesses.

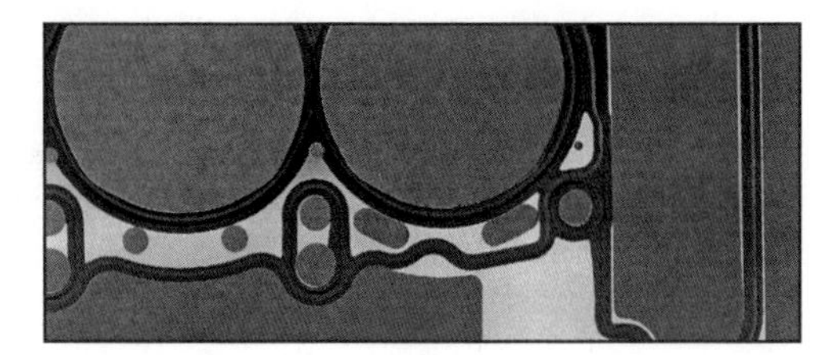

Fig. 7-296 Layered metal head gasket with partial coating.

Partial Elastomer Coating, Fig. 7-296

With partial coating, only the head gasket surface areas that are relevant to sealing are coated. This makes it possible to omit the coating on the sealing surfaces that extend into the coolant or the motor oil; thus, there is no coating there that could peel off under critical conditions.

Further advantages of this process are that, with the special application procedure, both the thickness of the coating layer and the coating medium can be selected to suit the application. The coating requirements in the combustion chamber and liquid areas, which differ in part, can thus be properly met. For coolant and oil sealing, a thicker layer and softer elastomer are beneficial if, for example, the mating surface is rough or porous. At the same time, thinner layers are necessary to contain the ignition pressure at the combustion chamber. These conflicting goals can be resolved by selective coating.

Dual Stopper Design for Use with a Cylinder Liner

A modified gasket design is required in many cases where a separate cylinder liner is used. To avoid plastic deformations and to keep the liners from being shifted downward, the necessary sealing and preload forces have to be introduced into the gasket system in a defined manner.

Force application at the liner is closely defined with the use of a so-called double stopper, Fig. 7-297. In this configuration one stopper is formed by a folded bead around the edge of the combustion chamber, while a second stopper is formed behind that bead by overlapping two sheets of metal. The two layers are joined in this overlapping area with a laser welded seam. To achieve ideal response during operations, the stopper force acting on the liner must cause no plastic depression of the liner.

By employing sheet metal of varying thickness, the distribution of the pressure to the two stoppers can be regulated individually. Thus, for example, the stopper at the outside may be thicker by 20 μm; consequently, the larger share of the preload is directed not to the liner but rather to the outside area of the cylinder tube. This stratagem ensures the required preload on the components while at the same time avoiding any displacement of the liner.

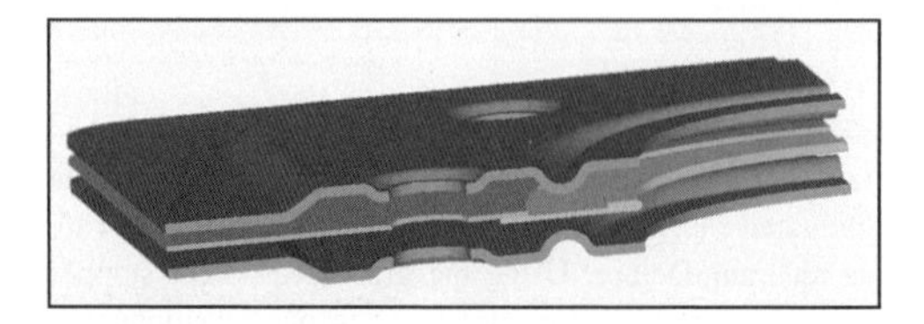

Fig. 7-297 A 3-D view of a head gasket with double stopper.

Stopperless Design

In gasoline engines, and particularly where aluminum engine blocks are used, it is possible under certain circumstances to do without the stopper.

In this way the elastic deformations of components caused by the head gasket are reduced dramatically. In addition to reducing cylinder deformation, deformations in the area around the valve seats can also be significantly reduced.

The implementation of this concept does, however, require that the bead geometries be matched exactly to the details of the mating components. In gaskets with stoppers, the deformation of the full beads are determined by the thickness of the stopper. Protecting the beads in this way creates ideal conditions with respect to durability and resilience.

Without stoppers, Fig. 7-298, the deformation of the beads depends largely on the stiffness of the components. This means that, depending on the stiffness of the cylinder head and the engine block, the beads are deformed to a greater or lesser extent. Attaining sufficient sealing pressures while achieving ideal durability requires individual adaptation to the conditions prevailing in the engine.

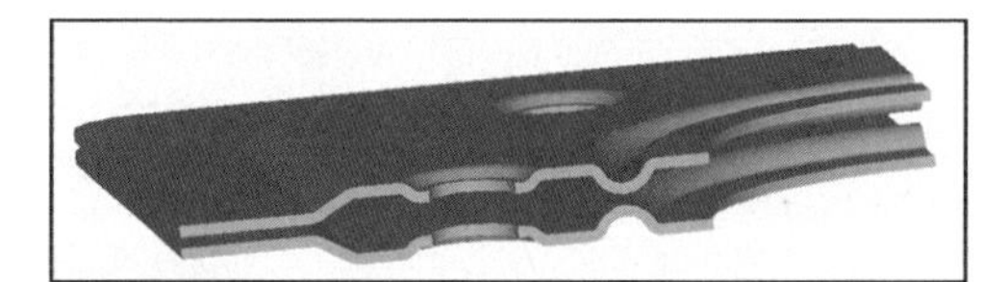

Fig. 7-298 A 3-D view of a head gasket without a stopper.

Integrated Supplementary Functions

Integrating a high-sensitivity sensor system directly into the head gasket provides for even more dependable monitoring of processes in the engine: integrated sealing gap sensors, Fig. 7-299.

Fig. 7-299 Head gasket with integrated sealing gap sensor.

The sensor system uses the enormous pressures created by combustion inside the cylinder. These pressures cause relative movement between the engine block and the cylinder head. The sensor registers this movement and is thus able to detect at an early date irregularities in the engine, such as misfiring or other ignition problems.

The measurement of coolant and component temperatures inside the engine is becoming even more significant since, in conjunction with cooling regulated according to the engine map, for instance, the values registered at the measurement points that were previously used are hardly representative. Particularly in operating ranges where there is little coolant flow or none at all, the temperature of necessity has to be measured at critical points in the engine.

7.21.1.4 Prospects for the Future

The requirements of future engine designs for the head gasket are characterized essentially by higher peak pressures, greater thermal loading, reduced component stiffness, and new materials.

With its modular construction, the layered metal head gasket offers every option for individual adaptation to the specific conditions prevailing in the engine. The engineering freedom offered by this system permits influencing component mounting and strains and the distribution of pressures in the engine. In this way, the available bolt force can be utilized efficiently while at the same time minimizing component deformations. The advanced Metaloflex® layered metal head gasket will continue to represent a reliable, durable, and economical sealing concept.

The metal-elastomer technology will also be the predominant head gasket design in the heavy utility vehicles sector. Separating the combustion chamber and liquid sealing functions enables ideal adaptation of the gasket, particularly in engines with "wet" cylinder liners.

7.21.2 Special Seals

7.21.2.1 Functional Description of the Flat Seal

Flat seals are highly effective, cost-favorable seals both for a number of liquid media and for gases. A broad range of pressure and temperature loads can be managed. The requirements on the flange surfaces at the components being sealed are low; surfaces machined with the milling head are sufficient. To achieve positive sealing for static, flat seals, sufficient surface pressure is guaranteed at all operating states. Influencing parameters such as operating media, fluctuations in temperature and operating pressure, engineering elements (such as bolts and sealing surfaces), the location of the gasket within the assembly, and the seal's long-term influence on the sealing assembly have to be taken into account during engineering.

Thus, the following requirements apply for the sealing element:

- Adaptation to component surfaces (microstructure—roughness/macrostructure—not plane)
- Pressure resistance (setting behavior) under the influence of heat and/or operating media
- Tightness across the entire surface of the seal
- Cross-sectional tightness in the seal material
- Mechanical stability (tensile strength)
- Elastic resilience properties
- Temperature resistance

Consequently, the ideal seal is an elastic rubberized metal with great strength and resistance to media and temperatures.

7.21.2.2 Elastomer Seals

Elastomer seals (Fig. 7-300) are employed in a broad spectrum of applications. They are made up of a composite material comprising fibers, fillers, and binders, Fig. 7-301. Since the end of the 1980s, rubber-asbestos elastomer seals have been replaced almost completely by asbestos-free qualities. In high-quality elastomer seals aramid fibers have largely been substituted for asbestos fibers. This material has superb mechanical and thermal properties. Cellulose and mineral fibers are used for economical gasket materials used in less critical areas.

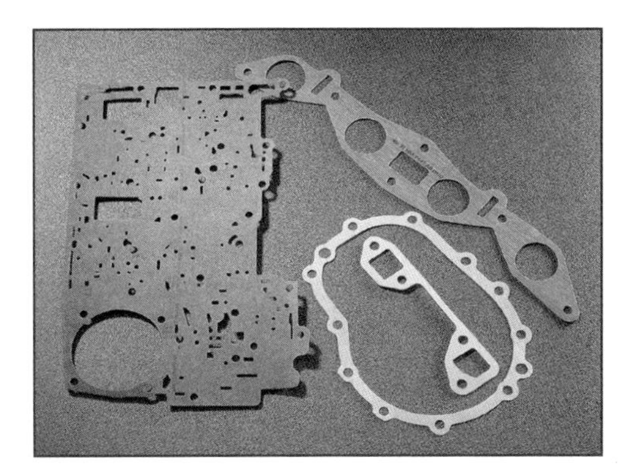

Fig. 7-300 Elastomer seals.

Fig. 7-301 Elastomer seals—composite structure.

The multitude of materials qualities available, such as EWP® sealing materials, makes it possible to select a suitable sealing material for almost every application. Elastomer sealing material is available in a range of thicknesses from 0.20 mm to over 2.5 mm. The choice of material thickness makes it possible to "tune" a gasket for adaptation capacity, mechanical stability, and setting properties. The performance capacities of the elastomer seal can be further improved by applying additional elastomer layers along a line. In these areas the prescribed preload force on the surface (low sealing pressure) is reduced to narrow linelike areas (high sealing pressure).

Elastomer seals are cut on modern CNC water jet machines. Gaskets are cut without conventional tools when this technology is employed.

The limits for the use of asbestos-free elastomer seals are found in areas that are subjected to severe thermal loads.

7.21.2.3 Metal-Elastomer Seals

Metal-elastomer seals, Fig. 7-302, differ from the elastomer seals described in the previous section in that they have a metal insert at the center of the material (Fig. 7-303). They are used primarily in automotive applications and are found in the coolant, oil, fuel, and exhaust areas.

The metal insert (substrate plate) is normally sheet steel that is toothed, perforated, or glued to a smooth surface.

Fig. 7-302 Metal-elastomer seals.

The metal insert provides a number of benefits:

- Great tensile strength
- Mechanical ruggedness
- Good dimensional stability
- Benefits in terms of process technology (coil manufacture)
- Cost reductions by lowering the fiber content
- Differing sealing materials on the substrate

The substrate lends the required tensile strength, and thus other specific properties of the sealing materials can be carefully optimized, as is shown in Fig. 7-304 below.

The specific properties of the materials listed in Fig. 7-304 are determined primarily by the composition of the sealing surface. Figure 7-305 indicates the most important adjustment parameters in the selection of the sealing layer.

The compound used for the sealing layers is determined most strongly by the thermal requirements. In a temperature range of up to 150°C, they are comparable with composite materials (Section 7.21.2.2). In exhaust system seals, graphite and mica materials, with their great resistance to high temperatures, are used. As was described in the previous section on elastomer seals, the performance capacity of the metal-elastomer seals can be further boosted by an additional elastomer coating applied along a line. The seal quality over broader surface areas, in particular, can be improved significantly in this way.

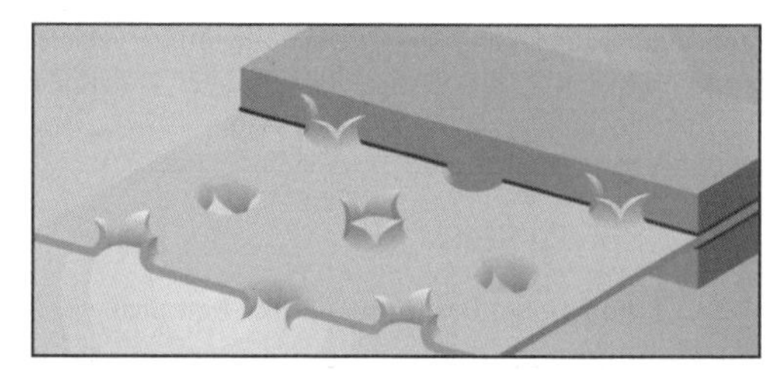

Fig. 7-303 Structure of the metal-elastomer seal.

Materials	Optimized properties	Application example
FW 522	Pressure resistance, cross-sectional tightness, resistance to media	Head gasket
FW 715	Adaptability, cross-sectional tightness	Oil pan
FW 520	Temperature resistance up to 450°C	Exhaust manifold
FW 501	Adaptability, temperature resistance up to 500°C	Exhaust gas return
FW 610	Temperature resistance up to 800°C	Turbocharger

Fig. 7-304 Survey of metal and metal-elastomer materials.

	Pressure resistance	Adaptability	Internal tightness	Resilience	Temperature resistance
Filler content	↑	↓	→	↓	↑
Fiber content	↓	↑	↓	↑	→
Share of elastomer	↓	↓	↑	↑	↓
Impregnating agent content	↓	→	↑	↓	↓
Sealing	↑	↓	↑	↓	→

Fig. 7-305 Materials parameters and their influence on functioning.

7.21.2.4 Special Metaloseal® Gaskets

The term Metaloseal® is derived from the words "metal sealing." The fundamental structure for metal seals, Fig. 7-306, is based on a metal substrate that is coated with an elastomer, usually on both sides. One of the major advantages is found in the fact that any of a variety of metals can be combined with varying elastomer compounds to suit the application at hand. Thanks to the beads that are formed in addition, the substrate material's properties can be matched perfectly to the sealing system, Fig. 7-307. As was already described in Section 7.21.2.1, the demands made on the sealing element can be satisfied only by metal gaskets that are coated and mechanically modified.

Substrate Materials

The choice of substrate materials has a direct influence on sealing properties. Ideal adaptability of the seal to the flange surfaces (macrosealing) can be achieved with two parameters: substrate material properties and bead geometry. The following table (Fig. 7-308) provides a survey of the various substrate materials available.

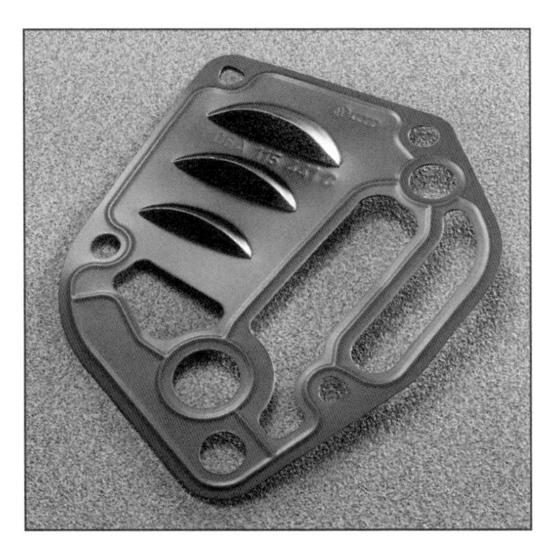

Fig. 7-306 Ribbed oil filter head gasket.

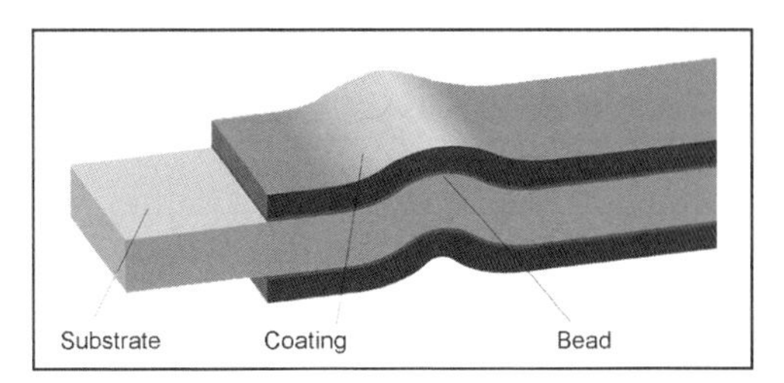

Fig. 7-307 Structure of metallic gaskets.

Standard material thicknesses lie between 0.20 and 0.30 mm. In special situations, thicker or multilayer gaskets can be used. This lets us achieve the best possible macrosealing properties for virtually every application, through the selection of suitable materials and bead geometries.

Coating

The selection of the elastomer is oriented primarily on the media to be sealed off and the prevailing operating temperature. One of the most important tasks in sealing is compensating for surface imperfections. Thus, the medium being sealed is kept from leaking across the surface area. The thickness at which each coating layer is applied can vary between 5 and 100 μm (on either side), depending on the particular situation. Listed in Fig. 7-309 are a few application examples for the various elastomer materials.

Functioning of a Metal Gasket

In the past it was necessary in some cases when using conventional elastomers to resort to engineering tricks in order to achieve a reliable seal. Thus, elastomer gaskets required exactly defined bolt torque in order to achieve sufficient surface pressures while at the same time avoiding excess pressure on the material, which would unavoidably damage the elastomer and result in leaks. In addition,

Substrate materials	Conditions for use
Cold-rolled strip	Standard design
Spring steel	Dynamic changes in sealing gap width, high pressures
Stainless steel	Aggressive media, corrosion protection, increased protection against frictional wear
Temperature-resistant steels	In exhaust systems or at temperatures between 400°C and 1050°C
Aluminum	To avoid contact corrosion when using magnesium, aluminum, or gray cast housings

Fig. 7-308 Metaloseal® substrate materials.

Elastomer materials	Conditions for use
NBR	Coolant, oil, air; fuel to a limited extent
FPM	Fuel
EPDM	Brake fluid, hydraulic fluid
Temperature-resistant coating	Exhaust system at flange temperatures <1000°C
Graphite coating	A slip-promoting coating with the ability to compensate for high relative motions between components

Fig. 7-309 Conditions for using the various elastomer materials.

there is always a conflict of goals among the various sealing properties when selecting the elastomer material (see Fig. 7-305, Section 7.21.2.3). This is where the advantages of metallic seals become apparent. A bead pressed into the substrate reduces the surface pressure to a line-shaped pressure. Thus, at the same bolt forces, higher surface pressure values can be achieved or, conversely, the same surface pressures can be achieved at lower bolt force. With the use of a metallic substrate, all of the physical properties of a metal can be exploited. In addition, one creates a further magnitude that can be adjusted as required: the bead force.

Bead force is influenced both by a certain ratio of the bead's height to width and by the shape of the bead itself—half or full bead—and can be modified individually to suit each application point. When the component is first tightened down, the elastomer coating is pressed into the surface by the force at the bead and closes any imperfections present there. In addition, the degree to which the seal adapts to the flatness of the component is determined by the bead. The bead functions, in the classical sense, like a spring that develops the required sealing force in response to deformation. Described in Fig. 7-310 is the interrelationship between the component's demands on

Requirements for the seal	Functional element
Ability to adapt to component roughness	Adaptable elastomer coating
Ability to adapt to component flatness	Beads
Cross section through the gasket	Nonporous elastomer coating
Pressure resistance (setting properties)	Metallic substrate, thin elastomer coating
Mechanical stability	Metallic substrate
Elastic resilience properties	Substrate (e.g., spring steel), bead
Temperature resistance	Substrate and coating material

Fig. 7-310 Each of the various functional elements in the Metaloseal® gasket responds to a specific requirement.

the seal and the capabilities that the various functional elements have to exert an influence.

Conditions for Use

Because of the multitude of options for combining metallic substrate materials and various elastomer compounds, almost all of the application points in the engine can be covered. Naturally every sealing system has to be analyzed, and the corresponding sealing properties, such as material structure and bead geometry, have to be defined. Figure 7-311 provides a survey of the wide range of applications for metallic gaskets.

7.21.2.5 Prospects for the Future

The ongoing increases in requirements for sealing systems give rise to new, innovative products. With the use of metallic substrates, the gaskets can assume additional functions. Oil splash plates or sensors for more efficient motor management are built into the gasket. Preassembly aids such as retainer clips or centering elements are created using bending processes.

7.21.3 Elastomer Sealing Systems

Greater performance at less weight, reduced fuel consumption, and lower emission levels—these central demands of the power plant engineer mean greater demands on sealing systems. Thus, engine components and attached assemblies are more frequently manufactured from plastic for weight and functional reasons. Reduced component stiffness (lower Young's modulus) in comparison to the aluminum and magnesium previously used are the consequences. When parts are clamped, greater deformation is encountered, and the sealing system has to compensate for this.

The elastomer-based sealing systems are superb in satisfying these exacting requirements. On the one hand, the sealing pressure required by elastomers is very low, and, on the other hand, their superior elastic properties enable tolerance compensation over a broad range. Because of elastomer materials' ability to resist extreme temperatures, these are used exclusively for containing liquids and gases. A metallic structure is used in elastomer gaskets to seal off the combustion chamber.

Suitable elastomers are selected to suit the medium to be sealed, the prevailing temperatures, and the requirements profile.

Figure 7-312 gives an overview of the elastomer compounds available and typical applications.

7.21.3.1 Elastomer Seals

Elastomer seals, Fig. 7-313, have no substrate. To prevent overloading the elastomer profile, these seals are, for instance, installed in a groove in the component. These components are always designed so as to eliminate any

Criteria	Application range
Temperature	−40°C to 1050°C
Pressure	Up to 350 bar
Media	Coolant, oil, exhaust gases, brake fluid, hydraulic fluid, air, fuel, biodiesel
Surface parameters	
Roughness	$R_{max} \leq 25\ \mu m$
Deviation from plane	≤0.30 mm

Fig. 7-311 Application range for metal gaskets.

Elastomer material							
Abbreviation (ISO 1629)	Chemical name	Applications at engine			Thermal application range		Application examples
		Fuel	Coolant	Oil			
FPM	Fluoro rubber	+	+	+	to −20	+230 °C	Head gasket, intake section
MVQ	Silicone rubber	–	o	o	to −50	+200 °C	Head gasket, special applications
MFQ	Fluorosilicone rubber	–	o	+	to −70	+180 °C	Head gasket, special applications
ACM	Polyacrylate rubber	–	–	+	to −30	+150 °C	Oil pan, valve cover
AEM	Ethylene-acrylate rubber	–	–	+	to −35	+160 °C	Oil pan, valve cover
EPDM	Ethylene-acrylate diene rubber	–	+	–	to −50	+130 °C	Water pump
ECO	Epichlorhydrin rubber	+	–	+	to −40	+120 °C	Special applications in fuel system
HNBR	Hydrated nitrile rubber	o	o	+	to −30	+150 °C	Special applications

+ Well suited / o Suitable / - Unsuitable

Fig. 7-312 Elastomer materials.

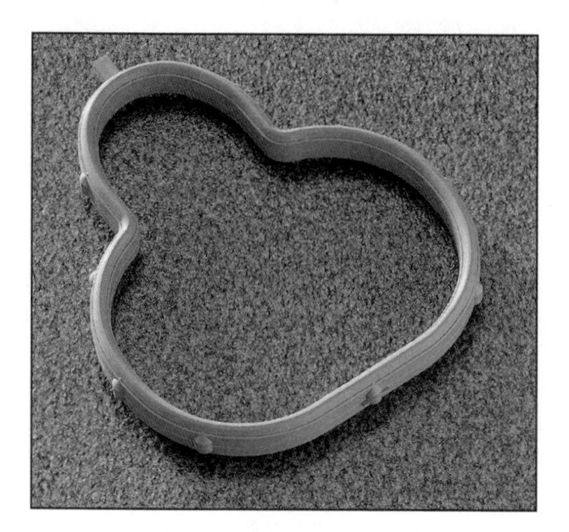

Fig. 7-313 Intake manifold gasket.

external deformation. The height-to-width ratio is characteristic for the design of this seal. The cross section is considerably thicker (higher) along the direction of the compression forces than it is wide. At compression of 20% to 30%, this gives a very broad working range for the seal, and also enables sealing plastic components, which are subject to severe deformation. This type of gasket is used, in particular, in combination with valve covers, intake manifolds, or water flanges made of plastic.

When sealing camshaft bearings and other three-dimensional passages in components, the elastomer gasket is the only option for sure management of the sealing point.

With special cross sections calculated using the finite element method (FEM), the gasket's profile is matched to the specific properties of the component being sealed. As a result of these calculations, a rectangular cross section is only seldom employed.

The T section is the preferred sealing profile for acoustic purposes. In combination with specially designed decoupling elements for the bolts, this design is used for valve cover seals that integrate acoustic decoupling. Since the components being sealed are pressed together by elastomer elements (see Fig. 7-314), this system can no longer be calculated with the engineering methods used in the past. To make these systems more functionally reliable, an analysis of the complete mounting system—comprising the seal, the decoupling element, bolt, and bushing—by FE calculations is unavoidable (see Figs. 7-320 and 7-322 below in Section 7.21.4.1).

Requirements made of acoustically decoupled systems include

- Decoupling structural sound
- Positive bolting of components
- Sealing
- Preassembly of individual parts

The interplay of finite-element calculations, laboratory simulation, and material development work is the basis for tailor-made, acoustically decoupled sealing systems.

7.21.3.2 Metal-Elastomer Gaskets

Since some components, for geometric or functional reasons, cannot use pure elastomer seals (they require a groove in the component), the metal-elastomer seal was developed, Fig. 7-315. In this type of seal the elastomer is vulcanized directly to an aluminum or steel substrate. The thickness of the elastomer is coordinated with that of the substrate but is always be considerably thinner than in solid elastomer seals. Here, like the pure elastomer seals, the elastomer is not installed in the path of primary force flow. No groove is required in the component since

Fig. 7-315 Metal-elastomer engine block gasket.

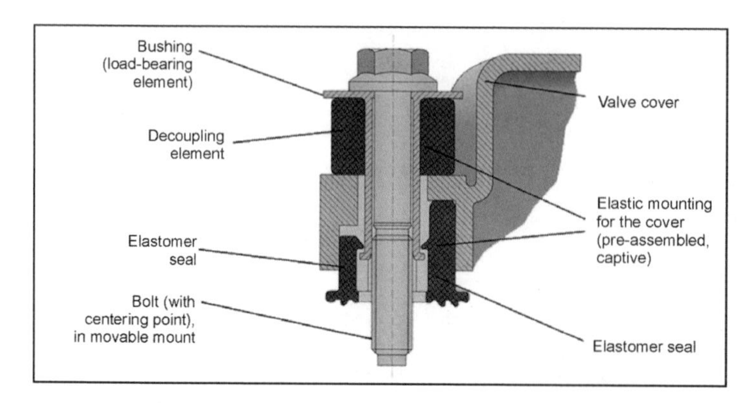

Fig. 7-314 Example of a decoupled cylinder head shroud system.

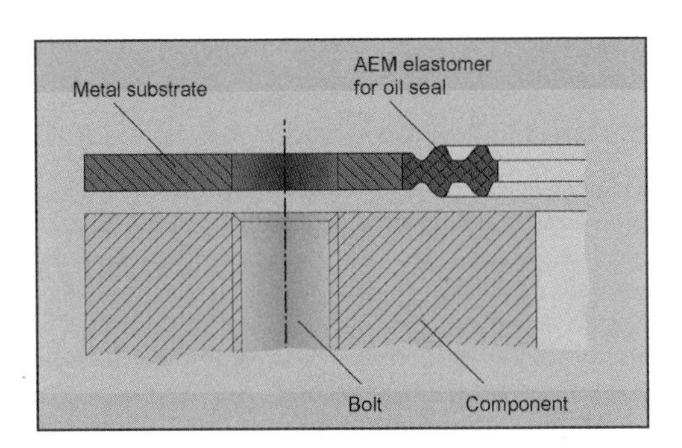

Fig. 7-316 Section through a metal-elastomer gasket.

the substrate, made of aluminum or steel, is at the main force transfer point, Fig. 7-316.

Design freedom in engine development is increased considerably with this concept by integrating supplementary functions into the substrate. In addition, the system is distinguished by great functional reliability and economy. Functions that are normally integrated in practice are

- Calibrating fluid flows
- Exhaust gas return
- Assembly aids
- Preassembly using clamps
- Cable grommets

The use of two-component injection machines makes it possible to vulcanize two different elastomers on a single substrate. The advantage is that the most suitable elastomer can be used for each of the media to be contained. Indispensable to this process are reliable coupling agents to ensure a good bond to the metal.

Metal-elastomer head gaskets made of metal substrates to which elastomer profiles are vulcanized are described in Section 7.21.1.2. This type of gasket is used in commercial vehicles and in the large engines found in ships and locomotives.

7.21.3.3 Modules

Important to achieving a properly functioning sealing system is not to see the system in isolation but rather to observe the complex interaction of all the individual systems involved. Consequently, seal and gasket manufacturers are now also developing other components and are offering them, together with the seals, as pre-assembled, multifunctional systems. These modules, ready for immediate installation, replace the previous individual parts to a greater extent. Here, every conceivable combination of sealing system and component (made of aluminum, magnesium, steel, or plastic) is possible. Lightweight designs are indispensable for reducing fuel consumption at rising engine performance. Plastics offer decisive advantages here and replace to a greater extent the materials used for engine components in the past. The know-how and the system competence arising from sealing technology, and from elastomer processing in particular, form the basis for developing innovative plastic modules. These are employed especially in the following areas:

- Cylinder head shrouds, Fig. 7-317
- Engine compartment capsules
- Oil separators
- Coolant flanges
- Intake manifolds

Depending on the requirements imposed on the plastic components, PA 6 is used for parts where appearance is important and PA 6.6 for components that have to introduce or transfer forces. To achieve the required strength and processing properties, one blends glass fibers and in some cases mineral fillers into the basic resin. Elastomer sealing systems are employed in modules with integral sealing functions since these can be ideally tuned to the medium to be contained and the requirements for component stiffness.

Numerous functions can be integrated into modules, both very efficiently and economically, due to plastics'

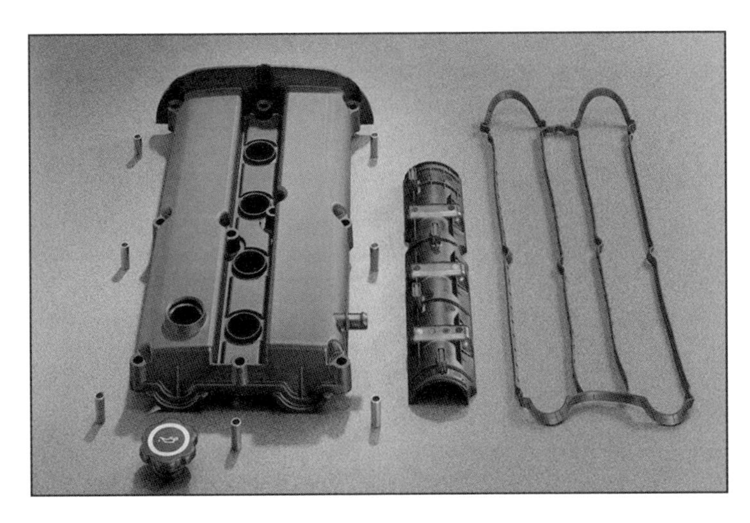

Fig. 7-317 Cylinder head shroud module with integrated gasket and oil separation.

processing properties. Major advantages are also found—as previously mentioned—in the amount of weight reduction and the manufacturing technology since plastic components make it possible to eliminate completely postinjection work such as deburring, tapping threads, or finishing surfaces.

Examples for the multifunctionality of modules include

- Acoustic decoupling of the component
- Integrating the blowby gas exit from the crankcase
- Integrating oil separator systems into a cylinder head shroud
- Integrating valves to regulate crankcase pressure
- Integrating cable passages from the cylinder head
- Providing a preassembled, complete system

In order to ensure reliable functioning of the module over the engine's entire service life, exhaustive tests of functions and geometries are conducted during the development phase. In addition, simulation tests are worked out that allow for depicting the loading conditions that occur during vehicle operation, making it possible to reduce testing times. When developing these tests, the experience and results drawn from practice are always taken into account.

The interplay of FE calculations, simulations, and engine testing makes it possible to prepare for mass production, in the shortest period of time, plastic modules that satisfy all the requirements for loading capacity and service life.

7.21.4 Development Methods

Engine tests continue to be a major factor in gasket testing. These tests, carried out on an engine on a test bed and driven by an external power source, are expensive and time-consuming, however. Since the trend is toward shorter development cycles, calculations for the sealing systems and laboratory testing under conditions that closely replicate those in the engine are moving further into the foreground. This is intended to enable fundamental assessments of the functioning of the gasket design even before actual testing in an engine, thus reducing to an absolute minimum the number of costly engine tests required. Preliminary examinations of seals without real-world engine components provide in-depth insights into the functional capacities of the product.

The finite element method is used as a calculation tool. This term describes the mathematical algorithm used to translate for computer processing a physical phenomenon at a section of the component being analyzed. A finite element model is the depiction of a geometry by a sufficient number of discrete elements.

7.21.4.1 Finite Element Analysis

The assignment for the person conducting the analysis is to identify the phenomena required to describe the problem at hand and to enter them into the calculation software. FE calculations are used to optimize the components both in the engineering phase and in the subsequent test phase. This preliminary selection lets us reduce the number of prototypes required.

Many of the engineer's problems can be converted directly into an FE calculation model right in the CAD program and, provided with the appropriate material behavior data and operating conditions, can be forwarded to an FE analysis program for calculation. The basis for this approach is linear calculation, using small component deformations, elastic material laws, and unequivocal mounting and loading. A further special field application is found for component calculations whenever one of the fundamentals for the linear approach is violated. Nonlinearity of a calculation problem (see Fig. 7-318) arises as a rule with major deformations of a component under load wherein, for example, the length of the lever arm used for chucking is shortened and a smaller flexural moment is created than what is defined in the basic dimension. If there are also path limitations for component deformation, then these are described as nonlinear contact conditions. The behavior of most technical materials is also linear only in a very limited range; there they adhere to Hooke's Law, which links tension and elongation, expressed as "Young's modulus." Optimization strategies lead to weight reduction or better utilization of the material in the spirit of a uniform strain level push to the boundaries of this range. If one departs from this linear range, then plastic deformations typically appear at metals, creep elongation at plastics, or stress-induced relaxation processes. Nonlinear responses in the tension-elongation function are always found in rubber materials.

There time factors—i.e., how quickly the load is applied and the effective time—play an essential part in the deformation response of a given body.

Product Calculations

Preliminary calculation and optimization of component properties require both detailed knowledge of material responses and a good understanding of the manufacturing path followed from the semifinished product to the part ready for shipment. At a full bead in a layered metal

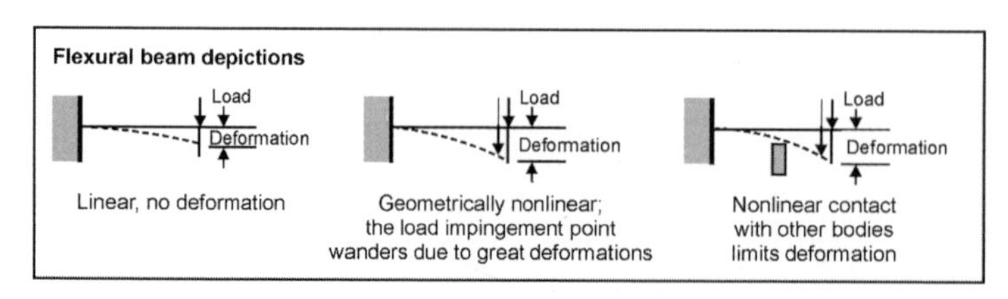

Fig. 7-318 Flexural beam: linear, nonlinear, and with contact.

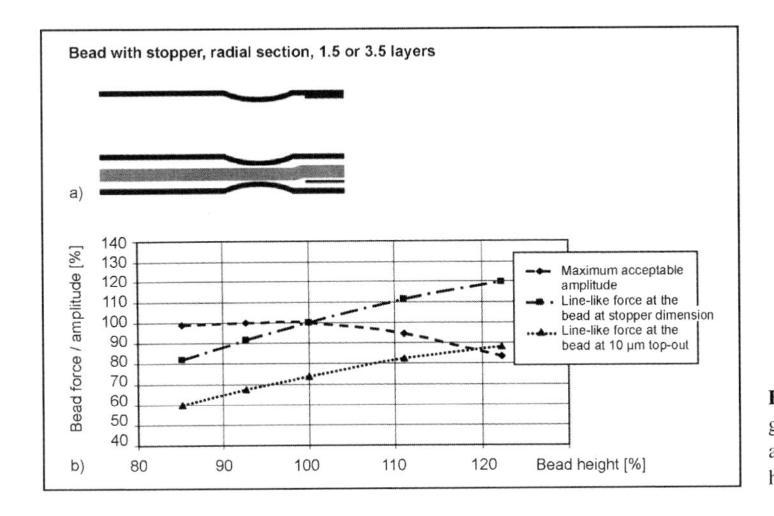

Fig. 7-319 (a) and **(b)** Sealing gap amplitude and line force along a bead as a function of bead height.

gasket (see Fig. 7-319, above) several reforming steps are carried out prior to final assembly in the engine. All the steps are "memorized" by structural changes in the metal and determine the bead's properties, spring characteristic, and tolerable sealing gap amplitude. With suitable tool dimensions the spring element can be designed for constant width at high force with appropriately smaller permissible sealing gap amplitude or for great sealing gap amplitude at lower force (see Fig. 7-319, below). The tuning required for the bead depends on the stiffness of the engine components and the ignition force.

Elastomer sections are frequently used at the engine to seal covers and shrouds, intake manifolds, and caps. They are characterized by great adaptability at the sealing surfaces and, at the same time, low preload force. A T section (see Fig. 7-320), installed between the valve cover and the cylinder head, is used to contain the oil splashed by the valve train. The vertical compression of the section generates the sealing pressure at the base of the groove and at the dual sealing lip toward the cylinder head. The section is designed for acoustically decoupled systems and has two blocks at the side that prevent direct contact between the cover and the head. The tension-induced relaxation of the elastomer material reduces the sealing force of the compressed profile over time; this has to be taken into account during design work.

Calculating the Component System

The cylinder head gasket forms the link joining the engine block and the cylinder head and, working in conjunction with the head bolts, forms the sealing system. To analyze the sealing system one requires—in addition to the geometric descriptions of the component in the form of an FE model, the material properties, and the sealing characteristics—information on the temperature distribution in the components and the ignition pressure in the combustion chamber. An engine runs under a wide range of load conditions and always has to be tight as regards gases and liquids. Extreme operational situations for the cylinder head gasket appear at full throttle with maximum coolant temperature and at cold start. Thanks to the bolt preload, the gasket is compressed to the height of the stoppers at the combustion chamber and, in other areas, locally to the thickness of the metal. The stopper acts like a wedge at the combustion chamber and places the components under elastic preload. The pressure on the stopper at the edge of the combustion chamber has to be greater than zero in order to ensure positive sealing in all operating states. In Fig. 7-321, one sees a raised area on the exhaust side when subjected to ignition pressure; it has to be corrected by adjusting stopper height in order to protect the combustion chamber bead against high sealing gap amplitudes. When the pressures at the stopper are too

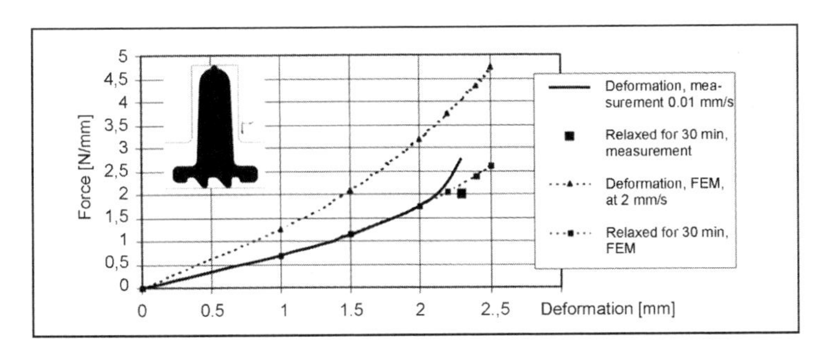

Fig. 7-320 Section through a T section in a groove. Calculating the force-deformation curve.

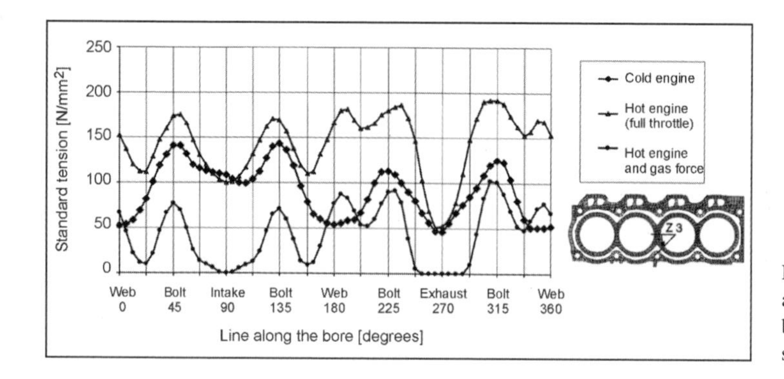

Fig. 7-321 Force distribution around the edge of the combustion chamber using a rigid stopper.

great, material overloads can occur at aluminum components, for example, causing damage to the component. The high temperature of the components at the combustion chamber further limits load-handling capacities.

Acoustic decoupling of a component interrupts mechanical transmission paths by elastic mounting, between elastomer elements, Fig. 7-322.

Impinging on a valve cover are, on the one hand, the sealing forces between the cylinder head and the valve cover, and, on the other hand, the forces at the bearing point, i.e., the decoupling element. The decoupling system (see Fig. 7-314, Section 7.21.3.1) comprising the cover, gasket, and several bearing points is preloaded with bolts and spacer bushings. If the deformation characteristics for the gasket and the decoupling element are known, then one can determine the working point at a given preload. Since all the components exhibit certain manufacturing tolerances, the actual preload in a system deviates from the design value. Calculations for the sealing profile ascertain the smallest permissible deformation at sure sealing; this is then specified as the minimum sealing pressure. In this way, the lowest seal compression force required for operations can be ascertained. The system's maximum preload force is limited by the load-bearing capacities of the decoupling elements; tolerable stress levels in the elastomer may not be exceeded. Within these limits the system is operationally reliable and can be fixed for a working range by tuning the preload and tolerance situation. The objective is to work with the lowest possible forces and thus to minimize deformations to the components.

7.21.4.2 Simulations in the Laboratory—Testing Functions and Service Life

Commonly used laboratory testing procedures employ servohydraulic testing equipment to perform hydraulic combustion pressure simulation to test head gaskets, shakers, and temperature controlled chambers for assembly tests. Hot gas generators are used to test items associated with the exhaust system.

Servohydraulic Testing Machinery

Servohydraulic testing equipment is employed for thermal quasistatic and dynamic testing. Quasistatic tests, which can also be conducted using electromechanical testing equipment, provide insights into the compression and resilience properties of seals and gaskets. Thermal quasistatic tests are used to examine the durability and creep characteristics of sealing materials when subjected to pressure and temperature.

Dynamic tests used to preselect and examine the seal design are of significance particularly for layered metal

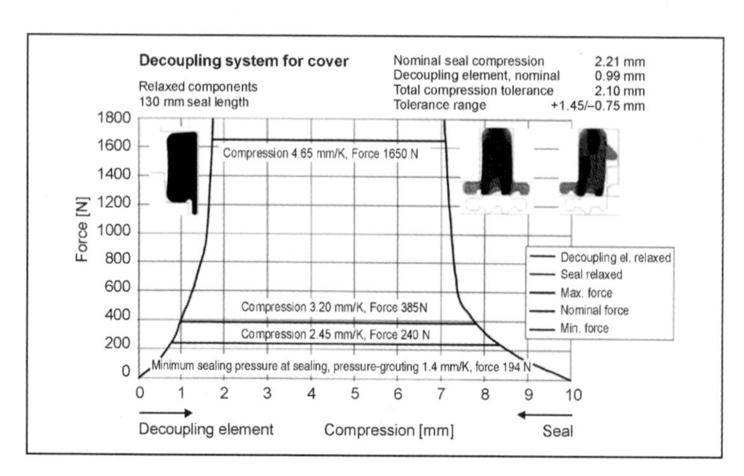

Fig. 7-322 Decoupling systems with a certain working range due to component tolerance.

gaskets. The gasket area surrounding the combustion chamber is chucked between metal flanges and is loaded repeatedly, for a prescribed number of cycles (e.g., 10^7) at a given frequency, at constant force amplitudes, or, preferably, at constant distance amplitudes. The objective is to determine the maximum permissible oscillation amplitude compatible with gasket durability. The clamping flanges can be formed to exhibit a defined surface quality (roughness, porosity) so that compression tests can be conducted to determine the minimum sealing pressure necessary to achieve a satisfactory seal.

Hydraulic Simulation of Internal Pressure

Operating on the basis of the tests conducted at the servohydraulic test stand, one uses dynamic internal pressure simulation, Fig. 7-323, to test the sealing system as a whole, under conditions closely approximating reality. For this purpose the head gasket is installed between the original mating components (engine block, cylinder head). The individual combustion chambers are then "fired" hydraulically, in the normal firing order, using fast-acting servo valves. Temperature cycles are run through, superimposed on the application of internal pressure, using a media circuit connected with the engine's water jacket. The interplay between component stiffness and gasket design is evaluated by measuring the dynamic variations in gap width that occur. Weak points in the components can be identified at an early stage in development; optimization work for the seal design can thus be conducted before the start of engine testing proper.

Fig. 7-323 Dynamic internal pressure simulation using original engine components.

Service Life Testing

These test procedures are used to examine the long-term properties of seals, seal materials, and modules. These are, in the main, tests of elastomer materials and plastics (modules). Exceptions are examinations of the setting behavior and pressure resistance of elastomer sealing materials.

In normal operations, elastomer seals and plastics are subject to aging over time, which does not occur during the pressure tests, brief thermal shock tests, and thermal conditioning procedures that are normally employed. To ensure full functioning of the module over the entire service life, simulation tests have been devised that both take account of the loading conditions found during vehicle operation and allow for reasonable testing periods. To do this, it is necessary to include temperature, media, and pressure loading in a testing program. This is done by connecting external media circuits (oil, coolant) to the test specimen and/or by exposure in a temperature chamber. With these tests, which imitate the engine operating states for temperature, one can simulate within a period of 2000 hours the loading corresponding to about ten years of vehicle operation. If the influences of oscillations and vibrations are to be assessed, then a test of this type may also be conducted using an appropriate vibration generator.

Vibratory Testing Systems

Engine components and modules are subjected during operation to mechanical vibration loads due to the influences of the road surface and direct vibration induced by the engine. Similar dynamic loads can be imposed upon the component examined using so-called "shaker" units. Both hydraulic and electrodynamic shakers are used, the latter being more common. A combination of a sliding-top table and a vibration loading along the vertical axis makes it possible to test horizontal vibration loading as well. Mechanical systems make it possible, where required, to impose vibration loads along several axes. Acceleration sensors are used to register component vibrations at the test specimen so that testing can be carried out specifically in the critical oscillation resonance range. In this way, fatigue phenomena at the test specimen can be examined with a considerable "time-lapse" effect.

Hot Gas Simulation

The thermal loading at the components, and thus at the sealing points in the exhaust system, can be simulated with hot gas generators. These deliver defined exhaust gas flows at constant temperature, burning heating oil, diesel fuel, or natural gas to do so. To achieve great component deformation, such as is found at the exhaust manifold during engine operation, the specimen is subjected to a thermal shock series in which hot gas and cold ambient air are passed through it alternately. The sealing function can be examined by pressure tests at room temperature (before and after the test series). This is, however, not a significant restriction for evaluating the gasket since it is particularly at low temperatures that the loss of bolt force due to thermal expansion in the mounting system comes fully to bear. When it is necessary to take account of dynamic influences, as well, the hot gas generator can be combined with a vibratory testing system. Either electrodynamic shakers or servohydraulic systems may be used, depending on the task at hand and the design of the specimen.

Bibliography

[1] "Integrierte Dichtspaltsensorik bei der Zylinderkopfdichtung," in MTZ, 5/2001, pp. 398–400.
[2] "Ventilhauben-Module," in ATZ/MTZ System Partners, April 2001, pp. 34–36.
[2a] "Dichter & Denker – Motordichtungen," in Motorsport und Business, 1/2001 (automotive industry insert), pp. 24–26.
[3] "Zylinderkopfdichtungskonzepte für zukünftige Motorgenerationen," in MTZ, 1/2001, pp. 30–35.
[4] "Zehn Jahre Audi TDI-Motoren mit Dichtungstechnik von Elring-Klinger," in MTZ special edition, 9/1999, pp. 78–80.
[5] "Dichtungstechnologie – kreative und innovative Entwicklungsleistungen für Meilensteine im Motorenbau," in special edtion, 60 Jahre MTZ, April 1999, pp. 59–61.
[6] "Neue Zylinderkopfdichtung mit integrierter Dichtspaltsensorik," in MTZ, 3/1999, pp. 148–151.
[7] "Zylinderkopfdichtungen, Spezialdichtungen, Module und Elastomer-Dichtsysteme," ElringKlinger AG.

7.22 Threaded Connectors at the Engine

7.22.1 High-Strength Threaded Connectors

Your basic modern engine contains between 250 and 320 threaded connections, which use from 80 to 160 different types of screws and bolts. The number of threaded connectors depends primarily on the engine configuration (e.g., four-cylinder inline or V-6 engine) and less on the combustion system (diesel or gasoline engine). Engines developed in Japan, when compared with European designs, have about 15% more threaded connectors per engine and, at the same time, fewer different screw designs. The size and number of bolts and screws rises with the displacement and number of cylinders.

Mass production among European car makers, in particular, has been heavily automated in the final assembly area since 1983. The front-runner here was VW with its "Hall 54" at the Wolfsburg assembly plant for the production of the GOLF III, which had just gone into production.[1] To accomplish this, it was necessary to design screws and bolts suitable for automatic feed, installation, and tightening.

Engine construction involves high-precision component manufacturing; the manufacturing tolerances for the basic units (e.g., cylinder block and head) are very close and the positioning accuracy for operating equipment and robots is better than 0.5 mm.

In fully automated assembly lines, the connector elements are moved by feed systems to the installation point; the bolts are screwed in and torqued down by a single or multiple power driver at an automated bolting station, necessary if only to absorb the reaction torque. Full automation does not make sense if many different engines are built on the same assembly line. With the further development of electrical control systems and ergonomic designs, hand-held power screwdrivers with integrated electronics (torque and rotation angle sensors) are used even more to monitor or control the tightening phase.[2a] This lowers the investment and maintenance costs for the assembly line and increases flexibility, moving toward "joint production systems."

7.22.2 Quality Requirements

If defects occurring while installing threads connectors are not detected, then there will be disturbances in the production process. One may count on malfunctions in the assemblies delivered to the customer. The screw or bolt is normally to blame for the disturbance, although, in addition to screw quality, the tolerances and properties of the components being joined and the threading in the nut as well as quality in assembly operations can have just as much influence on the connection.

Consequently, high-quality screws and bolts have to be used in automated systems. It is for this reason that reputed manufacturers not only make spot checks during manufacture but also often conduct a full test at the end of the manufacturing process, using automatic testing equipment. Thus, a full account is taken of the quality expectations held by screw and bolt users, with their "zero defects" targets. In practice, it is possible to achieve a reject ratio of less than 50 ppm, referenced to the major features examined, at screw sizes up to M 14; up to this size automated quality can be implemented without any technical problems. The most modern automatic machinery can process, depending on the scope and nature of testing, between 100 pieces per minute (mechanical testing) and 300 pieces per minute (optical testing). At larger dimensions, fully automated testing and the associated handling is often made uneconomical by the screw weight and size so that visual checks are made, usually combined with another step (such as hanging the parts on racks for surface finishing or when packing the parts). In conjunction with manufacturing using reliable processes, in which only random errors (occasional defects referenced to annual production volumes, at long intervals and at low rates) and no individual defective parts appear, defect rates of less than 50 ppm are achieved as a rule and otherwise less than 300 ppm.

This degree of process reliability has been achieved in recent decades due in no small part to consistent introduction of DIN EN ISO 9001 ff. and VDA Vol. 6.1[3] or even QS 9000 in the plants. Thus, the defect rate in manufacturing could be reduced from 2000 ppm to 600 ppm without undertaking any further efforts.

To avoid mixing parts later and to satisfy the demand for freedom from foreign parts, this test is made immediately before packing. The products are filled in special containers or in clear plastic bags and then sealed. Another option, even though seldom used, is to have the screws and bolts tested at the user's site.

A design proposal for screws amenable to assembly is shown in Fig. 7-324.

Experience has shown that there are difficulties if the bolts that are installed are drawn from a mixture made by different manufacturers unless exact specifications have been imposed in regard to the material, the 0.2% offset strength, and friction values. It is often necessary to set up the system anew following a change of suppliers.[4, 5, 6]

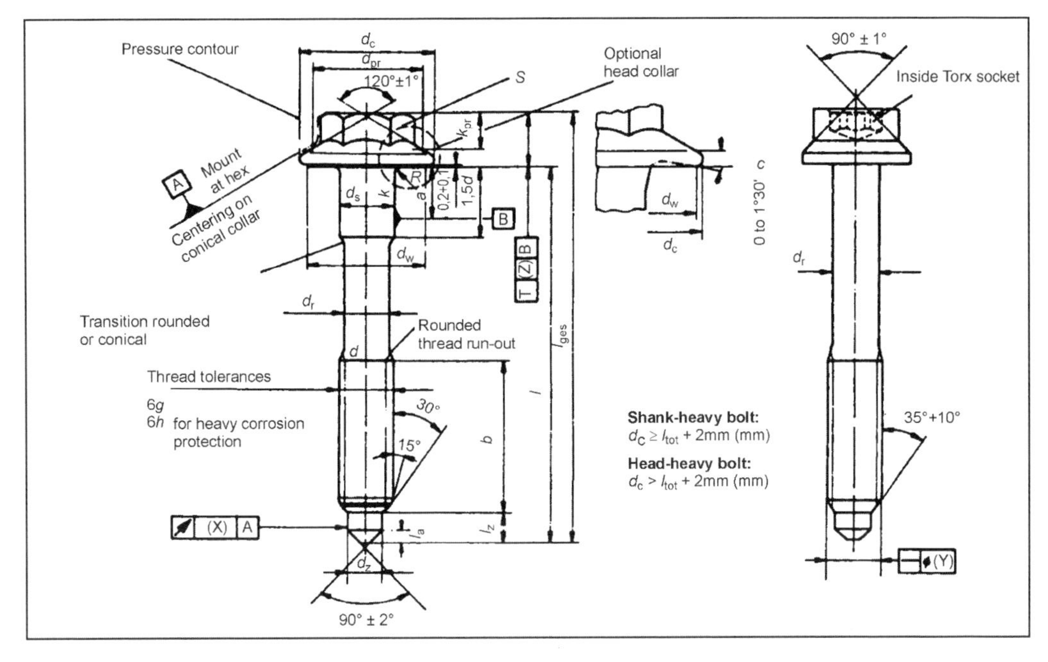

Fig. 7-324 Design suggestion for screws amenable to assembly.[1]

7.22.3 Threaded Connectors

At the engine there are generally five critical threaded connection areas; these are explained below:

- Head bolt
- Main bearing cap bolt
- Conrod bolt
- Belt pulley bolt
- Flywheel bolt

In addition, the following threaded connections can be problematic. They need not be characterized as critical from the applications technology viewpoint but may be among the major applications in the engine:

- Camshaft bearing cap bolt
- Oil pan fixing bolt, valve cover fixing bolt

Screw connections for subassemblies and flange mounting points are not discussed further at this point, with the exception of threaded connections for magnesium components. High-strength screws upwards of M 6 in size are used in most of these cases, and these are largely either standard designs or close to standard designs.

7.22.3.1 Head Bolt

The function of the head bolts is to make an operationally reliability connection for the complete system—comprising the cylinder head, head gasket, and engine block—over long-term operations, taking the maximum possible ignition forces into account. The primary goals are uniform, low component loading and tight seals against combustion gases, lubricants, and coolant.

While in the past, head bolts had to be retightened once or even twice to compensate for gasket setting, the cylinder head configuration requiring no retorqing is state of the art today.

This has been made possible by using waisted-shank bolts or waisted-thread bolts with great elasticity, closer tolerances for tensile strength and friction properties, cylinder head gaskets that resist setting (e.g., all-metal gaskets), and a tightening process with low scatter in the values for preload force. Rotation-angle controlled (turn-of-the-nut) tightening to beyond the elastic limit has established itself as the most common torquing process. Lightweight engineering is promoted more vigorously, and the resulting reduction in component stiffness at the engine block and cylinder head is normally compensated for by reducing the maximum screw strength. The minimum required screw force can be maintained only with a drastic reduction in the tolerances for tensile strength and friction values. When designing the cylinder head bolting constellation, it is necessary to understand the influence of temperature. It is conceivable that, while the engine is heating up, the head bolts heat up more slowly than the cylinder head and engine block that they join. There may be a considerable rise in the preload force if components such as aluminum, with higher coefficients of thermal expansion, are used for the latter. Considering this aspect, too, the use of waisted-shank bolts or waisted-thread screws (Fig. 7-325) is advantageous since, by virtue of the lower rise of the spring characteristics, the increase in screw loading is significantly less.[7, 8]

The expansion properties of steel can be influenced essentially fundamentally only by alloying with nickel.

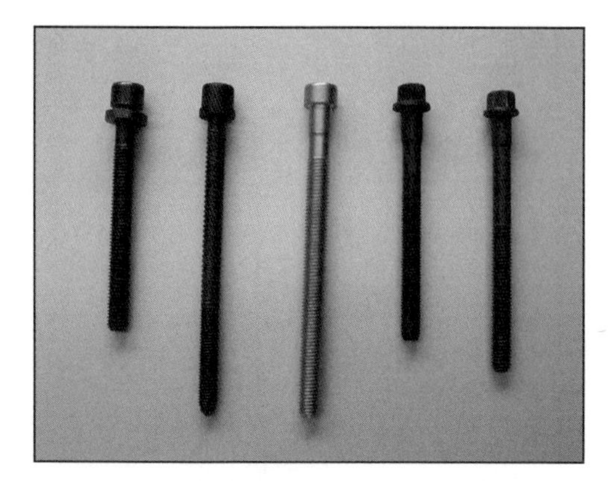

Fig. 7-325 Waisted-shaft or waisted-thread screws for head bolts (KAMAX Company).

Consequently, the latest developments provide for head bolts made of austenitic materials whose coefficients of thermal expansion are similar to that of aluminum. An as yet unsolved problem is the high degree of tool wear resulting from this material's great strength; consequently, economical manufacturing has not yet been implemented.

The constant need to reduce costs is responded to in two areas when optimizing the head bolts:

- Using waisted-thread screws as a compromise between sufficient elasticity and reduced manufacturing costs in comparison to waisted-shank bolts requiring a significantly more complex manufacturing process.
- Replacing the washer in aluminum cylinder heads by integrating its function into the screw head, in the form of a bolt with a flanged head. To avoid seizure during screw assembly, it is necessary to impose narrow limits on the geometry of the contact surface under the head and to select manufacturing technology that adheres to those limits. This includes surface treatment with extremely low variation in the friction values and excellent adhesion to the substrate material as is found, for example, in the thin-layer phosphating process with quasiamorphous crystal formation.

7.22.3.2 Main Bearing Cap Bolt

The main bearing cap bolts connect the main bearing caps with the engine block at the crankshaft bearings. As a rule, two such bolts are used for each main bearing cap; these are usually fully threaded, collared bolts and may be used with washers. Figure 7-326 shows the installation situation for such a connection and the associated force flows. Here

l_k = Clamping length
l'_k = Plate thickness
F_B = Operating force

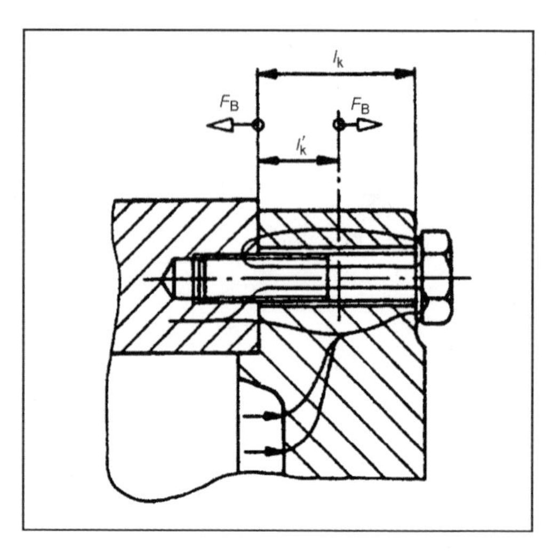

Fig. 7-326 Installation situation and force flow at the main bearing cap bolt.

The critical problem when designing this configuration is the tight installation space available for the bolt head in most instances. Very close attention must be paid to maintaining the permissible surface pressure for the rear of the bolt head and its mating surface. Every main bearing cap bolt is installed twice: the first time for machining the bearing shell seat to press-fit dimensions and then again after assembling the crankshaft and positioning the bearing caps. In the second assembly cycle, seizure may occur at the threads if the bolt exhibits damage such as impact dents at the tip or start of the threading. This is avoided preventively in screw engineering with ideal tip design and in manufacturing with the shortest possible drop heights (maximum 300 mm). The design of the tip is understood to include chamfering the start of the screw shaft before rolling the threads to ensure that the threads do not break out during rolling. At the start of the thread there appear only dull thread teeth that are not inclined to dent in response to impact.

To increase the engine block stiffness, the so-called ladder frames are used more frequently in engines to interconnect individual main bearing caps. In this way, the lower section of the engine can be stiffened to avoid twisting and warping. Usually the bearing caps are cast in place in the ladder frame made of aluminum. In this case the main bearing bolts are used to fix the complete unit in place.

Tightening processes using the 0.2% offset limit or the rotation angle as the lead variable have become the most common assembly techniques.

7.22.3.3 Conrod Bolt

The conrod bolt represents a typical case for a threaded connection subject to high dynamic loading. The range of sizes in passenger car engines is from M 7 to M 9, for utility vehicle engines from M 11 × 1.5, M 12 × 1.25, M 14 × 1.5, to M 16 × 1.5. To achieve correct dimensioning of the conrod bolt, one draws on data from the predecessor engine or for engines of similar design and size. Concerning the bolt for the large conrod eye, the operational loading on the bearing case due to the physical forces acting on the crankshaft system (masses and gas forces) are known.

Not known at the outset, however, are the operational loads by size, direction, and location, referenced to the bolt centerline in the parting plane and introduced into the individual threaded connection; this information is needed to ascertain the deformations and loads for the bolt. The professional literature[9, 10] mentions various analytical procedures used to calculate the axial force F_A, the transverse force F_Q (calculated magnitude derived from the friction value in the parting plane), and the eccentric distance a for the axial force from the screw centerline, dependent on the design parameters of the conrod bearing case. If these values are available, then it is possible, using the "KABOLT" program, which runs on a PC (screw calculations as per VDI 2230[11, 12, 13a]), to determine the preload value required to prevent partial liftoff and lateral shift of the connected components, and thus, to ascertain the appropriate thread dimensions and the strength class for the bolt. The determined values are used to designate the specifications for bolt tightening. Once the design calculations for the conrod joint have been concluded, pulser tests are used for the entire connecting rod to demonstrate durability. Subsequently, the calculated and laboratory results are verified with testing in the field. The calculation parameters for a conrod bolt connection are shown in Fig. 7-327. The example refers to a four-cylinder gasoline engine with displacement of 1996 cm^3.

The conrod bolt design is based primarily on the loading and on the assembly of the conrod. Depending on whether or not a nut is used, the bolts are equipped with heads shaped to accommodate torque transmission or with antirotation devices. The two halves of the conrod are centered with knurling, a fitting bushing, or a separate spline. Large conrods in utility vehicles often use interlocking areas following the tongue-and-groove principle.

The use of a sintered conrod makes good sense when a particular model is manufactured in medium-range numbers. While in conventional manufacture, the large conrod eye is cut away after machining in order to mount the conrod bearing shells. In recent years, "cracking" has established itself in large-volume manufacturing of sintered, cast, and forged conrods. Here the conrod end is separated from the conrod shaft in a device that applies a defined, external load to areas laid out to promote fracture. The advantage, in addition to eliminating the cutting work, is that the two halves of the conrod are self-centering. Then the fracture surfaces (postassembly) can be used to permit turning out the bearing shell seats. That is why a cracked conrod does not require an exact fit for the shaft at the conrod bolt. Here the screw diameter may exhibit a tolerance of 0.1 mm.

Each conrod is assembled twice after cutting. The first time is in preparation for machining the seats for the bearing shell. Here the preload force used for assembly must be similar to that found later during operation, so that similar deformations are induced in the conrod bearing housing. It is for this reason that the bolts are tightened to just below the 0.2% offset limit under torque or rotation angle control or under direct offset limit control. The conrod is disassembled after machining (to insert the bearing shells) and is then mounted on the crankshaft. Here a rotation angle controlled tightening process is used, which tightens the bolt into the range beyond the elastic limit; alternately, tightening under 0.2% offset limit control is employed. If one decides in favor of the rotation angle as the control magnitude, then it is necessary to conduct extensive laboratory trials in advance in order to formulate specifications for tightening. When using the 0.2% offset limit as the lead magnitude, it is sufficient, in a few tightening trials, to define the so-called "window."

Particularly because the conrod bolt, because of the manufacturing process for the conrod, has to be assembled twice and tightened into the offset limit range, one must ask which screws are particularly suitable for tightening beyond the 0.2% offset limit.[14]

When dimensioning threaded connections, it is necessary to remember that the threaded section, in the event of overloading due to static tensile forces, breaks at its weakest point. This is normally the case in the nonengaged threaded section or in the waisted-shank area. In the multiwaisted bolts recently developed, the failure is also in a waisted area. The conrod bolts shown in Fig. 7-328 are particularly suited for tightening into the range beyond the elastic limit.

When using bolts with a shaft (similar to DIN EN 24014), there should be at least six nonengaged turns in order to distribute plastic elongation over a larger area and, thus, to avoid the hazard of premature narrowing. The best tightening properties in the range beyond the elastic limit are demonstrated by waisted-shank bolts and screws that are threaded right up to the head (similar to DIN/EN 24017). The measured flexibility places the multiwaisted bolts between the waisted-shank bolt and the screw that is threaded along its full length.

The durability of threaded connections is determined exclusively by the magnitude of local stress concentrations. In bolt materials the fracture strength of the notched area compared to the smooth rod should, as a rule, be greater than 1, indicating a material with sufficient ductility. Permissible in high-strength bolts are durability values in the pulsating tensile range of $\sigma_A = \pm 55\ N/mm^2$.[15] Screw durability is increased if the threads are rolled after annealing. The additional dynamic forces resulting from dynamic operational forces (which are absorbed by the screw) are lowered (in connection with eccentric loading) as the preload force level gets higher. This, too, favors a tightening process that goes beyond the elastic limit.

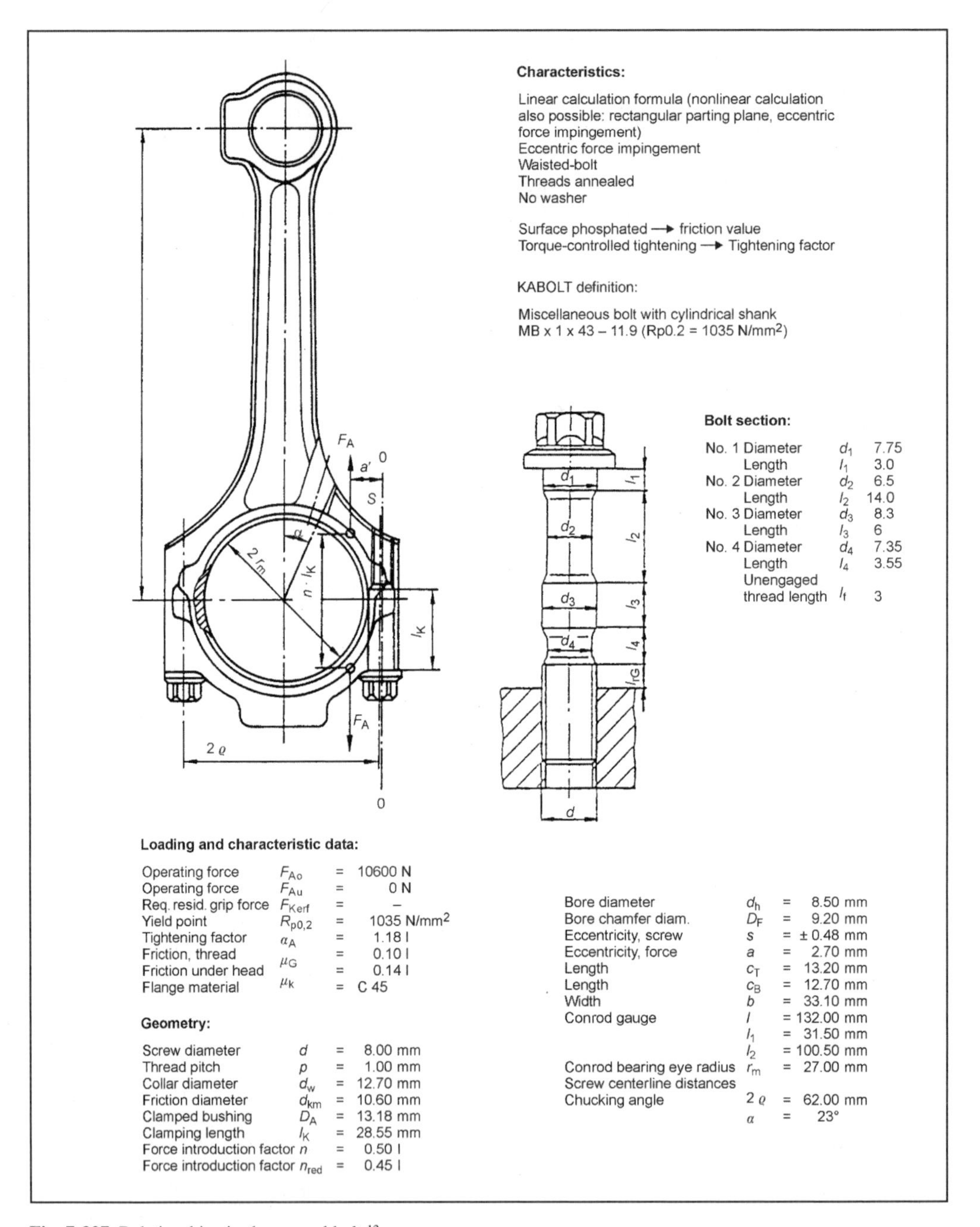

Fig. 7-327 Relationships in the conrod bolt.[12]

7.22.3.4 Belt Pulley Bolt

The belt pulley is secured with a bolt at its center. Often mounted on the crankshaft in addition to the belt pulley are a gear for the oil pump drive and possibly the vibration damper. The inside bore of the belt pulley is mounted on the crankshaft end journal. The large bore diameter in the belt pulley makes it necessary to create a positive connection between the bolt and the pulley with a large washer or a large-diameter bolt collar. Often an M 12 bolt is fitted with a washer or collar diameter of up to 38 mm (in gasoline engines) or an M 18 bolt with a collar diameter of up to 65 mm (in a diesel engine with 2.5 liter displacement, for instance). The pulley is press-mounted

Fig. 7-328 Conrod bolts for tightening beyond the elastic limit.[14]

separately on the crankshaft journal or is pulled onto the crankshaft by the screws at a previously defined tightening torque. In utility vehicle engines sizes of up to M 24 × 1.5 are used and the washer is positioned just before assembly. In large utility vehicle engines, the pulley is seated on the vibration damper and, passing through oversize bores, is bolted directly to the crankshaft with six or eight screws or lug bolts (e.g., M 10).

In the past, belt pulley bolts were tightened down with only a torque wrench. Today, the rotation angle technique has become more or less standard. Tightening takes place through a snug-tight fit until all mating planes are seated firmly one on the other. The bolt is then turned down further, the amount based on a measurement of the rotation angle. Extremely high ultimate tightening torques are achieved in this way. When using an M 12 × 1.5 − 10.9 bolt, torques of up to 260 Nm can be achieved, while the ultimate torque calculated theoretically lies between 120 and 150 Nm. The great spread in the ultimate torque results from the large head contact area, which causes "seizure" if there is even the slightest misalignment. A tightening technique based on the 0.2% offset limit cannot be applied for the pulley bolts when there is an extremely large screw head collar diameter or where several components have to be bolted together so that there is a large number of mating planes between the parts to be joined. Manufacturing inaccuracies and unavoidable grime results in greater setting and a connection that is so flexible that the 0.2% offset limit point is sensed not only for the bolt but also for the connection.

7.22.3.5 Flywheel Bolt

Because of the engineering design there is a relatively small pitch circle at the crankshaft. During assembly it is necessary to ensure that there is sufficient clearance between the bolts to accommodate the tightening tool. The bolts are all tightened simultaneously, using a multispindle tool and using the 0.2% offset limit as the control variable. This is also done because shorter clamping lengths (e.g., 7 mm) are present. Because of the narrow clearance between the crankshaft journal and the flywheel, the bolt heads are not as high as the standard heads. To be sure that the required torque can be applied safely and positively, a twelve-pointed (bihexagonal) head or a hexalobar head or, if necessary, an inside, multispline socket is used at the head. When oil is supplied by runners inside the crankshaft, the bolts used to seal against oil leaks are provided with a microencapsulated sealing adhesive or with an all-round nylon coating.

Some engine manufacturers still tighten down flywheel bolts under torque control, and then snug them down manually.

In the dual-mass flywheels that are used more frequently today, the module is delivered to the vehicle manufacturer complete with the bolts and is then assembled as a unit. The bolts are tightened down with a multispindle power driver, through bores in the clutch plate spring and the clutch disk.

7.22.3.6 Camshaft Bearing Cap Bolt

This threaded connection usually uses collared screws that are similar to the standardized styles, in sizes of M 6, M 7, and M 8 for passenger car engines and M 10 and M 12 for utility vehicle engines. Since during torque-controlled tightening, there is the hazard that differing clamping forces could be imposed on the camshaft bearings, the rotation-angle-controlled technique is used to a greater extent to achieve defined preload values. A special technique used by some car makers is to screw grub screws into the cylinder head; the bearing cap is then positioned and fixed with nuts.

7.22.3.7 Oil Pan Attaching Screws

The oil pan is also secured to the engine block with collared screws similar to the standardized styles, Fig. 7-329 (outside or inside socket wrench application, M 6, M 7, M 8 screws). To achieve complete freedom from leaks, the surface pressure must be uniform across the entire oil pan gasket. This is achieved with the smallest possible screw diameter that thus exhibits appropriate flexibility and with a suitably large collar diameter or a washer, where the screw forces are introduced uniformly. In addition, a large number of screws is needed so that when forces are introduced there are large overlaps in the "pressure cone" in the area of the seal. In spite of the great demands for tightness, this connection point is considered to be trouble-free. The screws are, as a rule, tightened under torque control (using a multispindle power screwdriver unit). To ensure that the oil pan is not canted, tightening is started at the middle of the engine block, continuing outward from there.

Ribs on the lower surface of the engine block reduce noise propagation. The oil pan itself is a source of high noise emissions because of its large surface area and low weight. Here the solution involving structural noise decoupling by the mounting screws used for the oil pan

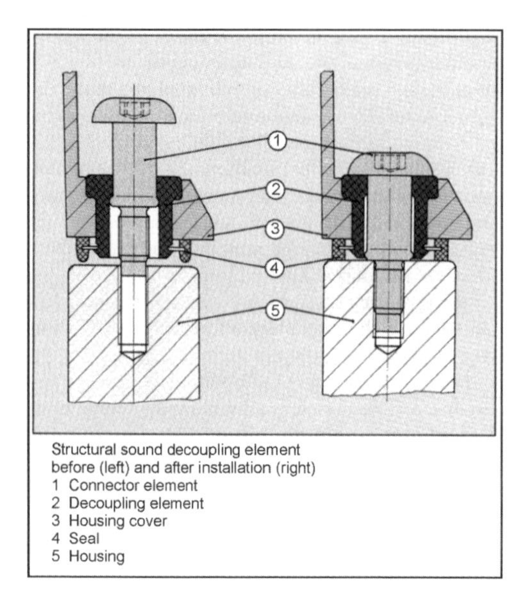

Structural sound decoupling element
before (left) and after installation (right)
1 Connector element
2 Decoupling element
3 Housing cover
4 Seal
5 Housing

Fig. 7-329 Oil pan screw with structural acoustic decoupling element (depiction before [left] and after [right] assembly) (KAMAX Company).

is a viable option. Widespread use has not been implemented because of the costs involved.

In order to cut costs, the market is exhibiting a trend to preassembling connector elements in system assemblies such as oil pans, valve covers, timing belt covers, etc. Here leading screw manufacturers are working on economical and space-saving preassembly solutions such as, for example, using self-tapping screws. Another solution, which at the same time could serve the interest of acoustic decoupling, employs a plastic bushing that is slipped over a special screw and is then premounted together with the component (Fig. 7-330).

7.22.4 Threaded Connections in Magnesium Components

The trend toward lightweight construction in automotive engineering, prevailing for years now, requires not only optimization of the components made from proven materials such as steel, aluminum, or plastic, but also the use of alternate materials such as magnesium. The advantage of magnesium is its relative stiffness even where cavity walls are thin. It is comparable to the density of plastic.

In engine design, this material is used only in secondary assemblies such as the cylinder head shroud for encapsulated engines or for the air filter body intake tube. Here magnesium replaces plastic. In the engine block itself, the thermal loads are too high for connection points as a whole. Thus, steel screws in conjunction with magnesium can be used only at room temperature due to the setting and relaxation properties. In the engine component area heat-treated aluminum screws are thus used, made from AL 6056 in conjunction with the die-cast magnesium

Fig. 7-330 Oil pan screw with inside hex lobes and collar (KAMAX Company).

alloy AZ 91, AS 21 up to 120°C (maximum temperature: 150°C), for example. When using magnesium components, contact corrosion properties in conjunction with steel or aluminum screws have to be taken into account.[16]

7.22.5 Screw Tightening Process

When selecting the assembly and tightening technique, one must remember that passenger car engines are manufactured in large numbers; utility vehicle engines, by contrast, are built in short production runs or even individually.[17]

7.22.5.1 Torque-Controlled Tightening

Torque-controlled tightening is normally used only for secondary applications (minimum preload force need not be exactly defined). It is employed only in demanding applications (such as mounting the belt pulley) in automated assembly lines. It continues to be used in service work. The problem is that the preload value applied under torque control has to be selected so that in the worst case (i.e., smaller actual coefficient of friction than what was estimated when establishing the torque level) the 0.2% offset limit is not exceeded as otherwise the screw would be stretched. Preload is the force that is present in the threaded connection after the completion of the assembly. At a very high actual coefficient of friction (higher than what had been assumed), the preload value is very low. Consequently, the properties of the screw cannot be fully exploited with this technique. Screw and bolt manufacturers and the automotive industry have agreed upon the coefficients of friction to be expected. They lie between $\mu_{Total} = 0.08$ and 0.14. They are a component in the quality agreement in each case and are spot checked for each batch of screws at a friction value test device.[14]

(Photo courtesy Atlas Copco Tools)

Fig. 7-331 Subassembly installation using a handheld power screwdriver with integrated torque and rotation angle transducers (Atlas-Copco).

A special form of torque-controlled tightening is the combination with "snugging down"; once the tightening phase is completed, the connection is retightened with a torque wrench (Fig. 7-331). This technique is used in mass production for all critical connector elements at those manual assembly stations that are still found in short production runs.

When manual connection is used, a torque-controlled pneumatic screwdriver is employed to tighten the screw or bolt down to the specified moment; then a torque wrench is used to retighten. The final position is normally marked with paint. The torque required to restart rotation is the snugging moment. Experience has shown that snugging usually goes beyond the adjustment value for the wrench so that an indirect, rotation-angle-controlled tightening process is often used.

7.22.5.2 Rotation-Angle-Controlled Tightening

When tightening the nut using the rotation angle as the lead variable, through to the 0.2% offset limit, the preload value is on average from 25% to 30% higher than for torque-controlled tightening. While in torque-controlled tightening the preload force varies by about ±25% (practically to the same extent as the friction), the preload force where the rotation angle or 0.2% offset limit is used as the lead variable varies by only about ±10%. When tightening using rotation control, the spread in preload is dependent on friction only in the range up to the snug-tight torque. The snug-tight torque is the moment that has to be applied until, by tightening the connection, all the mating surfaces are seated solidly one against another due to elastic and plastic deformation. The spread results primarily from the differing 0.2% offset limits for the bolts, provided that the required repetition accuracy when approaching the set angle is achieved. This is the case in today's pulse transducers. Beyond that, we see from the progress of the curves above the 0.2% offset limit that angular scatter has only a subordinate influence on assembly preload (Fig. 7-332). Torque monitoring is used to ensure quality in the connection.

We see that when tightening under torque control, the minimum preload force F_m lies between 48 and 57 kN. When working with yield point (0.2% offset limit) control, this value is between 67 and 85 kN while rotation angle control yields between 77 and 94 kN. Consequently, tightening under torque control gives the greatest spread in preload force at the smallest preload level. The preload force level when using turn-of-the-nut tightening is on average about 10% greater than that for the yield point technique.

The area around the $R_{p0.2}$ points represents the window for tightening under yield point control. The switch-off point for the power driver has to lie within this area so

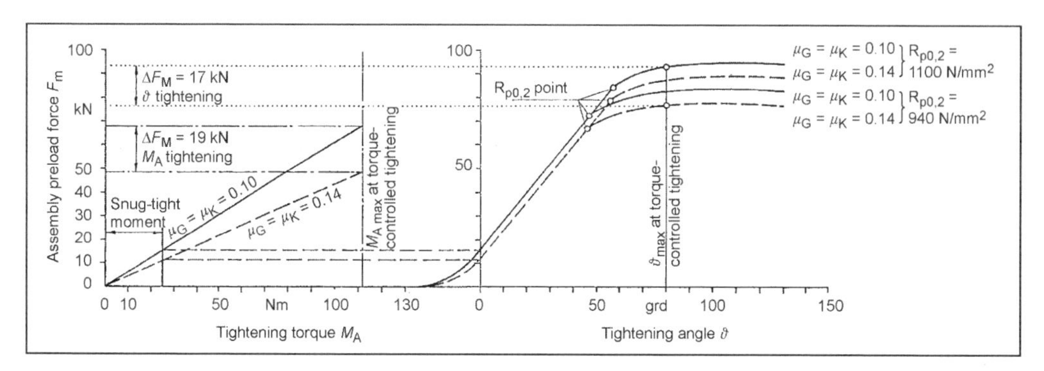

Fig. 7-332 Tightening curve for a screw as per DIN EN ISO 24014—M 12 × 1.5 × 70—10.9—for control by torque (left) and control with rotation angle and 0.2% offset limit (right), illustrating the influences of thread and head friction and screw strength.

that the threaded connection is registered as "OK" and can receive the paint marking, if that is specified.

When dimensioning threaded connections, it is necessary to remember that a threaded connector, in the event of overloading due to static tensile forces, breaks at its weakest point. This is normally the case in the nonengaged threaded section or in the waisted-shank area.

Using the turn-of-the-nut process (as a process that goes beyond the elongation limit) is not critical in screws and bolts where the shank length is greater than 2 × *d* or there are more than 10 turns of nonengaged threading. In that case, even tolerances as great as 20° are acceptable when specifying the rotation angle used for tightening. In a threaded connector with a pitch of $P = 1.5$ mm, turning the screw by 30° beyond the 0.2% offset limit induces plastic elongation of about 0.125 mm. Referenced to the 60 mm effective clamping length (grip) this represents permanent deformation of 0.21%. This value is not critical. Conversely, when using short screws ($<2 \times d$ shank length) the switch-off point has to be specified so closely that it is very near the yield point, particularly since today screws are often tightened into the offset limit range. The rule of thumb is that in these cases, referenced to the grip length, a maximum of 1% permanent deformation is acceptable. It is necessary to note here, however, that if the screw is tightened several times, then the head contact surface and the engaged threads can be damaged and thus tend to scuff and possibly seize. The required preload force cannot be attained in this event.

A further advantage of rotation-controlled tightening is its reproducibility even when using simple tools; consequently, it is a favorite technique for initial tightening on the assembly line and for service work.

7.22.5.3 Tightening under Yield Point Control

When compared with the rotation-controlled technique, this process offers the advantage that it always approaches the real 0.2% offset limit for the particular screw being installed. This process is used only to a very limited extent and in those situations where greater setting effects are expected during and shortly after tightening. The permanent elongation of the bolt each time it is tightened lies between 0.1% and 0.2% (the exact amount depending on the sensitivity of the power driver system) and thus below the yield point. Unacceptable permanent elongation of the screw or bolt beyond the offset limit is hardly possible. In comparison with tightening under rotation angle control, the mean preload force value is 4% to 7% lower. Quality assurance for the connection is affected by monitoring the window. This window specifies the power driver's switch-off point (defined by specifying maximum and minimum angle and torque values) within the tensile yield range of the bolt.

Bibliography

[1] Jende, S., "Robotergerechte Schrauben–Hochfeste Verbindungselemente für flexible Automaten," Techno TIP, 12/84, Vogel-Buchverlag, Würzburg.

[2] N.N., Industriewerkzeuge–Montagewerkzeuge, 2000–2001 catalog, Atlas Copco Tools GmbH, Essen.

[2a] N.N., Schraub- und Einpresssysteme. Firmenkatalog der Robert Bosch GmbH Automationstechnik, Edition 1.1, 2001, Murrhardt.

[3] VDA Publications, "Qualitätsmanagement in der Automobilindustrie," Qualitätsmanagement-Systemauditm Vol. 6, Part 1, 1998 edition, Verband der Automobilindustrie, Frankfurt.

[4] Jende, S., and W. Mages, "Roboterschrauben. Wie sollen Roboterschrauben gestaltet sein?" Schriftreihe Angewandte Technik, Verlag für Technikliteratur, 1990, pp. 12–18.

[5] Jende, S., "Automatische Montage hochfester Schrauben–Anwendungsbeispiele aus der Praxis," wt–Zeitschrift für industrielle Fertigung, Springer-Verlag, Berlin, Heidelberg, 1986.

[6] N.N., "Informations-Centrum Schrauben–Automatische Schraubmontage," Deutscher Schraubenverband e.V. [ed.], Hagen, 2nd edition, Mönning-Druck, Iserlohn, 1997

[7] Jende, S., and R. Knackstedt, "Warum Dehnschaftschrauben? Definition–Wirkungsweise–Aufgaben–Gestaltung," in VDI-Z, Vol. 128, 1986, No. 12.

[8] Illgner, K.H., and D. Blume, Schraubenvademecum, Bauer & Schauerte Karcher GmbH, 6th edition.

[9] Lang, O.R., "Triebwerke schnelllaufender Verbrennungsmotoren," Konstruktionsbücher, No. 22, Springer-Verlag, Berlin, Heidelberg, 1966.

[10] Grohe, H., Otto- und Dieselmotoren: Arbeitsweise, Aufbau u. Berechnung von Zweitakt- u. Viertakt-Verbrennungsmotoren, Kamprath–kurz und bündig series, Technology, 6th edition, Vogel-Buchverlag, Würzburg, 1982.

[11] VDI, Systematische Berechnung hochbeanspruchter Schraubenverbindungen, VDI Guideline 2230 (1986) and draft edition (1998).

[12] Jende, S., "KABOLT–ein Berechnungsprogramm für hochfeste Schraubenverbindungen, Beispiel: Die Pleuelschraube," in VDI-Z, Vol. 132, 1990, No. 7, pp. 66/78.

[13] PC-Bolt '98 (bolt calculation program), Institut für Maschinenkonstruktion/Konstruktionstechnik, Technische Universität Berlin, Berlin, 1998.

[13a] Esser, J., Ermüdungsbruch–Einführung in die neuzeitliche Schraubenberechnung, 23rd edition, TEXTRON Verbindungstechnik GmbH + Co., Neuss, 1998.

[14] Kübler, K.H., G. Turlach, and S. Jende, Schraubenbrevier, 3rd edition, KAMAX-Werke Rudolf Kellermann GmbH & Co. KG, Osterode am Harz, 1990.

[15] Scheiding, W., Verschrauben von Magnesium braucht mehr als Alltagswissen, Konstruktion und Engineering, Verlag Moderne Industrie, Landsberg/Lech, 2001.

[16] Kübler, K.H., and W. Mages, Handbuch der hochfesten Schrauben, 1st edition, KAMAX-Werke [ed.], Verlag W. Girardet, Essen, 1986.

[17] Jende, S., and W. Mages, "Schraubengestaltung für streckgrenzüberschreitende Anzugsverfahren–überelastische Grenzgänger," KEM, 9/1986 edition, Konradin Verlag, Leinfelden-Echterdingen.

7.23 Exhaust Manifold

Economical cast manifolds were the standard in vehicle engineering for many years. Only in sportier vehicles—in the interest of optimizing torque and performance—were single-walled tube-runner manifolds used. They enabled individualized runner lengths, diameters, configurations, and mounting. Combustion at full throttle was largely substoichiometric so that the exhaust temperatures were relatively low.

In the mid-1980s legislators in Europe imposed pollutant emission limits, making it necessary to equip the vehicles with catalytic converters. As emission laws became more stringent, exhaust pollutants following a cold start had to be reduced further and more quickly.

One of the options for rapid reduction was found by reducing the exhaust manifold's thermal mass (or capacitance). In the cast iron version the mass for a four-cylinder manifold, at from 4 to 8 kg, is quite high. If the exhaust manifold's thermal mass is low, then the heat in the exhaust can bring the catalytic converter up to the so-called light-off temperature more quickly. The light-off temperature is defined as the exhaust temperature at which half of the pollutants are converted. Options for reducing the mass are presented in Sections 7.23.2 to 7.23.4. Figure 7-333 shows the influence of manifold design on the temperature ahead of the catalytic converter when using a standardized test cycle.

A further aspect that has had a negative effect on traditional exhaust gas manifold design is the increase in exhaust temperatures, resulting from the increase in the power density and operating with a stoichiometric fuel-air mix across wide areas of the engine map. Whereas in the early 1980s we found exhaust temperatures of 850°C in gasoline engines and 650°C in diesel engines, these levels today have risen to beyond 1000°C in gasoline engines and as much as 850°C in diesel models.

Especially in gasoline engines, this fact has a significant influence on the selection of the casting material. Earlier cast manifolds using silicon-molybdenum (SiMo) alloys reached their application limits at exhaust temperatures of up to 900°C. Higher-quality gray casting qualities containing 20% to 36% nickel can be used up to about 1000°C. To handle even higher exhaust temperatures, it is necessary to resort to nickel- or cobalt-based alloys like those that are also used in turbine engineering. Cast manifolds, because of the typical wall thickness of from 4 to 6 mm (tube manifolds, by comparison, are 1.0 to 1.8 mm thick), generally operate in a temperature range that could have a negative effect on durability through time. The changes in microstructure occurring at these temperatures and the inadequate thermal strength result in plastic deformation.[6] During the cooldown phase microfissures appear, and these lead to manifold failure in the long run. Neither have extensive studies on the development of new manifold casting materials resulted in a sufficiently improved service life.[1] One solution is to assemble the manifold from steel sheet or steel tube components. This design is considered to be durable. Thus, there are examples in which cast SiMo manifolds were tested over 250 hours, while assembled manifolds for the same engine were tested under identical conditions for up to 500 hours.

For maximum exhaust temperatures, diesel engines offer a better operating environment for cast materials. In response to new legislation, however, there are trends toward replacing—in diesel engines, as well—cast components with those made of sheet metal.

Further arguments in favor of substituting sheet material for castings are efforts to reduce overall vehicle weight and ultimately to also reduce the great tendency for a cast manifold to heat up after the engine is shut down (Fig. 7-334).

The installation situation permits a very compact design using cast manifolds while sheet metal manifolds tend to take up more space, because of optimized runner lengths and minimum bending radii that have to be observed.

When the various manifold designs are heated and cooled, we find that cast materials, in comparison to tubing and sheet metal, involve a high degree of thermal lag. An assembled design with air gap insulation lies between casting and tubing in regard to this factor.

The need for heat shielding is determined primarily by the component's surface temperature, the postheating properties, and the proximity of nearby components. Since the energy transmitted in irradiated area rises with the fourth power of the surface temperature, it makes good sense to shield cast and tube manifolds that can reach surface temperatures of up to 800°C. One very good alternative is double-walled or jacketed manifold incorporating an air insulating air (AGI); here the tubing carrying the exhaust gas is separated from the supporting structure by an air gap. These manifolds, which, by their very nature, incorporate their own heat shield and exhibit maximum surface temperatures of from 450 to 500°C, generally do not require any additional shielding.

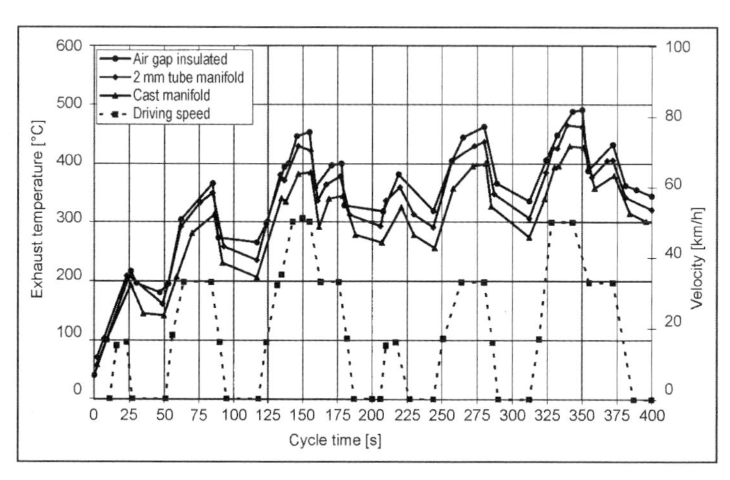

Fig. 7-333 Influence of manifold design on temperature ahead of the catalytic converter.

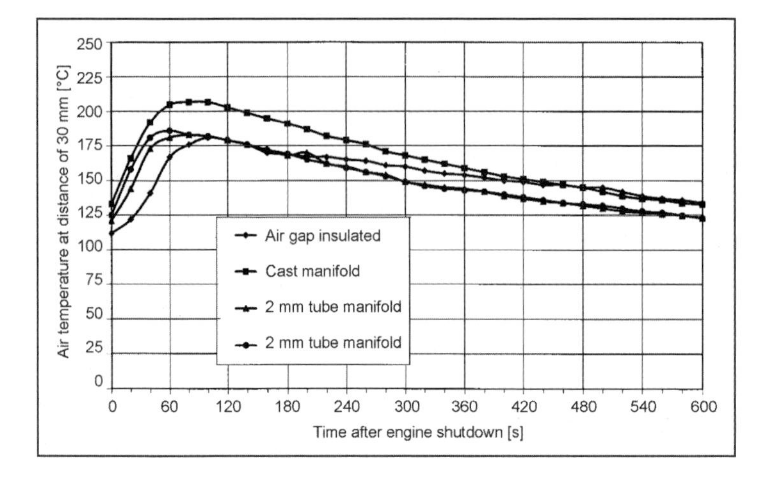

Fig. 7-334 Soaking behavior for various manifold designs.

7.23.1 Manifold Development Process

The essential steps in manifold development are listed below:

- Customer query for a desired manifold concept
- Customer specification of the available installation space (may also include the geometry for the draft concept as well as that for the cylinder head flange, exhaust flange geometry, space available for power driver use, surrounding engine compartment geometries, etc.)
- Specification of loading data (engine type and performance, vibration induction by the engine and/or road, exhaust gas temperature)
- Definition of the emission standard (EURO 3 or EURO 4 or some other norm)
- Development of a detailed concept and the design created using CAE tools including, for example, heat transfer calculations, calculations for flow mechanics, or FEM calculations[3, 4]
- Construction of samples with tooling similar to that to be used in mass production
- Certification testing at either the developer's or the customer's site
- Customer's production approval for the development
- Test with mass-production components to verify the design
- Construction of mass-production tooling
- Production launch

As a rule, the overall period between the inquiry and production launch is about two years. Development work today is carried out in only a 14-month period; eight months are consumed in pure development time, and the remaining six months are required to build mass-production tooling and set up the manufacturing lines.

7.23.2 Manifolds as Individual Components

7.23.2.1 Cast Manifold (Fig. 7-335)

Typical materials:

Nodular gray casting (GGG), SiMo gray casting: Nodular gray casting using silicon-molybdenum (GGG-SiMo), SiMo gray casting with vermicular graphite, austenitic cast iron (GGV-SiMo)[6]

Wall thicknesses: 7 to 8 mm for GGG manifolds
2.25 to 4 mm for chilled casting

Advantages:

- Compact design
- Wide latitude in designing the shapes
- Good acoustical properties with high material damping properties
- Economical ($15 to $18 for SiMo casting).

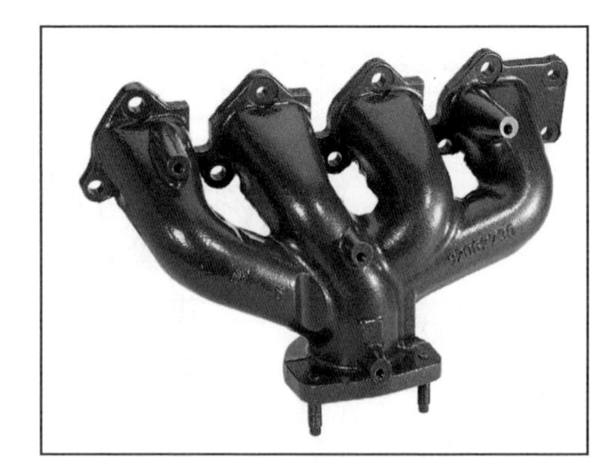

Fig. 7-335 Cast manifold for four-cylinder gasoline engine.

Disadvantages:

- Great weight.
- The maximum permissible exhaust gas temperatures for cast material are limited.
- If, because of the extreme temperatures, the use of nickel alloys is necessary, then the price will rise to between $35 and $40.
- Cast manifolds operate in a temperature range that can affect service life (bad for endurance, considering the engines' higher performance densities and resultant higher temperatures).
- High surface temperatures (heat shielding required).
- Critical in emissions following cold start because of the manifold's high thermal masses.
- Severe postheating properties because of great thermal mass.
- Any desired or optimized runner lengths can be implemented to only a limited extent with cast material (performance optimization is limited).

7.23.2.2 Tube Manifold (Fig. 7-336)

Typical materials:

Austenitic steels such as the type 1.4301, 1.4828, and 1.4841 alloys
Ferritic steels such as the 1.4509 alloy or newly developed ferritic steels containing up to 14% chrome along with titanium and niobium stabilization (examples being SUS 425 Ti, LR 429 EX, and F 14 Nb)
Wall thicknesses: 1.0 to 1.8 mm

Advantages:

- Performance-optimized design can easily be effected because of the greater options for selecting the shapes.
- Low weight.
- Standard steels that are readily available can tolerate high exhaust temperatures.
- Low postheating properties.

Fig. 7-336 Lightweight tube type manifold for a four-cylinder gasoline engine.

Disadvantages:

- More compact designs are possible but should not be implemented in four-cylinder engines because of performance considerations. Designs such as this are developed in some cases today to replace an existing cast design with a tube system occupying the same space. This, however, involved major problems in reaching the required durability levels, in addition to many other disadvantages.
- High surface temperatures (heat shielding required).
- When compared with the cast manifold, a tube type manifold is favorable for emissions at start-up. The situation can, nonetheless, remain critical if the manifold's thermal mass is still relatively high due to choosing an excessively thick wall, from 1.8 to 2.0 mm. This problem can be countered by reducing wall thickness to a typical value of 1.2 mm. Selected designs are being made up today at a wall thickness of from 0.8 to 1.0 mm.[5]
- Problematic acoustical properties due to low damping by the material. Additional efforts may be necessary under certain circumstances.
- Higher costs ($23 to $40).

7.23.2.3 Single-Wall, Half-Shell Manifold (Fig. 7-337)

Typical materials:

Austenitic steels such as the type 1.4301, 1.4828, and 1.4841 alloys
Ferritic steels such as the 1.4509 alloy or newly developed ferritic steels containing up to 14% chrome along with titanium and niobium stabilization (examples being SUS 425 Ti, LR 429 EX, and F 14 Nb)
Wall thicknesses: 1.5 to 1.8 mm

Advantages:

- Economical ($15 to $20).
- Low weight.
- Standard steels that are readily available can tolerate high exhaust temperatures.
- Low postheating properties.

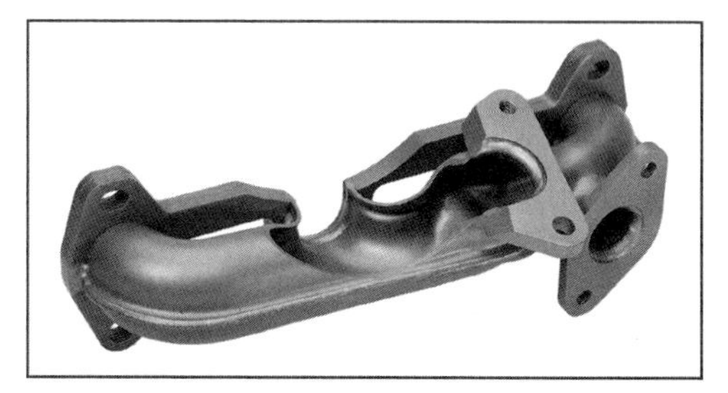

Fig. 7-337 Half-shell manifold for three-cylinder (diesel) engine.

Disadvantages:

- Only very short runner lengths can be realized in a four-cylinder engine; the geometry of such a manifold is then typically very limited.
- The shape involves a great deal of cutting loss.
- Very long welding seams are required.
- High surface temperatures (heat shielding required).
- Critical acoustic properties (additional efforts may be necessary under certain circumstances, in the form of double-wall shells).

7.23.2.4 Manifolds with Air Gap Insulation (AGI Manifold) (Fig. 7-338)

Separation of functions: Inside, there are lightweight components carrying the exhaust gasses; outside are the load-bearing elements with greater material thickness. These internal components are decoupled by floating seats. In this way, it is easy to achieve durability in such a manifold.

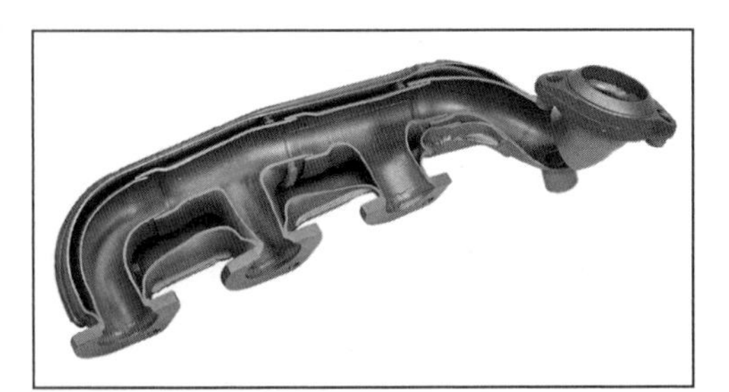

Fig. 7-338 Jacketed manifold with air gap insulation for V-6 gasoline engine.

Typical materials for the inside tube:

Austenitic steels such as the type 1.4301, 1.4828, and 1.4841 alloys
Typical materials for the load-bearing outside components:
Austenitic steels such as the 1.4301 alloy
Austenitic steels such as the type 1.4509 and 1.4512 alloys
Wall thicknesses: Interior components carrying exhaust gas 1.0 mm; load-bearing outside components 1.5 mm

Advantages:

- Relatively low weight and compact design.
- A design with optimized performance can be devised within a defined degree of latitude.
- Standard steels that are readily available can tolerate high exhaust temperatures.
- No high surface temperatures (thus nearby components can be positioned relatively close to the AGI manifold without further protective measures).
- Low postheating properties.
- Suitable for emission-optimized systems. The inner components carrying the exhaust gas have only a low thermal capacitance so that energy losses through to the catalytic converter are low; the outer components, with greater thermal mass, accept thermal energy only after the catalytic converter has reached full operating temperatures.
- A concept with a water-cooled outside jacket is even possible.[2]
- Good acoustic properties can be attained with moderate effort.

Disadvantages:

- High costs ($40 to $66)
- In some cases it is necessary to use high-pressure, internal reforming to achieve the complex geometries required while still taking up the least possible space; that means high costs and long lead times for the tools.
- Runners cannot be of any desired length.

7.23.3 The Manifold as a Submodule

7.23.3.1 Integrated Manifold and Catalytic Converter (Fig. 7-339)

Since a catalytic converter near the engine can be joined with the manifold using techniques such as welding or flanging, all the alternatives depicted at Section 7.23.2 are available for use in the manifold section.

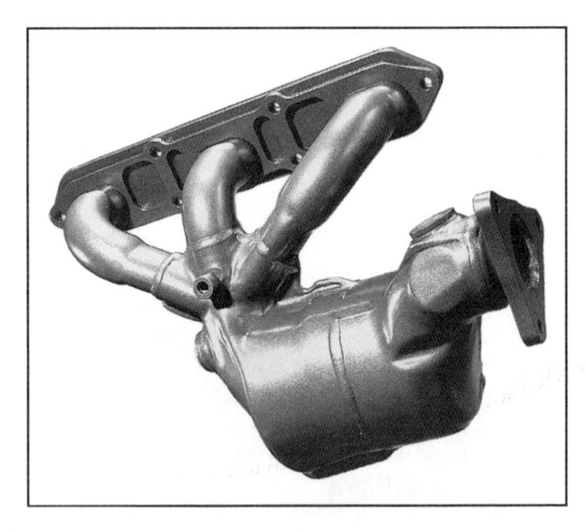

Fig. 7-339 Catalytic converter near the engine, with welded, cast manifold (six-cylinder boxer engine).

7.23.3.2 Integrated Manifold and Turbocharger

The manifold and turbocharger module shown in Fig. 7-340 is employed for both gasoline and diesel engines. Compared with an assembly made up of individual components, this module eliminates the masses of the flanges on the components while at the same time simplifying assembly. A clear disadvantage of this modular design is that the entire system has to be replaced even if just one

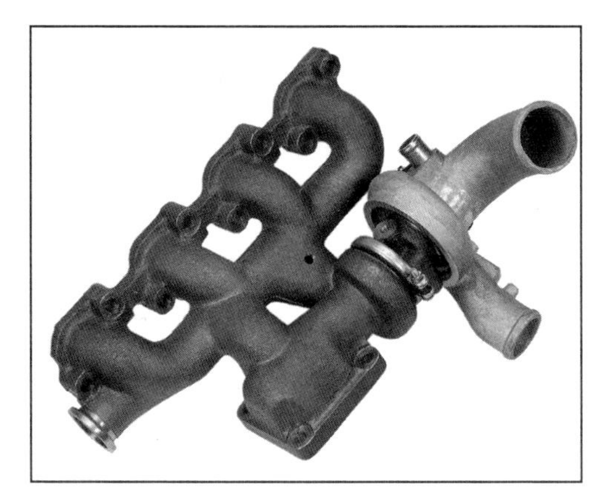

Fig. 7-340 Cast manifold with integrated, cast turbocharger (diesel engine).

of the components fails. Great costs are involved in unnecessary replacement of the turbocharger.

If in this area, too, for the reasons already discussed, one opts for other types of manifold, then it is necessary to provide additional support at the engine block for the heavy turbocharger unit.

Studies are currently being conducted to determine how a turbocharger housing made of sheet metal—to reduce thermal capacitance and weight—can be employed.

7.23.4 Manifold Components

Components such as connector nipples for the exhaust gas return system or the runners for secondary air supply, which until recently were contained in the manifold or the welded intake flanges, are more frequently being integrated into the engine block itself.

Flange concepts for tube manifolds are shown in Fig. 7-341.

Used here are flange designs ranging from complex, heavy cast flanges with integrated secondary air feed through to very simple, lightweight, deep-drawn flanges made of sheet metal. In some cases, deep-drawn flanges exhibit, while carrying out the same functions, up to a 50% reduction in mass when contrasted with a comparable cast flange. By raising the edge of the deep-drawn flange, for instance, one can achieve the same stiffness characteristics as in the cast flange. The seal is achieved with greater surface pressure induced by beads stamped into the metal around the entry openings.

Typically the thermal loads placed on the intake flange are low because of its contact with the relatively cool cylinder head. Consequently, economical, easily annealed standard steels such as the S3552J0 alloy can be used. Because of the higher temperatures at the exhaust ports, flanges of a similar design have to be made of higher quality ferritic or austenitic steels.[3]

Bibliography

[1] "Grenzen für Grauguss," Automobil-Produktion, Oktober 2000.

[2] Hein, M., Published patent application DE 4324458A1; German Patent Office, File No. P4324458.0, January 1994.

[3] Weltens, H., P. Garcia, and H. Neumaier, Neue Leichtbaukonzepte bei Pkw-Abgasanlagen sparen Gewicht und Kosten.

[4] Voeltz, V., A. Kuphal, S. Leiske, and A. Fritz, "Der Abgaskrümmer–Vorkatalysator für die neuen 1.0 l- und 1.4 l-Motoren von Volkswagen," in MTZ, Vol. 60, 1999, Nos. 7/8.

[5] Eichmüller, C., G. Hofstetter, W. Willeke, and P. Gauch, "Die Abgasanlage des neuen BMW M 3," MTZ, Vol. 62, 2001, No. 3.

[6] Hockel, K., "Der Abgaskrümmer von Personenwagenmotoren als Entwicklungsaufgabe," MTZ, Vol. 45, 1984, No. 10.

7.24 Control Mechanisms for Two-Stroke Cycle Engines

Characteristic of the principle behind the two-stroke cycle is that, in contrast to the four-stroke cycle, one complete working cycle is executed per crankshaft revolution; the expulsion of the burned charge from the cylinder and the introduction of fresh fuel and combustion air into the cylinder (scavenging process) takes place at crankshaft angles around BDC. The requirement here is that, with a suitable design of the mechanism controlling the change of gases, there is minimum mixing of fresh gas and exhaust gas (high scavenging efficiency) with a low required scavenging pressure gradient (low work expenditure for changing the charge), all this within the smallest possible crankshaft angle range around BDC (limited restriction on the useful piston stroke). There are several different scavenging processes available for the change of charges in two-stroke engines; these are explained in greater detail

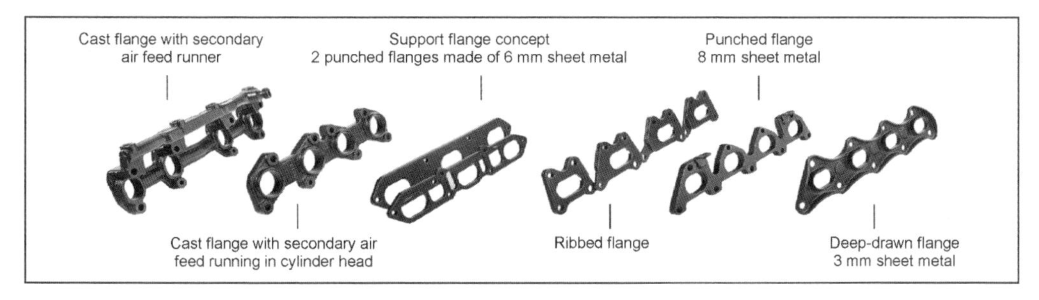

Fig. 7-341 Flange concepts for tube manifolds.

in Section 10.3 (see also Refs. [1, 2]). Their use requires a far different design for the drive components than what is found in four-stroke engines. Since the working cycle for the two-stroke engine transpires at the same frequency as crankshaft rotation, it is possible, in contrast to the four-stroke engine, to use the piston itself to control the gas flows.

Loop scavenging is used particularly in small engines and those running at high speeds; this principle is shown in Fig. 7-342. Here the piston controls the discharge of the exhaust from the cylinder through the exhaust slot(s), the inflow of fresh gas via the scavenging slots, and, when using the crankcase scavenging pump concept, the inlet of the fresh fuel-air mix into the crankcase as well. Because of the arrangement of the exhaust, intake, scavenging, and/or transfer passages at the cylinder, which penetrate the cylinder wall in the form of slots, the peculiarities described below result for the drive trains in two-stroke engines. The slots in the cylinder wall make it more difficult to achieve defined lubrication of the tribologic pair—the piston and cylinder. To ensure adequate lubrication and to avoid unacceptably high oil consumption, engineers must exercise great care when selecting the mating materials in regard to minimum lubricating oil requirements, metered lubricating oil feed, and/or sufficient oil stripper effect by the piston rings. To prevent the piston rings (and the ends in particular) from entering the exhaust, scavenging, and intake slots due to spring action, it is necessary to observe maximum slot widths (expressed as the ratio between slot width and cylinder diameter). This is explained in detail in Refs. [2, 3].

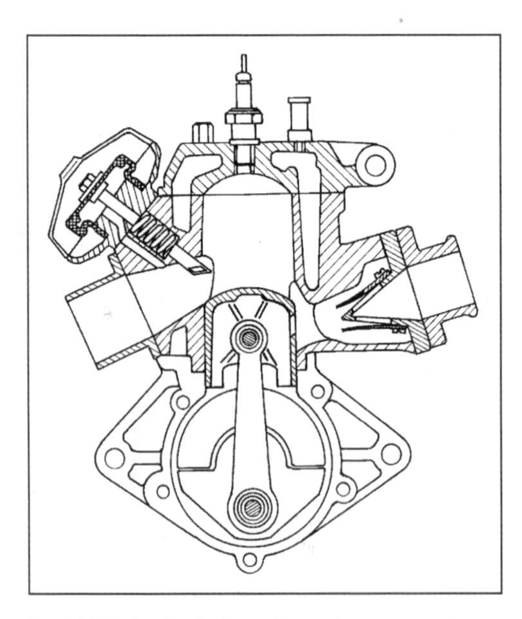

Fig. 7-342 Sectional view of a modern, two-stroke engine with loop scavenging, crankcase scavenging pump, reed valves at the intake system, and flat spool exhaust control.

In addition, the slots, normally rectangular in shape, have to be rounded at the corners at the upper and lower ends, and the transitions from the cylinder to the channel walls have to be rounded. Piston ring rotation in the piston grooves, accompanied by the hazard that the ends of the rings enter the slots in the cylinder walls under spring pressure, is prevented where required by pins pressed into the ring grooves.

The fact that firing is twice as often as in four-stroke engines and, above all, the piston controls fresh gas and exhaust flow, results in far higher thermal loads on the piston and cylinder in slot-controlled two-stroke engines when compared with four-stroke designs. This is discussed in Ref. [4]. This loading is seen as the essential cause for the limited service life often found in high-performance two-stroke engines. The situation is made all the more difficult where the incoming air or mix passes through the crankcase (crankcase scavenging pump). This largely eliminates effective cooling of the piston with splashed oil, a technique commonly used in higher-performance four-stroke engines. Among the strategies available to reduce the thermal load on the piston, piston rings, and wristpin boss are the following: Limiting individual cylinder volumes; careful designing of cylinder cooling (using water cooling if possible), particularly in the area around the exhaust slots; designing to reduce cylinder warping, which would make it more difficult to dissipate heat from the piston, through the piston rings, and to the cylinder walls; selecting a timing concept that prevents additional heating of the piston and fresh gas by exhaust blowback into the scavenging slots; selecting a scavenging process in which the exhaust flowing out of the cylinder is kept from coming into contact with any large surface area at the piston.

In modern loop scavenging cylinders for high-speed two-stroke engines, the fresh gas is generally introduced through between four and seven scavenging or transfer passages (in a mirror symmetrical arrangement to the exhaust channel), sweeping the wall at a shallow angle in the direction of the wall opposite the exhaust slot. This causes a rising stream of fresh gas to be formed along the cylinder wall. Near the cylinder head it reverses direction and forces the exhaust gas out of the cylinder. The transfer passages are located at the side of the cylinder and are tapered slightly along the direction of flow. This requires far more space between cylinders in multicylinder engines of this design when compared with similar four-stroke engines. The discontinuities in cylinder wall stiffness caused by the charge exchange runners results in more indirect force flow between the cylinder head and the crankshaft. Consequently, the highly asymmetrical thermal loading on the piston and cylinder due to the exhaust slots make it necessary to very carefully design the drive assembly and its cooling. It should be noted here that various strategies are used particularly in modern, two-stroke gasoline engines to increase the fresh gas fill efficiency, to influence fuel and air blending, and to avoid negative influences of gas pulsation in the intake and exhaust sections. Depending on the concept employed, these may involve rotary intake valves, reed valves (one-way valves), bypass reed controls, oscillation chambers, and, on the exhaust side, control spools or cylindrical valves. This may increase the complexity of the drive system considerably.

When using uniflow scavenging with exhaust valves, a concept employed particularly in diesel engines is used—the fresh gas enters the cylinder through scavenging slots under cylinder control while the exhaust gas flows out through several valves located in the cylinder head; their opening is synchronized with crankshaft rotation frequency. To achieve good scavenging efficiency, it is necessary that the intake runners or slots generally not impart any particular directional effect (aside from a slight tangential orientation to support gas blending); consequently, the volume of the intake plenum located upline from the scavenging slots, as shown in Fig. 7-343, in many cases adjoins the outside diameter of the cylinder sleeve (see also Ref. [5]).

Since the scavenging slots have to be covered by the piston skirt at TDC, long pistons are required particularly in long-stroke engines, resulting in a relatively large overall height for the engine. In contrast to loop scavenging, uniform-flow scavenging using exhaust valves causes somewhat less and more symmetrical thermal loading at the piston and cylinder. By contrast, the doubled actuation frequency for the exhaust valves (in comparison to four-stroke engines) and the high thermal loading on the cylinder head in fast-running engines places great demands on the design of cylinder head cooling and the kinematics of the valve train. In the design with four exhaust valves, often selected for high-speed engines, one objective in development is to achieve a shallow contour for the runners (small runner surface to be cooled, low exhaust heat losses where an exhaust turbocharger is used) so that the exhaust gas flow at the individual valves is hindered as little as possible. Aside from this, intensive cooling is necessary, particularly in the area around the injection nozzle to avoid carbonization problems. In order to exchange the charges—within the limited crankshaft arc available for this purpose—with the smallest possible amount of work, one must select a suitable valve train concept and valve train kinematics inducing minimum pressure loss as gases flow through the valves. Figure 7-344 shows the solution used in a 1.0-liter, two-stroke diesel engine currently being built by AVL. In this engine, the four exhaust valves

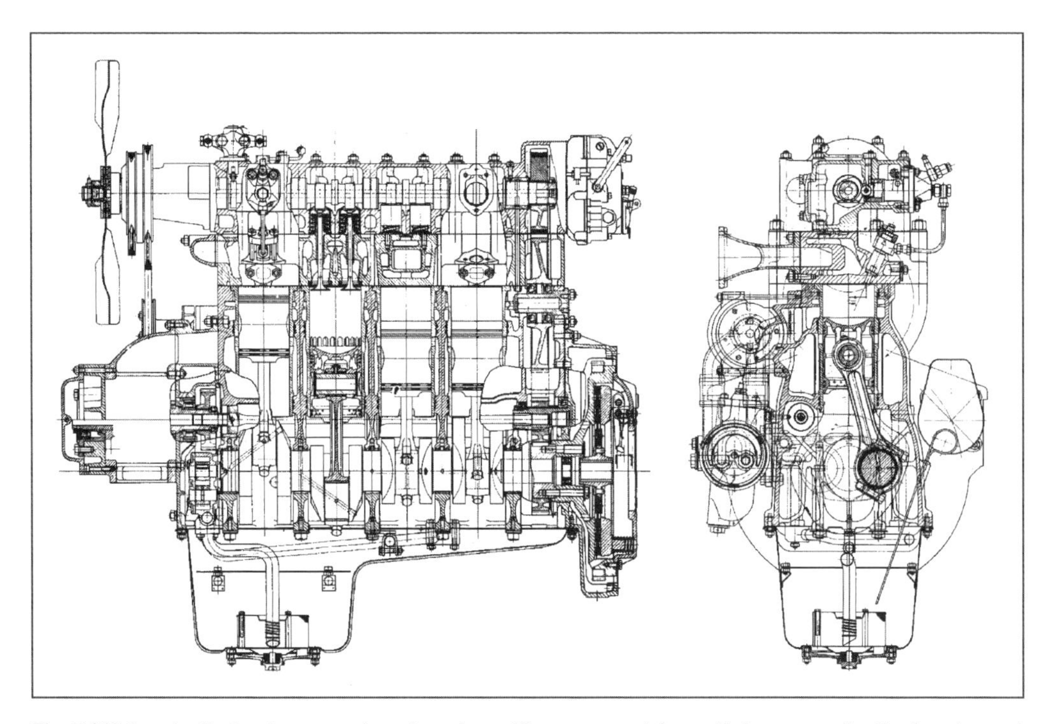

Fig. 7-343 Longitudinal and cross sections through a uniflow scavenged four-cylinder, two-stroke diesel engine made by Krupp.[5]

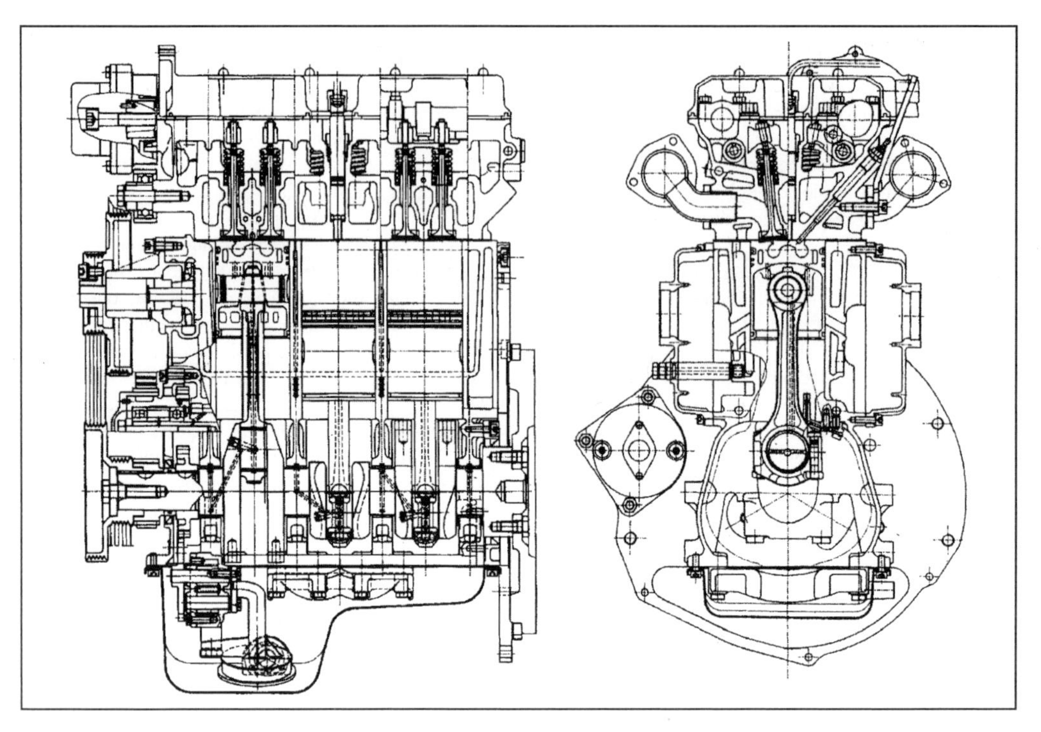

Fig. 7-344 Longitudinal and cross sections through a uniflow scavenged, two-stroke diesel engine made by AVL for passenger cars.[6]

Fig. 7-345 Illustration of the exhaust runner configuration and the valve train for a uniflow-scavenged, two-stroke diesel engine for passenger cars.

at each cylinder are activated by roller cam followers at two overhead camshafts. Figure 7-345 shows an alternate exhaust runner version to this concept.

Bibliography

[1] Venedinger, H.J., Zweitaktspülung insbesondere Umkehrspülung, Franckh'sche Verlagshandlung, Stuttgart, 1947.

[2] Bönsch, H.W., Der schnelllaufende Zweitaktmotor, 2nd edition, Motorbuch Verlag, Stuttgart, 1983.

[3] Küntscher, V. [ed.], Kraftfahrzeugmotoren–Auslegung und Konstruktion, 3rd edition, Verlag Technik, Berlin, 1995.

[4] N.N., Hütte; des Ingenieurs Taschenbuch IIA, 28th edition, Verlag Wilhelm Ernst & Sohn, Berlin, 1954.

[5] Scheiterlein, A., Der Aufbau der raschlaufenden Verbrennungskraftmaschine, 2nd edition, Springer-Verlag, Vienna, 1964.

[6] Knoll, R., P. Prenninger, and G. Feichtinger, "2-Takt-Prof. List Dieselmotor, der Komfortmotor für zukünftige kleine Pkw-Antriebe," 17th International Vienna Engine Symposium, 1996, VDI Fortschritt-Berichte Series 12, No. 267, VDI Verlag, Düsseldorf, 1996.

[7] Blair, G.P., Design and Simulation of Two-Stroke Engines, SAE International, Warrendale, PA, 1996.

[8] Meinig, U., "Standortbestimmung des Zweitaktmotors als Pkw-Antrieb," Parts 1 to 4, in MTZ, Vol. 62, 2001, Nos. 7/8, 9, 10, 11.

8 Lubrication

8.1 Tribological Principles

Engine technology is based on machine elements of different kinds that, linked by form and function, act on and influence one another, e.g., by

- Kinematics: Generation, transmission, and inhibition of movement
- Kinetics: Power transmission at boundary surfaces
- Transmission and transformation of mechanical energy
- Transport processes: Transportation of liquid and gaseous media

Tribology* plays an important role in these processes. According to DIN 50323, "Tribology . . . [is] the science and technology of surfaces influencing one another in relative motion. It covers the total area of friction and wear, including lubrication, and includes appropriate boundary surface reciprocal effects both between solids and between solids and liquids or gases."

Here lubrication permits, improves, and ensures the function, profitability, and service life of the components and functional groups of the engine and the complete powertrain.

In the field of their interactions, tribological systems can be reduced to a basic structure (system elements) (DIN 50320): Basic surface, mating surface, intermediate substance (particles, fluids, gases), and ambient medium (Fig. 8-1).

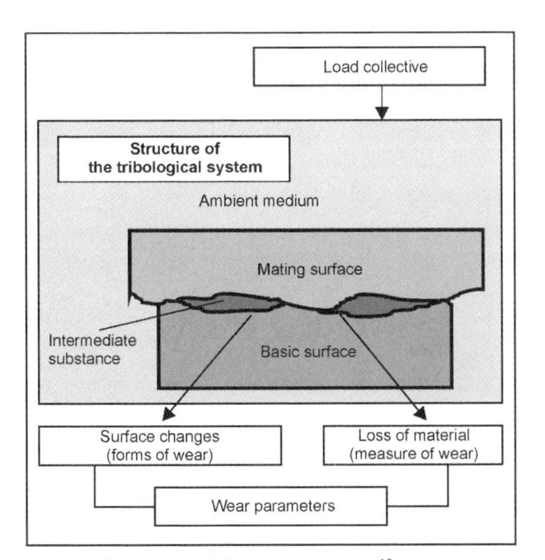

Fig. 8-1 Schematic: Tribological system.[12]

tribos (Greek) rubbing and *–logy* (Greek), a suffix of feminine nouns with the meaning of study of, knowledge, science.

Tribological stresses result from the movement process, effective forces (normal force), speeds, temperatures, and the duration of the load.

8.1.1 Friction

Friction is a complex phenomenon that is not easy to understand. It is ambiguous because it prevents movement as well as actually makes movement possible. There is no firm hold without friction—but also no movement away from the hold.

"Friction is an interaction between material areas of bodies in mutual contact. It opposes their relative movement. In the case of external friction, the areas of the substance in contact belong to different bodies, in the case of internal friction they belong to one and the same body." (DIN 50 323, Part 3).

Friction depends both on the state of movement of the friction partners, adhesive friction (static friction, striction) and motional friction (dynamic friction), and on the type of relative movement of the friction partners.

- Sliding friction: Sliding, translation in the contact surface, relative movement of the sliding partners
- Rolling friction: Rolling, rotation about an instantaneous axis in the contact surface
- Combined sliding and rolling friction: Rolling with microscopic or macroscopic proportions of sliding

Friction is also dependent on the condition of the substance areas involved:

- Dry friction
- Fluid or viscous friction
- Gas friction
- Mixed friction

In the engine, friction is undesirable because part of the mechanical energy already "generated" with poor efficiency is converted again into thermodynamically "lower valency" heat. By reducing the viscosity and load-bearing strength of the lubricant, this heat impairs the function of components. In extreme cases, damage can occur because of the bearings running warm or hot.

Dry friction is a result of several mechanisms:

- Adhesion and shearing: Formation and destruction of adhesive connections in the contact surfaces.
- Plastic deformation: Deformation due to relative tangential movement.
- Scoring: Sliding partners of different hardness, the rough peaks of the hard partner press into the surface of the soft partner and/or a hard particle between the sliding partners is pressed into the surface of one or both.
- Deformation: Elastic hysteresis and damping.
- Energy dissipation: Frictional energy (mechanical energy) is transformed into heat and is lost.

Static friction exists when a body is pressed onto its counterpart under the effect of a resulting force and adheres at rest. Static friction is the basis for the transmission of power between all parts of the engine permanently joined by bolts, clamps, or compression fits, such as crankcase and cylinder head, crankshaft and drive flange, or mounting bore and bearing. The critical coefficient of static friction μ_R for such connections depends on the material pairing, the surface condition, and the tribological conditions (lubrication); it is, therefore, a system property and not a material property.[1]

In the case of sliding friction (friction of movement), fluid friction, in particular, is of the greatest relevance to engine technology; it presupposes lubrication. The relevant friction conditions for machine parts are represented in the Stribeck curve named after Richard Stribeck (1861–1950) as

- Dry friction with direct metallic contact between the sliding partners.
- Boundary friction when the sliding partners are covered with traces of the lubricant.
- Mixed friction as a combination of dry and fluid friction when the lubricant film between the sliding partners is partially interrupted.
- Elastohydrodynamic lubrication: If high pressures exist between the sliding partners, the pressure in the oil film increases the viscosity of the oil. This is why—despite essentially unfavorable conditions—a sufficient minimum lubricant film thickness is obtained (for example, contraform contacts: gear pairs, cam/cam follower, etc.).
- Hydrodynamic lubrication: Fluid friction with complete separation of the sliding partners from one another by a lubricant film.

Losses because of friction are included in the mechanical efficiency. As the quotient of the effective power *Pe* and the indicated power *Pi*, the mechanical efficiency includes all the mechanical losses from the piston to the crankshaft flange. Furthermore, it also takes into account hydraulic losses (splash losses) and the drive powers of the ancillary machines necessary for operation of the engine. The mechanical efficiency of engines lies in the range from 75% to 90% at rated output and drops sharply at part load.

8.1.2 Wear

"Wear is a progressive material loss from the surface of a solid body caused by mechanical effects, i.e. the contact and relative movement against a solid, liquid or gaseous counterpart" (DIN 50320). Wear impedes functions and shortens service lives, but, as part of the gradual use, it is unavoidable in the operation of any machine.

Wear occurs when two friction bodies (basic and mating surfaces) are moved relative to one another under the effect of force—continuously, in oscillation, or intermittently. Here structural properties, strengths, hardness form, and surface geometry all have an influence on the wear. The wear process comprises several components that occur individually or in differing combinations with one another: Shearing, elastic and plastic deformations, as well as boundary surface processes. As a result, particles are released from the basic and mating surfaces and, in turn, increase the wear (Fig. 8-2).

For engine operation, it is the wear rate that is important, i.e., the speed at which the wear develops:

- Degressive: Running-in processes during which roughness unavoidable in production is smoothed out and the bearing surfaces of the partners are increased.
- Linear: Normal operation during which the wear increases steadily, but only slightly.
- Progressive: Self-propagating, the rate of wear accelerates so that functional faults quickly occur and lead to damage.

In engines, wear is predominantly caused by

- Sliding wear with dry contact and with boundary and mixed friction (incomplete separation of basic and mating surfaces).

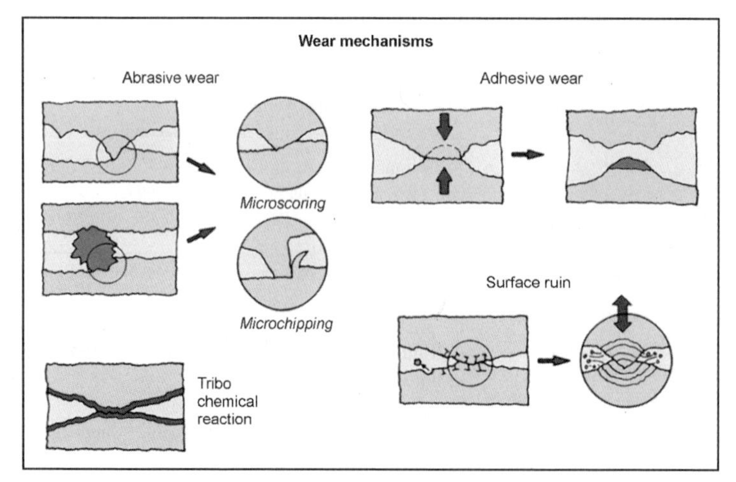

Fig. 8-2 Wear mechanisms.

- Vibrational wear. Typical: Fretting (rubbing oxidation, fretting corrosion).
- Fluid friction (complete separation of basic and mating surfaces).
- Cavitation: Formation of cavities because of localized low pressure in a fluid with subsequent implosion of the vapor bubbles. This causes damage to the adjoining surfaces; hydrodynamic properties deteriorate.
- Erosion: Exposure of solids to liquids containing particles [e.g., lubricants or fuels with foreign particles or gas streams with particles (exhaust gas with combustion residues)]; parts of the material surface are worn away.
- Wear due to impingement.
- Wear due to corrosion.

In the engine, wear expresses itself as a reduction in cross section, changes in surfaces, functional deterioration because of increased clearances, reduction in overlaps, and impairment of the geometry and kinematics. Consequences can be increased friction, seizing, and overload or vibration fractures. Wear in the engine is generally caused by

- Overloading
- Inadequate lubrication as a consequence of lack of lubricant and/or unsuitable or old oils
- Unfavorable operating conditions
- Malfunction or failure of engine components

Wear occurs predominantly in the following function groups:

- Engine: Pistons, piston rings, cylinders, bearings, and shafts
- Gearing drive: Gear wheels
- Control system: Cams and cam followers, valves, valve seats and valve guides, belt drives

8.2 Lubrication System

8.2.1 Lubrication

Lubrication[2,3] is the coating or wetting of sliding partners with a lubricant; this can be "liquids, gases, vapors, i.e. fluids, plastic substances and solids in powder form."

Functions of the lubrication are

- Power transmission.
- Reduction of friction and wear.
- Precision sealing: Parts sliding on and inside one another can, in principle, be sealed purely by means of a lubricant film.
- Damping of impact and vibration.
- Reduction of noise.
- Cooling: Dissipation of friction heat.
- Cleaning: Discharge of particles of all kinds.
- Corrosion protection.

The lubricant is a machine element; in the bearings it transmits the component forces by lubricant films with thicknesses of just a few thousandths of a millimeter. This ability is derived from the viscosity, i.e., the ability of the lubricant to resist a change in shape. The individual fluid particles rub together; tangential stresses (shear stresses) are created at their contact surfaces. The magnitude of these stresses is dependent on the shear rate perpendicular to the flow direction dv/dz and a material characteristic of the fluid, its kinematic viscosity η (viscousness) (Newton's shear stress). The kinematic viscosity, in turn, depends on the lubricant, its temperature, and pressure, as well as on the shear rate (Fig. 8-3).

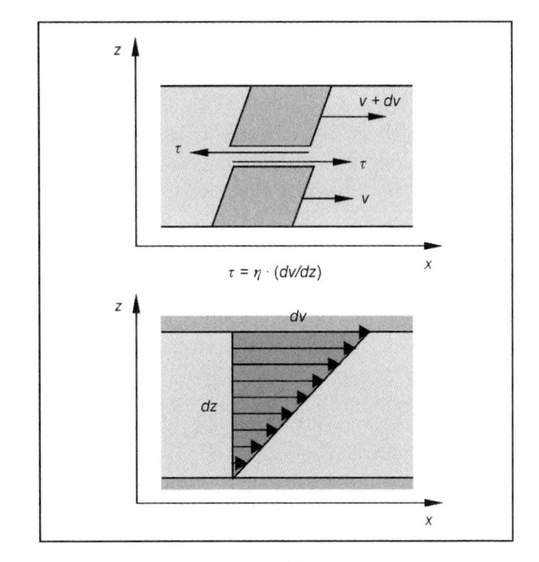

Fig. 8-3 Shear and shear rate.[13]

The shear stresses perform friction work (dissipation work) in sliding direction; this kinetic energy that is transformed into heat is "lost." In machine operation, the fluid friction has a disadvantageous effect: It costs mechanical energy and heats up the lubricant; that, in turn, reduces the load-bearing strength of the lubricant film. This heat of friction has to be dissipated, hence, necessitating additional design and operational measures. In the worst case, with mixed friction, it leads to wear of the sliding partners right up to seizure. But without inner friction, a fluid could not transmit forces.

8.2.2 Components and Function

A lubrication system consists of lubricant-conveying pipes, pumps, filters, heat transmitters, and the control elements in their arrangement relative to one another. Of particular note are the oil reservoir (oil sump), oil pump(s), oil heat exchangers, oil filters, control valves, filler neck, and the monitoring of the oil volume (oil level) and oil volumetric flow (oil pressure).

A distinction is made for the following:

- *Fresh oil or total-loss lubrication*: Here the oil is pumped from an oil reservoir to the individual consumers. It has to be ensured that clean, cool oil is delivered to the consumers at all times. With careful metering the oil consumption can be kept low. The fresh oil lubrication method is used in two-stroke SI engines with fuel injection.

- *Mixture lubrication*: This method of lubrication is used today predominantly for small two-stroke engines. The lubricating oil is added to the gasoline in a particular ratio (1:50 or 1:100) during refueling. The oil enters the cylinder together with the fuel on the intake stroke and into the crank chamber with the overflow. The discharged oil lubricates the bearings and the cylinder wall. Lubricating oil also enters the exhaust with the scavenging air, which increases the oil consumption and reduces the exhaust gas quality.
- *Forced-feed lubrication*: Four-stroke engines and two-stroke diesel engines are generally lubricated by this method. A pump delivers the oil from a tank via a system of pipes to the consumers, and from there it flows back pressure-free to the tank.
- *Dry sump lubrication*: Dry sump lubrication is used for conserving space (installation space) or for special operating conditions (off-road vehicles, sports cars). A suction pump draws the oil into a separate tank, and from there it is returned by a pressure pump to the oil system after cooling and filtration. The suction and pressure stages of the pump are often designed together.

Engine lubricating oil circuit[4,5,6]: The intake screen of the oil pump is located at the lowest point of the oil sump to ensure the oil supply even when the vehicle is at an angle. A positive-displacement pump—driven via gear wheel, chain, and toothed belt or mounted directly on the crankshaft—forces the engine oil through the filter and, depending on the design of the lubricating oil system, through a heat exchanger into the main oil line. A pressure relief valve located on the pressure side allows oil to bypass when the set pressure is exceeded. The control bores are designed to level out pressure peaks and suppress pressure fluctuations. The discharged oil either runs off freely or is returned to the intake side of the pump so that it does not become enriched with air.

From the pump, the oil passes through the filter. As protection against overloading because of excessive oil pressures, for example, during cold starting, the pump has a bypass valve; a nonreturn valve prevents the oil from running back when the engine is at standstill (Fig. 8-4).

The primary function of the oil filters is to protect the sliding partners from foreign particles in the oil. For this, the filter must be installed upline of the consumers so that the full oil flow passes through the filter (full-flow circuit). To relieve the full-flow filter and reduce its soiling, part of the oil is branched off from the main flow and is passed through a bypass filter—an oil centrifuge or a fine filter (Fig. 8-5).

Bypass filters are not, however, an alternative to oil changes as they can neither replace used additives nor filter fuel, water, and acids out of the lubricant.[7] If the engine oil is subject to high thermal loads, it has to be cooled separately, either with a water/oil or an air/oil heat exchanger. The oil heat exchanger is normally installed downline of the filter to minimize the pressure loss in the filter with the still warm and, therefore, low-viscosity oil. For optimum protection of the engine, however, the filter should be located downline of the heat exchanger, i.e., immediately in front of the oil consumers.

From the filter or heat exchanger, the oil passes via the main oil channel to the oil consumers. The engine is supplied with oil from the main oil channel through bores in the crankcase intermediate walls and in the main bearing shells. It passes through bores in the crankshaft to the connecting rod bearings and from there—depending on the design—through a bore on the connecting rod to the piston pin bearing (Fig. 8-6).

In order to deliver the oil to the main bearing journals, centrifugal force has to be overcome. On the other hand,

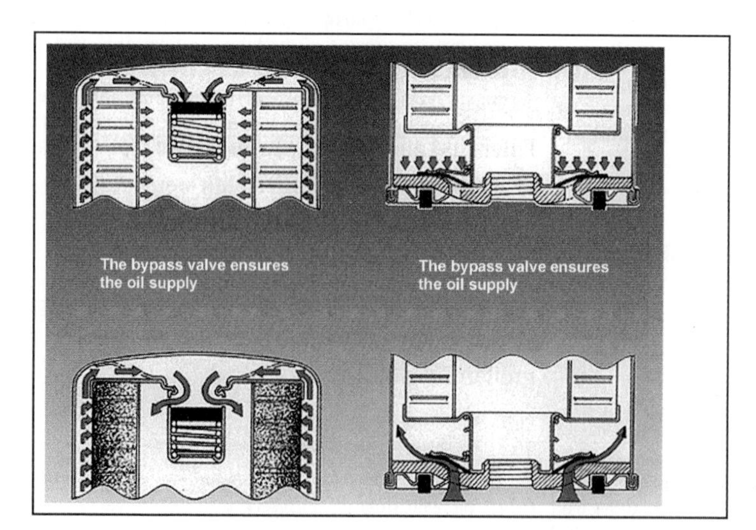

Fig. 8-4 Bypass valve and nonreturn valve for oil filters (Volkswagen).

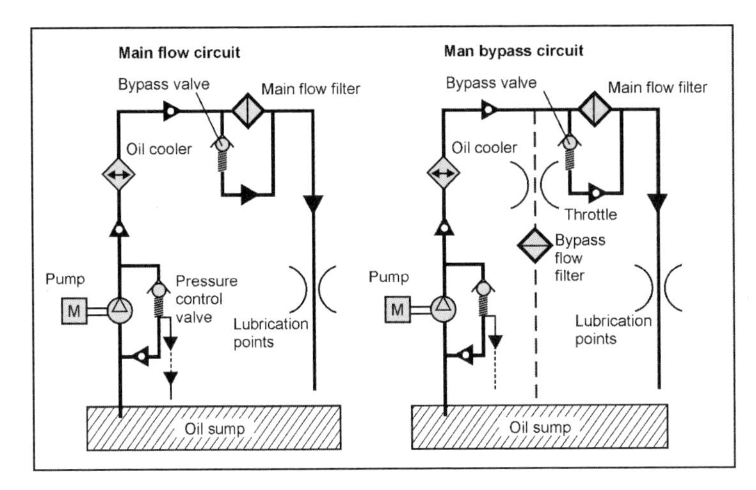

Fig. 8-5 Full-flow and full/bypass flow filtration.[14]

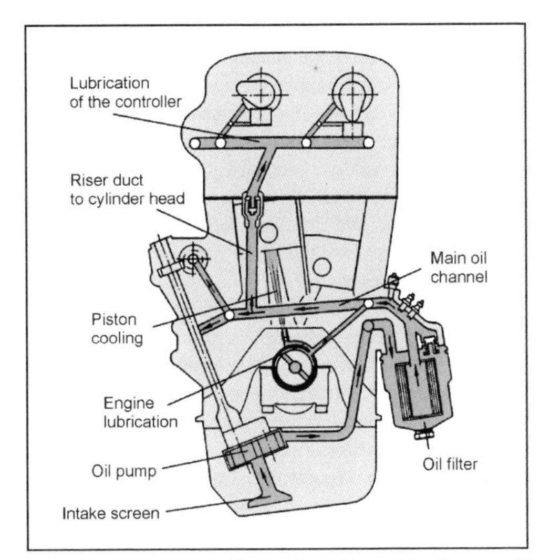

Fig. 8-6 Lubricating oil circuit (schematic) of a car SI engine (Volkswagen).

delivery from the bore in the main bearing journal to that of the cam journal or to the pinion pin bearing is enhanced by the centrifugal force or by the oscillating movement of the connecting rod. As a rule, one main bearing should supply only one cam journal with oil.

In high-performance engines, the oil circuit is split into two channels, one supplying the camshaft control with oil under high pressure, the other supplying the camshaft bearings and bucket tappets with oil under low pressure.[8] The oil supply to engine parts such as belt tensioner bearings and to engine accessories such as exhaust turbocharger, fuel injection pumps, etc., comes directly via oil channels. Components not connected to the oil supply system such as rocker arm contact surfaces or the flanks of gear wheels are lubricated indirectly by the spray oil in the crankcase. Under critical conditions, separate spray nozzles ensure an adequate supply of oil. The valve guides are also lubricated by sprayed oil, with the oil supply to the guides limited or metered by valve shaft seals. The trend today is towards more or less integrated oil lines and short oil paths with low pressure losses (hydraulic losses) (Fig. 8-7).

For engines with high specific output, piston cooling is now indispensable. Lubricating oil is diverted from the main flow and injected through injection nozzles against the underside of the piston or into piston cooling channels for the piston cooling. Pressure-controlled valves prevent

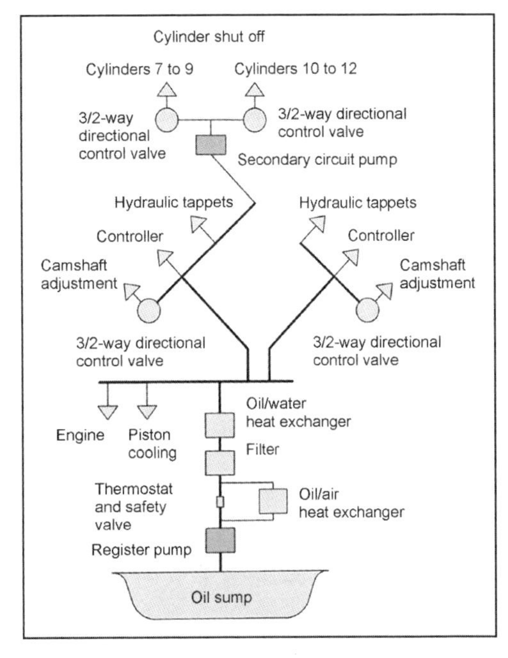

Fig. 8-7 Oil circuit of a V12 SI engine with cylinder shut-off (Mercedes-Benz).

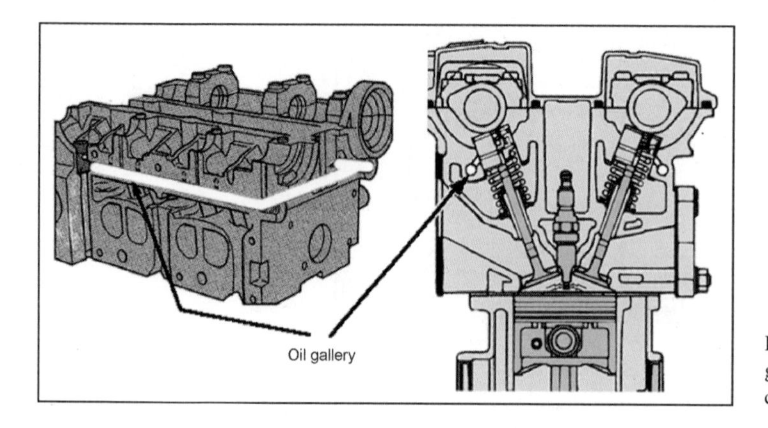

Fig. 8-8 Arrangement of the oil gallery in the cylinder head of a car SI engine (Ford).

heat being unnecessarily drawn from the piston when the engine, and hence the oil, is still cold. The spraying of the piston undersides through bores in the large connecting rod eye is a disadvantage because this cooling oil has to be additionally transported through the crankshaft.

As delivery begins only when the engine is started, there is a danger that the oil consumers receive no oil or too little oil during the first few revolutions of the engine. For this reason, nonreturn valves are fitted in risers and oil galleries in cylinder heads from which the collected oil can flow quickly to the consumers (Fig. 8-8). The electrically driven lubricating oil pilot pumps normally used on larger gasoline and large diesel motors cannot be used in motor vehicle engines because of the design complexity, the additional weight, and the cost.

Low oil levels and frequent oil circulation result in increased foaming of the oil. The upper limit for the gas content is considered to be 8%. Centrifugal separators and/or low-level oil return lines are used to counter foaming. As a result, the gas content can be reduced to below 4% (Fig. 8-9).

The oil in the sump is kept away from the engine by oil baffle plates so that the crankshaft cannot become immersed in oil because of the sloshing of the oil caused by the vehicle movement (Fig. 8-10).

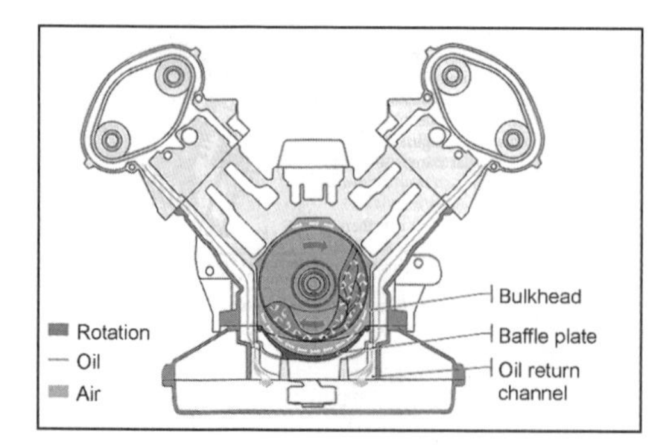

Fig. 8-9 Return oil passage from the cylinder heads of the Audi V6 Biturbo.

Oil pumps: Recirculating positive-displacement pumps—gear and ring gear pumps—of various designs are used for vehicle engines: External gear pumps, internal gear pumps (crescent pumps), and ring gear pumps (rotor pumps). These pumps are compact, have high efficiencies, exhibit a good intake behavior, and are suitable for a wide range of viscosities of the fluids to be pumped. The change in volume necessary for pressure boosting with positive-displacement pumps is affected by the meshing of the gear wheels. The displacement is calculated from the tooth geometry and the pump speed (Fig. 8-11).

Evaluation criteria for oil pumps are delivery characteristics, efficiency, sensitivity to cavitation, noise development, installation size, weight, and manufacturing costs. Important factors are a low intake head and a rapid pressure buildup in the oil circuit. The transport losses have to be covered, and the centrifugal force in the main bearing journals and the flow resistances of the oil consumers (bearings) have to be overcome. The pressure losses from the pump to the cylinder head lie in the order of approximately 1.5 to 2 bar. The flow velocity of the lubricating oil in the lines should not exceed 3 to 4 m/s.

Oil pumps are mounted on the crankshaft or engine block or in the sump. Mounting on the crankshaft permits an easier design and is cheaper (roughly 50% less expensive than installation in the sump), but it also forces larger impellers and higher pump speeds to be used than is really necessary. The power consumption is therefore significantly higher, irrespective of the pump type. Furthermore, the wobbling of the crankshaft has to be compensated, in ring gear pumps either by mounting the inner rotor in the pump housing or by centering the inner rotor on the crankshaft.[9]

If the pump is located in the sump, the intake head is lower and the pump draws in oil better during starting. In addition, lower pump speeds can be used (e.g., gear ratio 1:1.5), therefore reducing the drive power. One disadvantage here is the complexity of the drive with chain, toothed belt, gear, or worm drive.

The delivery characteristic of recirculating positive-displacement pumps is dependent on the pump speed.

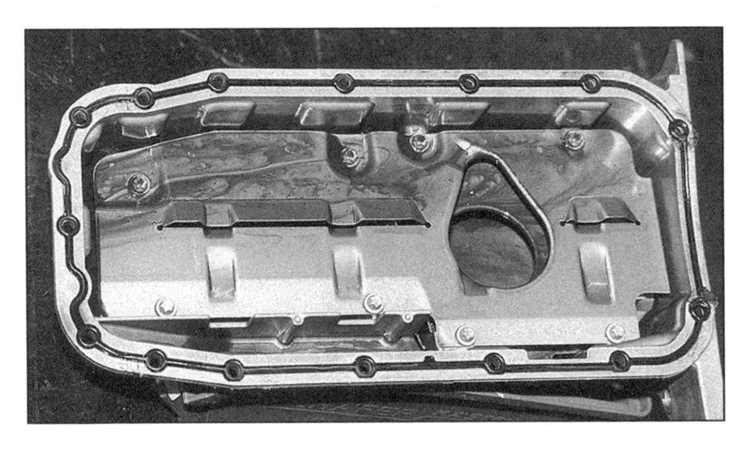

Fig. 8-10 Oil baffle plate of a four-cylinder car engine (Opel Ecotec).

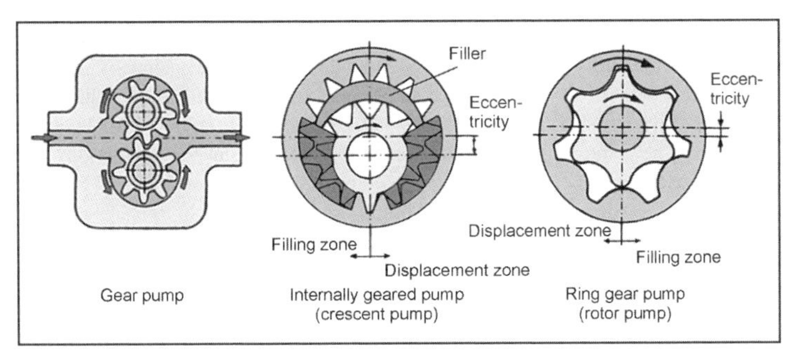

Fig. 8-11 Types of engine oil pump (schematic).

With increasing pump pressure, the volumetric efficiency drops because of the leakage losses. The oil demand of the engine, however, is more or less independent of the engine speed, so that with increasing engine revs the difference between delivery and demand becomes even larger. The individual oil consumers have different requirements: The bearings require a specific oil volumetric and hydraulic actuators a specific pressure. For the camshaft adjustment mechanism, for example, higher deliveries are required; a dedicated secondary pump is provided for the cylinder shutoff. The design of the pump for a minimum oil volumetric flow at (hot) idle speed—i.e., low engine revs and low viscosity of the oil—means that with increasing engine revs, oil has to be bypassed above a certain counterpressure so that roughly 50% of the hydraulic energy is transformed into heat.

A distinction is made between control valves that are controlled directly by the system pressure and valves that are controlled indirectly, i.e., by both the system pressure and a given pilot pressure (Fig. 8-12).

For additional consumers such as exhaust gas turbochargers, more oil has to be delivered. In addition, a reduction in the engine idle speed to lower the engine losses results in a significantly increased delivery at high engine revs. The disparity between the oil to be delivered at low engine revs and the oil volume actually required at high engine revs becomes even greater. For this reason, efforts are made to adapt the pump characteristic better to

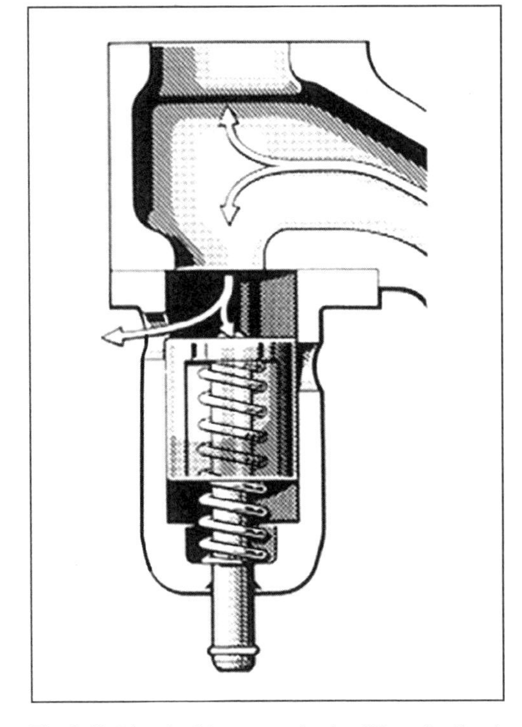

Fig. 8-12 Directly driven control valve (Mercedes-Benz).

the oil requirement of the engine by controlling the pump, by using register pumps, by varying the eccentricity on pumps with internal gearing, by using intake control for ring gear pumps, by axial shifting the secondary gear on pumps with external gearing, or by isolating the pump drive from the engine speed with the use of electric drives for the pump. However, such solutions demand a careful comparison of the design complexity and the additional weight and costs against the power savings that can be achieved.

For four- to six-cylinder engines, the oil demand is 40 to 100 l/min, and eight-cylinder engines require around 100 to 120 l/min. As a rough estimate the crankshaft main bearings of car engines require 3 l/min per bearing, the connecting rod bearings 4 to 5 l/min per bearing, the piston cooling 1.5 to 3 l/min per nozzle, the cylinder head about 12 l/min. However, 50% to 60% of the oil volumes are spilled off. Engines with aluminum crankcases require slightly more oil as the clearances increase with the temperature because of the greater thermal expansion. The delivery pressure is approximately 5 bar. The drive powers of oil pumps for four- to six-cylinder engines lie in the range from 0.5 to 2 kW, for larger engines up to 5 kW.

Oil monitoring: Because it is so vital for the engine, the oil supply has to be monitored.[10] As a rule, the pump counterpressure is used as the monitoring parameter. This is problematical in that it is not the physically relevant parameter, the oil volumetric flow, but a dependent parameter, the pump counterpressure, that serves as the monitoring parameter. On the one hand, this increases as a square of the flow velocity (in line with the volumetric flow), and, on the other hand, it is also dependent on the flow resistance. With increasing temperature, the viscosity (viscousness) of the lubricant decreases so that more oil has to be delivered to maintain the specified control pressure. If the line becomes clogged, the flow resistance increases so that, despite a lower oil volumetric flow, the pressure does not decrease. If, on the other hand, the coefficient of resistance drops because of an increase in bearing clearances, although more oil flows through the bearings, the pressure drops and incorrectly signals "low oil." For this reason, the oil pressure should be monitored at the end of the line, e.g., behind the last crankshaft bearing or in the cylinder head. Because the engine operator cannot keep an eye on the oil pressure gauge the whole time, he often notices a drop in the oil pressure only when it is too late, namely, from the generally disastrous consequences. For this reason, the drop in oil pressure should also be signaled acoustically.

Further monitoring parameters are oil temperature and oil level. Sensors are used for this purpose; it must also be possible to check the oil level manually using an oil dipstick with marks for the maximum and minimum oil levels.

Oil burden: The burden on the engine oil has increased continuously over the course of time: Because of smaller oil filling volumes, because of increasing powers as a result of higher engine revs and turbocharging, because of more compact engines (downsizing, particularly with the V-type engine) through more complex designs, longer inspection and oil change intervals, and because of widely (and frequently) changing engine loads and speeds. Furthermore, aerodynamically optimized body forms allow the temperature in the engine compartment to increase. The oil burden can be expressed in figures with various coefficients (Fig. 8-13), e.g., oil filling volume/swept dis-

Year Type	1937 Super 6	1940 Kapitän	1951 Kapitän	1960 Kapitän	1970 Commodore	1980 Commodore	1990 Omega	2000 Omega
Swept displacement dm^3	2.5	2.5	2.5	2.5	2.5	2.5	2.6	2.6
Power kW	40.4	40.4	42.6	66.2	88.2	110	110	110
Engine speed rpm	3600	3600	3700	4100	5500	5800	5600	5600
Oil filling l	5	4	4	4	4.5	5.75	5.5	5.5
Oil burden kW/l	8.1	10.1	10.65	16.55	19.6	19.1	20	20
Oil filling/ swept displacement l/dm^3	2	1.6	1.6	1.6	1.8	2.3	2.1	2.1
Oil change interval km	2000	2000	3000			7500	10 000	15 000

Fig. 8-13 Technical data of 2.5 l Opel engines.

placement or oil filling volume/power. More precise information is given by the oil burden coefficient:

Oil burden coefficient

$$= \frac{\begin{pmatrix} \text{Engine power}[\text{kW}] \\ \times \text{ Oil change interval}[\text{km}] \end{pmatrix}}{\begin{pmatrix} (\text{Oil volume } + \text{ Refill volume per} \\ \text{oil change interval}) \; [\text{L}] \cdot 1.000 \end{pmatrix}} \qquad (8.1)$$

Two such coefficients are compared in Ref. [11]:

Oil burden coefficient	KW · km/l
Ford Taunus 1949	11.5
Audi Quattro 1987	277.2

Oil consumption: The oil volume in the oil tank (sump) decreases during the course of the operating life because of oil losses and oil consumption. Oil losses occur when oil escapes between the rigid and moving parts of the engine. These can be the connection from the crankcase to the sump and cylinder head, the connection from the cylinder head to the cylinder head cover, the connections between oil filter and oil cooler, as well as leaking oil drain plugs and crankshaft seals.

The actual oil consumption results from internal leaks because of burning and/or evaporation of oil. Such leaks are caused by worn piston rings or piston ring grooves, mirroring in the upper area of the cylinder tracks, excessive clearance between valve stem, and valve guide or leaks in the turbocharger. The oil consumption can be estimated only roughly because it depends on a large number of parameters that change during the course of the engine life. "Normal" consumptions for car engines are 0.1 to 0.25 (0.5) l per 624 mi. A constant oil level does not always mean that no oil is being consumed because the oil consumption can—particularly in diesel engines—be "compensated" by the ingress of fuel into the oil system.

Oil change: The oil as a medium of lubrication is subject to a huge number of changes during the engine operation. These necessitate the periodic replacement of the oil fill (oil change). The oil change intervals have been significantly increased during the last decade. Criteria for the oil change are the content of liquid and solid foreign matter, the exhaustion of the additive effectiveness, and any impermissible changes in the viscosity. The filters have to be changed at the same time the oil is changed. The oil change intervals are specified by the engine manufacturers depending on the engine type (gasoline, diesel), engine model, service life in km or mi, operating time in months, and the respective operating conditions; they vary widely for car engines from (3000 mi), 9.300 to 2.400 mi (18.600 mi). These intervals must be strictly observed. The old oil must be disposed of in the prescribed manner.

More recently, the development is towards flexible, load-dependent oil change intervals from 2.400 to 25.000 mi, corresponding to 1 to 2 years of operation. The crucial factor for the oil change interval is the condition of the oil. It deteriorates during the engine operation because of oxidation, the formation of organic nitrates, reduction in the additive effectiveness, and, in diesel engines, additionally the incorporation of soot. Determining factors here are the engine size, i.e., the load on the engine, the operating conditions (cold start, hot running), and the oil grade. A sensor is used to monitor the operating temperature of the engine, the oil filling level, and the oil quality, where the dielectric constant is regarded as a criterion for the condition of the engine oil.[12]

Bibliography

[1] Czichos, H., and K.-H. Habig, Tribologie Handbuch, Vieweg, Wiesbaden, 1992.

[2] Affenzeller, J., and H. Gläser, Lagerung und Schmierung von Verbrennungskraftmaschinen, Die Verbrennungskraftmaschine-Neue Folge, Band 8, Springer, Wien, 1996.

[3] Fuller, D.D., Theorie und Praxis der Schmierung, Berliner Union, Stuttgart, 1960.

[4] Gläser, H., V. Küntscher, [ed.], Schmiersystem, in Kraftfahrzeugmotoren, 3. Aufl., Verlag Technik, Berlin, 1995.

[5] Reinhardt, G.P.u.a., Schmierung von Verbrennungskraftmaschinen, expert-Verlag, Ehningen, 1992.

[6] Treutlein, W., K. Mollenhauer [ed]., Schmiersysteme, in Handbuch Dieselmotoren, Springer, Berlin, 1997.

[7] Greuter, E., and S. Zima, Motorschäden, 2. Aufl., Vogel Buchverlag, Würzburg.

[8] Porsche 911, Sonderausgabe ATZ/MTZ.

[9] Eisemann, S., C. Härle, and B. Schreiber, Vergleich verschiedener Schmierölpumpensysteme bei Verbrennungsmotoren, MTZ 55 (1994) 10.

[10] Zima, S., Kurbeltriebe, 2nd edition, Vieweg, Wiesbaden, 1999.

[11] Eberan-Ebenhorst, C.G.A. von, Motorenschmierstoffe als Partner der Motorenentwicklung., in Schmierung von Verbrennungkraftmaschinen, TA Eßlingen, Lehrgang, 13–15.12.2000.

[12] Warnecke, W., D. Müller, K. Kollmann, K. Land, and T. Gürtler, Belastungsgerechte Ölwartung mit ASSYST, MTZ 59 (1998) 7/8.

[13] Standard DIN 50320 Wear (Terms).

[14] Motorenfilter, Die Bibliothek der Technik 31, Verlag Moderne Industrie, Landsberg/Lech, 1989.

9 Friction

9.1 Parameters

The useful power at the output shaft of the internal combustion engine (effective power P_e) is lower than the internal power at the piston (indicated power P_i). The difference is referred to as the friction loss P_r.

$$P_r = P_i - P_e \tag{9.1}$$

The friction loss includes the losses of the individual engine components such as the engine proper (crankshaft, connecting rods, pistons with piston rings), the valve train including the timing gear, and the requisite auxiliary drives. The internal power also allows for the losses due to the charge cycle, where the operating states and, consequently, the drive powers of the auxiliaries are often defined differently in the various standards.[1] The friction loss reduces the engine power available at the output shaft and, thus, also influences the fuel consumption of the engine.

Analogous to the effective and indicated mean pressure, the mean friction pressure p_{mr} is used to compare different engines with different swept volumes.

$$p_{mr} = p_{mi} - p_{me} = \left(\frac{P_i - P_e}{i \cdot n \cdot V_H}\right) = \left(\frac{P_r}{i \cdot n \cdot V_H}\right) \tag{9.2}$$

The friction of a complete engine includes the friction losses or drive powers of the individual components:

- Engine, consisting of
  (a) Crankshaft main bearing with radial shaft seal rings
  (b) Connecting rod bearings and piston group (pistons, piston rings, and piston pins)
  (c) Any mass balancing systems
- Valve train and timing gear
- Auxiliaries, such as
  (a) Oil pump, possibly with oil pump drive
  (b) Coolant pump
  (c) Alternator
  (d) Fuel injection pump
  (e) Radiator fan
  (f) Vacuum pump
  (g) Air conditioning compressor
  (h) Power steering pump
  (i) Air compressor

9.2 Friction States

Depending on the lubrication prevailing at the various friction points in the engine, different friction states occur. The most important are

- Solid friction (Coulomb's friction)
  Friction between solids without fluid intermediate layer.
- Boundary friction Friction between solids with an applied solid lubricant layer without a fluid intermediate layer.
- Mixed friction
  Fluid friction and solid friction or boundary friction occur simultaneously; the lubricant layer does not completely separate the two friction layers from one another, and a certain contact occurs.
- Fluid friction (hydrodynamic friction)
  A liquid (or gaseous) substance between the two friction layers completely separates the two from one another. In the internal combustion engine, the movement of the friction surfaces against one another creates the hydrodynamic supporting effect of the intermediate substance.

The occurrence of the different friction states is explained below using an example. In a hydrodynamic plain bearing, the different friction states occur as the engine passes through the engine rev band. The Stribeck curve in Fig. 9-1 shows the relationship between the coefficient of friction λ and the shaft speed n or the sliding velocity v at constant temperature (or constant viscosity η).

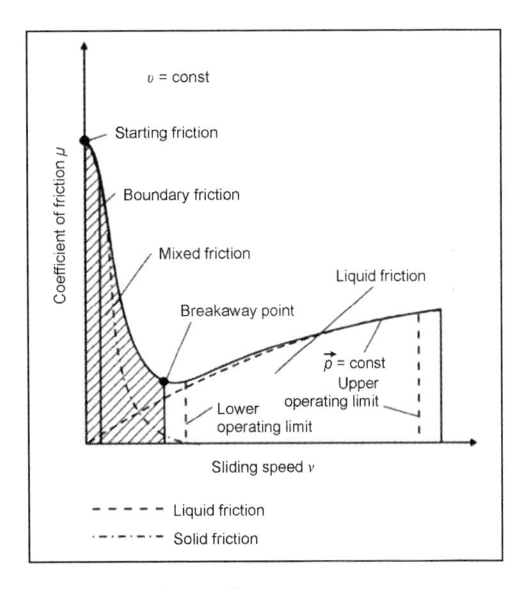

Fig. 9-1 Stribeck curve.[2]

The overall friction is made up of the two portions solid friction (or boundary friction) and fluid friction. At a standstill we have static friction. At low engine revs we first have solid friction and boundary friction, and then the mixed friction band occurs in which the friction decreases with increasing engine revs with a corresponding increasing buildup of a hydrodynamic supporting film. The breakaway point in this model representation is the point at which the hydrodynamic supporting film can completely separate the surface roughnesses of the two friction partners. The engine speed at which this state is achieved is also referred to as the transitional speed at which the minimum friction occurs. At engine speeds above the transitional speed, fluid friction occurs, and the friction increases again because of the increasing shear rates.

Increasing loads on the friction pair or decreasing viscosity of the fluid shifts the transitional speed upwards and extends the range of mixed friction.

Operating states on the left-hand branch of the Stribeck curve are unstable, as a brief variation such as an increase in engine revs or reduction in the load leads to a significant rise in the coefficient of friction and, hence, to an automatic amplification of the fault. For this reason, the operating point of a friction pair in continuous operation must be sufficiently far from the breakaway point on the right-hand branch of the Stribeck curve.

9.3 Methods of Measuring Friction

Exact calculation of the friction losses involves a great deal of work. There are various ways of determining the friction, although the majority of these exhibit significant inaccuracies. The following methods are commonly used for calculating the friction[3,4]:

- The rundown method: Here the engine is switched off after stabilization at an operating point, and the change in speed is measured as a function of time. The friction moment or mean friction pressure is then calculated using the moments of inertia of the moving masses.
- The shutoff method: On multiple-cylinder engines, the fuel supply to one of the cylinders is shut off, and this cylinder is then dragged along by the other working cylinders. The friction loss can be determined from the change in effective engine power before and after the fuel shutoff.
- The Willans lines: The fuel consumption of an engine is plotted on the Y axis against the mean effective pressure p_{me} for various engine speeds. The intersections with the negative p_{me} axis are then determined by linear extrapolation of the values down to fuel consumption zero; these can be roughly regarded as the mean friction pressures at the respective engine speeds.
- The motoring method: The engine is motored on a test rig by an external motor. The motoring power required to drive the engine is regarded as the friction loss. With this method either the engine can be motored at operating temperature and measured immediately after shutting off the fuel supply or it can be conditioned via external thermostat installations.
- The strip method: Strip measurement is a special form of motoring that is used to measure the friction losses of the various engine components, such as, the friction of the engine, the valve train, and the auxiliary drives. The designation derives from the method where the engine is dismantled (stripped) step-by-step on a motoring test rig. The friction losses of the individual components are determined from the difference between the measured values with and without these components. The total friction of the engine is obtained by addition of the values for the individual components.
- The indication method: This method can be used to determine the friction of an engine in motoring mode. Integration of the measured cylinder pressure over a working cycle gives the indicated work W_i which, referred to the swept volume, gives the indicated mean pressure p_{mi}. If the mean effective pressure p_{me} calculated from the torque measured at the drive shaft is subtracted from this, we obtain the mean friction pressure p_{mr}.
- Special measuring method: Apart from the friction measuring methods described above, there are a large number of other methods for determining, for example, the friction of individual components during operation. Torque measuring flanges can be used to carry out measurements on components driven by shafts.[2,4] For the piston group there are various facilities for measuring the piston frictional force.[5]

A crucial aspect for the precision and reproducibility of the individual methods and, hence, for the comparability of various measurements is strict compliance with the boundary conditions. For all these measurement methods, for example, the lubricating oil and coolant temperatures of the engine have to be set to less than ± 1 K. This is generally possible using only high-precision external thermostat installations.

Of the possibilities described for determining p_{mr}, the first three are subject to significant inaccuracies from the principle of the method alone and are therefore suitable only for the identification of trends.

With the motoring method, the problem is that the inertia moment of a complete engine includes not only the mechanical engine friction and the drive power of the auxiliary drives but also the charge cycle losses and that without additional indication no distinction can be made between the friction and the charge cycle losses. However, since the charge cycle losses react very sensitively to changes in ambient conditions on the test rig or to minor differences in the intake and the exhaust systems, the comparability of different engines is rather restricted with this method.

With the strip method, the boundary conditions can be set very accurately using external systems so that a good reproducibility and comparability of the results can be achieved. Characteristic for the strip method is the fact that the engine is always driven via the output shaft. This has the advantage over other measuring methods that the boundary conditions for the components under consideration are as close as possible to the conditions in the engine proper and a good transferability of the results is guaranteed. At the same time, this results in the limitation to the application of the strip method for determining the friction losses of any particular parts of the rotating engine: A functional (in the sense of the motoring operation) configuration of the engine must be possible *with* and *without* simultaneous movement of the parts under consideration. As a consequence, this means that the friction values measured for a component always also include the friction attributable to the drive; these are also eliminated when the components are removed. For example, the determination of the friction in the valve train also includes the friction generated in the timing belt or timing chain. This is also expedient in that the power losses can be allocated to the component in question and the load and any dynamics affect the level of the power losses.

The indication method demands a higher measuring complexity in order to obtain reliable results. A great influence comes from the fact that with multiple-cylinder engines, the individual cylinders can exhibit significant differences in their mean pressure. For this reason, a pressure measurement on all the cylinders at the same time is necessary. This causes considerable measurement complexity in practice. Furthermore, the complexity is increased by the fact that even minor errors in the TDC positioning and deviations in the pressure measurement from the calibration curve of the pressure sensors cause a significant difference in the p_{mi} value, and errors in the torque measurement distort the p_{me} value. Very great demands, therefore, have to be made on the accuracy of the indication and the torque measurement, as the result of the subtraction (of the mean friction pressure) is more than one power smaller than the initial parameters, so that the percentage errors are multiplied by a factor of ten.

Even minor deviations in the determination of the TDC of the piston therefore influence the calculation of the mean indicated pressure and, thus, also of the mean friction pressure. Fundamental studies have shown that an error of only 0.1° in the TDC position of the crankshaft can affect the calculated mean friction pressure by more than 10%, depending on the engine load.

A direct comparison of the different measurement methods is not possible, since the different boundary conditions influence the measurement results. This is illustrated as an example of a diesel engine with direct injection in Fig. 9-2. The fluid temperatures have been kept the same for the complete series of tests: 90°C oil temperature in the main gallery and 90°C coolant outlet temperature. A good correlation over the whole engine speed range is obtained between the results of the strip measurements and the motoring measurement (the charge cycle losses were determined by indication and deducted). The different friction values discovered with the motored engine are attributable to the following influences:

- The lubricant film temperatures in the engine are higher in spite of the same temperature in the main oil gallery.
- The combustion results in higher temperatures at the piston group and cylinder barrel.
- The lateral piston forces change due to the gas pressure.
- The load conditions of the injection pump change.

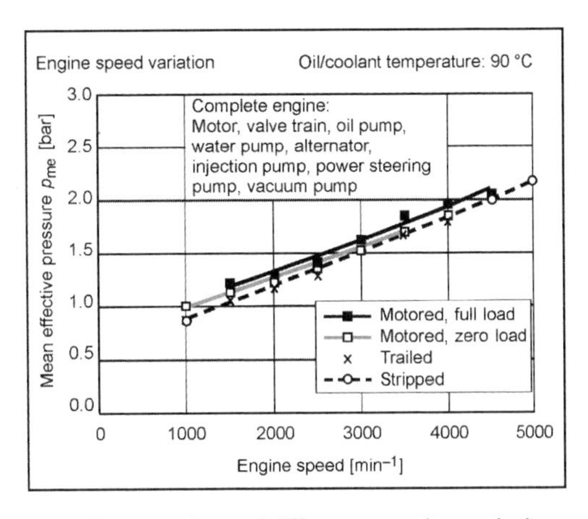

Fig. 9-2 Comparisons of different measuring methods on a car diesel engine with direct injection.

9.4 Influence of the Operating State and the Boundary Conditions

The operating state of the engine and the boundary conditions under which the engine is operated have a significant influence on the friction behavior. The most important parameters are described below.

9.4.1 Run-In State of the Internal Combustion Engine

In the first hours of operation, an adaptation of the friction partners takes place at the individual sliding points and with it a smoothing of the surface unevenness. This process involves a certain amount of wear and increases the friction loss of the engine. Thus, the running-in process takes place at different speeds for the different friction pairs and is completed in modern car engines after approximately 20–30 operating hours, but in individual cases only after more than 100 operating hours, so that the engine reaches a constant friction level. This remains more or less constant until engine components reach their service life limits, leading to an increase in the friction once again.

9.4.2 Oil Viscosity

Through the change in the shear forces, the viscosity of the lubricant has a significant effect on the conditions at the lubrication point. With otherwise unchanged boundary conditions, operation of the internal combustion engine with lubricating oils of different viscosities results in a change in the friction state. A lower viscosity of the lubricating oil means a lower load-bearing ability of the lubrication gap and, thus, a reduction in the lubricant film thickness. This is also associated with an increase in contact between solids in the mixed friction zone. Depending on the boundary conditions, the friction then drops if the hydrodynamic friction portion predominates or increases if the solid contact rises sharply. The behavior of different oils with different viscosities is illustrated in Fig. 9-3 for

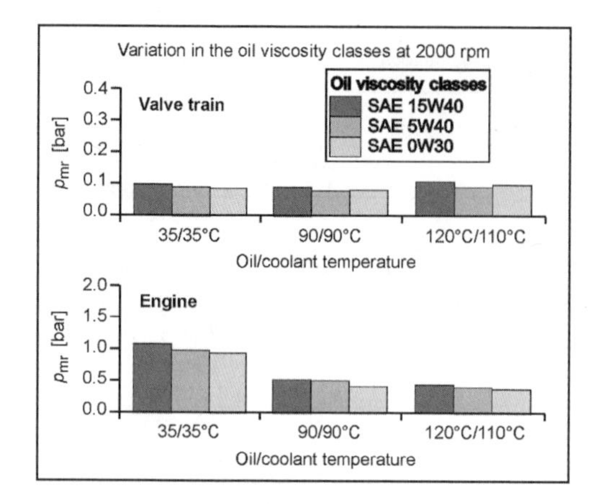

Fig. 9-3 Influence of the oil viscosity on friction.

a car SI engine at 2000 rpm. With the boundary conditions prevailing here, a reduction in friction with decreasing oil viscosity was observed in the engine. In the valve train, this reduction in friction is observed only at low temperatures. At higher temperatures, on the other hand, the friction increases because of the mixed friction conditions in the valve train caused by the lower oil viscosities. This change also has effects on the lubrication system and the oil pump drive power, as oil pressures and oil volumetric flows in the lubrication system are influenced by the various components and by the friction of the oil pump.

9.4.3 Temperature Influence

The operating temperature of the internal combustion engine, i.e., the temperatures of the components and the oil and coolant, influence the friction. The reasons for this are, first, the change in viscosity of the lubricant and, second, the change in the clearances in the various friction pairs. The effects of the changes in the fluid temperatures in the temperature range between 0 and 120°C are shown in Fig. 9-4. Even at fluid temperatures of approximately

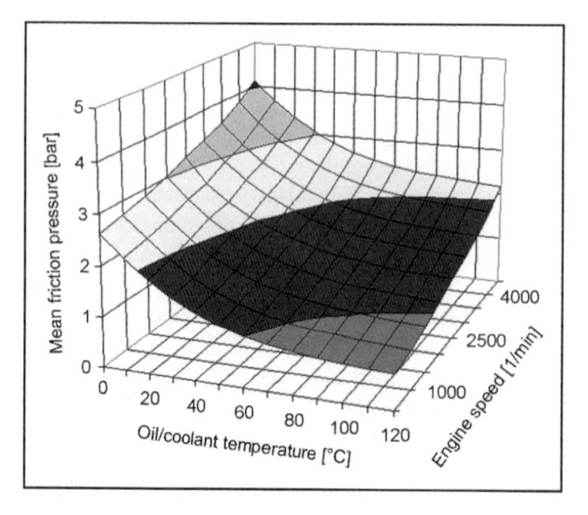

Fig. 9-4 Influence of the fluid temperatures on the friction.[6]

20°C, the friction losses are already doubled compared with an engine at operating temperature (90°C). This is one of the reasons for the increase in fuel consumption after a cold start and for short journeys when the engine is not at operating temperature.

9.4.4 Engine Operating Point

The engine operating point influences the friction both via the parameter "engine speed" and via the load. The influence of the engine speed is attributable to the increase in the sliding speeds at the friction points of the individual engine components. Increasing engine load has the following effects:

- Higher gas pressures and, thus, higher lateral piston forces, contact pressures of the piston rings, bearing loads, and forces for opening the exhaust valves
- Locally higher component temperatures and, hence, a possible increase in deformation
- Locally higher lubricant temperatures and, hence, a change in the friction state at the corresponding lubrication points
- Possibly modified drive power of the injection pump

The effect of the influences of engine load and engine speed on the friction behavior of a car SI engine is shown in Fig. 9-5. The measurements collected on the motored engine with loads between 0 bar (zero load) and full load are also compared with the results from drag measurements (p_{me} corresponds to the drag moment). The measurements in motoring mode at zero load show a good correlation with the measured values of the drag measurements at 0 bar.

The main influencing parameter is the engine speed: The engine friction increases at higher engine speeds. At moderate engine speeds, the engine load has only a very minor influence on the friction; i.e., the effects shown

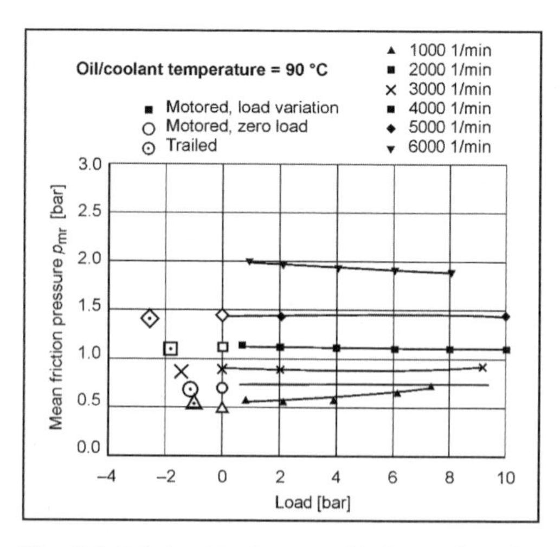

Fig. 9-5 Relationship between friction and engine load/engine revs.

have only a minor influence or compensate one another in this engine speed range. At an engine speed of 1000 rpm, the friction increases with increasing load. The friction of the piston increases at these low, sliding speeds because of the higher lateral piston forces. At high engine speeds, the friction decreases with increasing load. The reasons for this are the higher oil temperatures at the cylinder barrel at high engine powers, despite the same main oil temperature, and the partial compensation of the mass forces in the engine by gas forces.

9.5 Influence of Friction on the Fuel Consumption

The mechanical efficiency η_m of an internal combustion engine is defined as the ratio of mean effective pressure p_{me} to mean indicated pressure p_{mi}.

$$\eta_m = \left(\frac{p_{me}}{p_{mi}}\right) = \left(\frac{p_{mi} - p_{mr}}{p_{mi}}\right) \tag{9.3}$$

From this relationship, it is clear that at low engine loads, i.e., low mean effective and indicated pressure, the mechanical efficiency drops. The spreads of mean friction pressures of modern SI and diesel car engines are shown in Fig. 9-6. At an engine speed of 2000 rpm with values of 0.53–1.1 bar for SI engines and 1.02–1.4 bar for diesel engines including injection pump, the friction losses at full load are as high as 10% of the indicated power. In part-load operation, the mechanical efficiency drops so that the influence of friction on the fuel consumption continues to rise. A reduction in friction, therefore, offers a significant fuel savings potential and presents a worthwhile development objective. The span in each case between the engine with the highest and the engine with the lowest friction means not only an increase in fuel consumption, but also a reduction in the maximum power.

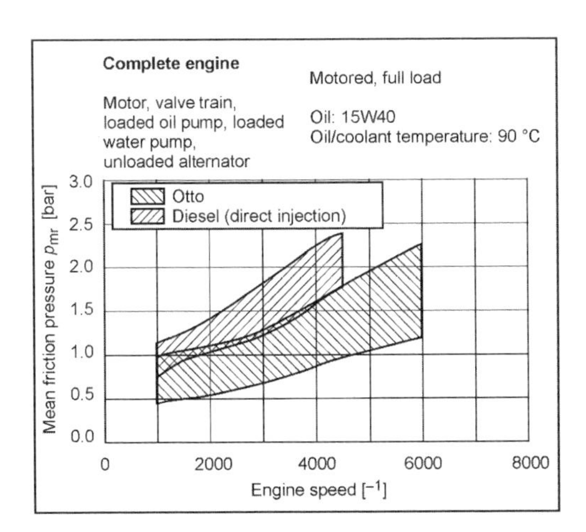

Fig. 9-6 Spread of the friction in motored engine mode (car engines).[4]

The development of friction over time is examined below, taking as an example the four-cylinder SI engine. Figure 9-7 shows the development of the mean friction pressure p_{mr} on the basis of studies in drag mode at 2000 rpm. The first thing of note is that the spread of the values has a very large bandwidth, although a downward trend is noticeable that is marked clearly by the regression line. The friction behavior of the SI engine, in particular, has been significantly improved in recent years. In purely statistical terms, the friction of a 2 l four-cylinder SI engine has been reduced by approximately 20% in roughly the last ten years. Extrapolation of the regression lines, however, results in an unrealistic reduction of the friction for the future.

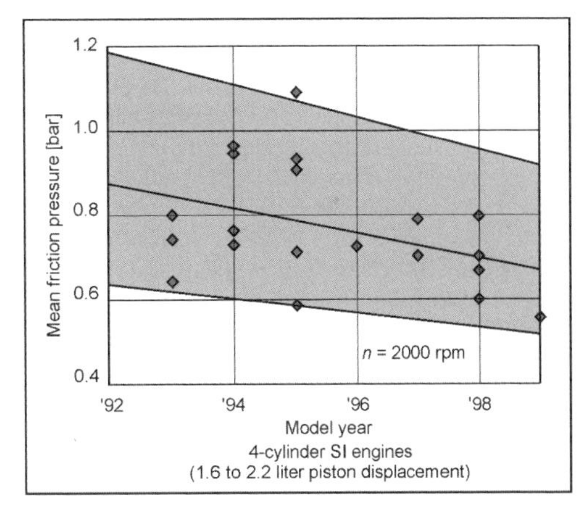

Fig. 9-7 Development of the friction in four-cylinder SI engines (1.6l–2.2l swept volume).[7]

The reduction of fuel consumption as a function of the mean friction pressure with the engine at operating temperature and an engine speed of 2000 rpm is shown in Fig. 9-8. The hypothetical case of the friction-free engine permits a reduction in fuel consumption of approximately 21% for the SI engine and of approximately 26% for the

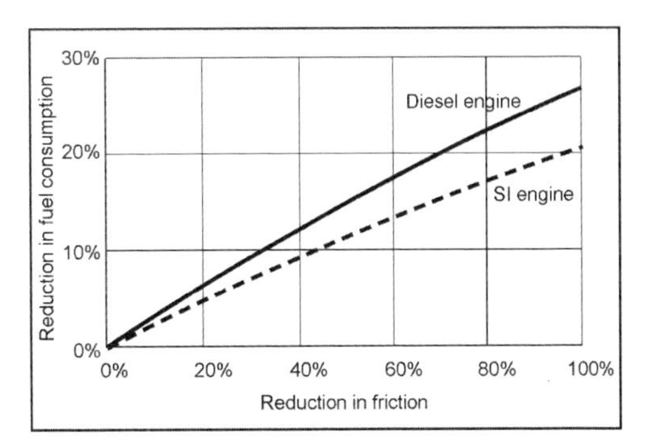

Fig. 9-8 Influence of the friction reduction on fuel consumption (friction considered at n = 2000 rpm).[8]

diesel engine. With the conventional measures used today (component optimization, roller valve train, modified pumps, heat management, etc.), it should be possible to exploit approximately 30% of this potential.

9.6 Friction Behavior of Internal Combustion Engines Already Built

9.6.1 Breakdown of Friction

When considering the friction losses of an engine, not only the total value but also the breakdown of the friction between the various components is of crucial importance. A common method employed for this is the "strip method" described in detail below.

Before the actual strip measurements, the complete engine with intake and exhaust sections is motored ("complete engine"). The drive torque measured here includes not only the mechanical engine friction but also the charge cycle losses. During this measurement, the oil pressures at the oil pump outlet, in the engine gallery, and, as much as possible, in the cylinder head together with the oil volumetric flows through the engine are recorded for every working point. The cooling system is subjected to external pressure for a constant pressure at the inlet to the coolant pump. The recording of these boundary conditions allows the boundary conditions on the complete engine to be set exactly later in the individual strip steps.

Following the recording of the boundary conditions on the complete engine, the measuring program for determining the friction of the individual components is performed. The stripping steps to be carried out are described below:

(a) The cylinder head is removed to determine the engine friction. To maintain the strain conditions of the engine block in the bolt area, the cylinder head is replaced by a plate with rounded cylinder openings. In this series of measurements, the gas chamber is therefore open, and the pistons are not subjected to gas forces. All the auxiliary drives are also removed. The oil pressure in the main gallery is set for the engine operation on the basis of the measurements performed for the complete engine or according to data from other sources using an external hydraulic oil supply.

(b) Removal of the pistons and connecting rods to determine the crankshaft bearing friction. The influence of the rotating masses is compensated by attaching "master weights" to the connecting rod bearing journals. The oil pressure in the engine gallery is set here again—as in (a)—using the external hydraulic oil supply.

(c) Measurement of the friction losses of crankshaft (including master weights) with valve train. The oil pressure in the engine gallery is set here again—as in (a)—using the external hydraulic oil supply.

(d) Measurement of the friction losses of crankshaft (including master weights) with oil pump. The oil pressure in the engine gallery is set here again—as in (a)—using the external hydraulic oil supply. The engine's own oil pump returns the oil directly back to the sump in a separate hose circuit via a variable throttle that regulates the oil pump pressure. The oil pump pressure is also set for the working point according to the pressures previously measured.

(e) Measurement of the friction losses of the crankshaft (including master weights) with coolant pump, alternator, power steering pump, and air conditioning compressor including tensioner and guide pulley(s). The oil pressure in the engine gallery is set here again—as in (a)—using the external hydraulic oil supply.

The friction losses of the pistons/connecting rod bearings, valve train, oil pump, and auxiliaries are determined from the differences between the results for the individual series of measurements. Furthermore, the sum of the determined values for the individual components gives a friction value for the whole engine referred to as "stripped complete engine." It describes the purely mechanical friction losses of the engine without the charge cycle losses.

A further detailing of the measuring program, for example, the determination of the friction of individual or all piston rings or the breakdown of the valve train friction between camshaft bearing friction and valve actuation is possible by including further stripping steps. On the other hand, measurement of all the components is not absolutely essential if only individual aggregates are to be considered.

The result of a strip measurement for a modern car SI engine is shown in Fig. 9-9. The percentage breakdown of the friction portions is shown in Fig. 9-10. The definition of the reference parameter total friction includes the

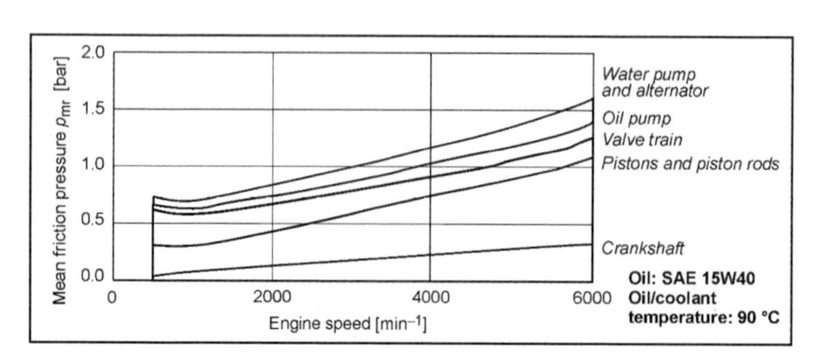

Fig. 9-9 Friction breakdown of a modern car SI engine.[8]

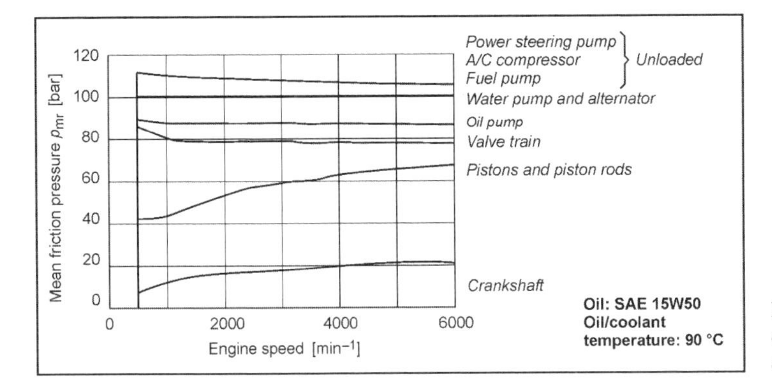

Fig. 9-10 Percentage breakdown of the friction in a modern car SI engine.[8]

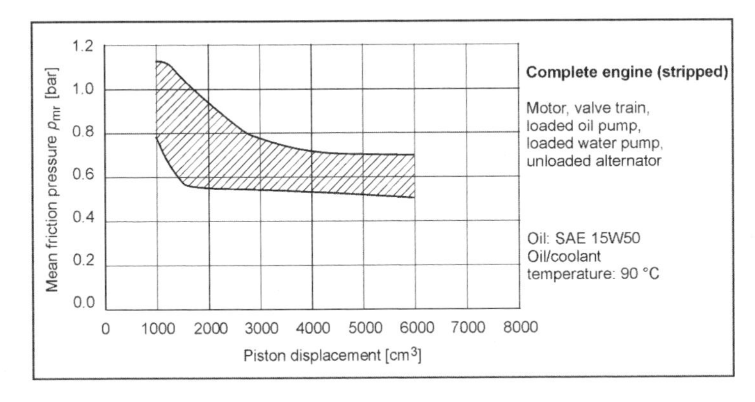

Fig. 9-11 Friction in car engines as a function of swept volume.

auxiliaries necessary for the engine operation—oil pump and coolant pump under load, alternator not under load, and components not purely for comfort such as power steering pump or air conditioning compressor.

Component-specific spreads can be elaborated in turn from the measurement results for the individual components. By comparing the measurement results for individual components with the corresponding spreads and, hence, with the state of the art, it is also possible to identify potentials for a reduction in the friction loss and to selectively exploit these potentials by optimization work.

Figure 9-11 shows the mean friction pressure of the stripped complete engine over the swept volume for an engine speed of 2000 rpm and an oil/coolant temperature of 90/90°C.

The spreads in these figures show that the swept volume above 1.5 l has practically no effect on the level of the mean friction pressure of a completely stripped engine. This is attributable to the fact that the power demand of various aggregates depends on the size of the vehicle and is not further reduced and also to the fact that the upper swept volume limit of the small car engine families lies at approximately 1.5 l. Because of the identical parts in the engine families, the engines are designed for this largest variant so that the smaller engines in the family have certain friction disadvantages.

9.6.2 Power Unit

The power unit of an internal combustion engine consists of the crankshaft including the radial shaft seal rings and of the piston group and connecting rods. Using the strip method, the engine can be further split into the friction of the crankshaft and the friction of piston group and connecting rods.

9.6.2.1 Crankshaft

The crankshaft friction is determined using master weights and includes the radial shaft seal rings. If we plot the mean friction pressure of the crankshaft against the engine speed and extrapolate the values up to a theoretical engine speed of 0 rpm, the Y branch received as a result can be roughly interpreted as the friction portion of the radial shaft seal rings that is relatively independent of the engine speed. The value obtained correlates with the measurement values from the separation of the radial shaft seal rings by stripping.

The friction moment of an individual main bearing referred to by its diameter shown in Fig. 9-12 for an engine speed of 2000 rpm can be calculated from the friction values for the crankshaft. It illustrates the measured values for a large number of engines and the regression lines for different engine concepts. The spread of the

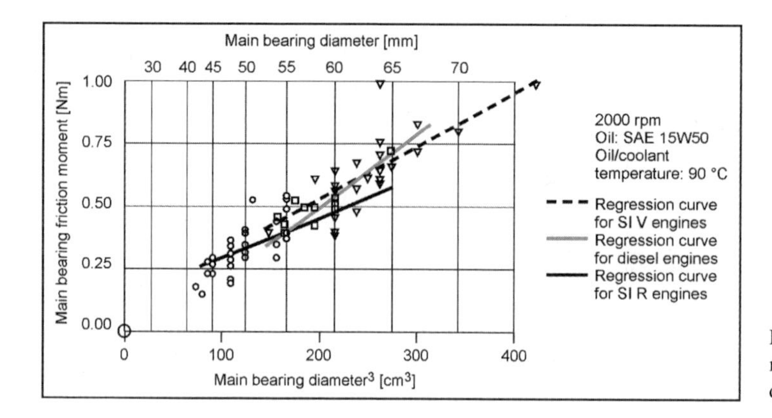

Fig. 9-12 Friction per crankshaft main bearing over main bearing diameter[3,4]

measured values around the respective regression lines shows that further parameters influence the friction in addition to the main bearing diameter. These include, e.g., the bearing geometry, the bearing clearances, deformations, or alignment deviations of the bearing race, as well as differences in the friction of the radial shaft seal rings.

9.6.2.2 Conrod Bearing and Piston Group

The friction of the piston group including connecting rod bearings can be determined by subtraction of the friction values for the crankshaft from the friction values for the engine. A further breakdown is difficult to achieve with the strip method as the connecting rods and piston group cannot be operated independently of one another. The friction of the connecting rod bearings can be determined using a practically friction-free aerostatic piston guide[9]; however, the work involved is enormous. The breakdown of piston and piston rings or the separation of single piston rings is possible, but it must be remembered that the removal of piston rings significantly changes the lubrication conditions of the piston and the other rings.

As shown above, the friction of the piston group has a very large proportion of the total friction in internal combustion engines. Great importance, therefore, has to be attached to its optimization in order to attain the goal of a low-friction engine. For this reason, a wide range of measuring systems have been developed for measuring the friction behavior of the piston group[5] or for monitoring the friction-influencing parameters, such as the cylinder deformation in motoring mode.[10]

The direct measurement of the piston friction forces in motoring mode provides the curve of the friction force over the crank angle, as shown in Fig. 9-13, which allows detailed conclusions to be drawn for the friction between piston and cylinder barrel, and in the event of force peaks occurring indicates possible wear. The influence of various parameters, such as piston micrograph, piston clearance, and piston ring pretension, can be examined in dragged and motoring mode. A variation in the piston ring surface pressure (piston ring tangential stress referred to the bearing piston ring surface) is shown in Fig. 9-14. The significant influence of the sum value on the measured

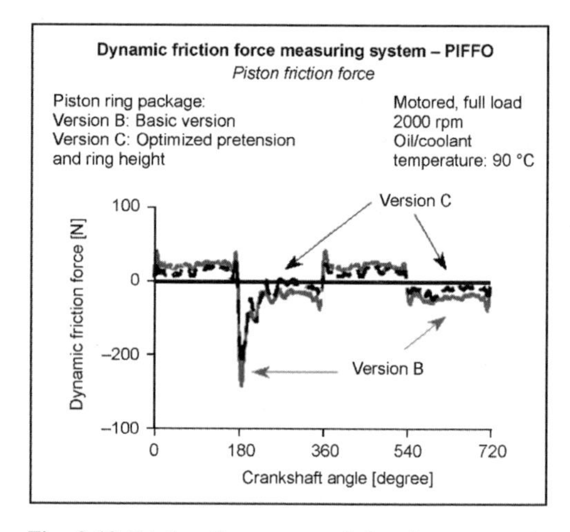

Fig. 9-13 Friction force curve of the piston group in motoring mode.[8]

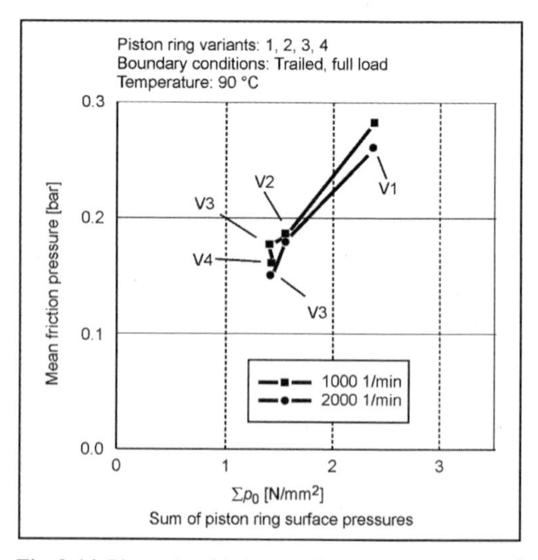

Fig. 9-14 Piston ring friction as a function of the preload.[5]

friction can be clearly seen. A comparison of a two-ring piston with the conventional three-ring piston of similar piston geometry and mass and the same sum value of the surface pressure, i.e., higher surface pressure of the individual rings for the two-ring piston, showed no significant differences in the mean friction pressure.

9.6.2.3 Mass Balancing

Mass balancing is the term used to refer to measures employed for partial or complete balancing of the mass forces and moments at crank drives. To improve comfort, an additional mass balancing is employed in many cases in car engines. The friction losses of the mass balancing gearing are affected by

- The order of the mass forces or moments to be balanced and, hence, the number and speed of the countershafts
- Number, design, and diameter of the bearing points
- Losses in the drive of the mass balancing elements

The balancing of the free second order mass forces in four-cylinder engines requires two countershafts that rotate at twice the crankshaft speed and, hence, exhibit unfavorable boundary conditions with respect to the friction behavior. Mass balancing gearings already built for four-cylinder engines exhibit friction values of 0.05–0.16 bar at 2.000 rpm; this can correspond to as much as 18% of the total friction in the engine.

9.6.3 Valve Timing (Valve Train and Timing Gear)

The friction of the valve train can be determined with the strip method from the difference between the measurement for the crankshaft with valve train and timing gear and the measurement for the crankshaft. A further separation, e.g., of the friction in the valve actuators or the camshafts, is possible, but in the analysis it has to be remembered that the timing gear dynamics are influenced and, hence, that the friction behavior changes.

Various valve train concepts are employed in modern car engines. Figure 9-15 illustrating the example of a multivalve engine shows that these concepts also have a considerable effect on the friction behavior of the valve trains. In valve trains with sliding tappets, the hydraulic valve lash adjustment increases the friction because of the additional friction caused by the pressure of the hydraulic element in the area of the cam base circle and the larger moving masses. Valve trains with roller tappets generally exhibit very favorable friction behavior. The unfavorable system dynamics of the timing gear associated with roller tappets, however, frequently necessitate higher pretensions in the timing gear. This can then lead to increased friction, particularly with chain drives.[11]

The breakdown of the friction within the valve train is necessary for the implementation of effective optimization measures. Figure 9-16 shows this breakdown for various valve train concepts. Sliding tappets exhibit the largest portion in the contact area of cam and tappet. This is because of the high contact forces and high relative velocities between cam and tappet. A reduction in friction can be achieved through a lowering of the contact forces by reducing the valve spring forces. With unchanged maximum engine speed, however, a reduction in the moving masses in the valve train is indispensable. The other possibility is to reduce the relative speeds by using a roller between the cam and the tappet.

9.6.4 Auxiliaries

In addition to the engine and the valve timing gear, a modern internal combustion engine also has a large number of auxiliaries. These are required for proper operation of the internal combustion engine and also to provide additional

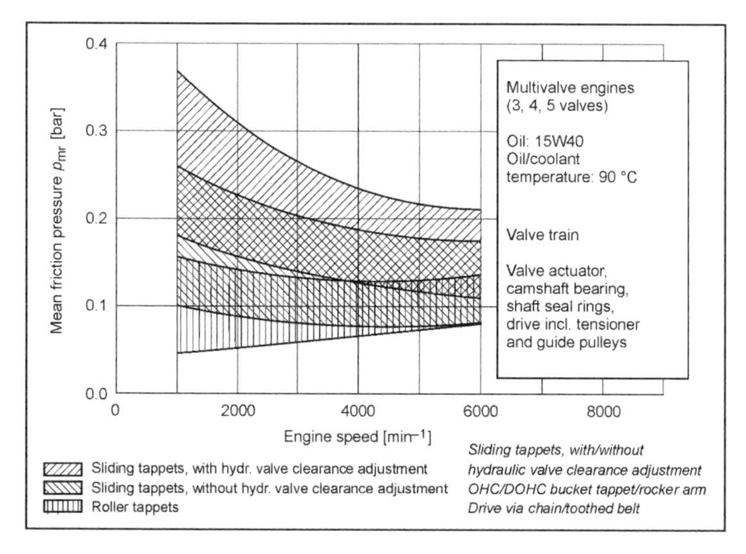

Fig. 9-15 Comparison of various valve train concepts.[4]

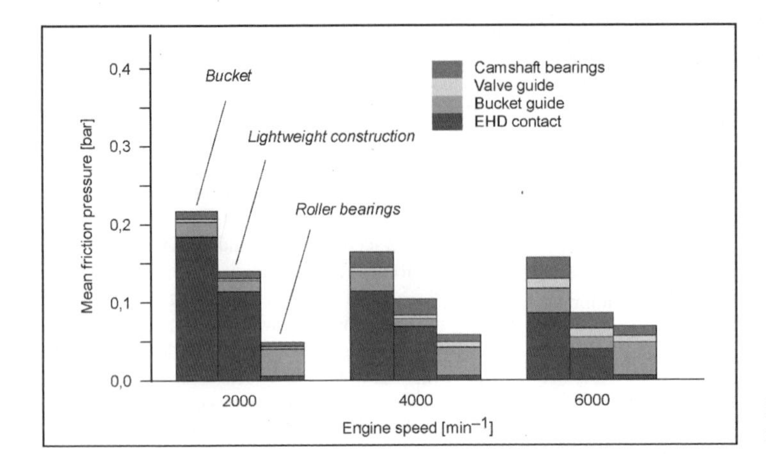

Fig. 9-16 Breakdown of friction in the valve train.[11]

functions such as safety and exhaust gas cleaning or meeting the rapidly growing comfort demands of the vehicle owners. Examples of the functions of the auxiliaries are

- Assurance of the engine's proper mechanical function in all operating states of the automobile: Lubrication oil pump, coolant pump, fuel supply system, radiator fan, mechanical turbocharger
- Assurance of proper supply of electrical energy to the engine and the automobile in all operating states using an alternator
- Creation of an additional exhaust gas cleaning facility: Secondary air pump, catalytic converter preheating
- Provision of auxiliary energies to cover enhanced passenger comfort and safety requirements: Power steering pump, air conditioning compressor, vacuum pump, starter, antilock brake system, traction control system, level control system.

Depending on the operating state, the drive of these auxiliaries consumes a large proportion of the mechanical energy provided by the internal combustion engine in the modern series application. The drive power required for these auxiliaries thus represents a mechanical loss and can be assigned to the friction loss. Various definitions make allowance for these auxiliaries in different ways. Our purpose here is not to consider the definitions but to look at the fundamental relationships with respect to the friction of the auxiliaries. As this plays a significant role in the fuel consumption of the vehicle, this aspect is becoming more and more important as a considerable increase in the energy demand is to be expected for the future due to additional or more powerful consumers.

This section gives an overview of the auxiliaries of a modern internal combustion engine. In view of the large number of such auxiliaries, we can look here only at the auxiliaries necessary for operation of the engine and the auxiliaries with the highest drive powers. We also take a brief look at the large number of components driven electrically and not by the internal combustion engine directly. The power supply to these components must not be neglected when considering the alternator drive power.

In modern engines, the auxiliaries are driven almost exclusively with a constant gear ratio to the crankshaft; this means that the speed of the individual auxiliaries is proportional to the crankshaft speed. The spread of the auxiliary speeds (ratio of maximum to minimum auxiliary speed) is defined by the spread of the speeds of the internal combustion engine due to the fixed transmission ratios. An adequate power output of the individual auxiliaries even close to the engine idle speed determines the transmission ratio. On the other hand, the power to be output by the crankshaft via the belt or chain drive increases with the engine speed, even if the power provided by the auxiliaries on the secondary side is not required. However, the individual power demands of the auxiliaries are not directly dependent on the engine speed. The direct drive thus represents a compromise between benefits and costs.

In the evaluations below, a distinction is made between the following definitions of power:

- Auxiliary power: Engine power required to drive the auxiliaries
- Power output: The power output by the auxiliaries (e.g., electrical energy or hydraulic energy)
- Power demand: Power output of the auxiliaries needed to cover the power demand of the engine or the automobile

Figure 9-17 shows the mean friction pressures of the auxiliaries necessary for the engine operation: Oil and coolant pumps deliver according to the engine operating point, the alternator is driven but does not output any electric power. The sum of the most favorable individual values for different engines shows that that there is still adequate optimization potential.

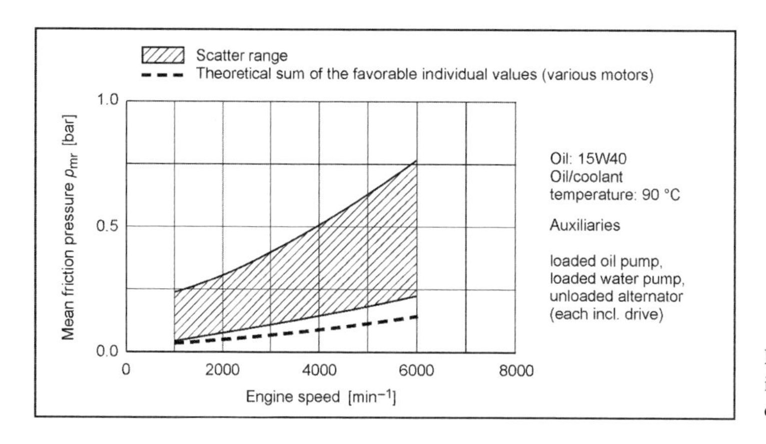

Fig. 9-17 Friction of the auxiliaries necessary for the engine operation.

9.6.4.1 Oil Pump

Modern four-stroke engines are lubricated by a forced-feed lubrication system. The following major components are supplied with lubricating oil by the oil pump:

- Crankshaft main and connecting rod bearings
- Piston spray nozzles
- Valve train and drive (camshaft, tappets, gear wheels, etc.)
- Turbocharger
- Other lubrication points according to engine form

The task of the engine oil circuit here is to

- Ensure a supporting oil film on all sliding surfaces under all operating conditions to effectively prevent mixed friction and the related wear
- Prevent localized overheating of components and the resulting damage by ensuring an adequate heat dissipation
- Pick up particles (soot and/or wear particles) and keep them in suspension
- Prevent or remove deposits
- Prevent corrosion

The oil pumps generally used in the motor vehicle engines are crescent-type or trochoid pumps driven directly by the crankshaft or externally geared pumps and trochoid pumps driven via a step-down gear and auxiliary drive. The drive power of the pumps differs significantly, depending on the drive system and pump type. The various optimization steps described in Refs. [12], [13], and [14] enable the pumps to be individually improved and adapted to the engine requirements. A feature common to all pump types is, as illustrated by the spread of the oil pump mean friction pressures in Fig. 9-18, the increase in the drive power at high engine speeds. In the majority of oil pump operating ranges, the widely used but energetically unfavorable bypass control results in lower efficiencies.

Adequate lubrication, in other words, a certain minimum oil pressure, must exist under all engine operating conditions because without this minimum pressure, engine damage can result within a very short time. The oil pump is, therefore, designed for the least favorable case, which means high oil temperature and an engine with a long service life and, therefore, large clearances. Further design criteria are low speeds to ensure the oil supply to hydraulic valve lash adjustment elements (fast idle) and to

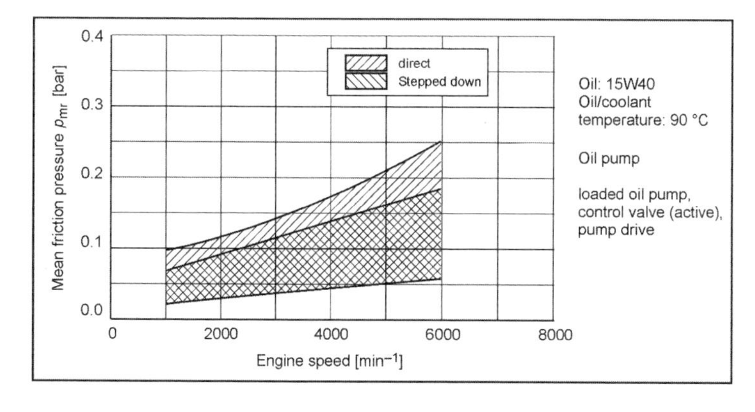

Fig. 9-18 Friction of various oil pumps.

oil pressure-controlled actuators (for example, camshaft adjusters), and high speeds for an adequate oil supply to the connecting rod bearings that are subject to high dynamic loads.[15,16] Figure 9-19 shows oil volumetric flows and oil pressures in a lubrication system. Here, it is essential that the necessary minimum oil pressure to ensure a demand-oriented lubricant supply is reached or exceeded during operation of the engine at all operating points, e.g., without tappet chatter and the risk of cavitation in the connecting rod bearings.

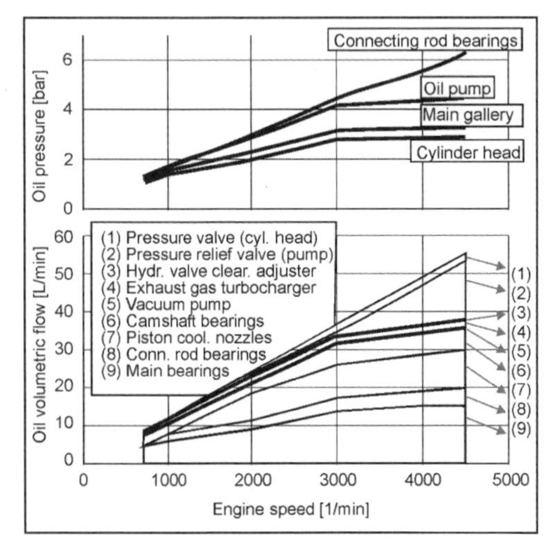

Fig. 9-19 Oil pressure and oil volumetric flow in the lubrication circuit.

The oil displacement of the engine increases less sharply with increasing engine speed than the delivery of the oil pump that increases more or less in proportion to the increase in engine speed. For this reason, part of the delivery is returned at medium and high engine speeds via a bypass valve, generally to the intake side of the pump.

Apart from the demand-oriented adaptation of the lubrication system and the detail optimization of the oil pump to the requirements of the engine, variable-displacement pumps have a great potential for reducing the drive power of the oil pump. Possibilities for adapting the delivery of oil pumps to the necessary demand are concepts with a variable delivery chamber volume that are, however, generally highly complex and expensive, as well as speed control by separation of the pump speed from the engine speed.

9.6.4.2 Coolant Pump

The coolant pumps for internal combustion engines are predominantly centrifugal pumps that are designed to provide an adequate coolant throughput to dissipate the heat both at low engine speeds and high engine load (for example, driving uphill with a trailer) and at rated output. With speed control of the pump dependent on the temperature of the components or of the coolant, e.g., via an electric drive, the temperature level of the walls surrounding the combustion chamber and, hence, the efficiency of the engine at part-load could be increased, the warm-up time of the engine shortened, and the drive power of the coolant pump reduced even at high engine speeds. The engine speed-proportional drive results in high deliveries at high engine speeds that, with an unfavorably designed coolant circuit, leads to high-pressure losses (and, hence, to high drive powers[17]). This offers an optimization potential in the design of the circuit with a correspondingly modified coolant pump.[18]

9.6.4.3 Alternator

High-performance and low-maintenance claw-pole generators (alternators) with a rated voltage of 14 V are almost exclusively employed to provide electrical energy in automobiles today. The efficiencies of the alternators are currently limited to maximum 60% to 70% and are achieved at low engine speed and with high loads on the alternator. Frequently, however, alternators are operated at high engine speeds and with low load and, hence, with low efficiencies of between 20% and 40%.

The electrical load requirements installed in the automobile have risen drastically in the last 40 years from around 0.2 to 2.5 kW and are expected to rise in the next 20 years to roughly 4 kW. The forecasts based on the Prometheus projects go even further. Here, around 8 kW of electric power will have to be provided up to the year 2010, when the power limits of the normal 14 V alternators of approximately 3 to 5.5 kW will be exceeded.[19] Figure 9-20 shows the friction of various alternators without electric power output. Starter-generator systems with higher electric power outputs and 42 V output voltages are expected for the vehicles of the future. Considerably higher electric powers are a precondition for the transition to electric motors for various components (e.g., oil pump, coolant pump, electromechanical valve train).

The power consumption of the consumers required to maintain the engine function is more or less independent of the vehicle operation. The electric power demand for all the other consumers, on the other hand, in particular for comfort functions, depends to a very great extent on the operating conditions (summer, winter, day, or night). Overall there is a spread in the total power demand from roughly 300 W up to 1200 W for a middle-class vehicle, depending on the operating conditions and frequency of operation.

Because of the physical relationships, the power output at engine idle speed and the maximum alternator output with a constant alternator weight cannot be defined independently of one another.[20] This unfavorable scenario is further aggravated by the growing electric power demand at idle speed and the desire to further lower the idle speed for fuel consumption reasons.

As a consequence of this, thought must be given to the alternator concept and to the drive management of the

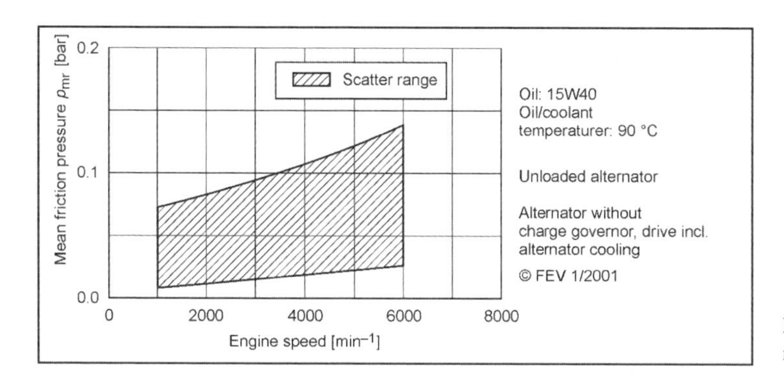

Fig. 9-20 Comparison of alternator drive powers.

alternator. A characteristic parameter in the alternator design is the 2/3 engine speed at which the alternator can output 2/3 of its maximum power. The transmission ratio between alternator and engine is normally selected so that the alternator runs at 2/3 of the engine speed at engine idle speed and thereby ensures the power supply to the engine and the automobile.

Alternator optimization goals are a high efficiency in all operating ranges, a low starting speed, and a high power output. The current should, therefore, rise sharply above the starting speed (1000 to 1500 rpm) so that a high power can be output to the running consumers even in the lower engine speed range. The largest proportion of the alternator losses during full load operation are, in particular, the iron and copper losses in the stator and the friction and fan losses, while the diode and excitation losses are relatively small.[21] Since the power output rises only slightly above an alternator speed of 5000 rpm, operation is to be recommended at alternator speeds of between 2000 and 5000 rpm.

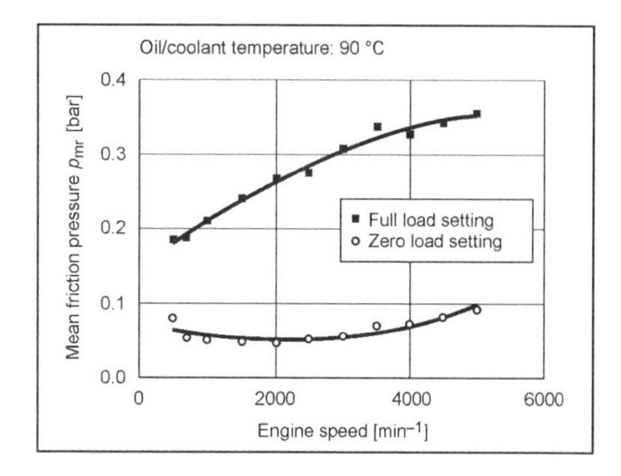

Fig. 9-21 Friction of injection pump (comparison of full load and zero load).

9.6.4.4 Fuel Injection Pump

The injection pump serves to inject the fuel through an injection nozzle directly into the combustion chamber towards the end of the compression stroke. Depending on the design of the injection volume and the engine operating point, the injection pressure lies between 50 and 200 bar for SI engines with direct injection and over 2000 bar for diesel engines.

Figure 9-21 shows the friction of a distributor injection pump of a diesel engine with direct injection. Between zero load and the maximum position of the fuel quantity positioner, the friction values are quadrupled. The friction values occurring at full load make up a major portion of the total friction in a diesel engine and are one of the main reasons for the increase in the engine friction between zero load and full load in diesel engines.

9.6.4.5 Air Conditioning Compressor

The development of air conditioning in automobiles began in the United States in the 1960s. Whereas only 20% of the automobiles in the North American market were equipped with air conditioning in 1965, this figure had already risen to 80% by 1980. The desire of the Japanese automobile manufacturers to conquer the North American market led to the Japanese discovering air conditioning for themselves so that, by as early as 1985, the percentage of automobiles in Japan with air conditioning was higher than that in the United States. A similar albeit delayed development has been observed since the end of the 1980s in Europe, too, but without the market penetration of the United States being achieved until now.

The cooling capacity demand for air conditioning in automobiles is dependent on the solar irradiation and the ambient temperatures. The average cyclic duration factor (running time) of the air conditioning in Europe is roughly 23% (United States, roughly 42%) and the average cooling capacity required 1 to 2 kW (United States, 4 to 5 kW).[22] Of all the auxiliaries, the air conditioning compressor has the highest power consumption that can be as high as 11 kW at high engine speeds, depending on the compressor design and operating condition. The average drive powers lie between 180 and 2000 W, depending on the cyclic duration factor.

Air conditioning compressors are generally driven by a belt at a speed proportional to that of the engine, thus creating a relationship between cooling capacity and engine speed, whereas the demand is practically independent of the speed. The air conditioning compressors are designed for maximum required cooling capacity that has to be available even at low engine speeds (when driving in town with a high percentage of engine idle time). At higher engine speeds, the compressors are, consequently, overdimensioned and have to be controlled. In many cases this is achieved with an electromagnetic clutch via which the compressor can be switched on and off.

Current developments are now moving away from the pressure-controlled compressor and increasingly toward volumetric flow-controlled compressors that reduce the excess power by varying their displacement. No noticeable improvement is to be expected in the energy balance, however, as the controlled compressor remains switched on for longer and because even with low cooling capacities the mechanical losses, particularly at high engine speeds, are considerable.[23]

In recent years more compact and lighter compressors (for example, vane-cell and spiral compressors) have been developed for use in compact vehicles, particularly in Japan.

9.6.4.6 Radiator Fan

The radiator fan has to ensure an adequate flow of air through the heat exchanger (radiator) to dissipate the heat at high load and low vehicle speeds.

Earlier the fan was driven directly by the engine at a speed proportional to the engine speed. In modern designs, temperature-controlled drive systems with a clutch (electric or hydrostatic drives) are used. These reduce the required drive power by comparison with the rigid drive by between 25% and 50%.

Electrically driven radiator fans are switched on when needed, depending on the coolant temperature. A switching hysteresis of around 10°C prevents continuous switching on and off. An electric fan is in operation when driving in town traffic for between 30% and 40% of the time. At higher vehicle speeds (main roads, highways), the air flow through the radiator is normally sufficient to dissipate the heat even without the fan.

At low engine speeds, viscous fans require less drive power than electric fans. This is because of the higher drive efficiency of the viscous fan at low engine speeds by comparison with the electric fan where the alternator efficiency also has to be taken into consideration. During the warm-up phase or part-load of the engine and at higher engine and vehicle speeds, the electric fan has the advantage over the viscous fan that it can be switched off when there is an adequate flow of air through the radiator. However, both drive systems still have a significant potential for reducing the transmission losses.

9.6.4.7 Power Steering Pump

Power-assisted steering systems that a few years ago were still reserved for the luxury-class automobiles are available today even for compact models. The trend towards broader tires and, hence, to increased steering effort, the more direct function of the power steering system, and the resulting improved handling of the vehicle have led to a significant increase in the market share of automobiles with power steering in recent years.

The steering assistance is provided by the oil pressure supplied by the power-steering pump and controlled at the steering gear according to the power assistance currently required. For cost reasons, vane cell pumps with bypass control are predominantly used as power-steering pumps in series-production vehicles.

The pressure requirement in the hydraulic system is dictated by the vehicle speed and the steering angle of the wheels. In present-day systems, maximum pressures of up to 130 bar occur in some cases at standstill with maximum steering angle. With increasing vehicle speeds, however, the required steering assistance drops sharply. The minimum pressure of the power steering system required to overcome the flow losses of the steering system when traveling straight ahead is vehicle and steering specific and lies in the order of 2 to 5 bar.

The displacement of a power steering pump has to be sufficiently high at low engine speeds and high steering speeds to ensure the steering assistance. For the design conditions, this means engine idle speed with the automobile at standstill and high steering speed on a dry road. These conditions occur during vehicle operation, particularly when parking or maneuvering. At higher engine speeds, a multiple of the useful oil flow is discharged as oil loss via the flow controller.

The driver power of the pump increases proportionally to the engine speed. The maximum possible drive power does not normally occur in practice, because high pressures in the steering system and high engine speeds do not occur simultaneously.

The required drive power of a power steering system is heavily dependent on the pump speeds and system pressures dictated by the vehicle operation. Typical drive powers of conventional power steering systems when traveling straight ahead average between 250 and 1200 W.

Use of variable-displacement power steering pumps such as variable-intake radial piston pumps allows the drive power to be significantly reduced. Great potential is offered here by electric power steering systems requiring average drive powers of only 100 to 200 W that have been used in recent years in small- and medium-sized series-production vehicles.

9.6.4.8 Vacuum Pump

On engines with throttle-free load control, a vacuum pump is employed to generate a vacuum (e.g., for the brake booster). The friction of normal vacuum pumps lies between 0.01 bar at low engine speeds and 0.04 bar at high engine speeds.

Bibliography

[1] Pischinger, S., Vorlesungsumdruck Verbrennungsmotoren, 21, Aufl., Selbstverlag, 2000.

[2] Affenzeller, J., and H. Gläser, Lagerung und Schmierung von Verbrennungsmotoren: Die Verbrennungskraftmaschine, Neue Folge Band 8, Springer-Verlag, 1996.

[3] Pischinger, R., G. Kraßnig, G. Taucar, and T. Sams, Thermodynamik der Verbrennungskraftmaschine: Die Verbrennungskraftmaschine, Neue Folge Band 5, Springer-Verlag, 1989.

[4] Koch, F., F.-G. Hermsen, H. Marckwardt, and F.-G. Haubner, Friction Losses of Combustion Engines—Measurements, Analysis and Optimization Internal Combustion Engines Experiments and Modeling, Capri, Italy, 15.–18.09.1999.

[5] Koch, F., U. Geiger, and F.G. Hermsen, PIFFO—Piston Friction Force Measurement During Engine Operation, SAE Paper 960306, 1996.

[6] Koch, F., F. Haubner, and M. Schwaderlapp, Thermomanagement beim DI Ottomotor-Wege zur Verkürzung des Warmlaufs, 22, Internationales Wiener Motorensymposium, Vienna, 26.04.–27.04.2000.

[7] Schwaderlapp, M., F. Koch, C. Bollig, F.G. Hermsen, and M. Arndt, Leichtbau und Reibungsreduzierung-Konstruktive Potenziale zur Erfüllung von Verbrauchzielen, 21, Internationales Wiener Motorensymposium, Vienna, 04.–05.05.2000.

[8] Koch, F., and U. Geiger, Reibungsanalyse der Kolbengruppe im gefeuerten Motorbetrieb-GfT Tribologie-Fachtagung, Göttingen, 5/6 November 1996.

[9] Haas, A., Aufteilung der Triebwerksverluste am schnellaufenden Verbrennungsmotor mittels eines neuen Messverfahrens, RWTH Aachen, Diss., 1987.

[10] Koch, F., E. Fahl, and A. Haas, A New Technique for Measuring the Bore Distortion During Engine Operation, 21st Int. CIMAC Congress, Interlaken, 1995.

[11] Speckens, F.-W., F. Hermsen, and J. Buck, Konstruktive Wege zum reibungsarmen Ventiltrieb, in MTZ 59 (1998) 3.

[12] Haas, A., T. Esch, E. Fahl, P. Kreuter, and F. Pischinger, Optimized Design of the Lubrication System of Modern Combustion Engines, SAE Paper 912407, 1991.

[13] Haas, A., E. Fahl, and T. Esch, Ölpumpen für eine verlustarme Motorschmierung. Tagung "Nebenaggregate im Fahrzeug," Essen, 1992.

[14] Fahl, E., A. Haas, and P. Kreuter, Konstruktion und Optimierung von Ölpumpen für Verbrennungsmotoren, Aachener Fluidtechnisches Kolloquium, 1992.

[15] Haas, A., P. Kreuter, and F. Maassen, Measurement and Analysis of the Requirement of the Dynamical Bearings in High Speed Engines, SIA Nr. 91191, Strasbourg, 1991.

[16] Esch, T., Luft im Schmieröl-Auswirkungen auf die Schmierstoffeigenschaften und das Betriebsverhalten von Verbrennungsmotoren, Lehrstuhl für Angewandte Thermodynamik, RWTH Aachen, 1992.

[17] Haas, A., R. Stecklina, and E. Fahl, Fuel Economy Improvement by Low Friction Engine Design, Second International Serninar "Worldwide Engine Emission Standards and How to Meet Them," London, 1993.

[18] Haubner, F., S. Klopstein, and F. Koch, Cabin Heating—A Challenge for the TDI Cooling System, SIA Congress, Lyon, 10.–11.05.2000.

[19] Bolenz, K., Entwicklung und Beeinflussung des Energieverbrauchs von Nebenaggregaten, 3, Aachener Kolloquium Fahrzeug- und Motorentechnik, 1991.

[20] Gorille, I., Leistungsbedarf und Antrieb von Nebenaggregaten, 2, Aachener Kolloquium Fahrzeug- und Motorentechnik, 1989.

[21] Henneberger, G., Elektrische Motorausrüstung, Vieweg Verlag, Wiesbaden, Braunschweig, 1990.

[22] Schlotthauer, M., Alternativantriebe für Nebenaggregate von Personenkraftwagen, in Antriebstechnik 24 (1985), Nr. 8.

[23] Fahl, E., A. Haas, and T. Esch, Tagung "Dynamisch belastete Gleitlager im Verbrennungsmotor," Esslingen, 1990.

10 Charge Cycle

The term "charge cycle" is understood as the exchange of the cylinder charge. In addition to the control elements in the cylinder head, the charge cycle is substantially influenced by the connected intake and exhaust system that determines the quality of the supplied fresh gas and removed exhaust.

The quality of this process is decisive for internal combustion engines since it substantially affects the maximum output and maximum torque as well as fuel consumption, exhaust quality, and running behavior.

Several factors influence the charge cycle such as the valve timing, valve lifting curves, design of the intake and exhaust systems, flow loss, wall temperatures in the ports and combustion chamber, environmental temperature, and pressure. The quality of the charge cycle can be described by the indices air expenditure λ_a and volumetric efficiency λ_l:

$$\lambda_a = \frac{m_G}{m_{\text{th}}} = \frac{m_K + m_L}{V_h \cdot \rho_{\text{th}}} \tag{10.1}$$

$$\lambda_l = \frac{m_{GZ}}{m_{\text{th}}} = \frac{m_{KZ} + m_{LZ}}{V_h \cdot \rho_{\text{th}}} \tag{10.2}$$

m_G is the quantity of mixture (fuel m_{KZ} and air m_{LZ}) fed to the cylinder, and m_{GZ} is the quantity of mixture remaining in the cylinder after the charge cycle. This stands in relationship to the mixture quantity m_{th} that could theoretically fill the cylinder. The air expenditure, therefore, provides more information on the intake system and the intake process, while the volumetric efficiency characterizes the fresh charge quantity actually remaining in the cylinder, i.e., the efficiency of the charge cycle, after the charge cycle has concluded [i.e., after IC (inlet closes)]. These two charge quantities differ by the amount of scavenging loss flowing into the exhaust from the intake during the valve overlap phase. When valves are actuated with a low valve overlap, the following approximation holds true: $\lambda_a \approx \lambda_l$; otherwise, $\lambda_a > \lambda_l$ holds true.

In the charge cycle, an important role is played both by the heat absorbed by the fresh charge in the intake system and cylinder and by the pressure loss. Assuming ideal gas, the following holds true for the volumetric efficiency λ_l:

$$\lambda_l = \frac{V_{GZ} \cdot \rho_{GZ}}{V_h \cdot \rho_{\text{th}}} = \frac{V_{GZ} \cdot T_{\text{th}} \cdot p_Z}{V_h \cdot T_Z \cdot p_{\text{th}}} \tag{10.3}$$

V_G and V_{GZ}, respectively, describe the supplied mixture volume and the mixture volume remaining in the cylinder after the charge cycle.

The effective output and, hence, the torque of an engine at a constant speed depends on the mean effective pressure. The formula for the mean effective pressure

$$p_{me} = \eta_{eZ} \cdot \lambda_l \cdot H_{GZ} \tag{10.4}$$

produces the effective output by taking into account the scavenging loss, pressure loss, and heat absorption during induction as follows:

$$P_e = i \cdot \eta_{eZ} \cdot \lambda_l \cdot H_{GZ} \cdot V_H \cdot n \tag{10.5}$$

The efficiency η_{eZ} and the lowest calorific value H_{GZ} refer to the composition of the cylinder charge after IC and EC (exhaust closes).

The following holds true for the torque:

$$M = \frac{1}{2 \cdot \pi} i \cdot \eta_{eZ} \cdot \lambda_l \cdot H_{GZ} \cdot V_H \tag{10.6}$$

The individual factors are mutually influential. The volumetric efficiency is greatly influenced by the speed. On the one hand, the throttling loss in the lines rises with the speed; on the other hand, gas-dynamic processes play a substantial role. The efficiency in the closed combustion chamber η_{eZ} increases with the volumetric efficiency since friction loss at constant speeds is constant, similar to throttling loss. For this reason, η_{eZ} also depends on the speed. In general, the maximum for the term $\eta_{eZ}\lambda_l n$ needs to be obtained for maximum output, and the maximum for the term $\eta_{eZ}\lambda_l$ needs to be obtained for maximum torque. This means that the two optimum values are separate from each other in two different, narrow speed ranges, which is why, with conventional engines (without variable valve actuation or a multistage manifold), a compromise always has to be made between torque and output.

10.1 Gas Exchange Devices in Four-Stroke Engines

In four-stroke systems, the charge cycles consist of expulsion and intake. These occur sequentially as a result of the displacement caused by the piston. The inlet and outlet of the cylinder must be periodically opened and closed by actuators.

The actuators must satisfy the following requirements:

- Large opening cross section
- Only a short time needed to open and close
- Flow-promoting design
- Strong seal during the compression, combustion, and expansion phases
- Great strength

Figure 10-1 shows two actuator designs of four-stroke engines. Lift valves provide a simple and secure seal, and the cylinder pressure reinforces the sealing effect. The fast acceleration and deceleration that occur during the stroke exert a great deal of stress on the valve gear from inertia. In addition, the grip can be lost at high speeds. Rotary-disk valves have quick opening and closing times and no inertia. However, their seal and operational safety

Fig. 10-1 Lift valve and rotary-disk valve timing.[1]

(jamming, seizing) are problematic because of the high temperatures and thermal expansion. Today, conventional timing systems use lift valves (Fig. 10-1, left).

10.1.1 Valve Gear Designs

To control the charge cycles of four-stroke engines, mushroom valves are used almost exclusively, and they are sometimes used for two-stroke engines. The required actuating mechanism, including the valves themselves, is termed the "valve gear."

A common feature of all valve gear arrangements is that they are driven via a camshaft that runs at half-crankshaft speed in four-stroke engines. The different valve gears can be distinguished by

- The number of valves per cylinder (Fig. 10-2)
- The position of the camshaft

Doubling the number of intake and exhaust valves to two is a sufficiently tried-and-true method to improve the volumetric efficiency and reduce the charge cycle work by providing larger flow cross sections. The advantages over a more complex valve gear are increased specific output, lower specific fuel consumption, and enhanced combustion. When this technical approach is pursued, we must ask if the conventional four valves per cylinder represent an absolute or relative optimum. In this regard, Aoi [SAE 860032] investigated four- to seven-valve arrangements. The following terms are defined in this context:

- Valve area: circular area of the valve openings per cylinder
- Valve opening area: lateral surface when the valves are open

Assuming the same cylinder diameter, the five-valve arrangement has the largest valve opening area, which at this juncture refers to the intake valves that have the predominant influence on the sought effect (Fig. 10-3). Given the same pressure ratio, this arrangement has the highest flow rate and best volumetric efficiency. Given equivalent valve opening areas, the cylinder diameter could be somewhat smaller for five valves than for four valves. The more compact combustion chamber of the five valves, therefore, has advantages for output.

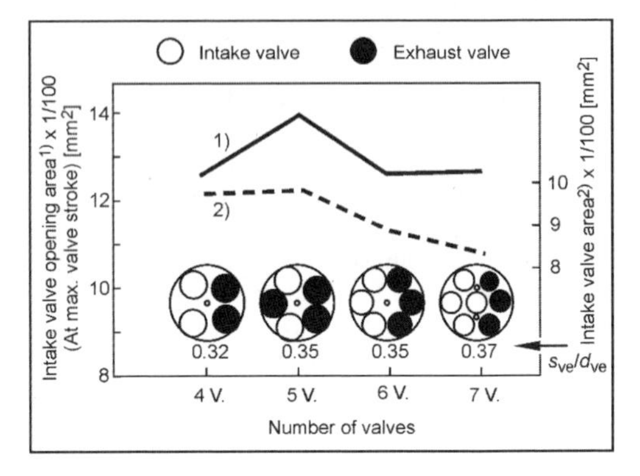

Fig. 10-3 Influence of the number of valves on the intake valve area and intake valve opening area.[9]

Nevertheless, four-valve spark-ignition engines have become widely accepted in passenger cars. This is primarily because the improvement attained with five over four valves is not worth the effort for most applications. This starts with the valve guide in the cylinder head and continues with the mechanical valve gear components. The lack of space in the cylinder head from new developments such as dual ignition or direct fuel injection also represents a problem that is difficult to solve. Figure 10-4 shows a four-valve engine with a radial valve arrangement, and a five-valve engine with a roofed combustion chamber.

Camshaft Position

- Bottom camshaft

The camshaft lies below the dividing line between the head and block (Fig. 10-5). Standing valves (Fig. 10-5, A) that can be actuated directly by tappets produce an inferior combustion chamber, however (knocking, hydrocarbon emissions); this design is antiquated. Overhead valves (Fig. 10-5, B and C) require a tappet, pushrod, and valve rocker to be actuated. The valves can be arranged in parallel (Fig. 10-5, B) or in a V (Fig. 10-5, C).

- Top camshaft

Number of valves per cylinder	2-valve	3-valve	4-valve	5-valve	6-valve	7-valve
Number of intake valves	1	2	2	3	3	4
Number of exhaust valves	1	1	2	2	3	3

Fig. 10-2 Valve arrangements.

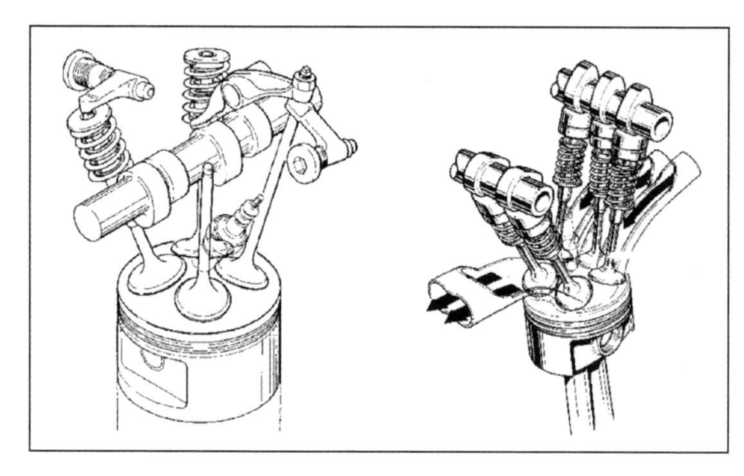

Fig. 10-4 The 4- and 5-valve engine.[1]

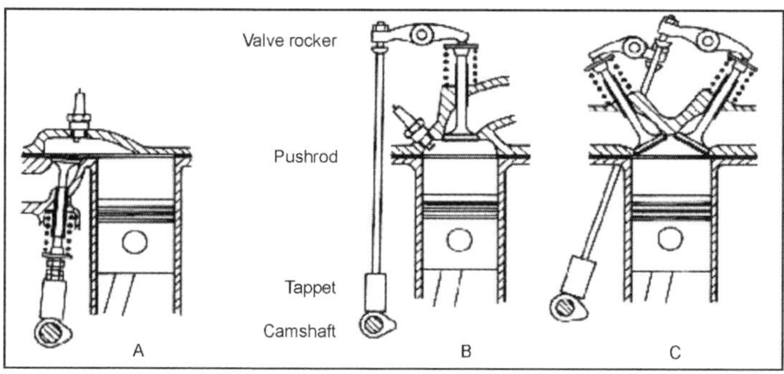

Fig. 10-5 Valve gears with a bottom camshaft—A: standing valves; B and C: overhead valves.

Camshafts above the head/block dividing line are usually used in modern, fast-running spark-ignition and diesel engines. The valves can be actuated via a valve lever or rocker arm, valve rocker, or tappet (Fig. 10-6). The advantage is that dispensing with the pushrod and tappet or valve lever or rocker arm reduces the unevenly moved mass and the elasticity of the valve gear.

In today's conventional valve gears, the transmission elements (valve rocker, valve lever, tappet, etc.) are pressed under spring force (valve spring) against each other or against the cam when the valve is open. This grip can be lost at high speeds. This does not hold true for desmodromic valves where lifting from the control cam is avoided by means of a second cam (Fig. 10-7); this makes valve springs unnecessary.

Valve clearance is also required. Because of the effort involved (manufacturing, servicing), this solution did not become popular.

10.1.2 Components of the Valve Gear

Camshaft

The camshaft transmits the torque introduced from the camshaft drive via the individual cams to the tappets. In

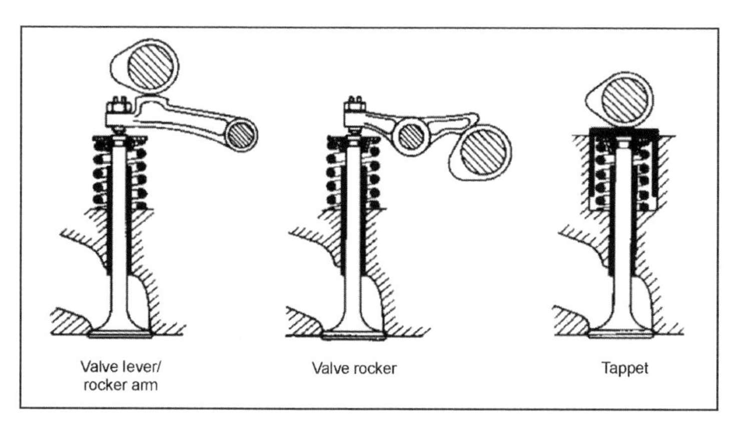

Fig. 10-6 Valve gears with an upper camshaft.[6]

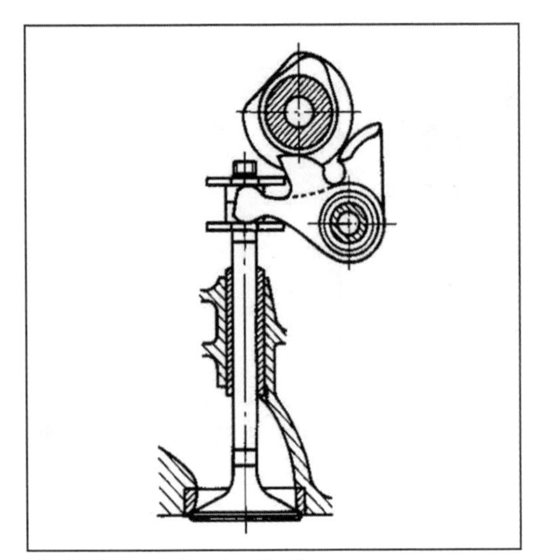

Fig. 10-7 Desmodromic valve.[5]

addition to the valve gear cams, the camshaft can have additional cams for actuating injection pumps (single pumps, pump-nozzle elements) or engine braking systems. Based on their manufacturing features, camshafts can be divided into cast, forged, and assembled camshafts.

Cast camshafts must be heat-treated after they are formed to give them the required strength and tribological properties. In the case of clear chill casting, the camshaft is hardened in one step by quickly cooling (quenching) the casting mold. In the case of centrifugal casting, the metal flows into a rotating permanent mold and hardens under the effect of centrifugal force. The camshafts are usually cast hollow to save weight.

In the case of assembled camshafts, the cams are manufactured separately from the shaft body and permanently joined later. Manufacturing them separately allows the materials to be adapted to function, manufacturing method, and stress. Cold-drawn structural steels (such as St52K) or alloyed steel (such as 100Cr6) can be used. For the cams, case-hardened steels (such as 16MnCr5) are used. The accepted forms of joining in series production are friction-lock connections by shrinkage or by hydraulic expansion of the tube from internal pressure, and keyed connections. With keyed connections, projections are created by roughing the tube at the attachment sites. The cam is given an internal splined profile and is pressed on with controlled force (KRUPP-PRESTA procedure). The additional advantage of assembled camshafts is the potential small cam spacing (multiple valves) and up to 40% less weight. However, the transmittable torque is limited by the joining method.

Multipart camshafts are frequently used for large engines. Individual camshaft segments are screwed together to create camshafts for engines with different numbers of cylinders. The bearing sites for the camshaft friction bearings used on all camshafts are ground directly on the tube in the case of assembled camshafts. The cam profile is also created by grinding. With a rolling contact, a negative radius of curvature (concave cam) of the cam profile is necessary to attain the desired valve gear kinematics. With fixed minimum grinding disk diameters, the negative radius of curvature can limit the valve gear kinematics. By belt sanding the cam profile, extremely small curvature radii can be created. The alternating loads from injection pumps and the valve gear generate flexural and torsional vibrations in the elastic camshaft. Torsional vibrations, in particular, generate angular deviations and, hence, deviations in the control and injection time between the first and the last cams. To minimize vibrations, the camshaft should be very rigid with comparatively low inertia (hollow shaft). Torsional vibration resonances can be calculated from the natural torsional frequency of the camshaft that arises within its speed range. Particular attention must be given to resonances that arise in low-seated engines with long camshafts. In certain instances (V-18/large engines), torsional vibration dampers must be placed on the free end of the camshaft.

Camshaft Drive

In addition to rare special designs (vertical shaft drive, pre-engaged drive), there are three conventional options for driving the camshaft with the crankshaft: gears, chain with a gear, and cogged belt. Gears are mainly used for bottom-mounted camshafts; the design becomes very complicated when gears are used for overhead camshafts.

Today, chains and gears as well as cogged belts are exclusively used for overhead camshafts (Fig. 10-8). A tensioning device is necessary for both types of drives. Cogged belts made of plastic with long fibers are less noisy and cheaper than chain drives. While chains have to be lubricated, the cogged belt needs to run in an oil-free area. Both drives must be encapsulated for protection as well as to avoid lubrication loss.

Valves

Figure 10-9 shows a valve with its installed elements. The seat surfaces and shaft ends of valves made of heat and wear-resistant alloys (such as Cr-Si or Cr-Mn steel) are

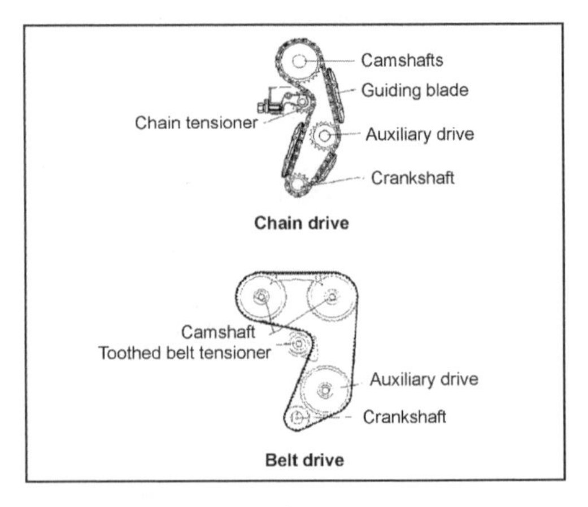

Fig. 10-8 Camshaft drives.[1]

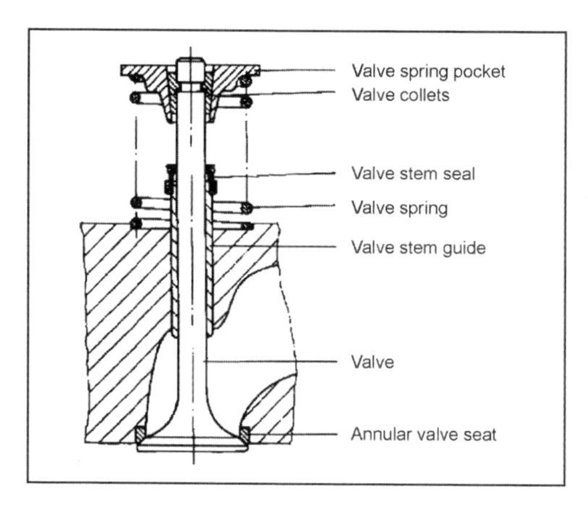

Fig. 10-9 Valve and valve components.[8]

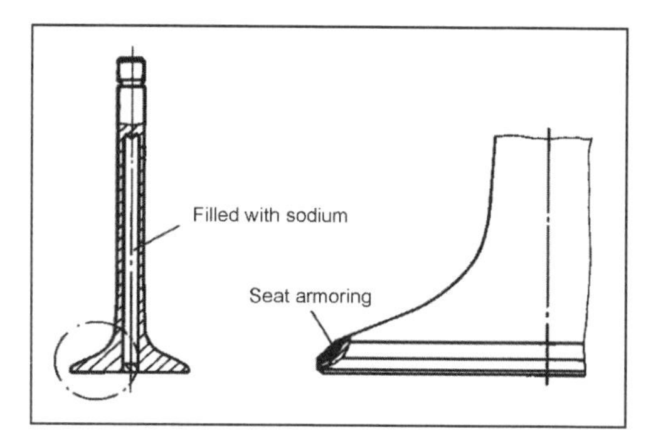

Fig. 10-11 Exhaust valve cooled with sodium.[1]

merely hardened or reinforced with hard metals. The valve shafts are chromed.

The valve shaft seals with spring-loaded elastomer sleeves must provide sufficient shaft lubrication and also prevent the penetration of excessive lubricating oil. Light-metal cylinder heads are provided with pressed-in valve guides and valve seat rings (made of special bronzes or alloyed cast iron) that are also frequently used in gray cast iron cylinder heads.

Valves are subjected to high thermal and mechanical loads as well as corrosion. The mechanical stresses arise from the valve head bending under pressure due to ignition and forceful contact when closing (impact). These stresses can be countered by providing the head with appropriate strength and shape. The valves with their large surface absorb heat from the combustion chamber. The top of the exhaust valve is also heated during opening by the exiting hot exhaust. In the valve, the heat primarily radiates to the valve seat, and a small part flows over the shaft of the valve guide. Intake valves reach temperatures of 300 to 500°C, and exhaust valves reach 600 to 800°C. A typical temperature distribution is shown in Fig. 10-10. If the seal on the valve seat is not perfect during the combustion phase, local overheating and melting occur that cause the valves to fail.

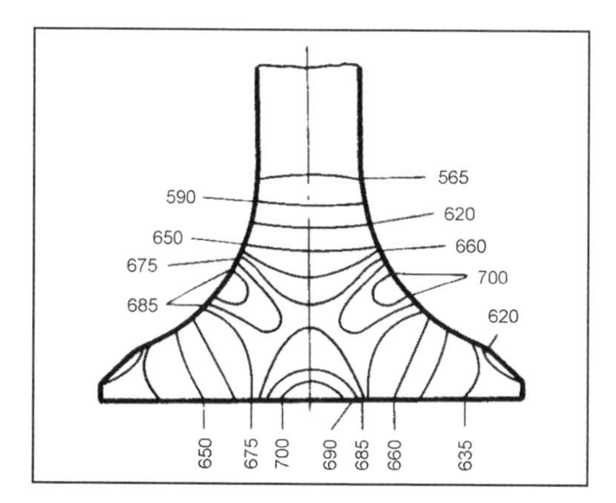

Fig. 10-10 Temperature distribution in the exhaust valve.[1]

To improve the heat conduction through the shaft, it is designed hollow and filled with sodium when it has to meet particularly high demands (Fig. 10-11, left). The movement of the sodium that is liquid at temperatures above 97.5°C enhances the transfer of heat. This lowers the valve temperature to 100°C. To reduce the wear, the seat can be reinforced by welding on stellite (Fig. 10-11, right).

The material of the valves must be very heat resistant and scale resistant. Special steels as well as titanium can be used.

Valve seat rings are installed in the cylinder heads to counter wear. A seat ring must always be provided for light metal cylinder heads (alloyed centrifugal cast metal and austenitic cast iron in special cases with heat expansion coefficients approximately as high as light metal). In the case of engines subject to high stress and also for exhaust valves in gray cast iron-cylinder heads, seat rings made of alloyed centrifugal cast metal are used.

The valve seat rings are either pressed in or shrunk on.

To avoid local temperature differences in the valve head as well as uneven wear, the valve should slowly rotate during operation. This movement can be supported by valve rotating mechanisms between the valve spring and cylinder head (rotovalves, rotocaps, and rotocoils) that convert the pulsing spring force into small rotary movements. The rotary movements are transferred via the valve spring and spring cap to the valve. The spring cap is affixed to the valve shaft with clamping cones (Fig. 10-12).

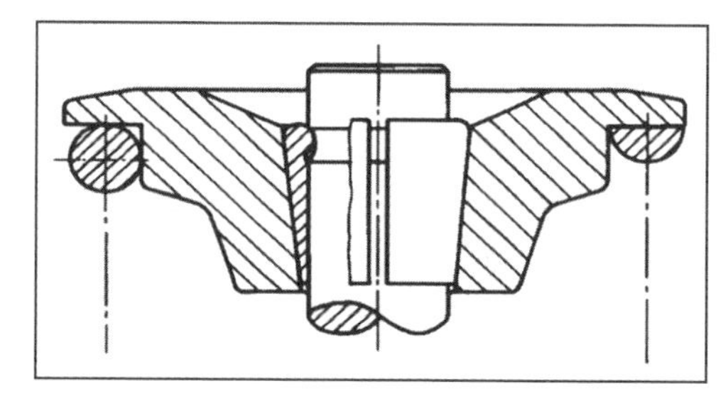

Fig. 10-12 Fixing the spring caps with clamping cones.[5]

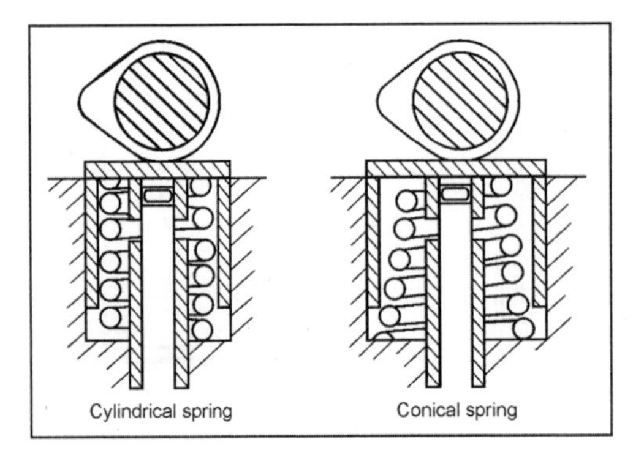

Fig. 10-13 Cylindrical and conical steel springs.

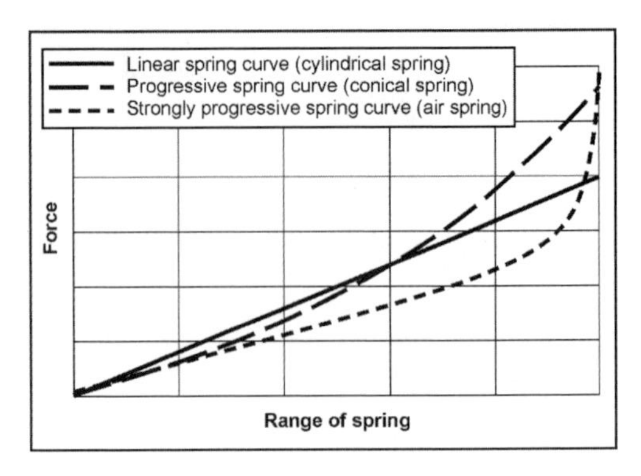

Fig. 10-15 Spring characteristic curves.

The shaft is thereby weakened only slightly since the round counterbore has a slight notch effect.

Valve Springs

Cylindrical or conical steel springs and pneumatic springs can be used as valve springs (Figs. 10-13 and 10-14). They primarily differ in the way they transmit force along the spring path. Whereas the cylindrical steel spring usually has a linear characteristic curve, the conical steel spring has a progressive characteristic, and the pneumatic spring has a strongly progressive characteristic curve (Fig. 10-15). The progressive characteristic curve is good for high speeds. Because of the expense and required supply of compressed air, pneumatic springs have been used only for motor sports.

Valve Rockers and Valve Levers

- Valve rockers

Valve rockers with pushrods are used for bottom-mounted camshafts and with valves arranged in a V shape for overhead camshafts.

Because of the strong contact pressure exerted on the pivot, the bearing must be especially rigid. For the valve rocker ratio $i = l_2/l_1$ (Fig. 10-25), values of 1 to 1.3 are recommended as a compromise for less surface pressure on the tappet, less moved mass, and high rigidity. The force of the valve rockers is transmitted to the valve along an axial path as much as possible to keep lateral forces from acting on the valve shaft and, thus, preventing increased wear of the valve guide. At one-half of the valve stroke, the center of rotation of the valve rocker should be perpendicular to the valve axis at the height of the shaft end to attain the least possible displacement of the valve rocker and valve in relation to each other (favorable sliding conditions). The force-transmitting spherical or cylindrical surface should be applied to the valve rocker and not the valve. For reasons of wear, the valve rocker end is hardened.

Figure 10-16 shows valve rocker designs. Valve rockers are usually cast or forged. Economical and light but less rigid are valve rockers pressed from sheet metal. It is advantageous to set the valve play at the resting rocker bearing. With forged valve rockers, the setting screw is normally on the rocker end, which increases the moved mass of the valve gear. Figure 10-17 shows a valve gear with hydraulic valve play compensation integrated in the valve rocker. The compensation element is supplied with lubricating oil via the valve rocker shaft and holes in the valve rocker.

- Valve levers (rocker arms)

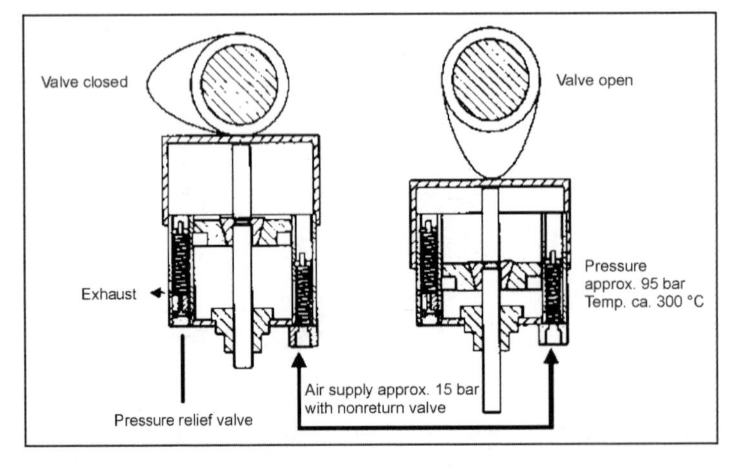

Fig. 10-14 Pneumatic spring.

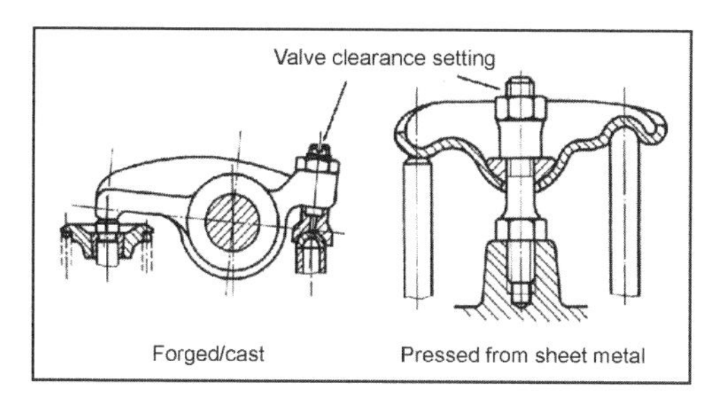

Fig. 10-16 Valve rocker.[5]

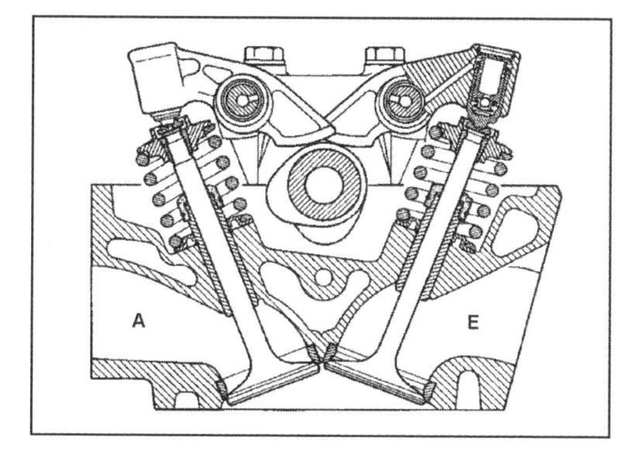

Fig. 10-17 Valve gear with valve rockers and hydraulic valve play adjustment.[10]

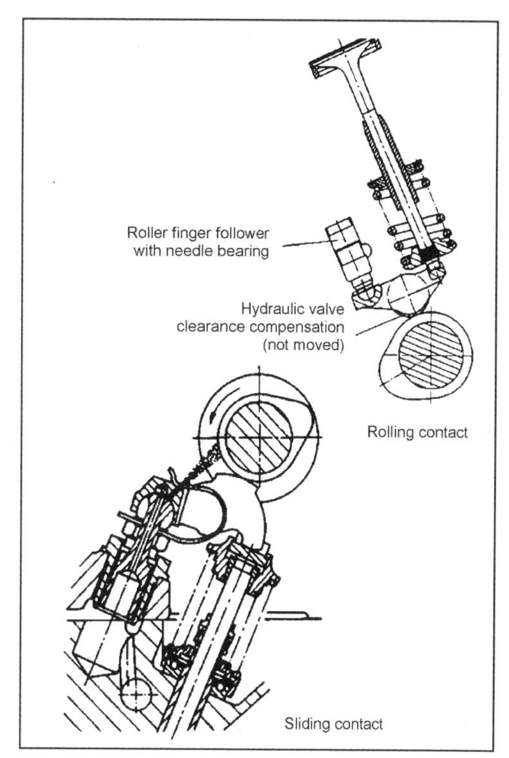

Fig. 10-18 Valve gear with valve lever (rocker arm).

The valve lever is exposed to a much lower degree of force than the valve rocker. The influence from changes at the bearing point is less; an automatic valve play adjustment system can be installed in the lever bearing in valve levers without substantially changing the overall elasticity of the valve gear. The designs of two valve levers are shown in Fig. 10-18.

It is possible to reduce friction loss, especially at low speeds, by using roller rocker arms. A roller finger follower on a needle bearing is used at the contact point between the rocker arm and camshaft. This can reduce the moment of friction of the valve gear by as much as approximately 30% in comparison to a sliding rocker arm arrangement (Fig. 10-18).

Figure 9-15 shows a spread that illustrates the advantages of the roller rocker arm in regard to reduced friction. The reduction of the valve gear friction, however, also reduces the damping of the oscillating torque introduced from the cam force and, hence, increases the load on the camshaft drive. Under certain circumstances, the subsequently required stronger chain or belt tensioning devices (tensioning pulley, tensioner blade, damping elements) can compensate for the friction advantages gained in the valve gear.

Tappets

Tappets in pushrod engines (Fig. 10-5, B) must guide the pushrods and absorb the transverse force that arises from the sliding of the cams. In overhead camshafts with a tappet drive (Fig. 10-6), the tappet has to keep the lateral force away from the valve guide. Normal tappet designs for pushrod engines are shown in Fig. 10-19. Flat-based tappets and bucket tappets can be removed both upward and downward. Roller tappets are used for maximum loads (diesel engines are subject to greater loads).

Figure 10-20 shows a bucket tappet that is almost exclusively used for overhead camshafts with a tappet drive.

The tappet diameter is determined by the maximum tappet speed. The surface pressure between the camshaft and tappet determines the cam width. Since the cam and tappet must glide on each other under high surface pressure, the materials of the two elements must be harmonized. The combination of hardened steel and white hardened

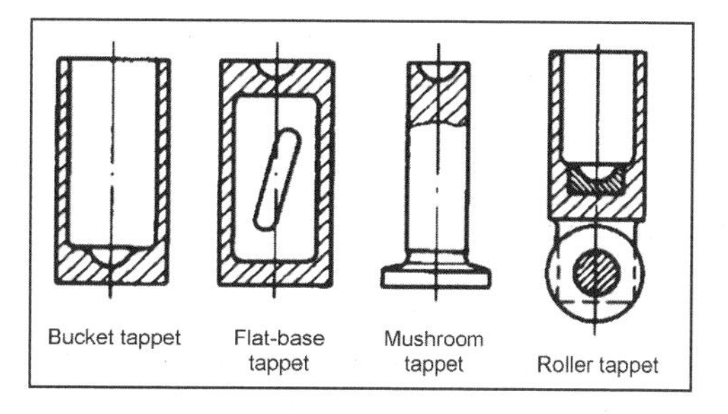

Fig. 10-19 Tappet for valve gear.[5]

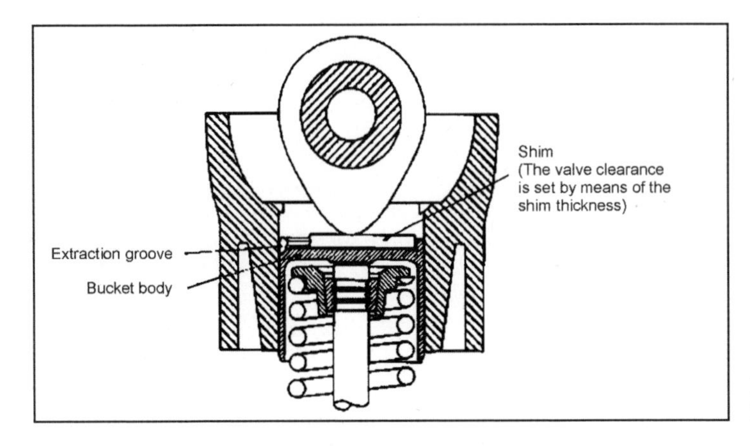

Fig. 10-20 Bucket tappet without hydraulic compensation.

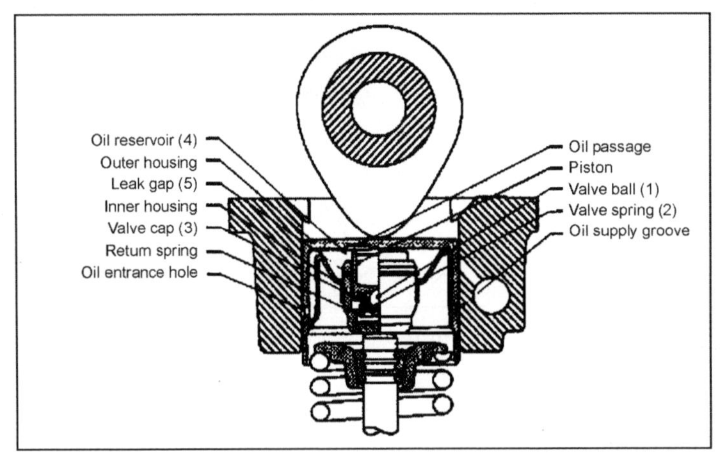

Fig. 10-21 Bucket tappet with hydraulic valve play compensation.

gray cast iron is quite suitable. Frequently, to avoid uneven wear, the tappet is rotated on its axis. For this reason, it is offset against the center of the cam by 1 to 3 mm. In addition to rigid tappets, there are tappets with automatic play adjustment (see Fig. 10-21). The play is kept constant with the amount of oil in the high-pressure chamber. If the valve play is too great, oil flows through the ball valve (1)–(3) from the reservoir (4); if it is too low, the excess oil exits via the leak gap (5). In addition to easier servicing by dispensing with the play setting, this system is also less noisy. The disadvantages are the large mass, low rigidity, and problems starting the engine after long periods of rest because of insufficient oil supply. Today, tappet engines almost exclusively use tappets with automatic play adjustment; in engines with valve levers, valve rockers, or rocker arms, the hydraulic valve play is adjusted with additional inserted elements.

10.1.3 Kinematics and Dynamics of the Valve Gear

For a good charge cycle, the valves must open and close quickly. However, the inertia of the valve gears needs to be taken into consideration in the design. Figure 10-22 shows the typical path of the cam stroke, cam speed $\dot{x}$, and

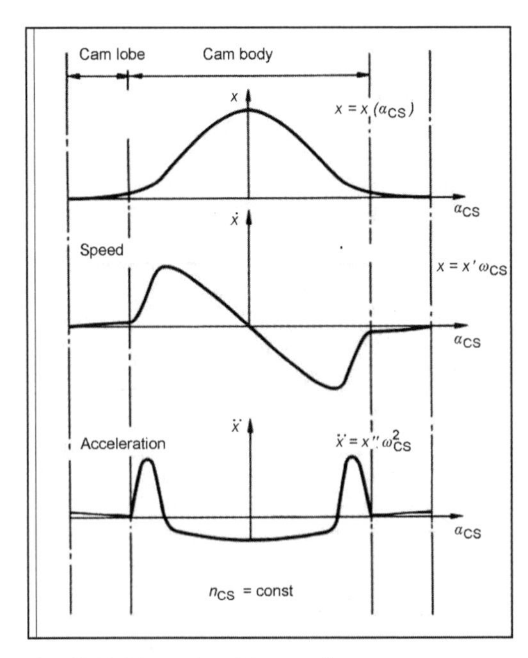

Fig. 10-22 Kinematics of the cams.[1]

cam acceleration $\ddot{x}$ over the angular displacement of the cam. These quantities correspond to the respective quantities of the valve movement.

The cam stroke or the cam contour is composed of the cam lobe and the cam body. At the cam lobe, the stroke speed $\dot{x}$ is slow so that common changes in the valve play do not generate strong impact pulses. The cam body determines the opening cross section for the charge cycle. The valve is closed by a deceleration corresponding to the cam lobe.

The stroke characteristic is a function of the camshaft angle α_{NW}. The following equation results for the stroke speed $\dot{x}$:

$$\dot{x} = \frac{dx}{dt} = \frac{dx}{d\alpha_{NW}} \cdot \frac{d\alpha_{NW}}{dt} = x' \cdot \omega_{NW} \tag{10.7}$$

with

ω_{NW} = Angular speed of the camshaft

At a constant camshaft angular speed, the following results for the stroke acceleration $\ddot{x}$:

$$\ddot{x} = \frac{d^2x}{dt^2} = \frac{d^2x}{d\alpha_{NW}^2} \cdot \frac{d\alpha_{NW}^2}{dt^2} = x'' \cdot \omega_{NW}^2 \tag{10.8}$$

In these equations, x' and x'' are speed-independent functions that are determined only by the geometry of the cams. The cam shape also influences the characteristic of the valve movement. Figure 10-23 shows the relationship between stroke characteristic and cam shape in connection with a flat-based tappet. In the figure, the rotation of the cam has been replaced by the swing of the tappet in the opposite direction with a standing cam. The cam shape is the envelope curve of the tappet-sliding surface. For kinematic investigations, the cam drive can be replaced by thrust cranks whose articulation corresponds to the curvature midpoint M of the cam contour belonging to the contact point B. x' (rotated vector) and x'' depend on the crank length (r_M) and position of the momentary thrust crank. We can see that the distance of the cam contact point B

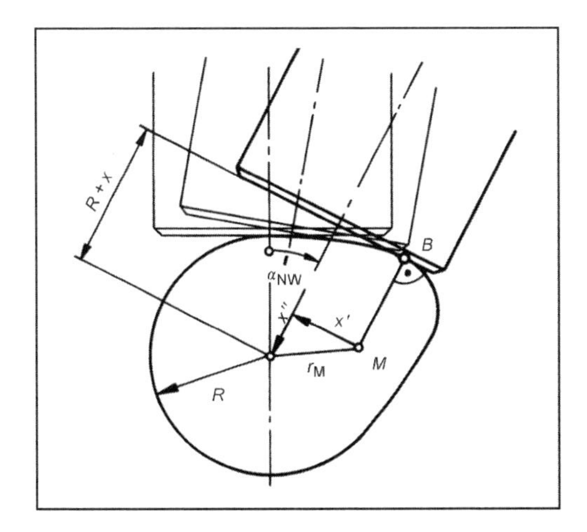

Fig. 10-23 Kinematics of the tappet stroke.

from the middle of the tappet is proportional to the speed. The tappet diameter must therefore be adapted to the maximum stroke speed.

It is important that friction is always present between the cam and tappet or rocker arm. In addition, there also must be friction between the valve and tappet or rocker arm for the valve to follow the cam stroke. The valve stroke may be recalculated corresponding to the valve rocker or valve lever ratio $i = l_2/l_1$. To test the grip, the force between the cam and tappet must be found. The inertia force and spring force must also be considered.

With a valve gear corresponding to Fig. 10-24, the following results for the force on the cam F_N:

$$F_N = F_F \cdot \frac{l_2}{l_1} + \left[m_{\text{Stö}} + m_{\text{St}} + \frac{J_K}{l_1^2} + m_V \left(\frac{l_2}{l_1}\right)^2 + \frac{m_F}{2} \cdot \left(\frac{l_2}{l_1}\right)^2 \right] \cdot \ddot{x} \tag{10.9}$$

with

F_F = Valve spring force
m_F = Mass of the valve spring (only one-half is used since it rests on one side against the cylinder head)
J_K = Moment of inertia of the valve rocker
$m_{\text{Stö}}$ = Tappet mass
m_{St} = Pushrod mass
m_{red} = Reduced mass
m_V = Mass of the valve
F_{red} = Reduced spring force

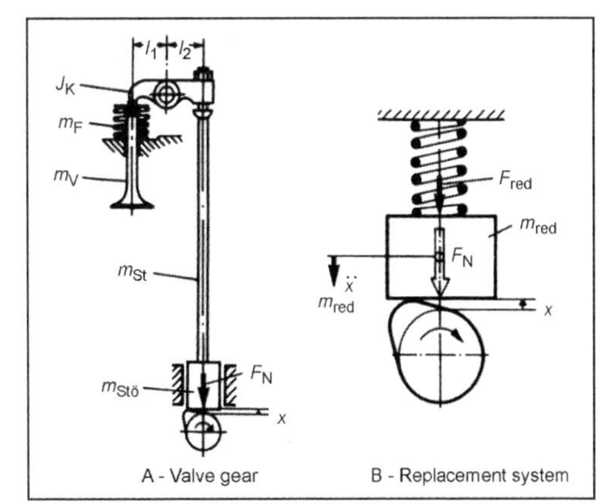

Fig. 10-24 Rigid valve gear.

If all the quantities on the cam side are "reduced," then the equation for the cam force is

$$F_N = F_{\text{red}} + m_{\text{red}} \cdot \ddot{x} \tag{10.10}$$

This equation corresponds to the replacement system in Fig. 10-25. The following condition must be fulfilled for the grip:

$$F_{\text{red}} + m_{\text{red}} \cdot \ddot{x} > 0 \tag{10.11}$$

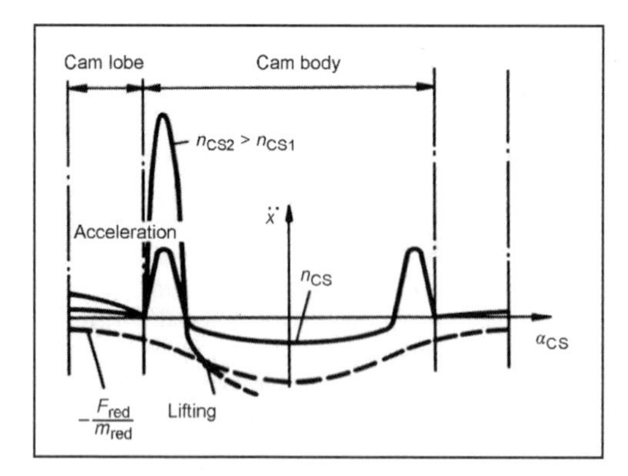

Fig. 10-25 Lifting conditions for a valve gear.

or expressed otherwise

$$\ddot{x} < \frac{F_{red}}{m_{red}} \tag{10.12}$$

The characteristic of the tappet acceleration depends on whether or not the valve is lifted. Figure 10-25 shows the acceleration of the cam stroke $\ddot{x}$ over the angular displacement of the cam at two camshaft speeds n_{NW1} and n_{NW2}. If the characteristic of $\ddot{x}$ intersects the curve $-F_{red}/m_{red}$, the grip is interrupted. This can occur only in the deceleration period of the main cam. There always exists speed above which lifting occurs.

The valve gear should be designed so that the valve is not lifted at maximum camshaft speed (= $\frac{1}{2}$ crankshaft speed). This is attained when the moved mass (m_{red}) is small and the spring force is high.

10.1.4 Design of Gas Exchange Devices in Four-Stroke Engines

Charge Cycle Energy Loss

The loss of charge cycle energy reduces the indicated work and, hence, the indicated output; this causes the specific consumption to rise. It occurs either as expansion energy loss at the beginning of the charge cycle [between EO (exhaust opens) and BDC] or as increased pumping work during the charge cycle. Pumping work represents the work required by the piston to draw a fresh charge during the intake cycle into the combustion chamber, and to expel the exhaust from the combustion chamber during the exhaust cycle. Correspondingly, the pumping work can be divided into intake work and exhaust work. In calculations, the pumping work is represented with the aid of the average effective pump pressure.

During intake, pressure losses occur at several locations, and this raises the intake work: Flow loss when the medium enters and leaves the intake system, pressure drop in the lines because of bends and rough surfaces, pressure loss in the air filter, at the airflow sensor, at the throttle valve, and loss at the valves. Assuming quasistationary flow, the overall pressure loss in the intake system toward atmospheric pressure can be described by the sum of the individual losses of the various components [Heywood]

$$\Delta p = \sum_i \Delta p_i = \sum_i \xi_i \cdot \rho \cdot v_i^2$$
$$= \rho \cdot \bar{v}_K^2 \cdot \sum_i \xi_i \cdot \left(\frac{A_K}{A_i}\right)^2 \tag{10.13}$$

where ξ_i equals the loss coefficient, v_i the local flow speed, and A_i the smallest flow cross section of the respective component. This makes it clear that to achieve less pump work in the charge cycle, greater flow cross sections are desirable, and that the pressure loss depends on the average piston speed $\bar{v}_K$ or the rpm; i.e., it increases with engine speed. The cross-sectional flow can be increased by increasing the size of the geometric opening cross section (valve stroke, valve seat ring diameter, number of valves).

Increased intake work primarily arises from throttle control while operating under a partial load. In spark-ignition engines, the amount of charge in the partial-load range required for the desired load is attained by adjusting the throttle valve, i.e., by changing the flow cross section. The piston must aspirate corresponding to the pressure loss at this point against a lower pressure than the atmosphere (the absolute pressure of the intake pipe drops). In the idling range, the increased intake work can be up to 30% of the work accomplished by the engine (Figs. 10-26 and 10-27).

The charge cycle energy losses are represented by the hatched area (without throttling) and by the hatched and dotted areas (with throttling). From the EO to the BDC, there is a loss of expansion work. The expulsion of the exhaust produces loss from exhaust work. Upon induction of the fresh charge at a vacuum, intake work is expended. Throttling loss occurs during throttling, in addition to the expansion and flow losses from the actuators. In a best-case scenario, the losses (hatched area) to the left of the compression line in Fig. 10-27 are avoided by controlled intake without throttling, such as by infinitely variable valve gears. In this case, the supplied amount of charge is controlled by adjusting the valve timing (IC is very important in this instance) or—depending on the variability of the system—by the variable stroke of the intake valve.

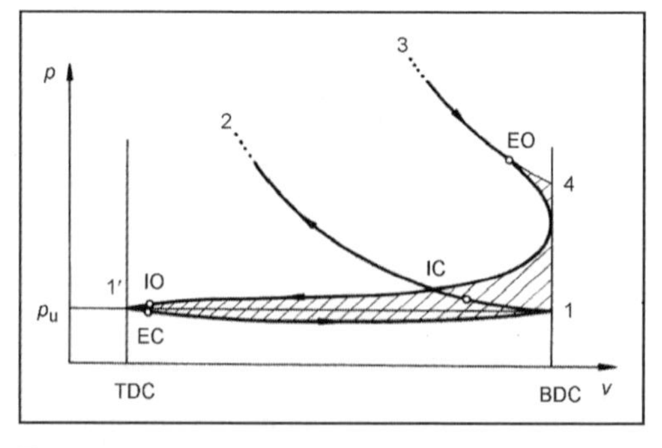

Fig. 10-26 Intake energy loss without throttling under a full load.[1]

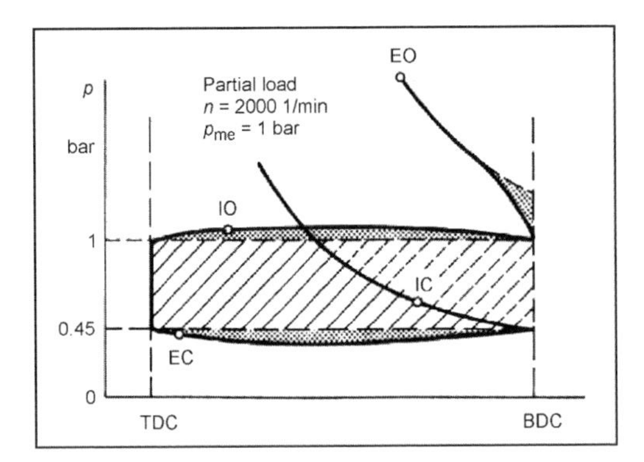

Fig. 10-27 Intake energy losses with throttling under a partial load.[1]

Increased pump work results not only from the intake of fresh air at a vacuum, but also during the expulsion of exhaust. Although the combustion gases are at a higher pressure than atmospheric pressure, they cannot leave the cylinder at the right time through the outlet and the exhaust system without work being done by the piston (i.e., before the end of the expulsion cycle). The exhaust counterpressure has a decisive influence on this process (Fig. 10-28).

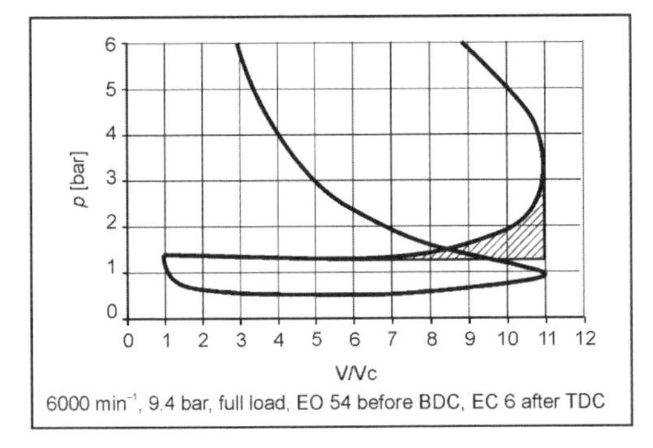

Fig. 10-28 Exhaust work.

For the charge cycle, the time of the EO is very important. This time always represents a compromise. When the EO is late, more expansion work is gained, and consumption is lowered. At higher speeds, however, greater exhaust work is required for the exhaust to leave the cylinder within the shorter period, which increases consumption. With an early EO, less exhaust work is necessary since the cylinders can be purged more easily and quickly. However, expansion work is lost, and the thermal load on the exhaust valve increases (Fig. 10-29).

Intake Systems

Both in the intake system and in the exhaust system, gas-dynamic processes occur that are based on the periodic excitation of the piston and natural frequency of the system. These can be used to improve the charge cycle process. These gas-dynamic effects in the intake system are

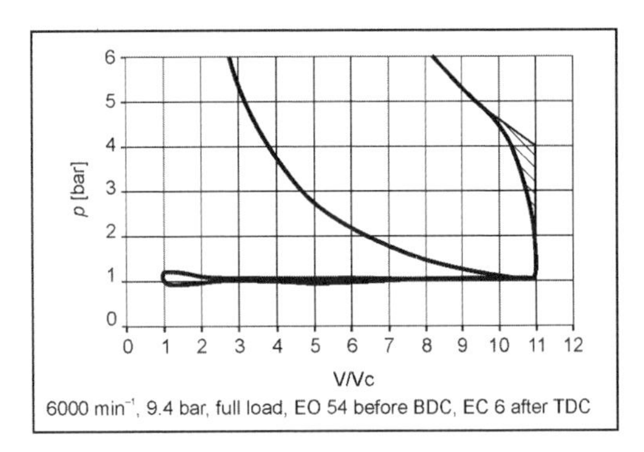

Fig. 10-29 Expansion energy loss.

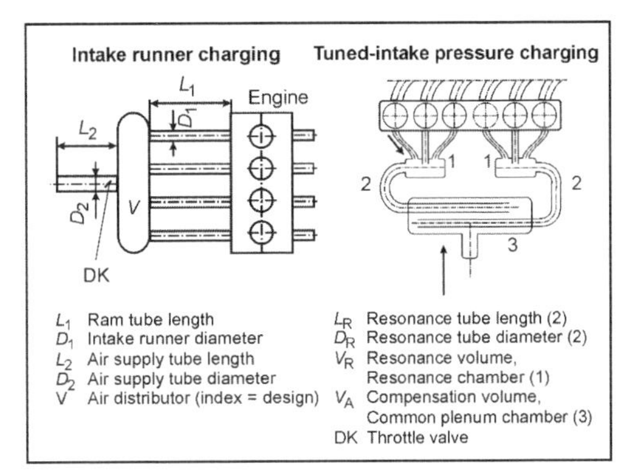

Fig. 10-30 Diagram of the ram tube and resonance

divided into ram tube and resonance effects. A schematic illustration of both intake systems is shown in Fig. 10-30.

Ram Tube Charging

The ram tube effect is based on the vacuum wave triggered by the descending piston that travels in the induction pipe opposite the direction of flow to the common plenum chamber and is reflected there at the open tube end. The overpressure wave that arises in this manner increases the cylinder charge by increasing the pressure gradient via the intake valve. This effect is particularly useful briefly before the intake valves are closed while the piston is ascending. The pressure wave prevents the expulsion of the fresh charge from the combustion chamber into the induction pipe and generates a charging effect.

Corresponding to the acoustic design, the pressure wave requires the following time at speed a to leave and return in the ram tube:

$$t = \frac{2 \cdot L_{\text{Intake}}}{a} \tag{10.14}$$

The inlet time (from IO to IC) should average one-third of the time required for an engine revolution at a given speed:

$$t \approx \frac{1}{3 \cdot n} \tag{10.15}$$

This allows the optimum length of the induction pipe to be determined at a given speed n:

$$L_{\text{Intake}} \approx \frac{a}{6 \cdot n} \tag{10.16}$$

Hence, the induction pipe length is the quantity that determines the ram tube effect. Corresponding to the acoustic design, there is a preferred speed for each induction pipe length at which there is maximum air expenditure. This has been demonstrated in engine tests in which only the induction pipe length was varied.[11] Figure 10-31 shows the influence of the induction pipe length on the maximum mean effective pressure. A shorter induction pipe shifts the torque peak in the direction of higher speeds and vice versa.

In real engine operation, however, the influence of the induction pipe length is more complex and partially overlaps with the influence of other intake-side parameters. For example, in addition to the pressure characteristic before the closing intake valve, the charge cycle is strongly influenced by the formation of a free vibration in the induction pipe in the period between IC and IO in correlation with the intake vibration that forms in the period between IO and IC.

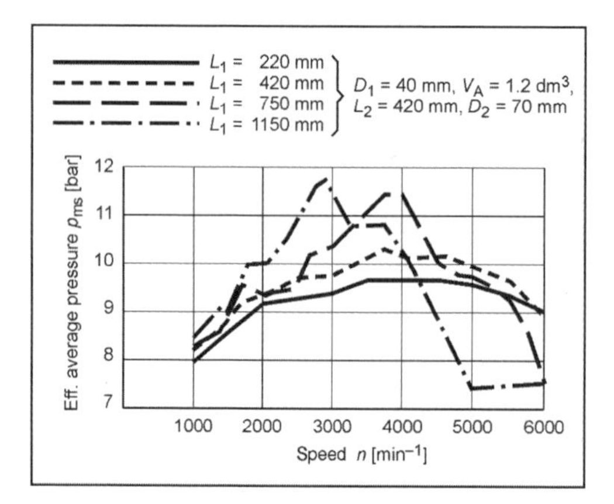

Fig. 10-31 Influence of the induction pipe length L_1 on the maximum mean effective pressure over the speed.

A fixed induction pipe length is therefore advantageous only within a specific range of speed. At higher speeds, a short induction pipe length is desirable, and, at slow speeds, a long pipe is desirable. Engines are therefore designed with a multistage manifold; i.e., the induction pipe length is adapted to the engine speed (Fig. 10-32).

When the throttle valve is open, the intake wave coming from the cylinder is reflected at this point (high speeds from 4000 min^{-1}). At speeds up to 4000 min^{-1}, the throttle valve is closed (long induction pipe). Figure 10-33 shows a further developed three-stage intake manifold. Recently, stageless variable induction pipes have also been used.

While the time of the waves depends on the induction pipe length, the amplitude of the wave is influenced by the induction pipe cross section. The flow speed in the induction pipe rises with the rpm so that the amplitude correspondingly rises as well (Fig. 10-34). Sufficiently high amplitudes to yield a corresponding recharging effect at low speeds can be created with a small induction pipe cross section. At high speeds, however, the cylinder charge falls with a small flow cross section. A good cylinder charge at high speeds, therefore, requires a large induction pipe cross section.

When there are several intake valves such as those used in four-valve engines, the induction pipe cross section can be adapted as a function of the load and speed by closing a port (Fig. 10-35). At low speeds and a low load, only the primary port is used. As the speed and load increase, the secondary port is added.

At lower speeds, the cylinder charge is better when the shutoff valve is closed (Fig. 10-36). In addition, a specific charge motion (swirling) can be generated with the inflow to improve the mixture. This increases the efficiency during partial-load operation, especially when the engine is operated with a lean mixture (lean engine).

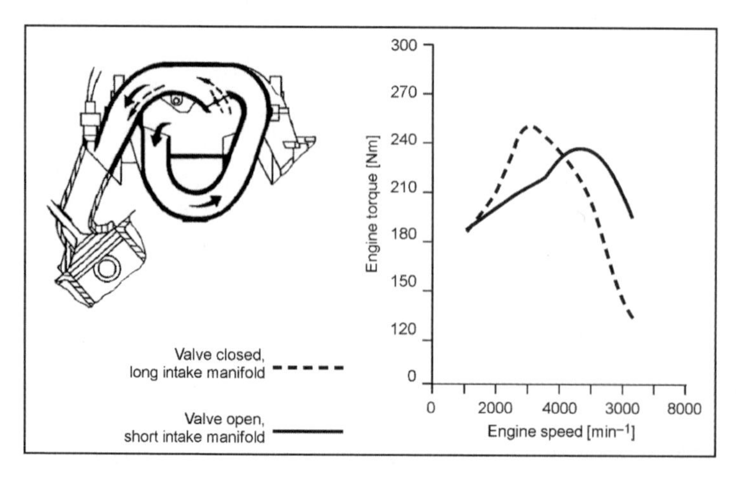

Fig. 10-32 Intake system with two-stage manifold; diagram (Audi V6).

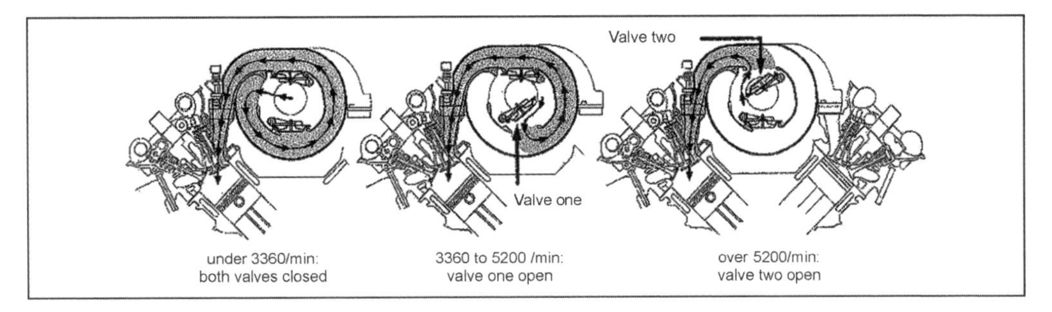

Fig. 10-33 Intake system with three-stage manifold.

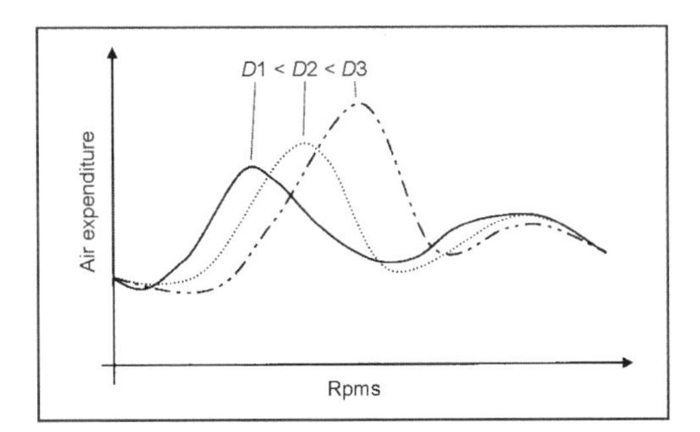

Fig. 10-34 Air expenditure as a function of the pipe diameter.[7]

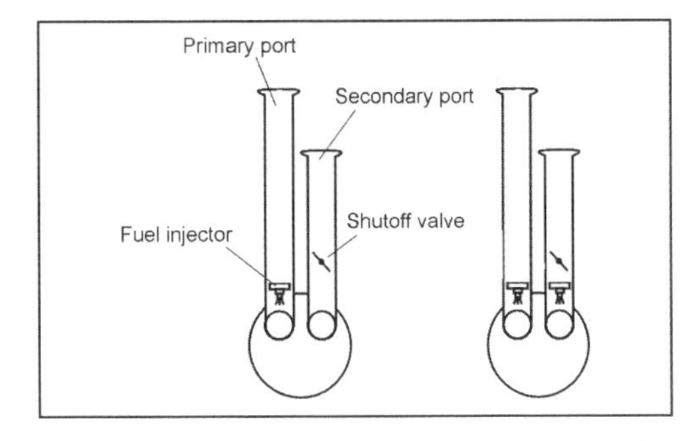

Fig. 10-35 Intake systems with channel closing.[1]

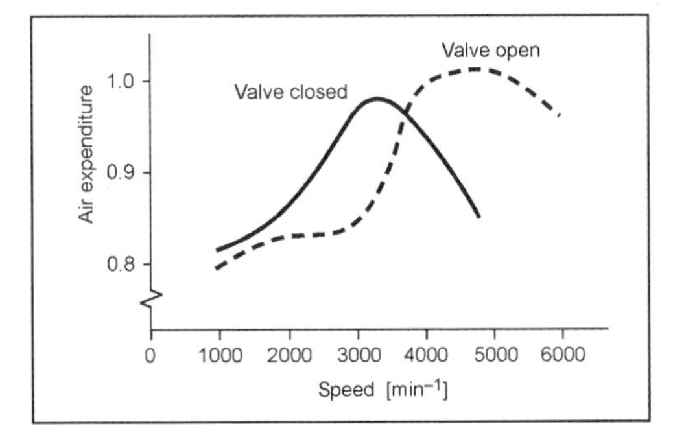

Fig. 10-36 Influence of closing a port on air expenditure.[1]

Resonance System

With a resonance charge, the charge effect is generated by an oscillating container-pipe system. The periodic intake cycles of the individual cylinders cause oscillating pressure through short induction pipes in the container that increase the pressure gradient between the inlet port and combustion chamber at the beginning and end of the intake phase.

This oscillating pressure that substantially increases the air expenditure has a definite maximum when the excitation by the cylinder corresponds to the natural angular frequency of the container-pipe system. An optimum condition for exciting oscillations is when the individual intake phase is offset by a 240° crankshaft angle, i.e., by three cylinders per resonance container.

When the intake valve is open, the system vibrates similar to a Helmholtz resonator. Vibrations arise when the air column in the inlet port moves against the "rigid" air in the cylinder, and the entire system functions like a spring mass system. The air in the cylinder can be viewed as the spring, and the air column can be viewed as the mass. The natural frequency of a Helmholtz resonator can be determined as follows:

$$f = \frac{a}{2 \cdot \pi} \sqrt{\frac{A_{\text{Intake}}}{L_{\text{Intake}} \cdot V_{BE}}} \qquad (10.17)$$

where A_{Intake} is the cross-sectional area of the induction pipe, and V_{BE} is the container volume.

In transferring the Helmholtz equation (10.17) to the internal combustion engine, Engelman used the compression chamber for the volume V_{BE} plus the half stroke volume of a cylinder and created the following simple relationship for the resonance speed n_{res} in a system consisting of a cylinder with an induction pipe:

$$n_{\text{res}} = \frac{15 \cdot a}{\pi} \sqrt{\frac{A_{\text{Intake}}}{L_{\text{Intake}} \cdot (V_c + 0.5V_h)}} \qquad (10.18)$$

This allows the natural frequencies of the Helmholtz resonator effect to be precisely described for a cylinder with an induction pipe. If there are several cylinders, the overlapping of the waves influences the results, and the phenomenon becomes very difficult to describe.

This vibration behavior is also noticeable with closed intake valves. The manifold volume acts as the resonance volume. With this design (volume), the natural frequency of the system can be varied so that it increases the air expenditure at certain speeds when an overpressure wave arrives in the intake port briefly before IC of the intake valve.

The resonance charge is particularly important in combination with turbocharging to compensate for the low torque at low speeds. In addition, it is useful to combine ram tube charge and resonance charge for six- and twelve-cylinder engines. At low speeds, the resonance vibration in the container is exploited, while short induction pipes at higher speeds contribute to the increase in air expenditure as a ram tube system. Figure 10-37 schematically illustrates a combined ram tube and resonance charge with a six-cylinder engine.

The adaptation is realized by opening or closing the resonance control valve. In the torque position, the resonance control valve is closed so that two "three-cylinder" intake systems with long pipes are active. In the output position, the resonance control valve is open, and the intake module works for all six cylinders as a ram tube system that is then fed from the entire upper manifold range with short ram tubes. The cross section and lengths can be tailored and optimized with one-dimensional calculations with these effects in mind. The air mass is controlled with the central throttle valve. The gain in torque from such a system is shown in Fig. 10-38.

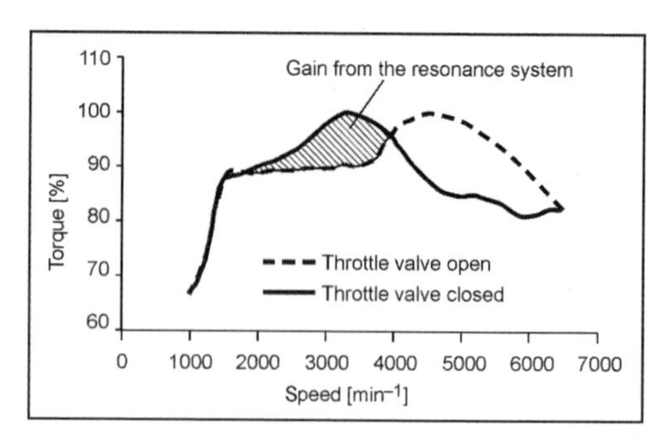

Fig. 10-38 Torque characteristic of an inline six-cylinder spark-ignition engine with a resonance system.[7]

Exhaust Systems

The exhaust system fulfills three tasks. It influences the power characteristic of the engine, it reduces exhaust noise, and it reduces the pollutants in the exhaust together with an installed catalyst. These tasks cannot be fully separated from each other. The noise damping always influences the power characteristic, generally in an undesirable manner; conversely, maximum performance exhaust systems are often too loud. The sound pressure at the exhaust valve lies between approximately 60 and 150 dB (A). This needs to be reduced to the legally prescribed value (Fig. 10-39).

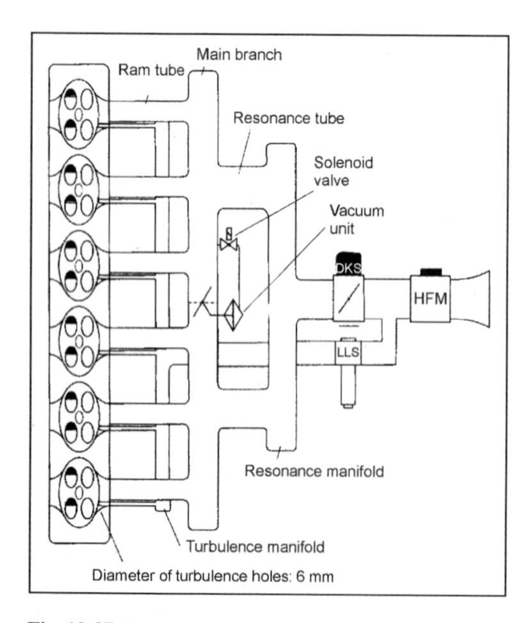

Fig. 10-37 Intake system of an inline six-cylinder engine.[7]

Similar to the processes at the fresh gas side of a reciprocating piston engine, transient flow behavior is also found in the exhaust system. When the exhaust valve is opened because of the overpressure in the cylinder, and later by the upwards-moving piston, an overpressure wave is induced that continues toward the tailpipe. Pressure and speed waves are reflected at the open pipe ends and are returned as an aspiration wave. This supports the charge cycle by lowering the exhaust counterpressure when the pipe lengths in the exhaust system are dimensioned correctly. Contrastingly, a returning overpressure wave can hinder the exit of fresh gas that is already in the cylinder. This mechanism is primarily exploited during the operation of two-stroke engines.[3]

Designs

There are two basic muffler designs: The resonator-type muffler and the absorption-type muffler. Frequently,

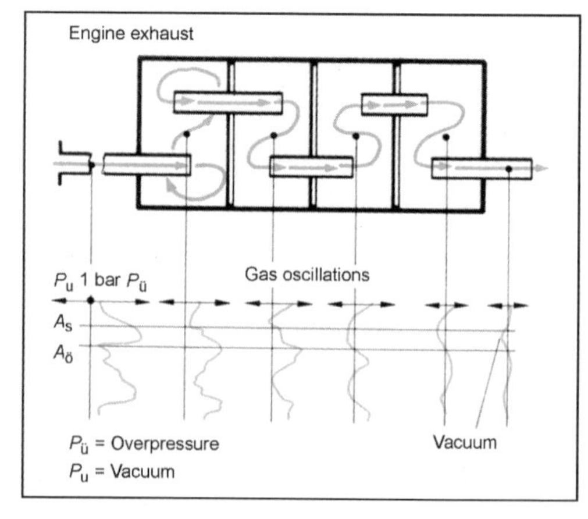

Fig. 10-39 Decrease of gas-column vibrations in mufflers.[2]

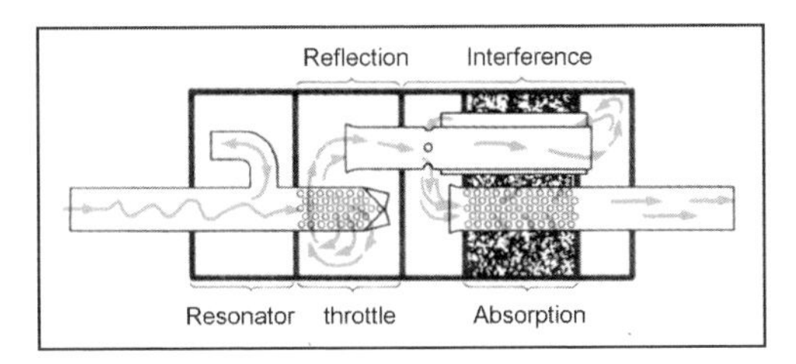

Fig. 10-40 Combined muffler system.[2]

combinations of the two types are used (Fig. 10-40), which reduce noise within the relevant range of 50 to 8000 Hz. Depending on the engine design (displacement, output, supercharging, number of valves and cylinders, etc.), a certain minimum volume is required for the reflection or absorption range (or several mufflers can be used: front, central, and rear mufflers).

With absorption-type mufflers, the flow of gas is guided through the muffler, and the gas-guiding pipe is perforated. The area between the jacket and perforated pipe is filled with absorbent material. The pulsing flow of gas can expand through the perforation into the area filled with absorbent material. A majority of the vibration energy is attenuated by friction and converted into heat. The flow of gas that leaves the muffler is largely pulse-free. The absorption-type muffler is especially distinguished by good sound suppression in the frequency range above 500 Hz and its low exhaust counterpressure.

In reflection mufflers (also termed interference mufflers), the sound is suppressed by being diverted, by changes in cross section, and by partitions inside the muffler. The corresponding chambers and changes in cross sections must be precisely harmonized with each other. Interference occurs when the sound waves extinguish each along two paths of different lengths (by being 180° out of phase). This principle is particularly effective in the range below 500 Hz.

Pressure peaks of extremely loud vibrations build in resonators (Fig. 10-40, left) that have a particularly low flow loss. The frequency at which a resonator is effective depends mainly on the dimensions (length l, diameter d, and cross-sectional area A) of the pipe extending into the resonator volume V. The resonance frequency f_0 can be calculated according to the following equation:

$$f_0 = \frac{c_0/2 \cdot \pi}{A/(l + 0,7 \cdot d) \cdot V} \tag{10.19}$$

A problematic side effect of reflection mufflers is the excitation of vibration that the wall structure of the muffler experiences from the pulsing exhaust flow. The resulting structure-borne sound can increase the noise emitted from the muffler. This can be countered by selecting sufficiently thick walls of the intermediate plates in the muffler, by using a sufficiently rigid construction of the overall muffler structure, and by using an outer double-layer jacket with or without an absorbent intermediate layer.

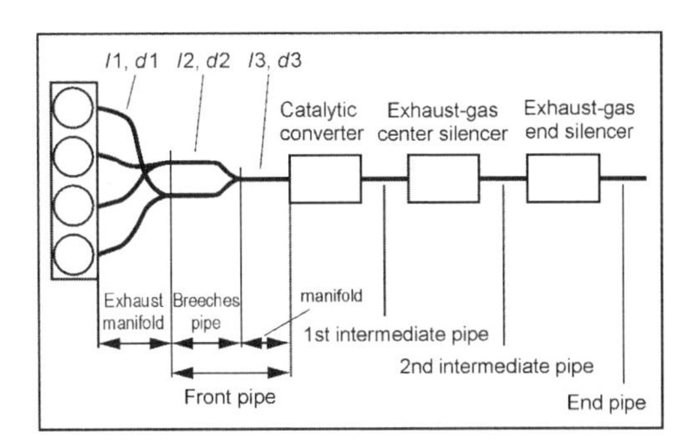

Fig. 10-41 Schematic construction of an exhaust system.[4]

Overall System

Figure 10-41 shows the basic construction of an exhaust system for a four-cylinder engine. When a single catalyst is used, it is necessary to connect the exhaust pipes of the individual cylinders. The exhaust from all the cylinders runs through the manifold that holds a central exhaust gas oxygen sensor to measure the integral air to fuel ratio.

The combined reflection/absorption-type muffler or combined reflection/branch resonator muffler are preferred to minimize exit noise. Based on the transient flow, the exhaust system can, given a suitable design, clearly improve the charge cycle similar to the intake system.[4]

The exhaust system largely affects charge cycle properties with three mutually influential factors:

- Gas-dynamic effects
- Exhaust work
- Residual share of gas in the exhaust

The exhaust work is determined by the flow properties of the exhaust system. The flow properties and the gas-dynamic effects in the exhaust system largely determine the residual exhaust gas of the cylinder charge when operating under a full load, which, in turn, strongly influences the combustion properties. The ignition conditions that change with the residual exhaust gas, the inner efficiency, and, hence, the torque behavior is significantly influenced by the adapted ignition points.

Design Criteria

In addition to the requirements for noise suppression and exhaust treatment, there are certain design criteria for the exhaust system related to the charge cycle.

Even Distribution

The exhaust pipes that can be assigned to the individual cylinders at the exhaust manifold must have pipes with equal lengths and cross sections. In view of the options within the vehicle interior, the elbows of the exhaust manifold and the pipe connections should be designed similarly. These requirements also apply to the Y pipe.

Exhaust Counterpressure Level

To achieve a low exhaust counterpressure, superior flow properties should be sought for the cylinder head exits and the exhaust system. The exhaust counterpressure cannot be reduced to zero because of the flow resistance of the catalyst and the basic function of noise suppression, since noise suppression always involves an irreversible conversion of energy that is manifested by the exhaust counterpressure behavior.

Gas-Dynamic Effects

The exhaust system should support the charge cycle in definite speed ranges for pipe length, cross section, and pipe branching.

Catalyst Operating Conditions

The installed position of the catalyst requires fundamentally contradictory design criteria. For superior starting when the engine is cold, the overall pipe length from the cylinder head to the catalyst should be as short as possible. In contrast, the temperature of the catalyst should be kept low during high engine performance to ensure a long life. This can be achieved when the pipe is as long as possible.

Gas-Dynamic Processes

The exhaust under high pressure in the combustion chamber causes a pressure wave when the exhaust valve is opened that makes the exhaust pulsate at a high amplitude. According to acoustic theory, the pressure amplitude advances at the speed of sound through the exhaust line and is reflected at the open pipe end as a negative pressure amplitude. If it is at the exhaust valve at the right time, the negative pressure amplitude can support the charge cycle by extracting residual gas from the combustion chamber.

Real exhaust systems have different reflection sites in the exhaust line from the cylinder head to the entrance in the catalyst housing because of the individualized pipe branching.

Figure 10-42 schematically illustrates the pressure wave for cylinder 1 in an exhaust system from Fig. 10-41. After passing along the exhaust path l_1, the positive pressure curve meets the first reflection site where the pressure pulse is divided according to the design of the pipe branches and the pipe cross sections of the exhaust manifold and Y pipe. At a correspondingly sharp branching angle, a small amount of the pressure pulse with a primarily positive amplitude passes through exhaust line l_1 of cylinder 4 and is reflected from the closed exhaust valve as a mainly positive pressure pulse. Another part of the pressure pulse is reflected from the pipe branch as a vacuum pulse and returns against the main direction of flow to cylinder 1. The majority of the original pressure passes along exhaust path l_2 of the Y pipe up to the pipe branch at the manifold where a division of the positive pressure pulse occurs similar to the transition from the exhaust manifold to the Y pipe. The remaining portion of the original pressure pulse that passes along exhaust path l_3 is reflected at the transition to the catalyst housing as a vacuum.

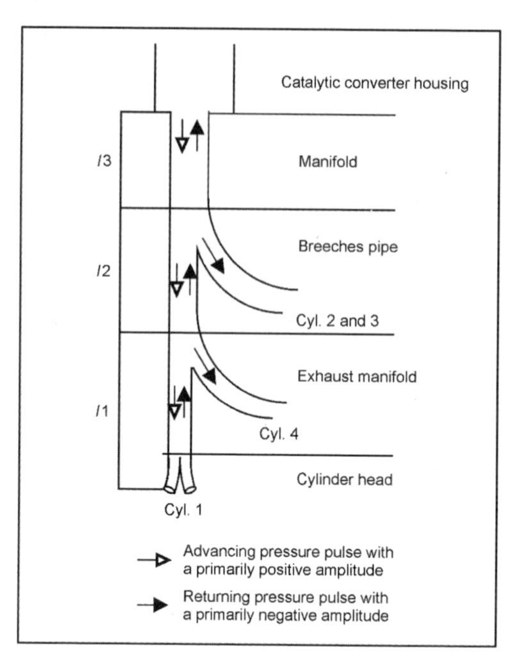

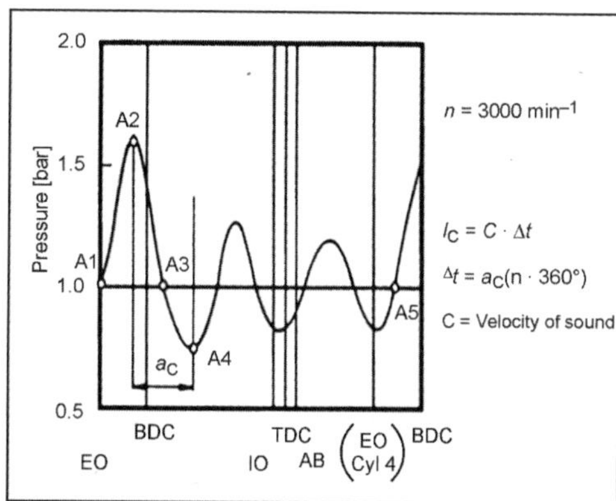

Fig. 10-42 Left: Schematic representation of the reflection sites. Right: Pressure characteristic in the exhaust manifold (100 mm after the exhaust valve).[4]

The rise of the positive pressure triggered by the opening exhaust valve starts at A1. The rise in pressure to the maximum A2 depends mainly on the function of the lifting valve. The further course of the pressure curve from A2 to A4, the maximum of the reflected vacuum characteristic, depends on the design of the exhaust system. The characteristic length l_C that remains constant for the respective exhaust system independent of the working point can be calculated from the crank angle a_C that extends from A2 to A4 by considering the rpm and speed of sound. The pressure curve from A4 to A5 is characterized by the overlapping wave movements in the exhaust system. The basic characteristic is similar for any respective exhaust systems and is nearly independent of the working point. At A5, the pressure of cylinder 4 starts to rise at the measuring sensor after passing through l_1 of cylinder 4 and l_1 of cylinder 1 up to the measuring sensor after EO of cylinder 4.[4]

The location and characteristic curve of the pressure from A_3 to A_5 and the characteristic length l_C strongly influence the engine properties. A minimum pressure during valve overlap is always advantageous.

The characteristic length l_C essentially depends on the exhaust pipe lengths l_1 and l_2, the ratio of the Y pipe diameter to the exhaust manifold diameter d_2/d_1, as well as on the design of the transition from the exhaust manifold to the Y pipe. As the sum of $l_1 + l_2$ increases and the diameter ratio d_2/d_1 decreases, the characteristic length l_C increases since the main reflection site is farther from the inlet port. This is also shown by the following experimentally determined pressure characteristics of three different exhaust system variants (Fig. 10-43).

Valve Timing

The valve timing is always a compromise since the engine operates within wide ranges of speed and load. Because of the factors described in Section 10.6.1, one cannot simultaneously optimize the charge cycle for maximum torque and maximum rated horsepower without additional features such as the camshaft adjustment system, the control cam system, or a multistage manifold. The offsetting of the valve timing is related to these factors. The terms "early" and "late" indicate a relative position to the basic control times that are indicated as the degree of crankshaft angle relative to the closer dead center.

- Exhaust opens (EO)

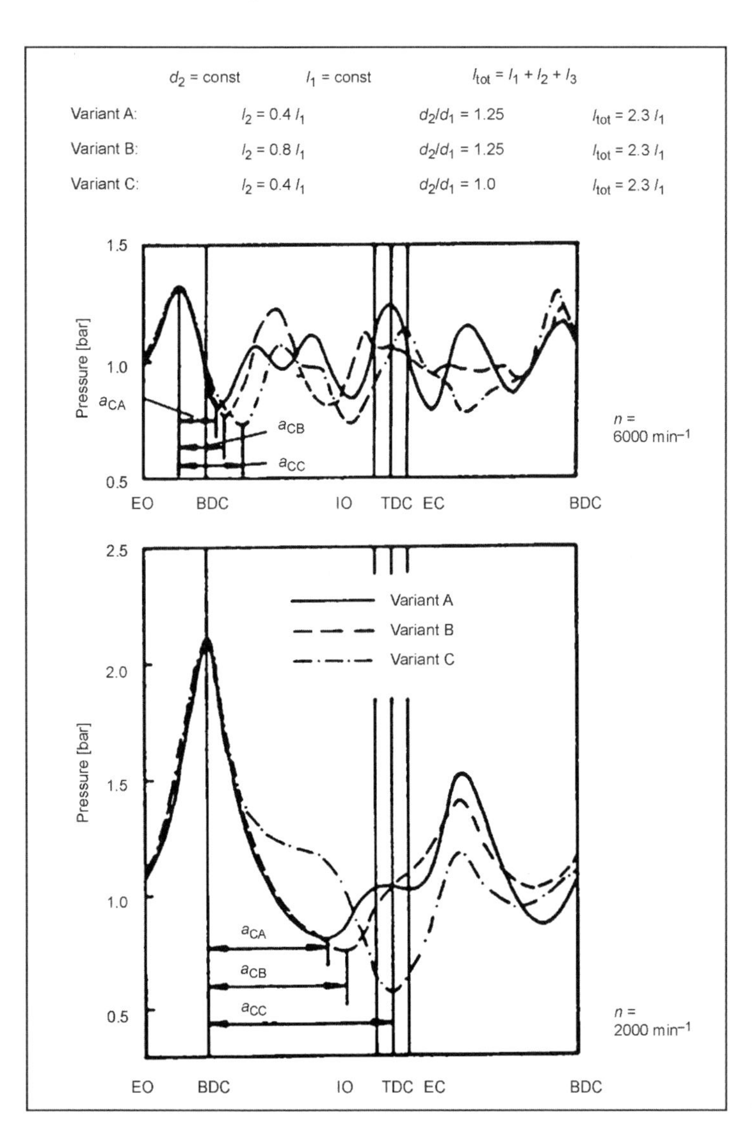

Fig. 10-43 Pressure characteristic in the exhaust manifold (pressure sensor 100 mm after the exhaust valve).[4]

The exhaust usually opens in spark-ignition engines at a 50°–30° crankshaft angle from BDC shortly before the end of the expansion cycle. This control time represents a compromise between a gain in expansion work and greater exhaust work.

If the EO is moved in the late direction (i.e., the EO occurs closer to the BDC), the working gas expands longer and exerts work on the piston, the thermal efficiency increases, and consumption falls. A longer expansion lowers the hydrocarbon emissions and the exhaust temperature. At greater speeds and loads, the exhaust work substantially increases at the start of the expulsion cycle, which, in turn, increases consumption. A late EO is primarily relevant for partial loads, and its influence on full loads is slight (Fig. 10-44).

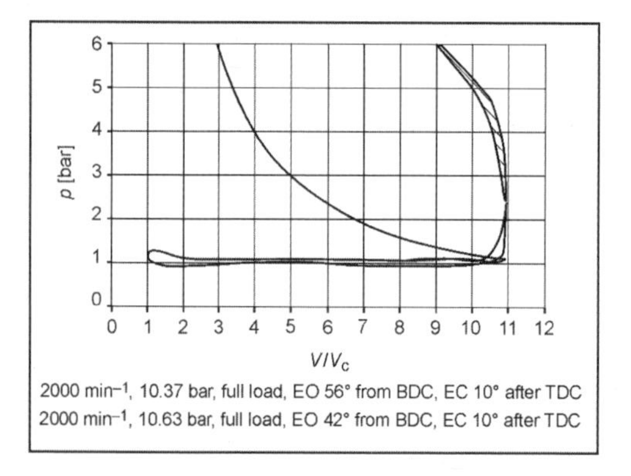

Fig. 10-44 Increase in expansion work by shifting the EO in the "late" direction.

When the EO shifts in the early direction, the opposite occurs: Expansion work is lost, the thermal efficiency drops, and the fuel consumption increases. The hydrocarbon emissions and exhaust temperature rise. However, less exhaust work is required since the cylinder pressure is always at a higher level, and the exhaust leaves the cylinder more quickly. An important factor is that the consumption increases at a partial load. Another fact is that the thermal load on the exhaust valve rises with an early EO and, hence, increases the material wear.

The pressure loss during expulsion also depends on the lifting curve of the exhaust valves. When the valve stroke rises strongly during opening, it is easiest for the exhaust to leave the cylinder. For this reason, the required compromises with two exhaust valves are less critical than with only one exhaust valve: When there are two exhaust valves, there is a more effective opening area available for expulsion at a faster rate. The exhaust can, therefore, leave the cylinder at the beginning of the expulsion cycle since it is at a higher pressure. There is, therefore, less exhaust work for the piston.

- Exhaust closes (EC)

A common approach to EC is an 8°–20° crankshaft angle after TDC, which indicates the end of the valve overlap phase. In addition to IO (inlet opens), EC is the control time that can be used to control the length of the overlap. At low speeds and load levels, the EC controls the amount of exhaust drawn back by the exhaust system, and at higher load levels and speeds, it controls the residual gas that can be expelled.

Under a full load, the cylinder can be thoroughly purged by a late EC, which increases the volumetric efficiency. This is used for engines with a higher rated horsepower such as sports engines. An increasingly greater portion of the fresh charge flows through the cylinder without participating in combustion (scavenging loss from short-circuit flow), which increases consumption and the hydrocarbon emissions.

Under a partial load, an increasingly greater portion of the exhaust is drawn back (internal exhaust recirculation) by the suction of the piston. This can yield substantial advantages for consumption and emissions. The last part of the exhaust is always relatively rich in uncombusted hydrocarbons since the combustion is incomplete of the cylinder charge zones close to the wall. This component is expelled relatively late. If this component in the exhaust is "recombusted," consumption is reduced, and there are fewer hydrocarbon emissions. Because of the diluted charge, the combustion temperature is lower, which reduces nitrogen emissions. Another consideration is that the fresh mixture becomes homogenized because of the hot residual gases and, hence, produces a better mixture. There is less intake work with a later EC. This occurs for two reasons: First, the drawn back exhaust component expands in the cylinder and supports expansion. Second, when there is more residual exhaust gas in the cylinder charge, less throttling for load control is required to compensate for this quantity while retaining the load. This further reduces consumption. The restriction on internal exhaust gas recirculation is determined by the residual gas compatibility during combustion.

With an early EC, the combustion gas cannot leave the cylinder at the right time (exhaust lockup) so that the residual exhaust gas in the cylinder rises. This causes the volumetric efficiency and the rated horsepower to drop. The scavenging loss is lower, which slightly lowers consumption. In this case as well, the last component of the exhaust is recombusted, which can have advantages for consumption and emissions under a partial load (the nitrogen oxide emissions are reduced because of the low combustion temperature). The exhaust remaining in the cylinder continues to flow (partially guided by the piston) very strongly into the induction pipe, which improves the mixture preparation. Since there is a continually smaller area for expelling the exhaust after a certain piston position, the exhaust work is increased. At the end of the expulsion cycle, the residual gas can be compressed by an early EC, which slightly increases consumption. An early EC is limited by the increased exhaust work, a fresh charge diluted with exhaust, and an inhomogeneous mixture from a strong inflow of exhaust into the induction pipe.

When dynamic effects in the exhaust system are optimized, the efficiency of the expulsion can be improved if a vacuum wave reduces the static pressure in the exhaust port shortly before EC and thereby sucks the exhaust out of the cylinder.

- Inlet opens (IO)

The control time IO is commonly set at 20°–5° crankshaft angle before TDC for spark-ignition engines. As the beginning of the valve overlap phase, it is also important like the EC for regulating the residual amount of gas in the fresh charge under partial loads and for scavenging the residual gas under full loads. As such, it has a substantial influence on idling quality.

The duration of the valve overlap phase is shortened with a late IO. Under a partial load, this produces a charge that is less diluted with exhaust, which increases the speed of combustion. Under such conditions, the rpm can be lowered during idling, which reduces consumption. Given the lower residual exhaust gas and the fast combustion, the combustion temperature increases, and emissions of nitrogen oxide increases. The hydrocarbon emissions can be lowered under the following conditions: Since the intake valve opens later, the flow in the cylinder is faster at a specific piston position, which increases the flow within the cylinder. This, in turn, improves the mixture preparation, and combustion is more thorough, which shortens the combustion or ignition delay as well as the length of combustion. When the IO is late, the intake work increases since a vacuum is generated in the cylinder in the first phase of intake. This increases consumption. Under a full load, the mean effective pressure is less since the air expenditure is lower.

With an early IO, the valve overlap phase is lengthened, and a particularly large amount of exhaust returns into the induction pipe under a partial load. This has a negative influence on combustion since the mixture becomes inhomogeneous and burns more slowly. However, this effect can also be put to positive use for induction pipe injection with throttle-free load control (variable valve actuation). Since there is no induction pipe vacuum with throttle-free load control, there is frequently insufficient mixture preparation, which causes the combustion to last longer and be incomplete. Fuel deposits can also form close to the valve. These deposits can be vaporized by the returning hot exhaust and be sucked back inside, which heats the induction pipe wall and improves the mixture. Investigations have shown [Göbel, MTZ] that this method can positively influence mixture preparation despite the reaction-inhibiting higher residual exhaust gas, which in the final analysis enhances the reactivity of the mixture.

- Inlet closes (IC)

The valve timing element that is primarily responsible for the torque and power characteristic is the IC. It usually lies at a 40°–60° crankshaft angle after BDC, and it influences the charging of an engine much more than the other control times. The characteristic quantities such as torque and output are primarily determined by the IC.

Offsetting the IC in the late direction to a time optimized for the maximum torque yields greater air expenditure and volumetric efficiency at higher speeds. A higher rated horsepower is correspondingly attained with a late IC. As illustrated in Section 10.6.2.1, the gas dynamic effects at higher speeds play the most important role (especially the recharging effects). When the IC is offset, the most important task is to exploit these effects by capturing the overpressure wave in the cylinder. At lower speeds and under a full load, a long opening time has a negative influence on the torque. Since the intake valve is closed later, a greater amount of charge is pushed back into the induction pipe by the piston. This is countered by a lower pulse because of the lower gas speed, which, in turn, lowers the volumetric efficiency. The influence of the IC control time on air expenditure under a full load is shown in Fig. 10-45. By offsetting the inlet camshaft by a 20° crankshaft angle toward late, the air expenditure is clearly reduced at low speeds. At the nominal rpm, the air expenditure is contrastingly increased by approximately 8% in an eight-cylinder spark-ignition engine with four valves per cylinder.

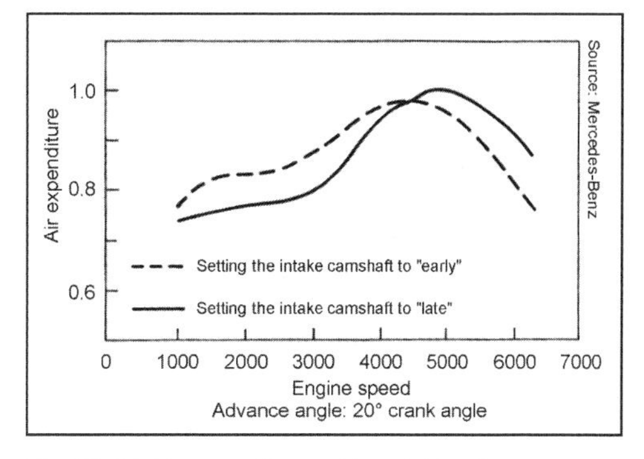

Fig. 10-45 Air expenditure with variable IC.

Under a partial load, a late IC lowers the intake work since the charge is aspirated with less throttling. This lowers the consumption. The thermal efficiency of the process is lower since the effective compression becomes increasingly low. The combustion temperature is reduced by the low peak pressure, which, in turn, reduces the nitrogen oxide emissions.

With variable valve actuation and a late IC, the engine can be operated without throttling. The goal is either to achieve a higher rated horsepower or to reduce the consumption under a partial load. Under a partial load, the excess charge is returned by the piston into the induction pipe during the compression cycle. Because of the throttle-free load control, less intake work is required. As described above, this reduces consumption, thermal efficiency, the consumption temperature, and nitrogen oxide emissions. The limit for a late IC is the drop in thermal efficiency and the worse mixture preparation in the intake port because of the lack of a vacuum (lower gas speed).

Given an early IC and conventional valve actuation, the intake phase becomes shorter, which reduces air expenditure. Under a full load and at higher speeds, this reduces the volumetric efficiency and yields a low rated horsepower. However, since less charge is returned into the induction pipe at low speeds, the volumetric efficiency and torque increase. Under a partial load, the required load can be attained with last throttling because of the shorter intake phase, which reduces the amount of intake work. This has a positive effect on consumption.

With variable valve actuation and an early IC, the load no longer has to be controlled by throttling; rather it can be regulated by the selected IC valve timing. The goal can be either to increase the torque under a full load or to reduce the consumption under a partial load. As soon as the amount of charge is in the cylinder that is required for the load, the intake valve is closed. In this phase, the piston is still moving toward BDC, and a vacuum is generated in the cylinder. Since the load control is throttle-free, the amount of intake work is much lower than when throttling is used to control the load, and this reduces consumption. The difference in pressure between the intake and exhaust systems is low, and only a slight amount of its exhaust is sucked back by the outlet. Assuming that the IO timing is at a conventional position and that the overlap phase is not long, an early IC produces stable combustion under low loads at slow speeds. The limits on a low IC is the mixture formation. Since the intake ends earlier than the BDC, there is frequently negligible charging movement in the cylinder during ignition, which can make combustion longer and incomplete after a long combustion delay. This can produce greater hydrocarbon emissions and increase consumption despite the low amount of work involved in the charge cycle. Furthermore, there is the danger of fuel condensation in the cylinder from the charge cooling because of the generated vacuum. As mentioned under the section "Inlet opens," the mixture is insufficiently prepared in the inlet port because of the absence of a vacuum, which makes the aspirated mixture inhomogeneous. Fuel deposits can form close to the valve.

Flow Cross Sections

For high volumetric efficiency and low work losses during a charge cycle, the large control valves must have a large geometric opening cross section. The characteristic curves of the opening cross sections of the intake and exhaust valves correspond to the valve lifting curves (Fig. 10-46).

The valve stroke and opening cross section for the intake valve are greater than for the exhaust valve. The opening cross section is made even greater since the intake valve is larger than the exhaust valve (intake valve diameter > outlet valve diameter).

The flow cross section at the valve strongly influences the charge cycle. The flow cross-section is smaller than the geometric cross section because of hydrodynamic processes (Fig. 10-47).

Both the geometric opening cross section and the flow cross section are ring areas that surround the valve axis

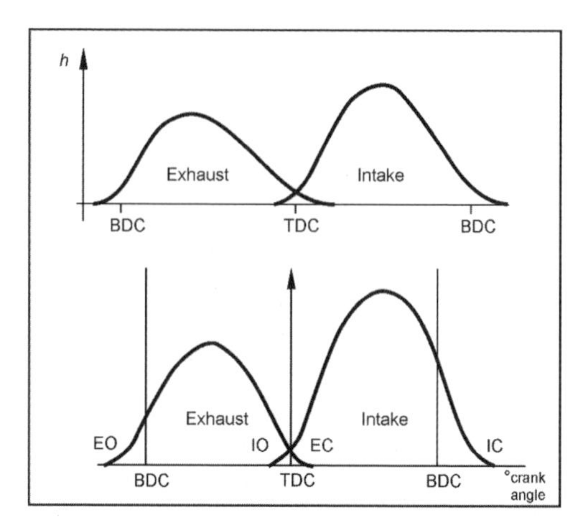

Fig. 10-46 Valve lifting curves and valve cross sections.[1]

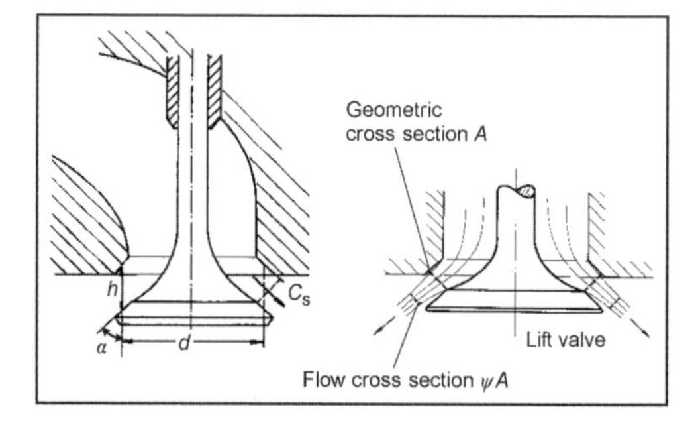

Fig. 10-47 Flow cross section and valve stroke.[1]

corresponding to valve seat angle α. The valve stroke is the perpendicular distance of the valve head to the valve seat.

Assuming an isentropic flow at the valve seat, the theoretical speed c_{is} results in flow cross section A_S. Because of friction, the actual speed c_s is less than c_{is}. The following holds true for the mass flow at the valve:

$$\dot{m} = \dot{V} \cdot \rho = A_S \cdot c_S \cdot \rho = \psi \cdot A \cdot \varphi \cdot c_{iS} \cdot \rho \quad (10.20)$$

with

ρ = Density in the flow cross section
ψ = Jet contraction (constriction number)
φ = Friction coefficient

The following holds true for the isentropic flow cross section A_{is}:

$$A_{iS} = \psi \cdot \varphi \cdot \frac{\rho}{\rho_{iS}} \cdot A \quad (10.21)$$

with

ρ_{is} = Density given an isentropic flow in the flow cross section

The following equation is thereby obtained for the mass flow:

$$\dot{m} = A_{iS} \cdot c_{iS} \cdot \rho_{iS} \quad (10.22)$$

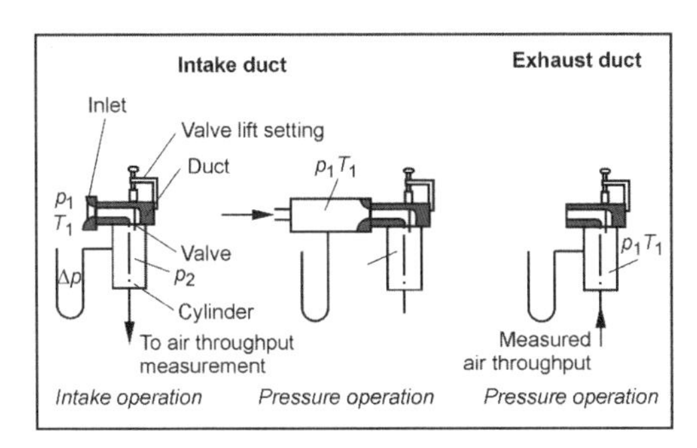

Fig. 10-48 Measuring setup to determine flow.[1]

The isentropic flow cross section A_{iS} of a valve as a function of the valve stroke is determined in a stationary flow test. A flow is guided through the cylinder head or a corresponding model, and quantities are measured for different valve strokes (Fig. 10-48).

T_1, p_1 = Thermal condition before measuring, e.g., in a collection tank

p_2 = Pressure in the cylinder

$\dot{m}$ = Mass flow, measured, for example, with an orifice

The measurement can be done using suction or pressure (compressed air). The isentropic flow cross section A_{iS} can be calculated from the recorded measured values. The following holds true:

$$c_{iS} = \sqrt{\frac{2 \cdot \kappa}{\kappa - 1} \cdot R_L \cdot T_1 \cdot \left[1 - \left(\frac{p_2}{p_1}\right)^{\frac{\kappa-1}{\kappa}}\right]} \qquad (10.23)$$

and

$$\rho_{iS} = \rho_1 \cdot \left(\frac{p_2}{p_1}\right)^{\frac{1}{\kappa}} \qquad (10.24)$$

$\kappa = 1.4$ for air

In terms of approximation, A_{iS} is independent of the set pressure ratio p_2/p_1 in the stationary flow test. In addition, A_{iS} can be transferred to real engines even though the flow is transient since a quasistationary calculation is permissible, given the short throttling sites in the direction of flow.

The flow factor of the valve α_V is used to evaluate the quality of the actuators:

$$\alpha_V = \frac{A_{iS}}{A_V} \qquad (10.25)$$

with

A_V = Valve area corresponding to the inner valve seat diameter

α_V does not provide any information on the quality of the charge cycle. A measure of the valve flow in a given engine and, hence, for the charge cycle is the flow factor α_K,

$$\alpha_K = \frac{A_{iS}}{A_K} \qquad (10.26)$$

with

A_K = Piston surface

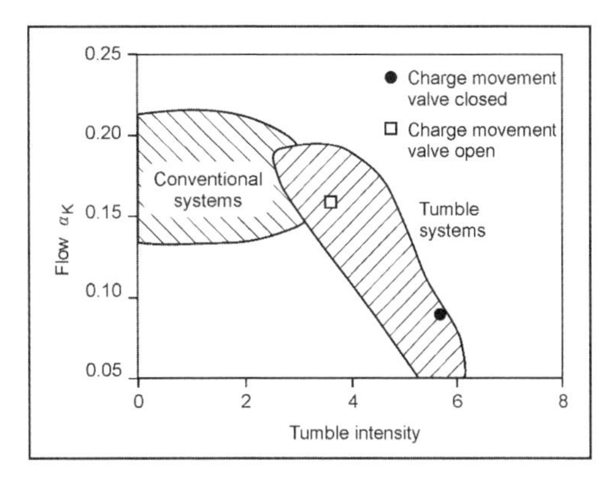

Fig. 10-49 Flow as a function of the tumble intensity.

Figure 10-49 shows the flow factor spread for modern engines as a function of the intensity of a tumble flow. The values for the VW FSI (1.4 l with direct gas injection) are shown as solid dots and open squares for open and close charge movement valves in the inlet port.

α_K is very suitable for comparing different engines with the same average piston speed. Reference values for the inlet-side flow speed of the engine given a maximum valve stroke $h_{V,\max}$ are

Spark-ignition engine	Two-valve: $\alpha_K = 0.09–0.13$
	Four-valve: $\alpha_K = 0.13–0.17$
Diesel engine	Two-valve: $\alpha_K = 0.075–0.09$
	Four-valve: $\alpha_K = 0.09–0.13$

10.2 Calculating Charge Cycles

The simulation of the engine combustion process, especially combined with a one-dimensional simulation of the gas dynamics in the intake and exhaust systems, is today a generally accepted tool for predicting output data of engines in the design phase or during construction. It is also used for analyzing the charge cycle and the thermodynamic process of engines running on a test bench. Especially for the last application it can, when used correctly, offer information that could not otherwise be experimentally determined except at great expense.

Because of the complexity of the charge cycle process, an enormous amount of effort goes into its theoretical analysis. Depending on the respective question, a certain amount of simplification is required. For this reason, various calculations for special applications have been developed for analysis and simulation. A distinction is drawn among purely thermodynamic zero-dimensional models, one-dimensional models that couple zero-dimensional analysis with gas dynamics in the intake and exhaust systems, and three-dimensional spatial models (CFD). Whereas a one-dimensional analysis makes it possible to describe the entire engine from the air filter to the exhaust system and thereby offers a temporal description and spatial (one-dimensional) description along the pipes of the processes, the three-dimensional CFD calculation is limited to spatial (three-dimensional) and temporal analysis of the processes in subsystems of the engine because of limited computer capacity.

The Filling and Emptying Method

The easiest way to describe the charge cycle in a real engine is the filling and emptying method. Since spatial gradients of the state variables are not covered by this method, the filling and emptying method belongs to the zero-dimensional methods of calculation. Despite this simplification, it is still sufficient in most cases for comparisons and an initial evaluation of the charge cycle.

In the filling and emptying method, the intake line and the exhaust line in the cylinder are viewed as containers whose contents are characterized by pressure, temperature, and material composition (Fig. 10-50).

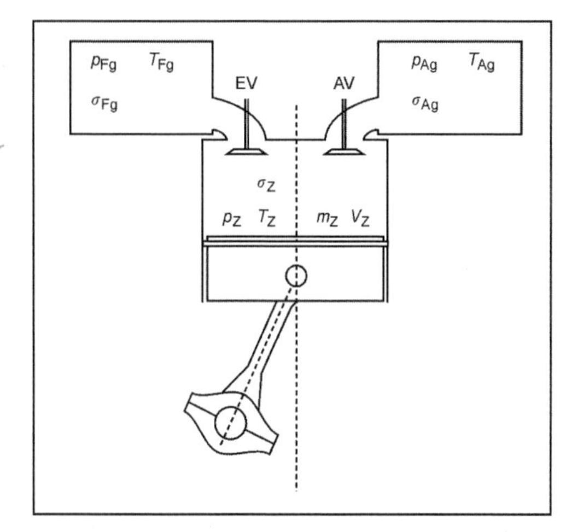

Fig. 10-50 Model of the filling and emptying method.

The filling and emptying method is based on the first law of thermodynamics:

$$\frac{d(m_z \cdot u)}{d\alpha} = -p_z \cdot \frac{dV}{d\alpha} - \sum \frac{dQ_w}{d\alpha} + \sum \frac{dm_e}{d\alpha} \cdot h_e - \sum \frac{dm_a}{d\alpha} \cdot h_a \tag{10.27}$$

To be able to determine the mass flows in the inlet and outlet, information must be obtained concerning the states at the inlet and outlet of the cylinder. Physically, strong three-dimensional flows occur through the valves that manifest jet disintegration and zones of turbulence. From a simplified perspective, the zero and one-dimensional models assume that the flow through these throttle sites is quasistationary. In this instance, "quasistationary" means that the state vector at the inlet and outlet areas of the throttle site (Fig. 10-51) does not change within a unit of time of the calculation, and that the change over time of

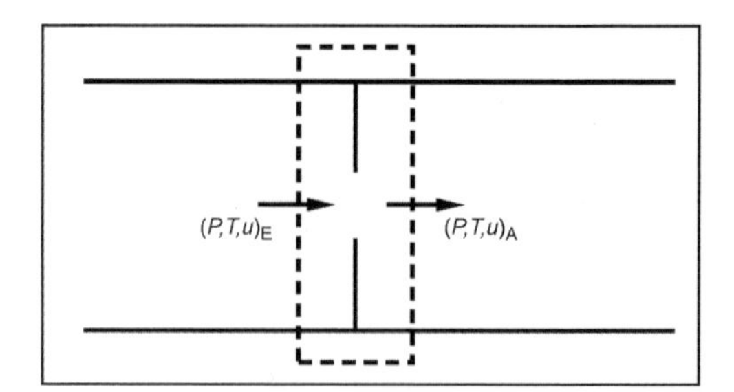

Fig. 10-51 State variables at a throttle site.

the vector results from the sequence of different stationary states. Since the throttle site does not extend infinitely, this analysis is more applicable for smaller throttle sites in the direction of flow in comparison to the connected pipes (Fig. 10-51).

Given these assumptions for this model, and the basic equations of the one-dimensional model, stationary flow can be used to calculate the state vectors $(\boldsymbol{p}, \boldsymbol{T}, \boldsymbol{u})_E$ and $(\boldsymbol{p}, \boldsymbol{T}, \boldsymbol{u})_A$ at the edges of the pipes. Using the continuity and energy equation for one-dimensional stationary flow, we obtain St. Venant's theoretical flow equation that holds true when an isentropic, loss-free change in state in a flow cross section arises after the inlet and outlet surfaces of the quasistationary throttle sites. Since, however, the change in state is not isentropic and the pulse attenuates, this approach must be corrected. A stationary measurement is required that quantifies the thermodynamic effect of the flow phenomena that causes the pulse to attenuate. This pulse attenuation manifests itself thermodynamically by an irreversible increase in the entropy of the fluid. The mass flow passing through the throttle site in an irreversible flow is smaller than the mass flow that would result with a loss-free flow. This loss is measured with the aid of the flow coefficient α that is defined as the ratio of the actual mass flow to the theoretical (isentropic) mass flow. The mass flows at the inlet and outlet are therefore calculated as follows:

$$\dot{m} = A_{\text{eff}} \cdot p_{01} \cdot \sqrt{\frac{2}{R \cdot T_{01}}} \cdot \psi \tag{10.28}$$

where

$$A_{\text{eff}} = \alpha \cdot \frac{d_{vi}^2 \cdot \pi}{4} \tag{10.29}$$

and the flow function ψ in the subsonic range is

$$\psi = \sqrt{\frac{\chi}{\chi - 1} \cdot \left[\left(\frac{p_2}{p_{01}} \right)^{\frac{2}{\chi}} - \left(\frac{p_2}{p_{01}} \right)^{\frac{\chi+1}{\chi}} \right]} \tag{10.30}$$

and in the transonic range

$$\psi = \psi_{\max} = \left(\frac{2}{\chi + 1} \right)^{\frac{1}{\chi+1}} \cdot \sqrt{\frac{\chi}{\chi + 1}} \tag{10.31}$$

The flow coefficient α changes with the valve stroke and is experimentally determined using stationary flow experiments.

Calculation Method

The goal of the calculation is to determine the characteristics of the pressure, temperature, mass, composition of the cylinder charge, and the characteristic curve of the mass change influenced by valves as a function of the characteristic of the crank angle during the charge cycle phase.

These quantities cannot be measured or can be measured only with great effort. Only the pressure can be indicated by means of a quartz sensor. The characteristic curves of these quantities are therefore calculated from a starting point by using numeric integration.

The initial values of the pressure, temperature, mass, and composition are determined at "outlet opens" by measuring or estimating, and their differential changes are calculated from this starting point using basic thermodynamic equations. On this basis, a suitable integration is applied step-by-step until all the values are known up to the time of "inlet closes."

One-Dimensional Gas Dynamics

The filling and emptying method is a quasistationary single-zone model. In this instance, quasistationary means that transient processes are viewed as stationary for short intervals; i.e., the individual quantities (pressure, temperature) are dependent only on time but not location. Dynamic influences such as pressure pulses that, for example, arise in the ram tube charge and resonance charge cannot (of course) be included. Amplitudes and phase angles of the oscillations can support the charge cycle at certain rpm and hinder them and other speeds. The characteristic of the volumetric efficiency is essentially determined by means of the rpm and torque characteristic of the engine.

These oscillations are excited by pressure waves that arise when the valves are opened and closed and when piston motion occurs. The following figure illustrates the pressure curve determined with the aid of induced low pressure in the intake pipes of a slow one-cylinder four-stroke engine at 3000 rpm. At the beginning of the intake process, the downward movement of the piston generates a vacuum wave at the intake valve. This vacuum wave advances to the air filter that acts as an open pipe end. It is reflected as an overpressure wave, returns to the intake valve, and reaches it at IC (Fig. 10-52).

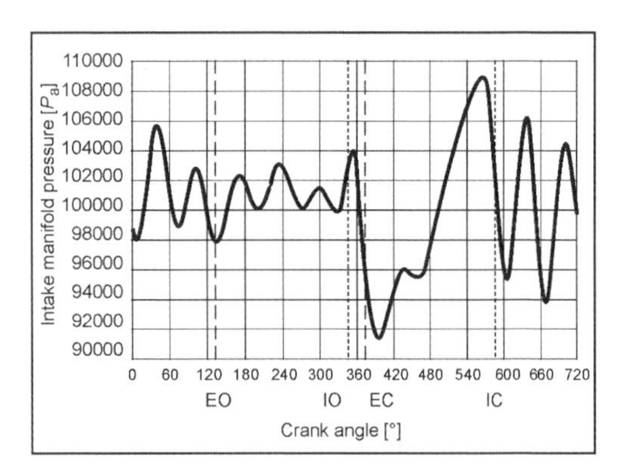

Fig. 10-52 Pressure characteristic in an induction pipe at 3000 rpm.

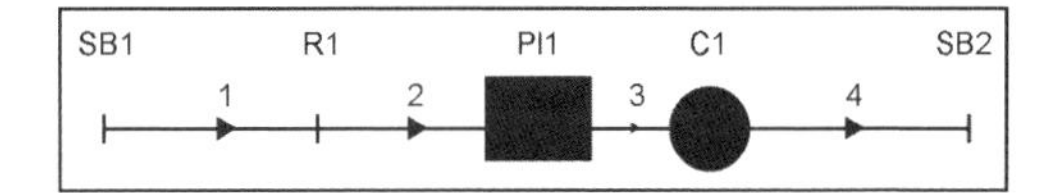

Fig. 10-53 Schematic representation of an entire engine system.

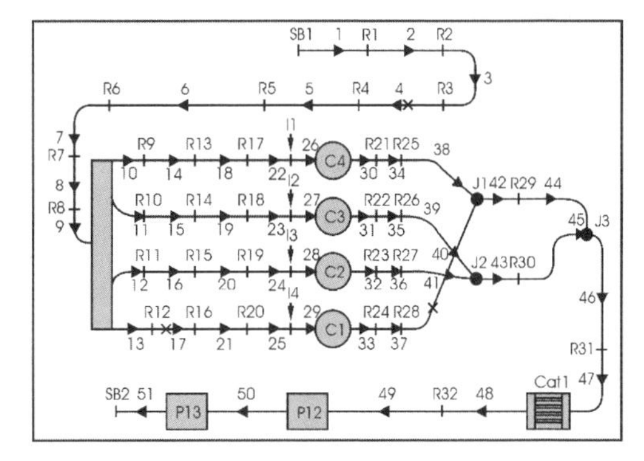

Fig. 10-54 Schematic representation of a four-cylinder spark-ignition engine.

In the one-dimensional simulation of the engine intake flow, the overall engine system is divided into individually abstract (i.e., simplified) elements such as the cylinder (C1), air filter (Pl1), orifices (SB1, R1, SB2), and pipes (1–4) (Figs. 10-53 and 10-54).

This is done assuming that the flow in the overall system can be described by a one-dimensional transient tubular flow in the pipe elements and by a one-dimensional quasistationary throttled flow in the components that connect the pipe elements.

The one-dimensional transient analysis within a pipe element assumes that the state quantities such as pressure p, density ρ, and speed u are sufficiently defined by averages in the individual pipe cross sections. Furthermore, it is assumed that there is no pulse loss because of internal friction in the flow. Only the friction of the flow against the pipe wall is covered. This means that the processes in a pipe element such as the conversion of pressure energy into movement energy are irreversible only as a result of the included wall friction. A nonlinear inhomogeneous differential equation system is accordingly created for a one-dimensional transient tubular flow within the flow plane (x,t plane) based upon the conservation equations for mass, pulses, and energy:

$$\frac{\partial \rho}{\partial t} = -\frac{\partial(\rho \cdot u)}{\partial x} - \rho \cdot u \cdot \frac{1}{A} \cdot \frac{dA}{dx} \tag{10.32}$$

$$\frac{\partial(\rho \cdot u)}{\partial t} = -\frac{\partial(\rho \cdot u^2 + p)}{\partial x} - \rho \cdot u^2 \cdot \frac{1}{A} \cdot \frac{\partial A}{\partial x} - \frac{F_R}{V} \tag{10.33}$$

$$\frac{\partial E}{\partial t} = -\frac{\partial[u \cdot (E + p)]}{\partial x} - u \cdot (E + p) \cdot \frac{1}{A} \cdot \frac{dA}{dx} + \frac{q_w}{V} \qquad (10.34)$$

where F_R is the wall friction, V is the volume, q_w is the flow of heat, and E is the total energy.

To solve this problem regarding initial values and boundary values, we need information concerning the state at the pipe edges. This state vector is determined by the flow in the components that connect the pipe ends with each other. Stated simply, the filling and emptying method assumes that the flow is quasistationary through these throttling sites.

Bibliography for Chapters 10.1 and 10.2

[1] Spicher, U., Verbrennungsmotoren A und B, Printed Lecture at the University of Karlsruhe (TH).

[2] Schwelk *et al.*, Fachkunde Fahrzeugtechnik, Holland and Jansen Verlag, Stuttgart, 1989.

[3] Stoffregen, J., Motorradtechnik, Vieweg-Technik, 3rd edition, 1999.

[4] Marquard, R., Konzeption von Ladungswechselsystemen für Pkw-Vierventilmotoren unter Fahrzeugrandbedingungen, Dissertation, TH Aachen, 1992.

[5] Pischinger, S., Verbrennungsmotoren I und II, Printed Lecture, FH Aachen.

[6] Jungbluth, G., *et al.*, Bau und Berechnung von Verbrennungsmotoren, Springer-Verlag, Berlin, 1983.

[7] Shell Lexikon Verbrenungsmotoren, Supplement to ATZ and MTZ.

[8] Köhler, E., Verbrennungsmotoren, 2nd edition, Vieweg-Verlag, Braunschweig, Wiesbaden, 2001.

[9] Aoi, K., K. Nomura, and H. Matsuzaka, Optimization of Multi-Valve, Four Cycle Engine Design: The Benefit of Five-Valve Technology, SAE Technical Paper 860032.

[10] Brüggemann, H., M. Schäfer, and E. Gobien, Die neuen Mercedes-Benz 2,6 und 3,0-Liter-Sechszylinder-Ottomotoren für die neue Baureihe W 124, MTZ 46 (1985).

[11] Duelli, H., Berechnungen und Versuche zur Optimierung von Ansaugsystemen für Mehrzylindermotoren und Einzylinder-Einspritzung, VDI-Fortschrittberichte, Series 12, No. 85, 1987.

10.3 The Charge Cycle In Two-Stroke Engines

10.3.1 Scavenging

The characteristic feature of different two-stroke engine designs is the respective type of cylinder scavenging and the related type of scavenging air supply. The selected scavenging approach greatly influences the complexity of the design, the component load, operating behavior, air/gas mixing conditions, fuel consumption, and the emissions of the engine.

When the cylinder is scavenged, the combusted mixture is displaced from the cylinder by fresh gas, without mutual mixing in the ideal exception of displacement scavenging. In contrast, when a cylinder is scavenged in a real engine, a mixture of fresh gas and exhaust occurs in addition to the displacement of the exhaust. As schematically illustrated in Fig. 10-55 (especially when there is a great deal of scavenging air such as at high load map points), a part of the scavenging gas mixed with the exhaust is expelled from the cylinder (loss of fresh gas). To evaluate the results or the efficiency of the scavenging procedure in two-stroke engines, the retention rate or air expenditure is used as an index in addition to volumetric efficiency (see also Refs. [2] and [3]).

Figure 10-56 shows an overview of the most important two-cycle scavenging procedures with methodically related advantages and disadvantages.

Loop scavenging: With loop scavenging (according to Schnürle), the fresh gas passes into the cylinder generally through two to six scavenging channels (overflow channels) that symmetrically mirror each other across the midaxis of the exhaust port and run in the opposite direction of the exiting exhaust. The scavenging streams align with each other and form an increasing stream of fresh gas on the side of the cylinder opposite the exhaust port. The fresh gas stream reverses direction at the cylinder head

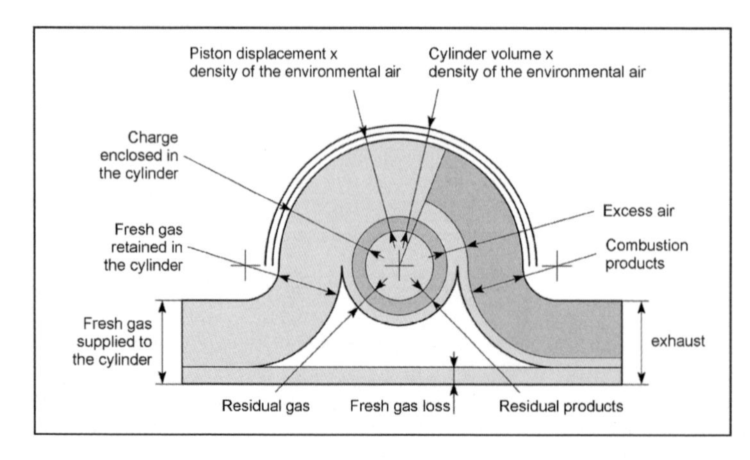

Fig. 10-55 Mass balance of the two-cycle scavenging process according to Ref. [1].

Scavenging approach	Advantages	Disadvantages
1. Loop scavenging 	• Compact design • High speeds are possible • The combustion chamber recess can be located in the cylinder head where it is well cooled • Simple design without a piston valve	• Asymmetrical timing diagram is possible only with additional components (piston valve) • Asymmetrical thermal load on the piston • The piston rings are especially endangered by the scavenging and exhaust ports • Comparatively difficult to generate charge turbulence
2. Uniflow scavenging with exhaust valves 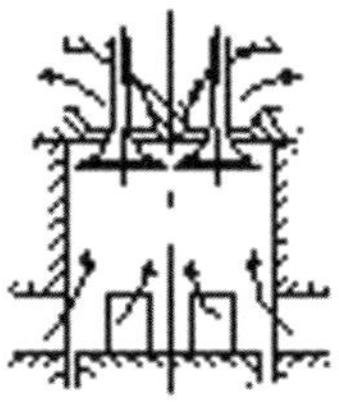	• Effective scavenging/low air expenditure • Easy to generate and influence the combustion chamber turbulence • The combustion procedure can largely be transferred to four-stroke engines • Asymmetrical timing diagram is possible without additional components	• Larger overall height in comparison to 1 • A more involved and optimized valve gear is required for large effective cylinder strokes and low consumption
3. Uniflow scavenging with opposed pistons 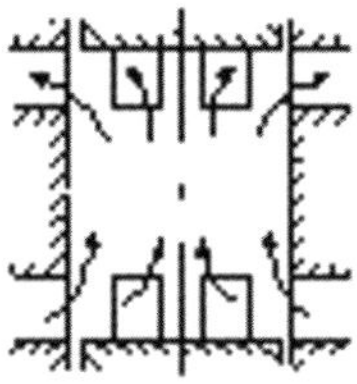	• Minimization of the combustion chamber surfaces heated in the high-pressure phase • Asymmetrical timing diagram can be achieved only by controlling the piston edges • Effective scavenging/low air expenditure	• More involved construction • Larger overall height (overall width) • Extreme thermal load on the piston controlling the exhaust ports • A conventional combustion method cannot be used due to the arrangement of the nozzle holder/spark plug
4. Reversed head scavenging 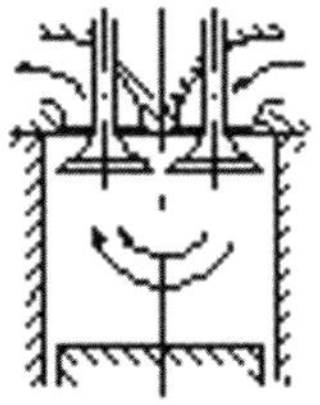	• Engine-transmission unit very similar to that of four-stroke engines • The piston rings are not endangered from scavenging and exhaust ports	• Low scavenging effect/large air expenditure • Because of the restricted opening time cross section, strong rise in the charge cycle work and consumption at higher speeds

Fig. 10-56 Comparison of different scavenging approaches.

and expels the exhaust from the cylinder. This type of scavenging that is particularly widespread in small engines and is suitable for high rpms is easy to design and results in compact engine dimensions. With direct injection (DI) diesel engines, the combustion chamber recess can be located in the cylinder head where it is well cooled. Disadvantages are the asymmetrical thermal load on the piston, the endangerment of the piston rings from scavenging and exhaust ports, and the fact that the oil consumption is difficult to control when pressure circulation lubrication is used. In addition, other technical measures are required to create the conventional charge turbulence for DI diesel engines and to create an asymmetrical control diagram.

Uniflow scavenging: With uniflow scavenging, the fresh gas passes into the cylinder through intake ports in the perimeter of the cylinder and displaces the exhaust through several exhaust valves in the cylinder head that are controlled with the crankshaft speed. A tangential arrangement of the scavenging channels makes it comparatively easy to generate and influence the turbulence that supports the mixture. This turbulence generally lasts the entire work cycle while attenuating, and it does not have to be completely regenerated in the following scavenging cycle. The advantages of uniflow scavenging are that it is comparatively effective (low air expenditure), asymmetrical timing can be achieved without additional constructive measures, and tried-and-true DI diesel combustion methods for four-stroke engines can be transferred largely unchanged to two-stroke engines. In contrast to loop scavenging, it is comparatively easy for the piston rings to freely rotate given a corresponding scavenging port design that increases their life. The overall height of a cylinder head with valves yields a taller engine compared to similar four-stroke engines, especially with oversquare stroke-to-bore ratios, since the scavenging ports are covered by the piston shaft and a collision of the connecting rod with the piston shaft must be excluded in the design. In addition, there are substantial demands on the design of the exhaust valve drive because of the double valve actuation frequency and the limited valve opening (crank) angle with the simultaneous requirement for large opening time cross sections.

Opposed piston uniflow scavenging: With opposed piston uniflow scavenging, two pistons move in the opposite direction in one cylinder, and their inner end position encloses the combustion chamber (TDC position).

In their outer end position (BDC position), one of the pistons opens the intake ports, and the other piston opens the exhaust ports so that the inflowing fresh gas expels the exhaust from the cylinder with the main direction of flow along the cylinder axis. The advantages are effective scavenging, minimization of the combustion chamber surface heated in the high pressure phase, and easily realizable asymmetrical timing. Serious disadvantages of this approach result from the complex construction, bulky engine dimensions, extreme thermal load on the exhaust-side piston (see also Ref. [4]), and the limited transferability of the combustion process to modern four-stroke engines.

Reversed head scavenging: With reversed head scavenging, the fresh gas generally flows through at least two or three valves actuated at crankshaft speed at BDC into the cylinder and displaces the exhaust from the cylinder through the simultaneously opened exhaust valves supported by a reversal of direction at the piston floor. The advantage of this type of scavenging is the design of the engine-transmission unit that largely corresponds to that of a comparable four-stroke engine. Furthermore, the absence of scavenging in the exhaust ports reduces the hazard to the piston rings. These advantages contrast with the great disadvantage that intake and exhaust valves must be located on the limited combustion chamber surface of the cylinder head. In contrast, for a comparable two-stroke engine with uniflow scavenging and, for example, with four exhaust valves, a basic approximation indicates that the available opening time cross sections are cut in half. At the same time, a great deal more scavenging air is required to introduce the same amount of fresh gas in the cylinder because of the less effective scavenging (mixture of fresh gas and exhaust from turbulence and contact of a large surface area of the gas stream) in reversed head scavenging. For this reason, the required charge cycle work and the resulting specific fuel consumption lies only within an acceptable range at low engine speeds. These restrictions of the nominal speed and consumption run counter to the requirements of designs of drives for future passenger cars. Apart from that, short-stroke, reversed head two-cycle diesel engines and possibly two-cycle spark ignition engines hold promise for low-speed airplanes (no intermediate transmission, high propeller efficiency).

We refrain from discussing other types of scavenging such as cross scavenging, fountain scavenging, reverse MAN scavenging, and the various dual-piston scavenging approaches (see also Refs. [5] and [6]) because of their limited efficiency, complicated design, or other disadvantages.

10.3.2 Gas Exchange Organs

As noted, the fresh gas stream entering the cylinder in the case of inflow scavenging and the exhaust stream leaving the cylinder in the case of loop scavenging are controlled by ports in the cylinder wall and the ascending and descending piston. A feature of port control is that a large flow cross section can be opened and closed within comparatively small crank angle ranges in comparison to conventional valve actuation in the cylinder head. High nominal speeds can, therefore, be obtained with port controlled two-stroke engines. A characteristic quantity used in designing and determining the gas flow rate through a port is the (opening) time cross section (see also Refs. [6] and [7]). This defines the time integral over the respective port cross-sectional area from the opening to the closing of the respective port. Without additional measures, symmetrical timing results for port-controlled two-stroke engines at the dead centers of the crankshaft. With the goal of improving the charging of the combustion chamber with an asymmetrical intake timing diagram, a series of two-stroke spark-ignition engines were equipped with tubular and roller rotary disk valves and then later with disk type rotary disk valves. With asymmetrical timing of the rotary-disk valves, the start of intake is substantially earlier than with port control. Since the vacuum in the crank chamber is comparatively low at this point in time, the air column in the intake tract is excited to form gas column oscillations comparatively less at low and average speeds. This produces more continuous torque characteristics and favorable conditions for the formation of a fuel-air mixture with a very constant air-fuel ratio in the carburetor. Instead of rotary-disk valves, modern two-stroke spark-ignition engines have frequently used reed valves in recent years (see also Refs. [7] and [8]). These act as nonreturn valves and automatically open given a specific pressure gradient toward the crank chamber, and they independently close given an opposite pressure gradient. Figure 10-57 shows the construction of a reed valve for two-stroke engines. The basic body (made of die-cast aluminum or plastic) in the form of a gable to reduce flow resistance is

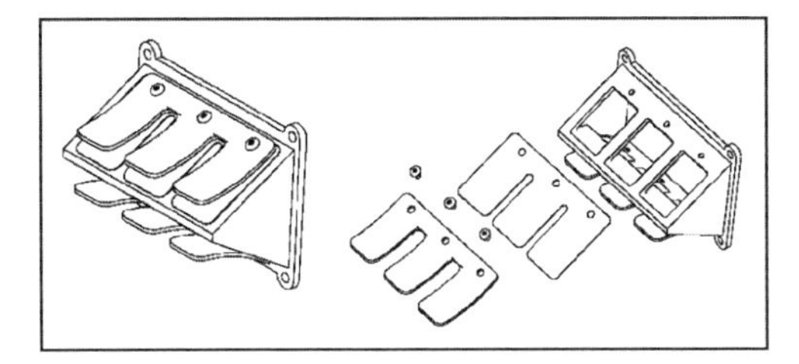

Fig. 10-57 Illustration of the constructions of a reed valve for use in an intake system of a two-stroke engine.

generally sprayed with a thin elastomer coating at the area where the reeds are contacted to reduce mechanical load and improve the seal and acoustics. The reeds fixed to one side of the basic body (mechanical replacement model: cantilever with a surface load) are made of either 0.15–0.2 mm thick Cr-Ni sheet steel or more recently 0.4–0.6 mm thick fiberglass-reinforced epoxide resin plates. Given the same length and width, the natural frequencies of steel and epoxide resin reeds are approximately the same since the quotients of their elasticity modulus and density are about the same.

Since the reeds open more as the pressure differential increases, a linear relationship between the pressure differential and mass flow results in a first approximation. To prevent the reeds from moving in an undefined manner (opening too wide with subsequent premature closing of the reeds, vibration in the second eigenform, etc.), reed valves with arched stops of sheet steel are provided that the reeds contact as they execute a rolling off movement when they open. The natural frequency of the reeds should be at least 1.3 times that of the opening frequency (intake frequency of the engine). Reed valves are placed either directly on the crank chamber or as shown in Fig. 10-58 and are used together with the piston intake control.

With the goal of compensating for the disadvantages of symmetrical timing of port-controlled loop scavenging,

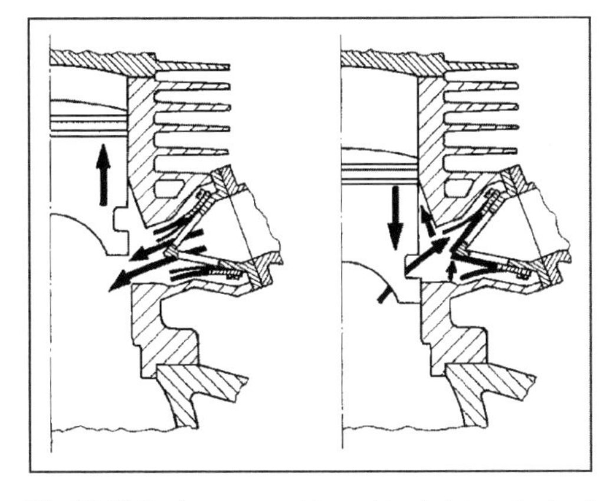

Fig. 10-58 Intake system with combined piston edge/reed valve control.

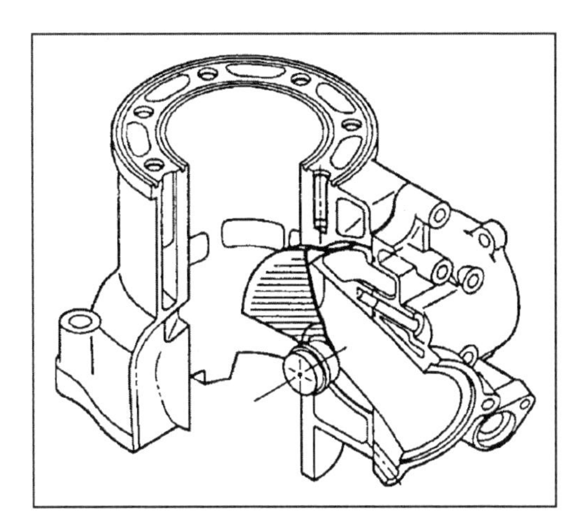

Fig. 10-59 Section of a loop-scavenged cylinder with an exhaust port pivot valve according to Ref. [9].

some modern high-performance spark-ignition engines use flat-seat valves, pivot valves, or rotary-disk valves. This can improve the fresh gas charging, the torque and performance curve, or, as is the case with the Honda AR combustion method (activated radical), the ignition of the air-fuel mixture. Figure 10-59 shows a section of such a cylinder.

10.3.3 Scavenging Air Supply

Whereas the pressure gradient for the charge cycle arises from the expulsion and intake process of the engine-transmission unit itself in four-stroke engines, the required scavenging pressure gradient for the charge cycle in two-stroke engines is generated by a separate scavenging blower (compressor). The cylinder can be scavenged only when the intake and exhaust organs are open simultaneously. The flow through the intake and exhaust organs can be described in simplified terms as a flow through two series-connected throttles (see also Refs. [10] and [11]) that can be replaced by an equivalent cross section. Since, apart from influences such as pressure pulsation, gas temperature, and exhaust counterpressure, it does not matter whether the ports or valves open and close a few times slowly or many times quickly within a given period of time, the air flow rate through a two-cycle engine to produce a respective scavenging pressure gradient is independent of the engine speed. In contrast, there is a quadratic relationship in the first approximation between the scavenging pressure gradient and the scavenging air quantity. At higher engine speeds, a much higher scavenging pressure is needed to attain the same result. The amount of scavenging air can be varied over wide ranges for a corresponding mapping point depending on the required engine temperature, exhaust temperature, emissions, consumption, and engine performance (supercharging), assuming that the scavenging blower is correspondingly flexible. A displacement-type compressor (reciprocating piston

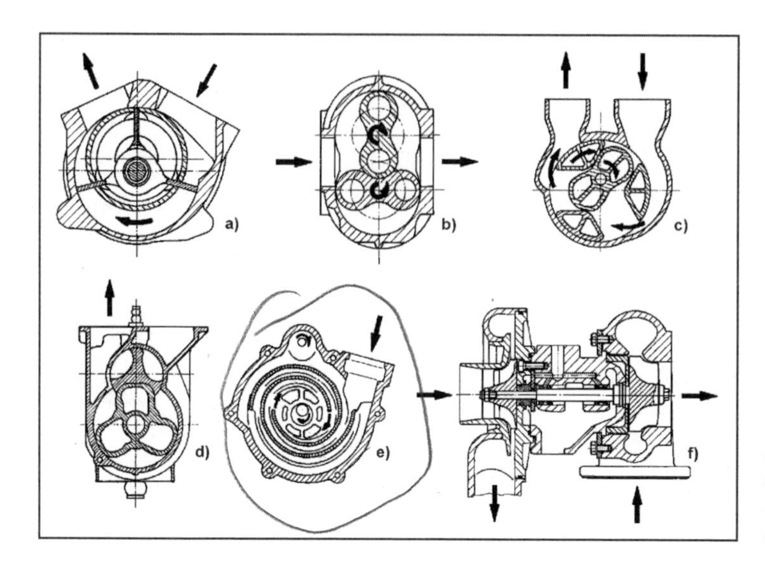

Fig. 10-60 Overview of the different designs of blowers and superchargers: (a) Vane-type superchargers, (b) Roots superchargers, (c) Rotary piston superchargers, (d) Screw compressors, (e) Spiral superchargers (G-superchargers), (f) Turbochargers.

compressor and rotary piston compressor) as well as flow compressors can be used for scavenging or possibly supercharging two-stroke engines (see also Refs. [10], [12], and [13]). Figure 10-60 shows an overview of different blowers and supercharging designs.

Reciprocating piston compressor: The simplest type of reciprocating piston compressor for two-stroke engines uses the crank housing and the bottom of the piston to enclose the working volume. With this design that is particularly widespread among small two-stroke spark ignition engines (the advantages are a compact design, low additional costs, steep compression curve, low additional drive power), the working gas generally flows through holes in the cylinder wall or piston shaft into the crank housing when the piston moves upward. When the piston subsequently executes a downward movement, the fresh gas is compressed and flows via overflow ports and from scavenging ports exposed by the piston head into the crank housing. In the following upward movement of the piston, the fresh gas is compressed and flows via overflow channels and scavenging ports exposed by the piston head into the cylinder. By using reed valves or rotary-disk valves, or by changing to a crosshead charging pump, the amount of scavenging air can be increased that is limited by the stroke-to-bore ratio and the dead space. In particular, given the limited scavenging efficiency of two-stroke engines and the fact that operating at a full load generally requires a substantial amount of excess air even in modern diesel combustion systems because of the smoke limit, the low volumetric efficiency of the crank housing scavenging pump is a profound disadvantage, apart from the complicated stepped piston design. Assuming that a highly effective, flow enhancing oil separator with low-pressure loss cannot be used for the scavenging air, the necessity of minimizing the lubrication oil in the scavenging air (problem: hydrocarbon and particle emissions, piston ring deposits, racing engine) means that the engine-transmission unit generally cannot be born on tried-and-true, low-noise, economical, and reliable friction bearings with oil spray cooling of the pistons. Another substantial disadvantage of crank housing scavenging pumps is that the crank chambers need to be sealed from each other in multicylinder engines. Using a separate, mechanically driven reciprocating piston compressor avoids some of the cited disadvantages; however, apart from the limited flexibility in adjusting the fuel delivery, substantial additional installation space is required, and major additional costs are involved.

Rotary compressor: Under the general term of the "rotary compressor" (rotary piston compressor), we find a series of compressors whose delivery or compression is determined by the compressing effect of rotating elements or pistons. The driveshaft is mechanically coupled to the crankshaft of the engine to scavenge or supercharge internal combustion engines. Belonging to this group of superchargers are Roots superchargers, vane-type superchargers (encapsulated blowers), rotary-piston superchargers, spiral-type superchargers (g superchargers), and screw compressors. Similar to reciprocating piston compressors, the delivered mass flow is approximately proportional to the drive speed and decreases slightly at higher pressures because of increasing leakage. In general, average compressor efficiency is attained. With an equivalent delivery rate, the dimensions of reciprocating piston compressors and radial compressors are approximately the same.

Flow-type superchargers: Of the flow-type superchargers, primarily radial compressors (turbocompressors) are used for vehicle engines. The delivered flow of radial compressors is approximately linear, and the pressure is approximately the square of the drive speed. Modern radial superchargers offer highly efficient compression. Since, in contrast to four-stroke engines, the two-stroke engine has a mass flow rate characteristic that is only more or less independent of the engine speed that can be defined as an opening (throttle) with a constant cross section, a radial blower mechanically coupled to the engine is a suitable scavenging blower. Corresponding to the goal of limiting the size of the radial supercharger, it is useful to drive the supercharger with a high-speed transmission. To optimally adapt the air mass flow delivered by the supercharger largely independent from the crankshaft speed for each mapping point to the desired scavenging or supercharging level of a two-stroke engine, it is desirable to drive the supercharger with a variable transmission ratio like the previously discussed displacement supercharger. Such a solution was, for example, used for the "ZF-Turmat" (see also Ref. [14]). Apart from high construction costs, problems with vibration, and the useful life of variable drive transmissions, a general disadvantage of mechanically driven superchargers is that a substantial amount of the effective output must be sent to the crankshaft to drive the supercharger. This correspondingly increases the specific fuel consumption.

Exhaust turbochargers: The exhaust turbocharger that has been successfully used for decades in four-stroke engines can also be used in two-stroke engines for passenger cars and trucks as a scavenging and supercharging blower. The advantage of turbocharging is that the exhaust energy converted in the turbine is used, which would otherwise largely be lost. According to Schieferdecker [15], a requirement for the use of freewheeling turbochargers in two-stroke engines is that the joint efficiency of the turbine and compressor must be at least 60%, which is more or less attained with modern turbochargers used in passenger cars and trucks. To utilize as much exhaust energy as possible in the turbine, it is also essential that the exhaust lines from the respective cylinder to the spiral housing of the supercharger be optimized for both good flow and minimal heat loss. In addition to a short, cramped port design, the air gap insulation, and possibly even the use of port liners, needs to be considered. To ensure a positive scavenging pressure gradient over as wide a mapping range as possible, superchargers should be used with variable turbine geometry [adjustable blades, sliding supercharger, double helix supercharger (Twin skroll, Aisin)] (see also Ref. [14]). An advantageous side effect of turbocharging and supercharging with superchargers that have an adjustable turbine geometry is that the backup of exhaust in front of the turbines allows highly effective charging even for scavenging approaches with symmetrical timing (such as loop scavenging). Such an approach, although in an extreme form, was used for the turbocompound airplane engine, the Napier Normad (see also Ref. [16]). To generate a positive scavenging pressure gradient when accelerating from a low load and low rpm and when starting an engine, you need a series-connected additional mechanically or electrically driven supercharger or a mechanical auxiliary turbocharger drive. An interesting alternative is an electrically supported turbocharger. With these types of superchargers, a part of the propulsion power for the compressor is supplied as needed by, e.g., an asynchronous electrical motor integrated in the supercharger (see also Ref. [17]).

For the thermodynamic conditions when coupling with two-stroke engines for the pressure wave supercharger (Comprex supercharger), the same observations apply that were made for turbochargers. A basic disadvantage is that the fresh gas is heated when it briefly and directly contacts the exhaust, and that mechanical or electrical support of the compressor output is impossible given the functional principle of the supercharger.

Bibliography

[1] Schweitzer, P.H., Scavenging of Two-Stroke Cycle Engines, Macmillan, New York, NY, 1949.
[2] Küntscher, V. (Pub.), Kraftfahrzeugmotoren–Auslegung und Konstruktion, 3rd edition, Verlag Technik, Berlin, 1995.
[3] List, H., Der Ladungswechsel der Verbrennungskraftmaschine, Teil II, Der Zweitakt, Springer-Verlag, Vienna, 1950.
[4] Gerecke, W., Entwicklung und Betriebsverhalten des Feuerrings als Dichtelement hoch beanspruchter Kolben, in MTZ, Vol. 14, No. 6, June 1953, pp. 182–186.
[5] Venediger, H.J., Zweitaktspülung insbesondere Umkehrspülung, Franckh'sche Verlagshandlung, Stuttgart, 1947.
[6] Bönsch, H.W., Der schnelllaufende Zweitaktmotor, 2nd edition, Motorbuch Verlag, Stuttgart, 1983.
[7] Kuhnt, H.-W., H. Budihartono, and M. Schneider, Auslegungsrichtlinien für Hochleistungs-2-Takt-Motoren; Speech at the 4th International Annual Symposium for the Development of Small Engines, Offenburg, March 16 and 17, 2001.
[8] Blair, G.P., Design and Simulation of Two-Stroke Engines, SAE-Verlag, Warrendale, Pa, 1996, ISBN 1-56091-685-0.
[9] Bartsch, Ch., Ein neuer Weg für den einfachen Zweitakter, Honda EXP-2 als Versuchsobjekt, in Automobil Revue, No. 5/1 February 1996.
[10] Zinner, K., Aufladung von Verbrennungsmotoren, Grundlagen–Berechnung–Ausführung, 3rd edition, Springer-Verlag, Berlin, Heidelberg, New York, Tokyo, 1985.
[11] Wanscheid, W.A., Theorie der Dieselmotoren, 2nd edition, VEB Verlag Technik, Berlin, 1968.
[12] Küntscher, V. [ed.], Kraftfahrzeugmotoren–Auslegung und Konstruktion, 3rd edition, Verlag Technik, Berlin, 1995.
[13] Zeman, J., Zweitaktdieselmaschinen, Springer-Verlag, Vienna, 1935.
[14] Hack/Langkabel, Turbo- und Kompressormotoren, Entwicklung, Technik, Typen, Motorbuchverlag, Stuttgart, 1999.
[15] N.N., Fahrzeugmotoren im Vergleich: Dresden Symposium on June 3–4, 1993, VDI Gesellschaft Fahrzeugtechnik, VDI Reports 1066, VDI-Verlag, Düsseldorf, 1993.
[16] N.N., Der Napier-Diesel-Flugmotor "Normad," in MTZ, Vol. 15, No. 8, August 1954, pp. 236–239.
[17] Huber, G., Elektrisch unterstützte ATL-Aufladung (euATL)–Schaffung eines neuen Freiheitsgrades bei der motorischen Verbrennung, 6th Conference on Charging Engineering, Dresden, 1997.

10.4 Variable Valve Actuation

Variable valve actuation can be used to positively influence the desired quantities for the combustion engine such as specific consumption, emission behavior, torque, and maximum output. Depending on their physical functional principle, variable valve actuation systems are divided into systems that are mechanically, hydraulically, electrically, and pneumatically actuated. Numerous such systems are known, and there is extensive research on both simple systems in which the control time can be varied between two positions and on more complex systems in which even the engine load can be controlled by variable control times. Figure 10-61 shows a detailed categorization of variable valve actuation. This categorization starts

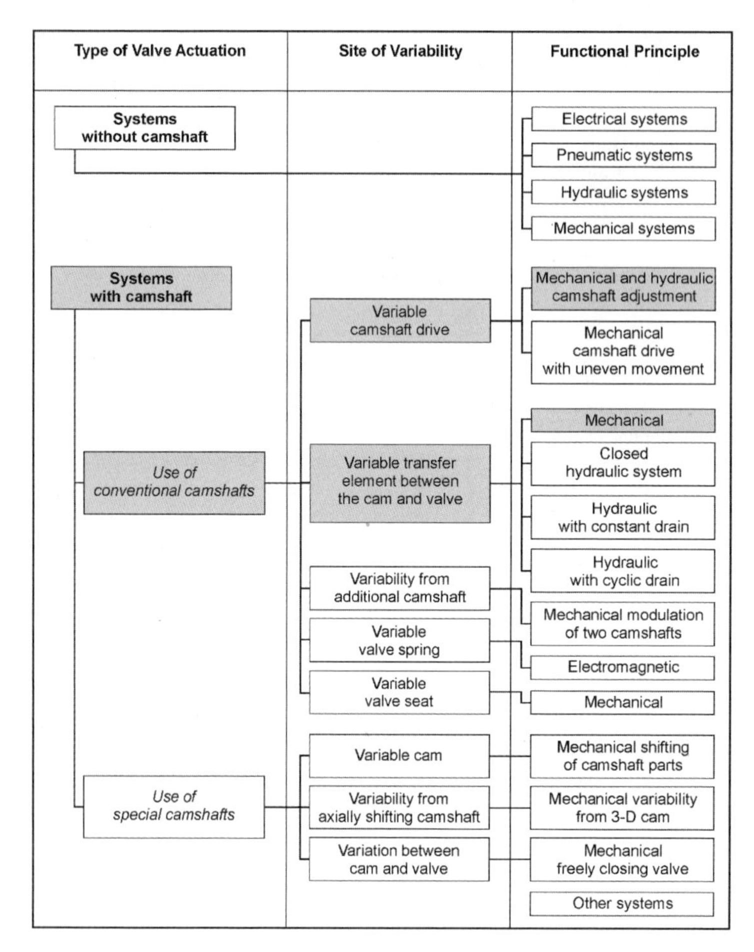

Fig. 10-61 Categories of variable valve control.[1]

with the component of the camshaft. The camshaft criterion is the first of three selected categorization levels. Systems, whose energy is provided for valve actuation without a camshaft, are categorized according to their physical functional principle. This accordingly yields electrically, pneumatically, hydraulically, and mechanically actuated systems. With systems that use a camshaft for control, a distinction is drawn between the use of conventional and special camshafts. Those camshafts are termed conventional that have a conventional cam geometry, use common materials, and are created using familiar manufacturing procedures. The second categorization level deals with the site where variability takes effect. The third categorization level that describes the operational and functional principle of variable valve actuation is divided into 17 groups.

In this section, we describe only the individual systems. The series-connected systems are of particular interest. In the categorization in Fig. 10-61, the groups that use series solutions have a gray background.

The numerous types of variable valve actuation make it difficult for a developer to select a suitable type of control for his application. Such a wide variety of systems is used for cylinder heads that substantial adaptations are required when variable valve control is used. A new cylinder head generation usually has to be developed for a system to be used in stock engines. Usually a more complex design is required for variable control times in contrast to conventional engines, and this is expressed in higher costs.

In the future, the option of using variable valve actuation to control engine load will gain in importance. A basic goal of varying the valve lifting curves is to lower charge cycle loss under partial loads and, hence, reduce fuel consumption. The goal of many developmental activities is to dispense with the throttle valve in spark-ignition engines to control load solely by varying valve lifting. In comparison to pure throttle control (TC) with conventional throttle valves, Fig. 10-62 shows four load control methods that vary intake valve lifting.

The load control method "early inlet closure" (EIC) limits the amount of fresh gas by early closure of the intake valve after charging based on the set load. When the engine idles, the intake valve opening time corresponds approximately to a 60° crankshaft angle. With the control mode "late inlet closure" (LIC), the part of the charge that is not needed for the specified output is

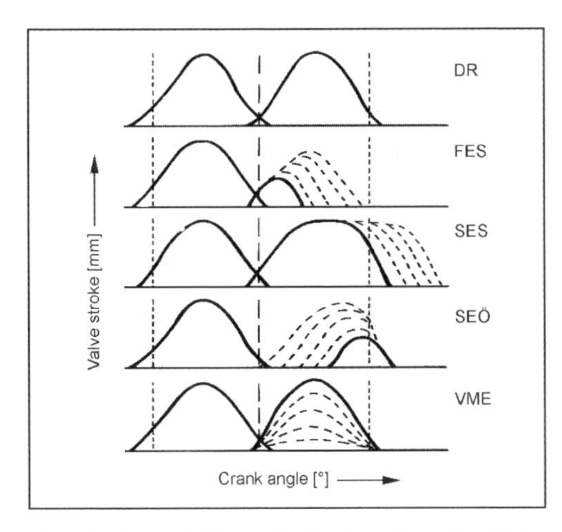

Fig. 10-62 Possibilities of adjusting the valve lifting curves with variable valve actuation.

expelled from the cylinder. This charge quantity passes through the throttle site of the valves twice with the corresponding loss. When the load is controlled using the "late inlet open" (LIO) method, the intake valve is opened only when the remaining opening time corresponds to the required amount of inflowing mixture. At the start of induction, a strong vacuum is in the cylinder that promotes mixture through turbulence. The cylinder charge is influenced by the control mode "variable maximum intake valve stroke" (VMI) by reducing the valve stroke with equivalent opening angles. Instead of the throttle valve, the valve acts as a throttle site that does not reduce the amount of charge cycle work. The valve friction can, however, be lowered, since the valve springs are only partially compressed.

The effects of the parameters on the valve lifting curve are familiar. An ideal valve gear is one that allows the valve lifting curves to be changed as freely as possible. It also makes sense to combine different load control procedures. Depending on the system, however, only a limited degree of freedom can be attained using the different types of variable valve actuation. In addition, a substantial amount of system engineering is required for valve actuation to approach the desired complete variability. When systems are used that turn the camshafts relative to the crankshaft position, the attainable improvements to the engine are substantial. These systems are widely used in stock engines, and we discuss them in great detail in the next chapter.

At this point, we can guess the degree to which variable valve actuation can improve consumption or emissions. In the professional literature such as Ref. [1], we find that improvements to consumption average between 5% and 15% within some engine mapping ranges. Frequently, however, the engines in the literature are optimized in other ways in addition to variable valve control so that it is difficult to directly identify the specific influence from variable valve actuation.

In comparison to spark-ignition engines, the potential improvement to diesel engines from variable valve actuation is limited. Relatively few investigations have been made into this.

10.4.1 Camshaft Timing Devices

10.4.1.1 Overview of the Functional Principles of Camshaft Timing Devices

As early as September 29, 1918, a patent was issued for the adjustment of a spark-ignition engine camshaft.[2] The desired variation during engine operation was attained with a sleeve with interior and exterior teeth and straight and helical teeth that moved axially between the camshaft and the drive wheel (Fig. 10-63). The angular position of the cam and camshaft were adjusted relative to the crankshaft.

The inventor of this patent, Samuel Haltenberger, intended the timing device for an airplane engine to adapt output to different flight heights. The helical-toothed sleeve (2) is moved in an axial direction by air pressure using a timing device linkage (4). The relative angular position of the camshaft (1) changes in relation to the driving bevel gear (3) that is linked to the crankshaft. Based on the same functional principle of a straight and helical-toothed sleeve, in 1983 Alfa Romeo started mass producing a camshaft timing device for a two-valve engine with dual camshafts (Fig. 10-64). The timing device is seated on the intake camshaft and enables the control times to be adjusted between two positions. While idling, the late control time position is held by a return spring (10), and an early control time is set depending on the oil pressure and speed. A solenoid (6) that actuates the control valve (5) applies the engine oil pressure to the helical-toothed piston (9). The adjusting element is the helical-toothed piston (9) that is moved by the oil pressure against the spring force. The helical teeth (3) on the piston and camshaft are used to rotate the camshaft relative to the driving sprocket (4) and, hence, to the crankshaft when the piston shifts axially.

The systems shown in Figs. 10-63 and 10-64 are all designs where a mechanical functional principle is used. This means that the force to actuate the valves flows only via components that are engaged by friction or are positively engaged. The adjusting elements such as the piston in the Alfa Romeo timing device in Fig. 10-64 can, however, be moved and held by oil pressure. For a camshaft timing device that operates based on hydraulics, a hydraulic component lies in the flow of force to actuate the valve. This is done using a quantity of oil that must be at a correspondingly high pressure to keep the positions of the adjusting elements stable.

The position of the camshaft timing device should logically be directly adjacent to the camshaft drive. The flow of force to drive the camshaft can be most easily interrupted here, and the camshaft adjustment can easily be varied by selecting the suitable adjusting element.

In researching the literature and known patent applications, one can find numerous different functional

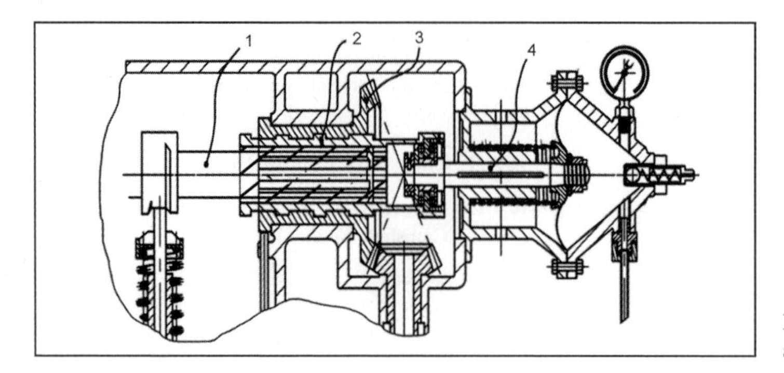

Fig. 10-63 Patent of a camshaft adjuster from 1918.[2]

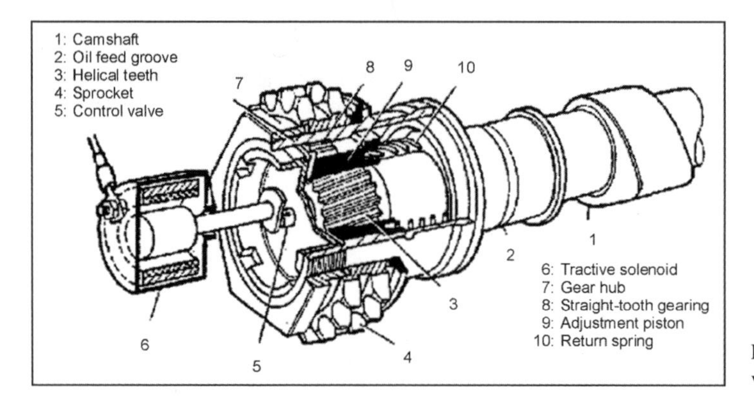

Fig. 10-64 Camshaft timing device by Alfa Romeo from 1983.[3]

principles for camshaft timing devices. In a patent search, we found approximately 800 different applications. When the application dates are plotted over the last 20 years, we can see a strong rise over time in activities in this field. After the Alfa Romeo timing devices started being mass produced, the number of patent applications began to rise drastically. In Fig. 10-65, which illustrates the situation up to January 2000, the number of applications is counted from 1979 to 1999 from a compiled database. For 1998 and 1999, not all applications could be entered since there are 18 months between the application date and date of publication.

The known timing devices can be categorized according to their different functional principles. Figure 10-66 shows these principles. Essentially, the timing devices are systems that are based on either a mechanical or a hydraulic functional principle. The most frequent solution is to axially shift a piston to change the angle by using helical teeth. Basically, only three principles are used for production engines (with gray backgrounds in Fig. 10-66). Belonging to the first group are systems that use helical toothing like Alfa Romeo's approach based on a mechanical functional principle. A second solution is the hydraulically actuated chain timing device where the camshaft is

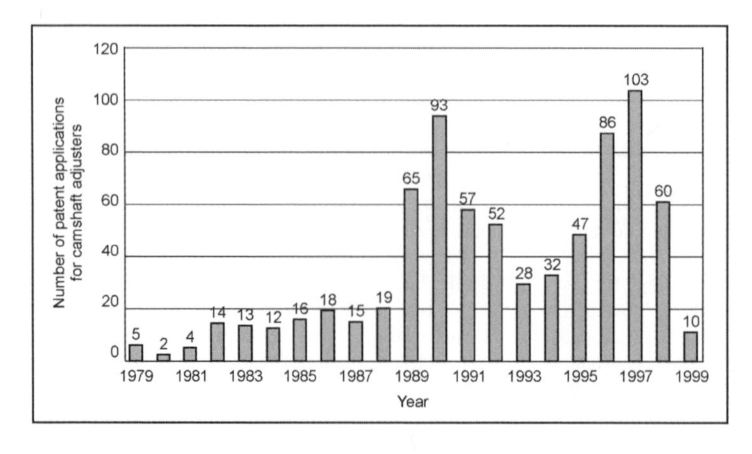

Fig. 10-65 Number of found patent applications and unexamined applications of camshaft timing devices from 1979 to 1999.

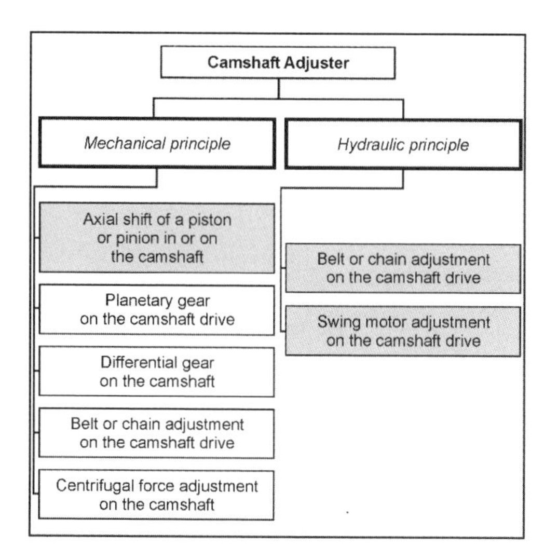

Fig. 10-66 Categorization of camshaft timing devices according to their functional principles.

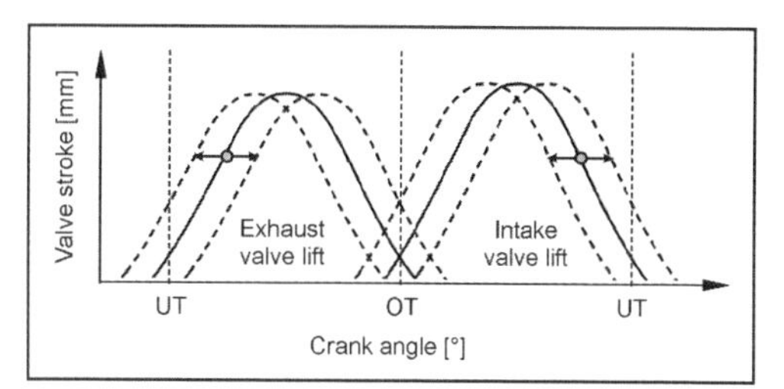

Fig. 10-67 Changing cam contours using continuously variable timing devices on intake and exhaust cams.

rotated to the desired extent by adjusting the chain sag. Belonging to a more current group are systems with hydraulically actuated swing motors on the camshaft drive. Individual descriptions of the systems can be found in Section 10.4.1.3.

All camshaft timing devices for production engines are on the camshaft drive. The timing devices do not affect the valve stroke or valve opening time. Numerous other known systems exist. The sites at which the valve stroke and valve opening time are adjusted are usually between the cam and valve. This allows the camshaft timing devices to be combined with these systems.

An example of a system to change the valve stroke or opening time is the so-called "VTEC" system by Honda.[4] This system allows different valve strokes and opening times by changing the transmission geometry between the cam and valve. These systems are used for many different engines (see also Section 10.4.2).

10.4.1.2 The Effects of Camshaft Timing Devices on Engines

The goals of camshaft timing devices can vary widely. The maximum output, the torque curve via the rpm, and the exhaust behavior can be positively influenced in passenger car spark-ignition engines by altering the relative angle of the camshaft to the crankshaft. Standard camshaft timing devices offer two angle positions and a variably changing angular position. Figure 10-67 shows the options for adjusting valve lifting curves from using two continuously variable camshaft timing devices. The curves in dashed lines represent the possible end positions of the control times.

Since camshaft timing devices are used only to change the position of the control times and not the valve lifting curves, the effects on the drive are limited. However, the potentially attainable improvement in engines is easier to estimate during development than, for example, infinitely variable valve actuation. To estimate the potential improvement, the charge cycles are calculated with numeric programs. The overall charge cycle of the engine can be estimated in reference to the torque and output behavior and the residual exhaust gas. All of the components participating in the charge cycle such as the induction pipe or exhaust system are parametrized and depicted in the calculation model.[1] The valve lifting curves are determined and included with the possible control times in the charge cycle calculations. This allows a reliable prediction of the engine output and torque characteristic. The parameters required to adjust the camshaft are roughly estimated and then refined in experiments.

The maximum torque or the maximum output can be positively influenced by using a camshaft timing device on the intake valve side, depending on the cam contour. Only a compromise is possible for output and torque for engines with fixed control times and cam contour positions. The position at which the intake is closed on the intake valve lifting curve has a decisive influence on maximum engine output. At higher speeds, the inlet is closed at later control times. The time is selected to optimize the cylinder charge and, hence, attain high volumetric efficiency. A return flow of the charge from the cylinder to the intake port can be avoided by adapting the speed of the inlet closing.

With camshaft timing devices, the valve overlap can be varied so that the residual gas in the engine exhaust can be controlled. Normally, the residual gas is supplied to the cylinder via an external exhaust return device. The temperature of combustion is restricted by the residual gas in the cylinder. This has a positive influence on the NO_x emissions. With continuously variable camshaft timing devices, internal exhaust return can be achieved by changing the valve overlap. This allows the exhaust to overflow from the exhaust port to the inlet port during the overlap phase at dead center during the charge cycle. The advantage of internal recirculation is attained with a short system dead time and more even distribution of the recirculated exhaust. Compromises always have to be made when designing the valve overlap. For example, the maximal possible valve overlap is limited by the position of the valves that collide with the piston when the overlap is too great.

Full load	Intake manifold position	Intake camshaft	Exhaust camshaft
2 ——	long	early	late
1 ·——	long	late	late
2 -----	short	early	late
1 — —	short	late	late
3 Partial load		early	late
4 idling		late	early

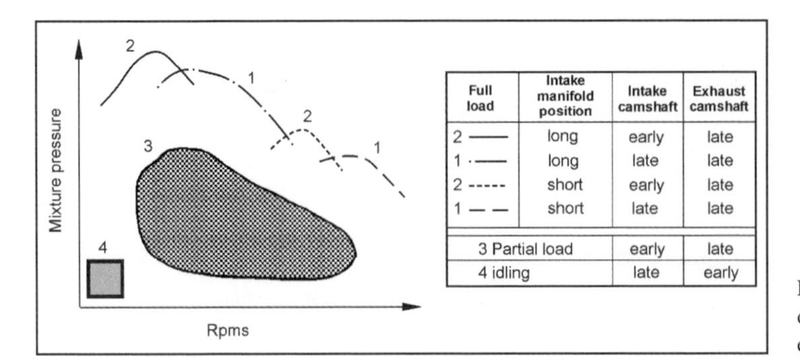

Fig. 10-68 Control strategy for dual cam adjustment of a VW V6 engine.[6]

An example is the control strategy of dual camshaft adjustment of VW engines.[5,6] Four basic positions are shown in Fig. 10-68 with the corresponding short or long induction pipe position for an intake engine with a multistage manifold and intake and exhaust cam adjustments.

This representation also shows the influence of different induction pipe lengths in combination with camshaft timing devices on the intake and exhaust valve side. Given the degrees of freedom that this enables, it is logical to work out a correspondingly appropriate adjustment strategy. The strategy can differ depending on the engine design. For example, to attain a high torque at average speeds, a long induction pipe channel is necessary. As speed increases, the intake control time is switched from "early" to "late" depending on the speed. At higher speeds, a short induction pipe channel is selected, and the intake camshaft is shifted in the direction of "late" to attain maximum output.

Figure 10-69 shows examples of control times of valve lifting curves for the individual camshaft and induction pipe positions of six-cylinder engines.

The first mass-produced camshaft timing devices with only two control time positions mainly sought to improve the output or the torque behavior. Today, the goal is also to control the inner exhaust recirculation by using continuously variable timing devices.[5] The intake camshaft is shifted for increasing torque, especially at low speeds, and for internal exhaust gas recirculation where the crank angle is offset from the "inlet open" output position toward "early" with a maximum 52° crank angle. The exhaust shaft can be adjusted from the output position "outlet close" toward "early" to optimize idling or toward "late" to attain maximum exhaust recirculation rates. A maximal 22° crank angle is sufficient for this. In comparing a conventional two-valve engine without camshaft adjustment to the four-valve engine described in Ref. [5] with camshaft adjustment, we can attain savings in consumption of 15.5% while idling and of 5.5% in the partial load range at 2000 min^{-1} and 2 bar. When using intake and exhaust valve time offsets, the specific consumption reduction is approximately 10%.

10.4.1.3 Camshaft Adjusters for Production Engines

After the start of mass production of the Alfa Romeo camshaft adjuster, other designs were used by other companies such as Mercedes-Benz, Nissan, and others.[7] Most of these systems used straight/helical teeth similar to Alfa Romeo as the functional principle.

A system that adjusts the control times by changing the chain side length is the camshaft chain timing device by the company Hydraulik-Ring.[8] The adjusting element is between the dual camshaft drive wheels, and the intake camshaft is driven by the exhaust camshaft. The timing device system combines a chain tensioner that is commonly

	Early position	Late position
Inlet open	26° before TDC	26° after TDC
Inlet close (long channel)	179° after TDC	231° after TDC
Inlet close (short channel)	184° after TDC	236° after TDC
Outlet open (short channel)	236° before TDC	214° before TDC
Outlet open (long channel)	231° before TDC	209° before TDC
Outlet close	26° before TDC	4° before TDC

Fig. 10-69 Control times for dual cam adjustment in a VW V6 spark-ignition engine with a 1 mm valve stroke.[6]

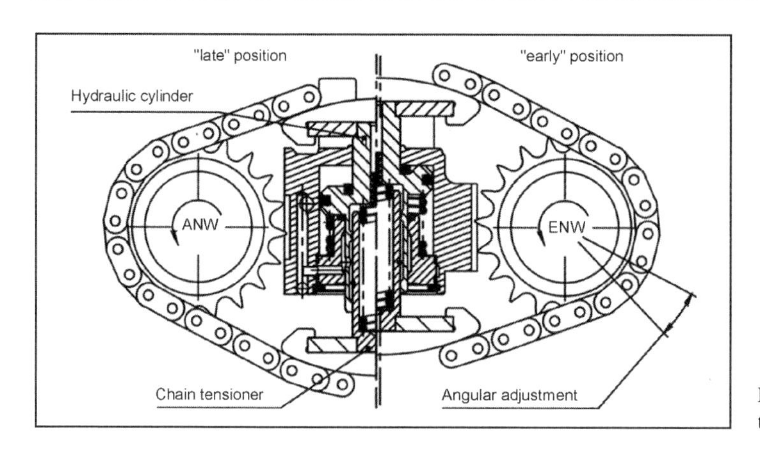

Fig. 10-70 Functional principle of the camshaft chain timing device.

used for such a short drive with a hydraulic cylinder to change the chain side length. The hydraulic cylinder under oil pressure on both sides is moved, depending on the desired control time position. In this manner, one chain side is lengthened, and the other is simultaneously shortened. This timing device provides two control time positions for the intake camshaft (Fig. 10-70).

During adjustment, the chain drive remains taut between the two drive wheels of the camshaft because of the chain tensioner integrated into the system. The adjustment cylinder of the timing device is controlled by an electronically controlled hydraulic 4/2-way valve. The timing device solution shown here uses hydraulically variable valve actuation since the end positions are held only by oil pressure. The design is such that the adjustment is made with the available engine oil pressure even under difficult conditions. A costly additional oil pump can be dispensed with. This timing principle is used in various series engines by Audi, Porsche (Fig. 10-71), and Volkswagen.[9–11]

Developments of the continuously variable adjustment of intake camshafts have enabled more than two camshaft positions to be held.

BMW was the first to use continuously variable adjustment of the camshaft in mass production (Fig. 10-72). First, this was used only for the intake camshaft, but it was followed later by continuously variable adjustment of the intake and the exhaust cams.[12]

A new generation of camshaft adjusters is represented by systems designed around the principle of swing motors.[13]

In this system, both the intake and the exhaust camshafts can be easily adapted to existing cylinder heads. Inside the timing device is a pivotable rotor that is firmly fixed to the camshaft. The outer part is driven either by a chain or a cogged belt. The connection between the outer and inner parts is formed by the oil space that is filled with engine oil pressure and that contains the pivotable rotor. Both sides of the blades of the rotor are supplied with oil pressure via an electronically controlled 4/2-way proportional valve.

The relative angular position of the camshaft is changed depending on the change in oil pressure on both sides of the rotor. The angular position of the camshaft measured with a sensor is compared to the position set by

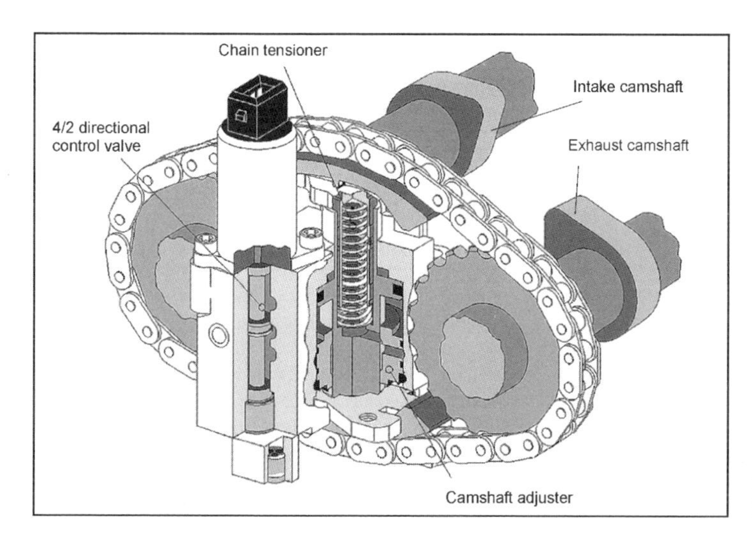

Fig. 10-71 Camshaft adjusters as a chain timing device for the Porsche Boxster.[14]

Fig. 10-72 Continuously adjustable camshaft adjustment for a BMW six-cylinder engine.[16]

the Motronic system. The desired position of the camshaft is permanently readjusted by controlling the proportional valve to hold stable intermediate positions of the rotor and, hence, of the camshaft. The oil is supplied only by engine oil without an additional pump. The system is controlled by the speed, load, and engine temperature. In comparison to conventional, toothed, continuously variable camshaft adjusters, these systems represent a much more economical solution so that one can expect their increasing use in series spark-ignition engines. The time and money spent on manufacturing the components can also be reduced when parts of the components are sintered and the seal of the oil chamber has a simple design. Timing devices of this type can be even more economical than toothed two-step timing devices. A more precise description of this system is found in Ref. [13].

The design of the dual camshaft adjusters with swing motors by Hydraulik-Ring for a six-cylinder engine is shown in Fig. 10-73.[5]

Figures 10-74 and 10-75 show the arrangement of dual camshaft adjusters with swing motors for the left cylinder bank of the 3.0 l Audi V6 engine. In this engine, a two-step timing device is used on the exhaust valve side, and a continuously variable timing device is used on the intake valve side. With this design of a cogged belt camshaft drive, the timing device housing needs to be encapsulated oil-tight.

In addition to Hydraulik-Ring series systems used by Audi and VW, similar systems with swing motors are used by Renault, Toyota, and Volvo.[14]

A wide variety of hydraulic valves are used for the hydraulic control of the camshaft adjuster.[14] Usually, directional controlled valves are used to control the oil flow. These can be subdivided into proportional and switching valves. Camshaft adjusters that hold only two end positions and, hence, can have only two different control times are equipped with 4/2-way valves.

Today, primarily 4/3-way proportional valves are used for continuously variable systems (Fig. 10-76). The

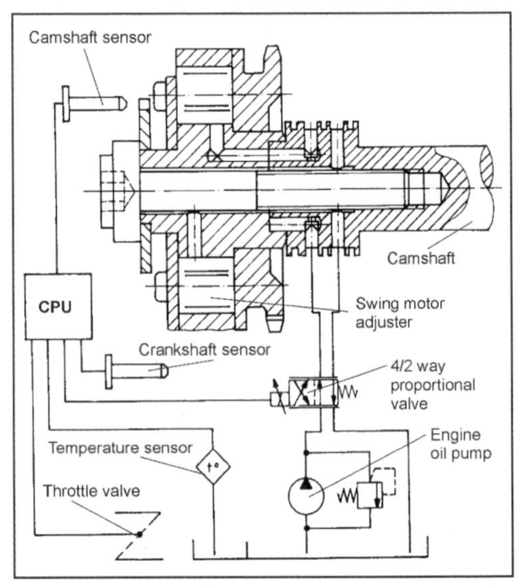

Fig. 10-73 Functional principle and control loop of a continuously variable camshaft adjuster designed with a swing motor.[13]

bulk of hydraulic valve engineering has less to do with manufacturing individual valves for small series than with implementing the technical requirements for economical large series production. The difficult problems of series production need to be dealt with such as dirty oil, engine vibration, high temperature fluctuations, or fluctuations in the vehicle power supply. Usually a special valve is used to adapt the valves to the individual engine. A well-thought-out modular system is useful to meet the primary requirement of economical mass production. Close collaboration between the developers of the variable valve actuation system and the developers of the engine is essential for successful mass production.

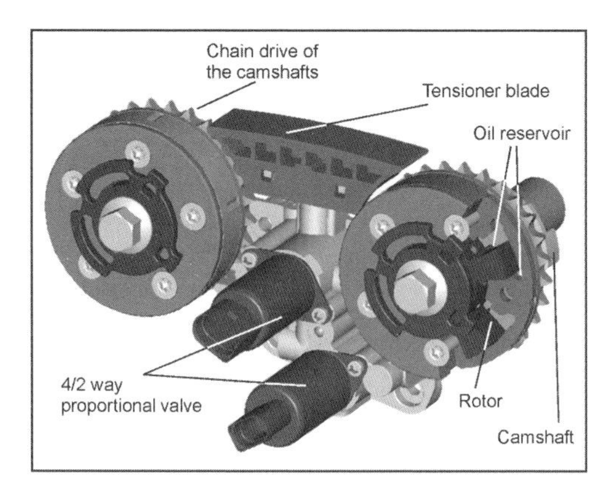

Fig. 10-74 Camshaft adjuster arrangement in a six-cylinder engine based on a swing motor principle.[6]

Fig. 10-75 Dual cam adjustment system in a 3.0 l V6 spark-ignition engine by Audi.[15]

10.4.1.4 Reflections about Camshaft Adjusters

An overview of the patent applications for camshaft adjusters and the number of different systems in production engines clearly illustrates that, in the future, probably all modern spark-ignition engines will use camshaft adjusters

We know of no series systems of production engines with only one camshaft in which the intake and exhaust cams can rotate in opposite directions. Perhaps it might make sense, however, to offset the entire camshaft via a camshaft adjuster, if only at narrow adjusting angles.

There are many reasons to use timing devices that allow the continuous variable offsetting of the camshaft. It is recommendable to also use these systems for multivalve engines with dual camshafts in which one system is affixed to a camshaft. In particular, the control of the internal exhaust gas recirculation with continuously variable systems can have a positive influence on the direct exhaust emissions.

To implement the functional principle, designers prefer timing devices with swing motors. The primary focus of these elements is on the development and use of light components and weight reduction. The swing motors can be economically and easily controlled by hydraulic direction control valves. The harmonization of the valves with the timing device is one of the essential engineering tasks during development. Here as well, the aim is to lower costs. In contrast to the other known systems for changing the control times during operation, camshaft adjusters have a simple design and are correspondingly economical. These systems should be integrated in the cylinder head design early in the initial development of a cylinder head. The oil circulation required for regulating the system can then be more easily harmonized with the hydraulic directional control valves.

Because of the potentially attainable improvement, new engines will probably increasingly use continuously

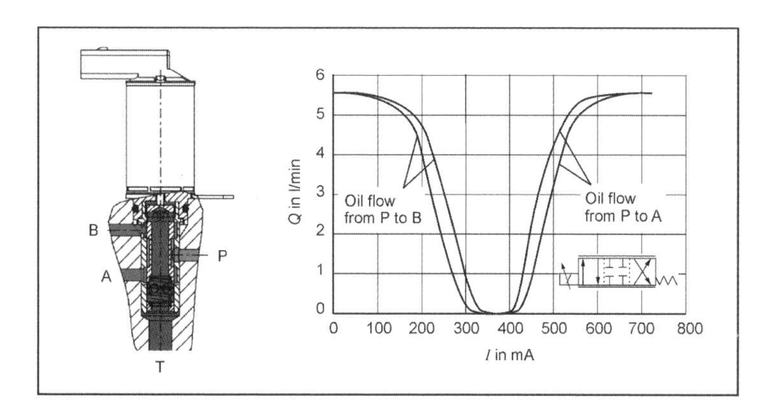

Fig. 10-76 A cross section, *Q-I* characteristic curve, technical data, and the hydraulic symbol of 4/3-way proportional valves.

variable timing devices. The control of inner exhaust gas recirculation requires cylinder head designs with at least two camshafts. The continuously variable camshaft timing device works similarly in spark-ignition engines with direct injection. In this case as well, the internal exhaust gas recirculation can be controlled by this system. Camshaft adjustment will also be used for this still-new combustion process.

Camshaft adjusters can be combined with variable valve actuation that allows the valve stroke or opening period to be varied. Porsche has used this approach in a series six-cylinder engine.[16] Numerous applications are possible for camshaft adjusters that offer substantial potential optimization of internal combustion engines. Infinitely variable valve actuation systems need to focus on the potential improvement that can be achieved with these measures.

10.4.2 Systems with Stepped Variation of the Valve Stroke or Opening Time

Honda used variable valve actuation for the first time in the mass production of spark-ignition engines with its VTEC system that influences the valve stroke or valve opening time.[4] The principle is based on a rocker arm solution in which, by moving small, hydraulically actuated pistons inside the rocker arm, different coupling states can be achieved to allow switching between different cam contours. Figure 10-77 shows an outline of the system used in a four-valve engine with dual camshafts. The right part of the figure shows an isometric representation of the valve and camshaft arrangement. For each cylinder, the camshaft has a central cam with a large valve stroke and opening time geometry. To the side is a cam profile with smaller cam contours. Inside the rocker arm module, a two-part piston is shifted by oil pressure parallel to the axis of the camshaft. This is done depending on the engine mapping as a function of the engine speed, the induction pipe pressure, the vehicle speed, or the coolant temperature. The oil is supplied for switching the cam contour through openings and channels in the bearing shaft on which the rocker arm module pivots. When operating at low speeds, the smaller cam contours act on the gliding surface of the rocker arm. The rocker arms are separated by precisely harmonizing the geometry of the two-part adjusting piston with the rocker arm width. A relative stroke is created between the central rocker arm and the individual rocker arms on the side. The central rocker arm is supported on a spring element. The space for this must be created in the cylinder head. In cylinder head designs with more than four valves, this is a particular challenge to the developer. When coupled as shown in Fig. 10-77, the central cam acts on the rocker arm module, and all components are moved simultaneously without a relative stroke. The two-part adjusting piston is reset with a small spring. The adjusting oil pressure is established by the engine oil circuit without an additional oil pump. The VTEC system is located on the intake and exhaust valve side.

For this and similar solutions, Honda developed numerous patent applications. By the number of different inventors alone of the patent applications, we can guess at the enormous amount of development. Four-valve solutions with one or two camshafts have been created for production engines. Both rocker rollers and rocker arms with sliding surfaces are used.[7] Up to three differently acting cam contours have been realized in this context.

Mitsubishi has also developed a similar series system for four- and six-cylinder engines based on the same functional principle.[7] In this solution, three cam contours are used, and one cam contour consists of a base circle to stop the valve. With both engines, two and three cylinders are stopped using this valve actuation system. To achieve this, Mitsubishi requires a small oil pump in the cylinder head.

DaimlerChrysler uses variable valve actuation to stop the cylinders in its stock V8 and V12 engines. The solution is based on a valve rocker module that is used with a central camshaft in a three-valve approach. Figure 10-78 shows the valve rocker module of this system without a camshaft. The functional principle is the same as the above-described Honda solution. Inside the rocker roller module, a two-part adjusting piston is moved electrohydraulically against a spring force. Depending on the coupling state, different cam contours are selected between

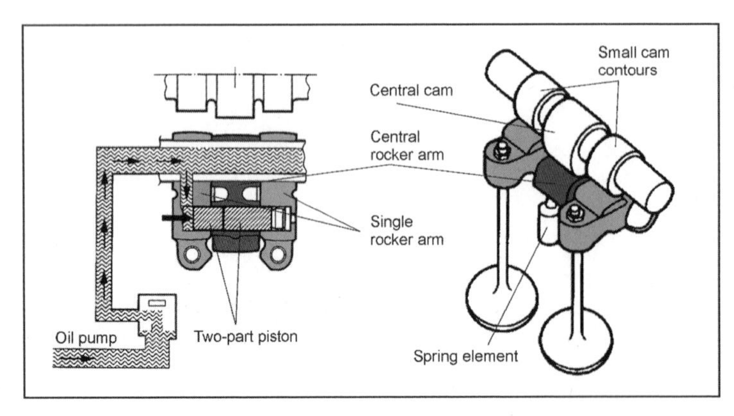

Fig. 10-77 Honda VTEC system.[4]

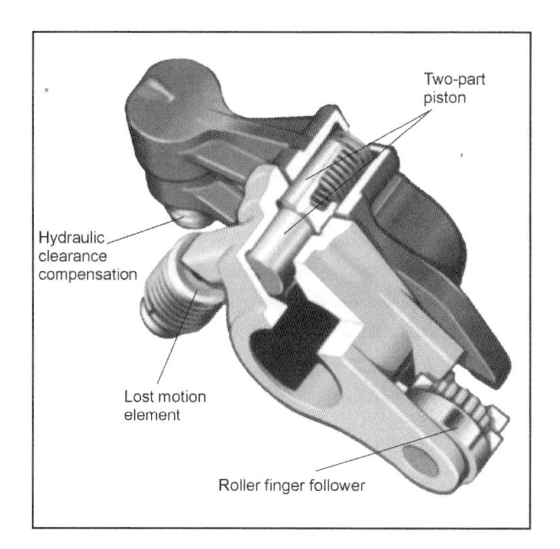

Fig. 10-78 Rocker roller module to stop the valves by DaimlerChrysler.[7]

valve strokes, except that one valve stroke is a zero stroke, and this stops the valves to shut off the cylinders.

The primary goal of the system used in this instance is to reduce fuel consumption during partial load operation by stopping the cylinders. This is particularly effective with high-displacement engines with many cylinders. There is little effect on the smooth running of these engines. These measures can reduce consumption by approximately 15% in contrast to conventional engines.

Similar to Mitsubishi and Honda, Toyota has also pursued solutions for mass production involving switching the valve contour on the intake and exhaust valve side. In this case as well, an adjusting piston in a rocker arm module is electrohydraulically pushed against a spring force (Fig. 10-79).

The interesting fact about this solution is that a rocker arm module is used in which a roller is the contact surface that faces the cam at low speeds, and a sliding surface is used at high speeds. At high speeds, the rocker arm module is pivoted by the sliding contact, and a stop below the lost-motion element provides coupling. The stop is held by oil pressure and is moved by spring force at low speeds toward the bearing center of the rocker arm module. At high speeds, the sliding surface lowers with the lost-motion element into the rocker arm module. The spring force of the lost-motion element can be minimal since the moved mass of the element is also low. For this solution, Toyota also uses a continuously variable camshaft adjuster on the intake valve side. This combination allows the valve lifting curve to be widely varied in contrast to engines with fixed control times.

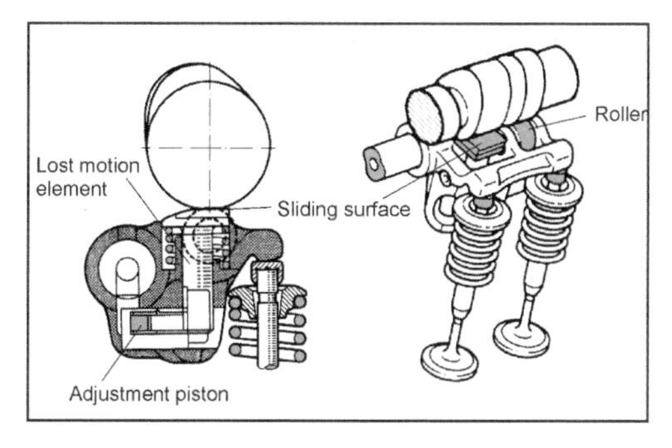

Fig. 10-79 Toyota valve actuation VVTL-i for different valve strokes.[17]

Porsche has traditionally used bucket tappet solutions for four-valve engines. In 2000, Porsche presented a turbo engine for the first time with variable valve actuation using different valve strokes with a switching bucket tappet.[16]

In addition, a camshaft control device was placed on the intake valve side to offer two control time positions. As with Toyota, a combination of two independently functioning systems of variable valve actuation is used. The switching bucket tappet can execute two valve strokes and consists of an inner and an outer tappet (Fig. 10-80). Its rotational position is oriented by a special guide in the cylinder head. The surface can be ball-shaped for correspondingly strong maximum strokes. Inside the tappet are small hydraulically actuated pistons that activate the inner or outer tappet for valve actuation depending on the position. In this case as well, the term "mechanically variable valve actuation" is appropriate since only the adjusting piston is electrohydraulically controlled, and the valves are actuated by the mechanical positive engagement of the components.

Usually a new generation of cylinder heads is used with this type of valve actuation. The geometry of the cam contours is conventional; i.e., they are smooth and can be manufactured by normal cam production systems. Corresponding to the categorization in Section 10.4, these

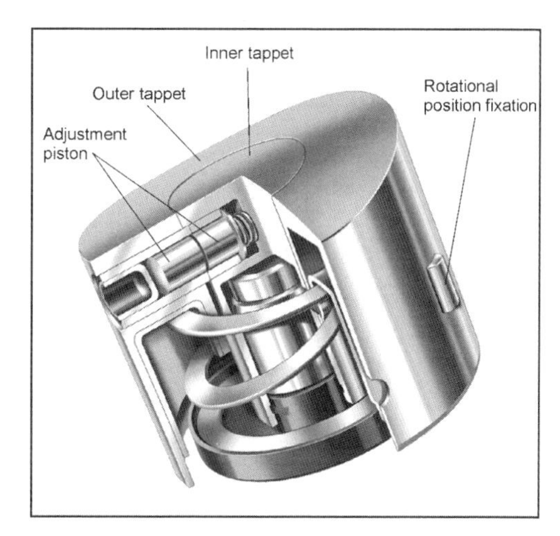

Fig. 10-80 Switching bucket tappet by Porsche.[16]

solutions represent systems with a variable transmission element between the cam and valve. The functional principle is mechanical since only mechanical actuators and contact elements are used in the flow of force toward valve actuation. The adjusting piston is controlled hydraulically via an electrically actuated directional control valve. It is anticipated that these or similar systems will become more widespread in production engines.

10.4.3 Infinitely Variable Valve Actuation

In the following, we briefly discuss a few systems that enable an infinitely variable valve lifting curve. These are systems that use camless controls as well as those with a mechanical functional principle that use a camshaft.

10.4.3.1 Mechanical Systems

With its Valvetronic system, BMW has created a continuously variable valve actuation system on the intake valve side. The load of the engine can be controlled solely by variable valve lifting. The system uses a special force transmission mechanism between the cam and valve and, corresponding to the categorization in Section 10.4, is classified as a mechanical variable valve actuation system. In Fig. 10-81 we see the Valvetronic system with the intake camshaft and intake valve module. In the flow of force between the camshaft and valve, there is a transmission mechanism that swings the roller rocker arm to actuate the valves. A setting shaft designed as an eccentric shaft and driven by an electrical control motor changes the lever geometry of the transmission mechanism. Valve strokes between 0.3 and 9.7 mm can be set. The entire adjustment processing can take place within 0.3 s. The conventional throttle valve can be dispensed with. The friction loss of the valve gear can be reduced during operation by the variable valve strokes in comparison to conventional valve gears since the valve springs are compressed less with smaller valve strokes.

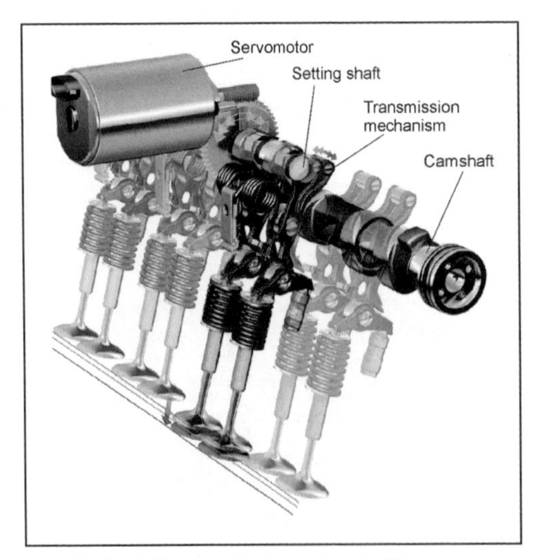

Fig. 10-81 "Valvetronic" system by BMW as a module with the valve gear components.[18]

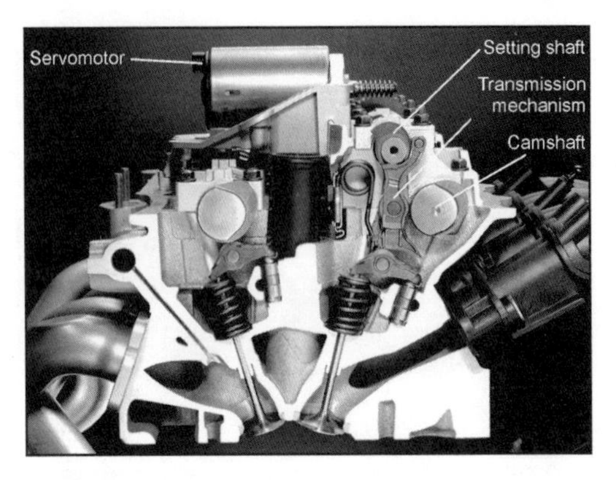

Fig. 10-82 Arrangement of the "Valvetronic" system in a cylinder head.[18]

Figure 10-82 shows an installed variable valve actuation system in a cross section of a cylinder head. The exhaust valve side remains conventionally actuated with a rocker arm. The required space for valve actuation is kept within limits. Space in the vehicle is required only for the control motor. The eccentric shaft, the transmission mechanism, the camshaft, and the control motor are premounted in a separate cast holder and attached as a module to the cylinder head.

Similar systems have also been developed. They have not yet been found in stock engines, however. BMW has also used continuously variable camshaft adjusters on both camshafts in addition to the variable stroke. The variation of the control times that can be attained is substantial. This combination was used first in a four-cylinder engine in compact passenger cars. In contrast to the predecessor model, savings in consumption of approximately 15% have been attained.

10.4.3.2 Hydraulically Actuated Systems

In the 1980s, there was a whole series of research efforts dealing with hydraulically variable valve actuation. The developmental goal was a freely settable valve actuation via the medium of oil. An example of a design to change the control times based on a hydraulic functional principle is the system by Fiat shown in Fig. 10-83. Developments toward similar solutions were made at many companies.

The intake valve is actuated via the camshaft and a hydraulic transmission mechanism. With the movement of the tappet in the tappet chamber, pressure builds that moves the piston above the valve and, hence, moves the valve. The oil pressure in the tappet chamber can be interrupted by a solenoid valve. This limits the valve stroke, and the engine load can be controlled without a throttle valve. Oil can be conveyed to the tappet chamber via a small pressure tank. The solenoid valve must be designed to switch extremely quickly. A problem with this type of valve actuation is the operating behavior at low temperatures

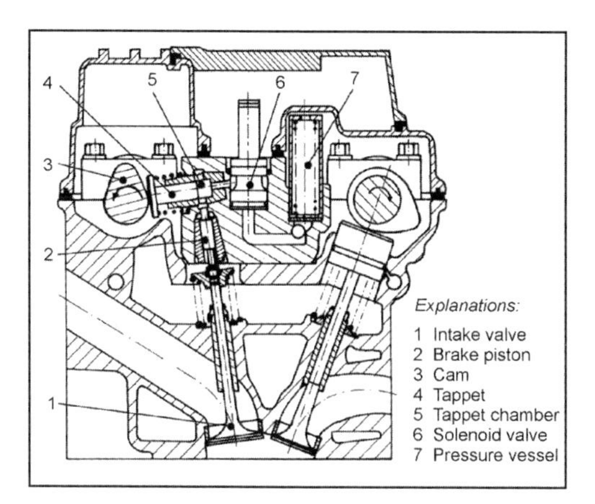

Fig. 10-83 Hydraulically variable valve actuation by Fiat.[19]

and the related strongly differing oil viscosities. A reproducible valve lifting curve is also difficult to obtain.

At present, there are few activities in the area of variable valve actuation. The only development of which we are aware is by Fiat. Whether these systems have a chance at being used in stock engines is difficult to determine.

10.4.3.3 Electromechanical Systems

Camless systems have the greatest potential for varying the valve lifting curve. They use valve actuators that can individually set any control time of any valve. Based on this idea, there have been investigations of valve actuation with an electromechanical functional principle for about 20 years. An armature between two coils alternately supplied with power is connected to the charge cycle valve via the armature guide. In addition, springs are used that actuate the armature and the valve. The armature is excited to vibrate when the bottom or top coil is supplied with power. The valve stroke can be set from 0 mm to a maximum stroke, and the load of the engine can be controlled by widely varying the valve lifting. Figure 10-84

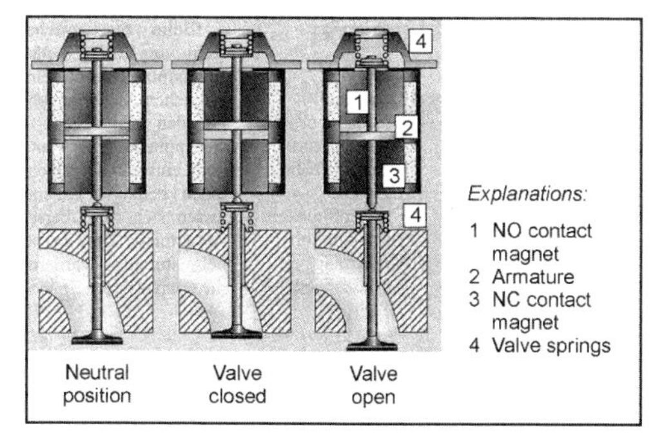

Fig. 10-84 Functional representation of an electromechanically variable valve gear.[20]

shows the basic construction of this type of control. To open the valves, the opener magnet is excited with current, and the closer magnet is excited to close it. When the coils are not excited with current, the armature and, hence, the valve remain in midposition between the coils. This position is held by the spring. In case of a system failure or engine stoppage, there is a corresponding clearance in the piston.

The future will show the degree to which this type of variable valve actuation will replace the purely mechanical valve actuation in stock engines. The additional potential for thermodynamic improvement in comparison to systems such as the Valvetronic system by BMW mentioned in Section 10.4.3.1 is limited.

Bibliography

[1] Hannibal, W., Vergleichende Untersuchung verschiedener variabler Ventilsteuerungen für Serien-Ottomotoren, Dissertation, University of Stuttgart, 1993.

[2] Haltenberger, S., Vorrichtung zur Ventilverstellung, Patent DE PS 368775, 1918.

[3] Bassi, A., F. Arcari, and F. Perrone, C.E.M.—The Alfa Romeo Engine Management System–Design Concepts–Trends for the Future, SAE Paper 85 0290, 1985.

[4] Inoue, K., R. Nagahiro, and Y. Ajiki, A High Power, Wide Torque Range, Efficient Engine with a Newly Developed Variable Valve-Lift and Timing Mechanism, SAE Paper 89 0675, 1989.

[5] Metzner, F.-T., and H. Flebbe, Doppelnockenwellenverstellung an V-Motoren, 8. Aachener Kolloquium Fahrzeug- und Motorentechnik, 1999.

[6] Ebel, B., and F.-T. Metzner, Die neuen V-Motoren von Volkswagen mit Doppelnockenwellenverstellung, in MTZ 61 Motortechnische Zeitschrift (2000) 12.

[7] Hannibal, W., and K. Meyer, Patentrecherche und Überblick zu variablen Ventilsteuerungen, Speech at Haus der Technik, March 2000.

[8] Ulrich, J., and O. Fiedler, Der Motor des neuen Porsche 968, in MTZ 52 Motortechnische Zeitschrift (1991) 12.

[9] Knirsch, S., M. Mann, H. Dillig, H.-J. Reichert, and T. Bartholmeß, Der neue Sechszylinder-V-Motor von Audi mit Fünfventiltechnik, in MTZ Motortechnische Zeitschrift 57 (1996).

[10] Metzner, F.-T., and P. Keiser, Der neue V6-4V-Motor von Volkswagen, 20th International Viennese Engine Symposium, 1999.

[11] Batzill, M., W. Kirchner, H. Körkemeier, and H.J. Ulrich, Der drive für den neuen Porsche Boxter, in special print der ATZ und MTZ, 1997.

[12] Braun, H.S.; R. Flierl, R. Kramer, R. Marder, G. Schlerf, and J. Schopp, Die neuen BMW Sechszylindermotoren, in special edition of ATZ and MTZ, 1998.

[13] Knecht, A., Nockenwellenverstellsystem, "Double-V-Cam," Ein neues System für variable Steuerzeiten, special print from Systems Partners 98, Vieweg Verlagsgesellschaft mbH, Wiesbaden, 1998.

[14] Wenzel, C., W. Stephan, and W. Hannibal, Hydraulische Komponenten für variable Ventilsteuerungen, Speech at Haus der Technik, Essen, 2000.

[15] Endres, H., H.-D. Erdmann, A. Eiser, P. Leitner, W. Kaulen, and J. Böhme, Der neue Audi A4, Der neue 3,0-l-V6-Ottomotor, in special print of ATZ and MTZ, 2000.

[16] Schwarzenthal, D., M. Hofstetter, H.-P. Deeg, M. Kerkau, and H.-W. Lanz, VarioCam Plus, die innovative Ventilsteuerung des neue 911 Turbo, Speech at the 9th Aachen Colloquium, October 4–6, 2000.

[17] N.N., Die variable Ventilsteuerung VVTL-i der Fa. Toyota, Press release from Toyota Cologne, January 2001.

[18] N.N., "Valvetronic" Information für den Kundendienst, BMW, Munich, March 2001.

[19] Hack, G., Freie Wahl, in Autor Motor Sport, 17/1999, S. 48–50.

[20] Koch, A., W. Kramer, and V. Warnecke, Die Systemkomponenten eines elektromechanischen Ventiltriebs, 20th Viennese Engine Symposium, May 6.

10.5 Pulse Charges and Load Control of Reciprocating Piston Engines Using an Air Stroke Valve

The air stroke valve represents an innovative approach for optimizing charge cycle processes in reciprocating piston engines. A substantial increase in torque is attained over the entire range of engine speed and especially at low speeds with an additional valve in the induction pipe that closes and opens the intake cross section extremely quickly during each induction stroke. By specifically controlling the flow and oscillation processes using this so-called air stroke valve, pressure waves are generated in the intake tract that are used to create charge effects.

With the air stroke valve, numerous additional effects can be created such as heating the aspirated air while cold starting an engine using thermodynamic state changes, or throttle-free load control in connection with a conventional nonvariable valve gear.

Mahle Filtersysteme GmbH, Stuttgart, designed an actively switching air stroke valve system and built it for testing in a single-cylinder engine. The results reveal the enormous potential of the air stroke valve for greatly influencing charge cycle processes in engines, especially to substantially increase the cylinder charge over the entire speed range.

10.5.1 Introduction

For internal combustion engines, an even, high torque characteristic from idling to high speeds is ideal.

Such behavior can be approximated only with a great deal of effort since many measures for optimizing torque work only within a narrow range of speed because of the oscillation processes in the induction pipe that produce corresponding resonance frequencies and speeds. Induction engine designs frequently work with a variable induction pipe length or resonance valve systems to enhance the torque characteristic extending to the partial load range. Another possibility for increasing torque is infinitely variable mechanical, electrohydraulic, or electromechanical valve gears where the charge cycle can be optimized by shifting the valve timing of the intake valve (Fig. 10-85, Refs. [1, 2]). A high torque at low speeds improves driving comfort, yields more agile engine behavior in transient driving conditions, and offers substantial potential fuel when the engine operates at low speed. A frequently used means of increasing torque below the nominal speed in spark-ignition engines, and even more so in diesel engines, is to convey more air into the combustion chamber by means of an exhaust turbocharger or mechanical supercharger. However, these systems, as illustrated in Fig. 10-85, have a markedly low torque below approximately 1500 to 2000 min^{-1}. In addition, there are limitations in the engine dynamics during transient driving since, during acceleration, the rotor of the turbocharger must be accelerated to create a flow of air mass corresponding to the set point.[3]

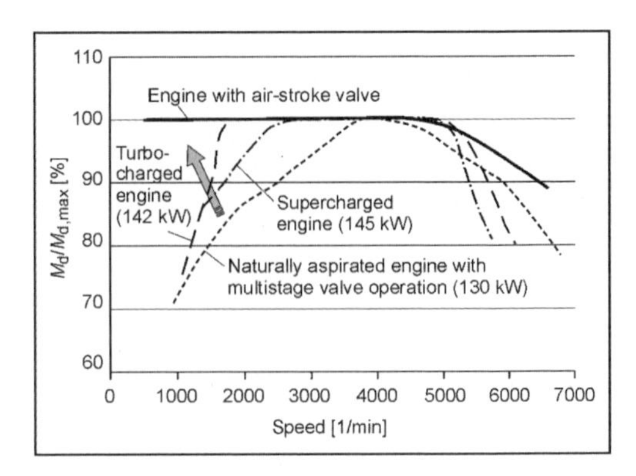

Fig. 10-85 Torque characteristics of spark-ignition engines with a multistage manifold, exhaust turbocharger, compressor, and air stroke valve.

An interesting alternative to increasing torque even at the lowest speeds is the air stroke valve that can theoretically be used to maintain the rated torque of the engines at idling speed without switching the length or resonance valves in the induction pipe. This approach can, in particular, be used to convey into the combustion chamber the air mass required upon a load increase in the next cycle without a response time or delay. With the air stroke valve, the length of the intake arms can be kept very short, and, thus, results in a substantial set of advantages over conventional ram tube systems. Combining an air stroke valve with an exhaust turbocharger for an immediate increase in the mass flow in response to the driver's input is a promising way to substantially increase vehicle dynamics, especially in down-sized approaches for supercharged, small-volume engines. In addition, consumption can be reduced, and the reduction of pollutant emissions can be anticipated.

However, the technical implementation of the air stroke valve poses extreme demands on the mechanical components, drive design, and control electronics since large changes in cross section are required with extremely short switching periods at specific times in the engine cycle.

10.5.2 Design and Operation of the Air Stroke Valve

The basic idea of the air stroke valve is to influence the charge cycle by placing a valve in the induction pipe upstream from the intake valve. This valve is used to apportion the air mass as needed within wide ranges and increase it above the "natural" intake. The increase in air mass, so-called dynamic supercharging, is achieved by controlling the opening and closing processes of the air stroke valve as a function of the engine operating parameters relative to the movement of the intake valve and piston. The engine intake valve opens, and the piston starts the intake stroke while the air stroke valve remains closed. The air in the intermediate area between the air stroke valve and intake valve is expanded in the combustion chamber. After a sufficient vacuum has built up, the air stroke valve is opened, and the fresh air enters at a high speed. A vacuum pressure wave runs from the valve to the manifold and is reflected at the induction pipe intake as an overpressure wave toward the combustion chamber as an effect that overlaps the transport of air mass. At the base of the piston, the inflowing air is delayed and reflected toward the induction pipe, which creates a backflow. The pressure increase from the conversion of kinetic energy into potential energy and the vibrations in the induction pipe are used to increase the air mass by either closing the air stroke valve at the proper time before the start of the backflow or timing the process so that the engine intake valve encloses the increased pressure in the combustion chamber. By using an air stroke valve, the inflowing air mass can be increased, and the backflow that arises over a wide range of speed toward the end of the intake process can be prevented.

The air stroke valve method was protected by patents in 1987[4,5] that generally describe two solutions. The simplest one is a controlled nonreturn valve where the energy is used to open the valve by the pressure difference between the intake system manifold and the combustion chamber with the aspirating piston (see Fig. 10-86). When the valve opens, a spring is tensioned that provides the

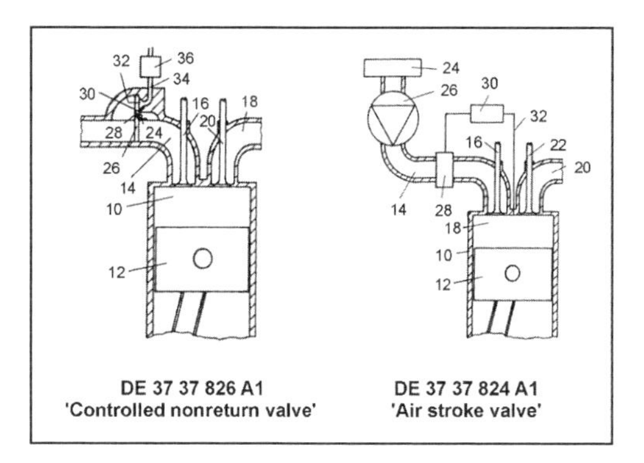

Fig. 10-86 Block diagrams from the patents.

energy to close the nonreturn valve. The arrangement can be controlled by the opening pressure difference and the closing spring force when the inflow terminates and when the backflow starts. Additional advantages are offered by electromagnetically controlling the nonreturn valve by using the above-described effects, such as creating a sufficient vacuum, at the right engine operating time. The results of investigations of a controlled nonreturn valve have been presented by Kreuter *et al.*[6] and illustrate the substantial possibilities for enhancing the charge with such a system.

The output of the system is enhanced by equipping the air stroke valve with an external drive as schematically illustrated in Fig. 10-86. The valve control then becomes independent of the current pressure difference between the manifold and combustion chamber, as well as the local gas speed at the nonreturn valve. This allows charge effects to be exploited at more opportune switching times, and it enables throttle-free load control of the engine by allowing the actuation time of the air stroke valve to be adjusted as desired in relation to the opening of the intake valve.

When a valve is used with freely timeable drive control, the torque can be increased by a more sophisticated variation: The double cycles of the inflow in the intake phase are described in Ref. [7]. The double triggering of intake and vibration processes of the air in the induction pipe while the intake valve opens further increases the cylinder charge in contrast to single cycles of the air stroke valve. In addition, such a valve system offers numerous other procedural possibilities that will be further described in the next section.

10.5.3 Options for Influencing the Charge Cycle

When operating a quickly switching valve with a freely controllable drive in the induction pipe, various functions can influence the charge. These concern the regulation of the air mass in the combustion chamber, and thermodynamic effects for specifically increasing the temperature of the inflowing fresh air.

10.5.3.1 Dynamic Supercharging in Induction Engines (Pulse Charge)

Dynamic supercharging increases the air mass in the combustion chamber by opening the air stroke valve once or twice during the opening phase of the intake valve. Using the above-described process, the air column in the intake tract is accelerated at specific times and then delayed to excite vibration. For this to work, the pressure waves induced by the vibration must be enclosed in the combustion chamber with the engine intake valves so that the resulting increase in density improves the cylinder charge (see Fig. 10-87). The dual operation of the air stroke valve during the intake stroke poses great demands for the dynamics of the valve and the reproducibility of the switching procedures.

Basically, the effect of the air stroke valve immediately starts upon load changes such as accelerations. Other advantages of this method are the usefulness of the dynamic supercharging when the engine starts. In addition, mixing is supported by faster charging because of the high inflow speed of the fresh air in the opening phases of the air stroke valve. The resulting homogenization of the mixture and the acceleration of combustion can potentially reduce the raw HC emissions of the engines during starting and regular operation.

10.5.3.2 Supporting and Recharging Supercharged Engines

When using mechanical supercharger or exhaust turbochargers, the air stroke valve can offer additional possibilities for enhancing the properties of the engine.[8] Particularly in the case of turbochargers, the increased

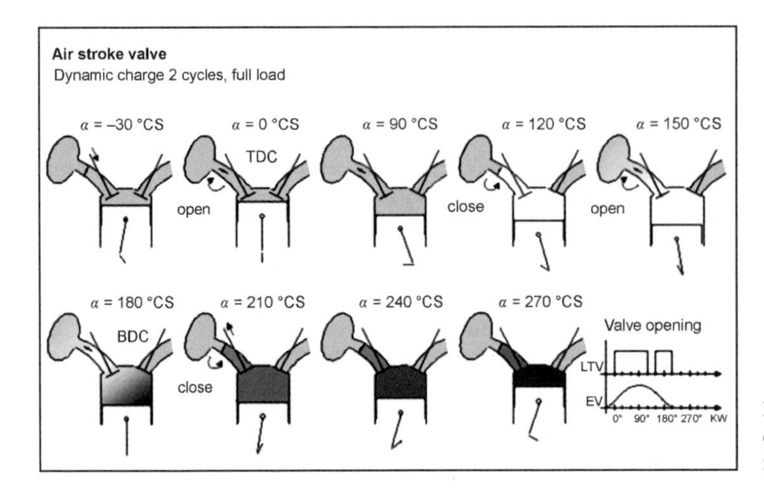

Fig. 10-87 Circuit diagram of dual air stroke valve timing in the intake phase to increase torque.

mass flow rate can greatly accelerate the response behavior of the supercharger. Especially when starting the engine, additional air mass is available in the combustion chamber that generates higher final compression temperatures and thereby improves cold start properties. Since the air mass flow is already accelerated at low rpm by the air stroke valve, the charging pressure can be lowered, which thereby reduces the specific load of the compressor. It is, therefore, conceivable to use smaller superchargers and dispense with turbochargers with a variable geometry and involved vane adjustments during partial loads.

From the perspective of energy, it makes sense to use the procedural variant where supercharged engines are "recharged." First, uncompressed air is aspirated by the piston. At bottom dead center, compressed air is supplied from the supercharger through a parallel tract of the air stroke valve system.[8] The air mass in the combustion chamber rises by approximately 50% from one cycle to the next. Only a part of the combustion air flows through the compressor, and dynamic supercharging effects can additionally reduce the compression in the supercharger. This allows the size of the supercharger to be reduced and lowers the energy required for the supercharger since only the air mass flow required for recharging must be compressed.

10.5.3.3 Throttle-Free Load Control

Throttle-free load control is an important step toward substantially reducing the consumption under partial loads in spark-ignition engines by minimizing the charge cycle loss. The opening time of the air stroke valve is adapted in relation to the required air of the engine. To achieve minimal air mass flow, a phase shift can be set between the air stroke valve opening and the intake valve opening so that both opening times only slightly overlap. This reduces the importance of a fast valve operating time in this method. In contrast, the valve leakage and the volume between the air stroke valve and intake valve gains in importance since the air mass enclosed there during the intake process is available in the combustion chamber.

When a valve is used that can be operated independent of the valve control time of the engine, different methods such as an early inlet close (EIC) or late inlet open (LIO) can easily be used in connection with a conventional mechanical valve gear to optimize engine operation under a partial load. The valve operation must be very precise to limit the aspirated air mass to that required by idling.

10.5.3.4 EGR Control

Similar to recharging supercharged engines, the exhaust gas recirculation cylinder can be selectively controlled so that exhaust is first aspirated, with a change to the aspiration of fresh air during the induction stroke. This yields a specified charge stratification in the combustion chamber that opens up additional possibilities for variation in terms of the inflow speed and charge movement.

10.5.3.5 Hot Charging

With hot charging, both an increase in the air mass and a rise in temperature of the aspirated air in the combustion chamber are sought to positively influence the mixing process in spark-ignition and diesel engines during cold starts and the warm-up phase. This process can be used by the starter during the first engine revolutions. If the air temperature is increased enough, the glow plugs in diesel engines can be dispensed with. In addition, the exhaust aftertreatment system starts much more quickly, which makes it easier to meet the D4 exhaust standard for diesel engines and to start the cabin heating system. Another possibility is to reduce the compression ratio in diesel engines for reasons of consumption without influencing their ability to start cold.

The increase in temperature results from a change in state in the aspirated fresh air. First, the pressure in the cylinder strongly drops when the air stroke valve is closed and the intake valve is opened by the movement of the piston. The enclosed mixture consisting of residual gas and fresh air is expanded. Since sufficient heat can be supplied through the cylinder walls, this process can be viewed as isothermic. After the air stroke valve opens, the cylinder fills very rapidly. The air is accelerated extremely fast by the strong vacuum. Thermodynamically, this process corresponds to the compression of the aspirated air in the cylinder that necessarily results in an increase in temperature. The amount of the attained temperature difference depends on the moment of opening and the length of opening of the air stroke valve, as well as the leakage of the closed valve.

The arising strong air expenditure, the high inflow speeds, and the higher final compression temperature allow more fuel to be injected with improved combustion quality. This leads to higher exhaust temperatures in a cold start, less combustion noise, faster warm-up, improved load assumption by the engine, and reduced cold start emissions. A prerequisite is also a very quickly operating valve in this method that allows a sufficiently fast intake of aspirated air into the cylinder and prevents backflow at the end of the intake phase.

10.5.3.6 Cold Charging Supercharged Engines

Like hot charging, the air stroke valve can also be used to specifically reduce temperature of the air in the combustion chamber of supercharged engines. By means of EIC (see Section 10.5.3.3), charging air compressed by the supercharger and precooled by the heat exchanger is enclosed in the combustion chamber, expanded by the piston movement, and thereby further cooled. This reduction of temperature lowers the temperature and pressure at the end of compression in the combustion chamber that reduces NO_x formation in spark-ignition and diesel engines and decreases the knocking tendency in spark-ignition engines. Alternately, a high charging pressure can be achieved with the resulting higher final compression temperature, which can be used to further increase the torque and output of the engine.

10.5.3.7 Cylinder Shutoff

Another variation is to alternately shut off the intakes to individual cylinders during partial-load operation, which is easily done by closing the air stroke valve during the intake process. The air supply of the shutoff cylinder can be periodically connected to prevent the combustion chamber from cooling as, for example, is possible with electromechanical valve gears (see Ref. [9]). This process can be easily realized in connection with a conventional mechanical valve gear. Primarily, the shift of the working cylinder to a higher load can be exploited. In contrast to electromagnetic valve actuation with the possibility of completely closing all valves, additional shifting work must be done by the engine since the exhaust valve remains open.

10.5.4 Prototype for Engine Tests

To demonstrate the functioning of the various air stroke valve methods, a prototype was designed and built that allows the system to be tested with an engine. The primary interest was to investigate the dynamic supercharging of the engine.

10.5.4.1 Parameters and Design

Very short operating times for the air stroke valve are required for dynamic supercharging. Extensive preliminary investigations using one-dimensional charge cycle calculations from the program GT-Power clearly show that valve operating times (the opening and closing process) of $\Delta t_S = 2$ ms are necessary to completely exploit the available potential. For this reason, a spring-mass oscillator was selected to be used as a direct valve drive that acts on the valve shaft. The movement of the oscillating system is controlled by two reverse solenoids that can be independently actuated, between which a hinged armature pivots supported by a spring. This has the advantage that oscillation can occur at the system resonance frequency upon starting. The energy required for valve actuation is then saved as spring energy and is immediately available for acceleration. In the actual polarity reversing process, only part of the loss arising during movement must be compensated, which means that the drive requires less energy. Another advantage is that the direction of the hinged armature can be reversed immediately after contact, which further increases the flexibility of the air stroke valve drive in regard to the crankshaft angle for the operating times.

10.5.4.2 Implemented Prototype

A Rotax BMW F650 single-cylinder engine with a stroke volume of 650 cm^3 was selected as a test engine. The engine had two symmetrical intake ports and was equipped with Otto direct injection, which allows unhindered access to the intake line and easy adaptation of the air stroke valve drive. For measuring, the engine was equipped with a very short induction pipe (280 mm). The air stroke valve (see Fig. 10-88), a symmetrical butterfly valve in a rectangular channel cross section of 30 mm × 60 mm, is in the two individual channels before the intake line branches. Because of the very long tubular section in the cylinder head, the ratio between the sum of all clearance volumes (combustion chamber, channels in the cylinder head, induction pipe up to the valve) and the stroke volume is

$$\sum V_c/V_H = 0.48$$

The natural frequency of the oscillated spring-mass system is determined by the torsion spring in the armature shaft and the moment of inertia of the hinged armature and valve. Since, in particular, the dynamic supercharging was to be investigated in detail at the required fast operating times, a design was created that permitted opening and closing times of $\Delta t_S = 2.1$ ms. The switching time can be varied by changing the moment of inertia by adding weight. An angular resolver was on the free end of the shaft to determine the current valve position.

The magnets to control the switching processes are designed as U-shaped magnets that are aligned with each

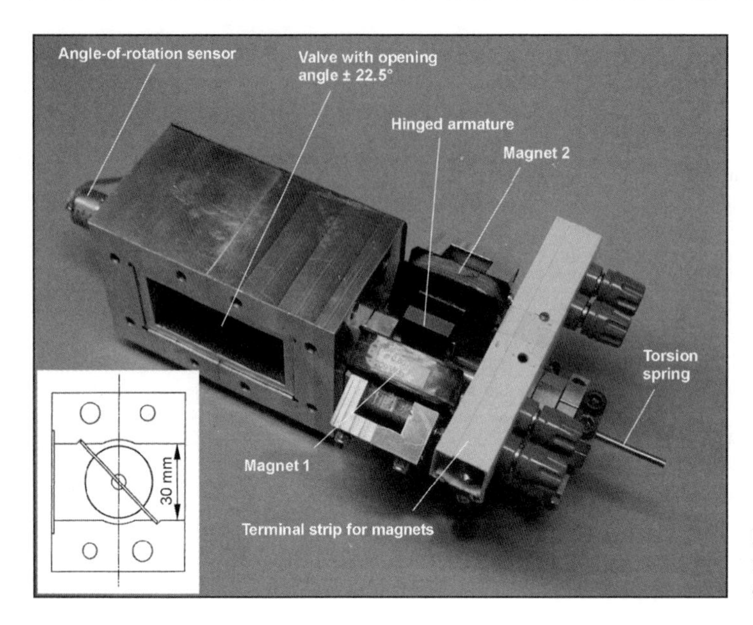

Fig. 10-88 Air stroke valve prototype for testing in a single-cylinder engine.

other at an angle of 45°. The alternating control of the actuator is carried out with the power electronics developed for this drive that controls the sequence of the output and holding current to the magnets. When the polarity of the armature is reversed, the holding magnet is switched off, and a high output current is initially fed to the opposite magnet. If the armature is lying on the holding magnet, the current is reduced over time to a lower holding current. The required output for this prototype is between 20 and 30 W, depending on the engine speed. If valve actuation is not desired, the armature can be held in open valve position.

In operating the air stroke valve, it is important to have a low amount of leakage from the induction pipe to the space between the air stroke valve and intake valve in closed valve position. The seal was attained by creating a 2-mm-deep cutout in the induction pipe wall (see Fig. 10-88) into which the valve swings with a small gap and then contacts. By correspondingly adjusting the drive, the valve edges lie on the contact edges in the induction pipe with a slight amount of force from the elasticity of the valve when in the closed position.

In the drive design, particular attention was paid to achieving very short operating times to especially reveal the effects of dynamic supercharging and hot charging. Other issues involving the durability of the overall system and noise emissions of the drive were largely left untouched in the development of this prototype.

10.5.5 Demonstration of Function in Single-Cylinder Engines

10.5.5.1 Increasing Air Expenditure by Dynamic Supercharging

In the following, air expenditure λ_L is used as a reference for dynamic supercharging. The air expenditure describes the ratio between the measured air mass and theoretical air mass flow rate that is formed from the stroke volume V_H and speed n. Since in the first approximation the torque rises in a linear relationship to the air expenditure in a faster-running engine given a stochiometric air ratio $\lambda = 1$, λ_L represents a suitable comparative quantity for evaluating charge cycle processes. Figure 10-89 shows λ_L in a slow-operating test engine as a function of the rpm. Because of the short induction pipe length, λ_L is quite low at low rpm. If the air stroke valve is opened twice in the intake phase given the same rpm, the air expenditure is substantially greater given optimal control times, and the increase rises strongly as the engine speed decreases. At $n = 1000\ \text{min}^{-1}$, the gain in air expenditure is approximately 13% in contrast to the pure induction without valve control. The rise in air expenditure predicted by charge cycle calculations can be almost completely achieved with this prototype.

To illustrate the charge cycle processes occurring in the intake system, Fig. 10-90 plots the pressure in the induction pipe measured with induced low pressure against the crank angle in a slow-running engine. One pressure measuring site was upstream from the air stroke valve and, hence, characterizes the pressure level of the induction pipes in the direction of the manifold. The second measuring site was in the volume between the air stroke valve and intake valve. To provide orientation in the engine cycle, the stroke curves of the intake and exhaust valves as well as the movement of the air stroke valves are also plotted. The pressure characteristics shown at a speed of $n = 1000\ \text{min}^{-1}$ clearly illustrate the physical processes when an engine is dynamically supercharged.

Because of the large valve overlap between the intake and exhaust valves, it is necessary to keep the air stroke valve closed when the intake valve starts opening to

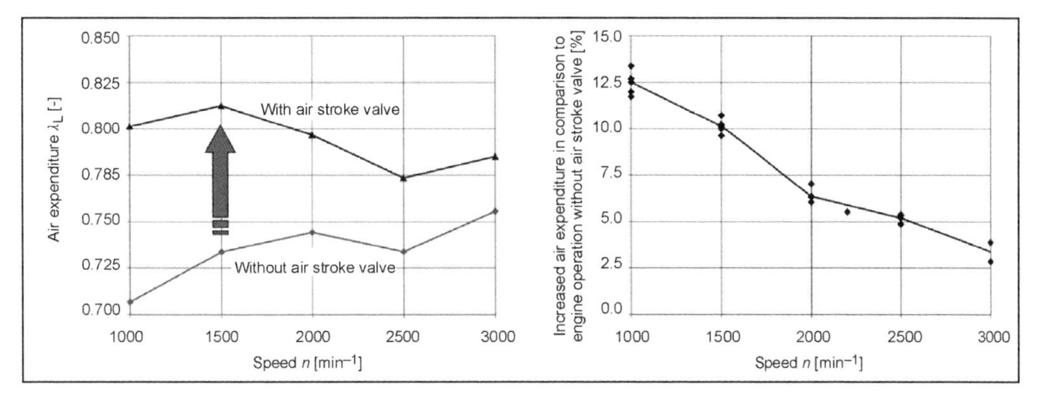

Fig. 10-89 Increase in air expenditure from an air stroke valve with two cycles in the intake phase in a slow-running engine.

prevent the inflow of exhaust. Only when the exhaust valve is largely closed is the intake cross section of the air stroke valve released and the delayed inflow process begun. During the intake valve stroke, the valve is closed. The pressure in the chamber between the air stroke valve and intake valve quickly drops to 150 mbar due to piston movement. Shortly before BDC in the charge cycle, the valve is briefly opened a second time. Because of the pressure difference, another air flow shoots into the combustion chamber, which causes a rise in pressure of nearly 100 mbar because of the reflection from the piston head in contrast to the induction pipe pressure. This pressure peak is retained in the cylinder chamber for dynamic supercharging, which significantly increases the density caused by the additional mass introduced into the combustion chamber. In Fig. 10-90, we see the pressure drop in the chamber between the air stroke valve and intake valve after the intake valve closes that arises from leakage across the air stroke valve.

10.5.5.2 Increasing Torque by Dynamic Supercharging

In Fig. 10-91, the measurements of air expenditure in the slow-running engine are compared with the data from a fast-running engine. The fast-running engine was operated with and without an air stroke valve at a constant stochiometric air-to-fuel ratio of $\lambda = 1$ and at the same ignition point to enable a direct comparison of the influence of the air stroke valve on engine operation. Two operating points of 1500 and 2200 min^{-1} were investigated. The results show that the air expenditure measurements of the slow-running engine can be transferred very easily to a fast-running engine.

It is notable that the torque at $\lambda = 1$ manifests a greater rise in comparison to the air expenditure. This indicates that the dual cycles of the air stroke valve with the resulting high air speed increase the charging movement in the combustion chamber and thereby positively influence the mixture.

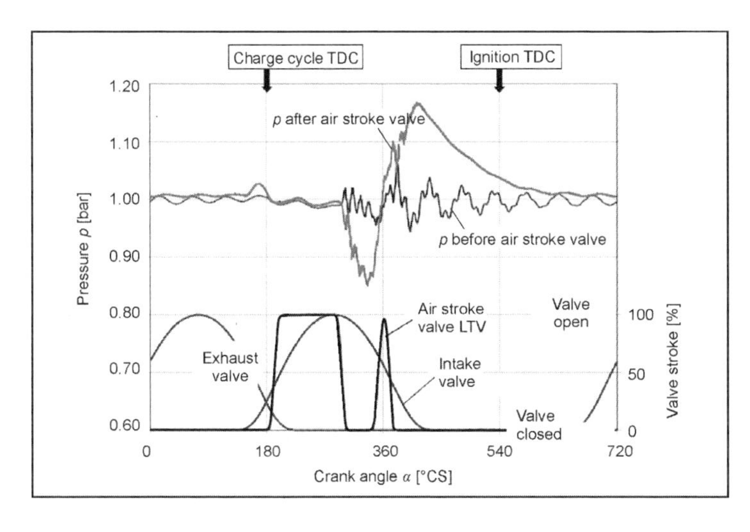

Fig. 10-90 Pressure characteristics before and after the air stroke valve with dual cycles in the intake phase in a slow-running engine.

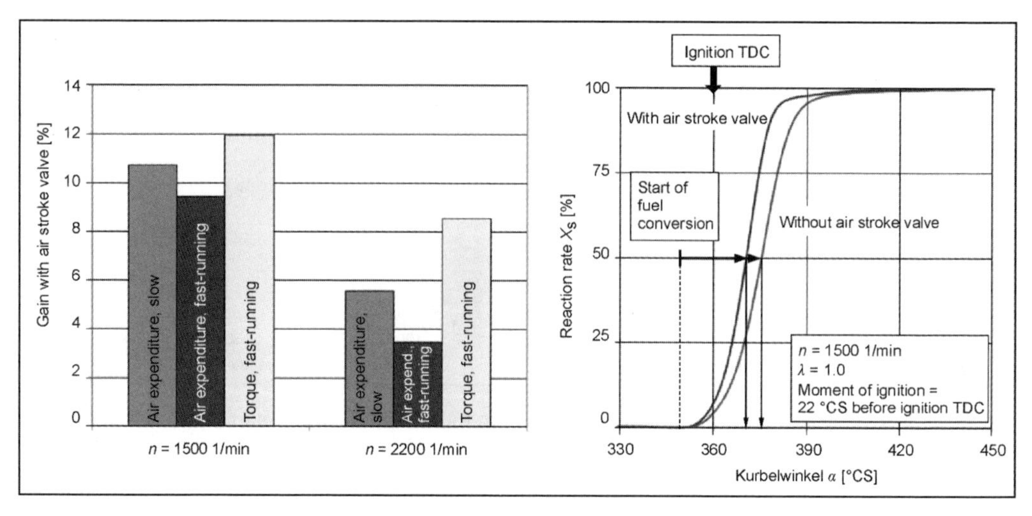

Fig. 10-91 Increase in air expenditure and torque in slow-running and fast-running engines.

A confirmation of this is provided by analyzing the combustion characteristic of the engine cycles with and without an air stroke valve as shown in Fig. 10-91 with $n = 1500\ \text{min}^{-1}$. Because of the tendency of the engine to knock, a late ignition is necessary when operating without an air stroke valve. This shifts the 50% conversion point away from the time of optimum efficiency that lies at 6°–8° after TDC ignition. When an air stroke valve is used, the mixture in the cylinder burns up 20% faster. The result is that the 50% conversion is very close to the optimum efficiency despite late ignition, and the state changes are, hence, closer to a constant volume cycle. The indicated mean effective pressure rises at the given speed from the air stroke valve by $\Delta p_{mi} = 9.8\%$. We note that the air expenditure does not completely reach that measured when the engine is running slowly. This is primarily because of the fact that the air stroke valve-control times could not be fully optimized due to the insufficient robustness of the drive in a fast-running engine.

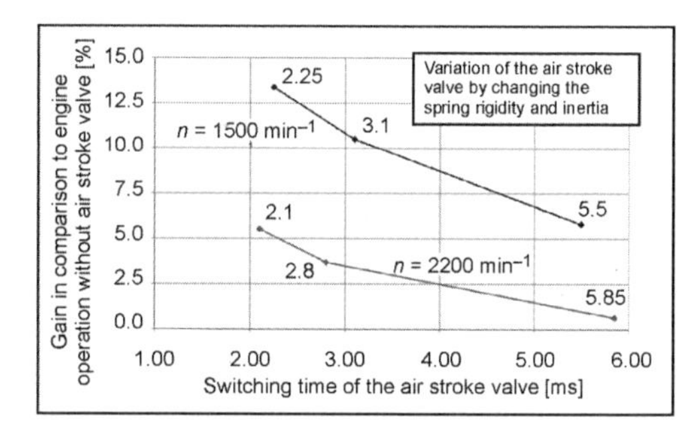

Fig. 10-92 Gain in air expenditure as a function of the air stroke valve actuation time Δt_S with two cyles in the intake phase in a slow-running engine.

10.5.5.3 Required Air Stroke Valve Operating Times in Dynamic Supercharging

An extremely important variable in using air stroke valves for dynamic supercharging is the operating time of the air stroke valve system. Figure 10-92 shows the gain from actuating the air stroke valve twice per cycle in contrast to an induction pipe without an air stroke valve represented in measurements at different valve speeds. The tests with different operating times were done by varying the moment of inertia and spring rigidity in the drive. This clearly illustrates that a fast valve with operating times of $\Delta t_S = 2$ ms is needed to realize the potential improvement. For example, the improvement from the air stroke valve at $n = 1000\ \text{min}^{-1}$ is cut in half when the operating time of the valve rises from $\Delta t_S = 2.25$ ms to $\Delta t_S = 5.5$ ms. At higher speeds, the improvement to the system falls almost to zero at operating times above $\Delta t_S = 5$ ms since dual actuation of the air stroke valve becomes increasingly useless to the charge cycle process.

The explanation can be found by analyzing the operating times optimized for air expenditure with dual actuation of the air stroke valve at various valve operating times Δt_S that are listed in Fig. 10-93. One can see that the first air stroke valve starts to open at the same time in each case when the exhaust valve is nearly closed. The complete closure of the air stroke valve the second time is also the same in every case and occurs briefly after bottom dead center. A very short second opening time of the air stroke valve has a positive effect on air expenditure. Theoretically, the optimum opening time of the second cycle can be determined from the induction pipe and inlet port lengths. When the valve operating time is long, the second cycle becomes too long. The time remaining for the intake valve to open is then insufficient to have a positive effect.

When the air stroke valve has longer operating times, the opening must be complete much earlier in the first

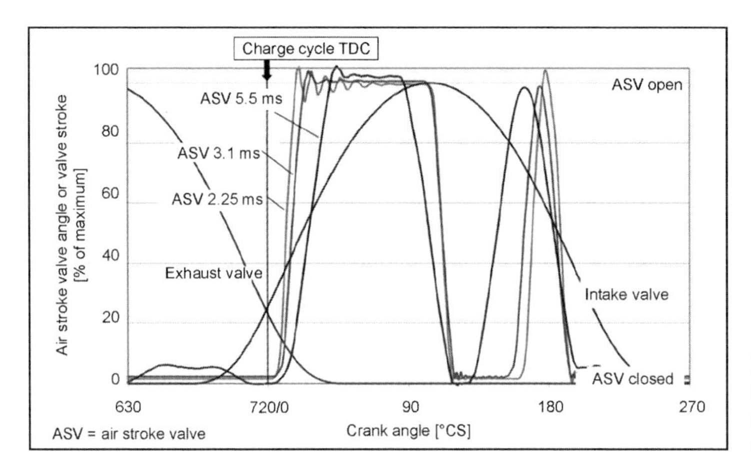

Fig. 10-93 Movement of the air stroke valve when varying the operating time Δt_S in a slow-running engine.

operating cycle, and the second operating cycle must begin immediately afterward. The inflowing air mass in the second operating cycle is thereby limited since a sufficient vacuum cannot build in the closing phase before the second opening. Furthermore, the backflow from the cylinder cannot be efficiently suppressed at the intake end when the air stroke valve operating times are long. The backflow occurs too quickly after the valve starts to open so that the air stroke valve cannot be reclosed at this time.

10.5.5.4 Hot Charging

In addition to dynamic supercharging, the feasibility was investigated of hot charging by opening the air stroke valve once briefly during the cycle while measuring the single-cylinder engine (see Section 10.5.3.5). A characteristic result of extensive measurements at $n = 1000$ min^{-1} is shown in Fig. 10-94. The cycle valve is opened only at a 30° crankshaft angle, which corresponds to the shortest possible time of 4.5 ms (= $2\Delta t_S$). The 30° crankshaft angle opening window of the air stroke valve was shifted over the opening phase of the intake valves. For the respective types of operation, the temperature increase in the cylinder chamber and the air mass in the combustion chamber were calculated from the measured pressure and temperature data.

It was shown that at the time of IC, temperature differences of up to $\Delta T_{IC} = 45$ K could be obtained in contrast to operating the engine without an air stroke valve, which is sufficient to preheat air for diesel engines in a cold start without glow plugs. The measured indices show that at these points, the combustion chamber pressure behind the closed valve falls to 0.46 bar up to the time at which the valve is opened. Because of the intense inflow from this pressure differential, the fresh air mass is increased to the level where it would be for pure induction operation due to the dynamic charging effects despite a very short valve opening time; there is therefore no limit to the air mass available during combustion when hot charging is used. The overall gas mass in the combustion

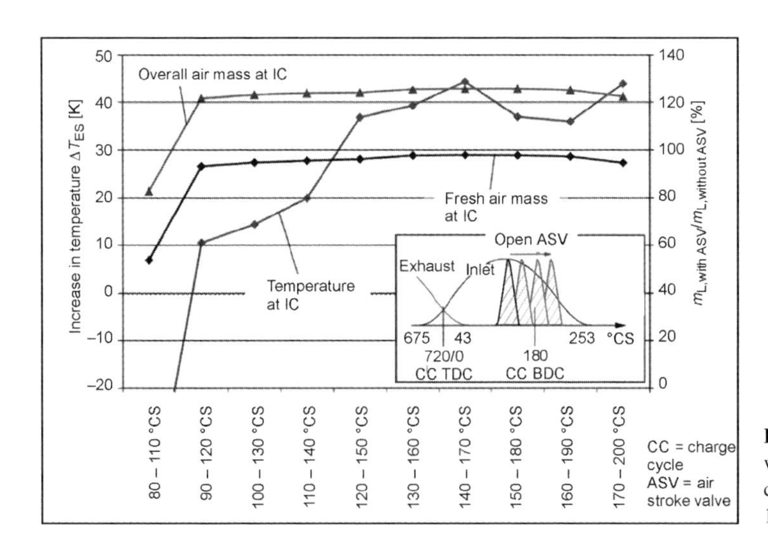

Fig. 10-94 Air temperature at IC with a gas mass in the combustion chamber with hot charging n = 1000 min^{-1}.

chamber rises by nearly 25% due to the residual exhaust gas. Additional increases in temperature are possible with systems that are ideally sealed.

10.5.6 Summary and Outlook

The air stroke valve represents an innovative system for optimizing the charge cycle. For spark-ignition and diesel engines, it allows numerous procedural variations to influence the air mass and temperature in the combustion chamber. However, it is very difficult to engineer such a system since extremely short operating times of around 2 ms must be attained for opening and closing the intake cross section.

The test of a prototype in a single-cylinder engine with Otto direct injection impressively demonstrated the performance of air stroke valves. In experiments to increase torque by dynamically supercharging the engine even at very low speeds, an increase in air expenditure, for example, of 13% at $n = 1000\ \text{min}^{-1}$ was obtained in contrast to mere induction-based operation. Furthermore, it has been demonstrated that the combustion of the charge is faster in a fast-running engine from the more intense charging motion. The results of hot charging experiments show that the temperature of the air aspirated into the cylinder chamber can be increased at IC by $\Delta T_{IC} = 45$ K with the same fresh air mass in comparison to induction without an air stroke valve. This reveals the substantial potential for improving the mixture in a cold start.

Some of the possibilities cited in Section 10.5.3 can theoretically be realized by infinitely variable mechanical, electromechanical, or electrohydraulic valve gears. Given the presently attainable operating times for electromechanical valve gears of approximately 3 ms for the opening and closing process, there are limitations to the possibility of throttle-free load control at high speeds and the effectiveness of dynamic supercharging. However, the serious disadvantage of these valve gears in comparison to the air stroke valve is the high power consumption of these systems of around 100 W per drive.[2] This comparatively large amount of power is required to move the large mass of the intake and exhaust valves.

At the moment, additional experiments are underway at Mahle Filtersysteme on air stroke valves in a four cylinder engine with a revised valve and drive design and reduced operating times of 1.8 ms. In contrast to the experiments presented here, the power consumption of the new drives has been reduced to 10 W per cylinder. The use of air stroke valves is therefore conceivable in combination with conventional, economical valve gears for highly dynamic charge control and supercharging—a possibility with a great deal of promise. This has a substantial set of advantages over conventional ram tube systems because of the short induction pipe length.

Bibliography

[1] Flierl, R., R. Hofmann, C. Landerl, T. Melcher, and H. Steyer, Der neue BMW Vierzylinder-Ottomotor mit VALVETRONIC, Teil 1: Konzept und konstruktiver Aufbau, in MTZ Motortechnische Zeitschrift 62 (2001) 6.

[2] Salber, W., H. Kemper, F. van der Staay, and T. Esch, Der elektromagnetische Ventiltrieb – Systembaustein für zukünftige Antriebskonzepte, Teil 1, in MTZ Motortechnische Zeitschrift 61 (2000) 12.

[3] Miersch, J., C. Reulein, and Ch. Schwarz, Rechnerischer Vergleich unterschiedlicher Motorenkonzepte zur Verbrauchsreduzierung und Dynamiksteigerung, 4th International Stuttgarter Symposium on Internal Combustion Engines, February 20–22, 2001, expert-verlag, Renningen, 2001.

[4] Schatz, O., Patent DE 37 37 828 A1, 1987.

[5] Schatz, O., Patent DE 37 37 824 A1, 1987.

[6] Kreuter, P., R. Bey, and M. Wensing, Impulslader für Otto- und Dieselmotoren, 22nd Viennese Engine Symposium, April 26–27, 2001 Progress Reports VDI: Series 12 Transportation Engineering, Vehicle Engineering, 45, VDI-Verlag, Düsseldorf, 2001.

[7] Schatz, O., Patent DE 43 08 931 C2, 1993.

[8] Schatz, O. and T. Steidele, Pulse charging—A New Approach for Dynamic Charging, 2nd International Conference on New Developments in Powertrain and Chassis Engineering, Strasbourg, 14.–16. 06. 1989, ImechE-Paper C382/116, 1989.

[9] Salber, W., H. Kemper, F. van der Staay, and T. Esch, Der elektromagnetische Ventiltrieb – Systembaustein für zukünftige Antriebskonzepte, Part 2, in MTZ Motortechnische Zeitschrift 62 (2001) 1.

11 Supercharging of Internal Combustion Engines

The major goals in the development of internal combustion engines, namely high efficiency, i.e., low fuel consumption and low emissions, have been discussed at length in the previous chapters. Another important point here is the increase in power concentration of an internal combustion engine.[1] It is, therefore, a question of obtaining as much power as possible from a defined engine volume and/or a given engine weight. Under certain circumstances, the increase in power concentration may also be linked to an improvement in efficiency.

The power output of an internal combustion engine is proportional to the mean effective pressure p_{me}, the speed n, and the total piston displacement V_H.

$$P_e = p_{me} \cdot n \cdot V_H \cdot \frac{1}{Z} \tag{11.0}$$

$$p_{me} = \rho_2 \cdot \lambda_L \cdot \eta_e \cdot \frac{H_u}{\lambda \cdot L_{\min}} \tag{11.1}$$

Z = 2 Four-stroke
Z = 1 Two-stroke
P_e = Effective power
p_{me} = Mean effective pressure
n = Speed
V_H = Piston displacement
ρ_2 = Density after charging
λ_L = Volumetric efficiency
η_e = Effective efficiency
H_u = Net calorific value
λ = Excess air factor
$L_{\min}$ = Minimum excess air factor

An increase in the piston displacement results not only in an increase in power but also in a significant increase in the engine weight and the necessary installation space as well as a deterioration in efficiency due to the increased friction loss. The friction losses increase disproportionately to the increase in engine speed with which an increase in power can also be achieved.

Calorific value H_u and minimum excess air factor $L_{\min}$ are fuel parameters and are assumed to be fixed.

$$p_{me} \sim \rho_2 \cdot \frac{1}{\lambda} \eta_e \cdot \lambda_L \tag{11.2}$$

The mean effective pressure is, therefore, proportional to the density of the air, the effective efficiency, and the volumetric efficiency and is inversely proportional to the excess air factor. The density of the air depends on the charge pressure and charge air temperature.

$$\rho_2 = \frac{p_2}{R \cdot T_2} \tag{11.3}$$

ρ_2 = Density behind the charger
p_2 = Charge pressure
R = Gas constant
T_2 = Temperature behind the compressor

The effective output of the engine is, hence, significantly increased with the increase in the air density. Today, up to 31 bar mean pressure is achieved, in particular, with diesel engines, and 19 bar mean pressure is already achieved with SI engines.

11.1 Mechanical Supercharging

During mechanical supercharging, the compressor is driven mechanically by the crankshaft (see Fig. 11-1) with the compression work having to be performed by the engine.

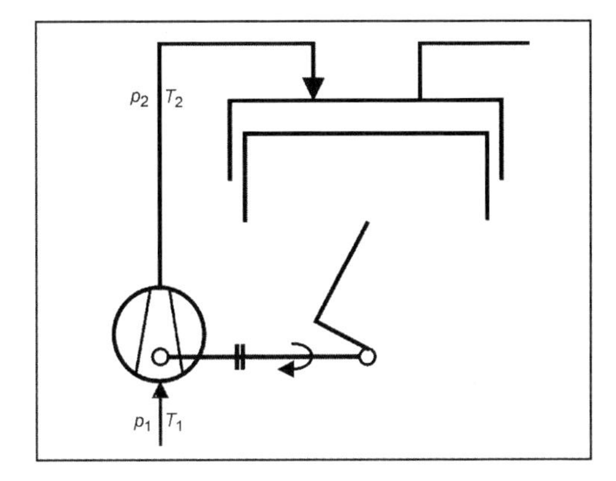

Fig. 11-1 Principle of mechanical supercharging.

The process now takes places at a higher pressure level. This results in a corresponding increase in the mean pressure, providing the air-fuel ratio remains constant. Mechanical supercharging initially results in deterioration in engine efficiency with the increase in output. If we compare it with a naturally aspirated engine of the same output, however, the mechanically supercharged engine produces a higher efficiency due to the lower mechanical and thermal losses. The compressors used are generally Roots blowers (Fig. 11-2), screw-type (Fig. 11-3) or spiral-type superchargers (Fig. 11-4), and, less frequently, radial compressors (with step-down gearing). Mechanical supercharging is predominantly used today in car SI engines where it has the benefit that during cold starting no heat is taken from the exhaust gas stream that is of great importance for the starting of the catalytic converter during the warm-up phase.

Fig. 11-2 Roots blower.[2]

Fig. 11-3 Screw-type supercharger.[3]

Fig. 11-4 Spiral-type supercharger.[4]

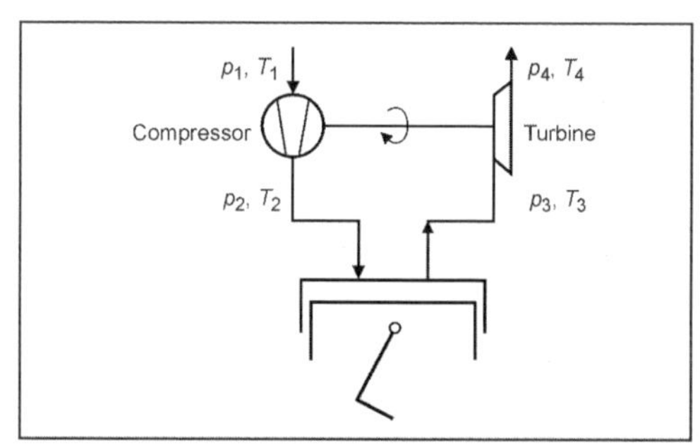

Fig. 11-5 Schematic representation of exhaust gas turbocharging.

11.2 Exhaust Gas Turbocharging

During exhaust gas turbocharging, the engine and the turbocharger (see Fig. 11-5) are linked thermodynamically and not mechanically. The compressor is driven by the turbine. The turbine receives the exhaust gas stream from the engine and, thus, covers the power requirement of the compressor.

Ram Induction

With ram induction, a large exhaust gas line volume is provided between turbine and compressor with the aim of reducing the pressure hammer of the individual cylinders and of pressurizing the turbine as continuously as possible, i.e., with a constant state p_3, T_3.

If we assume in an initial approximation that pressure p_3 is equal to pressure p_2, the engine will be operated at a high-pressure level without any change in the thermal efficiency. If we look more closely, however, we observe that a larger volume is relieved in the turbine so that a slight gain is possible. If $p_2 > p_3$, part of the turbocharger work will be output again to the crankshaft via the positive charge cycle loop.

Pulse Turbocharging

During pulse turbocharging, the kinetic energy of the exhaust gas is additionally used in the form of pressure waves. Figure 11-6 shows the pressure curve of a turbine.

Compared with the ram induction, this offers a gain as an isentropic expansion to the ambient state takes place instead of the irreversible throttling from the cylinder pressure to the exhaust gas counterpressure p_3. In fact, this gain cannot be completely exploited as a throttling takes place at the exhaust valves anyway, and because the turbine efficiencies with nonstatic charging are lower than with static charging. Compared with ram induction, pulse turbocharging has advantages especially in part-load operation and in the acceleration behavior.

Appropriate grouping of the cylinders with the given ignition sequence prevents exhaust gas from being pressed into a cylinder during the valve overlap, as this would result in an increase in the residual gas content. With turbocharged SI engines, the increased residual gas content results in a greater knock tendency; this, in turn, leads to a delayed ignition angle and, hence, to a loss of torque and increased fuel consumption.

The exhaust gas turbocharger consists of a compressor and a turbine (Fig. 11-7). The internals are shown in Fig. 11-8.

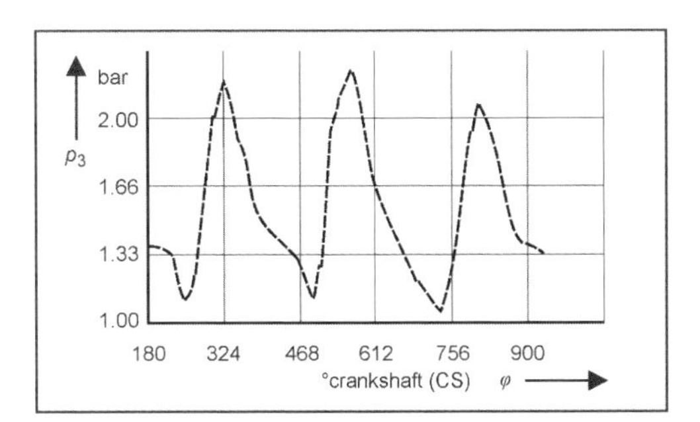

Fig. 11-6 Pressure wave during pulse turbocharging.

The operation of the compressor is imaged in a compressor map (Fig. 11-9).

Compressor speed and isentropic compressor efficiency are plotted against the volumetric flow $\dot{V}_1$ and pressure ratio p_2/p_1. If we follow a speed line to the left,

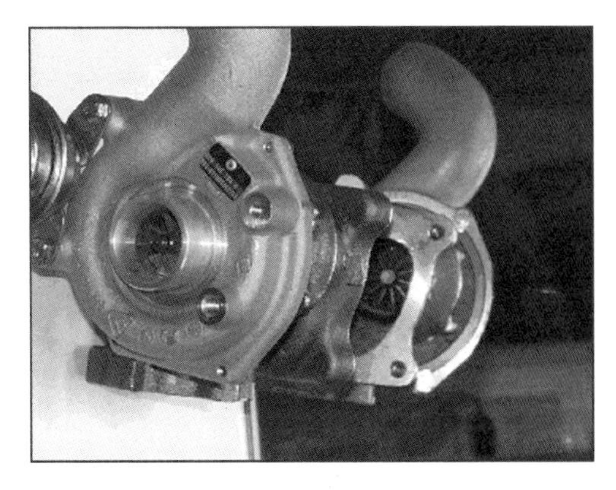

Fig. 11-7 Exhaust gas turbocharger K 03.[5]

Fig. 11-8 Turbocharger internals.[5]

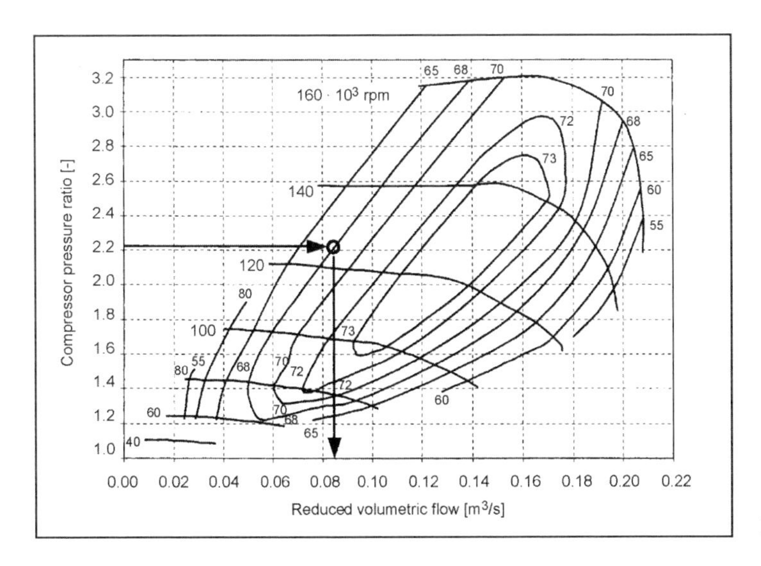

Fig. 11-9 Compressor map.

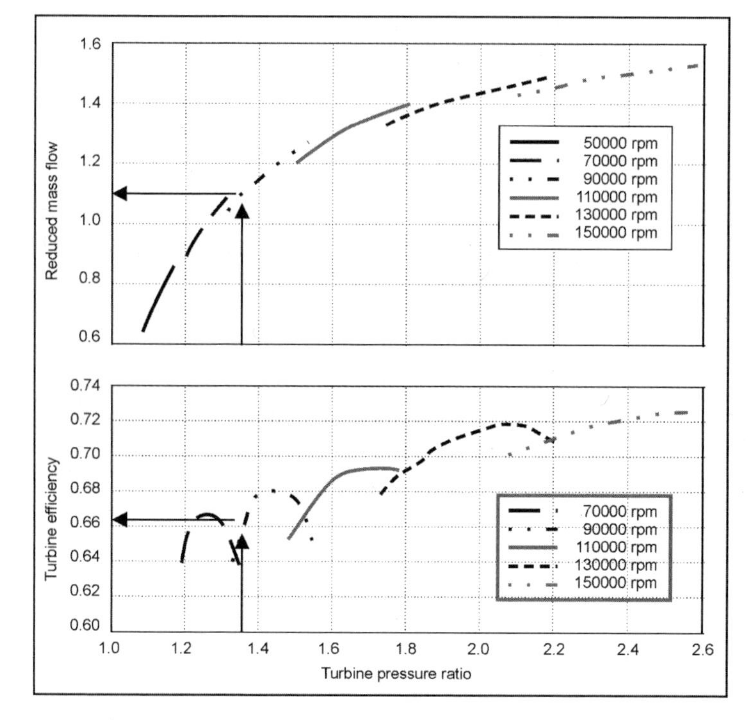

Fig. 11-10 Turbine map.

i.e., the compressor is increasingly throttled on the pressure side, we will reach the pump limit. This must not be reached during operation as otherwise the compressor would be destroyed.

For the presentation of the turbine behavior, the isentropic turbine efficiency and the flow coefficient are plotted against the turbine pressure ratio p_3/p_4 (Fig. 11-10) with the turbine speed as a parameter.

11.3 Intercooling

If we consider an isentropic compression process from 1 to 2 (Fig. 11-11), the temperature increases due to an isotropic compression as in Eq. (11.4).

$$\frac{T_2}{T_1} = \left(\frac{p_2}{p_1}\right)^{\frac{k-1}{\kappa}}$$

T_1 = Temperature upline of compressor
T_2 = Temperature downline of compressor
p_1 = Pressure upline of compressor
p_2 = Charge pressure
κ = Isentropic exponent

Since the compression is performed polytropically instead of isentropically, a further increase in temperature occurs [Eq. (11.5)].

$$T_2 - T_1 = \frac{(T_2 - T_1)_s}{\eta_{sV} \cdot \tau_K} \tag{11.5}$$

T_1 = Temperature upline of compressor
T_2 = Temperature downline of compressor
η_{sV} = Isentropic compression efficiency
τ_K = Cooling coefficient of the compressor

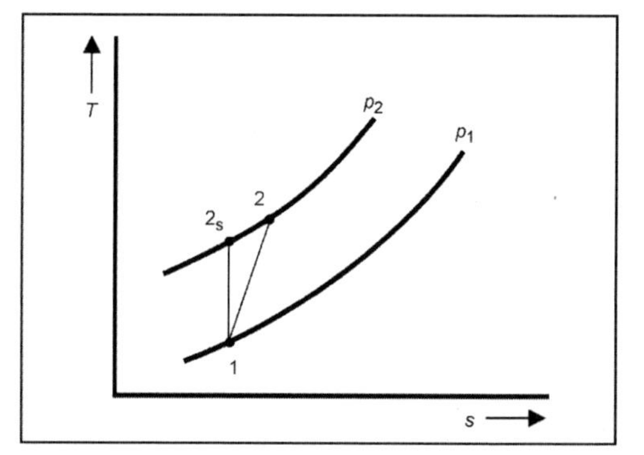

Fig. 11-11 Isentropic and polytropic compression.

The isentropic compression efficiency η_{sV} is calculated as

$$\eta_{sV} = \frac{h_2 - h_1}{h_{2s} - h_1} \approx \frac{c_p \cdot (T_2 - T_1)}{c_p \cdot (T_{2s} - T_1)} \tag{11.6}$$

h_1 = Enthalpy upline of compressor
h_2 = Enthalpy downline of compressor
h_{2s} = Enthalpy downline of compressor, isentropic
c_p = Specific thermal capacity for p = const

The cooling coefficient τ_K mentioned in Eq. (11.5) for turbochargers makes allowance for the heat dissipation via the compressor housing (particularly with large compressors) to the environment and lies in the range between 1.04 and 1.1. The temperature increase associated with the increase in pressure leads to a reduction in the density as shown in Eq. (11.3).

An intercooler allows the charge density and also the output to be increased as shown in Eq. (11.2).

Example:

$p_1 = 1$ bar; $T_1 = 293$ K (20°C)

Compressor: $\pi_V = p_2/p_1 = 2.5$

$\eta_{sV} = 0.85$

$T_2 = 313$ K (40°C)

Figure 11-12 shows a comparison between a naturally aspirated engine, a turbocharged engine, and a turbocharged engine with intercooling to 40°C. The same air-fuel ratio has been assumed for all three cases. This shows a direct relationship between the density and the output. The ambient state 1 bar and 20°C is assumed for the naturally aspirated engine. Comparison of the naturally aspirated engine with the turbocharged engine and a turbocharger compression ratio of 2.5 shows an increase in mean pressure to 187%, and for the turbocharged engine with intercooling to 40°C an increase in mean pressure to 234% is shown.

Figures 11-13 and 11-14 show the maps of turbocharged car and truck diesel engines.

Engine	$\rho_2 \left[\frac{\text{kg}}{\text{m}^3}\right]$	Mean pressure
Naturally aspirated engine	1.19	100%
Turbocharged engine	2.23	187%
Turbocharged engine with intercooling	2.78	234%

Fig. 11-12 Density and mean pressure of engines.

11.4 Interaction of Engine and Compressor

11.4.1 Four-Stroke Engine in the Compressor Map

Figure 11-15 shows the displacement lines of a four-stroke internal combustion engine. If the engine speed n is held constant, the volumetric flow $\dot{V}_1$ shows only a slight linear increase with increasing compression ratio p_2/p_1. The engine then operates as a volumetric displacement machine, and its throughput increases in relation to the increase in engine speed.

With increasing valve overlap and constant engine speed, the volumetric flow $\dot{V}_1$ increases less sharply with increasing compression ratio p_2/p_1.

Positive-Displacement Superchargers:
Some examples of positive-displacement superchargers are piston compressors (reciprocating piston and rotary piston), roots blowers, and screw-type superchargers.

From Fig. 11-16 it can be seen that the throughput increases with increasing compressor speed and drops slightly with increasing counterpressure. At constant speed we obtain the working points 1, 2, or 3, depending on the counterpressure.

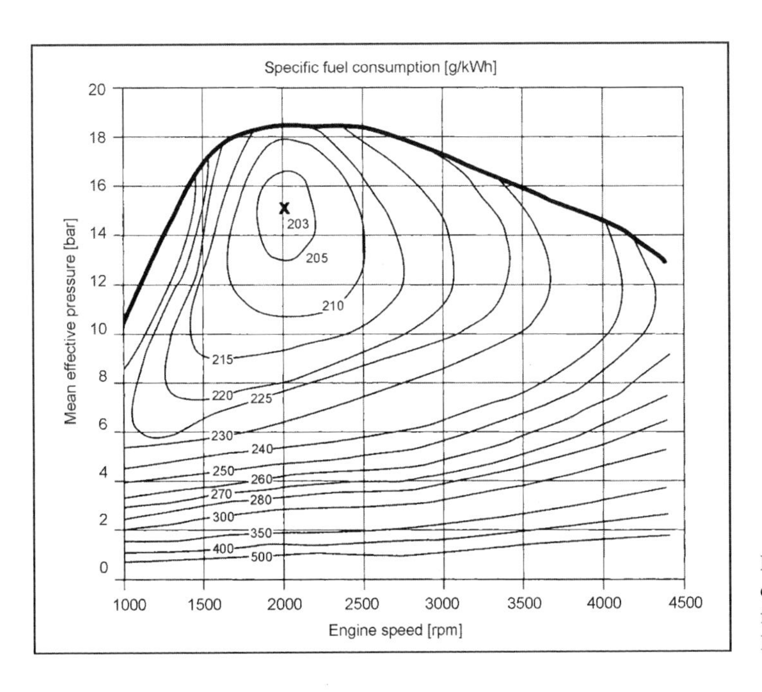

Fig. 11-13 Specific fuel consumption of the OM 611 four-cylinder engine from DaimlerChrysler.[6]

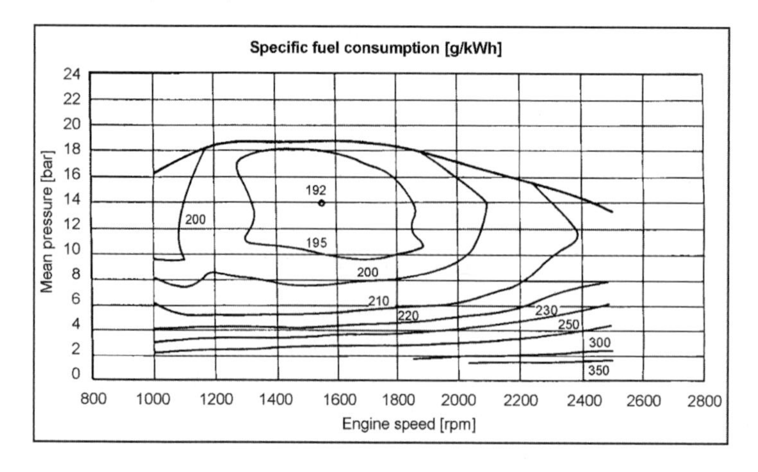

Fig. 11-14 Consumption map of OM 904 LA/125 kW from DaimlerChrysler.[7]

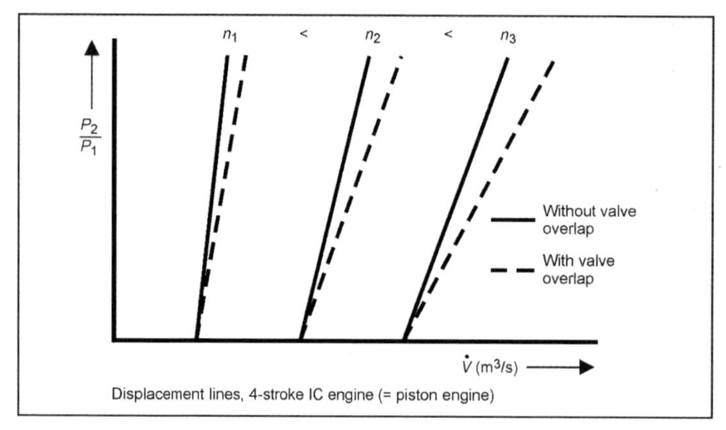

Fig. 11-15 Displacement lines.

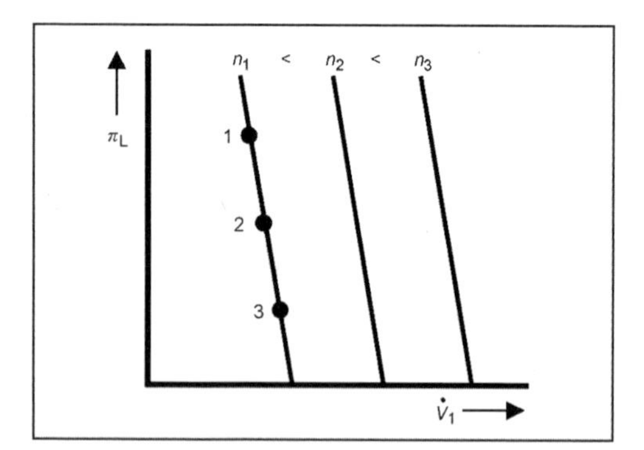

Fig. 11-16 Displacement lines of positive-displacement superchargers.

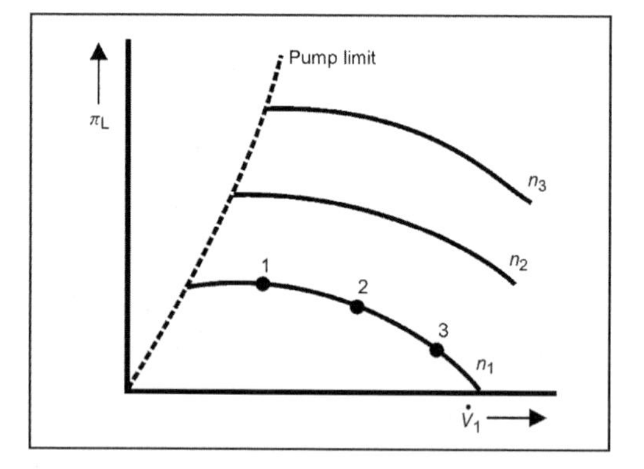

Fig. 11-17 Compressor map of radial compressor.

Radial Compressors:
The radial compressor operates on the centrifugal principle. The increase in pressure is created by the difference in circumferential speed between the inlet and outlet at the impeller. The kinetic energy thus admitted is converted into pressure in the diffuser. The compressor map shown in Fig. 11-17 is limited by the pump limit. To the left of the pump limit is an unstable compressor operation that starts by the breakaway of the flow at the inside of the compressor blades and results in extreme pressure fluctuations that under certain circumstances may destroy the compressor.

The speed lines drop slightly to the right of the pump limit; towards the displacement limit they drop more sharply. Depending on the counterpressure, this results in working points 1, 2, or 3 at constant compressor speed.

11.4.2 Mechanical Supercharging

Positive-Displacement Supercharger—Mechanically Linked to the Four-Stroke Engine (Fig. 11-18)

With a given transmission ratio, we obtain the operating curve 1-2-3-4 shown. By changing the transmission ratio, we can also create the operating curve 1′-2′-3′-4′ that leads to an increase in the mean working pressure.

Radial Compressor—Mechanically Linked to the Four-Stroke Engine

As shown in Fig. 11-19, air throughput and charge pressure increase at roughly the square of the rise in speed. This results in the mean pressure curve over speed shown in Fig. 11-20.

11.4.3 Exhaust Gas Turbocharging

During exhaust gas turbocharging, the engine and exhaust gas turbocharger are linked thermodynamically. The respective turbocharger speed is set depending on the power balance between compressor and turbine. If we consider the power balance at the turbocharger shaft, the change in the angular velocity can be calculated as

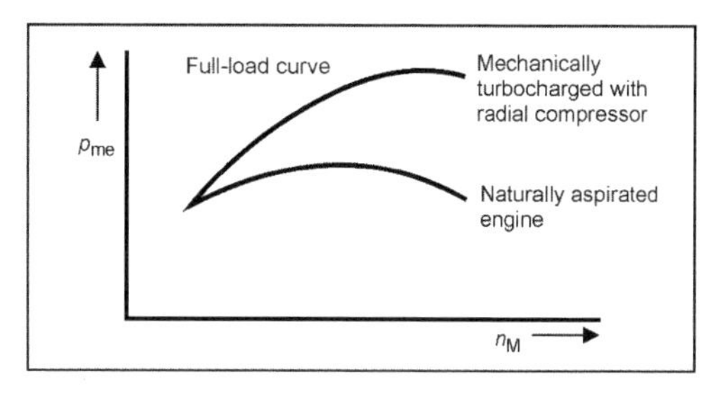

Fig. 11-20 Mean pressure curve over speed.

$$\frac{d\omega_{TL}}{dt} \cdot J_{TL} \cdot \omega_{TL} = P_V + P_T \tag{11.7}$$

$\frac{d\omega_{TL}}{dt}$ = Change in the heat propagation rate turbocharger
J_{TL} = Turbocharger polar inertia moment
ω_{TL} = Angular velocity turbocharger
P_V = Compressor power
P_T = Turbine power

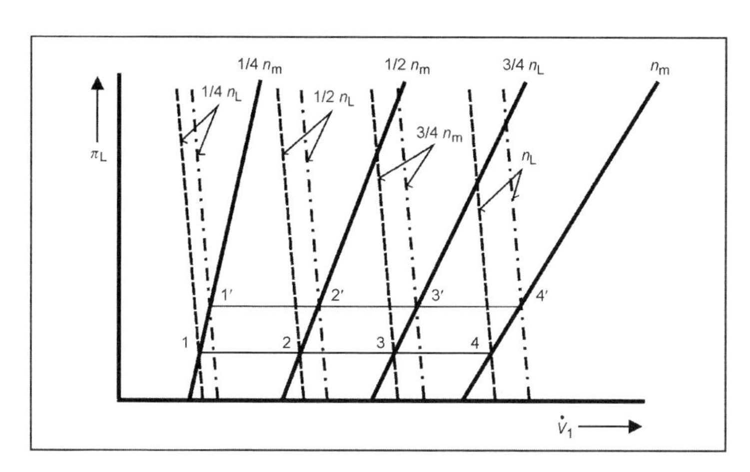

Fig. 11-18 Mechanical linking of positive-displacement turbocharger and four-stroke engine.

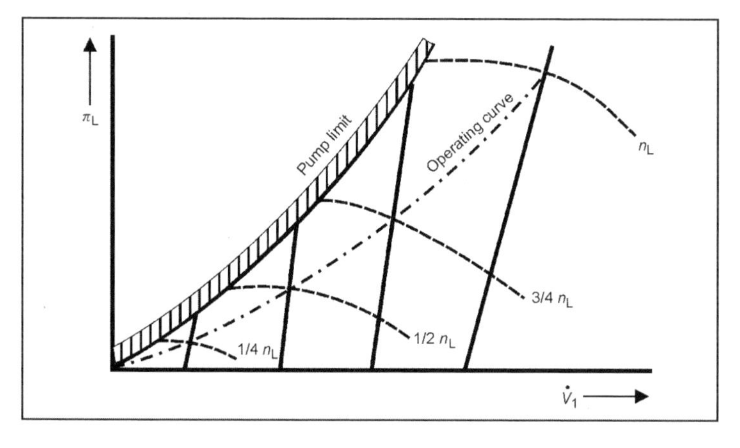

Fig. 11-19 Mechanical linking of radial compressor and four-stroke engine.

For the static state the left-hand side of the equation is 0

$$P_V + P_T = 0 \tag{11.8}$$

$$\dot{m}_V + \dot{m}_B = \dot{m}_T \tag{11.9}$$

$\dot{m}_T$ = Turbine mass flow
$\dot{m}_V$ = Compressor mass flow
$\dot{m}_B$ = Fuel mass flow

and the operating point lies on the engine displacement line. The power balance can thus be developed further.

$$P_V = \dot{m}_V \cdot \Delta h_{sV} \cdot \frac{1}{\eta_{sV} \cdot \eta_{mV}} \tag{11.10}$$

Δh_{sV} = Isentropic enthalpy gradient in compressor
η_{sV} = Isentropic compressor efficiency
η_{mV} = Mechanical compressor efficiency

$$P_T = \dot{m}_T \cdot \Delta h_{sT} \cdot \eta_{sT} \cdot \eta_{mT} \tag{11.11}$$

Δh_{sT} = Isentropic enthalpy gradient in turbine
η_{mT} = Mechanical turbine efficiency

$$\Delta h_{sV} = R_1 \cdot T_1 \cdot \frac{\kappa_1}{\kappa_1 - 1} \cdot \left[\left(\frac{p_2}{p_1}\right)^{\frac{\kappa-1}{\kappa_1}} - 1\right] \tag{11.12}$$

R_1 = Gas constant upline of compressor
T_1 = Temperature upline of compressor
κ_1 = Isentropic exponent upline of compressor
p_1 = Pressure upline of compressor
p_2 = Charge pressure

$$\Delta h_{sT} = R_3 \cdot T_3 \cdot \frac{\kappa_3}{\kappa_3 - 1} \cdot \left[1 - \left(\frac{p_4}{p_3}\right)^{\frac{\kappa_3-1}{\kappa_3}}\right] \tag{11.13}$$

Δh_{sT} = Isentropic enthalpy gradient in turbine
R_3 = Gas constant upline of turbine
T_3 = Temperature upline of turbine
κ_3 = Isentropic exponent of exhaust gas
p_3 = Exhaust gas counterpressure
p_4 = Pressure downline of turbine

The group efficiency η_{TL} is defined as the overall efficiency of the charge group:

$$\eta_{TL} = \eta_{mV} \cdot \eta_{sV} \cdot \eta_{mT} \cdot \eta_{sT} \tag{11.14}$$

η_{sT} = Isentropic turbine efficiency

Using Eqs. (11.9) to (11.13), the power balance solved for π_V is calculated as

$$p_V = p_2/p_1 \tag{11.15}$$

π_V = Compressor pressure ratio

and with

$\kappa_L = 1.4$

the turbocharger main equation is

$$p_V = \left[1 + \frac{\dot{m}_T}{\dot{m}_V} \cdot K_1 \cdot \frac{T_3}{T_1} \cdot \eta_{TL} \cdot \left(1 - \frac{p_4}{p_3}\right)^{\frac{\kappa_3-1}{\kappa_3}}\right]^{3,5} \tag{11.16}$$

K_1 = Constant [–]
η_{TL} = Group efficiency

If we assume $\dot{m}_T/\dot{m}_V \cong 1.03 - 1.07$, then the compressor pressure ratio is a function of the following factors:

$$p_V = p_V\left(\frac{T_3}{T_1}; \eta_{TL}; \frac{p_4}{p_3}\right) \tag{11.17}$$

The charge pressure p_2 thus increases with increasing exhaust gas temperature T_3 and increasing pressure in front of the turbine p_3 (where the change in group efficiency as a function of T_3 and p_3 has still been neglected).

The pressure p_3 is obtained with a given turbine as a function of the mass throughput and gas state and can be calculated for the ram induction as

$$\dot{m}_T = A_T \cdot \psi_T \sqrt{2 \cdot p_3 \cdot \rho_3} \tag{11.18}$$

$$\text{where } \psi_T = \sqrt{\frac{\kappa_3}{\kappa_3 - 1}} \cdot \sqrt{\left(\frac{p_4}{p_3}\right)^{\frac{2}{\kappa_3}} - \left(\frac{p_4}{p_3}\right)^{\frac{\kappa_3+1}{\kappa_3}}} \tag{11.19}$$

$A_{T\,\mathrm{red}}$ = Turbine equivalent cross section
ψ_T = Flow function
κ_3 = Isentropic exponent of the exhaust gas

If we consider the turbine as the throttle point (with p_3 upline and p_4 downline of the throttle point), we obtain the following relationship:

$$p_3 - p_4 = \frac{\rho_3}{2} \cdot v_3^2 \sim \frac{\dot{m}_T^2}{\rho_3^2} \cdot \frac{\rho_3}{A_T^2} \sim \frac{(n_M \cdot V_H \cdot \rho_2)^2}{\rho_3} \tag{11.20}$$

ρ_2 = Density downline of the turbocharger
ρ_3 = Density upline of the turbine
v_3 = Flow velocity, turbine
$A_{T\,\mathrm{red}}$ = Turbine equivalent cross section
n_M = Engine speed
V_H = Piston displacement

The mass throughput m_T through the turbine depends in a first approximation on the gas state at the intake organs (p_2, T_2), on the engine speed n_M (displacement line), and on the density ρ_3. The reduced turbine cross-sectional area $A_{T\,\mathrm{red}}$ has been assumed to be constant in this consideration. The following relationship thus exists:

$$\frac{p_3}{p_4} = \frac{p_3}{p_4}(p_2, T_2, n_M, T_3, A_T) \tag{11.21}$$

T_2 = Temperature downline of the compressor

Whereas in the case of an engine with a mechanically driven turbocharger and constant transmission ratio the charge pressure and, hence, the maximum torque are only a question of the engine speed, it is possible—as shown by Eq. (11.20)—to increase the exhaust gas counterpressure p_3 through a further reduction in the reduced turbine cross section $A_{T\,\mathrm{red}}$. As a result, the enthalpy gradient at the turbine increases. The turbocharger output and speed are increased, and, consequently, the charge pressure also increases.

Different operating points for the same $A_{T\,\mathrm{red}}$ fundamentally result in a different enthalpy gradient at the turbine and, thus, also a different charge pressure. This thermodynamic interaction of the engine and exhaust gas turbocharger is now discussed, taking three borderline cases as examples.

1. Generator Mode

In "generator mode," the speed n_M has to be kept as constant as possible in view of the high demands on the constant rotational frequency of the generator (Fig. 11-21).

For the engine with a mechanical turbocharger, we stay at one operating point, as n_M = const (Fig. 11-22).

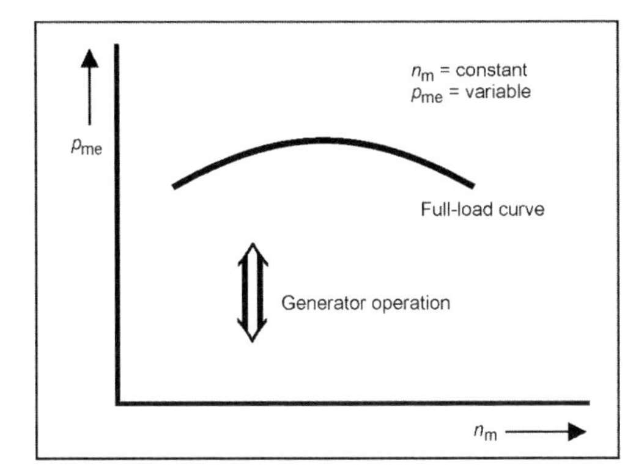

Fig. 11-21 Generator mode.

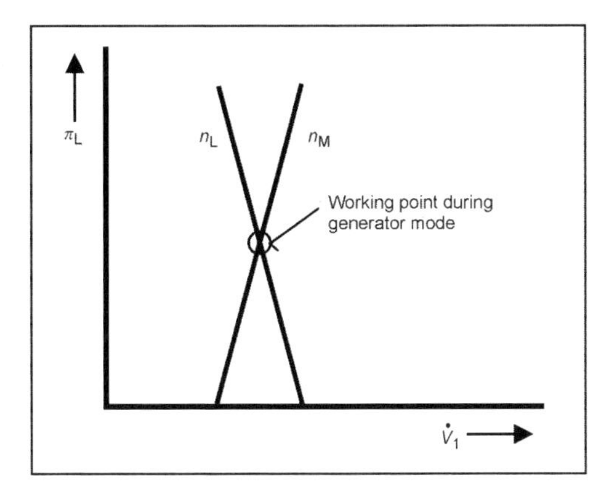

Fig. 11-22 Operating point in generator mode.

For the exhaust gas turbocharged engine, the change in load results in a different p_3 and T_3 and, hence, in a different turbine power and different charge pressure.

The operating points 1, 2, and 3 all lie on the engine displacement line that belongs to the generator speed (Fig. 11-23).

With an increase in load (increase in fuel injection), p_3, T_3, and, hence, the turbine power increase. The turbocharger speed increases, as do charge pressure p_2 and mass throughput.

2. Speed Reduction p_{me} = constant, n_M = variable

As illustrated in Fig. 11-24, the mean pressure moves along a horizontal line for different engine speeds. This results in a flatter operating line (a) in the compressor map

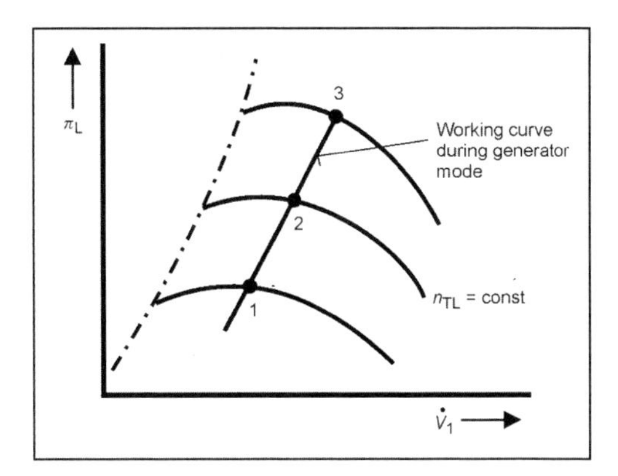

Fig. 11-23 Engine displacement line and generator mode.

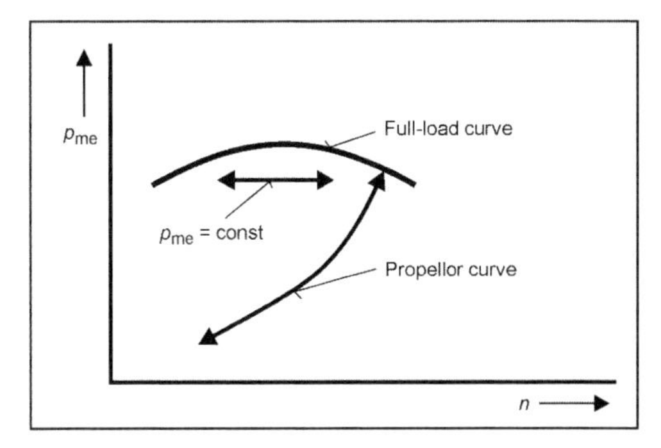

Fig. 11-24 Speed reduction.

(Fig. 11-25); i.e., with decreasing speed the operating point moves toward the pump limit (Danger!). This mode of speed reduction also occurs roughly in vehicle mode along the full-load line and makes the highest demands on exhaust gas turbocharging.

3. Propellor Mode n_M = variable, $p_{me} \sim n^2_M$

In ship drives with a fixed propeller, the propeller torque taken up depends on the square of the propeller speed. In

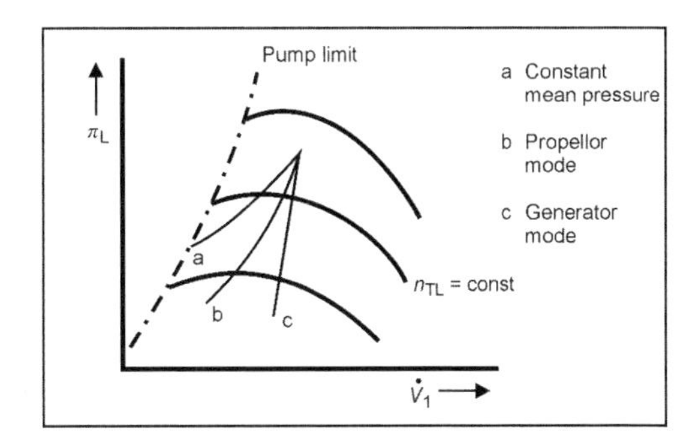

Fig. 11-25 Operating line between generator mode and speed reduction.

the compressor map, Fig. 11-25, the operating line lies between generator mode and speed reduction.

Figure 11-26 shows a superimposition of all lines of constant load and constant speed. In vehicle mode, the whole range is thus covered, which requires wider compressor maps. Figure 11-27 shows the mean pressure curve for the full-load line of naturally aspirated, mechanically turbocharged, and exhaust gas turbocharged engines. This latter line shows a highly unfavorable behavior as the torque also drops with decreasing speed. For good acceleration behavior in vehicle mode, however, a rise in the mean pressure curve is required with decreasing speed. This can be achieved by external controlling of the turbocharger.

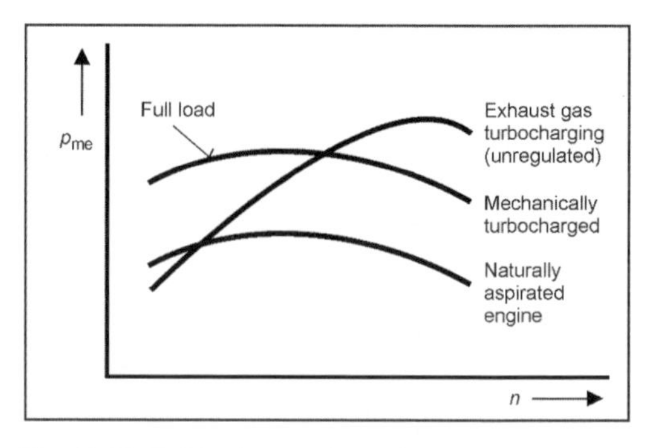

Fig. 11-27 Full-load curves for various engine variants.

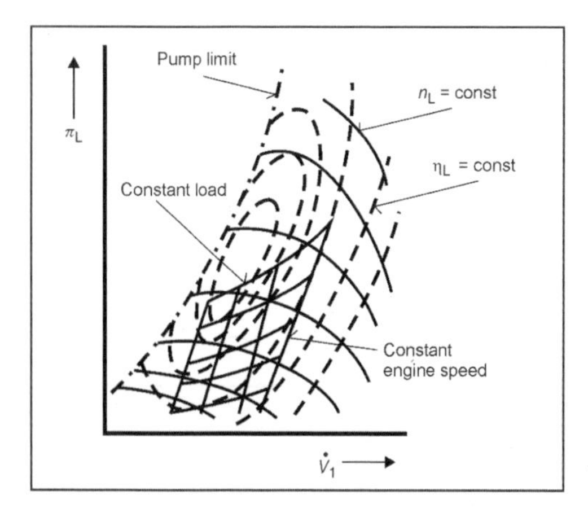

Fig. 11-26 Superimposition of maps.

Optimum Torque Curve by Adaptation of the Charge Pressure

In order to achieve a high charge pressure even at low engine speeds (SI < 2000 rpm, diesel car < 1800 rpm, truck diesel < 1100 rpm), a small turbine neck cross section $A_{T\,red}$ is chosen; this increases the pressure upline of the turbine. With increasing speed, however, the charge pressure increases also because of the increasing exhaust gas enthalpy stream so that the maximum pressure in the cylinder also rises. To limit the associated component load, the charge pressure is controlled to a constant value by allowing the excess exhaust gas enthalpy stream to bypass the turbine (waste gate) and, thus, to escape unused to the exhaust pipe (Fig. 11-28), representing a loss for the engine. The charge pressure curve along the full-load line and the effective mean pressure are shown in Fig. 11-29 for an Audi 2.7 l Biturbo engine.

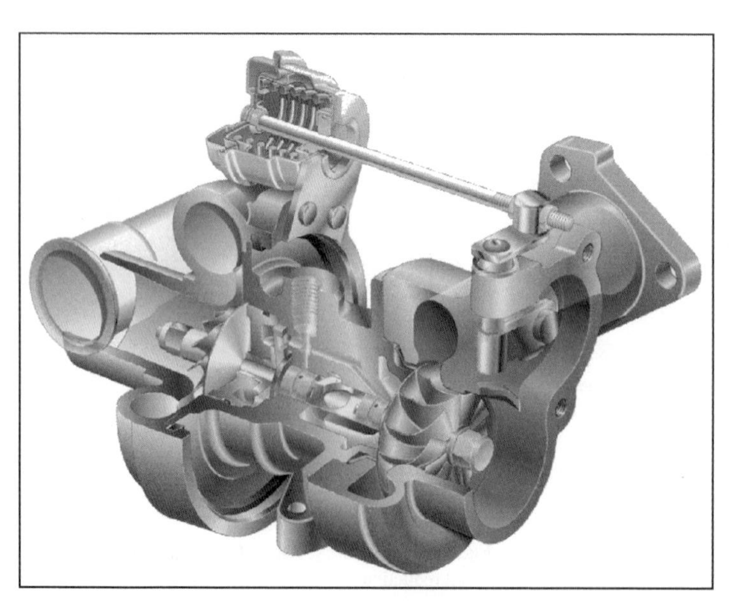

Fig. 11-28 Waste gate.[5] *(See color section.)*

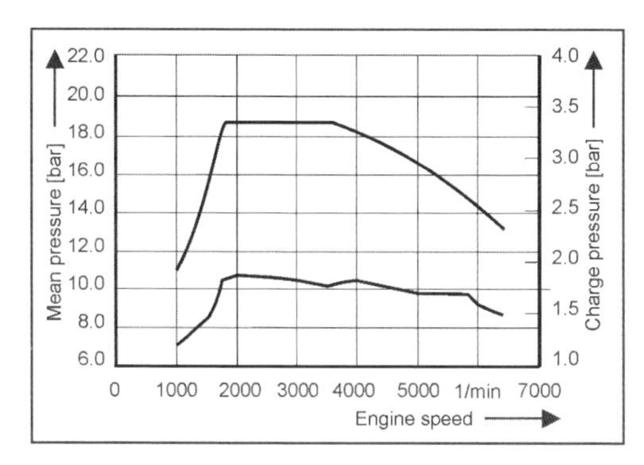

Fig. 11-29 Mean pressure and charge pressure curves of an AUDI V6 2.7 l Biturbo engine.[8]

Fig. 11-30 Variable turbine geometry, blade position open.[5]

With the variable turbine geometry (cf. Fig. 11-30), it is possible to set the reduced turbine cross section very small even at low speeds. This generates higher exhaust gas counterpressures, and a correspondingly higher charge pressure is achieved.

With higher speed and, thus, increasing mass throughput, the blades are turned in the direction of maximum contact cross section (blade position shown in Fig. 11-30).

A similar effect is achieved with the variable slide valve turbine (Fig. 11-31). A coaxial bush covers one channel of the two-stage turbine when moving towards the turbine wheel.

Figure 11-32 shows a turbocharged SI engine with two exhaust gas turbochargers and two intercoolers, while Fig. 11-33 shows a turbocharged medium-speed ship diesel engine, and Fig. 11-34 shows a corresponding exhaust gas turbocharger with axial turbine.

For the dual-stage turbocharging (Fig. 11-35), two exhaust gas turbochargers are connected in series, where the compressed air is aftercooled behind the first compressor and cooled again behind the high-pressure compressor. This dual-stage compression with aftercooling produces a good compression efficiency and with a compressor pressure ratio >5 also produces a correspondingly high mean pressure of up to 30 bar.

The high degree of integration of the turbocharger group can be seen in the boxed detail in Fig. 11-36.

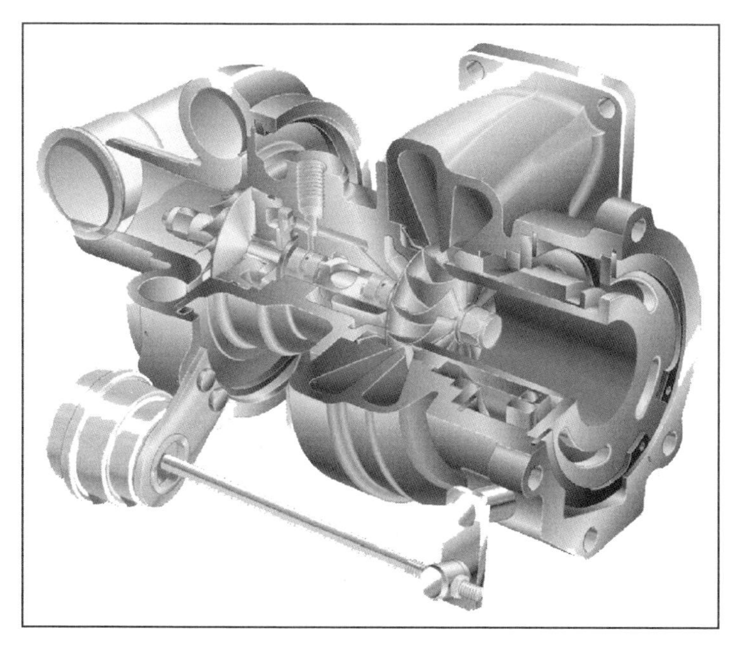

Fig. 11-31 Variable slide valve turbine.[5] *(See color section.)*

Fig. 11-32 AUDI RS4 engine.[9]

Fig. 11-33 Queen Elizabeth 2, diesel electric plant 9 × 9 L 58/64 95.5 MW.[12]

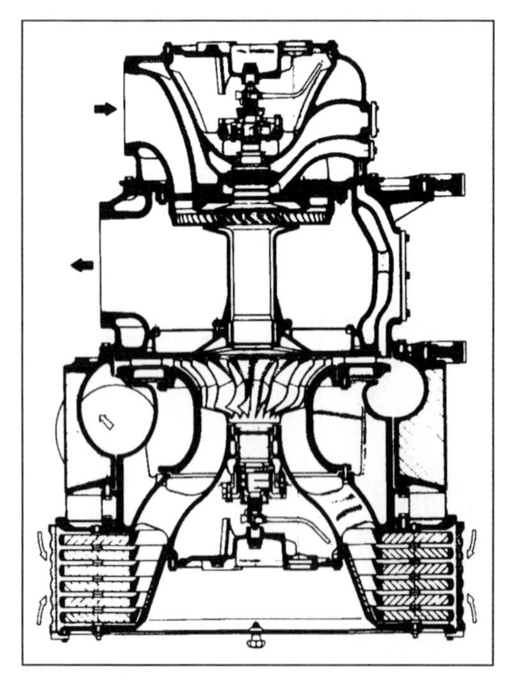

Fig. 11-34 MAN-exhaust gas turbocharger with axial turbine.[10]

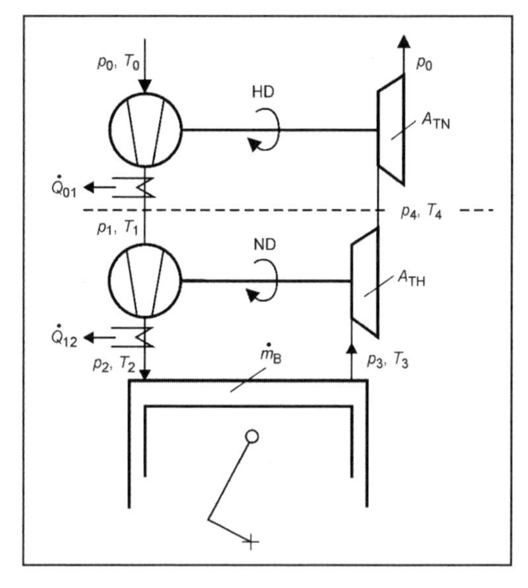

Fig. 11-35 Schematic of two-stage turbocharging.

11.5 Dynamic Behavior

The internal combustion engine forms part of a drive system from which a rapid response behavior is demanded.[11] This applies to all applications. Emergency power generators have to be able to assume the full power from a standstill within the shortest possible time (<15 s).

In vehicle operation, the internal combustion engine also has to spontaneously develop its maximum torque even under extreme load conditions (such as a car with a trailer starting in the mountains). Naturally aspirated engines control the torque more or less directly with the throttle plate angle (SI) or via the fuel injection volume (diesel engine).

If we calculate the twist equation for a roughly torsionally rigid drive system [Eq. (11.22)], then we see that with a given consumer torque M_V (= load) the effective engine torque M_{Me} and the polar inertia moment of the

Fig. 11-36 20 V 1163 TB 73 L, 6500 kW at 1250–1300 rpm.[12]

whole drive system $J_{ges\,A} = J_M + J_A$ significantly influence the gradient of the crankshaft angular velocity.

$$(J_M + J_A) \cdot \frac{d\omega_M}{dt} = M_{Me} + M_V \tag{11.22}$$

J_M = Polar mass moment of inertia of engine
J_A = Polar mass moment of inertia of drive
$\frac{d\omega_M}{dt}$ = Change in crankshaft angular velocity
M_{Me} = Effective engine torque
M_v = Consumer torque

Figure 11-37 shows an elasticity test for a vehicle with a turbocharged SI engine for acceleration from 60 to 100 km/h in 5th gear on the highly dynamic test rig.

It takes almost 3.5 s for the intake manifold pressure, and, hence, the mean pressure, to reach its static value.

Figure 11-38 shows further measurements for a load shift in a dynamic SI engine at constant engine speed (2000 rpm = const) on the highly dynamic engine test rig, where the mean pressure has been standardized to the static maximum value. The measured load signal rises rectangularly at 1 s to 100%. After a dead time, the naturally aspirated engine produces an equally spontaneous rise. The exhaust gas turbocharged SI engine rises with the same spontaneity up to approximately 55% of the achievable static mean pressure. The subsequent slow rise of 13%/s is attributable to the acceleration of the turbocharger internals. The engine reaches its maximum mean pressure after approximately 3 s. Before we proceed to discuss measures for improving the torque development in the exhaust gas turbocharged internal combustion engine, we see in Fig. 11-39 the acceleration behavior of a mechanically supercharged engine that achieves a significantly faster buildup of mean pressure compared with the exhaust gas turbocharger.

Improvement Measures:

Adjustment devices such as exhaust gas turbocharger with waste gate or variable turbine geometry enable the charge pressure to be built up significantly faster during an acceleration phase. In addition, the dynamic charge pressure buildup during nonstatic processes can be improved by using smaller impellers for turbine and compressor. The influence of the polar mass moment of inertia J_{TL} of the internals can be seen in the twist equation [Eq. (11.23)] for the exhaust gas turbocharger shaft.

In V-engines, for example, the dynamic behavior can be improved by grouping the cylinders on the exhaust gas side into a bank feeding two smaller turbines; on the air intake side, the two compressors are connected to a common intake pipe.

$$\frac{d\omega_{TL}}{d\varphi} = \frac{1}{\omega_{TL} \cdot J_{TL}} \cdot (P_T + P_v) \tag{11.23}$$

$\frac{d\omega_{TL}}{d\varphi}$ = Change in angular velocity ATL
ω_{TL} = Angular velocity ATL
J_{TL} = Polar mass inertia moment of turbocharger
P_T = Turbine output
P_V = Compressor output

Brief injection of additional air into the compressor means, on the one hand, that the internal combustion engine is adequately supplied with air immediately after a load demand and that the increased fuel injection volume

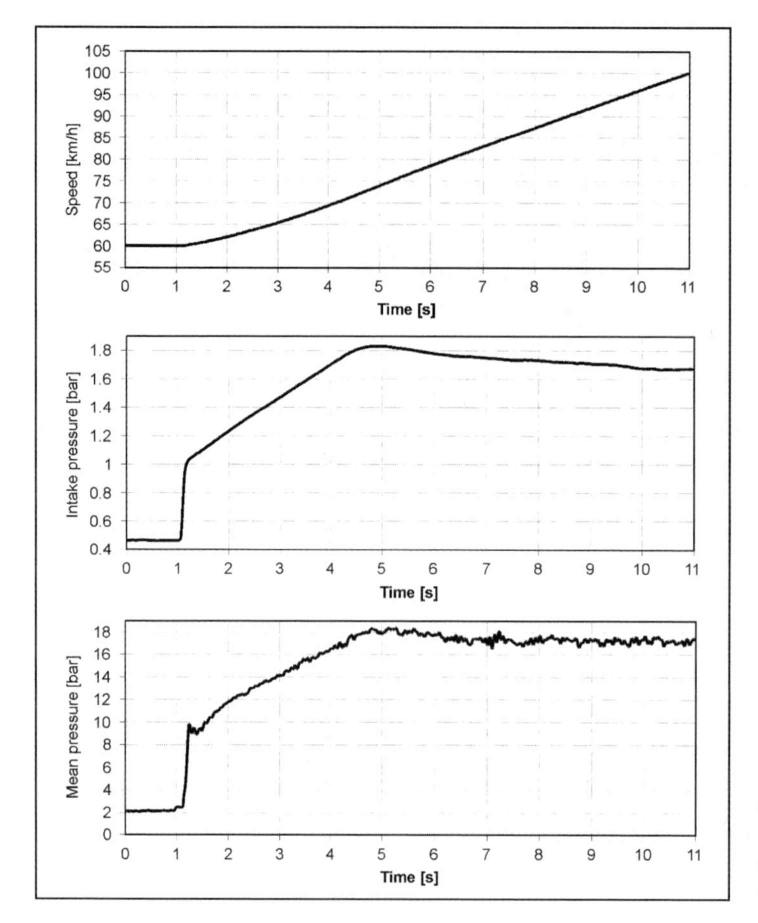

Fig. 11-37 Elasticity test (60–100 km/h in 5th gear) highly dynamic test rig and turbocharged SI engine.

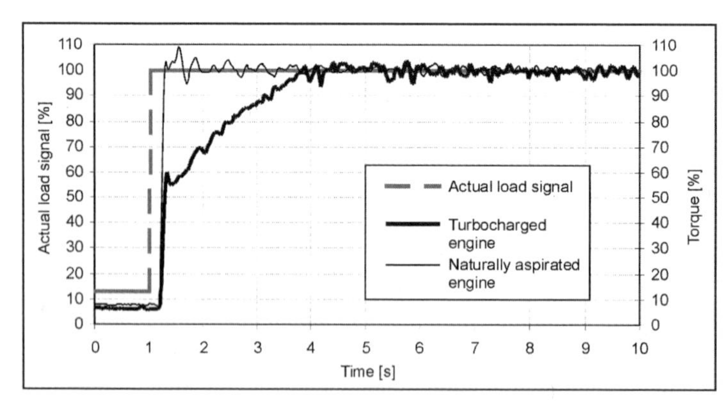

Fig. 11-38 Comparison of naturally aspirated and turbocharged SI engine; load shift at n = 2000 rpm = const.

corresponding to the limit air ratio provides a rapid increase in torque. On the other hand, the blow compressor wheel is accelerated so that the compressor delivers correspondingly more air with the increasing speed. The air injection is terminated when the turbine takes over the compressor work and the additional acceleration work required.

Electric Support for Exhaust Gas Turbocharging: Since the internal combustion engine does not spontaneously provide sufficient acceleration power for the turbocharger internals in response to a torque demand, it is expedient to use stored electrical energy to accelerate the turbocharger internals using an electric motor connected between compressor and turbine ("euATL") (Fig. 11-40).[13] The electric motor must also withstand the high turbocharger speeds when switched off and have sufficient torque for the acceleration of the internals (compressor and turbine wheel).

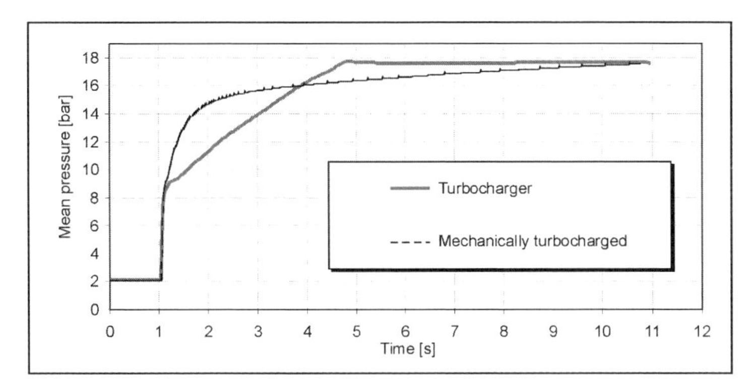

Fig. 11-39 Comparison of mechanical supercharging using a Roots blower and exhaust gas turbocharging on the vehicle acceleration process in the car SI engine (elasticity test).

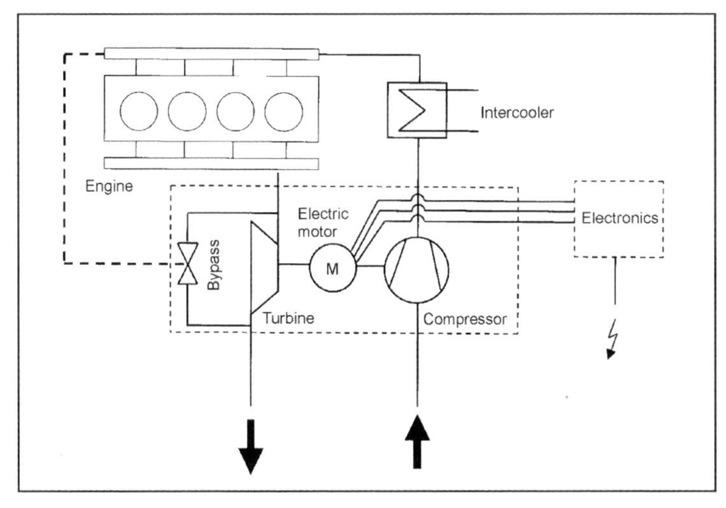

Fig. 11-40 Schematic diagram of the electrically supported exhaust gas turbocharger.

If an electrically driven compressor ("eBooster") is connected in series (Fig. 11-41) that briefly takes over the air supply to the internal combustion engine, the electric motor has to accelerate only the compressor wheel whose polar mass moment of inertia is only 1/3 of that of the turbine wheel. With an appropriate design of the eBooster compressor, the maximum speed is lower than with the euATL, hence, offering benefits for the design of the eBooster (Fig. 11-42). The wider compressor map of this two-stage controlled supercharging also offers the

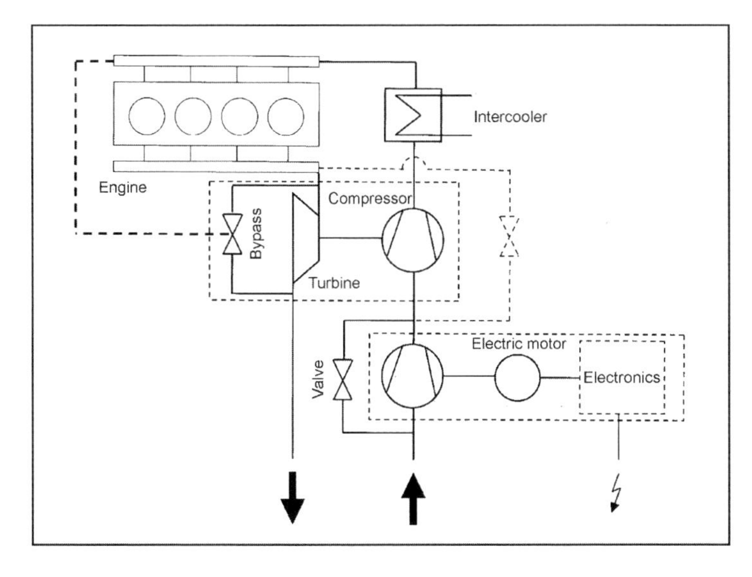

Fig. 11-41 Schematic diagram of the eBooster supercharging system.

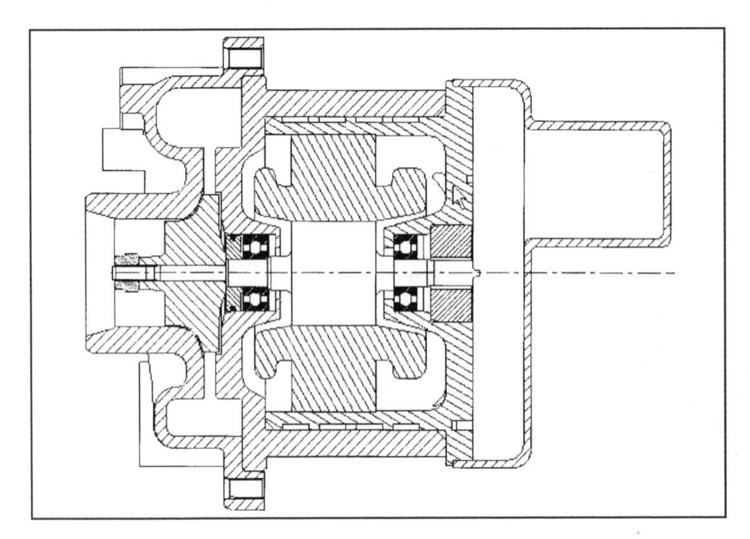

Fig. 11-42 Sectional view of the eBooster unit.

possibility of correspondingly raising the charge pressure and, hence, the torque of the internal combustion engine in the lower speed range, provided that sufficient electrical energy is available.

11.6 Additional Measures for Supercharged Internal Combustion Engines

11.6.1 SI Engines

With the turbocharged SI engines, the higher charge pressure results in higher ultimate compression temperatures. This increases the risk of autoignition and of knocking. For this reason, it can be necessary to lower the compression ratio. In any case, the start of ignition of the SI engine must be shifted towards "retard" in order to avoid impermissibly high ignition pressures and knocking combustion (Fig. 11-43).

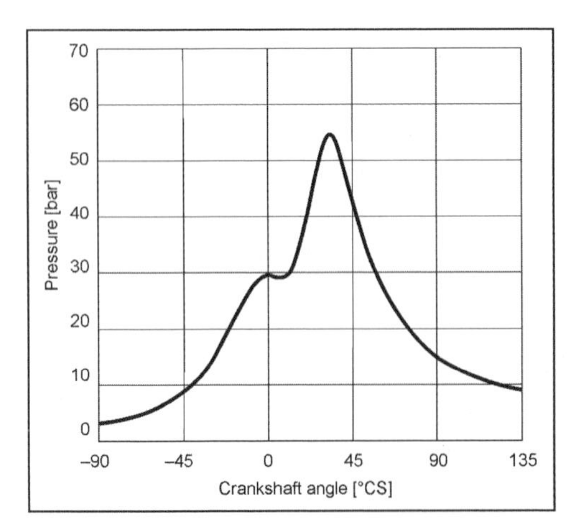

Fig. 11-43 Pressure curve of a turbocharged SI engine with retarded ignition angle.

High exhaust gas recirculation rates increase the risk of knocking, particularly if an unfavorable exhaust pipe design exists in front of the turbine inlet.

In part-load operation, the mass flow of turbocharged SI engines is throttled by the throttle plate positioned downline of the compressor. The open-air circulation plate creates a bypass around the compressor so that the mass flow not required by the engine (part load) is returned in front of the compressor. As a result, no pressure is built up behind the compressor. This is also used for exhaust gas turbocharged SI engines, but is more important for mechanically supercharged engines because of the displacement characteristic of the turbocharger.

The turbine materials used today are high temperature steels (NiCr steels). With an exhaust gas temperature of $T_3 > 950°C$, however, the strength drops sharply. Since the exhaust gas temperature of a turbocharged SI engine at full load can exceed 1000°C, the engine is enriched. This is performed in the engine controller using an additional control loop with exhaust gas temperature sensors. In the meantime, turbine materials are available that can withstand temperatures of up to 1050°C.

11.6.2 Diesel Engines

In diesel engines, the high charge pressure also results in very high ultimate compression pressures with compression ratios of $\varepsilon > 14$. Depending on the mechanical strength, the start of injection, therefore, has to be set very late in diesel engines so that under certain circumstances the compression pressure can be equal to or higher than the ignition pressure.

With medium-speed diesel engines, high charge pressures are used in conjunction with large valve overlaps (of up to 120° on the crankshaft) also to reduce the thermal load on the engine. The medium-speed engine is operated with high excess air factors ($\lambda \cong 2$).

With turbocharged diesel engines, the external exhaust gas recirculation demands additional measures in the form of a clocked control valve and software to control the charge pressure and the exhaust gas recirculation rate. Measures must also be taken to ensure that a negative scavenging gradient $(p_2 - p_3) < 0$ exists at all times.

Bibliography

[1] Zinner, K., Aufladung von Verbrennungsmotoren, Springer-Verlag, Berlin, Heidelberg, New York, 1980.

[2] EATON Corporation, Air Management Systems Division, Michigan, USA.

[3] LYSHOLM Technologies Schweden.

[4] SIG Schweiz-Industrie-Gesellschaft.

[5] 3K WARNER Turbosystems GmbH, Kirchheimbolanden.

[6] Naber, D., K.-H. Hoffmann, A. Peters, and H. Brüggemann, Die neuen Common-Rail-Dieselmotoren mit Direkteinspritzung in der modellgepflegten E-Klasse, in MTZ 60, Jahrgang (1999), Heft 9, S. 578–588.

[7] Bergmann, H., F. Scherer, and H. Ostenwald, Die Thermodynamik des neuen Nutzfahrzeugmotors OM 904 LA von Mercedes-Benz, in MTZ 57, Jahrgang (1996), Heft 4, S. 216–224.

[8] Technische Universität Dresden, 6, Aufladetechnische Konferenz, Dresden, 1997.

[9] Technische Universität Dresden, 7, Aufladetechnische Konferenz, Dresden, 2000.

[10] MAN B & W Diesel AG, Augsburg.

[11] Zellbeck, H., and J. Friedrich, Simulation des Beschleunigungsverhaltens von Pkw-Ottomotoren mit neuen Aufladeverfahren, 20, Internationales Wiener Motorensymposium, 6–7 Mai 1999.

[12] MTU Motoren- und Turbinen-Union Friedrichshafen GmbH.

[13] Hoecker, P., J.W. Jaisle, and S. Münz, Der eBooster – Schlüsselkomponente eines neuen Aufladesystems von BorgWarner Turbo Systems für Personenkraftwagen, 22, Internationales Wiener Motorensymposium, 26–27 April 2001.

12 Mixture Formation and Related Systems

In combustion, from a chemical perspective, the oxidation of fuel molecules requires that the oxidator (oxygen) have sufficient access to the fuel molecule. It is, therefore, necessary to prepare the fuel, i.e., transform it into a gaseous phase and mix it with air. This is normally done using mixture formation systems. A distinction is drawn between internal and external mixture formation during engine operation.

12.1 Internal Mixture Formation

Internal mixture formation occurs in the cylinder of an internal combustion engine. The air is inducted through the piston and compressed, and then the fuel is injected into the compressed air at a suitable time. The air-fuel mixture becomes an ignitable composition within certain ranges that leads to the ignition of the mixture at a corresponding temperature. Pronounced inhomogeneities arise with this type of mixture formation, and local air-fuel ratios of $\lambda = 0$ (pure fuel) to $\lambda = \infty$ (pure air) arise. A diffusion flame causes combustion. The utilized fuel must meet certain ignition quality criteria. The reaction occurs with prepared droplets, i.e., droplets surrounded by an ignitable mixture. Diesel engines have provided a typical example of internal mixture formation. Recently, there has been increasing development of spark-ignition engines that also use internal mixture formation, so-called spark-ignition engines with direct injection. The basic difference from the diesel engine is the use of gasoline and an external ignition source. In the future, we can expect that spark-ignition engines with internal mixture formation will represent a large share of all engines produced since their potential for reducing fuel consumption appears greater than with diesel engines with direct injection.

Whereas conventionally operating diesel engines use internal mixture formation that produces an inhomogeneous distribution of air and fuel within the cylinder, in the future, additional advantages may be achieved with homogeneous diesel combustion that allows the reduction of emissions and fuel consumption.

12.2 External Mixture Formation

External mixture formation is characteristic in conventional spark-ignition engines. The air and fuel are mixed before they enter the cylinder of the engine. A more-or-less homogeneous mixture consisting of air and fuel vapor is generated. This used to be found predominantly in engines that had a carburetor or single-point injection as the mixture forming mechanism. There was enough time available to mix air and fuel and transport this mixture to the intake valve. The danger of these mixture formers is that the fuel present in a vapor phase would condense on cold intake manifold walls, and the mixture would be unevenly distributed to the individual cylinders. The type of intake manifold injection used today eliminates these disadvantages; the fuel is injected directly before the intake valve and partially toward the open intake valve and into the cylinder. In this case as well, there is sufficient time to homogenize the mixture over the intake and compression phases.

12.3 Mixture Formation using Carburetors

With a few exceptions, mixture formation using a carburetor is no longer the preferred approach in passenger car engines. Large numbers of carburetors are used only for varieties in certain countries and for two-wheeled vehicle drives. In this section, we therefore cover only the basic details of mixture formation in relation to carburetors.

The task of a carburetor is to offer the required amount of fuel to the inducted air for the desired mixture ratio depending on the operating state of the engine. The throttle valve that governs the air and mixture flow is integrated into the carburetor.

The energy required for metering the fuel and conveying it within the carburetor is taken from the air stream.

The carburetor and the connected intake manifold with its branches to the individual cylinders that distribute the mixture generated by the carburetor are to be viewed as a functional unit. The operating behavior of the engine is greatly influenced by the precision with which the intake manifold was engineered to produce an even mixture distribution under all operating conditions.

12.3.1 Mode of Operation of the Carburetor

The functional principle of the carburetor is based on the fact that by reducing a cross section in an air-conducting channel, there is less pressure than in the larger cross section or than in the atmosphere due to the greater flow rate in the narrow cross section.

This pressure differential is used to supply fuel to the air through suitable cross sections (Fig. 12-1).

A characteristic feature of carburetors is the generation of a differential pressure signal from an air stream, and its direct conversion into a fuel stream. In principle, the air side and fuel side are designed identically and can be described with the Bernoulli equation for fluid mechanics.

Given a simplified assumption of an incompressible flow, the following equation results for the air mass flow:

$$\dot{m}_L = A_L \cdot \alpha_L \cdot \varepsilon \cdot \sqrt{2 \cdot \Delta p_L \cdot \rho_L} \tag{12.1}$$

A_L = Cross section of the air funnel,
α_L = Flow factor,

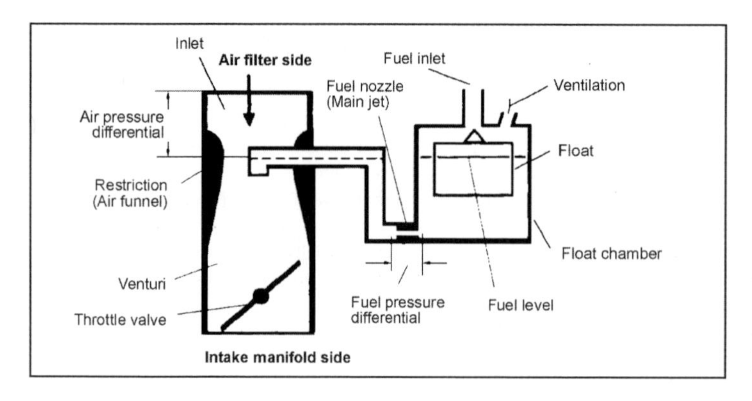

Fig. 12-1 Functional principle of the carburetor.

ε = Factor for air compressibility,
Δp_L = Pressure differential between the air funnel and the environment,
ρ_L = Air density in the air funnel

For the fuel mass flow, the following holds true:

$$\dot{m}_{Kr} = A_{Kr} \cdot \alpha_{Kr} \cdot \sqrt{2 \cdot \Delta p_{Kr} \cdot \rho_{Kr}} \qquad (12.2)$$

A_{Kr} = Cross section of the fuel nozzle,
α_{Kr} = Flow factor of the nozzle,
Δp_{Kr} = Pressure differential at the nozzle,
ρ_{Kr} = Fuel density

A carburetor has a fuel accumulator (float chamber) with a free fuel surface whose level is kept constant. A distinction is made between the following:

Constant air funnel cross section (fixed air funnel carburetor) (Fig. 12-1): Most carburetors are constructed according to this principle. In the intake air duct, there is a venturi-shaped air funnel with a fixed cross section to which a main nozzle is assigned. In small air streams, the pressure differential generated with the air funnel remains low. The pressure differential between the inlet and the intake manifold must, therefore, also be used to meter the fuel.

Carburetors with a constant air funnel cross section require several nozzle systems and an accelerator pump for an appropriate fuel supply in the engine map. To compensate for the influence of the different Reynold's numbers in the fuel and airflow, compensating air is mixed with the fuel.

Variable air funnel cross section: The intake air duct cross section is normally changed with a movable element. The following items are conventionally used:

- An air valve
- A piston that penetrates the channel
- A swiveling lever that constricts the channel

This allows a wide range of air streams to be controlled using a differential pressure that changes only slightly. For symmetry, a conical nozzle needle that extends into a needle jet is connected to a movable element to meter the fuel.

If the movable element also works while the engine is idling, the fuel can be dosed with the needle jet for the entire range of the air streams in an idling engine warm from operation. This is termed a constant vacuum carburetor.

When the movable element does not work while the engine is idling and rests on a stop, this is termed a "constant pressure stage." Constant pressure stages are frequently used as a second stage in multistage carburetors.

12.3.2 Designs

The designs can be categorized according to the number of intake air ducts and the spatial position.

12.3.2.1 Number of Intake Air Ducts

Single-barrel carburetor: This type of carburetor has an intake air duct with a throttle valve and is the most frequent design. Figure 12-2 shows an example of a downdraft carburetor on which several sizes and expansions of the carburetor are based for "beetle" engines. This is a fixed air funnel carburetor.

The system consists of a float chamber with a float and float needle valve, and internal float chamber ventilation in the inlet. It has a main system with an air funnel, a discharge arm, a compensating air nozzle with a venturi tube, and a main nozzle, as well as a dependent idling system with an idling nozzle, idling air nozzle, idle mixture-adjusting screw, and transition holes.

A throttle valve stop screw serves to set the opening of the throttle valve and, hence, control the air and mixture flow for idling. In a development, an idling fuel shutoff valve was added. The accelerator pump with a diaphragm, as well as an intake and pressure valve is actuated via a linkage together with the throttle valve. The orifice of the injection tube is calibrated. A pressure-tapping hole for the spark advance ends at the narrowest cross section of the air funnel. The single-barrel carburetor can be equipped with an automatic choke.

Dual carburetor: A dual carburetor is the combination of two single-barrel carburetors in one housing. Each of the intake air ducts is assigned an equivalent set of systems. Usually, there is only one float chamber and one

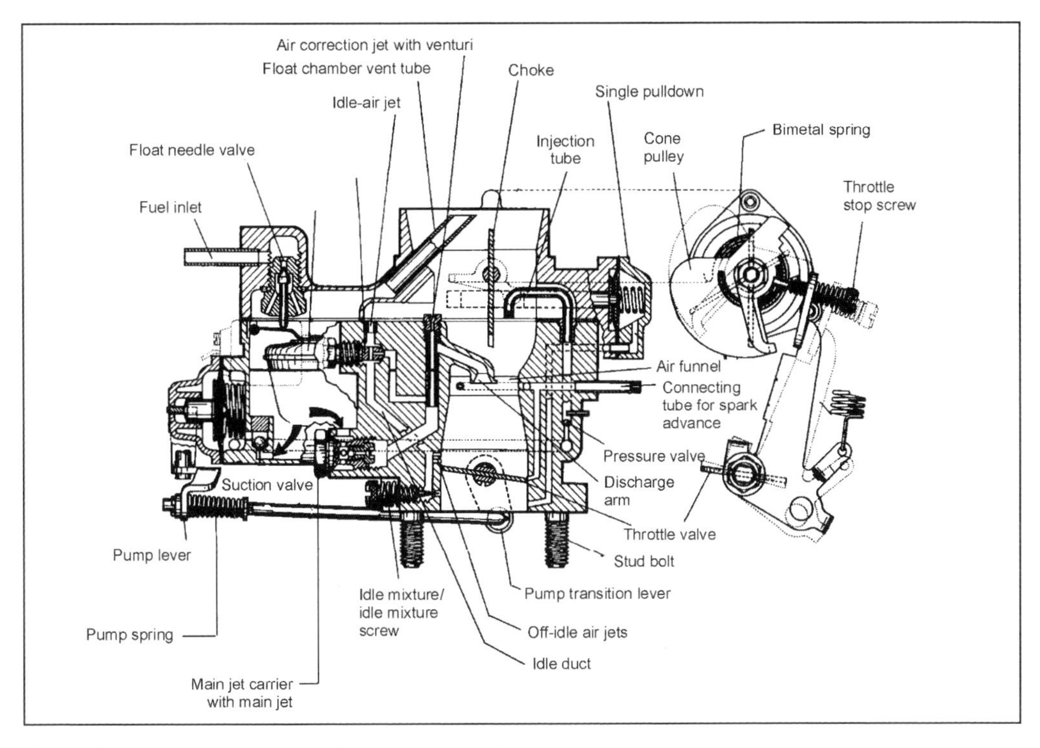

Fig. 12-2 Downdraft carburetor as a fixed air funnel carburetor.

accelerator pump. There is only one set of the required control organs for the starting control, for example.

The dual carburetor has two parallel intake air ducts in a common carburetor housing, each with a throttle valve that supplies the two separate intake manifolds. The throttle valves are activated simultaneously and can be either on a common shaft or on two parallel shafts. The same holds true for the chokes.

Triple-barrel carburetor: This is a combination of three fixed air funnels and has three inline parallel intake air ducts, each with a throttle valve in a common carburetor housing. The classic application was two triple-barrel carburetors on a six-cylinder boxer engine.

Two-stage carburetor: Over a long section, the air stream is divided into two stages. The first is used for the smaller air throughputs including idling and partial load, and the second stage, frequently with a larger cross section, is opened only to attain maximum output. The two-stage carburetor contains two parallel intake air ducts in a common housing that are both connected to an intake manifold. The two throttle valves in the carburetor are opened sequentially.

The first stage of a two-stage carburetor has all the additional necessary systems. The accelerator pump and starting control are contrastingly required only for the first stage. This is generally directly opened with the gas pedal; the second stage is linked to the opening of the first stage in the carburetor. Similar to single-barrel carburetors, a two-stage carburetor can be combined to form a double-barrel two-stage carburetor. Usually, double-barrel two-stage carburetors are supplied by means of a float chamber. Such carburetors are used in large volume six- and eight-cylinder engines.

12.3.2.2 Position of the Intake Air Duct

The following different air duct positions exist:

Downdraft carburetor: The intake air duct is at a right angle; the throttle valve is in the bottom part of the carburetor. The air flows from top to bottom.

Horizontal draft carburetor: The intake air duct lies horizontally in a flat stream or horizontal draft carburetor.

Semidowndraft carburetor: The intake air duct lies at an angle; the air stream can be directed upward or downward.

Updraft carburetor: In this case, the air stream rises from bottom to top. The throttle valve is in the top of the carburetor.

12.3.2.3 Designs for Special Applications

Pressurized carburetor: A pressurized carburetor is used on the pressure side of the supercharger of a supercharged engine sealed to the outside.

Carburetor for two-stroke engines: Given the limited intake of the generally used crankcase scavenging and the strong pulsations in the intake air duct, the cross section of the airflow is bigger than in carburetors for four-stroke engines.

12.3.3 Important Auxiliary Systems on Carburetors

Additional systems to the basic carburetor allow the carburetor to be used within the entire operating range of the engine with minimized emissions and fuel consumption as well as improved drivability. The following briefly lists the most important systems. It merely represents a selection; no claim is made for completeness.

(a) Acceleration enrichment. The preparation of the mixture that begins in the carburetor primarily occurs in the intake manifold. The boiling point of the fuel and the heating of the intake air and the intake manifold must be harmonized with each other to achieve the most homogeneous mixture in hot running engines. Yet even in hot running engines, high-boiling components of the fuel are still liquid in the intake manifold as film on the wall that is entrained by the air. The air enters the cylinder faster than the fuel. When the throttle valve is opened, the mixture becomes lean so that it must be temporarily enriched.

This is done by acceleration enrichment using an accelerator pump. Accelerator pumps are volume-displacing pumps with an intake and pressure valve. In the intake stroke, fuel streams from the float chamber through the intake valve into the pump interior. Upon the delivery stroke, fuel is generally delivered by the pressure valve to an injection device. This is calibrated and leads into the intake air duct. The delivery stroke is executed by a pump spring that is pretensioned in various ways. A distinction is drawn between

- Mechanically actuated plunger pump (Fig. 12-3)
- Mechanically actuated diaphragm pump
- Pneumatically actuated accelerator pump

(b) Actuating and constructing a second carburetor stage. In multistage carburetors, there are several systems for actuating the second stage with similar constructions. It is important for the second stage to always be reliably closed with the first stage.

A distinction is drawn between mechanically and pneumatically actuated systems.

Mechanically actuated second stage: In mechanically actuated systems, the second carburetor stage is primarily actuated by a trailing linkage. The opening is frequently restricted to prevent a temporary lean mixture.

The second stage frequently opens only when the first stage is already half open, and a catch of the throttle control lever is correspondingly positioned. Both valves reach the full opening position simultaneously. The design of the second stage largely corresponds to that of the first stage: There is a main system and transition system similar to the idling system in the first stage with a large fuel reserve. The transition system should ensure a smooth torque increase when the second stage opens quickly at low rpm with a correspondingly high pressure in the intake manifold.

For carburetors whose second stage is a constant pressure stage, the throttle valves on a common shaft actuate the second stage with a trailing linkage.

Pneumatically actuated second stage: In this design, both stages are constructed as fixed air funnel carburetors. A diaphragm unit actuates the second stage with a connecting rod articulated to the throttle control lever of the second stage. A corresponding design ensures that the second stage can open only when the first is almost completely open, and the second is closed with the first stage.

(c) Nozzles. These meter fuel, compensating air, and the premixture. To form an optimum flow and protect the actual calibrating section, they have an inlet cone and sometimes an outlet cone. Figure 12-4 shows a typical main nozzle. They can be used in both flow directions.

(d) Float chamber. A float chamber serves to control the fuel level in the carburetor, functions as a fuel accumulator, and contains a guided float that immerses in the fuel. The float actuates a needle, thus closing the body of the float needle valve that blocks the inflow of fuel when a set level has been attained.

There are carburetor designs that do not control the fuel level with a float chamber. The level in the fuel accumulator is controlled by means of the pressure in the fuel accumulator (diaphragm carburetor).

In order to work, the float chambers must be ventilated. A distinction is drawn between external and internal ventilation. With external ventilation, the gas area of the float chamber is directly connected to the environment surrounding the carburetor. This prevents problems when the engine is running hot that could arise from the fuel evaporating out of the float chamber.

With internal ventilation, a pipe leads into the inlet or pure air side of the air filter.

(e) Starting controls. In carburetor engines, special attention must be given to controlling the engine from a cold start through warm-up to a "hot" engine. In particular, the legal limits for exhaust emissions, problems associated with mixture distribution, fuel deposits on the wall, costs, and operating and driving comfort are requirements that must be taken into account.

The requirements, in particular, the enrichment of the mixture, result from

- Increased friction from unattained operating temperature
- Insufficient mixture preparation
- Increased power demand for auxiliary systems

In a stoichiometric mixture of air and conventional fuel (boiling curve), the dew point under environmental pressure is approximately 35°C. The homogenization of the mixture can generally happen only in the cylinder when the dew point temperature of the mixture in the

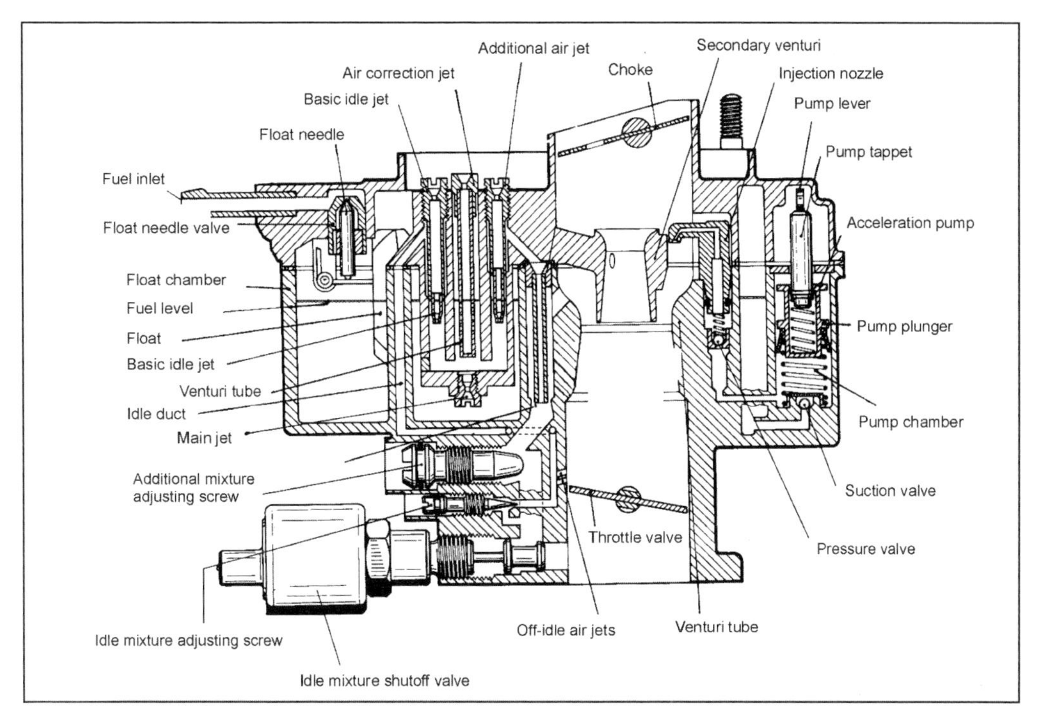

Fig. 12-3 Carburetor with mechanically actuated accelerator pump.

cylinder is exceeded during the compression stroke. Consequently, there is always a large amount of liquid fuel in the intake system in cold engines. The mixture must, therefore, be enriched in comparison to a warm operating state even when the engine is idling to compensate for deficient homogenization. In addition, the fuel must be enriched during acceleration. The amount of enrichment must be greater as the air and intake manifold wall temperature decreases. The main reference variable for enrichment is the temperature of the intake manifold.

The mixture formation is adapted to the following phases:

- Initial poststart phase. The engine starts as soon as there is an ignitable mixture. This process occurs faster at high temperatures than at low temperatures. In addition, a stationary mixture enrichment is required depending on the initial temperature and the time after the start.

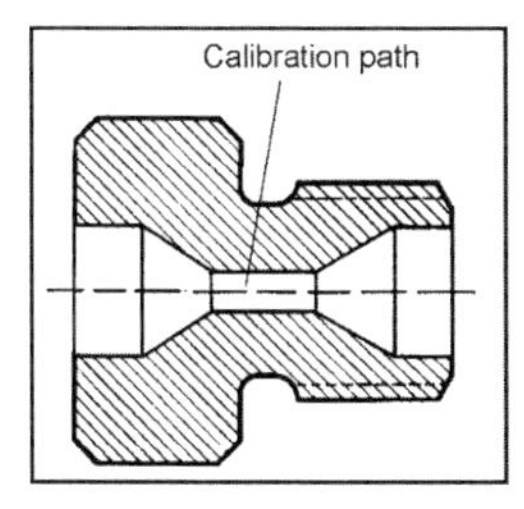

Fig. 12-4 Nozzle with a calibration section.

- Run-up to a stable idling speed. An initial, very strong enrichment must be attenuated for the mixture to remain flammable.
- Warm-up to operating temperature. The mixture stream and the enriching can be adjusted depending on the temperature of the intake manifold corresponding to the engine heat.

There are three basic systems for controlling starting:

- The manual starter, which is not discussed here since it is defunct.
- Automatic choke (Fig. 12-5). The mixture stream for starting and idling the cold engine is ensured by opening the throttle valve; the mixture is enriched with the choke. The temperature-dependent control element is a bimetal spring against which a choke can be drawn by the air stream. The functional relationship between the position of the choke and the position of the throttle valve is established by a cone pulley. In contrast to a fully automatic start, the functional process before the start must be triggered by a single depression of the gas pedal. The bimetal spring is heated to raise its temperature to that of the engine. As the temperature of the bimetal spring rises, the choke opens, and mixture enrichment is attenuated.
- Fully automatic start. The essential difference from an automatic choke is that this system is not triggered before starting. The control of the fuel mixture stream for idling a cold engine is separate from the mixture enrichment, depending on the inducted air stream.

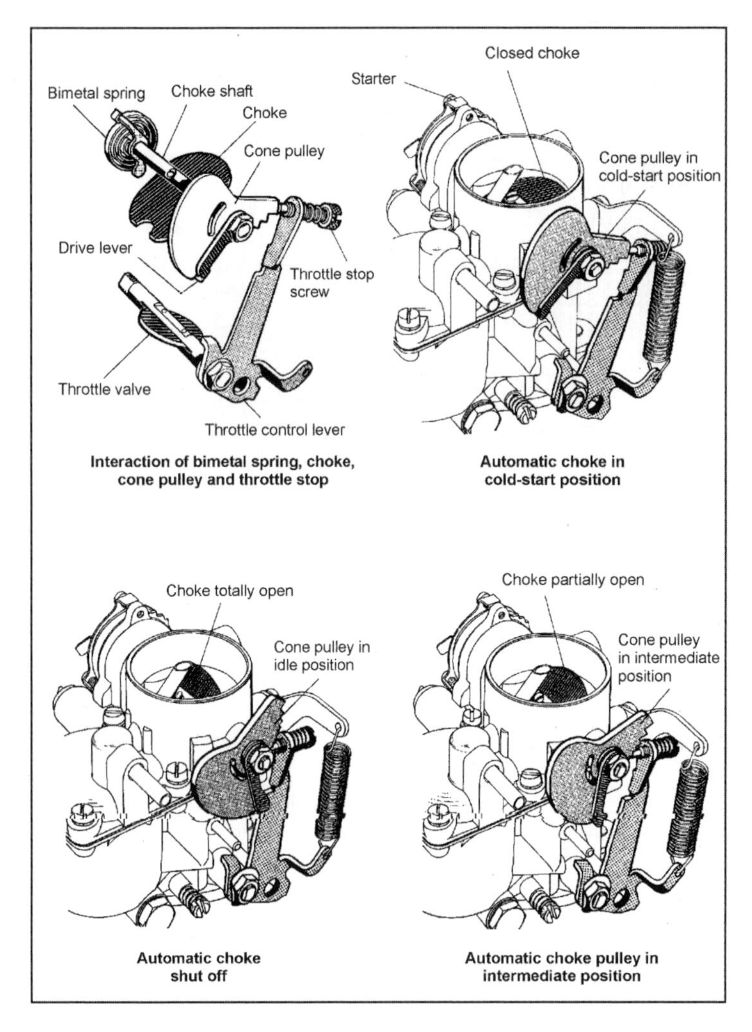

Fig. 12-5 Automatic chokes.

There are whole series of devices and additional functions for carburetors that improve their operating behavior such as the throttle valve actuator, pressure tapping holes, systems for compensating air, idling transition, additional mixture, and circulating air, to name only a few, which, however, are not discussed here.

12.3.4 Electronically Controlled Carburetors

Electronically controlled carburetors were developed to improve the adaptation of the mixture to all engine operating ranges and, hence, meet the demands for minimizing untreated emissions and lowering fuel consumption. A lambda closed-loop control was later adapted to the system. The mechanical design of the electronically controlled carburetor is basically identical to conventional carburetors. Additional features are the following: the throttle valve is actuated in the near-idling range, the mixture enrichment is influenced, and there are additional sensors and an electronic control unit (Fig. 12-6).

The electronically controlled carburetor is basically a fixed air funnel carburetor with a two-stage design with a pneumatically actuated second stage. The throttle valve for the first stage is actuated in the range close to idling with a continuously variable, position-regulated throttle valve actuator.

The continuously variable mixture enrichment over the entire range is carried out by means of the choke in the first stage that is actuated to restrict the air flow to enrich the mixture.

On the throttle valve shaft of the first stage is a throttle valve potentiometer that determines the throttle valve position or the change of the throttle valve position. Signals from the sensors consisting of the temperature sensor for the coolant and intake manifold, the idle switch on the throttle valve actuator and a speed tap are input variables for the electronic control that moves the throttle valve of the first stage as well as the choke. The following functions exist:

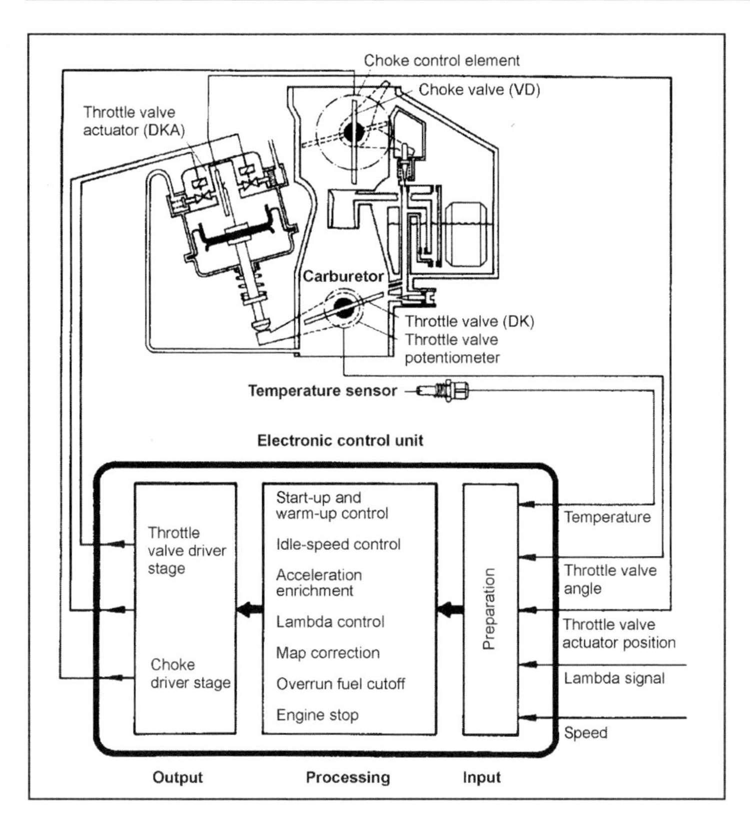

Fig. 12-6 Block diagram of an electronically controlled carburetor.

- Control of engine start and warm-up
- Acceleration enrichment
- Lambda closed-loop control
- Influence of the air-fuel ratio in the program map
- Idling speed control
- Overrun fuel cutoff
- Catalytic-converter protection function by shutting off the fuel

When starting with an electronic carburetor, the throttle valve of the first stage is actuated, and the choke is fully closed. During the run-up phase, the idling speed control takes over the control of the throttle valve with a set point depending on the coolant temperature of the engine. The mixture enrichment is controlled with the choke, and a basic enrichment is dictated as a function of the intake manifold temperature in the program map while stationary. This is added to acceleration enrichment when the load of the engine increases.

12.3.5 Constant Vacuum Carburetor

The functional principle of the constant vacuum carburetors is illustrated in Fig. 12-7 with the example of a Zenith-Stromberg CD carburetor. The system works with a variable air cross section in the form of a horizontal draft carburetor. The change in cross section in the intake air duct is caused by a plunger inserted from above. In addition to pressure from its weight, a spring presses the plunger against a bridge, filling the bottom section of the intake air duct. Above the plunger, an area is sealed with a diaphragm that abuts the bottom of the plunger and accordingly compensates the pressure. The bottom of the diaphragm is ventilated toward the inlet. Because of the spring, a linear rise in the pressure differential is required to lift the plunger.

Because of this pressure differential, the inducted air is accelerated between the inlet and the narrowest place between the plunger and the bridge. The position of the plunger is, hence, a measure of the inducted air stream.

The float chamber is below the bridge. Affixed in the center of the piston is a nozzle needle that extends into a needle jet and meters the fuel as a function of the plunger lift.

With the needle jet, the fuel can be measured on a characteristic curve depending on the air stream. A constant-vacuum carburetor is adapted so that the plunger is completely opened at a full load and approximately half maximum speed. With larger air streams, the constant-vacuum carburetor then operates like a fixed air funnel carburetor. By exploiting the pressure oscillations in the intake system, the mixture ratio of the air and fuel can be set over the entire engine operating range.

12.3.6 Operating Behavior

Hot operation: During engine operation, the carburetor is cooled by the inducted air and the initial evaporation of

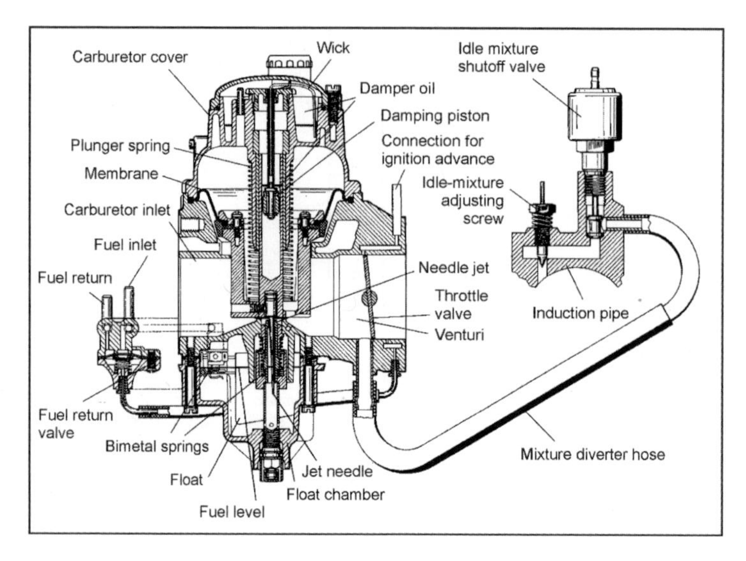

Fig. 12-7 Constant-vacuum carburetor.

the fuel. High environmental temperatures and subsequent heat after the engine is turned off can lead to operational problems.

The main problems are high component temperatures in the carburetor that lead to fuel evaporation, an engine with cylinder heads in a uniflow arrangement in which the intake manifold and exhaust manifold are stacked, and the boiling behavior of the fuel.

Improvements in relation to hot operation are possible with

- A hot idling valve to open a bypass for air to circumvent the carburetor. This is primarily used while the engine is idling hot to counter mixture enrichment.
- Fuel return. A valve connected parallel to the float needle valve returns unneeded fuel to the tank. This removes heat from the carburetor.
- Reduction of heat, e.g., by shields and an insulating flange.

Altitude compensation: Generally speaking, the mixture is enriched in carburetors proportional to the root of the environmental pressure. For example, the enrichment at 1600 m above sea level is approximately 10%. The following corrective options exist:

- Influence the pressure in the float chamber
- Influence the differential pressure signal for fuel metering
- Influence the air correction and the metering cross section for fuel

Icing: Because of the evaporation heat of the fuel, heat is drawn from the inducted air and the components of the carburetor, causing them to cool. This can cause water vapor in the air to freeze and, hence, cause problems in operation. The tendency toward ice formation is greatest at air temperatures of 5°C with high humidity, especially in fog.

Icing can be controlled very effectively by additives in the fuel (such as alcohols) that lower the freezing point.

Another possibility is to preheat the intake air and, hence, increase the air temperature.

Distinctions are drawn among the following:

Idling icing: This can arise when ice from the cooling mixture forms on the edge of the throttle valve at low engine loads. In addition, ice can form on the holes through which exists the premixture for idling, as well as on the transition holes, and can distort the differential pressure signal. In addition to using fuel with additives, this can, for example, be dealt with by increasing the amount that the intake air is preheated. The idling ice disappears as the heat increases.

Full-load icing: Full-load icing or air funnel icing arises in older carburetors in which the venturi tube of the main system is located in the middle of the air funnel, and the premixture enters the air over a short path.

This can be dealt with by increased preheating of the intake air or by fuel additives. In carburetors in which the venturi tube is not in the middle of the air funnel, carburetor icing rarely occurs.

Idling speed control: The idling speed can be indirectly controlled with mechanical means by controlling the intake manifold pressure. A separate idle controller is integrated in electronically controlled carburetors.

Engine afterrunning: To prevent the engine from "afterrunning" after being shut off due to the self-ignition of the mixture at hot areas in the combustion chamber, the supply of fuel is shut off along with the ignition.

Overrun: The transition from load operation to overrun must in certain cases be supported in two ways. Added mixture in overrun prevents an undesirable lean mixture. This is done with an additional mixture system. The other possibility is to shut off the fuel in overrun that is normally speed controlled and additionally serves to lower the fuel consumption.

The transition to overrun is accomplished with a throttle valve dashpot or an overrun air valve to minimize load change reactions.

12.3.7 Lambda Closed-Loop Control

There are two possibilities for a lambda closed-loop control:

- In a fixed air funnel carburetor, an additional possibility of intervention to achieve lambda closed-loop control is to provide valves that are in the main system and/or idling system, and that influence the air-fuel ratio. Alternately, an arrangement is used in the form of a partial load control in which a channel to the intake manifold is opened and closed with a solenoid valve.
- With electronic carburetors, the lambda closed-loop control is one of the features of an electronic control unit. The air-fuel mixture ratio is influenced with a choke.

Bibliography

[1] Pierburg, A., Vergaser für Kraftfahrzeugmotoren, Vertrieb VDI-Verlag, Düsseldorf, 1970.

[2] Löhner, K., and H. Müller, Gemischbildung und Verbrennung im Ottomotor, Vol. 6 in H. List, Die Verbrennungskraftmaschine, Springer-Verlag, Vienna, New York, 1967.

[3] Lenz, H.P., Gemischbildung bei Ottomotoren, Band 6, in H. List, A. Pischinger, Die Verbrennungskraftmaschine, reissue, Springer-Verlag, Vienna, New York, 1990.

[4] Behr, A., Elektronisches Vergasersystem der Zukunft, MTZ 44 (1998) No. 9, p. 344.

[5] Großmann, D., Lexikon Verbrennungsmotor, Vergaser, Supplement MTZ/ATZ, to appear in 2002.

12.4 Mixture Formation by Means of Gasoline Injection

12.4.1 Intake Manifold Injection Systems

The demands for low vehicle emissions and low fuel consumption are the primary influences on the design of modern intake manifold injection systems where fuel is injected through electronically controlled fuel injectors for individual cylinders into the intake arms of Otto engines. A typical configuration for fulfilling standards for minimum emissions is shown in Fig. 12-8.

The measures to reduce emissions beyond the basic functions of the engine control systems—injection and ignition—are mainly based on the required emissions standards, the untreated emissions of the combustion engine, and the vehicle weight class in the smog test. For example, in the aftertreatment of exhaust, measures such as secondary air injection in combination with retarded ignition are required to quickly heat the catalytic converter. These measures, as well as their diagnosis, increase the complexity of the engine control system with sensors, actuators, cables, and computer programming.

Typical functional features of modern engine control systems are

- Torque-based load control with an electronically controlled throttle valve [electronic throttle control (ETC)]
- Model-based functions such as model-based intake manifold filling with load detection via a hot film air mass meter or an intake manifold pressure sensor
- Control of the position of a continuously adjustable camshaft on the inlet and/or outlet side
- Control of various relays to turn on or shut off components (main relay, fuel pump relay, fan relay, starter relay, air-conditioner compressor relay, etc.)
- Active camshaft position sensor for quickly detecting the camshaft position and, hence, quickly synchronizing the engine control while starting the engine
- Cylinder-selective knock control based on a crankcase vibration sensor to provide optimum output and consumption control of the moment of ignition
- Control of the tank ventilation valves to regenerate the carbon canister while the engine is running
- Special catalytic converter heating function with secondary air system, retarded ignition, and transmission shifting point control
- Precise control of the mixture composition via an oxygen sensor ("lambda sensor") upstream from the catalytic converter, and a trim control via a second oxygen sensor downstream from the catalytic converter
- Onboard diagnosis (OBD) of all exhaust-relevant components and functions.

In the design of the fuel system, care is taken to ensure that the fuel in the fuel rail is only slightly heated. If the fuel is heated in the phase after shutting off the engine (hot soak), vapor bubbles can arise in the fuel rail that can lead to problems in a subsequent hot start.

A distinction is drawn between two basic fuel system designs:

1. Fuel system with a return system (Fig. 12-9): A characteristic of this fuel system is that the pressure regulator is directly on the fuel rail. The pressure diaphragm receives the intake manifold pressure on one side so that a constant differential pressure arises between the fuel in the fuel rail and the intake manifold. Given a constant fuel injector control time, this makes the injected fuel quantity independent of the intake manifold pressure.

The advantages of the fuel system with a return system are

- Favorable fuel pressure control dynamics
- Good hot start behavior from raising the fuel rail with cool fuel from the tank
- The injected fuel quantity is independent of the intake manifold pressure

A substantial disadvantage is that the fuel is heated in the tank (up to 10 K in contrast to systems without a return). This increases the fuel evaporation in the tank and the load on the carbon canister.

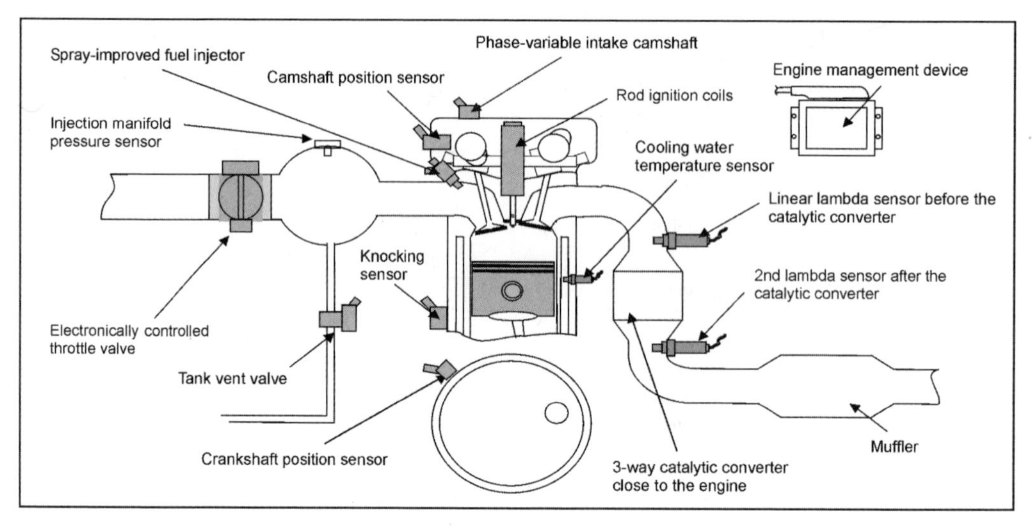

Fig. 12-8 Intake manifold injection system.

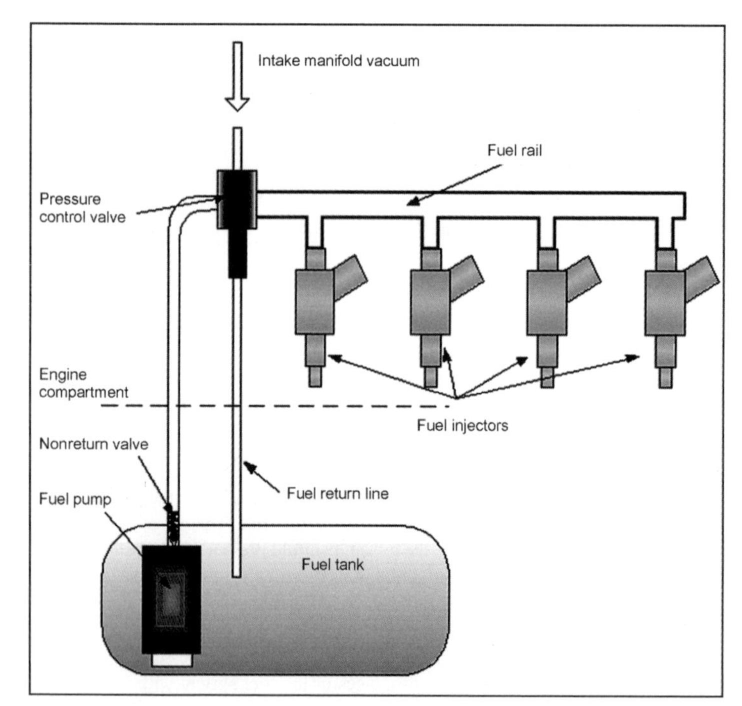

Fig. 12-9 Design of a fuel distribution system with return flow.

2. For this reason as well as to reduce system costs, return-free fuel systems were developed (Fig. 12-10). They are characterized by the integration of the fuel pump and pressure control valve in the tank or close to the tank.

The advantage of this design is that the excess fuel does not have to be first pumped into the engine compartment and then flowed via the pressure regulator back into the tank. The injection times are correspondingly corrected by the engine control system as a function of the constant fuel pressure of approximately 350 kPa (3.5 bar ±0.5 bar). To avoid large pressure fluctuations in the fuel rail that can lead to fluctuations in the injected fuel quantity, pressure pulsation dampers are used in return-free fuel systems.

12.4.2 Systems for Direct Injection

In addition to the described possibility of injecting liquid fuel into the intake manifold of the spark-ignition engine, direct injection systems have been developed in recent years. The fuel is directly injected into the combustion chamber from a central fuel rail under high pressure by

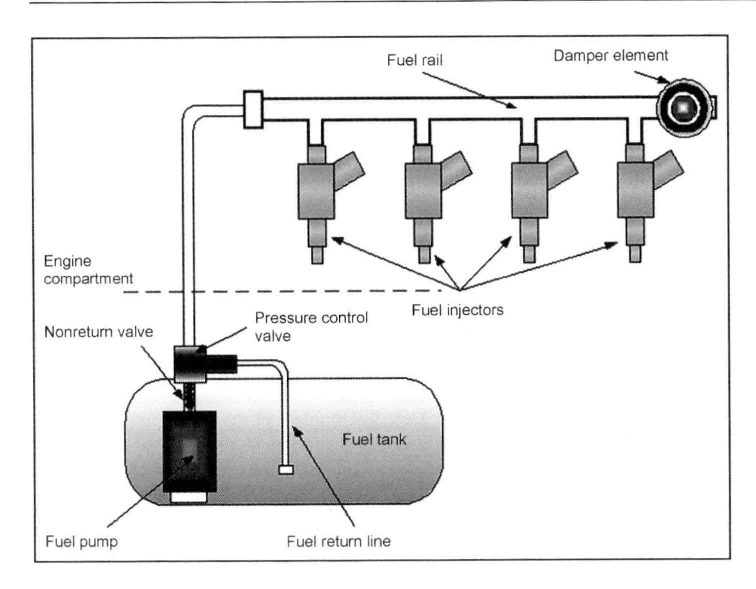

Fig. 12-10 Layout of a return-free fuel distribution system.

electronically controlled fuel injectors (the common rail principle).

By layering the fuel and air by injecting during the compression phase, a relatively lean mixture is produced with a relatively rich mixture cloud close to the spark plug that ensures reliable ignition. Because of the excess air in this mode of operation, there is a reduction in the charge cycle work and in the loss of wall heat during the high-pressure phase of combustion, which together substantially lowers the specific fuel consumption.

In addition, the direct injection cools the interior of the cylinder from the evaporation of the fuel that reduces knocking at full loads. This makes it possible to increase the compression by approximately one unit. The specific fuel consumption is also lowered. Stratified operation makes sense only within a limited range of operation under a partial load in a spark-ignition engine. In other ranges, the engine is operated homogeneously lean, stoichiometrically, or rich under a full load.

Depending on the installation site and position of the fuel injector and the arrangement of the air intake into the cylinder, a distinction is drawn between wall-directed, air-directed, and jet-directed combustion (Fig. 12-11):

- High-pressure direct injection (HPDI) with wall-directed combustion (Fig. 12-11, left):

  The fuel injector is on the side, and the fuel is sprayed onto the piston head. With the shape of the piston recess and the type of airflow, the injected fuel is directed toward the spark plug.

  Depending on the geometry of the intake ducts where the air streams in, a distinction is drawn between a reverse tumble procedure and a HPDI with air-directed combustion.

- HPDI with air-directed combustion (Fig. 12-11, middle):

  The fuel injector is also located on the side, but the fuel is injected into the air in the center of the combustion chamber in contrast to wall-directed combustion. A substantial amount of air movement is necessary, which is generated by a variable tumble.

- HPDI with jet-directed combustion (Fig. 12-11, right):

  This type of combustion has the greatest potential for lean engine operation and, hence, the greatest potential savings in fuel consumption—more than 30% at a low partial load. The fuel injector is located in the center of the combustion chamber, and the spark plug is close to it at a lateral angle. This prevents the fuel from contacting the piston or combustion chamber walls. This type of combustion places large demands on the preparation of the jet by the fuel injector. To ensure reliable ignition and low fouling of the spark

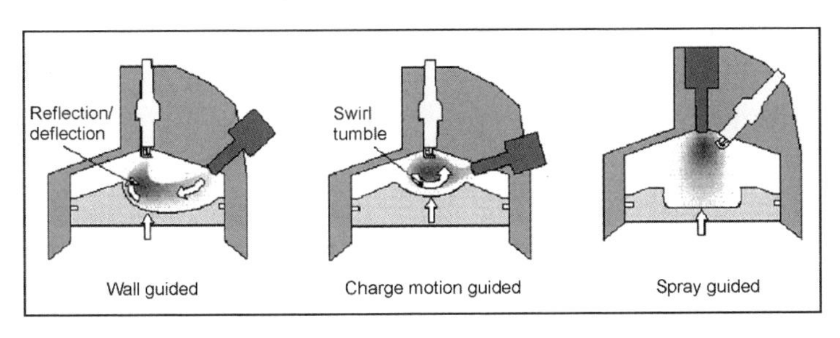

Fig. 12-11 Combustion.

plug, the fuel in the area of the spark plug must be finely atomized, and the spray pattern may not change substantially when the combustion chamber pressure changes.

In addition, there exists air-supported direct injection (OCP_{TM}—"orbital combustions system" which is a registered trademark of Orbital Corporation) in which the central injector position close to the spark plug has proven to be particularly advantageous.

The overall lean mixture during combustion in stratified operation creates a problem for exhaust aftertreatment: Conventional three-way catalytic converters cannot reduce NO_x emissions. Despite lower untreated NO_x emissions from exhaust gas recycling rates of up to 30%, untreated NO_x emissions must undergo special aftertreatment to meet the exhaust thresholds. Beyond selective NO_x reduction catalysts that have a low thermal stability, so-called NO_x storage catalytic converters are used. Storage catalytic converters absorb the NO_x emissions in lean operation and convert them in substoichiometric operation into N_2 and CO_2. A complex engine management function controls this process. Storage catalytic converters tend to experience "sulfur poisoning" and, therefore, require fuels with low sulfur content.

The exhaust aftertreatment measures reduce the savings in effective fuel consumption that result from direct injection.

12.4.2.1 Air-Supported Direct Injection

In addition to liquid high-pressure direct injection systems, there are air-supported direct injection systems such as OCP_{TM}. Air-supported direct injection systems enable stable combustion with good stratification that is compatible with large amounts of recycled exhaust gas because of the high quality of the mixture preparation. The main feature of air-supported direct injection is an arrangement of an electromagnetic fuel injector and an electromagnetically actuated air injection valve (Fig. 12-12) that injects finely atomized fuel into the combustion chamber.

Figure 12-13 provides an overview of the air-supported direct injection system. The injection system is divided into two subsystems, the compressed air path and the fuel path. A compressor driven by gears or a belt generates the required compressed air. The pressure level is set by a mechanical pressure regulator to a set point, and the fuel is conveyed by an electrical fuel pump. The fuel pressure is regulated to be at a constant differential pressure from the compressed air (approximately 0.7–1.5 bar).

The fuel is metered with a conventional fuel injector for intake manifold injection. The fuel is injected into a venturi in the air injection valve. By means of the air injector, a finely atomized mixture cloud is introduced into the combustion chamber that can be directly ignited. In stratified-charge operation, jet-directed combustion is, therefore, possible with low untreated emissions. By synchronizing the phase angle of injection (the actuation of the air injector) and ignition, an optimum air-fuel ratio in the mixture can be maintained at the spark plug for stable combustion under all operating conditions.

As Fig. 12-13 shows, the system consists of an air filter, an air mass meter with an integrated air temperature sensor, an electrical throttle valve actuator, and the intake manifold.

The externally recycled exhaust can be introduced into the individual intake tubes via the manifold or via a line in the cylinder head. In any case, a position-controlled EGR valve is necessary to provide precise metering. Optionally, internal exhaust recycling can be provided by adjusting the phase angle of the intake and exhaust camshaft, allowing the torque characteristic and output of the engine to be improved.

The exhaust aftertreatment subsystem consists of a three-way catalytic converter close to the engine and an underfloor NO_x storage reduction catalytic converter. A broadband oxygen sensor allows the air-fuel ratio to be controlled in lean operation, including the regeneration phase and the stoichiometric operation. The exhaust temperature sensor optimally controls the NO_x storage catalytic converter and trigger measures to protect it. A binary lambda sensor downstream from the NO_x storage catalytic converter is required for the control of the regeneration phase to work properly. Alternately, a NO_x sensor can be used.

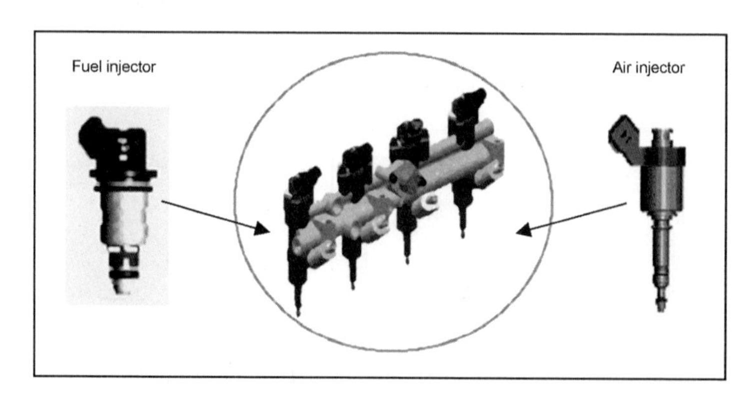

Fig. 12-12 Air and fuel rail with a fuel injector and air injection valve swirling method with duct closure.

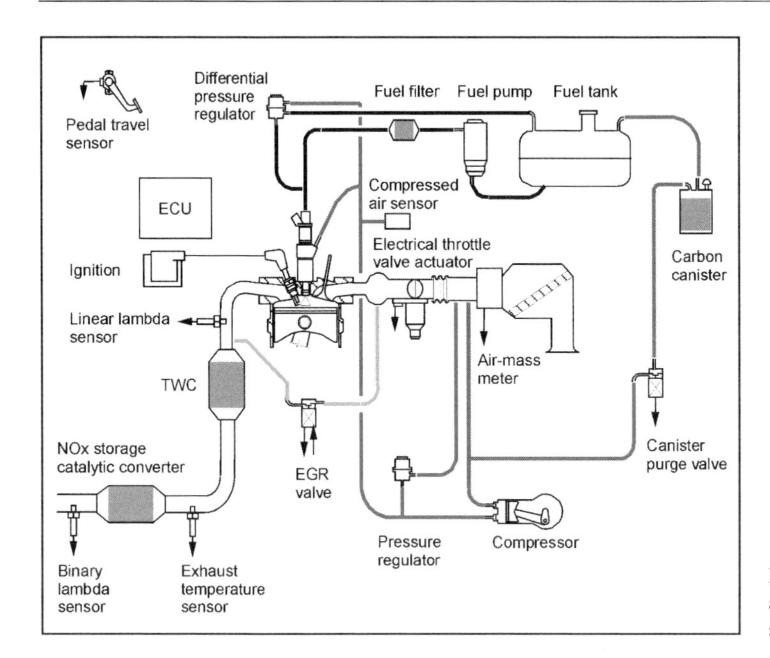

Fig. 12-13 System overview of an air-supported direct injection system.

The compressor in the system offers a new and interesting solution for scavenging the carbon canister. The air inducted by the compressor is guided through the carbon canister. This allows stratified-charge operation to also be used to attain a sufficient scavenging rate without the disadvantages of untreated emissions.

The process of fuel preparation in the OCP_{TM} injector differs from that in a high-pressure direct fuel injector: With a high-pressure injector, the jet decomposes primarily from turbulence and inertia in the liquid jet itself. Up to the end of the decomposition process, a distance of approximately 10–50 times that of the orifice diameter is necessary. In the case of an air-supported injector, the jet dissipates when the aerodynamic forces exceed the surface tension in the liquid. The pressure level in the air injector is such that the critical pressure ratio at the valve orifice is exceeded during injection. The resulting sound velocity of the airflow causes strong aerodynamic forces to be exerted on the fuel jet. The essential part of the atomization process is over directly at the valve outlet. Other thermodynamic effects such as evaporation play a special role particularly when injecting fuel into a high-temperature medium. This interaction of the fuel spray with the cylinder charge represents the interface between the fuel system and the combustion system. The atomization quality can be seen in Fig. 12-14. The average Sauter mean diameter (SMD) is 10.3 μm; only an insignificant number of droplets have a diameter over 40 μm.

The required air mass flow for the OCP_{TM} air injector in reference to the entire amount of air inducted by the engine varies from 15% in homogeneous idling to 1.5% in full-load operation. As an absolute quantity, this results in approximately 5–9 mg air per injected fuel per pulse for a 1.5-liter four-cylinder engine. The absolute pressure in the air rail is preferably set at 6.5 bar. Compressed air is generated by means of a water-cooled piston compressor that is driven by the engine via gears or a belt.

Because of the high turbulence and stratification in the air-supported OCP_{TM} direct-injection system, normally no additional measures are required to move the charge such as swirl valves. Because of the low sensitivity to internal cylinder flow, this injection system is particularly suitable for engines with different valve configurations without active or passive measures to move the charge.

12.4.2.2 High-Pressure Injection

Another possibility for directly injecting fuel into the cylinder is liquid fuel injection using the common rail principle (fuel injection from a common pressure line).

Figure 12-15 provides a system overview of high-pressure gasoline injection divided into an engine control system and a fuel system.

A spark-ignition engine with direct injection requires the use of an electronically controlled throttle valve for the various modes of operation with homogeneous charging and stratified charging. To control the mixture to produce lean mixtures with stratified charging requires a linear λ sensor that can also ensure this function for homogeneous operation with $\lambda = 1$.

The high-pressure pump is fed from the low-pressure system that has a pressure of 1–4 bar. In the mechanically driven high-pressure pump, the fuel pressure is increased to 120 bar. The high-pressure is controlled via an electrically controlled pressure control valve. The return from the high-pressure line ends directly from the pressure control

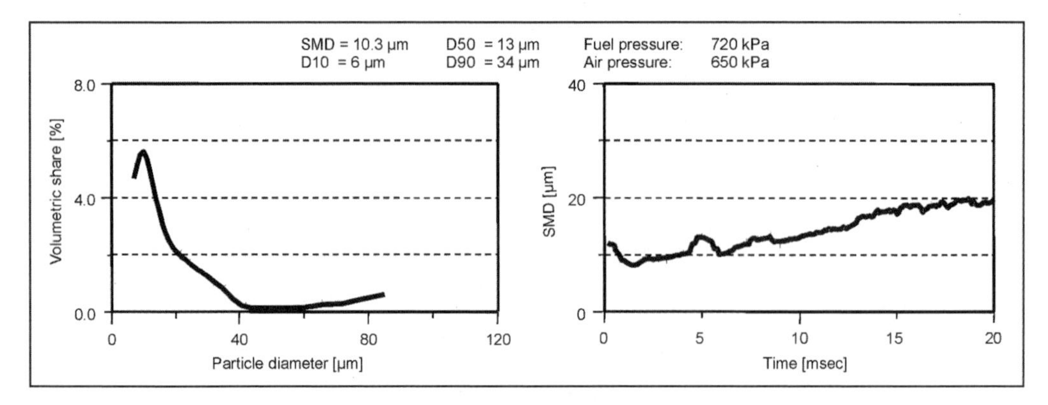

Fig. 12-14 Drop size distribution in relation to volume and time in air-supported direct injection.

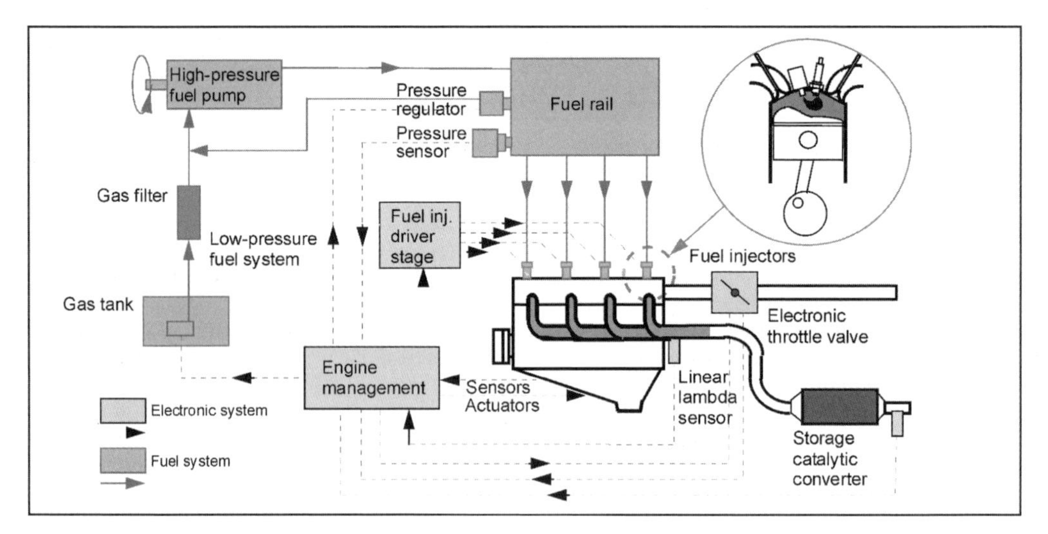

Fig. 12-15 Gasoline direct injection—system overview.

valve in the feed for the high-pressure pump. A pressure sensor serves to detect the pressure. For reasons of safety, an overpressure valve is integrated in the high-pressure circuit that limits the maximum fuel pressure. If the fuel pressure in the fuel rail is kept constant over the entire speed and load ranges, the electrical pressure regulator can, in principle, be replaced by a mechanical pressure regulator.

The fuel injector is located directly in the cylinder head. Because of the high fuel pressure, the magnetic force required to open the valve needle must be much higher than with low-pressure fuel injectors. In addition, the valve needle must be opened and closed extremely quickly for charge stratification and metering.

The fuel atomization quality strongly depends on the fuel pressure, the counterpressure, the flow calibration, and the spray dispersal angle.

Figure 12-16 shows the atomization quality of a high-pressure fuel injector in comparison to a low-pressure fuel injector and an air-supported fuel injector. Different types of combustion and combustion chambers require different flow calibrations and jet shapes.

The high-pressure pump provides the pressure (50 to 120 bar) for the fuel rail. The injection pressure has recently been approaching 200 bar, and no limit is yet in sight for jet-directed procedures. Given a multihole nozzle, this yields a stable spray dispersal angle, good evaporation, and mixture preparation. The high-pressure pump is driven directly by the engine camshaft and is, therefore, mounted on the engine. A distinction is drawn between radial and axial high-pressure pumps.

Figure 12-17 shows an axial plunger pump. A swash plate is rotated by a mechanically driven shaft that is responsible for the alternating stroke movement of the three pistons. Fuel passes into the cylinder through a groove in the swash plate and is ejected via a nonreturn valve in the outlet. Each piston is mounted to the swash plate by a ball-and-socket joint. The bearing and pump chambers are separated by a shaft-sealing ring. The combined materials and coatings are adapted to the wear and lubrication

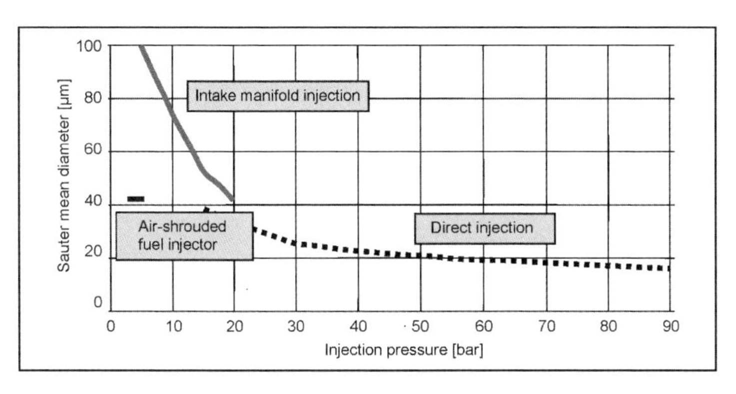

Fig. 12-16 Comparison of the atomization quality of high-pressure injection valves and intake manifold fuel injectors.

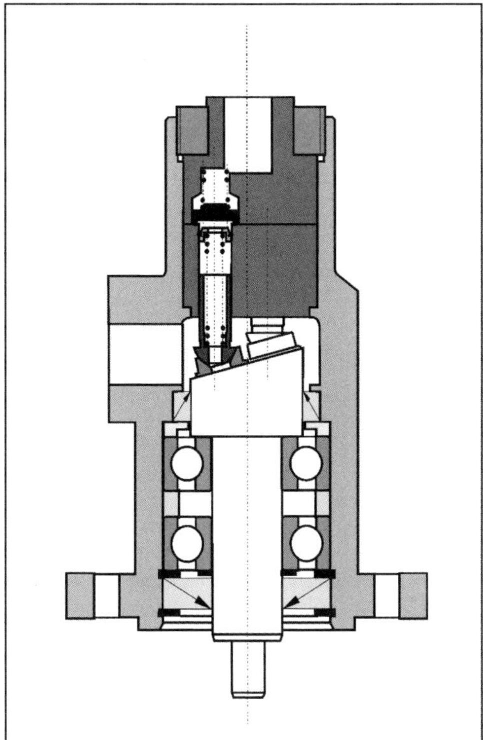

Fig. 12-17 High-pressure fuel pump.

requirements of gasoline operation. The extraordinarily narrow tolerances require the fuel to be finely filtered.

For high-pressure direct injection, there are numerous new requirements on the engine control system:

- The pressure in the high-pressure fuel system must be regulated.
- Lean operation in engines with direct injection requires a linear lambda sensor that covers the lean range and operating range where $\lambda = 1$.
- The high-pressure fuel injectors require a control system that is adapted to the special requirements of this technology. The high fuel pressure and greater demand for linearity and reproducibility from injection to injection make it necessary to alter the fuel injector control system. For the fuel injectors to open quickly, an increase in voltage and current to 80 V/10 A is required. In integrating the driver stages in the electronic control unit, the greater power loss of the driver stages must be taken into consideration.
- For lean engines, the use of an electrical throttle valve is essential. The control of this engine-operated throttle valve ensures that the pedal and throttle valve positions are completely independent.
- In unthrottled engine operation, there is no pressure differential for scavenging the carbon canister. To attain the necessary scavenging rates, a pump is required for scavenging the carbon canister in engine designs with a high-stratified charge component.

In contrast to conventional engines with intake manifold injection systems that work under nearly all operating conditions and a homogeneous stoichiometric mixture (i.e., the mixture is enriched only in special engine states such as cold start, warm-up, and full load), the Otto engine with direct injection is operated using different injection and combustion strategies.

The fuel preparation strategies that produce different homogeneous operating conditions and stratified charging are discussed below (Fig. 12-18).

The goal of stratified charging is to concentrate a well-prepared fuel-air mixture at the spark plug so that a locally limited, ignitable mixture arises ($\lambda \approx 1$) that creates favorable conditions for combustion despite the overall lean mixture. Because of the local concentration of the mixture in the center of the combustion chamber, stratified charge operation also allows high exhaust gas recycling rates. The stratification of the mixture around the spark plug is attained by late injection during the compression cycle. The throttle valve is opened completely for maximum air induction into the cylinder.

The jet direction, jet shape, jet penetration depth, and air flow in the cylinder are the decisive parameters for a successful stratification of the injected fuel near the spark plug.

The jet direction cannot be changed during engine operation. The penetration depth depends on the difference between the fuel jet speed and the air flow speed in the cylinder. The jet speed can be influenced by the injection

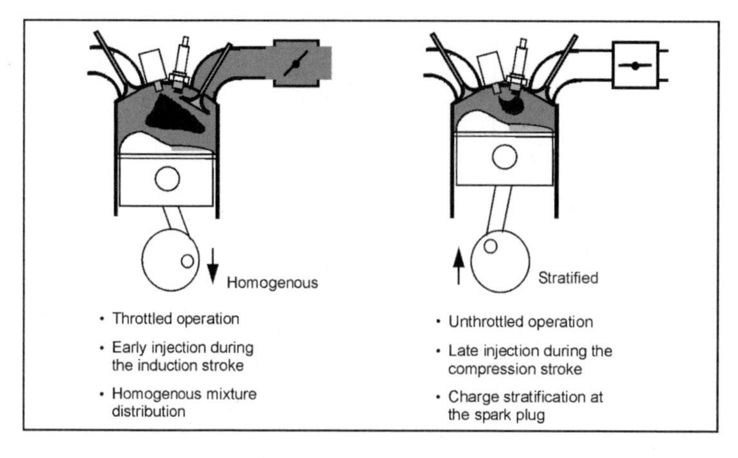

Fig. 12-18 Engine operation with homogeneous and stratified charging.

pressure. To attain a specific movement of air flow in the combustion chamber with turbulence near the spark plug (for good mixture preparation), the design of the intake duct and the combustion chamber (a key responsibility of the engine manufacturer) must be adapted and optimized. Present intake systems are constructed with a tumble or swirl design.

When the operating state of the engine is to be changed from stratified charging to homogeneous charging with the same engine output, the inducted air mass must be reduced by closing the throttle valve; simultaneously, the amount of injected fuel in the cylinder must be increased to compensate for the greater throttling loss. To produce a homogeneous mixture in the combustion chamber, the fuel is injected during the intake cycle at the point in time in which the air speed is at its maximum.

Problems with mixture preparation, drivability, and, above all, exhaust emissions prevent stratified charging from being used over the entire operating range of the engine.

The different operation states over the working range of the engine are shown in Fig. 12-19. The example also includes cooling water temperature to illustrate the influence of different environmental states.

The following combustion states exist:

- Homogeneous rich
- Homogeneous $\lambda = 1$ with or without exhaust gas recycling
- Homogeneous lean with exhaust gas recycling
- Stratified charging with a high exhaust gas recycling rate

At low and medium loads and speeds, the engine is operated with stratified charging and a high exhaust gas recycling rate. This yields lower fuel consumption. The exhaust temperature determines the operating range at low loads in which the engine can be operated unthrottled. For the catalytic converter to convert the pollutants, the catalytic converter temperature may not fall below 250°C. Completely unthrottled operation is, therefore, impossible while the engine is idling. Even in a cold start and during warm-up, the engine runs with a homogeneous, slightly lean mixture to quickly start the catalytic converter. A hot engine can function with throttled stratified charging.

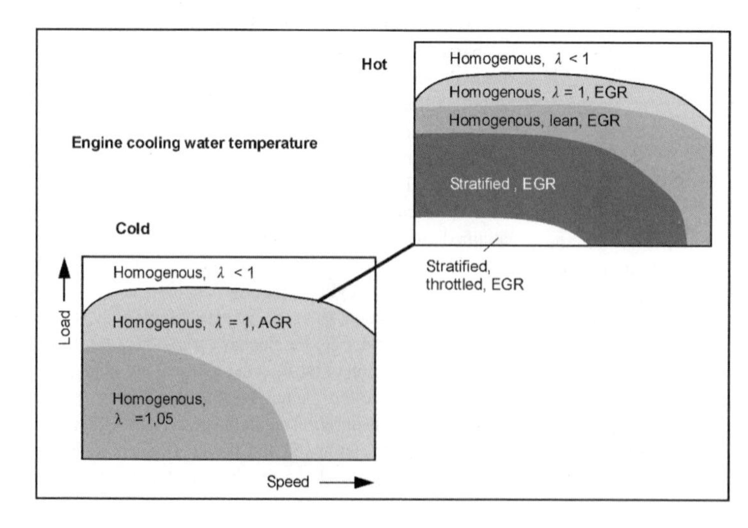

Fig. 12-19 Operating strategies in the engine map.

At a low partial load and high speeds, it is difficult to attain good mixture preparation with stratified charging because of the short mixture preparation time and the danger of soot formation. A homogeneous mixture with EGR is, therefore, preferable.

The NO_x emissions and the danger of soot formation pose limits to stratified charging in the upper partial-load range. In this range, operation with homogeneous charging and EGR has only a slightly negative effect on fuel consumption in comparison to stratified charging with EGR; however, the NO_x emissions are lower, and there is no danger of soot formation.

Homogeneous lean operation is limited by the exhaust temperature. At temperatures over 500°C, storage catalytic converters are no longer able to store nitrogen oxides so that the engine is operated with a stoichiometric mixture and high EGR rates to reduce the NO_x emissions and fuel consumption.

Exhaust gas recycling is not possible when operating at full load. The engine is controlled in the same manner as engines with intake manifold injection, i.e., with a mixture for maximum performance and optimal catalytic converter protection.

In addition to changing the operating states during the transitions between different load states (such as the transition from stratified charging at partial load to a homogeneous enriched mixture at full load during acceleration), a change between two operating conditions can also be necessary in an unchanging load for exhaust aftertreatment. The main requirement is a transition without a change in torque since this is perceived by the driver.

To regenerate the catalytic converter during lean operation, increased fuel consumption of up to 3% is to be expected.

12.4.2.3 Injected Fuel Metering

The fuel for injection is metered via electronically controlled fuel injectors (Fig. 12-20). The fuel in the annular orifice between the valve needle and needle seat is metered by the length of time in which the needle is opened. In principle, the design of the fuel injectors is the same for the intake manifold injection and direct injection. The differences lie in the design of the magnetic circuit and the maximum flow.

The needle is lifted by applying current to the solenoid coil when the magnetic force on the needle is greater than the force from the fuel pressure, the spring, and the friction. As soon as the flow of current is interrupted in the coil, the magnetic field starts to decay, and the needle closes the annular orifice supported by the spring force and the fuel pressure.

After the fuel leaves the fuel injector, a jet geometry arises that is a function of the geometry of the fuel injector after the metering annular orifice (above all the needle seat and orifice plate geometry). A distinction is drawn between the following valves:

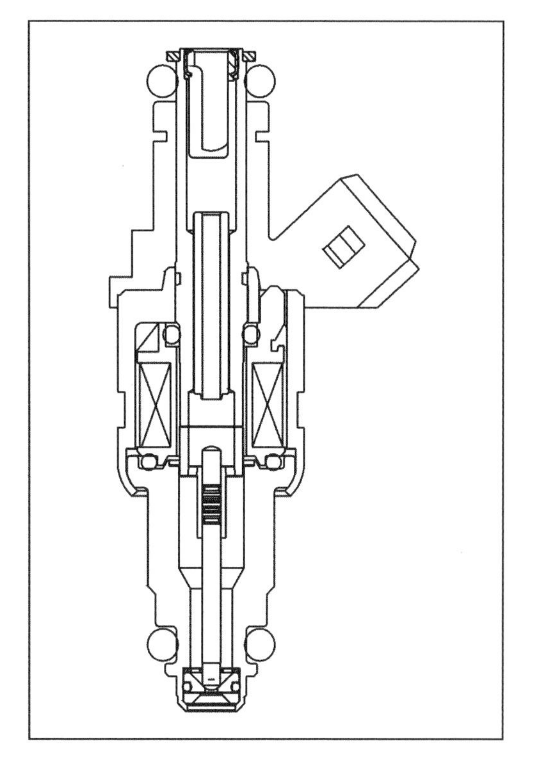

Fig. 12-20 Fuel injector.

- Pencil stream valve: The injection jet has a small maximum spray dispersal angle of 8°. This type of injector is primarily used for applications in which the fuel injector is installed relatively distant from the intake valve.
- Cone spray valve: The injection jet has a larger spray dispersal angle of 10° to 30°. This fuel injector is mainly used when the distance to the intake valve is relatively short. The droplet sizes are smaller than with the pencil stream jet valve.
- Split stream valve: The injected fuel is divided into two injection streams. The angle between both jet axes is normally 15° to 35°. This type of fuel injector is chiefly used for multiple-valve engines with two intake valves.

In addition to the jet geometry, there is a series of other quantities that must be determined for the fuel injectors to be used in an engine:

- *Static flow:* This is the maximum flow through a fuel injector when maximum current is applied to the coil. It depends on the fuel pressure, the diameter of the holes in the orifice plate at the fuel injector exit, and the needle stroke.
- *Dynamic flow:* This is the flow for a coil operating time of 2.5 ms.

- *Linear flow range:* A linear flow range (LFR) is the ratio of the maximum and minimum flows with a maximum 5% deviation from the linearity lines (lines through the characteristic of the injected fuel quantity over the operating time of the coil).
- *Droplet size:* This characterizes the atomization of the fuel injector. The droplet size of a cloud of droplets is usually indicated by the Sauter mean diameter that describes the ratio of the average droplet volume to the average droplet surface in a delimited measured volume. In addition to the average droplet size, the droplet size distribution in the injection jet also has a large influence on the emission behavior of the internal combustion engine. Furthermore, the droplet speed is important since it characterizes the penetration depth of the fuel jet when injected into the air, and it characterizes the secondary jet decomposition when the droplets contact a surface.
- *Leak rate:* Because of applicable legislation concerning evaporation emissions, the requirements are stringent. Since it is difficult to determine the leak rate with liquid media, it is done with nitrogen. The leakage may not exceed 1.5 cm^3/min.

12.5 Mixture Formation in Diesel Engines

Diesel engines operate using internal mixture formation (see Section 12.1). At the end of the compression cycle, liquid fuel is injected around the ignition TDC into highly compressed air. Directly after penetration of the fuel droplets whose average Sauter mean diameter is 2–10 μm, the physical and chemical preparation of an ignitable air-fuel mixture begins. The processes of fuel evaporation, the mixture with the air, and subsequent ignition and combustion occur simultaneously. The goals of mixture formation are to achieve the fastest possible ignition of the fuel vapor and to burn as completely as possible all the injected fuel. If these two basic conditions are fulfilled, the pollution generated by combustion is very low with the avoidance of extreme pressure peaks and, hence, loud combustion and a high mechanical and thermal load (see also Sections 14.3 and 15.1).

The air-fuel mixture in the combustion chamber is strongly inhomogeneous both in terms of location and time. The local air ratio in the combustion chamber ranges from 0 (in fuel droplets) to infinite (zones of pure air). The global air ratio, i.e., the ratio of the air mass actually in the combustion chamber to the air mass required for complete combustion of the injected fuel ranges from approximately 1.1 to 6 in practically designed diesel engines.

There is only a very short time available for mixture formation in diesel engines. Assuming an injection time of approximately 36° crankshaft angle, there is only 1.5 ms available at a speed of 4000 min^{-1}. In comparison, the mixture formation time for a conventional spark-ignition engine with intake manifold injection at a comparable speed is 15 ms. The time from the start of injection to the first ignition of an air-fuel mixture is much shorter. This time, termed the ignition lag, is up to 2 ms. Ignition lag is strongly influenced by the temperature and pressure conditions in the combustion chamber and the atomization of the fuel. After the first ignition, the additional mixture formation of the uncombusted hydrocarbons with the available air oxygen is accelerated by the beginning combustion and the related temperature increase as well as the arising turbulence. The energy required for the mixture formation comes from either the fuel system or the air movement and the starting combustion itself.

In engines with a divided combustion chamber (prechamber or whirl chamber diesel engines), the rich combustion starting primarily in the whirl chamber is responsible for the main energy for mixture formation in the main combustion chamber. Minimal requirements are placed on the injection system; depending on the whirl chamber procedure, the air movement participates to different degrees. In the direct injection system normally utilized today without a divided combustion chamber, the injection system contributes the majority of energy. In engines with a greater speed range or injection systems with comparatively low injection pressure, the air is guided so that turbulence arises in the combustion chamber that supports the mixture formation. The greater the influence of the air movement in mixture formation, the lower the injection pressure can be. It should be noted that the generation of swirling air is associated with greater losses in the charge cycle. The fuel injection in the combustion chamber is therefore of prime importance for mixture formation in diesel engines. In addition to other functions, the injection pressure plays the central role.

Whereas for whirl chamber engines, the injection pressure level has remained constant at approximately 300–400 bar, for engines with direct injection, the injection pressure has risen dramatically over the last 10 years (Fig. 12-21). This is largely related to the development

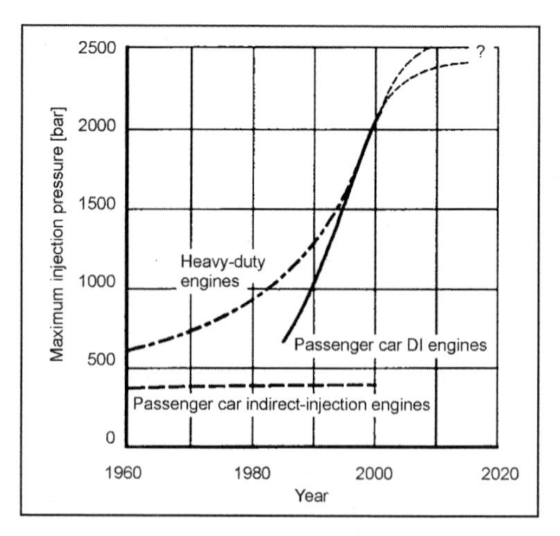

Fig. 12-21 Development of the maximum injection pressure in recent decades.

of high-speed diesel engines with direct injection for passenger cars. Since the available time is very short because of the high speeds, a correspondingly large amount of mixture formation energy must be made available by a high injection pressure.

When injecting the liquid fuel into the combustion chamber, it is important for the fuel to be distributed in many very small droplets and, hence, present a large surface for evaporation, and also to reach the air in the combustion chamber to prevent soot formation because of locally insufficient oxygen. This is accomplished by immediately adjusting the injection pressure, the nozzle aperture geometry, and combustion chamber recess, as well as the correct injection time. The fuel droplets must be prevented from reaching the cylinder wall beyond the combustion chamber recess and collecting in the fire land area between the piston and cylinder. The droplets would then escape combustion, evaporate, and leave the exhaust pipe as uncombusted hydrocarbons.

Figure 12-22 shows the penetration depth of the liquid and vapor fuel over the period after the start of injection as a function of the injection pressure and aperture geometry.[1] In the left picture, we can clearly see that the penetration depth of the liquid jet is independent of the pressure. The speed of the jet tip, however, is much faster at the maximum injection pressure. The stronger pulse ensures greater air entrainment in the injection jet and, hence, faster evaporation. The picture on the right shows that a larger nozzle aperture at the same injection pressure causes the liquid fuel to penetrate less deeply. It should be noted that, as the drop in size increases with the nozzle hole diameter, the aerodynamic resistance (increased by the square of the drop diameter and the speed) can rise so much that the penetration depth falls with larger holes. A balance must be sought between the nozzle hole cross section, injection pressure, and air movement.

In addition to this classic type of internal mixture formation in diesel engines with liquid fuels, there are various special types of diesel engine mixture formation such as in diesel or gas engines and homogeneous diesel combustion that have recently been the subject of intense investigation (see Section 15.1.2.4). A detailed description of mixture formation in connection with diesel combustion can be found in Chapter 15.

12.5.1 Injection Systems—An Overview

Tasks

The injection system is largely responsible for providing diesel engines with high exhaust quality, low fuel consumption, fast response behavior, and smooth running with little noise. Depending on the use of the diesel engine, the stated goals can assume different amounts of importance. The injection system and diesel engine are correspondingly adapted. The main tasks of the injection system are the following.[2,14]

Precise metering of the fuel mass per work cycle. The load of diesel engines is controlled by metering the injected fuel mass (quality control). As a result, this must be as precise as possible to attain maximum performance; on the other hand, when the stoichiometric air-fuel ratio is approached, there is a danger of excessive noise. The more precise and stable the metering of the fuel in a full load curve (over time) is, the smaller the safety margin from the smoke limit; i.e., the engine performance can be maxed out. The fuel quantity tolerances should be as low as possible under a full load and should not exceed approximately $\pm 2.5\%$. During idling and in the partial load range (in particular, in stationary operation, i.e., when there is no intentional control), the requirements on the stability of the fuel metering are very high from

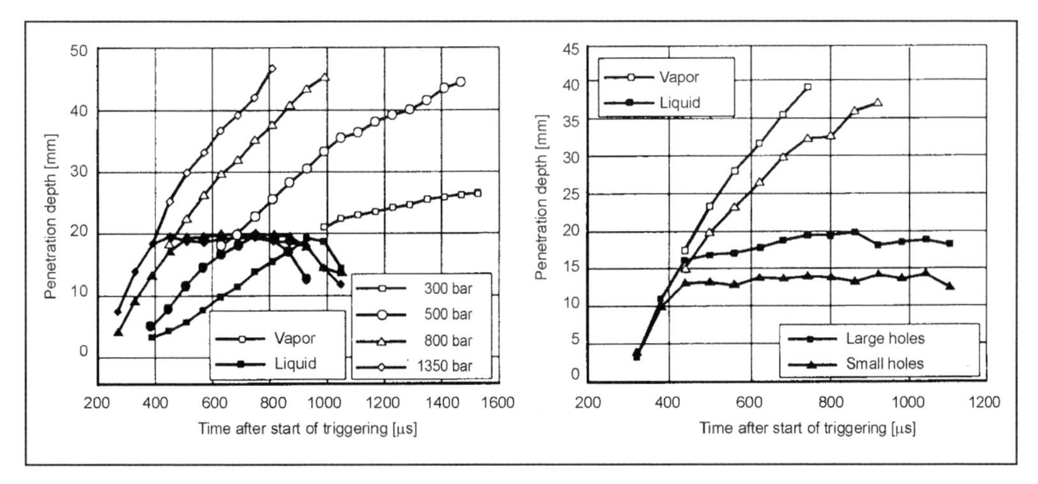

Fig. 12-22 Penetration depth of the liquid and vapor fuels in an injection chamber as a function of the injection pressure and the nozzle hole size.[1]

cylinder to cylinder as well as from injection to injection. The deviations should be less than 1 mg/injection. It may be necessary to adapt the injected amount of fuel to each cylinder to achieve the desired smooth running.

Adapting the injection rate to operation conditions. The injected fuel mass during the injection process per unit time [injection rate and its change: $dm/dt = f(t)$] decisively influences exhaust emissions, smooth operation and fuel consumption. In principle, the injection rate can be influenced by changing the injection orifice cross section of the nozzle and by changing the injection pressure. Despite substantial efforts, a reliable nozzle with a variable cross section has not been successfully created, which leaves only pressure modulation. The pressure can be modulated comparatively easily with a slight degree of variation by adjusting the cam shape and, hence, via the cam speed or plunger speed in the high-pressure injection pump. Pressure modulation in accumulator injection systems is presently in the developmental stage. However, injection rates can be changed to a certain extent by pressure stages in the nozzle holder. Figure 12-23 shows changed injection rates during the time for the main injection.[3,4]

In general, a high injection rate at the related large amount of injected fuel at the beginning of injection produces a strong burst of combustion with a high local temperature and, hence, high NO_x formation.

Multiple injections. Controlling the injection rate during an injection is frequently insufficient to meet requirements. Increasingly, multiple injections are required with different quantities depending on the operation point in the program map. Figure 12-24 shows a selection of conceivable multiple injection systems, some of which are in use.

Figure 12-25 shows multiple injections at the optimum operating point and their share of fuel at three places in the program map.

A small, reduced preinjection substantially reduces the ignition lag for the following main injection and can, hence, soften the combustion characteristic that reduces combustion noise. A secondary injection under high pressure and, hence, with a well-prepared jet is able to oxidize the soot generated during preinjection or increase the exhaust temperature given a corresponding exhaust aftertreatment system, e.g., to regenerate the particle filter. In addition, secondary injection can support the supply

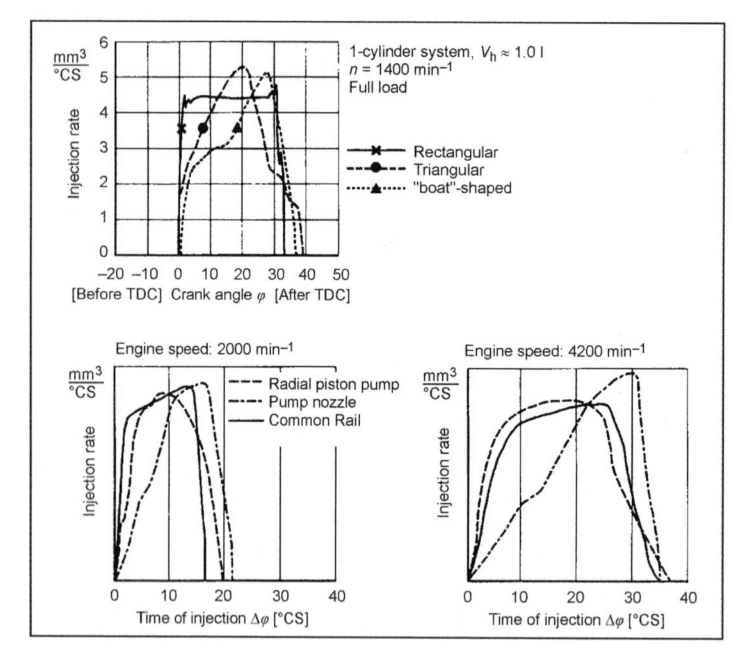

Fig. 12-23 Different injection curves [injection rate = $f(t)$] for the main injection, top[3] and bottom.[4]

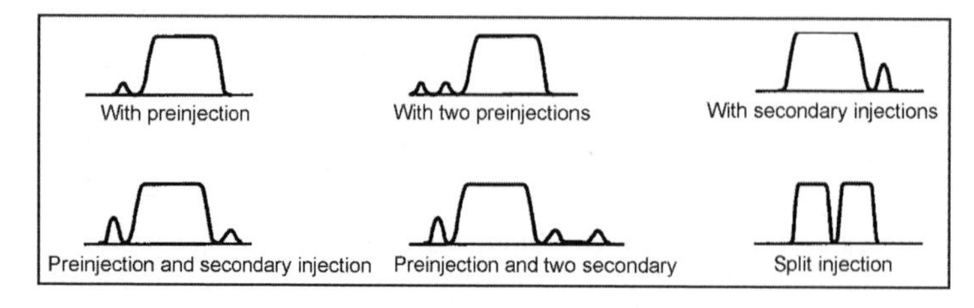

Fig. 12-24 Qualitative nozzle needle stroke characteristics for multiple injections.

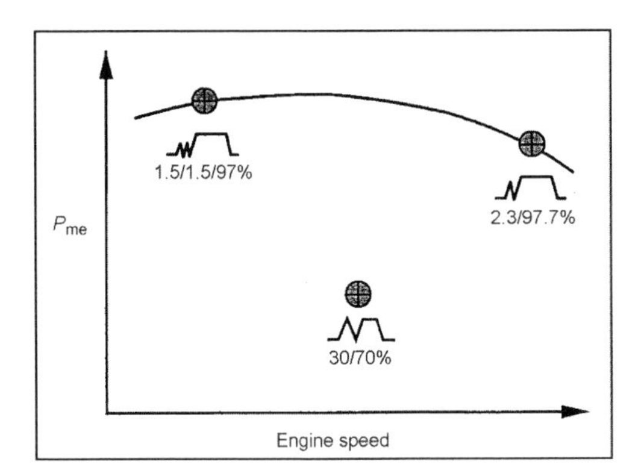

Fig. 12-25 Example of optimum, operation-point-dependent multiple injections in the engine map; quantity distribution in percent.[5]

of uncombusted hydrocarbons to subsequent exhaust aftertreatment.

Minimum injection. In connection with multiple injections, in particular, where the preliminary and secondary injections range from approximately 1 to 3 mg per injection, e.g., in passenger car engines, the need for precisely metering the very small quantities increases dramatically. The injected fuel quantity tolerances should be less than 0.5 mg per injection. Since the fuel injector needles always move within the so-called ballistic area for these minimum amounts—i.e., they do not reach the mechanical stop—all the manufacturing-related tolerances strongly affect the quantity. The result is a very strong quality requirement for the fuel injectors, in particular, the nozzle needle and the injection orifice.

Adapting the injection time. A purely speed-related adaptation of the beginning of fuel supply in systems with long fuel injection tubing is insufficient, as has been known for a long time. Even systems without injection tubing or with electronically controlled injectors require a freely settable injection start from early (as in cold starts) to late in certain areas of the program map to reduce nitrogen oxides. Within other operating ranges, the optimum setting for fuel consumption is desirable. In addition, the times between the start of injection can be individually adjusted for multiple injections. The precision of the beginning of injection should be $< \pm 1°$ crankshaft angle.

Flexible adaptation to operating and environmental conditions. In addition to the cited main tasks, a modern injection system reacts flexibly to dynamic processes depending on the air mass. For example, when accelerating under a full load, the fuel quantity should be adapted only to the dynamically measured air mass to avoid undesirable smoke emissions. When the engine speed has been reached, the quantity of fuel is reduced corresponding to the use of the diesel engine to protect it from excess speed (full-load speed regulation). Within the bottom operating range, the engine is operated at the lowest possible speed both stably and nearly load independent. The respective amount of required fuel also is adapted to quickly heat the engine depending on the environmental and fuel temperatures. The amount of fuel needs to be adapted to elevation as well. For example, at a high elevation above sea level, the amount of fuel under a full load should be reduced because of the lower air density to limit smoke. When the engine is overrunning, a ramplike increase in the amount of injected fuel when the speed drops below the idling speed is needed for the engine to "catch up" when idling. The injection masses need to be adapted to the respective operating conditions depending on the charge pressure and exhaust gas recycling.

These multifaceted and sometimes interrelated tasks and demands on injection systems can be handled only by electronically controlled systems. Mechanically controlled systems with edge-controlled fuel metering cannot meet these demands, or can only at the cost of substantial compromises. In uses where the cited greater demands are not as important, in particular on dynamic behavior, these mechanical, robust systems are still satisfactory. Edge-controlled systems have largely been abandoned in vehicle engines that have to meet stringent exhaust laws and simultaneously use as little fuel as possible under dynamic conditions.

Design and Parts

In a modern-day fuel injection system, the mechanical system, hydraulics, electrical system, and electronics work together. The overall system can be divided into five subsystems (Fig. 12-26).

Low-pressure system: The low-pressure system ensures that the fuel from the tank is supplied for high-pressure injection. Either the required pump can pump the fuel from the tank to the engine as a tank module or, as a pump integrated in the high-pressure pump, it can draw fuel for the tank and convey it at the necessary supply pressure to the high-pressure units. This feed pressure can be 1 to 15 bar depending on the injection system and rpm. In common rail systems, a rail-pressure-dependent, low-pressure-side control is used to reduce the power consumption of the high-pressure pump. When designing low-pressure pumps, one must keep in mind that in most cases, a substantial volumetric flow of fuel is required to cool the injection components and to compensate for leakage. The utilized filters are both large-pore prefilters as well as fine filters; the latter have an average pore size of 5 to 10 μm, depending on the injection system. The filters also have the task of removing any water that is in the fuel to prevent corrosion of the injection components. To ensure operation at extremely low temperatures, the filter frequently has an electrical fuel heater, or heated fuel from the high-pressure area is recycled to the fuel before the filter to prevent the filter from being blocked by paraffin deposits from the fuel.

High-pressure system: The high-pressure system is essentially characterized by the high-pressure pump. The

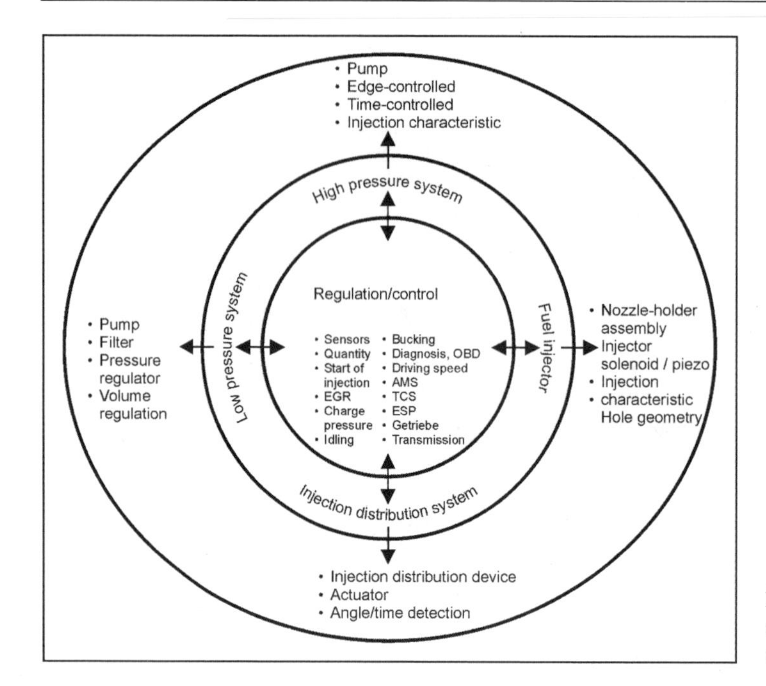

Fig. 12-26 The diesel injection system as the sum of different subsystems.

required high-pressure and injection output is also generated by plunger pumps. Internally or externally supported radial piston pumps and single-cylinder axial piston pumps are also used. Only these pumps are capable of generating pressures greater than 1000 bar stably over a long period of time and over metering the required amounts as needed

In conventional, so-called edge-controlled systems, the actual pump element also has the function of metering precise quantities in addition to the tasks of volume feed and increasing pressure. In modern systems with electronically controlled valves, the high-pressure pump exclusively serves to generate injection pressure and supply the fuel. The metering is assumed by an electronically controllable valve (normally a solenoid valve or piezoactuator).

To supply the fuel to the engine at the right time, a timing device system is necessary. In general, there are two basic systems. In conventional systems, because of mechanical or hydraulic forces, there is a phase displacement between the pump driveshaft and, hence, the camshaft or ignition distributor shaft of the high-pressure pump and the crankshaft of the internal combustion engine. In systems in which the metering is accomplished by electronically controllable valves, the required adaptation of the injection time can be achieved by a change of the operating time of the valves or by a combination of the operating times and phase displacement.

Fuel injector (injection nozzle): The fuel conveyed by the high-pressure pump is fed via the fuel injector to the combustion chamber of the engine. In addition to the injection time and precise metering, the main task of these valves is jet preparation for subsequent mixture formation and combustion. Either the fuel injector can be directly pressure controlled or, as is the case with standard common-rail systems, it can be hydraulically controlled by an electronically controllable valve.

The connection between the high-pressure system and the fuel injector is made with inline systems using high-pressure lines. These are steel lines with inner diameters of approximately 1.5 to 2.5 mm. To increase the fatigue limit, it is important for the inside of the pipes to be as smooth as possible without rough recesses, overlaps, and defects. Special procedures are required to achieve this, such as autofrettage, which is plastic smoothing of the inside under extreme pressure.

Regulation and control: The above-cited four subsystems are coordinated by a regulation and control system. While conventional systems are primarily controlled mechanically, hydraulically, and, in certain cases, even pneumatically, modern injection systems use electronic information acquisition and processing systems and electrically actuated control organs.

The electrical actuators are usually only for control; the displacement energy is normally generated hydraulically or pneumatically.

The electronic control and regulation system is incorporated into the overall engine and vehicle management system and is, hence, connected to all the subsystems from which the control proceeds to affect the engine torque or speed.

Figure 12-27 schematically illustrates the injection systems characterized by pumps that are used today; the edge-controlled systems are in the top row. The regulation and control can be both mechanical and electrical.

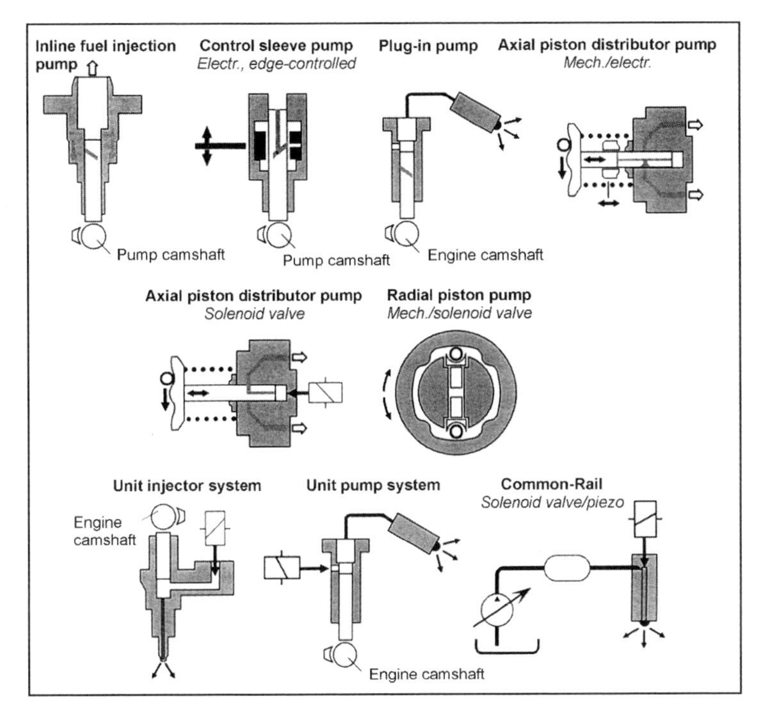

Fig. 12-27 Presently utilized injection systems: top, edge-controlled systems; middle, solenoid-valve-controlled compact systems (≤6 cylinder); bottom, individual cylinder systems that are solenoid-valve-controlled or piezocontrolled.

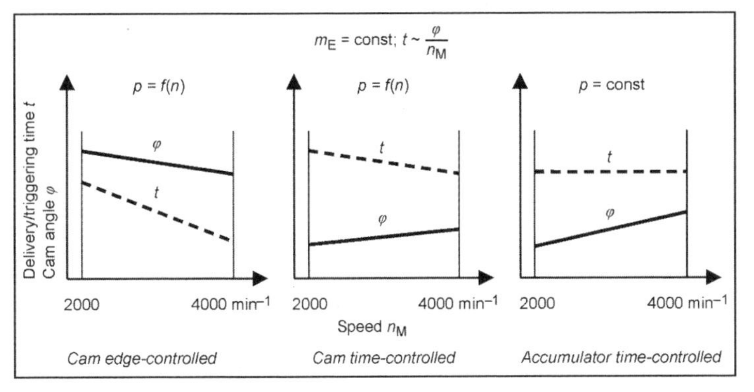

Fig. 12-28 Breakdown of diesel injection systems for pressure generation and the manipulated variable for metering.

In the middle and bottom rows are the systems in which the feed is controlled by electrically actuated control valves. With the exception of computer-regulated systems (CRS), injection can occur in the other systems only while volume is conveyed and pressure is generated, i.e., while the pump plunger is moving.

In general, the injection systems used today can be divided into three categories: these are the so-called cam-edge-controlled, cam-time-controlled, and storage-controlled systems (Fig. 12-28).

In cam-edge-controlled systems, the manipulated variable is the cam angle of fuel delivery or the delivery stroke of the piston. Since the piston speed and the pressure increase as the speed increases (also influenced by predelivery and postdelivery effects in the case of edge control), the delivery angle at high speeds is less than at low speeds assuming a constant quantity.

In cam-time-controlled systems, the manipulated variable is the control time. In this case as well, the control time is lower at high speeds than at low speeds because of the speed-related pressure. In the case of storage-time-controlled systems, the pressure can be constantly set by means of the speed and, hence, the control time. The delivery angle, therefore, doubles as the speed doubles.

12.5.2 Systems with Injection-Synchronous Pressure Generation

The main feature of injection systems with injection-synchronous pressure generation is that the pressure generation and the fuel delivery or injection occurs individually at the right time for each engine cylinder; i.e., the pressure is generated at the same rhythm as the engine ignition sequence. Individual pump systems, inline fuel-injection pumps, distributor injection pumps, and line-free pump

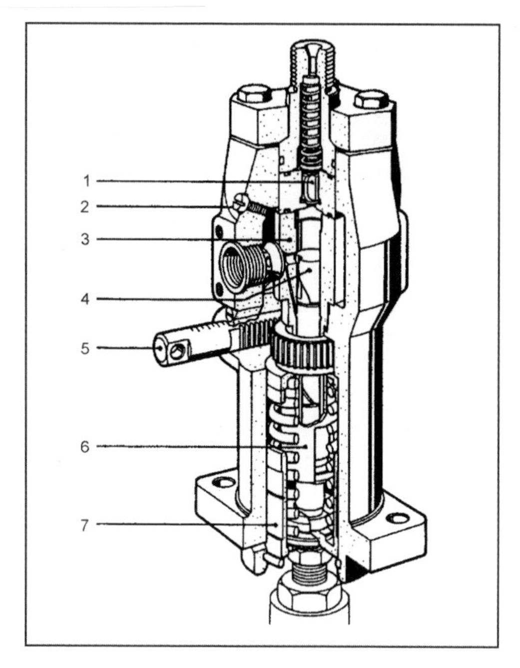

Fig. 12-29 Mechanically controlled single-plunger fuel injection, individual injection pump for large engines[6]: 1, Pressure valve; 2, Vent screw; 3, Pump cylinder; 4, Pump piston; 5, Control rack; 6, Control sleeve; 7, Guide bushing.

nozzle systems are based on this principle. The metering can be done by edge control (mechanically or electronically controlled) or by electrically actuated control valves (see also Section 12.5.1).

12.5.2.1 Individual Pump Systems with a Line

In addition to inline fuel injection pumps, the individual fuel injection pump system shown in Fig. 12-29 with mechanical control is one of the oldest diesel injection systems. This system is characterized in that the drive for the pump plunger is provided by special cams that are on the camshaft for the valve control of the engine. This design permits the use of individual injection pump systems (also frequently termed plug-in pumps) only for engines with bottom-mounted camshafts. The system is, therefore, not suitable for modern, high-speed passenger car diesel engines whose valves are controlled exclusively by overhead camshafts. The main areas of use for mechanically controlled individual pump systems are, therefore, small engines, engines for construction machinery, and stationary engines, as well as large engines for diesel locomotives or ship engines.

The attainable injection pressures for large engines range up to 2000 bar. For ship engines, special designs for heavy oil operation are available. The long life and reliability required in these cases lead to a very robust construction with sealed cylinders on one side, so-called blind-hole elements.

A great deal of engineering is required to freely adapt the start of delivery and, hence, fuel injection. Subsequent developments have therefore produced the solenoid valve controlled individual pump system, the pump line nozzle (PLN) system, and the unit pump system (UPS) or electronic unit pump (EUP). As a result of the short fuel injection tubing between the individual pump and the nozzle-holder assembly, the demands are slight regarding the adjustment of the beginning of delivery, and this can be flexibly accomplished by controlling the start for the solenoid valve on the delivery cam. These systems are, therefore, useful for high-speed commercial vehicle engines (Fig. 12-30). In addition, there is a greater need for freely adjusting the start of injection in large engines.[13]

Today, unit pump systems attain maximum injection pressures of approximately 1800 bar with a potential of 2000 bar. The unit pump system is a development of the edge-controlled plug-in pump for the cam-time-controlled electronic individual pump system.

12.5.2.2 Inline Fuel Injection Pumps

The elements of the inline fuel injection pump consisting of the pump barrel and pump plunger and corresponding to the number of available engine cylinders are in their own housing in high-speed aluminum engines. The pump plungers are moved by the pump's own camshaft that is driven by the timing gear drive of the engine. The fuel is metered exclusively through edge control by rotating the pump plunger. Each pump plunger bears an angled timing edge, so that a different delivery stroke and, hence, a different amount of injected fuel is delivered or can be set in connection with the cylinder-side, fixed spill port as a function of the angular position of the pump plunger. The entire plunger stroke remains constant and corresponds to the cam pitch. The plunger is rotated by a control sleeve that mates with a laterally movable control rack. The control rack is moved by the governor connected to the injection pump. The governor can be either a mechanical governor that primarily shifts the control rack depending on the speed and, in particular, provides full-load speed regulation or an electronic governor that acts on the control rack by using an electromagnetic actuator mechanism. To adapt the injected fuel quantity to the wide variety of operating conditions, add-on modules are required for mechanically controlled pumps such as a charge pressure-dependent full-load stop, and temperature and quantity-dependent adjustors of the injected fuel quantity.

To reliably supply the pump elements with fuel, a low-pressure supply pump is usually mounted to the inline fuel-injection pump that is actuated by a special cam on the pump's camshaft. The supply pump feeds fuel to the fuel gallery of the inline fuel injection pump at pressure of up to approximately 3 bar. At the high pressure exit of the inline fuel injection pump, a pressure valve separates the high-pressure area in the pump from the fuel injection tubing and the nozzle-holder assembly; therefore, after the line and nozzle system injects the fuel, the fuel in the system remains under pressure (i.e., a certain amount of static pressure

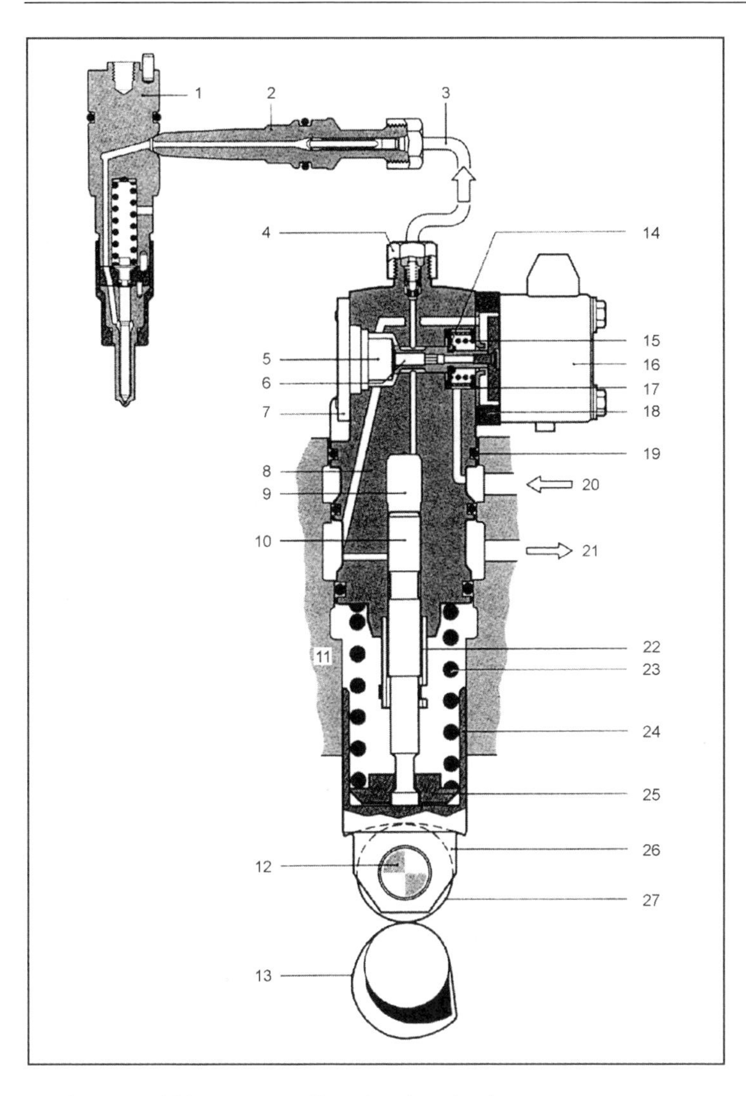

Fig. 12-30 Pump line nozzle (PLN) or unit pump system (UPS) for commercial vehicle engines.[7] 1, Injection nozzle holder assembly; 2, Delivery connection; 3, High pressure line; 4, Connection; 5, Lift stop; 6, Solenoid valve needle; 7, Plate; 8, Pump housing; 9, Hoch pressure chamber (element chamber); 10, Pump plunger; 11, Engine block; 12, Roller tappet bolt; 13, Cam; 14, Spring seat; 15, Solenoid valve spring; 16, Valve housing with coil and magnet core; 17, Armature plate; 18, Intermediate plate; 19, Seal; 20, Fuel feed (low pressure); 21, Fuel return; 22, Pump plunger retainer; 23, Tappet spring; 24, Tappet body; 25, Spring seat; 26, Roller tappet; 27, Tappet roller.

remains). In addition, a return-flow throttle valve is frequently integrated in this pressure valve that prevents undesirable secondary injections at only a slight injection pressure.

Similar to the situation with mechanically controlled individual injection pumps, inline fuel injection pumps require a comparatively great deal of engineering to freely adapt the start of delivery. The speed-related control of the start of delivery can be achieved with a front-end timing device. This functions with the help of flyweights and suitable kinematics to yield a phase displacement between the pump camshaft and the engine crankshaft. A simple load-dependent control of the start of delivery is occasionally enabled by a top timing edge on the pump plunger. Hydraulic front-end timing devices are also used that can change the start of delivery in relation to both load and speed within certain limits. Despite these various solutions, the absence of a freely adjustable start of delivery is a disadvantage in conventional inline fuel injection pumps. This led to the construction of the control-sleeve pump.

Figure 12-31 shows a section of a pump element of a control-sleeve inline fuel injection pump. Its main feature is that it can move in the area of the piston timing edge of the pump barrel (control sleeve). This permits the adjustment of the plunger lift to port closing, i.e., the piston path until the inlet passage for the fuel. A small plunger lift corresponds to an early start of delivery; a large plunger lift corresponds to a late start of delivery. The height of the control sleeves of the individual pump elements are changed by a common actuator shaft. The actuator shaft, as well as the control rack necessary for fuel metering, are activated by two separate, electromagnetic actuator mechanisms. The fuel is metered in control-sleeve pumps like conventional inline fuel injection pumps or conventional individual pump systems. In comparison to standard inline fuel injection pumps, varying the start of delivery by changing the plunger

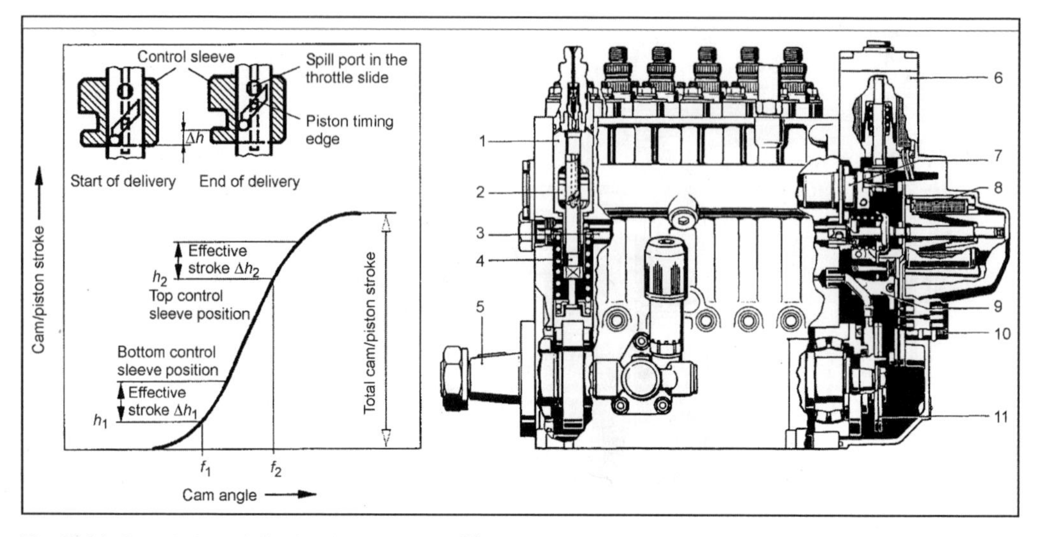

Fig. 12-31 Control-sleeve inline fuel injection pump.[2,6] Left: Functional principle for varying the start of delivery. Right: Pump with electromagnetic actuator mechanisms: 1, Pump barrel; 2, Control sleeve; 3, Control rack; 4, Pump plunger; 5, Camshaft; 6, Start-of-delivery actuator solenoid; 7, Control-sleeve setting shaft; 8, Rack travel actuator solenoid; 9, Inductive control rack travel sensor; 10, Plug-in connection; 11, Disk to block start of delivery and part of the oil return pump.

lift to port closing requires higher cams. For this reason and because two actuator mechanisms are required, this pump design is used only for commercial vehicle engines. Conventional inline fuel injection pumps with mechanical or electromagnetic control are contrastingly used in all engine sizes. The large inline fuel injection pumps for ship engines are presently controlled only mechanically. The pressure range of inline fuel injection pumps extends from approximately 550 bar for the small inline pump (M-type) to approximately 1350 bar for the control-sleeve pump.

Because of greater demands for lower exhaust emissions, fuel consumption, and the related demand for maximum injection pressure in the injection system, multiple injections, and freely variable start of delivery, the inline fuel injection pump system is increasingly being replaced by solenoid-valve-controlled injection systems.

12.5.2.3 Distributor Injection Pump

In addition to the inline fuel injection pump, the distributor injection pump is the second most familiar compact pump design. It consists of a low-pressure supply pump, a high-pressure supply pump, a timing device unit, a speed and fuel governor, and various mechanical and electrical components. The high-pressure pump can either be designed either as an axial piston pump or as a radial piston pump.

Figure 12-32 shows an axial piston pump with edge-controlled fuel metering and an electromagnetic actuator mechanism. In contrast to the inline pump, this pump is distinguished in that only one pump element is required for all engine cylinders. This is possible because the frequency of the stroke of the pump plungers corresponds to the ignition frequency of the internal combustion engine and not that of an individual engine cylinder. At the same time, the pump plunger rotates at the camshaft speed. The fuel is fed to the engine cylinders by the stroke of the plunger. The rotation distributes the fuel to the engine cylinders corresponding to the firing sequence. This dual function of the plunger allows the distributor injection pump to be used for engines with up to six cylinders. The applications for this pump are, in particular, for high-speed diesel engines for passenger cars and light commercial vehicles. In individual cases, they can also be used for medium-duty class engines. The nozzle-side injection pressure reaches values of 1200 to 1300 bar.

A particular advantage of the distributor injection pump is the integration of a solenoid-valve-controlled timing device. This makes it particularly suitable for engines with a large speed range. Instead of electromagnetic actuator mechanisms, an inertia-supported actuator assumed mechanical control of the timing of the control sleeve and, hence, the fuel metering mechanisms in the older conventional design. A further development of the axial piston distributor pump is shown in Fig. 12-33. This is a cam-time-controlled pump that has a solenoid valve in the high-pressure area that can control both the charging of the pump element and the start and end of delivery (hence, the delivery quantity). Because of the slight amount of dead space in the high-pressure area of the pump, pressures of approximately 1500 bar can be attained with this variation. An electronic control unit is on the pump body that assumes the pump control functions, in particular, the activation of the fuel solenoid valve and the solenoid valve for timing the start of delivery. The information for this is provided from the pump internal incremental pump angle time signal (increment angle time signal). This is generated by the speed or angle-of-rotation

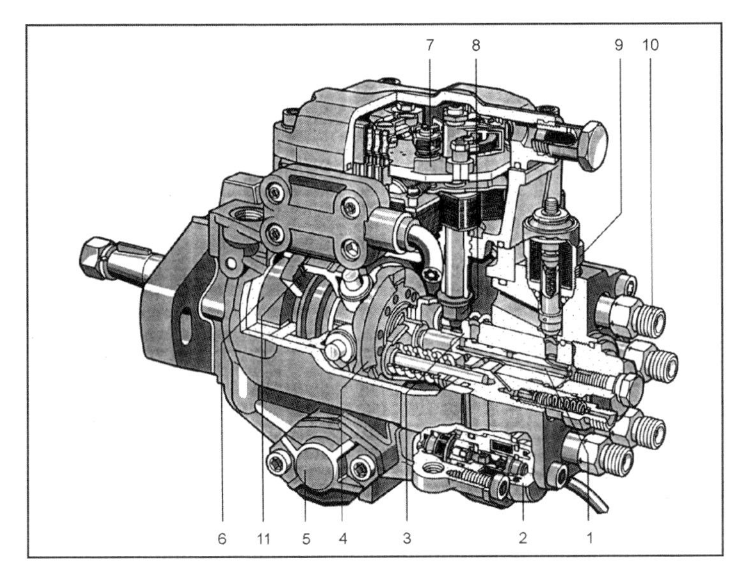

Fig. 12-32 Axial piston distributor injection pump, edge controlled with electromagnetic actuator mechanism designed by Bosch. 1, Distributor plunger; 2, Solenoid valve for timing; 3, Control collar; 4, Timing device; 5, Cam plate; 6, Supply pump; 7, Electrical fuel quantity positioner with feedback sensor; 8, Actuator mechanisms; 9, Electrical shutoff device; 10, Pressure valve holder; 11, Roller holder.

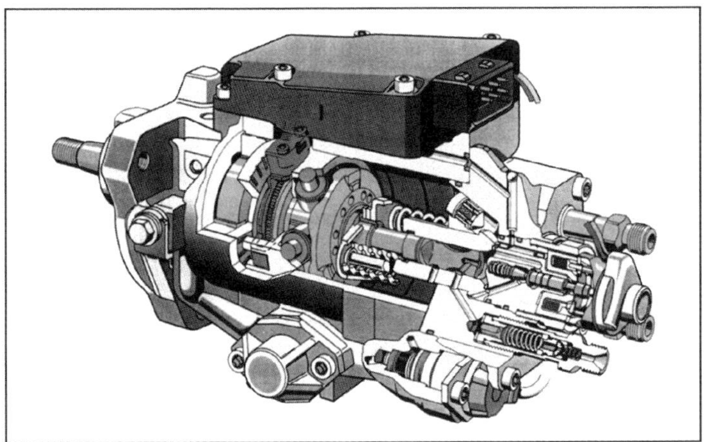

Fig. 12-33 Axial piston distributor pump, cam time controlled with solenoid valve; designed by Bosch.

sensor on the trigger wheel of the drive shaft. The position of the sensor is changed together with the adjustment of the roller ring necessary for timing the start of delivery. With the cam plate speed (camshaft speed) and the assignment of the delivery stroke to the TDC signal of the crankshaft speed sensor, the start of delivery can be precisely timed without a needle motion sensor in the nozzle-holder assembly having to provide the information at the beginning of injection. At the same time, the increment angle time signals can set the exact control time of the fuel solenoid valve and hence the fuel metering. The pump electronic control unit (ECU) is connected to the second ECU, the engine control unit, or it can contain its functions so that only one more electronic control unit is necessary. The drive and the timing devices are similar to the edge-controlled axial piston distributor injection pump.

In distributor injection pumps with a radial piston high-pressure pump, the pressure generation function and the distribution function are separate. As the name indicates, the pressure generation piston is in a radial position. The fuel metering can be controlled by either the cam pitch or the cam time via solenoid valves—similar to the solenoid-valve-controlled axial piston distributor pump.

Figure 12-34 shows a cam-pitch-controlled radial piston distributor pump. With this type of pump, the delivery stroke (that corresponds to the respective overall stroke) is changed by axially shifting the ignition distributor shaft and by conical surfaces. The axial position of the ignition distributor shaft is detected by an inductive travel sensor and represents a measure of the delivery stroke or the delivery quantity. This pump is also primarily used for passenger cars similar to the axial piston distributor pump.

To attain the maximum injection pressure, a radial piston pump was developed with a higher supply rate and solenoid valve control (Fig. 12-35).

With this pump, injection pressures of 1800–1900 bar can be attained. A cam ring bears inner radial cams whose stroke is transferred via rollers and slippers to the radial

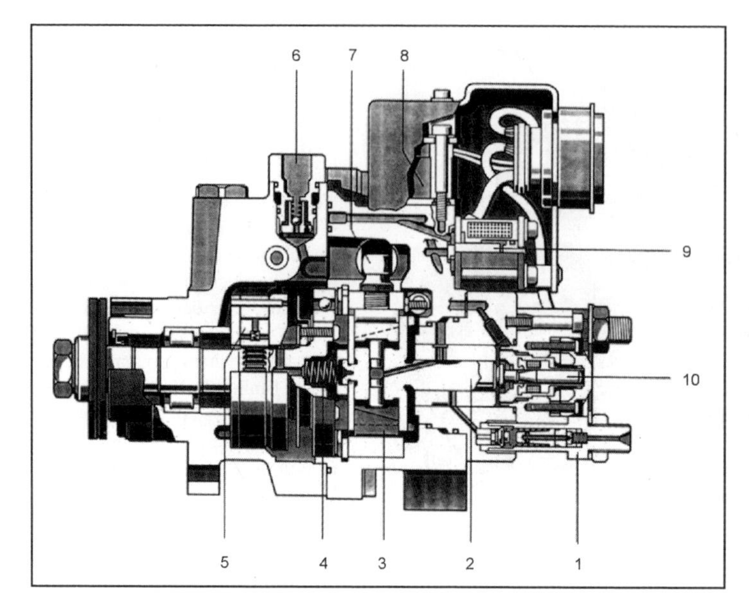

Fig. 12-34 Radial piston distributor injection pump, stroke controlled and electronically controlled.[8] 1, Delivery valve holder; 2, Ignition distributor shaft; 3, Slippers; 4, Return spring; 5, Supply pump; 6, Pressure holding valve; 7, Timing device; 8, Solenoid valves (timing device, return); 9, Solenoid valves (shutoff, charging); 10, Sensor for axial rotor position.

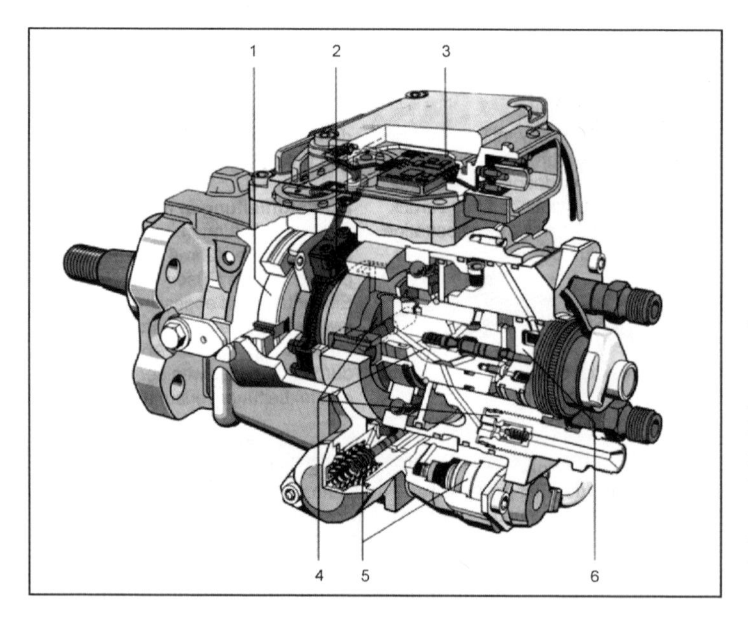

Fig. 12-35 Radial piston pump, cam time controlled with solenoid valve.[9] 1, Vane-type supply pump with pressure control valve; 2, Angle-of-rotation sensor; 3, Pump ECU; 4, Radial piston high-pressure pump with ignition distributor shaft and outlet valve (pressure valve); 5, Timing device, and timing device solenoid valve (pulse valve); 6, High-pressure solenoid valve.

delivery plungers that take over the supply of fuel and thereby generate high pressure. The diameter number of the delivery plungers determines the fuel delivery rate. The short, direct flow of force within the cam drive produces a certain amount of resilience and, hence, a very rigid system that allows high injection pressure.

The fuel is distributed to the engine cylinder via the rotating ignition distributor shaft in which the controlling solenoid valve needle is centrally located. The force-generating actuator solenoid lies stationary in the distributor head. The needle is, therefore, geometrically separated into two parts. The sealed seat and the sealing part of the needle are in the rotating ignition distributor shaft. The magnetic force is transmitted by the external needle part to the concomitantly rotating needle part. In a currentless state, the high-pressure valve is opened by spring force; the pump area can thereby fill the radial pistons with fuel via the low-pressure circuit. After current is applied, the valve closes and the high-pressure delivery begins. The timing device basically has a design similar to the axial piston distributor pump; however, the dimensions are adapted to the increased requirements. The solenoid valves for fuel quantity and start of delivery are controlled analogously to the axial piston distributor pump via the pump ECU that, in turn, receives the required information via the increment angle time signals as described above. Because of the high fuel delivery rate of this pump, a so-called return-flow throttle valve in the area of the pump outlet is usually sufficient; i.e., it is an open system. Hence, the pump and line do not have to be joined by a hermetic seal between the individual injections into the engine cylinder. The radial piston pump is used in engines ranging from passenger car engines to heavy-duty engines.

Whereas conventional distributor injection pumps with edge control are not suitable for generating preinjections by interrupting the supply phases, this is possible with solenoid-valve-controlled systems. Figure 12-36 shows the nozzle needle stroke of an injection system with a solenoid-valve-controlled radial piston pump when using a two-spring nozzle-holder assembly. At low and high speeds, we clearly see the preinjection generated by repeatedly switching the solenoid valve. In the case of low speeds, we see the rate-of-discharge curve formed by the two-spring nozzle-holder assembly.

Common to all distributor injection pumps is that the valve gear is exclusively lubricated by the fuel in contrast to individual injection systems and inline fuel injection pumps. In the latter two systems, the valve gear, i.e., the combination of the cams and tappets is lubricated by engine oil and is, hence, tribologically insensitive in comparison to the fuel-lubricated distributor pump valve gear. The utilized diesel fuel must, therefore, meet a minimum lubrication standard. The new fuel standard DIN EN 590[10] established in the 1990s requires this.

12.5.2.4 Pump Nozzle System

In pump nozzle units (PNU), also termed unit injector systems (UIS) or electronic unit injectors (EUI), the high-pressure-generating pump element and the fuel injector form an assembly. The pump plungers are driven via the engine's own overhead camshaft on which special injection cams are located. Figure 12-37 shows the PNU system in the cylinder head with the example of a passenger car engine and the design of the PNU system. Because of the absence of fuel injection tubing, the fuel volume to be compressed during delivery (dead space) is very small. This allows this system to attain maximum injection pressures in diesel engines. At present, the injection pressure is just over 2000 bar with a potential of 2500 bar. The system is used both in passenger car engines and in commercial vehicle engines. However, the engines must have an overhead camshaft. The arrangement of the pump nozzle element in the cylinder head requires a completely new cylinder head design with an integrated fuel supply and removal system, as well as a particularly rigid and robust camshaft drive. While in the past pump nozzle elements were also used with hydraulic and mechanical control in individual cases, today solenoid-valve-controlled systems predominate.

In the system in Fig. 12-37, preinjection can be used over a wide range of the engine program map via a small, hydraulically actuated bypass plunger. After the first time the nozzle needle is opened, pressure continues to rise, and a bypass plunger is actuated that interrupts the injection and simultaneously increases the nozzle needle opening pressure for the main injection. If the delivery generated achieves this increased opening pressure, the main injection starts. In the future, preinjection and secondary injection (in connection with exhaust aftertreatment) will be realized by repeated triggering of the solenoid valve.

Figure 12-38 gives a summary and qualitative evaluation of the attained injection pressure levels for the different injection systems.

12.5.3 Systems with a Central Pressure Reservoir

Injection systems with a central pressure reservoir are presently termed common rail injection systems.

The common rail injection system allows a freely selectable pressure within given limits. This gives the developer another degree of freedom for optimizing combustion in contrast to cam-controlled injection systems.

The flexibility of being practically unrestricted in setting essential injection parameters has always been a desired goal of diesel injection engineering and opens up new dimensions to combustion developers. In addition to the freely selectable injection pressure, there is, in principle, the possibility of multiple injections independent of a cam

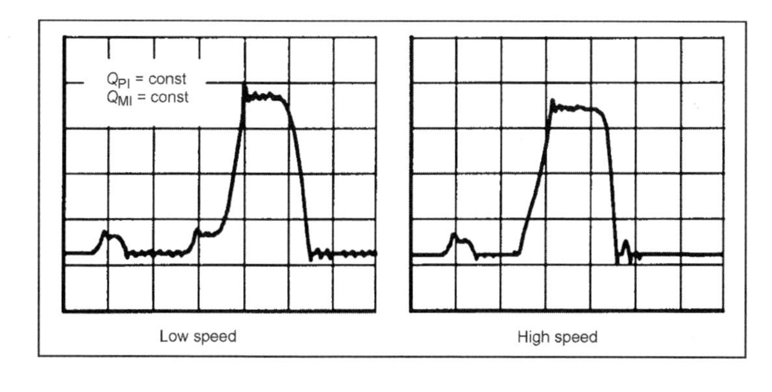

Fig. 12-36 Preinjection realized with a radial piston distributor injection pump in connection with a two-spring nozzle-holder assembly.

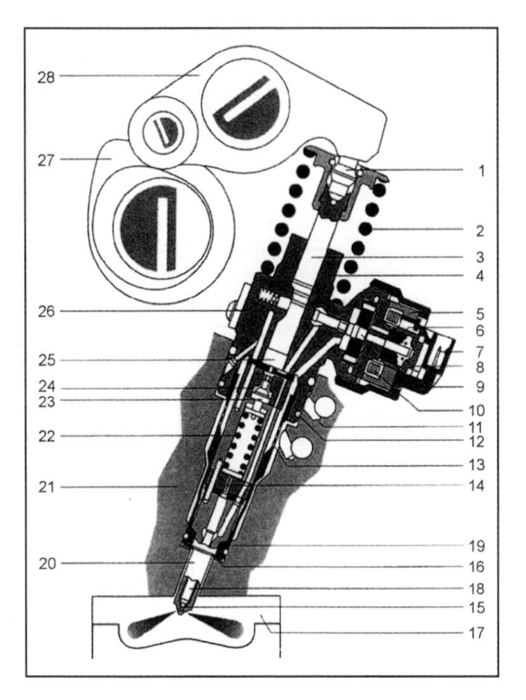

Fig. 12-37 Pump-nozzle unit for passenger car engines.[7] 1, Ball pin; 2, Return spring; 3, Pump plunger; 4, Pump body; 5, Connector; 6, Magnet core; 7, Compensating spring; 8, Solenoid valve needle; 9, Armature; 10, Coil of the electromagnet; 11, Fuel return (low pressure part); 12, Seal; 13, Inlet passages (approximately 350 laser-drilled holes as a filter); 14, Hydraulic stop (damping unit); 15, Needle seat; 16, Sealing washer; 17, Combustion chamber of the engine; 18, Nozzle needle; 19, Tensioning nut; 20, Integrated injection nozzle; 21, Cylinder head of the engine; 22, Compression spring (nozzle spring); 23, Storage plunger; 24, Storage chamber; 25, High pressure area (element chamber); 26, Solenoid valve spring; 27, Camshaft; 28 Rocker arm.

ramp or contour such as in distributor pumps and pump nozzle systems. Physically, the CR system permits injection at any time. The number of injections and the time at which they occur is essentially limited by the cost.

The possibilities for comfortably integrating the system into the engine design and the clearly reduced load on the pump drive in comparison with all other cam-controlled systems makes the common rail system (Fig. 12-39) an increasingly familiar injection system in diesel engines.

12.5.3.1 High-Pressure Pump

At present, two fuel metering approaches are used:

- High pressure blow-off
- The volumetric flow approach

In contrast to a high-pressure blow-off approach, the volumetric flow control combines lower drive power in the program map and a lower injection of heat into the system by returning the fuel into the tank. The necessity still exists of reacting quickly to transient pressure changes via a valve in the high-pressure range of the system. Figure 12-40 shows a high-pressure pump.

The volumetric-flow-regulated high-pressure pump has an effective energy use of 70% to 90% over the pump speed range and pressure range of 200–1500 bar at maximum delivery, and there are large areas above 80%. However, the advantages of this approach are manifested only in partial delivery. Over the entire range of the injection pressure and fuel delivery rate, the efficiency down to the lowest pump speeds is usually above 50%. The effective energy use of the high-pressure blow-off approach under the same conditions is contrastingly below 20%.

In volumetric-flow-regulated high-pressure pumps, the influence of the fuel delivery rate on torque fluctuations (Fig. 12-41) of the pump drive is an important factor. The figure shows rail pressures of 500 and 1500 bar at a pump speed of 1000 rpm. It is representative for the

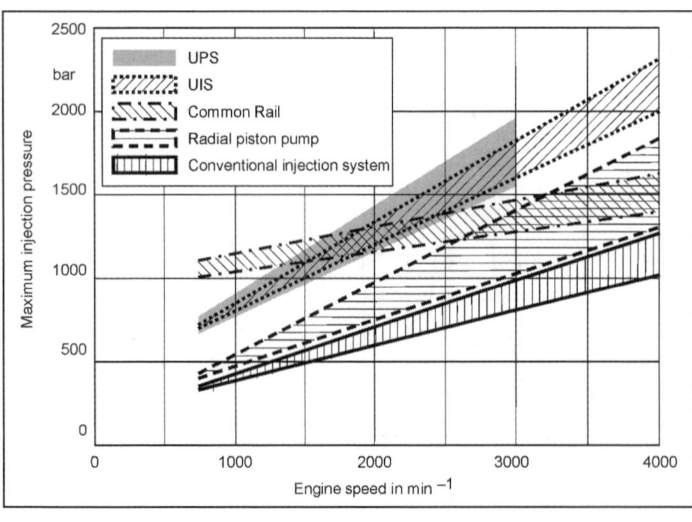

Fig. 12-38 Injection pressures for different injection systems.

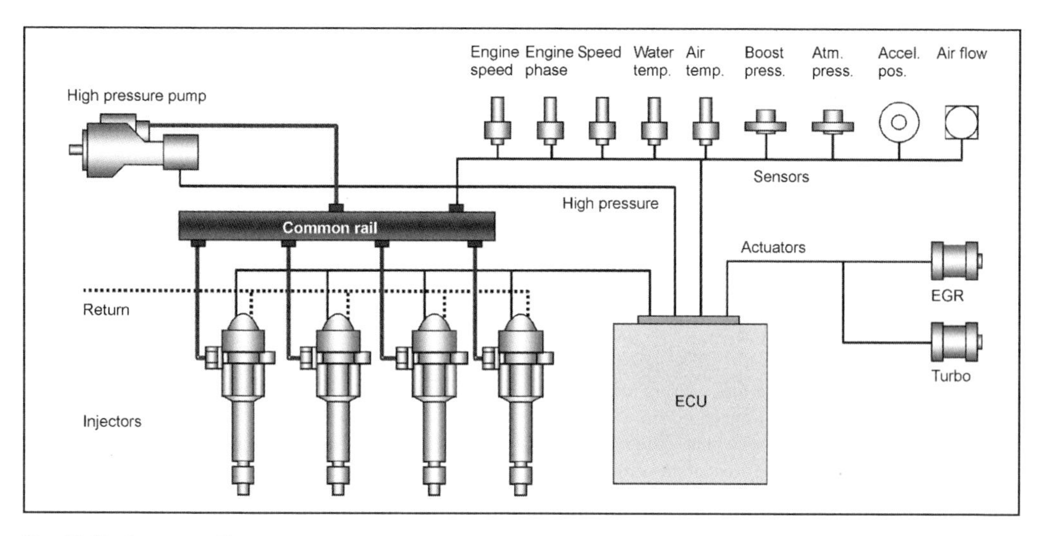

Fig. 12-39 Common rail system.

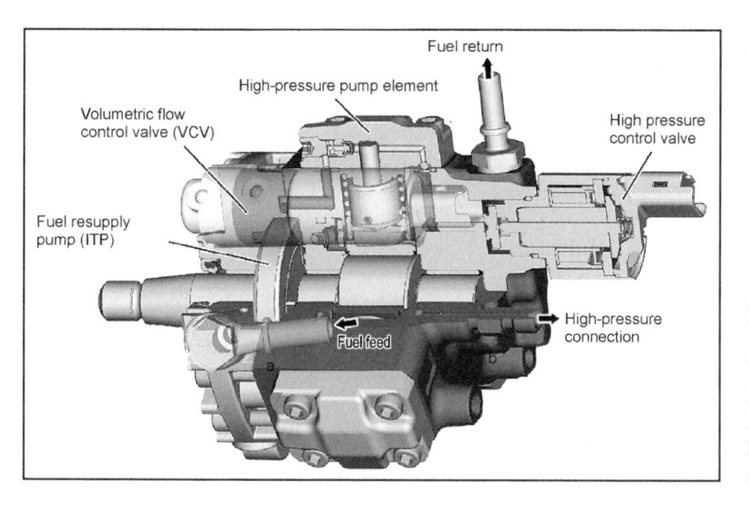

Fig. 12-40 High-pressure pump. 1, Fuel resupply pump; 2, Volumetric flow control valve; 3, High-pressure pump element; 4, High-pressure control valve; 5, a Fuel feed, b High-pressure connection, c Fuel return.

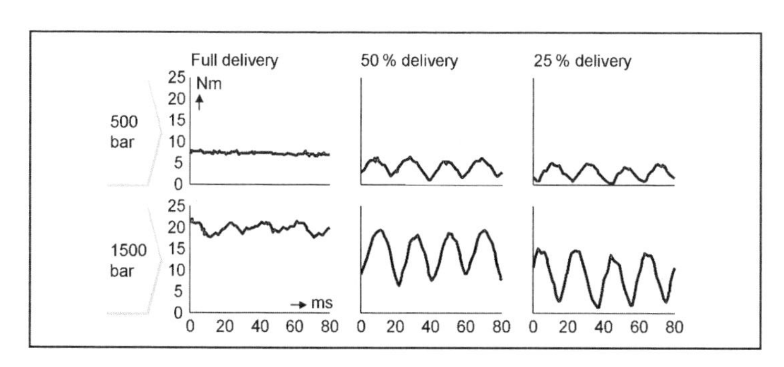

Fig. 12-41 Supply rates and torque fluctuations.

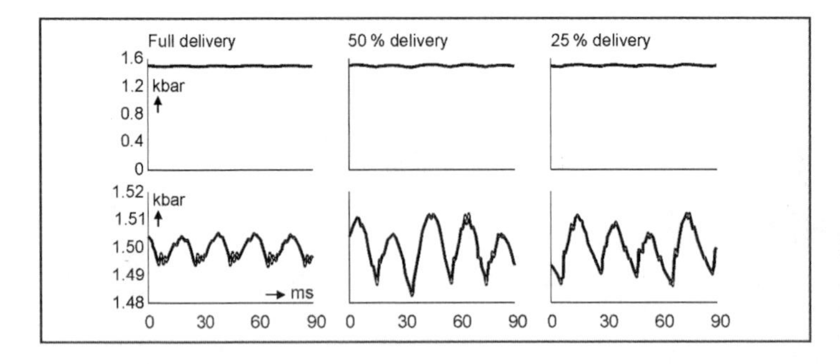

Fig. 12-42 Pressure pulsations in the rail with a different fuel delivery rate and volumetric flow control.

entire pressure range that, as the fuel delivery rate drops, the average torque falls with a moderate rise in torque fluctuations. Overall, we can state that, in comparison with cam-controlled systems, the drive of the common rail high-pressure pump is much more robust.

Figure 12-42 illustrates the influence of the volumetric flow control at 1500 bar on the pressure fluctuations in the rail that can impair the injected fuel quantity. Just as with torque fluctuations in the drive, the fuel delivery rate has practically no influence on the pressure pulsations in the rail. Given a rail pressure of 1500 bar and maximum delivery, pulsations rise of ±5 bar. Given a very low fuel delivery rate of 25%, the width of the fluctuations increases to approximately 15 bar.

As an example, we can compare common rail high-pressure pumps that control pressure via a high-pressure bypass and those that control pressure via a volumetric flow control valve in reference to pressure oscillation behavior. Both control approaches produce similar results. The fear that controlling pressure by means of a volumetric flow valve could induce impermissible pressure oscillations in the rail is, therefore, unfounded. Accordingly, the above-cited advantages of pressure control can be exploited with a volumetric flow control valve without disadvantages. A comparison of the pressure pulsations in the rail with high-pressure bypassing *and* volumetric flow control is shown in Fig. 12-43.

Proportional directional-control valves are used for volumetric flow control, and proportional pressure limiting valves are used for high-pressure blow-off.

The fuel can be predelivered to the high-pressure pump via an electrical presupply pump (perhaps integrated in the fuel tank) or a mechanical supply pump that is separate or integrated in the high-pressure pump. The advantage of the first solution is that after the fuel tank has been emptied, the system can be rapidly refilled, whereas the presupply pump integrated in the high-pressure pump housing has the advantage that there are fewer components in the injection system, and the overall fuel system can be cheaper.

12.5.3.2 Rail and Lines

The rail (Fig. 12-44) serves as a high-pressure reservoir for the fuel that is delivered by the high-pressure pump. Furthermore, it supplies the injectors with the necessary amount of fuel for all operational conditions. The rail is designed so that it can be filled quickly and so that the fluctuations arising during injection can be quickly suppressed. The length and the diameter of the lines between the rail and injectors are designed to promote this.

The goal of such a design is to have a similar pressure for each cylinder at the same engine operation time during each injection since otherwise the injector-to-injector spread can become too great due to the timing, which can cause problems with emissions and driving dynamics.

Rails are forged or welded from drawn steel. Special attention must be given to the high-pressure fatigue strength of the weld connections. The connection to the lines should be such that no tension arises in the weld seam.

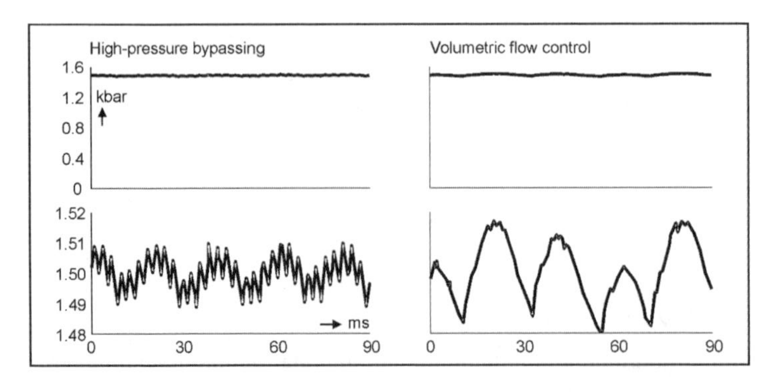

Fig. 12-43 Comparison of pressure pulsations in the rail with high-pressure bypassing (left) and volumetric flow control (right).

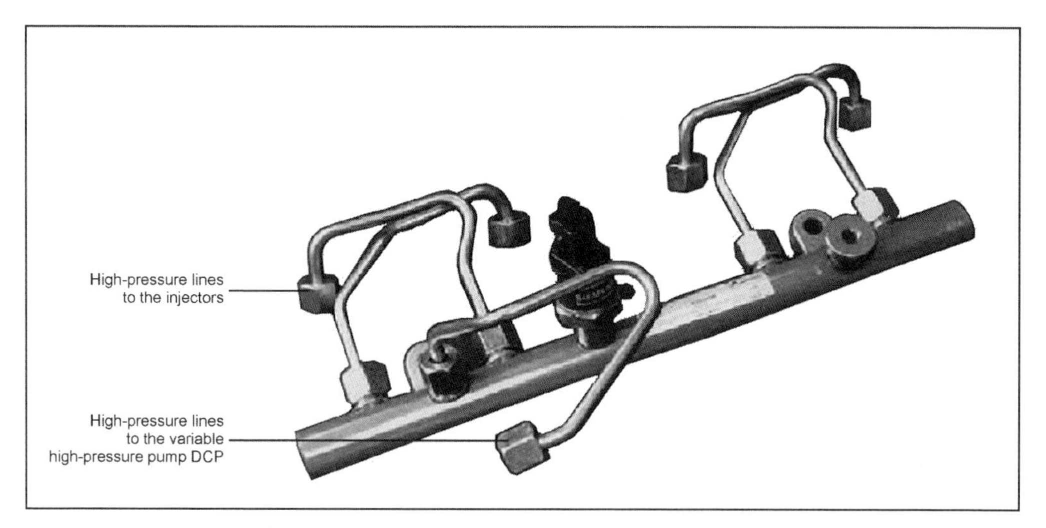

Fig. 12-44 High-pressure rail.

The lines between the pump and rail and between the rail and injectors are made of seamless drawn steel.

12.5.3.3 Injectors

Figure 12-45 shows the layout of the common rail injector using the example of a piezoinjector. The heart of the injector is a piezoactuator that allows relatively low electrical voltages, while, at the same time, satisfies the automotive requirements regarding temperature and vibration. The actuator is able to open or close the servovalve much faster than 100 μs. Together with the harmonized input and output throttle combination to the control area above the nozzle needle, the nozzle opening speed can be influenced and, hence, the rate-of-discharge curve and also the minimum injected fuel quantity that is determined by the minimum operating time. These processes are triggered with practically no response time. It becomes clear that piezoengineering enables highly reproducible fuel injection.

A realized example of such an injector is shown in the enlarged sectional view in Fig. 12-46. The piezoactuator (4) is a multilayer stack in which numerous individual ceramic plates are joined. Initial pressure is applied in a housing. A problem to be solved is temperature compensation. Because of the necessarily wide range of operating temperature in motor vehicles, the expansion of the ceramic plates from temperature is very large in relation to the lengthwise expansion in response to voltage. This temperature is compensated by selecting a suitable material for the pretension spring surrounding the piezostack together with the installation housing, as well as appropriately setting the play of the actuator. On the one hand, the injector cannot be open too long (too little play), which can cause engine damage, and, on the other hand, the

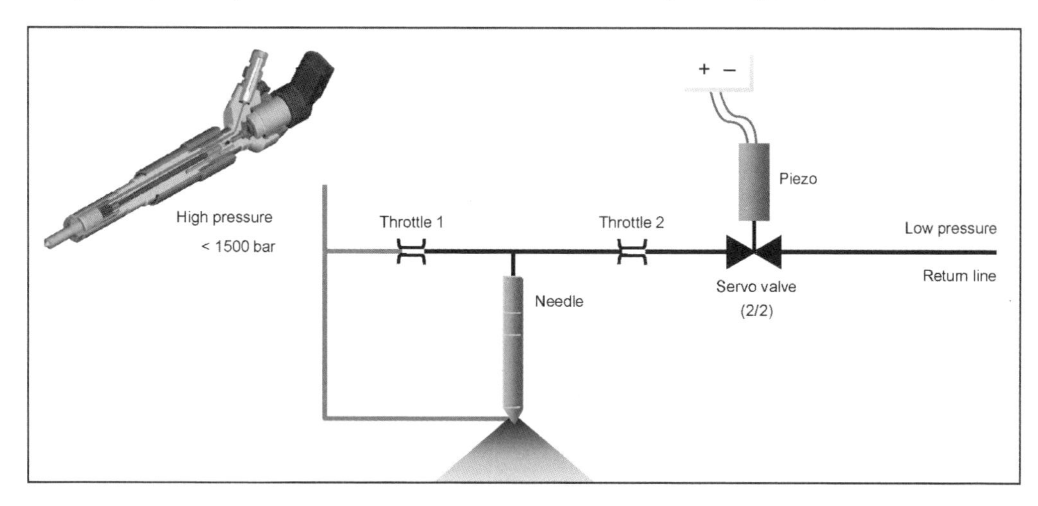

Fig. 12-45 Common rail piezoinjector.

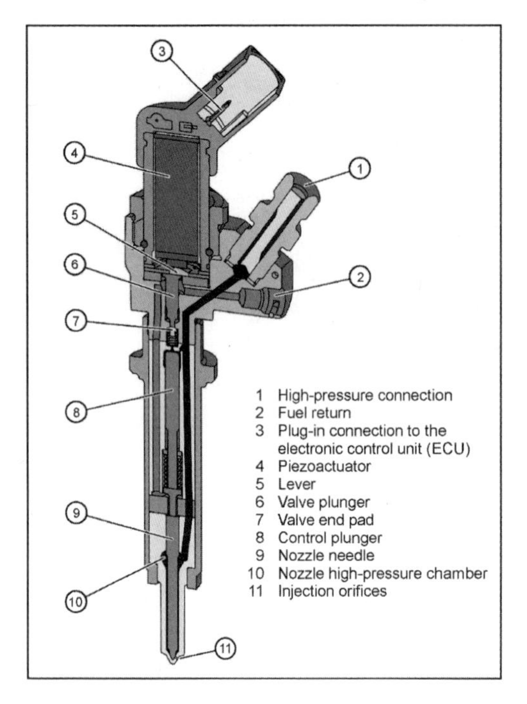

Fig. 12-46 Sectional view of a piezoinjector.

injector cannot remain closed during very short operating times (too little play), which notably increases the combustion noise in the absence of pilot injection. Another peculiarity is the servovalve that opens inward to the high-pressure area instead of outward in contrast to a magnet-actuated valve. The reason is that when voltage is applied, the piezoelement expands and also exerts a large outward force. This makes it more functionally appropriate for the piezoelement to open the valve against high pressure and yields a simpler injector design than if the movement is in the opposite direction when voltage is applied to the piezoelement when the servovalve is being closed.

Overall, this actuator design allows a control valve lift of approximately 30 μm to be maintained over the entire temperature range of a vehicle engine from −30 to +140°C.

The functioning of this construction can be seen in Fig. 12-47. If the injector is not controlled (left half of the figure), there is fuel at the high pressure of the rail both in the injector control area (2) and in the high-pressure chamber (3) of the nozzle. The hole for the fuel return (5) is sealed by the valve end pad (4) with a spring. The hydraulic force exerted by the high fuel pressure on the nozzle needle (6) in the control area (2) (F1) is greater than the hydraulic force acting on the nozzle tip (F2) since the area of the control plunger in the control area is greater than the free area under the nozzle needle. The nozzle of the injector is closed.

If the injector is triggered (right half of the figure), the piezoactuator (7) presses via the lever (8) on the valve plunger (9), and the valve end pad (4) opens the hole that connects the control area (2) with the fuel return. This causes the pressure in the control area to drop, and the hydraulic force acting on the nozzle needle tip (F2) is greater than the force acting on the control plunger (F1) in the control area. The nozzle needle (6) moves upward, and the fuel passes via the injection orifices into the combustion chamber of the engine.

When the engine idles, the valve that connects the control area with the fuel return is closed by spring force along with the injector nozzle.

Figure 12-48 shows the performances of second generation piezoinjectors. Assuming a 0.5 l single stroke volume, and given an advantageous overall adjustment of the injectors, i.e., with sufficiently fast opening and closing flanks, we achieve up to 1500 bar minimum injection quantities below 1.5 mm³. In the low-pressure range, we get as low as 0.7 mm³. At the same time, the distance of

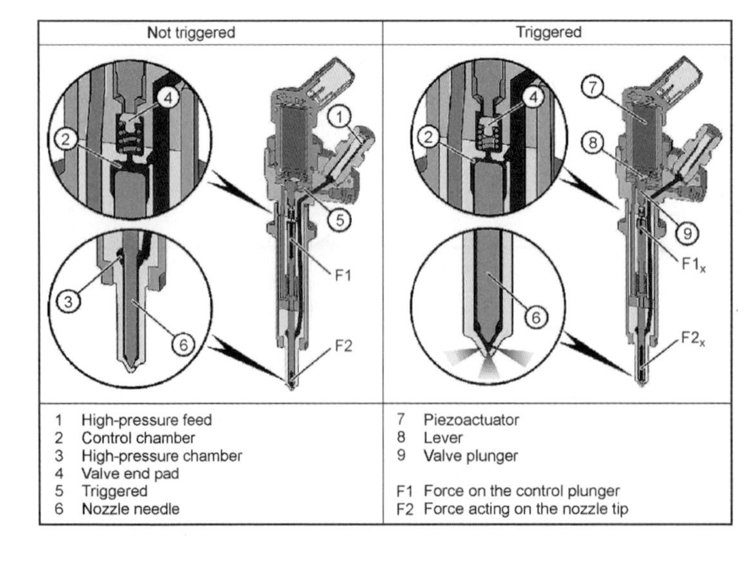

Fig. 12-47 Injector function.

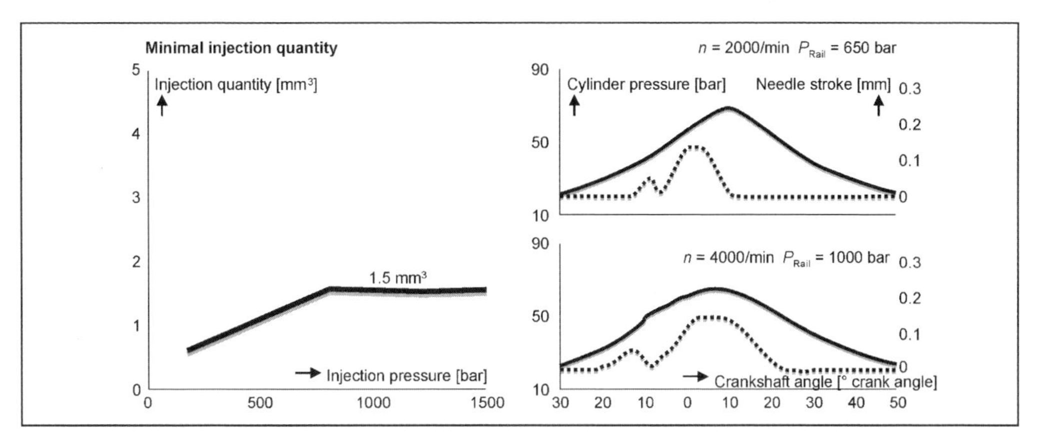

Fig. 12-48 Performance of 2nd generation CR injector with piezoengineering.

the start of injection from preinjection to the main injection can be very short. At a speed of 2000 rpm, the smallest distance is a crankshaft angle of less than 6°; at a speed of 4000 rpm, a distance of approximately 12° crankshaft angle is still possible. There is no restriction to larger intervals between the start of injection.

From both diagrams, we can see that preinjection is possible within the entire program map range, i.e., both within the entire pressure range and within the entire speed range.

12.5.3.4 Injection Nozzle

The task of the injection nozzle is to atomize and distribute the fuel to attain the desired micromixture and macromixture.

The injection nozzles that are used in the common rail injection system are sac-less nozzles and blind-hold nozzles (see Section 12.5.4).

12.5.3.5 Electronics

The system block diagram in Fig. 12-49 shows the sensors and actuators, illustrating the complete functional scope of the common rail injection system.

All driver stages, including energy recovery, are integrated within the engine electronics. In contrast to solenoids, piezoelements require a totally new driver stage approach. Whereas with solenoids current flows during the entire valve opening phase regulated by peak and hold, the piezoactuator is electrically similar to a capacitor. The piezoelement is charged and lengthens; at the end, it is

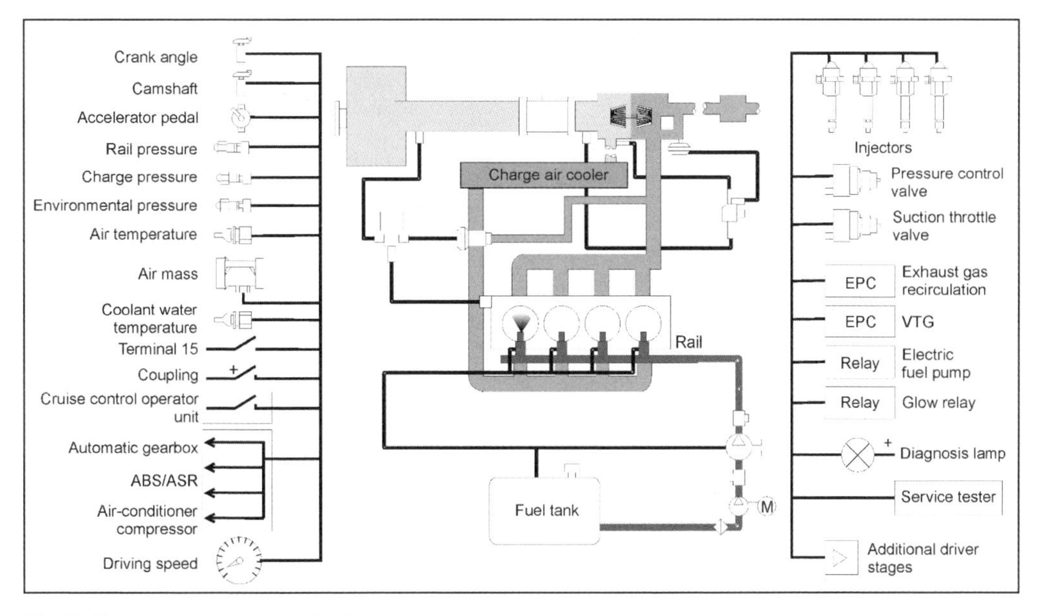

Fig. 12-49 System block diagram for CR.

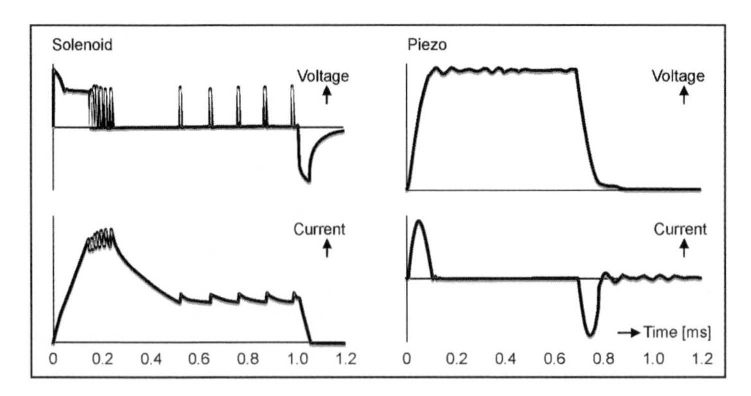

Fig. 12-50 Comparison of the electrical properties of solenoid and piezotechnology.

discharged and returns to its initial length. A comparison of the electrical properties of solenoids and piezoelements is shown in Fig. 12-50.

Another aspect of piezotechnology is electromagnetic compatibility. Strong voltage peaks are expected because of the fast switching times. However, this is not a major problem, since piezoactuators can be charged and discharged in a sinusoidal oscillation. This technology is, therefore, no more problematic for electromagnetic compatibility (EMC) than solenoid technology with cyclical peak and hold phases.

Furthermore, piezotechnology has the potential of energy recovery. For example, approximately 50% of the utilized energy can be recovered in piezotechnology given extremely fast switching.

The absence of magnetic remanence allows a high degree of repeatability in piezoactuators from shot to shot, and it allows individual injections to be rapidly fired to form an injection series, which allows the targeted control of combustion.

The intervals between injections are limited only by the speed of the driver stage.

12.5.3.6 Developmental Trends

General developmental trends for the common rail injection system of the future are the following:

- Increase of the injection pressure
- Flexible injection rate control
- Increased use of closed-loop control strategies
- Greater compactness of the components
- Reduction of tolerances
- Regulated exhaust aftertreatment

Above all, we can expect a further increase of the injection pressure for improved fuel preparation and combustion. While in second generation common rails with piezotechnology the injection rate is a question of the basic design, the next large developmental step of piezotechnology will permit flexible injection rate control. In addition, closed-loop control strategies will increasingly be used.

It is mainly the trend toward small, low-consumption cars that is driving the increased compactness of the injector and pump. This development will also further increase the demands made on the tolerances and the minimum injection (Fig. 12-51). With piezotechnology, given adapted injection orifices, preinjections of 0.5 mm^3 will be feasible.

Bibliography

[1] Schöppe, D., Anforderungen an moderne Dieseleinspritzsysteme für das nächste Jahrhundert, Conference on Diesel Engine Technology, Esslingen, 1997.

[2] Egger, K., and D. Schöppe, Diesel Common Rail II–Einspritztechnologie für die Herausforderungen der Zukunft, International Vienna Engine Symposium, 1998.

[3] Klügl, W., K. Egge, D. Schöppe, and H. Freudenberg, The Next Generation of Diesel Fuel Injection Systems Using Piezo Technology, FISITA World Automotive Congress, Paris, 1998.

[4] Piezo Common Rail PCR2 DW10TD, After Sales Documentation 7/2000.

[5] Egger, K., U. Lingener, D. Schöppe, and J. Warga, Die Möglichkeiten der Einspritzung mit einem Piezo-Common-Rail-Einspritzsystem für Pkw, Int. Vienese Engine Symposium, 2001.

12.5.4 Injection Nozzles and Nozzle-Holder Assemblies

The fuel delivered by the pump element is injected through the injection nozzle at a high pressure into the combustion chamber of the diesel engine and distributed very finely. The nozzle itself is mounted in a nozzle-holder assembly that is screwed or inserted into the cylinder head to form a seal. The high-pressure element, nozzle-holder assembly, and nozzle form a constructive unit in the pump nozzle unit. With common rail systems, the injector as a control element also serves the function of the nozzle-holder assembly. The main tasks of the nozzle in combination with the nozzle-holder assembly are to form the rate-of-discharge curve, atomize and distribute the fuel in the combustion chamber, and seal the hydraulic system from the combustion chamber. The nozzle construction and design need to be precisely harmonized with the different engine conditions. These are primarily:

- Combustion processes [direct injection (DI), indirect injection (IDI)]
- Geometry of the combustion chamber
- Number of injection jets, the spray shape, and spray direction
- Injection time
- Injection rate

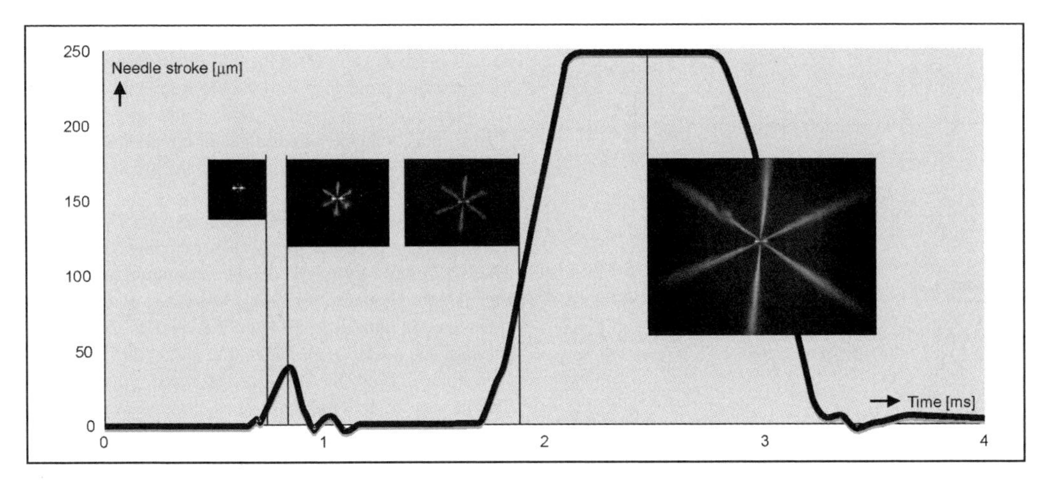

Fig. 12-51 Jet symmetry over the needle stroke at 100 bar.

Figure 12-52 shows a few basic designs of pintle nozzles (for IDI engines) and hole-type nozzles (for DI engines). In each case, the nozzles open inward. Outward-opening nozzles are no longer mass produced. The pintle profile in pintle nozzles can be used to adapt the nozzle-stroke-related opening cross section and, hence, the fuel flow or rate-of-discharge curve to engine requirements. In flattened pintle nozzles, a larger duct is released so that combustion-chamber-side coking of the nozzle needle can be reduced.

For reasons of strength, the design of the nozzle cone shape in hole-type nozzles is very important. In addition, the size of the residual volume between the tip of the nozzle needle and the inner contour of the nozzle body between the nozzle needle seat and the injection orifice is important because of the fuel volume inside that does not participate in combustion. The smaller this volume, the fewer hydrocarbons evaporate from this volume that can then be found in the exhaust as uncombusted HC emissions. Blind-hole nozzles usually have a greater cone strength and greater residual volume than sac-less nozzles in which the injection orifice is found in the area of the nozzle seat. The residual volume is thereby separated

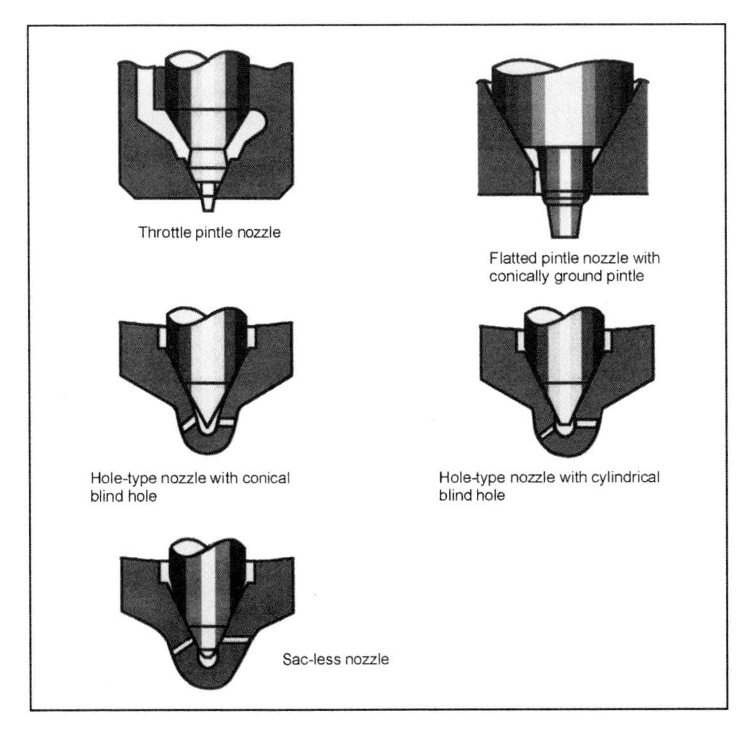

Fig. 12-52 Basic designs of injection nozzles.

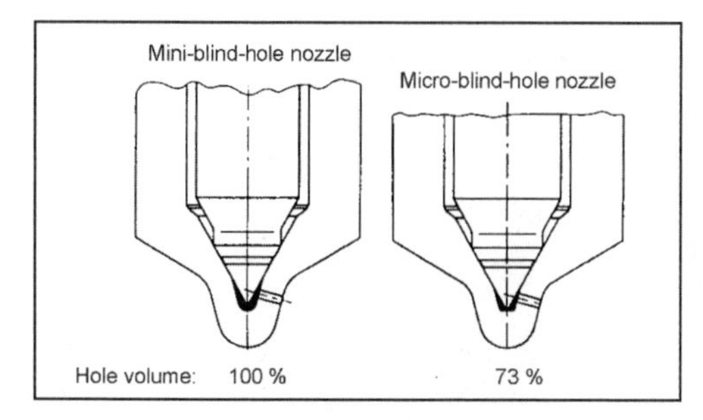

Fig. 12-53 Miniblind-hole and microblind-hole nozzles (Bosch).

from the combustion chamber, and only the fuel remaining in the individual injection orifices can evaporate. Most injection orifices themselves are presently made by electroerosion and are no longer mechanically drilled. Laser methods are now also being investigated.

Figure 12-53 shows new developments in nozzle needle and cone designs: the miniblind-hole nozzle and microblind-hole nozzle. The advantage of these nozzles in comparison to sac-less nozzles is a more even injection behavior of the individual nozzle holes with a low residual volume.

This is important, in particular, for the minuscule amounts involved in preinjection and postinjection. In sac-less nozzles, an uneven injection pattern of preinjected fuel (1–3 mm^3 per injection) can arise when the strokes are very small because of manufacturing-related tolerances. If the seat is moved back and viewed in the direction of flow, uneven cross sections in the seat area and, hence, uneven pressure do not as strongly affect injections in which the seat throttle predominates (minimum strokes) since the injection orifice starts a few millimeters from the seat. Figure 12-54 compares the

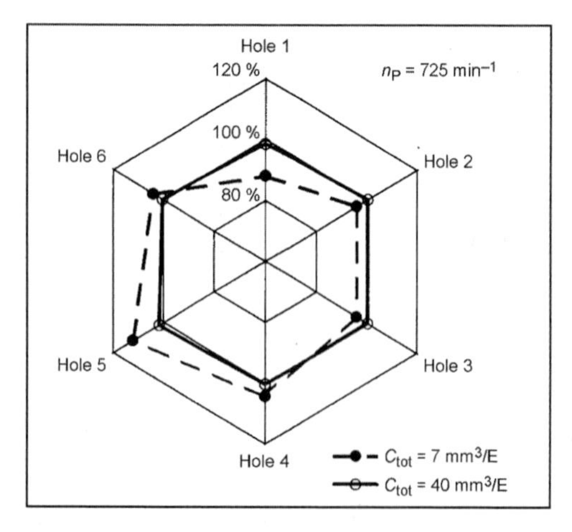

Fig. 12-54 Individual injected fuel quantities per injection orifice for a blind-hole nozzle at two different levels of fuel.

individual injection quantity per injection orifice with small and average injected fuel quantities determined with a measuring procedure according to Ref. [11].

The number of injection orifices strongly depends on the combustion behavior and the air circulation (swirl) in the combustion chamber. In general, the greater the swirl, the fewer injection orifices are necessary and vice versa. Normally, engines with direct injection have 5 to 12 injection orifices. In passenger car diesel engines, the number is 5 to 7 holes. Today, the standard minimum hole diameter is approximately 0.12 mm. More important than the diameter is the flow through the nozzle holes. From drilling the nozzle holes, burrs arise at the start of the nozzle hole, i.e., on the inside of the nozzle, that strongly influences the flow behavior of each individual hole. For this reason, nozzles are rounded with hydroerosion today after drilling. In addition to deburring, this manufacturing process also yields the conical shape of the injection orifice with a larger diameter at the start of the injection hole and a smaller diameter at the end of the injection orifice on the external contour of the cone. This evens the speed profile in the nozzle hole and prevents cavitation zones. Figure 12-55 depicts the beginning of the nozzle hole with and without hydroerosive rounding. This can more than halve the flow tolerance.

Another possibility of evening out the injection jets for each nozzle hole is to improve the guidance of the nozzle needle by means of a dual needle guide (Fig. 12-56).

Nozzle-holder assembly. The nozzle is, as mentioned, installed in the nozzle-holder assembly. The nozzle needle is closed by the initial pressure from the compression spring in the nozzle-holder assembly. If the hydraulic force [proportional pressure and ($d^2_{needle} - d^2_{seat}$)] exceeds the initial force, the nozzle opens. In principle, the fuel must overcome two throttling points: First, the stroke-related seat throttle (variable throttle), and then the fixed restriction characterized by the injection orifice geometry. For small strokes, the seat throttle predominates. When the nozzle is completely open and the nozzle needle lies on the mechanical stop, the injection orifice geometry determines the flow cross section. Figure 12-57 shows the characteristic nozzle flow of a hole-type nozzle. In the area of the smallest strokes, the seat throttle determines the flow. This steeply increases with the stroke. In this area, the so-called ballistic area of the nozzle needle movement, manufacturing tolerances and set tolerances play a very large role.

To shape the rate-of-discharge curve, particularly in edge-controlled injection systems, the two-spring nozzle-holder assembly is frequently used (Fig. 12-58). First, the initial force of the spring is overcome in the top part of the nozzle-holder assembly. The opening pressure is approximately 120 to 180 bar, and after passing through the plunger lift to port closing (a few hundredths of a millimeter), the initial force of the second (bottom) spring is overcome. The opening pressure of this second stage is then 250 to more than 300 bar. This makes it possible to have an additional preinjection in the low-speed range

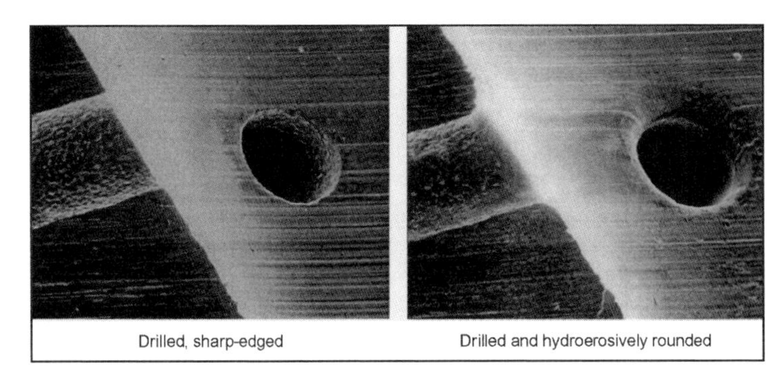

Fig. 12-55 Nozzle hole inlet with and without hydroerosive rounding (Bosch).

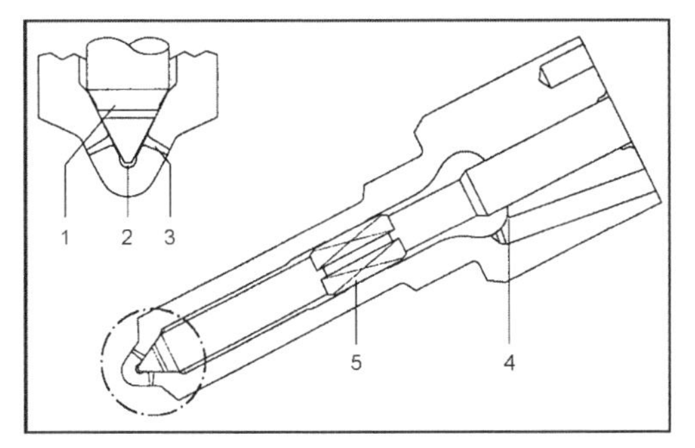

Fig. 12-56 Dual needle guide of a sac-less nozzle.[2] 1, Seat geometry, stable over the long term; 2, Minimal dead space; 3, Injection orifices, conically and hydro-erosively rounded; 4, High-pressure-resistant internal geometry; 5, Dual needle guide.

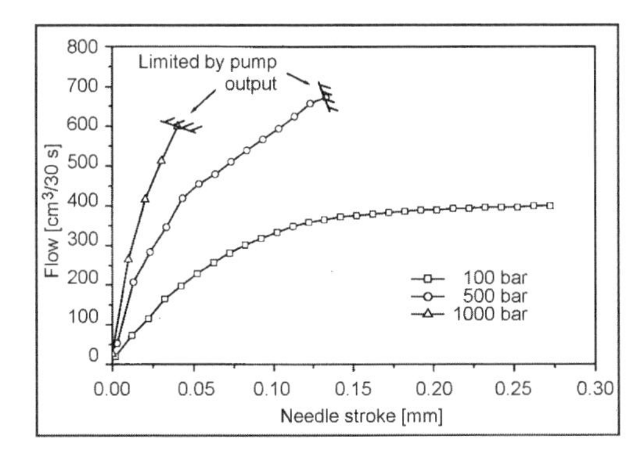

Fig. 12-57 Volumetric flow of a nozzle-holder assembly (sac-less nozzle) that depends on the nozzle needle stroke for three different pressures.

(see Fig. 12-36). At higher speeds, the pressure builds up so strongly and quickly that the first stage is immediately overcome, and a normal needle lift characteristic arises.

In the common rail injectors built today, the nozzle needle is opened and closed purely by hydraulic force (see Section 12.5.3). In the future, developments are conceivable in which pressure modulation during injection in connection with spring-loaded nozzle needles will enable rate-of-discharge curves with common rail injectors. In addition, developments are underway in which the nozzle hole cross section changes depending on the needle stroke, so-called Vario nozzles or register nozzles. These nozzles open outward. The advantage of these different constructions is that only a small cross section is exposed and the jet preparation is improved while the engine is idling or under a partial load.[2]

Left: 1, Edge filter; 2, Inlet passage; 3, Pressure pin; 4, Intermediate disk; 5, Nozzle-retaining nut; 6, Head thickness; 7, Nozzle; 8, Locating pins; 9, Compression spring; 10, Shim; 11, Leak fuel hole; 12, Leak fuel connecting thread; 13, Holding element; 14, Connecting thread; 15, Sealing cone.

Right: 1, Holding element body; 2, Needle movement sensor; 3, Compression spring 1; 4, Guide washer; 5, Compression spring 2; 6, Pressure pin; 7, Nozzle-retaining nut.

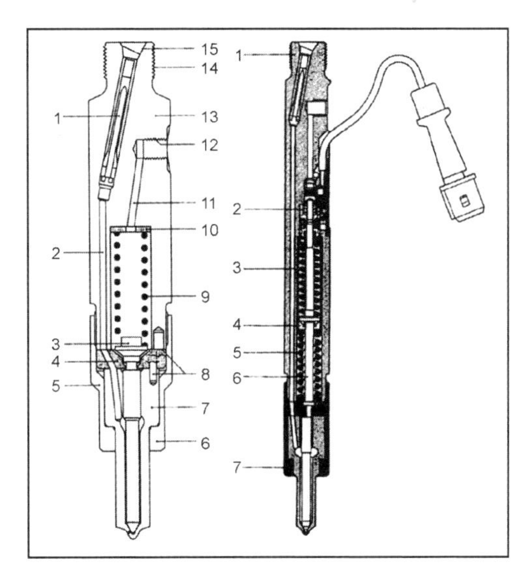

Fig. 12-58 Conventional nozzle-holder assembly (left) and two-spring nozzle holder assembly with an integrated needle motion sensor to determine the start of injection (right).[9]

12.5.5 Adapting the Injection System to the Engine

For diesel engines to provide the best results at every working point corresponding to requirements, the entire injection system must be exactly adapted to the engine. The phrase "the application of the injection system to the engine" is used.

To best solve the individual tasks defined in Section 12.5.1, numerous geometric parameters of the injection system components and working-point-related input variables of the injection system must be determined and implemented corresponding to the target values. The electronically controlled systems offer many more degrees of freedom and possibilities for optimization than conventional, mechanically regulated systems. For example, the following important engine and vehicle-related restrictions must be observed when determining the fuel mass to be injected:

- Smoke limit (especially at a full load)
- Maximum permissible cylinder pressure
- Exhaust temperature
- Speed of the engine
- Torque and speed limits

The required injection volume per work cycle and cylinder for the four-stroke engine is calculated using the following equation:

$$V_K = \frac{P_e \cdot b_e \cdot 2}{z \cdot n_M \cdot \rho_K} \tag{12.3}$$

P_e = Effective performance of the engine
b_e = Specific fuel consumption of the engine (mass/performance and time)
z = Number of cylinders
n_M = Engine speed
ρ_K = Fuel density

The implementation of these requirements by the injection system depends on the fuel metering principle. In conventional edge-controlled or directly lift-controlled pumps, the volume $V_{1\ \text{Stroke}}$ delivered by the pump per stroke depends only on the cross section of the piston and the size of the effective stroke:

$$V_{1\ \text{Stroke}} = A_{2\ \text{Piston}} \cdot h_{3\ \text{Efficiency}} \tag{12.4}$$

$A_{2\text{Piston}}$ = Cross-sectional area of the pump plunger
$H_{3\text{Efficiency}}$ = Effective delivery stroke of the pump plunger

In cam-time-controlled systems, the delivered quantity per injection depends on the closing time of the solenoid valve, the plunger cross section, and the plunger speed:

$$V_{1\ \text{Stroke}} = A_{2\ \text{Piston}} \cdot v_{2\ \text{Piston}} \cdot \Delta t_{SD} \tag{12.5}$$

$A_{2\text{Piston}}$ = Cross-sectional area of the pump plunger
$v_{2\text{Piston}}$ = Average speed of the pump plunger during delivery
Δt_{SD} = Closing time ($\hat{=}$ delivery time) of the control valve

The predelivery and postdelivery effects, as well as the delivery during opening and closing of the high-pressure solenoid valves, are not included. The fuel volume leaving the nozzle is contrastingly simplified by

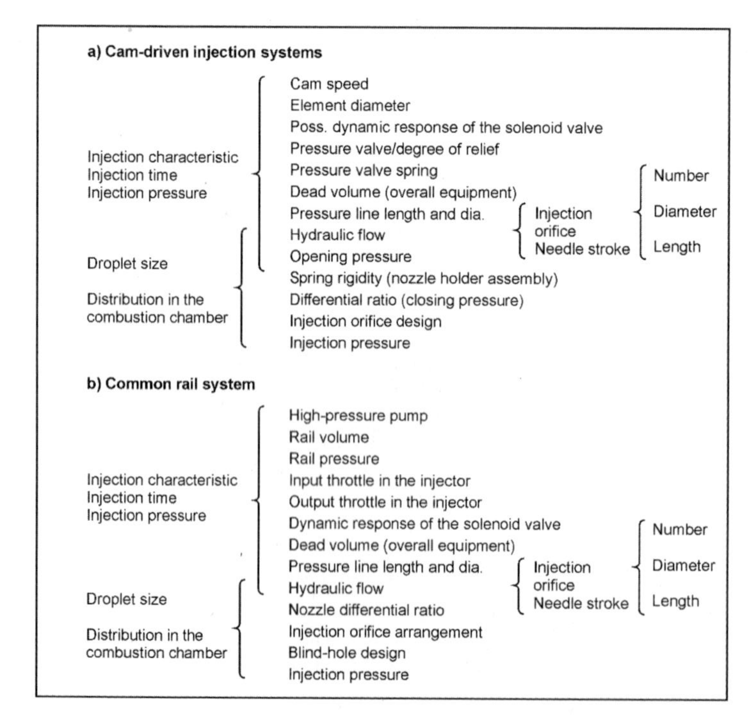

Fig. 12-59 Important constructive design and adaptation parameters of injection systems.[2]

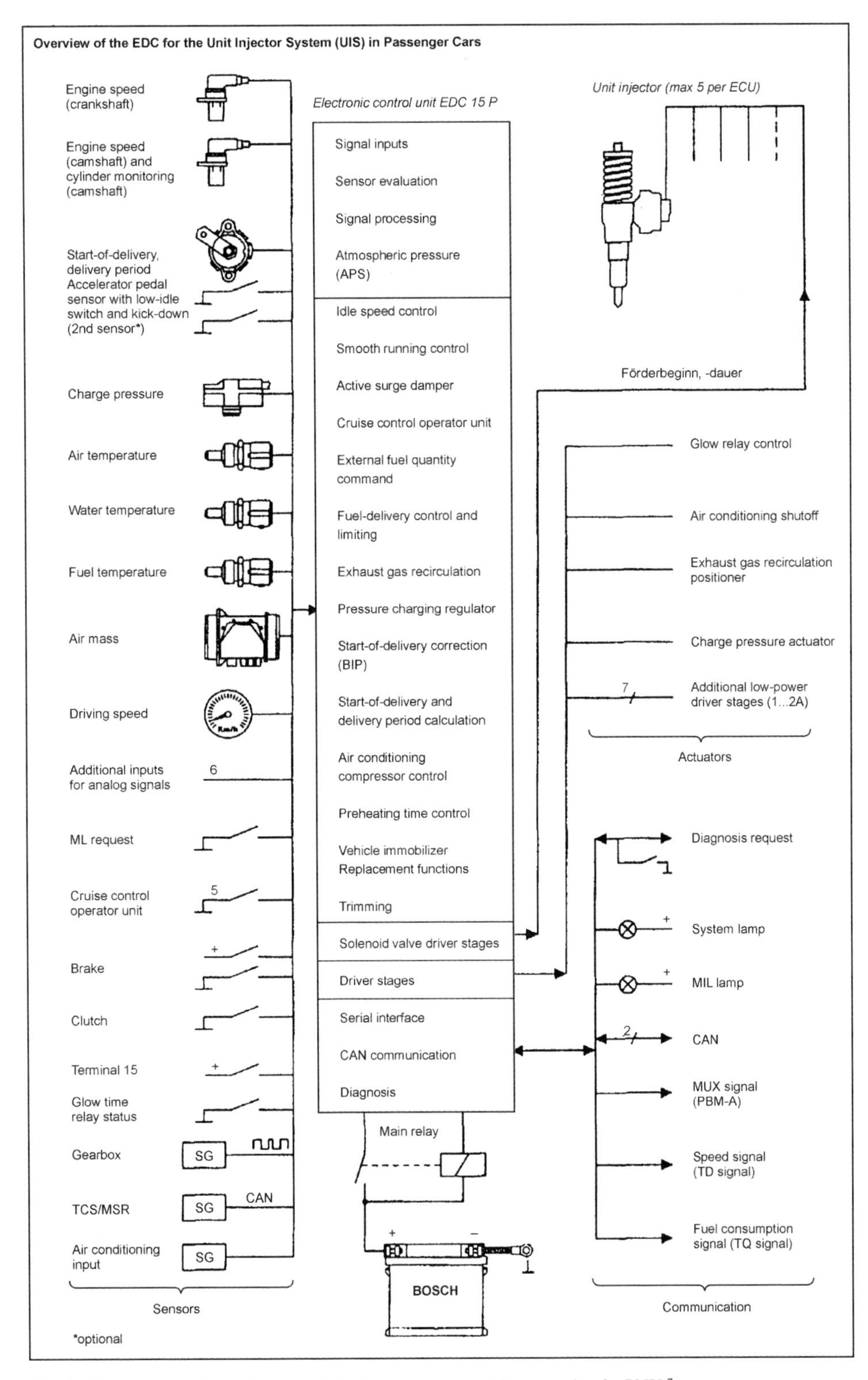

Fig. 12-60 System configuration of an injection system using the example of a PNU.[7]

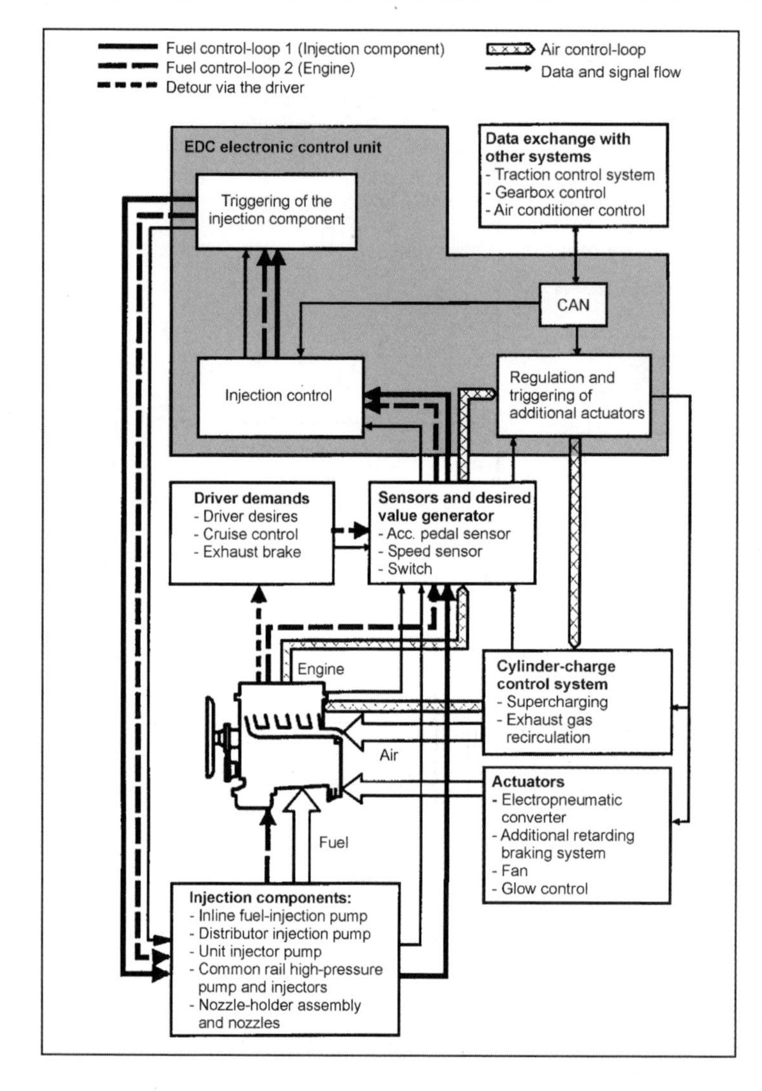

Fig. 12-61 Basic design of the electronic diesel control system.[12]

$$V_{\text{Aus}} = A_D \cdot \Delta t \cdot \alpha \cdot \sqrt{\frac{2}{\rho_K} \cdot \Delta p} \tag{12.6}$$

A_D = Geometric nozzle hole cross section
Δt = Injection time
α = Flow factor
ρ_K = Fuel density
Δp = Differential pressure (fuel side/combustion chamber side)

The injection time can be calculated in simplified form from the needle lift signal or the control time of the injector side, electrically actuated valve in common rail systems. It must be remembered that the metering cross section of the nozzle and the differential pressure between the inside of the nozzle and the combustion chamber during the injection phase are not constant. The same holds true for the flow coefficient. In addition, it must be remembered when designing the high-pressure pump that the fuel at high pressures above 1000 bar cannot be assumed to be incompressible. The high-pressure pump must, therefore, be larger in reference to its delivery capacity depending on the dead space and available pressure and temperature levels to cover the "storage behavior" of the fuel volume to be compressed. Powerful tools are available for the numeric simulation of injection systems and for determining processes that cannot be measured or can be measured only with difficulty in the pump, line, nozzle, solenoid valve, and injector.[2,13]

Figure 12-59 provides a general overview of the hardware-side adaptation parameters of the injection system for the engine application for cam-driven injection systems and for a common rail system. In addition, we have the operating parameters such as temperature, charge pressure, air pressure, exhaust-gas recycling, and information from other vehicle systems such as electronic stabilization program (ESP) or acceleration slip regulation (ASR), as well as the driver's wishes (gas pedal position, cruise control), and information from the sensors of the exhaust aftertreatment system.

In the past, all of these tasks were accomplished only by mechanical engine management in the form of the mechanical control of diesel injection. None of these requirements related to vehicle and engine operation, and the driver's wishes could be implemented. The mechanical control was restricted to the basic functions of operating the engine such as idling control, maximum speed governing, full-load torque control, manifold-pressure-dependent fuel quantity compensation, atmospheric-pressure-dependent full-load torque control, and temperature-dependent fuel quantity compensation (such as while starting). Only with the introduction of electronic diesel control could the above-cited requirements be comprehensively covered in the application.

Electronic diesel control (EDC) can be divided into three system blocks.

Sensors and desired value generators detect the operating conditions of the engine and the set points and convert physical quantities into electrical signals so that they can be processed in the second block, the electronic control unit.

In the *electronic control unit,* this information is processed according to mathematical guidelines (control and regulating algorithms). The electronic control unit also provides the electrical output signals for the actuators and is the interface for other systems and for diagnosis.

The third block consists of the *actuators.* They convert the electrical output signals of the electronic control unit into mechanical quantities such as triggering the solenoid valve for metering fuel.

Figure 12-60 shows an overview of the EDC system for a pump-nozzle unit in passenger cars.

The basic design of an electronic diesel control is shown in Fig. 12-61. Strictly speaking, "diesel control" covers both control and regulation since in many cases the actuators are activated based on input variables by predetermined data program maps or characteristics without the reaction directly being checked. On the other hand, in a series of cases, reactions such as the speed of the engine in the idling speed regulation system and the nozzle needle movement in the injection start regulation system are measured and used to activate the actuators.

The electronic control unit in the electronic diesel control system is, hence, strictly speaking a control and regulation unit. For further details concerning electronic engine management, see Chapter 16.

Bibliography

[1] Pauer, T., R. Wirth, and D. Brüggeman, Zeitaufgelöste Analyse der Gemischbildung und Entflammung durch Kombination optischer Messtechniken an DI-Dieseleinspritzdüsen in einer Hochtemperatur-Hochdruckkammer, 4th International Symposium for Combustions Diagnostics, Baden-Baden, May 18/19, 2000.

[2] Mollenhauer, K. (Pub.), Handbuch Dieselmotoren, 2nd edition, Springer, Berlin, Heidelberg, 2002.

[3] Härle, H., Einfluss des Einspritzverlaufs auf die Emissionen des Nkw-DI-Motors, Conference on Diesel and Gas Direct Injection, Berlin, December 9/10, 2000 (see also Ref. [13]).

[4] Eichlseder, H., Der Einfluss des Einspritzsystems auf den Verbrennungsablauf bei DI-Dieselmotoren für Pkw. 5, Tagung "Der Arbeitsprozess des Verbrennungsmotors," Graz, 09/1995.

[5] Chmela, F., P. Jager, P. Herzog, and F. Wirbeleit, Emissionsverbesserung an Dieselmotoren mit Direkteinspritzung mittels Einspritzverlaufsformung, in MTZ 60 (1999) No. 9, pp. 552–558.

[6] Robert Bosch GmbH (eds.), Technische Unterrichtung Kraftfahrzeugtechnik: Diesel-Reiheneinspritzpumpen PE, 1998/99 edition.

[7] Robert Bosch GmbH [eds.], Technische Unterrichtung Kraftfahrzeugtechnik: Diesel-Einspritzsysteme Unit Injector System/Unit Pump System UIS/UPS, 1999 edition.

[8] Lewis, G. R., Das EPIC-System von Lucas, in MTZ 53 (1992) No. 5, pp. 224–229.

[9] Robert Bosch GmbH [eds.], Technische Unterrichtung Kraftfahrzeugtechnik: Diesel-Radialkolben-Verteilereinspritzpumpen VR, 1998/99 edition.

[10] DIN Deutsches Institut für Normung [eds.], DIN EN 590 (Ausgabe 2000–02), Kraftstoffe für Kraftfahrzeuge.–Dieselkraftstoff.–Anforderungen und Prüfverfahren, Beuth, Berlin, 2000.

[11] Tschöke, H., A. Kilic, and L. Schulze, Messadapter für Mehr-lochdüse, Offenlegungsschrift DE 199 09 164 A1 of 9/7/00.

[12] Robert Bosch GmbH (Pub.), Technische Unterrichtung Kraftfahrzeugtechnik: Elektronische Dieselregelung EDC, 2001 edition.

[13] Tschöke, H., and B. Leyh, Diesel- und Benzindirekteinspritzung, expert-Verlag, Renningen-Malmsheim, 2001.

13 Ignition

13.1 Spark-Ignition Engine

13.1.1 Introduction to Ignition

In combustion engines (SI engines) with externally supplied ignition, the combustion process is triggered by an electrical discharge in the combustion chamber toward the end of the compression cycle. The required components are an ignition coil as the high-voltage source and a spark plug as the electrode in the combustion chamber. From the spark, a high-temperature plasma channel arises between the spark plug electrodes. An exothermic chemical reaction occurs in a thin reaction layer around this channel. This develops into a self-sustaining and expanding flame front.[1]

13.1.2 Requirements of the Ignition System

The ignition system must ensure a reproducible ignition process throughout all conceivable changes and dynamic fluctuations of the engine's operating states. For the spark to jump to the spark plug electrodes, the ignition system must have sufficient high voltage. The pressure, temperature, and density of the mixture at and between the ignition electrodes at the time of ignition influence the required voltage. These parameters vary widely over the speed and load. According to Paschen, the required ignition voltage increases linearly with the pressure and electrode spacing. The energy transferred to the mixture by the spark must suffice to trigger self-sustaining combustion. The optimum time of ignition plays a central role and is measured in the engine during the application phase and saved in a program map in the engine control unit as a function of the speed and load.

13.1.3 Minimum Ignition Energy

Homogeneous, stoichiometric fuel-air mixtures require energy of less than 1 mJ for ignition while idling. In richer or leaner mixtures, the required energy rises to 3 mJ.[2] In real engines, the conditions are substantially less favorable. The energy requirement rises sharply because of the inhomogeneous distribution of air, fuel, recycled exhaust gas, etc., between the cylinders, and because of inhomogeneous cylinder charging and transfer and heat losses to feed lines and electrodes. Conventional ignition systems provide approximately 40 mJ with a spark duration of 1 ms at the spark plug to ensure ignition.

13.1.4 Fundamentals of Spark Ignition

13.1.4.1 Phases of the Spark

The spark forming at the spark plug can be divided into three sequential types of discharge with very different energy and plasma physical properties (Fig. 13-1).[3–5]

Initially, the voltage at the spark plug rises sharply. As soon as the current charge forming in the field reaches the opposing electrode, breakdown occurs within a few nanoseconds. The impedance of the electrode path falls drastically, and the current rises quickly from the discharge of the leakage capacitance of the spark plug.

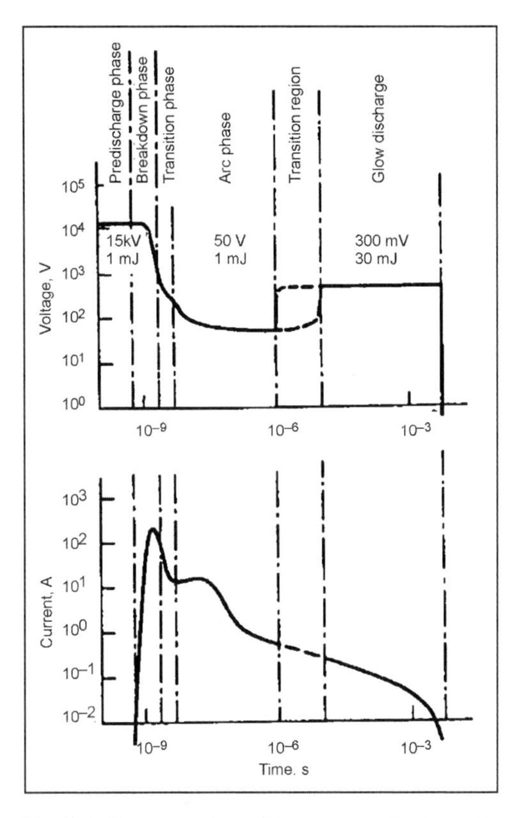

Fig. 13-1 Curve over time of the current and voltage of a transistor coil ignition (TCI).[4] Typical values of occurring voltages and energy transfer in the individual spark phases.

Because of the fast rise in voltage in the ignition coil, arcing does not occur upon reaching static breakdown voltage but upon overvoltage due to the ignition lag. Very high temperatures of 60 000 K arise in the conductive channel from the complete dissociation and ionization of the atoms and molecules. The pressure wave begins to propagate at supersonic speed.

The spark then transitions into the *arc phase* with very small voltages in which the current is determined by the discharge of the high-voltage-side capacitance. At the cathode, a hot spot (ignition spot) arises because of the strong emission of electrons; cathode material vaporizes and strongly erodes the electrodes. The temperature in the channel drops to approximately 6000 K. The plasma expands by thermal conduction and diffusion processes, and the exothermic reaction begins that produces an advancing flame front.

At currents below 100 mA there is a transition to *glow discharge*. Multiple transitions can occur between arc and glow discharge in a transition range depending on changes and movements of the mixture between the electrodes. In the glow discharge phase, the voltage rises again (the electron stream is supported by the contacting ions); the temperature in the channel is now only approximately 3000 K. This is below the melting temperature, and the electrodes are now primarily atomized by contacting charge carriers.[6]

The energy accumulator, the coil, fully discharges in the discharge channel. When the voltage falls below the threshold voltage necessary for maintaining the channel, the spark terminates. The residual energy decays in the secondary winding of the ignition coil.

13.1.4.2 Energy Transmission Efficiency

Figure 13-2 shows the amount of energy that can be sent to the mixture in the described phases of the spark.

The breakdown phase has the greatest ignition efficiency and causes faster energy conversion in the initial phase of the combustion process. By enlarging the spark plasma and increasing its propagation speed, the reliability of ignition can be improved.[4]

Because of the substantial heat loss from the electrode, the energy available in the spark plasma is much less than the electrical energy supplied to the spark plug. With conventional transistorized coil ignition, basically the glow phase stimulates ignition, and the ignition reliability increases with the peak current and the length of the discharge.[7]

A long spark duration promotes ignition. Even with lean mixtures ($\lambda = 1.5$) and a fast flow (>30 m/s), the long glow discharge of transistorized coil ignition is sufficient by itself to continually ignite a flammable mixture that is transported through the flow field into the electrode area.[8]

13.1.5 Coil Ignition System (Inductive)

Coils used in distributorless ignition systems switched with transistors are dry ignition coils cast with epoxide resin that consist of a closed magnetic circuit made of laminated low-loss electrical sheet steel with concentrically superposed primary and secondary windings (Fig. 13-3). When the primary current is turned on, energy is inductively stored in the air gap of the magnetic circuit. After the primary current is interrupted by the transistor (Fig. 13-4), a secondary-side voltage builds in the coil until breakdown at the spark plug. The maximum attainable voltage essentially depends on the cutoff voltage and the secondary/primary turns ratio in the coil.

After arcing, energy discharges in the spark via the secondary winding of the coil. During this glow phase (combustion time), the spark gap at the spark plug can be considered from an electrical point of view as being replaced by a zener diode gap that restricts the secondary voltage to the value of the firing voltage and keeps it constant until the spark breaks contact.

The definitions of the properties of such an ignition coil are uniformly governed by ISO 6518. The available voltage is defined as the maximum attainable voltage with substitutional resistance corresponding to the relevant obstruction. For example, 1 MΩ/25 pF of the electrical load corresponds to an ignition coil directly connected to the spark plug, and 1 MΩ/50 pF corresponds to an ignition coil that is connected to the spark plug via an ignition cable.

The output or combustion energy is determined by measuring the discharge time of the ignition coil ending with a zener diode circuit with 1000 V. By means of the turns ratio and the interrupting current of the coil, the maximum spark current (glow current) is set on the secondary side of the ignition coil. The spark duration can be varied within wide limits by setting the stored inductance and operating point of the magnetic circuit.

The coupling between the primary and secondary sides of the ignition coil is more than 90%. Of the electrical energy stored in the primary current circuit, only approximately 50% arrives at the spark plug because of the transmission loss and resistance in the circuit. The conditions in the combustion chamber (pressure, temperature, mixture movement, etc.) determine the firing voltage during spark ignition together with the electrode distance. The influence on the energy and spark duration is shown in Fig. 13-5.

Double spark ignition coils are used widely in which both ends of the secondary winding are series connected via ignition cables to the spark plugs that belong to cylinders whose firing sequence is shifted by a 360° crankshaft angle. When there are four cylinders, cylinders 1 and 4

	Breakdown, %	Arc, %	Glow, %
Radiation loss	< 1	5	< 1
Heat conduction at the electrodes	5	45	70
Overall loss	6	50	70
Plasma energy	94	50	30

Fig. 13-2 Energy balance of the three types of discharge.[3]

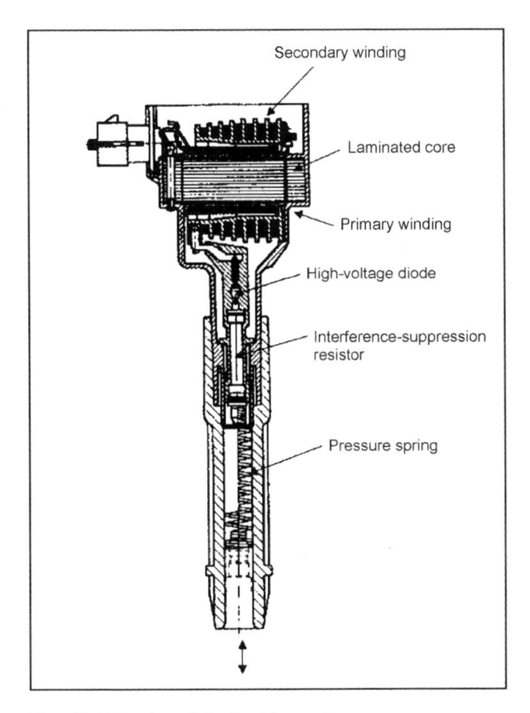

Fig. 13-3 Design of the ignition coil.

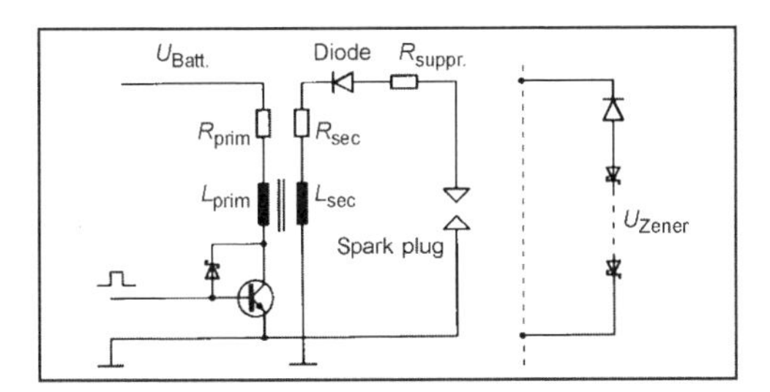

Fig. 13-4 Schematic layout of transistorized coil ignition.

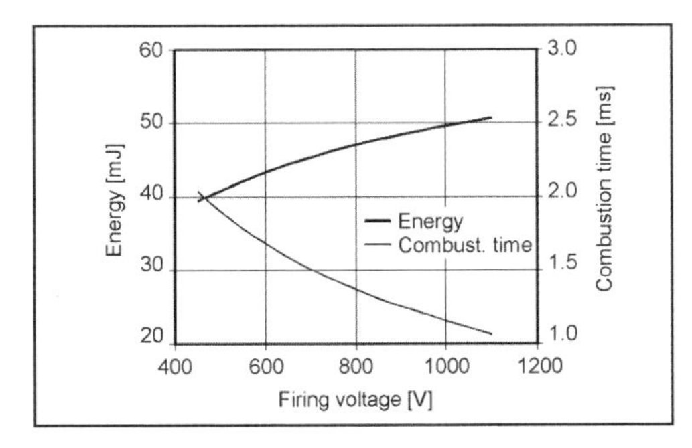

Fig. 13-5 Influence of the firing voltage on the energy and spark duration.

and cylinders 2 and 3 are connected to a coil. The series connection causes two spark plugs to fire simultaneously—one in a cylinder filled with a fuel-air mixture, the other in a cylinder in the exhaust cycle in which a support spark arises with only a small amount of additionally required voltage due to the pressureless state.

Because of the series connection, one of the two spark plugs ignites with a positive high voltage and the other with a negative high voltage. For ignition with a negative high voltage, the required voltage is slightly less (1 – 2 kV) than with a positive voltage because of the higher temperature of the middle electrode of the spark plug and the subsequently reduced work function of the electrons while the engine operates. At the same time, the electrode erosion at the spark plugs is strongly asymmetrical because of the different polarities of the ignition voltage.

Different arrangements are possible for ignition with double spark coils. On the one hand, the double spark coils can be combined into a block or packet, and the spark plugs can be connected via ignition cables; alternately, the ignition coil can be directly placed on or connected to a spark plug, and the connection to the spark plug in the correlating cylinder can be with an ignition cable.

In higher-end vehicles, single spark coils are used to better control ignition and problems with valve overlap, etc., where each cylinder is fired with its own ignition coil (Fig. 13-6).

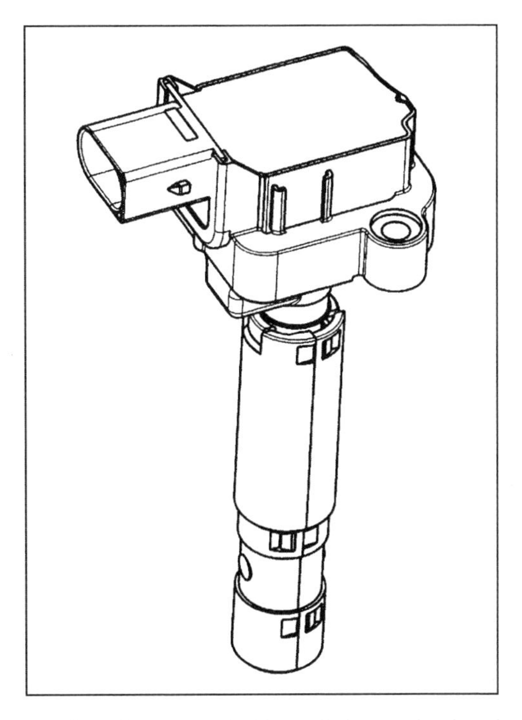

Fig. 13-6 Single spark ignition coil that can be shoved directly onto the spark plug with 70 mJ, 35 kV output voltage, and 2 ms combustion time.

The coils are mounted on the cylinder head and directly contact the spark plug or are combined into blocks with several individual spark coils and connected via ignition cables to the spark plugs.

In single spark ignition coils, a high-voltage diode is required in the secondary circuit to suppress the voltage pulse that arises at the inductance coil when the current is switched on since an ignitable mixture can be in the cylinder at this time at a low pressure and, hence, low required voltage.

By directly connecting these coils to the spark plug and, hence, dispensing with the interference-free ignition cables, the ignition coil itself must have the interference-suppression element such as a wound, inductive resistor to suppress high-frequency interference that arises from the flashover at the spark plug.

The use of ignition coils with or without ignition cables (built separately or plugged on directly) determines from the different external capacitive loads the optimum turns ratio with which the coil can provide the maximum output voltage (Fig. 13-7).

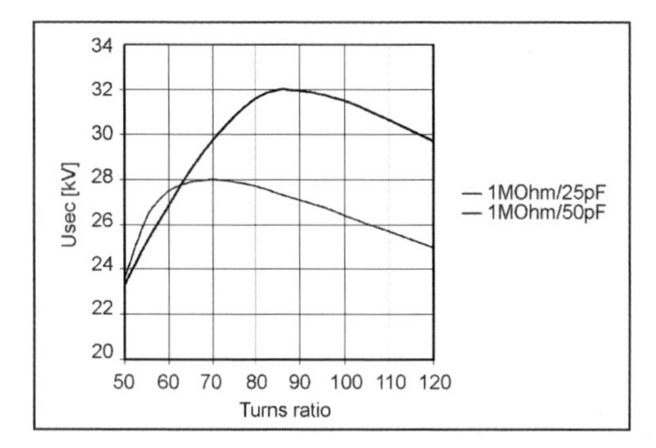

Fig. 13-7 Influence of the external load of the ignition coil on the optimum turns ratio.

Pencil coils are becoming more important (Fig. 13-8). Their design with an open, long magnetic circuit allows the size and diameter of the ignition coil to be reduced so that the coil can be mounted directly in the spark plug shaft. The component requirements for temperature resistance and insulation strength are, hence, greater.

The system that is chosen depends on the application, the special requirements, and cost. The same holds true for the integration of other components and intelligent functions in the ignition coil such as the installation of electronic semiconductor switches and/or the integration of diagnostic and self-protection tasks.

13.1.6 Other Ignition Systems

Despite repeated efforts at introducing alternative ignition systems (plasma ignition, laser ignition, and many others), the traditional coil ignition has become generally accepted because of its favorable cost-benefit ratio.[7]

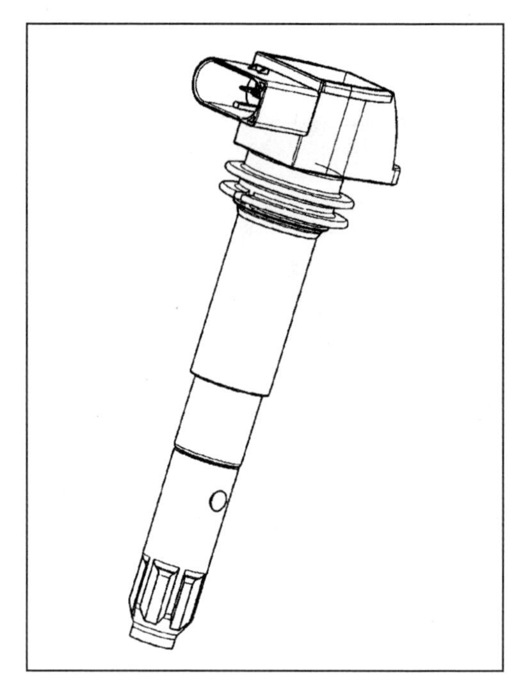

Fig. 13-8 Pencil coil with a diameter of 22 mm, 32 kV output voltage, and 60 mJ.

Only in exceptions (such as racecar engines) is capacitor discharge ignition (CDI) used. With CDI, the energy is temporarily stored in a capacitor, and the required high voltage is switched via a fast low-loss ignition transformer. These ignition systems have an extremely fast voltage rise (a few kV/λs) and, hence, effectively resist shunts from deposits on the spark plugs. A disadvantage is that the very short combustion time of approximately 100 λs can lead to misfiring when there are inhomogeneous mixtures, and the strong spark current can increase spark plug erosion.

An additional improvement is "AC ignition" as is used in Mercedes 12-cylinder (V-12) engines.[9] A capacitor functioning as an energy accumulator with a weakly coupled ignition transformer is connected to a resonant circuit with a resonance frequency of approximately 20 kHz. After flashover, energy is delivered in the spark from the secondary side of the coil while the capacitor is recharged (reverse converter principle). In contrast to CDI, the spark does not cease since enough energy remains in the system to maintain oscillation. The danger of misfiring from inhomogeneous mixtures is much less than CDI.

With AC ignition, a type of ignition is obtained in which the combustion time is freely settable independent of the provided ignition voltage in contrast to transistorized coil ignition. With a combustion time tailored to demand (energy controlled ignition), spark plug wear is less, and, e.g., the ionic current can be measured at the spark plugs to detect misfiring after the controlled end of the spark.[9]

All of the discussed types of ignition beyond transistorized coil ignition require additional components in addition to the coil such as capacitors and power supplies (100–800 V) for generating the required charge voltages, and this increases cost and limits the acceptance and use of such ignition systems.

13.1.7 Summary and Outlook

To increase operational reliability, ignition systems should have low source impedance and/or a fast voltage rise (shunt resistance).

Furthermore, ignition systems must provide sufficiently high voltage. In future ignition systems, we can anticipate a further rise in the demands on available voltage (lean operation, high EGR rates, turbocharging, Otto-DI). In particular, the required ignition voltage in a lean-running engine with direct fuel injection under a partial load in stratified-charge operation is higher than for a comparable engine in stoichiometric operation since the charge dilution from excess air and/or exhaust recycling increases the gas density in the cylinder and, hence, raises the breakdown voltage at the time of ignition.

However, the demand for maximum ignition voltage that is typically attained under a full load in homogeneous operation is comparable in both instances; the demands on an engine with direct fuel injection therefore remain unchanged in regard to maximum electrical insulation resistance of the ignition coil, wire, and spark plug in comparison to an engine with multipoint fuel injection.[10]

Only when the ignition system has a large capacity to store energy can a sufficiently large plasma channel be generated. The energy requirements are also higher for an engine with direct fuel injection under partial load in stratified-charge operation in contrast to an engine with intake manifold injection since more energy must be supplied to the mixture (70–100 mJ) because of the charge dilution from excess air or EGR to ensure repeatable and sufficient arc development.[10] One can assume that this value will fall with improvements in mixture control.

Bibliography

[1] Heywood, J.B., Internal Combustion Engine Fundamentals, McGraw-Hill, New York, 1989.
[2] Autoelektrik, Autoelektronik am Ottomotor, Bosch, VDI-Verlag, 1987.
[3] Albrecht, H., R. Maly, B. Saggau, and E. Wagner, Neue Erkenntnisse über elektrische Zündfunken und ihre Eignung zur Entflammung brennbarer Gemische, Automobil-Industrie (4), 45–50, 1977.
[4] Maly, R., and M. Vogel, Ignition and Propagation of Flame Fronts in Lean CH_4-Air Mixtures by the Three Modes of the Ignition Spark, Proceedings of the 17th International Symposium on Combustion, pp. 821–831, The Combustion Institute, 1976.
[5] Schäfer, M., Der Zündfunke, Dissertation, Universitat Stuttgart, 1997.
[6] Hohner, P., Adaptives Zündsystem mit integrierter Motorsensorik, Dissertation, Universitat Stuttgart, 1999.
[7] Maly, R., Die Zukunft der Funkenzündung, MTZ 59 (1998) 7/8.
[8] Herweg, R., Die Entflammung brennbarer turbulenter Gemische, Dissertation, Universitat Stuttgart, 1992.
[9] Schommers, J., U. Kleinecke, J. Miroll, and A. Wirth, Der neue Mercedes-Benz Zwölfzylindermotor mit Zylinderabschaltung, Part 2, MTZ 61 (2000) 6.
[10] Stocker, H., M. Archer, R. Houston, D. Alsobrooks, and D. Kilgore, Die Anwendung der luftunterstützten Direkteinspritzung für 4-Takt Ottomotoren-der ‚Gesamtsystemansatz', 7th Aachen Colloquium on Vehicle and Engine Technology, 1998, p. 711et seq.

13.2 Spark Plugs

13.2.1 Demands on Spark Plugs

The spark plug represents the electrode necessary for ignition in the combustion chamber and, hence, has to satisfy quickly changing engine requirements.

Electrically, the spark plug must ensure high-voltage transmission and isolate the required ignition voltages of over 30 kV, prevent arcing, and consistently resist dielectric loads from high field strengths and quickly changing fields over its life.

Mechanically, the spark plug should seal the combustion chamber against pressure and gas, and absorb the mechanical forces that arise when screwing in the plug.

Thermally, good heat conduction protects the spark plug against loads from small thermal shocks in each combustion cycle and keeps down the temperature of the spark plug.

Electrochemically, the spark plug must resist attacks from spark erosion, combustion gases, and residue such as hot gas corrosion, oxidation, and poisoning from sulfur in the fuel, and it must resist the formation of deposits on the insulator.

13.2.2 Design

Given the above requirements, the basic design of the spark plug has scarcely changed over the course of the development of the engine (Fig. 13-9). Nevertheless, primarily over the last 20 years, changes have been made in the form of constructive details and improved materials because of the increased need to adapt the spark plug to the specific conditions of each engine; this has led to a substantial increase in the change interval. Different surface ignition approaches became possible with the use of unleaded fuels.

The insulator of the spark plug consists of an aluminum oxide ceramic that provides strong electrical arcing resistance, and it is usually provided with a ribbed insulator flashover barrier on the insulator neck. Embedded in the insulator, the center electrode and igniter are connected gastight by a special electrically conductive glass seal. With corresponding additives, this conductive glass seal can be provided with a specific resistance to improve erosion resistance and interference suppression.

The gastight connection between the insulator and metallic body is created with an internal sealing ring, and the initial mechanical force on the sealing ring arises from the spark plug body that is first beaded onto the insulator and then electrocoated in a special heating procedure.

Welded onto the spark plug body are one or more ground electrodes that form the gas discharge path together with the center electrode. The various spark plug types are

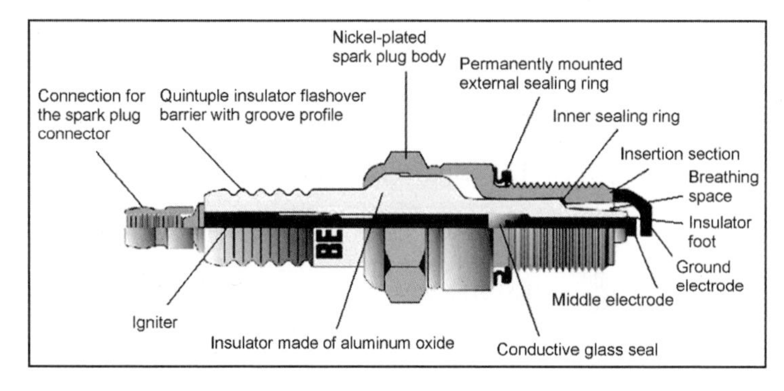

Fig. 13-9 Design of a spark plug.

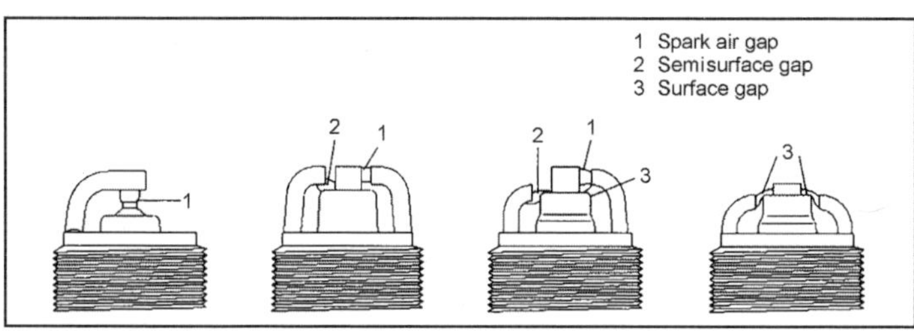

Fig. 13-10 Different spark paths.

distinguished according to their electrode or spark path (Fig. 13-10).

Air gap: In spark plugs with hook electrodes (J-type), good-to-optimum mixture accessibility is provided by the open spark path through the gas chamber (air).

Surface gap: If the spark glides across the insulator when arcing, deposits and combustion residue can be burned up. Electrical shunts are avoided, but the ignition spark must be energy rich to compensate for the cooling that arises while the spark glides over the insulator. Simultaneously, the lower required voltage sometimes permits a longer spark path and, hence, greater mixture accessibility.

Semisurface gap: By arranging the electrode, spark paths can be set that partially traverse the air and partially run across the insulator. By combining mutually independent air and surface gap paths, the rise in the required ignition voltage from electrode erosion can be reduced, which greatly extends the life of the spark plugs.

The electrode position determines the spark position in the combustion chamber (Fig. 13-11).

Fig. 13-11 Normal and advanced spark position.

13.2.3 Heat Range

The heat range is a measure of the thermal resistance of a spark plug and describes the maximum operating temperature that arises in the spark plug from the equilibrium between the absorption and release of heat.

After starting the engine, the spark plug should reach the “self-cleaning temperature” of >400°C as quickly as possible to oxidize (burn off) deposits on the insulator to prevent electrical shunts. At the same time, the heat conductivity must be sufficient so that the stationary end temperature does not exceed 900°C at any point on the spark plug that could produce uncontrolled autoignitions. In terms of the design, the heat range of the spark plug is controlled by the geometric shape of the insulator nose and the breathing space, as well as the electrode’s position, geometry, and heat conductivity (Fig. 13-12). Spark

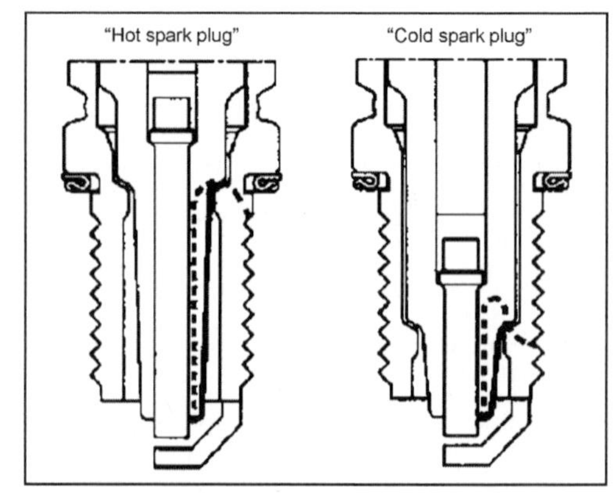

Fig. 13-12 Hot and cold spark plugs.

plugs with long insulator paths up to the internal seal and open breathing space form large heat absorbing surfaces with poor heat conduction. These spark plugs are termed "hot"; spark plugs with short insulator noses are correspondingly called "cold."

By using compound electrodes such as nickel electrodes with a copper core—copper is unsuitable to be used directly in the combustion chamber, but its heat conductivity is very good—the removal of heat from the electrode is substantially improved.

When the spark position is extremely advanced in the combustion chamber, specially adapting the cross section and the heat-absorbing surface of the insulator nose tip allows the self-cleaning temperature to be quickly reached and produces a more-or-less self-regulation of the upper temperature of the insulator below 900°C. This type of spark plug is, hence, suitable for use in combustion chambers with relatively low and also very high temperatures.[1]

Stratified combustion procedures usually require spark plugs that extend far into the combustion chamber.[2] This can increase the mechanical and thermal load on the electrode. To prevent vibration fractures, the thread insert is lengthened. This allows shorter and, hence, colder ground electrodes. All electrodes are also equipped with a copper core.

13.2.4 Required Voltage for Ignition

The difference between the high voltage offered by the ignition coil and the required ignition voltage (Fig. 13-13) defines the voltage reserve. The arising electrode erosion increases the electrode spacing and, hence, the required voltage (Fig. 13-14) and, together with the voltage reserve, determines the maximum possible life (length of use) of the spark plug. A one-sided increase in the available voltage of the ignition coil to allow the spark plug to be operated longer is counterproductive: It produces problems with the high-voltage capacity of the feed lines and increases electrode erosion because of the high ignition energy.

The required ignition voltage in Figs. 13-13 and 13-14 displayed according to the amount and frequency of occurrence is calculated in a mixture of overland travel and a circular track test with a high acceleration component. A clear rise in the required voltage over the operating time can be discerned at the spark plugs with two lateral Cr-Ni electrodes.

One of the tasks of the spark plug is to keep down the ignition voltage itself and the additional rise of the ignition voltage over the time of operation. The reduction of the electrode spacing to lessen the essential ignition voltage is subject to narrow restrictions because of the required mixture accessibility, in particular, with lean mixtures, and the occurrence of quenching, etc. If the spacing is too small, misfiring occurs from the combustion of a volume that is too small for initial ignition or is because of poor mixture accessibility.

The reduction of the electrode cross section increases the electrical field strength from a peak effect with less required ignition voltage. This necessitates the use of high-grade metal electrodes that reduce electrode erosion because of increased electron discharge work and higher material melting and boiling points (Fig. 13-15). At the same time, the heat-absorbing surface is reduced.

Because of the temperature of the electrode, it is preferable for the polarity of the ignition voltage to be negative since the hotter center electrode enhances electron discharge and, hence, lowers the required voltage.

13.2.5 Ignition Characteristic (and Mixture Ignition)

In addition to the cited features, spark plugs are also evaluated for their ability to reduce cyclic combustion fluctuations and to shift the lean limit to influence the smooth running of the engine as well as the exhaust gas and fuel consumption.

Spark plugs with small electrodes are optimally suitable for reducing the required ignition voltage and the contact surface of the flame with the electrodes to prevent heat loss. Large ignition gaps with favorable mixture accessibility can be advantageously realized in spark plugs with surface gaps that limit the rise in voltage resulting

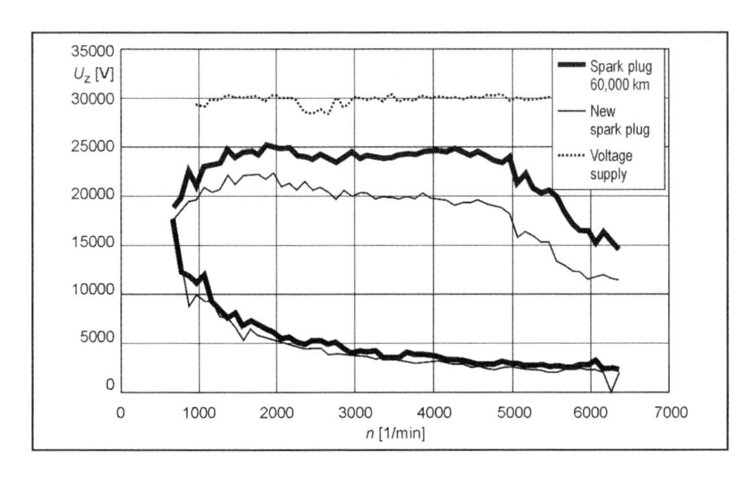

Fig. 13-13 Required voltage (min, max) and available voltage.

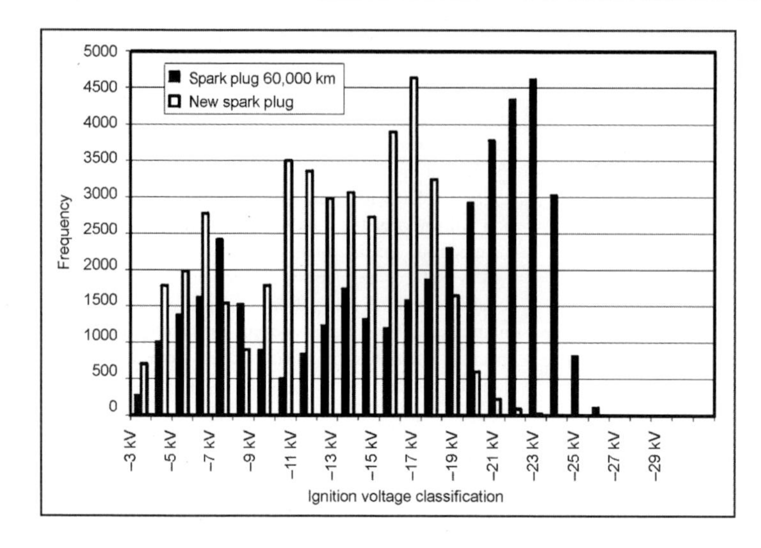

Fig. 13-14 Frequency distribution of the required ignition voltage.

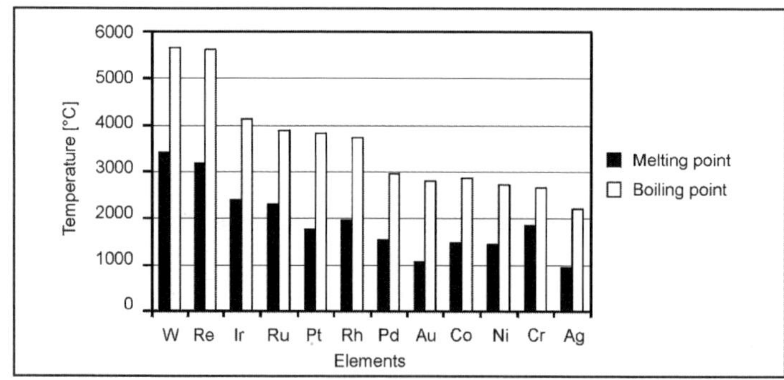

Fig. 13-15 Melting and boiling points of different metals.

from electrode erosion and offer a favorable electrode orientation at average flows of 2–5 m/s; the flame is moved away from the electrode but not extinguished.[3] The influence of these measures on electrode wear is not taken into consideration.

Surface gap spark plugs (including those with several electrodes) are, according to a set of investigations, not suitable for igniting leaner mixtures since the insulator heat loss and the mixture accessibility are inferior[4,5]; however, the sparking distance is greatly increased at the cost of a greater required ignition voltage. This improves mixture accessibility, and the shunt resistance of this type of spark plug is more effective.

In modern engines, flow speeds of more than 10 m/s are normal, and they can achieve 30 m/s in SI engines with direct fuel injection. The orientation of the electrodes in the combustion chamber becomes irrelevant because of turbulence, and the orientation does not influence the ignition behavior; however, the influence of the spark position is clearly recognizable.[6] According to these investigations, spark plugs with a spark position (spark air gap) that is advanced extremely far into the combustion chamber work better in engines with intake manifold fuel injection; in engines that operate using a stratified charge (FSI by VW),[6] the overall behavior of surface gap spark plugs is improved. The self-cleaning behavior on the insulator is probably the decisive factor.

Optimum arc formation increases the combustion speed, but the greater combustion chamber temperature enhances the formation of NO_x.

Cold starts are particularly demanding on spark plugs where the spark plug must ensure a faultless start without shunts and especially flawless engine acceleration (load assumption). More voltage is required for acceleration, and this can cause electrical shunts and, hence, misfiring when there are deposits on the spark plugs. Similar problems occur with repeated starts, continuous short-distance travel, or slow driving in which the spark plug does not become sufficiently hot. Technical assistance is provided by equipping the spark plugs with surface gaps on the insulator (cleaning from the surface spark) or providing corona edges facing the insulator on the high-voltage center electrode (cleaning from additional ionization). Sharp edged or pointed electrodes in the spark-over path reduce the required voltage and, hence, the tendency to shunt.

In summary, it must be noted that spark plugs need to be specially readapted to each engine and engine variation (supercharging, EGR rate, etc.). No general conclusion can be made regarding which spark plug type is best suited for each application. The best possible adaptation to the parameters of thermal behavior, spark geometry, and the required ignition voltage is required. At the same time, the spark plug should be located at the site of the most favorable flow conditions (advanced spark position—SI engines with direct fuel injection), which poses additional demands on the selected electrode material and the design of the insulator nose.

13.2.6 Wear

The spark plug electrodes are subject to several wear mechanisms.

1. The *thermal* stress from internal engine processes arising from compression and ignition wears the material of the electrode extending into the combustion chamber from hot gas corrosion and scaling.
2. Another cause of wear is *chemical* reactions such as oxidation of the electrodes triggered by fuel, additives, and combustion gases. Notable wear of the electrode material occurs at high temperatures from aggressive gases.
3. The *spark erosive* attack on the electrode causes the materials to partially melt and evaporate from the high temperatures in the plasma channel. This creates a demand for materials with high melting and boiling points.

Nickel is the primary electrode material that is alloyed with aluminum and chrome as oxide formers and manganese and silicon against sulfur in the oil and fuel to improve chemical resistance (Fig. 13-16). With a melting point of only approximately 1450°C, the material is not resistant against attack from hot gas and spark erosion (Fig. 13-17). Nevertheless, operating lives of 60 000 km and more are possible with optimized alloys and suitable geometric designs.

Platinum fulfills the demands for a high temperature and oxidation stability. Chemical attacks on the grain boundaries by the platinum poisons sulfur and silicon increase wear. The arc of the spark partially melts the electrode surface that can then more easily react with combustion gases.

Iridium has even higher melting and boiling points, but it is unsuitable as an electrode material when pure. To exploit its high temperature resistance, platinum, palladium, or rhodium are alloyed and form oxides that protect the surface of the iridium.[9]

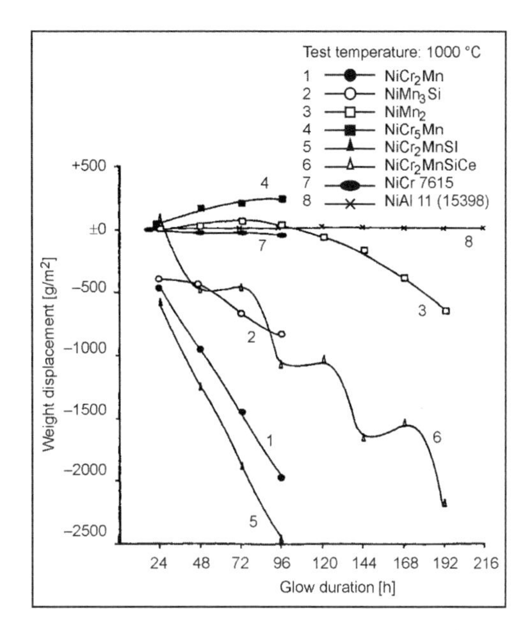

Fig. 13-16 Hot gas resistance of various Ni alloys.[7]

In high-performance spark plugs, precious metal electrodes are particularly suitable. However, because of the high cost of the precious metals, the amount of material is restricted, and chrome-nickel electrodes are used where only the areas that form the arcing path are reinforced with precious metal. With a suitable design, the demands can be combined for high performance (life) with nearly unchanged required ignition voltage, favorable mixture accessibility and idling stability, and reduced shunt sensitivity and superior cold start behavior.

Figure 13-18 shows the principle of flow guidance with two-material electrodes. Small anchoring sites (1) made of materials with high discharge and low vaporization rates (such as Pt) are on both electrodes with inverse properties (such as Cr-Ni), and they determine the required ignition voltage and arcing site. This geometry and the selected materials force the first spark to arc via the anchoring sites, but the discharge immediately transitions to the areas of the support electrodes formed as sacrificial

Spark phase	Duration	Energy	Spark erosion
Rise	60 μs		
Breakdown	2 ns	0.5 mJ	$12 \cdot 10^{-12}$ g/mJ
Arc	1 μs	1 mJ	$210 \cdot 10^{-12}$ g/mJ
Glow	2 ms	60 mJ	$3.5 \cdot 10^{-12}$ g/mJ

Fig. 13-17 Wear from different spark phases.[8]

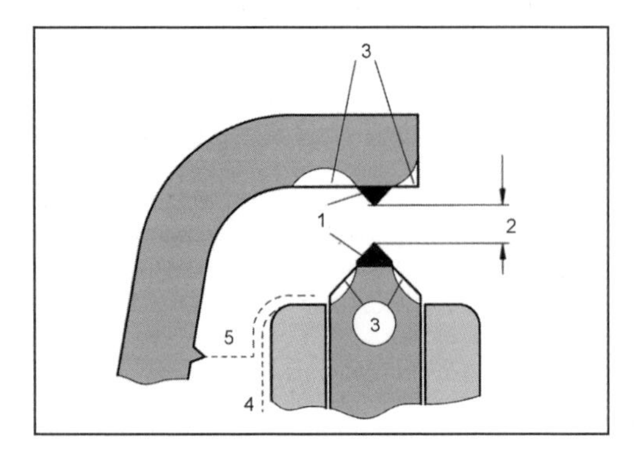

Fig. 13-18 Principle of flow guidance.[8]

zones (3). The erosion of the anchoring sites is minimal, and the electrode spacing (2) and required ignition voltage remain constant. The erosion (Figs. 13-19a and b) is shifted to specified regions of the base electrode; the effective spark length rises over time, enhancing ignitability.[8] Since the ignition voltage remains nearly constant over the life of the spark plug from the reduced electrode erosion, the electrode spacing can be larger, and a more favorable electrode geometry can be selected that improves ignitability and idling stability. A restriction of the life from deposits (4) on the insulator is prevented by an auxiliary spark path (5) that eliminates these deposits with occasional creeping discharges. At the same time, these additional creepage spark paths improve cold-start behavior and prevent misfiring in operating conditions with a very high required voltage.

Fig. 13-19 (a) Erosion behavior of a Cr-Ni electrode, standard spark plug after 28 000 km, change in the electrode spacing from 0.7 to 1.1 mm. (b) Erosion behavior of a platinum-reinforced electrode, long-life spark plug after 105 000 km, change of the electrode spacing from 1.00 to 1.05 mm.

13.2.7 Application

In principle, spark plugs must be redesigned for each engine since the requirements are very different because of the type of mixture guidance, the EGR rate, the position of the spark plug, etc. The thermal suitability of a spark plug is ideally evaluated in the original aggregate. The heat range is adapted by measuring ionic current in which the changes in combustion are observed, and the preignition and postignition can be observed by blanking individual ignitions (self-ignition). The postignition is uncritical for the engine.

In addition, investigations are carried out using thermocouples on the spark plugs in the engine under different speed and load conditions to determine the hottest cylinder, the maximum electrode temperature, and other component temperatures. The spark plug needs to be dimensioned such that preignition cannot occur under a full load.

From measurements of special "heat range measuring engines," the temperature of the spark plug can be clearly increased on the test bench by advancing the ignition angle, and the temperature of the individual spark plug components can be determined pyrometrically with an optical access to the cylinder, and the suitability for preignition can be checked. The heat range reserve can be indicated in degrees crankshaft angle to shift the ignition in an early direction without causing preignition.

Bibliography

[1] Meyer, J., and W. Niessner, Neue Spark plugntechnik für höhere Anforderungen, in ATZ/MTZ Special Edition, System Partners 97.

[2] Eichlseder, H., P. Müller, S. Neugebauer, and F. Preuß, Inner engineische Maßnahmen zur Emissionsabsenkung bei direkteinspritzenden Ottomotoren, TAE Esslingen, Symposium: Entwicklungstendenzen Ottomotor, December 7/8, 2000.

[3] Pischinger, S., and J.B. Heywood, Einfluss der Zündkerze auf zyklische Verbrennungsschwankungen im Ottomotor, in MTZ 52 (1991) 2.

[4] Lee, Y.G., D.A. Grimes, J.T. Boehler, J. Sparrow, and C. Flavin, A Study of the Effects of Spark Plug Electrode Design on 4-Cycle Spark-Ignition, Engine Performance, SAE, 2000-01-1210.

[5] Geiger, J., S. Pischinger, R. Böwing, H.-J. Koß, and J. Thiemann, Ignition Systems for Highly Diluted Mixtures in SI-Engines, SAE, 1999-01-0799.

[6] Kaiser, Th., and A. Hoffmann, Einfluss der Zündkerzen auf das Entflammungsverhalten in modernen Motoren, in MTZ 61 (2000) 10.

[7] Brill, U., Krupp-VDM, private communications, 1994.

[8] Maly, R., Die Zukunft der Funkenzündung, in MTZ 59 (1998) 7/8.

[9] Osamura, H., and N. Abe, Development of New Iridium Alloy for Spark Plug Electrodes, SAE, 1999-01-0796.

13.3 Diesel Engines

13.3.1 Autoignition and Combustion

Autoignition characterizes diesel engine combustion. Combustible fuel is injected toward the end of the compression cycle into the hot, compressed cylinder charge, mixed with it, and ignited. During the ignition lag (between injection and the start of autoignition), a series of complex physical and chemical subprocesses occur such as spray formation, vaporization, mixing, and chain branching (initial chemical reactions) without any notable conversion of energy.

The ignition depends on the starting conditions of mixture formation:

- The pressure and temperature of the charge
- The temperature, viscosity, vaporization characteristics, and ignitability of the fuel
- The pressure, time, and characteristic of injection, as well as the nozzle geometry that determines the spray formation (size, distribution, and pulse of the droplets)
- Charge movement
- Charge composition, i.e., the oxygen component and the changes in the specific thermal capacity from the EGR, etc.
- The combustion chamber geometry

Autoignition starts locally in the areas with completely evaporated fuel mixed with sufficient atmospheric oxygen. During this phase, injection typically continues, and combustion and mixture preparation occur simultaneously. The ignition process is strongly inhomogeneous since liquid and gaseous phases simultaneously exist with a complex dynamic interaction. The local temperature is the decisive factor in determining the ignition lag and related processes.

The fuel-air mixture prepared during the ignition lag burns quickly upon the onset of ignition. The combustion of the fuel prepared subsequently occurs with a slower diffusion combustion. The fuel preparation is further accelerated from the increasing release of energy. High conversion rates in autoignition generate high-pressure gradients and, hence, usually high noise emissions. To avoid this, the combustion of premixed components is limited as much as possible, for example, by introducing preinjection.

The start of combustion or the moment of ignition must be optimized in relation to exhaust gas emissions, fuel consumption, performance, and noise. Compromises are required since the measures taken within the engine are mutually influential.

In passenger cars, engines with direct fuel injection have predominated in recent years in contrast to approaches with a divided combustion chamber.[1] The injection engineering and the means used to support cold starts have become very sophisticated. We now can have several injections per work cycle at a high maximum injection pressure, a largely free start of injection, and injected fuel quantity to improve fuel consumption and smooth running. Components to support cold starts such as glow plugs were improved to offer reliable support at extremely low start temperatures, faster heating speed, less required energy, and longer life.

Passenger car diesel engines are equipped with electric engine starter systems whose design is oriented around the cold-start threshold temperature that the engine requires to reliably start.[2]

Ignition strongly depends on the initial conditions. In particular, during a cold start, these starting conditions are so poor that satisfactory ignition cannot take place without additional measures.

13.3.2 Diesel Engine Cold Starts

Cold starts include all those starting processes in which the engine and media are not at their operating temperature. At temperatures below +60°C, cold starts are supported by changing the injection time quantities. As the engine warms up, the smooth running, throttle response or load assumption are enhanced, and the pollutant emissions are reduced.

More extensive measures are necessary at temperatures below freezing since the starting quality worsens disproportionately until the temperature decreases so much that the engine cannot be started.

13.3.2.1 Important Influential Parameters

Diesel engine combustion is optimized for hot engine operation. The chosen external parameters substantially influence cold start quality:

- Engine construction (DI/IDI)
- Cylinder number
- Charge volume or surface/volume ratio
- Compression ratio
- Starter features (starter output, battery)
- Injection system
- Air guidance and charge
- Internal losses (oil viscosity, gearbox, auxiliary systems,…)

In contrast to combustion in a hot-running engine, the conditions during a cold start and the following warm-up of the engine are much poorer for autoignition and the subsequent complete combustion of the fuel. The most important influential parameters on the start behavior and the relationships of the parameters to each other are shown in the diagram in Fig. 13-20. Attention has, therefore, been given to the development of cold-start components and injection systems with more degrees of freedom. For the sake of clarity, the representation of additional relationships such as the direct influence of the temperature on the charge loss (gap dimensions/oil film) or the final compression temperature are not shown.

Low temperatures reduce the battery performance and increase the drivetrain friction so that the attainable starter speed is lower from the increase in required torque. This increases charge and heat losses because of the longer end phase of compression. The revolution speed of the engine decreases at very low environmental temperatures around the compression or ignition dead center so that the long dwell time of the hot compressed charge in the combustion chamber greatly decreases the temperature and charge pressure.[3] This dramatically worsens conditions for mixture formation and ignition, where the temperature has a much greater influence on start quality in contrast to pressure.[3–5]

As the temperature decreases, higher starter speeds are required to ensure a reliable cold start. The required minimum start speed and, hence, the cold-start threshold temperature can be greatly lowered by means of start aids that enable starting at temperatures around −20°C and below (Fig. 13-21). The output of the starter and battery is designed for the required cold-start threshold temperature, where a fully charged battery is assumed. If the battery is

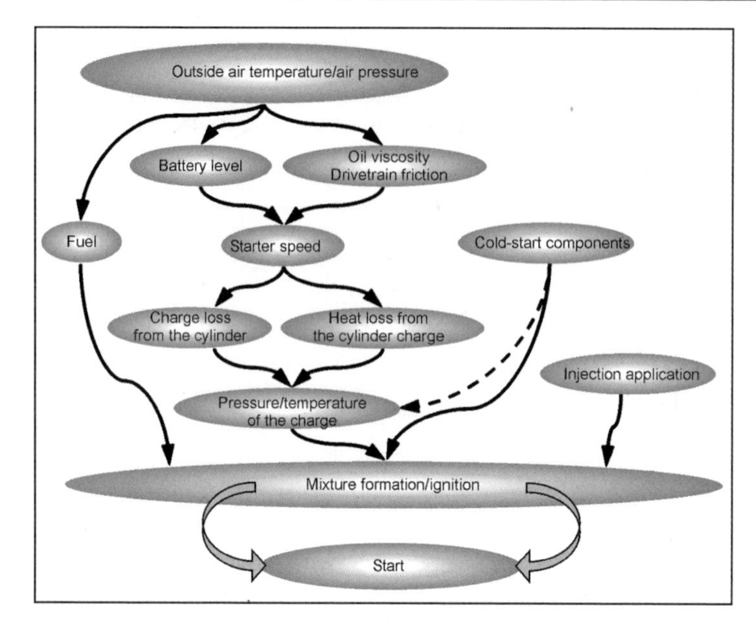

Fig. 13-20 Important influence parameter from cold start.[3]

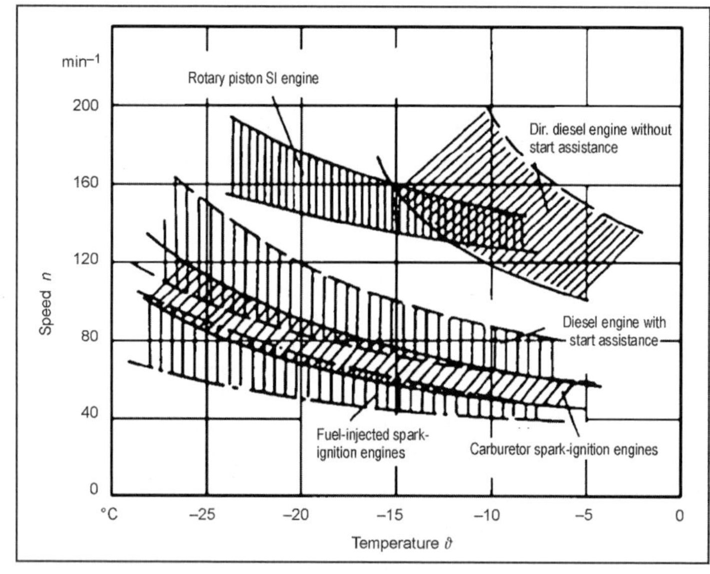

Fig. 13-21 Minimum start speed.[2]

only half-charged, the threshold temperature rises, e.g., from −24°C to −20°C.[2]

An important quantity in this context is ignition lag that describes the time from the beginning of injection to ignition. The beginning of injection is determined by the needle lift signal, the solenoid valve, and injector flow or, in the case of optically accessible aggregates, by the exit of the fuels from the injection orifice. The start of combustion can be obtained from the cylinder pressure signal, from an ionic current signal, or optically from light signals. The ignition lag increases exponentially as the charge temperature decreases,[6] and it has a minimum[3] at an average start speed of approximately 200 min^{-1} (Fig. 13-22).

This is explained by the overlap of the physical ignition lag with the chemical ignition lag. While the physical ignition lag decreases as the start speed increases because of improved mixture preparation, the chemical ignition lag increases.[7] The reason is the kinetics of the initial reactions whose duration is nearly constant. In the portrayed example, the chemical ignition lag at $\varphi_0 = -20°C$ above 200 min^{-1} is a nearly constant 6 ms. The chemical ignition lag in degrees crankshaft angle, hence, increases proportionally to rpm. In contrast to the inline fuel injection pump used in the cited investigations, modern injection techniques have further lowered the physical ignition lag and greatly improved mixture formation. At greater starter

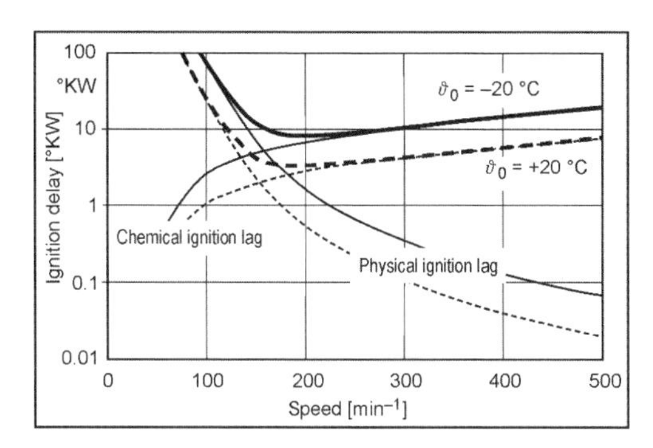

Fig. 13-22 Ignition lag.[3]

speeds, however, the chemical ignition predominates so much that the presented results are still applicable.

The longer ignition lag under cold-start conditions cannot be compensated at will by advancing injection. Fuel injected too early mixes with the slightly compressed charge until it falls below the ignition threshold or deposits on the combustion chamber walls. It is no longer available for combustion when the pressure and temperature necessary for autoignition are attained because of increasing compression.

13.3.2.2 Start Evaluation Criteria

In passenger cars, a reliable, independent start and subsequent stable and smooth engine running are required. We still have no regulation of the exhaust gas emissions in cold starts below the freezing point in passenger car diesel engines. In evaluating the starting quality, the impairment of driver comfort is the primary focus. This is based on the perception of noise or odor, visible exhaust gas clouds (soot, blue and white smoke), vibrations, the waiting time until the start, the starting time itself, and a poor reaction of the engine to acceleration. The quality of cold starts can be evaluated by measuring the noise level, smoke density, and other exhaust gas emissions—in particular, HCs—and the evaluation of the speed fluctuations during idling and increases in speed as a response to the quantity of injected fuel (Fig. 13-23).

Despite the possibilities for measuring cold starts, in the final analysis, the subjective impression of the driver is the decisive factor, which is much more complex, and widely varying importance is assigned the absolute measured quantities.

13.3.3 Components for Supporting Cold Starts

As the temperature decreases, the conditions for quick ignition and complete combustion worsen even under otherwise favorable conditions. Without a cold-start aid, the start quality decreases until the start becomes too long for the driver at temperatures below −10°C or even becomes impossible. Aids for supporting cold starts have the task of improving ignition conditions until the combustion in the cylinder is highly effective within the available time limits. The limits are set by the engine processes in the power cycle and are set, on the one hand, by the optimum beginning of injection so that the injected fuel can ignite before it deposits on the combustion chamber wall or mixes so thoroughly that it falls below the ignition threshold; on the other hand, the limits are set by the maximum available

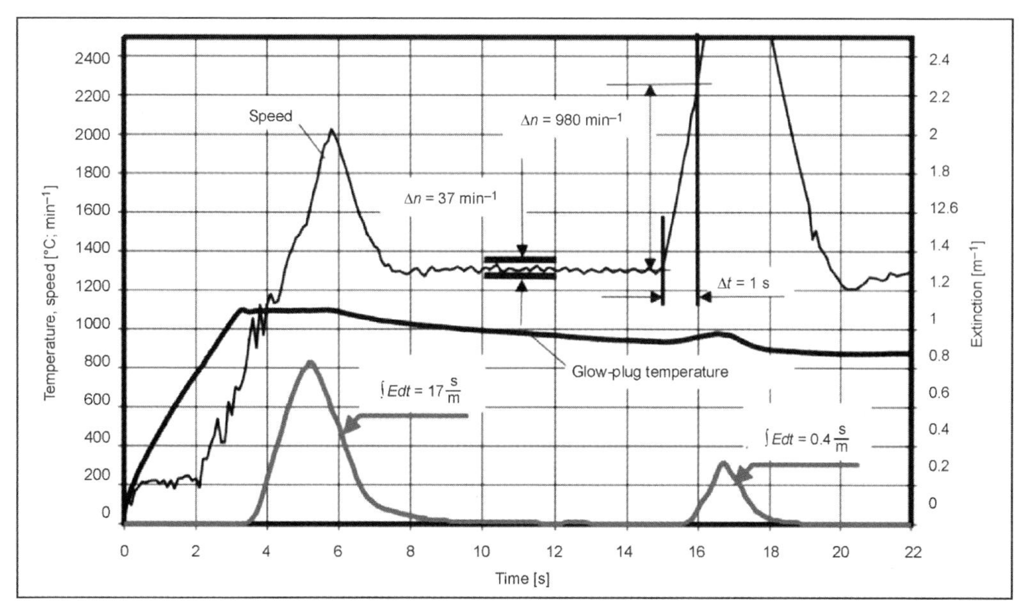

Fig. 13-23 Start evaluation conditions.

time for complete combustion. Furthermore, a sufficient amount of fuel must be converted to continue accelerating the engine by a release of energy that exceeds internal losses. To fulfill these requirements, the start of combustion or the maximum rise in pressure must be at top dead center. Typically, the combustion during a cold start is subject to strong cyclic fluctuations so that substantial instabilities arise, including misfiring.[8] The task of cold-start aids is to compensate for the worsening of start conditions, particularly in delayed mixture preparation, and to introduce smooth ignition at the right time for stable combustion.

This is done with glow plugs by electrically generated heat that is directly introduced into the combustion chamber and locally promotes mixture formation and ignition. Another approach especially targeted for engines with large displacement is to heat the intake air with flame glow plugs or electrical heating flanges that heats the entire air charge in the intake tract to a substantially higher level so that the injected fuel conditions correspond to those in a hot-running engine.

13.3.3.1 Glow Plug Systems

A heating system with glow plugs as the active heating elements in the combustion chamber and an electronic control interprets the commands of the engine management, prepares information on the state of the system, and returns it to the control unit. In modern passenger car diesel engines, the glow plug has become a standard component. For engines with a divided combustion chamber, it is an essential cold-start aid which ensures that starting also occurs within the frequently rising temperature range of 10–30°C. Because of the drastic worsening of the start quality below freezing, the glow plug as a cold-start aid is also used for diesel engines with direct fuel injection.[4]

Principle

The glow plug is typically close to the injection nozzle but is not directly positioned in the injection jet and extends approximately 3–8 mm into the combustion chamber. It offers directly to the combustion chamber a comparatively low amount of heat in the form of a hot surface. The power input depending on the design and size is 30–150 W in a state of equilibrium. Glow plugs attain surface temperatures of 800–1100°C. The physical and chemical ignition lag is reduced next to the hot glow element tip because of the accelerated vaporization of the fuel droplets and the initial chemical reactions that are faster at higher temperatures.[9] Subsequently, local combustion must be supplied sufficient energy to independently maintain the flame and ignite fuel injected by the injection jets far from the glow plug so that all the introduced fuel is completely combusted in the remaining time. The glow plug, therefore, acts as an indirect, local ignition aid; the energy for igniting the majority of the fuel originates from the fuel itself.

The glow plug continues to be supplied with current after starting for up to three minutes depending on the engine temperature (postglowing) to ensure favorable and constant ignition conditions during the engine warm-up phase.

The energy introduced into the combustion chamber by preheating the charge or the combustion chamber walls is not decisive, although it still must be taken into consideration. Good start qualities can also be obtained by using quickly heating glow plugs without additional preheating. In addition, the thermal mass of the metal combustion chamber walls is high enough so that a significant temperature increase cannot arise within the cited performance range over 3–15 s. Heat supplied to the air charge during the preheating phase is lost with the first gas exchange. The experience that long preheating phases improve the start quality is based on the fact that a self-regulating glow plug heats a greater area as the heating time increases and, hence, saves more heat energy. The glow plug cools less while the starter operates, as is the case with shorter preheating phases.

It is frequently assumed that a "hot spot," a comparatively miniscule hot point, is sufficient for ignition. Since the locations with favorable ignition conditions fluctuate strongly from cycle to cycle, in particular, during a cold start, and a large thermal mass reduces temperature fluctuations at the glow element, a "hot area" or "hot volume" is necessary in practice.

Requirements

The glow plug should provide a sufficiently high temperature over the shortest possible time to support ignition and maintain this temperature independent of the momentary conditions or even adapt it to the conditions.

The installation space available for the glow plug is especially limited in modern engines with a four-valve design with pump-nozzle elements or injectors; the glow plugs must, therefore, be as slim as possible, but they must also be very robust.[10] In addition, the installation situation makes exchanging the glow plugs expensive, so the glow plugs need to last the life of an engine.

Since the load on the vehicle electrical system is particularly critical during cold starts, the glow plugs require a minimal power input.

Legal regulations require a permanent monitoring system, OBD, for emissions-relevant components. This is realized in glow plug systems by monitoring each individual glow plug and providing feedback to the engine control unit. With electronic glow plug systems, there are other possibilities of influencing emissions within the engine. By intermediate heating, i.e., turning on the glow plugs again when the aggregates cooled during overrunning, controlled combustion is ensured with minimum emissions.

Design

Glow plugs consist of a metal resistance heating element wound into a coil that is protected from combustion chamber gases by a metal sheath resistant to hot gas corrosion. In this glow tube, the coil is embedded in compressed magnesium oxide powder that provides electrical isolation, good heat transmission, and mechanical stability. This component forms the heating element together with the supply of current to the helical heating wire. This is pressed into a body with a sealing seat, a thread, and a hex head that is used to screw the glow plug into the cylinder head and that creates the ground contact. The current is carried to the heating element with a threaded or plug-in connection. A standard size for a glow plug is an M10 thread and a 5-mm-diameter heating element. The length and the head shape vary depending on the requirements (Fig. 13-24).

a top constant temperature. By selecting the control material and the resistance division between the heating and control filament, various characteristics in the temperature curve can be represented.

The glow plug is controlled via a relay or an electronic switch, and its nominal voltage corresponds to the voltage offered by the vehicle electrical system when the engine idles. The glow plug is cooled by the movement of air during starting and while the engine is running. This is, however, compensated by the higher available vehicle electrical system voltage so that the desired temperature is maintained during postglowing (Fig. 13-25).

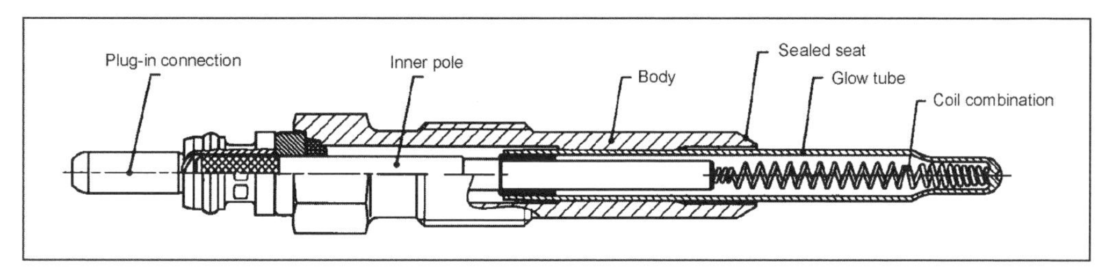

Fig. 13-24 Glow plug design.

(a) Self-Regulating Glow Plug

In self-regulating glow plugs, the coil consists of a combination of a helical heating wire and a control filament. The helical heating wire consists of a high-temperature-resistant material whose electrical resistance is largely independent of temperature, whereas the resistance of the control filament has a large positive temperature coefficient. With cold glow plugs, first a high current arises that quickly heats the helical heating wire. Through heat conduction and self-heating, the control filament subsequently becomes increasingly hot so that overall resistance increases and the current decreases. We, therefore, have a combination of fast heating with independent regulation at

(b) ISS Glow Plug System (Instant Start System)

One fast starting glow plug system consists of an electronic control unit and a fast starting glow plug.[11] The design is similar to that of the self-regulating glow plug, but the coil combination is substantially shorter, and the glowing area is reduced to approximately one-third. In diesel engines with direct injection, this corresponds to the part of the heating element extending into the combustion chamber. As a side effect, the power demand is 2 to 3 times less, which is particularly important in engines with eight or more cylinders.

The glow plug is designed for operation with a nominal voltage such as 5 V that is less than the vehicle system

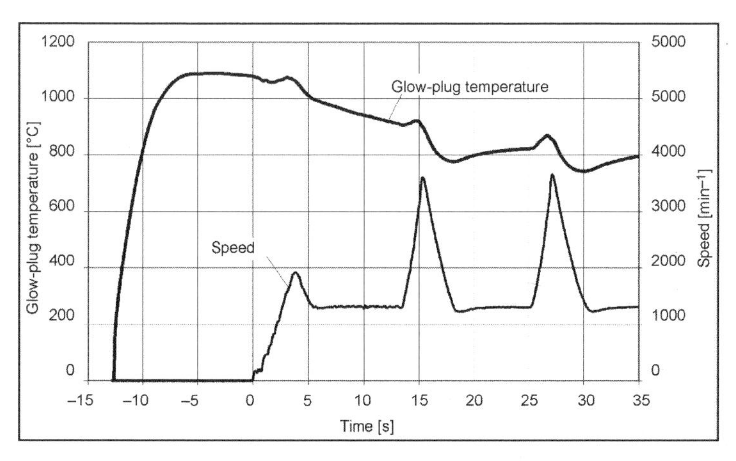

Fig. 13-25 Start with self-regulating glow plug.

voltage with which this glow plug attains a constant temperature of approximately 1000°C. Using the electronic control unit, the vehicle system voltage is clocked, and the voltage at the glow plug is effectively reduced to 5 V. The desired temperature at the glow plug can, therefore, be maintained as soon as a vehicle system voltage of more than 5 V is available. The glow plug temperature is, hence, independent from the vehicle system voltage that is frequently only 7–9 V, particularly during starter operation.

In running engines, the glow plug is cooled by the charge cycle and air movement in the compression phase. The temperature of the glow plug decreases as the rpm increase at a constant glow plug voltage and injected fuel quantity, whereas the temperature increases with increasing injected fuel quantity and constant glow plug voltage and rpm. These effects can be compensated for with the aid of the electronic control unit by always supplying to the glow plugs the optimum effective voltage required for the respective operating point. Other influencing variables are compensated in an analogous manner. The glow plug temperature can, hence, be applied depending on the operating state.

Furthermore, the combination of a low-voltage glow plug with an electronic control unit is used to heat the glow plug extremely quickly by applying the full vehicle system voltage to the glow plug for a predefined time and only subsequently cyclically applying the required effective voltage. The conventional preheating time is reduced to a maximum of 2 s down to the lowest temperatures, hence, enabling the same start times as SI engines.

Because of the high dynamics of the glow plug system, a start without preheating is also possible. At low temperatures, it is nevertheless logical to set short preheating times that can coincide with the required initializations, checks, etc. Given an already hot glow plug, substantially better ignition prerequisites exist from the beginning.

The electronics also assume protective functions for the glow plug and communicate with the engine control unit (for OBD). With its expanded degrees of freedom, the glow plug system will be used in the future in the application phase to optimize the internal engine combustion processes and the life of the glow plugs.

13.3.3.2 Heating Flange

Today, electrical heating flanges are primarily used for commercial vehicle engines with a piston displacement greater than 0.8 l per cylinder. They allow a reliable start at low temperatures and the reduction of smoke emissions.[12] As demands increase for the reduction of emissions during cold starts and the improvement of driving comfort, correspondingly adapted electrical heating flanges are becoming interesting for passenger car applications.

Principle

The heating flange with a 0.5–2 kW connecting cable is installed in front of or inside the intake manifold. The electrical output is converted into heat in the heating flange and released to the intake air.

Normally, metal heater elements do not have temperature-dependent resistance. In addition, there are heating flanges with a PTC characteristic that have metal or ceramic elements.[12] The optimum characteristic for good starting can be supported with correspondingly controllable power electronics.

The heating flange heats the intake air temperature to at least 30 K. Figure 13-26 shows the theoretical increase in the final compression temperature T_2 plotted against the intake air temperature T_1 for various compression conditions and a polytropic exponent of $n = 1.37$ that is calculated according to the relationship for polytropic compression $T_2 = T_1 \cdot \varepsilon^{n-1}$.

This shows that, e.g., at a compression ratio of $\varepsilon = 18.5$, an increase in the intake air temperature of $\Delta T_1 = 50$ K yields an increase in the final compression temperature of $\Delta T_2 = 147$ K.

The cited relation yields much higher values for the final compression temperature during a cold start. Given T1 = −25°C with the above data, T2 would theoretically = 457°C. This value would not be attained because of heat and charge losses and the reduced intake air density.

For the relative estimation of the "reinforcement" of the air temperature increase from polytropic compression, this relation is still valid despite the absolutely lower final compressions temperature. In the literature, polytropic exponents of $n = 1.2–1.3$ are frequently used to calculate the thermodynamic state during a cold start. These exponents result from pressure measurements, taking into consideration the heat and charge losses. However, in an integrating combustion chamber temperature measurement, Rau[3] has shown that the exponent for calculating the temperature is close to the theoretical exponent, i.e., $n = 1.38$ in the relevant range (250 K $< T_1 <$ 830 K).

Because of the globally higher charge temperature, the heating flange improves mixture preparation and substantially reduces ignition lag.

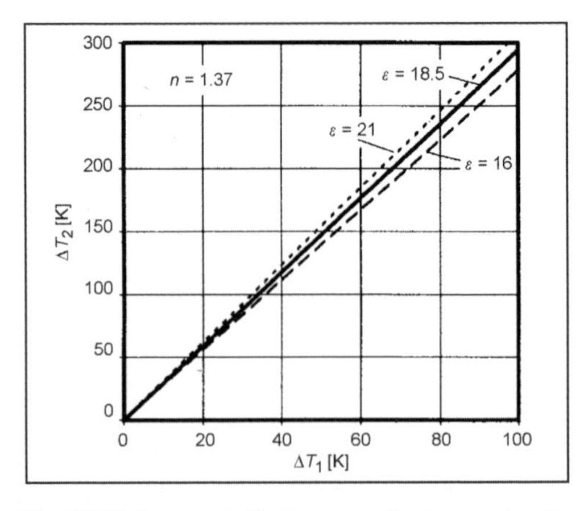

Fig. 13-26 Increase in final compression temperature by preheating intake air.

Requirements

For the electrical heating flange, a short heating time and good heat transfer from the heater element to the air are required with minimal flow resistance in the intake air duct. The electrical connecting cable must heat the intake air as much as possible without overloading the vehicle's electrical system. The available installation space is determined by the intake air cross section. The dynamic changes in the intake air flow speed require that the heating flange have sufficient thermal mass to prevent fast cooling and overheating of the element.

For the OBD, the function of the heating flange can be monitored with the aid of an electronic control unit and sent to the engine control unit.

Design

The electrical heating flange consists of an approximately 20-mm-wide frame or flange in the intake air guide. It assumes a sealing function, connects to and guides power, and holds power electronics and the heater element, including insulation. The heater element consists of one or more metal strips that typically meander with approximately five windings in a ceramic insulator and are connected at one side to the frame for a ground connection (Fig. 13-27).

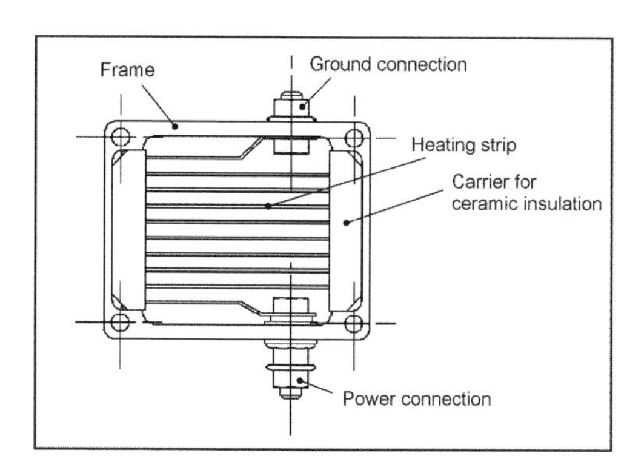

Fig. 13-27 Heating flange.

Function

When current is applied to the heating flange, the heater element reaches 900–1100°C and heats the resting, surrounding air. With the activation of the starter, preheated air is inducted and compressed. The higher global charge temperatures improve ignition conditions. The heating flange heats the flowing air in the intake tract by approximately 50°C and is thereby cooled to 500–600°C (Fig. 13-28).

The thermal mass of the heater element slows fast changes in the air flow, and slow changes are compensated by the self-regulating behavior of the heating tape or an electronic control. Because of the heat output of the heating flanges that is substantially greater than that of a glow plug, ignition conditions are achieved quickly in the entire combustion chamber that, together with the adaptation of fuel injection, substantially reduce smoke emissions in the warm-up phase[12] (Fig. 13-29).

13.3.4 Outlook

13.3.4.1 Combined Systems

The glow plug system is the suitable cold start aid for diesel engines in passenger cars to ensure the fastest start with a minimum drain on the vehicle electrical system. In contrast, electrical heating flanges have the potential to further reduce warm-up emissions, improve smooth engine running, and improve load assumption. It, hence, makes sense in view of increasingly stringent exhaust regulations to combine both systems to attain fast starts with minimum emissions and maximum smooth running. This solution is particularly recommendable when there are many cylinders and a large charge volume.

13.3.4.2 Measurement of Ionic Current

In SI engines, the measurement of ionic current is already used to obtain information about combustion directly from the combustion chamber.[13] So that no additional probes have to be introduced into the combustion chamber, glow plugs are recommendable for diesel engines

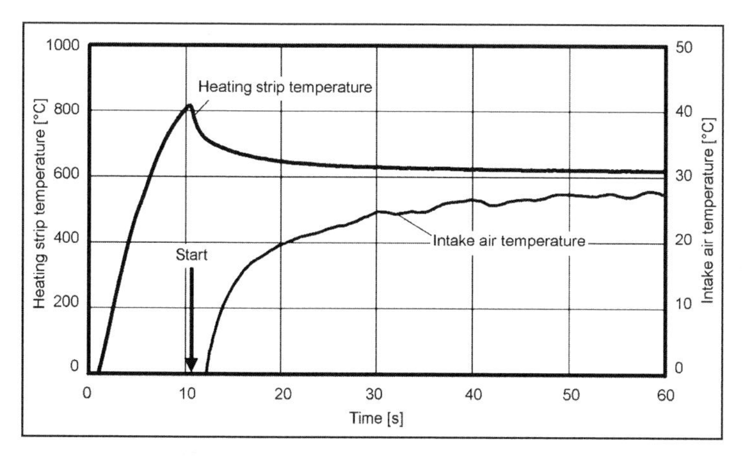

Fig. 13-28 Heating behavior of a heating flange.

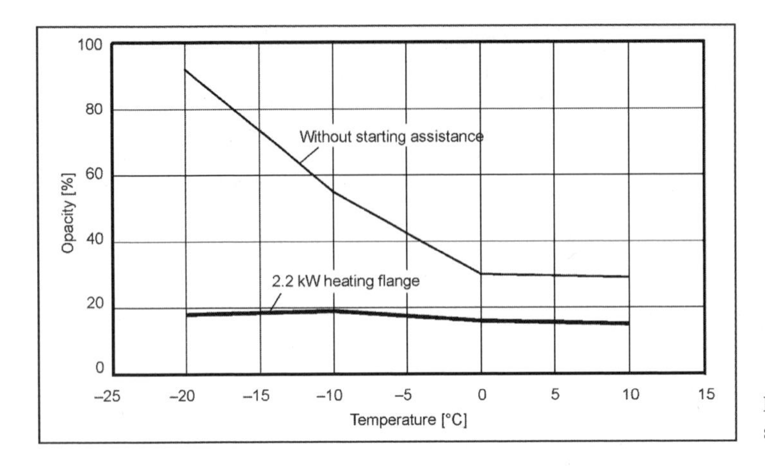

Fig. 13-29 Opacity 30 s after starting.[12]

with their favorable position[14] and the possibility of oxidizing soot on the electrode. If the heating element is isolated from the glow plugs and a voltage is applied, an electrical field forms in the combustion chamber around the glow plug tip. The charges of the particles in the field flow through the electrode. Current of a few microamperes to milliamperes can be measured by a suitable circuit, amplified, and possibly prepared and sent to the engine control unit (Fig. 13-30).

Diesel engine combustion is especially subject to significant local stochastic fluctuations.[15] Subsequently, in contrast to the integrating cylinder pressure signal, the ionic current signal measured at the glow plug can sometimes only indirectly determine thermodynamic information such as the combustion function, the location of main combustion, etc., with a great deal of calculation.

Measuring ionic current using glow plugs is cheaper in comparison to indicating cylinder pressure and represents a robust internal engine sensing mechanism that can be continuously evaluated. Potential applications of ionic current measurement are, for example,

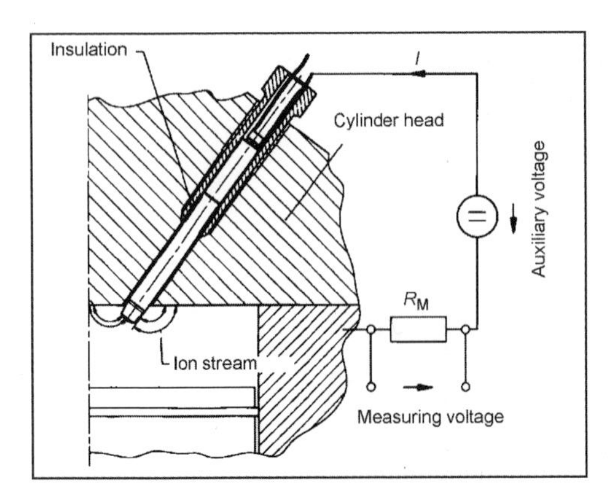

Fig. 13-30 Principle of ionic current measurement.

- Detection of misfiring
- Cylinder equalization at the start of combustion, balancing tolerances in the injection and intake system, etc.
- Fulfillment of OBD requirements by direct feedback from the combustion chamber
- Compensation for differing fuel quality

To realize "ionic-current-regulated diesel engines, substantial efforts are being made at present to develop corresponding evaluation algorithms and governor structures to correlate the measured signals with the processes in the combustion chamber. Furthermore, the position of the sensor and its design for long-term use must be optimized. To attain the isolation of the heating element from the cylinder head required for measuring ionic current, a separate ground connection is necessary between the heating element to the cylinder head that can be interrupted for the ionic current measurement. A circuit that does this is integrated in the glow plug so that the external design of the glow plug does not change.

13.3.4.3 Regulated Glow Plug Systems

The self-regulating glow plugs that are frequently used today will be increasingly replaced in the future by electronically controlled systems. The next goal is to develop regulated systems that do not require complex calculations of the control output depending on the engine parameters. Instead, a higher-level engine control unit should transmit only the required amount of heating in the form of a set point to the glow plug control unit that interprets the set point and correspondingly regulates the required voltage sent to the glow plug. To attain this goal, glow plugs must be developed that can return an easily evaluated and stable temperature signal to the glow plug control unit.

Bibliography

[1] Bauder, R., Die Zukunft der Dieselmotoren-Technologie, in MTZ 59 (1989) 7/8.

[2] Henneberger, G., Elektrische Motorausrüstung, Vieweg, Wiesbaden, 1990.

[3] Rau, B., Versuche zur Thermodynamik und Gemischbildung beim Kaltstart eines direkteinspritzenden Viertakt-Dieselmotors, Dissertation, Technical University of Hannover, 1975.

[4] Petersen, R., Kaltstart- und Warmlaufverhalten von Dieselmotoren unter besonderer Berücksichtigung der Kraftstoffrauchemission, in VDI-Fortschrittsbericht, Series 6, No. 77, 1980.

[5] Reuter, U., Kammerversuche zur Strahlausbreitung und Zündung bei dieselmotorischer Einspritzung, Dissertation, RWTH Aachen, 1990.

[6] Pischinger, F., Verbrennungsmotoren, Vorlesungsumdruck, RWTH Aachen, 1995.

[7] Sitkei, G., Kraftstoffaufbereitung und Verbrennung bei Dieselmotoren, Springer, Berlin, 1994.

[8] Zadeh, A., N. Henein, and W. Bryzik, Diesel Cold Starting: Actual Cycle Analysis under Borderline Conditions, SAE 909441, 1990.

[9] Warnatz, Technische Verbrennung, Springer-Verlag, 1996.

[10] Endler, M., Schlanke Glühkerzen für Dieselmotoren mit Direkteinspritzung, in MTZ 59 (1998) 2.

[11] Houben, H., G. Uhl, H.-G. Schmitz, and M. Endler, Das elektronisch gesteuerte Glühsystem ISS für Dieselmotoren, in MTZ 61 (2000) 10.

[12] Merz, R., Elektrische Ansaugluft-Vorwärmung bei kleineren und mittleren Dieselmotoren, in MTZ 58 (1997) 4.

[13] Schommers, J., U. Kleinecke, J. Miroll, and A. Wirth, Der neue Mercedes-Benz Zwölfzylindermotor mit Zylinderabschaltung, Teil 2, in MTZ 61 (2000) 6.

[14] Glavmo, M., P. Spadafora, and R. Bosch, Closed Loop Start of Combustion Control Utilizing Ionization Sensing in a Diesel Engine, SAE paper 1999-01-0549.

[15] Ernst, H., Zündverzug und Bewertung des Kraftstoffs, in Deutsche Kraftfahrtforschung No. 63, VDI-Verlag, 1941.

14 Combustion

14.1 Principles

14.1.1 Fuels

Fuels for SI and diesel (CI) engines are mineral oil based and consist of hundreds of individual components. This composition has a very great influence on the physical and chemical properties and, thus, on the combustion characteristics.

During the production of fuels from coal, a synthesis gas is first produced by hydrogenation or gasification that is subsequently converted using an appropriate synthesis method to produce an alternative fuel, e.g., methanol (CH_3OH) or methane (CH_4). Fuels produced from coal as well as nonfossilized fuels such as rape oil or rape methyl ester that are obtained from biomass are of only subordinate importance for motor vehicle applications and are used today at best in niche markets.[1]

This section focuses on the classification and chemical structure of simple hydrocarbons, the $C_xH_yO_z$ compounds, insofar as it is necessary for an understanding of the oxidation of hydrocarbons.

Hydrocarbon compounds are generally subdivided into alkanes (earlier, paraffins), alkenes (olefins), alkines (acetylenes), cycloalkanes (naphthenes), and aromatics.

The alkanes (paraffins) are chainlike or "aliphatic" hydrocarbons with purely single bonds (monovalent), while a distinction is made between the normal alkanes with a straight-chain structure and iso-alkanes with a branched-chain structure. The alkenes (olefins) are chainlike hydrocarbons with one or two double bonds, while alkenes (mono-olefins) have one and alkadienes (diolefins) two double bonds. The alkines (acetylenes) also have a chainlike structure and a triple bond. Figure 14-1 shows the structural formulas for these aliphatic hydrocarbons.

The structural formulas of the cycloalkanes (naphthenes) with their circular structure and purely single bonds and of the circular aromatics with their double bonds whose basis is the benzene ring are shown in Fig. 14-2.

Oxygenated hydrocarbons are chainlike compounds for which a distinction is made between alcohols, ethers, ketones, and aldehydes.

Alcohols contain a hydroxyl group (R-OH). The simplest alcohols are methyl alcohol (methanol: C_3H-OH) and ethyl alcohol (ethanol: C_2H_5-OH). Ethers are hydrocarbon residues linked together by an oxygen bridge (R_1-O-R_2) and ketones residues linked by a carbonyl group (R_1-CO-R_2).

Aldehydes contain a CHO group, e.g., formaldehyde (HCHO). The structural formulas of the oxygenated hydrocarbons are shown in Fig. 14-3, where the CHO groups should not be confused with the OH group (-COH) attached to the carbon.

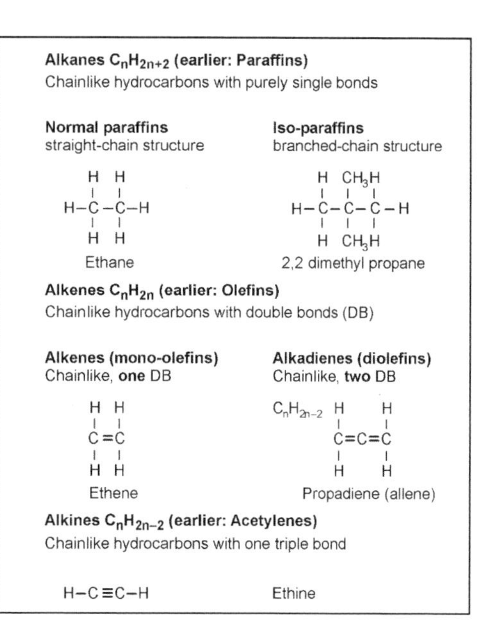

Fig. 14-1 Aliphatic hydrocarbon compounds.

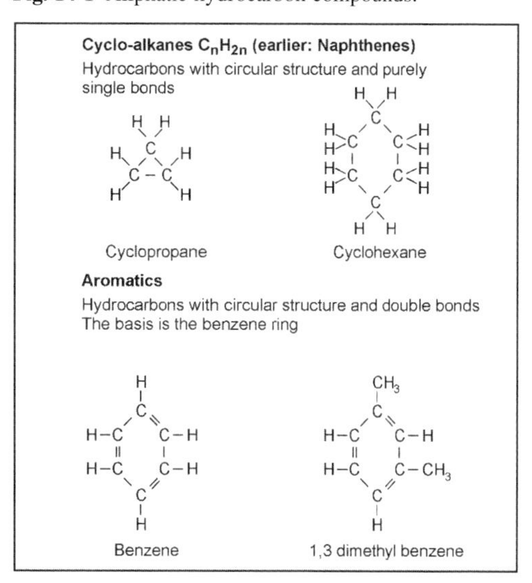

Fig. 14-2 Alicyclic (open-chain) and aromatic hydrocarbon compounds.

Two-component substitute fuels are used to determine the ignition quality of SI and CI engine fuels, namely, the substitute fuel consisting of

- *n*-heptane (C_7H_{16}) with the octane number OZ = 0
- Iso-octane (C_8H_{18}) with the octane number OZ = 100

for the SI engine, and the substitute fuel consisting of

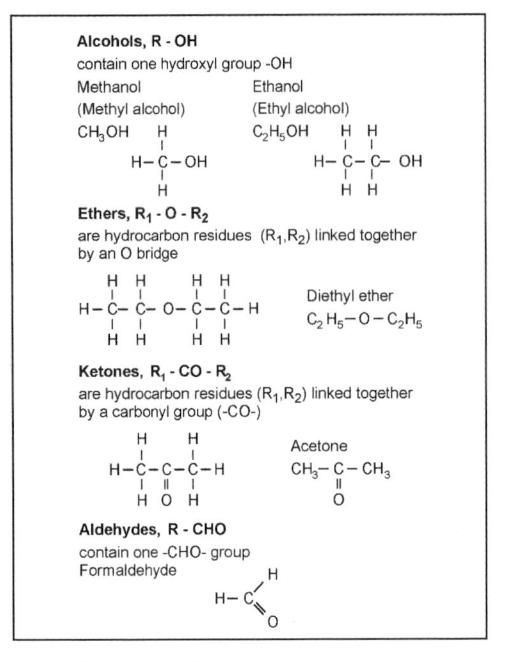

Fig. 14-3 Oxygenated hydrocarbon compounds.

- α-methyl naphthalene ($C_{11}H_{10}$) with the cetane number CZ = 0
- n-hexadecane (cetane: $C_{16}H_{34}$) with the cetane number CZ = 100

for the CI (diesel) engine, where the octane number is defined as the iso-octane content and the cetane number as the cetane content of the two component substitute fuel. The structural formulas of the components of the two substitute fuels are illustrated in Fig. 14-4.

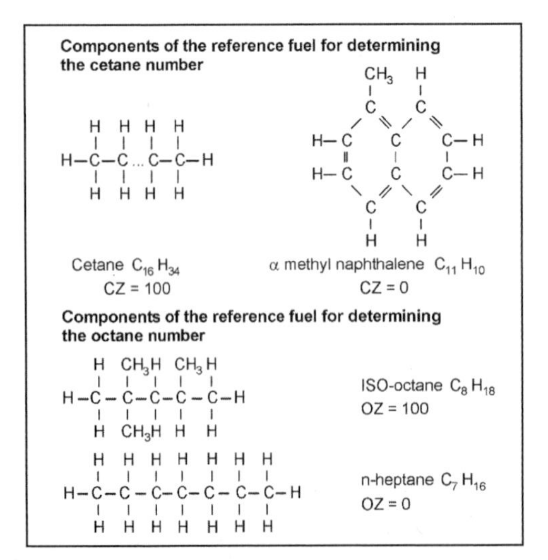

Fig. 14-4 Structural formulas for the components of the substitute fuels for the SI and CI engines.

While for SI engine fuels, a lower ignition quality and, hence, a higher knock resistance is required, the opposite is the case for CI (diesel) engine fuels.

The octane number drops with increasing numbers of hydrocarbon atoms for n-alkanes and alkenes while increasing the number of branches for iso-alkanes and the number of components with double bonds.

The net calorific value for the combustion of hydrocarbon compounds lies in the range

$$40.2 \text{ MJ/kg (benzene)} < H_u < 45.4 \text{ MJ/kg } (n\text{-pentane})$$

The maximum laminar flame speed of the liquid fuel components in air at 1 bar is only approximately 2 m/s; during combustion of these components in the engine, on the other hand, turbulent flame speeds of up to 25 m/s occur.

14.1.2 Oxidation of Hydrocarbons

With complete combustion, hydrocarbon compounds C_xH_y are converted into carbon dioxide CO_2 and water vapor H_2O. This reaction can be generally described by the gross reaction equation

$$C_xH_y + \left(x + \frac{y}{4}\right)O_2 \rightarrow xCO_2 + \frac{y}{2}H_2O + \Delta_R H \tag{14.1}$$

where the reaction enthalpy $\Delta_R H$ represents the heat released by the combustion process. In reality, however, the combustion does not take place as described by this gross reaction equation, but has a very complex reaction pattern based on elementary reactions that only today has started to be roughly understood and is illustrated schematically in Fig. 14-5.[1]

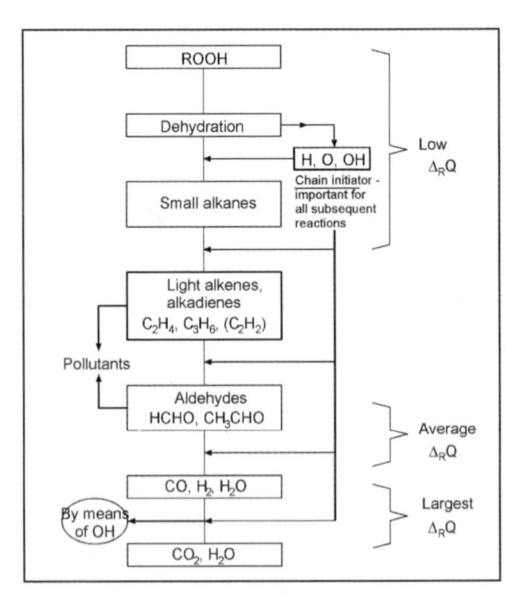

Fig. 14-5 Hydrocarbon oxidation process.

In a first reaction phase, hydrocarbon peroxides (ROOH) are formed that are broken down into small alkanes by dehydrogenization. The subsequent reactions with the radicals H$^\bullet$, O$^\bullet$, and O$^\bullet$H (chain propagators) first create light alkenes and alkadienes and ultimately aldehydes, such as acetaldehyde CH_3CHO and formaldehyde HCHO.

The formation of the aldehydes during which only about 10% of the released heat is produced is accompanied by the occurrence of a cold flame. In the following blue flame CO, H_2, and already H_2O are formed, and in the last stage, the hot flame, CO_2, and H_2O are ultimately formed. During the oxidation of the hydrocarbons to CO, roughly 30% and finally during the oxidation of the CO to CO_2 the remaining 60% of the thermal energy stored in the fuel are released. The main release of heat, therefore, takes place only at the end of the reaction process during the oxidation of CO to CO_2.

Figure 14-6 shows schematically a detail from the hydrocarbon oxidation process during which carbonyl compounds and formaldehyde are produced; Figure 14-7 shows qualitatively the temporal concentration and temperature curve during the hydrocarbon combustion.

In order to calculate the temperature and concentrations in the flame front it can be assumed that the eight components H, H_2, O, O_2, OH, CO, CO_2, and H_2O in the flame front are in partial equilibrium because of the high temperature prevailing there. This "OHC system" is therefore described by the five reaction equations

$$H_2 = 2H \tag{14.2}$$

$$O_2 = 2O \tag{14.3}$$

$$H_2O = \frac{1}{2}H_2 + OH \tag{14.4}$$

$$H_2O = \frac{1}{2}O_2 + H_2 \tag{14.5}$$

$$CO_2 = CO + \frac{1}{2}O_2 \tag{14.6}$$

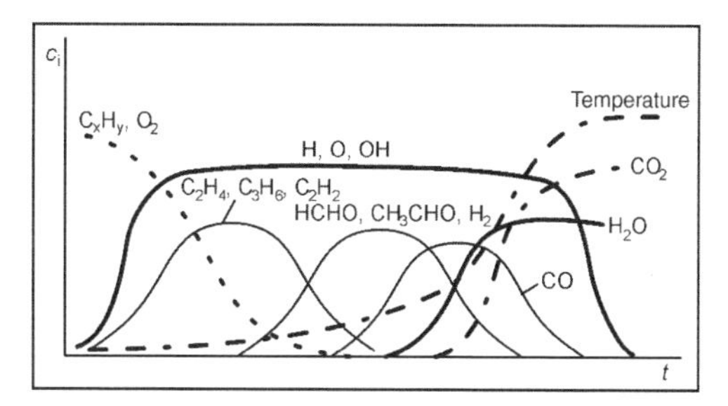

Fig. 14-7 Curve of temperature and concentration over time during hydrocarbon combustion.

while for the five equilibrium constants we have

$$K_1 = [H]^2[H_2]^{-1} \tag{14.7}$$

$$K_2 = [O]^2[O_2]^{-1} \tag{14.8}$$

$$K_3 = [H_2]^{\frac{1}{2}}[OH][H_2O]^{-1} \tag{14.9}$$

$$K_4 = [O_2]^{\frac{1}{2}}[H_2][H_2O]^{-1} \tag{14.10}$$

$$K_5 = [CO][O_2]^{\frac{1}{2}}[CO_2]^{-1} \tag{14.11}$$

Together with the atom balances for the atoms O, H, and C (better CO) and the condition that the sum of the partial pressures of all the components has to be equal to the total pressure, we ultimately obtain a nonlinear equation system that can be clearly solved using conventional numeric integration methods, e.g., the Newton-Kantorovitch method. Figure 14-8 shows as an example the concentration distribution of the OHC components as a function of the temperature for a total pressure of 1 bar.

If, for a subsequent calculation of the thermal nitrogen formation, only the oxygen atom concentration is required, this can also be approximately calculated according to Ref. [2] from the equation

$$[O] = 130[O_2]^{\frac{1}{2}} \exp\left(-\frac{29\,468}{T}\right) \tag{14.12}$$

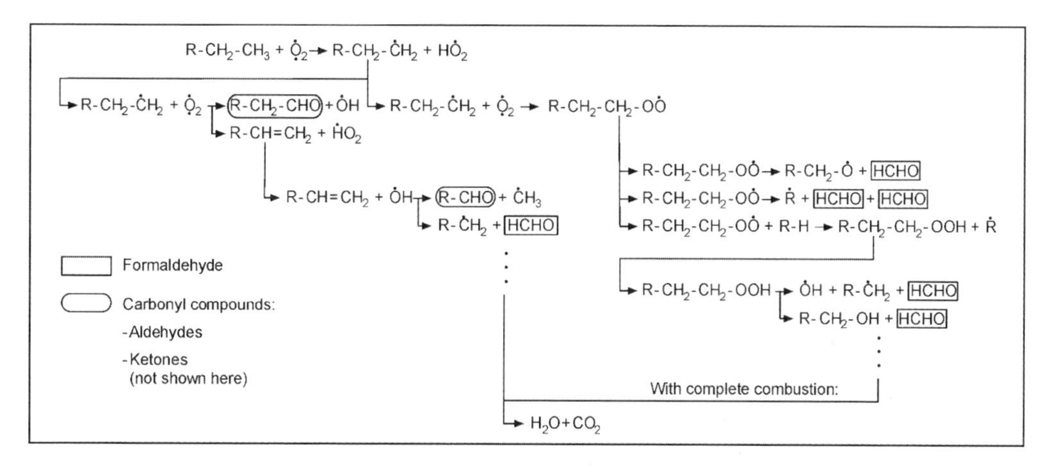

Fig. 14-6 Detail from the hydrocarbon oxidation process.

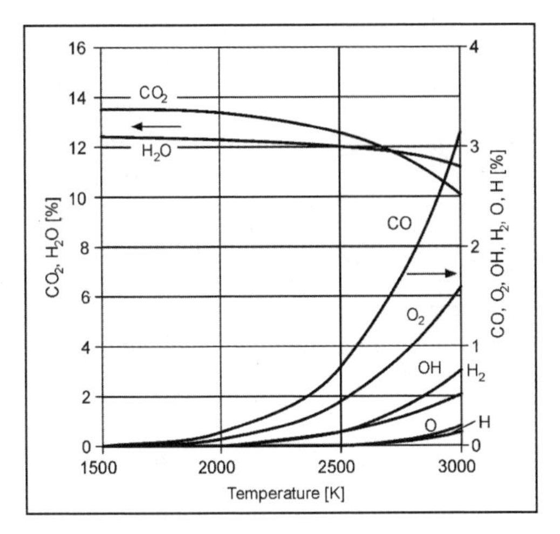

Fig. 14-8 Partial equilibrium of OHC components as a function of temperature for total pressure 1 bar.

14.2 Combustion in SI Engines

14.2.1 Mixture Formation

14.2.1.1 Intake Manifold Injection

In the SI engine, the fuel is generally mixed with air outside the combustion chamber, on older engines via the carburetor or by central injection upline of the air manifold and on newer engines by injecting the fuel into the intake manifold upline of the intake valve. On four-valve engines with two intake channels, an injection valve with a double jet toward the two intake channels is generally employed. Timing and volume of the fuel injected are determined by the start of opening and the duration of opening the injection valves, where these inject either once the full volume or twice half the volume of fuel. We speak of parallel injection when fuel is injected into all the cylinders simultaneously, irrespective of the momentary valve position. If, on the other hand, the fuel is injected into each cylinder separately and at an optimized moment before or even during the opening of the intake valve, then we speak of sequential injection. The qualitative differences for a four-cylinder engine are illustrated in Fig. 14-9.

The load of the SI engine is controlled quantitatively; i.e., the load is controlled by the volume of mixture via the throttle valve.

In the SI process, two working strokes are available for the mixture formation, the intake, and the compression stroke, while the flow characteristics in the combustion chamber and the fuel supply have a major influence on the mixture formation.

The admitted liquid fuel first has to evaporate completely before the fuel vapor subsequently mixes with the combustion air; the evaporation and mixing processes take place simultaneously. The fuel can burn completely only if the local excess-air factor of the air-fuel mixture in the combustion chamber is equal to or greater than one.

Mixture formation in the combustion chamber takes place roughly as follows:

- During the intake stroke, an extensive mixing of the air and fuel takes place, with small droplets of diameters $d_T \leq 20$ μm evaporating completely.
- During the compression stroke, intensive mixing takes place, where even the large droplets with diameters up to $d_T \approx 200$ μm evaporate completely.
- Although at ignition TDC droplets are no longer to be found, there is still, nevertheless, a large proportion of fuel and air still unmixed. The standard fluctuation of the fuel concentrations in the combustion chamber is roughly 10%–15%.

14.2.1.2 Direct Injection

The demand for a reduction in specific fuel consumption and, hence, in the CO_2 emissions led to the development of direct fuel injection also for the SI engine. Here the timing of the fuel injection determines the time available for the mixture formation. An early injection at the beginning of the intake stroke results in a fairly homogeneous mixture similar to that obtained with intake manifold injection. This early injection is used particularly in the upper load range. The late injection during the compression stroke, on the other hand, results in a favorable fuel-air ratio for the part-load range that is on average

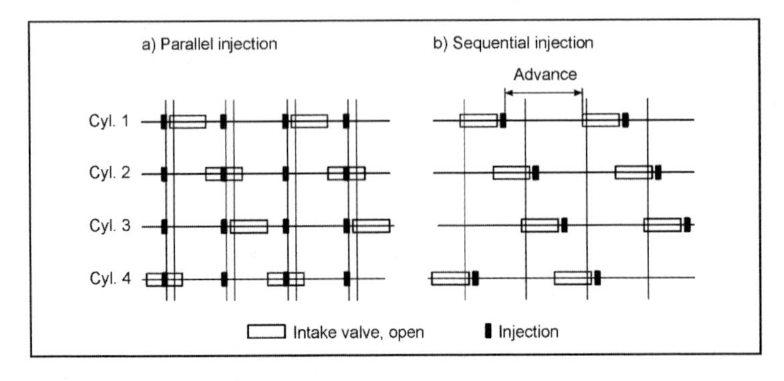

Fig. 14-9 Parallel (a) and sequential (b) injection.

exceptionally lean (lean-burn engine). The associated dethrottling of the engine, in particular, results in thermodynamic benefits for the combustion process and, hence, to a favorable fuel consumption. The situation is shown qualitatively in Fig. 14-10.

With later injection in the compression stroke, the throttle valve is completely open, and the excess-air factor decreases with the load.

With a changeover to an early injection timing during the intake stroke, the throttle valve is more or less closed and subsequently opened with increasing load so that the air-fuel ratio remains constant and equal to $\lambda = 1$. These two load sections are therefore referred to as the lean-burn concept and the Lambda = 1 phase. It should already be pointed out at this stage that the proven three-way catalytic converter cannot be used for exhaust gas treatment with the SI DI engine due to $\lambda \cong 1$ in lean-burn operation.

A distinction is made with respect to mixture formation in lean-burn operation between jet-directed, wall-directed, and air-directed combustion processes.

In the jet-directed combustion process, mixture formation and the development of the charge stratification are essentially attributable to the characteristics of the fuel jet. An ignitable mixture is formed at the outer edge of the jet, and the spark plug has to be positioned so that this ignitable mixture reaches it at the moment of ignition. Because of its function, this process is highly sensitive to the spray quality supplied by the injection system, and the assurance of an adequate mixture formation is problematical in the upper part-load range.

In the wall-directed combustion process, the fuel is first injected onto the surface of a specially formed piston recess where it evaporates and mixes with the combustion air. Formation of the mixture in good time is supported by a selective charge movement (tumble). This process also permits a high operational stability in the part-load range. The wall-directed mixture formation is significantly influenced, however, by the surface temperature of the piston recess; this can create problems during cold starting and in the upper part-load range.

In the air-directed combustion process, the mixture is formed by the interaction of fuel jet and a directed flow (swirl) of the cylinder charge. The combustion chamber must, therefore, be designed so that this swirl transports the resulting mixture to the spark plug and ensures that an ignitable mixture is available at the electrodes of the spark

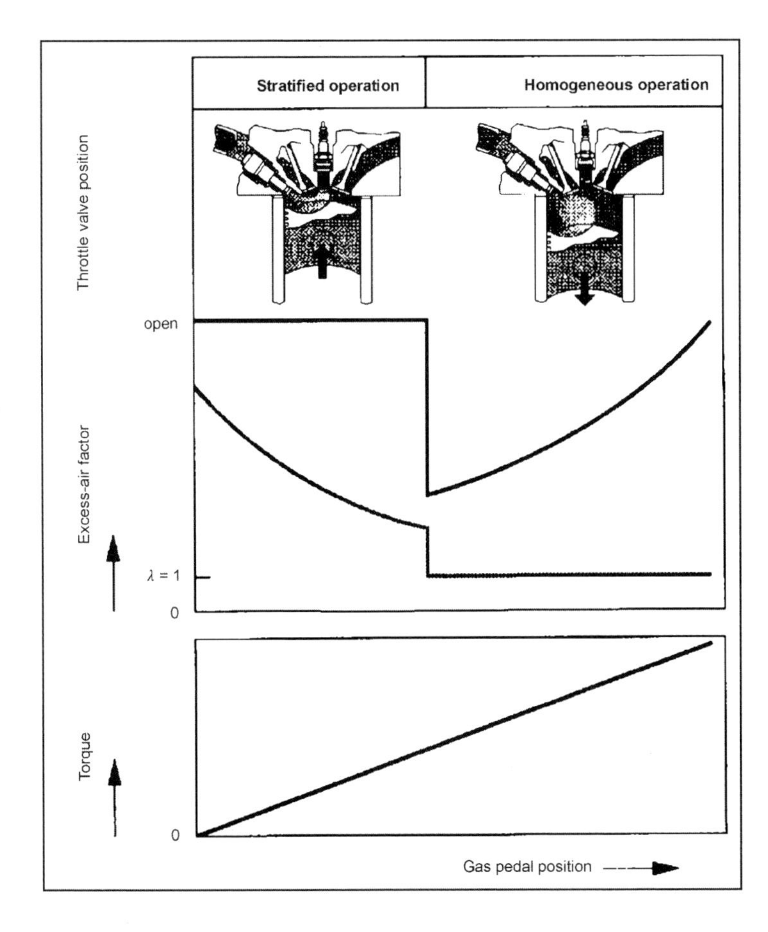

Fig. 14-10 Excess-air factor and throttle valve position in SI engine with direct fuel injection.

plug at the moment of ignition. This process ensures a stable operation with a stratified cylinder charge over a wide range of the engine map, and it offers the greatest potential for achieving a low HC emission.

The jet-, wall-, and air-directed combustion processes are illustrated qualitatively in Fig. 14-11, while a detailed description can be found in Ref. [3].

14.2.2 Ignition

The more or less homogeneous fuel-air mixture that exists at the end of the compression process is ignited by an electric ignition spark from a spark plug shortly before TDC. The actual moment of ignition is an optimization parameter; it is adapted to the engine operation so that an optimum combustion process is obtained.

In order to trigger the ignition, there has to be an ignitable mixture at the spark plug. Too lean and too rich mixtures do not ignite. From bomb trials we know the ignition limits $0.6 \leq \lambda \leq 1.6$ and from engine trials $0.8 \leq \lambda \leq 1.2$. The ignition temperature of the mixture has to be exceeded locally in the area of the ignition spark; from engine trials we know this temperature range is $3000\ \text{K} \leq T \leq 6000\ \text{K}$.

The ignition lag is roughly 1 ms and is the time between the moment of ignition and the start of combustion. The ignition voltage lies between 15 kV (normal) and 25 kV (cold start), the ignition energy lies in the range of 30–150 mJ (theoretically, 1 mJ), and the ignition duration is roughly 0.3 to 1 ms.

The ignition systems necessary for the ignition can basically be split into two groups, namely, into battery ignition systems and solenoid ignition systems. Battery ignition systems are the standard today, with solenoid ignition systems being used only in small and inexpensive engines and also where maintenance of the battery cannot be assured.

The conventional coil ignition system consists of battery, ignition switch, ignition coil, contact breaker, turnoff capacitor, ignition distributor, and spark plug. The current flowing through the primary winding of the ignition coil creates a strong magnetic field there. The current supply is interrupted by the contact breaker at the moment of ignition. The resulting rapid decay of the magnetic field induces the high voltage necessary for ignition in the secondary winding of the ignition coil. The induced primary voltage is approximately 350 V, while the secondary voltage is around 25 kV. The voltage causes a flashover at the spark plug.

In a transistor coil ignition system, a transistor is used as an electronic switch in the primary circuit instead of the cam-actuated contact breaker of the conventional coil ignition system. Compared with the conventional coil ignition system, the ignition voltage drops only slightly with increasing revs as a result, and the moment of ignition remains practically constant. Furthermore, greater ignition energy can be provided; the maximum number of sparks possible is roughly 30 000 per minute. In addition, the transistor coil ignition system is practically maintenance-free.

With the CDI system the ignition energy is stored in the electric field of a capacitor instead of in a magnetic field. The increase in the ignition voltage of approximately 300 V/μs um is roughly a power of ten larger than with the coil and transistor ignition systems.

With the conventional coil ignition system in which the ignition timing is adjusted mechanically in the ignition distributor, only simple spark advance curves can be produced. With the electronically controlled adjustment of the ignition timing, on the other hand, the ignition can be adapted much better to the engine operation, and the speed and load-dependent ignition map is far more differentiated than for the mechanical system. It consists

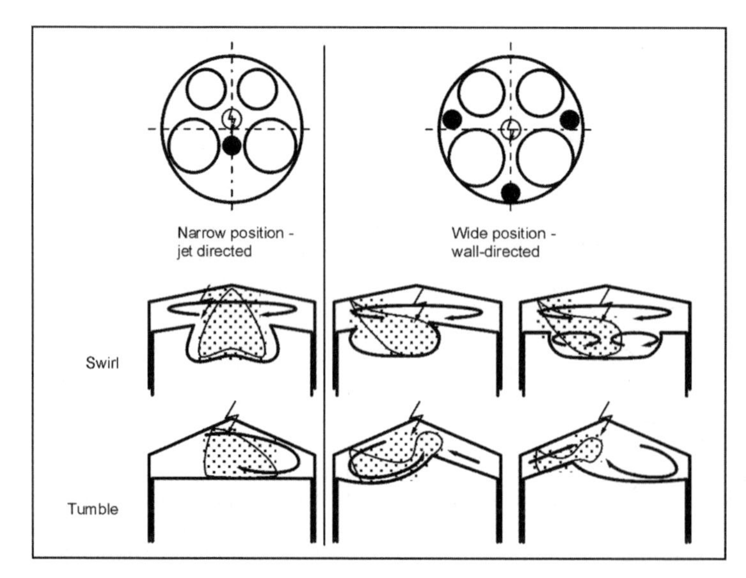

Fig. 14-11 Jet-directed, wall-directed, and air-directed combustion processes for SI DI engines.

of roughly 4000 individual values that are determined in trials and then stored electronically in the ignition map.

Figure 14-12 shows at the top a mechanical and at the bottom an electronic ignition map. The differences can be clearly and unambiguously seen.

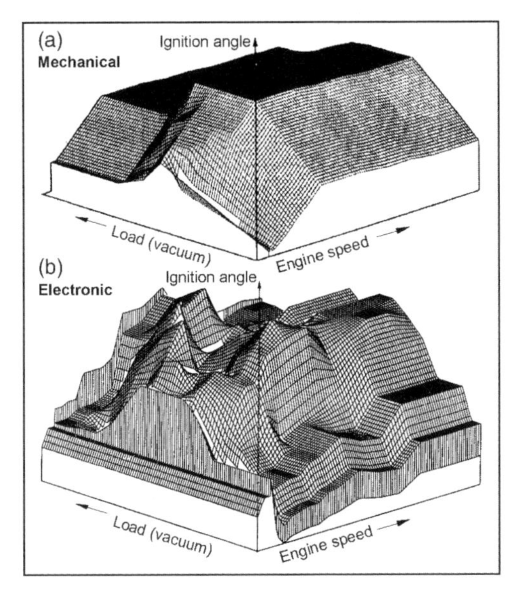

Fig. 14-12 Mechanical (a) and electronic (b) ignition maps.

14.2.3 Combustion Process

14.2.3.1 Flame Propagation

If, when considering the flame propagation, we assume an absolutely homogeneous fuel-air mixture in the combustion chamber, then we have the ideal case of a completely premixed flame. The chemical processes; taking place at the flame front are slow compared with the heat and material transport processes; the SI combustion is, therefore, chemically controlled.

Figure 14-13 shows qualitatively the pressure curve of the high-pressure process in dragged and motored mode. The terms "combustion time," "ignition lag," or "combustion lag" and "effective combustion time" are explained in the figure. The end of combustion is defined as the moment at which the fuel is "completely" combusted—as a rule, 99.9%. The figure also shows the ratio of combusted to admitted fuel, m_{BV}/m_B as a function of the crank angle that, under the assumption that the fuel is completely combusted, is identical with the burn-through function or the cumulative combustion cycle.

Figure 14-14 shows the position of the flame front at different crank angles for two working cycles each for intake manifold injection and direct injection. It can be seen that the differences occurring are relatively small and in the order of the cyclical fluctuations described in Section 14.2.3.3.

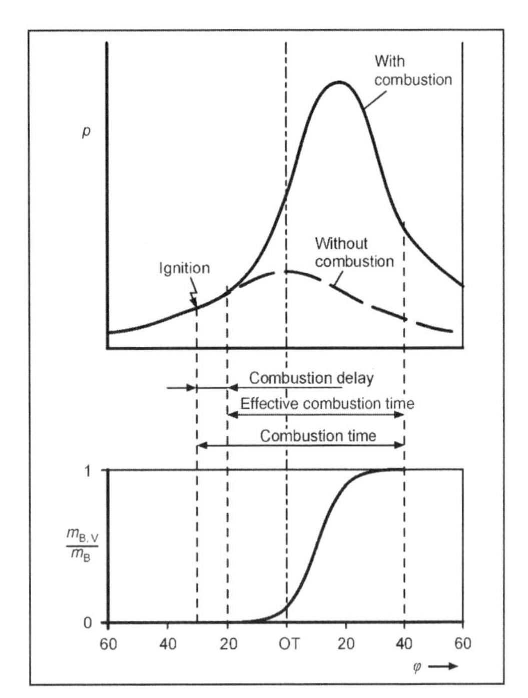

Fig. 14-13 Pressure and cumulative combustion curve as a function of the crank angle.

In order to achieve a faster burn-through of the mixture, three-valve engines with two decentrally positioned spark plugs (dual ignition) are now also being employed alongside the well-known four-valve engines with centrally positioned spark plug.[5] The flame propagation shown for these two engines in Fig. 14-15 shows clearly that the mixture burns through faster in the three-valve engine with dual ignition as the combustion paths are shorter; see also Section 15.2.

14.2.3.2 Mean Pressure and Fuel Consumption

The excess-air factor λ has a major effect on the flame speed and, hence, via the combustion and pressure curve, the achievable mean pressure, as well as the specific fuel consumption. For $\lambda > 1.1$ the combustion takes place increasingly slowly because of the lower combustion temperature caused by the heating of the excess air and the resulting lower flame speed. The minimum fuel consumption is achieved for air-fuel ratios of roughly $\lambda = 1.1$. On the other hand, because the mixture calorific value increases with decreasing air-fuel ratio, the maximum mean pressure is already achieved at roughly $\lambda = 0.85$. The optimum air-fuel ratio, thus, lies in the range $0.85 < \lambda < 1.1$. The "fish-hook curve" in Fig. 14-16 shows the development of the specific fuel consumption b_e as a function of the effective mean pressure p_{me}, where various

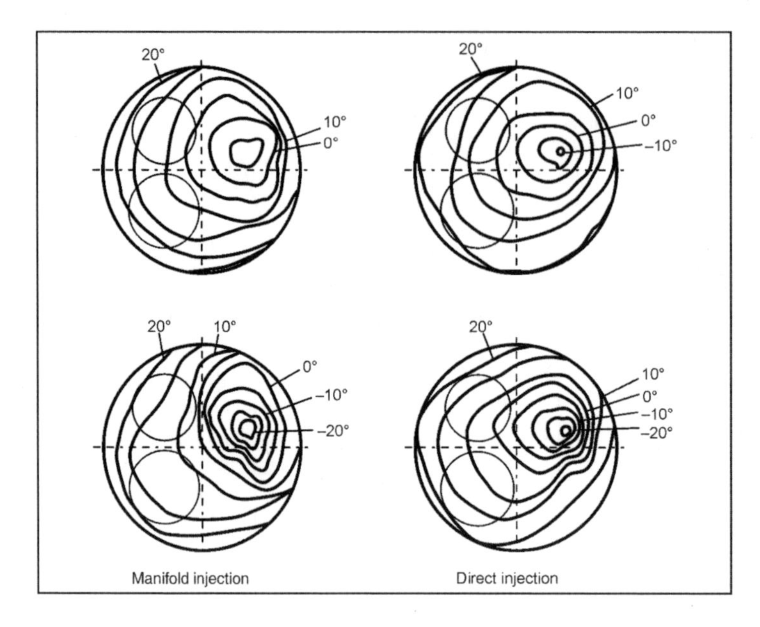

Fig. 14-14 Flame propagation for injection into the intake manifold upline of the intake valve and for direct injection during the intake stroke (Lambda = 1 mode).

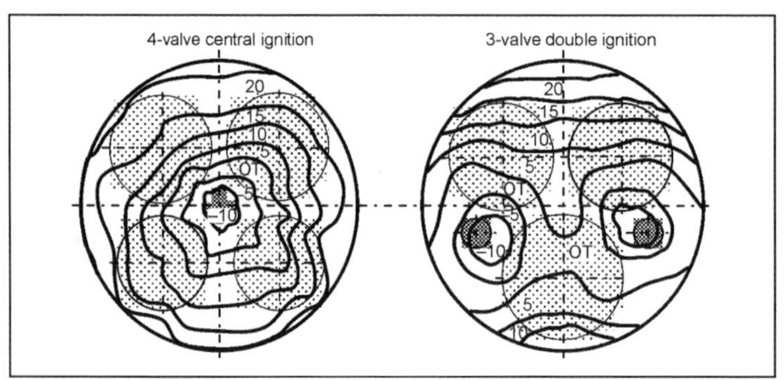

Fig. 14-15 Comparison of burn-through behavior for an engine with central ignition.

excess-air factors λ at constant engine revs and constant throttle valve position have been plotted.

When using the three-way catalytic converter for exhaust gas cleaning, the excess-air factor for SI engines must be controlled very precisely in the λ window $0.999 < \lambda < 1.002$. It can be seen from Fig. 14-16 that this value lies fairly exactly between $b_{e\,\min}$ and $p_{m\,\max}$.

14.2.3.3 Cyclical Fluctuations

Relatively large fluctuations in the pressure curve from working cycle to working cycle are a typical feature of the SI combustion. The reasons for these cyclical fluctuations are temporal and local fluctuations in the turbulent speed field and the mixture composition in the combustion chamber and, hence, also in the area of the electrodes of the spark plug. The resulting cyclical fluctuations in the ignition lag have a significant effect on the pressure curve in the combustion chamber and can result in a more or less complete combustion.

Figure 14-17 shows the effects of the cyclical fluctuations (top) and the influence of the ignition angle on the pressure curve during combustion of methanol. The comparison of the two diagrams shows that the cyclical fluctuations have a similar effect on the adjustment of the ignition angle by approximately 15° on the crankshaft. During thermodynamic evaluations of the tests, an average of the measured pressure curves, therefore, has to be taken, normally over 64 or 128 working cycles.

The reduction of the cyclical fluctuations during SI combustion by optimizing the mixture formation, the ignition map, and the flame propagation (dual ignition) is a worthwhile goal with a view to reducing the specific consumption and the HC emissions; see also Ref. [4].

14.2.3.4 Engine Knock

During normal combustion relatively "gentle pressure curves" with maximum pressure increase rates of approximately 2 bar/KW are to be observed. By comparison, sharp pressure fluctuations occur in the air-fuel mixture during knocking combustion that can be explained by the following process. After initiation of the combustion by the ignition sparks, the unburned residual mixture is

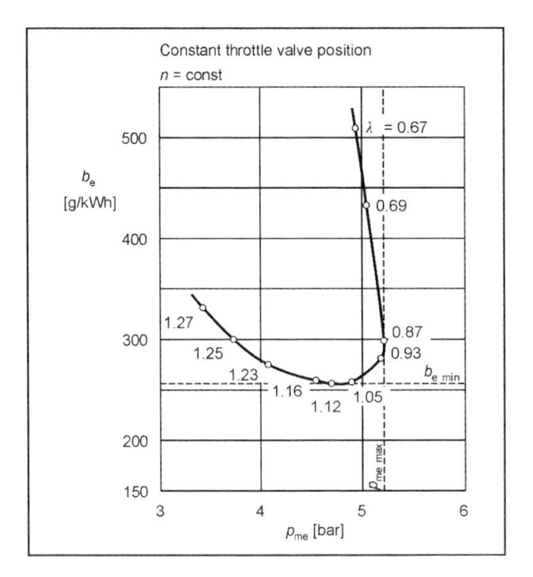

Fig. 14-16 Specific fuel consumption and effective mean pressure for different excess air factors.

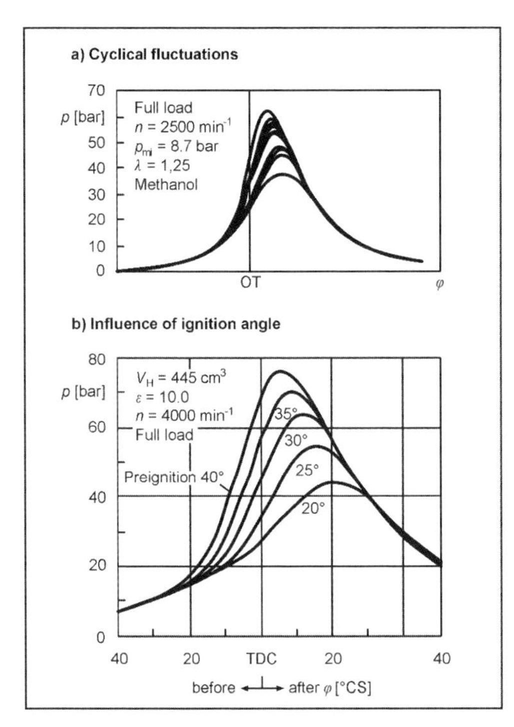

Fig. 14-17 Cyclical fluctuations and influence of ignition angle on pressure curve in the combustion chamber.

further compressed by the propagating flame front and, hence, additionally heated up. If the ignition limit is exceeded in the process, then a spontaneous autoignition takes place in the remaining mixture. This practically isochoric residual gas combustion results in steep pressure gradients that spread in the form of pressure waves in the combustion chamber, causing the familiar knocking or ringing noise.

The spontaneous autoignition is controlled almost exclusively by the chemical reaction kinetics. Trials with single-cylinder engines have shown that knocking occurs at temperatures of around 1100 K. The elementary chemical processes that take place here can be explained relatively well by the chain branching via the HO_2 radical as

$$HO_2^{\cdot} + RH \rightarrow H_2O_2 + R^{\cdot} \tag{14.13}$$

$$H_2O_2 + M \rightarrow O^{\cdot}H + O^{\cdot}H + M \tag{14.14}$$

On the other hand, however, the heat losses at the walls surrounding the combustion chamber in multicylinder series production engines are far higher, so that knocking occurs there at much lower temperatures in the range between 800 and 900 K. In this temperature range, the H_2O_2 degradation is relatively slow, and the autoignition is described by the considerably more complex low-temperature oxidation; see Section 14.3.3.

The spontaneous autoignition is essentially dependent on the composition of the fuel. For a qualitatively correct description of the autoignition process in SI engines, reduced reaction mechanisms were developed in Ref. [5] and [6]. While a four-step mechanism for the autoignition of *n*-heptane is described in Ref. [5], a formally similar three-step mechanism for two-component SI fuels consisting of *n*-heptane and iso-octane was developed in Ref. [6] as

$$F + \alpha O_2 \rightarrow P \tag{14.15}$$

$$F + 2O_2 \rightarrow I \tag{14.16}$$

$$I + (\alpha - 2)O_2 \rightarrow P \tag{14.17}$$

with the species involved

$$F = \frac{OZ}{100}(i - C_8H_{18}) + \left(1 - \frac{OZ}{100}\right)(n - C_7H_{16}) \tag{14.18}$$

$$I = \frac{OZ}{100} I_8 + \left(1 - \frac{OZ}{100}\right) I_7 + H_2O \tag{14.19}$$

$$P = \left[8\frac{OZ}{100} + 7\left(1 - \frac{OZ}{100}\right)\right] CO_2 + \left[9\frac{OZ}{100} + 8\left(1 - \frac{OZ}{100}\right)\right] H_2O \tag{14.20}$$

the intermediate products

$$I_7 = OC_7H_{13}OOH \quad \text{and} \quad I_8 = OC_8H_{15}OOH \tag{14.21}$$

the stoichiometric coefficient of oxygen

$$\alpha = 12.5\frac{OZ}{100} + 11\left(1 - \frac{OZ}{100}\right) \tag{14.22}$$

and the octane number OZ of the fuel.

Apart from the composition of the fuel, the geometry of the combustion chamber has a major influence on the knock tendency. Combustion chambers with a low knock tendency have

- Short flame paths thanks to a compact design and a centrally located spark plug
- No hot spots at the end of the flame path due to the positioning of the spark plug near the exhaust valve
- High flow velocities and, hence, a good mixture formation thanks to the swirl, tumble, and squeeze flows

The roof-shaped combustion chamber with its valves inclined at 20° to 30° to the cylinder axis and centrally located spark plug illustrated in Fig. 14-18 has proved to be very effective with four-valve cylinder heads. In the future, modeling of the detailed processes of combustion and 3D simulation may make major contributions to optimizing the combustion process in the light of the combustion chamber geometry.

A further type of undesirable combustion process is glow ignition. It is caused by extremely hot zones or "hot spots" at the walls surrounding the combustion chamber with temperatures of around 1200 K that lie well above the autoignition temperature. The most frequent hot spots are combustion residues that are deposited as hot scales on the walls, predominantly of the exhaust valve. Figure 14-19 shows qualitatively the pressure curve during knocking combustion, with the start of knocking combustion and that of glow ignition also plotted. Knocking combustion can occur only after initiation of the combustion by the ignition spark, while glow ignition can occur beforehand. The pressure waves occurring during both processes can result in mechanical material damage, and the increased thermal load can lead to fusion of elements on the piston and cylinder head.

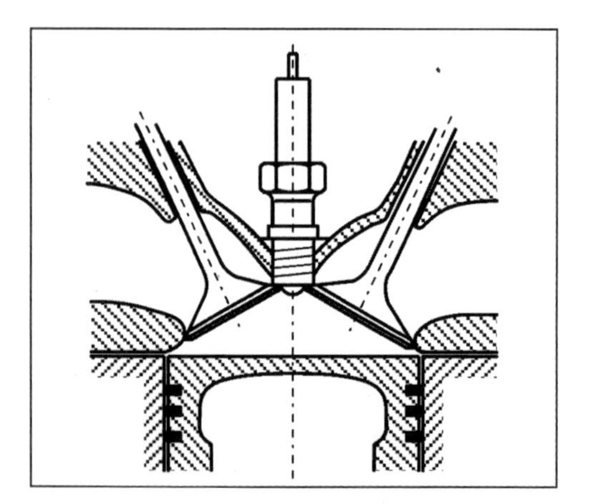

Fig. 14-18 Roof-shaped combustion chamber of four-valve cylinder head with centrally located spark plug.

A detailed description of the reaction kinetic processes during knocking can be found in Ref. [7]. The wide range of forms that glow ignition can take are described at length in Ref. [8].

14.3 Combustion in Diesel Engines

Combustion in the diesel engine is characterized by the following features. The fuel is injected under high pressure towards the end of the compression stroke, normally shortly before top dead center, into the main combustion chamber (direct injection) or, on older engines, into a prechamber (indirect injection). Injection systems used are distributor injection pumps with injection pressure of around 1450 bar, pump-nozzle systems with pressures over 2000 bar, and common-rail injection systems with pressures of approximately 1650 bar. These injection systems are described in detail in Chapter 12.5.1. The injected fuel evaporates, mixes with the compressed hot air, and ignites. By contrast with the SI engine, only a very short length of time is available for the mixture formation with the diesel engine. A fast injection and the best possible atomization are, therefore, essential for an intensive mixture formation.

Figure 14-20 shows qualitatively the part processes taking place during diesel engine mixture formation and combustion.

The individual part processes, in particular the spray formation, droplet evaporation, and mixture formation, interact with one another and can, therefore, not be considered in isolation. The process of mixture formation and

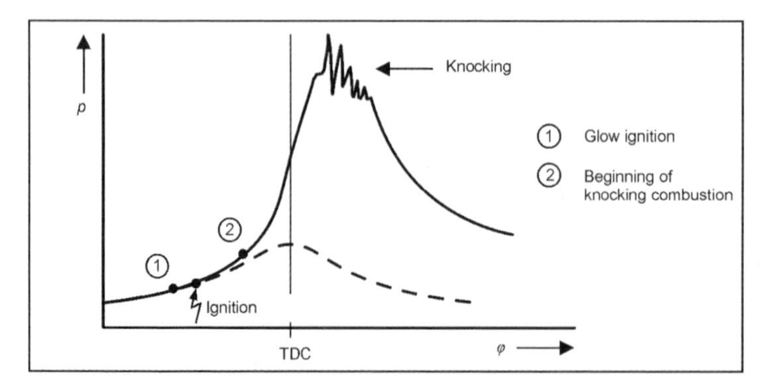

Fig. 14-19 Pressure curve during knocking combustion.

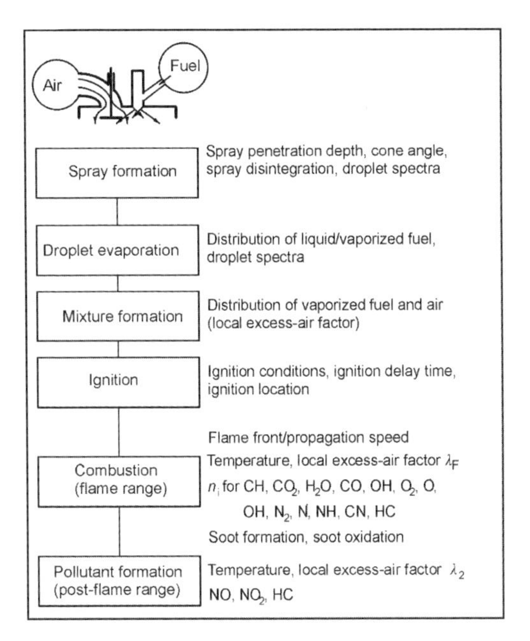

Fig. 14-20 Part processes of mixture formation and combustion in the diesel engine.

combustion in the diesel engine is therefore extremely complex.

14.3.1 Mixture Formation

14.3.1.1 Phenomenology

The injected fuel leaves the injection nozzle as a jet at high speed, disintegrates into small droplets because of its high velocity relative to the surrounding, highly compressed air and the high turbulence in the jet, and is atomized as it progresses further into the combustion chamber. Figure 14-21 shows a qualitative sketch of the injection jet leaving the injection nozzle.

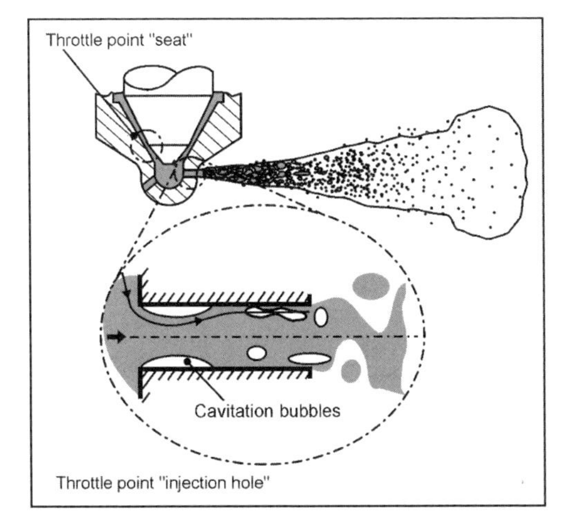

Fig. 14-21 Schematic diagram of jet propagation.

The propagation of the jet in the diesel engine and, hence, the mixture formation is essentially determined by the injection system and the injection parameters, but also by the flow map (swirl and turbulence) in the combustion chamber. The turbulent kinetic energy created by the injection jet into the combustion chamber, however, is at least one order higher than the kinetic energy of the combustion air, so that the flow map in the cylinder becomes significant only towards the end of the injection when the jet has already been sharply decelerated.

With very high injection pressures, the breakdown and evaporation of the liquid jet is already initiated in the nozzle bore by cavitation. Because of the extremely sharp pressure gradient through the injection bore, the pressure drops below the vapor pressure, and first vapor bubbles are formed. Implosion of these bubbles then creates pressure oscillations that accelerate the breakdown of the liquid fuel jet and the primary droplet formation. Both the deformation of these primary droplets because of their oscillation behavior and the surface forces caused by the high relative velocity between droplet and combustion chamber air lead to secondary droplet breakdown, and droplets of different sizes in the diameter range of 10 to 100 μm are formed.

This secondary droplet breakdown is described by the Weber number:

$$\mathrm{We} = \frac{\rho \cdot d_T \cdot w_{rel}^2}{\sigma} \tag{14.23}$$

that represents the dimensionless relationship between mass inertia force and surface tension force of the droplets. Depending on the value of the Weber number, the breakdown mechanisms illustrated in Fig. 14-22 are to be observed:

- Oscillation breakdown (1): We < 12
- Bubble (2) and lobe breakdown (3): We < 50

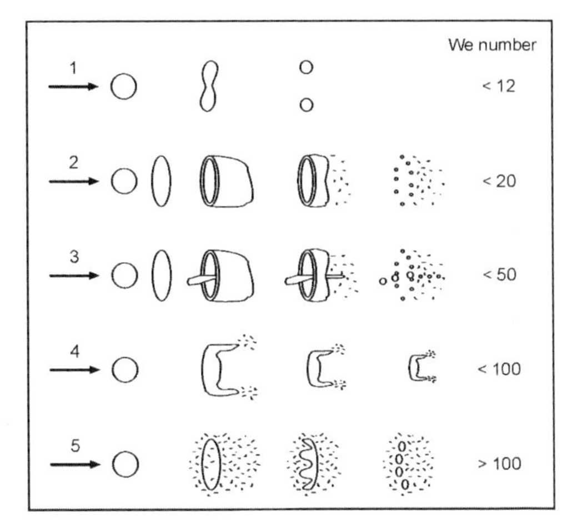

Fig. 14-22 Droplet breakdown mechanisms from Ref. [9].

- Shear breakdown (4): We < 100
- Catastrophic breakdown (5): We > 100

Apart from the droplet breakdown, collisions between several droplets also occur in the injection jet. Depending on the size, velocity, and impact angle, the droplets are repelled by one another, breaking down into smaller droplets or combining to form larger drops (drop coalescence).

At the edge of the jet, the fuel droplets mix with the hot air in the combustion chamber (air entrainment). This causes the droplets to heat up because of convective thermal transmission and temperature radiation of the hot combustion chamber walls until they ultimately evaporate. Apart from the temperature, the droplet evaporation rate is also influenced by the diffusion of the fuel from the droplet surface into the surroundings of the droplet so that heat and material transport take place simultaneously.

Jet propagation and mixture formation are at least qualitatively understood today and can be described relatively well using semiempirical models for the various part processes.

14.3.1.2 Fuel Jet Propagation

A simple correlation function is developed in Ref. [10] for the jet tip path and jet tip velocity that correlates relatively well with measured values. The correlation

$$s = C \cdot A_1^{\alpha_1} \cdot A_2^{\alpha_2} \cdot A_3^{\alpha_3} \cdot A_4^{\alpha_4} \cdot A_5^{\alpha_5} \cdot t^{\alpha_t} = K \cdot t^{\alpha_t} \tag{14.24}$$

for the distance that the jet tip travels within a time t is ultimately derived using a dimension analysis. For the velocity of the jet tip this means

$$\dot{s} = \alpha_t \cdot K^{\alpha_t - 1} \tag{14.25}$$

The dimensionless parameters A_i are calculated as

$$A_1 = \frac{\mathrm{Re}^2}{2}, \tag{14.26a}$$

$$A_2 = \frac{\mathrm{Re}^2}{\mathrm{We}} \tag{14.26b}$$

$$A_3 = \rho/\rho_g, \tag{14.27a}$$

$$A_4 = l/d_D, \tag{14.27b}$$

$$A_5 = \eta/\eta_g \tag{14.27c}$$

With the jet velocity w at the nozzle opening and the geometry of the nozzle bore l/d_D we obtain for the dimensionless parameters:

$$\text{Reynolds number Re} = \frac{w \cdot d_D \cdot \rho}{\eta} \tag{14.28}$$

$$\text{Weber number We} = \frac{w^2 \cdot d_D \cdot \rho}{\sigma}$$

where no index stands for "liquid fuel" and the index g is for "air." The exponents α_i are calculated by correlation of the measured values for s and $\dot{s}$ as $\alpha_1 = 0.3$, $\alpha_2 = -0.008$, $\alpha_3 = 0.2$, $\alpha_4 = 0.16$, $\alpha_5 = 0.6$, and $\alpha_t = 0.4$.

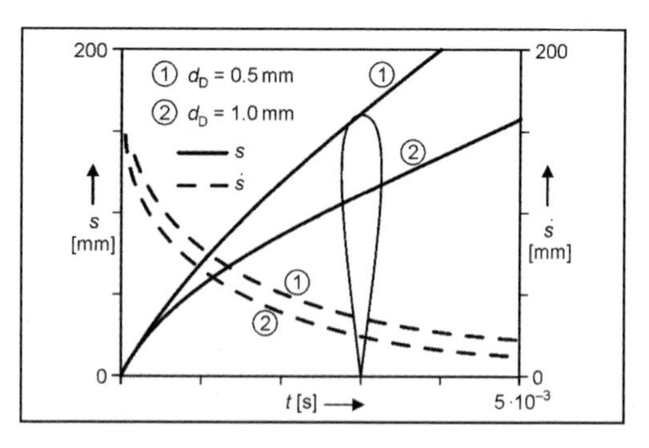

Fig. 14-23 Calculated jet tip path and velocity from Ref. [10].

Figure 14-23 shows the values for the jet tip path and jet velocity according to Ref. [10], calculated using this correlation function for two different injection nozzles.

By analogy we obtain the correlation for a jet angle Θ

$$\Theta = 2 \cdot \arctan\left[\frac{c_0 \cdot w \cdot d_D \cdot t}{\left(\frac{\rho_g}{\psi \cdot \rho}\right)^{\alpha_3} \cdot s^2}\right] \tag{14.29}$$

where $c_0 = 2.65$ and $\psi = 0.65$.

If the droplet breakdown does not start until after the injection process we obtain the following correlation for the mean Sauter diameter:

$$\frac{\mathrm{SMD}}{d_D} = k \cdot \mathrm{Re}^{\beta_1}\mathrm{We}^{\beta_2} \cdot A_3^{\beta_3} \cdot A_5^{\beta_5} \tag{14.30}$$

where $k = 4.12$, $\beta_1 = 0.12$, $\beta_2 = -0.75$, $\beta_3 = 0.18$, and $\beta_5 = 0.54$. The mean Sauter diameter is a characteristic parameter for describing the droplet distribution spectrum. It is defined as the ratio between the total volume and the total surface area of all the droplets and so reduces the droplet distribution spectrum to the equivalent droplet with the diameter

$$\mathrm{SMD} = \frac{\sum_{i=1}^{n} V_{T,i}}{\sum_{i=1}^{n} A_{T,i}} \sim \frac{\sum_{i=1}^{n} d_{T,i}^3}{\sum_{i=1}^{n} d_{T,i}^2} \tag{14.31}$$

Figure 14-24 shows the development of the Sauter diameter over time and the droplet distribution spectra measured at different times. The upper diagram shows that the Sauter diameter remains practically constant after a short starting phase and indicates a good match between the measured values and the correlation. The middle diagram shows the development of the droplet spectra over time. While at the beginning of the injection smaller droplets predominate, after a certain time larger droplets are observed (because of the breakdown of the liquid core and coalescence of smaller droplets) and then smaller droplets again (because of droplet breakdown and evaporation) the farther the jet moves away from the nozzle opening.

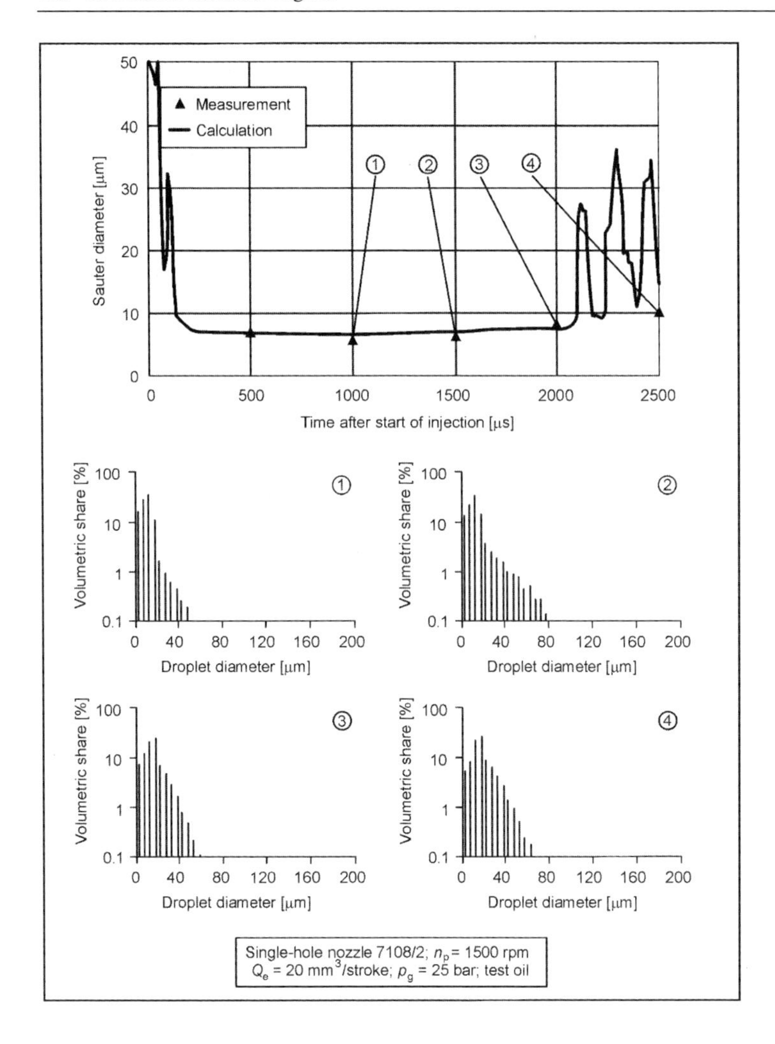

Fig. 14-24 Sauter diameter and droplet distribution spectra as a function of the injection time from Ref. [10].

For the diesel engine, the mixture formation cannot be considered independently of the jet propagation, on the one hand, and the combustion, on the other. It is a typical feature particularly of the diesel engine combustion that jet propagation, mixture formation, and combustion take place partially simultaneously. In continuation of the considerations above, the local excess-air factor in the jet can be developed from the correlation function

$$\lambda_l = \frac{\tan \Theta}{\sqrt{\psi} \cdot L_{\min}} A_3^{\alpha_3} \cdot \frac{s}{d_D} \qquad (14.32)$$

with the theoretical air expenditure $L_{\min}$ = 14.5 kg air per kg fuel. Figure 14-25 shows the comparison of measurement vs calculation for the jet tip path, jet angle, and local excess-air factor as a function of time.

When using this correlation function, however, we should be aware that it can provide only an integral description of the observed phenomena and that it allows nothing to be said about the convective transport processes for pulse and heat critical for the diffusion combustion taking place on the microscale; see also Ref. [11].

14.3.2 Autoignition

The physical and chemical processes taking place during the ignition lag time are very complex; the main physical processes are the atomization of the fuel, the evaporation, and the mixing of fuel vapor with air to create an ignitable mixture. The chemical processes taking place are the prereactions in the mixture taking place up to autoignition that occurs at a local excess-air factor of $0.5 < \lambda < 0.7$.

The start of oxidation of hydrocarbons can be seen as a branched chain propagation process whose reaction course is heavily dependent on the temperature and, according to Ref. [7], can be classified into the three temperature ranges described below.

At high temperatures above $T > 1100$ K, the chain branch

$$H^{\bullet} + O_2 \rightarrow O^{\bullet}H + O^{\bullet}$$

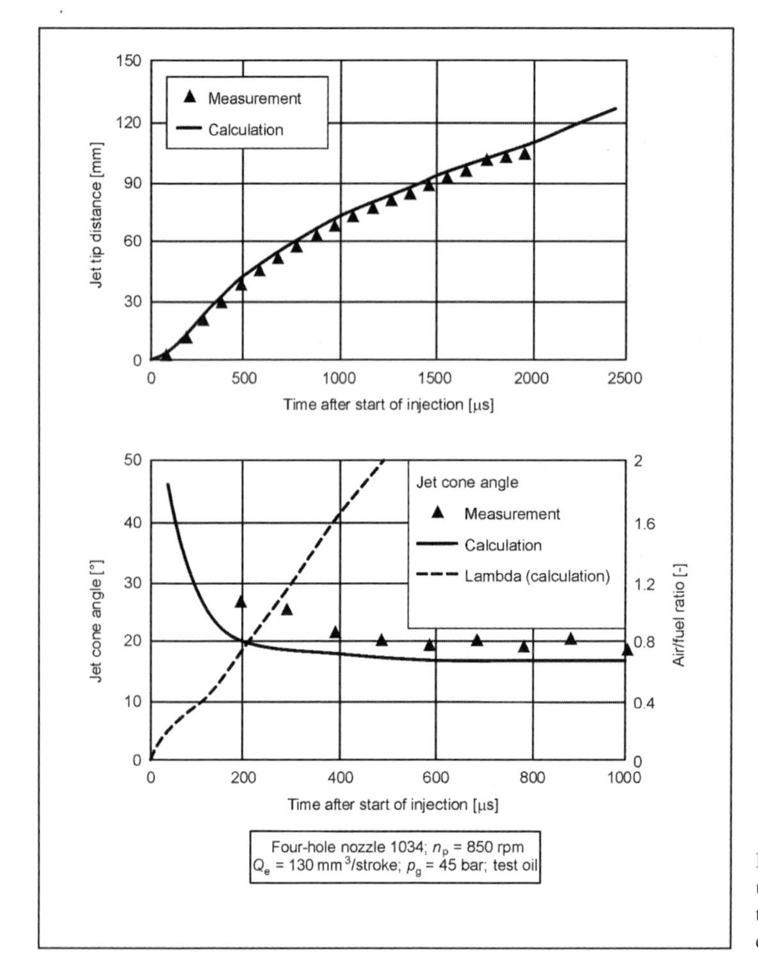

Fig. 14-25 Comparison of measurement and calculation for the jet tip path, jet angle, and local excess-air factor.

dominates. This reaction is, however, very dependent on temperature and rapidly loses significance at lower temperatures. In the middle temperature range of $900\ K < T < 1100\ K$, the additional branches

$$HO_2^{\bullet} + RH \rightarrow H_2O_2 + R^{\bullet}$$

$$H_2O_2 + M \rightarrow O^{\bullet}H + O^{\bullet}H + M$$

gain in importance, where the OH radicals are partially transformed back into the original HO_2 radical.

In the low temperature range up to $T < 900$ K, the H_2O_2 degradation is relatively slow, and degenerated branch reactions gain in importance characterized in that precursors of the chain branch (e.g., RO_2) break down again at higher temperatures. This leads to an inverse temperature dependence of the reaction rate that can be described as a two-step reaction mechanism. This two-step reaction mechanism that was originally developed to describe the knocking combustion in the SI engine[7] produces an extensive reaction pattern because the residual molecules created can have a large number of isomer structures, roughly 6000 elementary reactions with around 2000 species for the autoignition of n-$C_{16}H_{34}$.

Various autoignition models have been developed for use in the simulation of the engine combustion.

A model used today with 30 elementary reactions and 21 species is generally regarded as a basic model containing more or less the character of elementary kinetics.[12] Figure 14-26 shows that the results of this model correlate very well with measurements in stoichiometric n-heptane/air mixtures and at different pressures.

An eight-step reaction model is described in Ref. [13] that contains a chain propagation mechanism that has been expanded to include a degenerated branch process with two reaction paths and two chain termination reactions. This autoignition model requires the adaptation of 26 reaction parameters; it is used in the commercially available FIRE code and is described in detail in Ref. [14].

A "relatively simple" five-step reaction mechanism is described in Ref. [15] that illustrates very well the autoignition of mixtures of n-heptane and iso-octane in

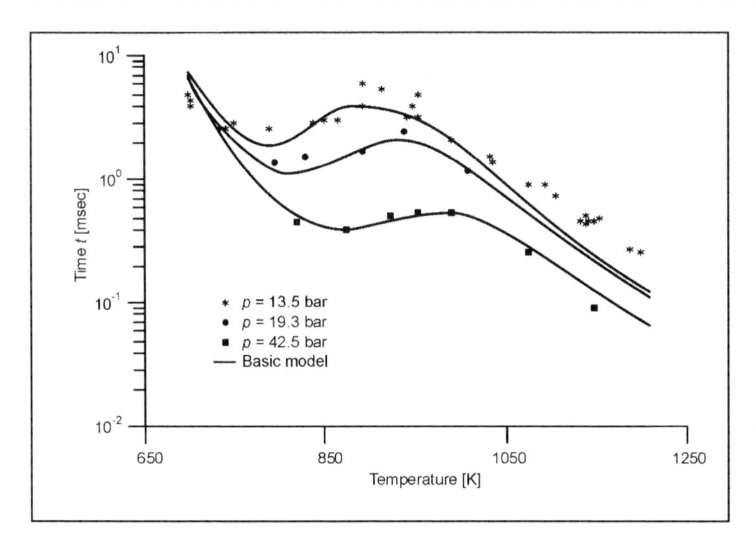

Fig. 14-26 Comparison of the basic model with measurements for the autoignition to that from Ref. [12].

the middle temperature range. This model is of great interest for simulation calculations because of its simplicity.

The complex models above are often not essential for autoignition in diesel engines that typically occurs at higher temperatures. For this reason, a one-equation model is used with good success in practice, which describes the ignition lag using just one Arrhenius equation as

$$\Delta t_{ZV} = A \cdot \frac{\lambda}{p^2} \exp\left(-\frac{E}{RT}\right) \tag{14.33}$$

as a function of pressure, temperature, and excess-air factor.[11]

In summary, Figure 14-27 shows the distance of the local ignition ranges from the nozzle opening as a function of time after the injection for the various injection nozzles described in Ref. [16].

It can be seen that the mixture ignites after approximately 1 ms and that the ignition range lies closer to the nozzle opening, the smaller the diameter of the injection bore.

14.3.3 Combustion Process

14.3.3.1 Phenomenological Description

The combustion process in the diesel engine can be roughly subdivided into three phases as shown in Fig. 14-28 and described below.

The fuel injected during the ignition lag time mixes with the surrounding air and forms a more or less

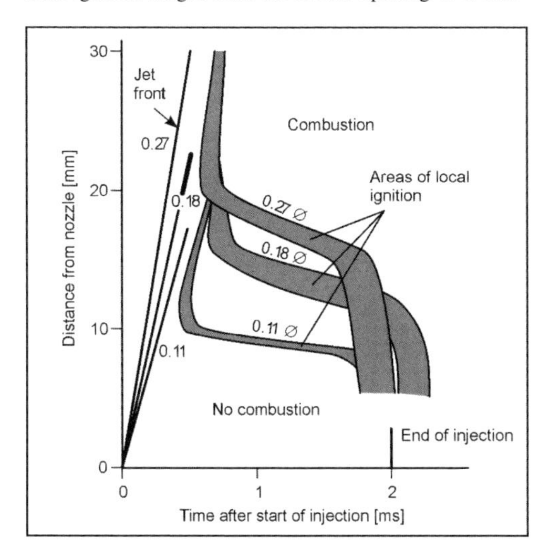

Fig. 14-27 Position of the local ignition ranges as a function of the time after the start of ignition for different injection nozzles according to Ref. [16].

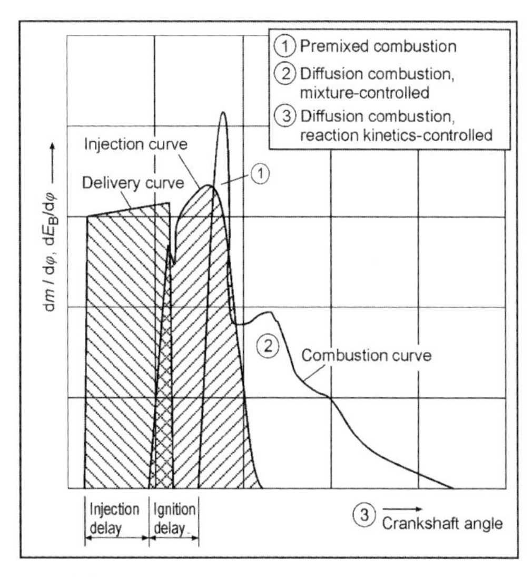

Fig. 14-28 Delivery, injection, and combustion curves in the diesel engine.

homogeneous and reactive mixture. After the ignition lag time that is controlled physically and chemically, this mixture burns very rapidly. This first phase, the "premixed" combustion, is therefore very similar to the SI engine combustion. The combustion noise typical of diesel engines is caused by the high-pressure rise speed $dp/d\varphi$ during this premixture combustion. This noise can be influenced by changing the moment of injection: advanced injection means "hard" and retarded injection "soft" combustion; see Fig. 14-29. More recently, the combustion noise is significantly reduced by the use of a preinjection of roughly 5% of the fuel volume.

The mixture formation processes continue during the main combustion. In this second phase, the chemistry is fast, and the combustion process is mixture controlled. We, therefore, speak also of mixture-controlled diffusion combustion. The end of this phase of the main combustion is characterized by the maximum temperature being reached in the combustion chamber.

Toward the end of the combustion, pressure and temperature in the flame front have dropped so far that the chemistry becomes slow by comparison with the mixing processes taking place at the same time. This third phase, therefore, increasingly becomes a reaction kinetically controlled by diffusion combustion.

The determining factor for the thermodynamic quality of the whole combustion process is the curve of the released thermal energy

$$\frac{dE_B}{d\varphi} = f(\varphi) \tag{14.34}$$

It results in a heating of the fuel-air mixture in the cylinder and, hence, in a significant rise in pressure. Figure 14-30 shows pressure and combustion curves (heat release rate) at full and partial loads in a high-speed high-performance diesel engine with relatively retarded injection.

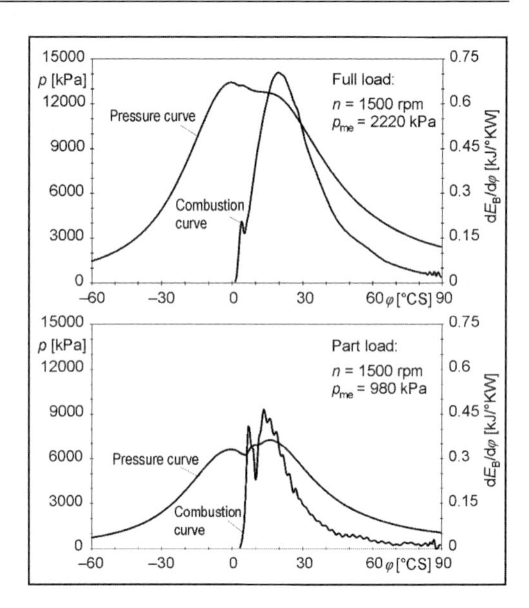

Fig. 14-30 Pressure and combustion curves in a high-speed diesel engine at full load and partial load.

14.3.3.2 Equivalent Combustion Curves

A series of models has been developed to simulate the heat release that are described in detail in Refs. [1] and [11]. In practice, on the other hand, semiempirical equivalent combustion curves are frequently used that have to be adapted to the engine in question and, therefore, cannot be extrapolated.

Starting from triangular combustion curves, Ref. [17] has developed the relationship

$$\frac{E_B}{E_{B,\text{tot}}} = 1 - \exp(-a \cdot y^{m+1}) \tag{14.35}$$

where

$$E_{B,\text{tot}} = m_B \cdot H_u \tag{14.36}$$

for the maximum releasable heat volume and

$$y = (\varphi - \varphi_{BB}) \cdot \Delta\varphi_{BD} \tag{14.37}$$

for the dimensionless crank angle with the combustion time

$$\Delta\varphi_{BD} = \varphi_{BE} - \varphi_{BB} \tag{14.38}$$

on the basis of reaction kinetics considerations where φ_{BE} = end of combustion and φ_{BB} = start of combustion.

$$\frac{dE_B}{d\varphi} = f(\varphi, m) \tag{14.39}$$

$$E_B = \int f(\varphi, m) \cdot d\varphi = F(\varphi, m) \tag{14.40}$$

Figure 14-31 shows the combustion curve and the cumulative combustion curve or burn-through function in relation to the dimensionless crank angle for different form parameters m.

At the end of combustion, i.e., at $\varphi = \varphi_{BE}$ or at $y = 1$, $\eta_{U,\text{ges}}$ percent of the total energy admitted with the fuel should have been released. We, consequently, have the relationship

$$\frac{E_B}{E_{B,\text{tot}}} = \eta_{U,\text{tot}} = 1 - \exp(-a) \tag{14.41}$$

and from this the numeric values

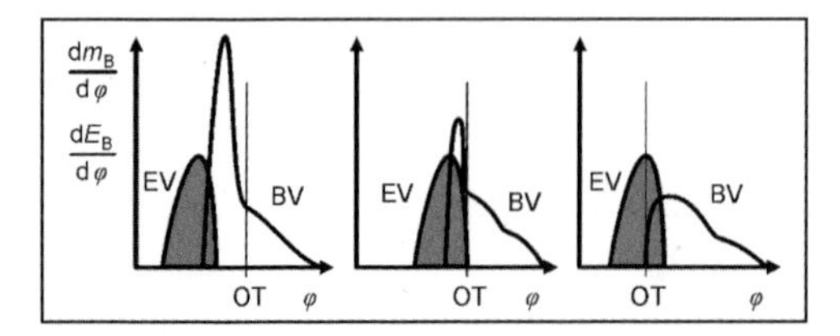

Fig. 14-29 Injection curve and combustion curve during hard and soft combustion.

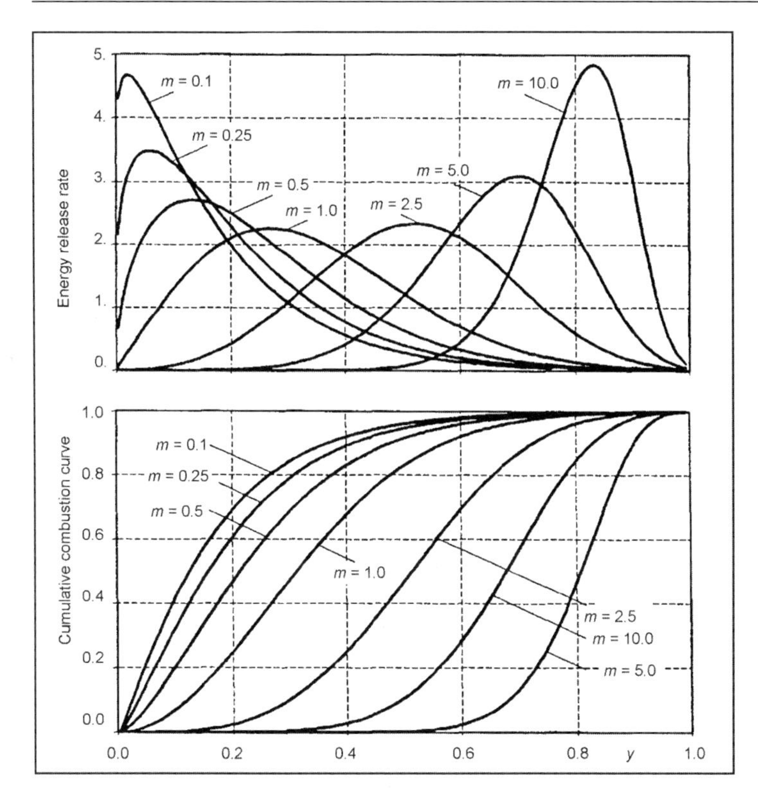

Fig. 14-31 Combustion curve and burn-through function according to Vibe.[17]

$\eta_{U,\,\text{tot}}$	0.999	0.990	0.980	0.950
a	6.908	4.605	3.912	2.995

For the degree of implementation, Ref. [18] has stated the empirical relationship

$$\eta_{U,\text{tot}} = \begin{cases} 1 : \lambda > \lambda_{RB} \\ a \cdot \lambda \cdot \exp(c \cdot \lambda) - b : 1 \le \lambda \le \lambda_{RB} \\ 0.95 \cdot \lambda + d : \lambda \le 1 \end{cases} \tag{14.42}$$

where

$$c = -1/\lambda_{RB}$$

$$d = -0.0375 - (\lambda_{RB} - 1.17)/15$$

$$a = (0.05 - d)/[\lambda_{RB} \cdot \exp(-1) - \exp(c)]$$

$$b = a \cdot \exp(c) - 0.95 - d \tag{14.43}$$

where λ_{RB} is the excess-air factor at which an exhaust gas blackening with the Bosch soot number $RB = 3.5$ is reached. The interval $1.17 < \lambda_{RB} < 2.05$ is stated as the validity range.

The VIBE equivalent combustion curve is defined by the three VIBE parameters: Start of combustion φ_{BB}, combustion time $\Delta\varphi_{BD}$, and form parameters m. Consequently, only three parameters can be adapted here for a given working point. The adaptation is performed such that the start of combustion φ_{BB}, the ignition pressure p_z, and the mean pressure $p_{m,\,i}$ correspond to those of the real engine process.

The VIBE parameters for *any* working points are converted using semiempirical functions as well as in relation to the main parameters: Excess-air factor λ, engine speed n, output, ignition lag $\Delta\varphi_{ZV}$, and start of combustion φ_{BB} using the formula

$$\frac{\Delta\varphi_{BD}}{\Delta\varphi_{BD,0}} = \left(\frac{\lambda_0}{\lambda}\right)^{0.6} \cdot \left(\frac{n}{n_0}\right)^{0.5} \cdot \eta^{0.6}{}_{U,\text{tot}} \tag{14.44}$$

$$\frac{m}{m_0} = \left(\frac{\Delta\varphi_{ZV,0}}{\Delta\varphi_{ZV}}\right)^{0.5} \cdot \frac{p \cdot T_0}{p_0 \cdot T} \cdot \left(\frac{n_0}{n}\right)^{0.3} \tag{14.45}$$

$$\Delta\varphi_{ZV} = 6 \cdot n \cdot 10^{-3} \cdot \left[0.5 + \exp\left(\frac{7800}{2 \cdot T}\right) \cdot \left(\frac{0.135}{p^{0.7}} + \frac{4.8}{p^{1.8}}\right)\right] \tag{14.46}$$

$$\varphi_{BB} = \varphi_{FB} + \Delta\varphi_{EV,0}\frac{n}{n_0} + \Delta\varphi_{ZV} \tag{14.47}$$

with the start of delivery φ_{FB} and the injection lag $\Delta\varphi_{EV,\,0}$. Further details can be found in Ref. [11].

14.4 Heat Transfer

14.4.1 Heat Transfer Model

The heat transfer from the hot exhaust gas in the combustion chamber is affected by convective heat transfer and by the heat radiation of the glowing soot particles. The description of the heat transport is made more difficult by the formation of soot layers at low load and by their burning off at full load. An overview of the state of our knowledge can be found in Ref. [11]. The heat transfer model described below is derived from Ref. [19] and is still state of the art even today.

A dimension analysis is performed for the dimensionless heat transfer coefficient, the Nusselt number, for a stationary and fully turbulent tubular flow

$$Nu = C \cdot Re^{0.8} \cdot Pr^{0.4} \tag{14.48}$$

with the

$$\text{Nusselt number } Nu = \frac{\alpha \cdot D}{\lambda} \tag{14.49}$$

$$\text{Reynolds number } Re = \frac{\rho \cdot w \cdot D}{\eta} \tag{14.50}$$

$$\text{Prandtl number } Pr = \frac{v}{a} \tag{14.51}$$

If we consider the mixture in the combustion chamber as an ideal gas with the thermal equation of state

$$p/\rho = R \cdot T \tag{14.52}$$

and, furthermore, assume the correlations

$$\frac{\lambda}{\lambda_0} = \left(\frac{T}{T_0}\right)^x, \ \frac{\eta}{\eta_0} = \left(\frac{T}{T_0}\right)^y, \text{ and } Pr = 0.74 \tag{14.53}$$

for the temperature dependence, then ultimately we obtain

$$\alpha = C \cdot D^{-0.2} \cdot p^{0.8} \cdot c_m^{0.8} \cdot T^{-r} \tag{14.54}$$

where $r = 0.8 \cdot (1 + y) - \alpha$ and the assumption that the characteristic speed w in the engine is equal to the mean piston speed c_m. By comparison with measured values, the exponent for the temperature dependence is determined as $r = 0.53$ and the constant as $C^* = 0.013$. For motored engines, a modification of the characteristic speed has to be introduced as

$$w = C_1 \cdot c_m + C_2 \cdot \frac{V_h \cdot T_1}{p_1 \cdot V_1} \cdot (p - p_0) \tag{14.55}$$

because the combustion drastically increases the turbulence and, hence, the heat transfer. The pressure curve in the motored engine is described by $p(\varphi)$, and V_1, p_1, and T_1 are the values at "Close inlet valve." For the constants C_1 and C_2 we obtain by adaptation to measure values

$$C_1 = \begin{cases} 6.18 + 0.417 \cdot c_u/c_m\text{:Charge cycle} \\ 2.28 + 0.308 \cdot c_u/c_m\text{:Compression/Expansion} \end{cases} \tag{14.56}$$

$$C_2 = \begin{cases} 6.22 \cdot 10^{-3}\text{m/K:Prechamber} - \text{engine} \\ 3.24 \cdot 10^{-3}\text{m/K:DI} - \text{engine} \end{cases} \tag{14.57}$$

where the validity range for the intake swirl c_u/c_m is given as $0 < c_u/c_m < 3$. The speed corrected with the "combustion element" provides values that are too low for dragged engines and in the lower load range. For this reason, the relationship

$$w = c_m \cdot \left[1 + 2 \cdot \left(\frac{V_c}{V}\right)^2 \cdot p_{m,i}^{-0.2}\right] \tag{14.58}$$

was recently proposed, and it was recommended that the largest numeric value be used for the characteristic speed. For diesel engines with direct injection, the constant C_2 must be corrected for higher wall temperatures as follows

$$C_2 = \begin{cases} 3.24 \cdot 10^{-3}\text{m/K:} & T_W < 550\text{K} \\ 5.0 \cdot 10^{-3} + 2.3 \cdot 10^{-3} \cdot (T_W - 550)\text{m/K:} & T_W > 550\text{K} \end{cases} \tag{14.59}$$

Further details can be found in the cited literature.

14.4.2 Determination of Heat Transfer Coefficients

Because of the temporal fluctuation in the gas temperature in the combustion chamber, temperature fluctuations occur in the walls surrounding the combustion chamber as shown in Fig. 14-32.

The energy transport by thermal conductivity in solids is described by the Fourier differential equation

$$\frac{\partial T}{\partial t} = a \cdot \frac{\partial^2 T}{\partial x^2} \tag{14.60}$$

with the thermal conductivity

$$a = \frac{\lambda}{\rho \cdot c_p} \tag{14.61}$$

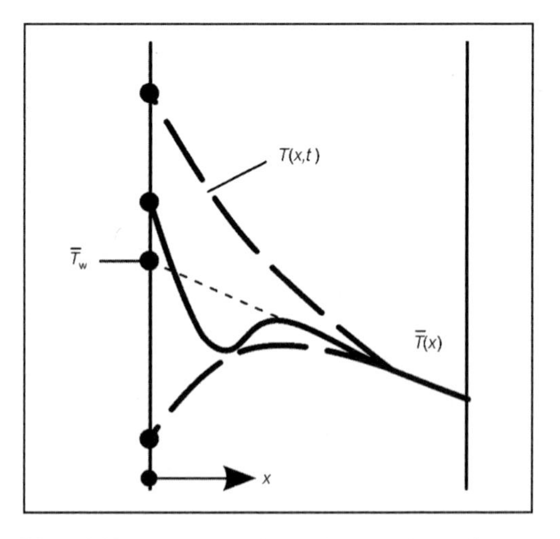

Fig. 14-32 Temperature fluctuations in the walls surrounding the combustion chamber.

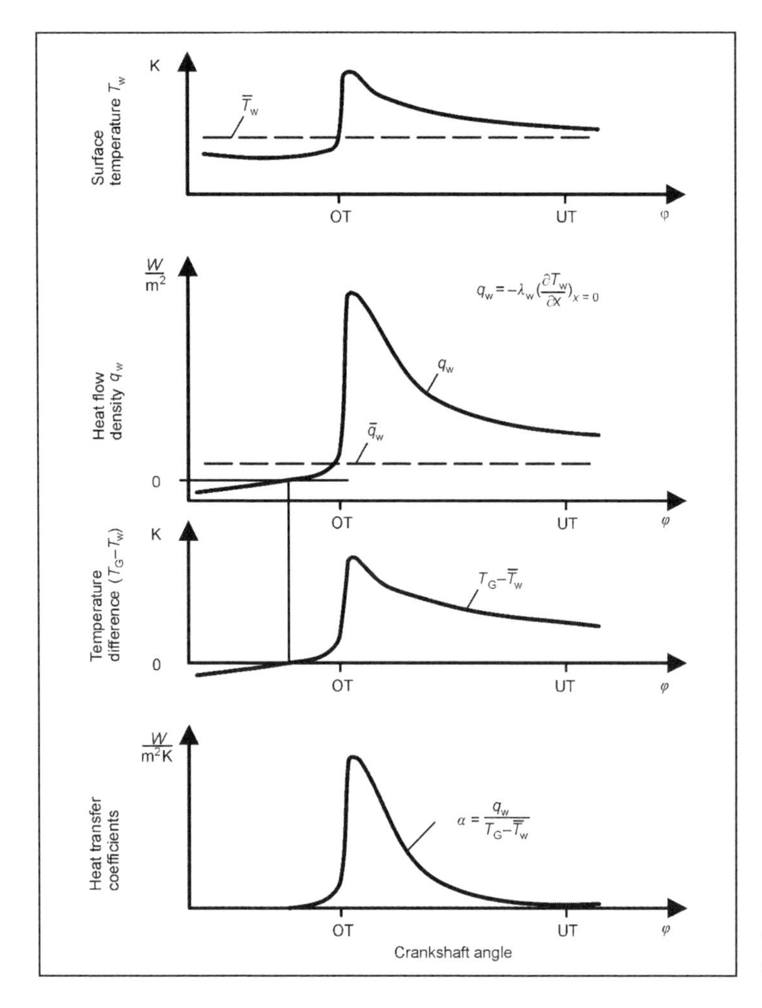

Fig. 14-33 Experimental determination of heat transfer coefficients.

If, as a boundary condition, we assume temperature fluctuations in line with the Fourier series on the gas side

$$T_W(t) = \overline{T}_W + \sum_{i=1}^{\infty}[A_i \cdot \cos(iwt) + B_i \cdot \sin(iwt)]$$

for $x = 0$ (14.62)

and above a certain wall thickness a constant heat flow density

$$\frac{\partial T}{\partial x} = -\frac{\overline{q}}{\lambda} \quad \text{for} \quad x \to \infty \tag{14.63}$$

then as a solution for the temperature curve in the wall we obtain the Fourier series

$$T(x,t) = \overline{T}_W \cdot \frac{\overline{q}}{\lambda} \cdot x + \sum_{i=1}^{\infty} \exp\left(-x \cdot \sqrt{\frac{i \cdot w}{2 \cdot a}}\right) f_i(iwt) \tag{14.64}$$

$$f_i(iwt) = A_i \cdot \cos\left(iwt - x\sqrt{\frac{i \cdot w}{2 \cdot a}}\right) + B_i \cdot \sin\left(iwt - x\sqrt{\frac{i \cdot w}{2 \cdot a}}\right) \tag{14.65}$$

By differentiation we obtain from this the heat flow density at the surface $x = 0$

$$q(0,t) = \overline{q} + \lambda \sum_{i=1}^{\infty} \sqrt{\frac{i \cdot w}{2 \cdot a}} F_i(iwt) \tag{14.66}$$

$$F_i(iwt) = (A_i + B_i) \cdot \cos(iwt) + (B_i - A_i) \cdot \sin(iwt) \tag{14.67}$$

To determine the heat coefficient

$$\alpha(t) = \frac{q_W(t)}{T_G(t) - T_W(t)} \tag{14.68}$$

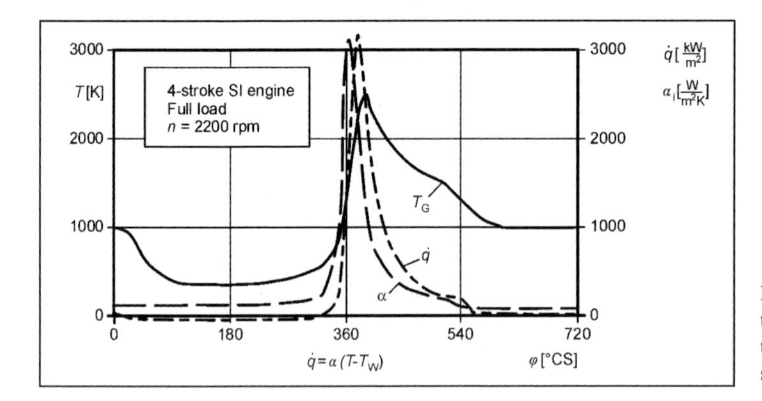

Fig. 14-34 Curves of gas temperature, heat flow density, and heat transfer coefficient for a four-stroke SI engine at full load.

the curve of the surface temperature $T_W(t)$ is now determined experimentally.

By setting $T_{W\,\mathrm{exp}}(t) = T_{W\,\mathrm{theo}}(t)$ we obtain by Fourier analysis the Fourier coefficients A_i and B_i, and, hence, from Eq. (14.67) the heat flow density at the surface

$$q_W(t) = q(0, t) \tag{14.69}$$

The fluctuations of the surface temperature $T_W(t)$ are relatively small so that, as a rule, the mean wall temperature T_W can be inserted into Eq. (14.69) instead of $T_W(t)$.

Figure 14-33 shows graphically the procedure for the experimental determination of the heat transfer coefficient, and Fig. 14-34 shows the curves of gas temperature, heat flow density, and heat transfer coefficient for a four-stroke SI engine at full load.

Further details can be found in the cited literature. This chapter is an abridged and in some areas completely revised version of the corresponding chapter in Ref. [1].

Bibliography

[1] Merker, G.P., and G. Stiesch, Technische Verbrennung–Motorische Verbrennung, B.G. Teubner-Verlag, Stuttgart, Leipzig, 1999.

[2] Bockhorn, H., A Short Introduction to the Problem-Structure of the Following Parts, in Bockhorn [ed.] Soot Formation in Combustion, Springer-Verlag, Berlin, Heidelberg, New York, 1994.

[3] Braess, H.-H., U. Seiffert [ed.], Vieweg Handbuch Kraftfahrzeugtechnik, Friedr. Vieweg & Sohn Verlagsgesellschaft mbH, Braunschweig, Wiesbaden, 2000.

[4] Bargende, M., H.-K. Weining, P. Lautenschütz, and F. Altenschmidt, Thermodynamik der neuen Mercedes-Benz 3-Ventil-Doppelzünder V-Motoren, U. Essers [ed.], in Kraftfahrwesen und Verbrennungsmotoren, 2. Stuttgarter Symposium, Expert-Verlag, Renningen-Malmsheim, 1997.

[5] Müller, U.C., N. Peters, and A. Liñán, Global Kinetics for n-Heptan Ignition at High Pressures, Twenty-Fourth Symposium (International) on Combustion, The Combustion Institute, 1992, pp. 777–784.

[6] Müller, U.C., Reduzierte Reaktionsmechanismen für die Zündung von n-Heptan und iso-Oktan unter motorrelevanten Bedingungen, Dissertation, RWTH-Aachen, 1993.

[7] Warnatz, J., U. Maas, and R.W. Dibble, Verbrennung, 2, Aufl., Springer-Verlag, Berlin, Heidelberg, 1997.

[8] Urlaub, A., Verbrennungsmotoren, 2, Aufl., Springer-Verlag, Berlin, Heidelberg, 1994.

[9] Mayer, W.O.H., Zur koaxialen Flüssigkeitszerstäubung im Hinblick auf die Treibstoffaufbereitung in Raketentriebwerken, Forschungsbericht DLR-FB-93-09, 1993.

[10] Renner, G., Experimentelle und rechnerische Untersuchungen über die Struktur technischer Dieseleinspritzstrahlen, Fortschritt-Berichte VDI, Reihe 12, Nr. 216, VDI-Verlag, Düsseldorf, 1994.

[11] Merker, G.P., and C. Schwarz, Simulation verbrennungsmotorischer Prozesse, Teubner-Verlag, Stuttgart, Leipzig, 2001.

[12] Fieweger, K., and H. Ciezki, Untersuchung der Selbstzündungs- und Rußbildungsvorgänge von Kraftstoff/Luft-Gemischen im Hochdruckstoßwellenrohr, SFB 224-Forsch, Bericht, 1991.

[13] Halstead, M.P., L.J. Kirsch, and C.P. Quinn, The Autoignition of Hydrocarbon Fuels at High Temperatures and Pressures—Fitting of a Mathematical Model, Combustion and Flame, 30, 45–60, 1977.

[14] Fuchs, H., K. Pachter, G. Pitcher, R. Tatschl, and E. Winkelhofer, Dieseleinspritzung, Modellierung der Gemischbildung im Dieselmotor, FVV-Bericht Nr. 613, Forschungsvereinigung Verbrennungskraftmaschinen e.V., Frankfurt/M., 1996.

[15] Schreiber, M., A. Sadat Sahak, A. Lingens, and J.F. Griffiths, A Reduced Thermokinetic Model for the Autoignition of Fuels with Variable Octane Rings, Twenty-fifth Symposium Combustion, The Combustion Institute, Pittsburgh, 1994, pp. 933–940.

[16] Winkelhofer, E., B. Wiesler, G. Bachler, and H. Fuchs, Detailanalyse der Gemischbildung und Verbrennung von Dieselstrahlen, Tagung "Der Arbeitsprozess des Verbrennungsmotors," Graz, 1991.

[17] Vibe, R.R., Brennverlauf und Kreisprozess von Verbrennungsmotoren, VEB-Verlag Technik, Berlin, 1970.

[18] Betz, A., Rechnerische Untersuchung des stationären und transienten Betriebsverhaltens ein- und zweistufig aufgeladener Viertakter-Dieselmotoren, Dissertation, TU-München, 1985.

[19] Woschni, G., Die Berechnung der Wandwärmeverluste und der thermischen Belastung der Bauteile von Dieselmotoren, in MTZ 31 (1970) 491–499.

15 Combustion Systems

15.1 Combustion Systems for Diesel Engines

15.1.1 Diesel Combustion

General overview. Combustion is to be understood as chemical reactions in which a substance releases heat (exothermic reaction) while bonding to molecular oxygen (oxidation). Combustion starts with ignition. This can take place only under certain conditions that may be described in a simplified manner as follows:

- The reaction partners must possess a minimum energy level, so-called activation energy. Only molecules that have reached this energy level can react with each other. The share of molecules in a mixture of reaction partners with a sufficiently high energy level increases exponentially as the mixture temperature rises.
- The reaction mixture must have a specific composition. When there is a large share of one or another reaction partner, the potential molecular collisions are insufficient to trigger a stabile, self-supporting reaction. Accordingly, ignition reliably occurs only within the ignition limits (an air-fuel ratio of approximately 0.7 to 1.3). These ignition limits expand as the mixture temperature rises. Inert gas components in the reaction mixture (such as exhaust) reduce the reaction speed similar to a shift in the mixture composition toward a "lean" ignition limit.

The mode of operation of diesel engines is based on autoignition of the fuel introduced into the combustion chamber. The fuel is supplied to the combustion chamber by injection with a suitable system, the injection system (see 12.5.1). To achieve reliable autoignition of the fuel, the air must be sufficiently hot in the combustion chamber. This is essentially attained by a correspondingly high engine compression ratio. Mixing the fuel with available air and, hence, creating optimum conditions for ignition of the forming air-fuel mixture is a necessary prerequisite for subsequent combustion.

In addition to the charge movement in the combustion chamber, the combustion chamber geometry, the thermal state of the cylinder charge, the walls neighboring the combustion chamber, and the type of fuel injection influence mixture formation. In conventional mixture formation and combustion processes in diesel engines, this is done in the combustion chamber itself. For this reason, the mixture formation in diesel engines is also termed internal mixture formation in contrast to classic spark-injection engines (mixture formation in the intake manifold). The degree of homogeneity of the concentrated field of oxygen and fuel (liquid and vapor) that forms in the combustion chamber during fuel injection and changes according to time and place is a measure of the mixture formation quality. The quality substantially depends on local and temporal processes and the completeness (pollutant formation) of combustion in the diesel engine. The measurable pollutant emissions in the engine exhaust arise from the interaction between pollutant formation and pollutant decomposition in the combustion chamber and the exhaust system. This is especially true for emissions of soot, hydrocarbons, and carbon monoxide.

The amount of heat released by combustion after ignition determines the gas pressure and temperature characteristics in the combustion chamber in conjunction with the exchange of heat between the fuel, the walls neighboring the combustion chamber, and the liquid fuel. This also determines the success of energy conversion (mean pressure and fuel consumption) and the mechanical and thermal loads on the engine components. In addition, the change over time of the gas pressure substantially influences the noise emitted by a combustion engine (combustion noise).

The injection of the fuel, the decay of the fuel jet into a spray, the evaporation of the fuel, the mixture of the fuel with the air, heat transfer between the working substance, combustion chamber walls and fuel, the air movement created or intentionally generated (swirl ducts) by the piston movement as physical processes, and the combustion (oxidation) of the fuel as a chemical process occur simultaneously to a degree under continually changing conditions and are mutually influential. These processes, therefore, must be considered in terms of their interrelationships. Figure 15-1 schematically illustrates the relationships and interactions of the processes occurring in the combustion chamber of a diesel engine.

Because of the complexity of mixture formation and combustion in diesel engines, their theoretical and experimental study is extremely problematic. An additional difficulty is that conventional engine fuels are not pure substances but are mixtures of different hydrocarbons that cannot be exactly defined. It is, hence, difficult to determine physical and chemical properties and chemical reactions under engine combustion conditions, and these properties sometimes can only be approximated. This is also the reason why a high level of understanding has been gained for a series of subprocesses, but all the details of the overall processes involved in diesel engine combustion have not yet been explained.

Fuel injection. The design and construction of the injection system including the injection nozzle determine how the fuel is fed to the combustion chamber. The injection itself can be essentially characterized by the following:

Given a single injection per power cycle:

- The time of the start of injection and the length of injection

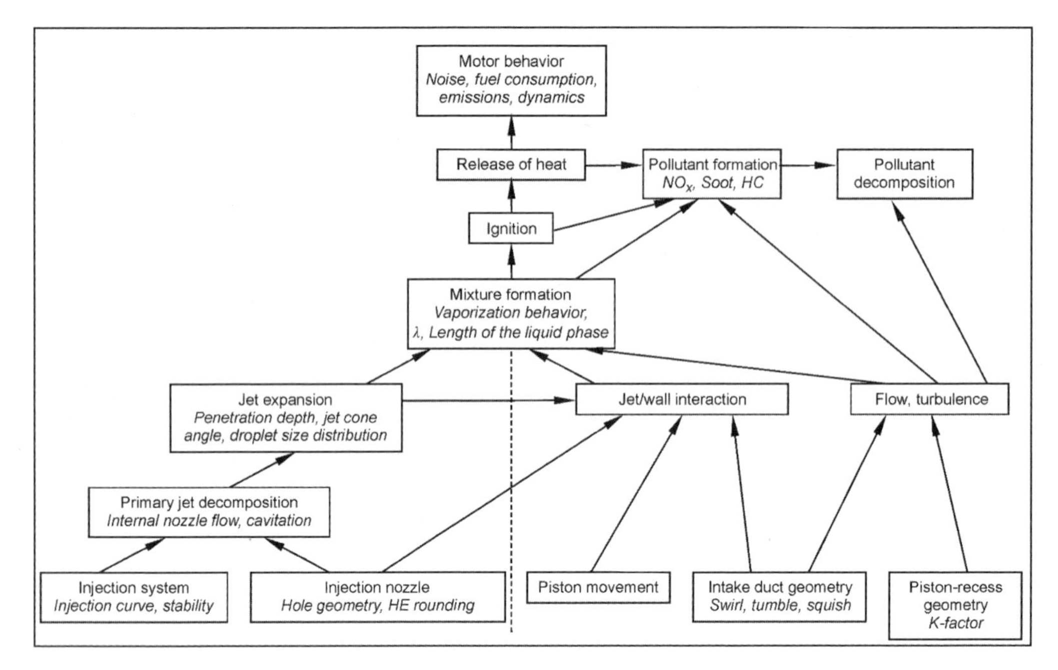

Fig. 15-1 Processes involved in mixture formation and combustion in diesel engines.[1]

Given divided injection:

- The point in time and the length of the individual injections
- The time characteristic of the injection rate
- The geometry, the number, and the alignment of the nozzle openings in relation to the combustion chamber (see Section 12.5.1).

Mixture formation. The goal of mixture formation is to generate the optimum local mixture of fuel and air (micromixture formation) and the optimum distribution of the air-fuel mixture to the combustion chamber volume (macromixture formation). The goal of optimization is to attain the maximum engine work with minimum fuel consumption and simultaneous minimum exhaust emissions. The limits to engine noise and the mechanical and thermal loads on components also need to be maintained. Since some of the measures to attain these goals counteract each other, optimization can be only a compromise between the individual demands. A notable example is the contradictory behavior of nitrogen oxide emissions and the specific fuel consumption. If we examine the different parameters (such as exhaust regulations) that exist for the individual types of engines such as large diesel ship engines as opposed to passenger car diesel engines, it becomes clear that there can be no general quantitative formulation of optimum mixture formation conditions for all diesel engines. Nonetheless, there are a few basic considerations that must be taken into account when designing and optimizing all diesel engines.

For normally used hole-type nozzles (in DI engines), the mixture formation process presently consists of the following steps: Mixture formation starts directly with fuel injection. Depending on the injection system, the fuel jet leaves the injection nozzle at different speeds (>100 m/s). Given the high relative speed of the exiting fuel in comparison to the surrounding air and the implosion of cavitation bubbles from the injection orifices immediately after they leave the nozzle, the fuel jet disperses almost immediately directly after the nozzle exit. Figure 15-2 shows a measurement-based model of this process.

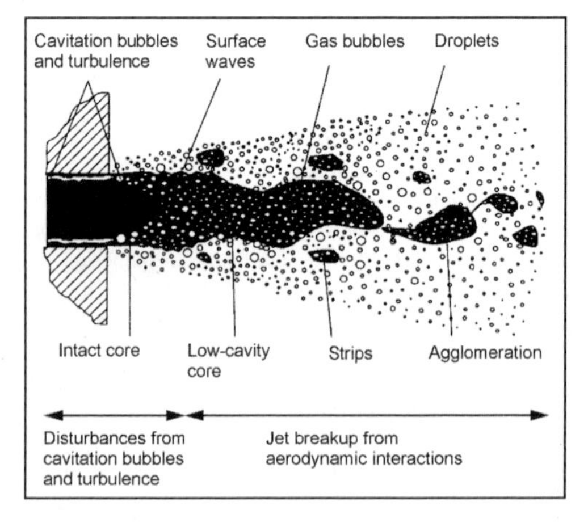

Fig. 15-2 Two-phase spray model and spray images (2-D scattered light technique) for two rail pressures during common rail injection in a chamber.[2,3]

The forming fuel jet consists of numerous individual fuel droplets of different sizes (2 to 30 $\times$ 10^{-3} mm) and shapes.[4] Depending on the parameters of the injection system and the gas state in the combustion chamber, each fuel jet has its own characteristic statistical distribution of droplet size. The droplet size essentially depends on the following influences. The arising droplets are smaller:

- The smaller the nozzle bore diameter
- The faster the exit speed from the nozzle
- The greater the air density in the combustion chamber
- The lower the fuel viscosity
- The smaller the surface tension of the fuel

Additional air movement in the combustion chamber increases the relative movement between fuel and air and, hence, the atomization quality, and the micromixture and macromixture formations.

A typical distribution of the droplet sizes in the fuel jet is shown in Fig. 15-3.

Since the fuel is injected at the end of the compression stroke, the fuel droplets in the injection jet are immediately exposed to the high gas temperature in the combustion chamber. This produces a strong exchange of heat from the heated combustion chamber air to the relatively cold fuel droplets. As the temperature exchange continues between the air and fuel, the evaporation at the droplet surface increases. The forming fuel vapor then mixes with the surrounding air.

Differences in concentration and temperature then form in the environment around the droplets (see Fig. 15-4) and, therefore, also in the entire fuel jet (heterogeneous mixture) that subsequently trigger diffusion processes near the individual fuel droplets and in the jet.[6] In the middle of Fig. 15-4, we see the change over time of the air conditions at the edge of a fuel jet approximately 26.5 mm from the nozzle exit for three different injection pressures.[2] The bottom of Fig. 15-4 shows a snapshot of the distribution of the air conditions in a fuel jet.[7] It becomes clear that the ignition conditions in a diesel fuel injection jet *(ignition lag)* can always attain

- A mixture composition within the ignition limits
- A sufficiently high mixture temperature

after a certain amount of time.

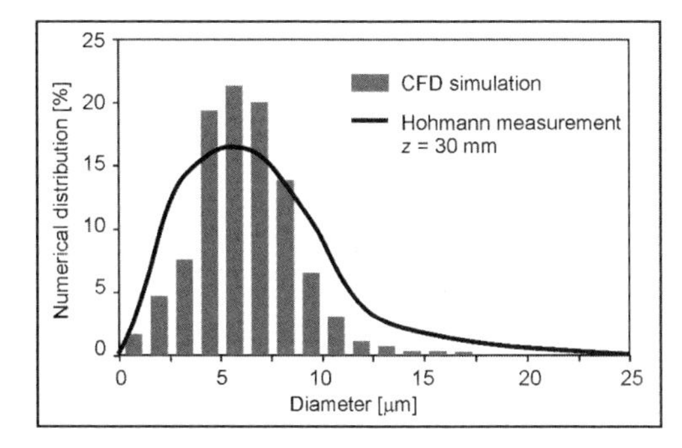

Fig. 15-3 Droplet size distribution in a fuel jet 30 mm from the nozzle.[5]

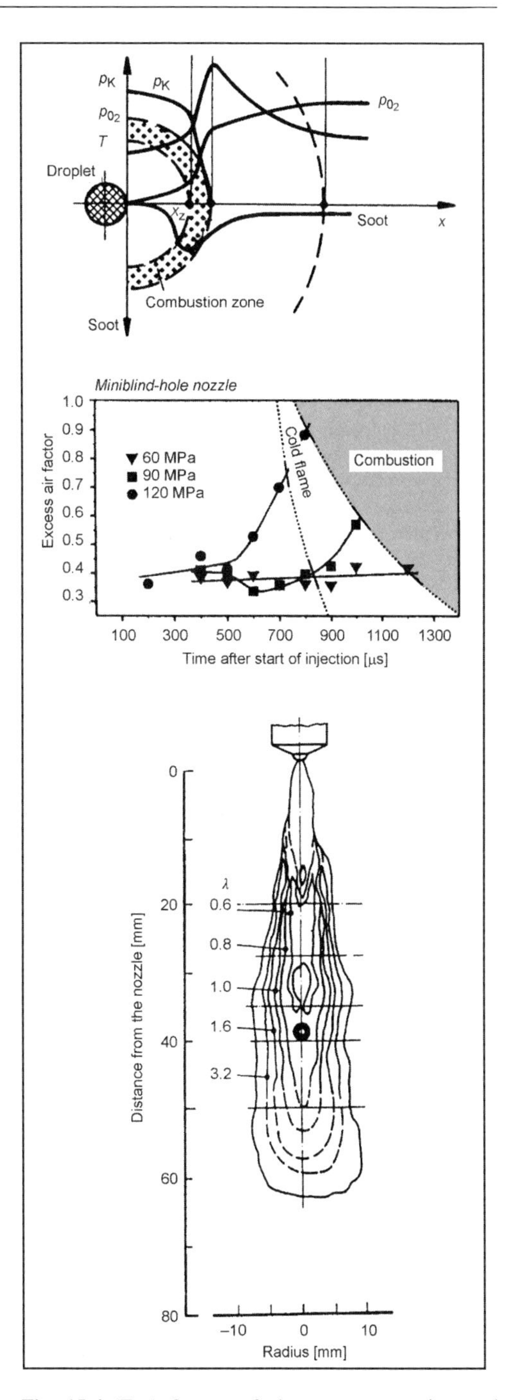

Fig. 15-4 (Top) Oxygen, fuel, soot concentration, and temperature in the environment of a burning individual droplet. (Middle) Change over time of the air-fuel ratio at a site in the injection jet up to the beginning of cold flame reactions and when ignition conditions are attained for different injection pressures. (Bottom) Momentary local distribution of the excess air factor in a fuel jet.

The primary influential parameters on the development of a fuel jet in diesel engines are schematically illustrated in Fig. 15-5.

Ignition lag, ignition, and combustion.[4] The physical and chemical processes that start in the combustion chamber at the beginning of fuel injection need time before ignition conditions are attained. This span of time, the ignition lag, extending from the start of injection to the first ignition is very important for the subsequent combustion process. Depending on the conditions existing in the engine upon fuel injection, the ignition lag is up to 2 ms.

With a short ignition lag, relatively little fuel is injected and physically and chemically optimized up to the start of combustion. After ignition, this produces a moderate rise in pressure and temperature in the combustion chamber. Since a rise in pressure and maximum pressure in the combustion chamber are major causes of combustion noise, they are set at a relatively low level. With a low maximum pressure, the mechanical component load is also lower. The maximum gas temperature, the formation of nitrogen oxides related to the high gas temperature, and the thermal component load are relatively low. On the other hand, combustion occurs at comparatively low pressures and temperatures, yielding a higher specific fuel consumption and increased soot formation. The latter is because of the relatively large amount of fuel that is injected into the developing hot flame after ignition, and the excessively slow mixture of the forming fuel vapor with the air. Local zones of insufficient air and high temperatures promote soot forming crack reactions. A comparatively long ignition lag produces the correspondingly opposite effects. The cited physical and chemical reasons for the ignition lag also provide indications of how to effectively influence it.

A short ignition lag is produced by

(a) Physical influences

- High gas temperature and high gas pressure at the beginning of injection
- Strong atomization of the fuel
- High relative speeds of the fuel and air

These influences cause the fuel to quickly evaporate, which permits the rapid distribution and mixture of the fuel with the air in the combustion chamber.

The gas pressure and temperature at the beginning of injection can be increased by the following constructive measures:

- High compression ratio
- Late injection time
- Supercharging
- High coolant temperature and suitable cooling channel
- Combustion chamber design (influence of the wall temperature)
- Use of ignition aids (glow elements, earlier glow plugs, intake air preheating)

The atomization of the fuel is mainly determined by the selected injection system as well as the injection time (gas state) and the temperature-related fuel properties (see above).

The relative speeds of the fuel and air can be influenced by the constructive design and harmonization of the

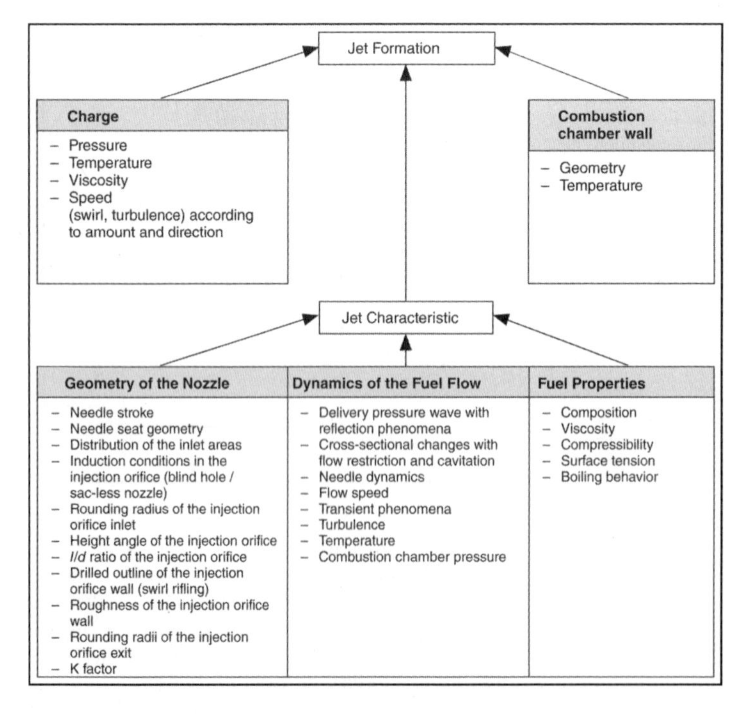

Fig. 15-5 Influences on the development of the injection jet.[8]

injection system, combustion chamber shape, and intake duct. The different approaches to the solution of this task include the essential differences between the diesel engine combustion processes.

(b) Chemical influences

- High ignitability of the fuel (high cetane number)
- High fuel temperature, high gas pressure, and temperature at the start of injection

ensure fast chemical preparation of the fuel.

The actual combustion in the diesel engine presently can be described as follows: The slowest processes control the combustion process. Directly after ignition, the fuel that was physically and chemically well prepared during the ignition lag burns quickly and with a high conversion of energy. This first phase is also termed premixed combustion and is mainly controlled by the still relatively slow chemical processes (low temperature). After this, combustion transitions into a second phase that is characterized by the continued injection of fuel into the existing flame, and, hence, by a strongly inhomogeneous charge composition and temperature. In this phase, combustion is again controlled by the slowest process, mixture formation (diffusion). The speed of the chemical reaction is much faster because of the quick rise in temperature in the first phase. As combustion progresses into the third phase, the conversion rate decreases because zones of decreasing oxygen and gradually sinking gas temperature from expansion reduce the reaction speed. In addition, the charge movement initiated by the intake process decreases in this phase. The phase is then controlled by slower mixture formation and the decreasing reaction speed. This produces a thermodynamically problematic delay of combustion far into the expansion stroke. Figure 15-6 shows the typical qualitative characteristic of the injection and combustion rate for a diesel engine with direct fuel injection.

There is much less time available for internal mixture formation in diesel engines in comparison to the classic spark-injection engine. In addition, the boiling curve of the diesel fuel is much higher. This disadvantages the diesel engine in comparison to the spark-injection engine when used in vehicles. With increasing speed, the time problem becomes more pronounced. The unavoidable inhomogeneity of the cylinder charge yields lower average pressures because of the less efficient exploitation of air (smoke limit). Both lower the power output per liter for diesel engines. Hence, today, diesel engines are usually supercharged. Compensating for this disadvantage remains an essential goal in the development of combustion processes for diesel engines.

Pollutant formation. In developing combustion processes for diesel engines, major considerations are particle and nitrogen oxide emissions. The exhaust particles largely consist of soot on which is deposited a greater or lesser number of hydrocarbon and/or sulfur compounds.

Hydrocarbon and carbon monoxide emissions play less of a role in diesel engine combustion. The formation of pollutants is directly related to local conditions of ignition, mixture formation, and combustion in the combustion chamber.

According to Ref. [9], soot and nitrogen oxide formation is strongly influenced by reaction kinetics. These processes are not completely understood, however. From numerous investigations of flames and shock wave tubes, we have gleaned the following picture (Fig. 15-7): The soot formation is limited by the temperature and the excess air factor. This restriction is independent from the predominant pressure. The so-called soot yield (soot mass/overall hydrocarbon mass) increases with pressure, however. At temperatures of approximately 1600 K and excess air factors <0.6, the soot yield approaches a maximum. For many hydrocarbons, the soot formation thresholds are very similar, allowing these considerations to also be applied to diesel engines. The previously described heterogeneous mixture formation in diesel engines means that despite an overall excess air factor >1, local excess air factors <0.6 can arise. As long as the mixture temperature remains below approximately 1450 K, soot formation is largely excluded. When a burning, relatively rich mixture cools (close to the wall, for example) or fuel is heated that is insufficiently mixed with air, intense soot formation occurs.

Because of the influence of the wall (quenching) and the fuel composition, the start of soot formation in diesel engines must be assumed when there is an excess air factor <0.8.

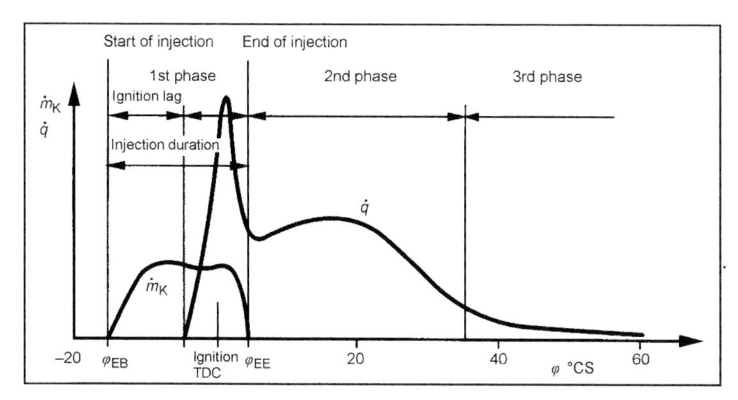

Fig. 15-6 Qualitative characteristic of fuel injection and heat release.[6]

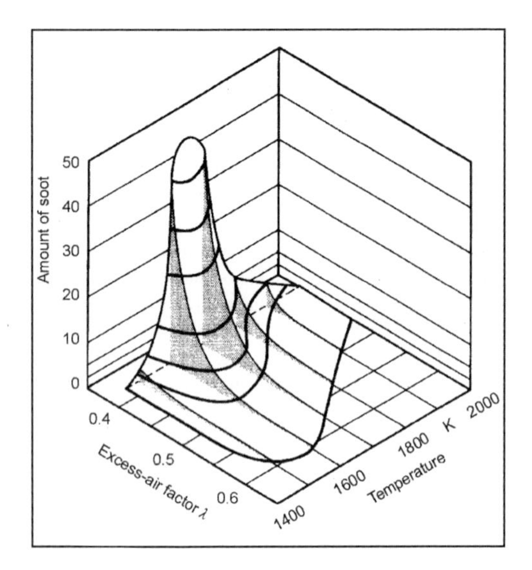

Fig. 15-7 Soot yield as a function of the temperature and excess air factor.[9]

Figure 15-8 shows a temperature/excess air factor diagram in which the states of the mixture and combusted material close to ignition TDC are plotted in addition to the soot formation range. In addition, the range of intense nitrogen oxide formation (components formed within 0.5 ms) is also portrayed. As we know, the highest nitrogen oxide formation rates occur with an excess air factor of 1.1. In this range, a part of the formed soot can be recombusted, as the reaction time of soot particles (d = 40 nm) in this range shows. As the excess air factor further increases toward the average combustion chamber value, there is a decrease in the combustion temperature and, hence, nitrogen oxide formation. The graph also provides an explanation of the typical contradictory behavior of soot and nitrogen oxide emissions in diesel engines. Relatively low temperatures and a deficiency of air promote soot formation and lessen nitrogen oxide formation. High temperatures and excess air have opposite effects. A clear reduction of both emissions is basically attainable only when the fuel-air mixture is "leaned off" or homogenized as quickly as possible at a very low temperature before ignition while avoiding "rich" areas, or the mixture is combusted at excess-air factors between 0.6 and 0.9.

15.1.2 Diesel Four-Stroke Combustion Systems

The combustion strategy that arose over the course of developing diesel engines can be explained and understood based on the above-described processes in the combustion chamber.

Rudolf Diesel did not have an opportunity to use industrially manufactured, highly developed injection techniques during his lifetime. His attempt to use the high-pressure fuel injection system with which we are familiar today failed because of the technical restrictions

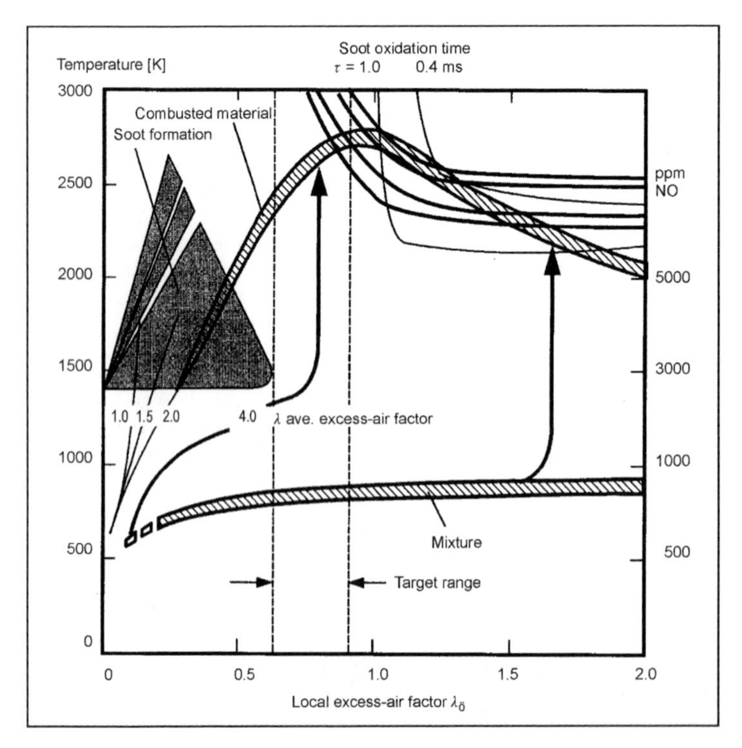

Fig. 15-8 States of the mixture and combusted material in diesel engine combustion.[7]

of his time. As an "emergency solution," he developed a method in which the liquid fuel was blown into the combustion chamber of the engine using compressed air. According to Ref. [10], this combustion process is distinguished by very even and "gentle" engine running. The exhaust was soot-free and odorless. Given today's knowledge, this result can be explained as follows:

- Blowing in the air produces a very fine fuel atomization (fast evaporation and mixing).
- Before the actual combustion phase, there is an intense mixture of fuel and air in the nozzle.
- Because of the cooling effect of the blown air and further cooling as the fuel flows into the cylinder (expansion), soot formation is largely prevented in the arising mixture.

The decisive disadvantage of the later developments of this procedure was the large amount of work to drive the required air compressor. This resulted in correspondingly high fuel consumption. Because of the direct introduction of the fuel into the combustion chamber, this method can be termed a direct injection (DI) method, although it is fundamentally different from the high-pressure fuel injection procedure used today. The direct injection of fuel is, therefore, the historically oldest combustion system. In the described form, it was limited only to relatively low-speed engines, i.e., engines with large combustion chambers. To make the diesel process useful for high-speed engines and vehicle use, additional developmental steps were necessary. An important prerequisite was the introduction of high pressure injection that became technically feasible at the beginning of the 1920s, which was cheaper and metered fuel better. The disadvantage of high-pressure injection in contrast to blowing in air was that without additional measures, the mixture was formed only by the injection nozzle (without air support). To a certain degree, the charge turbulence that rises in the combustion chamber as rpm increase is sufficient to compensate for the shorter time because of the increased mixing speed and, hence, increasing combustion speed. Given accelerations within ranges that are covered by today's medium-high speed and high-speed engines, solutions had to be found that permitted a corresponding acceleration of the mixture formation and combustion processes. The recognized important variables were the relative speeds of the fuel and air, and the influence of the combustion chamber wall on the speed of the mixture formation processes. The relative speeds of the fuel and air are most effectively influenced by

- The fuel speed in the combustion chamber (injection pressure)
- The air speed in the combustion chamber (combustion chamber and intake duct design)

The best results in engines are attained by the optimum harmonization of fuel injection and air movement in the combustion chamber.

The diesel engine combustion system arose against this background.

15.1.2.1 Methods using Indirect Fuel Injection (IDI)

In engines that use this method, the combustion chamber is divided. It consists of a main combustion chamber and a secondary combustion chamber. The secondary combustion chamber is in the cylinder head. The main combustion chamber is formed by the cylinder and a recess in the piston head that is adapted to the position of the chamber mouth. These engines are therefore also termed indirect-injection or divided chamber engines. The main and secondary combustion chambers are connected by one or more channels with a cross section that is narrower than that of the main and secondary combustion chambers. The secondary combustion chamber is designed as a whirl chamber or prechamber. Both methods have the following in common.

The fuel is injected under moderate pressure (<400 bar) by means of an inline or distributor injection pump into the secondary chamber. Throttling pintle nozzles (small injected quantity during the ignition lag) are used as the injection nozzles. After the first amount of fuel is quickly mixed with the air followed by a relatively short ignition lag (high chamber wall temperatures), ignition occurs in the secondary chamber. The air overflowing from the main combustion chamber at a high speed into the secondary chamber because of the displacement effect of the piston during the compression phase greatly supports mixture formation in the secondary chamber. Directly after ignition, the pressure and temperature quickly rise in the secondary chamber above the values in the main combustion chamber. The higher chamber pressure causes the air-fuel mixture forming and partially burning in the secondary chamber to intensely flow into the main combustion chamber. This causes the outflowing mixture to intensely mix with the sufficient air in the main combustion chamber.

Combustion processes that use indirect injection tend to form a greater amount of soot. The reason is the lack of air at relatively high temperatures in the secondary combustion chamber after ignition. Under high engine loads, a part of the soot formed in this phase can still burn in the main combustion chamber. Under a partial load, however, the temperatures for effective afterburning are too low. The formation of nitrogen oxides is largely suppressed in the indirect fuel injection method. The air deficiency in the chamber is then an advantage in this regard. When the mixture is expelled from the secondary chamber, it is quickly diluted so that high local temperatures are largely avoided along with excess air factors that support nitrogen oxide formation. The intense mixture formation in the indirect fuel injection method also produces favorable hydrocarbon and carbon monoxide emissions in these engines. Another advantage of the intense mixture formation is the resulting relatively low rise in cylinder pressure that produces correspondingly low noise. This combustion process also allows high air utilization (close to the stoichiometric mixture composition) under a full load and simultaneously at high engine speeds.

The described properties of the indirect fuel injection method secured indirect-injection engines a predominant position among high-speed engines for a long time, especially passenger car diesel engines. Even among medium-speed engines, indirect-injection engines were represented in the upper speed range.

Prechamber system.[11] This method arose in the 1920s, and its development is largely concluded. Figure 15-9 shows a prechamber according to Ref. [12].

In contrast to a two-valve design, the design shown here in a four-valve engine has greater potential for savings in fuel consumption and lower exhaust emissions because of the symmetrical and central arrangement of the prechamber in relation to the main combustion chamber.

All prechambers are rotationally symmetrical. The actual chamber area in which the fuel is injected can be spherical to egg-shaped. The chamber is connected to the main combustion chamber via a duct that ends in several combustion holes. The number, direction, and diameter of combustion holes must be optimized together with the piston recess. According to Ref. [4], the chamber volume should be approximately 40% to 50% of the compression volume. This percentage strongly influences soot and nitrogen oxide formation and should be correspondingly optimized. The optimum cross section of all combustion holes is 0.5% of the piston cross section. A greater number of holes reduces soot emissions. The direction of the combustion holes is set in relation to the thermal load of the piston heads. The injection nozzle is on the top of the chamber opposite the duct. The compression ratio of these engines is between 21:1 and 22:1. In the above example, the ratio is 21.7:1. The prechamber system is not very suitable for very small cylinder-stroke volumes. The mixture formation in the chamber can be optimized with a spherical pin whose geometry and position is adapted to the chamber (see Fig. 15-9). This pin is perpendicular to the direction of the fuel injection jet and supports the preparation of the contacting fuel jets and the fuel distribution in the secondary chamber. Because of the special shape of the bottom of the spherical pin, a slight swirl is generated in the secondary chamber during inflow of the air coming from the main combustion chamber in the compression phase that provides a more intense mixture formation in the secondary chamber. It is best for the injection nozzle to be at a slight angle in relation to the lengthwise chamber axis. Despite the relatively high compression ratio, the method does not work without ignition assistance. It is helpful to place a glow element or glow plug downwind from the air stream in the secondary chamber following the injection nozzle. To support the warm-up phase, the glow element can be operated for up to one minute.

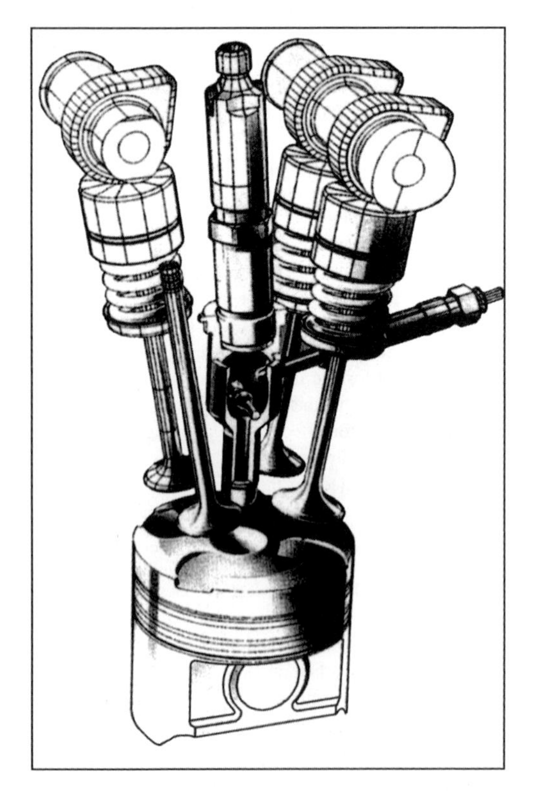

Fig. 15-9 Combustion chamber arrangement of a four-valve prechamber engine (DaimlerChrysler AG).

Whirl chamber system.[13] In this type of combustion system, the secondary chamber is in the cylinder head as is the case with the prechamber system (Fig. 15-10).

The secondary chamber can be disk-shaped, spherical, or oval. The main combustion chamber and secondary chamber are connected via a passage with a relatively large flow cross section. The transfer passage ends tangentially in the actual combustion chamber so that, as the piston moves upwards, the air flowing into the secondary chamber is subject to a forceful swirling movement. The ratio of the whirling speed to the engine speed is related to the operating state of the engine, especially the rpm, and is between 20 and 50. The whirling in the chamber at the beginning of compression initially corresponds to a solid swirling that becomes more similar to a potential

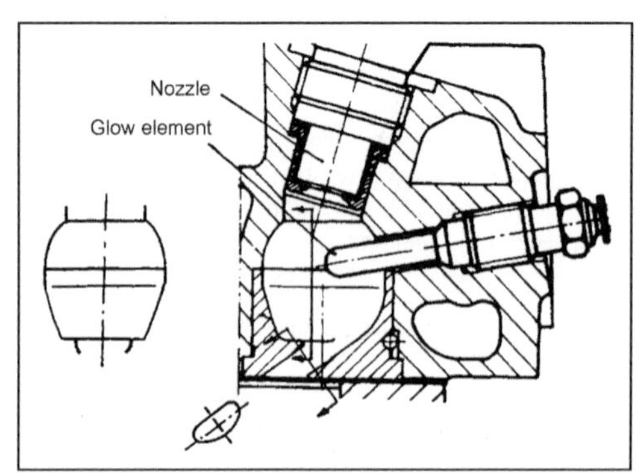

Fig. 15-10 Whirl chamber with an injection nozzle and glow element (Opel Omega 2,3 D).[13]

swirling in the last phase before TDC. The maximum circumferential speed increases with the rpm and always occurs in the range of 10° to 20° crank angle before TDC. The secondary chamber size and position and geometry of the transfer passage are to be optimized with the nozzle arrangement in the secondary chamber and the combustion chamber recess (usually shaped like goggles) to be provided in the piston head opposite the passage exit. The piston recess suppresses the burning flame at the edge of the recess and, hence, reduces the danger of incompletely combusted fuel being transported to colder regions of the piston head and enhancing soot formation there. At the present stage of development,[4] the optimum chamber volume is approximately 50% of the compression volume. The optimum overflow cross section is 1% to 2% of the piston cross section. The injection nozzle is in the top part of the secondary chamber so that the fuel jet enters the secondary chamber tangentially against the incoming air directed at the hot, opposing chamber wall so that the swirling air in the secondary chamber penetrates the jet at a right angle. As the fuel jet passes through the whirl chamber, a part of the fuel is quickly evaporated and becomes ignitable. The majority of the injected fuel reaches the 900 K chamber wall where it vaporizes relatively slowly. Ignition greatly accelerates this process. The forming fuel vapor is mixed quickly and intensely by the whirling motion of the air in the chamber. The remaining combustion process is the same as in a prechamber engine. The compression ratio of these engines is between 22:1 and 23:1. The stroke/bore ratio for these types of engines is between 0.95 and 1.05. The whirl chamber system can be used at up to speeds of approximately 5000 rpm (somewhat greater than the prechamber system) and is, hence, especially suitable for use in passenger cars. The combustion properties and attainable average pressure at the soot limit are comparable with those of prechamber engines. The whirl chamber system also does not work without ignition assistance. The position of the glow element or the glow plug in the secondary chamber strongly influences engine performance so that it requires a specially optimized design. The disturbing influence of the glow element on chamber flow can be compensated to a certain degree by reducing the cross section of the transfer passage (increase in speed).

15.1.2.2 Direct Fuel Injection Method (DI)

In the direct fuel injection method, the combustion chamber is not divided[4,14–16] (Fig. 15-11). The actual combustion chamber is formed by a recess in the piston head. Up to 80% of the compression volume can be incorporated in this piston recess. Diesel engines with a cylinder diameter larger than approximately 300 mm usually work without additional air movement in the combustion chamber. The mixture is formed exclusively by the injection system, especially the injection nozzle. Multihole nozzles are used as injection nozzles with up to twelve nozzle holes depending on the engine size. An engine with four valves allows the combustion chamber to be symmetrical (which has a positive effect on mixture formation and the thermal load) because of the centrally located injection nozzle aligned with the cylinder axis. Figure 15-11 shows an asymmetrical design with two valves.

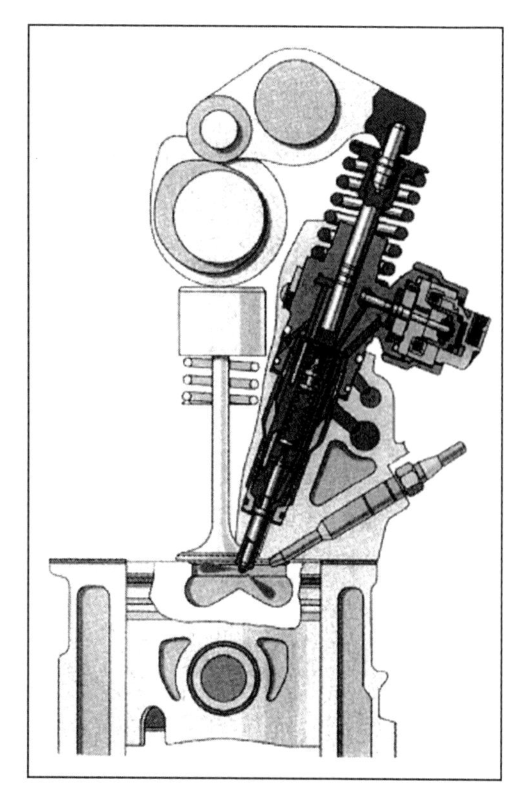

Fig. 15-11 Combustion chamber arrangement of a two-valve DI engine with a pump-nozzle injection system (Audi AG).

The injection pressure (1300 to over 2000 bar) and the nozzle bore diameter determine the size of the fuel droplets and the relative speeds of the fuel and air in the injection jet. The combustion chamber is largely open and adapted to the shape and position of the injection jets. In smaller, faster running engines, the air movement generated by the intake process, fuel injection, and piston movement is no longer sufficient for good mixture formation. Special measures must be taken to increase the relative speeds of the fuel and air in the combustion chamber. By designing the intake duct as a swirling and/or tangential duct, the air intensely swirls around the cylinder axis as it flows into the combustion chamber. This swirling overlaps the turbulence that already exists in the combustion chamber and causes the rapid distribution and mixture of the fuel vapor arising upon injection directly next to the injection jet with the air in the combustion chamber (macromixture formation). Another possibility of increasing the relative speeds of the fuel and air in the combustion chamber is to narrow the piston recess near the piston head.

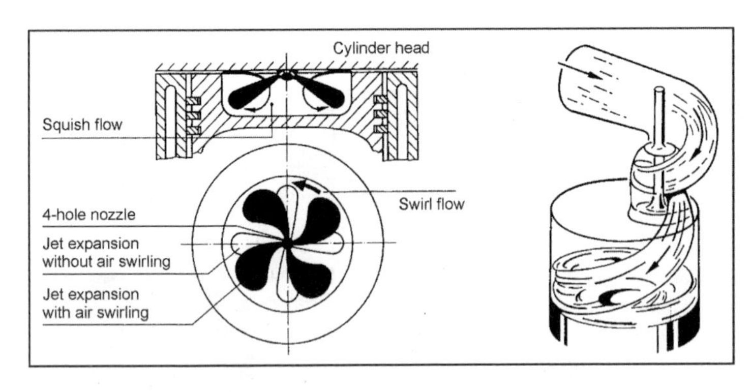

Fig. 15-12 Flow processes in the combustion chamber of a diesel engine that uses direct injection and mainly air-distributed fuel.[4]

During the compression stroke, the air above the piston head is thereby directed toward the piston recess. When the air flows into the recess, an intense whirling movement arises, the so-called squish (Fig. 15-12).

The advantage of the squish in contrast to swirling is that it increases in intensity as the piston approaches TDC (fuel injection phase) while the swirling generated during induction decreases. As the speed of the engine increases, both methods are combined. To attain the best values for fuel consumption and exhaust emissions, the intake ducts, the combustion chamber geometry, and the fuel injection must be optimized and harmonized with each other (Fig. 15-13).

A reduction of the number of nozzle holes requires an increase in the swirl and vice versa. If the swirl is too high, the fuel in the individual fuel jets becomes mixed. This produces local "over-enrichment" of the mixture and, hence, worse air utilization and high exhaust emissions. In vehicle engines, it is particularly difficult to optimally harmonize the mixture formation over the entire operating range. Simulation methods (3-D) and improved experimental possibilities (such as the transparent engine) help solve these tasks. Swirling must be adapted to the load and rpm for optimum engine operation. High-speed engines need compression ratios between 15:1 and 19:1 and, like engines with indirect fuel injection, must have glow elements as a starting aid for reliable cold starts and warm-up. Today, these engines attain maximum speeds of up to 4500 rpm and an efficiency of 43% with exhaust turbocharging at their best point. In large engines, compression ratios between 11:1 and 16:1 are necessary depending on the amount of turbocharging. Also, efficiencies of more than 50% are attained. The described relationship between engine speed (engine size), air movement, and combustion chamber shape is revealed in the comparison of typical combustion chamber shapes in engines that use direct injection with increasing speeds (Fig. 15-14).

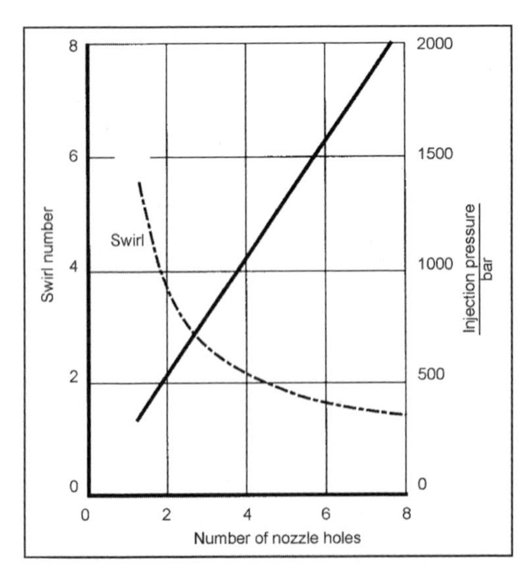

Fig. 15-13 Typical relationship between injection pressure, swirl number, and nozzle hole number.[16]

On the left side of Fig. 15-14 the typical combustion chamber of a medium-speed engine is shown, and on the right is a passenger car engine. We can clearly see the increasing narrowing and deeper piston recess as the speed increases. This increases the squish, and the swirl is sustained into the expansion stroke. The necessary swirl increases in the same manner (Fig. 15-15).

There is likewise a simultaneous reduction of the number of nozzle holes. It becomes difficult to optimally harmonize the combustion process as the engine speed rises because the system becomes more sensitive to the combustion chamber geometry. In passenger car combustion chambers, particular attention needs to be paid to the specific shape of the recess edge (turbulence ring).[17] The design with four valves and a central injection nozzle is becoming increasingly popular with smaller cylinder sizes.

15.1.2.3 Comparison of Combustion Systems

The previously cited combustion systems are primarily compared regarding their specific fuel consumption, exhaust emissions, and combustion noise.[4,18] They basically differ according to how they generate the relative speeds of the fuel and air that are required for mixture formation. The indirect injection methods work at low injection pressures and also relatively low fuel speeds and, therefore, require high air speeds. In direct injection methods, high fuel speeds are attained by high injection pressures. They can therefore get by with lower air speeds.

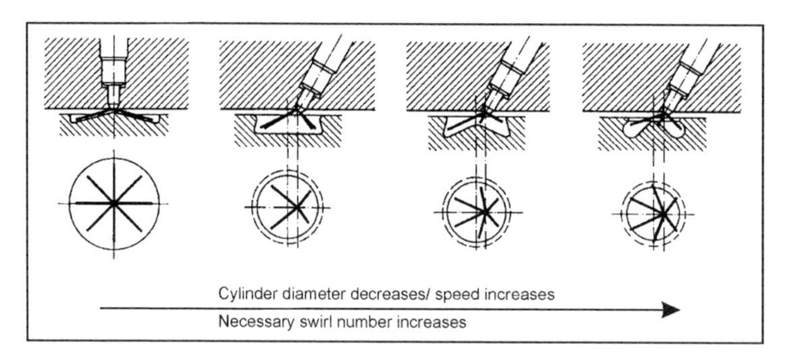

Fig. 15-14 Influence of engine size (speed) on the combustion chamber recess shape and required air movement in diesel engines that use direct injection as described by Ref. [4].

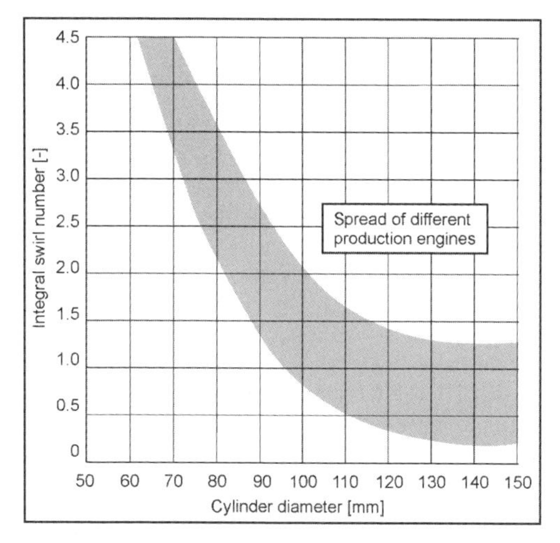

Fig. 15-15 Typical relationship of the required swirl number to cylinder diameter.[17]

The swirl ducts that are necessary in engines that use direct injection to generate air movement restrict the cylinder charge at high speeds and increase charge cycle losses. The necessary flow speed near TDC tends to be lower with direct injection, a bit faster in the whirl chamber system, and highest in the prechamber system. As flow speed in the combustion chamber rises, the flow losses increase. In addition, faster flow speeds cause a greater transfer of heat and, hence, a greater loss of wall heat and higher thermal load; this is further pronounced in indirect-injection engines in contrast to engines that use direct injection due to the larger combustion chamber surfaces. Because of the greater flow and heat transfer losses, combustion systems using indirect fuel injection have an approximately 15% higher fuel consumption than engines using direct injection. Because of the less favorable surface/volume ratio (30% to 40% higher than DI) of the combustion chambers, engines using indirect fuel injection manifest worse cold start behavior that cannot be fully compensated by a higher compression ratio. The higher injection pressures required in engines that use direct injection lead to more expensive injection systems subject to greater loads.

The higher charging speeds with the indirect fuel injection method allow better air utilization. This allows low air-fuel ratios at the smoke limit, which in turn compensates for worse delivery and fuel consumption in comparison to direct injection engines. More-or-less equally high average full load pressures are therefore attainable.

Black smoke emissions are worse for the indirect fuel injection method, particularly in the lower load range in comparison to direct injection. As the engine load increases, nitrogen oxide emissions are better for the indirect fuel injection method than for direct fuel injection. In terms of hydrocarbons, indirect fuel injection has advantages over direct injection (Fig. 15-16).

The enormous advances in the development of injection technology, especially increasing injection pressure, have reduced the emission advantages of indirect fuel injection. Only in the case of nitrogen oxides does direct injection need substantial improvement because of the higher constant volume component in the supply of heat, which is also the reason for the louder combustion

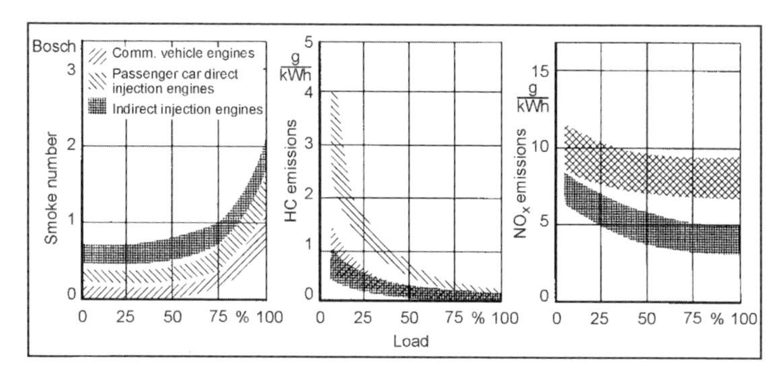

Fig. 15-16 Comparison of exhaust emissions in different combustion processs.[9]

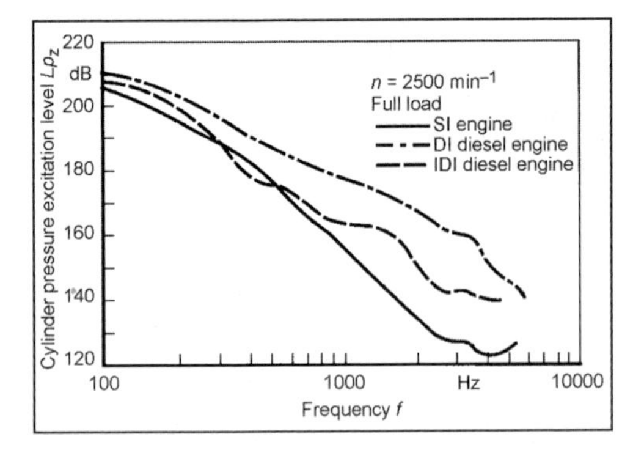

Fig. 15-17 Gas pressure excitation spectra of different diesel combustion systems in comparison to the spark-injection engine.[4]

(Fig. 15-17). Since DI engines have a substantially higher exhaust gas recirculation rate than indirect-injection engines, the emissions disadvantage can be compensated.

The present activities of injection system manufacturers toward fully manipulable injection time curves and any desired division of fuel injection offer promising improvements. At present, the future appears to lie with combustion systems with direct injection despite the existing deficits, especially because of the clear fuel consumption advantage. This positive development in favor of combustion systems with direct injection (Fig. 15-18) is supported by additional advantages.

Engines that use direct injection experience a lower thermal load. This makes them highly suitable for exhaust turbocharging that can be used to reduce exhaust emissions. The progress over recent years in turbocharger development (such as variable turbine geometry) make high-speed turbodiesel engines with direct injection serious competitors of spark-injection engines for use in passenger cars. In summary, it can be stated that combustion systems using direct injection have largely displaced combustion systems with divided combustion chambers independent of engine speed and size.

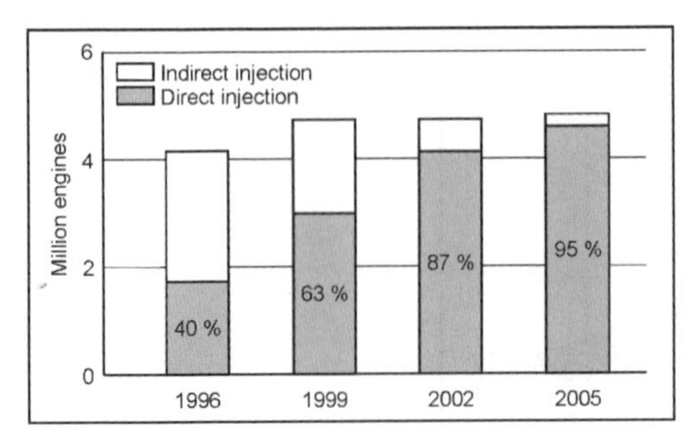

Fig. 15-18 Production development of a passenger car with a diesel engine in West Europe (Bosch GmbH).

15.1.2.4 Special Methods and Features

MAN-M method. This method completely breaks new ground. Whereas the rule of thumb is to keep the fuel away from the combustion chamber wall, in this case the fuel is directly applied to the wall. The spherical combustion chamber is in the piston head. This arrangement yielded the name middle sphere method (Fig. 15-19).

The fuel is injected with a one- or two-hole nozzle at a relatively low pressure tangential to the combustion chamber wall where it first spreads as a film. Only a small part of the fuel is distributed by the air to start ignition. Combustion systems with direct injection can, hence, be divided into wall distributing and air distributing systems. Neither method can be fully represented, however.

Hence, the designed combustion processes, especially for small cylinders, are better described as primarily wall or air distributing. In the MAN-M method, applying the fuel to the wall reduces the fuel speed to nearly zero, and the liquid fuel is not exposed to a high combustion chamber temperature (a wall temperature of approximately 340°C at full load). To attain a high relative speed of the fuel and air, we need a high air speed in the combustion chamber that is attained by means of swirl ducts. During the ignition lag, a slight amount of fuel evaporates from the combustion chamber wall. A correspondingly slight amount of fuel is prepared for combustion, which yields a very low pressure rise and combustion noise. After ignition, the fuel film rapidly evaporates off the wall because of the high gas temperature. The intensely swirling air quickly mixes the air and fuel. In the whirl itself, there is a separation of the hot combustion gases that travel along spiral paths to the center and the relatively cold air that is forced outward toward the fuel. Since the fuel is initially separated from the high gas temperature, the soot emissions are relatively low. This combustion system is therefore distinguished by good air utilization, and it attains high average pressures at the smoke limit. Disadvantages are the high flow and heat transfer losses that increase fuel consumption thermal load, especially on the piston and cylinder head. For this reason, this method is not particularly suitable for supercharging. In the partial load range, mixture formation worsens because of the falling temperature, which, in particular, leads to increased hydrocarbon emissions. The disadvantages of this method in comparison to the primarily air distributing combustion system with direct injection are the reasons it is no longer used today. The M method was chiefly used in commercial vehicles.

FM method. In developing the M method, people found that it was also good for burning low-boiling fuels (multifuel suitability). This led to the development of the FM method. Inner mixture formation, the combustion chamber shape, and load regulation were taken over from the M method. Ignition occurs with the aid of a spark plug [F = Fremdzündung (externally supplied ignition)] as in a spark-injection engine. The process closely corresponds to the constant pressure process. The behavior of the

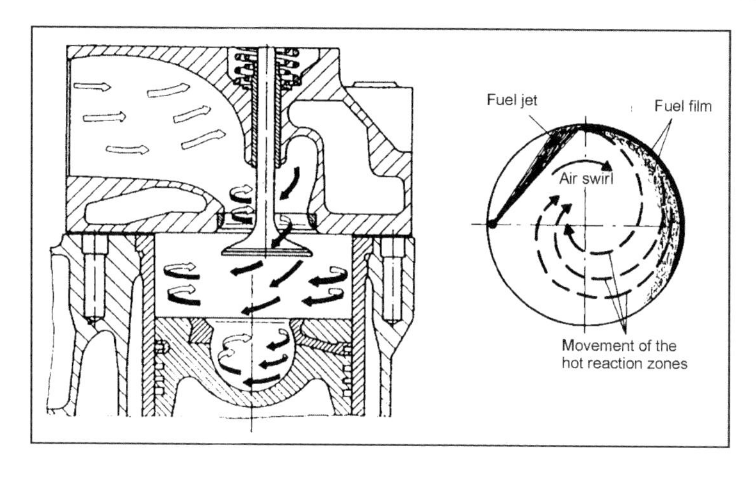

Fig. 15-19 Combustion chamber—fuel and air movement in the MAN-M method.[6,19]

exhaust emissions is somewhat better than with the M method. Because of the combination of features of the classic diesel and spark-injection methods, the FM method is counted among the hybrid combustion systems.

Ignition jet method. In this method, a small amount of diesel fuel (up to 10% of the full load heat consumption) is injected to ignite a fuel-air mixture that is homogeneously premixed outside of the engine cylinder and introduced into the combustion chamber. The homogeneously premixed fuels can be lean mixtures of diesel, alcoholic, or gaseous fuels. In practice, this method is mainly used in ignition jet gas-diesel engines. These engines can also be operated as fuel-changing engines; i.e., the quantity of diesel fuel can be increased from the ignition quantity to the full load quantity while simultaneously reducing the quantity of gas. The engine then runs completely on diesel. The advantage is that such engines can also be operated when a continuous gas supply cannot be ensured and/or when full diesel engine performance is needed such as when using weak gases.

Homogeneous compression ignition in diesel engines.[20] Continuously stricter exhaust regulations are motivating the search for better diesel engine combustion systems. A recent increase has been noted in research activities investigating autoignition of homogeneous diesel fuel-air mixtures. We know that homogeneous mixtures can be burned much more cleanly and with at least the same high level of efficiency as heterogeneous mixtures. Homogenizing the mixture prevents local temperature peaks that always arise in the combustion of heterogeneous mixtures and largely suppresses the formation of nitrogen oxides. Mixture compositions that lead to strong soot formation can be avoided (see 15.1.1 Pollutant Formation). The hydrocarbon and carbon monoxide emissions are higher than with heterogeneous diesel combustion. These can be effectively eliminated by means of oxidation catalytic converters. The problems associated with applying homogeneous combustion to diesel engines mainly have to do with the generation of a sufficiently homogeneous diesel fuel-air mixture during the time available in the engine, and in exactly controlling the moment of ignition. The time problem involved with homogenization limits the attainable speed. The engine load is restricted by the increasing danger of knocking. Precisely controlling the moment of ignition requires additional effort. Under discussion, for example, are variable intake air temperature and/or variable compression ratios. Figure 15-20 shows the present potential as well as

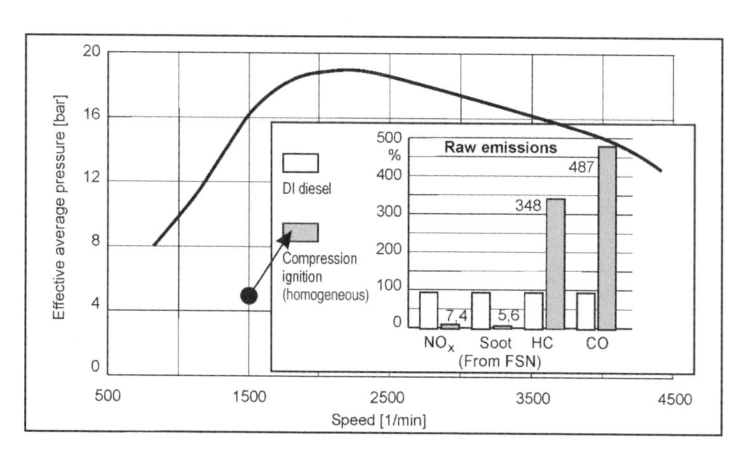

Fig. 15-20 Potential of homogeneous compression ignition in a DI-diesel engine.[2]

the load and speed limits of homogeneous diesel combustion. Internationally, this method is called HCCI (homogeneous charge compression ignition).

Gas-diesel engine. In this method, air is compressed as in conventional diesel engines. Highly compressed combustion gas is blown into the compressed air (internal mixture formation). While the combustion gas is being blown in, local regions with a combustible gas-air mixture arise in the combustion chamber. The autoignition temperature of the combustion gases is usually higher than that of diesel fuel. During the resulting longer ignition lag, a relatively large amount of ignitable gas-air mixture forms. This burns almost immediately upon autoignition. Impermissibly high-pressure gradients and maximum pressures arise in the combustion chamber. For this reason, a slight amount of diesel fuel is injected (pilot fuel). The gas is blown in only after the pilot fuel ignites. This allows the gas to burn almost immediately with an acceptable rise in cylinder pressure.

Special features of heavy oil operation. Heavy oil operation that is required today for medium-speed four-stroke diesel engines with piston diameters of approximately 200 to 600 mm and speeds between approximately 1000 to 400 rpm has a few particularities regarding combustion. The vanadium and sodium content of the heavy oil can produce deposits in the combustion chamber during combustion. The result is high-temperature corrosion. This impermissibly restricts the continuous operation of an engine or even makes it impossible. Together with the water rising during combustion, the sulfur in the heavy oil causes the formation of sulfuric acid and sulfurous acid when the oil falls below the dew point and, hence, causes low temperature corrosion. This makes it necessary to design the cooling system for the components forming the combustion chamber so that critical temperatures are not reached over the entire range of operation.

Bibliography

[1] Renner, G., and R.R. Maly, Moderne Verbrennungsdiagnostik für die dieselmotorische Verbrennung, U. Essers [ed.], in Dieselmotorentechnik 98, Expert-Verlag, Renningen-Malmsheim, 1998.

[2] Schünemann, E., C. Fettes, F. Rabenstein, S. Schraml, and A. Leipertz, Analyse der dieselmotorischen Gemischbildung und Verbrennung mittels mehrdimensionaler Lasermesstechniken, in IV. Tagung Motorische Verbrennung, Essen, March 1999, Haus der Technik, Essen, 1999.

[3] Fath, A., C. Fettes, and A. Leipertz, Modellierung des Strahlzerfalls bei der Hochdruckeinspritzung, in IV. Tagung Motorische Verbrennung, Essen, March 1999, Haus der Technik, Essen, 1999.

[4] Mollenhauer, K., Handbuch Dieselmotoren, 1st edition, Springer, Berlin, 1997.

[5] Adomeit, Ph., and O. Lang, CFD Simulation of Diesel Injection and Combustion, SIA Congress, "What Challenges for the Diesel Engine of the Year 2000 and Beyond" (Lyon, May 2000), SIA, Suresnes, France, 2000.

[6] Urlaub, A., Verbrennungsmotoren: Grundlagen, Verfahrenstheorie, Konstruktion, 2nd edition, Springer, Berlin, 1995.

[7] Dietrich, W.R., and W. Grundmann, Das Dieselkonzept von DEUTZ MWM, ein schadstoffminimiertes, dieselmotorisches Verbrennungsverfahren, Progress Reports VDI, Series 6: Energieerzeugung, No. 282, VDI Verlag, Düsseldorf, 1993.

[8] Heinrichs, H.-J. Untersuchungen zur Strahlausbreitung und Gemischbildung bei kleinen direkteinspritzenden Dieselmotoren, Dissertation, RWTH Aachen, 1986.

[9] Pischinger, F., H. Schulte, and J. Jansen, Grundlagen und Entwicklungslinien der dieselmotorischen Brennverfahren, in VDI Reports No. 714 (Conference: Die Zukunft des Dieselmotors, Wolfsburg, November 1988), VDI Verlag, Düsseldorf, 1988.

[10] Meurer, J.S., Das erstaunliche Entwicklungspotenzial des Dieselmotors, in VDI Berichte Nr. 714 (Conference: Die Zukunft des Dieselmotors, Wolfsburg, November 1988), VDI Verlag, Düsseldorf, 1988.

[11] Armbruster, F.-J., Einfluss der Kammergeometrie auf den Energiehaushalt und die Prozesssimulation bei Kammerdieselmotoren, Fortschrittberichte VDI, Reihe 12, Verkehrstechnik/Fahrzeugtechnik, Nr. 149, VDI Verlag, Düsseldorf, 1991.

[12] Fortnagel, M., P. Moser, and W. Pütz, Die neuen Vierventilmotoren von Mercedes-Benz, in MTZ 54 (1993) No. 9, pp. 392–405.

[13] Sun, D., Untersuchung der Strömungsverhältnisse in einer Dieselmotor-Wirbelkammer mit Hilfe der Laser-Doppler-Anemometrie, Dissertation, University of Stuttgart, 1993.

[14] List, H., and W.P. Cartellieri, Dieseltechnik – Grundlagen, Stand der Technik und Ausblick, in MTZ Special Edition "10 Jahre TDI-Motor von Audi," September 1999.

[15] Spindler, S., Beitrag zur Realisierung schadstoffoptimierter Brennverfahren an schnelllaufenden Hochleistungsdieselmotoren, Progress Reports VDI, Series 6: Energieerzeugung, No. 274, VDI Verlag, Düsseldorf, 1992.

[16] Dietrich, W.R., Die Gemischbildung bei Gas- und Dieselmotoren sowie ihr Einfluss auf die Schadstoffemissionen–Rückblick und Ausblick, Part 1, in MTZ 60 (1999) No. 1, pp. 28–38; Teil 2 in MTZ 60 (1999) No. 2, pp. 126–134.

[17] Thiemann, W., M. Dietz, and H. Finkbeiner, Schwerpunkte bei der Entwicklung des Smartdieselmotors, M. Bargende and U. Essers [ed.], in Dieselmotorentechnik 2000, Expert-Verlag, Renningen-Malmsheim, 2000.

[18] Kirsten, K. Vergleich unterschiedlicher Brennverfahren für kleine schnelllaufende Dieselmotoren, Dissertation, RWTH Aachen, 1986.

[19] Beitz, W., and K.-H. Grote, Taschenbuch für den Maschinenbau, 20th edition, Springer, Heidelberg, 2001.

15.2 Spark-Injection Engines

Combustion in spark-injection engines uses spark plugs for externally supplied ignition. The air-fuel mixture can be prepared in different ways:

- Homogeneous mixture preparation by external mixture formation [port fuel injection (PFI)]
- Homogeneous mixture preparation using fuel injected directly into the combustion chamber during the intake phase [direct injection spark ignition (homogeneous DISI)]
- Stratified mixture preparation using fuel injected into the combustion chamber toward the end of compression [direct injection spark ignition (stratified DISI)]

In homogeneous mixture preparation, the output is set by adjusting the charge (quantity control). In stratified mixture formation, the output is set by varying the excess air/fuel ratio (quality control), which enables throttle-free load control. In the following, we first discuss combustion by homogeneous mixture formation in PFI engines. The particular features of DISI engines are treated in the following chapter.

15.2.1 Combustion Processes in Port Fuel Injection (PFI) Engines

Combustion of Hydrocarbons

Normally, fuels for spark-injection engines consist of a mixture of approximately 200 different hydrocarbons (alkanes, alkenes, alcohols, and aromates). In PFI engines, a largely homogeneous air-fuel ratio occurs at the end of compression that is ignited shortly before TDC by an electrical ignition spark. There must be an ignitable mixture near the ignition spark. An air-fuel ratio of approximately $0.8 \leq \lambda \leq 1.2$ must exist near the ignition spark. To enable chemical reactions and, hence, combustion in the air-fuel mixture, the reaction partners must possess activation energy that is produced by the ignition spark. This required ignition energy is 30–150 mJ per combustion. The ignition spark generates local temperatures of 3000–6000 K. To ensure reliable ignition, an ignition voltage at the spark plug of 15–25 kV is required with a spark duration of 0.3–1 ms (related to the environmental state and charge movement). For flame reliable propagation, the energy released from combustion must be greater than the heat transported to the evaporating fuel and the walls delimiting the combustion chamber. The heat is released by the combustion of hydrocarbon compounds with oxygen according to the following overall reaction equation:

$$C_xH_y + (x + y/4)\,O_2 \rightarrow x\,CO_2 + y/2\,H_2O$$

Since the probability is slight of all the required reaction partners occurring simultaneously, the hydrocarbons are oxidized by numerous elementary reactions[1] in which alkanes rise in a first reaction phase via the dehydration of hydrocarbon peroxide, and the alkanes in turn form aldehydes by reacting with H, O, or OH radicals. The formation of the aldehydes requires approximately 10% of the entire released energy and is accompanied by cold flames. In the subsequent blue flame, CO, H_2, and H_2O (30% of the saved energy) are formed. In the following hot flame, CO_2 and H_2O rise, and 60% of the energy saved in the fuel is released.

Cylinder Pressure Characteristic, Indicated Thermal Efficiency, and Flame Propagation

The energy released during combustion leads to a rise in the temperature and pressure of the cylinder charge in the combustion chamber that is detected only in the cylinder pressure characteristic analysis after a delay following the start of ignition (Fig. 15-21). This is influenced by the local heating to ignition temperature of the mixture directly next to the spark plug and is approximately 1 ms independent of the rpm. The combustion time can be calculated with the aid of physical models for the conversion of energy and heat dissipation from the combustion chamber.[2] These produce the combustion function that indicates the ratio of combusted to utilized fuel mass as a function of the crankshaft angle. This allows us to evaluate the state and length of combustion as well as its thermodynamic effect. For homogeneous mixtures, the center of gravity of combustion is most efficient at 8° crank angle after TDC,

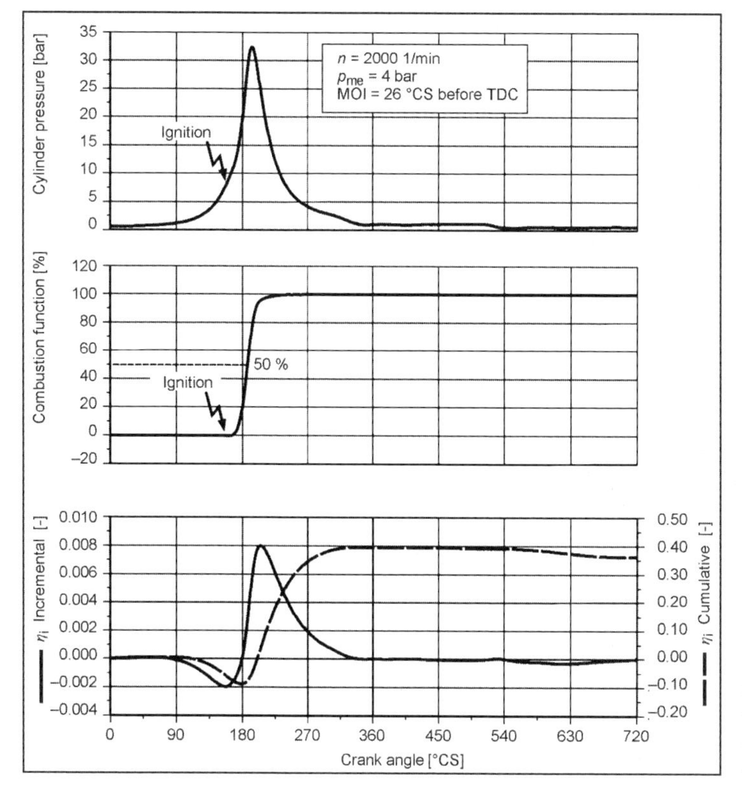

Fig. 15-21 Pressure characteristic and pressure characteristic analyses cylinder wall (flame quenching).

and the effective combustion period is 30°–50° crankshaft angle depending on the working point and combustion process.

We are limited in our ability to portray the detailed effects of changes in the engine parameters (such as variation of the control time, changes in charge movement) using the previously cited simulation models. To accomplish this, the DPA method (differential pressure characteristic analysis[3]) is suitable that evaluates the indicated work based on the measured pressure characteristic at each degree crank angle (Fig. 15-21, bottom). With the aid of the utilized fuel mass and a few simple calculations, the advantages and disadvantages of different engine configurations can be compared and optimized.

The flame front proceeding from the spark plug is thin and widens during normal combustion at a rate of approximately 20–25 m/s. Shortening the combustion period increases efficiency by approaching isochoric energy conversion and can be attained by the following:

- Fast flame front speed from greater charge movement (swirl, tumble, or squish)
- Shorter flame paths from a compact combustion chamber design with a central spark plug or several ignition sites
- Higher charge density from a higher compression ratio

Figure 14-15 shows the flame propagation in a 4V engine with a central spark plug in comparison to dual ignition in a 3V engine.[4] In dual ignition, the cylinder charge burns faster because of the shorter combustion paths, and the flame tends to reach the combustion chamber walls. This reduces the tendency of the flame to become extinguished before the cylinder and substantially reduces the amount of uncombusted hydrocarbons in the exhaust. The fast conversion of energy also reduces cyclic fluctuations in SI engine combustion.

Cyclic Fluctuations and the Influence of the Ignition Angle

The fluctuations in the cylinder pressure characteristic from power cycle to power cycle (Fig. 15-22) are typical for SI engine combustion and arise from the fluctuations in the turbulent velocity field and local charge composition that influence the propagation of the flame front and, hence, energy conversion. Figure 15-23 illustrates the substantial influence of the moment of ignition on the maximum cylinder pressure and efficiency-influencing position in reference to TDC.

Influence of the Compression Ratio

By increasing the compression ratio, we can partially compensate the combustion-inhibiting influence of low cylinder pressure under a partial load that arises when setting output by throttling the intake air. Figure 15-24 shows the increase and loss in fuel consumption that arises by changing the compression ratio starting from $\varepsilon = 10$.

Knocking Combustion

As the load increases, limits are set on increasing compression and advancing ignition because of the tendency of uncombusted residual mixture from the cylinder charge to self-ignite. Important properties beyond compression and the moment of ignition are the fuel properties, temperature of the combustion air, combustion chamber shape, component temperatures, and the charge state (composition, flow field). The theory preferred in Ref. [6] concerning the origin of engine knocking is based on the secondary ignition of the uncombusted mixture. The additional progress of engine knocking is determined by the propagation of these secondary reaction fronts triggered by these autoignition sites. Given extremely fast energy conversion, local changes in pressure can arise that generate pressure oscillations of the cylinder charge around

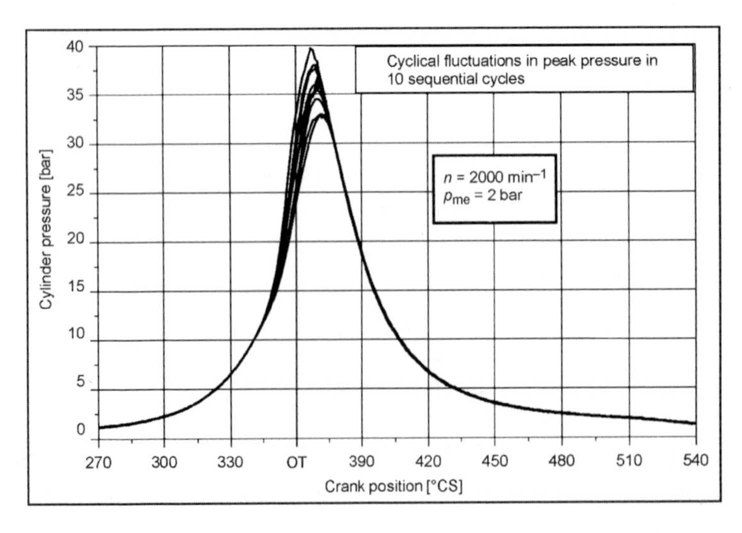

Fig. 15-22 Cyclic cylinder pressure fluctuations.

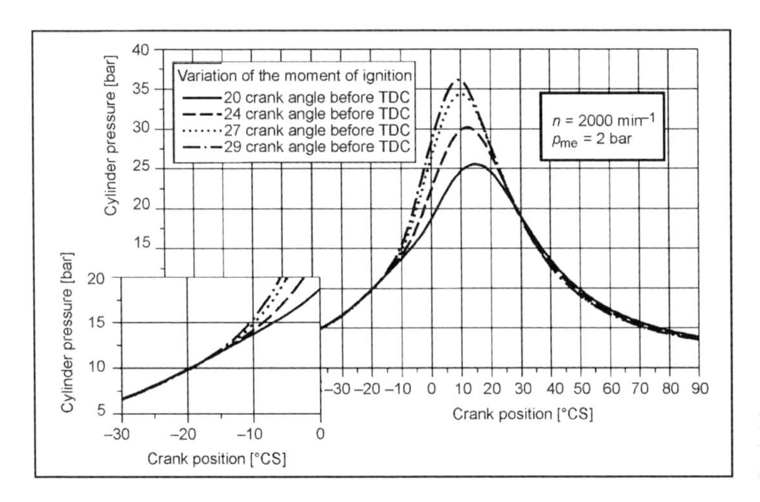

Fig. 15-23 Influence of the ignition angle on the cylinder pressure characteristic.

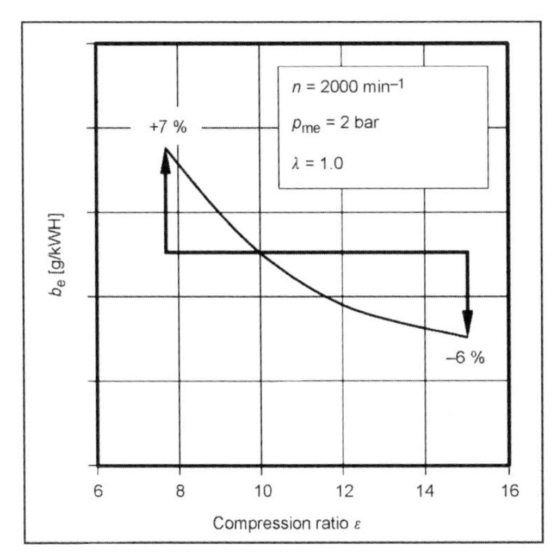

Fig. 15-24 Influence of the compression ratio on partial load consumption.[5]

5–20 kHz and that can be detected in the cylinder pressure signal detected (Fig. 15-25). The high-frequency oscillations die asymptotically and last up to 60° crankshaft angle.

The pressure waves occurring in knocking combustion excite the cylinder charge to form characteristic resonance oscillations that can be calculated by using the general wave equation applied to a hollow cylinder and by using the Bessel function.[7] Figure 15-26 shows a typical calculated resonance oscillation pattern in a hollow cylinder. The resonance oscillations are a function of the cylinder diameter. Figure 15-27 illustrates their influence on the frequency of the most important oscillation modes.

The spontaneous propagation of the reaction fronts is frequently inhomogeneous because of the sequential, apparently random ignition of neighboring mixture components at shock wave propagation speeds of up to 600 m/s that are, hence, close to the velocity of sound of the final gas, and can trigger thermal explosions that can damage the engine. If the final gas is burned by heat conduction and diffusion processes, many isolated, autoignition sites arise distributed over the end gas, and pressure waves are not

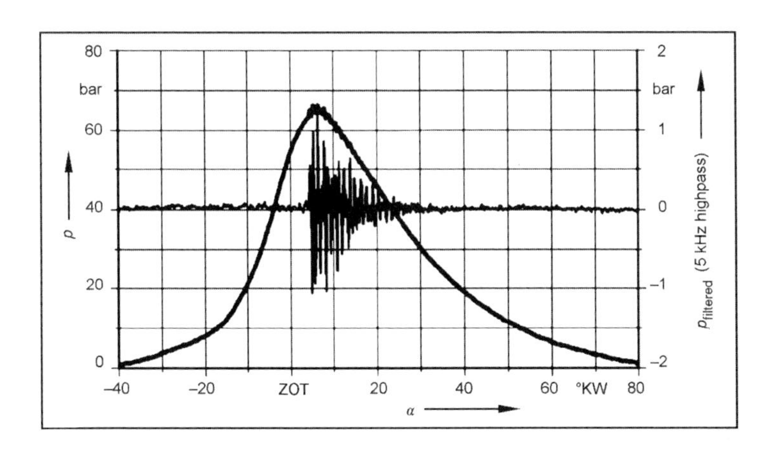

Fig. 15-25 Cylinder pressure characteristic and filtered cylinder pressure during knocking combustion.

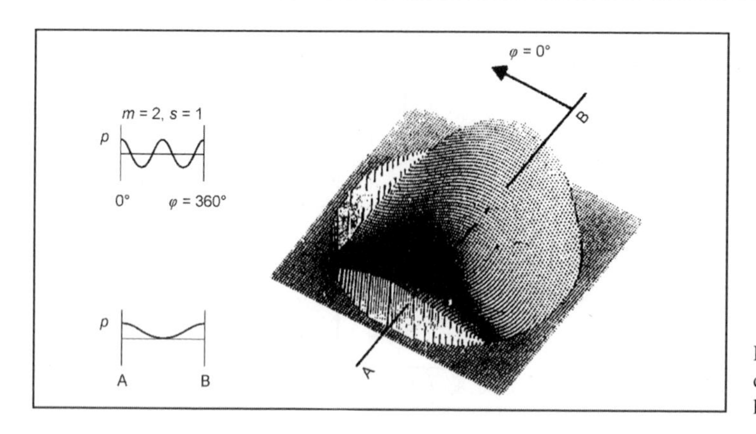

Fig. 15-26 Calculated pressure distribution of a resonance oscillation in a cylinder.

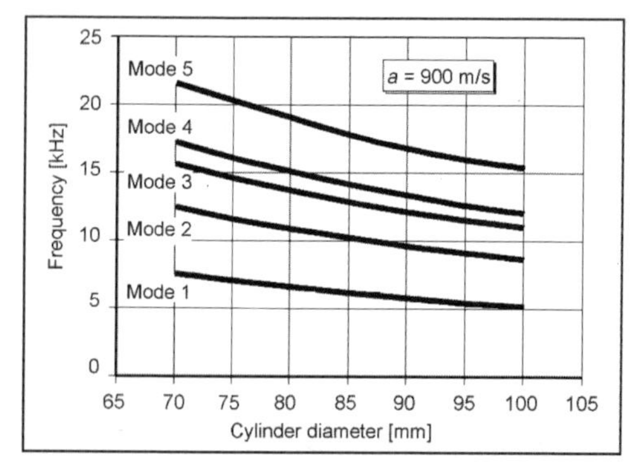

Fig. 15-27 Cylinder pressure resonance frequencies as a function of the cylinder diameter.

generated.[6] Figure 15-28 shows typical flame propagation during knocking combustion.

Flame Speed

The flame speed of normal combustion is the product of the combustion speed and transport speed of the local fresh gas. The combustion speed is strongly influenced by the local charge composition and increases with the charge turbulence in the combustion chamber. The transport speed is influenced by the piston movement, the squish flows, and charge movement triggered by the intake process (swirl, tumble).

Figure 15-29 shows the average flame speed in relation to the air-fuel ratio. The probability that the reaction partners will contact each other is greatest at $\lambda = 0.8$–0.9. The maximum work is accomplished at $\lambda = 0.8$–0.9 because of the fast combustion. The flame speed slows greatly with richer or leaner mixtures and must be corrected by advancing the ignition angle. Lean mixtures lower the energy remaining in the exhaust because of their low thermal capacity and the lower final combustion temperature resulting from the dilution of the charge.[9] The efficiency of the combustion engine therefore increases when the charge is diluted. Since the flame speed and charge dilution have a contradictory influence on the efficiency of real combustion, the optimum efficiency arises

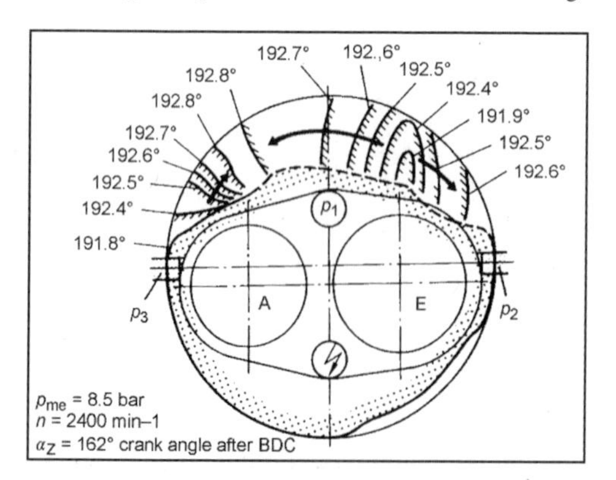

Fig. 15-28 Flame propagation during knocking combustion (light-guide measuring technique).[8]

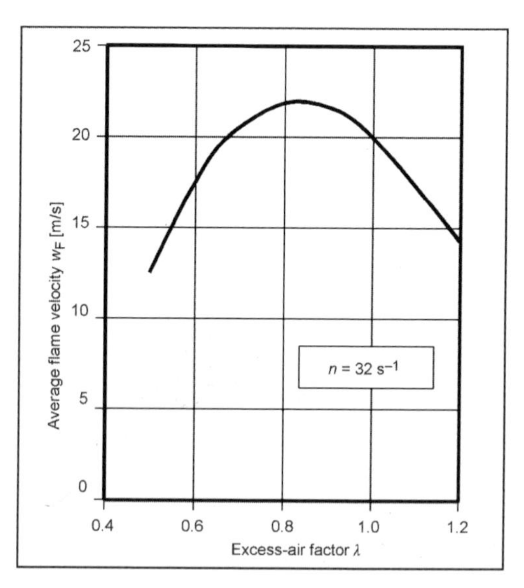

Fig. 15-29 Average flame speed in relation to the excess air factor.

at $\lambda = 1.1-1.3$ in conventional PFI spark-injection engines with a homogeneous mixture distribution.

Cylinder Charge Dilution

The charge can be diluted with environmental air or recycled exhaust. By increasing the air-fuel ratio or the share of residual exhaust gas, the share of components not participating in the chemical reaction is increased. To more easily compare the influence of exhaust gas recirculation and overstoichiometric charging, these inert components can be summarized as the parameter "inert gas component" or IG[10]:

$$m_{IG} = m_{N_2} + m_{RG} + m_{O_2,(\lambda>1)} + m_{H_2O,L} \quad (15.1)$$

$$IG = \frac{m_{IG}}{m_B \cdot L_{St}} \quad (15.2)$$

Figure 15-30 shows the influence of charge dilution on the combustion speed and the indicated thermal efficiency in the high-pressure phase, and on the overall process under a partial load.[11] The axes were scaled for the share of residual exhaust and excess air factor assuming an "equivalent inert gas component."

Diluting the charge lengthens the ignition phase. The combustion period remains initially constant, and efficiency increases. As charge dilution increases, the cylinder pressure necessary for ignition limits the advance of the ignition angle, which makes the reaction phase longer, cyclic fluctuations increase, and efficiency drop.

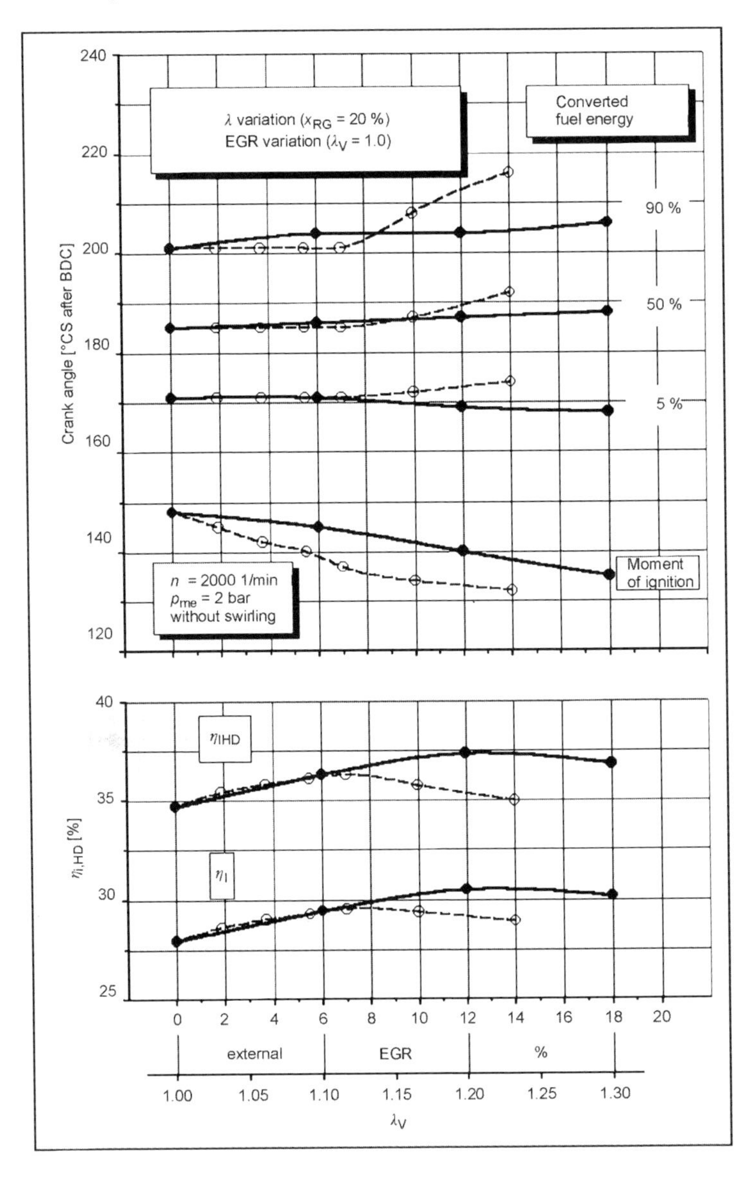

Fig. 15-30 Combustion characteristic and efficiency as a function of charge dilution.[11]

Given an equivalent IG component, the ignition phase is longer when exhaust is added than when air is added; when exhaust is added, the advance of the ignition angle is restricted, and the efficiency tends to worsen. When exhaust gas is recirculated, the oxygen partial pressure drops, which slows flame propagation.

With the recirculation of external exhaust gas, the indicated thermal efficiency η_i does not fall as much as the indicated high-pressure efficiency $\eta_{i,\,HD}$ while dilution increases. The reason for this is the thermal unchoking resulting from the high intake air temperature in the intake manifold, which reduces charge cycle loss. Despite the reduction of the charge cycle loss, combustion mixtures diluted with exhaust gas do not attain the internal efficiency of air-diluted mixtures since they enable greater dilution of the charge.

Exhaust gas recirculation is used to lower fuel consumption in $\lambda = 1$ approaches since this allows three-way catalytic converters to be kept for the treatment of exhaust gas. In addition to external exhaust gas recirculation, the exhaust gas recirculation rate can be internally controlled by means of variable valve timing. Figure 15-31 illustrates how, by varying the exhaust camshaft position by a 40° crankshaft angle, the valve overlap increases when ignition is retarded, and the efficiency changes. Differential pressure characteristic analysis shows that given an equivalent working point, compression work increases as the valve overlap increases. The reason is that the cylinder

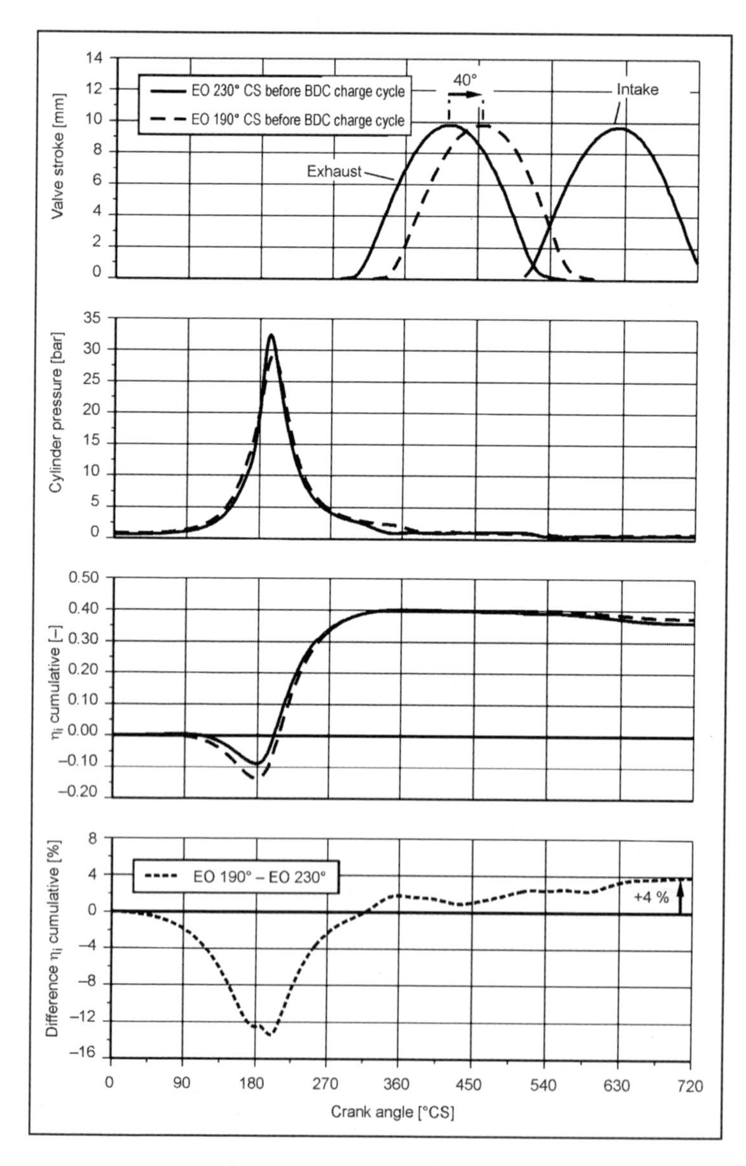

Fig. 15-31 Internal exhaust gas recirculation by varying the exhaust timing, moment of ignition = optimum fuel consumption, pressure characteristic analysis.

charge increases as the exhaust gas recirculation increases. During expansion, a higher share of residual exhaust gas produces a longer combustion period and lower cylinder peak pressures. However, because of the superior charge characteristics and later EO, there is an improvement in efficiency. The unchoking in the intake phase that occurs as the exhaust gas recirculation increases reduces the charge cycle losses and, accordingly, further increases the efficiency of variants with a greater valve overlap by an overall 4%.

In addition to efficiency, the level of untreated emissions is an important factor in evaluating combustion processes. Figure 15-32 shows the fuel consumption/emission tradeoff of 4V engines with intake valve shutoff when the charge dilution is varied at a stationary working point. In contrast to the starting point ($\lambda = 1.0$, no exhaust gas recirculation), increasing the exhaust gas recirculation to 17.5% lowers fuel consumption by 4% and HC + NO_x emissions by 50%. The alternative type of charge dilution using a lean mixture allows a maximum excess air factor of $\lambda = 1.4$. In contrast to the basic variant, this lowers fuel consumption by 9% and HC-NO_x emissions by 40%.

Charge Movement

Lean adjustment is primarily improved by increasing the charge movement. This can be achieved by giving the intake duct a special shape for the inflowing fresh cylinder charge. Swirl ducts or an intake duct shutoff in 4V engines generate a rotating vortex whose axis is parallel to the cylinder axis. Swirling flows remain during intake and compression and dissipate only during expansion. Turbulence ducts generate whirling in the cylinder whose axis is perpendicular to the cylinder axis. Whirling arises from a one-sided inflow through the intake valve as a result of the burbling in the flow in the intake duct. Tumble flows generally last up to the time of compression and decay into microturbulences close to ignition top dead center.

Figure 15-33 shows engine behavior when there is external exhaust gas recirculation of swirl and tumble flows in a 4V engine. The charge swirl was created by shutting off an intake valve. In contrast to the tumble approach, the swirl variation has a much shorter ignition lag in this type of engine. Because of the large charge movement, the flame core can more quickly reach a larger mixture area after the start of ignition and induce a detectable energy conversion. The combustion phase is also faster in the swirl approach than the tumble approach. The faster energy conversion during swirling means that less preignition is required, thereby resulting in more favorable ignition conditions at the moment of ignition. Cyclic fluctuations (σ_{pmi}) are much lower as the exhaust gas recirculation rate increases in the swirl variant. The improved combustion stability and the short combustion period give the swirl variant its advantage in fuel consumption.

Combustion Chamber Shape

The combustion chamber shape influences the following properties in SI engines:

- Inflow of the cylinder charge
- Charge movement in the cylinder
- Speed of energy conversion
- Untreated emission level
- Knocking

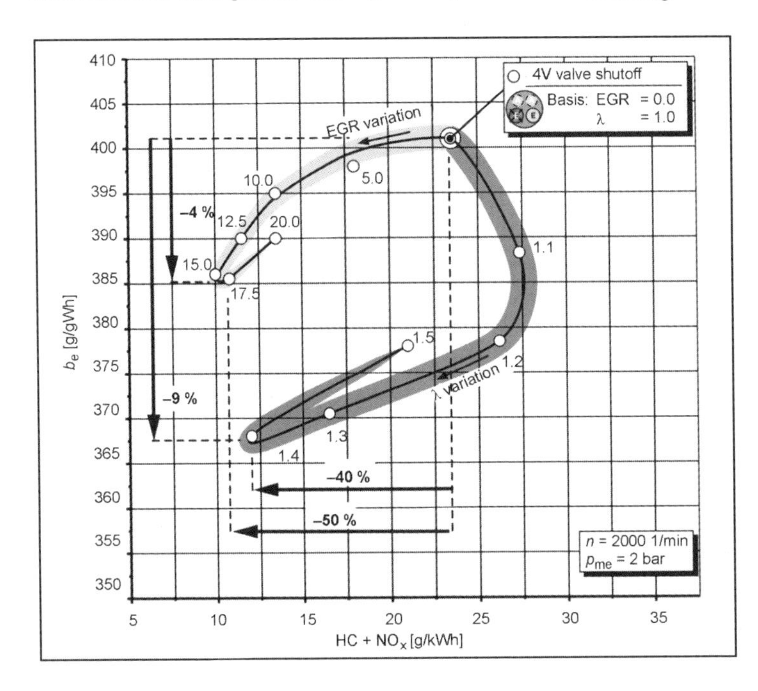

Fig. 15-32 Fuel consumption/emission trade-off from varying the charge dilution.[11]

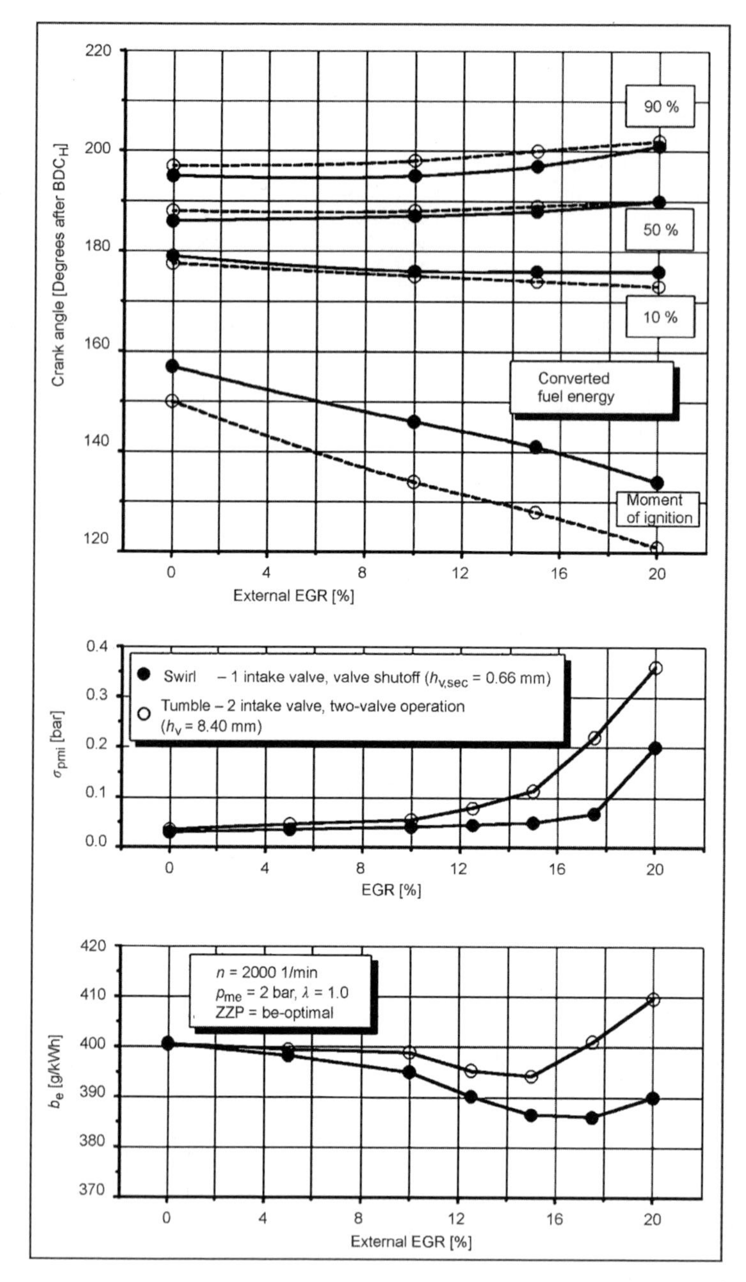

Fig. 15-33 Combustion characteristic of swirl and tumble flows in a 4V engine.[11]

The combustion chamber design must therefore meet the following requirements:

- Unhindered inflow to the valve seat
- High flow speeds for the cylinder charge at the ignition TDC
- Short flame paths arising from a centralized spark plug position and compact combustion chamber geometry
- Minimal dead space (fire land height, valve pockets)
- Avoidance of hot components

These requirements are met by roof-shaped combustion chambers with valves arranged in a V. Because of charging advantages, most current engines are four-valve engines with two intake and two exhaust valves. Figure 15-34 shows an example of a 4V standard combustion chamber.

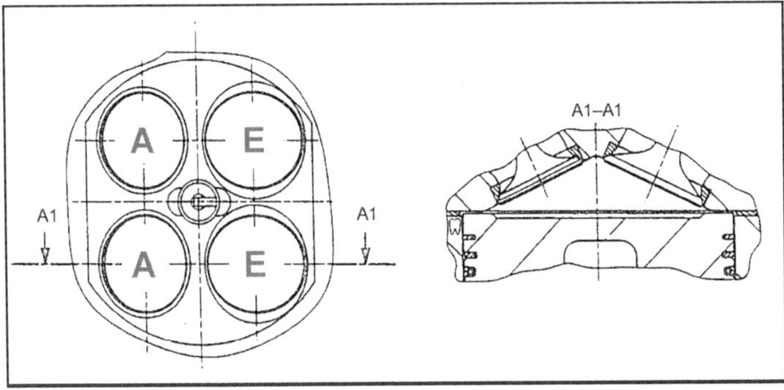

Fig. 15-34 The 4V combustion chamber of a series spark-injection engine.

Because of cost advantages, two-valve engines with parallel suspended valves and a camshaft are also used.

Influence of Load and Loss Analysis

The performance of PFI engines is set by throttling the intake air. The reduced density of the inducted fresh charge reduces the cylinder pressure that, in turn, reduces the flame speed. As shown in Fig. 15-35, this substantially lowers efficiency in the high-pressure phase under low loads. In addition to losses in the high-pressure phase, the efficiency of the engine under a partial load is worsened by the greater charge cycle losses arising from the throttling of the intake air and load-independent engine friction.

Figure 15-36 shows the energy loss distribution for a PFI engine under partial and full loads. Whereas approximately 30% of the used energy is available as useful work under a full load, only approximately 15% can be used under a partial load. At more than 50%, the majority of the used energy is lost as exhaust energy because of the selected process parameters (compression ratio, homogenous cylinder charge with $\lambda = 1$, combustion chamber shape). Under a partial load, loss components further rise to approximately twice that of full load operation because of engine friction, throttling, and too slow combustion.

15.2.2 Combustion Process of Direct Injection Spark Ignition (DISI) Engines

In comparison to a similar spark-injection engine with PFI, we can expect potential fuel consumption of 15%–20% with DISI (Fig. 15-37).

The cited reasons for this are a comparative

- Drastic reduction of throttling losses and reduction of wall heat loss in stratified combustion
- Higher effective and geometric compression ratio enabled by internal mixture cooling and corresponding knock behavior.[12]

A comparison of PFI and DISI engines is shown in Fig. 15-38 showing loss distribution with the speed at 2000 rpm and an effective mean pressure of 2.0 bar.

Decreasing throttling in the DISI engine reduces the charge cycle work by one-third in comparison to the PFI engine. Minimizing the wall heat loss in a DISI engine lowers the release of heat to the coolant water by

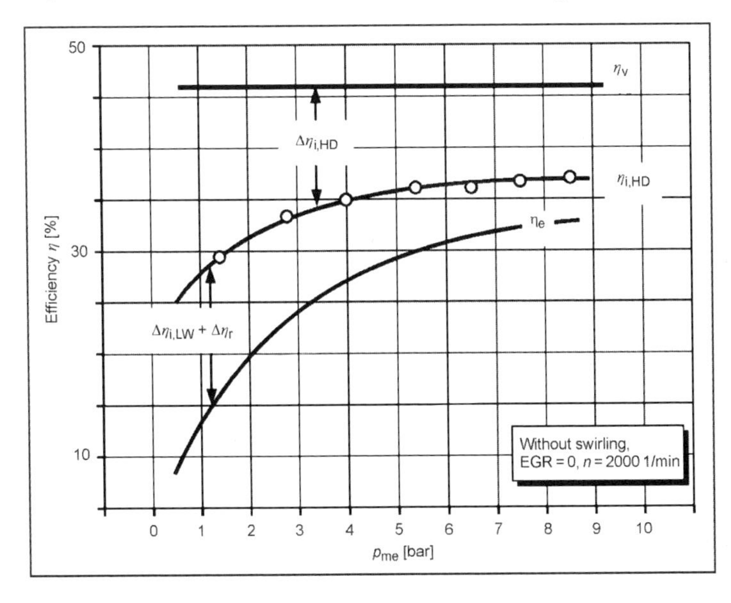

Fig. 15-35 Efficiency of PFI engines as a function of load.[11]

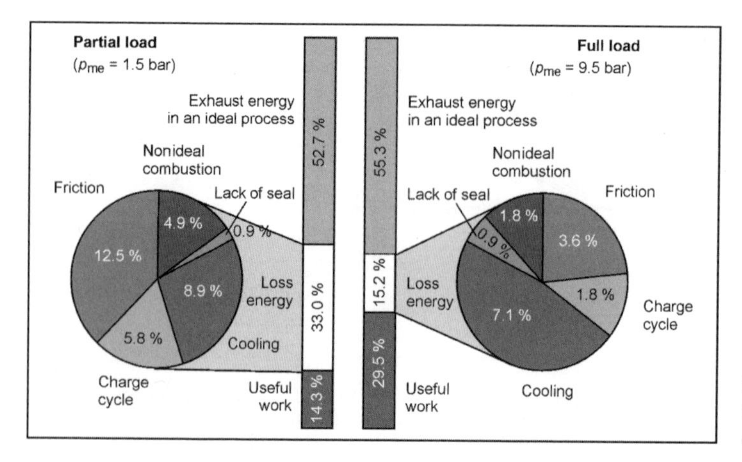

Fig. 15-36 Loss analysis of a four-cylinder PFI engine (1.6 l displacement).[11]

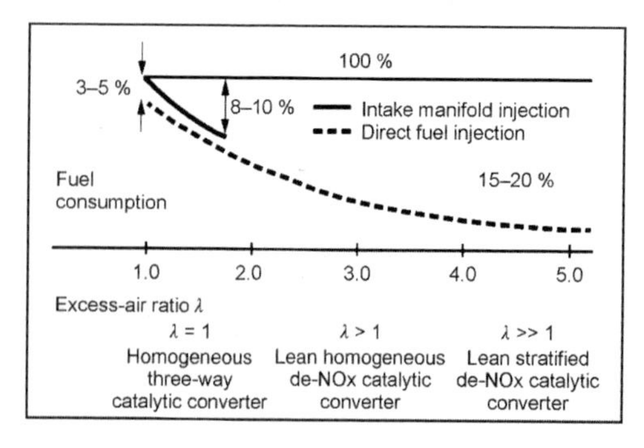

Fig. 15-37 Fuel savings, partial load operation.[12]

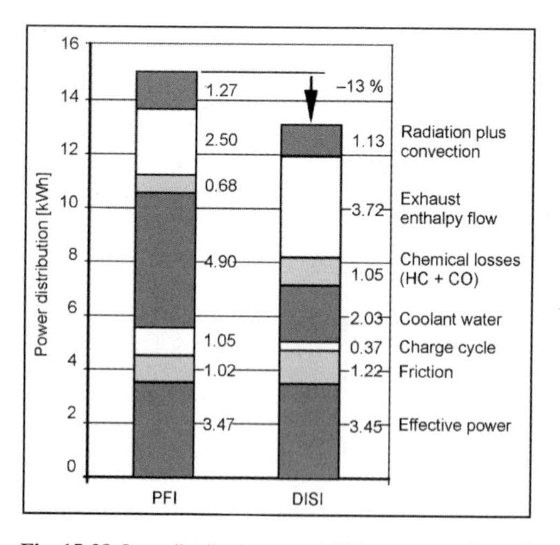

Fig. 15-38 Loss distribution, n = 2.000 rpm, p_{me} = 2 bar.[13]

approximately 60% because of lower process temperatures. The greater exhaust mass flow with slightly more chemical energy than a DISI engine and the greater friction work (piston group, high pressure pump) reduce the cited advantages of the DISI approach; however, an overall consumption advantage of 13% remains.

Types of Operation

In contrast to systems using ported fuel injection, several types of engine operation are possible with direct injection that we identify and qualitatively evaluate in the following.

(a) Stratified Operation

An essential parameter of mixture preparation is injection timing. Specifically, controlling injection during the compression stroke prevents the mixture from being completely mixed in the combustion chamber before the moment of ignition; stratification of the fresh charge results. This makes the mixture leaner than the introduced mixture mass. In certain methods, high air-fuel ratios (λ = 6) can be measured on the test bench.[14]

(b) Homogeneous Operation

Injection during the intake stroke is a characteristic of homogeneous operation. The fuel consumption and leaning as well as the throttling loss are comparable to values of spark-injection engines with port fuel injection. This type of operation is used in the full load range and for lean operation in the partial load range.

The advantages are the knocking behavior under a full load from higher internal cooling of the directly introduced fuel and the exhaust gas treatment at a stoichiometric air-fuel ratio from the use of a conventional catalytic converter system.[15]

(c) Dual Injection

To be understood as dual injection is the distribution of the injection event to two points in time. For example, during relatively lean operation after starting the engine, a high-energy level in the exhaust is possible with acceptable operating behavior. This energy is used to heat the catalytic converter in the exhaust gas treatment system. Another variant of dual injection using two injection events seeks to divide the charge in the compression stroke in stratified operation, which can reduce knocking during a full load.[16]

	Homogeneous operation	Homogeneous lean operation	Stratified operation	Dual injection
Injection timing	Intake stroke	Intake stroke	Compression stroke	Intake, compression, and exhaust cycle
Mixture	Homogeneous	Homogeneous	Stratified	Inhomogeneous
Air-fuel ratio	0.7 to 1.0	1.0 to 1.7	1.7 to 4.0	0.6 to 1.5
Exhaust temperature	High	Medium	Low	Medium–very high
Throttling	High	Medium	Low	Medium

Fig. 15-39 Features of the different types of operation.

Figure 15-39 presents the characteristic features of the different types of operation.

The time required for mixture formation in stratified operation is not available at higher speeds; in this instance, homogeneous operation is used with its early injection timing. Another limitation at higher loads is mixture formation with expanded zones of over-rich ignition. The operation strategy in Fig. 15-40 also includes lean operation with a homogeneous mixture in which the advantages of lean operation can be exploited. To lower the greater untreated NO_x emissions in comparison to the PFI approach and maintain emission thresholds with the exhaust treatment system during overstoichiometric engine operation, most DISI engines must operate with high exhaust recirculation rates that additionally promote fuel evaporation.

Comparison of Homogeneous and Stratified Operation

In addition to measuring the fuel consumption and pollutant emissions, analyzing the cylinder pressure characteristic is an important tool for comparing homogeneous and charge-stratified operation. In the following, we, therefore, consider cylinder pressure characteristics at a speed of n = 2000 rpm and effective mean pressure of p_{me} = 2.0 bar, and the development of the internal efficiency under the same load.

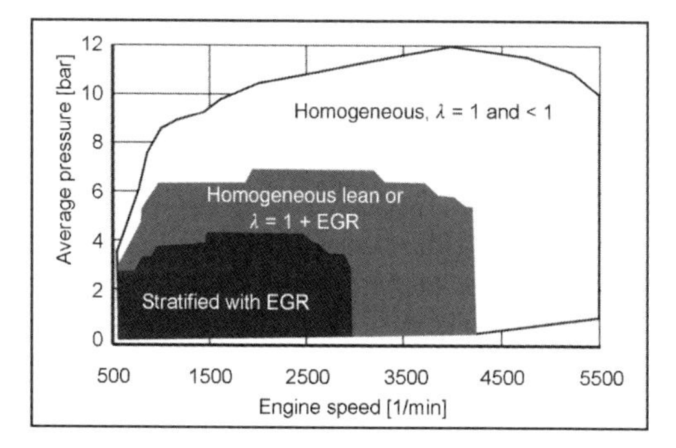

Fig. 15-40 Operation strategy.

Figure 15-41 shows the cylinder pressure characteristics of both types of operation that are typical for partial load operation. In stratified charge operation, the compression and combustion peak pressure are higher since substantially more air mass must be compressed due to the unthrottled operation. In addition, the charge cycle work is much lower in stratified operation.

Figure 15-42 shows the progress of the indicated thermal efficiency η_i. Starting at the charge cycle TDC upon induction, more work has to be expended in homogeneous operation up to the beginning of the compression stroke than in stratified operation (negative internal efficiency) due to throttling. Only upon the beginning of the compression stroke does the piston experience force in the direction of TDC in homogeneous operation because of the greater difference in pressure between the cylinder and the crankcase. The gradient in the indicated thermal efficiency curve is, hence, initially positive. Only after approximately one-half of the piston stroke is the cylinder pressure greater than the environmental pressure as can be seen in Fig. 15-41. We now have force acting against the piston stroke. The gradient of the indicated thermal efficiency curve again becomes negative. Toward the end of the compression stroke, the greater air mass is compressed in stratified operation, and more work is, therefore, expended.

During the power cycle, the curve for stratified operation has the greater positive gradient and attains greater indicated efficiency than homogeneous operation toward the end of the power cycle. The portrayed advantage is partially explainable by the lower wall heat loss and the superior charge properties in stratified operation. However, this positive difference in efficiency cannot be completely maintained up to the end of the emission cycle since in contrast to homogeneous operation a greater exhaust mass must be expelled. The indicated efficiency curve of stratified operation, therefore, has the greater negative gradient.

Combustion Approaches

According to publications, most current developmental approaches that use direct injection are designed for stratified operation. In this type of operation, late injection

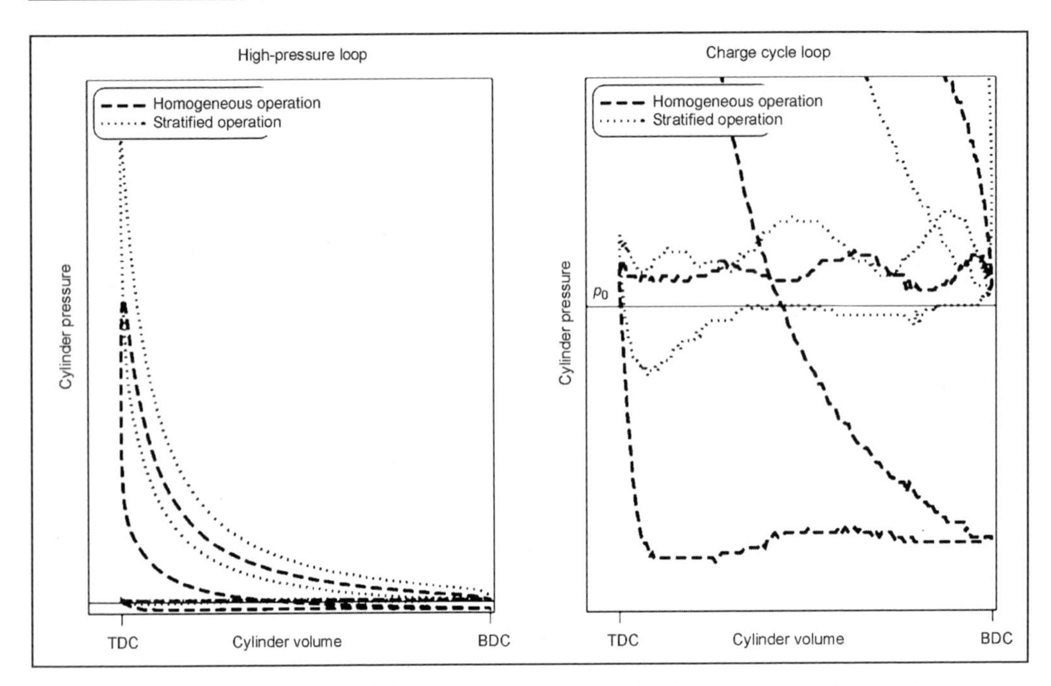

Fig. 15-41 Cylinder pressure characteristics of homogeneous and stratified charge operation in a DISI engine; $n = 2000$ rpm, $p_{me} = 2$ bar.

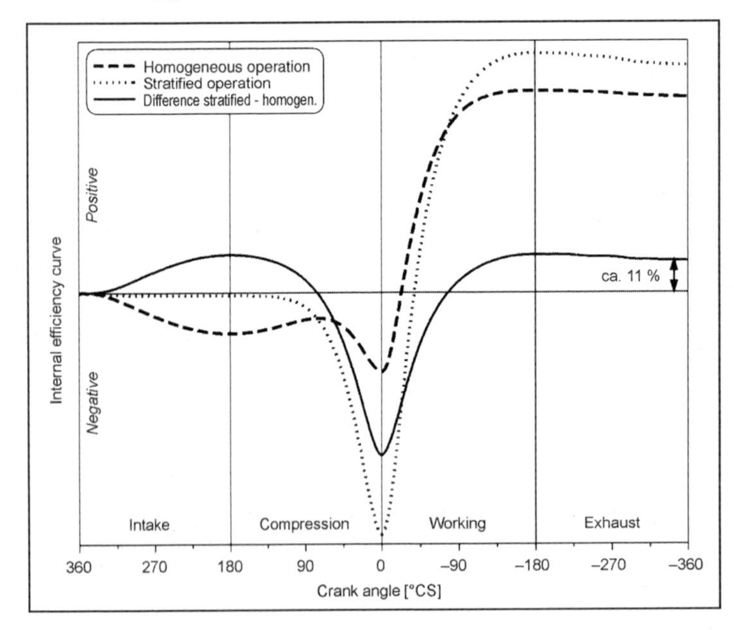

Fig. 15-42 Indicated thermal efficiency for homogeneous and stratified charge operation in a DISI engine; $n = 2000$ rpm, $p_{me} = 2$ bar.

directly into the combustion chamber poses special requirements on the selected combustion process:

- The mixture must be formed relatively quickly; liquid fuel or zones with overrich mixture need to be attenuated by the time of ignition.
- The mixture must be transported to the spark plug in a controlled and reproducible manner.
- A clear stratification of the ignitable mixture should be discernable with the goal of lower wall thermal losses using exhaust or air.
- Zones with a noncombustible lean mixture are to be avoided, as are overly rich zones.
- Avoid directly wetting the piston path and the spark plug with fuel.

The precise understanding of the injection process allows the selected method to be evaluated in reference to the cited requirements. Usually a specific charge movement is required for mixture formation and the transport of the mixture to the spark plug, and the charge movement must be precisely understood during the compression phase in order to design the method.[15]

The methods presently being developed can be classified into three combustion approaches and are characterized as follows (Fig. 12-11):

(a) Air-directed combustion method: The fuel is transported by a charge movement generated from the site of introduction to the spark plug. The combustion chamber walls are not wet when the method is properly applied. The precise timing of injection and a stable charge movement are decisive for the quality of the method. The quality of the mixture formation that is supported by the charge movement is high in corresponding designs.

(b) Wall-directed method: The fuel is guided by a correspondingly shaped combustion chamber wall (the piston in this instance) to the site of ignition. This method is associated with greater fuel deposits on the combustion chamber walls; evaporation before ignition cannot usually eliminate the entire fuel film. Since this method is based on uniform conditions, the process is stable. The higher untreated emissions and the comparatively low potential fuel consumption prevent this method from being widely used.

(c) Jet-directed method: Introducing the fuel directly adjacent to the site of ignition has the highest potential of the compared methods for stratifying the fresh charge. The corresponding benefit in fuel consumption is contrasted with a somewhat unsatisfactory mixture quality at the spark plug at the moment of ignition. Only a part of the fuel jet or spray consists of an ignitable mixture. The preparation of the mixture cannot be substantially supported by charge movement because of the danger of blowing the mixture away from the site of ignition. Since the spark plug experiences a substantially alternating thermal load from being occasionally wet with fuel, continuous operation is not ensured.

The quantities of injection timing, moment of ignition, and combustion period are different in stratified operation without exhaust recirculation in comparison to the process shown in Fig. 15-43.

The different times between the end of injection and moment of ignition are notable. The time is the shortest for the jet-directed method since local ignition starts after the end of injection. The time of mixture formation is contrastingly longest for the wall-directed method since the mixture is guided a relatively long way across the piston surface.

In the following ignition phase (time between ignition and the 5% conversion rate), the relatively poor mixture preparation of the jet-guided method is characterized by a long delay. The good mixture formation of the two other methods yields a similar combustion lag despite different modes of operation (length of mixture formation in the wall-directed method, intense charge movement in the air-directed method).

In the following reaction, the time until an 85% conversion rate is noted, which also takes into account the incomplete conversion of the wall-directed method. The unprepared fuel on the piston surface restricts the maximum conversion to 88%. The air-directed method has the shortest combustion period because of the favorable mixture quality and the intensity of the charge movement. This relatively fast conversion continues until the end of combustion. The two other methods have a longer combustion time because of the poor local mixture quality and the lesser charge movement.

An advantage of the wall-directed method is the relatively late and efficient location of combustion.

Target Quantities and Parameters

A basic motivation for developing DISI engines is the potential savings in fuel consumption. This potential is frequently reduced in the overall vehicle by requirements such as

- Exhaust gas treatment under current emission standards
- Thermal behavior

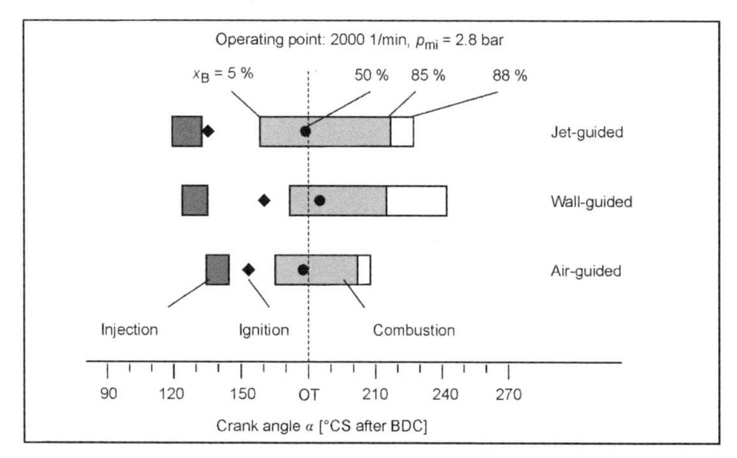

Fig. 15-43 Process characteristics of different DISI combustion methods at the working point of n = 2000 rpm, p_{mi} = 2.8 bar.[17]

- Diagnosis
- System costs
- Long-term stability

In evaluating a combustion process, however, these parameters should be taken into consideration since they may directly influence the combustion process. As an example, take NO_x emissions behavior in an active exhaust gas treatment system: because of the accumulation of untreated NO_x emissions in the catalytic converter storage system, a regeneration process is necessary. This occurs during rich engine operation. The fuel consumption behavior is influenced by the frequency of this process. A reduction of the NO_x emissions can, hence, indirectly influence the fuel consumption. As a means to lower untreated NO_x emissions, exhaust gas recirculation is frequently used. This can occur both externally and internally.[18]

Another task must be accomplished in a direct injection approach to ensure the long-term stability of the vehicle propulsion system that the customer takes as a given. Since no fuel film is introduced into the intake manifold, deposits that can arise in exhaust recirculation systems or from crankcase ventilation cannot be decreased from corresponding fuel additives.

The formation of deposits on the intake valves can cause the engine to stall. By selecting the right materials for the intake valves and reducing the substances that form deposits, this behavior can be clearly reduced.[16]

The goals of combustion process development are guided by several parameters, some of which have been discussed above. Some examples of essential developmental tasks are[5]

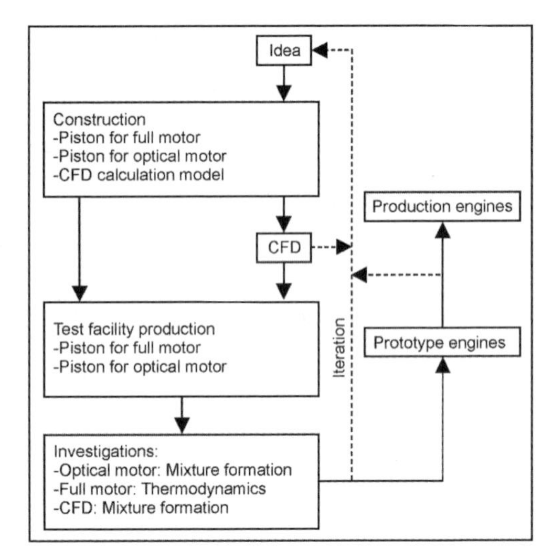

Fig. 15-44 Developmental tools.[19]

- Generating charge movement in air-directed systems
- Robust injection and spray formation, especially in jet-directed systems
- Wetting with fuel in wall-directed systems
- Dealing with the demands of exhaust gas treatment in stratified and lean operation

Measuring Technique	Used for	Dim.	Comments
Video stroboscopy	Injection (liquid phase)	2-D	+ Statistics easy to obtain; + Combinable with endoscopes; – Max. 1 picture per cycle
High-speed film	Injection (liquid phase)	2-D	+ Recording of overall cycle; – Picture processing, statistics
LIF	Injection (liquid and vapor phase)	2-D	+ Both phases may be visible – Difficult to quantify – Max. 1–2 analyses/cycle
Raman	Fuel concentration	0-D (1-D)	+ Quantitative – Low signals, distortable – Max. 1–2 analyses/cycle
LDA	Flow	0-D	+ Precise, statistics easy to obtain + (Record of overall cycle) + Established and well-developed – Linear information problematic
PIV	Flow (possibly simultaneous injection)	2-D	+ 2-D flow information + Statistics easy to obtain + Established and well-developed – Max. 1 analysis/cycle

Fig. 15-45 Overview of optical investigation methods.[19]

- Exhaust gas recirculation
- Stability of the combustion process/cyclic fluctuations
- Control unit functions
- Cold starts

Tools and Methods

Tightly coordinated, familiar developmental steps and tools are used to resolve the complex questions involved in developing approaches for direct injection. There exists a broad range of variations and possibilities available for the short time normally available for the development of spark-injection engines, especially in optimizing the combustion method. To restrict the amount of experimentation to reasonable limits, statistical models or CFD calculations must be used (Fig. 15-44). Other means for estimating the scope of experimentation are optical investigations. An exemplary overview of a few optical investigative methods is listed in Fig. 15-45.

To speedily accomplish these investigations, developers must deal with organizational problems in addition to technical demands. Specific tools are, therefore, used in the individual CFD and optical investigations that limit the normally time-consuming research to essential conclusions and substantially reduce the required time. The results of these investigations largely serve as a comparison with CFD calculations.

The tools used in the subsequent developmental process for harmonizing engine control unit (ECU) functions are also model supported. To work efficiently, the torque-based functional structure of the ECU requires the use of statistical experimental design [design of experiments (DOE)].[20]

Most models are not precise enough regarding stochastic behavior such as cyclic fluctuations in the combustion characteristic. The entire engine must, therefore, be represented in the testing of thermodynamic variables when developing combustion processes.

The actual fuel consumption potential of SI engines that use direct injection, therefore, can be realized only by closely linking the development of the combustion process, exhaust gas treatment, and operation strategy together with comprehensive models.[14,15]

Bibliography

[1] Merker, G.P., and G. Stiesch, Technische Verbrennung, Motorische Verbrennung, Teubner-Verlag, Stuttgart/Leipzig, 1999.

[2] Pischinger, S., Verbrennungsmotoren, Vorlesungsumdruck, RWTH Aachen, 19th edition, 1998.

[3] Wurms, R., Differenzierte Druckverlaufs-Analyse–eine einfache, aber höchst wirkungsvolle Methode zur Interpretation von Zylinderdruckverläufen, 3rd International Indiziersymposium, 1998.

[4] Niefer, H., H.K. Weining, M. Bargende, and A. Waltner, Verbrennung, Ladungswechsel und Abgasreinigung der neuen Mercedes-Benz V-Motoren mit Dreiventiltechnik und Doppelzündung, in MTZ (1997), pp. 392–399.

[5] Pischinger, F., and P. Wolters, Ottomotoren–part 2, H.-H. Braess and U. Seifert [ed.], Vieweg Handbuch Kraftfahrzeugtechnik, Friedrich Vieweg & Sohn Verlagsgesellschaft, Wiesbaden, 2000.

[6] Stiebels, B., Flammenausbreitung bei klopfender Verbrennung, Fortschr.-Ber. VDI Reihe 12 Nr. 311, VDI Verlag, Düsseldorf, 1997.

[7] Adolph, N., Messung des Klopfens an Ottomotoren, Dissertation, RWTH Aachen, 1983.

[8] Kollmeier, H.-P., Untersuchungen über die Flammenausbreitung bei klopfender Verbrennung, Dissertation, RWTH Aachen, 1987.

[9] Pischinger, R., G. Kraßing, G. Taucar, and T. Sams, Thermodynamik der Verbrennungsmaschine, Die Verbrennungskraftmaschine, New edition, Vol. 4, p. 99, Vienna: Springer-Verlag, Vienna, New York, 1989.

[10] Südhaus, N., Möglichkeiten und Grenzen der Inertgassteuerung für Ottomotoren mit variablen Ventilsteuerzeiten, Dissertation, RWTH Aachen, 1988.

[11] Fischer, M., Die Zukunft des Ottomotors als Pkw-Antrieb–Entwicklungschancen unter Verbrauchaspekten, Dissertation, TU Berlin, 1998, Schriftenreihe B–Fahrzeugtechnik–des Institut für Straßen- und Schienenverkehr.

[12] Eichlseder, H., W. Hübner, S. Rubbert, and M. Sallmann, Beurteilungskriterien für ottomotorische DI-Verbrennungskonzepte, Spicher, U.u.A.: Direkteinspritzung im Ottomotor, Expert Verlag, Renningen, 1998.

[13] Österreichischer Verein für Kraftfahrzeugtechnik u. Techn. Universität Wien: Krebs, R., and J. Theobald, Die Thermodynamik der FSI-Motoren von Volkswagen, 22., Internationales Wiener Motorensymposium, 2001.

[14] Österreichischer Verein für Kraftfahrzeugtechnik u. Techn. Universität Wien: Karl, Abthoff, Bargende, Kemmler, Kühn, Bubeck, Thermodynamische Analyse eines DI-Ottomotors, 17., Internationales Wiener Motorensymposium, 1996.

[15] Österreichischer Verein für Kraftfahrzeugtechnik u. Techn. Universität Wien: Zhang, H., K. Bayerle, G. Haft, D. Klawatsch, G. Entzmann, and H.P. Lenz, Doppeleinspritzung am Otto-DI-Motor: Anwendungsmöglichkeiten und deren Potenzial, 22., Internationales Wiener Motorensymposium, 2001.

[16] Haus der Technik: Beermann, H., B. Hanula, H. Hoff, A. Krause-Heringer, C. Glahn, and S. Limbach, Untersuchung von Ablagerungen an Komponenten von BDE-Ottomotoren. Leipertz, A.: Motorische Verbrennung, HdT Conference, March 2001.

[17] Grigo, M., Gemischbildungsstrategien und Potenzial direkteinspritzender Ottomotoren im Schichtbetrieb, Dissertation, RWTH Aachen, 1999.

[18] Dahle, U., S. Brandt, and A. Velji, Abgasnachbehandlungskonzepte für magerbetriebene Ottomotoren, Spicher, U.u.A.: Direkteinspritzung im Ottomotor, Expert Verlag, Renningen, 1998.

[19] Haus der Technik: Stiebels, B., R. Krebs, and M. Zillmer, Werkzeuge für die Entwicklung des FSI-Motors von Volkswagen, Leipertz, A.: Motorische Verbrennung, HdT Conference, March 2001.

[20] Fischer, M., and K. Röpke, Effiziente Applikation von Motorsteuerungsfunktionen für Ottomotoren, in MTZ (2000), et seq.

15.3 Two-Stroke Diesel Engines

Despite a certain popularity of two-stroke diesel engines in the 1950s and 1960s as small stationary engines, tractor engines (Lanz, Hanomag, F&S, ILO, Stihl, O & K, Hirth), as well as in commercial vehicles (Krupp, Ford) (see also Refs. [1] and [2]), two-stroke diesel engines are presently not used to power passenger cars or commercial vehicles. The reasons for the irrelevance of the two-stroke engine in this market segment are the increased requirements concerning engine life, lubricating oil consumption, and emissions that cannot be sufficiently met with conventional, simple engines (crankcase scavenging pump, symmetrical timing diagram, limitation to three moving parts). Other reasons are the limited developmental status of scavenging blowers that are an alternative to the crankcase scavenging pump, as well as problems with cooling, lubrication, and materials. The increasing use of exhaust turbocharging in four-stroke diesel engines also lessens the performance advantage of two-stroke diesel engines. The advantages of the two-stroke diesel engine for lower excitation of drive train oscillations in engines with few cylinders, the torque characteristic, the weight-to-power ratio, cold-start behavior, engine heating after a cold start, untreated NO_x emissions, and exhaust gas treatment conditions make the two-stroke diesel engine especially interesting for one- to three-cylinder engines in low-consumption passenger cars. In the 1990s, this led to related development projects by companies including Toyota,[3] AVL,[4] Yamaha,[5] and Daihatsu.[6]

In two-stroke diesel engines, the choice of the combustion process is strongly influenced by the scavenging approach. In two-stroke diesel engines with uniflow scavenging using intake ports and exhaust valves, it is fairly easy to generate a swirling flow in the cylinders by the design of the scavenging ports and intake ports. The swirl can be influenced as a function of the engine load and speed by placing valves in front of the scavenging ports. For this reason, similar mixture formation conditions can be generated, and comparable combustion processes can be used in comparison with the direct injection four-stroke diesel engines that are primarily used today. Section 7.24 shows a uniflow scavenging three-cylinder two-stroke engine by AVL for use in passenger cars (see also Ref. [4]). When cam-actuated injection pumps are used (distributor injection pump, pump nozzle), the injection frequency that is twice that of four-stroke engines must be taken into consideration when designing the pump and cam. In particular, the use of common rail injection allows the engine to be operated within certain ranges of the program map (such as low load at high speed) with four strokes. Independent of the selected injection system, effective cooling of the nozzle holder or injection nozzle must be provided when designing the cylinder head.

In two-stroke engines with loop scavenging (head loop scavenging or piston-controlled loop scavenging, e.g., according to Schnürle), a swirling flow does not form in the combustion chamber around TDC but rather a more-or-less pronounced tumble flow. For this reason, loop-scavenged two-stroke diesel engines in older vehicles used the indirect fuel injection method almost without exception (prechamber, whirl chamber). In small stationary engines with loop scavenging (F & S, ILO[1]), direct injection was contrastingly used, sometimes with a radial arrangement of the nozzle holder assembly. Modern direct fuel injection systems with injection pressures up to 2000 bar and a greater number of injection orifices need only comparatively low-swirling combustion air to provide satisfactory mixture formation so that a DI combustion method remains a possibility for loop-scavenged two-stroke diesel engines, in certain circumstances with a slightly radial squish flow at TDC.

Figure 15-46 shows an example of a diesel engine with loop scavenging with a lengthwise section and cross section of a two-cylinder two-stroke diesel engine with 1.0 liter displacement by Yamaha[5] that was intended for use in small cars. The stroke is 93 mm, and the bore is 82 mm. At 33 kW, the rated output is 4000 rpm. The maximum torque of 80 Nm is attained at 2500 rpm. The engine with an overall weight of 95 kg is designed for use in 3 liter vehicles and should meet the future Euro 4 thresholds. The cylinder crankcase made of an aluminum alloy with a Ni-P-SiC–coated cylinder barrel has four transfer passages per cylinder through which the fresh gas from the crankcase reaches the cylinder. The cylinder barrel and the crankshafts with roller bearings or connecting-rod bearing are intentionally provided with lubricating oil via map-controlled total loss lubrication to enable minimum consumption of lubricating oil. The outlet of the loop-scavenged cylinder is provided with two superposed exhaust ports. To improve the torque characteristic, the top exhaust port can be closed by means of a throttle valve, while the ratio in engine operation varies between 13:1 and 18:1. Apparently given the difficulties in generating a characteristic combustion chamber swirl for DI combustion, a chamber combustion method was used. In this type of combustion process (Fig. 15-47), strong swirling is generated via four tangential blow ports after the start of combustion in the chamber with low overthrust losses in the cylinder. According to Yamaha,[5] this allows complete combustion to occur with low fuel consumption and emissions.

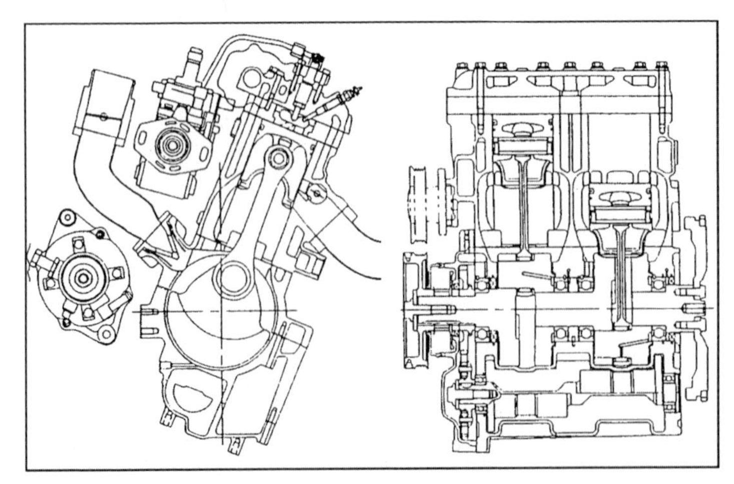

Fig. 15-46 Lengthwise and cross-sectional views of the 1.0 liter two-stroke diesel engine by Yamaha.[5]

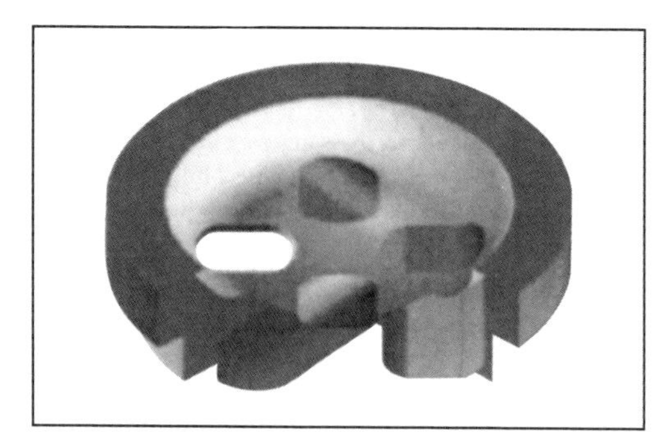

Fig. 15-47 Representation of a whirl chamber of the 1.0 liter two-stroke diesel engine by Yamaha.[5]

15.4 Two-Stroke SI Engines

In contrast to two-stroke diesel engines, the two-stroke SI engine for passenger cars enjoys a long tradition. In particular, the positive experiences in the development and production of two-stroke motorcycle engines formed the basis in the 1920s for the market introduction of passenger cars with two-stroke SI engines by the companies DKW, Aero, Jawa, and Ceskoslovensko Zbrojovka. The large demand for reasonably priced vehicles after the mass acquisition of cars following World War II provided the foundation, especially in Germany, for the development and production of numerous passenger cars with two-stroke SI engines. In addition to DKW, two-stroke passenger cars were produced by Lloyd, Goliath, Gutbrod, Glas, and others. The market share of two-stroke passenger cars by the end of the 1950s in West Germany was approximately 20%. In East Germany up to the cessation of production at the beginning of the 1990s, the brands Wartburg and Trabant achieved a maximum market share for two-stroke passenger cars of over 60%. Against the background of increasing customer and public sensitization to directly perceptible hydrocarbon emissions (blue smoke), rough idling, service life problems, and comparatively high fuel consumption under a full load, DKW in Ingolstadt stopped producing passenger car two-stroke engines in 1996 (Saab of Sweden stopped in 1968). Practically simultaneous to the cessation of production of passenger cars with two-stroke engines in Wartburg and Sachsenring at the beginning of the 1990s, publications and presentations by Orbital,[7,8] AVL,[9] Subaru,[10] Toyota, GM, and Ficht, etc. (see also Refs. [11] and [12]), reawakened interest in two-stroke SI engines. According to these publications, improving mixture formation (direct injection) and using alternative scavenging methods could overcome the specific disadvantages of traditional passenger car two-stroke engines and produce low emission and consumption in engines especially for small passenger cars.

The essential characteristic of the two-stroke method is that the engine undergoes one complete power cycle per rotation in contrast to the four-stroke method. The combusted charge is removed and the fresh gas is introduced (scavenging process) into the cylinder at the same time within a crankshaft angle range around BDC. Since the gas volume communicates with the atmosphere via the open exhaust organs, compression always begins after the intake and exhaust organs close (apart from gas-dynamic influences and supercharging and boosting effects—even when the intake air is throttled under a partial load) at a cylinder pressure that approximately corresponds to the atmospheric pressure. In contrast to throttle-controlled four-stroke SI engines, comparatively high final compressions also occur under a partial load. As shown in Section 10.3, there are various scavenging methods available for the charge cycle of two-stroke engines that are associated with respective advantages and disadvantages. Because of the simple and compact design and the demand for comparatively high nominal speeds, two-stroke SI engines were developed for passenger car use almost exclusively with loop scavenging and a crankcase scavenging pump. In contrast to throttle-controlled four-stroke SI engines, the charge cycle work falls in conventional two-stroke engines with a crankcase scavenging pump when approaching a partial load (see also Ref. [13]). This principle of load control yields a high exhaust component in the cylinder under a partial load due to the "open" gas exchange; during a charge cycle, only the amount of exhaust is expelled from the cylinder that corresponds to the fresh gas entering the cylinder as determined by the degree of intake port throttling. A high exhaust component in the cylinder lowers NO_x emissions and improves the physical preparation of the fuel because of the increased temperature under a partial load. On the other hand, the high inert gas component under a partial load and especially during idling drastically worsens ignition conditions. A large amount of residual exhaust gas in connection with a high final compression pressure under partial load gives rise to the demand for an ignition system with high ignition energy. If under these conditions the scavenging process does not place an ignitable mixture near the spark plug, misfiring occurs. In the following scavenging process, more of the air-fuel mixture is scavenged in the cylinder that improves ignition conditions. If ignition occurs after one or more scavenging processes, the subsequent combustion, as a result of the prior reactions in the mixture during preceding compression cycles, is characterized by high-energy conversion rates, pressure gradients, and peak pressures. This behavior of mixture-purged two-stroke SI engines leads to inferior tractability under a partial load and especially during idling. Furthermore, the expulsion of uncombusted mixture components increases fuel consumption and hydrocarbon emissions. Because of the influence of gas fluctuations in the intake system, especially in two-stroke engines with a crankcase scavenging pump, the cylinder charge changes with the rpm, and so does the mixture composition, particularly when mixture formation is external (in a carburetor). In addition to the residual exhaust gas, there are other influences on ignition, tractability, and emissions. As the load rises, the increasing fresh gas component in the cylinder produces smoother engine running. Experience shows that the fuel consumption is comparatively good in mixture-scavenged two-stroke engines under an average partial load at average speeds. As a full load is approached, the increasing amount of mixture scavenged in the cylinder depending on the gas-dynamic design of the intake and exhaust systems causes a more-or-less marked increase in the loss of fresh gas and, hence, an increase in fuel consumption and HC emissions.

According to the present state of technology, a prerequisite for maintaining strict current and future exhaust pollutant thresholds in passenger car four-stroke SI engines (at least in some areas of the program map) is the oxidation of uncombusted hydrocarbons (HC) and carbon monoxide (CO) as well as the simultaneous reduction of nitrogen oxides (NO_x) in three-way catalytic converters with a stoichiometric air-fuel ratio ($\lambda = 1$ control). The basic condition for the three-way catalytic converter to function in a two-stroke SI engine is for the uncombusted mixture leaving the cylinder during the charge cycle to have the same (stoichiometric) air-fuel ratio as the fresh charge. This state is obtainable in theory in two-stroke SI engines with external mixture formation. However, the described misfiring under a partial load and the related serious time-related changes in the exhaust composition produce regulatory difficulties in maintaining a narrow λ "window."

In the internal mixture formation process (direct injection), the cylinders are scavenged with air. Depending on the quality of the scavenging method, the mean pressure (load), and the amount of scavenging air that may be used to cool the cylinder, the oxygen directly introduced into the exhaust must be compensated by enriching the mixture remaining in the cylinder (increase of the injected quantity) for operation at $\lambda = 1$. A rise in the oxidizable exhaust components (HC, CO) from the enrichment of the mixture in the cylinder is undesirable from the perspective of fuel consumption, the thermal load on the catalytic converter, and the limited reduction of pollutants from the conversion rates of the catalytic converter.

In contrast to the use in passenger cars, the low weight, the small required area, the mechanical robustness, and the low-maintenance operation of two-stroke SI engines have led to their dominant position as outboard engines, jet ski and snowmobile engines, small two-wheeler engines, and power tool engines. The technical and environmental demands that are increasing in this market segment as well have led to the development and introduction to the market of numerous technical improvements, some of which have drastically reduced fuel consumption and/or pollutant emissions. This includes the fresh air supply in small stationary engines, the introduction of oxidation catalytic converters in connection with the optimization of loop scavenging, and a lean mixture adjustment in mopeds and scooters, the use of secondary air systems in small two-wheelers, and the mass production of electronic direct injection for outboard engines and two-wheelers. Above all, the strict pollution thresholds in the most important markets for passenger cars and the high demands on comfort and service life as presented in Ref. [14] that have not been sufficiently met in at least some new approaches represent a serious hurdle for the use of two-stroke SI engines in passenger cars. To successfully market two-stroke SI engines in automobiles, the following central design features must be taken into consideration along with the related developmental tasks (see also Ref. [15]):

- Use or optimization of a scavenging method that offers a reliably combustible mixture to the spark plug with minimal fresh gas losses, even under a partial load.
- Transition from external mixture formation (carburetor/port fuel injection) to a direct fuel injection system that ensures good mixture preparation and the presentation of a reliably combustible mixture to the spark plug even at low, partial-load operating points within the short time available for mixture formation. Air-supported direct fuel injection systems offer good mixture formation within the short time available, but they need to be optimized for the high system cost and high power input.
- Use and optimization of ignition devices and ignition methods that reliably, stably, and consistently ignite mixtures that are difficult to ignite under a partial load under real vehicle operating conditions.
- Develop and use scavenging and supercharging blowers that permit freely selectable cylinder scavenging and supercharging over the entire program map of the engine with minimal power input. Electrically supported turbochargers, perhaps with variable turbine geometry, offer the option of using a part of the otherwise unused exhaust energy and simultaneously compensating for the disadvantages of symmetrical timing diagrams (loop scavenging) because of the collection of exhaust in front of the turbine.
- Not using the crankcase as a scavenging pump makes it possible to use the crankshaft mounted on a plane bearing, has a better service life, cost, and acoustics, and effectively cools the strongly heated piston, perhaps with a cooling channel, by means of a forced-feed lubrication system and oil injection nozzles. The cylinder/piston stroke combination and the piston ring assembly are essential factors in optimizing the minimum required lubricating oil and maximum oil scraping effect of the piston rings and sufficient mechanical and thermal strength of these valve train components.
- Based on the technological knowledge of exhaust treatment in DI four-stroke SI engines, exhaust treatment systems need to be adapted to the requirements for two-stroke SI engines to allow fulfillment of the strict future pollutant thresholds, even when not operating the engine at a stoichiometric air-fuel ratio.

Figure 15-48 provides an exemplary view of a loop-scavenged three-cylinder two-stroke engine by Orbital. The engine has a stroke of 72 mm and a bore of 84 mm. At 58 kW, the rated output is 4500 rpm. The maximum torque of 130 Nm is attained at 3500 rpm. The overall

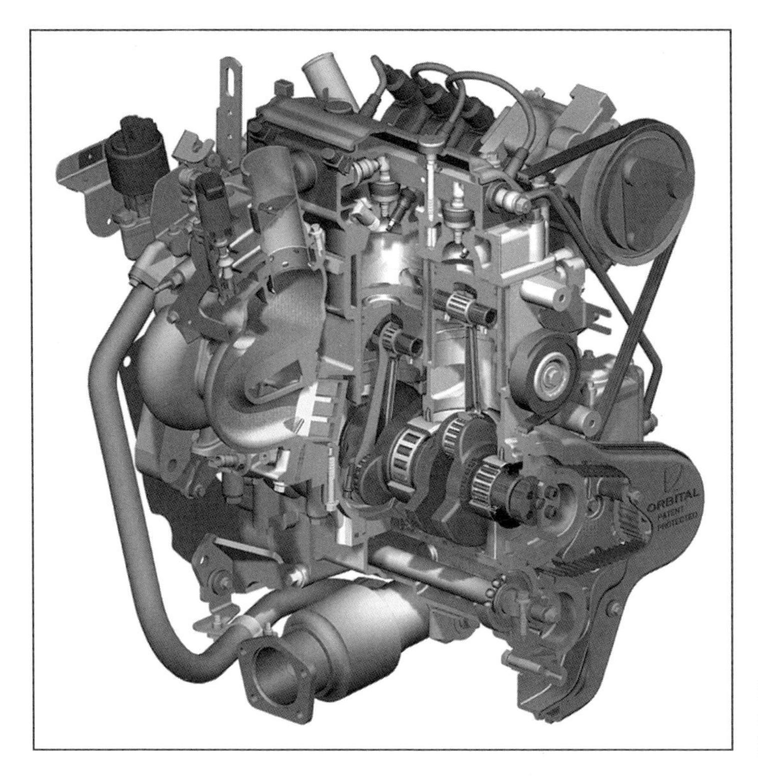

Fig. 15-48 Section of the 1.2 liter three-cylinder two-stroke engine by Orbital.[18] *(See color section.)*

weight of the engine is 85 kg. According to statements in Ref. [16], Euro III thresholds are maintained after running continuously for 80 000 km with a sufficient safety margin. The engine was and is intended for use in Indonesian passenger cars for the brands Maleo and Texmako.

The water-cooled cylinder crankcase made of an aluminum alloy has several overflow passages per cylinder and is divided in the crankshaft midplane. The forged crankshaft consists of a single piece and has divided roller bearings at the pins for the main and connecting rod bearings. When the piston moves upward, intake air is sucked into the respective crankcase through the intake manifold. Reed valves in front of the crankcase prevent a return flow of gas into the induction tract during compression in the crankcase. The crankshaft bearing and the cylinders are supplied with fresh oil by an electronically controlled lubricating oil pump. The fuel-oil mixture ratio is normally between 1:50 and 1:200. To attain a high torque over the entire speed range, there is a barrel controller in the exhaust ducts near the exhaust ports that can change the exhaust timing. The barrel controller is adjusted via a direct current motor. A particular feature of the engine is air-supported gasoline direct injection (see also Ref. [17]). The main element of this injection system is an electromagnetically controlled valve in the cylinder head for injecting an air-fuel emulsion into the combustion chamber. The liquid fuel is precisely metered by an injection valve of a conventional ported fuel injection system and injected into a mixture chamber. By injecting air compressed in a reciprocating piston supercharger into this chamber, an air-fuel emulsion forms that is finely atomized and blown into the combustion chamber. According to Ref. [17], an average Sauter diameter (SMD) of less than 8 μm is attained. A good mixture quality is attained (for example, at 3000 rpm at the end of injection around 25°–30° crank angle before TDC in the stratified air-fuel mixture) that permits a lean adjustment of the air-fuel ratio up to 100:1, given stable combustion under a partial load.

Bibliography for 15.3 and 15.4

[1] Frese, F., sowie A. Fuchs, in Bussien, Automobiltechnisches Handbuch, Bd. 1, 18th edition, Technischer Verlag Herbert Cram, Berlin, 1965, pp. 757–788 and pp. 789–791.

[2] Scheiterlein, A., Der Aufbau der raschlaufenden Verbrennungskraftmaschine, 2nd edition, Springer-Verlag, Wien, 1964.

[3] Nomura, K., and N. Nakamura, Development of a new Two-Stroke Engine with Poppet-Valves: Toyota S-2 Engine, in A New Generation of Two-Stroke Engines for the Future?, P. Duret [ed.], Editions Technip, Paris, 1993, pp. 53–62.

[4] Knoll, R., P. Prenninger, and G. Feichtinger, 2-Takt-Prof. List Dieselmotor, der Komfortmotor für zukünftige kleine Pkw-Antriebe, 17th International Vienna Engine Symposium, April 1996, in VDI Fortschritt-Berichte, Series 12, No. 267, VDI Verlag, Düsseldorf, 1996, and AVL Infounterlagen.

[5] http://www.yamaha-motor.co.jp vom März 1999, and information from N.N., Diesel Progress International Edition (ISSN 1091 3696), Volume XVII, No. 4, Skokie, IL, USA, July/August 1999, pp. 42–43.

[6] N.N., IAA, Motoren und Komponenten, in MTZ Vol. 60 (1999) No. 11, p. 719.

[7] Schunke, K., Der Orbital Verbrennungsprozess des Zweitaktmotors, Speech at the 10th International Vienna Engine Symposium, April 1989, Progress Reports, VDI Series 12, No. 122, VDI-Verlag, Düsseldorf, 1989.

[8] Cumming, B.S., Opportunities and Challenges for 2-Stroke Engines, Article at the 3rd Aachen Colloquium for Vehicle and Engine Technology, Aachen, October 1991.

[9] Plohberger, D., and L.A. Miculic, Der Zweitaktmotor als Pkw-Antriebskonzept-Anforderungen und Lösungsansätze, Speech at the 10th International Vienna Engine Symposium, April 1989, Progress Reports VDI Series 12, No. 122, VDI-Verlag, Düsseldorf, 1989.

[10] N.N., Neuer Subaru-Zweitaktmotor im Versuch, in MTZ Vol. 52 (1991) No. 1, p. 15.

[11] Appel, H. [ed.], Der Zweitaktmotor im Kraftfahrzeug, Abgasemission, Kraftstoffverbrauch, neue Konzepte; Conference Proceedings of the Joint Colloquium in February 1990, University of Berlin, ISBN 37893 13695.

[12] N.N., Fahrzeugmotoren im Vergleich: Dresden Conference, June 1993, VDI Gesellschaft Fahrzeugtechnik, VDI-Berichte 1066, VDI-Verlag, Düsseldorf, 1993.

[13] Groth, K., and J. Haasler, Gaswechselarbeit und Ladungsendzustand eines Zweitakt- und eines Viertaktottomotors bei Teillast, in ATZ Vol. 62 (1962) No.2, pp. 51–53.

[14] Braess, H.H., and U. Seiffert [eds.], Vieweg Handbuch Kraftfahrzeugtechnik, Friedrich Vieweg & Sohn Verlagsgesellschaft mbH, Braunschweig/Wiesbaden, 2000.

[15] Meinig, U., Standortbestimmung des Zweitaktmotors als Pkw-Antrieb, Parts 1–4, in MTZ Vol. 62 (2001) No. 7/8, pp. 9–11.

[16] Shawcross, D., and S. Wiryoatmojo, Indonesia's Maleo Car, Spearheads Production of a Clean, Efficient and Low Cost, Direct Injected Two-Stroke Engine, IPC9 Conference, Nusa Dua, Bali, Indonesia, November 1997.

[17] Stan, C. [ed.], Direkteinspritzsysteme für Otto- und Dieselmotoren, Springer-Verlag, Berlin, Heidelberg, 1999.

[18] N.N., Information from www.orbeng.com.au. April 2001.

16 Electronics and Mechanics for Engine Management and Transmission Shift Control

16.1 Environmental Demands

The demands on engine management and transmission shift control are mainly determined by the following environmental conditions: Temperature, vibration, and protection against certain media (pressurized and unpressurized liquids, solids, etc.). The environmental conditions are chiefly a function of the installation site (Fig. 16-1) in the passenger car and are categorized according to four main classes of installation:

- Passenger compartment or electronic box (E-box)
- Engine compartment (surface mounted on the chassis)
- Surface mounted on the aggregate
- Integration in aggregate

The definition of the different classes (Fig. 16-2) allows the development of specific housing approaches and, e.g., overall integration approaches for transmission shift controls. This standardization reduces project-related expenditures, i.e., chiefly the costs of materials and tools. Standardization also allows the manufactured units to be simplified and supports a global manufacturing strategy.

There are different reasons for selecting an installation site such as

- Reductions in cost of the wiring harness (engine compartment, surface mounted on the engine)
- Reduction of electromagnetic compatibility (EMC) from shorter wiring harness (engine compartment, surface mounted on the engine)
- Installation in the passenger compartment, concentration of the electronic control units (E-box)
- Engine test including the electronic control unit before installation (surface mounted on the engine)
- Integration options (system approach), e.g., intake module (mounted adjacent to the engine)

A general trend has been away from installation sites in the passenger compartment toward sites close to the engine.

Figure 16-2 shows the predominant environmental conditions at the installation sites.

The environmental conditions for the housing become increasingly harsher closer to the engine, which is reflected in the design of the device (selected materials, manufacturing principles, functioning, etc.).

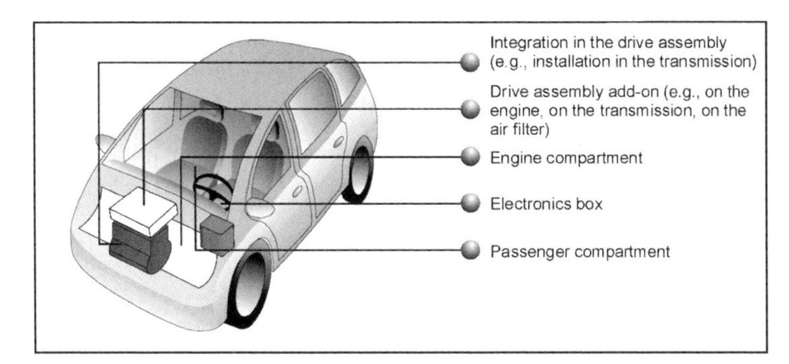

Fig. 16-1 Installation areas.

	Passenger compartment/ electronics box	Engine compartment	Assembly add-on *(e.g., on the engine, transmission, air filter)*	Integration in the assembly (e.g., installed in the transmission)
Vibration *(depending on the frequency)*	.... 5 g	... 16 g	... 28 g	... 40 g
Thermal class *(environmental temperature)*	... 80°C	... 105°C	... 125 °C	... 140 °C
Seal	Dust-tight	Dust-tight, jet-water-tight	Dust-tight, steam-jet-tight	Transmission-fluid-tight

Fig. 16-2 Classes of installation.

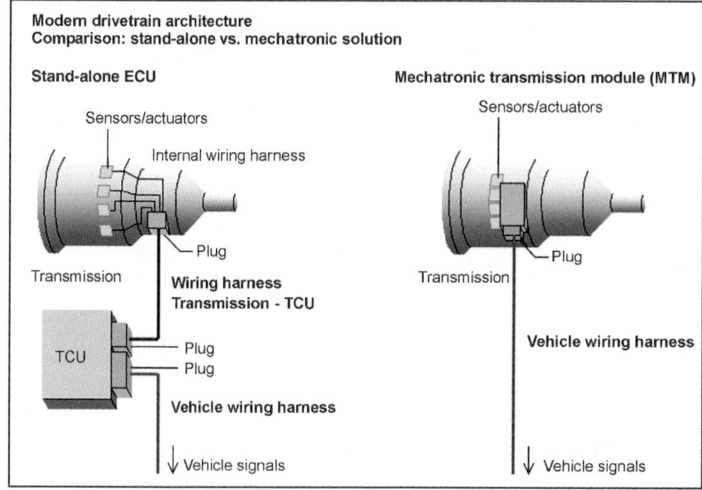

Fig. 16-3 Drivetrain; stand-alone ECU and integrated products.

In the case of electronic control units (ECU), a distinction is drawn between "stand-alone products" and "integrated products" (Fig. 16-3). To be understood as stand-alone products are engine management systems and transmission shift controls that are installed as independent units in passenger cars in contrast to integrated products that are combined with other functional units (e.g., the transmission). These two approaches are described in detail in the next sections. Figure 16-3 shows the different approaches using the example of a transmission shift control.

16.2 Stand-Alone Products (Separate Devices)

Whereas in the past a housing had the primary task of protecting the internal electronics from environmental conditions such as water, dust, and mechanical effects, thermal management is becoming increasingly important because of increased circuitry, electrical performance, and increased environmental temperatures since the classic location of the E-box has shifted toward the engine. Hence, the thermal parameters have become a decisive factor in the selected approach. In order to conduct outward losses from the electronics of up to 40 W, the results from thermal simulations and comparative measurements are referenced when designing cooling fins, thermal bridges, and circuit carriers.

The environmental temperature and the temperature at the attachment points frequently determine the circuit carriers and, hence, the manufacturing process.

The housing must also have a variety of fastening options (insertion, screwing, clamping) and resist local vibrations. The development team is also supported by stress analyses with the goal of optimizing weight (dimensioning parts to distribute stress under different load conditions). This is reflected in the trend toward light construction in automotive design.

Examples of individual housing types are shown and explained in Fig. 16-4 et seq.

The housing concept in Fig. 16-4 is a classic representative of an E-box installed under moderate conditions. The thermal design consists of a printed circuit board with a special layered construction that conducts the released heat from the electrical components to the metallic housing parts. To keep the path of transported heat as short as possible, the power semiconductors are usually placed on the edge of the printed circuit board.

Since normally more than one electronic control unit is installed in an E-box, the arising heat loss frequently cannot be removed only by free convection. In this case, additional air flow is necessary that is generated by an additional fan.

Figure 16-5 shows the housing type for installation in the engine compartment. This design fulfills the conditions that are required by today's European customers. The modular design permits adaptation to different types of fixation, including an integrated module built onto the intake system. Various plug connectors reflect the

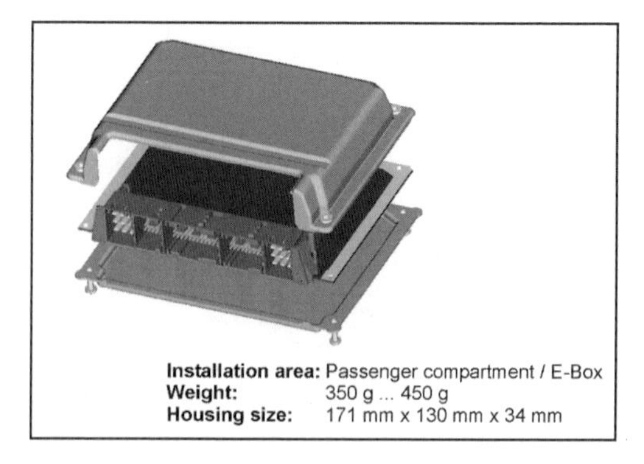

Installation area: Passenger compartment / E-Box
Weight: 350 g ... 450 g
Housing size: 171 mm x 130 mm x 34 mm

Fig. 16-4 Housing approach for the passenger compartment or electronics box.

Fig. 16-5 Housing approach for the engine compartment.

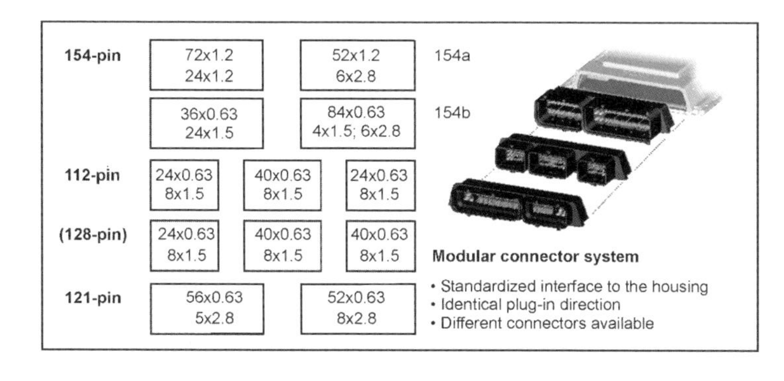

Fig. 16-6 Modularity of the plug connector.

appropriate functionality (corresponding to the number and type of connector pins) and wiring harness approach (number of plug-in modules) (Fig. 16-6).

Despite the numerous customer-specific design options, the technical design and, hence, the related manufacturing process remains nearly unchanged.

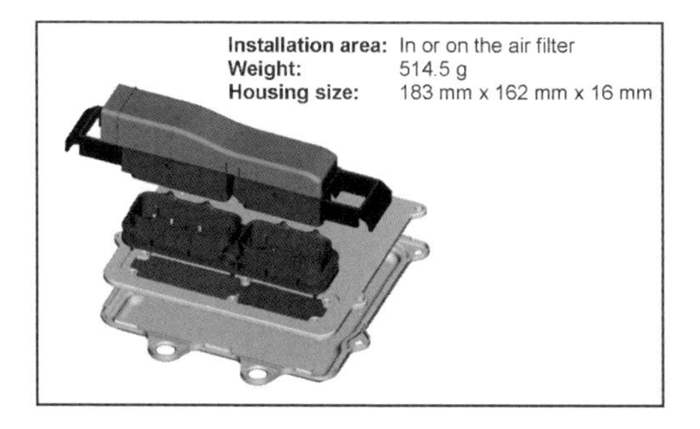

Fig. 16-7 Housing approach for externally mounting to the assembly (air filter, intake module).

The housing in Fig. 16-7 is specially designed in response to the demand for integrated solutions and for installation close to the engine on the air filter or intake module. It is particularly resistant to vibrations even as a printed circuit board since the printed circuit board is laminated to the aluminum holder. The holder is connected to the air-guiding part. This offers an effective form of thermal management.

Mounting devices on the engine (Fig. 16-8) poses the greatest demands on materials, design, and manufacturing, and is fundamentally different from the above-cited approaches. Ceramic materials are required as the substrate, and the ICs used as the electrical components are bare die or flip chip packages. The investment in the manufacture of these devices is substantial.

The advantages to the customer of this approach are the potential miniaturization and the possibility of integration in the engine or drivetrain (integrated transmission shift controls, smart actuators).

16.3 Connecting Approaches

The plug connector or "plug" is the result of extensive coordination with the customer and suppliers. The design of this part includes such topics as engine managment, customer system approach (distribution in the engine and chassis: wiring harness architecture), contact system (cross sections, surfaces), seal approaches (single core or collective seal), plug-in force, direction of installation, locking strategy, antitheft strategy, resistance to vibration, flexural strength, installation procedure (flow soldering, reflow soldering, bonding), and selection and combination of materials, to cite the most important.

The result is a part that is chiefly defined by the following criteria:

- Seal
- Number of pins
- Number of modules (chambers)
- Plug-in direction of the plug (perpendicular or parallel to the printed circuit board)

Since the developmental effort is substantial, automobile manufacturers are striving toward uniformity

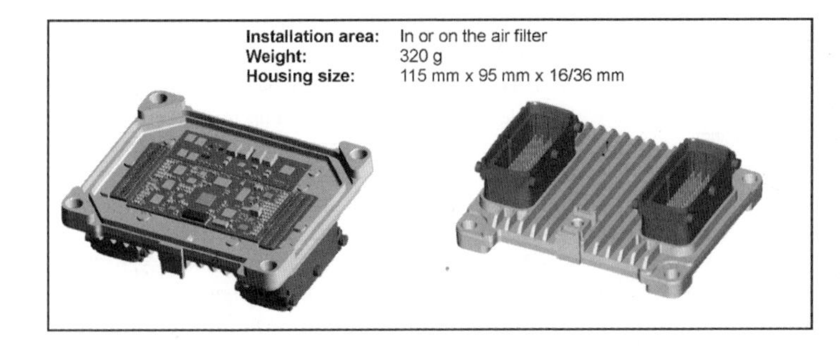

Fig. 16-8 Housing approach for aggregates add-on (surface-mounted on the engine).

in design and classes of requirements in collaboration with the suppliers of electronic control units (including SIEMENS) and connectors.

16.4 Integrated Products (MTM = Mechatronic Transmission Module)

In addition to the actual controlling of switching components, a transmission shift control includes the detection of the peripheral variables that influence path sensors, angle sensors, speed sensors, temperature sensors, and pressure sensors.

A MTM (mechatronic transmission module) integrated in the transmission covers the functions of electronic control and peripheral sensors as well as the subfunctions of potential and signal distribution, contacting the shifting components and other hydraulic and electronic interfaces. It ideally is a complete, independent module. In contrast to stand-alone transmission shift control units, electronic control units integrated in the transmission have maximum mechatronic potential since all important input and output components are directly in the transmission. If the transmission and the integrated control can be designed together in the concept phase, the maximum amount of integration can be achieved since the arrangement of all required components can be optimized in reference to their location, orientation, and technology (Fig. 16-9).

A control integrated in the transmission is usefully mounted to the hydraulic control plate, i.e., to the lowest site of the transmission in the oil pan, since the pressure control valves operate here and can be directly contacted and controlled. The MTM is subject to extremely harsh environmental conditions such as high, continuous vibration, continuously high temperatures, and very aggressive oils.

An MTM must offer a hermetically sealed holder for the electronics, long-lasting internal and external interface contacts, high mechanical strength and stability to survive the vibration and pressure, a high level of reproducible precision to provide an exact reference for attached sensors, and 100% compatibility with the used transmission fluid. In addition, the MTM must be manufacturable by means of automation and, of course, be economical.

We now discuss an example of the design and technology of a modern MTM (Fig. 16-10). Although a great deal of value is placed on reusable technology, customer desires and special demands always require adapted approaches.

An aluminum base plate serves as a base and interface element of the MTM to the hydraulic control plate of the transmission, and it also functions as a heat sink and heat transfer body for the ceramic substrate (LTCC) adhered to

Fig. 16-9 Modern MTM for lengthwise-installed CVT transmission.

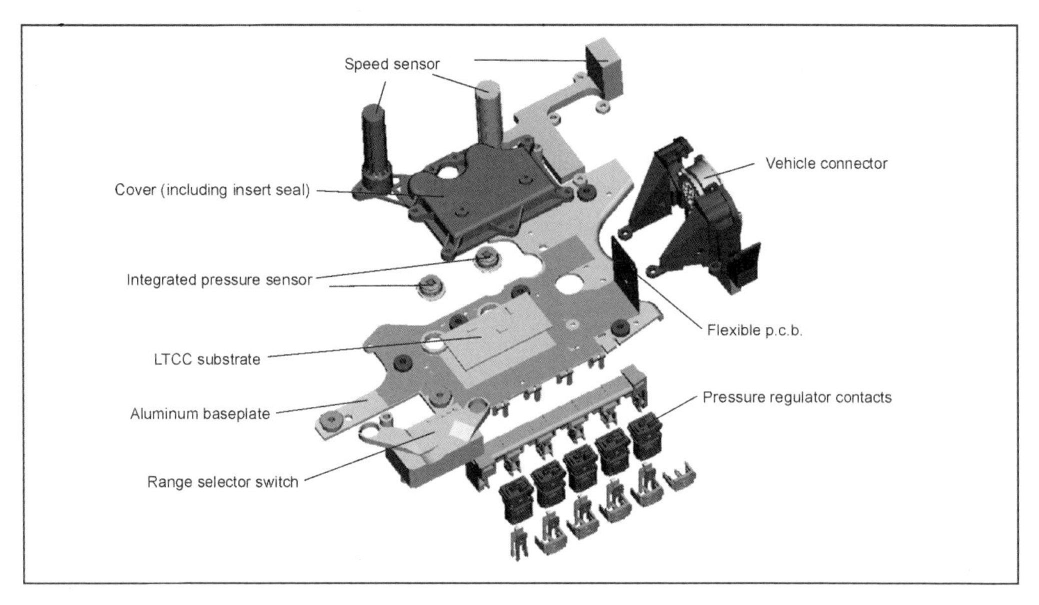

Fig. 16-10 Main components of a modern MTM CVT transmission.

the control plate with thermally conductive adhesive. Laminated to the base plate is a flexible printed circuit board that is the signal and potential distribution element of the transmission shift control electronics on the LTCC substrate to all peripheral sensing and signal detection components. Pressed into the aluminum base are pressure sensors with their own seals. The electrical contact of the substrate to the pressure sensors and flexible printed circuit board is made with 300 μm aluminum thick-wire bonds. The overall electronics and pressure sensors are hermetically sealed against the environment with a plastic cover and insertable seal to the laminated, flexible printed circuit and then riveted. The seal configuration and seal material must be tailored to provide a sufficient and permanent seal restoring force under all tolerances and extreme temperatures. To protect the electronic components and bonds against vibration, the electronics compartment must be filled with a silicone gel. Contacts of the flexible printed circuit board are created to the peripheral components such as speed sensors, control valve terminals, selector switch, vehicle plugs, and internal plugs via an approved laser welding process. When mounting the MTMs on the control plate, all pressure control valves are simultaneously contacted, the hydraulic interface to the pressure sensors is established, and all sensor reference positions are set. The design of the pressure control contacts can compensate for a 3D malpositioning of the valves in the hydraulic system.

To create another, higher integration step, the pressure control can be integrated as another component of a module in a future generation of integrated transmission shift controls to produce a so-called EHM (electro-hydraulic module).

16.5 Electronic Design, Structures, and Components

16.5.1 Basic Structure

Figure 16-11 illustrates the basic signal flow and the essential function blocks in a block diagram. The signals detected by the sensors are sent via an input filter structure to the computer. These signals are then converted and are sent via the driver stages to the actuators. By means of digital interfaces, contact can be made with other electronic control units or shop diagnosis devices. A voltage regulator ensures the required supply of voltage and current to the components. In addition, complex reset logic is required to ensure proper functioning.

16.5.2 Electronic Components

We now look at a few typical examples of electronic components in engine and transmission shift controls.

16.5.2.1 IC Knocking Input Filter Component

Up to two knocking sensors can be connected to this component. Their signals are processed by the filter in the component and sent to the microcontroller for evaluation. This is done via a serial interface. The serial interface is also used to program the variables that can be set in the component (Fig. 16-12).

16.5.2.2 Driver Stage Component

This component is a quadruple driver stage that is directly controlled by the microcontroller and that controls via the outputs up to four different actuators. A complex diagnosis

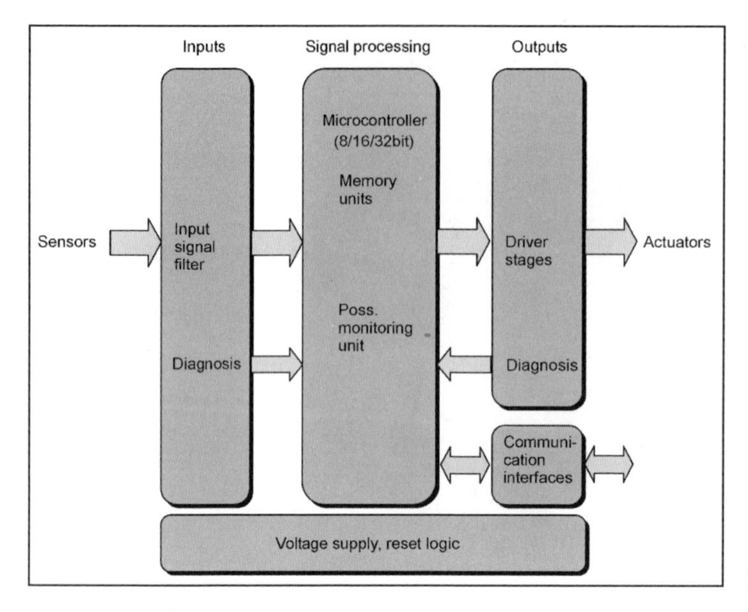

Fig. 16-11 Signal flow.

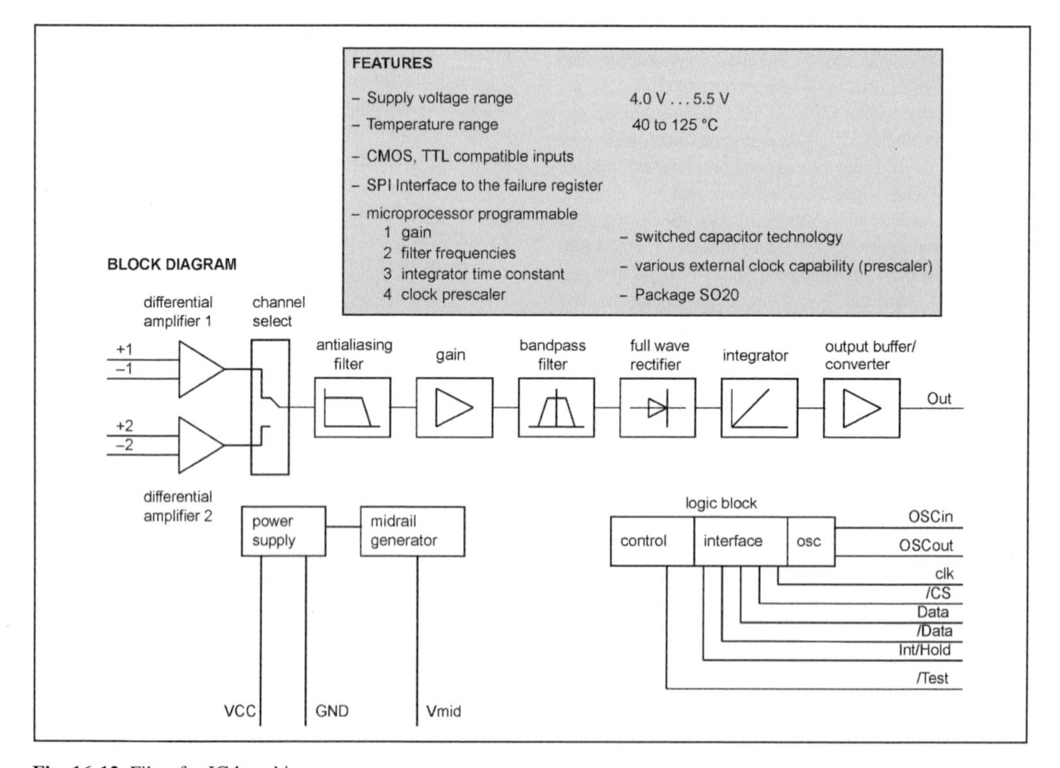

Fig. 16-12 Filter for IC knocking.

monitors the outputs for overcurrent, short circuit to ground or battery voltage, and excess temperature. Via a serial interface, these data can be retrieved by the microcontroller, evaluated, and saved (Fig. 16-13).

16.5.2.3 Microcontroller

The microcontroller was designed especially for applications in automotive technology. It combines high computational performance with a high degree of integration with peripheral components that are necessary for evaluating the input signals and controlling the driver stages (Fig. 16-14).

16.5.2.4 Voltage Regulator

This component comprises three regulators: The main regulator and two downstream regulators with substantially less power. The main regulator is responsible for the compo-

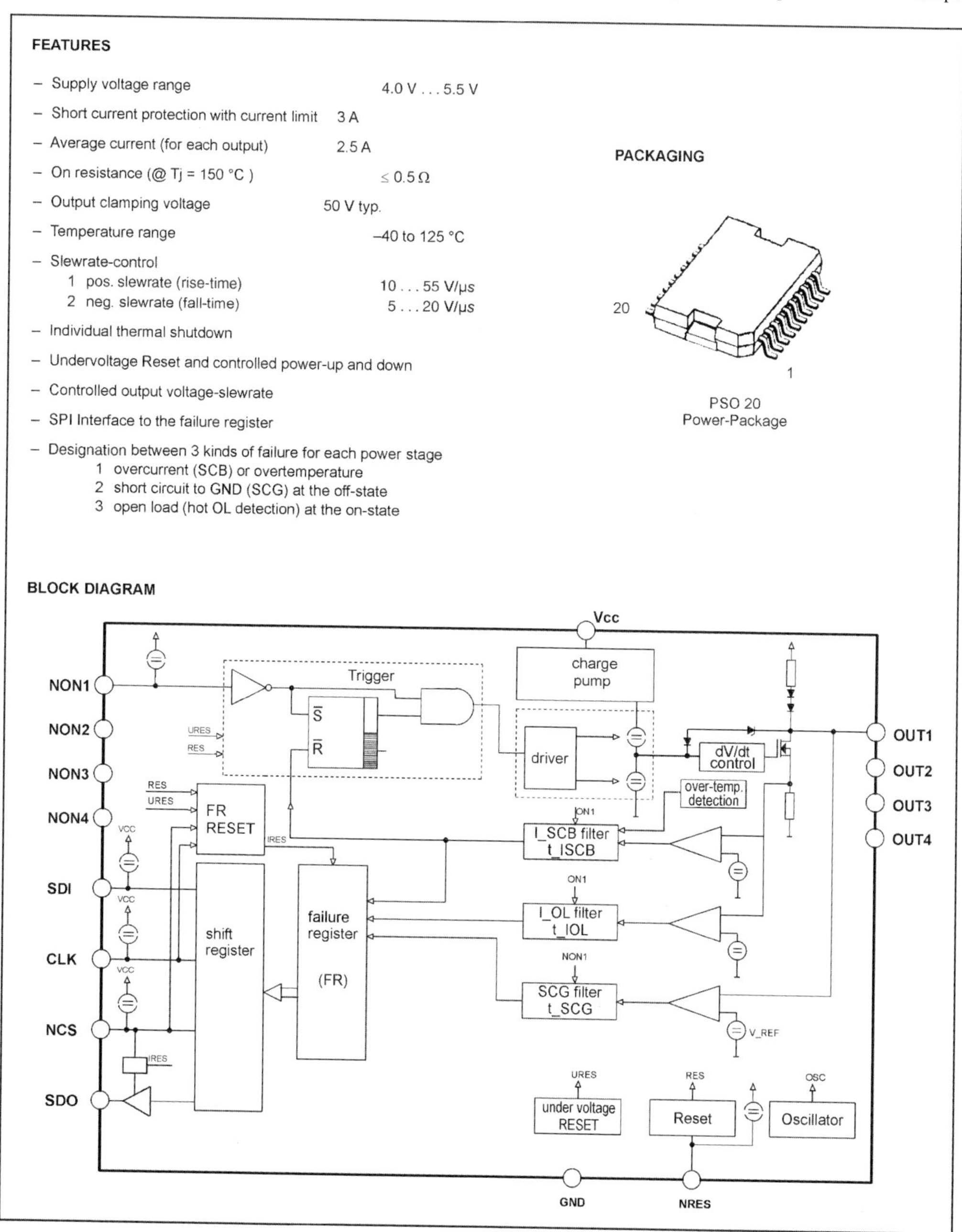

Fig. 16-13 Four-output driver stage.

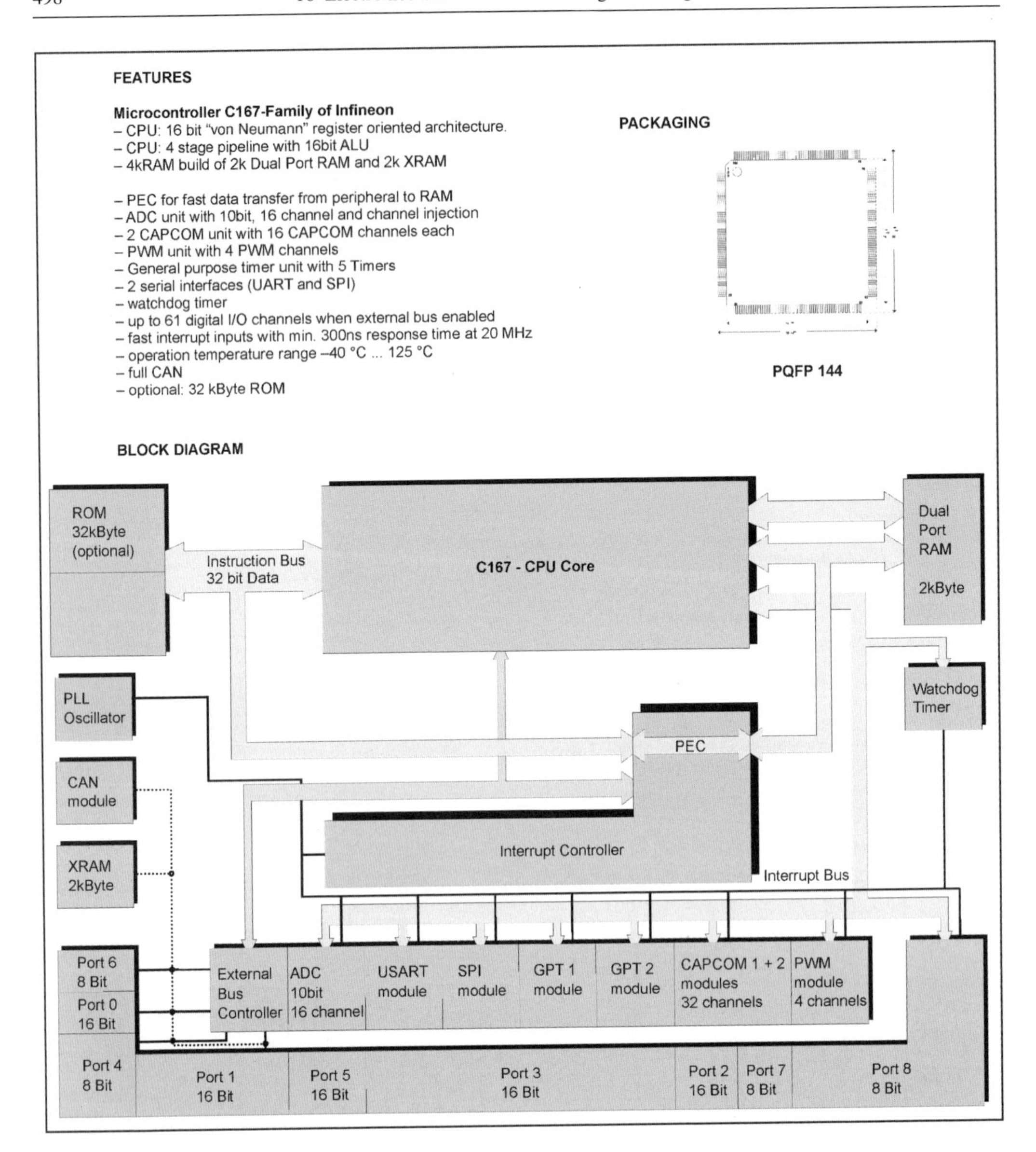

Fig. 16-14 Microcontroller.

nents in the electronic control unit, while the two downstream regulators can be used to control sensors outside of the electronic control unit. Furthermore, a monitoring unit and release logic are integrated in the component (Fig. 16-15).

16.6 Electronics in the Electronic Control Unit

16.6.1 General Description

The engine electronics assume the task of the central control unit of the combustion process. Depending on the engine and vehicle-side input signals, the integrated computer unit calculates the required variables of the actuators relevant to the engine functions.

The electronic control unit comprises the following main functional groups.

16.6.2 Signal Conditioning

Analog and digital signals pass from sensors, switches, and other control units throughout the engine compartment and vehicle interior via the wiring harness into the electronic control unit. Here the different types and amplitudes of signals are converted into digital voltages and

FEATURES

General
- Three 5V regulated outputs
- Three enable inputs
- Functional supply voltage range, VBD: 6 V to 27 V
- Load dump voltage ($t < 400$ms): 45 V
- Operating temperature range: –40 °C to 125 °C
- Overtemperature protection
- Protected from –45 V to +60 V input voltage
- Power SO20 SMD package
- Also available in Multiwatt-11 pin-through package

Main output
- Output voltage: 5V ±2%
- Current capability: 450 mA continuous
- Reset output with programmable delay time and adjustable thresholds

2 Tracking outputs
- ±1% precision referred to Main output
- Current capability: 100 mA and 50 mA respectively
- Protected from short circuits to GND and Battery voltage
- Diagnosis output for each tracking output
- Separate thermal shutdown protection

PACKAGING

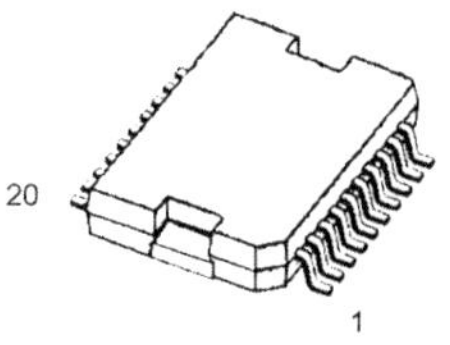

BLOCK DIAGRAM

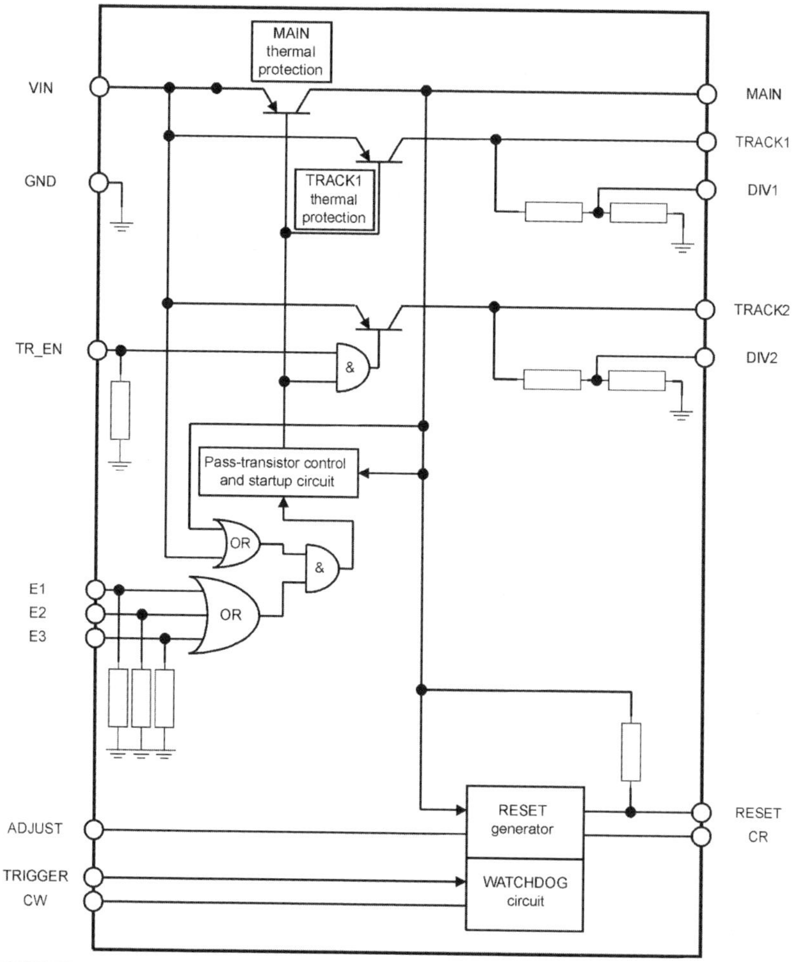

Fig. 16-15 Voltage regulator.

frequencies that represent information readable by the microcontroller.

The adaptation of the input signals from the knock sensor, lambda sensor, and induction-type pulse generator on the crankshaft is particularly complex.

In the stochastic signal from the knock sensor, the higher signal of the knocking engine is filtered out of the permanent engine noise level, amplified, rectified, and integrated. This is done with the aid of an integrated circuit that allows the midfrequency and amplitude to be preset to any desired level via a programmable register. Finally, a normalized signal is transmitted during a definable time slot to the A/D converter of the computer.

During operation, the lambda sensor yields a current proportional to the oxygen component in the exhaust gas. This current flow at the internal resistance of the sensor generates at the output a voltage that passes via an adapter circuit to the A/D converter.

The special feature of the crankshaft signal is the relation of the signal amplitude to the speed. It ranges from a few hundred millivolts at low speeds to several hundred volts. The signal is converted into the digital rectangular shape of the same frequency by zero transition point detection where interference is suppressed by a variable negative feedback.

16.6.3 Signal Evaluation

The computer unit itself comprises the CPU, the ROM for program code and parameters, the variable data memory, and the monitoring unit for the safety tests in E-gas systems.

The digitally prepared input signals serve as the variable actual values of the engine functions represented in binary code. Program maps and characteristics form the variable manipulated variables for the programmed arithmetic operations. The results of the numerous individual calculations are transferred in the form of level/time information to the output ports of the microcontroller.

A predefined safety approach is used in devices with throttle valves set by an electric motor. Algorithms are calculated simultaneously in the CPU and monitoring unit, and the results are exchanged via the serial interface and compared. In the case of deviations, the safety function is activated, which then turns off the redundant throttle valve, fuel injectors, and ignition to stop the vehicle.

16.6.4 Signal Output

The logical level of the output port driver of the controller is used directly as a control signal of the respective driver stage that in turn operates the actuators installed in the vehicle. The driver stages can be classified in three categories.

Low-side drivers control inductive and ohmic loads that are connected against the battery voltage such as valves, relays, and ignition coils, as well as heating resistors and the logical interface of other electronic controls.

High-side drivers connect the current flow for consumers that are connected to ground at one side.

In the case of bridge driver stages, the consumer is connected with both terminals of the electronic control unit. This type of connection is particularly for operating dc motors that require a continuous adjustment of forward and reverse movements.

All driver stages have a self-protection function that prevents the component from being destroyed in the case of electrical short circuits at the output to the battery or ground or a load short circuit. In addition, these interruptions are detected by circuitry and buffered in a fault register. The arithmetic-logic processor then can fetch the error code from the driver stage via the available serial interface and implement reactions such as limp-home functions, the triggering of an error light, and entry into the internal fault storage.

16.6.5 Power Supply

This circuit element draws the current for the electronic control unit from the vehicle system voltage. Depending on the level and load on the battery (e.g., the starter), 6 to 16 V are converted into a stable dc voltage of 5.0 V (also 3.3 V for new systems) to operate the electronics.

The suppression of interference from the vehicle electrical system (up to ± 150 V) by protective measure with semiconductors and capacitors contributes to flawless operation of the electronics.

In addition, this circuit block offers up to three stable 5 V voltages for supplying external potentiometers or sensors with current.

16.6.6 CAN Bus Interface

The controller area network (CAN) is a serial bus system. It was created especially for linking intelligent sensors, actuators, and electric motor/transmission controls (ECU/TCU) in the vehicle. A CAN is a serial bus system with multimaster properties. The CAN bus protocol was developed especially for applications affecting safety in the automobile industry. All CAN elements can transmit data; several nodes can simultaneously query the bus. The serial bus system has real-time properties. It was declared an international standard in ISO 11898. The object-oriented messages contain information such as speed and temperature and are available for all receivers. Each receiver independently decides, based on the transmitted identifier, if the message is to be processed or not. The arbitration between the bus elements is priority controlled by the identifier. The maximum data transmission rate is 1 Mbit/s.

16.6.7 Electronics for Transmission ECUs

The same applies for the function blocks of the CPU, power supply, driver stages, and interfaces as mentioned above.

However, there are a few function modules (Fig. 16-16) that were developed specifically for transmission shift controls:

(a) Current regulator for electromagnetic valves
(b) Redundant driver components: e.g., another high-side driver stage is added to a low-side driver stage

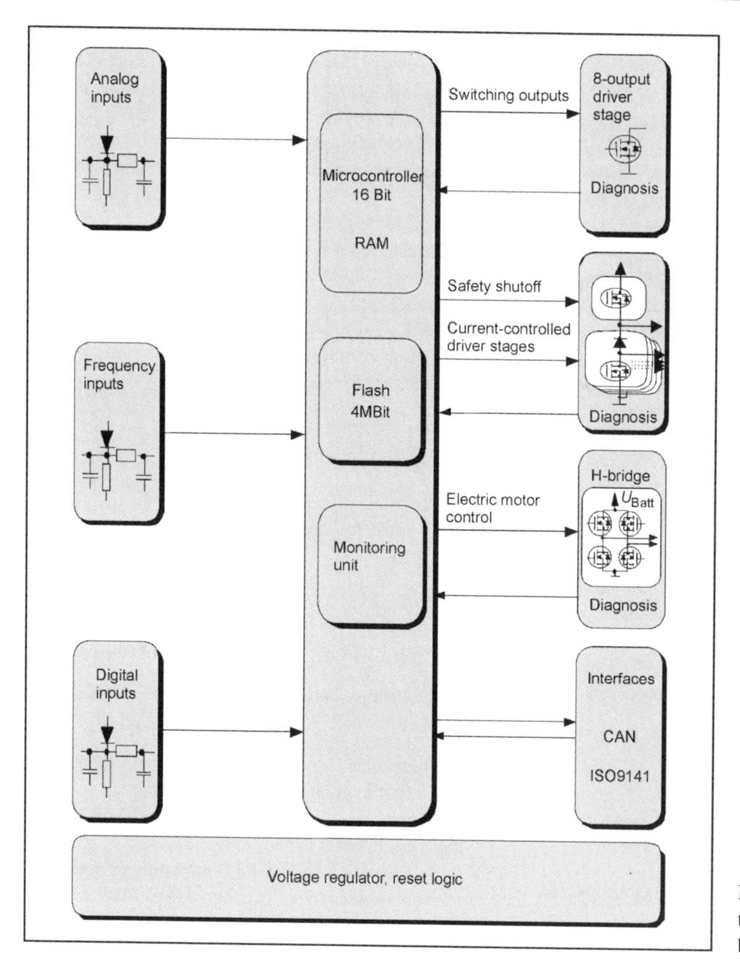

Fig. 16-16 Schematic representation of function modules in a block diagram.

to eliminate critical situations with the aid of the electronics

(c) Bridge circuits for electric motors with high current demands

(d) The control electronics are frequently integrated in the transmission. This produces much higher demands (temperature, vibration, seal, etc.) on the components and materials such as the housing and seal elements.

16.7 Software Structures

16.7.1 Task of the Software in Controlling Engines

Over the past 20 years, the importance of software has increased steadily and dramatically in all areas of automotive electronics, especially in engine management systems. On the one hand, functions have become more economical and effective than were accomplished beforehand with mechanical or electronic solutions, and, on the other hand, the potential of a freely programmable computer allows completely new and earlier unrealizable functions to be added. This includes the comprehensive self-diagnosis functions of modern electronic control units or the fine harmonization of the combustion process made possible by software that minimizes emissions and fuel consumption.

The amount of software (Fig. 16-17) in a typical engine management system has doubled about every three years in the past. Forecasts indicate that this trend will continue into the future.

At the same time, computing power has continuously increased. Whereas an 8-bit processor (e.g., Motorola 68HC11, Infineon 80C517) programmed in the assembler was sufficient for typical engine management in 1990, 16-bit processors (e.g., Infineon C167) were required in 1995 with the higher programming language "C," and in 2000, a 32-bit processor (e.g., the Motorola Black Oak) was used. A substantial part of the software (up to 50%) in a modern engine management system is not used for the

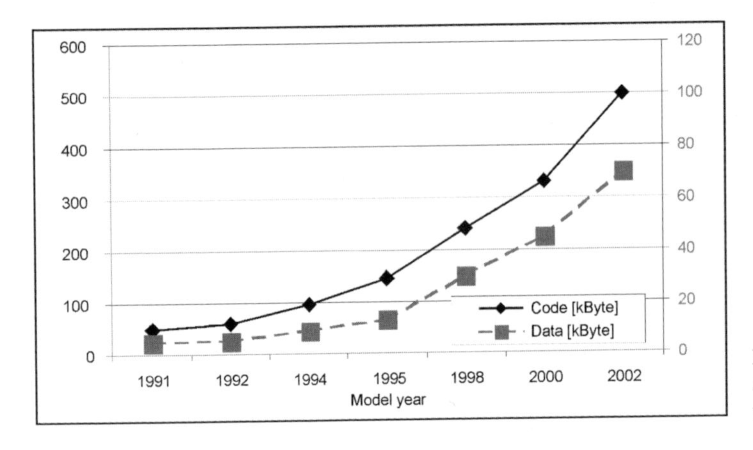

Fig. 16-17 Software of a typical six-cylinder engine management unit in kBytes.

actual functioning (control of the engines), but deals only with tasks surrounding the engine such as diagnosing the ECU and peripherals, OBD II, etc.

The growing amount of software and simultaneously shrinking developmental times have led to a sharp rise in the size of the software teams on a single project from two developers in 1990 to more than ten in 2000. This has, in turn, necessitated strict developmental processes with extensive quality controls.

16.7.2 Demands on the Software

What are the demands on the software structure for the electronic control of the drivetrain?

- To describe the required functions: Engine/transmission control, exhaust gas control, comfort functions, engine/transmission protection, self-diagnosis with extensive fault storage (EURO x, OBD), limp-home function, reprogrammability, communication with other electronic control units
- To provide fast reactions in real time: On the I/O level as low as 10 μsec; on function level 2, a simultaneous 1 msec to 1 sec, or 1.8 msec to 1.5 sec synchronized with the crankshaft
- To be largely independent of the hardware, especially the microcontroller to support a multisource strategy
- To cover different tasks (gas, diesel, automatic transmission, automatic-manual transmission, integrated starter generator, ...) with a high degree of reusability between the different areas
- To be integratable with software from customers and key-component manufacturers
- To use standard software components (OSEK operating system with communication and network services)
- To be compatible with function packages (aggregates; see Section 16.9) that represent a universal solution for the function (such as ignition) from the sensor, via the algorithm describing the function, to the actuator.

To economically meet such demands, the function and hardware must be largely separated. The solution is to use six software layers with specific responsibilities. In addition, a well-defined software development process is required to meet deadlines and ensure quality.

16.7.3 The Layer Approach to Software

The relationships between the layers are defined by a few rules:

- Each layer can use all the services of the subordinate layers, but none of the higher layers.
- Each layer can exchange data with the layers directly underneath, but it cannot jump a layer.
- Each layer can have control flows (without data transmission) to other layers. In control flows to higher layers, the layer may not recognize the recipient of the control flow.
- For each layer, only specific data types are allowed.
- Each layer is to be designed independently of the hardware, processor, and compiler. The two lowest layers form exceptions, but they need to follow this rule as much as possible.
- The six layers of software correspond to respective levels of abstraction of the "real world" (Fig. 16-18).
- The OSEK operating system is a standard software module and offers functions from different layers; whereas the communication and network services primarily act on the hardware abstraction layer, the operating system services are also available on the higher layers. The integration of the layer model is shown in Fig. 16-19.

Whereas a layered design is desirable from a software vantage point, from the viewpoint of system development, the formation of function packages termed aggregates is important. An aggregate includes everything that is required to fulfill a specific function (e.g., lambda control, ignition), from acquiring and preprocessing the relevant data in the layer model on the lower layers of BIOS, PAL, and HAL, to the actual regulating and control algorithms

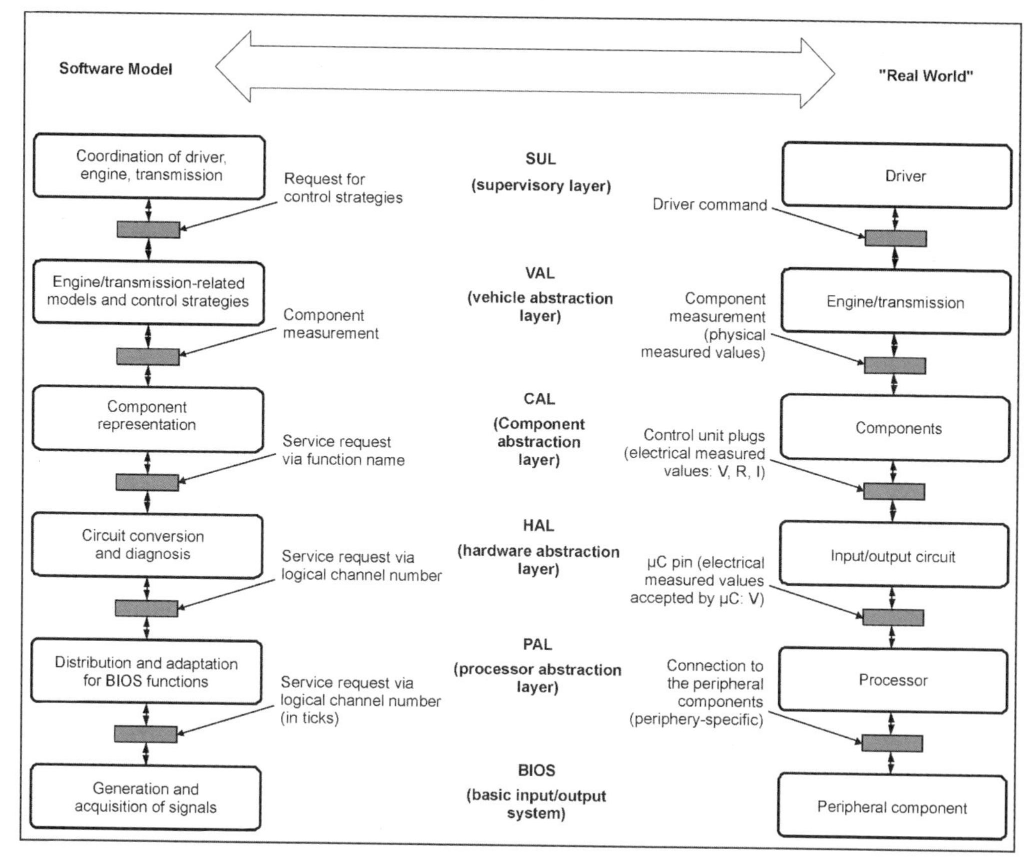

Fig. 16-18 The six layers of drive control software.

(located in CAL and VAL), to triggering corresponding actions. These are sent in the software via HAL, PAL, and BIOS to the actuators. Different function packets use the same sensor data and control the same actuators, which is made easy by the software layer model that provides standardized interfaces.

The modules that are assigned specific aggregates and sensors/actuators on a software layer then represent the preferred granularity of reuse.

The sought advantages are combined using the presented software layer model:

- A new microcontroller or a new input/output component affects only the BIOS and PAL layers or perhaps the HAL layer as well; the function algorithms remain the same.
- A function can assume the required data as given—the input/output of these data can be implemented or reused independently.
- Externally developed functions that adhere to the interface conventions can be easily integrated.
- The packaging and independence from the hardware is a step in the direction of "software as a product," i.e., the ability to sell and buy individual software functions.
- The model also increases reusability, which allows an exponential increase in software capacity and, hence, keeps the amount of effort involved in developing new software at a reasonable level.

16.7.4 The Software Development Process

Efficient and quality-oriented software development in large teams requires an appropriate, well-outlined developmental process that follows the CMM (capability maturity model). The V cycle (see Fig. 16-20) is widely used; this does a particularly good job of representing the interaction of task analysis across the levels of abstraction and the associated tests. The V is completely run through for each software delivery, usually at least five per project.

16.8 Torque-Based Functional Structure for Engine Management

Today, modern engine management systems fulfill more than just the legal requirements for exhaust gas emissions and fuel consumption; they also increase driving comfort. This expands the task of engine management for an SI engine far beyond the control of injection and ignition. In

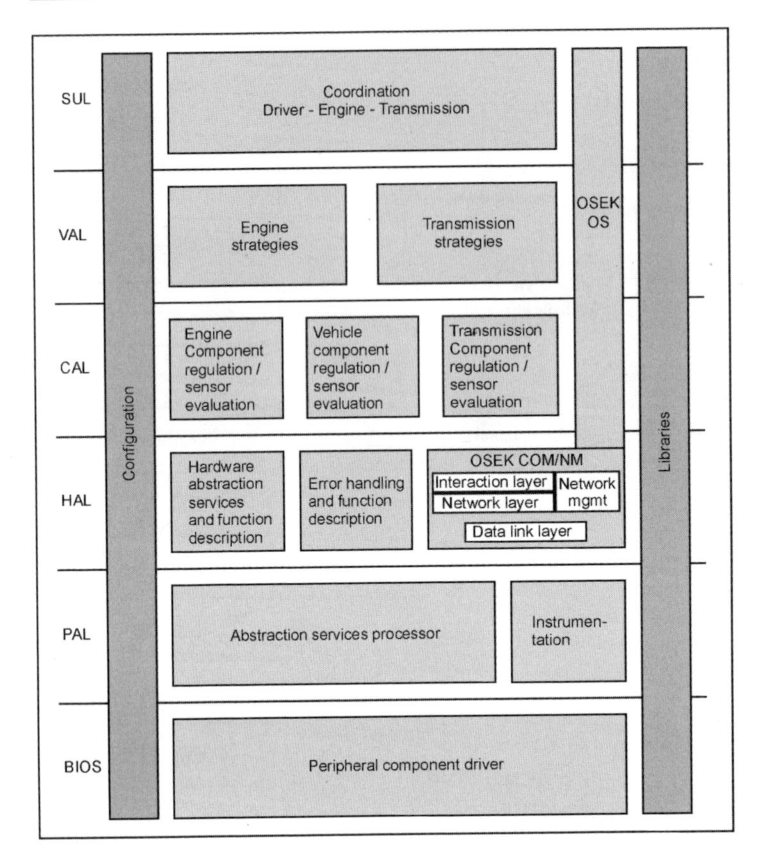

Fig. 16-19 Integration of the operating systems in the six software layers.

implementing these requirements, functional quality has been substantially increased by the drive-by-wire system (mechanical separation of the accelerator pedal from the throttle valve). This was accomplished by functionally influencing the cylinder charge simultaneous to the driver's input. This is especially necessary in SI engines with direct fuel injection and stratified engine operation under a partial load.

In a torque-based function architecture (Fig. 16-21), all demands that can be formulated as torque or efficiency are defined on the basis of these physical variables. This means that the interface between individual functions and the (sub)systems are defined as torque or efficiency. This produces a clear function architecture within an engine management system.

The goal is to attain the best possible compromise between drivability, fuel consumption, and exhaust gas emissions at all times.

Since many functions must be highly dynamic over time, the required nominal torque must be realized by using two paths. In addition to the charging path that undertakes all charge-influencing actuator judgments of the throttle valve, the crankshaft-synchronous ignition path assumes all interventions that directly influence combustion efficiency. All actuator judgments are undertaken in this path that influence the torque produced by the engine independent of the charge, i.e., the ignition and injection timing.

By coordinating the two paths, it is possible to quickly increase torque by adjusting the ignition angle. This is important for idle control, starting, transmission adjustments, and drive slip control. Efficiency can also be intentionally worsened by adjusting the ignition, e.g., a retard adjustment when heating the catalytic converter to improve exhaust. Also belonging to the fast torque path are the lambda and cylinder shutoff path that can be activated if necessary and also coordinated with the charging path and/or ignition path.

The main influencing variables such as fresh gas charging, the ignition angle, and lambda produce the inner torque (TQI) from combustion that, in contrast to the indicated torque, does not include the entire charge cycle. The torque produced by an engine (Fig. 16-22) results from indicated torque minus the torque lost from friction. The friction and charge cycle losses are combined in the lost torque TQ_LOSS. After subtracting the lost torque from the auxiliary systems, we have the clutch torque TQ_CLU. The drive torque available at the wheels is determined after taking into account the loss from the clutch and transmission.

The TQI (Fig. 16-23) is calculated as follows:

$$\text{TQI} = \text{TQ\_CLU} - \text{TQ\_LOSS}$$

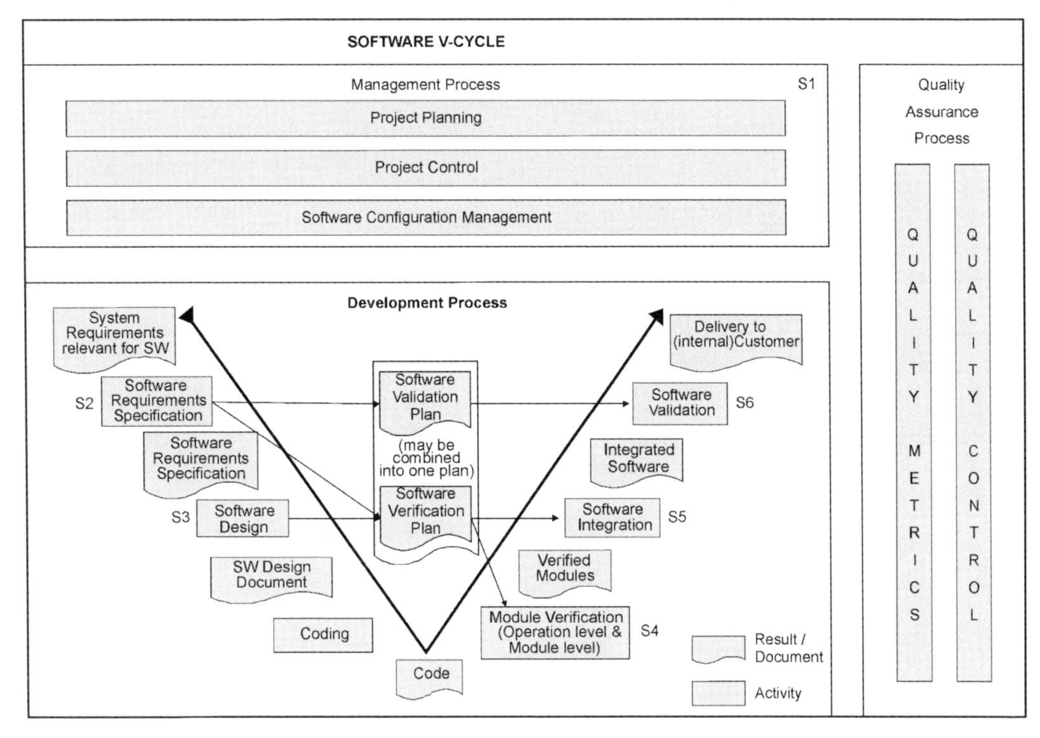

Fig. 16-20 Software developmental process.

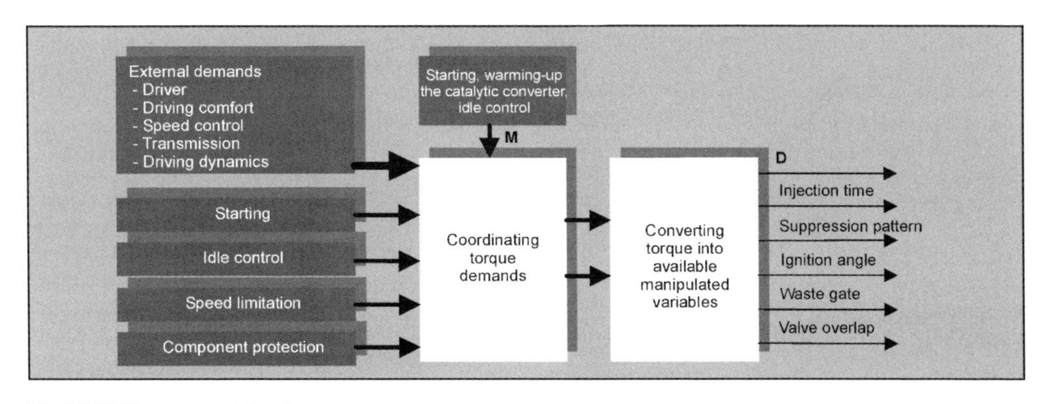

Fig. 16-21 Torque-based function structure.

The lost torque TQ_LOSS is always a negative torque since the engine is not fired and must be operated with a trailing throttle. In the equation, it is added to the clutch torque since the inner high-pressure torque TQI from combustion contains both the clutch torque measurable at the dynamometer and the lost torque TQ_LOSS. This means that the inner torque TQI must be at least as large as the lost torque TQ_LOSS to overcome the friction and charge cycle losses (TQ = 0). Only when the TQI is greater than the TQ_LOSS is there positive torque TQ at the clutch.

The driver command PV_AV is communicated to the system in E-gas systems (drive-by-wire) via a pedal travel sensor and interpreted in the torque structure as a torque request. This torque command can be changed by various actuator judgments such as the cruise control, load-reversal damping, or transmission adjustments. The resulting torque in the torque structure is then simultaneously fed into the two following paths: the slow torque adjustment path (charging path, slow air path), and then the fast torque adjustment path (ignition path, fast ignition path) where

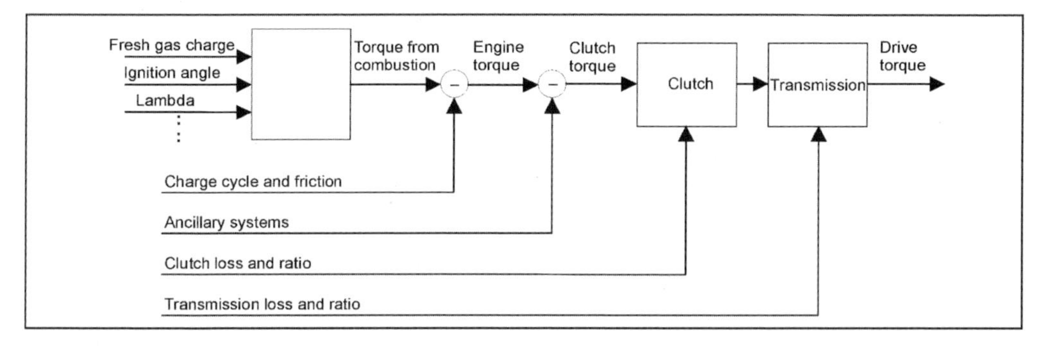

Fig. 16-22 Torque transmission.

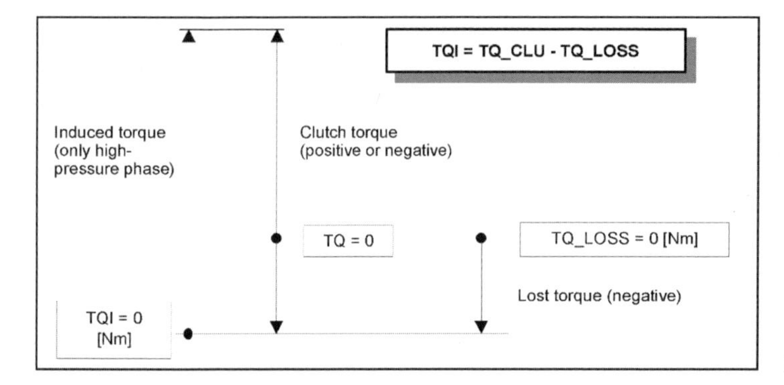

Fig. 16-23 Calculating the TQI.

the efficiency for lambda (injected fuel quantity) and cylinder shutoff are included in the ignition efficiency.

The charging path is therefore termed the slow path since its dynamics depend on the intake manifold volume, throttle valve cross section, engine speed, etc. Its time constant lies within the range of a few hundred milli-seconds. This is contrasted with the crankshaft-synchronous ignition path in which the time constant depends only on the engine speed. Whereas at an engine speed of 1000 rpm, a time constant of approximately 30 ms results; it is 5 ms at a speed of 6.000 rpm. This is one to two orders of magnitude smaller than the charging path time constant.

16.8.1 Model-Based Functions Using the Example of Intake Manifold Charging

Rising demands in the precision of metering the injected fuel mass per cycle, or the precision of the moment of ignition, have led to increasingly complex functions in engine management. The most frequently used multidimensional tables (in which the correction factors for the injection time are saved as a function of various input conditions) no longer suffice for attaining complex goals since, especially in transient engine operation, the detection of metering the fuel mass or the position of the moment of ignition is too imprecise. This yields deviations from the stoichiometric mixture that can produce problems with emissions and drivability.

Therefore, we seek to describe the physical relationships in models. This allows the set parameters such as the fuel mass for injection per cycle to be derived from physical formulas for the air mass inducted per cycle. A disadvantage is the increasing demands on processor performance and the increased memory arising from usually complex algorithms for the models.

This is illustrated using the example of the intake manifold charging model in Fig. 16-24. In the past, the injection time was calculated directly from the air mass signal (using a hot-film air-mass meter). In transient operation, i.e., when the throttle valve is opening or closing, the intake manifold is filled or emptied as a result of a change in the intake manifold pressure. The arising air mass error produces an incorrect injection time that is manifested by a rich mixture when accelerating (positive load jump) and a leaner mixture upon letting off the gas pedal (negative load jump).

Currently, with the aid of the intake manifold charging model, we are able to largely eliminate this mixture error. The air mass flowing out of the intake manifold into the cylinder is calculated from the absorption lines saved in the tables that depend on the modeled intake manifold pressure. The intake manifold pressure is calculated from the general gas equation using a differential equation. The air mass signal is initially not used. The air mass flowing via the throttle valve into the intake

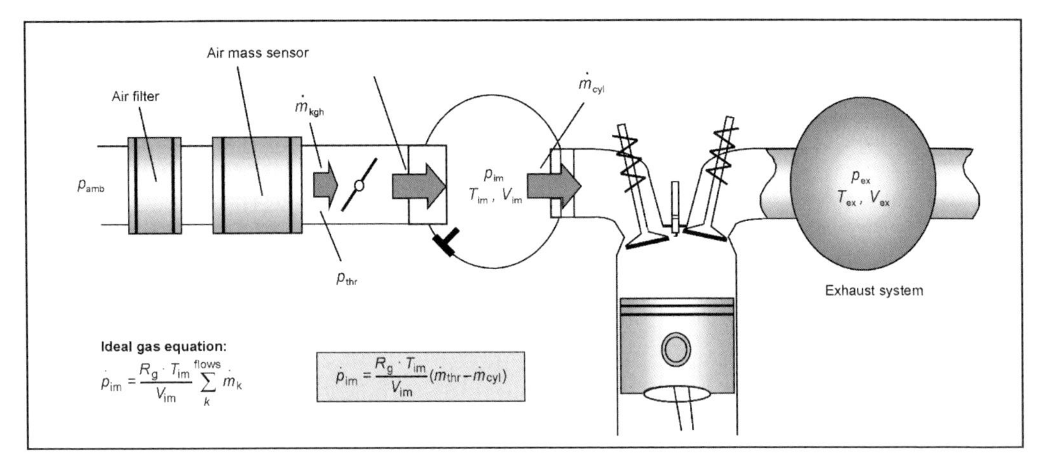

Fig. 16-24 Intake manifold model (without exhaust gas recirculation).

manifold is first determined only from the position of the throttle valve. Later, the parameters of the model such as the environmental pressure and reduced throttle valve cross section are corrected with the aid of adaptive methods.

To simplify the formulaic relationship of air mass through the throttle valve (Fig. 16-25) as a function of the differential pressure $p_{amb} - p_{im}$, a Ψ function is used (equation by St. Venant).

The air mass flowing into the cylinder is saved in the form of line equations as a function of the intake manifold pressure (Fig. 16-26).

The linearized first-order differential equation for the intake manifold pressure compiled from the individual

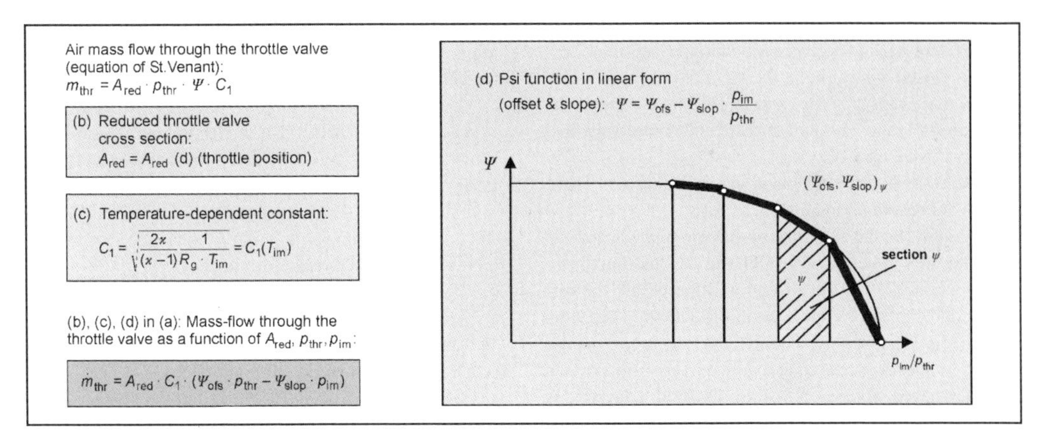

Fig. 16-25 Model for the air mass through the throttle valve.

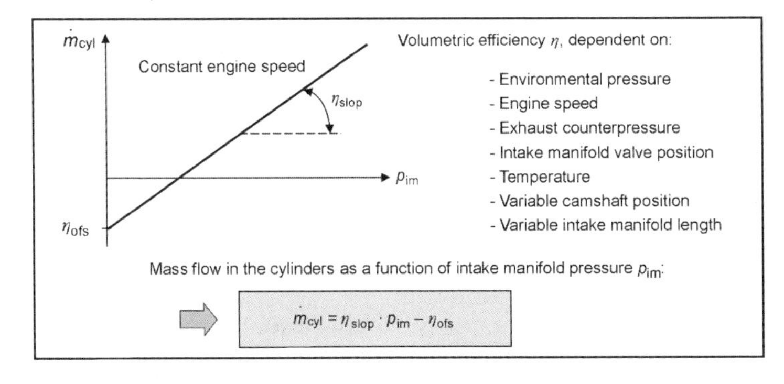

Fig. 16-26 Model of the air mass flow in the cylinder.

models can be numerically solved in the segment grid in the engine management system using a method of integration (e.g., trapezium rule).

The advantages of this model-based functional approach are

- Harmonization of a dynamic model largely by stationary determinable values. The parameter of the intake manifold volume primarily influences the dynamic behavior.
- Understandability and repeatability of the application since the function is chiefly based on physical variables.
- Independent structures of pressure and air mass guided systems.
- Simple diagnosis.
- Invertible function (important for determining the throttle valve set point in torque-guided systems and for determining the safety strategy).

16.9 Functions

16.9.1 λ Regulation

The three-way catalytic converter with λ regulation has become the most familiar exhaust gas treatment approach for SI engines with external mixture formation. λ regulation ensures that the pollutant components CO, HC, and NO are optimally converted. It is necessary to maintain a stoichiometric composition of the air-fuel mixture ($\lambda = 1$) within a very narrow λ range (λ window) (Fig. 16-27).

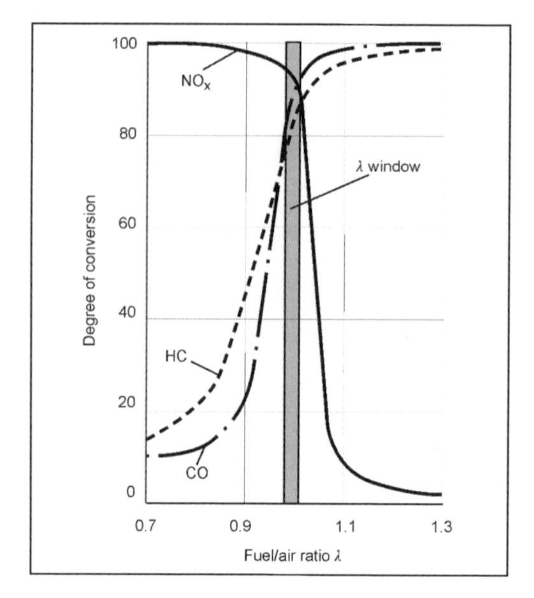

Fig. 16-27 The lambda window.

In a closed-control loop, the air-fuel ratio λ is measured by the λ sensor in the exhaust gas that compares the actual air-fuel ratio with the set point and corrects the fuel quantity if necessary.

There are binary and linear lambda sensors. These sensors are described in Section 18.3.1.

To optimize the function of the three-way catalytic converter, i.e., to best oxidize CO and HC and maximize the reduction of NO_x, the air-fuel mixture before the catalytic converter must have a certain fluctuation; i.e., the internal combustion engine must operate in a specific manner with both excess air and deficient air. This ensures that the oxygen storage unit of the catalytic converter is filled and emptied. When O_2 is stored, the NO_x level is also reduced; when the oxygen is released, oxidation is supported, which prevents stored oxygen molecules from deactivating sections of the catalytic converter.

The control algorithm (Fig. 16-28) for binary λ regulation is based on a PI controller. The P and I components in the program maps are saved via the engine speed and load. With binary regulation, the catalytic converter (λ fluctuation) is implicitly stimulated by the two-step control. The amplitude of the λ fluctuation is set at approximately 3%. A superposed trimming control via a binary after-converter sensor helps maintain the λ window before the catalytic converter.

In linear lambda control (Fig. 16-29), forced excitation is necessary to set the λ fluctuation. The diagram provides an overview of the structure of the linear λ regulation including forced excitation and trimming control.

At the actual λ set point, the forced excitation (Fig. 16-30) modulates a periodic deviation (λ pulse) to optimize catalytic converter efficiency.

The obtained signal is directly entered as a precontrol element in the fuel quantity correction; the signal may also be influenced by the secondary air and be further processed as a filtered λ set point taking into account the gas travel time and the delay behavior of the linear sensor.

The signal of the linear λ sensor is converted via a saved characteristic into a λ value. This characteristic can be corrected by the trimming control (Fig. 16-31). The trim controller is designed as a PI controller that uses the after-converter sensor signal, which is less exposed to cross sensitivity (preferably by a binary jump sensor).

The control deviation (Fig. 16-32) is then calculated from the corrected λ signal and the filtered λ set point as

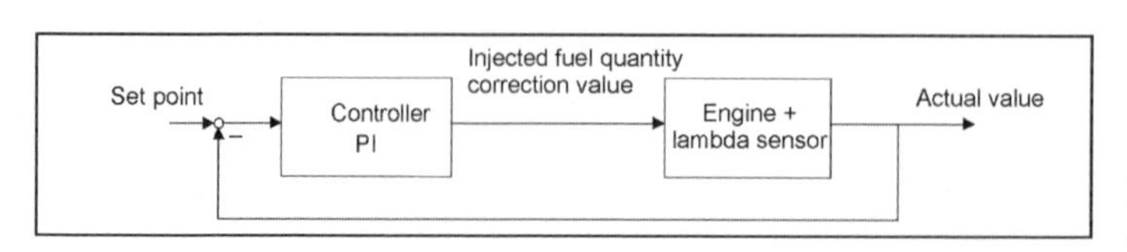

Fig. 16-28 Control algorithm for a binary lambda sensor.

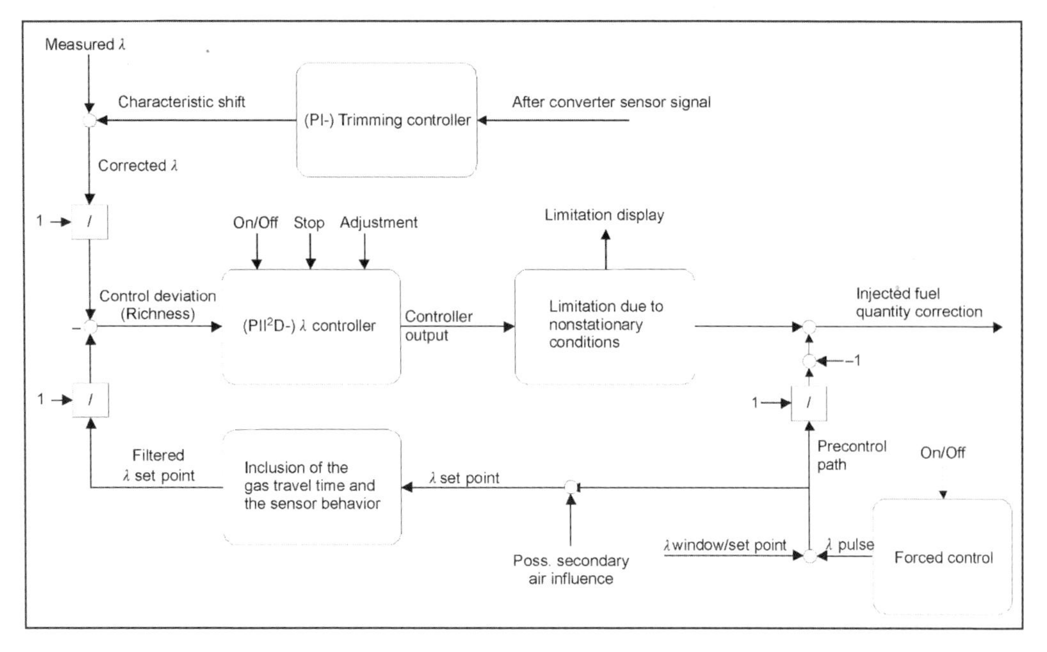

Fig. 16-29 Lambda control for a linear sensor.

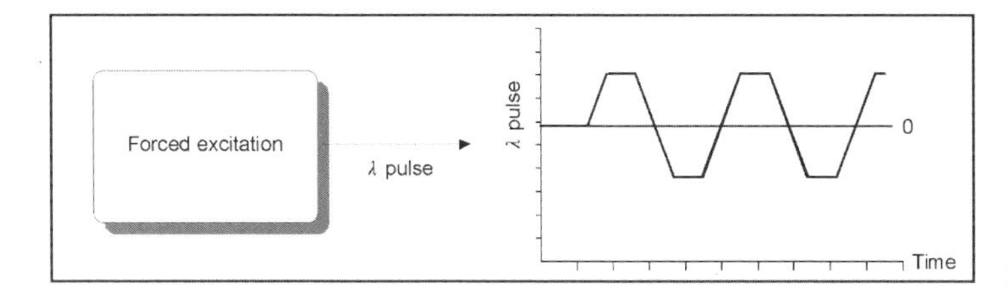

Fig. 16-30 Forced excitation.

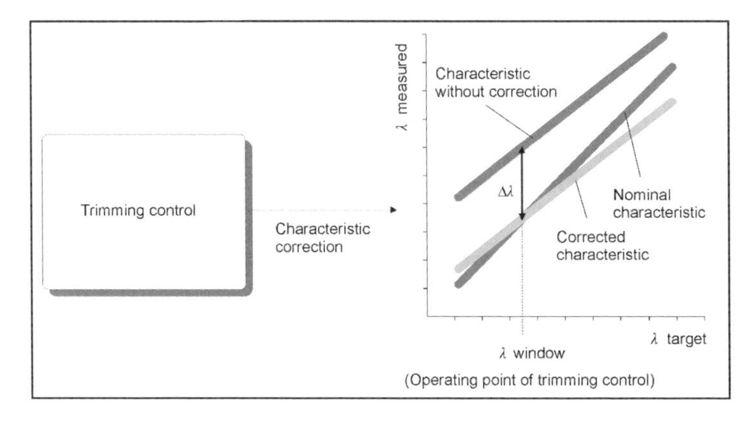

Fig. 16-31 Trimming control.

richness ($= \lambda^{-1}$) that serves as input to the actual λ controller, which is designed as a PII²D controller and is shown in Fig. 16-32. The I² component serves to balance the oxygen charge of the catalytic converter. The controller output can also be further limited under nonstationary operating conditions. The injected fuel quantity correction calculated in this manner is included together with the precontrol in the injected fuel quantity calculation.

Linear lambda control has the following advantages over binary lambda control:

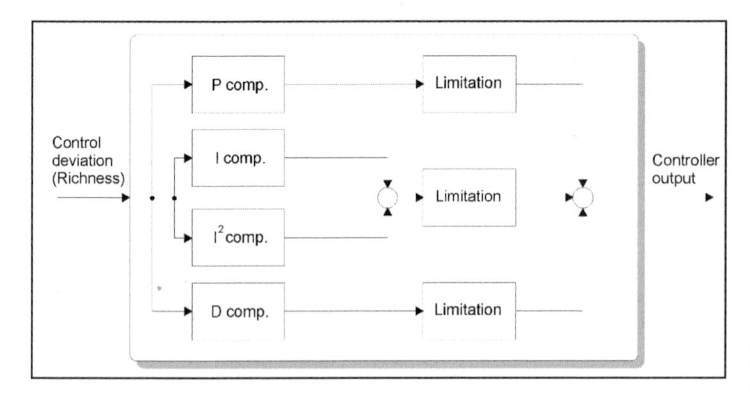

Fig. 16-32 Calculation of control deviation.

- Increased control dynamics and reduction of transient λ errors.
- Greater catalytic converter efficiency from settable forced excitation in a closed λ control loop.
- Ability to regulate $\lambda \neq 1$; this allows controlled warm-up or controlled catalytic converter protection.

16.9.2 Antijerk Function

Because of sudden changes in engine torque that can arise from acceleration or from letting off the gas pedal, the vehicle is excited to oscillate lengthwise. These perceptible changes in acceleration are very uncomfortable to passengers. The effect can be observed in nearly all passenger cars; its intensity depends on the type of construction of the drivetrain and its parameters (e.g., rigidity of the drivetrain). It is very important to reduce the effect by engine management since transient handling is an important parameter in the decision to buy a vehicle.

To develop the functions, a simple physical model is used that sufficiently describes the jerk effect. There are two functions in engine management for reducing lengthwise vehicle oscillations:

- Load-reversal damping (torque transient) based on control (driver command filter)
- Antijerk function (antijerk controller) based on a control loop

The drivetrain can be represented as a two-mass oscillator (Fig. 16-33). The mass m_1 represents the moment of inertia $J_1 = m_1 \cdot r^2_{rot}$ (rotating masses): The crankshaft and camshaft, pistons and connecting rod, flywheel and auxiliary systems.

The mass m_2 comprises the drivetrain masses (gears, propshaft, wheel masses) and the remaining vehicle mass. As torque increases in the engine, torsion is exerted on the drivetrain, and the stored energy acts on m_1.

If TQ_{engine} is a step function, the drivetrain oscillates at its natural frequency. The amplitude and frequency of this oscillation and its decay time are gear dependent. In a low gear, the amplitude and frequency are higher, and the decay time is longer than in a high gear.

To dampen load reversal, the following physical facts are exploited: Based on the model of the two-mass oscillator, we can show that the oscillation can be reduced depending on the excitation of the systems. Ramp-shaped signals are particularly suitable.

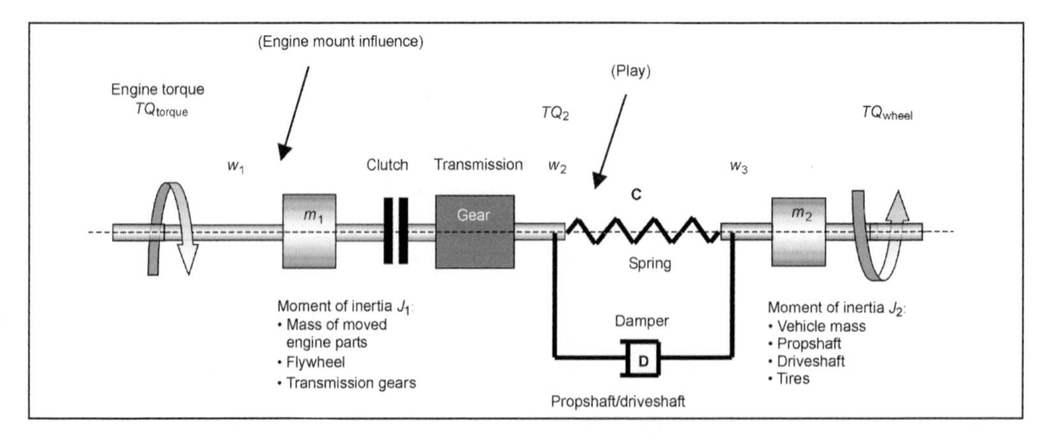

Fig. 16-33 Drivetrain as a two-mass oscillator.

The oscillation amplitude is smallest when the ramp rise time is the same as or a multiple of the period duration. This theory can actually be used only when the spontaneity of the vehicle is not impaired. A compromise between the comfort and the dynamics must be reached by shorter ramp times.

When the driver commands are filtered in this manner, oscillations remain in the drivetrain that must be compensated by the antijerk function.

The output value of load-reversal damping is a torque command of the driver or cruise control filtered by a ramp function. The ramp rise times are determined from the selected gear (Fig. 16-34).

Since the function is not only for damping oscillations in the drivetrain but also for tipping the engine on its bearings, a distinction is drawn between different torque ranges.

The core of the function is a variable calculation of the ramp rise. The speed gradient represents an additional condition for switching the torque range (Fig. 16-35).

The antijerk function and load-reversal damping work closely together. The control loop fights the remaining oscillations from the load-reversal damping. The load-reversal damping, hence, is applied for a high degree of spontaneity, whereas the antijerk function provides greater driving comfort.

The basic idea is to derive a correction signal from speed deviation that enters the torque set point in phase. Since the processes are very dynamic (typical frequencies for drivetrain oscillations range from 2 to 10 Hz), the torque command must be quickly implemented via the ignition.

Oscillations in linear vehicle acceleration can be detected from the engine speed that is very suitable for signal detection because of its resolution and updating properties.

The oscillations in the drivetrain are expressed as a speed differential based on the deviation of the actual speed from a reference speed. From this difference, the

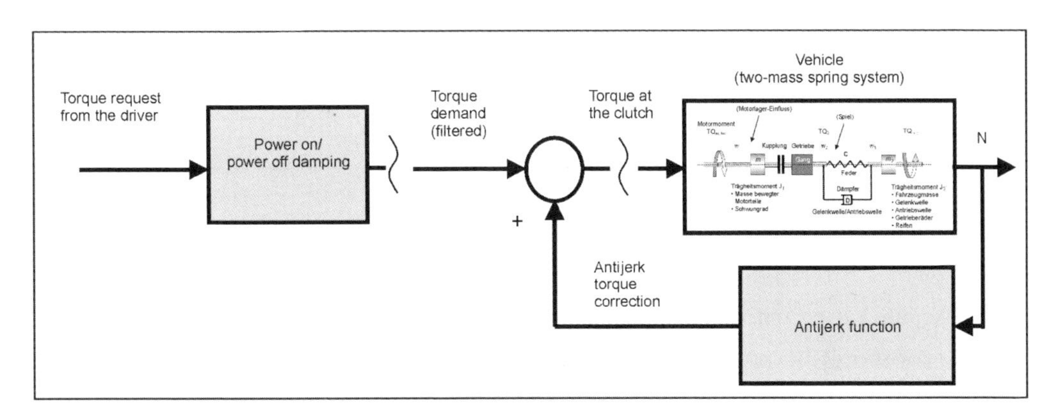

Fig. 16-34 Torque model.

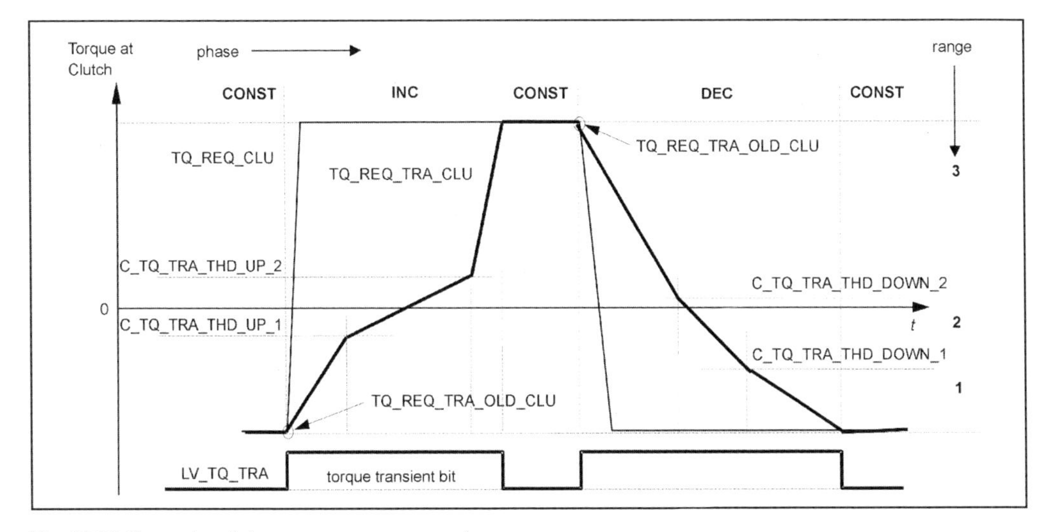

Fig. 16-35 Ramp rise of the torque.

correction signal is derived for the torque command. The phase and amplitude of the correction signal can be influenced by parameters.

The correction signal is active only during an applicable time frame. The antijerk function (Fig. 16-36) is triggered when oscillations in the speed occur.

In certain configurations of the drivetrain, the physical limits of an internal combustion engine are reached. The correction signal can no longer be implemented in phase.

The target position of the throttle valve is calculated from the torque command via various steps, and it is set by the throttle valve position controller (Fig. 16-38).

The goal of throttle valve control is to precisely match the actual air with the desired air mass (from the torque model). In the forward branch of the intake manifold charging model, the air mass flowing into the engine results from the throttle valve position and the speed. This relationship must be exactly invertible so that a throttle valve target position can be calculated in the reverse path from the charging set point.

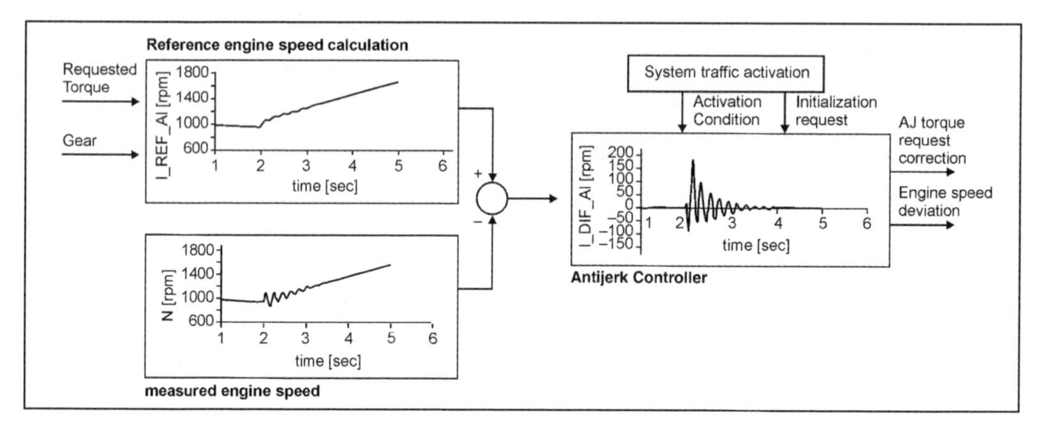

Fig. 16-36 Diagram of antijerk regulation.

16.9.3 Throttle Valve Control

In a torque-guided engine management system, the set point for setting the electronic throttle valve (ETC) results from the intended torque in an "inverse intake manifold charging model" or the reverse path of the intake manifold charging model (Fig. 16-37).

Figure 16-39 shows the structure of the throttle valve control loop. The input signal for the position controller is the difference between the actual and desired positions of the valve. Depending on this deviation, a control algorithm calculates a control signal [pulse width modulation (PWM)

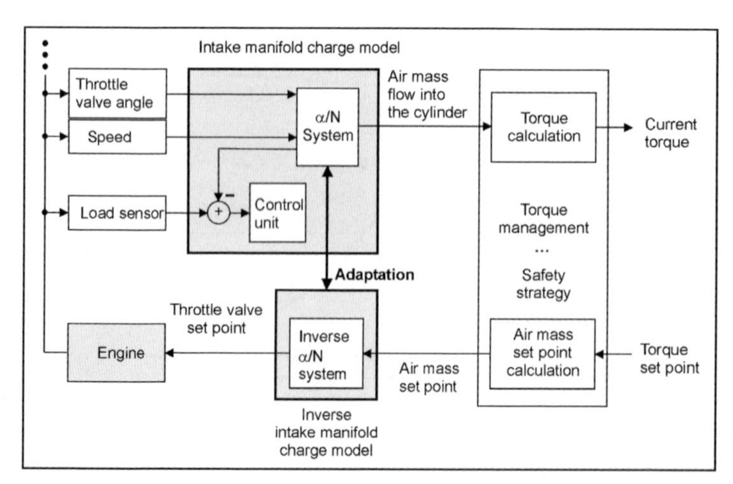

Fig. 16-37 Consistency of the forward and reverse air mass paths.

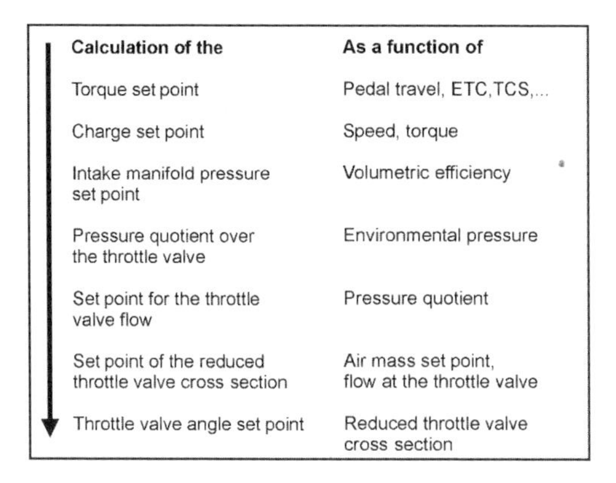

Calculation of the	As a function of
Torque set point	Pedal travel, ETC,TCS,...
Charge set point	Speed, torque
Intake manifold pressure set point	Volumetric efficiency
Pressure quotient over the throttle valve	Environmental pressure
Set point for the throttle valve flow	Pressure quotient
Set point of the reduced throttle valve cross section	Air mass set point, flow at the throttle valve
Throttle valve angle set point	Reduced throttle valve cross section

Fig. 16-38 Calculating the throttle valve set point from the target torque.

management system therefore intervenes at the moment of ignition in uncharged engines and influences the charge pressure and the moment of ignition in charged engines.

In a knocking control system, we make use of the noise arising from the pressure oscillations in the combustion chamber by tapping the structure-borne noise signals in the crankcase with the aid of a knock sensor. In the knock sensor, a seismic mass acts on a piezoceramic and induces a charge there proportional to the structure-borne noise oscillation of the installation site. The noise—typically within a frequency range of 5 to 15 kHz—arises as a resonance of the engine structure with the high-frequency components in the pressure characteristic that arise in the combustion chamber during knocking due to the turbulent flame velocities. Figure 16-41 shows a typical pressure characteristic and the structure-borne noise signal for normal and knocking combustion.

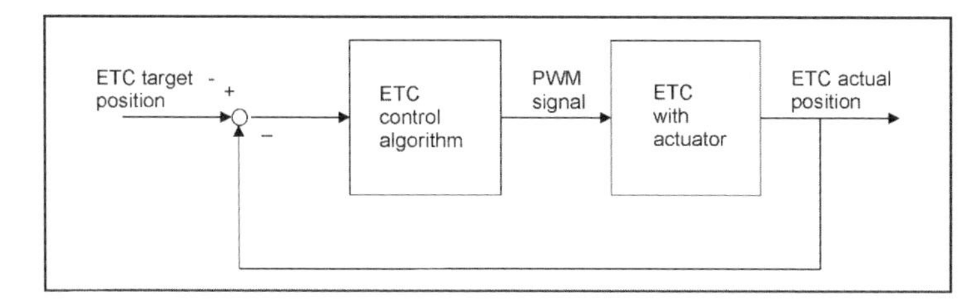

Fig. 16-39 Structure of the electronic position control of the throttle valve (ETC).

signal, Fig. 16-39] that influences the servomotor on the valve so that the actual throttle valve position is moved to the desired position.

16.9.4 Knocking Control

Knocking is uncontrolled, self-instigated combustion of normally inert gas components usually with high flame speeds around the velocity of sound, and also causes high-pressure peaks. Continuous knocking combustion damages the engine, primarily the pistons, cylinder head seal, and cylinder head.

Knocking can be reduced by the following measures:

- Later moment of ignition
- Higher octane number (RON) of the fuel
- Richer mixture
- Lower charge pressure
- Lower intake air temperature
- Reduction of deposits on the piston and valves
- Suitable construction of the combustion chambers

Knocking is problematic for engine efficiency since at today's conventional compression ratios of ca. 10–12, the most efficient moment of ignition is in the knocking range of the ignition characteristic (average pressure as a function of the moment of ignition) (Fig. 16-40).

If the engine is operated close to this optimum efficiency range, knocking control is required.

The goal of engine management is to operate the engine in a closed control loop at the knocking threshold to the extent that it is "before" the optimum moment of ignition. The engine

The engine management system detects knocking from the electrical knocking signal by first formatting the raw signal in an integrated circuit (IC) (Fig. 16-42).

The formatted raw signal is further processed in a microprocessor. The knocking event exists in a cylinder-selective form when the formatted raw signal exceeds the previously applied knocking limit adapted in engine operation. This evaluation occurs in a knocking time frame that is established for each cylinder by the crank angle of the engine. The energy calculated from the knocking signal determines in another block the extent of the ignition angle correction.

In case of error, i.e., when the knocking control cannot work properly because of a sensor error, the ignition angle is retarded for safety so that the engine operates reliably under all circumstances outside of the knocking

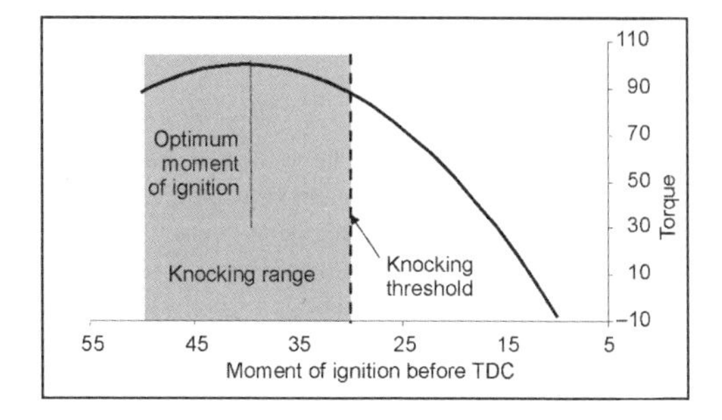

Fig. 16-40 Engine torque as a function of the moment of ignition.

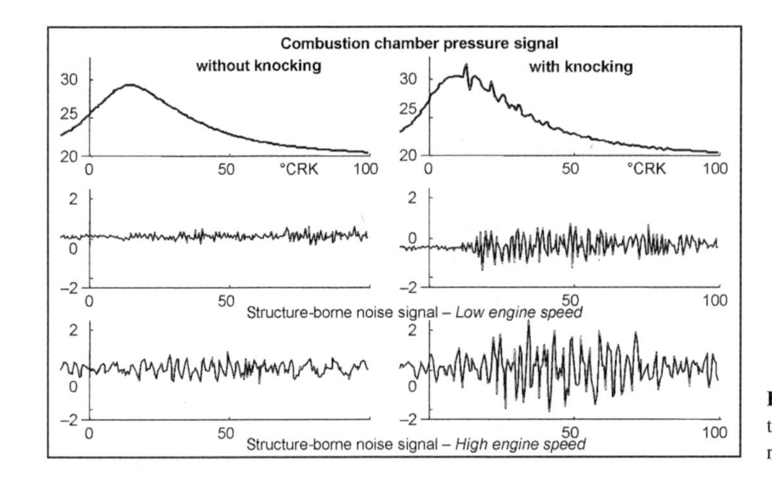

Fig. 16-41 Pressure characteristic and structure-borne noise recognition.

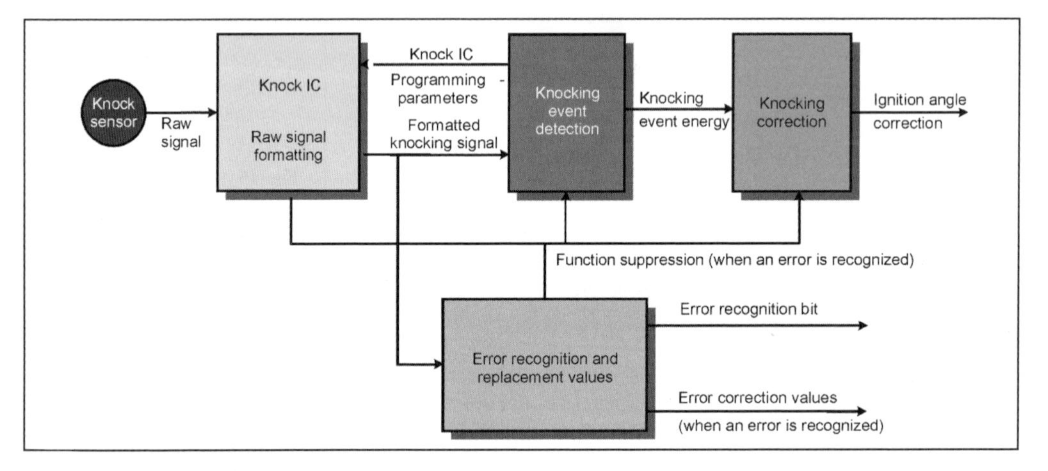

Fig. 16-42 Knocking signal processing.

range. Figure 16-43 shows the time characteristic of the knock control adjustments.

As can be seen in the figure, there is a fast and a slow ignition angle adjustment. This is because of the different phenomena that cause knocking. For example, deposits on the piston or the fuel quality are influences that only change slowly; in contrast, the intake air temperature and the engine operating point are influences that can change from power cycle to power cycle.

The position of the knock sensor should be set so that knocking is easily recognized while the engine is operating and can be clearly differentiated from other influences such as valve gear noise. The engine is extensively investigated in the engine development phase. Knocking is detected with combustion chamber pressure sensors, and the results are compared with the measured structure-borne noise signal.

In four-cylinder engines, a knock sensor is usually placed on the inlet side on the crankcase between cylinders 2 and 3. This allows the knocking noise of all four cylinders to be recognized. In six-cylinder inline engines, two knocking sensors are used. Two knocking sensors (one knock sensor per cylinder bank) are also used in V6 and V8 cylinder engines.

16.9.5 "On-Board" Diagnosis (OBD)

The air pollution from dense traffic in metropolitan areas led the United States in the 1960s to set legal limits of vehicle emissions. Today, the United States, especially California, has the world's strictest emission thresholds for passenger cars. This development has caused the exhaust gas purification systems for vehicle engines to become increasingly bigger and more complex.

These measures have substantially reduced pollutant emissions in new vehicles, yet simultaneously the portion of emissions has strongly risen from vehicles with defective exhaust gas purification systems. According to an estimation of the American Environmental Protection Agency (EPA) in 1990 (for example) approximately 60% of emissions came from uncombusted hydrocarbons from

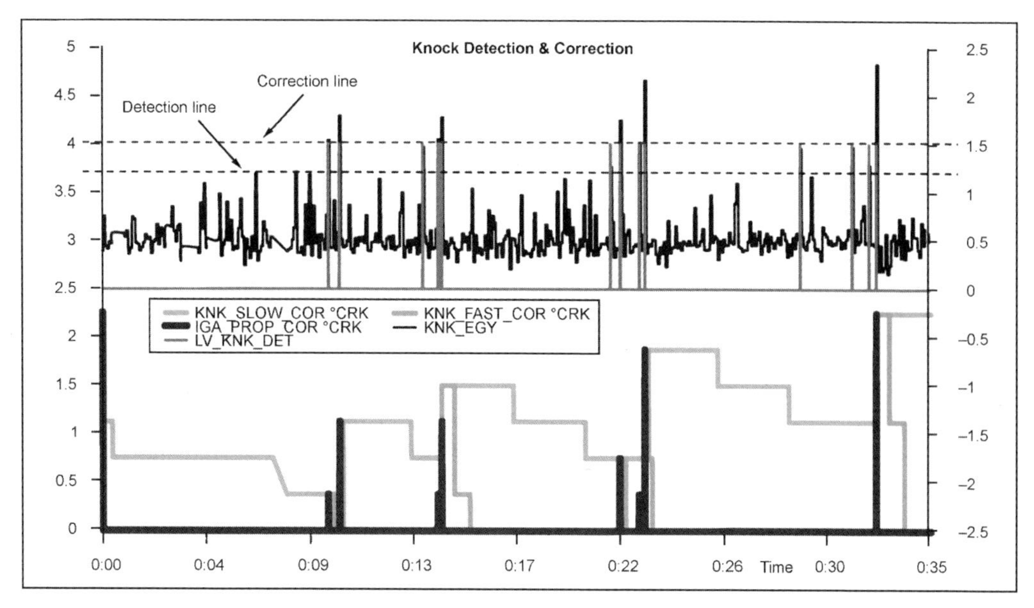

Fig. 16-43 Time characteristic of knocking adjustment.

vehicles with faulty exhaust gas purification systems. Because of this problem, the EPA has demanded that engine management systems provide vehicles with self-diagnosis systems that monitor all exhaust-influencing systems, functions, and components and inform the driver when these components are faulty.

The essential component for exhaust gas purification is the three-way catalytic converter. In the catalytic converter, the exhaust gas component carbon monoxide and the uncombusted hydrocarbons arising during engine combustion are oxidized into carbon dioxide and water. At the same time, the nitrogen oxides are reduced to nitrogen. For maximum conversion of all three exhaust gas components, the engine must operate with a stoichiometric mixture, i.e., with an air-fuel ratio of $\lambda = 1$. The mixture must, therefore, be precisely controlled.

To set the air-fuel mixture, the inducted air mass and the speed of the engine are measured. In the electronic control unit, these signals are used to calculate the opening time of the electrical fuel injectors and, hence, the fuel mass injected per power cycle to set a stoichiometric mixture. To adjust the mixture as precisely as possible to the required air-fuel ratio of 1, the so-called lambda control is superposed over this control process. The lambda sensor in the exhaust gas system determines if the mixture is set too rich or too lean. As a function of the sensor signal, a correction factor for the injection duration is calculated in the electronic control unit to produce an average air-fuel ratio of 1.

The catalytic converter starts operating only when its temperature lies above the so-called light-off temperature. Today, this temperature for catalytic converters is approximately 350°C. Fast heating of the catalytic converter during the warm-up phase can be attained by blowing secondary air into the exhaust gas system directly before the exhaust valves. The air-fuel mixture inducted by the engine is adjusted to be rich. The secondary air mass flow is adjusted so that the air-fuel ratio in the exhaust gas system is slightly lean. This causes the uncombusted hydrocarbons and carbon monoxide to be oxidized in the exhaust gas system. Since this reaction is exothermic, the exhaust gas temperature rises. This, in turn, causes the catalytic converter to quickly heat up.

In addition to the three-way catalytic converter, external exhaust gas recirculation is frequently used to lower nitrogen oxide emissions. Combusted exhaust gas is mixed with the combustion air. This causes the combustion temperature to fall, which consequently reduces nitrogen oxide emissions. The returned exhaust gas is metered by a valve in the return line.

In addition to exhaust gas emissions that arise from engine combustion, there are additional hydrocarbon emissions from the vaporization of the fuel in the tank. Tank ventilation systems are used to reduce these emissions.

These systems have the task of preventing the hydrocarbon vapors that arise in the vehicle tank from exiting into the atmosphere. An active charcoal filter that absorbs gaseous fuel is placed between the tank and the connection to the environment. This fuel is regenerated at specific intervals to prevent the filter from overloading. The tank ventilation valve between the active charcoal filter and the intake manifold of the engine is opened at specific intervals. The rising flow of air from the active charcoal filter causes desorption of the stored fuel. The rising fuel vapor-air mixture flows into the intake manifold and is burned in the engine.

16.9.5.1 Self-Diagnosis Tasks

The goal of self-diagnosis is to monitor the functioning of all the exhaust-relevant vehicle components and systems during normal driving conditions. If an error is determined, the problematic components are precisely located, and the type of error and location of the error and environmental conditions are saved in a memory. If the fault causes set exhaust thresholds to be exceeded, the driver is informed via a signal light in the dashboard and asked to bring the vehicle to a repair shop. In addition, suitable measures are undertaken to maintain driving safety, ensure continued operation, and eliminate subsequent damage. In the shop, it must be possible to read out the fault storage to allow the fault to be quickly found and fixed with the saved data.

The first thing that the California environmental authorities developed to attain this goal was a draft of a specific law concerning on-board diagnosis of engine management systems starting in model year 1988. All components had to be monitored that were connected to the electronic control unit of the engine management system. Starting in model year 1994, the expanded on-board diagnosis, abbreviated as OBD II, was required by law. For the first time, the monitoring of all exhaust-relevant vehicle components and systems was required. The requirements of the California environmental authorities were partially adopted by the other 49 states.

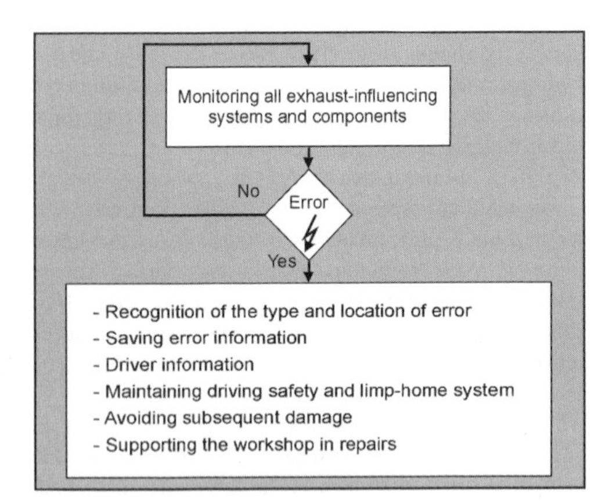

Fig. 16-44 Tasks of self-diagnosis.

In particular, the following main requirements exist (Fig. 16-44).for monitoring:

- The catalytic converter system
- The lambda sensor
- The entire fuel system including the fuel injectors, fuel pressure regulator, the fuel pump, and the fuel filter
- The secondary air system
- The exhaust gas recirculation system
- The tank ventilation system consisting of the active charcoal filter and tank ventilation valve
- Other systems held to be relevant to the exhaust that are not directly controlled by the engine management system, such as the transmission shift control for automatic transmissions.

Combustion misses should also be recognized.

In addition to monitoring these systems, there is a standardized fault light control and a standardized tester interface from which the fault memories can be read in the workshop.

16.9.5.2 Monitoring the Catalytic Converter

Monitoring the catalytic converter (Fig. 16-45) is one of the most important OBD II tasks. The catalytic converter is displayed as defective when the hydrocarbon emissions exceed a specific threshold in the U.S. FTP75 smog check. The respective threshold depends on the model year and the emissions rating of the vehicle.

When the diagnostic threshold is exceeded, the catalytic converter is displayed as defective.

For non-LEV (low-emission vehicle) certified vehicles, the diagnostic threshold is 1.5 times that of the hydrocarbon emissions threshold in the U.S. FTP75 smog test. For transitional low emission vehicles from model years 1996 and 1997, the diagnosis value is twice the exhaust gas threshold. For vehicles starting in model year 1998 and vehicles that are certified according to the low and ultralow emission thresholds, the diagnostic threshold is defined as 1.75 times the emissions threshold.

Based on the definition of the diagnostic thresholds, particularly low diagnostic thresholds result for vehicles that are certified according to the strict low emission and ultralow emission thresholds. For example, the maximum permissible HC emissions in the smog test for a ULEV (ultra low-emission vehicle) vehicle are 84% lower than for a vehicle not categorized as a low emission vehicle.

There are several processes for monitoring catalytic converters that all exploit the oxygen storage capacity of the catalytic converter. This storage ability correlates with the hydrocarbon conversion in the catalytic converter. Even a slight drop in the conversion rate leads to a clear reduction in the oxygen storage capacity of the catalytic converter.

The oxygen storage in the catalytic converter can be detected with a lambda sensor. In addition to the sensor in front of the catalytic converter, a second is installed after the catalytic converter, and the signals from the sensor after the catalytic converter are compared with the signals in front of the catalytic converter. With today's conventional lambda sensors, there is a low sensor voltage for lean mixtures and a high voltage for rich mixtures. Because of the design of the binary lambda control, there are rich/lean jumps in the sensor voltage with a relatively constant amplitude in the lambda sensor before the catalytic converter at $\lambda = 1$. With linear λ regulation, greater forced excitation is used for catalytic converter diagnostics. In a new catalytic converter with a relatively high oxygen storage capacity, these control fluctuations are

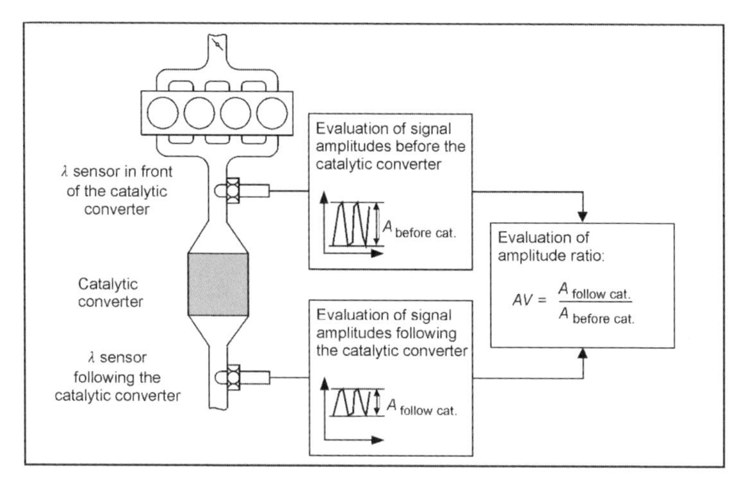

Fig. 16-45 Monitoring the catalytic converter.

strongly suppressed as shown by the sensor signal after the catalytic converter. A worn-out catalytic converter, as shown above, has a much worse storage capacity so that the control fluctuation in front of the catalytic converter influences the sensor following the catalytic converter.

The basic procedure for diagnosing the catalytic converter is as follows:

- The engine management system first determines the signal amplitudes of the lambda sensors before and after the catalytic converter. Then a quotient of the amplitudes is calculated. This amplitude ratio is used to evaluate the conversion rate of the catalytic converter.
- When the conversion rates are low, there is an average amplitude ratio of nearly 1. As the conversion rate rises, the ratio decreases.

For TLEV (transitional low-emission vehicle) vehicles and vehicles not categorized as low emission vehicles, this method represents a reliable way to diagnose catalytic converters.

In vehicles that are certified according to the strict LEV and ULEV thresholds, a worsening of the conversion rate of a few percent causes the diagnostic threshold to be exceeded. With these conversion rates, however, relatively low amplitude ratios are detected. It is very difficult to reliably distinguish between a defective and a functioning catalytic converter based on the amplitude ratio for these vehicles, especially taking into consideration the divergence within the product line.

To diagnose the catalytic converter efficiency of LEV and ULEV vehicles, a series of new methods have been developed. Let us consider two.

The majority of future LEV and ULEV vehicles will have a preliminary catalytic converter close to the engine in addition to the main catalytic converter. This preliminary catalytic converter has a relatively small volume, which, along with the fact that it is close to the engine, allows the operating temperature to be quickly reached and, hence, enables good exhaust conversion after a cold start.

One way to diagnose these catalytic converter systems is to monitor just the oxygen storage capacity of the preliminary catalytic converter with a downstream lambda sensor (Fig. 16-45). The assumption is that the preliminary catalytic converter ages much more quickly than the main catalytic converter. Since the volume of this catalytic converter is relatively small in comparison to the main catalytic converter, its maximum permissible drop in efficiency is much greater. Initial measurements show that the value to be diagnosed is a 30%–50% efficiency loss. A problem with this method is that the efficiency loss of the preliminary catalytic converter must correlate directly with the efficiency loss of the overall catalytic converter system. The suitability then strongly depends on the configuration of the catalytic converter system (Fig. 16-46).

In a second method, temperature sensors are before and after the preliminary catalytic converter in addition to the lambda sensors before and after the overall catalytic converter system. These additional sensors monitor the starting behavior and conversion in the preliminary catalytic converter. This exploits the effect that the reactions in the catalytic converter are exothermic, which increases

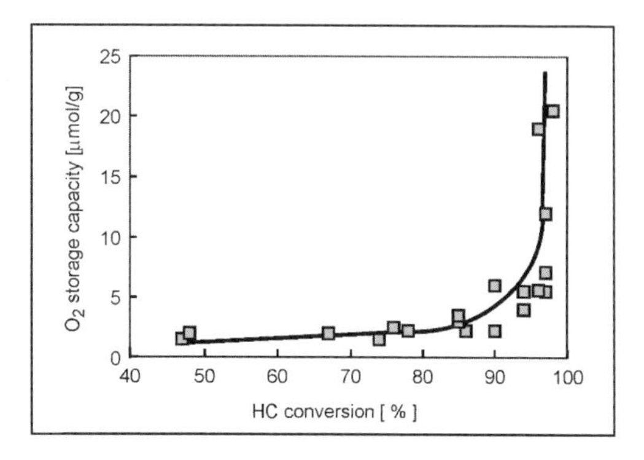

Fig. 16-46 Correlation between the oxygen storability and HC conversion.

the exhaust gas temperature following the catalytic converter. The increase in temperature, therefore, correlates with the efficiency of the catalytic converter. A disadvantage of this method is that, in addition to the second lambda sensor, precise and, hence, relatively expensive temperature sensors must be used. An overview of the diagnostic methods is shown in Fig. 16-47.

16.9.6 Safety Approaches

The law requires protective devices for systems that endanger life and property. When these devices fail, the remaining risk lies below a tolerable threshold. In complex systems with software, these are termed protective functions. In addition, the law states that safety-relevant systems must be state of the art. For so-called "drive by wire" engine management systems in which the throttle valve is not actuated directly via a Bowden cable but rather via an electrical drive independent of the gas pedal, a safety strategy is required in the engine management system.

This system eliminates hazardous situations for the driver. Such situations can be undesired acceleration (stepping on the gas), i.e., undesired starting of the vehicle, or an increase in engine speed. No engine output or only a low engine output is defined as a safe state. A distinction is drawn between individual errors (i.e., only one error) and multiple errors.

The engine management system must independently recognize individual errors and then be able to restore the vehicle to a safe state within 500 ms. Limited engine output is permissible that allows a "limp-home" function. In the case of multiple errors, it is permissible to include the reaction of the driver, e.g., brake actuation.

Monitoring	Technical Solution
Catalytic conversion system	Compare the signal amplitudes of the λ sensors before and after the catalytic converter For LEV/ULEV vehicles, also determine the starting temperature
λ sensor	Determine control frequency, signal range and heat resistance, overlapping control with the sensor following the catalytic converter
Misfiring	Calculate the uneven running from the angular velocity of the crankshaft
Tank ventilation system	Check the underpressure of the tank system
Exhaust gas recirculation system	Determine the intake manifold pressure with an active EGR
Secondary air system	Monitor the λ sensor signal
Fuel system	Monitor the λ control value

Fig. 16-47 Overview of diagnostic methods.

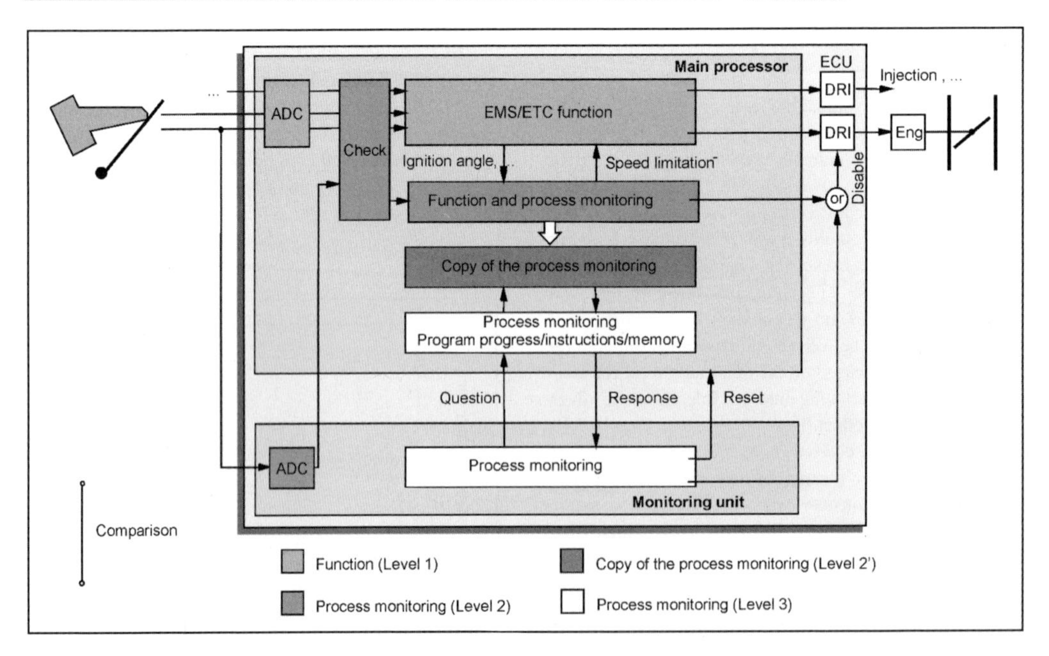

Fig. 16-48 Safety strategy in the engine management system.

To attain this goal, comprehensive changes are required in the engine management system:

- The pedal travel is detected by two independent position sensors.
- The throttle valve position is detected by two independent position sensors.
- The engine ECU contains a monitoring unit (usually a second processor) that operates independently of the main processor.
- The engine ECU contains extensive safety functions.

The safety functions (Fig. 16-48) are divided into several levels that realize different monitoring tasks. A distinction is drawn between the following levels:

- Level 1: Functions to control and regulate the engine including translating the position of the gas pedal into a throttle valve opening angle.
- Level 2: The process monitoring checks the control and regulation process of the engine (level 1) with a focus on all functions that could undesirably increase torque in case of an error.
- Level 2′: A copy of the code of the process monitoring system that is required for level 3 to function.
- Level 3: Processor monitoring.

Level 3 is executed partially by the CPU and partially by the monitoring unit. The monitoring module monitors the program's execution, the instruction sets, and the memory area. In addition, the monitoring unit gives the CPU arithmetic tasks and checks the function of the CPU by monitoring the responses. In addition, the monitoring unit also detects an analog input signal and makes it available for a plausibility check in the CPU, which supports the monitoring of the AD converter of the CPU.

The process monitoring on level 2 is designed so that some of the same functions as on level 1 are calculated, but not using the same data. To obtain consistency, the process monitoring must be precise. This leads to a redundant set point of the reference variable (e.g. indicated torque). In addition, the actual value of the reference variable is calculated on level 2 and compared with the redundant calculated set point. Actual values that are too high are thereby recognized, and corresponding error reactions are introduced that put the vehicle into a safe state.

17 The Powertrain

The integrated starter-motor/alternator (ISG) is examined in this chapter because it will play an important role in the future, inter alia, the conception of the powertrain.

17.1 Powertrain Architecture

Transmission of engine power output from the crankshaft to the drive wheels makes this output actually effective for the driver in the form of vehicle acceleration and deceleration. The torque-transmitting elements of an automobile powertrain are the following:

- The engine
- (Possibly) an integrated starter-motor/alternator (ISG)
- The gearbox, consisting of an initial movement element (e.g., a clutch) and the actual speed-reduction gearing system
- (Possibly) a power divider gearing system in the case of four-wheel drive
- The final drive gearing system(s) (the differential, possibly slip controlled)

(See Figure 17-1.) The powertrain's functions complementary to the combustion engine take the form of

- Initial movement (achieved with elements such as a clutch or a torque converter, or also with an integrated starter-motor/alternator)
- Possibly, the additional importation or recovery of electrical motive force using an integrated starter-motor/alternator, a battery, and/or large electrical capacities
- Balancing engine behavior and vehicle traction requirement (friction or geometrical-locking transmission elements such as a dual-shaft multistage transmission, a continuously adjustable variator, or a torque converter)
- Reduction of engine rotation irregularities (for example, damper elements in the clutch, a multimass flywheel, or a slip controlled torque converter)
- Distribution of power to the drive wheels (by distribution of torque between the front and rear axles, for example, and by subdifferentials between the left and right sides of the vehicle).

The transmission, in particular, combines the functions of an initial movement element and those of a power adjuster. In the latter case, the gearbox (in combination with the differential) adapts the engine operating characteristics to a significantly larger range of the torque/speed requirement at the gearbox output shaft (Figure 17-2, Ref. [1]).

17.2 The Motor-Vehicle's Longitudinal Dynamics

If the mass of a vehicle is concentrated at a single point in a model simulation, the acceleration and braking of this vehicle are then derived from the so-called "vehicle resistance equation" (moments listed here refer to summarized half-shafts):

$$m_{\text{vehicle}} a_{\text{vehicle}} = 1/r_{\text{wheel}} \, [i_{\text{tot}} M_{\text{motor-effective}} - M_{\text{vehicle resistance}}] \tag{17.1}$$

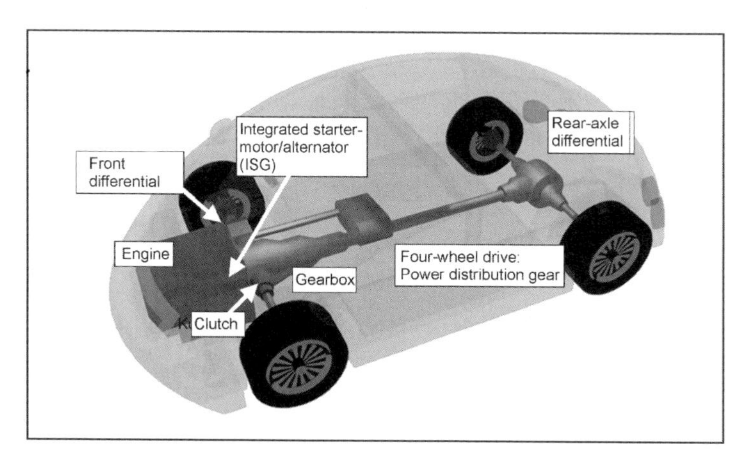

Fig. 17-1 Automobile powertrain.

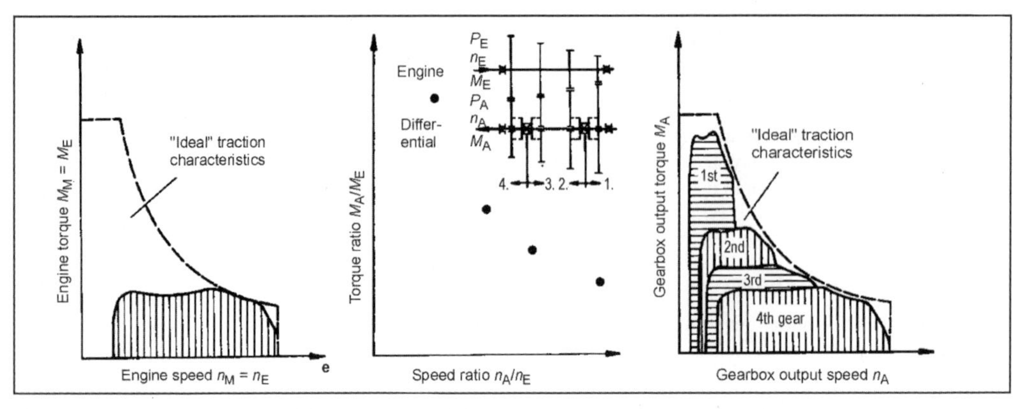

Fig. 17-2 Functions of an automobile transmission: Matching power requirement and engine output.[1]

in which

$$M_{\text{vehicle resistance}} = M_{\text{rolling resistance}} + M_{\text{climbing resistance}} + M_{\text{air resistance}} + M_{\text{braking}} \tag{17.2}$$

↑	↑	↑
Vehicle-mass, coefficient of rolling resistance (road surface and tire characteristics)	Vehicle mass and gradient	Vehicle speed, air resistance (air density, frontal cross section, coefficient of air resistance c_W)

m_{vehicle} = Vehicle mass
a_{vehicle} = Vehicle acceleration
r_{wheel} = Dynamic tire radius
i_{tot} = Overall transmission ratio
M = Moments

This equation can be used to determine vehicle-acceleration capability, and maximum vehicle speed (at $a_{\text{vehicle}} = 0$), as well as to calculate instantaneous climbing resistance in the context of real-time evaluation.

17.3 Transmission Types

Transmissions can be classified into the following types by the design of their transmission elements:

- Multistage transmissions as distinct from continuously variable transmissions
- Transmissions with natural axial eccentricity between the input shaft and the output or takeoff shaft (dual-shaft transmission) as distinct from transmissions with axial arrangement of the input shaft and output shaft (inline transmissions)

Multistage transmissions are based on geometrically locking transmission elements (e.g., sets of helical-toothed spur gears and planetary gears), whereas continuously variable transmissions are generally based on friction-locking functional principles. This friction-locking function necessitates additional auxiliary energy, with the result that continuously variable transmissions generally have poorer internal gearbox efficiency. These gearboxes balance out this disadvantage within the overall powertrain by their capability of adapting the engine working point optimally to road situations.

A further differentiating factor in automobile transmissions is their level of automation. Manually actuated ("stick-shift") transmissions (dual-shaft type), for instance, continue to play a significant role in Europe, whereas electrohydraulically actuated automatic transmissions (generally of planetary type) predominate in the United States and Asia. These types are increasingly being augmented with automated dual-shaft transmissions

- Based on an (electric-motor or electrohydraulically driven) dry clutch (automated stick-shift transmission)
- Based on a hydraulically actuated dual clutch (dual-clutch transmission)

and a continuously variable flexible drive transmission mechanism based on a so-called "push belt" or a chain.

Figure 17-3 shows a summary of various transmission types and a number of their characteristic features. Figure 17-4 shows the usual torque ranges for automobile applications.

Figure 17-5 shows an example of the DaimlerChrysler W5A 580 five-speed multistage transmission, which incorporates a slip controlled torque converter and three sets of planetary gears. This transmission is used in nearly all Mercedes-Benz standard (rear-wheel) drive automobiles.

Transmission type	Abbreviation	Transmission ratio	Weight	Noise	Consumption[a]	Gearshift comfort (ATZ value)[b]
Manual transmission (5-speed)	5MT	Dual-shaft transmission	Low	Low	−10.0%	—
Manual transmission (6-speed)	6MT	Dual-shaft transmission	Low	Low	−12.0%	—
Automatic multi-stage transmission (5-speed)	5AT	Sets of planetary gears	Medium	Low	−0.0%	9
Automatic multi-stage transmission (6-speed)	6AT	Sets of planetary gears	Medium	Low	−3.0%	9
Continuously variable transmission	S-CVT	Flexible transmission mechanism (based on so-called "push belt")	High	Medium	−5.0%	9.5
Continuously variable transmission	K-CVT	Flexible transmission mechanism (chain basis)	High	Medium	−5.0%	9.5
Toroidal drive	T-CVT	Friction-wheel transmission	Very high	Low	−7.0%	9.5
Automated manual transmission	E-AMT	Dual-shaft transmission with electro-mechanical actuation	Low	Low	−15.0%	6.3
Automated manual transmission	H-AMT	Dual-shaft transmission with electrohydraulic actuation	Low	Low	−14.0%	6.5
Dual-clutch transmission	DCT	Dual-shaft transmission with electrohydraulic actuation	Medium	Low	−8.0%	8.7

[a] Approximate consumption advantage compared to a five-speed automatic multistage transmission at 300 Nm operation and ungoverned gasoline engine.

[b] The ATZ value is a measure of the quality of the change of the transmission ratio. An ATZ value of 10 indicates an optimum (completely smooth) change of the transmission ratio, while a value of 1 indicates an extremely rough transition.

Fig. 17-3 Comparative assessment of a number of transmission types.[2–4]

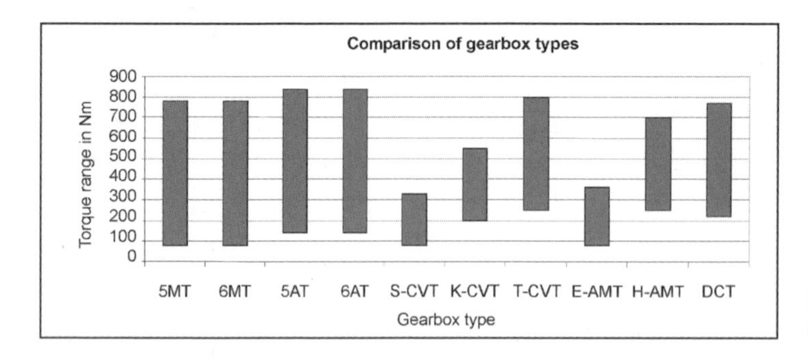

Fig. 17-4 Normal torque ranges for automobile gearboxes.

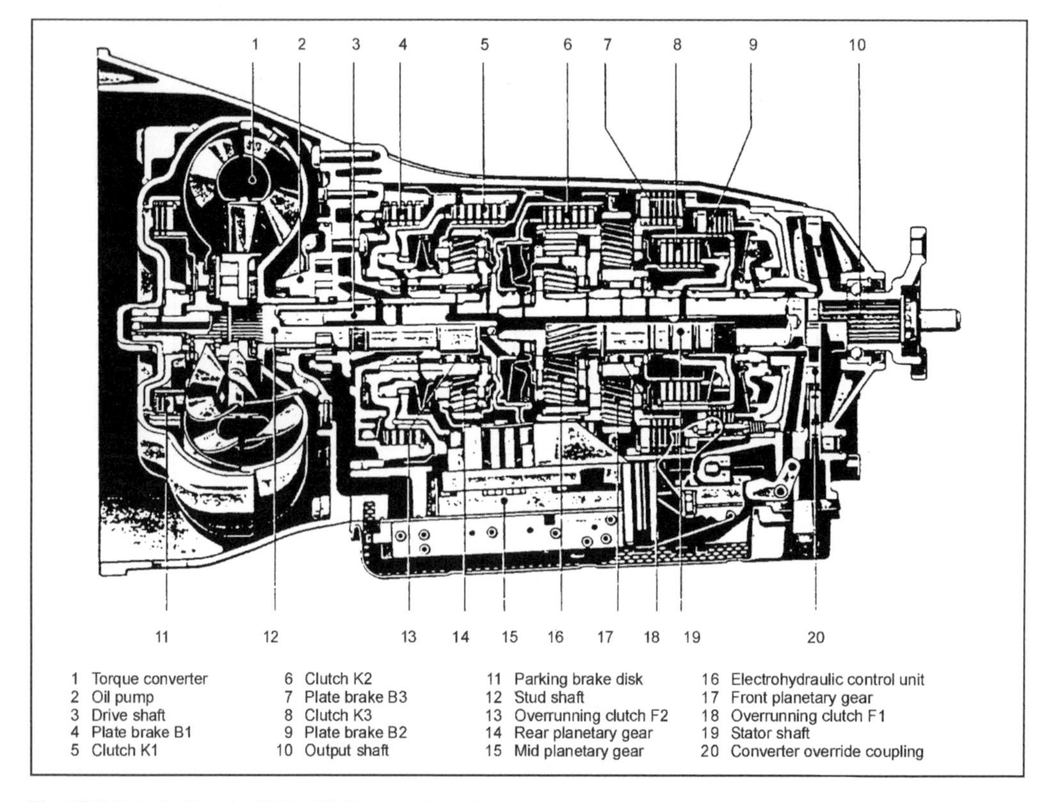

Fig. 17-5 DaimlerChrysler W5A 580 five-speed multistage transmission.

17.4 Power Level and Signal Processing Level

(a) Power level (Fig. 17-6):
This is the level of the actual torque-transmitting components.

(b) Signal level (Fig. 17-6):
Control and regulation of the entire powertrain is based on physical models of the individual components, which are functionally integrated by a torque-based model concept (starting from wheel torque and proceeding up to engine and transmission management).

(c) Links:

Modern powertrain architectures are characterized by a clear vertical correspondence between the power level and signal level components and their links. These links exist at the power level in the form of torque-transmitting shafts and in the form of communications channels at the signal level

Bibliography for Sections 17.1 to 17.4

[1] Mitschke, M., Dynamik der Kraftfahrzeuge, Band A, Antrieb und Bremsung, Springer, Berlin, Heidelberg, New York, 1995.
[2] Förster, H.J., Automatische Fahrzeuggetriebe, Grundlagen, Bauformen, Eigenschaften, Besonderheiten, Springer, Berlin, Heidelberg, New York, 1991.

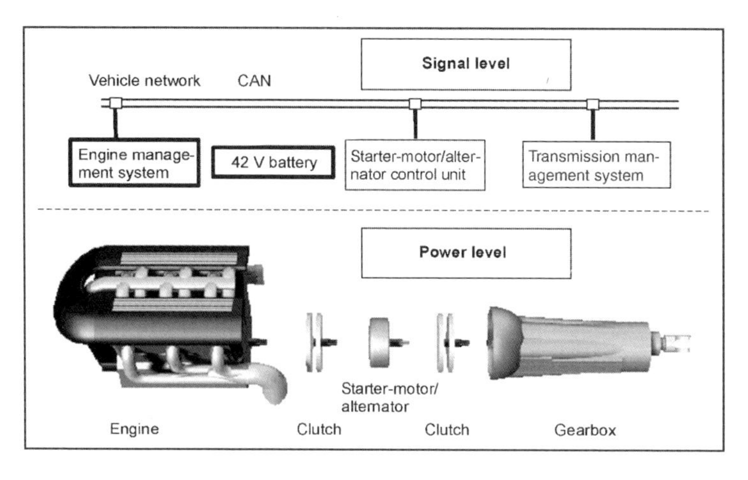

Fig. 17-6 Powertrain power and signal levels

[3] Lechner, G., and H. Naunheimer, Fahrzeuggetriebe: Grundlagen, Auswahl, Auslegung und Konstruktion, Springer, Berlin, Heidelberg, New York, 1994 (published in English in 1999: Lechner, G., and H. Naunheimer, Automotive Transmissions: Fundamentals, Selection, Design and Application, Springer, Berlin, Heidelberg, New York, 1999).

[4] Bock, C., Die ACEA-Vereinbarungen zur Flottenverbrauchsreduzierung und ihre möglichen Konsequenzen auf zukünftige Getriebekonzepte, Vortrag im Haus der Technik anlässlich der Tagung "CVT-Getriebe," Essen, 2000.

17.5 Transmission Management

17.5.1 Functions

17.5.1.1 Overview

The following functional groups can be defined for all transmission concepts:

- Gearshift strategy: Determines which target transmission ratio or which gear is selected
- Ratio transition: Management of the actual change of the transmission ratio
- Diagnosis functions: Achievement of a safe condition in case of component failure or emergency operation
- Special functions, such as control of the converter override function, gearshift lever disabler magnet actuation, and safety concept in "shift-by-wire" drive systems

These relationships are shown using the example of the automated manual transmission (AMT) in Fig. 17-7. The driving strategy in this context determines not only the target gear in automatic mode, but also checks manual gearshift commands by the driver (known, for instance, in the form of the so-called "Tiptronic" and "IntelligenTip®" systems). The subordinate level is responsible for the initiation and overall coordination of the gearshift sequence and, therefore, in the case of the AMT, for control of the engine (torque and speed), clutch torque, and logical gear position. The "actuator control" level is responsible for regulation of the appurtenant physical manipulated variables (path, pressure, angle, etc.).

17.5.1.2 Driving or Gearshift Strategy

Electronic transmission control systems, which for the first time enabled the driver to manually select a number of individual gear-changing programs such as "Economy," "Sport," and "Winter," were introduced during the 1980s. This proved not to be an optimum solution, however, since it at all times necessitated manual intervention by the driver in order to adjust the vehicle's gear-changing behavior to the road situations occurring.

Since, in addition, it could not be guaranteed that the driver would actually make a manual selection in every situation, it was ultimately necessary to make compromises in the case of these diverse gear-changing programs, too. For this reason, so-called "smart" driving and gearshift strategies, and gearshift strategies that automatically set the correct priorities on the basis of the prevailing conditions, have nowadays become an integral component of every automatic transmission system.

SAT (Siemens adaptive transmission Control)—see Fig. 17-8—the Siemens driving strategy for automatic multistage transmissions, is in successful use with a range of vehicle manufacturers, vehicle classes, and driving styles.[1]

Both global strategy criteria and short-term road situations are defined:

- Driver-type recognition
- Environment recognition: Road gradient
- Adjustment to low adhesion conditions (ice)
- Manual intervention (IntelligenTip®)
- Fast-off detection: Suppression of an upward shift if gas is reduced quickly (indicating the driver's intention to slow down)
- Bend recognition: Prevention of upward gear changes as a function of lateral acceleration
- Braking response: Additional change downs if the brake is actuated, taking account of engine speed limits and the road situation.

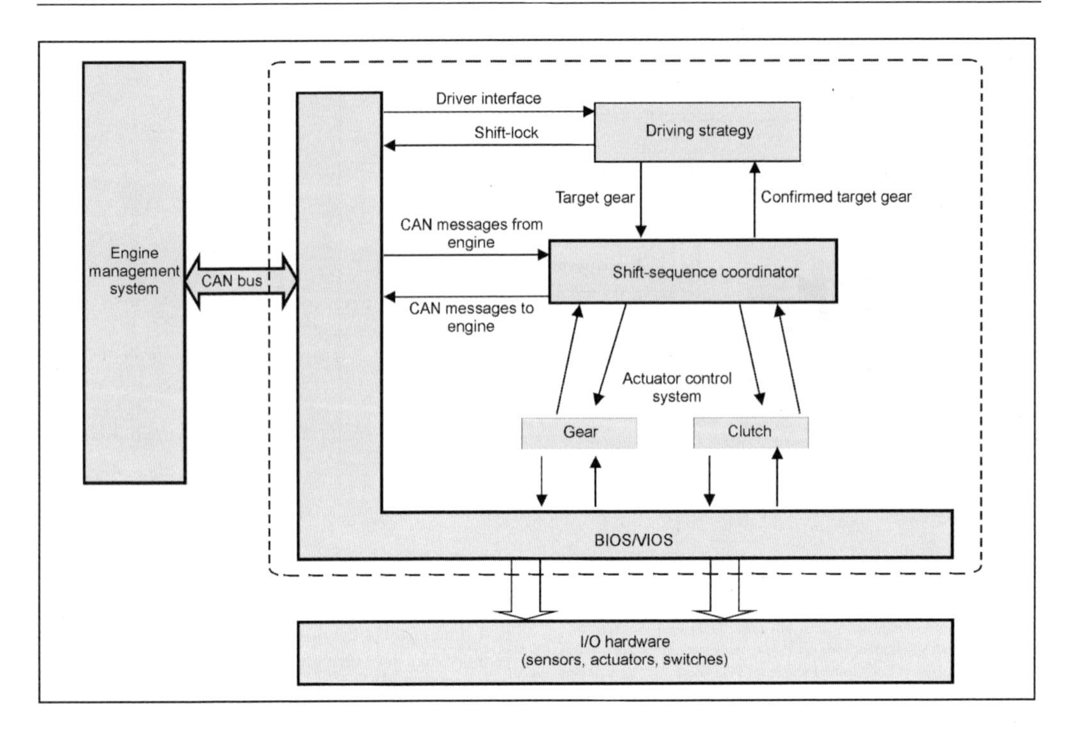

Fig. 17-7 Functional groups.

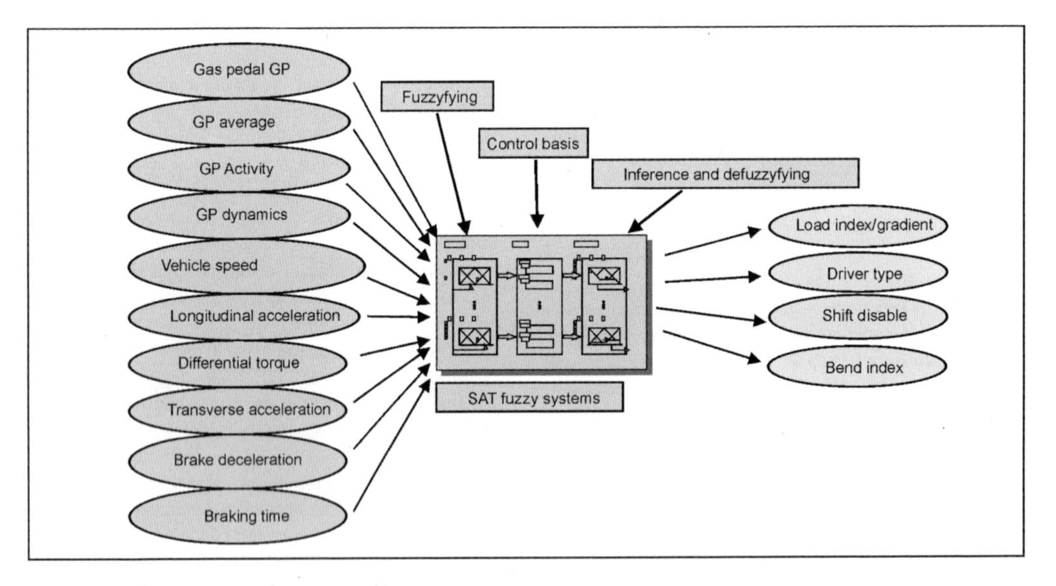

Fig. 17-8 Adaptive transmission control (Siemens).

A special feature of the SAT system is its comprehensive use of fuzzy logic, with 30 to 40 rules, depending on the expansion level. This permits achievement of high adaptation logic and dynamics. Thanks to an online learning component, conceptual future solutions, such as IntelligenTip®2 offer the driver more freedom to generate his own personal preferences in gearshift strategy; see Fig. 17-8.

17.5.1.3 Automatic Transmissions with Planetary Gears and Torque Converter

Present-day conceptual solutions give preference to direct single clutch management (see Fig. 17-9) using electro-hydraulic valves. This eliminates the need for a demultiplexer in the hydraulic system; the individual clutch pressures are calculated in the control unit software. This is also in line with the trend toward replacement of hydraulic functions by software functions, to achieve cost savings. Gear changing with the clutch management system is also made possible in principle.

17.5.1.4 Automated Stick-Shift Transmissions

The basic structure has already been explained in Fig. 17-7. Unlike automatic transmissions with planetary gears, explicit gear management is necessary for the dual-shaft transmission with automatic synchronization. The entire gearshift sequence is, thus, more subject to sequential control; i.e., clutch operation and gear-changing follow one another.

17.5.1.5 Continuously Variable Transmissions (CVT)

In the CVT (see Fig. 17-10), the variator and the individual thrust forces of the cones require continuous control (flexible drive CVT), whereas the gearshift sequence in multistage transmissions results in changing between discrete conditions. Principal attention is focused on minimization of thrust pressure to achieve the lowest possible fuel consumption and high ratio-change dynamics, but with dependable prevention of belt slipping. In addition, further

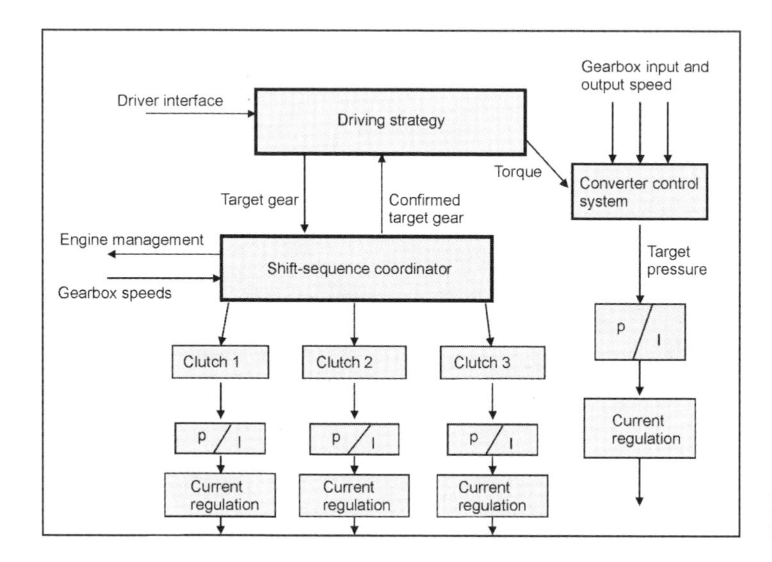

Fig. 17-9 Direct individual clutch management

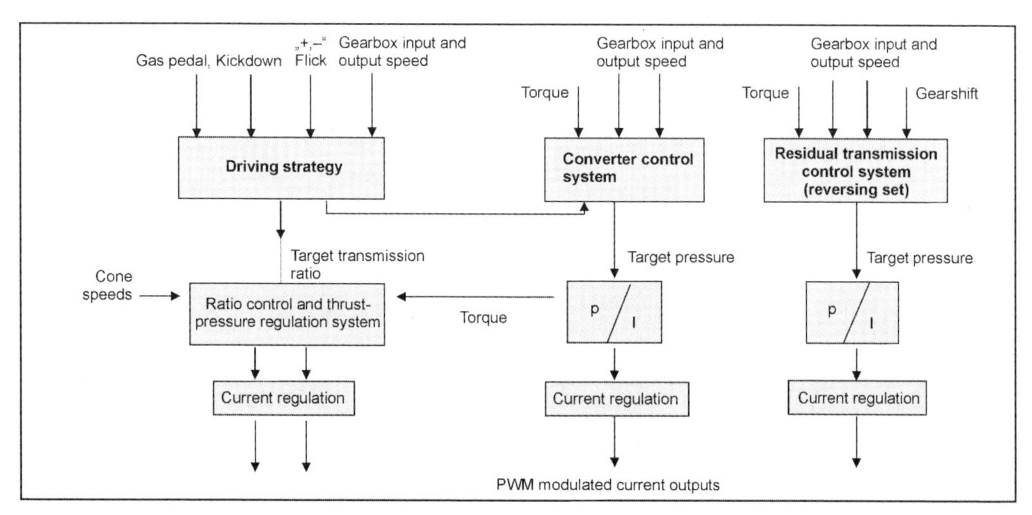

Fig. 17-10 Function diagram of the continuously variable transmission (CVT).

functions control converter override and the planetary gearing for vehicle reversing.

17.6 Integrated Powertrain Management (IPM®)

Future powertrain concepts consist of a number of individual subsystems: The engine, an electrical machine, and, in many cases, automated transmissions.

IPM® (integrated powertrain management)[3] has no direct influence on the process of conversion of the energy stored in the fuel (gasoline, diesel fuel, gas, or hydrogen), but instead attempts to optimize the working points for the energy converters (the engine and/or the electrical machine), the battery, as an energy storer, and the torque converter (transmission) on the basis of a holistic concept.

Because of the many degrees of freedom in such a system, it is important to control and coordinate downstream units optimally on the basis of central driver-intention interpretation and road-situation detection, taking account of the higher-level prioritization.

Integration in the sense of IPM® in this context covers the control and coordination of the entire system, but not design aspects such as system space requirement, installation, etc.; see Fig. 17-11.

An important feature of this concept is the introduction of a higher control level superimposed on the "component control systems." This guides the torque generator and/or converter through the relevant states and optimizes energy flow.

Integrated powertrain management is subdivided into three levels; see Fig. 17-12:

- Level 1 consists of driver and road-situation detection. Driver recognition includes instantaneous driver-intention interpretation and driver-type classification. The state of the powertrain (propulsion, braking, start/stop, coasting, boosting, recuperating, etc.) is determined at the second level on the basis of signals from Level 1 and other vehicle-sensor data.
- Level 2 is referred to as state management in the powertrain and performs the task, on the basis of the inputs from Level 1, of adjusting the powertrain to the

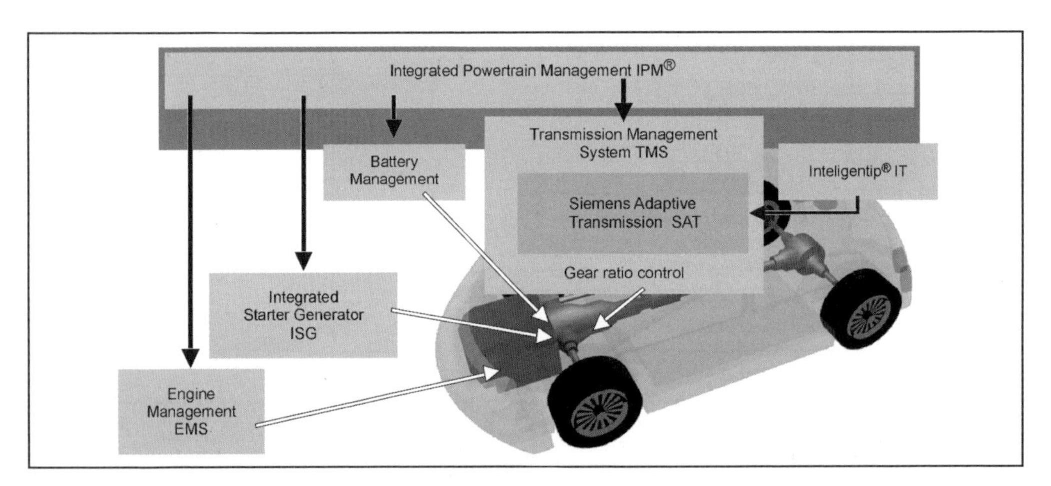

Fig. 17-11 Diagram of integrated powertrain management.

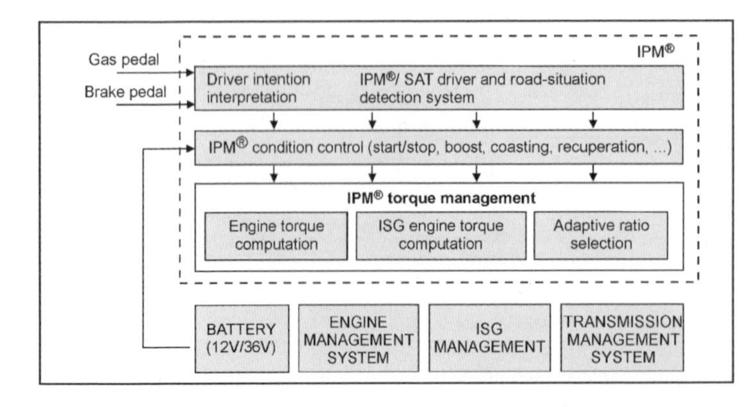

Fig. 17-12 Integrated powertrain management levels.

state that fulfills the currently prioritized optimization criteria.

- Level 3 supplies target data for the downstream units on the basis of physical variables. The engine and the electrical machine can, for example, each be operated at a defined working point by specifying a target torque or a target speed; a target transmission ratio is specified for the transmission system. In addition, certain driving conditions also access the clutch (so-called "free-wheeling," for example, disengaging the clutch if coasting in gear).

Bibliography for Sections 17.5 and 17.6

[1] Graf, F., F. Lohrenz, and C. Taffin, Industrialization of a Fuzzy Logic Transmission Controller, VDI Tagung: Getriebe in Fahrzeugen, Friedrichshafen, 1999.

[2] Heesche, K., F. Graf, W. Hauptmann, and M. Manz, IntelligenTip – eine trainierbare Fahrstrategie, VDI Tagung: Getriebe in Fahrzeugen, Friedrichshafen, 2001.

[3] Siemens-VDO Automotive AG.

17.7 The Integrated Starter-Motor/Alternator (ISG)

17.7.1 ISG: A System Overview

As a result of the continuously growing power demand in motor vehicles, alternators of outputs ranging from 4 to 10 kW will be needed in the future. The 14 V vehicle electrical system will no longer be adequate for distribution of this power and will be either replaced or augmented by a 42 V system.

The present-day solution of separate starter-motor and alternator will be fused into a single integrated system to achieve these ratings of as much as 10 kW. The structure in principle of an integrated crankshaft/starter-motor/alternator is shown in Fig. 17-13.

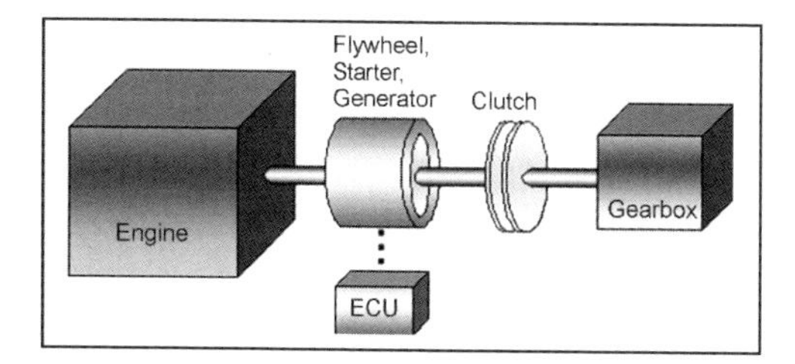

Fig. 17-13 Integrated starter-motor/alternator (ISG).

Additional system features, such as a boost function, retarder function, and start/stop function then become possible. Vehicle system simulations have demonstrated that integration of the starter-motor/alternator into the powertrain management system permits fuel savings of around 10% to 20 % in standardized driving cycles. The basic requirements made on a starter-motor/alternator can, therefore, be formulated as follows:

- High torque with low battery currents for starting and for "boosting" (assisted acceleration)
- Generation of the required output throughout the engine's speed range, with high efficiency and ultra-compact design
- Designs and structures suitable for use in automobiles

Asynchronous, synchronous, reluctance, and axial flux types are available for energy conversion. Current knowledge indicates that the asynchronous type, and also the permanently excited synchronous version, will best meet the requirements. The design of a suitable system is examined below, using the example of an asynchronous machine.

The following factors influencing the system as a whole must be taken into account:

- Dimensions and in-vehicle location
- Starting torque and alternator output
- Battery/starting current
- Electronics
- General design
- Noise
- Manufacturability
- Costs

17.7.1.1 Torque Structure in a Motor-Vehicle

The engine and starter-motor/alternator can be regarded as a torque source with the following features:

Engine

- Generation of torque in one direction only
- Torque adjustment time greater than 300 ms in the lower speed range
- Maximum torque not available in the lower engine speed range (particularly in the case of standard turbochargers)

Starter-Motor/Alternator

- Torque generation in two directions
- Torque adjustment times smaller than 5 ms achievable using field-oriented regulation
- Maximum torque developed only in the lower speed range

A suitable combination of both these torque sources make it possible to achieve a significant improvement in performance. A torque management system then permits rational distribution of the torque target data. A battery management system must also be integrated, since the battery's charge state must be taken into account for boost and retarder phases under all circumstances.

A brake management system must also be included if regenerative braking is to be possible.

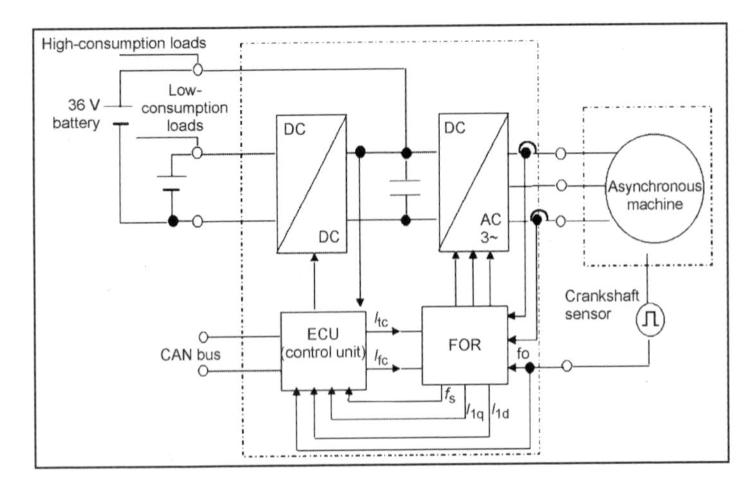

Figure 17-14 Starter-motor/generator structure.

17.7.1.2 Starter-Motor/Alternator Structure

As already mentioned, the starter-motor/alternator is linked via an interface (e.g., a CAN bus) to the torque management system for torque coordination.

The starter-motor/alternator unit is shown in the context of a dual-voltage vehicle system in Fig. 17-14.

17.7.1.3 Description of the Starter-Motor/Alternator's Most Important Modes of Use

Starter-Motor

This performs all the functions of a conventional Bendix-type starter, but can also accelerate the engine up to idling speed, permitting low-consumption, low-emission, and quiet starting. The start/stop function that can be optimally implemented in the ISG context permits achievement of entirely new dimensions, with starting times of less than 300 ms.

Booster

In this mode, the engine is assisted, particularly in its lower speed range. The relevant operating range extends from idling speed up to medium running speeds.

In order to avoid dips in performance during acceleration, it is necessary, as already mentioned, to know the battery's charge state, in order to permit derivation of potential torque assistance from this actual state value.

Alternator and Retarder

The starter-motor/alternator is capable of generating electrical power when the engine is running at idling speed or faster. The efficiency achieved in this mode is greater than 80% throughout the entire speed range and is practically independent of the electrical output yielded.

Recuperation of energy during braking can, of course, happen only if the battery is not already in its maximum permissible charge state. Charging current, which is dependent on a number of different parameters, can be determined using the battery management system and relayed via the interface to the starter-motor/alternator in the form of a "braking" target torque value.

17.7.2 Converters (Powertrain Management and Voltage Converters)

A suitable power-electronic controlling unit is required for regulation and control of this drive system. This unit must be capable of controlling and adjusting the necessary phase currents of the electrical machine and must provide adequate computing capacity for the necessary control algorithms.

Linking of the 14 V vehicle system, which will continue in use for many years, to the new 42 V system will be accomplished via a bidirectional DC/DC converter. It will also be a rational step to incorporate the DC/DC converter into the ISG control unit, since it will then be possible to multiply utilize important components with cost-optimizing effects.

17.7.2.1 Requirements Made on the Electronics from a System Viewpoint

The torque generated by the electrical machine can be regulated by controlling the voltage and current in the electric motor's stator.

The electrical machine's phase current is the central design criterion for the electronics. It, for its part, is determined mainly by the space available for the machine. Limited available space for the ISG motor means high phase currents.

Design of the ISG system is greatly complicated by the large number of differing requirements and operating states. These are examined in Fig. 17-15.

The driving cycle defined by the automobile manufacturer is a vital basis for the drafting of an optimized requirement profile. It indicates the duration of exposure of the power electronics to current under the applicable ambient conditions.

Operating condition	Phase current	With main effects on	Remarks
Cold starting	Maximum phase current	Power semiconductors, thermal design	Short-cycle operation, exploitation of thermal capacity
Hot starting	High phase current to achieve short starting times	Service life	Short-cycle operation in hot climates
Boosting	Medium to high phase current (depending on speed)	Cooling	Chronologically limited by battery capacity and heat generation (battery and ISG)
Alternator	Medium to high phase current (depending on speed and electrical load)	Cooling, selection of power semiconductors	Effects on thermal design
Typical driving cycle	Highly fluctuating phase current, depending on loads	Cooling, structure and connecting methods, intermediate circuit condensers	Decisive for design for optimized service life

Fig. 17-15 Effects of operating conditions on power electronics.

17.7.2.2 Function Groups and Design Criteria

The control unit's main function groups take the form of the DC/DC voltage converter for generation of the second system voltage, the DC/AC inverter, the field-orientated regulation (FOR) system for the powertrain and the regulation unit for the voltage converter, the communication system, activation of the end stages, and signal processing. This technology is designed around the following main parameters:

- Installation situation/cooling (flow rate and coolant pressure)
- Efficiency (loss-free phase-current measurement, cross sections, components)
- Phase current as a function of time
- Battery current (battery management)
- Necessary computing capacity (FOR)
- Automotive engineering necessities (tightness, volume, vibration)

The implementation of these functions necessitates several electronic components, the ultimate number of which can be drastically reduced by integrating device functions into user-specific integrated circuits (USC). This reduces unit volume and increases reliability.

17.7.2.3 Cooling

Water-based cooling has been used in the control unit under discussion to minimize its size and to make it possible to exploit the positive experience gained in electric vehicles. At present, restrictions imposed by the components used prevent the use of the engine cooling circuit because of the high temperatures that occur. A special cooling circuit, separate from the engine cooling circuit, is therefore installed in the vehicle. Any exceeding of permissible coolant temperature can, for the purpose of protection, be optionally counteracted with a power limiter. Air cooling is regarded as an alternative to liquid cooling, but is suitable only for smaller outputs, since the special location in the engine compartment does not permit the installation of larger equipment units close to the air inlet (accident performance). Figure 17-16 provides an overview of the zones in the electronic system in which heat losses may be anticipated.

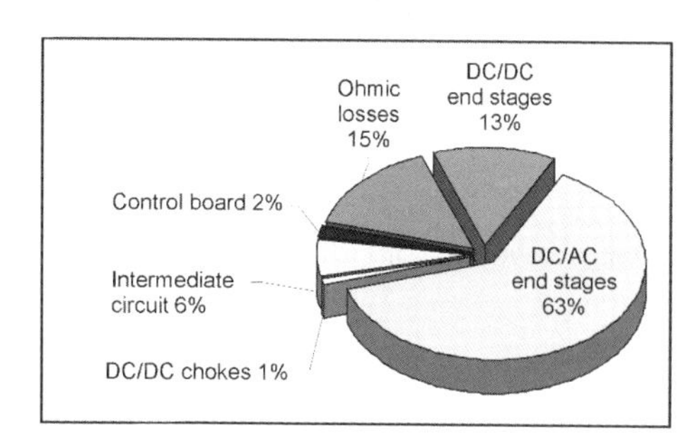

Fig. 17-16 Power loss distribution from control unit under cold starting.

The role played by free convection as a portion of total cooling can be ignored, since the control unit's power loss is around 500 W in continuous operation and actually rises to above 1.5 kW at starting.

The highly time-dependent load and temperature profiles encountered in automobiles differentiates this application from a conventional industrial plant, which is able to exploit forced convection.

Operating mode	Phase current	Power	Construction system
High power	$>150\ A_{rms}$	6–15 kW	Hybrid system
Low power	$<150\ A_{rms}$ (Cooling water: 85°C)	4–8 kW	Discrete MOSFET

Fig. 17-17 Classification of control units.

Fig. 17-18 Discrete technology ISG control unit.

7.7.2.4 Classification of the Converter's Power Electronics

Two different construction systems are used to meet the differing power requirements; see Fig. 17-17.

Between the two power classes, there is an overlap area essentially determined by the maximum cooling plate temperature and the available space.

Discrete Structure

Discrete construction technology utilizes individual regular power semiconductors and printed circuit boards (PCBs) employing conventional soldering methods for the mounting and connecting systems. The production of such electronic modules has proven extremely beneficial in automotive engineering. Figure 17-18 shows a solution for these design challenges, using individual semiconductors.

Hybrid Structure

The hybrid system is used in cases in which higher output currents are needed. In this case, power semiconductors are mounted without housings on, for example, ceramic substrates, which offer excellent thermal conductivity. The electrical connections are created directly on the ceramic surface, and the semiconductors are connected with adhesive bonding.

This method permits superior thermal connection of the transistors and the high-current substrates to the cooling system. The ohmic losses from the electrical connections are also reduced, since it is possible to achieve lower impedances. Figure 17-19 shows an example of a power stage incorporating this technology.

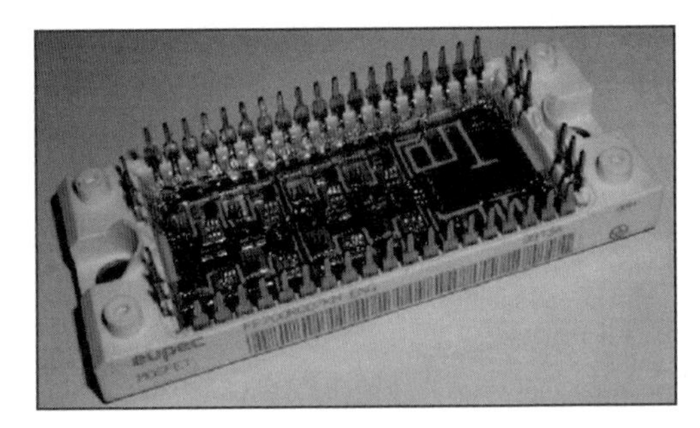

Fig. 17-19 Hybrid system end stages.

17.7.2.5 DC/DC Converters

Proposed architectures for dual 14 V/42 V vehicle systems incorporate a DC/DC converter, almost always of bidirectional type.

The rating desirable for such components is generally between 0.5 and 2.5 kW. Maximum efficiencies achievable for a reasonable economic input are generally stated to be around 90% to 95%. Innovative multistage topologies for the converter are notable for their extremely low input and output current ripple, which can actually reduce to zero.

Such components should, preferably, be integrated into the ISG's control unit. Installed components can then be multiply utilized, helping to minimize the space requirement. The unit volume increases practically fourfold compared to the integrated solution of the DC/DC converter, which is made as a separate control unit. The DC/DC converter end stage construction variants are identical to those of the inverter.

17.7.3 Electrical Machine

17.7.3.1 Design Criteria

Figure 17-20 shows an asynchronous machine with a stator bearer and rotor bearer for the 42 V vehicle electrical system.

A water-based cooling system is integrated into the stator bearer. This "drum rotor" design is a good solution if the inner boring of the rotor for mounting the clutch is also fitted with a dual-mass flywheel, and offers the following characteristics:

- Starting torque >200 Nm
- Battery current >500 A under cold-start conditions
- Alternator output ≥4 kW, with a system efficiency of >80% for all operating speeds and part electrical load

The volume per unit of output, and, therefore, initially, the rough dimensions of the machine, can be derived for a specified starting torque and alternator output. The design variable in this context is the starting torque.

The rotor's squirrel-cage winding is in the form of a die-cast aluminum component.

17.7.3.2 Simulation Tools

An economically optimum solution is needed for spatial and electromagnetic design if the starter-motor/alternator is to be produced on an industrial scale. The following knowledge is necessary for calculation and design of this machine:

- Electromagnetic design on the basis of the machine's fundamental and harmonic wave performance (analytical calculation)
- FEM computation for the fine contours
- Temperature distribution in the machine
- Mechanical strengths
- Noise excitation

In some cases, questions concerning component service life can be predictably answered in this context.

17.7.3.3 Thermal Simulation

Simulation of steady-state temperature distribution is necessary to permit the design of the cooling method and dimensioning of the ISG. Figure 17-21 shows a sectional view of a temperature distribution along path *a-b*, with boundary conditions as follows: Coolant temperature, 125°C; ambient temperature in the clutch housing, 130°C; temperature transmission via the clutch, 140°C; crankcase temperature, 150°C. The electrical machine's intrinsic heat generation in the rotor and stator derives from alternator output of, for example, 4 kW.

Design of the machine's insulation system also uses temperature distribution for dimensioning purposes.

Fig. 17-20 Asynchronous machine, "drum rotor" type.

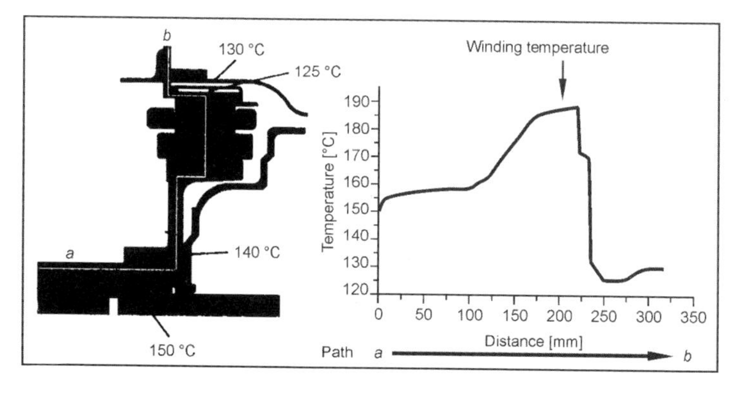

Fig. 17-21 Thermal simulation.

17.7.3.4 Mechanical Strengths

Calculation of the mechanical strengths of the entire design is necessary to permit development of an economically optimum design suitable for mass production. As a revolving component, the rotor is of particular importance in this context. To assure reliable operation throughout the targeted service life, it is necessary, for example, in the case of the rotor to calculate mechanical stresses, radial deformations, and centrifugal forces with and without imbalance from the various materials. The network model for FEM calculation of the rotor is shown in Fig. 17-22.

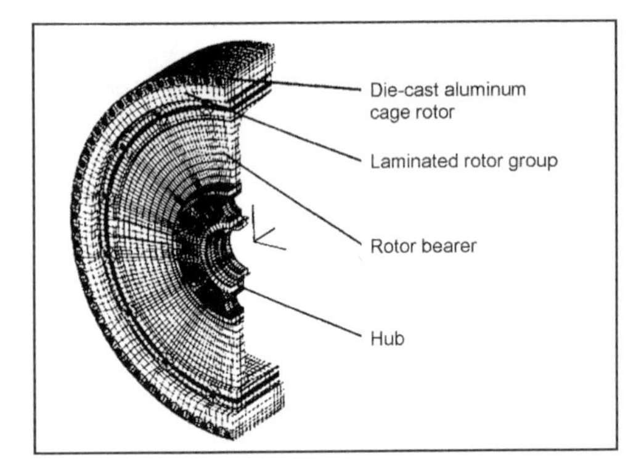

Fig. 17-22 FEM network model of rotor.

17.7.3.5 Requirements Made on the Electrical Machine

The requirements made on the electrical machine can be derived from the specification of the system as a whole. Machine type can be selected on the basis of the relevant background conditions.

Figure 17-23 selects a view that shows an initial, rough approximation of the features of the various machines. These machine types all offer specific advantages and disadvantages in practical service.

In the case of volume, starting current, and air gap, for example, the permanently excited synchronous machine exhibits advantages over its asynchronous cousin. The asynchronous machine, on the other hand, offers benefits in terms of costs, efficiency at high speeds, the position sensor, and robustness. A selection based on technical and economic criteria can be made when the variously weighted features are taken into account.

Figure 17-24 shows an actual example of a starter-motor/alternator system installed in an engine.

Demonstration of correct functioning in an engine is furnished here on the basis of dynamometer tests.

Fig. 17-24 Example of a starter-motor/alternator system installed on an engine.

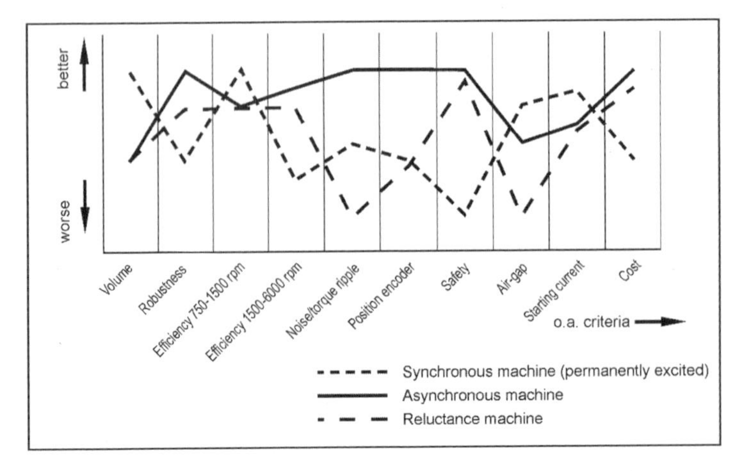

Fig. 17-23 Comparative view of synchronous, asynchronous, and reluctance machine.

17.7.4 Series Development

For industrial-scale production of prototypes whose functionality has been demonstrated, great demands are made in terms of suitability for series production. These requirements must be taken into account at an early stage when selecting the particular type of machine. The most important criteria are

- Spatial design
- Tolerances (stator, rotor, bearers, crankshaft, clutch housing)
- Residual imbalance in the completed rotor
- Validation and approval
- Vibration testing at up to 50 times gravitational acceleration
- Overspeeds of up to 13 000 rpm
- Temperature resistance, ambient temperatures greater than 150°C
- Cooling of the electrical machine
- Cost-efficient machine design

Bibliography

[1] Krappel, A., *et al.*, Kurbelwellenstartgenerator (KSG)–Basis für zukünftige Fahrzeugkonzepte, 2., erweiterte Aufl., Expert-Verlag, 2000.

[2] Vetter, A., *et al.*, High Current Module for the Automobile Industry, PCIM '99 Europe, Proceedings Power Conversion, Nürnberg, Germany, June 1999.

[3] Lovelace, E.C., *et al.*, Interieur PM Starter-Alternator for Automotive Applications, Proceedings ICEN '98, Instanbul, Turkey, September 1998.

[4] Miller, J.M., *et al.*, Starter-Alternator for Hybrid Electric Vehicle: Comparison of Induction and Variable Reluctance Machines and Drives, Proceedings IEEE '98.

[5] Altenbernd, G., H. Schäfer, and L. Wähner, Modern Aspects of High Power Automotive Starter-Alternator, Symposium on Power Electronics, Electronical Drives, Advanced Electrical Motors (speedam), Sorrento, Italy, June 1998, pp. 5–19.

[6] Altenbernd, G., *et al.*, Present Stage of Development of the Vector Controlled Crankshaft Starter-Generator for Motor Vehicle, Proceedings speedam, Ischia, Italy, June 2000.

[7] Schäfer, H., Starter-generator for Cars, Based on an Induction Machine with Field-Oriented Control SAE European Switched Reluctance Motors and Brushless Technology TOPTEC, September 1998.

[8] Rasmussen, K.S., and P. Thogersen, Model based Energy Optimisier for Vector Controlled Induction Motor Drives, EPE 1997 Trondheim.

[9] Späth, H., Steuerverfahren für Drehstrommaschinen, Springer, Berlin, Heidelberg, New York, 1983.

[10] Schäfer, H., Starter-Generator mit Asynchronmaschine und feldorientierter Regelung, in Sonderausgabe von ATZ und MTZ, Automotive Electronics, January 2000.

[11] Skotzek, P., *et al.*, High Performance Power Electronics for Integrated Starter/Generator Systems, Electronic Systems For Vehicles, 9th VDI Congress, Baden-Baden, Germany, October 2000

18 Sensors

18.1 Temperature Sensors

Most temperature measurements in the automobile utilize the temperature sensitivity of electric resistance materials with negative temperature coefficient (NTC).

The strong nonlinearity enables a large temperature range to be covered, Fig. 18-1.

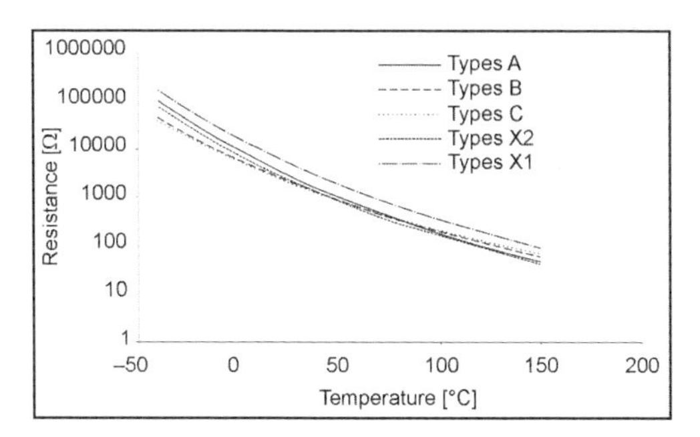

Fig. 18-1 Typical characteristics of temperature sensors (NTC).

For applications with very high temperatures (exhaust gas temperatures up to 1000°C), platinum sensors are employed.

The change in resistance is converted into an analog voltage by a voltage grading circuit with an optional parallel resistance to the linearization.

The sensors are employed for the following temperature ranges:

Application	**Temperature range**
Intake/charge air	− 40 to + 170°C
Ambient temperature	− 40 to + 60°C
Interior temperature	− 20 to + 80°C
Coolant temperature	− 40 to + 130°C
Engine oil	− 40 to + 170°C
Fuel	− 40 to + 120°C
Exhaust gas	+100 to +1000°C

Figure 18-2 shows various forms of temperature sensors for oil, coolant, and air temperature.

18.2 Knock Sensors

"Knock" is understood to mean an abnormal combustion in SI engines caused by a spontaneous ignition of the mixture in the cylinder. This undesirable combustion results in a significantly higher mechanical load on the engine. Continuous knock causes damage or even destruction, e.g., of the piston.

In an optimum combustion process, the zones of highest efficiency and of knock are very close together. Knocking generates oscillations with characteristic frequencies. These engine oscillations are registered by the knock sensor and transmitted to the engine controller where the signal is evaluated using corresponding algorithms to identify knocking. The engine controller then regulates the combustion process so that knocking no longer occurs (advancing the ignition timing by a few degrees on the crankshaft). Furthermore, knock control also permits operation with different fuel qualities.

Knock sensors are generally broadband knock sensors. These cover a frequency spectrum from, for example, 3 to over 20 kHz with an intrinsic resonance of over 30 kHz. The knock sensors are installed at appropriate positions on the engine block. In order to detect knocking for each individual cylinder, several knock sensors are employed on multicylinder engines (e.g., two sensors for six cylinders or four sensors for eight cylinders).

The functional principle of knock sensors is typically based on a piezoceramic ring that converts the engine vibrations into electrically processable signals using a superimposed (seismic) mass, Fig. 18-3.

Sensor sensitivity is expressed in mV/g or pC/g and is practically constant over a wide frequency range. The transmission behavior of the knock sensor can be adapted by the choice of seismic mass.

The resonance frequency can be increased by reducing the seismic mass. Since the tolerance band of the sensitivity of such sensors is approximately ±30%, the use of limit specimen sensors (of the sensitivity tolerance band) is important when tuning the engine controller.

Knock sensors with integral plugs are increasingly being used.

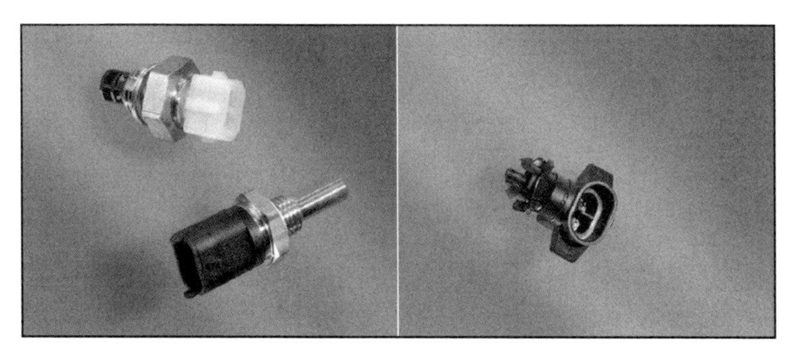

Fig. 18-2 Various forms of temperature sensors.

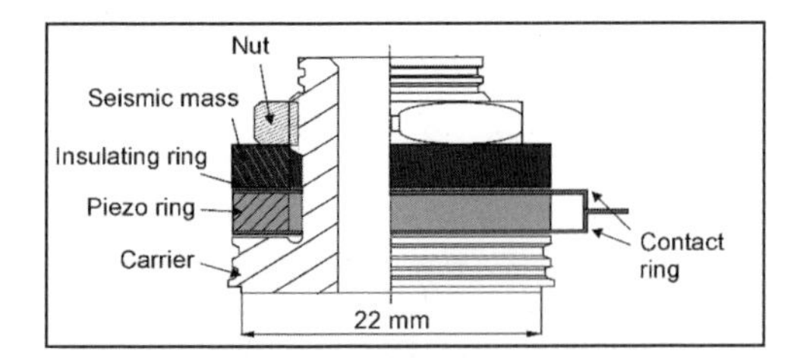

Fig. 18-3 Cross section through a knock sensor.

In some cases knock sensors are used today even in diesel engines to control the start of injection and the function of the injection nozzles (OBD).

18.3 Exhaust Gas Sensors

Installed directly downline of the manifold, they serve to control the fuel injection (λ control) in order to achieve the optimum conversion rate of a catalytic converter; installed downline of the catalytic converter they monitor its function and enable the OBD requirements to be satisfied.

Common to all the sensors employed today is that they consist of several layers of zirconium dioxide (ZrO_2) that conducts oxygen ions at temperatures above approximately 350°C and use the "Nernst equation": The voltage available above a ZrO_2 layer depends only on the difference in the oxygen partial pressures on each side of the layer.

18.3.1 Lambda Sensors

A distinction is made between the binary and linear lambda sensors. Binary sensors permit the control of the air-fuel ratio around the stoichiometric point $\lambda = 1$ and, hence, the setting of the fuel supply for an optimum conversion in the three-way catalytic converter. Linear sensors monitor the air-fuel ratio continuously between rich and lean and are particularly suitable for the control of lean engines, for example, SI engines with direct fuel injection.

With the binary lambda sensor, the Nernst voltage is measured between a catalytically active exhaust gas-side electrode and a reference electrode suspended in air. The voltage changes sharply around $\lambda = 1$, Fig. 18-4.

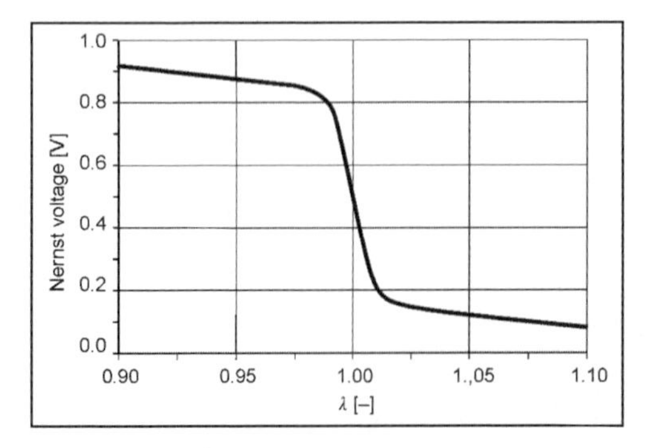

Fig. 18-4 Characteristic of a binary lambda sensor.

With the linear lambda sensor, the air-fuel mixture is controlled to a Nernst voltage corresponding to $\lambda = 1$ in a chamber inside the sensor by connecting an electric current referred to as the pump current.

The air reference is generated either via a channel in the ceramics or by a constant supply of oxygen to a cavity. The pump currents serve as a measurement signal and depend on the air-fuel ratio, Figs. 18-5 and 18-6.

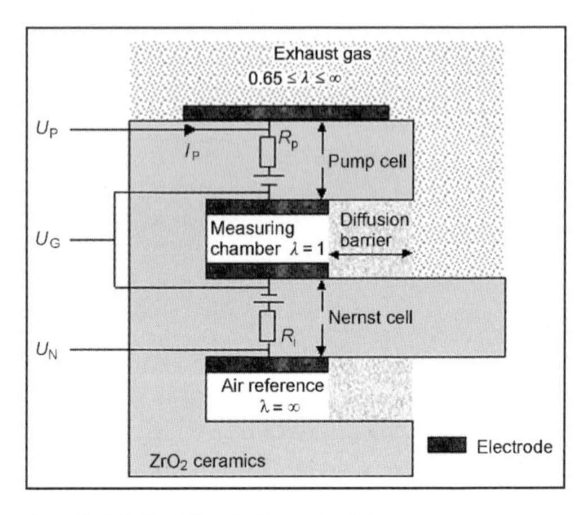

Fig. 18-5 Principle of a linear lambda sensor.

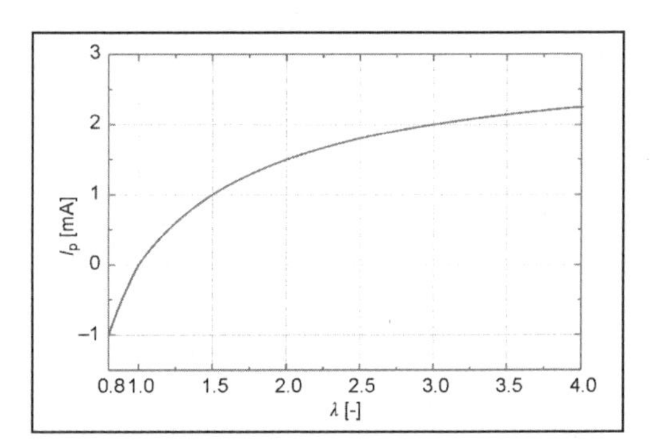

Fig. 18-6 Characteristic of a linear lambda sensor.

18.3.2 NO_x Sensors

The NO_x sensor permits the direct measurement of the nitrous oxide concentration in the exhaust gases of SI and diesel engines. It enables optimum control and diagnosis of NO_x catalytic converters by the engine controller [e.g., NO_x accumulator, selective catalytic reduction (SCR) catalytic converter] and compliance with OBD requirements for the checking of the three-way catalytic converter for low-emission concepts (SULEV, LEV 2).

The most promising functional principle of a NO_x sensor is based on the decomposition of nitrous oxides using a catalytically active electrode comprising a mixture of platinum and rhodium. The measurement of the oxygen produced in the process is well known from the amperometric linear lambda sensor. The structure of the multilayer ZrO_2 sensor ceramics comprises two chambers. In the first, the oxygen contained in the exhaust gas is reduced (lean exhaust gas) or increased (rich exhaust gas) to a constant partial pressure of 10–20 ppm by connecting a pump current. The necessary current is proportional to the reciprocal of the air-fuel ratio. The NO_x reduction at the measuring electrode takes place in the second chamber. The current necessary to keep the atmosphere around the electrode free from oxygen is proportional to the nitrous oxide concentration and forms the measurement signal, Fig. 18-7.

integration into the engine management system is possible only with electronic control of the sensor in its immediate vicinity. There are two ways of doing this, either with a stand-alone or "smart" NO_x sensor with the complete control system (heater control and pump current control) and digital communication to the engine controller or with a remote pump current controller with analog control.

18.4 Pressure Sensors

Different sensor types are employed in order to meet the various demands of the pressures to be measured.

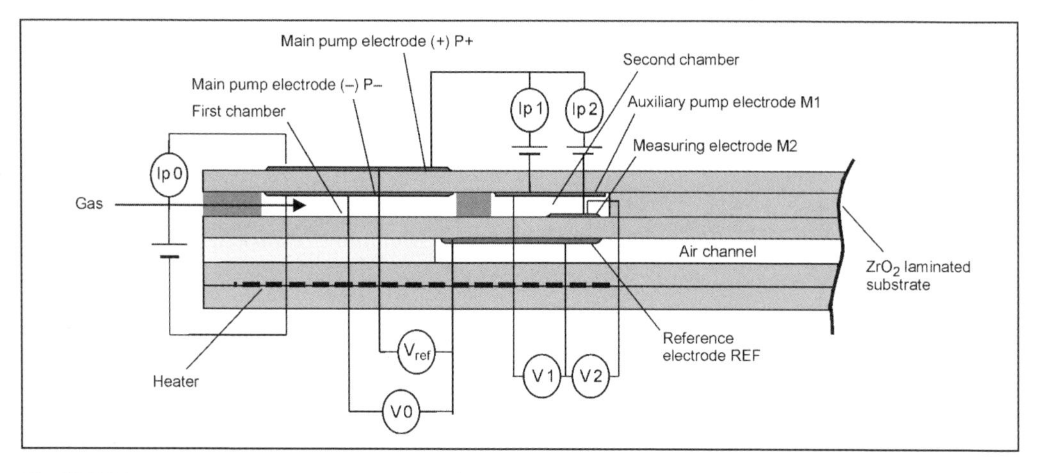

Fig. 18-7 Measurement principle of an NO_x sensor.[1]

A two-step setting of the residual oxygen in the first and second chambers using an additional electrode enables the cross sensitivity of the sensor to oxygen to be reduced. Knowledge of the air-fuel ratios also permits a numeric compensation of the NO_x signal. A disadvantage of such sensors is their high ammonia cross sensitivity, caused by oxidation of ammonia producing nitrogen monoxide in the first sensor cavity, Fig. 18-7.

The necessary currents at the NO_x measuring electrode are a few μA for a measuring range of several hundred ppm, Fig. 18-8. An electromagnetically safe

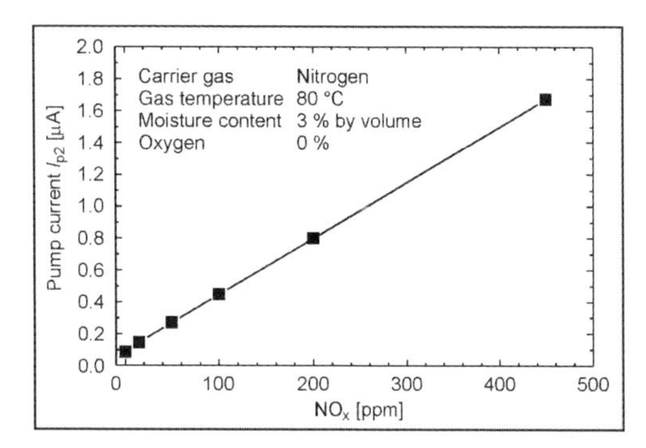

Fig. 18-8 Characteristic of an amperometric NO_x sensor.

These sensor types can be categorized as follows:

Sensor Type	Pressure range
Normal pressure sensors	ca. 0–5 bar
Medium pressure sensors	ca. 5–100 bar
High pressure sensors	ca. 100–2000 bar
Differential pressure sensors	ca. 0–1 bar

The fields of application and sensor principles for these sensor types are described in the following sections.

18.4.1 Normal Pressure Sensors

Normal pressure sensors can be subdivided into the following groups:

- MAP: Manifold absolute pressure sensor (intake air pressure sensor)
- BAP: Barometric absolute pressure sensor
- Turbo MAP: Manifold absolute pressure sensor for turbocharged engines

The MAP is used to determine the intake manifold vacuum pressure that exists downline of the throttle valve. The typical measuring range here is 0.2–1.1 bar.

Together with the temperature, this allows the intaken air mass to be calculated. Depending on the driver's wishes, this information forms the basis for determining the gasoline injection volume and the throttle valve position.

Using the lambda sensor signal, a closed control loop is set up that controls the air-fuel mixture in the range of $\lambda = 1$ in order to guarantee minimum exhaust gas emissions. The MAP is often used with an integral temperature sensor to reduce installation work, Fig. 18-9.

The BAP is used to determine the ambient pressure. The information received serves for compensation of the air pressure at different altitudes. The typical measuring range here lies in the order of 0.5–1.1 bar.

The Turbo MAP serves to determine the charge pressure. The typical measuring range here lies in the order of 0.5–2.5 bar.

The engine controller optimizes the combustion parameters using the charge pressure information. The charge pressure can also be used to control the turbocharger (VTG).

The following measurement principles are in use.

18.4.1.1 Piezoresistive Measurement Principle

Piezoresistive measuring cells are traditionally used. A piezoresistive measuring cell is a pressure cell consisting of a diaphragm with attached piezoresistors. The pressure acting on the cell causes the piezoresistors to expand. This results in a pressure-dependent change in resistance. These changes in resistance are transformed into voltages (typically between 0 and 5 V) by a separate electronic circuit.

In more recent versions, the pressure cell is integrated into the chip with "volume micromechanics."

18.4.1.2 Capacitive Measurement Principle

A fundamentally new development is the "surface-mounted micromechanical" pressure sensor, Fig. 18-10. Here the pressure cell and the associated evaluation electronics are manufactured on a chip using standard semiconductor processes (BiCMOS). This eliminates the bond wire connections between pressure sensor cell and evaluation electronics.

The pressure is measured by a special capacitor-like pressure cell. The pressure acting on the cell changes the distance between the two capacitor surfaces and results in a change in the capacitance. This is transformed into an output signal of 0–5 V in the integral electronics. The configuration is shown schematically in Fig. 18-10.

The desired characteristic for the different applications and pressure ranges mentioned above is set by calibration at the end of the sensor production.

18.4.2 Medium Pressure Sensors

These sensors are used, e.g., for engine and hydraulic oil pressure in automatic transmissions and applications outside the powertrain (e.g., A/C compressors). For fluids or aggressive media, a configuration is predominantly used that is similar to the high-pressure sensor described below.

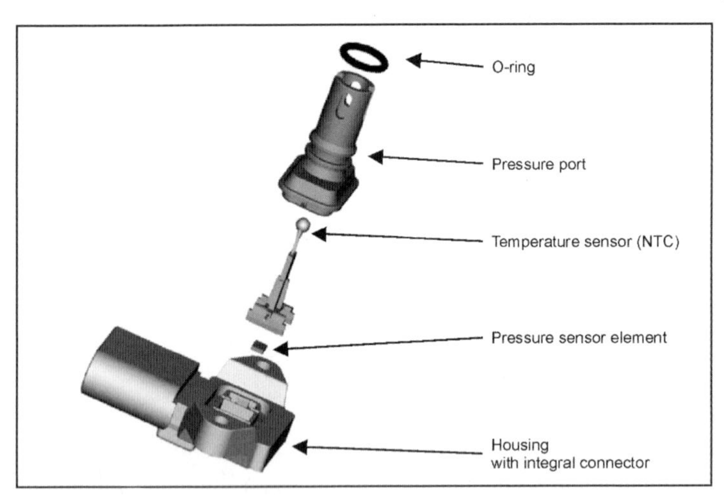

Fig. 18-9 Structure of a MAP with integral temperature sensor.

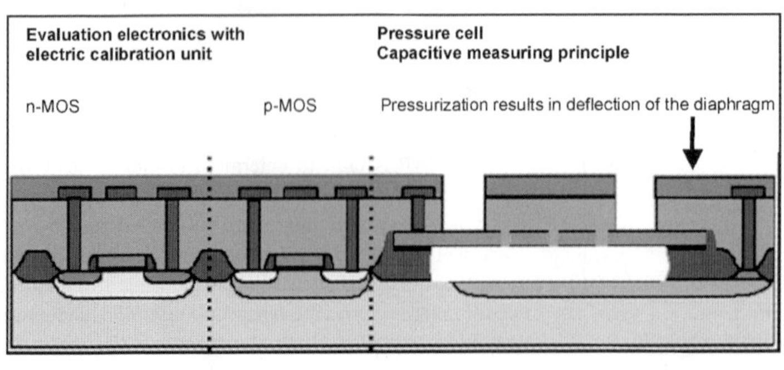

Fig. 18-10 Structure of a surface-mounted micromechanical pressure sensor.

18.4.3 High-Pressure Sensors

The pressure range for high-pressure sensors starts at approximately 100 bar, Fig. 18-11. The common design is of hexagonal form with an M12 screw thread connection. Plugs with three pins (supply voltage, ground, and output) are generally used. The calibration is also performed via the three pins or additional contacts.

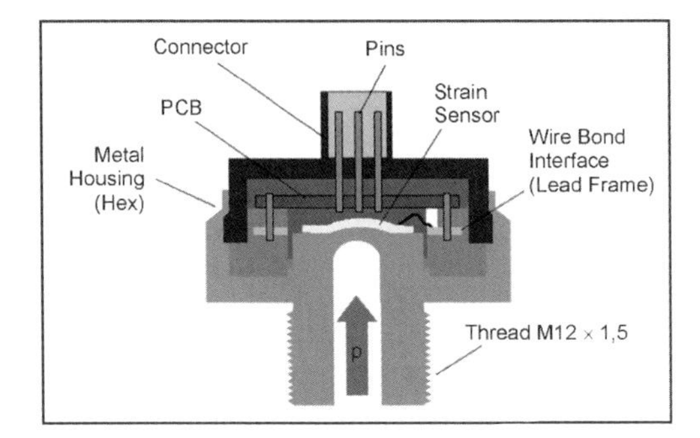

Fig. 18-11 Principle sketch of a high-pressure sensor.

The main fields of application can be defined as follows:

100–200 bar	HPDI gasoline direct injection systems
200–280 bar	Brake pressure sensors
1300–2000 bar	Common rail diesel injection systems

A fundamental distinction between the sensors is made on the basis of three different media separation concepts:

1. Hermetic separation of the measuring element by a steel housing and a steel diaphragm.
2. Hermetic separation by thin corrugated diaphragms, where the space for the measuring element (silicon pressure sensor) is filled with silicon oil under vacuum as a pressure transmission medium.
3. The measuring element (silicon chip) is exposed directly to the medium; voltage supply and signals are transmitted via hermetically sealed glass leadthroughs.

The above media separation concepts predominantly employ the following sensor principles or pressure measuring cells:

- Capacitive sensors with ceramic measuring cell (aluminum oxide) up to approximately 200 bar
- Thick film strain gauges on ceramic diaphragm up to approximately 200 bar
- Thin film strain gauges on steel diaphragm (100–2000 bar)

18.4.3.1 Technical Boundary Conditions

Of critical importance for the use of the sensors is the guarantee of the leak tightness over the service life. With specified operating temperatures from −40 to +140°C, enormous demands are made on the fatigue strength of the materials, as the loads on the material are extremely high, particularly in the 2000 bar range. The very large number of load cycles and temperature cycles demands the use of extremely reliable assembly and connection technologies as well as very extensive environmental testing.

In addition, a sealing concept for the thread connections tailored to the particular sensor is also needed. Tightening torques and clamping forces must not have any reactions (e.g., zero point shift) on the pressure measurement. Especially for the 2000 bar range, high-strength and, therefore, very cost-intensive materials have to be used. To minimize material costs, there are solutions in which the complex diaphragm body is welded to the housing. Furthermore, high-pressure sensors have to withstand a bursting pressure of roughly 1.5 to 2 times the rated pressure.

18.4.3.2 Signal Transmission

Customers demand almost exclusively ratiometric signal outputs. A "clamping" function of the output voltage is the standard in the meantime and, therefore, permits a diagnostic function. The measurement rate for the output signals normally lies between 1 and 5 ms.

18.4.3.3 Measuring Precision

High demands are also made on the measuring precision of the sensors. Whereas in the range up to approximately 200 bar an overall precision of approximately 2%–3% within a limited temperature range (e.g., 20–90°C) is demanded, the demands for common rail applications are higher and lie in the range of approximately 1.5%–2.5% overall precision. This precision can be achieved only if a time-consuming calibration for at least two load points and at least two temperatures is performed immediately after production. In view of the high pressures of more than 2000 bar, the setting up of the calibration stations and the performance of the calibrations is more difficult than with other sensors.

18.5 Air Mass Sensors

In order to be able to determine the air mass flow drawn in by the engine, either a manifold pressure sensor (MAP) or an air mass sensor is employed today. The output signal serves in the electronic engine controller as the basis for determining the load state.

In SI engines, the signal serves predominantly for controlling the fuel volume as an input parameter for the ignition map and for determining the exhaust-gas recirculation rate. In the interplay with the lambda sensor, the hot-film anemometer (HFM)/MAP forms a closed control loop.

Since with diesel engines there is no throttle valve and, hence, the intake manifold pressure is no measure for the intaken fresh air mass, an HFM has to be employed. Here the signal from the HFM serves as a control variable for the exhaust-gas recirculation (EGR); in newer systems it is also a control parameter for a map-dependent diesel injection pump.

Since the exhaust gas provides no feedback with diesel engines, the demands made on the measuring precision of the HFM are higher than with the SI engine.

The measurement of the intaken fresh air mass is therefore a precondition for reducing the pollutant emissions and for increasing the driving comfort.

18.5.1 Comparison of Air Mass-Controlled and Intake Manifold Pressure-Controlled Systems

By comparison with the intake manifold pressure-controlled system, the HFM offers the benefit of being able to measure the desired parameter directly and not having to rely on ancillary parameters such as, e.g., the intake air temperature and engine displacement maps. This reduces the application complexity and increases the precision as fewer variables have to be taken into consideration. With increasing performance of the engines through, e.g., turbocharging, direct injection, variable valve timing, etc., the complexity of the model calculation for an intake manifold pressure-controlled system increases, and, consequently, the precision decreases. The measuring precision of an air mass-controlled system remains more or less unaffected by these influencing factors.

18.5.2 Measuring Principles

Apart from the principles for determining mass flows employed earlier, sensor plate and hot wire, the following are predominant today.

Ultrasonic

With the ultrasonic run-time method, the time taken for a sound wave to travel from the transmitter to the receiver is measured. This parameter allows the flow velocity to be determined and, in combination with air density and temperature, the mass flow to be determined.

Hot-Film Anemometer

Practically all air mass sensors employed in the automobile today follow this principle. A heated element dissipates energy to the surrounding air. The dissipated heat energy is dependent on the air flow and can be used as a measurement parameter.

18.5.3 Hot-Film Anemometer

The HFM, Fig. 18-12, consists of a tubular housing with flow straightener (honeycomb-lattice combination), sensor guard, and the sensor module. The tube diameter is adapted to the air mass range required in each case.

Sensors, electronics, connecting elements, and plug are integrated into the sensor module.

Two temperature-dependent metallic-film resistors on a glass substrate (R_S and R_T) are positioned inside the tube in the direct intake air stream. These two resistors, in combination with R_1 and R_2, are linked in a bridge circuit, Fig. 18-13.

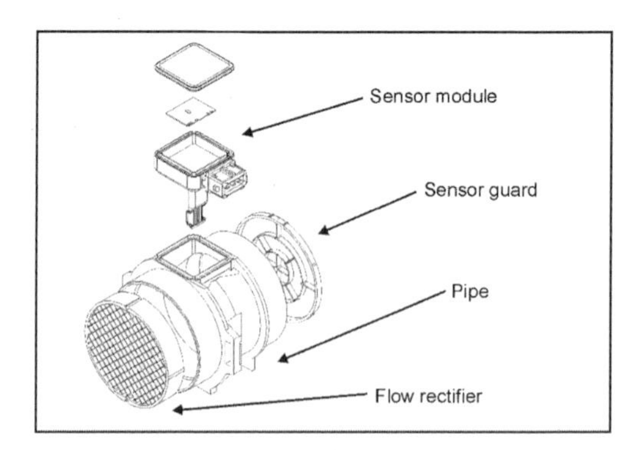

Fig. 18-12 Structure of a hot-film air mass meter.

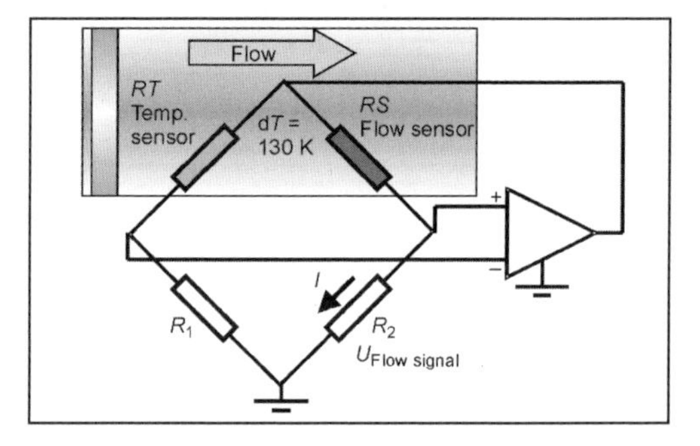

Fig. 18-13 Principle of a hot-film air mass meter.

The voltage at R_2 is a measure of the air mass flow rate. Depending on the intaken air mass flow, R_S is cooled more or less strongly.

The electronics control the necessary heating current through R_S so that there is always a constant temperature difference (e.g., 130 K) at R_S to the air temperature measured at R_T. The heating current is transformed into a voltage signal at resistor R_2.

The resistors R_S and R_T are matched to one another in such a way that the map is independent of the air temperature. The map also exhibits—thanks to the physics—an advantageous nonlinear characteristic permitting a practically constant proportional resolution.

Thanks to the use of materials specially adapted to the conditions in the automobile engine compartment, flow control, circuiting technology, and the mechanical configuration, the HFM signal is more or less independent of the temperature, pressure, and soiling.

Internal combustion engines with four or fewer cylinders create extreme pulsations in the intake manifold at wide-open throttle or in the case of no throttle valve (e.g., in diesel engines or SI engines with direct fuel injection). At certain engine speeds, at the resonance point, a pulsating return flow occurs that with conventional HFM results in a positive measurement error as the air passes over the sensor three times.

The air mass flow (Q) is calculated as a function of the degree of modulation (m) to

$$Q = Q_{\text{mean}} \cdot [1 + m \cdot \sin(\omega t)]$$

$$m = \frac{\hat{Q}}{Q_{\text{mean}}} \quad (18.1)$$

This effect can be compensated with an additional heating resistor R_H (booster).

Returning air is heated by the booster and passes over the heating resistor R_S. This prevents R_S from being cooled again by the returning air. Overheating the returning air produces an overcompensation that ensures that the air flowing toward the engine again is not measured a second time.

The return-flow compensation is independent of the resonance frequencies, temperature, and air pressure.

In many applications the temperature sensor (NTC resistor) for determining the intake air temperature is also integrated into the HFM.

18.5.4 Secondary Air Mass Sensors (SAF)

During the exhaust gas test cycles, a large proportion of the CO and HC emissions are produced during the start phase of an engine. In the first few minutes after the start, practically no conversion takes place because the temperature is still under the "light-off" temperature of approximately 350°C. In order to achieve the quickest possible heating of the catalytic converter, secondary air is blown into the exhaust gas line and the exhaust gases are enriched with additional hydrocarbons. This can be achieved by enriching the mixture or by subsequent injection of fuel into the exhaust system. The oxygen admitted via the secondary air creates a postcombustion of the rich mixture and, thus, permits a faster heating of the catalytic converter and a consequent significant reduction in the pollutant emissions. This is necessary in order to meet the strictest exhaust emissions requirements.

Based on the measurement principle of the main stream HFM, the secondary air mass sensor measures the fresh air mass admitted to the catalytic converter in addition to the exhaust gases during the start phase.

The advantages compared with an uncontrolled system are the independence from system tolerances and the possibility of also being able to carry out an extended diagnosis of the system during the secondary air phase.

18.6 Speed Sensors

Although one generally speaks of speed sensors; in this case we are dealing with incremental sensors. The following speed sensors are frequently used in the control of the powertrain:

- Crankshaft sensor
- Camshaft sensor
- Transmission speed sensor

Electronic controllers for internal combustion engines require information on the momentary position of the crankshaft and camshaft for exact control of ignition and fuel injection.

For the application on the crankshaft, high precision over the full range of functions (temperature, air gap, speed, mechanical tolerances) is demanded. Furthermore, the sensor should be able to detect even very low speeds in order to permit a rapid detection of the position during the start of the engine. For engines with up to eight cylinders, misfire detection uses the evaluation of the crankshaft sensor signal. A very high repetition accuracy ($<0.03°$) is therefore demanded.

The camshaft sensor is used for synchronization between the camshaft and crankshaft; this means an identification of the first cylinder. To achieve rapid synchronization, either specially encoded sensor wheels for the camshaft or camshaft sensors with a static function are used. In engines with variable valve timing, the camshaft sensors are also needed for control of the camshaft adjuster. One camshaft sensor is required for each adjustable camshaft. For the application as position sensor for the variable valve timing, the accuracy is of paramount importance.

Transmission speed sensors are used to measure the speed of the vehicle. Both the input and output speeds are required for control of automatic and CVT transmissions. The demands made on transmission speed sensors are considerably lower, but these sensors should be able to detect even very low speeds.

The measurement principles are based on passive and active speed sensors.

18.6.1 Passive Speed Sensors

Sensors used as passive speed sensors today are almost exclusively inductive sensors, also known as variable reluctance (VR) sensors.

Inductive sensors consist essentially of a coil around a magnetically precharged core. If the inductive sensor comes near a moving ferromagnetic sensor wheel, a voltage is induced. This voltage is evaluated in the controller. Each flank of the sensor wheel induces an electric voltage. With inductive sensors, the level of the induced voltage is dependent on the speed; there is, therefore, a lower speed/frequency limit for the function of the inductive sensor.

18.6.2 Active Sensors

Active speed sensors have integrated electronics for signal processing. Active sensors, therefore, transmit standardized signal levels that can be used in the electronic controller without additional processing.

Active sensors based on the Hall effect are the most widely used, but magneto-resistive (MR) sensors and giant MR (GMR) sensors are also being increasingly employed.

Of the Hall sensors, the differential Hall sensor is the most frequently employed.

The edge change of a ferromagnetic sensor wheel results in a difference in the magnetic field at the differential Hall element, Fig. 18-14. Thanks to the differential principle, these sensors are essentially insensitive to interferences such as temperature changes and external magnetic fields and are characterized by high precision. The differential principle allows sensors to be built with a lower speed limit of 0 rpm (zero speed). Because of the differential principle, however, these sensors can be used in only one installation position.

"Single-element" Hall sensors are used for static functions. These sensors allow tooth or gap to be detected

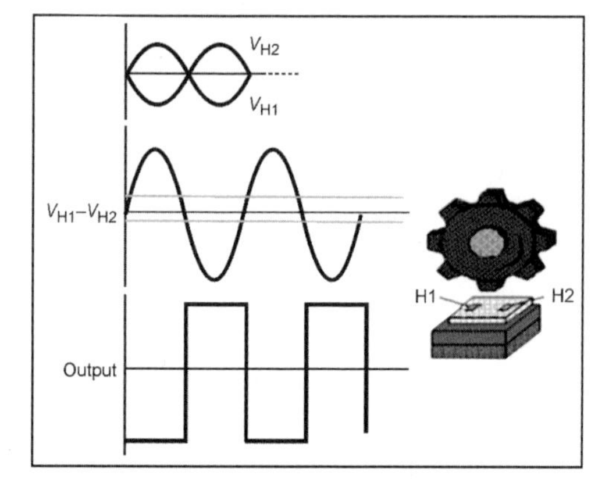

Fig. 18-14 Measurement principle of a differential Hall sensor.

Fig. 18-15 Various versions of active camshaft and crankshaft sensors.

without the sensor wheel moving (true power on). Thanks to the design as a single element, they can be positioned anywhere between Hall sensor and sensor wheel.

Figure 18-15 shows different forms of active sensors based on the differential Hall sensor.

Bibliography

[1] Bauer, H., Kraftfahrtechnisches Taschenbuch, 23, Aufl., 1999.
[2] Fiedeler, O., Strömungs- und Durchflussmesstechnik, Oldenbourg, 1992.
[3] Niebuhr, J., and G. Lindner, Physikalische Messtechnik with Sensoren, Oldenbourg, 1994.
[4] Tränkler, H.R., and E. Obermeier, Sensortechnik, Springer, 1998.

19 Actuators

19.1 Drives for Charge Controllers

Pneumatic and electric actuators are predominantly used as engine management actuators.[1–3] Figure 19-1 shows a comparison of the advantages and disadvantages of the most widely used drives.

19.1.1 Pneumatic Drives

Pneumatic drives, Fig. 19-2, as actuators are predominantly used as changeover contacts between two fixed positions. Pneumatic actuators consist of a vacuum unit with a diaphragm that is linked via a control valve to the vacuum supply of the vehicle. The element to be actuated is connected to the actuator either directly or via levers and cables.

The advantage of the pneumatic drive is its low price in combination with large actuating torques in relation to its size. A major disadvantage of pneumatic drives is the difficulty of achieving position control, therefore necessitating additional control elements for precise stopping in intermediate positions. This disadvantage has resulted in the widespread change from pneumatic drives to electric drives.

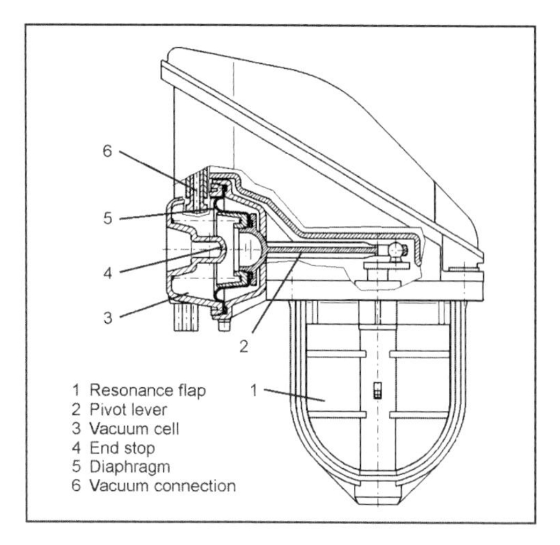

Fig. 19-2 Intake manifold resonance valve with pneumatic drive.

19.1.2 Electric Drives

19.1.2.1 Stepping Motor

The stepping motor, Fig. 19-3, is predominantly used as an actuator where low demands are made on the actuating force. The advantage of the stepping motor lies in its stepwise movement and the corresponding control. The counting of the positioning steps permits a determination of the relative position of the drive compared with its position at the start of the movement and, hence, in a simplified position control of the drive. For simpler demands, no additional sensor is required to determine the actual position, although an absolute determination of the current position is not possible.

The main disadvantage of the stepping motor is the low excess torque available to overcome any possible binding in the mechanism. Another related disadvantage is the possibility of an error left undetected in the position control as a result of an inability to make a required positioning movement.

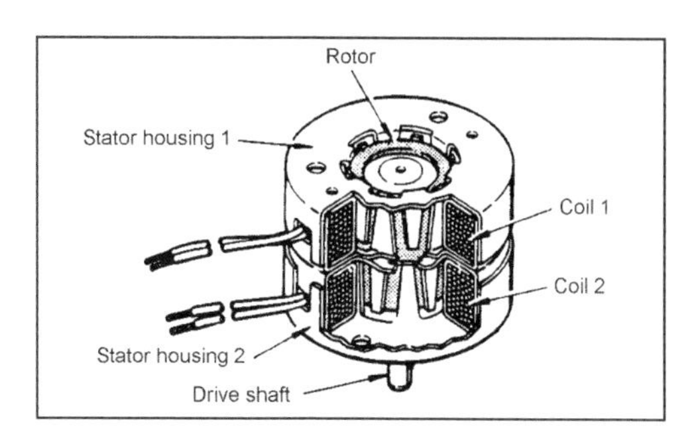

Fig. 19-3 Stepping motor.

Drive	Pneumatic drive	Stepping motor	DC motor with gearbox	Torque motor
Actuating torque	++	–	++	o
Actuating time	+	–	+	++
Position control	– –	++	+	+
Weight	+	–	o	–
Costs	++	+	o	+

++ very positive, + positive, o average, – negative, – – very negative

Fig. 19-1 Comparison of different drives.

19.1.2.2 DC Motor

The direct-current motor (DC motor)[4] is mainly used as an actuator in conjunction with gearings. The flexibility of the gearing designs and gear ratio here permits the use of the same DC motor for different actuating torque or actuating time requirements.

The main advantage of the DC motor and gearing combination is the high excess torque. This enables short actuating times to be achieved and allows brief binding to be overcome. Position control[5] of the DC motor drive is possible only in combination with a position sensor.

Disadvantages of the DC motor are its comparatively more complex construction and the wear behavior of the carbon brushes in the motor as well as in the gearing as compared with contact-free drives.

19.1.2.3 Torque Motor

The torque motor, Fig. 19-4, is used as an actuator where low demands are made on the actuating force in conjunction with the wish for short actuating times. The main advantages of the torque motor are the contact-free and, hence, wear-free drive and its simple construction. Disadvantages compared with the DC motor are the low excess torque and high weight in relation to the actuating torque. The torque drive also requires a position sensor for position control.

Fig. 19-4 Torque motor.

19.2 Throttle Valve Actuators

19.2.1 Key Function in SI Engines

The power output of a spark ignition engine is controlled quantitatively. This requires control of the intaken air mass. The most widely used technical solution for influencing air mass flows is the throttle valve actuator. The position of the throttle valve in the air duct determines the amount of air drawn in by the internal combustion engine and the pressure level in the intake manifold, Fig. 19-5.

19.2.2 Key Function in Diesel Engines and in Quality-Controlled SI Engines (Direct Injection)

The operation of the diesel engine and SI engine with direct fuel injection is regulated by quality control of the air-fuel mixture. Under ideal conditions, no throttling of the air mass flow is necessary. Nevertheless, throttle valve actuators are widely used in diesel and SI engines with direct fuel injection. The main function of these actuators is to create a vacuum to draw exhaust gases into the air taken in by the engine (exhaust gas recirculation) to comply with the strict legal requirements for low pollutant contents in the exhaust gases (see also Section 19.2.6).

19.2.3 Additional Functions

19.2.3.1 Idle-Speed Control of SI Engines

Apart from its main function of the cylinder charge control,[6] the throttle valve actuator together with its various attachments performs additional functions. The most important auxiliary function of the throttle valve actuator is to control the idle speed of the SI engine, Fig. 19-6.

The idle speed is normally controlled by influencing the air mass flow. In addition, "fine-tuning" is achieved by shifting the ignition timing.

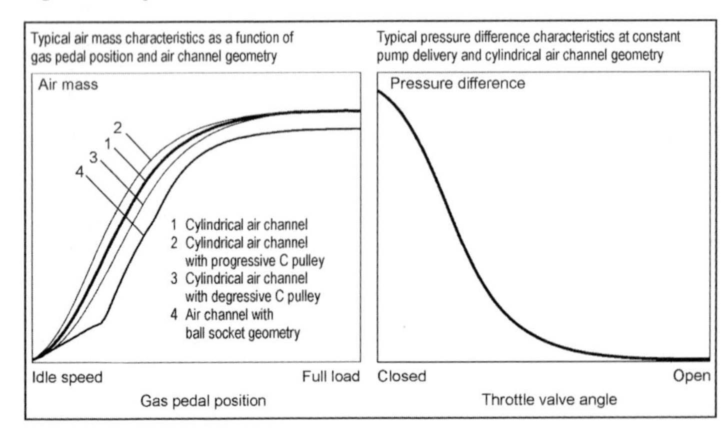

Fig. 19-5 Diagram of air mass and differential pressure characteristics.

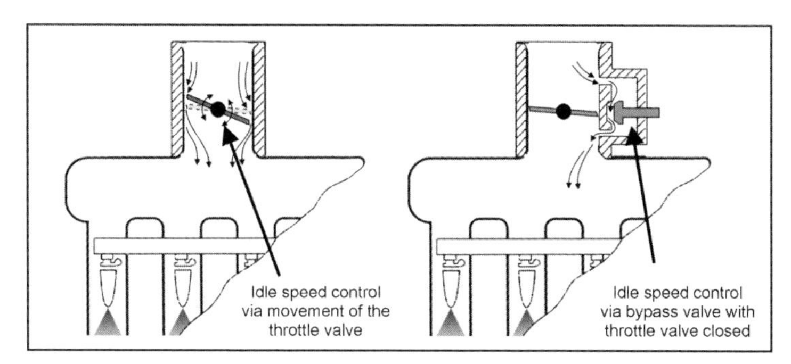

Fig. 19-6 Comparison of idle-speed control systems.

The air mass flow at idle speed can be controlled with an actuator in a bypass to the main air duct when the throttle valve is closed or by direct positioning of the throttle valve in the slightly open working range.

19.2.3.2 Position Signal

A sensor on the throttle valve actuator generates a position signal and transmits this to the engine control electronics. Potentiometers are widely used for this function.

The signal from the sensor also serves to distinguish between the part-load and idle-speed operation of the internal combustion engine. In some cases, switches on the throttle valve actuator are used in addition to the sensor signal.

With electrically powered throttle valves, the position sensor is generally integrated into the drive. To meet rising demands on the reliability of these systems, the potentiometers are increasingly being replaced by contact-free sensor systems.

19.2.3.3 Dashpot Function

The dashpot function provides a slower return of the throttle valve to the closed position when the gas pedal is suddenly relieved. Without the dashpot function, the throttle valve is slammed closed by return springs in such cases. This leads to an impact load and a sharp braking of the vehicle. To improve ride comfort, the impact load is damped. This is affected in some cases by a separate mechanical damper. Modern throttle valve actuators, however, perform this dashpot function by opening the bypass actuator for the idle-speed control or by a return of the throttle valve to the closed position at a slower speed and independently of the gas pedal return (see also Section 19.2.4).

19.2.3.4 Cruise Control Function

The speed control of a vehicle with SI engine (cruise control) is affected by actuation of the throttle valve independently of the driver. This can be affected by separate cruise control actuators linked to the throttle valve by means of cables or levers. Modern throttle valve actuators perform this function using a direct drive in the throttle valve actuator.[2]

The cruise control function requires a power source (e.g., electric or pneumatic drive) that can open the throttle valve independently of the gas pedal position.

19.2.4 "Drive by Wire"/E-Gas

By contrast with the cruise control function, it is necessary for the drive slip control or traction control system (TCS) and for the electronic stability program (ESP) to be able not only to open but also to close the throttle valve independently of the gas pedal position.

This function is difficult to achieve with mechanically linked systems. In some cases, a second throttle valve (open during normal operation) that is actuated independently of the gas pedal position is installed upline of the actual throttle valve to perform these functions. However, an engine drag-torque control is not possible with these systems. More widespread here is the use of "drive by wire" systems (also known as E-Gas systems), Fig. 19-7, that permit positioning of the throttle valve totally independently of the gas pedal position.

With these systems, a nominal position of the throttle valve is calculated in the engine controller on the basis of various key data and functions that is then implemented by the position control of the throttle valve actuator. The position is controlled by comparing the nominal and actual positions of the throttle valve and corresponding control of the actuator by the engine control electronics. In some cases the position control is also affected by

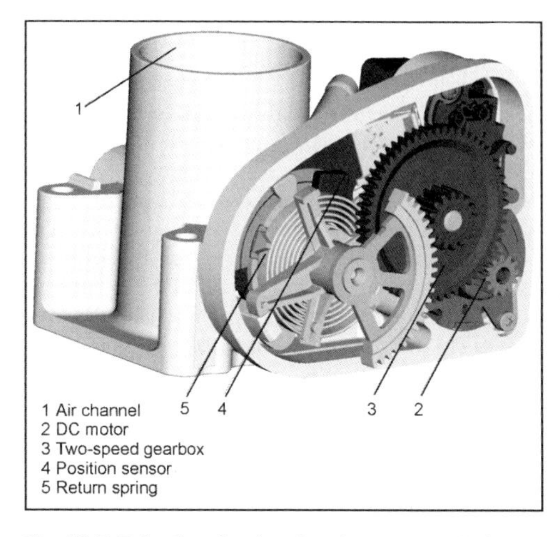

Fig. 19-7 Drive by wire throttle valve actuator (E-Gas 5).

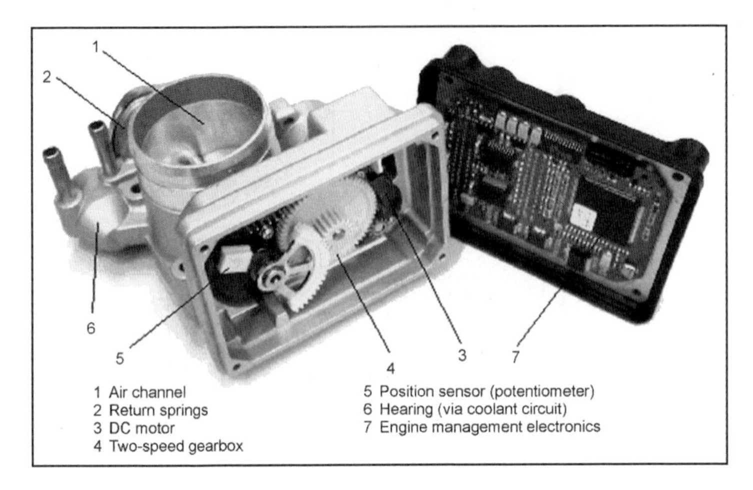

Fig. 19-8 Throttle valve actuator (E-Gas 7) with integral engine control electronics.

electronics in the throttle valve actuator, Fig. 19-8. In this case, only the nominal value is formed in the central engine control electronics and then transmitted to the throttle valve actuator. There are also applications in which the complete engine control electronics is integrated into the throttle valve actuator.

The positioning of the throttle valve independently of the gas pedal position also enables or simplifies other functions. The opening map of the throttle valve can be accelerated or decelerated in relation to the gas pedal position as desired. The idle-speed control, the cruise control function, and the impact load damping are all performed by the software in the engine control unit and require no additional mechanical parts.

Malfunctions or a failure of the throttle valve actuator can be detected by the engine control electronics and are not misinterpreted as drivers' wishes. For this, a safety concept has been integrated into the software of the engine control electronics. This is a significant benefit of the drive by wire compared with throttle valve actuators linked mechanically to the gas pedal.

19.2.5 Charge Pressure Control

Throttle valve actuators are also used to control or to limit the charge pressure of turbocharged engines. With mechanical turbochargers, another throttle valve actuator is installed in a bypass to the compressor in addition to the throttle valve that regulates the degree of cylinder charge of the internal combustion engine. If the charge pressure is too high in certain engine operating situations, the throttle valve in the bypass is opened and part of the compressed intake air escapes back into the area in front of the compressor. The charge pressure is reduced.

19.2.6 Vacuum/Prethrottle Actuators

Throttling of the air mass flow creates a vacuum in the intake manifold of internal combustion engines compared with the ambient pressure. This pressure difference is employed for various functions. It serves the brake booster as an energy medium. Controlled by external actuators (generally valves), the vacuum is used to feed the "blowby gases" (crankcase ventilation) and the air current for the regeneration of the carbon canister and for exhaust gas recirculation.

This function of the throttle valve actuator is also employed in diesel engines and quality-controlled SI engines (direct injection), Fig. 19-9.

The throttle valve of the prethrottle actuator is fully open during normal operation and closes only when a vacuum is required in the intake manifold, e.g., for EGR feeding. The demands made on such prethrottle actuators with respect to actuating torque and actuating time are typically slightly lower than the demands made on standard throttle valve actuators.

19.3 Swirl and Tumble Plates

Swirl plates are particularly employed for quality-controlled engines. They serve to create a swirling or tumbling motion of the airflow in the combustion chamber. This function permits the selective creation of different air-fuel ratios in the combustion chamber. Of particular importance here is the creation of an ignitable mixture near the spark plug in conjunction with lean mixtures in other areas of the combustion chamber during part-load operation.

Throttle valves are frequently used that change either the direction or speed of the airflow or both so that the air enters the combustion chamber with a swirling or tumbling motion. In some cases an undeflected airflow and an air flow diverted by a swirl plate actuator are mixed at a preset angle to also achieve this effect.

19.3.1 Swirl Plate Actuators (Swirl/Tumble Actuators)

Since it is necessary to influence the swirl motion in the airflow right into the combustion chamber, the swirl motion is created as close as possible to the combustion chamber, i.e., by a separate actuator as near as possible to the intake valves for each cylinder of the internal combustion engine.

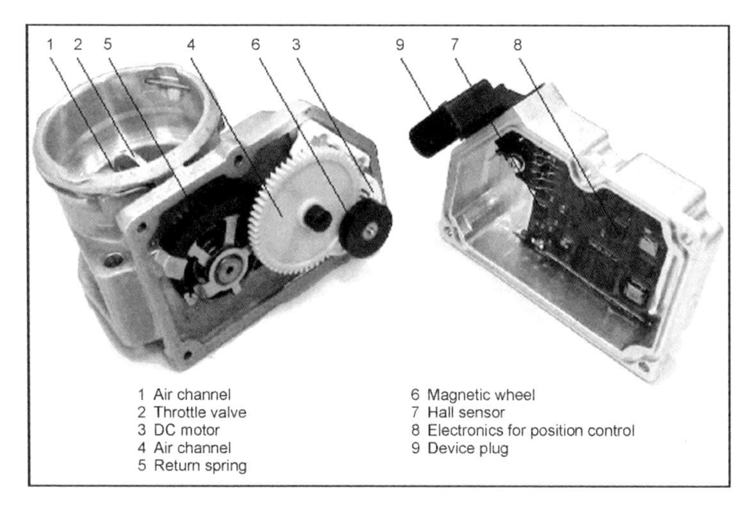

Fig. 19-9 Prethrottle actuator with integral electronics.

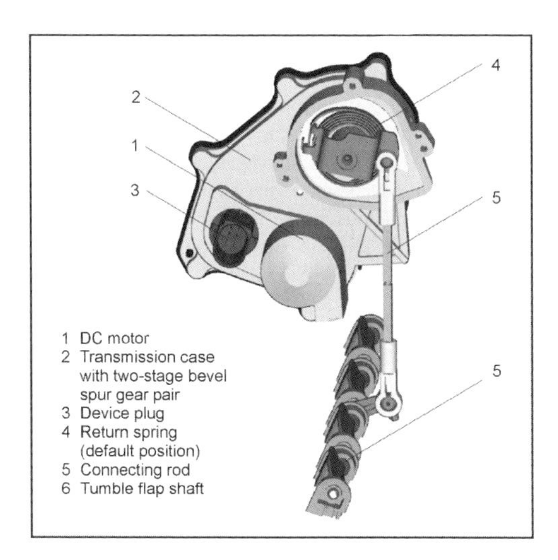

Fig. 19-10 Actuator with tumble plate system (four-cylinder engine).

As a cylinder-selective position control of the swirl plate is generally not necessary, the swirl plates of a cylinder bank are actuated by a common shaft that is positioned by an actuator. For the widely differing airflows in different driving situations, it is necessary to also be able to move to intermediate positions. The demands made on a swirl plate actuator with position control are thus comparable with those made on drive by wire throttle valve actuators. Because of the actuating torque and actuating time demands, DC motors with gearing drives are predominantly employed, Fig. 19-10.

19.4 Exhaust Gas Recirculation Valves

In the early 1970s, external exhaust gas recirculation was employed for the first time in series-production automobiles in North America. This technique was used in order to comply with the emission limit values of the time. During exhaust gas recirculation, part of the combusted exhaust gas is tapped at the exhaust manifold and returned via a pipe to the intake manifold where the combusted exhaust gas is mixed with the intaken mixture, Fig. 19-11.

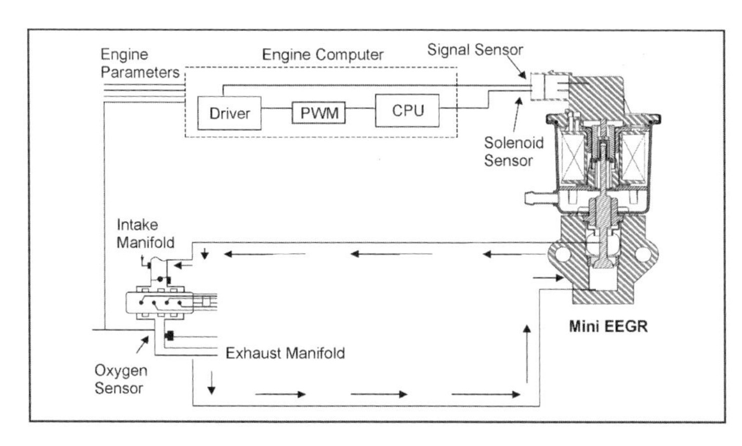

Fig. 19-11 Exhaust gas recirculation, schematic.

This admixing of the combusted exhaust gas lowers the peak combustion temperature and, hence, reduces the nitrous oxide emissions. In addition, exhaust gas recirculation can also help to reduce the fuel consumption in the part-load range. As the volume of recirculated exhaust gas has to be varied depending on engine load and engine revs, a corresponding control element is needed—the EGR valve.

In addition to external exhaust gas recirculation there is also an internal exhaust gas recirculation that exists on all four-stroke engines because of the overlap of the intake and exhaust valve systems. Internal exhaust gas recirculation has the same effect on the emissions, while the EGR volumes are relatively small because of the system design and can be influenced depending only on the load and engine revs on engines with variable valve train. Variable valve train systems are fundamentally employed with the aim of optimizing engine power and torque. Exhaust gas recirculation is just an additional benefit that alone would not justify the relatively high costs of these systems, and is, therefore, to be seen only as a bonus. In spite of the limited controllability of the internal exhaust gas recirculation volumes, additional external exhaust gas recirculation systems are normally not employed on engines with variable valve train.

Disk valves with a pneumatic drive (vacuum unit) were used for the first external exhaust gas recirculation systems. Here the vacuum unit was exposed to the pressure of the intake manifold, resulting in an adjustment of the EGR valve according to the operating point of the engine. Pneumatic deceleration valves, nonreturn valves, and pressure relief valves were installed in the system to limit the functional range to prevent negative effects of inappropriate exhaust gas recirculation volumes. Other control systems also use the exhaust gas backpressure as a control parameter for the vacuum unit. In some cases electric changeover valves were also integrated into the control line to shut off the exhaust gas recirculation at certain operating points. In the next stage of development, electropneumatic pressure transducers were used with which it became possible for the first time to control the position of the exhaust gas recirculation valve independently of the operating point of the engine. Nevertheless, the range of application of exhaust gas recirculation remained limited to operating points at which the level of the vacuum was sufficient to open the disk valve against the spring force or the pressures acting on the valve.

The wish to employ exhaust gas recirculation at higher load points and independently of the intake manifold vacuum triggered the development of electric exhaust gas recirculation valves, Fig. 19-12.

At the same time the demands made on the precision increased so that sensors that showed the position of the valve were integrated. By comparison with earlier generations, these exhaust gas recirculation valves permit a very accurate control of exhaust gas recirculation volumes with reduced actuation times at the same time. The integration of all the modules into one component simplifies the adaptation at the engine and reduces the function-relevant tolerances. Thanks to these functional benefits, electric exhaust gas recirculation valves are more and more widely used and are employed almost exclusively for new engine generations. Apart from stepping motors and lift and turn magnets, DC motors are now also more frequently used today as electric actuators.

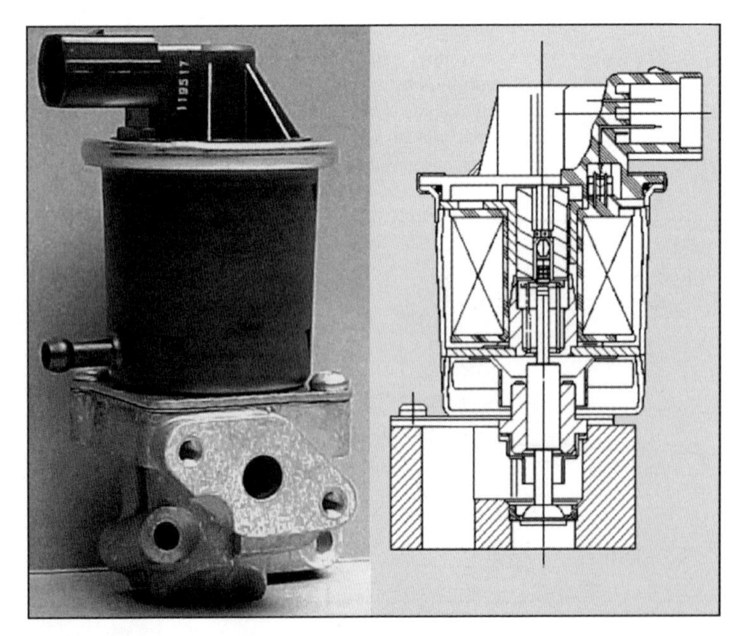

Fig. 19-12 Electrically controlled exhaust gas recirculation valve.

In addition to the further development of the actuators, the actual control valve has also been changed many times. Apart from disk and needle valves of different forms and dimensions, flap and rotary slide valves are also used today. Essentially, the valve should guarantee a constant function over the service life irrespective of the degree of soiling. Furthermore, the change in the differential pressure through the valve that occurs at every change in position should have the least possible influence on the set valve position. This is particularly important when moving from the closed to the slightly open position as the differential pressure acting here is subject to a large change. At the same time, the precision demands are very high in this working range. To improve the function in this range, there are valve developments with nonlinear opening characteristics. The valve design should also be as insensitive as possible to pulsing pressures. The best compromise at present appears to be flap valves, although, depending on the demanded EGR rate and the engine sensitivity to changes in volume, disk valves can also meet the requirements, Fig. 19-13.

For diesel engines, exhaust gas recirculation is a very effective method of complying with the demanded NO_x emissions and is used in Europe for all vehicles up to 3.5 t, and in some cases even above. The most widespread are controlled systems with pneumatically actuated valves, although electrically controlled valves are increasingly being employed for new applications, Fig. 19-14.

For conventional SI engines with intake manifold injection, electrically controlled systems are the most widespread, while EGR valves are currently installed in only approximately 50% of the vehicles as the emission values can also be achieved using alternative techniques.

For the SI engine with direct injection, a 100% EGR application can be assumed, as the benefits of this engine concept can be fully exploited only together with exhaust gas recirculation. In view of the high precision demands, flap valves with electric motors can be expected to become more widespread.

Fig. 19-14 Flap valve with electric motor.

19.5 Evaporative Emissions Components

19.5.1 Canister-Purge Valves

With the tightening of exhaust emissions legislation, the evaporative emissions of the tank system in vehicles with SI engines also came under closer scrutiny along with the combustion residues. This led to the tank now being vented via a "carbon canister" and no longer directly to the atmosphere. The activated charcoal contained in this

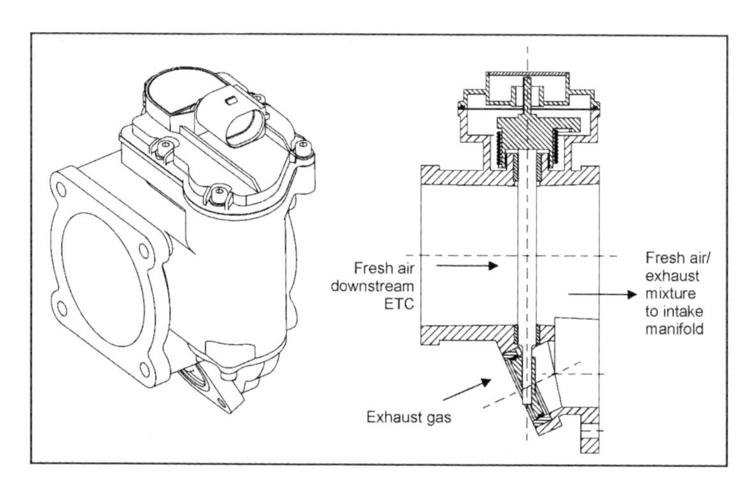

Fig. 19-13 EGR flap valve.

canister can bind large quantities of gasoline vapors that can be caused (for example) by parking in the sun, so that under normal circumstances no more gasoline vapors can escape into the atmosphere via the tank vent. At the same time, the carbon canister has to be regenerated at regular intervals so that the saturation limit is not exceeded. For regeneration, the absorbed gasoline vapors are drawn in by the engine and combusted.

This additional amount of fuel has to be precisely metered, however, so that it does not lead to an over-enrichment of the mixture. This is controlled via a "canister-purge valve." This is a clocked solenoid valve that is controlled by the engine control unit in relation to the lambda control. Essentially, the amount of fuel admitted to the engine via the injection system has to be reduced by the amount regenerated.

The function of the canister-purge valve is determined by the controllability of low throughputs and maximum throughput. The regeneration of the carbon canister should be possible even at engine idle speed, but because of the high-pressure differential and the small overall amount of fuel required, this necessitates a high control precision of the canister-purge valve. At the same time, however, large regeneration volumes are called for in the part-load and full-load ranges. Because of the small vacuum in this engine-operating mode, this necessitates a large flow cross section. Furthermore, the canister-purge valves should be compact and should create the least possible noise radiation. They are installed on the vehicle body, on the intake manifold, or even on the carbon canister, as long as this is located near the engine.

Because of the different emissions legislations as well as the function and application demands, a large number of different canister-purge valves have been developed, Fig. 19-15. A distinction is made between low-frequency valves (5 to 20 Hz) with pulsating throughput and high-frequency valves (>100 Hz) with continuous throughput.

Fig. 19-15 Types of canister-purge valves.

The low-frequency valves are generally cheaper, but the control precision is limited and a high level of noise can develop, particularly at temperatures below freezing. The valves with continuous throughput are more complex with corresponding disadvantages in the dimensions and costs, although fundamental functional and acoustic benefits are achieved. To decrease the sensitivity to pressure fluctuations, pressure-compensating valve seats or nozzles with ultrasonic flow are used in some cases.

19.5.2 Evaporative Emissions Diagnostics

With the introduction of the OBD II legislation (On Board Diagnose, 2nd generation) in North America, leak testing of the complete tank system became a legal requirement for the first time. This requirement was based on the observation that with a further reduction in the exhaust emissions, greater attention had to be paid to the evaporative emissions, as their share of the total emissions of the vehicle is very large. In particular, it was discovered that undetected leaks in the tank system and operating errors (e.g., lost or wrong tank filler cap) resulted in very high evaporative emissions over time. It, therefore, became a legal requirement for a diagnostic system to be installed on the vehicle that would detect all leaks greater than the throughput through a calibrated opening of 1 mm diameter. Here the system has to be able to distinguish between a normal leak (e.g., leaking hose connection, tank damage) and major leaks (missing tank filler cap).

When implementing this legislation in the vehicle, it was discovered that the technical complexity was far higher than had initially been assumed. In particular, the different climatic and operating conditions as well as the respective fuel level in the tank create a broad band of parameters that have to be adapted. In spite of the problems of implementation in the vehicle, the legislation was further tightened by reducing the diameter of the calibration opening from 1 to 0.5 mm.

The tank diagnostics can be performed with vacuum and pressure systems. Both of these system types exhibit fundamental advantages and disadvantages irrespective of the components employed. The legislation permits diagnostics both during vehicle operation and at standstill. The pressure systems offer a few benefits during vehicle operation while the vacuum method is favored for the vehicle standstill with the 0.5 mm legislation. Both technical (e.g., tank volume, tank form, space) and market-related aspects (e.g., the vehicle is sold only with the OBD II system, the vehicle is alternatively available also without the OBD II system, unit price per system in relation to the application costs, etc.) can influence the decision in favor of a particular diagnosis method. Furthermore, the experience gained to date and the strategy of the vehicle manufacturer greatly influences the choice of system. In Europe, on the other hand, leak diagnosis is no longer conducted because the associated cost is regarded as being disproportionately high. The only system that will be used in the future is one that will diagnose a correctly fitted tank filler cap, where a mechanical or electric switch contact is sufficient.

19.5.2.1 Tank Diagnostics with Pressure

With the Siemens leakage diagnosis pressure pump (LDP I), Fig. 19-16, a pressure of up to approximately 20 hPa is generated in the tank system using the intake manifold vacuum via a clocked three-way valve and a spring-loaded diaphragm.

Fig. 19-16 Tank diagnostics pressure pump.

Via the pump diaphragm, a switch detects the change in position, and the corresponding dropout time is compared with the default values stored in the control unit. A simple nominal and/or actual comparison then allows the leak tightness of the tank system to be evaluated.

If a leaking tank system is detected, the diagnosis is repeated to eliminate all ambient influences on the result. Only when the same fault is detected in two consecutive measurements is the OBD warning lamp switched on via the engine control unit. Additional software now makes reliable tank diagnosis possible with the LDP I even under the 0.5 mm legislation.

19.5.2.2 Tank Diagnostics with Vacuum

The Siemens NVLD (natural vacuum leak detection) system, Figs. 19-17 and 19-18, utilizes the ambient temperature influences allowing for the ideal gas law for tank leakage diagnosis (normal leak). The NVLD unit is directly connected to the tank or to the carbon canister. During engine operation, the vent is opened to the atmosphere via an electromagnetically activated valve.

When the vehicle engine is not running, the valve is closed and thus creates a tank system that is sealed to the atmosphere. The different operating states and ambient influences result in differences in temperature of the tank system and the fuel. Since the tank system is completely sealed to the atmosphere, these differences in temperature

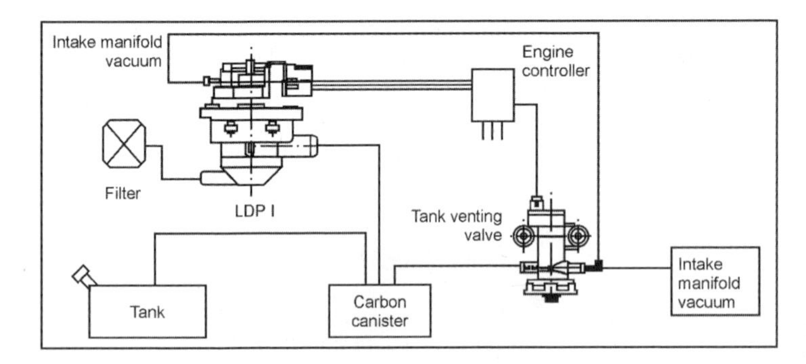

Fig. 19-17 Tank diagnostics with pressure (schematic).

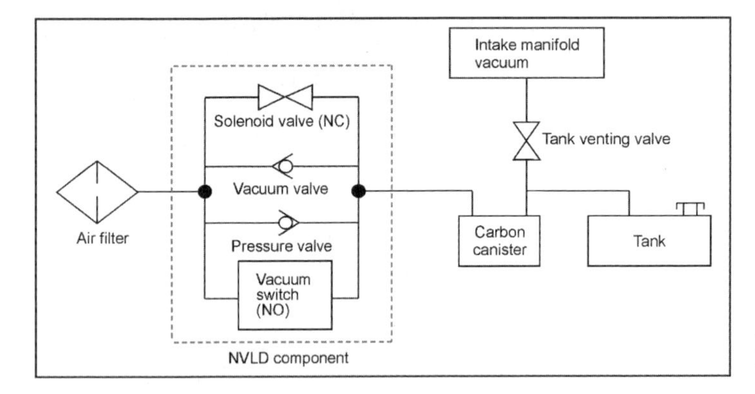

Fig. 19-18 Tank diagnostics with vacuum (Siemens NVLD system).

create changes in pressure in the tank. These changes in pressure also act on the diagnostic diaphragm that, in turn, is connected to a contact switch. When the tank system is leak tight, the changes in pressure trigger a switching signal that is registered by the vehicle electronics. If this switch signal is not received for a given period, the reverse conclusion is drawn that the tank has a leak. In addition, it is possible to detect major leaks during engine operation. Here the solenoid valve is closed and a vacuum is built up in the tank via the canister-purge valve. A possible major leak is then detected via the pressure diaphragm and the contact switch. Additional spring-loaded valves integrated into the NVLD unit ensure that certain threshold values for the pressure and/or vacuum level are not exceeded in the closed tank system.

Bibliography

[1] Moczala, H., *et al.*, Elektrische Kleinstmotoren und ihr Einsatz, Expert-Verlag, 1979.

[2] Richter, C., Elektrische Stellantriebe kleiner Leistung, VDE-Verlag, 1988.

[3] Kenjo, T., and S. Nagamori, Permanent Magnet and Brushless DC Motors, Oxford Science Publications.

[4] Vogt, K., Berechnung elektrischer Maschinen, VCH, 1996.

[5] Leonhard, W., Control of Electrical Drives, Springer, 1985.

[6] Luft, J., Elektromotorischer Systembaukasten Ansätze zur Gewichts- und Bauraumreduzierung, VDO, 1995.

20 Cooling of Internal Combustion Engines

20.1 General

The growing demands with respect to fuel consumption, exhaust gas emissions, service life, ride comfort, and package have led to modern cooling systems for internal combustion engines in the motor vehicle—with few exceptions—exhibiting the following characteristics:

- Water cooling of the engines with forced circulation of the coolant by a belt-driven centrifugal pump
- Operation of the cooling system at a pressure of up to 1.5 bar
- Use of a mixture of water and antifreeze, generally ethylene glycol, with a content of 30% to 50% v/v
- Aluminum in corrosion-resistant alloys as the dominant radiator material
- The coolants exhibit additional inhibitors for the corrosion protection of aluminum radiators
- Plastic as a dominant material for radiators, fans, and fan surroundings
- Temperature control via the fan drive and coolant thermostat
- Use of intercoolers, engine oil, transmission oil, hydraulic oil, and exhaust gas coolers, depending on engine type, engine output, and equipment features
- Preassembly of all front-end cooling components in one functional unit, the "cooling module"

Apart from the numerous development activities for even more compact, lighter, and more efficient components, the electronically controlled cooling system is increasingly gaining in significance for the demands outlined at the beginning.

20.2 Demands on the Cooling System

Peak temperatures of over 2000°C occur briefly inside the cylinders of an internal combustion engine. Charge cycles, expansion processes, etc., between the ignitions, however, result in far lower mean temperatures. Nevertheless, thermal overloading of the components exposed to the gas has to be prevented and the lubricating properties of the oil film between the piston and cylinder surfaces maintained by cooling.

In water-cooled internal combustion engines, around one-third of the admitted fuel energy is discharged via the cooling system, a further third is lost via the exhaust gas, and one-third is transformed into useful work, depending on the combustion process (Fig. 20-1).

Several thermally critical operating conditions are normally investigated during the design of cooling systems, such as "maximum speed on the flat," "fast hill climbing," or "slow hill climbing with trailer." A distinction is also made between operations in Europe and countries with hotter climates. Travel speed, ambient temperature, heat volumes to be dissipated, and the nominal values for maximum permissible coolant, charge air, and oil temperatures are always given. Typical rules of thumb and nominal values for the main methods of cooling are summarized in Fig. 20-2.

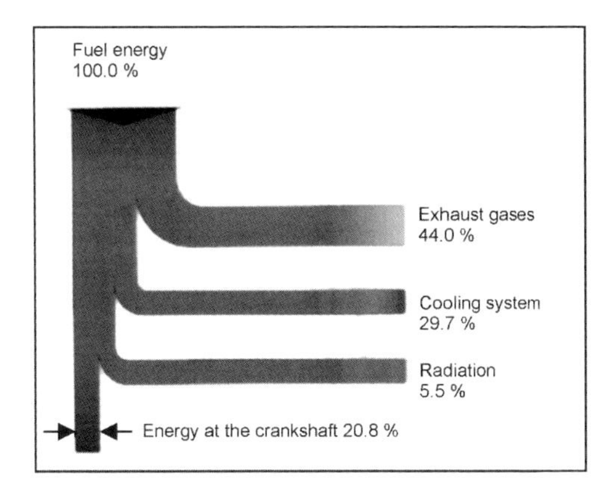

Fig. 20-1 Energy balance in a water cooled, 1.9 L SI engine at 90 km/h at constant speed in 4th gear.

Bandwidths for different operating conditions from the smallest car engine up to the most powerful truck engine are

Maximum coolant temperature	100°C ... 120°C
Maximum coolant throughput	5000 ... 25 000 L/h
Maximum charge air throughput	0.05 ... 0.6 kg/s
Maximum charge air intake temperatures (at 25°C ambient temp.)	110 ... 220°C

20.3 Principles for Calculation and Simulation Tools

In the radiators used in motor vehicles, heat is transferred from one flowing medium 1 via a fixed wall to a second flowing medium 2 from the higher to the lower temperature level, Fig. 20-3.

This heat volume is calculated using the parameters shown in Fig. 20-3 to

$$\dot{Q} = \alpha_1 \cdot A \cdot (t_1 - t_1') = \frac{\lambda}{\delta} \cdot A \cdot (t_1' - t_2')$$

$$= \alpha_2 \cdot A \cdot (t_2' - t_2)$$

$$(t_1 - t_1') = \frac{\dot{Q}}{\alpha_1 \cdot A}; \qquad (t_1' - t_2') = \frac{\dot{Q} \cdot \delta}{\lambda \cdot A};$$

$$(t_2' - t_2) = \frac{\dot{Q}}{\alpha_2 \cdot A};$$

$$t_1 - t_2 = \frac{\dot{Q}}{A} \cdot \left(\frac{1}{\alpha_1} + \frac{\delta}{\lambda} + \frac{1}{\alpha_2}\right) = \frac{1}{k} \cdot \frac{\dot{Q}}{A}$$

$$\dot{Q} = k \cdot A \cdot (t_1 - t_2) \tag{20.1}$$

	Cars	Commercial vehicles (Euro 3)
Maximum heat volume to be dissipated from the coolant SI engine IDI diesel engine DI diesel engine	 $Q_{KM} = (0.5 \ldots 0.6)P_{mech}$ $Q_{KM} = 1.0P_{mech}$ $Q_{KM} = (0.65 \ldots 0.75)P_{mech}$	 $Q_{KM} + Q_{LLK} = 0.65P_{mech}$
Maximum permissible values for the temperature difference between coolant at radiator inlet and ambient temperature	Approx. 80 K	Approx. 65 K
Maximum permissible values for the temperature difference between charge air at radiator outlet and ambient temperature	Approx. 35 K	Approx. 15 K

Fig. 20-2 Nominal values for cooling methods.

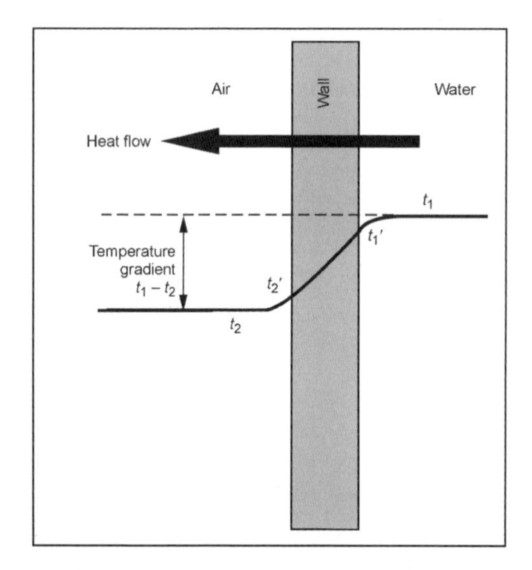

Fig. 20-3 Temperature curve for heat transfer from the water side with the higher temperature t_1 via a wall to the air side with the lower temperature t_2. t_1' and t_2' are the surface temperatures on either side of the partition wall.

The heat-transfer coefficients $\alpha_{1,2}$ can be increased by comparison with those of flat surfaces by finning. It has to be assured, however, that an advantage is actually achieved here in spite of the resulting increase in the flow resistance of the media and the necessary higher delivery energy.

The basic aim when designing a cooling system is to provide the demanded cooling capacities with the most compact, lightweight, and inexpensive radiators possible within the installation space available, so that an optimization process has to be carried out with respect to the arrangement and dimensioning of the heat transfer elements in the module, the choice of the fin or pipe geometry of the radiators, the power consumption of the fan, and the matching to the vehicle-specific boundary conditions, frequently also of the air drag coefficient value and crash behavior.

A common aid for the design is the use of analytical programs for heat-transfer calculation using the unidimensional flow filament theory. Given the radiator geometry, the heat transfer, heat transmission, and pressure drop conditions as well as the material streams, the parameters' pressure and temperature at the outlet from the heat transmitter can be calculated from the same inlet parameters. Backed up by empirical data from many years of measurement on a wide range of versions, fin or tube variants of different dimensions and for practically any operating points can be calculated very precisely in advance using these correlations within the framework of the similarity theory.

Designs of complete cooling modules with full and partial overlaps of heat-transfer media, fans, and fan housings are almost exclusively demanded today. As a result, "topology models," Fig. 20-4, with several flow paths are created for these modules of which each can be calculated in turn using the flow filament theory. The mutual influencing of the components is taken into account by the calculation codes.

Finally, this aid is complemented by elements such as headwind, fan, and all pressure boosting systems in the vehicle such as radiator grille and engine compartment ventilation. This permits the iterative calculation of the cooling air throughput in the vehicle and, consequently, of all thermodynamic parameters of the cooling system. Combined with broad experience from cooling capacity measurements in the wind tunnel, we obtain a reliable and quick simulation aid that significantly reduces the need for measurements on the vehicle.

The near future will bring the coupling of analytical unidimensional methods with the numeric three-

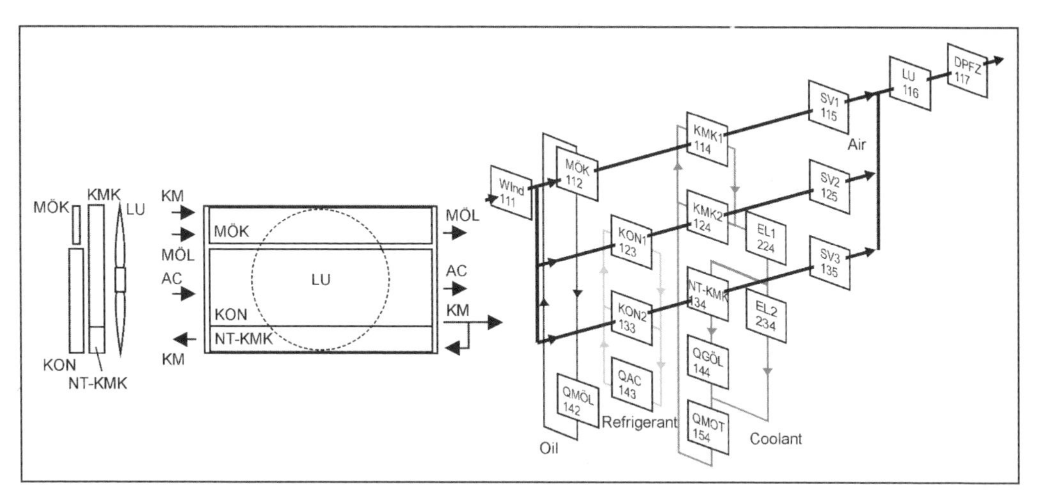

Fig. 20-4 Topology model for a unidimensional simulation of a cooling system in the vehicle.

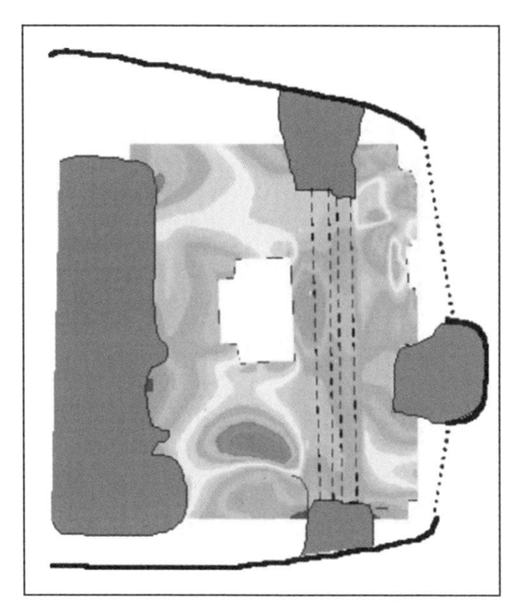

Fig. 20-5 CFD simulation of the cooling air flow in the front section of a car. (*See color section.*)

dimensional CFD methods as these can provide the detailed determination of the very complex cooling air flow in the engine compartment, Fig. 20-5.

20.4 Engine Cooling Subsystems

20.4.1 Coolant Cooling

The nonferrous metal radiators with copper fins and brass pipes common in early years have disappeared almost completely in Europe. They have been superseded in cars since 1975 and in commercial vehicles since 1988 by further-developed Al alloys offering a weight advantage of up to 30% with high-pressure resistance thanks to the brazing and a higher corrosion resistance.

Pipes and fins form the "radiator matrix." A distinction is made between the following:

Mechanically assembled rib and pipe systems of round or oval pipes and slot-fitted punched ribs linked to one another by expanding the pipes, Fig. 20-6. These systems typically cover the lower power segment, but thanks to improved expansion techniques and ever narrower oval pipes also achieve the power spectrum of soldered systems.

Soldered systems of flat pipes and rolled corrugated ribs. Today, these are generally manufactured with only one tube in the system depth; they can be ribbed to increase the strength, Fig. 20-7.

The system depths (extension in the cooling air flow direction) from the smallest car radiator to the largest commercial vehicle radiator range from 14 to 55 mm, with nonferrous metal radiators even to more than 80 mm, with cooling air surfaces ranging from 15 to 85 dm^2. Aluminum has more or less asserted itself as radiator material in Europe. In the United States and Japan, nonferrous metal systems are also still widespread. As further regional differences, the radiators for cars in Europe are mainly constructed in cross-flow design with the pipes running horizontally, Fig. 20-8, whereas in the United States and Japan they are frequently also built of downdraft design. In commercial vehicles, the arrangement in the downdraft, i.e., with the pipes running vertically, inside the vehicle frame are more widespread as then power variants can be formed simply via the pipe length with identical pipe shells and expansion tanks, Fig. 20-8.

The expansion tanks are always made of glass fiber-reinforced polyamide and are mounted on the radiator block with a gasket and a bead.

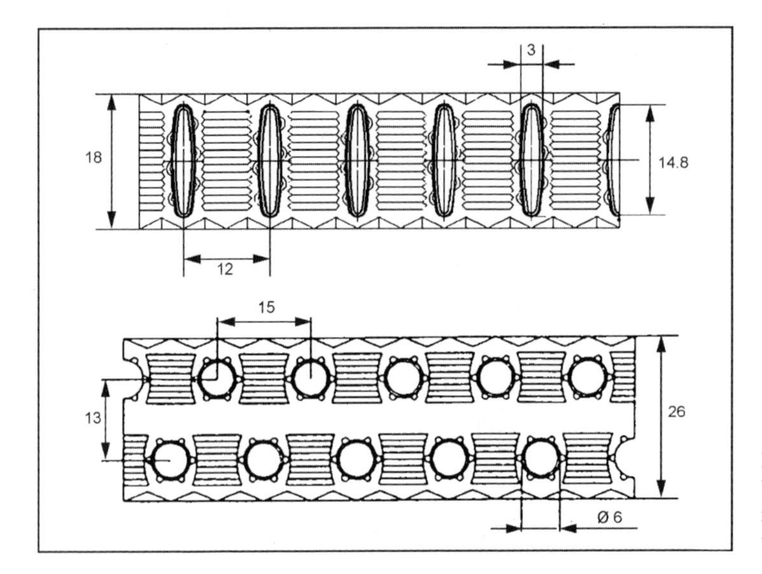

Fig. 20-6 Mechanically assembled rib and pipe systems for radiators with round and flattened oval pipes.

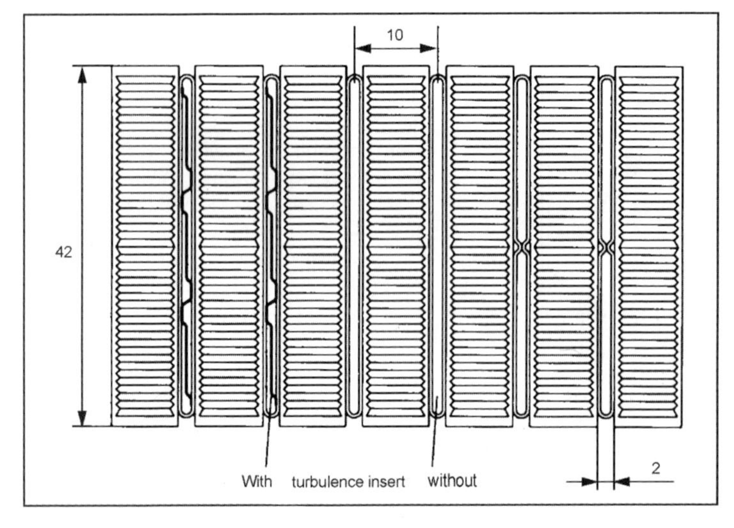

Fig. 20-7 Soldered flat tube system for radiators.

20.4.1.1 Radiator Protection Media

In a liquid-cooled internal combustion engine, the waste heat is dissipated to the environment to avoid overheating using a coolant. Coolants are operating media just like the lubricants and fuels and have to satisfy the following requirements:

- Optimum heat transmission properties
- High thermal capacity
- Low evaporation losses
- Good antifreeze properties
- Corrosion, erosion, and cavitation protection for all metallic materials
- Compatibility with elastomers, plastics, and coatings
- Prevention of deposits (fouling) and clogging
- Temperature stability
- Low maintenance requirements
- Long service life
- Simple handling
- Low operating medium costs
- Minimum environmental impact

The coolant generally consists of a mixture of tap water with a radiator protection medium tested and approved by the automobile and engine manufacturers, normally in a ratio of 50:50% v/v. Depending on the source, the tap water can exhibit significant differences in quality and have a considerable influence on the effectiveness of the coolant. For this reason, certain minimum demands are made on the quality of the tap water, Fig. 20-9.

The radiator protection medium consists of approximately 90% monoethylene glycol (1,2-ethanediol), 7% additives, and 3% water. The monoethylene glycol mixed with tap water results in a lowering of the freezing point of the coolant and protects the whole engine coolant circuit against freezing in the winter, for example, with

Fig. 20-8 Radiators for cars with cross-flow arrangement and commercial vehicle cooling module with radiator of downdraft design.

Property	Unit of measure	Demand
Appearance	—	Colorless, clear
Sediment	mg	0
*p*H value	—	6.5–8.0
Sum of the alkaline earths	mmol/L	0.9–2.7
Hydrogen carbonate	mg/L	≤100
Chloride content	mg/L	≤100
Sulfate content	mg/L	≤100

Fig. 20-9 Minimum demands on the quality of the tap water.

a common 1:1 mixture down to approximately 38°C. In some products the monoethylene glycol is replaced by monopropylene glycol (1,2-propanediol). The additives include substances for corrosion protection (inhibitors) and buffering, antifoaming agents, and pigments. The inhibitors here are of essential importance for the service life of the whole engine coolant circuit and effectively determine the quality of a radiator protection medium. The effectiveness of the inhibitors in the coolant provides additional protection for the materials in the engine coolant circuit against corrosion.

Before approval of a radiator protection medium, the corrosion protection capabilities, in particular, are assessed in extensive laboratory and technical tests. After successful completion of the most important tests such as the glassware test to ASTM D 1384, the knock chamber test to MTU (Motors and Turbines Union), the FVV (Research Association for Internal Combustion Engines) hot corrosion test, the FVV pressure ageing test, the FVV vibration test, the water pump test to ASTM D 2809, and the circulation test to ASTM D 2570, the vehicle fleet test critical for the product approval is finally carried out by the vehicle manufacturers.

During this practical test under the real conditions of road operation, the engine coolant circuits of the test vehicles are normally completely dismantled after running approximately 100 000 km and examined for signs of possible corrosion, erosion, and cavitations, and the results are evaluated. The compatibility with seal and hose materials and with plastics also plays an important role. This information together with the information gained on the test behavior of the coolant gives us a reliable overall view of the suitability of the radiator protection medium.

Depending on the operating conditions, the coolant is subject to a natural ageing. It is, therefore, essential to observe the service and maintenance instructions of the automobile and engine manufacturers. Complete changing of the coolant is normally carried out after 100 000 km or after two years for cars or after one year for commercial vehicles. New developments of organic inhibitor-based radiator protection media increase the service life of the coolant and contribute to reducing costs and conserving resources. Their share of the market is growing steadily.

20.4.2 Intercooling

Turbocharging with cooled charge air has become the general standard in the meantime for the commercial vehicle diesel engine and is almost always used for the car diesel engine with the aim of increasing the power density and of reducing fuel consumption and emissions. It is also finding greater attention than in the past during the course of the further development of SI engines. The increase in density achieved with the decreasing charge air temperature can be transformed into a higher output thanks to the improved cylinder charge. Furthermore, the lower temperature reduces the thermal load on the engine and results in lower NO_x contents in the exhaust gases.

Intercoolers are preferably soldered flat tube radiators of aluminum, cooled directly by the cooling air. The system depths range from approximately 30 mm to over 100 mm, the face areas from 3 dm^2 for cars up to 80 dm^2 for commercial vehicles. A large number of layouts are common in the car: Large intercoolers in front of the radiator, long and slim under or next to the radiator, or totally separate from the module, e.g., in the area of the wheel housing, hence, the large bandwidth in the system depths. The air receivers are made almost exclusively of plastic. In the commercial vehicle, large cross-flow arrangements in front of the radiator are most widespread, with the bracket for the whole module being preferably fastened to their air receivers, making the intercooler the supporting element for the module. The previously common cast aluminum construction of the air receiver is being replaced more and more by high-temperature-resistant plastics, Fig. 20-10.

Fig. 20-10 Intercooler with air shrouds of a high-temperature-resistant plastic for a lightweight commercial vehicle.

A current trend is charge air cooling using coolants, Fig. 20-11. Compared with the air-cooled systems used today, these systems exhibit a smaller pressure drop on the charge air side. Furthermore, valuable space is saved in the front section of the vehicle, and handling dynamics are improved. To date, this technology is mainly being employed in small numbers in high-performance engines' luxury class models, but it is to be expected that the use of coolants for cooling the charge air will become more prominent in future engine and vehicle developments.

20.4.3 Exhaust Gas Cooling

Diesel engines have to comply with ever stricter emission limits, Fig. 20-12. These limits, currently defined as "Euro3" level, can be achieved with low fuel consumption if the exhaust gas recirculation system familiar from the car is also cooled via an exhaust gas cooler. The admixing of noncombustible exhaust gas constituents to the cylinder charge and the cooling reduces the combustion temperature and, hence, the NO_x content of the exhaust gas.

As exhaust gas coolers are exposed to very high temperatures and extreme corrosion, especially in commercial vehicles, stainless steel is indispensable here as cooler material. Laser welding or nickel soldering are the most commonly used assembly methods. These coolers are designed as tube banks where the tubes containing the exhaust gas can be simple round tubes or tubes with special performance-enhancing but soiling-resistant measures.

The performances are sufficient for the statutory level Euro3 of approximately 2 kW for cars and up to 40 kW for commercial vehicles. The bandwidth of dimensions is, therefore, correspondingly large. The lengths alone vary from approximately 100 mm up to approximately 600 mm. These are already used in some series-production cars, and their number is increasing, Fig. 20-13. Widespread use in commercial vehicles is to be expected from 2002.

20.4.4 Oil Cooling

Part of the waste heat from the engine is absorbed by the lubricating oil. With some powerful engines, the cooling

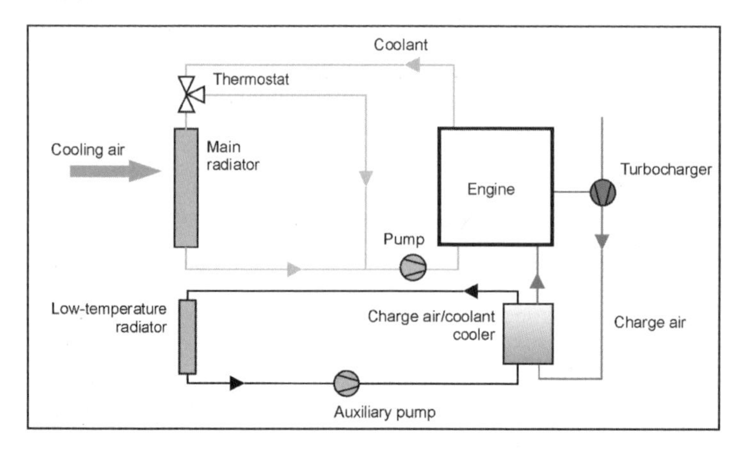

Fig. 20-11 Schematic of a coolant circuit for cars with indirect charge air cooling in a separate low-temperature circuit.

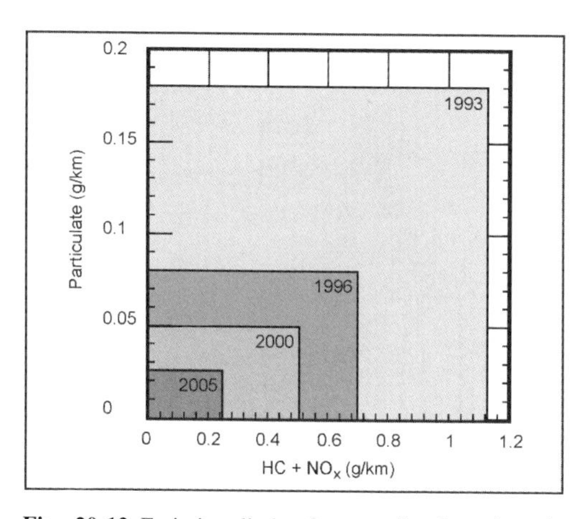

Fig. 20-12 Emission limits for car diesel engines in Europe from 1993 to 2005.

via the oil sump is no longer sufficient to maintain the maximum permissible oil temperature so that an engine oil cooler has to be employed.

Engine oil coolers in the car are preferably located near the engine, are of a round disc, disc stack, or flat tube design, and are made of aluminum, Fig. 20-14, so that cooling is performed indirectly with coolant.

Direct cooling with oil and air coolers is also common, with very high-pressure soldered flat tube designs in aluminum being arranged in the cooling module. In commercial vehicles, cooling is always performed with coolant, while the coolers are normally installed in an opening in the crankcase where they are exposed to the main flow of the coolant. The most widespread design is with plate-type coolers of stainless steel with turbulence inserts on the inside through which the oil flows. More recently, the use of aluminum coolers with higher capacities and comparable strengths but roughly half the weight has become possible.

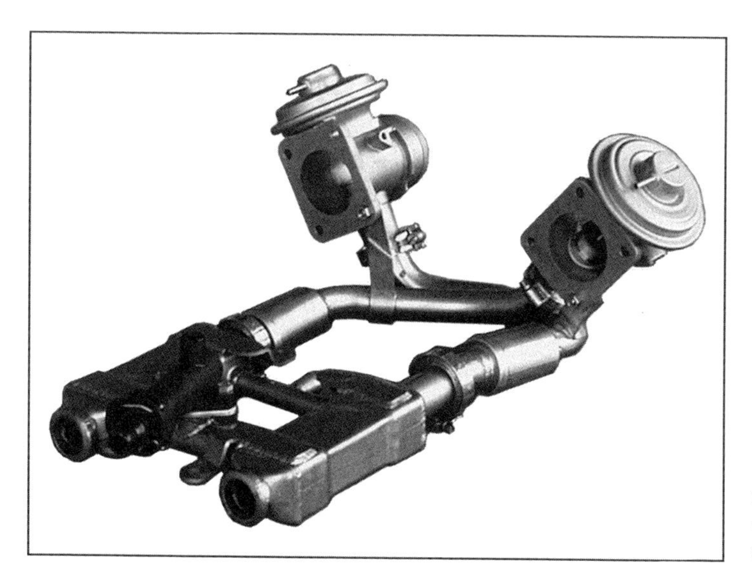

Fig. 20-13 Two-bore EGR cooler with EGR valves for a V8 car diesel engine.

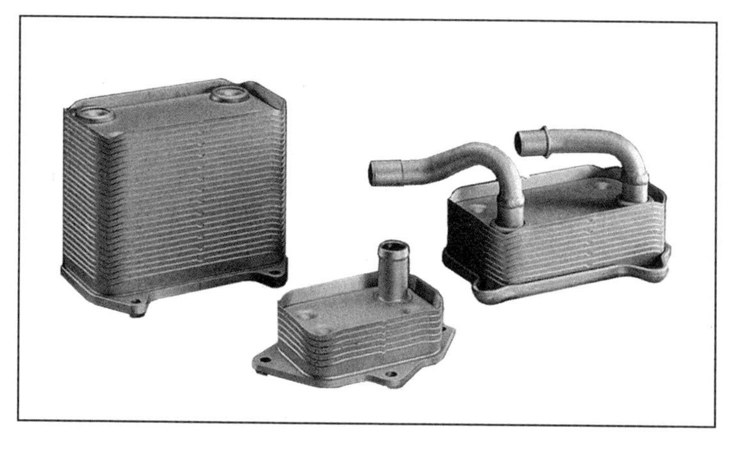

Fig. 20-14 Oil cooler of disc stack design.

Transmission oil coolers for cars with automatic transmissions can again be of air-cooled flat tube design or installed as very slim and long flat tube coolers in the expansion tank of radiators where they are cooled by the coolant. The latter form predominates today, although disc stack coolers installed in the module are also becoming more and more widespread.

Hydraulic oil used in power steering or other servo systems also has to be cooled. This is generally performed in simple pipe coils on a cooling module, and in rarer cases also with long tube yokes fitted with fin packs by mechanical expansion.

20.4.5 Fans and Fan Drives

Fans for engine cooling are manufactured today almost exclusively in plastic. In addition to the axial blades, there are, depending on the operating states in the vehicle, also hoop rings and inlet guides at the blade tips. Further typical fan characteristics can be curved blades and asymmetric blade pitches, Fig. 20-15. Such measures allow the fan efficiency and the noise emissions to be favorably influenced.

Fig. 20-15 Car fan with curved blades and hoop ring for drive with electric motor.

For cars, single or twin fans—generally suction fans—with maximum vane diameters of approximately 500 mm are used. With the exception of the most powerful engines, electric motors are employed as fan drives. They have an electric power consumption of up to 600 W with a stepped speed variation being provided via preresistors or an infinitely variable speed variation with brushless electric motors.

The upper power segment of cars and the full range of commercial vehicles are equipped with viscous couplings as fan drives, Fig. 20-16. Here a drive speed dictated by the crankshaft or an engine-side gearing—generally that of the coolant pump—is transmitted from the primary side by oil friction to a secondary side connected to the fan. A variable oil filling of the coupling allows the fan speed to be varied from an idle speed to just below the drive speed. The maximum fan diameter used in commercial vehicles is 750 mm with a power consumption of up to approximately 30 kW.

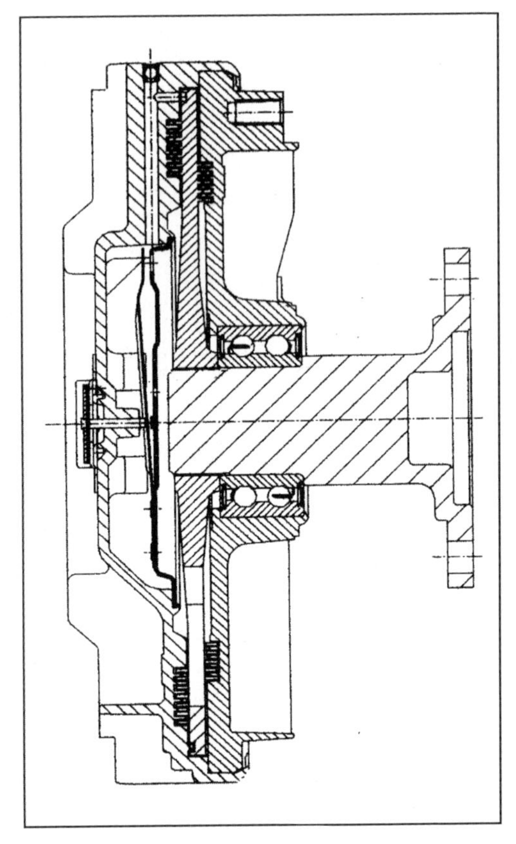

Fig. 20-16 Viscous coupling for the fan drive in commercial vehicles.

20.5 Cooling Modules

Cooling modules are units consisting of various components for cooling and possibly air conditioning of a vehicle that include a fan unit with drive, Fig. 20-17. The modular technique that has become more and more widespread since the end of the 1980s because it offers several fundamental technical and economic benefits:

- Optimum design and matching of the components
- Higher efficiency in the vehicle or smaller and less expensive components possible
- Less development, testing, logistics, and assembly costs for the vehicle manufacturer

In normal road vehicles, fixed cooling modules are almost exclusively used, attached to the longitudinal or transverse members of the vehicle. Generally, one of the heat transfer units serves as the module supporting element while the other components are snapped, clamped,

Fig. 20-17 Cooling module for use in cars with radiator, expansion tank, A/C condenser, refrigerant receiver, and electric fan with housing.

or clipped to its water tank or air shroud and side parts. The more components a cooling module contains, the more expedient is the use of a supporting frame to hold all the module elements.

20.6 Overall Engine Cooling System

The design of the cooling system is dictated by the operating states of the vehicle where cooling is of critical importance such as, e.g., driving at high speed or climbing hills with a large trailer load in the summer with the air conditioning system switched on. However, these critical cooling states occur only very seldom during a vehicle service life. This means that for the majority of the vehicle operating life either excessively high fluid flows are pumped for the engine cooling or that the temperatures in coolants or oils are too low or too high. This increases the fuel consumption and, consequently, the exhaust emissions, ride comfort is impaired, and the service life of the engine and attachments deteriorates.

The target for future cooling systems is to control all the fluid temperatures and media flows through a demand-oriented control of the engine cooling in such a way as to minimize energy demand and/or, depending on the priority, to achieve comfort, emissions, or service life advantages. For this, control interventions in the engine cooling will have to be possible in the future.

In present-day cooling systems, the following control possibilities for the fluid flows oriented to the cooling capacity requirements are already implemented:

- A thermostat whose wax element senses the temperature of the coolant flowing around it ensures that the coolant flow passes either through the radiator or past the radiator through a bypass line. Cooling of already very cold coolant can thus be more or less avoided or maximum cooling can be assured at very high temperatures.
- Electrically powered fans are switched on at various speeds or with an infinitely variable speed according to the coolant temperature in the expansion tank.
- On fans with viscous coupling, the oil filling and, hence, the fan speed is controlled depending on the cooling air temperature in front of the coupling. Hot cooling air is produced by flowing through hot heat-transfer elements. This is a sign of a high cooling requirement, and the fan is switched on via a bimetallic element.
- All other systems designed for critical operating conditions, however, are then operated without control. The coolant pump, for example, is driven via a belt by the crankshaft, charge air cooling is then almost always uncontrolled, and the oil cooling is only partially thermostatically controlled.

Such cooling systems were sufficient until now and were characterized by their very reliable operation. Future technologies for cooling systems, however, will be based on electronic control as are many other systems of the automobile. Via a network of sensors monitoring the thermal condition of the engine and cooling system, a control unit will trigger adjustments of delivery organs (fans, pumps) and control elements (valves, flaps, shutters) on the basis of the stored control concepts in order to save drive energy in auxiliary systems, to favorably influence exhaust gas and noise emissions, and to shorten heating-up phases in the sense of enhancing comfort and reducing wear by demand-oriented cooling. To achieve this, all the delivery and control elements must be controllable.

For the thermostat, this possibility has been created by using an electric heating of the wax element, Fig. 20-18. As a result, the thermostat position can be set independently of the current coolant temperature on the basis of set points from an engine map. The possibility of increasing the temperature during part-load operation of the engine reduces the fuel consumption.

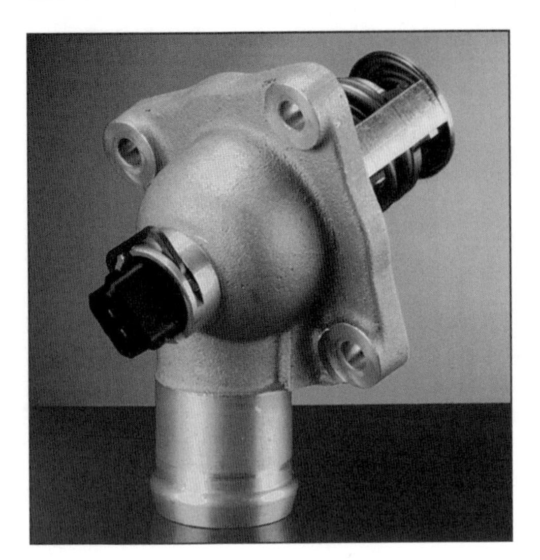

Fig. 20-18 Coolant thermostat with electric heating of the wax element.

In present-day vehicles, the coolant flow is generated by a coolant pump that is driven by a belt proportionally to the engine speed. In order to reduce the coolant throughput with low cooling requirements, but at the same time to be able to supply more coolant to the heater during the warm-up phase, the use of switchable or controllable pumps will be expedient in the future. In cars this can take the form of electric pumps. For more powerful engines, a 42 V onboard power supply is required. By being separated from the engine belt drive, the electric pump offers new design scopes. Alternatively, the coolant flow can also be influenced by the use of controllable throttle elements or switchable couplings in combination with the mechanical pump.

Apart from the control of the coolant flow in the main cooling circuit, there are also concepts for splitting the coolant stream into several circuits. One of these is the indirect charge air cooling in separate circuits or in low-temperature circuits connected to the main circuit described in Section 20.4.2. Circuits are also used for transmission oil temperature control in which the temperature transfer medium is supplied either with hot coolant to heat the oil during the warm-up phase or with cold coolant from a low-temperature section for cooling the oil. A thermostat ensures the switchover from heating to cooling.

The controlled delivery and throttling of the cooling air also offers a great improvement potential for the future. The stepped-speed electric fans in cars will be increasingly replaced by variable-speed fans with EC motors. Viscous couplings for commercial vehicles can now be electrically controlled because the oil filling is controlled by an electromagnetically actuated valve and no longer by a bimetallic element. This permits a control of the fan speed and a quick switching on and off of the fan.

In many driving situations, the fans are switched off. Nevertheless, a high cooling air flow is required at high travel speeds, hence, an increase in the drag of the vehicle. The use of aerodynamically optimized cooling air shutters here can reduce the fuel consumption and at the same time the noise emissions. In addition, a faster heating of the passenger compartment and of the engine is achieved in winter as heat losses can be reduced by isolating the engine compartment from the cold surrounding air.

Bibliography

[1] Kays, W.M., and A.L. London, Hochleistungswärmeübertrager, Akademie Verlag, Berlin, 1973.

[2] Eitel, J., Ladeluftkühlung mit Niedertemperatur-Kühlkreiskäufen für Kraftfahrzeug-Verbrennungsmotoren, in MTZ 53 (1992) Heft 3, pp. 114–121.

[3] Kern, J., and J. Eitel, State of the Art and Future Developments of Aluminum Radiators for Cars and Trucks, VTMS Conference, Columbus, OH, 1993.

[4] Ambros, P., Beitrag der Motorkühlung zur Reduzierung des Kraftstoffverbrauchs, Tagung Wärmemanagement HdT Essen, 1998.

21 Exhaust Emissions

Since the 1940s there have been systematic efforts in California to reduce the effects of mass transportation on air quality. In Europe, alarm was raised over automobiles in the 1960s because of carbon monoxide emissions directly harmful to humans. This led to the restriction of uncombusted exhaust components such as carbon monoxide and hydrocarbons. Because of the increase in trace gases from combustion and their dissemination to remote areas, trees became damaged in the 1970s and 1980s due to acid rain and photooxidants, among other things. Since nitrogen oxides and uncombusted hydrocarbons contribute to the formation of these substances, it became imperative to limit the amount of these emissions into the atmosphere. This need was responded to by the introduction of exhaust emission thresholds for street traffic in the United States starting in 1961, in Japan starting in 1966, and in Europe starting 1970.

Carbon monoxide emissions, which are directly harmful to humans, were reduced to a harmless level by legislation in industrialized countries. The drastic restriction of nitrogen oxides and hydrocarbon emissions began in the United States and Japan in the early 1980s. Central European countries were not far behind as they also strongly reduced these trace gases from passenger cars and power plants by the late 1980s.

By the beginning of the 1990s, it became clear that other exhaust emissions that were not harmful to humans could influence the earth's atmosphere. These effects that are summarized by the term "greenhouse effect" led people to focus on carbon dioxide emissions. Although the continuous reduction of fuel consumption by individual vehicles in the transportation sector has led to just a slight increase in fuel consumption by individual motorists, the strong increase in energy consumption for heating purposes and to produce electrical energy has led to a substantial increase in the CO_2 concentration in the atmosphere. The focus then became the more economical use of primary energy sources.

21.1 Legal Regulations

In this section, we address automotive exhaust emission thresholds for carbon monoxide (CO), hydrocarbons (HC), nitrogen oxides (NO_x), and particles (PM) for the European Union, the United States of America, and Japan. Since the statutory exhaust emission thresholds are indicated in different units [(g/km), (g/test), or (g/mile)], they were recalculated to (g/km) for comparison in this chapter. A direct comparison of exhaust emission thresholds is possible only when the emissions are measured using the same test cycle. However, this does not usually occur.

To measure exhaust emissions of passenger vehicles directly off the assembly line during type approval, there are numerous prescribed procedures worldwide. For passenger cars, we list the most important:

- U.S. procedure in its 1975 version (FTP 75) with the added test cycles SC03 (with air conditioning), and US06 (aggressive driving): The U.S. highway test cycle
- EC ECE 15/04 and EC MVEG-A test cycles
- Japanese 10.15-mode test, Japanese 11-mode cold test

21.1.1 Europe

The European emissions regulations for new passenger cars were originally specified in European Directive 70/220/EEC. This contains the thresholds (ECE R 15) defined by the United Nations Economic Commission for Europe (ECE). Changes to these regulations are found in the Euro 1 and 2 standards that became valid under directive 93/59/EC. The thresholds according to Euro 3 and 4 (2000/2005) published in Directive 98/69/EC were accompanied by an introduction of improved fuel qualities. A minimum diesel oil cetane number of 51 is stipulated, along with a clear reduction of sulfur both in gasoline and in diesel fuel.

The operation cycle for these regulations is found in ECE R83 (91/441/EEC). The test is carried out according to 98/69/EC. All emission thresholds are expressed in g/km. The development of standards over time for passenger cars is found in Fig. 21-1. The currently valid thresholds for gasoline and diesel are in Fig. 21-2.

In addition, an EEV (enhanced environmentally friendly vehicle) option is defined that reduces the Euro 3 thresholds for gaseous pollutants to approximately one-tenth and provides particle thresholds of 0.01 (g/km). The introduction of these vehicles is encouraged by tax incentives.

21.1.2 California, USA

Because of its special climate, the state of California has always taken the initiative in limiting emissions and, therefore, with the exception of CO, has always prescribed lower thresholds than the other states in the United States. National exhaust emission thresholds for vehicles in all of the United States were promulgated for the first time in the "Clean Air Act" of 1968. In 1977, new thresholds were set that brought about a 90% reduction in comparison to 1973. The mass emissions of exhaust have been measured since this regulation using the FTP-75 test cycle. The new thresholds led to the introduction of the three-way catalytic converter.

These standards were also sharpened in 1994 and 1998. The current status is portrayed in Fig. 21-3. A plan was created by the California "Air Resources Board" (ARB) in 1996 that stated that the exhaust emissions of passenger vehicles should be significantly reduced. These emission standards were incorporated into federal law as

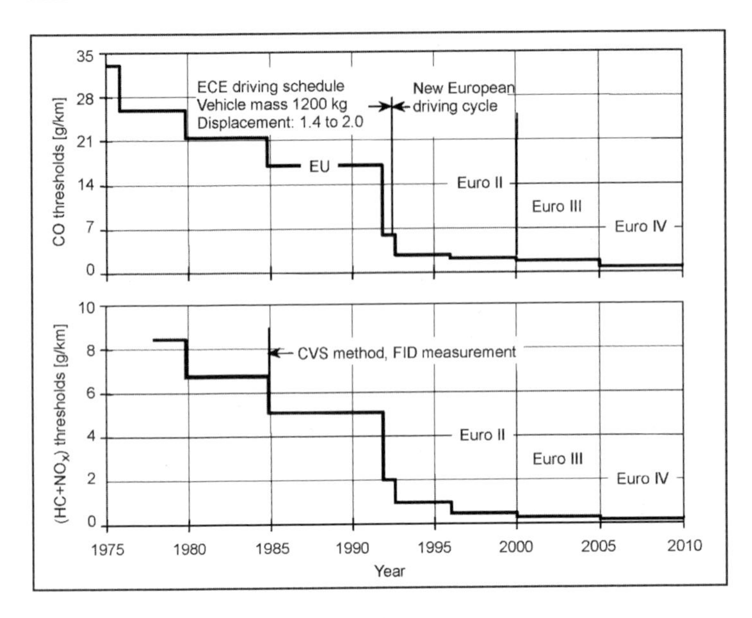

Fig. 21-1 Development over time of the exhaust emission standards for passenger cars with gasoline engines in the European Union.

Drive	Emissions category	CO	HC	$HC + NO_x$ (g/km)	NO_x	PM
Diesel	Euro 3	0.64	—	0.56	0.50	0.050
	Euro 4	0.50	—	0.30	0.25	0.025
Gasoline	Euro 3	2.30	0.20	—	0.15	—
	Euro 4	1.00	0.10	—	0.08	—

Fig. 21-2 Current exhaust emissions thresholds for passenger vehicles in the European Union.

Length of operation	Emissions category	THC	NMHC	NMOG	CO	NO_x	PM	HCHO
		g/mile						
5 years	Tier 0	0.41	0.34	—	3.4	1.0	0.20	—
50 000	Tier 1	0.41	0.25	—	3.4	0.4	0.08	—
miles								
10 years	Tier 0							
100 000	Tier 1	—	0.31	—	4.2	0.6	0.10	—
miles								

Fig. 21-3 U.S. federal certification exhaust emission standards for passenger cars.

the NLEV (National Low Emission Vehicle) standard (see Fig. 21-4) and the CFV (Clean Fueled Vehicle) standard (see Fig. 21-5). New vehicle categories were also defined according to emission categories TLEV (transitional low emission vehicle), LEV (low emission vehicle), ULEV (ultralow emission vehicle), and ZEV (zero emission vehicle) as shown in Fig. 21-6. The annual sales figures of automobile manufacturers must satisfy a specified percent of these categories.

The effort to introduce zero emission vehicles (ZEV) in California has been modified. The deadline has been extended to 2018, the percentage of PZEV (partial ZEV)

Length of operation	Emissions category	THC	NMHC	NMOG	CO	NO_x	PM	HCHO
		g/mile						
5 years 50000 miles	TLEV	0.41	—	0.125	3.4	0.4	0.08	0.015
	LEV	0.41	—	0.075	3.4	0.2	0.08	0.015
	ULEV	0.41	—	0.040	1.7	0.2	0.08	0.008
	ZEV	0.00	0.00	0.000	0.0	0.0	0.00	0.000
10 years 100000 miles	TLEV	—	—	0.156	4.2	0.6	0.08	0.018
	LEV	—	—	0.090	4.2	0.3	0.08	0.018
	ULEV	—	—	0.055	2.1	0.3	0.04	0.011
	ZEV	0.00	0.000	0.000	0.0	0.0	0.00	0.000

Fig. 21-4 U.S. national low emission vehicle emission (NLEV) standards for passenger cars.

Length of operation	Emissions category	THC	NMHC	NMOG	CO	NO_x	PM	HCHO
		g/mile						
5 years 50000 miles	LEV	0.41	—	0.075	3.4	0.2	—	0.015
	ILEV	0.41	—	0.075	3.4	0.2	—	0.015
	ULEV	0.41	—	0.040	1.7	0.2	—	0.008
	ZEV	0.00	0.00	0.000	0.0	0.0	0.00	0.000
10 years 100000 miles	LEV	—	—	0.090	4.2	0.3	0.08	0.018
	ILEV	—	—	0.090	4.2	0.3	0.08	0.018
	ULEV	—	—	0.055	2.1	0.3	0.04	0.011
	ZEV	0.00	0.000	0.000	0.0	0.0	0.00	0.000

Fig. 21-5 U.S. clean fueled vehicle (CFV) thresholds for passenger cars.

was increased, and the category of AT-PZEV (advanced technology PZEV) was added. The share of newly permitted ZEVs is supposed to rise from 2003 to 2018 from 10% to 16%, of which 50% of the vehicles must meet the AT-PZEV standard that contains the SULEV (super-ultralow emission vehicle) exhaust emission standard. These modifications primarily serve to lower costs while attaining environmental goals.

21.1.3 Japan

The first carbon monoxide emission restrictions for passenger vehicles were introduced in Japan in 1966 and use the four-mode test, which has since been discarded. In 1973, HC and NO_x were limited for the first time, and the ten-mode test started to be used. The thresholds introduced in 1975 lowered vehicle emissions of CO and HC by 90%. NO_x emissions were reduced by 90% by the regulations issued in 1976 and 1978. These thresholds are still valid today. Depending on the drive and engine designs, different thresholds apply to vehicles within the same category. For example, a distinction is drawn between vehicles fueled by gasoline and LPG. In passenger cars with diesel engines, a distinction is made according to combustion method (direct fuel injection or indirect injection engine) and the origin of the vehicle (Japan or imported).

Length of operation	Emissions category	THC	NMHC	NMOG	CO	NO_x	PM	HCHO
		g/mile						
5 years	Tier 0	—	0.39	—	7.0	0.4	0.08	0.015
50 000	Tier 1	—	0.25	—	3.4	0.4	0.08	0.015
miles	TLEV	—	—	0.125	3.4	0.4	—	0.015
	LEV	—	—	0.075	3.4	0.2	—	0.015
	ULEV		—	0.040	1.7	0.2	—	0.008
	ZEV	0.00	0.00	0.000	0.0	0.0	0.00	0.000
10 years	Tier 0							
100 000	Tier 1	—	0.31	—	4.2	0.6	—	—
miles	TLEV	—	—	0.156	4.2	0.6	0.08	0.018
	LEV	—	—	0.090	4.2	0.3	0.08	0.018
	ULEV	—	—	0.055	2.1	0.3	0.04	0.011
	ZEV	0.00	0.000	0.000	0.0	0.0	0.00	0.000

Fig. 21-6 California certification exhaust emission standards for passenger cars.

The presently valid emission standards and the proposed standards for 2002 are listed in Fig. 21-7. The current test method is the 10–15 mode cycle that has replaced the older ten-mode cycle since 1991, or since 1993 for imports. This test corresponds to the European ECE + EUDC (European Extra Urban Driving Cycle), but at a lower speed.

21.1.4 Harmonizing Exhaust Emission Regulations

To control the time and expense required to develop and permit vehicles, efforts are underway to recognize certifications of other countries (UN-ECE 1958 Agreement).

Vehicle weight	Emissions category	CO	CO	HC	HC	NO_x	NO_x	PM	PM
		Max	Mean	Max	Mean	Max	Mean	Max	Mean
		g/mile							
Diesel	1997	2.7	2.1	0.62	0.40	0.55	0.40	0.14	0.080
<1265 kg	2002[a]	–	0.63	–	0.12	–	0.28	–	0.052
Diesel	1997	2.7	2.1	0.62	0.40	0.55	0.40	0.14	0.080
>1265 kg	2002[a]	–	0.63	–	0.12	–	0.30	–	0.056
Gasoline	1997	2.7	2.1	0.39	0.25	0.48	0.25	–	–
	2002[a]	–	0.670	–	0.08	–	0.08	–	–

[a] Suggestions for 2002.

Fig. 21-7 Japanese exhaust emissions thresholds for passenger cars. The maximum limits apply to a production level of less than 2000 vehicles per year, and the mean limits apply to greater production volumes.

That this is a good idea can be seen by comparing the most recently valid NO_x + HC thresholds from the preceding sections. We note, however, the somewhat different requirements for exhaust purification systems since Europe and Japan tend to value a quick catalytic converter light-off after a cold start, and more weight is given to transient engine behavior in the United States.

21.2 Measuring Exhaust Emissions

21.2.1 Measuring Techniques for Certifying Automobiles

In general, the time and cost of these measuring procedures is very high for the reasons listed below. They all require a chassis dynamometer that must be calibrated to the respective vehicle, climatized test rooms to test specific cold-start conditions, and numerous highly sensitive exhaust emission measuring devices.

Here is a list of generally applicable features of type approval tests:

- Conditioning the vehicle at room temperature for approximately 12 h
- Cold start and measurement of starting emissions
- Dynamic test cycle with speeds from zero to 120 (km/h)
- Second test cycle with a hot start (U.S. FTP 75 Test)
- Precise measurements of exhaust emissions.

Figure 21-8 shows the basic arrangement of a vehicle chassis dynamometer for measuring exhaust emissions under conditions for certification. According to the presently valid laws governing approval, the dilution measuring technique must be used to determine mass emissions.

The molecule-specific absorption of bands of infrared light is used as a measuring standard for the exhaust emission components carbon dioxide (CO_2) and CO. To measure HC, flame ionization detectors (FID) are used. From a chemical viewpoint, the measuring technique of the flame ionization detector is based on the ionization of oxidizable hydrocarbon compounds in a hydrogen flame. Basically, the detector signal is proportional to the number of supplied hydrocarbon atoms. To detect nitrogen oxides (NO and NO_2), measuring devices are used that are based on chemoluminescence detectors (CLD). Particles are measured as mass emissions per kilometer using partial flow filtering and gravimetric evaluation.

To measure opacity in the recurring inspections of operated vehicles, partial flow opacimeters are used to determine the k value while operating the engine under "free acceleration."

Particularly exhaust thresholds that correspond to the ULEV or the Euro 5 standard place special demands on exhaust measuring technology since the concentrations in the exhaust must be measured when the engine is running hot and emission concentrations are very low. For this reason, new measuring methods have been suggested that deviate from the earlier used dilution measuring methods[1] and directly measure the concentrations in the exhaust emissions. An example of the direct evaluation of the exhaust mass emissions of nitrogen oxide is shown in Fig. 21-9.

21.2.2 Measuring Technology for Engine Development

The drastically stricter exhaust provisions require a detailed analysis of the origin of the exhaust components and the exploitation of every strategy to further reduce emissions using the latest measuring technology. A primary condition for future engine development is to lower vehicle fleet fuel consumption. In addition to the restricted pollutant components of total hydrocarbons, CO and NO_x, there are various other "unrestricted" pollutant components that arise during engine combustion such as benzene, toluene, xylene, aldehydes, or ammonia. Either these components are already in the fuel and pass uncombusted into the exhaust emissions or they are formed during combustion in the engine. Since specific components such as benzene are hazardous to health and smell bad, it is becoming increasingly important to detect these components.

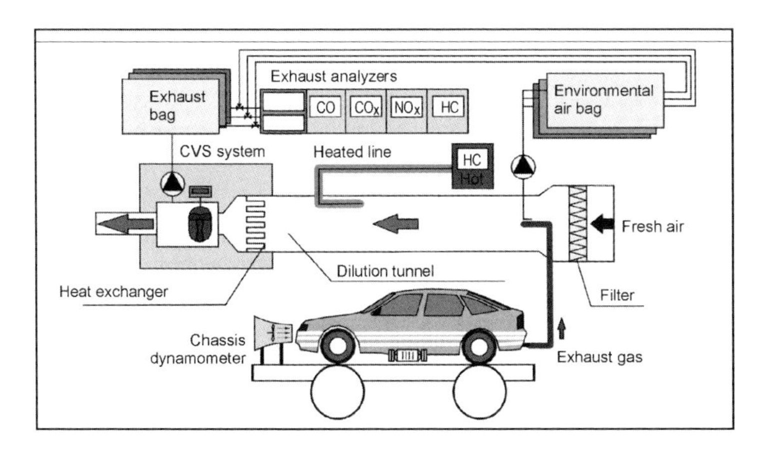

Fig. 21-8 Cycle of the chassis dynamometer for exhaust emission certification tests.

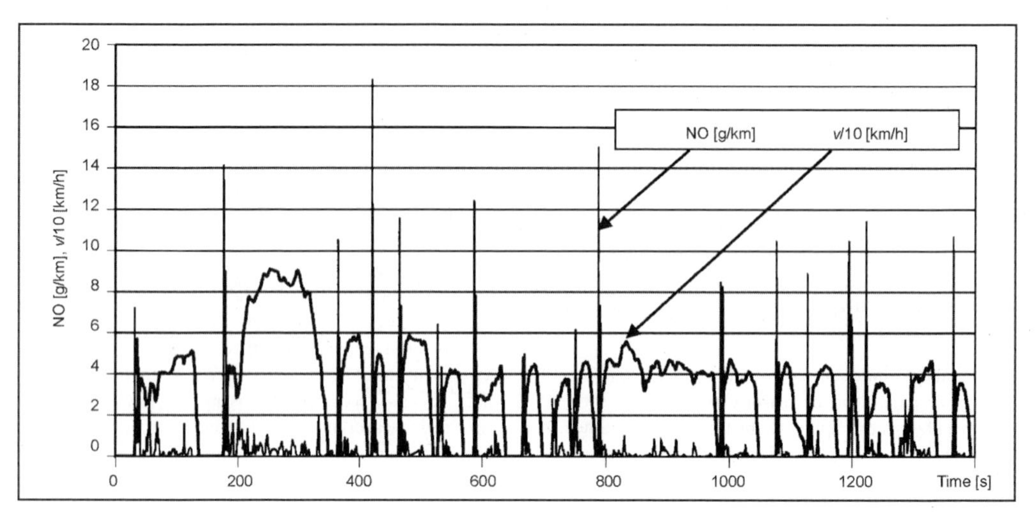

Fig. 21-9 Characteristic of the vehicle speed and nitrogen oxide mass emissions over the FTP 75 test cycle.[2]

In particular, there is great potential for improving exhaust emission values by optimizing transient engine operating conditions that make it easier to adapt to the operating conditions of the exhaust purification system. Fuel consumption can be influenced to a much greater degree, however, by other measures such as the combustion procedure and drivetrain management.

The goal of future research is, therefore, to develop engine management so that changes in load and speed do not cause a noticeable deviation from the optimum lambda characteristic that is determined by the functional principle of the catalytic converter. To accomplish this task, it is necessary to carry out high time resolution measurements in the combustion chamber and at specific points in the exhaust system to determine the precise sources of the emissions.

Given the dominance of transient operating conditions, it is also necessary when developing engines to carry out experimental investigations on the dynamic simulation test bench to adapt mixture formation and control systems. These efforts also alleviate the great amount of effort involved in preparing a complete vehicle for the chassis dynamometer. The goal of simulation on the dynamic engine test bench is to achieve the same speed and torque characteristic for the engine crankshaft as well as to attain equivalent temperatures, fuel consumption, etc., as experienced on the road or when operating the vehicle on a chassis dynamometer. The primary advantage lies in the high reproducibility of the individual test runs.

The illustration in Fig. 21-10 shows a measuring setup for a modern development test bench that also permits the measuring of the individual combustion cyles. The setup is typically divided into a simulation computer, highly dynamic electrical load machine, a crank-angle-related measured data memory, and high-speed exhaust measuring equipment.

The following emission measuring equipment for engine analysis is used on these test benches that can be categorized according to response speed:

- Standard measuring devices with a response time in seconds and above
- Transient measuring equipment with response times in the 100 ms range
- Measuring devices and methods for individual cycle analysis with response times around 1 ms

The measuring devices can also be categorized according to their sites of use on the engine:

- Measuring devices that analyze a sampled and conditioned partial flow of the exhaust emissions. The majority of measuring procedures fall under this category.
- Sensors and measuring devices that are used in the exhaust system in situ.
- Measuring methods for experimentally determining the gas composition in the combustion chamber.

Partial flow measuring devices:

A measuring setup for the engine with the most important exhaust emission measuring devices is seen in Fig. 21-11. Please refer to Section 21.2.1 for a discussion of the physical principles of the exhaust emission measuring devices for restricted exhaust components. Furthermore, the very important oxygen measuring method based on the paramagnetic properties of the oxygen molecule is discussed. In addition to the classic devices for measuring restricted exhaust emission components, mass spectrometers and particle size measurements are now used. Mass spectrometry determines the ratio of mass to charge for ions or their components. This is done by diverting the ions in magnetic and electrical fields or determining their movement

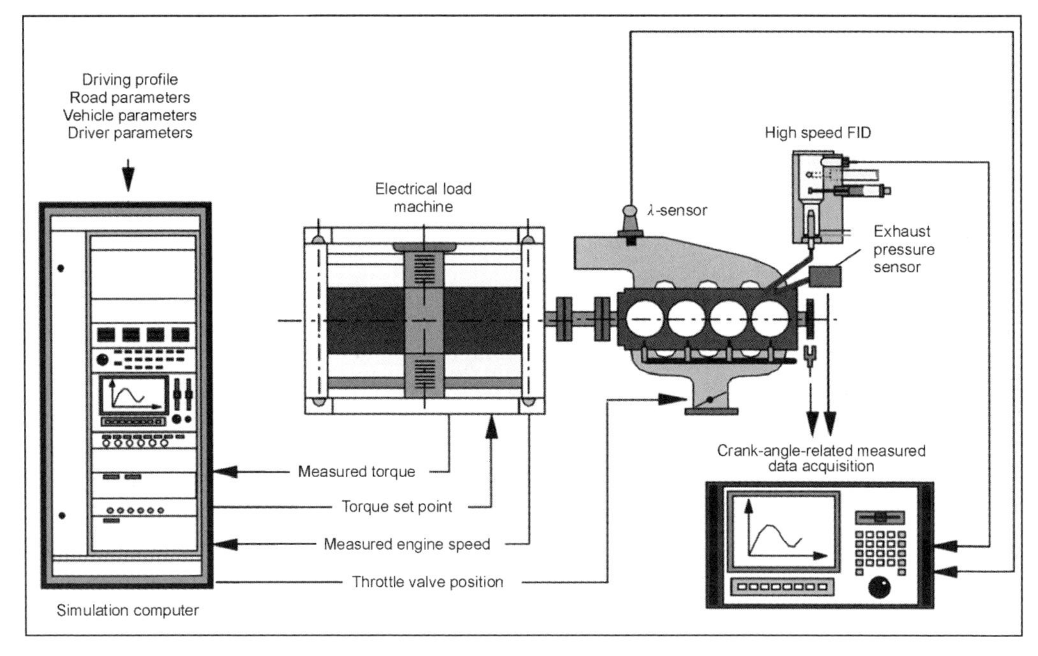

Fig. 21-10 Measuring setup for the crank-angle-related exhaust emission measurement.

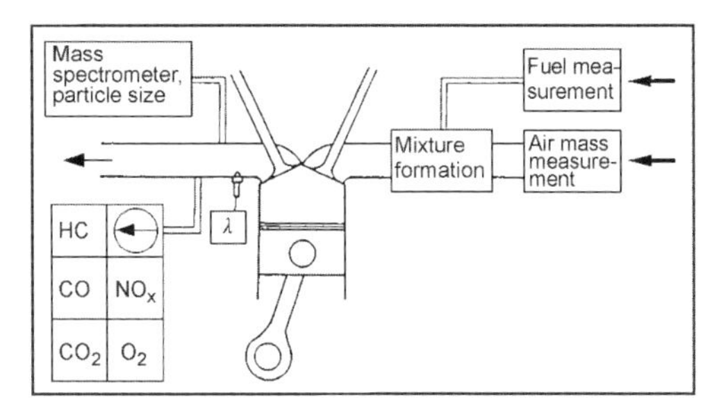

Fig. 21-11 Setup of exhaust emission measuring equipment for partial flow measurement on an engine test bench.

energy. Theoretically, each exhaust component or several can be determined at the same time by their number of moles. However, two effects stand in the way. On the one hand, there are several relevant exhaust emission components that have the same number of moles, and on the other hand, the simultaneous measurement of several components greatly increases the measuring time. Gas chromatography can selectively be used to detect nonrestricted exhaust emission components.

In particular, the detection of size-class-dependent particle emissions represents a very time-intensive test at present. A distinction is drawn between methods that work according to the impactor principle that allows a specific number of size classes to be gravimetrically determined at the same time based on the aerodynamic properties of the particles, and methods that are selective that can measure only one size class as shown in Fig. 21-12. These measuring devices combine several measuring principles. They distinguish between individual particle size classes by the variable electrical charge and subsequent aerodynamic evacuation. The particle fraction is then fed to a condensation nucleus counter in which the number of particles per volume unit is determined.

In situ exhaust measurement in the exhaust system:

To measure the air-fuel ratio at a high response speed both in the lean mixture range (λ greater than one) and in the rich range (λ lower than one), oxygen sensors are used that are mounted in the exhaust emission system. These "wide range sensors" use the oxygen ion pump principle and are available from various manufacturers. However, there are several factors that greatly influence the measuring precision of these sensors. Figure 21-13 shows the most important errors that can arise when these sensors are used. The typical error from excess exhaust counterpressure can be 20% of the measured value of $\lambda = 2$. These sensors are becoming increasingly important since they are also used in engine management as control sensors for lean engines.

The basic design and the output signal as a function of the lambda value can be seen in Fig. 21-14.

For individual cycle analyses with a high time resolution, sensors made of strontium titanate ($SrTiO_3$) are useful. These sensors have a response time of approximately 5 ms and permit the detection of the mixture composition of individual cylinders from the total engine exhaust. This represents a substantial advance, especially in the

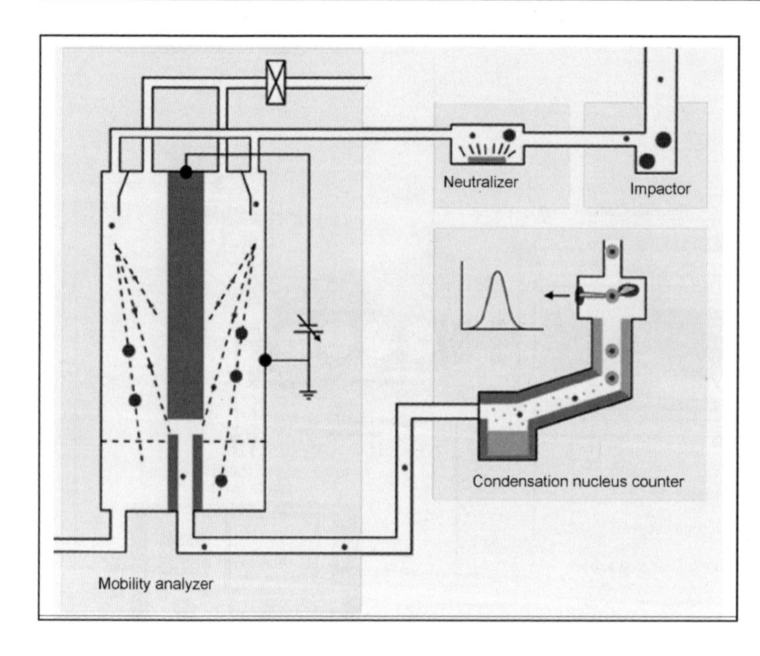

Fig. 21-12 Mobility analyzer to determine the size class distribution of particles.[3]

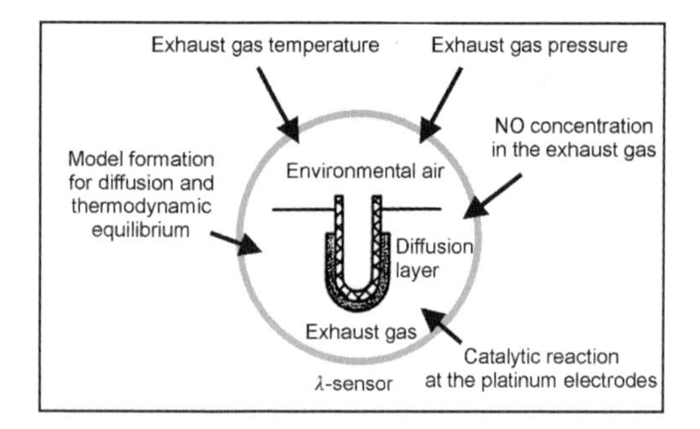

Fig. 21-13 Influence of errors when determining the air-fuel ratio with a lambda sensor.

cylinder-selective comparison of the fuel injection system while taking into account the charge of the individual combustion chambers.

Based on a similar technology as the lambda sensors in Fig. 21-14, NO_x sensors were made for in situ measurements[4] that offer useful support in the development and control of catalytic converter systems for lean engines.

Exhaust emission measurement in the combustion chamber:

The gas composition and individual gas components can be measured using different methods as shown in Fig. 21-15 that are divided into two main groups: Optical measuring methods that are applied directly in the combustion chamber and are usually used for experimental engines (such as "glass engines") and methods that withdraw gas directly from the combustion chamber.

Optical measuring uses different physical and quantum mechanical properties of molecules or atoms to determine the percentage of a specific gas component. These measuring methods are generally able to determine the distribution of numerous components at the same time.

The methods based on gas withdrawal work with either pulsed gas withdrawal valves or capillaries experiencing continuous flow. The measuring devices that are used in conjunction with gas withdrawal valves basically correspond to standard devices that are used for conventional exhaust emissions analysis. To obtain a sufficiently large gas volume for analysis, the average over numerous combustion cycles must be calculated. Individual cycle analyses, especially concerning cyclic fluctuations or transient effects during load changes can be done using only continuous gas withdrawal and gas analyzers that react very quickly for system-related reasons. Today, these are used for the exhaust emission components of hydrocarbons and nitrogen oxide.

The access to the combustion chamber is fairly easy when the continuous gas withdrawal method is used (Fig. 21-16). Since no mechanical actuation is required in the direct environment of the combustion chambers or the withdrawal site, the measuring position is less restricted. Another property of this sampling method is the favorable local resolution because of the lack of fuel condensation on the wall at the withdrawal valve and the simple geometric shape of the inflow cross section.

These measurements in the combustion chamber can provide much information about mixture formation and combustion up to the point at which pollutants arise. The results can be valuable for emission reduction and for optimizing various constructive details. A very interesting

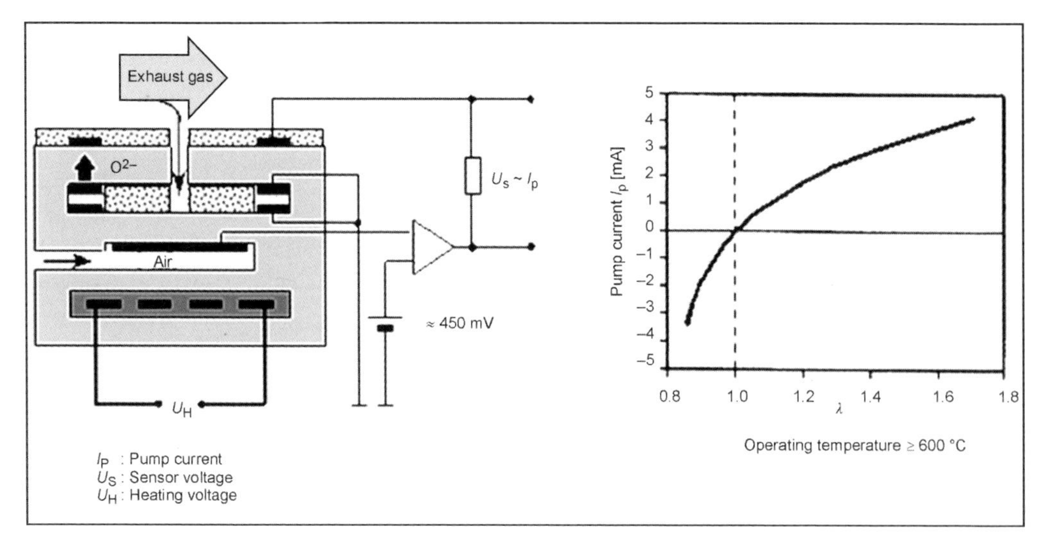

Fig. 21-14 Basic design and output signal of a lambda sensor according to Ref. [4].

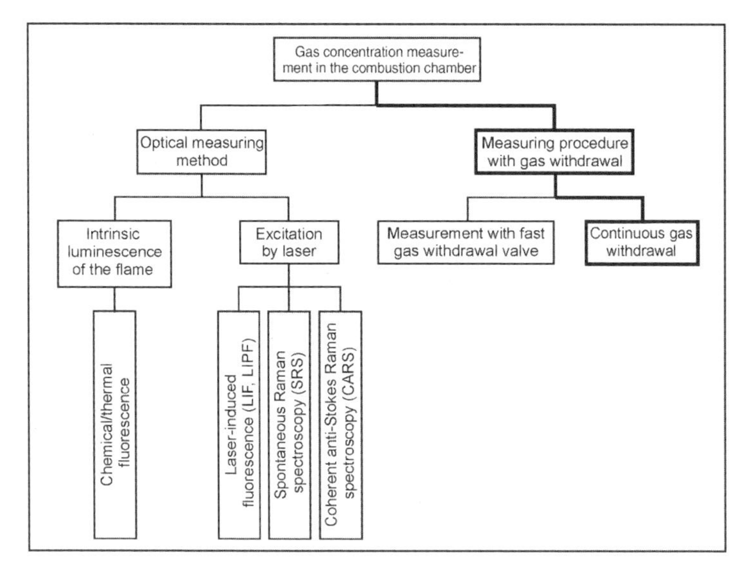

Fig. 21-15 Possible measuring methods for determining the gas composition in the combustion chamber.[5]

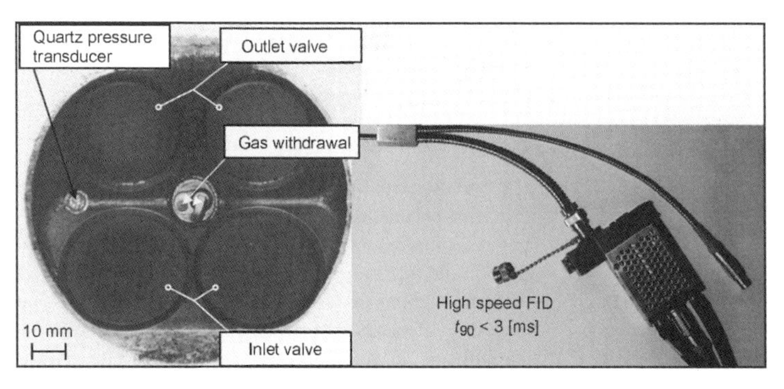

Fig. 21-16 Measuring setup for continuous hydrocarbon indication in a four-valve spark-injection engine.

use is in the development of combustion systems with direct fuel injection and lean designs in which the focus lies on the variable mixture at the spark plug and, hence, the cycle-selective lambda value.

In addition to engine experiments involving combustion, experiments with engines operating without combustion are also carried out. By definition, the HC concentration in the combustion chamber can be used in operation without combustion to calculate the local air-fuel ratio. The measurements also provide information about the process of mixture formation and the residual gas content in the area of the spark plug or the withdrawal site.

21.3 Pollutants and Their Origin

In the combustion of fuels with oxygen from the air that contain 21% by volume O_2, <1% by volume inert gases, and nitrogen N_2, energy is released as heat in an exothermic reaction. The release of heat by fuels based on hydrocarbons such as gasoline and diesel fuel is determined by numerous incomplete reactions depending on the composition of the hydrocarbons in the fuel. Important fuel components are paraffins, olefins, and aromates.

Given the complete combustion of hydrocarbons under ideal conditions or when there is excess air, theoretically only carbon dioxide and water arises as well as nitrogen from air, the carrier of the oxygen.

For this reason, the air-fuel ratio, lambda (λ), represents the most important parameter for the combustion process. "Lambda" is defined as the ratio of the actual air quantity relative to the ideal stoichiometrically required quantity.

$$\lambda = (m_L/m_K)/(m_L/m_K)_{\text{stoich}} = (m_L/m_K)/(m_{L,\text{th}})$$
$$= m_L/m_{L,\text{th}} \qquad (21.1)$$

m_L = Air quantity per unit time supplied to the engine
m_K = Fuel quantity per unit time supplied to the engine
$m_{L,\text{th}}$ = Theoretical air quantity required for the complete combustion of this fuel quantity

The following general reaction equation can be used for combustion in the operating range with excess air:

$$1\,[\mathrm{CH}\psi\mathrm{O}\varphi] + 4.762 \cdot (1 + \psi/4 - \varphi/2) \cdot \lambda\,[\text{air}] \qquad (21.2)$$

burns to form

$$1\,[CO_2] + (3.762 \cdot (1 + \psi/4 - \varphi/2) \cdot \lambda - n/2\,[N_2]$$
$$+ ((1 + \psi/4 - \varphi/2) \cdot (\lambda - 1) - n/2)\,[O_2] + n\,[NO]$$
$$+ \psi/2\,[H_2O]$$

[ ] = Component
n = Mole of nitrogen oxide
ψ = Hydrogen-to-hydrocarbon atomic ratio of the fuel
φ = Oxygen-to-hydrocarbon atomic ratio of the fuel

In addition to the main components of the exhaust such as carbon dioxide (CO_2 and water vapor are the main representatives of the pollutants), carbon monoxide, uncombusted and partially combusted hydrocarbons (HC-aldehydes, ketones, etc.), and nitrogen oxides (NO_x) are limited by the law. Pollutants primarily arise from the interruption of the reaction chain due to the short dwell time in the combustion chamber. The equilibrium, hence, no longer exists. Inhomogeneities in the mixture from different air-fuel ratios λ, combustion chamber wall effects, and impurities and additives in the fuel also contribute to the undesired by-products (Fig. 21-17).

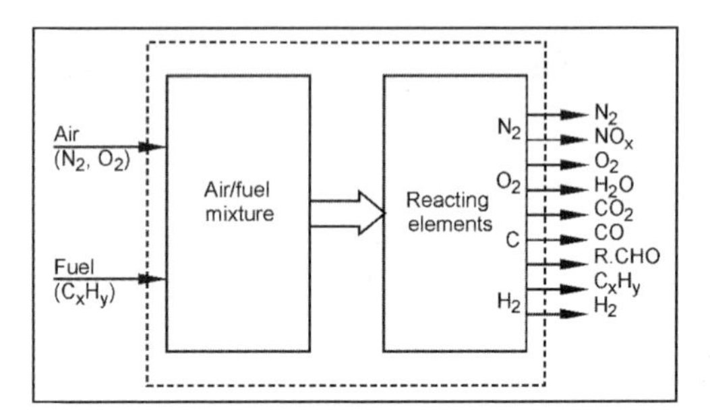

Fig. 21-17 Reaction mechanisms in the combustion chamber.[6]

In addition, depending on the type of fuel and combustion procedure, solids in the form of particle emissions are possible. Nonrestricted exhaust components that arise from thermally cracking the hydrocarbons and their by-products are gaining increasing attention since either they are potentially hazardous or they cause a noticeable odor.

21.3.1 Spark-Injection Engines

Combustion in spark-injection engines possesses the following general characteristics:

- Externally supplied ignition executed as single or multiple ignition
- A compression ratio of 8 to 14 depending on the used fuel
- Both four-stroke and two-stroke combustion are used

An important parameter is the air-fuel ratio λ that puts narrow limits on the combustion procedure. Depending on the approach to combustion and exhaust purification, a constant air-fuel ratio for the entire combustion chamber is selected, or stratified charging with varying ratios in the combustion chamber. Indirect fuel injection into the intake manifold and direct fuel injection into the combustion chamber are possible methods for mixture preparation.

21.3.1.1 Restricted Exhaust Emission Components

Carbon dioxide: In Europe, carbon dioxide emissions are restricted exhaust components although they are not toxic. Legal regulations are increasingly restricting CO_2 exhaust. Carbon dioxide arises from the complete combustion of the hydrocarbons in the fuel molecules. The CO_2 emission essentially depends on the fuel consumption and the fuel composition, and it attains its relative maximum given a

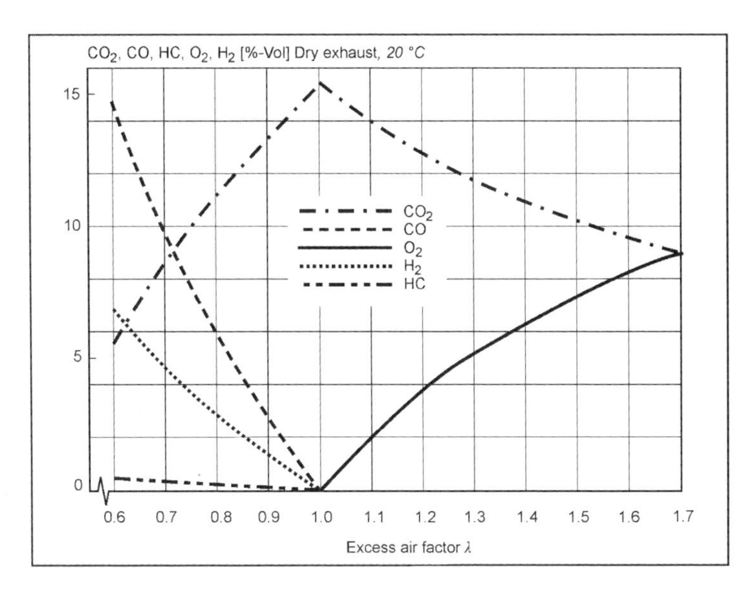

Fig. 21-18 Calculated concentration of exhaust emissions over the air-fuel ratio for a spark-ignition engine.

complete conversion at an air-fuel ratio of 1. Figure 21-18 shows the calculated ideal exhaust concentrations.

Carbon monoxide: CO arises as an intermediate step in the formation of carbon dioxide and from incomplete combustion under a lack of oxygen. This is characterized by the "water-gas equation."

$$CO + H_2O \Leftrightarrow CO_2 + H_2 \tag{21.3}$$

Essentially, carbon monoxide formation is determined by the local excess air factor and the temperature and pressure. When there is insufficient air, the CO emissions have a nearly linear relationship to the air-fuel ratio. The CO emissions result from a lack of oxygen. When $\lambda > 1$ (excess air), the CO emissions are very low and nearly independent of the lambda value. The CO emissions are also largely independent of other parameters such as the compression ratio, load, moment of ignition, and fuel injection law.

Hydrocarbons: HC emissions arise from uncombusted and partially combusted hydrocarbons and corresponding thermal cracking products. These components can originate from both the fuel and the lubricants. Various mechanisms are responsible for these emissions. For example, incomplete combustion of the hydrocarbons occurs because of partial ignition of the overall combustion chamber volume, and wall deposits of fuel. Other reasons are residual fuel in the dead spaces such as gaps in the cylinder head seal, valve seats, fire land, piston rings, spark plugs, and squish areas. Misfiring, emissions of hydrocarbons from the lubricant, and absorption of fuel molecules in the lubricant film of the cylinder barrel and at sites with impurities also increase hydrocarbon emissions. If one plots the mass-related HC emissions over the exhaust cycle, then we find increased HC emissions in the curve plotted against the crank angle shortly after the exhaust valve opens and before it closes that appear to be primarily determined by the previously cited wall phenomena (Fig. 21-19).

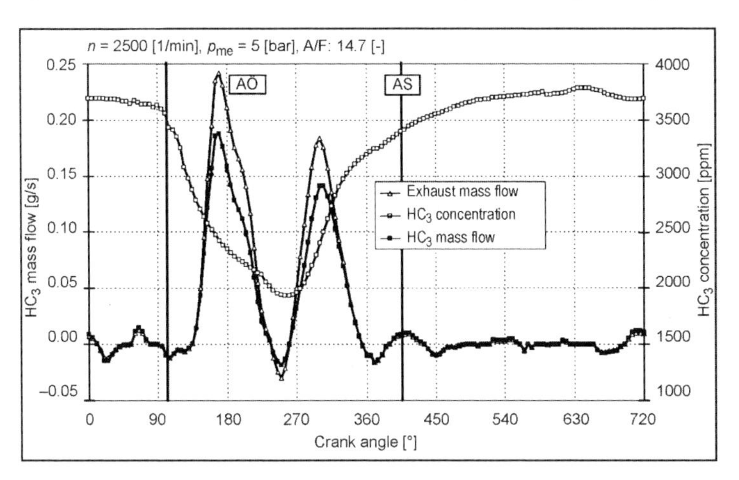

Fig. 21-19 Characteristic of mass-related hydrocarbon emissions after the exhaust valve as a function of the crank angle.[7]

When the flame front extinguishes as it contacts the cold wall (flame quenching), HCs are released since the mixture at the boundary layer cools to the wall temperature and the reaction is terminated. The origin of partially combusted hydrocarbons essentially depends on the temperature and oxygen content, and to a lesser degree on the molecular structure. With air-fuel ratios under 1, HC emissions strongly increase since there is too little oxygen for complete combustion in the combustion chamber. The same holds true as the air-fuel ratio increases and the ignition limit of the mixture is reached, which leads to misfiring when there is homogeneous mixture formation.

Nitrogen oxides: This generic term includes the seven oxides NO, NO_2, NO_3, N_2O, N_2O_3, N_2O_4, and N_2O_5. Nitrogen oxides arise from the nitrogen and oxygen in air during combustion. The processes are described by the expanded Zeldovich mechanism (1946). The most important representatives of these oxides are nitrogen monoxide NO and nitrogen dioxide NO_2. A general distinction is drawn between two important NO formation processes: The formation of thermal NO is influenced by the parameters of temperature, oxygen concentration, the air-fuel ratio, dwell time, and pressure. The maximum formation of NO occurs at approximately 2200–2400 K and quickly decreases at higher temperatures. Below 750 K, high activation energy is required for the decomposition of NO. NO arises quickly in a side reaction in the flame front from OH radicals that form other compounds with the nitrogen molecules. Because of the high temperatures, nitrogen oxides can also be formed from the nitrogen contained in the fuel. This formation process is, however, less important. The NO/NO_2 ratio of the untreated emissions in spark-injection engines is over 0.99. The maximum NO_x concentration occurs in the slightly lean range of $\lambda = 1.05–1.1$.

Spark-injection engines with direct fuel injection and charge stratification, in comparison to intake manifold injection, produce lower NO_x emissions due to the low average temperature. Charge stratification yields are more CO and HC emissions from local lean zones.

21.3.1.2 Unrestricted Exhaust Components

Particles: All components that can be separated by a filter below 51.7°C are considered particles. The particles consist of solid organic or liquid and soluble organic phases. This includes soot, various sulfates, ash, various additives from the fuel and lubricating oil, and abrasion and corrosion products. Abraded chrome arises as well as nickel aerosols from piston wear. The chrome aerosols have a particle size of 1.6–6.4 μm.[8] Condensated particle emissions play a rather subordinate role in spark-ignition engines. However, they are gaining more attention with fuel injection systems that use direct fuel injection.

Gaseous components: Of particular interest are aromates such as benzene, toluene, xylene, and polycyclic aromatic hydrocarbons (PACs) as well as aldehydes such as formaldehyde, acetaldehyde, acrolein, propionaldehyde, hexanal, and benzaldehyde. Aldehydes are intermediate products that rise in the oxidation of hydrocarbons, and their formation depends on temperature.[9] Of the BTEX (benzol, toluene, ethylbenzene, xylole) components, toluene occurs in the greatest quantity.[8] We can (in principle) draw a direct relationship between fuel composition, lubricant composition, and quality of the combustion characteristic with the generation of unrestricted components.

21.3.2 Diesel Engines

Diesel engines have the following characteristics:

- Inner mixture formation.
- Load control via the supplied fuel quantity with unthrottled, inducted air.
- Autoignition and a large amount of excess air. From an integral point of view, diesel engines operate with a load-dependent air-fuel ratio of between 1.2 (high load) and 7 (idling).
- Compression ratio of 14 to 22.
- Higher-boiling hydrocarbons as fuel.

Indirect injection engines (prechamber/whirl chamber) have favorable untreated emissions and noise emissions, but they are more commonly being replaced by engines with direct fuel injection for passenger cars because of CO_2 emissions that are up to 20% higher. In commercial vehicles, diesel engines with direct fuel injection are the main form of propulsion in Europe. Large engines that have the greatest efficiency of all heat engines also use this technique, however, usually with the two-stroke method. Different methods with different pressure generators are used for mixture formation such as inline fuel injection pumps, distributor fuel injection pumps, pump nozzles, pump-rail nozzles, and common rail systems. At present, the most important fuel injection method in passenger car engines uses air-distributing high-pressure fuel injection through multiple-hole nozzles.

21.3.2.1 Restricted Exhaust Components

Carbon dioxide: The clearly improved fuel consumption and the enhanced consumption in the partial load range leads to a real 20% reduction of CO_2 emissions per traveled kilometer.

Carbon monoxide: Because of inhomogeneity of the mixture from charge stratification, there are zones with air-fuel ratios that are less than one. In these areas, high concentrations of CO arise during the reaction that largely reoxidized to form CO_2. In contrast to spark-injection engines, this leads to substantially lower specific carbon monoxide emissions.

Hydrocarbons: The mechanisms and parameters are similar to spark-injection engines. In general, the HC emissions in diesel engines are much lower. Additional relevant variables are the mixture formation quality of the fuel injection system and metering precision. Postinjection increases HC emissions. The influence of minimized pollutant quantities with fuel injection nozzles is shown in Fig. 21-20.[10]

Nitrogen oxides: The formation processes are also comparable with those of SI engines. The NO-to-NO_2

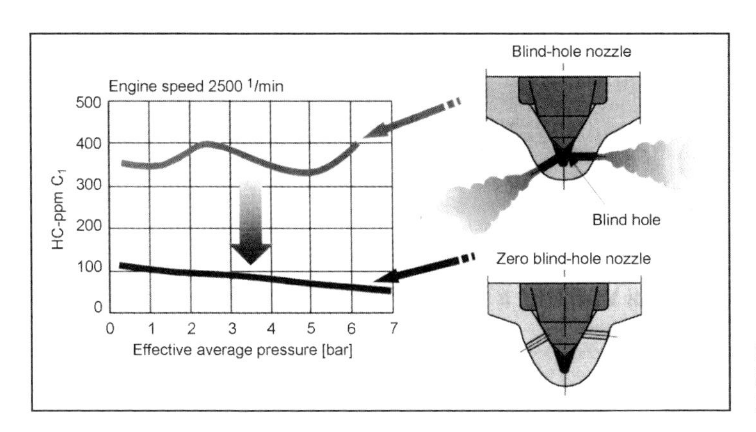

Fig. 21-20 Influence of the fuel injection nozzle construction on HC emissions.[10]

ratio for diesel engines is 0.6 to 0.9 depending on the load. Under a small load, more NO_2 is formed. This ratio is primarily determined by the oxygen concentration and the dwell time. NO_2 is formed at the flame front.

Diesel engines with divided combustion chambers have clearly lower NO_x emission than diesel engines with direct fuel injection. Given the extreme lack of air at high temperatures while fuel is injected into the prechamber, the NO_x formation rate is much lower. When the prepared mixture passes into the main combustion chamber, the opposite conditions predominate, i.e., a large amount of excess air at lower temperatures. Since the compatibility for exhaust recirculation is much higher with diesel engines with direct fuel injection than with indirect injection engines (by approximately a factor of two), the ratios are opposite.

Particles: For the most part, the particles from diesel engines consist of hydrocarbon particles. The remainders are hydrocarbon compounds (some of which are bound to soot), and a few are sulfates in the form of aerosols. In the combustion of different hydrocarbons, there are several intermediate substages in the individual stages such as cracking, dehydration, and polymerization. The creation of soot is largely determined by the local temperature (800–1400 K) and the oxygen concentration, and occurs in two phases.[9] The reactions in the primary formation phase occur almost exclusively from radical chain mechanisms in the core of the fuel jet behind the jet tips. O, H, and OH radicals are formed. Cyclic and polycyclic aromatic hydrocarbons form by polymerization and cyclization. From the addition of other units, relatively stabile intermediate products form from aggregation that become increasingly larger particles (so-called primary particles). As the primary particles coagulate to form large units, secondary particles arise. Uncombusted and partially combusted hydrocarbons, especially aldehydes can bond to the secondary particles due to their large specific surface. As combustion proceeds, the secondary formation phase is soot reoxidation that is governed by the dwell time and oxygen concentration.

The diameter of the particles varies between 1 and 1000 nm. For homogeneous mixtures, soot is found in the exhaust emissions at an air-fuel ratio below 0.5; at a λ above 0.6 under optimum conditions, there is no demonstrable soot formation.[11] In addition to soot formation as a source of particles, the lubricant is also an important source of particle emissions.

A particular problem area is the conflict between HC and NO_x particles. The conditions for low particle formation and low HC emissions contrast with the prerequisites for low nitrogen oxide emissions. Therefore, attention is especially given to the secondary formation phase, soot reoxidation. To support soot reoxidation, generally a large amount of mixture formation energy is required in the last phase of combustion that is attainable by a specific swirl and tumble in the combustion chamber, greater fuel injection pressure, a faster injection rate at the end of the fuel injection process, and more even distribution. These conditions are unfortunately favorable prerequisites for high NO_x emissions.

Figure 21-21 gives a qualitative summary of pollutant formation in diesel engines.[12]

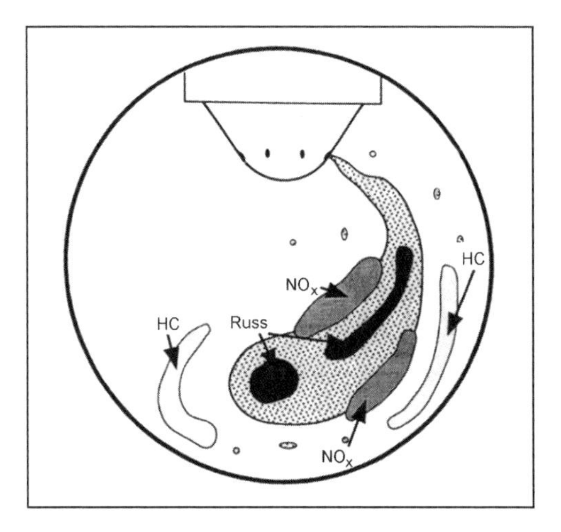

Fig. 21-21 Qualitative representation of diesel engine combustion and pollutant formation.[12]

21.3.2.2 Unrestricted Exhaust Emission Components

Important unrestricted components in untreated exhaust from diesel engines are cyanide, ammonia (NH_3), sulfur dioxide (SO_2), and sulfates. Of the specific hydrocarbons, methane, ethane, ethene, ethine, benzene, and toluene are of particular interest. Of the PACs, phenanthrene, pyrene, fluorene, fluoranthene, and anthracene predominate in descending order. The concentration of these parameters is at least six times greater than that of the other individual PAC substances, and forms approximately 90% of the PACs.[8] Phenols and different aldehydes such as formaldehyde, acetaldehyde, acetone + acrolein, and propionaldehyde are also being studied more closely.[13] The cited components are formed from trace substances in the fuel, in the lubricant, and, to a degree, from secondary reactions in the exhaust system.

If particle emissions are viewed according to their mass-related share of hydrocarbons, we find a ratio of 80% elementary hydrocarbons to 20% organic compounds. Chrome and nickel aerosols originate from abrasion, as is the case with spark-injection engines.

21.4 Reducing Pollutants

Basically, the measures to reduce pollutants can be divided into measures preceding and following the engine. In the first section, we address measures before and in the combustion chamber that play an important role in lowering the untreated emissions and fuel consumption.

21.4.1 Engine-Related Measures

21.4.1.1 Spark-Injection Engines

A majority of the approaches are both for engines with intake manifold fuel injection and for spark-injection engines with direct fuel injection. In general, it can be noted that a minimum of the untreated emissions generally does not yield the best overall result following exhaust treatment.

Mixture formation:

The air-fuel ratio of the mixture in the combustion chamber has the greatest influence on untreated emissions. The emissions of CO and HC are lowest in the slightly lean range of $\lambda = 1.05–1.1$; untreated NO_x emissions are highest in this range, however.

An equivalent air-fuel ratio for all cylinders is another prerequisite for low emissions. This requires highly precise metering of the fuel to all cylinders. A λ spread, in particular, strongly increases CO emissions and to a lesser extent HC emissions. NO_x emissions rise with a low λ spread. As the spread increases, NO_x emissions fall again. To measure the cylinder-selective λ differences for control purposes, refer to Section 21.2.2.

For a complete conversion of the fuel in the engine, the fuel must be well prepared. With intake manifold injection, normally the fuel is injected directly before the intake valves. When the intake manifold pressure and temperature are properly exploited, this position yields optimum fuel preparation with minimum wall film formation. Fuel preparation is further optimized by additionally surrounding the injected fuel jet with air, special jet geometries, fuel injection nozzles with a flash boiling effect, and piezoactuated injectors that ensure highly precise metering of very small quantities of injected fuel.

The preparation time is substantially shorter with air-atomizing direct fuel injection than with manifold injection (a duration similar to diesel DI). In addition, different fuel injection strategies with the fuel injection nozzles must be used for the respective operating modes (homogeneous or stratified charge).

Another approach for mixture preparation is to blow the mixture into the combustion chamber. For this, the fuel must be prepared outside of the combustion chamber. Particularly in the stratified charging of an extremely lean mixture, this method creates favorable conditions for unthrottled engine operation. Most pollutants can be thereby reduced. The ability to run on a lean mixture additionally enables fuel savings over a wide load and speed range.

Combustion characteristic and combustion procedure:

The combustion speed is essentially influenced by the fuel, the air-fuel ratio, the pressure and the temperature during conversion, and the flow state in the combustion chamber. The combustion characteristic for manifold injection largely depends on the intake valve lift and the start of fuel injection, the degree of mixture preparation, and the moment of ignition. Short dwell times at high temperatures reduce NO formation. Optimum combustion procedures do not exceed a maximum temperature of 2000 K.

Direct fuel injection also allows the fuel injection time to be freely selectable, but the mixture formation time is very short.[14] Charge stratification reduces untreated NO_x emissions and fuel consumption. However, we need to remember that the highly nonlinear NO_x formation does not yield flame front zones that are extremely hot. Directly adjacent to the spark plug, there is normally a rich mixture to ensure ignition. The majority of the surrounding charge is, however, lean. Very homogeneous combustion is to be ensured.

Valve control:

Valve gear/valve control times: The transition from two valves to four valves has a number of advantages, especially for engines with manifold injection. A central spark plug and the symmetrical four-valve combustion chamber form is optimum for low-pollutant combustion. Only higher hydrocarbon emissions can sometimes be determined. By means of the variable control times, fuel consumption and emissions can be influenced over a wide range. With an electromechanical valve gear[15] that allows a variation of many degrees of freedom, fuel consumption can be additionally lowered, particularly in engines with direct fuel injection. A small valve overlap, small valve stroke, and late intake time allow emissions to be greatly reduced in the partial load range. In addition, the engines can be operated unthrottled under a partial load that substantially lowers fuel consumption. Cylinder shutoff in engines with a large number of cylinders also decreases fuel consumption and, hence, CO_2 emissions.

Exhaust recirculation:

Exhaust emissions are guided from the exhaust section of the engine through an exhaust recirculation valve into the intake tract and replace a portion of the fresh charge. This gas mixture can absorb a large amount of heat under dissociation and, hence, lowers the temperature during combustion to prevent the formation of thermal NO. The related unthrottling of the engine also lowers fuel consumption. Variable inner exhaust recirculation can be attained by corresponding valve overlapping by means of phase shifters. When exhaust emissions are fed to the intake manifold, even distribution to all cylinders must be ensured. Exhaust recirculation rates greater than 15% yield somewhat higher HC emissions and poorer idling.

Compression ratio: High compression improves thermal efficiency. The peak combustion temperature is increased, which in turn produces higher NO_x emissions. Because of the high-pressure level, HC emissions also rise from the relative increase in the combustion chamber gap. CO emissions tend to fall as compression increases. Variable compression is under development with promising results, at least in terms of fuel consumption.

Combustion chamber design:

In addition to the geometry that affects the bore-to-stroke ratio, the surface, volume, and squish area, other parameters influence emission behavior. A central spark plug position for a shorter flame path, compact combustion chambers with a small surface, minimum dead volume of gaps, and specific squish areas especially lower HC emissions and fuel consumption. Measures to reduce the combustion process also somewhat lower NO_x emissions. An increase in the compression ratio reduces fuel consumption, but increases NO_x emissions. The most important measures influencing quality in spark-injection engines with manifold injection are summarized in the diagram in Fig. 21-22, and those for DI engines are shown in Fig. 21-23. Numbers were intentionally not used since the individual measures frequently can lead to substantially different results for different engines, and the intention was to show a generally applicable trend.

Other measures to lower untreated HC emissions are variable swirl formation in the intake port and regulated engine temperature control.

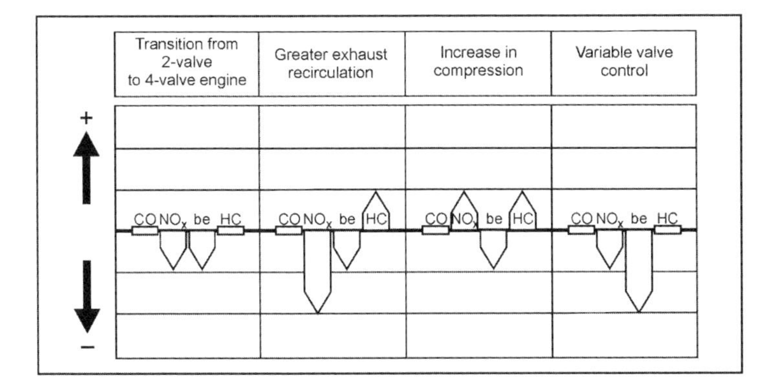

Fig. 21-22 Measures to reduce pollution in a spark-injection engine with manifold injection. (+) indicates a rise and (−) a drop in exhaust emissions.

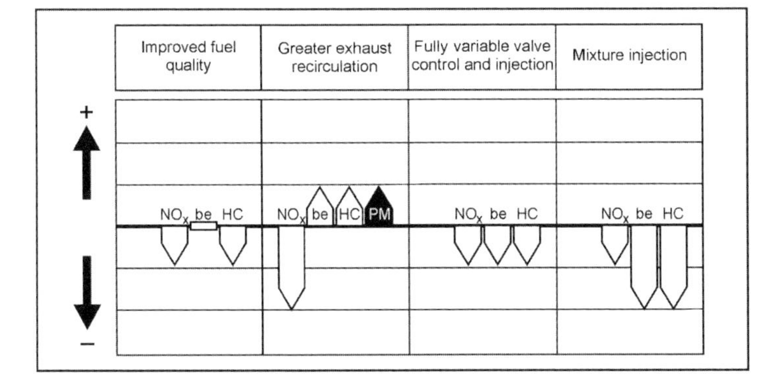

Fig. 21-23 Measures to reduce pollution in a spark-ignition engine with direct fuel injection. (+) indicates a rise and (−) a drop in exhaust emissions.

Ignition:

The primary method for externally supplied ignition is electrical ignition in the form of single or multiple ignition using spark plugs in the combustion chamber. The selected design influences the shape of the flame front and, hence, also the formation rate of nitrogen oxides. Another basic parameter is the moment of ignition in reference to TDC. As we know, late ignition produces low NO_x emissions. Moments of ignition optimized by means of adaptive control that covers all necessary parameters are possible with modern engine management systems. For the mixture to reliably ignite, sufficient ignition energy of 0.2–3 [mJ]c is required. A long spark duration with stabile, high combustion voltage supports the reliable and stable ignition of the mixture and produces low HC emissions. In other advances, the spark plug is used as a "combustion chamber sensor." By measuring the ion flow at the electrodes during combustion, the start of ignition (misfiring diagnosis/CH emissions) and the progress of ignition can be measured, and knocking can be monitored. In conjunction with electronic combustion control, this permits effective diagnosis and, hence, lower emissions over long periods.

Spatial ignition (to be understood as the simultaneous ignition of the mixture at a theoretically infinite number of locations in the combustion chamber), laser ignition (in which a laser beam spread by a suitable lens system ignites the entire contents of the combustion chamber with sufficient energy), and plasma ignition are under development and have advantages mostly for individual exhaust emission components. In particular, spatial ignition procedures have a very high potential to substantially lower untreated NO_x emissions. At the same time, fuel consumption under a partial load can be reduced.

Further improved fuel quality to prevent the coking of direct fuel injection systems produces more stabile emissions behavior, especially of jet-directed systems.

21.4.1.2 Diesel Engines

Approaches to optimize the emissions of diesel engines generally deal with the classic conflict between fuel consumption, nitrogen oxides, and particle emissions.

Combustion procedure and combustion characteristic:

The most important parameter for conventional nozzles for air-distributed, direct fuel injection is the start of injection in reference to the TDC. The ignition lag represents a relatively constant quantity. Standard approaches based on four-valve cylinder heads use a swirl and charge duct. There is a trend toward flatter and wider piston recesses that allow unhindered jet expansion. This development has been supported by multiple-hole fuel injection nozzles. This reduces the amount of fuel deposited on the wall. Charge losses can be reduced by increasing the swirl.

Homogeneous diesel combustion with lean premixture combustion and a faster injection rate toward the end of the reaction can produce nearly soot-free combustion with minimum NO_x emissions.[11,16] The practical implementation of this method is hindered at present by insufficient mixture formation and an inhomogeneous mixture distribution as well as the limited operating range in the load and speed program map. In addition, this combustion procedure requires a fully variable fuel injection system with optional preinjection, main injection, and secondary injection that can be varied over wide ranges. This permits the formation of nearly any fuel injection characteristic. This method competes with combustion chamber ignition in spark-injection engines.

Supercharging:

For diesel engines with direct fuel injection, turbocharging still represents the most efficient option for reducing all pollutant components. The most important parameters such as the charging pressure and charging air temperature must be varied for each load level. In addition, variable turbine geometries, sequential supercharging, and charge air temperature regulated according to load can reduce fuel consumption and especially NO_x emissions. Electrical secondary supercharging can substantially reduce particle emissions in transient processes such as starting from idling.

Fuel injection systems and fuel injection procedures:

High-pressure fuel injection systems such as the pump nozzle and common-rail system can optimize fuel preparation and especially lower particle emissions due to the high fuel injection pressure of 1500–2000 bar (over 2000 bar in the near future) in combination with new fuel injectors. Figure 21-24 shows the possible improvements as the fuel injection pressure rises. The central conflict of particles versus NO_x can be managed much better.[17]

The best strategies must be taken from the test stage for the most important parameters such as the moment of injection, fuel injection law, fuel injection pressure, nozzle shape (jet position, projecting mass, number of nozzles), fuel injection quantity, preinjection, injection spacing, secondary injection, injection time, and cutoff time. Figure 21-25[18] shows how to best circumvent the central conflict of particles vs. emissions by suitably harmonizing engine control in a common-rail fuel injection system.

A small preinjected quantity of less than 1 mm^3 minimizes NO_x and particle emissions when at a suitable duration from the main injection. Secondary injection reduces particle emissions while NO_x emissions remain the same. A very short cutoff time especially lowers HC emissions. Fuel injection systems that use piezoinjectors can approach the desired fuel injection curve because of the shorter operating times and permit extensive optimization.

Injecting water reduces NO_x by up to 25% in engines with an excessively high combustion temperature, but this also worsens CO and HC emissions.[11,16]

Valve control:

Greater charging is possible using multiple valves. This has a particularly positive effect on fuel consumption and general raw emission behavior. In particular, four valves now appear to be the best approach for diesel engines

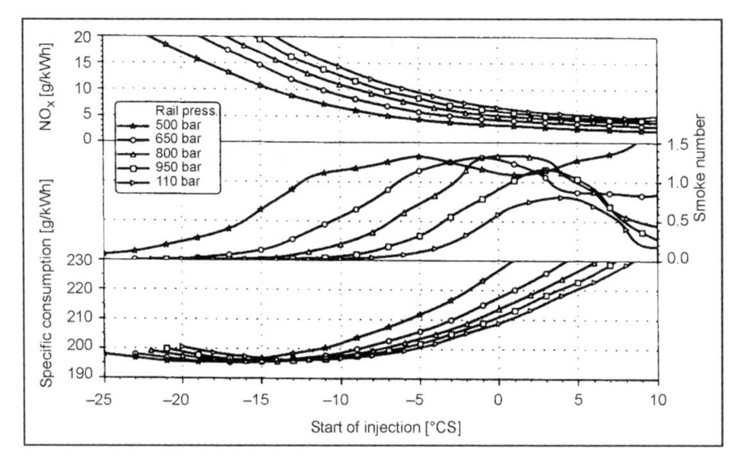

Fig. 21-24 Influence of rail pressure and start of injection on fuel consumption, smoke number, and NO_x.[17]

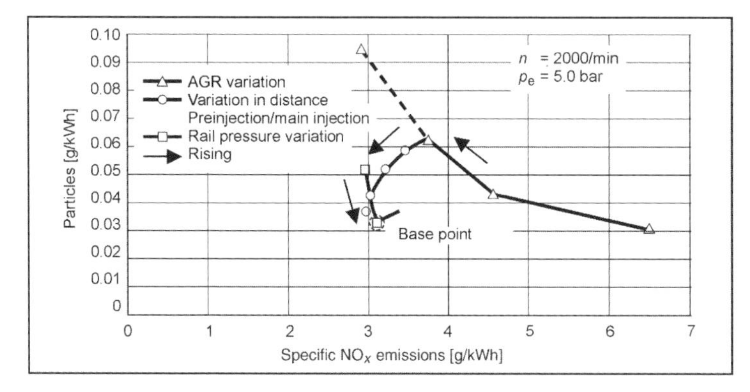

Fig. 21-25 Central conflict between particle and NO_x emissions in V8 TDI engines with common rail injection according to Ref. [18].

because of the injection nozzle position. Variable valve control and phase shifters are now used for passenger car diesel engines since turbocharging is so common, but they are presently not as important as they are in spark-injection engines.

Exhaust recirculation:

In contrast to spark-injection engines with externally supplied ignition, diesel engines can achieve much higher recirculation rates. However, as we can see in Fig. 21-26, there is a notable increase in the number and size of particles as the exhaust recirculation rate increases. Additionally cooling the recirculated exhaust lowers NO_x and particle emissions somewhat, but it increases CO and HC emissions. The conflict between NO_x and particles is improved by up to 15% with cooled exhaust recirculation.[16]

Combustion chamber design:

Basically, the same design rules hold true as for spark-injection engines. Depending on the injection method and the charge volume, different requirements arise for the

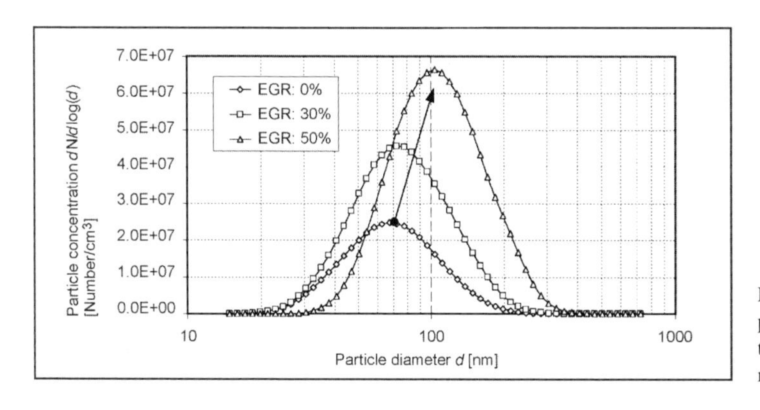

Fig. 21-26 Particle concentration plotted against the particle diameter as a function of the exhaust recirculation rate.[19]

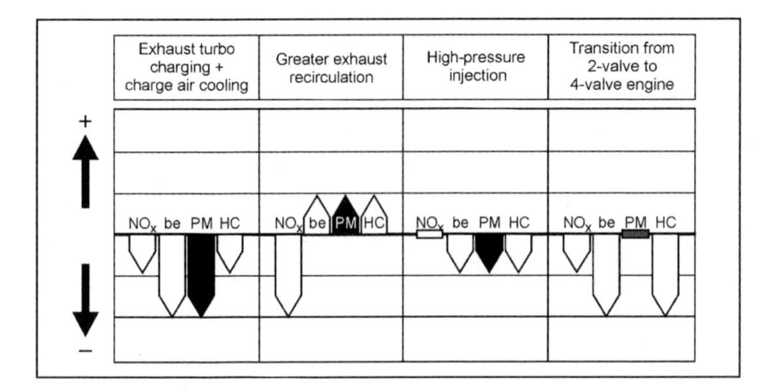

Fig. 21-27 Measures to reduce pollutants from diesel engines. (+) indicates a rise and (−) a drop in the exhaust emission level.

combustion chamber geometry. Wide, flat combustion chambers with a slow swirl are preferred when there is large piston displacement. Contrastingly, deeper recesses with faster swirling are used in combustion chambers of passenger vehicles with a cylinder volume of 450–550 cm^3.

Figure 21-27 summarizes the most important measures used to reduce diesel engine emissions.

The rapid development of proven technologies can bring about notable progress in the reduction of exhaust pollution. It appears possible that certain families of engines may attain particularly low exhaust thresholds such as the ULEV or Euro 4 without the use of fuel-consuming additional aggregates. At the same time, new steps should be taken to introduce new combustion procedures for the "pollutant free automobile that protects natural resources."

Bibliography for Sections 21.1–21.4

[1] Staab, J., Automobil-Abgasanalytik bei niedrigen Grenzwerten, in MTZ Motortechnische Zeitschrift, Vol. 58 (1997), No. 3, pp.168–172.

[2] Tauscher, M., Optimierung eines On-board Abgasmesssystems, Dissertation at the Technical University of Vienna, 2000.

[3] Mohr, M., Feinpartikel in Verbrennungsabgasen und Umgebungsluft, Internet publication: http://www.empa.ch/deutsch/fachber/abt137/motor/partikel.html, 1998.

[4] Neumann, H., G. Hötzel, and G. Lindemann, "Advanced Planar Oxygen Sensors for Future Emission Control Strategies," SAE Paper 970459, Detroit, 1997.

[5] Pucher, E., Ch. Weidinger, and H. Holzer, Kontinuierliche HC-Indizierung am 4-Ventil Ottomotor, Compendium of the 3rd International Indication Symposium, AVL Deutschland GmbH, Mainz, 1998.

[6] Pachta-Reyhofen, G., Wandfilmbildung und Gemischvertei-lung bei Vierzylinder-Reihenmotoren in Abhängigkeit von Vergaser- und Saugrohrkonstruktion, Dissertation at the TU of Vienna, 1985.

[7] Pucher, E., G. Schopp, and D. Klawatsch, Fast Response Cycle-to-Cycle Exhaust Gas Analysis for P.I. Engines, in 5th International EAEC Conference, Strasbourg, 1995.

[8] Puxbaum, H., *et al.*, Tauerntunnel Luftschadstoffuntersuchung 1997 – Results of the Measuring Session from October 2–5, 1997, Salzburg State, Department 16 Environmental Protection, Salzburg, 1998.

[9] Klingenberg, H., Automobile Exhaust Emission Testing: Measurement of Regulated and Unregulated Exhaust Gas Components, Exhaust Emission Tests, Springer-Verlag, Berlin, 1996.

[10] List, H., and W.P. Cartellieri, Dieseltechnik, Grundlagen, Stand der Technik und Ausblick–10 Jahre Audi TDI-Motor, in Special Edition of the MTZ, 1999, pp. 10–18.

[11] Spindler, P., Beitrag zur Realisierung schadstoffoptimierter Brennverfahren an schnelllaufenden Hochleistungsdieselmotoren, VDI Progress Report Series 6, Energy Generation No. 274, 1992.

[12] Boulouchos, K., *et al.*, Verbrennung und Schadstoffbildung mit Common rail Einspritzsystemen bei Dieselmotoren unterschiedlicher Baugröße, Internationale Konferenz "Common Rail Einspritzsysteme – Gegenwart und Zukunftspotenzial," ETH Zurich, 1997, p. 111.

[13] Kohoutek, P., *et al.*, Status der nichtlimitierten Abgaskomponenten bei Volkswagen, VDI Progress Reports, Vol. 348, Düsseldorf, 1998.

[14] K. Fraidl, P. Kapus, W. Piock, and M. Wirth, Fahrzeugklassen-spezifische Ottomotorkonzepte, VDI Progress Reports, Vol. 420, Düsseldorf, 2000.

[15] Langen, P., R. Cosfeld, A. Grudno, and K. Reif (BMW Group), Der elektromechanische Ventiltrieb als Basis zukünftiger Ottomotorkonzepte, 21st International Viennese Engine Symposium, 2000.

[16] Wirbeleit, C., *et al.*, Können innermotorische Maßnahmen die aufwändige Abgasnachbehandlung ersetzen? VDI Progress Reports, Vol. 420, Düsseldorf, 2000.

[17] Härle, H., Anwendung von Common Rail Einspritzsystemen für NKW-Dieselmotoren, Internationale Konferenz "Common Rail Einspritzsysteme–Gegenwart und Zukunftspotenzial," ETH Zurich, 1997, pp. 42–43.

[18] Bach, M., R. Bauder, H. Endres, H.W. Pölzl, and W. Wimmer, Dieseltechnik, Der neue V8-TDI-Motor von Audi, Part 3, Thermodynamik, 10 Jahre Audi TDI-Motor, in Special Edition of MTZ, 1999, pp. 40–46.

[19] Guber, M., D. Klawatsch, and E. Pucher, Comparative Measurements of Particle Size Distribution: Influences of Motor Parameters and Fuels, Proceedings Second International ETH-Workshop on Nanoparticle Measurement, ETH Zurich Laboratory for Solid-State Physics, 1999.

21.5 Exhaust Gas Treatment for Spark-Ignition Engines

21.5.1 Catalytic Converter Design and Chemical Reactions

The basic function of an automobile catalytic converter can be described by the following reaction equations [(21.4) to (21.9)]. The main pollutant components in the exhaust gas of spark-ignition engines are uncombusted hydrocarbons and carbon monoxide (these must be completely converted by oxidation after incomplete combustion), and nitrogen oxides, which are reduced to form nitrogen.

Oxidation of CO and HC into CO_2 and H_2O

$$C_yH_n + \left(1 + \frac{n}{4}\right)O_2 \rightarrow y\,CO_2 + \frac{n}{2}H_2O \tag{21.4}$$

$$CO + \frac{1}{2}O_2 \rightarrow CO_2 \tag{21.5}$$

$$CO + H_2O \rightarrow CO_2 + H_2 \tag{21.6}$$

Reduction of NO/NO_2 into N_2

$$NO\ (\text{or } NO_2) + CO \rightarrow \frac{1}{2} N_2 + CO_2 \tag{21.7}$$

$$NO\ (\text{or } NO_2) + H_2 \rightarrow \frac{1}{2} N_2 + H_2O \tag{21.8}$$

$$\left(2 + \frac{n}{2}\right) NO\ (\text{or } NO_2) + C_yH_n \rightarrow \left(1 + \frac{n}{4}\right) N_2 + y\, CO_2 + \frac{n}{2} H_2O \tag{21.9}$$

These reactions are catalyzed in the presence of precious metals Pt, Pd, and Rh. To attain the highest possible conversion rate, the precious metals are dispersed on a substrate oxide with a large surface. These substrate oxides are typically inorganic materials with a complex pore structure (such as Al_2O_3, SiO_2, TiO_2) on which the catalytic materials are applied together with promotors.

The catalytic substrate is produced in an aqueous solution with a solid content of 30%–50%, a so-called slurry. This slurry is coated onto honeycomb-shaped monoliths. Both ceramic and metal monoliths are used. The honeycomb structure ensures the greatest possible surface for the catalytic reaction in a small space. Figure 21-28 shows an example of a catalytic converter that consists of two ceramic monoliths.

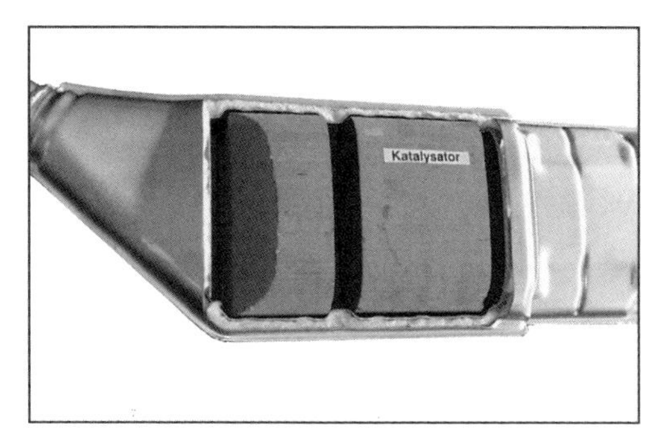

Fig. 21-28 Section of a catalytic converter.

21.5.2 Catalytic Converter Approaches for Stoichiometric Engines

21.5.2.1 Three-Way Catalytic Converter

The reactions discussed in the prior section to convert pollutant components in SI engine exhaust produce the conditions under which these reactions are possible. To oxidize HC and CO, excess oxygen is necessary, whereas to reduce nitrogen oxides, the presence of reducing components is required. Since all pollutant components must be converted when driving, there is a narrow operating window for exhaust composition and conversion in which combustion can occur.

By regulating the air-fuel mixture with the aid of a λ sensor within the narrow range around the stoichiometric ratio of $\lambda = 1$, it is possible for the oxidation reactions and reduction reactions to occur with a high conversion rate. Such catalytic converters are termed three-way catalytic converters since all three pollutant components are equally converted. The point at which both CO and NO_x are best converted as a function of the air-fuel ratio λ is the optimum operating point for the catalytic converter, termed the crossover point. Figure 21-29 shows the conversion of pollutant components HC, CO, and NO_x as a function of the air-fuel ratio λ. To maintain the presently very strict exhaust laws in Europe and the United States, these three-way catalytic converters are used for $\lambda = 1$ regulated spark-ignition engines.

In addition to conversion under hot operating conditions, the starting behavior (light-off) of the catalytic converter is a very important factor for characterizing its behavior. In addition to catalytic-converter-specific properties, substrate-specific properties play an important role in dynamic tests. They determine the thermal mass of the catalytic converter by using geometric parameters such as cell density and wall and film thickness, and they also influence the catalytic converter's thermodynamic properties of heat capacity and density. By reducing the thermal mass, the time before light-off in a cold start can be greatly reduced.

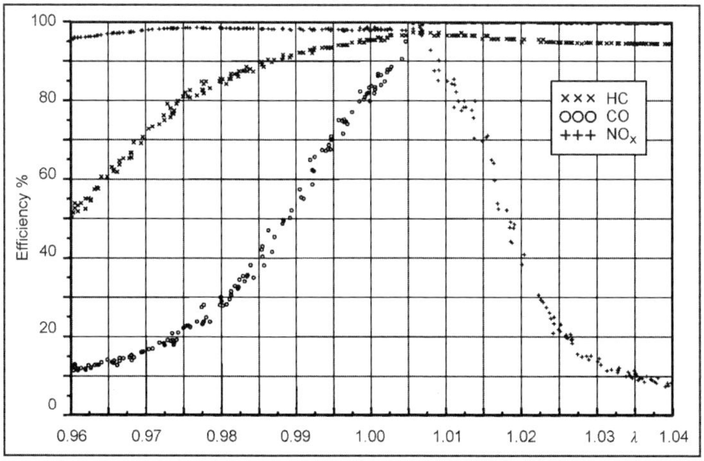

Fig. 21-29 The conversion of pollutant components as a function of the air-fuel ratio.

	Ceramic			Metal			
Cell density, cpsi	400	600	900	400	600	800	1000
Wall/film thickness, mil/mm	6.5	3.5	2.5	0.050	0.040	0.030	0.025
Geometric surface, cm^2/cm^3	27.3	34.4	43.7	36.8	42.9	51.6	56.0
Free cross section, %	75	80	86.4	89.3	89.8	93.7	91.4
Hydraulic diameter, mm	1.10	0.93	0.79	0.97	0.84	0.72	0.65
Density, g/cm^3	0.43	0.35	0.24	0.77	0.73	0.55	0.61

Fig. 21-30 Substrate parameter for high-cell and thin wall substrates.

At the same time, a large geometric surface must be made available for the catalytic reaction. Substrates with a high cell density and thin walls are suitable for improving the light-off behavior when the catalytic converter is heated quickly. Figure 21-30 shows the geometric parameters of selected standard substrates.

The temperature-related thermal capacity of ceramics is compared with that of metal substrates in Fig. 21-31.

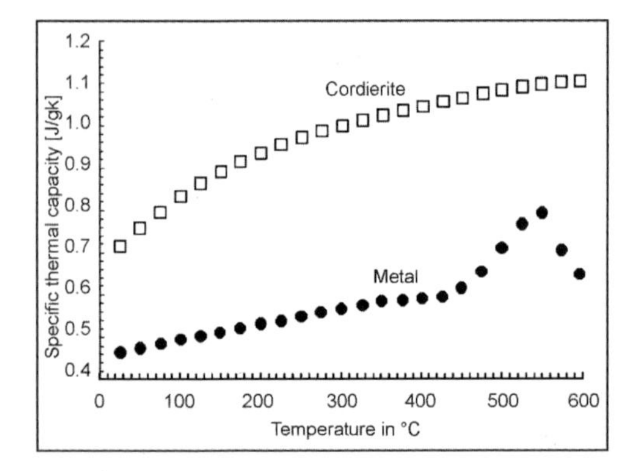

Fig. 21-31 Thermal capacity.

Different catalytic converter systems are distinguished based on the arrangement of the catalytic converter(s) in the overall vehicle (Fig. 21-32). On the one hand, a position close to the engine is possible (underhood main catalytic converter) that is subject to restrictions of thermal and mechanical stability and available installation space. On the other hand, there is the traditional underfloor position where the lower exhaust temperature is less supportive of catalysis. Frequently, a combination of an underhood preliminary catalytic converter and underfloor catalytic converter is used. This system combines the advantages of a fast heating curve of the underhood arrangement with the required installation space of the underfloor position for larger catalytic converters with the disadvantage of a higher system cost.

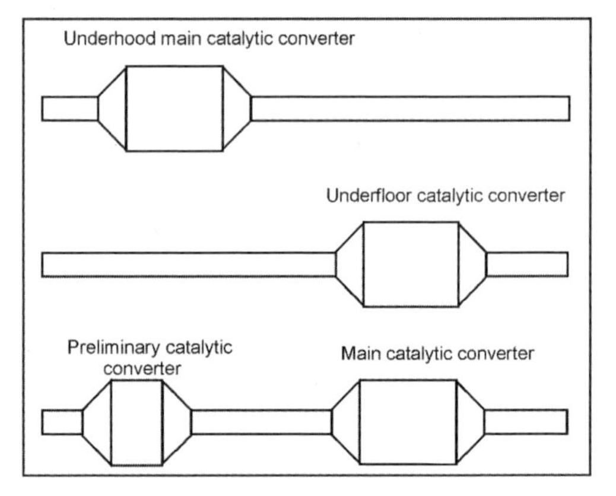

Fig. 21-32 Catalytic converter designs.

21.5.2.2 Oxygen Storage Mechanism

The chemical reactions of the three-way catalytic converter were discussed in detail in the above sections. It was also mentioned that the oxidation and reduction reactions can occur simultaneously with maximum conversion only when the air-fuel ratio is at the stoichiometric point, i.e., $\lambda = 1$. The engine control system solves this problem by continuously measuring a quantity that is proportional to the air-fuel ratio with the aid of a measuring sensor in the exhaust, the λ sensor, in a closed control loop. If the sensor measures exhaust that is too rich or too lean, the engine management makes a correction in one direction or the other. This means, however, that the air-fuel ratio is only stoichiometric on average over time; at specific engine loads, it can vary widely from one time to the next. Depending on the exhaust, i.e., rich or lean, the catalytic converter then produces HC, CO, and NO_x peaks in reference to the conversion. The surface chemistry of cerium offers a solution to this dilemma since it has the property of storing and then releasing oxygen.

The Basic Mechanism of Oxygen Storage

As mentioned, the air-fuel ratio (λ) is never exactly one, but rather fluctuates continuously around this point. This means that more oxygen exists in one-half of the fluctuation than is necessary for conversion, whereas there is a lack of oxygen in the other half. Although the amplitude and frequency of the oscillation can be controlled very precisely, the conversion of the exhaust can still suffer. For this reason, an element is incorporated in the catalytic coating that has the property of storing oxygen. When a half-cycle arises in which there is excess oxygen, the excess oxygen is stored and can be released for conversion in the following half-cycle in which there is a lack of oxygen. The following reaction equation applies:

$$Ce_2O_3 + 0.5\ O_2 \rightarrow 2\ CeO_2 \qquad (21.10)$$

where cerium is the element that forms the oxygen storage component. The unique surface chemistry of this element makes it able to store oxygen. Cerium can assume two different oxidation stages, and the mechanism functions according to the following intermediate step:

$$_{Ce^{4+}}O_{Ce^{4+}} + PM \rightarrow {}_{Ce^{3+}}\Delta_{Ce^{3+}} + PM - O \qquad (21.11)$$

If a carbon monoxide atom contacts the surface, it can bond with the stored oxygen, and the cerium oxide is thereby reduced as shown in the following reaction equation:

$$2\ CeO_2 + CO \rightarrow CO_2 + Ce_2O_3 \qquad (21.12)$$

which also includes an intermediate step that is catalyzed by the precious metal:

$$CO + PM + O \rightarrow PM + CO_2 \qquad (21.13)$$

The CO conversion can then be improved directly by the oxygen storage mechanism. The same also holds true for NO_x conversion that occurs according to the following pattern:

$$_{Ce^{3+}}\Delta_{Ce^{3+}} + NO \rightarrow {}_{Ce^{4+}}\Delta_{Ce^{4+}}\ 0.5N_2 \qquad (21.14)$$

It has been generally assumed that the conversion of hydrocarbons occurs via the same mechanism. However, this is not true. Hydrocarbon molecules require large precious metal surfaces to react and, hence, are not catalyzed like CO and NO_x by the cerium.

The Development of Oxygen Storage Mechanism

In the first three-way catalytic converters, cerium was incorporated in soluble form without special stabilizers. The advantage was a very large surface and, hence, oxygen storage ability when new. However, when these catalytic converters were exposed to high temperatures over a long period, this surface rapidly decreased. As a comparison, cerium when fresh has a surface of approximately 120 m²/g; after 4 h of ageing in an oven at 1050°C, it drops to less than 1 m²/g. Given increasingly tougher exhaust laws, catalytic converters are being placed increasingly closer to the engine exhaust outlet to support fast light-off. At these installation sites where consistently high temperatures predominate, the storability can drop very rapidly.

The development of stabilized cerium components was, therefore, a landmark invention for three-way catalytic converters. The stabilizer largely consists of zirconium, but it also includes other rare earth elements. Although the available surface when new is much less, approximately 80 m²/g, it is approximately 30–40 m²/g after ageing which is much greater than unstabilized cerium. Figure 21-33 shows the comparison of stabilized and unstabilized cerium after being aged 4 h at 1050°C. The different ageing behavior of the two materials is easy to see.

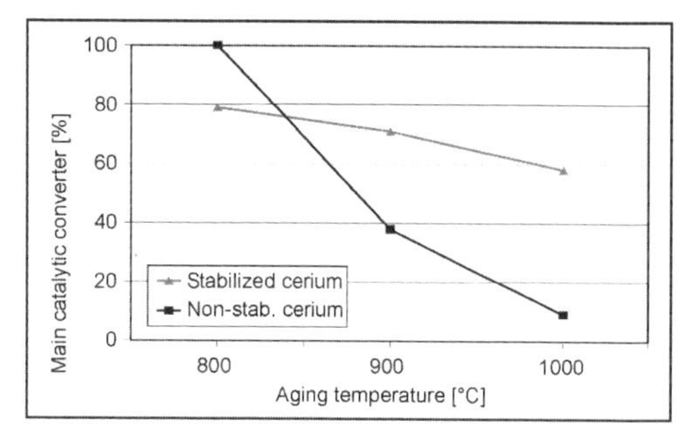

Fig. 21-33 Comparison of unstabilized cerium with stabilized cerium aged 4 h at 1050°C in an oven.

Only after this development could the life of underhood catalytic converters correspond to that of automobiles. Experience from approximately 30 years of catalytic converter use shows that the catalytic converter is no longer the weakest link in the chain of the exhaust gas purification system.

21.5.2.3 Cold Start Strategies

To maintain legally required exhaust standards, it is very important to bring the catalytic converter to operating temperature as soon as possible. The measures to heat the catalytic converters described below have been used in various series applications to accomplish this. A distinction is drawn between active and passive cold-start strategies.

Electrically Heated Catalytic Converters

One active measure is to electrically heat catalytic converters with electrical heating elements (Fig. 21-34). A high electrical output is required before or during the engine start that is supplied by the vehicle electrical system. This electrical output must be able to be generated when the engine is not very efficient (for example, during a cold start with typically poor generator efficiency and with an electrical system voltage of 12 V). For this reason, this system can achieve strict exhaust gas standards only for large engines.

Secondary Air

In addition to measures within the engine for increasing the exhaust gas enthalpy, injecting secondary air into the

Fig. 21-34 Electrically heated catalytic converter.

exhaust gas system is an effective variation for quickly heating the catalytic converter. Figure 21-35 illustrates heating catalytic converters upon a cold start with and without secondary air injection. The secondary air is injected by an electrical pump into the exhaust ports of the engine. The additional oxygen promotes exothermic oxidation so that the catalytic converter can be heated to the operating temperature in a few seconds. Since the fuel-air ratio of the engine can be set slightly rich during hot operation, the drivability is good in a cold start.

HC Storage Catalytic Converter

Another possibility of reducing HC emissions is to use HC storage catalytic converters (traps) (Fig. 21-36). The uncombusted hydrocarbons emitted during cold starts are adsorbed by a storage catalytic converter as long as the three-way catalytic converter does not work. After the three-way catalytic converter has light-off, the uncombusted hydrocarbons are released and then converted. For this system to work, the desorption temperature of the

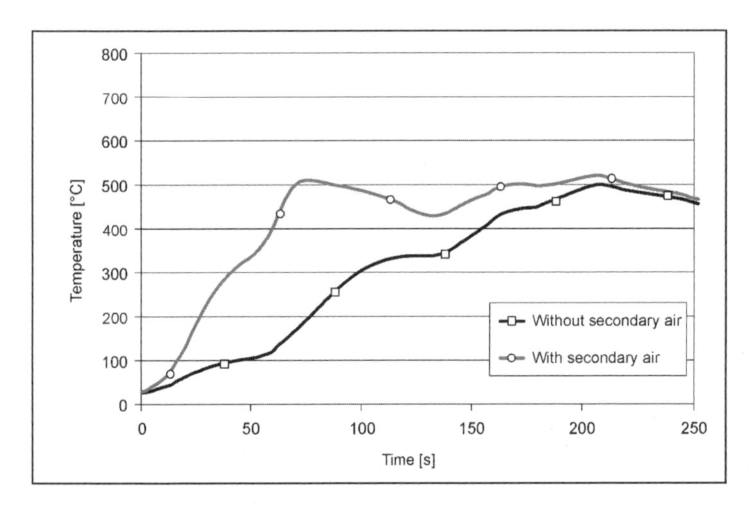

Fig. 21-35 Temperature characteristic in the catalytic converter bed with or without secondary air.

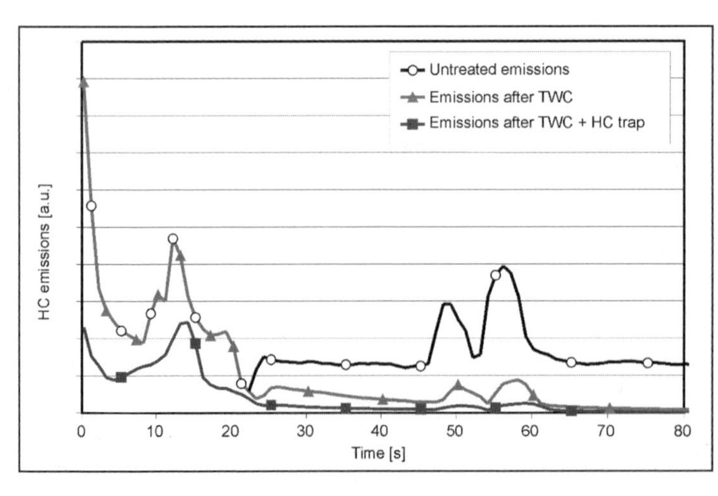

Fig. 21-36 HC emissions with an HC trap.

storage mechanism must be above the light-off temperature of the three-way catalytic converter so that the stored hydrocarbons can also be effectively converted and not just passed into the exhaust gas system at a later time. It is, therefore, necessary to have sufficient oxygen for oxidation at the moment the hydrocarbons are released. This can be accomplished by a suitable engine management strategy (precontrolled λ). At present, the use of storage catalytic converters is limited by the temperature stability of the storage materials. The temperature stability of the zeolites is far below that of a three-way catalytic converter.

21.5.2.4 Deactivation and Its Effect

One of the main causes of catalytic converter deactivation is the atmosphere to which the catalytic converter is exposed. Exhaust gas temperatures above 900°C are not a rarity and primarily rise in underhood installation sites. Additional deactivation effects are caused by fuel or motor oil in the exhaust and are irreversible with a few exceptions such as thermal ageing. For researchers, it is essential to understand the mechanisms in order to develop resistant materials or regeneration methods if possible.

Thermally Generated Deactivation

One of the most important goals in the manufacturing of catalytic converters is to ensure that the reactants have optimum accessibility to the active sites when the catalytic components are applied to the substrate. In a perfectly dispersed catalyst, each atom (or molecule) that participates in the conversion reaction is easily accessible as shown in Fig. 21-37.

A few catalysts are formed in this highly active state; however, they are extremely unstable since they easily grow into large crystals under heat. This growth reduces the catalytic surface. Furthermore, the aluminum oxide substrate with its enormous inner surface constructed of a network of pores is also subject to a sintering process. This results in a loss of inner surface. Another deactivation mechanism is described by the interaction of the catalytic species with the substrate material. Alloy formation reduces the active catalytic species. All of the previously described processes are influenced by the nature of the precious metals, by the substrate material, by the exhaust environment, and, above all, by high temperatures.

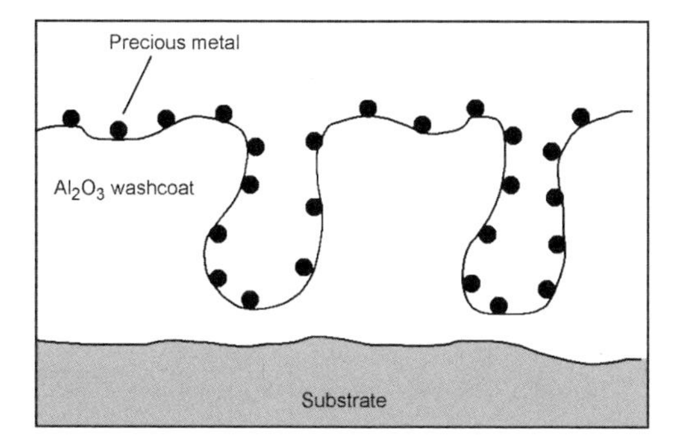

Fig. 21-37 Schematic sketch of an ideally dispersed catalyst on an aluminum oxide substrate.

Precious metal sintering:

Highly disperse catalytic species are subject to the natural tendency of combining into crystals under heat. In this process, the crystals grow, the ratio of surface to volume falls, and less catalytically active atoms or molecules are available on the crystal surface for the reactants. In other words, many of the active sites are buried in the crystal, and since fewer active sites participate in the reaction, the performance decreases. In Fig. 21-38, the phenomenon is illustrated with a simple sketch. The initially finely distributed precious metal forms crystals or agglomerates under heat.

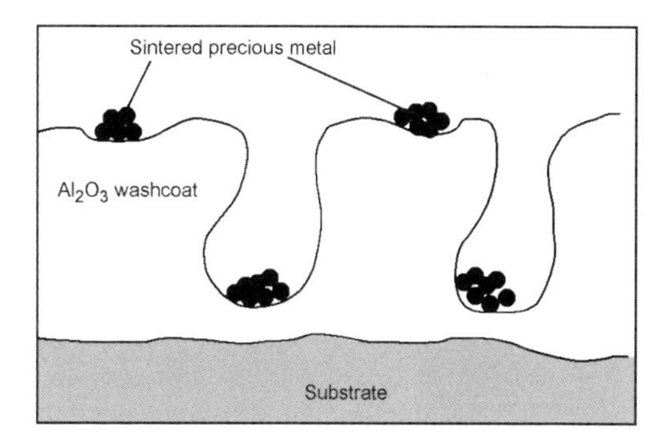

Fig. 21-38 Schematic sketch of precious metal sintering on a substrate.

The loss of performance due to precious metal sintering in catalytic converters is significant. A focus of catalytic converter research is, therefore, to precisely investigate the relationships of sintering and offer ways to stabilize finely distributed dispersions. Various elements from the group of rare earths have been successfully used in exhaust treatment to stabilize the precious metals. The precise mechanism of stabilization has not been completely researched; however, it appears as if the stabilizers fix the precious metal to the surface and, hence, reduce its mobility.

Substrate material sintering:

Within a given crystal structure (such as γ-Al_2O_3), the loss of surface is related to the loss of H_2O and a gradual loss of the pore structure as shown in Fig. 21-39. If the sintering process ceases, the pore openings gradually decrease, which raises pore diffusion resistance. A chemically controlled reaction could, hence, be gradually limited by pore diffusion. This phenomenon is decisively influenced by a progressive loss of activation energy of the corresponding reaction. In the conversion/temperature graph in Fig. 21-40, the slope of the curve gradually decreases.

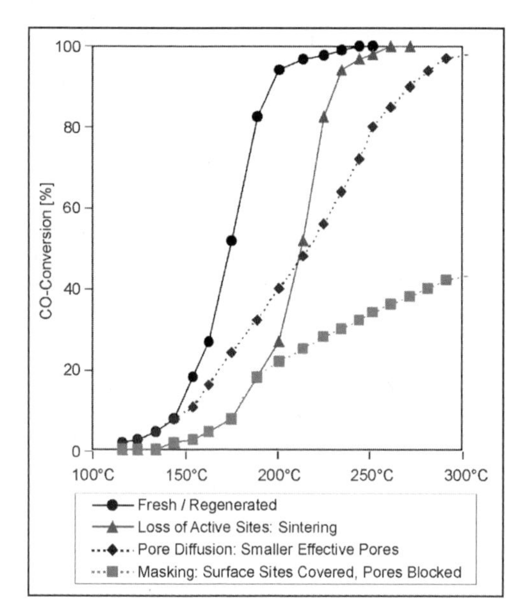

Fig. 21-39 Conversion as a function of the entrance temperature for different deactivation mechanisms.

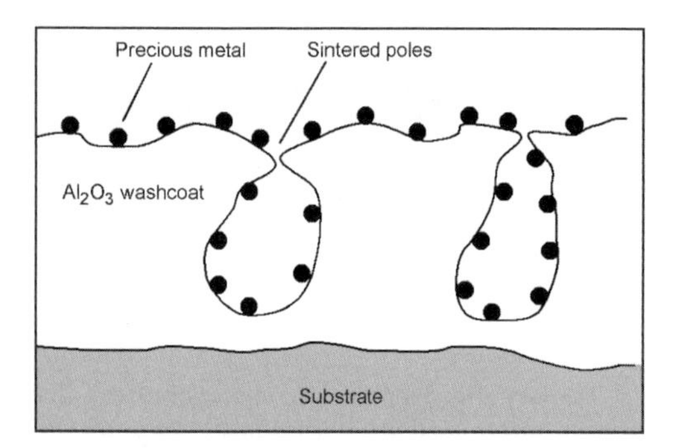

Fig. 21-40 Schematic sketch of sintered substrate material.

In an extreme case, the pores become completely closed, and the catalytically active sites inside the pores are no longer accessible to the reactants (Fig. 21-39).

Another mechanism for substrate oxide conversion is based on the transformation of the crystal structure. When, for example, γ-AL_2O_3 is converted to δ-AL_2O_3, there is a significant, stepwise loss of the inner surface of approximately 150 to $<50\ m^2/g$. The same thing is observed with TiO_2 transformed from an anatase structure to a rutile structure; the surface shrinks from approximately 60 m^2/g to $<10\ m^2/g$. The conversion/temperature graph in this case is usually subject to a loss of activity.

The sintering process of many substrate materials can be slowed in the presence of certain elements of the 3rd and 4th main groups in oxidized form. It is assumed that they form solid compounds with the substrate and, hence, reduce the surface reactivity largely responsible for sintering.

Interaction between precious metals and the substrate oxide:

The reaction of the catalytically active component with the substrate can be the reason for deactivation when the product has less activity than the originally finely distributed species. Under a high temperature and lean exhaust conditions, Rh_2O_3 reacts, for example, with the large highly active surface of the Al_2O_3 and forms an inactive mixed oxide. This process describes an important mechanism in the deactivation of the NO_x reduction activity. It is assumed that the reaction basically occurs according to the following process:

$$Rh_2O_3 + Al_2O_3 \xrightarrow{800^\circ \text{Air}} Rh_2Al_2O_4 \qquad (21.15)$$

Since the activity of the catalyst is impaired, the curve shifts toward a higher temperature with a significant change in the slope. This undesirable reaction causes the development of alternate substrate oxides such as SiO_2, ZrO_2, TiO_2, and their combinations. The negative interaction problem can be resolved by using these alternative substrate oxides; however, they are frequently not that resistant to sintering.

Deactivation from poisoning:

Another cause of catalyst deactivation is harmful substances from the exhaust gas or the machines that apply the catalyst layer. There are two basic poisoning mechanisms: Selective poisoning in which a chemical substance directly reacts with the active centers or the substrate material and, hence, impairs or stops activity, and nonselective poisoning in which impurities are deposited on or in the catalyst substrate material and close to the active centers and pores. The result is a decrease in the performance due to scarcely accessible centers.

Selective poisoning:

If a chemical species reacts directly with the active centers, the term "selective poisoning" is used. This process directly influences the activity or selectivity of a given reaction (Fig. 21-41). Some of these elements or molecules react with catalytic components by forming chemi-

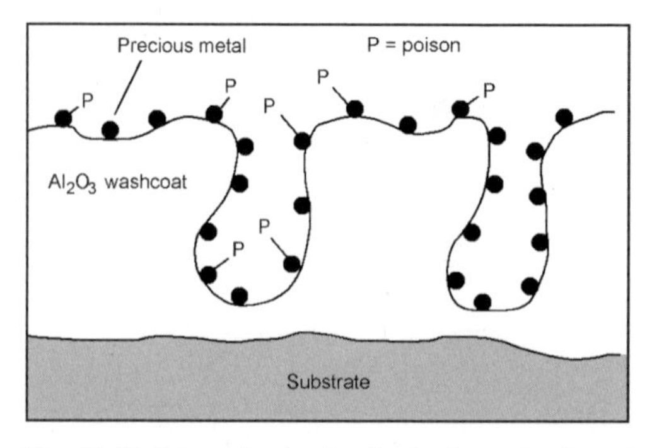

Fig. 21-41 Schematic sketch of selective poisoning of active centers.

cal compounds (such as Pb, Hg, Cd, etc.); an inactive alloy is formed. The process is irreversible and causes the permanent deactivation of the catalyst. Others only adsorb (or more precisely, chemisorb) the catalytic component (such as SO_2 on Pd) and block additional reactions. These mechanisms are reversible; desorption occurs by supplying heat, washing, or simply removing the harmful component from the process flow, and the catalytic activity is restored. If active centers are directly blocked, this always raises the light-off temperature; but the slope of the curve remains unchanged since the functioning of the remaining sites is retained and the activation energy is not changed. The conversion/temperature diagram looks similar to that of precious metal sintering.

If the substrate oxide reacts with a component from the gas stream and forms a new compound as is the case with $Al_2(SO_4)_3$, the pores are generally nearly blocked, which increases resistance to diffusion (Fig. 21-42). The activation energy falls, and the light-off curve shifts toward higher temperatures with a simultaneously reduced slope, which translates into poorer conversion (Fig. 21-39).

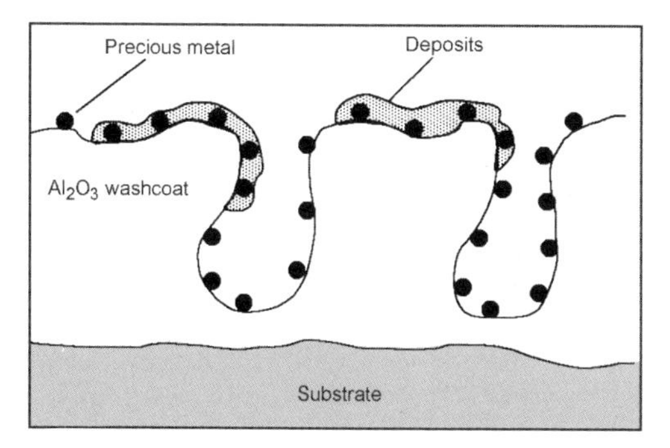

Fig. 21-42 Schematic sketch of nonselective poisoning of active centers.

21.5.3 Catalytic Converter Approaches for Lean-Burn Engines

Conventional spark-ignition engines are operated with a homogeneous fuel-air mixture that is created outside of the combustion chamber in the induction tract. In principle, the mixture supplied in such a spark-ignition engine must be throttled under a partial load. This throttle loss and the subsequent effects of the reduced cylinder charge on the thermodynamic process are the main causes why the efficiency of spark-ignition engines greatly decreases at lower engine loads. In spark-ignition engines, this is expressed by substantially higher consumption under a partial load in contrast to diesel engines.

The efficiency of spark-ignition engines under a partial load can be substantially increased by hyperstoichiometric engine operation. To get by without throttling, the mixture must be made very lean; i.e., the engine must be operated with a large amount of excess air. The ignitability of a homogeneous mixture formed outside the cylinder poses substantial restrictions to lean adjustment and, hence, unthrottling. Directly injecting the fuel into the combustion chamber together with charge stratification allows further, nearly complete unthrottling. The increase in efficiency related to lean operation directly leads to a reduction of fuel consumption.

Independent of whether the mixture is formed externally or internally (direct injection), excess oxygen is in the exhaust of lean-operated spark-ignition engines. This makes it substantially more difficult to convert pollutants in lean exhaust gas. In conventional spark-ignition engines with stoichiometric operation, there is a nearly complete conversion of pollutant components such as HC, CO, and NO_x the familiar three-way technology. The reaction kinetics of lean-operated spark-ignition engines are an obstacle to this, however. HC and CO are preferably converted in the catalytic converter as a result of the faster reaction speeds. The previously converted reaction partners are missing to reduce NO_x. For this reason, technologies are required that allow efficient exhaust treatment, especially nitrogen oxides, in a lean atmosphere. Lower exhaust gas temperatures pose an additional problem to exhaust treatment.

21.5.3.1 Options for NO_x Reduction in Lean Exhaust Gas

At present, there are different basic approaches to the conversion of nitrogen oxides from which various options can be derived for reducing NO_x in lean exhaust gas. The individual technologies can be divided into the following groups and have been discussed in Ref. [1].

- Direct NO decomposition
- Plasma technologies
- Selective catalytic reduction (SCR)
- NO_x storage catalytic converters

Direct NO Decomposition

The reaction steps of the direct decomposition of NO into nitrogen and oxygen are shown in Fig. 21-43.

Catalytic converters that can directly break down NO into N_2 and O_2 are the ideal product for use in SI lean-burn engines as well as in diesel engines. To convert these technologies into practical use will require a revolutionary invention. Although NO decomposition is thermodynamically preferred and the basic chemistry has been revealed in R&D laboratories,[2] transferring it to a real engine or vehicle operation has been unsuccessful to date.

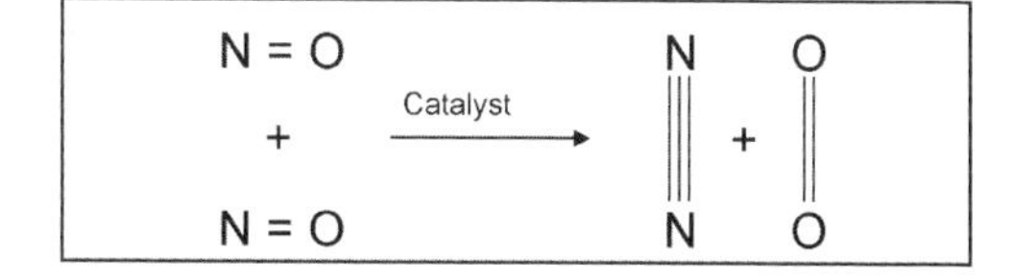

Fig. 21-43 Reaction producing direct NO decomposition.

Plasma Technologies

In its simplest form, the plasma system operates with AC voltage that is applied between two metal electrodes of which one is coated with a nonconductive material. Silent discharges consisting of microdischarges within the microsecond range occur which cause all rising reactive groups to decompose from chemical bonding and recombination processes. The rising plasmas are in a state of inner energy disequilibrium distribution with high electron temperatures of 10^4–10^6 K and a low-kinetic gas that typically ranges between 300–1000 K. The plasma consists of a series of electrons and excited radicals and ions as well as photons. Because of the inner energy disequilibrium distribution in these plasmas, chemical reactions can occur via nonthermal channels that permit strongly endothermic reactions.[3] The two undesired reactions for reducing NO in plasma occur there in addition to numerous other reactions.[4]

Reaction Partner Products

$$e + N_2 \rightarrow e + N + N \tag{21.16}$$

$$N + NO \rightarrow N_2 + O \tag{21.17}$$

Laboratory prototypes with heterogeneous catalysts in plasma fields were tested in engine exhausts with varying results. To what extent this technology will become a standard application in SI lean-burn engines is presently uncertain. An important criterion for the success of the plasma method is the required energy for generating the plasma and the related disadvantage in fuel consumption, as well as the reduction of NO_x at spatial velocities that are predominate in engines.

Selective Catalytic Reduction

NO_x conversion in "lean" atmospheres via specially tailored catalysts are termed "selective catalytic reduction." The required addition of suitable reducing agents yields the end products N_2, CO_2, and H_2O.

The term "passive SCR" stands for catalysts that use components that are exclusively in the exhaust for NO_x reduction; i.e., no subsequently added reducing agents are required (Fig. 21-44, top). To be understood as "active SCR" catalytic converters, the reducing agents must be introduced into the exhaust gas system before the catalytic converter (Fig. 21-44, bottom) following actual combustion.

Passive SCR catalytic converters:

These catalytic converters use the hydrocarbons available in the exhaust gas to reduce NO_x to produce the reaction products N_2, CO_2, and water. The basic work in this area is described in Refs. [2], [5], and [6]. The activity of fresh catalytic converters based on Cu-ZSM-5 zeolites is very good. The durability is problematic, however.[7,8] The worsening of NO_x conversion is primarily because of the sulfur in the fuel and thermal ageing in the presence of water.

Another example is a passive SCR iridium catalytic converter with a downstream three-way catalytic converter as schematically illustrated in Fig. 21-45.[9] It must be noted that when Ir catalytic converters are new, the NO_x conversion rates are lower than in storage catalytic converters. As compensation, the sulfur tolerance is much greater. Furthermore, when a passive SCR catalytic converter is used, a prior catalytic converter close to the manifold to reduce cold-start HC emissions cannot be used since this would also convert the hydrocarbons required to reduce NO_x in hot engine operation.[10,11] The HC emissions in the post-cold-start phase must, therefore, be countered with other suitable measures. Reducing the catalytic converter light-off time in the test cycle by putting it closer to the manifold is limited in practice by the temperature stability of the Ir catalytic converter.[12]

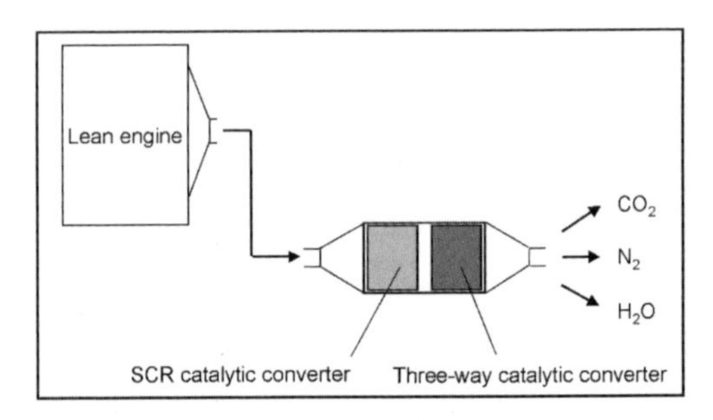

Fig. 21-45 Diagram of a passive NO_x SCR system.

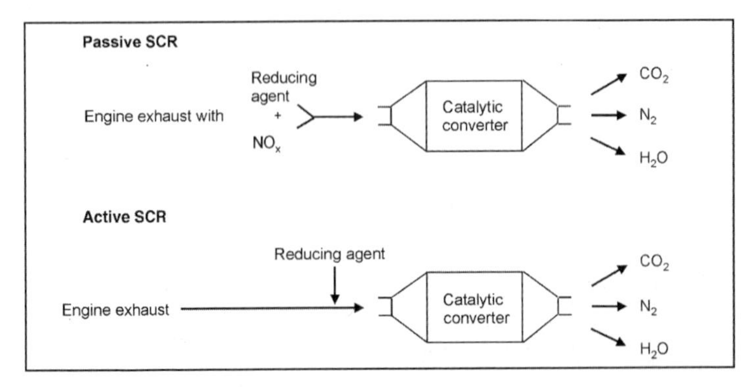

Fig. 21-44 The difference between passive and active SCR.

Active SCR catalytic converters:

Active selective catalytic reduction requires an efficient mixture of nitrogen oxides with the additionally introduced reducing agents before the catalytic converter inlet (Fig. 21-46). Reducing agents such as ammonia or urea are used. This technology is highly efficient in stationary applications such as energy generating systems in which the chemical reactions occur in a narrow window determined by temperature, flow speed, and NO_x concentration. In these applications, ammonia is the reducing agent that generates N_2 and H_2O.

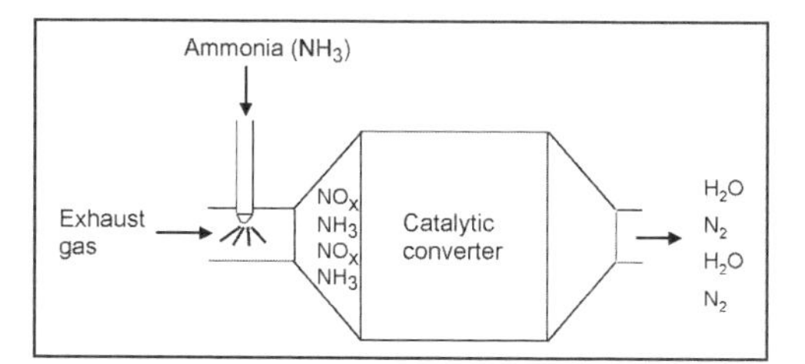

Fig. 21-46 Schematic sketch of an active SCR system.

The operating temperature range of the desired chemical reactions depends on the respective catalytic converter. Vanadium-titanium catalytic converters are the most efficient between approximately 200 and 450°C. At lower temperatures, the catalytic converter becomes impaired by ammonium sulfates, and, at higher temperatures, the catalytic converter oxidizes ammonia into NO. The top temperature limit when ammonia is used is approximately 600°C. For use in a spark-ignition engine with direct injection, urea (a NH_3 compound) is the most promising reducing agent that decomposes to form ammonia and carbon dioxide when injected into the exhaust gas. Urea has the advantage that no gaseous ammonia has to be stored in the vehicle.

Before SCR can be successfully used in passenger cars with lean SI engines, many problems have to be solved.[13] Under transient conditions, the proper amount of reducing agent must be provided by a control system without "NH_3 gaps." The injection of the reducing agent into the exhaust gas must be adapted to the strongly fluctuating quantity of NO_x, the flow rate, and the temperature, and may not contribute to vehicle emissions. The maximum temperature resistance of the catalytic converter is apparently insufficient for use in a lean-burn SI engine. For example, it is approximately 650°C for the cited vanadium-titanium catalytic converter. The cost of the overall system with the nozzles, storage container, tubing, onboard diagnosis, etc., must also be considered, as well as the yet-to-be-developed infrastructure for filling up the reducing agent. The prospects for a successful implementation in lean-burn gas engines are, therefore, rather low to moderate.

NO_x storage catalytic converters:

The most promising method for reducing NO_x emissions in lean-burn engine exhaust is to use NO_x storage catalytic converters, also termed NO_x adsorbers or NO_x traps.[14–17] Since initial series applications for[18,19] exhaust treatment in passenger cars with lean-burn SI engines are based on this technology, a discussion on NO_x storage catalytic converters with greater detail is in the following section.

21.5.3.2 The NO_x Storage Catalytic Converter

The functional principle is schematically illustrated in Fig. 21-47 and can be described by four basic steps involved in the conversion of NO_x into N_2.

During lean operation, the NO in the exhaust oxidizes at the precious metal in the catalytic converter by reacting with oxygen and forms NO_2.

$$NO + \frac{1}{2} O_2 \rightarrow NO_2 \tag{21.18}$$

The NO_2 then reacts with the metal oxides deposited in the catalytic converter that are used as storage materials, with the formation of a corresponding storage material nitrate.

$$NO_2 + MeO \rightarrow Me\text{–}NO_3 \tag{21.19}$$

Since this reaction is not catalytic, but rather stoichiometric, the storage material is "consumed." As the amount of stored NO_2 increases, the effectiveness of the nitrate formation decreases. A state of saturation is reached. To maintain a high storage efficiency, the storage material must be periodically regenerated. For this reason, the engine is briefly switched to substoichiometric (rich) operation. Under rich operating conditions, the temperature stability

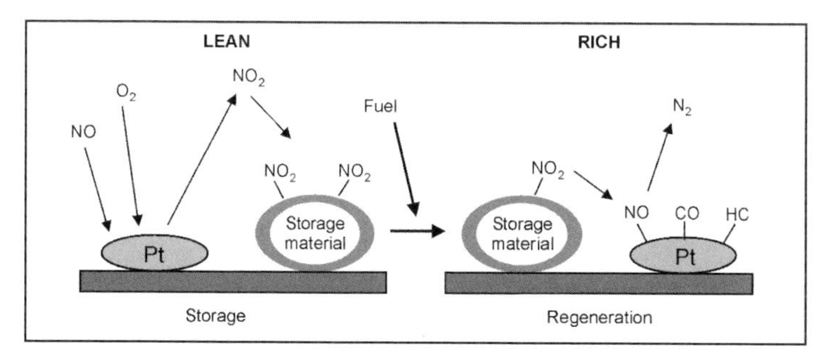

Fig. 21-47 Model example of NO_x storage and regeneration.

of the nitrate is lower than in lean operation so that the nitrate decomposes into NO and MeO.

$$Me{-}NO_3 \rightarrow MeO + NO + \frac{1}{2}O_2 \quad (21.20)$$

The released NO is then converted into N_2 with the aid of reducing agents HC and CO that are also present under rich operating conditions.

$$NO + HC/CO \rightarrow \frac{1}{2}N_2 + H_2O/CO_2 \quad (21.21)$$

For practical use in vehicles, we can determine the requirements on a NO_x storage catalytic converter that necessitate certain features. The essential criteria for evaluating the quality and usefulness of NO_x adsorbers are

- NO_x storability
- Ability to regenerate NO_x
- The operating temperature range for NO_x storage and regeneration
- HC/CO conversion in lean operation
- Conversions in $\lambda = 1$ operation
- Maximum temperature stability
- Sulfur resistance or ability to regenerate sulfur.

The NO_x storability, the ability to regenerate NO_x, the operating temperature range, etc., represent features of a NO_x adsorber that influence its conversion performance when new. The maximum temperature stability and sulfur resistance/regenerability are also features that affect durability.

NO_x Storability and Regenerability

Figure 21-48 shows the typical storage curve of two NO_x adsorber catalytic converters. After the stored material is emptied, a storage process starts with a high degree of efficiency that decreases as the catalytic converter fills. To permit low-consumption lean operation between two regenerations for as long as possible, the goal of development is the highest possible NO_x storability with a high efficiency. Figure 21-48 shows how catalytic converter B has a higher storage capacity than catalytic converter A.

Since efficiency above 90% is necessary to maintain Euro IV exhaust thresholds depending on the application, the NO_x storage potential cannot be fully exploited in practice, and the storage medium has to be regenerated before this point.

As mentioned in the discussion of the mode of operation, the engine is briefly run rich so that the NO_x arising during nitrate decomposition is converted into N_2 with the aid of the reducing agents HC and CO. To keep the fuel consumption as low as possible during rich operation, developers seek to efficiently exploit the regeneration material.

Figure 21-49 shows the NO_x efficiency of two catalytic converters in an engine test bench cycle with a 60 s lean phase and 2 s rich phase. An efficiently regenerating catalytic converter representing the current state of development is compared with an older, poorly regenerating catalytic converter. The efficiently regenerable catalytic converter can be completely regenerated with 2 s rich operation despite a high NO_x storability, recognizable by the greater efficiency in the initial lean cycle. In contrast, the "bad" catalytic converter that stores NO_x in the same manner is incompletely emptied during the rich peak, which decreases efficiency from cycle to cycle.

Temperature Range for NO_x Storage and Regeneration

When using an underfloor NO_x storage catalytic converter, inlet temperatures below 300°C (ECE range) to over 500°C (EUDC range) are expected in the European smog test depending on the application. To comply with Euro IV laws, a NO_x efficiency of more than 90% is required depending on the application. The temperature range in which the cyclic storage and regeneration attains such a level of efficiency in NO_x storage catalytic converters is the focus of particular interest. In addition to restrictions posed by the engine, it is restricted by the map

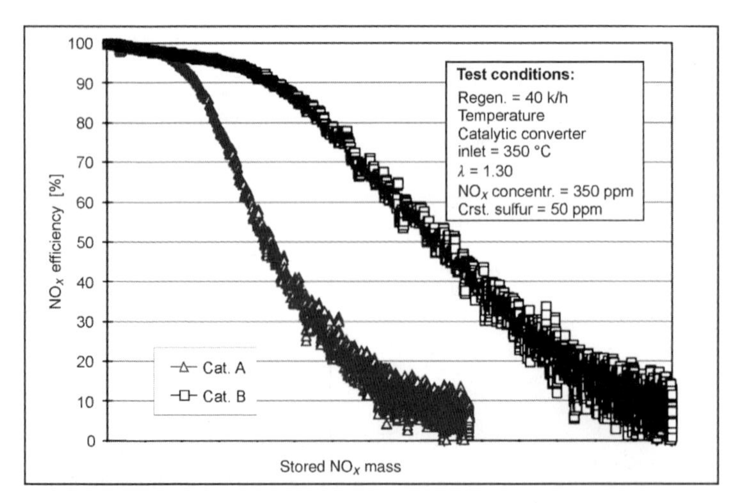

Fig. 21-48 NO_x storage curves for two catalytic converters at 350°C.

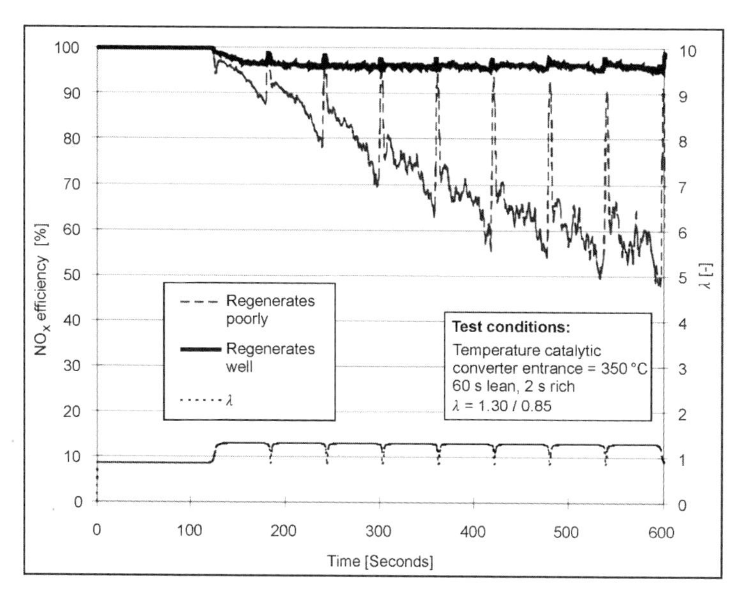

Fig. 21-49 NO_x storage regeneration of various catalytic converters.

range in which the engine can be operated lean with low consumption; from the user's perspective, the temperature range should, therefore, be as wide as possible.[18]

Figure 21-50 shows the different NO_x efficiency curves of two new catalytic converters (precious metal content: 125 g/cu.ft.) in an engine test bench test with a 60 s lean and 2 s rich phase plotted against the catalytic converter inlet temperature.

The efficiency at low temperatures is limited by the "light-off" of the catalytic converter in this case, the ability of the precious metals to oxidize NO into NO_2. The top temperature limit is essentially restricted by the stability of the formed nitrates, i.e., the ability of the storage material to form thermodynamically stabile nitrates even at high temperatures.[16] Since barium as a NO_x storage material does not form stabile nitrates like potassium, the efficiency of Ba catalytic converters drops at temperatures above 400°C, whereas the efficiency of a potassium catalytic converter at 500°C is still above 90% (Fig. 21-50).

Three-Way Characteristics and HC/CO Conversion in Lean Operation

Considering conventional three-way catalytic converters, NO_x storage catalytic converters have comparatively favorable three-way characteristics and HC/CO conversion rates in lean operation. The HC activity is negatively influenced when very basic materials such as potassium are used as the NO_x storage component.[20]

Figure 21-51 compares the conversion of a barium and potassium NO_x storage catalytic converter with that of a conventional three-way catalytic converter. In the left of the figure are the conversions in homogeneous, lean operation at λ 1.5 and a 350°C inlet temperature; on the right are conversions in controlled operation where λ = 1 at 450°C.

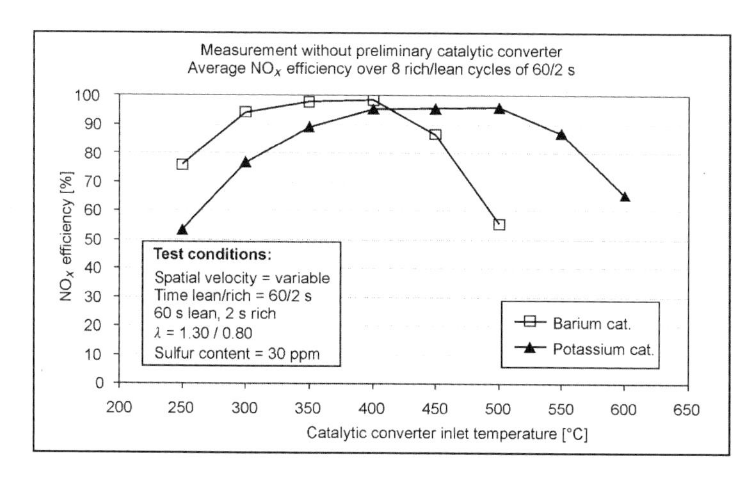

Fig. 21-50 Operating temperature range of new NO_x adsorbers.

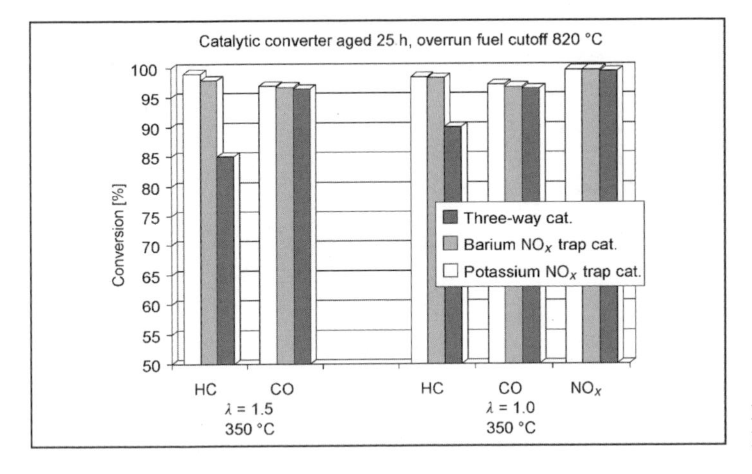

Fig. 21-51 HC/CO conversion in lean operation with $\lambda = 1$.

Temperature Stability

Figure 21-52 shows the influence of various ageing conditions on a NO_x adsorber catalytic converter with barium as the storage material. The reference point is 1 h at 650°C, with a catalytic converter stabilized at $\lambda = 1$. After 25 h of stoichiometric engine test bench ageing at 820°C in a catalytic converter bed, we see a reduction of the NO_x activity over the entire operating temperature range that, however, does not grow when extending the ageing to 50 h. The deactivation arises from the temperature-related sintering of the washcoat, the contained precious metals, and the NO_x storage component.

Ageing under stoichiometric conditions at the same temperature but with periodic overrun fuel cutoff phases substantially increases the deactivation of the catalytic converter that progresses with ageing. The causes are the increased sintering of the precious metals under lean conditions and the reaction of the barium with the aluminum oxide of the washcoat that also occurs under lean conditions at temperatures above approximately 700°C. The barium is irreversibly deactivated for NO_x storage. The speed of these effects increases with the temperature.[21]

One way to increase the maximum temperature stability is to use a NO_x storage material that does not produce this interaction with the washcoat. Results from using potassium as the storage material reveal a much greater ageing stability in comparison to barium catalytic converters. Figure 21-53 shows a comparison of both technologies under high-temperature ageing with an overrun fuel cutoff at 850°C catalytic converter inlet temperature.

The respective storage capacity of the individual technologies when new is set at 100% as a reference quantity. Based on a new state, the Ba technology shows a steady decrease in the remaining NO_x storage capacity as the ageing increases. After 50 h, the barium catalytic converter is largely deactivated. In contrast, the potassium technology shows a much higher remaining NO_x storability. Although the rate notably decreases after 25 h, the most striking factor is the substantial retention of the remaining storage capacity as ageing progresses.

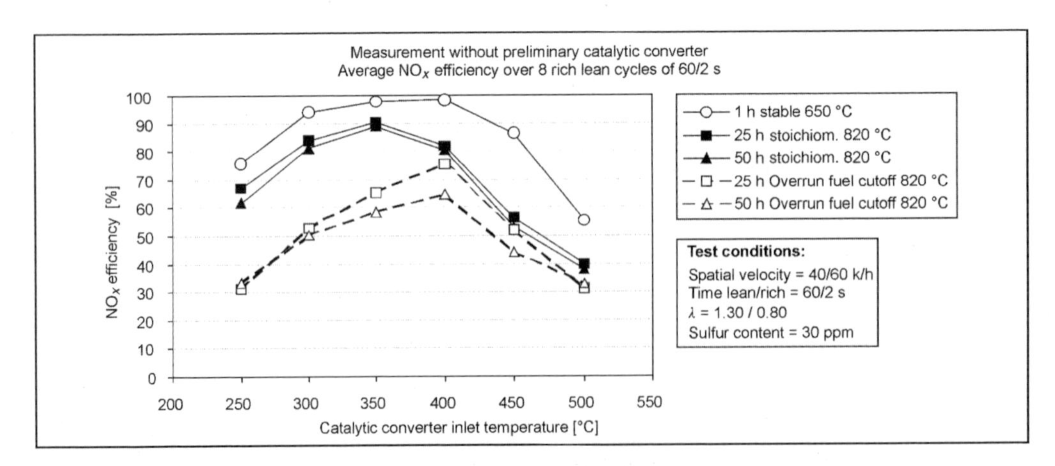

Fig. 21-52 Influence of various aging influences on the operating temperature range of a barium NO_x adsorber.

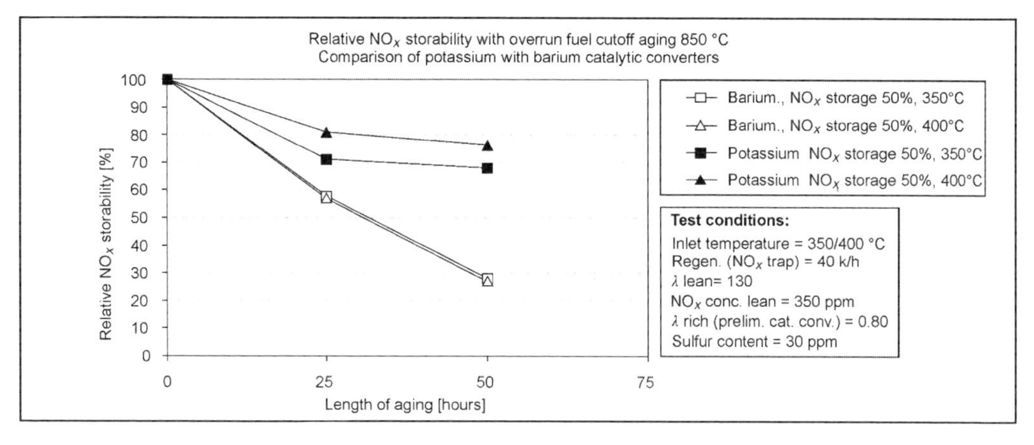

Fig. 21-53 Storability of NO_x adsorbers with potassium and barium after high-temperature ageing.

In addition to this clear advantage of maximum temperature stability, potassium catalytic converters, as mentioned in the section discussion on the temperature range, have a wider operating temperature window for NO_x storage and regeneration at higher temperatures.

This contrasts with disadvantages that are to be weighed depending on the system configuration, vehicle package, and cost. The following disadvantages can be cited:

- Lower HC conversion (see the section on HC and CO conversion)
- Higher desulfurizing temperature (see the section on sulfur poisoning)
- Incompatibility with certain substrate materials

Since the HC conversion of a potassium-containing NO_x adsorber can be much less than that of a barium catalytic converter, this fact needs to be taken into account in the corresponding system design. One option is to use larger preliminary catalytic converters that completely take over HC conversion.

The higher temperatures required to desulfurize potassium-catalytic converters due to more thermally stabile sulfates than in barium catalytic converters pose additional demands on the engine management system. These systems must be capable of providing catalytic converter inlet temperatures of approximately 750°C for desulfurization even at times in which lower temperatures would exist in normal engine operation.

A decisive disadvantage of potassium catalytic converters is the affinity of potassium with ceramic substrates used in series production. Potassium diffuses at temperatures above approximately 750–800°C into the ceramic substrate, is deposited there, and forms a compound with the ceramic. Two negative effects arise as a result. On the one hand, the storage component for NO_x storage becomes inactive. On the other hand, the mechanical stability of the ceramic substrate increases. This process is accelerated at temperatures above 800°C. With the ageing shown in Fig. 21-53, a potassium catalytic converter was used on a metal substrate where this affinity does not exist. Solutions are presently being developed that permit such a coating even on modified ceramic substrates.

Sulfur Poisoning and Regeneration

The problem with sulfur poisoning of NO_x adsorbers results from the fact that all materials that are suitable for NO_x storage also tend to store SO_2 by forming a corresponding sulfate. The reactions are analogous to those occurring with NO_x storage and are schematically represented in Fig. 21-54.

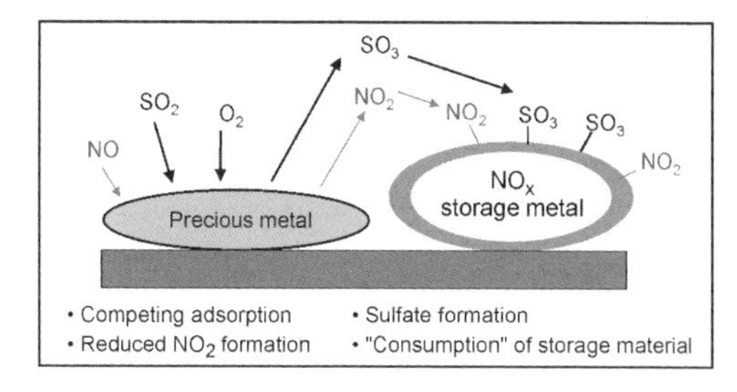

Fig. 21-54 Illustration of the sulfur poisoning of a NO_x storage catalytic converter.

In lean operation, the NO_x adsorber first oxidizes the SO_2 to form acidic gas SO_3. Just as is the case with NO_2, the SO_3 also reacts with the storage material to produce the corresponding sulfate. The storage material converted into a sulfate is, hence, no longer available for NO_x storage. The actual problem of sulfur poisoning is that these sulfates are more thermally stable than nitrates. Classic storage materials are, hence, incapable of sulfate regeneration under the same conditions as for NO_x regeneration.

Over time, the sulfate in the NO_x storage medium increases so much that the NO_x storage capacity falls to an unacceptable level.

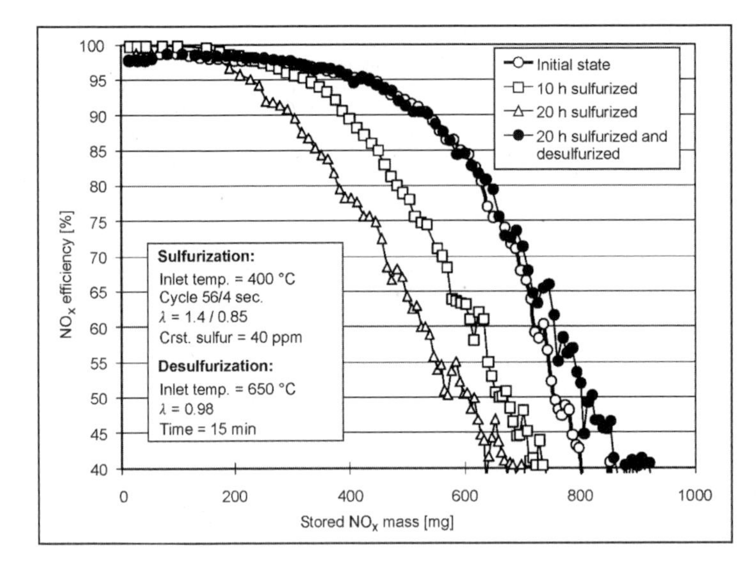

Fig. 21-55 NO_x storability during sulfurizing and desulfurizing.

In Fig. 21-55, we see the decrease in the NO_x storage capacity of a thermally aged barium catalytic converter with a 400°C inlet temperature not previously sulfurized after 10 and 20 h of sulfurization with 40 ppm fuel sulfur. The NO_x storability can be completely eliminated after further sulfurization.

The catalytic converter sulfurized for 20 h was then desulfurized. The conditions were as follows: 650°C catalytic converter inlet temperature, $\lambda = 0.98$, 15 min. constant operation. The NO_x storability of the catalytic converter was regenerated to its initial level as shown by the curves in Fig. 21-55.

The thermal stability of the sulfates is lower under rich conditions in contrast to lean conditions or when $\lambda = 1$, which causes the sulfate to decompose and accordingly regenerates the storage capacity. The sulfate decomposition accelerates the higher the temperature and the richer the exhaust. For barium-containing NO_x storage catalytic converters, temperatures of approximately 650°C are sufficient for sulfate regeneration. When more basic NO_x storage components are used that are able to form stable nitrates at higher temperatures than barium, higher temperatures are also required for desulfurization.

During sulfur regeneration under constantly rich engine operation conditions, sulfate decomposition produces SO_2 that is then converted in the catalytic converter into the undesired secondary emission product H_2S. The desulfurizing strategies that avoid the formation of H_2S are currently being developed. Because of space restrictions, however, we avoid giving a detailed discussion of the reactions during desulfurization.

Basically the loss of activity of a NO_x storage catalytic converter from sulfur poisoning accelerates as the amount of sulfur in the fuel increases.[22] The introduction of low-sulfur fuels reduces this problem, which in turn reduces the frequency of desulfurizing that is problematic for fuel consumption.

To date, 100% protection of the NO_x adsorber from sulfur poisoning by an upstream sulfur trap has been unsuccessful. An apparent benefit from sulfur traps is that the time can be increased between sulfur regenerations of the NO_x storage medium.[15,16]

21.5.3.3 System with a Precatalytic Converter and NO_x Adsorber

Based on the statements made concerning operating temperature range and temperature stability in Section 21.5.3.2, the NO_x adsorber needs to be placed under the floor. This means that an underhood precatalytic converter is required to deal with cold-start emissions. To maintain current and future exhaust laws dealing with lean SI engines, a system should be used with a precatalytic converter and a NO_x absorber. Figure 21-56 shows such a system.

In addition to the cited conversion of cold-start emissions, the precatalytic converter also assumes the tasks of three-way conversion under $\lambda = 1$ conditions. In addition, HC and CO are converted during lean operation. This feature is very helpful for the NO_x storage process in the adsorber. HC and CO molecules reaching the adsorber are converted by the adsorber in a reaction that competes with NO_x storage. As a result, less NO_x is stored with a higher efficiency, and the effectively exploitable storage capacity is less. Since it is under the hood, the precatalytic converter requires a minimum temperature stability of 950°C.

The target value for temperature stability of NO_x adsorbers is 900°C. This value has not been attainable with the present technology. For this reason, the maximum temperature load for initial series applications needs to be limited by cooling. Possible locations for cooling devices are shown in Fig. 21-56. The type of cooling that is used depends on the available space, the required cooling, and the cost. In conclusion, the effort involved in creating the system must be justified by the attainable practical advantage.

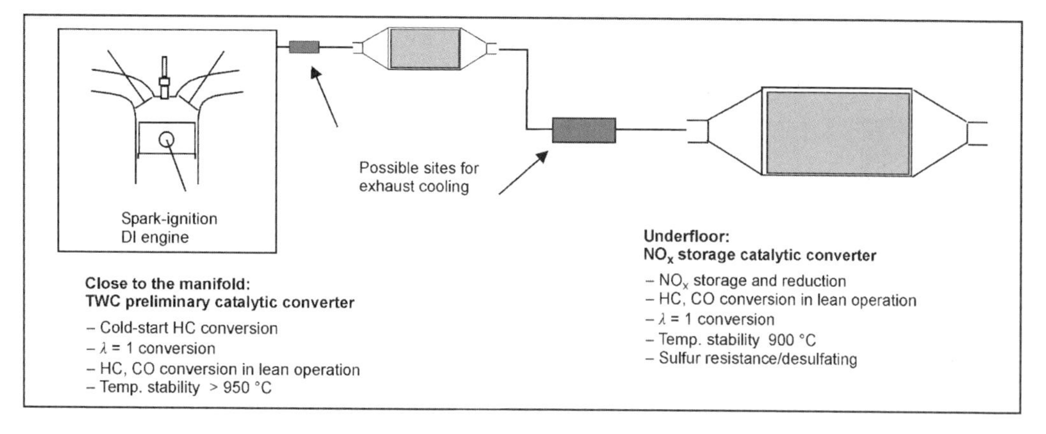

Fig. 21-56 Catalytic converter configuration for lean SI engines for EU IV applications.[23]

21.5.4 Metal Catalytic Converter Substrates

Since the initial development of catalytic converters in the early 1960s, there were efforts to use metal as a substrate in addition to the cordierite extrusions (Fig. 21-57).

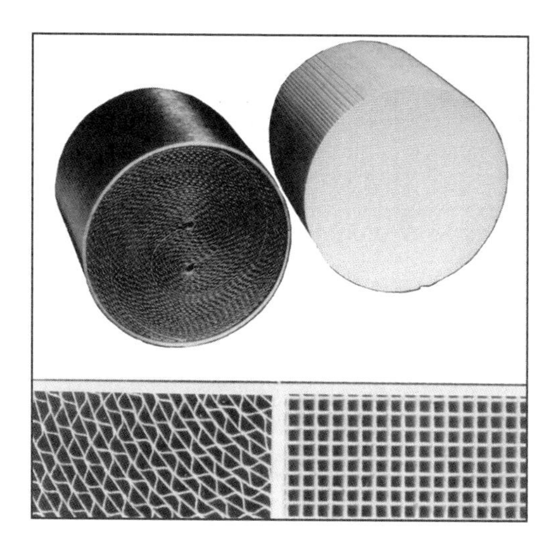

Fig. 21-57 Ceramic (left) and metal (right) honeycomb structures.

To produce a metal substrate, smooth and corrugated metal foils are wound to form honeycomb bodies and are introduced into a tube (Fig. 21-58). Over a period of approximately 20 years, it was rather difficult to maintain the necessary mechanical durability of metal substrates since spiral-wound substrates telescope under dynamic loads at high temperatures.

Only with the introduction of a high temperature soldering process to connect the individual foil layers and with the development of a new winding technique, were the obstacles to the use of metal substrate catalytic converters largely overcome.

Today, the utilized metal foils are 0.05 to 0.03 mm thick. The large-scale use of thinner foils is presently being worked on. The aluminum in the foils makes them very corrosion resistant, and the very thin aluminum oxide layer on the metal surface allows the oxide washcoats to adhere well with the substrate material.

The very thin metal cell walls only slightly raise the exhaust counterpressure (Fig. 21-59), which has a positive effect on fuel consumption and engine performance.

For effective exhaust purification, the time that passes until the operating temperature of the catalytic converter is reached is very important since approximately 70–80% of all pollutants formed during a test cycle are emitted during this period. Shortening this time is a focal point in the development of exhaust purification technology. The following constructive features should be incorporated to best exploit the exhaust energy to heat the catalytic converter:

- Low thermal capacity
- Large geometric surface of the substrate

Metal substrates are very suitable because of their physical features and large surface. The ratio of the substrate surface to substrate thermal capacity (which strongly influences the heating behavior) increases with the cell density, while the cell-wall thickness decreases as shown in Fig. 21-60.

Figure 21-61 demonstrates how the use of catalytic converter substrates with a higher cell density and equivalent dimensions decreases hydrocarbon emissions in a cold start.

The conversion of pollutants can also be enhanced by using substrates with a high cell density even after the operating temperature of the catalytic converter. An example is demonstrated in Fig. 21-62 by the illustration of hydrocarbon conversion in bag 1 of the FTP (Federal Test Procedure) cycle as a function of the cell density.[24] The effect from the increased conversion from the use of higher cell densities clearly exceeds the effect from enlarging the catalytic converter volume.

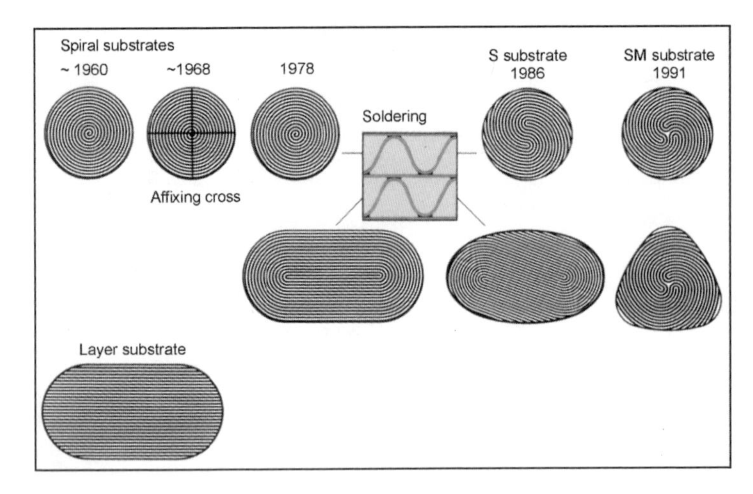

Fig. 21-58 Development of methods to produce metal substrates.

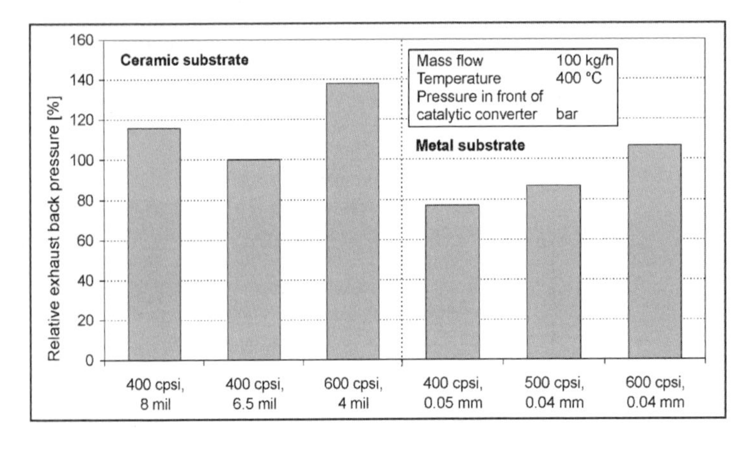

Fig. 21-59 Exhaust counterpressure of various catalytic converter substrates.

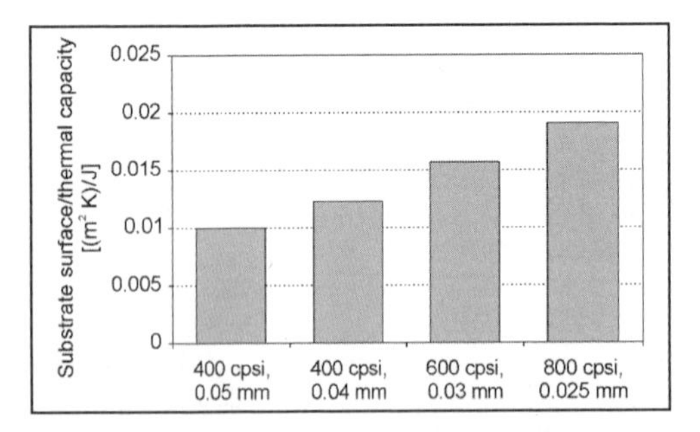

Fig. 21-60 Ratio of substrate surface/substrate thermal capacity for metal substrates of various cell densities.[1]

The increase in catalytic efficiency is not based only on the increase of the substrate surface as the cell density increases; when the channel diameter falls as the number of cells increase, the transfer of material from the gas phase to the channel wall also improves. This effect is illustrated in Fig. 21-63.

To further increase the catalytic conversion, especially of metal substrate catalytic converters, structures can be introduced in the channel walls as shown in Fig. 21-64. These microwaves are perpendicular to the gas flow and cause a strong transmission of material from the gas phase to the substrate walls from local turbulence.

By structuring the channels in this manner, the conversion can be increased by 10% to 15% percent in comparison to substrates with identical dimensions but with smooth channel walls. An alternative is to reduce the substrate volume or cell density without lowering the conversion rate, thereby decreasing the required installation space of the catalytic converter and minimizing its output loss.

The use of metals as substrate materials permits the cold-start phase particularly critical for pollutant emissions to be shortened by heating the catalytic converter to the necessary operating temperature. The metal monolith serves as a resistance heat conductor. Figure 21-65 illustrates how the overall emissions can be reduced with an electrically heated catalytic converter in comparison to an unheated system.

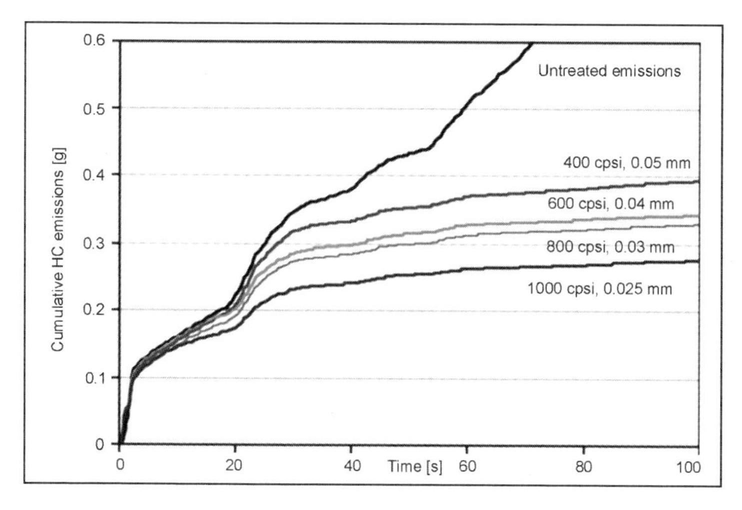

Fig. 21-61 Cumulative hydrocarbon emissions during the first 100 s of the FTP cycle (catalytic converter dimensions: 98.4 × 74.5 mm).[2]

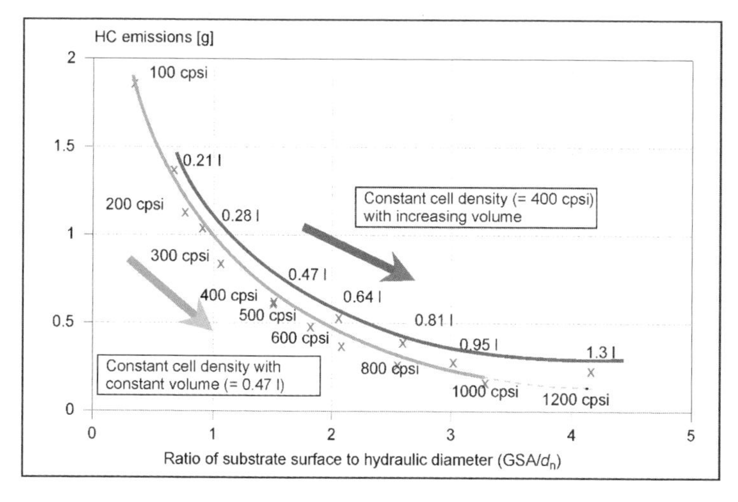

Fig. 21-62 Cumulative hydrocarbon emissions in bag 1 of the FTP cycle as a function of the ratio of the substrate surface/hydraulic diameter (GSA/d_h).

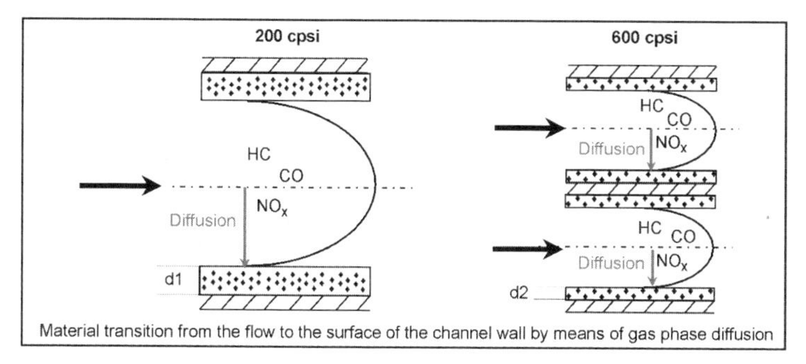

Fig. 21-63 Schematic illustration of the diffusion paths with different hydraulic diameters and cell densities.

Because of their construction, metal substrate catalytic converters can be directly welded to the exhaust gas system.

Recycling vehicle components has become more important in recent times. A special method developed for metal substrates allows the utilized materials to be almost completely recovered. The basic progression of this process is shown in Fig. 21-66.

Bibliography for Section 21.5

[1] Dahle, U., S. Brandt, and A. Velji, "Abgasnachbehandlungskonzepte für mager betriebene Ottomotoren," Conference "Direkteinspritzung im Ottomotor," Haus der Technik, Essen, 1998.

[2] Iwamoto, M., and H. Hamada, "Removal of Nitrogen Monoxide from Exhaust Gases through Novel Catalytic Processes," Catalysis Today, Vol. 10, 1991, pp. 57–71.

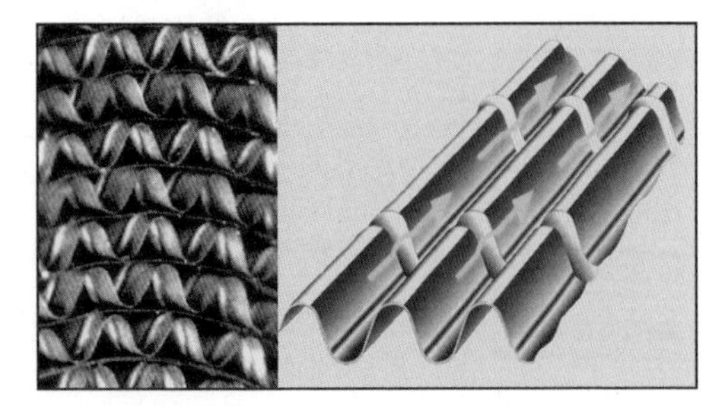

Fig. 21-64 View of the face of a metal substrate with structured channels (left), and schematic representation of the microwaves in the channel (right).

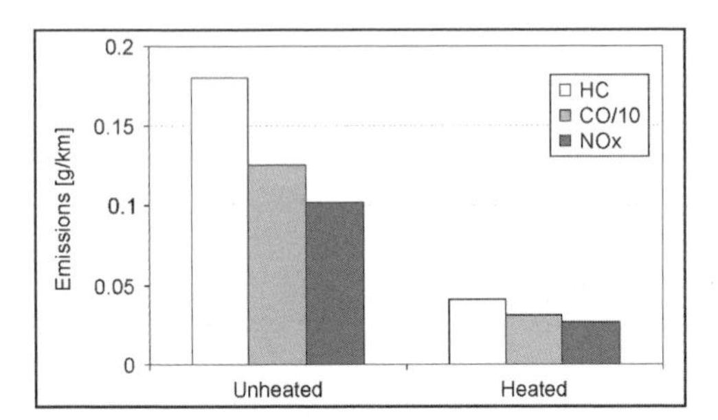

Fig. 21-65 Comparison of emission results for unheated and heated catalytic converters in the Euro II test for the BMW Alpina B12.[25]

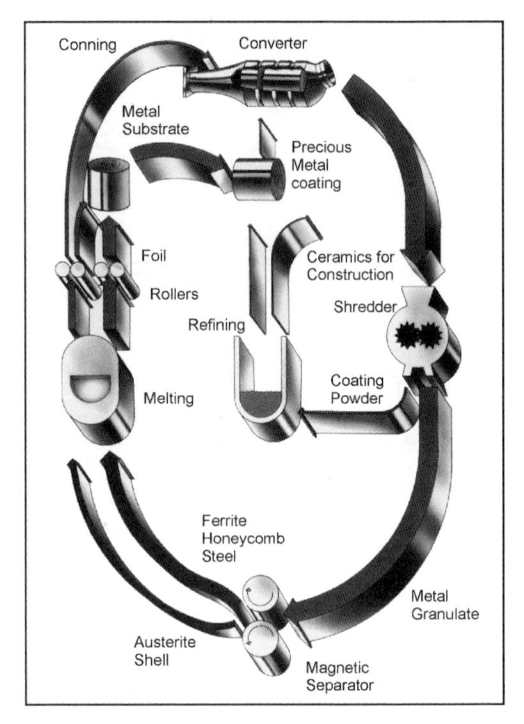

Fig. 21-66 Schematic representation of the recycling process developed for metal substrate catalytic converters.

[3] (a) Braun, D., U. Kuehler, and G. Pietsch, "Behavior of NO_x in Air-Fed Ozonizers," Pure and Applied Chemistry, Vol. 60, 1988, 741–746; (b) Oumghar, A., J.C. Legrand, A.M. Diamy, and N. Turillon, "Methane Conversion by Air Microwave Plasma," Plasma Chemistry and Plasma Processing, Vol. 15, 1995, pp. 87–107.

[4] Penetrante, B.M., M.C. Hsiao, B.T. Merritt, G.E. Vogtlinm, P.H. Wallmann, A. Kuthi, C.P. Burkhart, and J.R. Bayliss, "Electron-Impact Dissociation of Molecular Nitrogen in Atmospheric-Pressure Nonthermal Plasma Reactors," Appl. Phys. Lett., Vol. 67, No. 21, November 1995.

[5] Held, W., A. König, T. Richter, and L. Puppe, "Catalytic NO_x Reduction in Net Oxidizing Exhaust Gas," SAE 900496, 1990.

[6] Kawaki, and H. Muraki, "Exhaust Gas Cleaning Catalyst and Method," Japanese Patent Office Patent Journal, Kokai Patent No. Hei 2[1990]-265649.

[7] Kharas, K.C.C., H.J. Robota, and D.J. Liu, "Deactivation of Cu-ZSM-5 Lean-Burn Catalysts," Applied Catalysis B: Environmental, Vol. 2, 1993, pp. 225–237.

[8] Grinsted, R.A., H.W. Jen, C.N. Montreuil, M.J. Rokosz, and M. Shelef, "The Relation between Deactivation of Cu-ZSM-5 in the Selective Reduction of NO and Dealumination of Zeolite," Zeolites, Vol.13, 1993, pp. 602–606.

[9] Schreffler, R., "MMC, Japanese work to expand gasoline DI tech," Ward's Engine and Vehicle Technology Update, Vol. 24, No. 3, February 1998, pp. 2–3.

[10] Mitsubishi, "Global Standard Eco-engine, Mitsubishi Gasoline Direct Injection Engine," February 1998.

[11] Ando, H., K. Noma, K. Iida, O. Nakayama, and T. Yamauchi, "Mitsubishi GDI Engine, Strategies to meet the European Requirements," AVL-Tagung "Motor und Umwelt," 97, 1997.

[12] Hori, M., A. Okumura, H. Goto, M. Horiuchi, M. Jenkins, and K. Tashiro, "Development of New Selective NO_x Reduction Catalyst for Gasoline Leanburn Engines," SAE 972850, 1997.

[13] Schmelz, "Method and Apparatus for Controlled Introduction of a Reducing Agent into a Nitrogen Oxide Containing Exhaust Gas," U.S. Patent 5,628,186.

[14] Miyoshi, N., S. Matsumoto, K. Katoh, T. Tanaka, J. Harada, N. Takahashi, K. Yokota, M. Sugiura, and K. Kasahara, "Development of New Concept Three-Way Catalyst for Automotive Lean Burn Engines," SAE 950809, 1995.

[15] Strehlau, W., J. Leyrer, E.S. Lox, T. Kreuzer, M. Hori, and M. Hoffmann, "New Developments in Lean NO_x Catalysis for Gasoline Fueled Passenger Cars in Europe," SAE 962047, 1996.

[16] Hepburn, J.S., E. Thanasiu, A. Dobson, and W.L. Watkins, "Experimental and Modeling Investigations of NO_x Trap Performance," SAE 962051, 1996.

[17] Harada, J., T. Tomita, H. Mizuno, Z. Mashiki, and Y. Ito, "Development of Direct Injection Gasoline Engine," SAE 970540, 1997.

[18] Krebs, R., E. Pott, and B. Stiebels, "Das Emissionskonzept des Volkswagen Lupo FSI," Tagung "9th Aachen Colloquium on Vehicle and Engine Technology," Aachen, 2000.

[19] Krebs, R., E. Pott, H. Hahne, T. Kreutzer, U. Göbel, J. Höhne, and K.-H. Glück, "Die Abgasreinigung der FSI-Motoren von Volkswagen," MTZ 61 (2000).

[20] Matsumoto, S., N. Miyoshi, and Y. Ikeda, "NO_x Storage-Reduction Catalyst (NSR) for Automotive Engines: Sulfur Poisoning Mechanism and Improvement of Catalyst Performance," Conference: Zukünftige Abgasgesetzgebungen Europa und USA: Technische Lösungen, Ottomotoren, Haus der Technik, Essen, 1998.

[21] Strehlau, W., J. Höhne, U. Göbel, J.A.A. v.d. Tillaart, W. Müller, and E.S. Lox, "Neue Entwicklungen in der katalytischen Abgasnachbehandlung von Magermotoren," AVL Conference, "Motor und Umwelt," 1997.

[22] Dahle, U., S. Brandt, A. Velji, J.K. Hochmuth, and M. Deeba, "Abgasnachbehandlung bei magerbetriebenen Ottomotoren–Stand der Entwicklung," 4t Symposium on Developmental Tendencies in SI Engines, Esslingen, 1998.

[23] Faltermeier, G., B. Pfalzgraf, R. Brück, C. Kruse, and W. Maus, "Katalysatorkonzepte für zukünftige Abgasgesetzgebungen am Beispiel eines 1.8 l 5V-Motors," 17th International Viennese Engine Symposium, Vienna, 1996.
[24] Maus, W., R. Brück, P. Hirth, J. Hodgson, and M. Presti, "Potenzial von Katalysatorkonzepten zum Erreichen der SULEV-Emissionsgrenzwerte," 20th International Vienna Engine Symposium, 1999.
[25] Hanel, F.J., E. Otto, and R. Brück, "Electrically heated catalytic converter (EHC) in the BMW Alpina B12 5.7 Switch-Tronic," SAE 960349, 1996.

21.6 Exhaust Treatment in Diesel Engines

21.6.1 Diesel Oxidation Catalytic Converters

In contrast to spark-ignition engines, a diesel system is more thermodynamically efficient. This is expressed in lower fuel consumption and lower CO_2 emissions than its SI counterpart. For exhaust purification, oxidation catalytic converters (DOC) have been used for more than ten years in diesel passenger cars. Diesel engines are operated with excess oxygen. Their maximum exhaust temperature is approximately 800°C, and their average is clearly below that of comparable spark-ignition engines.

This means

- Better CO and HC emissions
- Higher NO_x emissions
- Much higher particle emissions
- More complex emissions control since gas phase reactions must be dealt with due to the share of particles in the exhaust, and less reducing agent is available due to lower HC concentrations.

For exhaust purification, these characteristics mean that exclusively oxidative reactions are used, and catalytically active components are used that chiefly have to cover the low temperature range (fast light-off).

21.6.1.1 Pollutants in Diesel Exhaust

Hydrocarbons and CO:

Even under excess oxygen, heterogeneous combustion chamber conditions can cause incomplete oxidation reactions and uncombusted or partially oxidated hydrocarbons in the exhaust in addition to CO. Some of these compounds are responsible for the typical smell of diesel exhaust. All engine parameters that improve the exploitation of the oxygen in the combustion chamber (such as swirling the mixture) or higher combustion temperatures can lower CO and HC emissions.

Particles:

Local rich conditions during combustion lead via the intermediate steps of acetyls and polycyclic hydrocarbons to the formation of graphite-like soot particles. By means of coagulation and agglomeration processes, soot particles approximately 100 to 300 nm in diameter (median) arise from these approximately 1–10 nm primary particles. Since the large surface area (up to 200 m^2/g) of these particles make them strongly adsorbent, a large share (greater than 50% by weight) of hydrocarbons, sulfates, water, and lubricating oil components can be demonstrated in diesel soot in addition to carbon.

Nitrogen oxides:

In oxidation reactions, the nitrogen oxides (NO and NO_2) occur in the presence of nitrogen. Since the concentration of both components and their ratio depend on the reaction temperature and the oxygen concentration during combustion, NO_x emissions can be reduced by suitable engine measures such as late fuel injection (gas temperature falls) and exhaust recirculation (oxygen concentration falls).

Sulfur oxides:

From the combustion of sulfur-containing fuel, primarily SO_2 rises that is further oxidized at temperatures greater than 300°C by precious metals to form SO_3, which reacts in the presence of water to form sulfuric acid (H_2SO_4). All three compounds can deactivate the catalyst; SO_2 and SO_3 by specific addition that blocks the precious metal, and H_2SO_4 by coating the washcoat surface and by producing condensation in the washcoat pores.

21.6.1.2 Characteristics of Diesel Oxidation Catalytic Converters

Design

Similar to three-way catalytic converters, DOCs consist of the following components:

- Honeycombs (monoliths) of ceramic or metal as a substrate for the catalytic coating
- Al_2O_3 for porous, thermally stabile coatings with large surfaces (100–200 m^2/g)
- Precious metal and promotors as catalytically active centers on whose surfaces the oxidation reactions occur

Manufacture

One possible manufacturing method includes the following steps:

- Precious metals and promotors are dissolved.
- This solution is applied to the Al_2O_3 surface (the arising suspension is termed a washcoat).
- The honeycombs are dipped in the washcoat.
- Subsequent drying and calcining processes remove the water from the honeycombs and fix the washcoat.

21.6.1.3 Deactivating the Catalyst Surface

While driving, the catalytic converter surface can be reversibly or irreversibly deactivated by chemical or physical influences. Figure 21-67 offers a few possible examples:

- The washcoat surface is coated by residues from oxidation or the further reaction of hydrocarbons (coking).
- Selective poisoning from coating and covering active centers such as the addition of sulfur compounds to precious metals.

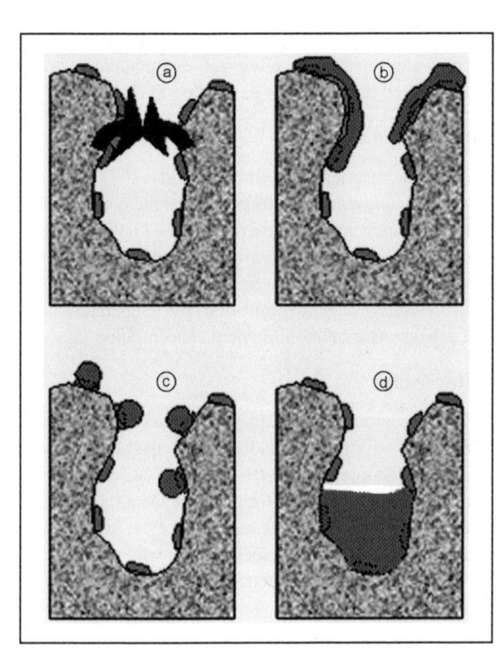

Fig. 21-67 Different types of precious metal and washcoat pore poisoning: (a) Sintering of pores; (b) Nonselective coating of the surface; (c) Selective poisoning of active centers; (d) Condensation of hydrocarbons.

- The restriction of pore openings hinders the accessibility of active centers (sintering). As shown in Fig. 21-68, "light-off curves" measured in model systems provide information on the respective deactivating mechanisms.

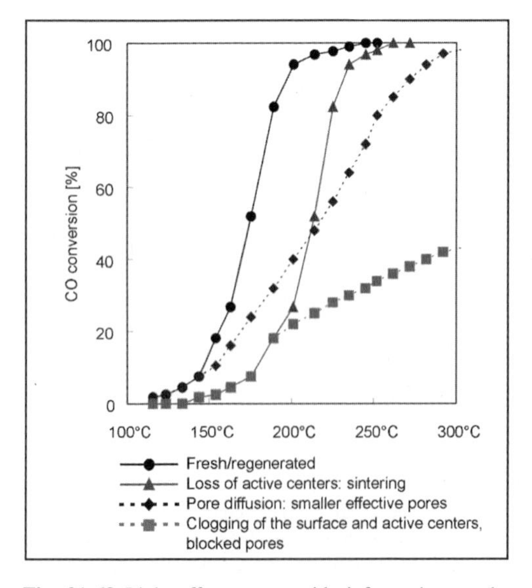

Fig. 21-68 Light-off curves provide information on the different deactivating mechanisms.

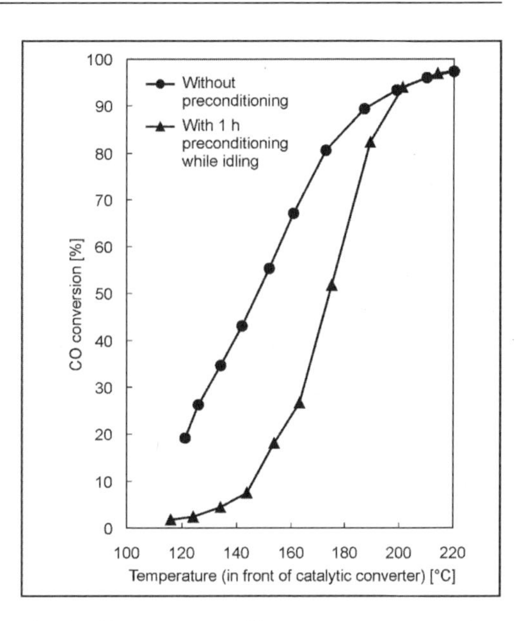

Fig. 21-69 The "light off" temperature occurs when 50% conversion is reached. Preconditioning during idling worsens light-off.

Deactivation under Low Temperature and Low Load Conditions

Temperatures up to approximately 250°C and low loads yield reversible poisoning of the surface from carbon-containing components. The worsening of the CO light-off curves in Fig. 21-69 after 1 h idling (<120°C, <20 Nm) can be reversed by briefly increasing the exhaust temperature (<1 min, <250°C); i.e., the activity can be completely recovered.

Deactivation from Sulfurization

An increase in the exhaust temperature beyond 300°C leads to the sulfurization of the catalytic converter. Regeneration under excess oxygen is possible; however, temperatures greater than 600°C are required. Alternately, the concentration of reducing agents (hydrocarbons) can be increased by enriching to $\lambda < 1$; this reduces the sulfur compounds adsorbed on the surface to form H_2S, which can be desorbed from the surface and detected in the exhaust. If we integrate over the area of the respective H_2S peaks, we obtain information on the sulfurization tendency of the respective catalytic converter. To protect from irreversible sulfurization, promoters can be added to the washcoat that suppress the affinity with sulfur compounds. Figure 21-70 shows a comparison of the H_2S signals of a standard washcoat and the correspondingly protected version. The corresponding vehicle test (MVEG) can be seen in Fig. 21-71. After ageing with 1000 ppm sulfur in the diesel fuel, the modified washcoat surface in the ECE part of the cycle has clearly less CO (g/km) than the standard version.

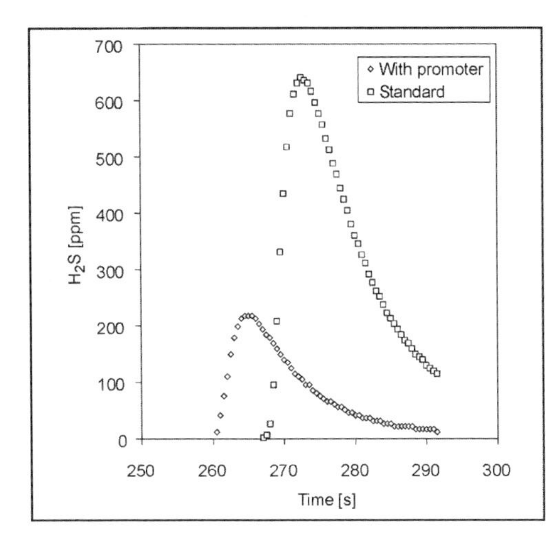

Fig. 21-70 A modification of the washcoat surfaces reduces the affinity for sulfur compounds.

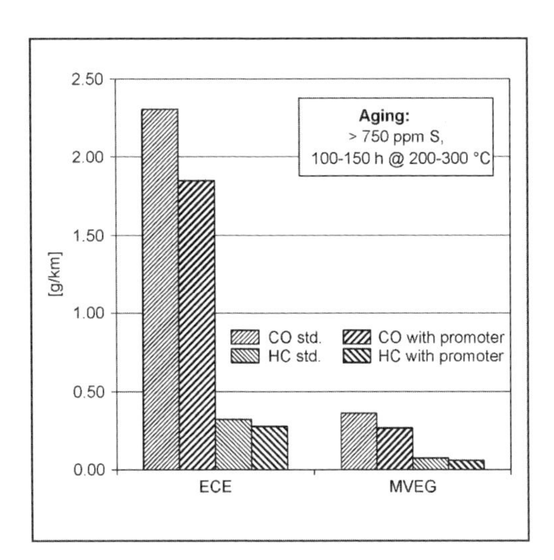

Fig. 21-71 Reduced affinity for sulfur compounds also improves CO and HC performance in the MVEG cycle.

Thermal Deactivation

Higher exhaust temperatures cause the precious metal to be sintered; i.e., small particles agglomerate into larger units. This translates into a loss of the available metal surface and, hence, lowers the oxidative effect. This irreversible process is portrayed in Fig. 21-72. Pt dispersion and CO light-off are inversely related, i.e., the lower the available Pt surface [measured as (%) dispersion], the lower the CO activity.

The temperature stability of all the other washcoat components must be checked in addition to that of the precious metal. Analyses of 50 h ageing at 700°C reveal both the favorable stability of Al_2O_3 surfaces (Fig. 21-73) as well

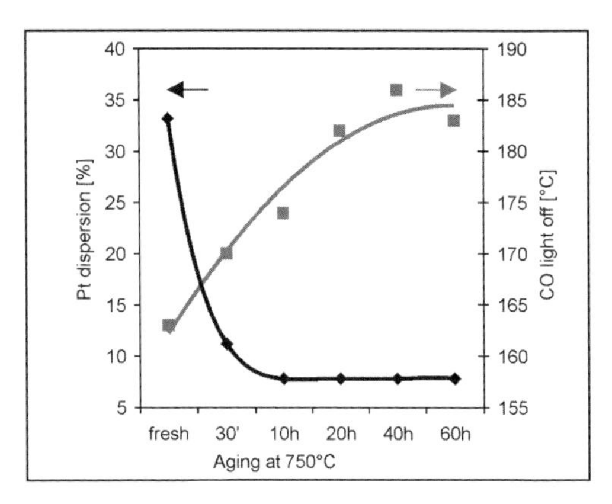

Fig. 21-72 High-temperature ageing reduces the dispersion of the Pt particles. This lowers the activity of CO oxidation.

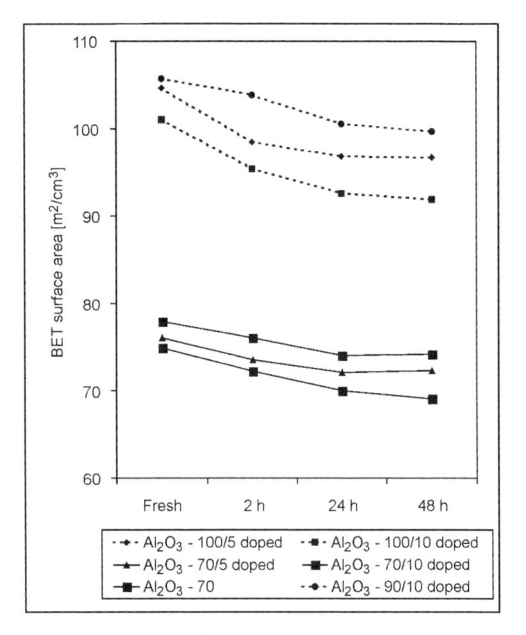

Fig. 21-73 Temperature stability of Al_2O_3 up to 900°C.

as the unrestricted functioning of a zeolite (Fig. 21-74) as indicated by its typical HC desorption curve.

21.6.1.4 Evaluating Diesel Oxidation Catalytic Converters

Light-Off

The activity of DOCs on engine test benches is primarily established by their so-called light-off. The conversion at specific temperatures and loads at the point at which 50% conversion occurs is termed the light-off temperature. Figure 21-75 shows the corresponding temperature and torque of a measured 1.9 l naturally aspirated engine,

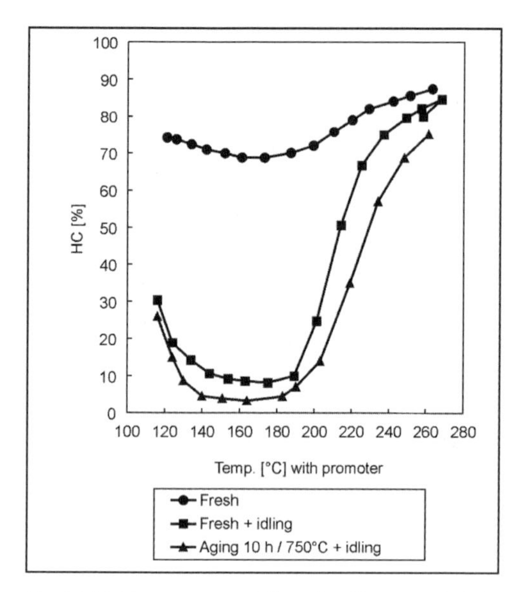

Fig. 21-74 Temperature stability of zeolites up to 850°C.

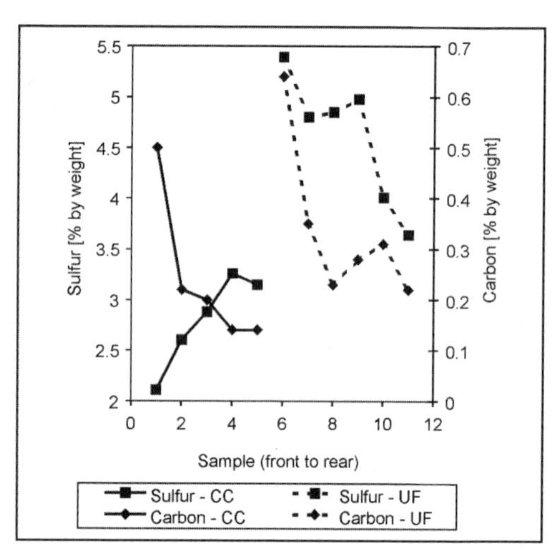

Fig. 21-76 Post mortem analysis: C/S profile of CC and UF catalytic converters.

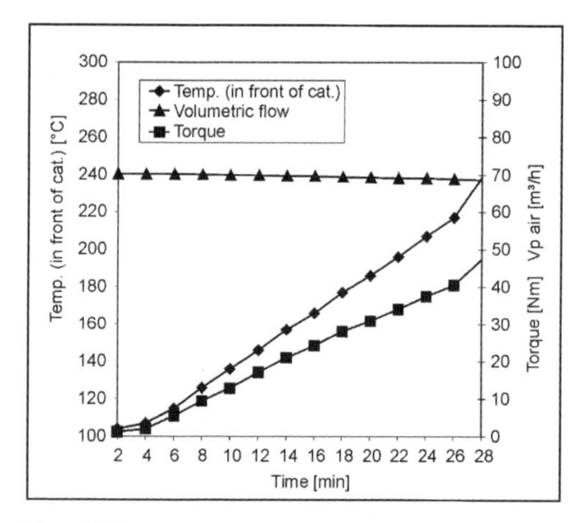

Fig. 21-75 Measured activity of catalytic converters in light-off tests: Temperature ramps and torque characteristic of a 1.9 l naturally aspirated diesel engine.

and Fig. 21-69 shows the associated curve. The CO light-off is approximately 175°C in this instance.

Postmortem Analyses of Deactivated Catalytic Converters

The physical chemistry of aged catalytic converters is investigated in postmortem analyses to determine the deactivation processes during ageing. Figure 21-76 shows carbon and sulfur gradients for a close-coupled (CC) catalytic converter and an underfloor (UF) catalytic converter. The characteristics indicate that more carbon-containing deposits are found in front in the cc arrangement, whereas there is strong sulfurization over the entire length of the monolith. In the UF catalytic converter, both gradients are parallel.

Endurance Tests

Endurance is determined both within specific driving cycles and in normal driving. Figure 21-77 shows an example of a DOC in the AMA cycle aged over 20 000 km. In this cycle, mainly low-temperature and low-load stability are checked. The curves of the catalytic converter measured after 5000, 10 000, 15 000, and 20 000 km in the MVEG cycle are horizontal for approximately 5000 km; i.e. deactivating ceases after this time. The catalytic converter reveals stabile conversion over the remaining life. The data in Fig. 21-77 originate from a vehicle that was regularly measured in normal operation over 80 000 km. In this case as well, we see a characteristic similar to that in Fig. 21-78. After initial ageing, the conversion remains stable over the entire duration of the test.

21.6.2 NO_x Adsorbers for Diesel Passenger Cars

The removal of nitrogen oxides from lean engine exhaust with the aid of NO_x storage catalytic converter technology can be used in spark ignition and diesel engine drive systems. In the following, we address the specific characteristics of exhaust purification of diesel passenger cars. The characteristic differences from spark ignition DI applications are the lower exhaust temperatures, higher soot emissions, and peculiarities of generating rich exhaust conditions that are necessary to regenerate the storage catalytic converter from adsorbed NO_x and SO_x.

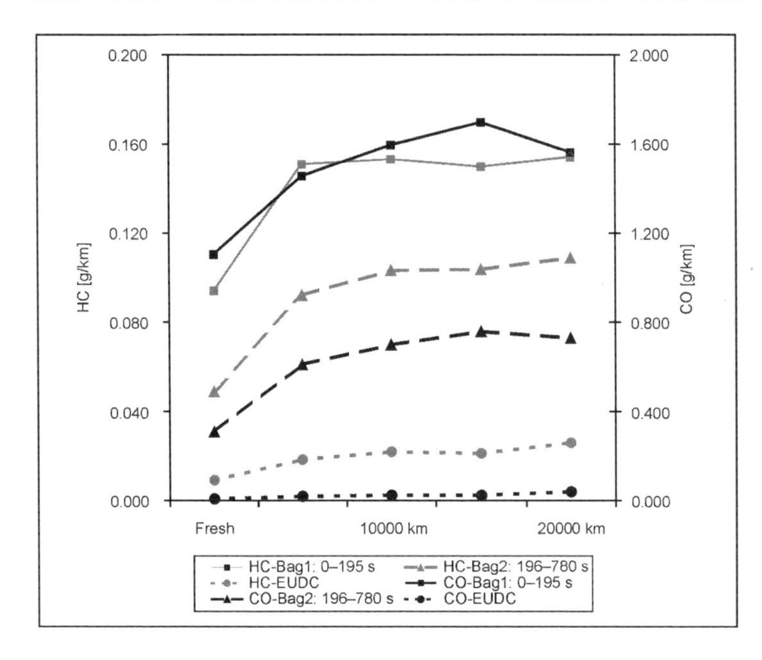

Fig. 21-77 Long-term operation under a weak load provides information on endurance (here: part of a test series that was run up to 80 000 km).

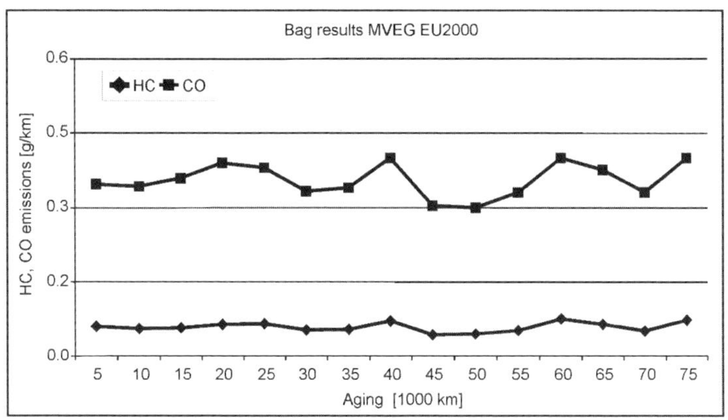

Fig. 21-78 Demonstration of endurance from 80 000 km in-field tests.

21.6.2.1 Operating Range of Storage Catalytic Converters

The low exhaust temperatures of diesel passenger cars mean that catalytic converters have a lower thermal peak load, and the window of operation is shifted toward lower temperatures. Figure 21-79 shows the typical NO_x emissions of a diesel passenger car as a function of catalytic converter bed temperature in the MVEG driving cycle.

Below 150°C, there is no NO_x conversion because the catalytic converter has not yet achieved light-off. Between 150 and 250°C, NO_x can be stored and reduced; however, because of the high spatial velocities and high NO_x concentrations, less NO_x is stored. The optimum efficiency of the catalytic converter lies between 250 and 300°C. At temperatures above 350°C, NO_x storage can be thermodynamically limited depending on the NO_x storage material. As can be seen in Figs. 21-79 and 21-80, the majority of NO_x emissions in city driving occur from 150 to 200°C. In this temperature range, the storage and reduction of NO_x is kinetically limited. The conversion of NO_x in city driving, hence, poses substantial demands on the low-temperature activity of the catalytic converter. The development of modern diesel engines with lower fuel consumption also further lowers the exhaust temperature in the ECE cycle.

21.6.2.2 Desulfurization

Another peculiarity in the operation of NO_x storage catalytic converters in diesel passenger cars is the restricted desulfurization temperature when regenerating SO_x. It is nearly impossible to drive at $\lambda < 1$ under a high load because of increased soot emissions and loss of torque. Typical desulfurization temperatures of diesel exhaust lie within the range of 500–550°C.

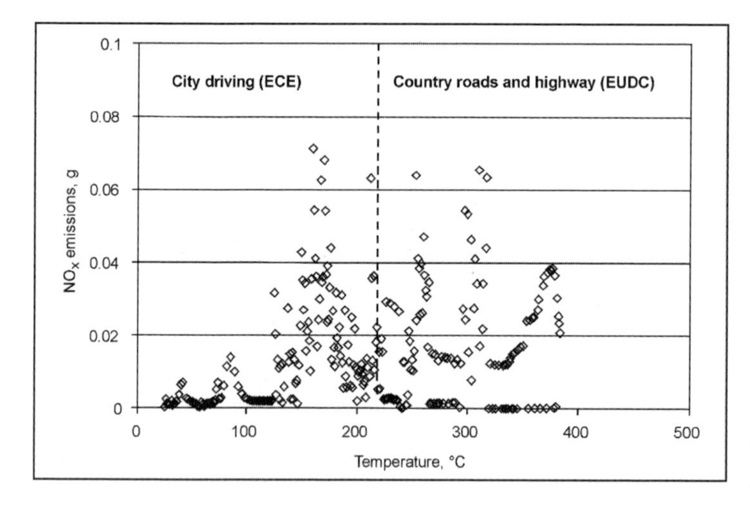

Fig. 21-79 NO_x emissions of a diesel passenger car with EU III motor calibration as a function of the catalytic converter bed temperature in the MVEG cycle.

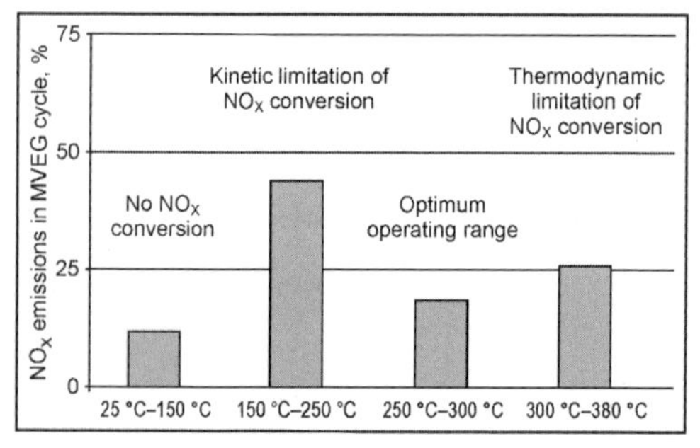

Fig. 21-80 Distribution of NO_x emissions in the MVEG cycle in four temperature ranges relevant to catalytic converters.

The desulfurization capacity of the catalytic converter greatly depends on the selected NO_x storage component (NSC). The tighter the NO_x is bound to the catalyst, the more efficient the storage of NO_x at high temperatures. A high NO_x bonding strength also reduces NO_x peaks during regeneration phases. The advantages of stronger NO_x adsorption are at the cost of higher desulfurization temperatures. Figure 21-81 illustrates the compromise between desulfurization capacity and the efficiency of NO_x storage and regeneration.

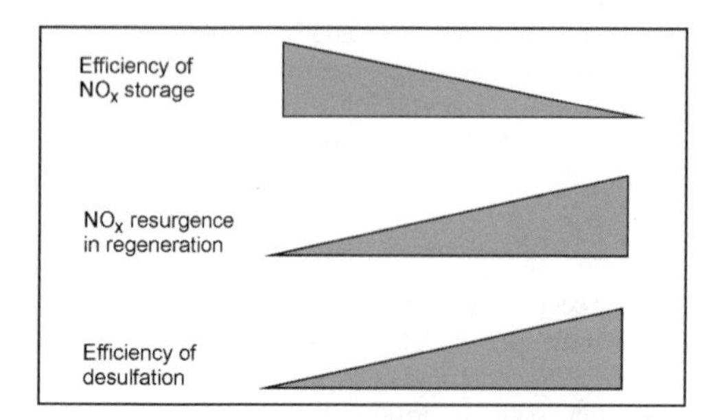

Fig. 21-81 Coupled characteristics of efficiency of NO_x storage and regeneration and the desulfurization capacity of NO_x storage catalytic converters.

Without measures to encourage desulfurization, the NO_x conversion rate of storage catalytic converters in the lean/rich cycle is linear as a function of the mileage or time. The mileage that can be traveled to attain a given NO_x conversion rate decreases inversely proportional to the sulfur content in the fuel. Figure 21-82 shows the relationship of the mileage as a function of the fuel sulfur content at a constant NO_x conversion rate.

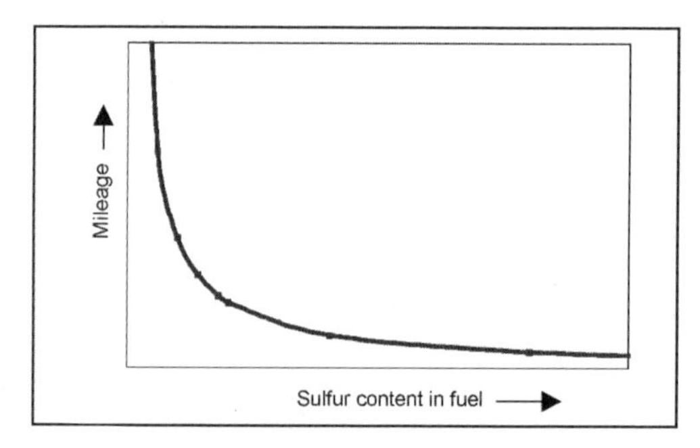

Fig. 21-82 Traveled mileage to attain a threshold of NO_x conversion as a function of the sulfur in the fuel.

The degree to which the NO_x conversion rate is reduced when operating with sulfur-containing exhaust also depends on the temperature and frequency of desulfurization. High exhaust temperatures and long periods between desulfurization leads to the formation of sulfates between the SO_x of the exhaust and the storage material of the catalytic converter that irreversibly damage the storage catalytic converter.

The development of new storage materials has lowered the desulfurization temperature and enhanced the long-term stability of NO_x conversion with the aid of periodic desulfurization in the lean/rich cycle of sulfur-containing synthetic exhaust. However, the required sulfur tolerance in all operating states of real driving remains insufficient. The long periods of constant lean driving at high exhaust temperatures are particularly critical.

21.6.2.3 Regeneration Methods

Both internal and external methods to regenerate catalytic converters from stored NO_x and SO_x are being investigated. In external regeneration, a reducing agent, usually diesel fuel, is sprayed in front of the storage catalytic converter. To reduce the oxygen mass flow, a partial stream of exhaust is guided over the storage catalytic converter.[7,8] This technique requires the use of exhaust valves that are problematic since they wear quickly. In addition, the injection of diesel fuel when the exhaust is below 250°C causes the fuel to condense. Alternative methods for onboard generation of gaseous reducing agents by reforming diesel fuel are being investigated; they are involved, however.

For internal or engine regeneration, reducing conditions are achieved in the exhaust by operating the engine with variable injection parameters. The "poorer" combustion of the fuel in the combustion chamber raises the exhaust temperature, and the throttling of the volumetric flow of intake air reduces the volumetric flow of exhaust. Both measures increase the efficiency of regeneration.

A disadvantage of internal regeneration is the increased soot emissions that arise during rich engine operation at higher loads and contribute to the deactivation of the storage catalytic converters. Within the framework of an integral approach to purifying exhaust of NO_x and particles, combined exhaust systems with soot filters and NO_x storage catalytic converters are under intense investigation.

Bibliography for Sections 21.6.1–21.6.2

[1] Heck, Farrauto, Catalytic Air Pollution Control, Van Nostrand, Reinhold, 1995.
[2] Farrauto, Voss, Monolithic Diesel Oxidation Catalysts, Applied Catalysis B: Environmental 10 (1996) 29.
[3] Bond, Heterogeneous Catalysis, Oxford University Press, 1990.
[4] Bode [ed.], International Conference on Metal-Supported Automotive Catalytic Converters (MACC 97), Wuppertal, Germany, 1997, Werkstoff-Informationsgesellschaft, Frankfurt.
[5] Kruse, Frennet, Bastin [eds.], 5th International Congress on Catalysis and Automotive Pollution Control (CAPOC 5), Brussels/Belgium, 2000, Universite Libre de Bruxelles, Vols. 1 and 2.
[6] Guyon, M., P. Blanche, C. Bert, and L. Philippe, Renault, Messaoudi, I.: Segime, NO_x-Trap System Development and Characterization for Diesel Engines Emission Control, SAE2000-01-2910, Baltimore.
[7] Patent DE 196 26 835 A1, Patent DE 196 26 836 A1.
[8] Beutel, T., U. Dahle, and A. Punke, "Euro 4 – Abgasnachbehandlungstechnologien für Magermotoren (Otto/Diesel)," VDA Technical Congress, IAA Frankfurt/Main, 1999.
[9] Cooper, B.J., and J.E. Thoss, Role of NO in Diesel Particulate Emission Control, SAE Technical Paper No. 890404, 1998.
[10] Mul, G., Catalytic Diesel Exhaust Purification, Proefschrift, Technische Universiteit Delft, 1997.
[11] Neeft, J.P.A., M. Makkee, and J.A. Moulijn, Diesel particulate emission control, Fuel Processing Technology, 47 (1996).

21.6.3 Particle Filters

Smoke, dust, and fog have always been important topics in job safety. In 1775, Pott reported on lung cancer in chimney sweeps. In 1868, Tyndall discovered the optical effect for measuring fine particles; in 1936, the newspaper *Dust* noted the importance of submicron particles. In 1959, the Johannesburg Convention determined the size of particles accessible to lungs. In 1980, particle filters were suggested for vehicle use, and, in 1983 (one year after the EPA's introduction of the first threshold of 0.6 g/mile), an SAE Congress was held that addressed this subject. The subject is, therefore, not new. What is new is that, after slow technological development over two decades in which many researchers and companies participated, more than 200 000 particle filters are being used, of which more than 25 000 reconditioned filters are in the workplace, several large bus fleets are using more than 1000 units, and initial large passenger car series are underway that collect over 99.9% of the solids (soot). This has led California to undertake a retrofitting project for millions of units. Such a project is being discussed in Japan. The use of particle filters will be required in new exhaust regulations in Europe, the United States, and Japan. Initial approaches are starting to include particle filters in the vehicle design and the engine management system. The technical foundation has been laid and has set the stage for emission laws worldwide.

21.6.3.1 Particle Definitions and Particle Properties

Different definitions of air pollution particles can be found in the regulations:

- According to the legally valid definition for street traffic, particle mass is everything that can be filtered and, hence, weighed at 325 K (gravimetric method) independent of the size of the particle and its chemical composition.
- In the workplace, most regulations count the overall mass of elementary carbon EC (soot) that is less than 5 μm; there are strong tendencies to shift this limit toward less than 500 nm.
- In environmental law, fine particles emissions are defined worldwide as the overall mass detected with high volume samplers of $<10\ \mu$m (PM10) and $<2.5\ \mu$m (PM2.5) independent of their chemical composition.

The information provided by these measuring guidelines does not give us a satisfactory physical definition of particles. These measuring procedures are also insufficient for toxicological evaluation since no information is provided on the size distribution of the particles in an aerosol or on their chemical composition and physical phase (solid or liquid).

The particles captured from the aerosol in diluted and cooled exhaust reveal an agglomeration structure under an electron microscope whose basic elements are nearly spherical and are quite dense (approximately 1.8 g/cm^3).[1]

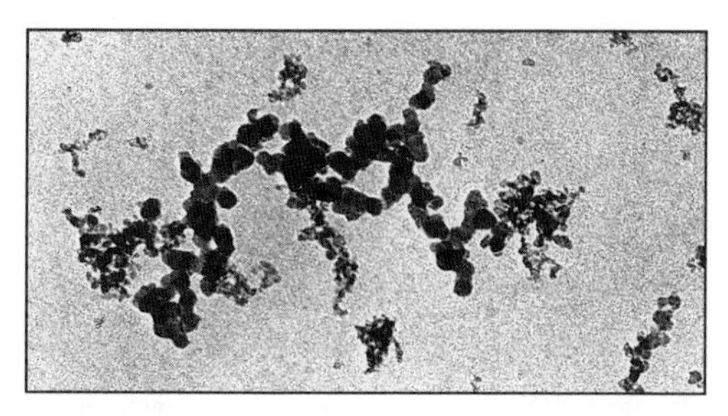

Fig. 21-83 Diesel particle agglomerations (Burtscher).[2]

These particles are created every time hydrocarbons are combusted as well as when the combustion of biomass and coal (Fig. 21-83) takes place.

These surface-rich agglomerations (BET surface: 100–200 m^2/g; see Ref. [1]) serve as condensation nuclei upon cooling; and capture films of hydrocarbons and sulfurous acid products that, in turn, can bind a large amount of water.

Substances that flow through the filter as a gas and can be filtered only upon further cooling as a result of condensation (droplet formation) and/or addition are not considered particles according to a physically correct definition of a hot gas filter. The definition must be limited to substances that have the characteristics of a particle when flowing through the filter, i.e., are basically solid particles such as soot, lubricating oil ash, abraded material, mineral particles that are not deposited in the intake filter of the engine, and sulfur products such as gypsum that can be formed with the calcium of the lubricating oil. Deposited and tightly bonded products at exhaust temperatures such as polycyclic hydrocarbons that are adsorbed during combustion must be included since they also remain bonded up to inhalation.

The size of these particles that come in a wide variety of shapes is difficult to describe. However, it must be defined since the actual geometric shape cannot be detected by any method in situ, i.e., as an aerosol. Comparative sizes are commonly used, such as the aerodynamic diameter for particles greater than 500 nm and the mobility diameter for particles less than 500 nm. The particles are, hence, not evaluated according to their actual geometric size but rather according to their characteristics in comparison to spherical particles with a density of one. Evaluating them according to their inertia (aerodynamic diameter) or their diffusion (mobility diameter) yields information on their "diameter." Since particles from industrial combustion are usually smaller than 500 nm and nearly all mechanisms of deposition deep in the lungs are related to diffusion, the definition of the mobility diameter is preferred in this context.[3,4] Another parameter frequently used for characterizing morphology is the fractal parameter that is usually far below 3 and frequently around 2, which indicates chain and flat structures.

The size distribution (Fig. 21-84) of the particles from engine combustion shows a logarithmically normal distribution around 60–100 nm at the engine exit that scarcely changes to the end of the tailpipe.

The majority of these solid particles lie within the invisible range (<400 nm). Visible smoke is caused by relatively few but very large agglomerations that primarily arise in older engines from combustion adjacent to the wall, not enough excess air, or deposits in the exhaust system.

Soot particles are largely inert and odorless and are insoluble in water and organic solvents.

If large amounts of ash, abraded material, or mineral particles arise, bimodal distributions frequently arise with a distinct second maximum around 20–30 nm.

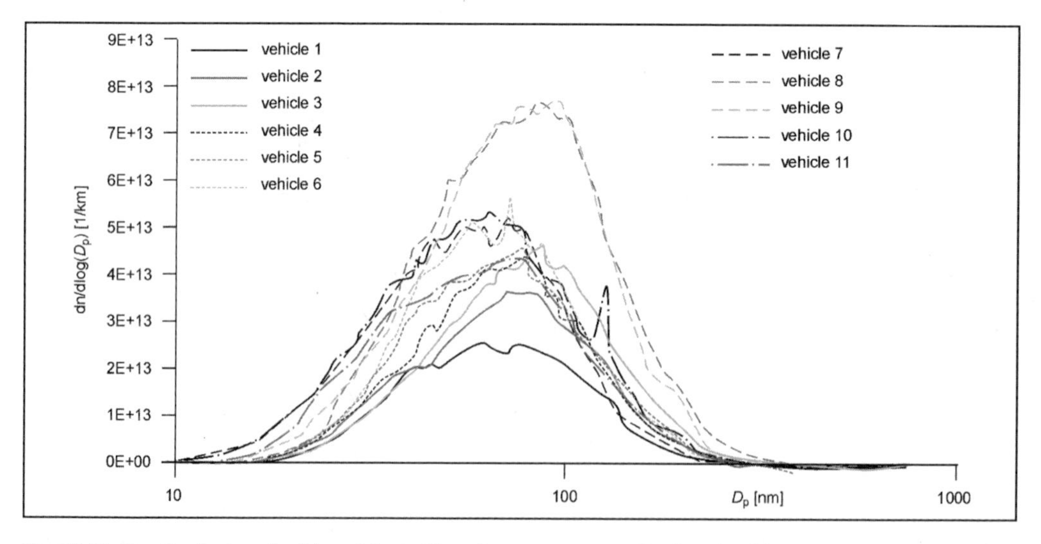

Fig. 21-84 Size distribution of solid particles in 11 modern passenger car diesel engines[5] D_p = mobility diameter (SMPS measuring procedure).[5]

21.6.3.2 Goals of Particle Filtration

The goals of filtration must be oriented around relevance to health and technical feasibility.

Relevant to health are particles that penetrate into the depths of the lungs, dwell there a long time, and cannot be fagocyted (digested by macrophages) or dissolved in bodily fluids.

Soot particles fulfill all these conditions. The maximum size of particles that are deposited in the alveoli of the lung is approximately 10–20 nm depending on the inspiration volume; smaller particles are deposited in the upper respiratory tract and reconveyed back to the throat by very efficient lung purification mechanisms (mucous layer, cilia).

The smaller the particles, the easier they leave the alveoli and pass into the blood vessels and then via the blood and lymph into the entire organism[6] (Fig. 21-85). In addition, these small solid particles transport adsorbed toxic substances (such as carcinogenic polycyclic aromatic hydrocarbons) into the organism. The goal must then be to efficiently collect particles in the range of 10–500 nm—preferably with a filtration rate that increases as the particles grow smaller. The deposited substances remain bonded in the filter matrix under all conditions, and particles and adsorbed substances are not released during the regeneration process.

The use of cutting-edge technologies, the second goal of preventive measures to protect health, is shown in Fig. 21-86.

In testing filters in the performance test of the Swiss Federal Agency for the Environment, Forest and Landscape (Bundesamt fuer Umwelt, Wald und Landschaft) BUWAL,[8] rates of deposition for solid particles were obtained in the critical range of 99%–99.9%—a common result for modern particle filters.

The concentration of the particles in undiluted pure gas downstream from such a filter lies approximately in the same range as today's atmospheric concentrations, and

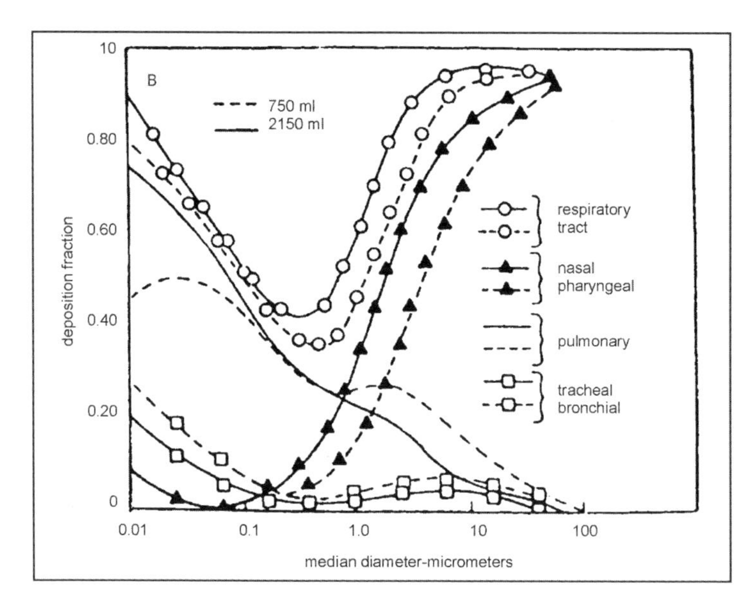

Fig. 21-85 Deposition of fine particles in the nose, bronchi, and alveolus.[3]

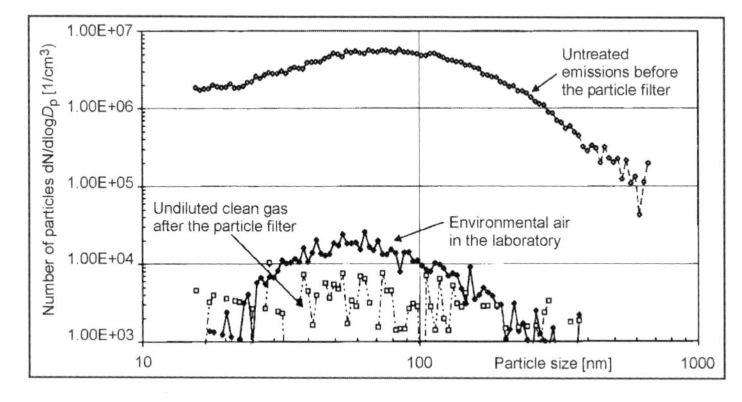

Fig. 21-86 Deposition rate of a ceramic cell filter in a commercial vehicle DI diesel engine[7] after field use of more than 2000 h of operation.

frequently lower. The nature of particle concentrations in environmental air lets us deduce that they are essentially determined by engine emissions into the atmosphere.

21.6.3.3 Requirements for Filter Media and Technical Solutions

The problems that diesel particle filters must overcome are quite challenging[9]:

- Exhaust temperatures up to 750°C, and temperature peaks during regeneration up to 1400°C
- High thermal and thermomechanical stress during quick temperature changes
- Danger of material damage from lubricating oil ash and additives
- Ability to store large amounts of soot and ash
- Low pressure loss that only slightly affects the turbocharger and engine
- Low thermal mass (quick light-off)
- Rates of deposition greater than 99% for particles 10–500 nm
- Insensitive to vehicle vibration especially when installed under the hood
- Insensitive to damage when cleaning inert ash components

In addition to all these requirements, the filter must be economical (for commercial vehicles, less than approximately \$10/kW; for passenger cars, less than approximately \$5/kW), have a small installation volume, and have a service life equivalent to the engine life.

Possible filter media are surface-rich structures made of high-temperature-resistant materials such as monolithic porous ceramic structures (wall flow) in the form of cells (Figs. 21-87 and 21-88) or foams (Fig. 21-89), high-alloy porous metal sintered structures and metal foams

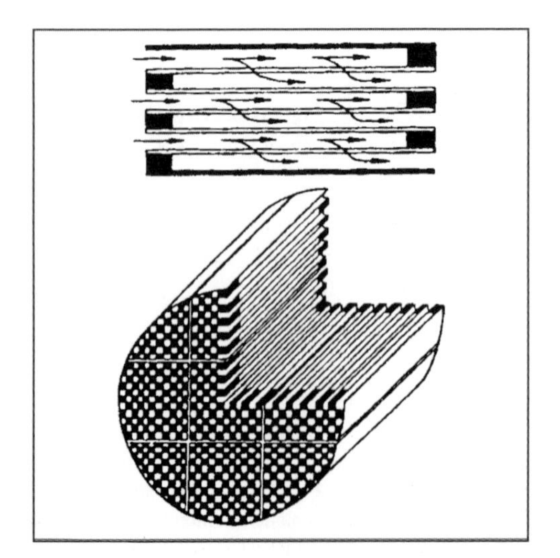

Fig. 21-87 Monolithic ceramic cell filter (NGK, Corning, etc.).[9]

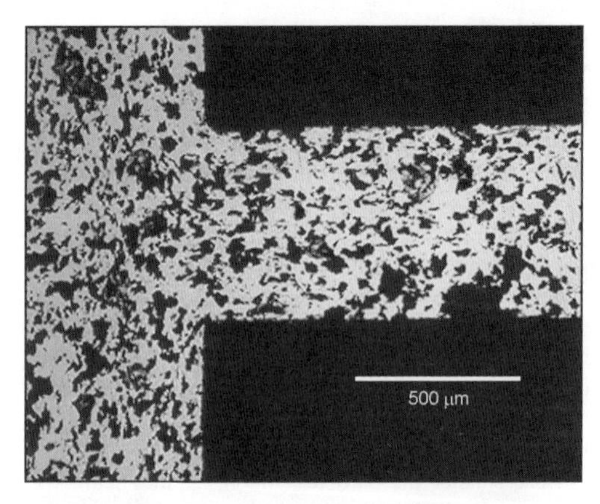

Fig. 21-88 Pore structure of a ceramic filter (Corning).[9]

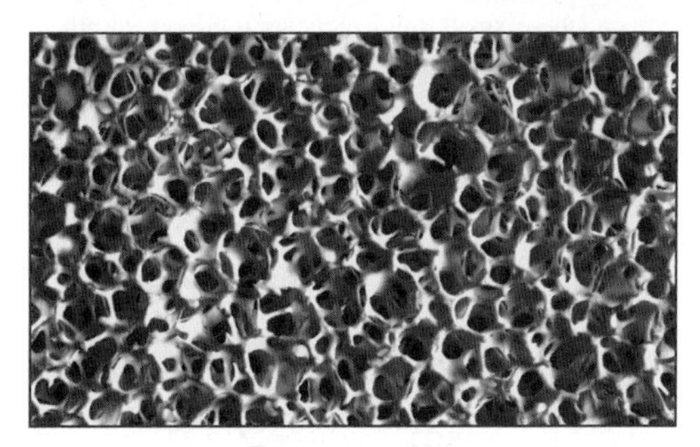

Fig. 21-89 Ceramic foam as the filter medium (Alusuisse).

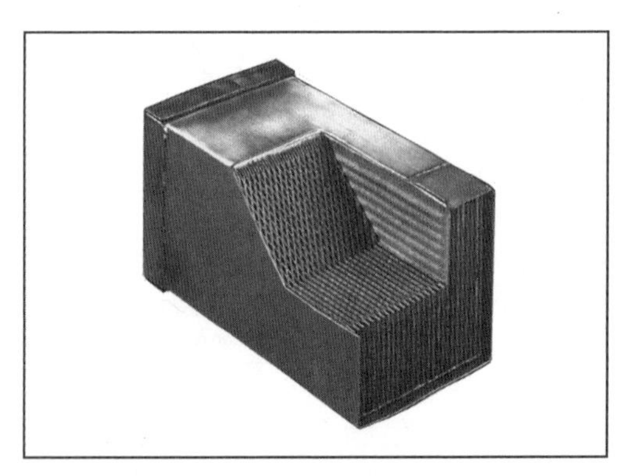

Fig. 21-90 Filter of porous sintered metal plates (SHW; HJS) formed into a cellular structure and welded.

(Fig. 21-90), and fiber structures such as cop or textile weaves (knits, plaits) (Fig. 21-91) and fleeces (Fig. 21-92), using ceramic or metal fibers. The pore size or the fiber diameter required for deposition must be around 10 μm or less.

A few examples are porous sintered ceramic (previously mostly cordierite, but recently also SiC), and alternately sealed cells with a flow through porous walls, and a large filter surface.

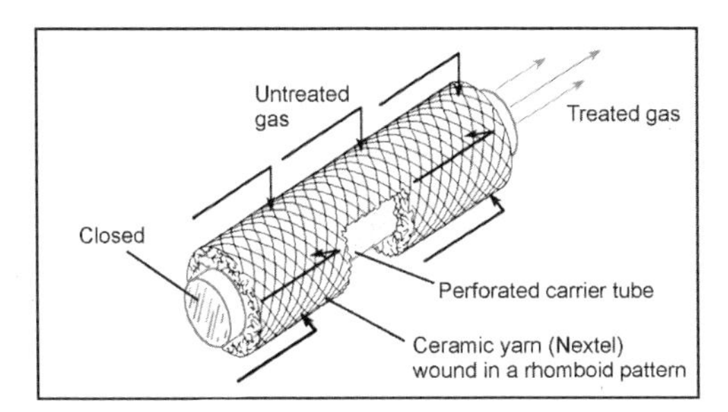

Fig. 21-91 Filter candle (3M, MANN & HUMMEL), in the form of a yarn coil; high-temperature ceramic yarn wound in a rhombic pattern on inner perforated sheet metal.[10]

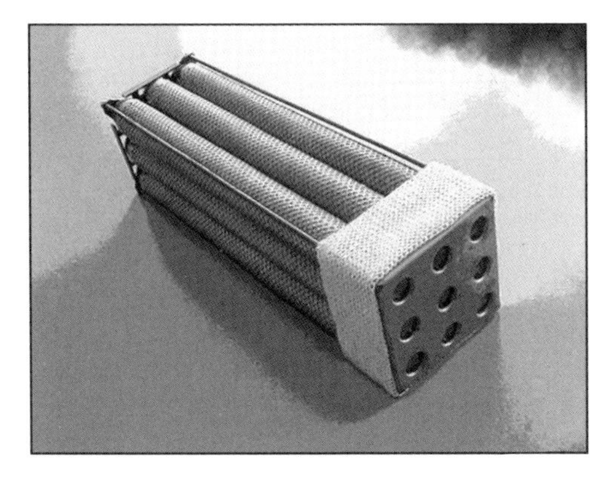

Fig. 21-92 Knit fiber filter (BUCK); a knit, pleated structure of high temperature fiber. Parallel filter candles.[11]

At present, the following filter systems are preferred:

- **Ceramic monolithic cell filter**
  A similar design to a cellular catalytic converter but with alternately sealed cells. This type of filter offers a large surface, a low structural volume (1–3 m^2/l), a low counterpressure, and a high deposition rate at low gas speeds through the walls (a few cm/s). The filters are usually made of cordierite by extrusion (NGK, Corning). Silicon carbide has recently been used (Notox, IBIDEN) as well as other ceramic materials. Intensive development of materials has led to thermo-shock-resistant structures. In particular with the material cordierite, experience has been gathered worldwide for two decades (more than 80 000 filters have been delivered).
- **Metal sintered filters**
  With a general structure similar to ceramic monoliths, SHW developed a filter based on metal materials. The basic element is a thin sintered plate (a few tenths of a millimeter) with a specific pore structure. These filters are relatively heavy in comparison to ceramics, but they are very robust. Their heat conduction is naturally very good. Recently, metal sintered filters have appeared in bag-like structures made of filter plates that weigh less (HJS, PUREM).
- **Fiber spiral vee-form filter**
  Yarns made of high-temperature fibers (Mullit) are wound in a special technique to create rhombic channel structures on a perforated substrate tube. Filter candles of this type have been developed by 3M and Mann & Hummel.
- **Fiber knit filters**
  Ceramic yarns are processed into round knit structures and shaped into deep structures by plating. The available macroscopic fiber surface is typically 200 m^2/l, whereas the microscopic surface of the fiber is 100–200 m^2/g. This filter type was developed by BUCK, and it comes with a catalytic coating and electrical internal heating. The flow preferably passes from the inside to the outside.
- **Fiber braided filter**
  High temperature fibers also come as plaits and can be used for filtration fixed on metal substrate structures. Such systems were developed by Hug and 3M.
- **Filter papers/filter felts/filter fleeces**
  Paper filters that have a similar structure to intake air filters are used only when the exhaust temperature can be kept reliably low (such as when cooling the exhaust of underground engines). In any case, paper can be used up to temperatures of approximately 300°C (DONALDSON, PAAS). These papers are basically used in fiber filters with short fibers that are in a random pattern and whose structure is held by binders. Felts and ceramic fibers can be used for higher temperatures as has been used for a long time in industrial hot gas filtration; fleeces made of resistance-welded metal microfibers are also used (BEKAERT).

In contrast to these surface-rich structures, flow-dynamic, electrostatic, and plasma methods have not become widely accepted. Exhaust scrubbers that were initially used frequently are now scarcely used except in underground coal mining; in any case, they are unsuitable for filtering nanoparticles.

This list is not exhaustive; there are numerous other technical solutions being developed. In addition to factors such as filter quality and pressure loss, designers are addressing size reduction, vehicle-tailored design, incorporation in the overall process, and linkage to other exhaust purification methods.

21.6.3.4 Deposition and Adhesion

In general, there are three areas in a filter with physically different deposition mechanisms as shown in the following figure with fiber deposition as an example (Figs. 21-93 and 21-94).

Larger particles are mostly captured by impaction from inertial force and a bit less from blockage. For the submicron particles (nanoparticles) at issue here, Fig. 21-97 below shows that diffusion is predominant.

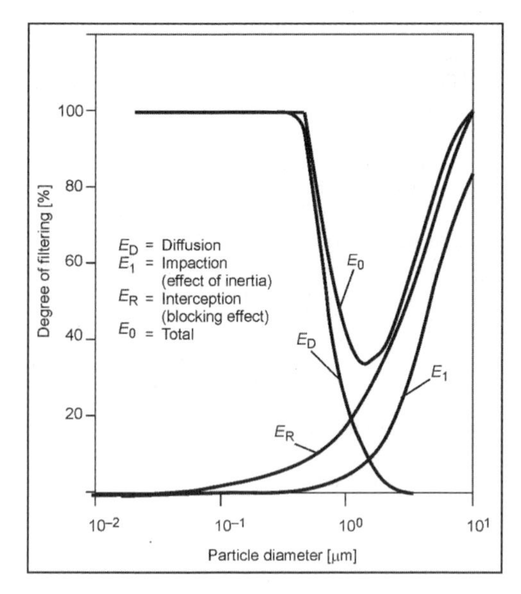

Fig. 21-93 The effects of deposition in a filter medium as a function of particle size (3M).

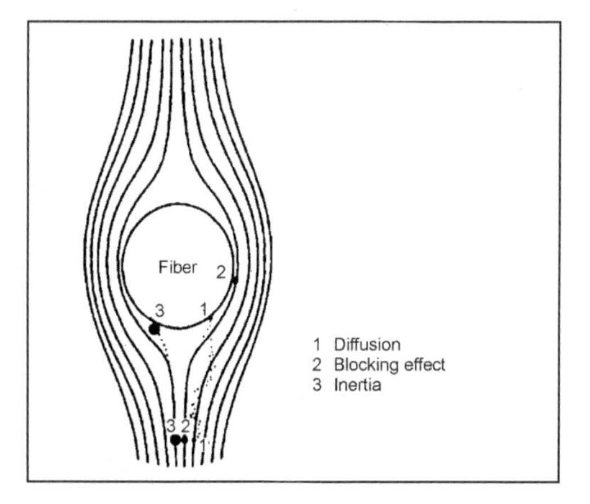

Fig. 21-94 Deposition mechanisms acting on an individual fiber (3M).

If we compare the deposition characteristic in Fig. 21-85, we find a similar deposition minimum around 1 μm where impaction becomes weak and the effect of diffusion starts to increase. Such particles are largely exhaled in contrast to the many larger ones that are deposited in the upper respiratory tract and the much smaller ones that tend to be deposited in the alveoli.

With these small particles, the ratio of drag force to inertial force in the Stokes' range (that does a good job describing the conditions in the filter[2] is large enough for the particles to follow the flow lines around any obstacle, even around very fine microscopic structures such as filter fibers.

$$\frac{\text{Drag force}}{\text{Inertial force}} \sim \frac{d \cdot \mu \cdot v}{d^3 \cdot \rho \cdot v^2} \tag{21.22}$$

d = Particle diameter
v = Speed
μ = Dynamic viscosity
ρ = Density

These small particles can be deposited only by means of diffusion. Diffusion takes time, however, which means that a sufficient filter depth L and low flow speed are required for successful deposition.

The process can be best described with reference to a channel flow (Fig. 21-95) where the channel diameter is 10 μm like the typical pore size of such very fine filters, i.e., 100 times greater than a typical soot particle.

L = Filter depth
B = Channel width (pore size)
v = Flow speed
c_D = Diffusion speed

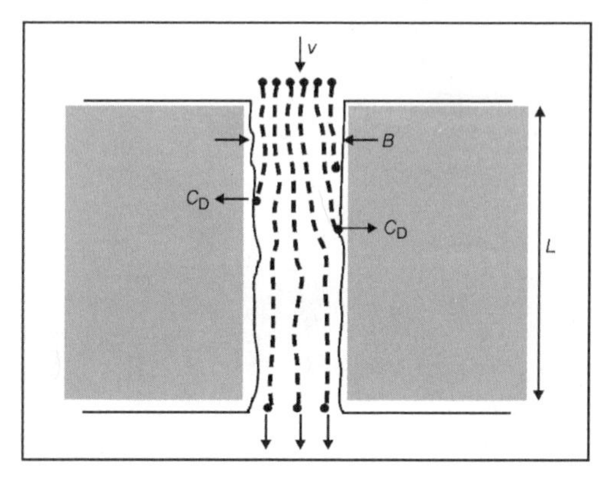

Fig. 21-95 Channel model for diffusion deposition of particles.

For a particle to reach the wall from the middle of a channel before it leaves the channel, the time required to flow through the channel $t_1 = \dfrac{L}{v}$ must be greater than the diffusion time from the middle to the wall $t_2 = \dfrac{B}{2 \cdot c_D}$.

In order to compare different geometries and flow conditions, we need the following condition: $B \cdot v/L =$ const ~ diffusion speed, a function of the particle diameter d and temperature T.

For typical ceramic cell filters, the filter depth (wall thickness) is 0.5 mm, the pore size is 10 μm, and the speed is a few cm/s. Fiber filters whose typical pore size is greater and that work at much faster speeds require a greater flow depth as indicated by this condition.

Since small particles have higher diffusion speeds or higher mobility b, we can expect that smaller particles are filtered better in such structures.

The terms "diffusion speed" and "mobility" are also equivalent in the sense of an Einsteinian relation:

$$D = k \cdot T \cdot b \tag{21.23}$$

D = Diffusion coefficient
T = Absolute temperature
b = Mobility
k = Boltzmann constant

From the integration of Brownian motion in a field with a varying particle density (strong decrease at the wall toward zero), we get the diffusion speed for particles of different sizes at room temperature as shown in Fig. 21-96. As a comparison, the theoretical sink speed is also given.

Particle size	Diffusion speed μm/s	Sink speed mm/h
10 nm	260	0.2
100 nm	30	3
1000 nm	5.9	126

Fig. 21-96 Numerical example according to Hinds.[3]

Since the time required to flow through the filter wall of a ceramic cell filter is usually around 0.01 s, the diffusion path of a 100 nm particle is only 0.3 μm; only particles close to the wall are deposited in such a channel, even at higher temperatures, although the diffusion speed increases with the temperature.

Filter structures, therefore, have to be much better than the channel model to ensure the deposition of fine particles. There are two ways to overcome this: The porous walls of the wall flow filter can be described as a flow model, and the fiber depth filter can be described as a circulation model (Figs. 21-97 and 21-98).[12]

In porous walls, the flow occurs in channels that run from pore to pore. There are numerous diversions, dwell times in pore caverns, and new walls when channels branch. Diffusion is substantially improved, and impaction is also enhanced. Such filters are distinguished by the superior deposition of coarser particles, but because of the channel-like nature of the passages, the rate of deposition tends to worsen for very small particles.

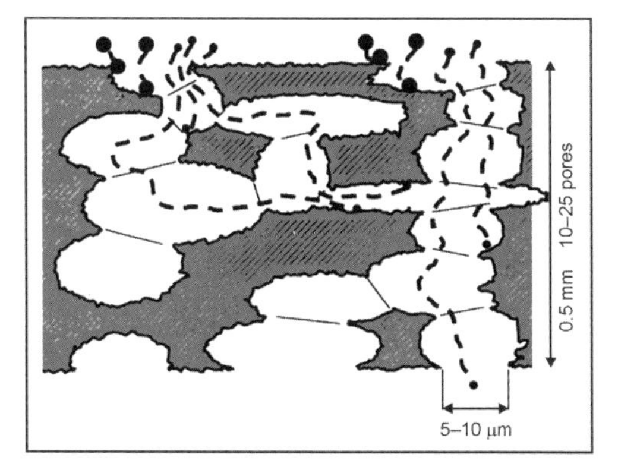

Fig. 21-97 Porous wall.

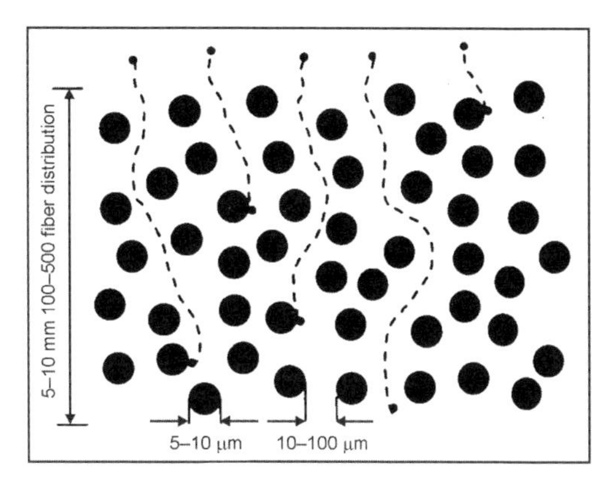

Fig. 21-98 Fiber bundle.

In the circulation model, new boundary layers are always being formed; i.e., the flow channel is continuously being divided so that particles frequently pass directly next to walls where they can be deposited by diffusion. With such pure depth filters, one can expect an increasing rate of deposition of small particles.

Definition of the Technical Rate of Deposition

The rate of deposition is defined by the overall mass or the number of particles. In the second case, the rate of deposition can be represented as a function of the particle size, the so-called dividing characteristic.

Rate of deposition according to particle mass PMD:

$$\mathrm{PMFR} = \frac{\mathrm{PM_{Before}} - \mathrm{PM_{After}}}{\mathrm{PM_{Before}}} \tag{21.24}$$

Rate of deposition according to particle count filtration rate (PCFR):

$$\mathrm{PCFR} = \frac{\mathrm{PZ_{Before}} - \mathrm{PZ_{After}}}{\mathrm{PZ_{Before}}} \tag{21.25}$$

$$\text{Penetration} = 1 - \text{rate of deposition} \tag{21.26}$$

When the sampling and measurement are designed so that only the solid mass is detected or only the solid particles are counted, both these definitions usually produce very similar values. This does not necessarily have to be the case. If a spectrum shifts toward very fine particles, the situation is better described by counting than by mass. Furthermore, in a particle size spectrum that is dominated by very fine particles, measuring by counting is a much more sensitive method. For health reasons, the number and surface are to be included in the measurement; the rate of deposition should be correspondingly defined.

Retention of Particles

In addition to deposition, the second component of filtration is reliable retention of the particles in the filter matrix, i.e., adhesion. If we discard the effects from the shape that can largely be ignored, the adhesion of a particle to a surface in hot gas filtration under dry conditions is determined according to Van der Waals by the following equation:

$$p = \frac{A}{6 \cdot \pi \cdot z^3} \tag{21.27}$$

p = Adhesive pressure
A = Constant
z = Contact distance

Small particles whose center of mass is very close to the surface adhere much better than larger particles. Furthermore, since the exposure to the flow of small particles that lie close to the boundary layer is slight, there is not much danger that submicron particles, especially once deposited on a surface, will be entrained by flow.

However, other particles can collect on deposited particles so that large agglomerations can form in a filter as illustrated in Fig. 21-99.

Fig. 21-99 Diesel soot, deposited as very fine particles on a ceramic fiber 10 μm in diameter and a large agglomeration that was formed in the filter.[11]

Such agglomerations offer a large area of attack for the flow. They can then be released and leave the filter, which is typical behavior of large-pore fiber depth filters that become overloaded. These are referred to as agglomerators.

Since very fine particles adhere well to surfaces, only filter cakes and large agglomerations are removed when trying to clean filters by blowing air through them. The very fine particles can be removed from the filter only by washing, which loosens the Van der Waals bonds.

21.6.3.5 Regeneration and Periodic Cleaning

The high deposition rate of solid particles causes them to quickly fill filters. Filters become filled with flammable components (soot) within a few hours, and they fill with inert solid particles (ash) after a few thousand hours. These figures can vary widely depending on the untreated emissions of the engine, mode of operation, lubricating oil consumption, fuel and lubricating oil properties, and filter characteristics. In every case, however, the flammable residue consisting of EC and organic hydrocarbon (OC) must be removed relatively frequently by combustion. This process is termed regeneration. For filters to be free of residue, regeneration should occur in such a manner that only CO_2 and water arise. This ideal is frequently not attained. The reasons, in addition to the CO/CO_2 equilibrium, are effects that occur during the heating phase in which the substances can leave the filter by evaporation, and low oxygen during heating that can lead to coking (pyrolysis) and, hence, nearly unregenerable residues.

The complex process of soot combustion that is not influenced by thermodynamic conditions but rather chiefly kinetic conditions can be described according to Ref. [13] by a reaction kinetic model as follows:

$$\frac{dM}{dt} = k_0 \cdot M^m \cdot p_{O_2}^n e^{\frac{-E}{RT}} \tag{21.28}$$

M = Relative soot mass
p_{O2} = Partial pressure of the oxygen
R = Gas constant
T = Absolute temperature
E = Activation energy

This relationship indicates the primary importance of temperature and sufficient oxygen. The activation energy E in filters without regeneration assistance is around 140 kJ/mole. This value can be reduced to 80–90 kJ/mole by catalytic measures. For soot to completely burn, temperatures above 600°C are required with an oxygen content above 7%; i.e., the conditions necessary for heating the filter system cannot be maintained in many vehicle conditions over the long term.

The combustion conditions can vary over a relatively wide range depending on the character of the soot and deposition history, on the absorption of hydrocarbons from the lubricating oil and fuel, and on newly formed substances.

To make matters more difficult, the emission of hydrocarbons and CO should not be too high during regeneration, and the stress from the released heat from soot combustion should be controlled as best as possible to not overtax sensitive structures such as ceramic monolithic cell filters.

Numerous regeneration methods have been developed to solve this problem that can be generally categorized as passive and active methods:

- "Active" when the regeneration is triggered by a controlled/regulated supply of energy
- "Passive" when catalytic measures lower the activation energy enough for the reaction to occur at the operating temperature.

In special cases (small engines, brief use, and operation in buildings) exchangeable filters are possible that are externally regenerated or disposed of after becoming filled.

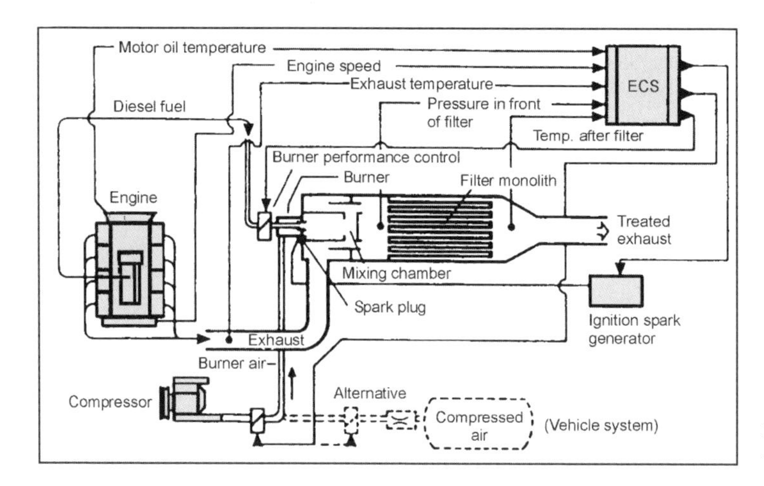

Fig. 21-100 Particle filter system with full-flow burner (Deutz).

Active systems include

- **Full flow and partial flow burners** (Figs. 21-100 and 21-101): There are numerous subvariants. In addition to the actual full flow burners that are regenerated under all operating conditions—a very difficult technical task—there are twin systems that are alternatingly regenerated under predetermined conditions, burners that are accessed only during idling, and burners that heat the filter element from the clean air side (PUREM).

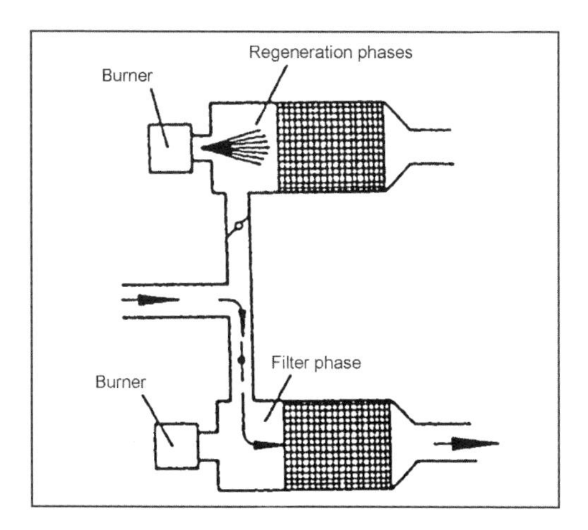

Fig. 21-101 Dual filter system with valve control (Eberspächer).

- **Electrical heating:** Numerous types of electrical systems have been developed, such as overall heating of the gas stream and, hence, the filter, or specific heating of the filter matrix using ohmic heat in electrically conductive materials (SiC), and sequential heating systems in which one filter candle after the other[11] or one filter channel after the other[14] is heated to the regeneration temperature. If a sizable soot cake has formed, it may be sufficient to merely ignite this coating; the fire then burns through the entire soot cake from the released heat.

  The main problem with electrical heating is the limited availability of the electrical energy in vehicles. For this reason, electrical methods have become popular only when the regeneration occurs at a standstill, and electrical energy is supplied from the outside (mains energy) (UNIKAT, HUSS, ERNST, JMC etc.). The process can be carried out slowly to spare the filter material. There is extensive experience with this method, especially with off-road vehicles and underground applications.

- **Raising regeneration energy from engine combustion:** Active systems include those in which the required energy is taken from the engine by special strategies during the regeneration periods. The normal measures are late injection, secondary injection, throttling, and exhaust recirculation. These strategies can raise the exhaust temperature by approximately 200°C, which in many cases, especially in combination with catalytic measures, are sufficient to regenerate the filter. All of these approaches raise fuel consumption, which, however, is not that problematic since the regeneration phases are short in relation to the operating time between regenerations, typically 1%–3%. Measures of this type are, however, usually possible only with original equipment, preferably in engines with electronic fuel injection systems.

Passive methods for regeneration assistance are just as varied.

- **Regeneration additives:** Those substances are added to the fuel at low concentrations (10–20 ppm) that can lower the soot burning temperature to approximately 300°C using a catalytic effect. Examples of such substances are cerium, iron, copper, and strontium. The end products of these additives surface in the exhaust as extremely small ash particles (around 20 nm)[15];

they may, therefore, be used only with corresponding particle filters.[8] The advantage is that the additives substantially lower raw soot emissions and, hence, relieve the filter.

- **Catalytic coating** (Fig. 21-102): The soot ignition temperature can be similarly lowered by coating the filter with transition metals.[13] A prerequisite is a very large specific surface of the filter material (greater than 100 m^2/g) and, hence, a very fine distribution of the active centers for the reaction.

Whereas massive soot deposits can be burned when additives are used (with the danger of generating high temperature peaks), coated filters avoid the formation of thick soot cakes since this can substantially restrict the effectiveness of the catalysts applied to the wall.

There are different functional variants where the catalyst is coated; on the untreated gas side that primarily enhances soot burning, and on the clean gas side where precious metals can be used to reoxidize CO and HC.

In CRT (continuously regenerating trap) systems[16] (Fig. 21-103), an oxidation catalyst coated with precious metal upstream from the particle filter generates more NO_2 from NO in the engine exhaust. NO_2 is not stable at these temperatures, however. In the downstream particle filter, the reverse process occurs; the released oxygen radical oxidizes the carbon at exhaust temperatures starting at approximately 230°C.

For this to work, however, sulfur-free fuel must be used to prevent the sulfating reaction ($SO_2 \rightarrow SO_3$) that is the preferred reaction to reduce NO_2 conversion.

There are many combinations of this procedure.[17,18] For example, there is the method developed by Peugeot (Fig. 21-104) for use in passenger cars.

In this system, cerium oxide is used as a fuel additive to lower the soot ignition temperature by approximately 200°C; however, this is grossly insufficient for passenger cars. In addition, the exhaust temperature is raised approximately 100°C by secondary injection, and the remaining uncombusted fuel is converted in a precatalytic converter, which further increases the temperature. Exhaust recirculation that lowers the combustion temperature is turned off in the regeneration phase, and the vehicle electrical system is tapped by electrical consumers. All of these elements are required to trigger regeneration even under adverse conditions when the counterpressure limit is reached.

The conditions under which the soot can be burned with sufficient oxygen are listed in Fig. 21-105.

Fig. 21-102 Schematic representation of a catalytically coated soot filter (Engelhard).

Fig. 21-103 CRT filter system (Johnson Matthey).

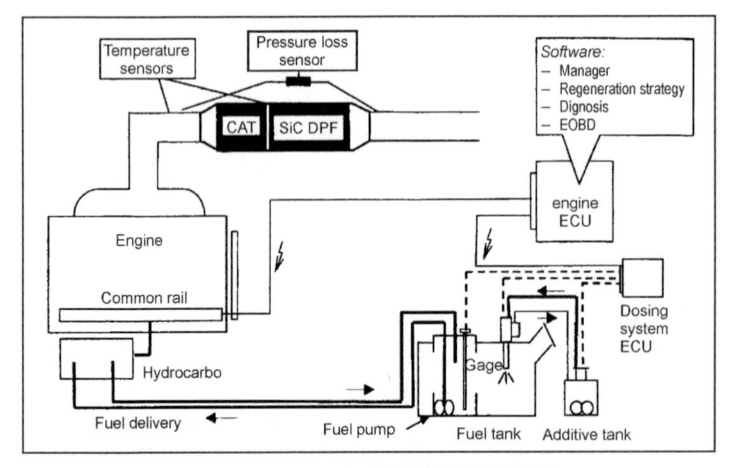

Fig. 21-104 Diagram of the Peugeot soot filter system for passenger cars[19].

AU: In Fig. 21-104 should "Hydrocarbo" be "Hydrocarbon"?

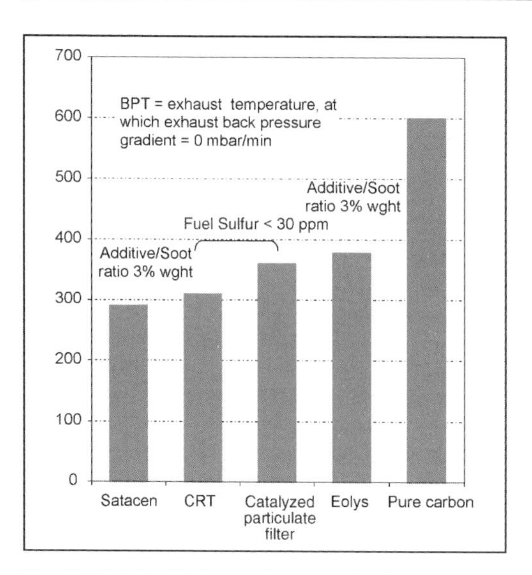

Fig. 21-105 Equilibrium temperature in filter regeneration with different catalytic aids.[20]

The filter must be purified at longer intervals from inert substances that primarily arise from lubricating oil additives. These substances include metal oxides from wear-preventing additives such as zinc oxide, and calcium as the main component of anticorrosion additives that can form gypsum with the sulfur from the fuel or lubricating oil. The inert substances become deposited in the particle filter and plug its pores. The filter is purified of these inert substances approximately every 2000 h (100 000 km). The filter must be removed and carefully washed. The ash substances need to be disposed of in an environmentally friendly manner. In addition to clogging, the disadvantageous effects of common lubricating oils on filters include filter damage, e.g., from the formation of glass phases. This has led to demands for new lubricating oils that have a low ash content, a low sulfur content, and less phosphorous and alkaline earths for the sake of filters. The goal is to significantly reduce the emission of lubricating oil ash to a maximum of 0.5 mg/kWh.[21]

A successful filter system and regeneration method largely depends on the knowledge of the operating behavior of an engine, i.e., the collective load that arises in typical use. The most important parameter in addition to the oxygen content is the temperature. Figures 21-106 and 21-107 illustrate the problem.

Figure 21-109 (below) in which the dwell times are cumulative within certain temperature windows illustrates an apparently sufficient level for several regeneration processes. In Fig. 21-107, we note that these temperature episodes are very short; only a filter system with a short light-off time can use these short phases of sufficient temperature to start regeneration.

21.6.3.6 Regeneration Emissions and Secondary Emissions

The loaded particle filter must be understood as a chemical reactor that, as an adsorber, has very large surfaces (50–100 000 m^2) and is operated over wide temperature ranges with pronounced adsorption or desorption phases that may include the formation of new substances. The wide variety of emissions from engine combustion allows chemical reactions that can lead to the emission of critical concentrations of toxic substances. By coating or incorporating catalytically active substances, the formation of such substances can be accelerated, and their concentration can increase. There are additional emission risks from the store and release behavior of such systems, as well as reactions that can arise when the soot is burning. Generally speaking, there are three groups of processes:

- During regeneration, emission peaks of HC and CO can arise: HC when adsorbed hydrocarbons evaporate in the heating phase and CO when regeneration is very quick or occurs with low oxygen.
- "Store and release" phenomena are always possible in adsorbing systems with fluctuating temperatures. When commercially available fuels are used, we note pronounced sulfate cycles, for example.

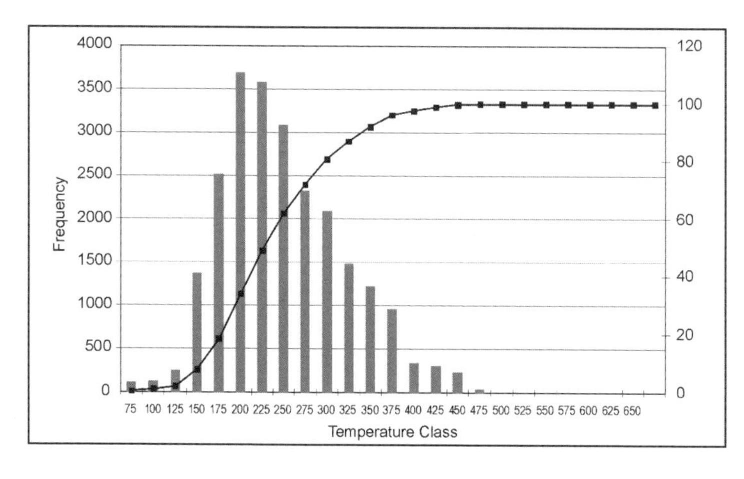

Fig. 21-106 Cumulative distribution of exhaust temperature in a tour bus; 268 kW.[22]

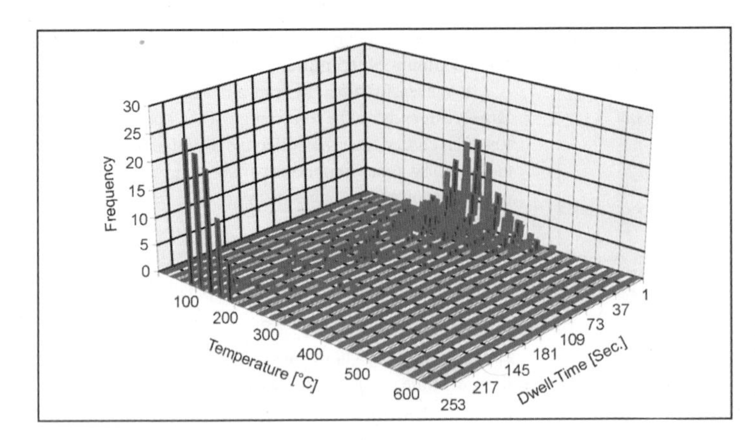

Fig. 21-107 Distribution of exhaust temperature according to periodic episodes in a tour bus; 268 kW.[22]

- Released substances that did not previously exist in the system are understood as secondary emissions (i.e., that are formed in the particle filter). Such reactions arise primarily with catalytic support, and the deposition of lubricating oil ash can produce noticeable catalytic effects. When precious metal coatings are used, a strong sulfate reaction is observed ($SO_2 \rightarrow SO_3$) as well as a substantial shift in the NO/NO_2 equilibrium.

In conjunction with copper additives, a massive increase in the emission of dioxins and furans of several orders of magnitude has been observed.[23] A shift in the polycyclic aromatic hydrocarbon spectrum is also conceivable, and nitro-polycyclic aromatic hydrocarbons can be formed; aldehydes, etc. It is, therefore, necessary to check secondary emissions when type testing catalytically coated or catalysis-supporting systems.

Normally, precious metals are required for the particle filter system to change the legally restricted gaseous pollutants CO, HC, and NO_x that are formed in the engine; for example, we see a reduction of CO and HC by approximately 90–95% with the CRT method when precious-metal-containing additives are used.

A generally valuable feature of particle filter systems is that polycyclic aromatic hydrocarbons, although volatile, are usually reduced equivalent to the rate of deposition of solid particles. This can be explained by the fact that only polycyclic aromatic hydrocarbons are adsorbed during the soot formation phase on the surface-rich structure, they remain fixed to the structure, and they are converted into the end products CO_2 and H_2O during regeneration. This process has also been demonstrated by the "time-of-flight" analytic method.[15]

21.6.3.7 Pressure Loss

Because of the unavoidable pressure loss when the exhaust flows through these fine-pore structures, particle filters always negatively influence the engine (supercharged engines more than naturally aspirated engines): Expulsion work increases as exhaust retention rises, combustion is influenced as the counterpressure increases, and component temperatures rise. The pressure loss of the particle filters increases as the filter becomes filled with soot and ash deposits. Interestingly, surface filters become progressively clogged from the formation of filter cake until completely blocked as the rate of deposition increases. With deep-bed filters, this process is revered according to the fiber growth model;[24] i.e., a certain threshold load is not exceeded,[11] and the rate of deposition simultaneously falls.

The pressure loss in the fine-pore filter element and soot cake follows a laminar law since the Reynolds numbers in reference to the pore size are less than 1.

The pressure loss is normally indicated as follows:

For fiber deep-bed filters[24]

$$\Delta p = K_1 \cdot L \cdot \left(\frac{1-\varepsilon}{\varepsilon}\right) v \cdot \mu \cdot \frac{1}{d^2} \tag{21.29}$$

L = Filter depth
ε = Porosity = Cavity volume $= \dfrac{\text{Pore volume}}{\text{Filter volume}}$
v = Approaching flow speed
μ = Dynamic viscosity
d = Fiber diameter
ρ = Density

For pore structures according to Ergug and Orning[25]

$$\Delta p = K_2 \cdot L \cdot \frac{(1-\varepsilon)^2}{\varepsilon^3} \cdot \mu \cdot v \cdot \left(\frac{O_p}{V_p}\right)^2 \tag{21.30}$$

O_p = Fiber surface
V_p = Pore volume

In addition, there is a significant turbulent component of flows in housing and filter feed channels that are proportional to $\rho\, v^2$.

The pressure loss of new filters at a nominal engine load is usually 20–40 mbar, i.e., similar to mufflers in commercial vehicles that are usually replaced by a filter. The conventionally accepted maximum permissible pressure loss is 200 mbar (in reference to the nominal speed and load).

This pressure loss has a negative effect on fuel consumption and performance arising from the expulsion work, where a proportional counterpressure of up to 200–300 mbar can be assumed. For an engine not subject to a load, the following holds true:

$$\frac{\Delta b_e}{b_e} = \frac{\Delta p}{p_e + p_r} \tag{21.31}$$

b_e = Specific fuel consumption
Δp = Filter pressure loss
p_e = Effective average pressure
p_r = Friction pressure

In a commercial vehicle operated under a relatively high load, the pressure loss accordingly reduces fuel consumption by approximately 2%. In vehicles with lower load factors such as passenger cars, the influence is greater, around 5%. If the pressure loss rises above this, combustion and supercharging are negatively influenced starting at approximately 400 mbar so that stronger, nonlinear effects arise as shown in Fig. 21-108 in the simulation of a modern DI commercial vehicle engine under a large load.

In this graph, the effects are actually too high since only the difference from the muffler should be indicated. Mufflers in commercial vehicles are designed with approximately 60 mbar pressure loss; in passenger cars, one frequently finds values above 200 mbar.

21.6.3.8 Installation Area and System Integration

The installation area required for a particle filter system is approximately 4 to 8 times that of the engine charge volume. It can be problematic to incorporate in the exhaust system such a component whose possible shapes are limited, especially when retrofitting.

In any case, a few manufacturers have been able to adapt their system designs to offer replacement filters in the dimensions of the muffler, even for CRT systems that contain a catalytic converter plus a filter element.

The problem is easier to solve with original equipment.

There are many interesting variations of structural and functional system integration:

- Optimum linkage of the filtration + catalysis + noise suppression.
- Linking the functions of filtration + denoxing.
- Placing the filter on the high-pressure side before the turbocharger to substantially enhance regeneration and reduce the effect of the pressure loss on the engine in relation to the turbocharger expansion gradient.[27]
- When a particle filter is used, there is no longer a need for a trade-off between NO_x and particle mass (PM) when harmonizing the engine's features. The engine can, therefore, be designed for lower NO_x emissions, taking into account slightly higher particle emissions.
- The exhaust after the filter is so particle-free that it can even be fed to the intake before the turbocharger. There is no need to fear engine wear and turbocharger compressor soiling from exhaust recirculation.
- The engine management system can briefly increase the exhaust temperature enough to regenerate the filter. The increased particle formation is not externally visible.
- Another facet of system integration is the use of fuels and lubricating oils that are tailored to operate with particle filters. The filtration of the intake air is improved to such a degree that the very fine-pore particle filters are not clogged by inducted mineral dust.

21.6.3.9 Damage Mechanisms, Experience

Monolithic ceramic cell filters, especially those made of cordierite, are sensitive components. This porous material has a flexural strength of only approximately 1 MPa and

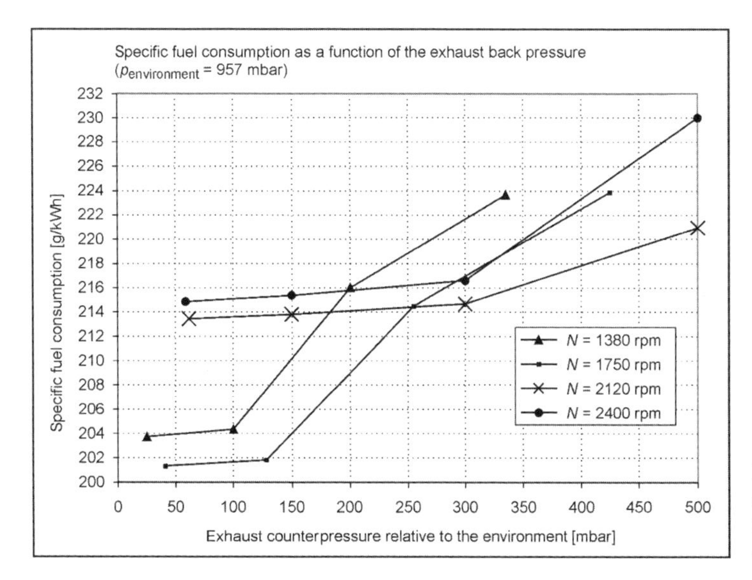

Fig. 21-108 Modeling the increase in fuel consumption at four full-load speeds as a function of increased exhaust counterpressure.[26]

very low heat conductivity. It is, therefore, sensitive to the thermomechanical stress that can arise in regeneration. Since the threshold temperature is approximately 1400°C, many filters have been damaged from uncontrolled regeneration.

Several paths have been pursued to overcome this problem:

- Development of harder and more temperature-resistant materials such as porous SiC with a flexural strength of 40 MPa
- Development of structures with a lower crack propagation tendency
- Optimizing the geometric design of the filter to minimize heat stress
- Measures to control regeneration to limit the temperature gradient and peak temperature

We now have experience with these approaches that have been incorporated into the exhaust systems. More than 10 000 filters have been used in buses with damage rates below 1%, and more than 20 000 h operating time in production machines.

Not to be underestimated is the load on fragile ceramic components from vibrations that can arise especially with underhood installation. The filter elements must be isolated with a ceramic mat because of the large differences in expansion and must be installed under initial stress in a metal housing. If the initial stress relaxes, damage is unavoidable. Damage of this type is not much of a problem with other filter structures, especially fiber filters or filters made of metal components such as sintered filters and metal fleece filters.

Another class of damage that chiefly arises with monolithic ceramics can be triggered by lubricating oil ash or additive ash: The end product of these substances is metal oxides such as the ceramic itself. There are many opportunities for these substances to form phases with the ceramic, and this usually causes weakening. Damage of this type is also possible with fibers, however, the fiber structures are less sensitive than a monolithic structure because of their high elasticity and redundancy.

When additives are used, filter damage causes these substances to be released into the atmosphere. This must be avoided; hence, filter damage must be recognized by the OBD, and the dosing of the additive must be immediately stopped.

For a long time, people feared that fuel additives would negatively influence engine combustion and wear. When added, these substances are metalorganic in nature and, hence, mix on a molecular basis. No negative effects in terms of wear have ever been noted. To the contrary, it is noted that the fewer deposits in the ring area have a positive effect on wear. In individual cases, primarily when copper is used, increased deposits have been observed on the injection nozzle that negatively influences the spray pattern. On the positive side, the effect of additives starts during combustion, which is partially explained by the catalytic coating of the combustion chamber surfaces; a stronger factor is probably the continuous supply with the injected fuel. In one case (cerium + Pt additive), a clear improvement of the fuel consumption was observed that indicates a change in the rate of heat release, i.e., accelerated combustion from the catalytic effect.[28]

21.6.3.10 Quality Criteria

In addition to economic criteria such as investment costs, infrastructure costs, and service, the following criteria must be included in the evaluation of a filter system:

- Rate of deposition, e.g., number-oriented PCFR, or penetration $P = 1 - \text{PCFR}$
- Overall relevant size range of 10–500 nm
- Pressure loss Δp, better in relation to average effective pressure of the engine $\Delta p/p_{me}$ for the load that best characterizes the application
- Volume throughput in comparison to size: $V/B =$ spatial velocity S $(1/s)$
- Thermal response time t_1, i.e., the time that passes after a sufficient rise in the exhaust temperature until the filter starts to regenerate and the pressure loss decreases
- Storage time for inert material until the filter must be cleaned: t_2

These quality parameters can be summarized in a multicomponent evaluation as a single filter parameter:

$$1/5 \cdot \left(\frac{P(-)}{0.01} + \frac{\Delta p_2/p_{me}}{0.02} + \frac{10}{S(1/s)} + \frac{t_1(s)}{100} + \frac{t_2(h)}{2000} \right) < 1$$

When this value is much greater than one, then at least one or more of the important parameters lie outside of the values that are now attainable.

21.6.3.11 Performance Test, Type Test, OBD, Field Control

- **Performance test**
  One can generally assume that a filter is capable of collecting a particular percentage of solid particles of a particular size independent of the precise composition of these particles.[29] The throughput, temperature, and load must be included as boundary conditions since impaction rises as the throughput increases; however, diffusion deposition decreases, the load has a varying effect (positive for a surface filter, negative for deep-bed filter) on the deposition rate, and the temperature can influence the diffusion constant and adhesion conditions.

  It is then sufficient to measure a filter at a maximum spatial velocity and maximum temperature when new to get a good idea of its rate of deposition. The filter is characterized by a single value for the rate of deposition per particle size class (separation curve); by multiplying this rate of deposition with solid particles in untreated emissions, we can determine the solid particles in the clean emissions of any engine.

  This test could, in principle, be applied to a filter test machine with a test aerosol as in many industrial filter applications (DIN 24184, 24185). It is preferable to carry out measurements on a representative diesel engine primarily because of the issue of adhesion conditions and agglomeration formation in the filter.

Transient processes such as the free acceleration of an engine from low idling to high idling, to cite an extreme example, do not produce any new information as expected since the very low speeds in the filter medium do not yield any flow-dynamic effects.

A filter can then be characterized by a simple test in reference to its basic function, namely, the retention of solid particles.

If, in addition to actual filtration, the filter must fulfill other functions such as catalysis, it must undergo additional tests.

- **Type test**
  The type test is done in driving cycles under transient conditions. The emissions factor (g/km or g/kWh) of the corresponding vehicle is measured as opposed to the rate of deposition. For regenerating systems, the type test must also be run during regeneration, and this result weighted over time must be included in the overall result.
- **On board control (OBD)**
  The load of the filter and pressure loss are measured continually as essential input for system control. These measurements are used further to monitor two critical pressure levels: An upper level that indicates the need to clean the filter from inert dust and a bottom level that signals damage to the filter.
- **Periodic field control**
  For engines up to EURO 2, control can be provided by the free acceleration method measuring opacity to yield reliable information on the proper functioning of the system. For engines with much lower untreated emissions and carefully controlled smoke emission during acceleration, the sensitivity of the simple opacimetric methods is no longer sufficient; however, there are sufficiently sensitive measuring devices available.[30]

21.6.3.12 Catalytic Soot Filter

One method to improve soot regeneration is to use catalytic soot filters that are an alternative for diesel passenger cars in addition to the systems discussed in Section 21.6.1 based on fuel additives. The catalyst is applied to the particle filter. A schematic representation of a wall-flow particle filter is shown in Fig. 21-102.

The catalyst is applied to the particle filter by a coating method similar to that used for diesel oxidation catalytic converters or three-way catalytic converters, and it permits catalytic reactions at the boundary layer between the solid soot particles and the catalytic coating. Differences in coating methods result from the alternately sealed channels that prevent linear flow as it occurs in uniflow structures and, hence, increase the manufacturing effort. It is possible to coat the inlet and outlet channels with different catalytic materials to obtain different functions.

Regeneration occurs from the oxidation of soot. The available oxidants are the residual oxygen from the engine exhaust as well as the oxygen bonded to the nitrogen as NO_2 that is in the engine exhaust as the pollutant NO_x. Figure 21-109 lists the most important chemical reactions that are relevant to soot burning.

The NO_2 available in small concentrations in the engine exhaust can be generated by a diesel oxidation catalytic converter before the filter as well as the catalytic coating in the soot filter. Another technical variation does not use an upstream oxidation catalytic converter. In this case, the reduction of the carbon monoxide and hydrocarbons required to satisfy exhaust laws occurs by means of the catalyst in the soot filter. The catalyst in the soot filter can also oxidize several times the NO that forms when the soot burns, which is also termed NO_2 turnover. The schematic representation in Fig. 21-110 illustrates such a cyclic process.

This multiple use of the NO can clearly improve the soot burning given the right system design. Another important function of the catalytic soot filter is the oxidation of the CO that rises in addition to the NO from the

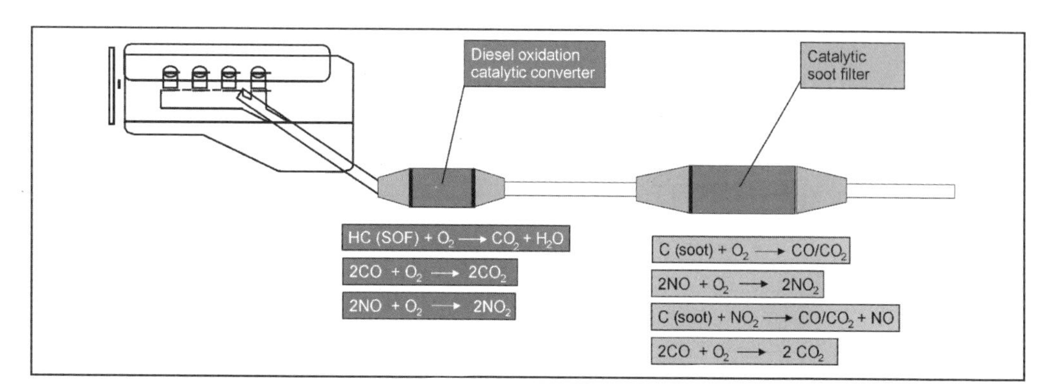

Fig. 21-109 Chemical reactions during soot burning in a catalytic soot filter (Engelhard).

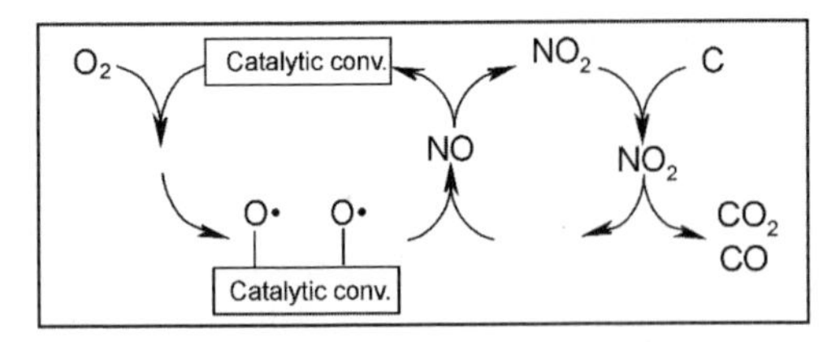

Fig. 21-110 Cyclical process of NO oxidation in soot burning (Engelhard).

incomplete burning of the soot. If the CO is not converted, it can lead to substantial pollutant emissions.

The relevant quantities in catalytic soot regeneration are the exhaust temperature and the concentration of the O_2 and NO_2. However, the soot regeneration can also be influenced by other parameters:

- Exhaust temperature
- Oxygen content of the exhaust (O_2, NO_2)
- Flammable residual components of the exhaust
- Exhaust mass flow
- Particle composition such as the mass of the deposited HCs
- Particle characteristic (such as the ability to form active O_2 centers)

Figure 21-111 illustrates the influence of the temperature and the use of hydrocarbons deposited on the soot to increase temperature.

The figure shows that the previously described burning to form NO_2 functions from only approximately 250 to 450°C. The bottom temperature threshold is set by the light-off behavior of the catalyst that catalyzes the NO oxidation. Between 250 and 450°C, the NO_2/C reaction determines the soot burning rate. At a NO_2:NO ratio of 1:1, the NO_x:C mass ratio in the exhaust must be at least eight corresponding to the stoichiometry of the NO_2/C reaction to attain quantitative soot burning. At 450°C, the NO_2/C and O_2/C reactions occur at the same speed (isokinetic point). Above 450°C, the more active O_2/C reaction dominates, and the soot burns independent of the NO_x concentration.

In the operation of a diesel passenger car, these conditions (NO_x:C, T) cannot always be maintained: Soot accumulates in the filter, the exhaust counterpressure rises, and the soot must be burned with oxygen. The required temperature can be reduced by approximately 150 K by the catalytic soot filter. The required temperature range of 450 to 600°C cannot be sufficiently ensured in practical driving so that active engine-side measures must be taken. The complete regeneration of the soot may occur at different speeds, taking into account the state of the soot in the filter, the O_2 concentration in the exhaust, and the volumetric exhaust flow. The speed of the soot burning rises with the temperature, the quantity of deposited hydrocarbons ("moist soot"), the O_2 concentration, and the decrease in the volumetric exhaust flow. Figure 21-111 shows, in particular, how available HCs that may not be in the liquid phase can enhance soot burning by exothermic reactions. In this case, the catalytic combustion of the SOF (soluble organic fraction) component provides the necessary activation energy for igniting the soot.

The catalytic soot filter must be protected from uncontrolled burning that can arise when the quantity of flammable components, i.e., soot and deposited HCs, becomes too great in the filter since the coating cannot withstand temperatures greater than 1000°C. However, the materials of the wall-flow filter are also subject to restrictions in view of potential local temperature differences that can easily arise from uncontrolled burning and that may cause the material to crack.

Another consideration in the use of catalytic soot filters is the exhaust counterpressure. The filter materials and catalytic coating must be harmonized: The primary conflict is between high filter efficiency with simultaneously effective regeneration and a minimum exhaust counterpressure. The filter efficiency can be influenced by the filter parameters of pore size, porosity, and pore structure as well as the type and mass of the applied catalytic coating.

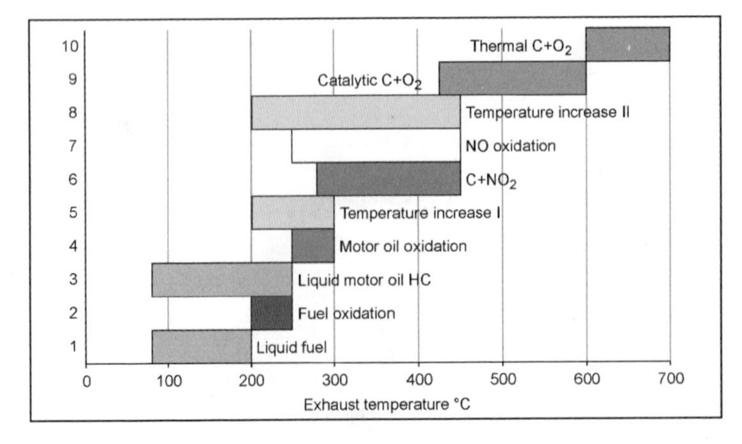

Fig. 21-111 Temperature ranges for soot regeneration (Engelhard).

During operation, the catalytic filter is loaded with ash from the engine oil. A positive factor promoting filter efficiency in comparison to fuel-additive-supported systems is that there is no ash from the fuel additive. Nevertheless, the system must be designed so that engine oil ash does not restrict the function of the catalytic converter over the desired operating time.

21.6.3.13 Particle Measuring

Since the harmful effects of very fine particles is related to their size, concentration, and substance, appropriate sampling and measuring techniques must be chosen to reliably characterize these quantities.

The measuring technique must also be suitable to ensure sufficient precision at the very low threshold level of EURO 4/5, and it should permit transient measurement in a manner similar to gas measurement with the introduction of the transient ETC cycle.

The gravimetric method used worldwide for determining the overall PM will not satisfy future requirements for precision, and it is not dynamic. However, there are major efforts underway to further develop this method.[31]

There is nothing objectionable about size-selective and substance-specific mass measurement. However, it may be very difficult to measure the very small masses such as <50 nm that occur in vehicles with a precision that is satisfactory for the type test or is sensitive enough to provide information on individual influential factors for engine development.

Conventional in situ measuring that has been used for quite a while in aerosol physics is much more sensitive and offers highly developed procedures for

- Separating according to phase during sampling (solid or liquid)
- Classifying according to mobility diameter or aerodynamic diameter
- Measuring the pollutant concentration as particle number or active surface

The methods are described in Refs. [2], [3], [32], and [33].

The SMPS (scanning mobility particle sizer) method is frequently used today for measuring in the engine exhaust that can be combined with upstream thermodesorbers for selectively separating the volatile substances from solid particles.

The measuring sequence is as follows:

1. Sampling probe with isokinetic suction.
2. Heated line made of electrically conductive material for preventing thermophoretic effects, subsequent condensation, and electrostatic deposition.
3. Dilution greater than 1:50 directly after sampling to prevent changes in the aerosol from agglomeration.
4. Electrical charge of the particles.
5. Classification of the particles in a differential mobility analyzer (DMA) (Fig. 21-112).
6. Count of the particles in a condensation nucleus counter (CPC) (not shown).

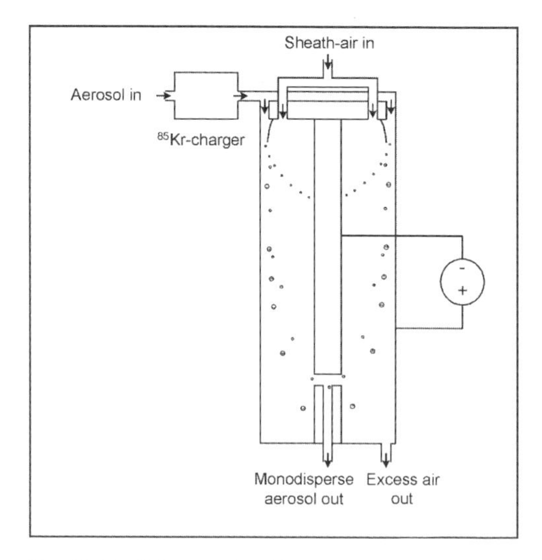

Fig. 21-112 Differential mobility analyzer (DMA).

The electrically charged particles drift along a trajectory that is determined by the ratio of aerodynamic drag to electrical field force in the annular area to the center electrode. At a specific throughput and set field strength, only a specific, very narrow-band class reaches the exit slot. By varying the voltage, approximately 60 size classes can be analyzed in 1–3 min.

For systemic reasons, this method is not suitable for dynamic measurements. The option remains of measuring one size class after the other or connecting several devices with set size classes.[34]

An interesting alternative to classifying sizes that has a much flatter separation curve in contrast to the DMA yet provides an online signal is the electrical diffusion battery (Fig. 21-113).

The particles are deposited in capture grids; the finest particles are deposited in step 1 with the lowest grid count where they discharge their electrical charge. The grid geometry physically determines the separation characteristic, i.e., the mobility diameter range of particles that are deposited in a specific grid geometry. The number of electrical elementary charges is determined by the particle size; it is a measure for the active surface of the particles. The measured current per step is, hence, an online signal for the overall surface of the deposited particles. Since the average diameter per step is known and the average number of elementary charges is known, the number of deposited particles per step can be derived.

Another measuring procedure that is frequently used is the electrical low-pressure impactor (ELPI) shown in Fig. 21-114.

This device also provides online information and is suitable for measuring during dynamic driving cycles. However, the classification, as in every impactor, is according to aerodynamic diameters, and the physical

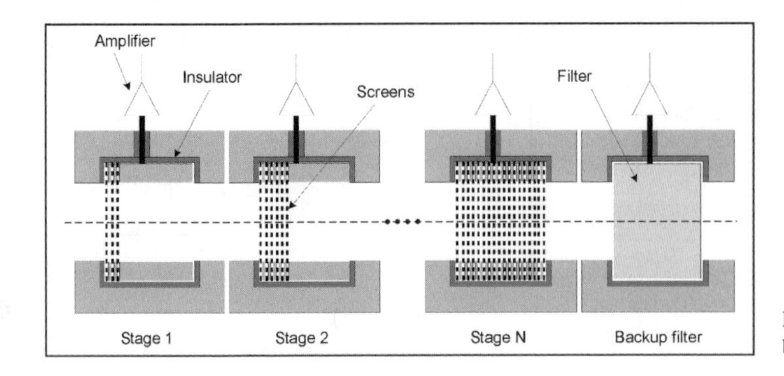

Fig. 21-113 Electrical diffusion battery.

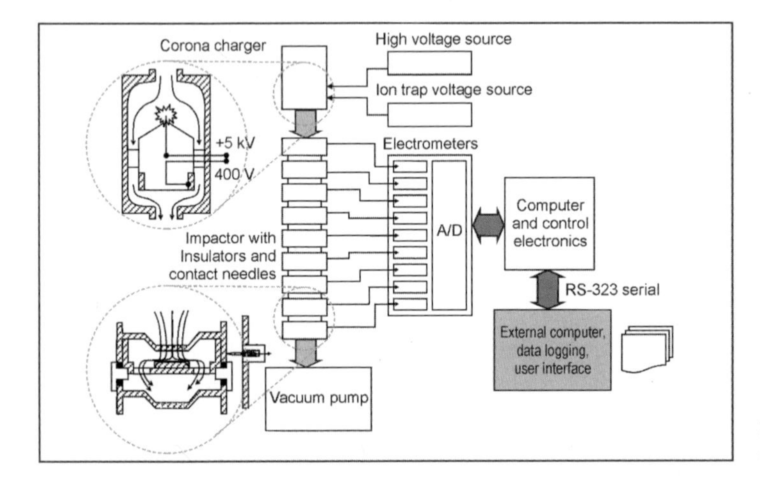

Fig. 21-114 Electrical low pressure impactor (ELPI).

effects of the greatly lowered pressure in the lower steps need to be taken into account. Compared with SMPS, ELPI offers a lower resolution for the smallest particle size, but the measuring range for large particles is greater.

Bibliography

[1] Pauli, E., Regenerationsverhalten monolithischer Partikelfilter, Dissertation, Fakultät für Maschinenwesen der Rheinisch-Westfälischen Technischen Hochschule Aachen, 1986.

[2] Burtscher, H., Partikelemissionen und Partikelfiltertechnik, Seminar Haus der Technik, Munich, 2000.

[3] Hinds, W.C., Aerosol technology, John Wiley, 1989.

[4] Siegmann, K., Soot Formation in Flames, J. Aerosol Sci. 31, Suppl. 1.

[5] ACEA-Programme on the Emissions of Fine Particles from Passenger Cars, 1999.

[6] Wichmann, E., and A. Peters, Epidemiological evidence of the effects of ultrafine particle exposure, The Royal Society, 10.1098/rsta.2000.0682, Phil. Trans. R. Soc. London A Vol. 358, 2000, pp. 2751–2769.

[7] Mayer A., *et al.,* Particulate Traps for Retro-Fitting Construction Site Engines VERT: Final Measurements and Implementation, SAE 1999-01-0116.

[8] Geprüfte Partikelfiltersysteme für Dieselmotoren, Vollzugsunterlagen Suva/BUWAL Schweiz, http.//www.BUWAL.ch/Projekte/Luft/Partikelfilter/d/Index.htm.

[9] Ebener, S., *et al.,* Parallele Reduktion von Partikel und NO_x – ein neues Abgasnachbehandlungskonzept, Viennese Engine Symposium, 2001.

[10] Hardenberg, H., Wickelrußfilter für Stadtomnibusse in der Erprobung im Verkehrsbetrieb, Der Nahverkehr 4/86.

[11] Buck *et al.,* Gestrickte Strukturen aus Endlosfasern für die Abgasreinigung, MTZ Motortechnische Zeitschrift 56, 1995.

[12] Baraket, M., Das dynamische Verhalten von Faserfiltern für feste und flüssige Aerosole, Dissertation ETH Zurich No. 9738/192.

[13] Mayer, A., *et al.,* Passive Regeneration of Catalyst Coated Knitted Fiber Diesel Particulate Traps, SAE 960138.

[14] Dürnholz, M., and M. Krüger, Hat der Dieselmotor im Pkw eine Zukunft? 6th Aachen Colloquium on Vehicles and Engine Technology, 1997.

[15] Kasper, M., Ferrocene, Carbon Particles, and PAH, Dissertation No. 12 725/1998 ETH Zurich.

[16] European Patent CRT EP 0835684.

[17] Hüthwohl, G., *et al.,* Partikelfilter und SCR, Abgasnachbehandlungstechnologien für Euro 4-Anforderungen, 4th Dresden Engine Colloquium, 2001.

[18] Toshiaki, Tanaka, *et al.,* Parallele Reduktion von Partikel und NO_x – ein neues Abgasnachbehandlungskonzept, Viennese Engine Symposium, 2001.

[19] Salvat, O., P. Marez, and G. Belot, Passenger Car Serial Application of a Particulate Filter System on a Common Rail Direct Injection Diesel Engine, SAE Paper 2000-01-0473, PSA Peugeot Citroen.

[20] Herzog, P., Exhaust Aftertreatment Technologies for HSDI Diesel Engines, Giornale della "Associazione Tecnica dell'Automobile," Torino, ATA Vol. 53, n. 11/12/2000, p. 389.

[21] Jacob, E., Einfluss des Motorenöls auf die Emissionen von Dieselmotoren mit Abgasnachbehandlung, Viennese Engine Symposium, 2001.

[22] Mayer, A., *et al.,* Particulate Trap Selection for Retrofitting Vehicle Fleets based on Representative Exhaust Temperature Profiles, SAE Paper 2001-01-0187.

[23] Heeb, N., Sekundäremissionen durch Abgasnachbehandlung, Seminar Haus der Technik/Essen, Partikelemissionen, 2000.

[24] Jodeit, H., Untersuchungen zur Partikelabscheidung in technischen Tiefenfiltern, VDI Progress Reports No. 108.

[25] Rausch, W., Untersuchungen am Sinterlamellen-Filtermedium, Aufbereitungs-Technik Mineral Processing, 6/1988.

[26] Partikelfilter für schwere Nutzfahrzeuge, herausgegeben vom Schweiz, Bundesamt für Umwelt, Wald und Landschaft, UM 130/12-2000.

[27] Mayer, A., Pre-Turbo Application of the Knitted Fiber Diesel Particulate Trap, SAE 940459.

[28] Fanick, E.R., and J.M. Valentin, Emissions Reduction Performance of a Bimetallic Platinum/Cerium Fuel Borne Catalyst with Several Diesel Particulate Filters on Different Sulfur Fuels, SAE 2001-01-0904.

[29] Mayer, A., *et al.*, Particulate Traps for Construction Machines Properties and Field Experience, SAE 2000-01-1923.

[30] 4. ETH-Konferenz "Nanoparticle Measurement" Zurich, 2000, Proceedings, Switzerland, Bundesamt für Umwelt, Wald und Landschaft.

[31] AVL-Forum Partikelemissionen 2000, Darmstadt.

[32] Matter, U., Probleme bei der Messung von Dieselpartikeln, Seminar Haus der Technik, "Feinpartikelemissionen von Verbrennungsmotoren," 1999.

[33] Kasper, M., U. Matter, and H. Burtscher, NanoMet: OnLine Characterization of Nanoparticle Size and Composition, SAE 2000-01-1998.

[34] Gruber, M., *et al.*, Partikelgrößenverteilung im instationären Fahrzyklus, Viennese Engine Symposium, 2001.

22 Operating Fluids

In automotive technology, the term operating fluids is used as a generic term for fuels, lubricants, coolants, and hydraulic fluids. In this chapter, we use it specifically in reference to application engineering. We do not, therefore, discuss the exploration, recovery, or processing of mineral oil and synthetic products.

22.1 Fuels

Let us first briefly note a few basic properties of fuels that do not appear in standards since these properties refer to the fundamentals of combustion. Nonetheless, an understanding of the related context is not without merit.

C/H Ratio, Air Requirement, and Air-Fuel Ratio

Fuels essentially consist of hydrocarbons composed of the elements C and H. The minimum amount of air (L) required for their complete combustion—the theoretical air requirement—can be calculated when the masses of carbon, hydrogen, and, possibly, oxygen are known from an elementary analysis of the relevant fuel. This is identified as L and is indicated in kg/kg. The calculation is carried out according to the following relationships:

$$L = \frac{\mathrm{O}}{0.23} \tag{22.1}$$

$$\mathrm{O} = 2.67 \cdot 0.01\,\mathrm{C} + 8 \cdot 0.133\,\mathrm{H_2} - 0.01\,\mathrm{O_2} \tag{22.2}$$

Example of a calculation of SuperPlus:

$$\mathrm{O} = 2.67 \cdot 0.01\,\mathrm{C} + 8 \cdot 0.01\,\mathrm{H_2} - 0.01\,\mathrm{O_2}$$

$$\mathrm{O} = 2.67 \cdot 0.847 + 8 \cdot 0.133 - 0.02$$

$$= 2.261 + 1.064 - 0.02$$

$$\mathrm{O} = 3.305\ \mathrm{kg}$$

$$L = 3.305 \cdot 0.23 = 14.369\ \mathrm{kg\ Air/kg\ Fuel}$$

(see Fig. 22-1).

Figure 22-1 lists the percent mass of C, H_2, and O_2 for a few important hydrocarbons and fuels, the resulting C/H ratio, and the theoretical air requirement.

Fuel	% (m/m)[a]				kg/kg
	C	H_2	O_2	C/H	L
Methane	75.0	25.0	—	3.0	17.4
Propane	81.8	18.2	—	4.5	15.8
Butane	82.8	17.2	—	4.8	15.6
n-Heptane	84.0	16.0	—	5.25	15.3
i-Octane	84.2	15.8	—	5.33	15.2
Cetane	85.0	15.0	—	5.67	15.1
Xylene	90.6	9.4	—	9.64	13.8
Toluene	91.3	8.7	—	10.5	13.6
Benzene	92.3	7.7	—	12.0	13.4
Gasoline[b]	~88.9	~11.1	—	~8.0	~14.1
Regular gasoline	~85.5	~14.5	—	~5.9	~14.9
Super gasoline	~85.1	~13.9	~1	~6.1	~14.6
SuperPlus	~84.7	~13.3	~2	~6.5	~14.4
Diesel fuel	~86.3	~13.7	—	~6.3	~14.8

[a] The % (m/m) corresponds to mass percent.
[b] ~70% benzene ~22% toluene ~8% xylene.

Fig. 22-1 C/H ratio and air requirement.[1]

The ratio of the actual air supplied for combustion to the theoretically required air is called the air-fuel ratio (λ). When there is excess air, i.e., $\lambda > 1$, the engine operates at a lean setting; given an air deficiency of $\lambda < 1$, the engine operates at a rich setting. When $\lambda = 1$, the air-fuel ratio is stoichiometric. For high performance, a spark-ignition engine operates with a fuel-rich air mixture around $\lambda = 0.90$ to 0.95. For low fuel consumption we can lean off up to $\lambda = 1.1$; in a spark-ignition engine with direct injection, the air-fuel ratio can be above $\lambda = 1.4$. In principle, diesel engines operate with excess air. Under a full load, they operate at $\lambda \approx 1.2$ and at $\lambda > 4$ while idling.

Today, a distinction is drawn between fuels that convert energy in internal combustion engines, aviation fuel for generating thrust in air travel, and fuels for heating purposes. Fuels can be liquid or gaseous. The energy chemically bound within the fuels is first converted by combustion into heat and then directly converted in the same machine into mechanical work. The term "sprit" frequently used in German media and in the German vernacular is insufficient since it actually refers only to ethyl alcohol. It originated from the economic crisis after World War I since fuel alcohol (Kraftspiritus) consisting of ethanol from potatoes (Sprit) produced by the administrative monopoly for distilled spirits increasingly had to be used as an additive to alleviate the drastic gas shortage. To further expand sales, "Reichskraftsprit GmbH, Berlin" was founded in 1925 whose products were called Monopolin. These were mixed with gasoline and/or benzene in quantities up to 65%.

22.1.1 Diesel Fuel

The boiling ranges of diesel fuels extend from approximately 180 to 380°C. Diesel fuels are used in high-speed diesel engines, especially automotive diesel engines (passenger cars and commercial vehicles). They consist of approximately 3000 hydrocarbons that are obtained in refineries using various methods for processing petroleum from a wide range of sources. Whereas earlier they consisted of relatively simple distillation products, in recent years they have become highly complex because of the much higher demands of engine manufacturers and developments in the mineral oil industry. Additives have been needed to greatly enhance the properties of the basic products that affect the engine. Since 1987, a few branded fuels have offered super diesel. The frequently used term, "gas oil," is now dated in the field of application engineering, although it is still used in refineries for middle distillates.

22.1.1.1 Diesel Fuel Components and Composition

Diesel belongs to the light middle distillates of petroleum. It is a mixture of primarily paraffinic hydrocarbons (alkanes) whose respective percentages influence engine behavior. Whereas primarily fractions from atmospheric distillation were used in an earlier period, today cracking components are more frequently used because of the continuously increasing demand for diesel fuel. Figure 22-2 lists the properties of typical refinery components of diesel fuel from distillation.

The properties of typical diesel refinery components that arise from today's cracking procedure are shown in Fig. 22-3.

The continuously increasing demand for diesel fuel has led to a shift in the types of fuel. Figure 22-4 shows the historical ratio of diesel consumption to gasoline consumption in Germany. The percentage of diesel in total mineral oil consumption in Germany is now approximately 22%.

Product description	Density (kg/m³)	Boiling range (°C)	Cetane number
Kerosene	805	150–260	45
Light gas oil	840	210–320	55
Heavy gas oil	860	200–400	55
Vacuum gas oil	870	250–400	56

Fig. 22-2 Diesel fuel components from distillation.[1]

Product description	Density (kg/m³)	Boiling range (°C)	Cetane number
Hydrocrackers	860	170–400	52
Thermal crackers	857	180–400	40
Catalytic crackers	953	195–410	40

Fig. 22-3 Diesel fuel components from cracking.[1]

Fuel	1975	1980	1985	1990[a]	1995	1996	1997	1998	1999
Diesel fuel	10.333	13.099	14.556	21.464	26.208	25.982	26.186	27.106	28.775
Gasoline	20.174	24.463	23.131	31.405	30.165	30.036	29.996	30.281	30.250
Diesel/gasoline	0.512	0.540	0.629	0.683	0.859	0.865	0.873	0.895	0.951

[a] Since 1990 in all of Germany.

Fig. 22-4 Ratio of diesel fuel consumption to gasoline consumption in Germany in millions of tons.[2]

In addition to conventional petroleum-based diesel fuel, there is a series of other, synthetically made substances that can be used for combustion in diesel engines. In 1925, Fischer-Tropsch synthesis was discovered in which synthetic gas was, for example, obtained from coal or natural gas. Synthetic hydrocarbons can be made from this gas with the aid of catalysts, and the hydrocarbons can be refined into gasoline or diesel fuel. This very inefficient method is rarely used today. Two synthetically manufactured products are at least interesting as mixed components for diesel fuel, even if they are not readily available, i.e., SMDS (Shell middle distillate synthesis) from natural gas, and XHVI (extra high viscosity index) that occurs in slight amounts as a by-product from the production of synthetic lubricants. Both products have a very high cetane number of greater than 70 (see "Ignitability") and are practically sulfur-free. Because of the high production costs and low availability, they can be used only as blend components for diesel fuel. Recently, biodiesel has been produced from biomass, primarily rapeseed oil. The percentage of biodiesel in overall consumption is less than 1%, however. This will be discussed further in Section 22.1.1.4. For diesel fuel, the demands of application technology and production are strongly contradictory. Figure 22-5 shows how the paraffin component, density, final boiling point, crack component, and sulfur content are advantageous or disadvantageous either in the engine or in production.

22.1.1.2 Characteristics and Properties

The minimum requirements for diesel fuel are found in DIN EN 590. They primarily concern density, ignition quality (cetane number), boiling curve, resistance to cold, and sulfur content. The standard characteristics of diesel fuels and their practical importance are shown in Fig. 22-6.

The compromise arrived at in 1998 between the automobile and mineral oil industries in the EU commission for the auto and oil program (EPEFE; see Section 22.1.2.2) produced the following change to a few environmentally relevant characteristics of DIN EN 590 (Fig. 22-7).

Density

Density is an essential parameter. As the density increases, the energy content increases per unit volume. For example, the calorific value in the standard permissible density range is 34.8 to 36.5 MJ/l (megajoule/liter). Given an unchanging injected quantity of fuel, the energy supplied to the engine increases with the density, which increases engine performance. However, the exhaust emissions and, especially, the particles increase under a full load due to the richer mixture. On the other hand, the volumetric fuel

Characteristic	Demands on production	Advantages for use	Disadvantages for use	Disadvantages for production
Paraffin component	High	Ignition quality	Low-temperature behavior	Cost
Density	Low	Exhaust emissions	Engine performance; Fuel consumption	Yield; cost
Final boiling point	Low	Exhaust emissions	Low-temperature behavior	Yield; cost
Crack components	Low	Ignition quality aging		Yield; cost
Sulfur content	Low	Emissions	Pump wear	Yield; cost
Characteristic bandwidth	Narrow	Harmonization		Yield; cost

Fig. 22-5 Contradictory demands between application technology and production.[3]

Characteristic	Unit	Demands	Influence on vehicle operation
Density at 15°C	kg/m³	820–845	Exhaust/performance/fuel consumption
Cetane number Cetane index	— —	Min. 51.0 Min. 46.0	Starting and combustion behavior, exhaust and noise emissions
Distillation Up to 250°C Up to 350°C 95% point	 % (V/V)[a] % (V/V) °C	 Max. 65 Min. 85 Max. 360	Exhaust emissions/deposits
Viscosity at 40°C	mm²/s	2.0–4.5	Evaporability/atomization/ lubrication
Flash point	°C	Over 55	Safety
Filterability (CFPP) 15. 04. to 30. 09. 01. 10. to 15. 11 and 01. 03. to 14. 04. 16. 11. to 28.(29.) 02.	 °C	 Max. 0 Max. –10 Max. –20	Low-temperature behavior
Sulfur content	mg/kg	Max. 350	Corrosion/particles/catalytic converter
Carbon residue	% (m/m)	Max. 0.30	Combustion chamber residue
Ash content	% (m/m)	Max. 0.01	Combustion chamber residue
Water content	mg/kg	Max. 200	Corrosion
Lubricity (WSD 1.4) at 60°C	μm	Max. 460	Wear

[a] The % (V/V) corresponds to percent by volume.

Fig. 22-6 Minimum diesel fuel requirements, according to DIN EN 590, and their importance (excerpt).[1]

Characteristic	Unit	DIN EN 590 (1993–1999)	Euro III (starting 2000)	Euro IV (starting 2005)
Sulfur (max.)	mg/kg	500	350	50
Cetane number (min.)	—	49	51	51
Density (max.)	kg/m³	860	845	845
T95 (max.)	°C	370	360	360
Polyaromates (max.)	% (m/m)	—	11	11

Fig. 22-7 Results of the EU Auto Oil program for diesel fuel.[1]

consumption increases as density decreases. Engine manufacturers would, therefore, like a further restriction of the density range in the standard. However, this would greatly restrict the use of crack components that are basically heavy, and this would restrict fuel availability in the face of continuously rising demand and would increase production costs. A helpful way out of this conflict would be to introduce a density sensor in the fuel tank to meter the fuel based on the measured density. The density of winter diesel fuel is lower than summer diesel fuel between five and ten units. The reason for this is discussed in the section on low-temperature behavior. In modern engine management systems, a density correction system is provided that is at least related to temperature.

Ignition Quality

This is characterized by the cetane number (CN). At present, it is set at a minimum of 51 in the standard. Engine manufacturers are requesting an increase to 58. In the market, it presently lies between 51 and 56 with a tendency toward higher values in summer fuel. In winter fuel, some of the higher-boiling components must be discarded to ensure sufficient cold resistance. Essentially, the CN of the individual fractions rises with the boiling temperature. A certain amount of time is required (ignition lag) to start the combustion of the fuel injected into the hot air. This variable depends on the engine construction and the operating conditions, but especially on the ignition quality of the diesel fuel. The primarily influential cetane number is the volumetric percentage of cetane $C_{16}H_{34}$ (*n*-hexadecane), the paraffinic reference fuel where CN = 100 (in a mixture with α-methylnaphtalene $C_{11}H_{10}$, an aromatic double-ring compound), and the reference fuel where CN = 0. The CN has a major influence on the combustion process and, hence, on exhaust and noise emissions. Figure 22-8 illustrates the improvement of combustion behavior from increasing the ignition quality.

A high CN also has a positive effect on the starting behavior and the emissions of uncombusted HC. Since the natural CN is frequently insufficient, it must be increased by the addition of organic nitrates, such as amyl nitrate or ethylhexyl nitrate (EHN). The required dose is usually under 0.2% (V/V), and an improvement of up to five units can be attained. Figure 22-9 shows the effect of EHN in different diesel fuel samples with the different response.

Diesel fuel sample	CN without EHN	CN with EHN	Gain
1	48.5	51.0	2.5
2	49.0	53.5	4.0
3	50.0	53.3	3.3
4	51.3	53.0	1.7
5	52.5	56.6	4.1
6	55.4	58.0	2.6

Fig. 22-9 CN increased by EHN.[12]

The cetane index (CI) indicated in the standard in addition to the cetane number is calculated from the density and boiling behavior as an alternative. It has only a limited correlation with the CN calculated in test engines since it cannot represent the universally used ignition accelerators. The CN is determined in a CFR or BASF test engine by changing the compression ratio ε, or by varying the throttling of the intake air. A high CN means that the compression ratio must be lowered or that the air flow rate must be reduced. For the standard value, the testing must be done with a coordinating fuel research (CFR) engine. The BASF engine evaluates 1.5 units higher than the CFR engine, so that the measured values must be correspondingly corrected.

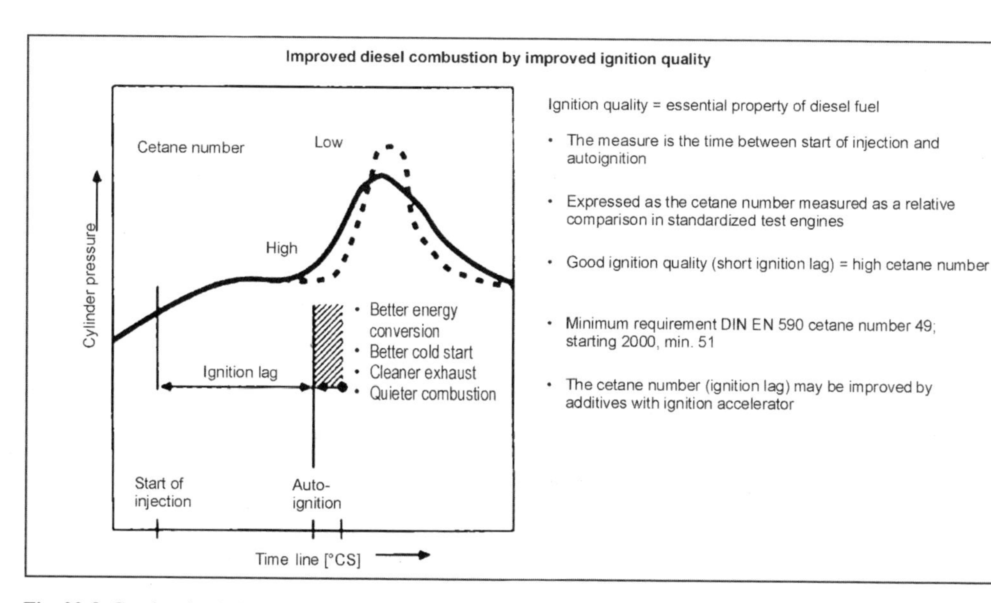

Fig. 22-8 Combustion behavior as a function of ignition quality.

Boiling Curve (Distillation)

Since fuels are a mixture of many hydrocarbons, they do not have an actual boiling point like pure hydrocarbons; rather, they have a boiling range. Diesel fuel starts to evaporate at approximately 180°C and stops at approximately 380°C. This behavior is not very important in comparison to gasoline since the mixture is prepared directly in the combustion chamber in diesel engines.

The three points established in DIN EN 590, i.e., 250 and 350°C and the 95% (V/V) point characterize only the top boiling range. Too large a portion of high boilers, especially aromates, i.e., an overly high final boiling point, enlarges the droplets in the injection jet. The resulting longer ignition lag has a negative influence on the progression of combustion, which in turn increases noise and soot. On the other hand, a slightly high volatility is advantageous for cold starts, whereas too great a percentage of low boilers causes evaporation directly at the injection nozzle, and this disturbs the intended distribution of the fuel in the combustion chamber. A restriction of the boiling curve sought by the automotive industry such as in Sweden for "Class 1" (200 to 290°C) would seriously limit the availability of diesel fuel, in Germany by approximately 40% (V/V). The desired lowering of the final boiling point would be primarily responsible, although it would understandably ease a few problems of the engine manufacturers.

Viscosity

The viscosity or inner friction of diesel fuel generally increases with the density. It must be prevented from falling below the set minimum value to ensure that there is sufficient lubrication between the sliding parts of the fuel injection system. If it is too high, the droplet size rises at the provided injection pressure. This results in poorer mixture formation and, hence, energy exploitation, lower performance, and higher soot emissions. The viscosity initially increases quickly as the temperature rises and then gradually decreases at a slower rate. Therefore the diesel fuel in the fuel tank, fuel lines, and fuel filter should, if possible, be prevented from becoming too hot by constructive measures.

Flash Point

The flash point is the temperature at which fuel vapors can be ignited by externally supplied ignition. It is important for determining the fire hazard and the subsequent safety measures in the storage and distribution systems. In the danger classification for this purpose, diesel fuel is rated A III—that is, less hazardous (gasoline is A I)—and must, therefore, have a flash point over 55°C. Even slight mixtures with gasoline cause this threshold to be exceeded impermissibly. For branded fuels, measures are taken to ensure that even a slight mixture with gasoline is impossible in storage and transport. The seriousness of a mixture with gasoline is illustrated in Fig. 22-10. When producing diesel fuel, the flash point also restricts the use of highly volatile components.

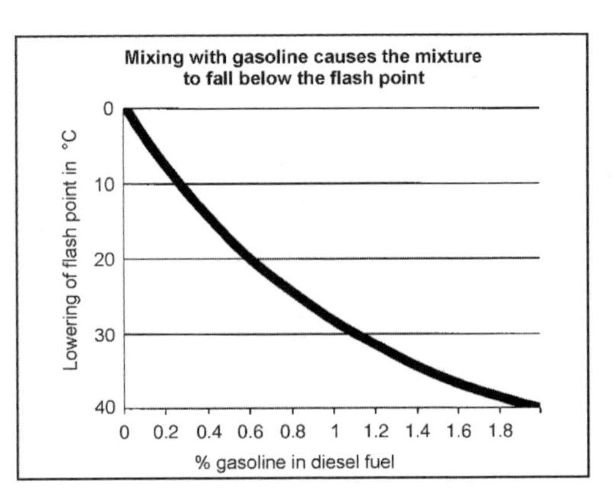

Fig. 22-10 Effect of gasoline in diesel fuel on the flash point.

Low-Temperature Behavior

This describes the flowability and filterability of diesel fuel. The paraffinic hydrocarbons that are particularly suitable for diesel fuel because of their favorable self-ignition behavior unfortunately form crystals as the temperature falls. They precipitate and clump into "slack wax." They accordingly impair the pumpability of the fuels and can plug the fuel filter. If this occurs, the engine can no longer be operated. The low-temperature behavior is substantially influenced by the fuel properties, the technical vehicle features, and the driving conditions. In DIN EN 590, the filterability in the cold filter plugging point (CFPP) test is a criterion for the resistance to cold of diesel fuel. In addition to the CFPP, brand manufacturers also use the criterion of the start of paraffin precipitation to determine the CP (cloud point), earlier also termed BPP (begin of paraffin precipitation). A distinction is, therefore, drawn between summer diesel fuel and winter diesel fuel. To produce a suitable winter quality, tailored additives are used. A combination of flow improvers and "wax antisettling additives" (WASA) has proven to be particularly effective. WASAs are also effective in storage and distribution systems where wax crystal clumping can be prevented. In winter traffic, super diesel achieves CFPP values at −33°C from the optimum combination of boiling range and additives, which is far below the limit of −20°C required in the standard. If vehicles have an installed fuel/filter heating system, further substantial improvements can be attained.

Sulfur Content

Petroleum naturally contains more or less sulfur depending on where it comes from, as illustrated in Fig. 22-11.

The sulfur is chemically bound, and more than 95% is converted into gaseous sulfur dioxide (SO_2) upon combustion. The remainder largely passes into the particle mass of the exhaust that contains sulfurous acids and sulfates. Corrosion and exhaust pollution arises. Whereas SO_2 emissions from diesel fuel no longer represent an environmental problem given the drastic desulfurization over the

Geographical location	Origin	Sulfur content % (m/m)
North Sea	General Brent	0.6–2.2 0.4
Middle East	Iran heavy Arabian light Arabian heavy	1.7 1.9 2.9
Africa	Libya light Nigeria	0.4 0.1–0.3
South America	Venezuela	2.9
Russia	Siberia	1.5

Fig. 22-11 Typical sulfur content of a few crude oils.[12]

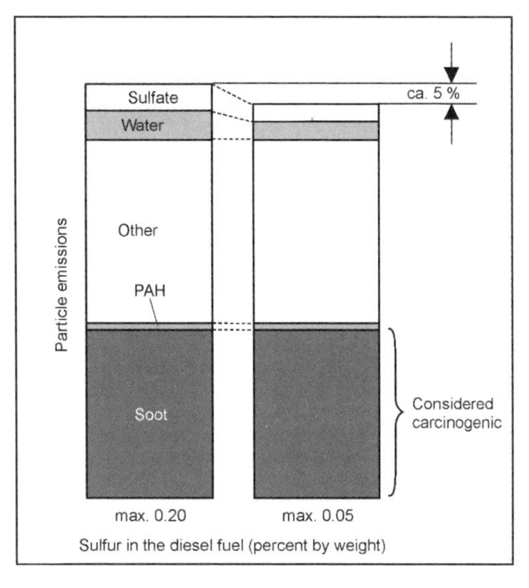

Fig. 22-12 Further desulfurization does not reduce a critical percentage of particles.[12]

last 30 years in refineries, the negative influence of the remaining sulfur [below 0.035% (m/m)] on particle emissions is now a focus of interest. In addition to the soot in diesel exhaust that is suspected of being carcinogenic, there are also polycyclic aromatic hydrocarbons (PAH). According to analysis specifications, these critical substances are measured as "particles" as well as sulfates and their adsorbed water. The goal is to reduce this component of the particles by further reducing the sulfur content. In contrast to the prior Euro II threshold valid since 1996 of a maximum 0.05% (m/m), the European standard requires a reduction to 350 mg/kg (350 ppm) in Euro III (which began in 2000), and to 50 mg/kg (50 ppm) in Euro IV starting in 2005. In any case, the reduction to the Euro II threshold in contrast to the pre-Euro I threshold of 3000 mg/kg has lowered particle emissions in the ECE R49 test by 9% to 18% as illustrated by numerous experiments.

Given the increasing use of oxidation catalytic converters, the percentage of sulfur converted to SO_3 (sulfate) has strongly increased so that particle emissions have also increased. For this reason among others, the European automotive industry (ACEA) seeks a further reduction of sulfur. On the other hand, it has been demonstrated that a few modern diesel passenger cars using existing fuel already comply with future exhaust requirements. Today's system of fuel production would have to be radically changed to comply with future highly stringent requirements, and this would send prices skyrocketing. Particles can only be clearly reduced by corresponding exhaust treatment systems such as particle filters since the critical percentage of particles cannot be effectively lowered just by limiting the sulfur content, as can be seen in Fig. 22-12.

Another problem is that the hydrogen treatment in the refinery required for desulfurization yields a welcome increase in the CN, but at the cost of decreasing density with the above-described consequences.

The NO_x storage catalytic converters that will be used in future approaches to exhaust treatment are extremely sensitive to sulfur. Engine manufacturers would like practically sulfur-free fuel containing less than 10 ppm sulfur. In this context, it may be appropriate to recall the experimental "smoke suppressors" that were introduced a few decades ago. These additives, primarily barium compounds (as well as manganese and calcium), could not reduce the particle emissions that were not measurable at that time, but gave only the visual impression of smoke reduction by brightening (masking) the particles. As an addendum to discussing the characteristics cited in DIN EN 590, let us briefly address a few other interesting characteristics.

Calorific Value

A distinction is drawn between the top calorific (TC) value that describes the combustion heat of the fuel including the condensation heat of water, and the bottom calorific (BC) value that indicates the actual useful amount of heat. In practice, only the BC is important (termed just the "calorific value" in the following). It provides information on the energy density. Whereas for scientific purposes the BC is generally expressed as MJ in reference to the unit of mass (kg), in practice the calorific value is expressed as MJ/l that refers to the volume.

Also of interest is the calorific value of the combustible air-fuel mixture that depends on the calorific value of the fuel and the air-fuel ratio. It is not the BC that primarily influences the performance of the engine; it is, rather, the calorific value of the combustible air-fuel mixture. Figure 22-13 compares the BC of diesel fuel with super gasoline, methanol, and RME (rapeseed methyl ester).

Fuel	Calorific value BC	
	MJ/l	MJ/kg
Diesel fuel	35.7 [a]	43.0 [a]
Rapeseed oil methyl ester	32.7	37.2
Gasoline (super)	30.8 [a]	41.0 [a]
Methanol	15.63	19.99

[a] Average.

Fig. 22-13 Comparison of calorific values of diesel fuel with super gasoline, methanol, and RME.[12]

We can see that diesel fuel has approximately 15% more energy than super gasoline, whereas RME has approximately 9% less than diesel fuel. In the case of methanol, we know that nearly twice the volume is consumed to produce the same amount of energy. It is also interesting to compare the calorific values and elemental analyses of three different diesel fuels as in Fig. 22-14. We can see that even with more-or-less large differences in density as is the case with fuels B and C, there is no notable difference in the calorific value.

Carbon Residue

This is determined by the last 10% of the distilled diesel fuel from low-temperature carbonization. It basically contains organic and a few inorganic components and provides information on the tendency of diesel fuel to coke the injection nozzles. Since ignition accelerators slightly increase the carbon residue, it makes sense to measure it only in diesel fuel without additives. Whereas DIN EN 590 permits a maximum of 0.3% (m/m), we find clearly lower values in commercial diesel fuels. The average is 0.03% (m/m).

22.1.1.3 Additives for Diesel Fuel

Additives are agents that improve the properties of fuel and lubricants and are normally added at concentrations in the ppm range. The goal of developing them, which is usually very expensive, is essentially to achieve a marked effect in the desired direction at the lowest dose without undesirable side effects. The additives useful for diesel fuel were already discussed with occasional details when describing the individual characteristics and their practical relevance. Some additional information is appropriate, however. Figure 22-15 shows different problems associated with diesel vehicles that can be solved with additives.

Detergent and Dispersant Additives

Detergents are soap-free, surface-active wetting and cleaning agents that reduce surface and interfacial tension. Usually they have a dispersant effect and can keep foreign materials in a liquid from clumping. A series of organic substances are suitable and proven as diesel fuel detergents and dispersants. These are amines, imidazolines, amides, succinimides, polyalkyl succinimides, polyalkyl amines, and polyether amines. Their task is to reduce or prevent deposits on the injection nozzles, especially throttling pintle nozzles, and in the combustion chamber. They are essential to ensure the operation of particularly fine direct injection nozzles and exactly maintain the pilot injection phase over a long period. Their efficiency with reference to the needle lift is particularly important. Also of interest is the positive effect of these additives on particle emissions during operation.

Corrosion Inhibitors

These use oxidation inhibitors and metal deactivators to ensure the aging stability of the diesel fuel that can be vary widely depending on the type of crude oil and manufacturing procedure. Oxidation inhibitors (antioxidants) prevent the corrosive attack of atmospheric oxygen. Together with metal deactivators, they form with the aid of organic compounds a catalytically inactive protective film that physically or chemically adheres to the metal surface.

Lubricity Additives

These are lubricity improvers that are added to the diesel fuel when, because of the strong drop in the sulfur content, the lubrication of the parts of the fuel injection pump under

Diesel fuel sample	Density/15°C (kg/m³)	Elemental analysis % (m/m)			Calorific value UC	BC	BC
		C	H	O	MJ/kg	MJ/kg	MJ/l
A	829.8	86.32	13.18	—	45.74	42.87	35.57
B	837.1	85.59	12.70	—	45.64	42.90	35.91
C	828.3	86.05	13.70	—	46.11	43.12	35.72
Average	831.7	85.99	13.19	—	45.84	42.96	35.73

Fig. 22-14 Calorific value and elemental analysis of commercial diesel fuel.[12]

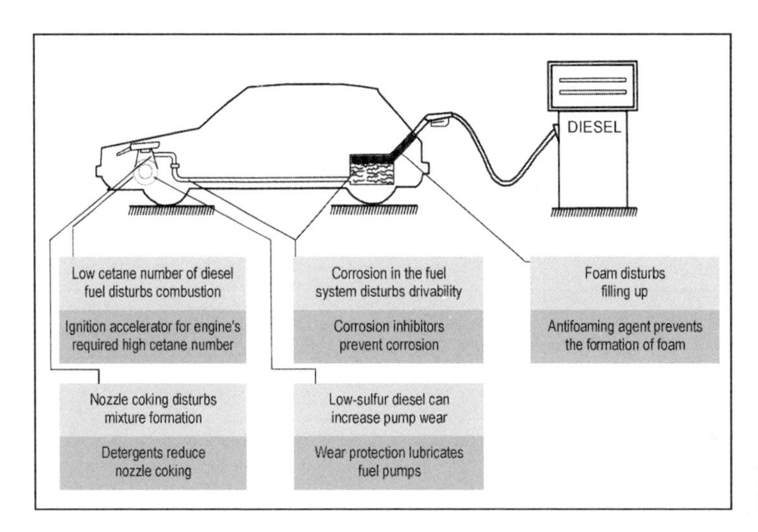

Fig. 22-15 Additives solve problems in diesel vehicles.[4]

high mechanical stress is no longer ensured by the fuel itself. Without an additive, high pump wear occurs after a short period of operation. This particularly affects distribution injector pumps and pump-nozzle systems. Long-term wear occurs even with a threshold of 0.05% (m/m) sulfur valid up to 1999. The HFRR test (high frequency reciprocating wear rig) is used to measure wear protection. It simulates sliding abrasion in the fuel injection pump: A sphere with a 6 mm diameter is rubbed under constant pressure on a polished steel plate under a liquid. In DIN EN 590, a threshold of 460 μm wear to the sphere diameter is prescribed at 60°C test temperature. Polar compounds are used as the high pressure additives.

Antifoaming Agents

The annoying foam of diesel fuel that arises while filling up can be largely suppressed by antifoam additives. They change the surface tension of the foam bubbles; i.e., they loosen or destroy the boundary layers between them. These are usually liquid silicones that are added to the diesel fuel in very slight amounts (~0.001%).

Odor Improvers

To reduce the penetrating smell of diesel fuel, especially the annoying smell when filling up diesel passenger cars, aromatic substances are used. However, there is no unanimity regarding the effectiveness of such measures.

Regeneration Aids for Particle Filters

To regenerate particle filters for future use, a new type of additive may be necessary to help burn the particles collected in the filter. In experiments, the iron compound ferrocene has proven to be particularly effective. A summary of the most important additives for diesel fuel and their purpose is in Fig. 22-16.

Super Diesel Additive Package

Leading market producers have introduced a new generation of the additive package for super diesel. The goal is to increase the cetane number beyond the new minimum limit of 51. In addition, a combustion improver is incorporated that greatly improves operation in cold starting, during warmup, and during typical daily driving. It also allows a reduction in fuel consumption. Furthermore, there is also greater protection against wear of the fuel injection system. Of course, the low-temperature behavior is exemplary and clearly exceeds minimum requirements. Diesel fuel from different manufacturers can be mixed, but the balanced effect of the additives may be lost.

22.1.1.4 Alternative Diesel Fuels

Every plant is a renewable raw material that is termed biomass. Some contain a particularly large amount of exploitable energy such as sugar beets, sugar cane, and rapeseed. By means of suitable conversion processes, these sun-fed primary energy sources can yield liquid secondary energy such as ethyl alcohol (ethanol) and rapeseed oil. In addition, biogas can also be generated. There are two basic reasons for the current interest in such biofuels that can be used in engines. Given the enormous overproduction in European agriculture, the present practice of fallow fields cannot be viewed as a final answer. Therefore, there are increasing efforts to dispose of suitable agricultural products in the energy sector. The EU is providing corresponding tax incentives for biofuels to support this process. There is also the increasing concern about greenhouse gases. These gases (primarily carbon dioxide—CO_2) are considered to cause climate changes. In addition to the natural emission sources for CO_2 that have existed forever, CO_2 emissions from the combustion of fossil energy carriers are under scrutiny. There is a worldwide effort to intentionally reduce them. A biofuel that is fundamentally suitable for operating diesel vehicles is rapeseed oil as an initial product for biodiesel.

Diesel fuel additives	Active ingredient	Improved characteristics	Advantage in use
Ignition accelerator, combustion improver	Organic nitrates such as ethylhexyl nitrate	Cetane number	Cold start, white smoke, combustion noise, exhaust emissions, fuel consumption
Detergents	Amines, amides, succinimides, polyetheramine		Clean nozzles, fuel consumption
Flow improvers	Ethylvinyl acetate	Low-temperature behavior	Reliable operation at low temperatures to allow the use of paraffinic components with a high CN
Wax antisettling	Alkylaryl amide	Low-temperature behavior	Starting, cold operation, storage
Lubricity	Fatty acid derivatives		Pump wear
Antifoaming agent	Silicone oils		Filling up
Corrosion protection	Allyl succinic acid ester/amine salts of alkenic succinimide acid		Protection of the fuel system in storage and in the vehicle

Fig. 22-16 Summary of the most important diesel fuel additives and their purposes.[4]

Biodiesel

Its use is based on the idea that it produces only as much CO_2 during combustion as is taken from the air when the rapeseed plant is growing. This is spoken of ideally as a closed CO_2 loop that does not raise the CO_2 concentration in the atmosphere. However, we must not forget that farming and converting the biomass also requires energy. In addition, biofuels are very expensive. When calculating the cost per ton CO_2 reduction for different biomass fuels or preventive measures, it becomes clear that other measures such as heat insulation and wind energy are much more economical. The suitability of rapeseed oil as a starting material for use in engines has been demonstrated in extensive experiments. However, it was quickly proven that pure rapeseed oil cannot be used without further modification. The fuel systems, engines, and engine oil have to be more or less extensively reconfigured. Investigations sponsored by the German Federal Ministry for Research and Technology have shown that most diesel engines in Germany cannot be directly operated with rapeseed oil. The technical problems primarily arise from the high viscosity. This causes the injection nozzles and piston ring grooves to coke, makes operation difficult at low temperatures, and worsens the atomization of the injected fuel. The poorer combustion causes a substantial increase in exhaust pollutants with the exception of just a slight increase of NO_x. In addition, there is the familiar "fritting" odor from the exhaust that can be reduced by catalytic converters. Furthermore, the emission of aldehydes and PAH is greater than with diesel fuel. Other problems are insufficient stability, low resistance to cold, and poor elastomer compatibility. In addition, the contained glycerides and glycerins can produce substantial deposits on the injection nozzle and in the combustion chambers. Rapeseed oil needs to be generally modified for it to be useful in modern engines. This can be done either by esterifying it into RME, or by hydrocracking in the refinery in a mixture with hydrocarbon refinery products. The most important general minimum requirements for diesel fuel made of VME (vegetable oil methyl esters) are presented in Fig. 22-17.

The transesterification of rapeseed oil is carried out with methanol. Basically, this conversion improves the cold resistance, viscosity, and thermal stability. In addition, undesirable minor constituents are removed. RME is a clearly better alternative fuel for diesel engines than pure rapeseed oil. However, additional energy must be used for conversion. In contrast to rapeseed oil where approximately 65% of the energy contained in the product is required to make it available, approximately 77% is required for RME. Even if RME meets the fuel requirements listed in a DIN draft (DIN 51606 E), elastomer compatibility must be ensured for the vehicle. In the relevant emissions tests, RME has lower particle, PAK, HC, and CO emissions than diesel fuel. The NO_x and aldehyde emissions are higher, though. Further disadvantages are low performance, higher volumetric fuel consumption, and clearly higher production costs. Accordingly, RME requires large state subsidies for it to be cost competitive at the pump. Without subsidies, the production costs of

Properties	Unit of measure	Threshold Min.	Max.	Test procedure
Density	kg/m³	875	900	ISO 3575
Kinematic viscosity	mm²/s	3.5	5.0	ISO 3104
Flash point	°C	100		ISO 2719
CFPP filterability 15. 04. to 30. 09. 01. 10. to 15. 11 and 01. 03. to 14. 04. 16. 11. to 29. 02.	°C	 0 –10 –20 –10		DIN EN 116
Sulfur content	% (m/m)		0.01	ISO 4260
Cetane number	—	49		ISO 5165
Ash content	% (m/m)		0.01	ISO 6245
Water content	mg/kg		300	ASTK D 1744
Neutralization number	mg KOH/g		0.5	DIN 51558/1
Methanol content	% (m/m)		0.3	Still open
Phosphorous content	mg/kg		10	Still open

Fig. 22-17 Diesel fuel from vegetable oil methyl ester (VME).[12]

RME are presently six times higher in comparison with diesel fuel. The most important characteristics of RME in comparison to typical diesel fuel are listed in Fig. 22-18. The high cetane number is a positive characteristic.

The characteristics can be notably improved to approach those of diesel fuel using the cited, alternative path of processing rapeseed oil by hydrocracking in the refinery mixed with hydrocarbon refinery products. Figure 22-19 shows the properties of pure rapeseed oil and diesel fuel in comparison to three fuels produced by different mixtures of rapeseed oil with vacuum gas oil with subsequent hydration in the hydrocracker. R10, R20, and R30 indicate the rapeseed oil percentage in the end product. Of note are the gradual reduction of the sulfur content and the improvement of the CN.

Likewise, rapeseed oil can also be added in the middle distillate desulfurization system. Figure 22-20 shows the properties of such fuels with 10%, 20%, and 30% rapeseed oil. In this case as well, there are advantages for the sulfur content and the CN, as well as disadvantages for the low-temperature behavior. We can see that the transesterification of rapeseed oil into RME produces an overall better end product.

It should also be mentioned that dimethyl ether (DME)$(CH_3)_2O$ is a suitable component for diesel fuel. It arises in another processing step from methanol or more recently, directly from natural gas or synthetic gas from other primary energies. DME is presently used as a liquid under pressure especially to replace fluorochlorohydrocarbons as a propellant gas in spray cans.

In summary, the technical feasibility of rapeseed oil as an alternative diesel fuel has been proven, although it requires a substantial effort. Production costs are, however, prohibitively high. It can become affordable only if it is

Fuel	Composition % (m/m) C	H	O	Density at 15°C (kg/m³)	Calorific value BC (MJ/l)	Cetane number CN
RME	77.2	12	10.8	822	32.8	51.0–59.7
Typical diesel fuel	86.6	13.4	0	830–840	35.5	51

Fig. 22-18 Comparison of characteristics of RME with diesel fuel.[12]

Characteristic	Unit of measure	Diesel fuel	Diesel fuel R10	Diesel fuel R20	Diesel fuel R30	Rapeseed oil
Density	kg/m³	841.5	835.7	830.5	824.9	920.0
Sulfur content	% (m/m)	0.19	0.13	0.09	0.04	0.01
CFPP	°C	–9	–7	–5	–2	16
Cetane number	—	54.5	59	63	66.5	41
Calorific value BC	MJ/kg	42.82	42.98	42.84	43.23	37.40
Viscosity/20°C	mm²/s	4.90	4.99	5.01	5.01	73.5

Fig. 22-19 Mixtures of diesel fuel and rapeseed oil (rapeseed oil in vacuum gas oil/hydrated).[12]

Characteristic	Unit	Diesel fuel R10[a]	Diesel fuel R20[a]	Diesel fuel R30[a]
Density	kg/m³	836.7	832.1	827.5
Sulfur content	% (m/m)	0.13	0.09	0.04
CFPP	°C	–5	–4	–2
Cetane number	—	58	63	69
Calorific value BC	MJ/kg	42.92	43.06	43.11

[a] The % rapeseed oil in middle distillate desulfurization.

Fig. 22-20 Diesel fuel-rapeseed oil mixtures after conversion in a middle distillate desulfurization process.[12]

supported by large state subsidies. In addition, the limited availability of vegetable oil-based fuels makes their replacement of diesel fuel improbable.

Alcohol-Diesel Fuel Mixtures

Methanol or ethanol alone as an alternative “diesel fuel” has basic, substantial disadvantages and requires substantial, expensive adaptations to the engine and fuel. To adapt diesel engines to pure alcohol requires, for example, a second injection system for dual-fuel operation. Diesel is used for cold starts, idling, and warm up, whereas alcohol is increasingly added as the load and speed increase. Other options are ignition assistance with glow plugs or spark plugs. Chemical ignition quality improvers as fuel additives have also been tested. They are expensive, however. Since alcohols have lower ignition quality and high evaporation heat, they must be correspondingly adapted as a fuel. A disadvantage for operation is the clearly lower calorific value (see Fig. 22-10), which translates into inferior performance and higher fuel consumption. Particularly advantageous are the lower particle and NO_x emissions. Mixtures of methanol or ethanol with diesel fuel are easier to use. Since methanol and ethanol are nearly impossible to mix with diesel fuel at room temperature, a large amount of solubilizers must be used such as ethyl acetate. The stabile mixture ranges can be determined from three-phase solubility diagrams for methanol with diesel fuel. Alcohols are technically feasible, but with today’s cost structure and tax load, they are not competitive.

Diesel and Water Emulsions

There are basic advantages to introducing water into the combustion process. In particular, nitrogen oxide formation is reduced from the decrease in the peak temperature as a result of internal cooling from water evaporation. The water can be introduced either through a second fuel injection system or by diesel and water emulsions. Whereas the first approach requires a substantial redesign of the engine and vehicle, diesel and water emulsions are much easier to realize. Tests have shown that with an increase in water content, NO_x emissions and black smoke strongly decrease as expected; however, HC and CO emissions increase. The higher HC emissions are particularly notable in the lower load range, which more than offsets the advantages gained in particle emissions. To attain realistic advantages for emulsions, a variable ratio of diesel fuel to water is required as a function of the working point, and this takes quite a bit of effort to realize. Hence, the use of emulsions is more successful in stationary engines. Diesel and water emulsions are also more costly since additional wear protection is required for the fuel injection system; modern high-pressure fuel injection systems are particularly sensitive. Furthermore, the insufficient long-term stability of the emulsion, especially at low temperatures, must be compensated by additives, and the attack of microorganisms must be countered. The required emulsifiers mean higher fuel costs that until now have inhibited the widespread use of such products.

CNG in Diesel Engines

CNG (compressed natural gas; methane) is natural gas compressed to 200 bar for vehicle use. Figure 22-21 shows the physical characteristics of CNG in comparison with diesel fuel.

One can see that even at 200 bar, the energy density in the gas tank is low. The low ignition quality of methane means that energy must be supplied to the diesel engine for ignition. The *jet ignition system* with two fuels can be used for this. Nevertheless, people prefer to convert diesel engines for city busses into spark-ignition engines to exploit the advantages of easier fuel storage (monofuel). The cylinder head and pistons are correspondingly altered, the injection nozzle is replaced with a spark plug, and high-tension ignition is used instead of the fuel injection pump. The compression ratio is reduced from 17.5:1 to 11.0:1. The use of such commercial vehicle approaches in metropolitan areas makes sense given the emissions advantages. It is questionable if this approach will gain wider acceptance given the higher energy consumption in comparison to diesel engines.

22.1.2 Gasoline

Gasolines have a boiling range of approximately 30 to 215°C and are provided for driving spark-ignition engines, primarily in the automotive sector. They consist of numerous hydrocarbons found in basic gasoline that is obtained in refineries by various processing methods from petroleum from a wide range of origins. They also contain slight amounts of other organic compounds and additives.

The term used for many decades, CF for carburetor fuel, is outdated given the general use of fuel injection.

Explosion Thresholds

The explosion threshold is always a factor in the consideration of gasolines. It describes the limits within which sudden combustion of an air-fuel vapor mixture occurs when an ignition source is activated. A distinction is drawn between a bottom threshold (slight amount of fuel vapor) and a top threshold (great deal of fuel vapor). At concentrations outside of these thresholds, no combustion can occur after ignition. Gasoline-air mixtures have a bottom explosion threshold of approximately 1% (V/V) fuel in air, and a top one of approximately 8% (V/V) fuel in air. When gasoline is stored, a very rich fuel-air mixture normally forms above the fuel that is far above the top threshold. Investigations have determined that fuels may exceed the top threshold given a minimum vapor pressure, low volatility, and low environmental temperature, which makes the fuel vapor-air mixture in the gas tank ignitable.

22.1.2.1 Gasoline Components and Composition

Gasoline is one of the low-boiling components of the petroleum. It is a mixture of reformates, crack gasolines (olefins), pyrolysis gasolines, isoparaffins, butane, alkylates, and replacement components such as alcohols and ethers. Figure 22-22 presents the basic characteristics such as density, octane number, and boiling behavior of the gasoline components in use today. The component methyl tertiary butyl ether (MTBE) is particularly important since it is required to manufacture SuperPlus. The alcohol mixture consisting of methanol and tertiary butyl alcohol (TBA) has favorable octane numbers. We do not address the antiknock lead compounds that busied fuel research departments for decades, since they are now forbidden in nearly every corner of the globe.

Figure 22-23 shows the elementary composition of the portrayed components according to their paraffins, olefins, and aromates determined by FIA analysis (fluorescence indicator absorption).

Figure 22-24 provides information on the quantity of the individual components in a typical gasoline from German refineries. The percentages of reformates and crack gasolines are approximately equivalent. The percentages of all the other components are much less, although they must all be present.

Replacement Components: Alcohols and Ethers

To compensate for lower octane numbers from the ban on lead, various ethers were discovered as essentially new gasoline components in addition to the further developed

Physical characteristics	Diesel fuel	CNG
Aggregate state in the tank	Liquid	Gaseous
Pressure in the tank	Atmosphere	200 bar
Density	830 kg/m^3	170 kg/m^3
Bottom calorific value; volume	34.7 MJ/l	7.2 MJ/l
Bottom calorific value; mass	42.0 MJ/kg	47.7 MJ/kg

Fig. 22-21 Physical characteristics of CNG in comparison to diesel fuel.[1]

Component	Density	Octane numbers		E 70[a]	E 100[b]
Unit of measure	kg/m^3	MON	RON	% (V/V)	% (V/V)
Distilled gasoline	680	62	64	70	100
Butane	595	87–94	92–99	100	100
Pyrolysis gasoline	800	82	97	35	40
Crack light gasoline	670	69	81	70	100
Catalytic crack light	685	80	92	60	90
Catalytic crack heavy	800	77	86	0	5
Hydrocrack light	670	64	90	70	100
Full range reformate 94	780	84	94	10	40
Full range reformate 99	800	88	99	8	35
Full range reformate 101	820	89	101	6	20
Isomerizate	625	87	92	100	100
Alkylate	700	90	92	15	45
Polymer gasoline	740	80	100	5	10
Methyl tertiary butyl ether	745	98	114	100	100
Methanol/TBE 1:1	790	95	115	50	100

[a] Evaporated quantity at 70°C. [b] Evaporated quantity at 100°C.

Fig. 22-22 The basic gasoline components.[12]

Component	Paraffins[a]	Olefins	Aromates
Unit of measure	% (V/V)	% (V/V)	% (V/V)
Distilled gasoline	94	1	5
Butane	100	—	—
Pyrolysis gasoline	Approx. 20	Approx. 10	Approx. 70
Crack light gasoline	Approx. 57	Approx. 40	Approx. 3
Catalytic crack light	61	26	13
Catalytic crack heavy	29	19	52
Hydrocrack light	100	0	0
Full range reformate 94	45	—	55
Full range reformate 99	38	—	62
Full range reformate 101	29	1	70
Isomerizate	98	—	2
Alkylate	100	—	—
Polymer gasoline	5	90	5

[a]Including naphthene.

Fig. 22-23 Elemental analysis of gasoline components.[12]

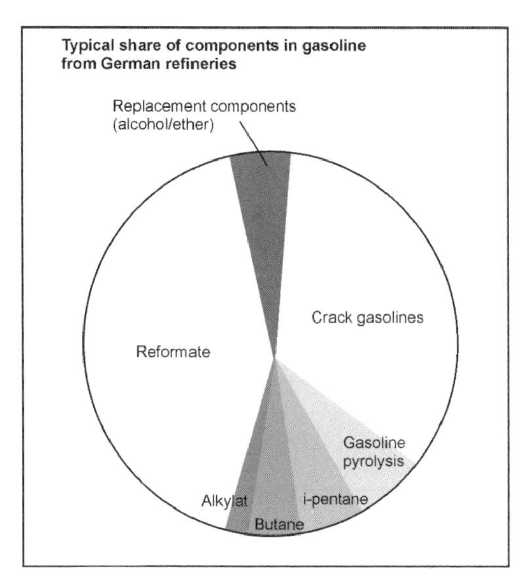

Fig. 22-24 Gasoline components in Germany.[4]

high-octane, classic components and alcohols. These are oxygen-containing hydrocarbon compounds in which a CH_2 group is replaced by an oxygen atom. Suitable for gasoline are ethers with at least five C atoms. Figure 22-25 shows are the most important physical characteristics of the alcohol components compared with those of super gasoline.

Methanol and ethanol have been used at different times and different places in the history of motorized travel. Their use as alternative fuels is discussed in detail in Section 22.1.2.3. Ethers are distinguished by their favorable miscibility with gasoline without an azeotropic increase of the volatility, and less sensitivity to water. The high octane numbers and low vapor pressure are noteworthy. Because of the lower oxygen content in comparison to methanol and ethanol, the reduction of the calorific value is within tolerable limits in contrast to normal fuel components. Today, MTBE is manufactured on an industrial scale. Figure 22-26 presents the most important physical characteristics for ether components with those of super gasoline. The other cited ethers with the exception of ETBE and TAME are scarcely used as fuel components because of their high production costs.

The use of particularly valuable, oxygen-containing replacement components, such as ethers and alcohols, is restricted to a relatively low amount given the EU limitation of the overall O_2 content in gasoline of 2.5% (m/m). For example, the ethanol content is limited to 5% (V/V), and the methanol content is restricted to a maximum 3% (V/V) if additional, suitable solubilizers—usually TBA—are added to the methanol. Figure 22-27 shows the EU directive on the use of oxygen-containing components. This limitation is to exclude undesirable side effects on elastomers, prevent separation at low temperatures, and prevent excessive lean adjustment that can disturb drivability.

MTBE has proven itself especially in SuperPlus as a replacement for lead-based antiknocking agents and

Name	Abbreviation	Boiling point	Density at 20°C	Vapor pressure	RON	MON	Bottom calorific value	Evaporation heat	O_2 content
Unit of measure		°C	kg/m³	kPa			MJ/l	kJ/kg	% (m/m)
Methyl alcohol	Methanol	64.7	791.2	32/81[a]	114.4	94.6	15.7	1100	49.93
Ethyl alcohol	Ethanol	78.3	789.4	17/70[a]	114.4	94.0	21.2	910	34.73
Isopropyl alcohol	Isopropanol	82.3	775.5	14/72[a]	118.0	101.9	23.6	700	26.63
Sec.-butyl alcohol	SBA	100.0	806.9				27.4		21.59
Isobutyl alcohol	IBA	107.7	801.6	4/63[a]	110.4	90.1	26.1	680	21.59
Tert.-butyl alcohol	TBA	82.8	786.6	7/64[a]			26.8	544	21.59
Super gasoline (up to 1999)	Super gasoline	30 to 215	725–780	S 60–70 W 80–90	95	85	Approx. 31	380–500	0–2

[a] As a mixture component in gasoline (10%).

Fig. 22-25 Most important physical characteristics for alcohol components in comparison to super gasoline.[12]

Name	Abbreviation	Boiling point	Density 20°C	Vapor pressure	RON	MON	Bottom calorific value	O_2 content
Unit of measure		°C	kg/m³	kPa			MJ/kg	% (m/m)
Methyl tertiary butyl ether	MTBE	55	740	48	114	98	26.04	18.15
Ethyl tertiary butyl ether	ETBE	72	742	28	118	102	26.75	15.66
Diisopropyl ether	DIPE	68	725	24	110	100	26.45	15.66
Tert.-amyl methyl ether	TAME	85	770	16	111	98	27.91	15.66
Isopropyl-tert.-butyl ether	PTBE	88.5	740	20			27.46	13.77
Super gasoline (typical 1999)	Super gasoline	30–215	725–780	60–90	95	85	Approx. 41	0–2

Fig. 22-26 Most important physical characteristics of ether components in comparison to super gasoline.[5]

Component	Upper limit for all EU countries	Upper limit for individual states	Upper limit for Germany
Methanol	3% (V/V)	3% (V/V)	3% (V/V)
Ethanol	5% (V/V)	5% (V/V)	5% (V/V)
IPA	5% (V/V)	10% (V/V)	10% (V/V)
TBA	7% (V/V)	7% (V/V)	7% (V/V)
IBA	7% (V/V)	10% (V/V)	10% (V/V)
Ether[a]	10% (V/V)	15% (V/V)	15% (V/V)
Other[b]	7% (V/V)	10% (V/V)	10% (V/V)
Mixtures	2.5% (m/m) O_2	3.7% (m/m) O_2	2.7% (m/m) O_2

[a] MTBE, TAME, and ETBE as well as others with min. 5 C atoms.
[b] Other monoalcohols.

Fig. 22-27 Maximum concentration of O_2-containing components (EU).[12]

benzene to increase the octane number. At present in Germany, SuperPlus contains an average of approximately 10% MTBE.

Gasoline Types

At present in Germany, there are three types of unleaded fuel: Regular since 1985, (Euro) Super since 1986, and SuperPlus since 1989. For older vehicles without a catalytic converter, unleaded super was available up to 1996, whereas leaded regular gasoline was removed from the market in 1988. In a few European countries, regular gasoline cannot be found since all automobile manufacturers are under pressure to offer minimum fuel consumption. Almost all of their engines are, therefore, designed to run on Super or SuperPlus. Figure 22-28 shows the percentages of the three types of gasoline on the German fuel market.

One can see that SuperPlus, after an initial success, probably because of greater cost, has not been widely

Fuel type	1991	1992	1993	1994	1995	1996	1997	1998	1999
Regular gasoline	39.2	39.6	38.8	39.4	38.4	37.6	36.9	35.3	34.3
Super	31.9	38.2	42.6	46.9	50.7	54.4	57.3	59.5	61.1
SuperPlus	6.8	7.2	7.3	6.0	5.4	5.3	5.8	5.2	4.6
Lead-free overall	77.9	85.0	88.7	92.3	94.5	97.4	100.0	100.0	100.0
Super leaded	22.1	15.0	11.3	7.7	5.5	2.6			

[a] In all of Germany.

Fig. 22-28 Percentages of gasoline types consumed in Germany[a] (%).[2]

accepted. In addition to conventional gasoline based on petroleum, a series of other synthetic substances are suitable for combustion in spark-ignition engines. In addition, there are several possibilities for using alternative gasolines. These will be discussed further in Section 22.1.2.3.

22.1.2.2 Characteristics and Properties

The minimum requirements for the three cited lead-free gasolines are found in DIN EN 228. As can be seen in Fig. 22-29, they primarily concern density, antiknock quality, the boiling curve, vapor pressure, benzene content, and sulfur content.

Because of the environmentally relevant importance of some fuel characteristics in practice, a compromise was reached to bridge the different perspectives of the automobile and mineral oil industries on a European basis (EU Commission) in 1998 within the framework of the Auto-Oil program that further stiffened the standards for environmentally relevant fuel characteristics in a first step in 2000 and in a second step in 2005. For gasoline, this primarily concerned the sulfur, benzene, and aromatic contents. The exhaust thresholds were established correspondingly as Euro III starting January 1, 2000 and as Euro IV starting January 1, 2005. Figure 22-30 shows the presently known changes in gasoline.

The changes to the European exhaust laws are shown in Fig. 22-31.

Density

The ranges of the density for all three unleaded gasolines are set to a uniform 720–775 kg/m³ at 15°C. Figure 22-32 shows averages and the ranges of density of commercial German gasoline for summer and winter.

As the density increases, the volumetric energy content of the fuel also generally raises, which is related to falling volumetric fuel consumption. Based on experience, a rise in density of 1% equals a drop in volumetric fuel consumption of 0.6%. One can see that the values in the summer are all higher, and that there are advantages in fuel consumption for super and especially for SuperPlus.

Antiknock Quality

The antiknock quality of gasolines is their ability to prevent undesired combustion, i.e., combustion not triggered by the spark plug or uncontrolled combustion of the uncombusted residual exhaust gas before it meets the flame front. Depending on the fuel composition and design, the flame front passes through the charge at a propagation velocity of more than 30 m/s. Figure 22-33 schematically compares normal combustion with knocking combustion. In knocking operation, the combustion speed is approximately 10 times faster, causes steep pressure peaks and cavitation-like pressure fluctuations, and is accompanied by a substantial increase in the combustion chamber temperature.

A simplified comparison of the corresponding pressure/time diagrams is shown in Fig. 22-34.

If knocking is continuous, the spark plugs, pistons, cylinder head seals, and valves become damaged or even destroyed, especially when preignition occurs. In Fig. 22-35 we see a piston that was destroyed from continuous knocking.

Modern engines are largely protected from such mechanical damage by the use of knock sensors—structure-borne sound sensors or ionic current meters. They delay the moment of ignition when knocking starts, reduce the charge pressure during charging, or throttle the intake air. In vehicles with knock control, the ignition map is electronically adapted to the fuel in the tank. However, when the antiknock quality is lower than that specified by the manufacturer, the later ignition setting causes a drop in performance, higher fuel consumption, and a higher thermal load on the catalytic converter. Conversely, a transition from Super to SuperPlus can increase performance with an earlier ignition setting in conjunction with lower fuel consumption and emission advantages. In determining the required antiknock quality of an engine, a distinction is drawn between acceleration knock and high-speed knock. Whereas acceleration knock as a transient

Characteristic	Unit of measure	Demands according to DIN EN 228		
		SuperPlus	Super	Regular
Density at 15°C	kg/m^3	720–775		
Antiknock quality RON MON		 Min. 98 Min. 88	 Min. 95 Min. 85	 Min. 91 Min. 82.5
Lead content	mg/1	Max. 5		
Boiling curve[a] Vaporized quantity (class A) At 70°C, E70 At 100°C, E150 At 150°C, E150 Vaporized quantity (Class D/D1) At 70°C, E70 At 100°C, E100 At 150°C, E150 Final boiling point FBP (Class A/D/D1)	%(V/V) °C	 20–48 46–71 Min. 75 22–50 46–71 Min. 75 Max. 210		
Volatility index VLI[b] (VLI = 10 × VP + 7 × E70) Class D1	 Index	 Max. 1150		
Distillation residue	%(V/V)	Max. 2		
Vapor pressure (DVPE) Class A Class D/D1	kPa	 45.0–60.0 60.0–90.0		
Evaporated residue	mg/100 ml	Max. 5		
Benzene content	% (V/V)	Max. 1		
Sulfur content	mg/kg	Max. 150		
Oxidation stability	min	Min. 360		
Copper corrosion	Degree of corrosion	Max. 1		

[a] Class A: 01. 05.–30.09. (summer).
Class D: 16. 11.–15. 03. (winter).
Class D1: 16. 03.–30. 04./01, 10.–15. 11. (transition).
[b] Vapor lock index.

Fig. 22-29 Gasoline characteristics according to DIN EN 228.[1]

condition is not that dangerous at a low speed and load, continuous high-speed knocking at a high speed and full load can be hazardous enough to cause engine damage.

Octane Number

The octane number is a measure of the antiknock quality of a gasoline. A distinction is drawn between the minimum requirements of RON (research octane number) and MON (engine octane number). Both terms are based on traditional names from American fuel research that do not have a logical correspondent. In practice, the SON (street octane number) is also relevant. For earlier carburetor engines, the FON (front octane number) or the RON 100 (corresponding to the RON of the fuel components boiling

Characteristic	Unit of measure	DIN EN 228 bis 1999	Euro III starting 2000	Euro IV starting 2005
Sulfur	mg/kg	500	150	50
Benzene	% (V/V)	5	1	1
Aromates	% (V/V)	—	42	35
Vapor pressure	kPa	70	60	?
Olefins	% (V/V)	—	(21) 18	?

Fig. 22-30 Results of EU automobile/oil program for gasoline.[1]

	Pollutant	91/441/EWG Euro I	94/12/EG Euro II	98/69/EG[a] Euro III	98/69/EG[a] Euro IV
Engine	In g/km	Starting 1992	Starting 1996	Starting 2000	Starting 2005
Spark ignition	CO	3.16	2.2	2.3	1.0
	HC + NO_x	1.13	0.5		
	HC			0.2	0.1
	NO_x			0.15	0.08
Diesel	CO	3.16	1.0	0.64	0.5
	HC + NO_x	1.13	0.7	0.56	0.3
	Particles	0.18	0.08	0.05	0.025

[a] Changed (stiffened) test procedures.

Fig. 22-31 Development of the European exhaust laws (passenger car).[1]

Density in kg/m³		SuperPlus	Super	Regular
Range	Summer	737–770	738–771	727–768
	Winter	728–765	733–764	725–751
Average	Summer	753	751	745
	Winter	745	743	735

Fig. 22-32 Density of commercial German gasoline.[1,12]

at 100°C) was used. Whereas RON and MON are measured in special CFR single-cylinder knock test engines by changing the compression ratio, the SON is determined in production vehicles by advancing the moment of ignition. The test according to the MON method is based on speed, moment of ignition, and mixture preheating under harder conditions; the MON is, therefore, always lower than the RON. In practice, this means that engines under high thermal stress—which today is practically every single one—have a minimum requirement for the MON of a fuel in addition to the RON. The RON-MON difference is termed the "sensitivity" and should not exceed the value of 10.

Figure 22-36 portrays the operating conditions when determining the RON and MON in the CFR test engine.

Octane Number Scale

The octane number scale extending from 0 to 100 is dimensionless. The 0 represents the particularly knock-susceptible reference fuel, regular heptane (C_7H_{16}), and 100 is the particularly knock-resistant reference fuel, isooctane (C_8H_{18}), also termed 2,2,4-isopentane [$C_5H_9(CH_3)_3$]. The ON of a fuel is determined in a comparative test between the fuel sample and *i*-octane/*n*-heptane mixtures. The compression ratio is first increased in the CFR test engine until the sample starts knocking. Then the associated ON is calculated by maintaining a constant compression ratio and changing the mixtures of *i*-octane and *n*-heptane until the engine starts knocking. The knock limit is determined with the aid of an electronic knock sensor. For example, RON 95 means that the gasoline measured in the CFR test engine using the research method acts like a mixture of 95% *i*-octane and 5% *n*-heptane when it reaches the knock limit.

Mixed Octane Number

The range of the ON scale ends by definition at 100. For fuels whose ON is greater than 100, a mixed octane number is determined with the aid of an added component with a known ON. When the result of the knock measurement is obtained for this mixture, the mixed octane number can be calculated with the following formula:

$$\text{Mixture ON} = \frac{100 \cdot M + b \cdot K}{a} \tag{22.3}$$

where

M = ON is the diluted mixture
K = ON is the added component
a = % M
b = % K

Example:

$$\text{Mixture ON} = \frac{100 \cdot 94 + 10 \cdot 80}{90} = \frac{10\,200}{90} = 113.3$$

with

$M = 94$
$K = 80$
$a = 90$
$b = 10$

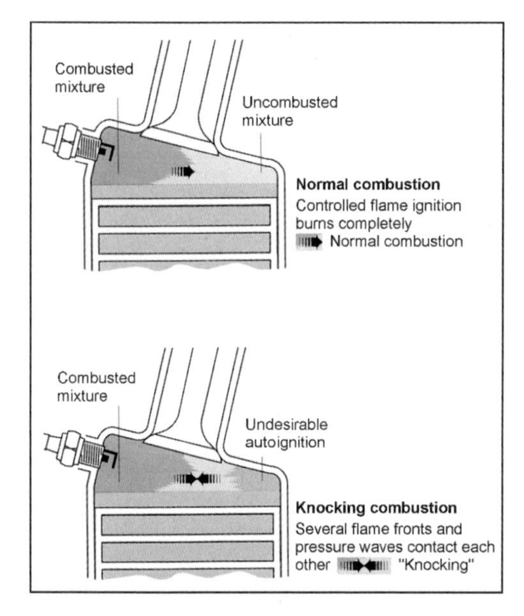

Fig. 22-33 Regular and knocking combustion.[5]

Fig. 22-35 Piston destroyed by continuous knocking.[5]

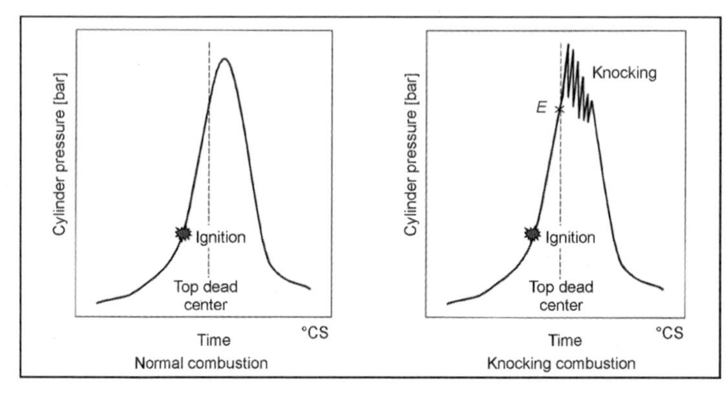

Fig. 22-34 Pressure/time diagram.[9]

	Engine speed (min^{-1})	Intake air (°C)	Mixture preheating (°F)	Moment of ignition °crank angle before TDC	Compression ratio
RON	600	51.7 ± 5	—	13	Variable 4 to 16
MON	900	38	Variable 285–315	Variable 14–26	Variable 4 to 16

Fig. 22-36 Operating conditions of the CFR test engine.[1]

Since this calculated value is actually valid only with a mixture of structurally similar hydrocarbons, its practical applicability is limited. Since 1956, the *Wiese scale* has been used as a practical method (DIN 51788). Starting with *i*-octane, increasing amounts of TEL (tetra ethyl lead) are added. The method corresponds to the performance number (PN) used for airplane fuels. Figure 22-37 shows the numeric values for the relationship between octane numbers greater than 100 and the respective TEL added to *i*-octane.

ON	% (V/V) TEL	ON	% (V/V) TEL
100	0.0000	111	0.0399
101	0.0020	112	0.0468
102	0.0042	113	0.0546
103	0.0066	114	0.0634
104	0.0092	115	0.0734
105	0.0124	116	0.0850
106	0.0158	117	0.0963
107	0.0195	118	0.1133
108	0.0238	119	0.1308
109	0.0285	120	0.1509
110	0.0338		

Fig. 22-37 Wiese scale for ON 100.[1]

Octane Number Requirement

The octane number requirement of an engine is measured on an engine test bench within the overall speed range under a full load. A knock limit curve map arises in which is entered the spark advance characteristic determined by the manufacturer. The octane number requirement then results from the intersections of the knock limit curves with the ignition map, which allows the maximum to be read immediately. Usually it lies around the maximum torque, i.e., the maximum average pressure.

Engine Design and Octane Number Requirement

From the vantage point of the engine, the octane number requirement is chiefly determined by the compression ratio. Given geometrically similar combustion chambers, an increasing piston displacement corresponds to a decrease in the knock limit compression. Hence, larger cylinders are more knock sensitive. To a certain degree, an oversquare cylinder ($s/D < 1$) has a higher octane number requirement with otherwise equivalent dimensions than an undersquare engine ($s/D > 1$). In both cases, the path traveled by a flame during combustion plays a role. The connnecting rod ratio r/l is important because a higher r/l maintains the efficiency of the piston overlap (squish area) while the overall combustion is approximately maintained. The end gas, therefore, has no occasion to assume a high temperature due to reduction processes without absorbing heat. The combustion that occurs at a nearly constant volume is also advantageous for the quality of the thermal efficiency. The general return to long-stroke engines improves the exhaust and further increases the compression ratio. For the combustion chamber design to have a very low octane number requirement with a high efficiency, the following points are observed:

- Compact combustion chamber with a very low surface to volume ratio (spherical cap or roof-shaped)
- Central location of the spark plug in the combustion chamber to attain equally long flame paths (four valves)
- Large squish area from piston overlap with minimum thickness (generates turbulence)
- Energetic charge movement
- Strong cylinder head cooling

In summary, it can be stated that the best results are obtained when the area filled by the mixture upon the moment of ignition is as close to the spark plug as possible. The valve timing also influences the relative knocking sensitivity of an engine. For example, a large valve overlap reduces knock due to its influence on the residual exhaust gas and mixture temperature. Early closing to increase torque in the lower speed range can raise the octane number requirement. Today's wide use of light metal and the practically uniform lack of air cooling have a positive effect.

Operating Conditions and Octane Number Requirement

The octane number requirement largely depends on the operating conditions. These are largely influenced by the state of the intake air, excess-air factor, speed, moment of ignition, volumetric efficiency, load, and the coolant temperature. When the pressure and temperature of the intake air rise, this raises the octane number requirement, whereas increasing humidity decreases the octane number requirement. The octane number requirement is highest at the stoichiometric excess-air factor. A richer or leaner mixture does not produce the pressure and temperature required for knocking due to the lower velocity of combustion. The temperature of the uncombusted fuel-air mixture is analogous. A rising speed generally translates into a rapid decrease of the octane number requirement since the piston is already moved away from the top dead center at a time critical to the end gas; the combustion chamber volume therefore increases, and the compression of the end gas correspondingly decreases. Furthermore, at high speeds, combustion is quicker because of the large turbulence generated in the combustion chamber. The throttling loss and lower compression end pressure also have the same effect. The moment of ignition naturally directly influences the octane number requirement. The earlier it occurs (far before TDC), the earlier combustion begins in the piston travel of the compression cycle which compresses the end gas.

In general, spark-ignition engines tend to knock, especially when the throttle valve is fully open, i.e., under a full load, since it is at this point that the maximum combustion pressure arises from the greatest cylinder charge. The maximum octane number requirement is usually at the speed where there is maximum torque (average pressure) since the volumetric efficiency, moment of ignition, and excess-air factor interact to promote knocking. As the coolant and oil temperature rise, the octane number requirement naturally increases since the critical conditions for spontaneous, undesired combustion of the residual exhaust gases are enhanced. On average, the octane number requirement increases by 1 for each 5°C rise in coolant water temperature. The influence of the oil temperature is somewhat less.

Combustion Chamber Deposits and the Octane Number Requirement

While an engine is operating, combustion chamber deposits form that cover its surface, the piston head, the valve head, and the spark plug. They arise both from the fuel and the lubricant. Soot arises from the fuel from incomplete combustion in the idling and warm-up phases. Cracked or coked oil components that remain on the top piston ring in the combustion chamber come from the lubricant or via the valve guides. Ash-forming additives, if not fully organic, can cause deposits. The deposits raise the octane number requirement by reducing the combustion chamber volume; i.e., they increase the compression ratio and insulate against heat. The knocking tendency rises quickly in a new, clean engine to a maximum until the deposit equilibrium is reached. In practice, this rather unstable equilibrium arises after approximately 10 000 to 20 000 km. The rise of the octane number requirement from deposits in city traffic has been greatly reduced from the transition to unleaded fuels. Of course, driving style also plays a large role. When all the factors combine, the octane number requirement can rise by 7 from a new engine to the deposit equilibrium, even when state-of-the-art operating fluids with additives are used. For example, an engine designed for regular gasoline operates without knocking only when it uses super.

Street Octane Number

Although the antiknock quality of a fuel provides information on the practical knocking behavior expected in a given automobile by determining the RON and MON, it is difficult to assign the laboratory octane number to actual street behavior. For example, different fuels with the same RON can produce widely different knocking behavior in one and the same automobile because of the numerous cited factors that can influence the octane number requirement. To precisely monitor these conditions, mineral oil researchers use test methods to determine the octane number that actually occurs on the street, the SON. In this case as well, the produced fuels are compared with the familiar reference fuels. The measurements are carried out either on suitable automobile test benches or on engine test benches. In comparison to the CFR test engine, the measuring range is greatly limited. Meaningful values can be measured only from approximately 10° to 15° crank angle around the basic spark adjustment set by the automobile manufacturer. This approximately corresponds to a bandwidth of 5 to 6 ON. For earlier carburetor engines, the CRC F-28 method was used, the "modified Uniontown method," that is basically the same as determining the octane number requirements for acceleration knock. As expected, the limit curves rise sharply because the knocking tendency falls quickly, and the octane number requirement drops rapidly with increasing speed. If the laboratory octane numbers RON and MON provide only a limited amount of information on the actual behavior of the fuel in practice, they are still a useful yardstick for indicating the interaction between the engine and fuel. The SON usually lies between the RON and MON. At low speeds it tends toward the RON; at high speeds and with a large amount of residual exhaust gas, it tends toward the MON. The SON-RON difference has proven useful for purposes of comparison. It is termed the street evaluation number (SEN). The advantage of this nomenclature is that the SEN is positive when the SON exceeds the RON which is usually the case, and it is negative when the SON is less than the RON. The positive and negative signs then can be used to directly evaluate a fuel in a given automobile or engine. A positive SEN also indicates that the relevant engine is evaluating a fuel with less "severity" than the CFR test engine in the RON method and vice versa: when the SEN is negative, the engine is providing a more severe evaluation than the CFR engine.

By establishing a minimum MON in addition to the RON in the standards, a large number of earlier-used mixture components with a low MON were excluded. In addition, the general use of off-center fuel injection has eliminated the earlier predominant sensitivity of engines to an uneven distribution of the octane numbers over the

Component	Property			Influence on SON	
	Octane numbers		Boiling behavior	Carburetor engine acceleration	High load and speed
	RON	MON			
Light distillate	Low	Low	Highly volatile	Negative	Negative
Butane *i*-pentane/*i*-heptane	High	High	Highly volatile	Positive	Positive
Light crack gasoline	High	Low	Highly volatile	Positive	Negative
Heavy reformate	High	Average/ high	Nonvolatile	Negative	Positive
Heavy crack gasoline	Average	Low	Nonvolatile	Negative	Negative

Fig. 22-38 Influence of a few fuel components on the SON.[5,9]

boiling range of the fuel. In general, this has eliminated the necessity of determining the SON, and it remains relevant only for research purposes.

The influence of a few fuel components on the SON is shown in Fig. 22-38. It can be seen that light distillate and light and heavy crack gasoline negatively influence modern fuel injection engines.

Front Octane Number

For the sake of completeness, we mention the front octane number that is no longer relevant today. It provides information on the RON of the components of the gasoline that boil below 100°C. It was especially useful for carburetor engines with long intake ports. Since only the light components reach the combustion chambers when the throttle valve is quickly opened, it had to be ensured that within this boiling range there were enough knock-resistant components available. In the low boiling range, with the exception of butane, the other light components such as distillate and reformate gasoline generally have a low ON. The antiknock quality of the front boiling range was, hence, too low in comparison to overall fuel. To deal with this fuel problem, high-octane light components such as isomerisates, catalytic crack gasoline, and alcohols are used. The earlier-used lead compounds were also adapted by introducing highly volatile lead *tetramethyl* instead of lead *tetraethyl*. Given the general transition from carburetor to individual cylinder fuel injection with the precise metering and preparation of the mixture under transient conditions, the FON became irrelevant and was, therefore, withdrawn from the standard.

Boiling Behavior (Distillation)

The boiling behavior or volatility is determined by the boiling curve and the vapor pressure. In addition to the antiknock quality, it is the most important evaluation criterion for gasolines that change into a vaporous state between 30 and 205–210°C.

Boiling Curve

When doing a boiling analysis according to DIN EN ISO 3405, the fuel specimen is evaporated and then condensed with variable heat output and a fixed temperature increase of 1°C/min. The resulting boiling curve contains a great deal of information for application engineering. Well-balanced boiling behavior is an essential prerequisite for operating automobiles with spark-ignition engines under every condition. The meaning of the boiling curve and its individual sections is shown in Fig. 22-39.

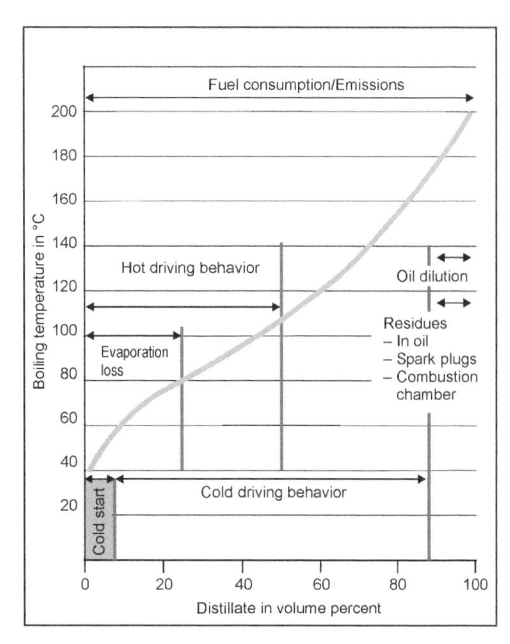

Fig. 22-39 Boiling curve and its influence on engine behavior.[5]

For example, light, i.e., low-boiling, components are responsible for fast starting cold engines, good response, and low exhaust emissions during the warm-up period. Too many can lead to vapor bubble formation and greater evaporation loss in the summer. In wet, cold weather, throttle valve icing can occur. Too many high-boiling components can lead to condensation on the cylinder walls in cold operation and dilute the oil film and oil supply. Too few components in the middle boiling range impairs drivability and may cause bucking during acceleration. After a hot engine is turned off and quickly restarted, the demands on the fuel are precisely the opposite. Under unfavorable conditions, parts of the fuel system can become so hot that a large portion of the fuel evaporates, which causes vapor bubbles in the fuel pump or vapor cushions in the fuel injection lines. Figure 22-40 shows the opposite requirements for cold starting and hot driving.

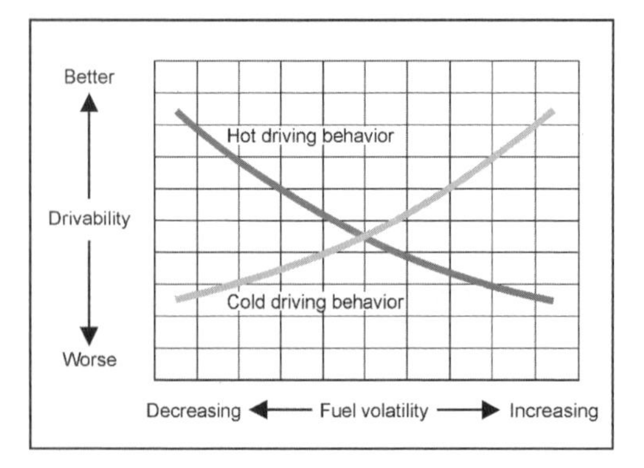

Fig. 22-40 Influence of volatility on cold-start and hot driving behaviors.[5]

In the EN standard (see Fig. 22-29), there are different volatility classes that cover geographic and yearly changes in the weather. Figure 22-41 shows examples of typical German winter values for E70, E100, and E150.

In addition, Fig. 22-42 gives a comparison of values for the end of boiling of German gasoline.

Vapor Pressure

The pressure that arises in a sealed container as a function of temperature from the evaporation of fuel is termed the vapor pressure. It influences (sometimes in connection with other volatility criteria) cold and hot starting, cold driving behavior, and evaporation loss. It is basically determined by the light components such as butane toward the initial boiling point. To determine the vapor pressure, the Reid method was incorporated in DIN EN 12 (RVP = Reid vapor pressure) up to 1999. The test temperature is 37.8°C (100°F) with a vapor-liquid ratio of 4:1. In general, the *wet* Reid method is sufficiently precise. For standard testing, different volatility classes are established depending on the environmental temperature. For the boiling temperature of E 70°C, eight classes are established for the vaporized quantity of fuel as a function of the vapor pressure, and this correlates with hot starting and hot driving behavior. Each volatility class is assigned a VLI (vapor lock index) value as a parameter. It is calculated from 10 RVP + 7 E70 and has been particularly useful for carburetor engines. Since the fuel in modern fuel injection engines is exposed to high temperatures particularly before and in the nozzles, an additional measuring method was worked out within a broader measuring range: 40 to 100°C (DIN EN 13016-2). With this dry method based on Grabner's test setup, the vapor-liquid ratio is 3:2. In particular, it indicates the azeotropic increase in vapor pressure

		SuperPlus % (V/V)	Super % (V/V)	Regular % (V/V)	Standard range class D% (V/V)
E70	Average Range	36 29–46	35 30–47	37 29–48	22–50
E100	Average Range	55 48–62	54 50–63	58 50–67	46–71
E150	Average Range	87 78–93	86 79–94	87 76–98	Min. 75

Fig. 22-41 Distillation values for German winter gasoline.[5]

	SuperPlus	Super	Regular
Range	176–210	172–210	162–208
Average	194	193	190

Fig. 22-42 Final boiling point of German gasoline.[5]

in the measurement of the vapor pressure of methanol-containing gasoline hotter than 38°C. The VP in this instance is clearly higher that gasoline without methanol, whereas adding TBA or ether yields a normal vapor pressure (wet). In general, a vapor pressure that is too low, i.e., a fuel that evaporates too slowly, results in unsatisfactory starting and cold driving behavior, whereas a vapor pressure that is too high produces problems in hot starting and hot driving behavior. In addition, the formation of an air-vapor mixture when storing fuel above the top explosion point requires a sufficiently high vapor pressure to be safe. The "true" vapor pressure at 50°C is found in the transport regulations. It applies to a vapor-liquid ratio of 0:1 and is calculated from the RVP.

With the amendment of DIN EN 228 on February 1, 2000, the method used to determine the vapor pressures also changed. The Reid method was replaced by the DVPE (dry vapor pressure equivalent) according to DIN EN 13016-1. The DVPE is calculated from the ASVP (air saturated vapor pressure), for example, using the Grabner equipment with a vapor-liquid ratio of 4:1. By changing DIN EN 228, the volatility classes also changed. For the first time, two transition periods were provided between winter and summer quality (see Fig. 22-29).

Benzene Content

Benzene (C_6H_6) is the basic element of aromatic hydrocarbons. Because of its high ON (RON and MON 100) and availability from coke manufacture, it was earlier used as an important component in super gasoline. However, this was engine benzene, a mixture of benzene, toluene, and xylene (see Fig. 22-1), the secret of the first super fuel of the world "ARAL" (aromates/aliphates) marketed in 1924, a product whose antiknock quality and other qualities was clearly superior to gasoline. After introducing catalytic reformers in the 1950s, the use of engine benzene in Germany became increasingly less important. After the health risks from handling benzene became known, benzene largely stopped being used as an additive, especially since we became aware of other ways of replacing undesirable lead-based antiknocking agents. However, other aromates continue to play a large role in modern gasoline. In EU standard 228 for gasoline, the benzene content was limited for a long time to a maximum 5% (V/V). In the market, it was an average of 2% (V/V) and in SuperPlus (since 1995) even 1% (V/V). Since January 1, 2000, the established maximum threshold for all gasoline qualities was 1% (V/V). Figure 22-43 shows the changing benzene content in German gasoline from 1986 to 1994.

However, numerous other aromates are also used in fuels. Figure 22-44 gives an overview of the aromates used in gasoline.

Aromates already exist in petroleum, but most are produced by catalytic reformers with the release of hydrogen. Figure 22-45 provides information on the aromate content in German gasoline.

It is interesting to compare the olefin content in German gasoline as shown in Fig. 22-46. We can see that it strongly decreases as the MON requirement increases.

Sulfur Content

Sulfur occurs in petroleum almost exclusively in bound form as mercaptan sulfur, disulfide sulfur, etc. Mercaptans (thioalcohols) are sulfur derivatives of alcohols in which the oxygen of the hydroxyl group OH is replaced by S. In petroleum, the S content ranges from 0.01% to 7.0%. In fuels, a high S content has always been undesirable, and sulfur has, therefore, been removed in the refineries as costs allow. Apart from SO_2 emissions, a few catalytic converters, especially uncontrolled catalytic converters,

Year	SuperPlus	Super	Regular	Super leaded	Average
1986	—	2.84	2.36	2.83	2.40
1987	—	2.69	2.17	2.89	2.50
1988	—	2.59	2.21	2.83	2.60[b]
1989	2.52	2.60	2.05	2.87	2.50
1990	2.59	2.76	2.16	2.73	2.40
1991	2.10	2.42	1.71	2.48	2.20
1992	2.42	2.49	1.78	2.49	2.20
1993	2.39	2.27	1.66	2.53	2.10
1994	2.04	2.08	1.57	2.26	1.90

[a] The % (V/V).
[b] Ban of regular leaded.

Fig. 22-43 Change of the benzene content.[a] [12]

Product	Empirical formula	Boiling point Boiling range	Mixed ON RON	Mixed ON MON
Toluene	C_7H_8	110°C	124	112
Ethyl benzene	C_8H_{10}	136°C	124	107
Xylene	C_8H_{10}	138–144°C	120–146	103–127
C_9 aromates	C_9H_{11}	152–176°C	118–171	105–138
C_9 + aromates (small quantities)	$C_{10}H_{12}$ $C_{11}H_{13}$	169–210°C	114–155	117–144

Fig. 22-44 Aromates used in gasoline.[12]

% (V/V)	SuperPlus	Super	Regular
Toluene	13.1	10.5	9.8
Xylene	12.7	11.0	11.4
C_8 + Ar	12.7	12.2	12.8

Fig. 22-45 Aromate content in German gasoline (1994 average).[12]

Olefin content % (V/V)	SuperPlus	Super	Regular
Range	0–17	1–22	1–37
Average	4	10	18

Fig. 22-46 Olefin content of German gasoline.[12]

tend to convert sulfur into hydrogen sulfide (H_2S) that smells under certain operating conditions. In addition, the catalytic converter efficiency decreases as the S content raises, which correspondingly increases the emission of CO, HC, and NO_x, which can have serious consequences, particularly in storage catalytic converters.

According to the quality standard DIN EN 228, gasoline can contain only 150 mg/kg sulfur as of January 1, 2000. Figure 22-47 shows the typical S content of German gasoline from 2000.

Figure 22-47 shows that the sulfur level is sometimes far below the present threshold of 150 mg/kg.

Reformulated Fuel

This is to be understood as a change in the composition and/or physical characteristics to reduce pollutant and evaporation emissions. Within the European auto-oil program (EPEFE), the influence on emissions of all essential gasoline parameters was investigated. Figure 22-48 shows the qualitative options and consequences of the different measures.

Apart from economic disadvantages, some of the possible measures have contrary effects on the individual types of emissions. As we can see, the only measure that reduces all types of pollutants in the exhaust gas is the continued reduction of the sulfur contents.

A few comments are made on the term "designer fuels" that has surfaced recently. These are tailored special fuels for the automobile industry that are used, for example, to meet special requirements for first tanking or for research purposes. They are not generally defined but are rather individually composed to achieve the desired special properties.

SuperPlus	Super	Regular
5–45 (mg/kg)	10–130 (mg/kg)	20–140 (mg/kg)

Fig. 22-47 Typical S content in German gasoline.[6]

Additives for Gasolines

We do not discuss the most important antiknock gasoline additive of an earlier era, lead, since it is no longer used. In addition, combustion chamber residue converters (scavengers) are no longer necessary. However, wear has arisen in a few older engines with "soft" valve seats when unleaded fuels are used under continually high loads, and this can be countered with special additives. In addition,

Fuel parameter	CO	HC	Benzene	NO_x	SO_2	CO_2	Possible refinery method	Economic disadvantage	Technical disadvantage
Sulfur reduction	⇓	⇓	⇓	⇓	⇓	—	Expand system Hydrodesul-furization	Increased costs	More CO_2
Lower final boiling point					—	⇓	No investments	Low availability Less profitable	Low density leads to higher volumetric consumption
Higher volatility within the average boiling range	⇑	⇓		⇑	—	—	Only some investments required	Low extra costs	Endangers MON level
Higher volatility within the lower boiling range	⇓	⇓			—	⇓	No investments		Evaporation losses Hot behavior
Reduce aromate content	⇓	⇓	⇓	⇑	—	⇑	Isomerization Alkylation	Substantial extra costs	Endangers RON level Refinery CO_2 rise
Lower benzene content	—	—	⇓		—	—	Various levels of investment	Moderate extra costs	
Alcohols/ethers as gasoline components	⇓	⇓	⇓ ⇑	⇓	⇓	⇓	Component tanks	Higher product costs	Higher volumetric consumption Hot behavior Corrosion

Fig. 22-48 Reformulated gasoline.[12]

additives against carburetor icing—earlier essential—are no longer used. Throttle valve icing that occasionally occurs can be countered with surface-active detergents. Today's additive packages in gasolines primarily prevent system-related deposits in the fuel and mixture formation systems, primarily on intake valves. It has been demonstrated by many investigations that the use of tailored additive packages are essential for the endurance and cleanness of the engines and its fuel systems, the maintenance of exhaust gas values of new engines, and attaining and maintaining overall favorable operation. They are also economical over the long term. These problems are not new. Modern and future high-performance engines have given rise to new problems. For example, different temperature and flow conditions exist at the intake valve. There is practically no flow of oil that earlier had a certain rinsing effect. The result is increased deposits on the rear of the valves. In regard to combustion chamber deposits, the crowded normal multivalve arrangement can substantially worsen emissions. A reduction of these deposits by 1 g can reduce NO_x emissions by 18%–19%. Depending on the engine and operating conditions, deposits of 4-8 g arise in practice without additives. The goal is to limit combustion chamber deposits to 1.3 g per cylinder. The problem in developing additives is optimizing a package to deal with the contradictory requirements of intake valve cleanness (requires thermostabile components) and combustion chamber deposits (lowest possible thermostability). Since oil consumption is nearly zero today, fuel is increasingly finding its way into the engine oil from fuel and additive condensate. This phenomenon can be very pronounced since today's oil-changing intervals are so long. Fuel and oil vapors accordingly pass through the enclosed crankcase ventilation system into the air intake system, into the combustion chamber, and up to the catalytic converter which can damage or destroy it. To reduce such problems, additive components should be prevented from entering the oil pan. This illustrates yet another conflict with the otherwise necessary thermostabile additives.

The transition to extremely sulfur-low fuels worsens the natural lubrication properties of the gasoline since surface-active components are removed in desulfurization.

The resulting higher pump wear must be counteracted by special antiwear additives. As a positive side effect, this can reduce fuel consumption by 3.5%. The spark-ignition engines with direct injection (DI engines) that will be common in the future have a series of particular problem zones that can be dealt with only by special additives. The anticipated consumption and emission advantages especially rely on the precise formation of a mixture cloud in time and space. This sensitive system can be destroyed by the slightest deposits with correspondingly negative effects. Deposits must, therefore, be avoided, particularly on the nozzles. The cleanness of the intake ducts is important to the generation of the required swirl. Fuel additives, however, may not enter the intake ducts of DI engines. The injection pressure is also much higher, and this increases fuel pump wear. The required protection from wear must be assumed by new "lubricity improvers" or "friction modifiers." The engine-damaging formation of black sludge that is caused in particular by the high NO_x formation in lean engines can be reduced by the fuel-additive entering the oil. Figure 22-49 provides an overview of the additives that are required today.

Gasolines from different manufacturers are all miscible, but the balanced effect of the additives can be lost, and in certain circumstances damage may arise.

22.1.2.3 Alternative Gasolines

Although numerous alternative fuels are mentioned in the media, only a few of them pose real alternatives to the familiar fuels based on fossil energy. To draw clear distinctions, we first need to define unrenewable and renewable or regenerative energy sources. The generated primary energies are not directly useful for powering vehicles in their normally available form and must first be converted by suitable methods into appropriate secondary energy. In addition, certain minimum requirements need to be posed on the secondary energies suitable as alternatives to today's gasoline such as technical feasibility and storage life in the distributor system, transportability, use of the gas station infrastructure, and storability with sufficient energy density in the automobile. Given these basic requirements, the following secondary energies are suitable in addition to today's gasoline types:

- LPG Liquefied petroleum gas. Pressurized, liquefied LPG based on propane and butane.
- CNG Compressed natural gas. Compressed natural gas based on methane.
- LNG Liquefied natural gas. Gas liquefied at low temperatures based on methane.
- MEOH Methanol. Alcohol, usually from natural gas (methane), also termed wood spirit.
- ETOH Ethanol. Alcohol from sugar-containing plants. Also termed spirit or sprit in German.
- GH_2 Gaseous hydrogen. Can be made from water and all hydrogen-containing energy carriers.
- LH_2 Liquefied hydrogen. Hydrogen that is liquid at a low temperature.

Strictly speaking, only those fuels that are not produced from the primary energies petroleum, natural gas, or coal can be considered true alternative fuels. They must also be available in a sufficient amount to supply a continuously

Component	Active ingredient	Improved	Comments
Antioxidants	Paraphenylene diamine Hindered alkyl phenols	Storage stability Polymerization	Improved stability of crack components
Metal deactivators	Disalicylide Propane diamine	Stops the catalytic effect of metals	Improved stability of crack components
Corrosion inhibitors	Carboxyl amine, ester amine compounds	Corrosion protection	Usually used together with detergents
Detergents	Polyisobutene amine Polyisobutene polyamide Carboxylic acid amide Polyether amine	Cleanness of intake and fuel system, prevents throttle valve icing Drivability Exhaust emissions	Most important gasoline additive used together with carrier oils
Lubricity improvers	Polyisobutene amine, etc.	Life of the fuel injection pump	Compensates for the lubricity loss in low sulfur gasoline
Wear protection	Organic potassium, sodium compounds	Protects exhaust valve seats	Lead replacement for old automobiles, usually a separate additive

Fig. 22-49 Overview of gasoline additives.[7]

growing percentage of the world's automobile population that is continuously growing itself. Accordingly, only hydrogen remains a real alternative fuel that is available over the long-term. Since the solution to the numerous related problems will take a great deal of time, we need to closely consider supplements to classic gasoline, i.e., LPG, CNG/LNG, methanol, and ethanol for the transitional period.

Gas Fuels LPG/CNG/LNG

Under the name of propellant gas or liquid gas, mixtures of the refinery gases propane and butane were used as emergency fuels, particularly in the initial period after World War II. They were primarily used in commercial vehicles with thin-walled steel tanks that were exchanged at filling stations. Today, LPG is sometimes used in dual operation with gasoline as a gas in pressure tanks that are filled at special LPG pumps. Special quality requirements are established in EU standard EN 589. The details are in Fig. 22-50.

We can see that the vapor pressure is much higher than with gasoline. The vapor pressure can be set by the ratio of propane to butane in the mixture. The maintenance of the vapor pressures in classes A to D is required to ensure cold starts. In addition, the manufacturer must ensure that a characteristic unpleasant smell is perceptible when 20% of the lower explosion threshold is reached (see 22.1.2.). Liquid gas is gaseous at normal pressure and temperature. Since gases in relation to their volume have substantially less energy than gasoline, they are liquefied under pressure for storage. At room temperature, they are liquid at 25 bar. Figure 22-51 shows a few interesting characteristics for liquid gas.

The use of liquid gas in spark-ignition engines has a few advantages such as clean combustion with high performance and low fuel consumption, and improved untreated emissions in the exhaust. Unfortunately, these advantages can be attained only in monovalent gas vehicles when the engine and automobile are set up for gas operation. Likewise, their high antiknock quality cannot be exploited without a clear increase in compression. Disadvantages are the increase in weight and reduced trunk area from the pressure tank. An important factor for economical use is the country-specific tax burden. In contrast to Holland and Italy where excess liquid gas from refineries is supported as a supplemental fuel by tax preferences, in Germany the sum of the tax burden and the money required to convert to dual operation because of an insufficient gas station network is prohibitive, even for high-mileage drivers. In addition, there are restrictive ordinances such as the ban on access to underground and above-ground garages. It is also difficult for engines to maintain the further-stiffened exhaust thresholds in dual operation. In summary, LPG can play only a subordinate role in Germany, even as a supplementary fuel.

CNG or natural gas—methane—under a high pressure (300 bar) was first used after World War II in the Ruhr for powering heavy commercial vehicles with spark-ignition engines from the Benzene Association of that period. Given the close collaboration with the mining industry at that time, methane from mines was compressed to 300 bar in a high-pressure ring main built in 1950, sent

Property	Unit of measure	Threshold		Test methods
		Min.	Max.	
MON	—	89		Calculated
Content of 1,3-butadiene	Molar %		0.5	ISO 7941
Hydrogen sulfide	Mg/m^3		<4	ISO 8819
Overall sulfur	Mg/kg		200	ISO 24260
Cu corrosion	Degree of corrosion	1		ISO 6251
Evaporated residue	Mg/kg		100	NF M 41-015
Absolute vapor pressure at 40°C	KPa		1550	ISO 2456
Absolute vapor pressure min. 250 kPa at temp.	°C			
Class A			−10	ISO 4256
Class B			−5	
Class C			0	
Class D			+10	

Fig. 22-50 Quality requirements of liquid gas (excerpt from DIN EN 589).[12]

Characteristic	Unit of measure	Propane	Butane	50/50
Empirical formula	—	C_3H_8	C_4H_{10}	—
Density of the gas at 15°C	kg/m³	1.81	2.38	2.06
Density of the liquid at 15°C	kg/m³	510	580	540
Boiling point	°C	–42	–0.5	–20.7
Volumetric calorific value	MJ/m³	93.45	108.4	101.9
Mass calorific value	MJ/kg	46.1	45.75	45.8
RON	—	111	94	100
MON	—	96	89.6	95

Fig. 22-51 Characteristics of liquid gases.[12]

via compressor stations to different distribution sites where high-pressure bottle batteries in heavy commercial vehicles were filled. This pioneering effort that was somewhat daring for its time was terminated in 1953 since a sufficient amount of gasoline again became available, and heavy trucks with spark-ignition engines fell out of favor. Based on the present consumption of natural gas, easily accessible supplies would last 60 to 65 years. Natural gas can, therefore, be counted on as a supplemental fuel over the long term. CNG is natural gas compressed to 200 bar for use in automobiles. It can be brought to this pressure at appropriately set-up gas stations. Figure 22-52 shows the design of such a station.

Natural gas that has a different composition depending on its origin, typically approximately 90% methane and 10% ethane, is primarily suitable for correspondingly adapted spark-ignition engines in monovalent and bivalent operation due to its high antiknock quality. In the United States and in Italy, corresponding tax incentives have led to the use of a substantial volume of natural gas, even in passenger cars. However, its use is primarily restricted to a small number of commercial vehicles and pickups due to the complex and large gas tanks. The substantial advantages in emissions are particularly favored in city busses. Figure 22-53 compares the values of CNG with gasoline.

In gas engines, the mixture is formed in a mixer that is similar to the earlier carburetor. The functional principle is that of a Venturi tube. Because of the vacuum that predominates at the narrowest site, the required amount of natural gas is sucked in through holes at the constriction and mixed with air. Since CNG is stored in automobiles under a pressure of 200 bar, a gas pressure controller is necessary to expand the natural gas to environmental pressure and supply it to the mixer. The performance is controlled via a throttle valve. In comparison to diesel engines, a drop in performance of approximately 5% must be taken into account. The reason for this is the lower air intake air volume corresponding to the amount of gas, and throttling arising from the throttle valve and Venturi mixer. In comparison to diesel engines, the volumetric fuel consumption

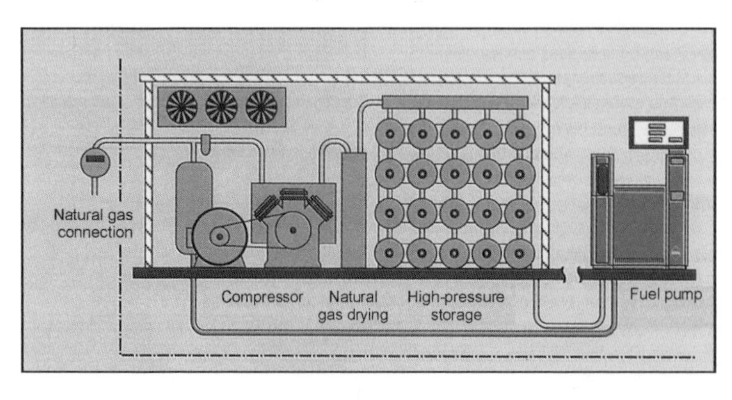

Fig. 22-52 Design of a CNG gas station.[8]

Physical characteristic	Super gasoline	CNG
Aggregate state in the tank	Liquid	Gaseous
Pressure in the tank	Atmospheric	200 bar
Density	751 kg/m^3	170 kg/m^3
Bottom calorific value (volume)	30.8 MJ/l	7.2 MJ/l
Bottom calorific value (mass)	41.0 MJ/kg	47.7 MJ/kg

Fig. 22-53 Physical characteristics of CNG in comparison to gasoline.[12]

following the full load curve shows losses of 8% to 10% from the reduced compression ratio. At a low load and speed, the situation is, however, much worse. For example, fuel consumption by municipal busses is 22% to 35% more depending on the use. However, a particular advantage in this type of use is the complete freedom from soot at high torques and low speeds, which makes the exhaust gas practically particle-free, and the engine is substantially quieter. The standard regulated three-way catalytic converter alsp ensures extremely low emissions. CNG can also be stored at –160°C at 2 bar, while natural gas is then in liquid form as LNG. The liquefaction of the gas requires additional energy and is done beforehand on a large industrial scale. A perfectly insulated cryotank must be used in the automobile in conjunction with the required control system. The expense is substantial. This technique has never been used in practice except for text purposes. CNG can likewise be used only to a limited extent as a supplemental fuel.

Hydrogen

This secondary energy can be obtained from numerous hydrogen-containing substances such as water, natural gas, methanol, or biomass with the expenditure of energy to release the H_2. The importance of hydrogen as an environmentally friendly cyclical system can be seen in Fig. 22-54.

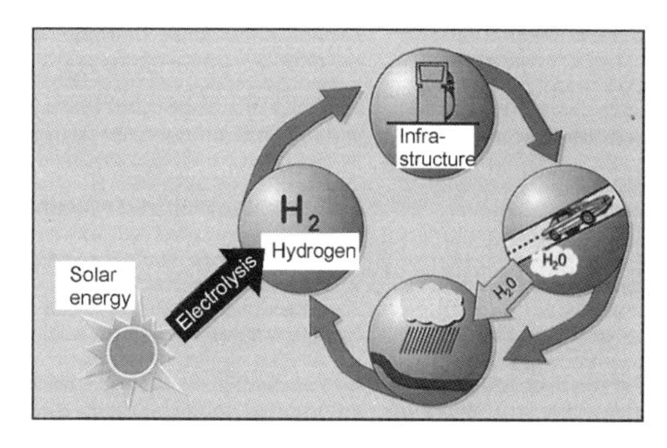

Fig. 22-54 Cycle of hydrogen technology.[8]

Ideally, but still extremely expensive, hydrogen is obtained by being regenerated, e.g., from water with the aid of solar energy, water power, wind energy, or electrolysis. When combusted in spark-ignition engines, NO_x arises from the air, but practically no pollutants or CO_2, just water in the form of vapor that can return into the cycle. Figure 22-55 lists the physical characteristics.

In terms of its mass, liquid hydrogen has approximately three times the energy of hydrocarbons and much wider ignition limits in air. There are theoretically three options for transport, storage, and the distribution system: High-pressure storage tanks, metal hydride storage tanks,

Description	Unit of measure	Physical characteristic
Density of the liquid at 20.3°K	kg/m^3	70.79
Density of the gas at 20.3°K	kg/m^3	1.34
Density of the gas at 273.15°K	kg/m^3	0.09
Evaporation heat	kJ/kg	445.4
Bottom calorific value	MJ/kg	119.97
Bottom ignition limit in air	% (V/V)	4.0–4.1
Top ignition limit in air	% (V/V)	75.0–79.2

[a] At absolute 1.013 bar.

Fig. 22-55 Physical characteristics of hydrogen.[a 12,13]

and liquid storage tanks. High-pressure storage tanks are unsuitable for transporting large amounts of hydrogen and for installation in passenger cars since they are unsafe, heavy, and very expensive. They are not completely out of the picture for commercial vehicles. Metal hydride storage tanks in which the hydrogen is adsorbed on metal alloys offer a high safety standard, but the amount of hydrogen they can store is limited despite the present level of development. For a passenger car range of 200 km, a tank weighing several hundred kilograms is required. In a liquid storage tank, the hydrogen is cooled to –253°C and stored in a cryotank using high-performance insulation (LH_2). The large amount of energy required for cooling to such low temperatures is substantial, however. In addition, when automobiles stand for a long period, hydrogen is lost from leakage. At the present level of development, a loss of more than 2% per day must be anticipated. Filling up requires an enormous effort since all moisture and air must be evacuated in addition to dealing with the low temperature. Basically, only fully automated fill-ups are possible without human interference. The long time that it initially took to fill up has been reduced to an acceptable level. Substantial developments are still required for commercial use. Possible energy converters in cars are both the spark-ignition engine and the fuel cell. The method of mixture preparation and combustion control would have to be redeveloped for the spark-ignition engine. Combustion occurs in a very lean range, which is optimum for controlling the combustion process and for NO_x emissions. This is associated with a performance loss, however. With less excess air, it may be necessary to inject water into the intake manifold since backfiring in the intake system could arise. It would be ideal to directly inject the liquid hydrogen into the combustion chamber. In addition, because of the high ignitability of hydrogen, the spark-ignition engine compression cannot be very high, which worsens the thermodynamic efficiency. The NO_x problem should take a while to solve. Given the worldwide success in developing fuel cells over recent years for electric engines in motor vehicles, one can assume that this system is more suitable for the use of hydrogen than the spark-ignition engine. However, the production methods for fuel cells must still be optimized. It is relatively expensive to generate hydrogen based on fossil energy carriers in terms of the overall energy balance. Finally, a suitable infrastructure must be developed for filling up such automobiles.

Alcohol Fuels

Alcohols are hydrocarbon-oxygen compounds whose particular feature is an OH group in the molecule instead of a hydrogen atom. Alcohols are, in principle, suitable for powering automobiles with spark-ignition engines. The techniques for producing them are known and well developed. They can be transported, stored, and distributed essentially in the existing system. The physical characteristics of the most important alcohols have already been portrayed in Fig. 22-25. In our discussion, only methanol and ethanol are of interest as supplemental fuels. Figure 22-56 again provides a comparison of their physical characteristics with super fuel.

Serious differences in practice are found in the antiknock quality, the calorific value, and the evaporation heat. A particular advantage is the high antiknock quality that can be used to improve efficiency through a correspondingly high compression ratio. Alcohols also burn faster, which means that the ignition map must be correspondingly adapted (see "racing fuels"). The substantially lower volumetric calorific value correspondingly yields a higher road fuel consumption. The much greater evaporation heat causes greater cooling of the fuel-air mixture which, because of the superior internal cooling, improves the charging and, hence, performance. The substantial increase in volume of the fuel-air mixture after fuel evaporation allows higher average pressures than gasoline and offers greater thermodynamic engine efficiency. In addition, the ignition range of an alcohol-air mixture is greater than that of gasoline, which allows more excess air under a partial load. The effects on untreated exhaust emissions are also positive. Special measures, such as preheating the air intake system, are required particularly at low temperatures to deal with the higher boiling point in comparison to the initial boiling point, and the lower vapor pressure in conjunction with the stronger cooling from the high evaporation heat. In comparison to gasoline, the drop in

Name	Abbreviation	Boiling Point	Density 20°C	Vapor pressure	RON	MON	Calorific value BC	Evaporation heat	O_2 content
		°C	kg/m^3	Hpa			MJ/1	kJ/kg	% (m/m)
Methyl alcohol	Methanol	64.7	791.2	32	114.4	94.6	15.7	1100	49.93
Ethyl alcohol	Ethanol	78.3	789.4	17	114.4	94.0	21.2	910	34.73
Super fuel	Super	30–215	725–780	S: 60–70 W: 80–90	95	85	Approx. 31	380–500	0–2

Fig. 22-56 Physical characteristics of methanol and ethanol in comparison to super gasoline.[8,12]

temperature of the theoretical mixture without preheating is 120°C for methanol and 63°C for ethanol. The aggressiveness of alcohols against metals and elastomers requires the use of special materials and special additives. Alcohols can be used in their pure form as well as in mixtures with hydrocarbons. Low concentrations as described in Section 22.1.2.1 do not require any changes to the automobile. Higher concentrations such as 15% (V/V) methanol in the gasoline require corresponding adaptations. The basic approaches for this were worked out 20 years ago in a joint project of the German automobile and mineral oil industries subsidized by the Federal Ministry of Research. The following facts are relevant concerning the production and use of methanol and ethanol:

The simple alcohol molecule CH_3OH (methanol) arises almost exclusively from the synthetic gases CO and CO_2 that can be obtained from every carbon-containing primary energy carrier, today preferably from natural gas. It has a large amount of H_2 (C/H ratio 4:1) which also makes it interesting as a starting product for fabricating hydrogen using the infrastructure of fuel cell technology, at least in the initial phase. It can be used as a gasoline-methanol mixed fuel, for example, M 15 (MEOH), or as methanol fuel (M 100). Since the danger of separation upon the addition of water increases as the methanol content decreases, the stability of methanol-gasoline-water mixtures must be monitored. The methanol fuel M 100 must contain HC components for cold starts and warm-up, as well as other substances such as additives to allow use in automobiles.

The addition of a certain amount of gasoline is also required for safety reasons since methanol burns with an invisible flame. The essential points of possible specifications for methanol fuel are shown in Fig. 22-57. The advantage of the high octane number, fast combustion, and larger volume expansion of the fuel-air mixture is contrasted with a consumption disadvantage of approximately 70% in an optimized methanol engine. In addition, methanol tends to preignite much more strongly than gasoline, which requires special modification of the engine such as cold spark plugs, etc. When methanol fuels are used, special engine oils without ash-free dispersants are required since they can form a sticky residue upon

Characteristic	Unit of measure	Summer	Winter
Methanol	% (m/m)	Min. 82	Min. 82
HC total[a]	% (m/m)	Min. 10, Max. 13	
Butane	% (m/m)	Max. 1.5	Max. 2.5
Density 15°C	kg/m^3	770–900	
Vapor pressure RVP	mbar	550–700[b]	750–900[b]
Water content	ppm	Min. 2000, Max. 5000[c]	
Higher alcohols	% (m/m)	Max. 5	
Formic	ppm	Max. 5	
Overall acid[d]	ppm	Max. 20	
Evaporation residue	mg/kg	Max. 5	
Chlorine	ppm	Max. 2	
Lead	ppm	Max. 30	
Phosphorous	ppm	Max. 10	
Sulfur	ppm	Max. 100	
Additives	%	Max. 1	

[a] Type of hydrocarbon, boiling behavior and quantity depending on use
[b] Example from central Europe
[c] With corrosion inhibitors
[d] Measured as acetic acid

Fig. 22-57 Specification for methanol fuel.[13]

contacting methanol. In addition, they require additives against corrosive engine wear.

Ethanol

C_2H_5OH is the second in the series of alcohols characterized by the hydroxil OH group. It can be obtained from biomass by fermenting agricultural products. Suitable starting products are all sugar, starch, and cellulose-containing raw materials. Figure 22-58 shows the possibilities for ethanol generation.

Glucose is converted into alcohol with yeast. To date, the generation of ethanol for fuel from sugarcane has the greatest economic importance in Brazil. To increase the yield and reduce the competition between foods and fuel production, it would be more advantageous to use primarily cellulose-containing plants. The fermentation must be preceded by a conversion process that transforms the different cellulose types into glucose depending on the plant type. From the user's perspective, cold-start problems arise unless special constructive measures are taken. The use of alcohols in diesel engines is discussed in the section "Alcohol-Diesel Mixtures." In summary, the alcohols methanol and ethanol are not viable alternative fuels, but they can serve a role as supplemental fuels in the (very) long transition period to hydrogen.

Racing Fuels

Of course, beyond the special constructive measures to attain maximum specific performance, advantages have also been sought from the fuel. As long as existing laws do not provide restrictive regulations, such as for commercial Super, performance enhancements from fuel have been substantial. The goal is to combine the maximum antiknock quality at a maximum compression ratio, the highest internal cooling for the best charging, and a high rate of combustion for maximum speeds with the greatest possible energy supply. Furthermore, the fuel volatility must be adjusted so that it meets the requirement for the greatest possible volumetric efficiency. Figure 22-59 presents the mixture components that are relevant for racing fuels with their particularly important characteristics (some of which are no longer used).

Performance can be increased directly by fuel with components with a high energy content and low stoichiometric air-fuel ratio. With a given air supply, this combination allows an increase in the actually supplied amount of energy. A typical example is nitromethane whose calorific value is clearly lower than gasoline; however, the much lower stoichiometric air-fuel ratio allows a more than twofold increase in the energy supply (specific energy). Its use is, however, limited by the high thermal

Sugar-containing		Cellulose-containing
Sugar cane Sugar beat Sorghum	Grain Corn Cassava Potatoes	Forest scrap wood, quickly growing trees Hemp, kenaf Bagasse, straw Stalks, husks, hulls Used paper

Fig. 22-58 Vegetable raw materials for generating ethanol.[8,12]

Components	Density at 20°C	RON	MON	Final boiling point	Evaporation heat	Bottom calorific value
	(kg/m³)			(°C)	(kJ/kg)	(MJ/l)
Acetone	791			56	524	
Diethyl ether	714			35	487	24.3
Ethanol	789	114.4	94.6	78.5	910	21.2
Methanol	792	114.4	94.0	64.7	1100	15.6
Benzene	879	99	91	80	394	34.9
Toluene	867	124	109	110	356	34.6
Nitrobenzene	1200			208	397	
Water	1000	—	—	100	2256	—

Fig. 22-59 Components of racing fuels.[1]

Characteristic	Unit of measure	Nitromethane	Methanol	Isooctane
Empirical formula	—	CH_3NO_2	CH_3OH	C_8H_{18}
O_2 content	% (m/m)	52.5	49.9	0
Evaporation heat	kJ/kg	560	1170	270
Bottom calorific value	MJ/kg	11.3	19.9	44.3
Stoichiometric air-fuel ratio	—	1.7 : 1	6.45 : 1	15.1 : 1
Specific energy[a]	MJ/kg	6.65	3.08	2.93

[a] Quotient from BC and stoichiometric air-fuel ratio.

Fig. 22-60 Specific energy of nitromethane in comparison to *i*-octane.[12]

and mechanical load on the engine-transmission assembly. Figure 22-60 provides information on these relationships for nitromethane and methanol in comparison to iso-octane.

Prestressed ring-shaped compounds such as quadrocyclane and diolefins such as diisobutylene yield measurable, direct increases in performance. Although they usually have a low ON and are unsuitable according to a conventional evaluation, they are much less knock sensitive than their ON seems to indicate due to a clearly higher rate of combustion with a corresponding adaptation of the ignition map. Another advantage of a high rate of combustion is that the extremely high speeds tend to shift the energy conversion toward the top dead center, which improves efficiency. This advantage also holds true for olefin-containing, conventional crack components. Given the sometimes contradictory properties of the fuel components, harmonizing the engine with the fuel is a delicate and, hence, involved procedure. The availability of the discussed special fuels is restricted, and they are all very expensive.

Figure 22-61 compares quadrocyclane with toluene, and diisobutylene with iso-octane.

Special racing fuels were already being used in the 1930s for Grand Prix racecars. Figure 22-62 presents the contents of the then strictly secret racing fuels of the prewar competitors Auto-Union and Mercedes-Benz with Alfa Romeo and Maserati.

22.2 Lubricants

Lubricants are design elements without which the internal combustion engine and transmission could not function. They were developed along with the automobile and continue to be developed because of a number of interrelated factors. The complex composition of modern lubricants allows them to satisfy even the highest requirements.

22.2.1 Types of Lubricants

The term "automobile lubricants" includes the following subcategories:

- Engine oils for four-stroke spark-ignition and diesel engines
- Two-stroke oils for motorcycles, scooters, and mopeds
- Universal oils for tractors

Characteristic	Unit of measure	Quadrocyclane	Toluene	Diisobutylene	Iso-octane
Empirical formula	—	C_7H_8	C_7H_8	C_8H_{16}	C_8H_{18}
Density	kg/m^3	919	874	719	699
RON	—	54[a]	124[a]	98[a]	100
MON	—	19[a]	112[a]	78[a]	100
Bottom calorific value	MJ/kg	44.1	40.97	44.59	44.83
Stoichiometric air-fuel ratio	—	13.43	14.70	13.43	15.05
Specific energy	MJ/kg	3.28	2.79	3.32	2.98

[a] Mixture ON.

Fig. 22-61 Specific energy of quadrocyclane and diisobutylene.[12]

Components	Auto-Union Mercedes-Benz	Alfa Romeo Maserati
	% (V/V)	% (V/V)
Ethanol	10	49.5
Methanol	60	34.5
Denaturing	—	0.5
Benzene	22	—
Petroleum ether	5	—
Water	—	0.5–3
Remainder	3[a]	12–15[b]

[a] Toluene/nitrobenzene/castor oil.
[b] Information not provided.

Fig. 22-62 Grand Prix racing fuels used before 1939.[1]

- Transmission oils
- Hydraulic oils
- Greases

In this section we discuss the first two categories, engine oils and two-stroke oils.

22.2.2 Task of Lubrication

The lubricant is to reduce friction between contact bodies, reduce their wear, remove any wear particles from the lubrication site, and prevent the penetration of foreign materials into the lubrication gap. They must also transfer force, for example, from the piston to the conrod, cool by transferring heat to the oil pan or oil cooler, provide a seal, for example, of the annular gap between the piston and cylinder, protect against wear, prevent deposits and corrosion, neutralize acidic combustion products, be compatible with the elastomers of the seals, be stable over time to allow long change intervals, have a low evaporation loss for low oil consumption, and manifest optimum viscosity temperature behavior to ensure easy cold starts and reliable hot operation. Recent design elements such as multivalve technology, hydraulic valve lash compensation, camshaft timing, and supercharging represent substantial and sometimes new challenges. The new generation of spark-ignition engines with direct injection can, in contrast to existing engines with manifold injection and catalytic converters, be operated in the lean range of combustion. This can produce a new set of problems for the engine oil. In addition, we are confronted with demands to reduce fuel consumption by reducing friction, not impair the exhaust purification systems in spark-ignition and (in the future) diesel engines (Euro IV), and maintain environmental compatibility.

22.2.3 Types of Lubrication

A distinction is drawn between full and partial lubrication (liquid friction and mixed friction). Liquid friction, the ideal state, predominates, for example, in plain bearings at a certain speed, or after the external application of oil pressure. However, mixed friction must be assumed when, for example, the crankshaft and bearing directly contact each other upon starting without the lubrication film that must first build up. It is also unavoidable that in certain component assemblies such as the valve gear with tappets and cams, or at the point where the pistons reverse in the cylinders, mixed friction predominates over long periods of operation. It is important to provide the lubricant with antiwear and antioxidation additives and to have the lubricant be sufficiently viscous to minimize wear.

22.2.4 Lubrication Requirements

Among the most important requirements for engine oil are the following.

Transmitting Force

From the connecting rod, the entire combustion pressure exerted on the piston is transferred to the crankshaft via the piston pin and connecting rod bearing only with the aid of the slight amount of oil in the lubrication gaps. The arising pressure in the thin lubrication gap can be as large as 10 000 bar.

Cooling

The engine oil plays a relatively minor role in removing heat, but its specific task is quite important, namely, piston cooling. On the one hand, the oil flying about within the engine conducts heat from the hot piston; on the other hand, particularly in highly supercharged diesel engines, usually additional oil is sprayed from below against the piston head or guided into a separate cooling channel to cool especially the upper piston ring area.

Sealing

Engine oil has the important task of providing a fine seal between the piston, piston rings, and cylinder barrel surface in order to transfer the high pressure from combustion with minimal loss to the piston head surface. Even when there is an optimum seal, approximately 2% of the combustion gases pass by the piston and enter the crankcase (blowby gas). This gas attacks the engine oil with aggressive reaction products from combustion.

Protection from Deposits

During combustion in spark-ignition and diesel engines, oil-insoluble residues necessarily arise in solid or liquid form. The residues must be prevented from agglomerating and collecting in the engine, for example in the piston ring grooves or in the oil pan. Oil-insoluble residues can form sludge under special circumstances. This problem is solved with detergent and dispersant additives.

Corrosion Protection

Mineral oil per se offers a certain amount of corrosion protection against slight amounts of water. This protection is, however, insufficient in the additional presence of aggressive combustion products. After the engine is turned off, humidity can condense into water inside the engine. Water also arises during combustion as a reaction product. One liter of fuel yields approximately 1 liter of water. Water leaves the hot engine largely in the form of steam together with the exhaust gases. A small part enters with the blowby gases into the crankcase and oil pan where it condenses when the engine cools. The engine oil can absorb only a certain amount of water so that corrosion can appear on unprotected metals if not counteracted by corrosion inhibitors.

Wear Protection

Mechanical and corrosive wear must be prevented primarily in the cylinder barrel, on the piston and piston rings, the bearings, and valve gear such as the cam, tappet, and rocker arm. In diesel engines, a special load arises in areas with mixed friction from soot formation. This especially includes the cylinder barrels. Mechanical wear can be effectively reduced by EP/AW additives (extreme pressure/antiwear), whereas corrosive wear can be kept under control by the neutralizing effect of corrosion inhibitors.

Seal Compatibility

The properties of radial shaft sealing rings used in engines, valve shaft seals, and other seals made of elastomer materials (elastic plastics) may not be altered by fresh or used oil. They may not become brittle, soften, or shrink and form cracks under stress. To ensure a lasting seal, a certain amount of swelling is desirable. To prevent or compensate for drying, i.e., the exchange of softener with polyalphaolefins (PAO) (for example) in the elastomers when certain synthetic basic liquids are used, "seal swell agents" are used.

Aging Stability

This is particularly important in view of the continual extension of the oil change intervals. At high operation temperatures, engine oils tend to "age" since oxygen bonds to the hydrocarbon molecules yielding acids, and resinous or asphalt-like components can form. The oil is continuously mixed with air as a thin film as it flows off, drips off, is flung off, and is sprayed inside the engine. This can make the engine oil thicker with the assistance of the blowby gases. To prevent this, antioxidants are used.

Evaporation Loss

The evaporation loss strongly depends on the viscosity and the type of basic liquid. Earlier, when only mineral oil raffinates were used as the basic oil, a general rule was the thinner the basic oil, the greater the evaporation loss. The basic liquids in high-performance engine oils of the "new technology" such as special raffinates, hydrocrack oils, synthetic hydrocarbons, and esters have the same viscosity yet different levels of evaporation loss. These are essential because of the increasing dwell times of the oil in the engine.

Viscosity Temperature Behavior

Today's assumptions that oils should be as thin as possible when cold and as viscous as possible when hot can be achieved only by multigrade oils with a high viscosity index (V.I.). In this case as well, modern basic liquids are far superior to earlier basic oil raffinates.

22.2.5 Viscosity/Viscosity Index (V.I.)

Viscosity is a measure for internal friction or the resistance a liquid offers to deformation. Assuming a laminar flow, the shear stress arising between two flowing layers is proportional to the speed gradients perpendicular to the direction of flow, according to Newton's shear stress law. The arising proportionality factor is termed dynamic viscosity or the absolute viscosity of the relevant liquid. It represents the force that counteracts the movement of a liquid layer in reference to 1 cm^2 that flows at the speed of 1 $cm \cdot s^{-1}$ parallel to a resting liquid layer at a distance of 1 cm. The unit of measure of the dynamic viscosity is 1 P (Poise) = 100 cP (centipoise; 1 cP = 1 $mPa \cdot s$; millipascal second). The term viscosity in the Newtonian sense is limited to the range in which the proportionality is retained independent of the gap width and shear speed. In lubricating oils, this proportionality can, for example, be lost by cooling when the original Newtonian liquid no longer follows the proportionality law from the precipitation of solid particles such as paraffins and the formation of a mixture of solids. This is particularly the case with artificially thickened oils such as multigrade oils. Instead of measuring dynamic viscosity, the easier solution in practice is almost always to measure the kinematic viscosity or relative viscosity. This results from the ratio of the dynamic viscosity to the density. The unit of measure is 1 St (Stokes) = 100 cSt (centistokes; cSt = 1 $mm^2 \cdot s^{-1}$). The viscosity is primarily influenced by the temperature and pressure and, in the case of non-Newtonian liquids, also by the shear speed.

22.2.5.1 Influence of Temperature on Viscosity

As the temperature rises, the distance of the molecules in the lubricating oil increases so that they move away from their mutual range of influence. This reduces inner friction and, hence, the viscosity. The dependence on temperature of the viscosity is particularly important in lubrication technology. It is evaluated and rendered comparable in an internationally uniform manner with the aid of the viscosity index (V.I.). The greater the V.I., the lower the temperature sensitivity of an oil. This relative identification is carried out using two extremely different temperature-sensitive reference oils. Both have the same viscosity at 100°C. With Pennsylvania reference oil, the viscosity increases very slowly as the temperature falls; with Gulf Coast reference oil, it rises quickly. The value V.I. = 100 has been assigned to the first reference oil, and V.I. = 0 to the second. The viscosity values of both reference oils are defined according to DIN ISO 2909 for the range of 2 to 70 $mm^2 \cdot s^{-1}$ at 100°C. For values above 70 $mm^2 \cdot s^{-1}$, equations are provided in the standard for calculation. The V.I. cannot be measured directly. It is calculated for a given oil by comparing it with the reference oils using the relationship in Fig. 22-63.

Figure 22-64 shows an example of graphically determining the V.I. for an oil with a viscosity of 8 $mm^2 \cdot s^{-1}$ at 100°C. For the L reference oil, 97 $mm^2 \cdot s^{-1}$ at 40°C was measured, and 57 $mm^2 \cdot s^{-1}$ at 40°C was measured for the H reference oil. These two values were assigned 0 and 100 V.I., and the differential range of 40 $mm^2 \cdot s^{-1}$ was divided into 100 increments. Since, for the oil sample P, a viscosity of 61 $mm^2 \cdot s^{-1}$ at 40°C was measured, the value 90 can be read on the V.I. scale.

With the development of multigrade oils with a V.I. > 100, a fundamental problem in determining the V.I. arose in the form of overvaluing the oils with a low viscosity. To solve the problem, a new calculation method was introduced that produced the $V.I._E$ (expanded V.I.). When the V.I. is particularly high, the $V.I._E$ allows a clear differentiation in evaluating the efficiency of V.I. improvers. The fully synthetic basic liquids in the engine oils of the new technology have a very high $V.I._E$ that is additionally raised by new V.I. improvers to cover the wide SAE ranges (see Section 22.2.8.1). The relationship in Fig. 22-65 is used to calculate the $V.I._E$.

$$V.I. = 100 \cdot \frac{L-P}{L-H} \tag{22.4}$$

H (High) ⇒ Viscosity of the reference oil with V.I. = 100 at 40°C

L (Low) ⇒ Viscosity of the reference oil with V.I. = 0 at 40°C

P (Sample) ⇒ Viscosity of the oil to be determined at 40°C

Fig. 22-63 Calculation of the viscosity index.

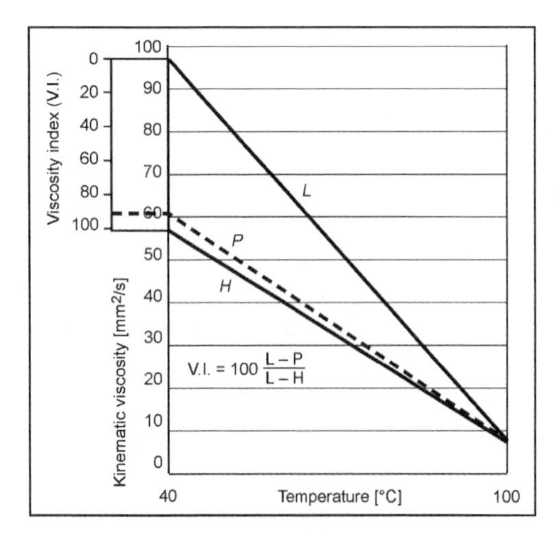

Fig. 22-64 Determining the V.I. in a graph.

$$V.I._E = 100 + \frac{G-1}{0.0075} \qquad G = \frac{\lg H - \lg P}{\lg Y} \tag{22.5}$$

H (High) ⇒ Viscosity of the reference oil with V.I. = 100 at 40°C

P (Sample) ⇒ Viscosity of the oil to be determined at 40°C

Y (Sample) ⇒ Viscosity of the oil to be determined at 100°C

Fig. 22-65 Calculating the $V.I._E$.

22.2.5.2 Influence of the Pressure on the Viscosity

When a lubricating oil is under a very high pressure, its viscosity rises strongly as under the influence of temperature because the now-denser molecules generate greater inner friction. In calculations of plain bearings, the influence from pressure is usually dismissed because it is assumed that the rise in viscosity from a rise in pressure is approximately compensated from the drop in viscosity due to the increase in temperature that always occurs. Accordingly, we know from substantial experience that a rise in pressure of approximately 35 bar has the equal but opposite effect as an increase in temperature of approximately 1°C. At very high pressures, for example, in roller bearings or geared transmissions (up to 15 000 bar), the rising pressure has to be taken into account. Roughly speaking, at room temperature the viscosity of most petroleum products doubles when the pressure rises by approximately 300 bar. Furthermore, the same rise in pressure within a high-pressure range increases the viscosity more than within a low-pressure range. Highly fluid oils are influenced less by a rise in pressure because of their viscosity than viscous oils. It is interesting to note that a rise in pressure can also increase the V.I. Naphthene-based oils respond more readily than paraffin-based oils. In general, like the influence of temperature, a change in pressure has less of an effect on paraffin-based oils than on naphthene-based oil. In Fig. 22-66, we see the influence of pressure on the viscosity and V.I.

22.2.5.3 Influence of Shear Speed on Viscosity

Because of V.I. improvers that can be used for a wide temperature range (see Section 22.2.8.3), Newton's shear stress law no longer applies to multigrade oils since the proportionality, i.e., the viscosity, is now determined by the gap

Pressure	Paraffin-based oil			Naphthene-based oil		
Bar	cSt[a]/40°C	cSt/100°C	V.I.	cSt/40°C	CSt/100°C	V.I.
1	52.5	6.8	90	55.4	5.8	16
1400	810	43.5	100	21.9	53.5	54
2500	8700	195	125	91 000	454	115

[a] 1 cSt = 1 $mm^2 s^{-1}$

Fig. 22-66 Influence of pressure on viscosity and V.I.[1,9]

thickness (film thickness) and the shear speed. This is a non-Newtonian liquid. Whereas with a Newtonian liquid the viscosity remains constant as the shear speed rises, it falls with a non-Newtonian oil. Since in this case an infinite number of viscosities are conceivable depending on the shear speed at which they were measured, the term apparent viscosity is used to draw a distinction; for the absolute viscosity in Poise, the corresponding shear speed is given in reciprocal seconds. One speaks of a shear gradient in this context. Under a high shear load, both the V.I. and viscosity can fall since long-chain polymer V.I. improvers can be broken up and lose some of their effectiveness. A loss in viscosity and V.I. with a high shear gradient can be permanent because of the mechanical or chemical breakup of the large polymer molecule into smaller molecules or temporary since the long-chain polymer molecules tend to align corresponding to the direction of flow that results in less flow resistance. When the V.I. loss is temporary and the shear stress decreases, the oil resumes its original viscosity. Permanent V.I. loss is undesirable in practice. In automobile engines, shear speeds of 50 000–1 000 000 s^{-1} arise. Earlier polymer V.I. improvers with a high molecular weight experienced both high temporary and permanent V.I. loss. Today's V.I. improvers are stable even under maximum shearing stress because of their moderate molecular weight and special structures, and they maintain the viscosity temperature behavior engineered for the fresh oil; i.e., the multigrade oil remains in the given SAE ranges (stay in grade).

22.2.6 Basic Liquids

Engine oils always consist of a basic liquid or a mixture of basic liquids and an additive package that has been harmonized in extensive experiments without which today's requirements cannot be met. The basic liquids, also termed basic oils, are mineral oils, synthetic oils, or a mixture of both (partially synthetic oils). The basic liquids influence important properties of engine oil such as viscosity, evaporation loss, and additive response. Different basic liquids react differently to the effect of the additives and can produce different engine test results. For this reason, they are divided into five groups according to ATIEL (Association Technique de l'Industrie Europeenne des Lubrifiants) as shown in Fig. 22-67.

In addition to the listed properties, there are other important criteria for selecting basic oils depending on the use.

22.2.6.1 Mineral Basic Oils

Mineral basic oils are still used today in most conventional lubricants; however, they are increasingly being replaced by synthetic basic liquids because of steadily increasing requirements. The raffinates from petroleum are obtained by atmospheric distillation, vacuum distillation, solvent refinement, deparaffination, and hydrofinishing. They consist of large molecules with infinite possibilities for branching, even with the same number of C and H atoms. Despite involved refining methods, there is no uniformly

Group	Composition	Sulfur content	Viscosity index
I	<90% (m/m) saturated hydrocarbons	>0.03% (m/m)	≥80 < 120
II	≥90% (m/m) saturated hydrocarbons	≤0.03% (m/m)	≥80 < 120
III	≥90% (m/m) saturated hydrocarbons	≤0.03% (m/m)	≥120
IV	Polyalphaolefins (PAO)		
V	All others not contained in groups I, II, III, or IV such as esters		

Fig. 22-67 Categorization of the basic liquids according to ATIEL.[1]

structured mineral basic oil from different crude oils. There are basic oils with different viscosities, from highly fluid spindle oil to highly viscous Brightstock for engine oils depending on the desired viscosity. In practice, generally mixtures of at least two basic oil components are used that lie between spindle oil and Brightstock. Adjacent distillation cuts are preferably used. The sulfur content of the crude oils suitable for producing lubricants is between 0.3% (m/m) (North Sea) and 2.0% (m/m) (Middle East). A distinction has always been drawn between paraffin-based and naphthene-based basic oils. The paraffin-based oils are preferred because of their better viscosity temperature behavior. Their V.I. is generally high, ranging between 90 and $\leq$100 (see Section 22.2.5). The viscosity of the mixture components at 100°C is between 3.7 and 32 $mm^2 \cdot s^{-1}$. For the high-performance oils in demand today, the properties of the best raffinates are no longer sufficient.

22.2.6.2 Synthetic Basic Liquid

Synthetic basic liquids are essential in engine oils when high-performance multigrade oils are required with minimum oil consumption, minimum residue, maximum wear protection, high fuel economy, and the potential for extremely long oil change intervals. The starting materials for synthetic engine oils are largely based on raw gasoline that exists in the form of ethene (ethylene) after cracking. The synthetic hydrocarbons PAO and PIB (polyisobutene) are made from this using various catalytic processes. If ethene is reacted with oxygen and hydrogen in the presence of a catalyst, synthetic esters or polypropylene glycols (PPG) and polyethylene glycols (PEG) arise in different steps. Another possibility for producing synthetic basic liquids is from vacuum residue. A hydrocrack oil and light gas or gasoline fractions arise from catalytic hydrocracking. For new technology engine oils, PAO, PAO plus esters, or PAO plus hydrocrack oil are used; the other synthetic basic liquids such as the above-cited polyglycols are used for hydraulic and industrial gear oils. Synthetic hydrocarbons such as PAO and hydrocrack oils have a very special molecular structure that does not exist in the starting products; they are more or less a tailored product. Mineral oils, esters, PIB, and PAO are all miscible with each other. In the transition period from mineral oils to today's fully synthetic engine oils, partially synthetic engine oils were introduced. These are still used today as a low-cost alternative for average loads. The advantages and extra costs of synthetic basic oils compared with raffinates are shown in Fig. 22-68.

22.2.7 Additives for Lubricants

Engine oils always consist of one or a mixture of several basic oils (see Section 22.2.6) and additive agents. Additives are oil-soluble agents of various types that are added to the basic liquids to produce properties that do not exist (sufficiently) in the basic oil, to reinforce positive properties, and to minimize or eliminate undesirable properties. Not all properties of the engine oil can be influenced by additives such as the thermal conductivity, the viscosity pressure dependence, the gas solubility, and the air releasability. Additives almost always work as mixtures and can have both synergistic and antagonistic effects. The additive content extends from a few ppm to 20% (m/m) and more in modern high-performance oils. Many lubricant additives are surface-active or interface-active materials whose structures can be compared with that of a match. The head is a functional chemical group that, for example, is "ignited" by water, acids, metals, or soot particles. It is also termed the polar group in which the actual active ingredients are concentrated. They can be organic (ash-free) or metallorganic (ash forming). The "stem" consists of a nonpolar hydrocarbon residue (radical), the oleophil (that is "drawn" by oil). It is primarily responsible for dissolving the additive in the oil. Many additive types have several stems on their polar group. Another important group of oil additives consists of high-molecular hydrocarbons with a special molecular structure that can also contain oxygen. Figure 22-69 lists the additive types for engine oils. In principle, transmission oils have the same composition. Like the basic liquids, the additives also need to be considered in terms of environmental compatibility. For example, chlorine-containing compounds are scarcely used nowadays. A measure of the amount of added metal-containing substances is the sulfate ash content in the fresh oil. In engine oils for spark-ignition and diesel passenger cars, it is 1.0% to 1.5%, and for European diesel commercial vehicles, 1.5% to 2.0%.

Component	Advantage	Reason	Extra cost (%)
Hydrocrack oil	High V.I. >110 Low evaporation loss Good viscosity temperature behavior	Molecular structure Uniform composition Low pour point	Approximately 50–300 (depending on the V.I. and quality)
Polyalphaolefin Polyisobutene	Very high V.I. <150 Very low evaporation loss Very good viscosity temperature behavior	Molecular structure Uniform composition Low pour point	Approximately 400 Approximately 200

Fig. 22-68 Advantages and extra costs of PAO and hydrocrack oils in contrast to raffinates.[10,12]

Additive type	Active ingredient	Function
V.I. improvers Dispersing or nondispersing	Polymethacrylate Polyalkylstyrenes Olefin copolymer (OCP) Star polymers PIB Styrene ester polymers	Improve viscosity temperature behavior
Detergents (basic)	Metal sulfonates Metal phenolates Metal salicylates (metal = Ca; Mg; Na)	Keep the inside of the engine clean Neutralize acids Prevent lacquer formation
Dispersants (ash-free)	Polyisobutene succinimides	Disperse soot, aging products, and other foreign materials Prevent deposits and lacquer formation
Oxidation inhibitors	Zinc dialkyldithiophosphates Alkylphenols Diphenyl amines Metal salicylates	Prevent oil oxidation and thickening
Corrosion inhibitors	Metal sulfonates (metal = Na; Ca) Organic amines Succinic acid half-esters Phosphorous amines, amides	Prevent corrosion
Nonferrous metal deactivators	Complex organic sulfur and nitrogen compounds	Prevent oxidation and oil thickening
Friction modifiers	Mild EP additives Fatty acids Fatty acid derivatives Organic amines	Reduce friction loss
Wear reducer (EP additives)	Zinc alkyl dithiophosphate Molybdenum compounds Organic phosphate Organic sulfur and sulfur phosphorous compounds	Reduce or prevent wear
Pour point depressant	Polyalkyl methacrylates	Improve flow properties at low temperatures
Foam inhibitor (defoamant)	Silicone compounds Acrylates	Reduce or prevent the formation of foam

Fig. 22-69 Typical engine oil additives.[11]

22.2.7.1 V.I. Improvers

Today's high-performance engine oils for passenger cars and commercial vehicles must function perfectly under all driving and weather conditions with extremely long oil change intervals. This means sufficiently low viscosity for reliable cold starts (even when the outside temperature is very low), immediate energy-saving lubrication, and sufficiently high viscosity for reliable lubrication under high thermal and mechanical loads. Only multigrade oils (see Section 22.2.8.3) meet these requirements. The given viscosity temperature dependence of the basic liquids is insufficient even when the V.I. is very high; suitable V.I.

improvers must therefore be used. These are polymers with a high molecular weight. Their mode of operation can be explained with reference to their dissolving behavior. At low temperatures, V.I. improvers are tightly bunched in the oil as solvents. Since they require little space, they have only a slight effect on the viscosity. As the temperature rises, the required space increases; the bunches unravel and, hence, increase the viscosity. When selecting a V.I. improver, we need to determine its sensitivity to high shear stress (for example, between the cam and tappet, in roller bearings, or between the gears of the oil pump) so that the desired V.I. is maintained even under high shear stress and after a long operating time. The response of basic oils to V.I. improvers is less favorable in basic oils with a high V.I. than in those with a low V.I., and it quickly falls as the added amount increases. Depending on the type, V.I. improvers usually have pour-point-lowering properties. Because of their molecular size, they form a disrupting site for crystal growth when incorporated in a paraffin crystal and accordingly give rise to small, separate crystals. Because of the marked influence of temperature on the viscosity of an oil (see Section 22.2.5.1), only multigrade engine oils with a V.I. greater than 100 are used. The top products that are designed as fuel economy and long-lasting oils for extremely long oil change intervals are all fully synthetic, and their basic liquids have a very high V.I., such as 130. These also receive additional VI improvers to create SAE classes 5W-50 or 0W-40. Styrene-diene, star copolymer, or olefin copolymer (OCP) type V.I. improvers are added predissolved in PAO or mineral oil. The amount of additive in market-ready oil is generally 1% to 10% (m/m).

22.2.7.2 Detergents and Dispersants

Numerous combustion products arise that damage the engine oil during combustion in spark-ignition and diesel engines. This includes oil-aging products, partially combusted and uncombusted fuel residue, soot, acids, nitrogen oxides, and water. These largely oil-insoluble solids or liquid foreign materials enter the oil circuit and have an undesirable or damaging effect. Resin- and asphalt-like oil aging products cause deposits on metal surfaces, oil thickening, and sludge deposits on engine parts. Acidic combustion products generate corrosion, catalyze oxidation, and can break down wear protection additives. Carbon and lacquer-like deposits cause the piston rings to cake tight in the ring grooves so that more blowby gas enters the crankcase and further burdens the oil. In addition, the seized piston rings cause bore polishing on the cylinder walls that can lead to premature wear. Sludge deposits can plug oil lines and the oil filter and cause scoring on the piston and cylinder barrels from insufficient lubrication. Detergent and dispersant additives are essentially soaps. They encase solid and liquid foreign particles and keep them suspended in the oil to prevent them from depositing on engine parts and agglomerating, which could lead to sludge formation. In addition, acidic products are neutralized. The mechanisms of the action of detergent and dispersant additives can be generally categorized as follows:

- Encase and wash
- Keep clean and keep solid foreign particles in suspension
- Encase liquid foreign particles and keep them in suspension (interface active)
- Chemically neutralize acidic components (basic)

Several specially harmonized, multifunctional agents are used as detergents and dispersants. They also protect against corrosion and neutralize acids. A long-lasting basic reserve needs to be provided because of the very long oil change interval. Usually metallorganic compounds such as phenates, phosphates, sulfonates, salicylates, and naphthenates are used that are made basic with excess metal carbonate. When they participate in combustion, they form ash in which calcium, magnesium, sodium, and zinc can be found, depending on the additive type. Polyisobutene succinimide is also an ash-free organic detergent that works well against cold sludge deposits that tend to arise during "stop-and-go operation." The amount of additive is generally 1% to 5% (m/m).

22.2.7.3 Antioxidants and Corrosion Inhibitors

Even the most superior lubricating oil tends to oxidize under the influence of heat and oxygen, i.e., age or become rancid. Acids form as well as lacquer deposits, resin deposits, and sludge-like deposits that are largely oil insoluble. The addition of antioxidants substantially improves the aging protection. Initially aging is very slow, and the oil scarcely changes. After the antioxidants are used up, the oxidation speed increases, and a rise in the oil temperature of 10°C doubles the reaction speed and cuts the useful oil life in half. This process can be accelerated by trace amounts of metals, in particular, copper and iron (whose activity increases as they grow finer) that enter the oil from abrasive and corrosive wear and substantially reduce the reaction temperature with oxygen. Water can also have this effect. Without highly effective antioxidants, today's conventional oil change intervals would be inconceivable. The mode of action of antioxidants primarily originates from free-radical scavengers and is supplemented by peroxide replacers and passivators. Radicals are hydrocarbon molecules in which free, highly reactive valences arose on the carbon from chain breakage. Oxygen or another radical immediately seeks to bond to the carbon. The free-radical scavengers saturate the free valence by transferring hydrogen from the additive. The peroxide replacers work only when oxygen-containing aging products have formed. They react with the oxygen and form nonreactive compounds. The nonferrous metal passivators are chemical substances of the triazole type that weaken the catalytic effect of copper and iron particles by encasing the metal ions in the oil. They accordingly also protect the surface of bearing materials from the corrosive attack of active sulfur, for example.

22.2.7.4 Friction and Wear Reducers (EP/AW Additives)

When parts gliding against each other under a high pressure and temperature are no longer fully separated by the lubricant, the surface of the sliding partners contact and increased wear occurs or, in an extreme case, seizure or even welding. EP/AW additives can be of assistance in this situation. They form very thin coatings on the sliding surfaces of the friction partners and are continuously consumed and renewed as needed. They are solid under normal conditions but can slide upon wear and prevent direct metal-to-metal contact. These are interface-active materials that contain zinc, phosphorous, and sulfur in the polar group. The most famous is ZnDTP (zinc dialkyldithiophosphate) that has been most successful in the area of mixed friction of the cam and tappet. Transmission oils use phosphoramines, thionates, and phosphorous-sulfur compounds in addition to sulfurized esters and hydrocarbons.

22.2.7.5 Foam Inhibitors

Air or other gases can be present as finely distributed bubbles or surface foam in the lubricating oil. The primary contributing factor is air from the swirling in the crankcase, as well as pressure and temperature. Surface foam can be dispersed by a special agent that reduces the surface tension between the oil and air. Foam suppressers must be largely insoluble in the oil and have a lower surface tension than the oil. Very low concentrations (0.01 g/kg oil) of silicone oils such as polydimethylsiloxane have been successful as the active ingredient. However, the removal of the dispersed air can become more difficult since the silicone oil prevents the recombination of small air bubbles into larger, easily rising bubbles.

22.2.8 Engine Oils for Four-Stroke Engines

The operating conditions to which engine oils are exposed extend from extremely short-distance use—50% of all travel in passenger cars is over distances less than 6 km—to extremely long-term loads over long distances. In addition, we now have engine oil change intervals of up to 30 000 km or a maximum two years for passenger cars with spark-ignition engines, and up to 50 000 km or a maximum of two years for passenger cars with DI diesel engines, as well as 100 000 km for commercial vehicles. The top-up requirement is also very low. By topping up the oil, not only is the missing amount of oil replenished, but fresh and unconsumed additive is supplied to the engine, which is more important. The oil volume in the oil pan of modern engines does not increase on par with the power density of the engine. The specific oil load, hence, rises continuously. To restrict weight, the amount of oil in the engine has tended to decrease. All of this has to be managed in engines whose specific performance averages 55 kW/L at speeds that may exceed 6000 rpm. In modern supercharged passenger car diesel engines with direct fuel injection, the average pressure as a measure of the engine load can exceed 20 bar. The engine oil is, hence, subject to substantial thermal and mechanical stress. In addition, the oil as a hydraulic fluid needs to efficiently accomplish many tasks in the engine such as hydraulic valve lash compensation, camshaft timing, and chain tightening under every operating condition during its entire time in the engine.

22.2.8.1 SAE Viscosity Classes for Engine Oils

In 1911, the SAE (Society of Automotive Engineers, USA) introduced a binding classification for engine oil viscosity that is still valid today after many adaptations. In its present version, a total of 12 classes are defined: Six for the winter (0W to 25W) and six for the summer (20 to 60). Figure 22-70 shows the viscosity grades for engine oils according to SAE J300 of 4/1997. They inform the user that he is using oil with the right viscosity stipulated by the engine manufacturer.

22.2.8.2 Single-Grade Engine Oil

Single-grade engine oils meet the viscosity requirements of only the individual SAE grades 0W to 60 shown in Fig. 22-70. In practice, we generally find only viscosities between SAE 20 and 40. These oils, therefore, have a low V.I. and are accordingly suitable only for undemanding engines that run primarily under unchanging conditions at practically the same temperature, such as engines for power generators. In automobile engines, single-grade engine oil has to be changed more frequently corresponding to the season and the operating conditions.

22.2.8.3 Multigrade Oils

The term "multigrade oil" means that such an oil covers the viscosity requirements of several SAE grades, for example, 5W-30. It covers the low W class in the low temperature range and ends with the high-temperature viscosity class at 100°C. Figure 22-71 shows in bold the most common combinations in Central Europe with other possible combinations that, however, are presently technically meaningless.

The combination 0W-40 can meet maximum demands on the viscosity temperature behavior and requires a fully synthetic basic liquid and a particularly shear-resistant V.I. improver (see Section 22.2.5.3) to maintain the physical requirements. The combination 15W-20 meets the lowest requirements and is made with a mineral basic oil and a relatively low V.I. improver, but it is technically irrelevant. In multigrade oils, the low temperature viscosity is determined by the basic liquid, whereas the high temperature viscosity is set by the thickening effect of the V.I. improver. The very high performance of modern high-performance multigrade oils derives primarily from combining synthetic basic liquids with highly effective additive packages plus temperature and shear-resistant V.I. improvers, without which today's extremely long oil change interval would not be possible.

SAE viscosity class	Maximum apparent viscosity at °C cP/°C	Low temperature pump viscosity cP/°C	Maximum pump threshold temperature °C	Kinematic viscosity at 100°C cSt		HTHS[a] viscosity in cP at 150°C and $10^6\ s^{-1}$ shear gradient
		Max.		Min.	Max.	Min.
0W	3250/–30	60 000/–40	–35	3.8	—	
5W	3500/–25	60 000/–35	–30	3.8	—	
10W	3500/–20	60 000/–30	–25	4.1	—	
15W	3500/–15	60 000/–25	–20	5.6	—	
20W	4500/–10	60 000/–20	–15	5.6	—	
25W	6000/–5	60 000/–15	–10	9.3	—	
20				5.6	< 9.5	2.6
30				9.3	< 12.5	2.9
40				12.5	< 16.3	2.9[b]
40				12.5	< 16.3	3.7[c]
50				16.3	< 21.9	3.7
60				21.9	< 26.1	3.7

[a] High-temperature high-shear viscosity.
[b] For 0W-40, 5W-40, 10W-40.
[c] For 15W-40, 20W-40, 25W40, and 40.

Fig. 22-70 SAE viscosity classes for SAE J300 engines-lubricating oils, as of April 1997.[1]

0W-40[a]	**5W-40**	**10W-40**	**15W-40**
0W-30	**5W-30**	**10W-30**	15W-30
0W-20	5W-20	10W-20	15W-20

[a] The viscosity combinations in bold are the most common.

Fig. 22-71 Multigrade oil viscosity classes.[1]

22.2.8.4 Fuel Economy Oils

Multigrade oils whose low-temperature viscosity lies between SAE grades 0W or 5W are termed LL engine oils in German (Leicht-Lauf oils) or FE engine oils (fuel-economy oils). Fuel consumption is clearly reduced in two ways:

- By lower viscosity during full lubrication (hydrodynamic lubrication)
- By friction-reducing additives for boundary lubrication (mixed friction)

Reducing viscosity has the greatest influence since primarily hydrodynamic lubrication predominates in the engine. The effect of friction modifiers for mixed friction is narrowly restricted. Overall friction losses are shown in Fig. 22-72:

It turns out that the largest reduction of consumption is within the partial load range close to idling. The first FE oils were SAE grade 10W-X, and then came 5W-X and 0W-X; as would be expected, the greatest savings are from 0W-X. However, all other requirements for hot operation must be covered, and the evaporation loss must remain low enough to keep oil consumption at a minimum. For example, VW standard 50 000 requires that it be ≤13% as per DIN 51581. According to the generally valid fuel consumption test based on Directive 93/116 EC, the FE oil is compared to a reference oil with a viscosity of SAE

Engine	Full load	Partial load
Spark ignition	3% to 5%	11% to 18%
Diesel	7% to 9%	13% to 14%

Fig. 22-72 Friction loss.[10,12]

15W 40. In addition, tractive torque measurements, for example, using the Mercedes-Benz OM 441 LA engine, can be used to evaluate the friction reduction from an FE oil for commercial vehicles. A typical measuring result of a comparison between the earlier used 15W-40 oil and a fuel economy oil 5W-30 is shown in Fig. 22-73. Both oils also had to fulfill all the other requirements from the other test runs (for example, according to MB 228.3).

Multigrade oil for commercial vehicles	Tractive torque (Nm)
15W-40 (MB 228.3)	285
5W-30 (MB 228.3)	244 (reduction of 17%)

Fig. 22-73 Tractive torque in the OM 441 LA engine.[12]

In addition, fuel economy oils also clearly improve the pumpability at low temperatures to ensure a much faster supply of oil to the engine after a cold start. This is true for both fresh oil and used oil, as can be seen in Fig. 22-74.

Also notable in these results is the low degree of thickening of the used oil as a result of the high quality of these oil types.

22.2.8.5 Break-In Oils

At an earlier time, it was necessary to fill new engines with a special break-in oil or initial operation oil that was drained after a relatively short dwell time of 1000 to 1500 km and contained the metal abrasion from this first operation phase. There were special break-in instructions, and the driver had to drive cautiously to prevent the thin initial operation oils with relatively few additives from being overly stressed. Frequently, they were designed as preservative oil when, for example, the vehicles were to be exported overseas. Because of the improved surface quality of the friction surfaces of the engine and the enormous strides in oil technology, the use of break-in oils in passenger cars and commercial vehicles is now restricted to a few special cases. The performance of initial operation oils practically corresponds to off-the-shelf engine oils since they also remain in the engine for a full change interval; however, they generally have somewhat higher corrosion protection.

22.2.8.6 Gas Engine Oils

The use of CNG/LNG in automobiles require special engine oils in monovalent operation since the higher combustion temperatures increase the tendency of deposits to form in the combustion chamber and on the pistons. These particularly hard deposits require low-ash additives. In comparison to gasoline, approximately twice the amount of water arises from the combustion of natural gas. Hence, the danger of corrosion is especially greater in short-distance driving. The absence of high-boiling hydrocarbons in natural gas also generates a greater need for lubrication at the intake valves, which could result in pocketing of the valve seat. In bivalent operation, these specific properties are not as predominant, and modern high-performance oils can cover nearly every requirement.

22.2.8.7 Methanol Engine Oils

In long-term experiments carried out in tandem by the automobile and mineral oil industries from 1977 to 1981, the special requirements were defined for engine oil in methanol engines. Less residue forms on the pistons, so the amount of added detergents and dispersants can be lower than in gasoline engines. In contrast, when contact occurs between methanol and engine oil from the blowby gas from crankcase ventilation, it was found that methanol is incompatible with long-chain V.I. improvers and dispersants, which increases residue formation in the intake system. As a solution, single-grade engine oil was first tried to avoid the negative influence of the V.I. improvers. The greater tendency for corrosion and cylinder wear in cold operation was found to be particularly unfavorable in long experiments on the test bench and the street. The suspected reason is the reduced wetting ability of the oils on metal surfaces in the presence of methanol. A possible solution is higher corrosion and wear protection.

22.2.8.8 Hydrogen Engine Oils

In the joint research on alternative fuels, the influence of hydrogen on engine oil was also investigated in hydrogen-operated spark-ignition engines. The different combustion processes in comparison to gasoline clearly affect the engine oil requirements. In the combustion of hydrogen, more than twice the amount of water arises as a combustion product in comparison to the combustion of conventional gasoline. Since additional water is injected into the combustion chamber in some engine designs for regulated

Multigrade oil for commercial vehicles	Fresh oil Oil pressure 2 bar (s)	Used oil Oil pressure 2 bar (s)
5W-30 (MB 228.3)	7	9
10W-40 (MB 228.3)	10	14
15W-40 (MB 228.3)	23	35

Fig. 22-74 Pumpability in the OM 441 LA engine 0°C.[12]

combustion, the related increased water vapor in the engine oil under cold driving conditions requires greater corrosion protection, stronger dispersability, and a greater ability to absorb and release water. On the other hand, there is much less combustion residue and resulting deposits so that the detergent content can be reduced. Recent engine designs for hydrogen operation get by without additional water injection. The risk from additional water to the engine oil has accordingly receded. These hydrogen engines can be operated with commercial engine oils. However, experience from operating spark-ignition engines with hydrogen are insufficient to draw any conclusions on the optimum composition of engine oils used in this context.

22.2.8.9 Performance Classes

Given the different conditions of use and demands on engine oils, numerous specifications have arisen over time concerning their composition and performance. These were usually worked out jointly by the engine and mineral oil industries with the collaboration of consumer organizations or military authorities. In addition, the individual automobile manufacturers keep publishing a growing quantity of brand-related specifications. Beyond worldwide approval requirements, there are also regulations limited to the regions of use. These describe both the physical properties and the performance behavior in engine tests. The specifications of the following associations and institutions describe performances behavior:

Usually, a distinction is drawn between passenger car spark-ignition engines, passenger car diesel engines, and commercial vehicle diesel engines. Since 1996, the European ACEA specifications have been the successor to the well-known, no longer current CCMC specifications that existed since the 1970s. In addition, API classifications are frequently required. The MIL specifications are now irrelevant in Europe. In recent years, the individual requirements and the related releases of the European automobile manufacturers have become the most important.

ACEA Specifications

These represent the current standards for engine oils for European automobile engines. In addition to a few selected American test engines, they cover primarily engines of European construction and design. The test conditions correspond to European driving conditions. They define the minimum requirements for physical and chemical laboratory tests and full engine test bench tests. Since a few American engine tests are also prescribed, there is a certain amount of overlap with American API classifications. The abbreviations for the ACEA specifications are no longer oriented around the previously conventional names of the outdated CCMC specifications. Whereas CCMC G (gasoline) was used for spark-ignition engines, CCMC D for commercial vehicle diesel engines, and PD for passenger car diesel engines, in the ACEA specifications these categories are now termed ACEA A, ACEA B, and AECA E. The group A concerns spark-ignition engines, B light or passenger car diesel engines, and group E concerns heavy commercial vehicle diesel engines. In addition to the number for the performance classes, they can also include the year when the respective specification took effect, for example, ACEA A3-98, ACEA B2-96, or ACEA E4-98. Figure 22-75 presents the ACEA specifications for service fill oils for spark-ignition engines with their most important features.

In contrast to the commercial vehicle sector, passenger car diesel engines require special additives that are very similar to those for spark-ignition engines because of the higher speed, higher specific performance, higher valve gear load, and more frequent short-distance use. Figure 22-76 shows the ACEA specifications for service-fill oils for passenger car diesel engines.

Commercial vehicle diesel engines are used under a wide range of operating conditions such as in government vehicles and municipal busses with a continuously changing load at low speeds, and in long-distance traffic with a continuous high load at higher speeds. Engine oils for commercial vehicles are different in many respects from

ACEA	Oil type	Important requirements
A1-98	Fuel economy engine oils	HTHS viscosity min. 2.9, max. 3.5 mPa · s, FE[a]
A2-96 Version 2	Standard engine oils	HTHS viscosity >3.5 mPa · s Higher requirements for evaporation loss, wear, cleanliness, black sludge, and oxidation stability
A3-98	Premium engine oils	HTHS viscosity >3.5 mPa · s Higher requirements for shear stability, wear, cleanliness, black sludge, and oxidation stability
A5-01	Premium fuel economy engine oils	HTHS viscosity <3.5 mPa · s on the performance level of ACEA A3

[a] Demonstrated fuel savings in the MB M111FE test in comparison to reference oil SAE 15W-40.

Fig. 22-75 ACEA specifications for spark-ignition engines.[1,12]

ACEA	Oil type	Important requirements
B1-98	Fuel economy engine oils	HTHS viscosity min. 2.9, max. 3.5 mPa · s, potential fuel savings.
B2-98	Standard engine oils	HTHS viscosity >3.5 mPa · s Higher requirements for evaporation loss, wear, cleanliness, oil thickening, oil consumption.
B3-98	Premium engine oils	HTHS viscosity >3.5 mPa · s Higher requirements for evaporation loss, stay-in-grade, wear, cleanliness, oil thickening, oil consumption.
B4-98	Premium engine oils for direct injecting diesel engines	Requirements higher than B2 in regard to piston cleanness in diesel engines with direct injection.
B5-01	Premium fuel economy engine oils for DI diesel engines	HTHS viscosity 3.5 mPa · s on a higher performance level than ACEA A4.

Fig. 22-76 ACEA specifications for passenger car diesel engines.[1,12]

those for passenger cars. For example, the requirements are particularly high wear protection for the cylinder barrel, very clean pistons over the long term, high dispersability of soot, reserves and high performance for extremely long oil change intervals, and low residue formation in turbochargers and intercoolers. Given the present level of technology for European commercial vehicle engines, these things can be achieved only with superior additives. Figure 22-77 shows the ACEA specifications for heavy commercial vehicle diesel engines.

Section 22.2.8.8 provides information on the engine tests stipulated for the individual classes.

ACEA	Oil type	Important requirements
E1-96[a]	Standard engine oils for naturally aspirated engines for undemanding conditions of use.	Basic requirements.
E2-96 Version 3	Standard engine oils with higher requirements.	Higher requirements for bore polish, piston cleanness, cylinder wear, oil consumption.
E3-96 Version 3	Premium engine oils for commercial vehicles with ATL engines; approximately corresponds to MB sheet 228.3 and MAN 271.	Higher requirements for oil consumption, sludge formation, rise in viscosity with a high soot content in the oil.
E4-99	Premium engine oils for commercial vehicles with ATL engines and an intercooler, approximately corresponds to MB sheet 228.5 and MAN 3277.	More demanding requirements for bore polish, piston cleanness, cylinder wear. Additional limitation of deposits in the turbocharger.
E5-99	Standard/premium engine oils for the combined requirements of European and American commercial vehicle engines.	Requirements less stringent than E4-99, additional tests in the Cummins M11 and Mack T9.

[a] Invalid since 1999.

Fig. 22-77 ACEA specifications for heavy commercial vehicle diesel engines.[1,12]

API Classifications

The description of performance requirements of engine oils for automobile engines was started much earlier in the United States than in Europe. In its engine oil classifications, API draws a distinction only between passenger car engines and commercial vehicle engines. Since the share of passenger cars with diesel engines in the United States is very low, they have no classification for light diesel engines. The engines used for the tests are made in the United States, and the test conditions favor American driving conditions. All the viscosity grades according to SAE are permissible. While the engine tests up to API SG can be done by the respective manufacturers, they need to be reported and registered since the introduction of API SH (CMA Code) when an API category for the product is claimed. At the same time, it is possible to obtain a license through API that allows an API label to be affixed to the packaging. Figure 22-78 shows the prior API classifications for passenger car engine oils.

These API classes have been on the engine oil packaging for decades in addition to the SAE grades and are listed in the operating instructions. The user can then check if the quality corresponds to that prescribed by the automobile manufacturer.

The API classification for commercial vehicles is more multifaceted since the design of commercial vehicle diesel engines made in America is sometimes substantially different from that of European models. In earlier classifications, they refer to the MIL specifications (see Section 22.2.8.3). In addition, we should note that two-stroke diesel engines can frequently be found in the American market, yet not in Europe. Among the test runs stipulated practically exclusively for the Caterpillar, single-cylinder diesel engine have been incorporated into later test runs for modern engines by Caterpillar, Cummins, Mack, and Detroit Diesel. Figure 22-79 illustrates the API classifications for commercial vehicles.

Parallel to the API classifications in the United States are the ILSAC certification (International Lubricant Standardization and Approval Committee) that uses the API classifications for passenger car engines in cooperation with the AAMA (American Automobile Manufacturers Association), and JAMA (Japan Automobile Manufacturers Association) to offer a classification of oil quality and usefulness for engine oil packaging that is more consumer oriented. The outdated ILSAC GF-1 corresponded to API SH, and the current ILSAC GF-2 corresponds to API SJ. Over the course of 2001, ILSAC GF-3 was introduced as a mirror of API SL.

MIL Specifications

Engine oil specifications have been established for automobiles in the U.S. Army since 1941. The requirements have been continuously adapted as these engines have

API class	Year introduced	Important requirements
SA[1]	1925	Unalloyed engine oils. Possible to add pour point improvers and foam suppressers.
SB	1930	Slightly alloyed engine oils with low wear, aging, and corrosion protection.
SC	1964	Engine oils with increased protection against scoring, oxidation, bearing corrosion, cold sludge, and rust.
SD	1968	Improvement of API SC with greater protection against scoring, oxidation, bearing corrosion, cold sludge, and rust
SE	1972	Improvement of API SD with greater protection against oxidation, bearing corrosion, rust, and lacquering.
SF	1980	Improvement of API SE with additionally improved protection against oxidation and wear.
SG	1989	Improvement of API SF with further improved oxidation stability and better wear protection.
SH	1992	Corresponds to API SG; however, the engine tests for API SH, in contrast to API SG, must be registered with a neutral institute.
SJ	1997	Corresponds to API SH with an additional laboratory test against high temperature deposits. Regulated exchangeability of basic liquids, stricter test instructions regarding read across.

[a] Service fill.

Fig. 22-78 API classifications for passenger car engine oils.[1]

API class	Year introduced	Important requirements
CA[a]	Mid 1940s	For naturally aspirated diesel engines and occasionally low-load spark-ignition engines. Protection against bearing corrosion and ring groove deposits.
CB	1949	For naturally aspirated diesel engines using poorer-quality diesel fuel with a high sulfur content. Occasionally also in spark-ignition engines. Protection against bearing corrosion and ring groove deposits.
CC	1961	For naturally aspirated diesel engines with a medium load, occasionally also in spark-ignition engines with a high load. Protection against high-temperature deposits, bearing corrosion, and cold sludge in spark-ignition engines.
CD	1955	For naturally aspirated diesel engines, supercharged and highly supercharged turbodiesel engines using diesel fuel with a very high sulfur content. Increased protection against ring groove deposits at high temperatures, and against bearing corrosion.
CD II	1985	For two-stroke diesel engines with increased requirements of wear protection and deposits.
CE	1984	For highly supercharged diesel engines under a high load at low and high speeds. Improved protection against oil thickening, piston deposits, wear, and oil consumption in comparison to API CD.
CF-4	1990	Improvement over API CE in terms of piston cleanness and oil consumption.
CF	1994	Like CD, but for indirect-injection diesel engines for a wide range of diesel fuels and a sulfur content over 0.5% (m/m). Improved control of piston cleanness, wear, and bearing wear.
CF-2	1994	For two-stroke diesel engines with increased requirements for cylinder and piston ring wear, as well as improved control of deposits.
CG 4	1994	For high-load, high-speed four-stroke diesel engines in street use, as well as off-road use, diesel fuel with a sulfur content of 0.5% (m/m) is especially suitable for engines that fulfill the emissions standards of 1994. Also covers API CD, CE, and CF-4. Additionally increased oxidation stability and protection from foaming.
CH-4	1998	In contrast to CG-4, enhanced requirements for diesel engines that correspond to the emissions standards of 1998. Sulfur content in the diesel fuel of up to 0.5% (m/m). Given a longer oil change interval, increased protection against noniron corrosion, thickening from oxidation and oil-insoluble soiling, foaming, and shear loss.

[a] Commercial.

Fig. 22-79 API classifications of commercial vehicle engine oils.[1]

developed. The term "HD oils" (heavy duty) arose in this context for oils under high stress in diesel engines, and it continues to be used by consumers to this day. This represented the transition from exclusively unalloyed mineral oils to alloyed oils that received chemical additives for the first time. The only slightly alloyed oils for spark-ignition engines were termed "premium oils" in contrast to the HD oils for diesel engines. Although the MIL specifications were originally intended only for military use, they were used throughout the world for a long time after World War II in the civilian sector in performance recommendations for engine oils. For the military, the specifications MIL-PRF-2104G have been valid since 1997. They allow single-grade engine oils SAE 10W, 30, and 40, and multigrade SAE 15W-40 as viscosities. The requirements of these specifications correspond to elements from API CF, CF2, and CG4. In addition to fulfilling chemical-physical requirements, the oils for tactical vehicles in the U.S.

Army must also fulfill special friction tests because of the specific construction of tactical military vehicles such as tanks. For a few years, the use of "MIL" to identify oil performance has been allowed only if the corresponding oil is permitted by the American military.

Automobile Manufacturer Specifications

Beyond the API classifications and ACEA specifications, European automobile manufacturers require special performance classes for the release of individual engine oils that must be fulfilled in addition to API and ACEA specifications whose requirements are sometimes clearly exceeded. Given the amazing progress in engine technology, the requirements are subject to a continually accelerating process of change. The fulfillment of the special requirements is confirmed by a written release. Some automobile manufacturers provide lists of the released oils. The most important automobile manufacturer requirements are compiled in Fig. 22-80. There are other special requirements, some of which are associated with the formal release, from other automobile and engines manufacturers such as Ford, Peugeot, Porsche, Renault, Rover (passenger cars and commercial vehicles), as well as DAF, Iveco, MTU, Scania, and Volvo (only commercial vehicle).

Engine Test Methods

To fulfill the requirements established in the individual engine oil specifications, binding engine tests are stipulated in addition to the usual physical and chemical requirements. They are updated from time to time as needed, usually every two years in the case of the ACEA. Some automobile manufacturers recognize only tests that are done at neutral, specially permitted test institutes.

But first, let us take a historical retrospective of the many outdated test methods of the last decades. By the end of the 1950s, spark-ignition engine oils had to undergo the MS test sequences in the framework of API classification for American V8 engines by General Motors (Seq. I/II/III), Chrysler (Seq. IV), and Ford (Seq. V). To be certified according to MIL specifications, diesel engines had to pass the Caterpillar single-cylinder test that runs over 480 h with L IA/E naturally aspirated engines and L-1H, L-1D, and L-1G ATL engines. In addition, oils underwent supplementary test runs L-38 and LTD over 40 or 180 h in the smaller CLR (coordinating lubricant research) Labeco single-cylinder engine.

In Germany, at the beginning of the 1960s, the suitability of alloyed oils was tested in the MWM KD 12E single-cylinder diesel engine in test methods A and later B over a 50 h transit period in reference to piston cleanness and ring clogging. In England, oils were tested in the Petter AV.1 single-cylinder diesel engine with a 120 h transit period in combination with the Petter W.1 single-cylinder spark-ignition engine with a 36 h transit period to attain DEF approval. Soon to follow was the certification test demanded by Daimler-Benz in Mercedes-Benz four-cylinder passenger car diesel engines. Figure 22-81 shows the first European engine tests standardized by the CEC that were done with single-cylinder test engines and a few full-size engines of the period.

Rapid engine development, the demand for further improvements in reliability, a longer service life, longer oil dwell times, and falling levels of oil consumption have required ever-new test engines and test methods to fulfill the requirements in the ACEA specifications. Today we have correspondingly suitable, Europe-wide specified motor oil tests that are listed in Fig. 22-82.

In addition, engine oil tests for API classification and particularly those of the European automobile manufacturers need to be observed. For API class SJ for spark-ignition engines valid since 1996, the sequence VI A test is also provided for determining the fuel economy of an engine oil. For diesel engines, API CH-4 became effective in 1998 with test runs in the CAT I K and Cummins NTC 400. It is informative to compare the engine oil test runs of ACEA and API. Figure 22-83 offers a comparison for passenger car spark-ignition engines, and Fig. 22-84 compares commercial vehicle diesel engines.

Test specifications and procedures are voluntarily observed, and the type and composition of lubricants being tested or developed are voluntarily maintained with the assistance of the EELQMS for engine oils (European Engine Lubricant Quality Management System), a joint initiative of ATC (Technical Committee of Petroleum Additive Manufacturers) and ATIEL. The European Technical Association of the European Lubricants Industry ATIEL and the European ATC have developed a fixed set of regulations (ATC Code of Practice and ATIEL Code of Practice) to which the member companies can voluntarily submit by submitting an annual written letter of conformance. They thereby certify that the performance classes of the oils that they manufacture or sell are based on exact and controlled tests corresponding to the prescribed conditions of the two Codes of Practice carried out in test facilities certified according to EN 45001. The tests are reported and registered at the ERC (European Registration Centre) that does not, however, publish any specific release lists. The list of the companies voluntarily participating in the quality assurance system is available to consumers and can be requested from ATIEL and ATC or viewed on the Internet.

22.2.8.10 Evaluating Used Oil

While oil is in the engine, numerous foreign materials collect in it, primarily residues from fuel combustion such as soot. In diesel engines these are especially uncombusted hydrocarbons, acidic reaction products, abraded elements from engine wear, and water. The load on the oil generated by these foreign materials consisting of liquid (low-molecular) and solid (high-molecular) aging and reaction products naturally changes the physical and chemical states of the used oils. Physically, the viscosity changes (usually by thickening but also by dilution with fuel condensate)

Manufacturer	Specification	Designation	Type of engine	Requirements
BMW	Special oils and long life oils	Special oil; long life oil	Passenger car SI and diesel	ACEA A3/B3 plus additional BMW engine and foam test, FE oils 0W-X and 5W-X. Long life oils for long oil change intervals.
MAN	MAN standards	270	Commercial vehicle diesel	ACEA E2, single-grade engine oil for normal requirements.
		271	C.V. diesel	ACEA E2, multigrade oils for normal requirements.
		M 3271	C.V. gas	CNG/LPG special oils.
		M 3275	C.V. diesel	ACEA E3, stricter physical requirements, high-performance oils.
		M 3277	C.V. diesel	ACEA E3 plus OM 441 LA corresponding to MB Sheet 228.5 plus deposit test, high-performance oils for max. oil change interval.
Daimler-Chrysler	Mercedes-Benz fuel regulations	MB Sheet 229.3	Passenger car SI and diesel	ACEA A3, B3, and B4 plus MB engine tests and special requirements, high-performance multigrade oils for very long intervals.
		MB Sheet 229.1	Passenger car SI and diesel	ACEA A2 or A3 and B2 or B3 plus MB engine tests, high-performance multigrade oils for long intervals.
		MB Sheet 227.0	Commercial vehicle diesel	ACEA E1 plus additional evaluation criteria in the OM 602A, single-grade engine oil for short oil change intervals.
		MB Sheet 227.1	Commercial vehicle diesel	ACEA E1 plus additional evaluation criteria in OM 602A, multigrade oils for short oil change intervals.
		MB Sheet 228.0	Commercial vehicle diesel	ACEA E2 plus additional stricter evaluation criteria in OM 602A, single-grade engine oil for normal oil change intervals.
		MB Sheet 228.1	Commercial vehicle diesel	ACEA E2 plus additional stricter evaluation criteria in OM 602A, multigrade oils for normal oil change intervals.
		MB Sheet 228.2[a]	Commercial vehicle diesel	ACEA E3 plus additional, further-stiffened evaluation criteria in OM 602A, single-grade engine oil for long oil change intervals.
		MB Sheet 228.3[a]	Commercial vehicle diesel	ACEA E3 plus additional, further-stiffened evaluation criteria in OM 602A, multigrade oils for long oil change intervals, SHPD[b] type.
		MB Sheet 228.5[a]	Commercial vehicle diesel	ACEA E4 plus additional further stiffened evaluation criteria in OM 602A, multigrade oils for max. oil change intervals, USHPD[c] type.
VW/Audi	VW Standard	501 01[d,e]	Passenger car SI and naturally aspirated diesel	ACEA A2 plus VW-specific engine and aggregate tests. Standard-multigrade oils.
		500 00[d,e]	Passenger car SI and naturally aspirated diesel	ACEA A3 plus VW-specific engine and aggregate tests. Stricter physical requirements. FE oils 5W/10W-30/40 for normal oil change intervals up to approx. the end of model year 1999.

Manufacturer	Specification	Designation	Type of engine	Requirements
		505 00[e]	Passenger car naturally aspirated and ATL diesel	ACEA B3 plus VW-specific diesel engine and aggregate tests. Standard or FE multigrade oils for normal oil change interval up to approx. the end of model year 1999.
		502 00	Passenger car spark-ignition	ACEA A3 plus VW-specific engine and aggregate tests under special inclusion of long-term stability. Standard or FE multigrade oils.
		505 01	Passenger car diesel	Special oil SAE 5W-40 for DI diesel engines with pump-nozzle fuel injection system, normal intervals.
		503 00[f]	Passenger car spark-ignition	ACEA A3 plus VW-specific engine and aggregate tests under special inclusion of long-term stability and fuel economy. HSHT viscosity reduced to ≥2.9 and <3.4 mPas. Comprehensive manufacturer tests. For automobiles starting approx. model year 2000 with long oil change interval. Not suitable for automobiles built beforehand.
		503 01	Passenger car spark-ignition	ACEA A3 plus specific manufacturer tests in ATL spark-ignition engines with high specific performance.
		506 00[f]	Passenger car diesel	ACEA B4 with stricter limits under special inclusion of long-term stability and fuel economy. HSHT viscosity lowered to ≥2.9 and <3.4 mPas. Comprehensive manufacturer tests. For automobiles with DI diesel engines that do not have a pump-nozzle fuel injection system, starting approx. model year 2000 with longer oil change intervals. Not suitable for automobiles built beforehand.
		506 01	Passenger car diesel	ACEA A1/B1 plus comprehensive specific manufacturer testing under special inclusion of long-term stability and fuel economy. HSHT viscosity lowered to ≥2.9 and <3.4 mPas. For automobiles with DI diesel engines with pump-nozzle fuel injection system and long oil change intervals.

[a] For multigrade oils XW-30 and 0W-40, additional testing in the OM 441LA test with premeasured bearings and tappets.
[b] Super high performance diesel oil.
[c] Ultra super high performance diesel oil.
[d] No new releases have been granted since 1997; extensions of releases are possible under certain conditions.
[e] Combinations are possible and common as 501 01 and 505 00, as well as 500 00 and 505 00.
[f] Only combined together.

Fig. 22-80 Important engine oil specifications of a few automobile manufacturers.[1,14]

particularly in cold seasons. The chemical changes particularly affect the alkalinity reserve as a measure of active ingredient consumption. These changes are evaluated and the abrasion elements in the used oil are determined in a used oil analysis. This is an important tool for determining the state of the oils while developing the engine and engine oil, and for evaluating the state of the engine and the engine oil in relationship to the dwell time in the engines of large fleets. In evaluating used oil, the effects of different operating conditions are monitored. Passenger cars, in particular, second cars, are generally operated under stop-and-go conditions with many cold

CEC method	Test name	Design	Requirements
L-01 A-69	Petter A.V. 1	1-cylinder prechamber diesel engine	Piston ring freedom, piston cleanness
L-13-T-74	Petter A.V. B	1-cylinder prechamber diesel engine	Piston ring freedom, piston cleanness under stricter operating conditions than Petter A.V. 1
L-02 A-69	Petter W.1	1-cylinder spark-ignition engine	High-temperature oxidation, bearing corrosion
L-03 A-70	Ford Cortina	4-cylinder spark-ignition engine	High-temperature oxidation, ring clogging
L-04 A-70	Fiat 600D	4-cylinder spark-ignition engine	Low-temperature sludge
L-05-T-70	MWM method A (DIN 51361)	1-cylinder prechamber diesel engine	Piston ring freedom, piston cleanness
L-06-T-71	Fiat 124 AC	4-cylinder spark-ignition engine	Piston ring freedom, cleanness
L-12 A-76	MWM method B (DIN 51361)	1-cylinder prechamber diesel engine	Piston ring freedom, piston cleanness under stricter operating conditions than the MWM A

Fig. 22-81 CEC-test methods.[1,12]

starts that are seldom interrupted by long drives. On the other hand, approximately 10% of all users primarily use their automobiles for long trips with a continuously high load. Hot operation and cold operation, of course, have radically different effects on the engine oil condition. The most conventional points for analyzing used oil are the following:

- Flash point ⇒ dilution by fuel
- Viscosity at 40 and 100°C
- Alkalinity ⇒ Active ingredient reserve; base number and acid number ⇒ TBN/TAN
- Dispersability
- Nitration ⇒ black sludge
- Overall soiling ⇒ solid foreign materials, oil-insoluble aging products
- Abraded elements and dirt ⇒ iron, copper, chromium silicon content
- Water and glycol content ⇒ leaky coolant circuit
- Spectrometric infrared analysis according to DIN 51451 ⇒ identity

Physical Changes

The oil can thicken, i.e., increase in viscosity during operation because of the evaporation of highly volatile oil components, from the increase in solid foreign materials from combustion, abrasion, and wear, and by oil aging from the oxidation and polymerization of oil components. Longer transit periods under a high load at a high speed promote a rise in viscosity. This makes cold starting and the supply of oil to critical lubrication sites more difficult and increases fuel consumption. Oil thickening is, therefore, one of several important criteria for establishing the oil change interval. Some of the engine oil testing methods cited in Section 22.2.8.8 serve as a standard; however, the tests required by the individual automobile manufacturers are the ones that are primarily used. The decrease in viscosity from oil dilution is primarily from fuel and water, in particular, in cold and short-distance driving. Uncombusted fuel components and water vapor from combustion condense in the cold engine, pass by the piston rings, and enter the oil pan. In modern, low-polluting engines that use electronically controlled mixture enrichment in cold operation, the condensation tendency is less. Oil dilution also rises when there is incomplete combustion in a cylinder, for example, from spark plug failure or damage to the nozzle. In modern engines, this is the exception since an increased service life, higher component quality, and electronically controlled ignition ensure reliable operation. Finally, multigrade oils can experience a permanent viscosity loss from the shearing of V.I. improvers when they are not sufficiently shear resistant (see Section 22.2.5.3).

ACEA	Method	Test name	Design	Primary criteria
A	CEC L-53-T-95	Mercedes-Benz M111SL	R4-cyl. SI engine	Black sludge, cam wear
	CEC L-54-T-96	Mercedes-Benz M111FE	R4-cyl. SI engine	Fuel economy
	CEC L-55-T-95	Peugeot TU-3M H	R4-cyl. SI engine	High-temperature deposits, ring clogging, oil thickening
	CEC L-38 A-94	Peugeot TU-3M S	R4-cyl. SI engine	Valve gear wear
	ASTM D-5533-93	Buick Sequence IIIE	V6-cyl. SI engine	High-temperature oxidation
	ASTM D-5302-95a	Ford Sequence VE	R4-cyl. SI engine	Low-temperature deposits, wear
B	CEC L-46-T-93	VW 1.6 TC D	R4-cyl. ATL diesel engine	Deposits, piston cleanness, ring clogging
	CEC L-78-T-99	VW TDI	R4-cyl. ATL diesel engine	Ring clogging, piston cleanness, oil thickening
	CEC L-56-T-98	Peugeot XUD11ATE (BTE)	R4-cyl. ATL diesel engine	Oil thickening, piston cleanness, sludge
	CEC L-51 A-98	Mercedes-Benz OM602A	R5-cyl. ATL diesel engine	Wear, oil consumption, oil thickening
E	CEC L-51 A-98	Mercedes-Benz OM602A	R5-cyl. ATL diesel engine	Piston cleanness, sludge, wear, oil consumption, oil thickening
	CEC L-42 A-92	Mercedes-Benz OM364A (LA)	R4-cyl. ATL diesel engine	Wear, piston cleanness, bore polishing
	(CEC L-42-T-99) CEC L-52-T-97	Mercedes-Benz OM 441LA	R6-cyl. ATL diesel engine	Piston cleanness, wear, bore polishing
	ASTM D 5967	Mack T-8E	R6-cyl. ATL diesel engine	Turbocharging pressure loss, oil consumption
	ASTM D 4485	Mack T-8	R6-cyl. ATL diesel engine	Relative viscosity, filter clogging
	M11 High Soot	Cummins M11	R6-cyl. ATL diesel engine	Rise in viscosity, oil consumption, oil filter clogging
		Mack T-9	R6-cyl. ATL diesel engine	Valve gear wear, sludge, cylinder wear, ring wear, bearing wear

Fig. 22-82 Engine tests for ACEA specifications.[1,12]

Figure 22-85 shows the substantial amount of oil dilution from extreme short-distance driving measured by the fuel in the used oil in fleet tests of SI engines in typical second car operation. The effect of a drive on the highway measured in a 1.4 l engine is notable. The oil dilution of 2.5% measured after evaporation of the fuel should correspond to the general dilution in alternating operation between short distances and long distances. We should not overlook that the oil dilution important for establishing the oil change interval can be masked by the opposite effects of oil thickening. In modern engines, the coolant water circuit thermostatically controlled by an oil-water heat exchanger usually quickly heats the cold oil by first heating the coolant water. The oil is then quickly heated to operation temperature, and condensation products can evaporate. If the engine oil temperature rises during operation above that of the coolant water, the oil is cooled via the heat exchanger by the coolant water or the radiator. This means that the oil dilution of modern engines is generally less critical.

The rise in the metal content from extreme short-distance driving in three automobiles can be seen in Fig. 22-86. Iron is used as an example that can originate from wear of the cylinder wall or the valve gear. The effect on engine wear of such engine oils strongly diluted by fuel is clear.

The investigation of the disassembled engines after traveling 10000 km under these conditions shows that sig-

ACEA	API	Test method	Conditions	Primary criteria
X		Peugeot TU 3M S	Cold/hot	Valve gear wear
X		Peugeot TU 3M H	Hot	High-temperature deposits, ring clogging, oil thickening
X		Mercedes-Benz 111SL	Cold/hot	Black sludge, cam wear
X		Mercedes-Benz 111FE	Cold/warm	Fuel economy
X	X	Buick Sequence IIIE	Hot	High-temperature oxidation
X	X	Ford Sequence VE	Cold/warm	Low-temperature deposits, wear
	X	L-38	Hot	Bearing corrosion, oil oxidation
	X	Sequence IID	Cold	Corrosion
	X	Sequence VI A	Cold/warm	Fuel economy
	ILSAC GF-3	Nissan KA 24	Cold/hot	Cam and rocker arm wear

Fig. 22-83 Comparison of ACEA and API engine oil tests for passenger car spark-ignition engines.[1,12]

ACEA	API	Test method	Conditions	Primary criteria
X		Mercedes-Benz OM 364° (LA)	Hot	Piston cleanness, wear, bore polish, turbocharger pressure loss, oil consumption
X		Mercedes-Benz OM 602A	Cold/hot	Piston cleanness, sludge, wear, oil consumption, oil thickening
X		Mercedes-Benz OM 441LA	Hot	Piston cleanness, wear, bore polish, turbocharger pressure loss, oil consumption
X	X	Mack T-8E	Hot	Relative viscosity
X	X	Mack T-8	Hot	Filter clogging, viscosity rise, oil consumption
X	X	Cummins M11	Hot	Oil filter clogging, valve gear-wear, sludge
X	X	Mack T-9	Hot	Cylinder wear, ring wear, bearing wear
X	X	Sequence III E	Hot	High temperature oxidation
	X	L-38	Hot	Bearing corrosion, oil oxidation
	X	Caterpillar I K/I N	Hot	Piston cleanness, wear, oil consumption
	X	GM 6.2L (RFWT)	Hot	Roller rocker arm wear

Fig. 22-84 Comparison of ACEA and API engine oil tests for commercial vehicle diesel engines.[1,12]

	2.0 l Engine	1.8 l Engine	1.4 l Engine
Distance (km)	Fuel content in %	Fuel content in %	Fuel content in %
1000	3.5	7.5	5.5
2000	7.0	18.0	15.0
4000	6.5	20.5	12.0
6000	12.5	19.5	10.0
8000	15.2	20.5	11.5
10 000	15.6	27.0	17.5
12 000	18.5		2.5[a]
14 000	17.0		2.5
16 000	17.5		

[a]After highway driving.

Fig. 22-85 Oil dilution from fuel in spark-ignition engines after extreme short-distance driving.[12]

	2.0 l Engine	1.8 l Engine	1.4 l Engine
Distance in km	Iron content in mg/kg	Iron content in mg/kg	Iron content in mg/kg
1000	10	7.5	20
2000	15	10	50
4000	25	45	75
6000	40	75	90
8000	80	110	100 (7500 km)
10 000	100	250	650
12 000	175		
14 000	400		
16 000	650		
18 000	800		

Fig. 22-86 Metal abrasion in spark-ignition engines from extremely short-distance driving.[12]

nificant wear arose on the cylinder barrels, piston rings, bearings, and valve gear.

Additional steps are required before the oil change interval can be substantially extended by additional bypass filters, both in spark-ignition and diesel engines. Spark-ignition engines have less filterable combustion products in the oil than diesel engines. Hence, bypass oil filters are not recommendable for spark-ignition engines. Diesel engines, especially commercial vehicle diesel engines with a large amount of oil can benefit from reducing the insoluble components in the oil. It has been shown that the majority of impurities remain in the oil, however. Most of the solid foreign materials in the oil have a particle size of 0.1 to 0.5 μm, yet the pore width of the finest oil filter is much greater. The dispersant effect of oil additives is much stronger than the adsorbing force of the filter medium.

The same holds true for low-molecular aging products. The filter is totally incapable of slowing the natural breakdown of additive efficiency. Only increasing the oil volume by installing an additional filter can help. A proportional extension of the oil change interval appears possible. An oil additive container that can be used for heavy commercial vehicles would produce a better result and help prevent additional special waste from the required disposal of the bypass filter cartridges that must be regularly changed. To measure the wide differences arising from strongly varying operating conditions and present the new, adapted oil change interval to the driver, various sensors have been used for a while to indirectly monitor the quality of the engine oil in automobiles. The number of cold starts, the engine speed, the performance, the oil dwell time, and the distance are factored into a flexible electronic oil change interval display. Recently, the electronics have even monitored the engine oil state in automobiles with sensors in the oil circuit that also register oil top-up. The quality of oil when it is exchanged in a garage can even be electronically entered to yield the fuel consumption for a certain engine oil quality and, hence, determine the length of the change interval.

Chemical Changes

The alkalinity reserve in fresh oil is defined by the TBN (total base number). This is a measure of the ability of the oil to neutralize acidic combustion products to reduce or prevent residue formation, corrosion, and wear. This is contrasted with the TAN (total acid number) that indicates the amount of weak and strong acids in the used oil. Both are used to evaluate used oil. Values beyond *p*H-9 (highest alloyed diesel engine oils) are termed SBN (strong base number), and those under *p*H-4 (need oil change) are termed SAN (strong acid number). Figure 22-87 presents the assignment of base numbers and acid numbers. A gradual decrease in the neutralization ability of used oil of up to 50% in comparison to the TBN of fresh oils is generally held to be acceptable.

A number of years ago with the spread of consumption-optimized spark-ignition engines before the introduction of engines with regulated catalytic converters that operate with a stoichiometric air-fuel ratio, black sludge formation was a serious problem. It arose from operating the engine with a lean air-fuel ratio that causes hotter combustion and, hence, produces more nitrogen oxides that enter the crankcase (and, hence, engine oil) with the combustion gases via the piston rings. The nitrogen oxides are transformed there into NO_2 either in the gas phase or by reacting with oil components. The NO_2 then reacts with polar additive components to form organic nitrates and, hence, the problematic black sludge. This process is also termed nitration. The amount of organic nitrates in used oil is an indicator of its continued usefulness. With the development of suitable engine oils and fuel additives, the formation of black sludge could be prevented. The introduction of catalytic converters and the related operation at $\lambda = 1$ attenuated the problem. In SI engines with direct injection that operate in the lean range depending on the design, steps were taken to keep this problem from repeating itself. Finally, we should note that modern passenger car engines generally consume 100 ml of oil every 1000 km; given the large amounts of oil circulating in the engine over conventional oil change intervals of 15 000 km, there is no real need for topping up. The oil may need to be replenished in engines with a long change interval that are increasingly entering the market. This is frequently determined by a sensor in the oil pan and signaled to the driver. In automobiles that are primarily driven long distances, it may be highly recommendable to top up the oil given the numerous cited negative influences on the reserve of agents to prevent the oil volume from dropping too far, and to refresh the reserves of chemically active ingredients. However, we must remember that excessively low oil consumption nearly always indicates harmful oil dilution from fuel. In larger commercial vehicle diesel engines, oil consumption of up to 400 ml/1000 km is common after the breaking-in phase.

*p*H	TBN	SBN	TAN	SAN
1 to 4				Decrease
4 to 9	Increase		Decrease	
9 to 11		Increase		

Fig. 22-87 TBN and TAN.[1]

22.2.8.11 Racing Engine Oils

Oils for engines in competitive vehicles must be optimized for their respective purposes. Let us consider the example of engine oils in modern Formula 1 racing engines in comparison to earlier Grand-Prix formulas in the 1930s. In earlier compressor engines with a very high specific performance (120 kW/l at 7000 rpm) a mixture of castor oil and synthetic esters was used primarily to prevent piston seizing. Castor oil is a vegetable oil from the seeds of the castor oil plant native to Brazil and India. It consists of 80% to 85% glycerides of ricinic acid and glycerides of other organic acids. A disadvantage is its insufficient oxidation stability and the formation of resin-like deposits that force the engine to be disassembled and cleaned almost every time it was used. For today's 3.0 l naturally aspirated engines that produce more than 200 kW/l at 18 000 rpm, frequently only fully synthetic, highly fluid oils are used that are optimized for lowest friction resistance with maximum shear resistance and high-temperature resistance. They must have high oxidation stability, high wear protection, and particularly effective foam suppression because of the extremely high speeds and oil movement in the dry sump oil tank and engine. Increased dilution of the oil with fuel is a given when there is particularly rich combustion during high performance. High dispersability must be ensured to avoid the ejection of foreign materials and additives. On the other hand, a racing oil of this kind does not have to be tailored for cold starts and must remain intact for only a very short life—just one race or approximately 300 km. Likewise, cost is not a consideration. For long-distance racing such as the 24-h Le Mans, the requirements are naturally stiffer; in addition to increased performance reserves, oil consumption and oil topping up are also factors.

22.2.8.12 Wankel Engine Oils

The same engine oils used for reciprocating piston engines are used to lubricate rotary engines. This is understandable for economic reasons since these engines are not very widespread, although the special features of the rotary engine would certainly be better served with tailored engine oils—preferably with low-ash additives. In rotary engines, some of the oil is used to lubricate the apex seal and continuously burned in the process. Because of the system-related high oil consumption of approximately 1 1/1000 km and the continuous need to replenish the oil supply, and because of the constructive features of the rotary engine, the otherwise applicable considerations such as low evaporation loss, high oxidation stability, high wear protection, etc., are not as important.

22.2.9 Engine Oils for Two-Stroke Engines

Given their design, two-stroke engines require a different type of lubrication than four-stroke engines since a forced-feed lubrication system cannot be used with crankcase scavenging. A distinction is drawn between conventional mixture lubrication in which a small concentration of special engine oil is premixed with the fuel and increasingly popular fresh oil lubrication that is load and speed-dependent where oil comes from a separate oil tank. Over the course of development of the two-stroke engine with the increase in environmental awareness, the mixture ratio was reduced from an initial 1:20 to 1:25, 1:50, 1:100, and finally to 1:150 with a concomitant profound increase in performance. Nevertheless, the oil consumption of the two-stroke engine has always been several times higher than the four-stroke engine. Given the continuous participation of the oil in combustion, the related deposits need to be dealt with that arise on the spark plug, in gas exchange openings, and in the exhaust system. Oils for two-stroke engines, therefore, require a clearly different lubricant technology than oils for four-stroke engines.

The basic requirements for two-stroke oils are the following

- High solubility in the fuel
- High corrosion protection since the crankshaft drive and bearing are continually exposed to environmental air
- Minimal residue formation during combustion (spark plug/exhaust outlet)
- Scoring and seizure protection for piston rings, piston skirt, and cylinder barrel
- Low-smoke and low-noise combustion

The dispersability and the viscosity temperature behavior so vital in four-stroke engine oils are of no importance. Multigrade oils are not used. The required performance is ensured by selecting suitable basic oils and special additives. Primarily SAE-30 basic oils are used. To deal with exhaust smoke, which is particularly critical today, polyisobutylene and synthetic esters have proven to be particularly suitable basic liquids for suppressing smoke in exhaust gas. Detergents, dispersants, and corrosion and rust protection additives are used to attain the previously cited properties. Primarily ash-free substances are used, especially since EP requirements need to be met. They are also advantageous in regard to environmental requirements.

22.2.9.1 Two-Stroke Performance Classes

The API classes TA to TC were used earlier to evaluate the quality of two-wheel, two-stroke oils; TA was used for mopeds, TB for scooters and motorcycles, and TC for high-performance engines. The required engine test runs are no longer feasible since the stipulated engines are no longer built. However, API TC (CEC TSC-3) still remains in effect. It was replaced by the JASO and ISO specifications (previously global). Given the dominance of Japanese two-stroke engine manufacturers, JASO (Japanese Automotive Standard Organization) specifications are the most referenced. The ISO (International Standard Organization) specifications that are valid worldwide differ only slightly. Figure 22-88 shows the JASO and ISO classes that have been introduced since 1996. They apply to air- and water-cooled two-stroke, two-wheel engines and evaluate the performance of the oils in terms of lubricity, engine cleanness, freedom of the exhaust system, and exhaust smoke. It has become increasingly important to avoid visible and smellable exhaust smoke. Today, practically every powerful two-stroke branded product must fulfill JASO-FC or ISO-L-EGD requirements. The specifications of the latter allow manufacturers to claim superior performance. The classification NMMA TC-W3 (National Marine Manufacturers Association) also covers the biodegradability of two-stroke oils for outboard engines. These oils can also be used in chainsaws. The classification TISI 1040 (Thailand Industrial Standards Institute) is not relevant in

JASO	ISO	Comments
FA	—	
FB	L-EGB	
FC	L-EGC	Low-smoke
—	L-EGD	Low-smoke

Fig. 22-88 JASO and ISO classes.[14]

Europe; it is applicable only for the Thai market and especially deals with smoke in the exhaust from oil.

In the initial developmental stage of two-stroke oil, the oil-fuel mixture had to be made in a mixer. "Self-mixing" two-stroke oil with a solubilizer soon became available in small containers and was added to the gasoline in the gas tank. The widespread autolube lubrication in modern two-wheelers with two-stroke engines makes it unnecessary to premix oil and fuel outside or inside the gas tank. The oil is contained in a separate tank and metered into the flow of the air-fuel mixture depending on the load and speed. Both the life of the oil and environmental requirements can be taken into account by specifically reducing or increasing the amount of oil in the fuel.

22.2.9.2 Two-Stroke Test Methods

Figure 22-89 provides the physical characteristics for oil specifications for two-wheel, two-stroke engines corresponding to international requirements (ISO) and Japanese requirements (JASO).

Figure 22-90 shows the engine tests for two-wheel, two-stroke oils by Japanese manufacturers.

22.3 Coolant

The coolant consists of water plus radiator protector. The radiator protector provides frost and corrosion protection. The radiator protector and water are usually mixed at a ratio of 1:1, which provides sufficient antifreeze protection in nonarctic areas, and the required corrosion protection. Water by itself is insufficient for today's cooling systems. Water for cooling systems should have the following optimum characteristics:

- Water hardness: 5° to 9° German hardness
- *p*H at 20°C 7 to 8
- Chlorine ion content max. 40 mg/l
- Total > chloride + sulfate max. 80 mg/l

22.3.1 Frost Protection

At temperatures below the freezing point, the coolant must be protected against freezing or it will expand and cause impermissibly high system pressure and can potentially destroy the cooling system, engine blocks, and cylinder head. Frost protection is achieved by adding glycols—multivalent alcohols—to the coolant water. Figure 22-91

Test purpose	Test of
Viscosity at operation temperature	Minimum viscosity at 100°C, 6.5 $mm^2 s^{-1}$
Spark plug gap bridging	Limit on sulfate ash content: ISO max. 0.18% (m/m) JASO max. 0.25% (m/m)
Life of oxidation catalyst	JASO: no phosphorous permitted
Safety during storage and transport	Flash point corresponding to national law

Fig. 22-89 Characteristics of two-wheel, two-stroke oils.[12]

Test purpose	Engine	Test conditions	Test criteria
Safety against piston seizure and scoring	Honda DIO AF 27	Alternating load at 4000 rpm; spark plug seat temperature 160–300°C; mixture ratio 50:1	Decrease in torque after cold start and at operation temperature
Cleanness of piston rings, piston skirt, combustion chamber residue	Honda DIO AF 27	Full load at 6000 rpm; mixture ratio 100:1, JASO 1 h	Evaluation of the engine parts after the end of the test
Smoke formation in the exhaust gas	Suzuki SX 800R	Partial load and idling at 3000 rpm; mixture ratio 10:1	Evaluation of visible smoke
Cleanness of the exhaust ports	Suzuki SX 800R	Load change at exhaust gas temperatures 330–370°C at 3600 rpm; mixture ratio 10:1	Threshold of the bottom pressure at the intake range

Fig. 22-90 Engine tests for two-wheel, two-stroke oils.[12]

	Monoethylene glycol	Monopropylene glycol	Diethylene glycol
Empirical formula	$C_2H_6O_2$	$C_3H_8O_2$	$C_4H_{10}O_3$
Density at 20°C kg/m³	1113	1036	1118
Boiling point in °C	198	189	245
Melting point in °C	–11.5	–60.0	–10.5
Specific heat at 20°C (kJ/kg)	2407	2460	2307

Fig. 22-91 Characteristics of glycols.[12]

	Density in kg/m³ at % (V/V) monoethylene glycol			
	50	40	30	20
10°C	1084	1073	1051	1035
30°C	1075	1063	1038	1030
50°C	1064	1049	1031	1022
70°C	1050	1037	1025	1015
90°C	1038	1025	1015	995

Fig. 22-92 Density of the coolant MEG.[1]

shows the characteristics of the three glycols suitable as radiator frost protectors. Monoethylene glycol (MEG) is the most used radiator frost protector. Measuring the coolant density provides a way to easily and quickly monitor its concentration. In Fig. 22-92, the measured density at the respective measured temperature is portrayed as an indication of the concentration.

As expected, the density increases with the concentration and falls as the temperature rises. A coolant concentrate based on MEG has a higher boiling point than water, which promotes engine efficiency. Today, we find coolant temperatures of up to 120°C at 1.4 bar system pressure. The boiling points for the respective MEG concentration are shown in Fig. 22-93, whereas Fig. 22-94 presents the behavior of water/glycol mixtures under cold conditions.

Concentration in % (V/V)	Boiling point in °C
0	100.0
10	101.5
20	103.0
30	104.5
40	106.5
50	109.0

[a] MEG/water mixtures at regular pressure.

Fig. 22-93 Boiling points.[a1]

% Monoethylene glycol (V/V)	Ice flaking points in °C	Pour points in °C
0	0	0
5	–2	–2.5
10	–4	–5
15	–6.5	–8.5
20	–9.5	–12
30	–17	–20.5
40	–27	–32
50	–37	–47

Fig. 22-94 Frost protection from monoethylene glycol.[1]

The specific heat of a coolant, i.e., its heat-absorbing property or ability to absorb and conduct engine heat, should be very high. It rises with the temperature, but drops with the MEG concentration.

22.3.2 Corrosion Protection

The coolant concentrate contains carefully harmonized agents that prevent corrosion from arising on the different metals that contact the coolant. Figure 22-95 provides information on the arising corrosive substances and the required inhibitors. Individual inhibitors can protect one metal, but may corrode others. The concentration of the individual agents is also a factor. If it is too high, it can be as problematic as if it were too low. The synergies between the individual components also need to be considered. A sufficient reserve alkalinity neutralizes the acidic substances that enter the coolant uncontrolled from the exhaust gas or oxidation products of glycol. The primary corrosion inhibitors are the following:

- Alkali metal phosphate
- Amine/phosphate
- Benzoate/nitrite
- Nitrite amine phosphate-free inhibitors (NAP)
- Silicate-free inhibitors (OAT)

Market-ready coolant concentrates usually contain approximately 93% (V/V) MEG and up to 7% (V/V) corrosion inhibitors. In addition to the corrosion inhibitors

Metal	Corrosion from	Inhibitor
Aluminum	Sodium nitrite, Borax	Sodium silicate, sodium nitrate, phosphates, sodium benzoate, benzotriazole, tolyl triazole
Cast iron	Sodium benzoate, sodium nitrate, glycolic acid	Phosphates, sodium nitrite, salts of carboxylic acids
Copper	Free amines, sodium nitrate	Benzotriazole, tolyltriazole, sodium mercaptobenzothiazole
Brass	Free amines, sodium nitrite	Benzotriazole, tolyltriazole, sodium mercaptobenzothiazole
Steel		Phosphates, sodium benzoate, sodium nitrite
Tin solder	Sodium nitrite, sodium nitrate, glycolic acid	Benzotriazole, Borax

Fig. 22-95 Corrosive substances and inhibitors.[12]

Property	Unit of measure	Characteristic	ASTM test method
Density at 15.5°C	kg/m^3	1110 to 1145	D 1172
Freezing point 50% (V/V) in distilled water	°C	Max. –37	D 1177
Boiling point (undiluted)	°C	Min. 163	D 1120
Boiling point 50% (V/V) in distilled water	°C	Min. 107.8	D 1120
Attacks automobile paints	—	No attack	D 1882
Ash content	% (m/m)	Max. 5	D 1119
*p*H 50% (V/V) in distilled water	—	7.5 to 11.0	D 1287
Chlorine content	mg/kg	Max. 25	D 3634
Water	% (m/m)	Max. 5	D 1123
Reserve alkalinity	ml	[a]	D 1121

[a] To be agreed upon between the manufacturer and user.

Fig. 22-96 ASTM standard D 3306 for coolant based on MEG (physical/chemical characteristics).[12]

listed in Fig. 22-95, there are small amounts of other additives such as antifoaming agent, sequestering agent for complexing calcium and magnesium ions in hard water, silicate stabilizers, denaturants, and dyes. This yields a complex mixture. To make sure that all requirements are fulfilled, the coolant should not contain less than 40% (V/V) coolant concentrate.

22.3.3 Specifications

Because of the complexity of the coolant concentrate, it must meet the characteristics in the corresponding specifications to be permitted. These describe the quality and performances behavior. Standardized methods are used to determine the described measured values. Figure 22-96 shows the ASTM standard D 330 for coolant based on MEG and the requirements for coolant performance. There are also numerous regulations provided by the individual automobile manufacturers for radiator protectors.

The following additional information regarding coolant requirements should also be noted.

Given the anticipated greater use of magnesium as a cast alloy component, precise investigations are required to see if the type and composition of presently used radiator protectors can satisfy the new requirements.

Deposits	No deposits may form in the cooling system since this would impair the removal of heat. If the water hardness is too high, lime and other minerals can precipitate starting at approximately 60°C and collect especially at locations that are critical to the transfer of heat.
Hot water corrosion	In today's high-performance engines, the temperature of the surfaces contacting the coolant is very high. The water reacts directly with the aluminum to produce gaseous hydrogen that can attack metal.
Surface corrosion	All metal surfaces are attacked by corrosive substances as a function of their relative roughness.
Contact corrosion	There are various metals in the cooling system. If, for example, entrained iron particles collect on an aluminum surface, a local element forms that can produce holes in the surface.
Gap corrosion	In gaps within the cooling system in which the coolant does not circulate evenly, entrained corrosive components can collect and cause increased corrosion.
Cavitation	From fluctuations in the system pressure of the cooling circuit, vapor bubbles can form in the cylinder head and water pump. They combine as the pressure rises. This jump in pressure removes material from the metal surface that can cause serious corrosion.

Bibliography

[1] Aral Research Archive.
[2] Aral [Pub.], Verkehrstaschenbuch 2000/2001, 43rd edition, Bochum, 2001.
[3] Aral [Pub.], Fachreihe Forschung und Technik – Kraftstoffe für Straßenfahrzeuge, Grundlagen, Bochum, 1998.
[4] Aral [Pub.], Fachreihe Forschung und Technik–Dieselkraftstoffe, Bochum, 2001.
[5] Aral [Pub.], Fachreihe Forschung und Technik–Ottokraft-stoffe, Bochum, 2001.
[6] Aral [Pub.], Fachreihe Forschung und Technik–Umweltfreundliche Kraftstoffe, Bochum, 1995.
[7] Aral [Pub.], Fachreihe Forschung und Technik–Kraftstoff-additive, Bochum, 1995.
[8] Aral [Pub.], Fachreihe Forschung und Technik–Alternative Kraftstoffe, Bochum, 2001.
[9] Waldmann, H., and G. H. Seidel, Kraft-und Schmierstoffe, Sonderdruck ARAL AG aus Automobiltechnisches Handbuch, 18th edition, 1965, Walter de Gruyter, Berlin, Supplementary volume, 1979.
[10] Aral [Pub.], Fachreihe Forschung und Technik–Schmierstoffe Grundlagen/Anwendung, Bochum, 1997/1998.
[11] Aral [Pub.], Fachreihe Forschung und Technik–Schmierstoffadditive, Bochum, 1996.
[12] Basshuysen, R.v., and F. Schäfer [Eds.], Shell Lexikon Verbrennungsmotor, Friedr. Vieweg u. Sohn Verlagsgesellschaft mbH AT2 u. MT2 (no date indicated since it has not been published in complete form).
[13] Menrad, H. [Ed.], Alkohol Kraftstoffe, Springer-Verlag, Wien, 1982.
[14] DEKRA [Pub.], Betriebsstoff-Liste, Motor-Presse-Verlag, Stuttgart, 1999.

23 Filtration of Operating Fluids

23.1 Air Filter

23.1.1 The Importance of Air Filtration for Internal Combustion Engines

In the early 1920s, automobiles frequently suffered from the very limited durability of the engines used. High levels of dust on the unsurfaced roads of the time caused severe piston-ring wear, leading to declining engine performance and ultimately resulting in the need to replace the old piston rings with new ones to restore the engine's vitality. The introduction of so-called "air-cleaners" [1] made it possible to double repair and servicing intervals to 4000 km.

Today, too, modern engines also draw in considerable quantities of dust while running, varying depending on the place of operation. These particles ultimately reach the oil circuit, where they are partly responsible for engine wear, the degree and intensity of component wear (e.g., big-end and little-end bearings) depending on the number of particles and their type, geometry, size, and hardness. If these particles are able to enter the bearing clearances, they produce scoring on contact surfaces or generate secondary particles, since they disintegrate between the contact surfaces as a result of the high mechanical loads occurring there. Also relevant in terms of wear are all sizes of particles up to the nominal gap width, since compound friction can occur in the connecting-rod bearings and these particles are able to enter the spaces there. It is the function of filters to remove such particles from the engine's intake air supply.[2]

Modern filters achieve particulate-removal efficiencies of up to 99.9%, signifying that, of 1000 particles of the specified diameter, at most one particle is able to pass through the filter element. The unremoved particles are able to reach the oil circuit, where they have to be removed on full-flow and bypass oil filters.

23.1.2 Impurities in Engine Intake Air

Atmospheric air contains dust particles, the diameters of which can range from 0.01 to 2000 μm. About 75% of airborne impurities are of a magnitude of between 0.1 and 100 μm, this range and its concentration distribution being highly geographically dependent.

23.1.3 Data for Assessment of Air-Filter Media

The function of modern air filters is that of reducing the pollution in the intake air to acceptable levels under the given operating conditions. Specific filtration values are available in the various specifications. These data include dust absorption capacity, removal efficiency, surge strength—i.e., no passage of dust even if surges occur in the flow of intake air—and stability, signifying retention of the filter element's pleated structure (see Fig. 23-1), even under wet conditions, such as driving in heavy spray and during heavy rain, for example. Adequately stable pleat geometry during operation and, in particular, in case of wetting is also important for trouble-free filter-element function, since maximum dust capacity can be maintained only if the pleated compartments remain uniformly arranged. In addition, the filter medium must also tolerate engine oils, fuels, and crankcase gases; i.e., it also needs to have chemical resistance.

The most recent developments in the field of filter media are orientated around the longer servicing intervals of up to 90 000 km for automobiles. Media with a pronounced "gradient structure" (as far as pore distribution is concerned) are currently used. As seen in Fig. 23-2, multilayer media consist of filter papers in combination with melt-blown layers (layers of plastic fibers). This increases dust storage capacity and thus service life by up to 30%.

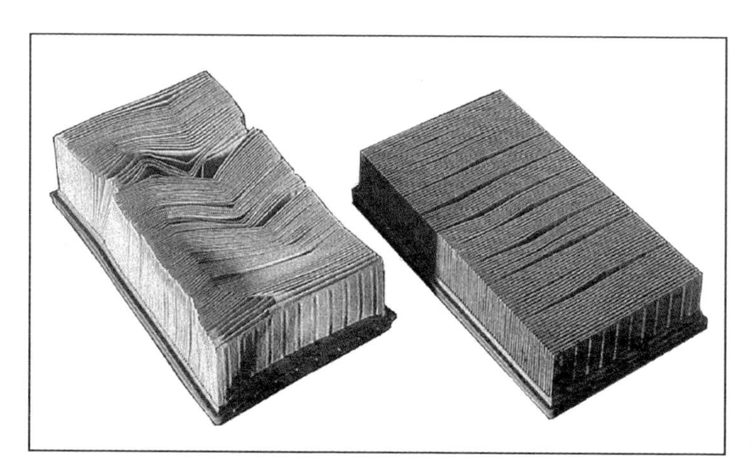

Fig. 23-1 Damaged pleated structure, causing reduced filter performance (at left), and original condition (right).

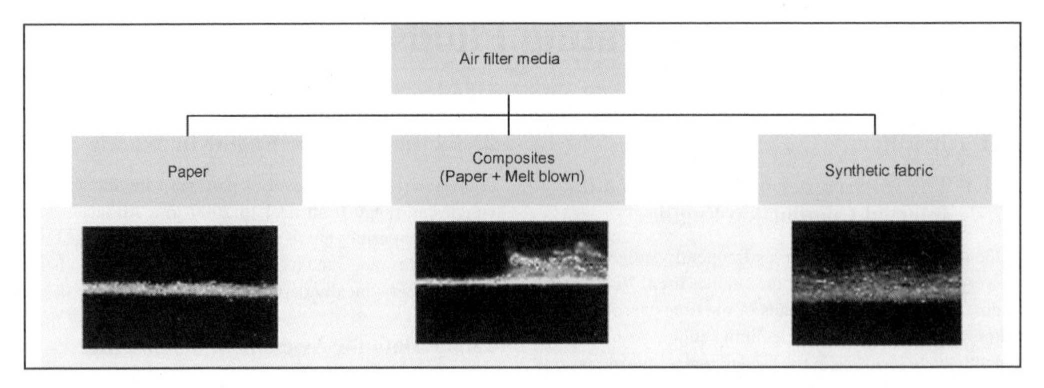

Fig. 23-2 Filter media for use with internal combustion engines.

Media consisting of pure plastic fibers (synthetic fabrics) (see Fig. 23-2) permit the generation of greater gradient structures and thus a dust capacity enhanced by as much as 50%.[3]

The materials must remain thermally stable up to engine compartment temperatures of about 85 to 90°C. Modern filter fabrics have a significantly lower pressure loss on the filter element, for the same size; see Fig. 23-3. Filter media that meet these requirements consist of cellulose or plastic fibers, or mixtures of the two.

23.1.4 Measuring Methods and Evaluation

Filter media are tested under standardized conditions specified in DIN/ISO 5001.[4] In conformity with the standard, removal efficiency can be determined using two methods, in the form of gravimetric removal efficiency, which is determined from the ratio of the increase in the mass of the filter medium under test to the total mass of dust fed during the test period.

To obtain detailed information on the performance range of filter media for the wear-relevant particles, it is necessary to measure removal efficiency as a function of particle size. This is shown as a function of particle diameters between 0.1 and 10 μm in Fig. 23-4. Removal efficiency increases steeply after only a slight pressure increase has occurred, and 100% of the particles are captured.

It is necessary to take both removal efficiency and dust absorption capacity into account to determine the efficiency of a filter medium.

The latter can be determined by feeding dust onto the filter element under load at a constant volumetric air flow until a pressure drop of 20 mbar occurs.

23.1.5 Requirements Made on Modern Air-Filter Systems

Modern air-filter systems, i.e., a system consisting of a "raw air" pipe with an air inlet, a filter housing, dampers to reduce intake noise, a filter element, a cleaned-air line with a hot-film air-mass sensor (HFM), and the engine intake manifold, must be geometrically adapted to the restricted space available under the hood. This results in the development of new filter media that require significantly

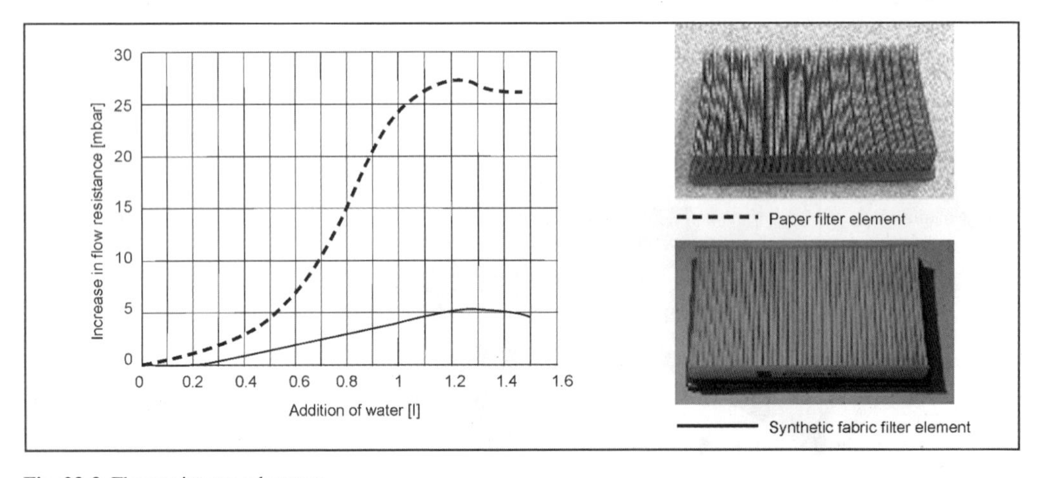

Fig. 23-3 Flow resistance when wet.

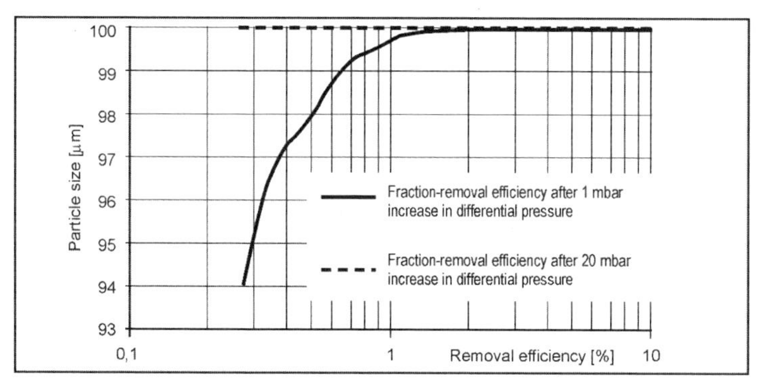

Fig. 23-4 Dust-fraction removal efficiencies for a standard filter medium.

less space without sacrificing filter performance. The filtration figures required for automobiles (diesel) and trucks are shown in Fig. 23-5.

The efficiency of the entire air-intake train depends on the performance of the individual components such as the location of the air-intake point on dust-intake and water-intake minimization criteria. The intake area should be located in sheltered, low-turbulence zones on the vehicle, for instance, under the covering in the wheel arch or in areas with no through flow in the engine compartment. Ideal placement in trucks is above the cab roof (overhead intake) or to the side of the cab, in order to achieve longer servicing intervals.

A flow-optimized filter housing assures complete exploitation of the medium's potential for dust absorption and removal efficiency, since the medium is not subjected at any point to excessive approach-flow velocities and, thus, high removal efficiency can always be guaranteed.

In addition to filtration of the combustion air, a further important function of the air-filter element is that of keeping dirt away from the hot-film air-mass sensor. Fouling this with particles affects the engine control system, clearly signifying that filtration does have a direct influence on driving comfort and convenience.

23.1.6 Design Criteria for Engine-Air Filter Elements

It is necessary for the design of an engine-air filter element to differentiate between the filter fineness required for automobiles with gasoline and diesel engines and trucks, as shown in Fig. 23-5.

The filter-media surface area is calculated on the basis of the engine's air-volume requirement per unit of time [$\mathring{V}$ (m³/min)].[5] Sufficient surface area is provided for cleaning this volumetric flow of air so that the air's flow velocity never exceeds a certain critical velocity range of $v < v_{crit}$. At excessively high filtration velocities (above 10 cm/s, for example), this results in the particles not being captured on the fibers during passage through the filter. The excessively high pulse results in rebounding off of the fibers, and the particles are able to pass through the filter medium. Permissible flow velocity v_{crit} may differ greatly, depending on the filter materials used, such as filter paper, multilayer media (composites), or synthetic fabrics.

The following critical velocities v_{crit} and dust-absorption capacities result for typical filter media when the removal efficiency demanded for gasoline and diesel engines is taken into account; see Fig. 23-5.

The necessary filter surface area is calculated in the following example: Surface area is calculated to achieve the required removal efficiency on the basis of volumetric air flow (e.g., 5 m³/min) and vehicle type (automobile/diesel), taking account of flow velocity. The specific dust capacities of the filter medium are exploited to achieve service intervals of, for example, 60 000 km (= 200 g of laboratory dust). A square meter of standard filter paper is needed to achieve the required dust capacity of 200 g, critical velocity then being exceeded, however, with the

Filter medium	v_{crit} (cm/s)	Gravimetric removal efficiency %			Spec. dust capacity (g/m²)	Weight per unit of area (g/m²)
		Automobile (gasoline)	Automobile (diesel)	Trucks		
Paper	10	>99	>99.8	>99.9	190–220	100–120
Multilayer medium (composite)	17	>99.5	>99.8	>99.9	230–250	100–120
Synthetic fabric	33	>99.8	>99.8	>99.9	900–1100	230–250

Fig. 23-5 Critical filtration velocity, specific laboratory dust capacities, removal efficiencies of filter media, and weights of filter media per unit of area.

result that the higher figure of 1.25 m^2 is selected for the ultimate design.

23.1.7 Filter Housings

23.1.7.1 Design of Filter Housings

In addition to the "available space" problem, ease of servicing, i.e., trouble-free filter-element changing, is of great importance in filter-housing design. This functionality is linked directly to the need for housing tightness. Practically all air-filter elements feature a polyurethane (PUR) seal.

Filter housings can be classified into flat and round-filter element types; see Fig. 23-6. In the case of rectangular filters, the PUR section seal is located in the housing groove and clamped axially by the filter-housing cover. Round filter elements predominate in the goods and utility vehicle sector, because of their greater simplicity of sealing and higher stability.

Fig. 23-6 Air-filter element types.

The design of filter housings is essentially determined by the requirements for homogeneous flow in the housing. This minimizes pressure drop and achieves uniform flow onto the surface of the filter element.

Bibliography

[1] Katz, H., Die Luft-, Brennstoff- und Ölreiniger im Kraftwagen, Autotechnische Bibliothek, Vol. 80, Berlin W62, Richard Carl Schmidt & Co., 1927.
[2] Affenzeller, J., and H. Gläser, Lagerung und Schmierung von Verbrennungsmotoren, Springer Wien, New York, 1996.
[3] Purchase, D.B., Handbook of Filter Media, Elsevier Science Ltd., 1997.
[4] DIN/ISO 5011.
[5] Lechner, F., University thesis, Technical University of Heidelberg, 2000.

23.2 Fuel Filters

Fuel-injection systems on modern gasoline and diesel engines react extremely sensitively to even the very smallest fouling in the fuel. Damage can be caused, in particular, by particulate erosion and, in diesel engines, also by corrosion resulting from water in the fuel. Such contamination is composed both of extremely hard mineral particles and of organic particles such as soot and tar. Diesel fuels sold under DIN EN 590 must achieve particulate contents of lower than 24 mg/l. International automotive industry associations recommend figures of below 24 mg/kg. Diesel fuels sold in Germany generally have particle contents of less than 10 mg/l, although the above-mentioned limit for particle concentration is in some cases greatly exceeded in the fuels available around the world. The fuel filter must be able to capture all larger particles (>15 μm) with certainty. Studies into wear on newer injection systems also document the "wear relevance" of the finer fraction, of around 5 μm. For this reason, a removal efficiency for the 3 to 5 μm particle-size fraction has become established as a characterizing dimension for filter fineness in recent years. As a result of their much greater injection pressures, diesel injection systems require greater protection against wear than gasoline injection systems, and thus finer filters. In addition, the fuel filter must also possess adequate capacity for storage of particles.

23.2.1 Gasoline Fuel Filters

Modern gasoline engines are equipped with electromagnetically operated injection valves. It is the fuel filter's task to prevent erosive wear on the electrical injection valve and the ingress of wear-causing particles into the engine's combustion chambers. Engines featuring gasoline direct injection (GDI) require significantly finer fuel filters for protection against wear than systems for injection into the intake manifold. The reasons are that, on the one hand, pressures higher by a factor of 30 occur at the injection valve and, on the other hand, other components, such as the pressure accumulator and the pressure-control valve, must also protect the injection system against the entry of particles.

The Necessary Filter Finenesses

A specified filter fineness (initial removal efficiency in accordance with ISO/TR 13353, Part 1, 1994) is the result of dynamometer and field tests performed by engine and injection-system manufacturers jointly with the filter producers. Figure 23-7 shows recommendations for minimum filter fineness for gasoline-engine injection systems.

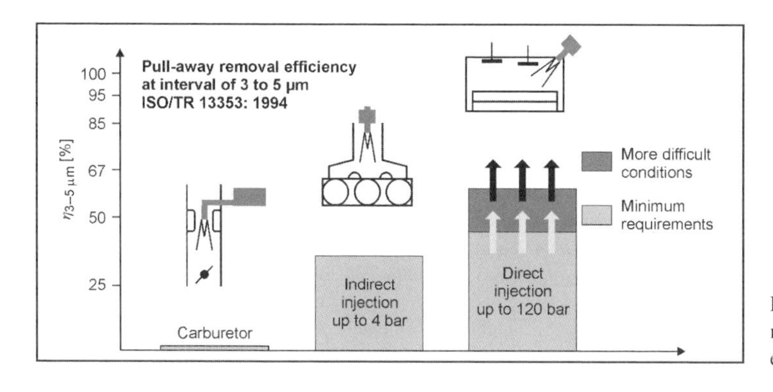

Fig. 23-7 Recommended minimum-filter fineness for gasoline-engine fuel filters.

Even finer fuel filters are necessary to achieve efficient protection against wear if the engine is to be operated with high particulate concentrations in the fuel or the outside air (more difficult conditions).

Types of Gasoline-Fuel Filters

The main form of gasoline-fuel filter is the inline filter. On some types, the pressure-control valve is also integrated into the filter head. The filter housings are made of plastic, aluminum, or steel, depending on their position in the engine compartment (crash safety) and the vehicle manufacturer's specification. Maintenance-free "vehicle lifetime" fuel filters are increasingly being demanded for gasoline-engine vehicles. A further trend arises as a result of the lowering of hydrocarbon emissions limits. This makes it necessary to integrate all external components of the low-pressure circuit, such as the injection pump, fine filter, and pressure-control valve into a single in-tank unit. Further elements such as the tank filling-level transducer, the swirl pot, which can be actively filled via an ejector, and the optional prestrainer, which serves to protect the pump, can also be integrated into the in-tank unit. Figure 23-8 shows a modern lifetime filter element. Such filter elements are also currently made in complex, noncylindrical geometries, in order to achieve maximum exploitation of available space in the in-tank unit.

Fig. 23-8 Spiral V type filter element for lifetime filtration in an in-tank unit.

Filter Element and Filter Medium Structure

The requirements currently made on filter fineness and dirt-storage capacity necessitate innovative filter concepts. With the same filter fineness, a simple strainer (surface-action filter) achieves only about a tenth of the filter service life (particle storage capability) of a modern deep-bed filter medium. The use of deep-bed filter media that, in addition, are also installed with an extremely high packing density is therefore necessary in order to meet the performance data specified for gasoline-fuel filtration.

This is achieved, in particular, through star-configuration pleating (see Fig. 23-8). The pleated construction, which possesses a large filter surface area, is supported on a pressure-proof central tube. The fuel flows radially from the outside to the inside of the filter. So-called spiral V filters, which consist of concentric filter-paper compartments, are also in use as an alternative. Deep-bed filters currently consist mainly of ultrafine cellulose fibers. More recently, spiral V elements are increasingly incorporating multilayer filter media with ultrafine synthetic fibers. The cellulose fibers of the filter medium are sheathed with special fuel-resistant impregnating agents.

23.2.2 Diesel-Fuel Filters

The rapid development of diesel-engine technology permits extremely high fuel efficiency in both automobile and commercial vehicle engines. In all modern diesel engines, the fuel is injected directly into the cylinder. The function of the fuel filter is that of protecting all the components of the high-pressure injection system. The filter can, for this purpose, be located in the low-pressure circuit, either on the pressure side in the feed line to the high-pressure pump or on the suction side in the inlet line to the fuel presupply pump (FSP). If on the pressure side, a differential pressure of up to 6 bar, depending on the design of the system, is available for fuel filtration, significantly more than with the suction-side arrangement. Installation on the pressure side, therefore, predominates on commercial vehicle engines and also increasingly on automobile engines.

Necessary Filter Finenesses

The introduction of modern, solenoid-valve-controlled diesel injection systems made it necessary to significantly increase filter fineness.[2,3] It quickly became apparent, particularly in the case of commercial vehicle pump-nozzle systems, that the use of the up to then finest filters (initial removal efficiency $\eta_{3-5\,\mu m} = 45\%$, up to 1997 the finest filter grade used in Europe for distributor injection pumps) resulted in wear of the solenoid valve's seats. This damage became apparent in practice in the form of declining engine output, rough running (caused by differing degrees of wear on the individual injectors), and greater generation of soot. The necessary filter fineness is determined primarily by assessment of wear following field testing of the vehicle. This time-consuming procedure is increasingly being superseded, or at least augmented, by special wear studies performed by the manufacturers of diesel engines and injection systems. A fuel specially blended with ultrafine particles (with ISO 12103 M1 or A2 test particulates, for example) is used on the injection-pump test bench. The injection-flow loss that occurs upon equipage with diverse filters is used as an indicator of wear and is then correlated with the filter's initial removal efficiency. Both in field tests and in the context of test-bench trials, indicators relevant for practice correlate extremely well with the initial removal efficiency determined in accordance with ISO/TR 13353 (1994) in the 3 to 5 μm particle-size range. Tests performed on pump line nozzle (PLD) engines (commercial vehicle, United States) also confirm the relevance of the ultrafine fraction about 5 μm particle size for wear.[4] Figure 23-9 shows recommendations concerning minimum filter fineness for diesel injection systems. As in Fig. 23-8, differentiation is made between normal and more difficult conditions.

Removal of Water

Damage caused by local impairment of lubrication and, in particular, by corrosion, can occur if water reaches the high-pressure side of a diesel injection system.[5] In many cases, the fuel filter, therefore, also performs the additional task of eliminating this free or emulsified water. Measures to prevent the ingress of impermissibly high quantities of water are generally required for distributor injection pump (VE) and common rail (CR) systems. Because of their short contact times, unit injector (PD) systems are relatively insensitive, but also require a system for removal of water if extremely high water levels are anticipated. Water is currently mainly eliminated by coalescence on special filter media. Filter media built up of fibers with a hydrophobic surface are particularly suitable for these applications. The water removed collects in a reservoir at the bottom of the filter and is discharged via manual or automatic systems when this water collecting space has filled. Testing for removal of water is performed by adding a 2% emulsion of water in diesel oil and is described in ISO 4020.

Diesel Fuel Filter Types

It is necessary to differentiate between diesel fuel filters that can be opened and filters that are replaced complete with the filter housing during servicing. This category of filters includes sheet-steel, aluminum, and plastic inline filters. Greater crash safety requirements have resulted in a significant revival of the sheet-steel type. Additional elements such as a water outlet, a water sensor (in the form of a conductivity sensor), a thermostatic valve (for return of hot fuel), and heating systems can also be integrated into inline filters. Another very widely used type of filter, one that cannot be opened, is the easy-change disposable fuel filter. The disposable filter is screwed onto a filter head featuring an external thread, and is sealed by an external elastomer element.

Particularly in the case of commercial vehicles, the demand for maximum filter fineness, and good water-removal performance combined with longer servicing intervals can be met only by multistage filtration concepts. Here, a prestrainer for removal of water and prefiltration of particles is first installed on the pressure or suction side. The fuel then passes through a pressure-side fine filter in which the ultrafine particles are removed. Figure 23-10 shows an easy-change disposable fuel prestrainer for commercial vehicles into which a large number of extra functions (e.g., a thermostatically controlled fuel return system, an electric heating system, a filling pump, and sensors for water-level and differential pressure) have been integrated.

Housing-type fuel filters are filters that can be opened. The housing cover is unscrewed for filter changing, and only the filter element is replaced. For easier servicing,

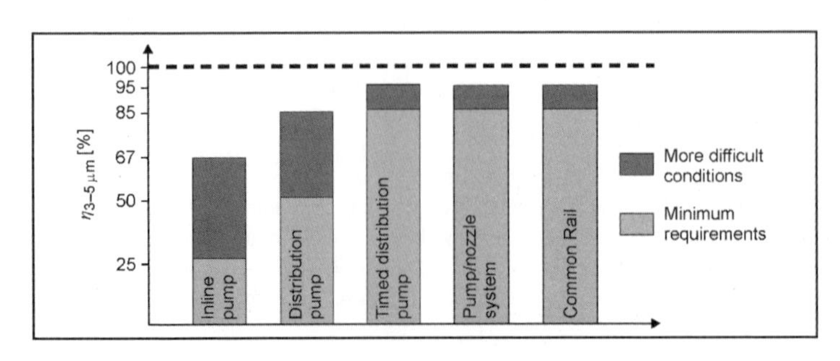

Fig. 23-9 Recommendation for minimum filter fineness for diesel-fuel filters.

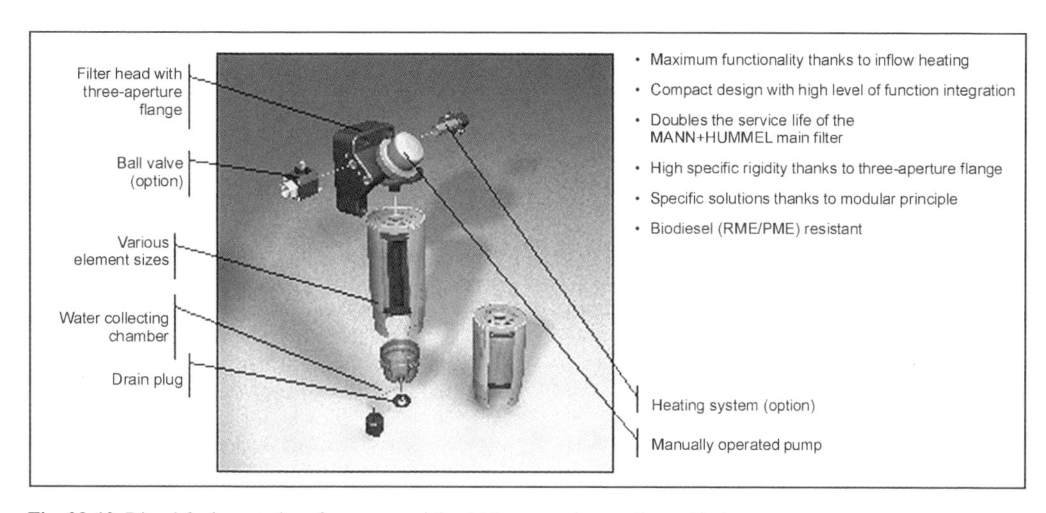

Fig. 23-10 Diesel fuel prestrainer for commercial vehicle, easy-change disposable type.

the housing cover is usually located at the top. Housing-type fuel filters may also contain additional elements such as a pressure-control valve, a thermostatic valve, an electric heating system, and water-level and differential-pressure sensors. In modern types, the filter element itself consists entirely of nonmetal, and therefore easily thermally recyclable, materials (Fig. 23-11).

Filter elements for diesel fuels are usually of the spiral V pleated type. The demand for high filter fineness combined with long servicing intervals necessitates the use of innovative filter media.

Structure of the Filter Element and Medium

Multilayer composite filter media (Mann + Hummel Multigrade) are capable of achieving ultrahigh filter finenesses with simultaneously high particulate storage capacities. Increased performance can be achieved only through systematic optimization of the composition and structures of the fibrous filter media. Ultramodern test methods based on measurements of removal efficiency using automatic particle counters are used for this purpose. Empirical development tools have recently been rationally augmented via the use of flow simulation (CFD) for calculation of particulate removal efficiency in fibrous deep-bed filter media.[6] Discoveries based on the optimization of the fiber structure have been integrated into modern composite filter media. Figure 23-12 shows the schematic structure of the MULTIGRADE filter media. Performance data for both particle-storage capacity (service life) and filter fineness increases significantly compared to conventional cellulose-based mixed fiber media.

Thanks to the hydrophobic properties of the base material and the ultrafine fiber diameter, the layer of polyester fibers on the approach side features extremely good water coalescence. Removal of water is, therefore, accomplished on the approach side. Performance data superior to single-layer filter papers can also be achieved by using so-called hybrid fiberglass papers. These filter media contain 5% to 20% microglass fibers with fiber diameters of about 1 λm. These filter media, which are used outside

Fig. 23-11 Metal-free filter element for a commercial vehicle housing-type fuel filter.

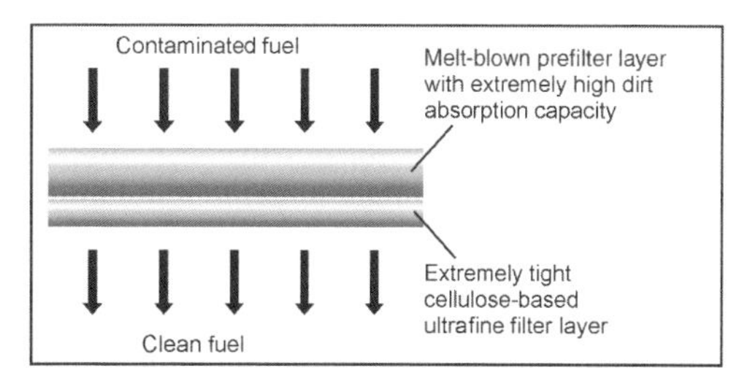

Fig. 23-12 Schematic structure of the MULTIGRADE filter media.

Europe, are not entirely undisputed, since the ultrafine, brittle glass fibers may detach.

23.2.3 The Performance Data of Fuel Filters

MULTIGRADE filter media, developed to achieve maximum particulate-storage capacity, are fuel filters suitable for use in in-tank units (lifetime use) for gasoline engines. The use of filter media tailored to the achievement of maximum filter fineness is necessary in the case of high-performance diesel-fuel filters. The MULTIGRADE concept achieves this without sacrificing service life. Figure 23-13 shows the service life and removal efficiency of a filter element containing MULTIGRADE F_PF for lifetime gasoline filters and the performance data for diesel filters incorporating the MULTIGRADE F_HC and MULTIGRADE F_HE ultrafine filter media compared to an early standard (cellulose/polyester hybrid fiber paper, up to 1997 the finest diesel filtration medium used in Europe).

Prospects

In the future, continuing advances in diesel and gasoline injection will also be accompanied by a continuous increase in the performance data of filter media and complete fuel filters. It will be possible to improve on the already extremely high standard only via the use of innovative and ever finer fibers as the filter-medium material and via optimization of the filter medium's microstructure. The focus of future refinements in filter media for diesel fuel will be on the increasing filter finenesses combined simultaneously with longer service intervals and greater compactness. Future diesel fuel filtration systems will include not only the integrated additional functions already used today, but also solutions for maintenance-free disposal of the water removed from the fuel. In the case of gasoline-engine fuel filters, the paramount technological challenge in the future will be the achievement in compact lifetime filter elements for in-tank units of the higher filter fineness required for DI engines.

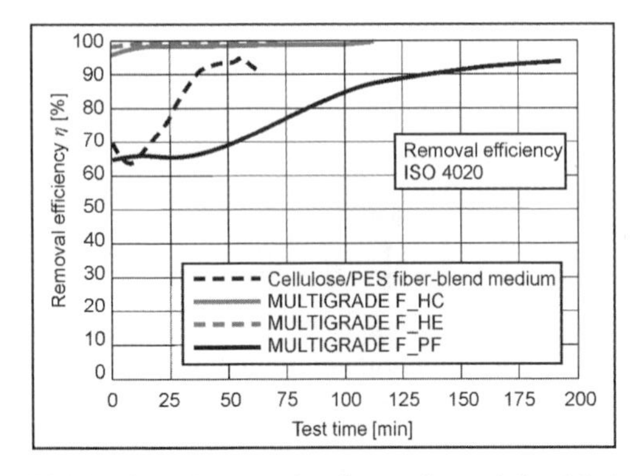

Fig. 23-13 Performance data for gasoline and diesel-fuel filters incorporating conventional and modern filter media.

23.3 Engine-Oil Filtration

23.3.1 Wear and Filtration

All the dirt that enters the engine from all the diverse possible sources collects in the engine oil. Dust particles from the atmosphere, allowed through even by good air filters, thus get into the oil. The remaining dirt from production and assembly of the engine enters the oil, as does metallic abrasion (wear particles) and soot from incompletely combusted fuel. Water (condensate) from combustion and uncombusted fuel (oil dilution), which form a complex mixture with the reaction products of the oil, such as oil-oxidation products and additive reaction products, for example, must also be added. The tendency for increasingly longer oil-changing—and therefore oil-filter-changing—intervals, must also be kept in mind.

Oil filters (see Fig. 23-14) are available in a selection of common types, and are a collection and accumulation point for the particles. This is true of both full-flow oil filters and bypass oil filters. These are deep-bed filters, which enable the oil to perform its intended functions during the period stated by the manufacturer. In addition to lubrication, i.e., the reduction by the oil of friction in all bearings and moving parts in order to reduce wear, these functions also include the cooling of hot parts of the engine and the sealing functions of thin oil films, and power transmission.

The oil-filter element can neither extend oil-changing intervals nor retard degradation processes. The filtration of ultrafine particles (above all, soot), which are able to pass through the full-flow oil filter but which can be

Fig. 23-14 A selection of different oil filters [disposable filter (with metal screw-on housing), metal-free elements for installation in oil-filter modules, in each case with paper, composite, and all-synthetic filter media].

successfully removed on bypass oil filters (ultrafine deep-bed filters) or centrifuges in the bypass oil circuit of diesel engines, prevents an increase in viscosity as a result of elevated particle concentrations. Failure to perform an oil change at the correct time can result in serious engine damage.[7]

Without appropriate filtration of the engine oil, wear caused by particles could be infinitely repeated, as a result of the continuous recirculation of the oil. The consequences are greater oil and fuel consumption, reduced engine performance, and significantly greater environmental damage as a result of poor exhaust figures.[8]

Wear relevance depends to a certain extent on the engine itself, i.e., on its tolerances and bearing clearances. Significant improvements have been achieved in these areas in recent years; i.e., production methods have been refined even further, and machining tolerances—and therefore lubrication gaps, too—have been reduced. Particles of a size of about 1 μm are, therefore, now critical—especially at high concentrations. Particularly dangerous are particle sizes ranging from about 8 μm up to about 60 μm, as is demonstrated by wear measurements. Figure 23-15 shows a typical wear diagram. The metallic abrasion caused on the engine was measured using tracer-marked metals, and illustrates the differing wear relevance of the various particle sizes. These statements can also be applied to other engines, although one must remember that greater abrasion, of course, also occurs with high concentrations of extremely small particles in the oil. Coarser particles are also highly dangerous, however, because they become comminuted (i.e., they disintegrate), and then form part of precisely the most dangerous particle-size fractions.

23.3.2 Full-Flow Oil Filters

Automobiles are equipped with a full-flow oil filter. This suffices provided high ultrafine particle levels are not anticipated between service intervals. Figure 23-16 shows a schematic view of a full-flow circuit and a bypass circuit. In the full-flow oil circuit, used in both gasoline and diesel engines in automobiles, the oil pump draws the oil in from the oil sump (nonpressurized); if necessary, the oil is cooled in the oil cooler and then passes through the full-flow oil filter. A pressure-control valve feeds the surplus mass flow of oil back into the sump.

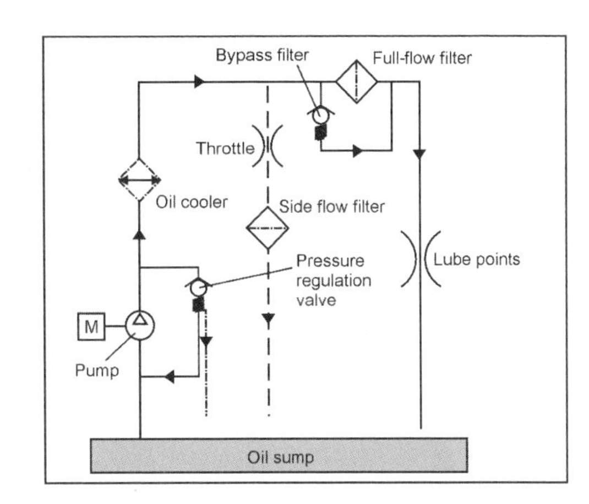

Fig. 23-16 Full-flow and bypass filter circuit (for commercial vehicles only) in the oil circuit.

The oil flowing to the engine must completely pass through the full-flow oil filter, generally resulting in a compromise between filter fineness (which generally signifies pressure losses in otherwise comparable filter media) and the size of the filter. The full-flow oil filter also features a bypass valve, which opens as a function of the differential pressure across the filter, and thus assures oil supply to the engine, which has even greater priority than filtration of the oil. This may be necessary, for example, in case of extremely low temperatures and highly viscous oil. It is, therefore, extremely important that the filter element is not expired, i.e., clogged.

Otherwise, this bypass valve remains at least partially open during normal operation and allows continuous passage of an unfiltered sidestream of oil, with all the consequences of elevated wear. The maximum differential pressure recommended by automobile manufacturers is between 1.5 and 2.5 bar. Filter media manage to achieve these requirements by using appropriate pleating and embossing methods. Figure 23-17 shows a section through a typical disposable filter. The spiral V filter element and the overflow valve (at bottom) can all be seen, as can the

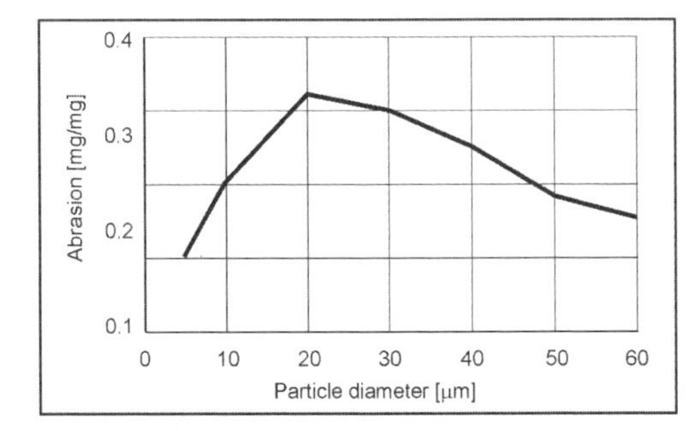

Fig. 23-15 Wear diagram showing the wear caused by particles of various sizes.

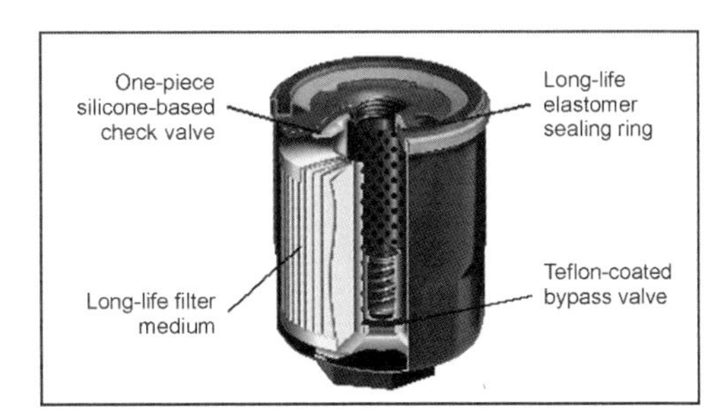

Fig. 23-17 Sectional view of a modern disposable filter with all-synthetic filter medium.

one-piece check valve, which prevents the filter element from running empty when the engine is stopped, and thus assures that the oil's full lubricating action is immediately available after starting.

A more complex solution is illustrated in Fig. 23-18, which shows a complete oil-filter module. The oil cooler is also directly integrated here, in addition to the metal-free filter element permitting environmentally friendly disposal. Also included are sensors for measurement of oil pressure, temperature, and (in the future) oil quality, permitting precise determination of the current condition of the oil. Such multifunctional oil modules are increasingly superseding conventional easy-change disposable filters, via the integration of even more functions, such as a heating system, an extra separator for crankcase venting, and mounting functions for other engine components.

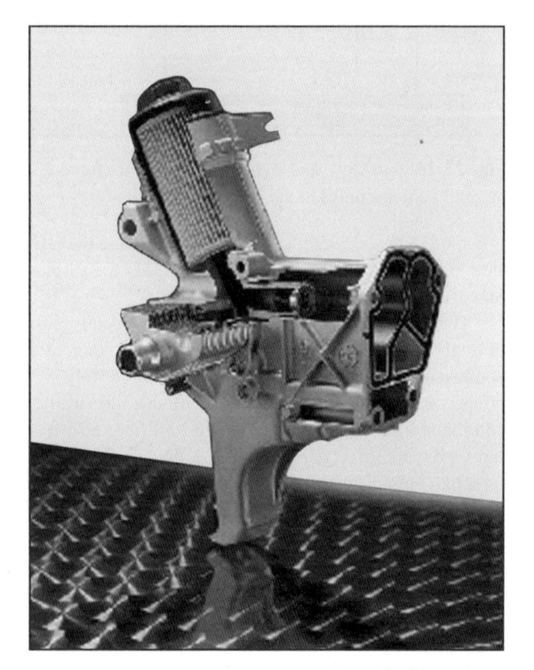

Fig. 23-18 Oil-filter module for a four-cylinder diesel engine with metal-free filter element and various integrated additional functions (oil-cooler, pressure and temperature sensors, alternator mounting, check and overflow valves, oil and water flow control).

23.3.3 Removal Efficiency and Filter Fineness

Unlike the situation with filter finenesses for engine intake air and fuel filtration, there is no specified minimum removal efficiency in the case of engine-oil filtration. The various manufacturers define the filter fineness for oil extremely diversely in their specifications; the separation curves specified are correspondingly diverse. Figure 23-19 shows a range of different separation curves as a function of particle size. The mass-average (i.e., the equivalent diameter referred to mass) 50% value of the particle is taken as the average filter fineness, i.e., not less than 50% of the particles of this particle size contained in the oil are removed; in other words, a maximum of 50% may pass through the filter. Full-flow oil filters that remove particles with diameters of 9 and 12 μm serve as the standard. However, more recently developed filter media that filter engine oils extremely finely, with a filter fineness of 4 μm, are now available.

So-called lifetime oil filters that assure the capture with certainty of relatively large particles but omit the filtration of small particles are also being tested. These filters have extremely good chemical resistance and more than 200% more dirt-storage capacity than standard oil-filter elements of the same size and, therefore, permit longer surface intervals.

New filter media are needed to achieve these figures, which diverge from the previous standard, i.e., from those achieved by filter media based on pure cellulose or on papers reinforced with synthetic fibers. These new media will, primarily, take the form of so-called composite materials, i.e., a harmonized combination of paper and melt-blown or all-synthetic fabrics. Fiberglass media, which could potentially be used because of their good filtration

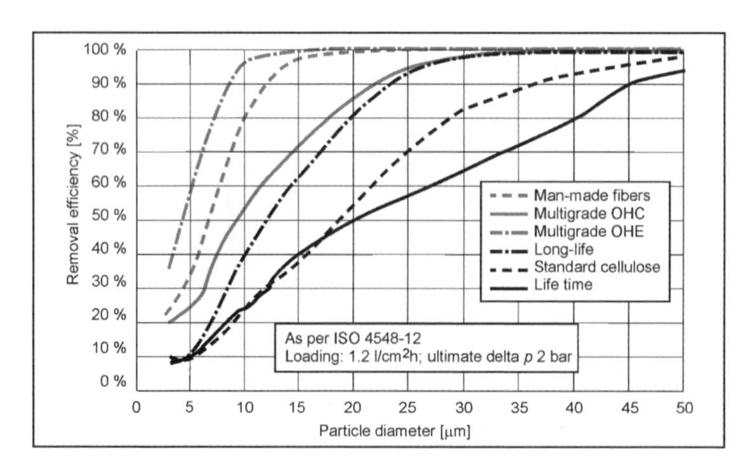

Fig. 23-19 Filter finenesses for six different oil-filter media, as a function of particle diameter.

properties, have up to now been rejected by the majority of manufacturers due to problems concerning fiber migration. Composites consist of at least two layers and possess graduated filter fineness and, because of the texture of the synthetic material, a higher dirt-storage capacity. These are also deep-bed filters.

The large range of media currently extends to oil filters consisting of man-made fibers that, just like lifetime filters, offer excellent resistance, particularly to modern, all-synthetic low-friction oils. In combination with the longer service periods, higher engine temperatures, and more refined additive packages for service life enhancement, these synthetic oils behave significantly more aggressively than their mineral and semisynthetic equivalents. Therefore, it must be assured that both the sealing materials and the filter media take account of these adverse conditions.

23.3.4 Bypass Oil Filtration

A high concentration of ultrafine particles can have a wear relevance similar to that of larger particles. Also much feared is so-called "bore polishing," an effect in which areas on the cylinder sliding surfaces become so extremely finely polished that the surface quality of the metal itself prevents the adhesion of oil films, with the result that the lubricating film breaks down. It is, therefore, a good idea to install an additional removing element for these ultrafine particles, particularly in high-mileage diesel engines, such as those installed in commercial vehicles and engines with a high soot level. Figure 23-16 also shows a bypass arrangement for the oil circuit. In this system, a small sidestream is diverted upstream the main filter, i.e., at a point at which maximum oil pressure is available, and then routed through a bypass filter. This sidestream is equivalent to about 5% to 10% of the total volumetric flow of oil. To achieve the corresponding filter effect of soot particles of <1 μm (primary soot particles are in the nanometer range, but agglomerate to form larger structures and can thus be mechanically removed using deep-bed filters), filtration velocity must be lower than in the case of the full-flow filter, and the filter medium must be correspondingly finer. This again uses up available space. The flow rate decreases as the deep-bed filter becomes increasingly clogged, but filter fineness increases.[9]

Another extremely effective method of removing ultrafine soot particles via a bypass flow is achieved via the use of centrifugal filters. A centrifugal filter consisting of metal or plastic (lower weight and more environmentally safe disposal) is installed in place of the larger bypass filter element. Only the oil pressure, which accelerates this centrifuge up to as much as 10 000 revolutions per minute, is needed to drive it. Figure 23-20 shows the plastic rotor of a centrifugal filter. The driving nozzles at the lower end of the rotor are easily visible. This figure also shows a top view of the sectioned rotor, in empty condition first, and then after use.

The high centrifugal forces achieve not only good removal efficiency for the particles that are relevant, the filter cake is also extremely compact. This signifies that such centrifugal filters are a genuine alternative to bypass filter elements. The rotor, filled with compact, ultrafine particles, is simply taken out and replaced with a new one at every service, which can be selected in accordance with the oil-changing and full-flow oil filter-changing intervals. There is no longer a need for time-consuming and complex cleaning, and there are also no imbalance problems.

Such compact centrifugal filters may also be of interest for diesel-engined automobile applications. The even more stringent legislation on exhaust-gas standards (EURO 4 and EURO 5) are in a conflict of aims between reduction of oxide of nitrogen and reduction of soot-particle concentrations. The combustion process is presently being optimized more toward lower NO_x concentrations, in order to eliminate SCR technology. The result, inevitably, is more intensive soot production, with the consequence of higher particle concentrations in the blowby gas and oil, too. It also must not be forgotten that oil changes occur at much longer intervals. Small, compact centrifugal filters are an interesting alternative to larger bypass flow oil filter elements when one remembers that there is also limited space available in modern automobiles.

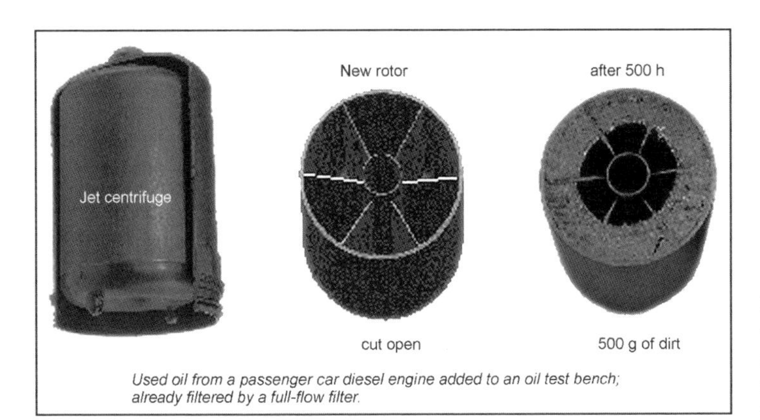

Fig. 23-20 Structure of a plastic centrifugal filter for bypass oil circuit removal of ultrafine particles. Sectional view of rotor, new and used.

Bibliography

[1] World-Wide Fuel Charter, Brochure by the International Associations of the Automotive Industry, ACEA, Alliance, EMA, JAMA, April 2000.

[2] Klein, G.-M., H. Bauer [Ed.], Kraftstofffilter, Kraftfahrtechnisches Taschenbuch, Bosch, 23rd edition, Vieweg, Braunschweig, Wiesbaden, 1999, pp. 436–437.

[3] Klein, G.-M., L. Bergmann [Ed.], Changes in Diesel Fuel Filtration Concepts, Proceedings of the 2nd International Conference on Filtration in Transportation, Stuttgart, Filter Media Consulting, LaGrange, USA, 1999, pp. 45–49.

[4] Bessee, G.B., *et al.*, High-Pressure Injection Fuel System Wear Study, SAE 980869.

[5] Projahn, U., and K. Krieger, Diesel-Kraftstoffqualität–Erkenntnisse aus Sicht des Einspritzlieferanten, Proceedings 9, Aachener Kolloquium Fahrzeug-und Motorentechnik, Pischinger [Ed.], Aachen, 2000, pp. 929–944.

[6] Klein, G.-M., H. Banzhaf, and M. Durst, Fuel Filter Solutions for Future Diesel Injection Systems, Proceedings World Filtration Congress 8, Brighton, U.K., 2000, pp. 887–890.

[7] Mach, W., and T. Trabandt, Auswirkungen fester Fremdstoffe in Gebrauchtölen auf das Verschleißverhalten von Dieselmotoren, Mineralöltechnik 10, 1998.

[8] Spanke, J., and P. Müller, Neue Ölwechselkriterien durch Weiterentwicklung von Motoren und Motorenölen, MTZ Motortechnische Zeitschrift 58 (1997) 10.

[9] Dahm, W., and K. Daniel, Entwicklung der Ölwechselintervalle und deren Beeinflußbarkeit durch Nebenstromfeinstölfilterung, MTZ Motortechnische Zeitschrift 57 (1996) 6.

24 Calculation and Simulation

The use of computer-simulation methods during development has become increasingly established in the industrial environment in recent years because of the significant contribution it makes to increased efficiency. Alongside expansions of the available technical and scientific software, and the rapid advances in computing power achieved, this success can essentially be attributed to the anchoring of the process chain concept in the CAE-assisted development process. This has made it possible to tailor the level of detail and informational output of the computation methods used to the engine's state of development and, therefore, to the problems requiring solution at that particular point. Also included is recognition of the fact that the value-creation factor in the computation procedures can be optimized only through a holistic view using design methods shaped by CAD practice.

The consequence is that the procedure of

- Provision of the necessary geometry, materials property data, characteristics data, etc., in a form suitable for calculation
- Definition of the load and boundary conditions for the particular problem
- Drafting of a computation model
- Evaluation and interpretation of the results

must be anchored in the design process, with due account taken of time and cost management factors.

24.1 Strength and Vibration Calculation

24.1.1 Procedures and Methods

Stress and vibration studies play a central role in component design. They are, on the one hand, a precondition for optimum materials efficiency and, thus, have a direct effect on manufacturing costs, and, on the other hand, it is possible in many cases to achieve functional improvements—a reduction in the weight of the crank web, for example, produces a direct reduction in vibration amplitudes of the machine as a whole. The calculation methods used in the development process are orientated around the problem itself, the accuracy levels required, and the time and resources allocated.

In many cases, component dimensioning can be accomplished using engineering concepts with simplified relationships. This makes it possible to achieve valuable information for support of the design process extremely quickly and efficiently. The use of more complicated procedures is necessary in the case of complex load states and of components, the geometry of which is defined by open-form surfaces. The finite element method (FEM) has proven to be the most effective instrument for this purpose; it permits the simulation of the loads resulting from static and dynamic forces, as well as from temperatures.

Figure 24-1 shows the finite element model of an engine. The geometry of the structure is simulated by using shell and spatial elements. In conjunction with a defined load, the deformations and loads are determined by the computer program within every element and, therefore, at every point in the component. The cost of these results is the input for the generation of this model. Despite the availability of high-performance software systems for support in model generation, a significantly greater time input is required than in the case of the use of the classical engineering formulas. It has been only the consistent use of CAD on a more widespread basis that has made it possible to integrate these simulation methods as a fully accepted

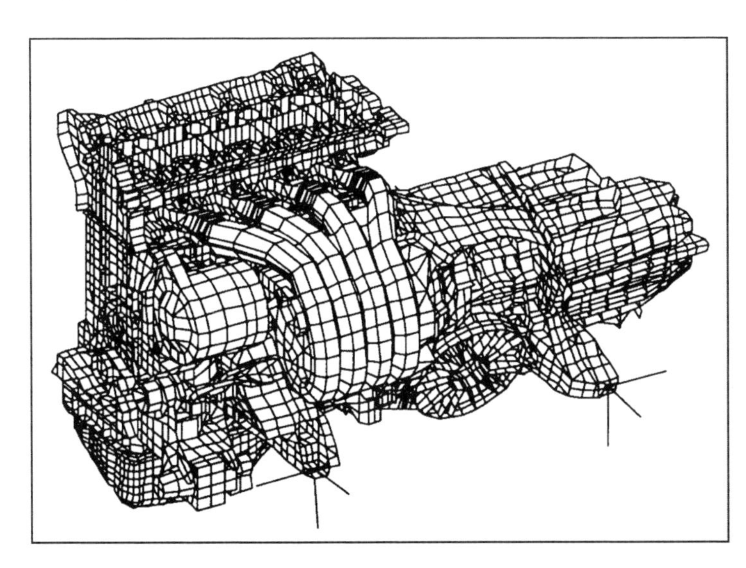

Fig. 24-1 Finite element model of an engine.

and acknowledged aid in the development process. The geometric data processed by CAD is adopted directly from computer systems that generate the required models largely automatically.

In addition to questions of component dimensioning on the criteria of strength, vibration problems also play a dominant role in the design of engines. Because of the physical phenomena involved, it is necessary from a simulation viewpoint to distinguish between nonlinear problems, the solution of which will be discussed at a later point, and linear problems. Finite element methods are used for the treatment of the latter. An important field here is calculation of acoustic performance deriving from gas forces and the oscillating masses involved. It is necessary to differentiate between structure-borne noise, which is essentially transmitted via the engine bearings and, thus, affects the noise level in the interior of the vehicle, the engine's acoustic radiation, and the muzzle noise emissions from the intake and exhaust systems, where attention is drawn to Ref. [1] in the context of the mathematical optimization of these latter.

A further step in the simplification of integration of calculation into the CAD environment can be found in a procedure derived from the finite element method theory, the so-called *p* method.[2] It is no longer necessary to generate a fine grid of the structure to be examined to permit the use of this method; the computation grids can, instead, consist of three-dimensional bodies of any shape and geometry design. The load factors and their distribution within the spatial elements are determined invisibly for the user internally by higher order polynomial formulations in accordance with the specified accuracy levels.

The boundary method should also be mentioned for the sake of completeness. This procedure also avoids the generation of a finite element model; it is necessary to define only the surface of the component to be examined using a grid consisting of triangular and rectangular elements. This method remains restricted in the field of structural calculation to special cases, however, for a number of different reasons (load data applicable only to the surface, high computing-time requirement, restriction to linear problems). The boundary method is, however, extremely well suited to the calculation of the acoustic field resulting from acoustic radiation from, for example, a complete engine.

The finite element method is suitable for study of the entire scope of structural calculation, whereas the boundary method and the *p* method are restricted to stress calculations.

These methods examine all questions of component dimensioning. Prime emphasis attaches in this context to strength functions involved in component design. Although the computation methods themselves are reliable, the simulation of the loads involved remains the basic problem, in a number of different forms. The load data for calculation of a connecting rod, for example, can be determined extremely easily; simulation of the complete engine, taking dynamic effects into account, is necessary for determination of forces and the resultant stresses in the crankshaft. Stress analysis of fittings such as alternator mountings, etc., are similarly complex, since their critical load states are the result of vibrational excitation. At least equally complicated is the investigation of loads caused by thermal exposure. These occur, most particularly, in the cylinder head as a result of the temperature gradients present there. As discussed at a later point, determination of the participating coefficients of heat transfer from flow calculation is an essential precondition here.

In addition to stress calculations, mathematical methods are also used to obtain information on component deformation. This includes oval distortion of cylinder sleeves as a result, for instance, of pretensioning of cylinder-head bolts and exhaust manifold expansion.

All these loads occur cyclically and can, therefore, be evaluated only in terms of a service life statement. More comprehensive methods make it possible to estimate component service- life from the stresses determined, on the precondition that the plot of load exposure is known.[3] Because of additional parameters such as materials properties, processing state, etc., such studies may involve high levels of uncertainty, however.

Nonlinear geometric problems associated with dynamic effects frequently occur; examples can be found in the crankshaft, the natural torsional frequency of which, inter alia, determines strength, and the intrinsic dynamic of a valve spring. It is necessary to use special methods for stress analysis of such phenomena; so-called multibody systems (MBS) have become established for this purpose. Unlike the finite element method, in which the components to be studied are broken down into individual cells, components are described in the MBS procedure in the form of bodies with inertia and flexibility properties. Solution supplies dynamic variables such as velocity, acceleration, forces, etc., for every body in the time range, taking account of all nonlinear effects. Model generation is usually complex and time consuming, since the nonlinearity of system parameters such as, for example, attenuation must also be stated in order to obtain dependable results. MBS methods are used primarily for the determination of forces under operating conditions as an input for stress studies using the finite element method. MBS are capable of simulating flexible component properties only to a limited extent, however. The two methods are, therefore, used in combination for special problems. Parts of the structure, whose flexibility properties are significant, are defined using finite element, and the modal parameters (frequencies and intrinsic geometries) determined from them become an element in the MBS model. The precondition is that the component described using finite element exhibits linear-elastic behavior. A typical example of such a procedure is the calculation of the acoustic radiation of the engine + transmission train. As a result of its elasticity, the cylinder and crankcase unit produces a vibrating surface with low amplitudes that can be determined extremely easily using finite element procedures, whereas excitation of forces is the result of rotating and oscillating masses that are modeled using MBS methods.

The finite element method is also used as the basis for mathematical optimization procedures.[4] These can be used extremely efficiently to perform special tasks such as minimization of component weight, for example. Specified parameters, such as material thickness or factors determining geometry, are calculated in such a way that the target quantity—e.g., weight—is minimized. Computation offers great potential in this area, since the results obtained in this way are generally better, and are available in a significantly shorter time, than those achieved using the classical "trial and error" procedure. Computing-time consumption is high, however, and restricts the number of parameters (also referred to as "design variables").

These optimization methods start from a given component geometry and vary only its parameters (e.g., the coordinates of significant points), whereas topology optimization, as a special variant of these methods, makes it possible to design the geometry of a not yet defined component in such a way that a specific variable (for instance, weight) is minimized. Although design work is also additionally necessary for ultimate definition of the component geometry, this optimization step generally supplies valuable information for its design.[5]

Such methods are used primarily for component dimensioning, i.e., for optimization of material exploitation for a given static load. Use for dynamic loads is essentially restricted to control of natural frequencies; problems associated with transmission patterns, acoustics, etc., must be left aside.

24.1.2 Selected Examples of Applications

Strength

Information on stresses and, therefore, on the "strength" of a component plays an essential role in the component's dimensioning. Methods of determining loads are, therefore, an important element in the design process. The finite element method offers ideal preconditions for the study of complex structures. Its integration into working procedures must be such that the results can be used to support the decision-making process, however. The closest possible linking with the CAD system, using the design as performed, has proven advantageous. The p method is particularly suitable in this context, since it sets only low restrictions concerning the size and shape of the elements, and since a computing network can, therefore, be automatically generated relatively simply. Figure 24-2 shows the networking and the stress-analysis results for a differential gear housing as an example. An element system would be too coarse and irregular for the obtainment of information on stresses using the classical finite element method. The shaded area represents the stress distribution; it is apparent that stresses vary even within the elements.

Strength of an Exhaust Manifold

The design of the exhaust manifold is one of the most difficult problems in component dimensioning for an engine.

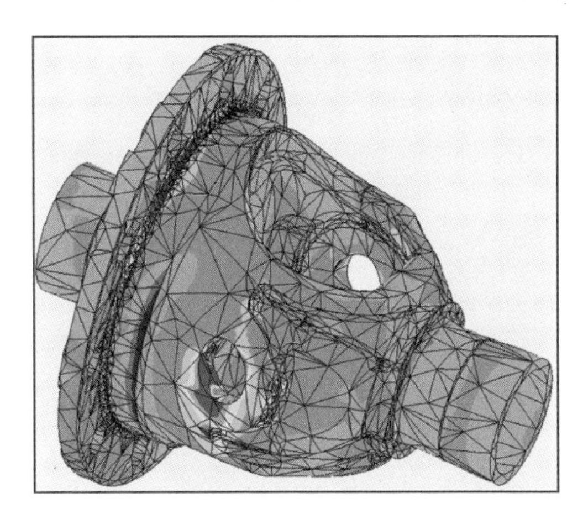

Fig. 24-2 Determination of stress using the p method.

Thermal load exposure and the high dependence of material properties on temperature necessitate intensive integration of experimental technology in layout and design; in many cases, adequate durability can be achieved only via the selection of a high-cost material. Simulation methods permit the optimization of a cost-intensive process to a particularly great extent.

The specified load consists of an accumulation of a number of heating and cooling phases. The loads to be determined occur, on the one hand, as a result of the temperature gradients in the manifold and as a result of the fact that the manifold's expansion is hindered because it is fixed to the cylinder head. The finite element method is used for the calculation of the relevant stresses and temperature distributions; determination of coefficients of heat transfer is accomplished using a three-dimensional simulation of flow within a manifold, as will be described at a later point. The dependence on temperature of all materials data must, of course, be taken into account.

The coefficients of heat transfer initially established and the exhaust-gas temperatures determined from one-dimensional computation are used to calculate the (non-steady-state) temperature field, from which nonlinear finite element calculation is used to produce the stress distribution, from which an estimation of the number of endurable load cycles can be made, using suitable materials data. Figure 24-3 shows the result of such a computation. The fracture-endangered points are indicated in the finite element computation model.

Acoustics

The acoustic radiation behavior of the engine is essentially determined by component geometry. Mathematical simulation makes it possible to obtain significant information for design and material selection at an extremely early stage of the design process. Generation of the model is extremely complex, however, since the crank web, an essential factor for simulation of the relevant excitation forces, must be additionally described using finite elements,

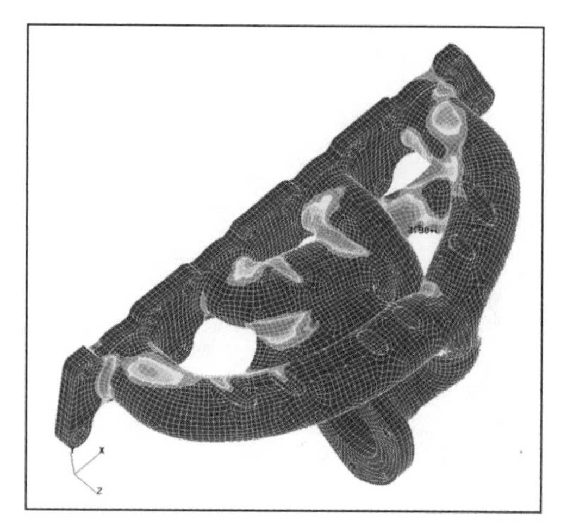

Fig. 24-3 Distribution of fracture-load cycles in an exhaust manifold. (*See color section.*)

in addition to a detailed simulation of the structure of the crankcase, gearbox, cylinder head, etc. The fact that the crankshaft bearings exhibit nonlinear behavior as a result of the hydraulic properties of the lubricating film must also be taken into account. For these reasons, a simplified method is used in practice, in order to obtain at least qualitative information for initial drafts. The so-called pulse echo method replaces the excitation forces caused by the crank drive by standard forces that are attached to the main bearings; this avoids the necessity of modeling the lubricating film and crank drive. The result is an evaluation of the transmission behavior of the structure from which, even at this early point, significant decisions can be derived. Complete modeling of the engine is then used for detailed optimization. In all cases, acoustic velocities on the surface are used for identification of those areas that exert the main influence on radiation. A practical and actual application is described in Ref. [6].

Figure 24-4 shows a calculated velocity distribution from which critical parts of the surface can be identified. The velocities found on the surface permit determination of the acoustic pressure in the immediate vicinity from a subsequent calculation. The data thus obtained are of only subordinate interest for automotive engines, however; behavior in the vehicle is of much greater significance. Simulation of this necessitates further "escalation" of the modeling process—a simulation of the vehicle components surrounding the engine, such as the body shell and suspension, is required here. Because of the time and financial input necessary, such a simulation cannot be integrated into the development process and can be used only to obtain information on the engine's acoustic performance in the vehicle in the context of basic research activities.

Valve Actuation Dynamic

Matching of valve-spring stiffness with the cam lift curve is accomplished from kinematic observations that also include the acceleration and deceleration of the valve. Superimposed on these movements are the vibrations of the valve and of the components of the entire camshaft drive and timing system. The resultant forces are definitive for the valve timing system's service life.

Multibody systems, using nonlinearities such as those of the chain drive system and damping mechanisms, for example, can be registered and are used for the registration of the relevant dynamic. Figure 24-5 shows the computation model of a valve timing system, using the intrinsic dynamic of the chain, and the valve springs and the camshaft are simulated. In order to register all the forces acting in the engine, the chain or belt used to drive the camshaft and the crank drive are depicted additionally in an engine model, from which the dynamic behavior of the entire engine is simulated.

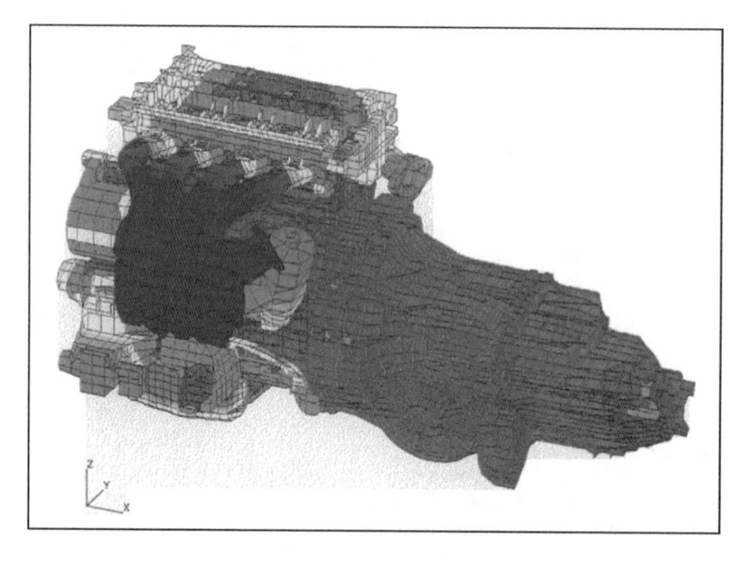

Fig. 24-4 Distribution of acoustic velocities on the surface of an aggregate. (*See color section.*)

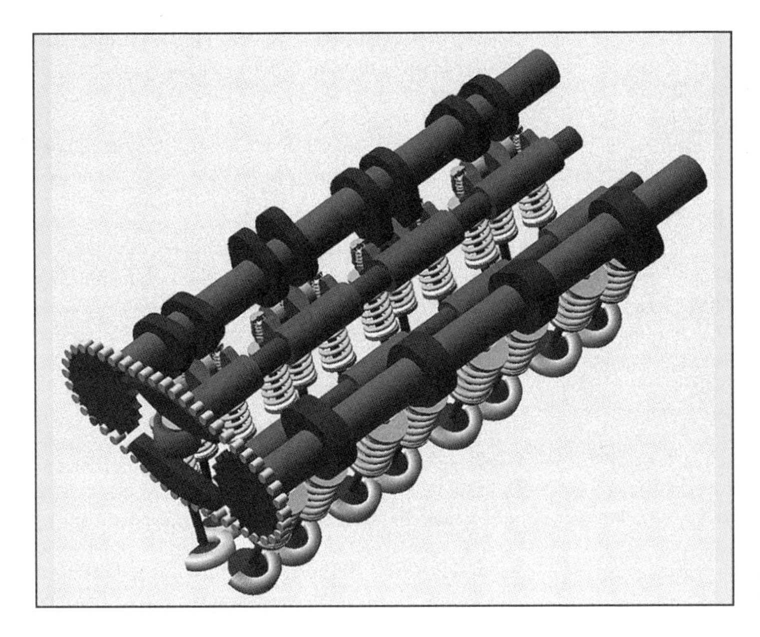

Fig. 24-5 MBS model of a valve timing system.

Component Optimization

One of the designer's essential tasks is that of determining that component geometry for which, with a minimum of material input, the permissible loading level is not exceeded. The finite element method provides valuable assistance in this process, since its use makes it possible to evaluate all possible variants and, thus, select the best. This target can be achieved much more efficiently via the utilization of optimization methods using—within a limited scope—component geometry that can be determined on the basis of an optimization strategy, taking certain background and boundary conditions into account. Weight was defined as the quantity to be minimized in the case of the connecting rod shown in Fig. 24-6, and permissible stress as the boundary condition. The starting point is shown in the left of the figure, the result on the right.

The shaded areas represent the stress distributions; an increase in weight of only six grams makes it possible to achieve a reduction of around 40% in maximum stresses.

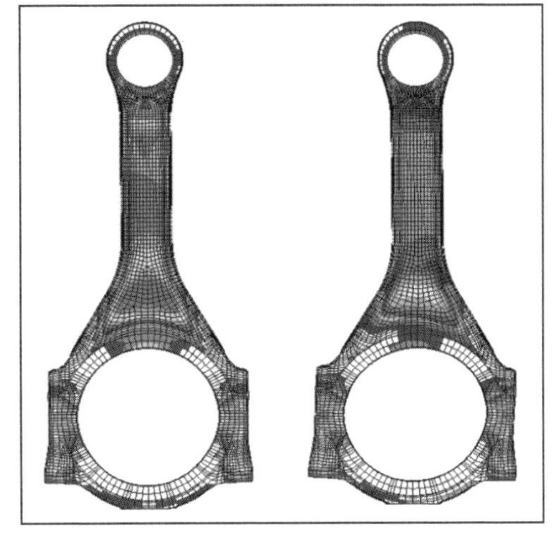

Fig. 24-6 Optimization of the geometry of a connecting rod: Plot of stress in initial state (left) and in optimized state (right). (*See color section.*)

24.1.3 Piston Calculations

As a result of the hot combustion gases (up to 2000 K), high cylinder pressures of up to 220 bar, and high acceleration forces caused by their oscillating motion, the pistons of internal combustion engines are subjected during operation to high thermal and mechanical loads, the magnitude of which is determined by the particular operating state (speed and part or full load). During the engine cycle, the piston undergoes a range of different stress states; exposure to gas forces predominates at TDC at ignition, whereas only exposure to centrifugal force takes place when there is no ignition. The term "high cycle" exposure is used, since this change occurs in every engine cycle. Transient thermal processes, such as the heating of the engine or the changeover from one operating state (idling) to another (full load) are correspondingly designated "low cycle" loads. Such low cycle thermal processes should not be confused with transient thermal loads on thin surface layers on the piston head, which occur as a result of the hot combustion gases in every engine cycle and, thus, give rise to a high cycle load exposure. Thin-walled designs are used to keep the oscillating masses in the engine small, resulting in high cyclic stress states in the component. This necessitates high calculation accuracy, in order to achieve the required service lives without complex, time-intensive, and costly experimental optimization procedures.

In view of the complex geometry and loads involved, FEM are used to determine temperatures and stresses as a function of operating states in the piston. In a subsequent computation, the results obtained are compared with the piston material's experimentally determined fatigue performance, in order to ascertain the number of endurable load cycles.

The dynamic performance of the piston in the cylinder is analyzed using secondary piston motion calculations. An important result of these calculations is the contact pressure between the piston and the cylinder as a function of crank angle. It permits the derivation of information on piston skirt wear and, in an extreme case, on the danger of piston seizure in the cylinder. In addition, the friction losses and kinetic energy of the piston upon impact with the cylinder is calculated, permitting assessment of the acoustic (NVH) behavior of the piston in the cylinder.

A further important aspect in piston calculation is the bearing system of the piston pin in the piston-pin holes. The contact pressure between the pin and the bearing system and the flexing and oval distortion of the pin are important criteria for assessment of whether the forces occurring can be transferred from the piston to the pin safely.

All such observations must never disregard that the interactions with other components have a decisive influence on the piston's operating performance. Deformations of the cylinder, for example, influence secondary piston motions and, therefore, the contact pressure between the cylinder and the piston. These secondary piston motions, for their part, have an influence on the dynamic behavior of the piston rings, and thus on blowby figures and oil consumption. The result ultimately is that only observation of the overall system (cylinder, piston rings, pistons, connecting rod, and bearings) can provide detailed information on operating performance.

Strength Calculations

Only FEM are nowadays used for calculation of complex stress states. With suitable discretization of the model, such methods permit relatively accurate prediction of the stress states in the component. These methods nonetheless naturally have their limitations. Because of the computing time necessary, for example, the number of elements for a half model of the piston is restricted to 50 000 to 80 000, resulting in individual details being depicted only approximately, or completely excluded ("defeaturing"). A further restriction on the accuracy of the calculation is the result of the continuum-mechanical concept and of the use of homogeneous material models. In actual practice, as a result of the processes used, the material features local heterogeneities, whose influence on the stress distribution and on fatigue performance can in many cases be included in the calculation only with empirically obtained influencing factors.

Materials Data

The allowable fatigue strength value of the materials as a function of temperature must be known to permit determination of piston service life. In addition, as a result, mainly, of thermal stresses, the mean stresses in the piston are generally not equal to zero, so the dependence of deflection stress on mean stress must also be taken into account.

Temperature-dependent Wöhler (stress-cycle) curves that are determined from the results of stress-controlled continuous vibration tests (10^5 to 10^8 vibration cycles) are used for calculation of service life (stress life) under a high number of cycles ("high-cycle fatigue"). Ideally, a whole group of Wöhler curves would be used for differing limiting stress ratios $s = \sigma_{min}/\sigma_{max}$, or mean stresses, to determine the influence of mean stress on material fatigue (fatigue strength diagram). Figure 24-7 shows a corresponding Smith fatigue strength diagram for a range of temperatures.

Strain life is an expansion of the basic stress life mechanism and permits better prediction for long cycle numbers, e.g., transient thermal events in which local material flow can occur. In addition to the "stress amplitude" damage criterion, material fatigue must also be registered in terms of plastic strain amplitude by strain-controlled vibration tests. The aluminum materials used for pistons may exhibit "cyclic hardening" or "cyclic softening," depending on their pretreatment and test temperature (see Fig. 24-8).

Other effects that occur under cyclic elastic-plastic loading and can play a significant role in assessment of component failure are "shakedown" (cyclic relaxation) and "ratcheting" (cyclic creep). Shakedown is the result of strain-controlled loading around a constant mean strain not equal to zero and ratcheting is the result of loading at a constant stress amplitude around a constant mean stress not equal to zero. These effects will not be examined in more detail here, however, despite the fact that they can most certainly play a significant role in assessment of component failure.

In order to determine service life, strain-controlled tests are continued up to failure of the specimen, and the results are plotted in Wöhler diagrams together with stress-controlled tests for high cycle numbers (see Fig. 24-9).

Tests in compression and in tension should always be performed, in principle, because of the asymmetrical mate-

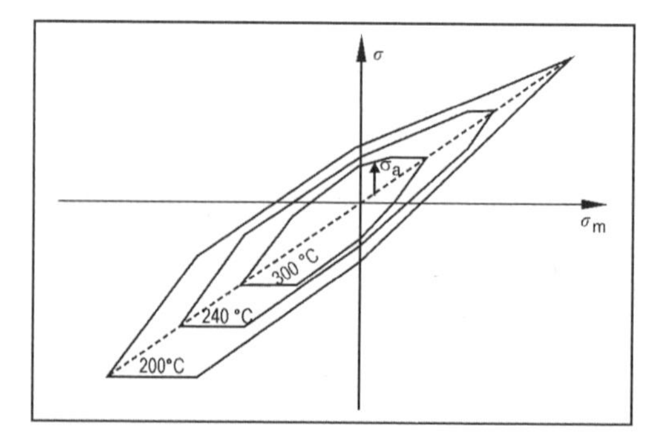

Fig. 24-7 Smith fatigue strength diagram for a piston material for various temperatures.

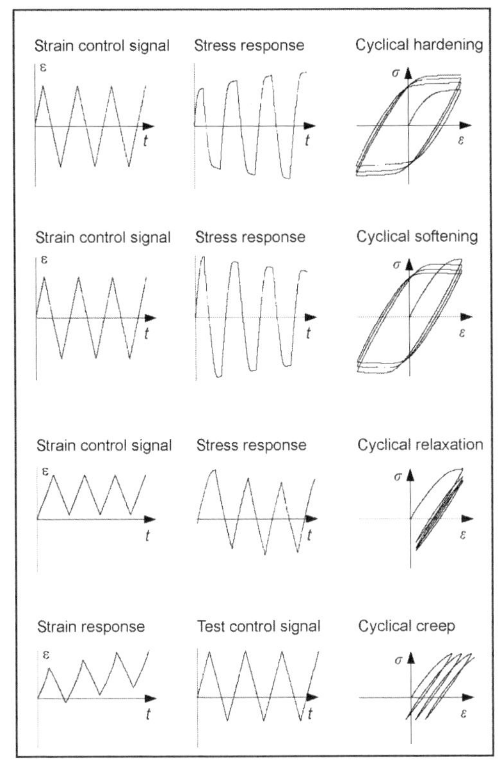

Fig. 24-8 Transient, cyclic material behavior (schematic view).

rials properties found in regular testing materials. The data for the field of compression are frequently not available, because of the large number of necessary tests (several temperature and mean stress levels, stress control and strain control, etc.); for this reason, it is necessary to use the tensile data with an appropriate correction for compressive states also. The calculation is generally excessively conservative if these data are used without correction.

Loads

(a) Thermal Loads

Determination of the correct thermal load is of decisive importance, because of the fact that the strength data of the aluminum materials used regresses severely at elevated temperatures and that the function of the piston rings is also no longer assured at excessive temperatures. It is generally assumed for piston temperatures that they remain constant throughout a working cycle and are not dependent on operating states. This assumption is justified for zones within the piston. Thin surface layers on the piston head are also subject to cyclic temperature exposure within the working cycle, however, resulting in thermally induced stresses that constitute a high cycle load on the material.

The temperature of the combustion gas of the cylinder and of the oil in the case of splash cooling (or otherwise of the oil mist) is specified together with the corresponding coefficients of heat transfer for calculation of temperature distribution in the piston. The gas data (temperature and coefficient of heat transfer) can be estimated from empirical data (databases) or, increasingly, calculated using combustion simulation programs.

This information is used to calculate the piston's temperature field by means of a finite element analysis (FEA). The maximum temperatures generally occur at the design output working point. Typical temperature fields for gasoline automobile engines are shown in Fig. 24-10, the same for automobile diesel engines in Fig. 24-11. The maximum figures on the piston head are about 320°C in the case of the gasoline engines, and about 380°C for the diesel engines.

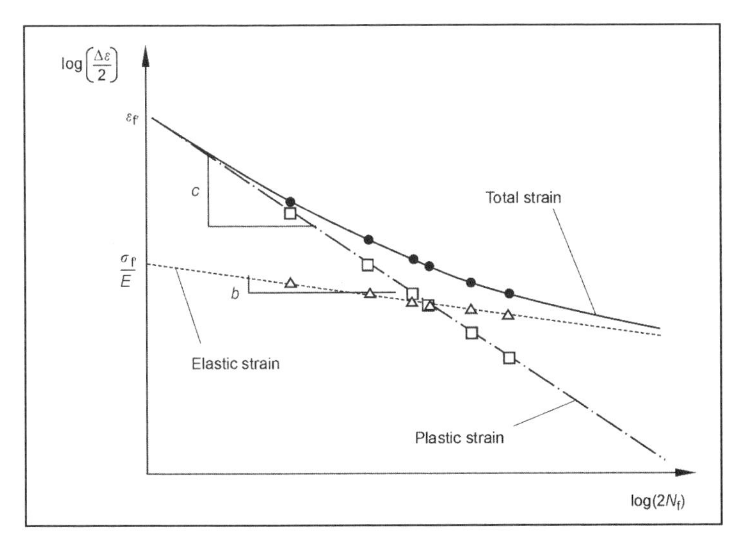

Fig. 24-9 Wöhler diagram.

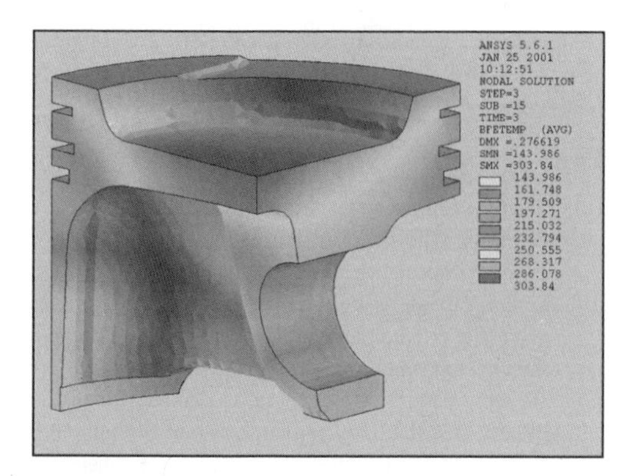

Fig. 24-10 Temperature distribution in an automobile gasoline engine at design output. (*See color section.*)

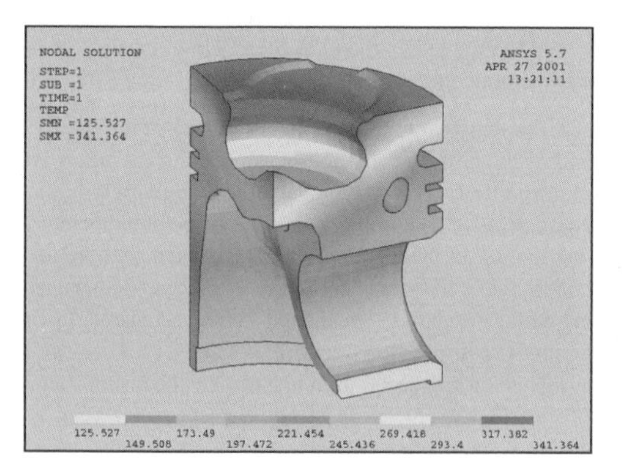

Fig. 24-11 Temperature distribution in an automobile diesel engine at design output. (*See color section.*)

The thermal stresses and deformations that occur without the inclusion of external loads (gas and/or centrifugal forces) can then be calculated from temperature distribution in a second computation. This step is generally omitted, however, and the temperature distribution also read in for the FE computation of mechanical stresses in the piston, to obtain the combination of thermal and mechanical stresses.

(b) Mechanical Loads

The piston is subjected during operation to loads exerted by gas F_G and masses F_{mK}.

$$F_K = F_G + F_{mK} \tag{24.1}$$

The gas forces result via multiplication of the cylinder pressure p_G acting on the piston head by piston head surface area A_K.

$$F_G = A_K \cdot p_G \tag{24.2}$$

Cylinder pressure is a factor that is periodically variable in the working cycle. It can be obtained either by measurement or by combustion simulation calculations. The latter practice is, naturally, frequently used during the conceptual phase for new engines, when corresponding measurements made on the actual engine are, of course, not yet available. As an alternative, data from comparable applications, or specified target data, are also frequently used for calculations.

Typical data may be

- Approximately 70 to 90 bar for nonsupercharged gasoline engines
- Approximately 100 to 120 bar for supercharged gasoline engines
- Approximately 180 bar for supercharged automobile diesel engines
- Up to 220 bar for supercharged truck diesel engines

The mass forces acting on the piston act in the direction opposite to piston movement. They are calculated from piston mass m_k, crank radius r, crank angle α, and connecting-rod ratio λ_p, at

$$F_{mK} = m_K \cdot r \cdot \omega^2 \cdot (\cos\alpha + \lambda_p \cdot \cos 2\alpha) \tag{24.3}$$

At top dead center, this formula reduces to

$$F_{mK} = m_K \cdot r \cdot \omega^2 \cdot (1 + \lambda_p) \tag{24.4}$$

Mass force increases quadratically with speed. To keep oscillating forces as low as possible, an attempt is made to keep piston mass as small as possible through appropriate design, particularly in the case of high-speed engines.

Figure 24-12 shows qualitatively the gas forces, mass forces, and total force acting on the piston as a function of crank angle. At TDC, the mass force acts contrary to gas force and, thus, reduces the total force acting on the piston. This observation should be used only to illustrate the force conditions, however. Within the FE analysis, each element is loaded with the appurtenant acceleration, with the result that the appurtenant mass force is calculated for each element. The mass forces resulting from secondary piston movement are frequently left out of the account in this context. These accelerative effects can also be calculated using appropriate programs. In a finite element analysis, such accelerations for each element can be superimposed on the mass forces from purely oscillating movement.

Stress Calculation

Because of their complex, three-dimensional structure and multilayer loading, pistons can be calculated numerically using only finite element methods. The temperatures, stresses, strains, and deformations in the component are obtained as a result of such calculations.

Correct transmission of forces into the piston is extremely important if realistic information is to be obtained. For this reason, the computation model usually consists of the piston itself, the piston pin, and the upper section of the connecting rod. The piston pin and the upper section of the connecting rod are modeled as authentically as possible, in order to reproduce the flexure and oval distortion of the piston pin as accurately as possible. The profile of any contoured boring of the piston-pin hole must

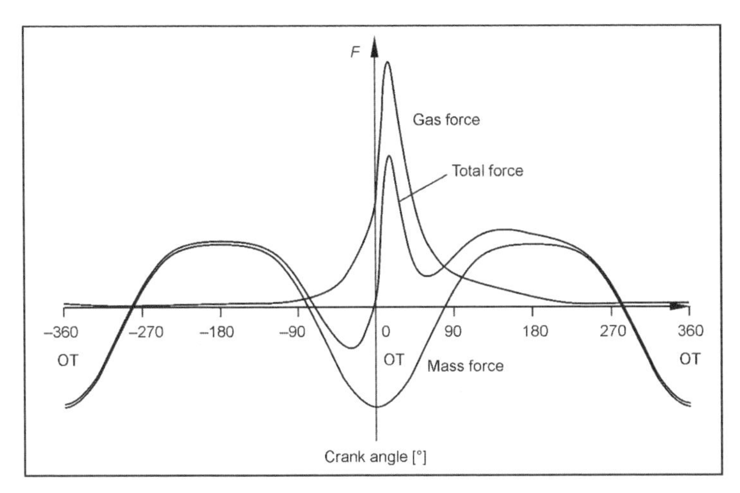

Fig. 24-12 Resultant piston forces in a four-stroke gasoline engine.

also be incorporated into the generation of the model, since the pressure distribution between the piston pin and piston pin boss is otherwise falsified. Where high contact pressures occur, an elastic-plastic calculation must be performed in order to take into account deformations and associated changes in stresses (generally a reduction in peak stresses). Figure 24-13 shows a typical model, a quarter model of the piston having been selected here for better comprehensibility, whereas half models are generally used for actual calculations.

A nonlinear problem arises from the necessary contact formulation between the connecting rod and the pin, and the pin and the little end boss.

The piston is usually networked with tetrahedral elements, since networking with tetrahedrons is simpler, compared to that with hexahedrons, and can thus be accomplished more quickly. Networking fineness is selected as a function of site, since the number of elements has an overproportionate effect on computing time. The difference between the stress averaged from a number of elements and nonaveraged stress at one node can be used as an indication of the quality of the grid. This difference should not be greater than 10% to 15%. A figure of 50 000 to 80 000 elements thus results for a complete half model.

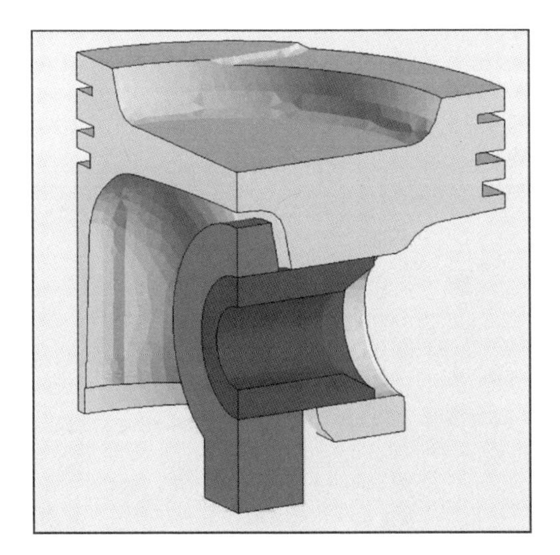

Fig. 24-13 Model for use for piston calculation. (*See color section.*)

Figure 24-14 shows the deformations and stresses occurring purely under gas force in the piston of a diesel engine. The compressive forces acting on the piston head cause the piston head to deflect inward. The force acting on the piston is transferred at the piston pin boss into the piston pin, causing the piston pin to flex and to become ovally deformed at its ends. The piston is bent downward around the piston pin on the pressure and counterpressure side simultaneously. This causes oval deformation of the piston shank, the diameter of which becomes smaller at the open end in the pressure/counterpressure direction. In addition to piston design, the design of the piston pin (internal/external diameter, length, etc.), supported length (total piston pin length, width of the little end boss), the design of the upper little end boss (parallel or trapezoidal), and the contact pressure between the piston and cylinder all have a significant influence on piston deformation.

The stresses and deformations occurring under purely thermal load in a diesel-engine piston are shown in Fig. 24-15. The bulging of the piston head and the increasing enlargements of diameter from the skirt end to the piston head are clearly apparent. The greater increase in diameter at the piston head is caused mainly by the higher temperature at the piston head compared to the piston shank.

The mechanical and thermal stresses and deformations in the piston are superimposed on each other to form the total load shown in Fig. 24-16.

Calculation of stresses and deformations can by approximation be performed on the assumption of purely elastic material behavior. More accurate information is obtained if the elastic-plastic performance of the material is known, and the FEA is taken as a basis. It must be noted

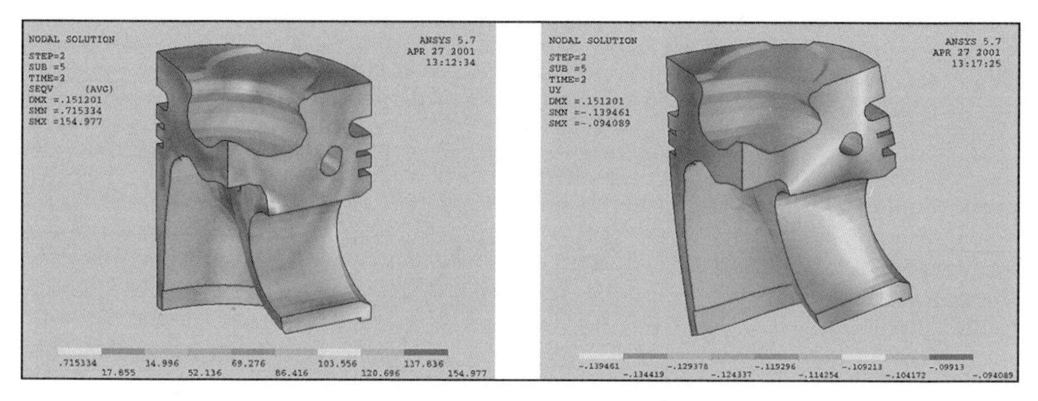

Fig. 24-14 Stresses and deformations in a piston under purely gas force (example). (*See color section.*)

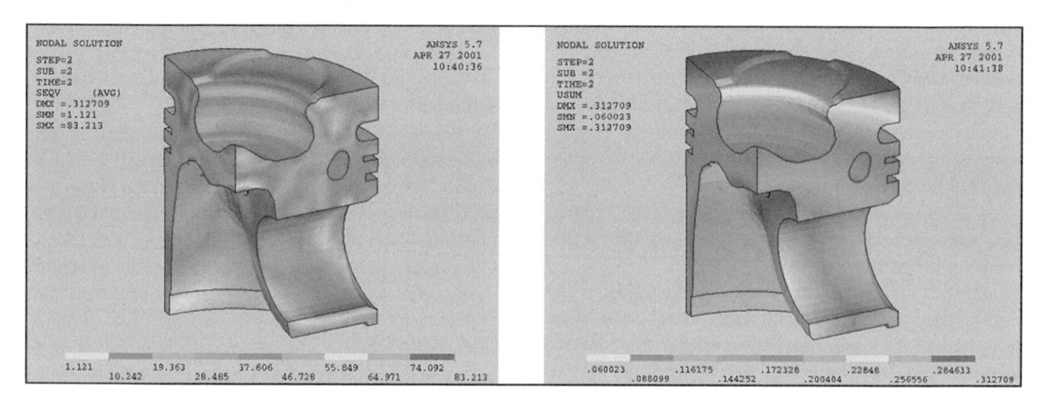

Fig. 24-15 Stresses and deformations in a piston under a purely thermal load (example). (*See color section.*)

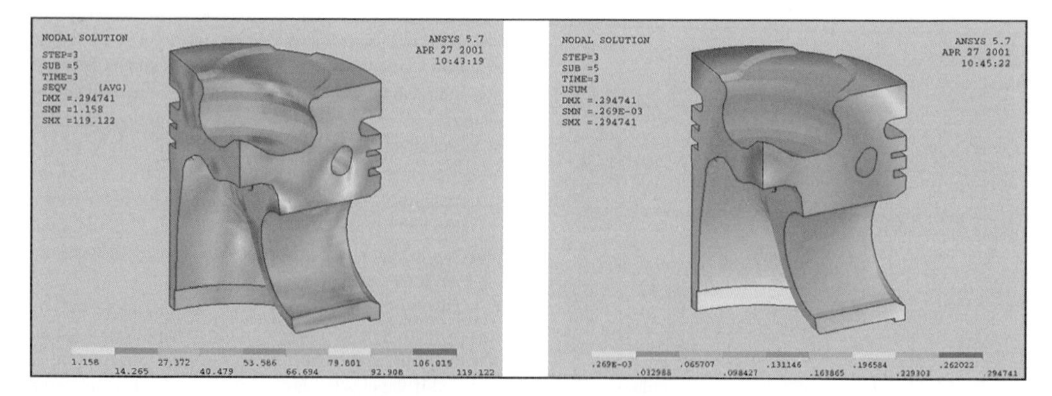

Fig. 24-16 Stresses and deformations in a piston (mechanical and thermal) example. (*See color section.*)

that the material model parameters for the piston's entire temperature range must be known. The inclusion of elastic-plastic material behavior results in a further nonlinearity in the FE computation.

The results obtained include, inter alia, the stress tensor and local temperature distribution. The complex geometry and loading of the piston generate triaxial, nonproportionate stress states in various zones of the piston.

The local stress tensors of the individual load steps can, as required, be depicted and evaluated using post-processors. It is useful, for example, to depict the radial stress on the pin hub, in order to permit assessment of the pressure distribution between the piston pin and the piston-pin hole. Also of interest in this context are the tangential stresses, since tangential tensile stresses contribute decisively to the occurrence of incipient hub cracking and crevice-induced fractures (see Fig. 24-17).

Alongside examination of maximum principal stresses, an important picture of piston loading can also be obtained from a study of stress intensity. The stress intensity in accordance with the energy of deformation hypothesis (von Mises' stress), which is proportional to the deviational strain energy (plasticity), is used, in particular, in the case of test piston materials. Stress intensity σ_v can be calculated to

$$\sigma_v = [\sigma_x^2 + \sigma_y^2 + \sigma_z^2 - (\sigma_x \cdot \sigma_y + \sigma_y \cdot \sigma_z + \sigma_x \cdot \sigma_z) + 3 \cdot (\tau_{xy}^2 + \tau_{yz}^2 + \tau_{xz}^2)]^{0.5} \qquad (24.5)$$

The stress intensity determined in this way for a piston under mechanical and thermal loads is shown by the example in Fig. 24-18.

It must be noted that information on the sign ("+" = tensile stress and "−" = compressive stress) and on the direction of the active stress is lost when the stress intensity is used. Observation of a range of stress evaluations is, therefore, unavoidable if a comprehensive picture of the stresses that prevail in the piston is to be achieved. Another possibility is examination of the stresses in the critical plane (see next section).

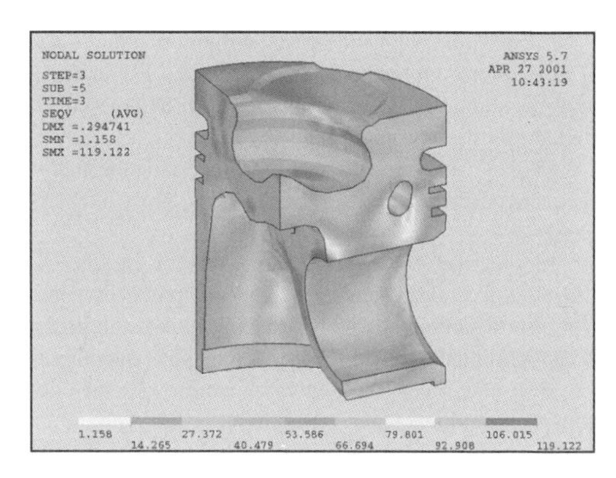

Fig. 24-18 Stress intensity on a piston under mechanical and thermal load (example). (*See color section.*)

Calculation of Service Life

To calculate service life, the temperature distributions and stress states for the individual working points are set against non-temperature-dependent material fatigue data in a second calculation. Examination here of only the oscillating load between two operating states makes it possible to derive the anticipated service life directly from comparison of the stress deflections and mean stress occurring against the permissible deflections. The individual load states must be linked by suitable damage accumulation hypotheses (see next section) if one wishes to predict the anticipated service life for a load group.

For one load point, e.g., maximum torque or maximum output, the stress states close to TDC with ignition load (TDCi) and at TDC without ignition pressure and exposure only to centrifugal force (TDC) are calculated. Under exposure to ignition pressure, the precise instant of TDCi is not observed, but instead a number of crank angle degrees after TDC, a point at which the maximum lateral force is acting on the piston.

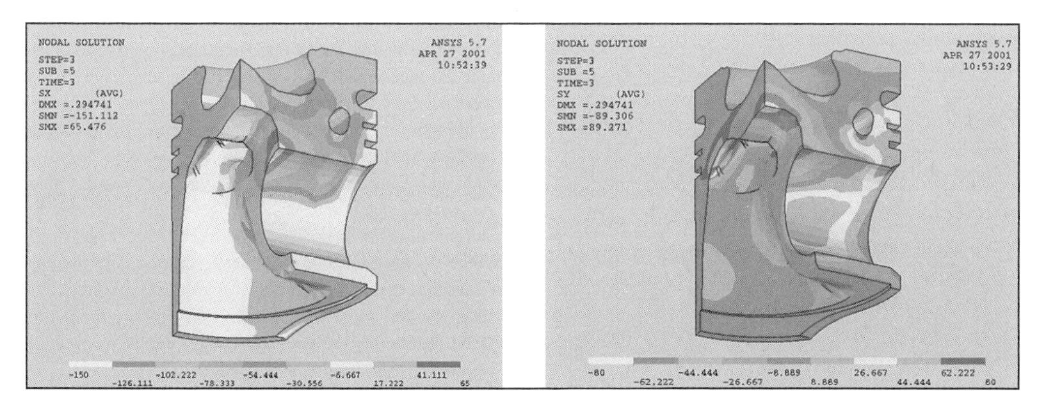

Fig. 24-17 Radial and tangential stresses in the hub boring (example). (*See color section.*)

Local temperature and stress distribution in the piston is known for both operating states from the FE analysis, and stress amplitude and mean stress can be determined directly for the stress invariants (hydrostatic pressure and "von Mises'" stress intensity). Care concerning the sign is recommended, however, since von Mises' stress intensity is signless.

In practice, the use of monoaxial or multiaxial, direction-dependent damage hypotheses produces superior information on operating behavior. As a result of the nonproportionality of the stresses acting, such hypotheses can be used correctly only in combination with the concept of the "critical plane."

Here, the mean stress and stress amplitudes for the two operating states are calculated in an initial step from the two local stress states for a sectional plane selected at will. The mean stresses and deflection stresses operating at this sectional plane permit, given the use of monoaxial or multiaxial damage hypotheses, determination of the damage acting on this sectional plane. The sectional plane is then rotated step by step about all three spatial axes and the deflection stresses (deflection stresses and mean stresses are calculated again together with the appurtenant damage for each new orientation. The plane with the highest damage is designated the critical plane and the appurtenant stresses (deflection stress, mean stress, and shear stresses) are designated critical stresses. The damage index of this critical plane is used for calculation of service life (Fig. 24-19).

Different planes, in which the critical stresses act, naturally generally a result for every element in the piston when this method is used; i.e., the critical stress vectors generally point in different directions, depending on the position in the piston. The critical stresses, for isolated areas, such as the boundary of the piston recess, or in the pin boring, naturally largely accord with the dominant stresses there, such as the tangential stresses.

Since materials characteristics data determined experimentally are available only in the form of monoaxial data, it is necessary to determine in the above observation a monoaxial stress intensity from the stress vector and the appurtenant shear stress using a suitable damage hypothesis for every plane. The normal stress hypothesis can be excluded for this purpose, because of the ductility of piston materials. The energy of deformation hypothesis involves the disadvantage that it always produces positive stresses, and that negative stress amplitudes are, therefore, not permitted by definition. The following linear combination of normal stresses and shear stresses has frequently proven its value in practice, the factors c_N and c_τ being material independent and generally obtained empirically:

$$\sigma_v = c_N \cdot \sigma_N + c_\tau \cdot \tau \tag{24.6}$$

Comparison of the deflection stress determined in this way and the appurtenant mean stress with the permissible stresses determined experimentally for each element, taking temperature into account, makes it possible to determine in advance the component's service life. The results are frequently stated in the form of safety factors or, directly, via statement of the foreseeably endurable number of cycles; their local distribution can be evaluated using postprocessors. The safety factors are defined as S_F = (allowable stress/actual stress) and should be greater than 1.0, so that the component does not fail before the expiry of the required service life.

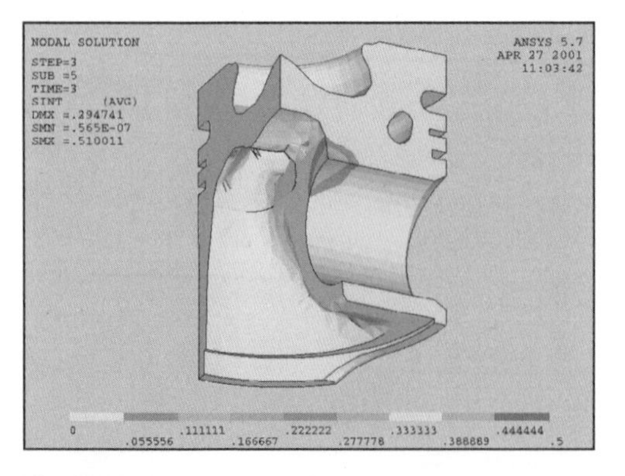

Fig. 24-19 Damage index using the critical plane method. (*See color section.*)

Damage Accumulation

The simple load case of an oscillating component loading between only two working points examined above is frequently used, but it does not, of course, represent the actual loads occurring under an extremely large number of differing operating steps. Suitable damage accumulation hypotheses, such as the Miner rule, in which the ratio of the required cycles to the endurable cycles is generated and summated across all operating states, for example, must be used if the influence of multiple operating states on service life is to be taken into account:

$$\text{Damage } D = \sum (\text{Required cycles})/(\text{Endurable cycles}) \tag{24.7}$$

The component fails prior to achievement of the required service life if the damage index is greater than 1.0.

The problem arises in practical application in the scope of FE analyses that an FE analysis would, in principle, need to be performed for every operating state, in order to determine the appurtenant stress states. The disadvantage here is, of course, the associated high computing input, with correspondingly long overall computation times. The piston is, therefore, subjected in many cases to a so-called "standard load" and a purely elastic FE analysis performed. Other, comparable, operating states are generated from this by multiplication of the stress tensors by the actual load. Where—as is generally customary—plastic deformations occur in the piston that necessitates an elastic-plastic calculation, the real stress states can be derived from the purely elastic determined stresses by using suitable conversion hypotheses (e.g., Neuber's or Glinka's).

It should be noted at this point, however, that this methodology necessitates two significant simplifications.

First, only purely elastic equations have been used, in order to adjust the elements in the FE model to a state of equilibrium, in terms of stresses and deformations. This equilibrium is no longer assured if the stresses are "retrospectively" converted to the plastic state by using approximations in accordance with Neuber or Glinka. Second, Neuber or Glinka approximations may be used only in cases in which plasticities are restricted to small, local areas.

24.2 Flow Calculation

The results of flow simulations provide the development engineer with a detailed insight into engine processes that, on the one hand, disclose potentials for improvement and, on the other hand, serve for virtual assurance of function. Nowadays, exhaust system concepts are evaluated using three-dimensional flow simulation. The potential of this "virtual product development" becomes rationally exploitable only in interaction with the greater use of optical measuring methods, since the tuning of numerical models on the basis of measured data is necessary in many applications.

The starting point for engine development is simulation of the gas charge exchange processes, which, taken together with a working-process calculation, supplies trend information on the engine's performance data, emissions behavior, energy consumption, mechanical loading, and acoustics. Although one- or quasidimensional computation methods are used here, the concepts utilized permit a spatially differentiated output of coefficients of heat transfer for finite element calculations in which information of component strength can be ascertained.

Following the initial phase of approximate component dimensioning, locally and chronologically resolved information on velocities, pressures, temperatures, and mixture compositions, which can be obtained only via three dimensional flow calculations, is needed to support the development process.

24.2.1 One- and Quasidimensional Methods

Calculation of the Charging and Working Process

A one-dimensional calculation of charging is a rational tool not only in the engine's conceptual phase, but also throughout the entire development process, and can be used for harmonization of components because of the interaction of a range of differing influences (flow, gas dynamics, energy conversion, etc.). Such a one-dimensional procedure has become established as an optimum compromise between development input and the accuracy required for definition of flow phenomena. The advantage of this method over the classical three-dimensional concepts can be found in its significantly simpler model generation and the much lower computing time requirement. These must be set against the disadvantage of the fact that physical variables, such as flow velocity, for example, are locally averaged, making it impossible to resolve local effects. It is, therefore, necessary, in order to obtain reliable results, to correlate the properties that cannot be depicted in the computing model empirically, where they are relevant to the result.

Figure 24-20 shows a schematic of the air and exhaust-gas routing of an eight-cylinder engine in which important elements of the mathematical model for the charging calculation are emphasized. One important result of simulation is cylinder volumetric efficiency. The results permit the dimensioning of the intake and exhaust systems (pipe lengths and cross sections) and optimization of the actuating times of the valve timing system.

The intake manifold tubes and exhaust system are represented by pipes in which gas flow is simulated one dimensionally with all the dynamic effects occurring. Both calculated and measured coefficients are used for inclusion of the pressure drop at pipe bends and branches. The exhaust turbocharger is defined by characteristics fuels.

A dominant factor in the calculation is the progress of combustion during energy conversion. The one-dimensional method does not make it possible to resolve the extremely complex combustion processes. The integration of a working-process calculation into the charging program, which calculates using a specified equivalent combustion progression, can be utilized here. A semiempirical equation, which was proposed by Wiebe,[8] is frequently used for determination of the rate of heat liberation. In addition to the empirically established parameters, it is also necessary to introduce a factor from the measured indicator diagram. This method ignores ignition retardation and is, therefore, used mainly for slow-running (low-revolution) gasoline engines. Modifications to an empirical model for precalculation of combustion progress need to be made here for the study of diesel engines and, in particular, of common rail diesel engines with direct injection in order to take into account the special circumstances involved such as multiple injection, variable rail pressure, and variable EGR rate (see Ref. [9], for example).

The same also applies to the generation of gasoline engines featuring direct gasoline injection. Rate of heat liberation and, therefore, combustion progress, can be directly used if they are known.

The two-zone model permits precise simulation of combustion. This assumes a homogeneous mixture and a rotationally symmetrical combustion chamber that is subdivided into a zone of combusted gas and a zone of uncombusted gas. The flame front, the propagation rate that is determined by the law of combustion, forms the dividing plane. Tuning the model is accomplished via assessment against measurements performed on the high-pressure part of the system. The advantage of this procedure is the fact that it permits at least approximate determination of coefficients of heat transfer on the combustion chamber wall, and these can be further utilized for calculation of stresses caused by temperature exposure. The two-zone model is suitable most especially for gasoline engines, because of the assumptions that are used to form the basis for the model.

Data from the engine's predecessor will continue to be needed for tuning mathematical models until it becomes

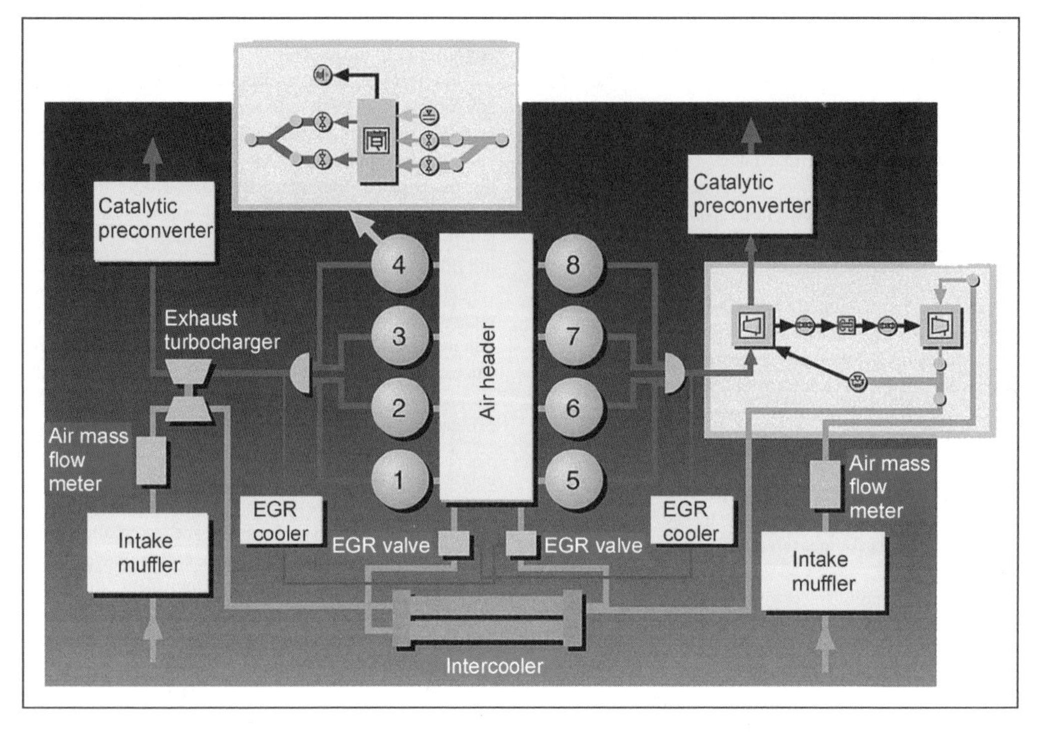

Fig. 24-20 Schematic view of air and exhaust-gas routing, showing elements of the charging calculation, from Ref. [7].

possible to develop generally applicable combustion progress models and to determine the range of parameters utilizable from existing models. Detail turbine and compressor characteristics fields will be required in the case of examination of supercharged engines, particularly for the study of their part-load operating ranges. The lack of such data complicates the use of the calculation procedure in the engine's conceptual phase.

Calculation of Vehicle Cooling Circuits

Simulation of the entire vehicle cooling circuit on the basis of a one-dimensional calculation method is a standard application in the design of the engine cooling system. In a procedure similar to the generation of charging models, all the components in the cooling circuit—essentially the water pump, engine, thermostat, radiator and oil cooler, hoses, and header tanks—are connecting to one another in a network and characteristics indices or fields assigned to them. Influencing factors are instantaneous volumetric flow, system pressures, and coolant temperatures. A useful step in the assessment of a range of differing cooling concepts is the simulation of the cooling circuit with one-dimensional simulation of the radiator/fan group, taking significant heat flows into account; see (for example) Ref. [10]. This makes it possible to calculate the effects of variations in fan and heat exchangers, or the influence of different operating states, on the vehicle cooling data achieved.

Calculation of Oil Circuits

It is also possible, using similar methods to those described above, to depict the oil circuit of the engine in the form of a one-dimensional mathematical model. Attention must be devoted here, in particular, to modeling the oil flow in the main bearings of the crankshaft and camshafts, in order to obtain realistic results.

Simulation of the Fuel Hydraulic Circuit

The fuel hydraulic circuit is a further example of an application for this type of calculation method. This involves simulation of highly dynamic processes, in which compressibility factors must also be taken into account, however. Particularly in high-pressure diesel injection systems with variable rail pressures, it is necessary to calculate the effects of fuel-line routings on the individual injectors in terms of the homogeneous distribution of fuel flow to the individual cylinders. Simulation of the effects of rail-pressure control on metering of fuel flow is a further application. The behavior of the injector itself need not be resolved in detail but can, instead, be described using characteristics fields.

Overall Process Analysis

In many cases, knowledge of the steady-state operating behavior of engines is not enough. Instead, it is necessary to simulate transient engine states, particularly in the field of the design and validation of electronic engine management system functions. Specific data on the vehicle itself, the power train, the transmission, engine electronics, ambient conditions, and driver behavior are, therefore, all included in the computation model in addition to precise thermodynamic modeling of the engine.[11] Simulation of vehicle acceleration for adjustment of the guide-vane mechanism of an exhaust turbocharger with variable turbine-blade geometry, taking various control strategies into account, may be mentioned here as a specimen application.

24.2.2 Three-Dimensional Flow Calculation

The Finite Volume Method

The computation methods mentioned above are not capable of describing flow processes with local resolution. Passage through extremely complex geometries, such as that of a water jacket or the inlet ports of the engine, for example, particularly necessitates the use of three-dimensional flow simulation. Figure 24-21 shows the mathematical model of the water jacket of a four-cylinder engine. It consists of around 500 000 volume cells that, taken together, fill out the structure through which flow occurs. Every geometric detail of the water jacket is resolved here. The finite volume methods have become established over finite difference and finite element procedures for industrial application. The conservation equations of mass and impulse are resolved in every individual volume unit. This set of equations is referred to as the Navier-Stokes equations; see (for example) Ref. [12]. Practically all technically relevant flow processes have a turbulent character. Further equations, so-called turbulence models, are needed for the description of flow turbulence. The classical route leads via chronological averaging of the Navier-Stokes equations. The additional terms generated in this averaging operation must also be modeled. Even today, the k-ε-model, which necessitates the solution of two additional equations for the characterization of a turbulent longitudinal and time scale, is a standard procedure.

Augmentations to this model, so-called "nonlinear turbulence models," are nowadays increasingly being used in applications in which the interaction of various turbulent eddy structures significantly determines flow topology.

The wall-boundary layer must be resolved, in order to permit adequately accurate treatment of detachments not attributable to discontinuities in the geometry and problems of heat transfer. So-called "low-returbulence models" are used then in these fields. The disadvantage of such methods can be found in the large number of computation cells and the resultant elevated level of need for resources such as hard-disk capacity and computing time.

The energy conservation equation must also be solved in cases in which additional thermodynamic problems, such as compressible flows, and flows involving energy input or heat transfer, are examined.

The system of partial, nonlinear differential equations must be discretized and linearized in an initial operation, in order to permit the finding of a solution for it. Significant difficulties are created here by the so-called convection term, which describes the transportation of a physical solution variable over the system boundaries of the individual volume units.

The commonly used discretization procedures differ in the accuracy with which they define the physical situation. This applies, in particular, to the depiction of steep solution variable gradients. The disadvantage of a complex

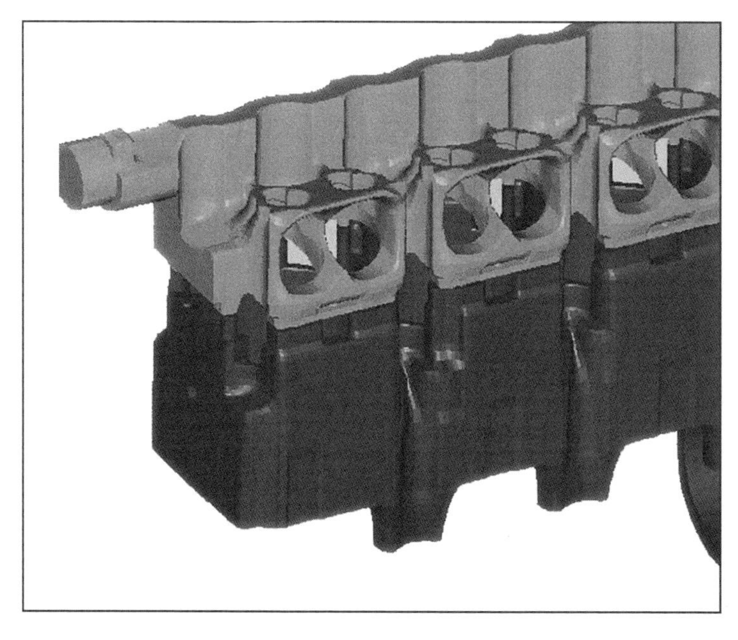

Fig. 24-21 Computing grid for the water jacket of a four-cylinder engine.

discretization procedure can generally be found in the associated difficulty of achieving a convergent solution.

A solution is generally achieved from an iterative procedure. In an industrial context, the solution of flow problems generally signifies the solution of a boundary value problem. This means that even areas of the solution zone located well downstream are capable of causing flow phenomena such as the occurrence of a detachment throughout the entire solution zone, even well upstream. Decisive importance, therefore, is attached to the definition of boundary conditions.

A description of the numerical methods of fluid mechanics can, for example, be found in Ref. [13].

The three velocity components, pressure, temperature, density, and characteristic turbulence data (degree and length of turbulence) are available as solution variables within each volume element. In addition, the coefficient of heat transfer and the heat flux transferred to the wall can be determined in every so-called "wall cell." These factors are in many cases further processed in finite element programs, with the aim of obtaining information on component strength.

Networking of the flow zone is an essential step on the road to the solution of fluid mechanical problems. It is here that the greatest advances have been achieved in the past five years. The precondition for "automatic" networking is a completely closed surface. The efforts being undertaken, therefore, shift toward CAD systems, in order to ensure the devotion of attention to the provision of the best-processed surface possible as early as the design stage.

There are, essentially, four trends:

- Automatic networking on the basis of a Cartesian computation grid, which is not tailored to the body but depicts the precise geometry using additional algorithms
- Automatic networking on the basis of tetrahedral grids (e.g., the Delauney method), in some cases offering the capability of converting tetrahedrons in the interior of the computation zone back to hexahedral elements
- Semiautomatic networking based on block-structured grids
- Semiautomatic networking on the basis of hexahedral elements, with toleration of a few tetrahedral or polynomial elements

Each of the methods discussed above has both advantages and disadvantages, depending on the required computation grid quality (particularly important in the solution of heat transfer problems), the necessary informational accuracy of the computation results, the available engineering hours, the quality of the CAD interface, the available computer capacity, and the total available time within which reliable results are to be produced.

The principle that the method must be holistically evaluated and is only as good as the weakest link in the chain applies here. A missing boundary condition cannot be made good by an extremely finely resolved computation grid. On the contrary, a computation grid containing a large number of distorted volume elements increases computing time and diminishes the accuracy of the results obtained.

Simulation Methods Based on the "Lattice Gas Theory"

A calculation method that does not solve conservation equations at the macroscopic level (in finite volumes), but instead resolves the processes on a microscopic scale (impact processes at the molecular level), has been becoming established for a number of years now, primarily in the field of simulation of flow on vehicle exteriors. This is a particle-based procedure, in which particles exist in time and space, and move in discrete directions at discrete velocities. All relevant flow variables can be determined using statistics based on discrete kinetic theory.

Turbulence modeling is accomplished here using VLES (very large eddy simulation).

Further information on this technique can be found, for example, in Ref. [14].

This method is currently undergoing expansion to further fields of application, such as flow through the engine compartment, brake cooling, and engine flows.

The benefits of the method can be found in the following:

- Automatic generation of the computation grid
- The absence of numerical errors
- The use of simple solution methods, since there is no need for solution of partial, nonlinear equations
- Explicit procedure, i.e., non-steady-state processes are reflected with chronological accuracy
- Short turnover times with a low engineering hour input

Its disadvantages are as follows:

- Absence of an experience base in engine applications
- Steady-state solutions can be depicted at present only via transient calculations and in the form of an asymptotic approximation
- Difficulty of treatment of turbulence
- High resource requirement

Combination of Calculation Methods of Differing Complexity

The increasing complexity of the problems requiring solution results in the combination of calculation methods with one another. This involves both the linking of various one-dimensional procedures with one another, the coupling of a 3-D technique with a 1-D code, and the combination of two 3-D methods.

Cylinder-charging calculations are, for instance, combined with 3-D CFD simulations to permit registration of gas-dynamic processes beyond the limitations of the 3-D calculation field. For reciprocation, the 3-D calculation supplies significantly better results in terms of the reflection of pressure waves at branch points, and in terms of the pressure loss in the passage of flow through complex components. Simulation of the recycling of exhaust gas into the intake system may be mentioned as an example here.

One example of the combination of two 3-D codes is the computation of component temperatures in the engine. It is necessary here to examine heat transfer on both the gas and the waterside. Coupling is accomplished as follows: Heat flows are transferred to the FE method from the CFD code, as the result of a calculation performed in isolation. The FE program, after computation, passes for its part the wall temperatures back to the CFD code. Following the exchange of a number of solutions between the two programs, the heat-flow distributions and wall temperatures will have reached a stable ultimate state.

24.2.3 Selected Examples of Application

Coolant Flow in an Engine's Water Jacket

The limits on performance-enhancing modifications to engines are increasingly defined by component strength. Efficient engine cooling under all engine-operating conditions is, therefore, a precondition for modern engine developments. The available water space is networked on the basis of CAD data and a fluid mechanical simulation is performed; see Fig. 24-21. The result is the ability to perceive zones of extremely low or extremely high velocities. Zones of low velocities are indicative of low convective cooling. Such situations must be avoided in the vicinity of heavily exposed components, such as the exhaust port, for example. High velocities in zones in which the pressure level has already dropped severely and coolant temperatures that are already extremely high result, on the other hand, in the danger of generation of cavitation bubbles as a result of local falls below vapor pressure.

An important target factor is the uniform cooling of all cylinders. Variations in the size of the passage apertures in the cylinder-head gasket usually make it possible to achieve this homogeneous distribution.

Figure 24-22 shows an example of the distribution of the coefficient of heat transfer around the heavily exposed exhaust ports of a five-valve engine. Systematic variation of the size and location of the passage apertures in the cylinder-head gasket has been applied to achieve optimization of flow through the zone between the two ports.

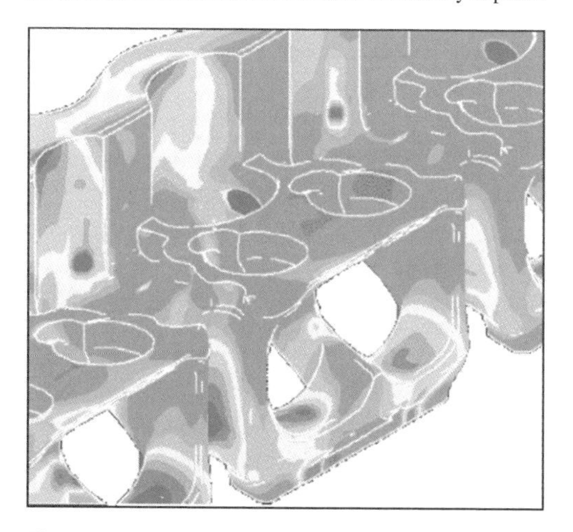

Fig. 24-22 Distribution of coefficients of heat transfer in the vicinity of the exhaust ports in a four-cylinder, five-valve engine. (*See color section.*)

Present-day development in computation methods includes solution of the energy equation in the coolant and the coolant's temperature-dependent physical data, and calculates the heat flow yielded by the component. The component temperatures necessary for this purpose are either adopted iteratively from an FE simulation or solved directly in the flow solver in the form of a so-called "conjugate heat transfer" problem. The computing grid around the component in which the heat-conduction equation is solved is expanded for this purpose. The convective flow of heat is corrected in both cases by a quantity containing the heat flux resulting from film and bubble boiling.

Charging

The targets of modern gasoline engine development are

- High torque throughout the entire speed range
- Low fuel consumption
- Low exhaust emission

Cylinder charging has a significant influence on these targets. The requirements listed above cannot be met with a fixed valve-timing setting. Only variation of the valve timing as a function of speed and load range permits attainment of the target criteria. Numerous simulations of the influence of timing settings on volumetric efficiency, residual gas content, and charging energy are necessary for this purpose. One possibility of rationally reducing the number of simulations to be performed can be found in "experiment planning" using ED (experiment design). Optimization of target criteria throughout the speed and load ranges can then be achieved with only a few simulation steps.

Mixture Generation and Energy Conversion

Charge Movement in Gasoline and Diesel Engines

The introduction of direct injection diesel and gasoline engines has heightened the importance of knowledge of internal engine processes. We should mention first the charge movement, which is optimized in terms of its alignment (swirl or tumble) and its intensity. Assessment of the inlet ports can be accomplished in a manner equivalent to testing using a steady-state flow simulation.[7] Figure 24-23 shows the iterative approximation of the integral values of swirl and flow through the inlet ports of a diesel engine to the target values, which envisaged simultaneous reduction of swirl intensity and flow resistance in the ports.

A subsequent simulation of transient internal cylinder flow additionally makes it possible to evaluate the influence of the integral values of swirl and flow on charge movement at the point of injection or ignition.

Such simulations, whether transient or steady state, have become routine. The 3-D flow simulation is dependent on boundary values of a charging calculation, however, since it is not yet possible to observe the entire engine

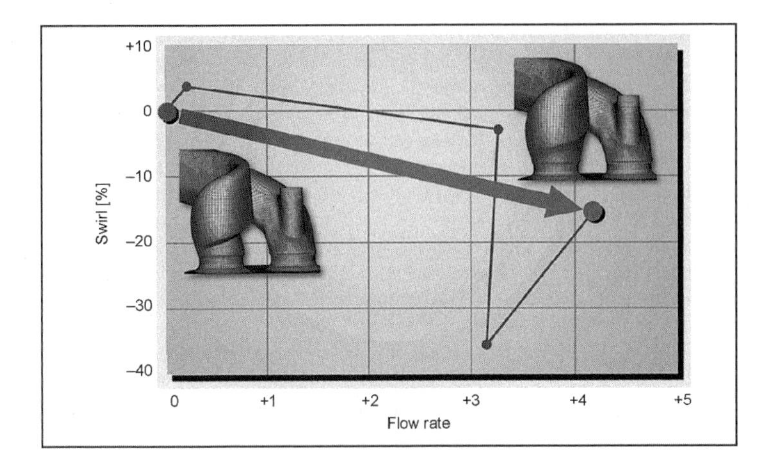

Fig. 24-23 Iterative approximation of the integral values of the inlet ports to the target values, from Ref. [7].

from the intake to the exhaust system within rational turnaround times. The quality of such boundary values is also determined by the quality of the subsequent, locally and chronologically highly resolved, three-dimensional flow computation.

Mixture Generation in Direct Injection Gasoline Engines

Mixture generation in the engine is subsequently examined on the basis of the simulation of charge movement. In simulation terms, this signifies that it is necessary to calculate a multiphase flow. Numerical treatment of a gas composed of a number of components (for description of an air-fuel mixture in an internal combustion engine, for example) can be performed with one additional equation per component (a type of mass conservation equation), whereas the input for examination of an additional phase in the same computation zone is considerably more complex. The second phase was long calculated in a LaGrange reference system (comotive reference system). The term "Euler/LaGrange observation" is used in this context. A motion equation taking mass, energy, and impulse exchange into account is solved for every particle and/or every droplet. The Euler/LaGrange formulation has the weakness of failing in case of high mass loadings (e.g., in so-called "dense" sprays, such as those generated by high-pressure diesel injectors).

Euler's formulation, which also regards the droplets of a spray as a continuum, contrasts with this. An Euler/Euler formulation necessitates extremely high input and encounters its limitations in extremely dilute sprays, for example. Attempts have recently been made to apply both formulations simultaneously in separate zones and, thus, achieve an ability to exploit the advantages of the respective method.[15]

Simulation requires as its input a so-called "spray model," which describes the injection process. Injection systems are nowadays usually first tested under controlled conditions (pressure and temperature) in a pressure chamber. These experimental data are used for calibration of the simulation.

Published concepts for description of the spray remain rare even today and are not keeping pace with the development of this type of injection system. The sprays requiring description can, on the other hand, be measured extremely well and fulfill essential criteria that are a precondition for the Euler/LaGrange observation mode.

Figure 24-24 illustrates, using the chronological development of the average-numbered droplet size and the (also average-numbered) droplet velocity, the quality achievable using a calibrated spray model. Two different radial positions 15 mm below the injector were selected for comparison of the calculation and the phase Doppler anemometer (PDA) measurement.

Validation of the calculation methods for specific applications, as performed above for spray calculations, for example, is an important step toward the acceptance of simulation results and toward the integration of calculation methods in the development process. Attention is for this reason drawn again and again to validation experiments in the examples presented below.

A spray model calibrated in this way simulates the injection process while the engine is turning. Significant target factors are

- Visualization of the fuel-air mixture
- Description of the chronological and local spread of the mixture "cloud"
- Analysis of the interaction between the injection jet and charge movement

Figure 24-25 shows an example of the droplets remaining in the combustion chamber at ignition TDC. This distribution shows that large droplets, in particular, accumulate underneath the exhaust valves. This, together with the distribution of the evaporated fuel in the combustion chamber, makes it possible to derive valuable information on necessary modifications to the engine concept.

Simulation permits the evaluation of the influence of engine speed, load point, combustion chamber geometry,

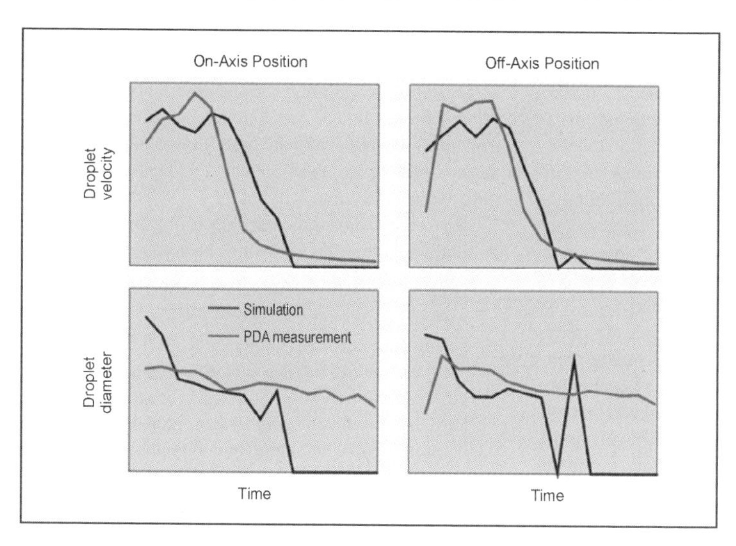

Fig. 24-24 Comparative assessment of calculated and measured data, shown in the form of development of droplet size values against time (top series) and against velocity (lower series). The left series in the figure applies to the jet axis, while the right series shows comparison with an off-axis position.[16]

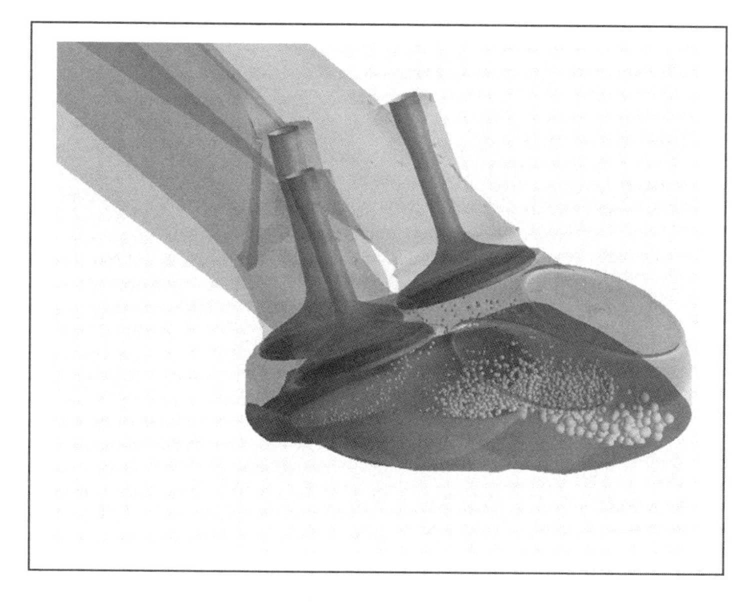

Fig. 24-25 View of fuel droplets remaining in the combustion chamber at ignition TDC.

injection characteristics (jet angle, etc.), and injection timing.

It must also be noted for qualification, however, that only a process averaged for a large number of cycles is depicted in the simulation. Effects caused by fluctuations in the charge movement or the injection process are not taken into account here. The field of engine fine-tuning thus (for the moment) remains closed to simulation.

Mixture Generation in Gasoline Engines with Inlet Manifold Injection

The wall film formed in the inlet port and on the valves necessitates complete observation of up to 20 engine cycles. Physical modeling of the droplet and wall interaction and of the secondary detachment of droplets out of the wall film is extremely difficult. Simulation of the problem necessitates extremely long computing times, with the result that only a few experiments in which the processes of inlet-manifold injection have been simulated with local and chronological resolution are known.

Mixture Generation in Direct-Injection Diesel Engines

Diesel injection jets have a significantly higher impulse than direct-injection gasoline engine systems. In a pump and jet system, injection pressures may rise above 2000 bar.

Cavitation phenomena that have effects on the nascent spray can occur within the injector.

As a result of the high impulse, a large number of extremely small droplets are fed into the combustion chamber in a compact spray-cone zone in an extremely short time. Droplet collisions thus occur as a consequence of the high droplet densities. The aerodynamic forces acting on the droplet result, on the other hand, in further disintegration of the droplets. These evaporate within a very short time. An intensive droplet and wall interaction nonetheless occurs as a result of the proximity of the injector and the piston at the point of injection.

Whereas the mixture generation and combustion phases in a gasoline engine occur at separate times, in a diesel engine the liberation of heat influences mixture generation as a result of the short ignition delay.

The first task in simulation is calibration of the spray model, a process that can be orientated essentially only around integral measured factors, such as the penetration depth of the jet of liquid. Mensurational registration of high-pressure diesel injection is extremely difficult, since conventional PDA measuring systems for the measuring of droplet sizes and velocities are unable to supply any information throughout large zones of the "dense" spray. Quantitative measurements of fuel-vapor concentration using Raman spectroscopy are, for their part, possible only in zones in which no further droplets exist. The consequence is that it is extremely difficult to assess the quality of a spray model intended to determine the spread behavior against time of both the liquid and the gas phases.

A further complicating factor is that droplet evaporation and, therefore, also mixture preparation, are decisively determined by so-called "entrainment" of the hot ambient air drawn into the jet of liquid. It is not possible to resolve such entrainment adequately on the basis of the commercial CFD codes currently used. Spatial resolution of the computer grid with edge lengths of around 0.01 mm are necessary in the vicinity of the injection nozzle in the case of injector hole diameters smaller than 0.2 mm. This results in an enormous number of computer cells and, therefore, in extremely long computing times. This weakness in the simulation of diesel-engine processes, which affects all the downstream and simultaneously occurring phenomena, such as combustion and droplet and wall interaction decisively, is the subject of intensive ongoing research.

Simulation of mixture generation in diesel engines has, therefore, not yet become a standard application. The existing computation results on diesel combustion were mainly achieved only after adaptation of model constants. The validity of these model constants for variation of the injection system or for other engine-relevant parameters is not estimable. Predictive calculations are, therefore, not possible on the basis of the current development status of simulation technology.

Simulation of Combustion in a Gasoline Engine

Procedures and methods for the experimental study of combustion processes in engines are developing simultaneously as well as in parallel with simulation methods. Potentials for the validation and coordination of numerical combustion models are therefore emerging. On the other hand, analysis of a combustion simulation permits a deeper insight into the processes occurring in the engine, and thus augments the knowledge gained from experimental studies.

A significant difficulty in the description of combustion phenomena is the fact that the instantaneous conversion rate depends both on local flow state and on the thermodynamic variables of state, i.e., pressure, temperature, and composition.

The combustion processes occurring in engines can be subdivided into the following categories:

- Homogeneously premixed combustion (full-load working point of a gasoline engine)
- Diffusion-controlled combustion (full-load working point of a diesel engine with direct injection, without preinjection)
- Partially premixed combustion (part-load working point of a gasoline engine with direct injection)
- Combination of diffusion-controlled and partially premixed combustion (part-load working point of a diesel engine with direct injection and preinjection)

Physical models have been developed for each of the above-mentioned categories of combustion processes. The first models included the so-called "eddy dissipation" models; see Ref. [17]. These were used, in particular, for simulation of diesel combustion. It is presupposed here that the chemical processes occur significantly more quickly than mixture of fuel and air at a microscopic level, for which a turbulent time scale constitutes the limiting factor. These conditions are fulfilled in diffusion flames.

This category of models has been and is being developed ever further. Such models are now capable of generating a link between the turbulence-dominated conversion rate and the chemical conversion rate times.

The so-called "flame Area" models[18] were developed specifically for homogeneous combustion processes (complete mixing of air and fuel). The flame front and its rate of propagation are calculated using an advance variable. Expansions of this model with an equation, which describes the mixing state, also permit application to partially premixed combustion, such as occurs in part-load ranges in gasoline engines featuring direct injection. "Flamelet" models are based on a detailed chemical reaction mechanism. Inclusion of the influence of turbulence is accomplished via the introduction of flame stretching factors, which are determined by the turbulent characteristics data. The solution of complex chemical conversion rates can be accomplished in advance and supplied to the simulation in the form of lookup tables.

The representative interactive flamelet (RIF) model reduces the complexity of including coupling between turbulence and chemistry, as a result of the fact that only between one and a maximum of 20 different turbulent characteristics are included in the calculation of instantaneous conversion rates.

The probability-density function (PDF), combustion models, in which the interaction between turbulence and chemistry is described by a multidimensional PDF, constitute a different category of combustion models. The chemical conversion rate is in many cases described here using a single-stage global reaction.

Detailed information on these combustion models can be found, for example, in Ref. [19].

Figure 24-26 shows an example of a comparative assessment of a calculated and measured development of combustion with a stratified charge in the AVL-DGI engine, shown in the form of a section through the cylinder and the injector axis. The PDF combustion model was used in this case. Formation of a flame core with a dominant direction of propagation in the zone of stoichiometric mixture composition initially occurs, immediately after initiation of flame generation.

Combustion simulations are currently used for comprehension of phenomena observed in actual engine operation. Because of the still unsolved problems of modeling, the solution of questions concerning ignition stability appears at present to be premature, and the predictive calculation of absolute pollutant concentrations in new engine concepts unrealistic.

Exhaust-Gas Aftertreatment

The subject of "numerical flow simulation" in the field of exhaust-gas aftertreatment is increasingly gaining in importance as demands for lower exhaust emissions from new vehicles multiply. There are two essential concerns that must be addressed:

- The engine's cold-start behavior
- The operational reliability of the catalytic converter

Simulation of the approach flow to a catalytic converter has become a standard application. A uniformity parameter is defined in order to permit classification of the quality of a particular variant; see (for example) Ref. [21]. The accuracy of the calculation method for a surging flow in the exhaust manifold has been demonstrated in the context of a validation study.[22] Figure 24-27 shows a comparative assessment of calculation against measurement for the plot against time of a mass-flow velocity at an engine speed of 2.000 1/min. The measurements were made using an optical measuring method in a manifold cross section located immediately downstream from the junction of the exhaust flows from cylinder 3 and cylinder 4.

It has now been demonstrated that calculations made on the basis of boundary conditions that do not change with time are adequate for the obtainment of trend statements concerning durability and quality even in the context of concepts with close-to-engine catalytic converters. The cylinder under examination in each case is charged with the mass flow that occurs briefly in engine operation during the so-called "pre-exhaust thrust."

The objective of mathematical optimization is uniform charging of the monolith, in order to reduce the vol-

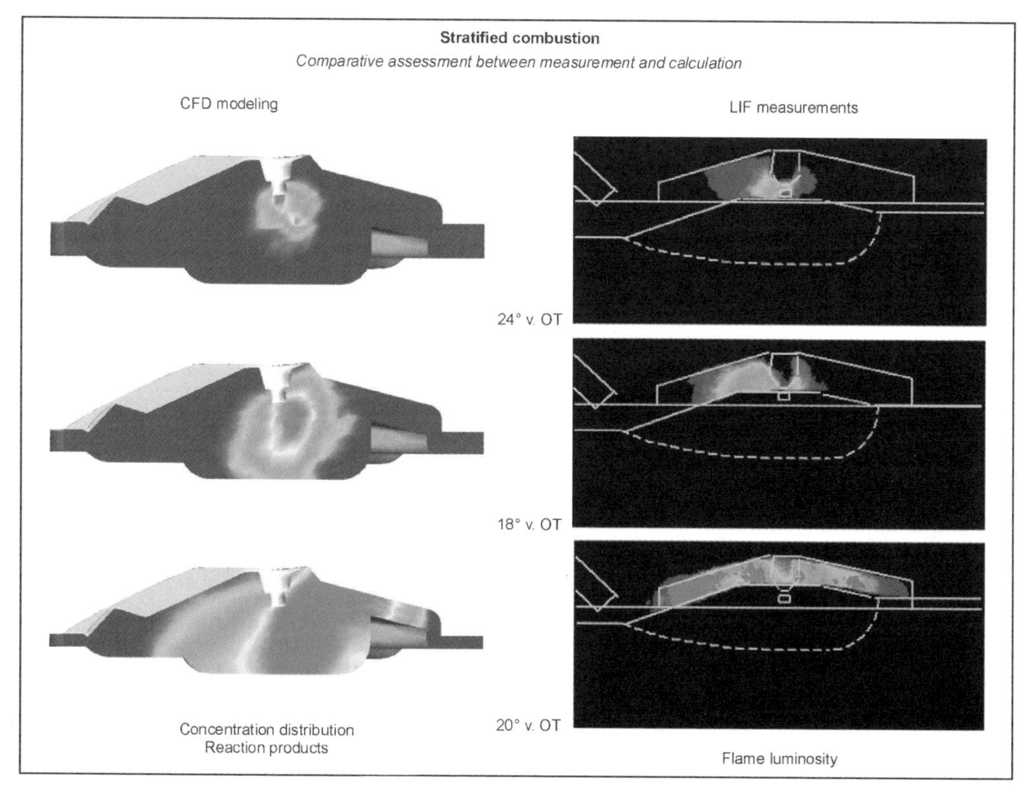

Fig. 24-26 Combustion with charge stratification; comparative assessment calculation vs. measurement.[20] (*See color section.*)

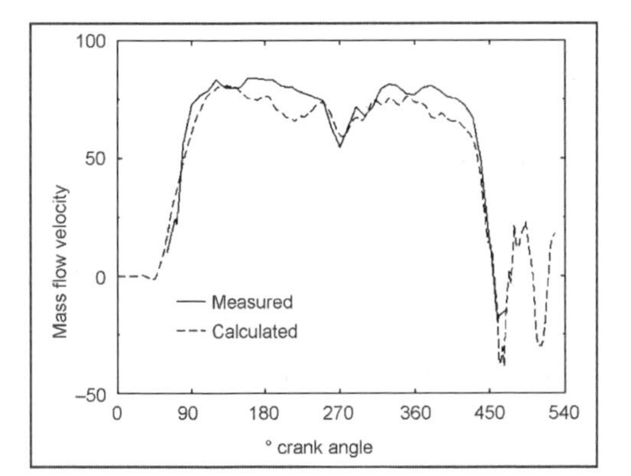

Fig. 24-27 Comparative assessment of calculated vs. measured mass-flow velocity in an exhaust manifold as a function of crank angle.

ume, and, therefore, the space required and costs for the catalytic converter.

In addition, a criterion for the durability of the coating of the catalytic converter is derived from the calculated pressure gradients in the first computer cell layer in the monolith.

Information on conversion performance cannot be derived from the computation results of a 3-D flow simulation. There are specially tailored program systems that are capable of describing conversion of pollutants in terms of trends for this purpose. These programs require an extremely large amount of prior information, which is not initially available, however. The combination of a detailed chemical* solver with a 3-D CFD code may provide a remedy in this field in the future. The decisive boundary conditions for all calculations of exhaust-gas conversion are statements of the engine's raw emissions, which must be stated in chronologically resolved form. This information can at present be obtained only from measurements.

Approach flow onto the λ probes can also be optimized by computer methods. The establishment of correlations between calculated and measured λ probe signals is still only at the preparatory stage, however. The question of whether steady-state computation operations will be adequate for the performance of trend analyses for cylinder-selective λ control has not yet been conclusively answered.

The combination of flow simulation in exhaust systems with a strength calculation has already become established, to the extent that the coefficients of heat transfer from the CFD computation are transferred to the FE program, in order to calculate there the thermomechanical load exposure of the component.

All the statements made above apply primarily to naturally aspired engines. Engines featuring exhaust supercharging require considerably more complex numerical description.

Good cold-start performance in an exhaust system does not necessarily correlate to optimum approach flow into the monolith. The limitations on correct calculation of heat transfer and the physical period of several seconds requiring simulation do not yet permit the use of 3-D flow simulation for clarification of cold-start performance.

Bibliography

[1] Enderich, A., and R. Handel, Mündungsschallprognose mit der Finiten Element Methode, MTZ 60 (1999) 1.

[2] Babuska, I., B. Szabo, and K.N. Katz, The p-Version of the Finite Element Method, SIAM J. Numer. Anal., Vol. 18, No. 3, June 1981.

[3] Haibach, E., Betriebsfeste Bauteile, Springer-Verlag, Berlin, 1992.

[4] Vanderplaats, G.N., Numerical Optimization Techniques for Engineering Design—with Applications, McGraw-Hill, Inc.

[5] Maute, K., E. Ramm, and S. Schwarz, Adaptive Topologie- und Formoptimierung bei linearem und nichtlinearem Strukturverhalten, NAFEMS Seminar zur Topologieoptimierung, 23 September 1997, Aalen.

[6] Fischer, P., P. Nefischer, and G. Kraßnig, Akustikberechnung lokal gedämpfter Motorstrukturen unter Einbeziehung stochastischer Erregungsprozesse, Kongress Berechnung und Simulation im Fahrzeugbau, VDI Berichte, Würzburg, 2000, p. 1559.

[7] Nefischer, P., P. Grafenberger, C. Hölle, and E. Kronawetter, Verkürzter Entwicklungsablauf durch Einsatz von CAE-Methoden beim neuen Achtzylinder-Dieselmotor von BMW, Teil 2, Thermodynamik und Strömung, MTZ 60 (1999) Nr. 11.

[8] Wiebe, I., Brennverlauf und Kreisprozess von Verbrennungsmotoren, VEB Verlag Technik, Berlin, 1970.

[9] Barba, C., C. Burkhard, K. Boulouchos, and M. Bargende, Empirisches Modell zur Vorausberechnung des Brennverlaufs bei Common-Rail-Dieselmotoren, in MTZ 60 (1999) Nr. 4.

[10] Betz, J., L. Beck, S. Micko, J. Hager, R. Marzy, and T. Gumpoldsberger, Entwicklung von Motorkühlsystemen bei erhöhten Anforderungen an transiente Betriebsbedingungen, in 22 Int. Wiener Motorensymposium, Wien, 2001.

[11] Ertl, C., E. Kronawetter, and W. Stütz, Simulation des dynamischen Verhaltens von Dieselmotoren mit elektronischem Management, in MTZ 58 (1997) Nr. 10.

[12] Zierep, J., Grundzüge der Strömungslehre, 4, Aufl., G. Braun Verlag, 1990.

[13] Ferziger, J.H., and M. Peric, Computational Methods for Fluid Dynamics, Springer-Verlag, 1996.

[14] Durst, F. [Ed.], Lattice Boltzmann Methods: Theory and Applications in Fluid Mechanics, LSTM Erlangen, KONWIHR Workshop, 2001.

[15] Wan, Y.P., and N. Peters, Application of the Cross-Sectional Average Method to Calculations of the Spray Region in a Diesel Engine, SAE 972866, 1997.

[16] Blümcke, E., T. Kobayashi and S.R. Ahmed [Eds.], Strategies for Simulating Transient In-Cylinder Flows Emphasizing Mixture Formation, in Proceeding of Workshop on CFD in Automobile Engineering, JSAE 2000.

[17] Magnussen, B., and B. Hjertager, On Mathematical Modelling of Turbulent Combustion, in 16th International Symposium on Combustion, 1976.

[18] Weller, H.G., S. Uslu, A.D. Gosman, R. Maly, R. Herweg, and B. Heel, Prediction of Combustion in Homogeneous-Charge Spark-Ignition Engines, in COMODIA 94, 1994.

[19] Peters, N., Fifteen Lectures on Laminar and Turbulent Combustion, ERCOFTAC Summer School Proceedings, Aachen, 1992.

[20] Tatschl, R., Ch. v. Künsberg Sarre, P. Priesching, and H. Riediger, Professor R. Pischinger [Ed.], Mehrdimensionale Simulation der Gemischbildung und Verbrennung in einem Ottomotor mit Benzin-Direkteinspritzung, in 7, Tagung "Der Arbeitsprozess des Verbrennungs-motors," 1999.

[21] Bressler, H., D. Rammoser, H. Neumaier, and F. Terres, Experimental and Predictive Investigations of Close Coupled Catalytic Converter with Pulsating Flow, SAE 960564, 1996.

[22] Bratschitz, B., E. Blümcke, and H. Fogt, Numerische und Experimentelle Untersuchungen an der Abgasanlage des Audi 1.8 l 5V-Turbomotors, in VDI Berichte Nr. 1559, 2000.

[23] Küntscher, V. [Ed.], Kraftfahrzeugmotoren Auslegung und Konstruktion, Verlag Technik, Berlin, 1995.

[24] Society of Automotive Engineers, Inc. [Ed.], SAE Fatigue Design Handbook AE-22, Society of Automotive Engineers, Warrendale, PA, USA, 3, Aufl., 1997.

[25] Draper, J. [Ed.], Modern Metal Fatigue Analysis, Safe Technology, Sheffield, U.K., 1999.

25 Combustion Diagnostics

25.1 Discussion

Combustion diagnostics is always applied in engine development when unexploited potential compared to thermodynamically possible targets is ascertained during measurement of consumption, output, and emissions. Given the high targets set for modern engines, thermodynamic combustion analysis using measurement of cylinder pressure is always a fixed element in the development sequence.

Measurements of cylinder pressure are augmented with a whole series of measured variables that define fluid state and component functions. Such "indicated data," which is generally registered in a form classified by cycle and crank angle, depending on the particular assignment, form the basis for thermodynamic evaluation of combustion and for optimization of the adjustment parameters for the engine. Comparative assessment of it against theoretically possible targets available from engine-simulation computations provides guides for appropriate development activity.

Such development activity is concerned essentially with charging, mixture generation, turbulent charge movement, and, ultimately, flame propagation. Within the scope of the thermodynamic information it supplies, engine indicating provides indications of deficiencies in these processes, but because of the characteristics of the sensor system used it is unable to provide any information on local events or on the component-related causes of the deficiencies.

The desire then arises in this context to determine through direct insight into engine processes precisely what it is that prevents the achievement of the theoretically possible potentials. This is accomplished using flow and combustion visualization methods.

The potentials for rendering internal engine flow, mixture generation, the combustion processes visible, and applying optical measuring methods are as numerous and diverse as the questions they are intended to answer. Of the large number of methods tested in the laboratory, however, only a very few are actually suitable for practical use on engines nearing maturity for series production. A number of these procedures exploit flame radiance as a signal source and, therefore, possess the potential of directly indicating the way in which changes in the engine affect local flame propagation processes. Such methods are described here in more detail under the subject of "Combustion Visualization." A number of other methods that can be used in subsectors of an operating characteristics field, given corresponding adaptation of the engine, are also examined here, in addition to this central aspect of combustion visualization.

25.2 Indicating

The term "indicating" is used to designate the measurement and depiction of the plot of cylinder pressure against time or crank-angle position.

Pressure indicating occupies a central ranking in combustion development[1,2] as a result of the great significance of cylinder pressure for thermodynamic comprehension of combustion in engines and is used for much more than just analysis of the plot of pressure.[3] The sensor systems, data-acquisition systems, and result analysis facilities necessary for this purpose have become widely used because of the availability of user-friendly measuring systems. The registration of supplementary characteristics variables such as measurement of the injection process, ignition current, and thermal variables proceed as a natural progression from pressure indicating.

In addition to high-pressure indicating for analysis of combustion, low-pressure indicating on the inlet side in the cylinder and on the exhaust side constitutes the precondition for analysis of charging and for determination of the masses of the gas available in the cylinder for combustion.

Figure 25-1 shows an example of measurement of cylinder pressure in a gasoline engine. The pressure signal acquired using the crank angle as the time base is used to generate the pV diagram or, given knowledge of the cylinder filling, the progress of combustion determined using a combustion model.

Low-pressure indicating, with measured data for intake-manifold pressure, cylinder pressure, and the pressure plot determined upstream from the turbine of the exhaust turbocharger, is shown in Fig. 25-2. The mass flows calculated from the measured data for low-pressure indicating using a charging model are also plotted. Calculated plots for pressure and mass flow that occur when the chronological outlet-valve lift sequence is changed in a charging simulation in order to achieve optimization of volumetric efficiency are shown in addition to these measured data.

A comparative analysis of combustion in a number of different combustion processes at a part-load working point is shown in Fig. 25-3. The comparison of the plots for pressure and the plots for combustion derived for their part using a simple combustion model very quickly provides the observer with an overview of the course, duration, and peak intensity of combustion, enabling him to judge the thermodynamic quality of the combustion process.

With other measured data, pressure indicating and mass balance also make a considerable contribution to the drafting of an energy balance and loss analysis for the various combustion processes. Figure 25-4 shows in this context a comparative assessment of the loss distribution in a

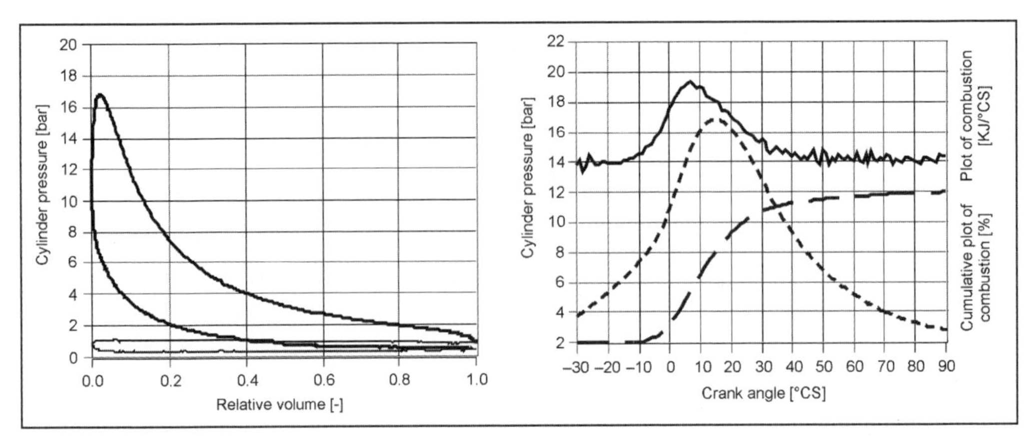

Fig. 25-1 Pressure indicating: *p-v* diagram and analysis of combustion plot.

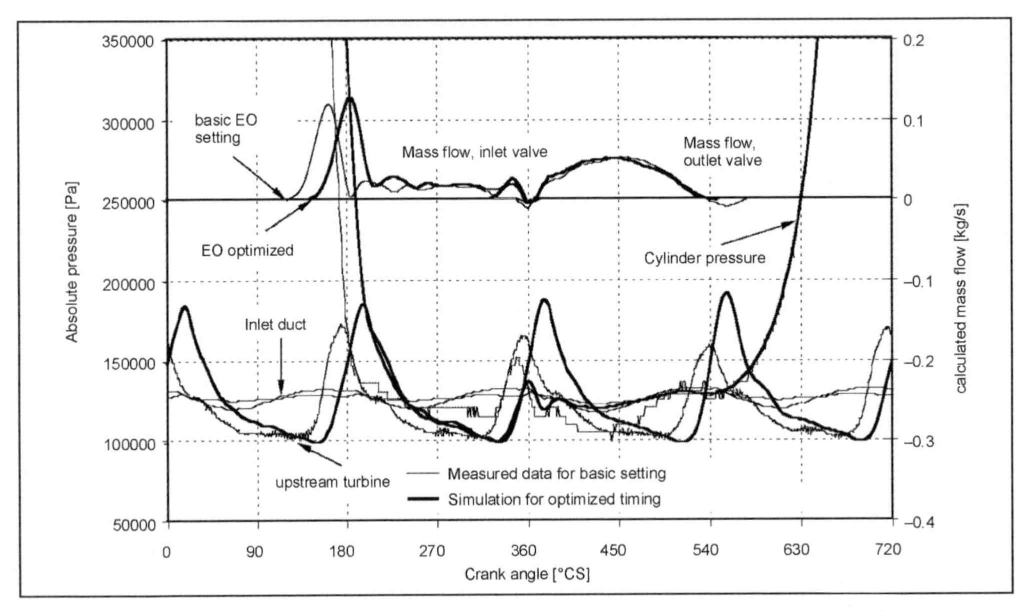

Fig. 25-2 Low-pressure indicating in the inlet, cylinder, and outlet train. Charging calculation for plot of mass flow, simulation of optimized mass flows, and plots of pressure.

DI gasoline engine when the engine is operated at the same working point in various ways.

25.2.1 Measuring Systems

The main structure of an indicating measuring system for measurement of pressure can be described as follows:

- Pressure transducer: This is installed either directly via a special boring in the combustion chamber or via special adapters in existing borings, such as those for the spark plugs or glow plug. Figure 25-5 shows examples of typical piezoelectric pressure sensors.
- Measuring amplifier: This amplifies the measuring signal received from the pressure transducer to a voltage range of a magnitude suitable for transmission across long cable distances to the data-acquisition unit while at the same time assuring a high signal-to-noise ratio. The length of cable between the pressure sensor and the amplifier is always kept as short as possible, in order to achieve high signal quality.
- Data-acquisition system: This is connected to the measuring amplifier and the crank-angle index mark generator, as well as to a PC for control of the entire system. Its principal function is that of recording the

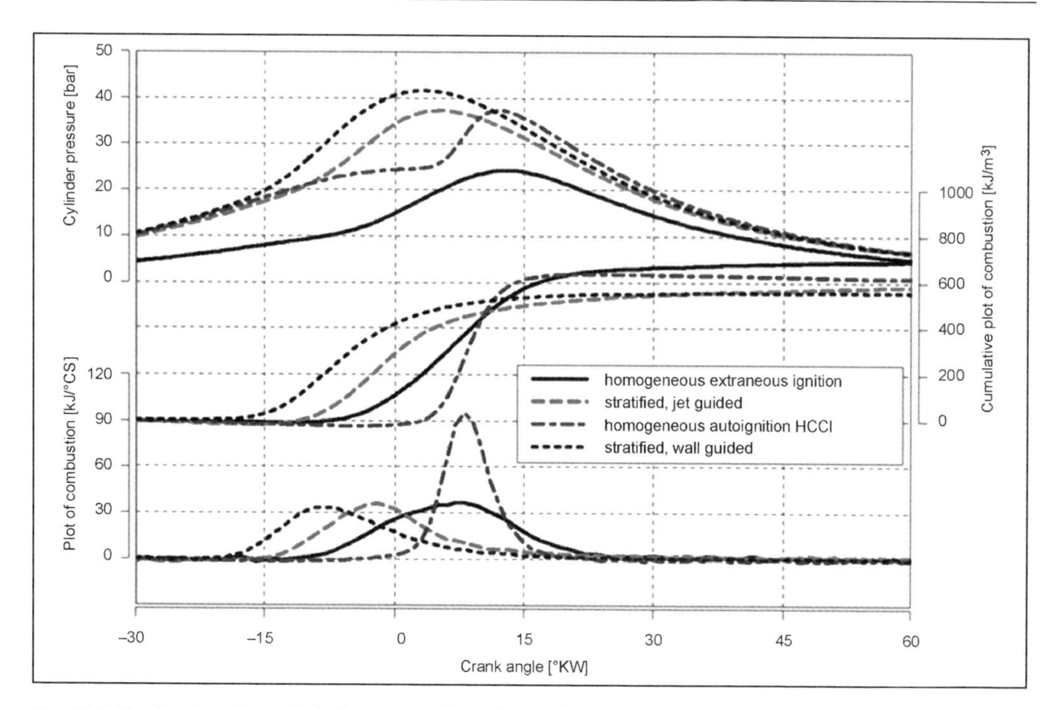

Fig. 25-3 Combustion-plot analysis for a range of gasoline-engine combustion processes, 2.000 rpm, pme = 2 bar.

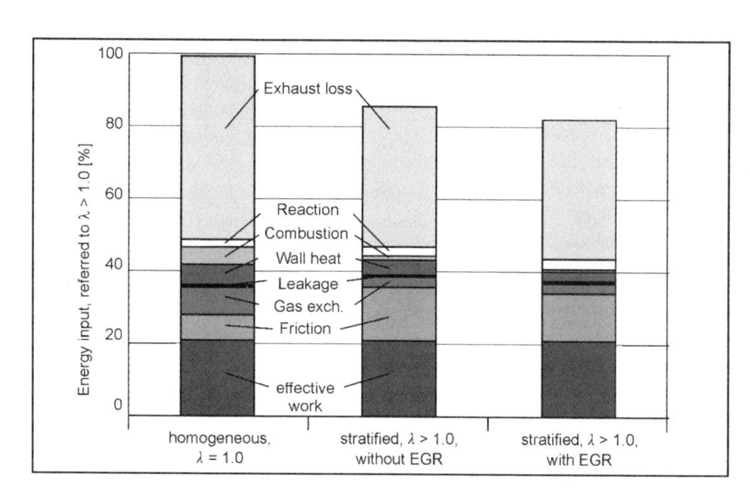

Fig. 25-4 DI gasoline engine: Loss distribution in stoichiometric and stratified operation, 2.000 rpm, pme = 2 bar

necessary measured data with the required measuring resolution. In addition to this basic function, result computations performed as early as during the measuring sequence, i.e., in "real time," are gaining increasingly in importance.

- System operation: This is accomplished via special PC software that makes it possible to parametrize the entire measuring system and the measurement itself, to obtain characteristics data, and calculations, and to define algorithms for the determination of characteristics data or the calculation of results from the measured data.
- Postprocessing: This is used for presentation and processing of the measured data. More complex computations, comparisons of results, and documentation procedures are performed here using corresponding graphical and computing aids. The scope is adjusted in each case by the user to match the need for performance of testing.

25.2.2 Quality Criteria

- Sensors: Sensitivity and signal dynamics are critical in this context in order to meet the demands of the

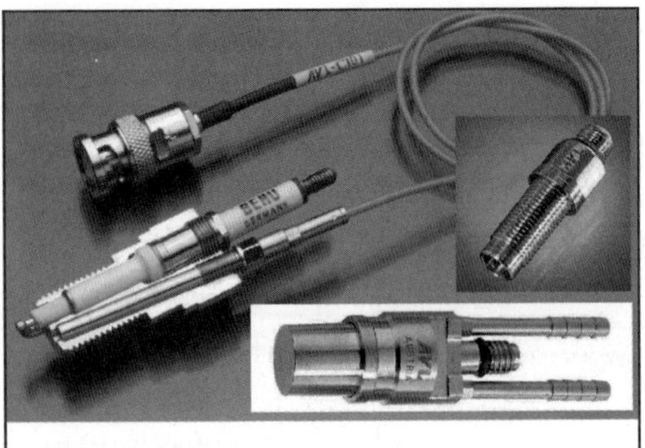

Applications: Combustion data, measurement of efficiency, energy balances, friction characteristics fields, limit overshoot monitoring, misfire detection, combustion noise, knocking detection, vibration excitation, tail gas measurement, EGR calibration, automatic characteristics field optimization, injection phase, mechanical loads

Fig. 25-5 Examples of piezoelectric sensors for measurement of cylinder pressure. (*See color section.*)

particular measuring task. In particular, sensors for practical use on the test bench must be insensitive to varying conditions of service and must reliably offer high long-term stability.

- Data-acquisition systems: The pure "measured-data acquisition" function occurs here immediately after real-time result analysis as early as during the measuring sequence. Direct indicated characteristics data for classification of the pressure plot is determined from the measured data itself by the statement of peak pressure p_{max}, location of peak pressure αp_{max}, pressure gradient $dp/d\alpha$, location of maximum pressure gradient $\alpha\ dp/d\alpha_{max}$, and maximum rate of pressure increase $dp/d\alpha^2$. Indirect indicated characteristics data are available in this form for indicated average pressures p_{mi}, $p_{mi\text{-}HD}$, $p_{mi\text{-}LW}$, friction agent pressure p_{mr}, onset of combustion, duration of combustion, and energy conversion points. Such real-time analyses are subject to continuous modification depending on needs and appropriateness, and as a function of computing potentials.
- Postprocessing: Measured data and the results of real-time analyses are supplied either locally or from databases with the result that they can be quickly integrated into any offline analyses defined by the user. The utilization of open-readable data formats that can be accessed via a whole series of user functions either defined or configured for the particular case is decisive in this context. This makes possible, from the indicating procedure, model functions or characteristics variables for quick and effective assessment of combustion on criteria that relate to topics such as

- Maximum component load exposure
- Noise generation caused by combustion
- Knocking and misfire detection
- Engine lean-running limits
- Optimum energy conversion

25.2.3 Indicating: Prospects

Under the clearly formulated preconditions of thermodynamic combustion analysis and as a result of technical advances in the fields of sensor systems and data acquisition, indicating has attained a central ranking in engine development. This established position has resulted in the desire to utilize cylinder-pressure measuring systems not only for the purpose of analysis in the context of combustion development, but also for the monitoring of engines during operational service. The introduction of such a functional diagnosis arrangement will be decisively orientated around the suitability of sensor systems for mass production and around its direct benefits in use.

The use of pressure indicating as a guide factor in combustion processes that are capable of realizing their potential in everyday use by precisely cycle-orientated regulation of combustion has a special technical attraction. Alongside the suitability of sensor systems for mass production, development in this field will be determined by the availability of actuators relevant to combustion.

25.3 Visualization

25.3.1 Functions and Discussion

The function of optical diagnosis methods in engine development is to provide insights into those flow, mixture generation, and combustion processes whose behavior cannot be adequately interpreted from the results of conventional indicating methods. In the development of combustion systems, questions concerning the detailed courses of the processes decisive for optimal combustion arise from normal pressure indicating in comparison to thermodynamic calculations and three-dimensional combustion modeling.

The main interest focuses here on the following topics in particular:

- The influence of flow within the engine on combustion
- Fuel-jet spread and mixture-generation processes
- Mixture state: The homogeneity or heterogeneity of the cylinder charge and of its temperature
- Combustion in the case of supplied ignition: Flame core formation, flame propagation, burnout of the end gas zones, spontaneous ignition of end gas, combustion anomalies
- Combustion in case of autoignition: Ignition sites, diffusion combustion, soot generation and burnoff, air efficiency, and flame temperature.

Such questions are studied at a number of levels:

- In the context of fundamental research: Research and measuring methods in which the automobile engine aspect determines the general orientation but in which the test apparatus may diverge extremely greatly from actual engine operation are used for basic functional analyses.
- Component tests: In this case, standardized test procedures are applied for the comparative evaluation of component properties.

- Actual engine operation: Optical sensor systems and measuring methods are orientated here specifically around the needs of engine operation uninfluenced by the measuring activity. Flame-observation methods under such real engine conditions are the central subject of this chapter.

25.3.2 Visualization Methods for Real Engine Operation

In the context of fundamental research, and in the arrangement of component tests, the study subject is always adapted to the specific question and to the needs of the particular test technology, whereas the emphasis is on undisrupted engine functioning in the case of visualization of actual engine operation. The limitations that result for the adaptation of the test technology to engine operation are, therefore, correspondingly restrictive in this field.

The following questions are presented: What results can be obtained using modern visualization methods under the restrictions imposed by real engine operation? How can they be utilized? What are the background preconditions and the necessary input?

25.3.2.1 The Radiant Properties of Gas, Gasoline, and Diesel Flames

The visible radiance that occurs in combustion of AC flames is the result of the chemoluminescence of the molecules formed during combustion and of the thermal radiation of soot. The spectral composition of the dominant contributory radiant ranges is shown in the emission spectra in Fig. 25-6; flame photographs are shown as an example in Fig. 25-7. The components fuel (CH), intermediate products (OH and CO), and radiant CO_2, H_2O, O_2 portions, along with other molecules and radicals, are generally found in the oxidation of CH molecules. The thermal radiation of soot particles also contributes to the flame's intrinsic luminosity in cases in which low-oxygen combustion results in the generation of soot. This particulate radiation can contribute to flame luminosity with significant intensity if local rich combustion occurs in a stratified charge; in diffusion combustion in diesel engines, flame radiance is massively dominated by such thermal soot radiance.

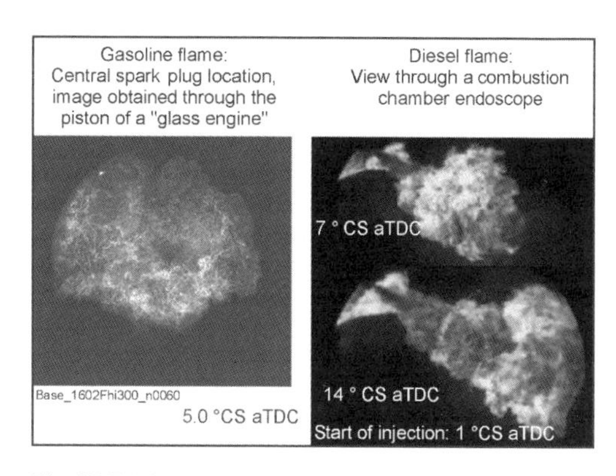

Fig. 25-7 Flame photograph. (*See color section.*)

25.3.2.2 Flame Spectroscopy

The spectral intensity distribution of the emission spectra contains information on the concentration of the radiant molecules and their initial components, on their temperature, and on the temperature of the radiant soot particles. Since the preconditions for thermal equilibrium are not present in many cases, the context of the transient processes occurring in combustion in engines, given the lifetimes of the radiant molecules and the fact that severe local gradients occur in the measuring volumes registered, exploitation of spectral radiance properties for quantitative measurements is limited to special cases (Lambda[4] OH

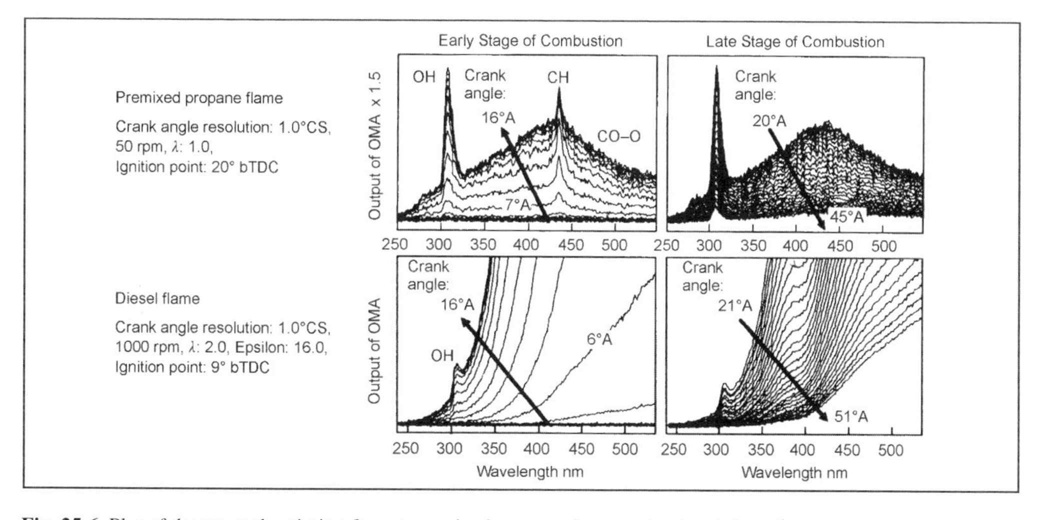

Fig. 25-6 Plot of the spectral emission from a premixed propane flame and a diesel flame.[4]

temperature,[5] soot temperature[6]). Standardized measuring methods are, therefore, restricted to only a few applications. In diffusion flames, for example, they exploit thermal soot radiation, either in spatially integral form[7] or by image evaluation methods applied to flame photographs for the determination of soot concentration and the temperature of the diffusion flame (Fig. 25-8, Ref. [8]). Development-relevant results for clarification of improved soot burnoff with divided injection are shown in this context in Fig. 25-9.[9]

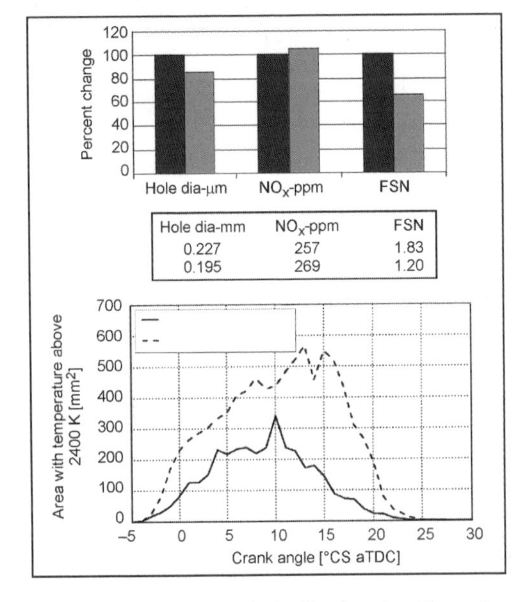

Hole dia-mm	NO_x-ppm	FSN
0.227	257	1.83
0.195	269	1.20

Fig. 25-8 Combustion analysis, diesel engine: Narrowing of the injection hole bore, effects on NO_x, and soot emissions. Also temperature-zone progress of diesel flames (Larsson[5]). Surface temperature analysis shows the mechanism of action for enhanced soot burnout.

25.3.2.3 Flame Propagation in Premixed Charges with Supplied Ignition

Following ignition and the formation of the flame core, the flame should propagate in such a way that chronologically optimum and locally uniform and complete combustion of the charge occurs. Flame propagation is driven by the advance of the flame front under the influence of turbulent charge motion. Figure 25-7 shows a flame image and the flame-front structure generated by charge turbulence.

An intensive interaction between flow and flame advance can occur across broad phases of combustion, because of the fact that the processes of turbulent flame propagation and directional flow within the engine may occur at comparable rates. This is the field for potential optimization of combustion within the engine.

The following development-relevant questions primarily arise in this context:

- How far is flame propagation as achieved at present removed from the ideal situation described above?
- What changes can be made to improve flame propagation?

25.3.2.4 Flame Propagation in Diffusion Combustion in a Diesel Engine

Ignition and combustion are determined here by gas state, and also decisively by the characteristics of the injection process. The spread of the diffusion flame (Fig. 25-7) is determined by injection and turbulent diffusion of the "cloud" of fuel vapor and its interaction with internal flow. Flame radiance immediately after autoignition is driven by the chemoluminescence of the reactants, but then very quickly becomes dominated by the thermal luminescence of the soot particles.

Primary development questions are

- How can air efficiency be raised by modifying the design of the injection system and charge movement?
- What changes to the injection system and charge movement would affect soot generation and burnoff and reduce soot emissions?
- How can excessively high flame-temperature peaks be avoided?

25.3.3 Visualization of Combustion in Real Engine Operation by the Flame's Intrinsic Luminescence

The engine is operated for this purpose on an engine test bench or in the vehicle in a dynamometer test. Analyses that augment standard engine indicating with information on the local and chronological progress of flow, mixture generation, and combustion, permitting the derivation of guidelines for systematic engine improvement, are of interest here for engine development purposes.

The flame's intrinsic radiance is used primarily here as the study subject, since it is accessible with the lowest technical input, thus very largely also avoiding disruption of combustion by the observation process. In an ideal case, the chronologically resolved, three-dimensional propagation of the flame in the combustion chamber should be registered by a visualization method in order to permit the ascertainment of deviations in combustion progress from the theoretical optimum. However, for technical reasons this rigorous requirement can be achieved only with great restrictions.

25.3.3.1 Technical Exploitation: Flame Propagation

Imaging flame-photographic and measuring methods that record the radiant intensity of the flame and derive from it information on the local and chronological progress of combustion are available for the technical achievement of flame observation.

Optical access in the case of flame photography is achieved via combustion-chamber windows. The combustion chamber is then depicted in the focal plane of a

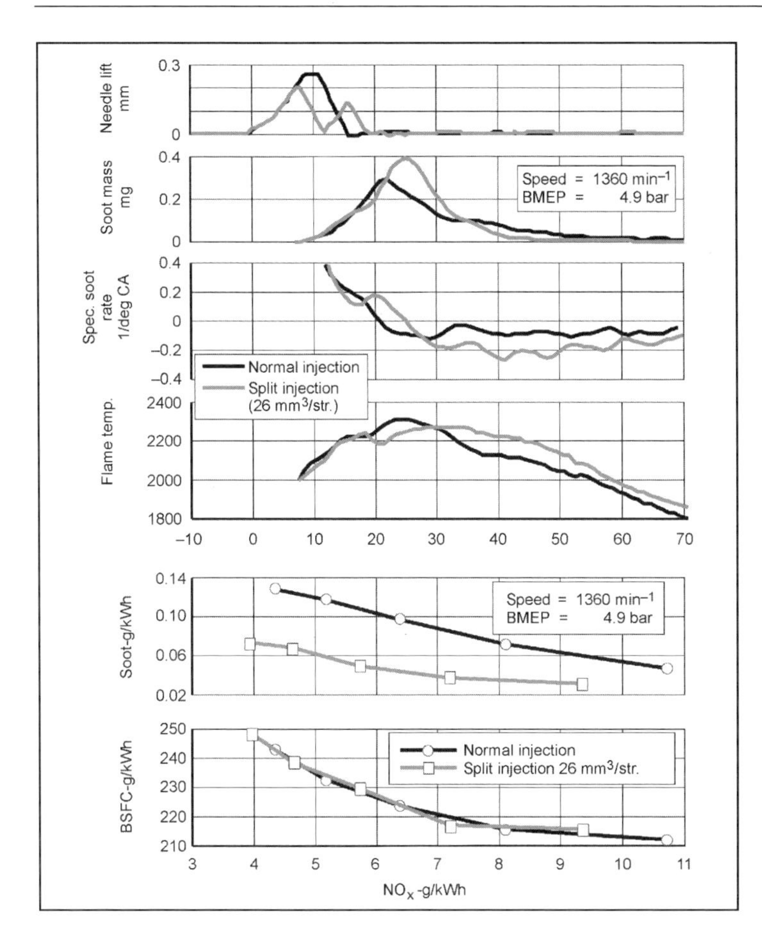

Fig. 25-9 Analysis of soot radiance in a diesel engine: Greater soot burnout is achieved because of split injection. The result reduced the No_x/soot trade-off and has no effect on consumption.

suitable camera via endoscopes. The system element relevant to engine operation in this context is the combustion-chamber window. This is either positioned on the combustion chamber via special borings[10] or used in specially adapted engine components.[11]

Flame photography: The advantage of direct flame observation via an endoscope is that picture capture using a camera immediately provides an image that can be interpreted. With appropriate selection of window position and angle of view into the combustion chamber, flame images that show extremely clearly the spread and turbulent structure of the diffusion flame, for example, in diesel engines, can be obtained (Fig. 25-7). The position of the flame in the combustion chamber can then be determined by image superimposition with reference images in a subsequent image-processing step.[8]

In flame photography, a luminescent, continuously changing cloud of gas with a heavily textured surface is depicted. The surface of the flame itself is registered in this process depending on the flame's transparency radiation from its interior. Given these properties in the image subject, image quality is influenced by the following factors:

- Motion-induced blurring: This can be minimized using correspondingly short camera shutter times.
- Variable subject distance: A well focused image is achieved because of the endoscope's high f-stop number and short focal length, provided the flame surface is sufficiently distant from the image-forming lens. The dimensions of the subject are distorted as a result of the short focal length, the expansive flame cloud, and the variable subject distance, however.
- Optically dense diffusion flames (diesel): Here, only thin surface layers contribute to the generation of the image. Because of the high level of absorption in the diffusion flame, the flame image is of low informational value only if the diffusion flame touches the viewing window.
- Optically thin (transparent) flames, premixed flames in gasoline and gas engines: In these cases, the highly textured flame surface and the diffuse underlying layers of combustion charge are imaged. The inner zone and the far boundary zone of the luminescent flame cloud dominate as soon as the flame touches the viewing window.[11]

Criteria for evaluation imaging systems include

- Combustion-chamber windows: Their size must not disrupt operation of the engine.
- Image transmission: Lens angle of acceptance, f-stop number, spectral transmission range.
- Camera characteristics: Local resolution (number of pixels), sensitivity (light yield), spectral sensitivity, signal dynamics, imaging frequency, exposure time, and shutter-induced attenuation.

Flame radiance: The advantage of high local resolution achievable in flame photography because of endoscope and camera characteristics is not exploitable for all problems relevant to engines and in many cases is actually problematical because of the high volumes of data involved. Fixation on a single window site for observation of important propagation processes can also be an excessive restriction. A remedy is provided here by observation methods in which flame propagation is reconstructed from measurement of flame radiance in limited areas of the combustion chamber volume.

This can be accomplished in a relatively simple arrangement by means of "photoelectric barriers," which, installed in the shell of a spark plug, detect the propagation of the flame core[12] or, in the form of a multichannel arrangement distributed throughout the combustion chamber, track flame advance.[13]

Combinations of small front-lens elements or "micro-optic" components and individual fiber-optics conductors produce a large number of potentials for the design of directional and locally delineated registration of flame radiance. The arrangement shown as an example in Fig. 25-10, for instance, registers visible radiance from five narrowly defined conical zones in the combustion chamber. A measuring signal typical of a single-channel system is also shown together with the pressure curve for combustion in Fig. 25-10. In addition to unequivocal localization, high signal quality (sensitivity, signal-to-noise ratio, and signal dynamics), intensity calibration of all measuring channels, and, particularly for evaluation of knocking combustion, high chronological resolution matching pressure-wave spread are definitive for utilization of such signals.

A sensor arrangement installed in the shell of a spark plug is shown in Fig. 25-11. As the installation diagram illustrates, growth of the flame core is tracked. The photographic image (obtained in a glass engine) and the signal are shown for comparison purposes. Since the spark-plug sensor records radiant intensities, intensity calibration of all measuring channels is always an element in the measuring procedure. The user must select intensity thresholds for evaluation of the results obtained. Reliability of results is achieved here by comparison of graduated threshold values.

Flame tomography: Maximum benefit is obtained from multichannel measurement of flame radiance if the geometrical arrangement of the combustion-chamber sections observed can be used for tomographic image reconstruction. This can be achieved using a sensor arrangement that superimposes an optical observation grid on the cross section of the combustion chamber.[14]

The arrangement of a number of observation cones is shown schematically in Fig. 25-12. Local flame intensity can be reconstructed from the measuring signals from all channels of the observation grid in a cylinder-head gasket and knowledge of the individual registration zones. Figure 25-13 shows examples of this from a DI gasoline engine. It is apparent with low swirl that intensive luminescent diffusion combustion occurs in the recessed zone of the piston, resulting in correspondingly high soot emissions. With swirling flow, the central recessed zone is obviously better mixed with air, with the result that no excessive diffusion combustion or soot emission occurs.

Local resolution in flame tomography is determined by grid density and is around 3 to 5 mm in technically practicable systems. This falls short by several orders of magnitude of the high local resolutions achievable using imaging cameras. Because of the distribution of the sensors around the entire circumference of the combustion chamber, however, the combustion-chamber cross section

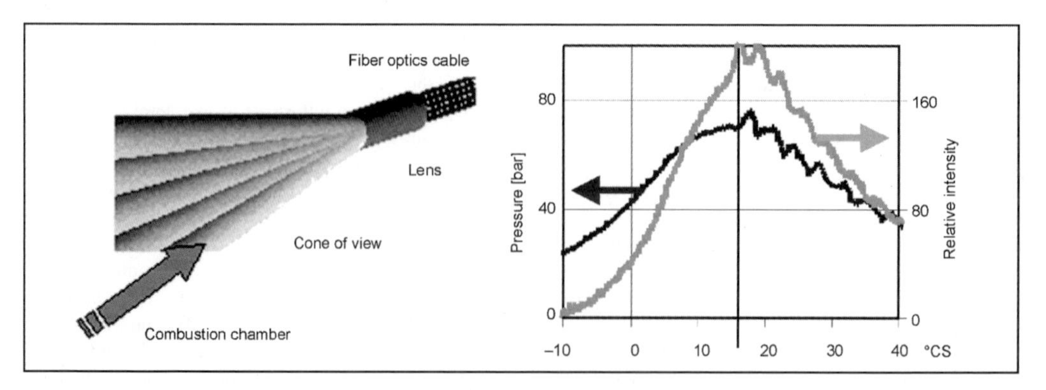

Fig. 25-10 Registration of flame radiance using micro-optic components, radiant intensity of the flame in the registration zone of a viewing cone, and, for comparison, the pressure signal for combustion.

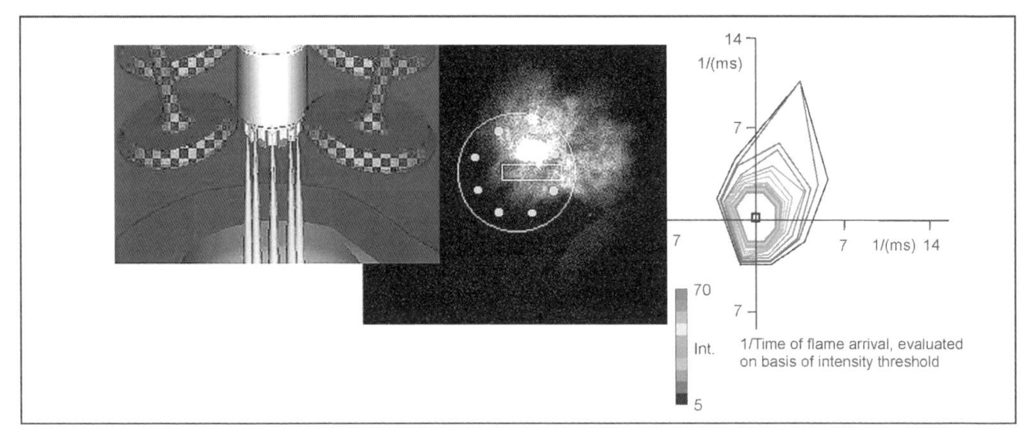

Fig. 25-11 Flame core generation, observation using a spark plug sensor. The result illustrates the symmetry or asymmetry of the flame core and its predominant direction of propagation. (*See color section.*)

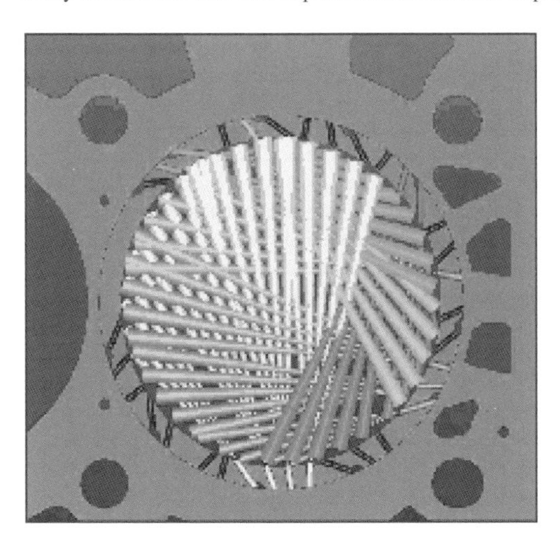

Fig. 25-12 Arrangement of a micro-optic sensor system in the cylinder-head gasket for tomographic flame reconstruction. (*See color section.*)

is registered uniformly and with no image-field distortion, with the result that flame propagation can be plotted clearly and with a homogeneously distributed resolution throughout the cross section.

Alongside depiction of intensity, flame propagation is also reproduced extremely conveniently in the form of progressive flame-front contours after image reconstruction and with the setting of a threshold value. Figure 25-14 shows propagation forms typical of certain flow conditions in modern four-valve engines. Knowledge of these and ascertainment of their dependence on operating conditions and/or the design of engine components can provide decisive indications for improvements. Examples of applications derived from development practice on extremely diverse engines can be found in various publications.[15,16]

The central advantage of measurement of flame radiance and reconstruction of flame propagation can be found in the arrangement of the sensor system specifically for each particular engine and flexibility in signal recording, the chronological resolution of which can be precisely adjusted to the requirements of the specific measuring task. Figure 25-15 illustrates this, using the example of a knocking spot distribution. The inadequate progress of the flame into the left side of the combustion chamber results here to a greater extent in end gas autoignition. Flame propagation, its one-sided retardation on the wall of the combustion chamber, and the resultant autoignition can be routinely registered by flame tomography.

Spark-plug sensor systems: The results obtained in observation of flames using tomographic sensors and model concepts of flame propagation processes also make it possible to study certain problems using simplified measuring methods. Because of the lower level of local resolution, the signal-acquisition systems of such methods must be precisely tailored to the signal patterns of individual combustion phenomena. The spark-plug sensors featuring

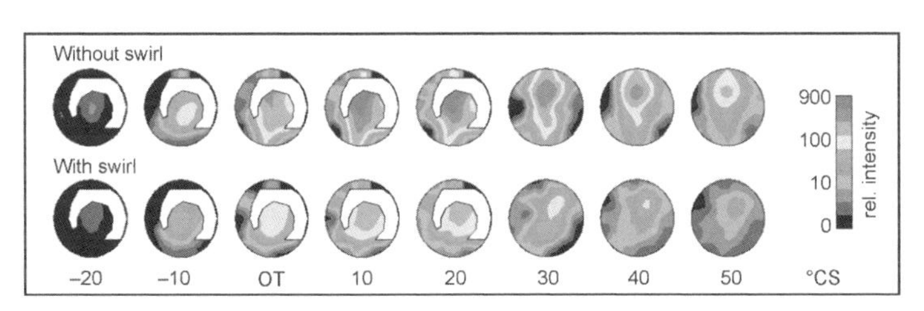

Fig. 25-13 DI gasoline engine: Flame tomography shows the local position of bright, soot-producing diffusion flames. Swirling flow produces a significant improvement. (*See color section.*)

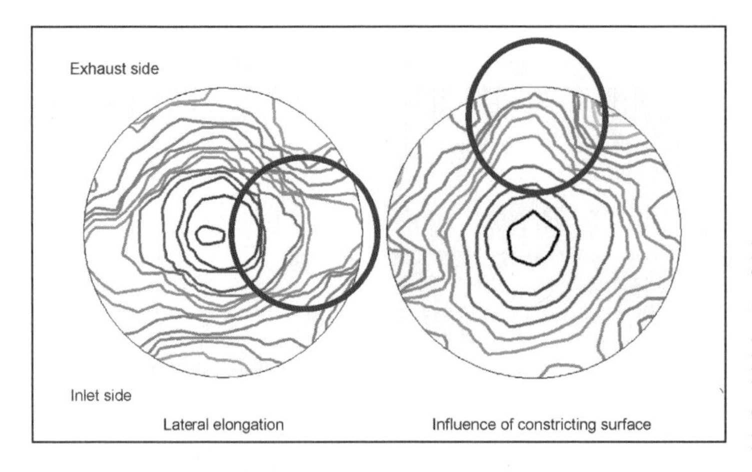

Fig. 25-14 Flame propagation: Tomograph with sensor system installed in the cylinder-head gasket. The isolines indicate the progress of the flame front against time. The influence of internal flow on flame propagation is clearly perceptible. (*See color section.*)

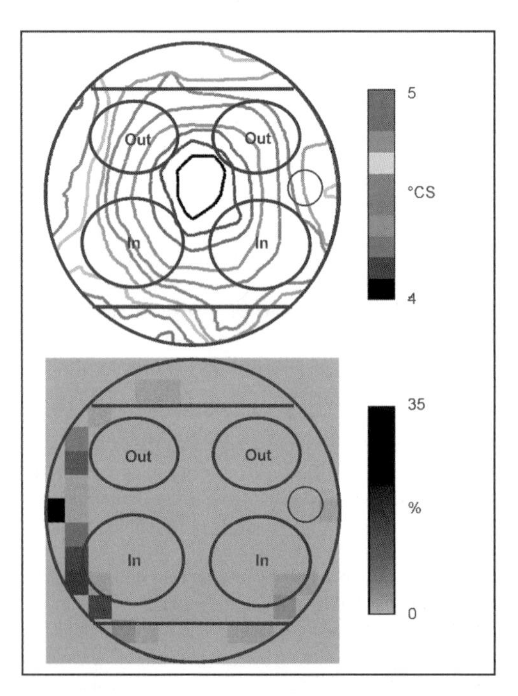

Fig. 25-15 Flame tomography supplies comprehensive documentation of flame propagation and knocking spot distribution. (*See color section.*)

built-in fiber optics already examined above have proven their capabilities for the study of flame core formation.

There are spark plug sensors specifically for observation of autoignition in knocking operation of the engine that register the engine's compression volume using a rotating fan-type sensor. Signal evaluation is matched to the specific propagation characteristics of the pressure and density wave occurring upon autoignition of end gas.[17] Figure 25-16 shows an overview of the sensor principle, signal pattern, localization, and result statistics available to the development engineer as an aid to decision making in the context of component modification.

The criteria for utilization of microsensor systems for flame radiance are

- Spectral sensitivity, signal sensitivity, and signal-to-noise ratio
- Signal dynamics, particularly at high signal amplitudes
- Localization of the individual channels, calibration procedures for multichannel systems
- Signal evaluation and data reduction

25.3.4 Visualization of Illuminated Processes

There is a whole series of questions in combustion development that can be studied only by active illumination of the processes involved. In the simplest case, the subject—a jet of fuel, for example—is diffusely illuminated and depicted using a suitable camera. Illumination and imaging technology make it possible to exploit properties of the subject by determining which velocity fields, fuel distribution, or distribution of specific combustion products can be rendered visible. Visualization of the fuel using the laser-induced fluorescence (LIF) method has become an indispensable aid in the development of direct injection systems, for instance.

The precondition for the use of subject illumination is always the optical access necessary for this purpose, which must be available simultaneously with the optical access for subject imaging. In special cases, a single window into the combustion chamber can be used for both tasks; the necessary flexibility and quality can in many cases be achieved only through separate access ways, however. In an extreme case, the engine is equipped with large windows for this purpose, or glass components are substituted for the piston and the cylinder sleeve (Fig. 25-17, Refs. 18,19]).

Engines like these can be operated under conditions closely approximating reality but within a restricted load and speed range, and provide the precondition for the application of appropriate visualization methods.

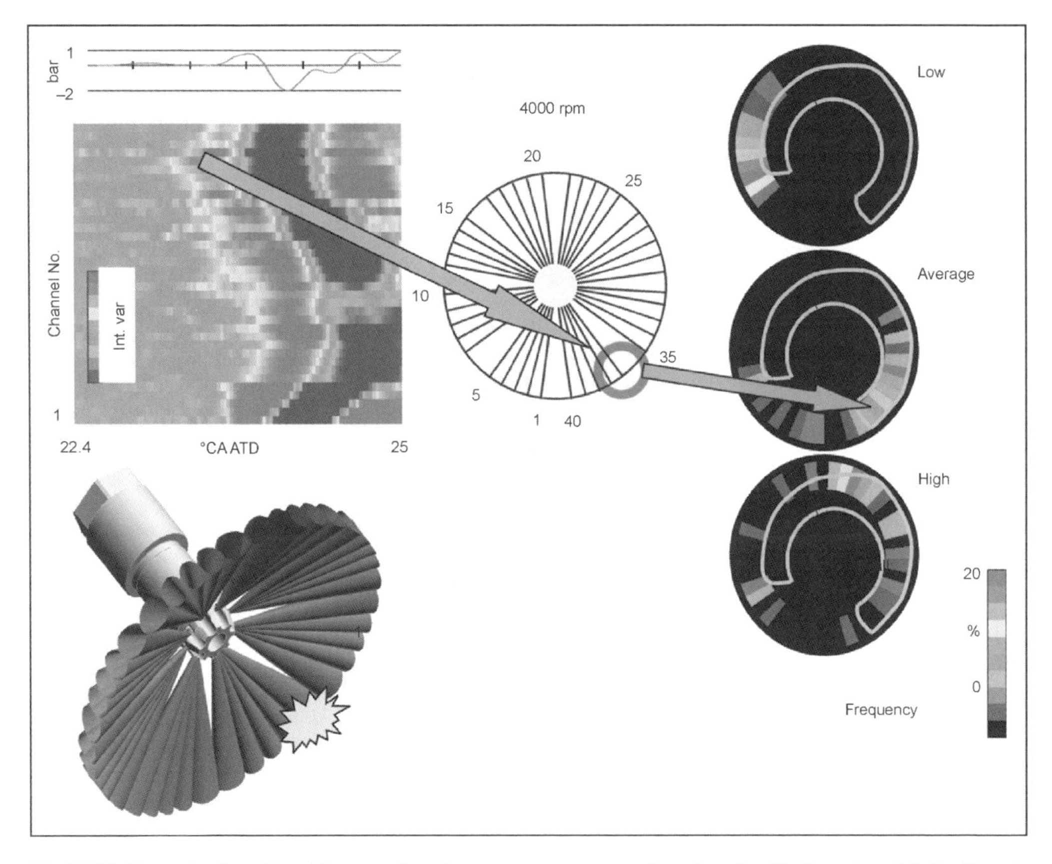

Fig. 25-16 Determination of knocking spots by a fan-type sensor, presentation of results: Single cycle and derived knocking spot statistics. (*See color section.*)

Fig. 25-17 Maximum optical access into the combustion chamber via the use of a glass cylinder and glass windows in the piston. The engine is operated for short periods with ignition for measuring purposes.

25.3.4.1 Visualization of Mixture Distribution

The development of DI gasoline engines, in particular, has accelerated the need for practicable methods for observation of charge stratification. Laser-induced fluorescence methods, in which molecules of fuel or of a tracer substance are made to fluoresce in a planar section of laser light, have proven their value for this. The resultant fluorescence is recorded using suitable cameras and, thus, initially supplies a qualitative image of mixture distribution.

Careful performance of these tests makes it possible to obtain a quantitative assessment of fuel concentration by applying calibration procedures and evaluation of pressure-dependent and temperature-dependent fluorescence yields to the intensity distribution of such images.[20] Significantly less effort and only relatively simple image-analysis methods are needed for the generation from groups of individual images of a probability analysis of a particular distribution recurring reliably at a certain crank angle in every cycle. These distribution statistics on the presence of mixture clouds meet the requirement for practically relevant and informationally useful visualization methods; examples are shown in Fig. 25-18.

25.3.4.2 Visualization of Velocity Fields

Particle Image Velocimetry (PIV): Here, the movement of scattered particles either naturally present in the flow field, such as droplets of fuel, or added to the flow as

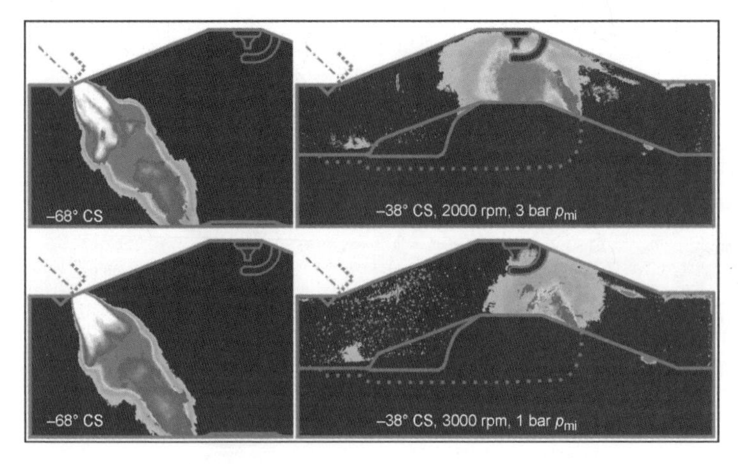

Fig. 25-18 Gasoline direct injection: Fuel distribution in the injection process and after deflection from the pistons. The stability of distribution states is determined from individual images using image statistics. Green/red: Fuel vapor with increasing stability; blue/white: Fuel droplets with increasing stability. *(See color section.)*

tracer particles is captured by double illumination or double exposure. The velocity field in the flow field observed is determined from evaluation of particle travel in the interval between making the two images.

Doppler Global Velocimetry (DGV): This is a method equivalent to PIV for area visualization of velocity fields. Here, too, tracer particles for scattering of the light beamed in are added to the flow field. The Doppler shift generated in the particles during the scattering process is evaluated as the velocity signal by an extremely narrow-band illumination and correspondingly modified spectral filters.[21]

25.3.5 Visualization: The Future

Methods for the visualization of internal engine processes have long been in use in basic research; the results are also used in the verification of computing processes for three-dimensional simulation of flow and of combustion in engines.

Visualization methods have come into more widespread use in engine development only since the development needs of modern combustion processes necessitated comprehensive detail understanding and optimization of processes within engines.

Unlike indicating methods in which measurement of cylinder pressure occupies a central position as a result of the thermodynamic significance of the pressure signal, capture or registration of flame propagation has achieved similar significance in the field of visualization. Here, too, theoretical understanding of optimum combustion provides clear guidelines for flame propagation. Its mensurational observation can then supply the development process with the necessary systems for component optimization.

Unlike the situation in indicating methods, however, visualization methods are only at the very start of their potential applications in engine development. The benefits demonstrated up to now must be consolidated by flexibility in sensor systems and precision in results. The large and diverse range of methods necessitates standardization of central measuring functions and the potential for uncomplicated integration of sensor and measurement technology innovations into open measuring systems. Since evaluation of combustion is performed ultimately on thermal dynamic criteria and on the basis of the results of emissions measurements, linking the results obtained from visualization, indicating, and waste-gas measurements is a central requirement for the development for combustion diagnosis systems.[22]

Bibliography

[1] Pischinger, R., G. Kraßnig, G. Tauçar, and Th. Sams, Thermodynamik der Verbrennungskraftmaschine, Springer, 1989.
[2] Heywood, J.B., Internal Combustion Engine Fundamentals, McGraw-Hill, 1988.
[3] Witt, A., W. Siersch, and Ch. Schwarz, Weiterentwicklung der Druckverlaufsanalyse für moderne Ottomotoren, Der Arbeitsprozess des Verbrennungsmotors, 7th conference, Graz, 1999.
[4] Kuwahara, K., and H. Ando, Time Series Spectroscopic Analysis of the Combustion Process in a Gasoline Direct Injection Engine, 4th Internationales Symposium für Verbrennungsdiagnostik, Baden-Baden, AVL Germany.
[5] Hirsch, A., H. Philipp, E. Winklhofer, and H. Jaeger, "Optical Temperature Measurements in Spark Ignition Engine," 28th EGAS conference, Graz, 1996.
[6] Gstrein, W., Ein Beitrag zur spektroskopischen Flammentemperaturmessung bei Dieselmotoren, Dissertation, Technical University of Graz, 1987.
[7] Hötger, M., Einsatzgebiete der Integralen Lichtleit-Messtechnik, in Motortechnische Zeitschrift (MTZ), 56/5, pp. 278–280, 1995.
[8] Larsson, A., Optical Studies in a DI Diesel Engine, SAE 1999-01-3650.
[9] Chmela, F., and H. Riediger, Analysis Methods for the Effects of Injection Rate Control in Direct Injection Diesel Engines, Thermo-fluidynamic Processes in Diesel Engines, CMT Valencia, 2000.
[10] Winklhofer, E., Optical Access and Diagnostic Techniques for Internal Combustion Engine Development, Journal of Electronic Imaging, Vol. 10 (3), 2001.
[11] Wytrykus, F., and R. Duesterwald, Improving Combustion Process by Using a High Speed UV-Sensitive Camera, SAE 2001-01-0917.
[12] Geiser, F., F. Wytrykus, and U. Spicher, Combustion Control with the Optical Fibre Fitted Production Spark Plug, SAE 980139.
[13] Spicher, U., G. Schmitz, and H.P. Kollmeier, Application of a New Optical Fiber Technique for Flame Propagation Diagnostics in IC Engines, SAE 881647.

[14] Philipp, H., A. Plimon, G. Fernitz, A. Hirsch, G. Fraid, and E. Winklhofer, A Tomographic Camera System for Combustion Diagnostics in SI Engines, SAE 950681.

[15] Liebl, J., J. Poggel, M. Klüting, and S. Missy, Der neue BMW Vierzylinder-Ottomotor mit Valvetronic, in MTZ Motortechnische Zeitschrift MTZ 62 (2001) 7/8.

[16] Grebe, U.D., P. Kapus, and P. Poetscher, The Three Cylinder Ecotec Compact Engine from Opel with Port Deactivation—A Contribution to Reduce the Fleet Average Fuel Consumption, 18th International VDI/VW.

[17] Philipp, H., A. Hirsch, M. Baumgartner, G. Fernitz, Ch. Beidl, W. Piock, and E. Winklhofer, Localisation of Knock Events in Direct Injection Gasoline Engines, SAE 2001-01-1199.

[18] Winklhofer, E., H. Fuchs, and G.K. Fraidl, Optical Research Engines—Tools in Gasoline Engine Development?, Proc. Instn. Mech. Engrs., Vol. 209, pp. 281–287, 1995.

[19] Gärtner, U., H. Oberacker, and G. König, Analyse der Brennverläufe moderner Nfz Motoren durch Hochdruckindizierung und Verbrennungsfilmtechnik, 3. Internationales Indiziersymposium, AVL Deutschland, 1998.

[20] Ipp, W., J. Egermann, I. Schmitz, V. Wagner, and A. Leipertz, Quantitative Bestimmung des Luftverhältnisses in einem optisch zugänglichen Motor mit Benzindirekteinspritzung, Motorische Verbrennung, BEV Edition 2001.1, pp. 157–172, A. Leipertz [Ed.], Erlangen, 2001.

[21] Willert, C., I. Röhle, M. Beversdorff, E. Blümcke, and R. Schodl, Flächenhafte Strömungsgeschwindigkeitsmessung in Motorkomponenten mit der Doppler Global Velocimetrie, Optisches Indizieren, Haus der Technik, Essen, Event No. H030-09-033-0, September 2000.

[22] Winklhofer, E., Ch. Beidl, and G.K. Fraidl, Prüfstandsystem für Indizieren und Visualisieren–Methodik, Ergebnisbeispiele und Ergebnisnutzen, 4. Internationales Indiziersymposium, AVL Deutschland, 2000.

26 Fuel Consumption

Reduction of fuel consumption and exhaust emissions has become one of the central tasks of vehicle development in recent years. The reasons for this can be found not only in legal requirements but also in the increased awareness in the handling of fossil energy source reserves and enhanced environmental awareness on the part of both customers and vehicle manufacturers. Continually increasing fuel prices, of course, also play a significant role.

Despite continually rising vehicle weights, it has proven possible to reduce consumption significantly in recent years; see Fig. 26-1.

Also directly dependent on fuel consumption are CO_2 emissions, a field in which the ACEA (Association of European Automobile Manufacturers) has submitted to the EU an undertaking to reduce values from the current average of 180 to 140 g/km (≅ 5.7 l/100 km fuel consumption) by the year 2008. This equates to a reduction in emissions and, therefore, in consumption, of nearly 23% (weighted for the vehicle fleet sold).

The present-day EU target is the achievement of CO_2 emissions of 120 g/km (≅ 4.9 l/100 km fuel consumption) by 2012.

26.1 General Influencing Factors

A certain quantity of energy, in the form of fuel, is required in order to overcome the various resistances to movement. The potentials for reduction of fuel consumption take the form of improvement of the efficiency of the power source and power transmission train and reduction of the vehicle's various resistances.

26.1.1 Air Resistance

Air resistance increases by the square of the resulting approach flow velocity, i.e., with longitudinal approach flow from ahead, at the square of travel speed.

Air resistance:

$$F_L = c_w \cdot A \cdot \frac{r_L \cdot v^2}{2} \qquad (26.1)$$

where

c_w = Coefficient of air resistance
ρ = Air density
v = Travel speed
A = Bulkhead surface area

The c_w value, as a shape factor, and bulkhead surface area A, as a size factor, are susceptible to design manipulations. The bulkhead surface area can be reduced only to a limited extent, since a certain size must be achieved for the passenger cell, and all the various equipment assemblies and modules must be accommodated. Figure 26-2 shows the trend in c_w values since 1900.

Limits are set on reduction of the c_w value as a result of design trends, vehicle complexity, the necessary flow through the engine compartment and interior, freedom of wheel movement, antilift provisions on both axles, flow through the wheel housings for cooling of the braking systems, underbody flow for cooling of the exhaust system, and necessary attachments (such as mirrors, windshield wipers, antennas, and handles). Figure 26-3 shows the influence of the c_w value on maximum speed and fuel consumption.

26.1.2 Weight

Vehicle weight becomes a factor in acceleration and uphill travel. Vehicle weight in this context plays a linear role in resistance.

Climbing resistance:

$$F_\alpha = m \cdot g \cdot \sin \alpha \qquad (26.2)$$

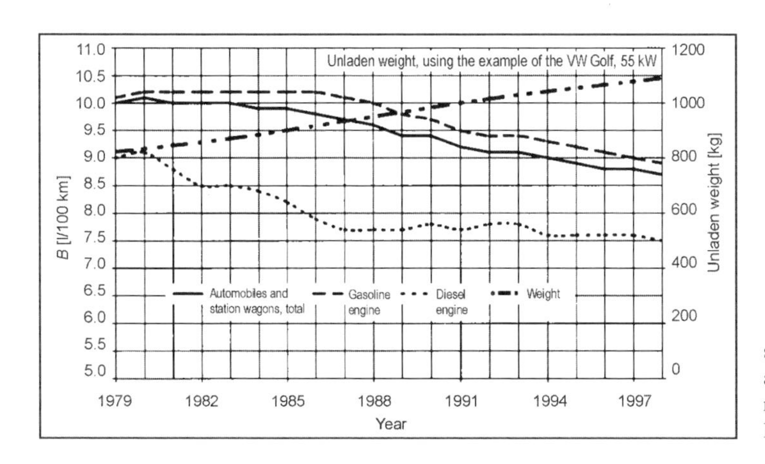

Fig. 26-1 The trend in fuel consumption and vehicle weight for automobiles and station wagons registered in Germany. Data from Ref. [1], augmented.

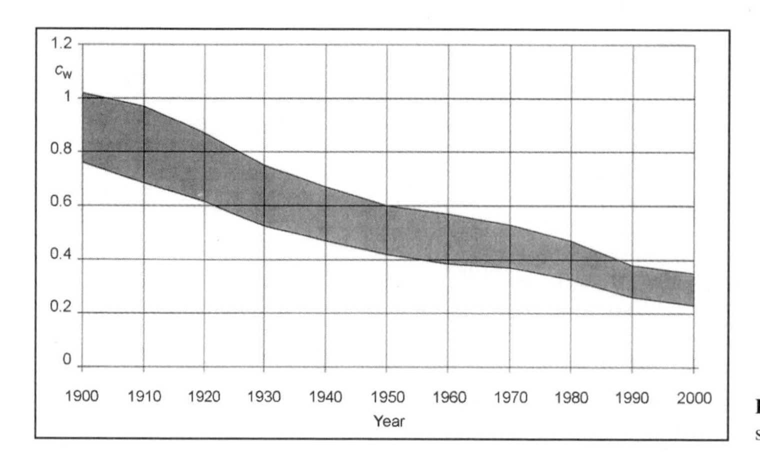

Fig. 26-2 Trend in c_w values since 1900.

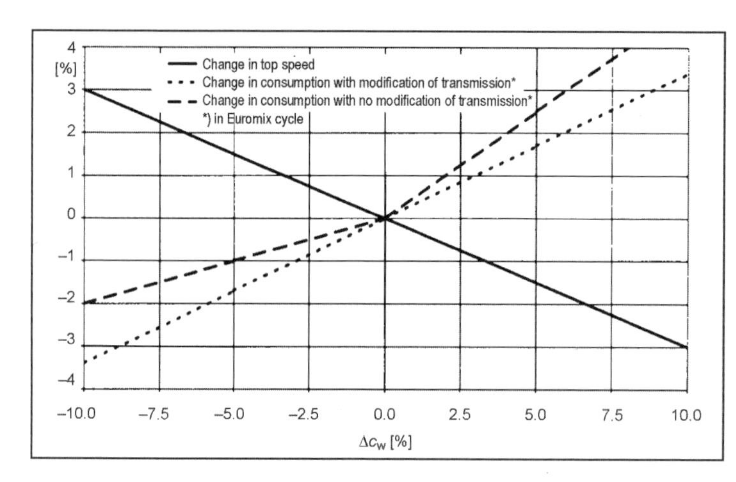

Fig. 26-3 The influence of c_w values on maximum speed and consumption. Data from Ref. [2].

Acceleration resistance:

$$F_a = e_i \cdot m \cdot a \tag{26.3}$$

Rotating mass factor:

$$e_i = \frac{\theta_{\text{Red}_i}}{m \cdot R_{\text{dyn}}^2} + 1 \tag{26.4}$$

where

g = 9.81 m/s²
m = Vehicle mass, including load
α = Road slope angle
θ_{Red} = Reduced rotating mass moment of inertia
R_{dyn} = Dynamic tire radius
i = Transmission ratio of observation

Average vehicle weight continues to rise at present times. The reasons for this can be found in increasing demands for convenience in the form of electric actuators for windows, sliding roofs, mirrors, seats, and in higher equipping specifications, air-conditioning, seat heating, and servo-steering systems. Additionally, the safety equipment developed in the past 20 years such as traction and brake control systems, electronic stability systems, active shock absorbers and transverse stabilizers, airbags, and seat-belt tensioners have also increased vehicle weight. The trend toward larger engines with the associated power transmission train components has a weight-increasing effect, as does the increasing selection of diesel engines. Further increases in weight are the result of enhanced crash safety and corresponding additional automobile body structures. The trend in the weight of various vehicle categories in recent years is shown in Fig. 26-4.

The increase in weight means that more powerful (and heavier) engines are necessary to achieve the same performance; weight then begins to escalate. A number of concepts exist to combat this weight spiral. A large portion of total body shell and suspension system weight can be reduced by intelligent lightweight structures and by the replacement of steel with lighter metals and alloys; aluminum and magnesium alloys as well as some plastics can be used as substitutes. Consistent vehicle conception should also use lower power and supercharged engines to reduce the weight of the entire power train and result in shifts in working points to the higher mean-pressure ranges. At

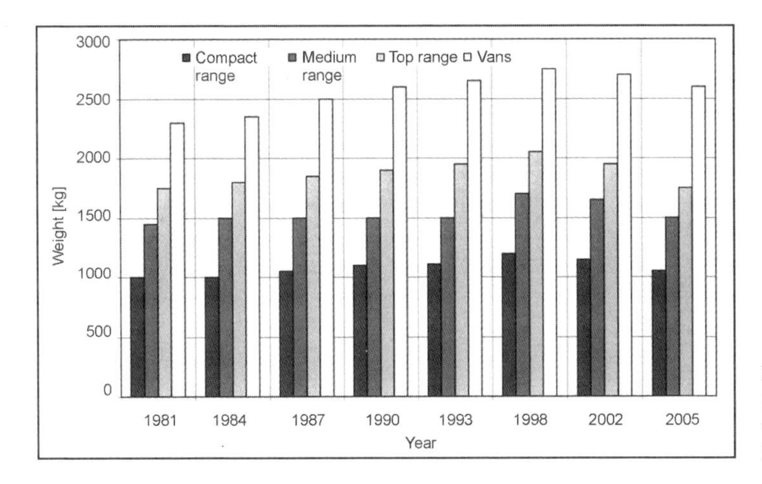

Fig. 26-4 Trend in weight since 1981 and forecast weights for various categories of vehicles. Data from Ref. [3].

present, however, this is the case only in the subcompact class of vehicles designed specifically for economy.

Every 100 kg of weight-saving reduces fuel consumption by around 0.2 l/100 km (Euromix).[4]

26.1.3 Wheel Resistance

Wheel resistance is made up of components resulting from the deformation of the tires upon contact with the road surface from bearing friction losses, from water-displacement resistances caused by displacement of water on wet road surfaces, and from toe-in resistances and lateral force resistances. The largest portion is made up of rolling resistance, which can be derived approximately from vehicle weight and the coefficient of rolling resistance, which summarizes the resistances generated at the tire.

Rolling resistance:

$$F_R = f_R \cdot F_G \tag{26.5}$$

where

f_R = Coefficient of rolling resistance
F_G = Vehicle weight

The coefficient of rolling resistance varies between orders of magnitude of 0.01 and 0.04, depending on tire type, road surface, and vehicle speed. Concepts for reduction of rolling resistance can be found in the subcompact cars mentioned above. It is not possible to completely exploit the consumption-reducing potential of low-rolling resistance tires since a drastic reduction in rolling resistance also incurs losses of drive smoothness, adhesion, and tire grip in wet conditions. It is necessary not only to select a different tire material mixture, but also to reduce tire width and increase tire pressure.

An increase in tire pressure also reduces the tire's flexing energy. The coefficient of rolling resistance of an automobile tire, for example, can be reduced by around 25% if tire pressure is increased by 0.5 bar.

The coefficient of rolling resistance increases with vehicle speed.

Coefficient of rolling resistance:

$$f_R = C_0 + C_1 \cdot v + C_2 \cdot v^4 \tag{26.6}$$

where

C_0, C_1, C_2 = Tire-specific constants
v = Vehicle speed

Tire rolling resistance and, therefore, consumption can be influenced via the particular rubber mixture selected. The greatest influence can be achieved in this context by varying the material of the running surface. Low absorption mixtures make it possible to reduce rolling resistance by up to 35%. The improvements achievable by means of variation of materials are only in the 1% to 5% range in the case of other tire elements, such as the sidewalls and bead.

26.1.4 Fuel Consumption

The various factors produce in total the following influence on distance-related fuel consumption ("mpg"):

$$B_e = \frac{\int b_e \cdot \frac{1}{\eta_{ü}} \cdot \left[\begin{array}{c} \left(m \cdot f_R \cdot g \cdot \cos\alpha + \frac{\rho}{2} \cdot c_w \cdot A \cdot v^2 \right) \\ + m \cdot (e_i \cdot a + g \cdot \sin\alpha) \end{array} \right] \cdot v \cdot dt}{\int v \cdot d} \tag{26.7}$$

The variables and units in the preceding equation are explained in Fig. 26-5.

When the vehicle is stationary, some other variables are idling consumption (around 0.5 to 1 l/h) and, either continuously or intermittently, the power consumption of electrical loads that may extend into the kW range of examples that can be found in Fig. 26-6.

Variable		Unit	Variable		Unit
B_e	Distance-based consumption	g/km	m	Vehicle mass	kg
b_e	Specific consumption	g/kWh	f_R	Coefficient of rolling resistance	—
$\eta_{ü}$	Efficiency Power train	—	g	Gravitational acceleration	m/s^2
α	Angle of road surface inclination	°	v	Vehicle speed	m/s
ρ	Air density	kg/m^3	e_i	Rotating mass correction factor in gear i	—
c_W	Coefficient of air resistance	—	a	Longitudinal acceleration	m/s^2
A	Bulkhead surface area	m^2	t	Time	s

Fig. 26-5 Variables and units.

Load	Power consumption	Load	Power consumption
Rear windshield heating	0.4 kW	Instrument panel	0.15 kW
Front windshield heating	0.7 kW	Stereo system	0.2 kW
Wiper motor	0.1 kW	Onboard computer	0.15 kW
Exterior lights	0.16 kW	Ventilation fan	0.1 kW
Control unit supply	0.2 kW	ABS/FDR pumps	0.6 kW
Fuel pump	0.15 kW	Total	2.91 kW

Fig. 26-6 Power consumption of electrical loads in automobiles.

The crankshaft-mounted combined starter-motor/alternator, the consumption potential of which is shown in Fig. 26-7, is an alternative to the three-phase generators currently widely used.

26.2 Engine Modifications

Characteristics data for modern gasoline and diesel engines are shown in Fig. 26-8. Specific fuel consumption in diesel engines is currently significantly better than that of gasoline engines.

Diesel engines generally have a slightly higher weight to power ratio than gasoline engines. Because of their various advantages and disadvantages, both processes continue to use those applications for which they are most suited. The modern diesel engine is superior in terms of fuel consumption to the gasoline engine, but the gasoline engine possesses a greater potential than the diesel engine. This is being exploited by fully variable valve timing systems and direct injection, downsizing, crankshaft-mounted combined starter-motor/alternators, reduction of friction losses, variable compression ratios, variable swept volumes

Function/Characteristic	Total potential savings
Start/Stop (ECE cycle)	15% to 25%
Efficiency boost/42 V vehicle system	
Braking-energy recovery	
Booster operation	

Fig. 26-7 Potential consumption reductions achievable with crankshaft-mounted combined starter-motor/alternator.[5]

Engine type		Max. speed (rpm)	Max. compression ratio ε	Max. mean pressure (bar)	Per liter output (kW/l)	Optimum fuel consumption point (g/kWh)
Gasoline engine, automobile	Naturally aspirated engine	5500 to 8500	Up to about 12	Up to about 13	80	Min. 225
	Super-charged	Up to 6800	Up to about 10	Up to about 19	Up to about 110	Min. 225
Motorcycle gasoline engine		15 000	Up to about 12	Up to about 12	50 to 110	—
Diesel automobile (direct injection)	Naturally aspirated engine	Up to 5000	Up to about 20	Up to 9	Up to about 35	About 210
	Super-charged	3500 to 4500	17 to 21	Up to about 25	Up to about 64	About 205

Fig. 26-8 Overview of characteristics data for gasoline and diesel engines.

(including cylinder shutoff and supercharging), and is estimated at 30% to 40% consumption reductions. Consumption figures similar to those for supercharged direct injection diesel engines are then achievable. The consumption potential for the diesel engine is estimated at 15% to 25%. The successful provisions here are, again, crankshaft-mounted combined starter-motor/alternators, minimization of friction losses, and improvement of mixture generation.

26.2.1 Downsizing

Increasing mean pressure makes it possible to increase effective output for the same engine capacity. With a smaller capacity, the same performance data as for a larger engine can be achieved. Smaller engines have lower absolute friction losses; because of the shift in the working point and the higher mean pressure, the same performance profile is achieved in sectors of better thermal efficiency. In diesel engines, increasing injection pressure increases mean pressure. According to Ref. [6], an increase in injection pressure from 600 to 1000 bar produces an increase in mean pressure of 17% with the same specific consumption. Modern injection systems already reach pressures of above 2000 bar. The mean-pressure increase thus possible is used mainly for obtainment of greater power from the same engine capacity.

In gasoline engines, thermal efficiency can be improved by raising the compression ratio. Figure 26-9 shows the resultant improvement in fuel consumption.

In order to avoid knocking, the compression ratio at full load is restricted to $\varepsilon \approx 11$. Considerably higher ratios, of up to $\varepsilon = 15$, are possible under part-load

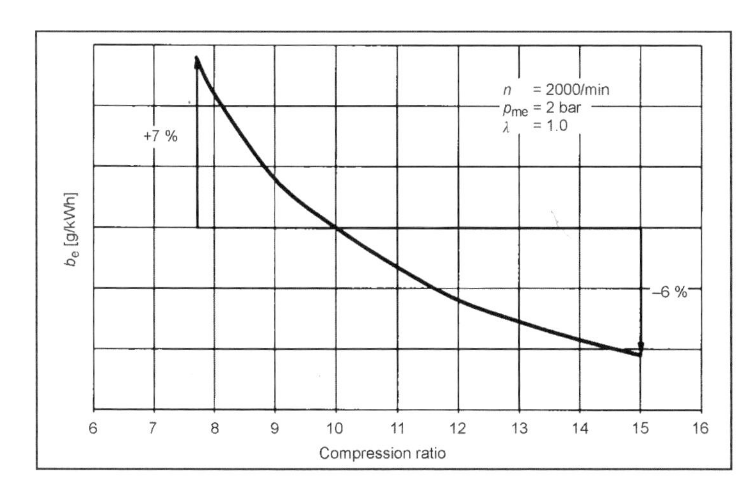

Fig. 26-9 The influence of compression ratio upon specific fuel consumption.[7]

operation. A variation of compression ratio is desirable for the purpose of optimization. This even makes it possible to avoid the efficiency loss in supercharge engines in the partial-load range, if the compression ratio is adjusted here. The high supercharging that is now possible produces a further improvement in thermal efficiency.

In summary, downsizing signifies the relocation of frequently traversed working points to ranges with lower specific fuel consumption. Since this range is in the high load region, the engine must be designed in such a way that it is operated under this high load for the major part of the load block used by the customer. The lower maximum power available has encountered little acceptance from customers despite the fact that the power reserves of a larger engine are used rarely in actual everyday driving.

26.2.2 Diesel Engine

The three diesel combustion systems, the prechamber, swirl chamber, and direct injection processes, were used alongside one another up to the late 1980s. The introduction of direct injection in automobile diesel engines in the late 1980s has resulted in increasing displacement of prechamber and swirl chamber engines. Virtually all of the new automobile diesel engines utilize direct injection.

The 15% to 20% lower specific fuel consumption of the direct injection diesel engine compared to indirect injection engines is primarily the result of the lower heat losses resulting from the undivided combustion chamber that is located in the piston head, and of the lower losses because of the obviation of flow between the subsidiary and the main combustion chamber. Indirect injection engines achieve an effective efficiency of around 36% compared to 43% in the case of direct injection. Disadvantages are the steep pressure rise (noise) and higher exhaust emissions compared to indirect injection engines. Softer, more homogeneous combustion is achieved by preinjection of small quantities of fuel and pulsed injection. The precondition is an injection system incorporating high-speed, largely free actuation of the injection nozzle, and complex tuning of the high-pressure fuel supply system up to the nozzle. This is achieved using common-rail, pump-nozzle systems and solenoid-valve actuated distributor pumps. The first two systems make use of the pressurized storage system for the fuel. The quantity required for injection is obtained from the pressure accumulator. In the case of the pump-nozzle system, storage is restricted to a crank angle window generated by the cam contour; the common rail system provides a continuous high system pressure. Common rail functions at peak pressures of up to 2000 bar. The high injection pressure compared to earlier systems provides the following potentials and effects:

- Smaller aperture diameter
- Higher jet velocity, greater penetration depth
- Earlier start of mixture generation because of greater reactive surface area and improved distribution
- Smaller average droplet diameters, more surface area for reaction
- More intensive mixture generation
- More rapid evaporation
- More rapid mixture distribution
- Higher conversion rates
- Higher degree of homogenization of the mixture "cloud"
- Shorter combustion duration
- Improved (internal) oxidation of soot, smaller particles

As a result of these advantages, the trend is now in favor of higher injection pressures. A compromise is necessary between the consumption advantages resulting from high injection pressures and the increased consumption deriving from the power necessary to drive a high-pressure pump or the camshaft and pump-nozzle elements. The common rail system possesses the lowest maximum drive power requirement. It needs only 40% to 50% of the drive power of a distributor pump and only 20% of the drive power of a pump-nozzle system.

The improved facilities for intervention to control mixture generation and reaction rate in the combustion chamber make it possible to achieve for the same specific consumption, using variable high-pressure injection systems, an increase in torque that more than balances out the losses resulting from the additional drive torque required.

Preinjection systems and injection rate adjustments can also be achieved by variable nozzle geometries, which are not yet ready for series production.

An electronic engine management system (digital diesel electronics), which controls the injection rate and time on the basis of the working point, is necessary to permit exploitation of the potential of high-pressure injection systems. Inlet duct geometry must also be tailored to the individual injection system and nozzle geometry in terms of defined flow, and, therefore, mixture-generation processes in the combustion chamber.

26.2.3 Gasoline Engine

26.2.3.1 The Lean-Burn Engine Concept and Direct Injection

The significantly higher specific fuel consumption of the gasoline engine in the part-load operating range compared to full load can be reduced by operating with an air surplus, i.e., so-called "lean-burn operation." The reasons for this are the following:

- Partial dethrottling as a result of higher air demand at the same mean pressure, and reduction of charging sequence energy
- Enlargement of thermal efficiency as a result of increase of the isentropic exponent
- Reduction of wall heat losses as a result of lower mixture densities in the wall zone

Limits are imposed on mixture leanness by the following:

- Ignition limit (lean-burn limit)
- Incomplete combustion as a result of locally differing mixture compositions

- Cycle fluctuations as a result of "wandering" of the combustion center, misfires
- Slowing of combustion rate

Thermal efficiency is

$$\eta_{\text{th}} = 1 - e^{1-k} \tag{26.8}$$

where
ε = Compression ratio
κ = Isentropic exponent

In lean-burn operation, overall charge mass increases as a result of the air surplus, with rising final compression pressures and temperatures as the consequence. The quantity of heat liberated is transferred into a greater charge, however, thus causing the average process temperature to fall. Both effects, i.e., the larger charge mass and the higher temperature spread, result in an increase in isentropic exponent κ, causing thermal efficiency to rise.

The reduced charging energy makes it possible to increase effective overall efficiency by up to 4%, depending on the working point and degree of leanness.[7]

The fuel-saving potential in the part-load range of the experimental engines constructed in the 1980s (normal FTP and ECE cycles) was up to 15%. These lean-burn concepts were not pursued further, however, due to the more stringent exhaust emissions legislation.

Catalytic aftertreatment causes no problems for lean-burn operation in the cases of hydrocarbon and carbon monoxide emissions.

The potentials of lean-burn operation can be better exploited in combination with gasoline direct injection. The advantages are as follows:

- Replacement of quantity regulation by quality regulation, reduction or elimination of throttling losses
- Charge stratification via corresponding injection-jet location and injection rate combined with reduced air flow
- Improved dynamics at load changes, delays caused by intake manifold filling, and creation of the wall fuel film are eliminated
- Internal cooling of the inducted air by evaporation of fuel in the cylinder, thus making higher compression ratios and, as a consequence, higher thermal efficiencies possible (shifting of the knocking limit)
- Adjustment of air ratio to various working points; see Fig. 26-10

Global statements concerning reduction of fuel consumption as a result of gasoline direct injection have little rationale, since they depend on the differing variants and working points. Significant consumption potentials exist in part-load operation, since engines can be run with high average air ratios in stratified charge operation. A reduction of 10% to 15% in consumption can be assumed in the test cycles currently widely used in Europe, the United States, and Japan. Significantly lower consumption benefits are achieved in full-load operation, as a result of stoichiometric operation.

A certain portion of consumption reduction is negated again by the regeneration phases necessary in part-load operation for reduction of oxides of nitrogen. For this purpose, the engine is operated for a short time (several seconds) at substoichiometric air ratios, in order to desorb the oxides of nitrogen fixed in the storage-type catalytic converter during lean-burn operation. During this phase, the catalytic converter operates like a conventional three-way catalytic converter.

26.2.3.2 Variable Valve Timing

Variable valve timing systems are a further method of influencing charging and exhaust recycling, and thus cutting consumption and pollutant emissions.

Matching of inlet valve lift and opening duration with the necessary flow of fresh gas makes it possible to achieve a partial dethrottling in the part-load range. Running is improved as a result of lower valve lifts in the idling and near-idling ranges. The reasons for this can be found in the improved mixing resulting from higher gas velocities (up to sound velocity) in the narrow valve gap, and the resultant more homogeneous and faster combustion. The reduction of idling speed thus possible offers an additional

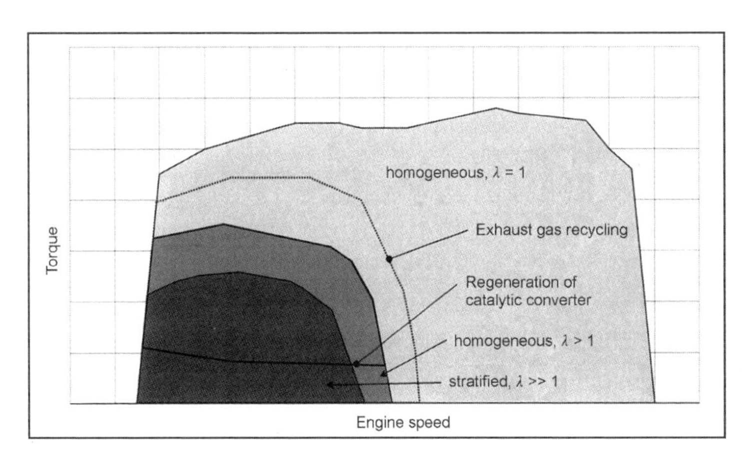

Fig. 26-10 Strategies in direct gasoline injection.

consumption-savings potential, because of the lower frictional losses.

There are two possibilities for adjusted opening duration: Early closure of the inlet when the necessary fresh-gas charge is reached, and delayed closure of the inlet; here, the inlet valve is open throughout the induction stroke and closes only during the compression stroke, when the unneeded mass of charge is transferred back into the intake system. Here, losses occur, compared to early closing of the inlet, as a result of the double movement of part of the charge mass; see Fig. 26-11.

In addition, variable valve timing and valve lifts make internal exhaust-gas recycling via shifting and prolongation of valve overlap possible. The advantages can be found in dilution and improved mixing of the fresh charge. Improved mixing produces the effects mentioned above. Dilution significantly reduces emissions of oxides of nitrogen, since the overall charge mass consists partially of inert exhaust gas. Combustion gas temperatures are thus lowered; less energy is available for the formation of oxides of nitrogen. The consumption benefits are to some extent negated by the lower combustion gas temperatures, resulting in slower and less homogeneous combustion.

Cylinder shutoff can be achieved via complete closure of the inlet valves of individual cylinders for a number of cycles; please see the corresponding section.

The greatest potentials for the future are possessed by gasoline engines in which the advantageous effects of variable valve timing and direct injection are combined with the stratified-charge concept. The consumption reduction potentials of a number of concepts are shown compared to the present series-production situation with camshaft spread in Fig. 26-12.

26.2.3.3 Ignition

Rapid and homogeneous ignition and high reaction rates are necessary for the achievement of a high mean pressure and a good thermal efficiency. Ignition depends, inter alia, on spark plug position and on the quality of spark transmission. A central spark plug location is generally desirable in order to achieve uniform combustion and short flame paths; see Fig. 26-13. Central spark plug location is relatively easily achieved with the use of quadruple-valve technology cylinder heads. Improved charge gas exchange conditions can thus be combined with the optimum spark plug position.

The combustion process in a real gasoline engine differs from the constant volume combustion assumed in the ideal engine. Observation of energy conversion against crank angle during the compression and power strokes generates a surface that expresses the progress of combustion. The concentration point of this area should be

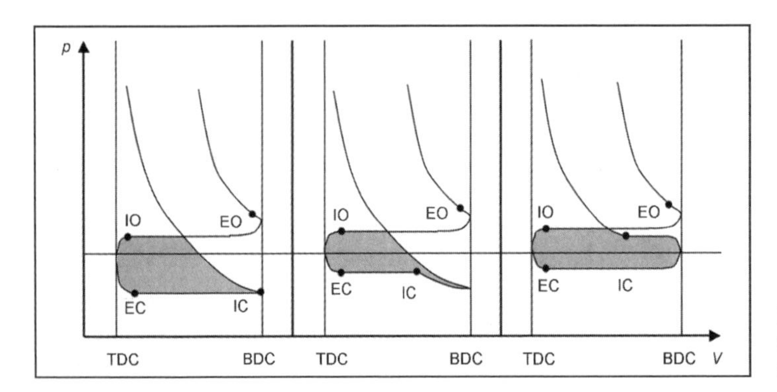

Fig. 26-11 Charging loops: conventional, early inlet closure, and delayed inlet closure.

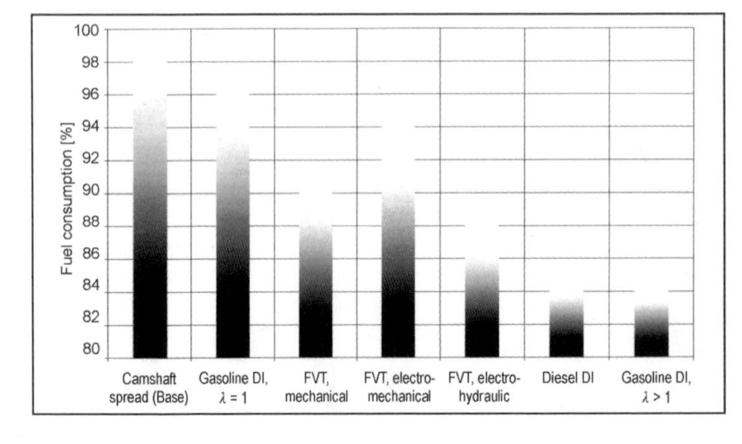

Fig. 26-12 Comparison of consumption-reduction potential in various gasoline engine concepts and in direct injection diesel engines, data from Ref. [9], augmented.

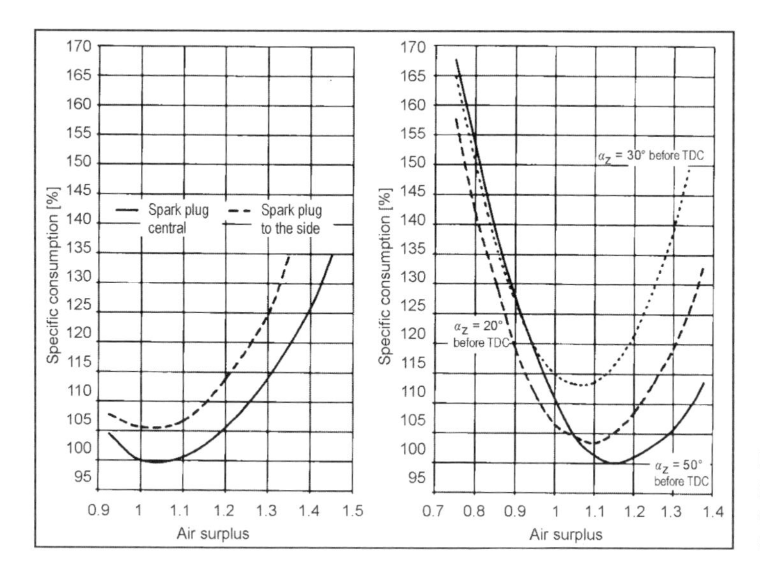

Fig. 26-13 The influence of spark plug position (at left), pinking, and air-fuel ratio upon fuel consumption, from Ref. [10].

located close to top dead center, to achieve good internal efficiencies.

Combustion progress and, therefore, the location of the combustion center, depend on ignition angle α_Z and air-fuel ratio λ; see Fig. 26-13.

Energy-optimized early ignition produces low specific consumption; compromises involving a later ignition angle are frequently necessary, however, in order to avoid knocking and to reduce exhaust emissions. With homogeneous operating strategies, minimum consumption occurs at slightly lean mixture ratios.

Ignition of the mixture can be further improved via the use of two spark plugs. More ignition energy is supplied, with the result that more homogeneous combustion and, therefore, lower cycle fluctuations are achieved even at critical working points, for instance, when idling and during exhaust-gas recycling operation. More rapid combustion is also achieved, improving thermal efficiency. Optimum combustion center positioning can be achieved with greater certainty. Series operation achieves reductions of about 2% in consumption compared to single plug ignition systems.[11] In addition, phase-displaced double ignition makes it possible to control pressure rise and plot of pressure in such a way that combustion noise can be reduced without delayed ignition and corresponding loss of efficiency. A reduction of 3 dB (*A*) has been achieved in this field in series operation.[11]

26.2.4 Cylinder Shutoff

Examination of the fuel-consumption characteristics field of a gasoline engine (Fig. 26-17 below) indicates that specific fuel consumption at low engine torques and low mean pressures, in particular, may be more than twice as high as at and around the optimum, high mean pressure point. This consumption sacrifice at low mean pressures is the result of, inter alia,

- A compression ratio selected for full load operation
- Low flow velocities at the inlet valve
- High throttling losses in gasoline engines as a result of the practically closed throttle butterfly
- Relatively high friction losses compared to the engine output
- High wall heat losses

In large automobile engines with a high power and torque output, only a small fraction of the engine's available power is required in in-town driving and on normal roads. The more power and torque an engine offers, the lower its engine working point drops into the part-load range. The result is high fuel consumption.

26.2.4.1 Concept for Reduction of Fuel Consumption

The basic concept of cylinder shutoff is that of increasing the torque of individual cylinders in the part-load range in order to achieve a working point with a better fuel consumption for these cylinders. Other cylinders are shut off for compensation. Eight- and twelve-cylinder engines are particularly suitable for the cylinder shutoff concept. In these engines, half the cylinders can be shut off in each case at low load and speed levels. The cylinders to be shut off are indicated by the engine's firing order on the basis of the rule of maintaining a constant firing order even in shutoff operation. In eight-cylinder V engines, an obvious tactic is to shut off two cylinders in each bank, whereas a complete bank can be shut down in a twelve-cylinder engine.

Comprehensive control algorithms and pilot control systems for throttle butterfly and ignition are needed to eliminate ride smoothness losses at gearshifts.

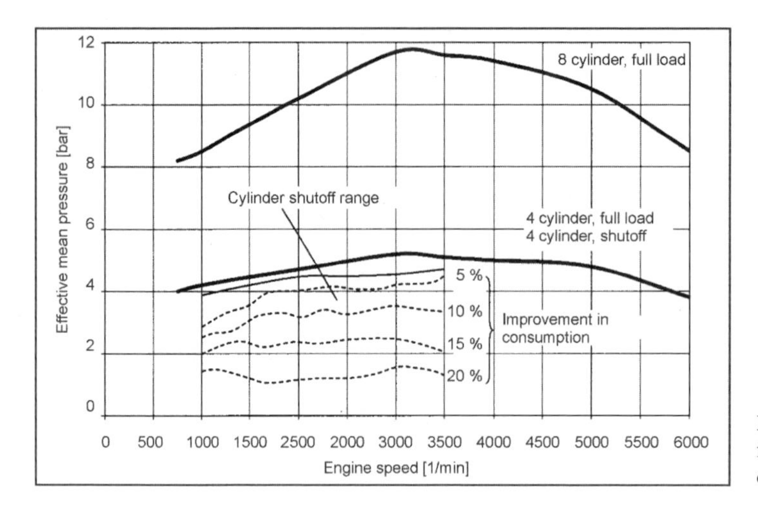

Fig. 26-14 Consumption benefits from cylinder shutoff in an eight-cylinder engine (from Ref. [12]).

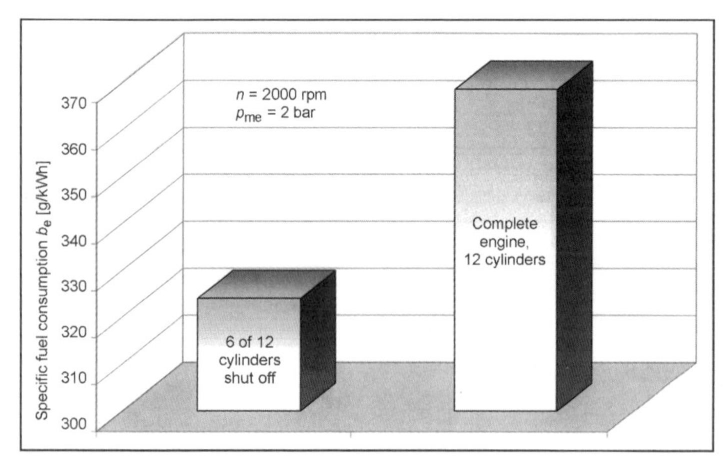

Fig. 26-15 Consumption benefit from cylinder shutoff in a twelve-cylinder engine (from Ref. [7]).

26.2.4.2 Consumption Benefits in the Part-Load Range

Figure 26-14 shows the plot of mean pressure for the eight-cylinder engine (operating on all cylinders) and for shut-off operation, as well as the cylinder shutoff range. Depending on the working point, the consumption benefits in cylinder shutoff are between 5% and 20%.

Figure 26-15 shows the consumption benefit for a twelve-cylinder engine at a part-load point with cylinder shutoff.

Improvements in consumption of 15% and 13% were determined for a vehicle with an eight-cylinder engine (W220 5.0 l) for constant travel at 90 and 120 km/h, respectively, and fuel consumption in the NEDC (New European Driving Cycle) was reduced by 6.5% as a result of cylinder shutoff.[12,13]

26.3 Transmission Ratios

Transmission ratios of the power train are determined in the multiratio gearbox, where the individual gear ratios are selected either by the driver (manually), or by an automatic system. In other systems, the gear ratio refers to the gearing in the differential gear (compensating gear), which features a fixed transmission ratio. The following describes the gear ratio from the transmission (from the engine) to the wheel:

- Transmission ratio in the selected gear i_{Gi}
- Differential gear ratio i_D

Overall power train transmission ratio in selected gear, i_A, is therefore

$$i_A = i_{Gi} \cdot i_D \qquad (26.9)$$

26.3.1 Selection of Direct Transmission

The engine's torque is converted via the multiratio transmission with its individual gear ratios and the transmission ratio of the downstream differential gear in accordance with the demand for tractive effort at the drive wheels. In engines installed longitudinally and featuring spur gear transmission, it is possible to assign a direct ratio to one of the gears (transmission ratio = 1:1). This achieves higher transmission efficiency, since no pair of gearwheels are engaged under load in such "direct drive" dimensions. The gear most frequently used on the road should be selected for the "direct transmission" stage to permit maximum exploitation of this consumption benefit. In automobiles, this is generally the highest gear, which may reach utilization rates of above 80%. In the lower gears, transmission efficiency is 95% to 96%, whereas 98% is achieved under 1:1 transmission.

The necessary overall transmission ratio is then a function, in this low-friction gear, of the differential gear.

26.3.2 Selection of Overall Transmission Ratio in the Highest Gear

The power train's smallest transmission ratio $(i_{Gi_{max}} \cdot i_D)$ influences maximum achievable speed, the vehicle's surplus power, its agility, and also fuel consumption, noise emission, and engine wear. Selection of overall transmission ratio is heavily dependent on the vehicle manufacturer's philosophy, and it is, therefore, not possible to provide any generally valid recommendations.

There are, in principle, three design possibilities; these are shown in Fig. 26-16.

Design for Maximum Speed

Here, the overall transmission ratio is selected in such a way that the resistance curve on the level (wheel resistance plus air resistance) intersects with maximum tractive effort at the wheel. Only this design achieves the highest possible maximum speed for the vehicle. At this working point, the engine is running at its design speed.

Overspeed Design

In this case, the power train's overall transmission ratio is higher than in the design for highest maximum speed. The intersection of tractive effort at the wheel with the resistance curve on the level is located beyond maximum power, i.e., at a correspondingly higher engine speed. This high engine speed level results in higher fuel consumption (see Fig. 26-16). The highest possible maximum speed is not achieved using this "short" transmission ratio, but high surplus power that can be used to overcome additional resistances is available at the wheel below maximum speed. This high surplus power produces an extremely agile vehicle.

Underspeed Design

Here, the overall transmission ratio is lower than in the design for highest maximum speed. The intersection of tractive effort at the wheel with the resistance curve occurs below the maximum tractive effort engine speed. Here, engine speed is lower than in the other two design modes and, thus, provides engine working points with favorable fuel consumption. The highest possible maximum speed is again not achieved in this layout, and surplus power at low speeds is extremely slight, with the result that the vehicle does not react so readily to a fully depressed gas pedal.

If an extremely pronounced underspeed arrangement is selected, it will be necessary to change down in gear as resistance rises since the surplus power available is not sufficient. This driving style sacrifices the consumption benefits of the underspeed design as a result of changing down a gear.

Of the extremely short and extremely long ratios in the highest gear shown in Fig. 26-16, a specific fuel consumption advantage of 16% results at the restrictive maximum speed for the highly underspeed design.

Selection of the Optimum-Consumption Gear

An unrestricted selection of transmission ratios are, of course, not available if the engine power is to be fully exploited. This is apparent in the case of a top speed trip at full power acceleration going up to top revs in all of the

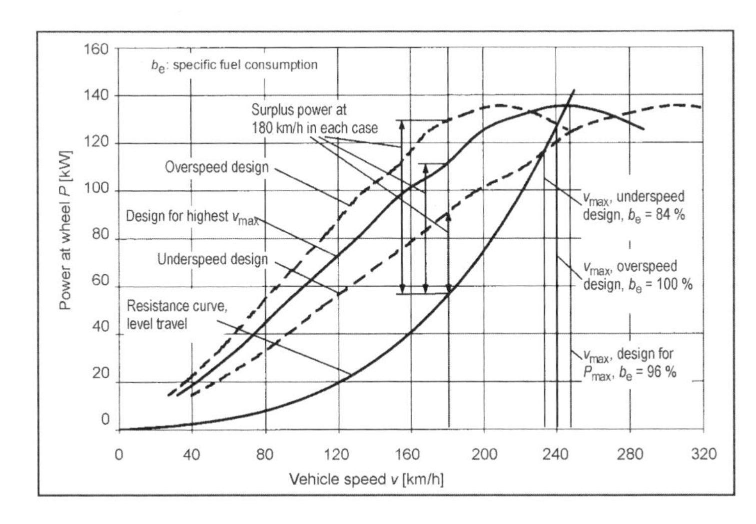

Fig. 26-16 Various designs of overall transmission ratio in top gear.

individual gears. The situation is different when travelling in the part-load range. The required tractive effort can then be provided in any one of several of the gears available. The driver can then select the optimum-consumption gear without having to accept any disadvantages, at least at constant travel speed.

Examination of a consumption characteristics field for an internal combustion engine (Fig. 26-17) demonstrates that there is only one working point at which the engine achieves its best specific consumption. The optimum point is in all cases located at high load and medium to low speeds. Specific consumption becomes greater, the further one departs from this point in this characteristics field. If unrestricted gear selection is available, the highest possible gear should always be selected, in order that engine load is high and that revs do not rise excessively. This is illustrated in an example in Fig. 26-17.

A constant power curve, such as is required for a travel speed of 100 km/h by the vehicle, has been plotted on the consumption characteristics. The engine is able to provide this motive power in gears 2 to 5. It is apparent that the engine reaches zones of lines with lower specific consumptions the higher the gear selected. Consumption at a constant 100 km/h is poorer by 60% if the vehicle is driven in second gear instead of fifth gear, for example. The best-consumption case would occur if it were possible to select the overall transmission ratio at this travel speed in such a way that the engine was operated at a speed of 1100 to 1200 rpm. Consumption savings of 15% would be possible compared to travel in fifth gear if this were the case.

The more gears that are available, the better one can target the engine's operating optimum for the particular power demand. This is extremely well implemented in the case of a fully variable transmission, selection of the transmission ratio being performed by an electronic system. In this case, it is possible to achieve the optimum working point for our 100 km/h travel speed in Fig. 26-17 and exploit the consumption advantage stated.

26.4 Driver Behavior

It is known from the characteristics fields for specific consumption against torque and speed that internal combustion engines achieve their best efficiency only within a narrow range of the overall characteristics field. This range is at low to average speed and high load, depending on engine design. The driver should remain within this range. This means

- Changing up at the lowest possible speed
- Accelerating at high load with low revs
- Driving smoothly in the highest possible gear, thinking ahead, and avoiding braking
- Using only 70% to 80% of the vehicle's maximum speed
- Switching the engine off if stationary for longer periods

The potential savings achievable via low gear change speeds in daily commuter traffic are shown in Fig. 26-18.

Consumption can be cut even further with an automatic stop/start system active when the engine is at operating temperature. Series-produced systems have been available for a number of years, but are used only in the small car sector. Consumption savings of 0.12 l/100 km are achieved.[15] The use of the integrated crankshaft-mounted combined starter-motor/alternator will permit the more widespread use of automatic stop/start systems in the future.

During the warming-up phase, fuel consumption is particularly high as a result of the high friction losses caused by the still cold lubricating oil and the low component temperatures. In addition, fuel consumption is also increased by a rich setting to avoid running irregularities (jolting, poor throttle response) and by heating strategies for the catalytic converter, such as delayed ignition, in the gasoline engine, for example.

Depending on initial temperature, the increase in fuel consumption during the warming-up phase may be up to 40% to 50% compared to the operating-temperature phase in the NEDC.

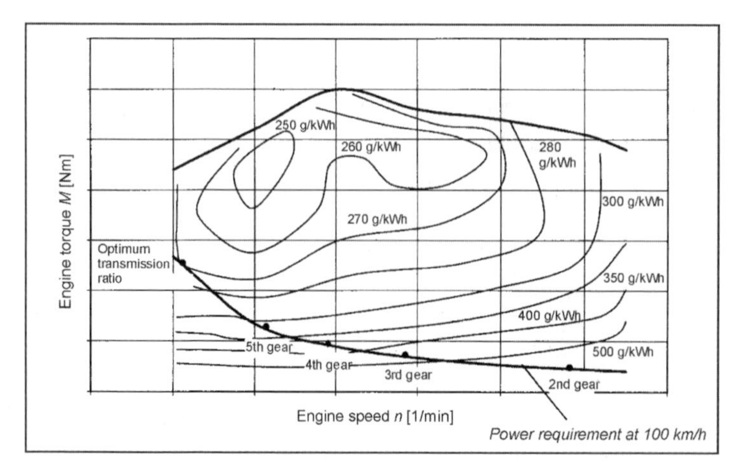

Fig. 26-17 Consumption characteristics field and the influence of selected transmission ratio at constant vehicle speed.

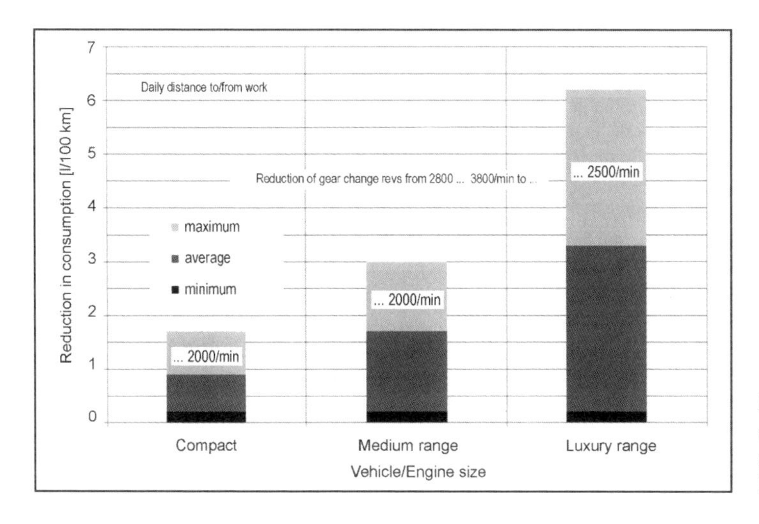

Fig. 26-18 Reductions in consumption as a result of cutting gear-shift engine speeds. From Ref. [14].

For driver behavior, this signifies making as few cold starts followed by a short journey as possible, and/or the use of engine preheating systems.

26.5 CO_2 Emissions

Emissions are classified by their origin. Flora and fauna, volcanoes, the oceans, and lightning are all natural emission sources. The emissions caused by humans ("anthropogenic" emissions) are the result of energy conversion processes, industry, transport, domestic heating, and waste incineration (for example).

Locally effective emissions include pollutants governed for motor vehicles, such as carbon monoxide (CO), hydrocarbons (HC), oxides of nitrogen (NO), and particulates.

Globally active are, above all, carbon dioxide emissions (CO_2), which are considered to be responsible alongside other greenhouse gases for global warming.

Total CO_2 emissions amount to some 800 Gt/year and originate from causes as shown in Fig. 26-19. About 3.5% of total CO_2 emissions are of anthropogenic origin (around 28 Gt/a). Road traffic accounts for about 11.5% of anthropogenic CO_2 emissions.

26.5.1 CO_2 Emissions and Fuel Consumption

The mass CO_2 emissions of a vehicle depend directly on its fuel consumption and can, as stated in Ref. [17], be calculated using the following formula:

$$m_{CO_2} = \frac{0.85 m_{Kr} \cdot 0.429\,CO - 0.866\,HC}{0.273} \qquad (26.10)$$

where m_{Kr} is the fuel mass, and CO and HC are emissions factors for carbon monoxide and uncombusted hydrocarbons.

Every provision for production of consumption, therefore, makes a direct contribution to the reduction of CO_2 emissions. The reduction of vehicle weight from 1500 to 1300 kg, for example, produces a decrease in CO_2 emissions of about 20 g/km.

It is apparent from the preceding equation that fuel composition also has an influence on CO_2 emissions. The higher carbon content, greater density, and lower calorific value of diesel fuel cause, for the same consumption, higher CO_2 emissions than gasoline. A diesel-engined vehicle emits 26.5 g/km CO_2 for every 1 l/100 km of fuel consumption, a gasoline-engined vehicle only 24 g/km CO_2.

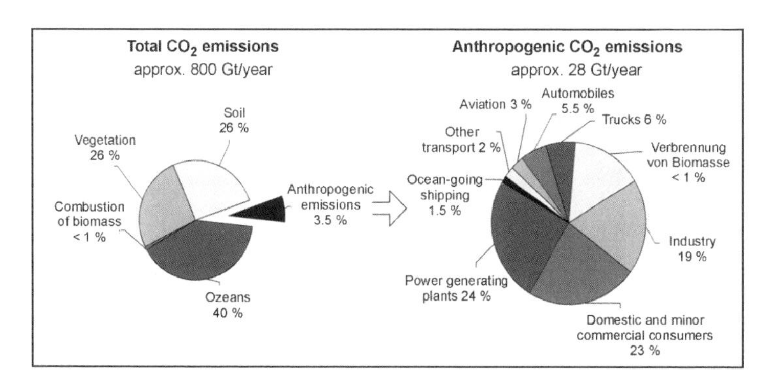

Fig. 26-19 Global annual CO_2 emissions and their causes.[16]

26.5.2 The Influence of Engine Use on CO_2 Emissions

The legally regulated exhaust emissions from a vehicle engine depend inter alia on application factors such as valve and ignition timing, ignition point, and fuel-air ratio λ. Nonregulated CO_2 emissions also manifest dependence on the fuel-air ratio, as is shown in Fig. 26-20.

Maximum CO_2 emission is reached at the stoichiometric air-fuel ratio ($\lambda = 1$). Gasoline engines equipped with a regulated three-way catalytic converter operate at this air-fuel ratio. In addition, CO_2 is also produced retrospectively from the CO emissions contained in the exhaust gas after a corresponding period of residence in the atmosphere, via the reaction of a portion of the carbon monoxide with air oxygen.

26.5.3 The Trend in Global CO_2 Emissions

Germany had around 45 million registered motor vehicles in the year 2000. The combined distance traveled by all these vehicles amounts to about 650 billion kilometers each year. This distance has approximately doubled over the last 20 years. This continuous rise has also been accompanied by a corresponding increase in the demand for fuel as well as in the amount of CO_2 released into the atmosphere; see Fig. 26-21.

A fall in the total demand for fuel and, therefore, in CO2 emissions is forecast in Ref. [16]. The prediction is based on the fact that the reduction of fuel consumption has become one of the automotive industry's highest priority targets. This is apparent, for example, in developments such as the direct gasoline injection engine, fully variable valve timing systems, DI diesel engines, and lightweight construction methods.

The use of hydrogen as an automotive fuel would eliminate CO_2 emissions. Hydrogen can be generated as a product of solar and nuclear energy, from biomass, and from water and wind energy.

In so-called closed circuits, alcohols are obtained from biomass as the fuel for the vehicle engine. The CO_2 emissions caused during combustion of the alcohols are degraded again by photosynthesis during the biomass's growth phase. The considerable energy input and soil occupation necessary for generation of the biomass should not be forgotten in the context of these so-called closed circuits, however.

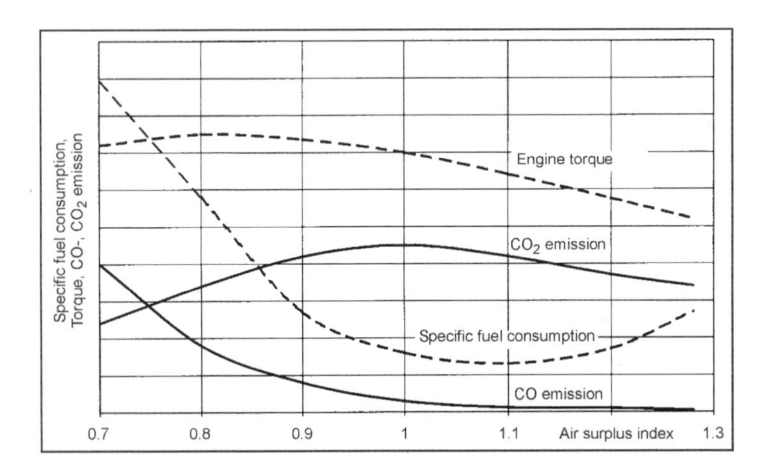

Fig. 26-20 CO_2 concentration, engine torque, and specific fuel consumption against air-fuel ratio.

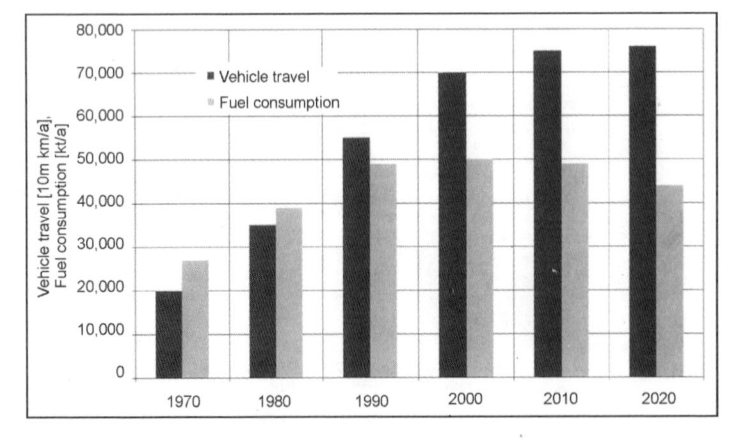

Fig. 26-21 Distance traveled and fuel consumption of automobiles and utility vehicles in Germany, 1970–2020.

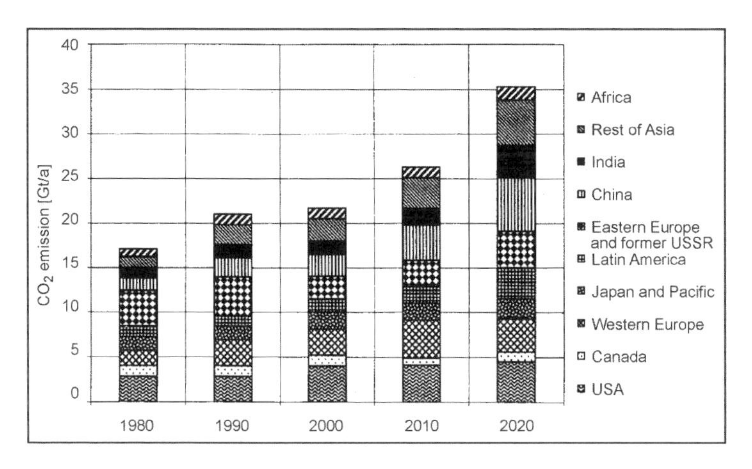

Fig. 26-22 Global CO_2 emissions from combustion of fossil fuels.[16]

The continued expansion of the global population and increasing industrialization in countries such as China and India are also causing an overall increase in world CO_2 emissions that cannot be balanced out by decreases in the CO_2 emissions of existing industrialized nations; see Fig. 26-22.

Bibliography

[1] Bundesministerium für Verkehr, Bau- und Wohnungswesen, Verkehr in Zahlen 1999, Deutscher Verkehrsverlag, Hamburg, 1999.

[2] Braess/Seiffert [Eds.], Vieweg Handbuch Kraftfahrzeugtechnik, Friedr. Vieweg & Sohn Verlags-gesellschaft mbH, Braunschweig/Wiesbaden, 2000.

[3] 20. Internationales Wiener Motorensymposium, VDI-Verlag, Düsseldorf, 1999.

[4] Hucho, W.H. [Ed.], Aerodynamik des Automobils, 3rd edition, VDI-Verlag GmbH, Düsseldorf, 1994.

[5] Kraftfahrwesen und Verbrennungsmotoren, 4. Internationales Stuttgarter Symposium, Expert Verlag, Renningen, 2001.

[6] Essers, U. [Ed.], Dieselmotorentechnik 98, Expert Verlag, Renningen, 1998.

[7] Bollig, C., K. Habermann, H. Marckwardt, K.I. Yapici, MTZ 11/97, Kurbeltrieb für variable Verdichtung, Verlag Vieweg, Wiesbaden.

[8] Carstensen, H., Systematische Untersuchung der Konstruktions- und Betriebsparameter eines Zweiventilmagermotors auf Kraftstoffverbrauch, Schadstoffemission und Maximalleistung.

[9] MTZ Motortechnische Zeitschrift 3/2000, Friedrich Vieweg & Sohn Verlagsgesellschaft mbH, Wiesbaden, 2000.

[10] Robert Bosch GmbH [Eds.], Ottomotor-Management, Vieweg, Braunschweig/Wiesbaden, 1998.

[11] 18. Internationales Wiener Motorensymposium, VDI-Verlag GmbH, Düsseldorf, 1997.

[12] Fortnagel, M., G. Doll, K. Kollmann, and H.K. Weining, Aus Acht macht Vier–Die neuen V8-Motoren mit 4,3 und 5 l Hubraum, ATZ/MTZ Jahresband 1988, Verlag Vieweg, Wiesbaden.

[13] Fortnagel, M., J. Schommers, R. Clauß, R. Glück, R. Nöll, CH. Reckzügel, and W. Treyz, Der neue Mercedes-Benz-Zwölfzylindermotor mit Zylinderabschaltung, MTZ 5/6/2000, Verlag Vieweg, Wiesbaden.

[14] Aral, A.G. [Ed.], Fachreihe Forschung und Technik–Kraftstoffe für Straßenfahrzeuge, 1998.

[15] MTZ Motortechnische Zeitschrift, Sonderheft 25 Jahre Dieselmotoren von Volkswagen, 5/2001, Friedrich Vieweg & Sohn Verlagsgesellschaft mbH, Wiesbaden, 2001.

[16] VDI-Bericht, Das Auto und die Umwelt, http://ivkwww.tu.wien.ac.at, 11/2000.

[17] Abthoff, J., C. Noller, and H. Schuster, Möglichkeiten zur Reduzierung der Schadstoffe von Ottomotoren. Fachbibliothek Daimler-Benz, 1983.

27 Noise Emissions

Anybody who has ever experienced a motor vehicle with a rigidly mounted engine or intake noise with no muffler—not to mention "naked" exhaust noise—will be in no doubt that of the many subdivisions of vehicle acoustics, engine acoustics was the first, and for a long time remained the most important. Passengers' demands for greater comfort and the concerns of the general public—represented by the law—drove developments in the engine-acoustics field, which has now achieved an extremely high level of sophistication despite the enormous increases in engine output. Scarcely anybody is now surprised at having to glance at the rev counter when the car is stationary to check that the engine is really still running. On the road, engine noise has been suppressed to such an extent that other sources such as tire and wind noise achieve equal and even dominant levels. A further indication of thorough mastery of engine acoustics can be found in the fact that it has now been possible for a number of years to conceive of "sound design."

Whereas the engine acoustics specialists' early tasks concentrated on the suppression of elemental exterior noise and vibration problems, aiming to find solutions in improvements in mufflers, internal engine mechanical balancing, and flexible mounting, the present-day field of engine acoustics has become significantly more diverse. This is true of both the nature of the noise sources involved—we may use the classifications of secondary noise radiation, generator noises, and timing mechanism noise—and the working methodology employed by engineers, which we can classify into transfer path analysis, noise-intensity measurement, holography, vibrometry, dummy-head technology, finite element method (FEM), boundary element method (BEM), and statistical energy analysis (SEA). Taken in conjunction with the enormous range of nonacoustic requirements (fuel consumption, emissions, heat balance, costs, package, etc.), the result is a complexity expressed within the manufacturing companies in the form of correspondingly large work groups and within supplier companies in the form of a high level of specialization. It is only necessary in this context to think, for example, of components such as exhaust systems, enclosure components, engine bearings, and dual-mass flywheels.

All of these branches, however, work on the basis of the same physical principles and methods, and utilize the same basic concepts of physical acoustics and psychoacoustics. We, therefore, summarize the essential concepts and terms in the following section, before going on to examine individual topics. Section 27.7 contains a short discussion of widely used analytical methods.

27.1 Basic Physical Principles and Terms

Despite the fact that the term "engine acoustics" does not really make it clear, it is not only "audible" phenomena that play a role here, but also the vibrations which can be felt by the occupants of vehicles and which, as in the case of so-called "idling shudder," can in many cases reach extremely low frequencies. So-called solid-borne noise is also important because only a very small range of noises actually originate as airborne noise (such as exhaust muzzle noise), most are instead generated in the form of vibration in solid bodies and then radiated by oscillating surfaces (such as mass forces, gas forces, and toothing forces) and/or need to pass through the body shell in the form of airborne noise on their way into the interior of the vehicle. If we think of piston canting noise, for example, this, too, in many cases has to traverse at least a short distance in the form of liquid-borne noise (in liquid-cooled engines) on its transmission route. This stage presents a considerable barrier to noise propagation, since neither liquids nor gases are capable of absorbing shear stresses. A significant difference compared to airborne noise can, however, be found in the considerably higher characteristic acoustic impedance (wave impedance), which signifies considerably greater coupling with the solid-borne sound perpendicular to the surface and results, inter alia, in it not being possible, because of the significant interactions that occur, to examine solid-borne noise and liquid-borne noise in isolation from one another.

The most widely used parameter for the description of solid-borne noise is acceleration that is "popular" as a result of the relative simplicity of measuring it. It must, however, be noted that, unlike airborne sound pressure, acceleration is a directional variable requiring measurements in three dimensions from a single point. The registration and quantification of solid-borne noise is, in general, significantly more complex, difficult, and diverse to measure than airborne sound because, as a result of their ability to also absorb shearing stresses, solid bodies generate a large range of different sound wave propagation forms (many of them simultaneously). The following can be mentioned as examples: Dilatational waves (valve stem); flexural waves (oil sump); torsional waves (crankshaft, camshaft). Because of the need not to influence the vibration system, as a consequence of constricted space or other restricting boundary conditions (temperature, pressure, tightness, etc.), and in particular, because of the desire to acquire information on the prevailing forces, a range of other measured variables such as noncontact path measurement (rotating and thin-shelled components) and measurement of strains (crankcase), pressure distributions (bearing seatings), and forces (engine bearings) are also used alongside acceleration. Another physical phenomenon, rotational vibration, is also of great importance, particularly in the field of engine and power train acoustics. The basic measured variable in this context is generally angular velocity, which can be determined from discrete angle pulses (gearwheels and incremental generators).

Rotary acceleration can, if required, then be determined by differentiation.

Compared to solid-borne noise, the registration and quantification of airborne noise, both in the vehicle interior and in the case of exterior noise, is initially relatively easy since space and temperature problems are encountered only rarely and the variable relevant to the human ear (i.e., sound pressure) can be measured directly using microphones. Sound pressure p is the term used to describe the amplitude of pressure fluctuation around static air pressure, with pressure amplitude cycles of 3 Pa, for example, being perceived as extremely loud (for comparison: 1 bar = 10^5 Pa). Although sound pressure is generally adequate for a description of airborne sound output, the most suitable parameter for quantification of airborne sound radiation (emission) is acoustic power P. This is the total sound wave power penetrating through an imaginary envelope and is calculated from the integral of acoustic intensity I across the envelope s:

$$P = \int_s \boldsymbol{I} \cdot ds \qquad (27.1)$$

Acoustic intensity represents mean energy transport per unit of area. It is a vector factor aligned parallel to the vector of acoustic velocity $\boldsymbol{v}$ and is calculated from

$$\boldsymbol{I} = \overline{p(t) \cdot v(t)} \qquad (27.2)$$

(averaging of time range) or from

$$\boldsymbol{I} = \tfrac{1}{2} \cdot \mathrm{RE}\{\tilde{p}\tilde{v}^*\} \qquad (27.3)$$

(frequency range), analogously to mechanical power $P = Fv$. In this context, acoustic velocity is the velocity of the local oscillating motion of the air particles. The vector character of acoustic intensity can be exploited mensurationally in the locating of sources in complicated sound fields and in the measurement of acoustic power, which also becomes possible in a reflecting environment.

Evaluation of noise signals is generally accomplished on two criteria. These are the most precise possible evaluations in terms of subjective human perception (we shall restrict our attention here to airborne noise) and in terms of the most efficient possible extraction of information on noise generation and the transmission route.

Concerning subjective perception, it is necessary first to note that the human ear is capable of registering acoustic pressures across a range of orders of magnitude of several powers of ten. For this reason, depiction of sound levels on a logarithmic dB scale has become established in acoustics, not only for sound pressure [sound pressure level (SPI)], but also, for example, for solid-borne noise variables, a variable proportional to energy being defined in every case:

$$L_x = 10 \cdot \log_{10} \cdot \left(\frac{x^2}{x_0^2}\right) \mathrm{dB} = 20 \cdot \log_{10} \cdot \left(\frac{x}{x_0}\right) \mathrm{dB}$$

or

$$L_x = 10 \cdot \log_{10} \cdot \left(\frac{X}{X_0}\right) \mathrm{dB} \qquad (27.4)$$

in which x is a field variable (e.g., sound pressure, acceleration, etc.), X an energy variable (e.g., acoustic intensity, acoustic power, etc.), and x_0 and X_0 their reference variables. In the case of sound pressure, $p_0 = 2 \cdot 10^{-5}$ Pa (root mean square value), in the case of acoustic power $P_0 = 10^{-12}$ W, and in the case of acoustic intensity $I_0 = 10^{-12}$ W m^{-2}.

Human hearing is not only nonlinear, but also frequency dependent. Its sensitivity declines significantly as low and as extremely high frequencies are approached. The fluctuation range becomes greater as the absolute sound pressure decreases. The various frequency components are, therefore, generally weighted with specified weighting curves (DIN IEC 651, Curve A for low, B for medium, and C for high sound volumes) for the sake of simplicity before being summated to an overall level, which is then correspondingly indexed [e.g., dB (A)]. Weighted levels, as a result of their simplicity, are used as the basis for a whole series of legal regulations but are suitable neither for diagnosis purposes nor for obtainment of information concerning the "quality" of a noise (see Sections 27.9 and 27.10), since they do not contain any information on the spectral or chronological structure of the noise.

Readily understandable designations that are associated with certain mechanisms, while at the same time provide implicit information on the frequency range involved or on the chronological structure of a noise, have become established in the automotive field as the conceptual basis for recurring acoustic phenomena. Figure 27-1 shows a number of typical examples concerning the engine.

An essential basis for the analysis of noises is provided, initially, by their spectrum, i.e., the classification of a signal into its frequency components. In practice, spectra are determined from digital signals using the so-called FFT (fast Fourier transformation), a highly efficient variant of the digital Fourier transformation. This generally supplies narrow-band spectra, i.e., a relatively high-frequency resolution, making it possible to calculate the classical, "coarser" subdivisions into octave (doubling of frequency) and third-octave spectra if required. Only high-frequency resolution permits detailed analysis of the noise. Sinusoidal noise components attributable to mass forces in the crank gear, for example, and therefore possessing defined frequencies, appear in the form of narrow peaks in the spectrum (linear spectrum). Such deterministic frequency components, which "wander" proportionally to speed, are referred to as "orders." A typical example is the 2nd order of the crankshaft in an inline four-chamber engine that dominates as a result of the unbalanced mass forces. The fundamental frequency, or 1st order, can be completely balanced out in a four-cylinder engine.

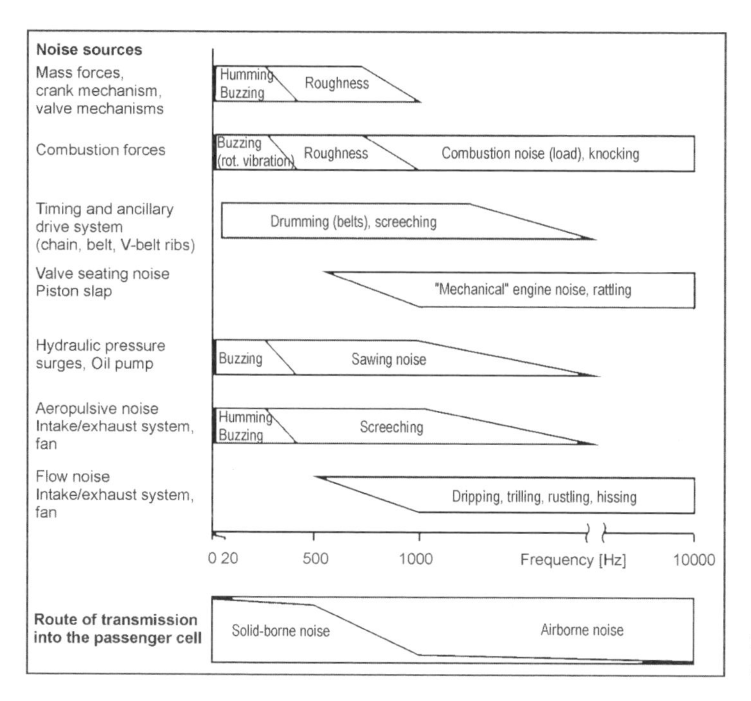

Fig. 27-1 Examples of noise sources.

Where the dominance of an order is known from the inception, a so-called "order filter" is used in many cases to examine only the level of this order. The higher orders, which definitively influence sound, also appear in the spectrum, however. Modulations, i.e., fluctuations in amplitude (or beat) or frequency significant enough for perception as noise or nuisance also become visible in the form of sidebands. These are peaks that occur at the interval of the modulation frequency adjacent to the midband frequencies.

An internal combustion engine is a diverse source of noise. If we leave aside the auxiliaries for the time being, not only aeropulsive and aeroacoustic sources, such as exhaust muzzle noise, intake noise, and fan noise, but also the noise radiation of the oscillating surface of the engine and attachment components are responsible for exterior noise. Their effectiveness can be characterized in the "degree of radiation," which is the ratio of actually radiated acoustic power P of a surface S to that of a large (significantly larger than the sound wave length) concurrent-phase oscillating plate with the same mean quadratic speed. If the interior zones oscillating in counterphase are significantly larger than the airwave length, the degree of radiation will be close to 1. The converse case is considerably more difficult to handle, but—simplified—the radiated acoustic power is smaller as the number of counterphase zones present becomes larger and more densely located relative to one another (hydrodynamic short circuit).

Once the airborne noise has been generated, it can be countered with insulation (energy reflection) and/or attenuation (energy dissipation), with at least a small amount of attenuation always being necessary. Insulation provisions (such as encapsulation) and combined provisions, such as mufflers, are evaluated specifically by degree of transmission or, in absolute terms, by the insertion insulation increment, the difference in level prior to and following the interposition of a provision

$$D_e = L_{\text{o.D.}} - L_{\text{m.D.}} \tag{27.5}$$

while the characterizing variable for purely airborne-noise-attenuating provisions such as noise-absorbing linings and claddings is their degree of absorption, i.e., the ratio of absorbed to incident intensity

$$\alpha = \frac{I_{\text{absorb}}}{I_{\text{einfall}}} \tag{27.6}$$

When interior noise is examined, a further component, actually dominant in the frequency range up to about 500 Hz, is added to the sources already mentioned, namely transmission of solid-borne engine noise via the body shell into the vehicle interior. In addition to the drive shafts relevant, in particular, in front-wheel drive vehicles, the essential transmission points are in this case the engine mountings. One development target of engine acoustics is, therefore, that of minimizing engine-side solid-borne noise amplitudes at the support points, since the degree of isolation of vibration achievable via the rubber bearings is limited.

On the body shell side, the corresponding development aim is minimization of "sensitivity" at the coupling points, which is quantified by the "acoustic transmission function."[1] This is the frequency-dependent ratio of sound pressure on an interior microphone and dynamic force at the point of excitation:

$$\tilde{H}_{ij} = \frac{\tilde{p}_j}{\tilde{F}_i} \tag{27.7}$$

It includes the complete transmission path, including radiation into the interior, absorption there, and the influence of resonance effects resulting from cavities in the passenger cell. So-called input inertance, on the other hand, discloses local weak points in the body shell in the form of the ratio of vibrational acceleration at the point of action of a force in the direction of the force and the force acting.[2–4]

$$\tilde{I} = \frac{\tilde{a}}{\tilde{F}} \tag{27.8}$$

27.2 Legal Provisions Concerning Emitted Noise

27.2.1 Methods of Measuring Emitted Noise

The level of noise and comfort in the interior of a vehicle is a matter of market forces, whereas the operating noise emitted into the environment ("emitted noise") became subject to official regulation at a very early stage. The measuring procedure applied is internationally standardized in ISO R 362. The measuring apparatus used is clearly the product of mensurational technology at the time of its origin and offers, in its simplicity, the indisputable benefit that this measurement can be performed identically and with little cost and complexity in all countries of the world.

The procedure, in principle, is shown in Fig. 27-2: At a constant 50 km/h, the vehicle approaches a 20 m long measuring section at the start of which the throttle is abruptly opened. The noise level during the subsequent (approximately) 1.3 s in duration full-throttle acceleration process (hence, the designation "accelerated drive past") is measured in each case by one microphone at a lateral distance of 7.5 m. The maximum dB (A) value achieved during this procedure on the right-hand or left-hand microphone is the result of the measurement. Since engine noise increases overproportionally with speed, the result is primarily dependent on the speed level achieved in this short period of time. A whole range of additional rules, therefore, specify the gear (in the case of stick-shift transmissions) or the selection (in the case of automatic transmissions) in which the accelerated drive past is to be performed. For five-gear stick-shift cars, for example, measurement is performed using 2nd and 3rd, and the arithmetical mean from the two measurements taken as the result. On automatics, measurement is performed at setting "D" with the kick-down function deactivated.[5,6]

27.2.2 Critical Evaluation of the Informational Value of the Emitted Noise Measuring Method

The disadvantages of the currently used method have been long known among specialists. A segment of a length of only approximately 1.3 s from a single operating state—full throttle—is capable of reflecting the contribution to traffic noise made by a vehicle under practical road conditions only extremely imperfectly. The "locality" background noise during morning cold starting of diesel engines, for example, is not registered at all. In addition, the measured result's dependence on speed means that

- Spontaneously reacting, rapidly accelerating vehicles reach a higher engine speed and, therefore, a higher noise level at the end of the short test length.
- More slowly accelerating vehicles achieve better measured results, since they only become noisy once they have already left the measuring length. The benefits of this deficiency are enjoyed, for example, by first generation turbocharged engines, which accelerated with elevated power production only after a certain time delay (the "turbo-gap").

Account is taken of the first point for vehicles with a rated output of >140 kW and a rated output/maximum permissible weight >75 kW/t using an additional rule referred to as the "Lex Ferrari": Only the value measured in 3rd gear is evaluated if a speed of 61 km/h or more is achieved at the end of the measuring length.

A further weakness can be found in the reduction of the measuring result to one single value, the dB (A) level, which in many cases fails to accord with subjectively perceived loudness or even with perceived unpleasantness. The A-weighting curve severely understates the role of

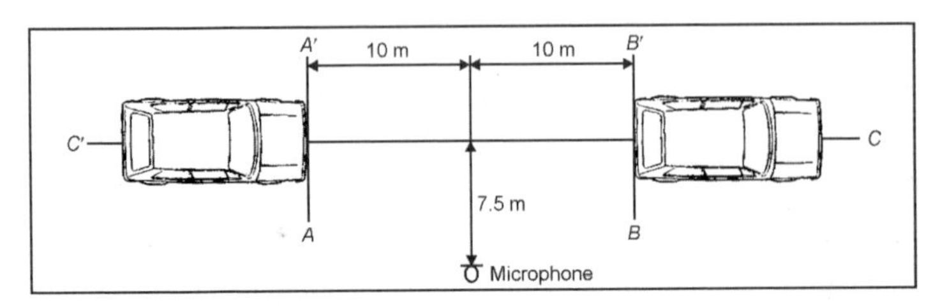

Fig. 27-2 Emitted noise measurement.[6]

low-frequency booming noises during starting of heavy freight vehicles, for example. For residents, they are precisely the main source of nuisance, since the noise-insulating effect of window glass is low in this frequency range. Pulsing noises, on the other hand, are perceived as significantly more unpleasant than "background noise" signals of the same sonic energy. Because of their knocking noise, diesel engines are considered significantly less pleasant than gasoline engines, despite having an identical or even a lower dB (A) value. The same also applies to the sputtering noise emitted by many two-stroke motorcycles.

27.2.3 Emitted Noise Limits, International Legislation; Future Trends

Exceeding—or rather nonexceeding—of the noise-emission limit must be demonstrated by the vehicle manufacturer before the "general roadworthiness permit" can be issued in the context of the vehicle type test. The emitted noise limits for automobiles applicable in Germany and most other European countries were lowered from 78 to 74 dB (A) between 1970 and 2000; an amount of −3 dB (A) corresponds to a halving of the sonic energy. The fact that the contribution made by tire noise is already as high as 68 to 70 dB (A) means that the engine's contribution cannot be much more.

Comprehensive surveys have ascertained that the noise nuisance experienced by residents has not declined to the same extent as permissible vehicle noise emissions. This is explained by the higher traffic levels of present time and by the fact that the ISO R 362 emitted noise measuring procedure is not capable of simulating actual on-the-road operation. Studies performed by the German "Umweltbundesamt" (Federal Office of the Environment, equivalent to FEA) demonstrated that vehicles spend a large percentage of their time in real traffic situations being operated at part load, whereas testing is performed only under full load in the ISO R 362 test.

For the future, the lowering of the limit for automobiles to 71 dB (A), i.e., another halving of noise emissions, is under discussion for the European Union in combination with a change in the measuring method used, toward "output-weighted-orientated vehicle acceleration." The proposal by the ACEA is that the measured result for a part-load drive past should be obtained via weighting of a measurement under full-throttle acceleration and a "no-load" drive past. This would ensure that greater importance is allocated to the tire/road-surface noise component.[7]

The emitted noise test in Japan is based on the EU guideline (there are differences concerning the vehicles' loading state and additional constant trips), whereas the procedure is performed in accordance with the SAE standard in the United States. The measuring method is similar, but the microphones are positioned at a distance of 15 m instead of 7.5 m. Practice has demonstrated that the European noise regulations are the more stringent; if a vehicle conforms with ISO R 362, for example, it will also meet the national-law regulations in the United States.

27.3 Sources of Emitted Noise

Depending on vehicle speed, the components making up traffic noise can, above all, be traced to two sources, the engine and the tires. Engine noise dominates at low speeds and road-surface/tire noise at freeway speeds. Wind noise, on the other hand, can be ignored.

A range of physically differing effects contributes to engine emitted noise:

- **Intake and exhaust muzzle noise.** The pressure surges excited in the intake and exhaust systems by the gas-charging sequences in the engine result in direct emission of airborne sound waves at the open end of the pipe system (the "muzzle"). Within the spectrum, ignition frequency, i.e., 2nd order in the case of four-cylinder engines, 3rd order in the case of six-cylinder engines, etc., dominates high-frequency, broadband flow noise resulting from flow velocity at the end tube that must also be added at high engine speeds and full-load operation.
- **Secondary radiation** from the intake and exhaust systems. The pressure surges in the interior of the system also cause the pipe and wall surfaces to oscillate, with the result that they also emit airborne noise to the exterior; this then is referred to as "secondary airborne noise."
- **Noise radiation from the engine structure**, i.e., the external surfaces of the engine and transmission unit. Combustion pressures and all nonregularly moving components in the engine, transmission, and auxiliaries result in dynamic forces acting on the housing structure and, thus, in motions and deformations of the outer walls, which then become the originating point for emission of pressure waves into the ambient air, i.e., radiation of noise. The spectrum of this airborne noise is dominated by high-frequency components above 500 Hz, which are subjectively perceived as "mechanical engine noise." The components of mechanical engine noise are, for example,
  - Pressure increase at compression and combustion, in the case of diesel engines, in particular
  - The impact of the valve disks on the valve seat
  - Piston canting
  - Tooth oscillation in the valve-timing system
  - Chatter of the floating wheels in stick-shift transmissions ("gearbox chatter")
  - Pressure surges from the hydraulic pumps

27.4 Emitted Noise-Reduction Provisions

27.4.1 Provisions on the Engine

Action taken at source, i.e., an engine design that produces the least possible noise, is always the most consistent and also the most efficient provision.[8] The design target must be

- **Maximum stiffness** in the force-transmitting housing structure and crank gear, in order that excitation of housing-wall-surface oscillation is minimized.
- **Oscillation-reducing design of engine exterior walls,** by

  (a) Ribs or fins (high dynamic stiffness, engine block, and gearbox casing, for example)
  (b) Isolation, e.g., flexible fixing of the cylinder-head cover or the intake manifold
  (c) Damping, e.g., a sheet-metal oil sump

  All of the listed alternatives conflict with the aims of cost optimization and functional requirements: Ribs and fins: Extra weight and space requirement; Isolation: Oil tightness, fixing of attachments and auxiliaries; Damping: Heat removal, extra weight. In the case of the oil sump, the ribbed aluminum oil sump has, in particular, become established over the damped sheet-metal oil sump in aluminum engines, since it is necessary as a load-bearing structural element to increase the stiffness of the engine plus transmission unit; see also Section 27.6.
- **Noise-absorbing attachment shells** as components of a "tight-fitting enclosure" are considered secondary provisions that prevent the radiation of noise. The cylinder-head claddings that are installed primarily for engine-compartment styling reasons (enclosure of cables and injection lines) also perform, as in-mold skinned plastic shells, an acoustic function.[9,10]
- **"Softer" combustion** is the name assigned to the task in the field of mixture generation, particularly in the case of diesel engines, to significantly reduce the subjectively perceived nuisance quality of combustion noise and improve both noise emissions and interior noise levels.
- **High-volume intake and exhaust mufflers** primarily constitute demands made in terms of space and cost on the overall vehicle conception. Muzzle noise must not supply any measurable contribution to total emitted noise if present-day limits are to be met. This provision is deemed achieved if an additional muffler, designated an "absolute muffler," achieves no further reduction in the dB (A) level in measurement of emitted noise. In the case of interior noise, however, an audible exhaust system component for the most pleasant possible engine noise is targeted, particularly at full-throttle acceleration. This is achieved, inter alia, with the support of analytical and simulation sound design methods during the development phase. Complex solutions involving exhaust butterfly valves that prevent low-frequency booming at low engine speeds and reduce flow noises (and exhaust counterpressure simultaneously) by opening an additional cross section at high engine speeds and exhaust flows are the result of such development activities.
- **Reduction of engine speed,** one of the most effective provisions for noise reduction, is also a matter of overall vehicle conception. In practical on-road operation, it achieves low noise emissions only if a "longer" transmission ratio is combined with high engine torque in the low rev range and, therefore, also offers acceptable driving characteristics.
- **Noise-optimized auxiliaries,** such as radiator fans and alternators, already play a role in achieving the low limits imposed for automobiles, despite their relatively small contribution to overall noise.

27.4.2 Provisions on the Vehicle

Secondary provisions implemented on the body shell to reduce engine noise emissions are referred to as "spaced" or "off-engine enclosures." The objective is that of using additional components installed in or on the body shell to make the engine compartment a largely sealed space from which only a small amount of engine noise can escape to the exterior. In addition, the peripheries are lined with noise-absorbing materials, such as foam or nonwoven cotton fabric (which, particularly in the lower engine compartment zones, require protection against absorption of oil and moisture, and must be of noncombustible type) in order to reduce the high noise level in the engine compartment. Practically all series-produced automobiles and those with diesel engines are equipped with a combination of the following enclosure elements (see Fig. 27-3):

- Absorbent engine-hood lining consisting of foam, nonwoven fiber fabric, of a cassette-module type in some cases, spaced from the sheet metal, in order that panel resonator effects achieve greater absorption at low frequencies, or of honeycomb type, with the effect of Helmholtz resonators.
- Undershield: This term identifies a plastic or metal shell that encloses the engine compartment at the bottom and is required even for reasons of air-resistance alone. Frequently featuring a cutout for the oil sump, in order to maintain ground clearance, it normally ends shortly before the firewall. In special cases, the front section of the transmission tunnel in acoustically critical vehicles is also closed on its underside, provided a solution can be found for the cooling problems that then arise. The undershield, too, features noise-absorbing lining on its inner side with safety against absorption of oil, fuel, etc., having priority over the best possible noise-absorbing properties. For this reason, skinned foam or cassette-type absorbers are used here instead of open-cell foam materials, which would be desirable from an acoustic viewpoint.
- Closure of the sides of penetrations to the wheel housing by rubber bellows for the track rods and, if necessary, provision of tunnel-like absorption sections lined with foam for the front-wheel drive shafts.
- Closure of cooling-air inlets at the front by, for example, thermally controlled slats; this is an extremely complex provision that is nonetheless practiced in top-

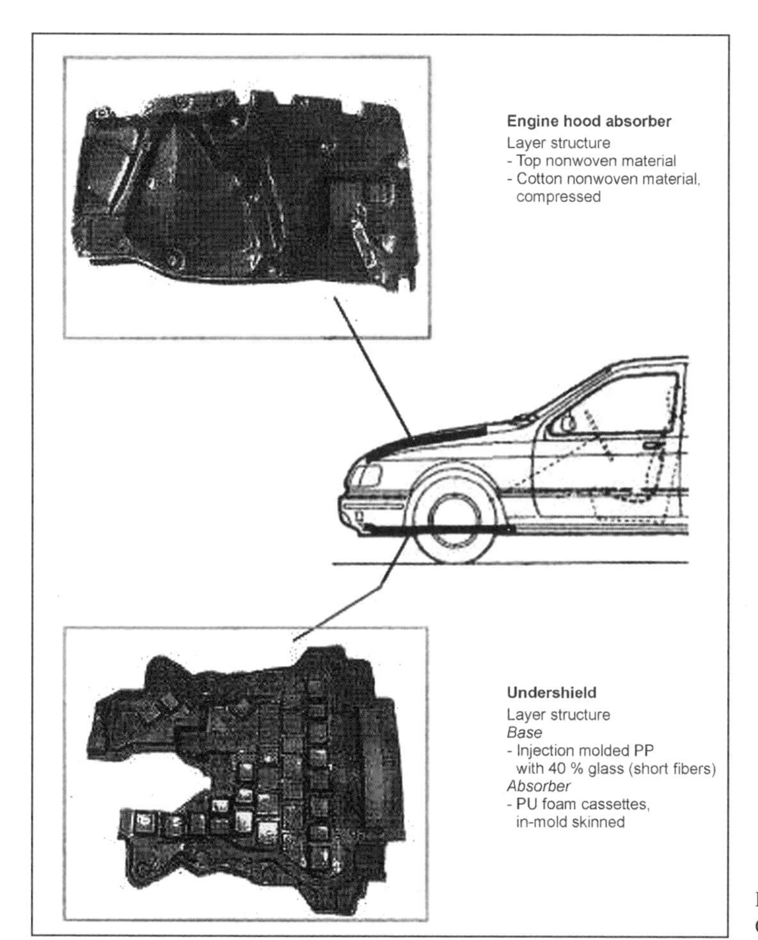

Fig. 27-3 Engine enclosure [HP-Chemie Pelzer].

end diesel engine vehicles. At cold starting, the slats are closed, reducing the diesel engine's cold "knocking." For functional reliability reasons (freezing in winter), such systems can be used only behind the radiator; i.e., there must be sufficient space between the radiator and the engine.

The reduction in emitted noise achievable with an engine enclosure is restricted by the size and number of the penetrations necessary for cooling the vehicle. The development of a complete engine enclosure is, therefore, more a problem of cooling than of acoustics. Vehicle studies incorporating completely enclosed engine compartments achieve spectacular publicity effects, but are generally a long way from any capability of surviving a trip through the passes of the Alps with a trailer in tow, or "hot and high" tests. It is necessary to differentiate between the following in the context of numerical data in dB (A), quantifying the measured noise reduction achieved with an enclosure:

- dB (A) value for reduction of radiated engine noise.
- Reduction of emitted noise emission factor, i.e., total vehicle noise in the legally prescribed drive past test.

In the drive-past test, 74.8 dB (A) is measured for an automobile, which is the result of the sum of the energy contents of 70 dB (A) tire noise and 73 dB (A) engine noise. A sophisticated engine enclosure would permit reduction of acoustic energy radiated by the engine to a half, i.e., to an acoustic pressure level of 70 dB (A). Tire noise and engine noise of 70 dB (A) each then produce a summated level of 73 dB (A). The reduction in emitted noise emission value achieved by the enclosure is, therefore, 1.8 dB (A).

27.5 Engine Noise in the Vehicle Interior

Although the provisions described in Section 27.4.1 also serve to reduce interior noise, the solid-borne sound paths, which dominate interior noise in the lower frequency range must be added here, which is not the case with emitted noise. All mechanical links between the engine and transmission unit and the body shell constitute potential transmission routes for solid-borne noise, particularly the engine mountings and the drive shafts, which in the case of front-wheel drive vehicles are connected with no intervening insulating elements to the suspension system from where the solid-borne noise can be transmitted via relatively stiff suspension attachment members into the body shell structure. Only the higher-frequency noise components above 500 Hz, so-called mechanical engine noise and combustion noises (Fig. 27-1), are transmitted via the airborne noise route from the engine compartment via the firewall and floor plates into the passenger cell.

The origins of solid-borne noise vibrations can be found primarily in oscillating mass forces because the low-frequency buzzing and booming noises are present in three- to five-cylinder engines (even when the engine is idling). Gas forces, the main causes of rotational asymmetry of the crank gear and of counterphase external reaction of the engine and transmission unit in the form of a rotary vibration about the principal axis of inertia in the crankshaft direction, are the second source of solid-borne noise. Their contribution to overall noise can be easily differentiated as a result of its nondependence on engine load and the increase in the degree of irregularity at low engine revolutions.

Excitation caused by free mass forces and torques can be controlled only by mass balancing; in terms of number and arrangement of cylinders, or additional balancing shafts, rotating at crankshaft speed (e.g., balancing of 1st order moments of inertia in three-cylinder, five-cylinder, and V6 engines), or at twice crankshaft speed (e.g., "Lancaster balancing" of 2nd order forces in the four-cylinder inline engine). The acoustic improvement must be set against higher costs and friction losses, however. Reduction of mass forces by lower piston and connecting rod masses is possible in theory, but these potentials have in most cases already been exhausted. The higher orders (>2nd order) are generally no longer the subject of mass-balancing studies, since excitation of them is significantly slighter and, in particular, since the precondition for mass balancing—rigid-body behavior in the crank gear and crankshaft—is no longer fulfilled at frequencies of above 250 Hz. In interior noise, the contribution from the higher orders is noticeable in the form of "roughness" in the engine noise and is a problem in the field of automobile engines of long-stroke (empirical value $H >$ approximately 80 mm), high-torque engines with a high connecting-rod ratio λ (because the higher orders increase superproportionally to λ). If two or more orders of approximately the same size are adjacent in the spectrum, for instance, 4th, 4.5th, and 5th, their superimposition results in a modulation, a surging fluctuation in noise level, which is perceived as an unpleasant noise characteristic. This "hammering" or "crankshaft rumble" appears at full-throttle operation since the "half orders" originate from combustion (ignition frequency of the individual cylinder in a four-stroke engine).

The second source of solid-borne noise, the non-steady-state character of torque demand, causes buzzing at ignition frequency (2nd order in four-cylinder engines, 3rd order in six-cylinder engines, etc.), particularly in the lower engine speed range, combined with occasionally severe vibration. A result of physics, and therefore not subject to influence by the designer, this problem is linked to torque at low revs; i.e., the better the engine's torque characteristic, the greater the problem of vibration and buzzing. This, for a long time, was the reason for the use of direct injection diesel engines only in the field of utility vehicles. Only years of development effort, including activity devoted to insulation of vibration, resulted in a breakthrough and use in automobiles, too. The rotary vibrations of the crankshaft itself can be largely isolated from the drive shaft by installing a heavy or dual-mass flywheel (Fig. 27-4). The reaction forces acting on the crankcase remain unaffected by this, however; it is, therefore, the rotary vibrations of the housing that make this problem so difficult to solve.

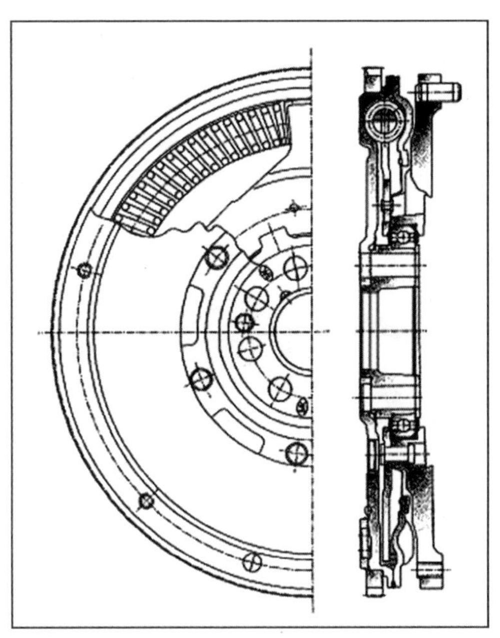

Fig. 27-4 Dual-mass flywheel [Luk].

The mass forces of the valve gear and intrinsic timing-system noise tend to play more of a role in four-cylinder engines for upper-end vehicles, since they impair the engine's "sonic impression" in a way similar to the noises from auxiliaries (alternator "whistling," "sawing" from hydraulic pumps, etc.). The mass forces of the valve gear occur as a reaction to the acceleration of the valves, tappets, rocker arms, etc., and the results are, in principle, the same as for the mass forces resulting from piston movement. Although they are one order of magnitude lower, they are perfectly capable of playing a codeterminant role in hum level in engines with good mass balancing, i.e., six- and eight-cylinder engines. The noise of the timing system, referred to as "timing-chain scream" or "timing-belt scream," is located in the medium-frequency range corresponding to the frequency of tooth engagement of the chain sprockets and belt pulleys. Similar to the situation in the case of cogs and gear wheels, it is the result of the periodic subjection of the teeth or the chain links to load and subsequent relief, and in the case of toothed belts, of expulsion of air upon tooth engagement ("air pumping"). These screaming noises become audible during idling phases, in particular, and in the lower engine rev range, where they are not yet screened by the increasing level of the other mechanical noises and combustion noises produced by the engine. In addition, low-frequency noises may also be generated by string oscillations of the chain or toothed belt.[11]

27.6 Acoustic Guidelines for the Engine Designer

The question of how the target of achieving low noise generation can be incorporated even at the engine design phase has already been the subject of numerous studies. As far as it is possible without knowledge of a particular design to provide any universally applicable guidelines, these tend toward maximum stiffness. This can be justified in physical terms by the fact that the deformations responsible for noise radiation and solid-borne noise intromission are, given identical forces, reduced and by the fact that the structural resonances are shifted toward higher frequencies where the amplitudes of the dynamic forces causing excitation become smaller.

The latter starts with the first bending mode of the engine and transmission unit, which in the case of a four-cylinder engine must be shifted under all circumstances into the frequency range significantly above the strongest vibration excitation of the second engine order (i.e., ≥ approximately 250 Hz in the case of a gasoline engine). The stiffness "weak point" is generally the bolted engine housing flange/clutch bell housing or torque converter bell housing joint, particularly if a non-load-bearing sheet-metal oil sump or a "short skirt" (engine side walls not extended down beyond the main bearing pedestals) prevents transmission of forces below the axis of the crankshaft. A remedy can be found in the use of an extrusion-molded aluminum oil sump with corresponding ribbing, and ribs or fins of the transmission bell housing. The aim of achieving the greatest level of "straight-line flow of forces" applies, i.e., any indentation or projection of the housing reduces the achievable stiffness. In the case of six- or eight-cylinder engines with no 2nd engine order excitation forces, a natural flexural frequency above the 1st order, i.e., ≥ approximately 120 Hz is sufficient. The greater masses of such generally high-capacity engines lower the natural frequency to such an extent that it is, nonetheless, necessary to design for high bending stiffness, particularly with the engine arranged longitudinally and long all-wheel-drive transmission systems. Static stiffness analyses or calculations are still sufficient for these ultralow oscillation modes, since the masses of the housing walls have only a little influence on the oscillation mode.

The first natural torsion frequency of a normal automobile engine is generally above the frequency of the most powerful rotational oscillation excitation. In special cases, however, this natural oscillation mode can also cause noise problems and necessitate structural stiffening provisions. Here, too, projections in the transmission bell housing are typical points of low stiffness.

Housing wall natural oscillations play an increasing role in the frequency range above about 500 Hz and can be reflected in the form of "hot spots" in the engine or transmission noise radiation. Such noise problems are so particular to the individual design that no precise recommendation (for ribs or fins or damping, for example) can be made without mensurational analysis or simulation. Larger flat, thin-walled zones should always be avoided, however, by cambering, ribbing, beading, divider walls, etc.

Maximum dynamic stiffness of the engine mountings, to which the flexible engine mountings are fixed, is also a design target of unconditional desirability. They should be regarded as cantilever beams with additional masses at the free end (covibrating engine mounting mass), the resonances of which increase the solid-borne engine noise transmitted via them into the body shell by as much as a power of ten. The designer should, therefore, always attempt to achieve more than 1000 Hz for the first natural support frequency, since the engine's excitation of solid-borne noise is then no longer dominant. This can be achieved in practice only provided the mountings are

1. Short; they do not project more than 100 to 150 mm beyond the engine wall.
2. Possess an adequate bolt-connecting base on the engine wall (square with approximately the mounting length as the side dimension), which must be correspondingly stiff.
3. Take the form of closed tapering hollow-section beams.

For this reason, sheet metal supports with an open sectional cross section are scarcely used in modern automobiles; they are being replaced by die-cast aluminum mountings, of which Fig. 27-5 shows an example.

Engine-end mounting length is a matter of the arrangement of the engine mounting in the overall vehicle concept, in which a whole range of functional requirements must be fulfilled. If, then, it is necessary to select between longer body-shell-side engine mounting brackets or longer engine-side mountings, the shorter engine-side mounting generally offers advantages from an acoustic viewpoint.

Similar stiffness criteria must also be taken into account in the arrangement and fixing of the auxiliaries. Their arrangement on the engine derives essentially from engine compartment packaging and the design of the belt-drive system. For vibration engineering purposes, they can be regarded as masses connected via a certain spring stiffness with the engine mass, vibrating in resonance at the corresponding frequency. Here again, the target under normal circumstances must be that of locating this resonance

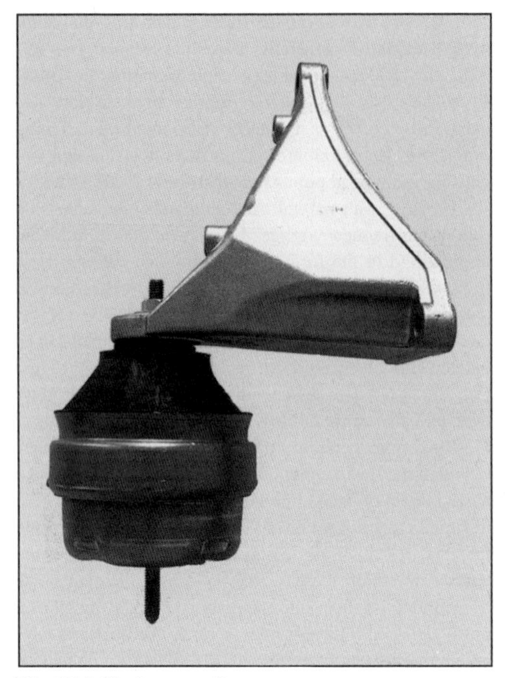

Fig. 27-5 Engine mounting.

frequency as high as possible, i.e., designing the stiffest mounting possible. The apparently elegant and lower-cost arrangement of two or more auxiliaries on one and the same mounting makes achievement of a high resonance frequency difficult because of the accumulation of masses, and also cannot be recommended for yet another reason: The solid-borne noise of an auxiliary unit, such as the servosteering pump, is transmitted directly into the neighboring auxiliary unit, the air-conditioning system compressor, and also (via its connecting elements) the air-conditioning system hoses into the body shell. It is also generally recommendable not to select a whole number ("integer") transmission ratio for the auxiliary unit, but 1.1 or 0.9, for example, instead, in order that the vibration excitation generated by the auxiliaries cannot coincide with the frequency of one of the engine orders. Unavoidable belt slippage will otherwise result in the superimposition of two oscillations of almost identical frequency, causing acoustically extremely unpleasant periodic fluctuations in level, so-called beats." [12–14]

27.7 Measuring and Analytical Methods

The complexity of the internal combustion engine and its peripherals as a noise source has resulted in the course of time in the development of a large range of experimental methods that would, applied together, provide an extremely detailed picture of the overall system. Because of the possible considerable cost and complexity involved, for measurement of the pressure distribution in a crankshaft bearing shell, for example, a standard range of diagnostic methods, using which the majority of problems can be solved, or at least identified and estimated, is initially applied in practice. It is necessary to differentiate from these methods for the testing of typical specification data or benchmarks, such as the acceleration level on the engine mountings or the engine's radiated acoustic power, for which simple prescribed procedures generally exist.

A universal tool that is generally utilized at the start of a diagnosis is the application of so-called signature analysis to an airborne noise signal, obtained either from individual microphones (close around the vehicle or in the interior) or from a dummy-head image (Fig. 27-6) of the noise. A large number of spectra are plotted in 3-D form as a so-called 3-D waterfall diagram (Fig. 27-7, left side) or in the form of a (generally color) coded 2-D figure, a Campbell diagram, across an engine-speed ramp (Fig. 27-7, right side). Such a depiction is extremely useful since, in particular, it is possible, on the one hand, to read off the excitation dimension in the form of its relevant orders as obliquely running curves and because, on the other hand, resonances in the transmission route are visible as a result of peaks with a fixed frequency. Where, in addition, individual orders of frequency ranges can be filtered out or elevated, it becomes possible to identify the components critical for the problem under study in an acoustic assessment.

Where only a few orders are relevant, and where precise quantitative information is required, one restricts oneself to the depiction of order curves against engine speed (Fig. 27-8). In this context, the use of variable frequency filters makes it possible to achieve data reduction even during the measurement. Typical order curves for an internal combustion engine are those of the largest unbalanced mass forces and moments and those of the ignition frequency, i.e., the 2nd engine order in a four-cylinder inline engine and the 1st, 2nd, and 2.5th engine order in a five-cylinder engine. It is normal practice in the higher-frequency airborne noise range (mechanical engine noise) to subdivide the noise into frequency bands (generally thirds or octaves) and to plot their level against engine speed. An increase in these levels may, for instance, indicate deficiencies in the acoustic insulation between the engine compartment and the vehicle interior.

Where a detailed analysis of noise radiation is required, the classical—but high-input—window method, in which small "windows" are opened point by point from a complete, tightly fitting insulating enclosure (consisting, for example, of mineral fiber and lead sheet), and their influence on radiation measured has been superseded (or at least augmented) by more modern procedures involving less feedback effect on the test object and the sonic field. In the case of the intensity method, the emitter is scanned point by point at a relatively small distance from the surface. The plotting of acoustic intensity above the projected surface then provides a good picture of the distribution of heavily and less heavily radiating zones, making it possible to determine the total radiated acoustic

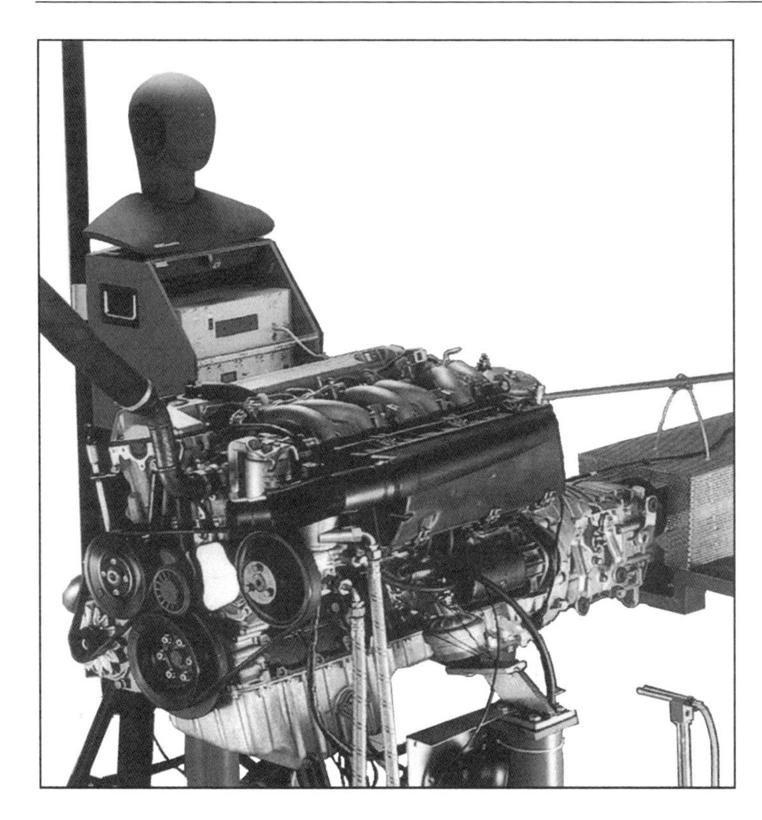

Fig. 27-6 Dummy head [Head acoustics].

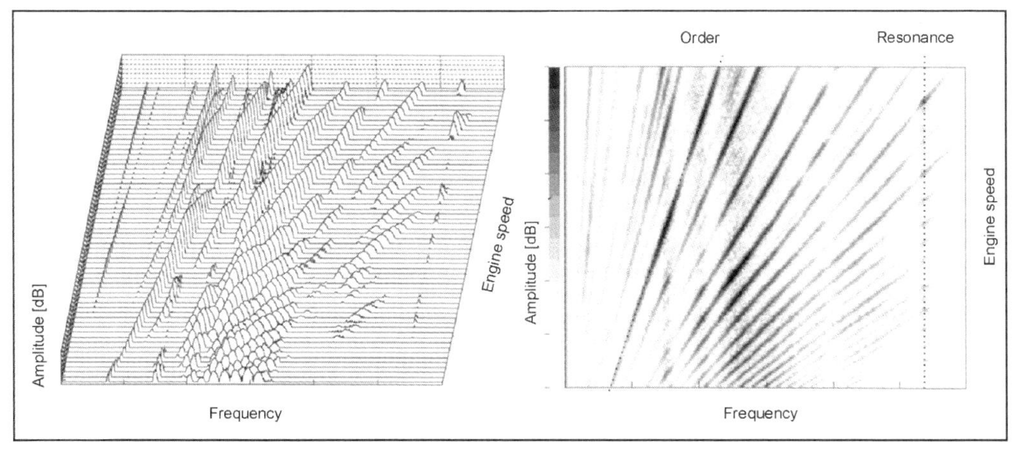

Fig. 27-7 Signature analysis: Campbell diagram (at right), 3-D waterfall diagram (left) [AFT Atlas Fahrzeugtechnik].

power simultaneously. One disadvantage is the long measuring time, throughout which a stable operating state must be maintained. In addition, an automatic system for movement of the intensity probe is necessary in many cases for safety and repeatability reasons. Acoustic near-field holography, on the other hand, is based on the signals from simple sonic pressure microphones arranged in a grid pattern. The use of computer-assisted signal evaluation based on the differential equations for sound wave propagation makes it possible—presupposing simultaneous measurement at a large number of points—to determine after a relatively short measuring time the intensity distribution on the radiating surface and in planes farther away.

So-called operational vibration analysis, which renders the motion modes occurring under actual operating conditions, is used where necessary in order to determine—and influence, where necessary—the oscillation modes relevant to radiation and/or solid-borne noise transmission.

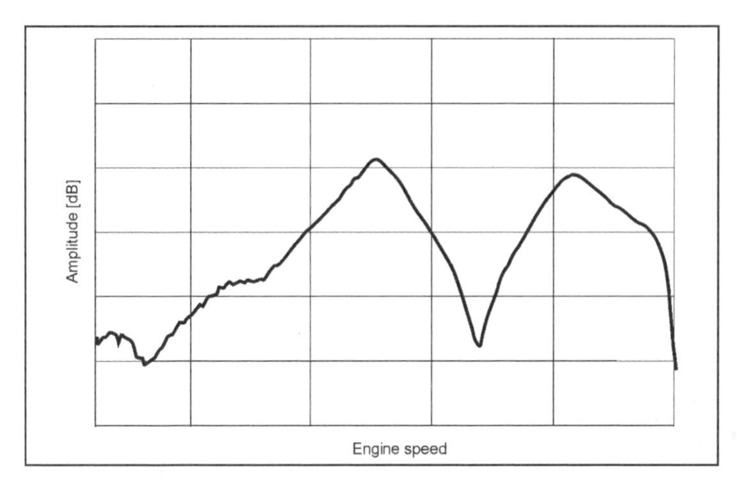

Fig. 27-8 Order level against engine speed, section from Fig. 27-7 [AFT Atlas Fahrzeugtechnik].

Experimental modal analysis, on the other hand, utilizes defined artificial excitation (e.g., an impact hammer) and is generally used for calibration of computation models or for checking whether certain natural frequencies and modes are within specified limits (e.g., the first natural flexural frequency of the engine plus transmission unit).[15] The method most frequently selected for modal analysis and operational vibration analysis is punctiform measurement of accelerations in three directions in each case. With the assistance of a computer, these can then be assigned to the nodal points of a wire-mesh model and frequency selectively animated in slow motion. Other solid-borne noise signals, such as inductively measured paths, are also suitable in principle for operational vibration analysis. Optical methods are frequently used where high temperatures, rotating, and/or thin-walled components are present. In laser vibrometry, surface velocity is measured point by point in one dimension as a time signal. This, it is true, does permit subdivision into frequency components, but necessitates, as a result of the point-by-point scanning method necessary, relatively long measuring times, and, therefore, stable, steady-state operating conditions. Laser double-pulse holography, on the other hand, supplies instantaneous images of the deformation of large areas (Fig. 27-9), which, however, represents the total deformation in the time between the two laser pulses and

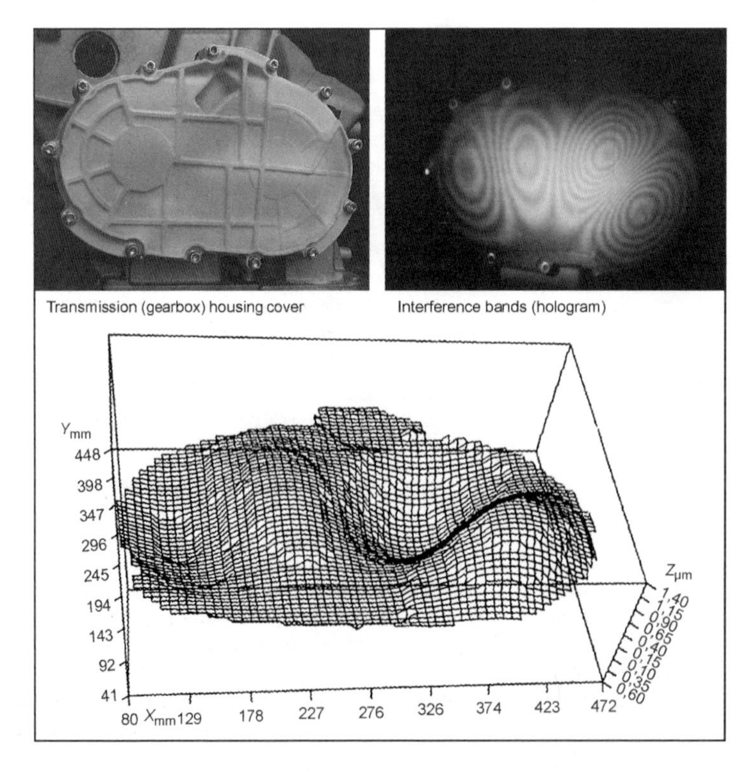

Fig. 27-9 Laser double-pulse holography, instantaneous images of the deformation of large surfaces areas [Laserlabor FHT-Esslingen].

permits "natural-mode-selective" evaluation only to a limited extent and only by skilled selection of the trigger point and time interval (typical: 0.8 msec).

Transfer path analysis is an extremely efficient tool for precise analysis of the solid-borne noise contributions to interior noise and, therefore, frequently forms the basis of an acoustic vehicle study.[16,17] It is essentially performed in three stages:

1. Direct (load cells, strain gauges) or indirect (bearing deformations, input impedances) determination of the cutting forces at the points considered relevant.
2. Determination, with a decoupled noise source, of the acoustic transmission functions from the points of intromission to the point of reception (e.g., the driver's ear) using artificial excitation.
3. Determination of the individual contributors to the noise by multiplication of the forces by the appurtenant transmission functions. It is possible, for instance, to determine for the 2nd engine order which engine mounting supplies in which direction the largest contribution to the buzzing noise perceived at the driver's ear.

27.8 Psychoacoustics

The weighting curves (e.g., A weighting) specified in DIN IEC 651 for sonic signals are an initial, approximate approach toward taking into account the nonlinear behavior of human hearing. A simple frequency weighting procedure does not permit the derivation of information on the subjectively perceived nuisance quality of a noise, however. Noises containing pulses, such as diesel knocking, are perceived as particularly unpleasant, whereas uniform sound at a constant level causes a low perception of unpleasantness, which is discernible less in the spectrum than in the time plot of such signals. The objective registration and quantification of these differences is an objective of so-called psychoacoustics, which defines for this purpose a range of psychoacoustic parameters on the basis of models and detailed audibility tests.

The basis for many psychoacoustic parameters is provided by a modified frequency scale, the "tonality scale" (0 to 24 Bark), which is based on the nonlinear frequency/site transformation of the basilar membrane and, thus, simulates the natural frequency discrimination faculty of the human ear. Using complicated algorithms in some cases, the psychoacoustic parameters can be calculated from measured signals, taking into account spectral and chronological concealment effects, as well as the sensitivities of the ear to fluctuations in amplitude and frequency.

The following are common psychoacoustic parameters:

1. Loudness: Linear variable for evaluation of perception of sound volume, using the "sone" (Reference: 1 kHz sinusoidal tone, 40 dB is equal to 1 sone). A computation method (after Zwicker) is standardized in ISO 532.
2. Sound volume: Level variable for assessment of perception of sound volume, using the "phon" unit; can be approximately calculated from loudness.
3. Severity: Weighting that emphasizes the high frequencies, which give a sound its severity (Unit: 1 acum).
4. Fluctuation amplitude: Assesses extremely low-frequency (<20 Hz) modulations in signal level, which are normally considered unpleasant.
5. Roughness: Assesses modulations in the 20 to 300 Hz frequency range that make a noise appear "rough," which is not always a negative characteristic ("sporty" sound).
6. Tonality: Used for classification of noises in terms of their pure tonal content in proportion to their noise content.[18]

These psychoacoustic parameters, taken together, characterize a noise significantly better than a weighted level, with the result that attempts are often made to derive by suitable combinations of the parameters characteristics data for the quality of a noise; however, these characteristics are restricted to certain contexts. We may mention, to illustrate the basic difficulty here, the exhaust noise from a Ferrari, which is considered "good" by a young man, but not by his grandmother. A large number of carefully tailored hearing studies with assessors drawn from vehicle purchasers and/or experts have been and still are performed in order to define the correct targets. The basis for this is provided, because of the high level of reproduction quality and manipulation potentials of the associated analysis systems, generally by dummy-head recordings (Fig. 27-6), actually reproduced in some cases in an original environment and including perception of low-frequency solid-borne noise (hand, foot, and seat vibration). This is also the interface between psychoacoustics and sound engineering.

27.9 Sound Engineering

The accepted tenets of vehicle acoustics have for some years now included recognition of the fact that simple "quietness" is, in many cases, no longer an appropriate target, since a certain level of acoustic feedback from the vehicle is, in fact, necessary and is expected. The function of sound engineering is now that of providing the desired acoustic information with a sound that is as pleasant and, where appropriate, also as typical of the marque as possible, and to strike certain characteristics, such as "sporty," "powerful," "dynamic," or "classy," depending on the vehicle type. Naturally enough, particular importance attaches in this context to the sound of the engine. The engine also incorporates at the same time the largest number of potentials for modification of noise, as a result of the large range of transmission routes and the composition of the noise in the form of a mixture of orders. If one sets aside for the time being purely electronic manipulations, using the spectacular but "artificial" results that can be achieved, it is possible to emphasize individual orders, for example, modifications to the intake and exhaust systems, and thus to generate the required sound, in many cases with a "sporty" or "dynamic" note.

Essential fundamentals for the quality of engine noise are set down, however, at the concept phase, where decisions are made concerning number and arrangement of cylinders, firing order, mechanical balancing, housing and crankshaft stiffness, air routing, etc. Provided the main criteria are taken into account at this stage, a good basic engine sound results almost automatically, signifying that "quietening" of undesirable, higher-frequency components resulting from gas forces, timing gear, and auxiliaries, which without doubt increase as engines and vehicles become more complex, is again needed to achieve a "perfect" acoustic whole. It is not always easy to decide what components must change, and to what extent, since people's subjective perception may, on the one hand, be extremely diverse, and, on the other hand, the costs involved in the development and implementation of remedies are not insignificant. The models, parameters, and methods of psychoacoustics are, therefore, increasingly used to facilitate this decision.[19]

27.10 Simulation Tools

The predictive calculation of the vibrations and radiated airborne noise from an engine that is still at the design stage even today remains a demanding target (Fig. 27-10).

In addition to the large number of degrees of freedom necessary, it is, in particular, the modeling of the charging and combustion processes, and also the nonlinear contact processes (impacts, lubricating films, friction, etc.) between the many moving and fixed components that make the achievement of an informationally useful overall model extremely difficult. The consequences have been a series of specialized models and methods, the results of which generally must first be compiled to obtain an overall statement.

The starting point on the structure dynamics side is an FE model of the engine plus transmission unit and an FE or multibody model of the rotating crank gear and also, where required, of the timing gear, which are sufficiently fine to permit registration of the complex oscillation modes in the acoustically relevant frequency range. Verification of them is normally performed in subsystems by calibration against experimental modal analyses. One of the main difficulties in this field is correct registration of attenuation. The most complex and difficult element, however, is calculation of the forces acting on the block structure during operation. It is necessary for this purpose to link the individual models, the linking procedure conditions through the hydrodynamic oil film of the bearings and the cylinder sliding surfaces being highly nonlinear. The gas forces acting on the crank gear can, for their part, either be taken from an indication measurement or be calculated using a complex charging and thermodynamic model. The overall result of the simulation is a velocity distribution on the surface of the engine, including the engine mountings.

Calculation of the radiated airborne noise, for the purpose of quantification of emitted noise, for example, can

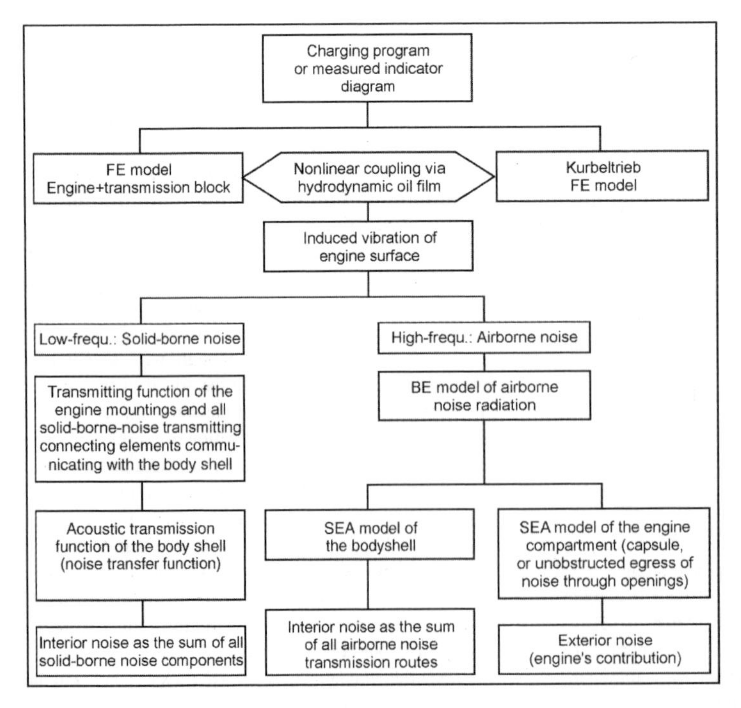

Fig. 27-10 Example of an engine noise simulation.

be accomplished relatively easily using FE or BE (boundary element) methods, since at this stage only a single, homogeneous medium is involved. It is necessary for calculation for vehicle interior noise to differentiate between the airborne noise passing through the body shell wall and the solid-borne noise transmitted into the body shell via the engine mountings, drive shafts, and various other connecting elements. The solid-borne noise transmission routes generally dominate at frequencies above approximately 500 to 1000 Hz. Here, the engine mounting oscillation amplitudes can be regarded as "path excitation" of the "Silentbloc" rubber mountings of the engine mounting system, because they are frequently used as reference factors that must not exceed certain limits. The solid-borne noise component of total interior noise can be calculated for every bearing using the dynamic stiffness of the engine mounting and the acoustic transmission function of the body shell at the bearing points, and these components are then summated with correct phasing for all bearings and all directions of oscillation. The acoustic transmission functions of the body shell can, for their part, be provided in the form of measured data or in the form of the results of a dynamic FE computation of body shell structure and cavity. Mathematical determination of body shell transmission functions is extremely difficult, however, because of the complex structure of a lined body shell incorporating a large number of difficult to define joints. In the higher frequency range, which is generally dominated by airborne noise excitation from the engine compartment, conditions are in some cases further simplified by the fact that it is possible as a result of a broadband noise character and the high density of body shell natural frequencies to observe energy flows for whole frequency ranges using statistical observation methods and ignoring phase relationships, and thus obtain simple algebraic equations (SEA). Because of the extremely high complexity and cost of an informationally useful acoustic model of the complete vehicle—even the development-phase updating of submodels constitutes a not insignificant organizational problem—"hybrid models" are increasingly used in practice, i.e., the combination of assemblies and modules or input variables registered using measuring means with computer models of new components. This route is open, in particular, in the case of new developments based on an existing vehicle floor group or an existing engine family.[20]

27.11 Antinoise Systems: Noise Reduction using Antinoise

The elimination of unpleasant noises by artificially generated antinoise is an option that became technically possible with the availability of high-speed digital control systems. The signal of the noise is registered as close as possible to its source, the opposite phase signal is generated in a computer, and then it is emitted via an amplification system. The "cancellation" of harmonic signal components functions best of all, e.g., one or more engine orders. The analogous mechanical principle is the familiar cancellation of mass crank gear forces by opposite-phase forces using balancing shafts.

Antinoise systems for four-cylinder vehicles, which capture engine noise in the vehicle interior using a number of microphones and reduce the buzzing of the 2nd engine order by more than 10 dB using systematically located loudspeakers, became available in the 1980s in the form of mature prototypes, such as those from Lotus Engineering. Practically all the major automobile manufacturers also had their own development programs in their research departments, which were unveiled to great publicity in the technical press in the form of experimental vehicles in which the buzzing noises could be switched off during travel at the push of a button. These initially impressive presentations at the same time disclosed a deficit: The vibrations generated along the same route by the engine were not susceptible to influence with antinoise by the loudspeakers, but do play a role in determining the subjective impression of comfort. Further developments were, therefore, logically enough, aimed at eliminating the solid-borne noise and, therefore, also the vibration at the points of transmission into the body shell by counterphase controlled vibration generators, in the form of piezoactuators, which further heightened the technical and cost input.

Complexity, and also the cost benefit ratio, was in the past also the main reason for antinoise not becoming established in mass-production automobiles. Put quite simply: "It is too expensive for the four-cylinder vehicles which need it—and the more expensive six- and eight-cylinder engines do not need it." The frequently expressed vision of "electronics instead of weight" (i.e., of balancing out the costs for the control system by saving on noise-insulation materials), was doomed to failure sheerly by physical principles: The sound package in the automobile body shell is installed almost entirely for reduction of the high-frequency noise components, the signals of which have a stochastic character and, therefore, no definable phase reference. Signal coherence is, however, the precondition for all stable interference phenomena and, therefore, also for noise cancellation.

Recent developments in higher-value four-cylinder engines indicate a pronounced trend toward the classical "balancing shaft" solution (Lancaster balancing), through which the mass forces of the 2nd engine order and the vibrations and buzzing noises caused by it can be completely eliminated. As a result of the entirely different cost framework, antinoise systems are in use in a series of applications in aerospace and plant engineering. Antinoise can ideally be emitted into a helicopter pilot's helmet close to his ear in order to cancel low-frequency rotor noise and, thus, improve ease of radio communication and general comfort. Antinoise systems for cancellation of low-frequency fan noises using a loudspeaker system installed directly on the fan outlet are employed in wind tunnel facilities. The main field of application is that of low-frequency noise problems, where the standard noise absorption and insulating materials offer little effective help.

Bibliography

[1] Bathelt, H., and D. Bösenberg, Neue Untersuchungsmethoden in der Karosserieakustik, in ATZ 78 (1976) 5, pp. 211–218.

[2] Heckl, M., and H.A. Müller, Taschenbuch der Technischen Akustik, Springer-Verlag, Berlin, Heidelberg, 1994.

[3] Henn, H., G.R. Sinambari, and M. Fallen, Ingenieurakustik, 2nd edition, Friedr. Vieweg & Sohn, Wiesbaden, 1999.

[4] Kremer, L., and M. Heckl, Körperschall, 2nd edition, Springer-Verlag, Berlin, 1996.

[5] Ehinger, P., H. Großmann, and R. Pilgrim, Fahrzeug-Verkehrsgeräusche, Messanalyse- und Prognose-Verfahren bei Porsche, in ATZ 92 (1990), No. 7/8, pp. 398–409.

[6] Klingenberg, H., Automobil-Messtechnik, 2nd edition, Springer-Verlag, Berlin, 1991, Volume A: Akustik.

[7] Betzel, W., Einfluss der Fahrbahnoberfläche von Geräuschmessstrecken auf das Fahr- und Reifen-Fahrbahn-Geräusch, in ATZ 92 (1990), No. 7/8, pp. 411–416.

[8] Basshuysen, R.v., Motor und Umwelt, in ATZ 93 (1991), No. 1, pp. 36–39.

[9] Albenberger, J., T. Steinmayer, and R. Wichtl, Die temperaturgesteuerte Vollkapsel des BMW 525 tds, in ATZ 94 (1992), No. 5, pp. 244–247.

[10] Eikelberg, W., and G. Schlienz, Akustik am Volkswagen Transporter der 4, Generation, in ATZ 93 (1991), No. 2, pp. 56–66.

[11] Geib, W. [Ed.], Geräuschminderung bei Kraftfahrzeugen, Friedr. Vieweg & Sohn, Braunschweig, 1998.

[12] Kollmann, F.G., Maschinenakustik, Springer-Verlag, Berlin, Heidelberg, 1993.

[13] Küntscher, V. [Ed.], Kraftfahrzeugmotoren, 3rd edition, Verlag Technik, Berlin, 1993.

[14] Mollenhauer, K. [Ed.], Handbuch Dieselmotoren, Springer-Verlag, Berlin, Heidelberg, 1997.

[15] Ewins, D.J., Modal Testing, Theory and Practice, Research Studies Press Ltd., Letchworth, 1984.

[16] Bathelt, H., Analyse der Körperschallwege in Kraftfahrzeugen, in Automobil-Industrie 1, März 1981, pp. 27–33.

[17] Bathelt, H., Innengeräuschreduzierung durch rechnergestützte Analyseverfahren, in ATZ 83 (1981) 4, pp. 163–168.

[18] Zwicker, E., and H. Fastl, Psychoacoustics, Facts and Models, Springer-Verlag, Berlin, New York, 1990.

[19] Quang-Hue, V. [Ed.], Soundengineering, Expert Verlag, Renningen-Malmsheim, 1994.

[20] Estorff, O.v., G. Brügmann, A. Irrgang, and L. Belke, Berechnung der Schallabstrahlung von Fahrzeugkomponenten bei BMW, in ATZ 96 (1994), No. 5, pp. 316–320.

28 Alternative Propulsion Systems

28.1 The Rationales for Alternatives

With only a few exceptions, modern vehicles are powered by gasoline or diesel engines and the appurtenant fuels. The exceptions take the form of vehicles fueled with CNG (compressed natural gas) or LPG (liquefied petroleum gas). Vehicles using electricity or hydrogen as their propulsion energy at present still retain the status of "niche" applications. Gasoline engine fuel currently occupies a world market share of somewhat more than 90%, diesel fuel around 10%. The rationales for alternative energy sources either can be found in local circumstances or are orientated around the availability of independent energy sources in a particular country or region. In the past, sufficient crude oil was available as the original energy form for gasoline and diesel fuel, but the search for alternatives has expanded enormously in recent years. The essential reasons for this are as follows:

- Many states are attempting to make their energy use less dependent on the dictates of the oil producing countries. In addition, the production of crude oil is becoming increasingly more complex and costly.
- Carbon dioxide emissions need to be reduced significantly. Part of the blame for global warming of the atmosphere is attributed to this gas. Carbon dioxide is produced in combustion of any carbon-containing energy source. The agreement between the ACEA and the Commission of the European Union specifies the reduction of production of CO_2 from new vehicles entering the European market to 140 g/km CO_2 by the year 2008. This signifies, given a 30% share of the European market held by diesel, an average consumption of 5.7 l/100 km.
- There is a desire for totally emissions-free vehicles.

The factors mentioned above must be taken into account in assessment of the use of alternative energies for vehicle propulsion systems. Figure 28-1 shows an overview of the arguments for and against the various energy sources.

The type of energy used is of decisive importance for the assessment of propulsion systems. It is necessary, in every case, to take into account the complete energy "chain." This will, of course, include original recovery of crude energy, preparation and processing (refining), transportation, and consumption in the vehicle.

28.2 The Wankel Engine

The Wankel, or rotary piston, engine takes the form of a special variant of the reciprocating piston engine and is, therefore, an intermediate link between the classical crankshaft engine and other types of propulsion systems. Figure 28-2 shows the structure and manner of operation of a Wankel engine.[1]

The advantages of this type of engine can be found in its complete balancing of moving masses, its compact design, the elimination of the valve drive and timing system, and its high torque output. Disadvantages include the geometrically unfavorable combustion chamber, negative quench effect, relatively high HC emissions, high fuel and oil consumption, and unsuitability for use as a diesel engine. Mazda, nonetheless, continued the development of this variant after its initial use in the NSU Ro 80.

28.3 Electric Propulsion

Electric propulsion systems were, in fact, used at an extremely early stage in the development of motor vehicles. As a result of the inadequate energy storage facility (the battery), they have up to now become established only in niche applications, unlike the gasoline and diesel engines. The electrically powered vehicle itself can, of course, be "emissions-free" in operation, but overall assessment must depend on the nature of generation of the electricity used. The propulsion system of an electrical vehicle comprises[2]

- Electric motor with electronic control system (thyristor) and cooling system
- Traction battery, including battery management system and the necessary charger
- Any necessary transmission gearing system, including differential gear

	CNG with catalyst	Gasoline	Diesel	Rapeseed oil methyl ester (REM)	LPD with catalyst	Methanol	Ethanol	Hydrogen	Electricity
Suitability	o	o	o	o	o	o	o	o	–
Availability	o	o	o	–	–	–	–	– – –	– –
Cost efficiency	–	o	+	–	o	–	–	– – –	– – –
Infrastructure	–	o	o	–	–	–	–	– – –	–
CO	+ + +	o	+	+	+	o	o	+ + +	+
HC	+ + +	o	+	+	+	o	o	+ + +	+
NO_x	+ +	o	–	–	+	+	o	+ +	+
Particulates	+ + +	o	–	–	+ + +	o	o	+ + +	+ +
Co_2	+ +	o	+ +	+ +	+ +	o	+ +	+ + +	–

Fig. 28-1 Propulsive energy sources suitable for use in motor vehicles.

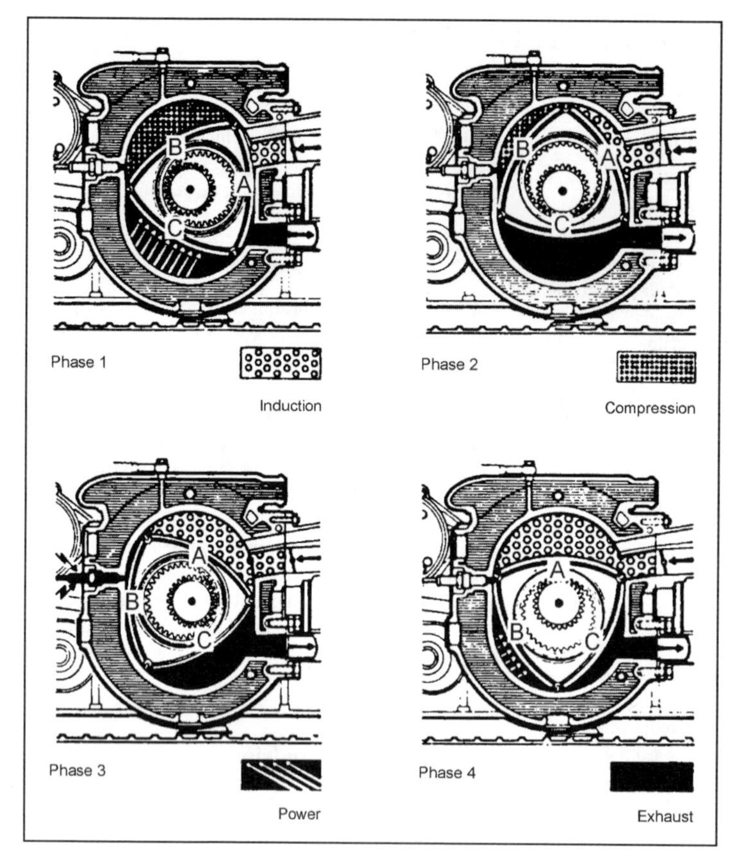

Fig. 28-2 Principle of the Wankel engine.

- System for transmission of power to the drive wheels
- Electrical steering and brake servosystem
- Heating and air-conditioning system
- Chargers (fixed or "on board")

A range of different electric motors are available for the propulsion system. The selection criteria are as follows: Low weight, high efficiency, compact design, low costs (manufacture and maintenance), and high torque across the broadest possible motor speed range. The possible variants can be seen in the following list:

- DC motors
  - Series-wound DC motors
  - Shunt-wound DC motors
- Three-phase motors
  - Asynchronous motors
  - Synchronous motors
  (a) Permanently excited synchronous motors
  (b) Separately excited synchronous motors
- Special motors
  - Brushless DC motors
  - Transverse flux motors
  - Switched reluctance motors

Figure 28-3 shows a comparative assessment of a number of electric motor types. The switched reluctance motor appears to be of particular interest.[2]

The front and/or rear wheels are frequently driven by the central electric motor for the purpose of transmission

	DCM	ASM	SSM	PSM	SRM	TFM
Efficiency	– –	+	+	+ +	+	+ +
Maximum motor speed	– –	+ +	+	+	+ +	– –
Volume	– –	+	+	+ +	+	–
Weight	– –	+	+	+ +	+	+
Cooling	– –	+	+	+ +	+ +	+
Manufacturing input	–	+ +	–	–	+ +	– –
Cost	–	+ +	–	– –	+ +	– –

DCM: DC motor; ASM: Asynchronous motor; SSM: Separately excited synchronous motor
PSM: Permanent-magnet excited synchronous motor; SRM: Switched reluctance motor
TFM: Transverse flux motor

Fig. 28-3 Comparative assessment of electric motors for electric vehicles.

of the torque to the drive wheels; in individual cases such as buses the electric motors are, in fact, installed in the wheels. A single-stage gearing system with a fixed transmission ratio is generally sufficient for the power transmission train, as a result of the high torque generated by the electric motor and because the electric motors can be overloaded for short periods.

The central reason for the low numbers of electric vehicles in use is the limited performance of batteries. The traction battery is the most important component in an electric propulsion system since the vehicle's range depends on its energy concept. The available electrical output determines vehicle performance. Figure 28-4 provides an overview of possible traction batteries.

It can be seen that the Californian demand for zero emissions vehicles has driven battery developments, particularly in the field of nickel/metal-hybrid and lithium/ion batteries. The range of niche products available will most certainly increase as battery performance rises.

Figure 28-5 shows a recent Ford electric vehicle variant, the "Think City Car." This vehicle is characterized by the following data:

Think City: Technical data

Seats	2
Dimensions	Length 2.99 m, Width 1.60 m, Height 1.56 m
Unladen weight	940 kg
Permissible laden weight	1130 kg
Trunk capacity	350 l
Max. payload	115 kg
Maximum speed	90 km/h
Acceleration	0 to 50 km/h in 7 sec
Range	85 km

Chassis and body shell

Chassis	Steel, zinc-plated
Superstructure	Extruded and welded aluminum
Body shell	Thermoplastic (polyethylene)
Roof	ABS plastic

Batteries and motor

Batteries	19 NiCd (nickel/cadmium) batteries, weight approximately 250 kg, water cooled
Output	11.5 kWh, 100 Ah
Charger (internal)	220V–16A/10 A (32.2/2 kW)
Charging time	4 to 6 hours (80% of battery rating), can be charged from any 220 V power socket
Motor type	Three-phase, water cooled
Max. motor output	27 kW
Voltage	114 V
Tires	115/70 R13

Battery type	Energy density		Power density		Service life		Cost
	Wh/kg	Wh/L	Wh/kg	Wh/L	Cycles	Years	EUR/kWh
Lead	30–50	70–120	150–400	350–1000	500–1000	3–5	100–150
Nickel/Cadmium	40–60	80–130	80–175	180–350	>2000	3–10*	225*–350
Nickel/Metal hydride	60–80	150–200	200–300	400–500	500–1000	5–10*	225*–300
Natrium/Nickel chloride	80–100	150–175	155	255	800–1000	5–10*	225*–300
Lithium/Ion	90–120	160–200	ca. 300	300	1000	5–10*	275*
Lithium/Polymer	150	220	ca. 300	450	<1000	—	<225*
Zinc/Air	100–220	120–250	ca. 100	120	—	—	60*
Target figures	80–200	135–300	75–200	250–600	600–1000	5–20	90–135

Fig. 28-4 Battery systems for electric vehicles, comparative assessment.

Fig. 28-5 The "Think City Car" electrical vehicle.[16]

28.4 Hybrid Propulsion System

As a general definition, hybrid propulsion system vehicles are those with two different propulsion systems and two different energy storage facilities. Appropriate electronic management systems make it possible to achieve a number of advantages over a conventional propulsion system. One such advantage is the reduction of fossil energy consumption, because the internal combustion engine is operated at or very close to its maximum efficiency working point, the acceleration is boosted by the electric motor, and/or the electric motor is used for initial movement. "Downsizing" the internal combustion engine element, thus, becomes possible in principle. Partial recovery of braking energy helps improve overall efficiency. Emissions in use are extremely low in many cases and close to zero in electric mode. A further advantage, particularly in electric operating mode, is the low noise output.

Like purely electric vehicles, hybrid propulsion systems must, at present, be regarded only as a niche specialty. Hybrid vehicle types can be classified as shown in the schematic in Fig. 28-6.[3,4]

The parallel hybrid makes it possible to provide traction from both propulsion units simultaneously. The serial hybrid always has an electric motor propulsion system; i.e., electric power must be generated on board. This can be achieved using an extremely diverse range of machines such as gasoline or diesel engines, gas turbines, Stirling engines, fuel cells, etc. The advantage is the fact that the electricity generating machine can be operated at the point that achieves optimum efficiency and optimum emissions. Sales have not yet been convincing, despite the fact that numerous companies have marketed corresponding series-manufactured vehicles.

The best known hybrid vehicle at present is the Toyota Prius, which is already on the market in Japan and is also available in modified form in Europe and the United States. The Prius is, in fact, more of a "mixed hybrid" and should really be regarded as a precursor of the fuel cell propulsion system. Figure 28-7 shows the layout in principle,[5] while Fig. 28-8[5] illustrates the CO_2 emissions-saving effect, but on the basis of the Japanese driving cycle, which includes an extremely high proportion of stationary vehicle times. Thirty thousand vehicles had been sold in Japan alone by the end of the year 2000. In addition to an optimized gasoline engine that permits achievement of the Californian SULEV (super ultra low emission vehicle) exhaust standards, numerous other interesting solutions, such as improvement of overall power train efficiency, stop/start function, and optimization of the heating and air-conditioning systems, have been incorporated into the latest versions.

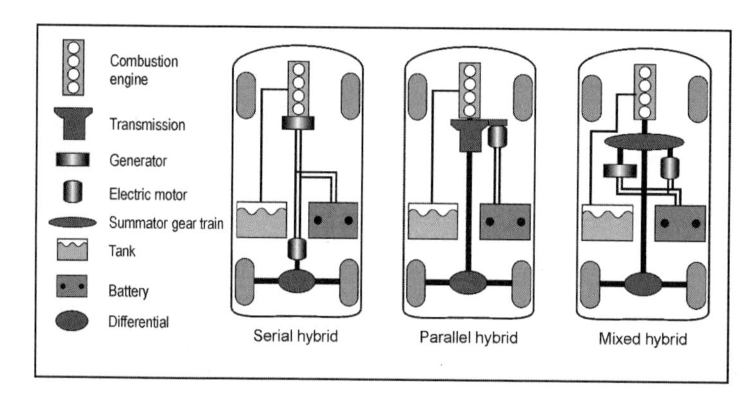

Fig. 28-6 Schematic of hybrid vehicles.

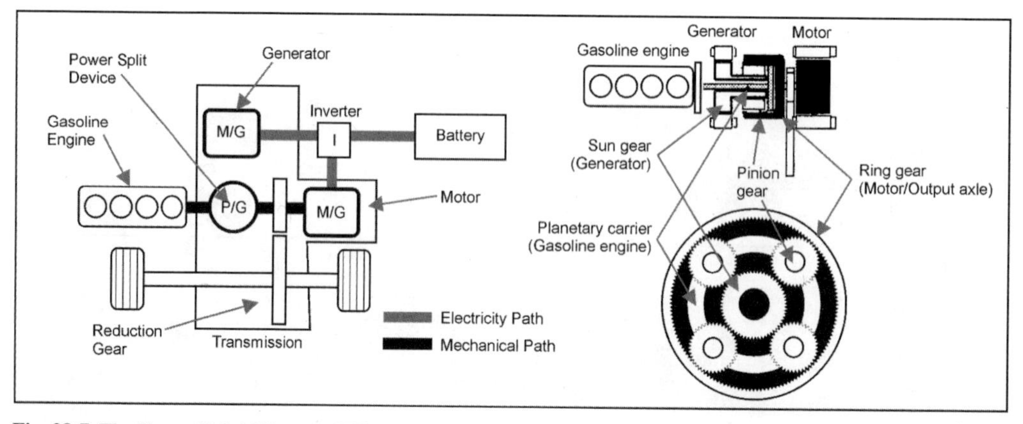

Fig. 28-7 The Toyota Hybrid System (THS).

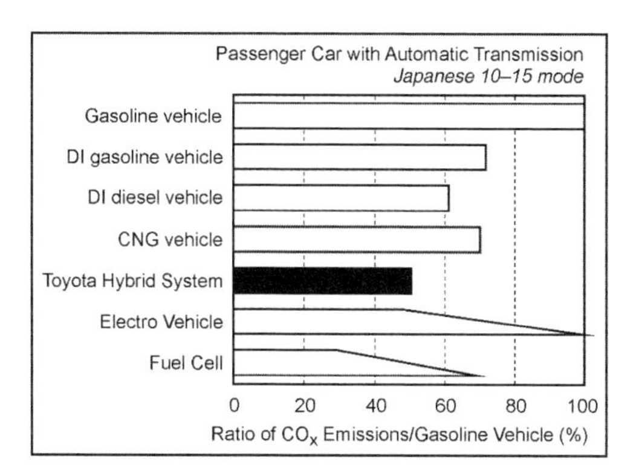

Fig. 28-8 Comparative assessment of CO_2 emissions from the Toyota Hybrid System and competing systems.[5]

28.4.1 Storage Systems

Flywheel energy storage systems and ultracaps (i.e., supercapacitors) are regarded as potential energy storage systems, particularly in conjunction with hybrid propulsion systems. Flywheel energy storage systems are currently used mainly in larger vehicles (buses and trams),[3,6,7] whereas the use of supercapacitors (high-performance capacitors) is also conceivable in automobiles. Numerous potential applications exist, as a result of the fact that supercapacitors are capable of generating extremely high powers for short periods.

These uses range from catalytic converter preheating, initial movement boosting, and power storage in electric and hybrid vehicles up to and including use in partial braking-energy recovery. Figure 28-9 shows a comparative technological assessment of the performance of a number of storage systems.[7]

28.5 The Stirling Engine

The Stirling engine (invented as early as 1918) has again and again been discussed as a potential power unit for motor vehicles. It functions on the basis of continuous external combustion or heat input. This thermal energy is transferred via a heat exchanger to the working gas in the cylinder. Using a displacer piston, the gas is transferred backward and forward between a chamber with a constant high temperature and a chamber with a constant low temperature. Internal pressure changes periodically as a result. The pressure changes are converted to mechanical energy by a working piston and a corresponding crankshaft mechanism. As shown in Ref. [8], the process's theoretical cycle (closed-circuit process with continuous heat input) can be described using two isotherms and two isochores.

The theoretical cyclical process of the Stirling engine is shown in the form of a *p-V* and *T-s* diagram in Fig. 28-10.

In the engine process, the cycle is implemented on a clockwise basis, and in the refrigeration set or heat pump, on a counterclockwise basis. The individual elements in the theoretical cycle process are

1 to 2: isothermal compression; following adiabatic compression, the working gas is recooled to its initial temperature in a cooler, the heat being yielded to the environment or to a fluid requiring heating.

2 to 3: isochoric heat absorption; absorption of heat in a regenerator.

3 to 4: isothermal expansion; after adiabatic expansion, the working gas is reheated in the heater to its initial state, input of heat from an external continuous combustion process being necessary: useful work is yielded in this subcycle.

4 to 1: isochoric heat removal; removal of heat in the regenerator.

The ideal process described can be achieved only if the working and displacer pistons move discontinuously. The efficiency of the ideal process equates to the Carnot efficiency. The Carnot efficiency forms the basis for the assessment of the efficiency of combustion engines:

$$\eta = 1 - T_1/T_3 = 1 - T_{min}/T_{max} \tag{28.1}$$

The most important advantages of Stirling engines are low emissions, the ability to use any suitable heat sources

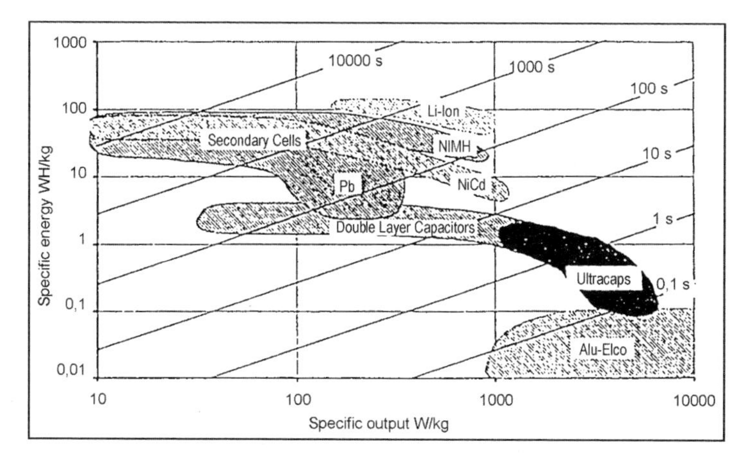

Fig. 28-9 Technological assessment of energy injection systems.

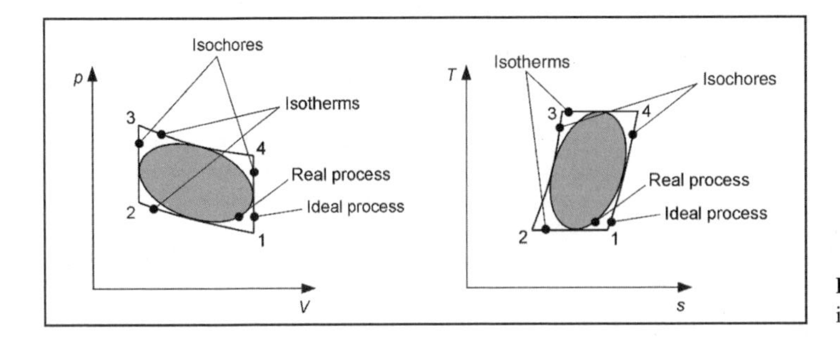

Fig. 28-10 The Stirling process in the *p-V* and *T-s* diagram.

(which can be generated using a range of different energy sources), an extremely good efficiency at the optimum working point (even in the part-load operating range, given regulation of swept volume), low vibration, and low noise; Fig. 28-11 shows a design provided by the STM Company.

The disadvantages are as follows: Poorer throttle reaction, higher load regulation complexity, greater space requirements (because of the size of the heat exchangers), and high production costs, with the result that, unlike the situation in stationary machines, the Stirling engine has up to now failed to become established in mobile applications.

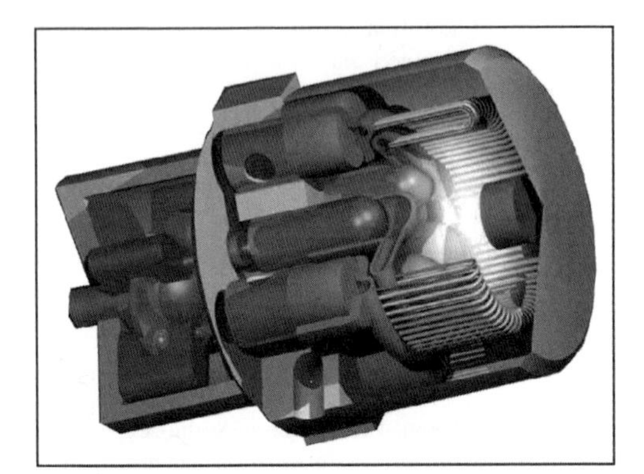

Fig. 28-11 STM Stirling engine (25 kW).

28.6 Gas Turbines

High-temperature gas turbines are propulsion systems that can be operated using an extremely large range of different fuels, i.e., different forms of energy. In some cases, chemical conversion processes that would otherwise be necessary for use in conventional engines can actually be omitted. Gas turbines powering motor vehicles have, for example, been operated directly using pulverized coal. Advantages for the overall efficiency of the conversion chain from the primary energy source up to the vehicle drive system derive from this.

The structure of a motor-vehicle gas turbine derives from the special requirements of automobile operation. Figure 28-12 shows the principle of this in schematic form.[9] The twin-shaft design, in which a separate work turbine is installed downstream of the gas generator set consisting of the compressor and compressor turbine, produces the elevation of torque needed for initial movement in a vehicle. Adjustable guide vanes upstream from the work turbine are also indicated in Fig. 28-12.

During operation, positioning of these vanes can be used to modify the passage cross section and, thus, vary mass flow in such a way that the required output is achieved at maximum permissible turbine inlet temperature. Minimum fuel consumption is thus attained. Reaction time at acceleration can be shortened by opening the vane cross section briefly, while, in coasting operation, the gas flow can be directed onto the rotor blades by turning the guide vanes in the opposite direction so that a braking torque is generated.

Alongside the magnitude of operating temperatures, the heat exchanger is the most important element in the achievement of good fuel consumption. The most common types of gas turbine (open design) that are conceivable for use in motor vehicles differ in the number of shafts and stages used (two). The following are possible:

- Single-shaft turbines (gas generator set and work turbine on a single shaft)
- Twin-shaft turbines (gas generator shaft and drive shaft are separated)
- Three-shaft turbines

The twin-shaft gas turbine constitutes a good compromise between complexity and performance. The plot of torque, for example, is significantly better than in the case of the single-shaft machine; load regulation is accomplished via regulation of working fluid temperature and/or via adjustable guide vanes on the turbine and compressor. Despite the emissions advantages, the multifuel capability, the lower vibrations, and the relatively good torque, excessively restricted suitability for lower power ranges, nonetheless, derives from the higher fuel consumptions, the high-temperature-proof ceramic systems necessary, and the high level of noise generation. Poorer throttle reaction compared to the reciprocating piston engine has also resulted in the lack of use of gas turbines as a direct propulsion form in series-produced motor vehicles. Use in a series-produced hybrid vehicle could be of interest, however.

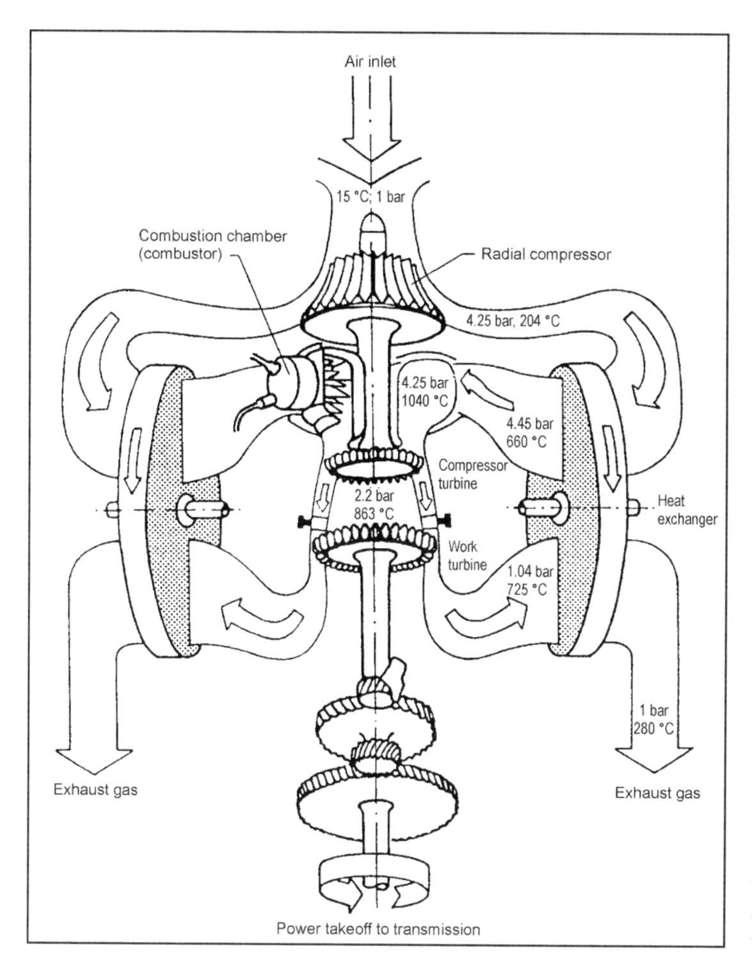

Fig. 28-12 Schematic structure of a twin-shaft automotive gas turbine.

28.7 The Steam Motor

The first steam motor for a vehicle propulsion system was produced in France in 1769. The steam age began in England in the 1830s.

Around 1900, large numbers of automobiles in France and the United States were equipped with steam motors. These were soon replaced by the new internal combustion engine.

A further alternative for vehicle propulsion was unveiled by a research and development corporation[10] in the year 2000: the steam motor. This propulsion unit is capable of fulfilling the Californian exhaust figures with a rational level of complexity and technology. Its structure, in principle, can be seen in Fig. 28-13.

The special feature here is the optimum design of the steam generation system, making it possible, as is not the case with the classical process, to achieve significant efficiency benefits. The consumptions achieved at present can be seen in Fig. 28-14.

The current state of the art propulsion system is, however, the gasoline engine with direct injection achieving conformity with the SULEV emission requirements. The use of the steam motor for the generation of electrical power as an auxiliary power unit in automobiles, for example, is also conceivable. Decisions concerning series-based use remain to be determined.

28.8 The Fuel Cell as a Vehicle Propulsion System

Fuel cells, as essential components of a vehicle propulsion system, have undergone particularly positive development in recent years. One of the reasons for this has been the high level of commitment by subsuppliers and one large automobile company, in particular.[11] In the meantime, all the main automobile manufacturers, numerous subsuppliers, and also scientific institutions are now working on the further development of the fuel cell and the improvement of its functionality in vehicles. The reasons for these activities include the following arguments:

- Dependence on crude oil as the main source of energy can be reduced, since various energy sources are

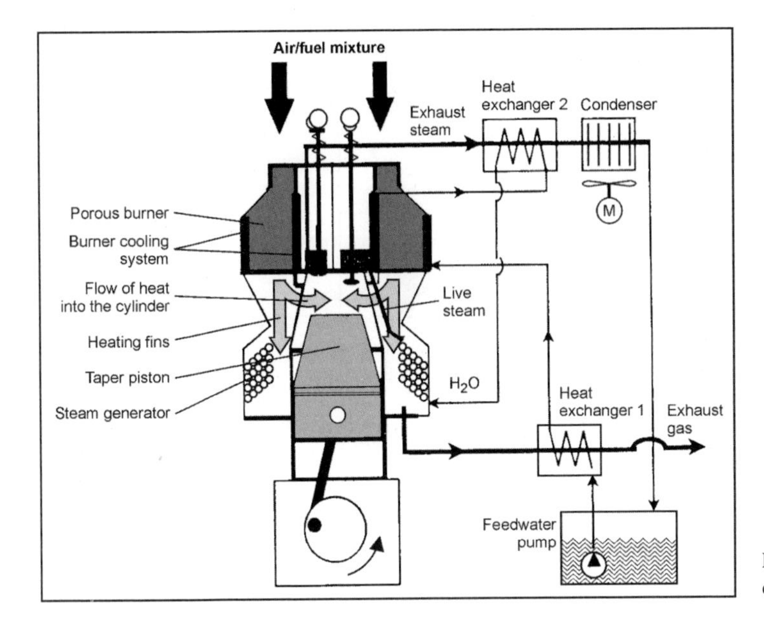

Fig. 28-13 The steam motor produced by IAV.

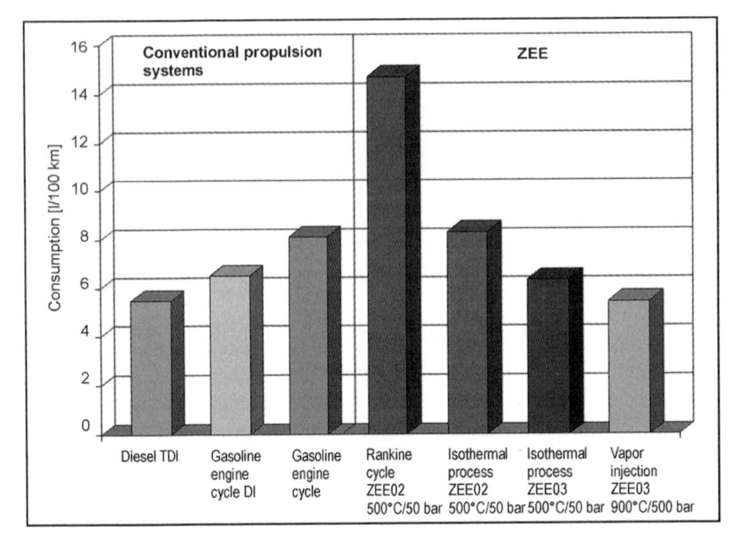

Fig. 28-14 Comparative assessment of consumption data, US FTP 75 driving cycle.

suitable for generating of hydrogen and since, given successful further development of the reformer and the fuel cell, efficiency is also higher when a special gasoline engine is used.

- The fuel cell can be operated with or without a reformer with a range of different energy sources, depending on the particular system and configuration. These energy sources range from hydrogen, CNG, and methanol up to and including a "fuel cell capable" gasoline engine fuel, i.e., a gasoline engine fuel with a low sulfur content, which can more easily be used for generation of hydrogen via a reformer.
- Depending on the particular system, the vehicle's operating emissions of HC, CO_2, NO_x, and particulates are zero or very close thereto. However, the complete emissions chain must be taken into account in assessment of overall emissions.
- If all development targets are achieved, the overall efficiency of the fuel cell vehicle, from the "energy source to the wheel," may be better, again with observation of the entire energy chain, than that of the future diesel engine vehicle. However, the efficiencies of the diesel engine and the fuel cell are at present identical.

Once one has recognized that modern vehicles featuring gasoline engines and equipped with the necessary exhaust aftertreatment systems are capable of conforming to the SULEV requirements of California, the reduction of carbon dioxide emissions remains as the most significant argument. Diesel-engined vehicles are also approaching this emissions level in research. Independence from crude oil is a further reason; i.e., the propulsion energy is not of fossil type and is, therefore, "CO_2 neutral." The necessary improvement in efficiency in gasoline and diesel engines and their power transmission trains, which is currently being achieved, is of particular significance as a "benchmark" for the fuel cell.

The PEM (polymer electrolyte membrane) has proven to be the currently most promising technology for use in mobile systems. The polymer membrane takes the form of a proton-conducting polymer film with a high power density and a working temperature of below 100° C. Alternative systems are at present already in industrial use for stationary applications. An alternative to the PEM is the AFC (alkaline fuel cell), which is used primarily in space projects.[12] With all the advantages of the fuel cell, there are also a whole series of disadvantages to which special attention must be devoted and intensive work performed for their elimination. These are

- Relatively high purchasing costs due to the use of expensive, noble metal, catalysts (high requirement for platinum)
- Relatively low specific power density of the cell
- High costs for fuel storage and vehicle fueling in some cases
- At present, still relatively long heating-up times, particularly where reformers are used
- Poor cold-start consumption levels

28.8.1 The Structure of the PEM Fuel Cell

The cell consists of the electrolyte membrane, the catalysts applied to it by coating, and the bipolar plates; see Fig. 28-15.

These are necessary for the supply of gases and takeoff of the electricity generated. The membrane itself consists of a 1/10 mm thick film (sulfonated fluorocopolymer), which separates the oxygen and hydrogen reaction gases and permits diffusion only of the protons. These are the result of the oxidation of hydrogen at the anode. On the cathode side, air oxygen evaporates to form steam with the protons that have migrated through the electrolyte. The difference in potential can be converted into electrical work in an external circuit.

The individual cells are then arranged in groups forming fuel cell stacks of, for example, 25 kW. Multiple stacks are capable, without further complication, of yielding 150 kW. The design, in principle, of a fuel cell system for a range of different energy forms is shown in Fig. 28-16.

The reformer is omitted if hydrogen is stored directly. The same applies in the case of the use of a direct methanol

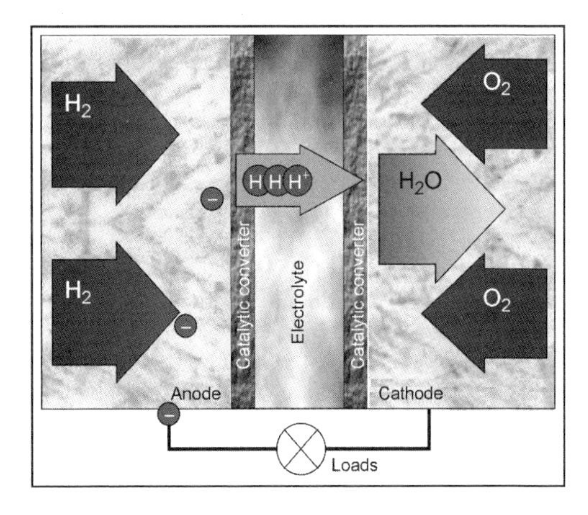

Fig. 28-15 The principle of the fuel cell.

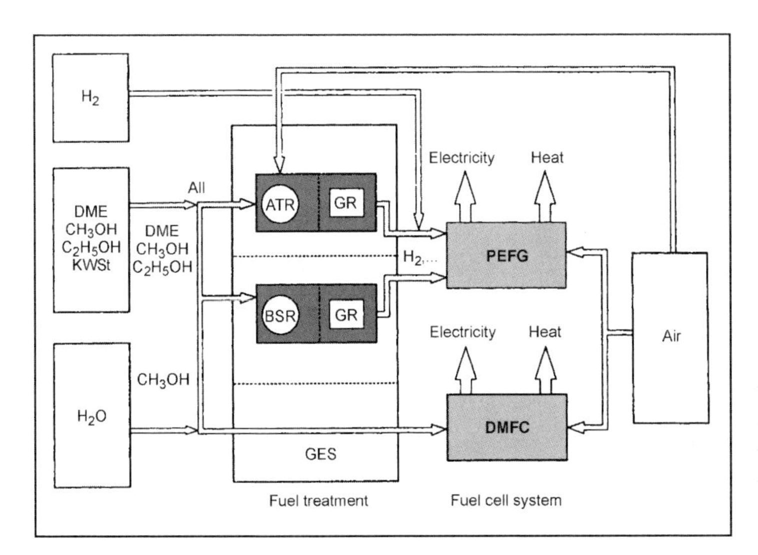

Fig. 28-16 Utilization of a range of different energy sources for fuel cell systems (ATR: autothermal reformer; HSR: heated steam reformer; GR: gas cleaning; PEFC: polymer electrolyte fuel cell).

fuel cell, in which the fuel (methanol) is directly oxidized in the fuel cell.[13] The following energy sources are suitable in principle:

- Hydrogen (H_2)
- Methanol (MeOH)
- Ethanol (EtOH)
- Dimethyl ether (DME)
- Diesel fuels (reference mixture: $C_{12.95} H_{24/38}$)
- Modified gasoline engine fuel

The systems exhibit differing levels of complexity, depending on the original energy source.

28.8.2 Hydrogen as the Fuel

Hydrogen and oxygen are used as the input substances in the supply of the fuel cell with hydrogen from a storage system; i.e., two storage systems are necessary. Here, too, an appropriate peripheral system is necessary for gas storage, supply, and cooling. The storage of hydrogen constitutes a particular challenge, however. Equally important is the generation of the hydrogen by, for example, electrolysis involving cracking of water or, the lower priced variant, production from natural gas. In terms of hydrogen storage, none of the processes currently used (pressurized gas storage, cryogenic storage, and metal hybrid storage) is suitable for use in private automobiles. Oxygen storage is a simpler problem and can be solved via the use of air. There are at present three alternative routes available to achieve greater independence from the performance of the storage systems: methanol with reformer, direct-methanol fuel cells, and gasoline with reformer.

28.8.3 Methanol as the Fuel

The use of methanol as the fuel means that the methanol must be converted to hydrogen via a reformer; see Fig. 28-17.

The methanol itself can, for its part, be recovered from a large range of substances such as natural gas, coal, and biomass, for example. This entire process chain must also be carefully examined to establish the overall CO_2 balance, however. One subvariant is the DMFC (direct methanol fuel cell) in which direct oxidation of the methanol is to take place in the fuel cell.

This technology is still a long time from series use, however.

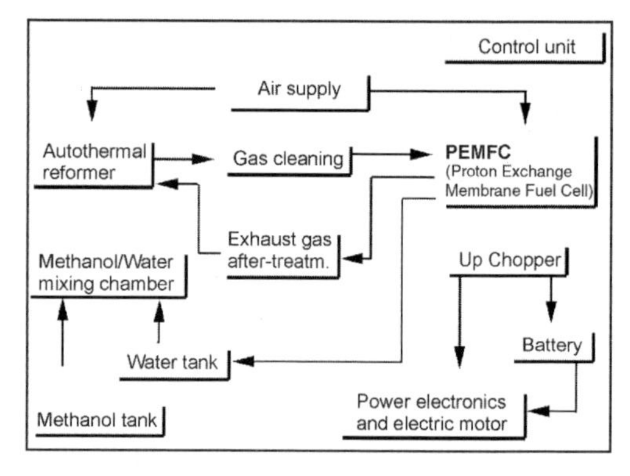

Fig. 28-17 Structure in principle of a methanol fuel cell powered vehicle.

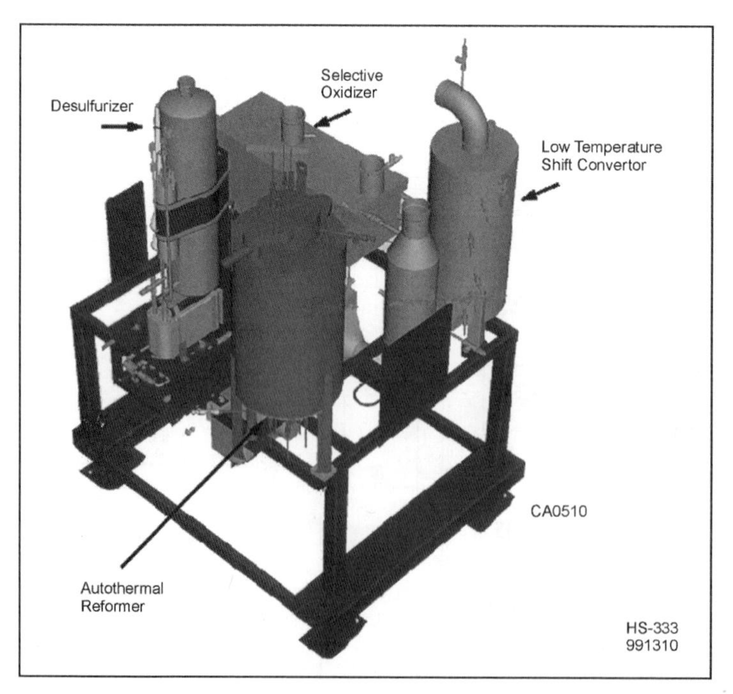

Fig. 28-18 Gasoline reformer for generation of hydrogen.

		Necar I	Necar II	Necar 5
Fuel cell system	Power Power density Voltage level	50 kW from 12 stacks 21 kg/kW 48 W/kg 130–230 V	50 kW from 2 stacks 6 kg/kW 167 W/kg 180–240 W	75 kW from 1 stack 15 kg/kW 66 W/kg 240–250 V
Tank system	Type Volume Pressure	Pressurized hydrogen tank, glass-fiber sheathed aluminum tank, 150 l 300 bar	Pressurized hydrogen tank, carbon-fiber-reinforced plastic tanks, 2 × 240 l 250 bar	Methanol tank 38 l
Propulsion system	Electric propulsion Maximum speed Range	30 kW 90 km/h 130 km	33 kW continuous output 45 kW maximum 110 km/h >250 km	33 kW continuous output 45 kW maximum 150 km/h 400 km
Permissible total weight		3500 kg	2600 kg	1450 kg

Fig. 28-19 Further development of fuel cell powered vehicles at DaimlerChrysler.

28.8.4 Gasoline Engine Fuel

The IFC Company[14] recently unveiled a reformer for a special gasoline engine fuel. Once operating temperature has been reached, the reformer's efficiency with an extremely low-sulfur fuel is better than 90%. Figure 28-18 shows a corresponding view of the subsystem.

28.8.5 The Fuel Cell in the Vehicle

As can be seen from Fig. 28-19, the progress made in using fuel cell systems is considerable.[11,15]

The latest automobile from DaimlerChrysler is already an extremely compact unit, incorporating a methanol reformer; see Fig. 28-20.

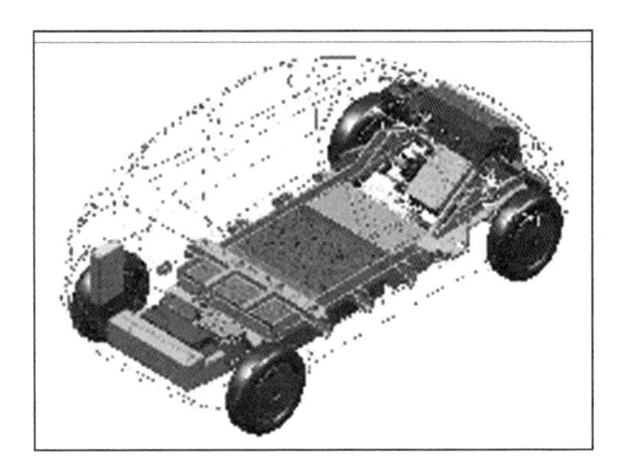

Fig. 28-20 Packaging of the Necar 5 propulsion system.[15]

Despite all the advances made, it must, nonetheless, be noted that considerable amounts of work on technology, performance, size, weight, and reliability—and also on cost—remain to be performed before this technology is ready for series use in private automobiles.

28.8.6 Evaluation of the Fuel Cell vis-à-vis Other Propulsion Systems

As already noted, it is most important that the fuel cell be assessed in conjunction with all other factors such as energy generation, transportation, and vehicle efficiency, including the reformer. Figure 28-21 provides information on the efficiency chains of individual systems.

The best efficiencies in the field of conventional propulsion systems are held by the diesel engine, the potential of which has been defined, given appropriate further development. These must be set against the figures for the fuel cell, for instance, hydrogen, with up to 40%, and for a modified gasoline engine fuel. In the latter case, efficiency even now is as good as that of the future diesel engine. Other studies also indicate with certainty overall efficiencies greater than 32.5% for the "gasoline reformer fuel cell" route. The development of the fuel cell and the appurtenant components on a massive scale will be worthwhile under all circumstances.

Fuel	Gasoline	Diesel	CNG	H_2	MeOH	MeOH	Gasoline
Conversion					Reformer		Reformer
Powertrain	Piston engine	Piston engine	Piston engine	Fuel Cell	Fuel Cell IMFC	Fuel Cell DMFC	Fuel Cell
Efficiency %							
Vehicle	18 – 24	24 – 28	18 – 24	30 – 40	26 – 36	24 – 34	21 – 29 [a]
Fuel	85 – 90	90 – 91	84 – 90	61 – 65	62 – 70	62 – 70	85 – 90 [a]
Vehicles and fuels	15 – 22	22 – 25	15 – 22	18 – 26	16 – 25	15 – 24	18 – 26 [a]
Vehicles and fuels	22 – 23	24 – 27	21 – 23	32.5 – 40	29 – 34	24.7 – 31	21 – 27 [b]

[a] Source: Prof. Höhlein, Jülich Research Center
Dr. Isenberg, Daimler-Benz 1998
[b] Source: FVV Fuel Cell Study
High values after 2005

Fig. 28-21 Efficiency of various energies and propulsion systems.

28.9 Summary

In the meantime, alternative energies and propulsion systems have the potential to achieve larger market shares. The precondition for this is that they achieve equivalent vehicle performances for the customer and can be sufficiently optimized for the other criteria of complexity, convenience, and cost to permit their acceptance by buyers. Another fact that cannot be forgotten is that present-day vehicle concepts (engine, transmission, power train management) are also undergoing continuous improvement. It will be necessary to become established against this future efficiency and performance level. The propulsion system distribution for Europe shown in Fig. 28-22 indicates an order of magnitude of up to 4% for alterna-

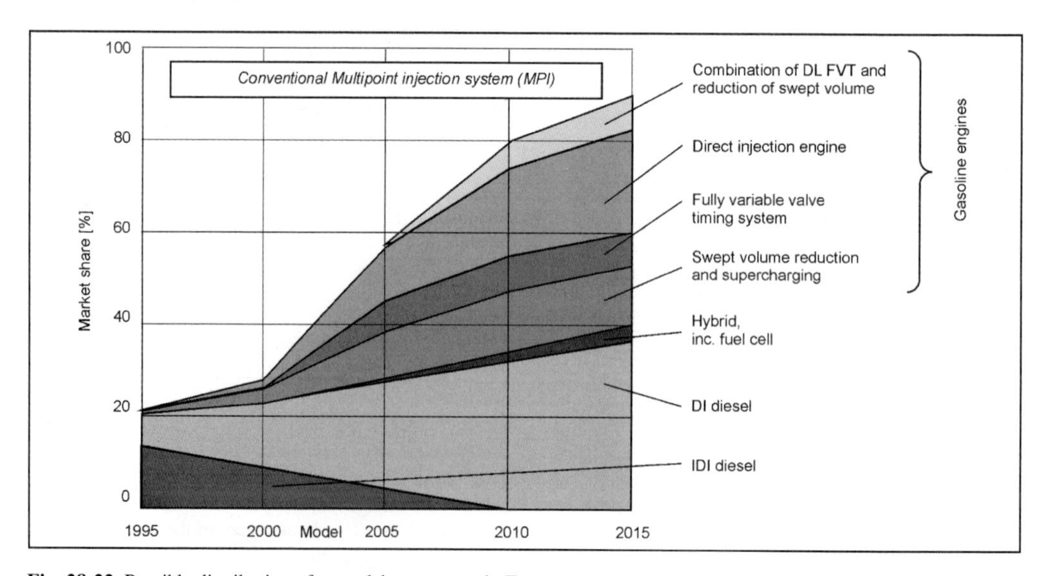

Fig. 28-22 Possible distribution of propulsion systems in Europe.

tive propulsion systems by the year 2015, i.e., more than 500 000 vehicles per year.

This category also includes the fuel cell, which will most certainly be used initially in buses and subsequently in series-produced automobiles. The fuel cell has the greatest potential for capturing a major share of the market from gasoline and diesel engines, which will, however, most certainly remain in series use for a long time. The fuel cell is at present subject to severe competition from hybrid vehicles designed for reduction of fuel consumption. Many developers expect that the fuel cell will come into use in large numbers in automobiles only when problems such as production, transportation, and storage of hydrogen have been solved.

Bibliography

[1] Bosch, Kraftfahrzeugtechnisches Taschenbuch, 23rd edition.

[2] Wüchner, E., Handbuch Kraftfahrzeugtechnik, Vieweg, Wiesbaden, 2000, ISBN 3-528-03214-X.

[3] Noreikat, K.E., *et al.*, Hybride Fahrzeugantriebe, Die Evolution zum Mehrwerthybrid, VDI Report 1565, Düsseldorf, 2000, ISBN 3-18-091565-X.

[4] Abthoff, I., Handbuch Kraftfahrzeugtechnik, Vieweg, Wiesbaden, 2000, ISBN 3-528-03214-X.

[5] Harada, I., Entwicklung eines neuen Toyota Hybrid Fahrzeuges, Proceedings Motor und Umwelt, AVL Graz, 2000.

[6] Täubner, F., *et al.*, Ergebnisse aus Prototypen neuer Schwungradspeicher, VDI Report 1565, Innovative Fahrzeugtechnik, Düsseldorf, 2000, ISBN 3-18-091565-X.

[7] Dietrich, T., Ultracapacitors–Power für innovative Automobilanwendungen, VDI Report 1565, Innovative Fahrzeugtechnik, Düsseldorf, 2000, ISBN 3-18-091565-X.

[8] Noreikat, K.E., Handbuch Kraftfahrzeugtechnik, Vieweg, Wiesbaden, 2000, ISBN 3-528-03214-X.

[9] Seiffert, U., *et al.*, Automobiltechnik der Zukunft, VDI Verlag, Düsseldorf, 1989, ISBN 3-18-400836-3.

[10] Mayr, B., *et al.*, "Zero Emission Engine ZEE," Der isotherme Dampfmotor als Fahrzeugantrieb, VDI Report 1565, Innovative Fahrzeugtechnik, Düsseldorf, 2000, ISBN 3-18-091565-X.

[11] Panik, F., *et al.*, Handbuch Kraftfahrzeugtechnik, Vieweg, Wiesbaden, 2000, ISBN 3-528-03214-X.

[12] Meitz, K., *et al.*, Alkalische Brennstoffzellensysteme für Fahrzeugantrieb, VDI Report 1565, Innovative Fahrzeugtechnik, Düsseldorf, 2000, ISBN 3-18-091565-X.

[13] Menzer, R., *et al.*, Potenzial unterschiedlicher Kraftstoffe bei ihrer Nutzung in Brennstoffzellensystemen, VDI Report 1565, Düsseldorf, 2000.

[14] Kelly, D., *et al.*, Development and Evaluation of Multi-Fuel-Cell-Power-Plant for Transportation Applications, VDI Report 1565, Düsseldorf, 2000.

[15] Daimler/Chrysler, Press folder, 2000.

[16] N.N.: Press release, Ford AG, Cologne.

29 Outlook

Automobiles have existed for more than one 100 years, and they have chiefly been driven by reciprocating piston engines. Both spark-ignition and diesel engines developed rapidly, and much developmental potential remains. Upon examination, we find that the rate of development has clearly accelerated in recent years.

Competitors such as the Stirling engine, the gas turbine, the Wankel engine, the steam engine, and the electric engine never had a serious chance at displacing the reciprocating piston internal combustion engine.

A while back, fuel cells joined the race.[1] To evaluate their prospects, we should not compare them directly with today's reciprocating piston engines; this is frequently done. Instead, we need to examine the developmental potential of both systems. Reciprocating piston engines have potential for improvement in the areas of fuel consumption and pollutant emissions, performance and torque, the aggregate weight and required installation space (packaging), as well as cost.

Now we address potential fuel savings of the competing spark-ignition and diesel engines.

Spark-Ignition Engine

The introduction of direct fuel injection in diesel passenger car engines in 1989 enlarged the diesel's lead in fuel consumption by 15% to 20% to approximately 35%. After direct fuel injection was successfully introduced into series production for the first spark-ignition engines a few years ago, this procedure has yielded savings of 10% to 17% in fuel consumption in the EUDC test, depending on the engine size.

The standard combustion methods for spark-ignition engines with direct fuel injection are wall directed. Jet-guided methods, which utilize multihole nozzles and higher injection pressures of more than 200 bar, are under development. This permits improvements of over 20% in fuel consumption in a smog test from improved charge stratification in reference to engines with external mixture formation. The first vehicles to use the jet-directed method will be mass-produced in 3 to 5 years. This will lower the advantage in fuel consumption held by diesel engines with direct fuel injection to 10%. Other developmental products that positively influence fuel consumption, some of which are in series production are[2,3]

- Fully variable valve gears with potential savings in consumption of up to 10%. The additional cost for mechanical systems is, however, approximately 15% (BMW). Electrohydraulic and electromechanical systems are also being developed. Camshafts may be dispensed with in electromechanical systems.
- Pulse turbocharging using an air pulse valve. Similar effects can be obtained as with fully variable valve control, but with less effort, and multistage manifolds can be dispensed with because of the increased charge over the entire speed range. Other functions such as hot and cold charging, cylinder shutoff and throttle-free load control allow the dramatic reduction of fuel consumption and pollutant emissions. The estimated fuel savings are 20% to 30%.
- Variable stroke volume, e.g., by cylinder shutoff, reduces fuel consumption by 8%–20% in the partial load range. Engine-internal mechanical systems are also in development.
- Fuel consumption can be reduced by approximately 5% with a dual spark plug engine and sequential dual ignition.
- Crankshaft starter generators with a 14/42V vehicle electrical system. In connection with downsizing, savings in fuel consumption of 15% to 22% are possible.
- Variable geometric compression ratio in conjunction with supercharging yields up to 30% reduction in fuel consumption.
- Downsizing with highly supercharged engines with over 25 bar average pressure and decreasing the stroke volume by reducing the number of cylinders from, e.g., six to four or three. Such approaches also reduce the engine weight and required space (packaging). The potential savings in fuel consumption is 20% to 25%.
- Turbocharging as a fuel consumption approach; i.e., relatively small superchargers allow fuel consumption to be reduced by approximately 15%.
- Reducing friction loss. Over the last 10 to 15 years, friction loss has been reduced by a remarkable 20%. Nevertheless, substantial potential remains since the theoretical case of a frictionless spark-ignition engine would yield an approximately 20% gain in efficiency. If we could attain one-half of this, fuel consumption would be reduced by approximately 10%.
- Thermomanagement with hot cooling under a partial load. Savings of 5% to 10% are conceivable.
- Fuel and lubrication oil optimization can yield potential savings of 5% to 10%.
- Drastically reducing the warm-up time. Fuel consumption savings of up to 5% are conceivable.

If all the cited improvements in fuel consumption are realized, the spark-ignition engine would actually *produce* fuel. Realistically, it is clear that aggressively combining complementary measures could likely reduce fuel consumption by 50% by 2010. This does not include savings in fuel consumption from advances in transmission technology. Depending on the comparison, this could lead to a further reduction of fuel consumption of 15% for both SI and diesel engines.

Diesel Engines

The potential for lowering fuel consumption in diesel engines is much less since direct fuel injection has already been adopted for nearly every engine in Europe within only 10 years. There are, however, still many possibilities for improvement[3,4]:

- Variable valve control could also be adopted for diesel engines. The gain in fuel consumption would be less, however, since diesel engines use quality control and, therefore, are already unthrottled under a partial load. The reduction of fuel consumption would be 5% to 10%.
- Pulse turbocharging. The potential savings in fuel consumption is also substantial in this instance: Approximately 20%. It would permit scrapping the glow plug or reducing the compression ratio, which permits a higher supercharging ratio and, hence, performance. When using hot charging and cold charging, pollutant emissions are reduced.
- Cylinder shutoff is possible; the reduction of fuel consumption is, however, less than in spark-ignition engines.
- Crankshaft starter generators have effects similar to those in spark-ignition engines.
- A variable geometric compression ratio would not be recommendable since it would be difficult to realize because of the stronger forces that occur.
- Reducing friction loss. Friction loss lowers efficiency in diesel engines by approximately 26%. To cut this in half would reduce fuel consumption by 13%, which is slightly greater than with spark-ignition engines.
- Downsizing as in spark-ignition engines is also possible. Under development are peak pressures of more than 200 bar in passenger cars and 250 to 300 bar in commercial vehicles for greater specific performance. To counter the weakness in starting, electrically supported turbochargers are being developed that increase the average pressure at starting speed. The gain in fuel consumption would be 20% to 25% as in spark-ignition engines.
- Thermomanagement, hot cooling, and fuel and lubrication optimization would lower consumption by 5% to 10%.
- Thermodynamic optimization could reduce fuel consumption by 5%.
- Turbocompound supercharging. By exploiting the exhaust energy better than pure exhaust-gas turbocharging, the fuel consumption can be lowered by 5%, and performance can be increased by 10%.

Even taking into account the imprecision of the above assumptions, we can still conclude that the fuel consumption of SI and diesel engines will grow increasingly closer in the future under the evaluated conditions. It can be expected that in the final analysis, the diesel engine will have a volumetric fuel consumption advantage of only approximately 10%. This means that the present importance of diesel engines will decrease since manufacturing costs are much higher than those of the spark-ignition engine.

The savings in fuel consumption of SI and diesel engines in mobile applications will be so high that, in comparison to fuel cells that naturally also have a certain amount of developmental potential, the lead will continue to widen. The relatively high manufacturing costs of fuel cells must also be taken into consideration. The argument that fuel cells are free of pollution in contrast to reciprocating-piston engines no longer applies to spark-ignition engines, and the problem will be solved at the latest for diesel engines with the introduction of homogeneous combustion (HCCI).[1]

We can, therefore, confidently conclude that the reciprocating-piston engine will continue its interesting and successful career for decades.

Bibliography

[1] Proceedings Verbrennungsmotor versus Brennstoffzelle—Potenziale und Grenzen für den Automobilantrieb, 13th International AVL Congress, 2001.

[2] Takimoto, M., Ausblick auf zukünftige Fahrzeugantriebe und deren Herausforderung für Toyota, 10th Aachen Colloquium Vehicle and Engine Technology, 2001.

[3] Heil, B., H.K. Weining, G. Karl, D. Panten, and K. Wunderlich, Verbrauch- und Emissionen-Reduzierungskonzepte beim Ottomotor, in MTZ 62 (2001), No. 11, pp. 900–915, Part 1.

[4] Heil, B., H.K. Weining, G. Karl, D. Panten, and K. Wunderlich, Verbrauch- und Emissionen-Reduzierungskonzepte beim Ottomotor, in MTZ 62 (2001), No. 12, pp. 1022–1035, Part 2.

Index

D

J

K

T

U

V

W

About the Editors

Dr.-Ing. E.h. **Richard van Basshuysen** VDI, was born in 1932 in Bingen/Rhein, Germany. From 1953 to 1955, after completion of training as an automotive mechanic and university admission, he studied mechanical engineering at the Wolfenbüttel University of Applied Sciences and graduated with an engineering degree. He earned his master's degree in engineering in 1982. From 1955 to 1965, Richard van Basshuysen worked at Aral AG, Bochum, in a scientific capacity. From 1965 to 1971, he was deputy head of overall experimental testing and director of power plant development at NSU in Neckarsulm (later Audi), where, among other things, he was among those responsible for development of the NSU RO 80 (Wankel engine) and Volkswagen K70. Additional career highlights at Audi include the following: department head, RO 80 engine testing and development (1971 to 1973); reciprocating piston engine development and development of new engine components and fuel metering systems (1973 to 1976); head of section, engine and vehicle development (1976 to 1988); overall head of power plant development and head of Audi luxury vehicle development, simultaneously member of the management board (1988 to 1990). Since 1990, independent consultant to the international automobile industry. In addition to his work as an engineer, Richard van Basshuysen has been active in various automotive engineering associations and has gained a reputation as an author of technical works. He was a member of the board of trustees of the Research Institute of Automotive Engineering and Vehicle Engines at the University of Stuttgart (FKFS) and has been a member of the advisory board and a specialist advisor to the Verein Deutscher Ingenieure (VDI, German Association of Engineers). In 2001, the Association awarded Dipl. Ing. van Basshuysen its highest honor, the VDI Benz-Daimler-Maybach Medal of Honor, for meritorious contributions to motor vehicle technology and to the VDI; in the previous year, he was awarded the highly endowed Ernst Blickle Prize for development and series introduction of direct injection passenger car diesel engines. Since 1991, Richard van Basshuysen has been editor of the international engineering and scientific journals ATZ (*Automobiltechnische Zeitschrift*) and MTZ (*Motortechnische Zeitschrift*), and has been a member of the editorial board of the trade journal *Automotive Engineering Partners*. His numerous publications encompass 47 scientific and technical articles and the technical books *Schadstoffreduzierung und Kraftstoffverbrauch von Pkw-Verbrennungsmotoren* (co-author, Wiesbaden, 1993); its English-language version, *Reduced Emissions and Fuel Consumption in Automobile Engines* (co-author, Society of Automotive Engineers, 1995); *Shell-Lexikon Verbrennungsmotor* (*Shell Lexicon of Combustion Engines*, Wiesbaden, 1996/2003, co-editor); and co-editor (with Fred Schäfer) of this volume, which first appeared in its German-language edition in 2002.

Prof. Dr.-Ing. **Fred Schäfer** VDI, was born in 1948 in Neuwied, Germany. From 1968 to 1974, he studied mechanical engineering at the Koblenz University of Applied Sciences and the University of Kaiserslautern. From 1974 to 1980, he worked as a scientific associate and docent at the University of Kaiserslautern. Fred Schäfer earned his engineering doctorate in 1980 with *An Investigation of the Addition of Hydrogen to Methanol on the Operation of an Unthrottled Otto Engine*. From 1980 to 1990, he worked in industry at Audi AG, Neckarsulm. Initially involved with engine development, Fred Schäfer advanced to head of the Engine Design department, with emphasis on production development of engines for the U.S. market, advanced development, and development of competition engines. He has been a professor of combustion engines and flow machinery at the Südwestfalen University of Applied Sciences, Iserlohn, Germany, since 1993. Fred Schäfer is a member of SAE and is the author of numerous technical and scientific papers, technical reports, and lectures. Within the framework of his activity as a university professor, he is director of the Institute for Reciprocating Engines at the Südwestfalen University of Applied Sciences, and conducts research and development projects on behalf of the automotive and supplier industries. In association with Richard van Basshuysen, he is co-author of the engineering books *Schadstoffreduzierung und Kraftstoffverbrauch von Pkw-Verbrennungsmotoren* (Wiesbaden, 1993); its English-language version, *Reduced Emissions and Fuel Consumption in Automobile Engines* (Society of Automotive Engineers, 1995); co-editor of *Shell-Lexikon Verbrennungsmotor* (*Shell Lexicon of Combustion Engines*, Wiesbaden, 1996/2003); and co-editor of this volume, which first appeared in its German-language edition in 2002 (second edition, Wiesbaden, 2003), encompassing contributions by 96 renowned authors from science and industry working under his coordination.

Color Section

The page number given in each figure caption refers to the location in the text where the figure is discussed.

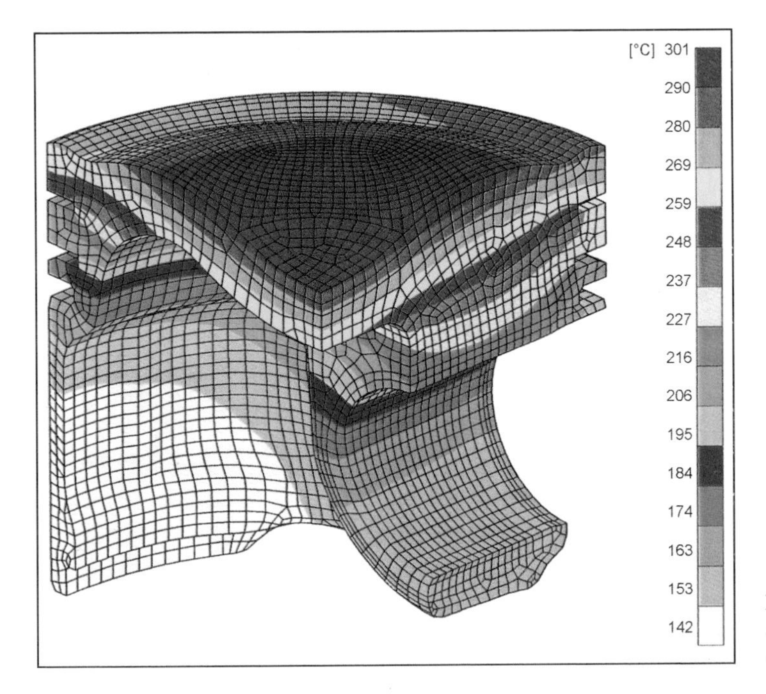

Fig. 7-8 Temperature distribution at a piston for a gasoline engine. *(See page 83.)*

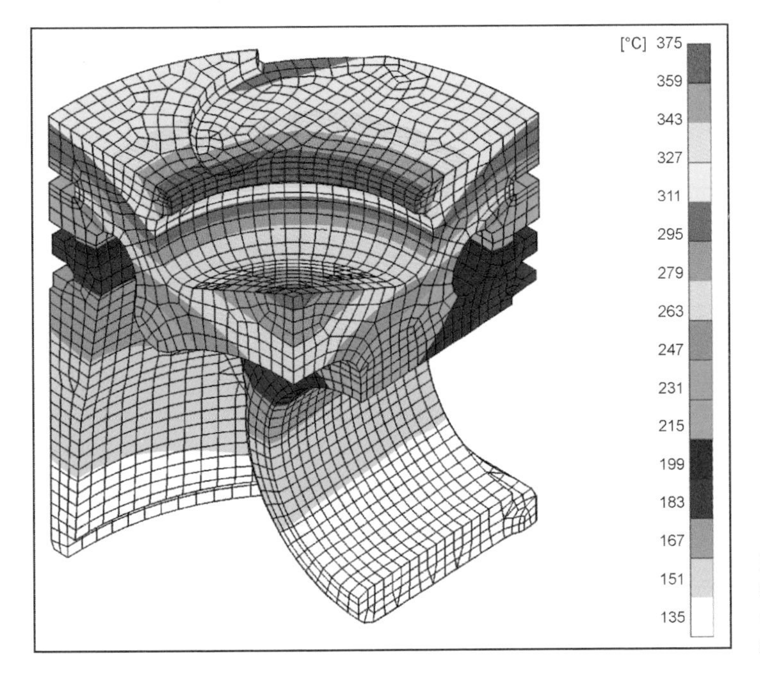

Fig. 7-9 Temperature distribution at a piston with cooling channel for a diesel engine. *(See page 83.)*

Fig. 7-33 Stress analysis for a conrod with an angular split, with a trapezoidal small end (half model) (Federal Mogul). *(See page 96.)*

Fig. 7-79 Section of the water jacket for coolant flow simulation.[9] (*See page 136.*)

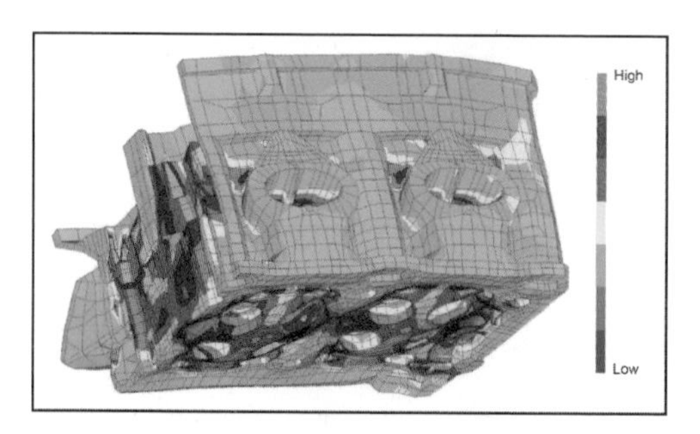

Fig. 7-80 Strength analysis at the cylinder head.[12] (*See page 136.*)

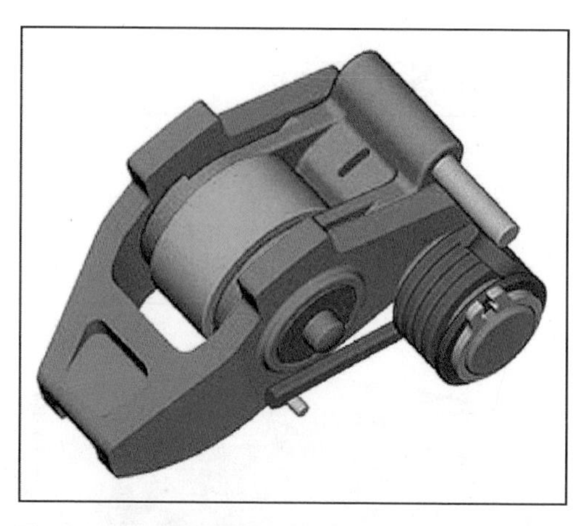

Fig. 7-120 Switchable cam follower. (*See page 157.*)

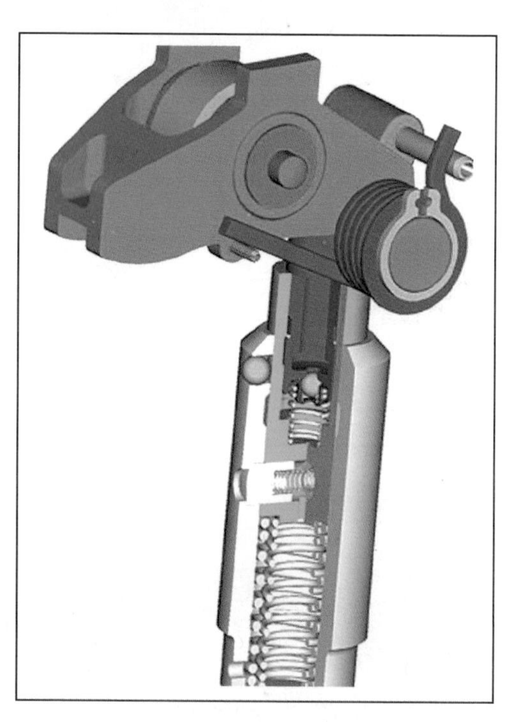

Fig. 7-123 Switchable support element and switchable cam follower. (*See page 159.*)

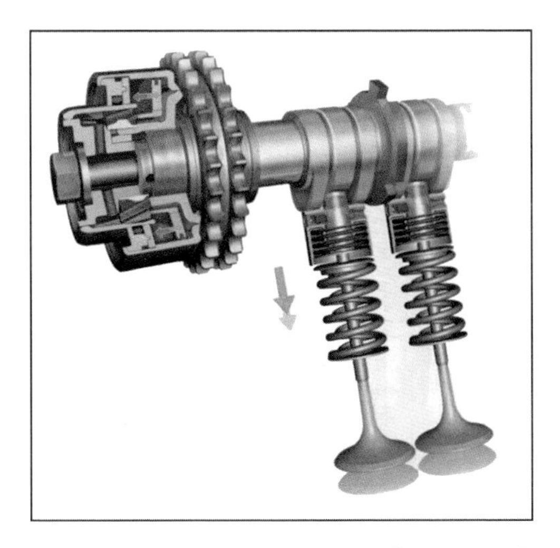

Fig. 7-124 Porsche VarioCam Plus System.[2] (*See page 159.*)

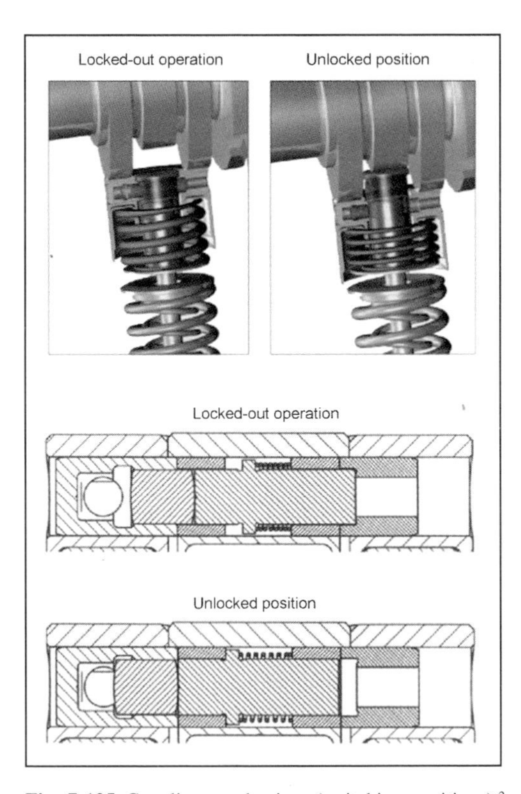

Fig. 7-125 Coupling mechanism (switching positions).[2] (*See page 159.*)

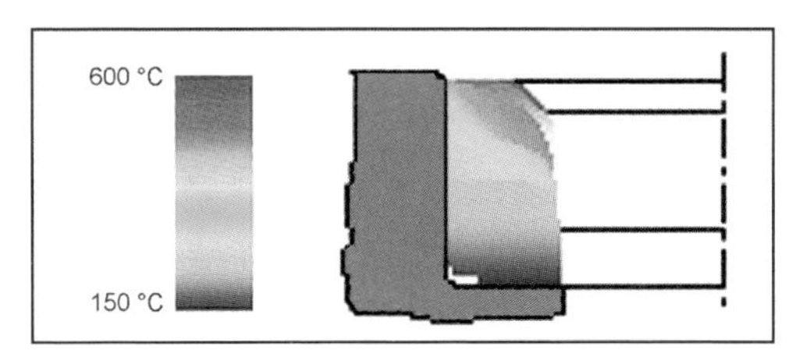

Fig. 7-168 Temperature distribution inside a valve seat insert at the exhaust port. (*See page 181.*)

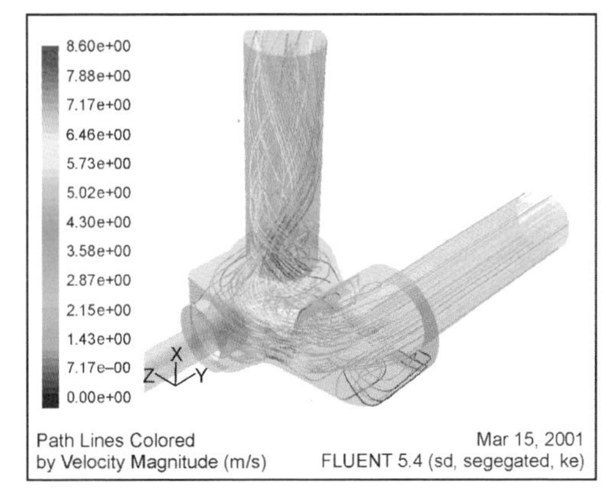

Fig. 7-209 Flow simulation near the valve using FLUENT V5. (*See page 201.*)

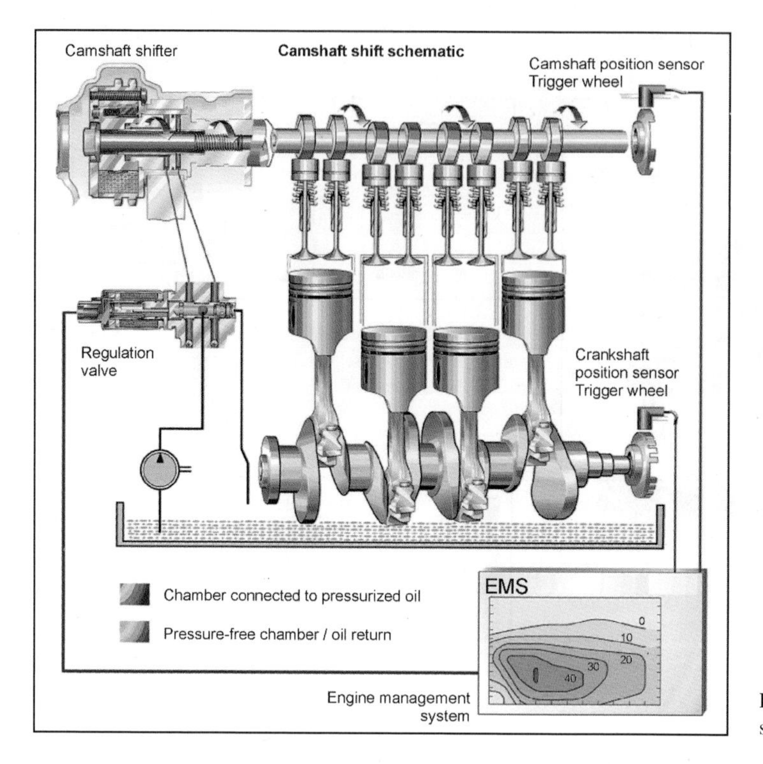

Fig. 7-224 Continuous camshaft shifting. (*See page 211.*)

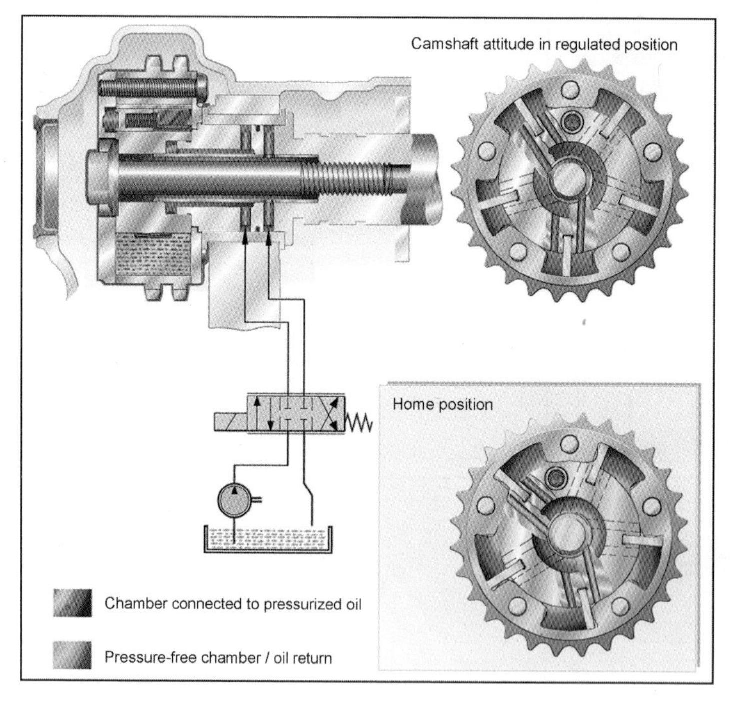

Fig. 7-225 Slewing motor or vane shifter. (*See page 212.*)

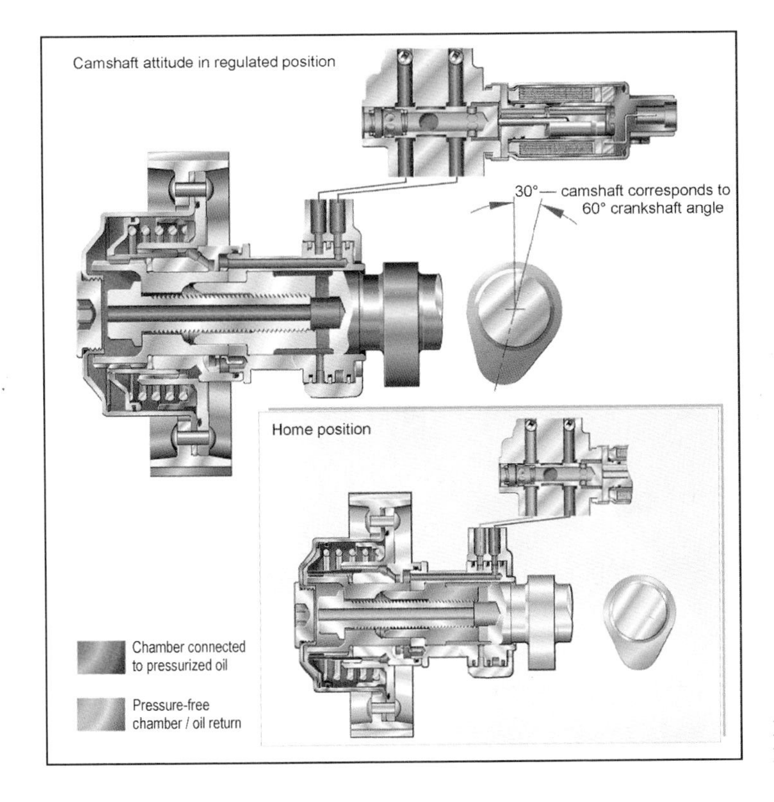

Fig. 7-226 Camshaft shifter with helical toothing. (*See page 213.*)

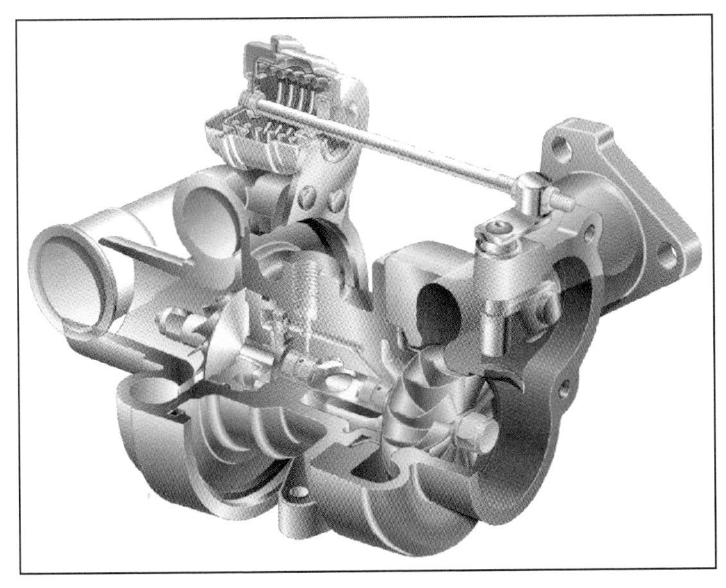

Fig. 11-28 Waste gate.[9] (*See page 364.*)

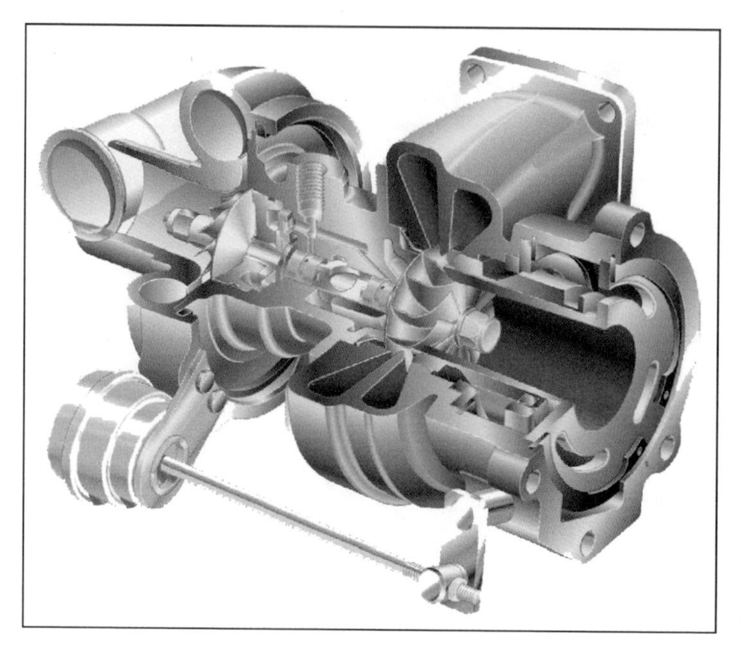

Fig. 11-31 Variable slide valve turbine.[9] (*See page 365.*)

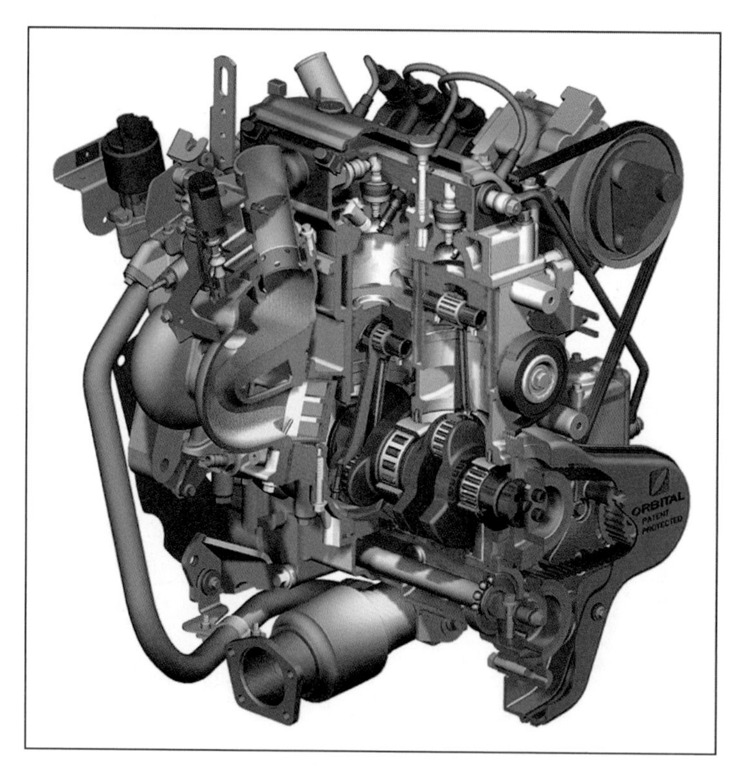

Fig. 15-48 Section of the 1.2 liter three-cylinder two-stroke engine by Orbital.[18] (*See page 489.*)

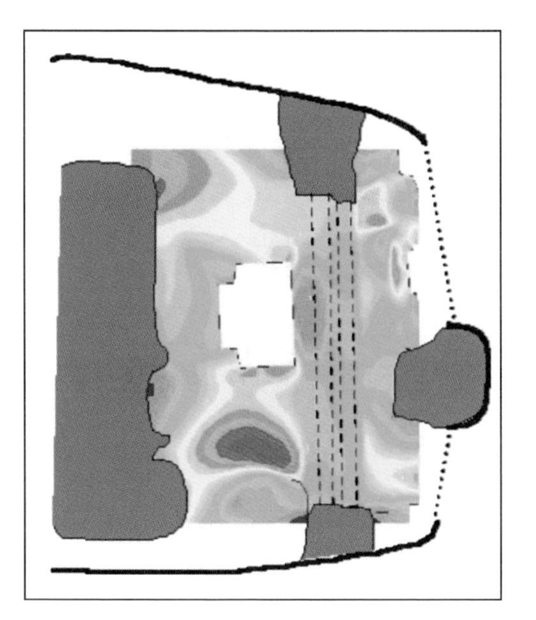

Fig. 20-5 CFD simulation of the cooling air flow in the front section of a car. (*See page 557.*)

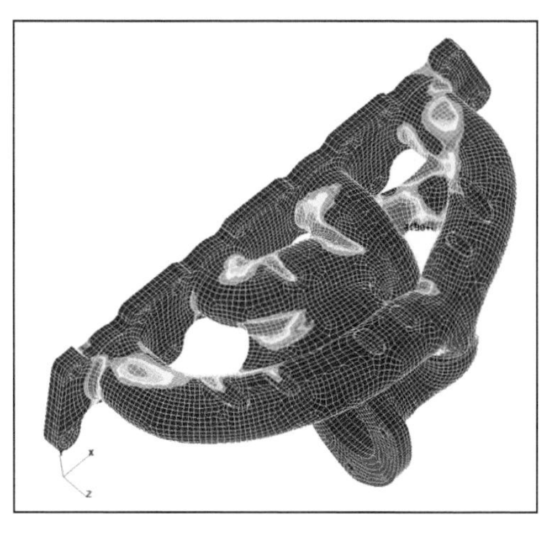

Fig. 24-3 Distribution of fracture-load cycles in an exhaust manifold. (*See page 704.*)

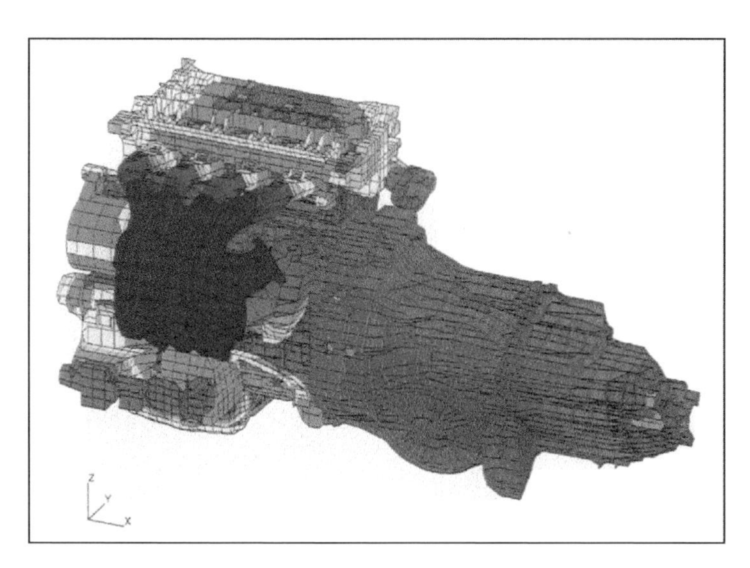

Fig. 24-4 Distribution of acoustic velocities on the surface of an aggregate. (*See page 704.*)

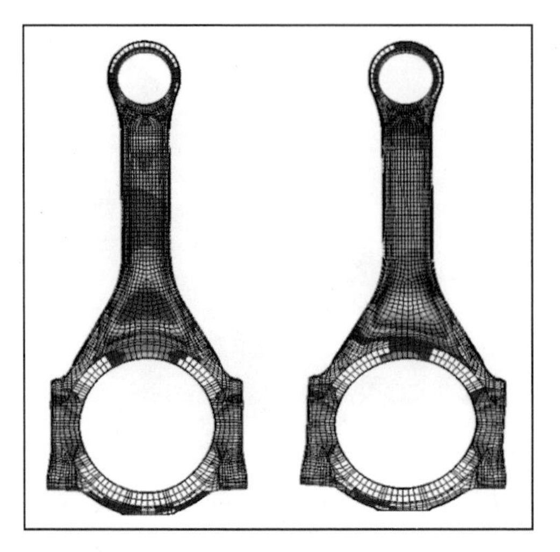

Fig. 24-6 Optimization of the geometry of a connecting rod: Plot of stress in initial state (left) and in optimized state (right). (*See page 705.*)

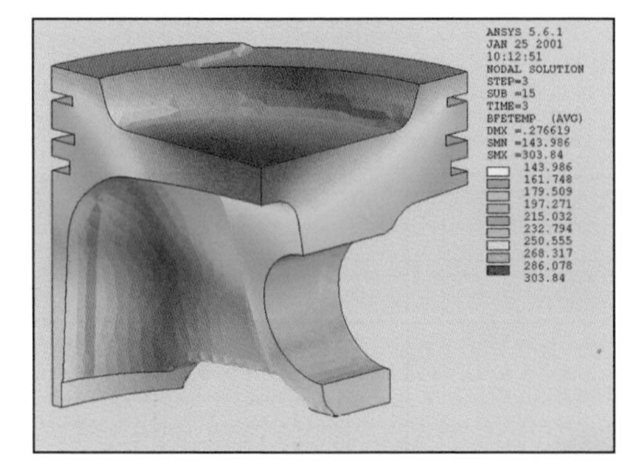

Fig. 24-10 Temperature distribution in an automobile gasoline engine at design output. (*See page 708.*)

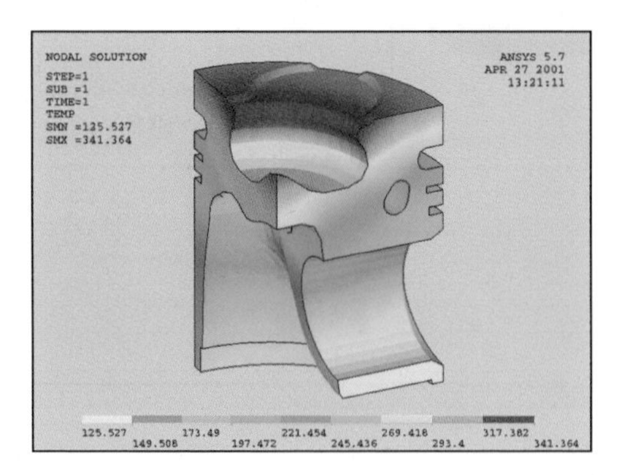

Fig. 24-11 Temperature distribution in an automobile diesel engine at design output. (*See page 708.*)

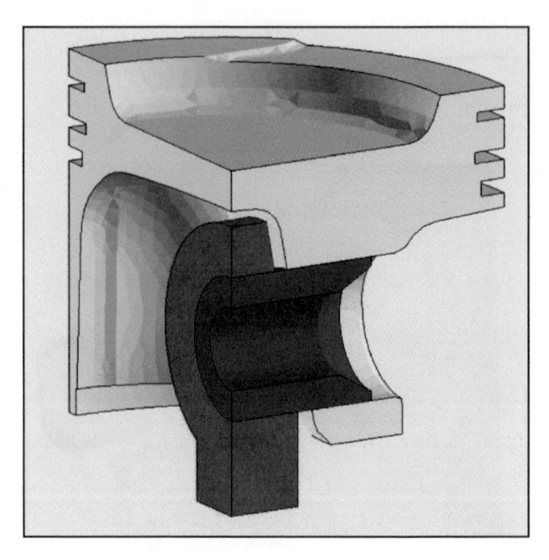

Fig. 24-13 Model for use for piston calculation. (*See page 709.*)

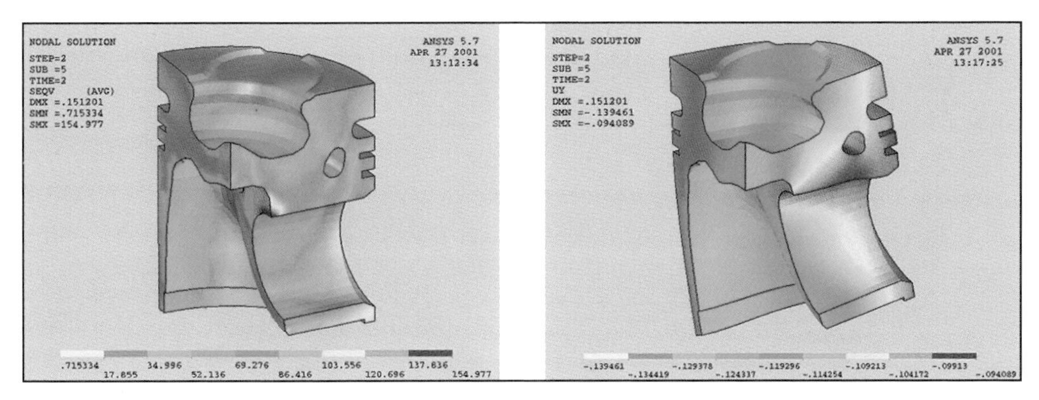

Fig. 24-14 Stresses and deformations in a piston under purely gas force (example). (*See page 710.*)

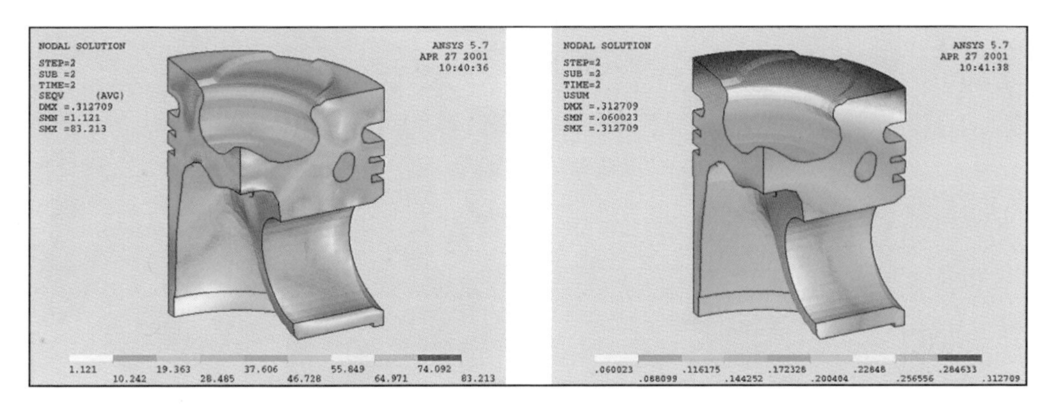

Fig. 24-15 Stresses and deformations in a piston under a purely thermal load (example). (*See page 710.*)

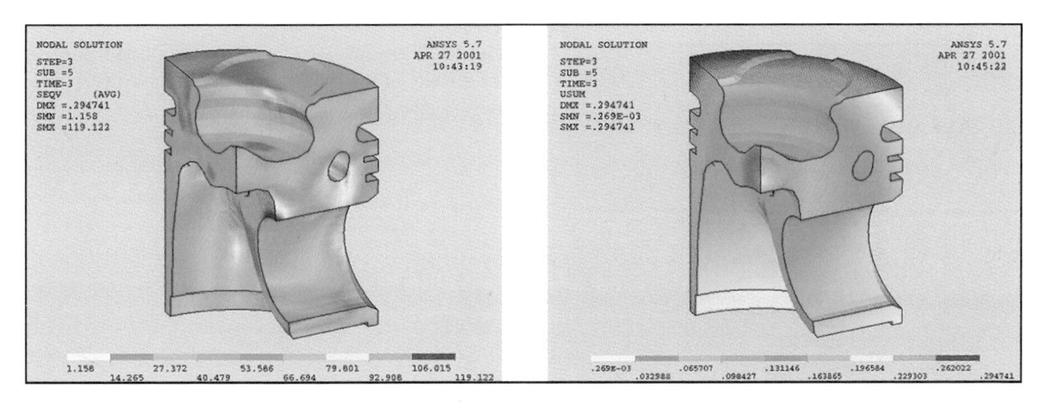

Fig. 24-16 Stresses and deformations in a piston (mechanical and thermal) example. (*See page 710.*)

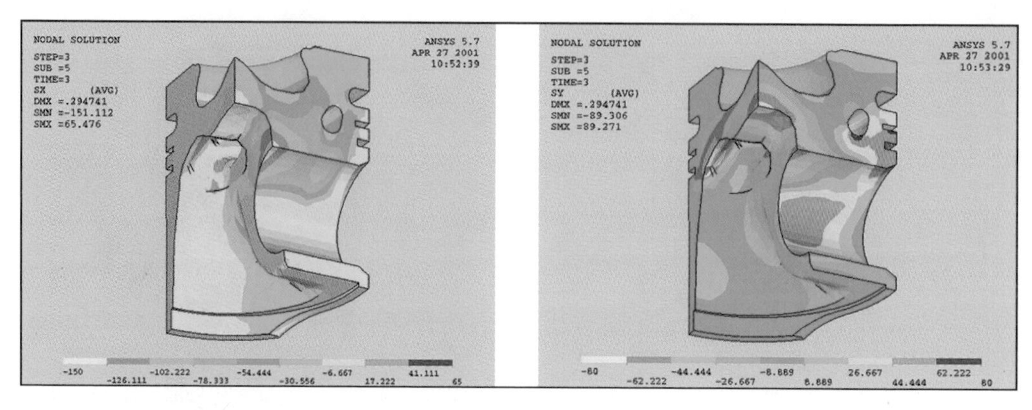

Fig. 24-17 Radial and tangential stresses in the hub boring (example). (*See page 711.*)

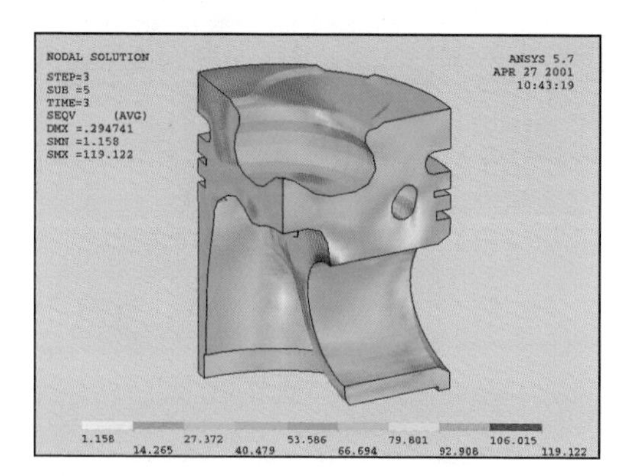

Fig. 24-18 Stress intensity on a piston under mechanical and thermal load (example). (*See page 711.*)

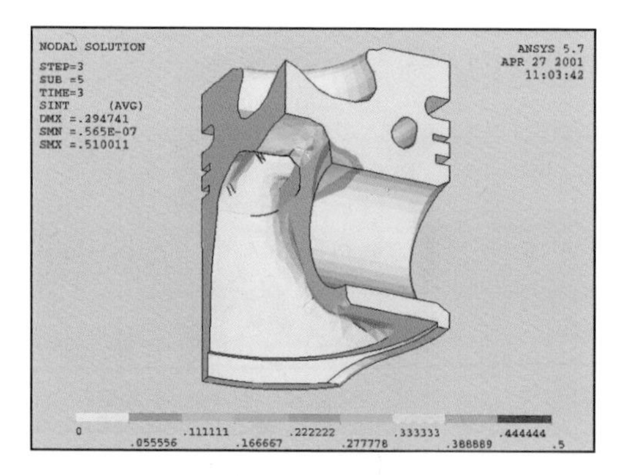

Fig. 24-19 Damage index using the critical plane method. (*See page 712.*)

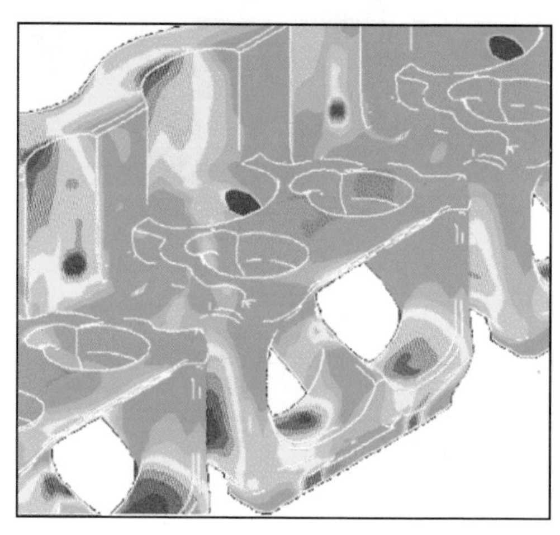

Fig. 24-22 Distribution of coefficients of heat transfer in the vicinity of the exhaust ports in a four-cylinder, five-valve engine. (*See page 717.*)

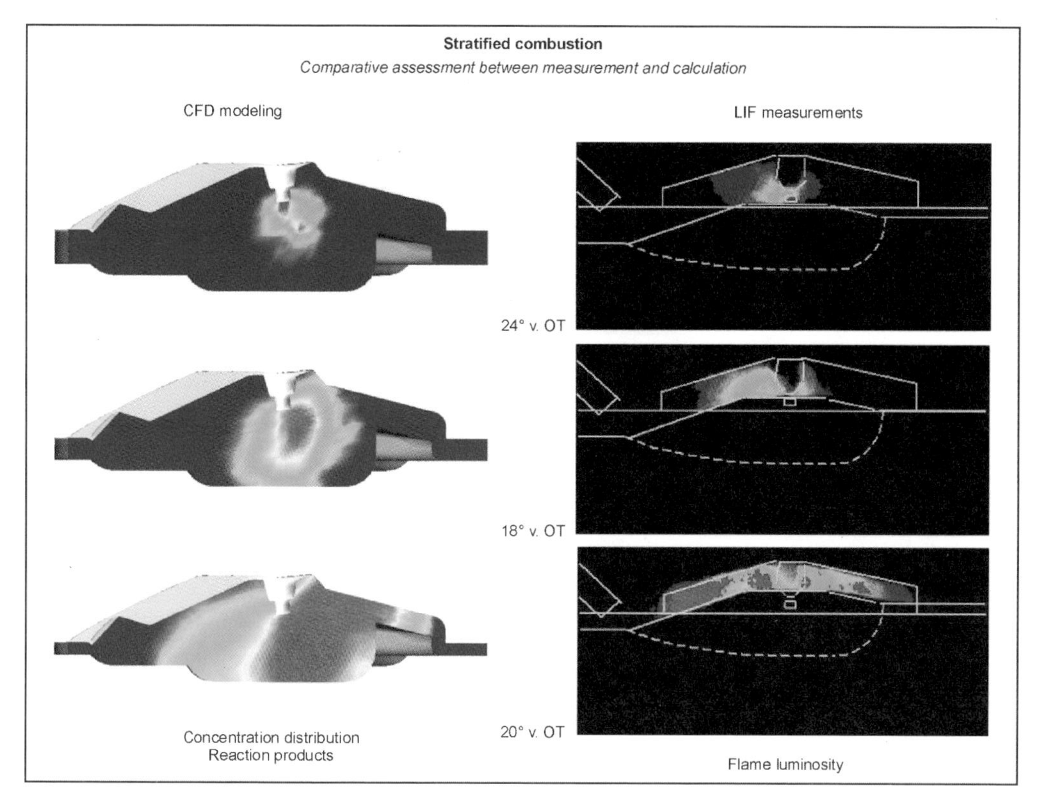

Fig. 24-26 Combustion with charge stratification; comparative assessment calculation vs. measurement.[20] (*See page 721.*)

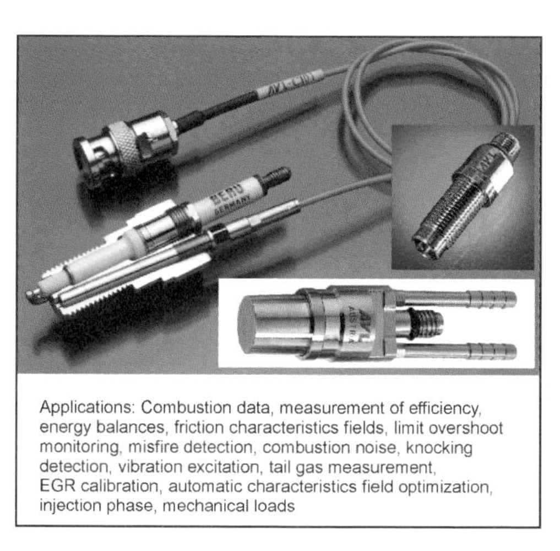

Fig. 25-5 Examples of piezoelectric sensors for measurement of cylinder pressure. (*See page 726.*)

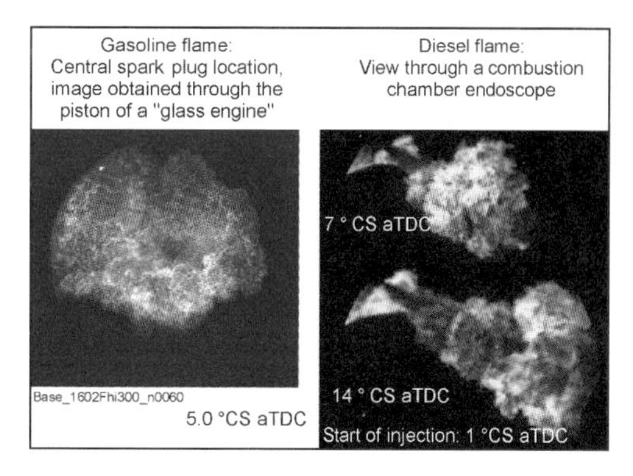

Fig. 25-7 Flame photograph. (*See page 727.*)

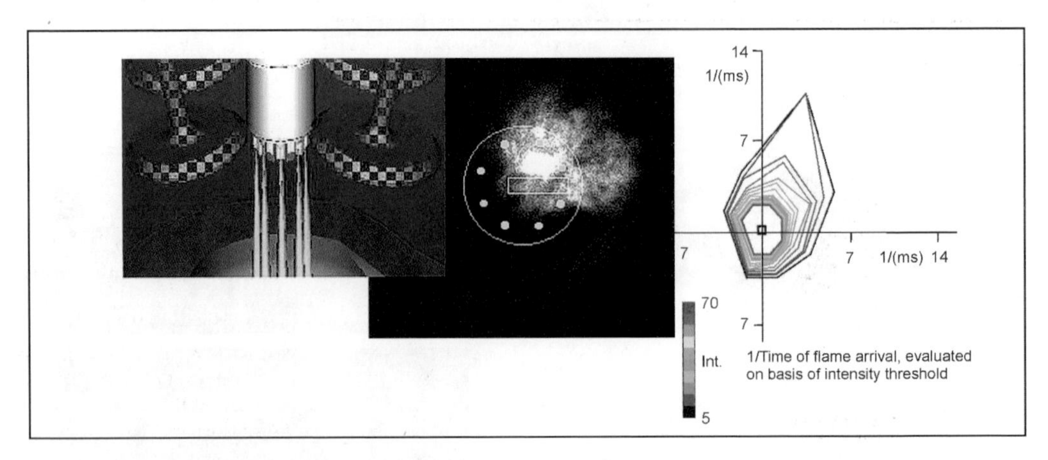

Fig. 25-11 Flame core generation, observation using a spark plug sensor. The result illustrates the symmetry or asymmetry of the flame core and its predominant direction of propagation. (*See page 731.*)

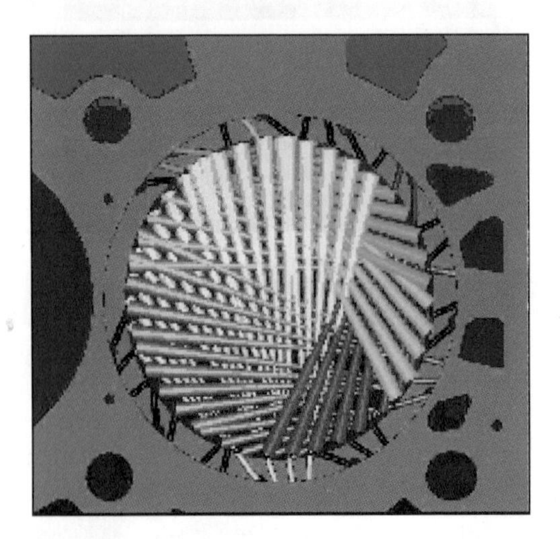

Fig. 25-12 Arrangement of a micro-optic sensor system in the cylinder-head gasket for tomographic flame reconstruction. (*See page 731.*)

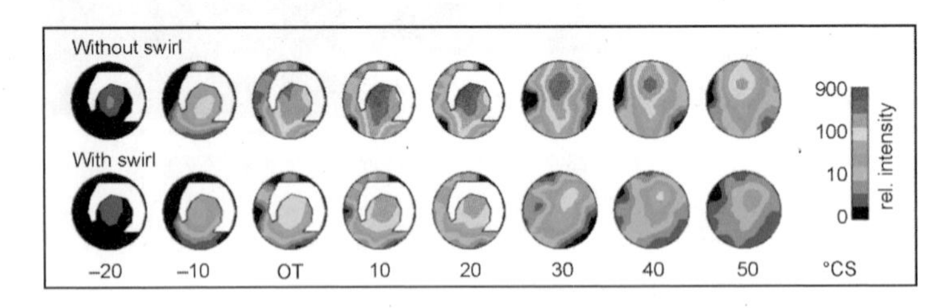

Fig. 25-13 DI gasoline engine: Flame tomography shows the local position of bright, soot-producing diffusion flames. Swirling flow produces a significant improvement. (*See page 731.*)

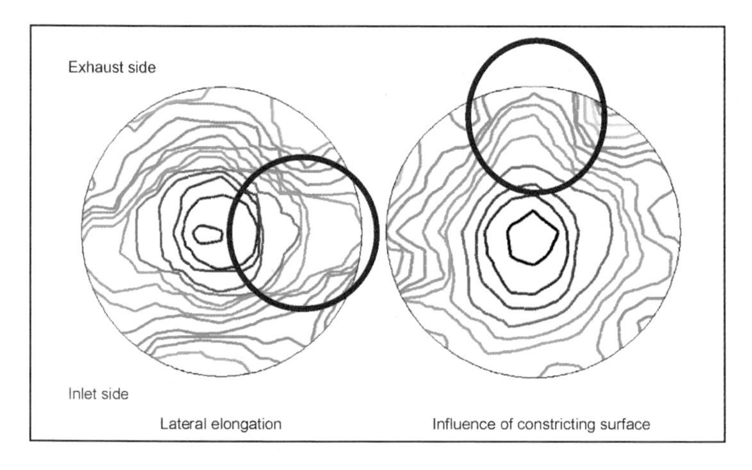

Fig. 25-14 Flame propagation: Tomograph with sensor system installed in the cylinder-head gasket. The isolines indicate the progress of the flame front against time. The influence of internal flow on flame propagation is clearly perceptible. (*See page 732.*)

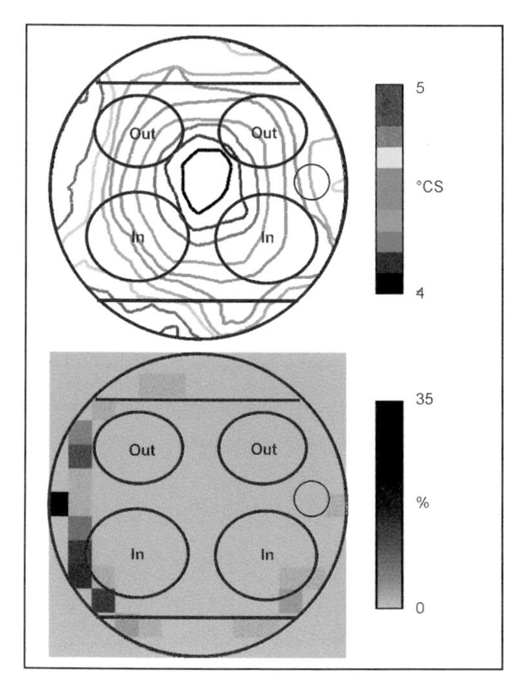

Fig. 25-15 Flame tomography supplies comprehensive documentation of flame propagation and knocking spot distribution. (*See page 732.*)

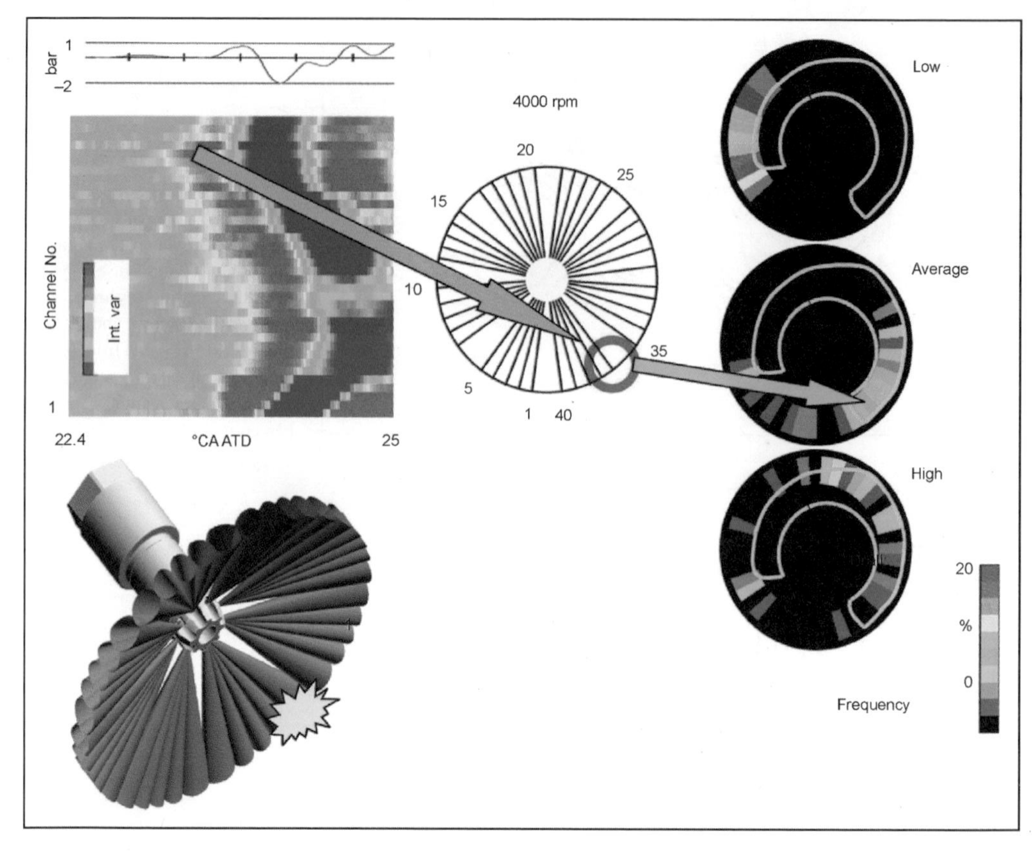

Fig. 25-16 Determination of knocking spots by a fan-type sensor, presentation of results: Single cycle and derived knocking spot statistics. (*See page 733.*)

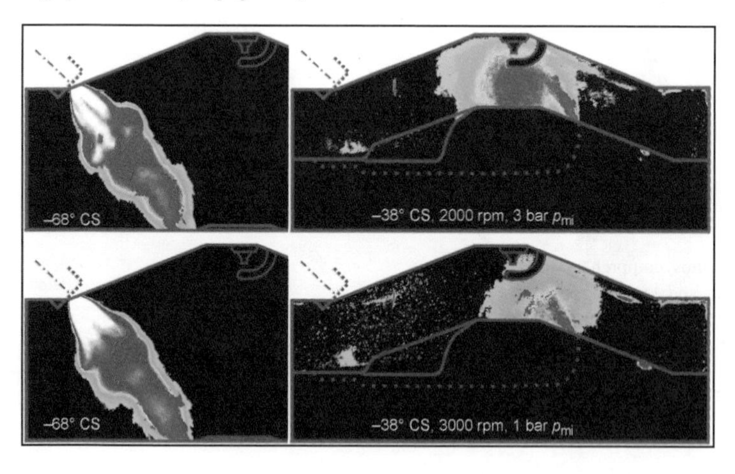

Fig. 25-18 Gasoline direct injection: Fuel distribution in the injection process and after deflection from the pistons. The stability of distribution states is determined from individual images using image statistics. Green/red: Fuel vapor with increasing stability; blue/white: Fuel droplets with increasing stability. (*See page 734.*)